PROCEEDINGS OF THE AMERICAN SOCIETY FOR COMPOSITES

HOW TO ORDER THIS BOOK

BY PHONE: 800-233-9936 or 717-291-5609, 8AM–5PM Eastern Time

BY FAX: 717-295-4538

BY MAIL: Order Department
Technomic Publishing Company, Inc.
851 New Holland Avenue, Box 3535
Lancaster, PA 17604, U.S.A.

BY CREDIT CARD: American Express, VISA, MasterCard

BY WWW SITE: http://www.techpub.com

PROCEEDINGS OF THE AMERICAN SOCIETY FOR COMPOSITES

TWELFTH TECHNICAL CONFERENCE

Edited by
RONALD F. GIBSON and GOLAM M. NEWAZ

Co-Sponsored by
WAYNE STATE UNIVERSITY
MICHIGAN MATERIALS AND PROCESSING INSTITUTE
AUTOMOTIVE COMPOSITES CONSORTIUM

October 6–8, 1997
The Dearborn Inn
Dearborn, Michigan

TECHNOMIC PUBLISHING CO., INC.
LANCASTER · BASEL

Proceedings of the American Society for Composites

a TECHNOMIC® publication

Published in the Western Hemisphere by
Technomic Publishing Company, Inc.
851 New Holland Avenue, Box 3535
Lancaster, Pennsylvania 17604 U.S.A.

Distributed in the Rest of the World by
Technomic Publishing AG
Missionsstrasse 44
CH-4055 Basel, Switzerland

Printed in the United States of America

10 9 8 7 6 5 4 3 2 1

Main entry under title:
Proceedings of the American Society for Composites—
Twelfth Technical Conference

A Technomic Publishing Company book
Bibliography: p.

ISSN No. 1084-7243
ISBN No. 1-56676-606-0

Table of Contents

Education

Joints and Fasteners

Preface

We are delighted to present this archival record of the Twelfth Annual Technical Conference of the American Society for Composites, which was held in Dearborn, Michigan on October 6–8, 1997. These proceedings contain technical papers that were presented in 35 sessions over the three days of the conference. Session tracks included Analysis and Design, Damage, Fracture and Fatigue, Dynamic and Time Dependent Behavior, Education, Joints and Fasteners, Manufacturing, Military Applications, Nondestructive Evaluation, and Thermal and Mechanical Characterization. On the basis of the number of papers presented in the different areas, composites manufacturing proved to be the dominant topic of the conference. Coincidentally, the theme of a National Science Foundation-sponsored workshop immediately following the conference at the same location was "The Future of Composites Manufacturing," so the two events complemented each other very nicely. The current interest in composites manufacturing stems from the need to make these materials more attractive in medium-to-high volume manufacturing environments by reducing costs and process cycle times. A relatively large percentage of the papers was also presented in the Dynamic and Time Dependent Behavior sessions and in the Damage, Fatigue and Fracture sessions, as research in these areas continues to build up the much-needed database on durability of composite materials. Other topics such as Joints and Fasteners, Military Applications, Nondestructive Evaluation, Thermal and Mechanical Characterization, and Education also attracted considerable attention at the conference. Finally, a technical poster session which is not recorded here allowed authors to present current research findings to conference attendees on an individual basis.

Many individuals worked very hard before and during this conference, and we would be remiss if we did not mention at least some of them. The enthusiastic participation of the keynote speakers, authors, chairs and co-chairs was obviously the major reason for the success of the conference. The efforts of Audra Duffer, the conference secretary, and Sandy Carpenter, who prepared the conference programs and made the local arrangements, are particularly noteworthy. The encouragement and support of our colleague and department chair, Ken Kline, is also gratefully acknowledged. Carl Johnson of Ford Motor Company is to be commended for his help in arranging the laboratory tour and the reception at Ford Research Laboratory. Financial support from the Michigan Materials and Processing Institute and MTS Systems was particularly important in helping to defray the costs associated with the conference. Finally, the faculty and students in the Advanced Composites Research Laboratory at Wayne State University are to be congratulated for their able assistance during the conference and the associated laboratory tours.

RONALD F. GIBSON AND GOLAM M. NEWAZ
CO-CHAIRMEN

Analysis and Design

Buckling Analysis of Thick Laminated Plates: Higher-Order Theory with Rotatory Moments

C. W. BERT AND M. MALIK

ABSTRACT

The higher-order shear deformable plate theory presented here concerns the buckling of laminated composite plates. The theory extends Reddy's higher-order theory (1984) to include the effect of rotatory moments sometimes called curvature terms. General equations of equilibrium in terms of stress resultants and boundary conditions are derived for plates laminated of orthotropic layers. Subsequently, the equations of equilibrium and boundary conditions are also obtained in terms of kinematic variables for the special case of symmetrically laminated cross-ply plates.

INTRODUCTION

It is now a well known fact that in many situations, the classical thin plate theory is quite inadequate for prediction of flexural behavior of plates. The theory grossly overestimates the buckling loads of the plates. This inadequacy results essentially from the neglect of transverse shear deformation; the effect of neglect of transverse shear deformation is most exemplified in composite plates in which, very often, the shear moduli happen to be quite small in comparison to the in-plane elastic moduli. Within the confines of applied elasticity (or mechanics of materials) approaches, a first-order shear deformable plate theory (FSDPT) offers the simplest way of accounting for the effect of transverse shear deformation. The assumed in-plane displacement functions with only first-order terms of the through-thickness coordinate in the FSDPT lead to uniform transverse shear strains through the plate thickness. The effect of nonuniformity of transverse shear strains is therein incorporated through the use of so-called shear correction factors. In the higher-order shear deformable plate theories (HSDPT's), the in-plane displacement functions (and sometimes the out-of-plane displacement function as well) are assumed with higher order terms of the through-thickness coordinate. Consequently, in the HSDPT's the nonuniformity of transverse shear strains is built in by itself and one doesn't need the shear correction factors. However, the HSDPT's are considerably more complicated than the FSDPT particularly in terms of cumbersome algebra. Of the various available higher-order theories, Reddy's theory (1984) is possibly the one most widely used.

Charles W. Bert, School of Aerospace and Mechanical Engineering, The University of Oklahoma, Norman, OK 73019-0601
Moinuddin Malik, School of Aerospace and Mechanical Engineering, The University of Oklahoma, Norman, OK 73019-0601

While including the effects of transverse shear deformation by the first- and higher-order theories, one finds that the rotatory inertia terms due to cross-sectional rotations come into effect naturally in the vibration analysis of plates. However, in the shear-flexibility-based buckling analysis, the moment terms due to cross-sectional rotations appear when one considers second-order strains in the potential energy due to in-plane stresses. These terms, introduced by Sun (1972, 1973) and later elaborated upon in the works of Dawe and Craig (1986), Whitney (1987), and Bert (1995) have been referred to as 'curvature terms'. These terms, in analogy with the rotatory inertia terms of the vibration problem may possibly be termed more appropriately as the 'rotatory moments' (Bert and Malik, 1997). Interestingly, the rotatory moment terms have not been considered in any of the available higher-order theories.

The motivation to the present work comes from a recent work of Bert and Malik (1997). Using the FSDPT, this work provided a comparison of the relative effects of transverse shear deformation and rotatory moments on buckling characteristics of symmetrically laminated cross-ply plates with various types of boundary conditions. Interestingly, the results also indicated that by including rotatory moments, the first-order theory yields critical load values of simply supported plates very close to those from the theory-of-elasticity solutions (Noor 1975); in fact, these comparisons were better than those of critical load values based on Reddy's theory (Reddy, 1984; Khdeir, 1988). Accordingly, the present work offers a higher-order theory for buckling of thick laminated plates wherein the effect of rotatory moments is included.

The present theory is developed within the framework of Reddy's theory (Reddy, 1984). The readers will find, admittedly, much similarity in the development with Reddy's work. Nevertheless, the present authors believe that this is the first higher-order theory of buckling of laminated plates in which rotatory moments are considered.

DEVELOPMENT OF THE THEORY

KINEMATICS OF DEFORMATION

The starting point of Reddy's theory is the hypothesis on the kinematics of deformation. The theory sets out the displacement field in a manner that the resulting stress field satisfies the conditions of vanishing transverse shear stresses on the surface of a plate laminated of orthotropic layers. Thus, the displacement components u, v, w in the x, y, z directions, respectively, are given in the form (Reddy, 1984)

$$\begin{aligned} u &= u_o + z\{\phi - \tfrac{4}{3}(\tfrac{z}{h})^2(\phi + w_{,x})\} \\ v &= v_o + z\{\psi - \tfrac{4}{3}(\tfrac{z}{h})^2(\psi + w_{,y})\} \\ w &= w(x, y) \end{aligned} \tag{1}$$

Here, in the above displacement field, u_o, v_o are the in-plane (x, y) displacement components of any point on the mid-plane $z = 0$; ϕ and ψ are the rotation components of a normal on the mid-plane in the y and x directions, respectively; and h is the thickness of the plate. It may be noted that the displacement field is described in terms of five independent kinematic variables: u_o, v_o, w, ϕ, and ψ. Also, as a usual notation, the subscripts of a variable following a comma indicate spatial derivatives with respect to that variable.

The displacement field (1) leads to the following strain field

$$\left.\begin{array}{ll}\epsilon_1 = u_{,x} = \epsilon_1^o + z(\kappa_1^o + z^2\kappa_1^2), & \epsilon_2 = v_{,y} = \epsilon_2^o + z(\kappa_2^o + z^2\kappa_2^2),\\ \epsilon_3 = w_{,z} = 0, & \\ \epsilon_4 = v_{,z} + w_{,y} = \epsilon_4^o + z^2\kappa_4^2, & \epsilon_5 = u_{,z} + w_{,x} = \epsilon_5^o + z^2\kappa_5^2,\\ \epsilon_6 = u_{,y} + v_{,x} = \epsilon_6^o + z(\kappa_6^o + z^2\kappa_6^2) & \end{array}\right\} \tag{2}$$

where

$$\left.\begin{array}{lll}\epsilon_1^o = u_{o,x}, & \epsilon_2^o = v_{o,y}, & \\ \epsilon_4^o = \psi + w_{,y}, & \epsilon_5^o = \phi + w_{,x}, & \epsilon_6^o = u_{o,y} + v_{o,x},\\ \kappa_1^o = \phi_{,x}, & \kappa_2^o = \psi_{,y}, & \kappa_6^o = \phi_{,y} + \psi_{,x},\\ \kappa_1^2 = -\frac{4}{3h^2}\epsilon_{5,x}^o, & \kappa_2^2 = -\frac{4}{3h^2}\epsilon_{4,y}^o, & \\ \kappa_4^2 = -\frac{4}{h^2}\epsilon_4^o, & \kappa_5^2 = -\frac{4}{h^2}\epsilon_5^o, & \kappa_6^2 = -\frac{4}{3h^2}(\epsilon_{4,x}^o + \epsilon_{5,y}^o)\end{array}\right\} \tag{3}$$

EQUATIONS OF EQUILIBRIUM

Consider a rectangular laminate subjected to in-plane compressive loading p_x on the sides $x = 0$ and a and p_y on the sides $y = 0$ and b, where the loadings are per unit side lengths. The equations of equilibrium of the plate in the state of neutral stability are set out via the principle of virtual displacements. The principle of virtual displacements may be stated as

$$\begin{aligned}&\int_{z=-h/2}^{h/2}\int_{A_{xy}} (\sigma_1\delta\epsilon_1 + \sigma_2\delta\epsilon_2 + \sigma_4\delta\epsilon_4 + \sigma_5\delta\epsilon_5 + \sigma_6\delta\epsilon_6)dA_{xy}dz\\ &\quad - \frac{1}{h}\int_{z=-h/2}^{h/2}\int_{A_{xy}} (p_x\delta\epsilon_1^{\text{NL}} + p_y\delta\epsilon_2^{\text{NL}})dA_{xy}dz = 0\end{aligned} \tag{4}$$

where

$$\int_{A_{xy}} (\,)dA_{xy} = \int_{x=0}^{a}\int_{y=0}^{b} (\,)dx\,dy$$

In (4), the second integral term comes from virtual work due to the nonlinear strains; it is in the inclusion of this term that the above equation differs from that of Reddy (1984). The nonlinear strain terms are given by

$$\epsilon_1^{\text{NL}} = \frac{1}{2}(u_{,x}^2 + v_{,x}^2 + w_{,x}^2), \quad \epsilon_2^{\text{NL}} = \frac{1}{2}(u_{,y}^2 + v_{,y}^2 + w_{,y}^2) \tag{5}$$

In order to proceed further, it is convenient to define the following stress resultants

$$\left.\begin{array}{l}(N_i, M_i, P_i) = \int_{-h/2}^{h/2} \sigma_i(1, z, z^3)dz; \;\; i = 1,2,6\\ (Q_1, R_1) = \int_{-h/2}^{h/2} \sigma_5(1, z^2)dz\\ (Q_2, R_2) = \int_{-h/2}^{h/2} \sigma_4(1, z^2)dz\end{array}\right. \tag{6}$$

The two kinds of resultants P_i and R_i arise as a consequence of the higher-order theory.

The equations of equilibrium are now obtained by integrating (4) by parts and using (5) and (6), and subsequently, collecting the coefficients of virtual displacements. Thus, the equations of equilibrium in terms of stress resultants are

$$\left.\begin{array}{ll}
\delta u_o: & N_{1,x}+N_{6,y}=p_x u_{o,xx}+p_y u_{o,yy} \\
\delta v_o: & N_{6,x}+N_{2,y}=p_x v_{o,xx}+p_y v_{o,yy} \\
\delta w: & \frac{4}{3h^2}(P_{1,xx}+2P_{6,xy}+P_{2,yy})+Q_{1,x}+Q_{2,y}-\frac{4}{h^2}(R_{1,x}+R_{2,y}) \\
& =p_x w_{,xx}+p_y w_{,yy}-\underline{\frac{4h^2}{315}[(p_x\phi_{,xx}+p_y\phi_{,yy})_{,x}+(p_x\psi_{,xx}+p_y\psi_{,yy})_{,y}]} \\
& -\frac{h^2}{252}[p_x w_{,xxxx}+(p_x+p_y)w_{,xxyy}+p_y w_{,yyyy}] \\
\delta\phi: & M_{1,x}+M_{6,y}-\frac{4}{3h^2}(P_{1,x}+P_{6,y})-Q_1+\frac{4}{h^2}R_1 \\
& =\underline{\frac{17h^2}{315}(p_x\phi_{,xx}+p_y\phi_{,yy})}+\frac{4h^2}{315}(p_x w_{,xx}+p_y w_{,yy})_{,x} \\
\delta\psi: & M_{6,x}+M_{2,y}-\frac{4}{3h^2}(P_{6,x}+P_{2,y})-Q_2+\frac{4}{h^2}R_2 \\
& =\underline{\frac{17h^2}{315}(p_x\psi_{,xx}+p_y\psi_{,yy})}+\frac{4h^2}{315}(p_x w_{,xx}+p_y w_{,yy})_{,y}
\end{array}\right\} \quad (7)$$

The rotatory moment terms in (7) are underlined. It is noted that the rotatory moment terms appear in the third (δw) equilibrium equation, as well as the fourth ($\delta\phi$) and fifth ($\delta\psi$) equations. This indicates that there is a coupling which is not present in FSDPT (Dawe and Craig, 1986, etc.).

In the above process, one also obtains the boundary conditions. The conditions on the $x=0$ and a sides are described by the following:

$$\left.\begin{array}{ll}
(1) & N_1-p_x u_{o,x}=0 \ \underline{\text{or}}\ \delta u_o=0 \\
(2) & N_6-p_x v_{o,x}=0 \ \underline{\text{or}}\ \delta v_o=0 \\
(3) & M_1-\frac{4}{3h^2}P_1-\frac{h^2}{315}p_x(17\phi_{,x}+4w_{,xx})=0 \ \underline{\text{or}}\ \delta\phi=0 \\
(4) & M_6-\frac{4}{3h^2}P_6-\frac{h^2}{315}p_x(17\psi_{,x}+4w_{,xy})=0 \ \underline{\text{or}}\ \delta\psi=0 \\
(5) & \frac{4}{3h^2}P_1+\frac{h^2}{63}p_x(\frac{4}{5}\phi_{,x}+\frac{1}{4}w_{,xx})=0 \ \underline{\text{or}}\ \delta w_{,x}=0 \\
(6) & \frac{4}{3h^2}P_6+\frac{2h^2}{315}(p_x\psi_{,x}+p_y\phi_{,y})+\frac{h^3}{504}(p_x+p_y)w_{,xy}=0 \ \underline{\text{or}}\ \delta w_{,y}=0 \\
(7) & \frac{4}{3h^2}(P_{1,x}+P_{6,y})+Q_1-\frac{4}{h^2}R_1-p_x w_{,x} \\
& \frac{h^2}{63}[p_x(\frac{4}{5}\phi_{,xx}+\frac{2}{5}\psi_{,xy}+\frac{1}{8}w_{,xyy}+\frac{1}{4}w_{,xxx})+p_y(\frac{2}{5}\phi_{,yy}+\frac{1}{8}w_{,xyy})]=0 \ \underline{\text{or}}\ \delta w=0
\end{array}\right\} \quad (8)$$

and, the conditions on $y=0$ and b sides are described by the following:

$$\left.\begin{array}{ll}
(1) & N_6-p_y u_{o,y}=0 \ \underline{\text{or}}\ \delta u_o=0 \\
(2) & N_2-p_y v_{o,y}=0 \ \underline{\text{or}}\ \delta v_o=0 \\
(3) & M_6-\frac{4}{3h^2}P_6-\frac{h^2}{315}p_y(17\phi_{,y}+4w_{,xy})=0 \ \underline{\text{or}}\ \delta\phi=0 \\
(4) & M_2-\frac{4}{3h^2}P_2-\frac{h^2}{315}p_y(17\psi_{,y}+4w_{,yy})=0 \ \underline{\text{or}}\ \delta\psi=0 \\
(5) & \frac{4}{3h^2}P_6+\frac{2h^2}{315}(p_x\psi_{,x}+p_y\phi_{,y})+\frac{h^2}{504}(p_x+p_y)w_{,xy}=0 \ \underline{\text{or}}\ \delta w_{,x}=0 \\
(6) & \frac{4}{3h^2}P_2+\frac{h^2}{63}p_y(\frac{4}{5}\psi_{,y}+\frac{1}{4}w_{,yy})=0 \ \underline{\text{or}}\ \delta w_{,y}=0 \\
(7) & \frac{4}{3h^2}(P_{2,y}+P_{6,x})+Q_2-\frac{4}{h^2}R_2-p_y w_{,y} \\
& \frac{h^2}{63}[p_x(\frac{2}{5}\psi_{,xx}+\frac{1}{8}w_{,xxy})+p_y(\frac{2}{5}\phi_{,xy}+\frac{4}{5}\psi_{,yy}+\frac{1}{8}w_{,xxy}+\frac{1}{4}w_{,yyy})]=0 \ \underline{\text{or}}\ \delta w=0
\end{array}\right\} \quad (9)$$

STRESS RESULTANTS

In order to apply the equations of equilibrium (7) and the boundary conditions (8) and (9) to a laminate having variously oriented orthotropic layers, it is convenient to describe these equations in terms of the five independent kinematic variables. For this purpose, one actually needs to express the stress resultants in terms of u_o, v_o, w, ϕ, and ψ. Consequently, the stress field in a layer of the laminate is expressed in terms of the strain field (2) through the following constitutive equations of a typical layer (Whitney, 1987)

$$\begin{Bmatrix} \sigma_1 \\ \sigma_2 \\ \sigma_6 \end{Bmatrix}^\ell = \begin{bmatrix} Q_{11} & Q_{12} & Q_{16} \\ Q_{21} & Q_{22} & Q_{26} \\ Q_{61} & Q_{62} & Q_{66} \end{bmatrix}^\ell \begin{Bmatrix} \epsilon_1 \\ \epsilon_2 \\ \epsilon_6 \end{Bmatrix}^\ell \tag{10}$$

and

$$\begin{Bmatrix} \sigma_4 \\ \sigma_5 \end{Bmatrix}^\ell = \begin{bmatrix} Q_{44} & Q_{45} \\ Q_{54} & Q_{55} \end{bmatrix}^\ell \begin{Bmatrix} \epsilon_4 \\ \epsilon_5 \end{Bmatrix}^\ell \tag{11}$$

where Q_{ij} are a set of symmetric plane-stress transformed-stiffness coefficients. In the above equations, the superscript ℓ refers to a layer of the laminate.

The stress resultants are obtained by using (10) and (11) in (6). Thus, the stress resultants due to the in-plane stresses σ_1, σ_2, and σ_6 are

$$\begin{Bmatrix} \{N\} \\ \{M\} \\ \{P\} \end{Bmatrix} = \begin{bmatrix} [A] & [B] & [E] \\ [B] & [D] & [F] \\ [E] & [F] & [H] \end{bmatrix} \begin{Bmatrix} \{\epsilon^0\} \\ \{\kappa^0\} \\ \{\kappa^2\} \end{Bmatrix} \tag{12}$$

whcre, in the above, the sub-columns and sub-matrices are of orders 3×1 and 3×3, respectively, comprising the elements formed by subscripts 1, 2, and 6. Also,

$$(A_{ij}, B_{ij}, D_{ij}, E_{ij}, F_{ij}, H_{ij}) = \int_{-h/2}^{h/2} Q_{ij}^\ell(1, z, z^2, z^3, z^4, z^6)dz \quad (i, j = 1, 2, 6) \tag{13}$$

are the plate stiffnesses. The first three stiffnesses are the usual ones and the last three are consequences of the higher-order theory.

The stress resultants due to the out-of-plane transverse shear stresses σ_4 and σ_5 are

$$\begin{Bmatrix} Q_2 \\ Q_1 \\ R_2 \\ R_1 \end{Bmatrix} = \begin{bmatrix} A_{44} & A_{45} & D_{44} & D_{45} \\ A_{54} & A_{55} & D_{54} & D_{55} \\ D_{44} & D_{45} & F_{44} & F_{45} \\ D_{54} & D_{55} & F_{54} & F_{55} \end{bmatrix} \begin{Bmatrix} \epsilon_4^0 \\ \epsilon_5^0 \\ \kappa_4^2 \\ \kappa_5^2 \end{Bmatrix} \tag{14}$$

where

$$(A_{ij}, D_{ij}, F_{ij}) = \int_{-h/2}^{h/2} Q_{ij}^\ell(1, z^2, z^4)dz \quad (i, j = 4, 5) \tag{15}$$

are the remaining plate stiffnesses. It needs to be noted that in the right sides of (13) and (15), integration in each layer and subsequent summation for all layers of the laminate is implied. Further, the plate stiffnesses, defined in (13) and (15), are sets of symmetric coefficients.

BUCKLING OF A SYMMETRIC CROSS-PLY LAMINATE

Here, as a matter of practical interest, expressions for stress resultants, the equations of equilibrium, and boundary conditions for the commonly used configuration of symmetrically laminated cross-ply plates are given. For this type of configuration, it may be seen easily from (13) and (15) that

$$\begin{aligned} &B_{ij} = E_{ij} = 0 \quad (i,j = 1,2,4,5,6) \\ &A_{i6} = D_{i6} = F_{i6} = H_{i6} = 0 \quad (i = 1,2) \\ &A_{45} = D_{45} = F_{45} = 0 \end{aligned} \tag{16}$$

and the expressions for the stress resultants in terms of the kinematic variables may be obtained as

$$\left.\begin{aligned} N_1 &= A_{11}u_{o,x} + A_{12}v_{o,y} \\ N_2 &= A_{12}u_{o,x} + A_{22}v_{o,y} \\ N_6 &= A_{66}(u_{o,y} + v_{o,x}) \\ M_1 &= (D_{11} - \tfrac{4}{3h^2}F_{11})\phi_{,x} + (D_{12} - \tfrac{4}{3h^2}F_{12})\psi_{,y} - \tfrac{4}{3h^2}(F_{11}w_{,xx} + F_{12}w_{,yy}) \\ M_2 &= (D_{12} - \tfrac{4}{3h^2}F_{12})\phi_{,x} + (D_{22} - \tfrac{4}{3h^2}F_{22})\psi_{,y} - \tfrac{4}{3h^2}(F_{12}w_{,xx} + F_{22}w_{,yy}) \\ M_6 &= (D_{66} - \tfrac{4}{3h^2}F_{66})(\phi_{,y} + \psi_{,x}) - \tfrac{8}{3h^2}F_{66}w_{,xy} \\ P_1 &= (F_{11} - \tfrac{4}{3h^2}H_{11})\phi_{,x} + (F_{12} - \tfrac{4}{3h^2}H_{12})\psi_{,y} - \tfrac{4}{3h^2}(H_{11}w_{,xx} + H_{12}w_{,yy}) \\ P_2 &= (F_{12} - \tfrac{4}{3h^2}H_{12})\phi_{,x} + (F_{22} - \tfrac{4}{3h^2}H_{22})\psi_{,y} - \tfrac{4}{3h^2}(H_{12}w_{,xx} + H_{22}w_{,yy}) \\ P_6 &= (F_{66} - \tfrac{4}{3h^2}H_{66})(\phi_{,y} + \psi_{,x}) - \tfrac{8}{3h^2}H_{66}w_{,xy} \\ Q_1 &= (A_{55} - \tfrac{4}{h^2}D_{55})(\phi + w_{,x}) \\ Q_2 &= (A_{44} - \tfrac{4}{h^2}D_{44})(\psi + w_{,y}) \\ R_1 &= (D_{55} - \tfrac{4}{h^2}F_{55})(\phi + w_{,x}) \\ R_2 &= (D_{44} - \tfrac{4}{h^2}F_{44})(\psi + w_{,y}) \end{aligned}\right\} \tag{17}$$

Using the plate constitutive equations (17) in the general equilibrium equations (7), one obtains the five equilibrium equations of symmetrically laminated cross-ply plates in terms of the five independent kinematic variables. These are

$$\left.\begin{array}{l}
\delta u_o: A_{11}u_{o,xx}+A_{66}u_{o,yy}+(A_{12}+A_{66})v_{o,xy}=p_x u_{o,xx}+p_y u_{o,yy}\\[2ex]
\delta v_o: (A_{12}+A_{66})u_{o,xy}+A_{66}v_{o,xx}+A_{22}v_{o,yy}=p_x v_{o,xx}+p_y v_{o,yy}\\[2ex]
\delta w: \frac{4}{3h^2}\{(F_{11}-\frac{4}{3h^2}H_{11})\phi_{,xxx}+[(F_{12}+2F_{66})-\frac{4}{3h^2}(H_{12}+2H_{66})](\phi_{,y}+\psi_{,x})_{,xy}\\
(F_{22}-\frac{4}{3h^2}H_{22})\psi_{,yyy}-\frac{4}{3h^2}H_{11}w_{,xxxx}-\frac{8}{3h^2}(H_{12}+2H_{66})w_{,xxyy}-\frac{4}{3h^2}H_{22}w_{,yyyy}\}\\
(A_{55}-\frac{8}{h^2}D_{55}+\frac{16}{h^4}F_{55})(\phi+w_{,x})_{,x}+(A_{44}-\frac{8}{h^2}D_{44}+\frac{16}{h^4}F_{44})(\psi+w_{,y})_{,y}\\
=p_x w_{,xx}+p_y w_{,yy}-\frac{4h^2}{315}[(p_x\phi_{,xx}+p_y\phi_{,yy})_{,x}+(p_x\psi_{,xx}+p_y\psi_{,yy})_{,y}]\\
-\frac{h^2}{252}[p_x w_{,xxxx}+(p_x+p_y)w_{,xxyy}+p_y w_{,yyyy}]\\[2ex]
\delta\phi: (D_{11}-\frac{8}{3h^2}F_{11}+\frac{16}{9h^4}H_{11})\phi_{,xx}+(D_{66}-\frac{8}{3h^2}F_{66}+\frac{16}{9h^4}H_{66})\phi_{,yy}\\
[(D_{12}+D_{66})-\frac{8}{3h^2}(F_{12}+F_{66})+\frac{16}{9h^4}(H_{12}+H_{66})]\psi_{,xy}\\
-\frac{4}{3h^2}\{(F_{11}-\frac{4}{3h^2}H_{11})w_{,xxx}+[(F_{12}+2F_{66})-\frac{4}{3h^2}(H_{12}+2H_{66})]w_{,xyy}\}\\
-(A_{55}-\frac{8}{h^2}D_{55}+\frac{16}{h^4}F_{55})(\phi+w_{,x})=\frac{h^2}{315}[17(p_x\phi_{,xx}+p_y\phi_{,yy})+4(p_x w_{,xx}+p_y w_{,yy})_{,x}]\\[2ex]
\delta\psi: [(D_{12}+D_{66})-\frac{8}{3h^2}(F_{12}+F_{66})+\frac{16}{9h^4}(H_{12}+H_{66})]\phi_{,xy}\\
+(D_{66}-\frac{8}{3h^2}F_{66}+\frac{16}{9h^4}H_{66})\psi_{,xx}+(D_{22}-\frac{8}{3h^2}F_{22}+\frac{16}{9h^4}H_{22})\psi_{,yy}\\
-\frac{4}{3h^2}\{[(F_{12}+2F_{66})-\frac{4}{3h^2}(H_{12}+2H_{66})]w_{,xxy}+\frac{4}{3h^2}(F_{22}-\frac{4}{3h^2}H_{22})w_{,yyy}\}\\
-(A_{44}-\frac{8}{h^2}D_{44}+\frac{16}{h^4}F_{44})(\psi+w_{,y})=\frac{h^2}{315}[17(p_x\psi_{,xx}+p_y\psi_{,yy})+4(p_x w_{,xx}+p_y w_{,yy})_{,y}]
\end{array}\right\} \tag{18}$$

Further, using (17) in (8) and (9), one obtains the boundary conditions. Thus the boundary conditions on the $x=0$ and a edges are specified as follows.

$$\left.\begin{array}{ll}
(1) & A_{11}u_{o,x}+A_{12}v_{o,y}=p_x u_{o,x}\ \underline{\text{or}}\ \delta u_o=0\\
(2) & A_{66}(u_{o,y}+v_{o,x})=p_x v_{o,x}\ \underline{\text{or}}\ \delta v_o=0\\
(3) & (D_{11}-\frac{8}{3h^2}F_{11}+\frac{16}{9h^4}H_{11})\phi_{,x}+(D_{12}-\frac{8}{3h^2}F_{12}+\frac{16}{9h^4}H_{12})\psi_{,y}\\
& -\frac{4}{3h^2}[(F_{11}-\frac{4}{3h^2}H_{11})w_{,xx}+(F_{12}-\frac{4}{3h^2}H_{12})w_{,yy}]\\
& =\frac{h^2}{315}p_x(17\phi_{,x}+4w_{,xx})\ \underline{\text{or}}\ \delta\phi=0\\
(4) & (D_{66}-\frac{8}{3h^2}F_{66}+\frac{16}{9h^4}H_{66})(\phi_{,y}+\psi_{,x})-\frac{8}{3h^2}(F_{66}-\frac{4}{3h^2}H_{66})w_{,xy}\\
& =\frac{h^2}{315}p_x(17\psi_{,x}+4w_{,xy})\ \underline{\text{or}}\ \delta\psi=0\\
(5) & -(F_{11}-\frac{4}{3h^2}H_{11})\phi_{,x}-(F_{12}-\frac{4}{3h^2}H_{12})\psi_{,y}\\
& +\frac{4}{3h^2}(H_{11}w_{,xx}+H_{12}w_{,yy})=\frac{h^4}{84}p_x(\frac{4}{5}\phi_{,x}+\frac{1}{4}w_{,xx})\ \underline{\text{or}}\ \delta w_{,x}=0\\
(6) & -(F_{66}-\frac{4}{3h^2}H_{66})(\phi_{,y}+\psi_{,x})+\frac{8}{3h^2}H_{66}w_{,xy}\\
& \frac{h^4}{210}(p_x\psi_{,x}+p_y\phi_{,y})+\frac{h^4}{672}(p_x+p_y)w_{,xy}\ \underline{\text{or}}\ \delta w_{,y}=0\\
(7) & -\frac{4}{3h^2}\{(F_{11}-\frac{4}{3h^2}H_{11})\phi_{,xx}+(F_{66}-\frac{4}{3h^2}H_{66})\phi_{,yy}\\
& +[(F_{12}+F_{66})-\frac{4}{3h^2}(H_{12}+H_{66})]\psi_{,xy}\}-(A_{55}-\frac{8}{h^2}D_{55}+\frac{16}{h^4}F_{55})(\phi+w_{,x})\\
& +\frac{16}{9h^4}[H_{11}w_{,xxx}+(H_{12}+2H_{66})w_{,xyy}]=\frac{h^2}{63}[p_x(\frac{4}{5}\phi_{,xx}+\frac{2}{5}\psi_{,xy}\\
& +\frac{1}{8}w_{,xyy}+\frac{1}{4}w_{,xxx})+p_y(\frac{2}{5}\phi_{,yy}+\frac{1}{8}w_{,xyy})]-p_x w_{,x}\ \underline{\text{or}}\ \delta w=0
\end{array}\right\} \tag{19}$$

The boundary conditions on the $y = 0$ and b edges are specified as follows.

$$\left.\begin{array}{ll}
(1) & A_{66}(u_{o,y} + v_{o,x}) = p_y u_{o,y} \ \underline{\text{or}}\ \delta u_o = 0 \\
(2) & A_{12}u_{o,x} + A_{22}v_{o,y} = p_y v_{o,y} \ \underline{\text{or}}\ \delta v_o = 0 \\
(3) & (D_{66} - \frac{8}{3h^2}F_{66} + \frac{16}{9h^4}H_{66})(\phi_{,y} + \psi_{,x}) - \frac{8}{3h^2}(F_{66} - \frac{4}{3h^2}H_{66})w_{,xy} \\
 & = \frac{h^2}{315}p_y(17\phi_{,y} + 4w_{,xy}) \ \underline{\text{or}}\ \delta\phi = 0 \\
(4) & (D_{12} - \frac{8}{3h^2}F_{12} + \frac{16}{9h^4}H_{12})\phi_{,x} + (D_{22} - \frac{8}{3h^2}F_{22} + \frac{16}{9h^4}H_{22})\psi_{,y} \\
 & -\frac{4}{3h^2}[(F_{12} - \frac{4}{3h^2}H_{12})w_{,xx} + (F_{22} - \frac{4}{3h^2}H_{22})w_{,yy}] \\
 & = \frac{h^2}{315}p_x(17\psi_{,y} + 4w_{,yy}) \ \underline{\text{or}}\ \delta\psi = 0 \\
(5) & -(F_{66} - \frac{4}{3h^2}H_{66})(\phi_{,y} + \psi_{,x}) + \frac{8}{3h^2}H_{66}w_{,xy} \\
 & = \frac{h^4}{210}(p_x\psi_{,x} + p_y\phi_{,y}) + \frac{h^4}{672}(p_x + p_y)w_{,xy} \ \underline{\text{or}}\ \delta w_{,x} = 0 \\
(6) & -(F_{12} - \frac{4}{3h^2}H_{12})\phi_{,x} - (F_{22} - \frac{4}{3h^2}H_{22})\psi_{,y} \\
 & +\frac{4}{3h^2}(H_{12}w_{,xx} + H_{22}w_{,yy}) = \frac{h^4}{84}p_y(\frac{4}{5}\psi_{,x} + \frac{1}{4}w_{,yy}) \ \underline{\text{or}}\ \delta w_{,y} = 0 \\
(7) & -\frac{4}{3h^2}\{[(F_{12} + F_{66}) - \frac{4}{3h^2}(H_{12} + H_{66})]\phi_{,xy} \\
 & +(F_{66} - \frac{4}{3h^2}H_{66})\psi_{,xx} + (F_{22} - \frac{4}{3h^2}H_{22})\psi_{,yy}\} \\
 & -(A_{44} - \frac{8}{h^2}D_{44} + \frac{16}{h^4}F_{44})(\psi + w_{,y}) + \frac{16}{9h^4}[H_{22}w_{,yyy} + (H_{12} + 2H_{66})w_{,xxy}] \\
 & = \frac{h^2}{63}[p_x(\frac{2}{5}\psi_{,xx} + \frac{1}{8}w_{,xxy}) + p_y(\frac{2}{5}\phi_{,xy} + \frac{4}{5}\psi_{,yy} + \frac{1}{8}w_{,xxy} + \frac{1}{4}w_{,yyy})] - p_y w_{,y} \\
 & \underline{\text{or}}\ \delta w = 0
\end{array}\right\} \quad (20)$$

CONCLUDING REMARKS

In the present work, a higher-order theory for buckling of thick laminated plates, was presented. The theory was based on Reddy's higher-order theory. The novelty of the present theory is that it includes the effect of rotatory moments (curvature terms). Due to page limitations, numerical results could not be included here. However, such results are expected to be published in some future research work.

REFERENCES

Bert, C. W. 1995. "Shear Deformation and Sandwich Configuration," Chapter 5 in *Buckling and Postbuckling of Composite Plates*, by G. J. Turvey and I. H. Marshall (editors), Chapman & Hall, London.

Bert, C. W. and M. Malik. 1997. "On the Buckling Analysis of Symmetrically-Laminated Cross-Ply Plates," *Mechanics of Composite Materials and Structures*, 4(1):39-67.

Dawe D. J. and T. J. Craig. 1986. "The Vibration and Stability of Symmetrically-Laminated

Composite Rectangular Plates Subjected to In-Plane Stresses," *Composite Structures*, 5(2):281-307.

Khdeir, A. A. 1988. "Free Vibration and Buckling of Symmetric Cross-Ply Laminated Plates by an Exact Method," *Journal of Sound and Vibration*, 126(3):447-461.

Noor, A. K. 1975. "Stability of Multilayered Composite Plates," *Fibre Science and Technology*, 8(1):81-89.

Reddy, J. N. 1984. "A Simple Higher-Order Theory for Laminated Composite Plates," *ASME Journal of Applied Mechanics*, 51(4):745-752.

Sun, C. T. 1972. "On the Equations for a Timoshenko Beam under Initial Stress," *ASME Journal of Applied Mechanics*, 39(1):282-285.

Sun, C. T. 1973. "Incremental Deformations in Orthotropic Laminated Plates under Initial Stress," *ASME Journal of Applied Mechanics*, 40(1):193-200.

Whitney, J. M. 1987. "Curvature Effects in the Buckling of Symmetrically-Laminated Rectangular Plates with Transverse Shear Deformation," *Composite Structures*, 8(1):85-103.

Whitney, J. M. 1985. "Structural Analysis of Laminated Plates," Technomic Publishing, Lancaster, PA.

Use of Effective Engineering Constants for Composite Laminates Optimized for Buckling Loads

S. VENKATARAMAN, R. T. HAFTKA AND T. F. JOHNSON

ABSTRACT

Optimum designs of composite laminates in structural optimization are often found to be poor in performance as the optimization often involves a large number of poorly understood failure modes. Additionally the designs are highly dependent on assumed loading directions. Industry often reduces the risk of failure by employing sub-laminates (often chosen to have fixed ratios of ply orientations) as building blocks for their laminate designs. These sub-laminates have been well understood through testing and analysis. Since the ratios of ply orientations is predetermined, the in-plane stiffnesses of the sub-laminates have invariant properties and hence designers often simplify their modeling by replacing the composite laminate by a homogenized orthotropic material with "*effective engineering constants*". For a stiffened composite panel it is shown while the error for initial designs incurred by using effective properties is only 20%, the error of optimized panels using such approach can be as high as 40%. An anti-optimization of laminate lay-ups, performed to obtain an estimate of the maximum error, revealed a maximum error of about 180% for optimized 16 ply laminate .The results indicate that the use of effective material properties can result in large errors in buckling loads. Errors are particularly large for high aspect ratio laminates. For such laminates a minimum of three repeating sub-laminates should be present before the error becomes negligible.

INTRODUCTION

The design of composite laminates often involves a number of poorly understood failure modes. Additionally, optimum designs are highly dependent on assumed loading directions. Consequently, when laminate lay-ups are designed, there is substantial risk of

---

Satchi Venkataraman, Graduate Student, Department of Aerospace Engineering, Mechanics and Engineering Science, University of Florida, Gainesville, FL 32611-6250.
Raphael T. Haftka, Professor, Department of Aerospace Engineering, Mechanics and Engineering Science, University of Florida, Gainesville, FL 32611-6250.
Theodore F. Johnson, Aerospace Technologist, Thermal Structures Branch, NASA Langley Research Center, Hampton, VA 23681-0001.

failure due to poorly modeled failure modes and/or off-design loads. Industry often reduces the failure risk by employing as building blocks sub-laminates which have been well understood through testing and analysis. These standard sub-laminates, while usually not quasi-isotropic, also possess substantial strength for every possible loading direction. This provides insurance against off-design conditions. The authors have often had to use such lay-ups in their work in the design of a liquid hydrogen tank (LH2) for a reusable launch vehicle on the recommendation of their industrial collaborators (Rockwell and McDonnell Douglas).

Often sub-laminates used for design purposes are chosen such that they have fixed ratios of various ply orientations. Since the stiffnesses of a sub-laminate with chosen rations of ply orientation have invariant properties, designers simplified their modeling by replacing the composite laminate (Schmidt , 1977 and Vitali, 1996) by a homogenized orthotropic material whose properties are obtained from the in-plane stiffness terms (see Appendix-A) of the laminate. The effective engineering constants for the laminate or sub-laminate are used in calculations of the stiffness matrix (ABD matrix) of laminates constructed from the sub-laminates. While this does not cause any error in the in-plane stiffness (A-matrix) it can cause substantial errors in the calculation of bending stiffness (D-matrix) terms when only few sub-laminates are used. While it may be evident that this simplifying assumption introduces some errors, the magnitude of such errors is not universally appreciated. The present paper discusses some of the errors and the worst case estimates that arise from using such an approximation in calculating buckling loads for composite laminate plates.

MOTIVATION

Interest in the effect of using equivalent or effective stiffness of laminates arose from an analysis and optimization of a J-stiffened panel of a LH2 tank construction. Figure 1 illustrates some stiffener geometries used in our design studies and some of the associated terminology. The details of the panel geometry and optimization problem are discussed in the following section.

PANEL GEOMETRY AND LOADING

The panel optimized was a 180 degree span 192 inch radius cylindrical panel stiffened longitudinally (stringers) and circumferentially (rings). The axial stiffeners or stringers were J type stiffeners (Figure 1) attached internally to the cylinder walls. The ring stiffeners were T type stiffeners. T stiffeners are the same as I-beam stiffeners except for the fact the top and bottom flanges are of different width. The stiffener geometry for the J-stiffener is characterized by different segments as illustrated in Figure 2. The panel is designed to withstand buckling failure under an axial compression of 1741 lb/in and a safety factor of 1.4. Details of the material properties and stacking sequence of the panel used for the analysis and design are shown in Tables A1 and A2, in Appendix - A. The optimization of the panels did not include any parameters associated with the ring stiffeners such as ring spacing, ring sizing or ring segment ply lay-ups. However, the rings were included in the model so as to account for their strengthening effect.

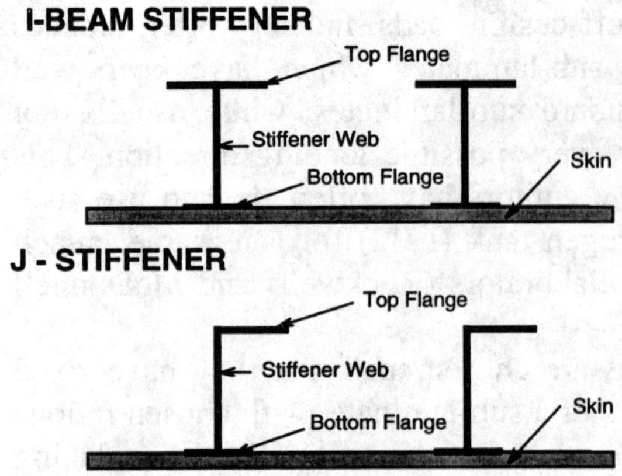

Figure 1. Stiffener geometry and associated terminology

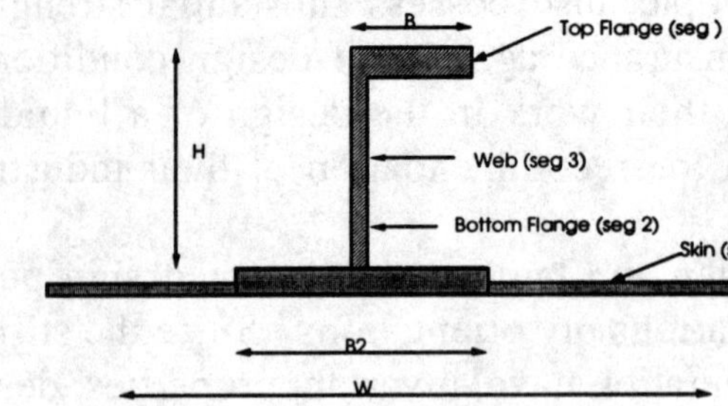

Figure 2. J-Stiffener geometry with terminology

Optimization results of the stiffened panel using the PANDA2 program (Bushnell, 1987), with stiffener geometry, spacing is presented in Table 1. The original sub-laminates are replaced by laminates with optimized lay-ups. This reduced the weight from 1484 lb to 1257 lb but resulted in laminate lay-ups with all plies oriented in the direction of loading. It is common knowledge that such a laminate would perform very poorly with hoop tension or shear loading. Also such thick unidirectional laminates are prone to matrix cracking.

TABLE 1 - PANEL DESIGN OPTIMIZED FOR STRINGER SIZE AND SPACING.

Description		Value
W	Stringer (STR) spacing	4.557 in
B	STR bottom flange width	1.165 in
B2	STR top flange width	0.581 in
H	Stringer height	1.326 in
Weight	Panel weight	1484 lb

In order to study the effect of using equivalent engineering constants for the laminates, two different approaches were used in the PANDA2 modeling of the panels. In the first approach the orientation and the properties of each ply were taken into account. The second approach used the equivalent laminate properties. Analysis of the initial design (Table A-2) with the two different approaches revealed a significant difference in the buckling load factor and buckling mode as can be seen from results in Table 2.

TABLE 2 - ANALYSIS OF INITIAL DESIGN USING EFFECTIVE STIFFNESS AND EXACT MODELING

Comparison of buckling load factor and failure modes	
Method used	Critical buckling load factor and failure modes
Exact laminate lay-up model	1.2528 (axial torsional buckling of skin-stringer segment)
Equivalent stiffness model	0.994 (simple support local buckling of panel) (error = -21.1%)

The panel modeled using exact ply lay-up details was found to be critical under torsional buckling of the skin stringer segment, and the model with equivalent stiffness was critical in local buckling. The two panel models were also optimized with the stiffener parameters such as stringer spacing, stiffener height and stiffener flange widths as design variables. The optimal designs obtained in the above two cases are presented in Table 3.

TABLE 3 - OPTIMIZATION RESULTS OF PANELS USING EFFECTIVE STIFFNESS AND EXACT MODELING

Panel Optimization Results		
Geometry variable	Using model with exact laminate lay-up	Using effective laminate properties
Stiffener spacing (in)	4.557	4.823
Stiffener height (in)	1.326	2.035
Bottom flange width (in)	1.165	1.324
Top flange width (in)	0.5806	0.683
Panel weight (lb)	1484	1665

For the panels modeled with effective material properties the stringers are less closely spaced, have a greater height and a larger top flange width as well. The panel optimized with equivalent stiffness is found to be heavier by about 180 lb. The comparison of these two cases strongly indicates that this is an effect of the reduced bending stiffness due to using the laminate equivalent stiffness calculated based on in-plane stiffness terms. The results obtained using the two approaches were analyzed using the opposite approach and are compared (Table 4). The optimization of the structure has accentuated the modeling errors. The errors for the buckling load have increased from 20% for the initial design to almost 40% for the optimized panels. A common occurrence for optimization program is to exploit errors in modeling to produce grossly wrong results (e.g. Haftka and Gürdal, 1992, pp 4-5).

TABLE 4 - ANALYSIS OF OPTIMIZED PANELS USING ALTERNATE APPROACH.

MODELING APPROACH USED FOR THE OPTIMIZATION	BUCKLING LOAD FROM ANALYSIS OF OPTIMIZED PANEL USING	
	EXACT APPROACH	EFFECTIVE STIFFNESS
EXACT APPROACH	1.4	0.93 (error of -33.6%)
EFFECTIVE STIFFNESS	2.38	1.4 (error of 41.2%)

EFFECT OF EQUIVALENT LAMINATE PROPERTIES ON BUCKLING OF AN UNSTIFFENED PLATE

In order to understand the effect of using equivalent engineering constants better, a simple rectangular unstiffened plate geometry was considered. The effect of equivalent stiffness in buckling load prediction of the plate at different aspect ratios was investigated. For comparison purposes two laminates with identical in-plane stiffness were chosen. The lay-up of the two laminates were $[+45/-45/0/90]_s$ and $[90/0/-45/+45]_s$. The two laminates have identical plies but different stacking sequence. The terms of the bending stiffness matrix (D-matrix) of the two laminates are compared with the D-matrix calculated for the laminate using its equivalent engineering constants that were calculated from the in-plane stiffness terms (Table-4). As expected, the largest difference is in D_{66} which is mostly influenced by the ±45° plies. For the [±45/0/90]s these plies are far from the center, leading to a large D_{66}, while for the [90/0/-45/+45]s they are close to the center, thus leading to small D_{66}. The D_{22} term is mostly influenced by the 90° plies so that it is much larger when these plies are on the outside. However there is also substantial contribution from the ±45° plies, which mitigates the difference between the laminates. Finally the D_{11} term is mostly influenced by the 0 plies which change position by the least amount between the two laminates. Furthermore, the effect of small move in position of 0° ply is counterbalanced by the opposite move of the ±45° plies which also contribute to D_{11}. Consequently there is very little difference in D_{11} between the laminates.

The buckling loads of a rectangular plate 100 inches in width (Figure 3) and loaded in biaxial compression of 100 lb/in and having simply supported boundary conditions were calculated for the two laminates and compared with the buckling load obtained from using the effective engineering constants (identical in-plane stiffness). The expression for the critical buckling load factor is given by:

$$\lambda_{cr} = \frac{\pi^2\left[D11\left(\frac{m}{a}\right)^4 + 2(D12+2D66)\left(\frac{m}{a}\right)^2\left(\frac{n}{b}\right)^2 + D22\left(\frac{n}{b}\right)^4\right]}{\left[N_x\left(\frac{m}{a}\right)^2 + N_y\left(\frac{n}{b}\right)^2\right]} \quad (1)$$

where,

λcr is the critical buckling load factor.
D_{ij} are the bending stiffness matrix terms.
N_x and N_y are the loads in the axial and transverse directions respectively.
m, n are the number of half waves in the direction of load N_x and N_y respectively.

The error in buckling loads due to equivalent laminate properties is defined as,

$$\%Error = 100\frac{(\lambda cr_{approx} - \lambda cr_{exact})}{\lambda cr_{exact}} \quad (2)$$

where,

λcr_{appro} is the buckling load factor calculated using equivalent laminate properties
λcr_{exac} is the buckling load factor calculated using exact ply representation.

The buckling loads of the rectangular plate were calculated for aspect ratios of 0.1 to 100 and the error in the buckling load calculated using the D-matrix obtained from equivalent stiffness are plotted (Fig. 4). For very large aspect ratios the D_{22} terms determine the buckling load, so that the [+45/-45/0/90]s laminate which overestimates D_{22} also overestimates the buckling load. For a square or near square plate, the D_{66} is the most important term (because of it coefficient of 4) and this is underestimated for the above laminate and hence the buckling load is also underestimated. The trends are reversed for the other laminate. For very small aspect ratios D_{11} dominates, and since it is the most accurately estimated bending stiffness coefficient, the error is small.

TABLE 5 - COMPARISON OF D-MATRIX TERMS FOR LAMINATES WITH IDENTICAL IN-PLANE PROPERTIES WITH D-MATRIX CALCULATED USING EFFECTIVE PROPERTIES.

Comparison of D-matrix terms for laminates with identical in-plane properties with D-matrix calculated from effective engineering constants for laminate (lb-in)		
Ply lay-up [+45 / -45 / 0 / 90] s	Ply lay-up [90 / 0 / -45 / +45]s	Laminate modeled with equivalent properties
42.711 24.639 7.692	41.830 5.007 2.564	47.399 14.823 0.0
24.639 32.455 7.692	5.007 72.599 2.564	14.823 47.399 0.0
7.692 7.692 26.104	2.564 2.564 6.472	0.0 0.0 16.288

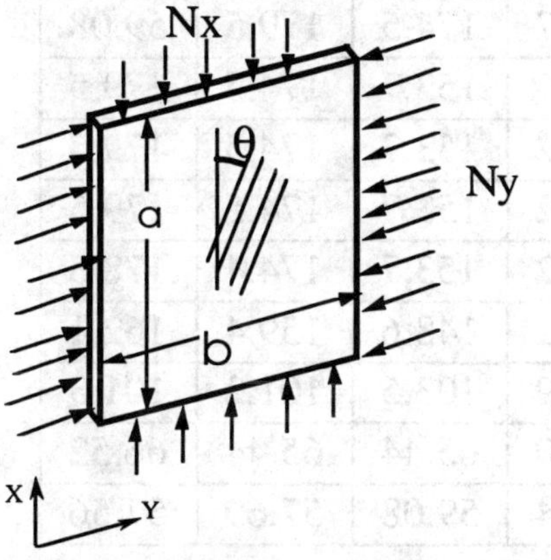

Aspect ratio = a/b

FIGURE 3. Rectangular plate in biaxial compression .

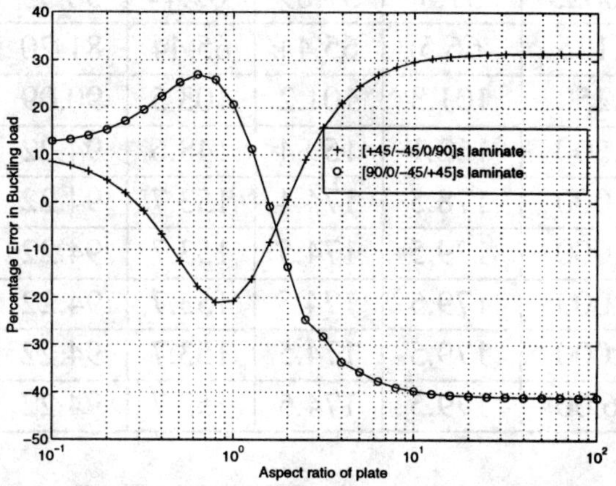

FIGURE 4. Error in buckling load estimation due to using effective material properties for quasi-isotropic laminates, Nx =Ny=100lb/in

ANTI-OPTIMIZATION OF LAY-UPS FOR MAXIMIZING ERROR FROM USING EFFECTIVE PROPERTIES

Next, the error in buckling load was calculated for the laminates for varying aspect ratios (a/b) and loading ratios (Ny/Nx). The aspect ratio and loading ratio were varied from one sixteenth to sixteen, with increments by a factor of 2.0 giving a possibility of 81 cases in all. However the symmetry of the loading and aspect ratio make some of these cases repetitive. For each geometry and loading, the worst errors were obtained by optimization of the stacking sequence. This process of optimizing for highest error is also known as anti-optimization (van Wamelen, 1993). The anti-optimization of the laminate lay-ups was performed using a Genetic Algorithm (GA). The design variables of the genetic algorithm were ply-orientations that are restricted to a discrete set of angles namely, ±45°, 0°, and 90°. The total number of plies in the laminate were a constant. Furthermore the basic unit of coding in the genetic algorithm represented a stack of two plies 0_2, 90_2 or ±45. This was used as a convenience to ensure all laminates obtained were balanced. The objective was to maximize the error in buckling load (Eqn. 4) due to the use of effective engineering constants.

The maximum value of the errors in buckling load factor obtained using effective engineering constants for the optimized 16 ply laminate ($[\theta_1 / \theta_2 / \theta_3 / \theta_4]_s$, where θ_i is a 0_2, 90_2 or ± 45 two ply stack) are shown in Table 6. For this 16-ply (4 design variables) laminate the maximum value of the errors are found to be as high as almost 180% and the minimum value of errors in this worst case scenario is still near 20% which is the case for an aspect and loading ratio of unity.

TABLE 5. PERCENTAGE ERROR IN BUCKLING LOAD ESTIMATE FOR WORST CASE LAMINATE OPTIMIZED FOR ONE SUB-LAMINATE IN THE TOTAL THICKNESS.

Load	Aspect Ratio (a/b)								
Nx/Ny	0.0625	0.125	0.250	0.500	1.000	2.000	4.000	8.000	16.00
0.0625	57.5	57.69	65.44	50.55	94.22	153.7	174.5	179.5	59.08
0.125	66.5	65.44	65.44	81.20	50.55	94.22	153.7	174.5	179.5
0.250	101.3	101.2	103.5	90.09	40.57	94.22	153.7	174.5	179.5
0.500	162.1	159.4	148.6	94.22	24.03	94.22	153.7	174.5	179.5
1.000	178.5	174.4	153.7	94.22	21.50	94.22	153.7	174.4	178.5
2.000	179.5	174.5	153.7	94.22	24.03	94.22	148.6	159.4	162.1
4.000	179.5	174.5	153.7	94.22	40.57	90.09	103.5	101.2	101.3
8.000	179.5	174.5	153.7	94.22	50.55	81.20	65.44	65.44	66.52
16.00	179.5	174.5	153.7	94.22	50.55	65.44	59.08	57.69	57.56

The large errors shown in Table 6 (as well as in Figure 4) are for the extreme case where the laminate has only one sub-laminate unit. In most practical situations the total thickness is a multiple (larger than 1) of the basic sub-laminate thickness. A select number of cases from Table 6 were chosen for further investigation of the effect of total thickness on the error. The cases chosen were such that, in the laminate obtained from optimization, the percentage of plies in any given direction did not exceed 50%. This was to avoid cases where a majority of plies were in one direction as they are not useful for practical purposes. The lay-ups of three such cases are shown in Table 7. The error in buckling load for a laminate with n sub-laminate (n=1, 2, ..., 5)are calculated for the chosen cases. It is seen for thin laminates (n=1) the errors are very high but as soon as they are doubled (n=2) or tripled (n=3) the error reduces by an order of magnitude for each increment (cases 1, 2). The case of a quasi-isotropic laminate is also presented in Table 7. Case 3 was the worst case laminate obtained for an aspect ratio and load ratio of unity. This laminate that resulted in only a 20% error for the given aspect ratio was found to be much worse (error of 70%) for an aspect ratio of 16.0.

Hence it is observed that the use of effective material properties obtained by homogenizing the lay-up of a composite laminate or sub-laminate can result in significantly large errors (Table 5). However when the number of sub-laminates are increased the plies are more evenly distributed through the laminate thickness and hence the errors are small when modeled with effective constants (which were obtained from homogenizing the laminate).

TABLE 7. EFFECT OF INCREASING THICKNESS (MULTIPLES OF SUB-LAMINATES) OF LAMINATE ON BUCKLING LOAD PREDICTION ERROR USING EFFECTIVE PROPERTIES.

Case No.	N_x/N_y	a/b	Worst case sub-laminate (16 plies)	% Error in buckling load for multiple sub-laminates				
				n=1	n=2	n=3	n=4	n=5
1	0.5	1/16	$[90_4 / \pm45 / 0_2]s$	162.1	16.54	6.57	3.55	2.24
2	0.5	0.5	$[90_4 / 0_4]s$	94.22	13.80	5.70	3.21	1.98
	1.0	1.0		21.50	4.63	2.00	1.11	0.71
3	1.0	16.0	$[90_4 / \pm45_2]s$	8.28	8.34	3.88	2.22	1.44
	1.0	1/16		68.32	11.29	4,72	2.60	1.65
5	1.0	10	$[\underline{+}45_2 / 0_2 /90_2]s$	34.34	6.83	2.92	1.62	1.03
6	1.0	10	$[90_2 / 0_2 /\underline{+}45_2]s$	27.88	9.92	4.67	2.68	1.73

TABLE 8. PERCENTAGE ERROR IN BUCKLING LOAD ESTIMATE FOR WORST CASE LAMINATE OPTIMIZED FOR 2 UNITS OF THE SUB-LAMINATE IN THE TOTAL THICKNESS.

Load	Aspect Ratio (a/b)								
Nx/Ny	0.0625	0.125	0.250	0.500	1.000	2.000	4.000	8.000	16.00
0.0625	9.64	9.15	10.52	10.97	9.16	13.80	17.85	18.90	18.35
0.1250	10.65	10.97	10.97	12.62	9.16	13.80	17.85	18.90	19.13
0.2500	11.34	11.45	12.62	13.56	4.63	13.80	17.85	18.90	19.13
0.5000	16.54	16.53	17.57	13.80	4.63	13.80	17.85	18.90	19.13
1.0000	18.73	18.82	17.85	13.80	4.63	13.80	17.85	18.82	18.73
2.0000	19.13	18.90	17.85	13.80	4.63	13.80	17.57	16.53	16.54
4.0000	19.13	18.90	17.85	13.80	4.63	13.56	12.62	11.45	11.34
8.0000	19.13	18.90	17.85	13.80	9.16	12.62	10.97	10.97	10.65
16.000	19.13	18.90	17.85	13.80	9.16	10.97	10.52	9.65	9.64

The worst case laminates with two units of the basic sub-laminate were obtained for the aspect ratios and load ratios as before by optimization of sub-laminate stacking sequence. Unlike before where the lay-up was optimized for a single sub-laminate and then error calculated for a laminate with two such units, the sub-laminate is optimized for the error in buckling load of the laminate with two units. The maximum value of error corresponding to the given aspect ratio and loading ratio for the laminates are shown in Table 8. It was found the magnitude of error (a maximum of about 20%) was very similar to the case when it was optimized for a single sub-laminate and analyzed for a multiple of two such units in the laminate.

CONCLUSION

In this work we demonstrated that using equivalent or effective material properties for laminates can lead to large errors in buckling loads, both conservative and non-conservative. For single sub-laminates at high aspect ratios these errors can be as high as

180%. The error decays rapidly as the number of sub-laminates are increased, with a maximum error of 20% for two stacks and 7% for three stacks. Hence the approximation error with effective material properties is a more significant for thin laminates. However, using such approximations in optimization for minimum weight structures under buckling constraints could lead to serious design errors, as optimization programs are notorious for exploiting model errors.

REFERENCES

Bushnell D., (1987), "PANDA2 - Program for Minimum Weight Design of Stiffened Composite Locally Buckled Panels," *Computers and Structures*, Vol. 25, No. 4, pp. 469-605.

Haftka, R.T. and Gürdal, Z. (1992), Elements of Structural Optimization, 3ed., Kluwer Academic Publishers.

Jones, R. M., (1975), Mechanics of Composite Materials, Hemisphere Publishing Corporation.

Schmidt, L.A. and Farshi, B., "Optimum Design of Laminated Fibre Composite Plates," *Intl. Journal for Numerical Methods in Engineering*, Vol. 11, 1977, pp. 623-640.

van Wamelen, A., Haftka R.T. and Johnson, E.R., "Optimum Lay-ups of Composite Specimens to Accentuate the Differences between Competing Failure Criteria," presented at the 8th Technical Conference on Composite Materials, American Society of Composites, Cleveland, OH., October 19-21, 1993.

Vitali, R., Park, O., Haftka, R.T., Sankar, B. and Rose, S., "Optimization of a Hat Stiffened Panel by Response Surface Techniques," Paper no. 1997-1151, to be presented at the 38th Structures, Structural Dynamics, and Materials Conference, April 7-10, 1997, Kissimee, FL.

ACKNOWLEDGMENT

This work was supported by the NASA Langley Research Center under grants NAG-1-1669 and NAG-1-1808. The authors would like to express their thanks to Dr. David Bushnell of Lockheed Palo Alto Research Labs for his constant support with using the PANDA2 program.

APPENDIX - A

DATA FOR COMPARISON OF EQUIVALENT PROPERTIES AND EXACT LAY-UP MODELS FOR J-STIFFENED PANELS AND QUASI-ISOTROPIC PLATES.

TABLE A-1. LAMINA PROPERTIES USED FOR J-STIFFENED PANEL.

Parameter	Lamina type	
	Tape	Fabric
Thickness of single ply (in)	0.005	0.0143
Elastic modulus in fiber direction, E_1 (psi)	2.15E07	1.16E+07
Elastic modulus in transverse direction, E2 (psi)	1.08E06	1.09E+07
Poisson ratio ν_{12}	0.3	0.06
Shear modulus G_{12} (psi)	6.0E05	1.12E6
Thermal expansion coef. in fiber dir α_1 (in/in/•F)	-1.0E-7	1.12E-06
Thermal expansion coef. in trans dir α_2(in/in/•F)	1.7E-05	1.73E-06
Max. allowable tensile stress in fiber dir. (psi)	222300	113700
Max. allowable tensile stress in trans. dir. (psi)	176300	74800
Max. allowable comp. stress in fiber dir. (psi)	7200	103900
Max. allowable comp. stress in fiber dir. (psi)	36000	75600
Max. allowable shear stress (psi)	10900	16320

TABLE A-2. PLY STACKING SEQUENCE OF DESIGN USED FOR ANALYSIS AND OPTIMIZATION

	Description	Value
seg- 1	Skin	$[+65_{3T}]_S$ (12 plies)[1]
W_1	Stringer (STR) spacing	5.0266 in
B_1	STR bottom flange width	1.500 in
$B2_1$	STR top flange width	0.600 in
	Stringer height	1.500 in
seg-2	Stringer bottom flange	$[0_{2F}]_S$ (2 plies)
seg-3	Stringer web	$[0_{1F}\ 0_{4T}\ 45_{1F}\ 0_{4T}\ 45_{1F}\ 0_{4T}\ 0_{1F}]_S$ (16 plies)
seg-4	Stringer top flange	$[0_{1F}\ 0_{4T}\ 45_{1F}\ 0_{4T}\ 45_{1F}\ 0_{4T}\ 0_{1F}]_S$ (16 plies)
W2	Ring spacing	31.447 in
B_2	Ring bottom flange width	1.0 in
$B2_2$	Ring top flange width	2.0 in
H_2	Ring height	9.0 in
seg 5	Bottom flange of ring	$[45_{1F}\ 0_{4T}\ 0_{1F}\ 0_{2T}]_S$ (16 plies)
seg-6	Web of ring	$[45_{1F}\ 0_{2F}\ 45_{1F}]_T$ (4 plies)
seg-7	Ring top flange	$[45_{1F}\ 0_{4T}\ 45_{1F}\ 0_{2T}]_S$ (16 plies)
Weight	Weight (lb.)	1496 lb

[1] Subscript T and F denote pre-preg Tape lamina and woven Fabric lamina, respectively.

APPENDIX - A

CALCULATION OF EFFECTIVE ENGINEERING CONSTANTS FOR COMPOSITE LAMINATES

Effective engineering constants are obtained by making the assumption that the composite laminate is a homogeneous orthotropic material and equating the actual in-plane bending stiffness matrix terms (A_{ij}) to the in-plane stiffness constants for a single layered equivalent material laminate. For a symmetric balanced laminate the engineering constants are obtained from the expressions,

$$\frac{E_x t}{1-\upsilon_{xy}\upsilon_{xy}} = A_{11}, \qquad \text{(A.1)}$$

$$\frac{E_y t}{1-\upsilon_{xy}\upsilon_{xy}} = A_{22}, \qquad \text{(A.2)}$$

$$\frac{\upsilon_{yx} E_x t}{1-\upsilon_{xy}\upsilon_{yx}} = A_{12} \qquad \text{(A.3)}$$

$$G_{xy} t = A_{66} \qquad \text{(A.4)}$$

where,

E_x, E_y are the effective elastic modulus of laminate in two orthogonal directions x,y
ν_{xy}, Poisson's ratio of the laminate
G_{xy} the in-plane shear modulus of the laminate and,
t is the total thickness of the laminate.

The in-plane laminate stiffness A_{ij} of the laminate are calculated from the classical lamination theory as follows.

$$A_{ij} = \sum_{k=1}^{n} [\overline{Q}_{ij}]_k t_k, \qquad (2)$$

where $[\overline{Q}_{ij}]_k$ and t_k are, respectively the transformed stiffness matrix and thickness of the kth ply

Elastic Buckling of Orthotropic Plates under Varying Axial Stresses

A. BADIR, H. HU AND A. DIALLO

ABSTRACT

The elastic buckling load of simply supported rectangular orthotropic plates subjected to a second degree parabolic variation of axial stresses in the longitudinal direction is calculated using analytical methods. The variation of axial stresses is equilibrated by nonuniform shear stresses along the plate edges and transverse normal stresses. The influence of the aspect ratio is examined, and the results are compared with plates subjected to uniform axial stresses.

INTRODUCTION

In most studies of stability of plates, the axial stress has been taken as uniform compression throughout flat rectangular plates (Timoshenko and Gere, 1961; Allen and Bulson, 1980; Whitney, 1987). Buckling of isotropic plates under a compressive stress that varies linearly from one loaded edge to the other has been studied by Libove, Ferdman and Reusch (1949). Cases of practical interest exist, however, in which the axial stress is not uniform but varies from tension at both loaded edges to compression in the middle. An example is the stability of the crown of the hat stiffened panel, a candidate configuration of the upper and lower skin of the Blended Wing Body (BWB) Aircraft. The BWB Aircraft is an advanced long-range ultra-high-capacity airliner with the principal feature being the wide double-deck body which is blended into the wing (Popular Science Magazine, 1995).

In the present paper, analytical methods are used to investigate the local stability of the hat crown plate in order to minimize its weight while optimizing its buckling strength. A varying tension-compression-tension stress is induced in the crown of the stiffeners. The axial stresses vary longitudinally due to bending, and the change in stresses is equilibrated by nonuniform shear stresses along the plate edges and transverse normal stresses as shown in Fig. 1.

ANALYTICAL MODEL

The distribution of the axial stresses is given by

Ashraf Badir, Assistant Professor, Department of Engineering, Clark Atlanta University, Atlanta, GA 30314
Hurang Hu, Research Associate, Department of Engineering, Clark Atlanta University, Atlanta, GA 30314
Abdouramane Diallo, Graduate Student, Department of Mathematics, Clark Atlanta University, Atlanta, GA 30314

$$\sigma_x(x) = \left[1 - 6\left(\frac{x}{a}\right) + 6\left(\frac{x}{a}\right)^2\right]\sigma_{x}\Big|_{x=0} \tag{1}$$

and the expressions for the shear and transverse normal stresses $\tau_{xy}(x,y)$ and $\sigma_y(y)$, are obtained by integrating the plane stress equilibrium equations

$$\frac{\partial \sigma_x}{\partial x} + \frac{\partial \tau_{xy}}{\partial y} = 0 \tag{2}$$

$$\frac{\partial \sigma_y}{\partial y} + \frac{\partial \tau_{xy}}{\partial x} = 0 \tag{3}$$

and are found to be

$$\tau_{xy}(x,y) = -3\frac{b}{a}(1 - \frac{2x}{a})(1 - \frac{2y}{b})\sigma_{x}\Big|_{x=0} \tag{4}$$

$$\sigma_y(y) = 6\left(\frac{b}{a}\right)^2 \frac{y}{b}\left(\frac{y}{b} - 1\right)\sigma_{x}\Big|_{x=0} \tag{5}$$

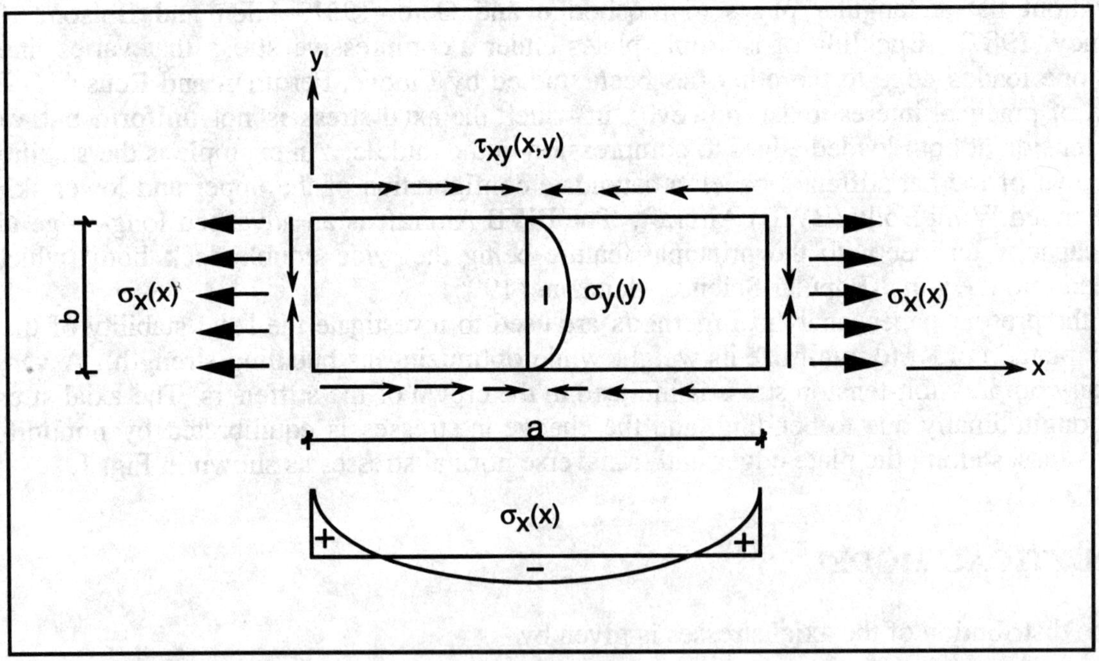

Figure 1. Plate subjected to tensile/compression/tensile longitudinal stresses $\sigma_{xx}(x)$, equilibrated by shear stresses $\tau_{xy}(x,y)$ and transverse stresses $\sigma_{yy}(y)$

The differential equation of the deflection surface of the orthotropic plate can be written as follows:

$$D_{11}\frac{\partial^4 w}{\partial x^4}+2(D_{12}+2D_{66})\frac{\partial^4 w}{\partial x^2 \partial y^2}+D_{22}\frac{\partial^4 w}{\partial y^4}-N_x\frac{\partial^2 w}{\partial x^2}-2N_{xy}\frac{\partial^2 w}{\partial x \partial y}-N_y\frac{\partial^2 w}{\partial y^2}=0 \tag{6}$$

where D_{ij}, w, N_x, N_{xy}, and N_y are the bending stiffnesses of the orthotropic plate (Jones, 1975), vertical displacement, axial, shear and transverse stress resultants, respectively. A solution to Eq. (6) that satisfies the boundary conditions cannot be obtained in closed form. Thus, approximate solutions are sought. Two different displacement fields are considered: (1) Double Fourier series, using Rayleigh-Ritz method, and (2) a polynomial in the longitudinal direction, using Galerkin method. A summary of the analytical models is described in the next sections, the influence of the aspect ratio is examined, and the results are compared with plates subjected to uniform axial loads.

DOUBLE FOURIER SERIES (RAYLEIGH-RITZ METHOD)

The total potential energy functional of the orthotropic plate is given by

$$U=\frac{1}{2}\int_0^a\int_0^b\left\{D_{11}\left(\frac{\partial^2 w}{\partial x^2}\right)^2+2D_{12}\frac{\partial^2 w}{\partial x^2}\frac{\partial^2 w}{\partial y^2}+D_{22}\left(\frac{\partial^2 w}{\partial y^2}\right)^2+4D_{66}\left(\frac{\partial^2 w}{\partial x \partial y}\right)^2+\sigma_x t\left(\frac{\partial w}{\partial x}\right)^2+\sigma_y t\left(\frac{\partial w}{\partial y}\right)^2+2\tau_{xy}t\frac{\partial w}{\partial x}\frac{\partial w}{\partial y}\right\}dy\,dx \tag{7}$$

where $t, a,$ and b represent the thickness, length and width of the plate, respectively.

Of all conceivable buckling patterns satisfying boundary conditions, the actual buckling pattern is that for which the potential energy U, as given by Eq. (7), is a minimum. The deflection surface of the simply supported skin in bending may be represented by a Double Fourier series as

$$w=\sum_{m=1}^{M}\sum_{n=1}^{N}a_{mn}\sin\frac{m\pi x}{a}\sin\frac{n\pi y}{b} \tag{8}$$

which satisfies the simply supported boundary conditions. Rayleigh-Ritz method requires that the coefficients a_{mn} be chosen such that U is a minimum. That is, they must satisfy the equations

$$\frac{\partial U}{\partial a_{ij}}=0 \quad (i=1,2\ldots M;\quad j=1,2\ldots N) \tag{9}$$

The deflection expression (8) can be splitted into two independent subsets, one corresponding to even values of n and the other, to odd values of n. The first subset corresponds to buckling that is antisymmetric about the line $y = b/2$ and the second, to buckling that is symmetric about that line. The second subset gives lower buckling loads, and attention is hereinafter confined to it. Similar argument can be made to the buckling pattern in the longitudinal direction (subscript m). Therefore, the buckling pattern is chosen as

$$w = \sum_{m=1,3,5\ldots}^{M} \sum_{n=1,3,5\ldots}^{N} a_{mn} \sin\frac{m\pi x}{a} \sin\frac{n\pi y}{b} \tag{10}$$

Substituting Eqs. (1), (4-5), and (10) into the energy expression (7), while using Eq. (9), we obtain the following relations

$$A_{mn}a_{mn} + \sum_{i=1,3,5\ldots}^{M} B_i a_{in} + \sum_{j=1,3,5\ldots}^{N} C_j a_{mj} + \sum_{i=1,3,5\ldots}^{M} \sum_{j=1,3,5\ldots}^{N} D_{ijmn} a_{ij} = 0 \qquad \begin{cases} m = 1,3,5\ldots M \\ n = 1,3,5\ldots N \end{cases} \tag{11}$$

where

$$A_{mn} = \frac{D_{11}\pi^2}{N_0 b^2}\left(\frac{m^4}{\beta^2} + \frac{2(D_{12} + 2D_{66})}{D_{11}} m^2 n^2 + \frac{D_{22}}{D_{11}} n^4 \beta^2\right) + m^2$$

$$\beta_i = mi \begin{cases} -1, & m = i \\ \dfrac{12}{\pi^2}\left[\dfrac{1}{(m+i)^2} + \dfrac{1}{(m-i)^2}\right], & |m \pm i| = even \end{cases}$$

$$C_j = nj \begin{cases} -1, & n = j \\ \dfrac{12}{\pi^2}\left[\dfrac{1}{(n+j)^2} + \dfrac{1}{(n-j)^2}\right], & |n \pm j| = even \end{cases}$$

$$D_{ijmn} = \frac{1}{\pi^2} \begin{cases} -6, & m = i, n = j \\ \dfrac{96mnij}{\left(m^2 - i^2\right)\left(n^2 - j^2\right)}, & m \neq i, n \neq j, |m \pm i| = even, |n \pm j| = even \end{cases}$$

and

$$\beta = \frac{a}{b} \quad \text{and} \quad N_0 = t\sigma_x\big|_{x=0} \tag{12}$$

β and N_0 represent the aspect ratio and buckling load, respectively. A nondimensional buckling coefficient k may be expressed as

$$k = \frac{N_0 b^2}{D_{11}\pi^2} \tag{13}$$

Relations (11) represent a set of $(M+1)(N+1)/4$ linear equations. the smallest eigenvalue of Eqs. (11) gives the buckling load.

POLYNOMIAL (GALERKIN METHOD)

We seek a nontrivial solution to the deflection w which satisfies equation (6) and the boundary conditions. For simply supported plates, the solution of Eq. (6) may be represented by

$$w = \sum_{m=1}^{M}\sum_{n=1}^{N} q_{mn}\left(\xi^{m+3} + A_{m1}\xi^{m+2} + A_{m2}\xi^{m+1} + A_{m3}\xi^{m} + A_{m4}\xi^{m-1}\right)\sin(n\pi\eta) \tag{14}$$

in which $\xi = x/a$; $\eta = y/b$ and the coefficients $A_{m1} \sim A_{m4}$ are determined by satisfying the four simply supported boundary conditions at the two loaded sides.

The expression for the Galerkin equations is given by

$$\int_{\eta=0}^{1}\int_{\xi=0}^{1}\left\{D_{11}\left(\frac{\partial^2 w}{\partial x^2}\right)^2 + 2D_{12}\frac{\partial^2 w}{\partial x^2}\frac{\partial^2 w}{\partial y^2} + D_{22}\left(\frac{\partial^2 w}{\partial y^2}\right)^2 + 4D_{66}\left(\frac{\partial^2 w}{\partial x \partial y}\right)^2\right.$$

$$\left. + \sigma_x t\left(\frac{\partial^2 w}{\partial x^2}\right)^2 + \sigma_y t\left(\frac{\partial^2 w}{\partial y^2}\right)^2 + 2\tau_{xy} t\frac{\partial w}{\partial x}\frac{\partial w}{\partial y}\right\}\frac{\partial w}{\partial q_{mn}} d\xi\, d\eta = 0 \qquad (m = 1,2\ldots M;\ n = 1,2\ldots N) \tag{15}$$

Theses relationships represent a set of $M \times N$ homogeneous algebraic equations. In order to avoid a trivial solution ($q_{mn} = 0$), the nondimensional buckling load k is chosen such that the determinant of the coefficient matrix vanishes. The minimum value of k that satisfies the requirement is the buckling load.

TABLE I. MATERIAL PROPERTIES

E_{11}	E_{22}	G_{12}	υ_{12}
9.25 Msi	4.67 Msi	2.27 Msi	0.397

TABLE II. CONVERGENCE OF BUCKLING COEFFICIENT *k* FOR A SQUARE PLATE.

M/N	1	3	5	7
1	1.75	1.69	1.65	1.65
3	1.72	1.65	1.65	1.65
5	1.72	1.65	1.65	
7	1.72	1.65		
9	1.72			

RESULTS

First, a convergence study for Rayleigh-Ritz method is performed. Orthotropic plates with material properties as listed in Table I have been studied. For a square plate, buckling coefficients k versus increasing values of M and N are listed in Table II. It can be seen that when M=N=3, k reaches its final value. Convergence of the solution in this case is faster than in the case of long strips. For example a four-term series ($M = N = 3$, note that only odd M and N are considered) converges to the final k value compared to a 12-term solution ($M = 23$, $N = 1$) for a long strip with $\beta = 20$.

The buckling coefficient k versus aspect ratio β is plotted in Fig. 2. Figure 2 shows that for plates with aspect ratio less than six, the buckling coefficient k increases as β increases. However, for long strips ($\beta > 6$), k does not significantly change and takes a value of approximately 5.75.

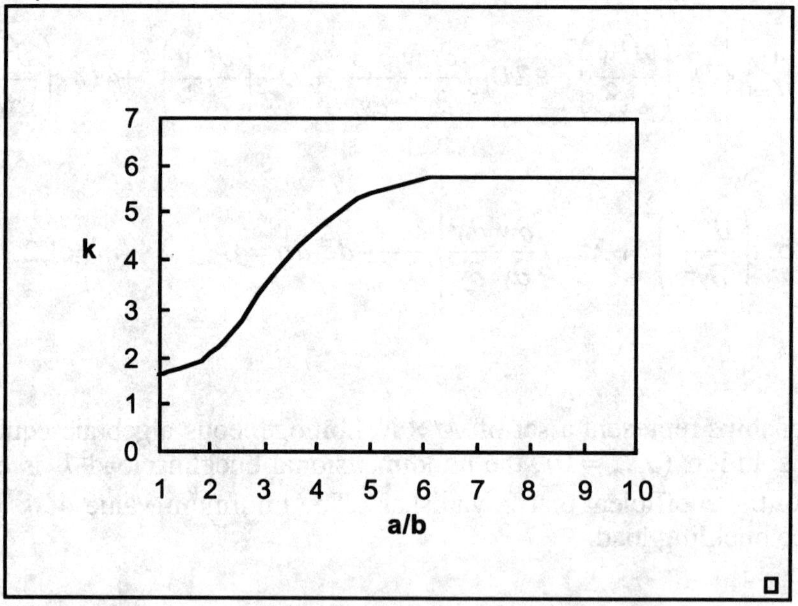

Figure 2. Buckling coefficient versus aspect ratio

TABLE III. CONVERGENCE OF BUCKLING COEFFICIENT FOR A SQUARE PLATE USING GALERKIN METHOD.

M	k
1	2.54
2	1.96
3	1.79
4	1.68
5	1.66

Galerkin method is also used to calculated the buckling load. The accuracy of the polynomial representation of the displacement field in the longitudinal direction, given by Eq. (14), is examined for a square plate. The values of the buckling coefficient k are listed in Table III, converging to a value of 1.66 (M=5, N=1) compared to 1.65 as obtained using Fourier Series (Rayleigh-Ritz method, Table II). For M=5 and N=1, the displacement w, Eq. (14) takes the following form

$$w = \Big[q_{11}\left(\xi^4 - 2\xi^3 + \xi\right) + q_{21}\left(3\xi^5 - 10\xi^3 + 7\xi\right) + q_{31}\left(7\xi^6 - 12\xi^5 + 5\xi^3\right) + q_{41}\left(\xi^7 - 6\xi^4 + 5\xi^3\right) + q_{5m}\left(4\xi^8 - 22\xi^5 + 18\xi^4\right)\Big]\sin(\pi\eta) \tag{16}$$

The buckling load intensity (load per unit width) calculated based on a parabolic and uniform load distribution is listed in Table IV. The parabolic load values represent the buckling load intensity at mid-span of the plate (i.e., $t\sigma_x\big|_{x=a/2}$ or $kD_{11}\pi^2/2b^2$).

Table IV shows that the buckling load intensity is influenced by both the load distribution and the aspect ratio. For a square plate, the buckling load calculated based on a uniform stress distribution is 244% higher than the buckling load calculated based on a tensile/compressive/tensile parabolic load distribution. The difference is only 5% (reduction)

TABLE IV. COMPARISON OF BUCKLING COEFFICIENTS FOR VARYING AXIAL STRESS AND UNIFORM STRESS

	Parabolic Load* (lb/in)		Uniform Load§ (lb/in)	% diff.
	Fourier Series	Polyno-mial	Exact	
a=b=150 in (a/b=1)	1.35†	1.36†	4.64	+244%
a=150 in. b=4.3 in. (a/b=35)	5773†	...	5482	-5%

* Associated with shear and transverse stresses as given in Eqs. (4) and (5).
§ No shear and transverse stresses.
† Buckling Load intensity at mid-span (x=a/2).

for long strips. One may expect that a tensile stress at the edges as shown in Fig. 1 will tend to restore the plate to its stable position and therefore increases its buckling load. However, the parabolic stress distribution is associated with both compressive and shear stresses. The transverses compressive stresses $\sigma_y(y)$ have a detrimental effect on the buckling load for square plates, while their effect is negligible for long strips.

CONCLUSIONS

The elastic buckling load of orthotropic plates subjected to a parabolically varying tension-compression-tension longitudinal stress, equilibrated by nonuniform shear stresses and transverse compressive stresses is analyzed. It is shown that the buckling load depends on the aspect ratio for plates with aspect ratio less than six. For square plates, the buckling load due to a uniform stress distribution is 244% higher than the buckling load calculated based on a tensile/compressive/tensile parabolic load distribution. However, the difference is only 5% (reduction) for long strips.

ACKNOWLEDGMENT

This work is supported by the NASA Langley Research Center under grant NAG-1-1714. This support is gratefully acknowledged.

REFERENCES

Allen, H. G., and P. S. Bulson, 1980. "Background to Buckling," McGraw-Hill Book Company.

Jones, R. M. 1975. "Mechanics of Composite Materials," McGraw-Hill, New York.

Libove, C., S. Ferdman, and J. Reusch. 1949. "NACA Tech. Note No. 1891."

Popular Science Magazine, April, 1995.

Timoshenko, S. P. and J. M. Gere, 1961. "Theory of Elastic Stability," 2nd ed. McGraw-Hill Book Company.

Whitney, J. M. 1987. "Structural Analysis of Laminated Anisotropic Plates," Technomic Publishing Company.

Skin-Stiffener Stresses in Composite Plates and Shallow Shells

M. E. ROBESON

ABSTRACT

The development of an approximate analysis for predicting stresses in the skin-stiffener interface region of composite plates and shallow shells is presented. The analysis determines interlaminar normal stress and transverse shear stresses in the direction of the stiffener axial coordinate and the direction of the inplane coordinate perpendicular to the stiffener axis. The analysis accounts for skin-stiffener interfaces with nonzero thickness and for curvature in the skin and stiffener. The interlaminar normal stress and transverse shear stresses at the interface are applied to the lower face of the skin and the upper face of the stiffener flange as unknown functions, and the displacements in the skin and flange are related through the deformation of the interface bond layer. The Rayleigh-Ritz method is used to determine the unknown functions that yield the interlaminar stresses at the skin-stiffener interface. Distributions of the interface stresses are presented and these results show that the stresses are greatest at the edge of the stiffener flange. In addition, results from the skin-stiffener analysis have been compared with results from other analyses, and all of these results were found to be in close agreement.

INTRODUCTION

Because of great potential for weight savings, advanced fiber-reinforced composite materials have become serious candidates for primary structural components of transport aircraft. To fully exploit the weight savings of these materials, the postbuckling strength of stiffened composite panel structures must be utilized and included in the design process. Thus, the postbuckling response of stiffened-panel structures under realistic combined loading conditions must be thoroughly understood for a wide range of design parameters. Achieving this goal requires the development of special purpose analyses that accurately predict behavioral trends. In these analyses, behavior associated with the strength limits and failure characteristics of stiffened composite panels should be considered in addition to the geometric nonlinearities that are required for postbuckling studies of thin-walled aircraft shell structures. Moreover, the behavior at the interface between the panel skin and attached stiffener is especially important.

Mark E. Robeson, Department of Mechanical Engineering, Old Dominion University, Norfolk, VA 23529

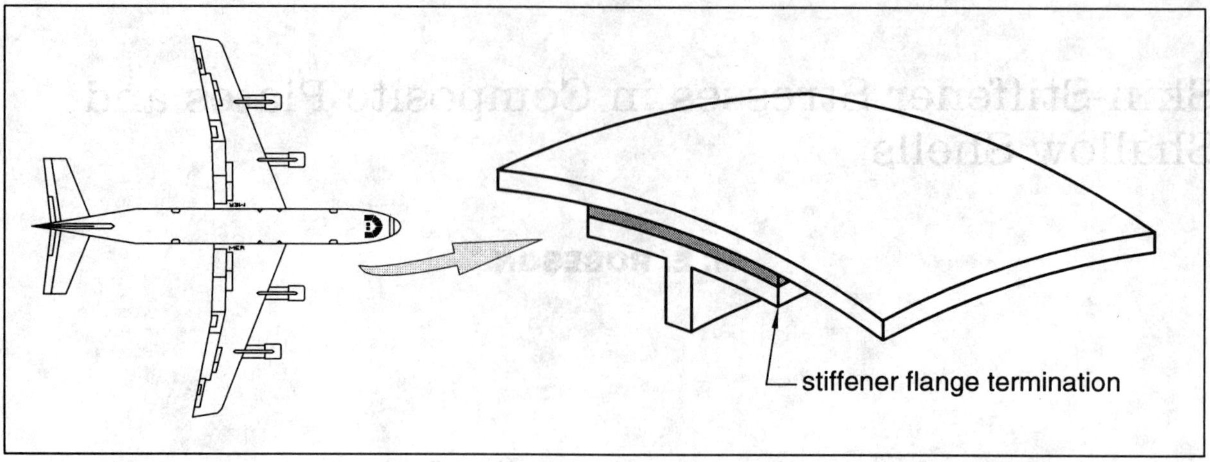

Figure 1. Generic transport aircraft and stiffened doubly-curved panel

Stiffened panels are used in composite aircraft structural components as a primary load-carrying mechanism. Figure 1 shows a representation of a generic modern transport aircraft and a corresponding generic stiffened doubly-curved panel. An important concern in the design of these panels is the failure of co-cured or bonded skin-stiffener interfaces, particularly when the panel is designed to buckle during service. Skin-stiffener separation in stiffened composite structures often initiates at the stiffener flange termination (see Figure 1) because of high interlaminar stresses that are caused by the abrupt change in local geometry. Other factors that affect these stresses include material discontinuities that exist where the stiffener flange is attached to the panel skin and the bending gradients in a buckled skin panel adjacent to the skin-stiffener interface. Additionally, if a distinct bondline is present, its thickness and material properties may significantly affect the interface stresses.

Many of the skin-stiffener interface analyses that have been developed use the assumption of generalized plane deformation; that is, instead of accounting for local deformation gradients in three mutually orthogonal directions, only two are considered. For example, a two-dimensional finite element analysis (Cohen and Hyer, 1987, 1992; Hyer and Cohen, 1988) was developed for a large-scale structure consisting of a plate with an attached stiffener. Displacements that were obtained from this global structural analysis were used as input for a localized finite element analysis. By using a stress-based boundary collocation procedure, this localized finite element analysis was coupled to a two-dimensional generalized plane deformation elasticity solution in the region of the stiffener flange termination. In the elasticity analysis, each component (the stiffener flange and the skin) was treated as having homogeneous material properties by "smearing" the properties through the thickness of the component. In addition, the stresses were assumed not to vary in the direction of the stiffener axial coordinate (generalized plane deformation), and the analysis neglects the effects of skin-stiffener interfaces with interfacial bond layers having nonzero thickness. This analysis later served as the basis for another analysis (Kan, Mahler, and Deo, 1990; Paul et al., 1989).

A computationally efficient, closed form method for calculating stresses at the skin-stiffener interface was developed by Kassapoglou (Kassapoglou, 1994). This method involves minimizing the complementary potential energy of a stiffened composite panel. The generalized plane deformation assumption was used in deriving equations that were solved in each orthotropic layer, subject to the stress boundary conditions and stress continuity conditions at the interface between layers. The skin and stiffener were modeled as homogeneous orthotropic materials.

Several skin-stiffener interface analyses have been developed utilizing beam theory. For example, Lopez (Lopez, 1989) developed an analysis based on beam theory and used it to investigate plate-stiffened flat panels in bending. In addition, a skin-stiffener analysis program called PSTRESS (Barkey, Madan, and Sutton, 1987) was developed utilizing beam theory. Two additional skin-stiffener analyses (Agarwal, 1982; Tsai, 1983) that were developed utilize beam theory to simulate the physical behavior of the skin-stiffener interface in composite tension field panels.

Physical evidence has caused some researchers (Hyer, Loup, and Starnes, Jr., 1990; Hachenberg and Kossira, 1990) to believe that a three-dimensional analysis is necessary to accurately determine the nature of the stress state in composite stiffened panels that are buckled. An example of this type of approach is the analysis developed by Wang and Biggers (Wang and Biggers, 1984), which is a skin-stiffener interface analysis based on linear plate theory and the principle of virtual work. This analysis accounts for local deformation gradients in three mutually orthogonal directions, and treats the skin and flange as separate orthotropic plates held together by unknown interface forces. Continuity is enforced through the adhesive interface deformation, and the principle of virtual work is used in conjunction with the Galerkin method to obtain three out-of-plane stress components. The interfacial bond layer is assumed to be isotropic and linear elastic.

The present analysis for determining stresses in the skin-stiffener interface region is significant because it fulfills a unique combination of needs. The analysis determines interlaminar normal stress and transverse shear stresses in both the direction of the stiffener axial coordinate and the direction of the inplane coordinate perpendicular to the stiffener axis, avoiding the assumption of generalized plane deformation. In addition, the analysis is capable of analyzing skin-stiffener interfaces having nonzero thickness (bonded stiffeners). The change in stiffness due to the presence of the stiffener web is also included in the analysis. Furthermore, the analysis accounts for curvature in skin and stiffener geometry. At present, the skin-stiffener analysis does not include geometrically nonlinear deformations.

SKIN-STIFFENER ANALYSIS

The present approximate analysis is based on thin shell theory (Dym, 1990) and is formulated for a section of stiffened panel that is comprised of a region of doubly-curved skin of thickness t^s, an attached stiffener flange of thickness t^f, and an adhesive interface of thickness h between the skin and flange. This section is illustrated in Figure 2. The stiffener web is rep-

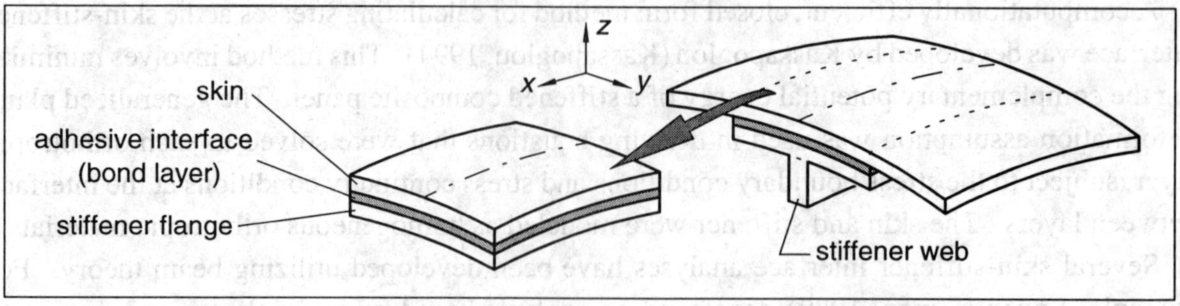

Figure 2. Representative section of stiffened panel under consideration in the analysis

resented by an extensional spring of stiffness k_z and a rotational spring of stiffness k_r. In addition, an extensional spring of stiffness k_y is present along the edges $y = \pm b$ of the skin to represent the y-direction membrane stiffness of the skin between stiffeners. For clarity, the extensional and rotational springs that act upon the representative local analysis region are not shown in Figure 2.

Figure 3 shows the details and dimensions of the representative local analysis region, including the system of arc-length coordinates (x, y, z). The effective transverse shearing stress resultants V_y^* and the moments M_y^* are used to represent the skin between stiffeners. These shear stress and moment resultants are used as known quantities that can be obtained from any generic global analysis of a stiffened panel. Their positive directions are shown in Figure 3. Membrane stress resultants N_x^* and N_y^* are present in the skin, while membrane stress resultants $(N_x^f)^*$ are present in the flange. Positive directions of these membrane stress resultants are also shown in Figure 3, while the applied forces and resultants that act on the negatively-ori-

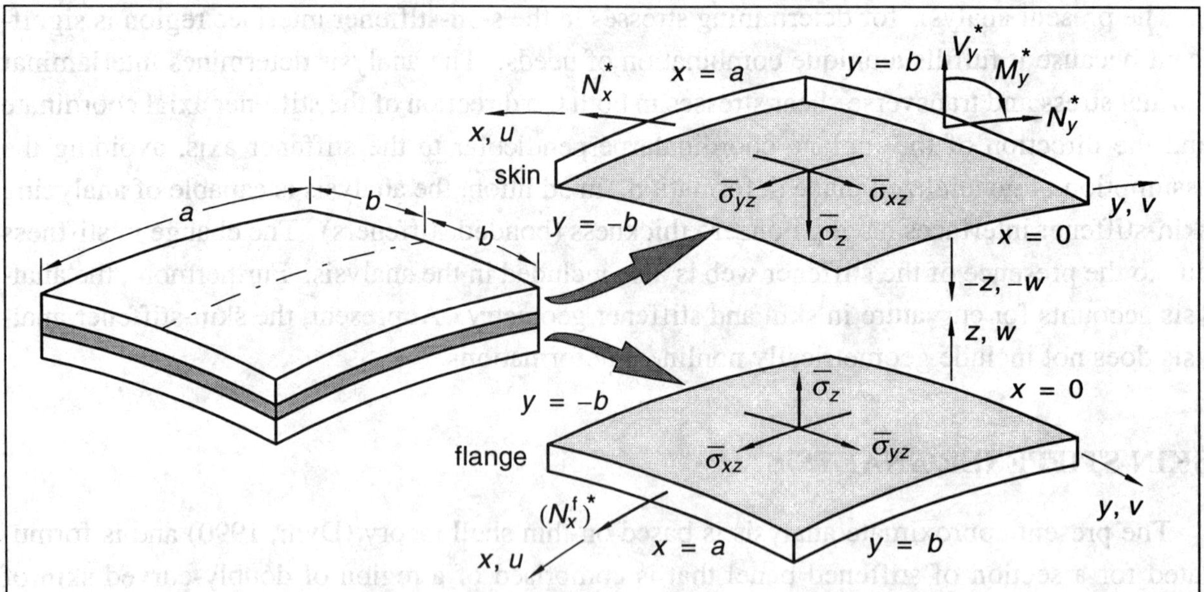

Figure 3. Coordinate system, loads, and dimensions of skin and flange

ented edges of the representative local analysis region are not shown. The interlaminar normal stress $\overline{\sigma}_z$ and the transverse shear stresses $\overline{\sigma}_{xz}$ and $\overline{\sigma}_{yz}$ at the skin-stiffener interface act upon the lower face of the skin and upon the upper face of the flange, as shown in Figure 3.

In the present analysis, the linear strain-displacement relations are given by

$$\begin{aligned} \varepsilon_x(x,y,z) &= \varepsilon_x^\circ(x,y) + z\kappa_x^\circ(x,y) & \gamma_{xy}(x,y,z) &= \gamma_{xy}^\circ(x,y) + z\kappa_{xy}^\circ(x,y) \\ \varepsilon_y(x,y,z) &= \varepsilon_y^\circ(x,y) + z\kappa_y^\circ(x,y) & \gamma_{xz}(x,y,z) &= 0 \\ \varepsilon_z(x,y,z) &= 0 & \gamma_{yz}(x,y,z) &= 0 \end{aligned} \tag{1}$$

where
$$\varepsilon_x^\circ = \frac{\partial u}{\partial x} + \frac{w}{R_x}, \quad \varepsilon_y^\circ = \frac{\partial v}{\partial y} + \frac{w}{R_y}, \quad \gamma_{xy}^\circ = \frac{\partial u}{\partial y} + \frac{\partial v}{\partial x},$$

$$\kappa_x^\circ = -\frac{\partial^2 w}{\partial x^2}, \quad \kappa_y^\circ = -\frac{\partial^2 w}{\partial y^2}, \text{ and } \kappa_{xy}^\circ = -2\frac{\partial^2 w}{\partial x \partial y}. \tag{2}$$

The symbols R_x and R_y are the principal radii of curvature along the x and y coordinate directions, respectively.

The constitutive equations used in the present analysis are for thin specially orthotropic shells and are given by

$$\begin{aligned} N_x &= A_{11}\varepsilon_x^\circ + A_{12}\varepsilon_y^\circ & M_x &= D_{11}\kappa_x^\circ + D_{12}\kappa_y^\circ \\ N_y &= A_{12}\varepsilon_x^\circ + A_{22}\varepsilon_y^\circ & M_y &= D_{12}\kappa_x^\circ + D_{22}\kappa_y^\circ \\ N_{xy} &= A_{66}\gamma_{xy}^\circ & M_{xy} &= D_{66}\kappa_{xy}^\circ \end{aligned} \tag{3}$$

where the A_{ij} are extensional stiffnesses and the D_{ij} are bending stiffnesses of classical plate theory.

The flange displacements and the skin displacements are related through the deformation of the interfacial bond layer. The bond layer is assumed to be isotropic and linear elastic, and stresses in this layer are assumed uniform through its thickness, h. Throughout the remainder of this paper, displacements, strains, extensional stiffnesses, bending stiffnesses, and stress resultants with the f superscript represent values in the stiffener flange, while these variables without superscripts represent values in the skin. For clarity, skin thickness t^s and flange thickness t^f retain the appropriate superscripts.

The skin and the stiffener flange of the present analysis are subjected to in-plane and bending strains. In addition, the interface layer is acted upon by surface shearing tractions $\overline{\sigma}_{xz}$ and $\overline{\sigma}_{yz}$ and by surface normal traction $\overline{\sigma}_z$ in the x, y, and z directions, respectively. Considering non-negligible effects of in-plane strain, bending strain, shear stress, and normal stress, the displacement continuity equations for the skin and the stiffener flange are given by

$$\begin{aligned} u^{\mathrm{f}} &= u + \frac{ht^{\mathrm{f}}}{2E}\frac{\partial \bar{\sigma}_z}{\partial x} - \frac{h}{G}\bar{\sigma}_{xz} + \frac{(t^{\mathrm{f}}+t^{\mathrm{s}})}{2}\frac{\partial w}{\partial x} \\ v^{\mathrm{f}} &= v + \frac{ht^{\mathrm{f}}}{2E}\frac{\partial \bar{\sigma}_z}{\partial y} - \frac{h}{G}\bar{\sigma}_{yz} + \frac{(t^{\mathrm{f}}+t^{\mathrm{s}})}{2}\frac{\partial w}{\partial y} \\ w^{\mathrm{f}} &= w - \frac{h}{E}\bar{\sigma}_z \, . \end{aligned} \tag{4}$$

For the present analysis, the total potential energy of the skin, Π^{s}, is given by

$$\begin{aligned} \Pi^{\mathrm{s}} = \int_{-b}^{b}\int_{0}^{a}\Big[\frac{1}{2}(N_x\varepsilon_x^{\circ} + N_y\varepsilon_y^{\circ} + N_{xy}\gamma_{xy}^{\circ} + M_x\kappa_x^{\circ} + M_y\kappa_y^{\circ} + M_{xy}\kappa_{xy}^{\circ}) \\ + \bar{\sigma}_{xz}u + \bar{\sigma}_{yz}v + \bar{\sigma}_z w\Big]dxdy - \int_{-b}^{b}(N_x^{*}u)\Big|_{x=0}^{x=a}dy - \int_0^a(N_y^{*}v)\Big|_{y=-b}^{y=b}dx \\ - \int_0^a\left(M_y^{*}\frac{\partial w}{\partial y}\right)\Bigg|_{y=-b}^{y=b}dx - \int_0^a(V_y^{*}w)\Big|_{y=-b}^{y=b}dx - \int_0^a\left(\frac{1}{2}k_y v^2\right)\Bigg|_{y=-b}^{y=b}dx \, . \end{aligned} \tag{5}$$

In addition, the total potential energy of the stiffener flange, Π^{f}, is given by

$$\begin{aligned} \Pi^{\mathrm{f}} = \int_{-b}^{b}\int_{0}^{a}\Big[\frac{1}{2}(N_x^{\mathrm{f}}(\varepsilon_x^{\circ})^{\mathrm{f}} + N_y^{\mathrm{f}}(\varepsilon_y^{\circ})^{\mathrm{f}} + N_{xy}^{\mathrm{f}}(\gamma_{xy}^{\circ})^{\mathrm{f}} + M_x^{\mathrm{f}}(\kappa_x^{\circ})^{\mathrm{f}} + M_y^{\mathrm{f}}(\kappa_y^{\circ})^{\mathrm{f}} + M_{xy}^{\mathrm{f}}(\kappa_{xy}^{\circ})^{\mathrm{f}}) \\ - \bar{\sigma}_{xz}u^{\mathrm{f}} - \bar{\sigma}_{yz}v^{\mathrm{f}} - \bar{\sigma}_z w^{\mathrm{f}}\Big]dx\,dy - \int_{-b}^{b}[(N_x^{\mathrm{f}})^{*}u^{\mathrm{f}}]\Big|_{x=0}^{x=a}dy \\ - \int_0^a\left[\frac{1}{2}k_r\left(\frac{\partial w^{\mathrm{f}}}{\partial y}\right)^2\right]\Bigg|_{y=0}dx - \int_0^a\left[\frac{1}{2}k_z(w^{\mathrm{f}})^2\right]\Bigg|_{y=0}dx \, . \end{aligned} \tag{6}$$

RAYLEIGH-RITZ SOLUTION

The Rayleigh-Ritz method is used herein to generate a system of algebraic equations, which are solved to determine the displacement field, and more importantly, the interlaminar stresses, at the skin-stiffener interface. The exact solution is represented approximately by series representation of the unknown functions that are chosen by using physical insight of the structure being analyzed. These series must satisfy the kinematic boundary conditions, and they typically contain a finite number of terms. Series for the displacements *u*, *v*, and *w* and for the interface stresses $\overline{\sigma}_{xz}$, $\overline{\sigma}_{yz}$, and $\overline{\sigma}_z$ were assumed and are given by

$$u = \sum_{i=1}^{m}\sum_{j=0}^{n} U_{ij}\cos\frac{i\pi x}{a}P_j\left(\frac{y}{b}\right) \qquad \overline{\sigma}_{xz} = \sum_{i=1}^{m}\sum_{j=0}^{n} (S_{xz})_{ij}\cos\frac{i\pi x}{a}P_j\left(\frac{y}{b}\right)$$

$$v = \sum_{i=1}^{m}\sum_{j=0}^{n} V_{ij}\sin\frac{i\pi x}{a}P_j\left(\frac{y}{b}\right) \qquad \overline{\sigma}_{yz} = \sum_{i=1}^{m}\sum_{j=0}^{n} (S_{yz})_{ij}\sin\frac{i\pi x}{a}P_j\left(\frac{y}{b}\right) \tag{7}$$

$$w = \sum_{i=1}^{m}\sum_{j=0}^{n} W_{ij}\sin\frac{i\pi x}{a}P_j\left(\frac{y}{b}\right) \qquad \overline{\sigma}_{z} = \sum_{i=1}^{m}\sum_{j=0}^{n} (S_{z})_{ij}\sin\frac{i\pi x}{a}P_j\left(\frac{y}{b}\right)$$

where U_{ij}, V_{ij}, W_{ij}, $(S_{xz})_{ij}$, $(S_{yz})_{ij}$, and $(S_z)_{ij}$ are the unknown coefficients, and $P_n(z)$ are Legendre polynomials. With the interface stresses expected to reach their respective maximums at the edge of the stiffener flange, Legendre polynomials were selected not only because of their orthogonality, but also because they should provide a good approximation to the stress distributions by representing the high stress levels at the stiffener flange edge.

The applied moment and shear loads on the longitudinal edges of the skin are also represented by series and are given by

$$M_y^* = M_0 + \sum_{i=1}^{m} M_i \sin\frac{i\pi x}{a} \quad \text{and} \quad V_y^* = V_0 + \sum_{i=1}^{m} V_i \sin\frac{i\pi x}{a} \tag{8}$$

With these series, different loads can be specified at each of the edges $y = b$ and $y = -b$.

Applying the Rayleigh-Ritz method, a system of algebraic equations is obtained, given by

$$\frac{\partial\Pi^s}{\partial U_{ij}} = 0,\ \frac{\partial\Pi^s}{\partial V_{ij}} = 0,\ \frac{\partial\Pi^s}{\partial W_{ij}} = 0,\ \frac{\partial\Pi^f}{\partial U_{ij}} = 0,\ \frac{\partial\Pi^f}{\partial V_{ij}} = 0,\ \text{and}\ \frac{\partial\Pi^f}{\partial W_{ij}} = 0 \tag{9}$$

where $i = 1, 2, ..., m$ and $j = 0, 1, 2, ..., n$ in each of equations (9). Solution of this system of equations determines the unknowns in the series given by equations (7).

RESULTS AND COMPARISONS

Use of the present skin-stiffener analysis is illustrated by considering the example of a bonded skin-stiffener combination. For this example, the length *a* is 5 inches and the stiffener flange width *b* is 0.75 inches. The interfacial adhesive bond layer is 0.01 inches thick, with a

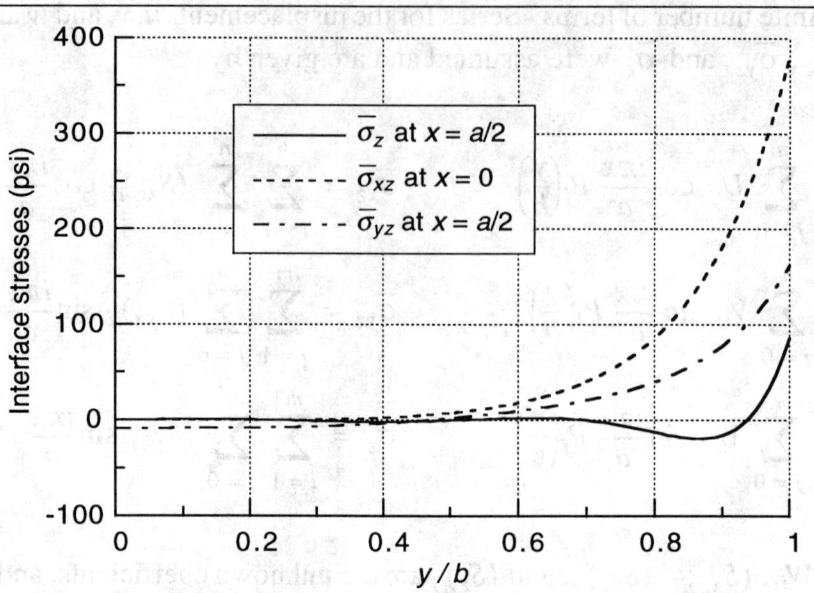

Figure 4. Distribution of interface stresses across the width of a skin-stiffener interface

Young's modulus of 245,180 psi and a shear modulus of 93,400 psi. Both the skin and stiffener flange are graphite-epoxy. The skin panel is 0.08 inches thick and has a stacking sequence of $[90/\pm45/0_2/\mp45/90]_S$. The stiffener flange panel is 0.04 inches thick and has a stacking sequence of $[90/\pm45/0]_S$. The loading for this example is an antisymmetric unit applied moment given by $M_y^* = (1)\sin(\pi x/a)$ in.-lb., and the extensional stiffness k_z and rotational stiffness k_r are zero, representing the absence of a stiffener web.

Results obtained from the present analysis for this example are shown in Figure 4. The interface stresses $\bar{\sigma}_{xz}$, $\bar{\sigma}_{yz}$, and $\bar{\sigma}_z$ are shown as a function of distance across the width of the skin-stiffener interface, from centerline to edge. Each stress component is shown in Figure 4 for the value of x that yields the maximum value over the length a. The results predict that each stress component reaches its respective maximum at the edge of the stiffener flange, which is consistent with the aforementioned observation that skin-stiffener separation often initiates at the stiffener flange termination because of high interlaminar stresses that occur there.

Accuracy of the present skin-stiffener analysis is illustrated by converting predicted interlaminar normal stress results to strain ε_z and comparing these results with results presented by Lopez. Lopez considered an example of a bonded skin-stiffener combination to compare his analysis with the analysis program PSTRESS. For this example, the length a is 1 inch and the stiffener flange width b is 1.25 inches. The interfacial adhesive bond layer is 0.038 inches thick, with a Young's modulus of 362,980 psi and a shear modulus of 83,920 psi. Both the skin and stiffener flange are aluminum, and the stiffener has no attached web. The skin panel is 0.123 inches thick, while the stiffener flange panel is 0.248 inches thick. The loading for this case is a symmetric applied moment given by $M_y^* = 29.8$ in.-lb., and the extensional stiffness k_z and rotational stiffness k_r are again zero, corresponding to the absence of a stiffener web.

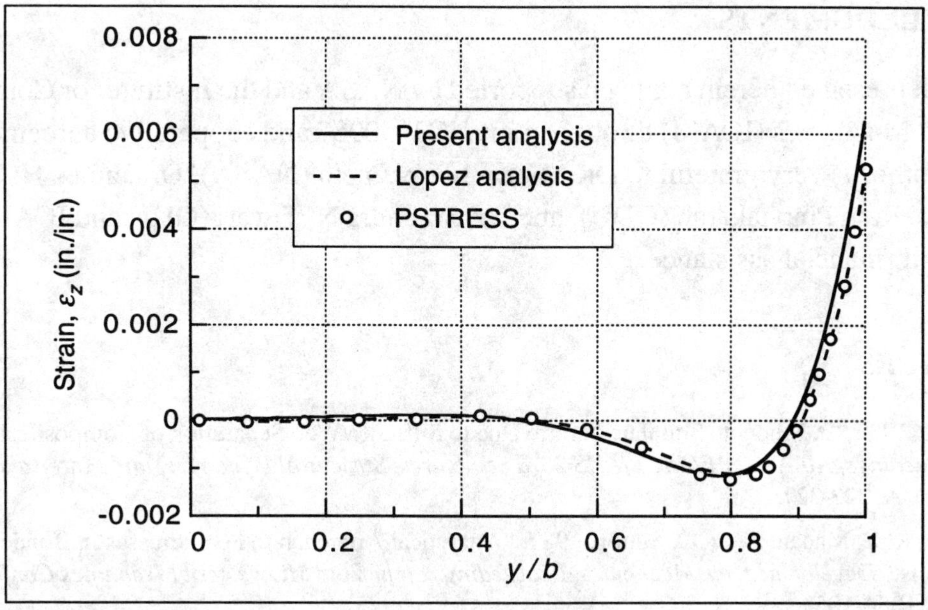

Figure 5. Comparison of present analysis with Lopez analysis and PSTRESS

The strain ε_z calculated by the present analysis, the Lopez analysis, and PSTRESS is shown in Figure 5 as a function of distance across the width of the skin-stiffener interface. As with the previous example (see Figure 4), the strain reaches a maximum at the edge of the stiffener flange, again corresponding to the observation that skin-stiffener separation often initiates at the stiffener flange termination. Most importantly, results from all three analyses are in close agreement.

CONCLUDING REMARKS

An analysis for determining stresses in the skin-stiffener interface region of composite plates and shallow shells has been developed. This analysis is significant because it accounts for curvature in skin and stiffener geometry, the change in stiffness due to the presence of the stiffener web, the presence of skin-stiffener interfaces having nonzero thickness, and local deformation gradients in three mutually orthogonal directions. Typical distributions of interlaminar normal stress, as well as transverse shear stresses in both the direction of the stiffener axial coordinate and the direction of the inplane coordinate perpendicular to the stiffener axis, have been presented. These results show that the interface stresses are greatest at the edge of the stiffener flange. In addition, results from the skin-stiffener analysis have been compared with results from other analyses, and all of these results were found to be in close agreement. Future work on the analysis includes providing more complete validation, incorporating nonlinear strain-displacement relations, and combining the determined stress fields with a failure theory.

ACKNOWLEDGMENTS

The work presented herein has been supported by NASA and the Institute for Computational and Applied Mechanics (ICAM) through grant NGT-50987 and cooperative agreement NCC1-232. The author is very grateful to Dr. Michael P. Nemeth (NASA), Dr. James H. Starnes, Jr. (NASA), Dr. Ram Prabhakaran (ODU), and Dr. Surendra N. Tiwari (ODU and ICAM) for their technical and financial assistance.

REFERENCES

Agarwal, B. L. 1982. "A Model to Simulate Failure Due to Stiffener/Web Separation of Composite Tension Field Panels," *Proceedings, AIAA/ASME/ASCE/AHS 23rd Structures, Structural Dynamics, and Materials Conference*, New Orleans, LA, 323-327.

Barkey, D. A., R. C. Madan, and J. O. Sutton. 1987. "Analytical Approach to Peel Stresses in Bonded Composite Stiffened Panels," *Developments in Mechanics, Proceedings of the 20th Midwestern Mechanics Conference*, West Lafayette, IN, 1054-1059.

Cohen, D. and M. W. Hyer. 1987. "Calculation of Skin-Stiffener Interface Stresses in Stiffened Composite Panels," NASA Contractor Report 184682, Virginia Polytechnic Institute and State University Center for Composite Materials and Structures Report Nos. VPI-E-87-32 and CCMS-87-17.

Cohen, D. and M. W. Hyer. 1992. "Influence of Geometric Nonlinearities on Skin-Stiffener Interface Stresses," *AIAA Journal*, 30(4):1055-1062.

Dym, C. L. 1990. *Introduction to the Theory of Shells*, Hemisphere Publishing Corporation, New York.

Hachenberg, D. and H. Kossira. 1990. "Theoretical and Experimental Investigation of Stringer Peeling Effects at Stiffened Shear Loaded Composite Panels in the Postbuckling Range," *Proceedings, 17th Congress of the International Council of the Aeronautical Sciences*, Stockholm, Sweden, 511-521.

Hyer, M. W. and D. Cohen. 1988. "Calculation of Stresses in Stiffened Composite Panels," *AIAA Journal*, 26(7):852-857.

Hyer, M. W., D. C. Loup, and J. H. Starnes, Jr. 1990. "Stiffener/Skin Interactions in Pressure-Loaded Composite Panels," *AIAA Journal*, 28(3):532-537.

Kan, H.-P., M. A. Mahler, and R. B. Deo. 1990. "Prediction of Stiffener-Skin Separation in Composite Panels," *NASA Conference Publication 3104, Pt. 2, Proceedings, 1st NASA Advanced Composites Technology Conference*, Seattle, WA, 939-952.

Kassapoglou, C. 1994. "Stress Determination at Skin-Stiffener Interfaces of Composite Stiffened Panels under Generalized Loading," *Journal of Reinforced Plastics and Composites*, 13:555-572.

Lopez, O. F. 1989. "The Effects of Bending on the Strain Distributions in a Stiffener-Skin Interface Region," M.S. Thesis, Old Dominion University, Norfolk, VA.

Paul, P. C., C. R. Saff, K. B. Sanger, M. A. Mahler, H.-P. Kan, and E. F. Kautz. 1989. "Out of Plane Analysis for Composite Structures," *Proceedings, 8th DOD/NASA/FAA Conference on Fibrous Composites in Structural Design*, Norfolk, VA, 263-279.

Tsai, H. C. 1983. "Approximate Solution for Skin/Stiffener Separation Including Effect of Interfacial Shear Stiffness in Composite Tension Field Panel," Naval Air Development Center Report NADC-83131-60.

Wang, J. T.-S. and S. B. Biggers. 1984. "Skin/Stiffener Interface Stresses in Composite Stiffened Panels," NASA Contractor Report 172261.

Beam on Elastic Foundation Model to Predict the Crush Behavior of Polymer Composite Tubes

A. G. CALISKAN AND M. M. CHADWICK

ABSTRACT

The crush behavior of polymer matrix composite (PMC) tubes were predicted using a simple beam on elastic foundation (BEF) model. It was observed that, in general, four dominant types of failure mechanisms could occur during crush; crack/delamination initiation, crack growth/arrest at the corners of the tubes, elastic bending/fracturing of the fronds, and delamination of the fronds. Due to differences in failure mechanisms among various fiber architectures and trigger geometries, significant changes in the load-displacement behavior were observed. To better understand the influence of these failure mechanisms, a simple model was constructed for these mechanisms, and the contribution of each to the overall energy absorption was calculated. The analytical results proved to be close to the experimental data, however in most cases predicted lower loads than the experimental data since the influence of friction and other nonlinear phenomena were not taken into account. Further work is required to assess whether or not these phenomena are significant to the overall crush behavior.

INTRODUCTION

Among many of the issues concerning the use of composite materials in automobiles is the energy absorption capabilities of composite structures. Of particular interest is the crashworthiness of a vehicle during a front end collision. Recently, a composite front apron has been built and tested for a Ford Escort by the Automotive Composites Consortium (ACC), and has passed a 35 MPH barrier test (Botkin et al, 1997). However, there is still a lack of understanding as to which factors control the energy absorption of composite structures. Some of these variables are constitutive materials, geometry of the structure, processing methods, rate of crush, direction of impact, frictional processes, and the use of triggering devices to initiate progressive crush. To better understand the effect these and other variables have on crush behavior, a simplified geometry (i.e a circular or square cross-section tube) is crushed. By controlling some of the variables listed above, it is hoped that a better understanding of the macroscopic and microscopic mechanisms can be achieved

To date, many of the analysis on the crush behavior of composite structures have not taken into account any of the microscopic details. Most of the previous work has involved

---

Ari G. Caliskan and Margaret M. Chadwick, Ford Motor Co., 20000 Rotunda Dr., P.O. Box 2053, M.D. 3182, Dearborn, MI 48121

using FE methods to predict overall crush behavior (Agaram et al, 1997). However, these models do not account for the different failure mechanisms. Instead, a continuum damage scheme is employed which mimics the effect of damage as a localized stiffness reduction. This smearing effect results in unsatisfactory prediction of energy absorption. In addition, the approach requires one to guess at the constitutive behavior of the damaged material. Simple models have also been constructed, however they do not show good correlation with experimental results (Hulway, 1991; Tao et al, 1994). To better understand crush behavior, microscopic examinations of the crush zone have been attempted to identify any distinct failure mechanism (Farley et al, 1992; Hull, 1991). The overall conclusions showed that different failure mechanisms occur due to changes in constitutive materials, trigger geometry, tube geometry, and crush rate. Moreover, as will be shown later, there is a direct correlation between the different modes of failure and energy absorption capabilities of a given material system.

The purpose of this study is first to identify the various macroscopic and microscopic failure mechanisms during the crush of square cross-section tubes. To achieve this the crush process of these tubes were recorded by video during the tests to observe macroscopic processes. In addition, regions of the crushed tube were sectioned and polished to investigate any microscopic failure modes which occurred during crush. Once these failure modes were identified, simple models were constructed to assess the contribution of each process to the overall energy absorption of the tube. It is hoped that the results of this work can shed some light as to the influence of the variables mentioned above to energy absorption and be incorporated into design codes to predict response of composite structures.

EXPERIMENTAL

Hollow square cross section tubes of 2"x2"x8" dimensions were made with a wall thickness of both 0.1" (thin walled) and 0.2" (thick walled). The corners of the tubes are rounded with an inner radius of $a=0.37''$ and $a=0.27''$ for the thin and thick tubes, respectively. All tubes were fabricated by SRIM using GMR-5000 (ICI), a polyurethane resin, reinforced with continuous strand mat $(CSM)_n$ E-glass fiber layers, where n is 3 for the thin tubes and 6 for the thick tubes. In addition, all tubes were postcured for 1 hour in a 150°C oven to insure full cure of the resin. The material properties of the tubes are listed in the following table.

Table I Material properties of the $(CSM)_n$ E-glass fiber/polyurethane tubes.

	Axial Stiffness E_{zz}	Hoop Stiffness $E_{\theta\theta}$	Tensile Strength X_t	Failure Strain ε_f
$(CSM)_n$	7.8 GPa	7.8 GPa	131 MPa	2.3%

All tubes were crushed with either a 6 mm or 12 mm radius trigger using a MTS system at a rate of 2"/min at ambient temperature. Tubes were crushed to a total displacement of 5" and the load was recorded. In addition, the tests were video taped in order to investigate macroscopic failure mechanisms during crush. Finally, selected regions of the tubes were sectioned and prepared using standard metallographic techniques. These sections were examined using an optical microscope.

MACROSCOPIC & MICROSCOPIC OBSERVATIONS

In the crush process there are significant macroscopic events controlling the overall load-displacement behavior of the tube. Initially, the tube is forced into the trigger and deforms elastically at the lip. This deformation causes stresses to build up at the top of the tube and cracks to initiate at the corners. The initial elastic portion of crush occurs in all of the material systems.

As the trigger is forced further into the tubes, the cracks at the corners of the tube grow. In the case of the $(CSM)_n$ tubes, the crack grows ahead of the trigger and arrests. In addition, the fronds that are formed in the $(CSM)_n$ tubes have fractures which produce distinct segments along the frond. This can be seen schematically in Figure 1. The microscopic observations supported many of the macroscopic results. The $(CSM)_n$ segment fractures showed fiber/matrix cracks with no damage in between the fractures.

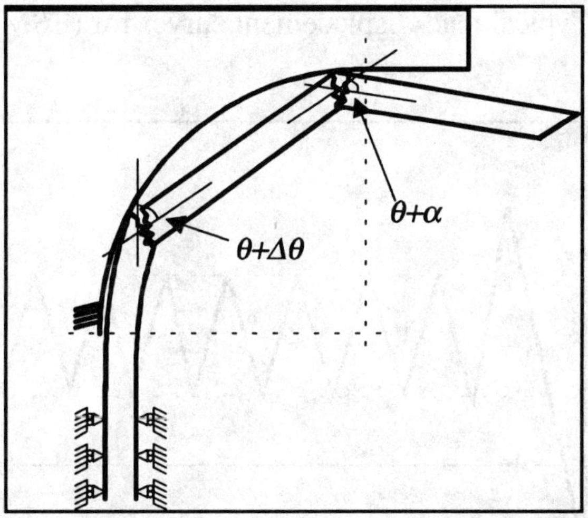

Figure 1 Schematic of frond segmentation in $(CSM)_n$ tubes.

ANALYTICAL MODEL

In this section, the failure mechanisms observed in square cross-section tubes crushed with a trigger will be analyzed. Each of these failure mechanisms will be viewed separately and the contribution of each to energy absorption will be calculated. In Figure 2, we see the load-displacement curves for a typical $(CSM)_n$ tube.

There are basically two events which characterize the basic shape of the load-displacement curves (see Figure 3). In the first region, the tube deforms elastically as it is being forced into the trigger. The extent to which this elastic region occurs is a function of material properties, geometry of the tube and trigger, crush speed, and frictional effects. The end of the elastic region is marked by crack initiation at the corners of the tube and/or delamination initiation at the tube walls. At this time it is unclear whether a delamination would cause an initial load drop since the majority of the stresses at the lip of the tube are in the hoop direction. In the analysis, we assume that the initial load drop is due to cracks forming at the corners.

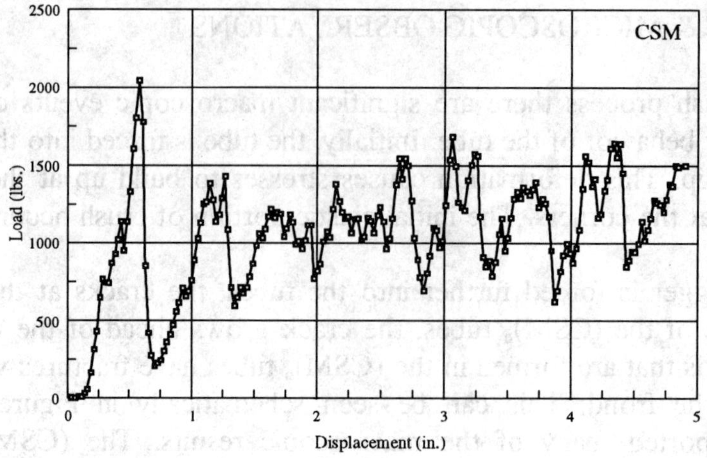

Figure 2 Typical load-displacement curves for $(CSM)_n$ systems.

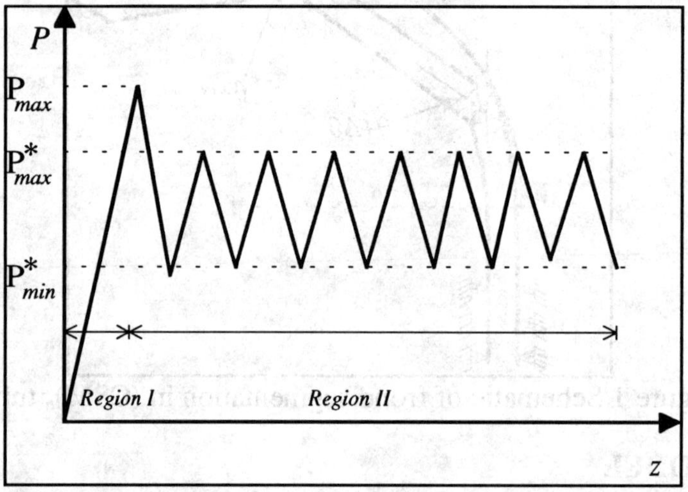

Figure 3 Generic Load-Displacement curve with regions of I. elastic and II. steady-state crush.

Once the crack is initiated, the load drops significantly. In this region of the curve, depending on the material system, any number failure mechanisms could occur and dominate the load-displacement behavior. For the $(CSM)_n$ tubes, it was observed that the cracks on the corners of the tube grow and arrest as the trigger is forced into the tube. This crack growth mechanism is due mainly to the build up and release of elastic strain energy and crack bridging by the fibers which is common in many composites (Caliskan, 1996) When the corner crack grows ahead of the trigger, the load decreases and then increases once the crack is arrested. When the fronds cannot support any more load, they fracture. Then, the whole cycle repeats itself. As a result, the load-displacement curve of the $(CSM)_n$ tubes exhibit a somewhat sinusoidal behavior. In addition, depending on the tube thickness and trigger radius, delaminations occurred in some of the tubes and not others. An explanation of these phenomena will be given in the following sections.

Since there are distinct failure modes in the crush process of these tubes, each failure mode will be examined independently with the appropriate assumptions and boundary conditions. There are basically 4 observed failure modes in the crush process of these tubes--crack/delamination initiation, crack growth/arrest at the corners of the tubes, elastic bending/fracturing of the fronds, and delamination of the fronds.

CRACK/DELAMINATION INITIATION

Prior to crack initiation, elastic energy is built up in the tube as the trigger is forced into the tube. To model this initial behavior, we first look at how the geometry of the trigger changes as the tube moves into it. In Figure 4 the trigger geometry is shown. The trigger radius is ρ, β is the crack length on the corners of the tube ahead of the trigger, ξ is the distance the tube has moved into the trigger, and δ is the end displacement of the lip of the tube. The effect of the trigger is essentially to deform the corners of the tube. As a result, initial loading in the radial direction is akin to internal pressure loading. Since the highest stresses are at the corners of the tube, we will model the problem as a circular tube with an internal pressure at the lip.

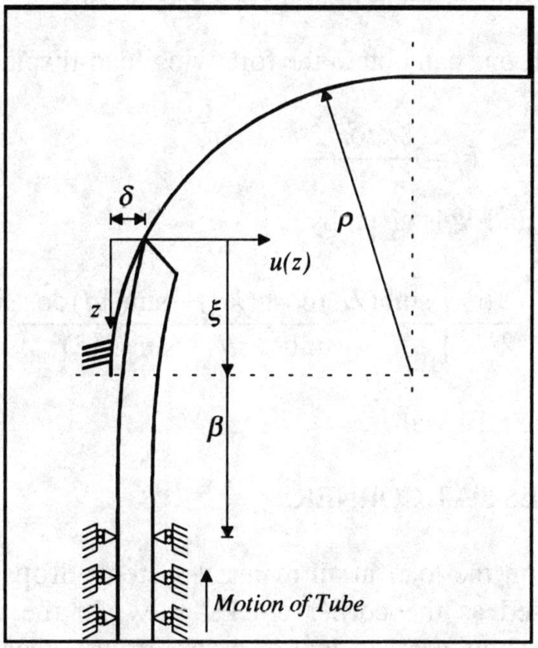

Figure 4 Trigger cross-section showing a tube wall being crushed through a trigger.

Consider a slender section of a circular tube constrained elastically by the rest of the tube (see Figure 5).

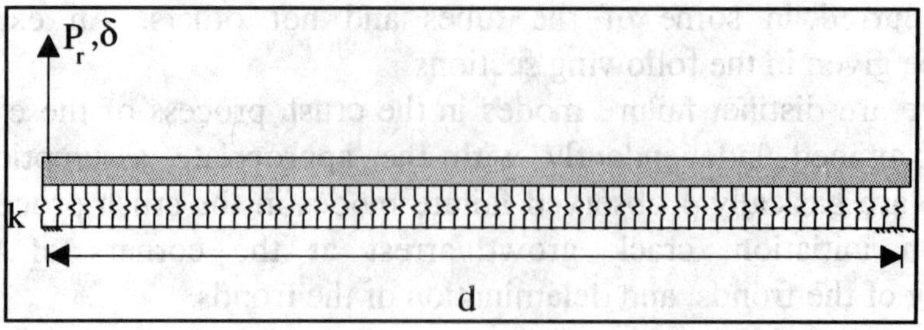

Figure 5 Beam on elastic foundation model of tube displaced at the lip

Where d is the length of the tube, k is the foundation stiffness given by $k = \frac{2\pi E_{\theta\theta} h}{a^2}$, and a is the radius of the cylinder. The governing equation for the radial displacement, u, of the tube is given by

$$\frac{d^4 u}{dz^4} + 4\lambda^4 u = 0 \tag{1}$$

where $\lambda^4 = \frac{k}{4E_{zz} I}$, E_{zz} is the modulus in the axial direction, I is the moment of inertia. After solving for u in Equation 1, one can obtain the following load-displacement relationship.

$$P_r = \frac{2\pi a \delta E_{zz}}{\Psi} \tag{2}$$

where P_r is the radial load and Ψ is given by

$$\Psi = \frac{6}{\lambda^3 h^3}\left[\frac{\sinh(\lambda d)\cosh(\lambda d) - \sin(\lambda d)\cos(\lambda d)}{\sinh^2(\lambda d) - \sin^2(\lambda d)}\right] \tag{3}$$

and h is the thickness of the tube wall.

CRACK GROWTH/ARREST AT CORNERS

After crack initiation the load in all material systems drops. This is due to the elastic strain energy being released as the corner cracks grow. In the case of $(CSM)_n$ materials, experimental observations show a great deal of crack growth ahead of the trigger, especially for the thinner tubes. As mentioned previously, the stresses at the corners are mainly in the hoop or in the θ-direction. So, the cracks in the corners are mainly Mode I cracks.

Once again we consider only the corners of the tube. We assume, since the loading is Mode I, that the corner of the tube acts like curved beam with a width of $2b = \frac{\pi}{2}a$ and thickness h (see Figure 6).

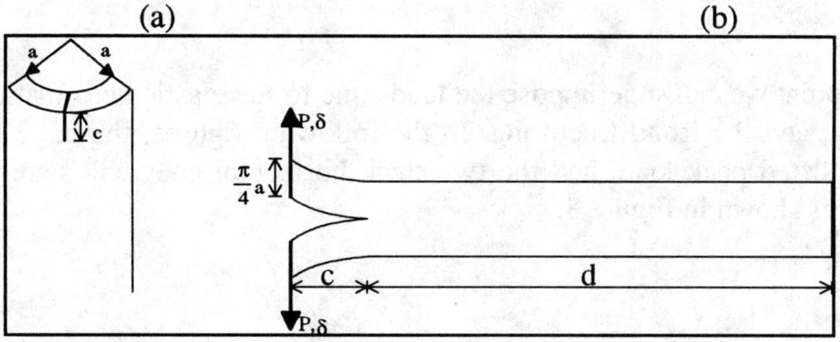

Figure 6 (a) Corner of a tube with a crack of length *c* (b) model of Mode I loading at the tube corner.

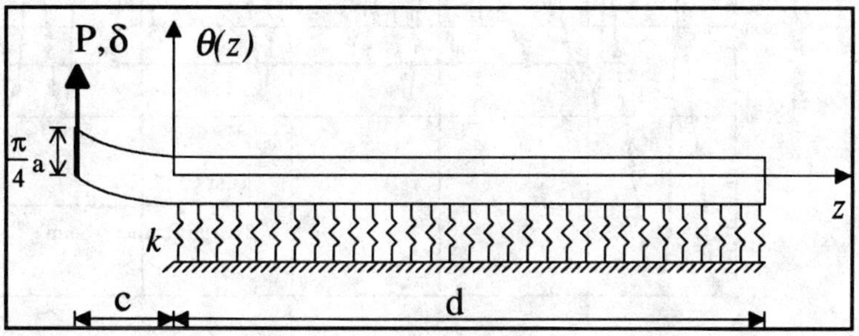

Figure 7 Beam on elastic foundation model of a beam with an unsupported portion.

From Figure 7, we see that this model is a generalization of the BEF model shown previously. Using a similar analysis, we arrive at the same load-displacement relation as before with Ψ being

$$\Psi = \frac{2\sinh(\lambda d)\cosh(\lambda d)}{\lambda^3 b^3(\sinh^2(\lambda d) - \sin^2(\lambda d))}\left[3 + 6\lambda^2c^2 + (6\lambda^2c^2 - 3)\frac{\sin(\lambda d)\cos(\lambda d)}{\sinh(\lambda d)\cosh(\lambda d)} + (2\lambda^3c^3 + 6\lambda c)\frac{\sinh^2(\lambda d)}{\sinh(\lambda d)\cosh(\lambda d)} - (2\lambda^3c^3 - 6\lambda c)\frac{\sin^2(\lambda d)}{\sinh(\lambda d)\cosh(\lambda d)}\right] \tag{4}$$

ELASTIC BENDING/FRACTURE OF FRONDS

In addition to the corners, the fronds also provide some resistance as the tube is being crushed. In the $(CSM)_n$ systems, the fronds bend and fracture at regular intervals. To model this behavior, the fronds are considered to be cantilever beams with an applied end displacement (see Figure 4). The length of the tube into the trigger, ξ, is increased until the stress at the root of the beam reaches either the tensile or compressive strength of the material, whichever one is lower. In the case of $(CSM)_n$, the tensile strength was lower. At this point, the load to initiate the fracture of the frond is recorded.

RESULTS

At this point we can superimpose the loads due to the elastic deformation at the lip, the corner cracking, and the frond fracturing. In the following figures, the large black circle will signify the calculated peak load, and the two thick horizontal lines will signify the calculated P^*_{max} and P^*_{min} ,as shown in Figure 3.

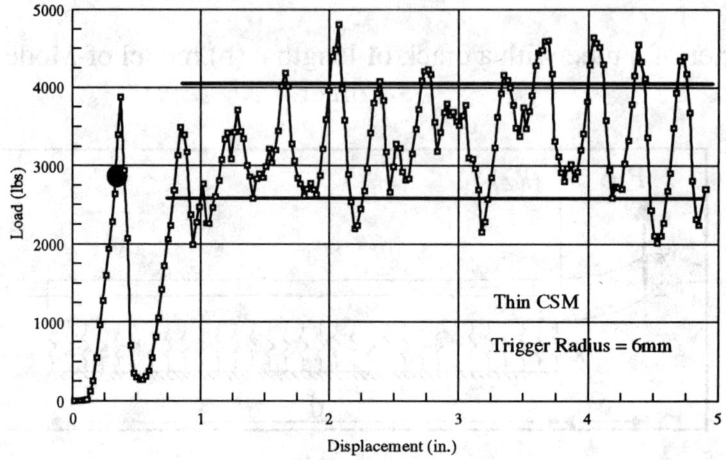

Figure 8 Load-Displacement curve and analytical prediction for a thin $(CSM)_3$ tube with a 6 mm trigger.

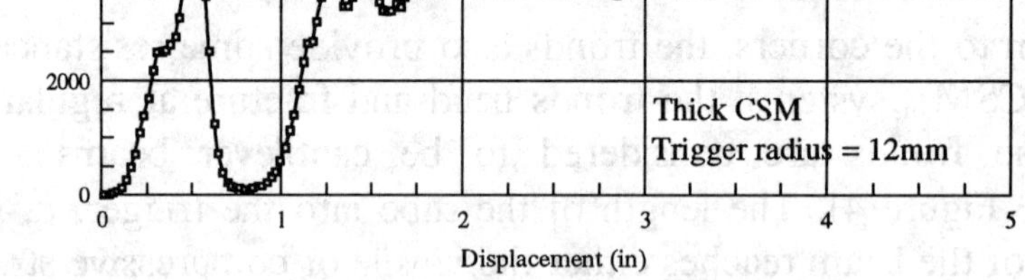

Figure 9 Load-Displacement curve and analytical prediction for a thick $(CSM)_6$ tube with a 12 mm trigger.

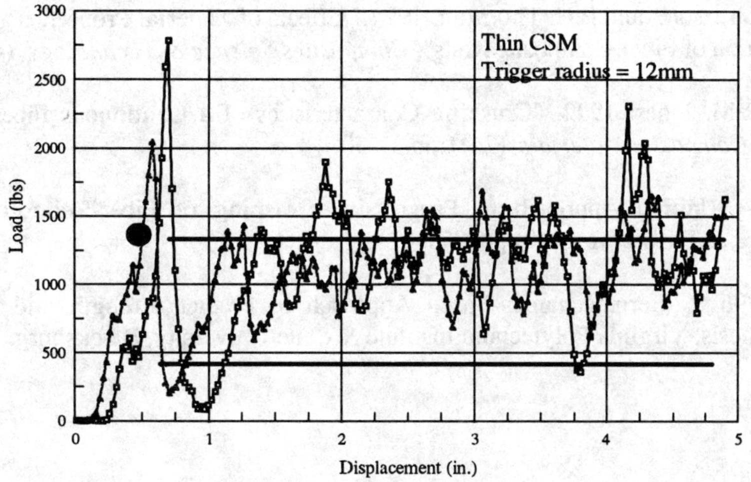

Figure 10 Load-Displacement curve and analytical prediction for a thin $(CSM)_3$ tube with a 12 mm trigger.

The results show fairly good agreement between the experimental and analytical results. Although, overall, the predicted loads are slightly lower than shown experimentally. This could be due to material variations as well as the effects of friction which were not considered in the model.

CONCLUSIONS

The crush behavior of PMC tubes were analyzed. The experimental results showed that in the case of the $(CSM)_n$ tubes the dominating failure mechanisms were the corner crack growth/arrest and frond deformation/fracture. In order to determine the effects of these failure mechanisms, a simple beam on elastic foundation model was constructed to relate the effect of deformation and fracture to the overall energy absorption of each material system. The calculations were in good agreement with experimental results. It is hoped that such simple models can provide an insight into the dominating failure mechanisms in PMC tubes and guide further development.

REFERENCES

Botkin M., S. Fidan, and R.A. Jeryan. 1997. "Crashworthiness Development of a Composite Front Structure for a Production Vehicle", *ACCE Conference Proceedings, Dearborn, MI, April 7-10, 1997.*

Agaram, V., S. Bilkhu, S. Fidan, M. Botkin, and N. Johnson. 1997."Simulation of Controlled Failure of Automotive Composite Structures with LS-DYNA3D", *ACCE Conference Proceedings, Dearborn, MI, April 7-10, 1997.*

Hulway, J.A. 1991."A Model for the Crush of Polymer Matrix Composite Tubes", Ph.D. Dissertation, University of Michigan, Ann Arbor, MI.

Tao, W.H., R.E. Robertson., and P.H. Thornton. 1994." Effects of Material Properties and Crush Conditions on the Energy Absorption of Fiber Composite Rods", *Composites Science & Technology*, (47): pp. 405-418.

Farley, G.L. and R.M. Jones. 1992. "Crushing Characteristics of a Continuous Fiber-Reinforced Composite Tubes", *Journal of Composite Materials*, (26)1:pp.37-50.

Hull, D. 1991. "A Unified Approach to Progressive Crushing of Fiber-Reinforced Composite Tubes", *Composites Science & Technology*, (40):pp. 377-421.

Caliskan, A.G. 1996 " Micromechanics-Based Approach to Predict Strength and Stiffness of Composite Materials", M.S. Thesis, Virginia Polytechnic Institute & State University, Blacksburg, VA.

An Analytical Model for Stitched Double Cantilever Beam Specimens

B. V. SANKAR AND S. M. DHARMAPURI

ABSTRACT

A simple analytical model is presented to describe crack propagation in stitched laminated DCB specimens. The stitches are smeared and modeled as continuous cohesive springs, which are assumed to be linear-elastic until final failure. Timoshenko beam theory is used to model the DCB. As the specimen is loaded, the stitches are assumed to break when the strain in the stitches reaches the ultimate strain. The crack tip is assumed to advance as the energy release rate at the crack tip reaches the Mode I fracture toughness of the parent laminate. A closed form solution is obtained for the problem of beam on elastic foundation and it is used to compute the deflections, stitch strain and strain energy release rate. A computer program is used to simulate the DCB test and crack propagation. Results presented include load-deflectoion diagrams, crack bridging length and apparant fracture toughness for various stitch parameters. A semi-cmperical formula is derived for the bridging length as a function of the stitch ultimate strain and fracture toughness of the parent laminate. The results are compared with experimental results. It is found that for the case Kevlar stitches the linear elastic model of the stitch yarn is not adequate. It is shown that inelastic behavior of the stitches may play a significant role in increasing the fracture toughness due to stitching.

INTRODUCTION

Laminated fiber composites that lack through-the-thickness reinforcements have poor interlaminar strength and fracture toughness. Translaminar reinforcement can be provided by inserting pins in the thickness direction (z-pinning) of the laminate or by stitching the layers with suitable yarns before resin impregnation. A number of experimental studies have shown that stitching can be effective in containing free edge delamination, and improving Mode I fracture toughness, Mode II fracture toughness and Compression After Impact Strength (Sankar and Sharma, 1992). Recently, there has been interest in modeling the effects of stitching. Jain & Mai, 1994 have developed analytical models to predict the effects of stitching on Mode I, Mode II and sublaminate buckling behavior. A comprehensive review of

---

B.V.Sankar, Professor and S.M.Dharmapuri., Graduate student, Dept. of Aerospace Engineering, Mechanics and Engineering Sciences, University of Florida, PO Box 116250, Gainesville, FL-32611-32650.

effects of stitching on delamination can be found in Dransfield et al. 1994. Jain and Mai, 1994. used a micromechanical approach in conjunction with the beam theory to study stitched DCB specimens. The developed a simple force-displacement relation for the stitch, and used the information to predict the crack growth resistance. Chen et al. 1990, used a plane strain finite element model to study the effects of stitching. They introduced the concept of overstep length and its role in effective strain energy release rate of the stitched composite.

In this paper we propose a simple model for studying the load-deflection behavior and apparent fracture toughness of stitched DCB specimens. The stitches are smeared over the entire delaminated interface and modeled as linear cohesive springs (Winkler type elastic foundation). Thus the load-deflection relationship of the stitch is linear. It is assumed that the stitches break when the strain reaches an ultimate value ε_u. Further we assume that the stitches debond completely from the surrounding material before they become effective and hence their length is equal to the laminate thickness *2h*, where *h* is the thickness of the sublaminate - either top or bottom - of the DCB specimen. The modeling of stitches as an elastic foundation needs some justification. In reality as the delamination propagates past the stitches, the stitches will debond from the surrounding composite. If they do not, then the stitches will possess infinite axial stiffness because of their infinitesimally small length. This will cause severe stress concentration, which will result in debonding of the stitches. As the stitches debond their stiffness will reduce relieving the interfacial shear stresses. We assume that this debonding is instantaneous and complete through the thickness of the laminate. The results for increase in fracture toughness due to stitching compare well with some experiments done on specimens with glass yarns. However for Kevlar stitch yarns the model greatly under-predicts the fracture toughness. When the inelastic behavior of Kevlar yarns were included the model results were closer to the test results indicating the nonlinear behavior of stitch yarns may be crucial in modeling the fracture process.

ANALYTICAL MODEL

We propose to use the Timoshenko beam theory to determine the deflections and also the force and moment resultants in a stitched DCB specimen. Then from the deflections we can determine the forces in the stitches. We shall also derive a method for computing the strain energy release rate (SERR) at the crack tip from the force and moment resultants acting on the beam. The problem we need to solve is depicted in Fig. 1. However it is sufficient to solve the modified problem as shown in Fig. 2. In Fig. 2 we consider only one half of the beam (top portion) because of symmetry. The beam is fixed at the right end, and a transverse force *F* and a couple *C* act at the left end. The stitches that bridge the crack are modeled as an elastic foundation which is described by the smeared stiffness *k*. The unit of *k* is force/length^3. The foundation constant *k* can be related to stitch parameters as shown in Eq.(1). where A_s and E_s are the cross section area and Young's modulus of the stitch, respectively, and *N* is the number of stitches per unit area of the laminate. The Equilibrium equations of the beam on elastic foundation (Sankar & Dharmapuri, 1997) can be derived as shown in Eqs. (2) & (3).

$$k = NA_sE_s / h \tag{1}$$

$$GA\frac{d\psi}{dx} + GA\frac{d^2 w}{d x^2} - kbw = 0 \tag{2}$$

$$EI\frac{d^2\psi}{d x^2} - GA\psi - GA\frac{dw}{dx} = 0 \tag{3}$$

The solution of Eqns. (2) & (3) can be expressed as:

$$\psi(x) = \sum_{i=1}^{4} a_i \, e^{\lambda_i x} \tag{4}$$

$$w(x) = \sum_{i=1}^{4} b_i \, e^{\lambda_i x} \tag{5}$$

where λ's are the characteristic roots of the characteristic equation and a_i and b_i are constants to be determined from the boundary conditions. The details of the solution can be found in (Sankar & Dharmapuri, 1997). Once the coefficients are determined, the solution of the problem depicted in Fig. 2 is complete. Obtaining the deflection curve, stitch forces and the shear force and bending moment resultants at any cross section is a matter of straight forward calculation. The distributed stitch force at any location x is given by $kbw(x)$, and it acts downward on the top half of the beam. It may be noted that the roots λ_i may be complex depending on the beam properties EI and GA, and the stitch stiffness k. Hence we used complex notations in the computer program used in the present study.

CRACK PROPAGATION CRITERIA

As the DCB specimen is loaded the crack tip will advance leaving the stitches behind. These stitches will bridge the crack surfaces until they are broken by excessive tensile stresses. A typical crack bridging is shown in Fig. 1. The actual crack tip where the delamination ends is shown in the figure as *C*. Thus the delamination or crack length is denoted by $(a+c)$, where a is the distance between the loading point *A* and the first unbroken stitch *B*, and c is bridging length (length *BC*). Both a and c continuously vary as the loading is increased. We have to define criteria for the crack propagation, i.e., the advancement of the crack tip, as well as for the stitches to break. The latter is quite straight forward and simple. We assume that the stitches will break as the strain in them exceed a certain ultimate strain, ε_u. The crack tip *C* will propagate when the strain energy release rate (SERR) reaches a critical value, G_{Ic}, of the composite material system. In the context of beam the strain energy density refers to strain energy per unit length of the beam, which is given by:

$$U_L = \frac{1}{2}\left(\frac{M^2}{EI} + \frac{V^2}{GA} + \frac{P^2}{EA}\right) \tag{6}$$

where *M, V* and *P* are the bending moment, shear force and axial force resultants, respectively. In the case of DCB specimen $P \equiv 0$. Further $U_L \equiv 0$ ahead of the crack tip as the force and moment resultants are zero. Accounting for the top and bottom laminates, the expression for SERR takes the form:

$$G_I = \frac{1}{b}\left(\frac{M^2}{EI} + \frac{V^2}{GA}\right) \tag{7}$$

where *M* and *V* are the moment and force resultants just behind the crack tip, which can be easily calculated using the constitutive relations and the solution for $w(x)$ and $\psi(x)$ given in (4) and (5).

A detailed explanation of the Numerical simulation of DCB Test is given in (Sankar & Dharmapuri, 1997).

NONDIMENSIONAL PARAMETERS

Since a large number of parameters are involved in describing the stitches and the delaminated beam, it will be convenient to introduce a set of nondimensional parameters to present the results - especially the effects of stitch parameters on the apparent increase in fracture toughness of the beam. The length dimension is normalized with respect to the laminate thickness *h*, and all the stiffness terms, e.g., Young's modulus and shear modulus, are normalized by dividing by the equivalent beam Young's modulus E_b, which can be calculated from the relation :

$$(EI)_{eq} = E_b B h^3 / 12 \tag{8}$$

Thus the laminate thickness and Young's modulus will be equal to unity in all the examples. Similarly, other parameters are nondimensionalized, details of which can be found in (Sankar & Dharmapuri, 1997).

RESULTS AND DISCUSSION

This section is divided as: (a) verification of our method of computing G_I ; (b) effects of ε_u , *k,* and G_{Ic} on the load-deflection diagrams, bridging length and apparent fracture toughness.

In order to verify our method of computing the SERR at the delamination front Eq. (7) we considered a case with very high ε_u so that the stitches would not break, but the crack would propagate. Thus there should not be any increase in the apparent fracture toughness G_{app}. In fact it can be seen from Fig. that as the crack propagates the nondimensional G_{Ic-app} (G_{app}/G_{Ic}) remains at unity for various stitch parameters. This validates Eq.(7) for computing G_I for a stitched laminate.

The effect of increasing the breakage strain of stitches on the load deflection behavior is shown in Fig. 4(a). It may be noted that the stitch strength does not have any significant effect on the load at which the delamination begins to propagate. This is indicated by the fact that the initial load drop occurs at the same load for all cases. However after the crack propagates and stitches come into effect via crack bridging, the load for a given deflection is higher in the case of strong stitches, i.e. stitches with higher strain to break.

In Fig. 4(b), the bridging length c is plotted as a function of crack length l for stitches with various strain to failure. The effect of ultimate strain of stitches is in general to increase the crack bridging length. An equation relating c and ε_u is presented at the end of this section. The apparent fracture toughness also increases with stitch ultimate strain as presented in Fig. 4(c).

The effect of stitch stiffness on the DCB behavior was very much dependent on the G_{Ic} of the laminate. For large values of G_{Ic} ($10^{-6} < G_{Ic} < 10^{-4}$) there was no significant change in the load deflection behavior, crack bridging length and the apparent fracture toughness due to changes in k_s. When the value of G_{Ic} was reduced (e.g., $G_{Ic} = 10^{-7}$), the effects of k_s could be realized as shown in Figs. 5(a) through 5(c). The crack begins to propagate initially at the same load level , see Fig. 5(b). but the load attains a second peak for higher k, thus increasing the load carrying capacity of the beam. The bridging length c decreases with increase in stitch stiffness Fig. 5(b). The apparent fracture toughness increases with k Fig. 5(c)

Obviously higher the G_{Ic}, larger the load at which the delamination begins to propagates Fig. 6(a). The load for a given deflection increases with G_{Ic}. From Fig. 6(b). it may be noted that increase in G_{Ic} decreases the bridging length. But the nondimensional apparent fracture toughness G_{Ic-app} decreases with increase in G_{Ic} indicating the stitches are not as effective in increasing the fracture toughness in laminates that already have higher fracture toughness. The effects of stitch parameters on the bridging length is summarized in Figs. 7(a) & 7(b). From the log-log plot an empirical formula was derived for the bridging length as:

$$\bar{c}=c/h=0.76\sqrt{\varepsilon_u}/G_{Ic}^{1/4} \quad \textbf{(9)}$$

A detailed comparison with experimental results has be presented in (Sankar & Dharmapuri, 1997). It is shown that the model does not capture the inelastic behavior of the stitch after debonding. Introducing plastic strain strain energy helps in comparing the results from this model and the experimental results.

CONCLUSIONS

A simple analytical model has been developed for simulating the load-deflection behavior of stitched double cantilever beam specimens. A simple method has been proposed to compute the strain energy release rate due to delaminations in stitched beams. The method is verified by the simulations performed using the present model. It is found that the maximum attainable increase in fracture toughness can be easily computed by the energy absorbed by the stitches in the breaking process. The crack bridging length behind the crack tip varies with the stitch parameters. A closed form expression has been derived for the nondimensional crack bridging in terms of the fracture toughness of the parent laminate and the ultimate strength of the stitches.

ACKNOWLEDGMENTS

The funding for this research was provided by the Florida Space Grants Consortium and the Technological Research and Development Authority of State of Florida.

REFERENCES

Dexter. H.B , & Funk J.G. 1986. Impact resistance and interlaminar fracture toughness of through-the-thickness reinforced graphite-epoxy. *AIAA paper 86-CP.*

Dransfield, K., Baillie, C.A. & Mai, Y-W.1994. Improvement of Delamination resistance in CFRP by stitching- A review. *Comp. Sci. & Technol.* ***50****. pp. 331-345.*

Jain L.K & Mai Y-W.1994 On the effect of stitching on mode I Delamination toughness of laminated composites. *Comp. Sci. & Technol.* ***51****. pp. 331-345.*

Mignery, L.A., Tan T.M. & Sun C.T.1985. The use of stitching to suppress delamination in laminated composites, ASTM STP 876, American Society of Testing Materials, Philadelphia, USA, *pp 371-385.*

Ogo, Y.1987. The effect of stitching on in-plane and interlaminar properties of carbon fiber/epoxy fabric laminates. MSc. thesis, University of Delaware.

Pelstring R.M. & Madan R.C.1989. Stitching to improve damage tolerance of composites., ***34th International SAMPE symposium.***

Sankar B.V.1991. A finite Element for modeling delamination in composite beams., ***Computers & Structures, 38,*** *No 2, pp-239-246*

Sankar.B.V. & Dharmapuri. S.M. 1997. Analysis of Stitched DCB Specimens. Report to *Florida Space Grants Consortium.* (to be submitted).

Sankar.B.V. & Sarma K.S. 1992. Effects of through-the-thickness stitching on impact and interlaminar fracture properties of textile graphite / epoxy laminates. *NASA CR - 195042.NASA Langley, Hampton, VA.*

Sankar. B.V. & Sonik V.1997. Pointwise Energy Release Rate in Delaminated plates., *AIAA,33,No7.*

Shu D. & Mai Y-W.1993. Effect of Stitching on Interlaminar delamination extension in composite laminates. *Comp. Sci. & Technol.* ***49****. pp. 165-171.*

Su. K.B.1989. Delamination Resistance of stitched thermoplastic matrix composite laminates. ***Advances in Thermoplastic Matrix composite materials***, ed. F.M. Newaz. ASTM STP 1044. American Society of Testing Materials, Philadelphia, USA, *pp 279-300.*

Ye. L.1992. Evaluation of mode I interlaminar fracture toughness for fiber-reinforced composite materials for fiber-reinforced composite materials *Comp. Sci. & Technol.* ***43****. pp. 49-54.*

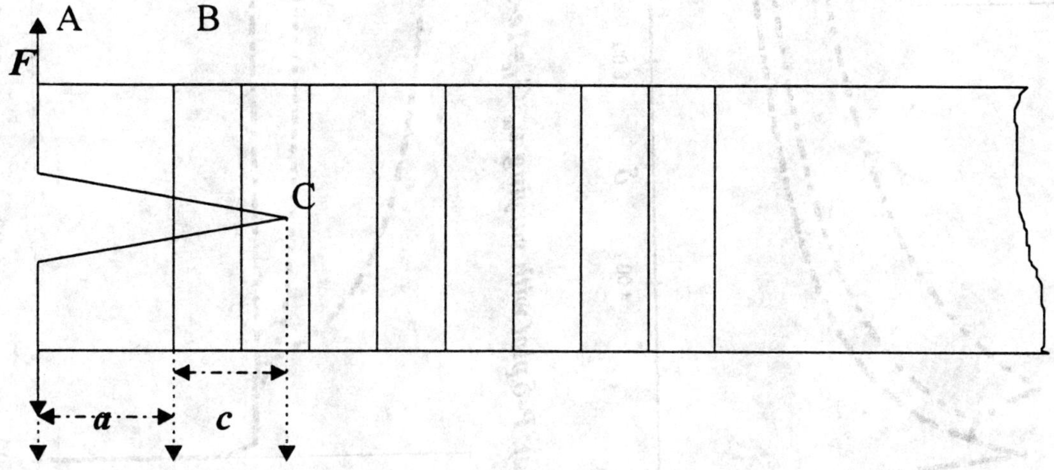

Fig. 1 Typical Stitched Double Cantilever Beam Specimen.

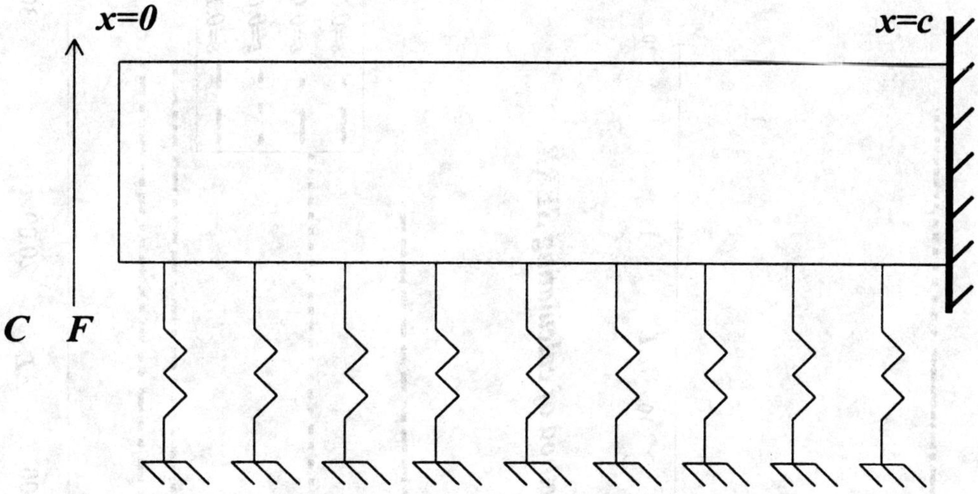

Fig 2. Analytical Model of DCB

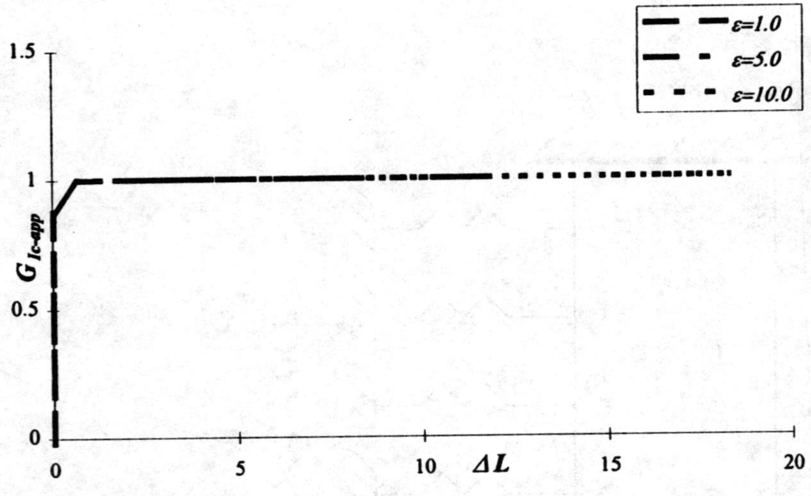

Fig. 3. Verification of method of calculating SERR.

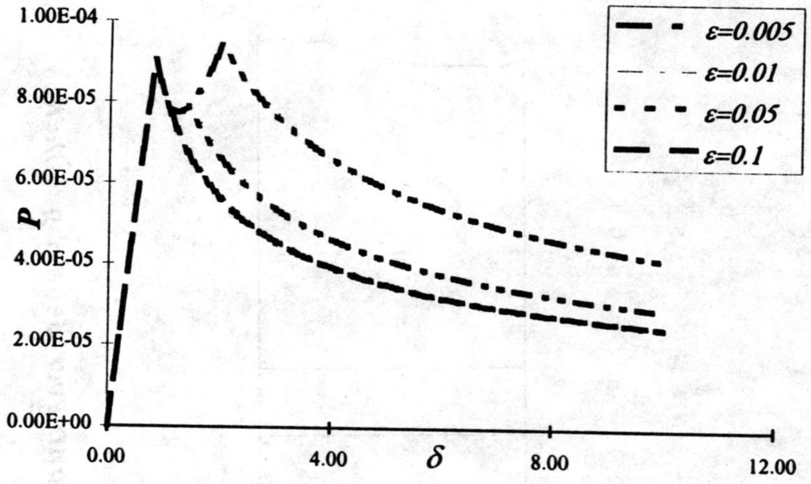

Fig. 4(a). P-δ plot with varying ε_u. G_{Ic}=1e-5, k=1e-3.

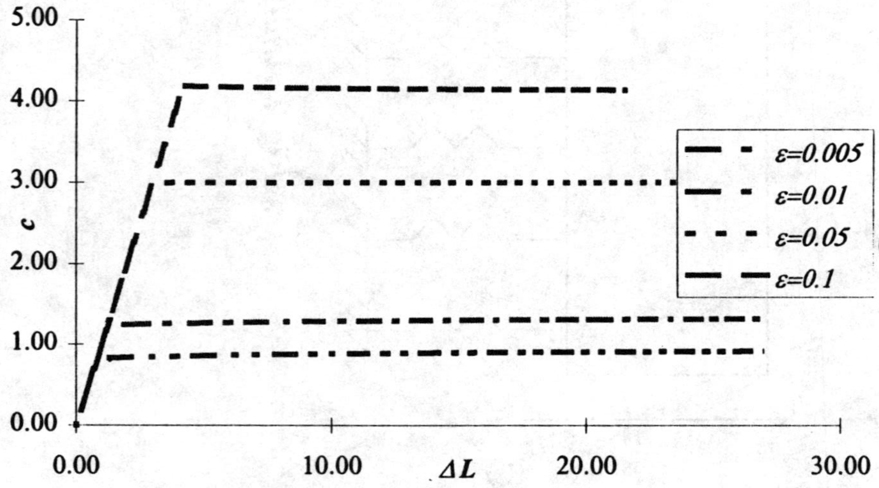

Fig.4(b). c-ΔL, G_{Ic}=1e-5, k=1e-5.

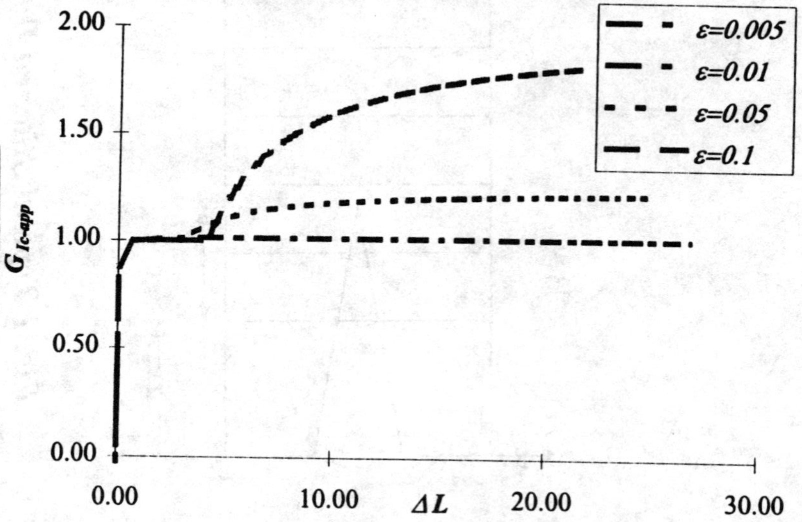

Fig. 4(c). G_{Ic-app}-ΔL, G_{Ic}=1e-5, k=1e-5

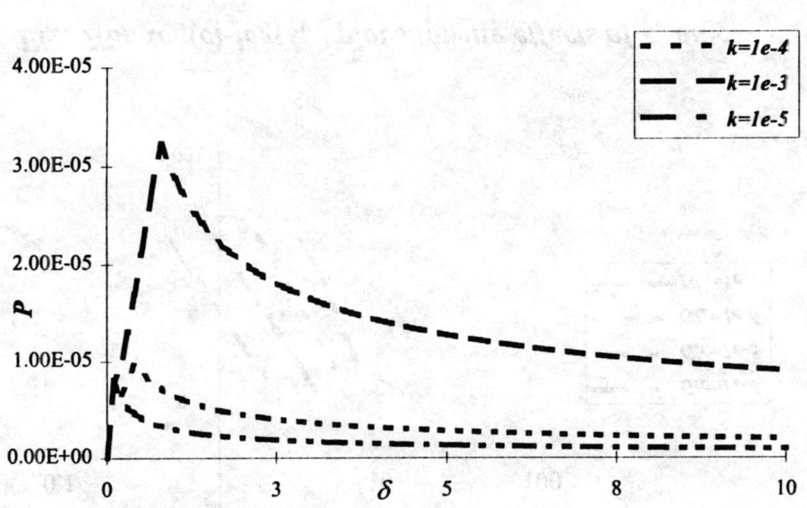

Fig. 5(a). P-δ plot G_{Ic}*=1e-7,* ε_u*=0.05*

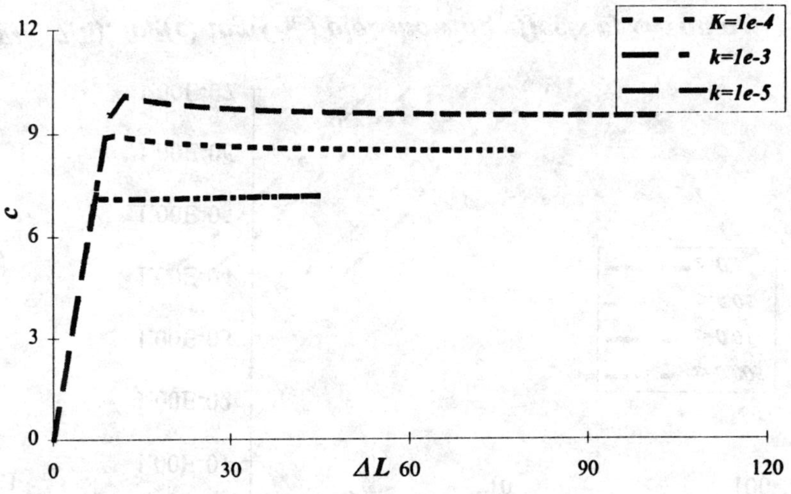

Fig. 5(b). c-ΔL plot G_{Ic}*=1e-7,* ε_u*=0.05*

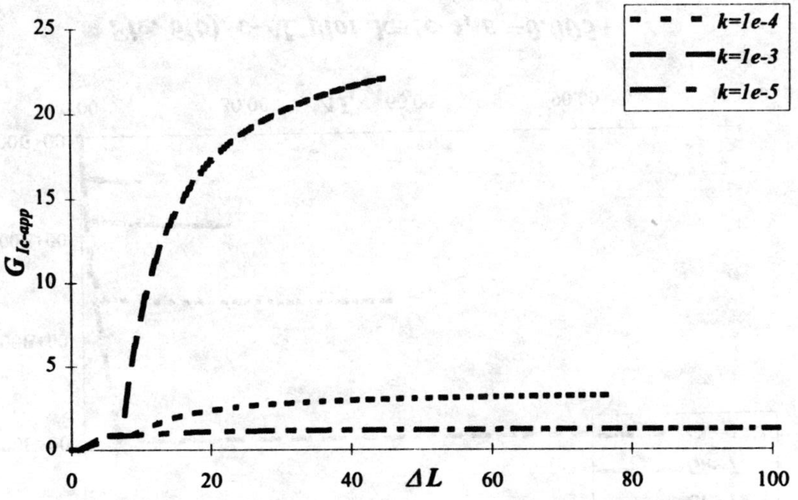

Fig. 5(c). G_{Ic-app}*-ΔL plot* G_{Ic}*=1e-7,* ε_u*=0.05*

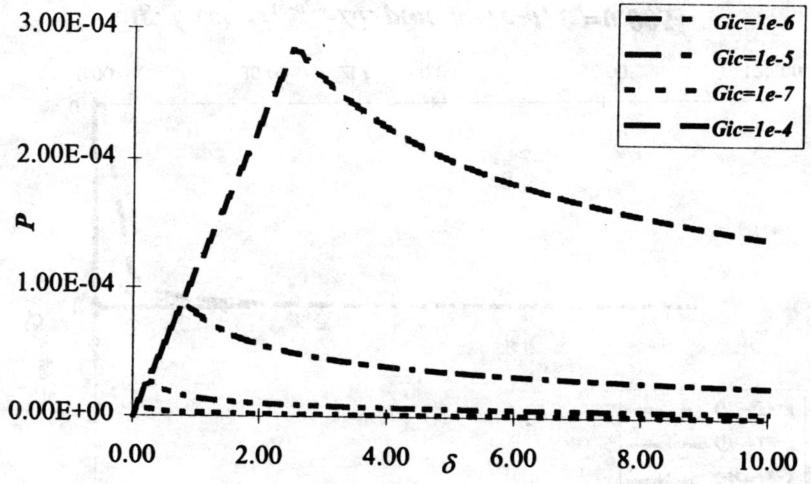

Fig. 6(a). P-δ plot k=1e-4, ε_u*=0.005*

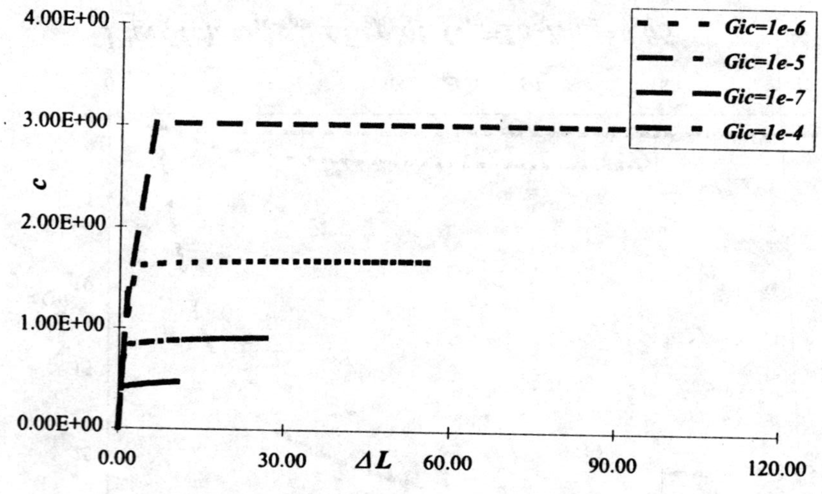

Fig. 6(b). c-ΔL plot k=1e-4, ε_u=0.005

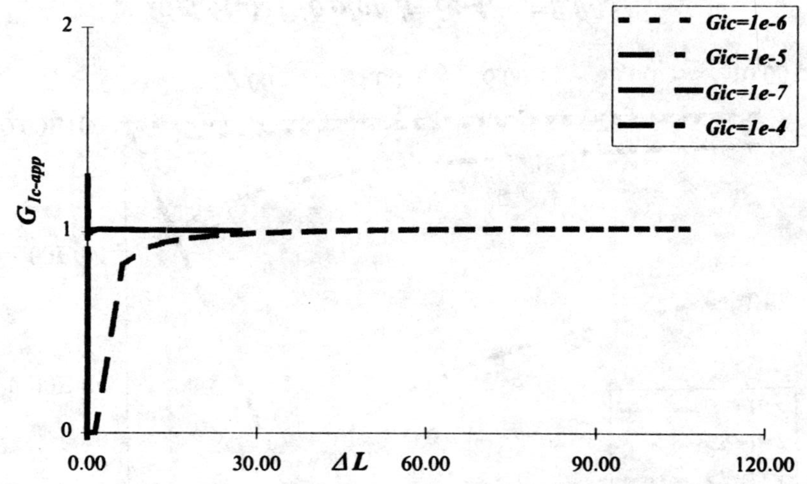

Fig. 6 (c). G_{Ic-app}-ΔL plot k=1e-4, ε_u=0.005

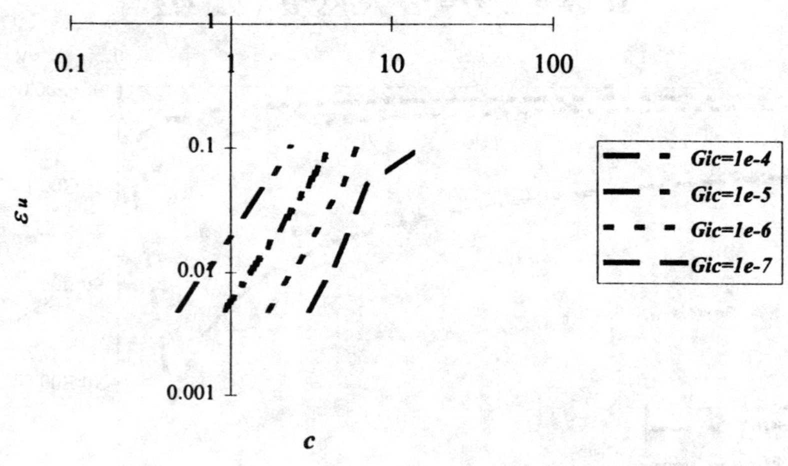

Fig. 7(b). log(c)-log(ε_u) plot showing effects of ε_u on c

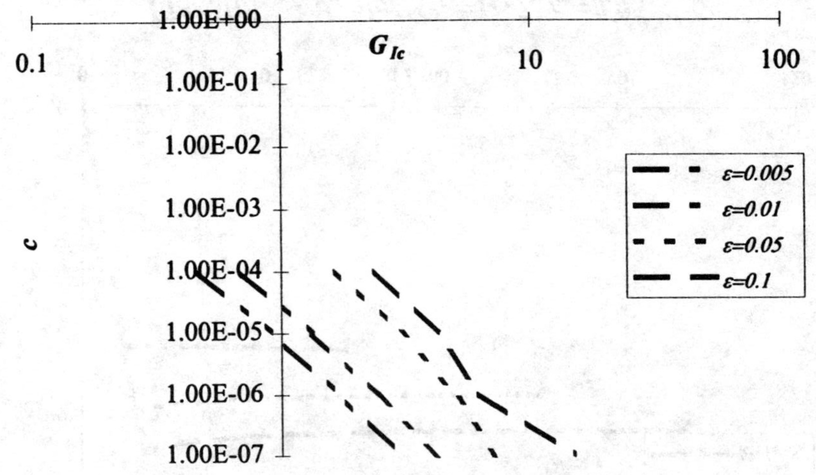

Fig. 7(b). log(c)-log(G_{Ic}) plot showing effects of G_{Ic} on c

Composite Beams on Elastic Foundations

N. NAIK

ABSTRACT

The behavior of deep composite beams on elastic foundation under transverse concentrated load has been evaluated using higher order shear deformation theory. A trignometric term is utilized to accurately describe shear deformation. The deformation and the induced stress state as a function of beam mechanical properties have been investigated.

INTRODUCTION

Portable bridges on marshy land for military applications and stiffeners of the pressurised cabins are the typical examples of beams on elastic foundations. In these applications, the structural components are normally made using polymer matrix composties. Reinforcing strands/yarns/fibers in polymer matrix composites are the other examples. Some studies are available on the behavior of beams/plates on elastic foundation under transverse loading (Hetenyi, 1961; Huang and Zou, 1994; Kobayashi and Sonoda, 1989; Voyiadjis and Kattan, 1990; Yokoyama, 1996). For the analysis of deep composite beams on elastic foundation, consideration of higher order shear deformation is necessary.

In the present analytical component of the work, the behavior of deep composite beam on an elastic foundation under transverse concentrated load and uniformly distributed load has been evaluated using higher order shear deformation theory. In this theory, a trignometric term is utilized to accurately describe shear deformation. In the analysis,

---

(Victoria Jubilee Technical Institute, Bombay University)
Quarter B-7, Central Area
IIT Campus, Powai
Mumbai - 400 076., India.

Winkler elastic foundation has been considered. The deformation and the induced stress state are calculated using higher order shear deformation theory, and the results are compared with those using Euler-Bernouli beam theory. The deformation and the induced stress state as a function of beam mechanical propeties have been investigagated. Predicted values are compared with the experimental results.

ANALYTICAL FORMULATION

Analytical formulation is presented for deflection and stress state considering shear deformation effects. A beam of length L and rectangular cross section with width b and height h, resting on Winkler elastic foundation is shown in Figure 1. Here, u and v are displacemetns of a beam element in the X and Y directions respectively, and x is the distance of the element from the origin along the X axis and y is the distance from the neutral axis along the Y axis.

Under the influence of vertical forces in the principal plane, the beam deflects vertically and the foundation offers reaction forces along the entire span. The reaction at any point is Kv, where, $K = K_o\, b$ and K_o is the elastic foundation modulus. The effect of friction between the beam and the foundation has been ignored.

Considering an element of the beam at a distance y from the neutral axis and a distance x from the origin, and expressing shear deformation as a trignometric function,

$$u = -y\frac{dv}{dx} + \frac{h}{\pi}\phi \sin\left[\frac{\pi y}{h}\right]$$

where, v and ϕ are functions of x.

Normal strain in the X direction, ε_x and shear strain in the XY plane, γ_{xy} are given by the following expressions:

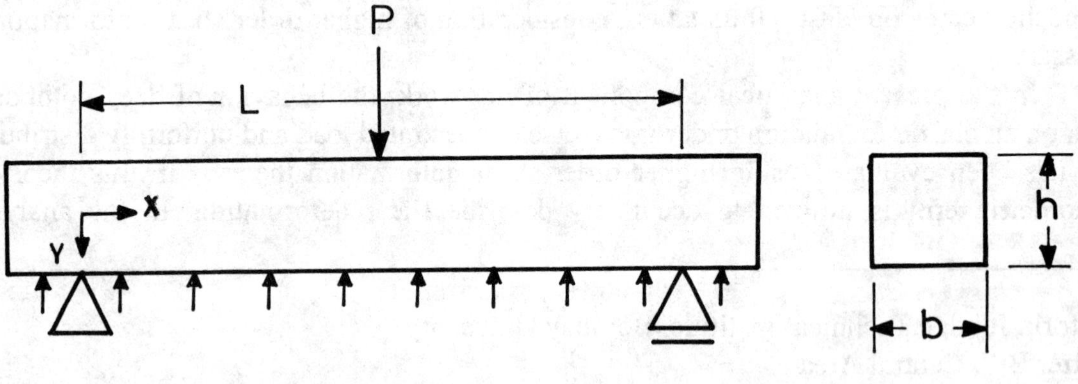

Figure 1. Schematic arrangement of composite beam on elastic foundation.

$$\varepsilon_x = \frac{\partial u}{\partial x} = -y\frac{d^2v}{dx^2} + \frac{h}{\pi}\sin\left[\frac{\pi y}{h}\right]\frac{d\phi}{dx}$$

$$\gamma_{xy} = \frac{\partial u}{\partial y} + \frac{dv}{dx} = \phi\cos\left[\frac{\pi y}{h}\right]$$

Normal stress, σ_x and shear stress, τ_{xy} are given by the following expressions :

$$\sigma_x = E\,\varepsilon_x = -E\,y\frac{d^2v}{dx^2} + \frac{E\,h}{\pi}\sin\left[\frac{\pi y}{h}\right]\frac{d\phi}{dx}$$

$$\tau_{xy} = G\,\gamma_{xy} = G\,\phi\cos\left[\frac{\pi y}{h}\right]$$

where, E is the modulus of elasticity in the X direction, and G is the modulus of rigidity in the XY plane.

For a portion of the beam between two cross sections at a distance dx apart and subjected to uniformly distributed load of magnitude q, if dx is very small, then vertical displacement of the element is v. Thus, total external force on this portion is given by [(q-Kv) dx]. If δv is virtual displacement in the Y direction given to this element, then external work done is given by [(q-Kv) dx δv].

Applying the principle of virtual work to an infinitesimal element of the beam,

$$\int_o^L\int_{-h/2}^{h/2}\int_{-b/2}^{b/2}\left(\sigma_x\,\delta\varepsilon_x + \tau_{xy}\,\delta\gamma_{xy}\right)dz\,dy\,dx$$

$$= \int_o^L (q - Kv)\,\delta v\,dx$$

where, $\delta\varepsilon_x$ is the virtual normal strain along the X axis, and $\delta\gamma_{xy}$ is the virtual shear strain in the XY plane.

Integrating with respect to z and y and bringing all terms to the left hand side,

$$\int_0^L\left[E\,I\frac{d^2v}{dx^2}\frac{d^2\delta v}{dx^2} - \frac{24\,EI}{\pi^2}\frac{d^2v}{dx^2}\frac{d\delta\phi}{dx}\right.$$

$$-\frac{24\,EI}{\pi^3}\frac{d\phi}{dx}\frac{d^2\,\delta v}{dx^2} + \frac{6\,EI}{\pi^2}\frac{d\phi}{dx}\frac{d\delta\,\phi}{dx}$$

$$\left.+\frac{6\,GI}{h^2}\,\phi\,\delta\,\phi\right]dx - \int_o^L (q - Kv)\;\delta v\,dx = 0$$

where, I is the moment of inertia of the beam cross section about the neutral axis.

Integrating with respect to x and grouping together similar terms,

$$\left[\left(EI\frac{d^2v}{dx} - \frac{24\,EI}{\pi^3}\frac{d\phi}{dx}\right)\frac{d\delta\ v}{dx}\right]_o^L$$

$$+\left[\left(-EI\frac{d^3v}{dx^3} + \frac{24\,EI}{\pi^3}\frac{d^2\phi}{dx^2}\right)\delta\ v\right]_o^L$$

$$+\left[\left(-\frac{24\,EI}{\pi^3}\frac{d^2v}{dx^2} + \frac{6\,EI}{\pi^2}\frac{d\phi}{dx}\right)\delta\ \phi\right]_o^L$$

$$+\int_o^L\left(EI\frac{d^4v}{dx^4} - \frac{24\,EI}{\pi^3}\frac{d^3\phi}{dx^3} - q + Kv\right)\delta\ v\ dx$$

$$+\int_o^L\left[\frac{24\,EI}{\pi^3}\frac{d^3v}{dx^3} - \frac{6\,EI}{\pi^2}\frac{d^2\phi}{dx^2} + \frac{6\,GI}{h^2}\phi\right]\delta\phi\ dx = 0$$

Differential equations for the given problem are as follows :

$$EI\frac{d^4v}{dx^4} - \frac{24\,EI}{\pi^3}\frac{d^3\phi}{dx^3} - q + Kv = 0$$

$$\frac{24\,EI}{\pi^3}\frac{d^3v}{dx^3} - \frac{6\,EI}{\pi^2}\frac{d^2\phi}{dx^2} + \frac{6\,GI}{h^2}\phi = 0$$

Solutions for the above equations under given loading for a simply supported beam are as follows :

$$v = \sum_{m=1}^{\infty} C_m \sin\left[\frac{m\pi\ x}{L}\right]$$

$$\phi = \sum_{m=1}^{\infty} D_m \cos\left[\frac{m\pi\ x}{L}\right]$$

C_m and D_m are Fourier constants given by the following equations.

$$C_m = \frac{q_m S_m}{(P_m S_m - R_m^{\ 2})}$$

$$D_m = \frac{q_m R_m}{(P_m S_m - R_m^{\ 2})}$$

$$\text{where, } S_m = 6\,EI\left[\frac{m}{L}\right]^2 + \frac{6\,GI}{h^2}$$

$$P_m = EI\left[\frac{m\pi}{L}\right]^4 + K$$

$$R_m = 24\,EI\left[\frac{m}{L}\right]^3$$

$$q_m = \frac{2}{L}\int_0^L Q \sin\left[\frac{m\pi\,x}{L}\right] dx$$

Q is the given loading condition as a function of x.

If the entire beam is subjected to a uniformly distributed load of magnitude q, then,

$$q_m = \frac{2\,q}{L}\int_0^L \sin\left[\frac{m\pi\,x}{L}\right] dx$$

$$q_m = 0 \qquad \text{(For even m)}$$

$$= \frac{4\,q}{m\pi} \qquad \text{(For odd m)}$$

If the beam is subjected to a single concentrated load, P, at a distance a from the origin, then,

$$q_m = \frac{2\,P}{L} \sin\left[\frac{m\,\pi\,a}{L}\right]$$

EXPERIMENTAL DETAILS

Experimental studies were carried-out on different composite beams to study the deflection and failure behavior. Coil spring (K_o = 14.63 x 10^6 N/m^3), rubber with large depth of 185 mm (K_o = 9.34 x 10^6 N/m^3) and rubber with shallow depth of 50 mm (K_o = 10.50 x 10^6 N/m^3) were used as elastic foundations. Thicknesses of composite specimens were 8.4, 7.3, 4.9 and 5.1 mm respectively for UD Glass/Epoxy specimen 1, UD Glass/Epoxy speciment 2, WF Glass/Epoxy and WF Carbon/Epoxy.

Mechanical properties of the compostie beams are given in Table I. Here, F_t, F_c and F_s are respectively tensile, compressive and shear strength values. The properties are with repsect to X direction. Experiments were conducted with simply supported ends and free ends. The simply supported length was 180 mm and free length was 255 mm. Same values were used for analytical predictions. Width for WF Glass/Epoxy specimen was 20 mm and for the others was 10 mm.

The experimental setup is shown in Figures 2 & 3. Experimental results at the center of the beam under central concentrated load are presented in Table II. It was observed that failure initiated with compressive failure on the upper surface of the beams.

RESULTS AND DISCUSSION

Using the analytical methodology presented earlier, deflection at failure, corresponding stress state and the load at the center of the beam were predicted and are presented in Table III. Here, σ_{bf}, σ_{sf}, Y_f and P_f are the bending stress, shear stress, deflection and the concentrated load at the central point at failure. It can be seen from Tables II and III that there is a good match between the predictions and the experimental values for UD Glass/Epoxy specimens. For the other two specimens, experimental deflections and the corresponding loads are significantly higher than the predictions. Possibly, the micro-damage had taken place much earlier than observed by experiments.

Effect of composite beam thickness was investigated analytically and the results are presented in Table IV. Predictions considering shear deformation are compared with those of simple bending theory (Hetenyi, 1961). It can be seen that the deflections are higher when shear deformation is considered. As the thickness of the beam increses, the difference increases. For a typical beam with thickness of 10 mm, the difference is 4.3% whereas for a thicker beam with thickness of 25 mm, the difference is 24.2% (Table IV). Hence, for the analysis of deep composite beams on elastic foundations, consideration of higher order shear deformation is necessary. It can be seen that the stress state is nearly identical.

Further, the effect of elastic foundation modulus on deflection and stresses was studied. Here, elastic foundation modulus was considered to be zero. For beams with lower thickness, the difference in deformation between considering shear deformation and considering simple bending theory remained nearly the same. As the thickness increased, the percentage difference increased compared to the case where K_o = 14.63 x 10^6 N/m^3.

TABLE I - MECHANICAL PROPERTIES OF THE COMPOSITE BEAMS

Material	E (GPa)	G (GPa)	F_t (MPa)	F_c (MPa)	F_s (MPa)
UD Glass / Epoxy	35	3.0	800	600	36
WF Glass / Epoxy	20	3.2	320	200	26
WF Carbon / Epoxy	90	3.2	440	400	42

TABLE II - BEHAVIOR OF COMPOSITE BEAMS : EXPERIMENAL RESULTS

Material	SIMPLY SUPPORTED ENDS		FREE ENDS	
	P_{max} (N)	y_{max} x 10^{-3} m	P_{max} (N)	y_{max} x 10^{-3} m
SPRING AS ELASTIC FOUNDATION				
UD Glass / Epoxy 1	1400	9.3		
UD Glass / Epoxy 2	1035	10.3	602	20
WF Glass / Epoxy	890	20.6		
WF Carbon / Epoxy	750	7.7	517	18
RUBBER AS ELASTIC FOUNDATION (SHALLOW DEPTH)				
UD Glass / Epoxy 1	1000	10.8		
UD Glass / Epoxy 2	1000	10.0	230	23
WF Glass / Epoxy	870	20.8		
WF Carbon / Epoxy	760	7.5		
RUBBER AS ELASTIC FOUNDATION (LARGE DEPTH)				
UD Glass / Epoxy 1	1500	10.0		
UD Glass / Epoxy 2	1020	10.6		
WF Glass / Epoxy	706	19.5		
WF Carbon / Epoxy	805	8.2		

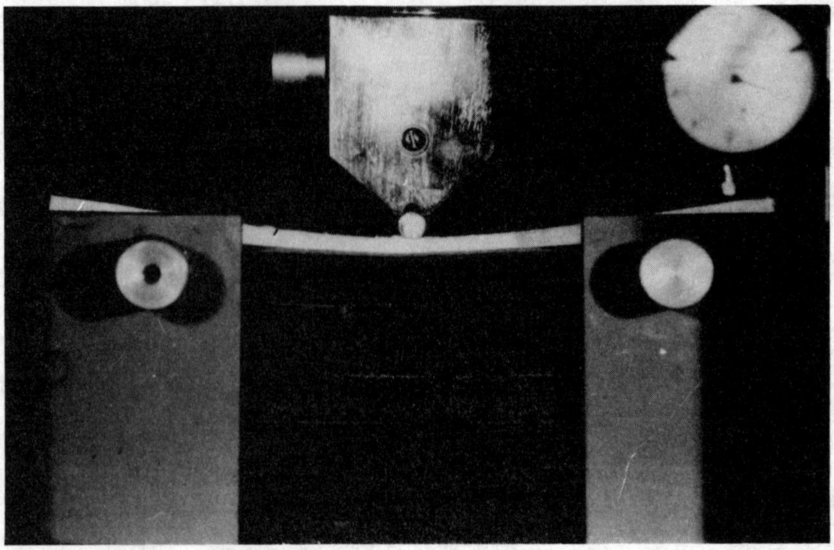

Figure 2. Experimental set-up : composite beam on rubber of large depth as foundation.

Figure 3. Experimental set-up : composite beam on spring as foundation.

TABLE III - PREDICTION OF FAILURE INITIATION WITH RUBBER (LARGE DEPTH) AS FOUNDATION

Material	σ_{bf} (MPa)	σ_{sf} (MPa)	y_f x 10^{-3} m	P_f (N)
SIMPLY SUPPORTED ENDS CONSIDERING SHEAR DEFORMATION				
UD Glass / Epoxy 1	602	-9.82	11.3	1640
UD Glass / Epoxy 2	602	-8.70	12.9	1260
WF Glass / Epoxy	201	-2.53	10.3	510
WF Carbon / Epoxy	408	-4.16	4.9	460

TABLE IV - PREDICTION OF DEFLECTION AND STRESSES : EFFECT OF BEAM THICKNESS

UD Glass/Epoxy, K_o = 14.63 x 10^6 N/m^3

b = 10 mm, L = 180 mm, P = 1000 N

h x 10^{-3}m	y x 10^{-3}m	σ_b (MPa)	σ_s (MPa)
A) SIMPLY SUPPORTED ENDS CONSIDERING SHEAR DEFORMATION			
10	4.12	258	-5.0
15	1.33	118	-3.3
20	0.60	67	-2.5
25	0.33	45	-2.0
B) SIMPLY SUPPORTED ENDS CONSIDERING SIMPLE BENDING THEORY			
10	3.95	259	-5.0
15	1.21	118	-3.3
20	0.52	67	-2.5
25	0.26	43	-2.0

CONCLUSIONS

An analytical formulation is presented for the behavior of composite beams on elastic foundations considering higher order shear deformation.

It is observed that for the analysis of deep composite beams on elastic foundation, consideration of higher order shear deformation is necessary for the accurate prediction of the deflection. But, there is no significant difference in the stress state.

In general, a good match is observed between the predictions and the experimental results.

ACKNOWLEDGMENT

I would like to thank Prof. R.P. Shimpi of Aerospace Engineering Department, IIT, Mumbai, for the useful discussion and guidance.

REFERENCES

Hetenyi, M. 1961. "Beams on Elastic Foundation", The University of Michigan Press, Ann Arbor.

Huang, W. and Y. Zou. 1994. "Linear Analysis Method for the Dynamic Response of a Winkler Foundation Supported Elastic Plate Subjected to Low Velocity Projectile Impact", Journal of Vibration and Acoustics, 116 : 67-70.

Kobayashi, H. and K. Sonoda. 1989. "Rectangular Mindlin Plates on Elastic Foundations", Journal of Mechanical Sciences, 31:679-692.

Voyiadjis, G.Z. and P.I. Kattan. 1990. "Bending of Thick Plates and Shells", in Advaces in the Theory of Plates and Shells, eds, Voyiadjis, G.Z. and D. Karamanlidis, Elsevier, Amsterdam, pp 87-121.

Yokoyama, T. 1996. "Vibration Analysis of Timoshenko Beam-Column on Two-Parameter Elastic Foundations", Computers & Structures, 61:995-1007.

Analysis and Design of Composite Materials for Simultaneous Heatsink Component and Thermal Expansion Constraint of Electronic Modules

B. J. SULLIVAN

ABSTRACT

Candidate composite material heatsinks designs for the electronic modules of avionics are reviewed for their ability to simultaneously act as a thermal plane for dissipation of heat and as a core for constraining the thermal expansion of the top surface of the printed wiring boards (PWB) to the coefficient of thermal expansion (CTE) of the surface mounted electronic chips. Specific designs are recommended based on conductance and mass. Stresses at the operating temperature extremes were also a factor in the screening study. The performance of the selected composite material is compared to that of the baseline heatsink design relative to heat transfer performance and vibration characteristics.

INTRODUCTION

Composite materials are being developed and qualified for use in electronic modules of many different applications. The properties which enhance their attractiveness in these applications include high thermal conductivity, low mass, high stiffness and very low or even negative coefficient of thermal expansion (CTE). This combination of properties makes them attractive for use as heatsink components which simultaneously constrain the thermal expansion of the top surface of printed wiring boards (PWB) to the CTE of the electronic chips mounted there. The lifetime of the chip to PWB joints can be directly correlated with the CTE mismatch between the electronic chips and the top surface of the PWB. Electronic module reliability is known to be improved with reductions in peak temperatures on the boards. Therefore composite materials offer improvements in two different areas which can dramatically improve the lifetime of electronic modules.

A screening study was performed first in order to identify the optimum composite material for use as the thermal plane/constraining core. Metal matrix composites including K-800X pitch fiber reinforced aluminum and K-800X pitch fiber reinforced copper matrix composites were considered. Organic matrix composites including K-800X pitch fiber reinforced cyanate ester (CE) and Amoco's Thermalgraph material impregnated with cyanate

Brian J. Sullivan, Materials Research & Design, Inc., 1024 E. Lancaster Avenue, Rosemont, PA 19010

ester resin were also considered. In addition, laminated structures consisting of copper or aluminum facesheets with K-800X/CE and Thermalgraph/CE at various orientations were examined.

The baseline design against which the composite heatsink must be compared is a thin aluminum panel with a thermal conductivity in the vicinity of 200 W/mK and a weight of just under one pound. Design requirements of the composite heatsink include maintaining the top surface of the PWB in the PWB/heatsink assembly to within a range of CTE's, e.g. 5 to 7 ppm/°C, keeping the weight of the composite heatsink less than that of the aluminum baseline while exceeding the conductance of the baseline design. Thickness limitations on the new heatsink design also exist for geometric and transverse vibration displacement constraints. Other design requirements include specified base excitation random vibration power spectral density, dynamic shock, and static equivalent "g" loads. Minimum and maximum temperatures expected to be experienced by the electronic modules were also considered as part of the screening and detailed design analyses relative to the residual stresses placed on the constituents of the laminated PWB/composite heatsink assemblies.

APPROACH

Analytical tools used in the screening and design analyses include classical lamination theory for property prediction of the assembled electronic module, finite element analysis of residual stresses arising from fabrication, assembly, and cooling of the electronic module, and closed form expressions for prediction of the natural frequency of the electronic modules. The screening study is conducted by determining the thickness of the composite material together with an appropriate fiber volume fraction which results in a heat sink capable of constraining the assembly to the 5 ppm/°C to 7 ppm/°C range, at the same or lower mass as the baseline design, and with the same or higher conductance.

At the conclusion of the screening study, finite element analyses were performed for the heat transfer analyses of both the baseline aluminum design and the selected composite material heatsink modules. Analysis of the random vibration, shock, and static equivalent loads on the baseline and selected composite design is also performed using finite element methods.

RESULTS OF SCREENING STUDIES

Since mechanical properties of the PWB material were not measured at the initiation of the screening study, properties were predicted based on the relative percentages of glass fibers, polyimide resin, and copper foil layers. A laminate model of the PWB material was constructed using the actual layup sequence and ply thicknesses. The in-plane elastic modulus of the PWB was calculated to be approximately 5.6×10^6 psi and the in-plane CTE was computed to be approximately 13×10^{-6} in/in/°C. The hard bond adhesive joining the PWBs to the heatsink was treated as a standard epoxy resin with a modulus of 500,000 psi and a CTE of 55×10^{-6} in/in/°C.

Figure 1 shows the CTE versus 0/90 K-800X/Al thickness for the $v_f = 0.43$ and $v_f = 0.47$ Gr/Al MMCs. This plot shows that the maximum thickness heatsink for the $v_f = 0.47$ composite is just over 0.1 inch and the maximum thickness heatsink for the $v_f = 0.43$ composite is 0.155 inch. Figure 2 shows the 0/90 K-800X/Al heatsink mass versus thickness for all permissible thicknesses (as determined from Figure 1) of both the $v_f = 0.43$ and $v_f = 0.47$ composites. Figure 2 shows that all thicknesses of the $v_f = 0.47$ Gr/Al composites which result in module CTE's in the 5-7 ppm/°C range will have a lower mass than that of the current baseline heatsink. Figure 2 also shows that thicknesses of the $v_f = 0.43$ Gr/Al composites up to 0.151 inch will have a lower mass than that of the current baseline.

Figure 3 shows that the Gr/Al MMC design with the highest possible conductance will be obtained using the 0/90 K-800X/Al at $v_f = 0.43$ and at the 0.151 inch thickness. At this thickness, the 0/90 K-800X/Al composite of $v_f = 0.43$ has the same mass as the baseline (310 grams) but has a conductance of 16.2 W cm/°K, a 102% increase over the calculated conductance (8 W cm/° K) of the current baseline.

Due to a design thickness limitation at the rail, another viable Gr/Al design would be to use a 0.130 inch uniform thickness Gr/Al composite heatsink to allow the MMC to be cast using the simplest possible mold at a thickness equal to the maximum allowable thickness at the rail. The mass of the 0.130 inch 0/90 K-800X/Al heatsink would be about 267 grams (~14% less than the baseline mass) and the conductance of the 0.130 inch Gr/Al heatsink would be about 14 W cm/°K (~74% greater than the baseline).

The same curves, i.e. assembled module CTE, heatsink mass, and heatsink conductance, were next generated for 0/90 K-800X/Cu composites. For $v_f = 0.47$ Gr/Cu composites, thicknesses of the 0/90 K-800X/Cu heatsinks in the approximate range of 0.05 to 0.16 inches will result in a CTE in the 5 to 7 ppm/°C range at the top surface of the PWB. The mass calculation indicates that the maximum allowable Gr/Cu thickness corresponds to the $v_f = 0.47$ 0/90 K-800X/Cu composite at a thickness of 0.065 inch. At this thickness, the conductance of the Gr/Cu design is approximately 25% greater than that of the baseline design.

The CTE, heatsink mass, and heatsink conductance curves for 0/90 K-800X/CE composites were calculated next. The results for these composites demonstrate that appropriate thicknesses of K-800X/CE heatsinks will be too thin to provide a conductance equivalent to the baseline heatsink. A similar finding holds for heatsinks of the ThermalGraph/CE composites.

Laminated structures composed of copper or aluminum facesheets in combination with K-800X/CE tape oriented at 90° and SRG/CE oriented at 0° were examined next. Design calculations performed on laminated structure heatsinks with copper facesheets were found a) to provide no conductance advantage over aluminum facesheet designs, b) to have greater mass than those with aluminum facesheets, and c) to experience yielding in the copper facesheets thus limiting the range of linear CTE response at the PWB surfaces provided by these heatsinks. Two laminated structure heatsink designs aluminum facesheets which are viable candidate heatsink designs are listed in Table 1.

Figure 4 details the specifics of each candidate composite heatsinks and illustrates the conductance for each. The designs with the highest conductance are the 0.15 inch thick 0/90 K-800X/Al and the 0.173 inch thick aluminum facesheet laminated structure. Both of these designs have a mass equal to that of the baseline. Of the two 0.130 inch thick candidates, the 0/90 K-800X/Al MMC has a higher conductance and a higher mass than the same thickness aluminum facesheet laminated structure.

While there is no minimum natural frequency requirement imposed on the heatsink, since random vibration is a design loading with significant non-zero acceleration power spectral density contributions from 10 to 2000 Hz, root mean square and peak stresses will be

Table I. Summary of aluminum facesheet laminated structure heatsink designs.

Constituent	Thickness (inch)	Thickness (inch)
Al facesheet	.035	0.024
90° K-800X/CE	.022	.017
0° SRG/CE	.053	.042
90° K-800X/CE	.022	.017
Al facesheet	.035	.024
Total Thickness (inch)	.173	.130
Total Mass (grams)	310.2	229.1
Total Conductance (W cm/°K)	17.5	13.2

lowest in the electronic modules with heatsinks providing the highest natural frequencies of vibration. Figure 5 displays the first natural vibration frequency of each candidate heatsink design. The 0.15 inch thick 0/90 K-800X/Al heatsink has the highest natural frequency, followed closely by the 0.173 inch thick aluminum facesheet laminated structure.

The final step in the screening study was to calculate stresses in the candidate composite heatsinks at the operating temperature extremes (-40°C and 100°C). The hard bond adhesive joining the PWB to the heatsink can be cured as low as 40°C. The cure temperature for the laminated structures would also need to be as low as possible to minimize the high in-plane compressive stresses in the unidirectional layers. The stress-free temperature for all constituents was assumed to be 35°C. Tables II and III contain the stresses at -40°C for the 0.15 inch thick 0/90 K-800X/Al design and the 0.173 inch thick aluminum facesheet laminated structure design, respectively. Stresses for the 0.130 inch thick Gr/Al and aluminum facesheet laminated structure designs are very close to their maximum mass counterparts and are not listed here. The findings of these stress analyses indicate that the Gr/Al stresses are quite manageable. However, in the aluminum facesheet laminated structure designs, the in-plane compressive stresses in the 0° SRG/CE layer may exceed its compressive strength, estimated to be approximately 35 ksi.

Table II. Peak stresses in 0.150" 0/90 K-800X/Al $v_f = 0.43$ candidate heatsink at -40°C for stress-free temperature of 35°C.

Constituent	σ_{xx} (ksi)	σ_{xx}^{fail} (ksi)	σ_{yy} (ksi)	σ_{yy}^{fail} (ksi)	σ_{zz} (ksi)	σ_{zz}^{fail} (ksi)	σ_{xy} (ksi)	σ_{xy}^{fail} (ksi)
Gr/Al	-4.0	-20.5	1.9	1.5	-4.0	-20.5	-1.2	N/A
Adhesive	2.5	5.0	1.0	5.0	2.5	5.0	-1.5	2.5
PWB	3.8	55.7	0.6	10.0	3.8	55.7	-1.2	5.0

Table III. Peak stresses in 0.173" Al facesheet laminated structure candidate heatsink at -40°C for stress-free temperature of 35°C.

Constituent	σ_{xx} (ksi)	σ_{xx}^{fail} (ksi)	σ_{yy} (ksi)	σ_{yy}^{fail} (ksi)	σ_{zz} (ksi)	σ_{zz}^{fail} (ksi)	σ_{xy} (ksi)	σ_{xy}^{fail} (ksi)
0° SRG/CE	-39.5	~ -35.0	3.52	1.4	1.20	1.4	-2.68	1.5
90° K-800X/CE	1.4	10.0	1.87	5.0	-42.8	-45.0	-3.13	8.0
Adhesive	2.85	5.0	1.31	5.0	2.81	5.0	-3.13	2.5
Aluminum	20.2	21.0	1.31	21.0	19.4	21.0	-2.68	12.0
Adhesive	2.85	5.0	0.8	5.0	2.81	5.0	1.39	2.5
PWB	3.63	55.7	0.8	10.0	3.16	55.7	0.94	5.0

CONCLUSIONS AND RECOMMENDATIONS OF SCREENING STUDY AND COMPARISON TO BASELINE DESIGN

The main findings of the screening study were as follows:

1) The 0.150 inch thick 0/90 K-800X/Al at $v_f = 0.43$ appears to be the best composite heatsink design. It has a high conductance, the highest natural frequency, the lowest stresses, and likely the lowest cost of all its composite heatsink competitors. Due to manufacturability issues, the 0.130 inch thickness heatsink is also an excellent candidate. The 0.130 inch thickness 0/90 Gr/Al design still ranks high in the screening study.
2) The 0.173 inch thick aluminum facesheet laminated structure design may also be a viable heatsink design. This design had the highest conductance and a high natural frequency; however, there is concern over the stresses predicted at -40°C. Its cost would also likely be higher than the Gr/Al heatsinks.
3) The Gr/Cu design is not competitive with the Gr/Al or aluminum facesheet laminated structure designs, since its higher density limits the allowable thickness, thus curtailing the conductance.
4) Designs fabricated with K-800X or SRG reinforcement in organic matrices are unable to match the conductance of the baseline heatsink.

As a result of these design calculations, it was recommended that heatsink qualification design analyses and tests proceed with the Gr/Al MMC heatsinks. The 0.130 inch thick design was selected for fabrication. Prior to testing, predictive analyses are being performed relative to the vibration response. Thc finite element heat transfer analyses of the 0.190 inch thick aluminum baseline and 0.130 inch thick Gr/Al heatsink module were completed. Three dimensional heat transfer finite element analysis models of the modules were constructed including the circuit board, adhesive, and heat sink. Analyses were conducted to simulate the conditions of actual CDI tests; specifically, all board and heatsink exposed surfaces were assumed to be insulated to eliminate convection heat transfer from these surfaces. Analyses for the current 30W heat load condition and for a projected 60W heat load case were each performed. Data provided by the vendor were used to determine the heat loads per unit area for both the 30W and 60W heat load analyses. Table IV provides the complete list of peak temperatures calculated for all of the MR&D heat transfer analyses.

Table IV. Summary of heat transfer analysis results for 10SU heatsink avionics modules.

Heatsink Model	Heatsink Thickness (inch)	Total Heat Load (Watts)	Minimum Rail Temperature (°C)	Maximum PWB Temperature (°C)
Baseline	0.19	30	76.0	88.95
Baseline	0.19	60	76.0	99.41
Gr/Al	0.13	60	76.0	95.68

It is apparent from this table that the Gr/Al heatsink, even at the reduced thickness of 0.13 inch from the baseline 0.19 inch thickness, offers a considerable improvement in peak board temperature over the CDI baseline design.

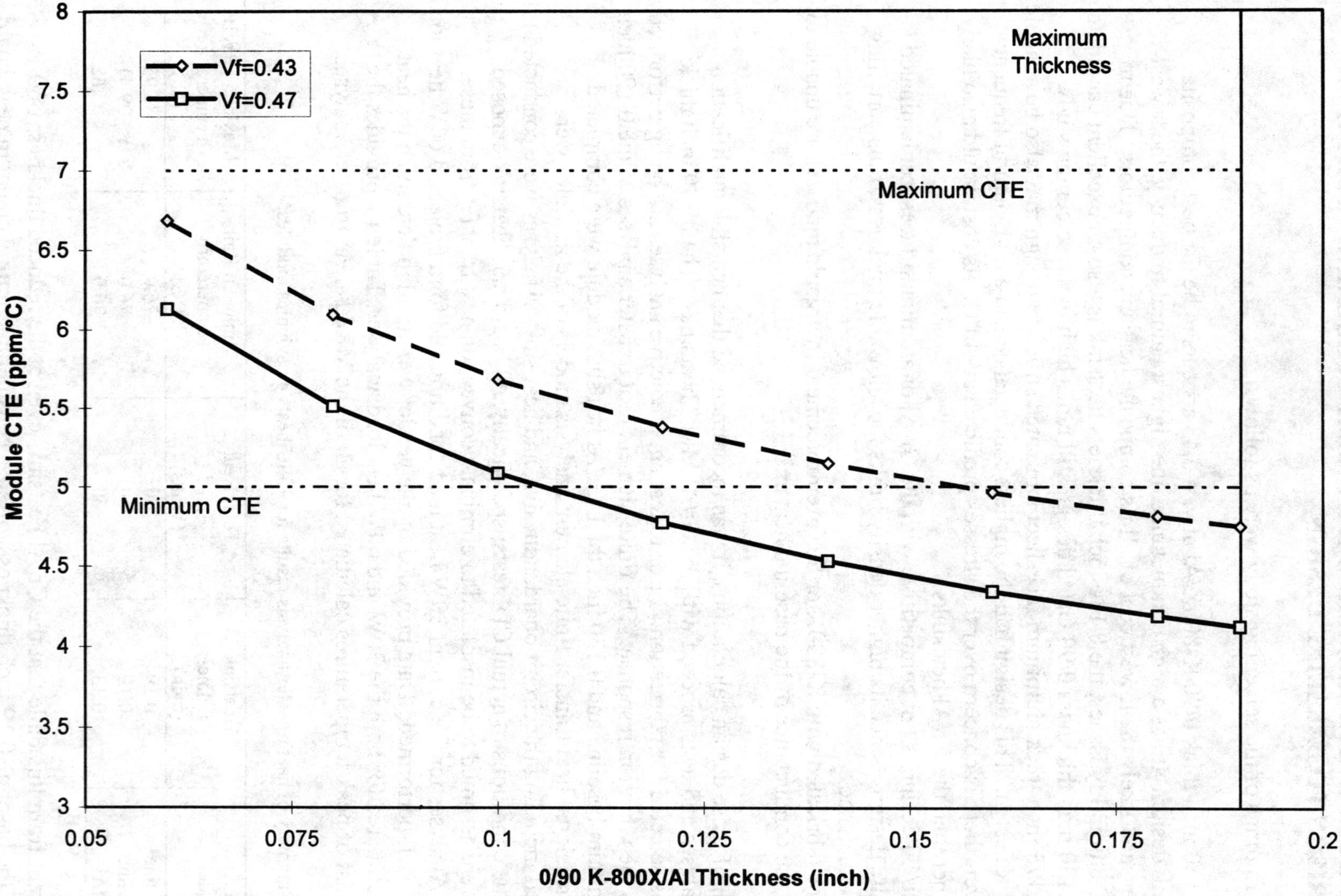

Figure 1. Effect of fiber volume fraction and Gr/Al thickness on electronic module coefficient of thermal expansion.

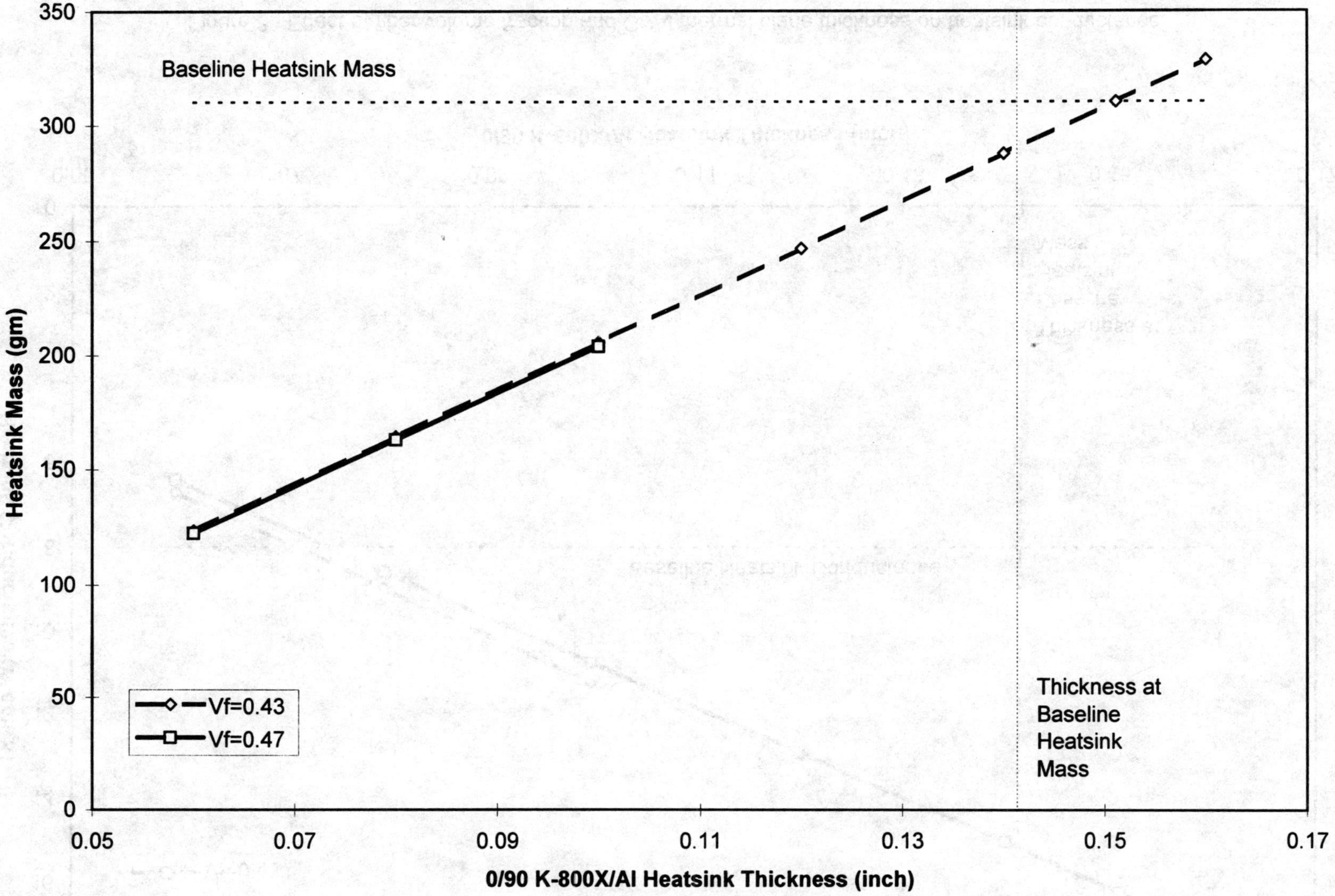

Figure 2. Effect of fiber volume fraction and Gr/Al thermal plane thickness on heatsink mass.

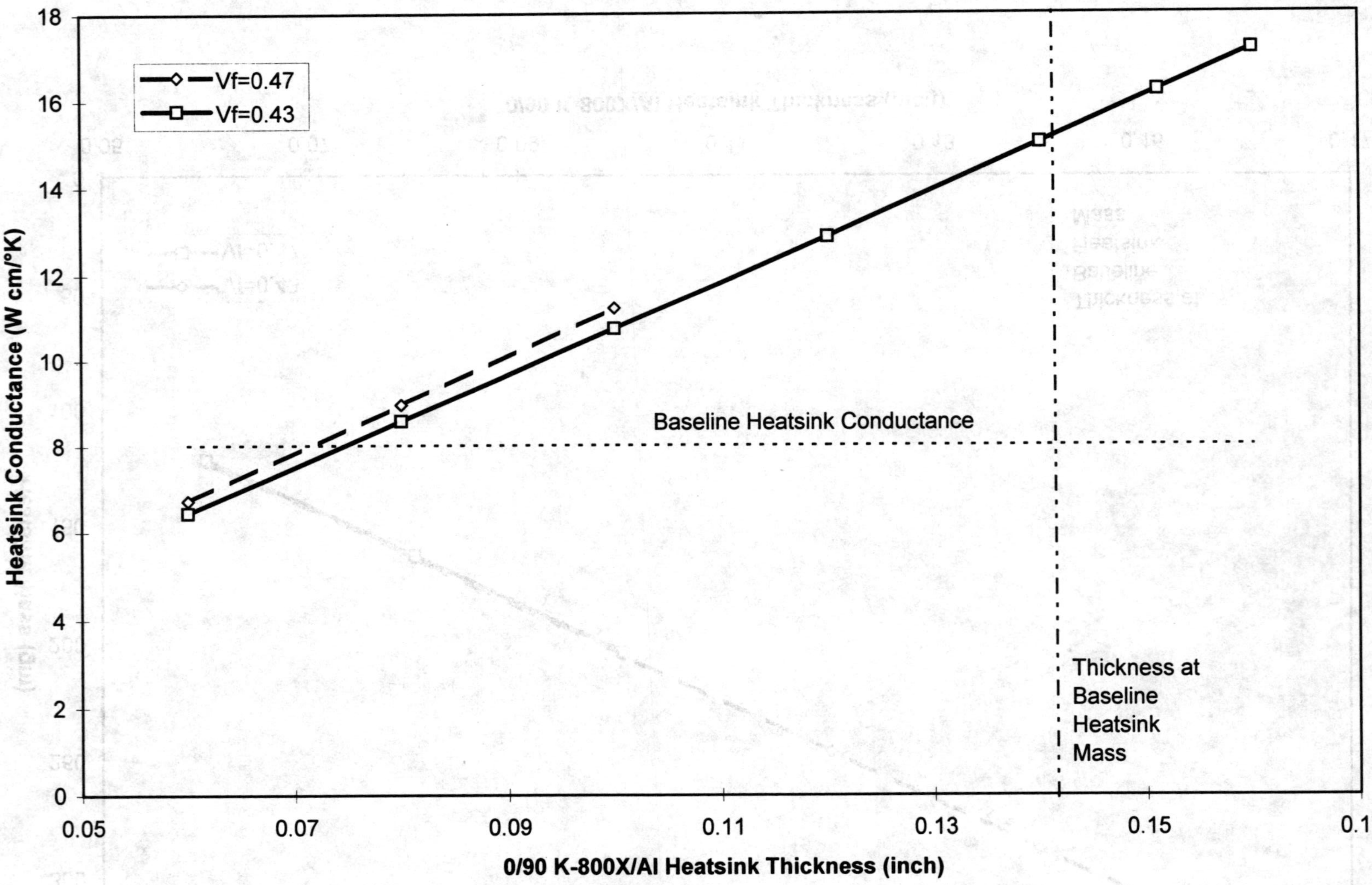

Figure 3. Effect of fiber volume fraction and Gr/Al thermal plane thickness on heatsink conductance.

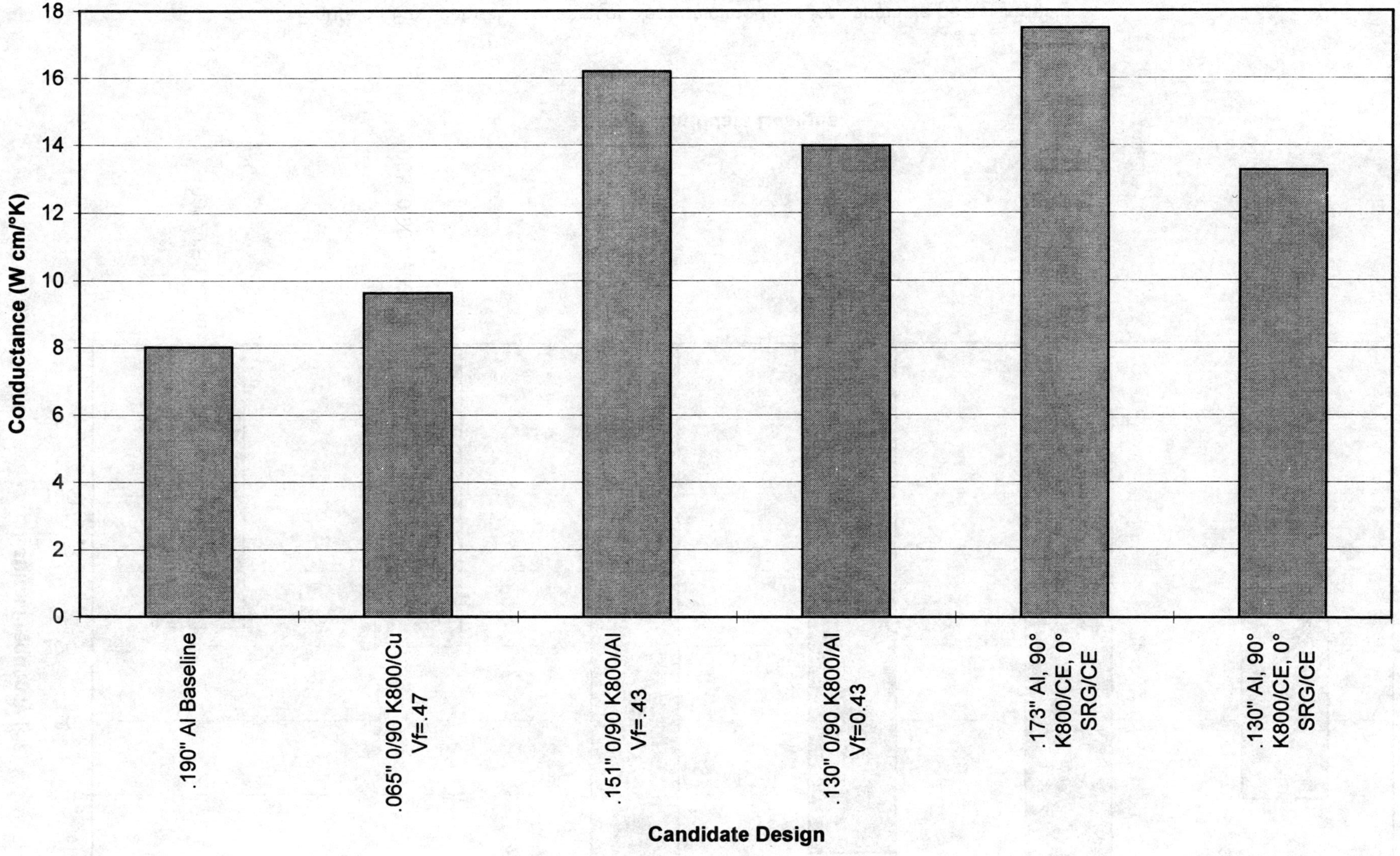

Figure 4. Conductance of candidate heatsink designs.

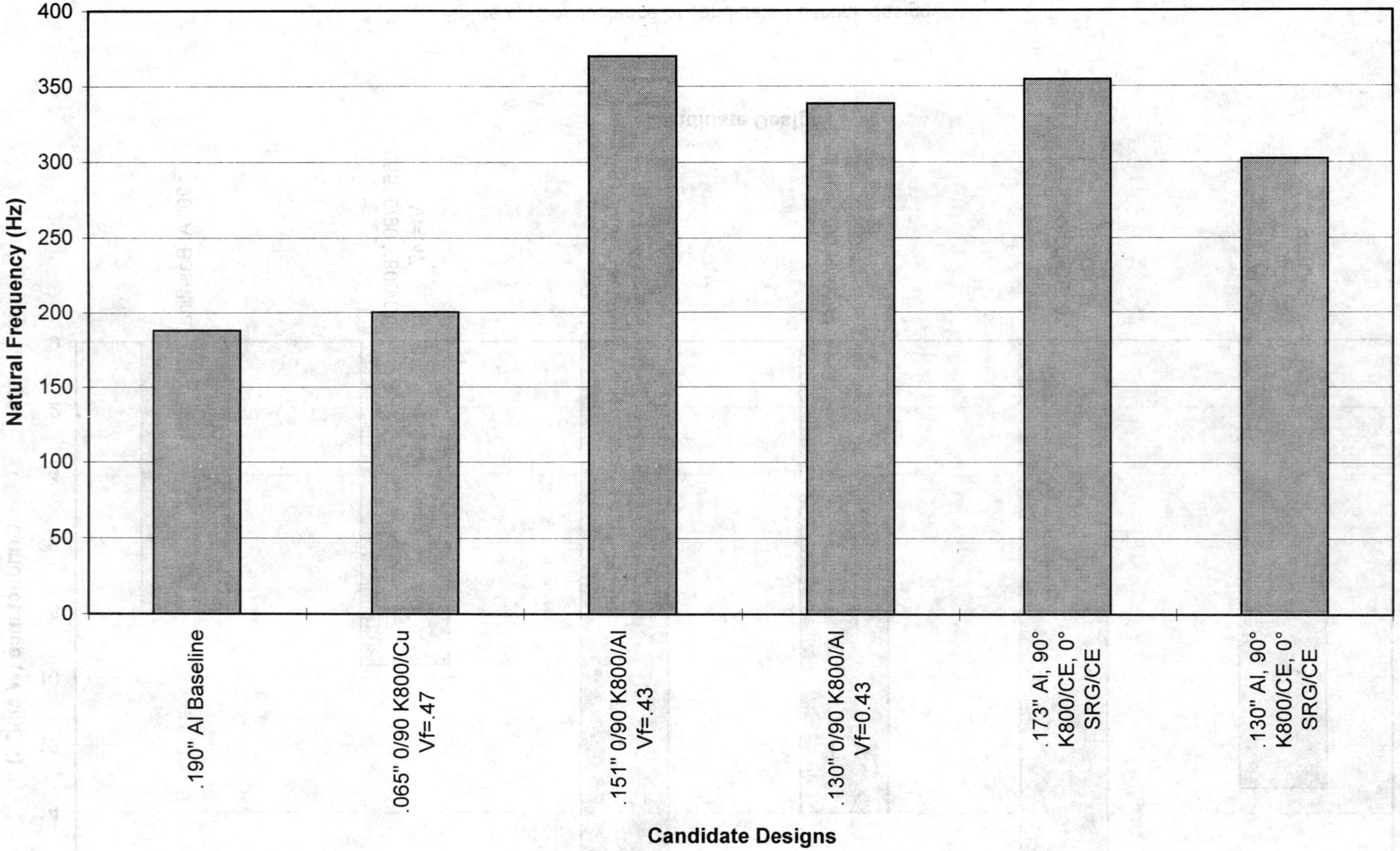

Figure 5. First natural frequency of electronic modules for candidate heatsink designs.

Analysis of a Fiber Composite Cylinder Laminated with an Elastic-Plastic Aluminum Liner under Internal Pressure Loading

S. R. SWANSON AND M. OFFOLTER

ABSTRACT

The basic equations are given for the analysis of a fiber composite cylinder wrapped over a ductile liner. The stiffness of the laminated cylinder is based on classical lamination theory along with thin-shell assumptions, in the linear range. After yielding of the liner, an incremental tangent matrix is employed, based on classical plasticity theory. Comparisons of the stiffness and residual strain predictions with experiments are given, and show good agreement.

INTRODUCTION

Lightweight pressure vessels are being used or developed, based on a fiber composite overwrap laminated to the outside of a metal or polymeric liner. The liner is intended to provide a gas seal as well as to serve as a mandrel for the composite manufacture, while the fiber composite provides strength with light weight. These fiber overwrapped cylinders are currently in use for portable gas cylinders such as oxygen tanks, and are also being considered for compressed natural gas (CNG) tanks for motor vehicles. If CNG becomes a more popular fuel, e.g. for pollution abatement purposes, this could potentially be a very large market for these tanks.

The liner and fiber composite are quite different materials, and the different properties can be used effectively in the design to maximum advantage. An effective design and manufacturing technique is to subject the pressure vessel to an initial pressure cycle, in which the ductile liner is strained beyond the yield point. This produces a favorable state of stress in the liner, much as in autofrettage of an all-metal thick-walled pressure vessel. This favorable state of stress improves the resistance of the liner to fatigue loading, as well as ensuring good contact between the liner and fiber overwrap over the range of temperatures needed for the application.

The present investigation considers the design of the cylindrical section of a pressure vessel with a ductile liner and a fiber composite overwrap. The liner is considered to be elastic-strain hardening plastic, and a simpler analysis with an elastic-perfectly plastic assumption is also considered. Experiments were carried out to verify the analysis, and these results are also presented.

The following will give the elastic and elastic-plastic analysis procedure for the composite overwrapped cylinder with ductile liner. Comparisons with experiments on composite cylinders with aluminum liners will then be shown, and a discussion given of factors in design of these pressure vessels.

Stephen R. Swanson and Matt Offolter, Department of Mechanical Engineering, University of Utah

ANALYSIS

The pressure vessel under consideration consists of a cylindrical section and end closures. One popular fabrication technique is to employ filament winding to apply the fibers over the liner. Although other forms of winding are possible, one method is to wind the "helical" windings in a $\pm\theta$ pattern over the cylindrical section, and continue the helicals over the end closures, or "domes." Additional reinforcement is then applied in the cylindrical section in the form of hoop windings, which are placed at an angle close to 90°, the hoop direction. It is typical practice to design the domes to be stronger so as to force the ultimate rupture of the tank to take place in the cylindrical section, as this simplifies the design without significant weight or material cost penalty. This is typically accomplished by requiring the helical fibers to carry a lower strain (or stress) relative to the hoop fibers in the cylindrical section. There is typically a stress concentration at the dome-cylinder junction, and the lower stress in the helical windings helps to prevent this from being a weak point.

The present paper considers only the cylindrical section of the pressure vessel. The geometry considered can thus be characterized by the internal radius, the thickness of the liner, the helical angle and thickness of the helical layers, and the thickness of the hoop windings. The material properties needed are the elastic and plastic properties of the liner, and the orthotropic properties of the fiber composite overwrap.

An analysis of the liner and composite cylinder will first be given that utilizes an elastic-plastic model with strain hardening. A simpler approximate analysis using an elastic-perfectly plastic model will also be considered.

Elastic Analysis - Standard lamination theory is used to construct the A matrix for the laminated aluminum and composite cylinder. The stress resultants are then related to the strains common to the liner and fiber cylinder by

$$\{N\} = [A]\{\varepsilon\} \tag{1}$$

where the stress resultants N are given by

$$\{N\} = \begin{Bmatrix} pr \\ pr/2 \\ 0 \end{Bmatrix} \tag{2}$$

For convenience, the first coordinate is taken in the hoop direction, so that the laminate is taken as a [0/±(90-θ)] layup. The stresses in the fiber composite and the aluminum liner can then be obtained from these strains by multiplying by the material stress-strain matrix. For example, the stresses in the aluminum liner are then obtained from

$$\{\sigma\} = [Q_{al}]\{\varepsilon\} \tag{3}$$

The pressure required to cause yielding in the liner can be obtained from the above. A convenient procedure is to calculate the stresses in the liner for a unit pressure. These stresses are then substituted into the von Mises yield criterion, as

$$f_1 = \frac{1}{\sqrt{2}}\left\{(\sigma_h - \sigma_a)^2 + (\sigma_a)^2 + (\sigma_h)^2\right\}^{1/2} \tag{4}$$

where the radial stress and shear stresses are taken as 0. The pressure at yield, p_y, is then given by

$$p_y = \sigma_{yp} / f_1 \tag{5}$$

After the liner yields, the elastic-plastic stress-strain relationship for the aluminum liner must be considered.

Elastic-strain hardening plastic analysis - After yielding, the stress-strain relationship for the liner is nonlinear, and must be analyzed on an incremental basis. This can be accomplished by using the usual assumptions of plasticity theory, which are the von Mises (distortional strain energy) yield function, an associated flow rule, and a hardening curve in the form of effective stress as a function of effective plastic strain (Mendelson, 1981). The yield function has already been shown in Eqn 4, specialized to the present plane stress case. The associated flow rule is given by

$$d\varepsilon_{ij}{}^{p} = \lambda \frac{\partial f}{\partial \sigma_{ij}} \tag{6}$$

which indicates that the plastic strain increments are normal to the yield surface. The function f is the yield function, taken here as the von Mises form. The magnitude of the plastic strain increments are determined by the hardening rule. The assumption of isotropic hardening gives the yield condition as

$$f = k(\varepsilon_{eff}{}^{p}) \tag{7}$$

where the value of k (and thus the value of f at yield) increases with plastic straining. Taking increments of f and k gives

$$df = \frac{\partial f}{\partial \sigma_{kl}} d\sigma_{kl} = dk = \frac{\partial k}{\partial \varepsilon_{eff}{}^{p}} d\varepsilon_{eff}{}^{p} = \frac{\partial k}{\partial \varepsilon_{eff}{}^{p}} \frac{\partial \varepsilon_{eff}{}^{p}}{\partial \varepsilon_{mn}{}^{p}} d\varepsilon_{mn}{}^{p} \tag{8}$$

where repeated subscripts imply summation. Substituting for the plastic strain increments from Eqn 6 gives

$$\frac{\partial f}{\partial \sigma_{kl}} d\sigma_{kl} = \frac{\partial k}{\partial \varepsilon_{eff}{}^{p}} \frac{\partial \varepsilon_{eff}{}^{p}}{\partial \varepsilon_{mn}{}^{p}} \lambda \frac{\partial f}{\partial \sigma_{mn}} \tag{9}$$

Solving for λ gives

$$\lambda = \frac{\dfrac{\partial f}{\partial \sigma_{kl}} d\sigma_{kl}}{\dfrac{\partial k}{\partial \varepsilon_{eff}{}^{p}} \dfrac{\partial \varepsilon_{eff}{}^{p}}{\partial \varepsilon_{mn}{}^{p}} \dfrac{\partial f}{\partial \sigma_{mn}}} \tag{10}$$

This expression for λ is then substituted into the flow rule of Eqn 6.

Expressions for the derivatives of the yield function with respect to the various components of stress can be established. Differentiating with respect to a normal stress, e.g.

$$\frac{\partial f}{\partial \sigma_a} = \frac{3}{2}\frac{S_a}{f} \tag{11}$$

where S_a is the deviatoric stress given by

$$S_a = \sigma_a - \sigma_m \tag{12}$$

The effective plastic strain is defined as

$$\varepsilon_{eff}{}^{P} = \frac{\sqrt{2}}{3}\left\{\left(\varepsilon^{P}{}_{h} - \varepsilon^{P}{}_{a}\right)^2 + \left(\varepsilon^{P}{}_{a} - \varepsilon^{P}{}_{r}\right)^2 + \left(\varepsilon^{P}{}_{r} - \varepsilon^{P}{}_{h}\right)^2\right\}^{1/2} \tag{13}$$

where the shear terms are not shown as they vanish in the present problem. Both the effective stress and effective plastic strain are defined so that they reduce to just the uniaxial stress and plastic strain in a uniaxial tensile test. The incompressibility condition for the plastic strains is also invoked and follows from the flow rule employed.

With the above, rules for differentiating the effective plastic strain are given by

$$\frac{\partial \varepsilon_{eff}{}^{P}}{\partial \varepsilon_x{}^{P}} = \frac{2}{3}\frac{\varepsilon_x{}^{P}}{\varepsilon_{eff}{}^{P}} \tag{14}$$

From the above, the incremental relationship between the plastic strain increments and the increments of stress can be formulated as

$$\begin{Bmatrix} d\varepsilon_h{}^{P} \\ d\varepsilon_a{}^{P} \\ d\varepsilon_r{}^{P} \end{Bmatrix} = C \begin{bmatrix} S_h{}^2 & S_h S_a & S_h S_r \\ S_a S_h & S_a{}^2 & S_a S_r \\ S_r S_h & S_r S_a & S_r{}^2 \end{bmatrix} \begin{Bmatrix} d\sigma_h \\ d\sigma_a \\ 0 \end{Bmatrix} \tag{15}$$

and the term C is defined as

$$C = \frac{9}{4 f^2 \phi B} \tag{16}$$

and

$$B = \frac{\partial \varepsilon_{eff}{}^{P}}{\partial \varepsilon_{mn}{}^{P}}\frac{\partial f}{\partial \sigma_{mn}} = \left(\frac{1}{f \varepsilon_{eff}{}^{P}}\right)\left(S_h \varepsilon_h{}^{P} + S_a \varepsilon_a{}^{P} + S_r \varepsilon_r{}^{P}\right) \tag{17}$$

The term ϕ is just the slope of the effective stress vs effective plastic strain hardening curve.

The above equations give a tangent modulus relationship between the plastic strain increments and the stress increments. By proceeding in increments of loading, is it straightforward to solve for the overall response of the laminated cylinder. The procedure is to use the above to obtain a tangential A matrix. The inverse of this matrix then gives a relationship between the increments of strain and the increments of pressure loading. The state of stress and strain in the aluminum liner must then be updated, and the pressure loads increased in another step.

Updating the plastic strains in the liner requires that the total strain, calculated from

$$\{\Delta\varepsilon\} = \left[A_t^{-1}\right]\{\Delta N\} \tag{18}$$

be separated into plastic and elastic components. This is accomplished by updating the strains from

$$\{\varepsilon\}_{n+1} = \{\varepsilon\}_n + \{\Delta\varepsilon\} \tag{19}$$

then calculating the stresses in the fiber composite cylinder from these strains, using equilibrium to determine the hoop and axial stresses in the aluminum liner, and then calculating the elastic strains that would result from these stresses. Subtracting the elastic strains from the total strains gives the plastic strains. The effective plastic strain required for the next increment of loading is then calculated.

Aluminum liner only - Pressure tests were performed on the aluminum liner alone, without the composite overwrap. This test was used to verify the yield stress and the hardening curve for the aluminum that had been characterized in uniaxial tensile tests. The analysis follows that given above, but is simpler as the liner stresses are determined from equilibrium and are in a fixed ratio of 2 to 1.

Analysis with elastic-perfectly plastic liner - An obvious simplification to the above can be obtained if the liner is considered to be elastic-perfectly plastic. A simple approximate solution can be obtained by assuming that the stresses in the liner don't change past the initial yield, and that the pressure drop across the liner also doesn't change. Thus the further response of the outer composite cylinder can be calculated as if it were acting alone, with an internal pressure given by the actual internal pressure minus the constant pressure drop across the liner. The stresses in the liner stay constant, and the total stresses and strains in the composite are the sum of the stresses at initial yield of the liner, plus the stresses from increasing the pressure past that causing initial yield.

The additional response of the composite to the overpressure can be calculated from

$$\begin{Bmatrix}\Delta\varepsilon_h\\ \Delta\varepsilon_a\end{Bmatrix} = A_c^{-1}\begin{Bmatrix}\Delta pr\\ \dfrac{\Delta pr}{2}\end{Bmatrix} \tag{20}$$

It should be noted that the A_C matrix here is just that of the composite. The total strain in the composite (and the liner) is then given by the strains at yield plus the additional strains from the above. The term Δp is defined as the internal pressure minus the pressure at initial yield of the liner.

COMPARISON OF EXPERIMENT WITH ANALYSIS

Tests were carried out on cylinders that were intended to simulate the cylindrical section of an composite overwrapped pressure vessel. The cylinders were fabricated with a 6061-T6 aluminum liner, with composite overwrap of AS4/3501-6 carbon/epoxy prepreg. Instead of the usual domes, these cylinders had the end closure provided by external hardware. The cylinder is coupled to the external hardware with tapered regions of fiber glass cloth reinforcement. This method of end closure has been used previously in biaxial tests for materials characterization (Smith and Swanson, 1997). The specimens were constructed by winding continuous strips of prepreg tape over the aluminum liner, first with the $\pm\theta$ helical winding and then with the hoop fibers. The specimens were instrumented with three strain gages in each of the axial and hoop directions, spaced around the circumference at the axial center, and then subject to various internal pressure cycles.

Pressure cycling tests were first carried out on the 6061-T6 aluminum liner material, without the fiber composite overwrap. A comparison of the strain response with the measured values for two aluminum cylinders is shown in Fig 1. It can be seen that the comparison is good, and that the aluminum is well characterized by the strain hardening plasticity model.

These pressure tests were continued until rupture of the aluminum liner. An unexpected result is the low strain levels at which failure of the liner occurred. More discussion of this will be given subsequently.

Tests were also carried out on four cylinders with the AS4/3501-6 fiber composite overwrap. Typical results are shown in Figs 2-4 for the hoop and axial strain response for two typical cylinders, for a number of pressure cycles. These loadings were applied to display the fundamental behavior of the elastic-plastic liner and composite overwrap. As shown in the figures, good agreement is seen between the analysis and experiments.

The hoop strain response is compared with the elastic-plastic model in Fig 2, and the hoop and axial strains with the simpler elastic perfectly-plastic model in Figs 3 and 4. It can be seen that the assumption of elastic perfectly-plastic response for the liner gives a good representation of the behavior. Because the total response of the liner and composite is being considered, the differences between the strain hardening plasticity model and the perfectly plastic model are small.

DISCUSSION

The comparison of the model predictions with the experiments indicates that the response of the composite overwrap cylinder with a ductile liner can be represented by the elastic plastic model given. Also, at least for the aluminum liner used in the present experiments, the elastic- perfectly plastic model gives sufficient accuracy, and is much simpler than the strain-hardening plasticity model.

As outlined in Swanson (1997), it is straightforward to use the above equations to relate the composite and liner properties to the desired characteristics of the cylinder. The desired parameters of the design include the rupture pressure, the pressure at first yield, the peak pressure employed to create favorable residual stresses, and the state of residual stress. All of these values can be obtained from the above equations.

The state of residual stress after cooldown from elevated temperature cure of the composite overwrap, and prior to the first pressure cycle used to yield the liner, is a complicated issue. If the aluminum liner and the composite overwrap are taken to be stress free at the cure temperature of 177 C, conventional lamination theory will predict tensile stress in the radial direction. However this is an over-simplified model for cylinders with hoop fibers, because of tension induced in the fibers by the mandrel expansion during the cure cycle. Ultrasonic inspection did reveal small areas of debond between liner and composite near the ends of several cylinders after cooldown, and one of the tests exhibited anomalous strain gage readings in the initial portion of the first pressure cycle which might be attributed to debond areas. Thus the tight mechanical fit introduced by the residual radial compression is a significant advantage of using a ductile liner with an overpressure cycle.

The relatively low failure strains of the 6061-T6 aluminum liners tested alone without the composite overwrap should be noted. While the nominal failure strain in tensile tests is listed as 17%, the strain gage readings at the time of brittle fracture of the liners were on the order of 1.8 %. As discussed by Mielnik (1991), typical forming limit diagrams, (biaxial strain state plots at instability or failure) show that the state of strain in the cylinders would tend to promote fracture, relative to a uniaxial tension test. Also, the ductility depends on the prior history of working. In any case, the low ductility exhibited in the experiments could potentially cause problems in the use of 6061-T6 aluminum as a liner material.

The thrust of the experiments was to compare with the model predictions for the pressure-strain response, and a number of pressure load and unload cycles were employed with each test specimen to determine the behavior. However the four specimens were also taken to failure. The failure values of the hoop fibers in the cylinder pressure tests are somewhat lower than exhibited by AS4/3501-6 in previous tests, and a value of 1.25 % appears to be representative, rather than a nominal value of 1.4 % seen in previous work (Swanson and Christoforou, 1986). No explanation for these somewhat lower failure values is available. Clearly further work is required to characterize the ultimate failure of the cylinders.

As discussed by Swanson (1997), a simple approximate theory for design of the cylindrical section of the composite overwrapped cylinder can be established if the pressure drop across the ductile liner is taken as constant during plastic straining of the ductile liner. The present analysis and experiments show that this is a good approximation.

SUMMARY AND CONCLUSIONS

Tests and analysis have been carried out for pressure tests of cylinders consisting of AS4/3501-6 carbon/epoxy wrapped on the outside of 6061-T6 liners. The analysis is based on elastic-plastic theory for the liner with strain-hardening, and conventional lamination theory. The results show that the strain response and residual strains can be well predicted by the model, and also that a simplified elastic-perfectly plastic model gives adequate accuracy.

REFERENCES

Mendelson, A., 1983, Plasticity: Theory and Applications, Krieger Pub. Co., Malabar, Florida.

Mielnik, E.M., 1991, Metalworking Science and Engineering, McGraw-Hill, New York.

Smith, L.V., and Swanson, S.R., 1997, "Design of a Cylindrical Specimen for Biaxial Testing of Composite Materials," J. Reinforced Plastics and Composites,16: 550-565.

Swanson, S.R., 1997, Introduction to Design and Analysis with Advanced Composite Materials, Prentice-Hall, Upper Saddle River, NJ.

Swanson, S. R., and Christoforou, A.P., 1986, "Response of Quasi-Isotropic Carbon/Epoxy Laminates to Biaxial Stress," J. Composite Mtls., 20: 457-471 .

Table 1. Parameters for composite cylinder pressure tests

aluminum liner thickness (6061-T6)	1.1176 mm (0.044 in)
inside radius	48.3 mm (1.900 in)
fiber composite thickness	1.707 mm (0.067 in)
composite layup	$[90_6/\pm30_3]$
composite properties (AS4/3501-6):	
E_{11}	131.0 GPa (18.4 Msi)
E_{22}	11.2 GPa (1.6 Msi)
G_{12}	6.55 GPa (0.95 Msi)
ν_{12}	0.28

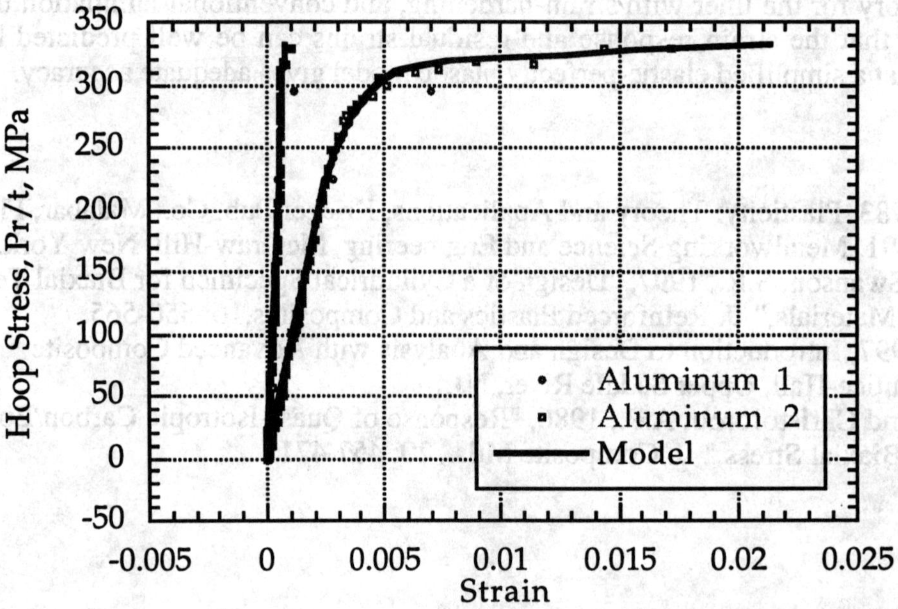

Fig 1. Comparison of model with experiment for axial and hoop strains in internal pressure loading of aluminum cylinders.

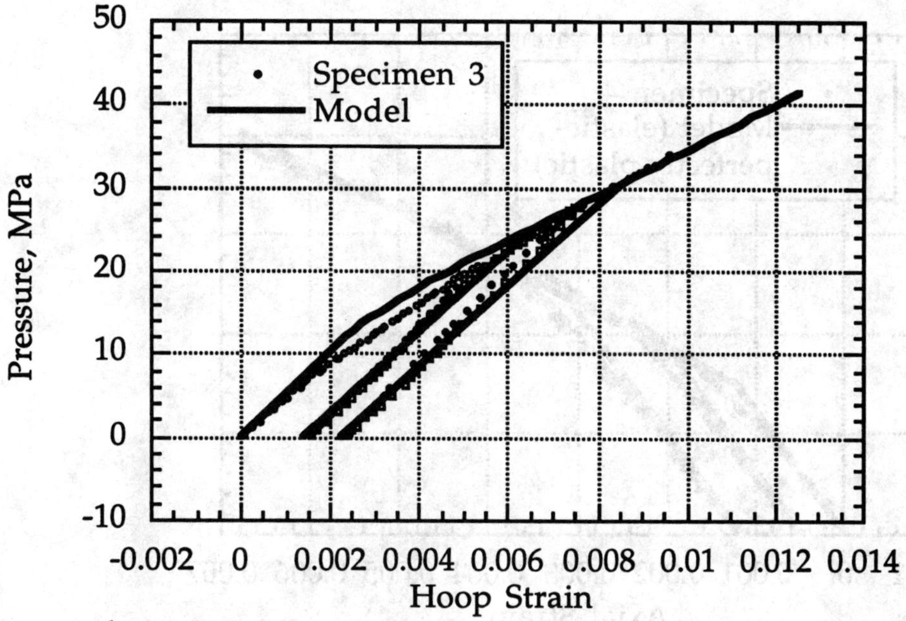

Fig 2. Comparison of measured hoop strain with model prediction in pressure cycling of aluminum cylinder with carbon/epoxy overwrap.

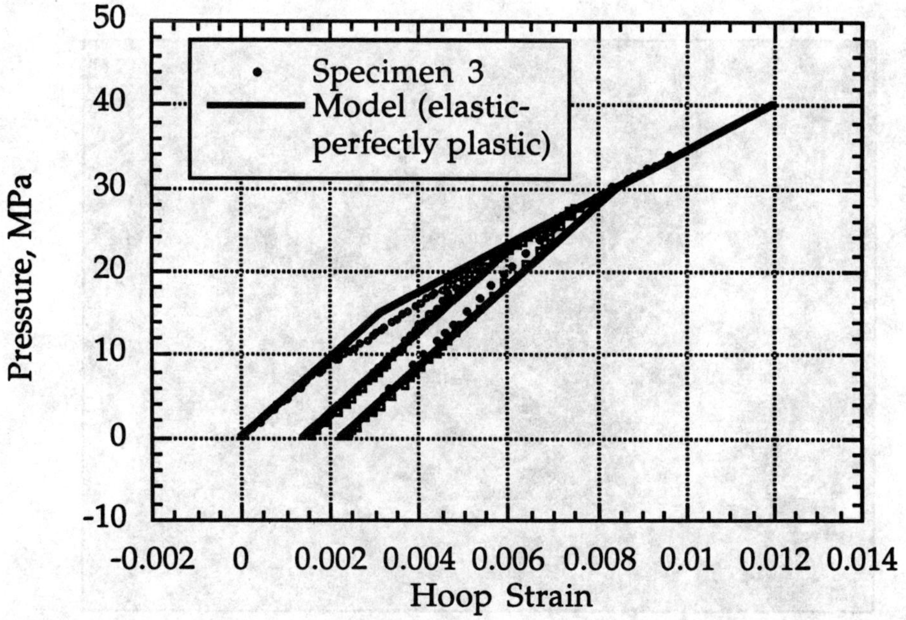

Fig 3. Comparison of measured hoop strain with prediction of elastic-perfectly plastic model in pressure cycling of aluminum cylinder with carbon/epoxy overwrap.

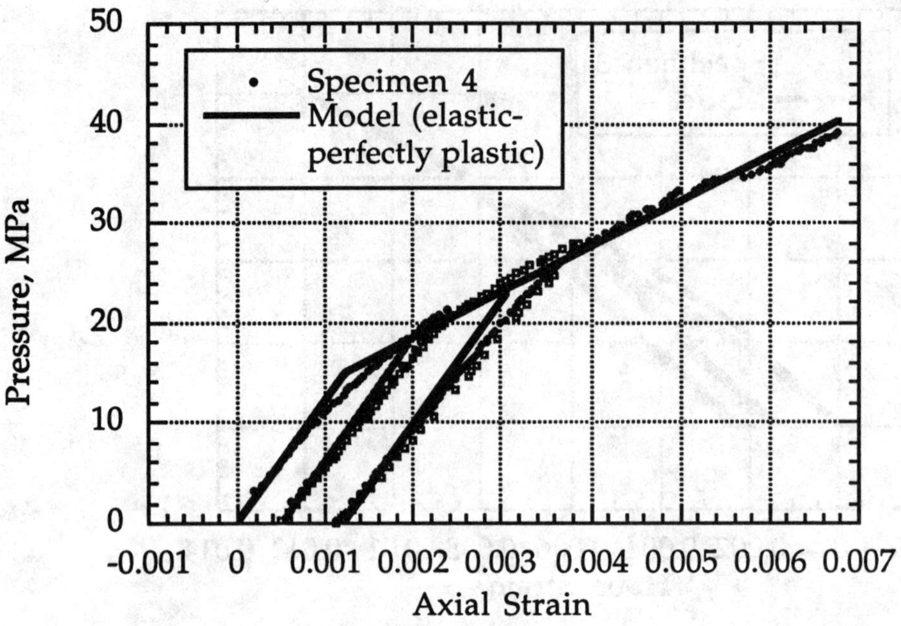

Fig 4. Comparison of measured axial strain with prediction of elastic-perfectly plastic model in pressure cycling of aluminum cylinder with carbon/epoxy overwrap.

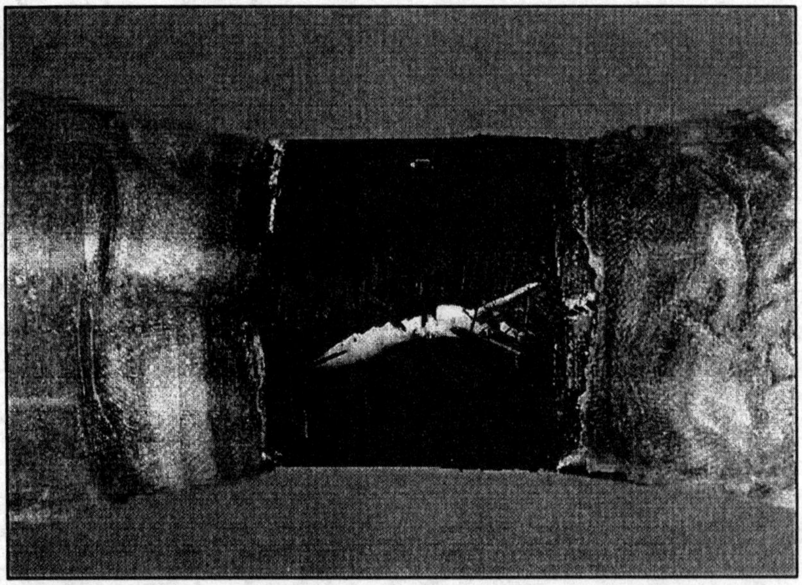

Fig 5. Typical failure of AS4/3501-6 carbon/epoxy wrapped over 6061-T6 aluminum liner under internal pressure loading.

Why Design Criteria for Laminates Are Excessively Conservative

M. R. PIGGOTT AND S. R. DAI

ABSTRACT

Design criteria for laminates are based on early work dating back to the 1960's. They involve, for elastic constants, the summing of transformed lamina elastic constants with very little attention to interaction effects. For strength they rely either on the maximum stress concept or a rounded off version of this (Tsai-Hill). Underpinning all this were some experiments on very long and narrow specimens. Recent work with wider specimens shows that the narrowness of the early specimens severely restricted the loading of the fibres. With wide and short $[\pm \phi]_s$ angle ply specimens the fibres can be broken, even with $\phi = 60^o$. The strengths can be five times higher than suggested by conventional theory. Furthermore, interaction effects are observed which do not occur with narrow specimens. These give much higher stiffnesses; perhaps as much as five times higher at $\phi = 15^o$ for carbon fibre laminates. Thc much greater strengths and stiffness are confirmed by literature data from tests on glass-epoxy filament wound tubes. These agree very closely with our $[\pm \phi]_s$ angle ply results for glass-epoxy. The conventional approach is well established, so design criteria are probably unnecessarily conservative. Angle ply laminates seem to be avoided, at all costs. This is unfortunate, as the tests described here show that angle ply laminates can be about as strong as standard $0^o/90^o/\pm 45^o$ structures, and have the advantage of being more resistant to premature cracking. A design procedure for angle ply laminates for aerospace, based on principal stresses, is proposed.

INTRODUCTION

Our theoretical basis for design criteria for higher performance composites rests on early work in the 1950's and 1960's. The most important criteria relate to the stiffnesses. If we take a regular balanced angle ply laminate for example, i.e. $[\pm \phi]_{ns}$ where ϕ can be any angle, it is assumed that we can sum the transformed compliances $\overline{S}_{ij}$, or the reduced transformed stiffnesses $\overline{Q}_{ij}$. Thus

$$\overline{S}_{11} = \frac{1}{4n} \sum_{k=1}^{4n} \overline{S}_{11k} \qquad (1)$$

M.R. Piggott, Department of Chemical Engineering and Applied Chemistry, University of Toronto, Toronto, Ontario M5S 3E5, Canada

gives the compliance in the major symmetrical direction, which we can write as $1/E_x$ using Engineering constants. (In the regular laminate, all the laminae have the same constitution and thickness; the laminate in equation 1 has 4n layers.)

Using Engineering units equation 1 gives

$$\frac{1}{E_x} = \frac{\cos^4\phi}{E_1} + \left(\frac{1}{G_{12}} - \frac{2\nu_{12}}{E_1}\right)\sin^2\phi\cos^2\phi + \frac{\sin^4\phi}{E_2} \qquad (2)$$

where E_1, E_2, G_{12} and ν_{12} are the Young's moduli, shear modulus and Poisson's ratio for each lamina.

The use of these equations assumes that putting the laminae together in this way does not involve any interactions. Tsai (1965) examined this experimentally and theoretically and showed that equation 2 was adequate for his glass-epoxy samples. (Although he also took thermal stresses into account, their effects were quite small.)

With respect to strength, the original approach used was the maximum stress theory of Stowell and Liu (1961). This gave three separate criteria leading to a strength vs ϕ plot which had cusps which were not observed experimentally. Tsai found a way of smoothing this out using a criterion due to Hill. The resulting equation, which fitted the results of Tsai's (1965) tests very well, was

$$\frac{1}{\sigma_x^2} = \frac{\cos^4\phi}{\sigma_{1u}} + \left(\frac{1}{\tau_{12u}^2} - \frac{1}{\sigma_{1u}^2}\right)\cos^4\phi\sin^2\phi + \frac{\sin^4\phi}{\sigma_{2u}^2} \qquad (3)$$

where σ_x is the strength in the direction corresponding to E_x, and σ_{1u}, σ_{2u} and τ_{12u} are the tensile and shear strengths of the laminae. (Again thermal stresses made little difference.) As can be seen, this is very closely analogous to the stiffness equation.

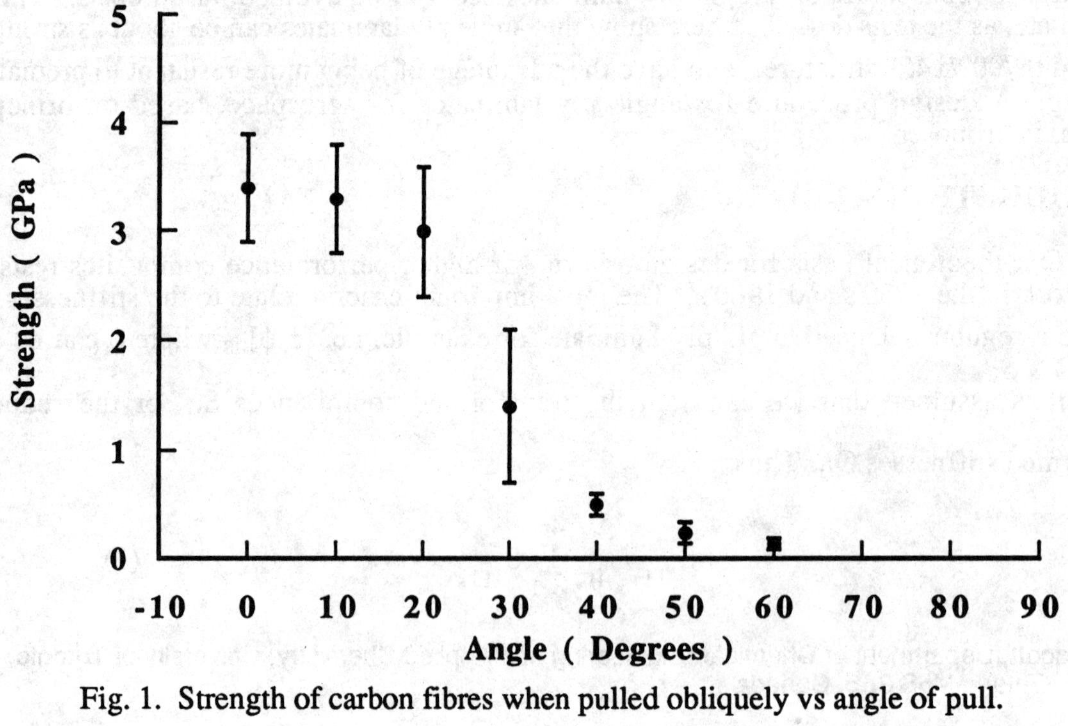

Fig. 1. Strength of carbon fibres when pulled obliquely vs angle of pull.

Equations 1 to 3 appear in pretty well all standard texts and are widely used in design software.

Analogous equations are used for randomly oriented fibre laminates, such as those made from sheet molding compounds. The appropriate equations for them are obtained by summing (i.e. integration) for all possible angles, ϕ.

The first published doubts about the use of these equations appears to have arisen when attempts were made by Piggott et al (1993) to account for the properties of composites containing imperfectly aligned short fibres. The strengths were too high. (In addition, the stress-strain plots for these were straight lines, contrary to the well known Kelly-Tyson theory. An alternative theory has now been accepted (Piggott, 1994) based on fracture failure at the fibre-matrix interface, rather than the benign yielding process of the prior theory.)

The unexpectedly high strengths suggested that it might be worth while to test individual fibres, pulled obliquely. When this was done, the tests showed that fibres, crossing cracks obliquely in reinforced polymers, could retain their strength to a remarkable degree: see fig. 1. (Khatibzadeh and Piggott , 1996a). There was also a fibre effect. The results could be fitted very roughly to the following equation:

$$\sigma_{fu\phi} = \sigma_{fu\phi}\left(1 + (1-\phi/\phi_h)^{1/3}\right) \quad (4)$$

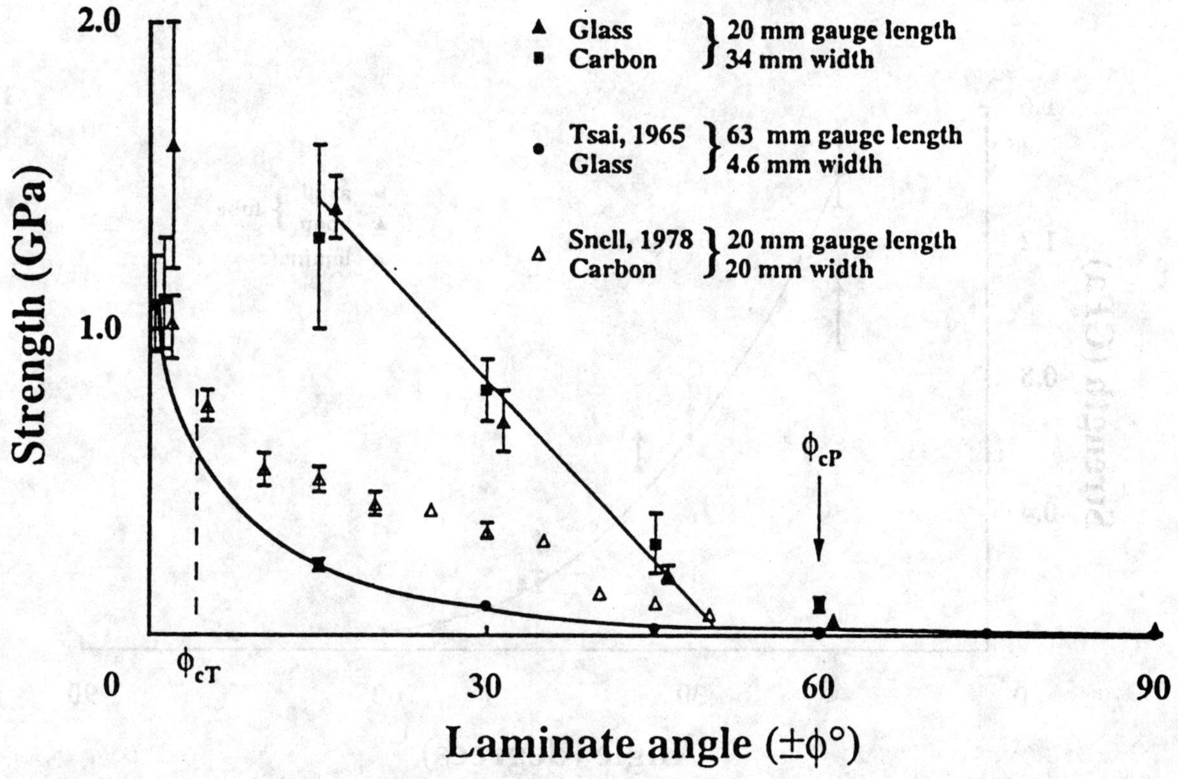

Fig. 2. Strengths of wide $[\pm\,\phi]_{ns}$ laminates compared with Tsai (1965) and Snell (1978) results.

where $\sigma_{fu\phi}$ is the strength of a fibre pulled at an angle of ϕ to its axis, σ_{fuo} is the normal strength (fibre pulled at zero degrees) and ϕ_h is the half strength angle, i.e.

$$\sigma_{fu\phi}(\phi = \phi_h) = \sigma_{fu\phi} / 2 \tag{5}$$

The next step was to investigate wider laminates. The early workers designed their test specimens so that no fibre went from one grip to the other. Thus they were very long and narrow, so the fibres were only stressed slightly and none failed. The recent work went to the opposite extreme of wide, short specimens, effectively made even shorter by the use of notches (Khatibzadeh and Piggott, 1996b). These gave the results shown in fig. 2. The Tsai (1965) results are also shown, together with some later carbon work of Snell (1978). It can be seen that using wider specimens increases the strength many fold, especially at the smaller angles.

(Interestingly, builders of fibreglass pressure vessels don't seem to have been strongly influenced by this work. Companies in many places were producing filament wound vessels which were effectively angle ply laminates, but with a cylindrical geometry, back in the 1960's and 1970's. These were three or more times stronger than Tsai could achieve (Hull et al, 1978). A well known expert apparently explained to Hull that he didn't appear to understand laminates when he pointed out that his results indicated that they could be very strong (Hull, 1996). Furthermore, software exists for making cross ply filament windings, with axial and circumferential fibres, presumably because of doubts about angle ply laminates.)

Well made filament windings have very high strength, and perform better under biaxial stress than under simple hoop stresses, (Soden et al, 1993). Fig. 3 shows some results from glass filament windings. Our glass results are also plotted for comparison. The agreement can be seen to be moderately good.

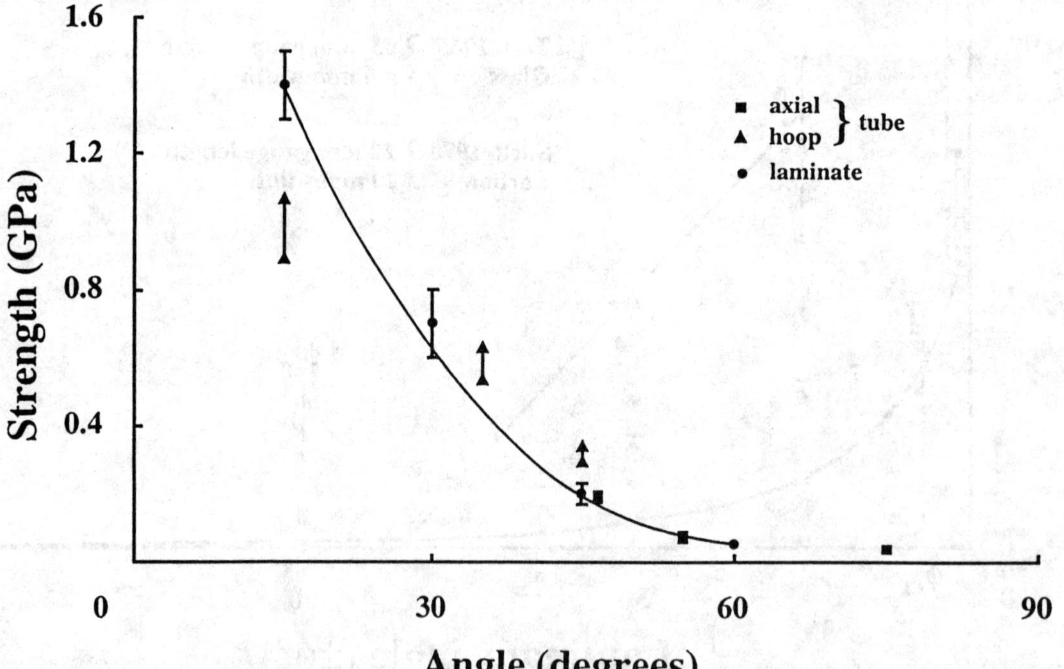

Fig. 3. Strength of wide [± ϕ]$_s$ glass-epoxy laminates compared with Soden et al's (1993) results for filament wound tubes.

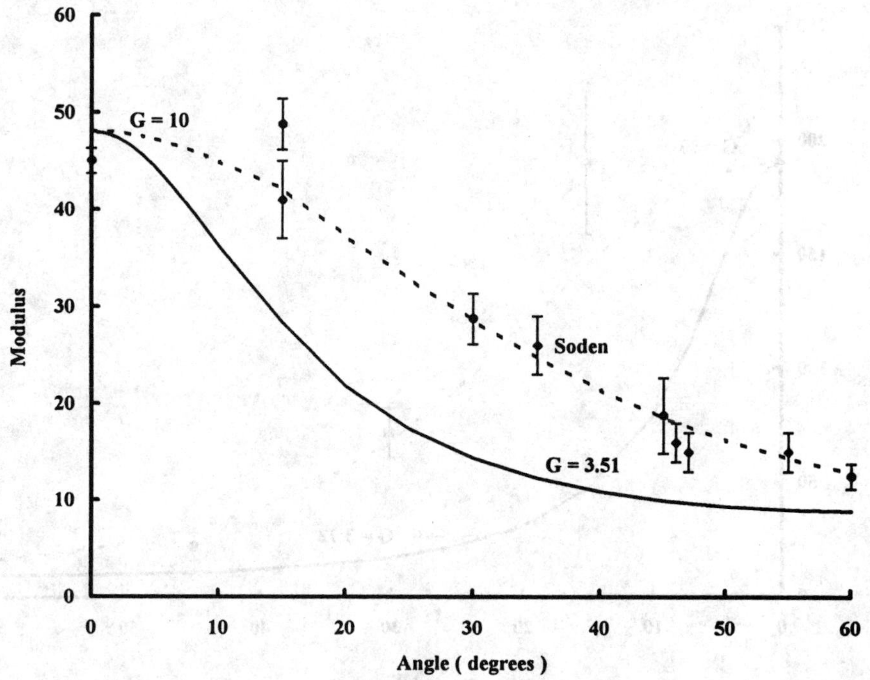

Fig. 4. Nominal stiffness of $[\pm \phi]_s$ glass-epoxy laminates compared with Soden et al's (1993) values for filament wound tubes. Curves are given by equation 2 with shear modulus values indicated.

Moduli were also measured on the filament windings, and can be seen to be in good agreement with our glass results; see fig. 4, (Khatibzadeh and Piggott, 1997). Shown also on the plot is equation 2 with literature values for E_2 and G_{12} and the Rule of Mixtures value for E_1. It seems clear that the effective value of G_{12} in the laminate is greater than the literature value (3.51 GPa) which was for a unidirectional sample. The curve that approximately fits the results uses equation 2 with G = 10 GPa. Thus some interaction appears to be taking place when laminae are put together in angle ply laminates. This must take the form of a constraint which curtails the shear strains.

In the case of carbon, the constraint seems to be much greater: see fig. 5, (Khatizbadeh and Piggott, 1997) However, the strains were estimated from cross head movements, and thus need to be verified.

Present designs for reinforced plastic aeroplanes are based on combinations of 0°, 90°, and $\pm 45^o$, with weighting factors 1, 0.1 and 0.1 respectively (Hart-Smith, 1996). These may not be as good as angle ply laminates because the 90° layers can split very easily, leading to cracking and loss of integrity. This paper describes some results of tests to investigate the relative resistance to splitting of the standard design and of angle ply laminates, using wide test specimens.

EXPERIMENTAL METHOD

Glass-epoxy prepreg from Ciba was used (S2-R7376). It was laid up in various patterns and hot pressed following the recommended temperature profile. This specified heating at 2^oC/min to 180^oC, holding for 150 min, then cooling at about 8^oC/min to 60^oC.

Each molding had eight layers as follows: $[\pm 30]_{2s}$, $[\pm 40]_{2s}$ and $[90 \pm 45\ 0]_s$. Samples 89 x 100 mm were prepared, and tested in the MTS machine using special grips shown in fig. 6. The cross head rate was 2 mm/min.

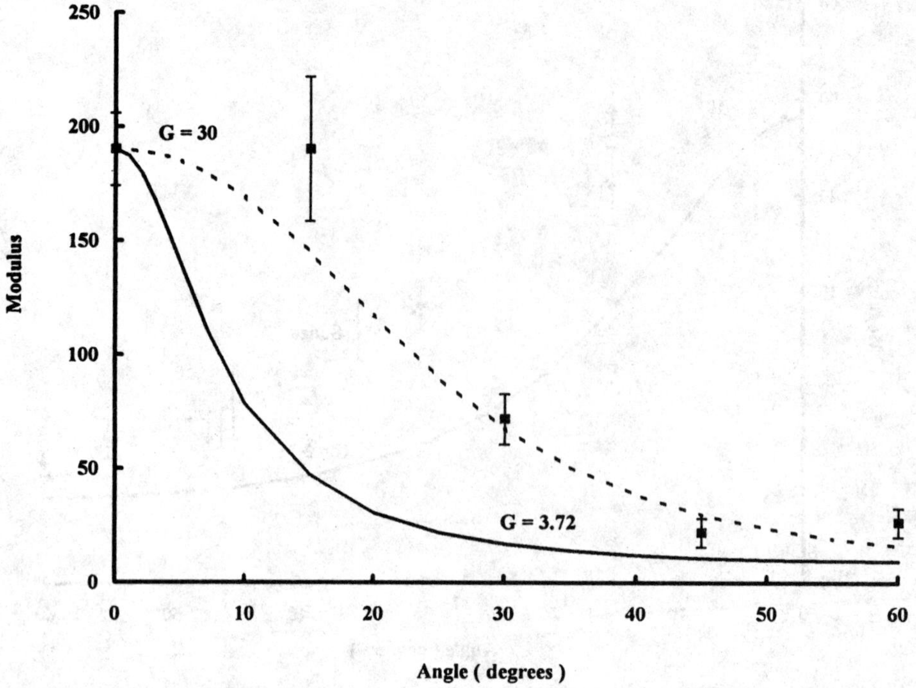

Fig. 5. Nominal stiffnesses of [$\pm\phi$]$_S$ carbon-epoxy laminates. Curves are given by equation 2 with shear modulus values indicated.

During the tests the samples were carefully observed to determine the onset of cracking. Five replicate tests were carried out for each laminate configuration.

EXPERIMENTAL RESULTS AND DISCUSSION

Table 1 lists the mean stresses and standard deviations (+/-) for the onset of cracking and final failure. It is clear that the conventional design cracks at relatively low stress, i.e. 71% of the failure stress. The angle ply laminates are very resistant to cracking, which did not occur until 90% [± 40]$_{2S}$ and 97% [± 30]$_{2S}$ of the final failure stress.

TABLE 1 - RESULTS OF TENSILE TESTS ON LAMINATES

Laminates	First crack stress	BreakingStress	Stress Ratio
[90 ± 45 0]$_S$	320 ± 90	450 ± 30	0.71
[± 30]$_S$	380 ± 50	390 ± 50	0.97
[± 45]$_S$	250 ± 30	270 ± 30	0.93

If we use the standard weighting factors, we can estimate the maximum design stress for the conventional design. This is (1.2/4) x 2.02 GPa using the Ciba data (2.02 ± .09 GPa) for the unidirectional strength, which comes to 0.6 GPa. Our failure stress (450 MPa) is somewhat low. The [± 20]$_S$ performed nearly as well (390 MPa), the difference being barely significant. The specimens were not notched, as in the earlier work so edge effects were much larger: the notched $[\pm 30]_s$ samples had a strength of about 700 MPa, (Khatibzadeh and Piggott, 1996b). With respect to the onset of cracking, the $[\pm 30]_{2s}$ outperformed the conventional design.

It thus seems worth while to propose that angle ply laminates replace standard designs.

SUGGESTED NEW DESIGN APPROACH

If angle ply laminates are to be used, we need to determine the angle. For this we may be able to adopt a criterion which has already been used, so that there is a basis of experience with it. For filament wound tubes such a criterion has been established on the basis of resolved stresses. Thus for a pressure vessel the stress ratio, hoop to axial is 2:1. Hence

$$\tan^2\phi = 1/2 \quad (6)$$

where ϕ is the equivalent angle ply laminate angle, i.e. 35° 16'. (The filament winding angle is 90-ϕ, i.e. 54° 44').

This criterion works well for filament windings, and they withstand biaxial stresses very well. Thus, it is recommended that the following procedure be used for aerospace and other structures:

1. Determine the principal stresses, σ_1 and σ_2
2. Determine ϕ using

$$\tan^2\phi = \sigma_2 / \sigma_1 \quad (7)$$

3. Prepare a $[\pm \phi]_{ns}$ laminate with the zero degree direction in the major principal stress direction.

For structures with loadswhich vary from place to place, both n and ϕ can be varied appropriately in step with the stresses.

CONCLUSION

Angle ply laminates have great potential, despite their bad reputation based on earlier work. They are widely used in filament wound structures, and have excellent properties, especially when subject to biaxial stress. Furthermore, flat angle ply laminates are markedly more resistant to premature cracking than lay ups which contain 90° layers. Thus $[\pm \phi]_{ns}$ angle ply laminates should be investigated as a possible replacement for the 0, 90 ± 45 lay ups presently being used in aerospace. A very simple design criterion, based on filament winding procedures should be used. This is the resolved stress criterion, which gives $\tan^2\phi = \sigma_2/\sigma_1$ where σ_1 is the maximum stress and σ_2 is the minimum stress (i.e. σ_1 and σ_2 are the principal stresses). The zero degree direction should correspond to the direction of σ_1.

ACKNOWLEDGEMENT

Financial support from NSERC (Canada) is gratefully acknowledged.

REFERENCES

Hull, D. and M.J. Legg and B. Spencer. 1978. *Composites*, 9, 17-24 and 263-71.

Hull, D. 1996. *Personal Communication..*

Hart-Smith, L.J. 1996. *Comp. Sci. Tech*, 56, 107-122.

Khatibzadeh, M. and M.R. Piggott, 1996a, *Comp. Sci. Tech* , 56, 1435-42.

Khatibzadeh, M. and M.R. Piggott, 1996b, *Comp. Sci. Tech,* 56, 1443-51.

Khatibzadeh, M. and M.R. Piggott. 1997. *Comp. Sci. Tech.* accepted for publication.

Piggott, M.R. and M. Ko and H.Y. Chuang. 1993. *Comp. Sci. Tech.,* 48, 291-9.

Piggott, M.R. 1994. *J. Comp. Mater.* 28, 588-606.

Snell, M.B. 1978.*Composites*, 9, 167-76.

Soden, P.D., R.Kitching, P.S. Tse and Y. Tsavalas. 1993. *Comp. Sci. Tech.* 46, 363-78

Stowell, E.Z., and T.S. Liu. 1967. *J. Mech. Phys. Solids,*, 9, 242-60.

Tsai, S. 1965. *NASA CR 224..*

Response of Woven Ceramic Matrix Composite Plates under Bending

A. ABDUL-AZIZ, L. GHOSN AND D. BREWER

ABSTRACT

Analytical studies are conducted to determine the behavior of ceramic matrix composite circular plates under bending. Stress and strain distributions are determined for four different woven composite systems at various loads. As a first approximation, the Von-Mises incremental plasticity theory is postulated to model damage growth in these composite systems. The material data used in the present study is based on actual experimental unidirectional stress-strain tensile data. Plate deflection as well as stress and strain distributions are calculated at various applied loads. Good agreement is observed between the analytically determined plate deflections and experimental results.

INTRODUCTION

Ceramic matrix composites (CMC) materials are candidate materials for combustor liner in future aircraft engine designs. They have shown to exhibit an increase in the strain to failure as compared to monolithic ceramics, and can withstand higher temperature than conventional metal systems. Furthermore, their low density is shown to decrease the thrust to weight ratio that will guaranty an increase in fuel efficiency. However, before these ceramic matrix composite materials can be used, the exhibited nonlinear damage needs to be modeled. As a first approximation, the classical plasticity theory is assumed to model the damage in woven systems. As a check to this hypothesis, the behavior of four woven ceramic matrix composites under biaxial loading are calculated using a nonlinear finite technique. This present study uses the MARC general purpose finite element program (MARC, 1995). The biaxial bending is induced by a ring on ring flexure apparatus. This loading technique was previously applied successfully by Shetty et al (Shetty, et al., 1980) on monolithic ceramics. Tsangarakis and his co-workers (Tsangarakis et al., 1990, 1992, 1995) extended the testing to cross-ply and unidirectional aluminum matrix composites as well as graphite/epoxy polymer composites.

Ali Abdul-Aziz, NYMA, Inc., Brookpark ,OH 44142
Louis J. Ghosn, Case Western Reserve University, Cleveland, OH 44106
Dave N. Brewer, U.S. Army Laboratory, NASA Lewis, Cleveland, OH 44135

Thus, the present work evaluates the biaxial stress-strain fields of ceramic matrix composite disk specimens using the finite element technique. Load-deflection curves are used for comparison with experimental data. The material data implemented in the biaxial analysis is based on actual experimental unidirectional stress-strain curves.

ANALYTICAL PROCEDURE

A finite element model representing the schematic of the experimental ring on ring setup is shown in Figure 1. The loading fixture consists of an inner-loading ring and an outer-support ring made of 301 stainless steel. A circular composite plate is situated between the two rings for testing. This test configuration provides a biaxial bending stress state within the area of the plate bounded by the inner loading ring. A typical diameter of the tested composite plate is about 2.05 in. The thickness is about 0.08 in. The loading ring diameter is 0.945 in and the support ring diameter is 1.653 in. The height of both rings is around 0.59 in. The contact radius of both rings mating the plate is 0.111 in.

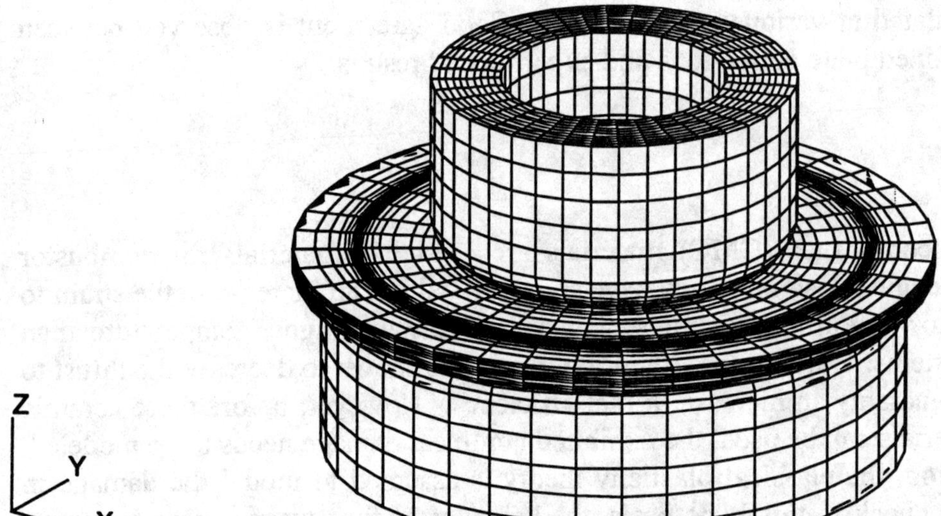

Figure 1. Schematic of the Ring-on-Ring Setup.

A three-dimensional finite element model of the test fixture is shown in Figure 2. Due to symmetry, only one quarter of the fixture is modeled. The mesh consists of 7722 nodes and 6040 eight-noded brick elements. Eight layers through the plate thickness are generated to capture the bending response. A higher mesh density at the contact regions between the plate and the rings is also required to capture the sharp stress gradients at these locations. The plate is allowed to slide relative to the rings. Hence, duplicate nodes are defined along the contact regions. One set of these nodes is associated with the plate and the other set is associated with the mating ring. Only, the axial degrees of freedom between two opposing nodes are tied to simulate the contact. No friction is assumed along the contacting areas. A true contact condition requires almost 6 times more computational time than the present simple tying option. Other boundary conditions are shown in Figure 2. The bottom surface of the outer supporting ring is fixed in the axial direction (z-direction). The external load is applied as an incremental axial displacement along the top nodes of the inner loading ring. The appropriate standard symmetry boundary conditions are also applied along the x=0 and y=0 planes.

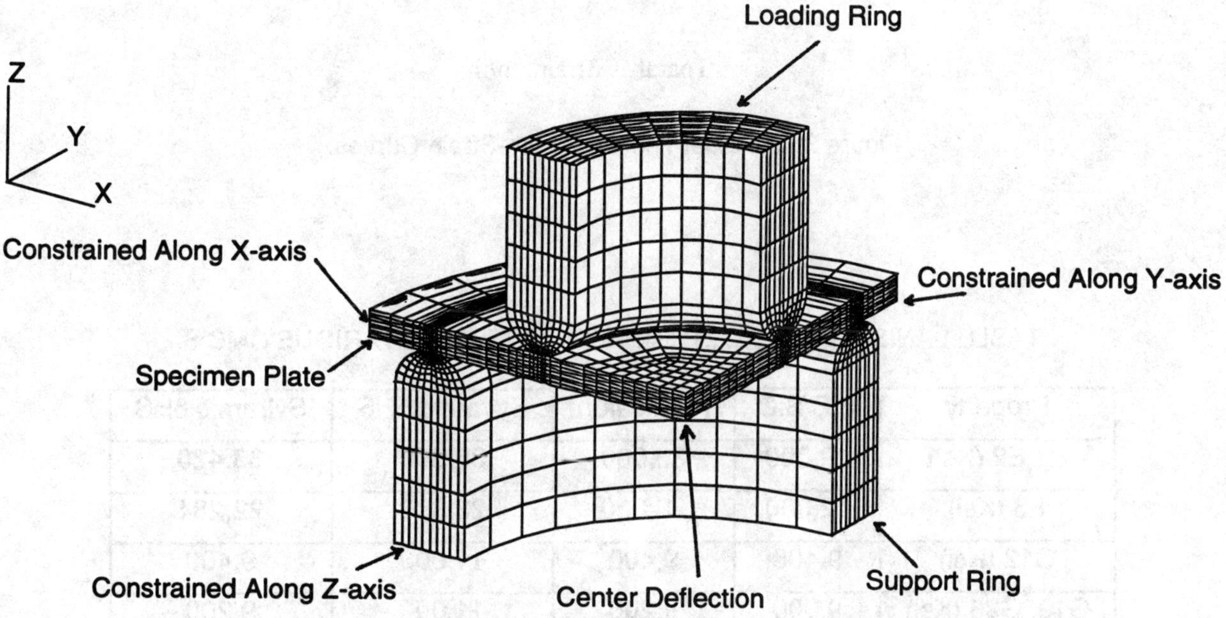

Figure 2. Finite Element Mesh and Applied Boundary Conditions.

The induced damage in the composite is simulated through an elastic-plastic analysis. The Von-Mises yield criterion is assumed to trigger damage initiation and propagation. The yield stress and the plastic strain are calculated from actual experimental unidirectional tensile stress-strain curves. The unidirectional stress-strain curves for the four material systems analyzed are shown in Figure 3. Four different types of woven silicon carbon based CMCs are considered, a baseline CG Nicalon /CVI SiC and Hi-Nicalon/MI SiC in 8 ply plain weave lay-up and Sylramic/MI SiC in 5 harness satin weave (5HS) and 8 harness satin weave (8HS) 8 ply configuration. The baseline material will be referred to simply as CG SiC/SiC. The Hi-Nicalon five weaves shows a higher ultimate stress compared with the commercial grade SiC/SiC and both Sylarmics 5HS and 8HS. However, the yield stress for the Sylarmics and the Hi-Nicalon are about the same but 50% higher than the commercial grade SiC/SiC system. The anisotropic elastic properties used in the present analysis are shown in Table I for the four CMC's woven composite systems.

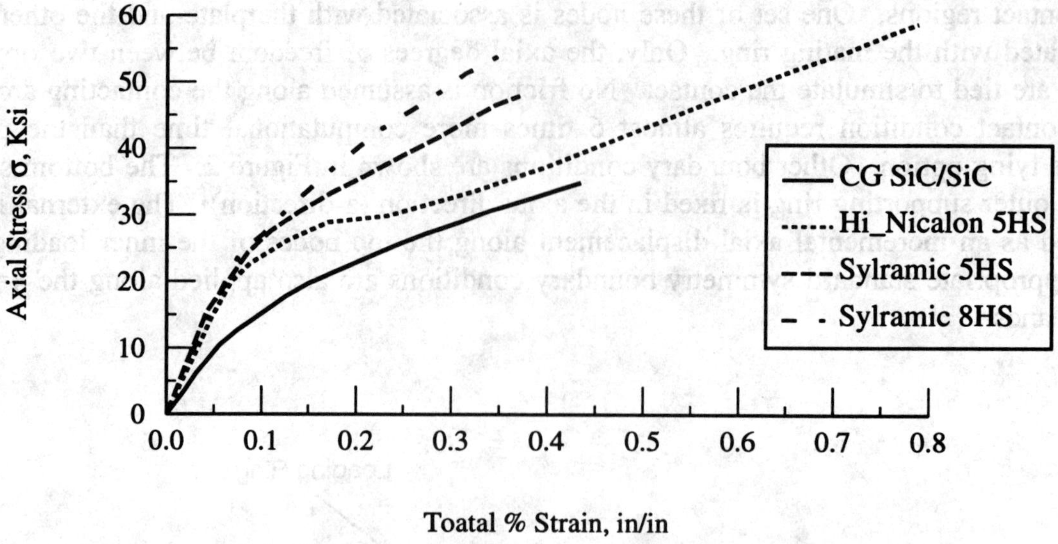

Figure 3. Uni-axial Tensile Stress-Strain Curves.

TABLE I-ANISTROPIC ELASTIC PROPERTIES FOR VARIOUS CMC'S

Property	SiC/SiC	Hi-Nicalon	Sylramic 5HS	Sylramic 8HS
E1,E2 (Ksi)	19,000	30,000	35,370	33,420
E3 (Ksi)	12,000	20,000	23,330	22,284
G12 (Ksi)	9,400	9,400	11,000	9,400
G13, G23 (Ksi)	9,000	9,200	11,000	9,200
ν12	0.14	0.14	0.083	0.083
ν13, ν23	0.18	0.16	0.099	0.089

RESULTS

A typical deformed mesh is shown in Figure 4. The maximum deflection occurs at the center of the plate. The axial deflection of the plate center relative to the displacement of the

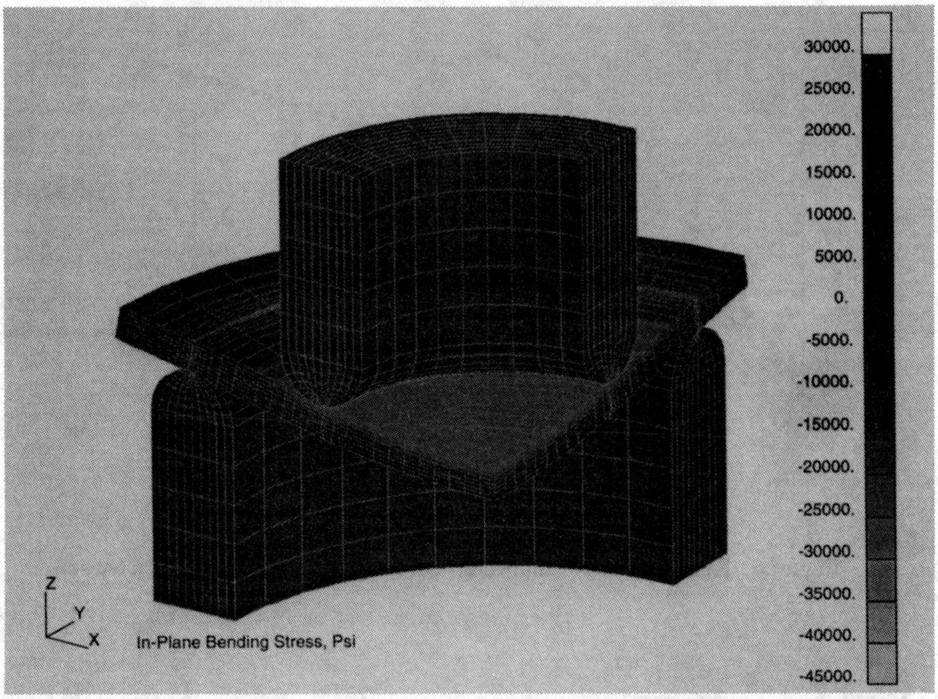

Figure 4. Deformed Mesh with in-Plane-Bending Stresses at 575 lbf.

inner ring is shown in Figure 5 for the commercial grade SiC/SiC woven system as a function of the applied external load. The experimental load deflection curve is also shown in the Figure 5. As seen from the figure, a good agreement is observed between the two curves given the crude assumption of the Von-Mises damage model. Experimental data were not available for both Sylramics and Hi-Nicalon systems. For completion, the relative center deflections versus the applied load are shown in Figure 6 for all four composite systems. The commercial grade SiC/SiC system shows more flexibility and earlier onset of non-linearity than the other systems.

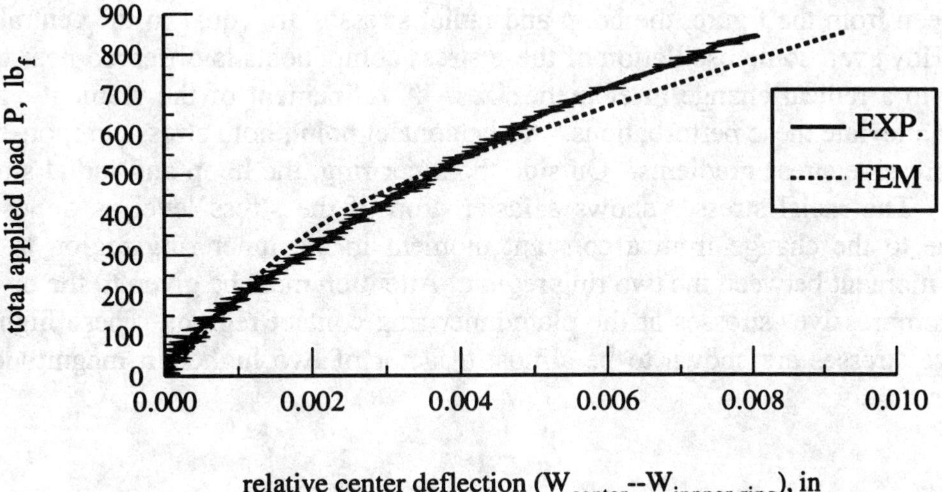

Figure 5. Relative Plate Center Deflection for the Commercial Grade SiC/SiC.

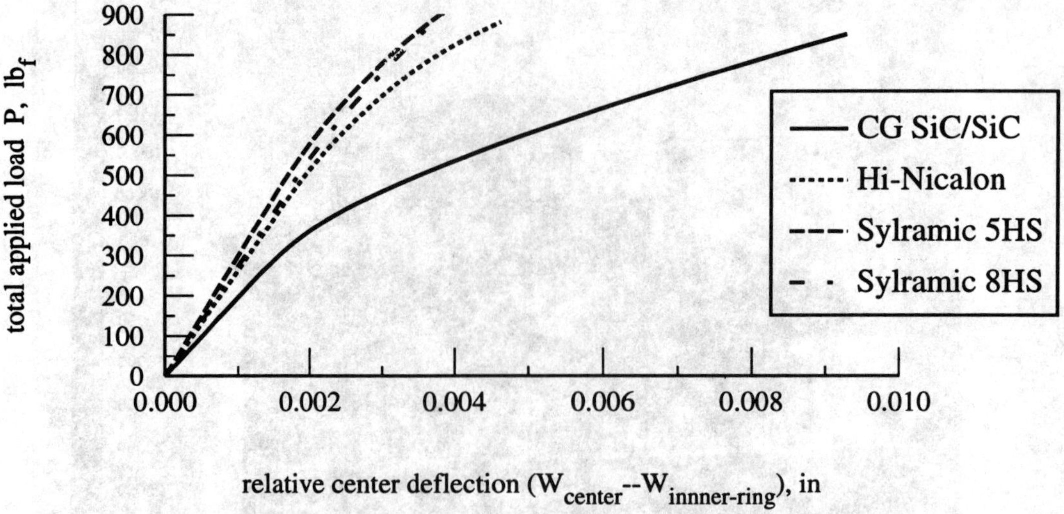

Figure 6. Relative Plate Center Deflection for four CMCs Woven Systems.

Figure 4 also shows a typical fringe plot of the in-plane stresses for the plate and both rings. The contour plot represents the X-direction stress distribution for the commercial grade SiC/SiC composite at a 575 Lbf load level. The stress distribution is symmetric and highly uniform throughout the gage area, showing as expected high tensile stresses at the bottom side of the plate and high compressive stresses at the top side. Also, a high stress gradient is clearly noted at the contact region between the plate and rings. The actual X-direction stresses at the center of the plate along the plate thickness are shown in Figure 7 for two loads (575 and 850 Lbf). The expected bending behavior is observed giving rise to a symmetric stress profile with tensile stresses on the bottom half and compressive stresses along the top half. The stress profile is far from linear given a low yield stress of the commercial grade SiC/SiC system. On the other hand, the strain distribution for the same center elements is plotted in Figure 8. The strain distribution reveals a linear profile as opposed to the stress profile. The calculated radial and hoop stresses of the near surface plate top and bottom elements are plotted in Figure 9 as a function of the radial distance, for a 850 lb_f applied load. As seen from the figure, the hoop and radial stresses are equal in the central region of the plate. However, some oscillation of these stress components is observed near the contact region due to a radical change in element sizes. A refinement of the element sizes in that region may alleviate these perturbations. At the contact point, both stress components display steep compressive stress gradients. Outside the inner ring, the hoop and radial stresses start to diverge. The radial stresses shows a faster drop of the stress level as oppose to hoop stresses due to the change from a constant moment in the inner ring region to a linearly decreasing moment between the two ring region. Attention must be given to the obvious high localized compressive stresses at the plate/inner ring contact region. These high localized compressive stresses are shown to be almost a factor of two higher in magnitude than the tensile stresses.

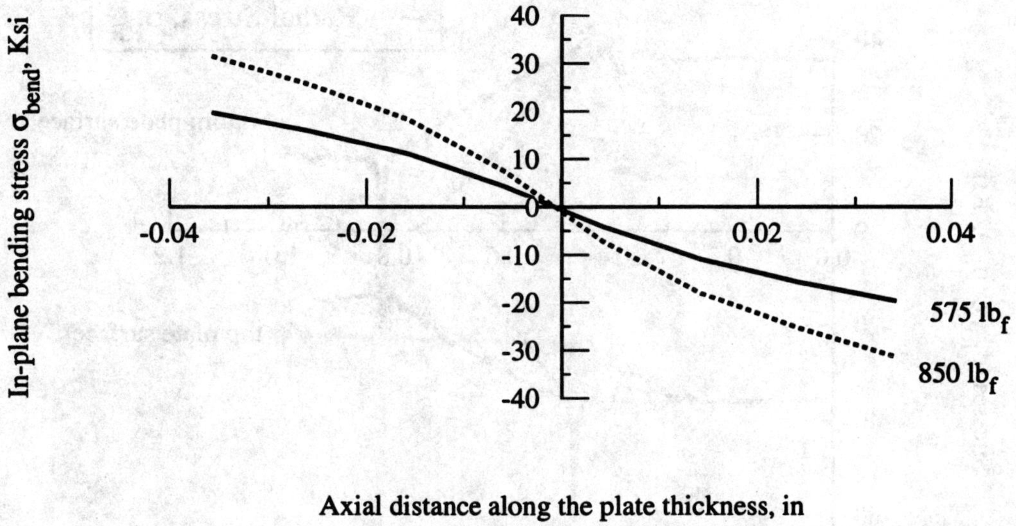

Figure 7. In-Plane bending Stresses through the Plate Thickness.

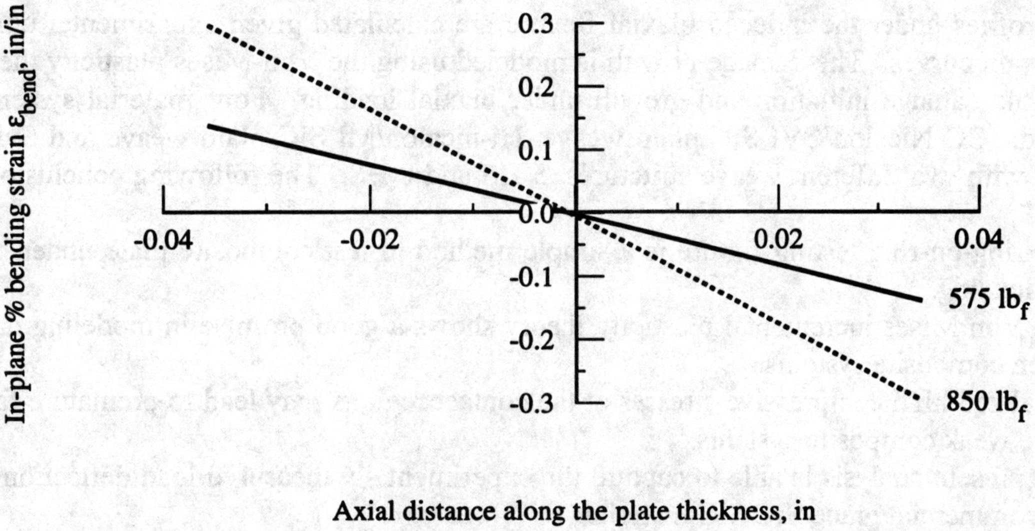

Figure 8. In-Plane Strain Profiles through the Plate Thickness.

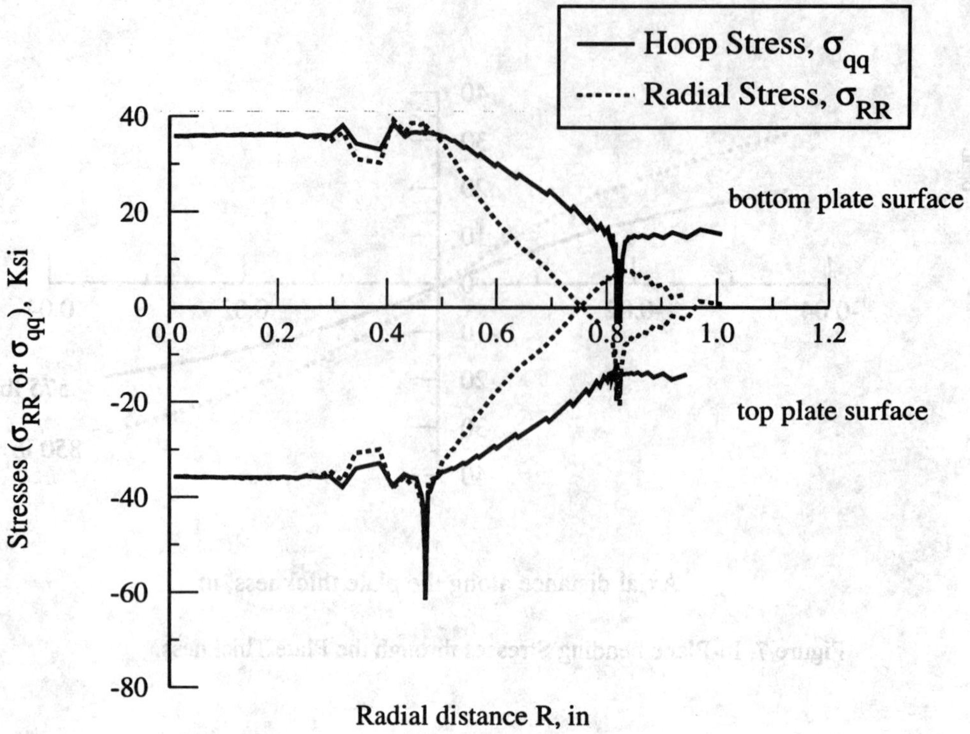

Figure 9. Radial and Hoop Stresses of the Top and Bottom Plate Elements for the CG SiC/SiC System.

SUMMARY

A stress analysis was conducted on the ring on ring loading fixture. The plate stress and strain profiles under the induced biaxial flexure are calculated given experimental uni-axial stress strain curves. The damage growth is modeled using the Von-Mises plasticity theory to predict the damage initiation and growth under biaxial loading. Four material systems are analyzed: CG Nicalon/CVI SiC plain weave, Hi-nicalon/MI SiC plain weave and Sylramic /MI Sic with two different weave structures, 5 HS and 8 HS. The following conclusions are drawn:

1. The ring-on-ring testing fixture is a simple method to test composite plate under biaxial flexure loading.
2. The Von Mises incremental plasticity theory shows a good promise in modeling damage of woven composite systems.
3. Localized high compressive stresses at the contact regions may lead to premature failures for some weak composite systems.
4. The present analysis is able to capture the experimentally measured load deflection curve for the commercial grade SiC/SiC material.
5. Experimental load deflection curves for the other composite systems are needed for model verifications.

REFERENCES

MARC general purpose Finite Element Analysis Program. 1995, Vol. A: user Information Manual; Vol. F: Theoretical manual. MARC Analysis Research Corporation, Palo Alto, California.

Tsangarakis, N. and M. Pepi. 1990. "Biaxial Flexing of a Fiber Reinforced Aluminum Composites," Journal of Composite materials, Vol. 24, No. 7, PP. 770-785.

Tsangarakis, N. and M. Pepi. 1992. "Biaxial Flexing of Unidirectionally Reinforced Aluminum Composite," Journal of Composite Materials, Vol. 26, no. 14, pp. 2060-2075.

Tsangarakis, N. and B. Taleghani. 1995. "Biaxial Flexural Testing of a Carbon Fiber Reinforced Epoxy Composite," Journal of Composite Materials.

Shetty, D.K et al., 1980. "Biaxial Flexure Tests for Ceramics," Ceramics Bulletin, Vol. 59, No. 12, pp. 1193-1197.

Meso-Mechanics and Meso-Structures: A Matter of Scale

C. C. CHAMIS

ABSTRACT

Meso-mechanics and meso-structures are described in terms of the scales at which they are observed and formulated. Select composite examples are presented to illustrate that both meso-mechanics and/or meso-structures are meaningful only when they are referred to a specific scale in a hierarchical scale observation/simulation. These examples include different types of composite unit cells, woven fabrics unit cells and progressive fracture as a composite enhanced infrastructure made from reinforced concrete. The results from the select examples indicate that meso-mechanics and meso-structures are elusive terms and depend mainly on the investigators knowledge and available information.

INTRODUCTION

The terms meso-mechanics and meso-structures are used to describe composites formulations which are based on some intermediate scale between the elemental level (micromechanics) and on the observation scale (macromechanics). The contention for meso-mechanics formulations is that micromechanics formulations may be either too difficult or credible testing cannot be performed to justify the relevant assumptions. On the other hand, formulations based on the observation scale, say for example laminates, usually are material and/or structural configuration dependent which must be repeated for every material/laminate. This latter approach is obviously time consuming and costly. For example, if the laminate response may be viewed as the observation scale, then the laminate theory based on known unidirectional composite (ply) properties is the meso-mechanics scale; while micromechanics will be the micro scale where the formulation starts from constituent properties and incorporates composite processing variables as well. Another example, formulations of structural response based on simulated component information may be viewed as meso-structural mechanics; while those based on the variables that describe the simulated component response will be viewed as micro-structural mechanics. The two examples mentioned above are generic and apply to other situations as well as such as (1) coupon testing versus granular structure, (2) granular structure versus metallurgical formulations, (3) polymer chains versus physical chemistry formulations, (4) structural system response versus finite element formulations etc. An alternate recently emerging view,

---

Christos C. Chamis, NASA Lewis Research Center, 21000 Brookpark Road, OH 44135.

which is by far more generic, is that of telescoping scale mechanics formulations (Chamis and Mital, 1996) based on elemental level variables and on subsequent propagation of that information to any desired progressively higher observation scale. A representative example of telescoping scale mechanics is composite mechanics which is performed by repeated application of laminate theory from fiber substructuring (where environmental and fiber architecture effects are included), ply substructuring and laminate substructuring. The objectives of the proposed paper are: (1) To describe telescoping scale mechanics and, (2) to demonstrate their generic features by applying them to select scales that may be viewed as meso-scales and meso structures.

COMPOSITE MECHANICS AND STRUCTUAL MESO-SCALES: SOME FUNDAMENTALS

Composite scale mechanics, as defined herein (author's rational), refers to both substructuring and telescoping. Composite scale substructuring refers to substructuring a laminate progressively to lower consistent scales through-the-thickness: mult-fiber ply, single fiber ply through-the-thickness of a ply, unit cell, and single fiber slicing (fiber substructuring). Scale telescoping reverses the process of substructuring. Both of these scales can be readily simulated by recursive application of laminate theory. The concept of composite scale mechanics by recursive application of laminate theory is illustrated schematically in Figure 1.

When we start with laminates we can progressively substructure (decompose) to lower scales of laminate behavior in terms of stress and strains. On the other hand it is not easy to substructure laminate behavior in terms of mechanical, thermal or any other properties. We can also substructure any behavior from the highest scale to the lowest by using consistent formal composite mechanics methods. Obviously then formal scalc substructuring methods are the inverse of those for scale telescoping.

We see in Figure 1 that: (1) our lowest scale is the slice with a scale a fraction of the fiber diameter; (2) the next scale in the telescoping sequence is a single fiber embedded in a matrix (typical cell) with a scale equal to fiber diameter plus some matrix; (3) the single fiber typical cell telescopes into a multi-fiber ply with scale of ply thickness; and (4) the multi fiber ply telescopes into multi-ply laminate with a scale of laminate thickness. Other scales in the telescoping sequence up to the structural system scale (not shown in the Figure 1) include: (5) multi-laminate finite element; (6) multi-finite-element subcomponent; (7) multi-subcomponent component and (8) multi-component composite structural system. The author classifies scales (2), (3) and (4) above meso composite mechanics scales, while scales (5), (6) and (7) are meso structural scales and, therefore, represent meso structures.

By the above classification: a meso scale can then be defined as a scale from which we choose information to predict behavior for the next scale in the hierarchical composite structure simulation; provided that formal methods exist to predict the information we choose from lower level scales. The following are typical examples: (1) starting with ply properties to predict laminate properties the ply is a meso scale because the ply properties can be predicted by micromechanics; (2) using the single fiber cell properties to predict ply properties, the single cell scale becomes a meso scale because the single cell properties can be predicted by the primitive - unit slice; (3) using laminate properties in a finite element, the laminate becomes a meso-mechanics scale, because laminate properties can be predicted by ply properties; (4) using finite element properties to predict composite subcomponent properties, the finite element becomes a meso structural scale because the element properties can be predicted by laminate properties; (5) using subcomponent properties to predict component properties, the subcomponent becomes a meso-structural scale because the subcomponent properties can be predicted from finite-element information. The above five examples illustrate that a meso-scale is a floating definition depending very much on the users knowledge, available information to him and time constraints. Below we discuss further in separate sections composite meso-scales and structural meso scales.

COMPOSITE MESO SCALES

We describe two levels of meso-scales in composite mechanics. These meso-scales are based on the schematics of the unit cells described in Figure 2. The ICAN Unit Cell has only two regions (A for matrix and B for fiber) and matrix. Region B then becomes a meso-scale. The METCAN Unit Cell has three regions (A- matrix only, B - matrix and interphase and C - matrix interphase and fiber). Regions B and C are meso-scales. The CEMCAN Unit Cell has also three regions (A - slices 1 and N_l, B - slices 2 and N_{l-1} for matrix and interphase and C - slices 3 to N_{l-2} for matrix, interphase, and fiber). Again regions B and C are meso-scales. In addition, C includes several meso-scales each with a different fiber volume ratio depending on the fiber part in that slice. Below we present equations for the transverse module for each unit cell to illustrate specific features in association with meso-scales. The transverse modulus equation for the ICAN Unit Cell (Chamis, 1986) is:

$$E_{\ell 22} = (1 - \sqrt{k_f})E_m + \frac{\sqrt{k_f}\,E_m}{1 - \sqrt{k_f}\left(1 - \dfrac{E_m}{E_{f22}}\right)} \tag{1}$$

The second term is the meso-scale contribution since it contains both fiber and matrix.

The equation for the transverse modulus for the METCAN Unit Cell (Hopkins and Chamis 1985), is:

$$E_{l22} = E_m\left\{(1 - k_f) + \frac{\sqrt{k_f}\left[1 - \left(\dfrac{D}{D_o}\right)\right]}{1 - \sqrt{k_f}\left[1 - \left(\dfrac{E_m}{E_d}\right)\right]} + \frac{\sqrt{k_f}\left(\dfrac{D}{D_o}\right)}{1 - \sqrt{k_f}\left[1 - \left(1 - \dfrac{D}{D_o}\right)\left(\dfrac{E_m}{E_d}\right) - \left(\dfrac{D}{D_o}\right)\left(\dfrac{E_m}{E_f}\right)\right]}\right\} \tag{2}$$

The second term is a meso-scale term since it contains both matrix and interphase. The third term is a higher level meso-mechanics term since it contains two meso-mechanics scales: (1) matrix and interphase (the first product term in the denominator), and (2) matrix,interphase and fiber (third product term in the denominator. Note the hierarchy of the meso-mechanics scales in the third term denominator.

The equation for transverse modulus for the CEMCAN Unit Cell (Murthy and Chamis, 1992) is:

$$E_{\ell 22} = \frac{1}{s} \sum_{l=1}^{N_l} \left\langle \left(\frac{E_{m22} E_{i\,22} E_{f\,22}}{K_m E_{f\,22} E_{\,i22} + K_i E_{\,f\,22} E_{m22} + K_f E_{\,i\,22}\, E_{m22}} \right) d_s \right\rangle_i \quad (3)$$

Note that the i term must be consistent with what the slice contains and satisfies the conditions of $k_f = 0$, and k_f and $k_i = 0$ as can be observed by inspection. Note also that equation (3) contains both a single level and two level meso-mechanics scales: single level when k_i or $k_f = 0$. Note also that equation (3) is more inclusive than equation (2) and equation (2) is more inclusive than equation (1). It should be apparent that more local details can be included in lower level scales obtained by progressive substructuring.

Some property comparisons (Mital, et al., 1993) from the different scale formulations are instructive. Comparisons for moduli are shown in Figure 3. The comparisons for the METCAN and CEMCAN Unit Cells are almost identical while those from the ICAN are somewhat higher especially for $E_{\ell 22}$ and $E_{\ell 33}$. Similar comparisons for the thermal expansion coefficients are shown in Figure 4. The comparisons are practically indistinguishable. Another interesting comparison is property prediction comparisons from the CEMCAN Unit Cell and 3-D finite element. The results from these comparisons (Mital et al., 1993) are shown in Figure 5. The comparisons are practically identical for both mechanical and thermal properties.

Another interesting case that contains a large number of composites meso-scales is simulating plain woven fabrics by using progressive laminate substructuring. A schematic of woven fabric unit cell is shown in Figure 6 (Mital, et al., 1995). A schematic of a typical meso-scale cell within the woven fabric unit cell is shown in Figure 7. Note that this meso-scale cell includes (1) fiber-tows in the plan, (2) fiber tows at an angle to that plane to represent the weave angulation and, (3) a resin rich area. It is important to note that the meso-scale shown in Figure 7 varies along the weave angulation. The properties of the woven fabric unit cell are predicted by two-tier composite scale telescoping. The first tier consists of single fiber ply telescoping to get the meso-scale cell properties while the second tier consists of meso-scale ply telescoping to get the woven fabric cell properties. Comparisons of predicted results obtained from meso-scale telescoping with 3-D finite element (Mital et al.,1996) are shown in Table I. Predictions obtained by using a 0/90 laminate to simulate the woven fabric unit cell are also shown in Table I in order to illustrate that a 0/90 laminate can be considered as a meso-scale in a plain weave woven fabric laminate. The results show that meso-scale predictions and those from the 3-D finite element are in very good agreement. Even those from the 0/90 are acceptable except for the values for E_{xx}. Side note: These good comparisons gain considerable significance when the computer times required are taken into consideration. For example: 0/90 takes about a small fraction of the time of the meso-scale cell; and the meso-scale scale cell takes a small fraction of the time of the 3-D finite element. Couple of other examples of laminate meso-scales that come to mind are: (1) intraply hybrid mechanics and smart composite structures with embedded sensor/actuators of the fiber or ribbon type.

From the above discussion of several composite meso-scales, it is concluded that meso-mechanics is a floating term and only makes sense when it is referred to a specific scale. The select cases used illustrate that it is very much the particular investigator's choice which will depend on the circumstances at the time of his investigation. The important observation/conclusion is one investigator's meso-mechanics is more likely another investigator's different mechanics scale in the simulation hierarchy.

STRUCTURAL MESO-SCALES AND MESO STRUCTURES

A very good example to describe composite structural meso-scales and meso structures is the hierarchical simulation of composite enhanced reinforced concrete infrastructures (Chamis and Gotsis, 1997). A typical cross-sections of two different reinforced concrete structural members are shown in Figure 8. Note the fiber composite laminate is placed at the bottom in order to restore or enhance the tensile strength of those sections. The properties of these sections required for finite element structural analysis are obtained by composite scale telescoping of the layered section. The scale hierarchy includes the fiber composite enhancement, the single or two-way steel reinforcement with concrete and several layers of concrete. The concrete properties are also simulated by using a local hierarchy of scale telescoping - from cement through sand, to gravel (Mital and Chamis, 1994). These concrete properties are then used with the reinforced steel bars to simulate that layer as a ply in the composites scale telescoping hierarchy.

Side Comments - The local hierarchical simulation is very interesting for the following reasons. If we consider the reinforcing bars as fibers and the concrete as matrix, then progressive scale substructuring will infer that formulations based at the sand grains scale are minimechanics while those based at the cement powder scale are nanomechanics. Conversely, during scale telescoping: (1) cement with sand is a first level of formulation and can be considered as micromechanics; (2) cement with sand and gravel is, then, as first level meso-scale and, (3) cement with sand, gravel and the reinforcing bars is a second level meso-scale. This second level meso-scale is the highest in the local reinforced concrete ply simulation. However, it is only the micromechanics scale in the formulation which considers concrete as a homogeneous matrix. The local concrete scale telescoping hierarchy is a very good example that illustrates the floating definition of meso-scale which, as was stated previously, depends mainly on the investigators knowledge and available information.

The next scale in the structural hierarchy is the finite element. A schematic of a finite element model for a reinforced concrete arch with composite enhancements is shown in Figure 9. (Chamis and Gotsis, 1997). Note the three different structural sections, note also that the arch tapers in width. Therefore, each finite element will have different stiffness properties and constitutes a structural meso-scale and also a different meso structure - consistent with the previously stated definition. The assembly of all the finite elements constitute the structure and, therefore, the structural scale which is the highest scale in the structural scale telescoping hierarchy.

Progressive fracture is employed in order to evaluate the strength enhancement of the arch due to composite layers. A schematic of progressive fracture simulation is shown in Figure 10. Note that hierarchical scale telescoping starts with the cement-sand nanoscale and ends at the structural scale (left side of figure). Note also that hierarchical scale telescoping is essential to synthesize the structural properties. Progressive hierarchical scale substructuring (decomposition - right side of the figure) starts from the structural scale and ends at the nanoscale where damage initiation, growth and progression are identified and tracked. Typical results obtained are shown in Figure 11. It is concluded that progressive fracture to evaluate the strength enhancement of composite repaired/enhanced reinforced concrete

Hopkins, D. A. and Chamis, C. C. 1985. "A Unique Set of Micromechanics Equations for High Temperature Metal Matrix Composites." NASA TM 87154.

Mital, S.K., Murthy, P.L.N. and Chamis C.C. 1993. "Ceramic Matrix Composites Properties/Microstresses with Complete and Partial Interphase Bond." NASA TM 106136.

Mital, S.K., and Chamis, C.C. 1994. "Thermal and Mechanical Behavior of Particulate Composite Materials." ASME PD-Vol. 62, pp. 275-283.

Mital, S.K., Murthy, P.L.N.,and Chamis C.C. 1995. "Simplified Micromechanics of Two-Dimensional Plain Woven Fabric Composite." Proceedings of the ASME Materials Division, pp. 829-840.

Mital, S.K., Murthy, P.L.N. and Chamis C.C. 1996. " Simplified Micromechanics of Plain Weave Composites." NASA TM 107165.

Mital, S.K., Gotsis, P.K. and Chamis, C.C. 1997. "Laminate Analogy for Composites Application to Infrastructures." Proceedings of 42nd International SAMPE Symposium, pp. 947-957.

Murthy, P.L.N. and Chamis, C.C. 1992. " Towards the Development of Micromechanics Equations for Ceramic Matrix Composites via Fiber Substructuring." NASA TM 105246.

TABLE I - COMPARISONS OF DIFFERENT SCALE SIMULATIONS

Property	Finite Element Analysis Results	0/90 Laminate	Micromechanics Fiber Slice
E_{xx}, GPa	49.0	72.5	54.5
E_{zz}, GPa	8.8	9.9	8.3
ν_{zx}	0.09	0.05	0.07
ν_{zy}	0.08	0.05	0.07
ν_{yx}	0.04	0.03	0.04
ν_{yz}	0.53	0.38	0.5
G_{xy}, GPa	4.1	4.6	4.4
G_{yz}, GPa	3.9	3.6	3.5

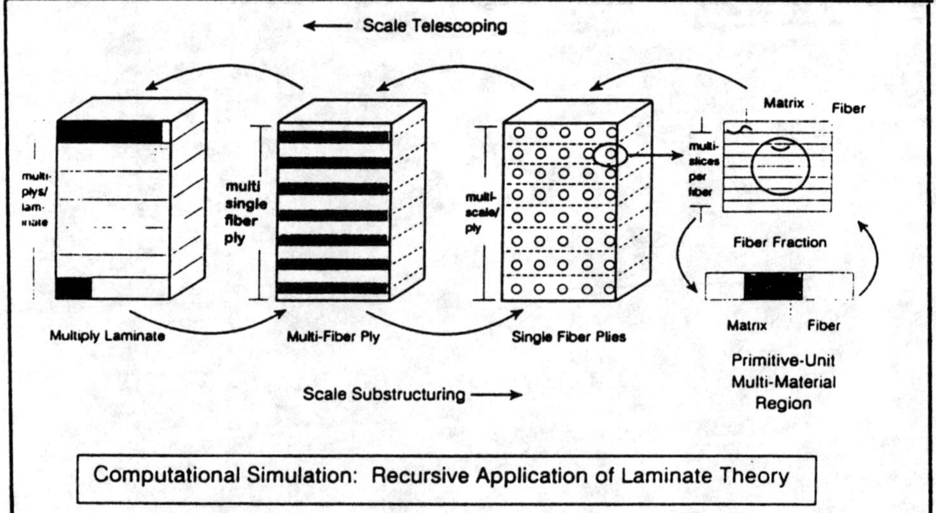

Figure 1 - Multi-scale Hierarchical Simulation

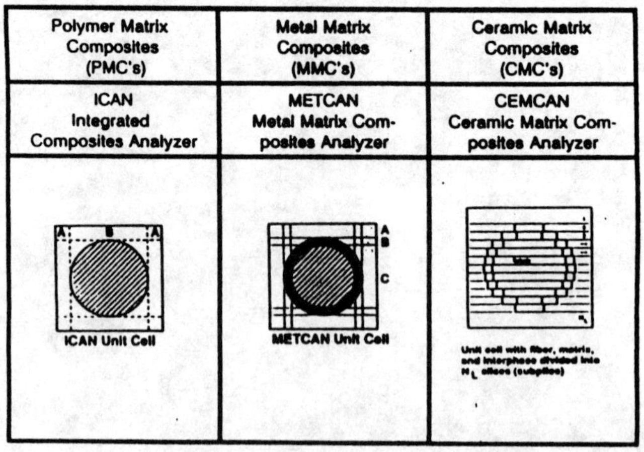

Figure 2 - Different Scale Unit-cell Simulations

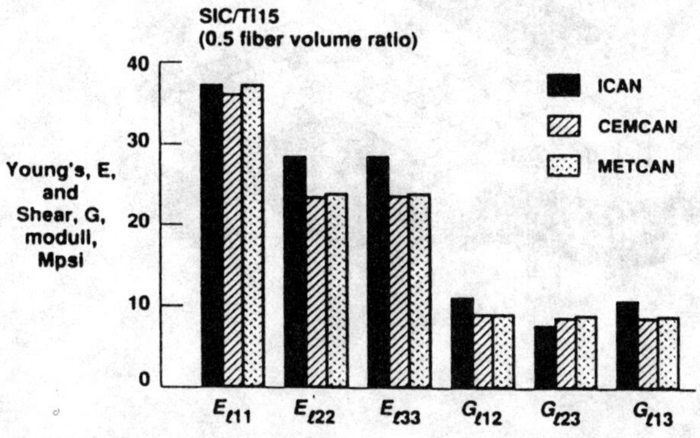

Figure 3 - Comparisons of Ply Mechanical Properties from Different Scale Simulations

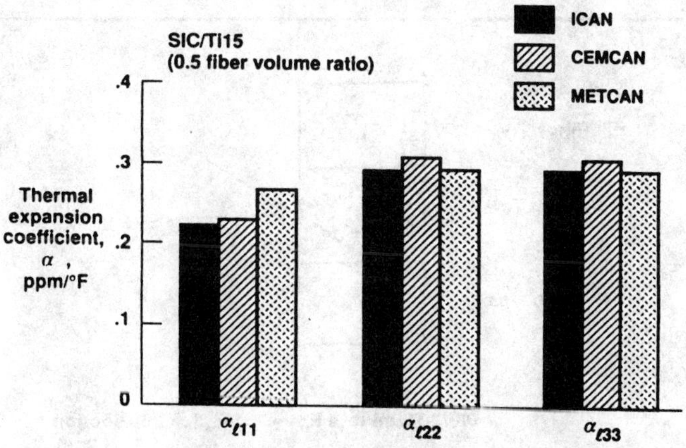

Figure 4 - Comparisons of Ply Thermal Properties from Different Scale Simulations

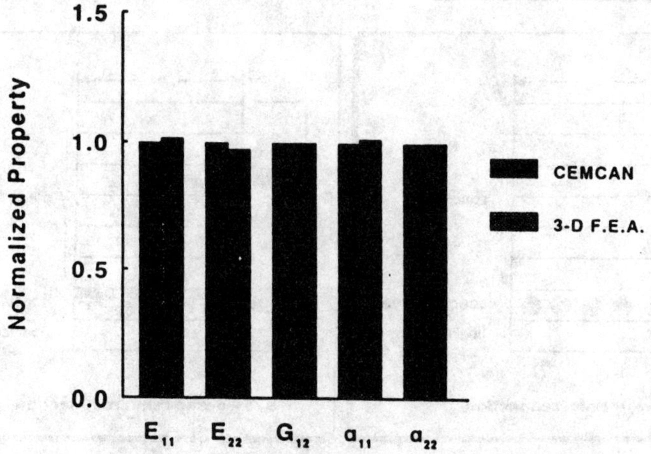

Figure 5 - Comparisons of Ply Properties from Slice-scale and 3-D Finite Element Simulations

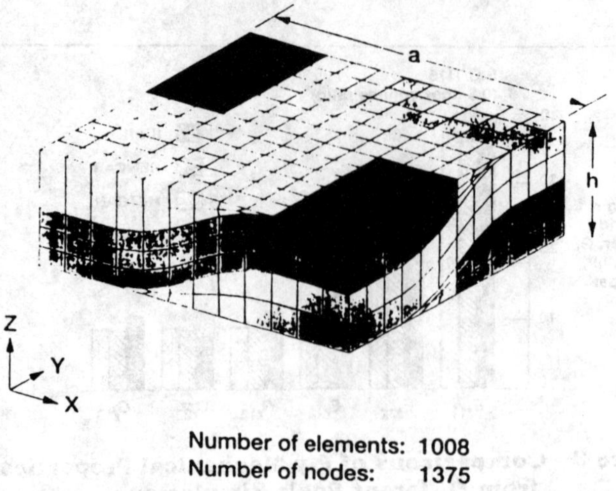

Figure 6 - Schematic of 3-D Finite Element Model of Composite Woven Fabric Unit Cell

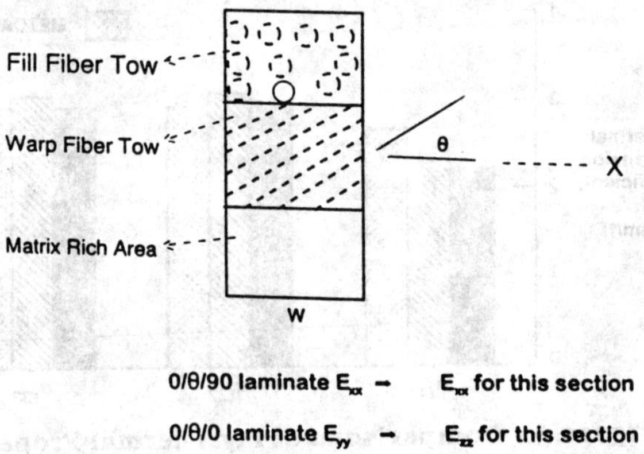

Figure 7 - Meso-Scale Cell for Simulating Woven Fabric Composites

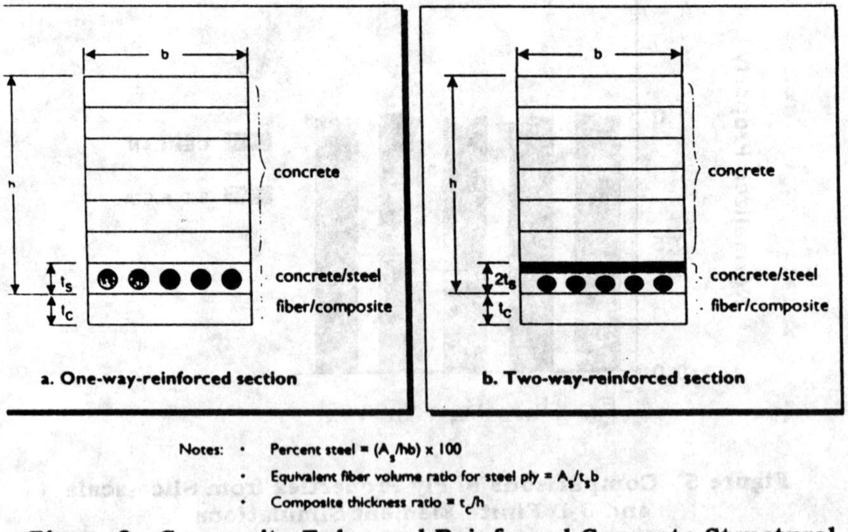

Figure 8 - Composite Enhanced Reinforced Concrete Structural Sections with Several Meso-Scales

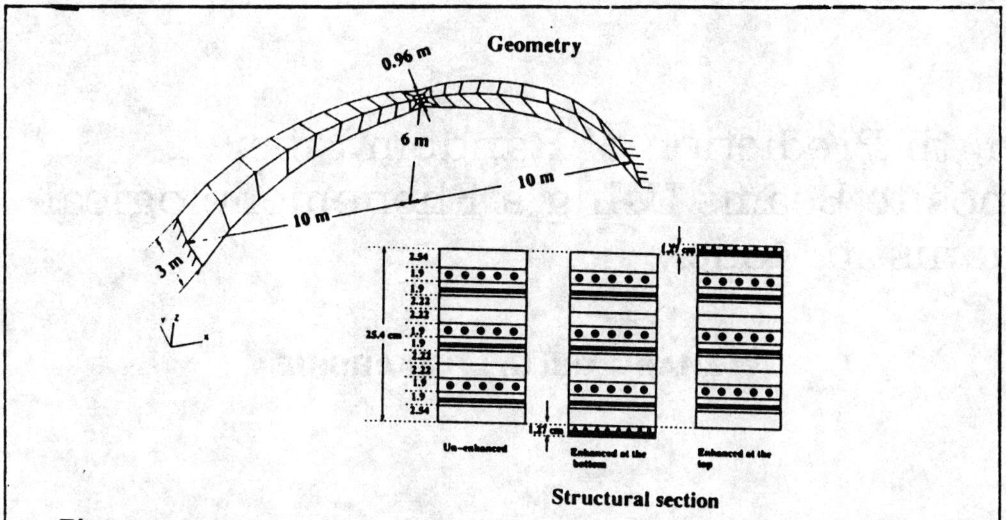

Figure 9 -Reinforced Concrete Arch with Composite Enhancements

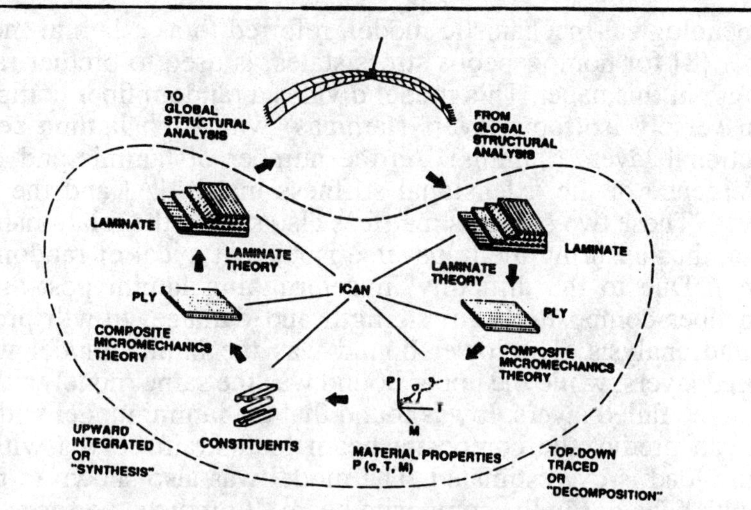

Figure 10 - Schematic of Scale Telescoping and Scale Substructuring for Structural Progressive Fracture

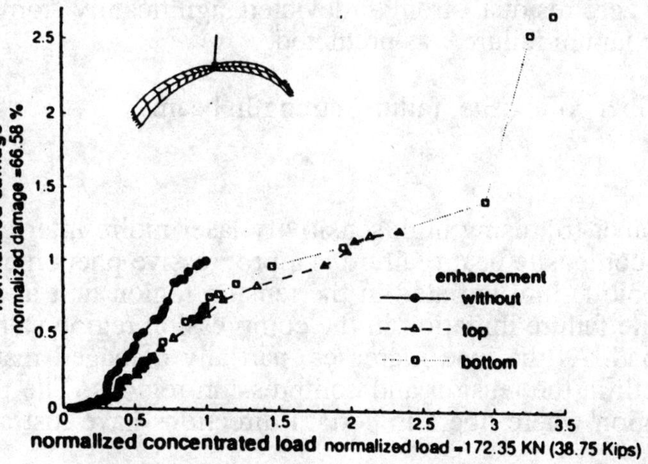

Figure 11 - Progressive Structural Fracture of a Concrete Reinforced Arch with and without Composite Enhancements

Strength Prediction of Random Fiber Composite Beams Using a Phenomenological-Mechanistic Model

Y. ZHANG AND D. L. SIKARSKIE

ABSTRACT

A phenomenological-mechanistic model, referred to as a lamini model and reported in a previous paper [8] for homogeneous stress states, is used to predict random fiber composite beam strength in this paper. This model divides a random fiber composite panel into a stack of thin transversely isotropic layers (laminas) with each lamina constructed from a series of unidirectional layers (laminis). As the number of laminis and laminas approach infinity, the convergence of the extensional stiffness matrix [A] and the bending stiffness matrix [D] is proved. These two stiffness matrices also satisfy the relationship established for isotropic materials, thus ensuring the transverse isotropy typical of random fiber composite plates (and beams). Due to the difficulty in determining lamini post-failure stress-strain behavior, random fiber composite beam strength and damage growth predictions actually come from a bound analysis. The lower bound was the lamini model with zero residual strength in the failed layers, while the upper bound was the same model with perfectly plastic residual strength in the failed layers. It was found that the lamini model with perfectly plastic residual strength can predict the composite beam load-strain curves with great accuracy, although the failure load is overestimated. The model was also shown to predict the failure initiation and growth behavior in the composite beam. It predicts the formation and development of the partially failed zones in both the beam tension and compression regions which were observed in the previous high sensitivity laser moire interferometry experiments [6]. The lower bound analysis underestimated the beam failure load. The load-strain curves predicted by the model with zero residual strength deviated significantly from the experimental results as soon as the first lamini failure was predicted.

KEYWORDS: random fiber composite, failure, strength, beam

INTRODUCTION

In our previous paper [6], using high sensitivity laser moire interferometry, we have shown that random fiber composite beam failure is a progressive phenomenon. It was found that observable microscopic failure initiated in the tension region first at about 40% of the ultimate failure load, while failure initiation in the compression region started at about 50% of the ultimate failure load. As the load increases, partially damaged material zones were formed and expanded both in the tension and compression regions. The partially damaged region is defined as a region where fibers in certain directions have lost all or a portion of

---

Yi Zhang, Ford Motor Company, Suite 155, Schaefer Court II, 14555 Rotunda Drive, Dearborn, MI48120.
David L. Sikarskie, Department of Mechanical Engineering and Engineering Mechanics, Michigan Technological University, Houghton, MI49931.

their load carrying capability, while the rest still retain their original integrity. In these regions, the composite material has partially lost its load carrying capability and also has a reduced stiffness. Unexpected was that the formation of these partially damage zones did not significantly affect the beam stiffness. This was attributed to the different load carrying capabilities of the chopped fibers oriented in different directions. Fibers oriented at 90° to the load direction start to fail at low stress levels, however, their failure does not have a significant effect on either composite stiffness or strength due to their limited reinforcement function.

In this paper, our aim is to develop a viable method for random fiber composite beam strength prediction. Due to the microstructural complexity of random fiber composites, it is extremely difficult, if not impossible, to develop a mechanistic model which is capable of predicting failure initiation and growth on a single fiber level. However, a purely phenomenological approach which completely ignores microstructural damage growth mechanisms is not able to achieve a good beam strength prediction either [5]. As a result, a phenomenological-mechanistic model which can partially capture the microscopic failure processes, but in a phenomenological manner becomes an appealing way for random fiber composite strength prediction. Such a model, referred to as a lamini model has been reported in a separate paper [8] for random fiber composite stiffness and strength prediction in homogeneous stress fields. The lamini model divides a random fiber composite panel into a stack (*n* layers) of thin transversely isotropic layers (laminas) with each lamina constructed from a series (*2m* layers) of unidirectional layers (laminis) having the same material constituents and processing conditions as the random fiber composite panel itself, see Figure 1. The laminis have their fiber

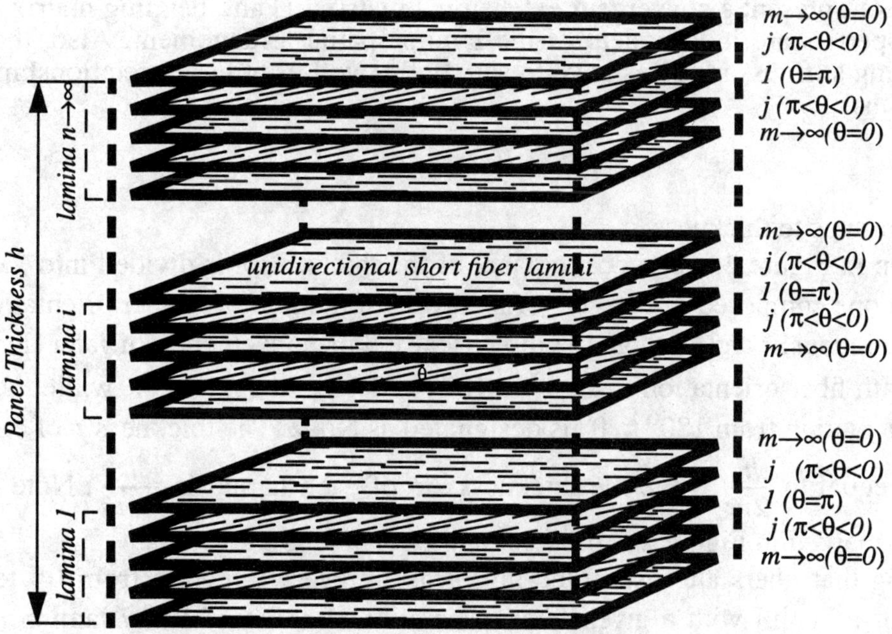

Figure 1 Schematic illustration of lamini model.

directions uniformly oriented from 0° to 180° and are symmetrically arranged to achieve the transversely isotropic nature of each lamina. Each lamini represents all fibers oriented in same direction within a lamina and with the same fiber volume fraction as the lamina. In the case where a random fiber composite panel has a fiber direction preference, i.e. the composite has more fibers in a certain direction, the lamini thickness can vary in different direction to give a representative behavior.

Collecting all fibers with the same orientation within a lamina into one lamini enables us to define composite failure on the lamini level, i.e. only the laminis reaching the stress state defined by a failure criterion fail whereas the other laminis remain intact. Contrast this with a fully mechanistic approach in which failure is defined at a single fiber level. As a result, the lamini model is able to track damage growth both through the thickness and in the plane of the structure (plate) while still being numerically efficient. The lamini model accurately predicts the Young's and shear moduli of a random fiber composite and provides close bounds on the experimental in-plane strengths, namely uniaxial tension, uniaxial compression, and shear[8]. The upper bound analysis assumes a failed lamini has a perfectly plastic residual strength, while the lower bound analysis assumes a failed lamini has zero residual strength. The reason for performing the bound analysis instead of an exact analysis is due to the difficulty of experimentally establishing the post-failure stress-strain behavior of the laminis.

Compared to existing laminate simulation models [1,2,3,4], one of the major advantages of the lamini model is its capability of predicting the stiffness and strength properties of a general random fiber composite structure such as a beam or a plate. The reason is that only the lamini model can achieve a global transverse isotropy under both in-plane and bending loading, essential for the bending strength prediction. In this paper, we will first show this through mathematically proving the convergence of both the extensional and bending matrices which also satisfy the relationship established for isotropic materials. Then, a bound analysis will be performed to model the damage growth process in the random fiber composite beam and to predict its failure load.

CONVERGENCE OF [A] AND [D] MATRICES

To represent the transversely isotropic nature of the random fiber composites, the lamini model has to present a convergent extensional matrix $[A]$ and bending matrix $[D]$, i.e. the stiffness properties are independent of the lamini spatial arrangement. Also, the extensional and bending stiffness matrices have to satisfy the well-established relationship for the isotropic materials,

$$[D] = \frac{h^2}{12}[A] \tag{1}$$

where h is the composite thickness.

As shown in Figure 2, a composite panel of the thickness h is divided into $2n$ layers of laminas, each one composed of two sets of m layers of laminis with fiber orientation from 0° to 180° continuously and arranged symmetrically. For convenience of later derivation, the lamini set with fiber orientation from 0° to 180° is designated as No. 1, while the lamini set with fiber orientation from 180° to 0° is designated as No. 2. The thickness t of each lamina is therefore equal to $\frac{h}{2n}$, while the thickness Δt of each lamini is $\frac{h}{4mn}$. Note that the fiber orientation marked in Figure 2 is the fiber angle relative to the y-axis.

Assuming that fibers are distributed uniformly in every direction from 0° to 180°, the position z of a lamini with a given fiber orientation θ in an arbitrary lamina i can be expressed as

$$z = \left[\frac{\theta}{\pi} + 2(i-1)\right]\frac{h}{4n} \tag{2}$$

in the lamini set No. 1, and

$$z = \left[\frac{\pi - \theta}{\pi} + (2i-1)\right]\frac{h}{4n} \tag{3}$$

in the lamini set No. 2, where $i = 1, 2, \ldots, n$.

Following classical lamination theory [9], the resultant in-plane forces for an $4mn$ layered laminate with in-plane load can be written as

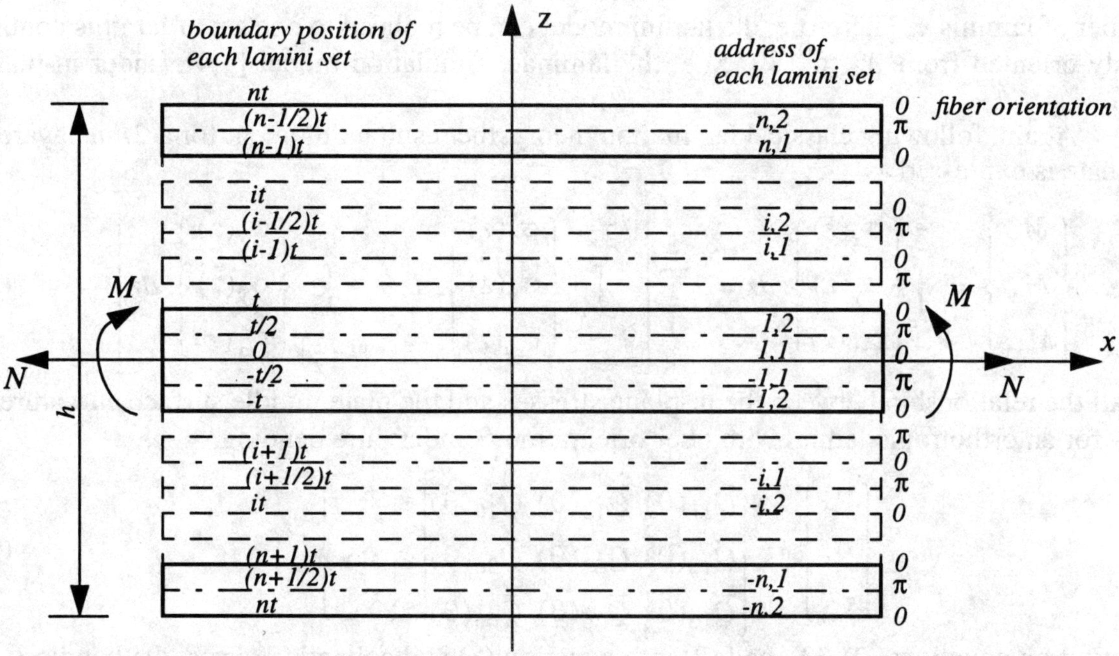

Figure 2. Layout of lamini model for failure load prediction of composite beam.

$$\begin{Bmatrix} N_x \\ N_y \\ N_{xy} \end{Bmatrix} = \int_{-\frac{h}{2}}^{\frac{h}{2}} \begin{Bmatrix} \sigma_x(z) \\ \sigma_y(z) \\ \tau_{xy}(z) \end{Bmatrix} dz = 2\sum_{1}^{n} \left(\int_{(i-1)t}^{\left(i-\frac{1}{2}\right)t} \begin{Bmatrix} \sigma_x(z) \\ \sigma_y(z) \\ \tau_{xy}(z) \end{Bmatrix} dz + \int_{\left(i-\frac{1}{2}\right)t}^{it} \begin{Bmatrix} \sigma_x(z) \\ \sigma_y(z) \\ \tau_{xy}(z) \end{Bmatrix} dz \right) \quad (4)$$

Note that each integral is a lamini set or $\frac{1}{2}$ of a lamina. Recall the constitutive equation for an orthotropic lamini with fiber orientation θ,

$$\begin{Bmatrix} \sigma_x \\ \sigma_y \\ \tau_{xy} \end{Bmatrix} = \begin{bmatrix} \overline{Q}_{11}(\theta) & \overline{Q}_{12}(\theta) & \overline{Q}_{16}(\theta) \\ \overline{Q}_{12}(\theta) & \overline{Q}_{22}(\theta) & \overline{Q}_{26}(\theta) \\ \overline{Q}_{16}(\theta) & \overline{Q}_{26}(\theta) & \overline{Q}_{66}(\theta) \end{bmatrix} \begin{Bmatrix} \varepsilon_x \\ \varepsilon_y \\ \gamma_{xy} \end{Bmatrix} \quad (5)$$

where $[\overline{Q_{ij}}(\theta)]$ is the transformed reduced stiffness matrix [9].

Substituting equations (2), (3), and (5) into equation (4), the resultant in-plane forces become

$$\begin{Bmatrix} N_x \\ N_y \\ N_{xy} \end{Bmatrix} = \frac{h}{2n\pi} \sum_{i=1}^{2n} \int_0^{\pi} [\overline{Q}(\theta)] d\theta \begin{Bmatrix} \varepsilon_x \\ \varepsilon_y \\ \gamma_{xy} \end{Bmatrix} \quad (6)$$

$[A]$ thus converges to

$$[A] = \frac{h}{\pi} \int_0^{\pi} [\overline{Q}(\theta)] d\theta \quad (7)$$

Equation (7) indicates that the extensional stiffness matrix $[A]$ is in fact independent of the number of laminis n. Therefore, the lamini model can be reduced to one set of laminis continuously oriented from $0°$ to $180°$, i.e. the laminate simulation model [1,4], under in-plane load.

Again, following classical lamination theory, the resultant moment for a $2mn$ layered laminate is expressed as

$$\begin{Bmatrix} M_x \\ M_y \\ M_{xy} \end{Bmatrix} = \int_{-\frac{h}{2}}^{\frac{h}{2}} \begin{Bmatrix} \sigma_x(z) \\ \sigma_y(z) \\ \tau_{xy}(z) \end{Bmatrix} z dz = 2\sum_{1}^{n} \left[\int_{(i-1)t}^{\left(i-\frac{1}{2}\right)t} \begin{Bmatrix} \sigma_x(z) \\ \sigma_y(z) \\ \tau_{xy}(z) \end{Bmatrix} z dz + \int_{\left(i-\frac{1}{2}\right)t}^{it} \begin{Bmatrix} \sigma_x(z) \\ \sigma_y(z) \\ \tau_{xy}(z) \end{Bmatrix} z dz \right] \quad (8)$$

Recall the relationship between the in-plane stresses and the plate middle surface curvature $\{K\}$ for an orthotropic lamini with fiber orientation θ under pure bending,

$$\begin{Bmatrix} \sigma_x \\ \sigma_y \\ \tau_{xy} \end{Bmatrix} = \begin{bmatrix} \bar{Q}_{11}(\theta) & \bar{Q}_{12}(\theta) & \bar{Q}_{16}(\theta) \\ \bar{Q}_{12}(\theta) & \bar{Q}_{22}(\theta) & \bar{Q}_{26}(\theta) \\ \bar{Q}_{16}(\theta) & \bar{Q}_{26}(\theta) & \bar{Q}_{66}(\theta) \end{bmatrix} z \begin{Bmatrix} K_x \\ K_y \\ K_{xy} \end{Bmatrix} \quad (9)$$

Substituting equations (2), (3), and (9) into equation (8) and collecting terms, the bending stiffness matrix [D] is obtained as

$$[D] = \frac{h^3}{16n^3\pi^3}\int_0^{\pi}[\bar{Q}(\theta)](\theta^2 - 2\pi\theta + 2\pi^2)d\theta - \frac{h^3}{8n^3\pi}\int_0^{\pi}[\bar{Q}(\theta)]d\theta \sum_{i=1}^{n} i + \frac{h^3}{4n^3\pi}\int_0^{\pi}[\bar{Q}(\theta)]d\theta \sum_{i=1}^{n} i^2 \quad (10)$$

Recall the identities

$$\sum_{i=1}^{n} i = \frac{n(n+1)}{2} \quad (11)$$

$$\sum_{i=1}^{n} i^2 = \frac{n(n+1)(2n+1)}{6} \quad (12)$$

Substituting equations (11) and (12) into equation (10), the bending stiffness matrix $[D]$ becomes

$$[D] = \frac{h^3}{16n^3\pi^3}\int_0^{\pi}[\bar{Q}(\theta)](\theta^2 - 2\pi\theta + 2\pi^2)d\theta + \frac{h^3}{8\pi}\left(\frac{2}{3} + \frac{1}{2n} - \frac{1}{6n^2}\right)\int_0^{\pi}[\bar{Q}(\theta)]d\theta \quad (13)$$

Obviously, the bending stiffness matrix $[D]$ is dependent on the number of laminas n, implying the limitation of the laminate simulation model [1,4]. In that model, basically only one layer of lamina was used, i.e. $n = 1$, resulting in a bending stiffness matrix $[D]$ as calculated from equation (13) for $n = 1$. This does not satisfy equation (1), the required relationship between matrices $[A]$ and $[D]$ for isotropic materials. Thus, the laminate simulation model reported in [1,4] contradicts the transverse isotropy and can not be used for beam or plate bending applications.

However, the present model, i.e. the lamini model, for $n \to \infty$, yields a [D] matrix

$$[D] = \frac{h^3}{12\pi}\int_0^{\pi}[\bar{Q}(\theta)]d\theta \tag{14}$$

Comparing equation (7) to equation (14), it is easily shown that equation (1) is satisfied for the lamini model. Thus, the lamini model in the limit is globally transversely isotropic as random fiber composite materials suggest. The lamini model also reflects, to a certain extent, the anisotropic nature of random fiber composites at a microscopic scale. The difference is that the anisotropy in the real material is random while the anisotropy in the model is now deterministic, i.e. an unstructured material is being modeled as a material with order such that both materials have the same global properties.

METHOD FOR BEAM FAILURE LOAD PREDICTION

It is assumed in this study that each lamini, to a first order, is an orthotropic linear elastic layer to failure. Also, Tsai-Wu failure criterion is used to define the lamini failure, although any other valid failure criterion could be equally applied. For a two dimensional stress state, the Tsai-Wu criterion can be expressed as

$$F_1\sigma_1 + F_2\sigma_2 + F_{11}\sigma_1^2 + F_{22}\sigma_2^2 + 2F_{12}\sigma_1\sigma_2 + F_{66}\sigma_6^2 = 1 \tag{15}$$

where $F_1 = \frac{1}{\sigma_x^t} + \frac{1}{\sigma_x^c}$, $F_2 = \frac{1}{\sigma_y^t} + \frac{1}{\sigma_y^c}$, $F_{11} = -\frac{1}{\sigma_x^t\sigma_x^c}$, $F_{22} = -\frac{1}{\sigma_y^t\sigma_y^c}$, $F_{66} = \frac{1}{\tau_f^2}$, σ_x^t, σ_x^c, σ_y^t, and σ_y^c are tensile and compressive strengths of the unidirectional short fiber lamini in the two principal material directions respectively. Note that the compression strength σ^c is negative. τ_f is the composite shear strength. The stress tensor σ_i is defined in the material coordinate system. The term F_{12} can be approximated as $-\frac{1}{2}\sqrt{F_{11}F_{22}}$ instead of running an additional biaxial test for determining it [8]. Again, because of the difficulty to experimentally establish the lamini post-failure stress-strain behavior, a bound analysis is performed to predict the beam failure load. The upper bound analysis assumes the failed lamini has a perfectly plastic residual strength, while the lower bound analysis assumes the failed laminis has a zero residual strength [8].

Assume that k layers of laminis have failed before the bending moment is increased to M_{k+1} and the addresses of these failed laminis are registered in a vector $\{f\}$. The extensional stiffness matrix [A], the extension-bending coupling matrix [B], and the bending stiffness matrix [D] are respectively reduced to

$$[A]_k = \Delta t\left\{\sum_{i=-2mn}^{2mn}[\bar{Q}]_i - \sum_{j=f_1}^{f_k}[\bar{Q}]_j\right\} \tag{16}$$

$$[B]_k = -\frac{1}{2}\sum_{j=f_1}^{f_k}[\bar{Q}]_j\left[\left(z_i + \frac{\Delta t}{2}\right)^2 - \left(z_i - \frac{\Delta t}{2}\right)^2\right] \tag{17}$$

$$[D]_k = \frac{1}{3}\left(\sum_{i=-2mn}^{2mn}[\bar{Q}]_i\left[\left(z_i + \frac{\Delta t}{2}\right)^3 - \left(z_i - \frac{\Delta t}{2}\right)^3\right] - \sum_{j=f_1}^{f_k}[\bar{Q}]_j\left[\left(z_j + \frac{\Delta t}{2}\right)^3 - \left(z_j - \frac{\Delta t}{2}\right)^3\right]\right) \tag{18}$$

in which z_i is the center line position of the corresponding laminis in thickness direction and Δt is the lamini thickness. Note that the extension-bending coupling stiffness [B] is zero before any lamini failure occurs, due to the symmetric nature of the lamini model. However, as soon as the laminis start to fail, the lamini model becomes unsymmetrical and $[B]$ becomes non-zero. Also, as the load increases, the number of damaged laminis increases also, and the extensional and bending stiffness matrices $[A]$ and $[D]$ gradually decrease by eliminating the contribution of the damaged laminis. When all laminis become damaged, the matrices $[A]$ and $[D]$ should be reduced to zero.

The resultant force and moment for a $2mn$ layered laminate with k layers of failed laminis can be expressed as

$$\begin{Bmatrix} \{N\}_{k+1} \\ \{M\}_{k+1} \end{Bmatrix} = \begin{bmatrix} [A]_k & [B]_k \\ [B]_k & [D]_k \end{bmatrix} \begin{Bmatrix} \{\varepsilon^\circ\}_{k+1} \\ \{K\}_{k+1} \end{Bmatrix} + \begin{Bmatrix} \{N\}_k^f \\ \{M\}_k^f \end{Bmatrix} \tag{19}$$

$\{\varepsilon^\circ\}_{k+1}$ and $\{K\}_{k+1}$ are the middle surface strains and the middle surface curvatures. $\{N\}_k^f$ and $\{M\}_k^f$ are the residual force and bending moment carried by all the failed laminis,

$$\{N\}_k^f = \Delta t \sum_{j=f_1}^{f_k} \{\sigma\}_j^f \tag{20}$$

$$\{M\}_k^f = \frac{1}{2} \sum_{j=f_1}^{f_k} \{\sigma\}_j^f \left[\left(z_j + \frac{\Delta t}{2} \right)^2 - \left(z_j - \frac{\Delta t}{2} \right)^2 \right] \tag{21}$$

$\{\sigma\}_j^f$ is the residual stress state in each failed lamini, which depends on the lamini post-failure residual strength assumption. For the upper bound analysis, i.e. perfectly plastic residual strength, the stress state $\{\sigma\}_{k+1}^f$ in each failed lamini remains constant at the failure stress, whereas for the lower bound analysis, i.e. zero residual strength, the stress state in the failed lamini is always reset to zero. The resultant internal force and moment on the left side of equation (19) should be equal to the external in-plane force and bending moment, which for the pure bending case

$$\{N\}_{k+1} = \{0\} \text{ and } \{M\}_{k+1} = \begin{Bmatrix} 1 \\ 0 \\ 0 \end{Bmatrix} M_{k+1} \tag{22}$$

where M_{k+1} is the bending moment causing the $(k+1)$ layer of lamini failure, and is the unknown to be solved. The solution sequence is as follows. Solve equation (19) for $\{\varepsilon^\circ\}_{k+1}$ and $\{K\}_{k+1}$ in term of M_{k+1} and express the stress state in each intact lamini as a function of M_{k+1}. Then, substituting the stress state in each intact lamini into Tsai-Wu failure criterion, equation (15), the bending moment M_{k+1} required to fail each intact lamini can be solved. The smallest one among the solutions is the bending moment which will cause the next lamini failure, whose address can also be registered. Having the failure bending moment solution, the stress and strain states in each lamini can also be calculated.

PREDICTIONS AND EXPERIMENTAL RESULTS

The random fiber composite panel (0.125 inch nominal thickness) used in this study was a structural injection reaction molding (SRIM) panel provided by the Technical Center of the General Motors Corp. The matrix polymer is a DOW MM364 isocyanate thermoset resin having a 5% elongation and the reinforcing fibers are 2 inch long chopped Owens Corning E-glass fibers. A mat preform was made in house by G.M. staff by chopping a glass roving and spraying it onto a screen. An epoxy binder was sprayed with the glass to give some stability to the preform. The preform was then placed in a mold, the resin injected and resultant plaque formed/molded [8].

To perform numerical calculations on the lamini model, both elastic and failure characterization data for a unidirectional short fiber composite panel having the same fiber and matrix material, and the same volume fraction as the material tested, are needed. Ideally, the required characterization data should be established from the unidirectional short fiber composite coupons of the same matrix, processing condition, and fiber volume fractions as the SRIM panel. This proved to be extremely difficult to accomplish and in the final analysis the required characterization data for the unidirectional lamini was estimated using techniques outlined in the separate paper [8]. The same stiffness and strength properties of the unidirectional chopped fiber laminis are used in this study. They are summarized as follows, the lamini Young's moduli along and perpendicular to the fiber direction are respectively $E_1 = 25.8$ GPa and $E_2 = 6.1$ GPa. The shear modulus is $G_{12} = 2.2$ GPa, the Poisson's ratio is $\nu_{12} = 0.35$. The lamini tension and compression failure stresses in the fiber direction are respectively $\sigma_1^t = 636.7$ MPa and $\sigma_1^c = 365.4$ MPa, the lamini tension and compression failure stresses perpendicular to the fiber direction are respectively $\sigma_2^t = 91.0$ MPa and $\sigma_2^c = 340.6$ MPa, and the lamini shear failure stress is $\tau_{12} = 202.7$ MPa.

The load strain curves predicted with perfectly plastic residual strength (upper bound) and zero residual strength (lower bound) are illustrated in Figure 3, superimposed with the

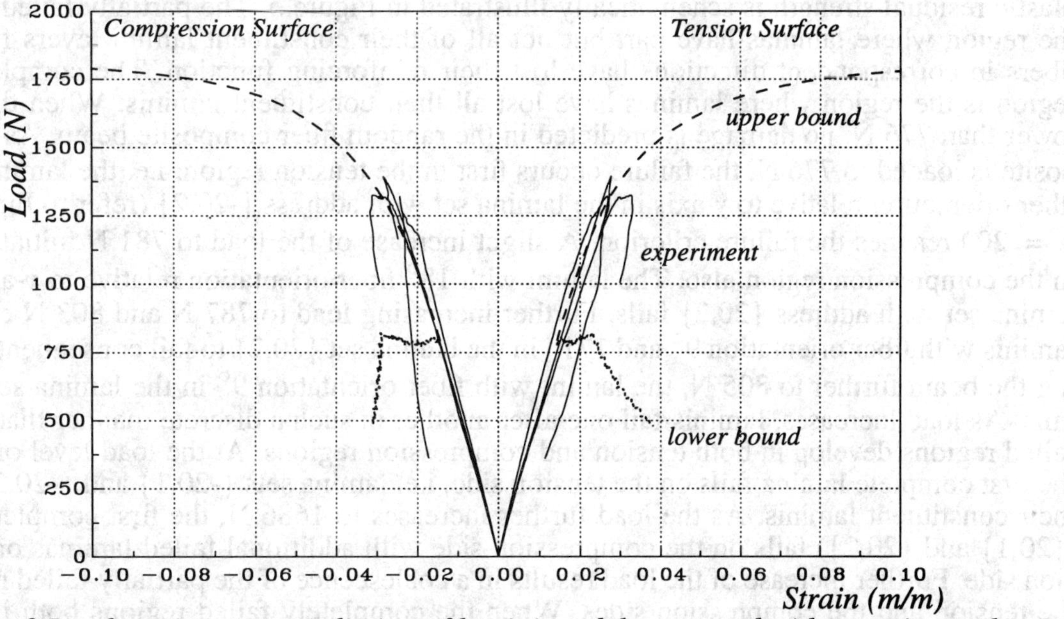

Fig. 3 Load strain curve prediction of lamini model compared with experimental results for four point bending.

experimental results. Random fiber composite beams were tested in an equidistant four point bending test fixture schematically illustrated in Figure 4. The strains were measured on the beam tension and compression surfaces. For detailed test procedures refer to reference [7]. The numerical computations assumed 40 ($n = 20$)layers of laminas and 40 ($m = 20$) layers of laminis per lamina. This subdivision assured convergence of the failure load calcula-

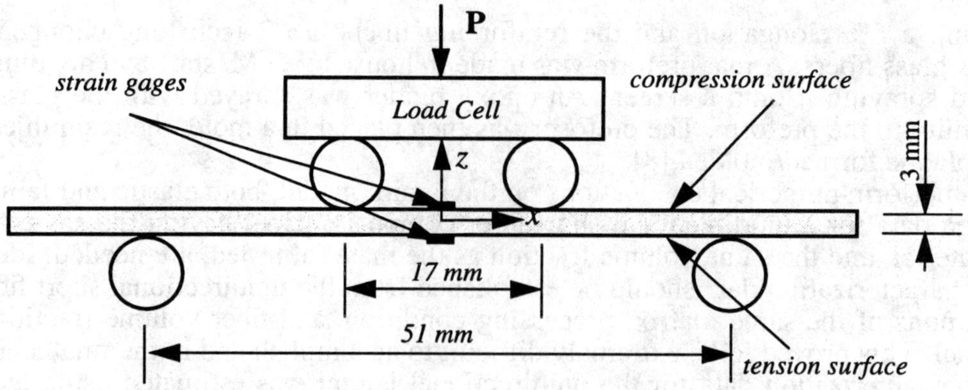

Figure 4. Schematic illustration of four-bend-test fixture.

tion. The lamini model with perfectly plastic residual strength is able to predict the non-linear load strain relations of the random fiber composite beam with great accuracy, for both the tension and compression surfaces. However, the assumption of perfectly plastic residual strength overestimates the beam failure load (1837 N), about 37% higher than the experimental data (1348 N). This assumption also results in an unrealistic strain prediction, i.e. infinite surface strains. This is, of course, due to the assumed perfect plasticity. Obviously, the lamini model with zero residual strength significantly underestimates the load carrying capability of the random fiber composite beam. The model predicts an 817 N load, about 39% lower than the experimental results (1348 N). The predicted load-strain curves also deviate significantly from the experimental data, as soon as the first lamini failure is predicted. The model predicted an ultimate failure load almost the same as the first lamini failure load.

The failure initiation and growth process predicted by the lamini model with perfectly plastic residual strength is schematically illustrated in Figure 5. The partially failed region is the region where laminas have part but not all of their constituent lamini layers failed, i.e. fibers in correspondent directions have lost their reinforcing function. The complete failed region is the region where laminas have lost all their constituent laminis. When the load is lower than 776 N, no damage is predicted in the random fiber composite beam. As the composite is loaded to 776 N, the failure occurs first in the tension region, i.e. the lamini with 0° fiber orientation relative to *y*-axis in the lamina set with address {-20,2} (refer to Figure 2 for $n = 20$) reaches the failure criterion. A slight increase of the load to 781 N initiates failure in the compression region also. The lamini with 18° fiber orientation relative to *y*-axis in the lamina set with address {20,2} fails. Further increasing load to 787 N and 803 N causes the laminis with fiber orientation 9° and 171° in the lamina set {20,2} to fail consequently. Loading the beam further to 805 N, the lamini with fiber orientation 9° in the lamina set {-20,2} fails. As load increases, laminis fail one after another in such a discrete manner that partially failed regions develop in both tension and compression regions. At the load level of 1284 N, the first complete lamina fails on the tension side, i.e. lamina sets {-20,1} and {-20,2} lost all their constituent laminis. As the load further increases to 1686 N, the first complete lamina ({20,1} and {20,2}) fails on the compression side with additional failed laminas on the tension side. Further increase of the load results in a coalescence of the partially failed regions in the tension and the compression sides. When the completely failed regions both in tension and compression sides coalesce, ultimate failure of the beam occurs. The ultimate failure load is predicted at 1837 N. Qualitatively, the lamini model prediction of failure initiation and growth is fairly consistent with the experimental observation using high sensitivity laser moire interferometry [6].

Note that the perfectly plastic and zero residual strengths in the lamini model serve well as the upper and lower bounds respectively for random composite beam failure load predictions. From the load-strain behavior and damage growth point of view, the model with

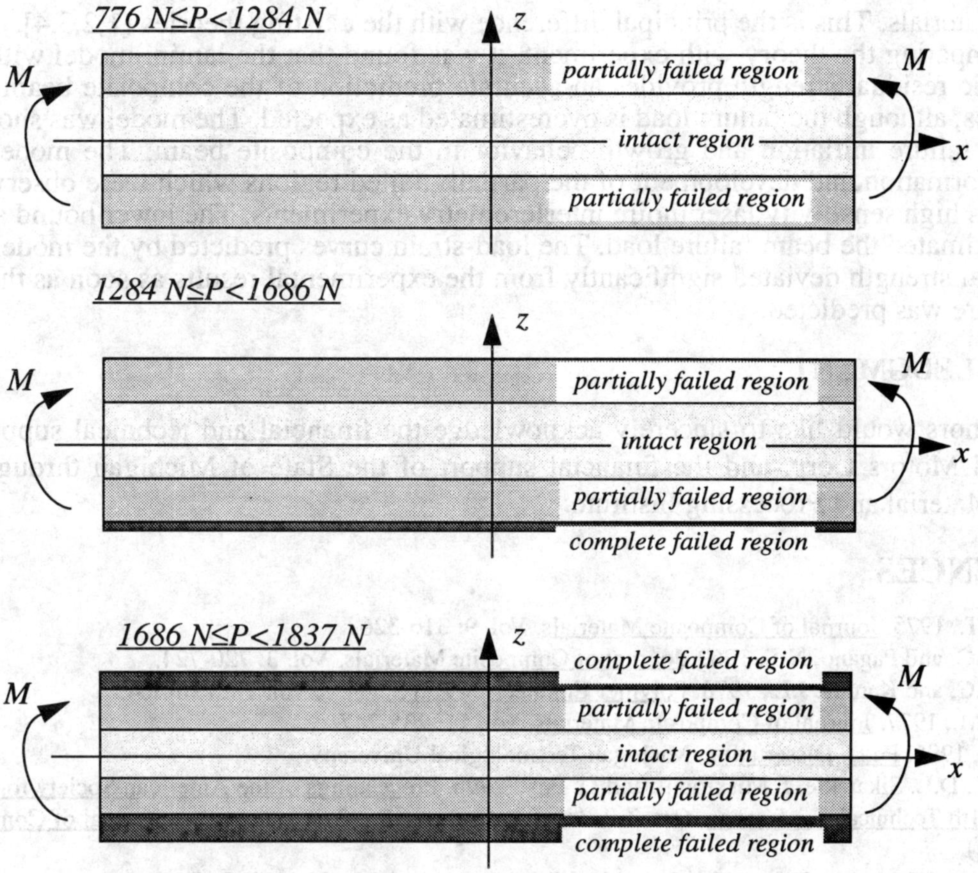

Figure 5. Schematic illustration of predicted damage growth process.

perfectly plastic residual strength gives a more reasonable description of the random fiber composite beam than the model with zero residual strength. The overestimate of the failure load and failure strain may be corrected by defining a more accurate beam ultimate failure condition than the one used in this study, i.e. the beam ultimate failure occurs at the time all the laminis fail. From the moire interferometry observation, the damage regions in the tension and compression regions did not coalesce at the time of beam collapse [6]. Establishing a more realistic lamini post-failure stress-strain behavior in the future will also improve the prediction accuracy.

SUMMARY

A phenomenological-mechanistic model, referred to as the lamini model, has been developed and used to predict the random fiber composite beam strength in this paper. To ensure that the lamini model represents a transversely isotropic material, it was mathematically proved that the extensional stiffness matrix $[A] = \frac{h}{\pi}\int_0^{\pi}[\bar{Q}(\theta)]d\theta$ and the bending stiffness matrix $[D]$ converges to $[D] = \frac{h^3}{12\pi}\int_0^{\pi}[\bar{Q}(\theta)]d\theta$ as the number of laminas approach infinity. The two stiffness matrices also satisfy the relationship $[D] = \frac{h^2}{12}[A]$ required for isotropic materials. This is the principal difference with the existing theories [1,2,3,4].

Comparing the theory with experiment, it was found that the lamini model with perfectly plastic residual strength provides an accurate prediction of the composite beam load-strain curves, although the failure load is overestimated as expected. The model was shown to predict the failure initiation and growth behavior in the composite beam. The model predicted the formation and development of the partially failed regions which were observed in the previous high sensitivity laser moire interferometry experiments. The lower bound analysis underestimated the beam failure load. The load-strain curves predicted by the model with zero residual strength deviated significantly from the experimental results as soon as the first lamini failure was predicted.

ACKNOWLEDGMENT

Authors would like to sincerely acknowledge the financial and technical support of the General Motors Corp. and the financial support of the State of Michigan through the Michigan Material and Processing Institute.

REFERENCES

1. Hahn, H.T., 1975. Journal of Composite Materials, Vol. 9: 316-326.
2. Halpin, J.C. and Pagano, N.J., 1969. Journal of Composite Materials, Vol. 3: 720-724.
3. Halpin, J.C. and Kardos, J.L., 1978. Polymer Engineering and Science, Vol. 18: 496-504.
4. Manera, M., 1977. Journal of Composite Materials, Vol. 11: 235-247.
5. Zhang Y., 1996. Ph.D. Dissertation, Michigan Technological University.
6. Zhang, Y., D.L. Sikarskie, I. Miskioglu, and D. Post. 1996. Proceedings of the American Society for Composites 11th Technical Conference, *Oct. 7-9, Atlanta, Georgia, P697-706.(in review,* Journal of Composite Materials*)*.
7. Zhang, Y. and Sikarskie, D.L., 1996. Journal of Composite Materials, Vol. 30: 966-983.
8. Zhang, Y. and Sikarskie, D.L., 1997. *Accepted for publication on* 11th International Conference on Composite Materials, *July 14-18, Gold Coast, Australia.*
9. Jones, R.M., 1975. Mechanics of Composite Materials, Hemisphere Publishing Company.

A Numerical Procedure for Elasticity and Failure Behaviour Prediction of Textile Reinforced Composite Materials

P. HAMELIN AND D. BIGAUD

ABSTRACT

A computer aided design tool and a numerical procedure allowing the prediction of textile reinforced composites behaviour are presented in this paper. Specifying a geometrical description, the software used allows to rebuild both simple or sophisticated textiles structures. It also aims to a pre-processing work for the subsequent mechanical analysis. Fabrics unit cells are regarded as a three-dimensional aggregate of subcells on which we apply a multi-scale energetical approach. First elements to validate the elastic prediction are provided by results obtained on a woven fabric material. The progressive failure procedure is validated by comparisons between simulation and experiments on 3D braided composites.

INTRODUCTION

The potentialities of woven, non-woven, knitted or braided fabrics used as reinforcements for polymer matrix composite materials increase. However the multi-functional character of those materials perfectly adapted to the structural element design is not thoroughly used. This is essentially the result of the difficulties of predicting their performances. The main problem of modelling the textile composites mechanical properties consists in dealing in an accurate way the large stress and strain variations within a textile. A textile composite is a very heterogenous structure. As a consequence, the stresses and strains fields differents elements greatly depend on the constituents repartition, on their size and on their spatial arangement. In other words, it depends on the textile architecture.

Many approaches exist to describe the global geometry of a composite. A simplifying description - we can speak about geometrical homogeneization - is necessary in order to develop a mechanical model. Many of them only provide good results for specific textile geometry, according to a certain kind of loading. Most of the existing models seem to be

---

Patrice Hamelin, Mechanical engineering and Material Laboratory, I.U.T. A Génie Civil, Université Claude-Bernard, 43 Blvd du 11 Novembre 1918, 69622 Villeurbanne Cx, France.
David Bigaud, Mechanical engineering and Material Laboratory, I.U.T. A Génie Civil, Université Claude-Bernard, 43 Blvd du 11 Novembre 1918, 69622 Villeurbanne Cx, France.

inaccurate because of the introduction of new reinforcements such as multi-directional or three-dimensional fabrics. This paper has two objectives. The first one is to describe a computer program developped in order to model the textile architecture geometry. The second objective is to present a numerical procedure allowing to estimate the elastic and failure behaviour of composites with textile reinforced polymer matrix. Finally, the model is compared with results provided from experimentations carried out on woven and braided fabric composites.

PRESENTATION OF TIS3D

The graphical technics implicated in computer aided design tools find an appropriate field of applications in the description of textile structures. The computer aided design software developed by the Mechanical engineering and Material Laboratory is called TIS3D. One of its main feature is to reconstitute basic volumes that represent usual textile structures. The geometrical description according to a reduced number of parameters allows to study woven, knitted, braided, 2D or 3D reinforcements of homogeneized geometries. Once the unit cell is defined, a regular volume elements meshing is superposed to the ''real'' geometry (Figure 1). Each element is specified as an intermediate element and is characterized by an individual material orientation.

Finally, the analysis of the unit cell mechanical behaviour consists in studying that of an aggregate of multi-scale elements with individual properties. For the mechanical prediction we must then consider three analyses scales:

- a microscopic scale at the level of which we have to determine the tow properties,
- an upper scale at the level of which we define properties of a intermediate element (8 meso-elements),
- finally the macroscopic scale at the level of which we consider the mechanical properties of the unit cell (aggregate of intermediate-elements).

THEORETICAL ASPECTS

GENERAL ASPECT

An energetical approach is used for each analysis level. At the microscopic scale, we idealize the fibre+matrix mixture by a bi-dimensional network of cells. This modelling has already been proposed by (Aboudi,1989). The main idea of the method consists in minimizing the complementary energy U_{cj} from the *j*-assembly defined as the average of energies in each *i*-cell (i=fibre or matrix).

$$U_{cj} = \frac{1}{2}\sum_{i=1}^{i=4} v_{i-micro}\{\sigma\}^T_{i-micro}[s]_{i-micro}\{\sigma\}_{i-micro}$$
$$= \frac{1}{2}\{\sigma\}^T_{j-meso}[s]_{j-meso}\{\sigma\}_{j-meso} \qquad (1)$$

v_i is the volume ratio of the i-cell. $[s]_\alpha$ represents the compliance matrix of the α entity.

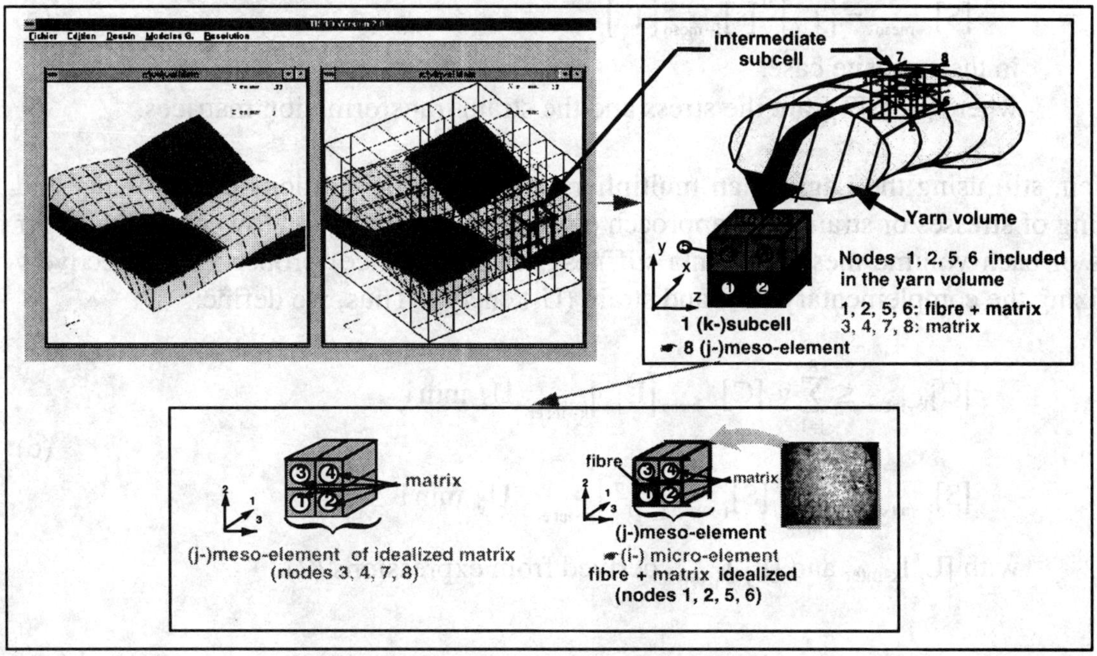

Figure 1. Unit cell multi-scale meshing.

Through Lagrangian multipliers (Chen and Cheng,1993) and imposing the conditions of continuity and averaging of stresses, it is possible to determine linear relations between the local tensors $\{\sigma\}_{i\text{-micro}}$ and the tensor of meso-elements stresses $\{\sigma\}_{j\text{-meso}}$ as matrices of stresses localization $[L_i^{\sigma}]_{j\text{-meso}}$ (eq.(2)).

$$\{\sigma\}_{i-\text{micro}} = \left[L_i{}^{\sigma}\right]_{j-\text{meso}} \{\sigma\}_{j-\text{meso}} \tag{2}$$

The compliance matrix $[s]_{j\text{-meso}}$ of the *j*-assembly is then easily explained:

$$\begin{aligned} [s]_{j-\text{meso}} &= \sum_{i=1}^{i=4} v_{i-\text{micro}} [s]_{i-\text{micro}} \left[L_i{}^{\sigma}\right]_{j-\text{meso}} \\ &= V_{fl} [s]_f \left[L_1{}^{\sigma}\right]_{j-\text{meso}} + \sum_{i=2}^{i=4} v_{i-\text{micro}} [s]_m \left[L_i{}^{\sigma}\right]_{j-\text{meso}} \end{aligned} \tag{3}$$

with

m: matrix, f: fibre.

$v_2 = v_3 = \sqrt{V_{fl}} - V_{fl} \quad v_4 = 1 - 2\sqrt{V_{fl}} + V_{fl}$

V_{fl} is the fibre packing ratio.

At the mesoscopic scale, we consider a intermediate volume element composed of eight subdivisions of identical volumes. The subdivisions properties are then defined as:

$$[S]_{j-\text{meso}} = [s]_{\text{matrix}} \tag{4}$$

if the subdivision does not belong to the yarn volume,

$$[S]_{j-meso} = [T_\sigma]^{-1}[s]_{j-meso}[T_\varepsilon] \tag{5}$$

in the opposite case.
where $[T_\sigma]$, $[T_\varepsilon]$ are the stress and the strain transformation matrices.

Then, still using the Lagrangian multipliers method and conditions of continuity and averaging of stresses or strains, the approach consists in supplying the lower and the upper bounds of each volume meso-elements stiffness (or compliance) properties, respectively minimizing the complementary (U_c) and strain (U_s) energy. Thus, we define:

$$\begin{aligned} [C]_{k-inter} &\leq \sum_{j=1}^{8} v_j [C]_{j-meso} [L_j^{\varepsilon}]_{k-inter} \quad U_s \text{ mini} \\ [S]_{k-inter} &\leq \sum_{j=1}^{8} v_j [S]_{j-meso} [L_j^{\sigma}]_{k-inter} \quad U_c \text{ mini} \end{aligned} \tag{6}$$

with $[L_j^{\varepsilon}]_{k\text{-inter}}$ and $[L_j^{\sigma}]_{k\text{-inter}}$ derived from expressions (7).

$$\begin{aligned} \{\varepsilon\}_{j-meso} &= [L_j^{\varepsilon}]_{k-inter} \{E\}_{k-inter} \quad U_s \text{ mini} \\ \{\sigma\}_{j-meso} &= [L_j^{\sigma}]_{k-inter} \{\Sigma\}_{k-inter} \quad U_c \text{ mini} \end{aligned} \tag{7}$$

where $\{\varepsilon\}_{j\text{-meso}}$, $\{\sigma\}_{j\text{-meso}}$, $\{E\}_{k\text{-inter}}$, $\{\Sigma\}_{k\text{-inter}}$ are the meso-element and intermediate element strains and stresses.

At the macroscopic level, the latter approach is extended to a three-dimensional assembly of regular volume intermediate-elements to eventually estimate the bounds of the unit cell elastic properties.

$$\begin{aligned} [C]_{macro} &\leq \sum_{k=1}^{N_x N_y N_z} v_k [C]_{k-inter} [L_k^{\varepsilon}]_{macro} \quad U_s \text{ mini} \\ [S]_{macro} &\leq \sum_{k=1}^{N_x N_y N_z} v_k [S]_{k-inter} [L_k^{\sigma}]_{macro} \quad U_c \text{ mini} \end{aligned} \tag{8}$$

where $v_k = \dfrac{1}{N_x N_y N_z}$; $N_xN_yN_z$ represents the number of intermediate elements.

with

$$\begin{aligned} \{E\}_{k-inter} &= [L_k^{\varepsilon}]_{macro} \{E\}_{macro} \quad U_s \text{ mini} \\ \{\Sigma\}_{k-inter} &= [L_k^{\sigma}]_{macro} \{\Sigma\}_{macro} \quad U_c \text{ mini} \end{aligned} \tag{9}$$

where $\{E\}_{macro}$ and $\{\Sigma\}_{macro}$ are the unit-cell strains and stresses.

RESULT OF THE METHOD

The key point of the method which consists in explaining the relations between internal and external stresses (or strains) through localization matrices, allows us, first, to determine the bounds for the elastic characteristics of a textile reinforcement unit cell. Then, it is possible to estimate the internal stress and strain states according to an imposed external stress $\{\Sigma\}_{macro}$ or an imposed external strain $\{E\}_{macro}$ up to the microscopic scale.

We propose in equations (10) the bounding of stress and strain tensors at the mesoscopic scale (yarn scale) for an imposed external stress. The knowledge of meso-element stress and strain states is directly exploitable to anticipate the composite failure behaviour. At this scale, we can include criteria taking into account the matrix and the fibre failure. As each of the criteria are applied to the assembly of subcells which compose the unit cell, progressive damage functions can be determined.

DETAILS ON NUMERICAL PROCEDURE FOR PROGRESSIVE FAILURE SIMULATION

The method used to simulate progressive failure of textile composites is based on the relations (10). In the present paper, the maximum stress criterion will be used to study the failure behaviour of 3D braided composites. However, it is possible to use a maximum strain or a quadratic criterion. The procedure for progressive failure simulation according to an imposed uniaxial stress and using a maximum stress criterion is detailed below.

First of all, each stiffness $[C]_{j\text{-micro}}$ or compliance $[S]_{j\text{-micro}}$ matrix is calculated by a first elastic analysis on the undamaged composite. Once this calculation achieved, we consider a macroscopic stress in the studied direction. The meso-elements internal stresses are then deduced in the local axis associated with the fibre direction. The minimal tension, compression or shear normalized stress (strength on real stress) is investigated in order to give the macroscopic stress value that would yield to the first meso-element failure. The constitutive matrix of the meso-element is then reduced by the selective RC method (rows and columns of the matrix $[C]_{macro}$ set to zero according to the stress allowable which was exceeded). The elastic analysis is carried on taking into account of the change of $[C]_{j\text{-micro}}$ (or $[S]_{j\text{-micro}}$).

According to an imposed macroscopic stress $\{\Sigma\}_{macro}$

$$[C]_{j-meso}\left[L_j^{\varepsilon}\right]_{k-inter}\left[L_k^{\varepsilon}\right]_{macro}\left[\sum_{k=1}^{N_xN_yN_z} v_k\left\{\sum_{j=1}^{j=8} v_j[C]_{j-meso}\left[L_j^{\varepsilon}\right]_{k-inter}\right\}\left[L_k^{\varepsilon}\right]_{macro}\right]^{-1}\{\Sigma\}_{macro} \le \{\sigma\}_{j-meso}$$

$$\{\sigma\}_{j-meso} \le \left[L_j^{\sigma}\right]_{k-inter}\left[L_k^{\sigma}\right]_{macro}\{\Sigma\}_{macro} \qquad (10a)$$

$$\left[L_j^{\varepsilon}\right]_{k-inter}\left[L_k^{\varepsilon}\right]_{macro}\left[\sum_{k=1}^{N_xN_yN_z} v_k\left\{\sum_{j=1}^{j=8} v_j[C]_{j-meso}\left[L_j^{\varepsilon}\right]_{k-inter}\right\}\left[L_k^{\varepsilon}\right]_{macro}\right]^{-1}\{\Sigma\}_{macro} \le \{\varepsilon\}_{j-meso}$$

$$\{\varepsilon\}_{j-meso} \le [S]_{j-meso}\left[L_j^{\sigma}\right]_{k-inter}\left[L_k^{\sigma}\right]_{macro}\{\Sigma\}_{macro} \qquad (10b)$$

PREDICTION OF ELASTIC BEHAVIOUR OF A WOVEN FABRIC

The validation of the numerical procedure in case of elastic prediction is provided through experimentations carried out on woven fabric composites. We study the cases of a plain-woven and five-harness-satin-woven fabrics. Both textile composites are glass/epoxy materials. The constituents properties and the main geometrical characteristics of the unit cells are given in tables I and II. The table III gives comparisons between the results obtained by the experimentations (only for the plain woven fabric), our numerical procedure (average of results provided by both the U_s and U_c approaches) and an exact method (F.E.M.) (Lene, Hassim and Paumelle,1991; Chouchaoui,1995). We can see that the results obtained by the numerical procedure can be accepted. Indeed, the difference from the exact solution does not exceed 1% as for the E_x and E_z moduli and 11% as for the E_y modulus. We have to be cautious as far as the Poisson ratios are concerned (24% for ν_{xz}, 11% for ν_{yx}). The satin-woven fabric can be considered in the same way. The results confirm that the moduli predictions can be accepted (1% as for the E_x and E_z moduli and 6% as for the E_y modulus). However, the discrepancies observed for the Poisson ratios are lower (about 18% for ν_{xz} and 9% for ν_{yx}).

TABLE I - GLASS AND EPOXY PROPERTIES [MPa]

	Glass	Epoxy
$E_{11}=E_{22}=E_{33}$	*73000*	3130
$G_{12}=G_{13}=G_{23}$	30000	1170
$\nu_{12}=\nu_{13}=\nu_{23}$	0.20	0.34

TABLE II - GEOMETRICAL DATAS [mm]

	Unit cell						Tow	
	t_m^*	H*	D_{weft}^*	D_{warp}^*	V_{fl}	V_{fg}	a*	b*
Plain	0.005	0.205	0.65	0.65	0.81	0.514	0.3	0.05
Satin	0.005	0.205	3.25	3.25	0.81	0.538	0.3	0.05

* t_m: matrix thickness, H: height of the unit cell, D_{weft} and D_{warp}: Distance along weft and warp directions, V_{fl}: tow fiber packing ratio, V_{fg}: composite fiber volume ratio,a and b: yarn geometric parameters (Figure 2).

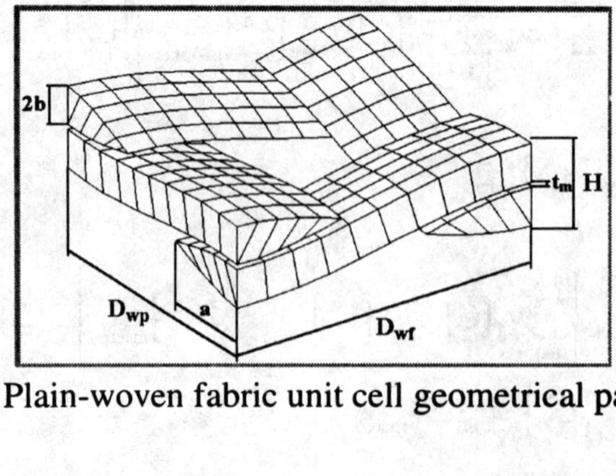

Figure 2. Plain-woven fabric unit cell geometrical parameters.

PREDICTION OF PROGRESSIVE FAILURE OF A 3D COMPOSITES

We propose to test the validity of our numerical procedure for the failure behaviour from experimentations carried out on Carbon/PA12 3D rectangular cartesian braided fabrics. The carbon yarns are spun and coated with the matrix.

MATERIAL PROPERTIES

The mechanical properties of the constituents are summed up in table IV. The fibre packing ratio is equal to 0.7. Datas used to study the braided composite failure behaviour are voluntarily restricted (in a first approach and considering the tensile loading) to axial tension and compression and braiding yarns shear strengths at the micro-elements scale. These datas are presented in table V. This restrictive choice is deduced from experimental observations usually done concerning the 3D composites failure. Referring Cox (Cox,1996), their failure results from phenomena combining axial tension and compression, and, yarn shear (kink band formation) failures, which the resin resistance does not affect.

The tensile strength is described through a three-parameter Weibull. The determination of this three parameters is carried out by testing four yarn groups of different lengths consolidated at 220°C and submitted to a 7-bar pression during one hour. This is a simulation of the consolidation cycle of the 3D braided composites. Each group are composed of 15 to 23 specimens with a gauge length of 35 to 140 mm. In figure 3a, we present the curves c.d.f. F versus σ/σ_0 allowing to determine two parameters of the Weibull law. Figure 3b shows the evolution of $(\sigma_0(L)/\sigma_0(L_o))$ ratio (normalized scale parameter) according to the gauge length ratio L/L_o. It is then possible to determine the third c.d.f. parameter, giving the tensile strength distribution law whatever long the specimen is. As for shear strength, we observe few variations since it is impossible to include scale phenomena (tests are carried out on tows with an identical section).

GEOMETRICAL PROPERTIES

In order to determine the unit cell geometry of the 3D rectangular cartesian braided composites, we make a sectioning of specimens (Figure 4). It allows us to rebuild the path of each braiding yarn along a repetition unit and thus to identify ten constitution subcells from the braided composite unit cell. The orientations of each of yarn elements within the subcells as well as the composite fibre volume ratio can be determined from the process parameters.

PROGRESSIVE FAILURE ACCORDING TO A TENSILE LOADING

43-yarn-braided-fabrics are loaded to failure (along the z-axis direction). The dimensions of the specimens are 100 mm x 10 mm x 2 mm (L x l x e). The average braiding angle is equal to 27°. The average fibre volume ratio , determined according to the ASTM D 3171 norm, is 0.517. The figure 5 shows a comparison between five experimental and two simulated progressive damage functions.

TABLE III - COMPARISONS BETWEEN EXPERIMENTAL AND NUMERICAL RESULTS

	Plain			Satin	
	Exact	Tis3D	Exp.	Exact	Tis3D
$E_x = E_z$	25200	24700	24860	25800	25540
E_y	11000	12380	14000	11100	11800
G_{xz}	5230	5140	6140	5700	5560
$G_{yx}=G_{yz}$	3260	3640	4000	3400	3740
ν_{xz}	0.140	0.178	0.194	0.150	0.181
$\nu_{yx}=\nu_{yz}$	0.310	0.348	0.360	0.320	0.352

TABLE IV - CARBON AND PA12 PROPERTIES

	Carbon	PA-12
E_{11}	230000	1300
E_{22}	15000	1300
E_{33}	15000	1300
G_{12}	50000	480
G_{13}	50000	480
G_{23}	12000	480
ν_{12}	0.33	0.36
ν_{13}	0.33	0.36
ν_{23}	0.33	0.36

TABLE V - TOW MAXIMUM STRENGTHS

Axial tensile strength	Shear strength
σ_o=912 MPa	90 MPa
s=6.612	
L_o=35 mm	
α=0.610	

s: shape parameter, σ_o: scale parameter, L_o: gauge length reference, α describes the effect of the gauge length on the c.d.f.

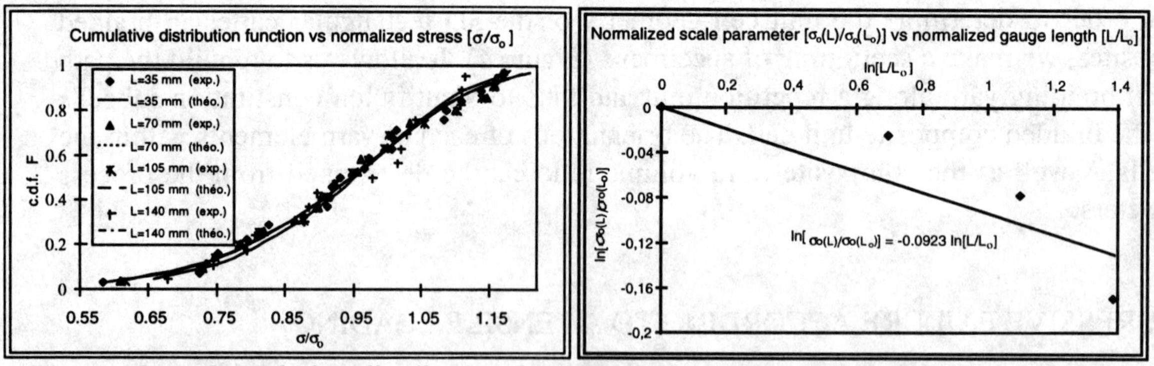

Figures 3. a- Cumulative distribution functions of single Carbon/PA12 tows made for various gauge length. b- Normalized scale parameter $[\sigma_o(L)/\sigma_o(L_o)]$ versus normalized gauge length $[L/L_o]$ - Determination of the third Weibull distribution parameter α.

The simulations are derived from the equation (10a) (maximum stress criterion) and considering that the macroscopic strain $\{E\}_{macro}$ according to the imposed macroscopic stress $\{\Sigma\}_{macro}$ is directly given by

$$\{E\}_{macro} = [S]_{macro}.\{\Sigma\}_{macro} \tag{11}$$

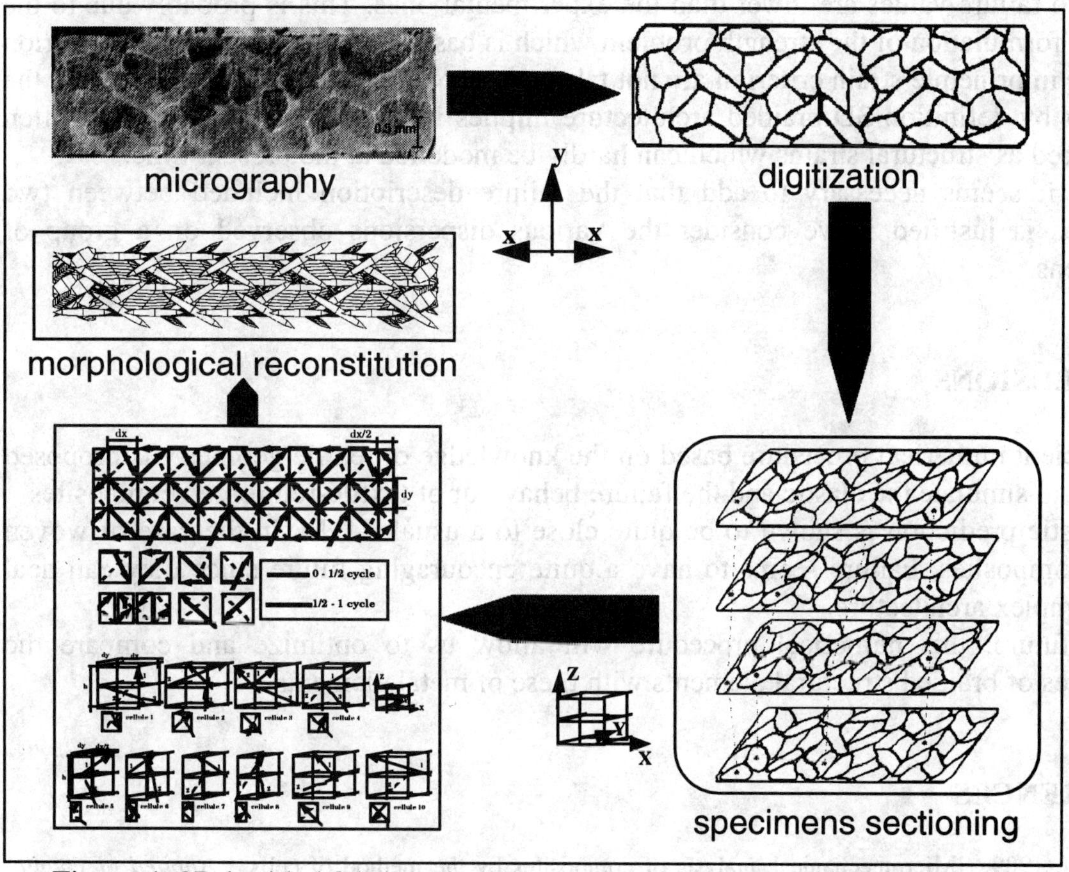

Figure 4. 3D braided fabric unit cell construction by specimens sectioning.

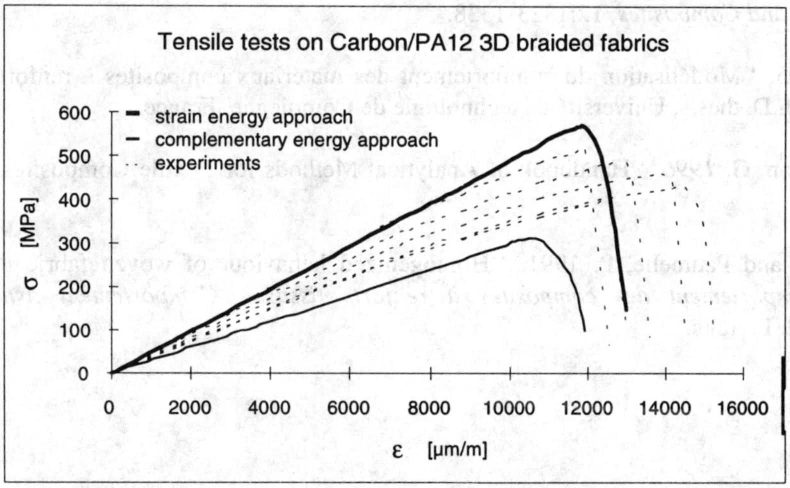

Figure 5. Comparisons between experimental and simulated damage functions.

We note that in the case of complementary energy approach the first simulated micro-element failure occurs more rapidly in terms of stress (310 MPa according to U_c and 560 MPa according to U_s). The failure of the whole braided composite is then predicted before that determined from the strain energy approach which seems rather logical according to expressions (10). We also can notice that strains to failure are almost similar (10%).
As far as the experimental curves are concerned we remark that strength values are included between the bounds predicted by our both approaches. However the simulated strains to failure values are lower than the experimental ones. This is probably due to the original formulation of the strength problem which is based on a maximum stress criterion and that in principle strain criterion are not taken in account. We can also explain it by the fact that by definition 3D braided architecture implies important strains to failure, often considered as structural strains which can hardly be modelled at the present time.
Finally, it seems necessary to add that the failure description included between two functions is justified if we consider the various dispersions observed on a group of specimens.

CONCLUSIONS

An efficient numerical procedure based on the knowledge of textile geometry is proposed in order to simulate the elastic and the failure behaviour of textile-reinforced composites. The elastic prediction is shown to be quite close to a usual F.E.M. in the case of woven fabric composites. Failure seems to have a quite encouraging future since they can deal with complex architecture.
In the future, this numerical procedure will allow us to optimize and compare the properties of braided structural elements with these of metal elements.

REFERENCES

Aboudi, J. 1989. ''Micromechanical analysis of composites by the method of cells,'' *Applied Mechanics Science*, 47(7).

Chen, D. and Cheng, S. 1993. ''Analysis of Composite Materials: A Micromechanical Approach,'', *Journal of Reinforced Plastics and Composites*, 12:1323-1338.

Chouchaoui, C.S. 1995. ''Modélisation du comportement des matériaux composites à renforts tissés et à matrice organique,'' Ph.D. thesis, Université de technologie de Compiègne, France.

Cox, B.N. and Flanagan, G. *1996*. ''Handbook of Analytical Methods for Textile Composites,'' Rockwell Science Center.

Lene, F., Hassim, A. and Paumelle, P. 1991. ''Homogenized behaviour of woven fabric composites,'' *Proceedings of «Comportement des composites à renforts tissus - Comportement dynamique des composites»*:69-82, Ed. Pluralis.

New Results in Development of Mathematical Methods for Prediction of Composite Effective Properties

N. S. BAKHVALOV, M. E. EGLIT AND A. E. YAKUBENKO

ABSTRACT. [1] Mathematical homogenization theory is used to calculate the effective moduli of composites. The following materials are considered: 1) composites reinforced by periodic system of plates and bars; 2) composites reinforced by an arbitrary system of high moduli fibres; 3) media with long thin pores; 4) elastic composites with small ratio of typical values of shear and bulk moduli; 5) viscous mixtures with small ratio of typical properly scaled viscosity and compressibility coefficients; 6) thermoconductive materials with small ratio of typical thermoconductivity coefficient and specific heat capacity.

INTRODUCTION

Composites are strongly inhomogeneous multiphases materials. A direct numerical calculation of strains, stresses, temperature, etc. in composites is practically impossible due to the rapid variations on the micro-scale level.

Fortunately composites behave as homogeneous media when the ratio ε of the typical size d of the inclusions to the size L of the whole sample and/or the characteristic wavelength of external fields is small. Then their properties can be described by the effective equations and the effective moduli (effective heat transfer, electric conductivity, magnetic permeability, elasticity, etc.) that are obtained by some type of averaging if the physical properties of each phase are known.

A way to obtain the effective equations and the effective moduli is to use the mathematical homogenization theory, developed in the studies of partial differential equations with rapidly varying coefficients.

There is a group of researchers in Moscow State University working actively in this field. The paper gives a brief review of some recent results of this group.

Let us describe briefly the homogenization algorithm. Let we have the exact equations for an inhomogeneous material. The aim of mathematical homogenization theory is to obtain the averaged equations that describe a homogeneous material and whose solutions are close in some sense to solutions of original exact equations.

---

[1]N.S. Bakhvalov, M.E. Eglit and A.E. Yakubenko, Math. and Mech. Department, Moscow State University, Moscow 119899, Russia. E-mail: bakhvalo@mech.math.msu.su, eglit@mech.math.msu.su, yakub@inmech.msu.su

The homogenization algorithm for periodic media includes the following steps (Bensoussan, Lions, Papanicolaou, 1978; Sanchez-Palencia, 1980; Bakhvalov, Panasenko, 1984).

1) Dimensionless slow variables $x_i = \bar{x}_1/L$ and fast variables $y_i = \bar{x}_i/d = x_i/\varepsilon$ are introduced. Here $\bar{x}_i$ are spatial coordinates, d is the size of a periodicity cell.

2) Solution u of the original problem is tried in the form of asymptotic series

$$u = u(t, x_i, y_i, \varepsilon) = \sum \varepsilon^n u_n, \qquad u_n \in M,$$

where M is the space of 1-periodic functions of y_i.

3) the series are substituted into original equations, boundary and initial conditions; coefficients at all powers of ε in the resulting equations are equated to 0. Thus the so-called cell-problems appear that should be solved to calculate the coefficients of the series. Averaging over the periodicity cell is applied to obtain the averaged (effective) equations.

4) The closeness of the averaged and original equations solutions is proved.

Cell-problems are usually complicated. But they often depend only on the structure of the material so after they are once solved they can be used to describe different processes in the material under consideration.

Many approximate formulae for effective moduli of composites are proposed basing on intuitive mechanical assumptions concerning the form of stress-strain state in the material. The advantage of the described method is that the closeness of the averaged and original equations solutions is strictly proved and the error is estimated.

Still the explicit approximate formulae for effective moduli are very useful, especially in optimal design of materials and constructions. Sometimes they can be derived mathematically; in particular it is possible if there are some additional small parameters besides the dimensionless size of the cell ε.

SMALL PARAMETERS CONNECTED WITH THE MATERIAL STRUCTURE

Small parameters may be connected with special geometrical properties of the structure or with very large range of moduli of different phases forming the composite. The following materials are considered in this paper: i) composites reinforced by periodic system of plates and bars; ii) composites reinforced by an arbitrary system of high moduli fibres; iii) media with long thin pores.

Let us consider constructions and composites reinforced by a periodic system of mutually orthogonal plates and bars. One can imagine them as large buildings consisting of identical rectangular rooms filled by soft and weakly conducting material (matrix).

Plates (walls) and bars (intersections of walls) are assumed to be relatively thin. Materials can be anisotropic. The properties of matrix, plates, bars and nodes (i.e., intersections of the bars) can be different. Moduli of plates, bars and nodes are assumed to be larger then the matrix moduli.

Denote the period along x_j by ε_j, the length of the sample by L, the thicknesses of plates by d_j; $\nu_j = d_j/\varepsilon_j$; $\bar{\nu} = \max(d_1, d_2, d_3)/\min(\varepsilon_1, \varepsilon_2, \varepsilon_3)$. It is assumed that $d_j << \varepsilon_j << L$.

Let us consider the equations

$$\frac{\partial}{\partial x_i} a_{kilj} \frac{\partial u_l}{\partial x_j} = f_k, \quad k = 1, \ldots, n; \tag{1}$$

They describe, for example, displacements in an elastic composite or stationary heat conduction obeying the Fourier law; a_{kilj} are supposed to be 1-periodic functions of $y_j = x_j/\varepsilon_j$; n is the dimension of the vector $\mathbf{u}$ with components u_l ($l, k \equiv 1$ if u is scalar, e.g., temperature).

At the condition $\varepsilon_j << L$ the equations with constant coefficients $\bar{a}_{kilj}$ can be constructed so that their solutions are close to the solutions of original equations (1). Coefficients $\bar{a}_{kilj}$ are called the effective moduli of periodic medium.

To calculate the effective moduli one should find vectors $\mathbf{m}^{qp}$ with components m_l^{qp} ($q = 1, 2, 3; p, l = 1, ..., n$) satisfying the equations

$$\frac{\partial}{\partial x_i} a_{kilj} \frac{\partial m_l^{qp}}{\partial x_j} = 0, \quad k = 1, \ldots, n \tag{2}$$

and conditions

$$\mathbf{m}^{qp} - x_q \mathbf{e}_p \in \mathbf{M}. \tag{3}$$

Here $\mathbf{e}_p$ is a unit vector of coordinate axis x_p ($\mathbf{e}_p$ should be replaced by 1 if $n = 1$), $\mathbf{M}$ is a subspace of functions from $H^{1,loc}(R_s)$ that are periodic over x_j with a period ε_j. The subspace of functions satisfying the conditions (3) is denoted further by $\mathbf{M}^{qp}$.

Let

$$U(\mathbf{u}, \mathbf{v}) = a_{kilj} \frac{\partial u_l}{\partial x_j} \frac{\partial v_k}{\partial x_i}, \qquad \Lambda(\mathbf{u}, \mathbf{v}) = \langle U(\mathbf{u}, \mathbf{v}) \rangle .$$

Here $< \cdot >$ means averaging over a periodicity cell

$$< f >= \frac{1}{\varepsilon_1 \varepsilon_2 \varepsilon_3} \int_0^{\varepsilon_1} (\int_0^{\varepsilon_2} (\int_0^{\varepsilon_3} f(x_1, x_2, x_3) dx_3) dx_2) dx_1.$$

Coefficients $\bar{a}_{kilj}$ are calculated by the formulae

$$\bar{a}_{kilj} = \Lambda(\mathbf{m}^{jl}, \mathbf{m}^{ik}) \tag{4}$$

The equations (2) are the Euler equations for a functional $\Lambda(\mathbf{m}, \mathbf{m})$;

$$\bar{a}_{qpqp} = \inf_{\mathbf{m} \in \mathbf{M}^{qp}} \Lambda(\mathbf{m}, \mathbf{m}), \tag{5}$$

infimum is reached by functions $\mathbf{m}^{qp}$ that are solutions of problems (2), (3).

From (5) it follow the estimates

$$\bar{}_{qpqp} \leq \inf_{\mathbf{m}(x_q) \in \mathbf{M}^{qp}} \Lambda(\mathbf{m}, \mathbf{m}). \tag{6}$$

Right parts of these inequalities can be found in quadratures for composites under consideration because corresponding Euler equations are ordinary equations of special form.

Lower bounds for $\bar{}_{ijkl}$ are possible to be obtained by different ways.

Consider for example heat conduction in a composite made of isotropic components. Let k_0, k_1, k_2 and k_3 be the thermoconductivity coefficients for matrix, plates, bars and nodes, respectively. The effective material is orthotropic. In a coordinate system with axes directed along the bars the coefficients $\bar{k}_{ii} \equiv \bar{a}_{i1i1}$ only can differ from zero. The following inequalities are valid

$$\inf_{\mathbf{m}\in\mathbf{M}^{q1}} \left\langle k \left(\frac{\partial m}{\partial x_q} \right)^2 \right\rangle \leq \bar{a}_{q1q1} \equiv \bar{k}_{qq}. \tag{7}$$

Solving the Euler equations one obtains the estimates (6) and (7) for the coefficient $\bar{k}_{11}$ in the form (Bakhvalov, Panasenko, 1984):

$$k_{11}^- \leq \bar{k}_{11} \leq k_{11}^+,$$

where

$$k_{11}^- = \left\langle \left\langle \left\langle (k_{11})^{-1} \right\rangle_1^{-1} \right\rangle_2 \right\rangle_3, \qquad k_{11}^+ = \left\langle \langle\langle k_{11}\rangle_2\rangle_3^{-1} \right\rangle_1^{-1}.$$

Here $\langle . \rangle_j$ means integration over y_j in the limits $[0, 1]$. Analogous estimates can be written for $\bar{k}_{22}$ and $\bar{k}_{33}$.

Fortunately for composites under consideration the upper and lower estimates are close to each other. Calculating k_{11}^- k_{11}^+, we obtain the formulae for $\bar{k}_{ii}$. We write them here for the simplest case. Let all plates have the same relative thickness ν. Then (Bakhvalov, Panasenko, Eglit, 1996)

$$\bar{k}_{11} = \bar{k}_{22} = \bar{k}_{33} = k_1 \frac{k_0 + 2k_1\nu + k_2\nu^2}{k_1 + k_0\nu}(1 + O(\Delta)),$$

where $\Delta = (k_1/k_2 + k_2/k_1 + k_2/k_3 + k_3/k_2)\nu$.

To obtain formulae for effective elastic moduli and for effective thermoconductivity coefficients for locally anisotropic materials we constructed approximate solutions $\tilde{\mathbf{m}}^{qp} \in \mathbf{M}^{qp}$ of the problems (2), (3). We will present here solutions used to calculate the elastic moduli.

It is useful to note that at $p = q$ the problems (2), (3) describe a unit extension of a periodicity cell along the x_q - axis at the following conditions: body forces are absent; $\mathbf{u}$ and its derivatives are periodic over the other space variables. At $p \neq q$ they correspond to a shear of a periodicity cell along the x_p-axis in the planes that are orthogonal to the x_q - axis.

For composites under consideration $\tilde{\mathbf{m}}^{qp}$ are constructed as follows. At the first step the displacements in the plates and bars that are orthogonal to x_q-axis are certain constants. In the plates that are parallel to x_q-axis they are equal to solutions of the extension or shear problems at the conditions for surfaces of plates to be free of forces and for displacements and their derivatives to be periodic on the side faces of plates. In the bars that are parallel to x_q-axis the displacements are those at simple extension or simple shear of isolated bars.

At the second step the displacements are spliced specially to satisfy the conditions (3).

The following estimates were shown to be valid for $\tilde{\mathbf{m}}^{qp}$

$$|\Lambda(\tilde{\mathbf{m}}^{qp}, \varphi)| \leq r\sqrt{\Lambda(\varphi, \varphi)} \quad \forall \quad \varphi \in \mathbf{M},$$

As

$$\Lambda(\mathbf{m}^{qp}, \varphi) = 0 \quad \forall \quad \varphi \in \mathbf{M},$$

then

$$|\Lambda(\tilde{\mathbf{m}}^{qp} - \mathbf{m}^{qp}, \varphi)| \le r\sqrt{\Lambda(\varphi,\varphi)} \quad \forall \quad \varphi \in \mathbf{M}.$$

With the use of $\varphi = \tilde{\mathbf{m}}^{qp} - \mathbf{m}^{qp}$ we obtain

$$\sqrt{\Lambda(\tilde{\mathbf{m}}^{qp} - \mathbf{m}^{qp}, \tilde{\mathbf{m}}^{qp} - \mathbf{m}^{qp})} \le r.$$

If r is small then $\tilde{\mathbf{m}}^{qp}$ are the approximate solutions of (2), (3). Substituting them into (4) we obtain approximate formulae for the effective moduli.

Various particular cases of two-dimensional and three-dimensional structures were considered. Limiting cases of porous media and frame structures were included as well.

The formulae for the simplest case of isotropic components have the form

$$\bar{a}_{1111} \approx \frac{4\mu_1(\lambda_1+\mu_1)}{\lambda_1+2\mu_1}(\nu_2+\nu_3) + (2\mu_2 + \frac{\mu_2\lambda_2}{\lambda_2+\mu_2})\nu_2\nu_3,$$

$$\bar{a}_{1212} = \bar{a}_{1221} \approx \mu_1\nu_3, \qquad \bar{a}_{1122} \approx \frac{4\mu_1^2}{\lambda_1+2\mu_1}\nu_1\nu_2,$$

Here λ_1, μ_1, λ_2, μ_2 are Lame coefficients of the plates and bars, respectively.

Formulae for the other nonzero coefficients can be obtained by change of independent variables. For example

$$\bar{a}_{1313} \approx \mu_1\nu_2.$$

It is proved that the relative error of the formulae for $\bar{a}_{qpqp}$ is not larger than

$$C(\beta_1, \beta_2)(1 + \mu_1/\mu_2)\bar{\nu}.$$

Here $\beta_i = \sqrt{(2\mu_i + 3\lambda_i)/(2\mu_i)^{\text{sign}\,\lambda_i}}$.

Let us consider composites reinforced by fibres (Bakhvalov, Eglit, 1995b). The fibres can be inhomogeneous and curvilinear; they can be located not periodically; the small break of fibres is permitted. At certain conditions effective characteristics of such composites are additive functions of characteristics of fibres. These conditions are the relations that contain ratios of fibres and matrix moduli, the fibres diameters and distances between fibres.

We consider the boundary value problem for the equations (1) in the domain G at the condition $\mathbf{u}|_{\partial G} = \mathbf{g}$. As usual the solution is understood in the sense of integral identity

$$\mathbf{u} \in H^1(G), \quad \Lambda_G(\mathbf{u}, \varphi) = 0 \quad \forall \quad \varphi \in H_0^1(G).$$

Here

$$\Lambda_G(\mathbf{u}, \varphi) = \int_G U\,dx$$

Let V_m $(m = 1, ..., M)$ be fibres, a_{kilj} be their moduli; $V_0 = G \setminus (\bigcup_{m=1}^{M} V_m)$ be matrix and $dist(V_i, V_j) \ge \alpha_0 > 0$ for i, $j > 0$. We consider the case of small ratios ε and α of the typical fibre diameter and distance α_0 to the macroscale L. Besides we assume that the ratio ω of the fibres moduli (thermal conductivity, elasticity) to matrix moduli is large. Let γ be a volume concentration of fibres.

The main idea is to assume that the influence of matrix on heat conduction and elastic behavior of the composite in the first approximation can be neglected. Then, e.g., heat can propagate only along fibres and the total heat flux equals the sum of fluxes through fibres. On this way it is easy to calculate approximately the effective moduli. More difficult problem is to obtain estimates of the error.

To calculate the effective thermoconductivity coefficients we should at first find the relation between the heat flux $\mathbf{q}^m$ along the fibre V_m and the temperature gradient along it. If V_m is homogeneous, straight and oriented along the x_1-axis then the temperature $v^{(1)}$ in V_m is given by the relation

$$v^{(1)} = v + N_j^m \frac{\partial v}{\partial x_j},$$

where v is the averaged temperature and N_j^m are linear functions of x_2, x_3 that can be found from the conditions $q_1^m = 0$, $q_2^m = 0$ at linear $v(x_1)$. Then for linear $v(x_1)$

$$q_1^m = k_{1j} \frac{\partial v^{(1)}}{\partial x_j} = c_{11}^m \frac{\partial v}{\partial x_1}.$$

The coefficient c_{11}^m can be called the effective conductivity of V_m.

Let V_m be inclined to x_1-axis and directed along x_1'. Then its effective conductivity tensor components c_{ij}^m equal $c_{11}^m g_{i1} g_{j1}$ where g_{ij} are components of the orthogonal matrix transforming coordinates x_j' into x_i.

If fibres are curvilinear (e.g., circular) they can be transformed into straight ones by a certain nonlinear transformation of coordinates. After that we apply the described algorithm and return to old variables.

Introduce $\hat{k}_{ij}$ by the relations $\hat{k}_{ij} = 0$ in V_0, $\hat{k}_{ij} = \sum_{i=1}^{M} c_{ij}^m$ where $c_{ij}^m = 0$ outside of V_m. Suppose the existence of $\bar{k}_{ij}(x)$ such that

$$\max_{i,j} \left\| \frac{\widehat{k}_{ij}}{\gamma} - \bar{k}_{ij} \right\|_{(-1)} \le \delta_A \ll 1,$$

As usual, here

$$\|\mathbf{f}\|_{(-1)} = \sup_{\varphi \in H_0^1(G)} \frac{(\mathbf{f}, \varphi)_G}{\|\nabla \varphi\|_{L_2(G)}}.$$

If the material is homogeneous in average, then $\bar{k}_{ij}$ are the average values of $\widehat{k}_{ij}/\gamma$.

The solution of original problem is near to the solution of the averaged problem for the homogeneous material with moduli $\bar{k}_{ij}$. The following inequality is proved

$$\int\limits_{G \setminus V_0} |\nabla(\mathbf{u} - \mathbf{v}^{(1)})|^2 \, dx \le b(R_1 + R_2), \tag{8}$$

where

$$R_1 = \varepsilon^2 + \omega^{-1} + \omega \gamma^2 \delta_A^2, \qquad R_2 = \varepsilon / (\omega \alpha).$$

There is an additional term R_3 in the right part of (8) if the fibres have "breaks"; it is equal to a common measure of all breaks.

So the effective moduli $\bar{k}_{ij}$ are additive functions of c_{ij}^m if R_1, R_2 and R_3 are small.

As an example let us consider a material that can be obtained from a material reinforced by periodic system of homogeneous straight fibres by the following way. Original fibres are replaced by fibres with different diameters not exceeding $O(\varepsilon)$ and located at a distance not larger than $O(\varepsilon)$. Let the ratios of the thermal conductivities of the new and old fibres are inversely proportional to the ratios of their cross-section areas. Then

$$\delta_A^2 = O\left(\varepsilon^2 \, ln \frac{1}{\varepsilon} + \frac{1}{\omega^2}\right)$$

and the values of $\bar{k}_{ij}$ for the new material equal those for original one and therefore can be calculated by the formulae given in (Bakhvalov, Panasenko, 1984).

Let us study a material with periodic system of long pores. The period of the material is different in two directions and the ratio of these two periods is an additional small parameter.

The aim is to obtain the explicit formulae for effective thermoconductivity coefficients for two-dimensional problems. Let the material be isotropic, with symmetric holes. The dimensionless period is ε_1 in the x_1 direction and ε_2 in the x_2 direction with $\varepsilon_2 << \varepsilon_1 << 1$. Let the length of a pore in x_1 direction be a and its thickness in x_2 direction be $g(x_1)$. Introduce $\bar{g}(x_1)$ such that $\bar{g}(x_1) = 0$ outside of pores, $\bar{g}(x_1) = g(x_1)$ in pores.

Denote the domain occupied by the material by G, its solid part by G_s, the external boundary of G by Γ and the boundary of pores by S.

We consider the boundary value problem

$$\frac{\partial}{\partial x_i}\left(k \frac{\partial u}{\partial x_i}\right) = f \quad \text{in} \, G_s, \quad k = const,$$

$$u|_\Gamma = g, \qquad \frac{\partial u}{\partial n}|_S = 0.$$

Here $\mathbf{n}$ is the normal to the boundary of pores.

It is proved (Bakhvalov and Saint Jean Paulin, 1996) that if $\varepsilon_1 + \varepsilon_2/\varepsilon_1 \to 0$ then the solution u is well approximated by solution v of the averaged problem

$$\bar{k}_{ij} \frac{\partial^2 v}{\partial x_i \partial x_j} = f \quad \text{in } G, \quad v|_{\partial G} = g,$$

where

$$\bar{k}_{11}^0 = k \left\langle \frac{1}{1 - \bar{g}(x_1)} \right\rangle_1^{-1}; \quad \bar{k}_{22}^0 = k(1-a); \quad \bar{k}_{12}^0 = \bar{k}_{21}^0 = 0; \tag{9}$$

$$\bar{k}_{ij} = \bar{k}_{ij}^0 / \mu_0$$

Here μ_0 is the area of the solid part of a periodicity cell.

Formulae (9) give approximate values of effective thermoconductivity coefficients at small ε_1 and small $\varepsilon_2/\varepsilon_1$.

SMALL PARAMETERS CONNECTED WITH THE SMALL INFLUENCE OF A CERTAIN PROPERTY OF MATERIALS

Consider now composites that are characterized by small parameters connected with the weak influence of different physical properties of components. The examples are

i) elastic composites with small ratio γ_1 of typical values of shear and bulk moduli;

ii) viscous mixtures with small ratio γ_2 of typical properly scaled viscosity and compressibility coefficients;

iii) thermoconductive materials with small ratio γ_3 of typical thermoconductivity coefficient and specific heat capacity.

It is found that the effective properties of these materials depend essentially on the relations between the values of additional parameters γ_i and the value of dimensionless length of inhomogeneity ε. We studied the process of sound propagation in composites (Sandrakov, 1987; Bakhvalov, Eglit, 1993, 1994, 1995a). Parameters γ_i are

$$\gamma_1 = \sqrt{\frac{\mu_0}{\lambda_0}}; \qquad \gamma_2 = \sqrt{\frac{\nu_0}{L\sqrt{\lambda_0/\rho_0}}}; \qquad \gamma_3 = \sqrt{\frac{\chi_0}{L\sqrt{\lambda_0/\rho_0}}}.$$

Here λ_0, μ_0, ν_0, χ_0 and ρ_0 are typical values of the compressibility coefficient, the shear modulus, the viscosity coefficient, the temperature conductivity coefficient and the density, respectively.

The simplest behavior is observed when the structure is very fine, i.e., ε is much less then γ_i. At $\varepsilon << \gamma_1 << 1$ and at $\varepsilon << \gamma_2 << 1$ the effective compressibility coefficient $\bar{\lambda}$ and density $\bar{\rho}$ are determined by the formulae

$$\bar{\lambda} = \langle 1/\lambda(x) \rangle^{-1}, \qquad \bar{\rho} = \langle \rho(x) \rangle.$$

If $\varepsilon >> \gamma_1$, $\varepsilon >> \gamma_2$, $\varepsilon >> \gamma_3$ and $\gamma_i << 1$, then shear elasticity, viscosity and thermoconductivity do not influence on the sound propagation. The sound speed is in general different in different directions and depend not only on mean density and compressibility, but on the details of material structure.

If the orders of parameters ε and γ_2, γ_3 are equal then the behavior of the material is the most complicated: it is a material with memory or with additional internal parameters.

REFERENCES

Bakhvalov N.S. and M. E. Eglit. 1993. "Averaging of the equations of the dynamics of composites of slightly compressible elastic components." *Comp. Maths Math. Phys.*, **33**(7): 939-952.

Bakhvalov N.S. and M. E. Eglit. 1994. "An estimate of the error of averaging the dynamics of small perturbations of very inhomogeneous mixtures". *Comp. Maths Math. Phys.*, **34**(3): 333-349.

Bakhvalov N.S. and M.E. Eglit. 1995a. "Homogenization of dynamic problems singularly depending on small parameters". Proceedings of Second Workshop on Composite Media and Homogenization Theory (Trieste, 1993) World Scientific, Singapore: 17-35.

Bakhvalov N.S. and M.E. Eglit. 1995b. "Explicit Calculation of Effective Moduli for Composites Reinforced by an Irregular System of Fibres" transl. in *Doklady Mathematics*, **51**(1): 46-501.

Bakhvalov N.S., G.P. Panasenko and M.E. Eglit. 1996. "Effective properties of constructions and composites with inclusions in the form of walls and bars" *Comp. Maths Math. Phys.*, **36**(12): 73-79.

Bakhvalov N.S. and G.P. Panasenko. 1984. "Homogenization. Averaging Processes in Periodic Media. Mathematical Problems in Mechanics of Composite Materials", Nauka, Moscow, ; Kluver Academic Publishers, Dordrecht-Boston-London, 1989.

Bakhvalov N. S. and J. Saint Jean Paulin. 1996. "Homogenization for thermoconductivity in a porous medium with periods of different orders in the different directions". *Asymptotic Analysis*, 13: 253-276.

Bensoussan A., Lions J.L., Papanicolaou G. 1978. "Asymptotic Methods in Periodic Structures", North Holland, Amsterdam.

Sanchez-Palencia E. 1980. "Non Homogeneous Media and Vibration Theory", Lecture Notes in Physics, 127, Berlin, Springer-Verlag.

Sandrakov G. M. 1987. "Averaging of a linearized system of hydrodynamics with low viscosity and the velocity of sound in mixtures". Preprint No. 178, OVM Akad. Nauk SSSR, Moscow (Russian)

An Improved Shear Deformation Theory for Laminated Composite Structures

R. C. AVERILL

ABSTRACT

In the present paper, a new laminated plate theory called MRZZ3 is developed that retains some of the improved predictive capabilities of higher-order zig-zag laminated plate theories yet is computationally equivalent to first-order shear deformation theory (FSDT) both in terms of its primary degrees of freedom and its finite element formulation. In particular, the present theory is capable of accurately predicting transverse shear interlaminar stresses and does not require an *ad hoc* shear correction factor. The theory is based on Reissner's mixed variational principle, but is cast in a form that is identical to FSDT.

INTRODUCTION

Laminated construction of panels and other structural elements is either being considered or implemented for many applications in the automotive, aerospace, civil, and defense industries. Advantages of laminated construction over single-layer construction include energy absorption (for noise reduction and/or crashworthiness), weight reduction, stiffness and strength tailoring, corrosion and fatigue resistance, manufacturability, and cost efficiency. However, the mechanical response of composite materials and structures is often enormously more complex than that of homogeneous and isotropic materials and structures. Thus, there is a great need for advanced theories and modeling techniques that can accurately simulate these complex responses in order to facilitate analysis and design activities as well as to provide greater understanding of the mechanics of composite materials and structures.

At the same time, there exists an enormous computational software infrastructure that was developed for the primary purpose of simulating the response of structures made of homogeneous and isotropic materials. In the case of shell-type structures (e.g., automotive body panels and airplane wing and fuselage skins), thin shell theories and associated three- and four-noded finite element models have been used very successfully for analysis of aluminum and metallic structures. A great deal of experience in developing and using such

R.C. Averill, Materials Science and Mechanics Department, Michigan State University, East Lansing, MI 48824-1226 models now exists. In fact, to most analysts and designers, the term "shell element" almost universally implies a three- or four-noded element with six degrees-of-freedom (three translations and three rotations) per node. It should be no surprise, then, that the typical analyst and designer now expects to be able to model composite structures using composite "shell elements" that look and behave in a manner identical to their isotropic counterparts. To date, attempts to develop such composite shell models have been only modestly successful.

The prediction of structural response in laminated composite structures is complicated by the extreme inhomogeneity that exists both at the material and at the laminate levels. Models that account explicitly for the microstructural or layerwise details are usually very accurate, but also are computationally expensive and complex (see, for example, Reddy, 1987; Toledano, 1987). On the other hand, simple models that account for only first-order effects (Mindlin, 1951; Yang, Norris, and Stavsky, 1966) are not able to capture the dominant kinematic modes of deformation, and thus often do a poor job of predicting structural response. A new class of theories, often called zig-zag theories, appears to have an excellent combination of both accuracy and simplicity (DiSciuva, 1985, 1987; Cho and Parmerter, 1993; Averill, 1994; Averill and Yip, 1996a, 1996b). The primary drawback of zig-zag theories is that the transverse deflection degree of freedom is required to be C^1 continuous, making the development of simple finite element models a difficult task. The author and his colleagues have recently developed a new family of zig-zag sublaminate models that seems to contain the requisite accuracy and simplicity needed to effectively model most laminated composite structures (Averill and Yip, 1996b; Yip and Averill, 1996; Cho and Averill, 1996). These zig-zag models have been cast in the form of eight-noded brick-type finite element models for plates and four-noded quadrilateral elements for beams, and contain only engineering-type nodal degrees-of-freedom (i.e., translations and rotations). These models are capable of predicting complex structural behavior using a very small and fixed number of degrees-of-freedom (DOFs), yet they allow for through-thickness refinement of the model, when necessary. While these models are effective for modeling very complex structural responses, they are not the most convenient choice for all composite structure analysis and design activities due to their brick-type element topology.

In order to take advantage of the available computational software infrastructure as well as the current user knowledge base, a laminated composite shell model is needed that conforms to the topological and behavioral norms established by homogeneous thin shell models, yet has greatly extended predictive capabilities that account for dominant kinematic behaviors associated with inhomogeneities in composite laminates. In the present paper, such a model is developed using Reissner's mixed variational principle (Reissner, 1984) to combine an accurate assumed transverse shear strain field based on cubic zig-zag theory with a simple first-order displacement field. The resulting theory provides improved accuracy, especially in the prediction of transverse shear stresses, and it can be cast in a form that is computationally equivalent to first-order shear deformation theory (FSDT) (Mindlin, 1951; Yang, Norris, and Stavsky, 1966).

FORMULATION OF THE THEORY

The plates to be considered here are composed of N layers perfectly bonded together, where the thickness and constitutive properties may vary arbitrarily from layer to layer. The reference surface of the plate is assumed to be coincident with the mid-plane, as shown in Figure 1.

The displacement field is assumed to have the following form, identical to FSDT:

$$\begin{aligned} u_x(x, y, z) &= u_0(x, y) + z\theta_y(x, y) \\ u_y(x, y, z) &= v_0(x, y) - z\theta_x(x, y) \\ u_z(x, y, z) &= w_0(x, y) \end{aligned} \tag{1}$$

where u_x, u_y, u_z are the displacements of an arbitrary point in the x,y,z directions, respectively, u_0, v_0, w_0 are the displacements of the reference surface (located at $z=0$) in the x,y,z directions, respectively, and θ_x, θ_y are rotations about the x and y axes, respectively. Based on Eqs. (1), the kinematically consistent strain fields are (with $\varepsilon_z = 0$):

$$\begin{aligned} \varepsilon_x &= \frac{\partial u_0}{\partial x} + z\frac{\partial \theta_y}{\partial x} \\ \varepsilon_y &= \frac{\partial v_0}{\partial y} - z\frac{\partial \theta_x}{\partial y} \\ \gamma_{yz} &= -\theta_x + \frac{\partial w_0}{\partial y} \\ \gamma_{xz} &= \theta_y + \frac{\partial w_0}{\partial x} \\ \gamma_{xy} &= \left(\frac{\partial u_0}{\partial y} + \frac{\partial v_0}{\partial x}\right) + z\left(\frac{\partial \theta_y}{\partial y} - \frac{\partial \theta_x}{\partial x}\right) \end{aligned} \tag{2}$$

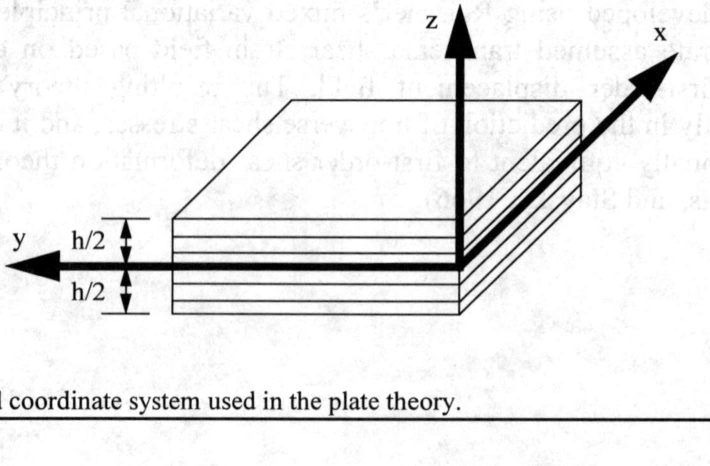

Figure 1. Geometry and coordinate system used in the plate theory.

The poor predictive capabilities of FSDT when applied to laminated composite structures can be attributed primarily to the form of the transverse shear strain fields γ_{yz}, γ_{xz} which are continuous through-the-thickness of the laminate. It is thus impossible in general to satisfy continuity of transverse shear stresses at ply interfaces, a necessary condition for internal equilibrium. In order to satisfy shear stress continuity, discontinuous shear strains are required. The magnitude of the actual discontinuity in transverse shear strains at ply interfaces is proportional to the mismatch in the transverse shear stiffnesses of the adjacent plies. Hence, the error in FSDT increases as this mismatch increases. Laminates with severe mismatches in ply shear stiffnesses include sandwich panels, adhesively bonded laminates, and laminates containing damage.

A much improved form of the transverse shear strains is obtained from cubic zig-zag theory (Cho and Parmerter, 1993; Averill and Yip, 1994; Yip and Averill, 1996), wherein the shear strains in the k-th layer vary according to a quadratic polynomial with a piecewise constant function superposed:

$$\begin{aligned} \hat{\gamma}_{yz}^{(k)} &= \phi_y + z\alpha_y + z^2\beta_y + \sum_{i=1}^{k-1} \eta_i \\ \hat{\gamma}_{xz}^{(k)} &= \phi_x + z\alpha_x + z^2\beta_x + \sum_{i=1}^{k-1} \xi_i \end{aligned} \tag{3}$$

In Eqs. (3) the variables $\alpha_y, \alpha_x, \beta_y, \beta_x, \eta_i, \xi_i$ can be eliminated by enforcing continuity of transverse shear stresses at ply interfaces:

$$\tau_{yz}^{(k)} = \tau_{yz}^{(k+1)} \qquad \tau_{xz}^{(k)} = \tau_{xz}^{(k+1)} \qquad \text{at } z = z_k,\ k = 1 \text{ to N-1}, \tag{4}$$

and satisfying the homogeneous shear traction conditions at the top and bottom surfaces of the laminate:

$$\tau_{yz}^{(1)}\Big|_{z=-\frac{h}{2}} = \tau_{xz}^{(1)}\Big|_{z=-\frac{h}{2}} = \tau_{yz}^{(N)}\Big|_{z=\frac{h}{2}} = \tau_{xz}^{(N)}\Big|_{z=\frac{h}{2}} = 0. \tag{5}$$

Using Eqs. (4,5), $\alpha_y, \alpha_x, \beta_y, \beta_x, \eta_i, \xi_i$ can be determined analytically in terms of only the two remaining variables ϕ_y, ϕ_x. After simplification, the assumed shear strain fields take the form:

$$\begin{aligned} \hat{\gamma}_{yz}^{(k)} &= (A_k + z\bar{A} + z^2\tilde{A})\phi_y + (B_k + z\bar{B} + z^2\tilde{B})\phi_x \\ \hat{\gamma}_{xz}^{(k)} &= (C_k + z\bar{C} + z^2\tilde{C})\phi_y + (D_k + z\bar{D} + z^2\tilde{D})\phi_x \end{aligned} \tag{6}$$

or:

$$\begin{aligned} \hat{\gamma}_{yz}^{(k)}(x,y,z) &= F_{A_k}(z)\phi_y(x,y) + F_{B_k}(z)\phi_x(x,y) \\ \hat{\gamma}_{xz}^{(k)}(x,y,z) &= F_{C_k}(z)\phi_y(x,y) + F_{D_k}(z)\phi_x(x,y) \end{aligned} \tag{7}$$

The constants $A_k, \bar{A}, \tilde{A}, \ldots \tilde{D}$ are lengthy expressions, and are omitted here for brevity.

The assumed form of the transverse shear strain fields in Eqs. (7) can be related to the kinematically consistent transverse shear strain fields in Eqs. (2) through Reissner's partially mixed variational principle (Reissner, 1984). In this principle, only the transverse strain fields are assumed independently of the displacement field, while the inplane strains are as given in Eqs. (2). The principle can be written in the following form (assuming $\sigma_z = 0$):

$$\begin{aligned}\iiint_V \Bigg\{ & -\left(\frac{\partial \sigma_x}{\partial x} + \frac{\partial \tau_{xy}}{\partial y} + \frac{\partial \tau_{xz}}{\partial z} + f_x\right)\delta u_x \\ & -\left(\frac{\partial \tau_{xy}}{\partial x} + \frac{\partial \sigma_y}{\partial y} + \frac{\partial \tau_{yz}}{\partial z} + f_y\right)\delta u_y \\ & -\left(\frac{\partial \tau_{xz}}{\partial x} + \frac{\partial \tau_{yz}}{\partial y} + f_z\right)\delta u_z \\ & +\left[\left(\frac{\partial u_y}{\partial z} + \frac{\partial u_z}{\partial y}\right) - \hat{\gamma}_{yz}\right]\delta \tau_{yz} \\ & +\left[\left(\frac{\partial u_x}{\partial z} + \frac{\partial u_z}{\partial x}\right) - \hat{\gamma}_{xz}\right]\delta \tau_{xz} \Bigg\} dV \\ & + \text{Boundary Terms} = 0 \end{aligned} \tag{8}$$

where f_x, f_y, f_z are body forces in the x,y,z directions, respectively.

In order to simplify the theory, the fourth and fifth terms in Eqs. (8) can be satisfied identically as follows:

$$\begin{aligned} \sum_{k=1}^{N} \int_{z_{k-1}}^{z_k} \left(F_{A_k}\phi_y + F_{B_k}\phi_x + \theta_x - \frac{\partial w_0}{\partial y}\right) dz &= 0 \\ \sum_{k=1}^{N} \int_{z_{k-1}}^{z_k} \left(F_{C_k}\phi_y + F_{D_k}\phi_x - \theta_y - \frac{\partial w_0}{\partial x}\right) dz &= 0 \end{aligned} \tag{9}$$

Carrying out the integration in Eqs. (9) and solving for ϕ_y, ϕ_x yields the following expression for the transverse shear strain fields:

$$\begin{aligned} \hat{\gamma}_{yz}^{(k)} &= G_{A_k}\left(-\theta_x + \frac{\partial w_0}{\partial y}\right) + G_{B_k}\left(\theta_y + \frac{\partial w_0}{\partial x}\right) \\ \hat{\gamma}_{xz}^{(k)} &= G_{C_k}\left(-\theta_x + \frac{\partial w_0}{\partial y}\right) + G_{D_k}\left(\theta_y + \frac{\partial w_0}{\partial x}\right) \end{aligned} \tag{10}$$

In Eqs. (10) $G_{A_k}, G_{B_k}, G_{C_k}, G_{D_k}$ are functions of z and of the layer thicknesses and transverse shear stiffnesses. These analytical functions are thus laminate dependent, and require a very modest amount of computation to be performed in order to cast them in their simplest and most usable form.

The above theory is called MRZZ3, where the M stands for Mindlin, the R stands for Reissner, and ZZ3 stands for cubic zig-zag theory.

NUMERICAL IMPLEMENTATION

The above theory can be implemented in one of several forms. Given that most finite element codes already contain one or more versions of FSDT or Mindlin-type elements, an implementation that requires the least amount of modification to these available element routines would be favorable. In fact, the above theory can be cast in a form that is identical to FSDT with a laminate-dependent shear correction factor, as shown below.

The variation of the internal shear strain energy in the laminate can be expressed as:

$$\delta U_S = \int_{\Omega} \sum_{k=1}^{N} \int_{z_{k-1}}^{z_k} \left(\tau_{yz}^{(k)} \delta\hat{\gamma}_{yz}^{(k)} + \tau_{xz}^{(k)} \delta\hat{\gamma}_{xz}^{(k)}\right) dz\,dx\,dy \tag{11}$$

Substituting Eqs. (10) into Eqs. (11), carrying out the through-thickness integration, and simplifying, one obtains:

$$\begin{aligned} \delta U_S = \int_{\Omega} \Bigg\{ & \left(-\delta\theta_x + \frac{\partial \delta w_0}{\partial y}\right)\left[\hat{A}_{44}\left(-\theta_x + \frac{\partial w_0}{\partial y}\right) + \hat{A}_{45}\left(\theta_y + \frac{\partial w_0}{\partial x}\right)\right] \\ & + \left(\delta\theta_y + \frac{\partial \delta w_0}{\partial x}\right)\left[\hat{A}_{45}\left(-\theta_x + \frac{\partial w_0}{\partial y}\right) + \hat{A}_{55}\left(\theta_y + \frac{\partial w_0}{\partial x}\right)\right]\Bigg\} dx\,dy \end{aligned} \tag{12}$$

Eq. (12) is of exactly the same form as that in FSDT, the only difference being the values of the laminate shear stiffness terms $\hat{A}_{44}, \hat{A}_{45}, \hat{A}_{55}$. In FSDT, $\hat{A}_{44}, \hat{A}_{45}, \hat{A}_{55}$ would be replaced by A_{44}, A_{45}, A_{55} (see Reddy, 1984 for a definition of these terms). Hence, the following shear correction factors can be defined:

$$k_{44} = \frac{\hat{A}_{44}}{A_{44}} \qquad k_{45} = \frac{\hat{A}_{45}}{A_{45}} \qquad k_{55} = \frac{\hat{A}_{55}}{A_{55}} \tag{13}$$

The shear correction factors in Eqs. (13) can be calculated during the pre-processing stage of a structural analysis and can be used with the standard form of FSDT. The advantage of this form of the present theory is that no modifications of existing implementations of FSDT models are required, yet the improved transverse shear strain fields are effected through simple shear correction factors.

Perhaps the most important benefit of the current theory is its ability to predict the through-thickness variation of transverse shear strains (and subsequently stresses) directly from Eqs. (10). This is accomplished without resorting to integration of equilibrium equations or other techniques that require the accurate calculation of stress gradients.

SUMMARY

A new shear deformation theory based on Reissner's partially mixed variational principle has been developed. The new theory contains some of the improved predictive capabilities of cubic zig-zag laminated plate theories, yet it is computationally equivalent to FSDT. The present theory is capable of accurately predicting transverse shear interlaminar stresses, does not require a shear correction factor, and can be cast in a form that is identical to FSDT with an appropriately calculated shear correction factor. The latter property of the current theory allows it to be implemented into available finite element codes with FSDT type elements without any modification to the existing element formulation.

REFERENCES

Averill, R.C. 1994. "Static and Dynamic Response of Moderately Thick Laminated Beams with Damage," *Composites Engineering*, **4**, p. 381-395.

Averill, R.C. and Yip, Y.C. 1996a. "Development of Simple, Robust Finite Elements Based on Refined Theories for Thick Laminated Beams," *Computers and Structures*, **59**, 529-546.

Averill, R.C. and Yip, Y.C. 1996b. "Thick Beam Theory and Finite Element Model with Zig-Zag Sublaminate Approximations," *AIAA Journal*, **34**, 1627-1632.

Cho, M. and Parmerter, R.R. 1993. Efficient Higher Order Composite Plate Theory for General Lamination Configurations, *AIAA J.*, **31**, 1299-1306.

Cho, Y.B. and Averill, R.C. 1997. "An Improved Theory and Finite Element Model for Laminated Beams Using First-Order Zig-Zag Sublaminate Approximations," to appear in *Composite Structures*.

DiSciuva, M. 1985. Development of an Anisotropic, Multilayered, Shear-Deformable Rectangular Plate Element, *Comp. & Struct.*, **21**, 789-796.

DiSciuva, M. 1987. An Improved Shear-Deformation Theory for Moderately Thick Multilayered Anisotropic Shells and Plates, *J. Appl. Mech.*, **54**, 589-596.

Mindlin, R.D. 1951. Influence of Rotary Inertia and Shear on Flexural Motions of Isotropic, Elastic Plates, *J. Appl. Mech.*, **18**, 31-38.

Reddy, J.N. 1984. *Energy and Variational Methods in Applied Mechanics*, John Wiley & Sons, Inc., New York.

Reddy, J.N. 1987. A Generalization of Two-Dimensional Theories of Laminated Composite Plates, *Comm. Appl. Num. Meth.*, **3**, 173-180.

Reissner, E., 1984. "On a Mixed Variational Theorem and a Shear Deformable Plate Theory," *Int. J. Num. Meth. Engr.*, **23**, 193-198.

Toledano, A. and Murakami, H. 1987. A Composite Plate Theory for Arbitrary Laminate Configurations, *J. Appl. Mech.*, **54**, 181-189.

Yang, P.C., Norris, C.H., and Stavsky, Y., 1966. "Elastic Wave Propogation in Heterogeneous Plates," *Int. J. Solids and Structures*, **2**, 665-684.

Yip, Y.C. and Averill, R.C., 1996. "A 3D Laminated Plate Finite Element with Zig-Zag Sublaminate Approximations -- Part I: Formulation and Application to Thick Plates and Sandwich Panels," submitted for publication.

Shear Deformable User Finite Element for MARC to Model Composites with Taper

B. B. HOFMAN AND O. O. OCHOA

ABSTRACT

A shear deformable composite element is implemented as a user element in MARC®. The element accommodates variable element thickness and hygrothermal load options. Example problems, a simply supported tapered plate, and a cantilevered tapered plate, both at various aspect ratios, are analyzed to benchmark the element and demonstrate the user element's ability to model tapered laminated plates and hygrothermal loads.

INTRODUCTION

In recent years extensive research has been done in the area of finite element analysis of composites. Research areas that have been of primary interest are higher-order theories, evaluation of three-dimensional stress fields, transverse shear effects and failure progression prediction. These features are of significant interest in the design of complex shape composite components such as engine blades.

A finite element, QHD40, that has the aforementioned features was developed by Ochoa. The QHD40 element is an eight-node, 40 degree-of-freedom, two-dimensional plate/shell element with seven degrees-of-freedom at each of four corner nodes, and three degrees-of-freedom at each mid-side node. The element employs total Lagrangian formulation for the nonlinear geometry option, treats each layer individually without smearing properties, has a damage progression option, and uses reduced integration for the shear terms to reduce hourglass and shear lock phenomena. New algorithms are added to facilitate tapered element thickness capability and hygrothermal analysis. The improved element is utilized as a user element in MARC® finite element code to perform analyses and demonstrate the element's capabilities.

Brad B. Hofman, Ford Motor Company, Wixom Assembly Plant, Wixom, MI 48393
Ozden O. Ochoa, Department of Mechanical Engineering, Texas A&M University, College Station, TX 77843-3123

TAPER FORMULATION

Almost all commercial finite element analysis codes currently have tapered plate/shell element capabilities which are suitable for homogeneous materials, but are not acceptable for heterogeneous materials such as layered composites. Commercial codes generally control element taper with the element thicknesses at the four corner nodes. To generate a tapered element, the user merely prescribes different thicknesses at the corner nodes. The problem with applying this technique to layered composites is as element thickness decreases, individual layer thicknesses also decrease, which is unrealistic from a manufacturing perspective.

QHD40 ELEMENT TAPER

In manufacturing complex composite parts, tapering is maintained by layer drop-off. That is, as the composite thickness decreases, some layers terminate, or drop-off. The remainder of the layers retain their constant thickness. This results in resin rich areas caused by the terminated layers. The resin rich areas that are present in an actual tapered laminate are difficult to model with finite elements due to the difference in the material properties between the layers and resin areas. In the QHD40 element, the drop-off layers are tapered gradually and terminated at the midplane (Figure 1). Additionally, to aid in formulation, the layer drop-off is assumed to be symmetric about the element mid-plane, which is realistic in the modeling of fan blades, which have symmetric layer drop-offs. Symmetric layer drop-off at the midplane is the only type of drop-off that the QHD40 element is formulated to perform, and this approach negates the presence of small resin pockets.

For every integration point in an element, the taper algorithm calculates the element thickness at that point by interpolating linearly from the element corners. The algorithm then calculates the distance from the element midplane to the midplane of each of the element layers. The distance from the element midplane to the surface of the element is calculated as the element thickness divided by two, denoted by distance x on Figure 1. Then half the thickness of the current layer ($t_1/2$) is subtracted from x, which is the distance from the element midplane to the center of the current layer, denoted by d_1 on Figure 1. Once this distance is known, the element stiffness is calculated for the current layer at that integration point. This process is repeated at every integration point in the layer.

For other layers in the element, the dimensions to the previous layer are used to calculate the distance to the current layer. In Figure 1 d_1, t_1, and t_2 (thickness of layer 2) are used to calculate d_2 as follows.

$$d_2 = d_1 - t_1/2 - t_2/2 \tag{1}$$

This procedure is repeated for every layer on the lower half of the element, starting at layer 1 and working toward the element centerline. Once the centerline is reached, the top half of the element is calculated, starting at layer n and similarly working toward the element centerline.

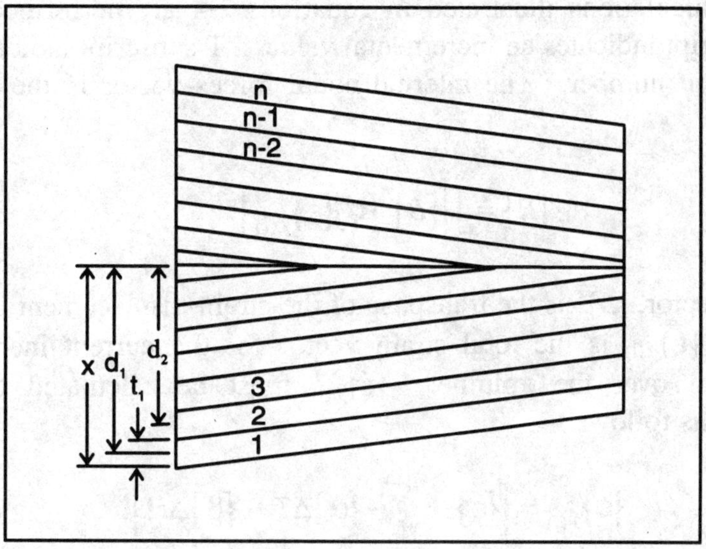

Figure 1. QHD40 element layer drop-off

The present taper formulation assumes that the layers on either side of the mid-plane remain parallel to the element surface. As the element tapers, the thickness decreases, thereby forcing the plies that contact the element centerline to terminate at the centerline. The QHD40 element assembles one half of the element at a time, starting at the element surface and moving toward the center. Once the center is reached, the algorithm then moves over to the other side of the element and repeats the process. This whole procedure is repeated for every integration point in the element. At a given integration point the element layers are assumed to be parallel to the element midplane, which means that the layers are not rotated to account for the angle between the layers and the element midplane. This same algorithm is used to track layer location at integration points during stress and strain recovery after MARC® solution.

HYGROTHERMAL FORMULATION

Hygrothermal loads are induced upon a structure by the service environment, of which temperature and moisture gradients are most prevalent. These hygrothermal loads have the potential to cause failure in laminated composites. Temperature changes cause irregular expansion due to the layered, anisotropic nature in laminated composites. This expansion induces stresses, which may cause failure. Moisture similarly causes composites to swell, leading to residual stresses. Analysis of the effect of these and other hygrothermal loads is critical when working with composites in structural applications. Several methods, both experimental and computational, have been developed to perform this analysis. This superposition technique, developed by O'Brien, Raju, and Garber, is one type of experimental hygrothermal analysis. The QHD40 element hygrothermal formulation is similar to a computational formulation developed by Wang and Crossman.

The following calculations illustrated in equations 2-4 are performed on a per layer basis, and an I subscript indicates an incremental value, a T subscript indicates a total value, and i is the increment number. The internal nodal forces vector in the finite element is treated as:

$$\{R\} = \int \left[[B]^T [C] \{\varepsilon\}_{T(i)} \right] dV \qquad (2)$$

$\{R\}$ is the residual vector, $[B]^T$ is the transpose of the strain-displacement matrix, $[C]$ is the constitutive matrix, $\{\varepsilon\}_{T(i)}$ is the total strain vector for the current increment and their product is integrated over the volume. $\{\varepsilon\}_{T(i)}$ must be calculated to accommodate environmental loads as follows:

$$\{\varepsilon\}_{T(i)} = \left[\{\varepsilon\}_{T(i-1)} - \{\alpha\}\Delta T - \{\beta\}\Delta H \right] \qquad (3)$$

$\{\alpha\}$ is the vector of thermal expansion coefficients, $\{\beta\}$ is the vector of moisture expansion coefficients, ΔT is the temperature change, ΔH is the percent moisture change by weight, and $\{\varepsilon\}_{T(i-1)}$ is the total strain vector from the previous increment.

The composite material was assumed to obey generalized Hooke's Law, so that the lamina stresses can be represented as:

$$\{\sigma\}_{I(i)} = [C]\left(\{\varepsilon\}_{I(i)} - \{\alpha\}\Delta T - \{\beta\}\Delta H \right) \qquad (4)$$

where $\{\sigma\}_{I(i)}$ is the incremental stress vector and $\{\varepsilon\}_{I(i)}$ is the incremental strain vector.

In the QHD40 element the hygrothermal strains that are subtracted in equations 3 and 4 are assumed to be linear. Additionally, the QHD40 element only accounts for the in-plane hygrothermal strains because the linear component of the transverse normal strain is assumed to be zero. The hygrothermal capability is incorporated very well into the QHD40 element when used with MARC®. Thermal and/or moisture expansion coefficients and temperature changes may all be entered through MENTAT® (MARC®'s pre- and postprocessor), and this information is all passed from MARC to the QHD40 element during the run. The QHD40 element then performs the above outlined calculations for every integration point in the element.

EXAMPLE PROBLEMS: TAPERED PLATE

Several problems are solved with the QHD40 element to verify and demonstrate the element's capabilities and accuracy. A simply supported tapered plate problem is solved and the results are compared to a Navier solution to benchmark the element. A tapered cantilevered plate is analyzed to demonstrate both the hygrothermal and taper capabilities of the QHD40 element.

SIMPLY SUPPORTED PLATE

A tapered square plate, L=4 inches, simply supported on all edges, was loaded uniformly across the entire plate surface. The plate was steel (E=30.0e7 psi, ν=0.30), and loaded with 25 lbf/in^2. The analysis was conducted for plates with aspect ratio S, S =10,20,50 and 100 (S is the smallest side dimension divided by the plate thickness). The results were compared to a Navier series solution that was modified to accommodate the variable plate thickness, which was obtained from Reddy. The modification of the Navier solution consisted of defining the plate stiffness as being variable, instead of constant as in a flat plate. TABLE I presents the calculated maximum deflection from the QHD40 element and the series solution for the aforementioned aspect ratios.

The QHD40 element results compared well with the Navier solution results at all aspect ratios. The higher aspect ratios produced better results, which is expected, however the lower aspect ratios produced good results also. The QHD40 element's ability to calculate stresses and strains in a tapered plate will be investigated with the cantilevered plate problem.

CANTILEVERED PLATE

The cantilevered plate problem consisted of a square plate clamped on one edge and transversely displaced on the opposite edge. The plate was 10 inches square, and 0.10 inch thick along the constrained edge. One plate analyzed was of constant 0.05 inch thickness, and the second plate tapered linearly from 0.10 inch thick at the constrained edge to 0.05 inch thick at the opposite edge.

The plate consisted of 16 layers of equal thickness, $[(0,45,90,-45)_2]_s$, and each lamina was astroquartz. The material properties are presented in TABLE II. The tapered plate

TABLE I - MAXIMUM DEFLECTION OF A TAPERED, SIMPLY SUPPORTED PLATE

S	w (QHD40) in.	w (Navier) in.
10	0.000471	0.000443
20	0.00362	0.000354
50	0.0554	0.0554
100	0.443	0.441

TABLE II - MATERIAL PROPERTIES OF ASTROQUARTZ

Elastic Constant	Value
E_1	6.41 Msi
E_2	2.20 Msi
E_3	2.20 Msi
G_{12}	0.23 Msi
G_{13}	0.23 Msi
G_{23}	0.23 Msi
ν_{12}	0.25
ν_{13}	0.25
ν_{23}	0.25

dropped from sixteen layers at the constrained edge to eight layers at the opposite edge, while the uniform thickness plate remained 16 layers thick over the entire plate area. A cross-sectional view of the tapered plate in Figure 2 illustrates the layer drop-off at the centerline of the laminate.

The tapered cantilevered plate was analyzed by displacing the free edge opposite the clamped edge transversely 0.25 inch in a series of increments. The flat plate of 0.05 inch thickness was also displaced 0.25 inch and used for comparison purposes. Figure 3 shows a side view of the transverse displacement of the flat and tapered plates to illustrate the curvature difference due to the taper.

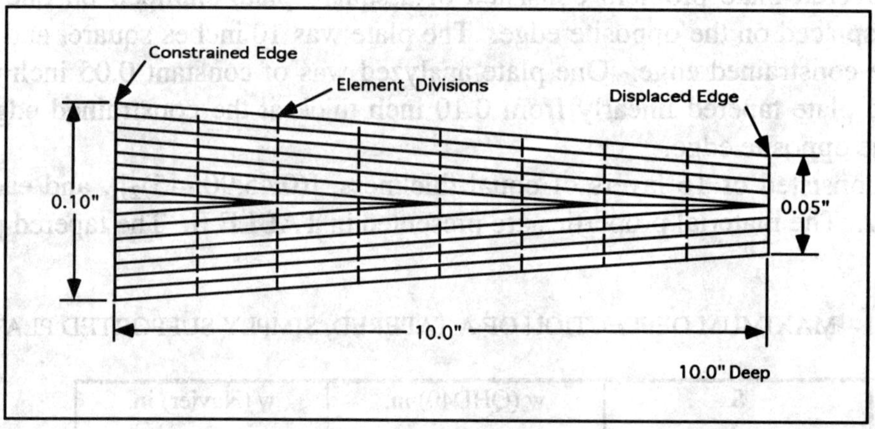

Figure 2. Cross-sectional view of tapered laminate

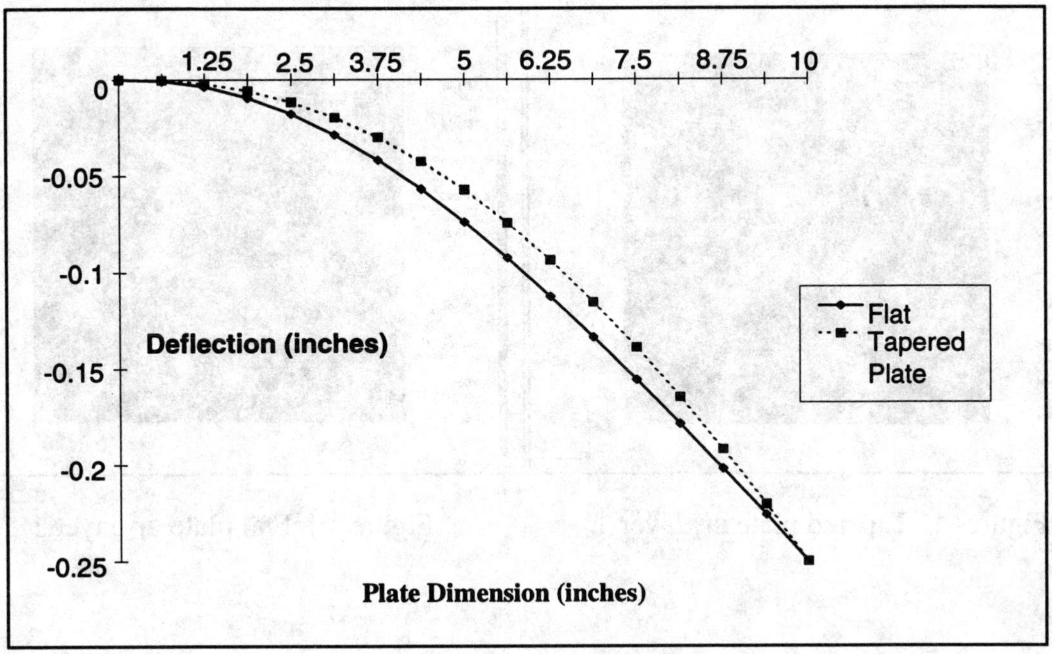

Figure 3. Tapered vs. flat plate deflection

A stated earlier, the tapered plate thickness is reduced from 16 layers to 8 layers, while the uniform plate is 16 layers throughout the plate. In the tapered plate, stress and strain plots in layers that don't drop off are qualitatively similar to the stresses and strains in the same layers in the flat plate. The stresses and strains in the tapered plate are somewhat larger that in the flat plate for a couple of reasons. First, the tapered plate is twice as thick at the constrained end as the flat plate, resulting in higher stresses and strains due to bending in layers further from the plate midplane. Second, as Figure 3 shows, the tapered plate has more curvature toward the displaced edge of the plate than the flat plate due to the variable plate stiffness, resulting in slightly higher stresses and strains due to bending along the displaced edge. Figures 4-5 are a set of strain plots of the tapered and flat plates to illustrate these observations.

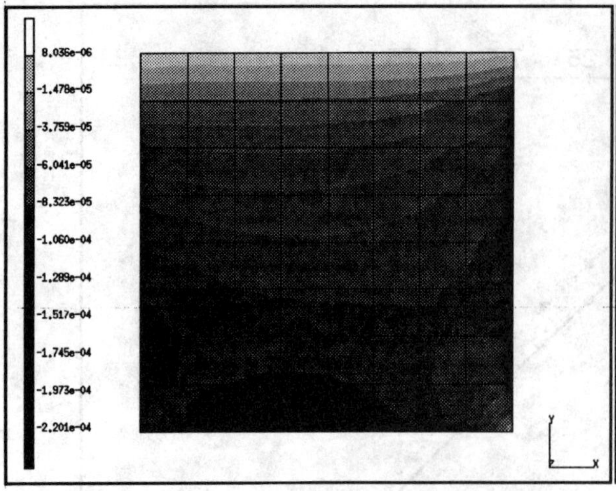

Figure 4. Tapered plate ε_{11} layer 1

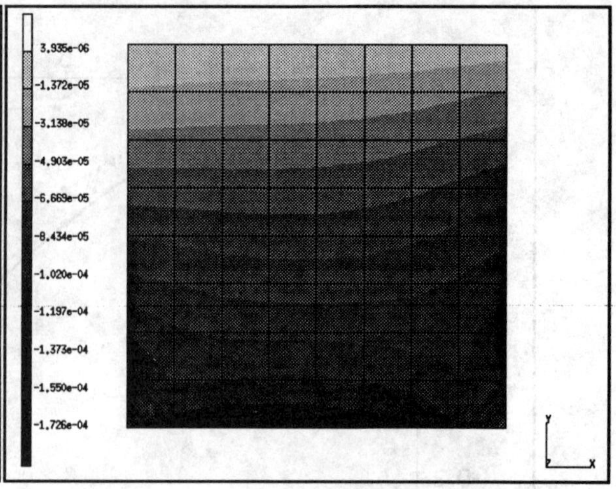

Figure 5. Flat plate ε_{11} layer 1

The QHD40 element calculates stresses and strains in layers that drop off in the same manner as in complete layers, except once a layer has dropped off, zeros are used as place markers in stress and strain arrays. This method facilitates the viewing of stress and strain plots in MENTAT®. Additionally, the QHD40 element accounts for dropped layers when calculating transverse stresses with the equilibrium equations in a tapered laminate.

Figures 6-7 are plots of σ_{11} in layer 7, which drops off in the middle of the tapered plate, and similar plots from the flat plate for comparison. Layer 7 drops at the horizontal line labeled A in the middle of the plate in Figure 6, and all following tapered plate plots are consistent in nomenclature. In Figure 6 zero stress indicates the area where the layer no longer exists due to drop-off. MENTAT® uses one element distance to drop the stress from the magnitude at the drop point to zero, which explains the gradual drop in stress from the layer drop point shown in these plots. Figure 8 shows the transverse stresses calculated by the QHD40 in a dropped layer. σ_{23} in layer 7 is minimal with the high and low points caused by the interaction of the plate corners with the clamped boundary condition.

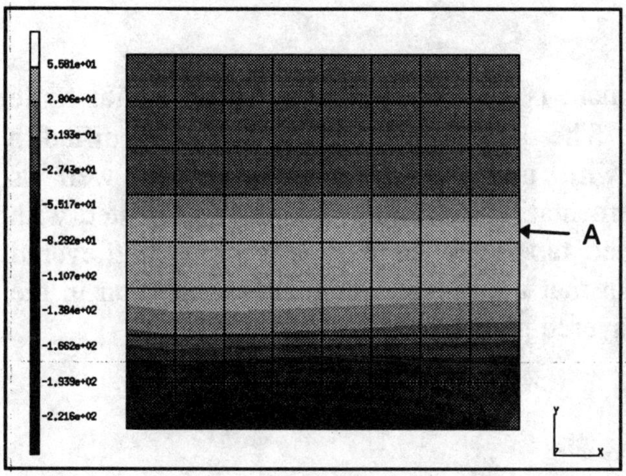

Figure 6. Tapered plate σ_{11} layer 7 (psi)

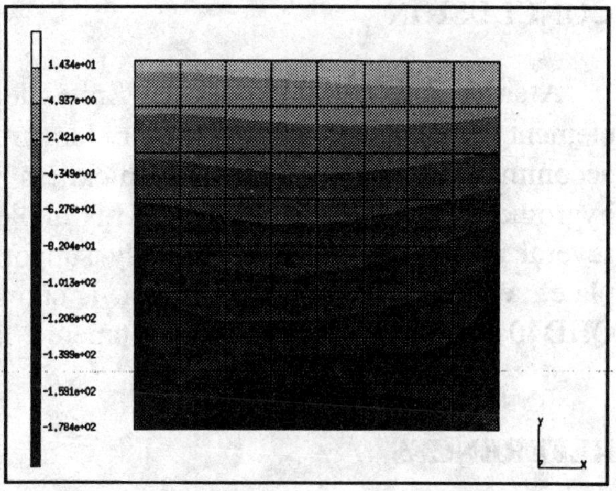

Figure 7. Flat plate σ_{11} layer 7 (psi)

The plots in Figures 6-7 are similar, however there are some differences which can be attributed to several factors. In the tapered plate (Figure 6) the abrupt end of the interior layers in the middle of the plate affects the stress distribution. Additionally, the variable plate stiffness in the tapered plate affects the resulting displacements (see Figure 3), which also affect stresses. Finally, the dropped layers vary in distance from the plate midplane, whereas in a flat plate a layer is always a constant distance from the plate midplane. This variable distance in the tapered plate results in varied bending effects on stresses. These factors will have a similar effect on strains.

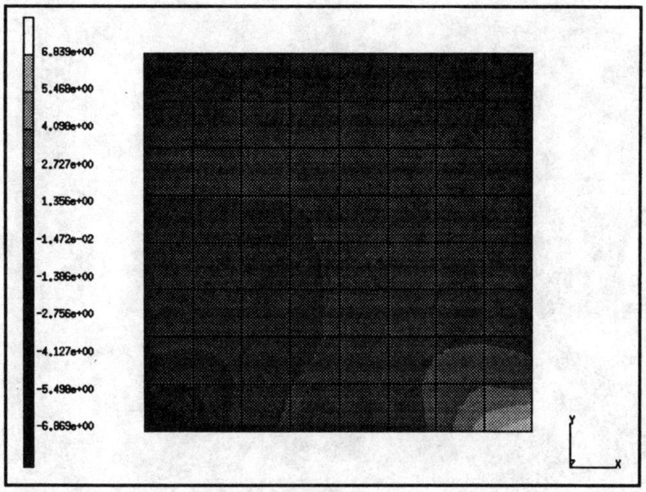

Figure 8. Tapered plate σ_{23} layer 7 (psi)

CONCLUSION

A shear deformable composite finite element is incorporated into a commercial finite element program, MARC®, as a user element. The shear deformable element is modified to accommodate variable element thickness with midplane layer drop-off, as well as hygrothermal analysis capability. This implementation and modification is presented with several example problems, a simply supported tapered plate, and a tapered cantilevered plate. With this example problem it is shown that the tapered element formulation in the QHD40 element adequately models tapered, layered plates, including layer drop-off.

REFERENCES

Ochoa, O. O. 1990. "Developing a Focused Design Methodology for Composite Material Systems," *Journal of Composite Structures*, 16:193-208.

Hofman, B.B. 1996. "Incorporation and Modification of a Shear Deformable Finite Element into MARC®," M.S. thesis, Texas A&M University, College Station, Texas.

O'Brien, T.K., I.S. Raju, and D.P. Garber. 1985. "Residual Thermal and Moisture Influences on the Strain Energy Release Rate Analysis of Edge Delamination," NASA TM-86437, Langley Research Center, Hampton, Virginia.

Wang, A.S.D., and F.W. Crossman. 1977. "Edge Effects of Thermally Induced Stresses in Composite Laminates," *Journal of Composite Materials*, 11:300-312.

Reddy, J.N. 1984. *Energy and Variational Methods in Applied Mechanics*, John Wiley & Sons, New York, New York.

®ABAQUS is a registered trademark of Hibbitt, Karlsson & Sorenson, Inc.
®MARC is a registered trademark of MARC Analysis Research Corporation, Palo Alto, California.

Unified Constitutive Modeling from Viscoelasticity to Viscoplasticity of Polymer Matrix Composites

Z. QIAN AND S. LIU

ABSTRACT

A unified constitutive model for polymer matrix composites is proposed with viscoelastic characterization at small deformation and viscoplastic characterization at large deformation, based on molecular chain deformation mechanisms. The evolution equations and the temperature dependence of drag stress and back stress are well established from molecular network theories. The material constants are then determined by master curves and a consistent procedure. The excellent agreement between experimental data of polycarbonate film tests and model predictions is achieved.

INTRODUCTION

Thermoplastic materials are being increasingly used for structural parts in the automobile industry, chemical industry, etc. The polymer serves as the matrix phase of "tough" advanced composites. Its super mechanical behaviors make it be easily processed such as stamp forming and molding. An important part of the study of deformation and failure mechanisms in polymer based composite systems consists of a study of the properties of the polymer matrix. The matrix is usually subjected to complicated strain states due to the presence of the reinforced phases such as fibers and ceramic particles (Mark, 1996) and due to complicated thermal loading during manufacturing processes such as molding and hot forming. The large deformation can be expected within interfaces near fibers and particles and at crack tip, where strains and stresses are locally higher. Recent work by Waas and Wineman (1996) shows that this can lead to the development near a fiber of small regions of large shear strains and shear strain gradients, in effect a shear strain boundary layer. This may play a significant role in fiber-matrix failure processes.

The polymer matrix is commonly molded as being viscoelastic. The molecular chain orientation hardening of polymers during large deformation results in the viscoplastic behaviors. Therefore, a constitutive model unified viscoelasticity with viscoplasticity is

Zhengfang Qian and Sheng Liu, Department of Mechanical Engineering, Wayne State University, Detroit, MI 48202 highly expected for mechanical behaviors of the polymer matrix, which is the focus of this paper.

The viscoelastic behaviors and models of polymers are summarized in textbooks such as Ferry (1980). Polymer drawing induces necking formation and its steady propagation, following by orientation hardening of molecular chains (Carmak and Lee, 1995). Hutchinson and Neale (1983) presented an early theoretical analysis for the necking propagation of a bulk polymer based on classical J2 flow theory and J2 deformation theory. Recent researches on constitutive modeling of glassy polymers mostly started from the point of view of rubber-like elasticity, where many theories addressed the behavior of networks of chain molecules (Boyce, Arruda, and Jayachandran, 1994; Wu and van der Giessen, 1993). These research works also incorporated the viscoplastic large deformation framework into finite element analysis. However, the constitutive models proposed by those authors have some shortages. For instance, the models are lack of ability to describe cyclic loading. The temperature dependence of mechanical behaviors is not well established. They are not so successful to describe shear deformation (Tomita and Tanaka, 1995). This deformation mode is particularly important for shear banding (localization), crazing, and necking formation of polymers, which can result in local failure and delamination. Therefore, further modification is desirable to deal with different deformation modes.

A unified constitutive model from viscoelsticity to viscoplasticity was proposed in this paper. It is able to describe both small deformation which is dominated by viscoelasticity and large deformation which is dominated by orientation hardening induced viscoplasticity. The temperature dependence both in small deformation range and in large deformation range was well established based on different deformation mechanisms. The material constants were then determined by master curves and a consistent procedure.

MODEL FORMULATION

RATE EQUATION OF AN INELASTIC STRAIN

The thermal activation theory (Tobolsky and Eyring, 1943) has been widely applied to metals and polymers. The thermally activated movements of local dislocation kinks by passing through dislocation obstacles produce inelastic flow (Frost and Ashby, 1982). Similar deformation mechanism operates in polymer materials, where thermal activation processes result in local molecular kinks rotation by the operation of disclinations (Argon, 1973). The inelastic shear strain rate can be written as a generalized form (Frost and Ashby, 1982)

$$\dot{\gamma}^p = \dot{\gamma}_0 \exp\left\{ -\frac{F_0}{kT}\left[1 - \left(\frac{\tau}{\tau_0} \right)^p \right]^q \right\} \tag{1}$$

Where the exponents p and q are around 1.0 for polymers, indicating a weak strain rate dependence. And the k is the Boltzmann constant, T the temperature, F_0 the activation energy of the process. τ_0 the reference shear stress, $\dot{\gamma}_0$ the reference shear strain rate.

Hasan and Boyce (1995) have evaluated the thermal activation model by using the parameters 1) p=q=1; 2) p=5/6, q=1 (Argon, 1973); 3) p=2/3, q=1 (Frost and Ashby, 1982). They found little effect on the predictions of the rate for the three parameter sets. Therefore, the p=q=1 is adopted here for simplification. In addition, in the range of power-law, above equation can be simplified as

$$\dot{\gamma}^P = \dot{\gamma}_0 \frac{\tau - \tau_1}{D_0(T)} \tag{2a}$$

with the relationship

$$D_0(T) = \tau_0 kT / F_0, \tau_1 = \tau_0(1 - kT / F_0) \tag{2b}$$

By introducing a back stress σ_B to replace the stress τ_1 and to consider long range interaction of polymer molecule, the generalized rate equation can be written as, keeping the basic mathematical structure of equation (2a).

$$\dot{\varepsilon}^P = \dot{\varepsilon}_0 \frac{\sigma - \sigma_B}{D(T)} \tag{3}$$

Where the D is the generalization of the $D_0(T)$, so-called drag stress. It has linear dependence on temperature according to eq.(2b). The $\dot{\varepsilon}_0$ is the reference strain rate, a material constant. And the back stress σ_B is not obtained in current form. Its evolution equation will be given by a suitable molecular network theory.

EVOLUTION EQUATION OF DRAG STRESS D

Drag stress D is used to describe short range interaction among molecular chains, i.e., the deformation resistance from surround molecule against an active molecular chain rotation. It is linearly dependent on deformation temperature according to eq.(2b). The viscoelastic behavior of polymers should be included in its evolution equation. One general form of the evolution equation can be written as

$$\frac{dD}{d\zeta} = g(D, T, \zeta) \tag{4}$$

Where the material time ζ is defined as

$$d\zeta = \frac{dt}{h(t)} + \frac{dp}{f(p)} \tag{5a}$$

$$p = \int \left|\dot{\gamma}^{P}\right| dt \tag{5b}$$

Where h(t) is time hardening function, and f(p) the deformation hardening function. ζ is, actually, a kind of combination Newtonian time t with deformation path length p.

A simple choice of function g expression is its linear term with D. The specific evolution equation is then given by

$$\frac{dD}{d\zeta} = \frac{1}{b}(D_{SS} - D) \tag{6}$$

For uniaxial loading, its integration form is obtained, which is

$$D = D_{SS} + (D_{SS} - D_0)\exp(-b\zeta) \tag{7}$$

Where $D_0(T)$, $D_{SS}(T)$ are the initial and saturated values of drag stress D, respectively They depend on temperature linearly. And the b in eq.(7) is a material constant. When the Newtonian time t is chosen as the material time ζ , from eqs. (6) and (7) we have

$$\frac{dD}{dt} = \frac{1}{b}(D_{SS} - D) \tag{8}$$

$$D = D_{SS} + (D_{SS} - D_0)\exp(-b\,t) \tag{9}$$

Considering the elastic Hooke's law, we obtain

$$\dot{\varepsilon} = \frac{\dot{\sigma}}{E} + \dot{\varepsilon}_0 \frac{\sigma - \sigma_B}{D} \tag{10a}$$

Neglecting the back stress in the case of small deformation, one has

$$\dot{\varepsilon} = \frac{\dot{\sigma}}{E} + \dot{\varepsilon}_0 \frac{\sigma}{D} \tag{10b}$$

The set of equations (8), (9) and (10b) just represents well-known Maxwell model of viscoelasticity with the nonlinear viscosity coefficient D(T, t). Therefore, the deformation mechanism of drag stress D is clarified in this special case. This Maxwell unit can be shown in Fig. 1. The test data of polymer films will approve the suitability of above equations. On the other hand, the model characterization in small deformation stage will be very useful for the determination of material constants, distinguished from material deformation characterization in large deformation stage.

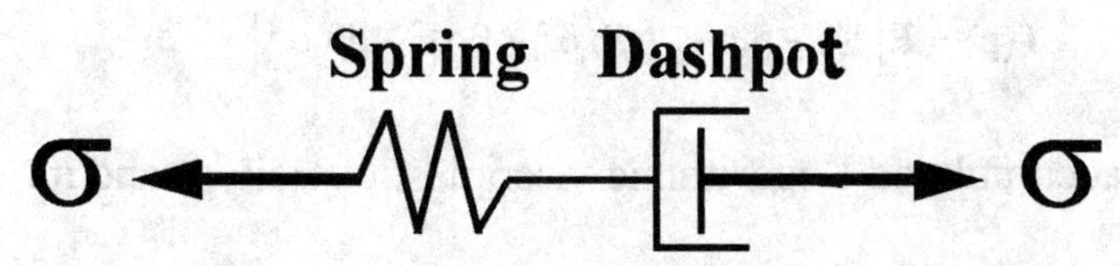

Figure 1, Maxwell Model of Viscoelasticity

EVOLUTION EQUATION OF BACK STRESS σ_B

The back stress is introduced to describe the orientation hardening of molecular chains during large deformation stage. This characterizes viscoplastic behavior of polymers. Molecular chains approach to their limited length, which results in so-called orientation hardening. The orientation hardening is not recoverable during the unloading of polymers after large deformation due to the configuration entropy changing of a molecular chain network. The derivation of the back stress evolution equation comes from molecular network theories, which were originally developed for the theories of rubber elasticity. Two of them are well known phantom network model (Flory and Rehner, 1943) and affine network model (James and Guth, 1943). The constrained junction network theory, proposed by Flory (1977), is modified to describe different deformation characteristics of polymers, including the behaviors of phantom network and orientation hardening of molecular chains as well as Mooney-Rivlin equation. The theoretical details were published in a recent paper (Qian, Lu and Liu, 1997). This integration form of the back stress for uniaxial loading was derived as (Qian, Lu and Liu, 1997)

$$\sigma_B = E(T)\left(\beta + \beta_1\lambda_P + \beta_3\lambda_P^3\right)\left(\lambda_P - \frac{1}{\lambda_P^2}\right) \tag{11}$$

Where the λ_p is the plastic stretch ratio. E(T) the Young's modulus, and β, β_1, β_3 are the material constants. The temperature dependence of the back stress is same as the temperature dependence of Young's modulus, which is a basic outcome of the Gaussian statistical theory of cross-linked molecular networks (for instance, Treloar, 1976).

The differential form of the back stress evolution equation is obtained from eq.(11)

$$\frac{d\sigma_B}{dt} = f(p)\dot{\varepsilon}^p - \sigma_B \dot{p} \tag{12a}$$

with the function expression of f(p)

$$f(p) = E\left[2(\beta - \beta_3)\alpha_p + 3\beta_1\alpha_p^2 + \beta\alpha_p^{-2} + 5\beta_3\alpha_p^4\right] \tag{12b}$$

and the definition of accumulated Logarithmic viscoplastic strain p, and its α_p

$$\dot{p} = \left|\dot{\varepsilon}^p\right| = \left|\dot{\lambda}_p\right| / \lambda_p \tag{12c}$$

$$\alpha_p = \exp(p) \tag{12d}$$

The evolution equation (12) makes the unified model be able to describe viscoplastic material behaviors during loading/unloading and cyclic loading. However, some recently proposed models (Boyce, Arruda and Jayachandran, 1994; Wu and van der Giessen, 1995) are still lack of the ability.

In summary, the proposed constitutive model consists of the equation set of eq.(8) (9) (10a) combining with eq.(11) and (12), which describes the viscoelastic behavior with viscoplastic behavior in a unified way.

THE DETERMINATION OF MATERIAL CONSTANTS

The Young's modulus E can be taken from the linear range of initial deformation stage of polymer testing in different temperatures. For the most polymer materials under glass transition temperature, the Young's modulus is linearly dependent on temperature, and weakly dependent on strain rates. The reference strain rate $\dot{\varepsilon}_0$ can be determined by initial yielding points at different strain rates. The constants $D_0(T)$, $D_{SS}(T)$ also have linear relationship with temperatures. After $\dot{\varepsilon}_0$ is determined, $D_0(T)$, $D_{SS}(T)$ and constant b can be determined by a master curve which is obtained by testing data at small deformation stage at different temperatures. Finally, the constants β, β_1, β_3 in back stress σ_B can be obtained by a least square program from a stress strain curve at large deformation stage. The master curve of the back stress can be also found out to verify the temperature dependence of the back stress.

MODEL VERIFICATION BY POLYCARBONATE FILM TESTS

The polycarbonate film strips with 22 millimeter in length, 6 millimeter in width, and 50 micron thickness were tested by a 6-axis mini fatigue tester (Lu, Qian and Liu, 1997) at different strain rates and different temperature range from 25°C to 80°C. The elastic modulus E decreases linearly with the increase of test temperature T. The relationship was found as following

$$E(T) = 3692 - 5.64T\ (^\circ K),\ (MPa) \tag{13}$$

Constant $\dot{\varepsilon}_0$ was 0.18 (1/sec), determined from tensile tests with two strain rates. The $D_{SS}(T)$ then was determined by the procedure described in above section. It also decreases with the increase of temperature. The relationship was

$$D_{SS}(T) = 3680.0 - 8.012T\ (^\circ K),\ (MPa) \tag{14}$$

The curve of $D_0(T)$ with temperature is parallel to that of $Dss(T)$ in current case. It was

$$D_0(T) = 6302.5 + D_{SS}(T) \tag{15}$$

The master curve of D(T, t) with the change of Newtonian time t is shown in Fig. 2 for different testing temperatures. The slope of the curve gave the constant b=0.569 (1/sec). The master curve of back stress σ_B was also obtained, as shown in Fig. 3. The material constants β, β_1, β_3 were determined from the curve via a least square program

$$\beta = 1.5235E-3, \quad \beta_1 = 0 \quad ,\beta_3 = 3.1087E-3 \tag{16}$$

The model prediction is presented in Fig. 4 for the small deformation stage. The excellent agreement between model predictions and experimental data was achieved for four testing temperatures. The full range deformation behaviors are shown in Fig. 5. The model predicts the experimental data well, including strong orientation hardening in large deformation stage. A little difference between experimental results and model predictions lies in the region of necking formation, where the calculation of stress and strain from experimental data is not accuracy. The reason is that different material points within the necking region are subjected to different deformation paths during the necking formation. And the stress-strain curve in the region is a kind of averaging of non-uniform deformation within the region. Therefore, no further attempt was made to improve the model prediction in the region.

SUMMARY AND CONCLUSIONS

A unified constitutive model is proposed with small deformation viscoelastic behavior and large deformation viscoplastic behavior, which is based on molecular chain deformation mechanisms. The temperature dependence of drag stress and back stress is well established from the same deformation mechanism. The material constants are then determined by master curves and a consistent procedure. The excellent agreement between experimental data of polycarbonate film tests and model predictions is achieved. The three dimensional formulation is complicated due to large deformation framework and orientation averaging approach, which will be given in the conference presentation.

ACKNOWLEDGMENT

The support of the SRC (Semiconductor Research Corporation) and the NSF PFF (Presidential Faculty Fellow) Award to S. Liu is gratefully acknowledged.

REFERENCES

Argon, A. S. 1973 " A Theory for the Low-Temperature Plastic deformation of Glassy Polymers", Phil. Mag., 28, 839-865.

Boyce, M. C., Arruda, E. M., and Jayachandran R. 1994. "The Large Strain Compression, Tension, and Simple Shear of Polycarbonate", Poly. Eng. Sci., 34, 716-725.

Carmak, M. and Lee, S. W. 1995. "Molecular Mechanism of Neck Formation in Uniaxial Stretched Poly (ethylene naphthalate) Films", Polymer, 36, 4039-4054.

Ferry, J. D. 1980., Viscoelastic Properties of Polymers, 3rd Edition, John Wiley & Sons.

Frost H. J. and Ashby M. F. 1982. Deformation Mechanism Maps, Oxford Press, London, 1-12.

Flory P. J. and Rehner, J. Jr. 1943. "Statistical Mechanics of Cross-Linked Polymer Networks. I. Rubberlike Elasticity", J. Chem. Phys., 11, 512-520.

Flory, P. J. 1977. "Theory of Elasticity of Polymer Network. the Effect of Local Constraints on Junctions", J. Chem. Phys., 66, 5720-5729.

Hasan O. A. and Boyce M. C. 1995. " A Constitutive Model for the Nonlinear Viscoelastic Viscoplastic Behavior of Glassy Polymers", Poly. Eng. Sci., 35, 331-344.

Hutchinson, J. W. and Neale, K. W. 1983. "Neck Propagation", J. Mech. Phys. solids, 31, 405-426.

James, H. M., and Guth, E. 1943. "Theory of the Elastic Properties of Rubber", J. Chem. Phys., 11, 455-481.

Lu M., Qian Z., and Liu S. 1997. "Thermo-Mechanical Testing of Thin Films", Proc. of SEM Spring Conference, Bellevue, WS

Mark J. E. 1996. "Ceramic-Reinforced Polymers and Polymer-Modified Ceramics", Poly. Eng. Sci., 36, 2905-2920.

Tobolsky A. and Eyring H. 1943, J. Chem. Phys., 11, p125.

Tomita Y. and Tanaka S. 1995. "Prediction of Deformation Behavior of Glassy Polymers Based on Molecular Chain Network Model", Int. J. Solids Structures, 32, 3423-3434.

Treloar, L. R. G. 1976. "The Mechanics of Rubber Elasticity", Proc. R. Soc. Lond., A351, 301-330.

Waas A. M. and Wineman A. 1996. "Failure Criteria for Composites", Air Force Ann. Report, AFOSR 49620, 1-30., Ann Arbor, Michigan.

Wu, P. D. and van der Giessen, E. 1993. "On Improve Network Models for Rubber Elasticity and Their Applications to Orientation Hardening in Glassy Polymers", J. Mech. Phys. Solids, 41, 427-456.

Zhengfang Qian, Minfu Lu, and Sheng Liu 1997. "Constitutive Modeling of Polymer Films Based on A Molecular Chain Network Theory", ASME InterPACK'97, 960-968, 1997.

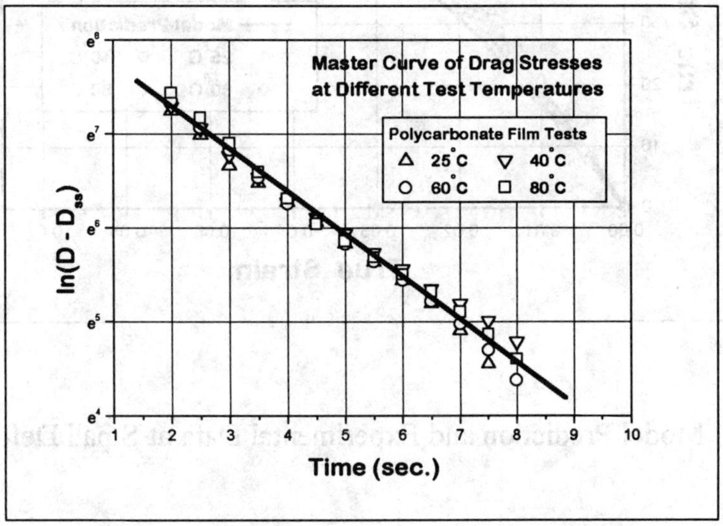

Figure 2. Master Curve of Drag Stress

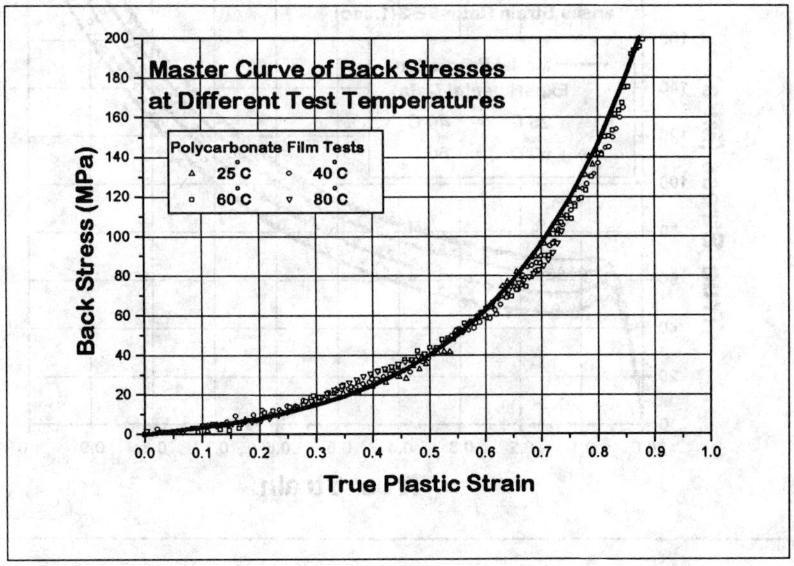

Figure 3 Master Curve of Back Stress

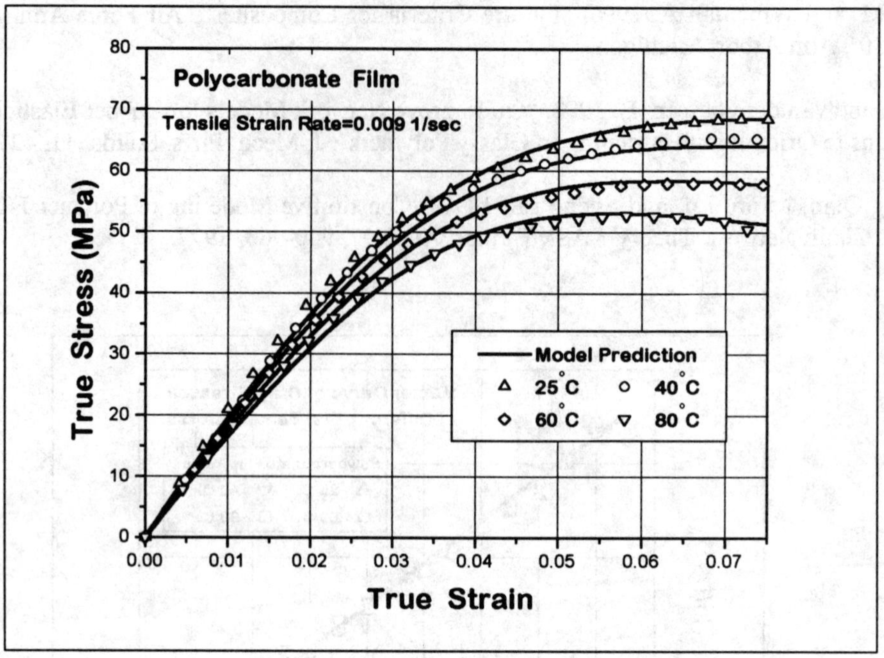

Figure 4. Model Prediction and Experimental Data at Small Deformation Stage

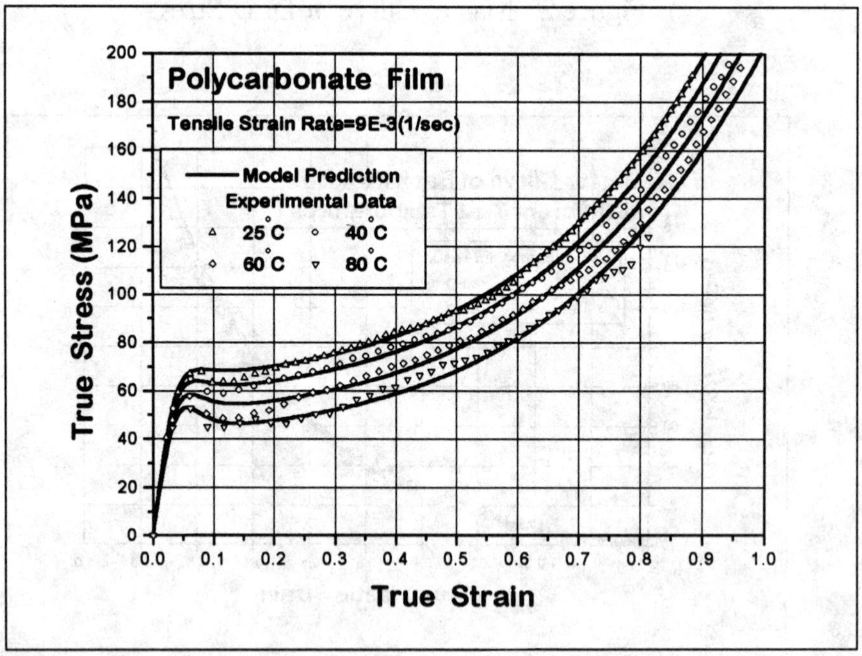

Figure 5 Model Prediction and Experimental Data for Tensile Tests

Effect of Interphase on the Transverse Properties of Unidirectional Fiber Reinforced Composites

S. FREEMAN AND P. K. MALLICK

ABSTRACT

This paper presents an analysis on the effect of interphase characteristics, such as interphase modulus and interphase thickness, on the transverse modulus, maximum principal stress and maximum strain energy in unidirectional continuous fiber composites subjected to loads normal to the fiber direction. In addition to the interphase characteristics, the fiber volume fraction and fiber-matrix modulus ratio are also varied. The transverse properties of the composite are calculated using the ABAQUS finite element software. A design of experiments approach is used to investigate the most important parameters affecting the transverse properties.

INTRODUCTION

Many investigators have noted the presence of a thin but distinct interphase at the fiber-matrix interface in fiber reinforced composites. The interphase surrounds the fibers and its properties may be different from the matrix properties. In polymer matrix composites, the difference in properties may be due to the difference in crosslink densities if the matrix is a thermoset polymer and due to the difference in crystallinity if the matrix is a thermoplastic polymer. Interphase material found in metal matrix composites is due to the reaction that takes place between the fibers and the matrix at the high processing temperatures used and it can be significantly different from the matrix in its alloy composition and physical characteristics.

While an interphase may naturally arise in many fiber reinforced composites, its thickness, composition and properties cannot be easily controlled. On the other hand, there had been several attempts to coat the fibers with a material that acts as an interphase and improves, most notably, the fracture toughness of the composite. The interphase material properties, in this case, can be controlled to produce properties that are desirable for a particular application.

S. Freeman, Altair Computing, Troy, MI 48084

P.K. Mallick, Department of Mechanical Engineering, University of Michigan-Dearborn, Dearborn, MI 48128

In recent years, a number of papers have been published addressing the thermo-mechanical stress analysis of fiber reinforced composites with coated fibers. In these composites, the coating acts as an interphase between the fibers and the matrix. Low et. al. (1995a) have shown that an elastomeric interphase can significantly reduce the tensile residual thermal stresses in a carbon fiber/epoxy composite. In another study, Low et.al. (1995b) have evaluated the residual thermal stresses as well as the effective composite properties as a function of fiber volume fraction, interphase thickness and interphase modulus. Carman et. al. (1993) used a concentric cylinder model to analyze the stress state in a coated fiber reinforced composite subjected to transverse loading. Their analysis indicates that, for a given fiber volume fraction, there is an optimum interphase modulus which produces the minimum stress variation around the interface between the matrix and the interphase. For very small fiber volume fractions (typically less than 20 percent), the optimum interphase modulus is close to that of the matrix modulus. For large fiber volume fractions (typically greater than 50%), the optimum interphase must be very compliant and its modulus should be less than one-third of the matrix modulus; however, the optimum value, in this case, increases with increasing interphase thickness.

The objective of this study is to examine the effects of fiber volume fraction, fiber modulus, interphase modulus as well as interphase thickness on the maximum principal stress, maximum strain energy and transverse modulus of a unidirectional continuous fiber composite subjected to transverse loading. A unit cell approach is used for the arrangement of fibers in the composite. Unlike the concentric cylinder model, the unit cell approach allows for the fiber-fiber interaction. The maximum principal stress and the maximum strain energy are calculated, since these two parameters are of interest in failure prediction of composite materials.

FINITE ELEMENT MODELING

Following assumptions were made in the finite element modeling :

(1) fibers in the composite are continuous, unidirectionally oriented and arranged in a regular square unit cell
(2) perfect bonding exists between the fibers and the interphase as well as between the interphase and the matrix
(3) the composite is free of any residual thermal or mechanical stresses
(4) fibers have a round cross section of diameter d_f , and,
(5) the interphase surrounding the fibers has a hollow round cross section of thickness t_i
(6) the interphase properties are constant throughout the interphase thickness.

Since a square unit cell is assumed, the unit cell dimension a can be calculated from the fiber volume fraction v_f using the following equation

$$a = \frac{d_f}{2}\sqrt{\frac{\pi}{v_f}}$$

In this analysis, a unit fiber diameter was assumed and the fiber volume fraction was varied.

Finite element models were created using Hypermesh, a finite element pre/post processor, and the analysis is performed using the ABAQUS finite element solver. Since the packing arrangement is symmetric, only one quarter of a representative unit cell was modeled. A sample of the finite element mesh is shown in Fig. 1. Standard 3-dimensional 8-node brick elements with reduced integration and hourglass control were used. In some locations, 6-node triangular prism elements were used to adjust the mesh density. Depending on the fiber volume fraction, the number of elements varied between 975 to 1180 and the number of nodes between 1290 to 1554.

The fibers in the unidirectional fiber reinforced composite are oriented in the z direction and the fiber cross section lies in the xy plane. The loading is in the x direction, which is transverse to the fiber direction. A unit thickness was assumed in the z-direction. A plane strain condition was assumed . The displacement boundary conditions used in the analysis are $u_x = 0$ at $x = 0$, $u_x = 0.1$ at $x = a$, and $u_y = 0$ at both $y = 0$ and $y = a$. Thus , the normal displacement in the transverse direction, which in this case is the x direction, is 0.1 units.

The matrix modulus was assumed to be 3.45 GPa, which represents a typical epoxy matrix. The fiber-to-matrix modulus ratios (E_f / E_m) are 5, 20, 60 and 200, representing polyester fiber, E-glass fiber, high strength carbon fiber and high modulus carbon fiber, respectively. The interphase-to-matrix modulus ratio (E_i / E_m) is either 0.02, 0.3, 1 or 4. The Poisson's ratio of the fibers is 0.2 and that for both the matrix and the interphase material is 0.35.

The finite element analysis produced the stress distribution, maximum stresses and strains, maximum principal stress and maximum strain energy. It also produced the normal force in the x-direction that creates the prescribed displacement of 0.1 units at x = a. The remote normal stress σ_o in the x-direction was calculated by dividing this normal force with the cross sectional area. The transverse modulus E_T of the composite was calculated by dividing the remote normal stress σ_o with the normal strain ε_o in the x-direction. Finally, the remote strain energy SE_o was calculated using the following equation

$$SE_0 = \frac{1}{2}\sigma_o \varepsilon_o$$

The maximum principal stress and the maximum strain energy were normalized by dividing them with the corresponding remote values and the transverse modulus was normalized by dividing it with the matrix modulus. These values are referred to as normalized maximum principal stress, normalized maximum strain energy and normalized transverse modulus in the following sections.

RESULTS

NO INTERPHASE

The first set of results refer to a condition that includes no interphase. In this set, three different volume fractions are considered, namely 25, 40 and 50%. The pertinent output values are shown in Table I. Both σ_{xx} and σ_{yy} are maximum at the fiber-matrix interface at 0° location relative to the loading direction. The shear stress τ_{xy} is also maximum at the interface, but at a 50° angle relative to the loading direction. The shear stress is zero at the location of the maximum normal stresses. The maximum principal stress occurs at the interface at the 0° location. The maximum strain energy occurs in the matrix, also at the 0° location, but near the end of the unit cell.

TABLE I - TRANSVERSE PROPERTIES - NO INTERPHASE

v_f	E_f/E_m	Maximum Principal Stress (σ_{max})	Max. ε_{xx}	Maximum Strain Energy (SE_{max})	σ_{max}/σ_o	SE_{max}/SE_0	E_T/E_m
0.25	5	1000	0.2	91.7	1.26	2.85	2.04
	20	1210	0.24	135	1.35	2.82	2.37
	60	1270	0.26	149	1.38	2.87	2.36
	200	1300	0.26	154	1.40	2.93	2.39
0.40	20	2130	0.44	431	1.44	4.08	3.01
	60	2320	0.49	512	1.47	4.56	3.20
	200	2400	0.50	546	1.49	4.75	3.27
0.50	5	2090	0.42	403	1.38	3.32	2.75
	20	3240	0.67	997	1.56	5.99	3.77
	60	3680	0.77	1300	1.61	7.12	4.14
	200	3870	0.81	1430	1.64	7.56	4.29

The results in Table I show that the maximum principal stress, the maximum strain energy as well as the transverse modulus of the composite increase with increasing fiber modulus as well as fiber volume fraction. Similar results were obtained for the maximum principal stress and the transverse modulus by Adams and Doner (1967), who used finite difference analysis to determine the effect of transverse normal loading on the stresses in a unidirectional continuous fiber composite. The composite configuration and boundary conditions are the same as those assumed in this analysis. However, it is noted that the normalized maximum principal stress is somewhat lower than that reported by Adams and Doner. The ratio of transverse modulus to matrix modulus, E_T/E_m, shows a better agreement, particularly at high volume fractions.

WITH INTERPHASE-DESIGN OF EXPERIMENTS

The objective of this part of the study is to determine the relative influence of the following factors on the maximum principal stress, maximum strain energy and transverse modulus of the composite under transverse tensile loads:
(1) Fiber volume fraction (v_f = 0.25, 0.40, 0.50)
(2) Fiber modulus to matrix modulus ratio (E_f/E_m = 20, 60, 200)
(3) Interphase modulus to matrix modulus ratio (E_i/E_m = 0.02, 0.3, 4)
(4) Interphase thickness (t_i/d_f = 0.05, 0.10, 0.20)

Since there are four factors and three levels for each factor, a full factorial analysis involves 3^4 or 81 simulations. Instead of conducting a full factorial analysis, it was decided to use a Design of Experiments approach that reduces the number of simulations needed to extract the main effects of each factor. The statistical design chosen was L27 and involves only 27 simulations. The results of these simulations are summarized below:

(1) Fiber volume fraction has a very significant effect on all three responses of interest (Fig. 2a). Both normalized maximum principal stress and the normalized transverse modulus behave in a nearly linear fashion with respect to the changes in fiber volume fraction. The maximum strain energy shows a non-linear trend , increasing moderately as fiber volume fraction is increased from 25 to 40 percent and increasing rapidly as the fiber volume fraction is increased from 40 to 50 percent.

(2) The effect of the fiber modulus is not very significant at high fiber-to-matrix modulus ratios (60 and above). Its effect becomes more significant at low fiber-to-matrix modulus ratios, particularly on the maximum strain energy (Fig. 2b).

(3) The effect of the interphase-to-matrix modulus ratio is not significant at high interphase modulus values (Fig. 2c). Its effect becomes more significant at low interphase modulus values, particularly on the maximum strain energy and the normalized transverse modulus.

(4) Interphase thickness has a significant effect on the normalized maximum principal stress (Fig. 2d). Its effect on the maximum strain energy and the normalized transverse modulus is relatively small.

EFFECT OF INTERPHASE MODULUS AND INTERPHASE THICKNESS

The effects of the interphase modulus on the maximum principal stress, transverse modulus and strain energy were investigated for an interphase thickness of 0.05 units, i.e., $t_i / d_f = 0.05$. The ratio of fiber modulus to matrix modulus, i.e., E_f / E_m, was either 5, 20, 60 or 200 and the ratio of interphase modulus to matrix modulus, i.e., E_i / E_m, was either 0.02, 0.3, 1 or 4. The results shown in Fig. 3 are for 25% fiber volume fraction. Similar results are obtained for 50% fiber volume fraction. The following observations are made from this study:

(1) The maximum principal stress increases with increasing E_i / E_m ratio (Fig. 3a). If the interphase modulus is lower than the matrix modulus, the maximum principal stress with the interphase is lower than that without the interphase. If the intephase modulus is greater than the matrix modulus, the maximum principal stress with the interphase is greater than that without the interphase.

(2) In this study the load was applied to create a normal displacement of 0.01 units transverse to the fiber direction in all the cases. Since the remote stress, σ_o , required to create this normal displacement also depends on the E_i / E_m ratio, a better comparison is the ratio of the maximum principal stress and the remote normal stress, σ_{max} / σ_o. This normalized maximum principal stress has the highest value at $E_i / E_m = 0.02$. It decreases to a much lower value at $E_i / E_m = 0.3$ and then increases; however, the increase is relatively small when the interphase modulus exceeds the matrix modulus (Fig. 3b).

(3) The normalized transverse modulus or E_T/E_m increases rapidly as the E_i / E_m ratio is increased from 0.02 to 0.3 (Fig. 3c) ; however, for interphase modulus greater than the matrix modulus, there is only a moderate increase in the normalized transverse modulus. Also to be noted is that the higher the fiber modulus, the higher is the E_T / E_m ratio.

(4) The normalized maximum strain energy or SE_{max} / SE_o has the maximum value at the lowest interphase modulus and decreases with increasing interphase modulus (Fig. 3d). For interphase modulus greater than the matrix modulus, there is a small increase in the normal-

TABLE II - EFFECT OF MATRIX MODULUS ON THE TRANSVERSE PROPERTIES (E_i/E_m = 0.3, t_i/d_f = 0.05)

E_f (GPa)	E_m (GPa)	E_f/E_m	σ_{max}/σ_o	SE_{max}/SE_o	E_T/E_m
207	1.035	200	1.37	8.44	2.17
207	3.450	60	1.36	8.26	2.15
207	10.350	20	1.33	7.80	2.10
690	3.450	200	1.37	8.46	2.17
69	3.450	20	1.33	7.76	2.10

ized maximum strain energy. The normalized maximum strain energy ratio also increases with increasing fiber modulus.

The next study examined the effect of interphase thickness on the normalized maximum principal stress, normalized transverse modulus as well as the normalized maximum strain energy. The fiber volume fraction was 25% and the fiber modulus to matrix modulus ratio was 60. The ratio of the interphase thickness to fiber diameter was either 0.05, 0.1 or 0.2. For E_i/E_m = 0.02, the normalized maximum stress ratio increases significantly with increasing interphase thickness; however, for the other two E_i/E_m ratios, including 0.3, there is no significant effect of the interphase thickness. The modulus change with increasing interphase thickness follows the predictable trend , i.e., with interphase modulus lower than the matrix modulus, the transverse modulus of the composite decreases with increasing interphase thickness and with interphase modulus greater than the matrix modulus, it increases with increasing interphase thickness. Similar trend is also observed with the normalized maximum strain energy.

For v_f = 50%, the normalized maximum principal stress also increases significantly with increasing interphase thickness for E_i/E_m = 0.2 and does not change much for either E_i/E_m = 0.3 or 4 (Fig. 5). The change in normalized modulus follows the predictable trend. The most interesting result is found in the effect on the normalized strain energy for E_i/E_m = 0.02. In this case, the lowest value occurs at an interphase thickness-to-fiber diameter ratio of 0.1.

In the above studies, the matrix modulus was 3.45 GPa and the fiber modulus was varied to obtain various E_f/E_m ratios. Three analyses were performed in which the fiber modulus was 207 GPa and the matrix modulus was either 1.035, 3.45 or 10.35 GPa, thus producing E_f/E_m ratio of either 200, 60 or 20. The interphase-to-matrix modulus ratio was 0.3 and the interphase thickness-to-fiber diameter ratio was 0.05. The results indicate that the matrix modulus has a small effect on the three transverse properties considered (Table II).

CONCLUSIONS

This study has shown that the two most important factors in controlling the maximum principal stress, maximum strain energy and transverse modulus of a unidirectional continuous fiber composite under uniaxial transverse tensile loading are the fiber volume fraction and the interphase modulus. The effect of interphase thickness becomes important at very low interphase modulus values.

REFERENCES

Low, B.Y., S.D.Gardner, C.U.Pittman and R.M. Hackett. 1995a. "A Micromechanical Characterization of Residual Thermal Stresses in Carbon Fiber/Epoxy Composites Containing a Non-Uniform Interphase Region," Composites Engineering, 5(4):375-396.

Low, B.Y., K.L. Anderson, , M. Vincent, S. Gardner, C.U. Pittman and R.M. Hackett. 1995b. "Toughened Carbon Fiber/Epoxy Composites: The Relative Influence of an Elastomer Interphase and Elastomer Dispersed in the Matrix," Composites Engineering, 5:437-457.

Carman, G.P., R.C.Averill, K.L.Reifsnider and J.N.Reddy. 1993. "Optimization of Fiber Coatings to Minimize Stress Concentrations in Composite Materials," J. of Composite Materials, 27 (6):589-612.

Adams, D.F.and D.R. Doner. 1967. "Transverse Normal Loading of a Unidirectional Composite," J. of Composite Materials, 1:152-163.

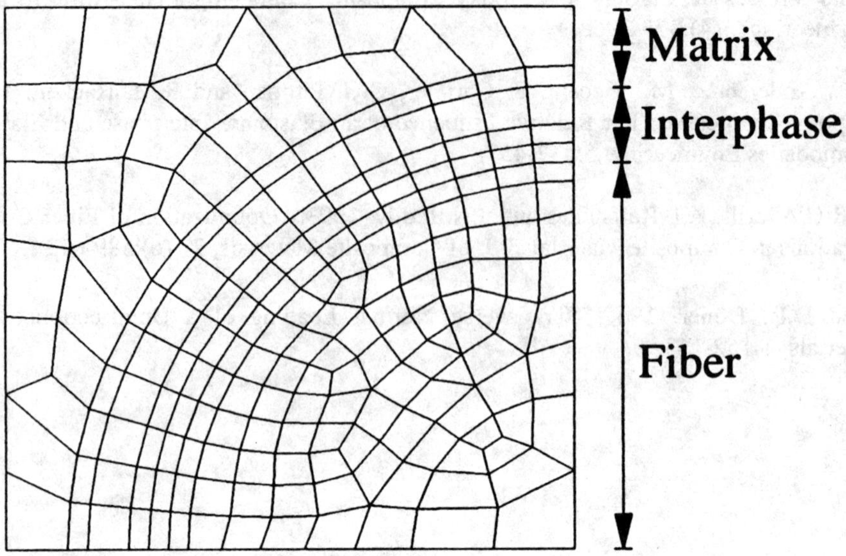

Fig. 1: Finite element model of the quarter unit cell

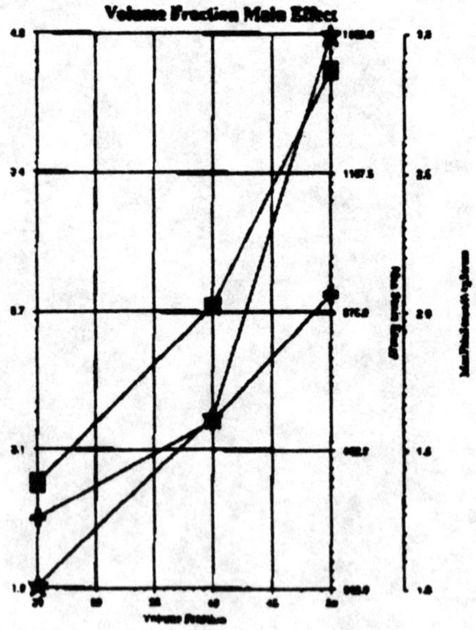

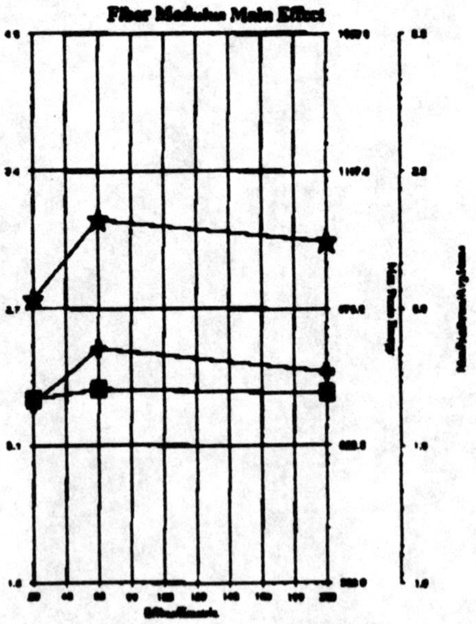

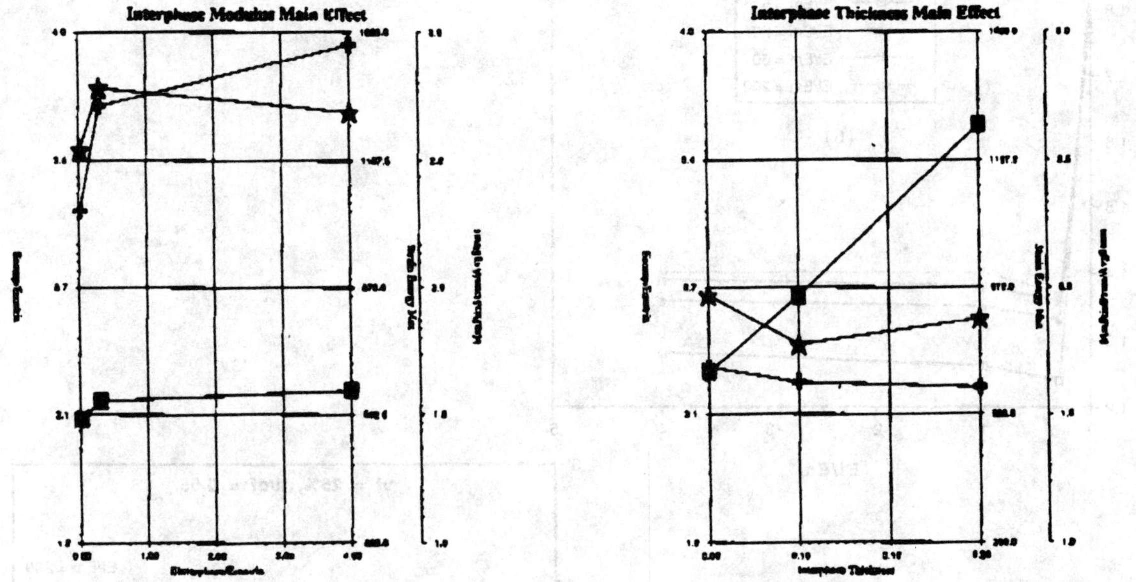

square = MaxPrinStress/AvgStress, diamond = Ecomp/Ematrix, star = MaxStrainEnergy

Fig. 2: Results of the Design of Experiments showing the main effects of fiber volume fraction, fiber modulus, interphase modulus and interphase thickness on the normalized maximum principal stress, normalized transverse modulus and maximum strain energy

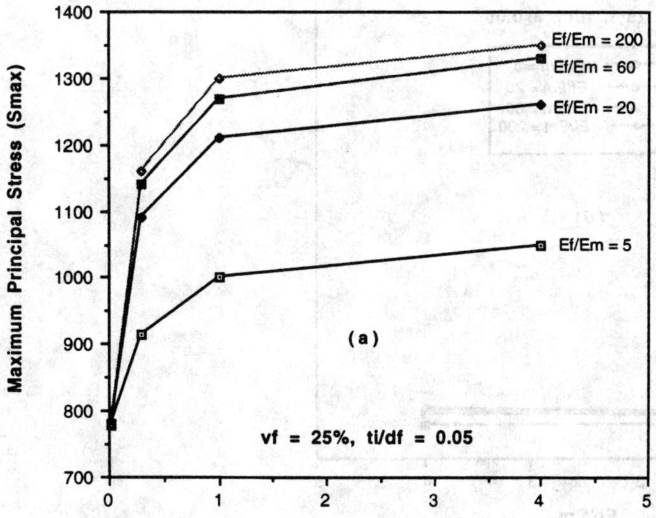

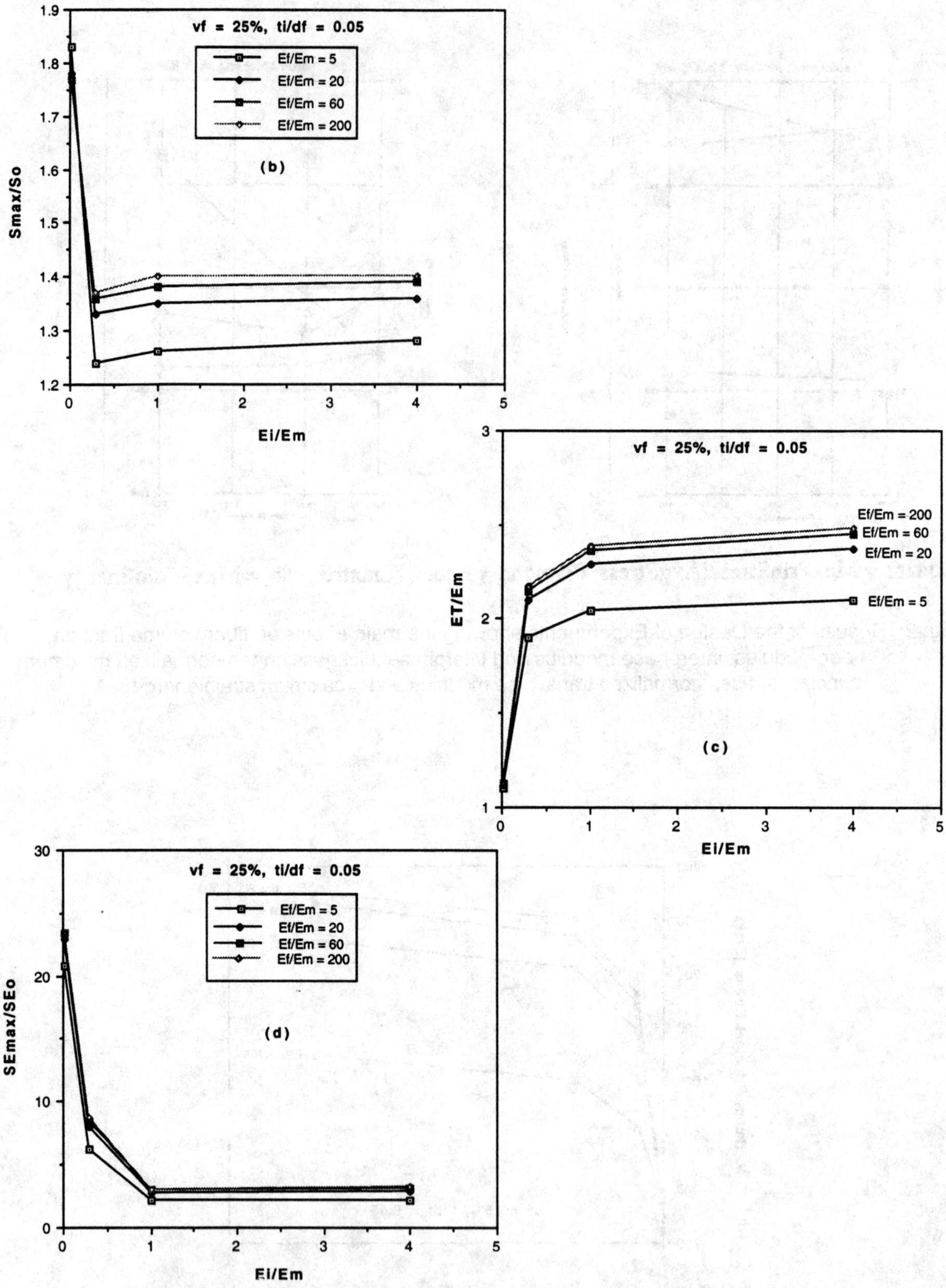

Fig. 3: Effect of interphase modulus on the transverse properties (fiber volume fraction = 25%, interphase thickness/fiber diameter = 0.05)

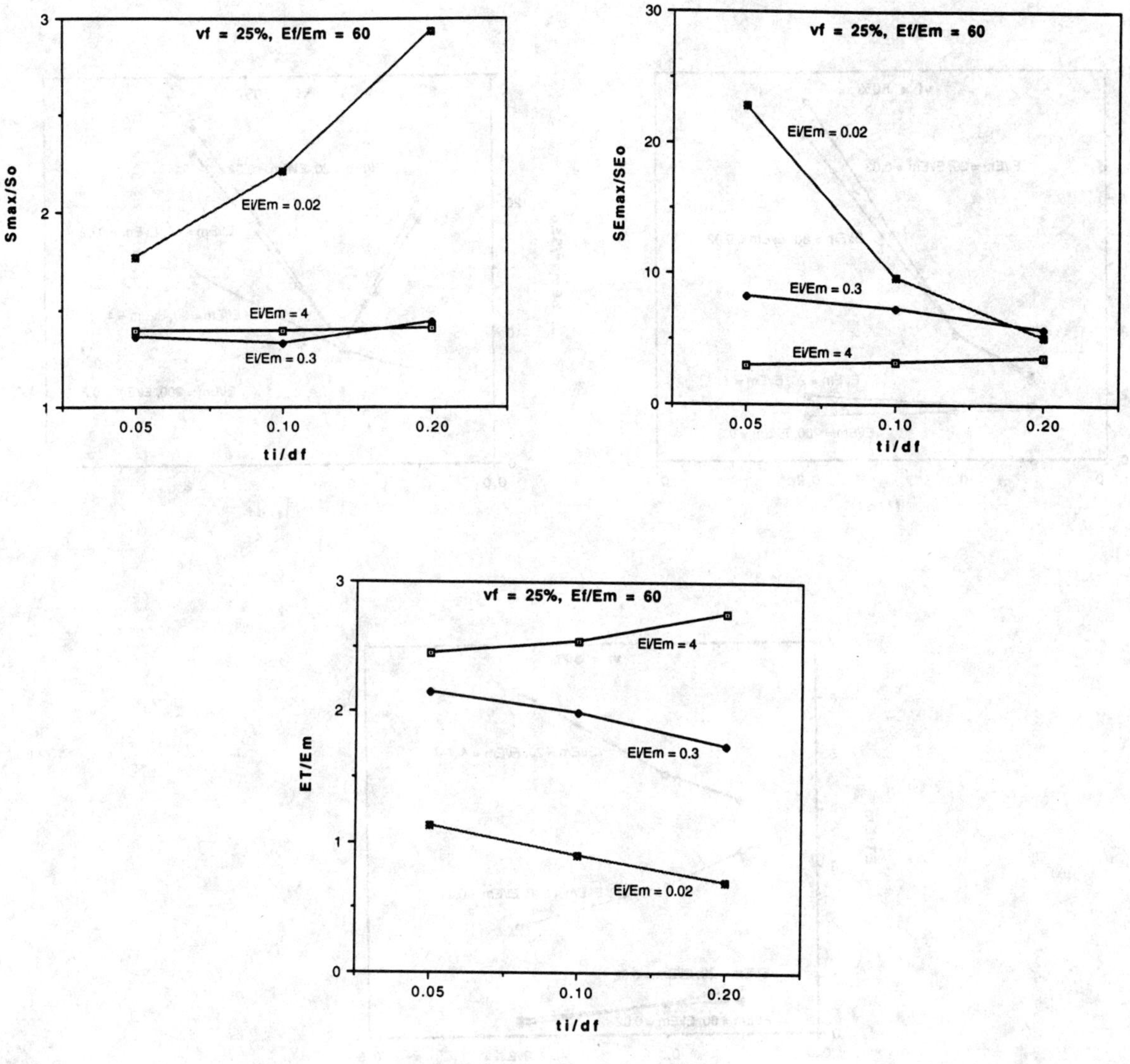

Fig. 4: Effect of interphase thickness on the transverse properties (fiber volume fraction = 25%, fiber modulus/matrix modulus = 60)

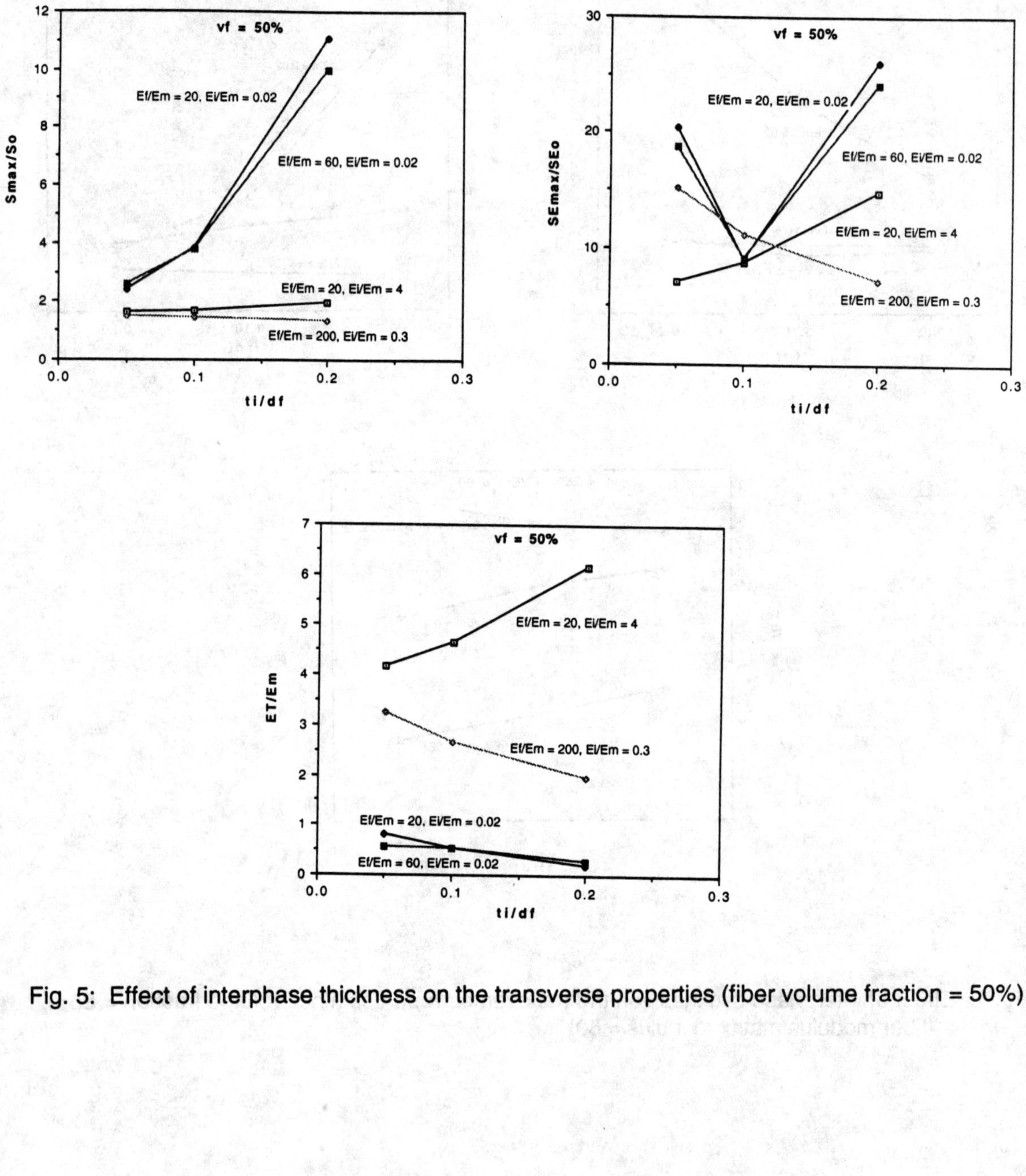

Fig. 5: Effect of interphase thickness on the transverse properties (fiber volume fraction = 50%)

Damage, Fracture and Fatigue

Continuum Description of Anisotropic Damage in Particle Composites

S. YAZDANI AND S. KARNAWAT

ABSTRACT

A damage constitutive model is developed for particle composites within the framework of continuum thermodynamics. It is assumed that the composite material consists of elasto-plastic inclusions within a brittle matrix phase. The model captures an anisotropic degradation of the material elasticity due to cracking and permanent deformations due to plastic flow in the presence of lateral pressure. The model is also capable of predicting the preloading effects on the strength behavior of the material and can address stiffness recovery in load reversal paths.

INTRODUCTION

Particle composites have been studied and researched extensively in the past few years. These composites are characterized as particle inclusions within a surrounding matrix. The overall behavior of the particle composites are dictated not only by the far field stress conditions but also by the mechanical behavior and properties of the composite constituents.

In this study we will focus on a class of particle composites whose inclusions are elasto-plastic and whose matrix is brittle. Using the continuum approach, the effects of size, shape, distribution, and spacing of the inclusions on the overall response of the material will be addressed through the material compliance tensor and its kinetic equation. In doing so, the model is capable of addressing material response under some complicated loading paths including nonproportional stress paths, load reversal paths, and preloading paths.

S. Yazdani, Associate Professor, Department of Civil Engineering and Construction, North Dakota State University, Fargo, ND 58105

S. Karnawat, Engineering Instructor, Turtle Mountain Community College, P. O. Box 340, Belcourt, ND 58316

Alternatively, one might adopt a meso-structural approach to the modeling of particle composites where the response of an inclusion within its surrounding matrix is studied and is subsequently homogenized to a continuum level. In spite of its elegance and direct relevance to the physics of the problem, it has been shown that the applicability of the meso-structural approach is limited to some non complicated loading history, and at present it requires intensive computational efforts.

The mechanics of the two phase composites has been studied by Stankowski (1990) where the inclusions were modeled as an elastic material embedded in an inelastic matrix phase. The inelastic response of the matrix was addressed by using the Drucker-Prager type plasticity model, and an incremental elasto-plasticity scheme was used for the inclusion-matrix interface to account for the adhesion, fracture, slip, and mobilized friction for normal and tangential tractions. In the Stankowski's numerical simulation, the particle composites were first subjected to uniaxial compression up to the prescribed level of loading which was followed by an unloading to the zero stress level. The specimen was subsequently given an abrupt 90 degree rotation and loaded in compression up to failure. A similar sequence of loading was conducted in tension. The study reported that the preloading reduces the load carrying capacity of the material when it is subjected to a new set of loading paths. This simulation was supported by the experimental study of van Mier (1984) on concrete (aggregate-mortar composites) cubes.

In this paper, a general continuum theory of particle composites is given. Utilizing the dissipation inequality, a flow potential incorporating effective damage parameters and a plasticity parameter are developed. The formulation is limited to small and isothermal deformations and to rate-independent processes. Isotropic or anisotropic behavior can be modeled within this formulation utilizing the current compliance tensor, which stores the components of induced damage.

GENERAL THEORY

The general formulations derived in this paper come from the widely accepted approach of thermodynamics with internal variables. A number of authors have taken this approach for the modeling of damage and damage evolution in brittle solids (Yazdani and Karnawat, 1996; Hansen and Schreyer, 1995; Yazdani and Schreyer, 1990; Ortiz, 1985; Ortiz and Popov, 1982).

With the assumptions of a purely mechanical theory and isothermal, rate-independent, and infinitesimal deformations in the stress space, the Gibbs free energy per unit volume, G, is given as

$$G = G(\sigma, \tilde{k}_I, \tilde{k}_{II}, \tilde{\gamma}) \tag{1}$$

where $\tilde{k}_I$ and $\tilde{k}_{II}$ are the effective damage variables in modes I and II, respectively, and $\tilde{\gamma}$ is referred to as an effective plasticity parameter. The second law of thermodynamics can be expressed by the Clausius-Duhem inequality as

$$\dot{G}(\sigma,\tilde{k}_I,\tilde{k}_{II},\tilde{\gamma})-\dot{\sigma}:\varepsilon\geq 0. \tag{2}$$

Equation (2) can be rewritten in an expanded form as

$$\left(\frac{\partial G}{\partial \sigma}-\varepsilon\right):\dot{\sigma}+\frac{\partial G}{\partial \tilde{k}_I}\dot{\tilde{k}}_I+\frac{\partial G}{\partial \tilde{k}_{II}}\dot{\tilde{k}}_{II}+\frac{\partial G}{\partial \tilde{\gamma}}\dot{\tilde{\gamma}}\geq 0. \tag{3}$$

Utilizing the standard thermodynamic arguments, two relations are obtained as

$$\varepsilon=\frac{\partial G}{\partial \sigma} \tag{4}$$

and

$$d_s=\frac{\partial G}{\partial \tilde{k}_I}\dot{\tilde{k}}_I+\frac{\partial G}{\partial \tilde{k}_{II}}\dot{\tilde{k}}_{II}+\frac{\partial G}{\partial \tilde{\gamma}}\dot{\tilde{\gamma}}\geq 0 \tag{5}$$

where $\frac{\partial G}{\partial \tilde{k}_I}$, $\frac{\partial G}{\partial \tilde{k}_{II}}$, and $\frac{\partial G}{\partial \tilde{\gamma}}$ are the thermodynamic forces, and $\dot{\tilde{k}}_I$, $\dot{\tilde{k}}_{II}$, and $\dot{\tilde{\gamma}}$ define the conjugate effective fluxes. For all admissible processes, the inequality presented by Equation (5) must be satisfied.

To progress further, a particular class of material has to be specified. A constitutive relationship that relates the elastic component of deformation to the stress tensor by means of the compliance tensor, **C**, can be written as

$$\varepsilon=\frac{\partial G(\sigma,\tilde{k}_I,\tilde{k}_{II},\tilde{\gamma})}{\partial \sigma}=\mathbf{C}(\tilde{k}_I,\tilde{k}_{II}):\sigma+\varepsilon^i(\tilde{k}_I,\tilde{k}_{II},\tilde{\gamma}) \tag{6}$$

where ε^i represents the component of the strain tensor that does not vanish upon unloading. Therefore, ε^i is referred to as an inelastic strain tensor. The form of Equation (6) also implies the dependence of the material compliance on the state of microcracking in both tension and compression.

In this work, we assumed that $\tilde{k}_I$ and $\tilde{k}_{II}$ are uncoupled from the outset, which means $\tilde{k}_I$ has no direct influence over mode II terms and $\tilde{k}_{II}$ does not affect mode I terms. With this, we proposed the following additive decomposition of the material compliance:

$$\mathbf{C}(\tilde{k}_I,\tilde{k}_{II})=\mathbf{C}^0+\mathbf{C}_I^c(\tilde{k}_I)+\mathbf{C}_{II}^c(\tilde{k}_{II}) \tag{7}$$

in which $\mathbf{C}_I^c$ and $\mathbf{C}_{II}^c$ denote the added flexibility tensor due to active microcracks in modes I and II, respectively. Similarly, an inelastic strain tensor can separated into three parts as

$$\varepsilon^{i}(\tilde{k}_{I},\tilde{k}_{II},\tilde{\gamma})=\varepsilon^{i}_{I}(\tilde{k}_{I})+\varepsilon^{i}_{II}(\tilde{k}_{II})+\varepsilon^{p}(\tilde{\gamma}) \tag{8}$$

where ε^{i}_{I} and ε^{i}_{II} denote the inelastic strain tensors due to inelastic damage processes in modes I and II, respectively, and ε^{p} represents the plastic strain tensor. Integrating Equation (6) with respect to σ, then substituting Equations (7) and (8), the general form of the Gibbs potential is obtained as

$$\begin{aligned} G = & \frac{1}{2}\sigma:C^{0}:\sigma+\frac{1}{2}\sigma:C^{c}_{I}(\tilde{k}_{I}):\sigma+\frac{1}{2}\sigma:C^{c}_{II}(\tilde{k}_{II}):\sigma+\varepsilon^{i}_{I}(\tilde{k}_{I}):\sigma \\ & +\varepsilon^{i}_{II}(\tilde{k}_{II}):\sigma+\varepsilon^{p}(\tilde{\gamma}):\sigma+A^{i}(\tilde{k}_{I},\tilde{k}_{II},\tilde{\gamma}). \end{aligned} \tag{9}$$

The constant of integration, term A^{i}, is decomposed into its additive terms as $A^{i}(\tilde{k}_{I},\tilde{k}_{II},\tilde{\gamma})=A^{i}_{I}(\tilde{k}_{I})+A^{i}_{II}(\tilde{k}_{II})+A^{p}(\tilde{\gamma})$, which represents an inelastic component of the Helmholtz free energy. For infinitesimal deformations and uncoupled theory, such additive decomposition of terms of any order is permissible. Substituting Equation (9) into Equation (5), then taking respective derivatives, the following form of the dissipation inequality is obtained:

$$\begin{aligned} d_{s} = & \frac{1}{2}\sigma:\dot{C}^{c}_{I}(\tilde{k}_{I}):\sigma+\frac{1}{2}\sigma:\dot{C}^{c}_{II}(\tilde{k}_{II}):\sigma+\dot{\varepsilon}^{i}_{I}(\tilde{k}_{I}):\sigma+\dot{\varepsilon}^{i}_{II}(\tilde{k}_{II}):\sigma \\ & +\dot{\varepsilon}^{p}(\tilde{\gamma}):\sigma-\dot{A}^{i}_{I}(\tilde{k}_{I})-\dot{A}^{i}_{II}(\tilde{k}_{I})-\dot{A}^{p}(\tilde{\gamma})\geq 0. \end{aligned} \tag{10}$$

To proceed further, let the fluxes in Equation (10) be represented by the following rate-independent linear evolution laws:

$$\dot{C}^{c}_{I}=\dot{\tilde{k}}_{I}\,\mathbf{R}_{I}(\sigma),\ \dot{C}^{c}_{II}=\dot{\tilde{k}}_{II}\,\mathbf{R}_{II}(\sigma) \tag{11}$$

and

$$\dot{\varepsilon}^{i}_{I}=\dot{\tilde{k}}_{I}\,\mathbf{M}_{I}(\sigma),\ \dot{\varepsilon}^{i}_{II}=\dot{\tilde{k}}_{II}\,\mathbf{M}_{II}(\sigma),\ \dot{\varepsilon}^{p}=\dot{\tilde{\gamma}}\,\mathbf{N}(\sigma) \tag{12}$$

where $\mathbf{R}_{I}(\sigma)$ and $\mathbf{R}_{II}(\sigma)$ are the fourth-order response tensors, which determine the direction of damage in modes I and II, respectively. Similarly, inelastic deformations in modes I and II are represented by the second-order response tensors, $\mathbf{M}_{I}(\sigma)$ and $\mathbf{M}_{II}(\sigma)$, respectively, and $\mathbf{N}$ is the second-order response tensor for the plastic flow. Substitution of Equations (11) and (12) into Equation (10) yield the dissipation as

$$d_s = \dot{\tilde{k}}_I\left(\frac{1}{2}\sigma:\mathbf{R}_I:\sigma + \mathbf{M}_I:\sigma - A^i_{I},_{\tilde{k}_I}\right) + \dot{\tilde{k}}_{II}\left(\frac{1}{2}\sigma:\mathbf{R}_{II}:\sigma + \mathbf{M}_{II}:\sigma - A^i_{II},_{\tilde{k}_{II}}\right)$$
$$+\dot{\tilde{\gamma}}\left(\mathbf{N}:\sigma - A^p,_{\tilde{\gamma}}\right) \geq 0. \tag{13}$$

Since $\tilde{k}_I$ and $\tilde{k}_{II}$ are the measures of energy dissipation in modes I and II, then by definition, irreversible character of damage implies

$$\dot{\tilde{k}}_I \geq 0, \text{ and } \dot{\tilde{k}}_{II} \geq 0. \tag{14}$$

Further, in the absence of any internal constraints, as indicated by Ortiz (1985) and Ortiz and Popov (1982), the coefficients of $\dot{\tilde{k}}_I$ and $\dot{\tilde{k}}_{II}$ must be nonnegative, which opens up two possibilities: (1) two independent damage surfaces, one for mode I and one for mode II, can be developed, and a numerical algorithm which invokes two surfaces simultaneously can be constructed (see, e.g. Yazdani and Schreyer, 1990; Ortiz, 1985); and (2) a combined surface can be formed by establishing a functional relationship between $\dot{\tilde{k}}_I$, $\dot{\tilde{k}}_{II}$, and the rate of total effective damage parameter, $\dot{\tilde{k}}$. The benefits of the later approach are well-documented and discussed by Yazdani and Karnawat (1996) thus, considered for the present theory.

With a closer look at the dissipation inequality, given by Equation (13), and for an uncoupled theory, it is plausible to assume that under tensile loading paths, $\dot{\tilde{k}}_{II}$ should be zero, whereas for compressive stress paths, $\dot{\tilde{k}}_I$ should vanish. To accommodate these considerations, simple equations that relate $\dot{\tilde{k}}$ to $\dot{\tilde{k}}_I$ and $\dot{\tilde{k}}_{II}$ are proposed as follows:

$$\dot{\tilde{k}}_I = b_1(\sigma)\dot{\tilde{k}} \text{ and } \dot{\tilde{k}}_{II} = b_2(\sigma)\dot{\tilde{k}} \tag{15}$$

which are constrained by $b_1(\sigma) + b_2(\sigma) = 1$ so that

$$\dot{\tilde{k}} = \dot{\tilde{k}}_I + \dot{\tilde{k}}_{II} \text{ or } \tilde{k} = \tilde{k}_I + \tilde{k}_{II} \tag{16}$$

Although the proposed form of Equation (15) appears to be arbitrary and other forms may equally be valid, its simplicity is nonetheless appealing and physically plausible. The scalar stress-based coefficients, $b_1(\sigma)$ and $b_2(\sigma)$, are responsible for smooth transition from the tensile behavior to compressive behavior as the external loads change from tension to compression. Substitution of Equation (15) into Equation (13) gives

$$d_s = \dot{\tilde{k}}\left\{ b_1\left(\frac{1}{2}\sigma:\mathbf{R}_I:\sigma+\mathbf{M}_I:\sigma - A^i_{I,\tilde{k}_I}\right)+b_2\left(\frac{1}{2}\sigma:\mathbf{R}_{II}:\sigma+\mathbf{M}_{II}:\sigma - A^i_{II,\tilde{k}_{II}}\right)\right\}$$

$$+\dot{\tilde{\gamma}}\left(\mathbf{N}:\sigma - A^p{}_{,\tilde{\gamma}}\right)\geq 0. \tag{17}$$

The irreversibility constraint on damage variables in modes I and II (given by Equation (14)) necessitates that the rate of an effective damage parameter must be nonnegative too, i.e., $\dot{\tilde{k}}\geq 0$.

It is experimentally shown that at low levels of lateral pressure the response of particle composite is controlled by the damage variable, whereas in the presence of large lateral pressure the particle composite behavior is governed by the plasticity parameter. So it is reasonable to assume that at an intermediate level of lateral pressure both damage and plastic flow are present. To reflect the dependence of lateral pressure on the material behavior, the rate of the total effective inelasticity parameter $\tilde{q}$, where

$$\dot{\tilde{q}}=\dot{\tilde{k}}+\dot{\tilde{\gamma}} \text{ or } \tilde{q}=\tilde{k}+\tilde{\gamma} \tag{18}$$

assumes the following relationship with $\dot{\tilde{k}}$ and $\dot{\tilde{\gamma}}$:

$$\dot{\tilde{k}}=a_1(P)\dot{q}, \qquad \dot{\tilde{\gamma}}=a_2(P)\dot{q} \tag{19}$$

with a restriction

$$a_1(P)+a_2(P)=1, \tag{20}$$

so that the total dissipation can split into its damage and plasticity counterparts. After substituting Equation (19) into Equation (17), the following form of the dissipation is obtained:

$$d_s = \dot{\tilde{q}}\left\{a_1\, b_1\left(\frac{1}{2}\sigma:\mathbf{R}_I:\sigma+\mathbf{M}_I:\sigma - A^i_{I,\tilde{k}_I}\right)+\right.$$

$$\left. a_1\, b_2\left(\frac{1}{2}\sigma:\mathbf{R}_{II}:\sigma+\mathbf{M}_{II}:\sigma - A^i_{II,\tilde{k}_{II}}\right)+a_2\left(\mathbf{N}:\sigma - A^p{}_{,\tilde{\gamma}}\right)\right\}\geq 0. \tag{21}$$

The irreversible nature of both $\dot{\tilde{k}}\geq 0$ and $\dot{\tilde{\gamma}}\geq 0$ necessitates that $\dot{\tilde{q}}\geq 0$ (because of Equation (18)), which translates Equation (21) into

$$\left\{ a_1\, b_1 \left(\frac{1}{2} \sigma : \mathbf{R}_\mathrm{I} : \sigma + \mathbf{M}_\mathrm{I} : \sigma - A^{i}_{\mathrm{I}, \tilde{k}_\mathrm{I}} \right) + \right.$$

$$\left. a_1\, b_2 \left(\frac{1}{2} \sigma : \mathbf{R}_\mathrm{II} : \sigma + \mathbf{M}_\mathrm{II} : \sigma - A^{i}_{\mathrm{II}, \tilde{k}_\mathrm{II}} \right) + a_2 \left(\mathbf{N} : \sigma - A^{p}_{, \tilde{\gamma}} \right) \right\} \geq 0. \tag{22}$$

Now, one can define the flow potential or flow surface by simply introducing a positive valued function to force the right-hand side of Equation (22) to 0. Let that positive function be $H(\sigma, \tilde{q})$. Based on this information, flow surface Ψ takes on the following form:

$$\Psi(\sigma, \tilde{q}) = a_1\, b_1 \left(\frac{1}{2} \sigma : \mathbf{R}_\mathrm{I} : \sigma + \mathbf{M}_\mathrm{I} : \sigma \right) + a_1\, b_2 \left(\frac{1}{2} \sigma : \mathbf{R}_\mathrm{II} : \sigma + \mathbf{M}_\mathrm{II} : \sigma \right)$$

$$+ a_2 \left(\mathbf{N} : \sigma \right) - \frac{1}{2} t^2(\sigma, \tilde{q}) = 0 \tag{23}$$

where

$$t^2(\sigma, \tilde{q}) = 2 \left(a_1\, b_1\, A^{i}_{\mathrm{I}, \tilde{k}_\mathrm{I}} + a_1\, b_2\, A^{i}_{\mathrm{II}, \tilde{k}_\mathrm{II}} + a_2\, A^{p}_{, \tilde{\gamma}} + H(\sigma, \tilde{q}) \right) \tag{24}$$

is identified as the hardening-softening function. After employing the standard Kuhn-Tucker loading-unloading criteria (i.e., $\dot{\tilde{q}} \geq 0$, $\Psi \leq 0$, and $\dot{\tilde{q}}\, \Psi = 0$), Equation (23) essentially completes the general formulation of particle composites; however, details must be worked out to fully describe a particular solid. Specific expressions for response tensors $\mathbf{R}_\mathrm{I}$, $\mathbf{R}_\mathrm{II}$, $\mathbf{M}_\mathrm{I}$, $\mathbf{M}_\mathrm{II}$, $\mathbf{N}$; and particular forms for a_1, a_2, b_1, b_2; and the hardening-softening function, t, must be identified.

REFERENCES

Hansen, N. R. and Schreyer, H. L. (1995). “Damage Deactivation,” *J. Appl. Mech.*, ASME, 62, 450.

Ortiz, M. (1985). “A Constitutive Theory for the Inelastic Behavior of Concrete,” *Mech. Mater.*, 4(1), 67.

Ortiz, M. and Popov E. P. (1982). “A Physical Model for the Inelasticity of Concrete,” *Proc. R. Soc. London*, A383, 101.

Stankowski, T. (1990). "Numerical Simulation of Progressive Failure in Particle Composites," Ph.D. dissertation, University of Colorado, Boulder.

van Mier, J. G. M. (1984). "Strain-softening of Concrete Under Multiaxial Loading Conditions," Ph.D. dissertation, University of Eindehoven, The Netherlands.

Yazdani, S. and Karnawat, S. (1996). "A Constitutive Theory for Brittle Solids With Application to Concrete," *Int. J. Damage Mech.*, 5, 93.

Yazdani, S. and Schreyer, H. L. (1990). "Combined Plasticity and Damage Mechanics Model for Plain Concrete," *J. Engng. Mech.*, 116(7), 1435.

Notch Toughness of Fiber Mat Composites

J. LINDHAGEN AND L. A. BERGLUND

ABSTRACT

The term fiber mat composites is meant to describe any composite material based on discrete fibers, often with an orinetation distribution of the fibers which is random-in-the-plane. Glass mat thermoplastics based on polypropylene belong to this group of materials and are becoming increasingly used. In a simplistic way of reasoning, PP is a tough polymer and is expected to show good performance as a composite matrix. In the present work, unnotched and notched strengths at room temperature and at -30°C were investigated. Increases in strength and stiffness with lowered temperature are caused by increased matrix modulus and yield stress. The damage zone at the notch root plays a key role in determining composite toughness. Its size and ability to carry tractions determines the notch toughness of the composite.

INTRODUCTION

The materials treated in the present study, GMT materials (glass mat thermoplastics), are commonly based on inexpensive thermoplastic materialssuch as polypropylene or polyamide reinforced with randomly oriented glassfibers. A review of the mechanical properties of GMT materials has been published by Berglund and Ericson (1995).Low price and rational component production have secured GMT a growing market in recent years, especially in the automotive industry. Large, near-net-shape components with intricate geometries and good mechanical properties can be produced at cycle times of typically 30 seconds. Automobile parts such as seat frames, bumpers and front ends are manufactured from GMT. Ongoing development work aims to improve mechanical properties so that future applications may include structural components subjected to higher loads. Unreinforced polypropylene, with aglass transition temperature of -10°C may show embrittlement at low temperatures. It is therefore of interest to study the low-temperature properties of PP based GMTs.

The objective of this study was to study notched and unnotched strength of fiber mat composites exemplified by two structurally different polypropylene based GMT materials.

EXPERIMENTAL

Two different commercial GMT materials (glass mat thermoplastics) with polypropylene matrix were used. Ahlström RTC, by Ahlström Glassfiber Ltd of Finland, is manufactured by a slurry deposition process similar to paper-making. Its reinforcement consists of evenly distributed

---

J Lindhagen and LA Berglund, Divison of Polymer Engineering, Luleå University of Technology, S-971 87 Luleå, Sweden

single fibers of 12 mm length, with a rather narrow fiber length distribution and with random fiber directions in the plane. The Ahlström material will be referred to as the short fiber material. Symalit FA8 is produced by Symalit AG of Switzerland. It is manufactured by a melt impregnation process in a band press where polypropylene films are laminated with layers of glass-fiber mat of the continuous swirled mat-type. It has continuous looped fiber bundles of varying diameter. The Symalit material will be referred to as the swirled mat material.

Tensile specimens were fabricated from circular plates. Each plate was manufactured from a square blank of raw material in an industrial press, with the use of a purpose-made tool. The following processing parameterswere employed: Heating of the blank: 4 min in a hot-air oven at 250°C. Tool temperature: 60°C. Press force: 883 kN, which corresponds to 18 MPa. Press time: 25 s. The plate diameter was 250 mm.Two tensile specimens were cut from each plate, with their straight edges parallell to one of the edges of the square blank. The full diameter of the plate was used to obtain the longest possible specimens. Specimen width was 40 mm and the length was 250 mm. Specimen thicknesses were 1.7 and 1.9 mm for the swirled mat and the short fiber material, respectively. Most specimens were notched with machined circular center holes of 5, 10 and 15 mm diameter. For each material, five specimens of each notch size were tested together with five unnotched specimens. A servohydraulic tensile testing rig was employed. Tensile data were sampled with a frequency of 2 Hz and stored and treated in a computer. The crosshead speed was 2 mm/min.

A full test series was also performed at a temperature of -30°C. A plywood box, isolated with 10 cm thick expanded polystyrene and with holes top and bottom for the grips, was mounted in the test rig. Liquid nitrogen was let in through an electronically controlled valve. It vapourised immediately, and the tests were thus performed in a liquid nitrogen vapour. Data from a standard PT100 temperature probe mounted inside the box were used to control the valve and hold the temperature at a steady level. All specimens were placed inside the box at least one hour in advance of being tested, after having been kept at -18°C in a freezer overnight. After specimens had been exchanged and the door of the box had been opened, the temperature was allowed to stabilize for at least twenty minutes.

An extensometer could not be used inside the isolating box. Therefore the recorded crosshead movement of the tensile rig must be used to calculate strain for the cold-temperature tests. The crosshead movement was divided by the specimen free length between grips. This method is less accurate, but for consistency it was used also for the room-temperature specimens. The free length of the specimens was 150 mm. Both nominal stresses, calculated from the net cross section area, and remote or gross section stresses will be discussed in this article. The tensile moduli are secant moduli calculated from the stress-strain plots as least-square linear curve fits between 0 and 0.2% strain.

Fiber volume fractions in the specimens were determined for both materials. 40x40 mm pieces were cut from the middle of five tensile specimens of each material. The pieces were weighed, placed in an oven at 500°C for four hours in order to burn off the matrix, and subsequently weighed again. The fiber volume fraction was then calculated based on the densities of fiber and matrix.

RESULTS AND DISCUSSION

STIFFNESS

In the low temperature tests, the tensile moduli of the two materials were higher than at room temperature, see Table I. At -30°C the Young's modulus increased by 32% and 25% for the short fiber and the swirled mat material, respectively. The changes can be explained by the stiffening of the matrix. Polypropylene has a glass transition temperature (Tg) of about -10°C. At -30°C the polypropylene matrix material is therefore in the glassy state. This increases the matrix Young's modulus, which in turn increases the modulus of the composite. The Young's modulus of polymers show a general increase with a decrease in temperature. For polypropylene the change is substantial from room temperature to temperatures below its Tg. Other common fiber composite matrices such as epoxy or polyester resins have a Tg above room temperature. These also show increased modulus at lower temperatures but the effect is smaller. Thomason and Vlug (1996) demonstrated that the modulus is independent of fiber length above a certain length. The present Young's moduli are similar to the data presented by Ericson and Berglund (1992).

STRAIN TO FAILURE

The strains to failure of the two materials are presented in Table I. For both materials, the strain to failure is almost independent of temperature. The small increase at -30°C is not statistically significant even on an 80% confidence level. The swirled mat material showed a lower strain to failure than the short fiber material at both testing temperatures.

Thomason et al. (1996) tested random polypropylene short glass fiber composites at room temperature. An important observation was a strong correlation between average fiber strength and the tensile strength of materials with random-in-the-plane fibre orientation. The data also showed a slight decrease in strain to failure with increased fiber content. However, this decrease can not explain the difference between the materials in the present study. At a given fiber content, the strain to failure was roughly independent of fibre length, in the interval 2.5-12 mm. Interestingly, Gaggar and Broutman (1976) found strain to failure in flexure to be independent of fiber volume fraction and matrix ductility in swirled mat GF-reinforced polyesters. These results indicate that the strain to failure of the present kind of composites is controlled by the fiber strain to failure. Furthermore, neat polypropylene is known to show decreased strain to failure with decreased temperature. In contrast, glass fibres are virtually unaffected by the temperature decrease of interest here. Since our data show the composite strain to failure to be practically insensitive to temperature, the strain to failure of the fibers appears to be the primary parameter controlling ultimate failure.

In view of the discussion above the lower strain to failure of the swirled mat material compared to the short fiber material is probably related to a difference in fiber strain to failure (see also next section).

STRENGTH AND NOTCH TOUGHNESS

Table I presents the strength of the two materials, notched and unnotched, at both testing temperatures. The strengths are presented as net section strengths, since this shows the ability of a material to withstand stress concentration and as remote (gross section) strengths.

At -30C, the unnotched strength increased for both materials. The increase was 19% for the short fiber material and 33% for the swirled mat material. Since the fiber strain to failure is unaffected by a decrease in temperature and the same was observed for the investigated GMT materials, the change in unnotched strength is related to the increase in matrix Young's modulus.

A higher relative increase in strength and work of fracture (area under curve) was observed for the swirled mat material, which is interesting. In the low strain region, the modulus-increase with lowered temperature is not very different for the two materials. However, at higher strains approaching failure, some interesting differences between the two materials develop. The swirled mat material shows a much steeper slope in the high strain region.

In another study damage mechanisms in thin transparent specimens of GMT model composites were studied by transmission optical microscopy, Lindhagen and Berglund (1997). At high strains, GMT materials exhibit considerable material damage. The damage consists of matrix plasticity, debonding at transversely oriented fibres and fiber fractures with associated debonding. The larger low temperature increase in strength and work of fracture of the swirled mat material may be related to the longer efficient fiber lengths. The continuous bundles of the swirled mat is expected to carry tractions more efficiently across the softened damage zone than short fibers.

In unnotched condition, the short fiber material exhibited higher strength than the swirled mat material. In the study by Ericson and Berglund (1992) the relation was the opposite. The present strength of the short fiber material (83 MPa) is similar to the previous study (80 MPa). However, the swirled mat material showed a marked decrease in unnotched strength (70 MPa) compared to the earlier study (92 MPa), although the fiber contents were the same. The strength of GMT with random in-the-plane fiber orientation is controlled by fiber strength, Thomason et al (1996b). The difference must therefore be attributed to a change in the grade of E-glass used in the commercial swirled mat material. Optical microscopy of this material also revealed the presence of very large diameter fibers, which are expected to have low strength.

The strength of the short fiber material decreased with the introduction of circular holes. This was also true for the swirled mat material, although the change was smaller, demonstrating lower notch sensitivity. At -30C the notched and unnotched strengths increased for both materials. However, the increase in notched strength was relatively larger than the increase in unnotched strength, which means that the materials became less notch sensitive at low temperature. At -30C, the swirled mat material was close to insensitive to notches of the size investigated in the present study.

Low notch sensitivities of random glass mat composites at room temperature have been reported elsewhere (Mallick,1992, Shirrell and Onachuk, 1986, Mallick, 1988). Materials with similar types of glass fiber reinforcement but different matrix materials compared to the present study were investigated. Swirled mat composites both with tough thermoplastic and brittle thermosetting matrices exhibited very low notch sensitivity.

MODELING OF NOTCH SENSITIVITY

In a previous study, notched strength predictions at room temperature for GMT materials were made although we were unable to present a satisfactory mechanism-based interpretation of our results, Lindhagen and Berglund (1997b). Four different failure criteria were applied. The advantage with these criteria is that for simple geometries, quantitative predictions of notched strength can be done easily. A disadvantage is that there is no straightforward route to physical interpretation of the results on the micromechanical level. The material constant used to characterize the notch sensitivity, a characteristic length, differs between different criteria. Most importantly, it is not clear how this parameter is related to microstructural parameters and constituent properties of the material.

Due to the limitations with the previous models, this section discusses a route which allows for incorporation of crack-bridging models at the micromechanical level. The notch in the specimen is assumed to lead to the formation of a localized damage zone in the material. This damage zone may be modeled as a localized inelastic band, where parallel, nonlinear springs act between the surfaces of an elastic material. The application of this concept to fracture in continuous fiber ceramic-matrix composites was reviewed by Bao and Suo (1992) and further discussed by Suo with coworkers (1993).

We consider a strip of material with a hole of radius a. The strip is loaded by the remote tensile stress s. Due to the stress concentration at the hole a damage zone is formed at the notch root. It is modeled as a damage band of a certain length L and leading to an opening displacement ∂. The limiting displacement of the springs in the band is ∂_0, which equals the separation of the damage band at failure. σ_0 is the maximum tension in the springs, and is assumed to be the material failure stress for unnotched specimens. Each individual spring is characterized by this behaviour, although different springs may have different displacement at the same level of remote stress. Failure of the spring closest to the notch is usually used as the criterion for total specimen failure.

Cottrell (1963) was able to correlate notch brittleness of materials to the ratio of a material length to notch size. This material length d is

$$d = (k\,\Gamma E)/\sigma_o^2$$

and is a measure of the size of the inelastic deformation region. Γ is the energy required to create a unit area of new fracture surface (work of fracture), and under plane stress conditions, E is the Young's modulus of the elastic substrate, in this case the composite modulus. The prefactor k varies within order of magnitude unity, depending on which bridging law is used.

A material can be notch-ductile (large inelastic effects) or notch-brittle (elastic). The ratio of notch length a to material length d is the important parameter which determines the maximum load at failure for a notched material:

$$a / d = 0, \text{ notch-ductile, } \sigma_n/\sigma_o = 1$$

$$a / d = \infty, \text{ notch-brittle, } \sigma_n/\sigma_o = 1/3$$

where σ_n is the notched strength of the specimen, measured on the net section.

The concept of a damage band and application of a crack-bridging model seems highly relevant in the present study. In our experiments, the black GMT material was observed to change to a grey appearance close to the hole and a band extended perpendicular to the loading direction, away from the hole, just prior to fracture. In addition, in-situ microscopy observations during loading in the region close to a hole were performed in another study (Lindhagen and Berglund, 1997a). Crack-bridging was an important mechanism. Although debonding of transverse fibers and matrix plasticity occurred, fiber fracture and debonding of bridging fibers were the critical events.

PREDICTION OF NOTCHED STRENGTH

We now consider our data within the presented framework of a localized damage band described by large-scale fiber-bridging. Figure 1 is from the article by Suo et al. (1993). Curves have been computed by FEM to show the dependence of notched strength on the ratio a/d for different ratios of notch length to specimen width, a/w. Our values of σ_n/σ_o are in the Figure for the three different hole sizes a/w. The values of σ_n/σ_o for specimens with a 10 mm hole were used to calculate a/d for each material at each different temperature. This value was then used to predict the strength of specimens with 5 and 15 mm holes. The results are presented in Table II. The errors of the predictions are less than 10% for all values except one, where the error is 14%. The model thus gives predictions with fair accuracy.

The value of d is of interest to the designer of components since it is a measure of the ductility of the material. It also scales with the largest hole radius that the material can sustain with low strength reduction. Values of d can be calculated for the different materials, temperatures and notch sizes from the estimated values of a/d from Figure 1. The results are presented in Table III. The obtained values of d range from about 5 to 40 mm. They are generally larger at low temperature and for the swirled mat material. Thus the investigated materials generally show low notch sensitivity with the present notch sizes and especially so at low temperatures. Although the parameter d should ideally be a material constant for a given material and temperature, Table III demonstrates some variability. Part of this, the two extreme values for instance, is caused by the low resolution of the graphical method at low values of a/d. We also expect data from larger holes to show reduced scatter from variations in microstructure and constituent properties.

The original physical interpretation of d was a damage zone size. The damage zone sizes of materials similar to the present were investigated by Karger-Kocsis and coworkers (1994, 1995). The experimental damage zone sizes from compact-tension specimens were 5-15 mm which supports the physical realism of the estimated range of values of d in the present study.

As the damage zone is formed at the notch, loads are redistributed so that the stress concentration is reduced. The effectiveness of load redistribution in the damage zone is determined by micromechanisms which permit large local displacement without complete loss of strength. Figure 2 shows a schematic sketch of a damage zone as inferred from the micromechanism observations of the previous study, Lindhagen and Berglund (1997a). The earliest damage is debonding of transversely oriented fibers. Debonds are accompanied by local matrix plasticity. This damage dominates at the furthest distances from the notch. Since these mechanisms occur at relatively low stress levels, their contribution to notch toughness is expected to be small. Closer to the notch, multiple debonds join to form transverse cracks. Fibers predominantly oriented in the loading direction still carry significant loads. These mechanisms are limited to the damage zone and contribute to a specific bridging law behaviour. Fiber fractures at the notch root determine the maximum load carried by the damage zone. After the occurence of critical fractures of fibers at the notch root, load is still carried across the fracture plane as the fibers are pulled out of the matrix, finally leaving a traction free crack. In the swirled mat material, some differences are observed in mechanisms. The longer fibers are better able to bridge the softened damage zone. Most importantly, the observations are consistent with a damage zone in which the load carrying capability is provided by bridging fibers.

In the fiber bridging model, the parameter ∂_o should be taken as the total amount of inelastic displacement to failure, assuming that all inelastic displacement takes place in the localized band adjacent to the notch. For the sake of discussion, calculated values of ∂_o for our materials are presented in Table III. They were calculated from the values of d also presented in Table III. A linear bridging law was used which leads to $d = (k\ \Gamma E)/\sigma_O^2$ and is a measure of the size of the inelastic deformation region. $\Gamma = \partial_o \sigma_O/2$ and, correspondingly, $d = \partial_o E/2\sigma_O$. The ∂_o values are in the range of 0.1 to 0.5 mm. As a reference, the amounts of inelastic displacement in the test specimens were roughly estimated from the recorded test data by comparing the load-elongation curves of notched specimens with the curves of unnotched specimens. The results show an inelastic displacement in the range of 0.1 to 0.4 mm, which is in the expected order of magnitude and in support of the physical realism of the model.

CONCLUSIONS

At lowered temperature, the Young`s modulus, tensile strength and notch toughness increased significantly for glass mat polypropylene. Data for the ratio of notched to unnotched strength of one hole size was used to predict the notched strength of other hole sizes with fair accuracy. The increased notch toughness at lowered temperature was interpreted based on failure of a fiber bridged damage zone. As matrix modulus and yield stress increased, so did the notch toughness. The work of fracture and the limiting opening displacement in the damage zone and damage zone size are suggested as important parameters which need to be characterized in continued experimental work.

REFERENCES

Bao, G. and Suo, Z. (1992). Remarks on crack-bridging concepts. Applied Mechanics Reviews, 45(8):355--366.
Berglund, L. and Ericson, M. (1995). Glass mat reinforced polypropylene. In Karger-Kocsis, J., editor, Polypropylene: Structure, Blends and Composites, volume 3: Composites, pages 202--227. London: Chapman & Hall.
Cottrell, A. (1963). Mechanisms of fracture. In Tewksbury Symposium on Fracture, pages 1--27. University of Melbourne 1965.
Ericson, M. and Berglund, L. (1992). Deformation and fracture of glass-mat-reinforced polypropylene. Composites Science and Technology}, 43:269--281.
Friedrich, K. and Karger-Kocsis, J. (1990). Fracture and fatigue of unfilled and reinforced polyamides and polyesters. In Schultz, J. and Fakirov, S., editors, Solid state behaviour of linear polyesters and polyamides, chapter5. Englewood Cliffs,N.J.: Prentice~Hall.
Gaggar, S. and Broutman, L. (1976). Effect of matrix ductility and interface treatment on mechanical properties of glass fiber mat composites. Polymer Engineering and Science, 16:537--543.
Karger-Kocsis, J. and Fejes-Kozma, Z. (1994). Failure mode and damage zone development in a GMT-PP by acoustic emission and thermography. Journal of Reinforced Plastics and Composites, 13:768--792.
Karger-Kocsis, J., Harmia, T., and Czigany, T. (1995). Comparison of the fracture and failure behavior of polypropylene composites reinforced by long glass fibers and by glass mats. Composites Science and Technology, 54:287--298.
Lindhagen, J. and Berglund, L. (1997a)Microscopical damage mechanisms in glass fibre reinforced polypropylene. Submitted to Journal of Applied Polymer Science.
Lindhagen, J. and Berglund, L. (1997b). Notch sensitivity and damage mechanisms of glass mat reinforced polypropylene. Polymer Composites, 18:40--47.
Mallick, P. (1988). Effects of hole stress concentration and its mitigation on the tensile strength of sheet moulding compound (SMC-R50) composites. Composites, 19:283--287.
Mallick, P. (1992). Effect of holes and notches on the tensile strength of random mat reinforced thermoplastic composites. In Proceedings of the 8th Advanced Composites Conference, pages 203--207, Chicago, USA. Materials Park: ASM International.
Shirrell, C. and Onachuk, M. (1986). Influence of mold coverage upon the notch strength of R25 sheet molding compunds. In Hahn, H., editor, Composite Materials: Fatigue and Fracture,ASTM Special Technical Publication (STP) 907, pages 32--50. American Society for Testing and Materials.
Suo, Z., Ho, S., and Gong, X. (1993). Notch ductile-to-brittle transition due to localized inelastic band. Journal of Engineering Materials and Technology, 115:319--326.
Thomason, J. and Vlug, M. (1996a). The influence of fibre length and concentration on the properties of glass fibre reinforced polypropylene: 1)Tensile and flexural modulus.Composites, part A, 27A:477--484.
Thomason, J., Vlug, M., Schipper, G., and Krikor, H. (1996b). The influence of fibre length and concentration on the properties of glass fibre reinforced polypropylene: 3)Strength and strain at failure. Composites, part A, 27A:1075--1084.
Thomason, J. and Vlug, M. (1997). The influence of fibre length and concentration on the properties of glass fibre reinforced polypropylene: 4)Impact properties. Composites, part A}, 28A:277--288.

Table I Mechanical properties of glass mat polypropylene.

	Notch size		Short fibre		Swirled mat	
Fibre weight fraction, $W_f \pm s$ [%]			35.4 ± 0.16		43.0 ± 0.95	
Fibre volume fraction, $V_f \pm s$ [%]			16.1 ± 0.10		21.0 ± 0.64	
Temperature			RT	-30°C	RT	-30°C
Young's modulus, $E \pm s$ [GPa]			5.00 ± 0.78	6.64 ± 0.37	6.18 ± 0.83	7.73 ± 0.55
Failure strain, $\varepsilon^* \pm s$ [%]			2.65 ± 0.08	2.76 ± 0.21	1.61 ± 0.19	1.75 ± 0.20
Failure stresses, $\sigma^* \pm s$ [MPa]:	UN	σ_0	82.9 ± 14.5	98.4 ± 10.2	70.0 ± 13.5	93.2 ± 19.2
remote (gross section) stresses, σ_n^r	5 mm	σ_{n5}^r	51.3 ± 7.9	73.0 ± 7.8	54.6 ± 3.3	78.8 ± 15.8
	10 mm	σ_{n10}^r	49.0 ± 5.9	60.2 ± 7.3	45.5 ± 4.2	65.2 ± 6.9
	15 mm	σ_{n15}^r	36.4 ± 3.3	50.9 ± 3.3	38.2 ± 3.6	52.6 ± 2.6
net section stresses, σ_n	5 mm	σ_{n5}	58.6 ± 9.0	83.4 ± 8.9	62.4 ± 3.8	90.1 ± 18.1
	10 mm	σ_{n10}	65.3 ± 7.9	80.2 ± 9.7	60.7 ± 5.6	86.9 ± 9.2
	15 mm	σ_{n15}	58.2 ± 5.2	81.5 ± 5.2	61.1 ± 5.7	84.2 ± 4.2

Table II Predictions of notched strength and deviations from data.

Material	Temp.	Notch size, *a*					
		2.5 mm			7.5 mm		
		rel. strength	σ_n^r, pred. [MPa]	deviation [%]	rel. strength	σ_n^r, pred. [MPa]	deviation [%]
Short fibre	RT	$0.65\ \sigma_0$	53.9	+5.0	$0.50\ \sigma_0$	41.5	+14
	-30°C	$0.68\ \sigma_0$	66.9	-8.3	$0.52\ \sigma_0$	51.2	+0.53
Swirled mat	RT	$0.73\ \sigma_0$	51.1	-6.4	$0.55\ \sigma_0$	38.5	+0.79
	-30°C	$0.80\ \sigma_0$	74.6	-5.4	$0.59\ \sigma_0$	55.0	+4.6

Table III Material parameters estimated from model and data.

Material	Notch size, *a* [mm]	Temp.	*a/d*	*d* [mm]	δ_0 [mm]
Short fibre	2.5	RT	0.54	4.63	0.154
		-30°C	0.21	12.2	0.361
	5	RT	0.42	11.9	0.395
		-30°C	0.32	15.6	0.463
	7.5	RT	0.85	8.82	0.293
		-30°C	0.30	25.4	0.754
Swirled mat	2.5	RT	0.14	17.8	0.405
		-30°C	0.01	250	6.03
	5	RT	0.21	24.4	0.553
		-30°C	0.07	71.4	1.72
	7.5	RT	0.27	27.8	0.629
		-30°C	0.20	38.5	0.927

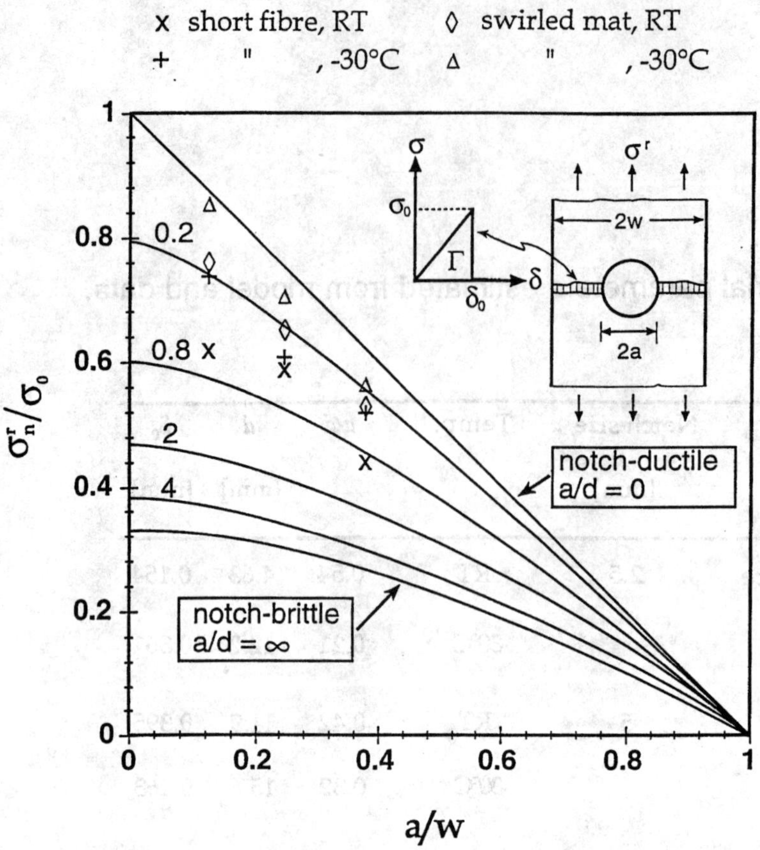

Figure 1 Notched strength for different notch sizes and material lengths. Experimental data are plotted in the diagram (curves from Suo et al, 1993).

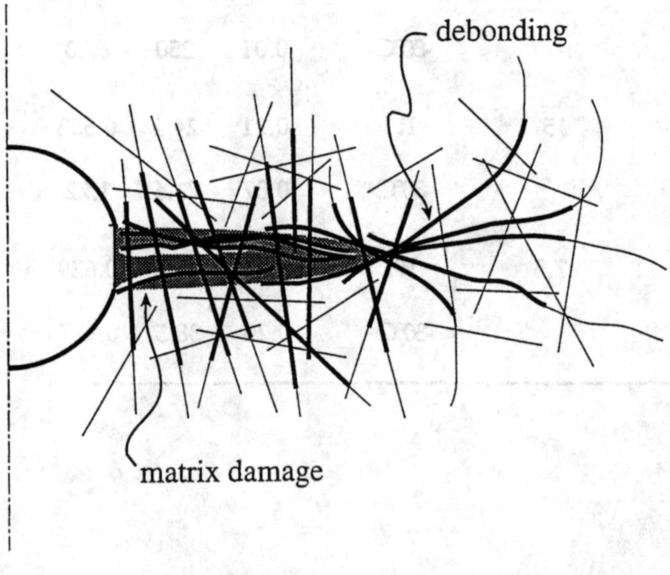

Figure 2 Damage zone in GF/PP fiber mat composite.

Frictional Interfacial Crack under Combined Shear and Compressive Loading

W. QIAN* AND C. T. SUN**

ABSTRACT— A closed form solution is presented for a bimaterial medium with an interfacial crack under combined shear and compressive loading based on the completely closed crack assumption. The solution indicates that the stress singularity at one tip is weaker than and at the other, stronger than $r^{-1/2}$. Numerical solutions show that if the crack opening zone is small due to large compressive loads, the crack tip stress field has the characteristic of a field with a strong singularity. A fracture criterion is proposed for interfacial cracks with weak as well as strong singularities. Finally the fiber pull-out problem is investigated.

INTRODUCTION

When crack surfaces are under combined compression and shear, the effect of sliding frictional stresses on the crack surfaces cannot always be ignored. Only when the near tip contact zone is extremely small does the K-dominated field agree well with that of the open crack solution (Sun and Qian, 1997). On the other hand, if the contact zone is finite, frictional contact of the crack surfaces can no longer be neglected. Most importantly, the frictional contact alters the stress singularity to become either weaker or stronger than the inverse square root singularity. Consequently, the strain energy release rate as conventionally defined either vanishes or becomes unbounded and, thus, cannot be used as a fracture parameter.

Many researchers have attempted to extend the fracture mechanics approach to treat various engineering problems involving interfacial cracks with frictional contact. Fiber pull-out and push-out problems have attracted the most attention (Gao et al, 1988; Sigl and Evans ,1989; Kerans and Parthasarathy, 1991; Hutchinson and Jensen 1990). Delamination in thin film/substrate micro-electronics system (Stringfellow and Freund, 1993) and composite laminates (Buchholz et al, 1995) also have been investigated. Various methods were proposed by those authors to derive the strain energy release rate or to employ different parameters for characterizing the interfacial toughness. Many of these approaches suffer deficiencies mainly due to the issue of

---

*W. Qian, formerly graduate student, Purdue University, West Lafayette, IN 47907-1282, currently at Performance Friction Corp., P.O. Box 819, Clover, SC 29710.

**C.T. Sun, School of Aeronautics & Astronautics, Purdue University, West Lafayette, IN 47907-1282.

weak or strong stress singularity and the energy dissipation associated with the sliding of crack surfaces.

Sun and Qian (1996) have shown that a new fracture criterion is needed to account for the weak singularity associated with interfacial cracks with friction. They also suggested that the oscillatory (open crack) model be used for an interfacial crack tip with small contact zone, and the contact (closed crack) model be used when the contact zone is finite. However, the loading in their study was confined to pure shear. In the present study, a combined compression and shear loading is considered. The crack tip stress singularity is examined for different crack surface movements.

A CLOSED FORM SOLUTION FOR A CLOSED CRACK PROBLEM

Consider a medium with a crack of length 2a lying along the interface between two dissimilar isotropic materials as shown in Fig. 1. The medium is under a combined remote compressive and shear loading, resulting in a partially closed crack; i.e., the right crack tip is closed, and the left crack tip is "open" if material "1" is stiffer than material "2" (Comninou and Schnueser, 1979). It can be shown that with additional compressive loading, the crack opening becomes smaller and smaller, and eventually the crack may be closed under very large compressions. Comninou and Dundurs (1980) solved a similar problem with only pure shear loading by using the elastic dislocation approach. However, their solution is not a closed form solution since it involves singular integral equations that require numerical solutions.

Following the same argument for adopting the oscillatory solution for small contact zone in interfacial cracks, we may adopt a solution for the completely closed crack to approximate the near tip solution for an interfacial crack which is almost closed. This model is not suitable for the partially closed crack (Comninou and Dundurs, 1980). However, it should be a good approximation as long as the crack opening zone is very small. For the closed crack problem, it is possible to obtain a closed form solution.

As mentioned by Comninou and Dundurs (1980), the only tractable formulation of interface cracks involving contact zones is through singular integral equations. The integral equations can be written by viewing the crack as an array of distributed edge dislocations and taking advantage of the known stress fields. Suppose that a discrete edge dislocation is placed at the point $(\xi, 0)$ of the interface. An edge dislocation with a Burgers vector in the y-direction (interface climb dislocation) b_y induces the following traction components at the interface (Comninou,1977).

$$\sigma_{yy}(x,0;\xi) = \frac{b_y\gamma}{\pi(x-\xi)} \tag{1}$$

$$\sigma_{xy}(x,0;\xi) = b_y\gamma\beta\delta(x-\xi) \tag{2}$$

If the Burgers vector b_x is in the x-direction (interface glide dislocation), the traction components are

$$\sigma_{yy}(x,0;\xi) = -\beta b_x\gamma\delta(x-\xi) \tag{3}$$

$$\sigma_{xy}(x,0;\xi) = \frac{b_x\gamma}{\pi(x-\xi)} \tag{4}$$

In the above equations, δ denotes the Dirac delta function; β is one of the Dundurs parameters defined as

$$\beta = \frac{\mu_1(\kappa_2 - 1) - \mu_2(\kappa_1 - 1)}{\mu_1(\kappa_2 + 1) + \mu_2(\kappa_1 + 1)} \tag{5}$$

and γ is defined as

$$\gamma = [(\kappa_1(1-\beta) + (1+\beta))/2\mu_1] + [(\kappa_2(1-\beta) + (1+\beta))/2\mu_2] \tag{6}$$

In (5) and (6), subscripts 1 and 2 stand for upper and lower materials, respectively (see Fig. 1), μ_i is the shear modulus, and

$$\kappa = \begin{cases} 3 - 4\nu & \text{for plane strain} \\ \dfrac{3-\nu}{1+\nu} & \text{for plane stress} \end{cases} \tag{7}$$

where ν_i is Poisson's ratio.

The stress field resulting from a series of n discrete dislocations along the interface are

$$\sigma_{yy}(x,0) = \sum_{i=1}^{n} \frac{B_y(\xi_i)\gamma\Delta\xi_i}{(x-\xi_i)\pi} - \beta\sum_{i=1}^{n} B_x(\xi_i)\gamma\Delta\xi_i\delta(x-\xi_i) \tag{8}$$

$$\sigma_{xy}(x,0) = \beta\sum_{i=1}^{n} B_y\gamma\delta(x-\xi_i)\Delta\xi_i + \sum_{i=1}^{n} \frac{B_x(\xi_i)\Delta\xi_i\gamma}{\pi(x-\xi_i)} \tag{9}$$

The two unknown densities of the dislocation distributions are $B_x(x)$ and $B_x(y)$. The stress fields resulting from a continuous array of dislocation or a crack in terms of $B_x(x)$ and $B_x(y)$ and remote combined loading p (assumed positive if compression) and τ are

$$\sigma_{yy}(x,0) = -p - \frac{\gamma}{\pi}\int_{-a}^{a} \frac{B_y(\xi)}{(\xi - x)} d\xi - \gamma\beta B_x(x) \tag{10}$$

$$\sigma_{xy}(x,0) = \tau - \frac{\gamma}{\pi}\int_{-a}^{a} \frac{B_x(\xi)}{(\xi - x)} d\xi + \gamma\beta B_y(x) \tag{11}$$

Assuming that the crack is closed and the crack surfaces are completely sliding, we have the boundary conditions

$$\Delta u_y(x,0) = 0 \tag{12}$$

$$\sigma_{xy}(x,0) = -\mu\sigma_{yy}(x,0) \tag{13}$$

or, in terms of $B_x(x)$ and $B_y(x)$,

$$B_y(x,0) = 0 \tag{14}$$

$$\tau - \frac{\gamma}{\pi}\int_{-a}^{a} \frac{B_x(\xi)}{(\xi - x)} d\xi = -\mu[-p - \gamma\beta B_x(x)] \tag{15}$$

In deriving (15), equations (10-11) and (13-14) have been used.

Solving the above singular equation (15), we obtain (Mikhlin, 1964)

$$B_x = \frac{\mu p - \tau}{(\mu^2\beta^2+1)\gamma} \times \left[-\mu\beta + \frac{1}{\pi(x-a)^{1-\lambda}(x+a)^{\lambda}} \int_{-a}^{a} \frac{(\xi-a)^{1-\lambda}(\xi+a)^{\lambda}}{(\xi-x)} d\xi\right] + \frac{c}{(x-1)^{1-\lambda}(x+1)^{\lambda}} \tag{16}$$

and

$$\cot \lambda\pi = \mu\beta \tag{17}$$

in which λ is the singularity index, and c is an unknown constant to be determined later by the single value displacement condition.

Using the hypergeometric functions and Gauss's recursion function (Gradshteyn and Rhyzik, 1965), we have the following integral identity (see Appendix) :

$$\int_{-a}^{a} \frac{(\xi-a)^{1-\lambda}(\xi+a)^{\lambda}}{(\xi-x)} d\xi = (x+a)^{\lambda}(x-a)^{(1-\lambda)}\pi \cot \lambda\pi - \frac{2\lambda\pi}{\sin \lambda\pi}\left(1+\frac{x-1}{2\lambda}\right)(-1)^{(1-\lambda)} \tag{18}$$

Using the above integral identity together with (17), $B_x(x)$ is finally obtained as

$$B_x(x) = \frac{\sin \lambda\pi}{(a+x)^{(1-\lambda)}(a-x)^{\lambda}} \left\{c + [x-(2\lambda-1)a]\frac{\tau-\mu p}{\gamma}\right\} \tag{19}$$

For single-valued displacements on circuits surrounding the crack, the density $B_x(x)$ must satisfy the conditions

$$\int_{-a}^{a} B_x(x) = 0 \tag{20}$$

Substituting equation (19) into (20), we have

$$c = 0 \tag{21}$$

Thus, the density $B_x(x)$ is

$$B_x(x) = \frac{\sin \lambda\pi}{(a+x)^{(1-\lambda)}(a-x)^{\lambda}} [x-(2\lambda-1)a]\frac{\tau-\mu p}{\gamma} \tag{22}$$

The relative sliding crack surface displacement $\Delta u_x(x)$ is obtained as

$$\begin{aligned} \Delta u_x(x) &= u_x^{(1)} - u_x^{(2)} = \int_{-a}^{x} B_x(x)dx \\ &= \sin \lambda\pi \left[\frac{\tau-\mu p}{\gamma}\right](a+x)^{\lambda}(a-x)^{(1-\lambda)} \qquad |x| < a \end{aligned} \tag{23}$$

The stress field can also be obtained with known density B_x as

$$\sigma_{yy}(x,0) = \begin{cases} -p - \beta \sin \lambda\pi(\tau-\mu p)\frac{x-(2\lambda-1)a}{(a+x)^{(1-\lambda)}(a-x)^{\lambda}} & |x| < a \\ -p & |x| > a \end{cases} \tag{24}$$

$$\sigma_{xy}(x,0) = \begin{cases} -\mu\sigma_{yy}(x,0) & |x| < a \\ \mu p + (\tau-\mu p)\frac{x+(1-2\lambda)a}{(x-a)^{\lambda}(x+a)^{1-\lambda}} & |x| > a \end{cases} \tag{25}$$

In deriving equation (25), the following integral identity is used (Gradshteyn and Rhyzik 1965).

$$\int_{-a}^{a} \frac{\xi}{(a-\xi)^{\lambda}(a+\xi)^{(1-\lambda)}(\xi - x)} d\xi = \frac{\pi}{\sin \lambda\pi}\left[1 - \left(\frac{a+x}{a-x}\right)^{\lambda} - \frac{a}{a-x}\left(\frac{a+x}{a-x}\right)^{(1-\lambda)}\right] \tag{26}$$

The same solution for the closed interfacial crack problem was also obtained by Kurtz (1993) using the complex function.

NEAR TIP SOLUTION

The value of the singularity index λ should satisfy (17). For a bimaterial medium loaded as shown in Fig. 1 with material 1 being stiffer than material 2, we have $\beta > 0$ and, thus, $\lambda < 0.5$. From (25) we can see that, in this case, the right crack tip ($x = a$)has a weak singularity, and the left crack tip has a strong singularity. On the other hand, if $\beta < 0$, then $\lambda > 0.5$ and the natures of singularity at the two crack tips are reversed. Figure 2 presents a simple way to judge weak and strong singularities at the crack tip based on the crack surface movements.

Without loss of generality, we consider an interfacial crack with crack surfaces situated to the left of the crack tip. The neartip stress field ahead of the crack tip along the bimaterial interface and the relative crack surface sliding displacement can be expressed in the general form

$$\sigma_{xy}(r,0) = K_{II}(2\pi r)^{-\lambda} \tag{27}$$

$$\sigma_{xy}(r,\pm\pi) = K_{II}\cos\lambda\pi(2\pi r)^{-\lambda} \tag{28}$$

$$\sigma_{yy}(r,\pm\pi) = -K_{II}\beta\sin\lambda\pi(2\pi r)^{-\lambda} \tag{29}$$

$$\Delta u_x(r) = u_x(r,\pi) - u_x(r,-\pi) = \left[K_{II}\gamma\sin\lambda\pi/2(1-\lambda)(2\pi)^{\lambda}\right] r^{1-\lambda} \tag{30}$$

where the generalized stress intensity factor K_{II} is defined as

$$K_{II} = \lim_{r\to 0}(2\pi r)^{\lambda}\sigma_{xy}(r,0) \tag{31}$$

In this form, the value of λ is less than 0.5 for a tip with weak singularity, and larger than 0.5 for a tip with strong singularity.

Following Irwin's crack closure integral approach, the strain energy release rate corresponding to a finite crack extension Δa can be obtained. We have

$$\begin{aligned}\hat{G}(\Delta a) &= \frac{1}{2\Delta a}\int_0^{\Delta a}\left[\sigma_{xy}(r,0) - \sigma_{xy}(\Delta a - r,\pi)\right]\Delta u_x(\Delta a - r)dr \\ &= \frac{K_{II}^2\gamma\sin\lambda\pi}{4(1-\lambda)(2\pi)^{2\lambda}}\Delta a^{1-2\lambda}\left[\frac{\Gamma(2-\lambda)\Gamma(1-\lambda)}{\Gamma(3-2\lambda)} - \frac{\cos\lambda\pi}{2(1-\lambda)}\right]\end{aligned} \tag{32}$$

where Γ is the gamma function. It is clear that for $\lambda = 0.5$, the above formula reduces to that for a frictionless interfacial crack, while a $\lambda < 0.5$ would cause $\hat{G}$ to diminish as $\Delta a \to 0$, and a $\lambda > 0.5$ would cause $\hat{G}$ to become unbounded as $\Delta a \to 0$. Consequently, the conventionally defined strain energy release rate does not exist if $\lambda \neq 0.5$. For this reason, Sun and Qian (1996) proposed using $\hat{G}$ in conjunction with a critical Δa_o to characterize the interfacial fracture toughness when friction is present; i.e., fracture occurs if

$$\hat{G}\left(\Delta a_o\right) = G_c \tag{33}$$

Under general loading conditions, an interfacial crack may be "open" or closed. Moreover, a closed crack may be naturally closed (weak) crack or a closed crack forced to close by compressive loads. To obtain these three types of crack tip stress fields, the corresponding stress intensity factor(s) must be obtained first.

Sun and Qian (1997) developed a couple efficient finite element based method to calculate the stress intensity factors or the open interfacial crack problem. For closed interfacial cracks with contact and friction, Sun and Qian (1996) employed the finite extension strain energy release rate $\hat{G}(\Delta)$ given by (32) to calculate K_{II}. They developed a finite element technique to perform a nodal release procedure to calculate the strain energy release rate $\hat{G}(\Delta a)$ for a finite crack extension Δa as well as dissipation energy rate G_d during the assumed crack extension. This technique is followed in the present study.

CENTER CRACKED BIMATERIAL MEDIUM

A center cracked bimaterial medium (see Fig. 1) under combined shear and compressive loading is selected to study the strain energy release rate and the singularity nature of its stress field. The medium is 20m × 20m with a 2m crack. The two materials are chosen so that $\beta = 0.49$. For numerical examples, we assume $\mu_1 = 35$MPa, $\mu_2 = 0.35$MPa, $\nu_1 = \nu_2 = 0$, and a positive shear $\tau = 0.12$MPa.

NEAR TIP STRESSES

The above interfacial crack problem is modeled by 4-noded plane strain finite elements using the commercial code ABAQUS. For $\beta > 0$ and $\tau > 0$, the left crack tip is "open". By increasing the compression load p, the open crack zone r_f decreases. For $\tau = 0.12$MPa, and $\mu = 0.5$ the finite element result yields

$p(MPa)$	γ_f/a
0.03	75.0%
0.06	31.0%
0.12	6.2%
0.16	1.9%
0.22	0.29%

The shear stress distribution ahead of the left crack tip is examined. Figure 3 shows the results for $p = 0.16$MPa, $\tau = 0.12$MPa and $\mu = 0.5$. The result directly from the finite element analysis (fea) is considered as the bench mark. The closed crack solution is obtained from (25) by assuming that the bimaterial medium is infinite. The oscillatory solution is the solution based on the open crack assumption and is obtained using K_I and K_{II} obtained from the finite element solution (Sun and Qian, 1997). The neartip (strong) solution is obtained using the aforementioned technique developed by Sun and Qian (1996) in conjunction with the assumption of a closed crack (of strong singularity).

The comparison of Fig 3 indicates that for a small open crack zone of $r_f/a = 1.95\%$ the near tip stress field agrees well with the near tip stress field with a strong singularity. Additional numerical results indicate that the open crack model yields reasonably accurate solutions only for $r_f/a > 6\%$.

Figure 4 shows the near tip shear stress distributions according to various models for $p = 0.22$MPa which produces an extremely small open crack zone ($r_f/a = 0.29\%$). It is evident that in this case the model of a closed crack with a strong singularity is an excellent representation. Also note that the closed crack model with a weak singularity cannot describe the near tip stress distribution.

ENERGY RELEASE AND DISSIPATION RATES

Using the finite element technique developed by Sun and Qian (1996), the strain energy release rate and energy dissipation rate can be calculated easily for either an open or a closed crack. For loading with $p = \tau = 0.12$MPa and $\mu = 0.5$, the interfacial crack stays open at the left crack tip in a small region of $r_f/a = 6.25\%$. In Fig 5, individual strain energy release rates, $\hat{G}_I$, and $\hat{G}_{II}$, total strain energy release rate $\hat{G}$ as well as dissipation energy rate G_d are plotted against various crack extensions Δa. Since the open crack region is not that small, the near tip stress field may agree more with the open crack solution. Thus, $\hat{G}_I$ and $\hat{G}_{II}$ may exhibit oscillatory behavior as $\Delta a \to 0$. This behavior is clearly shown in Fig 5. It is noted that $\hat{G} = \hat{G}_I + \hat{G}_{II}$ and the energy dissipation rate G_d are independent of the crack closure distance Δa.

Since the near tip stress field is dominated by the oscillatory stress field based on the open crack assumption, the fracture criterion can be established using the total strain energy release rate $G = \hat{G}$ ($\Delta a \to 0$) together with the mode mixity factor (Sun and Qian, 1997). For $p = 0.22$MPa, $\tau = 0.12$MPa, and $\mu = 0.5$, the open crack region r_f reduces about 0.3% of half crack length a. The whole crack is almost closed. As discussed earlier, the stress field (see Fig 4) near the left crack tip is dominated by the strong singularity solution, and $\hat{G}_I$ is almost vanishing. The energy release rate and dissipation rate are presented in Fig 6. The fact that $\hat{G}$ tends to increase as $\Delta a \to 0$ confirms that the near tip field is dominated by a strong singularity. It is interesting to note that the variation of $\hat{G}$ with respect to Δa is small except for extremely small Δa. This means that choosing the critical distance Δa_o is not a critical issue in the fracture criterion given by (33).

FIBER PULL-OUT SIMULATION

Fiber pull-out tests are often used to measure the fiber/matrix interfacial properties of fibrous composites. Once debonding initiates, an interfacial crack between the fiber and matrix appears. Due to the mismatch of coefficients of thermal expansion between the fiber and matrix, thermal residual stresses result from the manufacturing process (e.g. curing for polymeric matrices) of the composite. As a result, high compressive residual radial stresses may exist along the fiber/matrix interface.

Consider a concentric composite model with three phases, i.e., a circular fiber of radius R_f at the center surrounded by the matrix of radial thickness $4R_f$; and the outer ring represents the surrounding composite. The initial debonding length between the fiber and matrix is $a = 2R_f$. Figure 7 shows the axisymmetric section of the model.

In the numerical example, the elastic properties of the three phases of the model are assumed to be isotropic, but their thermal conductivities are assumed to be transversely isotropic:

$$\begin{aligned} &Fiber : E_f = 10GPa, \nu_f = 0.3, R_f = 0.064mm, \alpha_r^f = \alpha_\theta^f = 2 \times 10^{-6}/°C, \alpha_z^f = 0 \\ &Matrix : E_m = 0.1GPa, \nu_m = 0.3, \alpha_r^m = \alpha_\theta^m = 4 \times 10^{-6}/°C, \alpha_z^m = 0 \\ &Composite : E_c = 3.6GPa, \nu_c = 0.3, \alpha_r^c = \alpha_\theta^c = 4 \times 10^{-6}/°C, \alpha_z^c = 0 \end{aligned}$$

Assume a manufacturing temperature of $350°C$ above room temperature.. The thermal residual stresses are calculated using ABAQUS by assuming a temperature drop of $\Delta T = -350$ $°C$. The resulting radial stress distribution along the fiber/matrix interface is plotted in Fig 8.

The fiber is subsequently pulled in tension to a final level of 39.05MPa. The coefficient of friction $\mu = 0.5$ is assumed. Figure 9 shows the crack opening displacement (Δu_r) profiles at three different pull-out loads. At 18% load level (not shown), the debonding crack is completely closed and the crack surfaces slide against each other. The crack surfaces near the tip start to open at 43% load level. At 100% load level (i.e., fiber stress = 39.05MPa) the entire debonding crack is open.

From the result of Fig 9, we see that the fiber pull-out process may involve different fracture processes depending on the load level. Using the rule suggested in Fig 2, we note that the closed debonding crack at low loading levels has a strong singularity. If the debonding crack extends at these load levels, the fracture criterion (33) should be used. On the other hand, at higher load levels greater than 40% load, the crack tip opens substantially and the near tip stress field is characterized by the classical open crack oscillatory solution. If the debonding crack grows when the crack tip opens for a finite region, then the fracture criterion for open interfacial cracks should be used.

Table 1 lists the strain energy release rates associated with $\Delta a/R_f = 0.01$ for different load levels

CONCLUSION

A closed form solution is presented for a center cracked bimaterial medium with a frictional interfacial crack under combined shear and compressive loading based on the completely closed crack assumption. The solution indicates that one of the crack tip has a weak singularity and the other has a strong singularity. The corresponding strain energy release rate either varnishes or becomes unbounded and, thus, cannot be used as a fracture criterion. A new fracture criterion based on strain energy release rate for a critical crack closure distance is thus proposed.

From the numerical investigation, we find that the near tip stress field of an interfacial crack may be characterized by a strong singularity even if the crack is open, as long as the open zone is small. In such case, the result for the oscillatory open crack solution should not be used for fracture prediction.

For some interfacial crack problems such as fiber pull-out, we have demonstrated that the near tip crack surfaces may experience different amounts of frictional contact and , as a result, different crack tip stress singularities at different load levels. Therefore the selection of a proper fracture criterion may depend on the fracture load.

ACKNOWLEDGMENT

This work was supported by the Office of Naval Research through grant No. N00014-96-1-0822 to Purdue University. Dr. Y.D.S. Rajapakse was the technical monitor.

REFERENCES

Buchholz, F.G., Wang, H., Ding, S. and Rikards, R., 1995, 'Delamination analysis for cross-ply laminates under bending with consideration of crack face contact and friction', *ICCM-10* **1**, 141–148.

Comninou, M., 1977, 'A property of interface dislocations', *Philosophical Magazine*, **36**(5), 1281-1283.

Comninou, M. and Dundurs, J., 1980, 'Effect of friction on the interface crack loaded in shear', *Journal of Elasticity* **10**, 203–212.

Comninou, M. and Schmueser, D., 1979, 'The interface crack in a combined tension-compression and shear field', *Journal of Applied Mechanics* **46**, 345–348.

Gao, Y.C., Mai, Y.W. and Cotterell, B., 1988, 'Fracture of fiber reinforced materials', *Journal of Applied Mathematics and Physics(ZAMP)* **39**, 550–572.

Hutchinson, J.W. and Jensen, H.M., 1990, 'Models of fiber debonding and pullout in brittle composites with friction', *Mechanics and Material* **9**, 139–163.

Gradshteyn, I.S. and Ryzhik, I.N,. 1965, 'Table of integrals, series, and products', *Academic Press,* Fort Worth, TX.

Kerans, R.J. and Parthasarathy, T.A. 1991, 'Theoretical analysis of the fiber pullout and pushout tests', *Journal of American Ceramic Society* **74**, 1585–1596.

Kurtz, R.D., 1993, *Analysis of Bimaterial Crack Problems in the Presence of Friction,* Ph.D. thesis, Purdue University, West Lafayette, Indiana.

Mikhlin, S.G., *Integral Equations*, 1964, The Macmillan Co., New York NY.

Sigl, L.S. and Evans, A.G., 1989, 'Effects of residual stress and frictional sliding on cracking and pullout in brittle matrix composites', *Mechanics and Materials* **8**, 1–12.

Stringfellow, R.G. and Freund, L.B . 1993, 'The effect of interfacial friction on the buckle-driven spontaneous delamination of a compressed thin film', *International Journal of Solids and Structures* **30**, 1379–1395.

Sun, C.T. and Qian, W., 1996, 'The role of friction in the fracture of interfacial cracks', *Proceedings of the 14th U.S. Army Solid Mechanics Symposium*, Myrtle Beach, SC, October 16-18, 1996.

Sun, C.T. and Qian, W., 1997,"The Use of Finite Extension Strain Energy Release Rates for Fracture in Interfacial Cracks," to appear in *International Journal of Solids and Structures.*

APPENDIX

The integral identity in (18) can be derived as follows. First, from the integral table by Gradshteyn and Rhyzik (1965), the following integral can be represented by hypergeometric function $F(i,j,k,l)$ as

$$\int_{-1}^{1} \frac{(x+1)^{\lambda}(1-x)^{1-\lambda}}{(x-c)} dx = \pi(c+1)^{\lambda}(1-c)^{1-\lambda}\cot\lambda\pi - \frac{2\lambda\pi}{\sin\lambda\pi}F(-1,1,\lambda,\frac{1-c}{2}) \tag{A.1}$$

Using the following Gauss's recursion function and some known hypergeometric functions (Gradshteyn and Rhyzik, 1965) as

$$\lambda F(-1,1,\lambda,\frac{1-c}{2}) + (1-\lambda)F(0,1,\lambda+1,\frac{1-c}{2}) - (1-\frac{1-c}{2})F(0,2,\lambda+1,\frac{1-c}{2}) = 0 \tag{A.2}$$

we have

$$F(-1,1,\lambda,\frac{1-c}{2}) = 1 + \frac{c-1}{2\lambda} \tag{A.3}$$

Using (A.3), we obtain from (A.1) the following integral identity:

$$\int_{-1}^{1} \frac{(\xi-1)^{1-\lambda}(\xi+1)^{\lambda}}{(\xi-x)} d\xi = (x+1)^{\lambda}(x-1)^{(1-\lambda)}\pi\cot\lambda\pi - \frac{2\lambda\pi}{\sin\lambda\pi}\left(1+\frac{x-1}{2\lambda}\right)(-1)^{(1-\lambda)} \tag{A.4}$$

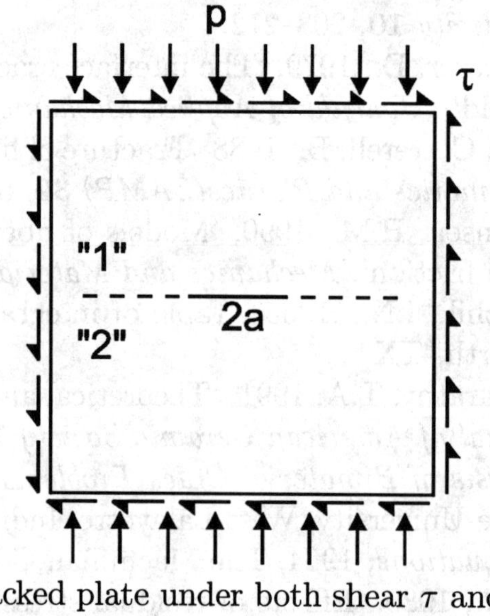

Figure 1. A center cracked plate under both shear τ and compressive p loading

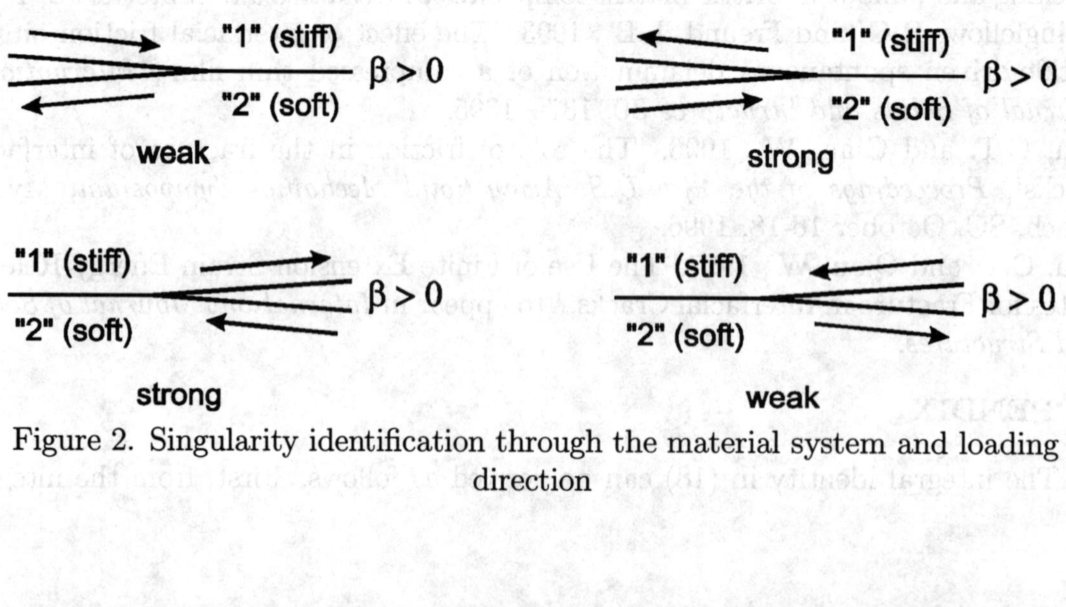

Figure 2. Singularity identification through the material system and loading direction

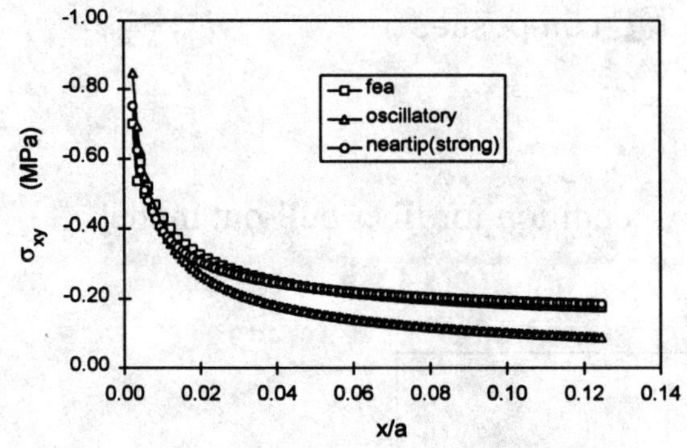

Figure 3. Comparison of shear stresses near left crack tip for p=0.16MPa τ=0.12MPa, and μ=0.5 (r_f/a=1.95 %).

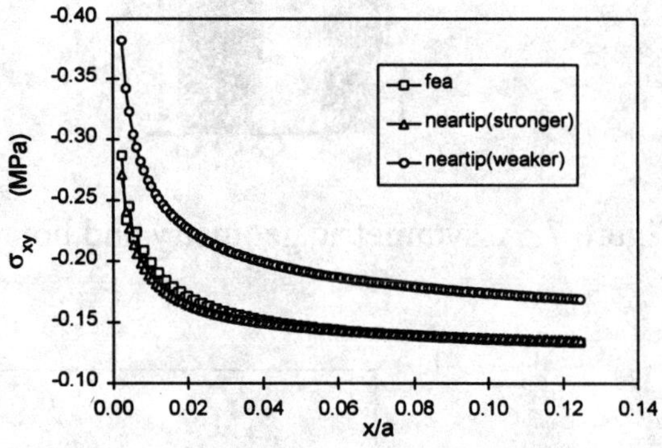

Figure 4. Comparison of shear stress ahead of open left crack tip, p=0.22MPa, τ=0.12MPa and μ=0.5 (r_f/a=0.29%).

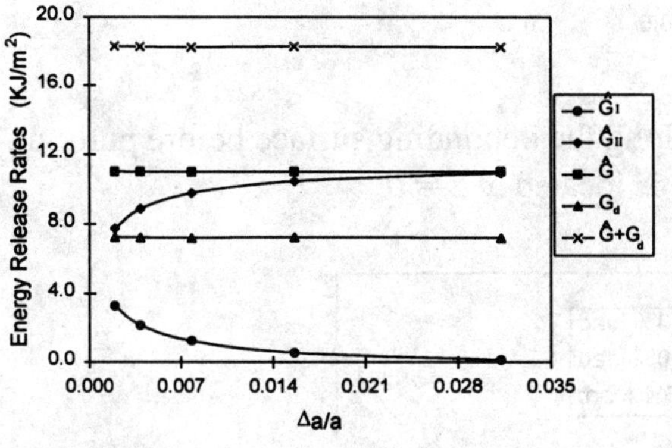

Figure 5. Energy rates under different crack extensions Δa at the open left crack tip, p=τ=0.12MPa , μ=0.5 (r_f/a=6.25%).

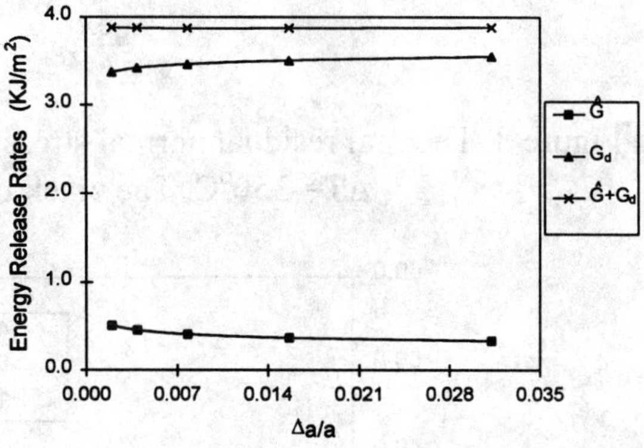

Figure 6. Energy rates under different crack extensions Δa at the open left crack tip, p=0.22MPa ,τ=0.12MPa , μ=0.5 (r_f/a=0.29%).

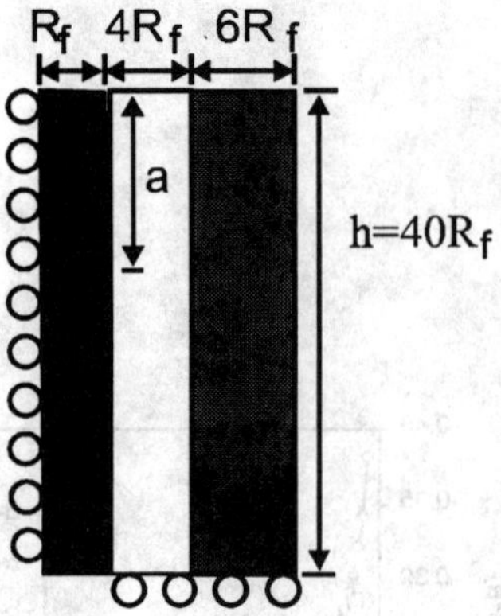

Figure 7. Assymmetric geometry and boundary condition for fiber pull-out model.

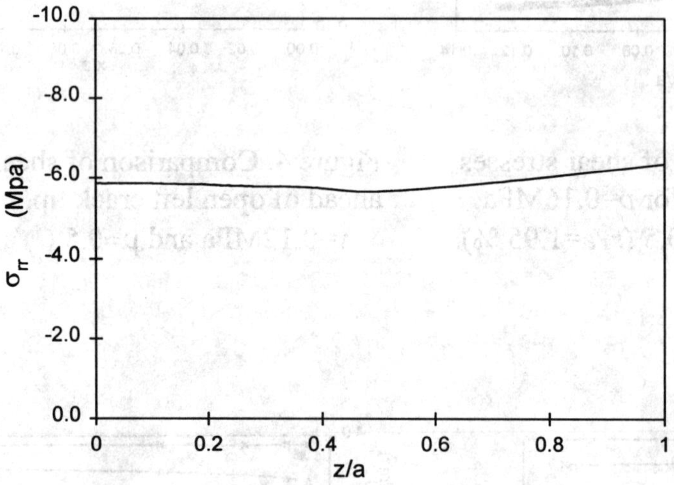

Figure 8. Thermal residual normal stress along the debonding surface before pull-out, ΔT=-350°C. The crack tip is located at $z = 0$.

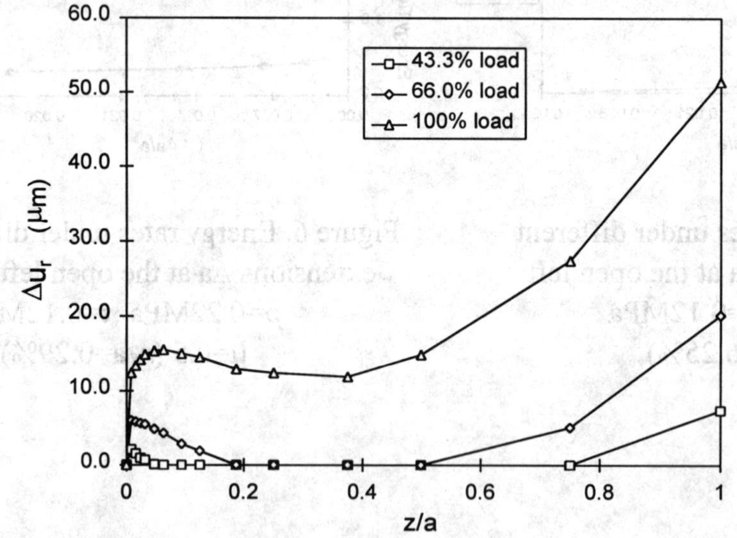

Figure 9. Crack opening displacement for fiber pull-out at different loading stages. The crack tip is located at $z = 0$.

A Model for Compressive Failure of Notched Composite Laminates under Multiaxial Loads

A. M. WAAS* AND J. H. AHN†

ABSTRACT

A finite element analysis (FEA) is used to develop a micromechanics based failure initiation predictive capability for analyzing notched composite laminates loaded in multiaxial compression. Results from a previous experimental study on compression failure mechanisms in composite laminates were used as motivation. Experimental results showed failure to initiate in the form of a narrow kink band near the hole edge and confined in extent to a distance within one half of the hole radius for uniaxial loading and by fiber/matrix debonding and kink banding for remote biaxial loading. A finite size rectangular region adjacent to the cutout edge corresponding in size to the area in which the kink banding occurs is analyzed. The agreement between experimentally measured far-field failure initiation load and analytical prediction is found to be good.

INTRODUCTION

Fiber reinforced composites are known for their superior tensile strength. Impact damage intolerance and poor compression strength have been the Achilles heel in Aerospace Structural applications. Laboratory coupon tests on carbon fiber/polymer matrix composites have indicated tensile strengths that are about twice the reported compressive strength values, Matthews and Rawlings(1994). A comprehensive survey of compression failure mechanisms is given in Waas and Schultheisz(1995).

In this study, a new micromechanics based non-empirical model to predict failure initiation in notched composite laminated plates loaded in compression (Waas and Babcock, 1986, Waas, 1988, Waas et al., 1990) is introduced. The development of the model relies on previous experimental work described in Khamseh and Waas(1992, 1997), that provides experimental data for compressive strength as a function of cutout size. Experimental results have shown the compression failure initiation mode for the

*Anthony M. Waas, Assoc. Professor, Composite Structures Laboratory, Dept. of Aerospace Engineering, Univ. of Michigan, Ann Arbor, MI 48109-2118, email:dcw@umich.edu

†J.H. Ahn, Graduate Research Assistant, Dept. of Aerospace Engineering, Composite Structures Laboratory, Univ. of Michigan, Ann Arbor, MI48109-2118, email: doolyii@umich.edu

laminates to be kink banding at the hole edge *followed* by delamination near the failure site, resulting in the propagation of the damage into the interior of the specimen (uniaxial). For remote biaxial loading, post-experiment microscopic examination revealed, both, fiber/matrix debonding and a shifted kink band (Khamseh and Waas, 1997). Thus, from the photomicrograph alone it cannot be conclusively stated whether kink banding proceeds or precedes fiber/matrix debonding. It is expected that the analytical model reproduce the salient features associated with the physics of the failure process, in particular, the objective is to construct an analysis model that takes as input measured constitutive and geometric data and outputs important quantities such as failure initiation load and corresponding failure mode.

ANALYTICAL/NUMERICAL MODELING

A finite element mesh of a region adjacent to the hole edge was generated using the SDRC-IDEAS Pre/Post processing software as shown in Figure 1. In choosing this microregion, observations from experiment, and symmetry with respect to loading and geometry were used to reduce the meshsize. The experimental results that were discussed earlier showed that the initially kinked region extended to a small fraction of the hole radius in the plane of the zero degree ply. Furthermore, surface strain gages placed remotely from the hole indicated no change in value corresponding to the inception of kink banding. Thus, it is possible to assume that kinking is a localized event, not affecting the strain state of the laminate in the far-field (regions that are removed form the location of the cutout).

For the pre-buckling analysis, material properties obtained from the experiment were used to determine the stresses and displacements using the solution as described in Lekhnitskii [1], for an equivalent homogeneous and orthotropic plate of infinite extent remotely loaded by uniform tractions corresponding to the experimental configuration, under conditions of plane stress. On the boundaries of the mesh, displacement fields obtained from the pre-buckling analysis were enforced as boundary values. The response analysis was split into two parts; Kink banding that is symmetric about the y-axis, i.e. kinking displacements that satisfy $u(\mathrm{x})=-u(\mathrm{-x})$ and $v(\mathrm{x})=v(\mathrm{-x})$, and antisymmetric kink banding, i.e. $u(\mathrm{x})=u(\mathrm{-x})$ and $v(\mathrm{x})=-v(\mathrm{-x})$, where u and v are the displacements in the x and y directions, respectively. For the symmetric case, the following boundary values were enforced (refer to Figure 3): edge AB, $u=0$, along edges BC, CD, and DA, $u=u_p$, $v=v_p$ (where the subscript p refers to the values for the displacements obtained from the pre-buckling analysis). For the antisymmetric case, the boundary conditions were the same along edges BC, CD, and DA, but along edge AB, $v=0$.

The loads on the remote boundaries were applied in an incremental fashion using the Riks algorithm and incorporating geometrical and material non linearity. The matrix was modeled as a elastic-plastic isotropic solid obeying $J2$ incremental theory of plasticity. The stress vs. strain curve for the matrix was generated by employing a standard uniaxial compression test of the pure matrix. The boundary loads were increased incrementally via the Riks method, while keeping track of the resultant horizontal force (in the x-direction) as a function of remote boundary displacement on edge BC. A plot of this is indicated in Figure 2 for uniaxial loading.

RESULTS AND DISCUSSION

As can be seen from the plots of Figure 2, the calculated resultant force in the x-direction (RF1) versus load proportionality factor, LPF (i.e. the factor by which the boundary displacements applied in the first step of the incremental Riks analysis is multiplied by) curves all display similar features. Initially, the curve is linear, followed by an unstable unloading path from a maximum load, that eventually asymptotes to a constant value. The maximum load is referred to as a limit load while the constant value is called a plateau load. If small initial geometrical imperfections are included, then a family of curves can be generated as a function of initial imperfection magnitude. This trend is shown in Figure 2 where curves generated with three imperfection magnitudes are shown along with the response obtained in the perfect case, for uniaxial loading. As the imperfection size increases, the limit load decreases and eventually disappears. Then the RF1-LPF curve shows a smooth increase from zero to the same plateau load value that is obtained with the perfect case. This is an important result, namely that the plateau load is *independent* of the imperfection size and can be obtained by analyzing the case corresponding to the perfect mesh. This finding is true, regardless of whether the external loads are uniaxial or biaxial.

Notice that the nonuniformity in the prebuckling stress field coupled with a progressively decreasing instantaneous shear modulus of the matrix material provides the necessary synergy to initiate the evolution of kink banding. Beyond the limit load, both RF1 and LPF decrease (the Riks method in Abaqus enables us to capture this) and this decrease as can be seen for the deformed configurations of the mesh is due to deformation localization, close to the edge of the hole. The remote (far-field) loads on the specimen corresponding to the limit load and the plateau load are tabulated in Table 1, as a function of hole size for uniaxial loading, while Table 2 provides a similar comparison for biaxial loading. It must be noted that, despite fiber/matrix debonding occurring simultaneously with kink banding for biaxial loading, it appears that predictive far-field loads for kink banding (without regard to modeling fiber/matrix debonding) are in good agreement with experiment suggesting that the energy release associated with fiber/matrix debonding is small compared to the plastic energy associated with kink banding. Furthermore, whether fiber/matrix debonding precedes kink banding cannot be conclusively stated via post-experiment microscopic examination of failed specimens that showed a fiber/matrix debond and a kink band on the same photomicrograph. This suggests that kink banding occurring at a location slightly skewed with respect to the x-y axes at the hole edge (i.e.not at 90^0 as would be the case with uniaxial loading), could drive fiber/matrix debonding failure and not vice-versa. This aspect warrants further investigation.

In Table 1 and Table 2, the measured experimental result for the far-field stress corresponding to failure initiation is indicated. Figure 5 and Figure 6 provide a comparison of the measured and predicted far-field failure initaiation stresses as a function of hole size for remote uniaxial and remote biaxial loading. The agreement between experimentally measured far field failure initiation values and predicted values based on the plateau load are seen to be very good over a significant range of hole sizes, although the FEA predictions are seen to be higher throughout. It must be noted that throughout this paper, all the results presented herein correspond to converged results as far as mesh size is concerned. That is, for a given geometry, the mesh density was progressively increased in both the planar directions (x and y), until the change in the obtained limit load and plateau load was less than 1% of the previous value. It is not at all surprising to obtain fast convergence in the present problem because the event that we are studying is truly local and scales with the size of the cutout, which is the only structural length scale of importance.

The presence of the cutout introduces a natural length scale that is reflected in the gradients associated in the prebuckling stress field. Therefore, even without an initial geometrical imperfection, we are able to simulate kink banding. The fact the the plateau load is only influenced by the cutout size and no other length scale is a very important finding of the present work. The present problem can be contrasted against the uniform prebuckling deformation situations that have been analyzed earlier (Kyriakides et al., 1994). In that investigation and the present study, an essentially three dimensional situation has been successfully modeled using a two dimensional approach. Thus, it appears that other possible modeling omissions, such as the presence of fiber ends within the matrix, the three dimensional random packing of the fibers and the presence of a fiber/matrix interphase, may have only a weak influence on the overall outcome of the localization process. Soutis, Fleck and Smith(1991), Soutis, Curtis and Fleck (1991) and Soutis and Fleck (1990), collectively referred to as SFS and Guynn and Bradley (1989), have previously introduced predictive models for failure initiation in notched composites. To use the SFS, one needs as input the unnotched strength of the laminate (obtained via tests) and also the fracture toughness associated with kinking(also obtained form suitable fracture tests). Unlike SFS, the present work directly models the salient features of the deformation localization; the only parameters that are needed as input are the material properties of the constituents, the geometry and the stacking. The effect of stacking (i.e plies other than zero plies) is accounted for via the calculation of the prebuckling displacement field which is influenced by the number, type, constitutive property and thickness of the layers in the laminate.

In the SFS, the unnotched compressive strength of the laminate is needed as input, and it is not entirely clear that a macroscopic value of laminate strength may capture the same features of the failure mechanisms near a cutout edge. For example, it is quite possible for an unnotched laminate to fail by a mechanism other than kink banding, while the corresponding notched laminate may in fact still fail by kink banding owing to the presence of the notch that introduces a well defined location for the failure to initiate. Furthermore, it is not clear that the mechanics of a kinked zone can be modeled as a crack with a linear traction-crack closure law. Notwithstanding these issues, the agreement between experiment and analysis that is reported in Soutis, Fleck and Smith (1991), is impressive.

CONCLUDING REMARKS

A new nonempirical procedure is introduced for predicting the initiation of failure in a laminated fiber composite plate that contains a circular cutout and loaded in multiaxial compression. Based on previous experimental work by the authors and a number of other investigators, it is now established that fiber composites containing cutouts fail by kink banding occurring close to the cutout edge when subjected to compression (uniaxial), while fiber/matrix debonding is seen to occur with kink banding in specimens that are biaxially loaded. Exploiting the fact that this initial manifestation of damage is a *localized* event (i.e. in experiments surface strain gages placed remote from the cutout, are unaffected by the initial kink band that forms near the cutout-see Khamseh and Waas (1992, 1997)), we have employed the commercial Finite Element package Abaqus to analyze this local event. Material properties of the constituents and the full uniaxial stress-strain curve of the matrix material is needed as input. Following the pioneering work of Argon (1972), Budiansky et al. (1993) and Kyriakides et al. (1994), the combination of a progressively decreasing modulus for the matrix and a nonuniform prebuckling stress field whose gradients are uniquely related to the cutout size are sufficient to fully characterize the kink banding and, leads to accurate predictions for the load carrying capacity of composite plates containing cutouts when loaded in multiaxial compression. The finite element predictions of the salient features of the failure initiation event is found to compare favorably with experimental data. The present work is non-empirical and provides a useful predictive tool for design engineers.

ACKNOWLEDGMENTS

We are grateful for financial support from the AFOSR under grant F49620-95-1-0326, with Dr. Walter Jones as the Scientific Officer, and from the Michigan Materials Processing Institute, with Dr. Paul Duvall as the program monitor.

REFERENCES

1. Matthews, F.L. and Rawlings, R., 1994, *Composite Materials: Engineering and Science*, Chapman and Hall, London.

2. Waas, A.M. and Schultheisz, C.R., 1995, "Compressive Failure in Composites, Part II", *Progress in Aerospace Sciences*, vol. 32, pp43-78.

3. Waas, A. and Babcock, C.D., Observation of the Initiation and Progression of Damage in Compressively loaded Composite Plates containing a Cutout, GALCIT SM Report 86-34(1986). See also GALCIT SM Report 85-12, by the same authors, 1985.

4. Waas, A., 1988, "Compression Failure of Fibrous Laminated Composites in the

Presence of Stress Gradients: Experiment and Analysis", Ph.D. thesis, California Institute of Technology.

5. Waas, A.,Babcock, C.D. and Knauss, W.G. 1990, "An Experimental Study of the Compression Failure of Fibrous Laminated Composites in the Presence of Stress Gradients", *International Journal of Solids and Structures*, Vol. 26, No. 9/10, pp. 1071-1098.

6. Guynn, E. G., and Bradley, W. L., 1989, "A Detailed Investigation of the Micromechanisms of Compressive Failure in Open Hole Composite Laminates", *Journal of* Composite Materials, vol.23, May, pp. 479-504.

7. Soutis, C., and Fleck, N.A., 1990, "Static Compression Failure of Carbon Fibre T800/924C Composite Plate with a Single Hole", *Journal of* Composite Materials, Vol. 24, pp. 536-558.

8. A. Khamseh and A. Waas, Failure Mechanisms in Uniply Composite Plates under Uniaxial Compression, ASME J. Materials and Technology, October, 1992.

9. A. Khamseh and A. Waas, Failure Mechanisms of Composite Plates with a Circular Hole under remote Biaxial Planar Compressive Loads, ASME J. Materials and Technology, vol. 119, pp56-64, 1997

10. Soutis, C., Fleck, N. and Smith, F., Failure Prediction Technique for Compression Loaded Carbon Fiber-Epoxy Laminate with an Open Hole, J. Composite Materials, 25, pp1476-1498, 1991

11. Soutis, C., Curtis, P. and Fleck, N., Compressive Failure of Notched Carbon Fiber Composites, Proc. Royal Soc. London A, 440, pp241-256, 1991.

12. Tsai, S.W., and Hahn, H.T., 1980, "Introduction to Composite Materials", Technomic Publishing Company, Connecticut.

13. S. Kyriakides, R. Arseculeratne, E.J. Perry and K.M. Liechti, "On the Compressive Failure of Fiber Reinforced Composites", *Int. J Solids Structures Vol.32, No. 6/7,pp.689-738*, 1995.

Table 1 Comparison of FEA Predictions with Experiment (Uniaxial Loading).

Specimen	Hole Size $(\frac{d}{r}, r)$	Avg. Exp. Far field stress (Psi)	FEA Prediction Based on Limit Load (Psi)	FEA Prediction Based on Plateau Load (Psi)
1	0.062,0.0520	6600	8294	6646
2	0.103,0.0864	4833	8136	5320
3	0.144,0.1207	4275	7774	4859
4	0.198, 0.166	3800	7240	4525

Table 2 Comparison of FEA Predictions with Experiment (Biaxial Loading).

Specimen	Hole Size $(\frac{d}{r}, r)$	Exp. Far field stress in x direction (psi)	FEA Prediction based on Plateau Load (Psi)
1,2	0.062,0.0520	10400,10800	10507
3,4	0.103,0.0864	9600,10000	9846
5,6	0.144,0.1207	9000,5800	9215
7,8	0.167, 0.140	7500,7000	7730

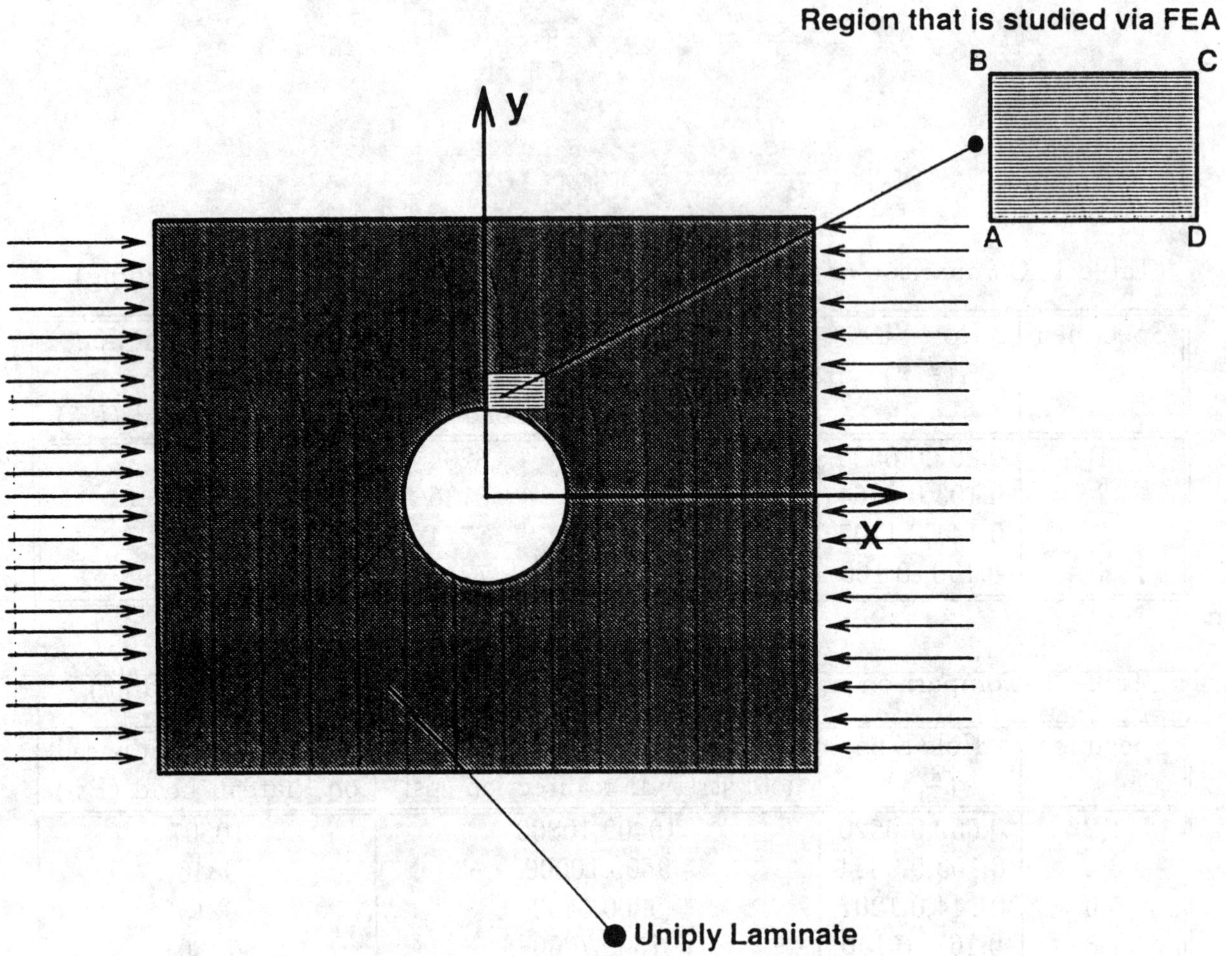

Figure 1 Microregion analyzed via FEA.

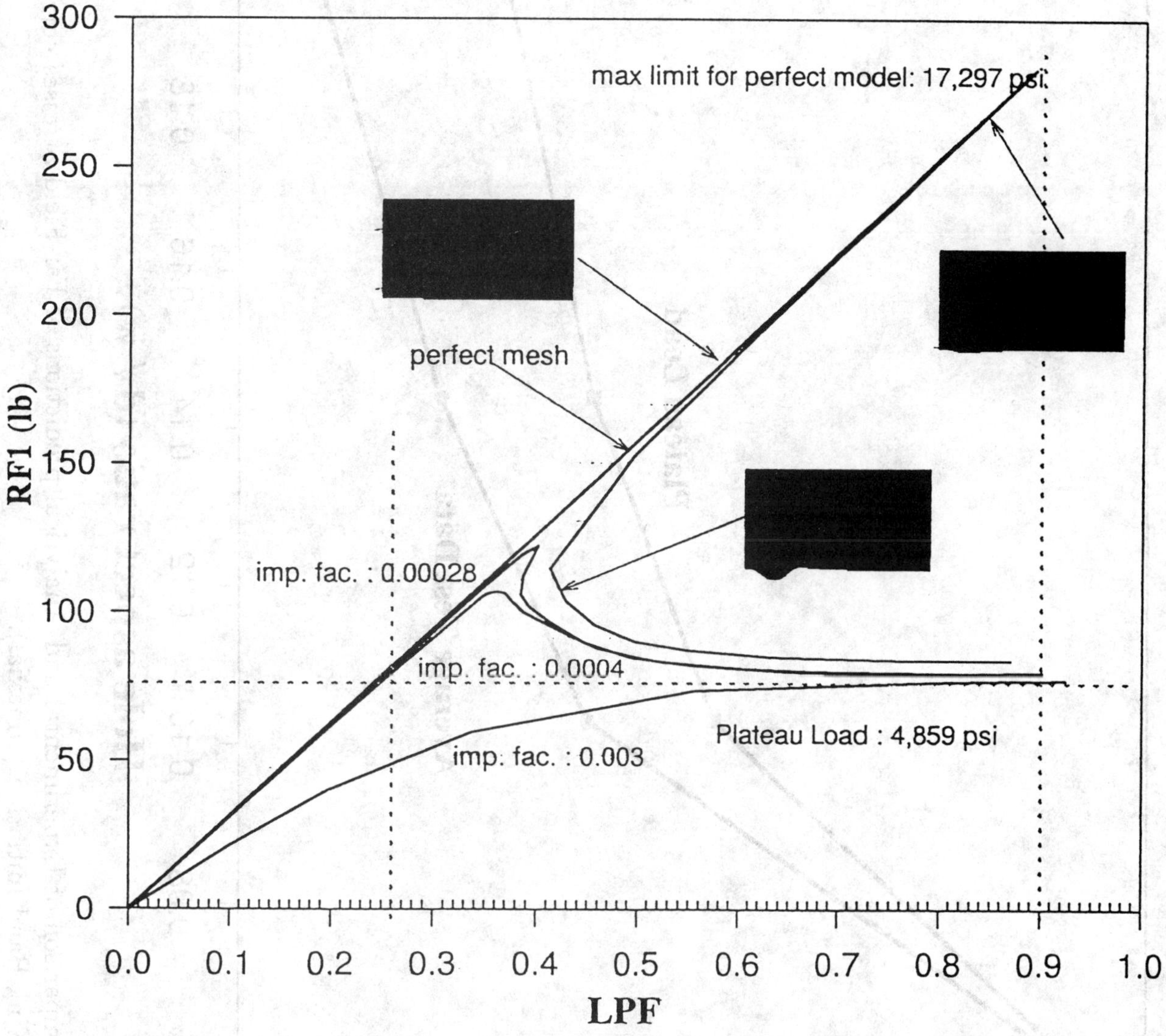

Figure 2: Response comparison between perturbed model and perfect model.

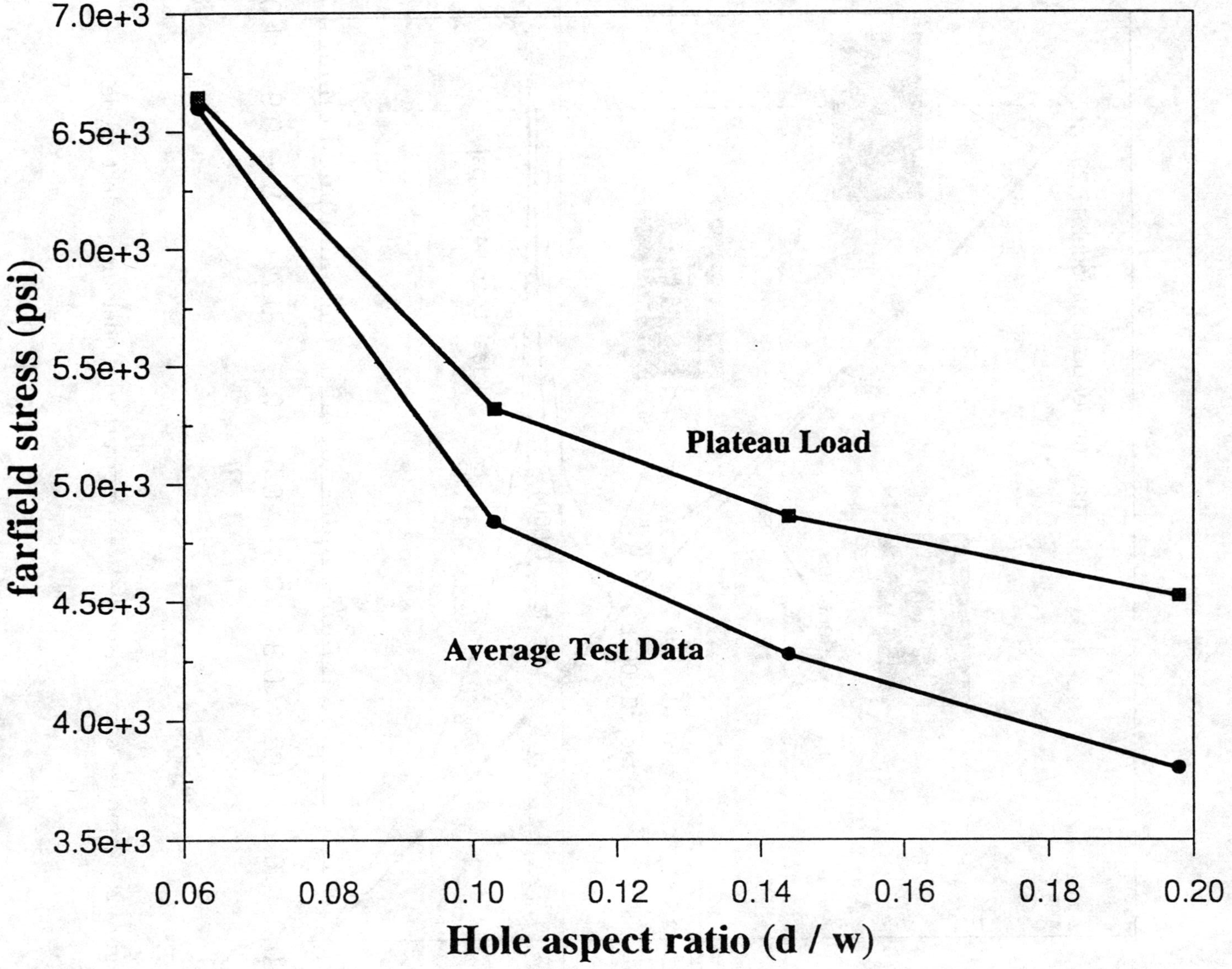

Figure 3: Comparison of Experimental and Analytical predictions of Far-Field Stresses corresponding to Kink Band Initiation (Uniaxial).

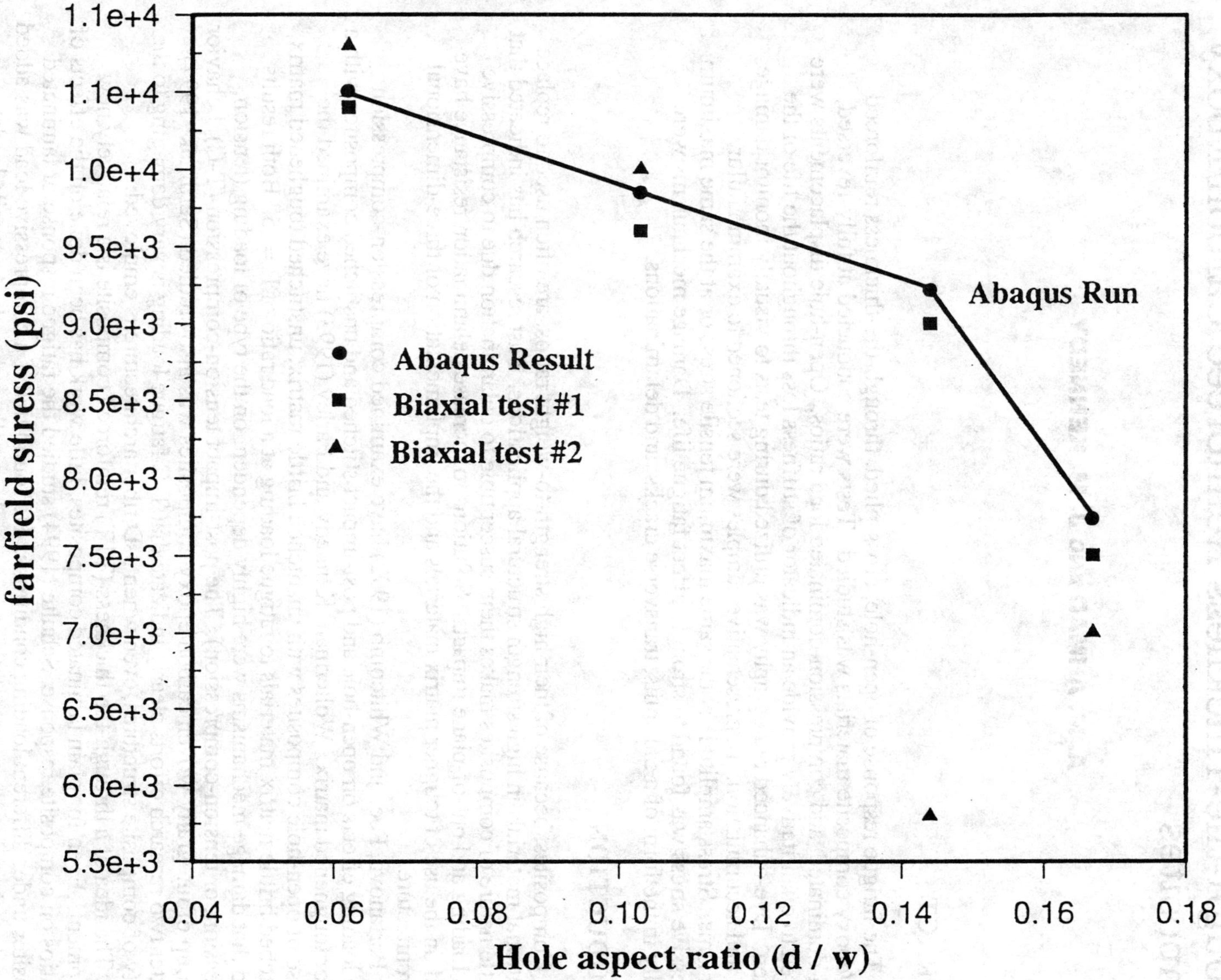

Figure 4: Comparison of Experimental and Analytical predictions of Far-Field Stresses corresponding to Kink Band Initiation (Biaxial).

The Effect of R-Ratio on the Fatigue Response of Through-the-Thickness Reinforced Carbon/Epoxy Composites

A. M. AHMAD AND J. M. KENNEDY

ABSTRACT

The fatigue response of open-hole, cross-plied, through-the-thickness reinforced carbon/epoxy composite materials was studied. Tests were conducted at fully-reversed, tension-dominated, and compression-dominated load ratios. Open-hole displacements were measured to quantitatively provide an indicator of stiffness loss throughout the life of the specimen. Dye-enhanced radiography was utilized during tests to visually monitor damage growth and accumulation. Representative samples were sectioned to examine failure mechanisms. Stress amplitude at the same maximum tensile stress or at the same maximum compressive stress was found to strongly affect fatigue life. Damage mechanisms were generally in the form of axial splits, transverse cracks, and delaminations.

INTRODUCTION

Composites, because of their high strength-to-weight ratios, are often used to replace conventional materials in high strength structural applications. Past research has indicated that the brittle nature of composites makes them susceptible to delamination due to compressive fatigue loading and out-of-plane impacts. Solutions to increase delamination resistance have focused on the use of tougher matrix materials and the implementation of three-dimensional fiber architecture.

Portanova, Poe, and Whitcomb (1992) have examined compression-compression fatigue loading effects on open-hole and post-impact stitched and unstitched composites with a brittle or toughened matrix. Wolterman, Kennedy and Farley (1993) have examined the response of open-hole composites with unstitched brittle matrix, unstitched toughened matrix, and stitched brittle matrix materials to fatigue loading at a load ratio of $R = -5$. Both results showed that damage mechanisms were highly dependent on the type of loading (tension, compression, or tension-compression). The post-impact tension-compression (T-C) behavior of a group of integrally woven carbon/epoxy composites was also studied (Serdinak, 1994). The latter two research efforts clearly indicate that T-C fatigue loading is very damaging to the integrity of composite structures, even when 3-D fiber architecture is employed.

The idea of a through-the-thickness (TTT) reinforced composite is a relatively new development. Even for plain laminated composites, little work has been done on the effects of load ratios on composite response. Schulte (1994) studied the fatigue response of laminated composites under different loading conditions. He found that if a compressive load was added to a tensile fatigue load cycle, the fatigue life was drastically reduced. As tensile loads increased, the damage mode became distributed over the entire specimen, in comparison to the

Atiya M. Ahmad, Graduate Student, and John M. Kennedy, Associate Professor, Department of Mechanical Engineering, Box 340921, Clemson University, Clemson, SC 29643-0921

local buckling experienced under increased compressive fatigue loading. It was found that the tensile load excursions caused delamination growth, whereas additional compressive excursions accelerated damage development. But the question remains on how the benefits of TTT-reinforcement (i.e. delamination suppression) affect the fatigue damage response under T-C loading.

The purpose of this research was to examine the damage mechanisms that develop when open-hole, TTT-reinforced carbon/epoxy composites undergo fatigue loading. Both carbon-type and non carbon-type TTT reinforcing fibers were examined. Four main load ratios were tested: a ratio heavily weighted in compression, a ratio moderately weighted in compression, a ratio moderately weighted in tension, and a ratio heavily weighted in tension. Intermediate load ratios were also tested to develop a correlation between predominant loading modes and resulting damage mechanisms.

MATERIALS AND EXPERIMENTAL PROCEDURE

The lay-up was a nine-layer preform, $[(0/90)_4/0]_T$, with five layers in the warp direction and four layers in the fill direction. The inplane fiber was a 21K filament count AS4 carbon fiber served with a fine denier PVA monofilament. A catcher yarn of AS4 carbon was located in the center of the preform and TTT yarns were introduced in such a manner that those in the warp direction looped around catcher yarns in the fill and vice versa, providing TTT reinforcement in both directions. Row spacing was 0.252 inches and penetration spacing was 0.126 inches. Hercules 3501-6 resin was infiltrated using a RTM process. Each sample had only one of the following TTT-reinforcing fibers: Kevlar 29, 1500 denier; Toray carbon, T-1000; or S2 75 1/3 glass. A 0.6 inch diameter open-hole was centrally located in a 5-inch gage length specimen. Anti-buckling supports were not necessary for the compression tests.

One static tension, one static compression, and a multitude of constant amplitude fatigue tests ($\omega = 4$ Hz) were conducted at room temperature for each TTT-reinforcing fiber type. The test program was designed to reveal the effects of load ratio changes on life. The first fatigue test was always fully-reversed, constant amplitude loading at some percentage of the static compressive strength. Thus, the allowable maximum and minimum stresses were fixed for the test program (see Figure 1). Tests on the remaining samples were conducted at different load ratios (recall, $R = \sigma_{min}/\sigma_{max}$) but with either the same maximum stress or minimum stress as the $R = -1$ test. All compression-dominated fatigue tests had the same σ_{min}, and all the tension-dominated fatigue tests had the same σ_{max} (see Figure 1a, 1b). The R-ratio varied from test to test, but not during a test. A variety of load ratios were examined to develop a correlation between predominant loading modes and resulting damage mechanisms.

Damage growth was quantitatively monitored by measuring displacements using extensometers with a 2-inch gage length that were mounted on both surfaces over the open-hole. Far-field stress versus displacement data were examined. Damage growth was also periodically recorded during the test using dye-enhanced radiography techniques. Lastly, representative samples fatigue loaded to failure were sectioned to examine failure modes.

RESULTS AND DISCUSSION

Kevlar TTT-reinforced fiber type samples were tested under fully-reversed ($R = -1$) fatigue loading conditions to generate an S-N curve. The tests ranged from 48% to 64% σ_c. Damage growth was observed and quantified by monitoring secant modulus, a continually changing parameter under fatigue loading (Stinchcomb, et al, 1975). Based on the load levels tested, secant modulus exhibited a gradual decrease during the majority of life, and dropped off sharply when failure became imminent. Figure 2 is a representative graph of a full cycle stress-strain curve and its evolution with damage accumulation. The material response became nonlinear over time, as signified by the changing shape of the stress-strain curves. As damage accumulated, the material stiffness decreased, allowing larger displacements at given stress levels. Radiographs showed axial splitting tangent to and in the vicinity of the open-hole as the dominating damage mechanism, followed by localized delamination and transverse cracks originating from the open-hole surface. Inter-ply delamination and ply pull-out were common failure mechanisms.

Fatigue tests for the Toray carbon TTT-reinforced fiber type were conducted under a variety of loading situations. Figure 3 shows results for secant modulus versus cycles under both fully-reversed (R = -1) and compression-dominated (R = -5) loading conditions at two different stress levels. For tests at 45% σ_c, R = -1 and R = -5 loading conditions had large and distinct differences in secant modulus degradation. The compression-dominated samples had a slower rate of stiffness loss than the fully-reversed samples, and thus, the compression-dominated samples had the longest life. This is not as clearly indicated for the samples at 58% σ_c. Both had similar secant modulus behavior and life (the R =-5 test 58% σ_c was stopped prior to failure). Recall the test sequence for compression-dominated loading presented earlier in Figure 1a. The first test was always fully-reversed loading at some percentage of static compressive strength. The remaining tests had the same minimum stress as the R = -1. The results presented for 45% σ_c in Figure 3 show that a smaller stress amplitude, with the same maximum compressive stress, led to a longer life. The same behavior was observed between the R = -1 test and the R = -5 tests at 58% σ_c. Further, though the R = -5 test at 58% σ_c had a larger maximum compressive stress, the R = -1 test at 45% σ_c exhibited the same life and stiffness loss behavior at a 30% larger stress amplitude than the R = -5 test at 58% σ_c.

Figure 4 shows one radiograph near failure for each sample tested. The dark rectangles at the top and bottom of all radiographs are extensometer tabs. Both the fully-reversed and the compression-dominated tests developed similar damage modes: longitudinal cracks, transverse cracks and delamination. The intensity of the damage modes differed. The R = -1 samples had damage accumulation in areas local to the open-hole, approximately within one hole diameter in any direction. The R = -5 samples, on the other hand, had axial splits and transverse cracks growing as far as two and a half hole diameters away. Transverse splits were scattered throughout the gage section and did not develop into the larger transverse cracks as shown for the R = -1 samples. The delaminated regions also remained distinct and did not converge as they did for the R = -1 specimen. In fact, for the R = -1 specimen, small squares bounded by the TTT-reinforcement (indicated by the circled regions in the radiographs) were observed to bulge under the compressive loading excursions. The R = -5 samples never developed localized intense damage. Damage developed more slowly at the smaller stress amplitude than at the larger one, even though the maximum compressive stress was the same. Another interesting observation centered around the stress amplitude: the larger the stress amplitude, the more likely it was for damage to develop local to the hole. This is evident when comparing the samples run at 58% σ_c with those run at 45% σ_c. Even under R = -5 loading conditions (58% σ_c), where axial splits and transverse cracks were very scattered throughout the gage section, a large transverse crack developed near the open-hole and a localized delamination developed.

The effect of mean stress on damage development was also examined. A compression-dominated R-ratio (R = -5) and a tension-dominated R-ratio (R = -0.2) were compared to each other and to the fully-reversed loading condition (R = -1). Results for secant modulus versus cycles are shown in Figure 5. The R = -5 sample and the R = -0.2 sample had the same stress amplitude but the limits were set by the R = -1 test. A comparison of secant modulus degradation for the different load ratios showed that a larger stress amplitude led to a more pronounced decrease in secant modulus. The mean stress, in this case, had little effect in secant modulus degradation; both compression- and tension-dominated samples had about the same stiffness loss for one million cycles of loading.

The radiographs for the R = -1 condition included, again, axial splitting along the open-hole boundary followed by transverse cracking perpendicular to the loading direction and delamination local to the hole. Radiographs at approximately 70% of life for the R = -5 and R = -0.2 conditions are shown in Figure 6. The damage modes are similar: dispersed axial splitting and transverse cracks. Some delamination along the axial splits tangent to the hole are observed in both cases. However, in the R = -5 tests near the hole, there was a fairly intense local delamination that was bounded by horizontal stitch rows.

The third TTT-reinforcement fiber tested was S2-Glass. Seven tests were conducted, each at a different R-ratio. The R-ratios tested were: -0.2, -0.6, -0.8, -1, -1.25, -1.67, and -5. Tests were run at 50% of the ultimate static compressive strength, with the tension-dominated tests all having the same maximum stress and the compression-dominated tests all having the same minimum stress. The fully-reversed test, R = -1, had the largest stress amplitude and the shortest life. Figure 7 shows effects of stress amplitude and R-ratio on secant modulus degradation for tension-dominated tests. Again, the larger the stress amplitude, the shorter the life, even though all tests had the same maximum tensile stress.

Radiographs for selected samples near failure are shown in Figure 8. The R = -0.2 sample showed primarily axial splitting and transverse cracks in an area of about one hole diameter in all directions. Delamination are not readily evident from the radiograph. For the R = -0.6 test, the damage zone decreased and became more local to the hole and, specifically, in the area tangent to the hole where delaminations started. For the R = -1 samples, in addition to the delaminations, a transverse crack was observed to have started at the hole boundary and moved transverse to the loading direction to the outer edge. In short, an increase in stress amplitude produced a damage zone more local to the hole. This was the same phenomenon was observed on tests of Toray carbon TTT-reinforced samples.

The samples tested under compression-dominated R-ratios exhibit similar results (see Figure 9). The larger the stress amplitude, the shorter the life, even though all tests had the same maximum compressive stress. Also, as shown in Figure 10, damage zone increased as stress amplitude increased. Again, even under compression-dominated loading at R = -5, the primary damage mode is axial splitting. Ply-cracks and delamination are also observed.

As for mean stress effects, a close examination of Figures 7 and 9 indicate that secant modulus degradation was independent of mean stress. For instance, the R = -0.2 and R = -5 tests had similar secant modulus curves, as did the pairs R = -0.6, R = -1.67 and R = -0.8, R = -1.25.

Figure 11 shows a cross-section of a sample that underwent R = -1 loading. The outer 0^0 ply is rich with matrix cracks and has delaminated. A matrix crack can be seen in the second 0^0 ply also (the third ply from the bottom), although there is significantly less damage than that observed in the outer ply. The 90^0 ply has also delaminated from the 0^0 plies. This behavior was common under the R = -1 condition. Figure 12 shows a cross-section of a sample that underwent compression-dominated loading (R = -1.67). Delaminations between all plies are observed. A compression failure in both the 90^0 ply and the 0^0 ply is observed (left-center of picture). Photomicrographs for tension-dominated loading were not available at the time of this publication.

CONCLUSIONS

The objective of this research was to examine the effect of R-ratio on failure modes for TTT-reinforced composite materials. It was anticipated that as the R-ratio moved from one that was compression-dominated to one that was tension-dominated, the failure mode would at some point transition respectively. Tests were run in the range from predominantly tension to predominantly compression, and all samples failed with some form of longitudinal splitting around the open-hole, transverse cracks growing from the open-hole region to the outer boundaries, and delamination in the region local to the hole. Photomicrographs from a compression-dominated test showed local compression failures in plies and delamination

As expected, the TTT-reinforcement decreased the intensity of the delaminations that occurred. Damage growth was more dependent on stress amplitude than on the maximum tensile or the maximum compressive stress. For instance, the fully-reversed tests, with a larger stress amplitude, always failed much earlier than the predominantly tension-dominated test, with a smaller stress amplitude, even though the maximum tensile stress was the same. Similar behavior was observed with the compression-dominated tests; the fully-reversed tests failed much earlier than the predominantly compression-dominated test, even though the maximum compressive stress was the same. A larger stress amplitude caused localization of damage mechanisms around the open-hole area, whereas smaller stress amplitudes caused a more dispersed region of longitudinal and transverse splitting, sometimes as far as two hole diameters away.

Lastly, secant modulus degradation proved to be a good indicator for estimating fatigue life. Although all samples exhibited a gradual decrease in secant modulus, failure was imminent when a large and sudden decrease in secant modulus was observed.

ACKNOWLEDGMENTS

This work was conducted under NASA Grant 13100-Z125. Samples were supplied by the NASA Langley Research Center.

REFERENCES

Portanova, M.A., Poe, C.C., and Whitcomb, J.D., *ASTM STP 1120,* ASTM, Philadelphia, 1992, pp. 37-53.

Schulte, K., *ASTM STP 1185,* ASTM, Philadelphia, 1994, pp. 278-305.

Serdinak, T.E., *Post-Impact Fatigue of Cross-Plied, Through-the-Thickness Reinforced carbon/Epoxy Composites,* Thesis, Clemson University, 1994.

Stinchcomb, W.W., Reifsnider, K.L., Marcus, L.A., and Williams, R.A., *ASTM STP 569,* ASTM, Philadelphia, 1975, pp. 115-129.

Wolterman, R.L., Kennedy, J.M., and Farley, G.L., *ASTM STP 1156,* ASTM, Philadelphia, 1993, pp. 473-490.

FIGURES

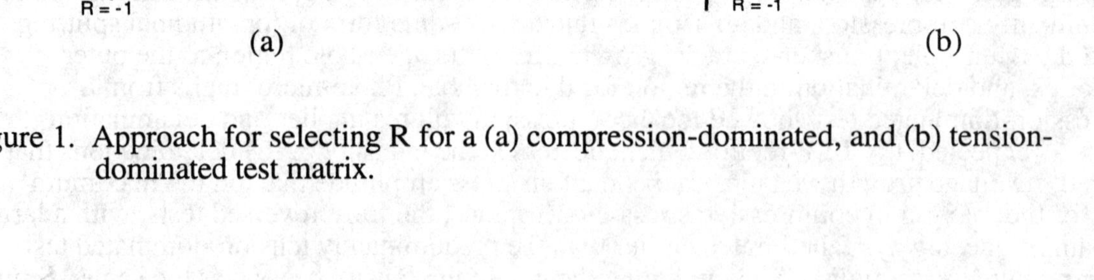

Figure 1. Approach for selecting R for a (a) compression-dominated, and (b) tension-dominated test matrix.

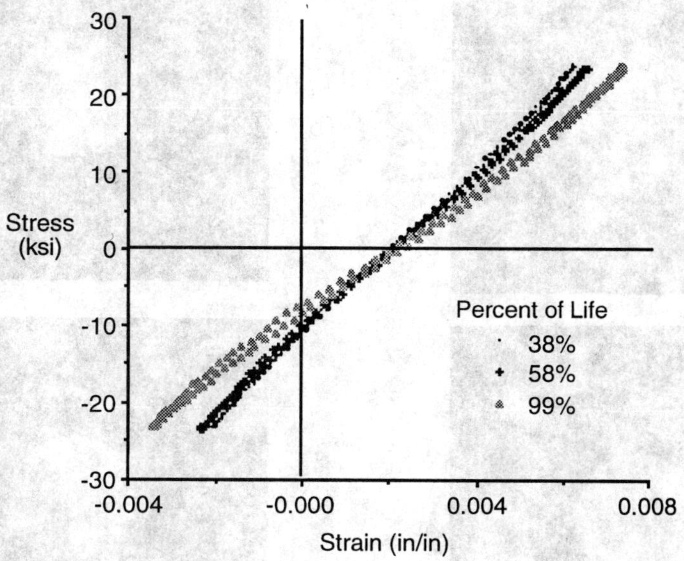

Figure 2. A representative diagram showing far-field stress versus average strain across the open-hole at selected life intervals for fully-reversed loading. The TTT-fiber type is Kevlar, and the stress level was 55% σ_c.

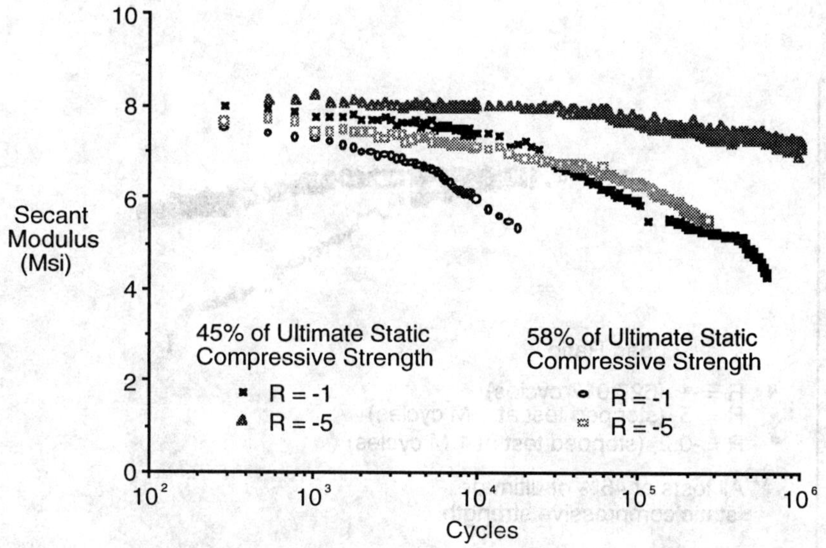

Figure 3. Secant modulus versus cycles for Toray carbon TTT-reinforcement fiber type at different R-ratios and stress amplitudes.

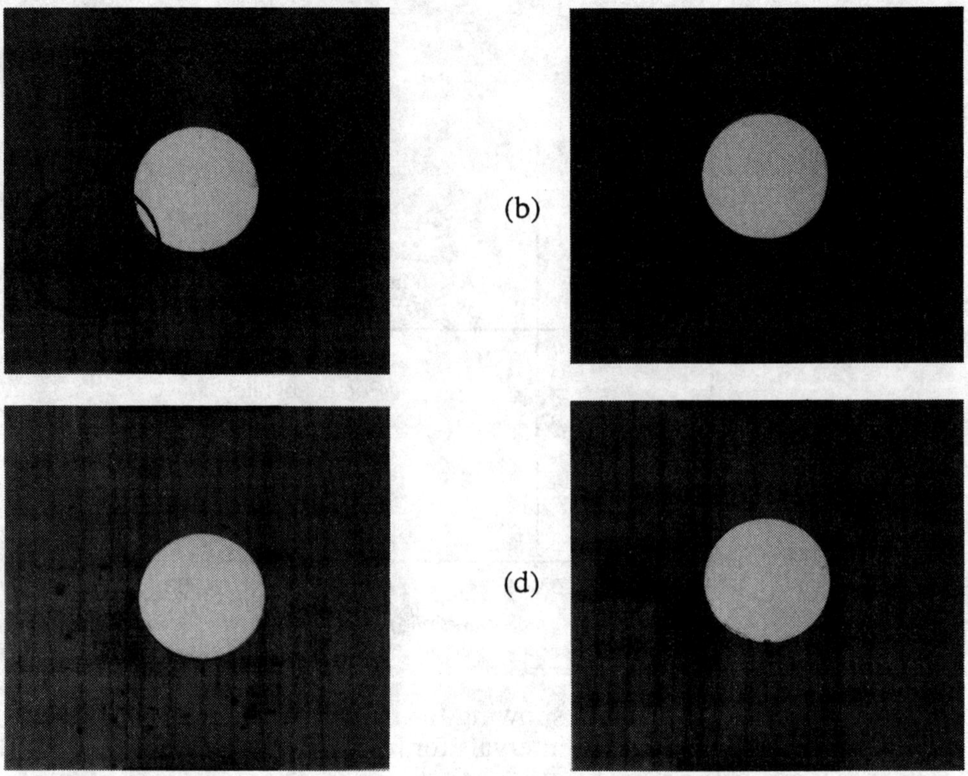

Figure 4. Damage near failure for Toray carbon TTT-reinforcement at (a) R = -1, 45% σ_c, ~79% of life, (b) R = -1, 58% σ_c, near failure, (c) R = -5, 45% σ_c at 1 M cycles, (d) R = -5, 58% σ_c at ~91% of life.

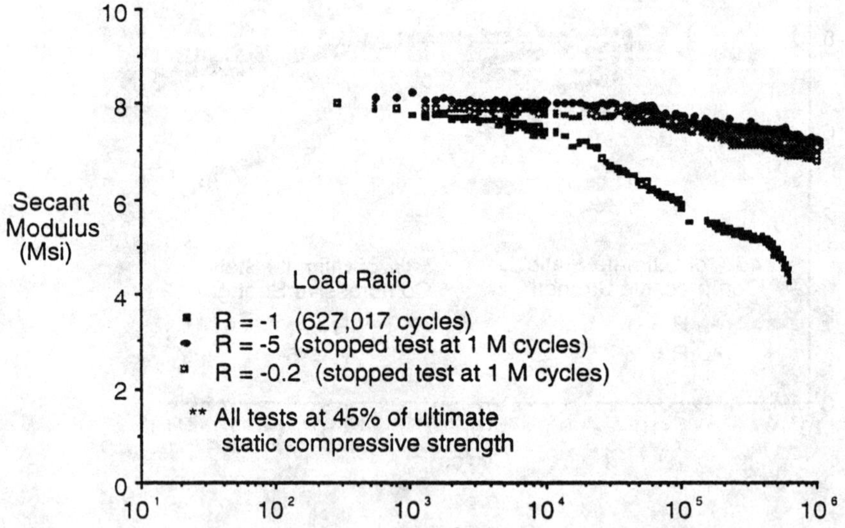

Figure 5. Secant modulus versus cycles for compression-dominated loading (R = -5) of Toray carbon TTT-reinforcement

(a) (b)

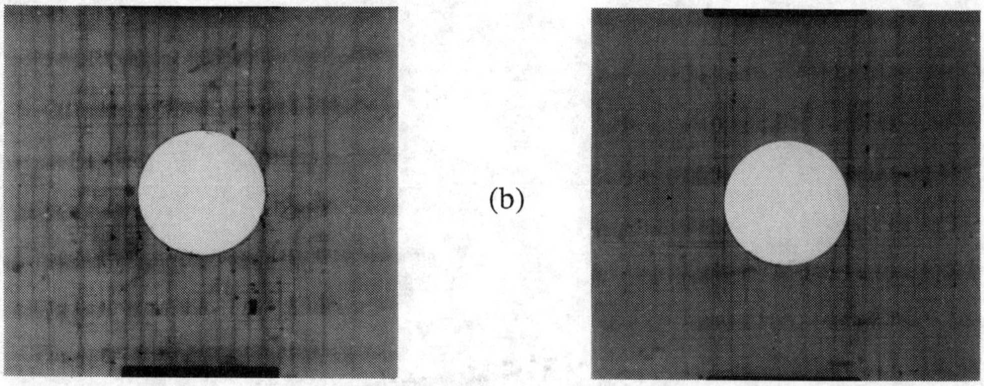

Figure 6: Radiographs at 70% of life for (a) R= -5, (b) R= -0.2 at 45% σ_c for Toray carbon TTT-reinforcement fiber type.

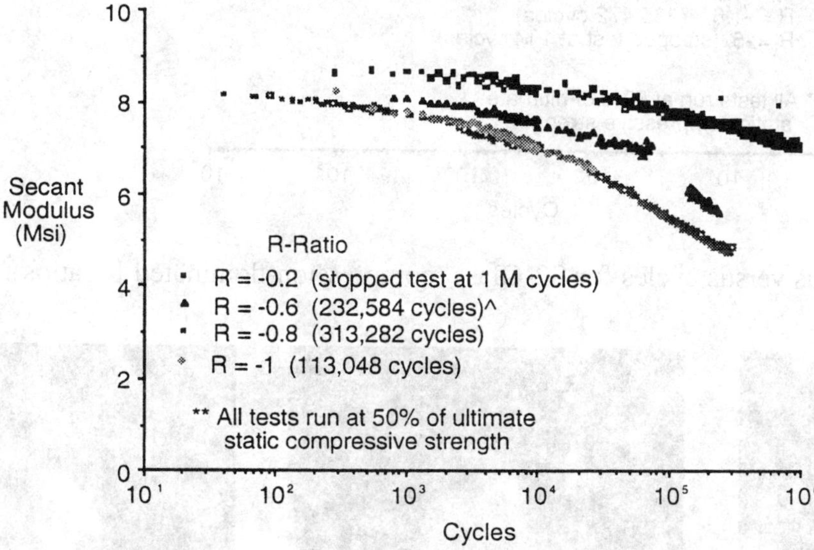

Figure 7: Secant modulus versus cycles for S2-Glass (tension-dominated R-ratios).

(a) (b) (c)

Figure 8: Damage near failure for S2-Glass TTT-reinforced: (a) R = -0.2 at 1M cycles, (b) R = -0.6 near failure, (c) R = -1 at 93% of life. Tests at 50% σ_c.

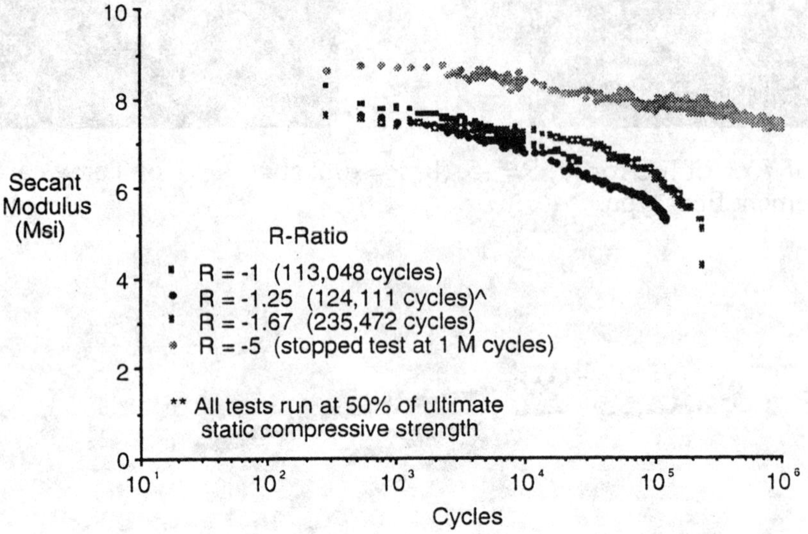

Figure 9: Secant modulus versus cycles for S2-Glass (compression-dominated R-ratios).

(a) (b) (c)

Figure 10: Damage near failure for S2-Glass TTT-reinforcement (compression dominated R-ratios): (a) R = -5, at 1 M cycles, (b) R = -1.67, at ~97% of life, (c) R = -1, at ~93% of life. Tests at 50% σ_c.

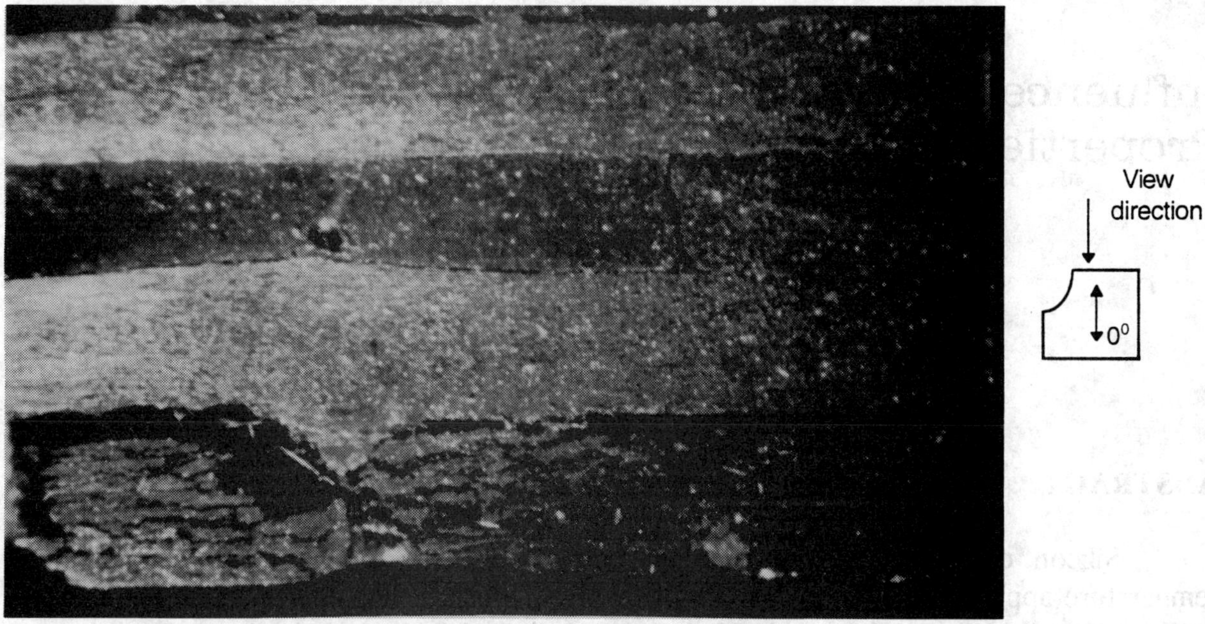

Figure 11. Photomicrograph of cross-section for S2-glass TTT-reinforced sample after R = -1 loading, 50% σ_c.

Figure 12. Photomicrograph of cross-section for S2-glass TTT-reinforced sample after R = -1.67 loading, 50% σ_c.

Influence of Interfacial Debonding on the Properties of Metal Matrix Composites

J. ZHU AND S. LIU

ABSTRACT

Silicon carbide reinforced titanium composites are primary candidates for high temperature applications. The properties of these composites, however, may be significantly affected by fabrication and operation induced defects and inelastic behaviors of matrices so that it is essential to investigate the influence of these factors on the material properties. In this study, three micromechanics models are used to investigate the behaviors of a SCS-6/Ti-15-3 composite under transverse tension. The effect of thermal residual stress on material behaviors is evaluated. Other factors, such as interfacial friction, matrix plasticity, and different interface conditions, are also considered in the models. The results obtained from these models are compared with the experimental data.

INTRODUCTION

Metal Matrix Composites (MMCs), such as titanium matrix reinforced by silicon carbide fibers, are primary candidates for future high temperature applications, for example, gas turbine, hypersonic flight vehicles, microelectronic packaging heat sinks, and space structures. These new materials are lighter than the monolithic metallic materials and may have acceptable strength and fracture toughness. However, these materials exhibit nonlinear macroscopic stress-strain responses prior to failure. Inelastic deformation (usually attributable to plasticity, creep, or viscoplasticity of matrix materials) gives rise to the macroscopic nonlinear response. Thermal mismatch between fiber and matrix microcracking, and discrete fiber breakage), may present a reliability concern to the applications of these advanced materials (Majumdar and Newaz, 1991; Binenda, *et al.*, 1990; Gabb, *et al.*, 1991; Ghosn, *et al.*, 1990, 1991; to name a few). In order to predict the life and reliability of the components made of these materials, all these effects have to be evaluated.

Two approaches are popular in obtaining the effective macroscopic constitutive properties of composites: the phenomenological approach and the micromechanics approach.

Jiansen Zhu, Nokia Research Center, 3500 W. Balcones Center Drive, Suite 2.9134, Austin, TX 78759-5398

Sheng Liu, Mechanical Engineering Department, Wayne State University, 5050 Anthony Wayne Drive, Detroit, MI 48202

The former utilizes the experimental data from standard testing coupon specimens to build constitutive equations, while the latter typically uses the unit cell technique combined with homogenization schemes analytically or computationally in a microscopic scale. The latter method can handle the complications such as local damage and its evolution in fibers, in matrix, and at interfaces. Material inelasticity can be implemented systematically in numerical schemes such as finite element methods.

The effect of debonding on the mechanical behaviors of composites can be studied by introducing the models which simulate failure modes within the constituent phases and at their interfaces. Discussion of several popular models has been presented in recent papers, such as in Tandon and Pagano (1993), in terms of modeling debonding. Both analytical and finite element numerical models have been proposed. In most cases, only linear elastic properties for both fiber and matrix are considered. Most models have been limited to fully debonded and frictionless interfaces between the fiber and matrix (Tandon and Pagano, 1993).

For interfacial delamination modeling, the inherit difficulty in implementing contact constraints was discussed in details (Liu, *et al.*, 1991, 1993; Liu, 1992). It is even more difficult if friction is considered at the debonded interface when arbitrary sliding is involved (Simo and Laursen, 1992; Liu and Zhu, 1994b, 1995b).

The current work is aimed at understanding the actual interface condition for metal matrix composites, A robust contact model is utilized to consider partial or full debonding, friction effect, and arbitrary slippage. The materials investigated in this study are composites made of Ti-15-3Cr-3Al-3Sn alloy and silicon carbide (SCS-6) fibers. The material properties are listed in Table 1. This titanium alloy is a high strength, metastable beta strip alloy with good weldability and corrosion resistance. The composites were fabricated by consolidating thin sheets of the matrix interspersed by SCS-6 fibers, using the hot isostatic pressing (HIP) technique. The consolidation temperature was 815°C. The composites were tested at room temperature (25°C) and high temperature (538°C) by (Santhosh, *et al.*, 1992). Therefore, thermal mismatch was induced by the temperature difference between consolidation and testing. Flat tension test specimens were machined from rectangular panels by the electric discharge machining (EDM) technique. The tests were carried out on an MTS 810 tensile testing machine with a 25 kN load cell and a 1.5 kW heater. Strain measurements were conducted using an MTS quartz rod extensometer. The test results are used here to check the suitability of the results obtained from the finite element simulations in the following section.

MODELING

The unit cell model technique is adopted in this investigation to study the micromechanics properties of metal matrix composites. The unit cell model is a popular technique which has been widely used in micromechanics study of composites (Allen and Boyd, 1993; Aboudi, 1990; Liu and Zhu, 1994b, 1995; for example) and recently in other fields, such as powder forming (Ren,*et al.*, 1994; Zhu, *et al.*, 1996).

It is assumed that fibers are uniformly distributed in matrix and have the same radii. Therefore, a unit cell square model containing a quarter of one fiber, Model *I*, is extracted from metal matrix composites. The top edge is free or is constrained to move parallelly in order to consider different locations of fibers. Furthermore, two models containing three fibers, Model *II* and Model *III*, are constructed to study the interaction between the fibers. For Model *I*, the lower and left boundaries are restrained to be able to move only vertically and horizontally, respectively. For Model *II*, the left boundary is restrained to be able to move vertically and the top and lower boundaries are free. For Model *III*, the left and lower boundaries are restrained to be able to move vertically and horizontally, respectively. The top edge is constrained to move parallelly. A controlled displacement is applied to the right boundary to simulate transverse tension for all models (Fig. 1).

The large deformation theory is used in this study (Bathe, 1985; Fung, 1965; Marsden and Hughes, 1981). The high temperature ductile material behavior is represented by the J_2 flow theory, using the finite strain generalization(Hutchinson, 1973).

Deformable solid contact elements are used to model the contact if there is debonding between fiber and matrix (Simo and Laursen, 1992; Liu, 1992; Liu and Zhu, 1994a), which are used to keep the fiber and matrix away from overlapping. The conditions can be expressed as

$$\begin{aligned} \lambda_N &\le 0 \\ g(u_i) &\ge 0 \\ \lambda_N g(u_i) &= 0 \\ \lambda_N g(u_i) &= 0 \end{aligned} \tag{1}$$

where λ_N is the contact force and g is the gap between two surfaces. If the friction is considered, the following Kuhn-Tucker conditions are used if the coulomb friction law is adopted,

$$\begin{aligned} \Phi = \|\lambda_T\| - \mu\lambda_N &\le 0 \\ u_T = \xi \frac{\partial \Phi}{\partial \lambda_T} & \\ \xi &\ge 0 \\ \xi\Phi &= 0 \end{aligned} \tag{2}$$

where λ_T is the tangential force along the interface; $\mu \ge 0$ is the coefficient of friction; and ξ is the norm of slip rate.

Figure 2 shows a typical mesh for Model *I*, which contains 925 nodes and 876 4-node plane strain elements. Different meshes, although not shown here, have been generated for studying the effect of mesh sizes. In both Model *II* and Model *III*, there are 2129 nodes and 2064 4-node plane strain elements (Fig. 3). If debonding is included in modeling, double nodes and deformable solid contact elements are added along the debonded interface.

RESULTS AND DISCUSSION

The results of finite element simulations will be presented and discussed in this section. First, it is interesting to investigate the thermal residual stress induced by cooling from the consolidation temperature to the testing temperature. Figure 4 and Figure 5 present the results of tension with and without considering thermal residual stress at different temperatures. It is shown that the thermal residual stress has a significant effect on the material properties when the interface between the fiber and matrix is debonded. When the interface is debonded, the initial part of homogenized stress-strain curves cannot match the experimental data if the thermal residual stress is not considered. When the interface is bonded, however, the influence of the thermal residual stress is not so significant. From these two figures, it is also shown that the actual interface should be somewhere between a bonded interface and a fully debonded interface.

When the thermal residual stress is considered, the fiber/matrix interface is in compression because of the thermal mismatch induced by temperature drop. A fine mesh is needed to obtain a smooth curve. Otherwise, there is a "overshot" in the homogenized stress-strain curves when the material enters the stage of plasticity, as shown in Fig. 6. The coarser the mesh, the larger the "overshot" is. The explanation is that the composite stiffness drops suddenly when the applied load opens the compressed interface. When the mesh is sufficiently fine, the "overshot" is almost eliminated. If the friction is considered between the debonded interface, however, the phenomenon of "overshot" may also be eliminated (Fig. 7).

Figure 8 and Figure 9 show the deformed meshes with both the bonded interface and the fully debonded interface under tensile loading at room temperature. Figure 10 shows the equivalent plastic strain for the debonded interface model under tension at room temperature. It was observed that there is a strain concentration in the matrix near the interface in the direction of about 30° measured from horizontal axis. This is due to the combined effect of tension and shearing.

It was observed that the fully debonded interface does not represent the true picture of the interface, as the homogenized stress-strain curves lie below the experimental data. If the top edge is constrained to move parallelly to simulate a fiber within the composite, the homogenized stress-strain response will be enhanced. This is demonstrated in Fig. 11, which shows the results at room temperature. At high temperature, constraining the top edge to move parallelly exhibits the same enhancement. Figure 12 shows a deformed mesh under tension at high temperature. It is interesting to note that the interface is fully separated and there is no contact between the fiber and matrix. In this case, the fiber has no contribution to the material properties of transverse tension.

It was observed that for Model *I*, the experimental data can only be matched by the partially debonded interface model. Figure 13 shows the results obtained from a partially debonded model. It is shown that 41.7% bonded interface model presents a reasonable result as compared with experimental data. We would like to point out that this partial debonding is not unique as other distributions of debonding could also match the experimental data. Figure 14 shows a deformed mesh of partially debonded model at room temperature. The same conclusion can be drawn for the case of high temperature. Figure 15 shows the results of a partially debonded interface model under tension at high temperature. These results show again that the actual interface condition is somewhere between a bonded interface and a fully debonded interface. It is also interesting to note that the same debonding model with 41.7% bonded interface fits both room temperature and high temperature experimental data well.

Model *II* and Model *III* are used to consider the interaction between fibers. Figure 16 shows the results from different models with all fiber/matrix interfaces fully debonded. It is noted that Model *III* presents the highest homogenized stress-strain curve and Model *I* with free top edge shows the lowest response. The reason is that Model *III* has more restraints than Model *II* and Model *I* has the least restraint along the boundary. Figure 17 presents the deformed mesh of Model *II* with all three fiber/matrix interfaces debonded under tension at room temperature. It was observed that the debonded interface is in partial contact.

If the middle fiber/matrix interface is bonded and other two interfaces are fully debonded, Model *II* and Model *III* present different results. The stress-strain responses obtained from Model *II* is higher than that of experiments (Fig. 18). However, the Model *III* shows a result which is very close to the experimental data (Fig. 19). This illustrates again that the actual interface condition is somewhere between a bonded interface and a fully debonded interface. Figure 20 shows a deformed mesh of Model *III* under tension at room temperature, in which the debonded interface is fully separated. This result is the same as that obtained from Model *I*.

CONCLUSIONS

Three models using the unit cell technique are presented here to investigate the transverse tensile behaviors of unidirectional SCS-6 reinforced titanium composites under transverse tension at different temperatures. The results obtained from finite element simulations are compared with the experimental data. The following conclusions can be drawn from this study:

(1) in order to avoid the "overshot", a sufficiently fine mesh should be used to study the coupling between thermal and mechanical responses;

(2) thermal residual stress is important in studying the behaviors of metal matrix composites; and

(3) the actual interface condition between fiber and matrix is somewhere between a fully bonded interface and a fully debonded interface.

REFERENCES

Aboudi, J., 1990, "Micromechanical Prediction of Initial and Subsequent Yield Surfaces of Metal Matrix Composites," *International Journal of Plasticity, Vol. 6,* pp. 471-484.

Allen, D.H. and J.G. Boyd, 1993, "Convergence Rates for Computational Predictions of Stiffness Loss in Metal Matrix Composites," *Composite Materials and Structures, AMD-Vol. 179/ AD-Vol. 37,* ed. by C.W. Bert, V. Birman, and D. Hui, pp. 31-45.

Bathe, K.J. and A. Chaudhary, 1985, "A Solution Method for Planar and Axisymmetric Contact Problems," *International Journal for Numerical Methods in Engineering*, 21, pp. 65-88.

Binenda, W.K., D.N. Robinson, S.M. Arnold, and P.A. Bartolotta, 1990, "A Creep Model for Metallic Composites Based on Metal Testing: Application to Kanthal Composites, " *NASA TM 103172.*

Fung, Y.C., 1965, "Foundations of Solid Mechanics," Prentice-Hall, Inc., New Jersey.

Gabb, T.P., J. Gayda, and R.A. Mackay, 1990, "Isothermal and Nonisothermal Fatigue Behavior of a SiC/Ti Composite," *Journal of Composite Materials, Vol. 24*, pp. 667-686.

Ghosn, L.J., P. Kantzos, and J. Telesman, 1991, "Modeling of Crack Bridging in a Unidirectional Metal Matrix Composite," *NASA TM 104355*, Lewis Research Center.

Hutchinson, J.W., 1973, "Finite Strain Analysis of Elastic-Plastic Solids and Structures," *Num. Solu. Nonlin. Struct. Prob.*, (ed. by R.F. Hartung), 17, ASME, New York.

Lerch, B.A., D.R. Hull, and T.A. Leonhardt, 1990, "Microstructure of a SiC/Ti-15-3 Composite," *Composites, Vol. 21*, pp. 216-224.

Lerch, B.A. and J.F. Saltsman, 1991, "Tensile Deformation Damage in Sic Reinforced Ti-15V-3Cr-3Al-3Sn," *NASA TM 103620*, Lewis Research Center.

Liu, S., 1992, "Damage Mechanics of Cross-Ply Laminates Resulting from Transverse Concentrated Loads," Ph.D. Thesis, Mechanical Engineering Dept., Stanford University .

Liu, S., A. Kutlu, and F.K. Chang, 1991, "Matrix Cracking and Delamination in Laminated Polymeric Composites Resulting from Transverse Concentrated Loading," *Proceedings of the 1st International Conference on Deformation and Fracture of Composites*, Manchester, England, March 25-27, pp. 30/1-30/7.

Liu, S., Z. Kutlu, and F.K. Chang, 1993a, "Matrix Cracking and Delamination Propagation in Laminated Composites Subjected to Quasi-Static Transverse Concentrated Loading," *J. of Composite Materials, Vol. 27, No. 5*, pp. 436-470.

Liu, S. and J.S. Zhu, 1994a, "Behavior of Debonded Thin Film Substrate Systems Subjected to Microindentation," *Recent Advances in Structural Mechanics, ASME 1994, PVP-Vol. 295/NE-Vol. 16*, pp. 177-189.

Liu, S. and J.S. Zhu, 1994b, "Micromechanical Behavior of Metal Matrix Composites Containing Cracking and Debonding," *Durability of Composite Materials, ASME 1994, MD-Vol. 51* , pp. 191-206.

Liu, S. and J.S. Zhu, 1995, "Micromechanics of High Temperature Composites Containing Cracking and Debonding," *Journal of Damage Mechanics, Vol. 4, No. 4,* Oct., pp. 380-401.

Majumdar, B.S. and G.M. Newaz, 1991, "Thermomechanical Fatigue of A Quasi-Isotropic Metal Matrix Composite," *Composite Materials: Fatigue and Fracture, Third Volume, ASTM STP 1110*, T.K. O'Brien, Ed., ASTM 1991, pp. 732-752.

Marsden, J.E. and T.J.R. Hughes, 1981, "Mathematical Foundations of Elasticity," Academic Press, Inc., New York.

Ren, F., A. Chandra, and V. Tvergaard, 1994, "A Unit Cell Analysis of Ti-Al Powder Compaction at Elevated Temperatures," *Mechanics in Materials Processing and Manufacturing, ASME 1994, AMD-Vol. 194*, pp. 239-260.

Santhosh, U., J. Ahmad, and A. Nagar, "Nonlinear Micromechanics Analysis Prediction of the Behavior of Titanium Alloy Matrix Composites," *Fracture and Damage, ASME 1992, AD-Vol. 27*, pp. 65-76.

Simo, J.C. and T.A. Laursen, 1992, "An Augmented Lagrangian Treatment of Contact Problems Involving Friction," *Computers and Structures, Vol. 43, No. 1*, pp. 97-116.

Tandon, G.P. and N.J. Pagano, 1993, "Effective Moduli of Unidirectional Fiber Composite Containing Interfacial Arc Microcracks," *Thermomechanical Behavior of Advanced Structural Materials, AD-Vol. 34/AMD-Vol. 173*, pp. 113-128.

Zhu, J.S., S. Liu, and A. Chandra, 1996, "Study of Powder Sintering-Compaction by A Micromechanics Model," *J. of Engineering Manufacture, Proceedings Part B*.

TEMPERATURE (°C)	25	482	538	566	650
TITANIUM ALLOY (Ti-15-3)					
TENSILE MODULUS (GPa)	86.3	72.2	77.8	64.4	53.0
POISSON'S RATIO	0.36	0.36	0.36	0.36	0.36
YIELD STRENGTH (MPa)	763	577	447	287	198
TANGENT MODULUS (GPa)	3.20	3.50	2.60	2.30	1.10
CTE (μm/m/°C)	8.48	9.71	9.87	9.98	10.26
SILICON-CARBIDE (SCS-6)					
TENSILE MODULUS (GPa)	393	376	374	373	370
POISSON'S RATIO	0.25	0.25	0.25	0.25	0.25
CTE (μm/m/°C)	3.56	3.99	4.07	4.12	4.27

Table 1. Material Properties

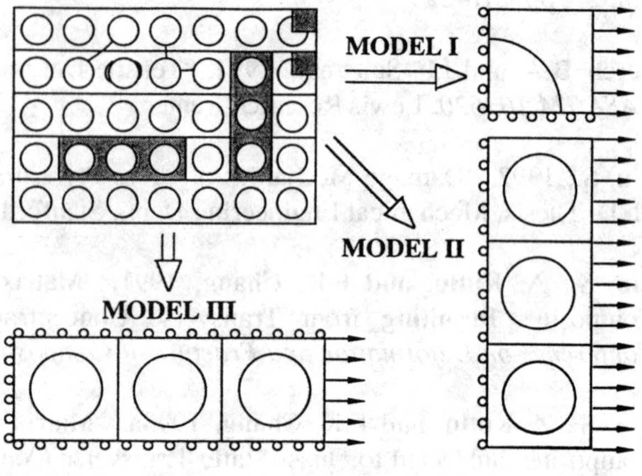

Figure 1. Schematic of three models using unit cell technique.

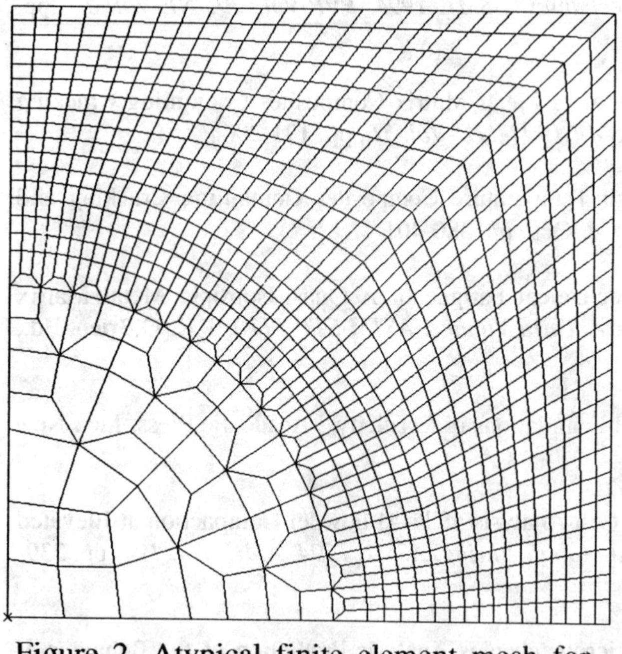

Figure 2. Atypical finite element mesh for Model I containing a quarter of a fiber.

Figure 3. A typical finite element mesh for both Model II and Model III containing three fibers.

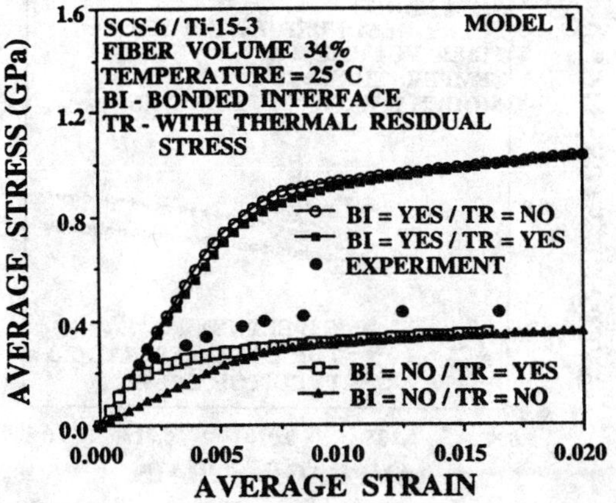

Figure 4. Homogenized stress-strain relationship obtained from different interfaces with and without thermal residual stress at room temperature.

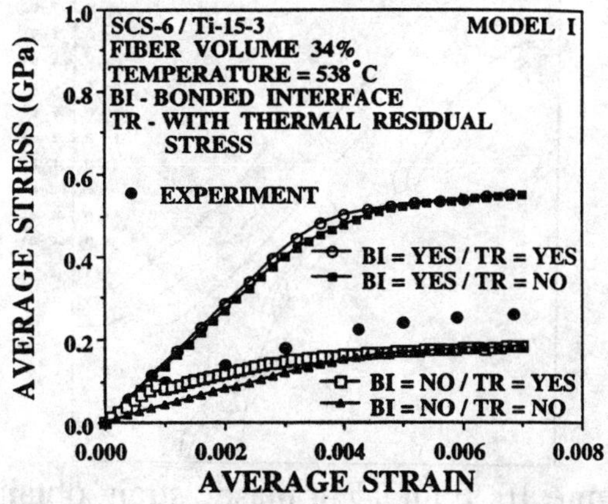

Figure 5. Homogenized stress-strain relationship obtained from different interfaces with and without thermal residual stress at room temperature.

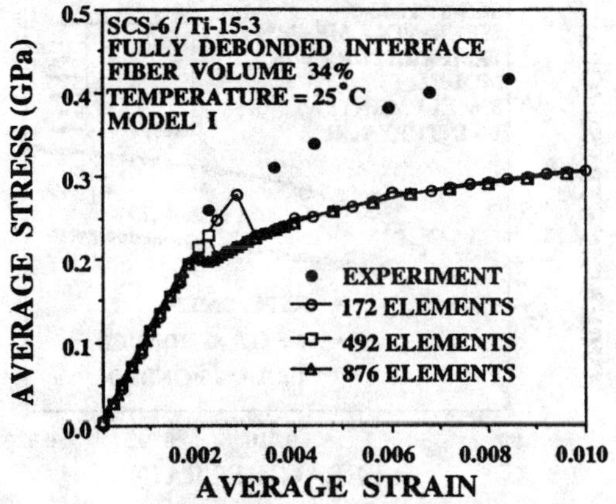

Figure 6. Homogenized stress-strain relationship obtained from different meshes at room temperature.

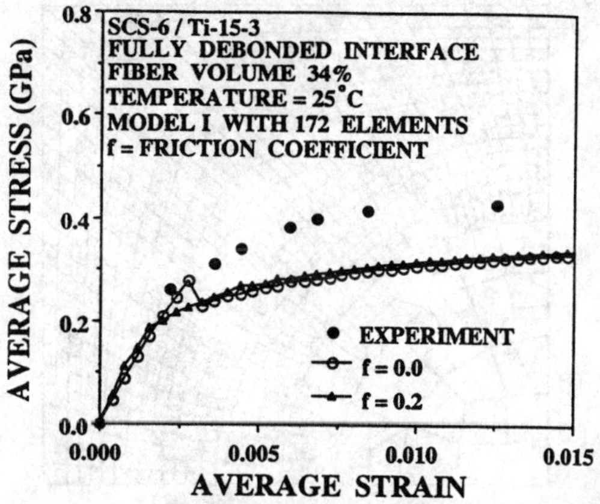

Figure 7. Homogenized stress-strain relationship obtained from fully debonded interface with and without interface friction at room temperature.

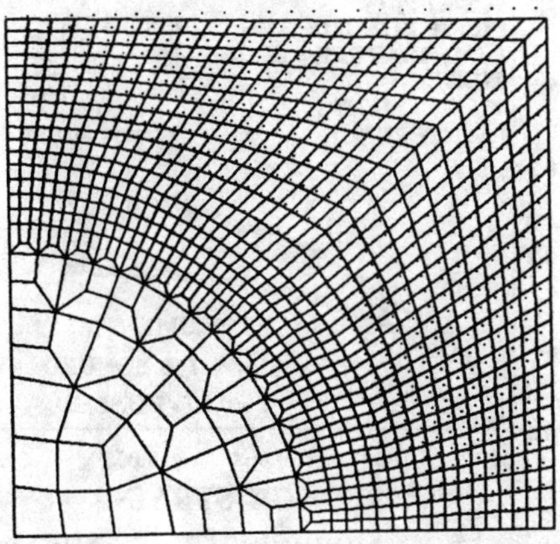

Figure 8. A deformed mesh without debonding at room temperature.

Figure 9. A deformed mesh with fully debonded interface under tension at room temperature.

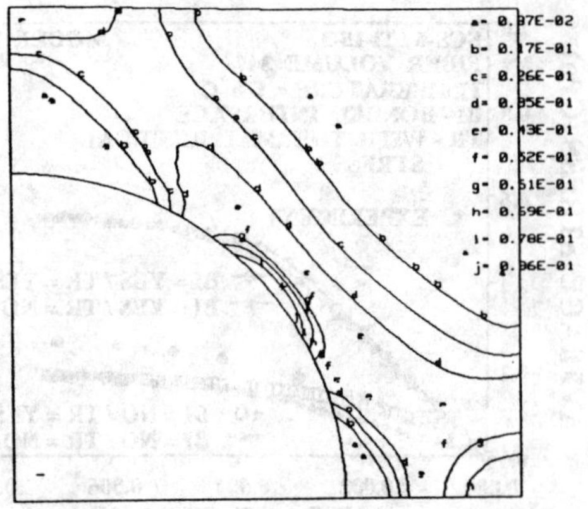

Figure 10. Equivalent plastic strain obtained from the model without debonding under tension at room temperature.

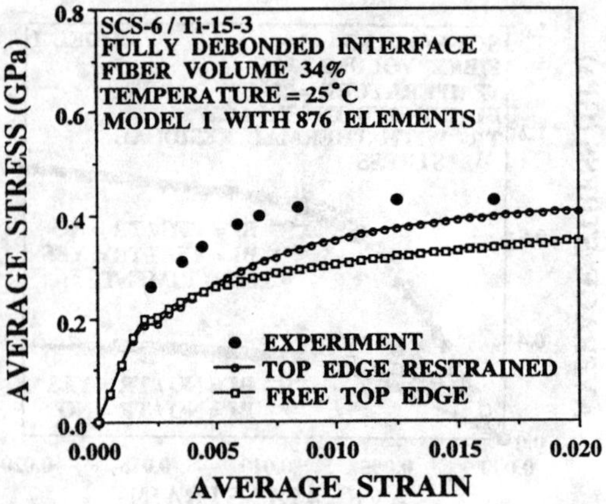

Figure 11. Homogenized stress-strain relationship obtained from fully debonded interface model with top edge restrained to move parallelly.

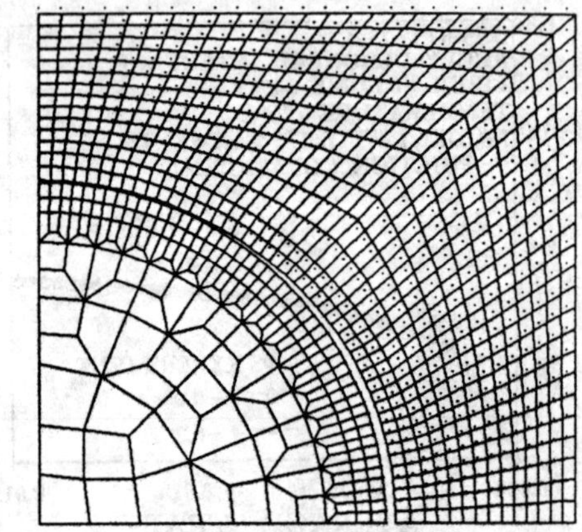

Figure 12. A. deformed mesh of fully debonded interface model with top edge restrained to move parallelly.

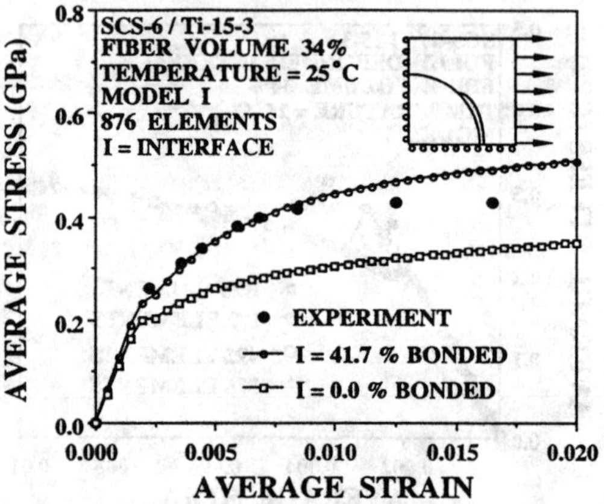

Figure 13. Homogenized stress-strain relationship obtained from Model I with partial debonded interface at room temperature.

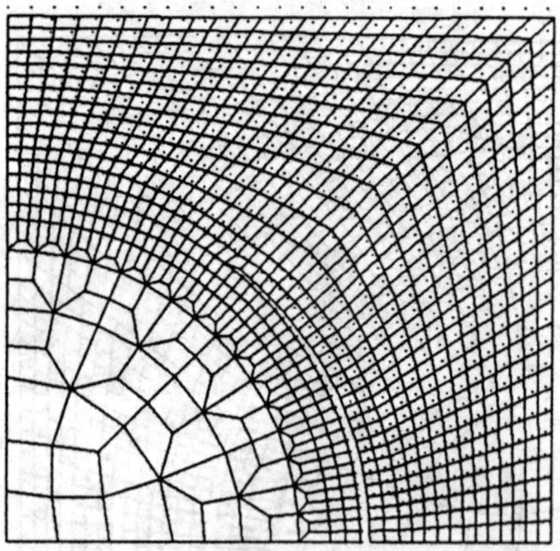

Figure 14. A deformed mesh of Model I with partial debonded interface at room temperature.

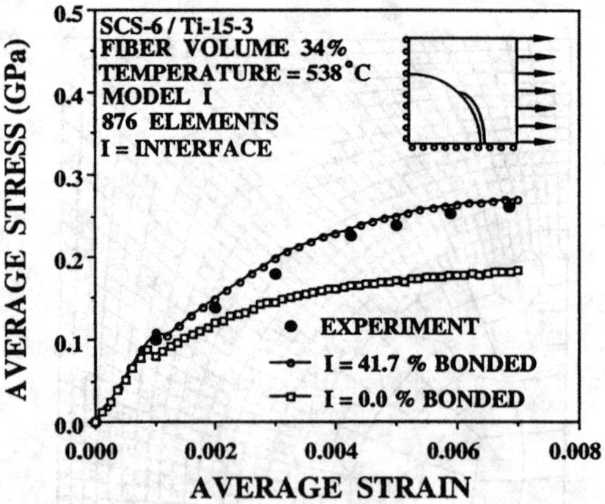

Figure 15. Homogenized stress-strain relationship obtained from Model I with partial debonded interface at high temperature.

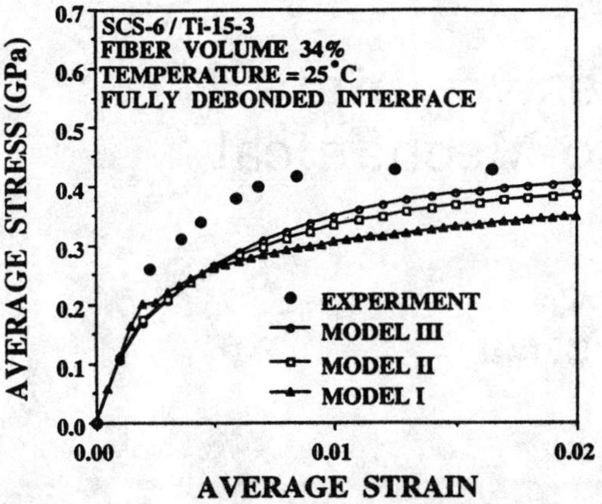

Figure 16. Comparison of homogenized stress-strain relationship obtained from different models with fully debonded interface at room temperature.

Figure 17. A deformed mesh of Model II with all three fiber/matrix interfaces fully debonded at room temperature.

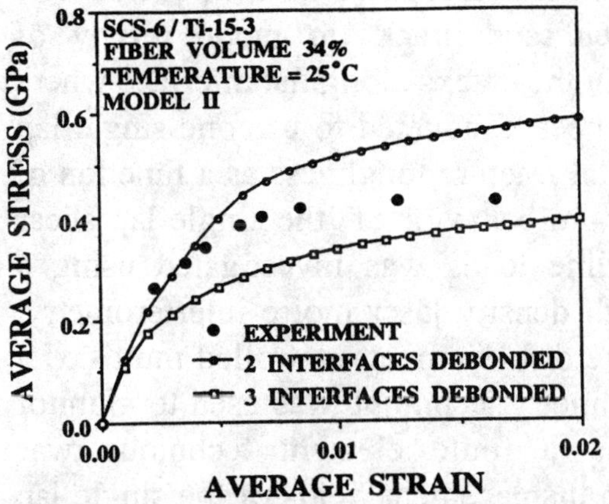

Figure 18. Homogenized stress-strain relationship obtained from Model II with middle fiber/matrix interface bonded and other two fully debonded.

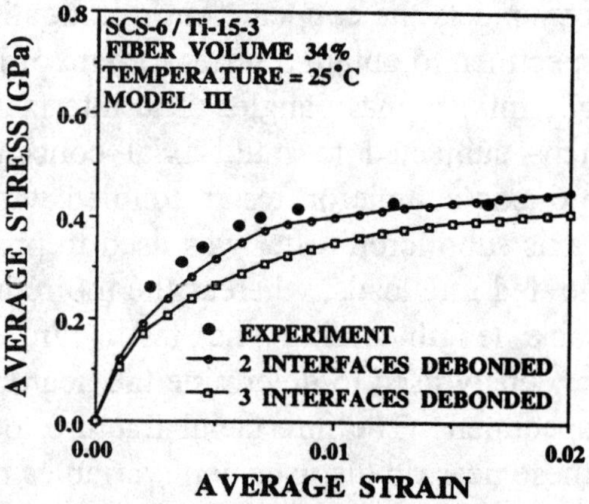

Figure 19. Homogenized stress strain relationship obtained from Model III with middle fiber/matrix interface bonded and other two fully debonded.

Figure 20. A deformed mesh of Model III with middle fiber/matrix interface bonded and other two fully debonded at room temperature.

Studying of Delaminated Adhesive Joints by Using a Multi-Axial Thermo-Mechanical Fatigue Tester

M. LU, J. WANG AND S. LIU

ABSTRACT

It is very difficult to measure the multiple points of the interfacial fracture toughness with a single specimen by using conventional experimental facilities. In this paper, a new technique was introduced to control the far field and crack tip mode mixity of copper/epoxy/copper single lap shear specimens with the cracks along the interface where the epoxy meets the copper. An investigation has been conducted to use one single lap shear specimen to obtain a series of data of interfacial fracture toughness as a function of the mode mixity phase angle. The interfacial fracture behavior of the single lap shear specimens subjected to multi-axial concentrated line loads was investigated using a unique 6-axis submicron tester coupled with a high density laser moiré interferometry. The 6-axis submicron tester was used to provide the displacement-controlled multi-axial concentrated line loads, whereas the moiré interferometry technique was used to monitor the crack length during the test. In addition, a finite element technique was simultaneously used to determine the near crack tip displacement fields of the single lap shear specimens. The interfacial fracture toughness and phase angle were computed by using these near tip displacement variables through the analytical energy release rate and phase angle expressions derived by authors. The results show the great advantages of using the new technique for the measurement of mixed-mode interfacial fracture toughness.

INTRODUCTION

The mechanical integrity of layered structures such as composite materials and electronic packages is directly determined by the adhesive strength of the interfaces between these dissimilar materials [Liu *et al*, 1993, 1994, 1995]. Therefore, experimental measurement of the adhesive strength between interfaces in layered structures is of critical importance for the determination of the overall layered structure reliability. In particular, increasing the adhesive strength of the interfaces may significantly help

---

Minfu Lu, Jianjun Wang and Sheng Liu, Department of Mechanical Engineering, Wayne State University, Detroit, MI 48202

improve the life of the layered structures. Therefore, precisely measuring and evaluating the adhesive strength of the layered structures has become a major concern.

It is common knowledge that one of the most usually encountered failure modes in layered structures is delamination. Unlike crack problems in homogeneous media, delamination cracks found in layered structures are often subjected to mixed-mode loading situations. Application of fracture mechanics concepts in the measurement of interface bond strength is now commonly recognized as the preferred methodology for characterizing interface integrity. It is also commonly recognized that the fracture conditions of interest at the bond interface are mixed mode in nature, i.e., conditions with combined K_I and K_{II} stress intensity factors. However, only a qualitative assessment of crack resistance can be measured using Mode I tests and there definitely exists a level of uncertainty as to how to apply such data in the design of new layered structures.

A failure envelope containing interfacial fracture toughness as a function of mode-mixity would allow a more quantitative assessment of the interfacial crack resistance and facilitate the design of new layered structural devices. It is common practice to use a number of different specimen shapes such as double cantilever beam and edge notched flexure. It is obvious that such a strategy can be time-consuming and expensive. Moreover, it is often difficult to produce the different specimen shapes using identical processing conditions. Therefore, it would be useful to evaluate test methods that use the same specimen geometry, but vary the geometry of the test fixture or change the ratio of the displacement-controlled multi-axial concentrated line loads to control the mode-mixity conditions at the crack tip.

In an effort to perform fracture tests under mixed-mode conditions, Reeder and Crews [Reeder *et al*, 1990] have developed a mixed-mode bending test to measure interlaminar fracture toughness as a function of mode-mixity for carbon reinforced epoxy composites. Such a test method uses the same specimen geometry and varies the mode-mixity at the crack-tip by changing the fixture settings. Other investigators [Goodle *et al*, Brandenburger *et al*, 1995] have also developed various test fixtures that permit control of the mixed-mode conditions, through control of the ratio of applied shear stress to normal stress. However, the mentioned test methods only provide one value of interfacial fracture toughness with one specimen during each test. Since the interfacial fracture toughness of the interface between dissimilar materials is related to the phase angle, a number of tests need to be conducted in order to acquire the relationship between the interfacial fracture toughness and the phase angle for a bi-material system. Such a strategy can also be time-consuming and expensive. Therefore, in order to shorten testing time and reduce the cost of test, a new facility and methodology need to be developed. They should be capable of providing a series of data of interfacial fracture toughness which are related to different phase angles with one specimen during each test. To fulfill these commissions, the objective of the current study is to introduce a new technique which can be used to control the far field of the specimen tested and the mode mixity conditions at the crack tip by changing the ratio of the displacement-controlled multi-axial concentrated line loads. In this research, the 6-axis submicron tester was used to provide the displacement-controlled multi-axial concentrated line loads, whereas the moiré interferometry technique was used to monitor the crack length during the test. In addition, a finite element technique was simultaneously used to determine the near crack tip displacement fields of the tested specimens. The interfacial fracture toughness and phase angle were computed by using these near tip displacement variables through the analytical energy release rate and phase angle expressions derived by authors.

As a demonstration, the new technique was used to measure the interfacial fracture toughness and the phase angle of the single lap shear specimen. Test setup and experimental results will be described in detail in the coming sections of this paper.

EXPERIMENTS

The key problem in the mixed-mode fracture test is to control the far field of the specimen tested and the mode mixity conditions at the crack tip by changing the ratio of the displacement-controlled multi-axial concentrated line loads. In doing so, the 6-axis submicron tester which was recently designed and built at Wayne State University [Lu, 1996] was utilized. The mini-tester is a general purpose thermal-mechanical fatigue testing machine and fully controlled by a PC computer with a user-friendly interface. This machine can apply a load to a specimen that is to be tested along six degrees of freedom, i.e.: three orthogonal translations and three rotations. The displacement resolution of each closed-loop dc-motor-drive stage is kept at 0.1 micron in translation and 0.001° in rotation. A set of integrated high temperature/cryogenic multi-grips and stages are connected to a 6-axis load cell, which will enable the user to perform tests along six axes. With other optional accessories, three-point bending, four-point bending and Brazil mixed-mode fracture tests can be implemented. A small environment chamber, which can be easily mounted, is capable of surrounding the grips to allow for testing at a ramming temperature to both extreme hot and extreme cold. This can be achieved by connecting an electric heater and bleeding liquid nitrogen through a solenoid valve. In order to achieve a desired temperature, a temperature controller should be used. The machine can work in two positions; vertical and horizontal. The vertical position is suitable for compression and three-point or four-point bending. The horizontal position is suitable for applications where the temperature chamber is being used. It is strongly recommended that the horizontal position be used for all tests except for compression testing. Figure 1 shows a typical configuration of the mini tester.

The mini tester is fully controlled by a PC. The software associated with the mini tester works well in an MS™ Windows 3.1 environment. It has a user-friendly interface. With the use of the control panel, prompted on the screen of the PC, the mini tester can run automatically and can be easily interrupted by the user at any time necessary. When using this software one can perform various material structure tests. For example, one can test the tension, compression, and torsion individually or in combinations. Using this software can also test low cycling fatigue, creep, and relaxation. Different curve fitting techniques are introduced by this software and can be used in a real-time environment.

Since the fixture of the machine can freely move along six degrees of freedom, the alignment of the tested specimen can be guaranteed. Moreover, the tester is of high precision and resolution so that it can be used to easily control the far field of the specimen tested and the mode mixity conditions at the crack tip by changing the ratio of the displacement-controlled multi-axial concentrated line loads and to precisely measure the variation of the load line deflection with time automatically. Therefore, the 6-axis mini-fatigue tester is suitable for the study of layered structure such as electronic packages, composite materials and their material behavior at an ambient temperature, ranging from -128°C to 400°C.

The high density laser moiré interferometry is a whole-field in-plane displacement measurement technique, featuring both high displacement sensitivity and high spatial resolution. It is especially effective for the non-uniform in-plane deformation measurements and has been used in the research and development for microelectronic packages [Guo *et al*, 1993, 1994, 1995]. It was, therefore, adopted to monitor the crack propagation and measure the crack length during the test.

The demonstrated experimental work was conducted to evaluate the interfacial fracture toughness of a single lap shear specimen by using this 6-axis submicron tester coupled with a high density laser moiré interferometry. The specimen is composed of two strips of copper and a layer of adhesive with an initial crack length a=3.0mm at the edge of the interface. The dimensions and geometry of the specimen considered are shown in Figure 2 (not to scale).

The grating was replicated onto the surface of the single lap samples at room temperature. Figure 3 presents the image picture of the specimens taken by a CCD camera. A portable moiré interferometer was adopted to acquire in situ displacement fringe patterns. The fringe patterns were recorded by a CCD camera and stored into a PC computer in a digital format. In addition, the computer image processing technique was used to analyze the fringe patterns.

The specimen was horizontally gripped into the fixture in order to capture the fringe patterns of the tested specimen at certain loading conditions by using the moiré interferometry facility. One end of the specimen was clamped by the grip of the mini tester, while the other end of the specimen was properly touched onto a 6-axis load cell to provide both load in x-direction and load in z-direction simultaneously as shown in Figure 4.

The test setup of the single lap shear specimen in conjunction with the high density laser moiré interferometry measurement system is presented in Figure 5, while the schematic drawing of the test setup is illustrated in Figure 6.

In the test, a single lap specimen was first gradually preloaded to a proper compression force in z-direction, and then a x-direction load was applied to the specimen as a function of time. The stage speed with respect to the loading rate is 0.001mm/s. The far field of the tested specimen and the mode mixity conditions at the crack tip can be controlled by changing the ratio of the multi-axial concentrated line loads both in x-direction and in z-direction. The load line displacement (deflection) and the values of load with time were automatically recorded by a computer when the crack propagated.

FINITE ELEMENT MODELING

In order to evaluate the energy release rate G and the phase angle φ corresponding to various crack lengths and to determine the critical energy release rate G_C caused by the given loads, the same single lap specimen was taken into consideration. The boundary conditions of the specimen considered are shown in Figure 2. The problem is idealized by a finite element mesh shown in Figure 7. A close-up view of fine mesh is plotted in Figure 8. It is noted that a fine local mesh is arranged along the interface of a copper strip/adhesive to handle crack initiation and crack growth and to accurately calculate the near tip displacement fields.

Due to the fact that the single lap shear specimen was subjected to line loads across the width of the sample, plane strain conditions were assumed with the clamped boundary conditions at the fix end in the analysis. The nodes at the fix end were set to u=v=0.

The properties of the materials involved were assumed to be linear elastic. The Young's modulus and Poisson ratio of the copper strips are 129.9GPa and 0.41 respectively, while the Young's modulus and Poisson ratio corresponding to the adhesive are 2.59GPa and 0.4 respectively, which were measured by authors

In the analysis, the commercial code ABAQUS [*ABAQUS User's Manual*, 1996] was used for all calculations. The 4-noded element, CPE4 in ABAQUS, was selected.

STRAIN ENERGY RELEASE RATE

Performing mixed-mode fracture tests on bonded materials generally requires computation of the total strain energy release rate G and determination of the K_{II} / K_I ratio (phase angle). According to both the U-field and the V-field obtained from the captured fringe patterns under the given loads and the measured crack length, the near tip displacement fields can be obtained from the finite element analysis (global/local solution). Once the u and v fields around the crack tip composed of a bi-material system are determined, the energy release rate and phase angle can similarly be derived as [Sun *et al*, 1996]:

$$G = \frac{1}{4}\left[\frac{D_{11}}{\cosh^2 \pi\varepsilon} K_I^2 + \left(D_{11} - \frac{W_{21}^2}{D_{11}}\right) K_{II}^2\right] \quad (4.1)$$

$$\varphi = \tan^{-1} \frac{K_{II}}{K_I} \quad (4.2)$$

The above parameters D_{11}, K_I, K_{II}, W_{21} and ε are defined as follows:

$$D_{11} = \frac{1-\nu_1}{G_1} + \frac{1-\nu_2}{G_2} \quad (4.3)$$

$$K_I = \frac{\Delta u_1 \, \mathrm{Im}(\zeta_2) + \Delta u_2 \, \mathrm{Re}(\zeta_2)}{\mathrm{Im}^2(\zeta_2) + \mathrm{Re}^2(\zeta_2)} \frac{1}{D_{11}} \sqrt{\frac{\pi}{2r}} \quad (4.4)$$

$$K_{II} = \frac{\Delta u_1 \, \mathrm{Re}(\zeta_2) + \Delta u_2 \, \mathrm{Im}(\zeta_2)}{\mathrm{Im}^2(\zeta_2) + \mathrm{Re}^2(\zeta_2)} \frac{1}{D_{11}} \sqrt{\frac{\pi}{2r}} \quad (4.5)$$

$$W_{21} = \frac{1-2\nu_1}{2G_1} - \frac{1-2\nu_2}{2G_2} \quad (4.6)$$

$$\varepsilon = \frac{1}{2\pi} \ln \frac{G_1 + G_2(3-4\nu_1)}{G_2 + G_1(3-4\nu_2)} \quad (4.7)$$

The parameter ζ_2 is of the form

$$\zeta_2 = \frac{1}{(1+4\varepsilon^2)\cosh \pi\varepsilon}(1-2i\varepsilon)\left\{\cos\left[\varepsilon \ln\left(\frac{r}{2a}\right)\right] + i\sin\left[\varepsilon \ln\left(\frac{r}{2a}\right)\right]\right\} \quad (4.8)$$

where Δu_1 and Δu_2 stand for the near-tip displacement field in both opening and shearing directions respectively. G_1 and ν_1 represent the shear modulus and Poison ratio of material I respectively. G_2 and ν_2 are representative of the shear modulus and Poisson ratio of material II respectively. It can be seen from the Equation (4.1) to Equation (4.8) that the energy release rate and phase angle used in this study are only the near tip displacement-based functions. That is, the energy release rate and phase angle can easily be calculated through Equation (4.1) to Equation (4.8) if the near tip displacement fields are previously determined. The near tip displacement fields can be obtained through either micromoiré interferometry or numerical work such as finite element analysis. Finite element analysis was used to determine the near tip displacement fields in this research.

RESULTS AND DISCUSSION

VERIFICATION

In order to verify Equation (4.1) to Equation (4.8), numerical calculations based on these equations were compared with the available analytical results. Numerical calculations were performed to determine the strain energy release rate of a double cantilever beam subjected to a pair of loads at the free ends. Two properties of the double cantilever beam were considered. These include: (1) double cantilever beam with isotropic material (shown in Figure 9); (2) double cantilever beam with a bi-material system (shown in Figure 10).

According to the classical beam theory [Williams, 1988], the energy release rate for the double cantilever beam with isotropic material, in which the transverse shear is not considered, can be derived as:

$$G = \frac{12P^2a^2}{EB^2H^3} \tag{5.1}$$

Similarly, the energy release rate for the double cantilever beam with a bi-material system is defined as :

$$G = \frac{6P^2a^2}{B^2H^3}\left(\frac{1}{E_1} + \frac{1}{E_2}\right) \tag{5.2}$$

where **P** stands for concentrated load; **a** crack length; **B** the width of the beam; **H** the half height of the beam; **E** the Young's modulus of isotropic material; E_1 and E_2 are the Young's moduli of material I and material II respectively. It is noted that the transverse shear force is neglected for slender beams.

The near tip displacement fields for the double cantilever beams considered were calculated using finite element analysis. In the calculations, **B**, **H,** and **P** were chosen as one unit. The crack length, **a,** was assumed to be 20 units long. The Young's modulus of isotropic material, **E,** was assumed to be 11900 units. The Young's moduli of material I and material II, E_1 and E_2, were assumed to be 11900 units and 169540 units respectively. In addition, the beam length, **L,** was assumed to be 40 units long. Meanwhile, the unit system selected in the numerical calculations were assumed to be compatible.

The results of the energy release rates calculated from Equation (4.1) to Equation (4.8) and the classical beam theory (Equation (5.1) to Equation (5.2)) are listed in Tables 4 and Table 5. It is found that the relative error between the energy release rates obtained from both Equation (4.1) to Equation (4.8) and Equation (5.1) for the double cantilever beam with isotropic material is only 1.6%, while the relative error between the energy release rates computed from both Equation (4.1) to Equation (4.8) and Equation (5.2) for the double cantilever beam with a bi-material system is less than 5%. It is obvious that the energy release rates predicted from Equation (4.1) to Equation (4.8) agree with the classical beam theory.

EXPERIMENTAL RESULTS

Figure 11 shows a typical line load-deflection curve recorded by the 6-axis tester. It is noted that there exists a drop phenomenon on the curve. It indicates that the crack propagation in the tested specimen occurred during the test. Since the increase of the crack length causes the reduction of the specimen stiffness, the load will drop corresponding to the same displacement. Such a phenomenon may occur several times with one specimen during the test. Therefore, a series of values of the interfacial fracture toughness corresponding to different phase angles can be obtained by using one specimen. In addition, since the moiré interferometry technique was used to monitor the crack length during the test, a series of values of the interfacial fracture toughness related to different phase angles can also be obtained according to various crack lengths captured by the CCD camera corresponding to different loading conditions.

The typical displacement fields of the single lap shear specimen by using 1200l/mm grating moiré measurement in both U and V directions are presented in Figure 12. Each fringe in the Figure represents 0.417 μm deformation. He-Ne laser moiré interferometer was used to obtain these fringe patterns. It is noted that there clearly exists discontinuities along the interface in the u-field and v-field fringe patterns. Therefore, the crack length can be more precisely measured from the discontinuities of the fringe patterns.

INTERFACIAL FRACTURE TOUGHNESS

As mentioned above, the energy release rate and phase angle for a single lap shear specimen corresponding to a certain crack length at the edge of the interface can be computed through Equation (4.1) to Equation (4.8) when the near tip displacement fields of the specimen are obtained by using the finite element method. The interfacial fracture toughness for a single lap shear specimen in the current study is determined by the fact that when crack propagation occurs after the ratio of the displacement-controlled multi-axial concentrated line loads reaches a certain value, the energy release rate at this moment is defined as the interfacial fracture toughness of the specimen under this given loads. Due to the utilization of the moiré interferometry technique to measure the crack length and the application of the 6-axis machine to record the relationship between crack length and the ratio of the multi-axial concentrated line loads, a series of values of interfacial fracture toughness and phase angles by using one specimen can be calculated in terms of Equation (4.1) to Equation (4.8). As an demonstrated example, Figure 13 shows the relationship between the interfacial fracture toughness and the phase angle obtained by using one specimen during the test. It indicates that the value of the interfacial fracture toughness increases as the phase angle increases. Meanwhile, the results shows that the facilities and methodology used in the current study can indeed be proved to possess the ability to control the far field and the mode mixity conditions at crack tip, and to efficiently do the mixed-mode fracture test.

SUMMARY

An investigation has been conducted to use one single lap shear specimen to obtain a series of data of interfacial fracture toughness as a function of the mode mixity phase angle. The interfacial fracture behavior of the single lap shear specimens subjected to multi-axial concentrated line loads was investigated using a unique 6-axis submicron tester coupled with a high density laser moiré interferometry. The 6-axis submicron tester was used to provide the displacement-controlled multi-axial concentrated line loads, whereas the moiré interferometry technique was used to monitor the crack length during the test. In addition, a finite element technique was simultaneously used to determine the near crack tip displacement fields of the single lap shear specimens. The interfacial fracture toughness and phase angle were computed by using these near tip displacement variables through the analytical energy release rate and phase angle expressions derived by authors. The results show the great advantages of using the new technique for the measurement of mixed-mode interfacial fracture toughness. Meanwhile, the results indicates that the facilities and methodology used in the current study can indeed be proved to possess the ability to control the far field and the mode mixity conditions at crack tip, and to efficiently do the mixed-mode fracture test. Moreover, in order to verify the equations derived by the authors for the calculation of the energy release rate, numerical calculations based on these equations were compared with the results of the double cantilever beam obtained from the classical beam theory. It is found that the relative error between the energy release rates obtained from both the equations and the classical beam theory for the double cantilever beam with isotropic material is only 1.6%, while the relative error between the energy release rates computed from both the equations and the classical beam theory for the double cantilever beam with a bi-material system is less than 5%. It is shown that the energy release rates predicted from these equations agree very well with the classical beam theory.

ACKNOWLEDGMENT

NSF Support through Research Initiation Award and Presidential Faculty Fellows Award to Dr. Sheng Liu is gratefully acknowledged. The authors also wish to express their gratitude to the SRC for the partial support.

REFERENCES

S. Liu, Z. Kutlu, and F. Chang, "Matrix cracking and delamination propagation in laminated composites subjected to a transverse concentrated line loading," J. of Composite Materials, Vol. 27, No.5, 1993, pp.436-470.

S. Liu, "Delaminations and matrix cracking of cross-ply laminates due to spherical indenter," Composite Structures, Vol. 25, 1993, pp.257-265.

S. Liu and F. Chang, "Matrix cracking effect on delamination growth in laminated composites induced by a spherical indenter, " J. of Composite Materials, Vol. 28, No.10, 1994, pp.940-977.

S. Liu, "Quasi-impact damage initiation and growth of thick-section and toughened composite materials," Int. J. of Solids and Structures, Vol. 31, No. 22, 1994, pp.3079-3098.

S. Liu *et al.*, "Investigation of Crack Propagation in Ceramic/Adhesive/Glass System," IEEE Trans., CHMT, Sept., 1995, pp.627-633.

S. Liu and Y. Mei *et al.*, "Bimaterial Interfacial Crack Growth as A Function of Mode-Mixity," IEEE Trans., CHMT, Sept., 1995, pp.618-626.

S. Liu and Y. Mei, "Behaviors of Delaminated Plastic IC Packages Subjected to Encapsulation Cooling, Moisture Absorption and Wave Soldering," IEEE Trans., CHMT, Sept., 1995, pp.634-645.

[8] S. Liu, "An Investigation to Popcorning Mechanisms for IC Plastic Packages: Defect Initiation," Int. J. of Microcircuits and Electronic Packaging, (to be published).

J. R. Reeder and J. H. Crews, "Mixed-Mode Bending Method for delamination," AIAA Journal 28, 1990, pp.1270-1276.

J. P. Goodle, R. A. Pearson and T. Y. Wu, "Interlaminar Fracture Toughness of a Glass-Filled FR-4 Epoxy Composite as a Function of Mode-Mixity," Application of Fracture Mechanics in Electronic Packaging and Materials, ASME EEP-Vol. 11, 1995, pp.163-169.

P. D. Brandenburger and R. A. Pearson, "Mixed-Mode Fracture of Organic Chip Attachment Adhesives," Application of Fracture Mechanics in Electronic Packaging and Materials, ASME EEP-Vol. 11, 1995, pp.179-185.

M. Lu, W. Ren and S. Liu, "A Multi-Axial Thermo-Mechanical Fatigue Tester for Electronic Packaging Material, " ASME Symposium of Sensoring, Monitoring, Modeling and Verification of Electronic Packaging, Nov. 17-21, Atlanta, 1996, pp.87-92.

Y. Guo, C.K. Lim, W.T. Chen, and C.G. Woychik, "Solder Ball Connect (SBC) Assemblies Under Thermal Loading: I. Deformation Measurement via Moiré Interferometry and Its Interpretation, " IBM J. of Research and Development, Vol. 37, No. 5, Sept. 1993, pp.635-647.

Y. Guo and C.K. Lim, "Hybrid Method for Strain/Stress Analysis in Electronic Packaging Using Moiré Interferometry and FEM, " SEM Spring Conference, Baltimore, Maryland, June 6-8, 1994

Y. Guo, "Experimental Determination of Effective Coefficients of Thermal Expansion in Electronic Packaging, " ASME International, Intersociety Electronic Packaging Conference & Exhibition, Lahaina, Hawaii, March 26-30,1995.

ABAQUS User's Manual, Version 5.5, Hibbitt, Karsson & Sorensen, Inc., 1996.

C.T. Sun and W. Qian, "The Use of Finite Extension Strain Energy Release Rates for Fracture Interface Cracks, " International J. of Solids and Structures, Vol. 000, No. 0, pp.1-15, 1996.

J.G. Williams, "On the Calculations of Energy Release Rates for Cracked Laminates, " International J. of Fracture, Vol. 36, 1988, pp.101-119.

TABLE 1 COMPARISON BETWEEN THE ENERGY RELEASE RATES OBTAINED FROM THE FEA AND THE CLASSICAL BEAM THEORY
(DOUBLE CANTILEVER BEAM WITH ISOTROPIC MATERIAL)

FEA result (J/m^2)	Beam theory (J/m^2)	Error (%)
37.84	37.24	1.6

TABLE 2 COMPARISON BETWEEN THE ENERGY RELEASE RATES OBTAINED FROM THE FEA AND THE CLASSICAL BEAM THEORY
(DOUBLE CANTILEVER BEAM WITH BI-MATERIAL SYSTEM)

FEA result (J/m^2)	Beam theory (J/m^2)	Error (%)
205.82	215.83	4.6

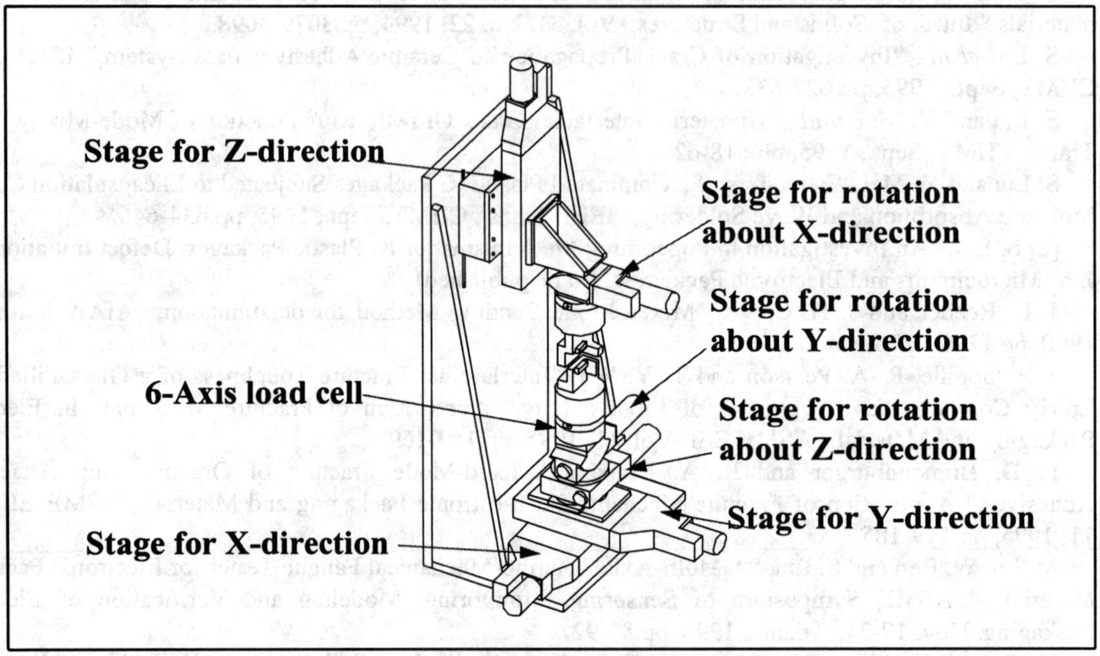

Figure 1 Schematic Drawing of The Mini Tester

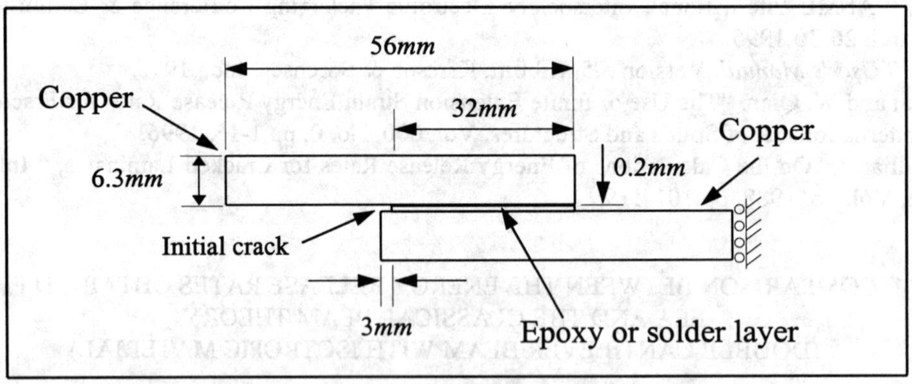

Figure 2 Dimensions, Geometry And Boundary Conditions of Single Lap Shear Specimen

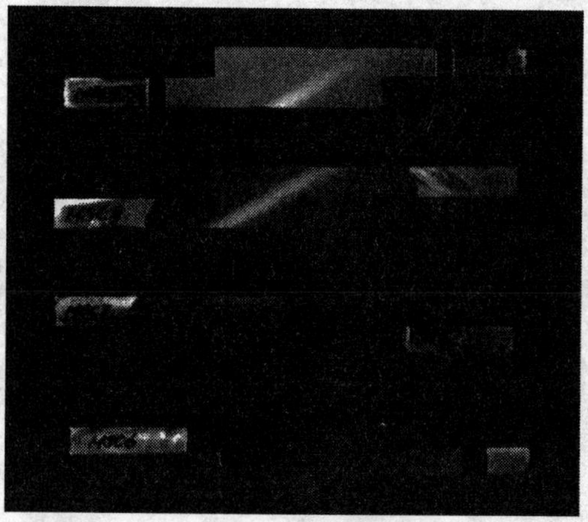

Figure 3 Image Pictures of Specimens Taken By A CCD Camera

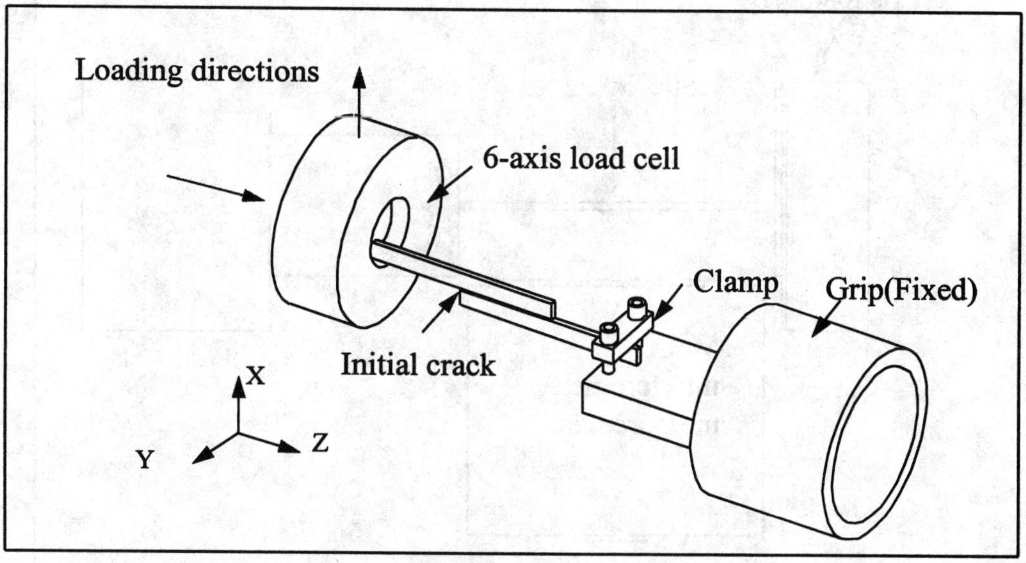

Figure 4 Fix of Specimen By The 6-Axis Machine

Figure 5 Test Setup of Single Lap Shear Specimen in Conjunction with The High Density Laser Moiré Interferometry Measurement System

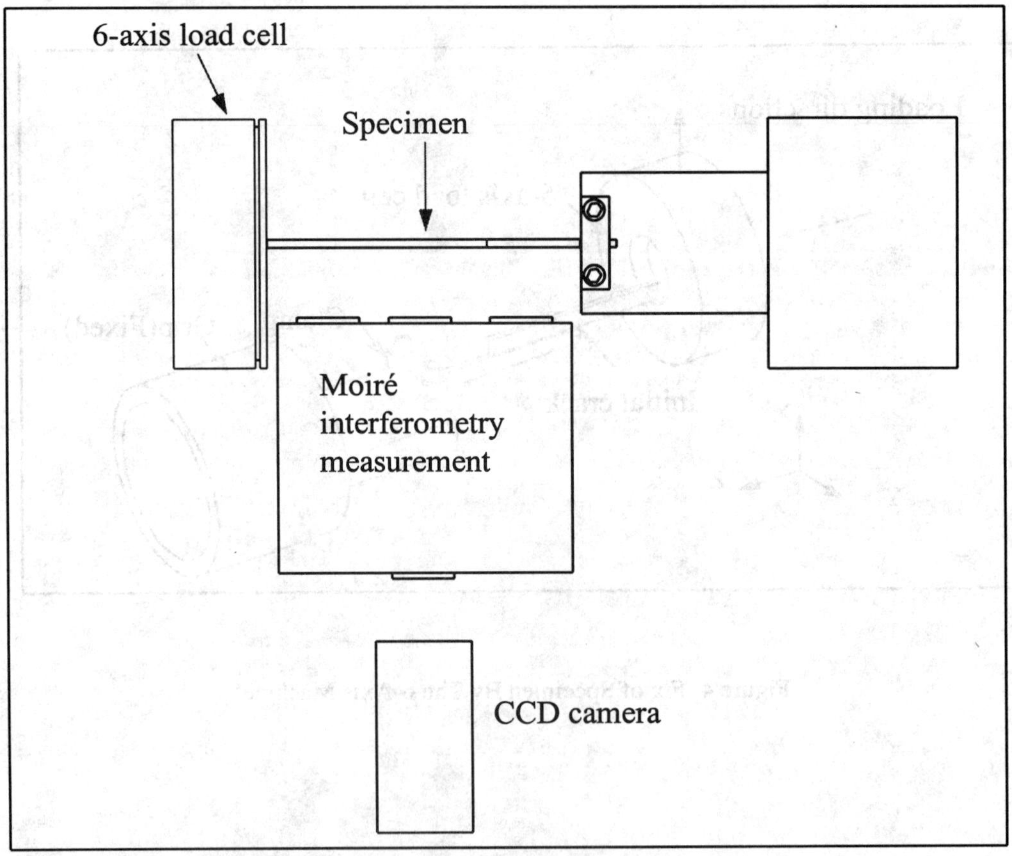

Figure 6 Schematic Drawing of The Test Setup

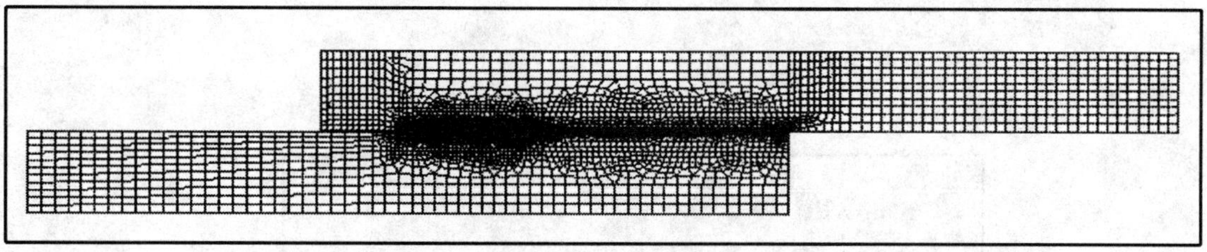

Figure 7 Finite Element Mesh for Single Lap Shear Specimen

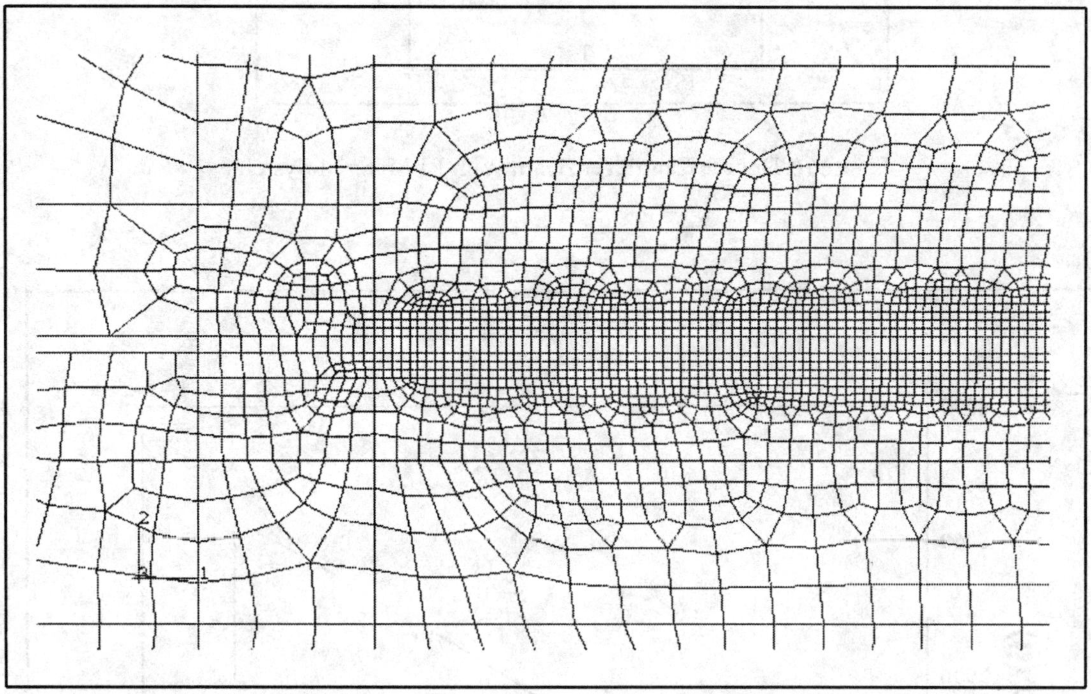

Figure 8 Local Finite Element Fine Mesh for Single Lap Shear Specimen

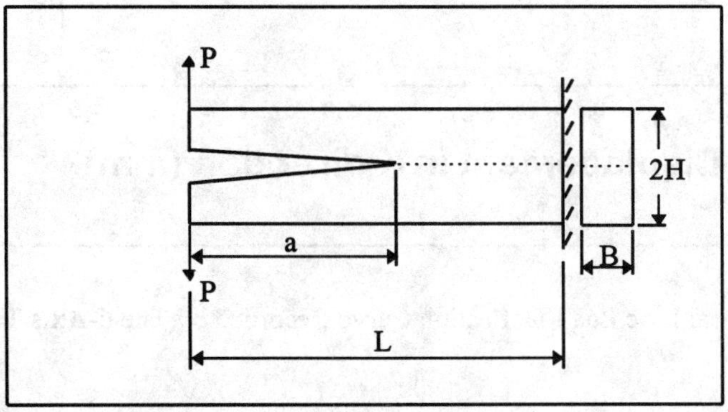

Figure 9 Double Cantilever Beam With Isotropic Material

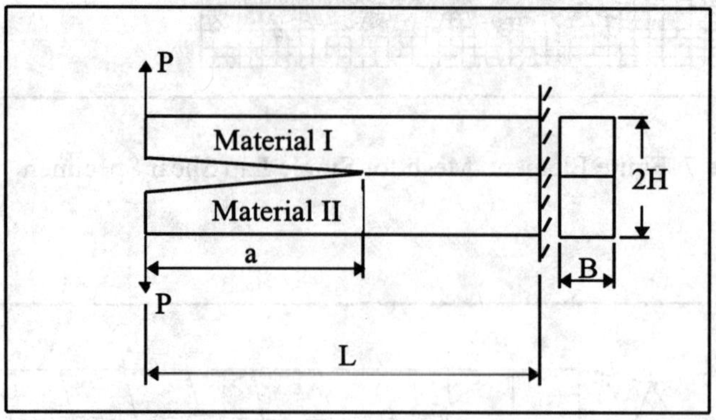

Figure 10 Double Cantilever Beam With Bi-Material System

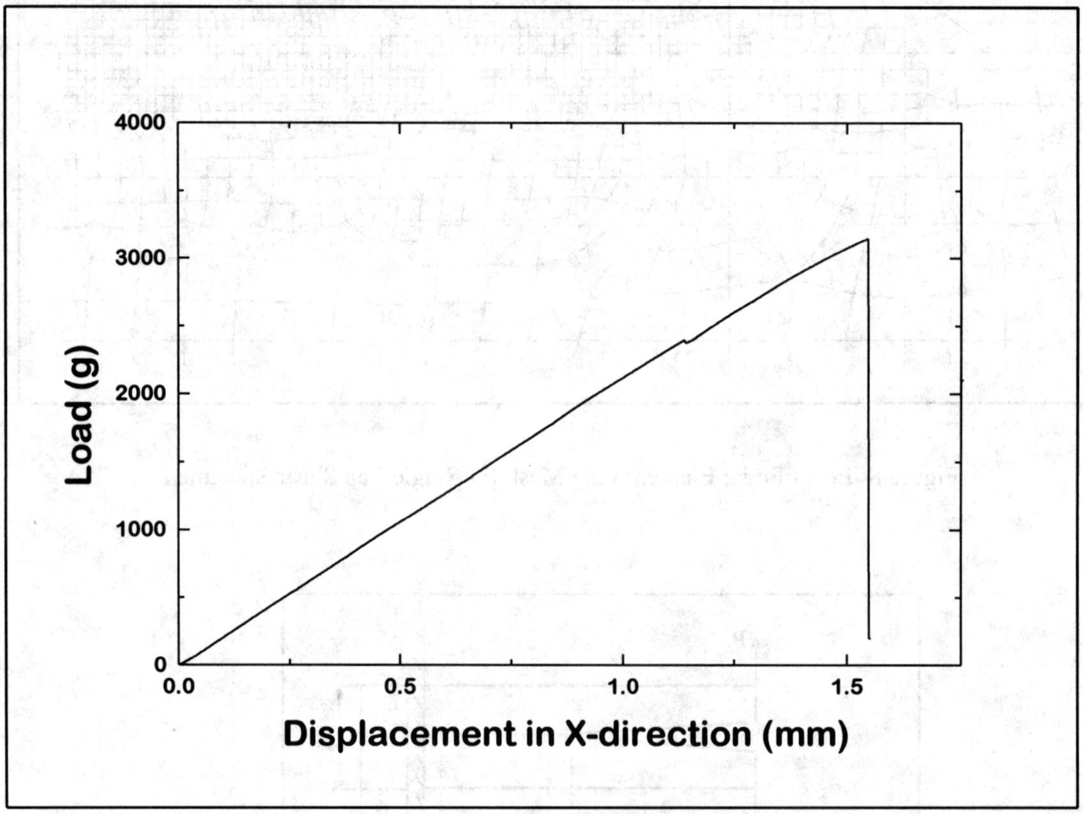

Figure 11 Typical Line Load-Deflection Curve Recorded by The 6-Axis Tester

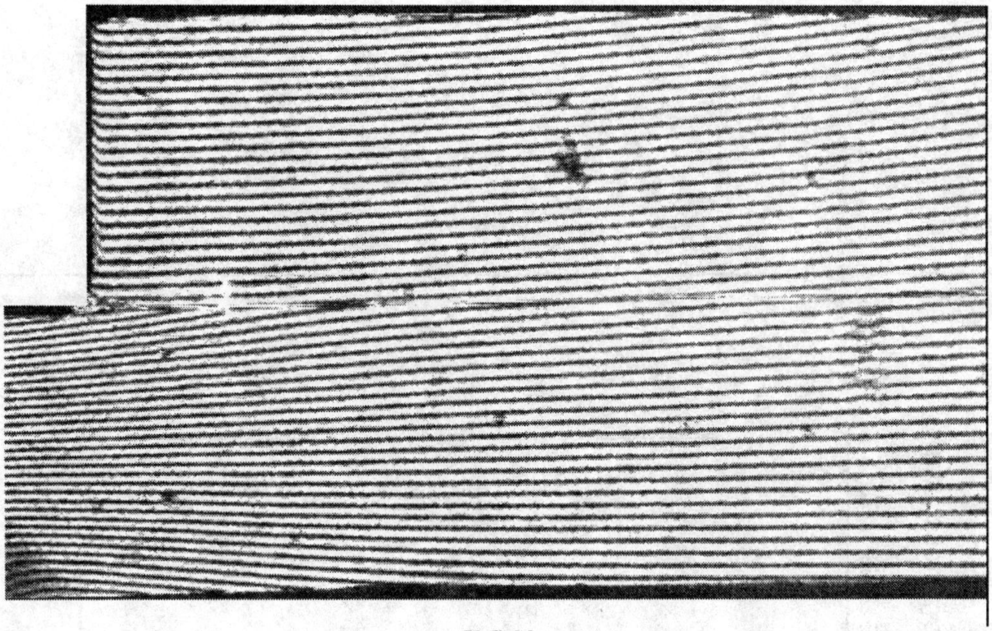

U-field

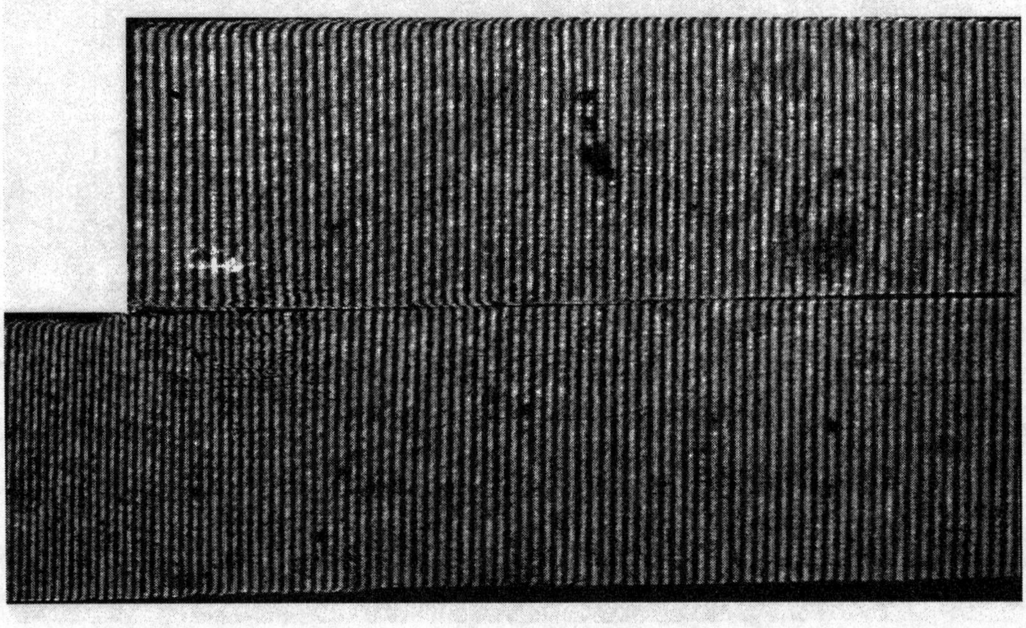

V-field

Figure 12 Fringe Patterns of Specimen during Test

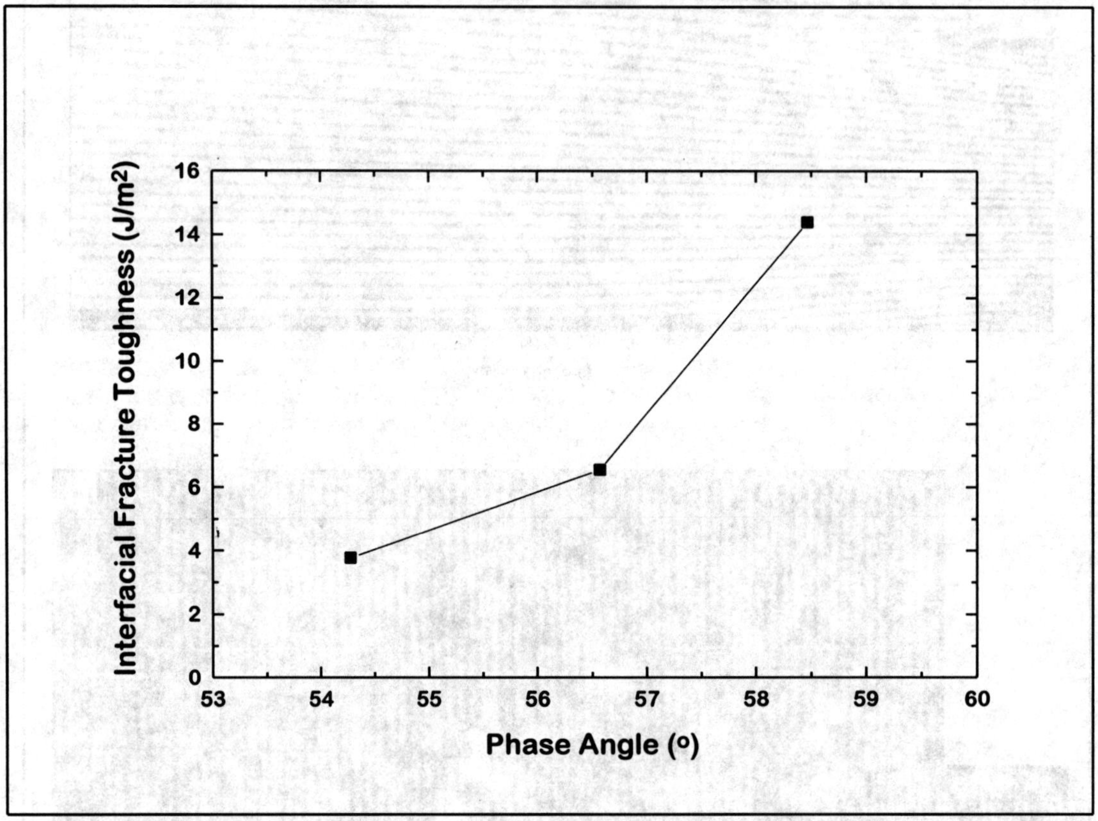

Figure 13 Relationship Of The Considered Specimen Between Interfacial Fracture Toughness And Phase Angle

Delamination in Polymer Matrix Composites

U. SANTHOSH,* J. AHMAD,* S. CHANDU* AND G. M. NEWAZ†

ABSTRACT

In this paper we postulate that delamination in polymer matrix composites (PMCs) is predominantly matrix cracking and that the 'property' that characterizes delamination toughness of the composite *is* the matrix toughness. We present experimental data from double-cantilever-beam (DCB) tests of a cross-ply PMC and single-edge notch tests of the constituent resin material that was taken through the same processing cycle as the composite. The tests are analyzed using a finite element model with a very detailed mesh near the crack tip. The fiber and matrix regions near the crack plane are modeled individually without 'smearing' their mechanical properties. Results are compared with the classical beam theory approach used to analyze DCB specimens. Preliminary results show that while the different analysis methods yield similar values for the critical strain energy release rate of the composite, the critical stress intensity factor, which is more representative of the local stress state around the crack tip is sensitive to the analysis method. In addition, modeling the true state around the delamination crack tip yields a value of the fracture toughness that is consistent with that of the *in-situ* matrix material thus supporting our postulate.

INTRODUCTION

Delamination and transverse cracking are two matrix dominated failure modes seen in laminated composites. These failure modes have been well studied in the literature in polymer matrix composites (PMCs). The Mode I delamination toughness and crack growth resistance is measured in the laboratory In these studies the basic approach is to measure the change in the compliance of the specimen with crack growth using either the double cantilever beam test or some variation of this test. The compliance change is then converted to a toughness value, usually the critical Mode I strain energy density (G_{Ic}) using a simple beam approximation. Several researches [1-3] have found correlation (Figure-1) between the Mode I critical strain energy release rates (G_{Ic}s) of the neat resin as measured either by the compact tension (CT) or the edge notch panel (ENP) test and the interlaminar G_{Ic}s of their composites as measured by the double cantilever beam (DCB) test. In brittle resin systems (G_{Ic} less than 200 J/m^2) the

* Research Applications, Inc., 7026 Corporate Way, Suite 211, Centerville, OH 45432

† Dept. of Mechanical Eng., 5050 Anthony Wayne Dr., 2100 Engineering Bldg., Detroit, MI 48202

composite G_{Ic}s are reported to vary from slightly to about three times higher than that of the corresponding matrix materials, while for some very tough resins the composite Mode I toughness is seen to be lower than that of the neat resin. The scatter in the data has been attributed to a number of reasons that include the stress distribution in the composite, fiber bridging, local yielding and viscoplasticity.

Transverse cracking is another of the matrix dominated failure mode often seen in composites. In this mode cracking occurs in the plies due to loads that are transverse to the direction of the fibers. Thus this mode is different from delamination which is caused by out-of-plane stresses at the ply interface. A comparison (Figure-2) of the transverse cracking toughness of PMCs to their delamination toughness (G_{Ic}) has shown that the two values are comparable [3-5]. This correlation between the two in PMCs indicates that they are governed by the same fundamental matrix property, most likely the fracture toughness of the matrix.

A lot of work has been done to improve the resistance of polymer matrix composites (PMCs) to these types of failure by using matrix materials with increased fracture toughness. However the 'apparent' increase in fracture toughness has been seen to be limited and no clear understanding exists as to the reasons. There is also a wide scatter in the fracture toughness values of PMCs as determined experimentally from different specimens and geometries. One reason for this can be attributed to the way the experimental data, i.e., the change in compliance of the specimen, is used to determine the delamination toughness of the composite: Most of the analysis of delamination in PMCs is based on modeling the laminate as a collection of well defined homogeneous, orthotropic plies with the delamination crack growth occurring at the distinct interface between the laminae. While this type of modeling is useful in characterizing the global structural properties of the composite such as the various modulii and Poisson's ratios, cracking is a local phenomenon that is governed by the local stress state around the crack tip. Within the very small range of dimension near the crack tip where the delamination (or crack opening) stress is significant, the material can no longer be view as homogeneous but instead should be viewed as comprising of distinct fiber and matrix regions. Another reason is that the data available in the literature is for different material systems, and often assumptions have to be made while comparing their relative toughness values. The composite toughness is usually compared to that of the constituent resin system that has not gone through the same processing cycle as the resin in the composite, i.e., not with that of the *in-situ* matrix material.

In this paper we postulate, then set out to demonstrate, that a fundamentally different way of interpreting test data can lead to a consistent and widely applicable set of rules that dictate the nature of delamination growth and failure of composites. A description of this work, which involved laboratory experiments, numerical analyses, and fracture mechanics considerations, is presented.

A POSTULATE FOR DELAMINATION IN PMCs

The correlation in Figure-2 can easily be explained once it is understood that at the local level of the crack, the two composite failure modes, transverse cracking of the matrix and delamination, are similar: growth of the crack through the matrix region. Thus the basic property which can be expected to characterize the toughness values of these two modes of crack growth is the toughness of the matrix material. We therefore postulate that *delamination in PMCs is predominantly matrix cracking and that the 'property' that characterizes delamination toughness of the composite is the matrix toughness*. Similarly, *the 'property' that determines the composite delamination growth resistance under fatigue and/or creep is the crack growth resistance of the matrix material under fatigue and/or creep*. The experimental observation discussed earlier in this paragraph thus directly supports the postulate.

However, the data available in the literature is for different material systems, and often assumptions have to be made while comparing their relative toughness values. The composite toughness is usually compared to that of the constituent resin system that has not gone through the same processing cycle as the resin in the composite. No data has been found in the literature where the toughness of a PMC has been compared to that of the *in-situ* matrix material. Thus while there is some preliminary evidence to support our postulate the evidence is neither sufficient nor complete in proving the validity of the postulate. So in this study we tested and compared delamination toughness of a cross-ply composite with that of the resin material which was taken through the same processing cycle as the composite.

EXPERIMENTAL EFFORT

The material used in this study was woven cross-ply IM7-6/5250 graphite/epoxy composite. The 3mm thick (2h) 16-ply laminate was fabricated at Pratt & Whitney from Hercules IM7-6K graphite yarn with tensile modulus of 300 GPa, tensile strength of 5.3 GPa and ultimate elongation of 1.81%. The weave was 4 Harnes Satin with a count of 11.5 x 11.5 and area weight of 0.205 kg/m^2. The neat matrix material was 5250 epoxy that was processed under the same conditions as the composite.

The delamination of the cross-ply composite was characterized using the Double Cantilever Beam (DCB) Test specimen is shown in Figure-3. The specimen was 311.8mm long (L) and 36mm wide (B). The initial crack length (a_o) was 75mm. Teflon strips were inserted between the plies before consolidation to create initial delamination. Hinges were glued to the specimen with epoxy and the load was applied to the hinges. Figure-4 shows the single edge notch panel (ENP) specimen used to determine the fracture toughness of the matrix. The specimen had an initial crack length of 6.35 mm. The load was applied at the center of the width of the specimen at a point 6.35 mm from the ends. The notches were introduced in the specimens using a diamond saw. The notch required special attention as the tip of the initial notch could be either blunted during the saw cut or could subsequently be made sharp using a blade cut after the notch was introduced. Both approaches were studied.

All fracture tests were conducted under fixed-grip conditions. Monotonic response of matrix was evaluated only in the linear regime to estimate the Young's modulus.

The Mode I delamination tests were conducted at room temperature using a servo-hydraulic MTS machine with a load cell of capacity of 4500 N. The specimens were tested by applying a tensile load to the attached hinges at a constant displacement rate of 12.7 mm/min. At each loading step, the test is stopped and a crack length is measured using the marks initially affixed along the specimen using liquid paper or whiteout. An accurate scale was marked at the edge of the specimen. All crack length measurements were done visually. The load and displacement were measured at this instant using data acquisition system. The specimen was then unloaded to the initial position for each crack length increment and reloaded with the same displacement rate to create a larger crack length. This procedure was repeated for a convenient number of loading-unloading steps. A typical load-deflection response is shown in Figure-5. For each specimen, using this procedure, a number of sets of crack length, deflection and loads were recorded for estimation of energy release rate.

The tensile load carried by the ENP specimen at fracture was relatively high, e.g. approximately 750 N, when the blunted crack introduced by the diamond saw was used. This specific specimen design was loaded in one end through a pin in a hole. However, for specimens with a sharp knife-edge introduced with a surgical blade, the load at fracture had dropped to approximately 390 N indicating that a sharper crack tip lowered the fracture toughness substantially. These specimens had exactly the same dimension but were gripped in wedge-action grips at two ends and did not have a hole. These two methods of gripping itself have no effect on the measured fracture toughness. All tests were conducted at room temperature under displacement control condition. The crosshead rate was 12.7 mm/min.

The Young's modulus of the 5250 resin was measured using a 127mm x 25.4 mm x 3.2 mm rectangular specimen. The specimen was loaded only in the elastic regime. The test was conducted at room temperature with a crosshead speed of 12.7 mm/min. The strain in the gage section was measured using an axial extensometer with a fixed gage length of 12.7 mm. The matrix Young's modulus was estimated from the stress-strain curve in Figure-6 to be 3.5 GPa.

The DCB experimental data obtained from the IM7-6/5250 composite specimens are shown in Figures 7a and 7b. In these figures a is the total length of the delamination crack, P is the applied load and delta (δ) is the total crack opening displacement. The critical loads at which the crack grew for the different crack lengths are also given in Table-I.

Table-I: DCB Test Data for IM7-6/5250

Test-1			Test-2		
a (mm)	P (N)	δ (mm)	a (mm)	P (N)	δ (mm)
90	31	64	89	46	21
100	28	73	102	42.5	30
110	27	84	117	38	42
120	26	105	127	33	58
130	23	110	141	30	64
140	20	115	154	24	70
145	19	120	169	23	89
			180	22	109
			188	20	117

Similarly the load displacement data for the ENP matrix specimens with the as-cut and sharp notches are shown in Figures 8 and 9. The data shows the load up to the point the crack grew which corresponded to a load of 752 N for the as-cut and 395 N for the sharp notch; as expected the sharp notch results in larger stress concentration at the crack tip.

ANALYSIS OF TEST DATA

The crack growth behavior observed in the DCB (Test-1) and ENP tests were studied by doing finite element stress analysis of the test specimens under the test conditions. The ABAQUS general purpose finite element code was used in the study. The elastic analysis used isoparametric eight noded plane elements and the region around the crack tip was modeled with sufficient detail to yield a convergent solution.

The delamination fracture toughness of the composite was characterized by the critical stress intensity factor, K_{Ic}, which can be related to the critical, or the failure stress, σ_f, by

$$K_{Ic} = \alpha\sigma_f\sqrt{\pi a} \tag{1}$$

Here α is a factor associated with the geometry of the flaw and that of the specimen. As will be explained later, choosing K_{Ic} has some distinct advantage over choosing G_{Ic} as the fracture toughness parameter. K_{Ic} can be obtained from the results of the finite element analysis by either of two methods: (i) the J Contour Integral or (ii) the Crack Opening Displacement (COD). The COD approach is the more direct approach since it relates the stress intensity factor of the crack to the displacement near the crack tip. However the path independent J integral approach is the preferred method since a path can be chosen sufficiently far from the crack tip and still a good estimate of the integral can be obtained, even if there is considerable yielding near the crack tip. Thus a much coarser mesh is acceptable near the crack tip and this makes the analysis more computationally efficient. In the linear elastic regime the J integral is simply the strain energy release rate G_I.

ANALYSIS OF THE DCB SPECIMEN

Mode I fracture toughness values are usually obtained from the critical crack growth load in the DCB specimen by using beam theory. The formula derived in literature [5] is:

$$G_{Ic} = \frac{3}{2}\frac{P_{cr}\delta}{B(a_o + \Delta a)} \tag{2}$$

Table-II gives the G_{Ic} values obtained from the experimental data using the above equation. The K_{Ic} values listed in the table were obtained assuming plane strain condition:

$$K_{Ic} = \sqrt{\frac{G_{Ic}E}{1-\upsilon^2}} \tag{3}$$

Table-II: DCB Composite G_{Ic} Determined Using Beam Approximation

a (mm)	G_{Ic} (kJ/m^2)	K_{Ic} (MPa$\sqrt{\text{m}}$)
90	0.9185	4.24

The Young's modulus of the 5250 matrix material was obtained from the experimental measurements (E=3.5 GPa) while the Poisson's ratio was assumed to be typical of other epoxies (ν=0.35). The composite was assumed to have 0.55% volume fraction and the elastic properties of the cross-ply composite were determined. Static large deformation finite element analysis of the DCB was then done to determine the load point displacement. The composite modulus was then scaled so that the displacement obtained from the stress analysis matched the displacement measured in the experiment. Thus for the cross ply composite we had

$$E_1 = E_2 = E_3 = 18.6 \text{ GPa}, \nu_{12} = \nu_{13} = \nu_{23} = 0.22, G_{12} = G_{13} = G_{23} = 7.6 \text{ GPa}$$

The DCB specimen was modeled using plane strain elements. The accuracy of the analysis was verified by also constructing a three-dimensional model and comparing the results of the J integral calculation. The contour values agreed to within 5% and from a computational viewpoint it was decided to use the two-dimensional model. Two models of the DCB specimen were used in this study. In the first model the specimen was treated as homogeneous with the effective properties given above. This model was studied using both small and large deformation analyses; the later was considered because the test involved relatively large transverse deflections (20-40 times the thickness). The results from the large deformation analysis are therefore expected to better predict the actual behavior in this test. The results of the analyses are given in Table-III. The large deformation analysis is seen to give a value of critical strain energy release rate that is about 20% lower than that obtained from the small deformation analysis. The results of the small deformation analysis however compare quite well with the simple beam solution in Table-II that also makes the small deformation assumption.

Table-III: DCB G_{Ic} Determined Using Homogeneous Composite Properties

a	Small Deformation Analysis		Large Deformation Analysis	
(mm)	G_{Ic} (kJ/m^2)	K_{Ic} (MPa$\sqrt{m}$)	G_{Ic} (kJ/m^2)	K_{Ic} (MPa$\sqrt{m}$)
90	0.9830	4.39	0.8282	4.03

Another model of the DCB specimen in which the region around the crack plane was modeled as matrix material was also studied. This small region of matrix material had a thickness of the order of the distance from the outer surface of the fiber to the ply surface. The remaining region was modeled as a homogeneous material with effective properties that were obtained as before by doing a large deformation stress analysis of the DCB specimen and matching the load point displacement. The composite material properties used were:

$$E_1 = E_2 = E_3 = 19.0 \text{ GPa}, \nu_{12} = \nu_{13} = \nu_{23} = 0.22, G_{12} = G_{13} = G_{23} = 7.6 \text{ GPa}$$

Again, both small and large deformation stress analyses were done. The values of the critical strain energy release rate (G_{Ic} or J integral) for the different crack lengths are shown in Figure-10 and in Table-IV.

Table-IV: DCB G_{Ic} Determined Assuming Crack in the Matrix

a	Small Deformation Analysis		Large Deformation Analysis	
(mm)	G_{Ic} (kJ/m^2)	K_{Ic} (MPa$\sqrt{m}$)	G_{Ic} (kJ/m^2)	K_{Ic} (MPa$\sqrt{m}$)
90	1.0270	2.01	0.8553	1.83

The table also gives the critical stress intensity factor computed from the J integral using the elastic properties of the matrix, since in this model the crack is in the matrix and not in the homogeneous composite.

Comparing the large deformation analysis results from Tables III and IV shows that the G_{Ic} values do not change much whether the crack is considered to be in the homogeneous composite or whether it is in the matrix material. Thus if G_{Ic} is used as the composite toughness criteria, then the critical toughness value will be relatively insensitive to the type of the matrix in the composite. This is because by changing the matrix property we could still have about the same value for the composite G_{Ic} as long as the global properties of the composite does not change much. However K_{Ic} is more susceptible to the local stress state around the crack tip and its sensitivity to the material property close to the crack is seen to offer a better discriminator of the composite fracture toughness.

In summary it can be stated that since the material surrounding the delamination is most likely the matrix component of the composite, the K_{Ic} determined by large deformation analysis of the crack in the matrix material is most likely the best estimate of the composite fracture toughness.

ANALYSIS OF THE ENP SPECIMEN

The experimental data from the ENP specimen of the resin was analyzed in order to determine the Mode I fracture toughness of the matrix material. Unlike in most other tests of this kind the matrix material that was tested was exposed to the same processing cycle as the composite. This 'fiberless composite' therefore represents the *in-situ* matrix material in the composite of the DCB specimens. Our postulate, stated earlier in this section, would be validated if the fracture toughness (K_{Ic}) of the *in-situ* matrix was determined to be equal to the delamination toughness (K_{Ic}) composite as obtained from the large deformation analysis of the DCB specimen with the crack assumed to be in the matrix.

Handbook solutions are available for the isotropic edge notch panel specimen. Assuming plane stress condition, since the thickness of the specimen is only one-eighth of the specimen width, we have [6]

$$K_{Ic} = \frac{P_{cr}}{B\sqrt{W}} \frac{\sqrt{2\tan(\pi a/2W)}}{\cos(\pi a/2W)} \left[0.75 + 2.02\left(\frac{a}{W}\right) + 0.37\left(1 - \sin(\pi a/2W)\right)^3\right] \tag{4}$$

and $$K_{Ic} = \sqrt{G_{Ic} E} \tag{5}$$

Plane stress finite element analysis of the model of the specimen was done and the critical stress intensity factor obtained from the value of the J contour. The fracture toughness as determined from the handbook solution and from the finite element analysis are given in Table-V. The two are in good agreement with each other.

Table-V: Fracture Toughness of the Matrix

K_{Ic} (MPa$\sqrt{m}$)	
Handbook Solution	FEA Using J Integral
1.97	1.96

DISCUSSION

Comparison of Tables IV and V show that the initial delamination toughness of the composite, is in excellent agreement with the fracture toughness of the *in-situ* matrix material: the difference being 2% in the case of small deformation case and 6% in the case of large deformation analysis. Thus the limited amount of data available in this Phase I study strongly suggest that the 'postulate' *is valid.* Further tests involving other materials and test geometries and detailed analyses of test data should be performed for further validation. If the postulate is proven valid then major advance in damage tolerant design and life prediction of PMC components will be possible. For example it will help explain the 'apparent' dependence of the composite fracture toughness on the specimen geometry and loading. The analysis technique will also have to address issues such as fiber bridging, crazing in the matrix and mixed mode crack growth to better explain the observed crack growth phenomena in PMCs. We propose to address these issues in the future.

CONCLUSIONS

After reviewing the available data we postulated that delamination in PMCs is predominantly matrix cracking and that the 'property' that characterizes delamination toughness of the composite is the matrix toughness. Large deformation finite element analysis of results from crack-growth experiments done on cross-ply graphite/epoxy laminates and on *in-situ* matrix material showed excellent agreement between the initial delamination toughness of the composite and the fracture toughness of the *in-situ* matrix material. Albeit based on only limited amount of test data (that could be generated thus far), our investigation demonstrated that knowing the matrix toughness and the elastic properties of the matrix and the composite, one could predict delamination toughness. Further tests involving other materials and test geometries, and detailed analyses of test data that includes the effect of fiber bridging, crazing in the matrix and mixed mode crack growth should be performed for further validation. Once further validated and refined to include general loading situations, our finding can profoundly simplify the task of establishing a fundamentally based and pragmatic methodology for damage tolerant design and life prediction of composites.

REFERENCES

1. Hunston, D. L., Moulton, R. J., Johnston, N. J., and Bascom, W. D., "Matrix Resin Effects in Composite Delamination: Mode I Fracture Aspects", *Toughened Composites, ASTM STP*

937, N. J. Johnston Ed., ASTM, Philadelphia, 74-94, (1987).
2. Bradley, W. L., "Relationship of Matrix Toughness to Interlaminar Fracture Toughness", in *Application of Fracture Mechanics to Composite Materials*, K. Freidrich, Ed., Elsevier Science Publ., 159-187, (1989).
3. Bradley, W. L., and Cohen, R. N., "Matrix Deformation and Fracture in Graphite-Reinforced Epoxies", *Delamination and Debonding of Materials, ASTM STP 876*, W. S. Johnson, Ed., ASTM, Philadelphia, 389-410, (1985).
4. Lee, S. M., "A Comparison of Fracture Toughness of Matrix Controlled Failure Modes: Delamination and Transverse Cracking", *J. Material Science*, Vol. 20, 185-196, (1986).
5. Whitney, J. M., Browning, E., and Hoogsteden, W., "Double Cantilever Beam Tests for Characterizing Mode I Delamination of Composite Materials", *J. Reinforced Plastics and Composites*, Vol. 1, 297-313, (1982).
6. Sih, G., Handbook of Stress Intensity Factors, Institute of Fracture and Solid Mechanics, Lehigh Univ., PA, pp. 5.5.1.7-5.5.1.10.

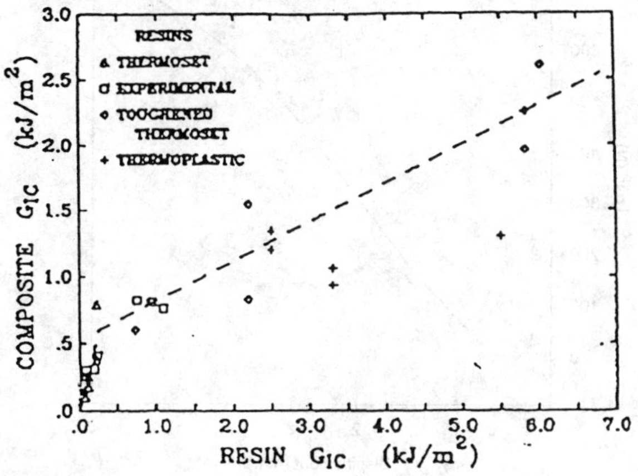

Figure-1: Composite G_{Ic} vs. Matrix G_{Ic} [3] in PMCs

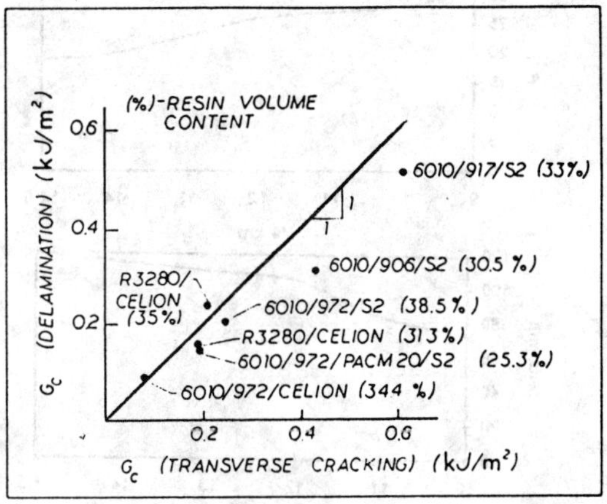

Figure-2: Delamination Toughness vs. Transv. Cracking Toughness [4]

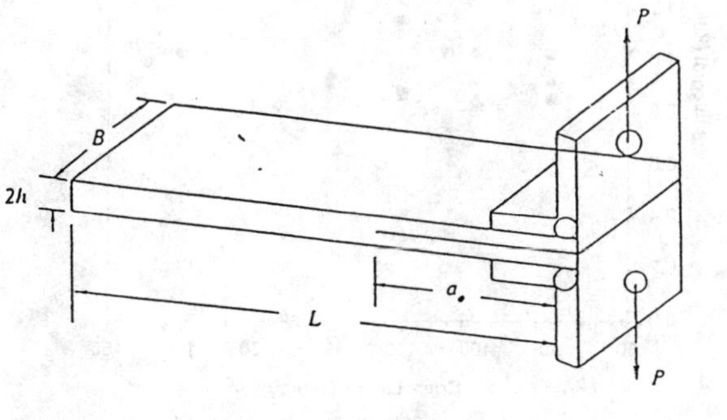

Figure-3: Double Cantilever Beam Specimen

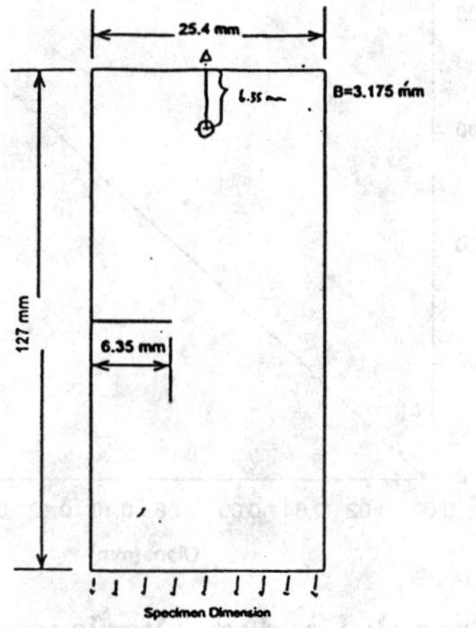

Figure-4: Edge Notch Panel Specimen

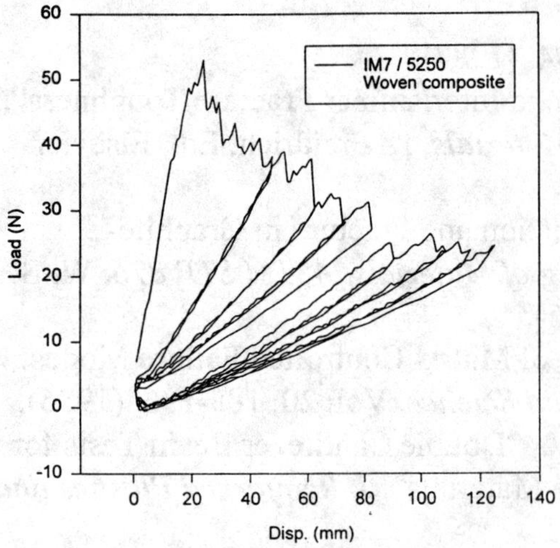

Figure-5: Typical Load-Deflection Curve in Mode I Delamination Test

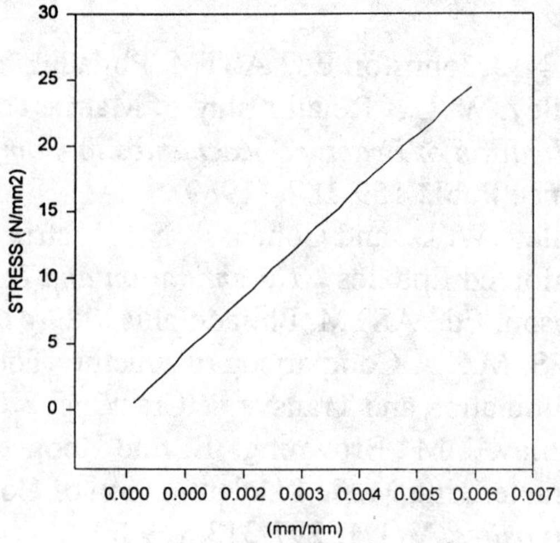

Figure-6: Stress vs. Strain Response of 5208 Resin

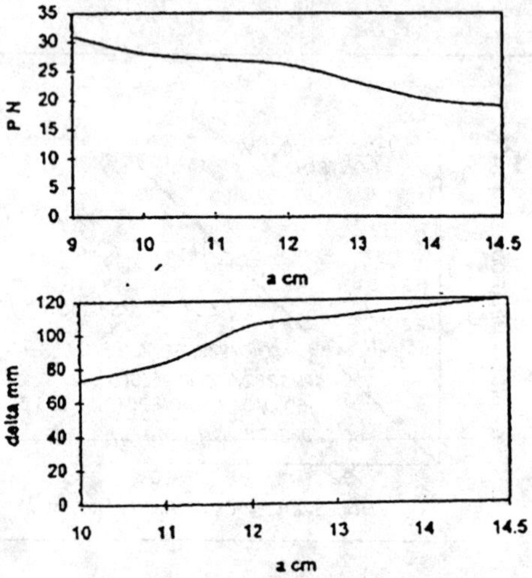

Figure-7: DCB Specimen; (a) Load vs. Crack Length & (b) Displ. vs. Crack Length

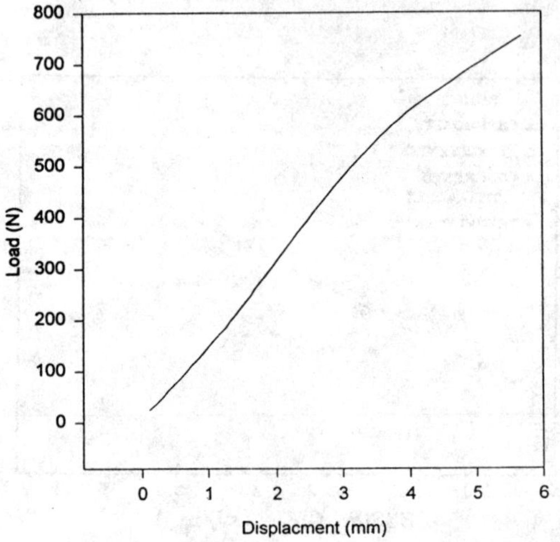

Figure-8: Load vs. Displacement in the ENP Specimen (As-Cut Notch)

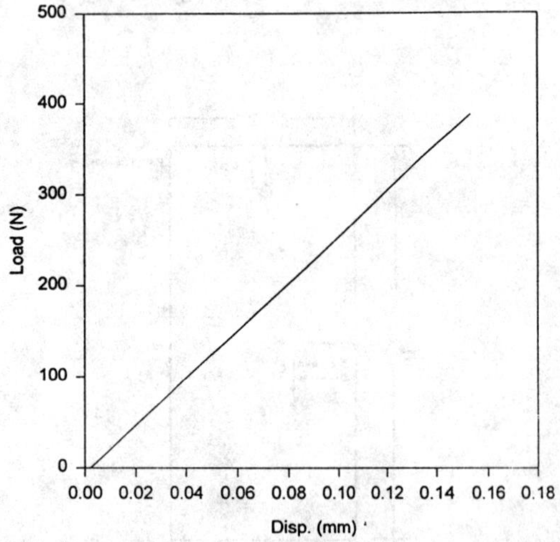

Figure-9: Load vs. Displacement in the ENP Specimen (Sharp Notch)

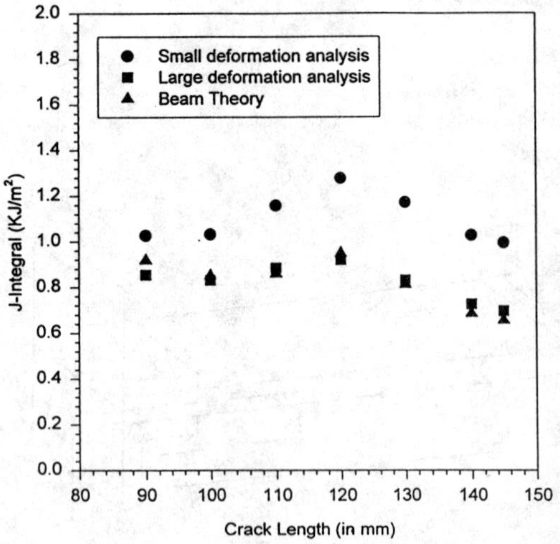

Figure-10: J Integral vs. Crack Length

On Multiaxial Fatigue Damage Parameters for Particle Reinforced Metal Matrix Composites

F. ELLYIN, Z. XIA AND C. LI

ABSTRACT

Appropriate multiaxial fatigue damage parameters for particle reinforced metal matrix composites (PMMCs) are investigated by numerical and experimental studies. The correlation of macroscopic with microscopic parameters is first explored through a three dimensional finite element analysis on a representative unit cell model. Multiaxial fatigue test data of an alumina reinforced 6061 aluminum alloy is then presented and correlated with different macroscopic mechanical parameters. Both the empirical correlation and the physical understanding of the damage mechanisms indicate that a strain energy parameter is most suitable parameter for PMMCs undergoing multiaxial fatigue damage.

INTRODUCTION

Particle reinforced metal matrix composites (PMMCs) are a new class of advanced engineered materials which are gaining wide acceptance for critical applications in industries such as automotive, electronics, heavy transport equipment and aerospace. Development of these composites attempts to combine attractive properties of metals (e.g. ductility) with those of ceramics (e.g. high modulus, abrasion and corrosion resistance) to achieve light, stiff and strong materials.

Since some applications are in load bearing components, these components are subjected to loads of multiaxial and cyclic nature. Therefore, the fatigue damage and life prediction are important topics to be investigated. Due to introduction of the reinforcing particles, the damage initiation and short crack growth of PMMCs could be considerably different from those of unreinforced alloys. It has been identified that interface debonding (between the matrix and particles) and particle fracture are two main modes of damage initiation in PMMCs (Xia and Ellyin, 1997; Mochida, 1991). Clusters in PMMCs are preferred sites for particle fracture, interface debonding and short crack growth (Li and Ellyin, 1996). Furthermore, it is found that micro-damage of PMMCs is dependent on the macroscopic multiaxial stress state, for example, Xia and Ellyin (1997) reported that the alumina reinforced 6061 aluminum alloy could show a significant degradation in elastic modulus under biaxial

---

F.Ellyin, Z. Xia & C. Li, Department of Mechanical Engineering, University of Alberta, Edmonton, AB, Canada, T6G 2G8

cyclic loading. Ravichandran and Liu (1995) have reported that the damage-induced dilatation of particle reinforced polymer is a function of the hydraulic pressure.

It should be noted that almost all experimental studies on fatigue behavior and fatigue life of PMMCs are limited to the uniaxial cyclic loading. Under the uniaxial fatigue, either stress range, strain range or product of the stress and strain ranges can be used to correlate the fatigue life data (Wang, et al. 1996; Hurd, 1988, Ellyin and Li, 1996). There are extensive studies on the multiaxial fatigue theories for metals and alloys, e.g. see Ellyin (1997). However, multiaxial fatigue behavior and appropriate multiaxial fatigue parameters for PMMCs are still topics requiring further investigation.

In this paper, we will first explore the correlation of macroscopic mechanical parameters with the microscopic damage-related parameters in PMMCs. This is achieved through a finite element analysis on a body centered cubic (BCC) unit cell as a representative volume element of alumina reinforced 6061 aluminum alloy composite undergoing damage. Next, multiaxial fatigue test data of the above composite recently obtained in our laboratory will be used to correlate with different multiaxial fatigue criteria. We conclude that a macroscopic strain energy parameter is a suitable multiaxial fatigue damage parameter for the PMMCs.

CORRELATION OF MICRO AND MACRO DAMAGE PARAMETERS

Numerical analysis based on a unit cell model has been employed in predicting overall properties of composite materials (Meijer, et al. 1997) or in studying the local stress-strain field of heterogeneous materials (Wienecke et.al. 1995). This method is also very useful for exploring relations between macro (overall) and micro (local) parameters.

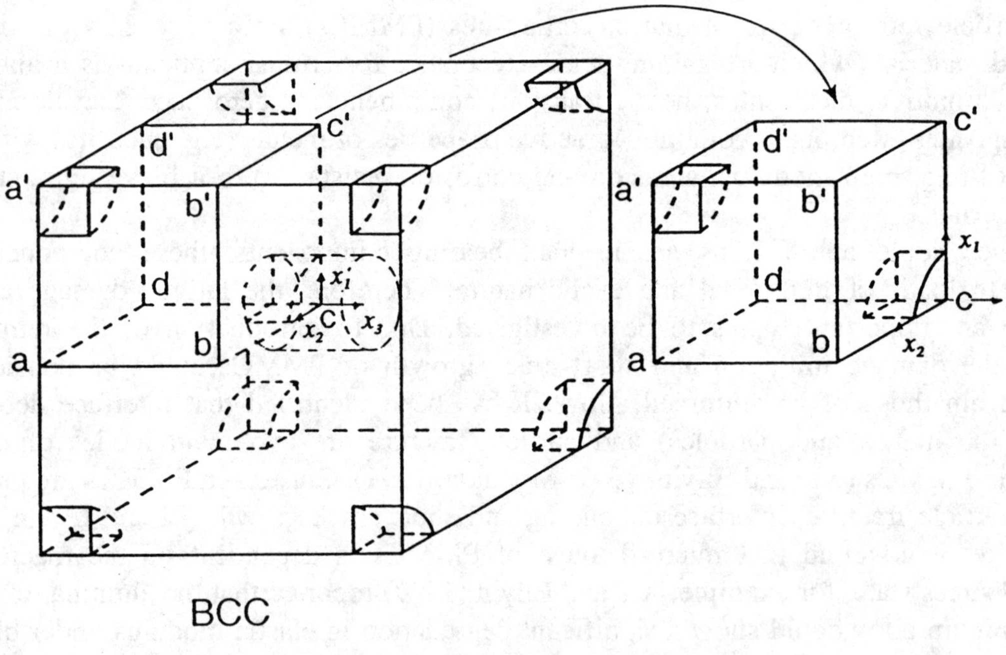

Fig. 1 One eighth of a BCC unit cell is required according to the symmetry conditions.

A three dimensional, body centered cubic (BCC) unit cell model was used in the present study. The particle (Al_2O_3) was assumed to be a short cylinder. One eighth of the BCC model was required according to the symmetry conditions (see Fig.1). On the three symmetric planes, the normal displacements were assumed to be zero. On the other three boundary surfaces, the normal displacements in each plane was assumed to be the same (coplanar conditions). The finite element mesh consisted of 1520 elements. Before applying a uniaxial or an equi-biaxial tensile loads, the model was assumed to experience a temperature drop from 500° C to 20° C. Thus, a residual compressive stress of about 400 MPa in the particle was introduced. This value is approximately equal to the measured value by Shi et al. (1993).

As mentioned in the previous section, interfacial debonding and particle fracture are two main local damage modes in PMMCs. We thus take the maximum normal stress at the interface, $\sigma^{i}_{n,max}$, and the maximum principal stress in particle, $\sigma^{P}_{1,max}$, as the microscopic damage parameters. Obviously, these two parameters can be thought as being the driving forces causing debonding of interface and fracture of particles, respectively. Two macroscopic mechanical parameters , the applied equivalent stress (see eq. 2) and the applied strain energy , $W = \int \sigma_{ij} d\varepsilon_{ij}$, were chosen to correlate with the micro damage parameters. Two loading cases, i.e. uniaxial and equi-biaxial loading were considered.

Figures 2a,b show the correlation of the applied equivalent stress with the $\sigma^{i}_{n,max}$ and $\sigma^{P}_{1,max}$, respectively. One can see that at low and intermediate stress range, for the equi-biaxial loading, the interfacial normal stress and the principal stress in particles are higher than that for the uniaxial loading. This indicates that the equi-biaxial loading is more damageable than the uniaxial loading. The difference between the two loads decreases as the applied equivalent stress increases, especially at a high applied stress level. This is because macroscopic plasticity is induced at such high stress levels.

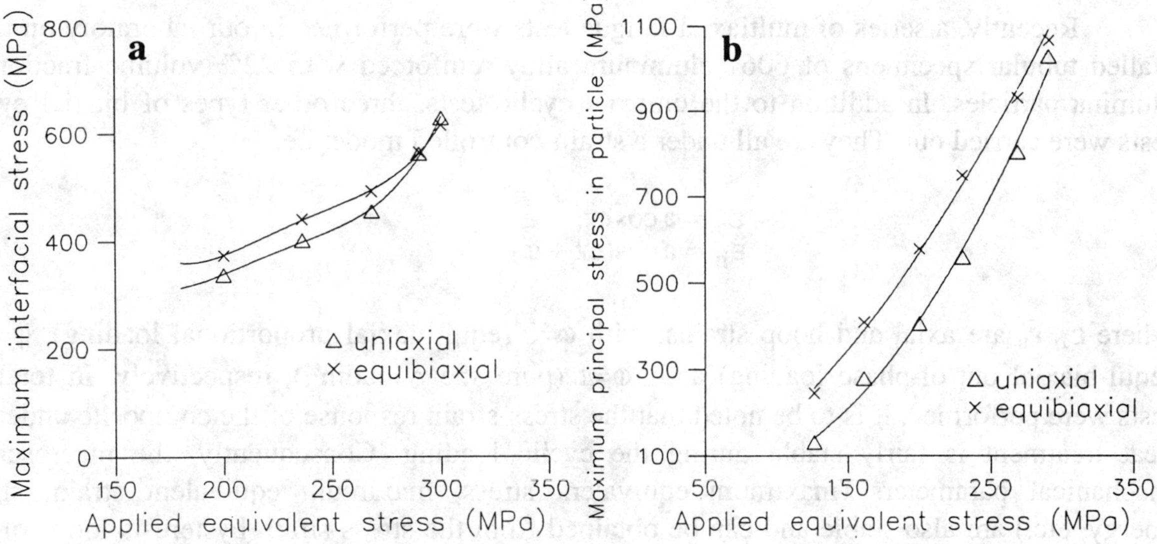

Fig. 2 Correlation of macro- and micro-damage parameters. **a**. Maximum interfacial normal stress versus applied equivalent stress; **b**. Maximum principal stress in particle versus applied equivalent stress.

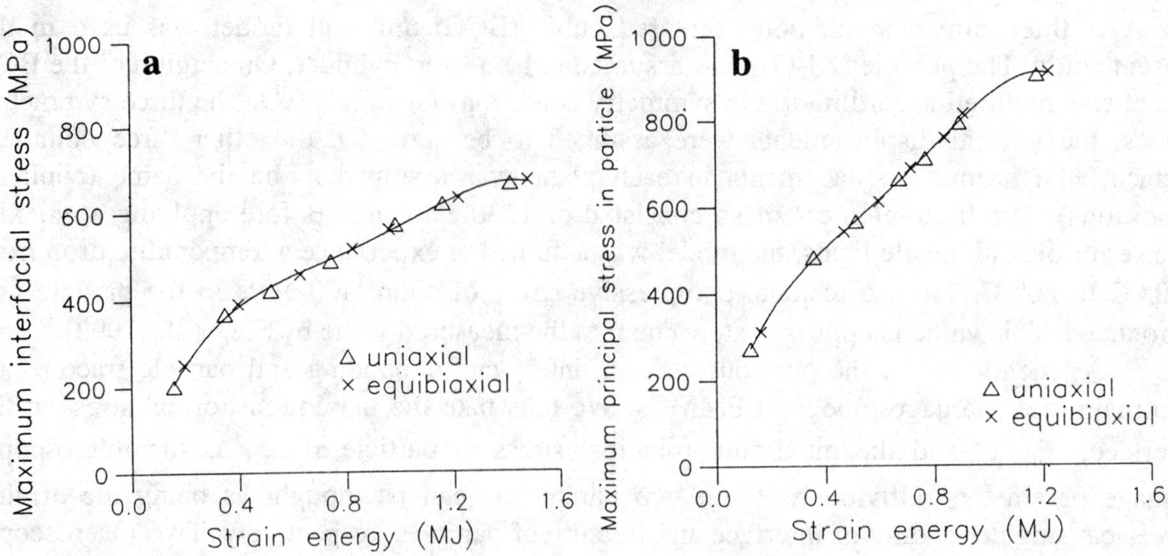

Fig. 3 Correlation of macro- and micro-damage parameters: **a.** Maximum interfacial normal stress versus applied strain energy; **b.** Maximum principal stress in particle versus applied strain energy.

The correlation of applied strain energy with the two micro damage parameters is shown in Figs. 3a,b .It is interesting to note that the distinction between the uniaxial and the biaxial loading disappears when the applied strain energy is used as a correlation parameter. For either low, intermediate or high stress range, a single curve can fit a functional relation for both loading cases. Therefore, the strain energy parameter is a better indicator in correlating the multiaxial loading with that of the uniaxial for the PMMCs

ANALYSIS OF MULTIAXIAL FATIGUE TESTS DATA

Recently, a series of multiaxial fatigue tests were performed in our laboratory on thin-walled tubular specimens of 6061 aluminum alloy reinforced with 22% volume fraction of alumina particles. In addition to the uniaxial cyclic tests, three other types of biaxial cyclic tests were carried out. They are all under a strain-controlled mode, i.e.

$$\begin{aligned} \varepsilon_a &= a\cos\omega t \\ \varepsilon_h &= a\cos(\omega t+\varphi) \end{aligned} \qquad (1)$$

where ε_a, ε_h are axial and hoop strains, with $\varphi=0$ (equi-biaxial proportional loading), $\varphi=\pi/2$ (equi-biaxial out-of-phase loading) and $\varphi=\pi$ (pure shear loading), respectively. In total 36 tests were performed. It is to be noted that the stress-strain response of the composite under T6 heat treatment is fairly stable during the cyclic loading. Consequently, the macroscopic mechanical parameters (maximum equivalent stress, maximum equivalent strain, strain energy, etc.) are also stable and can be obtained from the stress-strain hysteresis loops of the cyclic loading. In this study the stress-strain loops at two-hundreds cycles were used.

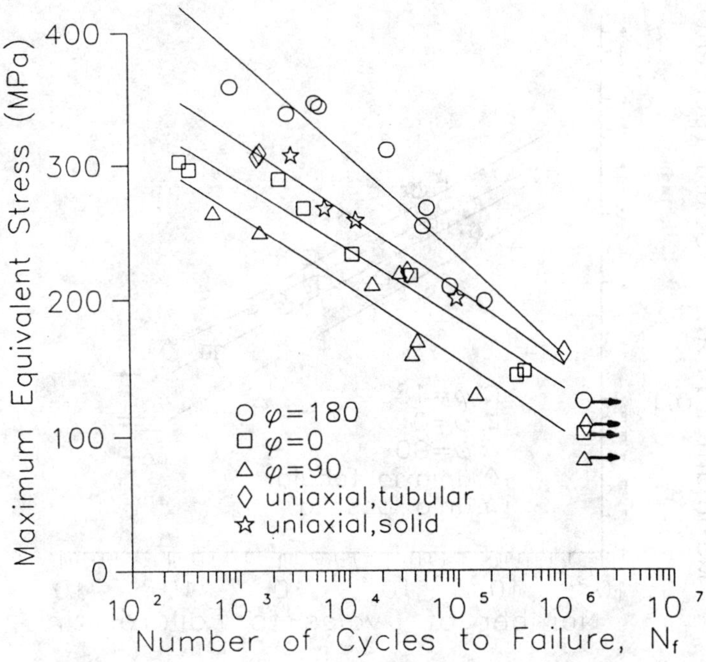

Fig. 4 Correlation of the maximum equivalent stress with the Multiaxial fatigue life data of Al_2O_3/6061-Al composite.

First, we use the conventional stress parameter , the maximum equivalent stress,

$$\sigma_{eq,max} = (\sigma_a^2 + \sigma_h^2 - \sigma_a \sigma_h)_{max}^{1/2} \tag{2}$$

to correlate with the fatigue life data. The result is shown in Figure 4. It is seen that the data points are dispersed over a relatively wide range in the figure, and they cannot be correlated by using a simple curve. However, for an individual type of loading path the data points can be approximately fitted by a linear curve in the log N_f vs $\sigma_{eq,max}$ coordinates. It is also noted that the curve for the out-of-phase loading path indicates the lowest life. The damage sequence in order of severity is: non-proportional, and proportional equi-biaxial, uniaxial and pure shear loadings.

Next, we use a strain parameter, the maximum equivalent strain,

$$\varepsilon_{eq,max} = (2e_{ij}e_{ij} / 3)_{max}^{1/2} \tag{3}$$

where e_{ij} is the deviatoric strain components. As shown in Fig. 5. four different lines are obtained for the four individual types of loading paths in the $\log N_f$ vs $\log \varepsilon_{eq,max}$ coordinates. However, this parameter also cannot correlate all data in a single curve.

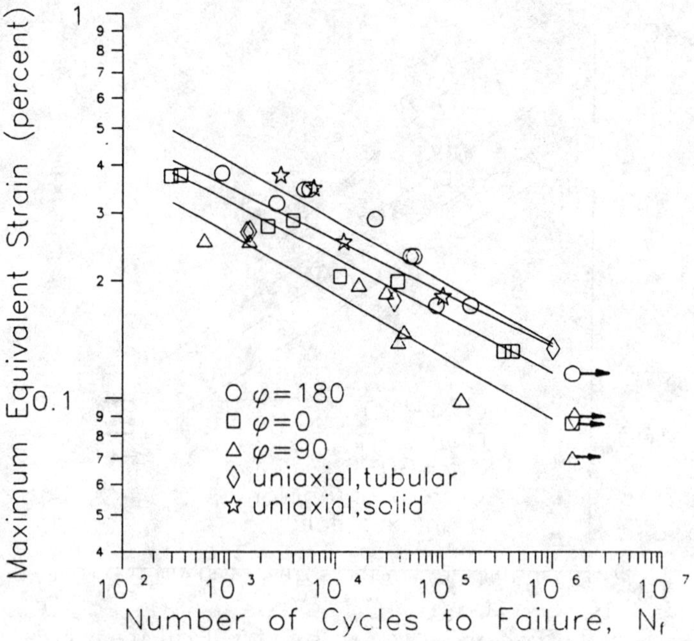

Fig. 5 Correlation of the maximum equivalent strain with the Multiaxial fatigue life data of Al_2O_3/6061-Al composite.

Ellyin et al. (1993) have suggested a strain energy parameter as a criterion in predicting multiaxial fatigue life of materials. This energy parameter can be expressed as

$$\Delta W^t = \Delta W^p / \bar{\rho} + \Delta W^{e+} \tag{4}$$

In the above, ΔW^p is the plastic energy per cycle and is calculated by

$$\Delta W^p = \int_{cycle} \sigma_{ij} d\varepsilon_{ij}^p \tag{5}$$

ΔW^{e+} is the elastic energy associated with positive stress and is determined from,

$$\Delta W^{e+} = \int_{cycle} H(\sigma_i) H(d\varepsilon_i^e) \sigma_i d\varepsilon_i \tag{6}$$

where σ_i (i=1,2,3) are principal stress, ε_i^e are the elastic part of the principal strains. H(x) is the heaviside function, i.e., H(x)=1 if x>0 and H(x)=0 if x<0. $\bar{\rho}$ is a multiaxial constraint factor, it specifies the surface strain constraint condition of different applied stress states.

Here we simply assume $\bar{\rho}=1$, and then eq. 4 reduces to:

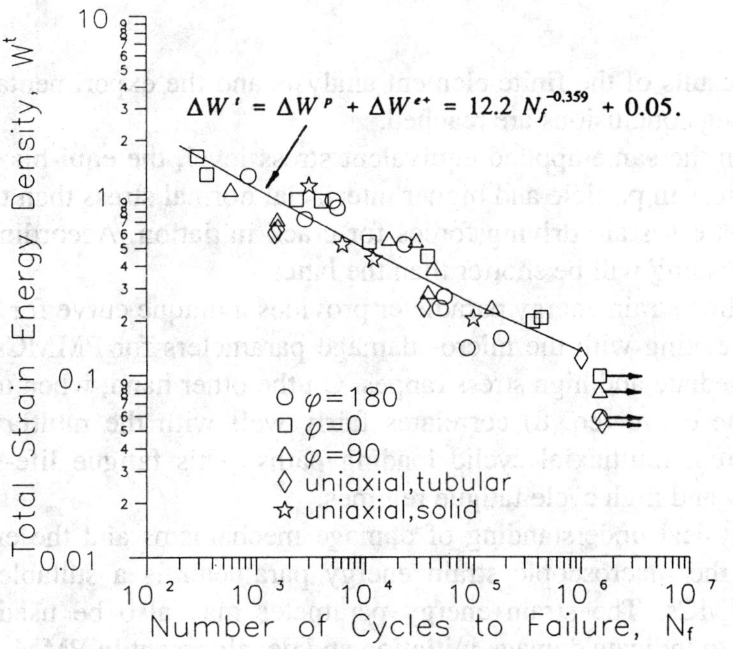

Fig. 6 Correlation of the total strain energy density with the Multiaxial fatigue life data of Al_2O_3/6061-Al composite.

$$\Delta W^t = \Delta W^p + \Delta W^{e+} \tag{7}$$

The correlation of this strain energy parameter with the multiaxial fatigue life data is shown in Fig. 6. One can see that all the life data points fall within a narrow band. Moreover, the points for different loading paths are inter-dispersed, and thus no clear lines can be drawn to separate the four groups of data. A single best fit curve,

$$\Delta W^t = 12.2 N_f^{-0.359} + 0.05 \tag{8}$$

correlates fairly well with the data, and the coefficient of correlation is 0.909 for the above best fit equation.

The above comparison once again demonstrates that the strain energy parameter is the most suitable one to correlate with the multiaxial fatigue data of the PMMCs. Such a correlation is not surprising because the essential interaction between stress and strain and hence the path dependence, is inherently included in the strain energy parameter. The assumption of $\overline{\rho} = 1$ may also have an inherent reason. As mentioned before, the damage process has a highly localization characteristic due to the existence of reinforcing particles. The fatigue crack initiation takes place at particle/matrix interfaces and particles with relatively large size or clusters of particles. Therefore, the influence of the surface constraint may not be as determinant as noted in isotropic materials due to the localized characteristic of damage within the interior clusters of PMMCs.

CONCLUSIONS

From the results of the finite element analysis and the experimental data of multiaxial fatigue the following conclusions are reached.

1. Based on the same applied equivalent stress level, the equi-biaxial loading induces higher principal stress in particle and higher interfacial normal stress than the uniaxial loading. These two micro stresses are driving forces for crack initiation. Accordingly, the fatigue life under the former loading will be shorter than the latter.

2. The applied strain energy parameter provides a unique curve for biaxial and uniaxial loading when correlating with the micro- damage parameters for PMMCs. This single curve covers low, intermediate and high stress ranges. On the other hand, when using a strain energy parameter a unique curve (eq. 8) correlates fairly well with the multiaxial fatigue data of PMMCs for different multiaxial cyclic loading paths. This fatigue life-strain energy curve applies to both low and high cycle fatigue regimes.

3. Both physical understanding of damage mechanisms and the empirical correlation have proven that the macroscopic strain energy parameter is a suitable multiaxial fatigue parameter for PMMCs. The strain energy parameter may also be used as a macroscopic damage parameter to indicate damage initiation and development in PMMCs.

ACKNOWLEDGEMENT

This research is supported by the Natural Science and Engineering Council of Canada (NSERC strategic grant STR 0149082).

REFERENCES

Ellyin, F. 1997, **Fatigue Damage, Crack Growth and Life Prediction**, Chapman & Hall, London.

Ellyin, F., Kujawski, D. and Xia, Z. 1993, "Multiaxial and Mixed-Mode Fatigue", in: **Fatigue 93, Proceedings of the Fifth International Conference on Fatigue and Thresholds**, Bailon, J.P. and Dickson, J.I. eds., Engineering Materials Advisory Services Ltd., Vol. III:1504-1514.

Ellyin, F. and Li, C. 1996, "The Role of Cyclic Plasticity in Crystallographic Crack Growth Retardation", Materials Science Research International, **1:** 137-143.

Hurd, N.J. 1988, "Fatigue Performance of Alumina Reinforced Metal Matrix Composites", Material Science and Technology,: **4:** 513-517.

Li, C. and Ellyin, F. 1995, "Short Crack Growth Behavior in Particulate Reinforced Metal Matrix Composites", Metall. Mater. Trans., **26A:** 3177-3182.

Meijer, G., Xia, Z. and Ellyin, F. 1997, "Biaxial Cyclic Analysis of Al_2O_{3p}-6061 Al Composite", Acta Materialia, (in press).

Mochida, T., Taya, M. and Lloyd, D.J. 1991, "Fracture of Particles in a Particle/Metal Matrix Composite under Plastic Strain and Its Effect on the Young's Modulus of the Composite", Mater. Trans., JIM, **32:** 931-942.

Ravichandran, G. and Liu, C.T. 1995, "Modeling Constitutive Behavior of Particulate Composite Undergoing Damage", Int. J. Solids Struct., **32:** 979-990.

Shi, N. et al. 1993, "Deformation-Induced Residual Stress Changes in SiC whisker-Reinforced 6061 Al Composites", Metall Trans., **24A:** 187-196.

Wang, L. et al. 1996, "Cyclic Deformation and Low Cycle Fatigue Behavior in a 6061 Al/22% vol. SiC Whisker Composite", Mater. Trans. JIM, **37:** 762-768.

Wienecke H.A., Brockenbrough, J.R. and Romanko, A.D. 1995, "A Three-Dimensional Unit Cell Model with Application Towards Particulate Composites", ASME J. Appl. Mech., **62:** 136-140.

Xia, Z. , Ellyin, F. and Meijer, G. 1997, "Mechanical Behavior of Al_2O_3 particle-reinforced 6061 aluminum alloy under uniaxial and multiaxial cyclic loading", Composite Science and Technology, **57:** 237-248.

The Durability of Graphite-Bismaleimide Aerospace Composites Bonded Using a Modified Epoxy Adhesive

L. M. BUTKUS[1] AND W. S. JOHNSON[2]

ABSTRACT

Adhesive bonding is a common method used in the fabrication of composite-composite and composite-metal joints, and it has become especially popular in the aerospace industry. The advantages of this technique include the elimination of stress concentrations inherent in riveted designs, improved sealing and aerodynamic properties, and, often, reduced costs. To realize the full potential of adhesively bonded composites, the mechanical properties of the adhesive layer must be fully understood. The behavior of adhesives in terms of stress and strain has been well documented. However, as with metallic components, stress-strain data should be supplemented with information regarding the fatigue and fracture characteristics of the adhesive bondline. To ensure reliability, environmental durability must also be addressed. This paper will present results from research on the durability of a bonded composite system intended for use on advanced fighter aircraft. The bonded system consists of graphite-bismaleimide adherends joined using a modified epoxy adhesive. Experimental data will describe the Mode I fracture toughness and fatigue crack growth behavior of the adhesive bondline in terms of the strain energy release rate (G). Prior to testing, selected specimens were subjected to conditions based upon projected aircraft operating environments: a) 100 thermal cycles between -54°C (-65°F) and 104°C (220°F), or b) 5,000 hours of isothermal exposure at 104°C (220°F). Although the thermal cycling had no appreciable effect on the fracture toughness or fatigue crack growth behavior of the system, the isothermal exposure did degrade the monotonic fracture toughness. The relationship between the environmentally-induced changes in the bondline mechanical properties and the chemical and/or physical aging of the epoxy adhesives was investigated using infrared spectroscopy, thermogravimetric analysis, and differential scanning calorimetry. Results from this project have particular significance to the design, durability, and damage tolerance of composite joints found in aircraft structures and other bonded components.

[1] L.M. Butkus, Graduate Student, School of Mechanical Engineering, Box #1001, 801 Ferst Drive, Georgia Institute of Technology, Atlanta, GA 30332-0405

[2] W.S. Johnson, Professor, Schools of Materials Science & Engineering and Mechanical Engineering, 778 Atlantic Drive, Georgia Institute of Technology, Atlanta, GA 30332-0245

BACKGROUND

Adhesively bonded structures in aerospace applications have multiplied dramatically during the past two decades. Progress in this field has been made in joint designs and in the development of materials, including advanced composites and toughened epoxy adhesives.

Previous research in the area of bonded joint analysis and design may be grouped into two major areas of emphasis. The first, a stress-based approach, was initiated by Goland & Reissner (1944), and used extensively by Hart-Smith (1983) and Hart-Smith & Thrall (1985) in the U.S. Air Force's Primary Adhesively Bonded Structures Technology (PABST) program. A more recently developed approach based upon fracture mechanics principles was first proposed by Ripling, Mostovoy, & Patrick (1963). Subsequent research by Shaw (1983), Johnson & Mall (1985), Mall & Johnson (1986), Brussat *et al.* (1987), Fernlund, *et al.* (1994), and others has shown that the strain energy release rate (G) may be used to describe the fatigue and fracture behavior of adhesively bonded joints.

The research described in the following article has used fracture mechanics to investigate the durability of bonded aircraft structures. The specific focus of this paper will be on the Mode I fracture and fatigue behavior of a bonded composite system used on the U.S. Air Force's new fighter aircraft, the F-22 *Raptor*. The particular bonded system consists of graphite-bismaleimide adherends bonded with a modified epoxy adhesive. Due to the stability of the bismaleimide matrix, this system is intended for use in sections of the outer skin exposed to high temperatures.

Fatigue and fracture results are presented in terms of the strain energy release rate (G_I) and will be discussed with respect to their influence on bonded composite joint design.

EXPERIMENTAL PROCEDURES

SPECIMEN GEOMETRY

Tests were performed on double cantilever beam specimens fitted with bonded hinges. (Fig. 1) Specimens were nominally 25 mm (1 in) wide and 300 mm (12 in) long.

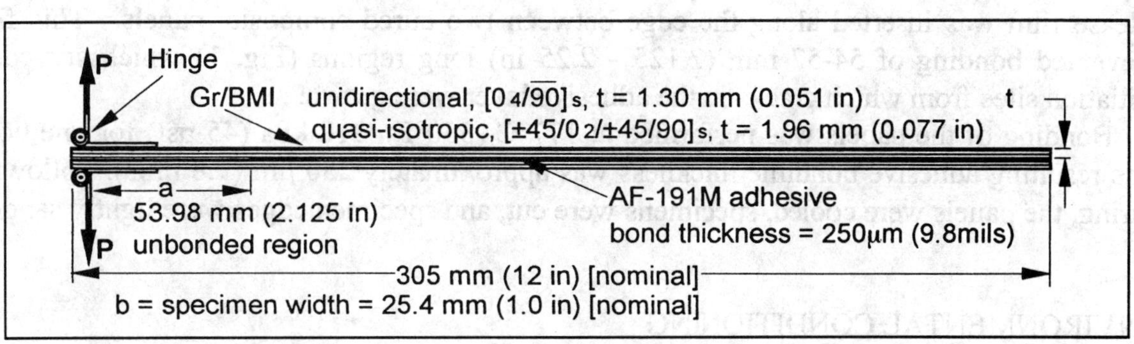

Figure 1. Double Cantilever Beam Specimen Geometry

SPECIMEN MATERIALS AND FABRICATION

Fabrication of the bonded graphite-bismaleimide (Gr-BMI/AF-191M/Gr-BMI) specimens employed manufacturing processes currently used in the aerospace industry.

The adhesive used was AF-191M, a toughened epoxy manufactured in film form (1.42 kg/m^2 [0.06 lb/ft^2]) by 3M Corp. The adhesive was supported by a non-woven nylon carrier or "scrim" cloth. (Fig. 2) This scrim cloth was located near one face of the film rather than along the centerline and comprised approximately 4% of the adhesive film by volume.

Composite adherends consisted of BASF IM7/5250-4 graphite-reinforced bismaleimide with an approximate 65% fiber volume fraction.

Figure 2. Non-Woven Nylon Scrim Cloth Contained in the AF-191M Film Adhesive

Composite panels were cured prior to bonding at 191°C (375°F), a vacuum pressure of 75 kPa (22 in. Hg), and an applied pressure of 655 kPa (95 psi) for 20 minutes. Two adherend layups were investigated: 1) a unidirectional 9 ply $[0_4/\overline{90}]_s$ layup with a nominal thickness of 1.30 mm (0.051 in.), and b) a quasi-isotropic 14 ply layup $[\pm45/0_2/\pm45/90]$, with a nominal thickness of 1.96 mm (0.077 in.). The quasi-isotropic layup is typical of ply stacking sequences used for composite components. The unidirectional layup was chosen to compare the behavior of bondline cracks in the vicinity of 0° plies with their behavior in the vicinity of the ±45° plies in the quasi-isotropic adherends.

Pre-bond surface preparation of pre-cured composite adherends consisted of peel ply removal followed by light hand sanding with 180 grit abrasive paper. Panels were then wiped with methanol and dried. Prior to bonding, a 102 μm (4 mil) thick Wrightlon™ release film was inserted along the edge between two cured composite panels. This film prevented bonding of 54-57 mm (2.125 - 2.25 in) long regions (Fig. 1) which served as initiation sites from which cracks in the adhesive layer were grown.

Bonding of the panels was performed at 177°C (350°F), 310 kPa (45 psi) for one hour. The resulting adhesive bondline thickness was approximately 250 μm (9.8 mils). Following curing, the panels were cooled, specimens were cut, and specimen edges were lightly sanded.

ENVIRONMENTAL CONDITIONING

Specimens were tested in three conditions: 1) as-received, 2) following isothermal exposure to the maximum operating temperature for bonded BMI composites, and 3) following exposure to a thermal cycling profile simulating temperature excursions between high altitude sub-sonic cruise and maximum aircraft performance conditions.

Specimens identified for testing in the as-received condition were stored in laboratory conditions (22°C [72°F], 50±5% rh).

Specimens (with unidirectional adherends, only) selected for testing following long-term isothermal exposure in a "hot/dry" atmosphere were exposed to 104°C (220°F) for 5,000 hours in a circulating air oven. Following exposure, these specimens were stored in a desiccator at room temperature prior to testing.

Specimens selected for thermal cycling were initially exposed to a "hot/wet" environment (71°C [160°F], >90% rh) for 320 hours. This simulated service conditions and introduced moisture into the bondline. Following "hot/wet" exposure, the specimens were sealed in foil tape and subjected to 100 cycles between -54°C (-65°F) and 104°C (220°F) using an average ramp rate of 6°C/min (11°F/min). (Fig. 3) Cycling was performed in a chamber using a pneumatic cage to shuttle specimens between hot and cold compartments. A thermocouple placed between two adherends monitored bondline temperatures. Prior to testing, cycled specimens remained sealed and were stored in laboratory conditions.

TESTING

All testing was performed at room temperature under laboratory conditions using computer-controlled test systems. Loads, displacements (from machine LVDT readouts), and cycle counts were collected by a digital data acquisition system. Crack growth within the adhesive layer was measured using a traveling microscope (mag.: 20X) or a Questar long focal length microscope (mag.: 200X). To further assist in tracking crack growth, one edge of each specimen was painted white and imprinted with a scale having 0.5 mm gradations. Crack length was monitored on the painted edge.

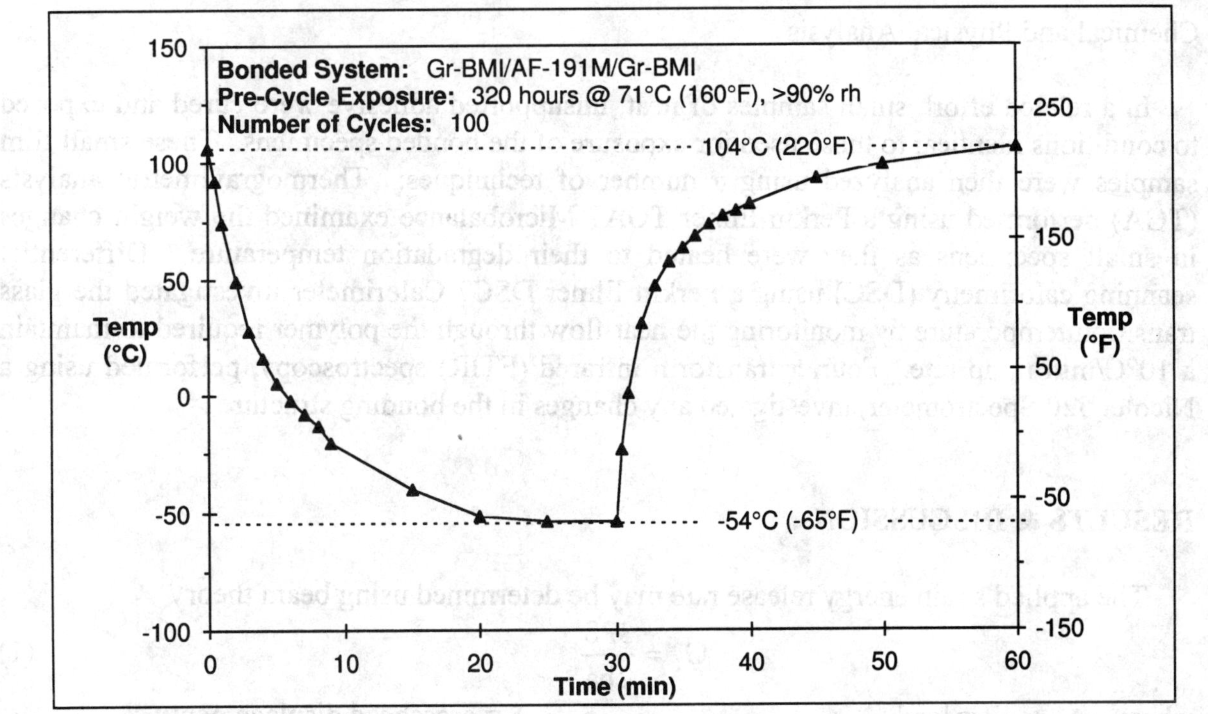

Figure 3. Thermal Cycling Profile for Gr-BMI/AF-191M/Gr-BMI Bonded Specimens

Monotonic Testing Procedures

Monotonic testing was conducted on a 4.5 kN (1 kip) screw-driven test frame to obtain the Mode I fracture toughness (G_{IC}) of the Gr-BMI/AF-191M/Gr-BMI systems. A crosshead displacement rate, equal to a crack mouth opening rate, of 1.0 mm/min (0.04 in/min) was used, consistent with procedures outlined in ASTM D3433-75 and D5228-94a. Deviation from linearity of a load versus displacement trace indicated the onset of crack growth which was confirmed by optical observations. Several loading runs were performed on each specimen permitting the calculation of multiple G_{IC} values. Each run proceeded until the load reached a maximum and then decreased during crack extension. The crosshead was returned to zero displacement after a minimum crack growth of 5 mm (0.20 in).

Fatigue Testing Procedures

Fatigue testing was performed on the unidirectional Gr-BMI/AF-191M/Gr-BMI system. Test equipment consisted of a servohydraulic test machine fitted with a 2.2 kN (500 lb) load cell operating under displacement control at a frequency of 1 Hz. Periodic compliance cycles at 0.1 Hz captured peak/valley load and displacement values subsequently used to calculate compliance and to estimate crack length and applied G_I. Tests used a maximum displacement of 10 mm and a displacement R ratio ($\delta_{min}/\delta_{max}$) of 0.1. Displacement control permitted load shedding and facilitated the identification of threshold levels of the applied strain energy release rate. A threshold crack growth rate of 1×10^{-6} mm/cycle (4×10^{-8} in/cycle) was chosen based upon studies by Marceau, *et al.* (1978) and Mall, *et al.* (1982).

Chemical and Physical Analysis

In a related effort, small samples of neat, unsupported adhesive were cured and exposed to conditions identical to those used for exposure of the bonded specimens. These small film samples were then analyzed using a number of techniques. Thermogravimetric analysis (TGA) performed using a Perkin Elmer TGA7 Microbalance examined the weight changes in small specimens as they were heated to their degradation temperature. Differential scanning calorimetry (DSC) using a Perkin Elmer DSC7 Calorimeter investigated the glass transition temperature by monitoring the heat flow through the polymer required to maintain a 10°C/min ramp rate. Fourier transform infrared (FTIR) spectroscopy, performed using a Nicolet 520 Spectrometer, investigated any changes in the bonding structure .

RESULTS & DISCUSSION

The applied strain energy release rate may be determined using beam theory

$$G_I = \frac{3P\delta}{2ba} \tag{1}$$

where P = load, δ = crosshead displacement, b = specimen width, a = crack length

Equation (1) may be further modified (ASTM D5228-94a) to account for the relationship between specimen compliance and observed crack length using

$$G_I = \frac{3P\delta}{2b(a+|\Delta|)} \tag{2}$$

where Δ is the a-axis intercept of the linear relationship between the cubed root of the compliance, $(\delta/P)^{1/3}$, and the crack length, a.

Mode I Fracture Toughness

Critical strain energy release rates (G_{IC}) from the monotonic fracture toughness tests were obtained using the Modified Beam Theory, equation (2), the visually observed crack length, and the load, P, at which crack growth began and the load versus displacement trace deviated from linearity. Figure 4 compares the G_{IC} values for the Gr-BMI/AF-191M/Gr-BMI specimens tested in the three conditions examined in this project.

The baseline G_{IC} of 1700 J/m² is relatively high compared to the Mode I toughness of several adhesive and matrix systems examined by Johnson & Mangalgiri (1987). Based on a statistical analysis using a 95% confidence interval, thermal cycling did not appreciably affect G_{IC}. However, 5,000 hours of exposure to 104°C (220°F) did reduce the toughness of the system with unidirectional adherends by approximately 28%.

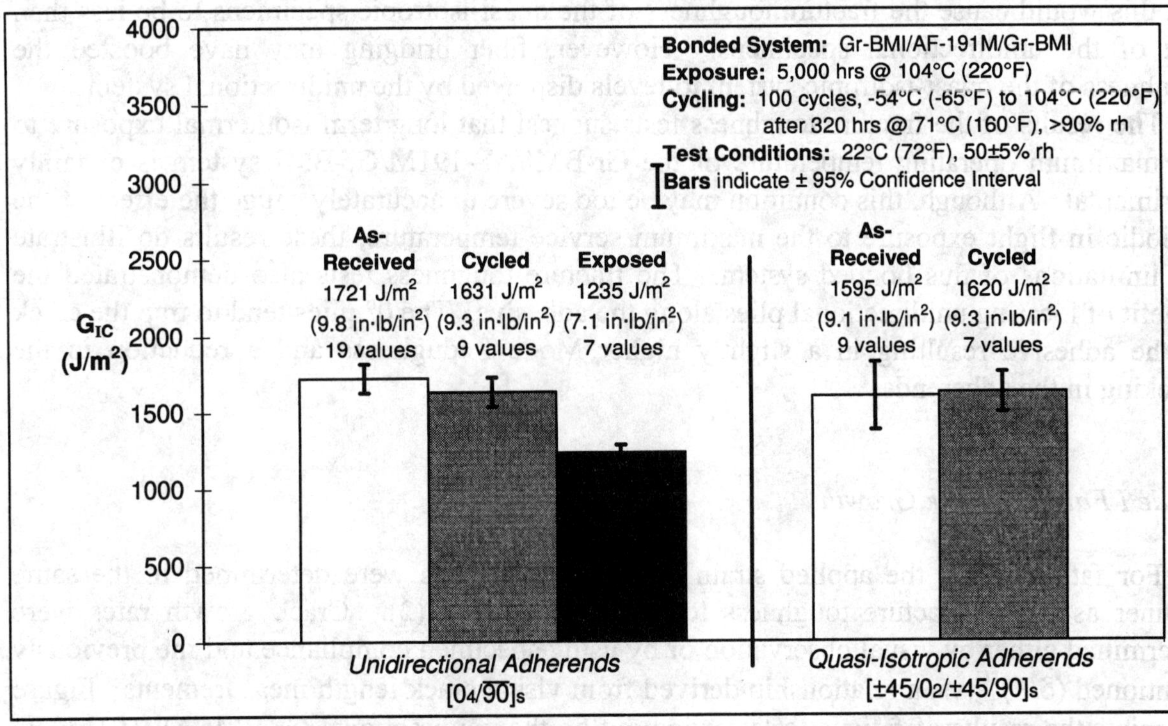

Figure 4. Comparison of the Effects of Environmental Exposure on the Mode I Fracture Toughness of the Gr-BMI/AF-191M/Gr-BMI Bonded System

Figure 5. Fracture Surfaces of the Gr-BMI/AF-191/Gr-BMI Bonded Systems(As-Received)
Upper Photo: Unidirectional Adherend, Lower Photo: Quasi-Isotropic Adherend

Isothermal exposure also resulted in darkening of the normal pale yellow AF-191M along the specimen edges, suggesting a physical or chemical change occurring at during exposure.

The crack path for specimens with unidirectional adherends was cohesive. (Fig. 5) However, because the scrim cloth was located closer to one face of the adhesive film, the crack path was offset towards one adherend and followed the plane of the scrim cloth. The crack front was nearly straight across the adherends with little evidence of "tunneling" in which the interior portion of the crack grows more rapidly than that located near the edge.

The crack path for specimens with quasi-isotropic adherends often departed from the bondline. This resulted in interlaminar cracking within the ±45° plies located near the adhesive, a behavior previously observed in bonded graphite/epoxy composites by Johnson & Mall (1986). It is likely that the BMI matrix is less tough than the AF-191M adhesive, and this would cause the fracture toughness of the quasi-isotropic specimens to be less than that of the unidirectional specimens. However, fiber bridging may have boosted the toughness of the quasi-isotropic system to levels displayed by the unidirectional system.

The results of the fracture toughness tests suggest that long term isothermal exposure to the maximum operating temperature of the Gr-BMI/AF-191M/Gr-BMI system is certainly detrimental. Although, this condition may be too severe to accurately gauge the effect of the periodic in-flight exposure to the maximum service temperature, these results do illustrate the limitations of this bonded system. The fracture toughness tests also demonstrated the benefit of locating unidirectional plies along the adhesive. The 0° plies tend to trap the crack in the adhesive resulting in a slightly higher Mode I toughness and a reduction in the cracking in the adherends.

Mode I Fatigue Crack Growth

For fatigue tests, the applied strain energy release rates were determined in the same manner as for he fracture toughness tests using equation (2). Crack growth rates were determined either by visual observation or by using specimen compliance and the previously mentioned $(\delta/P)^{1/3}$ vs. a relationship derived from visual crack length measurements. Figure 6 depicts the results of fatigue tests carried out on the unidirectional Gr-BMI/AF-191M/Gr-BMI specimens. (Specimens with quasi-isotropic adherends were not tested in fatigue.)

In contrast to the toughness behavior exhibited by the Gr-BMI/AF-191M/Gr-BMI system, fatigue testing indicated this bonded system is quite resistant to environmental exposure. No significant differences were observed in the threshold level or growth rate sensitivity (slope of the da/dN vs. ΔG_I line) among the three conditions tested.

Despite the lack of contrast in fatigue behavior among the specimens subjected to various environments, two observations may still be made regarding the general characteristics of fatigue crack growth in this system. First, $G_{I,max}$ at threshold was approximately one order of magnitude lower than the fracture toughness (G_{IC}). Thus, components that use this material system and which will be subjected to cyclic loading, must be designed with consideration given to fatigue and monotonic behavior.

Second, the crack growth sensitivity of the Gr-BMI/AF-191M/Gr-BMI system is extreme compared to that of metals. An indication of this is the steep slope (approx. 4-6) of the fatigue data. In contrast, fatigue data from aluminum and titanium alloys usually displays a slope of 2. Thus, small changes in the applied strain energy release rate or miscalculations of applied loads may cause crack growth rate differences of an order of magnitude or more.

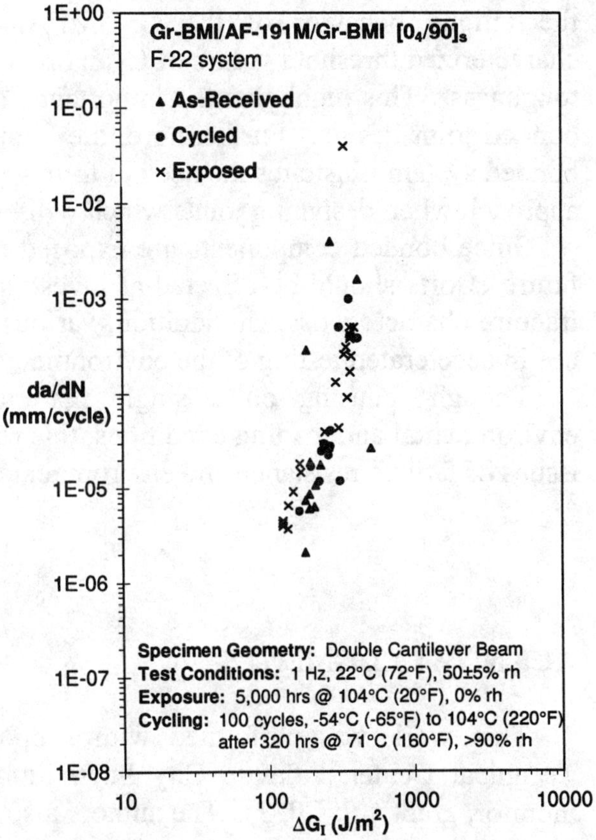

Figure 6. Comparison of the Effects of Environmental Exposure on Mode I Fatigue Crack Growth in the Unidirectional Gr-BMI/AF-191M/Gr-BMI Bonded System

Chemical And Physical Analyses

The chemical and physical analyses of the AF-191 epoxy adhesive revealed only slight changes caused by isothermal exposure or thermal cycling. No change in degradation temperature was identified by the TGA results. Nor were any significant changes noted in the chemical bonding structure revealed by results of the FTIR spectroscopy. Only T_g, the glass transition temperature, shown by DSC analysis, differed among the adhesive specimens that were subjected to the various environmental conditions. The glass transition temperature decreased from approximately 105°C (221°F) in the as-received condition to 102°C (216°F) for the thermally cycled material and 101°C (213°F) for the material subjected to isothermal exposure. Because T_g depends upon a polymer's resistance to motion within its structure which is enhanced by molar mass, branching, and crosslinking, this decrease may indicate a reduction in the amount of crosslinking. This slight change may, in turn, be the cause of the reduction in the fracture toughness of the exposed bonded specimens. However, the changes indicated by the chemical and physical analyses are extremely subtle and may not be significant. Additional work is required to draw parallels between mechanical behavior and the environmental effects on the polymer structure of the AF-191 adhesive.

CONCLUSIONS

This study investigated the Mode I fatigue and fracture properties of a bonded graphite-bismaleimide/modified epoxy (Gr-BMI/AF-191/Gr-BMI) system used on high performance military aircraft. These properties were examined in specimens in the as-received state as well as in those subjected to long-term isothermal exposure and thermal cycling indicative of the service conditions. The fracture toughness of this system was high compared to other adhesive and composite resins but was reduced by long-term exposure to the maximum operating temperature. However, neither long-term exposure nor thermal cycling affected the fatigue behavior of the Gr-BMI/AF-191M/Gr-BMI. This fatigue behavior was characterized threshold values of G_I an order of magnitude lower than the monotonic fracture toughness. This highlights the importance of including cyclic and monotonic properties in bonded joint design. Furthermore, the high degree of crack growth rate sensitivity in this bonded system illustrates the need to follow a conservative philosophy such as the "safe-life" approach when designing joints which will be subjected to cyclic Mode I loading.

Since bonded components are exposed to modes of loading other than simply Mode I, future efforts should be directed at investigating the Mode II and mixed-mode fatigue and fracture characteristics. In addition, various extreme environments should be evaluated for use in accelerated testing of the environmental durability of this and other bonded systems.

Though focusing on a single material system subjected to a select number of environmental and loading conditions, this research highlights the need to address the crucial issues of fatigue resistance and environmental durability in bonded composite joint design.

ACKNOWLEDGMENTS

This work was performed with support from the Federal Aviation Administration Technical Center, Atlantic City International Airport, NJ, (Dr. Don Oplinger, technical monitor, grant # 95G023). The authors also wish to thank Mr. Scott Reeve of the Lockheed Martin Aeronautical Systems Co. for providing specimens and personnel from the Technology & Engineering Sciences Division (WR-ALC/TIECD, Robins AFB, GA) for assistance with the thermal cycling experiments.

REFERENCES

ASTM D3433-75, "Standard Practice for Fracture Strength in Cleavage of Adhesives in Bonded Joints," Annual Book of ASTM Standards, American Society for Testing and Materials, Philadelphia, 1994.

ASTM D5528-94a, "Standard Test Method for Mode I Interlaminar Fracture Toughness of Unidirectional Fiber-Reinforced Polymer Matrix Composites," Annual Book of ASTM Standards, American Society for Testing and Materials, Philadelphia, 1994.

Brussat, T.R., Chiu, S.T., and Mostovoy, S. 1987. *Fracture Mechanics for Structural Adhesive Bonds - Final Report*, Lockheed Co., Burbank, CA, for the Air Force Materials Laboratory, AFML-TR-77-163.

Fernlund, G., Papini, M., McCammond, D., and Spelt, J.K. 1994. "Fracture Load Predictions for Adhesive Joints," *Composite Science and Technology*, Elsevier Science Publishers, London, 51: pp. 587-600.

Goland, M. and Reissner, E. 1944. "The Stresses in Cemented Joints," *Journal of Applied Mechanics*, 11, American Society of Mechanical Engineers, New York, NY, 11: pp. A-17 to A-27.

Hart-Smith, L.J. and Thrall, E.W. 1985. "Structural Analysis of Adhesive-Bonded Joints," Adhesive Bonding of Aluminum Alloys, E.W. Thrall and R.W. Shannon, eds., Marcel Dekker, Inc., New York: pp. 241-322.

Hart-Smith, L.J. 1983. "Adhesive Bonding of Aircraft Primary Structures," in High Performance Adhesive Bonding, G. DeFrayne, ed., Society of Manufacturing Engineers, Dearborn, MI: pp. 99-113.

Johnson, W.S. and Mall, S. 1985. "A Fracture Mechanics Approach for Designing Adhesively Bonded Joints," in Delamination and Debonding of Materials, ASTM STP 876, W.S. Johnson, ed., American Society for Testing and Materials, Philadelphia: pp. 189-199.

Johnson, W.S. and Mall, S. 1986. "Influence of Interface Ply Orientation on Fatigue Damage of Adhesively Bonded Composite Joints," *Journal of Composites Technology and Research*, 8(1): pp. 3-7.

Johnson, W.S. and Mangalgiri, P.D. 1987. "Influence of the Resin on Interlaminar Mixed-Mode Fracture," Toughened Composites, ASTM STP 937, N.J. Johnston, ed., American Society for Testing and Materials, Philadelphia.

Mall, S., Johnson, W.S., and Everett, J.A., Jr. 1982. "Cyclic Debonding of Adhesively Bonded Composites," in Adhesive Joints: Formation, Characteristics, and Testing, K.L. Mittal, ed., Plenum Press, New York.

Mall, S. and Johnson, W.S. 1986. "Characterization of Mode I and Mixed-Mode Failure of Adhesive Bonds Between Composite Adherends," in Composite Materials: Testing and Design (Seventh Conference), ASTM STP 893, J.M. Whitney, ed., American Society for Testing and Materials, Philadelphia: pp. 322-4.

Marceau, J.A., McMillan, J.C., and Scardino, W.M. 1978. "Cyclic Stress Testing of Adhesive Bonds," *Adhesives Age*, April: pp. 37-41.

Ripling, E.J., Mostovoy, S. and Patrick, R.L. 1963 "Application of Fracture Mechanics to Adhesive Joints," in Adhesion, ASTM STP 360, American Society for Testing and Materials, Philadelphia: pp. 5-19.

Shaw, S.J. 1983. "Adhesive Joint Failure - A Fracture Mechanics Approach," in Adhesion 7: Proceedings of the 20th Annual Conference on Adhesion and Adhesives, K.W. Allen, ed., Elsevier Applied Science Publishers, London: pp. 173-196.

Investigation of the Durability Characteristics of Fibre Reinforced Plastic (FRP) Materials in Concrete Environment

A. COOMARASAMY AND S. GOODMAN

ABSTRACT

Fibre reinforced plastic (FRP) materials have been successfully used in place of traditional materials such as metals, wood and ceramics in many commercial and industrial applicatios including aerospace industries, for many years. In recent years, the potential use of these materials as concrete reinforcement has attracted much attention. Commercially available FRP materials (mostly glass fibre reinforced plastic- GFRP materials) from six different manufacturers were used in this investigation. Accelerated laboratory tests were performed to evaluate the long term durability characteristics. Most of the tests were performed on samples conditioned in simulated concrete pore solution (0.6 M KOH + 0.2 NaOH + sat'd $Ca(OH)_2$) at 60° C. This solution was selected as a replacement for concrete exposure as it allows for the results to be generated in relatively short time periods. GFRP bars were tested under tension after conditioning to evaluate the changes in tensile strength and stiffness with respect to the unconditioned samples. Experiments were also performed on small samples of GFRP rebars and FRP grids to determine mass uptake/absorption of alkaline solution and for microscopic studies to evaluate morphological changes. The results obtained show that some of the products tested have retained three quarters of the original strength after 25 weeks of exposure to the simulated pore solution at 60°C, whereas some of the other products have found to loose the structural integrity and morphological characteristics within a short period of exposure. The results also show that the durability of these composites in alkaline environment is strongly dependent on the matrix resin, the fibre composition and the quality of the manufacturing process. The results of this study also show that the acceleration test regime used in this investigation can be used as a method for determining the durability characteristics of FRP materials in concrete environment and for the selection FRP products for demonstration projects, before finalizing the design codes and specifications. A simple mass uptake measurement under accelerated test conditions can be used as a first step screening procedure to eliminate poor quality products, from further testing.

Arunasalam Coomarasamy, Research Scientist, Ministry of Transportation Ontario, 1201 Wilson Avenue , Downsview, Ontario, M3M 1J8.

Stephen Goodman, Co-Op student (From the University of Waterloo) Ministry of Transportation Ontario; Presently, Junior Engineer, ITX Stanley Ltd, Main Street, Cambridge, Ontario, N1R 6R1.

INTRODUCTION

Fibre Reinforced Plastics (FRP) have been successfully used in place of traditional materials such as metals, wood and ceramics in many commercial and industrial applications including aerospace industries, for many years. Recently, the potential use of these materials as concrete reinforcement has attracted much attention from the civil engineering profession. The main advantages of FRP over steel for this application are that the FRP materials do not corrode, have better electromagnetic properties and have higher ratio of strength to mass density. The disadvantages of FRP are that the modulus of elasticity and the fracture strain are low when compared to steel.

Fibre Reinforced Plastics are advanced composite materials (ACM) made up of continuous fibres impregnated in a resin matrix. The plastic resins used as matrix include thermosetting resins such as epoxies, polyesters, vinyl esters and thermoplastic resins such as poly ether ether ketone (PEEK). Glass fibres, aramid fibres and carbon fibres made from pitch as well as from poly acrylonitrile (PAN) are being used as continuous fibres in these composites. Though FRP could be produced by several different manufacturing processes, products that are produced by pultrusion process (eg rebar, cable, grid and sheet) are of main interest for use in highway structures. The pultrusion process makes it possible to use higher fibre to resin ratio to attain higher strength properties. FRP products are inert with respect to corrosion in a traditional sense. However the non-corrosiveness is by no means a guarantee for the durability of FRP in the concrete environment. The long term performance and durability data for FRP products in highway applications are not available. The main concerns still are the durability and structural integrity of FRP products in the concrete environment. The Canadian Bridge Design Code which is being revised specifies a design life of 75 years for highway structures. Hence there is a need for a suitable accelerated test procedure, to evaluate the performance and durability of the new materials such as fibre reinforced plastic composites, in a reasonable time frame. Also there is a need for appropriate design code and specifications for the use of these new composite materials in highway structures.

The main objective of this study is to determine the effect of highly alkaline concrete environment on the critical strength characteristics such as the tensile properties of a selected range of commercially available FRP materials, particularly the glass fibre reinforced plastic rebar and grid samples. Another objective is to study the effect of accelerated tests on the physical and the morphological characteristics of the FRP materials selected.

MATERIALS AND METHODS

Commercially available Glass Fibre Reinforced Plastic (GFRP) rebar samples of 12- 13 mm (half an inch) nominal diameter from five different manufacturers and FRP grids samples containing basalt fibre, glass fibre and carbon fibre from a single manufacturer were used in this investigation. The codes used for the different samples are given below; the second letter in the codes refers to the type of resin matrix. P refers to polyester and V refers to vinyl ester.

CODE	AP	AV	BP	BV	CP	CV	DV	EP	EV	FV - G	FV - B	FV - C

Note: FV- G - glass fibre grid, FV-C Carbon fibre grid, FV-B Basalt fibre grid.

Accelerated laboratory tests were performed to evaluate the long term durability characteristics. Most of the tests were performed on samples which were conditioned in a simulated concrete pore solution prepared by mixing potassium hydroxide (KOH), sodium hydroxide (NaOH) and saturated calcium hydroxide $Ca(OH)_2$) at the following concentrations in distilled water: 0.6 M KOH + 0.2M NaOH + sat'd $Ca(OH)_2$) (Walzak, 1994). The pH of this solution was approximately 13.5. The treatment temperature was 60° C and the duration of the treatment ranged from 0 to approximately 6 months. The above solution was selected as a replacement for concrete exposure as it allows for the results to be generated in relatively short time periods. However it is recognised that there are limitations to the comparisons that can be made as pore solution can be viewed as a model system and not a completely accurate representation of the conditions that would prevail within the concrete.

GFRP bars and sections of grids were tested under tension before and after conditioning in the alkaline solution to evaluate the changes in tensile strength and stiffness with respect to the unconditioned samples. Experiments were also performed on small samples to determine mass uptake/absorption of alkaline solution and for microscopic studies to evaluate morphological changes.

PREPARATION OF TENSILE TEST SPECIMENS

FRP rebar specimens for tensile testing were cut to 1 metre lengths. Since the brittle nature of FRP does not allow direct clamping of the test specimens using the standard grips for tensile testing, both ends of the rebar specimens were bonded to steel tube anchors 20 cm long with an internal diameter of 15.9 mm, using Sikadur 35 high strength two component epoxy adhesive system. The adhesive system needed a minimum of a 4 week cure period at ambient temperature for adequate strength development. The anchored ends were also sealed using an industrial bridge deck sealant system to prevent any attack of the alkaline solution on the epoxy bond between the rebar and the steel anchors.

TEST APPARATUS AND PROCEDURE

The testing machine used was a Satec UTM model M 120HVL with computer controlled load-rate and data acquisition. A Satec Epsilon series extensometer model 3542SP-0800-200 was used to measure the elongation and the strain of the specimens over a gauge length of 200 mm. The load cell of the Satec UTM and the extensometer were linked to a computerized data acquisition system which scanned for data at a rate of 1 scan/second (Goodman and Coomarasamy, 1997).

One of the anchored ends of the test specimen was clamped-on first to the upper grip of the testing machine followed by attaching the extensometer at the middle section of the specimen. Then the bottom grip of the testing machine was engaged which applied a small compressive load on the specimen. The test procedure was initialized and the specimen was loaded at a rate of 17,800 N/min (400 lbf/min). This load rate produced failure usually between 2 - 5 minutes, which was within the recommended time of 1-10 minutes as described in the draft ASTM Standard which is under development by ASTM D 20 Committee for determining the tensile properties of Fibre Reinforced Plastic Rods for Concrete Reinforcement. The computer program allowed loading by the UTM to cease automatically once the specimen had failed.

EXPERIMENTS IN SMALL ALKALINE BATH

In order to investigate the effect of aqueous alkali on the physical and morphological characteristics of FRP materials, small rebar specimens of 8 -12 cm length and FRP grid specimens were immersed in a small bath containing an alkaline (lime) solution of pH 12.5 maintained at 60° C, for several weeks (Goodman and Coomarasamy, 1997). The initial mass of the specimens were first recorded. Every seven days, the specimens were removed from the bath and allowed to air dry for approximately 5 minutes. The mass of each specimen was recorded and compared to the unconditioned values. Morphological changes were also observed and noted. Some of the samples were examined in more detail by scanning electron microscopy (SEM). Similar experiments were also performed at 60° C in the alkaline solution of pH 13.5 described earlier.

RESULTS

THE EFFECT OF AQUEOUS ALKALI ON MASS UPTAKE AND MORPHOLOGICAL CHARACTERISTICS

The percentage increase in mass of FRP samples in the alkaline solution of pH 12.5 at 60° C, after 49 days of immersion are given in Table 1.

TABLE 1 - INCREASE IN MASS OF FRP SAMPLES IN AN ALKALINE SOLUTION FOR 49 DAYS

a) REBAR SPECIMENS			b) GRID SPECIMENS		
Bar Number	% Increase in Mass	Average % Increase	Grid Number	% Increase in Mass	Average % Increase
AV1	1.28				
AV2	1.26	1.3			
AV3	1.44		FV-G1	-3.19	
AP1	1.28		FV-G2	-3.49	-3.3
AP2	1.26	1.3			
AP3	1.36				
DV1	0.31				
DV2	0.22	0.3			
DV3	0.26				
CP1	2.16		FV-C1	2.17	
CP2	2.20	2.4	FV-C2	1.88	2.4
CP3	2.79		FV-C3	3.18	
CV1	0.47				
CV2	0.67	0.6			
CV3	0.53				
EV1	1.54				
EV2	1.35	1.5			
EV3	1.49		FV-B1	1.22	
EP1	1.81		FV-B3	1.30	1.3
EP2	1.91	1.8			
EP3	1.76				

In general, the samples containing vinyl ester type of resin showed lesser extent of degradation than the samples containing polyester type of resin system (Table 1a). Of all the rebar samples examined, the types DV and CV showed small mass increases indicating that these products were least affected by the alkaline treatment. It has also been observed that the outer surface of the samples containing polyester resin matrix, showed higher degree of degradation by the formation of gel like material accompanied with swelling, followed by blistering and eventual dissolution of the resin in some cases.

Figure 1 illustrates the effect of the alkaline solution of pH 13.5 on the resin system of type BP FRP composite rebar sample after approximately 9 weeks immersion at 60° C. The resin system, in this case a polyester has completely disintegrated leaving only the glass fibres intact. Some of the FRP rebar samples containing vinyl ester matrix too suffered severe degradation, probably due to some deficiencies in the manufacturing process. The rebar samples which showed extensive degradation in this alkaline immersion test were not selected for the evaluation of tensile properties.

Among the three different types of grid samples studied, the sample containing glass fibre showed a decrease in mass (Table 1b), due to the dissolution of the fibres and disintegration of the matrix resin system. The other two types of grid samples showed moderate mass increases. The edges of grid samples were somewhat uneven, with insufficient coverage of the fibres. This allows easy ingress of treatment solution into the sample with an easy attack of the components of the composite and the fibre-matrix interface. SEM pictures of thin sections of grid samples which were subjected to the alkaline treatment, showed disintegration of the interfacial bonding, especially in the samples containing glass and the basalt fibres. The results clearly show that, careful consideration should be given in the selection of the material types as well as manufacturing process for the production of composite materials.

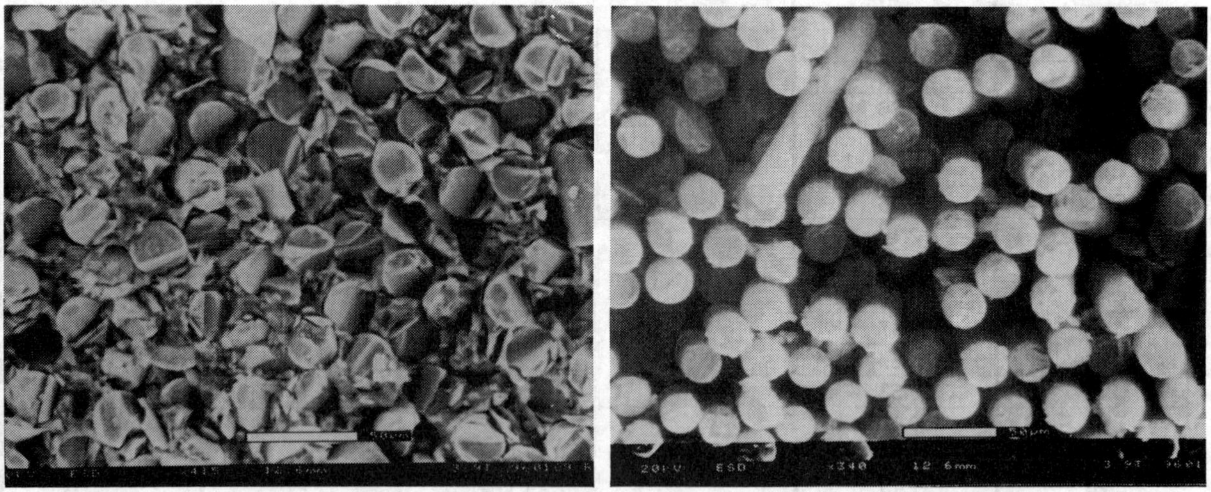

Figure 1 - FRP Rebar (Type BP) Cross-sections Before (left) and After (right) Treatment in alkaline solution of pH 13.5 at 60° C for 9 weeks, indicating dissolution of matrix resin. Magnification: x135

TENSILE TEST RESULTS

The stress vs. elongation curves obtained for the sample type AP before and after conditioning in the alkaline solution of pH 13.5 at 60° C for 17 weeks, are given in Figure 2. It can be seen that the material had lost it's strength characteristics substantially within the treatment period.

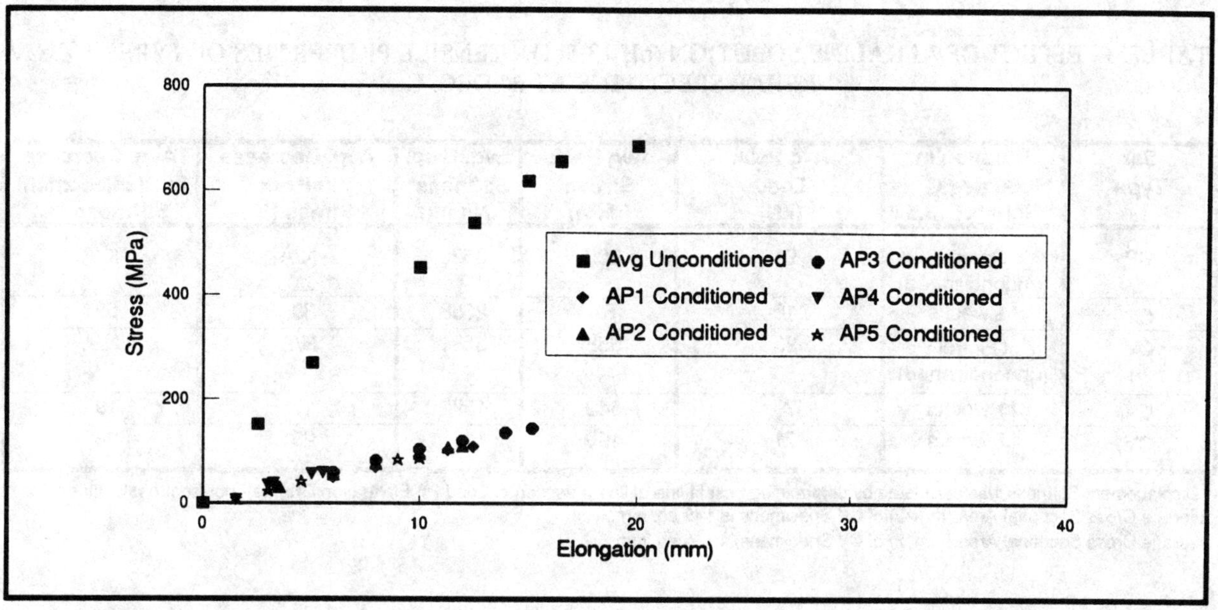

Figure 2 - Stress vs. Elongation Graphs for Unconditioned and Conditioned AP Rebars

The tensile test results (obtained from the stress /elongation graphs) for the sample types AP and AV before and after treatment in the alkaline solution of pH 13.5 are given in Table 2.

TABLE 2 - EFFECT OF ALKALINE SOLUTION (pH 13.5) ON TENSILE PROPERTIES OF TYPE AP & AV REBAR SPECIMENS AT 60 DEG. C

Bar Type	Duration in 60 deg C Soln pH 13.5	Avg Peak Load (kN)	Avg Peak Stress (MPa)	Avg. Disp. Stiffness* (N/mm)	Avg. Total Decrease in Peak Stress (%)	Avg. Total Decrease in Disp. Stiffness (%)
AP	Control (unconditioned)	90	691	4748	N/A	N/A
AP	78 days (approx. 11 wks)	17	131	1477	81	69
AP	118 days (approx. 17 wks)	13	102	1412	85	70
AV	Control (unconditioned)	57	403	5735	N/A	N/A
AV	78 days (approx. 11 wks)	31	222	3650	45	36
AV	118 days (approx. 17 wks)	31	218	3477	46	39

* Displacement Stiffness was calculated by determining best fit line of linear section of Load vs. Elongation Graphs - not from peak values
Average Cross Sectional Area (mm^2) of AP Specimens is 130 sq. mm.
Average Cross Sectional Area (mm^2) of AV Specimens is 142 sq. mm.

The tensile properties given in the table include peak load, peak stress and displacement stiffness. The displacement stiffness has been defined as P/d, where P is the load and d is the displacement (Mochizuki, Matsuzaki and Sugita, 1993). More details of this procedure are found in the reference cited.

The tensile test results for CP and CV types of test specimens before and after 11 weeks treatment are given in Table 3.

TABLE 3 - EFFECT OF ALKALINE SOLUTION (pH 13.5) ON TENSILE PROPERTIES OF TYPE CP & CV REBAR SPECIMENS AT 60 DEG. C

Bar Type	Duration in 60 deg C Soln pH 13.5	Avg Peak Load (kN)	Avg Peak Stress (MPa)	Avg. Disp. Stiffness* (N/mm)	Avg. Decrease in Peak Stress (%)	Avg. Decrease in Displacement Stiffness (%)
CP	Control (unconditioned)	91	608	5683	N/A	N/A
CP	11 weeks	10	67	2388	89	58
CV	Control (unconditioned)	95	658	5681	N/A	N/A
CV	11 weeks	79	543	4592	17	19
CV	17 weeks	71	490	4117	26	28

* Displacement Stiffness was calculated by determining best fit line of linear section of Load vs. Elongation Graphs - not from peak values
Average Cross Sectional Area (mm^2) of CP Specimens is 149 sq. mm.
Average Cross Sectional Area (mm^2) of CV Specimens is 145 sq. mm.

The results for types DV and EV before and after alkaline treatment are presented in Tables 4 and 5 respectively. The results presented are the average of 5 tests for each type of sample/treatment.

TABLE 4 - EFFECT OF ALKALINE SOLUTION (pH 13.5) ON TENSILE PROPERTIES OF TYPE DV REBAR SPECIMENS AT 60 DEG. C

Bar Type	Duration in 60 deg C Soln pH 13.5	Avg. Peak Load (kN)	Avg Peak Stress (MPa)	Avg Disp. Stiffness* (N/mm)	Avg. Total Decrease in Peak Stress (%)	Avg. Total Decrease in Displ. Stiffness (%)
DV	Control (unconditioned)	133	717	6215	N/A	N/A
DV	78 days (approx. 11 wks)	123	668	5880	7	5
DV	118 days (approx. 17 wks)	110	595	6207	17	0
DV	(approx. 25 wks)	92	495	6329	26	-8

* Displacement Stiffness was calculated by determining best fit line of linear section of Load vs. Elongation Graphs - not from peak values
Average Cross Sectional Area (mm^2) of DV Specimens is 185 sq.mm.

TABLE 5 - EFFECT OF ALKALINE SOLUTION (pH 13.5) ON TENSILE PROPERTIES OF TYPE EV REBAR SPECIMENS AT 60 DEG. C

Bar	Duration in 60 deg C Soln pH 13.5	Avg Peak Load (kN)	Avg Peak Stress (MPa)	Avg. Disp. Stiffness* (N/mm)	Avg. Decrease in Peak Stress (%)	Avg. Decrease in Displacement Stiffness (%)
EV	Control (unconditioned)	41	260	2920	N/A	N/A
EV	11 weeks	14	73	522	72	82

* Displacement Stiffness was calculated by determining best fit line of linear section of Load vs. Elongation Graphs - not from peak values
Average Cross Sectional Area (mm^2) of EV Specimens is 157 sq. mm.

Figure 3 shows the stress /elongation plots obtained for DV type FRP material.

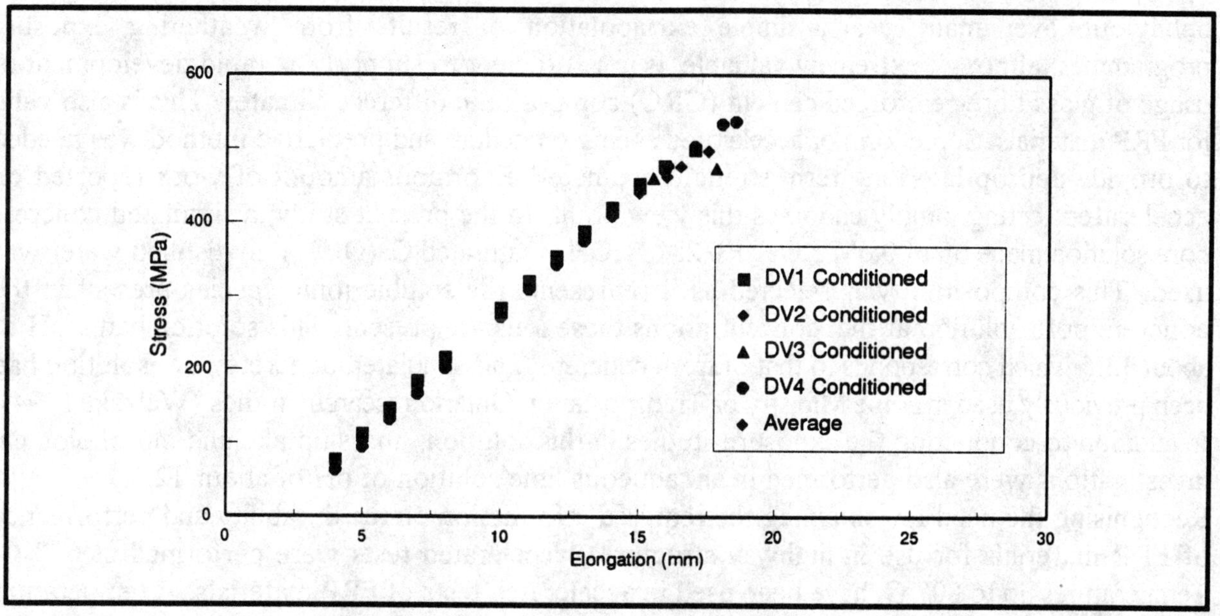

Figure 3 - Stress vs. Elongation Graphs for Unconditioned and Conditioned DV Rebars

The results in Table 2 show that the sample containing the polyester type of resin matrix had lost about 80 % of it's strength within 11 weeks of treatment at 60° C whereas the equivalent rebar with a vinyl resin had lost about 45% within the same time frame. The displacement stiffness of the materials have also decreased substantially under the test conditions used. The results in the Tables 3 and 5 also show similar decreases after 11 week's treatment for the product types CP and EV. The product type CV which contains a vinyl resin matrix showed a decrease of about 26% in strength and about 28 % decrease in displacement stiffness. These results are much better when compared to those of the other samples except sample type DV. The sample type DV showed a decrease of 7%, 17% and 26% of strength after 11 weeks, 17 weeks and 25 weeks treatment in the alkaline solution of pH 13.5 at 60° C (Table 4). The changes in displacement stiffness were also small. The plots in Figure 3 where the results of individual tests for the type DV after 25 weeks treatment are plotted, shows that the consistency of this material even after 25 weeks treatment in the alkaline simulated pore solution is in deed good, which suggests that a quality manufacturing process has been used in making this product.

DISCUSSION

One of the main reasons for considering non-steel elements for concrete reinforcement, is that steel corrodes in concrete, resulting in loss of strength and structural integrity. Highway structures are particularly prone to this corrosion, as they are exposed to the outdoor environment and to the deicing salts in colder climates. Substantial amount of money is being spent by highway authorities for rehabilitation of the corrosion induced damaged structures. It is essential that all tensile reinforcing elements including FRP materials to be used in concrete structures, to retain sufficient strength capability during the normal life of the concrete structure.
As it had been previously stated the concrete is highly alkaline having a pH of about 13.5, and the alkalinity decreases with maturation and carbonation.

The authors of a study on glass fibre-reinforced cements (Proctor, Oakley and Litherland, 1982) state that, in attempting to provide the necessary information on expected long term behaviour over many years, simple extrapolation of results from weathering exposure programmes, although extremely valuable, is not sufficient to support the rapid development of usage of glass fibre-reinforced cement (GRC) composite in different climates. This is also valid for FRP materials. Some form of accelerated ageing procedure and predictive method was needed to provide appropriate long term strength estimates. Enormous amount of work reported on accelerated testing, amply endorses this view point. In the present study, a simulated concrete pore solution made up of 0.6M KOH + 0.2M NaOH + saturated $Ca(OH)_2$) in distilled water was used. This composition was selected as it represents the soluble ionic species present in the concrete pore solution at the concentrations these ions are present. This solution had a pH of about 13.5 which corresponds to that of fresh concrete. This simulated concrete pore solution had been previously used in some Ministry of Transportation Ontario research studies (Walzak, 1994). In addition to conducting the exposure studies in this solution, mass uptake and morphological investigations were also performed in an aqueous lime solution of pH of about 12.5.
Recognising the need for obtaining the required information on the durability and performance of FRP materials for use in highway structures, accelerated tests were performed at 60° C. Temperatures up to 80° C have been used in accelerated tests of FRP materials. A temperature of 60° C was chosen, as it was felt that this temperature would not initiate degradation mechanisms that would not develop during normal service level exposure and would not significantly alter the integrity of the product , while providing the required information in a relatively short time frame. Proctor, Oakley and Litherland have worked out the acceleration factor for GRC. The acceleration factor was 272 times for the GRC system, if the accelerated test was performed at 60° C, when the mean annual service temperature is 10.4° C. Using a similar method, Gerritse (Gerritse, 1992) has worked out the acceleration factor for FRP in concrete environment. He concluded that the acceleration factor of more than 50 could be applied when the temperature is increased from room temperature to 60° C and even a factor of more than 200 when the temperature is increased to 80° C. The mean annual temperatures for Toronto, Sudbury, and Timmins (Ontario, Canada) were found to be 9.2, 3.6 and 1.1° C respectively (Based on the weather data for 1995, from the Environment Canada). Therefore the acceleration factor for the three different locations would be different. Experiments are in progress to determine the acceleration factor for the systems under investigation. However, as a rough estimate, an acceleration factor of about 100 may be applied for the tests conducted at 60° C, when a service temperature of 9.2° C is considered..

The results obtained show that some of the products tested have retained three quarters of the original strength after 25 weeks of exposure to the simulated pore solution at 60°C, whereas some of the other products have found to loose the structural integrity and morphological characteristics within a short period of exposure. From the results, it is evident, as expected (Gentry et al, 1996) the FRP containing vinyl ester resin in general was found to be more durable than the FRP systems with polyester resin matrix. However the results also show that durability and structural integrity of these materials are also heavily dependent on how well these products are manufactured. The simple mass uptake results and morphological studies indicated this pattern qualitatively and therefore could be used as a screening procedure to eliminate poor quality products without conducting extensive testings on them.

CONCLUSIONS

The results of this study show that the accelerated test procedure (conditioning at 60° C in a simulated concrete pore solution of pH of 13.5) used in this investigation can be used as a method for determining the durability characteristics of FRP materials in concrete environment and for the selection FRP products for demonstration projects, before finalizing the design codes and specifications. A simple mass uptake measurement carried out under accelerated test conditions at 60° C in an alkaline solution, can be used as a first step screening procedure to eliminate poor quality FRP products, from further testing.

REFERENCES

Gentry, T. R., L. C. Bank, A. Barkatt, and L. Prian. 1996. "Accelerated Test Methods to Determine the Long-Term Behaviour of Composite Highway Structures Subject to Environmental Loading" Paper ID 866 Symposium On Composite Materials in Non-Aerospace Applications, ASTM STP 1323, ASTM, West Conshohocken, PA.

Gerriste, A. 1992. "Durability Criteria For Non-Metallic Tendons in an Alkaline Environment," Proceedings of the 1st International Conferance on Advanced Composite Materials in Bridges and Structures, Sherbrooke, Quebec, Canada. Edited by K. W. Neale and P. Labossire, Canadian Society of Civil Engineering, Montreal, Canada.

Goodman, S. and A. Coomarasamy. 1997. "Basalt Fibre Reinforced Plastic Grids as Potential Concrete reinforcement," Research and Development Branch Report (to be Published), Ministry of Transportation Ontario, Ontario, Canada.

Mochizuki, S., Y. Matsuzaki, and M. Sugita. 1993. "Evaluation Items and Methods of FRP Reinforcement as Structural Elements," International Symposium on Fiber-Reinforced-Plastic Reinforcement for Concrete Structures, Edited by Antonio Nanni and Charles W. Dolan.,American Concrete Institute, Detroit, MI.

Proctor, B. A., D. R. Oakley and K. L. Litherland. 1982. "Developments in the Assessment and Performance of GRC over 10 years." Composites, 13(2), pp.173-179.

Walzak, T. L. 1994. "Adhesion Loss Mechanisms of Epoxy Coatings on Rebar Surfaces," Final Report for the Ministry of Transportation Ontario and Concrete Reinfocing Institute, Ontario Canada.

Constraint Effects on Energy Absorption in Unidirectional PMC Tubes

M. ABDEL-HAQ, G. B. BROGGIATO AND G. M. NEWAZ

ABSTRACT

An investigation was carried out to understand the effect of an external constraint on the failure modes and energy absorption capability of unidirectional polymeric composite (PMC) tubes. The constraint was a steel ring that was tightened on the outside of the tube. This ring could slide during tube crushing. The tightness of the ring was controlled using a strain gage positioned on the ring. The effect of the constraint was shown to depend on the initial position of the ring along the tube and its tightness. The study was performed on fiberglass/polyester tubes and the results showed that the external constraint not only can improve the specific energy absorption of the tubes, but also can reduce significantly the undesired high initial peak loads. A nonlinear finite element approach was used to model the fundamental failure modes that dominate the energy absorption of the tubes with an external constraint. These modes including delamination and longitudinal crack propagation (splitting), were modeled using the finite element code ABAQUS. A user friendly FORTRAN subroutine was implemented to the code to incorporate the material failure model for predicting the fiber/matrix failures that occur as crushing is experienced by the tube. The predicted model results compared well with the experimental data for the various crushing stages of the polyester tubes.

INTRODUCTION

Because of their high specific strength and stiffness, polymer matrix composites(PMC) have been introduced in many automotive applications to reduce weight and improve fuel economy. Unlike metals that absorb energy by plastic deformation, PMC tubes absorb energy by multiple microfracture processes started with a crush zone that propagates through the material in a stable manner and has the tendency to absorb a significant amount of energy in the case of a vehicular crash. The crush zone can be initiated by using a trigger at one end of the tube (Thornton, 1979).

M. Abdel-Haq, Mechanical Engineering Department, Wayne state university, Detroit, MI 48105
G. B. Broggiato, Dept. of Mechanics and Aeronautics, University of Rome "La Sapienza", Rome, Italy
G. M. Newaz, Mechanical Engineering Department, Wayne state university, Detroit, MI 48105

The failure modes of composite tubes are shown to depend on a large number of different sets of variables such as fiber type, matrix type, fiber architecture, testing speed, geometry, ect. which affect the energy absorption performance. However, most of these variables have been investigated in details (Hull, 1991; Czaplicki and Robertson and Thornton, 1991; Ramakrishna and Hull, 1993; Thornton, 1986, 1990; Hamada and Coppola and Hull, 1995; Farley, 1986)

Hull (1991) reviewed in detail the role of hoop constraints e.g. due to hoop reinforcement in cross-ply tubes regarding their position, distribution, and proportion on the nature of crushing. It was concluded that it is possible to control the load-bearing capacity of the tubes during crushing by changing the distribution of the fibers, but the choice of fiber architecture is often restricted by the suitable fibers and the manufacturing methods.

Although extensive experimental work has been conducted on the energy absorption of composite tubes, little efforts have been devoted to model the crushing process. Farley and Jones (1989) used the finite element method to predict the energy absorption capability of axially crushed graphite/epoxy and kelver/epoxy tubes. However, the models were reported as being computationally intensive. A finite element approach was used by Bolukbasi and Laananen (1995) to model the sustained crushing of the flat plate taking into consideration the different failure modes such as delamination and material failure. Mamalis et al. (1996a) carried out a theoretical analysis of the failure mechanisms of the stable mode of collapse of thin-walled tubes under static axial compression based on experimental observations. Mamalis et al. (1996b) analyzed theoretically the collapse mechanisms of fiberglass composite automotive frame rails tested under axial compression. This analysis showed to be as a good approximation of the energy absorption values, however, the peak and mean loads were not compared. Wolf and Baslid and Bunsell (1994) developed a model that describes the interaction between geometry and material properties that occurs during crushing. The crack length and the crushing stress values were compared with the experimental results. The primary objectives of the current program were

1) to introduce, for the first time, the concept of external constraint to control the progressive failure modes and improve the energy absorption capability of unidirectional PMC tubes. The constraint was presented by a steel sleeve that was tightened at the outer wall of the tube and which was able to slide as crushing occurred supporting the splaying of the 0° fibers. The main features desired by using the external constraint is to obtain a load-displacement curve for unidirectional tubes that is close to ideal curve for an energy absorber where after a small crush displacement, a stable crush zone is formed which then propagate at constant load that is approximately at the same level of the peak load. A higher peak load leads to high acceleration levels which are undesirable in the case of crash as the peak load is only a small portion of the total amount of the absorbed energy, and

2) to use a nonlinear finite element approach to model the crushing process of unidirectional composite tubes with an external constraint. The nonlinear finite element program ABAQUS was used and a user friendly FORTRAN subroutine was implemented to the code to incorporate the material failure model for predicting the fiber/matrix failures that occur as a result of large deformation. The model was mainly based on the observations of the failure modes that control the crushing process. Thus, a mechanism-based modeling approach was undertaken.

MATERIAL AND GEOMETRY

The tubes used in the experiments are standard pultruded fiberglass/polyester tubes produced by Morrison Molded Fiber Glass Company (MMFG) using the pultrusion process. The properties of this material system are given in Table I.

All the tubes were 204 mm long with internal diameter of 44 mm and 3 mm thickness. The Aluminum initiator radius was 12. mm and its base diameter was 100 mm. A 45 ° chamfer was lathe-machined at one end of the tube in contact with the initiator.

EXPERIMENTS

The tests were conducted using a servo-hydraulic MTS machine with an automated data acquisition system connected to a time-based medium speed image acquisition system. The image acquisition system that consists of a CCD camera, a frame grabber board and a PC took a signal from the MTS machine at the start of the test and collected 240 images for each experiment. A strain gage was attached to the constraint to control its tightness and a value of 1320 micostrains was used to define the initial pressure of the constraint for each test. The tube is then centered on the bottom test platen and the initiator was placed on the chamfered end of the tube. The lower platen was then raised until it touched the upper face of the initiator. The crushing rate was 50 mm/min. for all tubes and three channels of data were recorded for each test e.g., time, load and displacement. The images were then analyzed to define accurately the position of the constraint throughout the test. The different failure modes such as the length of the longitudinal crack, the position and time of occurrence of the transverse crack and delamination were recorded. Finally, these analyses were used to explain the behavior of the load-displacement curve for each tube.

TABLE I- MATERIAL PROPERTIES FOR FIBERGLASS/POLYESTER TUBES

Material property	
Longitudinal Tensile Strength, $S1_T$	204 MPa
Transverse Tensile Strength, $S2_T$	48 MPa
Longitudinal Tensile Modulus, $E1_T$	17.0 GPa
Transverse Tensile Modulus, $E2_T$	5.4 GPa
Longitudinal Compressive Strength, $S1_C$	204 MPa
Transverse Compressive Strength, $S2_C$	102 MPa
Longitudinal Compressive Modulus, $E1_C$	17.0 GPa
Transverse Compressive Modulus, $E2_C$	6.8 GPa
Longitudinal Shear Strength, S_{12}	50 MPa
Shear Modulus, G_{12}	2.9 GPa
Poisons Ratio, ν_{12}	.33
Density, ρ	1900 Kg/m^3

FINITE ELEMENT ANALYSIS

A quasi-static finite element analysis based on the nonlinear ABAQUS code was performed to model the fundamental mechanisms in the crushing process of composite tubes with an external constraint. The fundamental processes of interest included delamination, material failure and constraint sliding. The computational studies were performed using 400 two-dimensional 8-node plain strain elements to represent one frond. It was shown that axial splitting of the tubes by itself is not a dominant parameter in terms of energy absorption and it may contribute with less than 5% of the total absorbed energy (Mamalis et. al., 1996b) However, the length of the longitudinal crack is so essential since the fronds act as rigid beams and the load bearing ability of these fronds decreases as the longitudinal crack length increases. The initiator was modeled as a rigid body that has only one degree of freedom to move toward the frond. The crushing process was simulated by applying known levels of displacements to the rigid body that predicted the associated crushing forces.

DELAMINATION AND MATERIAL FAILURE MODEL:

Crack propagation analysis was carried out on a nodal basis. The crack-tip node debonds when the fracture criterion, f, reached the value 1.0 where f is defined as:

$$f = \sqrt{\left(\frac{\sigma_n}{\sigma_f}\right)^2 + \left(\frac{\tau_1}{\tau_{1f}}\right)^2 + \left(\frac{\tau_2}{\tau_{2f}}\right)^2} \tag{1}$$

where σ_n is the normal component of stress across the interface, τ_1 and τ_2 are the shear stress components at the interface, and σ_f, τ_{2f}, τ_{1f} are the normal and shear failure stresses. After debonding, the delaminated surface behave as frictional contact surface and prohibit any penetration between the two layers.

As crushing proceeds, local fiber/matrix failures developed in the fronds depending on the stress state in the different sections. The maximum stress criterion based on Chang and Lessard (1991) model was used for prediction of such failures. In this analysis three failure modes were considered.

- Matrix tensile/compressive failure can be occur due to a combination of transverse tensile/compressive stress (σ_2) and the shear stress (σ_{12}). The failure index , e_m, is defined as:

$$e_m = \sqrt{\left(\frac{\sigma_2}{S2_T}\right)^2 + \left(\frac{\sigma_{12}}{S_{12}}\right)^2} \quad \text{(Tensile)} \tag{2}$$

$$e_m = \sqrt{\left(\frac{\sigma_2}{S2_C}\right)^2 + \left(\frac{\sigma_{12}}{S_{12}}\right)^2} \quad \text{(Compressive)} \tag{3}$$

Failure is assumed to occur once em ≥ 1.0

- Fiber-matrix shearing failure occurs due to a combination of fiber compressive stress (σ_1) and the matrix shear stress (σ_{12}). The failure index, e_{fs}, is defined as:

$$e_{fc} = \sqrt{\left(\frac{\sigma_1}{S1_C}\right)^2 + \left(\frac{\sigma_{12}}{S_{12}}\right)^2} \quad (4)$$

Failure is assumed to occur once $e_{fc} \geq 1.0$

- Fiber buckling failure occurs when the maximum compressive stress (σ_1) in the fiber direction exceeds the fiber buckling strength.

$$e_b = \left(\frac{\sigma_1}{S1_C}\right) \quad (5)$$

After e_b exceeds 1.0, it is assumed that the material can no longer support any load.

The failure criteria, Eqns. 2-5, are applied to each finite element each state. If failure due to matrix tension/compression is detected within any element, the modulus in the transverse direction in that element (E2) is reduced by 95% and the Poisson's ratio (ν_{12}) is set to zero. If a fiber-matrix shear failure is detected, then the shear modulus (G_{12}) is set to zero in addition to the Poisson's ratio (ν_{12}). However, if any failure was detected, the program iterates the current solution with the degraded properties until the nonlinear equations of equilibrium are satisfied.

RESULTS AND DISCUSSION

The tests were performed on fiberglass/polyester tubes. Figure 1 shows the load-displacement curves with and without external constraints. The load-deflection response without using a constraint can be characterized as follows; axial compression of the fronds up to a peak load at a crushed distance of 13.66 mm. Then a significant load drop occurred due to transversal cracking of the first section of the fronds. The following section is then axially compressed in a lower rate than the first due to dalamination up to a constant load level. The progressive crushing then become almost uniform due to the steady state speed of the longitudinal crack combined with delamination. The experiments showed that the first transversely cracked section was delaminated. Experimental tests were performed to determine the initiation of delamination in the polyester tubes. No delamination was noticed when the tubes were crushed up to 12.5 mm after which the delamination started and the transverse cracks occurred at 13.66 mm.

The effect of the constraint on the failure modes is shown in Figures 2-3. Two positions of the constraint were evaluated; (1) exactly at the chamfer end of the tube, and (2) at 12 mm from the chamfer end. For the first case (Figure 2), no transverse cracking was observed and delamination started at the early stages of the crushing. The ring started to slide along the tubes reaching a steady sliding speed that is equal to the constant speed of the initiator on the stable crush region. It is important to mention that the position and the speed of the ring in this region is an exact representation of the position and the speed of the longitudinal crack that controls the length of the frond exposed to crushing at any instant. The effect of the constraint for this case can be characterized by a significant reduction in the peak load and significant improvement in the sustained crushing of the tube. Figure 3 shows the constraint effect when it was positioned at 12 mm from the chamfer end. The main difference between this case and the previous one is that a transverse crack occurred prior to the constraint sliding at a crushed distance of 10 mm causing a significant drop in the load. The constraint speed was constant and the same as the crushing speed causing a stable crushing process. However, the constraint was able, as in the previous case, to reduce the peak load and improve the average load levels of the tubes. Figure 1 shows a comparison between the two cases. One main advantage of the first constraint position is that the stable crushing zone started in the early stages of the crushing without a significant jump in the load levels. Table II summarized the effect of the constraint and its position on the performance of fiberglass/polyester tubes. The specific energy absorption (SEA) has been used as a measure of the performance of the composite tubes. It is defined as the ratio of the energy dissipated during the stable crushing zone to the crushed tube weight.

$$SEA = \frac{F_{avg}(\delta_c - \delta_{avg})}{\delta_c A_s \rho} \tag{6}$$

Where δ_c is the crushed distance, δ_{avg} is the crushed distance where the load started to be constant, A_s is the cross sectional area of the tube and ρ is the material density. Table II indicates that the case where an external constraint located exactly at the outer surface of the tube provides best performance. The constraint at this position eliminated the occurrence of the transverse crack which is a low performance failure mode (Farley, 1989) and created an early progressive failure process that looks like an ideal energy absorber where the peak and sustained loads are of the same level with no sudden drop in load in between.

The finite element approach was used to study the fundamental mechanisms that affect the energy absorption capability during the crushing process in PMC tubes with an external constraint. Only one frond was model assuming that the axial splitting by itself is not a dominant mechanism in the crushing process. The frond was assumed to contain an initial central crack and the crushing process was simulated by applying specific displacements to the rigid body that represents the initiator, and predicting the associated crushing forces by finite element analysis. Frictional sliding interfaces were modeled between the surface frond and the surface of the initiator and between the frond's layers immediately after debonding. A coefficient of friction of 0.25 was assumed for both interfaces. The constraint was modeled using two-dimensional plain strain elements positioned at the outer surface of the frond. Frictional sliding interface is modeled between the constraint and the frond. The motion history of the constraint was prescribed to the model based on experimental observations.

TABLE II- EFFECT OF THE CONSTRAINT AND ITS POSITION ON THE PERFORMANCE OF FIBERGLASS/POLYESTER TUBES.

TUBE	Constraint * Position (mm)	F_{peak} (KN)	δ_{peak} (mm)	$F_{avg.}$ (KN)	$\delta_{avg.}$ (mm)	$F_{peak} / F_{avg.}$	SEA KJ/Kg
Polyester	---	26.50	13.66	6.84	28.2	3.87	5.84
Polyester	12	11.10	10.00	12.60	20.0	0.88	12.00
Polyester	0	11.00	6.00	11.85	9.92	0.92	12.70

* From the edge of the chamfer.

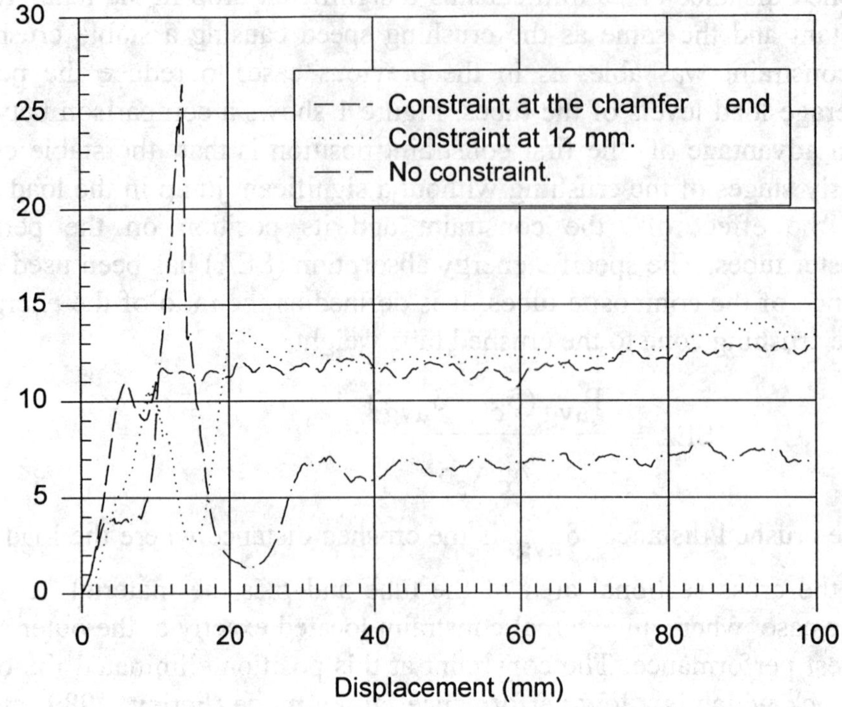

Figure 1- Effect of the constraint and its position on the crushing performance of fiberglass/polyester tubes

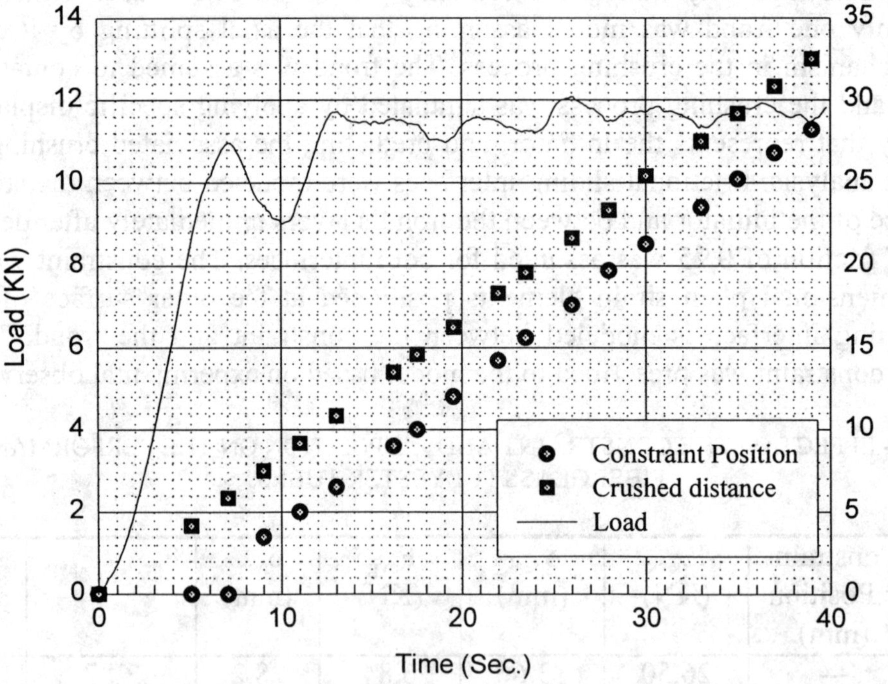

Figure 2- Load respond of a polyester tube with a constraint placed at 0 mm from the edge of the chamfer.

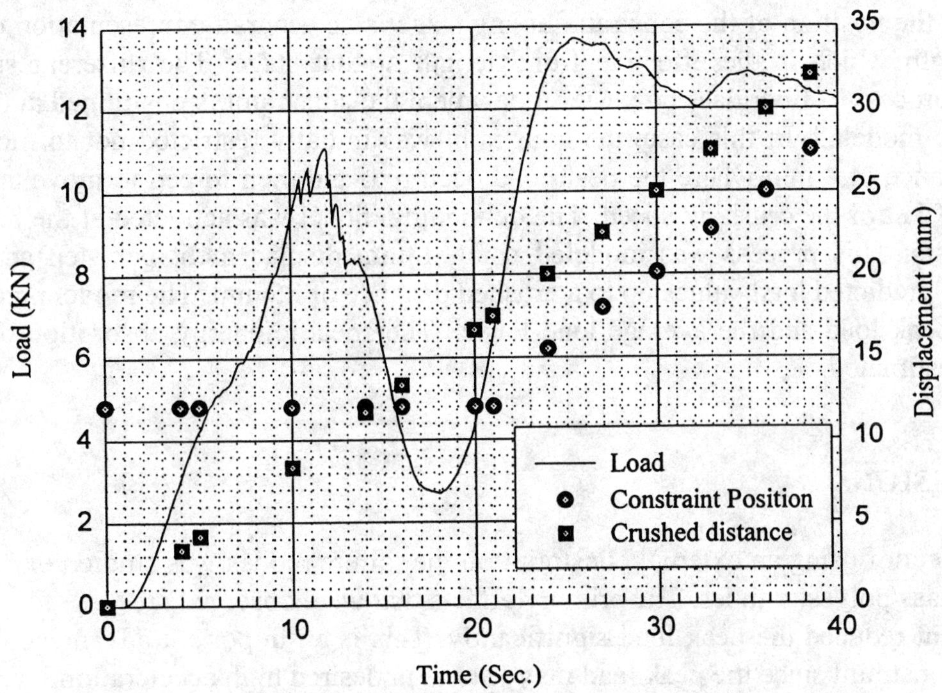

Figure 3- Load respond of a polyester tube with a constraint placed at 12 mm from the edge of the chamfer.

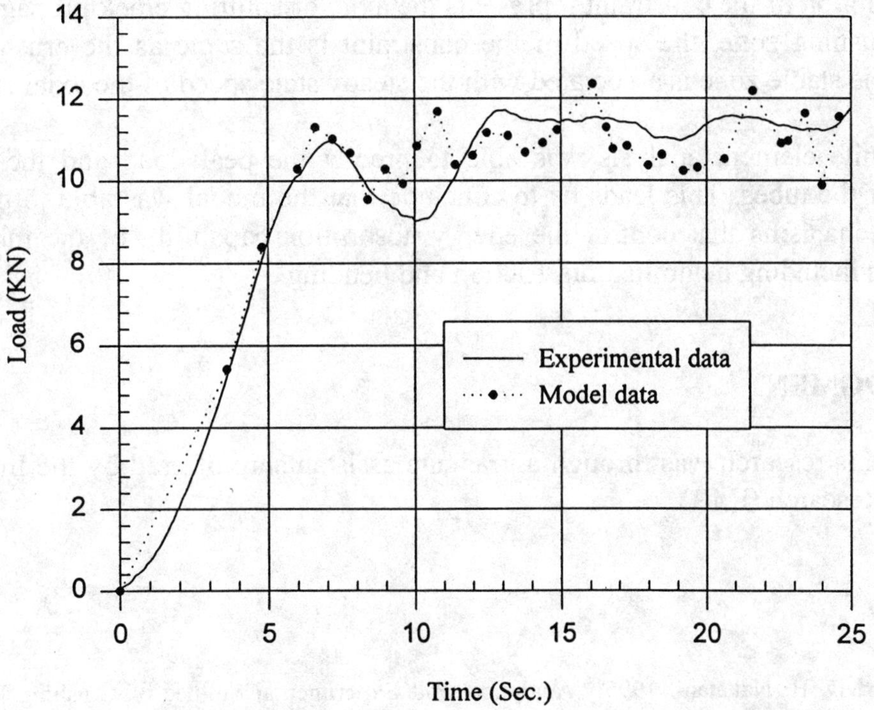

Figure 4- Comparison between the predicted results and the experimental data for polyester tube with constraint at its chamfer end.

However, the position of the constraint at any stage is an accurate representation of the axial crack length which is the effective frond length at that stage. The different steps of the crushing process of fiberglass/polyester tube where the constraint is positioned at the chamfer edge were modeled. In this case, the constraint was initially restricted not to move up to a crush distance of 4 mm where the peak load occurred, and then forced to move at a constant rate equal that of the crushing speed. The debonding model was activated at the beginning of the analysis. This process was modeled using 1 mm displacement per step and Figure 4 shows the predicted load values up to a crushed distance of 25 mm. The model predicted well both the peak load and the average load levels. The specific energy absorption of the tubes was also estimated.

CONCLUSIONS

1) The effect of using an external constraint on the failure modes was studied experimentally for fiberglass/polyester tubes. The primary effects that were observed are
- Constraint reduced the peak load significantly. This is an important advantage of using an external constraint since the peak load may lead to undesired high acceleration levels.
- The sustained crushing load improved in all cases by using a constraint.
- The start of this stable crushing zone was advanced when the constraint is used.
- Constraint eliminated the occurrence of the transverse cracks when it was placed at the edge of the chamfer. No sudden load drop has occurred for this case.

2) The sliding motion of the constraint represents the axial or splitting crack propagation rate. In the stable crushing zone, the speed of the constraint is the same as the crushing speed indicating that the stable zone is associated with the steady state speed of the axial crack.

3) Nonlinear finite element analysis was able to predict the peak load and the sustained crushing load of the tubes. This leads us to conclude that the model was able to predict the fundamental mechanisms that control the energy absorption capability of the tubes in the crushing process including delamination, friction and bending.

ACKNOWLEDGMENT

Funding of this research was through a graduate assistantship offered by the Institute for Manufacturing Research (IMR).

REFERENCES

Bolkbasi, A. O. and D. H. Nakatani. 1995. "Analytical and Experimental Studies of Crushing Behavior in Composite Laminates, " *Journal of Composite Materials*, 29(8):1117-1139.

Chang, F. K. and L. B. Lessard. 1991. "Damage Tolerance of Laminated Composites Containing an Open Hole and Subjected to Compressive Loading: Part I - Analysis, " *Journal of Composite Materials*, 25: 2-43.

Czaplicki, M. J., R. E. Robertson and P. H. Thornton. 1991. "Comparison of Bevel and Tulip Triggered Pultruded Tubes for Energy Absorption," *Composite Science and Technology*, 40: 31-46.

Farley, G. L. 1986. "Effect of Fiber and Matrix Maximum Strain on The Energy Absorption of Composite Materials, " *Journal of Composite Materials*, 20:322-334.

Farley, G. L. and R. M. Jones. 1989. "Energy Absorption Capability of Composite Tubes and Beams, " NASA Technical Memorandum 101634, AVSCOM Technical Report 89-B-003.

Hamada, H., J. C. Coppola, and D. Hull. 1995. "Effect of surface Treatment on Crushing Behavior of Glass-Cloth Epoxy Tubes, " *Composites*, 23(1):245-252.

Hull, D. 1991. "A Unified Approach to Progressive Crushing of Fibre-Reinforced Composite Tubes," *Composite Science and Technology*, 40:377-421.

Mamalis, A. G., D. E. Manolakos, G. A. Demstheneous, and M. B. Ioannidis. 1996a. "Analysis of Failure Mechanisms Observed in Axial Collapse of Thin-Walled Circular Fiberglass Composite Tubes," *Thin-Walled Structures*, 24:335-352.

Mamalis, A. G., D. E. Manolakos, G. A. Demstheneous, and M. B. Ioannidis. 1996b. "The Static and Dynamic Axial Collapse of Fiberglass Composite Automotive Frame Rails," Composite Structures, 34(1):77-90.

Ramakrishna, S. and D. Hull. 1993. "Energy Absorption Capability of Epoxy Composite Tubes with Knitted Fibre Fabric Reinforcement," *Composite Science and Technology*, 49:349-356.

Thornton, P. H. 1979. "Energy Absorption of Composite Structures," *Journal of composite Materials*, 13:247-262

Thornton, P. H. 1990. "The Crush Behavior of Pultruded Tubes at High Strain Rates," *Journal of Composite Materials*, 24:549-615.

Thornton, P. H. 1986. "The Crush Behavior of Glass Fibre Reinforced Plastic Sections," *Composites Science and Technology*, 27:199-224.

Wolf, CH., P. Bastid, and A. R. Runsell. 1994. "Relation of Energy Absorption of Composite Structures to Material Strength," *Composite Engineering*, 4(2):195-218

ABAQUS/Standard User's Manual. 1995. Version 5.5, Hibbitt, Karlsson and Sorensen, Inc.

Characterization of Fatigue Damage Development and Progression in Composite Laminates Using Stochastic Modelling and Analysis of Test Data

R. GANESAN, M. EL KARMALAWY AND S. V. HOA

ABSTRACT

In the present paper, a stochastic approach is developed for modelling and analysis of fatigue response test data of composite laminates. The damage development is identified and quantified through the correlation analysis. The stochastic analysis results are related to the microstructure observations. The fatigue behaviour of AS4/3501-6 Graphite Epoxy cross ply lay up $[0/90]_{4s}$ with 60% volume fraction is described through this stochastic approach.

INTRODUCTION

The long-term fatigue response of composite laminates has to be predicted based on short-term laboratory tests for the purposes of failure prediction and design of laminates subjected to dynamic loadings. The elastic modulus of a composite laminate has long been used as a parameter for quantifying the fatigue response of the laminate [Talreja (1987)]. Laminated composites are known for the high degree of variability in their mechanical properties. This variability can be attributed to (i) the variations due to manufacturing, and (ii) mechanical behaviour variations. The relative occurrences and the contribution to the overall failure of various damage mechanisms and failure modes have a random variation from one laminate specimen to another, and they can not be precisely quantified. Recognizing the random nature of fatigue process in laminates, a stochastic approach is developed and deployed in the present work for modelling and analysis of test data.

STOCHASTIC APPROACH

For quantifying the fatigue behaviour of composite laminates in terms of strength and modulus reduction, so far only a deterministic approach has in most cases been deployed. This deterministic approach has resulted in disagreement between theory and practice.

Rajamohan Ganesan, Moataz El Karmalawy and Suong van Hoa, Concordia Centre for Composites, Department of Mechanical Engineering, Concordia University, Montreal, Quebec, Canada, H3G 1M8.

The Weibull and Gaussian distribution functions have been used to model the fatigue life data of composites (Hahn and Kim, 1975; Ramani and Williams, 1977; Talreja, 1987). The approach has been to model the test data on life times of the laminates and to develop stress-life or strain-life models. This approach has severe limitations. Moreover, (i) the damage development and progression in the form of various failure mechanisms including matrix cracks, delamination and fibre fracture can not be quantified, (ii) the dominant damage modes during various stages of fatigue cycling can not be identified, (iii) the probability distribution function of life times has not been related to the fatigue damage process, and (iv) the correlation characteristics of test data are not accounted for. A rigorous stochastic approach can take into account all sources of variability in material response and lead to a reliable prediction of material response and damage development.

ANALYTICAL PROBABILITY DISTRIBUTIONS

In the present work, the actual probability distribution of fatigue response is determined from the test data through the damage descriptor, rather than assuming a Gaussian or Weibull distribution for the life times and failure strengths thereby assuming without a proper basis that this distribution is the actual distribution of fatigue response. As can be seen from the numerical results presented later, the true distribution of the damage descriptor is far different from Gaussian or Weibull distribution. It may be noted here that the type of distribution function has a direct bearing on the accuracy of failure prediction and hence, establishing the true distribution of test data is crucial for successful failure prediction and design. The probability density function of the compliance (which is the reciprocal of modulus) at load cycle n_{k-1}, which is denoted here as $f_{Ck-1}(c_{k-1})$ is determined based on the Maximum Entropy Method (Siddall, 1983). Based on this method, the analytical form of the probability density function is taken to be

$$f_{Ck-1}(c_{k-1}) = \exp\left[\lambda_0 + \sum_{i=1}^{p} \lambda_i c_{k-1}^{\,i}\right] \tag{1}$$

The values of the parameters λ_i , i = 0, 1, 2,..., p are determined such that the entropy of the random variable c_{k-1} given by

$$S = -\int_R f_{Ck-1}(c_{k-1}) \ln\left[f_{Ck-1}(c_{k-1}) \right] dc_{k-1} \tag{2}$$

where R denotes the range of the random variable c_{k-1} , is maximum, subject to the following constraints: (i) the total probability is unity, i.e.

$$\int_R f_{Ck-1}(c_{k-1})\, dc_{k-1} = 1 \tag{3}$$

and (ii) the probabilistic moments of the random variable c_{k-1} are equal to the corresponding statistical moments obtained from the compliance data, i.e.

$$\int_R c_{k-1}^{\,i} \, f_{Ck-1}(c_{k-1}) \, dc_{k-1} = M_i \, , \quad i = 1, 2, ..., q \tag{4}$$

where M_i is the i-th statistical moment about the origin of the compliance values c_{k-1} . The above problem should be solved based on the techniques of optimization. The optimization technique that is best suited to this problem is the Nonlinear Programming Technique proposed by Jacobson and Oksman (Siddall, 1983).

AUTOCORRELATION FUNCTIONS

The autocorrelation function is a stochastic measure that quantifies the physical relationship between two or more sets of data. This relationship is extracted in terms of the Autocorrelation Function or equivalently its Fourier transform called the Power Spectral Density Function (Bendat and Piersol, 1986). Here, the set of specimen compliance values at a load cycle value n_{k-1} and that at a load cycle value n_k are considered. From these two sets of data, the autocorrelation function $R(n_{k-1}, n_k)$ is determined:

$$R\,(n_{k-1}\,,\, n_k\,) = (1/N) \sum_{i=1}^{N} c_{k-1}\, c_k \tag{5}$$

In the above equation, N denotes the number of laminate specimens tested. Since the compliance process is a non-homogeneous stochastic process, the autocorrelation function is not unique. So, a parameter known as the "Parameter of Correlation Function" is introduced along with the "Lag Distance" which constitutes another parameter. Here, the parameter of correlation function $R(n_{k-1}, n_k)$ is n_{k-1} and the lag distance is the difference $n_k - n_{k-1}$, which is denoted by τ. Based on these parameters and from the autocorrelation function, the correlation coefficient $r(n_{k-1}, n_k)$ is defined such that the changes in mean values and variances corresponding to changes in the parameter of correlation function are accounted for.

$$r\,(n_{k-1}\,,\, n_k\,) = \{\, (1/N) \sum_{i=1}^{N} [c_{k-1} - \langle c_{k-1}\rangle]\,[c_k - \langle c_k\rangle] \,\} \,/\, \{\sigma_{k-1}\, \sigma_k\,\} \tag{6}$$

In the above equation, $\langle\ \rangle$ denotes the ensemble mean value, σ_{k-1} is the standard deviation of compliance values at the load cycle n_{k-1} , and σ_k is that corresponding to the load cycle value n_k. In addition another parameter is also introduced, which is the "Length of Correlation". This is the value of lag distance at which the correlation coefficient drops to a constant value or becomes asymptotic.

EXPERIMENTAL INVESTIGATION

The composite material considered is AS 4/3501-6 Graphite Epoxy cross ply lay up $[0/90]_{4s}$. The manufacturing conditions are as follows: Each plate is first heated for 1 hour at a temperature of 116° C and at a pressure of 586 kPa (and full vacuum in the bag), then it is heated for 2 hours at a temperature of 176° C and at a pressure of 655 kPa (and full vacuum in the bag). Tests were conducted at room temperature on an MTS hydraulic machine. Specimens were clamped to the machine grips using Al tabs as outlined in the ASTM D 3039/D 3039-93 test procedure. An on-axis T-T fatigue load at 5 Hz frequency and with a load ratio R of 0.1 was applied. The maximum load is chosen to be that required to induce gradual failure to the specimens. The value chosen corresponds to 70% of the average strength. Further details are available in Ganesan et al (1997). The values of fatigue compliance at different load cycles were determined and plotted in Figure 1.

FRACTOGRAPHY INSPECTION OF DAMAGE DEVELOPMENT AND PROGRESSION

The microstructure of the laminate has been monitored at different stages of fatigue loading using the standard fractography technique. The detailed results are presented elsewhere [El Karmalawy (1997)]. The following observations were made: (i) The first damage mechanism upon fatigue loading the specimens, was matrix cracking within the transverse plies, more often in plies that were farther away from the axis of symmetry than those closer to it; (ii) These matrix cracks occurred at random spatial locations in the laminate and further, the crack size as well as the crack density (i.e. number of cracks within a specified area) increased with increasing number of load cycles (Figures 2 and 3); (iii) The cracks started either at material defects such as voids, and most often spontaneously at random locations within the plies; (iv) Upon further loading, cracks extended throughout the transverse plies reaching the two adjacent longitudinal ply borders on each side of the laminate; (v) During propagation, the cracks favoured paths that turn around the graphite fibres; (vi) The crack propagation became unpredictable once the crack reaches an interface between the transverse and longitudinal plies; (vii) The cracks also moved along the interface causing local delamination (Figure 4); (viii) The interfacial crack suddenly jumped to the longitudinal plies causing one or more fibres to break; (ix) Once the fibre cracking occurred in longitudinal plies, it started propagating across the fibres as shown in Figure 5; (x) Again the crack moved parallel to the fibres until it reaches another jumping point across the next fibre and so on; (xi) Dislocations within the fibres were caused due to further fatigue loading, and a complete separation between fibre bundles was also present; (xii) The surface texture of the specimens became "rough" after fatigue loading.

STOCHASTIC CHARACTERISTICS OF TEST DATA

The analytical probability density function for the distribution of compliance values when the number of load cycles is 1×10^5 has been obtained and is shown in Figure 6. The mean value and standard deviation for this random variable have been determined and they were equal to 1.635×10^{-11} Pa^{-1} and 0.01154×10^{-11} Pa^{-1} respectively. The signal-to-noise ratio of compliance, which is defined as the ratio of the mean value of compliance to the standard deviation of compliance is calculated for this case (i). It is equal to 142. In a similar manner, the distributions of compliance values when the number of load cycles is (ii) 341 and (iii) 3.5 $\times 10^4$ were analyzed, and the corresponding probability density functions were determined. The mean value and standard deviation for the case (ii) are 1.658×10^{-11} Pa^{-1} and $0.0208 \times$

10^{-11} Pa^{-1} respectively. The corresponding values for the case (iii) are 1.678×10^{-11} Pa^{-1} and 0.0251×10^{-11} Pa^{-1} respectively. The signal-to-noise ratio for case (ii) is equal to 80 and that for case (iii) is equal to 67.

The autocorrelation functions have been determined next. First the values of compliance when the number of load cycles is (i) 86 were considered. The initial value of the parameter for this case is 86. Using this initial value and different values of lag distances, the autocorrelation function has been determined and is shown in Figure 7. In a similar manner, the compliance values when the number of load cycles is (ii) 316 and (iii) 1.001×10^5 have been considered. Correspondingly, the initial values of load cycles (which are the parameters of the autocorrelation functions) were (ii) 316 and (ii) 1.001×10^5 respectively. The corresponding autocorrelation functions have been determined.

From the autocorrelation function shown in Figure 7 and that for the cases (ii) and (iii), it has been observed that (i) the autocorrelation functions are not symmetric when the negative and positive lag distances are considered, thus indicating the strong non-stationarity present in the compliance values, (ii) in the case when the number of load cycles is 316, for most part of the positive lag distances, the correlation value does not change significantly. This has been the case for most of the compliance distribution in the second stage of the Figure 1. This indicates that the same physical process i.e. the same damage mechanism controls this part of the damage accumulation. The second stage corresponds to the extension of delamination (90°) damage in the composite. Also, it has been observed that the correlation values decrease for negative lag distances, when significant fatigue damage has set in the composite material. In the first two cases, the correlation values drop for larger values of positive or negative lag distances thereby indicating that the fatigue damage leads to the so-called "Wide Band" randomness in the compliance values. In other words, as more fatigue damage sets in the composite, different failure mechanisms become active.

DETECTION AND QUANTIFICATION OF DAMAGE MECHANISMS

The autocorrelation function can be utilized to detect and quantify the dominant damage mechanism. The parameters, the length of correlation and the parameter of correlation function are considered for this purpose. It has been observed from the fractography inspection that around a load cycle value within the range of 360-400, the saturation state is reached as for the matrix cracks. Within this range of load cycles, the autocorrelation functions were determined at load cycle values of 76, 86, 101, 151, 251 and so on. Two sample autocorrelation functions are shown in Figures 8 and 9. From these autocorrelation functions it has been observed that (i) the correlation coefficient is equal to unity corresponding to the value of the parameter of correlation function, (ii) in each case, the correlation coefficient drops to an almost constant value after a certain load cycle value, and (iii) when the parameter of correlation increases, the length of correlation decreases (Figure 10) such that an almost constant value of correlation coefficient is obtained when the parameter of correlation function is approximately equal to 320.

Consider an autocorrelation function say for instance, Figure 8. The physical relationship between the compliance values at a load cycle value of 76 and that at a load cycle value of 350 is quantified by the correlation coefficient value of 0.45. A similar physical relationship exists between the compliance values at load cycles 76 and 400, since the correlation coefficient remains almost at the same value. This points to the presence of two separate dominant physical processes, i.e. damage mechanisms, at these two load cycle values in each case. The same damage mechanism is dominantly present in the range of load cycle values corresponding to which the correlation coefficient values remain the same. Hence, it can be observed that the load cycle value that is approximately equal to 320 separates two different damage mechanisms. The microstructure observations can now be considered. Before the load cycle value of 350 matrix cracks development has been dominant. This confirms the observation deduced from the autocorrelation function.

CONCLUSION

An analytical method based on the Maximum Entropy Method has been developed for determining the true probability distribution of fatigue test data of laminates rather than assuming a Gaussian or Weibull distribution as in the existing works. This distribution has been shown to be superior to the standard distributions in that it leads to accurate predictions of fatigue response (Ganesan et al, 1997). A systematic way of distinguishing and quantifying various dominant damage mechanisms has been developed based on the correlation analysis.

REFERENCES

Bendat, J. S. and A. G. Piersol. 1986. *Random Data: Analysis and Measurement Procedures*, Wiley Interscience, New York.

El Karamalawy, M. 1997. Fatigue Damage Characterization and Prediction, and Reliability Evaluation for CFRP Composites. Ph. D. Thesis, Department of Mechanical Engineering, Concordia University, Montreal, Canada.

Ganesan, R., S. V. Hoa, S. Zhang, and M. El Karmalawy. 1997. "A Stochastic Cumulative Damage Model for the Fatigue Response of Laminated Composites," Presented at the International Conference on Composite Materials, ICCM-11, Melbourne, Australia, July 11-14, 1997.

Hahn, H. T. and R. Y. Kim. 1975. "Proof Testing of Composite Materials," *Journal of Composite Materials*, 9: 297-311.

Ramani, S. V. and D. P. Williams. 1977. "Notched and Unnotched Fatigue Behaviour of Angle-Ply Graphite/Epoxy Composites". In: *Fatigue of Filamentary Composite Materials*, ASTM STP 636. Eds.: K. L. Reifsnider and K. N. Lauraitis, pp. 27-46.

Siddall, J. N. 1983. *Probabilistic Engineering Design: Principles and Applications*. Marcel Dekker, New York.

Talreja, R. 1987. *Fatigue of Composite Materials*. Technomic publishing Company, Lancaster.

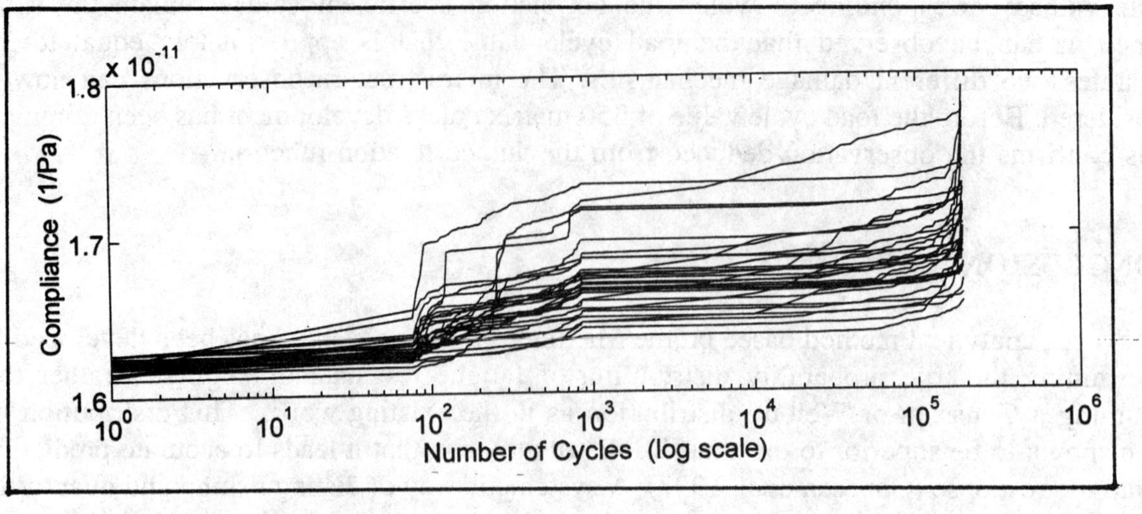

Figure 1. Different Realizations of Fatigue Modulus versus Number of Cycles Curves

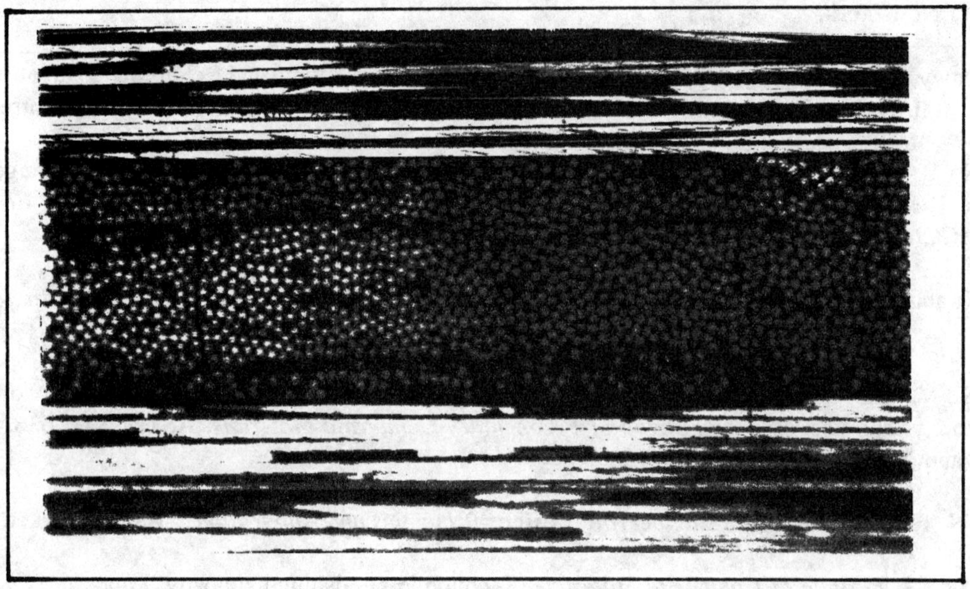

Figure 2. Micrograph Showing the Cracks Development in Transverse Plies During Fatigue (x 200), (at Mid Length of Specimen and at 200 Cycles)

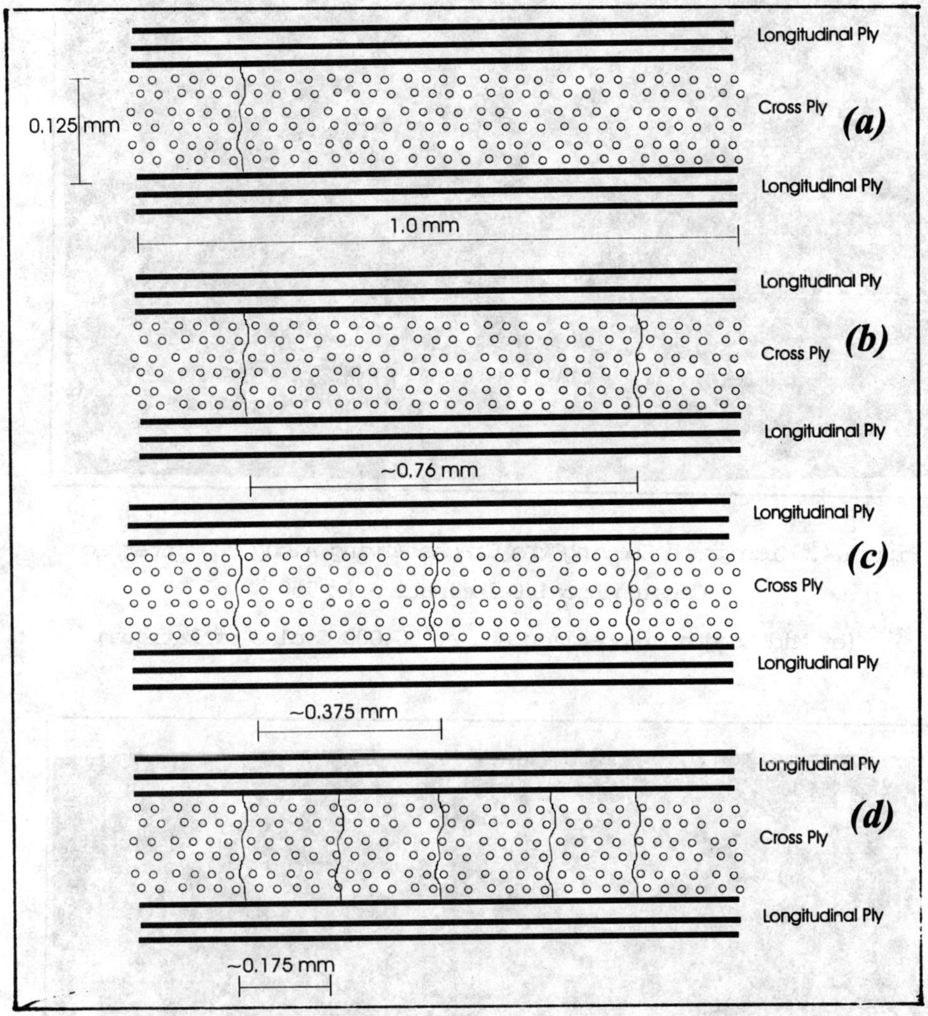

Figure 3. Schematic Showing the Cracks Development and Progression in Transverse Plies During Fatigue

(a) Crack Initiation During the First Few Cycles of Fatigue

(b) Multiple Cracks Development During the First 100 Cycles

(c) Cracks Density Increasing to 1.7 cracks/mm at 200 Cycles of Fatigue

(d) Saturation State, 5.7 Cracks/mm, at 400 Cycles of Fatigue

Figure 4. Micrograph Showing the Cracks Moving Along the Interface Causing Local Delamination (x 560)

(at Mid Length of Specimen and at 30,000 Cycles of Fatigue)

Figure 5. Micrograph Showing Crack Propagation Across the Fibres (x 560)

(at Mid Length of Specimen and at 160,000 Cycles of fatigue)

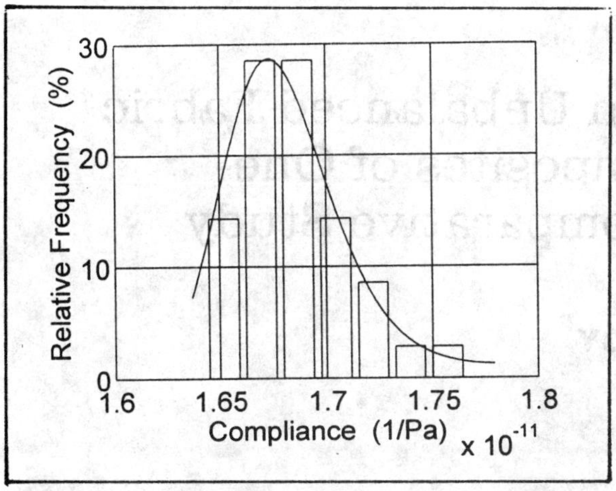

Figure 6. Frequency Histogram and Analytical Probability Density Function (Number of Cycles - 100,000)

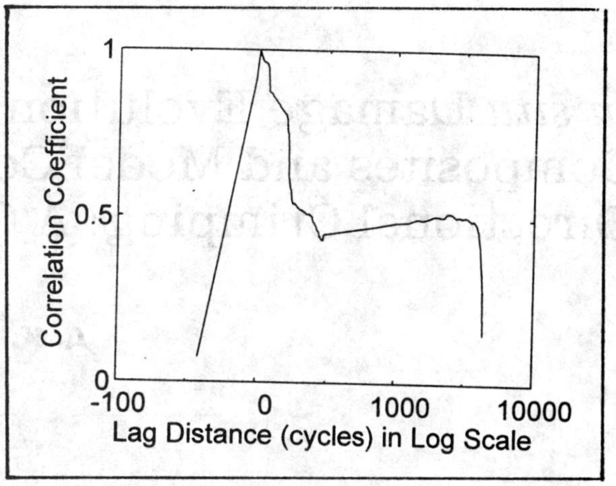

Figure 7. Autocorrelation Function (at Load Cycle 86)

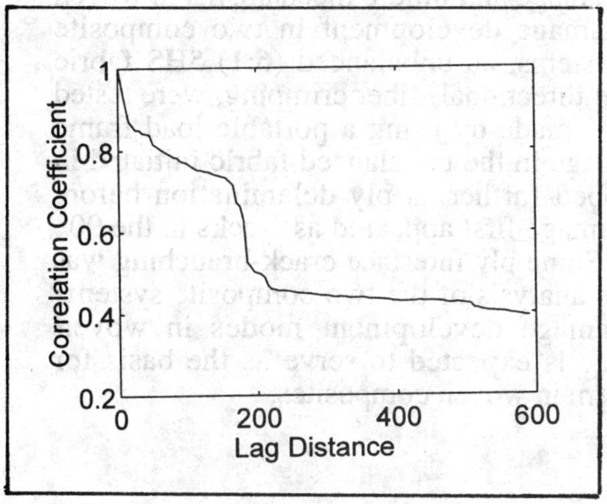

Figure 8. Autocorrelation Function Parameter of Correlation - 76 Cycles

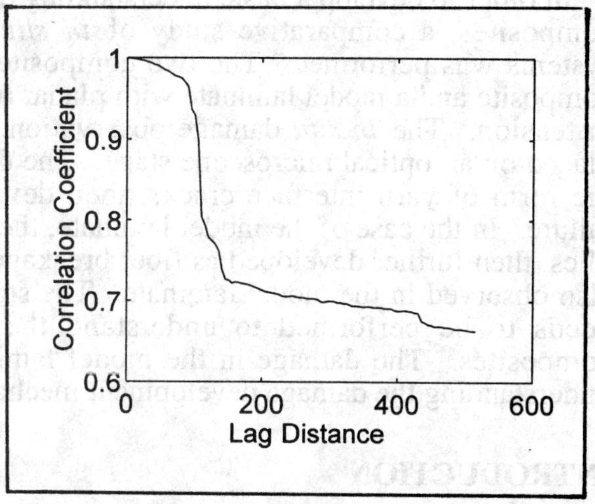

Figure 9. Autocorrelation Function Parameter of Correlation - 151 Cycles

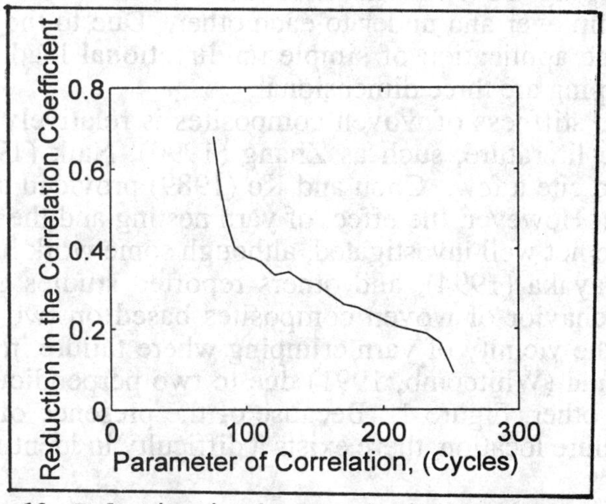

Figure 10. Reduction in the Correlation Coefficient After 320 cycles for different Parameters of correlation

In situ Damage Evolution in Unbalanced Fabric Composites and Model Composites of One-Directional Crimping: A Comparative Study

A. K. ROY

ABSTRACT

In order to develop a basic understanding of damage and failure mechanisms in woven composites, a comparative study of *in situ* damage development in two composite systems was performed. The two composite systems, an unbalanced (6:1) 5HS fabric composite and a model laminate with planar (one-directional) fiber crimping, were tested in tension. The *in situ* damage observation was made by using a portable load frame placed on an optical microscope stage. The damage in the unbalanced fabric initiated in the form of yarn interface cracks, then developed further as ply delamination before failure. In the case of the model laminate, the damage first appeared as cracks in the 90° plies, then further developed as fiber breakage. Some ply interface crack-branching was also observed in the model laminate. The stress analysis of the two composite systems needs to be performed to understand the damage development modes in woven composites. The damage in the model laminate is expected to serve as the basis for understanding the damage development mechanism in woven composites.

INTRODUCTION

The failure mechanism of woven (textile) composites is not well understood. In the vicinity of yarn crimping of two-dimensional woven fabric architecture, two perpendicular yarns crimp over and under to each other. Due to the perpendicular yarn crimping, even under the application of simple unidirectional load, the stresses in the vicinity of the yarn crimping are three dimensional.

The prediction of the stiffness of woven composites is relatively easy and has been done extensively in the literature, such as Zhang (1990), Naik (1992), Hahn (1994), Yurgartis (1993), etc., to cite a few. Chou and Ko (1989) provided a broad overview of this class of composites. However, the effect of yarn nesting and the failure mechanisms of woven composites are not well investigated, although some work has been done in this area. Naik (1994), Karayaka (1994), and others reported studies of the yarn nesting effects on the failure behavior of woven composites based on two-dimensional stress analysis. However, in the vicinity of yarn crimping where failure in general occurs, the stress is three-dimensional (Whitcomb, 1991) due to two perpendicular yarns crimping over and under to each other, Figure 1. Because of the presence of two perpendicular yarns crimping at the failure location, there exists a difficulty in identifying

Ajit K. Roy, University of Dayton Research Institute, 300 College Park, Dayton, OH 45469-0168, e-mail: royak@ml.wpafb.af.mil

experimentally the influence of the stress components on the failure of two-dimensional woven composites, as observed by Roy (1994, 1996).

To optimize the material properties and improve the performance of woven composites, we need to understand the physics and identify the influence of the stress components on the failure mechanism. A model composite is fabricated to simulate the fabric yarn crimping only in one direction (in a plane), Figure 2. In the vicinity of the planar yarn crimping, the stress field is two-dimensional (except for edge effects). Thus the model composite, to some extent, simplifies the process of understanding the influence of the stress components on the failure of the composite in the vicinity of the yarn crimping. For the convenience of defining the waviness in the model laminates, the ratio of the yarn (or lamina in this case) waviness amplitude, a, to the length of the wave, λ, is defined as the lamina waviness ratio. In addition to the damage observation in the model laminate, an unbalanced (6:1) commercial woven composite was also tested in tension in the warp direction. The objective of this work is to compare the damage evolution process in these two systems with a view toward using the data in understanding the failure mechanism in realistic woven composites.

MODEL COMPOSITE LAMINATES

The yarn crimping in woven (textile) composites, as discussed before, occurs in two perpendicular directions as shown in Figure 1, while the model system with simulated planar yarn crimping is shown in Figure 2. The laminates were made from AS4/3501-6 graphite/epoxy unidirectional prepregs. The waviness in the composite laminates was produced by inserting strips of unidirectional plies perpendicular to the longitudinal direction of the laminates. Four different ratios of lamina waviness were produced by selecting two different lay-up sequences of cross-ply laminates and two different strip widths (1.63 and 3.25 mm). However, only one of the model laminates of lamina waviness of 0.026 was used in this study. The cross-sectional micrograph of the model laminate used in this study, before testing, is shown in Figure 3. The geometrical configuration of the model laminates was characterized by the lamination sequence away from the wavy region, lamination sequence at the mid-section of the wavy region, and the lamina (yarn) waviness ratio. Consequently, a new laminate notation was used for the model laminates to indicate these characteristics. For example, the notation used for the model laminate shown in Figure 3 was ($[90_2/0_2]_s$, $[90/0_2/90]_s$, 0.026), indicating the lamination sequence away from the wavy region, mid-section lamination sequence of the wavy region, and the lamina waviness ratio, respectively. In the 5HS unbalanced woven composite tested, the yarn waviness ratio, a/λ, is found to be approximately 0.03. The cross-sectional micrograph of the unbalanced woven composite before testing is shown in Figure 4.

EXPERIMENTAL PROCEDURE

The objective of the present work, as stated before, was to generate experimental data on damage development in the two composites stated above, which will later be correlated with stress analysis for assessing damage and developing failure criteria for woven composites. Among the available damage detection techniques, e.g., acetate film replication, fluorescent dye resin impregnation, etc., the *in situ* microscopic observation of damage under load was found to be more promising than any other damage detection techniques. The specimens were tested in tension. In order to observe and record *in situ* damage development through a microscope, a portable load frame was used for this study, Roy (1996). The far-field strain of the model laminate was measured with a 0.0625 inch strain gage installed at 0.5 inch away from the middle of the wavy region, and the strain on the 5HS woven composite specimen was measured by a 0.125 inch strain gage installed at the mid-gage length.

IN SITU DAMAGE OBSERVATION

The specimens were loaded in tension in the portable load frame placed on a microscope stage. The specimens were loaded incrementally by increasing the gas pressure to the inflatable diaphragm of the load frame through a pressure regulator. The load and strains were recorded at each load step, and the edges of the specimens were simultaneously observed through the microscope for damage. Photographs of the specimen edge were taken using a 35-mm camera connected to the microscope. The image is also stored digitally for later retrieval.

OBSERVED DAMAGE

(a) Unbalanced (6:1) Woven Laminate

The number of yarns in the warp direction of a (6:1) unbalanced woven composite is approximately six times that of the fill direction. Further, the yarn waviness ratio of the unbalanced woven composite laminate along the warp direction was measured to be 0.03. The cross-sectional micrograph of the composite laminate before testing is shown in Figure 4. Since the waviness ratio of the woven composite along the warp direction was close to that of the model laminate, and to compare the damage evolution process with that of the model laminate, the woven laminate was tested in the warp direction. Before any load application, the cross section of the specimen was inspected with an optical microscope. No process-induced cracks were visible on the specimen cross section. Upon application of incremental loading, first cracking occurred at a far-field strain level of 0.35 percent (3550 $\mu\varepsilon$) and 65 percent of the failure load. The crack appeared as an interface crack near the tip of the fill yarn bundle, as shown in Figures 5 and 6. With further increasing load, the yarn interface crack propagated along the interface from the tip of the yarn to the midsection of the yarn. At 98 percent of the failure load (far-field strain of 5690 $\mu\varepsilon$) intra-yarn delamination occurred at multiple locations, shown in Figures 7 and 8.

(b) Model Laminate ($[90_2/0_2]_s$, $[90/0_2/90]_s$, 0.026)

The experimental procedure for the model laminate was the same as that of the unbalanced woven composite laminate. The first cracking in the model laminate occurred as a crack in the 90° ply near the wavy region, Figure 9, at a far-field strain of 0.37 percent (3700 $\mu\varepsilon$). With increasing load, the cracks in the 90° plies grew further and started to branch out as interface cracks at the 0°/90° ply-interface, Figures 10 and 11, at a far-field strain of 44 percent (4360 $\mu\varepsilon$). This far-field strain value of the interface crack initiation should provide a basis for developing interface crack growth criterion, which later potentially can be used to understand the interface damage development in woven composites. Significant fiber breakage in the wavy region was visible at a far field strain of 69 percent (6920 $\mu\varepsilon$), Figure 12.

CONCLUSIONS

In order to develop an understanding of damage mechanism in woven composites, a comparative study of damage evolution in a commercial 5HS unbalanced (6:1) woven composite and a model laminate with planar crimping subjected to tensile loading was performed. The yarn waviness of the unbalanced woven composite along the warp direction was close to that of the model laminate. The damage development in the case of the woven composite was primarily in the form of cracking at the yarn interface and intra-yarn delamination. Whereas, in the model laminate, the damage initiation was as cracks in the 90° plies, which further grew as ply-interface cracks. Some fiber breakage was also observed in the wavy region. No large-scale ply-interface crack was observed near the wavy region in the model system. An appropriate stress analysis should be performed to analyze the differences in the damage development in these two composites, which may aid in developing an understanding of damage mechanism in woven composites.

ACKNOWLEDGMENTS

The author wishes to express his appreciation to Mr. J. Camping of the University of Dayton Research Institute for designing the portable load frame. The work was performed at the Materials Directorate, Wright Laboratory, under Contract Number F33615-95-D-5029. Encouragement from Dr. N. J. Pagano, Wright Laboratory, is greatly appreciated.

REFERENCES

Chou, T. and F. K. Ko, 1989, Textile Structural Composites, Chapter 7, Elsevier, New York, USA.

Zhang, Y. C. and J. Harding, 1990, A Numerical Micromechanics Analysis of the Mechanical Properties of a Plain Weave Composite, Computers and Structures, Vol. 36, No. 5, pp 839-844.

Naik, N. K. and V. K. Ganesh, 1992, Prediction of On-axes Elastic Properties of Plain Weave Fabric Composites, Composites Science and Technology, Vol. 45, pp 135-152.

Hahn, H. T. and R. Pandey, 1994, A Micromechanics Model for Thermoelastic Properties of Plain Weave Fabric Composites, Journal of Engineering Materials and Technology, Vol. 116 (4), pp 517-523.

Yurgartis, S. W. and J. P. Maurer, 1993, Modeling Weave and Stacking Configuration Effects on Interlaminar Shear Stresses in Fabric Laminates, Composites, Vol. 24, pp 651-658.

Naik, N. K. and V. K. Ganesh, 1994, Failure Behavior of Plain Weave Fabric Laminates Under In-Plane Shear Loading, Journal of Composites Technology & Research, Vol 16, No 1, pp 3-20.

Karayaka, M. and P., 1994, Deformation and Failure Behavior of Woven Composite Laminates, Journal of Engineering Materials and Technology, Vol. 116 (2), pp 222-232.

Whitcomb, J. D. , 1991, Three-Dimensional Stress Analysis of Plain Weave Composites, Composite Materials: Fatigue and Fracture (3rd Vol), ASTM STP 1110, T. K. O'Brien, Ed., American Society for Testing and Materials, Philadelphia, pp 417-438.

Roy, A. K., 1994, Experimental Observation of Damage in 8HS Fabric Composites in Tensile Loading, Proc. 9th Technical Conference, American Society for Composites, Univ. of Delaware, Newark, USA, Technomic Publishing, Lancaster, PA, pp 1273-1280.

Roy, A. K., 1996, In Situ Damage Observation and Failure in Model Laminates Containing Planar Yarn Crimping of Woven Composites, Mechanics of Composite Materials and Structures, Vol. 3, pp 101-117.

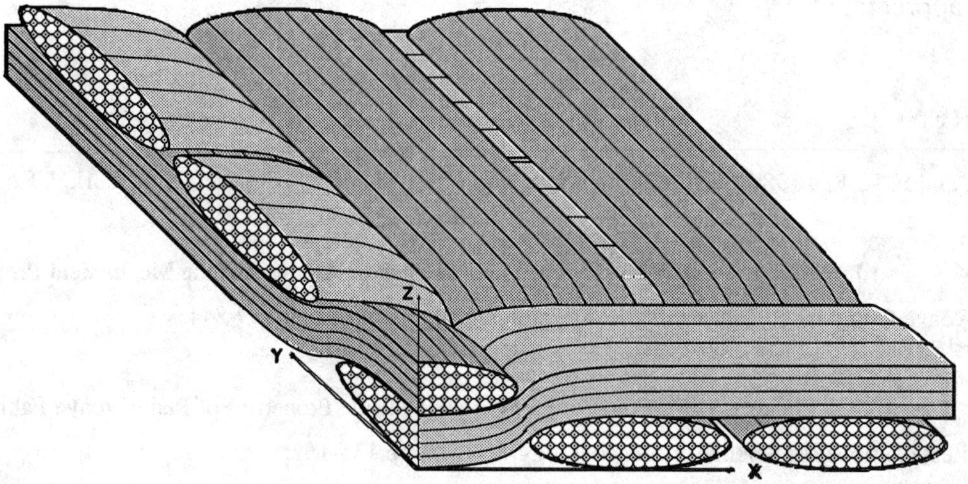

Figure 1. Schematic configuration of 8HS fabric composites.

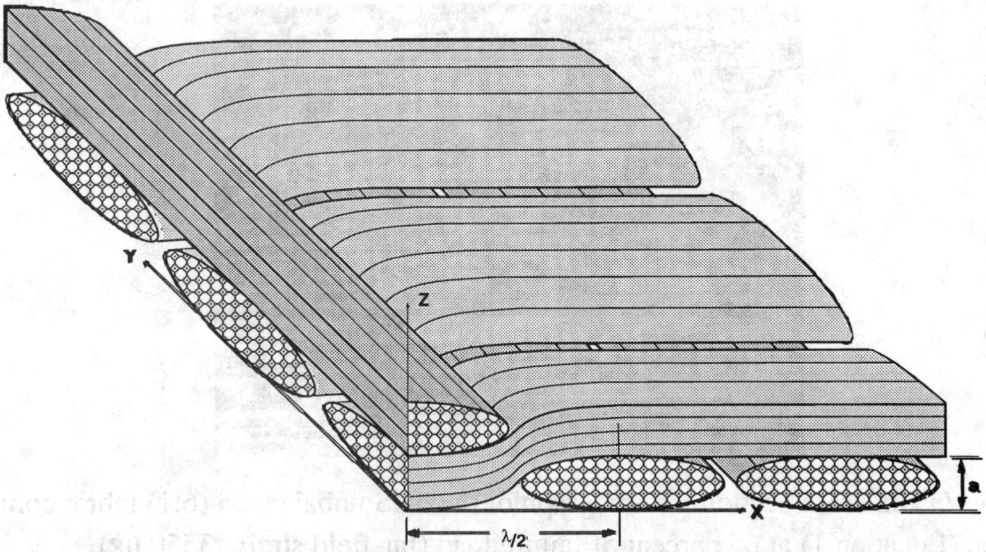

Figure 2. Stress-strain curves of a specimen of $[0]_{4T}$ laminate during incremental loading.

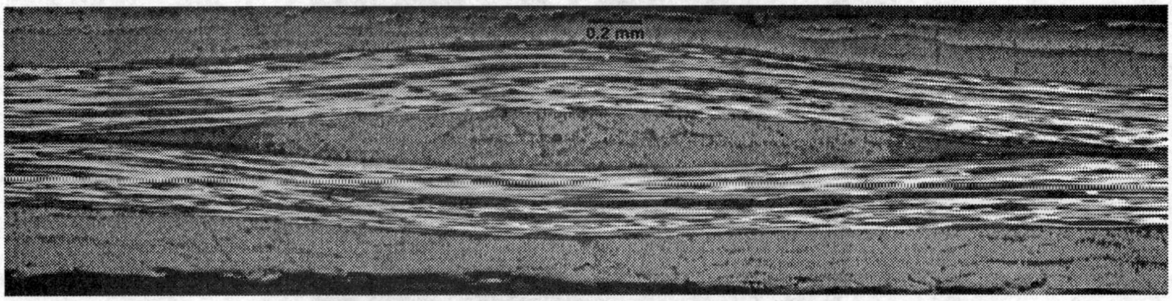

Figure 3. Cross-sectional micrograph of the model laminate.

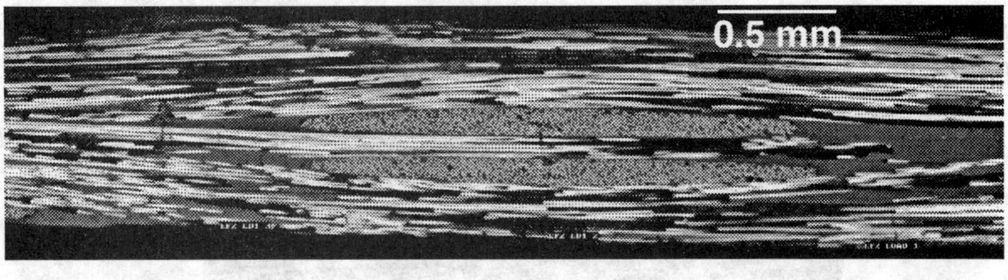

Figure 4. Cross-sectional micrograph of the unbalanced (6:1) woven laminate shown along the warp direction.

Figure 5. *In situ* cross-sectional micrograph of the 5HS unbalanced (6:1) fabric composite specimen (Location 1) at 65 percent of failure load (far-field strain: 3550 $\mu\varepsilon$).

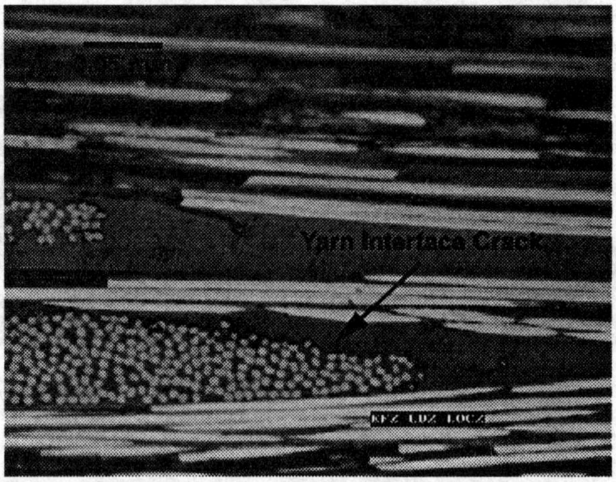

Figure 6. *In situ* cross-sectional micrograph of the 5HS unbalanced (6:1) fabric composite specimen (Location 2) at 65 percent of failure load (far-field strain: 3550 $\mu\varepsilon$).

Figure 7. *In situ* cross-sectional micrograph of the 5HS unbalanced (6:1) fabric composite specimen (Location 1) at 98 percent of failure load (far-field strain: 5690 $\mu\varepsilon$).

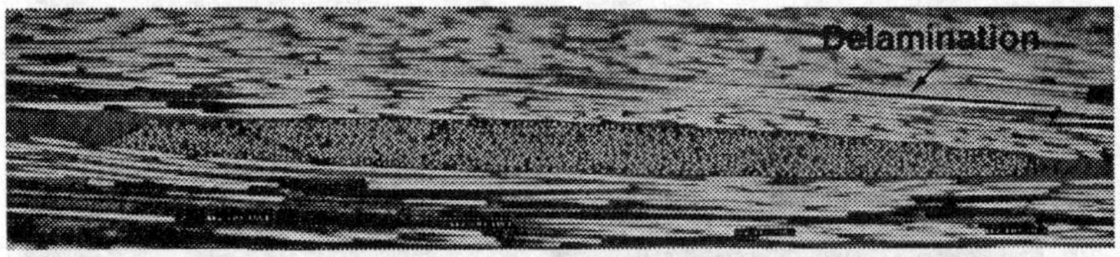

Figure 8. *In situ* cross-sectional micrograph of the 5HS unbalanced (6:1) fabric composite specimen (Location 2) at 98 percent of failure load (far-field strain: 5690 με).

Figure 9. *In situ* cross-sectional micrograph of the model laminate at 37 percent of far-field strain (3700 με).

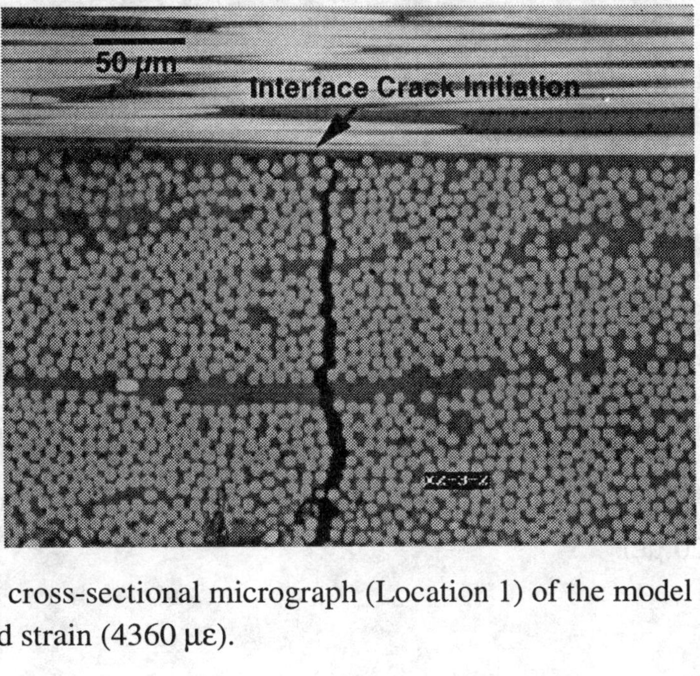

Figure 10. *In situ* cross-sectional micrograph (Location 1) of the model laminate at 44 percent of far-field strain (4360 με).

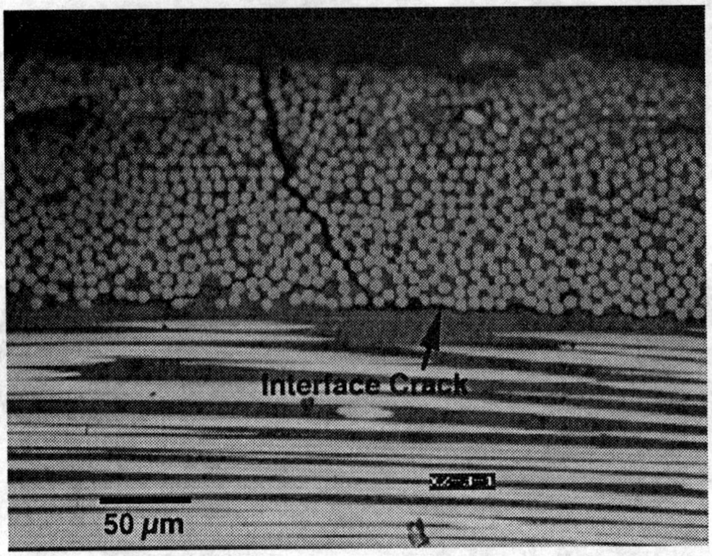

Figure 11. *In situ* cross-sectional micrograph (Location 2) of the model laminate at 44 percent of far-field strain (4360 με).

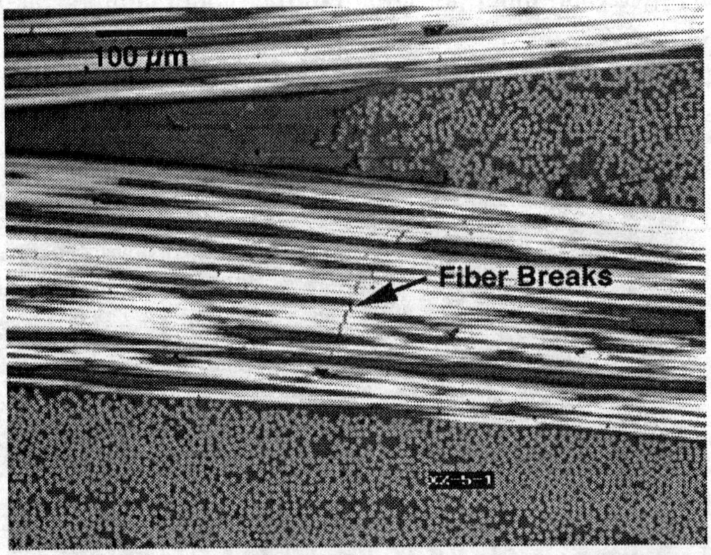

Figure 12. *In situ* cross-sectional micrograph of the model laminate at 69 percent of far-field strain (6920 με).

Damage Growth Sequence of SiC Fiber Reinforced Glass Matrix Cross-Ply Laminates

N. TAKEDA AND M. KIRIYAMA

ABSTRACT

Damage initiation and growth behavior was studied for SiC fiber reinforced glass matrix composite cross-ply laminates. The replica technique was used to observe the matrix cracking sequence in $[0_3/90_m/0_3]$ (m=1, 2, 3) laminates, and the effect of the number of 90° plies was clarified experimentally. A novel in-situ SEM (scanning electron microscope) observation was also conducted to measure the fiber/matrix interfacial debonding and sliding in 0° plies using specimens with parallel micro-lines printed on their surfaces. Interfacial debonding was found to grow intermittently as the applied stress increases. The debond length distribution depended on the number of 90° plies as well as the damage modes. Based on the above microscopic observation, an analytical model for damage growth was proposed using the energy balance calculation. The model was based on Kuo and Chou, but modified to include the effects of the interaction between different damage modes as well as the Poisson contraction at debonded regions. The predicted damage sequence agreed reasonably well with the experimental one.

INTRODUCTION

Understanding of damage initiation and growth behavior is very important for the development of fiber reinforced ceramic matrix composites (CMC) (Marshall and Evans, 1985). In particular, it is necessary to study the sequence of such microscopic damage process up to the final fracture to ensure the damage tolerance capability.

Many studies have been devoted to model the damage process in unidirectional, cross-ply and woven CMC. For example, some tensile mechanical properties of glass or glass-ceramic matrix composite laminates were experimentally obtained (Prewo, 1986, 1988). The matrix crack growth criterion was studied for SiC/LAS unidirectional laminates using fracture mechanics approach (Budiansky, Hutchingson and Evans, 1986). The fiber/matrix interfacial properties were also studied quite extensively (Marshall and Oliver, 1987; Hsueh, 1995). A full statistical treatment of the matrix crack evolution and associated stress-strain behaviors has been conducted recently (Ahn and Curtin, 1997). The damage evolution in cross-ply lamina-

---

Nobuo Takeda, Manami Kiriyama, Center for Collaborative Research, The University of Tokyo, 4-6-1 Komaba, Meguro-ku, Tokyo 153, Japan nates is more complex than in unidirectional ones since the matrix cracks occur in both 0° and 90° plies (Kuo, 1992; Karandikar and Chou, 1993; Kuo and Chou, 1995).

Although cross-ply or woven laminates are most important in practical applications of CMC, the damage microscopic initiation and growth behavior has not been well characterized yet. The purpose of the present study is to characterize such damage sequence in SiC fiber reinforced glass matrix cross-ply laminates both experimentally and theoretically from the micro-mechanical viewpoint. The replica technique was used to measure the overall matrix cracking sequence, and a novel in-situ SEM/micro-line technique was used to quantify the fiber/matrix debonding and sliding behavior in 0° plies. Then, a modified analytical model based on Kuo and Chou was proposed to compare with the experimental results.

EXPERIMENTS

MATERIALS

SiC (Nicalon™, Nippon Carbon Co. Ltd.) fiber reinforced borosilicate glass (Pyrex™ 7740, Corning Co.) matrix composites were fabricated by Nippon Carbon Co. Ltd. Fibers coated with approximately 70 nm-thick carbon layer were used to fabricate cross-ply laminates by glass powder slurry impregnation and hot pressing at an optimized process condition (1100°C, 10min). The properties of constituent materials are summarized in Table 1. The stacking configuration was $[0_3/90_m/0_3]$ (m=1, 2, 3) and the nominal fiber volume fraction V_f was 0.36. The average thickness of a unidirectional ply is 0.2 mm, which results in the total thickness ranging from 1.4 mm (m=1) to 1.8 mm (m=3).

REPLICA TECHNIQUE

Tensile tests were conducted at the cross-head speed of 0.4 mm/min using polished specimens with 5 mm in width and 25 mm in gage length. Tests were stopped intermittently at several loading stages to take replicas of the damaged gage section by using polyacetate films. Matrix crack initiation and multiplication behavior in both 0° and 90° plies was quantitatively measured by observing replicas through an optical microscope after the tests.

IN-SITU SEM/MICRO-LINE TECHNIQUE

The in-situ/micro-line technique has been established recently to quantitatively measure the microscopic deformation and interfacial damage evolution in fiber reinforced composites under loading (Takeda et al., 1997), and is used here for brittle matrix composites. Parallel micro-lines were printed on the parallel-sided specimen surfaces polished with 1μm diamond pastes, using photo-lithography technique. Micro-line patterns with 1 μm in width and 3 or 5 μm in gap were used on specimens in the same size as those used in the replica observation. The printed specimens were loaded at the same cross-head speed inside a SEM with a servo-hydraulic loading stage as well as a temperature control unit. Two-dimensional distribution of fiber/matrix interfacial debonding and sliding in 0° plies, in particular, was observed and measured quantitatively as a function of the stress level. Photos were taken at various loading stages and fed into a personal computer as digital images, and then the relative displacement were digitally measured. The resolution in the displacement is 0.02 μm in the present study.

EXPERIMENTAL RESULTS

MATRIX CRACK INITIATION AND GROWTH

Figure 1 shows photomicrographs of replicas showing matrix cracks under tensile load. Characteristic damage patterns in CMC cross-ply laminates are matrix cracks generating in both 0° and 90° plies. Matrix cracks penetrating into both 0° and 90° plies are dominant in $[0_3/90_1/0_3]$ laminates. However, matrix cracks existing only in 0° plies can also be observed more often from the earlier loading stage as the number of 90° plies increases.

TABLE 1---MECHANICAL PROPERTIES OF FIBER AND MATRIX

	Nicalon	Borosilicate Glass
Young's Modulus (GPa)	196	64
Poisson's Ratio	0.20	0.30
Thermal Expansion Ratio (10^{-6}/K)	3.1	3.25

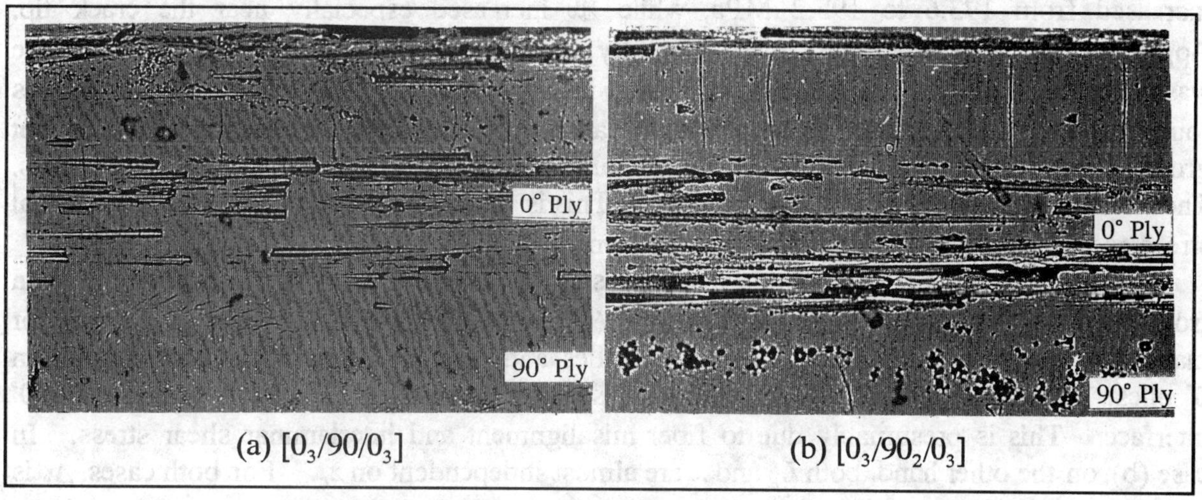

Figure 1 Photos of Replicas Showing Matrix Cracks Under Tensile Load

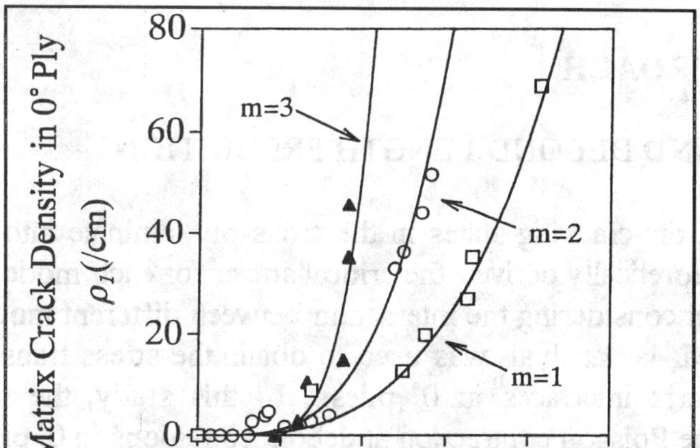

Figure 2 Relationship between Matrix Crack Densities in 0° and 90° Plies

Figure 2 shows the relationship between matrix crack densities in 0° and 90° plies for all laminate configurations. The following experimental formula can be obtained by fitting all experimental data,

$$\rho_1 = \rho_2^{(1.390m + 1.889)} \times 10^{-4} \tag{1}$$

where the matrix densities in 0° and 90° plies are denoted by ρ_1 and ρ_2, respectively.

IN-SITU OBSERVATION OF INTERFACIAL DEBONDING AND SLIDING

Figure 3 shows in-situ SEM photos of a damage state near a 0° matrix crack among matrix cracks penetrating into both 0° and 90° plies. The position of fiber/matrix debonding tip can be observed where the interfacial relative displacement or sliding, Δx, becomes zero.

Figure 4 shows some representative results of Δx as functions of x (distance from the crack tip) and the stress level at the 0°/90° interface in $[0_3/90_3/0_3]$ specimens. Let us explain the sequence of debonding and sliding at $x<0$. The debond length L_d, which is defined as a distance where Δx can be observed, was almost the same even when the applied stress σ increased from 173.6 to 195.2 MPa, while Δx increased especially near the crack tip. Approximately 3μm increase in L_d was found by the stress level of 213.1 MPa, but no increase was observed at 230.1 MPa while Δx increased near the crack tip. Much increase in L_d was found again at 247.3 MPa. In summary, it can be concluded that the debonding does not grow gradually as the applied stress increases, but discontinuous or intermittent growth occurs. The slope $\Delta x/x$ increases near the debond tip. This is due to the relaxation in thermal residual stress as well as the decrease in interfacial friction.

Figure 5 shows the distribution of Δx at several distances z_0 in the thickness direction from the 0°/90° interface. Both cases (a) and (b) are shown in Fig. 5, where case (a) is for matrix cracks penetrating into both 0° and 90° plies, and case (b) is for cracks only existing in 0° plies. In case (a), both L_d and Δx decrease with increasing z_0, except near the 0°/90° interface. This is presumably due to fiber misalignment and interlaminar shear stress. In case (b), on the other hand, both L_d and x are almost independent on z_0. For both cases, Δx is almost zero at the specimen surface (z_0=0.6 mm).

THEORETICAL APPROACH

STRESS ANALYSIS AND DEBOND LENGTH PREDICTION

Kuo has classified the cracking states in the cross-ply laminate into 5 damage modes as shown in Fig. 6, and theoretically derived the critical stress for each mode based on the energy balance criterion, without considering the interaction between different modes (Kuo, 1992; Kuo and Chou, 1995). Shear-lag analysis was used to obtain the stress transfer both at the 0°/90° interfaces and fiber/matrix interfaces in 0° plies. In this study, the stress distribution is modified, considering the Poisson contraction at debonded regions in 0° plies as noted in the Δx measurement. Thus, at debonded regions, the relationship between the interfacial shear stress τ_0 and the fiber axial stress σ_f is given by

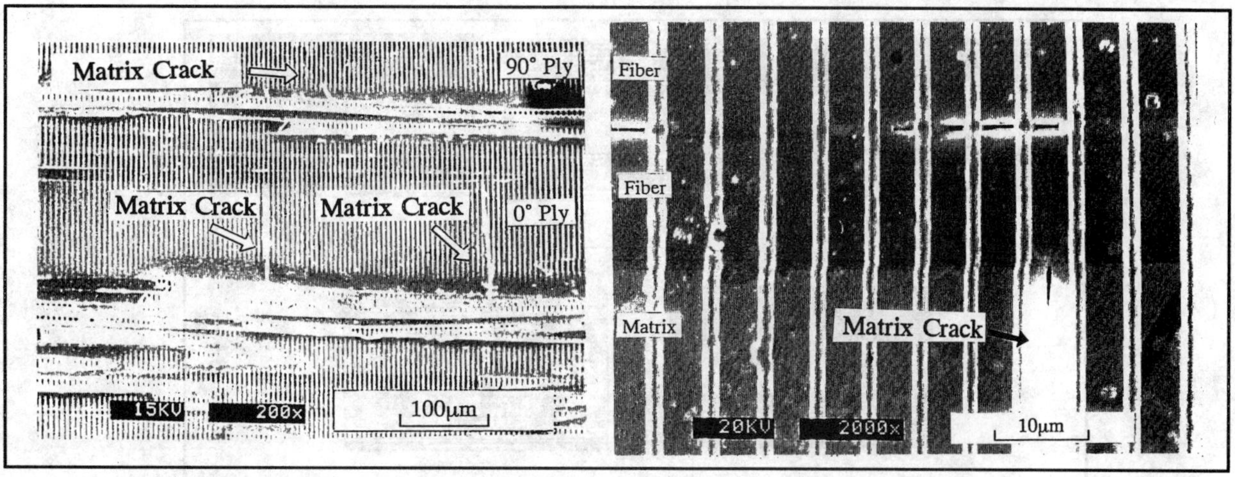

Figure 3 SEM Photos of Fiber/Matrix Sliding on Specimen Surfaces with Printed Micro-lines

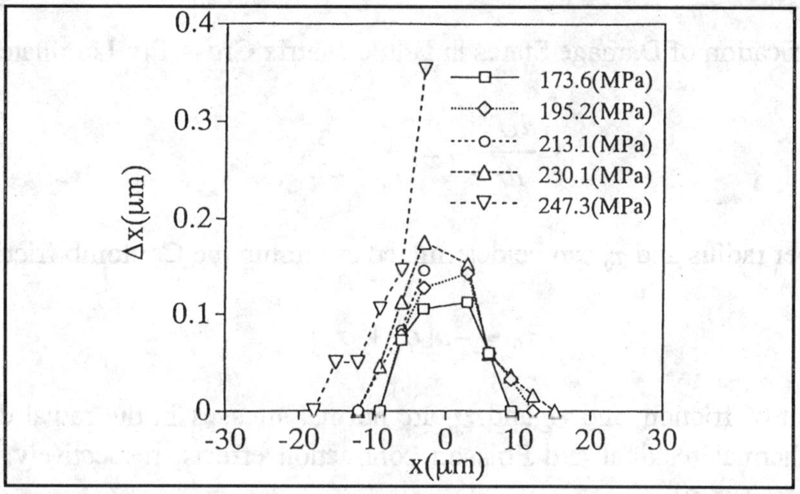

Figure 4 Relative Interfacial Displacement as a Function of Distance from Crack Tip at 0°/90° Interface ($[0_3/90_3/0_3]$)

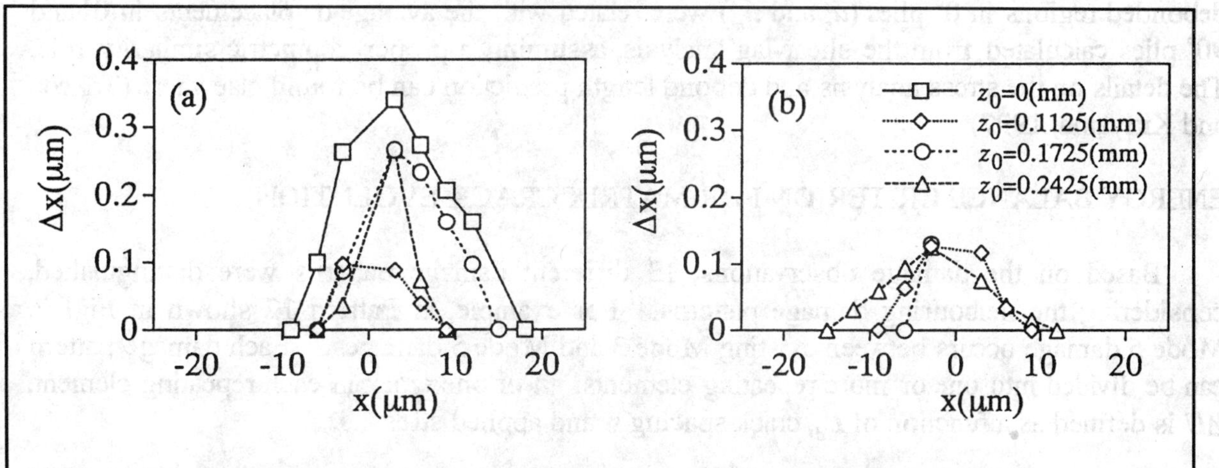

(a) Crack Penetrating into Both 0° and 90° Plies (b) Isolated Crack in 0° Plies

Figure 5 Distribution of Relative Interfacial Displacement at Different Distances z_0 in the Thickness Direction from 0°/90° Interface. ($[0_3/90_3/0_3]$, σ_{CP}=183.9 MPa)

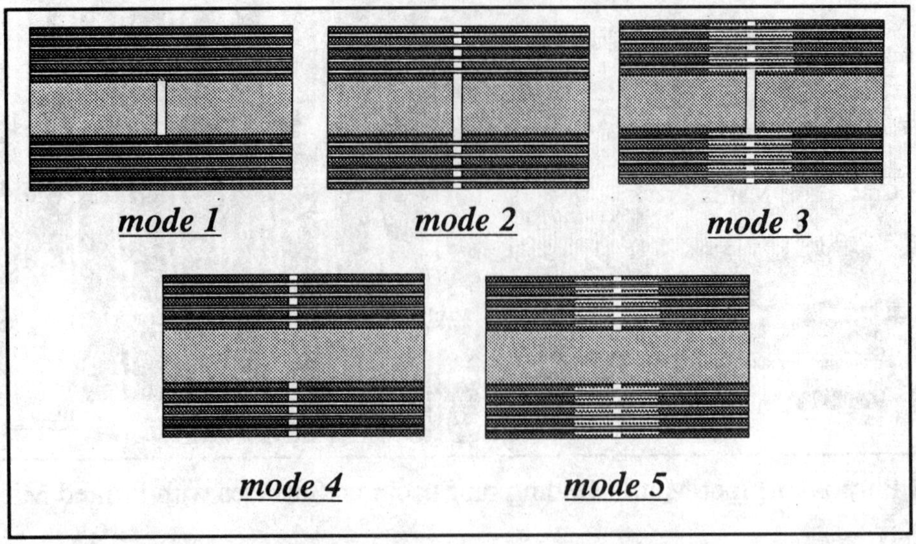

Figure 6 Classification of Damage States in Brittle Matrix Cross-Ply Laminates (After Kuo)

$$\frac{d\sigma_f}{dx} = -\frac{2\tau_0}{r_0} \tag{2}$$

where r_0 is the fiber radius and τ_0 can be determined assuming the Coulomb friction as

$$\tau_0 = -\mu\left(\sigma_c + \sigma_p\right) \tag{3}$$

μ is the coefficient of friction, and σ_c and σ_p are normal stresses in the radial direction at the interface due to thermal residual and Poisson contraction effects, respectively. The friction occurs only when $\sigma_c + \sigma_p < 0$.

The debond length L_d in mode 3 and 5 was predicted assuming the shear stress distribution in the thickness in 0° plies is linear. Displacements of fibers and matrix at debonded regions in 0° plies (u_f and u_m) were related with the average displacements in 0° and 90° plies calculated from the shear-lag analysis, assuming a proper geometric similarity rule. The details on the stress analysis and debond length prediction can be found elsewhere (Takeda and Kiriyama, 1997).

ENERGY BALANCE CRITERION FOR MATRIX CRACK EVOLUTION

Based on the damage observations, 13 different damage patterns were distinguished, considering the neibouring damage patterns. For example, in Pattern 12 shown in Fig. 7, Mode 5 damage occurs between existing Mode 3 and Mode 5 damages. Each damage pattern can be divided into one or more repeating elements. For one crack in each repeating element, ΔU is defined as a function of L_d, crack spacing s, and applied stress σ_{CP}.

$$\Delta U\left(\sigma_{CP}, L_d, s\right) = W - U_{all} \tag{4}$$

where W is work done by external load and U_{all} is the sum of energy terms.

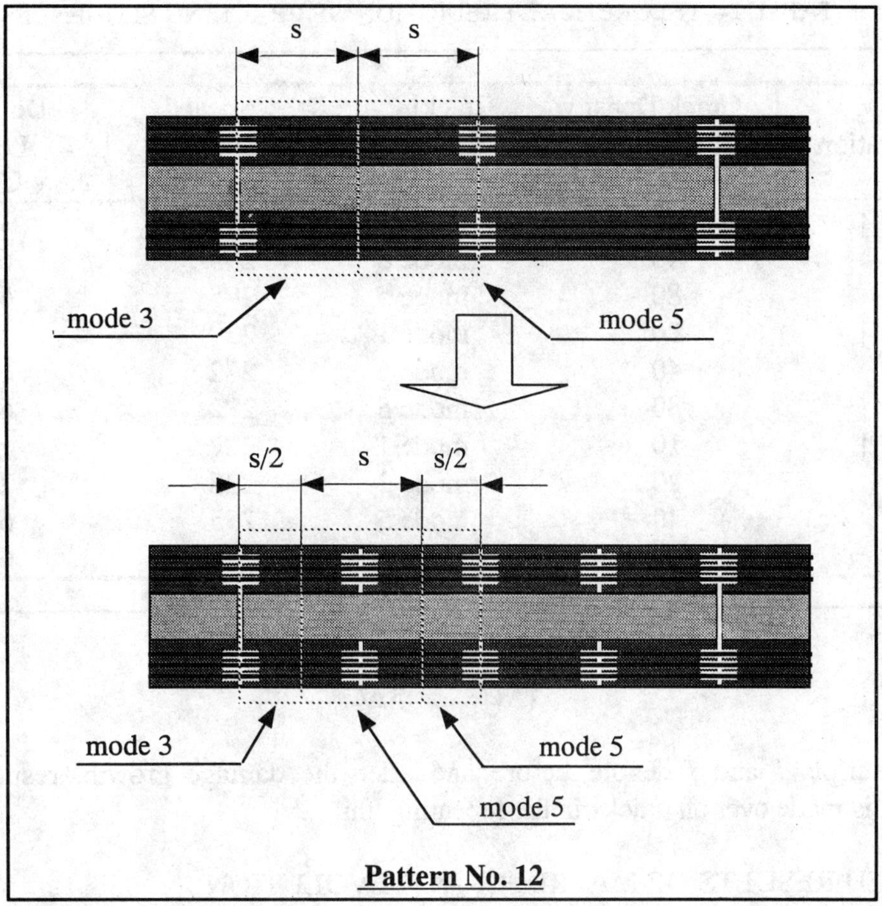

Figure 7 An Example of Schematic Representation of Cracking Sequence (Pattern #12)

$$W = (b+d)\frac{\sigma_{CP}}{E_f}\int_0^s \sigma_f dx \qquad (5)$$

$$U_{all} = U_c + U_{db} + U_s + U_f + U_m + U_2 \qquad (6)$$

U_c is crack surface energy, U_{db} interfacial debonding energy, and U_s energy loss due to interfacial friction. ΔU_f and ΔU_m are strain energies in fibers and matrix in 0° plies, respectively. ΔU_2 is strain energy in 90° plies. All terms in Eq. (6) can be written as,

$$U_{db} = \frac{2bV_f}{r_0}L_d\gamma_{db}, \quad U_s = \frac{2bV_f}{r_0}\int_0^{L_d}\tau_0(\Delta u_f - \Delta u_m)dx, \quad U_f = \frac{bV_f}{2E_f}\int_0^s(\sigma_f)^2 dx$$

$$U_m = \frac{bV_m}{2E_m}\int_0^s(\sigma_m{}^f)^2 dx, \quad U_2 = \frac{d}{2E_2}\int_0^s(\sigma_2{}^f)^2 dx \qquad (7)$$

where $2b$ and $2d$ are total thickness of 0° and 90° plies, respectively. E_f, E_m, and E_2 are Young's modulus of fiber, matrix, and 90° ply, respectively.

From the energy balance within each repeating unit before and after a damage growth such as in Fig. 7, the critical stress for such a growth can be simulated using the following formula,

TABLE 2---THEORETICAL PREDICTION OF CRACKING SEQUENCES

Stacking Configuration	Crack Density After Cracking (/cm)	Cracking Mode	Critical Stress (MPa)	Debonding Length of New Cracks (μm)
$[0_3/90_1/0_3]$	20	mode 1	254	-
	40	mode 5	295	65.60
	80	mode 5	295	65.60
$[0_3/90_2/0_3]$	20	mode 1	239	-
	40	mode 5	272	65.59
	80	mode 5	272	65.59
$[0_3/90_3/0_3]$	10	mode 1	227	-
	20	mode 3	228	25.77
	40	mode 5	255	65.47
	80	mode 5	255	65.47

$$\sum(\Delta U)^i \leq \sum(\Delta U)^f \tag{8}$$

where subscripts i and f denote before and after the damage growth, respectively. The summation is made over all cracks in the repeating unit.

PREDICTED RESULTS OF MATRIX CRACK EVOLUTION

The above energy balance criterion can predict matrix crack evolution in cross-ply laminates. Because matrix cracks normally initiate at yarns in 90° plies (yarn density: 1 yarn/mm), the initial crack density was assumed to be 10/cm, while γ_m and γ_{db} were to be 2.5 J/m^2 and 0.8 J/m^2, respectively. Table 2 shows the predicted matrix crack evolution for each laminate configuration. The initial crack was always Mode 1, and several Mode 1 cracks followed at the earlier loading stage. The following damage evolution depended on the laminate configuration, although the final saturated crack density were similar.

As obtained experimentally in the microscopic observations that revealed the interfacial debonding always occurred when cracks appeared in 0°plies, the present analysis predicted only Modes 1, 3, and 5 could occur in the present material system.

COMPARISON BETWEEN THEORETICAL PREDICTION AND EXPERIMENT

Figure 8 shows the overall crack density as a function of the applied stress. The theoretical prediction agrees well with the experimental results, although the former provides more steep increase than the latter at the large crack density region.

Figure 9 shows the crack density in both 0° and 90° plies as a function of the applied stress. The agreement between the theory and the experiment is reasonably good. However, the theory predicts the occurrence of matrix cracks in 90° plies at higher stress level than the experiment, and also provides more Mode 5 cracks than Mode 3 ones at lower stress level. The predicted debond length in Mode 5 cracks is also larger than the experimental one, especially at lower stress level. This is probably due to the difference in the matrix constraint

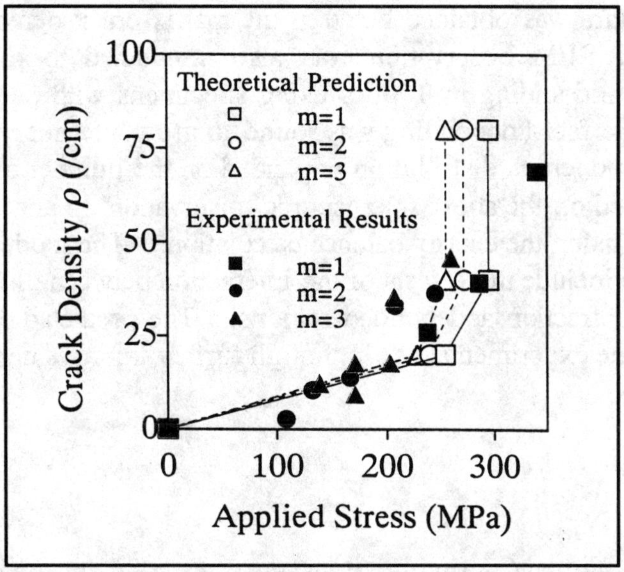

Figure 8 Comparison of Predicted and Experimental Overall Crack Densities

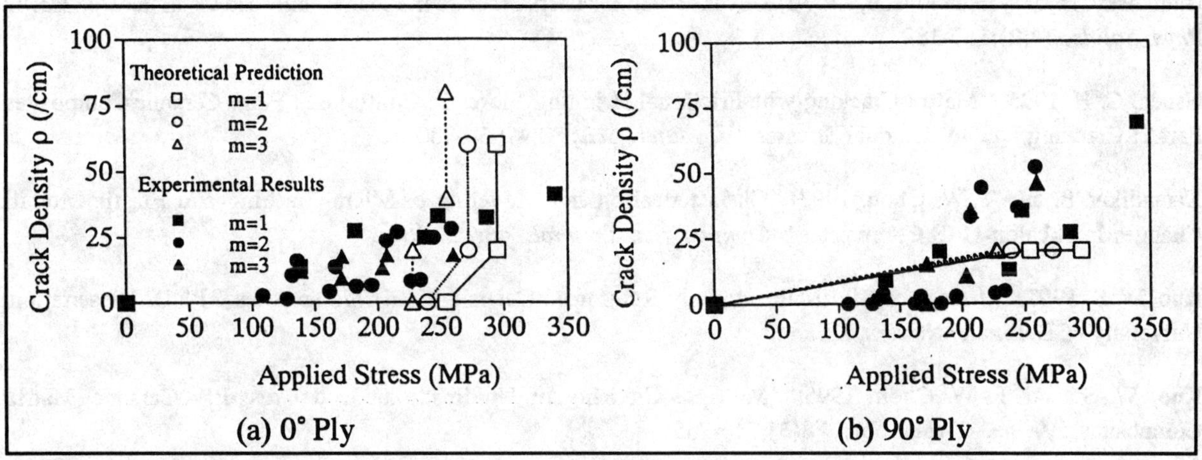

Figure 9 Comparison of Predicted and Experimental Crack Densities in 0° and 90° Plies

applied to fibers between the theory and the experiment. That is, the experimental debond length was measured at the specimen surface where fibers were exposed, while the theoretical one was for the fiber wholly surrounded by matrix and other fibers.

CONCLUSIONS

Damage initiation and growth behavior was studied for SiC fiber reinforced glass matrix composite $[0_3/90_m/0_3]$ (m=1, 2, 3) cross-ply laminates. The replica technique was used to observe the matrix cracking sequence. Characteristic damage patterns were matrix cracks generating in both 0° and 90° plies. Matrix cracks penetrating into both 0° and 90° plies were dominant in $[0_3/90_1/0_3]$ laminates. However, matrix cracks existing only in 0° plies could also be observed more often from the earlier loading stage as the number of 90° plies increased.

An experimental formura was obtained between the matrix crack densities in 0° and 90° plies. Then, a novel in-situ SEM observation was also conducted to measure the fiber/matrix interfacial debonding and sliding in 0° plies using specimens with parallel micro-lines printed on their surfaces. Interfacial debonding was found to grow intermittently as the applied stress increased. The debond length distribution depended on the number of 90° plies as well as the damage modes. Based on the above microscopic observation, an analytical model for damage growth was proposed using the energy balance calculation. The model was based on Kuo and Chou, but modified to include the effects of the interaction between different damage modes as well as the Poisson contraction at debonded regions. The predicted damage sequence agreed reasonably well with the experimental one, although further improvement is necessary for better prediction.

REFERENCES

Ahn, B. K. and W. A. Curtin. 1997. "Strain and Hysteresis by Stochastic Matrix Cracking in Ceramic Matrix Composites," *J. Mech. Phys. Solids*, 45:177-209.

Budiansky, B., J. W. Hutchinson, and A. G. Evans. 1986. "Matrix Fracture in Fiber-Reinforced Ceramics," *J. Mech. Phys. Solids*, 34(2):167-189.

Hsueh, C. H. 1995. "Matrix Cracking with Frictional Bridging Fibres in Continuous Fibre Ceramic Composites Part 2: Cracking Due to Residual Stresses," *J. Mater. Sci.*, 30:4793-4800.

Karandikar, P. and T. W. Chou. 1993. "Characterization and Modeling of Micro-Cracking and Elastic Moduli Changes in Nicalon-CAS Composites," *Compos. Sci. Technol.*, 46:253-264.

Kuo, W. S. 1992. "Damage of Multi-Directionally Reinforced Ceramic-Matrix Composites," Ph.D. Dissertation, University of Delaware, Newark, DE.

Kuo, W. S. and T. W. Chou. 1995. "Multiple Cracking of Unidirectional and Cross-Ply Ceramic Matrix Composites," *J. Am. Ceram. Soc.*, 78(3):745-755.

Marshall, D. B. and A. G. Evans. 1985. "Failure Mechanisms in Ceramic-Fiber/Ceramic-Matrix Composites," *J. Am. Ceram. Soc.*, 68(5):225-231.

Marshall, D. B. and W. C. Oliver. 1987. "Measurement of Interfacial Mechanical Properties in Fiber-Reinforced Ceramic Composites," *J. Am. Ceram. Soc.*, 70(8):542-548.

Prewo, K. M. 1986. "Tension and Flexural Strength of Silicon Carbide Fibre-Reinforced Glass Ceramics," *J. Mater. Sci.*, 21:3590-3600.

Prewo, K. M. 1988. "Carbon Fibre Reinforced Glass Matrix Composite: Tension and Flexural Properties," *J. Mater. Sci.*, 23:2745-2752.

Takeda, N. and M. Kiriyama. 1997. "Matrix Crack Evolution in SiC Fiber Reinforced Glass Matrix Cross-Ply Laminates," (Submitted for Publication).

Takeda, N., H. Niizuma, S. Ogihara, and A. Kobayashi. 1997. "Application of Micro-Line/Grid Methods to Temperature-Dependent Microscopic Deformation and Damage in CFRP Laminates," *Exp. Mech.*, 37(2):(in press).

A Fatigue Life Model for Polymeric Composites

N. BONORA AND G. M. NEWAZ

ABSTRACT

Damage in composite material is quite complex. Several different damage mechanisms occur under continuous straining. Matrix crack is the first damage mechanism that develop in the composite and it can be responsible for the largest reduction of the material stiffness along the direction normal to the cracks. For brittle composite system such as many practical polymeric matrix composites (PMC), the repetition of load in fatigue should not introduce any new damage mechanism with respect to the monotonic case. As a consequence of this, fatigue life of PMCs can be addressed relating fatigue damage evolution with quasi-static damage measurements. In the present paper, a phenomenological damage model, originally proposed for ceramic matrix composite, is reformulated for PMCs. The comparison with the experimental measurements relative to several PMCs confirmed the model proposed and the necessity to address to damage in terms of stiffness loss versus effective imposed strain. If the basic features of the evolution of a given damage state with cycles is known, the monotonic damage information can be used to predict the composite material fatigue life.

INTRODUCTION

Failure mode in composite materials is not unique. Many damage processes, such as transverse matrix cracking, fiber breaking, fiber-matrix debonding, interlaminar delamination, fiber pull-out, may take place before final failure. Each of this processes is responsible for the reduction of material capabilities to sustain load and the decrease of the material structural performance and properties such as strain to failure, composite strength, thermal properties and so on. The damage processes mentioned above develop in the material according to a proper order. Usually, matrix cracking is the first damage process to take place as a result of the fact that matrix has the lowest strain to failure of all the composite constituents. Transverse matrix cracks in composite laminates is the principal responsible for the stiffness reduction. Many attempts to model the loss of stiffness in composite laminates have been made. Either shear lag analyses, micromechanical models and continuum damage models

Bonora, N., Industrial Engineering Dept., University of Cassino, Via G.Di Biasio 43, 03043 Cassino, Italy
Newaz G.M., Mechanical Engineering Dept., Wayne State University, 5050 A. Wayne Dr., Detroit, MI-48105

have been proposed by many authors, (Daniel and Lee, 1990; Stecenko and Stefanovic, 1990; Lee and Daniel, 1990; Talreja, 1995; Allen et al.,1987a-b). In many of these models, stiffness reduction is always related to the applied stress experienced by the composite.

It has been observed that for those composite that can be considered brittle, i.e. the only energy dissipation is the one spent to generate new fracture surfaces and the residual strain after unloading is the result of the incomplete closure of fracture surfaces (Figure 1), stress-strain description cannot be sufficient to describe uniquely the material constitutive response. In fact, as it has been shown for a pure brittle composite system such as Nicalon-LAS (Lithium Alumina Silicate), traditional stress-strain curves including damage processes, obtained on several identical specimens, show a poor reproducibility and a scattered behavior. Bonora and Newaz (1996), have shown that it is more appropriate to address the composite behavior in term of effective applied strain instead of stress. Strain is always monitored directly in the composite while stress is derived on the assumption that the load is carried uniformly by the reference net section.

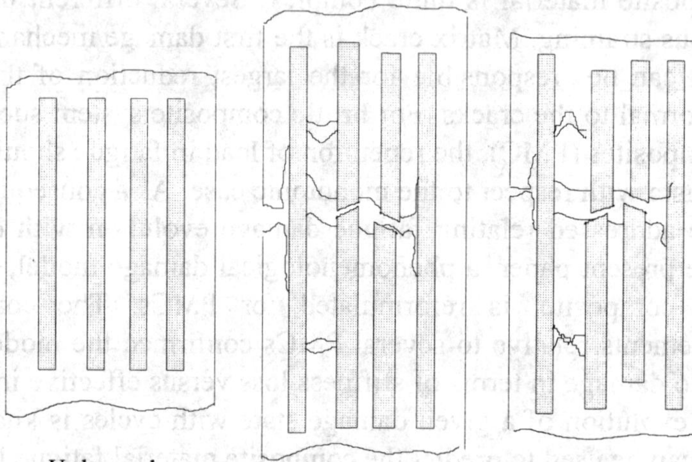

Figure 1- Sketch of the residual strain after unloading due to damage

The experimental tests performed on Nicalon-LAS composite systems showed that experimental data scatter in stress-strain curves disappear if the material tensile behavior is given in terms of damage versus imposed strain where damage variable quantifies the reduction of the composite stiffness (Newaz and Bonora, 1996). For pure brittle composite system, it has been observed that fatigue loading does not introduce any new damage mechanism in the overall complex damage process (Karandikar and Chou, 1993). Newaz and Bonora (1996) demonstrated that if fatigue does not result in new mechanisms, fatigue damage curve has to coincide with the static one. They found for Nicalon-LAS systems that:

a) the applied strain in the first cycle is responsible for the initial amount of damage that will evolve in the following cycles;
b) in constant stress fatigue loading, the effective strain experienced by the material increases as a result of the "*mechanical ratcheting*" consequent to the incomplete closure of the fracture surfaces during the unloading;

In the present paper, the concepts developed by the authors on CMC composite are successfully applied to PMCs and confirm that fatigue behavior of brittle composite systems can be easily approached from the engineering point of view in terms loss of stiffness using the model proposed. The current authors extended the stiffness loss concept into a damage-life relationship to develop the framework for life prediction model.

STATIC AND FATIGUE DAMAGE MODELING

Brittle composite systems, such as PMCs, are significantly affected due to applied strain field. Matrix is the first material component that fails under straining and matrix cracking is the key damage process that first occurs in the material and that is responsible for most of the loss of stiffness in most practical composites.

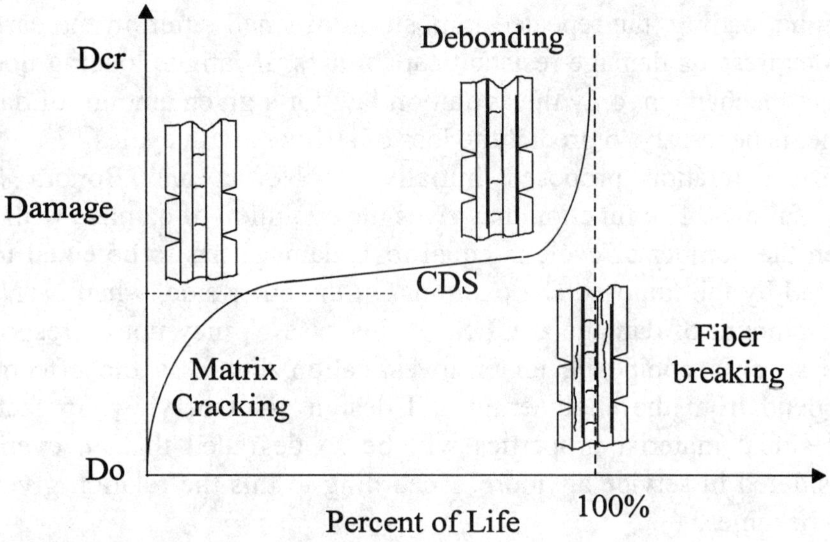

Figure 2 - Damage evolution scheme for PMC under fatigue loading, (Reifsnider at alii,1984)

Reifsnider and Jamison (1982) and Reifsnider et al. (1984) showed that matrix cracking is the dominant mechanism in the early stage of life and produces most of the damage accumulation, as depicted in Figure 2. The remaining life is dominated by other processes such as delamination. Anyway, when the composite has reached the saturation of matrix cracks, known as the Characteristic Damage State (CDS), composite stiffness is often reduced up 40-60% of the initial value. Even if the composite has not failed, from a functional point of view, performance requirement of the component during service may have to be compromised.

Bonora and Newaz (1996), following the work of Lemaitre and Chaboche (1980) on damage models for brittle materials, proposed a damage law in order to describe the effect of progressive cracking in brittle composites. Here damage is defined as the reduction of the stiffness along the direction normal to the matrix cracks. It is well known that a more rigorous damage model requires a tensor formulation for damage variable. However, in many practical applications, for instance for structural components carrying loads along the fiber direction as in ring structures with hoop loading where continuous fibers are used as hoop reinforcement, a single and most important damage vector can be identified and approached using a single scalar variable.

The damage law given by Bonora and Newaz (1996) can be rewritten for polymeric matrix composites as:

$$D(\varepsilon) = D_0 + \left[D_{cr} - D_0\right]\left[\frac{\ln(\varepsilon / \varepsilon_{th})}{\ln(\varepsilon_{CDS} / \varepsilon_{th})}\right]^{\frac{1}{m}} \quad (1)$$

where D_0 is the initial damage in the laminate, D_{cr} is the value of damage at CDS, ε_{th} is the threshold strain for which damage process, i.e. matrix cracking, is activated; ε_{CDS} is the strain at CDS and m is a material constant. During quasi-static tensile load ε_{CDS} can be very close to strain at failure because the material cannot respond to progressive straining due to the short time involved.

Under fatigue loading, the repeated imposition of small strain in the early cycles allows the material to express its damage resistant capabilities. If fatigue loading does not introduce any new damage mechanism, only the evolution law for a given amount of damage, produced in the first cycle, is necessary to predict the loss of stiffness with cycles.

The empirical relation proposed initially by Newaz and Bonora (1996), can be reformulated as follows. The function that gives the evolution of damage with cycles has to be bounded. When the number of cycle is equal to 1, damage has to be equal to the amount of damage produced by the imposed static strain. On the other side, when $N=N_f$ damage has to be equal to the amount of damage at CDS. In this case N_f may not correspond to the life at failure for the specific composite under investigation, but it is the effective life that the material can spend from the engineering and design point of view. In fact, after $N=N_f$ is reached, the residual material properties will be so degraded that an eventual component cannot be considered in service anymore. According to this the relation given in Newaz and Bonora (1996) becomes,

$$D(N,\varepsilon) = D_{CDM} + \left[D(\varepsilon) - D_{CDM}\right]\left[1 - \frac{\ln(N)}{\ln(N_f)}\right]^{\frac{1}{\beta}} \quad (2)$$

that, recalling Eqn. (1) and re-arranging can be written as follows,

$$D(N,\varepsilon) = D_{CDM} - (D_{CDM} - D_0)\left\{1 - \left[\frac{\ln(\varepsilon / \varepsilon_{th})}{\ln(\varepsilon_f / \varepsilon_{th})}\right]^{\frac{1}{m}}\right\}\left[1 - \frac{\ln(N)}{\ln(N_f)}\right]^{\frac{1}{\beta}} \quad (3)$$

where the material exponent β can be determined as described in (Newaz and Bonora, 1996).

In order to apply the model proposed for PMC laminates, the definition of damage as a variable needs to be discussed in detail. Damage variable, in the definition of Kachanov, indicates the material loss of stiffness as a result of the irreversible microstructural changes. In composite laminates, damage growth can be quite significant and the appearance of matrix cracks very often correspond to structural changes that involve macroscopical stress redistribution. A direct estimation of damage along the direction normal to the matrix crack can be given through the measure of the crack density. Using a vector damage formulation, Talreja (1981, 1982 and 1985) proposed the following expression that related damage with the crack density and the dimensions of the array of cracks,

$$D = \sqrt{(\eta_c \cdot l_c \cdot w_c \cdot f_c)} \tag{4}$$

where η_c is the crack density per unit length (number of cracks /mm), l_c is the average crack length that for non woven composites with 90° plies can be taken as the specimen width, w_c is the average crack width (i.e. equal to nt_0 where n is the number of cracked plies and t_0 is the ply thickness).

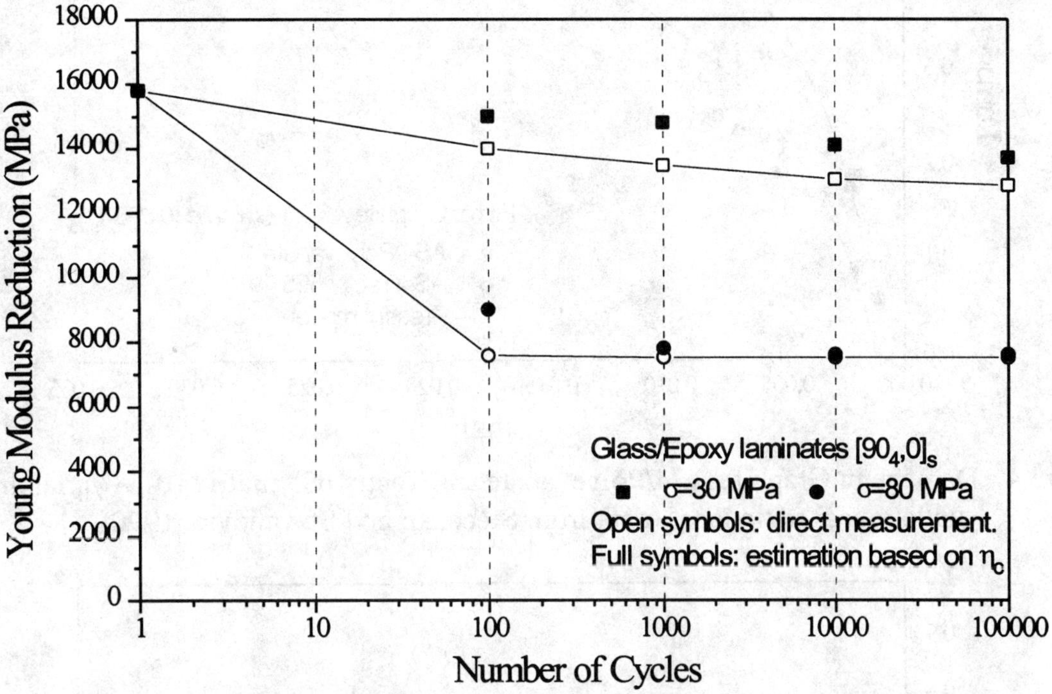

Figure 3 - Comparison between direct measurements of Young's modulus reduction and estimation based on crack density, (data from Bonora et al., 1992).

This expression has been verified on several laminated composites; the estimation of the reduction of the elastic modulus, in the direction normal to the crack array, is in very good agreement with direct stiffness measurements. In Figure 3, a comparison between the calculated damage and the direct measurements is given for a Glass/Epoxy $[90_4,0]_s$ laminate under fatigue loading, (Bonora et al., 1992).

MODEL VERIFICATION

The verification of the proposed model for PMC was performed with particular care to the following issue: a) to verify the model with quasi-static damage data and to identify the damage parameters; b) to verify that fatigue damage data can be reduced to the same static damage curve if the mechanical ratchetting is taken into account; c) to use this information to predict the loss of stiffness under fatigue loading.

The effective damage evolution law under quasi-static load was determined for several PMCs and the damage parameters were identified. In figure 4 damage evolution law for a two Graphite based PMCs, as a function of the imposed strain (Yalvac et al, 1991), is given.

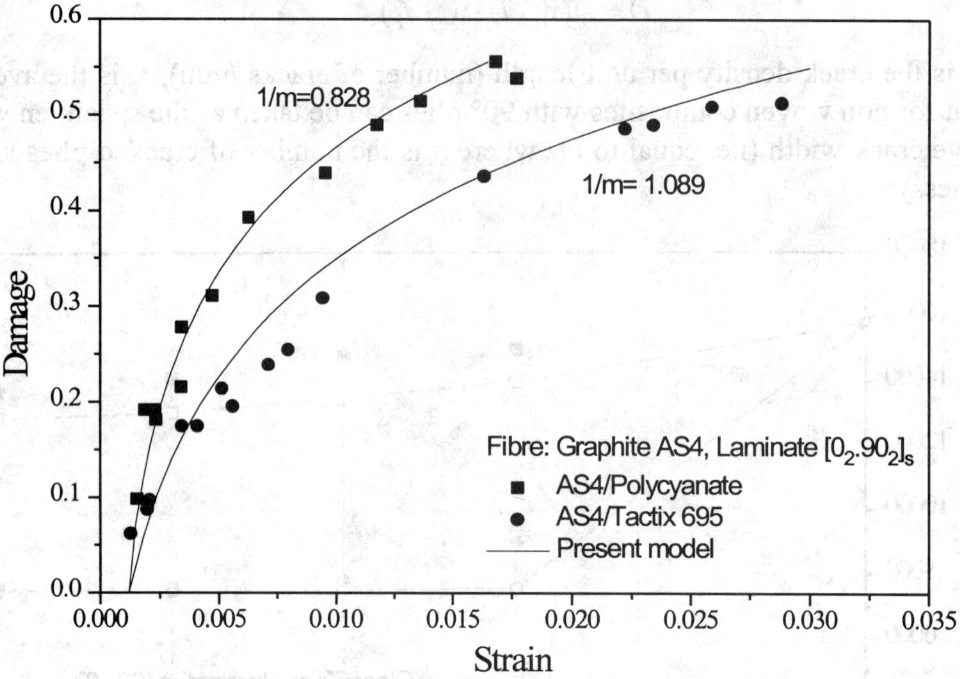

Figure 4 - Damage in Graphite AS4/Polycyanate and Tactix 695 matrix [0_2,90_2]$_s$ laminates under quasi static load, (data from Stecenko and Stevanovic, 1990).

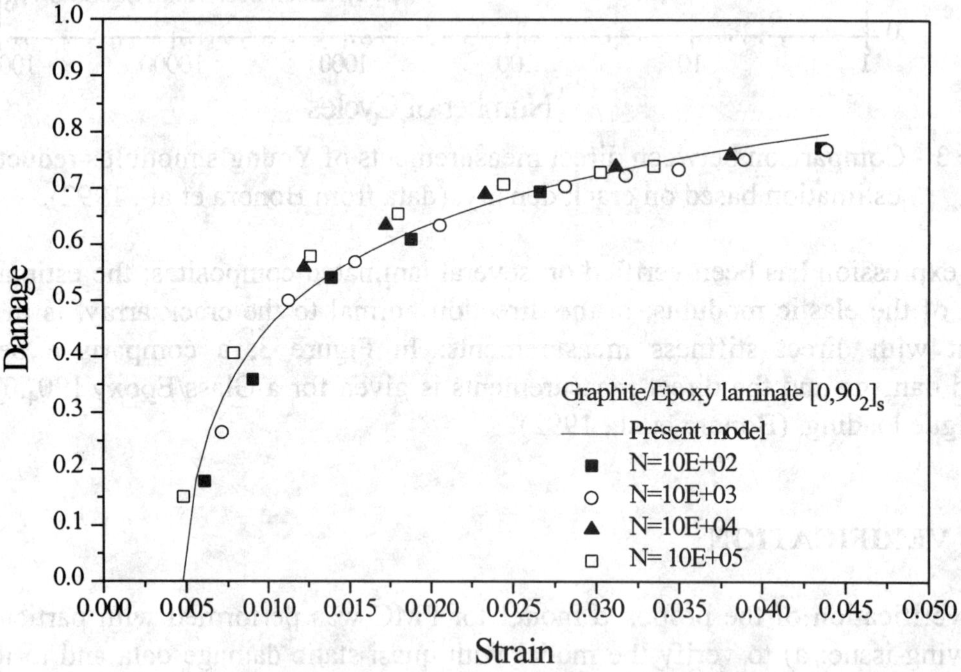

Figure 5 - Damage in Graphite/Epoxy [0,90_2]$_s$ laminate under fatigue loading: comparison between experimental data and the present model, (data from Charewicz and Daniel, 1986)

The statement that fatigue damage data should fall on the static damage curve was verified using data for a glass/epoxy (Bonora et al., 1992) and Graphite/epoxy laminates (Charewicz and Daniel, 1986). In these cases, for both materials, the crack density as a function of the number of cycles was available for several stress levels.

Thus, damage has been estimated using the relation proposed by Talreja and given in Eqn.(4). This information, if available for several stress levels, allows to construct complete damage curve at several number of cycles.

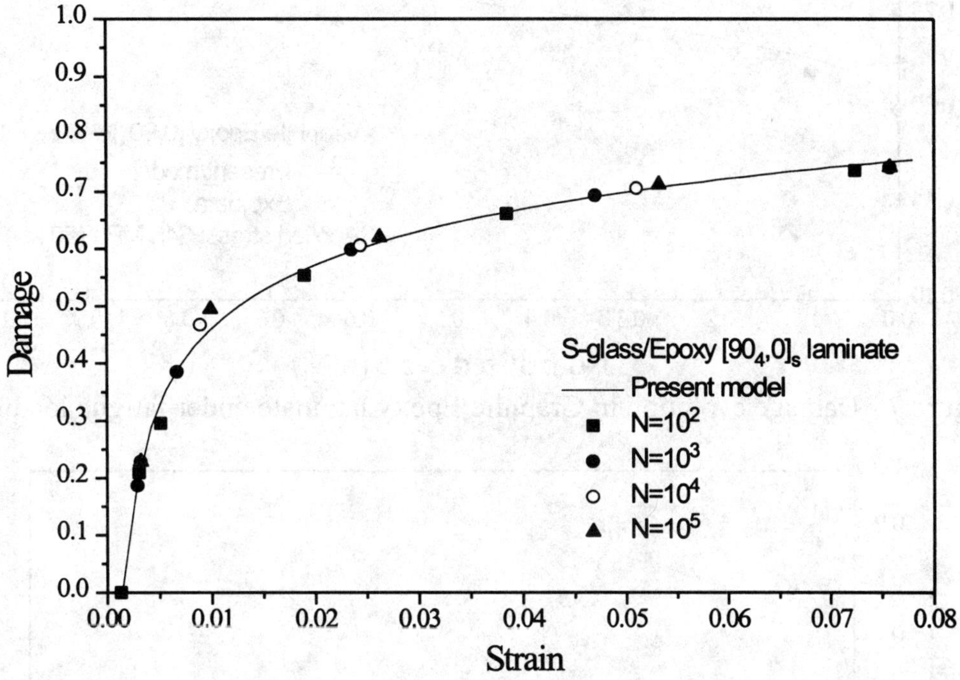

Figure 6 - Damage in S-Glass/Epoxy [90_4,0]$_s$ laminate under fatigue loading: comparison between experimental data and the present model, (data from Bonora et alii, 1992).

In figure 5, the damage curves obtained at 10^2, 10^3, 10^4 and 10^5 cycles for a graphite/epoxy composite are compared together with the model proposed. In figure 6 the same plot is given for a S-Glass/Epoxy laminate. Once the static damage law was assessed and the correlation with fatigue damage was established, the information from static tests were used to predict the evolution of damage with fatigue loading. In order to completely identify the material parameters needed in the fatigue damage evolution law given in Eqn. (3), information relative to the number of cycles that leads to the CDS need to be known through fatigue tests. The parameter that bring the information on the evolution of damage under cyclic loading is the damage exponent β. The authors have found, normalizing the fatigue curves relative to different stress levels respect to N_f, that this parameter seems to be characteristic of the material under investigation and independent from the stress level used in cycling. The β exponent was determined using fatigue data relative to high stress levels. In figure 7 the comparison between the present model and life data for a graphite/epoxy laminate is given. In figure 8, the comparison of the decay of the Young's modulus versus the number of cycles is given for two different stress levels with the predictions using the proposed model.

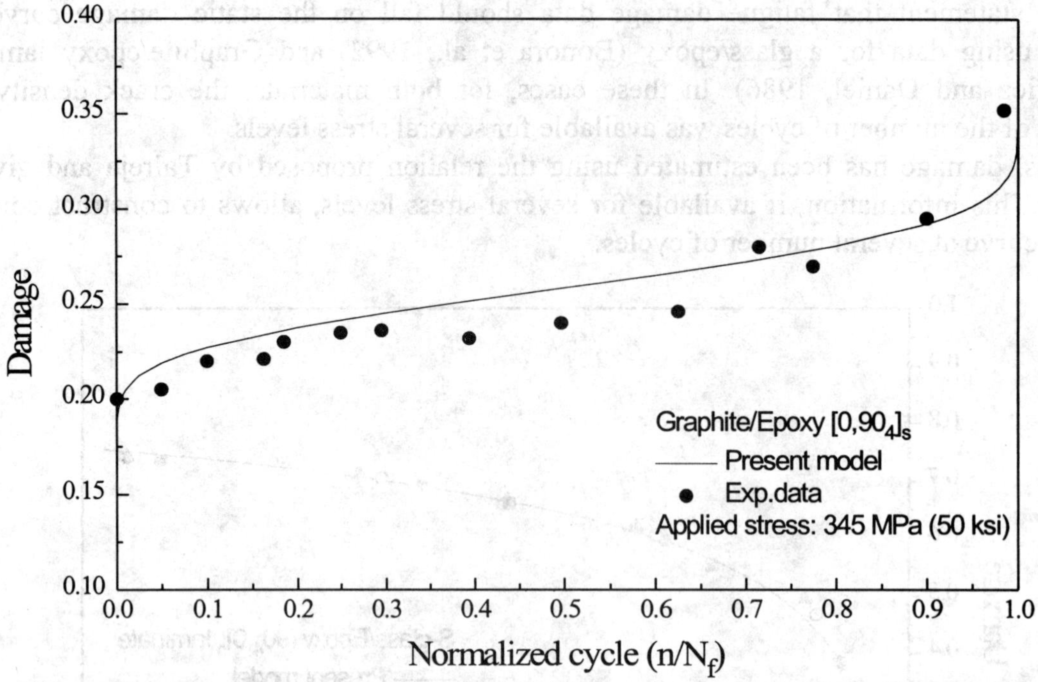

Figure 7 - Damage evolution in Graphite/Epoxy laminate under fatigue loading.

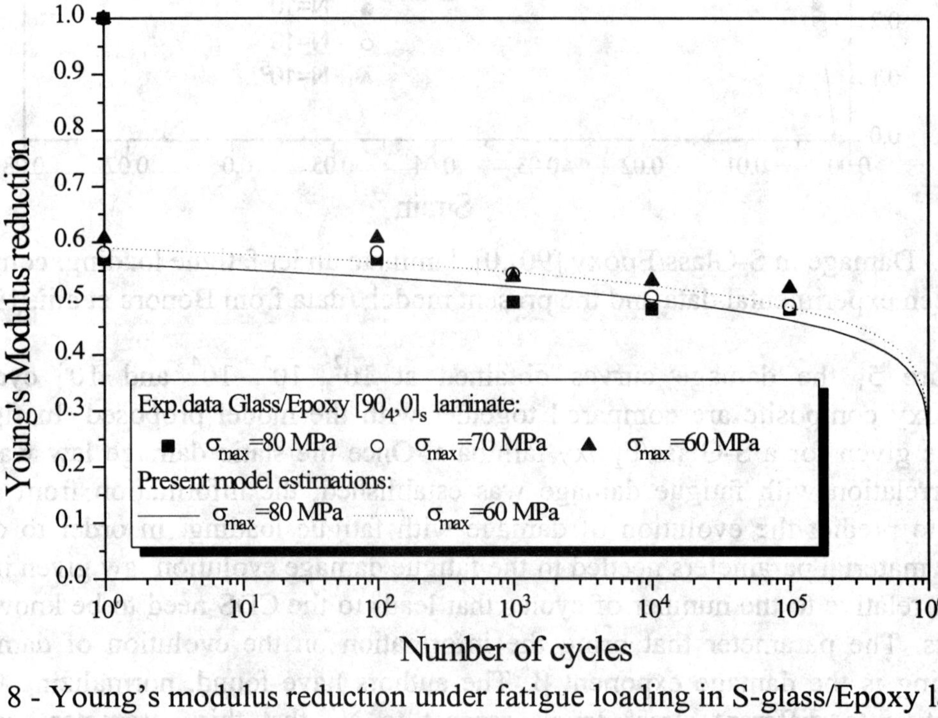

Figure 8 - Young's modulus reduction under fatigue loading in S-glass/Epoxy laminate: comparison between experimental data and the present model.

CONCLUSIONS

Damage is an essential part of composite response subjected to in-service loading and it needs to be taken into account in order to fully understand the material behavior and to predict composite material fatigue life. All brittle composites develop damage as soon as the applied strain in service is sufficient to initiate cracks in the matrix or fail a few fibers. The same concept of failure needs to be reformulated for these materials because PMCs loose their structural integrity and their capability to carry loads well before the specimen or component break up.

In addition, a stress based approach to constitutive modeling including damage for composite material can be misleading due to the "*composite*" nature of these materials for which local effective stresses change from point to point due to the local inhomogeneity, fiber misalignment, local fiber density, etc.

Brittle composite systems, such as many practical PMCs, seem to be very sensitive to the effective applied strain. In this paper, it is shown that stiffness loss versus the imposed strain identifies a characteristic material behavior specifically related to the effective damage process that takes place due to progressive straining. It has been shown that if fatigue loading does not introduce a different damage mechanism with respect to monotonic tensile loading, then the damage curve, as a function of strain, has to be the same for the material under investigation for both static and fatigue loading.

Thus, once the quasi-static damage curve is determined, and if the evolution of an assigned damage state with cycles is known, provided damage mechanisms in static and fatigue loading are similar, fatigue life of PMC can be successfully estimated.

REFERENCES

Allen, D.H., Groves , S.E. and Harris, C.E., 1987a, "A Thermomechanical Constitutive Theory for Elastic Composites with Distributed Damage-Part 1: Theoretical Development", *Int.J.Solids Struct.*,Vol.23, No 8, pp.1301-1318.

Allen, D.H., Groves, S.E. and Harris, C.E., 1987b, "A Thermomechanical Constitutive Theory for Elastic Composites with Distributed Damage-Part 1: Application to Matrix Cracking in Laminated Composites", *Int.J.Solids Struct.*,Vol.23, No 9, pp.1319-1338.

Bonora, N. and Newaz, G., 1997, "Modeling Damage Evolution in a Hybrid Ceramic Matrix Composite Under Static Tensile Load",to appear in ASME, *J. of Engn. Mat. and Technology*.

Bonora, N., Marchetti, M., Milella, P.P., 1992. "Theoretical Forecasting and Experimental Validation of Damage Tolerance and Accumulation in Glass/Epoxy Laminates", in *J. of Reinforced Plastics and Composites*, Vol. II , pp. 56-81.

Charewicz, A. and Daniel, I. M., 1986, " Damage Mechanisms and Accumulation in Graphite/Epoxy Laminates", *Composite Material: Fatigue and Fracture*, ASTM STP 907, H.T. Hann Ed., pp.274-297.

Daniel, M.I. and Lee, J.W., 1990, "Damage Development in Composite Laminates Under Monotonic Loading", *J. of Comp. Tech. and Research*, Vol. 12, No2, pp.98-102.

Karand

Life Prediction in Continuous Fiber Metal Matrix Composites

J. W. FOULK, K. L. E. HELMS AND D. H. ALLEN

ABSTRACT

The authors have in recent years been developing a computational algorithm for predicting damage evolution in continuous fiber metal matrix composites subjected to environmental degradation. Due to multiple nonlinearities, the model utilizes a finite element framework derived from thermodynamics and continuum mechanics. Matrix inelasticity, damage evolution, and degradation of mechanical properties due to oxidation-related effects are also included within the model. Matrix inelasticity is modelled with the Bodner anisotropic hardening viscoplastic model. Damage growth such as surface cracking, matrix cracking, fiber-matrix debonding, and fiber cracking is modelled via the inclusion of cohesive zone elements in the unit cell. The locations of these elements are chosen to correspond with experimentally observed damage. Environmental degradation is accounted for by including degraded material properties which result from oxygen-induced changes in microstructure.

The current paper discusses experimental results that have recently been obtained by the authors for a four-ply unidirectional SCS-6/Ti-β21S composite subjected to elevated temperature isothermal monotonic and fatigue loading cases. The results of these experiments help provide information that is critical for accurate deployment of the damage evolution algorithm.

INTRODUCTION

Over the last three decades metal matrix composites (MMC's) have been considered as candidate materials for a variety of advanced technological applications. MMC's have the potential to provide improved elevated temperature performance as compared to the capabilities of monolithic metals or polymer matrix composites. Unfortunately, continuous fiber MMC's have found only limited applications due to a variety of shortcomings that

---

James W. Foulk, Kayleen L. E. Helms, David H. Allen*, Center for Mechanics of Composites, Texas A&M University, College Station, TX 77843-3141

* Author to whom all correspondence should be directed.

generally lead to low fracture toughness and limited life (Larsen, et al., 1993).

Titanium metal matrix composites (TMC's) have been developed for use in a variety of elevated temperature aerospace applications. These include the National AeroSpace Plane (NASP), Integrated High Performance Turbine Engine Technology (IHPTET), and the High Speed Civil Transport (HSCT). TMC's are being considered because they may provide performance improvements over monolithic metals at elevated temperatures (Ashley, 1993). Unfortunately, in most circumstances, TMC's have failed to reach production primarily because they have been found to have a relatively short life in elevated temperature environments.

RECENT RESEARCH

Due to environmental considerations, Ti-β21S was developed as an oxidation resistant titanium beta alloy for the fabrication of TMC's. (Larsen, et al., 1993). Although Ti-β21S has superior oxidation resistance when compared to other titanium matrix materials, rapid degradation of mechanical properties has been observed at elevated temperatures. This degradation of mechanical properties can be indirectly attributed to oxidation dissolution (Wallace, et al., 1992; Ignatov, et al., 1973). Several experimental research efforts have noted the impact of environmental degradation on fatigue life (Neu, 1995; Ghonem, et al., 1994; Castelli, 1994; Mall and Portner, 1992; Revelos, et al., 1995).

A significant amount of research has been devoted to understanding and predicting the behavior of MMC's. Several micromechanics-based modelling approaches have been suggested (Coker et al., 1993; Neu, 1995; Newaz, 1995; Tamin, et al., 1994). In conjunction with these approaches, several computer codes have been developed to aid in life prediction of MMC's. A full review of many of these codes was recently completed (Kroupa et al., 1994). As a result of much study, it has been determined that essential components for successful modelling include the incorporation of: 1) a constitutive model accounting for matrix viscoplasticity; 2) a damage evolution model to accurately predict energy dissipation due to microfracture; 3) a method of accounting for environmental degradation; and 4) a failure criterion based on maximum allowable fiber stress (Lissenden, et al., 1993; Aboudi, 1990; Allen, et al., 1994; Nicholas, 1995). Detailed literature searches on TMC's are given in two recent publications (Allen, et al., 1995; Nicholas, et al., 1995).

EXPERIMENTAL OBSERVATIONS

In order to better understand the mechanisms of damage evolution in SCS-6/Ti-$^{\beta}$21S $[0]_4$ composite laminates, the authors have undertaken a limited experimental program. The following issues have been studied experimentally: 1) the effects of oxidation on the microstructure of Ti-ß21S; 2) the initiation of surface cracks and typical crack spacings in oxidized specimens; and 3) stress-strain ratchetting and energy. These experimental studies were carried out by thermomechanically testing four ply unidirectional specimens and then performing destructive examinations of the specimens by various techniques including optical and scanning electron microscopy, as well as an electron microprobe.

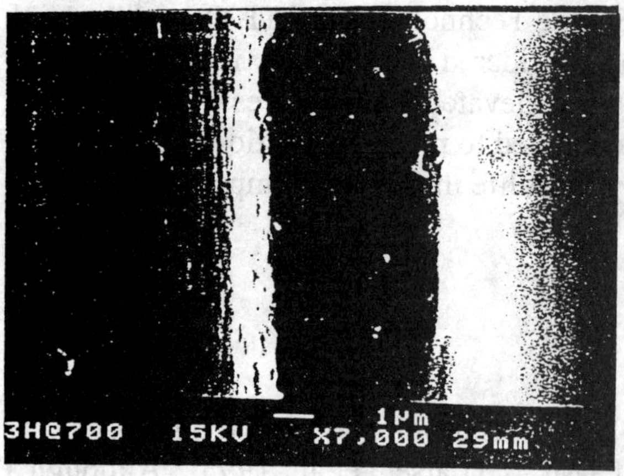

Fig. 1a

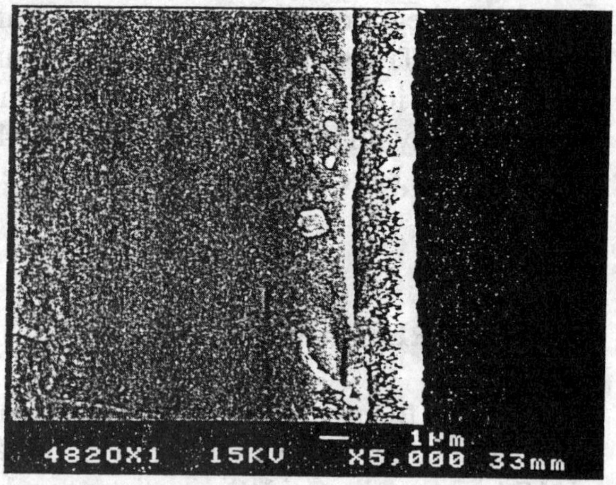

Fig. 1b

Figure 1. Comparison of Oxide Layer Development in a) Ti-15-3 Composite Soaked at 700°C for 3 Hours, and b) Ti-$^{\beta}$21S Composite Soaked at 700°C for 24 Hours

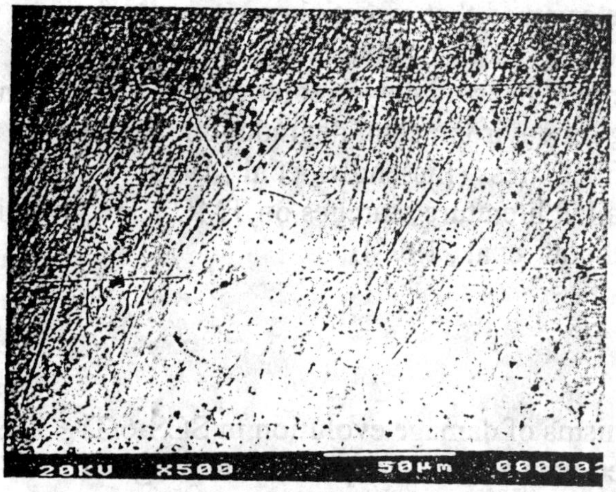

Fig. 2a

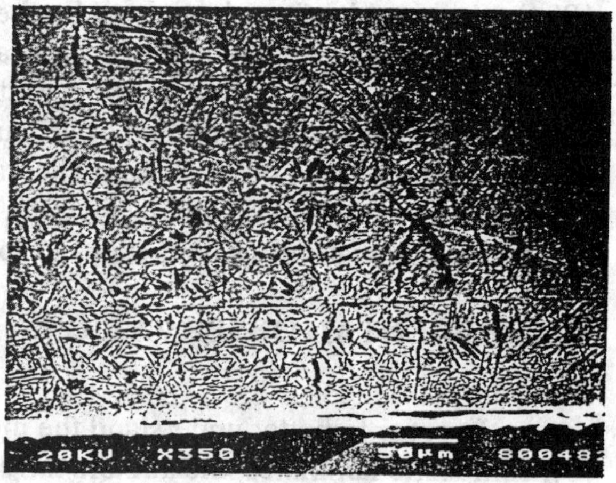

Fig. 2b

Figure 2. Edge View of Ti-$^{\beta}$21S a) Undegraded, and b) Soaked at 700°C for 24 Hours

As shown in Fig. 1a, when Ti-15-3 is heated for 3 hours at 700°C it develops a rutile oxide layer on the surface that is several microns in thickness. By contrast, when Ti-$^{\beta}$21S is heated for 24 hours at 700°C very little surface oxidation is observed, as shown in Fig. 1b.. This apparently led the research community to adopt this material over less oxidation resistant alloys such as Ti-15-3, which oxidizes very rapidly (Foulk, et al., 1996). However, closer comparison of the microstructure in the undegraded specimen, as shown in Fig. 2A, compared to that of the degraded specimen, as shown in Fig. 2b, indicates that a very fine grained alpha phase has formed along the beta grain boundaries, and the volume fraction of alpha phase is a function of the distance from the free surface, thus indicating that oxygen, which serves as an alpha stabilizer, has diffused into the interior of the composite. This contention is supported by the experimentally measured content of the surface layer after the heat treatment, as shown in Fig. 3. Wallace, et al. (1992) have shown that this development of alpha phase inherently embrittles the matrix material. These observations suggest that the fracture toughness of the sembrittled surface layer should be degraded as a function of the distance from the free surface of the coupon.

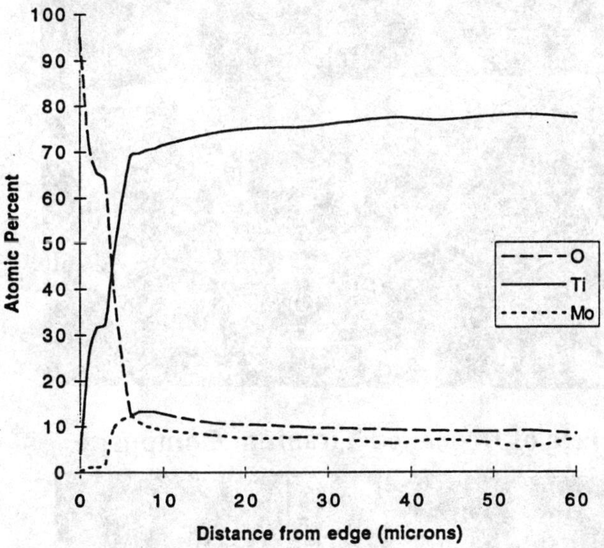

Figure 3. Atomic Percent of Oxygen, Titanium, and Molybdenum as a Function of Distance From the Free Surface

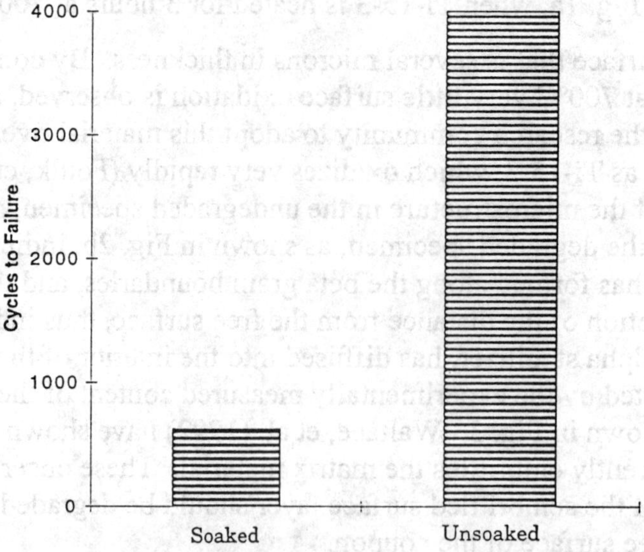

Figure 4. Comparison of Life of Degraded and Undegraded Titanium Composite

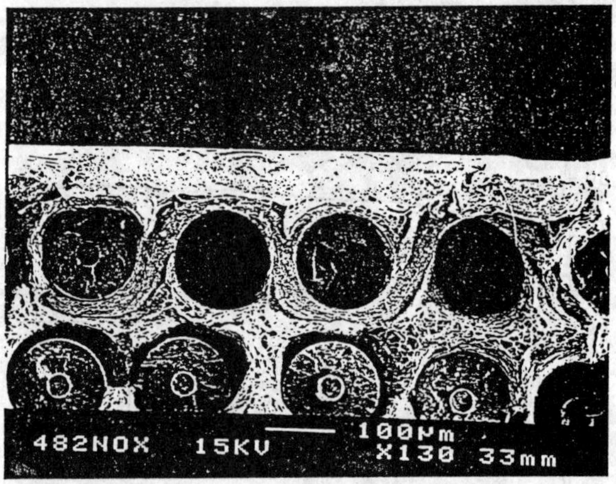

Figure 5. Fracture Surface of Degraded Titanium Composite

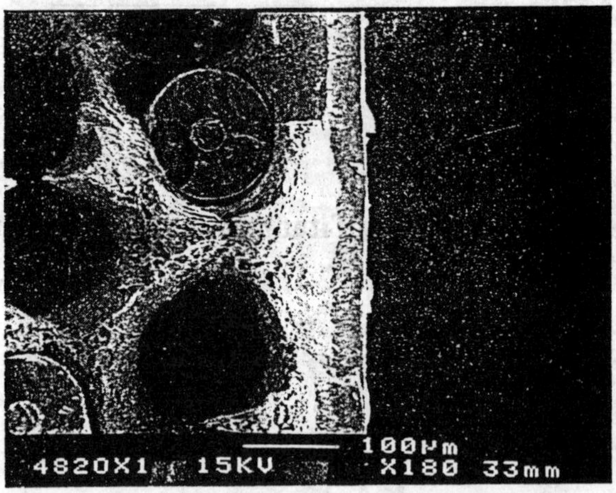

Figure 6. Closeup of Fracture Surface of Degraded Titanium Composite

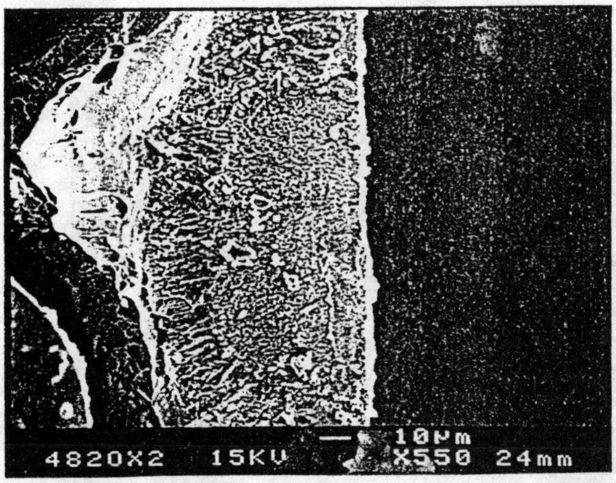

Figure 7. Mist and Velocity Hackles on the Fracture Surface of Degraded Titanium Composite

In order to examine the effects of environmental degradation, SCS-6/Ti-ß21S $[0]_4$ specimens were fatigued in air at 482 °C. One specimen was soaked in air at 700 °C for 24 hours, while the other was not. Figure 4 illustrates the difference in life. In addition, scanning electron microscopy illustrates the extremely brittle nature of the fracture surface for the material tested in air, as shown in Fig. 5. Note also the molybedenum interweave in the top row of fibers. This interweave is placed in the laminate prior to processing in order to minimize fiber swimming. However, it appears to have the adverse effect that it encourages crack growth. A closer inspection of the fracture surface, shown in Fig. 6, indicates that there appear to be three distinct regions of fracture. First, there is a very thin layer near the free surface that appears to coincide with the oxide layer, and this fracture surface appears to be brittle in nature. Second, a much thicker layer that also appears to be brittle in nature occurs under the surface layer. Finally, there is a tertiary layer that appears to be much lighter in color than the secondary layer. As shown in Fig. 7, a still closer inspection of the fracture surface indicates fracture patterns such as mist and velocity hackles in the secondary layer that are indicative of a brittle rather than a ductile material. These observations provide justification for modeling the embrittled region as elastic rather than viscoplastic.

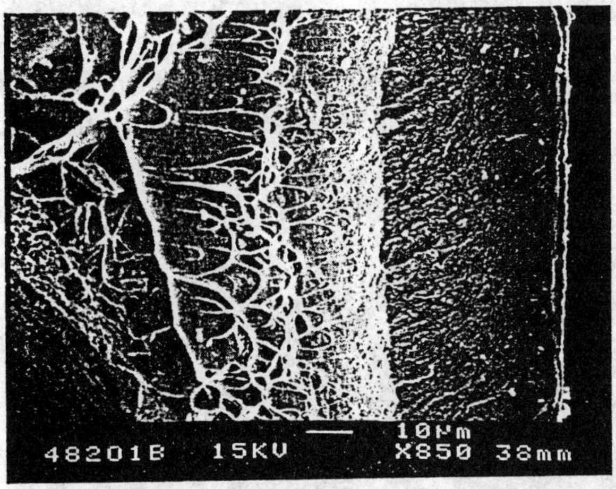

Figure 8. Brittle to Ductile Transition Region on the Fracture Surface of Degraded Titanium Composite

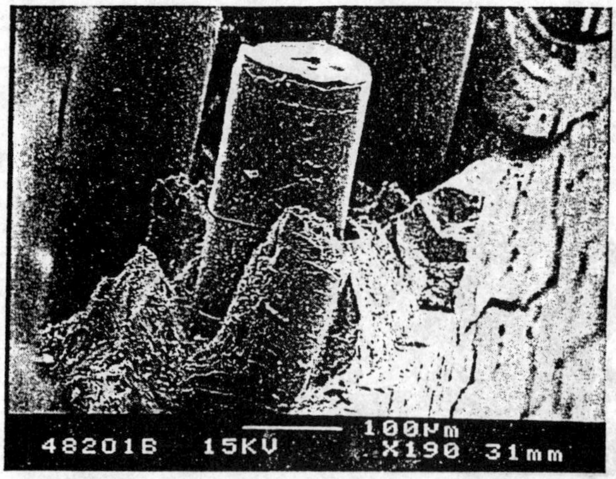

Figure 9. SEM Photo Showing Inclined Fracture Surface of Degraded Titanium Composite

Figure 10. Closeup of Inclined Fracture Surface in Degraded Titanium Composite

By contrast, further inspection of the tertiary layer, as shown in Fig. 8, indicates that the tertiary layer fails by ductile cleavage. This is supported by Figs. 9 and 10, illustrating the fracture surface in the tertiary layer is located on a plane that appears to be about forty-five degrees to the loading (fiber) direction. These observations suggest that a viscoplastic constitutive model should be utilized to analyze the response of the tertiary layer.

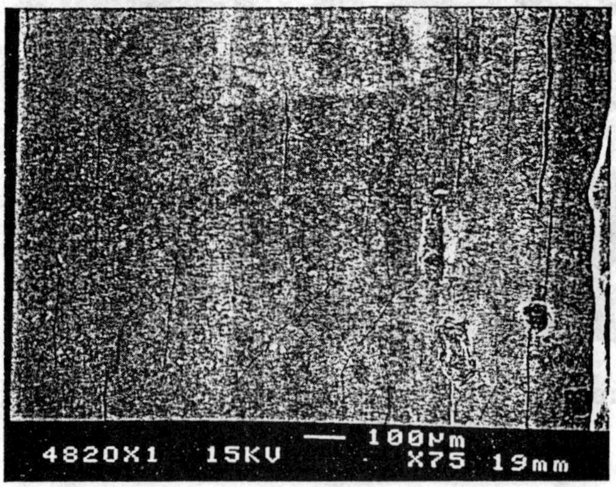

Figure 11. Surface Cracking on the Free Surface of Degraded Titanium Composite

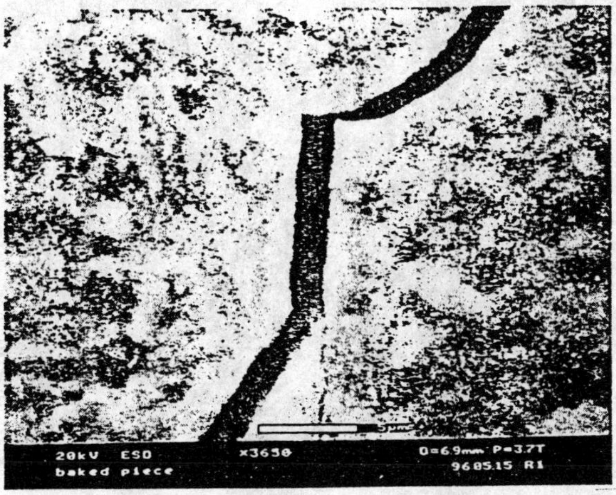

Figure 12. Closeup of Surface Crack in Degraded Titanium Composite

SEM inspection of the free surface of the pre-degraded specimen after failure indicates the presence of surface cracks (See Fig. 11). However, the locations of these cracks do not coincide with the cracks in the secondary layer, as shown in Fig. 12. This suggests that cracks are initiated in the secondary layer, followed by subsequent crack propagation in the surface and tertiary layers. Note that these surface cracks were not observed in the specimen that was not pre-degraded.

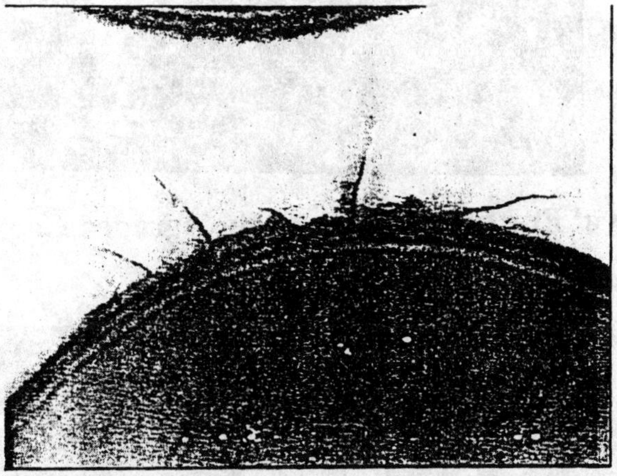

Figure 13. Radial Cracks Emanating From Fiber-Marix Interface in Degraded Titanium Composite

As shown in Fig. 13, radial cracks develop at the interface between the matrix and fibers. This damage mode appears to occur in both degraded and undegraded specimens. Although not shown in the photo, it is also observed that the fiber and matrix debond in both coupons as well.

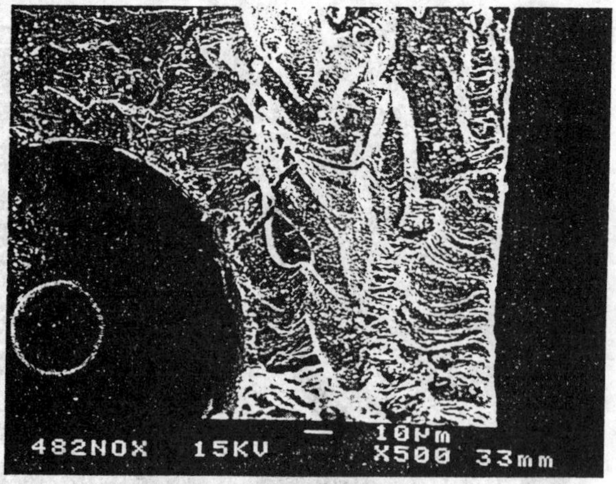

Figure 14. Fracture Surface of Undegraded Titanium Composite

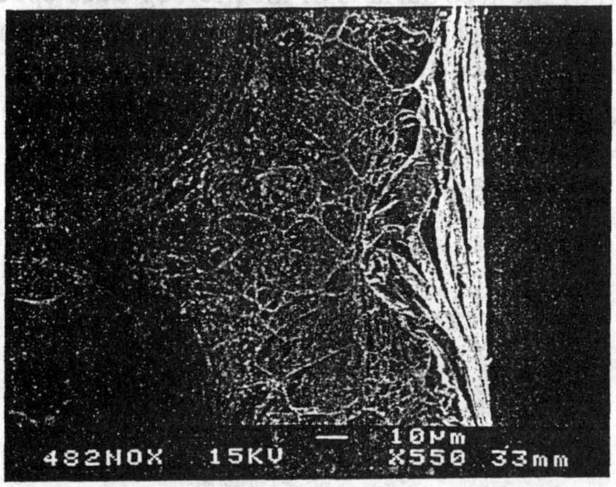

Figure 15. Closeup of Fracture Surface of Undegraded Composite

Figure 16. SEM Photo Showing Inclined Fracture Surface in Undegraded Titanium Composite

The undegraded specimen exhibits a significantly different fracture surface, as shown in Fig. 14. It is apparent that fracture mode in the matrix is everywhere ductile. Note also the presence of a rather large radial crack in the photo. Further evidence of the ductile nature of the fracture mode in the undegraded material is shown in Fig. 15, wherein ductile cleavage is prevalent. Finally, Fig. 16 shows that the failure surface in the undamaged specimen is on an inclined plane to the loading (fiber) direction all the way to the free surface of the coupon. These observations suggest that when modelling the undegraded specimen no gradation of the fracture toughness as a function of depth from the free surface should be included, and that the matrix material should be modelled as viscoplastic throughout the specimen.

CONCLUSION

Failure of the $[0]_4$ composite has been observed to be caused by a succession of separate but related damage events. In the undegraded specimen, damage initiates in the form of matrix radial cracks emanating from the fibers
(Foulk, et al., 1996). Continued loading causes fiber/matrix debonding and small surface cracks. Eventually, fiber cracking will result in the failure of the part. In the degraded specimen, damage is first observed in the form of large substrate cracks which appear along grain boundaries. These cracks form on planes perpendicular to the loading direction. As the cracks propagate through the embrittled layer near the surface, fiber/matrix debonding and crack bridging occurs. As the fiber bridges the crack, the stress in the fiber exceeds the maximum allowable fiber stress and, in turn, causes failure. Although fiber cracking is the final failure mechanism for both the degraded and undegraded specimens, failure occurs much sooner in the oxidized specimens.

The authors are in the process of incorporating the experimentally observed phenomena reported herein in a computationally based damage evolution model currently under development. Some preliminary model predictions based on these findings can be found in (Foulk, et al., 1997a, 1997b).

ACKNOWLEDGEMENT

The authors are grateful to the Air Force Office of Scientific Research for providing funding for this research under grant no. F49620-94-1-0341. We also wish to thank Lockheed-Martin Aerospace Corporation - Fort Worth Division for providing the SCS-6/Ti-β21S specimens used in this research.

REFERENCES

Aboudi, J. 1990. "Micromechanical Prediction of Initial and Subsequent Yield Surfaces of Metal Matrix Composites," *International Journal of Plasticity*, 6:471-484.

Allen, D.H., Eggleston, M.R., and Hurtado, L.D., "Recent Research on Damage Development in SiC/Ti Continuous Fiber Metal Matrix Composites," *Fracture of Composites*, E.A. Armanios, Ed., in *Key Engineering Materials* series, Vols. 120-121, Transtec Publications

Ltd., pp. 257-284, 1996.

Allen, D. H., Jones, R. H., and Boyd, J. G. 1994. "Micromechanical Analysis of a Continuous Fiber Metal Matrix Composite Including the Effects of Matrix Viscoplasticity and Evolving Damage," *Journal of Mechanics and Physics of Solids*, 42(3):502-529.

Ashley, S. 1993. "Boeing 777 Gets a Boost from Titanium," *Mechanical Engineering*, July 1993:60-65.

Castelli, M. G. 1994. "Thermomechanical Fatigue Damage/Failure Mechanisms in SCS-6/Timetal 21S $[0/90]_S$ Composite," submitted to *Composites Engineering*.

Coker, D., Ashbaugh, N. E., and Nicholas, T. 1993. "Analysis of the Thermomechanical Behavior of [0] and [0/90] SCS-6/Timetal®21S Composites," in the proceedings of the *ASME Winter Annual Meeting*.

Foulk, J.W., Allen, D.H., and Helms, K.L.E., "A Model for Predicting the Damage andOxidation Dependent Life of SCS-6/TI-$^{\beta}$21S $[0]_4$ Metal Matrix Composite," to appear, 1997.

Foulk, J.W., Allen, D.H., and Lagoudas, D.C., "A Micromechanical Model for Predicting Fatigue Response of Metal Matrix Composites Subjected to Environmental Degradation," *Proc. IUTAM Symposium on Transformations in Composites*, Cairo, Egypt, 1997.

Foulk, J.W., Helms, K.L.E., and Allen, D.H., "Computational Finite Element Analysis for Predicting the Effects of Environmental Degradation on Life in Metal Matrix Composites," *Damage and Interfacial Debonding in Composites*, Studies in Applied Mechanics, Vol. 44, Elsevier, New York, pp. 29-44, 1996.

Ghonem, H., Wen, Y, and Zheng, D. 1994. "An Interactive Simulation Technique to Determine the Internal Stress States in Fiber Reinforced Metal Matrix Composites," *Materials Science and Engineering*, A177:125-134.

Ignatov, D. V., Model, M. S., Sokyriansky, L. F., and Shinyaev, A. Ya. 1973. *Titanium Science and Technology, Vol. 4*, Jaffee, R. I. and Borte, H. M., eds., Plenum Press, New York.

Kroupa, J. L., Neu, R. W., Nicholas, T., Coker, D., Robertson, D. D., Mall, S. 1994. "A Comparison of Analysis Tools for Predicting the Inelastic Cyclic Response of Cross-Ply Titanium Matrix Composites," in the proceedings of the *ASTM Symposium on Life Prediction Methodology for Titanium Matrix Composites*, Hilton Head, South Carolina, March 22-24, 1994.

Larsen, J. M., Russ, S. M., Jones, J. W. 1993. "Possibilities and Pitfalls in Aerospace Applications of Titanium Matrix Composites," in the proceedings of the *NATO AGARD Conference on Characterization of Fibre Reinforced Titanium Metal Matrix Composites*,

Bordeaux, France, September 1993.

Lissenden, C. J., Herakovich, C. T., and Pindera, M.-J. 1993. "Inelastic Deformation of Metal Matrix Composites," Applied Mechanics Program Report Am-93-03, School of Engineering and Applied Science, University of Virginia, Charlottesville, VA.

Mall, S. and Portner, B. 1992. "Characterization of Fatigue Behavior in Cross-Ply Laminate SCS-6/Ti-15-3 Metal Matrix Composite at Elevated Temperature," *Journal of Engineering Materials and Technology,* 114(3):409-415.

Neu, R. W. 1995. "Thermomechanical Fatigue Damage Mechanism Maps for Metal Matrix Composites," *Thermo-Mechanical Fatigue Behavior of Materials: 2nd Volume, ASTM STP 1263*, M. J. Verrilli and M. G. Castelli, Eds., American Society for Testing and Materials, Philadelphia, PA.

Newaz, G. 1994. "Evaluation and Modeling of Mechanical Response and Strength of MMC in Compression," private correspondence.

Nicholas, T. 1995. "Fatigue Life Prediction in Titanium Matrix Composites," *Journal of Engineering Materials and Technology*, 117:440-447.

Nicholas, T., Zuiker, J., and Pernot, J. 1995. "Characterization of Titanium Matrix Composites," NASP Technical Memorandum 1199, Vol. I.

Revelos, W. C., Jones, J. W., and Dolley, E. J. 1995. "Thermal Fatigue of a SiC/Ti-15 Mo-2.7 Nb-3 Al-0.2 Si Composite," *Characterization of Titanium Matrix Composites: Volume II--Mechanical Behavior and Damage Tolerance of TMC's*, NASP Technical Memorandum 1199, Larsen, J., Sanders, B., Gambone, M. L., and Khobaib, M., eds.

Tamin, M. N., Zheng, D., and Ghonem, H. 1994. "Time-Dependent Behavior of Continuous-Fiber-Reinforced Metal Matrix Composites: Modelling and Applications," submitted to the *Journal of Composites Technology and Research*.

Wallace, T. A., Wiedemann, K. E., and Clark, R. K. 1992. "Oxidation Characteristics of Beta-21S in Air in the Temperature Range 600 to 800°C", NASA Technical Memorandum 104217.

FIGURE TITLES

Figure 1. Comparison of Oxide Layer Development in a) Ti-15-3 Composite Soaked at 700°C for 3 Hours, and b) Ti-$^{\beta}$21S Composite Soaked at 700°C for 24 Hours

Figure 2. Edge View of Ti-$^{\beta}$21S a) Undegraded, and b) Soaked at 700°C for 24 Hours

Figure 3. Atomic Percent of Oxygen, Titanium, and Molybdenum as a Function of Distance From the Free Surface

Figure 4. Comparison of Life of Degraded and Undegraded Titanium Composite

Figure 5. Fracture Surface of Degraded Titanium Composite

Figure 6. Closeup of Fracture Surface of Degraded Titanium Composite

Figure 7. Mist and Velocity Hackles on the Fracture Surface of Degraded Titanium Composite

Figure 8. Brittle to Ductile Transition Region on the Fracture Surface of Degraded Titanium Composite

Figure 9. SEM Photo Showing Inclined Fracture Surface of Degraded Titanium Composite

Figure 10. Closeup of Inclined Fracture Surface in Degraded Titanium Composite

Figure 11. Surface Cracking on the Free Surface of Degraded Titanium Composite

Figure 12. Closeup of Surface Crack in Degraded Titanium Composite

Figure 13. Radial Cracks Emanating From Fiber-Marix Interface in Degraded Titanium Composite

Figure 14. Fracture Surface of Undegraded Titanium Composite

Figure 15. Closeup of Fracture Surface of Undegraded Composite

Figure 16. SEM Photo Showing Inclined Fracture Surface in Undegraded Titanium Composite

History Dependent Load Effects in Composites

F. W. BRUST, R. MOHAN AND P. KRISHNASWAMY

ABSTRACT

There are a number of features associated with history dependent loading of composites that have been observed in the laboratory, including increased failure rates under high temperature cyclic loading as well as creep ratcheting. In this paper, the effect of damage development in metal matrix composites (MMC) subjected to history dependent loading is investigated through numerical simulations. The effects of void growth in the MMC (near the fiber) are investigated as well. The results of the analyses shed some light on the nature of damage development and the occurrence of creep ratcheting in MMCs. Some implications of the simulations are discussed.

INTRODUCTION

In order to successfully utilize titanium-based metal matrix composites (MMC) for high temperature structural applications, a through understanding of the mechanisms and mechanics of deformation and damage in these composites under cyclic loading is needed. This is because these materials are intended for use in structural applications which experience repeated mechanical and in many instances repeated thermal loading as well. Such combined loading situations induce complex interactions of the constituents that affect the overall deformation and damage response of titanium matrix MMCs.

Detailed investigations of the deformation and failure of titanium matrix MMCs under isothermal cyclic mechanical loading as well as thermo-mechanical loading conditions have been conducted by Gayda and Gabb (1991), Russ *et al* (1991), Nicholas *et al* (1993), Majumdar and Newaz (1995), and Newaz and Majumdar (1995), among others. Under isothermal slow loading conditions the mechanisms of deformation and failure and thus fatigue lives depend strongly on the maximum-to-minimum load ratio (R-ratio). Similarly, the characteristics of deformation and failure are significantly affected by the phase difference between thermal and mechanical loadings. Owing to the high temperature and slow loading conditions creep deformation of the Ti-based matrix appears to be an

---

Battelle Memorial Institute
Columbus, OH 43201

important consideration. The creep behavior of the matrix is strongly influenced by the cyclic loading conditions.

The purpose of this paper is to report the results of some numerical simulations of a titanium matrix-based composite subjected to constant load creep as well cyclic creep conditions. In the case of cyclic loading conditions, we consider both isothermal and thermo-mechanical conditions. Qualitative and quantitative comparisons are made between the present results and the detailed experimental investigations of Majumdar and Newaz (1995) and Newaz and Majumdar (1995).

NUMERICAL APPROACH

The unit cell model of the composite considered in this work is similar to that studied by Brust *et al.* (1993). The model is illustrated in Figure 1. Note the load sequence required in Figure 1. The HIP fabrication pressure must be accounted for in order to predict off-axis response accurately. The HIP pressure on the thin sections (1.99 mm thick) is neglected. The finite element model itself is generalized plane strain, with temperature-dependent inelastic properties for the matrix (plasticity, steady state creep, viscoplasticity, etc.) and temperature dependent elastic properties of the fiber. The interface is modeled with interface elements, which can account for frictional contact between the fiber and the matrix after debond is predicted locally. Additional details of the modeling procedure may be found in Brust *et al.* (1993). Although the 90-degree and the 0-degree loading cases are indicated in Figure 1, in this paper we consider only the 0-degree loading case. For this later case, the load is applied in the generalized plane strain direction, i. e. out of plane of paper. This two-dimensional analysis performs quite well since interface friction has a very small effect for the 0-degree case.

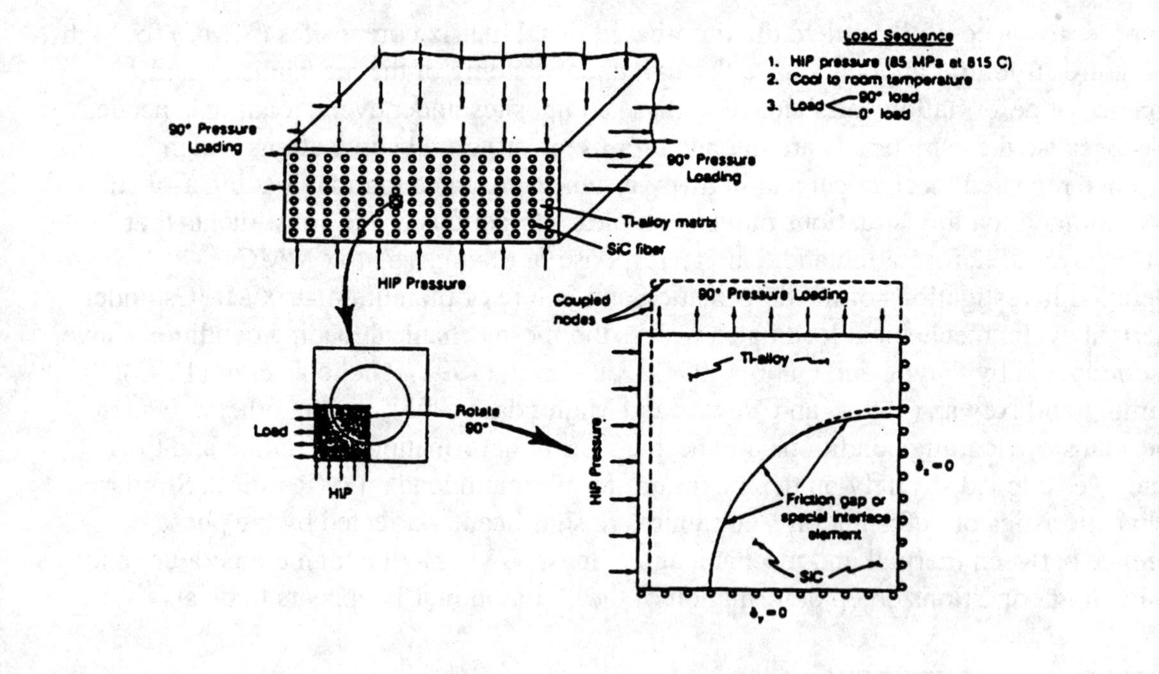

Figure 1. Unit Cell used in Numerical Simulations of Titanium MMC.

RESULTS AND DISCUSSIONS

In this section, the results of the simulations of a unidirectionally reinforced titanium matrix composite subjected to high temperature loading conditions are presented. The volume fraction of the SiC fiber is 0.35. The temperature dependent elastic properties of the SiC fiber are identical to that used by Brust *et al.* (1993) and Newaz and Majumdar (1995). The titanium matrix is considered to be governed by a nonlinear viscous law. In the case of constant load creep the constitutive law used for the matrix is due to Murakami and Ohno (1982). This law, though more complex, does not require abundant experimental tests to obtain the necessary constants. The material constants may be obtained from monotonic creep tests. However, in the case of cyclic loading we consider a simpler primary creep law, primarily due to the lack of creep data at various temperatures. Additionally, the main purpose of the simulations of cyclic loading was to understand the fundamental mechanisms that lead to failure. In the following the results of the simulations are discussed.

CONSTANT LOAD CREEP

The 0-degree specimen was loaded to 800 MPa parallel to the fiber direction and held at constant load for four hours. For the creep problem, the load sequence is as shown in Figure 1, except that after Step 2, a uniform temperature of 538 C (from room temperature) is modeled before applying the load in Step 3. It is interesting to observe the stresses in the matrix and fiber as creep evolves. Figure 2a shows the residual stress before load is applied in the out-of-plane direction. Figure 2b shows the stress after a load of 800 MPa is applied before creep occurs. At this stage, the average fiber stress is about 1550 MPa and the average matrix stress is about 450 MPa. Figure 2c shows the stresses after about one hour of creep hold. Here the stress in the matrix has relaxed to an average value of less than 50 MPa, while the fiber stress has increased to an average value of greater than 2200 MPa. Finally, Figure 2d shows stresses after five hours of hold time. Here the matrix stress is less than 10 MPa while the fiber stress is greater than 2300 MPa.

Here it is seen that, for this combination of load, temperature, and fiber-volume fraction, most of the creep has been exhausted from the matrix, and the fibers hold most of the load. Since the fiber failure stress is in excess of 3000 MPa (at this temperature), we do not expect a creep rupture in this case, and failure did not occur experimentally.

The accuracy of this analysis may be seen in Figure 3 where the average fiber-direction creep strain is compared to the experimental data. The predicted and experimentally observed longitudinal creep strain is illustrated in Figure 3a. The predicted results compare well with experimental data. The specimen was then unloaded. A comparison of the width direction total strain versus longitudinal (applied) is shown in Figure 3b. As mentioned above the comparisons between experimental measurements and analytical predictions for off-axis strains are poor if the HIP pressure is not included in the model (Load sequence 1 in Figure 1). Here good comparison is obtained. Note that very little difference is observed between assuming a perfect bond or permitting fiber-matrix interface debonding to occur for loading parallel to the fiber direction.

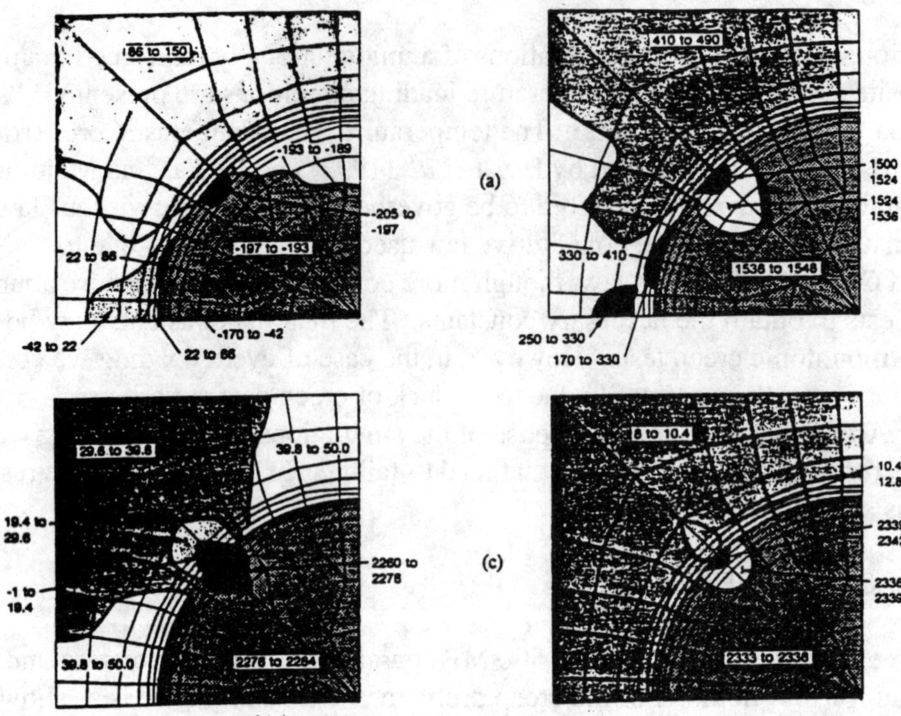

Figure 2. Axial stress in MMC at 538 C: (a) residual stress, no load, (b) at 800 MPa, instantaneous with load application, (c) at 800 MPa after 1 hour, (d) at 800 MPa after 5 hours. The numbers represent stresses in MPa.

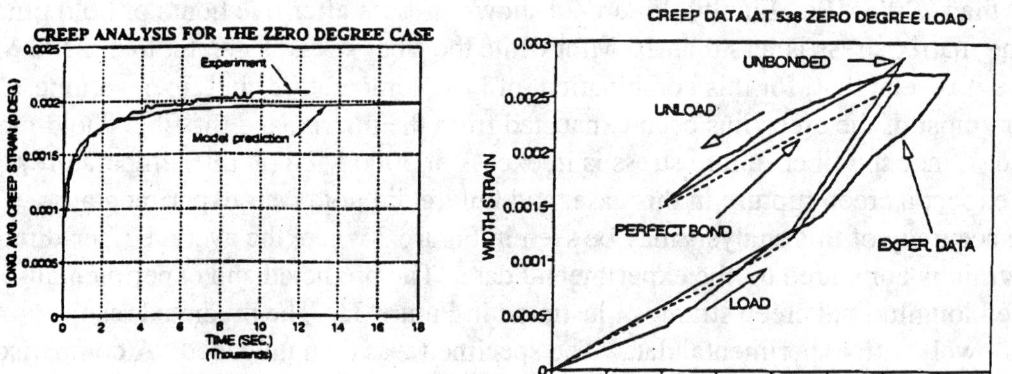

Figure 3. Creep strains: Predictions compared with experimental data. These are average (global) strains.

ISOTHERMAL CYCLIC LOADING

In this section, we consider the case of slow cyclic loading (0-degree loading) of the MMC under isothermal conditions. Majumdar and Newaz (1995) conducted detailed investigations of deformation and damage in titanium matrix MMCs loaded under isothermal fatigue conditions. They considered slow rates of loading and examined the response of the composite at several different maximum-to-minimum load ratios (R-ratio). At high temperatures, they observed creep ratcheting, for R-ratio above 0.0, while little or no ratcheting for R-ratio of -1.0. Another significant finding of Majumdar and Newaz (1995) was that at elevated temperatures matrix cracking was much more predominant mechanism of damage for R= -1 (Figure 12 in Majumdar and Newaz, 1995) compared with the predominance of fiber cracking for R = 0.1 (Figure 13 in Majumdar and Newaz, 1995). The predominance of fiber cracking for R = 0.0 was reported by them.

As mentioned previously the matrix is assumed to obey a primary creep law ant the fiber is considered to be elastic. Owing to the assumed matrix behavior, the rates of loading considered in the analysis are much slower than those considered by Majumdar and Newaz (1995). However, the results should shed some light on the deformation behavior of the MMC under high temperature cyclic loading conditions. Additional details of the analysis will be published elsewhere (Mohan and Brust, 1997).

The case examined here is the effect of hold time on the deformation of the MMC for an R-ratio of 0.0, under load-controlled loading. Figure 4 shows the far field average strain as a function of time in the first few cycles. Although the increase in the minimum and maximum strain levels is small, there is a definite trend toward ratcheting behavior, consistent with the findings of Majumdar and Newaz (1995). More conclusive evidence can be shown when these calculations are performed for large number of cycles. However, this would involve long hours of compute CPU time. Not surprisingly, the ratcheting behavior is likely to be more pronounced for longer hold times.

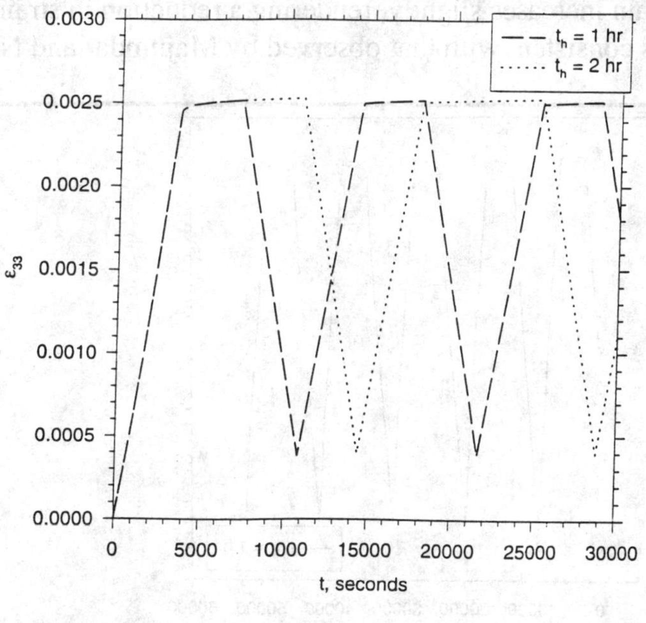

Figure 4. Variation of far field strain with time for R=0.0 under isothermal cyclic loading.

The temporal variation of axial stress in the fiber and the matrix in the vicinity of the interface between them is shown in Figure 5. It can be seen that the matrix mean stress decreases slowly as a result of time-dependent relaxation, thereby causing the fibers to assume increased stress. This appears to the primary reason for the ratcheting behavior at R = 0, supporting the conclusion of Gayda *et al.* (1990).

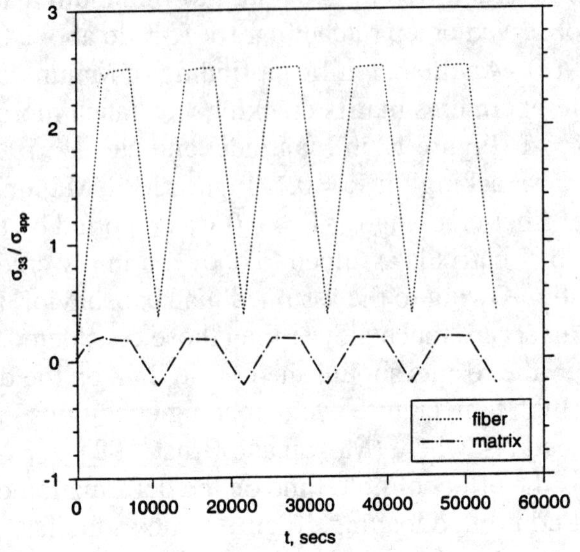

Figure 5. Normalized axial stress versus time for R=0.0 under isothermal cyclic loading.

The other case considered here involves fully reversed loading (R = -1.0). The variation of far field strain with time for R = - 1.0 is shown in Figure 6. Interestingly, little evidence of ratcheting is observed. In fact, the maximum strain decreases slightly over time while the minimum strain increases slightly, rendering a reduction in strain range over time. Again this behavior is consistent with that observed by Majumdar and Newaz (1995).

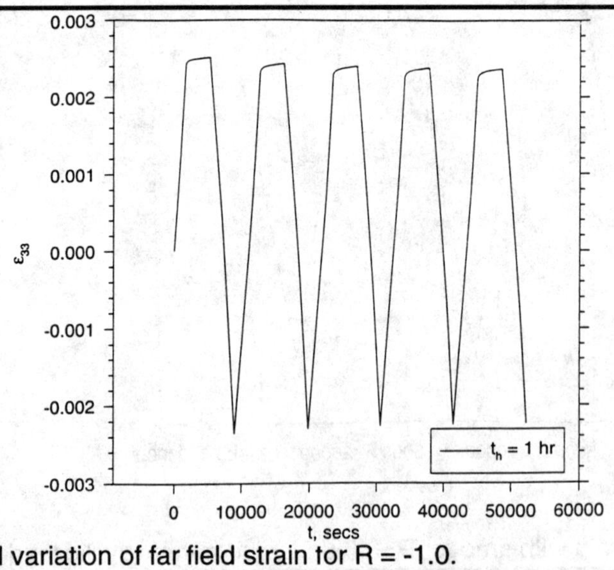

Figure 6. Temporal variation of far field strain for R = -1.0.

Figure 7 shows the temporal variation of axial stress in the matrix and fiber for R = -1.0. In contrast to the behavior at R = 0.0, the maximum stress in the matrix increases while it decreases in the fiber for the case of R = - 1.0. This result offers some explanation for the predominance of matrix cracking and less evidence of fiber cracking observed for R = -1.0 by Majumdar and Newaz (1995).

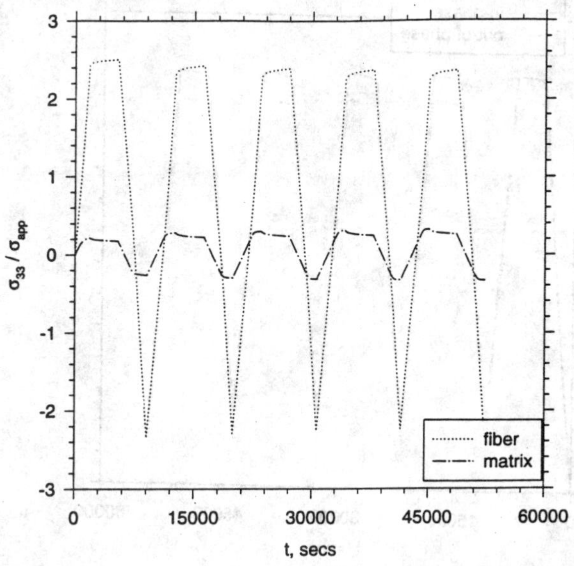

Figure 7. Normalized axial stress versus time for R=-1.0 under isothermal cyclic loading

THERMO-MECHANICAL CYCLIC LOADING

It is well known that MMC component life can be limited by damage caused by thermo-mechanical fatigue loading at high temperatures (Russ *et al.*, 1991; Mirdamadi and Johnson, W. S., 1991; Castelli *et al.*, 1992; Newaz and Majumdar, 1995). Newaz and Majumdar conducted detailed studies on in-phase thermo-mechanical loading of titanium matrix composite and concluded that distributed and banded fiber cracking appears to be the predominant mechanism of failure. The investigations of Castelli *et al.*(1992) indicated that in-phase thermo-mechanical loading generally leads to shorter lifetimes than corresponding out-of-phase loading. On the other hand, Mirdamadi and Johnson suggest that the thermo-mechanical lives for these composites for out-of-phase and in-phase loadings depend on a single correlating parameter, namely, the fiber stress in the unidirectional ply. It must be mentioned that detailed mechanism-based study of out-of-phase loading effects on damage is not available.

Results are presented here for in-phase as well as out-of-phase loadings of a 0-degree titanium matrix composite. A load ratio of 0.0 with a hold time of 1 hour and a temperature range of 170 C (480 C to 650 C) is considered. The variation of normalized axial stresses in the fiber and the matrix is shown in Figures 8 and 9, respectively. It can be seen that the maximum axial stress in the fiber is higher for in-phase loading than that for out-of-phase loading. In addition, the axial stress in the fiber is decreasing slightly with time for both the loading cases. The maximum stress in the matrix is higher for out-of-phase loading, as one would expect, than that for in-phase loading (see Figure 9). Interestingly, the rate of increase in matrix maximum stress is higher for the out-of-phase loading case than the corresponding increase for the in-phase loading case.

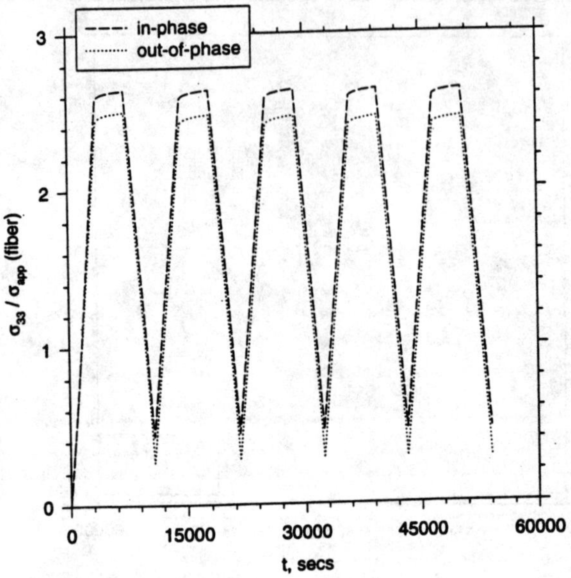

Figure 8. Normalized axial stress versus time in fiber under thermo-mechanical cyclic loading.

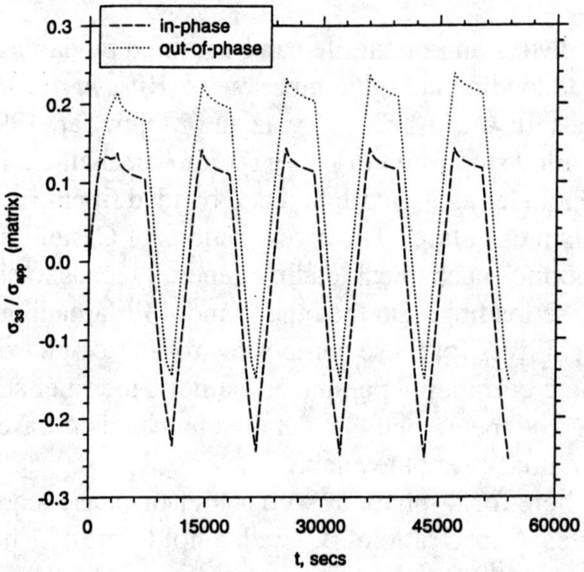

Figure 9. Normalized axial stress versus time in matrix under thermo-mechanical cyclic loading.

The results shown in Figures 8 and 9 suggest that the predominant mechanism of damage for in-phase loading may be due to fiber failure, while for out-of-phase loading may be due to matrix cracking. The former inference is supported by the findings of Newaz and Majumdar (1995). The results cannot offer conclusive evidence as to why the in-phase loading can be more damaging than the out-of-phase loading as evidenced by Castelli et al. (1992). It may be possible that the higher rate of decrease in the axial stress in the fiber for out-of-phase loading mitigates fiber cracking compared with the in-phase loading case. It was seen earlier that for a load ratio of 0.0 ratcheting was observed under isothermal conditions. Interestingly, for in-phase as well as out-of-phase thermo-mechanical loading (with R = 0.0) no ratcheting was observed, as evidenced in Figure 10. Notice a slight decrease in the strain range with time for both types of thermo-mechanical loading.

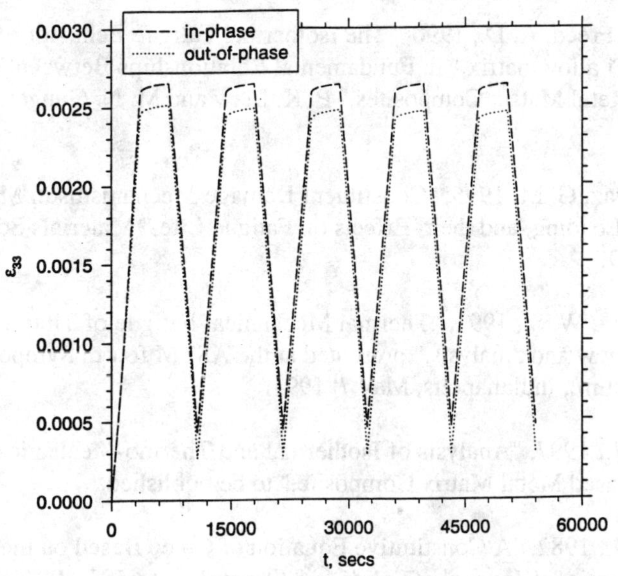

Figure 10. Temporal variation of average axial strain of MMC under thermo-mechanical cyclic loading.

CONCLUSIONS

Micromechanical analysis of history dependent high temperature loading of MMCs has been conducted. The temporal deformation of MMC under constant load as well as varying load histories have been studied. The varying load histories involved load cycling under isothermal as well as thermo-mechanical conditions. The analytical results for constant load creep agree well experimental measurements. The temporal variation of stresses and strains under cyclic loading shed light on the mechanisms of failure observed by Majumdar and Newaz (1995) for isothermal loading and by Newaz and Majumdar (1995) for thermo-mechanical loading.

REFERENCES

Brust, F. W., Newaz, G. M., and Majumdar, B. S., 1993, "A Study of Tension, Compression and Creep Deformation of Metal Matrix Composites," in The Proceedings of the American Society for Composites, 8th Technical Conference, 734-741.

Castelli, M. G., Barlotta, P. A., and Ellis, J. R., 1992,"Thermo-Mechanical Testing of High Temperature Composites: Thermo-Mechanical Fatigue Behavior of SiC(SCS-6)/Ti-15-3," in Composite Materials: Testing and Design (Tenth Volume), G. C. Crimes, ed., ASTM STP 1120: 70-86.

Gayda, J., and Gabb, T. P., 1991, "Isothermal Fatigue Behavior of a $[90]_8$ SiC-Ti-15-3 composite at 426 C," NASA Technical Memorandum No. 103686.

Gayda, J., Gabb, T. P., and Freed, A. D., 1990, "The Isothermal Fatigue Behavior of a Unidirectional SiC/Ti composite and the Ti alloy matrix," in Fundamental Relationships Between Microstructure and Mechanical Properties of Metal Matrix Composites," P. K. Liaw and M. N. Gunger, The Metallurgical Society, 497-514.

Majumdar , B. S., and Newaz, G. M, 1995, "Constituent Damage Mechanisms in Metal Matrix Composites under Fatigue Loading, and their Effects on Fatigue Life," Materials Science and Engineering, A200:114-129.

Mirdamadi, M., and Johnson, W. S., 1991, "Thermal Mechanical Fatigue of Titanium Metal Matrix Composites: Literature Review and Analysis," presented at the ASTM fourth Symposium on Composite Materials: Fatigue and Fracture, Indianapolis, May 7, 1991.

Mohan, R., and Brust, F. W., 1997, "Analysis of Isothermal and Thermo-Mechanical Fatigue Deformation of Titanium-based Metal Matrix Composite," to be published.

Murakami, S., and Ohno, N., 1982, "A Constitutive Equation of Creep Based on the Concept of a Creep Hardening Surface," *International Journal of Solids and Structures*, 18:597-609.

Newaz, G. M., and Majumdar , B. S., 1995, "In-phase Thermo-Mechanical Fatigue Mechanisms in a Unidirectional SCS-6/Ti 15-3 MMC," NASA Contractor Report No. NAS3 - 26822

Nicholas, T., Russ, S., Schehl, N., and Cheny, 1993,"Frequency and Stress-Ratio Effects on Fatigue of Unidirectional SCS-6/Ti-24Al-11Nb," in J. P. Bailon and Dickson, J. I., (eds.), Fatigue 93, II:995-1000, Engineering Materials Advisory Services Ltd., West Midlands, UK.

Russ, S. M., Nicholas, T., Bates, M., and Mall, S., 1991, "Thermomechanical Fatigue of SiC/Ti-24Al-11Nb Metal Matrix Composite," in Failure Mechanisms in High Temperature Composite Materials, G. K. Haritos, G. Newaz and S. Mall (eds.) AMD- Vol. 22 and AD-Vol. 22, 37-43, ASME, New York.

Investigation of the Fatigue Behavior of SCS-Ultra/Ti-6-4

J. R. CALCATERRA[1] AND S. MALL[2]

ABSTRACT

Titanium Matrix Composites (TMCs) have been the subject of numerous studies over the past two decades. The reinforcement of choice for these composites has normally been silicon carbide, in either continuous or particulate form. SCS-Ultra is a new fiber, developed for use in TMCs, with expected properties that are stronger and stiffer than current reinforcements. The focus of this study is to evaluate the fatigue behavior of SCS-Ultra/Ti-6-4 at 427°C. The improvement in ultimate strength over SCS-6/Ti-6-4 was less than expected and the fatigue lives of the new material were much shorter at all stress levels. Post-test evaluation with a scanning electron microscope revealed several processing flaws in the new composites. Because of these flaws, the SCS-Ultra fibers failed at a lower composite stress than predicted and there were numerous crack initiation sites to greatly accelerate the progression of damage in fatigue tests. Despite the poor fatigue performance of the SCS-Ultra/Ti-6-4 specimens, this material shows promise for greatly increased ultimate strength. Both the stress and strain to failure are greater than SCS-6/Ti-6-4, even with the previously mentioned processing defects.

INTRODUCTION

Titanium Matrix Composites (TMCs) have been the subject of numerous fatigue studies over the past decade [1]. In the majority of these studies, Silicon Carbide (SiC) has been the reinforcement of choice for continuous fiber composites. Typical SiC fibers are rather large, with diameters ranging from 100 to 140 μm, and have ultimate tensile strengths near 3500 MPa. A new type of silicon carbide fiber has recently been developed which promises to improve the strength and stiffness of TMCs. This fiber has been designated SCS-Ultra. The objective of the current research is to determine basic material properties, such as fatigue life and static response, for composites with the new fiber. The study will also compare the fatigue behavior of the SCS-Ultra/Ti composites to more conventional TMCs.

1. Post-Doctoral Research Associate, Department of Aeronautics and Astronautics
2. Professor and Head, Department of Aeronautics and Astronautics

MATERIAL

This material used in this study was SCS-Ultra/Ti-6Al-4V. SCS-Ultra is a new silicon carbide fiber manufactured by Textron Specialty Materials, Inc. This fiber was developed to have greater stiffness and strength than SCS-6, which is currently one of the strongest SiC fibers available. The increase in strength is due to a finer SiC grain size in the SCS-Ultra fibers. The nominal modulus of the SCS-Ultra fiber is reported to be 410 GPa (60.5 Msi) with a average ultimate strength of 6480 MPa (940 ksi). This can be compared to a modulus of 390 GPa (56.5 Msi) and a strength of 3500 MPa (507 ksi) for SCS-6. The strain to failure shows a similar increase, from 0.9% to 1.55%. The manufacturing process of these fibers is similar to SCS-6. They are fabricated through chemical vapor deposition of silicon carbide onto a 32 μm carbon substrate. The resulting fiber diameter is very large and can vary between 140 μm (5.6 mil) and 178 μm (7 mil). The SCS-Ultra fibers used in the current research have a nominal diameter of 178 μm (7 mil). Other than the increase in size, these fibers were very similar in appearance to SCS-6. When used in a titanium matrix, SCS-6 fibers are coated with a carbonaceous outer layer to protect the fiber from detrimental reactions caused during laminate manufacture. However, this coating would not accommodate the larger strain to failure of the SCS-Ultra. Because of this, a new coating system was developed to not only protects the fiber, but also permits greater elongation. Ti-6Al-4V, hereafter referred to as Ti-6-4, is one of the most popular $\alpha + \beta$ titanium alloys currently in use. It is useful up to 427°C (800°F), is weldable and machinable, and has outstanding creep properties.

SCS-Ultra/Ti-6-4 laminates were produced using the foil-fiber-foil method. In this method, mats of fibers are sandwiched between foils of the matrix material. The mats can be oriented in any direction to produce the final composite lay-up. A titanium-niobium (Ti-Nb) wire was used to ensure proper spacing and alignment of the fiber mats. The composites are produced by stacking the desired number of foil and fiber layers and consolidating the laminate with a hot isostatic pressing cycle. The cross-weave material is absorbed into the matrix during processing. The current material had a foil thickness of 100 μm (4 mils) and an orientation of $[0]_4$. This resulted in a very thin material with a fiber volume fraction (V_f) of 0.42.

Six dogbone specimens were obtained from one 89 x 133 mm (3.5 x 5.25 in) plate. The width and length of the grip area were 12.7 mm (0.5 in) and 25.4 mm (1 in) respectively. The gage section width was 7 mm (0.28 in) and the transition radius was 152.4 mm (6 in). Composite thickness was nominally 0.71 mm (0.028 in). Specimens were machined using wire electric-discharge machining. Visual inspection of the plate and specimens revealed large amounts of bending in the coupons. Non-destructive evaluation of the material revealed that there were numerous twisted and crossed fibers in the composite prior to testing.

EXPERIMENTAL PROCEDURES

One tensile strength and four fatigue tests were conducted in this investigation. All tests were completed at an elevated temperature of 427°C (800°F) in a laboratory air environment. Test temperature was maintained with radiant heaters that were controlled by two K-type thermocouples. The thermocouples were spot welded to a thin stainless steel plate which was then bolted to the specimen. The plate was positioned so that the thermocouples were located on either end of the 12.7 mm (0.5 in) gage section. Controllers kept the temperature to within 1°C of the set point. Strain was measured by a high-temperature, ceramic-rod, contact extensometer that was held onto the specimen by a spring loaded mounting arm. The gage length of the extensometer was 12.7 mm (0.5 in) with a range of +20% to - 10%. The thinness of the specimens necessitated the use of V-notched ceramic rods to prevent slippage. Specimens were gripped with water-cooled, hydraulic friction grips and load was applied with a servo-hydraulic test frame capable of loads of 100 kN (22 kips).

All tests were conducted in load control. The static test was performed with a loading rate of 5.1 N/sec (1.15 lb./sec). This corresponds to an initial stress rate of 1 MPa/sec (145 psi/sec). Fatigue tests were all taken to failure where failure was considered to be separation of the specimen into two pieces. Cyclic tests all had R_σ=-0.05 and a frequency of 1 Hz. The maximum stress of the tests ranged from 500 to 1050 MPa (72.5 to 152.3 ksi). After testing, the fracture surfaces of all specimens were examined using a Scanning Electron Microscope (SEM). All relevant test parameters can be found in Table 1.

Table 1. All Relevant Test Parameters

Specimen No.	Cross Sectional Area (mm^2)	R ratio	Frequency (Hz)	Maximum Applied Stress (MPa)	Cycles to Failure (Failure Stress)
Ultra 1*	5.10	na	na	na	(1554 MPa)
Ultra 2	5.02	0.05	1	800	34390
Ultra 3	5.08	0.05	1	1050	5477
Ultra 4	5.06	0.05	1	600	64109
Ultra 5	5.13	0.05	1	500	86948

* Statically tested to failure

RESULTS AND DISCUSSION

MECHANICAL RESULTS

The tensile test's stress-strain response can be seen in Figure 1. Two monotonic stress-strain responses from previously tested SCS-6/Ti-6-4 [3] specimens are included for comparison purposes. The SCS-6/Ti-6-4 data has been shifted to have the same initial strain as the SCS-Ultra composite. As can be seen from the graph, the initial thermal strain is 0.00260 mm/mm, which is quite high. Unidirectional TMCs typically have initial thermal strains closer to 0.002 mm/mm at 427°C (800°F) [4]. The modulus of the new composite is 182.4 GPa (26.5 Msi) and the yield stress is about 600 MPa (87 ksi). These quantities are both less than their SCS-6/Ti-6-4 counterparts. The failure stress is 1554 MPa (225.38 ksi), due mainly to a large increase in the strain to failure. These are both larger than the typical SCS-6/Ti-6-4 values, but not as large as predicted based on the virgin strength values of the new fiber. Figure 1 suggests that the increase in strain to failure of the SCS-Ultra fibers is very beneficial but there have been some detrimental effects during processing that decreased both fiber stiffness and strength.

A fatigue life (S-N) diagram for SCS-Ultra/Ti-6-4 is shown in Figure 2. This figure also includes previously generated SCS-6/Ti-6-4 data [5]for comparison purposes. As can be seen in this plot, the fatigue life of the SCS-Ultra/Ti-6-4 is approximately one fifth of the life for the SCS-6/Ti-6-4 composite. The gap between the two curves seems to grow as maximum applied stress is decreased. This is most likely due to the processing defects in the SCS-Ultra material. These defects would not cause much of a fatigue life change at higher values of applied stress, where composite damage is assumed to be dominated by the statistical failure of the fibers [6].

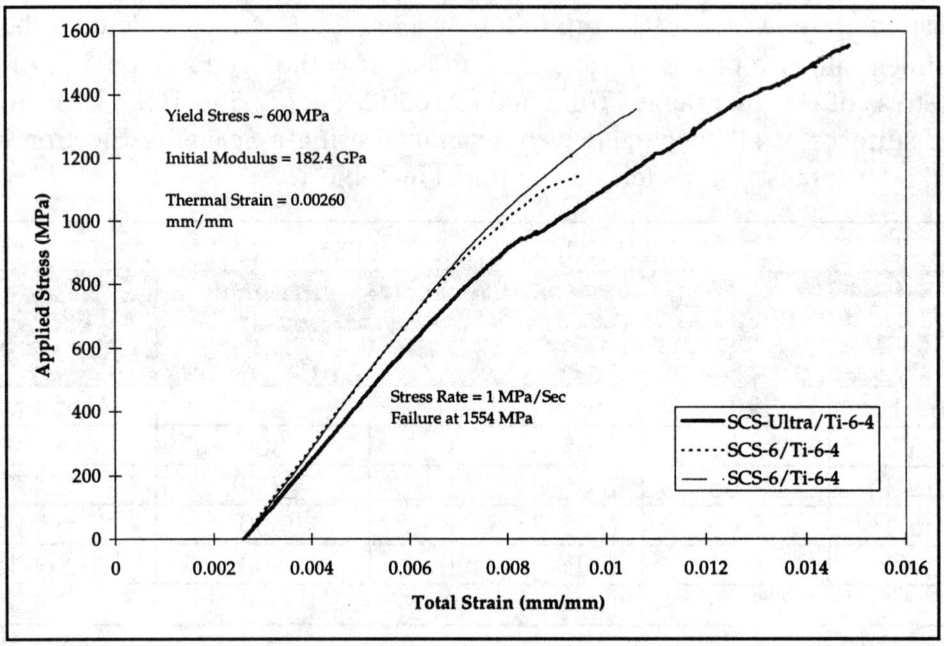

Figure 1. Tensile response of SCS-Ultra/Ti-6-4 compared to SCS-6/Ti-6-4.

However, at lower values of applied stress, where composite damage is dominated by matrix cracking, these defects would become much more influential since they would serve to accelerate crack initiation.

Typical hysteresis loops during cycling are shown in Figure 3. It is apparent from this plot that the SCS-Ultra composite undergoes significant strain rachetting over the cyclic life. It can also be seen that the modulus decreases as cycle count increases. This is similar to the tension-tension fatigue behavior of other titanium alloy composites [7]. The amount of decrease in the composite modulus prior to failure changes with applied stress. The modulus decreases to 90.4% of its original value at an applied stress of 1050 MPa (152.2 ksi), but only decreases to 97.7% of its original value at 500 MPa (72.51 ksi).

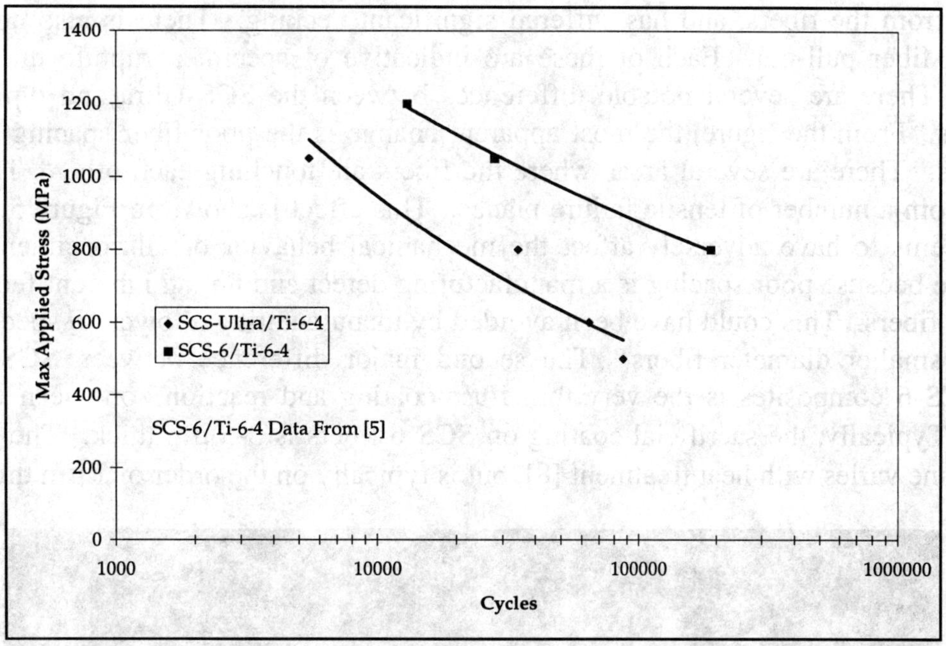

Figure 2. Comparison of the fatigue lives of SCS-Ultra/Ti-6-4 and SCS-6/Ti-6-4.

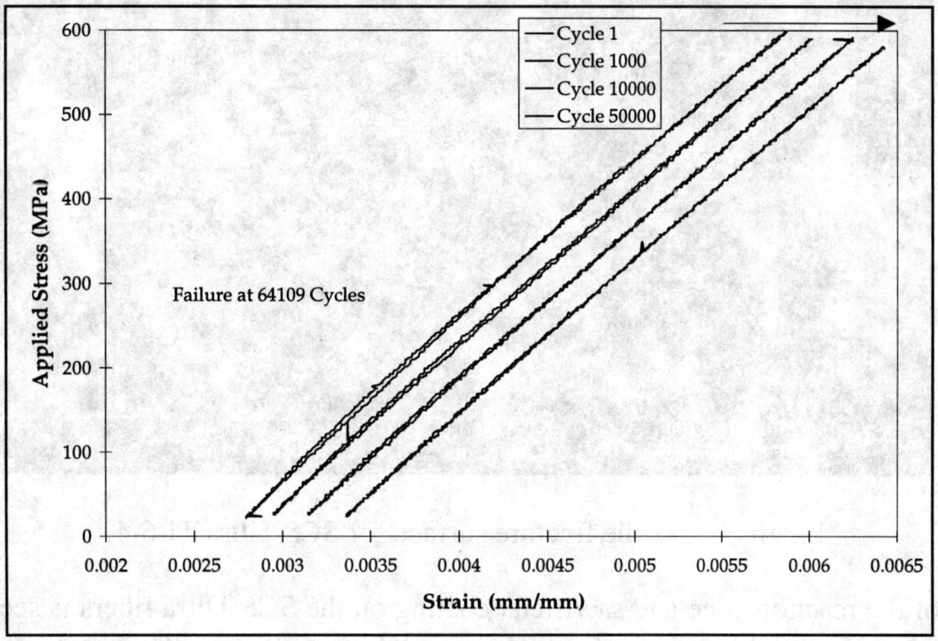

Figure 3. Typical hysteresis loops for SCS-Ultra/Ti-6-4

MICROSCOPIC RESULTS

Each specimen was investigated with a Scanning Electron Microscope (SEM) after failure. The failure surface of the ultimate tensile strength specimen is shown in Figure 4. The matrix on the exposed fracture surface is completely covered with ductile dimples, has debonded from the fibers, and has suffered significant necking. There is also an extensive amount of fiber pull-out. Each of these are indicative of specimen rupture due to tensile overload. There are several notable differences between the SCS-Ultra and typical SCS-6 composites. From the figure, the most apparent change is the poor fiber spacing in the new composites. There are several areas where the fibers are touching each other. These fibers served to join a number of tensile failure planes. This effect is shown in Figure 5. The fiber spacing seems to have adversely affect the mechanical behavior of all specimens. This is unfortunate because poor spacing is a manufacturing defect and not an inherent feature of the SCS-Ultra fibers. This could have been avoided by manufacturing a lower V_f specimen or by using the smaller diameter fibers. The second major difference between SCS-Ultra and typical SCS-6 composites is the very thin fiber coating and reaction zone seen in the new material. Typically, the sacrificial coating on SCS-6 fibers is 3-5 μm thick. The size of the reaction zone varies with heat treatment [8], but is typically on the order of 1 μm thick.

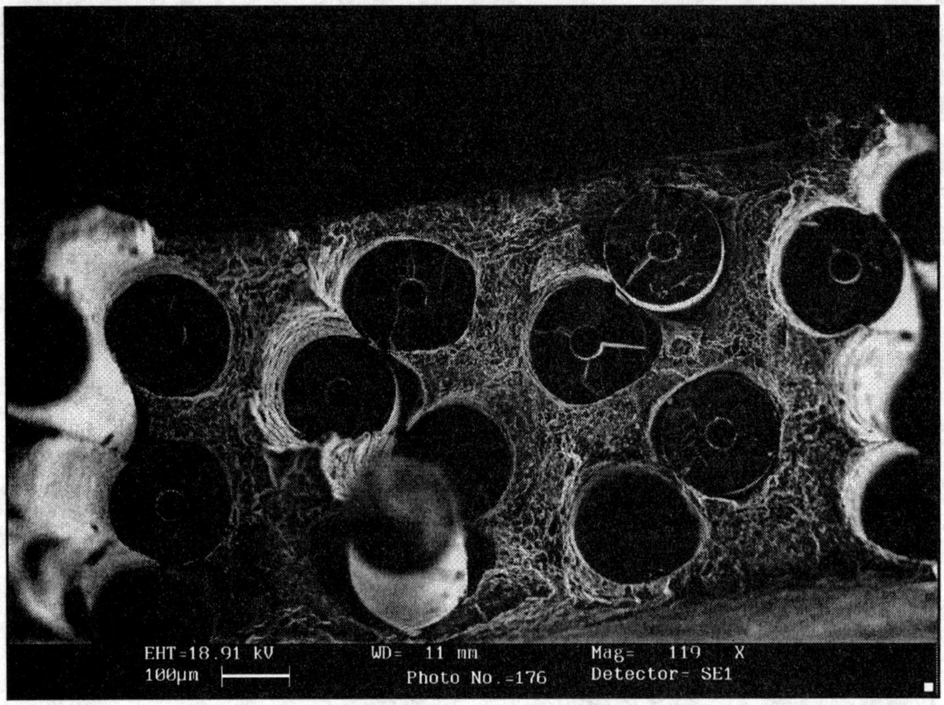

Figure 4. Tensile fracture surface for SCS-Ultra/Ti-6-4.

A portion of the reaction zone and sacrificial coating on the SCS-Ultra fibers is seen in Figure 6. The unreacted coating in Figure 6 is approximately 1 μm thick. The reaction zone in the titanium matrix is approximately 780 nm thick.

Both the touching fibers and the thin reaction zone in the titanium matrix served as excellent crack initiators. Although cracks did not have enough time to grow in the tensile test, they severely affected the lives of the fatigued specimens. In general, the amount of matrix cracking increased with decreasing maximum stress. The fracture surface of the specimen with a maximum applied stress of 1050 MPa had approximately 20% of its area

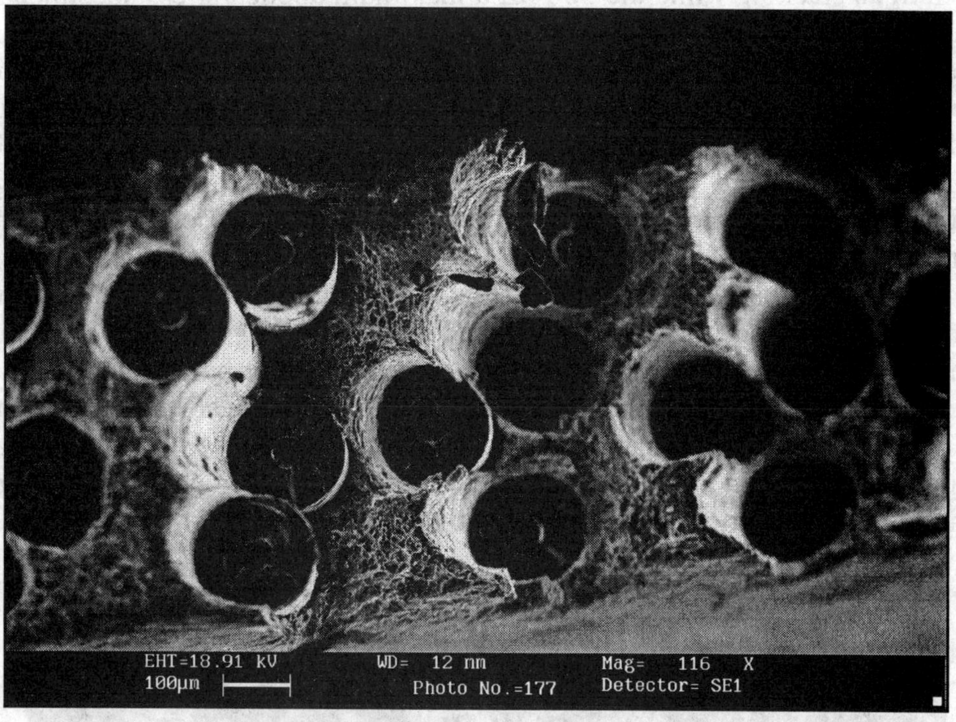

Figure 5. Tensile failure planes joined by touching fibers.

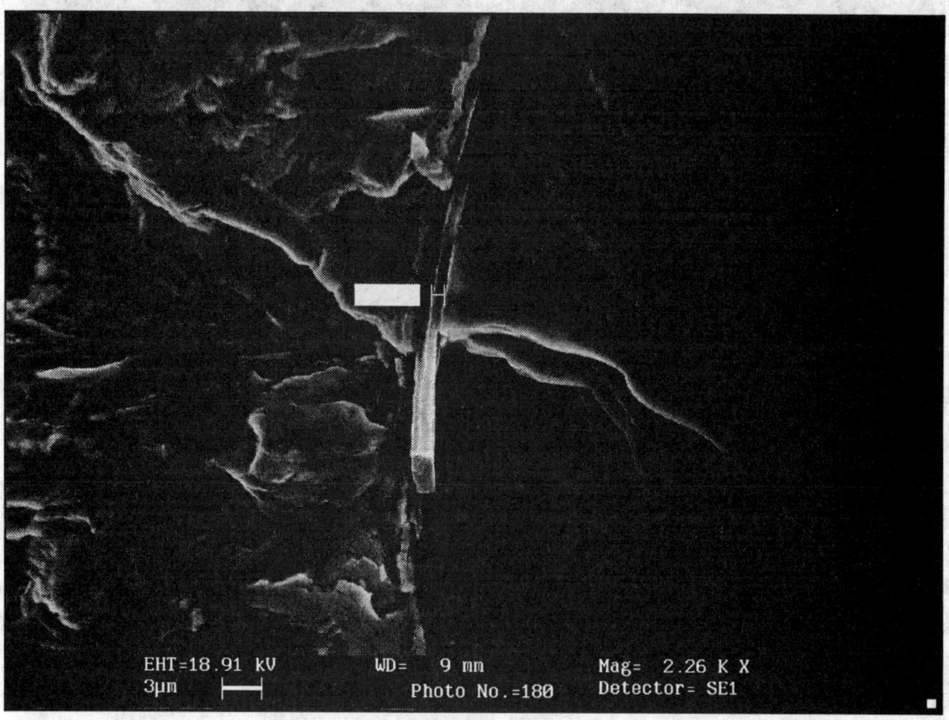

Figure 6. Close-up of coating and reaction zone. White box is next to distance indicator. Indicated distance is 1 µm.

covered by fatigue cracking while the 500 MPa maximum applied stress specimen had about 70% fatigue cracking. Besides the amount of cracking, the morphology of the failure surface also changed with applied stress. At higher stress levels, there was a significant amount of fiber pull-out, as shown in Figure 7. In all specimens, the majority of pull-out was surrounded by matrix that suffered ductile failure. The amount of fiber pull-out decreased with decreasing applied stress. This, coupled with the increasing amount of matrix fatigue cracking at lower stresses, suggests that the controlling damage mechanism in these composites changes from fiber dominated to matrix dominated as the stress decreases. The transition between damage modes has been widely discussed in TMCs [9,10]. In the current study, the shift from fiber to matrix dominated failure occurs near 800 MPa. However, the large number of manufacturing flaws in the material make this value far from certain. It may be that the flaws facilitate matrix cracking as a damage mechanism or that the thin coating leads to reduced strength in the SCS-Ultra fibers. The latter supposition is supported by the odd fracture surface of the 600 MPa specimen shown in Figure 8. At first glance, the large amount pull-out suggests that this portion of the failure surface was dominated by fiber damage.

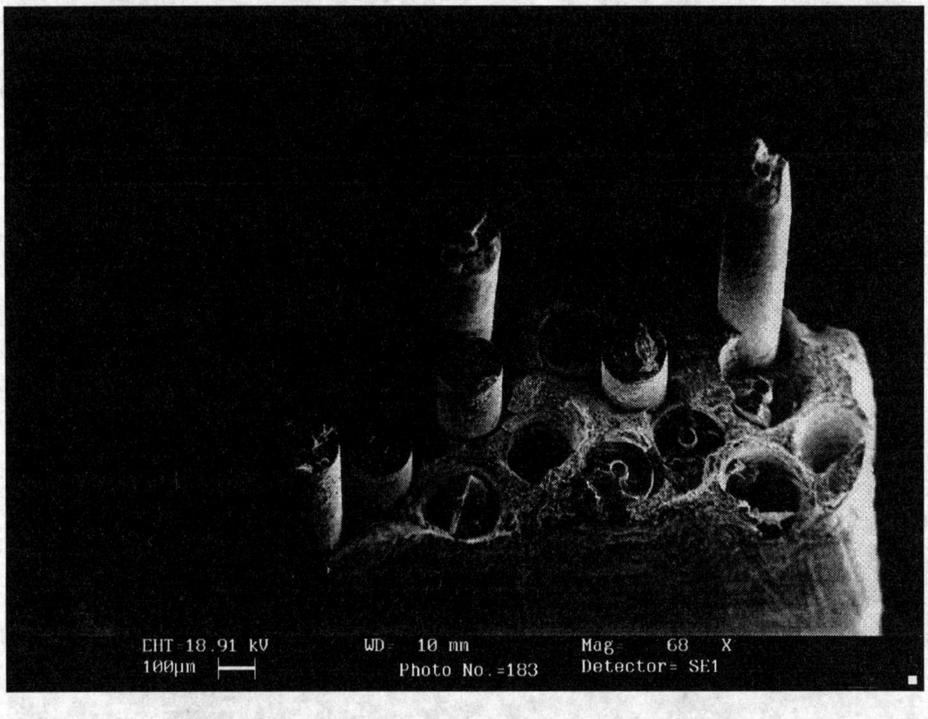

Figure 7. Fiber pull-out evident in 1050 MPa maximum applied stress specimen.

However, upon closer inspection, the entire matrix suffered from fatigue cracking, there is no ductile necking and the matrix has not significantly debonded from the fiber.The morphology of the matrix suggests that fatigue cracking was the controlling damage form in this area of the specimen. It is possible to have a cracked matrix bridged by unbroken fibers. However, upon final failure of the composite, the fibers typically break very close to the matrix fatigue plane due to the crack-induced stress concentration. The large pull-out lengths imply that the fibers broke away from the plane of final failure and were not significantly affected by the dominant crack. There are two possible explanations for this. First, there was a stress concentration at the point of fiber failure that was greater than the

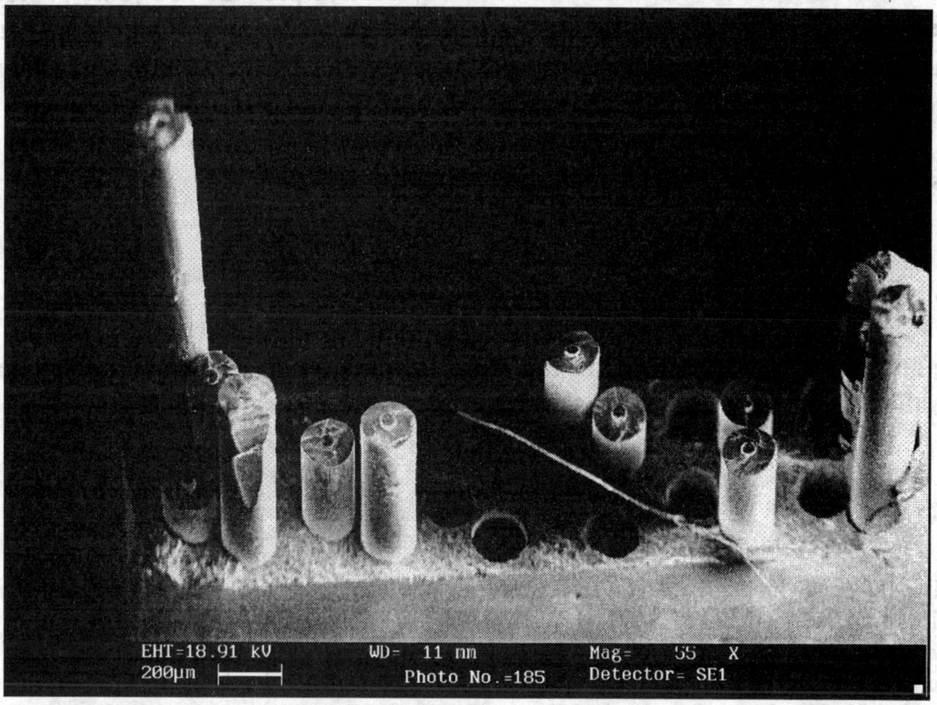

Figure 8. Fracture surface of 600 MPa maximum applied stress specimens showing fiber pull-out and matrix cracking.

induced stress concentration caused by the dominant crack. This is unlikely since the size of the stress concentration is proportional to the square root of the crack length. The second explanation is that the fiber was weak enough at the point of final failure to cause breakage. Significant strength drops have been observed is SCS-6 [11] when exposed to a surface flaw. It is very possible that the thin fiber coating did not protect the constituents from detrimental reactions during processing that, in turn, reduced the local strength of the fiber. This would also help explain the shorter fatigue lives in the fiber-dominated damage regime and the lower-than-predicted ultimate tensile strength. The final cause of the unusual fiber breakage was probably a combination of the two possibilities, with a stress concentration from a remote matrix crack interacting with an already weakened fiber.

CONCLUSIONS

The tensile strength of SCS-Ultra/Ti-6-4 showed an increase over other TMCs, though the increase was not as large as predicted. Despite the higher strength, the fatigue life of the new material is approximately 5 times shorter than SCS-6/Ti-6-4. Damage mechanism trends were similar to those seen in other TMCs. Microscopic inspection of the SCS-Ultra composite revealed that its poor mechanical properties were caused by a large number of manufacturing flaws and were not due to the fibers themselves. These manufacturing flaws included bad fiber spacing, not typically seen in TMCs, and a thin fiber coating. The fiber spacing could be corrected by producing material with smaller V_f or by using the smaller diameter fibers. The interfacial coating must be adjusted to provide adequate protection to the fibers while reducing the crack initiation sites in the reaction zone. Once the manufacturing problems are corrected, the composite should be re-evaluated for a better estimate of fatigue life and a clearer picture of damage progress. In short, SCS-Ultra fiber composites offer great promise in the future thanks to their increased strength and strain to failure. This should not be overlooked due to problems with material processing.

ACKNOWLEDGMENT

The authors would like to thank the Aero Propulsion and Power Directorate of Wright Laboratory (Ted Fecke) for supplying the material and sponsoring this research.

REFERENCES

[1] Castelli, M.G., and Gayda, J., "An Overview of Elevated Temperature Damage Mechanisms and Fatigue Behavior of a Unidirectional SCS-6/Ti-15-3 Composite," *DE-Vol. 55, Reliability, Stress Analysis, and Failure Prevention,* American Society of Mechanical Engineers, 1993, pp. 213-221.

[2] Calcaterra, J.R., and Johnson, W.S., "A Comparison of Damage Mechanisms for SCS-6/Timetal 21-S Subjected to Various Thermomechanical Fatigue Cycles," *Proceedings of the American Society for Composites - Tenth Technical Conference,* October 18-20, 1995, pp. 205-214.

[3] Lee, S.M., "Residual Strength and Fatigue Characterization of SCS-6/Ti-6-4," *Master's Thesis,* Air Force Institute of Technology, December 1996.

[4] Johnson, W.S., Mirdamadi, M., and Bahei-El-Din, Y.A., "Stress-Strain Analysis of a $[0/90]_{2s}$ Titanium Matrix Laminate Subjected to a Generic Hypersonic Profile," *Journal of Composites Technology & Research,* JCTRER, Vol. 15, No. 4, Winter 1993, pp. 297-303.

[5] Pittman, R.N., "Frequency Effects on Fatigue Behavior of a Unidirectional Metal Matrix Composites at Elevated Temperature," *Master's Thesis,* Air Force Institute of Technology, December 1995.

[6] Talreja, R., "A Conceptual Framework for the Interpretation of MMC Fatigue," *Materials Science and Engineering,* A200, 1995, pp. 21-28.

[7] Castelli, M.G., "Isothermal Damage and Fatigue Behavior of SCS-6/Timetal 21S $[0/90]_s$ Composite at 650°C," *NASA Contractor Report 195345,* June 1994.

[8] Kagawa, Y., Masuda, C., Fujiwara, C., and Fukushima, A., "Effects of Reaction Layer on Interfacial Shear Properties and Strength of Fiber in Silicon-Carbide (SiC) Fiber-Reinforced Titanium Alloy Composite," *Life Prediction Methodology for Titanium Matrix Composites, ASTM STP 1253,* W.S. Johnson, J.M. Larsen, and B.N. Cox, Eds., American Society for Testing and Materials, 1996, pp. 26-42.

[9] Jeng, S.M., Wang, P.C., and Yang, J.M., "Fatigue Damage Evolution and Degradation of Mechanical Properties in Silicon-Carbide (SiC) Fiber-Reinforced Titanium Matrix Composites," *Life Prediction Methodology for Titanium Matrix Composites, ASTM STP 1253,* W.S. Johnson, J.M. Larsen, and B.N. Cox, Eds., American Society for Testing and Materials, 1996, pp. 377-394.

[10] Neu, R.W., "Thermomechanical Fatigue Damage Mechanism Maps for Metal Matrix Composites," *Thermo-Mechanical Fatigue Behavior of Materials: 2nd Volume, ASTM STP 1263,* M.J. Verilli and M.G. Castelli, Eds., American Society for Testing and Materials, 1995.

[11] Gambone, M.L., Gundel, D.B., and Wawner, F.E., "The Effects of Isothermal Exposure on SiC Fibers," *The 7th Conference on Metal Matrix, Carbon, and Ceramic Matrix Composites,* J.D. Buckley, Ed., NASA Conference Publication 3235, 1994, pp. 235-252.

Analysis of Quasi-Isotropic MMC Using Iosipescu Shear Test

K. ZHANG, G. NEWAZ, S. CHANDU AND J. AHMAD

ABSTRACT

The shear constitutive response and deformation mechanisms under Iosipescu test for a quasi-isotropic, $[0/+45/-45/90]_s$ laminate made from TMIETAL 21S/SCS-6 MMC were investigated using a combination of mechanical measurement, microstructure analysis and finite element method. The objectives were to evaluate the material response using the Iosipescu test and attempt to establish Iosipescu test as a reasonable routine measurement for properties under shearing loading for Quasi-isotropic MMCs. Shear induced inelastic deformation is dominant due to plasticity which is associated with slip bands and damage which includes matrix grain boundary cracks, reaction zone cracks, debonding and fiber cracks. The prediction for the constitutive response using finite element analysis fits test data very well. The correlation of predicted results with experimental data shows that Iosipescu shear test is a valid method to investigate shear properties of quasi-isotropic MMC laminates.

INTRODUCTION

Considerable research to date has been conducted in characterizing silicon-carbide (SiC) fiber-reinforced titanium composites to understand their monotonic behavior (Sun and Chen 1989; Pindera, 1989; Sun *et al.* 1990; Newaz and Majumdar, 1991; Majumdar and Newaz, 1992; Newaz and Majumdar, 1993; Brust, Majumdar and Newaz, 1993; Newaz and Majumdar, 1994; Gao and Zhao, 1995; Allen Jones and Boyd, 1994; Newaz, Majumdar and Brust, 1996, Newaz and Zhang,1996). One important aspect is a continuing effort to understand and model the composite's constitutive response and failure conditions.

Constitutive modeling and inelastic deformation mechanisms have been studied extensively for longitudinal and transverse MMC laminates. However, little attention has been paid to investigate the constitutive response of pure shearing loading on MMCs, especially for the differently oriented ply-stacked laminate, its inelastic deformation mechanisms and the evolution of damage to failure. Furthermore, properties and constitutive response under pure shearing loading for MMCs using a reliable test technique has not been identified to date.

A variety of test methods are available for introducing shear stresses in unidirectional

---

Ke Zhang and Golam Newaz, Mechanical Engineering Department, Wayne State University, Detroit, Michigan
S. Chandu and J. Ahmad, Research Applications Inc., Dayton, Ohio

fiber MMCs. However, all of the tests have inherent deficiencies or disadvantages. The most uniform shear strain-stress state can be achieved in a MMC by applying torsional loading to a thin-walled tube specimen. However, such specimens are expensive requiring special fixtures and process techniques. The Iosipescu Test standard as explained in ASTM D 5379M-93 is recommended for in-plane shear testing specimens with unidirectional, cross-ply, woven composite and short-fiber composites. Tests on MMCs of continuous fibers are not within the scope of the standard. Iosipescu test has been conducted with various composites (Adams and Lewis 1995; Pindera, Ifju and Post, 1990; Swanson, Messick and Toombes, 1985; Morton *et al.*, 1992). For continuous fiber MMCs, especially oriented ply-stacked MMCs, some limited analysis has been performed as well (Santhosh and Ahmad, 1992; Chandu, Ahmad and Newaz, 1997).

For metal matrix composites, the tubular specimens can be quite expensive. Testing of tubes also requires specially designed fixtures (Swanson, Messick and Toombes, 1985). Off-axis tensile tests are not ideal for generating shear properties for titanium based metal matrix composites because of the relatively strong interaction of the transverse normal stress and shear stress at all possible angles due to the nature of orthotropy in a single ply. Also, off-axis tests are not applicable for multi-ply oriented laminates. The most reasonable approaches for shear testing of MMCs appear to be the Iosipescu test and $[\pm 45]_s$ tension test. The specific advantage of the Iosipescu shear test is the ability to develop pure shear stresses in a line zone between the two notches in the specimen. For ± 45 tension test, the shear properties are estimated after manipulating tension strain data in the normal and transverse directions. In this test, the plies in the lamina are subjected to both transverse normal and shear stresses.

EXPERIEMENTAL APPROACH

The material for testing was Ti 15-3/SCS-6 composite of 8-ply unidirectional for off-axis tension test and 8-ply quasi-isotropic $[0/+45/-45/90]_s$ lay-up for Iosipescu shear tests, approximately 0.00785-inch thick each ply with a fiber volume fraction of approximately 0.34. The SCS-6 (SiC) fiber diameter is approximately 140 μm.

The test specimens were cut from an electric-discharge machining (EDM) technique and then mechanically polished after EDM machining to remove any damage associated with the machining. Monotonic loading was applied at room temperature. All specimens were tested in the as-fabricated condition, i.e., no heat-treatment was performed prior to the testing.

The dimensions of a specimen with a test fixture and a strain gage (Micro-measurements, A2P-08-C085C-500) designed for pure shear strain measurement is shown in figure 1. The tests were performed at MTS test machine with an Iosipescu fixture. The strain data came from the strain gage and the stress data from the loading value divided by the cross section area between the notch tips. The specimens were loaded to a different deformation levels then unloaded to zero to investigate the constitutive response in both elastic and inelastic regimes and to determine how consistent are the elastic shear modulus, yielding and hardening tendency. Unloading from different loading levels, the zero-load strain offset, plasticity versa damage effects in nonlinear regime can be analyzed. In order to obtain information about shear failure of 8-ply quasi-isotropic $[0/+45/-45/90]_s$ Ti 15-3/SCS-6 composite, one of the specimens was loaded until its

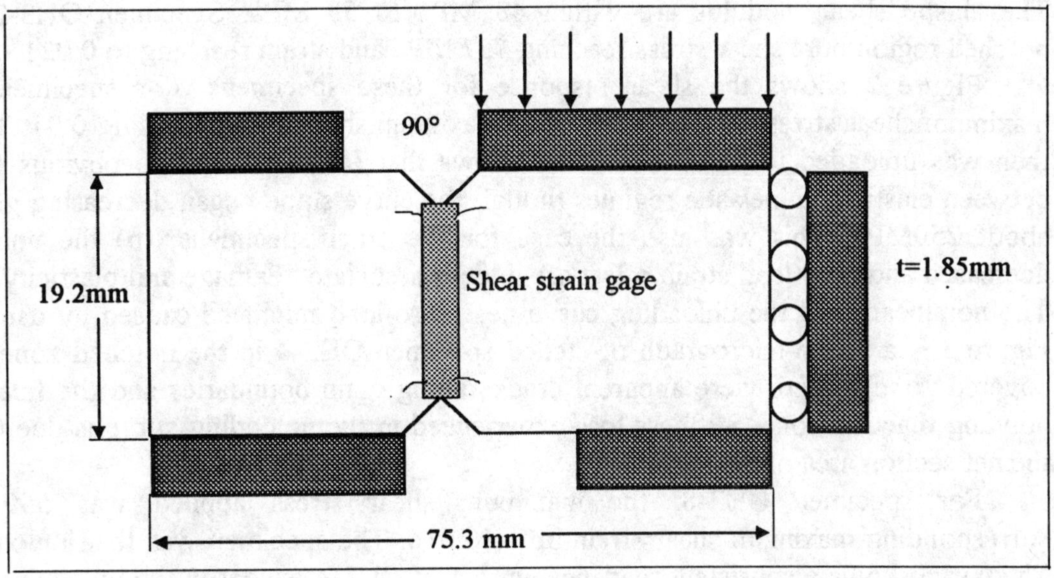

Figure 1. Dimensions of an Iosipescu test specimen with strain gage under loading fixture.

failure. Following mechanical testing, specimens were sectioned using a diamond wafering blade; debonding and fiber cracking can be significant during machining if adequate care is not taken during sectioning. Specimens were metallographically polished up to the first set of fibers. Specimens were etched using Kroll's reagent, which was found to be effective in revealing matrix damage and slip bands

RESULTS AND DISCUSSION

MECHANICAL BEHAVIOR

TABLE I shows the Iosipescu test results for specimens OH3-2, OH3-4,OH3-8 and OH3-12. It can be seen that the apparent yield stress was in the range of 200 MPa to 270 MPa and the corresponding yielding strain ranged from 0.0038 to 0.005.

TABLE I IOSIPESCU SHEAR TEST RESULTS

Specimen Number	Elastic Shear Modulus (GPa)	Maximum Stress (MPa)	Maximum Strain	Apparent Yield Stress (MPa)	Apparent Yield Strain	Unload Strain Offset	Failure Stress (MPa) / Failure Strain
OH3-2	55	757	0.021	270	0.0045	X	757/0.021
OH3-4	48	434	0.01	250	0.005	0.0004	X
OH3-8	54	559	0.014	200	0.0038	0.0017	X
OH3-12	48	659	0.017	250	0.005	0.0028	X

The elastic shear modulus are within 48 MPa to 55 MPa. Specimen OH3-2 failed with notched region pure shear stress reaching 757 MPa and strain reaching to 0.021.

Figure 2 shows the shear response for these specimens. For specimen OH3-4 the maximum shear stress was 434 MPa with maximum shear strain reaching 0.01. The specimen then was unloaded to zero. The curve shows that (a) there was no obvious turning point between elastic and inelastic regimes though the curve slope began decreasing gradually after about 250MPa. This was also the case for the other specimens; (b) the unloading slope decreased and zero-load strain offset are due to associated damage and plasticity respectively. The nonlinearity of the unloading curve near zero-load might be caused by damage closure. Figure 3 is a photo-micrograph of etched specimen OH3-4 in the notched zone (strain gage covered zone). There were apparent cracks along grain boundaries and the fiber and matrix bonding reaction zone. Stiffness loss experienced in the unloading curve is due to damage in the net section area of the specimen.

For specimen OH3-8, the maximum shear stress applied was 559 MPa with corresponding maximum shear strain up to 0.014. The specimen was then unloaded to zero. The curve shows a consistent tendency as that of OH3-4 except in the unloading period. The zero-load strain offset means that plasticity took a dominant role in the inelastic deformation regime. Figure 4 illustrates obvious matrix slip bands in etched OH3-8 specimen. These slip bands are associated with the plastic deformation mechanism.

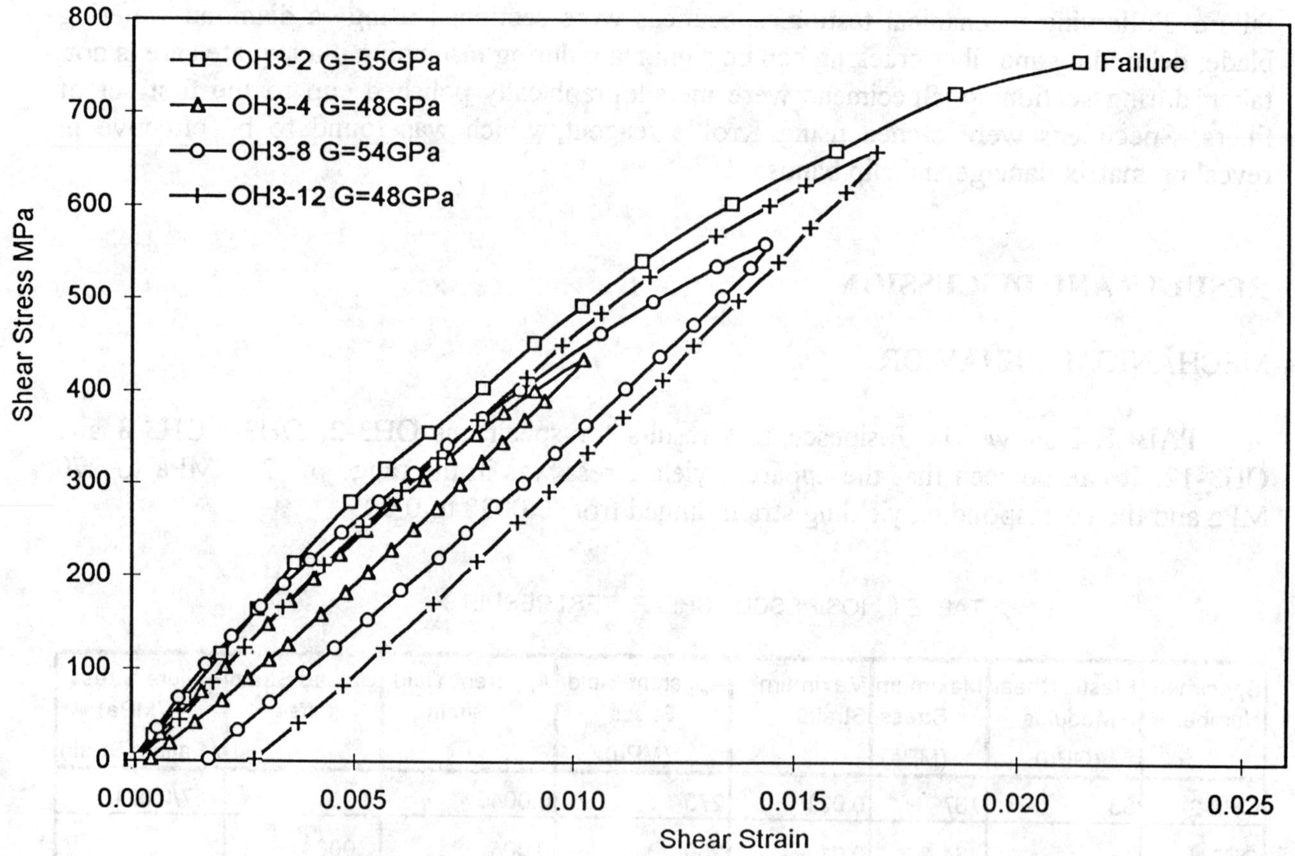

Figure 2. Mechanical response for Iosipescu test specimens of Quasi-isotropic laminate made of TMIETAL 21S/SCS-6 MMC.

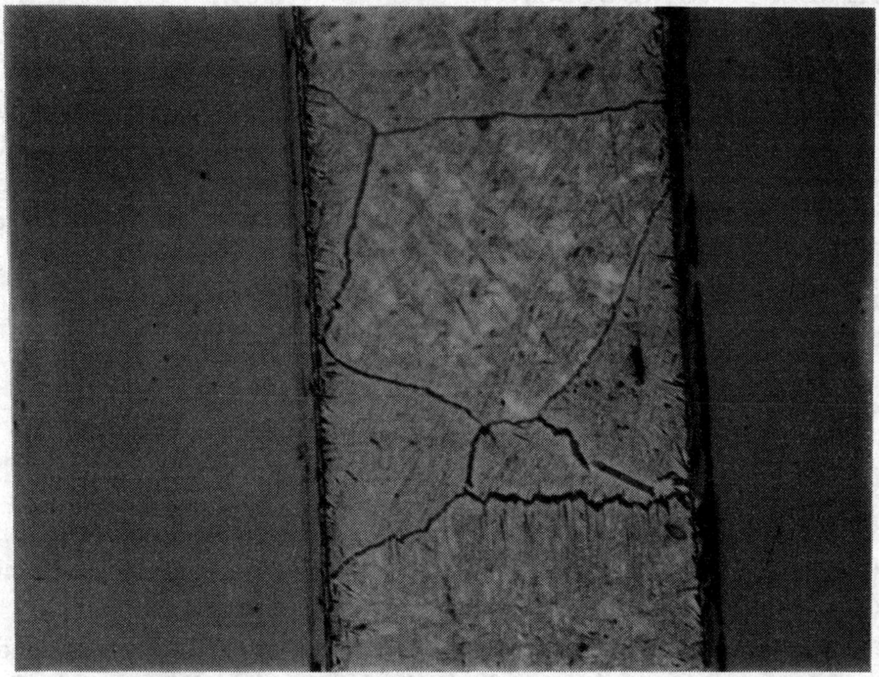

Figure 3. Etched microstructure (x500) of specimen OH3-4 in strain gage covered zone, cracks along grain boundaries and bonding reaction zone.

Figure 4. Etched microstructure (x1000) of specimen OH3-8, slip bands initiated from grain boundaries and bonding reaction zones.

CONSTITUTIVE RESPONSE MODELING

For the $[0/+45/-45/90]_s$ laminate, each laminae of the TMIETAL 21S/SCS-6 MMC is treated as an anisotropic ply. The laminate properties depend on the individual plies. In the model, each ply is modeled as a separate layer and connected to other plies through common nodes. The model is based on a plane stress analysis. In the inelastic range, the deformation becomes much more complex and the hardening effects are significant. It is necessary to determine inelastic equivalent stress, σ_e, like what we have in the case of homogenous plastic deformation. With a formulation regarding σ_e, we can relate our test data (monotonic tension test, for instance) to our analysis. Here we have according to Sun (1989):

$$\sigma_e = [\frac{3}{2}(\sigma^2_{22} + 2\, a_{66}\, \sigma^2_{12})]^{1/2} \tag{1}$$

where, σ_e represents equivalent inelastic stress and 1 represents fiber direction, 2 is 90° of the fiber direction (transverse). σ_{22} represents the transverse stress of a MMC laminae. σ_{12} represents the in-plane shear stress. σ_{11} represents stress in the fiber direction.

Sun (Sun *et al.* , 1990) assumed that only the transverse normal stress σ_{22} and in plane-shear stress σ_{12} of composite contribute to inelastic deformation. Whereas the longitudinal (fiber direction) σ_{11} stress does not affect inelastic deformation though it contributes to elastic deformation. This is in fact not true. The nonlinear deformation starts after the tension stress reached approximately 1100 MPa (Majumdar and Newaz 1992). But here we can take Sun's assumption as an approximation to simplify our calculation. With Sun's equation if we relate it to the 90° tension test data (Newaz and Zhang, 1997) as yield and hardening reference in an analysis, then for this situation (transverse loading), $\sigma_{12}=0$, and Sun's formula becomes:

$$\sigma_e = [\frac{3}{2}\sigma^2_{22}]^{1/2} \tag{2}$$

This case, the apparent yield stress is

$$\sigma_e = \sigma_y = 250 \text{ MPa (test data)} \tag{3}$$

For off axis monotonic tension test.

$$\sigma_{11} = \sigma_x \cos^2\theta \tag{4}$$

$$\sigma_{22} = \sigma_x \cos\theta \sin\theta \tag{5}$$

$$\sigma_{12} = \sigma_x \sin^2\theta \tag{6}$$

at $\theta = 10°$ we have

$$\sigma_{11} = 0.970\sigma_x \tag{7}$$

$$\sigma_{22} = 0.030\sigma_x \tag{8}$$

$$\sigma_{12} = -0.171\sigma_x \tag{9}$$

Putting these value into Sun's formula, we get:

$$\sigma_e = [\frac{3}{2}((0.970\sigma x)^2 + 2\, a_{66}(-0.171\sigma x)^2)]^{1/2} \tag{10}$$

where x is loading direction, and θ is the angle between fiber direction and loading direction.

From the test data (Newaz and Zhang, 1997), the apparent yield stress for 10° tension loading at x direction is σ_x= 850 MPa. Whereas with transverse tension test we already know that yielding equivalent stress is 250 MPa. With this information, we can get a_{66} which is approximately about 1.46. For MMC inelastic deformation includes not only plastic but damage - matrix grain boundary cracks, fiber and matrix bond separation and fiber cracks- effect too. Therefore, we should take both effects into the account in determining parameter a_{66}. We can modify Sun's model by choosing a suitable value of a_{66}. In Sun's paper (Sun *et al.*, 1990) a_{66} = 0.5 was chosen to fit the yield stress test curve of different angles of off-axis monotonic tension tests for SCS-6/Ti-6-4 MMC. But for [0/+45/-45/90]s laminate, shear test model in which the plies stacked have different orientation, the situation is more complicated. Analysis also shows that in different elements of the model either transverse tension stresses or transverse compression stresses is in effect in addition to shear stresses. And different stress state combination has different yield and hardening condition. From test data the transverse compression yield stress (750 MPa in Newaz, Majumdar and Brust, 1996) is much higher than the transverse tension yield stress (250 MPa in Newaz and Zhang, 1997). Here we can still use equation (1) for both tension and compression occasions, but choose a_{66} = 0.4, σ_y = 250 MPa (yield) for a case when $\sigma 22 > 0$ in an element; choose b_{66} = 0.5, σ_y= 750 MPa (yield) for a case when $\sigma 22 < 0$ in an element. During the analysis we use FORTRAN subroutine to check each element in different plies to see whether $\sigma 22 > 0$ or $\sigma 22 < 0$ to determine the field parameter and then apply tension hardening condition or compression hardening condition to the element accordingly. We used the transverse (90°) monotonic tension and compression test data as the hardening reference for the analysis.

FINITE ELEMENT ANALYSIS RESULTS

We used ABAQUS code to simulate a monotonic loading Iosipescu shear test for TMIETAL 21S/SCS-6 MMC $[0/+45/-45/90]_s$ laminate. Figure 5 shows the comparison of the constitutive responses between the FEA predication and the tested specimen OH3-8. The calculated response comes from the center portion elements of the notch zone (strain gage covered zone) of the surface layer, it is found that if the elements were taken from the 80% of the central portion between two notch tips, then the calculated curves can fit the test results well, otherwise the calculated response may be lower than the measured value. It can be seen that the calculated result fits the test data very well. Unlike the testing response, the calculated unloading stiffness, remained the same as the loading one. This is because the FEA model did

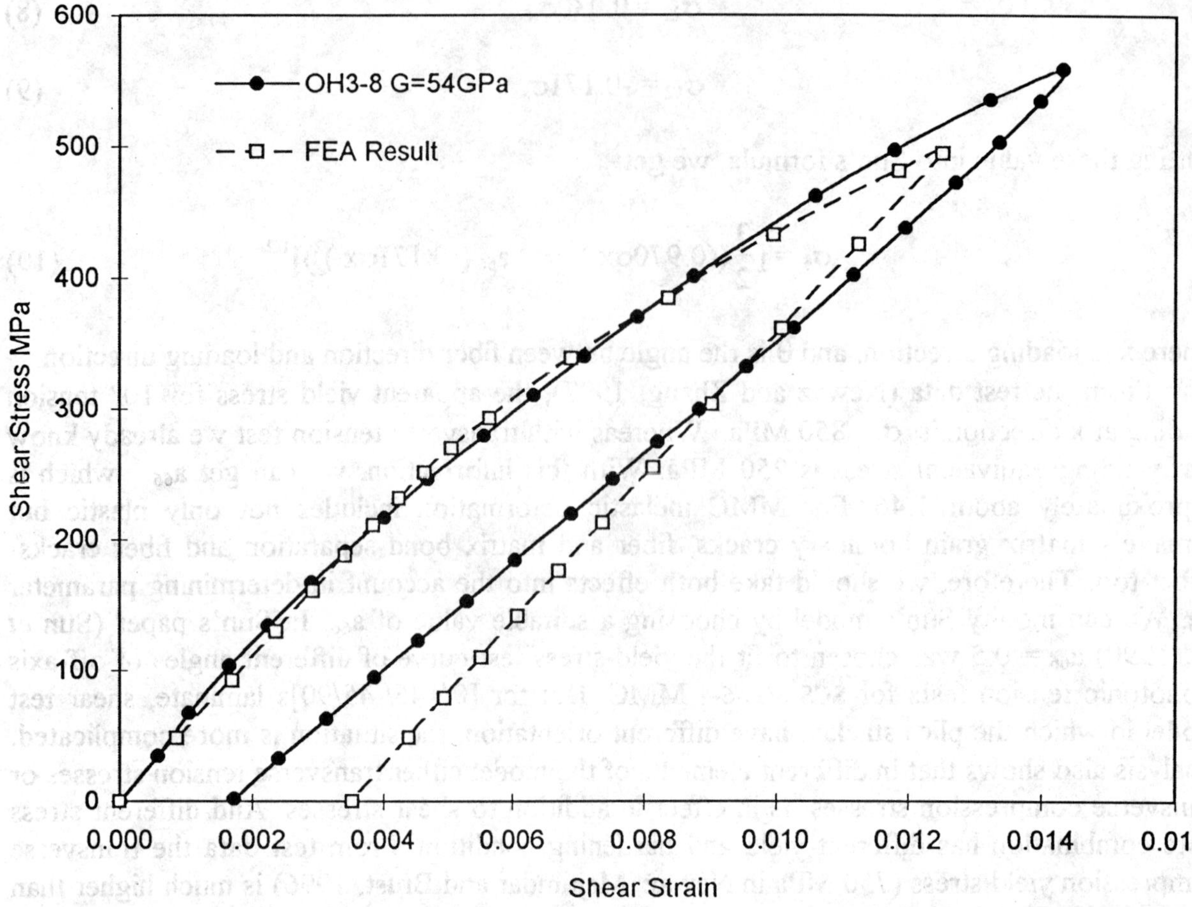

Figure 5. Comparison of analysis prediction and test measured constitutive shear response for Quasi-isotopic laminate made of the TMIETAL 21S/SCS-6 MMC.

not take the damage closure effect into the unloading analysis.

With the model surface color stress contours (they can not be clearly shown in white and black) for σ_x and τ_{xy} (x coincident with fiber direction) show that there is a small red color stress concentrated area on the edge next to a notch tip, its location is approximately the same as it was mentioned previously which caused the surface fiber fractures in specimen OH3-4 and further initiated failure in specimen OH3-2. The color contour also shows σ_x along fiber direction within the area exceeded 1900 MPa, Referring to Majumdar and Newaz, 1991, specimen under 0° tension loading failure at about 1500 MPa to 1800 MPa. This can explain why there were fiber fractures in the area of the surface layer of specimen OH3-4 and OH3-8.

It should be pointed out that Sun did not take the fiber direction inelastic deformation and neither the damage effects into his model. Therefore, when finite element model for MMC based on this assumption is limited to shear strain < 0.01, the calculation may provide good prediction. Errors may increase when the shear strain is larger than this limit.

CONCLUSIONS

1. Iosipescu shear test provides a reasonable method for measurement of shear constitutive response of MMC material.
2. Iosipescu shear test has very good reproducibility in terms of overall properties and mechanical behavior. Iosipescu test can be recommended as a routine measurement to obtain MMC material linear and nonlinear shear properties at room temperature. Finite element analysis model for MMC Iosipescu test using Sun's assumption as an approximation plus a subroutine to check the stress state to determine the transverse tension or compression stress and then applying different yielding and hardening criterion can obtain satisfactory result for the case of shear strain<0.01.
3. Sun's effective stress model for plastic deformation was successfully used for prediction of experimental data. The basic input for finite element analysis required 0° tension, 90° transverse tension and compression data. Shearing coefficient a_{66} was assumed to be generated by manipulating off-axis tension test data. However a_{66} obtained by this way included neither damage effect nor fiber direction inelastic effect, and the subsequently determined a_{66} can not perform a good enough prediction in the analysis. Furthermore one parameter a_{66} is not enough to interpret the difference between transverse tension and compression constitutive responses. So a better way is to assume two trial shearing parameter a_{66} and b_{66} to interpret transverse tension and transverse compression of a laminae respectively. The values of a_{66} and b_{66} are dependent of the property of the material.
4. At an early stage of shear inelastic deformation (strain <0.01), damage effect play an important role. This can be seen by the compliance change in the unloading curve. The damage mechanism was associated with matrix grain boundary crack and cracks along the fiber and matrix reaction zone. When shear deformation increases, plasticity becomes a primary inelastic deformation mechanism which is associated with slip bands in the matrix.

ACKNOWLEDGEMENT

Funding for this research was obtained through an Air Force Phase II subcontract from Research Application, Inc., Centerville, OH.

REFERENCES

Adams, D.F., and Lewis, E.Q.,1995, "Experimental Strain Analysis of The Iosipescu Shear Test Specimen," Experimental Mechanics, 352-360.

Allen, D. H., Jones, R. H., and Boyd, J. G., 1994, "Micromechanical Analysis of A Continuous Fiber Metal Matrix Composite Including The Effects of Matrix Viscoplasticity and Evolving Damage," J. Mechanics And Physics of Solids, 42: 505-29.

Brust, F. W., Majumdar, B. S., and Newaz, G. M., 1993, "Constitutive Response Analysis of Metal Matrix Composites via the Unit Cell Model," ASTM Symposium on Composite Materials, Pittsburgh.

Chandu, S., and Ahmad, J., and Newaz, G. M., 1997, "Detailed Interpretation of Off-Axis and Iosipescu Test Data On Metal Matrix Composites," to appear in the J. of Reinforced Plastics & Composites.

Gao, Z., and Zhao, H., 1995, "Life Predictions of Metal Matrix Composite Laminates Under Isothermal and Nonisothermal Fatigue," J. Composite Materials 29 No. 9:1142-68.

Majumdar, B. S., and Newaz, G. M., 1992, "Inelastic Deformation of Metal Matrix Composite: Plasticity and Damage Mechanisms," Philosophical Magazine A, Vol. 66, No. 2, 187-212.

Morton, J., Ho, H., Tsai, M.Y. and Farly, G. L., 1992, "An Evaluation of the Iosipescu Specimen for Composite Materials Shear Property Measurement," J. Composite Materials, 26 (5), 708-750.

Newaz, G. M., and Majumdar, B. S., 1991, "Deformation and Failure Mechanism in Metal Matrix Composites," The Winter Annual Meeting of The America Society of Mechanical Engineers, Atlanta, Georgia, AD Vol. 22, 55-65.

Newaz, G. M., Majumdar, B. S., and Brust, F. W., 1992, "Thermal Cycling Response of Quasi-isotropic Meal Matrix Composites," ASME J. Eng. Materials and Technology, 114, 156-61.

Newaz, G. M., and Majumdar, B. S., 1994, "A Comparison of Mechanical Response of MMC At Room Temperature and Elevated Temperature," J. of Composites Science and Tech., 50.

Newaz, G. M., Majumdar, B. S., and Brust, F. W., 1996, "Inelastic Deformation Mechanisms in SCS-6/Ti 15-3 Metal Matrix Composite (MMC) Lamina Under Compression," Life Prediction Methodology for Titanium Matrix Composites, ASTM STP 1253, W. S. Johnson, J. M. Larsen, and B. N. Cox, Eds., American Society for Testing and Materials, 208-230.

Newaz, G. M. and Zhang, K.,1997 "Inelastic Response of Off-axis MMC Lamina," American Materials Science of Technology , submitted to ASME J. of Eng. Mat. & Technology.

Pindera, M. J., 1989, "Shear Testing of Fiber Reinforced Metal Matrix Composites," Metal Matrix Composites: Analysis, and Failure Modes ASTM STP 1032, W. S. Johnson, Ed., American Society for Testing and Materials, Philadelphia, 19-42

Pindera, M.J., Ifju, P. and Post, D., 1990, "Iosipescu Shear Characterization of Polymeric and Metal Matrix Composite," Experimental Mechanics, 30 (1), 101-108.

Santhosh, U. and Ahmad, J., "Metal Matrix Composite Response Under Biaxial Loading," Constitutive Behavior of High Temperature Composites, MD-Vol. 40, Majumdar, B. S., and Newaz, G. M., and S. Mall editors, ASME 1992.

Sun, C.T., 1989, "Modeling Continuous Fiber Metal Matrix Composite as an Orthotropic Elastic-Plastic Material," Metal Matrix Composites; Testing, Analysis, and Failure Modes, ASTM STP 1032, W.S. Johnson, Ed., American Society for Testing and Materials, Philadelphia, 148-160.

Sun, C. T., Chen, J. L., Sha, G. T., and Koop, W. E., 1990, "Mechanical Characterization of SCS-6/Ti-6-4 Metal Matrix Composite," J. of Composite Materials, Vol. 24, 1029-1059.

Swanson, S.R., Messick, M. and Toombes, G.R., 1985, "Comparison of Torsion Tube and Iosipescu In-Plane Shear Test Results for Carbon Fiber-Reinforced Epoxy Composite," Composites, 16 (3), 220-224.

Dynamic and Time-Dependent Behavior

The Effect of High Strain Rate on the Compressive Mechanical Properties of Unidirectional E-Glass/Epoxy (3M Scotchply 1003)

A. T. DEE, J. R. VINSON AND R. F. GIBSON

ABSTRACT

Polymer matrix composites offer excellent mechanical properties such as high specific strength and stiffness, making them attractive for applications in several naval, aerospace, automotive, and recreational-sports structural components. Although many composite materials are selected for applications where high strain rate loading is probable, little is known of their responses to shock loading. Mechanical properties can vary significantly with strain rate. Therefore, the use of static properties in the analysis and design of structures which undergo dynamic loading may adversely lead to very conservative, overweight designs or lead to designs which fail prematurely and unexpectedly. The use of dynamic material properties ensures that designs are optimized for structural integrity and weight efficiency when subjected to dynamic loading.

In this study, a Split Hopkinson Pressure Bar is used to obtain high-strain-rate, compressive mechanical properties of a unidirectional E-Glass/epoxy (3M Scotchply 1003) laminate. While E-glass/epoxy has many uses, its properties are of special interest to the Office of Naval Research, where the material is applied to ship and submarine hulls. For the 1-direction (parallel to the fibers) and the 3-direction (thickness direction), the compressive mechanical properties of yield stress, yield strain, ultimate strength, ultimate strain, and modulus of elasticity are determined for strain rates varying from 183 sec^{-1} to 653 sec^{-1}. The data from the 48 dynamic tests are statistically analyzed, represented by equations, compared to quasi-static material properties, and discussed in some detail. Also, an examination of the fracture surface is performed using a scanning electron microscope (SEM).

INTRODUCTION

The Split Hopkinson Pressure Bar (SHPB) facility at the University of Delaware is used to determine high strain rate material properties. Impact of the specimen is initiated by releasing nitrogen gas from a pressurized chamber which causes the specimen to be compressed between the two bars. The striker bar is displaced on Teflon rings through the barrel where it

---

Alex Dee, Univeristy of Delaware, Dept. of Mechanical Engineering, 126 Spencer Lab, Newark, DE 19716
Jack R. Vinson, H. Fletcher Brown Professor of Mechanical & Aerospace Engineering, University of Delaware.
Ronald F. Gibson, Professor of Mechanical Engineering, Wayne State University, Detroit, Michigan.

strikes the incident bar (bar 1). The velocity of the striker bar, V_{sb}, just prior to impact is measured using two infrared beams set four inches apart.

Upon impact with the striker bar, the incident bar receives an elastic compressive strain wave with a wave velocity C_B and a wave shape $\varepsilon_I(t)$ and is displaced along its length axis. This bar is guided by Teflon bearings which are bolted to a steel I-beam and secured on a reinforced concrete base to minimize any vibration. When the wave reaches the bar 1-specimen interface, a portion of the incident wave is reflected back into bar 1 as a tensile wave with a wave shape $\varepsilon_R(t)$. The remaining portion is transmitted into the specimen as a compressive wave. The wave transmitted into the specimen travels through the specimen and reaches the specimen-bar 2 interface, where a portion of the wave is reflected back into the specimen. The remaining portion is transmitted into bar 2 as a compressive wave with a wave shape $\varepsilon_T(t)$. Bar 2 is displaced similarly to bar 1, but is arrested after the test by a simple dash pot.

Because the specimen length is short compared to the bar lengths, the initial stress wave in the specimen undergoes numerous internal reflections during the test. It is calculated that the initial wave duration in a typical specimen is approximately two microseconds. Within a composite material specimen, there may be differences in the wave speed in the fiber and matrix materials. However, if a minimum of four wave reflections within the specimen is achieved prior to specimen failure, the stress distribution along the specimen length is smoothed out and the specimen is assumed to be in a uniform state of stress at any instant. At the University of Delaware SHPB, the wave pulse time window (defined as the time it takes a wave to travel the length of the striker bar and back) is approximately 290 microseconds. Thus, to record information accurately at specimen failure, the specimen must fail within 290 microseconds after the start of the test and also after a time such that four internal reflections in the specimen has been achieved.

Strain gages are mounted on bars 1 and 2 equidistant from the specimen interfaces and are connected to an oscilloscope. The oscilloscope records and stores the output from the strain gages. Using this data, the stress and strain versus time of the specimen during the test can be calculated for the strain rate of that particular test.

MATERIAL DESCRIPTION

The unidirectional laminate is an E-Glass/epoxy (3M Scotchply 1003) with a fiber volume fraction about 0.5. The laminate was made by autoclave-style press molding at Wayne State University, using heated platens in an hydraulic vacuum to produce autoclave-quality material.

SPECIMEN PREPARATION

Compression specimens in the shape of right circular cylinders (0.2 inches in diameter and 0.3 inches in length) were core drilled from the laminate and prepared for testing by the senior author (Figure 1). Since the laminate is transversely isotropic (2- direction equivalent

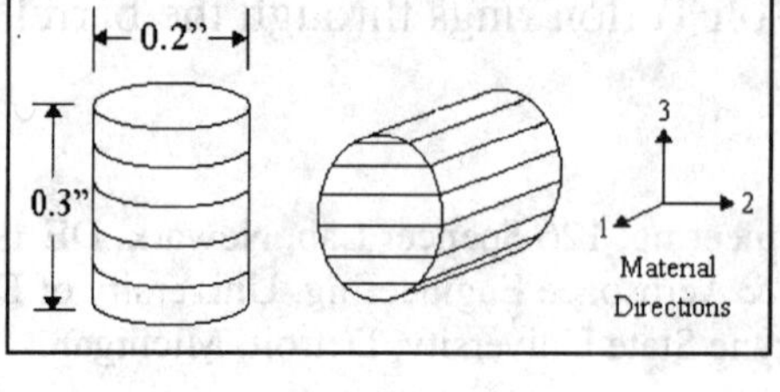

FIGURE 1. ILLUSTRATION OF A 1-DIRECTION SPECIMEN (RIGHT) AND A 3-DIRECTION SPECIMEN (LEFT).

to the 3-direction), only specimens in the 1 and 3 directions were prepared. The 1-direction refers to the in-plane direction parallel to the fibers, while the 3-direction denotes the through-thickness direction perpendicular to the fibers. Prior to testing in the SHPB, specimen ends were lubricated to minimize any radial forces. Tested specimens that were viewed under SEM were sputter coated with gold to reduce charging.

TEST RESULTS

The 1-direction specimens fractured in the fiber direction, most likely due to fiber buckling and interfacial debonding. Significant brooming was noticed at one end of failed specimens tested at the average 183 sec^{-1} strain rate. However, brooming became less evident with increasing strain rate.

The 3-direction specimens fractured in shear. At lower strain rates at and below 557 sec^{-1}, the shear plane was at a 45° angle to the specimen axis. Above the 557 sec^{-1} strain rate, the shear plane angle was observed to be a slightly steeper 40° angle to the specimen axis.

For completeness, all the data from the 19 tests in the 1-direction and the 19 tests in the 3-direction are presented in Tables I and II. Moreover, 6 quasi-static strength tests were performed (3 tests in each direction) on an Instron universal machine at a rate of 0.05 inches/minute. Quasi-static strength data is provided in Table V. These data will allow other researchers to study and analyze the tests independently. These data are then presented statistically in Tables III, IV, and VI which provide mean values and standard deviations.

TABLE I. E-GLASS/EPOXY (3M SCOTCHPLY 1003): 1-DIRECTION TESTS

Group	Strain Rate (/sec)	Yield Strength (MPa)	Ult Strength (MPa)	Yield Strain (%)	Ult Strain (%)	Modulus (MPa)	Notes
A1	187	410.1	1.07E+03	0.74%	2.35%	3.97E+04	Fracture predominately
A2	196	477.7	1.07E+03	0.74%	2.19%	4.02E+04	in the fiber direction.
A3	163	373.0	1.02E+03	0.50%	2.08%	3.92E+04	Large fragments. Brooming
A4	185	507.0	1.10E+03	0.74%	2.14%	4.10E+04	predominant at one end.
B1	354	589.0	1.08E+03	0.93%	1.10%	3.79E+04	Fracture predominately
B2	379	627.0	1.06E+03	1.21%	1.20%	3.67E+04	in the fiber direction.
B3	357		1.07E+03		1.10%	6.13E+04	Large Fragments and smaller
B4	353		9.57E+02		1.17%	6.01E+04	splinters.
C1	434	-	9.42E+02	-	1.13%	4.98E+04	Fracture predominately
C2	441	-	1.01E+03	-	1.17%	5.09E+04	in the fiber direction.
C3	452	-	9.65E+02	-	1.19%	6.01E+04	Splinters and dust.
D1	490	-	9.73E+02	-	1.24%	5.38E+04	Fracture predominately
D2	474	-	1.01E+03	-	1.28%	6.25E+04	in the fiber direction.
D3	496	-	9.59E+02	-	1.11%	5.80E+04	Splinters and dust.
D4	518	-	9.13E+02	-	1.31%	5.49E+04	
D5	520	-	9.56E+02	-	1.30%	5.32E+04	
E1	584	-	9.34E+02	-	1.27%	6.07E+04	Fracture predominately
E2	560	-	9.86E+02	-	1.65%	6.37E+04	in the fiber direction.
E3	568	-	1.02E+03	-	1.42%	6.85E+04	Splinters and dust.

TABLE II. E-GLASS/EPOXY (3M SCOTCHPLY 1003): 3-DIRECTION TESTS

Group	Strain Rate (/sec)	Yield Strength (MPa)	Ult Strength (MPa)	Yield Strain (%)	Ult Strain (%)	Modulus (MPa)	Notes
F1	286	-	2.98E+02	-	3.04%	9.98E+03	Fracture at a 45 degree shear plane. Large debris of intact specimen.
F2	285	-	3.03E+02	-	2.50%	9.90E+03	
F3	277	-	2.75E+02	-	3.03%	1.31E+04	
F4	373	-	2.85E+02	-	3.09%	1.15E+04	
G1	559	2.39E+02	2.88E+02	1.61%	2.60%	1.47E+04	Fracture at a 45 degree shear plane. Large debris of intact specimen. Some splinters.
G2	562	2.27E+02	2.90E+02	1.66%	3.01%	1.68E+04	
G3	557	-	2.78E+02	-	3.19%	1.34E+04	
G4	572	-	3.10E+02	-	2.92%	1.64E+04	
G5	537	-	2.80E+02	-	3.47%	1.34E+04	
H1	617	-	3.20E+02	-	3.35%	1.53E+04	Fracture at a 40 degree shear plane (w.r.t. specimen axis). Many smaller fragments. Some splinters.
H2	627	-	2.94E+02	-	2.71%	1.54E+04	
H3	620	-	2.98E+02	-	2.48%	1.63E+04	
H4	617	-	2.93E+02	-	2.59%	1.53E+04	
I1	640	-	2.95E+02	-	2.92%	1.57E+04	Fracture at a 40 degree shear plane (w.r.t. specimen axis). Many smaller fragments. Some splinters.
I2	649	-	3.02E+02	-	2.92%	1.42E+04	
I3	652	-	2.88E+02	-	3.26%	1.38E+04	
I4	678	-	2.95E+02	-	2.81%	1.50E+04	
I5	646	-	3.08E+02	-	2.71%	1.54E+04	
I6	652	-	2.93E+02	-	2.75%	1.55E+04	

STATISTICAL ANALYSIS, RESULTS, AND DISCUSSION

It is interesting to see that with the complexities of these high strain rate tests, and the small sample sizes, how small the coefficient of variance (COV), (i.e. the ratio of standard deviation to mean value, given as a percentage), is for the various mechanical properties as shown in Table VII. The highest COV value is 18.7% (on a yield strain) while the lowest COV is 2.0% (on a strain rate).

To analyze the trends from Tables III, IV, and VI, it is important to define significant differences. For the present purpose, a quantity P is said to be significantly different from Q if the mean value of P falls outside the region of the mean $\pm$ 3 standard deviations of Q. The $\pm$ 3 standard deviations are plotted as error bars for each relevant dynamic material property in Figures 2-5 as a visual means of determining significant differences.

From Table III, it is seen that significant differences are predominant between the average groups of strain rates in the 1-direction, with only the group C average strain rate not being significantly different from group D. Failure of the specimens was achieved at each strain rate. Yielding was found in the 1-direction at strain rates equal to and below 361 sec^{-1}. Therefore, in the strain rate range between 361 sec^{-1} and 442 sec^{-1}, a transition from ductile to brittle behavior must occur. This trend from ductile behavior to brittle behavior has been noticed in other polymer matrix composite materials. Concerning ultimate strength, Figure 2 shows that these values do not differ significantly with strain rate. Only the group D average ultimate strength is significantly different from the group A average ultimate strength. Ultimate strain values plotted in Figure 2 show no strain rate dependence. Modulus values show strain rate sensitivity with the group A modulus values being significantly different from all other groups, and groups C, D, and E being significantly different from group A (Figure 6).

From Table IV, it is seen that average strain rate groups in the 3-direction are generally significantly different except that the average strain rates of groups G and I are not significantly different from the average strain rates of group H, and group I is not significantly different from group H. Yielding was found in the 3-direction only for the group G average strain rate. Therefore, in the strain rate range between 305 sec^{-1} and 557 sec^{-1}, a transition from brittle to ductile behavior must occur, and a transition from ductile to brittle behavior must occur between 557 sec^{-1} and 620 sec^{-1}. Ultimate strength, and ultimate strain in the 3-direction show no strain rate dependence (Figures 3 and 5, respectively). Modulus of elasticity values in the 3-direction showed little strain rate dependence with only the average modulus values of group F being significantly different from average modulus values of groups H and I.

It is seen that significant differences are prevalent when comparing dynamic material properties to the quasi-static material properties in Table VI. The average quasi-static ultimate strength values in the 1 and 3 directions are significantly different (weaker) from all their respective average, dynamic ultimate strength groups and vice versa. Yielding was not present in quasi-static tests for either the 1 nor the 3-direction.

TABLE III. E-GLASS/EPOXY (3M SCOTCHPLY 1003): 1-DIRECTION STATISTICS

Group		Strain Rate (/sec)	Yield Strength (MPa)	Ult Strength (MPa)	Yield Strain (%)	Ult Strain (%)	Modulus (MPa)
A	Average	183	4.42E+02	1.06E+03	0.68%	2.19%	9.18E+04
	St Dev	14	6.13E+01	3.17E+01	0.12%	0.12%	8.31E+03
B	Average	361	6.08E+02 *	1.04E+03	1.07 % *	2.34%	6.70E+04
	St Dev	12	2.69E+01 *	5.67E+01	0.20% *	0.12%	7.82E+03
C	Average	442	-	9.72E+02	-	2.36%	5.36E+04
	St Dev	9	-	3.51E+01	-	0.11%	5.67E+03
D	Average	499	-	9.62E+02	-	2.47%	5.65E+04
	St Dev	19	-	3.46E+01	-	0.13%	3.84E+03
E	Average	571	-	9.80E+02	-	2.08%	6.43E+04
	St Dev	12	-	4.37E+01	-	0.08%	3.94E+03

*B1-2 used for average and standard deviation

TABLE IV. E-GLASS/3M SCOTCHPLY 100 EPOXY 3: 3-DIRECTION STATISTICS

Group		Strain Rate (/sec)	Yield Strength (MPa)	Ult Strength (MPa)	Yield Strain (%)	Ult Strain (%)	Modulus (MPa)
F	Average	305	-	2.90E+02	-	2.92%	1.11E+04
	St Dev	46	-	1.24E+01	-	0.28%	1.51E+03
G	Average	557	2.33E+02 *	2.89E+02	1.64% *	3.04%	1.50E+04
	St Dev	13	7.92E+00 *	1.28E+01	0.04% *	0.32%	1.62E+03
H	Average	620	-	3.01E+02	-	2.78%	1.56E+04
	St Dev	34	-	1.27E+01	-	0.39%	4.79E+02
I	Average	653	-	2.97E+02	-	2.90%	1.49E+04
	St Dev	13	-	7.13E+00	-	0.20%	7.53E+02

*G1-2 used for average and standard deviation

TABLE V. E-GLASS/EPOXY (3M SCOTCHPLY 1003): QUASI-STATIC TESTS

Direction	Number	Yield Strength (MPa)	Ult Strength (MPa)	Notes
1	1	-	6.45E+02	Fracture by brooming at
1	2	-	6.99E+02	one end of specimen and
1	3	-	6.13E+02	some longitudinal splitting.
3	1	-	1.67E+02	Fracture at a 40-45 degree shear
3	2	-	1.77E+02	plane (w.r.t. specimen axis).
3	3	-	1.77E+02	Specimens remained in one piece.

TABLE VI. E-GLASS/EPOXY (3M SCOTCHPLY 1003): QUASI-STATIC STATISTICS

Direction		Ult Strength (MPa)
1	average	6.53E+02
	st dev	4.35E+01
3	average	1.74E+02
	st dev	6.21E+00

TABLE VII. COEFFICIENT OF VARIANCE OF THE DYNAMIC MECHANICAL PROPERTIES

Property	1-Direction	3-Direction
Strain Rate	2.1 - 7.7%	2.0 - 14.9%
Yield Strength	4.4 - 13.9%	3.40%
Ultimate Strength	3.0 - 5.4%	2.4 - 4.4%
Yield Strain	17.6 - 18.7%	2.20%
Ultimate Strain	3.8 - 5.3%	6.9 - 14.0%
Modulus	6.1 - 11.7%	3.1 - 13.6%

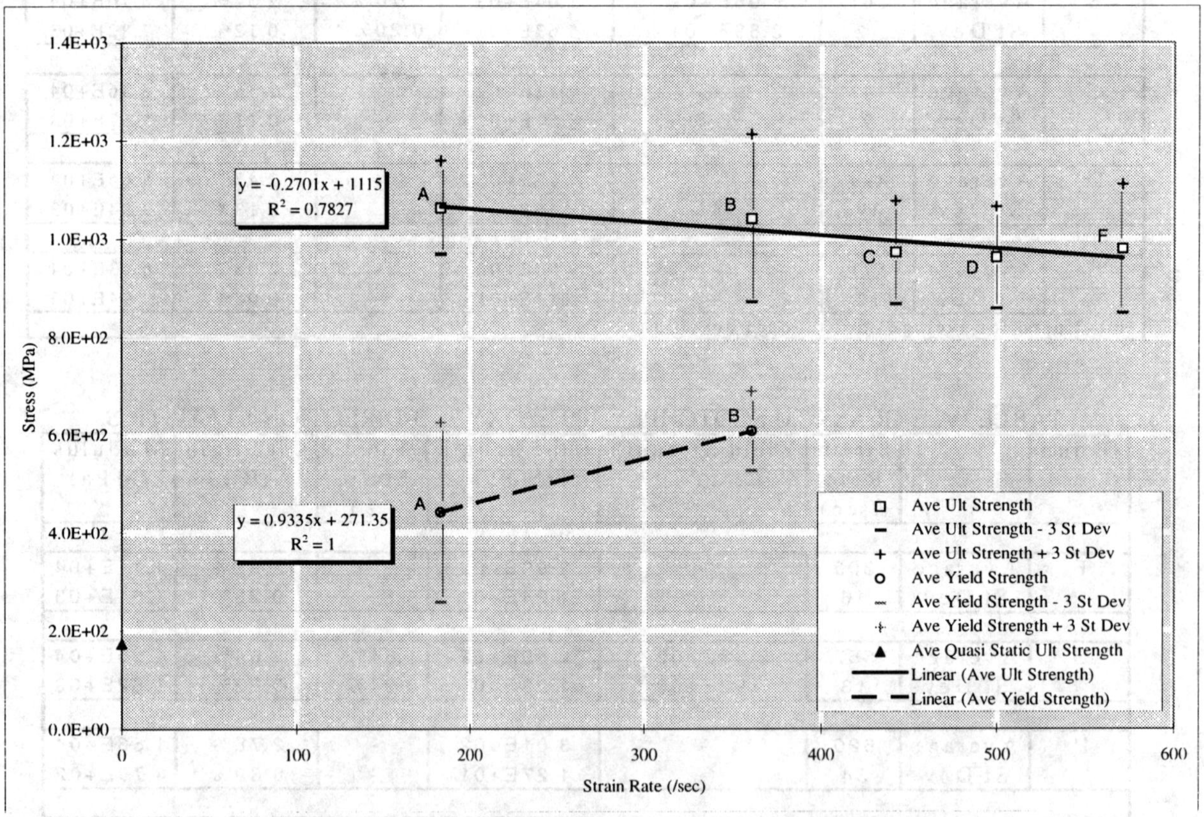

FIGURE 2. ULTIMATE AND YIELD STRENGTH AS A FUNCTION OF STRAIN RATE - E-GLASS/EPOXY (3M SCOTCHPLY 1003), 1-DIRECTION

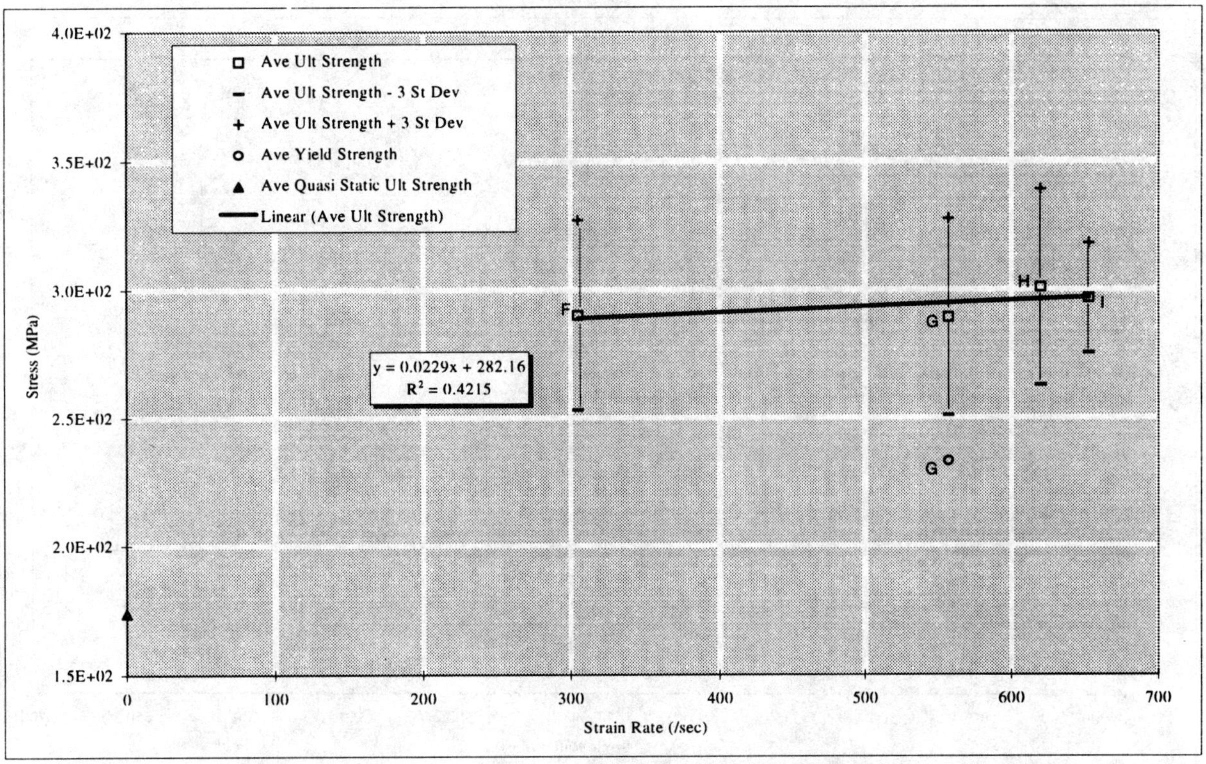

FIGURE 3. ULTIMATE AND YIELD STRENGTH AS A FUNCTION OF STRAIN RATE - E-GLASS/3M SCOTCHPLY 1003 EPOXY , 3-DIRECTION

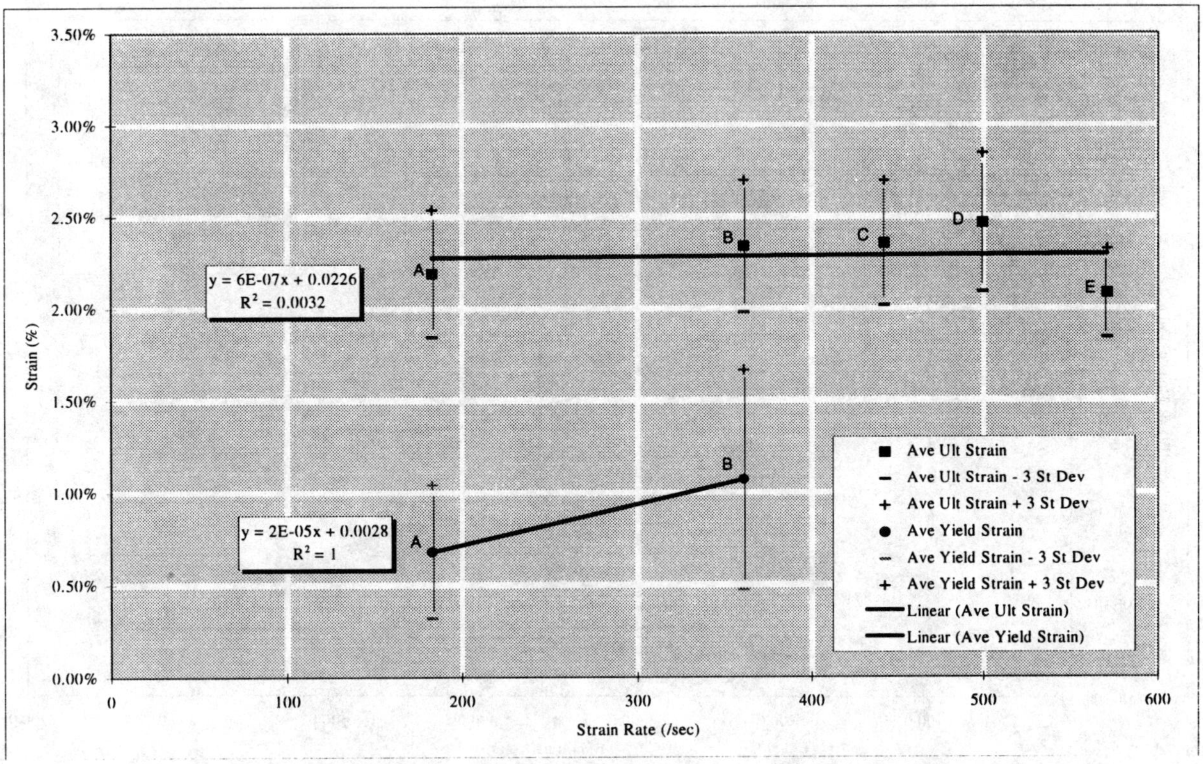

FIGURE 4. ULTIMATE AND YIELD STRAIN AS A FUNCTION OF STRAIN RATE: E-GLASS/EPOXY (3M SCOTCHPLY 1003), 1-DIRECTION

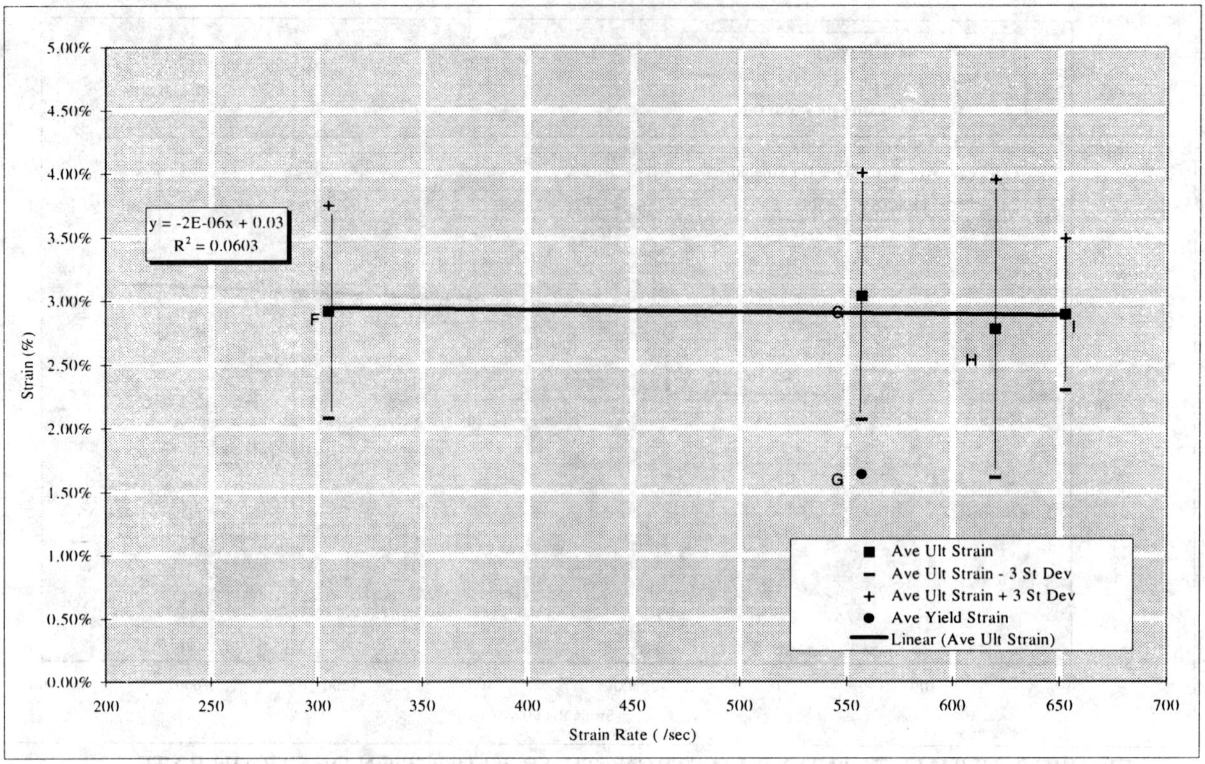

FIGURE 5. ULTIMATE AND YIELD STRAIN AS A FUNCTION OF STRAIN RATE: E-GLASS/EPOXY (3M SCOTCHPLY 1003), 3-DIRECTION

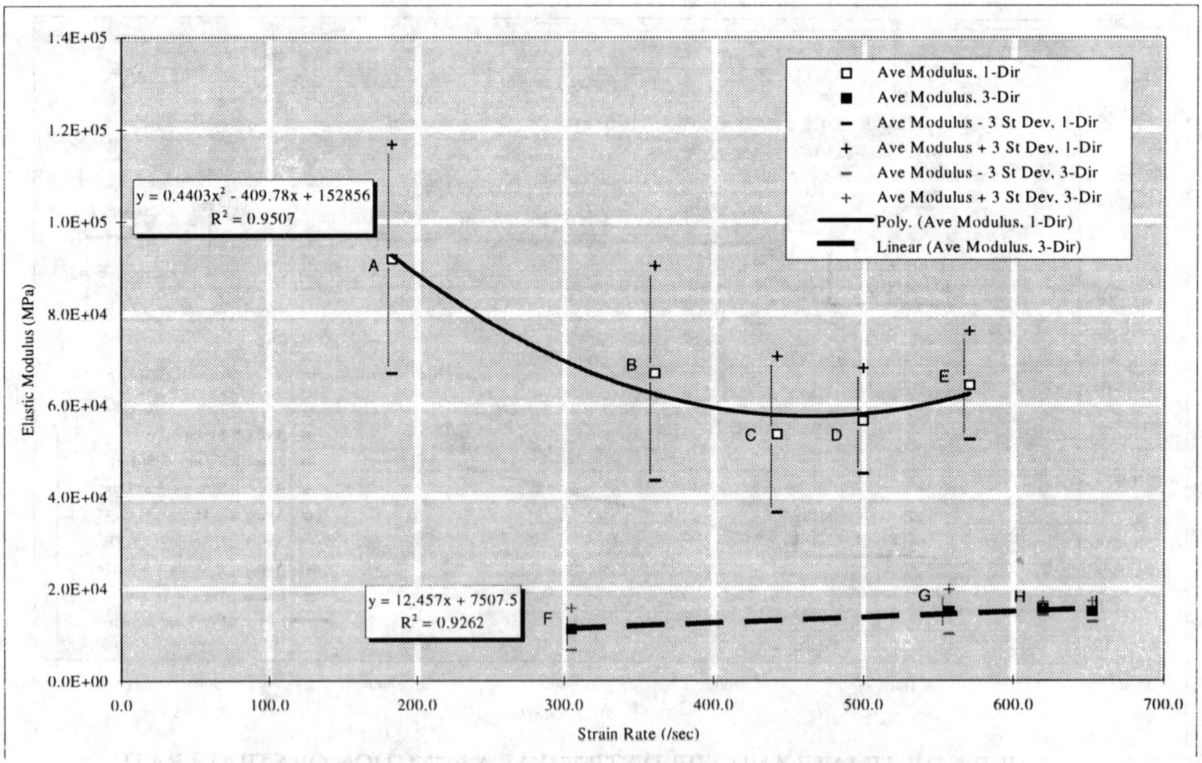

FIGURE 6. MODULUS OF ELASTICITY AS A FUNCTION OF STRAIN RATE: E-GLASS/EPOXY (3M SCOTCHPLY 1003), 1 AND 3-DIRECTIONS

FRACTOGRAPHY

Figure 7 shows SEM photos of 1 and 3-direction fracture surfaces tested at 361 sec^{-1} and 557 sec^{-1} respectively. The 1-direction failure surface shows interfacial debonding, indicating weak resin/fiber adhesion in the failure mode of longitudinal splitting (Figure 7B). Resin cusps or hackles in resin rich areas are also clearly visible (Figure 7A). These cusps are associated with brittle fracture and believed to be induced by tearing resulting from interlaminar shear. The 3-direction fracture surface shows fractured fibers coated with resin, indicating good resin/fiber adhesion in the shear failure mode (Figure 7C).

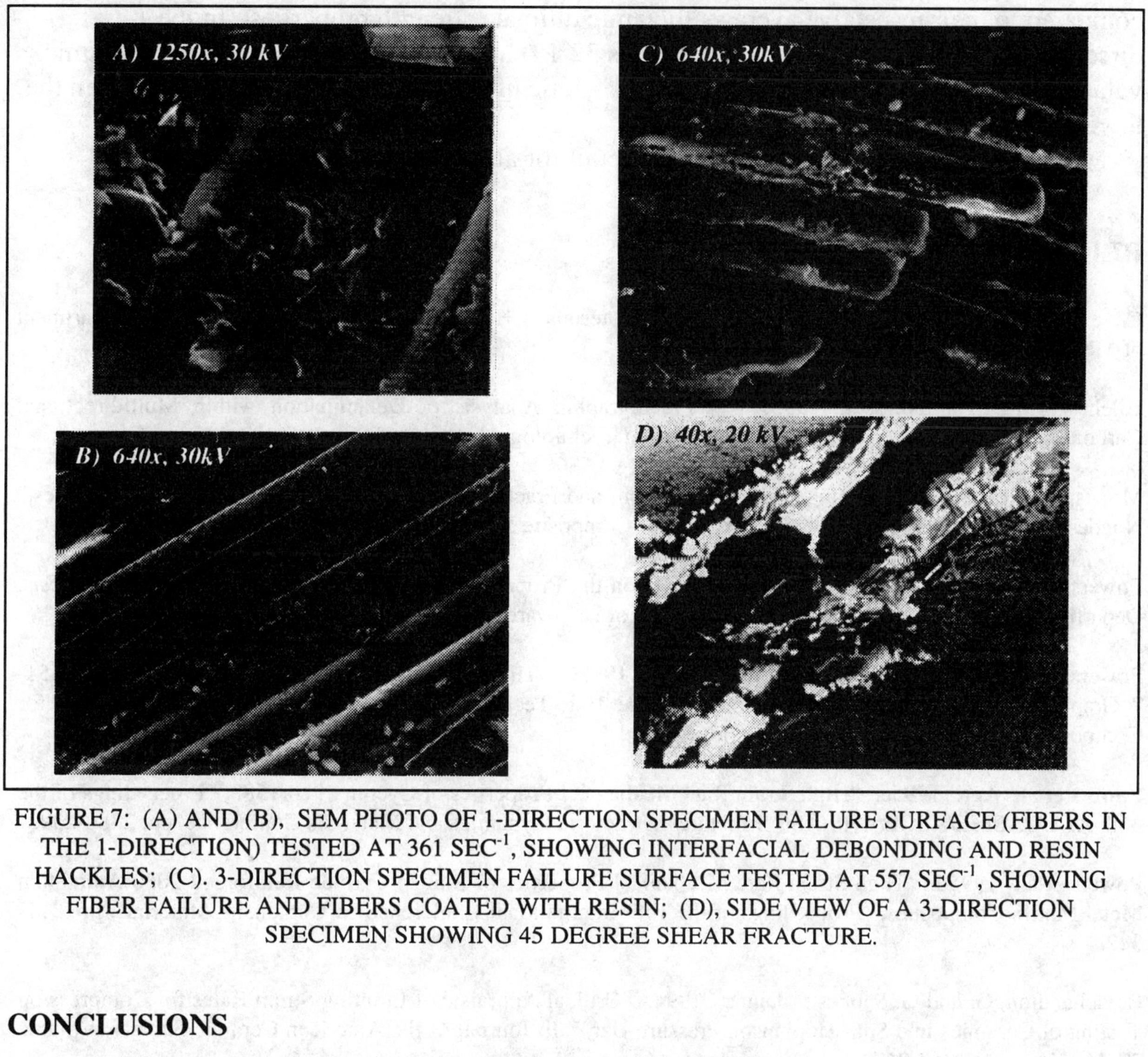

FIGURE 7: (A) AND (B), SEM PHOTO OF 1-DIRECTION SPECIMEN FAILURE SURFACE (FIBERS IN THE 1-DIRECTION) TESTED AT 361 SEC^{-1}, SHOWING INTERFACIAL DEBONDING AND RESIN HACKLES; (C). 3-DIRECTION SPECIMEN FAILURE SURFACE TESTED AT 557 SEC^{-1}, SHOWING FIBER FAILURE AND FIBERS COATED WITH RESIN; (D), SIDE VIEW OF A 3-DIRECTION SPECIMEN SHOWING 45 DEGREE SHEAR FRACTURE.

CONCLUSIONS

The data from these 38 experiments for the E-Glass/epoxy (3M Scotchply 1003) in the 1 and 3 directions provides a data bank for this material at room temperature for strain rates between 183 sec^{-1} to 653 sec^{-1}.

In the 1-direction, average ultimate strength displays little strain rate dependence, but does decrease 8.5% over the tested strain rates. Average yield strength increases 72.7% between strain rates of 183 sec^{-1} and 361 sec^{-1}. Average modulus values decrease about 42.8% over the tested strain rates.

In the 3-direction, ultimate stress and strain show no strain rate dependence, however, the average elastic modulus increases 34.2% over the tested strain rates. Yielding was found only at the average strain rate of 557 sec^{-1}, indicating material transitional behavior near this strain rate. The 3-direction ultimate strength is approximately 30% weaker compared to the 1-direction. However, the 3-direction ultimate strain is approximately 25% greater than the 1-direction ultimate strain.

Considering quasi-static strength values, they are significantly lower in both directions compared to their respective average, dynamic ultimate strength properties. In the 1-direction, the quasi-static ultimate strength is 32.1% lower then the lowest average dynamic value (group D). In the 3-direction, the quasi-static ultimate strength is 39.8% lower then the lowest average dynamic value (group G).

More experiments may provide greater detail for all of the above conclusions.

REFERENCES

Choe, G.H. 1989. "Impact Testing of Composite Materials at High Strain Rates," Master's Thesis, Department of Mechanical Engineering, University of Delaware.

Gilchrist, M.D., N Svensson. 1995. "A Fractographic Analysis of Delamination within Multidirectional Carbon/Epoxy Laminates," Composites Science and Technology, Volume 55, pp. 195-207.

Morris, G.E. 1979. "Determining Fracture Directions and Fracture Origins on Failed Graphite/Epoxy Surfaces," Nondestructive Evaluation and Flaw Criticality for Composite Materials, ASTM STP 696, pp. 274-297.

Powers, B.M. 1995a. "High Strain Rate Effects on the Properties of Composite Materials," Master's Thesis, Department of Mechanical Engineering, University of Delaware.

Powers, B.M., J.R. Vinson and I.W. Hall. October 1995b. "High Strain Rate Mechanical Properties IM7/8551-7 Graphite Epoxy Composite," Proceedings of the 10th Technical Conference of the American Society for Composites.

Powers et al. April 1995c. "High Strain Rate Mechanical Properties of Cycom 5920/1583," Proceedings of the 36th AIAA/ASME/ASCE/AHS/ASC Structures, Structural Dynamics and Materials Conference, pp. 2386-2392.

Powers et al. 1995d. "High Strain Rate Mechanical Properties of Silicon Carbide Reinforced 2080 Aluminum Metal Matrix Composites," Proceedings of the 10th International Conference on Composite Materials, pp. 317-322.

Ravichandran, G. and G. Subhash. January 1994. "Critical Appraisal of Limiting Strain Rates for Compression Testing of Ceramics in a Split Hopkinson Pressure Bar," In Journal of the American Ceramic Society, Volume 77, Number 1, pp. 263-267.

Sierakowski, R.L. 1985. "Dynamic Measurements of the Mechanical Properties of Filamentary Composites," In Dynamic Response of Composite Materials, SEM.

Zukas, J. (ed.). 1982. Impact Dynamics, John Wiley and Sons, New York, N.J.

Effect of Fiber Waviness on High Strain Rate Behavior of Composites

H. M. HSIAO AND I. M. DANIEL

ABSTRACT

The effect of fiber waviness on the high strain rate behavior of carbon/epoxy composite materials was investigated. Falling weight impact and Split Hopkinson Pressure Bar systems were developed for basic characterization of unidirectional aligned composites in compression at strain rates up to 1800 s^{-1}. Seventy-two and forty-eight ply unidirectional carbon/epoxy (IM6G/3501-6) laminates loaded in the longitudinal and transverse directions were characterized. Compression tests of 45° off-axis laminates of the same unidirectional material were also conducted to obtain the in-plane shear stress-strain behavior. Strain rates over a wide range, from quasi-static up to 1800 s^{-1}, were recorded. An incremental analysis was then used to predict the high strain rate behavior of composites with fiber waviness using the results obtained from the strain rate characterization. It is shown that, under longitudinal compressive loading, the nonlinear stress-strain behavior shows significant stiffening as the strain rate increases due to the shear component involved.

INTRODUCTION

Fiber waviness is a type of manufacturing defect that occurs particularly during the filament-winding process. Inspection of cylinder microstructures often reveals localized regions of layer waviness in the hoop direction resulting from post consolidation and cure shrinkage. Fiber/layer waviness represents one major factor that may partially explain why failure analysis of hydrostatically loaded composite cylinders based on properties obtained from flat and autoclave-cured specimens do not correlate well with cylinder test results.

The degrading effect of fiber waviness on the stiffness and strength of composites has been studied by various investigators (Kuo, Takahashi and Chou, 1988; Telegadas and Hyer, 1992; Bogetti, Gillespie and Lamontia, 1994). Analytical models were developed by the authors to predict the elastic properties, stress-strain behavior and compressive strength degradation as a function of fiber waviness, which were compared favorably with their experimental results (Hsiao and Daniel, 1996a, 1996b, 1996c). In this paper these models were extended to incorporate the effect of fiber waviness on the dynamic behavior of composites based on strain rate data obtained from unidirectional aligned composites through the application of drop weight and Split Hopkinson Pressure Bar (SHPB) techniques.

H. M. Hsiao and I. M. Daniel, Robert R. McCormick School of Engineering and Applied Science, Northwestern University, Evanston, IL 60208

HIGH STRAIN RATE CHARACTERIZATION

Seventy-two and forty-eight-ply unidirectional laminates were selected for basic high strain rate characterization. The material used was IM6G/3501-6 carbon/epoxy composite. Falling weight impact and SHPB systems were developed for dynamic characterization of composite materials in compression at strain rates up to 1800 s^{-1} (Hsiao, Daniel and Cordes, 1997). Strain rates below 10 s^{-1} were generated using a servohydraulic testing machine. Strain rates between 10 s^{-1} and 500 s^{-1} were generated using the drop tower apparatus. Strain rates above 500 s^{-1} were generated using the Split Hopkinson Pressure Bar.

TRANSVERSE COMPRESSIVE BEHAVIOR

Transverse stress-strain curves to failure under quasi-static and high strain rates are shown in Fig. 1. This comparison shows a significant strain rate effect. The transverse strength, which is a matrix dominated property, shows a nearly twofold increase from the quasi-static value. The initial modulus follows a similar trend, although not as pronounced, with an increase of up to 37%. The ultimate strain shows no strain rate effect at all, which implies that it can be used as a failure criterion in analysis under dynamic loading. Figure 1 also shows that the stress-strain behavior is a function of strain rate. The material stiffens as the strain rate increases. This stiffening behavior is very significant in the nonlinear region between quasi-static and 1 s^{-1} strain rates. Above the 1 s^{-1} strain rate, the material continues stiffening until it behaves almost linearly at a strain rate of 1800 s^{-1}. Two possible reasons for this phenomenon are proposed. The first one is the viscoelastic nature of the polymeric matrix itself and the second one is the time-dependent damage accumulation. At lower rates, damage accumulates more gradually such that a well defined nonlinear region occurs near the end of the stress-strain curve. At higher rates, however, damage does not have enough time to develop and thus the damage accumulation process has a diminishing effect on the stress-strain curve as the strain rate increases.

Figure 2 shows the variation of strength with strain rate. Two strain rate regions can be distinguished, which implies that the deformation mechanisms of the polymeric composite could change as the strain rate increases. In the first region, corresponding to strain rates between quasi-static and approximately 10 s^{-1}, there is a linear dependence of strength on the logarithm of strain rate. In the second region, corresponding to higher strain rates, the strength seems to be a power function of strain rate. Polymer viscoelasticity suggests that shear flow is highly impeded at high strain rates and, correspondingly, the relaxation modulus has an exponential-like increase for very short loading durations.

IN-PLANE SHEAR BEHAVIOR

The 45° off-axis compression test was used to obtain the in-plane shear behavior of the unidirectional composite. Figure 3 shows the shear stress-strain curves obtained under quasi-static and high strain rates of loading. This comparison reveals a strong strain rate effect. The shear stress-strain behavior, which is also matrix-dominated, shows high nonlinearity with a plateau region at a stress level that increases significantly as the strain rate increases. The "yield point" of the curve also increases with increasing strain rate. The strength increases sharply with strain rate from the quasi-static value by up to 80%. The initial modulus follows a similar trend, although not as pronounced, with an increase of up to 18%. The shear strength-strain rate relationship follows a similar trend as the one observed under transverse compression (Fig. 2).

LONGITUDINAL COMPRESSIVE BEHAVIOR

Figure 4 displays the longitudinal stress-strain curves to failure under quasi-static and high strain rates of loading by the pure end loading method. The stress-strain curve stiffens as the strain rate increases, although the magnitude of the change is much smaller compared to the transverse and shear behavior. The initial modulus shows only a slight increase with strain rate. The strength and ultimate strain are significantly higher than the static values by up to 79% and 74%, respectively. The increase in strength and ultimate strain observed may be related to the shear behavior of the composite and the change in failure modes. It is known that longitudinal compressive failure is intimately related to and governed by the in-plane shear response of the composite, even in the presence of the slightest initial fiber misalignment (Budiansky, 1983; Daniel, Hsiao and Wooh, 1996). Any compressive failure observed follows some form of initial shear failure of the composite. The longitudinal compressive strength can be predicted by the relation

$$F_{1c} = \left. \frac{\tau(\gamma)}{\varphi + \gamma} \right|_{max} \qquad (1)$$

where φ is the initial fiber misalignment and $\tau(\gamma)$ the nonlinear shear stress-strain relation

The longitudinal compressive strength is the maximum of the above function. The strength value corresponds to the tangent on the shear stress-strain curve intersecting the abscissa at $-\varphi$. It is apparent from this equation that if the in-plane shear behavior stiffens substantially with increasing strain rate, then, this would lead to the increase in longitudinal compressive strength. Figure 3 gives a good illustration of the increase in longitudinal compressive strength and ultimate strain observed during the tests. This relationship can also be used to explain the significant increase observed in the compressive strength of woven composites. It is found that a woven composite is more sensitive to strain rate than a unidirectional one, and that different woven composites have different strain rate effects due to the reinforcement structure. Plain-weave composites show a higher strain rate effect than satin-weave, unidirectional or multidirectional ones. This is attributed to the higher shear-dependence in the plain-weave case than in the other cases.

EFFECT OF FIBER WAVINESS

Figure 5 shows an ultrasonic C-scan image of a 72-ply $[(0_8/90_8)_2/\bar{0}_{8w}]_s$ crossply carbon/epoxy composite with embedded layer waviness. A 5MHz unfocused transducer was used with the gate placed on the back surface of the plate. It is shown that the period of layer waviness can be observed clearly by the conventional pulse-echo technique.

An incremental analysis was used to predict the dynamic behavior of wavy composites using the basic strain rate characterization data. A representative volume encompassing one period of waviness (Fig. 6) was divided into infinitesimally thin slices, which were analyzed as off-axis laminae. Successive load and time increments were applied with continuous monitoring of the resulting stress-strain behavior. The load increment can be controlled in such a way that either constant strain rate or constant loading rate can be achieved. The incremental analysis algorithm used consists of the following steps:

(1) Increase the applied stress and time by small increments.

(2) Calculate the stresses and the corresponding loading rates in the principal material directions for each dx segment.

(3) Calculate incremental strains for each segment.

(4) Update the effective (average) strain and strain rate for the entire representative volume by integration.

(5) For each segment use the stresses and loading rates obtained in step 2 to find the corresponding instantaneous tangential stiffnesses in its principal material directions from the basic strain rate characterization curves by interpolation/extrapolation on the $\log S_{ij}(\sigma_{ij}) - \log \dot{\sigma}_{ij}(\sigma_{ij})$ scale to update the material properties for the next increment.

(6) Update the effective strain and strain rate (or loading rate) at every step to obtain the cumulative stress-strain behavior of the wavy composite.

(7) Obtain the Young's modulus by curve-fitting the initial part of the stress-strain curve.

Figure 7 illustrates the predicted major Young's modulus of IM6G/3501-6 material as a function of strain rate and waviness parameter A/L. It is shown that the major Young's modulus degrades seriously as the fiber waviness increases. It increases moderately as the strain rate increases for the same waviness parameter A/L.

Figures 8 and 9 illustrate the predicted stress-strain curves (A/L=0.075) under uniaxial compressive loading σ_x for constant strain rate and constant loading rate, respectively. It is shown that strong nonlinearity exists due to fiber waviness. The nonlinear stress-strain curve stiffens significantly as the strain rate increases due to the shear component involved.

CONCLUSIONS

Thick unidirectional laminates were used for basic high strain rate characterization. Falling weight impact and SHPB systems were developed for dynamic characterization of composite materials in compression. The transverse compressive strength increases sharply with strain rate to nearly double the quasi-static value at the highest rate. The initial modulus follows a similar trend, although not as pronounced, with an increase of up to 37%. The ultimate strain shows no strain rate effect at all, which implies that it can be used as a failure criterion for analysis under dynamic loading. The shear stress-strain behavior shows high nonlinearity with a plateau region at a stress level that increases significantly as the strain rate increases. The shear strength increases sharply with strain rate from the quasi-static value by up to 80%. The longitudinal compressive strength and ultimate strain are significantly higher than the static values by up to 79% and 74%, respectively. The increase in longitudinal strength and ultimate strain observed may be related to the stiffening of the composite in-plane shear behavior under dynamic loading and the change in failure modes.

An incremental analysis was used to predict the dynamic behavior of wavy composites using the basic strain rate characterization data. Successive load and time increments were applied to a representative volume with continuous monitoring of the resulting stress-strain behavior. Results show that the major Young's modulus increases moderately as the strain rate increases for the same waviness parameter A/L. It is also shown that strong nonlinearity occurs in the stress-strain behavior due to fiber waviness with significant stiffening as the strain rate increases due to the shear component.

ACKNOWLEDGMENTS

The work described in this paper was sponsored by the Office of Naval Research. We are grateful to Dr. Y. D. S. Rajapakse of ONR for his encouragement and cooperation.

REFERENCES

Bogetti, T. A., Gillespie, J. W. and Lamontia, M. A. 1994, "The Influence of Ply Waviness with Nonlinear Shear on the Stiffness and Strength Reduction of Composite Laminates," *Journal of Thermoplastic Composite Materials*, 7:76-90.

Budiansky, B. 1983, "Micromechanics," *Computers and Structures*, 16:.3-12.

Daniel, I. M., Hsiao, H. M. and Wooh, S. C. 1996, "Failure Mechanisms in Thick Composites under Compressive Loading," *Composites Part B: Engineering*, 27B(6):543-552.

Hsiao, H. M. and Daniel, I. M. 1996a, "Nonlinear Elastic Behavior of Unidirectional Composites with Fiber Waviness under Compressive Loading," *Journal of Engineering Materials and Technology, Trans. ASME*, 118(4):561-570.

Hsiao, H. M. and Daniel, I. M. 1996b, "Effect of Fiber Waviness on Stiffness and Strength Reduction of Unidirectional Composites under Compressive Loading," *Composites Science and Technology*, 56(5):581-593.

Hsiao, H. M. and Daniel, I. M. 1996c, "Elastic Properties of Composites with Fiber Waviness," *Composites Part A: Applied Science and Manufacturing*, 27A(10):931-941.

Hsiao, H. M., Daniel, I. M. and Cordes, R. D. 1997, "Strain Rate Effects on the Transverse Compressive and Shear Behavior of Unidirectional Composites," submitted to *Journal of Composite Materials*.

Kuo, C.-M., Takahashi, K. and Chou, T.-W. 1988, "Effect of Fiber Waviness on the Nonlinear Elastic Behavior of Flexible Composites," *Journal of Composite Materials*, 22,:1004-1025.

Telegadas, H. K. and Hyer, M. W. 1992, "The Influence of Layer Waviness on the Stress State in Hydrostatically Loaded Cylinders: Failure Predictions," *Journal of Reinforced Plastics and Composites*, 11(2):127-145.

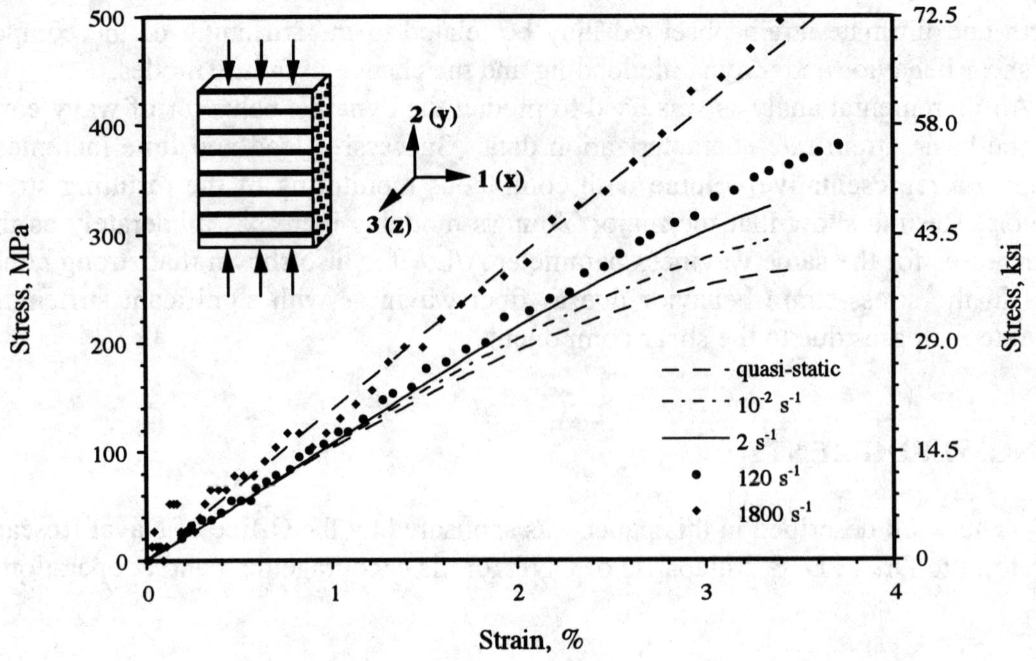

Figure 1 Transverse Compressive Stress-Strain Curves for Unidirectional IM6G/3501-6 Carbon/Epoxy under Quasi-Static and High Strain Rate Loading.

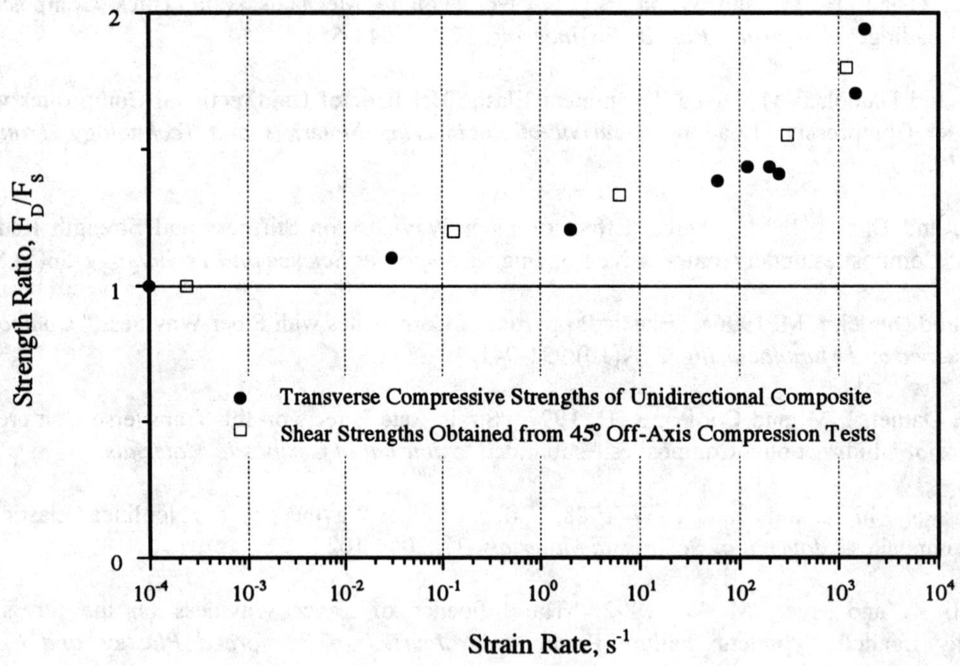

Figure 2 Variation of Transverse Compressive Strength and In-Plane Shear Strength with Strain Rate.

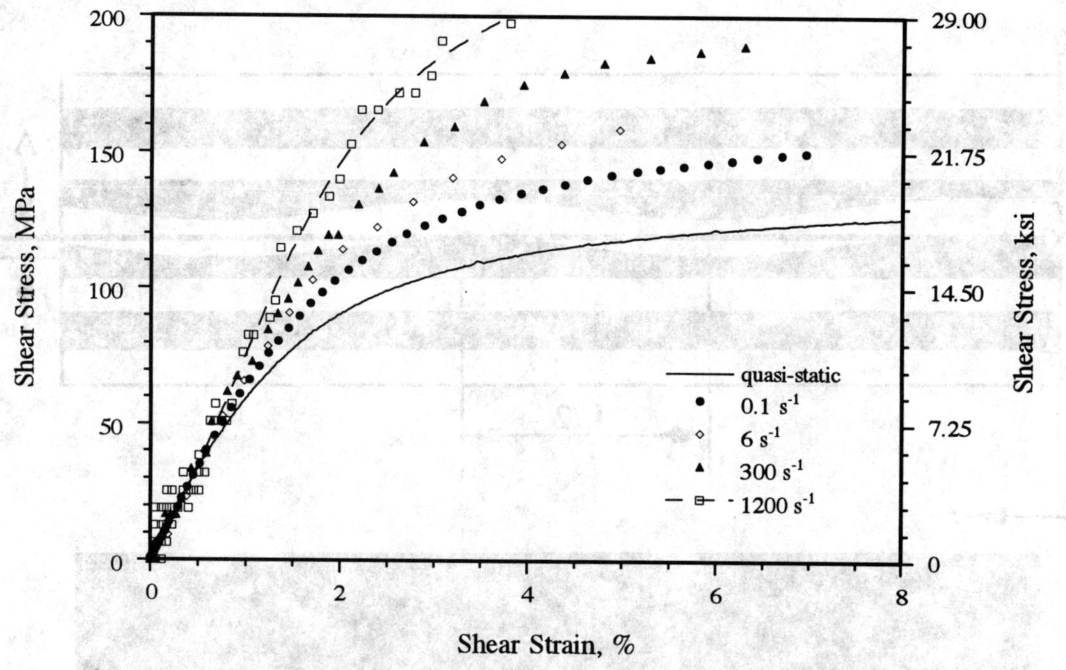

Figure 3 Shear Stress-Strain Curves for 45° Off-Axis Unidirectional IM6G/3501-6 Carbon/Epoxy under Quasi-Static and High Strain Rate Loading.

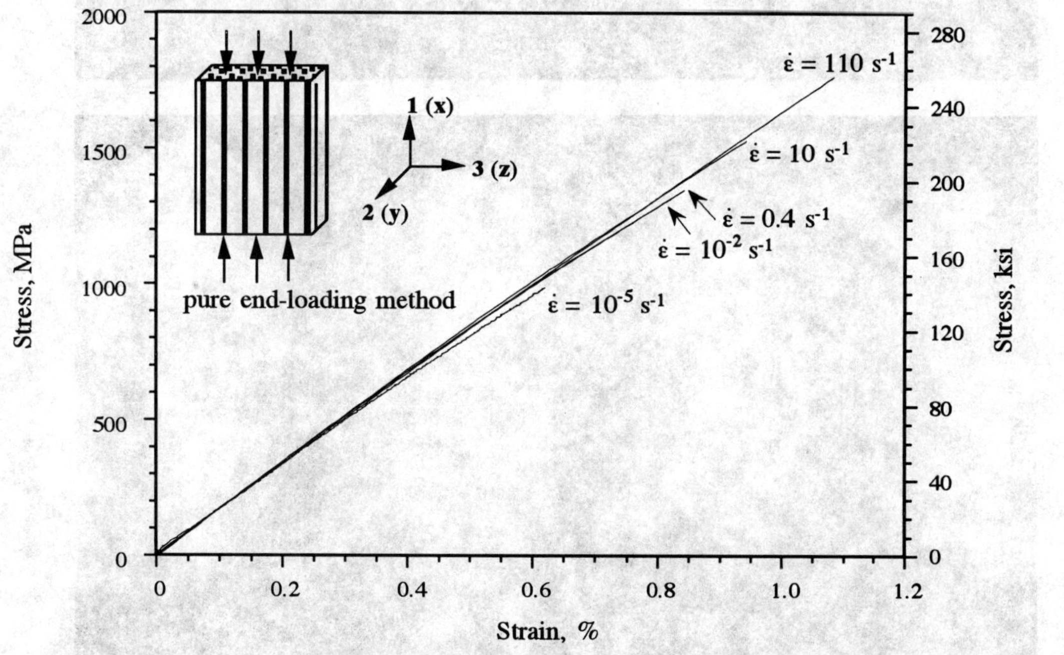

Figure 4 Longitudinal Compressive Stress-Strain Curves for Unidirectional IM6G/3501-6 Carbon/Epoxy under Quasi-Static and High Strain Rate Loading.

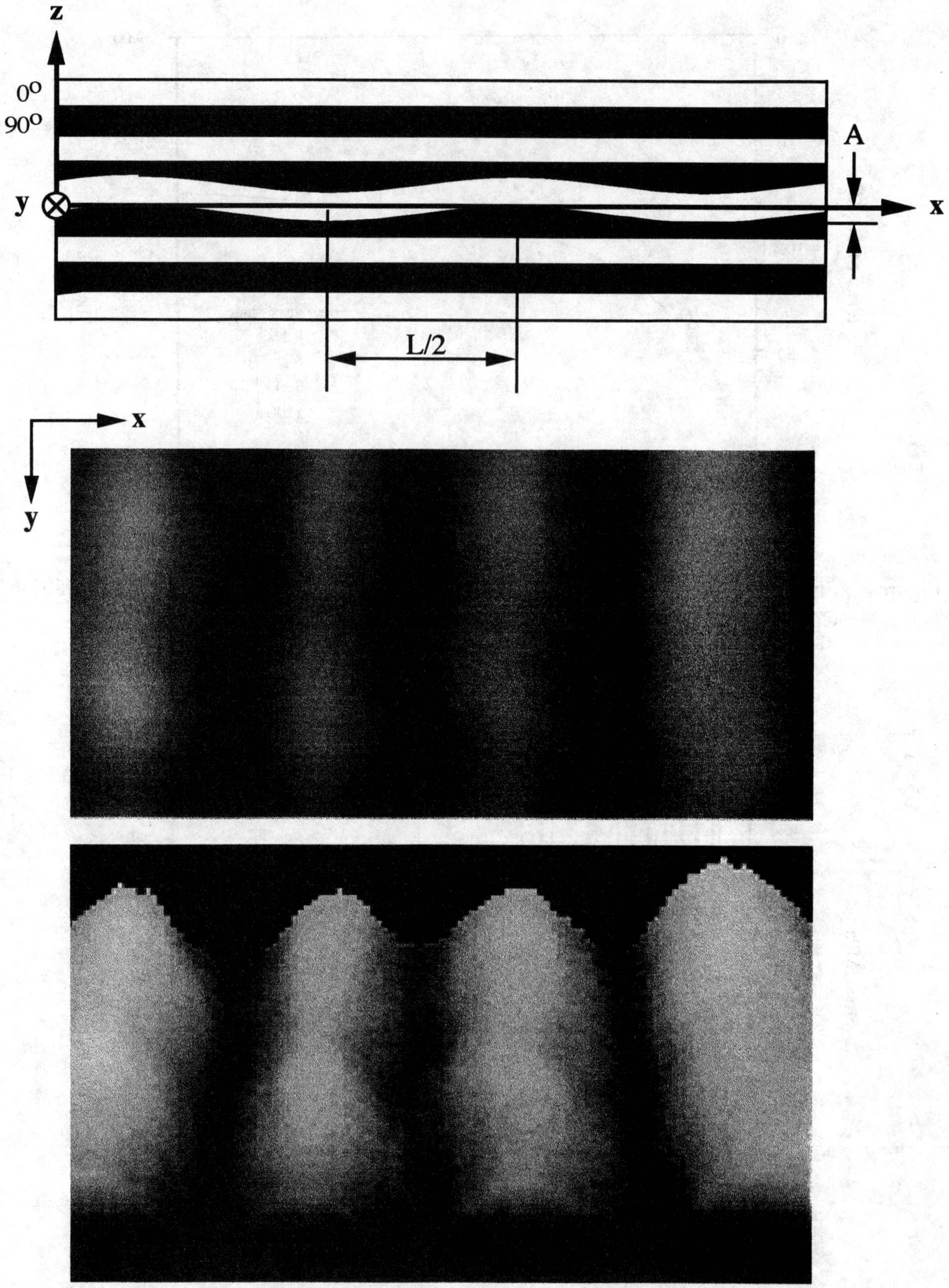

Figure 5 Ultrasonic C-scan Image of a 72-ply Crossply Composite with Embedded Layer Waviness.

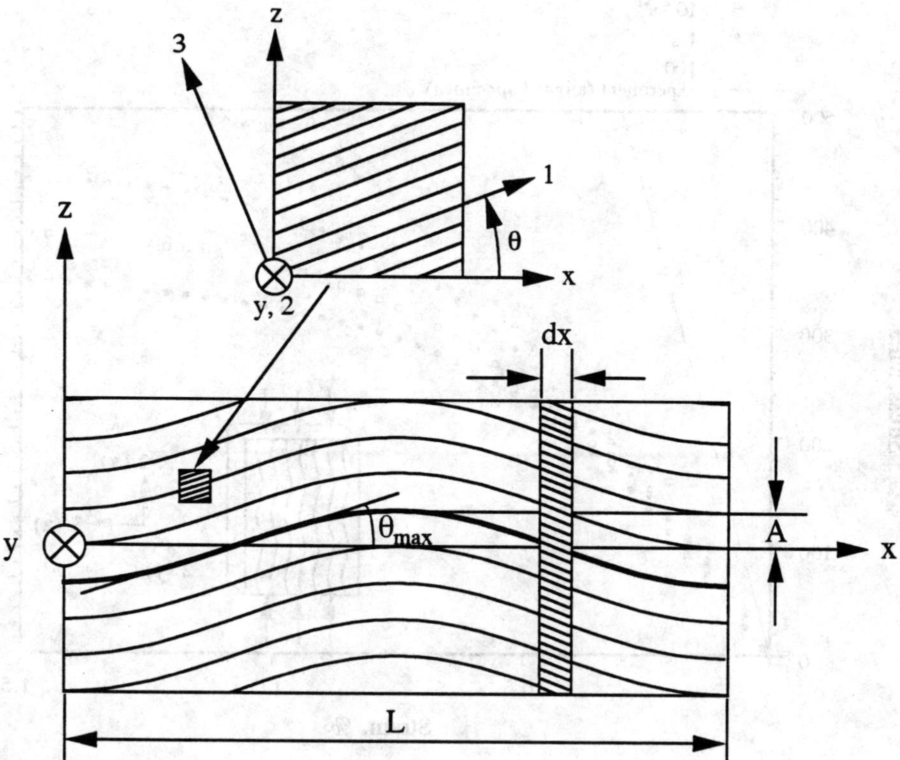

Figure 6 Representative Volume and Coordinates for a Unidirectional Composite with Uniform Waviness.

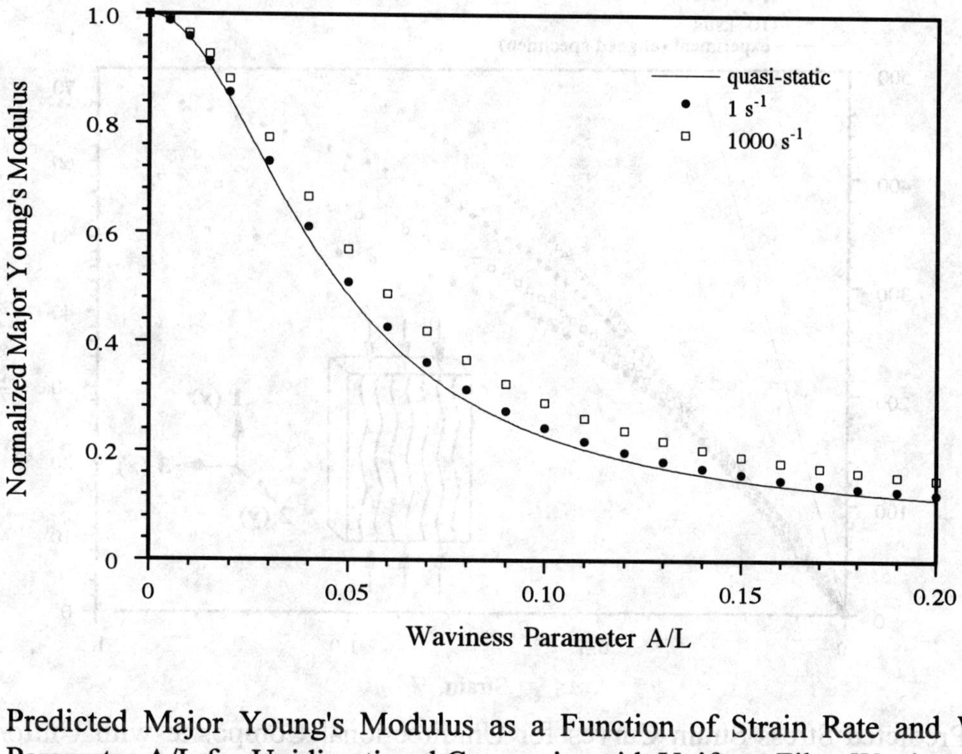

Figure 7 Predicted Major Young's Modulus as a Function of Strain Rate and Waviness Parameter A/L for Unidirectional Composite with Uniform Fiber Waviness.

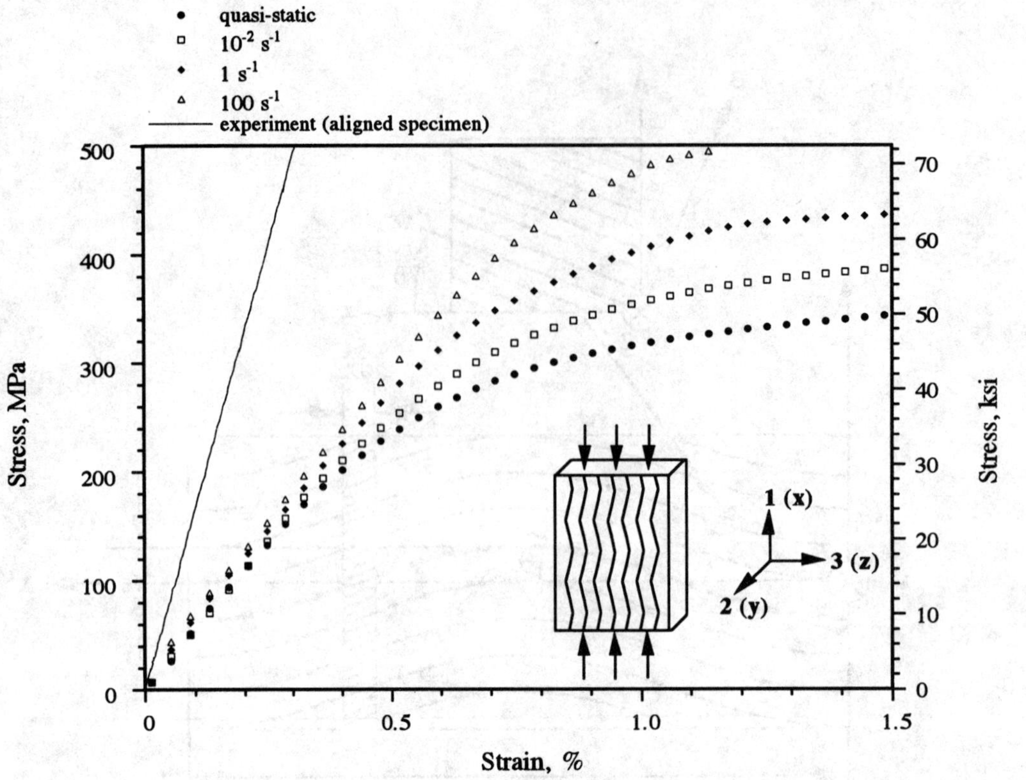

Figure 8 Predicted Stress-Strain Curves for Unidirectional Composite with Uniform Fiber Waviness under Uniaxial Compression in the x Direction (Constant Strain Rate).

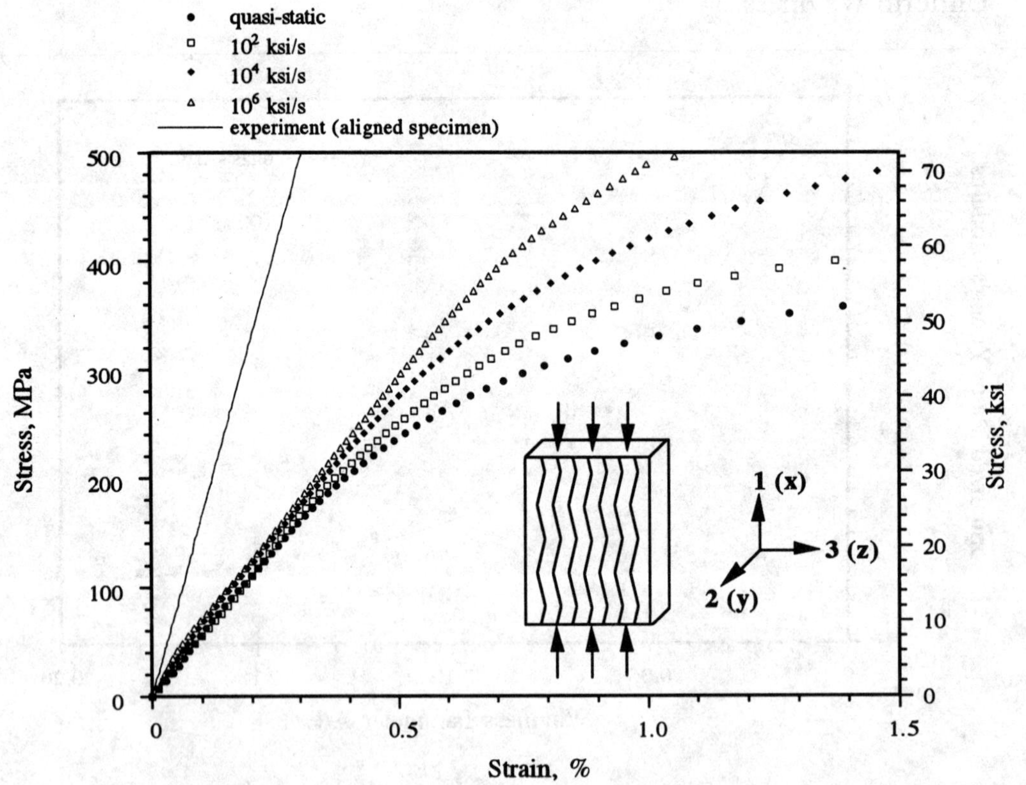

Figure 9 Predicted Stress-Strain Curves for Unidirectional Composite with Uniform Fiber Waviness under Uniaxial Compression in the x Direction (Constant Loading Rate).

Ultrasonic and Vibration NDE of Thick Composite Structures

Z. ZHENG, E. O. AYORINDE AND R. F. GIBSON

ABSTRACT: A method of obtaining the three-dimensional elastic constants of a completely free orthotropic thick plate from experimental plate vibration data, using an optimized three-mode Rayleigh formulation that incorporates through-the-thickness shear and rotary inertia, Timoshenko beam mode solution library and inverse technique with nonlinear program, is presented in the paper. The elastic constants of an aluminum plate and a glass/epoxy plate were successfully extracted, and they are shown to be consistent with results of ultrasonic tests and static test values.

INTRODUCTION:

Composite materials are becoming more attractive to many sectors of industry, such as the aerospace and automotive industries, having better performance than traditional materials in the areas of high stiffness-to-weight ratio, high strength-to-weight ratio, and really good corrosion resistance. It is therefore necessary to develop methods for reliable determination of their elastic properties. The most commonly applied composite materials consist of polymer resins reinforced with high-strength, high-stiffness fiber. Many thin laminated composite plates and panel structures in various industries are essentially orthotropic, and their elastic behavior may be adequately described by four elastic constants. For thicker structures of such materials nine independent elastic constants are required for complete elastic definition, but when they also manifest transverse isotropy only five independent elastic constants, such as the longitudinal and transverse Young's moduli, the shear moduli in two orthogonal planes, and the in-plane major Poisson's ratio, are needed to elastically define the material.

Zhiyang Zheng, Emmanuel O. Ayorinde and Ronald F.Gibson, Advanced Composites Research Laboratory, Wayne State University, Detroit, MI 48202

Early work in using vibration testing to determine elastic constants focused on isotropic materials (e.g. Zelenev and Electrova, 1973). Later work extended this to thin anisotropic plates (e.g. Deobald and Gibson , 1988; Ayorinde & Gibson, 1993). Most of the works are based on thin plate theory using the finite element or Rayleigh-Ritz method with various assumptions of vibration displacements. The method developed by Ayorinde & Gibson (1993) used the classical lamination theory and an optimized three-mode Rayleigh formulation with a suitably formed least-squares objective function in the inverse analysis.

The thin plate theory is inadequate for thick plates because of through-the-thickness shear and rotary inertia effects. Rather few methods (e.g. Mota Soares et al, 1993, and Ayorinde 1995} have been developed for thick plates with completely-free edge conditions (which can be

simulated best in the experimental testing) only a couple of approximate solutions based on thick plate theory with finite element methods or Rayleigh methods are found. Ayorinde (1995) developed an optimized three-mode Rayleigh formulation which decreases the items in the assumption of general displacement without sacrificing the precision of natural frequencies but the method is still of low efficiency compared to the thin plate theory, since where solving the highly transcendental frequency equations of Timoshenko beams takes much more time. The work of this paper improves the efficiency of the Ayorinde method.

THEORETICAL ANALYSIS

FLEXURAL VIBRATION OF RECTANGULAR THICK PLATE

We consider a rectangular plate lying at equilibrium in the x-y plane. The plate is of uniform thickness h and is composed of a number of layers, each consisting of unidirectional fiber-reinforced composite material. The material of each layer is assumed to possess a plane of elastic asymmetry parallel to the xy plane, and axes of material asymmetry also exist parallel and normal to the fiber direction. In general, of course, the material axes of an individual layer do not coincide with the plate axes.

The vibration theory of symmetrically laminated rectangular plates considering the effects of both transverse shear deformation and rotary inertia was given by Craig & Dawe (1986). For plates which are symmetrically laminated with respect to the median surface there is no bending-stretching coupling, and the displacements u, v and w at a general point in the plate can be assumed as

$$u=z\,\psi_x(x,y,t),\qquad v=z\,\psi_y(x,y,t),\qquad w=w\,(x,y,t)$$

here ψ_x and ψ_x are the total rotation along the x and y directions, respectively, of an initially straight vertical line. The constitutive equations for a symmetric laminate consisting of NL orthotropic layers are

$$\begin{Bmatrix} M_x \\ M_y \\ M_{xy} \\ Q_y \\ Q_x \end{Bmatrix} = \begin{bmatrix} D_{11} & D_{12} & D_{16} & 0 & 0 \\ D_{12} & D_{22} & D_{26} & 0 & 0 \\ D_{16} & D_{26} & D_{66} & 0 & 0 \\ 0 & 0 & 0 & A_{44} & A_{45} \\ 0 & 0 & 0 & A_{45} & A_{55} \end{bmatrix} \begin{Bmatrix} \partial\psi_x / \partial x \\ \partial\psi_y / \partial y \\ \partial\psi_x / \partial y + \partial\psi_y / \partial x \\ \partial w / \partial y + \psi_y \\ \partial w / \partial x + \psi_x \end{Bmatrix} \quad (1)$$

here M_x, M_y and M_{xy} are the bending and twisting moments per unit length, and Qx and Qy are the transverse shear forces per unit length.

The potential energy of a vibrating rectangular plate lying in the x-y plane may be written (Dawe, D.J. and Roufaeil, O.L. 1980) as:

$$\begin{aligned} \text{P.E.}_{\max} = \frac{1}{2}\int_0^a\int_0^b \Big[& D_{11}\left(\frac{\partial\psi_x}{\partial x}\right)^2 + D_{22}\left(\frac{\partial\psi_y}{\partial y}\right)^2 + 2D_{12}\frac{\partial\psi_x}{\partial x}\frac{\partial\psi_y}{\partial y} \\ & + D_{66}\left(\frac{\partial\psi_x}{\partial y} + \frac{\partial\psi_y}{\partial x}\right)^2 + 2D_{16}\frac{\partial\psi_x}{\partial x}\left(\frac{\partial\psi_x}{\partial y} + \frac{\partial\psi_y}{\partial x}\right) + 2D_{26}\frac{\partial\psi_y}{\partial y}\left(\frac{\partial\psi_x}{\partial y} + \frac{\partial\psi_y}{\partial x}\right) \\ & + A_{44}\psi_x + \frac{\partial W}{\partial y})^2 + A_{55}(\psi_x + \frac{\partial W}{\partial x})^2 + 2A_{45}(\psi_x\psi_y + \psi_x\frac{\partial W}{\partial y} \\ & + \psi_y\frac{\partial W}{\partial x} + \frac{\partial W}{\partial x}\frac{\partial W}{\partial y}) \Big] dydx \end{aligned} \quad (2)$$

The kinetic energy is

$$\text{K.E.}_{\max} = \frac{1}{2}\int_0^a\int_0^b \Big[m_1\left(\frac{\partial W}{\partial x}\right)^2 + m_2\left\{\left(\frac{\partial\psi_x}{\partial t}\right)^2 + \left(\frac{\partial\psi_y}{\partial t}\right)^2\right\} \Big] dydx \quad (3)$$

where

$$m1 = \int_{-h/2}^{h/2} \rho dz = \sum_{i=1}^{NL} \rho_l (h_l - h_{l-1}) \quad (4)$$

$$m2 = \int_{-h/2}^{h/2} \rho z^2 dz = \frac{1}{3}\sum_{l=1}^{NL} \rho_l (h_l^3 - h_{l-1}^3) \quad (5)$$

With these the beam frequency of each mode is obtained according to the Rayleigh formula ,as

$$\omega^2 = \text{P.E.}_{\max} / \text{K.E.}_{\max} \quad (6)$$

Taking into account the transverse shear and rotary inertia, the Timoshenko motion equations for translation in the z-direction and rotation about the y-axis of the length dx of beam becomes

$$EI\frac{\partial^2\psi}{\partial x^2}+\kappa^2 AG(\frac{\partial W}{\partial x}-I\rho\frac{\partial^2\psi}{\partial t^2})=0 \qquad (7)$$

$$\rho A\frac{\partial^2 W}{\partial t^2}-\kappa^2 AG(\frac{\partial^2 W}{\partial t^2}-\frac{\partial\psi}{\partial x})=0 \qquad (8)$$

where E =modulus of elasticity, G =modulus of rigidity, I =area moment of inertia of cross section, A =cross-sectional area, ρ =material mass density, κ^2 =shear coefficient.The Huang (1961) solution to this is

$$2-2\cosh b\alpha\cos b\beta+b\frac{[b^2r^2(r^2-s^2)^2+(3r^2-s^2)]}{[1-b^2r^2s^2]^{1/2}}\sinh b\alpha\sin b\beta=0 \qquad (9)$$

where

$$b^2=\omega^2\rho AL^4/EI \qquad (10)$$

$$\xi=x/L \qquad (11)$$

$$r^2=I/EI \qquad (12)$$

$$s^2=EI/\kappa^2 AGL^2 \qquad (13)$$

$$K_1=b(\beta^2-s^2)/\beta L \qquad (14)$$

$$K_2=b(\beta^2+s^2)/\alpha L \qquad (15)$$

FOUNDATION OF THICK BEAM MODE SOLUTION LIBRARY

Practical computation of plate frequencies shows that directly solving the equation (9) is most time consuming, especially with the higher frequencies of beam . Decreasing the time of this process means greatly saving the time of extracting the elastic constants in the inverse program since where the beam frequency equation needs solving many times. Here the library method is developed, and verified as a very efficient way to approach the purpose.

Two condensed factors r and s thst control the parameter b from which the beam frequency ω can be obtained in the equation (9). The r and s refer to the geometry and material properties of beams. For a beam with rectangular cross-section , the r and s can be easily found as:

$$r = \frac{1}{12}\left(\frac{h}{L}\right)^2 \tag{16}$$

$$s = \frac{E}{\pi^2 G}\left(\frac{h}{L}\right)^2 \tag{17}$$

For a given r and s, we can find some particular r1,r2 , s1,s2, which make

$$r1 \le r \le r2 \quad \text{and} \quad s1 \le s \le s2$$

and frequencies $F_i(r1,s1)$, $F_i(r1,s2)$, $F_i(r2,s1)$ and $F_i(r2,s1)$, where subscript i means the ith beam mode. The ith beam frequency according to r and s can be written as:

$$F_i(r,s) = \frac{FS2 - FS1}{r2 - r1}(r - r1) + FS \tag{18}$$

where

$$FS1 = (F_i(r1,s2) - F_i(r1,s1)) \times (s - s1)/(s2-s1) + F_i(r1,s1) \tag{19}$$

$$FS2 = (F_i(r2,s2) - F_i(r2,s1)) \times (s - s1)/(s2-s1) + F_i(r2,s1) \tag{20}$$

ELASTIC CONSTANTS FROM FREQUENCY INFORMATION

An iterative method minimizing the errors between frequencies from free vibration theory of plate and experimental frequencies is used. This process may be constructed as an optimum problem

$$\begin{cases} \min \; F(x) \\ s.t. \; G_i(x) \ge 0 \end{cases} \tag{21}$$

where $F(x)$ is the objective function, and in the thesis it is assumed to have a form

$$F(x)=\sum_{i=1}^{n}\left(1-\frac{f_{ic}}{f_{ie}}\right)^2\left|\frac{f_{ic}}{f_{ie}}\right| \tag{22}$$

f_{ic} and f_{ie} are natural frequency with same mode index coming from forward problem computation, and experimental test, respectively, and n represents total number of the experimental frequencies which are used.

$G_i(x)$ is the constaint function. It provides some limits in the optimum process.

The constrained optimum problem (20) is changed to a unconstrained optimum problem by the penalty function method as:

$$\min\ \psi(x) = F(x) + \lambda\,(G_i(x))^2 \tag{23}$$

where λ is called the penalty factor. Its value is determined as:

$\lambda = 0$, when $G_i(x) \geq 0$

$\lambda =$ large values, when $G_i(x) < 0$

This is solved by the Powell method.

ULTRASONIC METHOD

The ultrasonic immersion method of Rokhlin and Wang (1992) was used to obtain elastic constants of the samples. In this method, the sample is placed between the ultrasonic transducer and an aluminum reflector. The locations are fixed. The times of flight for ultrasonic waves to reach and arrive back from the sample and also the aluminum reflector are monitored as the sample is angularly rotated. The phase velocities are obtained from the geometry and speed parameters, and the elastic constants are derived from these.

RESULTS AND DISCUSSION

The natural frequencies of flexural vibration of aluminum and graphite/epoxy plates with various thickness ratio were investigated by using the theory, NASTRAN and ABAQUS in the research, and were found to be in excellent agreement.

The beam mode library developed was applied to the plates above. Both the computer time of the forward program and the accuracy with the library and without the library were investigated. The accuracy and the computer time of the forward program with and without the library are given in Table 1 and Figure 1 for the graphite/epoxy plate. The results show that using beam mode library can save the time of forward problem about 30% to 80% from thickness ratio 0.075 to 0.15. It is predicted to save even more time in higher thickness ratios. Irrespective of thickness ratio (in the range investigated) the frequencies of plates calculated with library are of very high accuracy (the errors are less than 0.1%).

We investigated the use of the vibration method for identification of elastic constants of unidirectional glass/epoxy composite using a plate (126.9 mm × 127.0 mm × 13.28 mm). The plate was fabricated from prepreg Scotchoply layers in an autoclave-style press at the Advanced Composite Material Laboratory of Wayne State University. An aluminum plate (127×127×12.7 mm) with the density of 2770 kg/m^3, and Young's modulus of 72.4 GPa and Poisson's ratio of 0.33 was also investigated. The experimental mode indices and corresponding frequencies of the glass/epoxy and the aluminum plates were obtained by vibration testing, and then the experimental modal information were input into the inverse program to extract the elastic constants. For the aluminum plate, the results show good agreement with the static values. For the glass/epoxy plate, the Young's moduli E_1 and E_2, and the shear modulus G_{12} are consistent with the results of the ultrasonic and static tests. Thickness direction shear data could not be obtained because of insufficient thickness for specimen preparation.

CONCLUSION

The vibration method presented in this research, which uses the experimental plate vibration modal data, forward program with beam mode library and inverse program with nonlinear program, successfully extracts the elastic constants of the glass/epoxy plate and the aluminum plate with great efficiency. It provides an effective, efficient and nondestructive way for the identification of elastic properties of thick composites. The rsults are comparable to ultrasonic, impact (Split Hopkinson Bar), and staic results.

ACKNOWLEDGMENTS

The authors are grateful to the Office of Naval Research and Program Officer Dr. Yapa D.S. Rajapakse for support for this research, and to Professor Jack Vinson, University of Delaware for the Split Hopkinson Pressure Bar tests on our specimens.

Table 1. Modal frequencies of Graphite/Epoxy plate with and without beam library solution

Mode Indices	Natural Frequencies of Plate (Hz) Thickness Ratio=0.085		Thickness Ratio=0.13	
	Without Lib.	With Lib.	Without Lib	With Lib
(2,2)	741.2875	741.2820	1104.0587	1104.0515
(1,3)	859.5015	859.5013	1280.9271	1280.9270
(2,3)	1696.7078	1696.6931	2457.1233	2457.1102
(1,4)	2294.6044	2294.6016	3303.7508	3303.7506
(3,1)	2845.3857	2845.3843	3969.4178	3969.4171
(2,4)	3083.6574	3083.5507	4335.8628	4335.8007
(3,2)	3142.3587	3142.4245	4352.1499	4352.1964
(3,3)	3982.7981	3982.9349	5414.1593	5414.2496

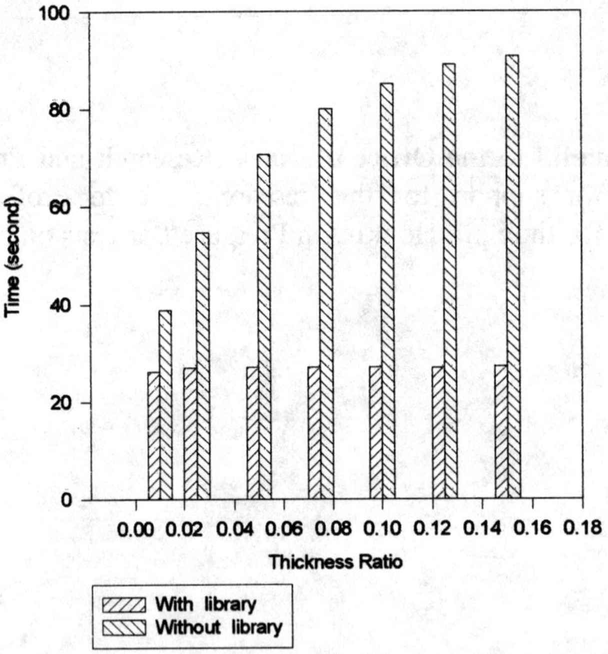

Fig. 1. Time of Forward Problem - Graphite/epoxy

Table 2. Elastic constants of glass/epoxy.

	Vibration	**SHP B**	**Ultrasonic**	**Static**
E_1(GPa)	43. 61	39.78-64.3	47.06	43.93
E_2(GPa)	13.72	11.12-14.9	13.28	12.36
G_{12}(GPa)	4.87		5.00	4.66
G_{23}(GPa)	3.61		3.51	3.58#
ν_{12}	0.30		0.29	0.26#

(SHPB= Split Hopkinson Pressure Bar - Low to High strain rates)
From C_{ij} values for typical glass/epoxy (Zimmer and Cost, 1970)

Table 3. Elastic constants of aluminum from vibration and static tests

	Vibration	**Static**
E_1(GPa)	70.13	72.40
E_2(GPa)	70.66	72.40
G_{12}(GPa)	27.20	27.20
G_{23}(GPa)	25.64	27.20
ν_{12}	0.36	0.33

References

Ayorinde, E.O., "Elastic Constants of Thick Orthotropic Composite Plates", *Journal of Composite Materials*, Vol. 28, No. 8, pp. 1025-1039, (1995).

Ayorinde, E.O., and Gibson, R.F., "Elastic Constants of Orthotropic Composite Materials Using Plate Resonance Frequencies, Classical Lamination Theory and an Optimized Three-mode Rayleigh Formulation", *Composites Engineering*, Vol. 3, No. 5, 395-407, (1993).

Craig, T.J., and Dawe, D.J., "Flexural Vibration of Symmetrically Laminated Composite Rectangular Plates Including Transverse Shear Effects", *International Journal of Solids and Structures*, Vol. 22, No. 2, 155-169, (1986).

Dawe, D.J., and Roufaeil, O.L. , "Rayleigh-Ritz Vibration Analysis of Mindlin Plates", *Journal of Sound and Vibration*, **69**(3), 345-359, (1980).

Deobald, L. R., and Gibson, R.F., "Determination of Elastic Constants of Orthotropic Plates by a Modal Analysis/Rayleigh-Ritz Technique," *Journal of Sound and Vibration*, Vol. 124, No.2, pp. 269-297 ,(1988).

Huang, T.C., "The Effect of Rotatory Inertia and of Shear Deformation on the Frequency and Normal Mode Equations of Uniform Beams With Simple End Conditions", *Journal of Applied Mechanics*, Trans. ASME, December, 579-584, (1961).

Rokhlin, S.I., and Wang, W.,"Double through-transmission bulk wave method for ultrasonic phase velocity measurement and determination of elastic constants of composite materials," *Journal of the Acoustic Society of America*, Vol. 91, No. 6, pp. 3302-3312, (1992).

Soares, C.M. et al. ,"Identification of Material Properties of Composite Plate Specimens," *Composite Structures*, Vol. 25, pp. 277-285.

Zelenevv Yu. V., and Electrova, L.M., "Determination of the Elastic Constants of a Unidirectional Fiber Composite Using Ultrasonic Velocity Measurements," *Soviet Physical Acoustics*, Vol. 18, No. 3, pp. 339 - 341.

Zimmer, J.E., and Cost, J.R., "Determination of the Elastic Constants of a Unidirectional Fiber Composite Using Ultrasonic Velocity Measurements," *Journal of the Acoustical Society of America*, Vol. 47, No. 3, pp. 793-803.

A Finite Deformation Anisotropic Plasticity Model for Fiber Reinforced Composites

H. D. ESPINOSA, H. C. LU, S. K. DWIVEDI AND P. D. ZAVATTIERI

ABSTRACT

A total lagrangian explicit formulation is presented to model the large deformation response of fiber reinforced composite material to impact and penetration. The inelastic response of each ply is described by an anisotropic rate dependent plasticity model. The parameters of the anisotropic constitutive model are determined experimentally using the off-axis and the Arcan shear tests. The interface between plies is treated separately through a contact/interface algorithm which takes into account delamination, opening and subsquent closing of the interface and large sliding between plies. The response parameters of the interface elements are modelled in terms of the actual materials properties. Hence, attempt has been made to model the pre- and post-failure response of fiber reinforced composite targets based on exprimentally observed failure mechanisms. The formulation is thought to simulate the penetration of GRP composite plates at impact velocities in the range of 180-200 m/s, Espinosa et al. 1996.

INTRODUCTION

Fibre reinforced composite materials have gained prominence as advanced materials owing to their high stiffness/density and strength/ density ratio. Specially, the glass fibre reinforced plastic (GRP) has become the potential material for designing weight efficient armor systems and combat vehicles. It becomes compulsary to evaluate their structural integrity under dynamic conditions at high strain rates before making full utilization in such applications.

The overall energy absorbing capability of fibre reinforced materials depend upon their consituents which include fibers, matrix, fibre-matrix interface and interface between plies. There have been a spate of studies over the last decade to gain insight into the failure mechanisms induced by impact damage in fiber- reinforced composite materials. Different failure modes including delamination, interlaminar matrix cracking, fiber/matrix debonding, fiber breakage and fiber pull-out have been observed. At low velocity impact, matrix fracture and interply delamination have been found to govern the damage of composites. While, extensive fiber fracture and microbuckling have been observed at high velocity

[0]Horacio Dante Espinosa, Associate Professor; Hung Chen Lu, Reseaech Assistant; Sunil Kumar Dwivedi, Post-Doctoral Researcher; Pablo D. Zavattieri, Research Assistant; School of Aeronautics and Astronautics, Purdue University, West Lafayette, IN 47907-1282

impact. In the latter case, the magnitude and extent of damage is a function of the penetrator shape and mass, striking velocity, plate thickness, and composite constituents' properties.

For instance, higher fiber strength provide better impact resistance to composites (Broutman and Rotem, 1975). According to Beaumont, 1979, the post-debond fiber sliding is the primary energy absorbing mechanism in glass fiber-reinforced composites, whereas fiber pull-out is responsible for much of the toughness in a carbon fiber composite. On the other hand, the tensile strength of the matrix material has a significant effect on impact behavior (Williams and Rhodes, 1982, Hunston, 1984, Elber, 1985). The matrix properties govern the damage threshold and determine the extent of impact damage while the fibre properties govern the resistance to penetration. The toughness of matrix is fully transferred to the toughness of composite for brittle polymers, while only partial transfer takes place for tougher polymers resins. Similarly, the interface between matrix and ply and those between plies affect the impact resistance of composites. The transverse fracture energy of a composite directly depends on the matrix-ply bond strength (Dorey, 1980, Ying, 1983). Measures to increase the interply bond strength have been suggested to enhance the impact resistance of composites (Sun and Rechak, 1988, Pelstring and Madan, 1989).

Additionally, numerical techniques have been used to study the impact resistance of fiber- reinforced composites. Lee and Sun, 1993a and b, studied the damage mechanisms of AS4/PEEK composites under quasi-static penetration experiments and established the load-deflection relation. This was used in a dynamic analyses to determine the ballistic limit of graphite/epoxy laminates. A comprehensive review of experimental as well as numerical work can be found in Abrate 1991, 1994 and Cantwell and Norton, 1991.

The response of fiber reinforced composite materials is a complex function of many factor, e.g., strength of matrix vis-a-vis fiber, fiber orientation, failure mechanism such as delamination, fiber breakage, fiber pull out, and the material post-failure response. One type of composite material may behave quite differently from another under the same load. Furthermore, recent experimental findings have been found to be contradictory to some of past experimental results, which in part may be due to the advancement of the measurement techniques. While there have been very few attempts to simulate the penetration of composite targets through modelling of non-linear phenomena, e.g., constitutive behavior, failure, delamination, contact/friction, etc. These new findings prompted the need for experimental as well as numerical investigations dealing with damage resistance and damage tolerance of fiber composites.

The approach is to understand the response of composite materials to impact loading through experiments and translate the findings to numerical simulations through proper modelling of inelasticity and pre- and post-failure constitutive behavior for full scale design and analyses. With this in view, Espinosa et al., 1996, devised a new technique for recording projectile tail velocity histories using a normal velocity interferometer (NVI) and target back surface velocity with multi-point interferometry. The technique has been used to study the dynamic failure mechanisms of woven fiber-reinforced plastic laminates (GRP) made of S-2 glass fibres embedded in a polyester resin matrix with approximately 60% fiber by volume. The penetration experiments have been carried out with the conical shaped penetrator having rounded head in the velocity range of 180-200 m/s. Three

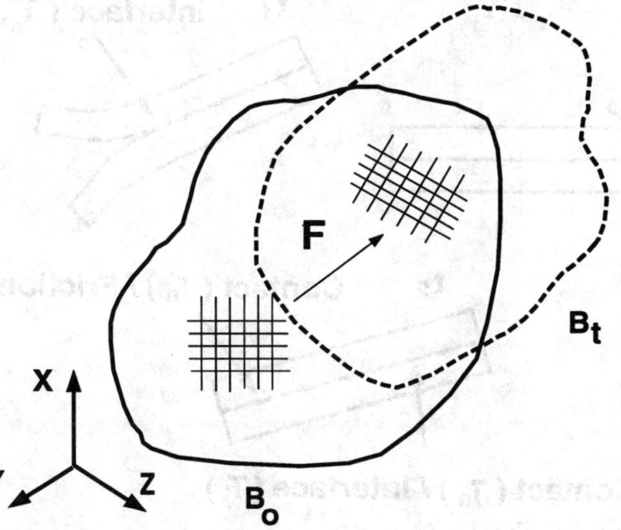

Figure 1: **Total Lagrangian Continuum Model**

distinctive zones of failure have been observed which includes extensive delaminations and fiber shearing, tensile fiber failure and large fiber deflection and lastly fiber microfracture and buckling.

In parallel, Espinosa and co-workers developed a multiple-plane microcracking model to describe the inelastic response of ceramics and a versatile contact algorithm with cohesive interface to deal with the complex situations involving debonding, large sliding and subsquent closing and opening of interfaces, Espinosa et al., 1996, 1997. The contact-interface model has been applied to simulate the delamination/contact of GRP plates as studied experimentally. A finite deformation anisotropic plasticity model was formulated and experiments carried out to determine the anisotropic model parameters of GRP. The mathematical formulation and experimental results are given in the following sections.

FORMULATION

It is evident from the above that the deformation of composite materials during impact and penetration is predominantly large involving large rotations and failure through various mechanisms. A total lagrangian continuum model as shown in Figure 1 has been adopted in the present formulation.

The large deformation and rotations of material are accounted through the energy conjugate second Piola Kirchoff's stress and Green Lagrnage's strain tensors in the initial undeformed configuration B_0. The constitutive behavior of the material is determined experimentally and included in the model. Additionally, a contact/ interface model is developed to deal with the phenomena of delamination during the deformation. As shown in Figure 2, the interface may be intact at time t_0. It may break and the plies separate at time t_1. As a result, one may have the situation of local delamination and/or a progressing front of delamination. The delaminated plies may come in contact as shown at time t_2

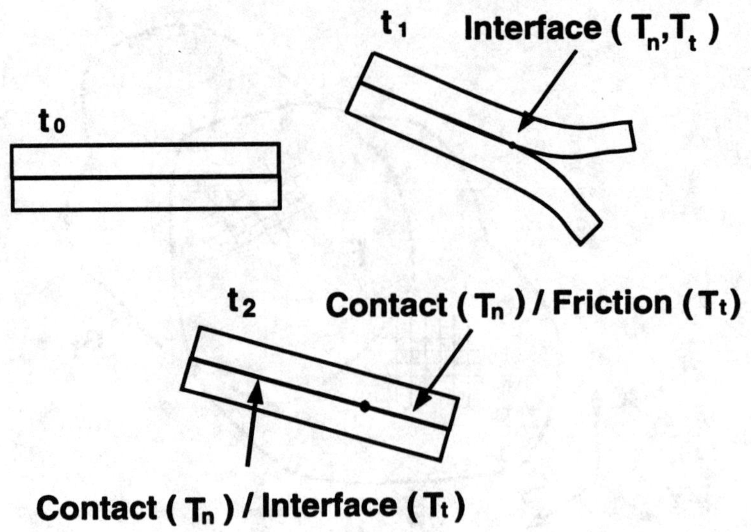

Figure 2: **Interface/Contact Model with Finite Kinematics**

and slide under compressive tractions involving frictional effects. The large deformation contact/interface model takes into account all these phenomena to predict the response of composite materials closer to reality.

FIELD EQUATIONS AND FINITE DEFORMATION ANISOTROPIC PLASTICITY MODEL

The rate form of virtual work equation in total lagrangian formulation, describing the motion from time step n to $n+1$, is given as,

$$\int_{B_0} \Delta\epsilon_{ij} L_{ijkl} \delta\epsilon_{rs} dB_0 + \int_{B_0} S_{ij(n)} \delta\eta_{ij} dB_0 - \int_{B_o} \rho_o(\mathbf{b_i} - \mathbf{a_i})\delta u_i dB_o - \int_{S_{o\sigma}} \mathbf{t}\delta u_i dS_o = 0 \quad (1)$$

where, S_{ij} is the second Piola Kirchoff's stress components, ϵ_{ij} and η_{ij} are the linear and non-linear components of the Green lagrange strain rate $\dot{E}_{ij}$, b_0, a and t are the body force vector, acceleration vector and boundary traction vector and superscript n indicates the value at time step n. In the absence of the body force and boundary traction, the above equation can be written at time step n for the explicit integration as,

$$M\ddot{U} = -F_{int} \quad (2)$$

where M is the global mass matrix, $\ddot{U}$ is the global acceleration vector and F_{int} is the internal force vector. The impactor as well as all the layers of composite target are discretized spatially into six node quadratic triangular elements. The mass of triangle is lumped at their nodes to get the global lumped mass matrix. Details on the numerical integration of the above equations can be found in Espinosa et al., 1996, 1997.

Assuming linear variation from time step n to $n+1$, the Green lagrange strain rate at new time step $n+1$ is given as,

$$\dot{E}_{ij} = \frac{1}{2\Delta t}[(F_{ki}F_{jk})_{n+1} - (F_{ki}F_{jk})_n] \tag{3}$$

where F_{ij} is the deformation gradient and Δt is the incremental time step. All quantities in the above equation are defined in the global co-ordinates. The second Piola Kirchoff's stress rate tensor is given by the constitutive law,

$$\dot{S}_{ij} = L_{ijkl}\dot{E}^e_{kl} \tag{4}$$

where L_{ijkl} is the elastic anisotropic material stiffness constitutive tensor in the global co-ordinates and $\dot{E}^e_{ij}$ is the elastic component of the Green lagrange strain rate tensor. In the case of anisotropic materials, the elastic constitutive matrix (L_l) is defined in the local co-ordinate system of the ply (Cook, R.D.,1974). It is tranformed to $L = T^T L_l T$ to obtain the constitutive matrix in the global co-ordinates following standard transformation proceedures.

The elastic components of the strain rate tensor are obtained by an additive decomposition of the total strain rate, namely,

$$\dot{E}^e_{ij} = \dot{E}_{ij} - \dot{E}^p_{ij} \tag{5}$$

In the above equation, $\dot{E}^p_{ij}$ is the plastic strain rate which is given based on the associative flow rule as,

$$\dot{E}^p_{ij} = \dot{\lambda}\frac{\partial f}{\partial S_{ij}} \tag{6}$$

Here, f is the flow potential and $\dot{\lambda}$ is the plastic rate proportionality factor. The inelastic behavior of the composite is modeled based on the small deformation yield function, quadratic in stresses, as proposed by Sun and Chen, 1989, i.e., Hill's type potential.

$$\begin{aligned} 2f(S_{ij}) = {} & a_{11}S_{11}^2 + a_{22}S_{22}^2 + a_{33}S_{33}^2 \\ & +2a_{12}S_{11}S_{22} + 2a_{13}S_{11}S_{33} + 2a_{23}S_{33}S_{22} \\ & +2a_{44}S_{23}^2 + 2a_{55}S_{13}^2 + 2a_{66}S_{12}^2 \end{aligned} \tag{7}$$

Experimental evidences show that the fiber composites behave linearly up to failure if the load is applied in the fiber direction. Hence, for orthogonal fibers oriented along direction 1 and 2, it is assumed that $E^p_{11} = E^p_{22} = 0$. The above flow potential then reduces to,

$$f(S_{ij}) = \frac{1}{2}S_{33}^2 + a_{44}S_{23}^2 + a_{55}S_{13}^2 + a_{66}S_{12}^2 \tag{8}$$

For equal fiber volume fractions in the principal 1 and 2 directions, the additional constraint of $a_{44} = a_{55}$ is obtained which further reduces the yield function to,

$$f(S_{ij}) = \frac{1}{2}S_{33}^2 + a_{44}(S_{23}^2 + S_{13}^2) + a_{66}S_{12}^2 \tag{9}$$

Defining an effective stress as,

$$\bar{S} = \sqrt{3f} \tag{10}$$

and using equations (7) and (10) and the rate of plastic work given by,

$$\dot{W}^p = S_{ij}\dot{E}^p_{ij} = \bar{S}\dot{\bar{E}}^p . \tag{11}$$

The proportionality factor of equation (6) is obtained as

$$\dot{\lambda} = \frac{3\dot{\bar{E}}^p}{2\bar{S}} \tag{12}$$

where $\dot{\bar{E}}^p$ is the effective plastic strain rate. The effect of strain rate and temperature can be modeled by defining the material strength in terms of an effective stress which includes temperature and rate terms, namely,

$$\dot{\bar{E}}^p = \dot{\bar{E}}_0^p[\frac{\bar{S}}{g(\bar{E}^p, T)}]^m \quad \text{if} \quad \bar{S} > g(\bar{E}^p, T) \tag{13}$$

$$g(\bar{E}^p, T) = \bar{S}_y[1 - (\frac{T - T_0}{T_m - T_0})^\alpha] \tag{14}$$

$\bar{S}_y$ in the above equation is the flow stress and is defined by an experimentally found power law, viz.,

$$\bar{S}_y = (\frac{\bar{E}^p}{A})^{1/n} \tag{15}$$

A summary of the constitutive equations is given in Table 1.

EXPERIMENTAL DETERMINATION OF THE ANISOTROPIC MODEL PARAMETERS

The anisotropic yield function given in equation (9) involves two coefficients which need to be determined experimentally. Following Sun and Chen, 1989, the coefficient a_{66} is determined from off-axis tensile tests to obtain a master effective stress-effective plastic strain curve.

As shown in Figure 3, let X axis be the uniaxial loading direction which makes an angle θ with respect to the fiber direction X_1. The stress referring to the material principle axes are

$$\begin{aligned} S_{11} &= cos^2\theta S_x \\ S_{22} &= sin^2\theta S_x \\ S_{12} &= -sin\theta cos\theta S_x \end{aligned} \tag{16}$$

where S_x is the applied axial stress. For the case of 2-D plane stress parallel to the X_1-X_2 plane, the plastic potential function can be reduced to,

$$f(S_{ij}) = a_{66}S_{12}^2 \tag{17}$$

Table 1: **Summary of constitutive equations**

$$\dot{S}_{n+1} = L\dot{E}^e_{n+1}$$
$$\dot{E}_{n+1} = \dot{E}^e_{n+1} + \dot{E}^p_{n+1}$$
$$\dot{E}^p_{n+1} = \dot{\lambda}\frac{\partial f}{\partial S_{n+1}} = \frac{3\dot{\bar{E}}^p}{2\bar{S}}\frac{\partial f}{\partial S_{n+1}} = \dot{\bar{E}}^p_{n+1}N_{n+1}$$
$$N_{n+1} = \frac{3}{2}\frac{\frac{\partial f}{\partial S}_{n+1}}{\bar{S}_{n+1}} = \frac{3}{2\bar{S}}\begin{bmatrix} 0 & 2a_{66}S_{12} & 2a_{55}S_{13} \\ 2a_{66}S_{12} & 0 & 2a_{44}S_{23} \\ 2a_{55}S_{13} & 2a_{44}S_{23} & S_{33} \end{bmatrix}_{n+1}$$
$$\bar{S}_{n+1} = \sqrt{3f_{n+1}(S_{ij})_{n+1}}$$
$$g(\bar{E}^p_{n+1}, T) = \bar{S}_{y,n+1}[1 - (\frac{T-T_0}{T_m - T_0})^\alpha]$$
$$\bar{S}_{y,n+1} = (\frac{\bar{E}^p_{n+1}}{A})^{\frac{1}{n}}$$
$$\dot{\bar{E}}^p_{n+1} = \dot{\bar{E}}^p_0[\frac{\bar{S}_{n+1}}{g(\bar{E}^p_{n+1}, T)}]^m$$

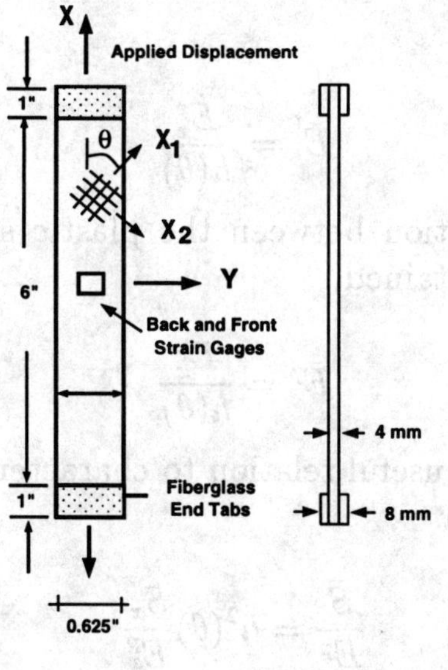

Figure 3: **Off-axis tensile test specimen**

Substituting the above equation into the effective stress equation, the effective stress can be derived in term of the stress S_x and the fiber direction θ, namely,

$$\bar{S} = \sqrt{3a_{66}cos^2\theta sin^2\theta S_x} = h(\theta)S_x \tag{18}$$

where, $h(\theta) = \sqrt{3a_{66}cos^2\theta sin^2\theta}$ is a function of the fiber orientation θ. The relationship between the incremental plastic strain and stress in term of the current state of stress reduces to,

$$\dot{\bar{E}}^p = \frac{2}{3}\bar{S}\dot{\lambda} = \frac{2}{3}h(\theta)S_x\dot{\lambda}. \tag{19}$$

The inelastic strain components are derived by substituting the yield function into the associated flow equation and are given by,

$$\begin{bmatrix} \dot{E}^p_{11} \\ \dot{E}^p_{22} \\ 2\dot{E}^p_{12} \end{bmatrix} = \begin{bmatrix} 0 \\ 0 \\ 2a_{66}S_{12}\dot{\lambda} \end{bmatrix} \tag{20}$$

Using coordinate transformations, equation for the strain in the X direction can be derived as,

$$\begin{aligned} \dot{E}^p_x &= cos^2\theta\dot{E}^p_{11} + sin^2\theta\dot{E}^p_{22} - 2cos\theta sin\theta\dot{E}^p_{12} \\ &= 2a_{66}cos\theta sin\theta S_x\dot{\lambda} = \frac{2}{3}h^2(\theta)S_x\dot{\lambda} \end{aligned} \tag{21}$$

Comparing the above equation with the equation of the effective strain, the relationship between the effective strain and the plastic strain component in X direction is obtained as,

$$\dot{\bar{E}}^p = \frac{\dot{E}^p_x}{h(\theta)} \tag{22}$$

Integrating both sides, the relation between the plastic strain in the X direction and effective plastic strain can be obtained,

$$\bar{E}^p = \frac{E^p_x}{h(\theta)} \tag{23}$$

The above equations provide an useful relation to characterize the ratio between effective stress and effective plastic strain.

$$\frac{\bar{S}}{\bar{E}^p} = h^2(\theta)\frac{S_x}{E^p_x} \tag{24}$$

Hence, determining the variation of S_x v.s. E^p_x experimentally , the variation of $\bar{S}$ v.s. $\bar{E}^p$ can be obtained from the above relation.

The effective stress-effective plastic strain relation needs to be unique in monotonic loading for a given material and strain rate. The parameter a_{66} is chosen so that the

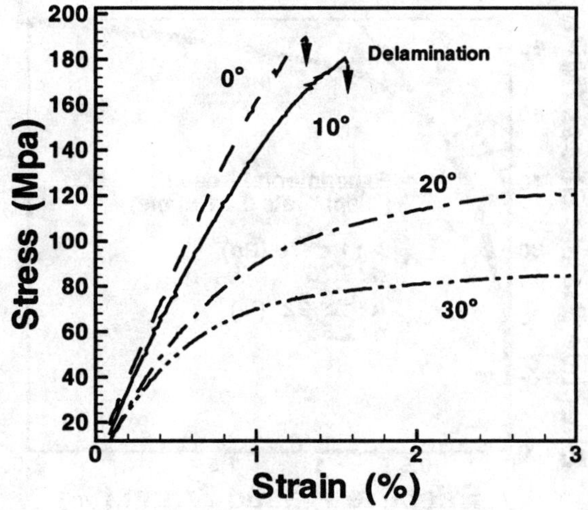

Figure 4: **Experimental stress-strain curves from off-axis tension tests**

resulting effective stress- effective plastic strain is independent of the off-axis angle θ. Experimental results have indicated that thermoplastic composites do not have a well-defined yield point so that the master effective stress-effective plastic strain curve can be fitted with a power law

$$\bar{E}^p = A(\bar{S})^n \tag{25}$$

where the power law coefficient **A** and **n** can be obtained by the curve fitting.

Experiments have been carried out at four different values of θ as 0, 10, 20 and 30. The S_x v.s. E_x curve obtained are shown in the Figure 4. Figure 5 shows the variation of $\bar{S}$ v.s. $\bar{E}^p$. Values of **A** and **n** have been obtained by fitting two curves as shown in the figure with their values.

The other coefficient, a_{44}, is related to the behavior of the composite when subjected to out-of-plane shearing. Its value can be identified from a pure shear loading obtained with an Arcan loading fixture, Arcan et al., 1978. The potential function for a pure shear test in the **1-3** direction can be simplified as,

$$f(S_{ij}) = a_{44}S_{13}^2 \tag{26}$$

since the rest of the stress components are equal to zero. By using the definition of effective stress, the relation between the effective stress and applied stress can be found as,

$$\bar{S} = \sqrt{3a_{44}}S_{xz} = \sqrt{3a_{44}}S_{13} \tag{27}$$

Applying the associated flow rule, the strain component in the **1-3** direction can be calculated in terms of the applied stress S_{xz},

$$\dot{E}_{13}^p = 2a_{44}S_{13}\dot{\lambda} \tag{28}$$

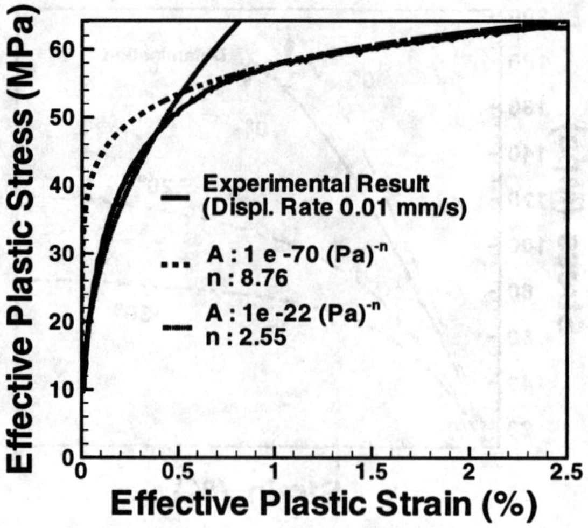

Figure 5: **Effective stress vs. effective plastic strain**

with

$$\dot{\lambda} = \frac{\dot{E}^p_{13}}{2a_{44}S_{13}} \tag{29}$$

The effective plastic strain rate can be related to the strain component S_{13} by the following expression,

$$\dot{\bar{E}}^p = \frac{2}{3}\bar{S}\dot{\lambda} = \frac{\dot{E}_{13}}{3a_{44}} \tag{30}$$

The relationship between the effective plastic strain and the experimentally measured strain can be found to be,

$$\bar{E}^p = \frac{E_{13}}{3a_{44}} \tag{31}$$

The stress and strain relation between the effective values and experimental values can thus be built as

$$\frac{\bar{S}}{\bar{E}^p} = \frac{3a_{44}S_{13}}{E^p_{13}} \tag{32}$$

Based on the above equations, the coefficient a_{44} can be obtained by matching the power law curve dentified through off-axis tests. Figure 6 shows the Arcan fixture used in our experiments.

CONTACT / INTERFACE MODEL

The above model is integrated with a dynamic contact/interface model. A versatile multi-body contact model for explicit dynamic analysis have been developed. The penetration of nodes of one body, i.e., slave nodes into the surface of other body, i.e., master

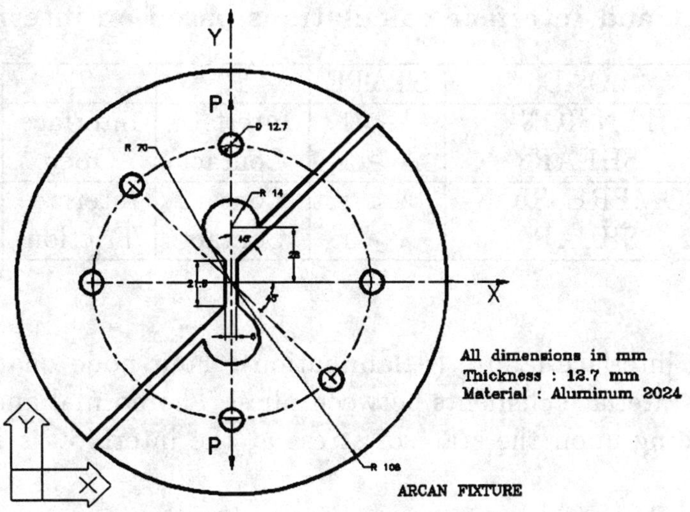

Figure 6: **Arcan's experiment for pure shear loading, Arcan et al., 1978.**

surface is checked. If found penetrating, then the force required to impose the impenetrability condition is applied to the slave nodes. The equal and opposite force is applied to master surface and properly distributed to its nodes to maintain the conservation of force. Tangential forces arising from Coulomb's friction, in which a friction coefficient is taken as the function of the relative sliding velocity, are also applied on the surface. A detailed description of the contact model can be found in Espinosa et al., 1996, 1997.

The contact model is integrated with an interface elements to simulate the delaminations and subsquent large sliding, opening and closing of the interface. The model is based on the interface model proposed by Tvergaard, 1990 for quasi-static calculations. It assumes that the interface carried forces that oppose separation and shear between two surfaces until debonding. The magnitude of these forces are function of the relative separation and shear displacement between the two surfaces. The normal and tangential tractions are given as,

$$T_n = \frac{u_n}{\delta_n} F(\lambda_c), \quad T_t = \alpha \frac{u_t}{\delta_t} F(\lambda_c) \tag{33}$$

$$F(\lambda_c) = \frac{27}{4} T_{max}(1 - 2\lambda_c + {\lambda_c}^2), \quad \text{for} \quad 0 \leq \lambda_c \leq 1 \tag{34}$$

where, λ_c is a non-dimensional parameter given as,

$$\lambda_c = \sqrt{\left(\frac{u_n}{\delta_n}\right)^2 + \left(\frac{u_t}{\delta_t}\right)^2} \tag{35}$$

In the above equation, u_n and u_t are the normal and tangential displacement jumps at the interface. δ_n and δ_t are critical values at which delamination takes place. T_{max} and δ_n and δ_t can be easily determined from energy arguments based on the mode I critical strain energy release rate of the interface. It is evident that the value of λ_c varies from **0** to **1**

Table 2: **Contact and Interface calculations based on interface traction**

LOAD	STATE	T_n	T_t
TENSION-	$\lambda < 1$	Interface	Interface
SHEAR	$\lambda \geq 1$	Contact	Open
COMPRESSION-	$\lambda < 1$	Contact	Interface
SHEAR	$\lambda \geq 1$	Contact	Friction

with $\lambda_c = 1$ defining interface failure (delamination). Four node quadrilateral elements are embedded as the interface elements between plies. The normal and tangential forces are computed depending upon the state of stress at the interface as shown in Figure 2 and Table 2.

As long as the value of λ_c remains less than unity, the interface normal traction is computed from contact model in the case of compression-shear and from interface model in the case of tension-shear. The shear traction is computed from the interface model. For values of λ_c larger than unity, delamination case, only the contact model is used to compute the interface tractions in both the states. The dependance of the interface model on the materials properties and swithching from interface to contact and vise versa make the approach realistic to simulate the delamination and subsqunet large sliding at the interface between plies. Rate effects in the delamination process can be easily incorporated in terms of the λ_c parameters as given by,

$$T_{max} = T^o_{max}(1 + \beta \ln[\frac{\dot{\lambda}_c}{\dot{\lambda}_{c,o}}]) \quad \text{with} \quad T^o_{max} = \frac{48 G_c}{27 \delta_n} \tag{36}$$

ACKNOWLEDGMENTS

This research was supported by the National Science Foundation through Grants Nos. MSS-9309006 and MSS 9311289, and by the Army Office of Scientific Research under Grant DAAH 04-95-1-0168 and Purdue-MURI'96.

REFERENCES

Abrate, S., 1991, "Impact on laminated composite materials" *Applied Mechanics Review, American Society of Mechanical Engineers* 44(4):155-189.

Abrate, S., 1994, "Impact on laminated composites: Recent Advances" *Applied Mechanics Review, American Society of Mechanical Engineers* 47(11):517-544.

Beaumont, P.W.R., 1979, "Fracture Mechanisms in Fibrous Composites," Fracture Mechanics, Current Status, Fracture Prospects edited by R. A. Smith (Pergamon Press), 211-233.

Broutman, LJ., and Rotem, A., 1975, "Impact Strength and Toughness of Fiber Composite Materials," Foreign Object Impact Damage to Composites, American Society for Testing and Materials, ASTM STP 568:114-133.

Cantwell, W.J., and Norton, J.. 1991. "The Impact Resistance of Composite Materials - A Review" *Composites,* 22(5):347-362.

Cook, R. D., 1974, "Concpets and Applications of Finite Element Analysis", John Willey and Sons Inc., New York.

Dorey, G., 1980, "Relationship between Impact Resistant and Fracture Toughness in Advanced Composite Materials," Effect of Service Environment on Composite Materials, *AGARD CP 28*

Elber, W., 1985, "Failure mechanics in Low Velocity Impact on Thin Composite Plates," *NASA Techical Paper 2152.*

Espinosa, H. D., Lu, H-C. and Xu, Y., 1996, "A novel technique for penetration velocity measurement and damage identification in ballistic penetration experiments." *to appear in J. Composite Materials.*

Espinosa, H.D., Emore, G.L., and Zavattieri, P.D., 1996, "Computational Modeling of Geometric and Material Nonlinearities with an Application to Impact Damage in Brittle Materials," *in Advances in Failure Mechanisms in Brittle Materials,* edited by R.J. Clifton and H.D. Espinosa, AMD-Vol. 219, MD-Vol 75, pp. 119-161.

Espinosa, H.D., Zavattieri, P.D. and Emore, G.L., 1997, "Adaptive FEM computation of geometric and material nonlinearities with application to brittle failure." *to appear in the Special Issue of Mechanics of Materials.*

Hunston, D.L., 1984, "Composite Interlaminar fracture ; Effect of matrix Fracture Energy," *Composites Tech Review 6*:176-180.

Lee, S-W. R., and Sun, C.T., 1993a, "A Quasi-Static Penetration Model for Composite Laminates", *J. of Composite Materials*, 27:251-271.

Lee, S-W. R., and Sun, C.T., 1993b, "Dynamic Penetration of Graphite/Epoxy Laminates Impacted by a Blunt-Ended Projectile," *Composites Science and Technology*, 49:369-380.

Williams, J.G. and Rhodes, M.D., 1982, "Effect of resin on Impact Damage Tolerance of Graphite/Epoxy Laminateds" Composite Materials: testing and Design (sixth Conference), American Society for Testing and Materials, ASTM STP 787, edited by Daniel, I.M., 450-480.

Ying, L., 1983, "Role of Fiber/Matrix Interphase in Carbon Fiber Epoxy Composite Impact Toughness," *SAMPE Quart.*, 14(3):26.

Effect of Moisture on High Strain Rate Properties of Graphite/Epoxy Composites

E. WOLDESENBET[1] AND J. R. VINSON[2]

ABSTRACT

Many applications of structural materials involving composites include impact or dynamic loading in a humid environment. Because mechanical properties vary with strain rates, it is necessary to know the properties of composites at the strain rates used in the application to ensure that designs are optimized for structural integrity and weight efficiency including when the structure is subjected to dynamic loading. In addition, composite materials are known to be subjected to degradation under humid conditions, and this adds to the complexity of determining the high strain rate behavior of composites. Exposure to moisture may lead to several undesirable effects, such as matrix degradation and fiber/matrix debonding. Several researchers have found that water absorption by composites causes degradation of matrix dominated quasi-static properties. However, very little is known of the effect of absorbed moisture on the high strain rate properties of polymer matrix composites. These dynamic properties are very useful in the automotive, aerospace, and naval applications of composite structures.

The Split-Hopkinson Pressure Bar facility of the University of Delaware is used herein to study the effect of absorbed moisture in high strain rate tests of composites. The material used in all tests is a unidirectional IM7/8551-7 graphite/epoxy composite. Cylindrical specimens were fabricated with a length to diameter ratio (L/D ratio) of 1.5. The study includes dry, medium, and saturated moisture conditions of test samples which have been immersed in water kept at room temperature. The specimens were kept in water up to 559 days. The strain rate values obtained in these tests vary from 200 - 1200/s. The tests show an overall increase in the high strain rate properties of the unidirectional IM7/8551-7 graphite/epoxy composite with moisture. In addition, a better understanding of the effect of other factors, such as matrix property and fiber/matrix interface, on the high strain rate properties are achieved.

---

[1]Doctoral Candidate, Department of Mechanical Engineering, University of Delaware.

[2]H. Fletcher Brown Professor of Mechanical & Aerospace Engineering, University of Delaware.

INTRODUCTION

The Split Hopkinson Pressure Bar facility at the University of Delaware used in this study is described in several papers (Powers et al, 1995; Woldesenbet and Vinson, 1997), and the reader is asked to refer to these papers . The Split Hopkinson Pressure Bar (SHPB) has been accepted as the test method for high strain rates since Kolsky first presented the test methodology for compression in 1949, (Kolsky, 1949). The method was developed based on wave propagation theory in elastic bars and the interaction between a stress wave and a short specimen. After Kolsky's introduction, the method was expanded to tensile testing (Nicholas, 1980) and shear testing (Duffy et al., 1971). Zukas has shown that for many metals, the mechanical properties vary significantly with strain rate, (Zukas, 1982). This new data will add to what is now available for other materials. At Delaware, material systems tested to date include glass/epoxy, glass/polyester, graphite/epoxy, carbon/metal matrix, and carbon/ceramic matrix, and the most recent papers regarding these materials are given in the References, (Powers and Vinson, 1995; Powers et al., 1995, Powers et al.,1996).

The effect of moisture on the static properties of composites have been studied for a while and there are numerous published reports, (Silverman et al, 1992; Shen and Springer, 1981; Barker and Balasundaram, 1987). This study is undertaken to study the effect of moisture on the dynamic properties of composites.

SAMPLE PREPARATION

The IM7/8851-7 carbon/epoxy specimens were made at the University of Delaware from continuous fiber unidirectional pre-impregnated tape. The tape was 16 in. wide and was cut into 48 in. strips. A 75 ply unidirectional laminate was fabricated by hand lay-up to achieve the desired thickness after processing. The laminate was processed in a Thermal Equipment Corporation computer controlled autoclave. The finished laminate was then sectioned using a water-cooled masonry saw equipped with a diamond grit blade, which usually produces no delamination and rough surfaces. Cylindrical specimens were fabricated with a length to diameter ratio (L/D ratio) of 1.5 from the sectioned pieces. The diameter and length of the samples were approximately 1/2 in. and 3/8 in. respectively.

Recently, a study was conducted using a graphite/epoxy composite that examined various L/D ratios from 0.5 to 2.0 (Woldesenbet and Vinson, 1997). Within the range of variables examined, statistically, no significant differences could be found in the material properties and this provided the basis for making cylindrical samples with the specific dimensions used in this study. The cylindrical samples were then immersed in a distilled water. The weight of five specimens with length being in each direction, that is, the fiber, transverse and thickness directions, was measured periodically to determine how much moisture they gained. A scale (a Mettler H80) was used to measure the weight of the samples which can accurately measure up to 0.0001 grams.

RESULTS & DISCUSSIONS

Figure 1 shows the moisture absorption of the IM7/8551-7 graphite/epoxy composite samples. The maximum moisture intake was lower for the samples with the fibers across the length than for the samples with the fibers across the diameter. However, the slopes of all the graphs which is an indication of the rate of absorption are about the same. Dynamic tests were performed on the SPHB after 64 days of immersion in distilled water, which was equivalent to 0.25 and 0.38% moisture in the 1 and 2(or 3) direction samples respectively, and after 559 days which was equivalent to 0.89, 1.06, and 1.02% moisture intake in the 1, 2, and 3 direction samples respectively.

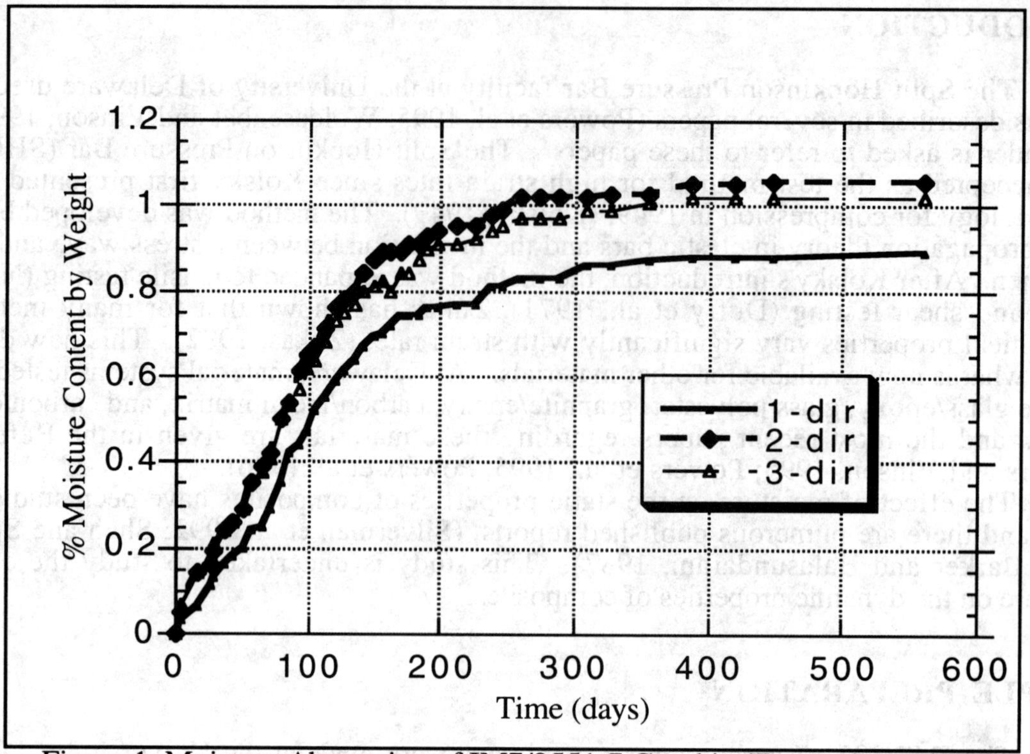

Figure 1. Moisture Absorption of IM7/8551-7 Graphite/Epoxy Composite

Figure 2- 4 show the stress vs strain rate for dry (d), wet (w), and saturated wet (sw) samples. In all tests, the wet samples which represent the specimens that were tested after 64 days of moisture absorption show the highest stress values for all the high strain rates used in this study.

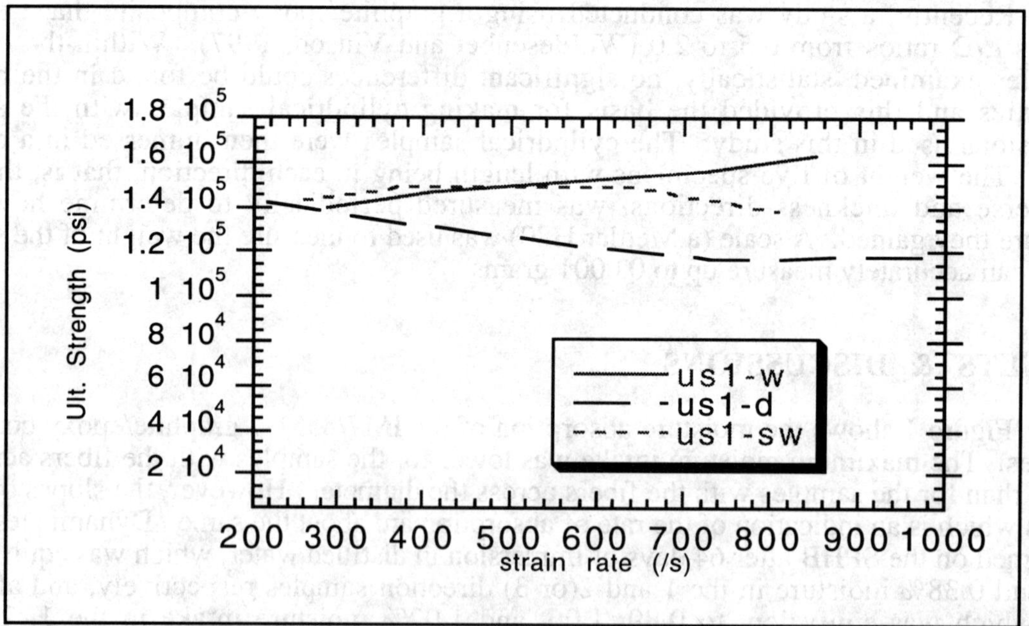

Figure 2. Comparison of 1-dir.dynamic ultimate stress of dry and wet IM7/8551-7 composite.

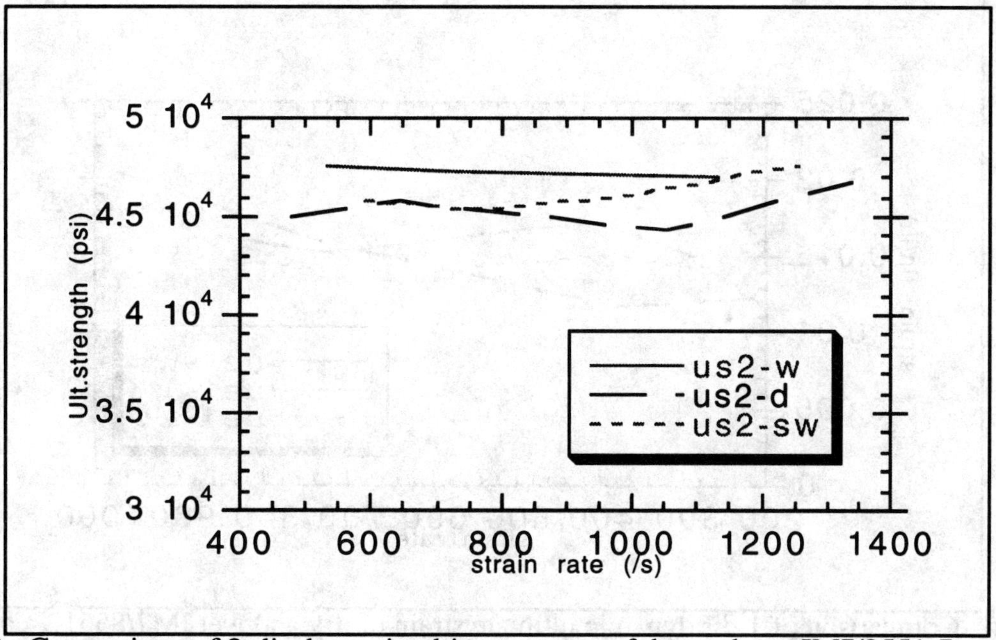

Figure 3. Comparison of 2-dir.dynamic ultimate stress of dry and wet IM7/8551-7 composite.

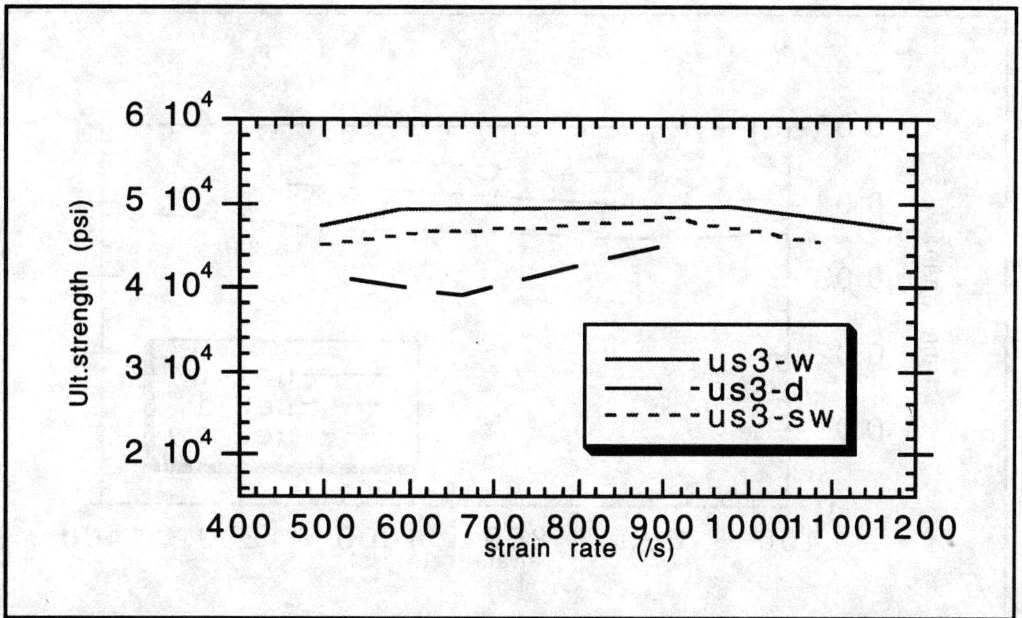

Figure 4. Comparison of 3-dir.dynamic ultimate stress of dry and wet IM7/8551-7 composite.

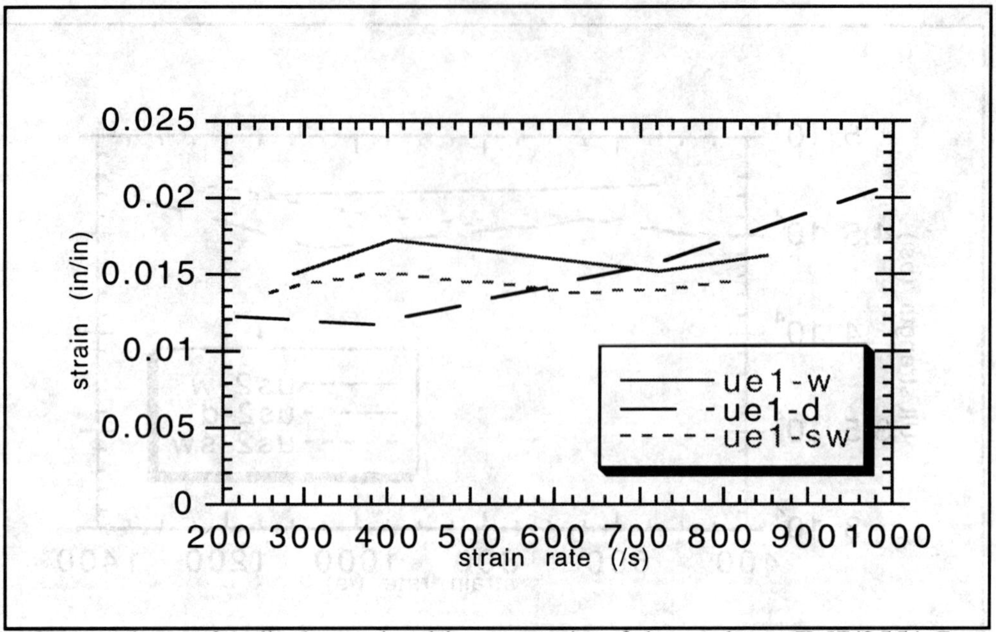

Figure 5. Comparison of 1-dir.dynamic ultimate strain of dry and wet IM7/8551-7 composite.

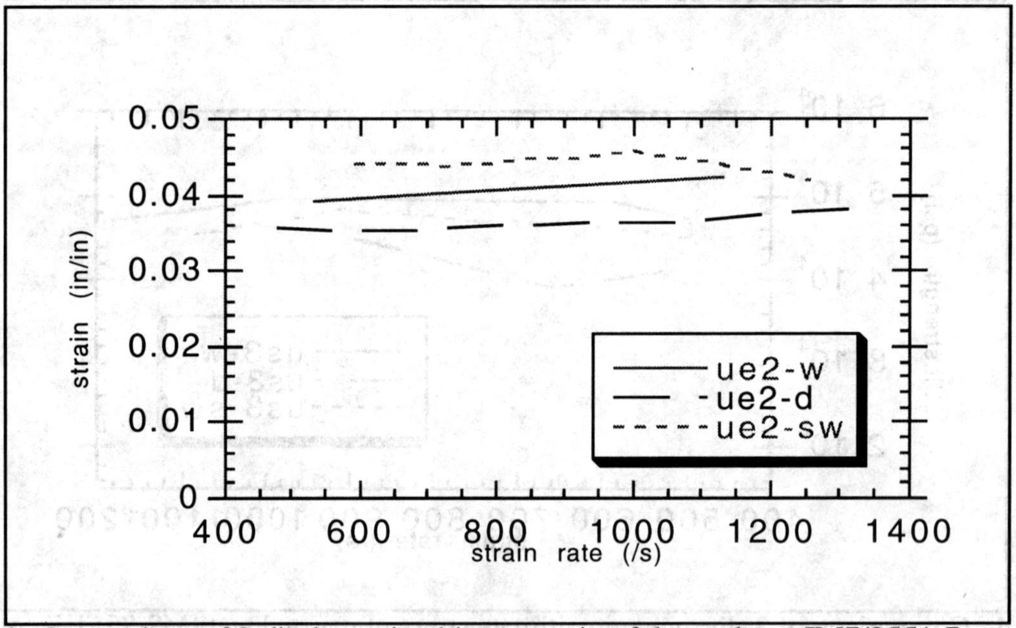

Figure 6. Comparison of 2-dir.dynamic ultimate strain of dry and wet IM7/8551-7 composite.

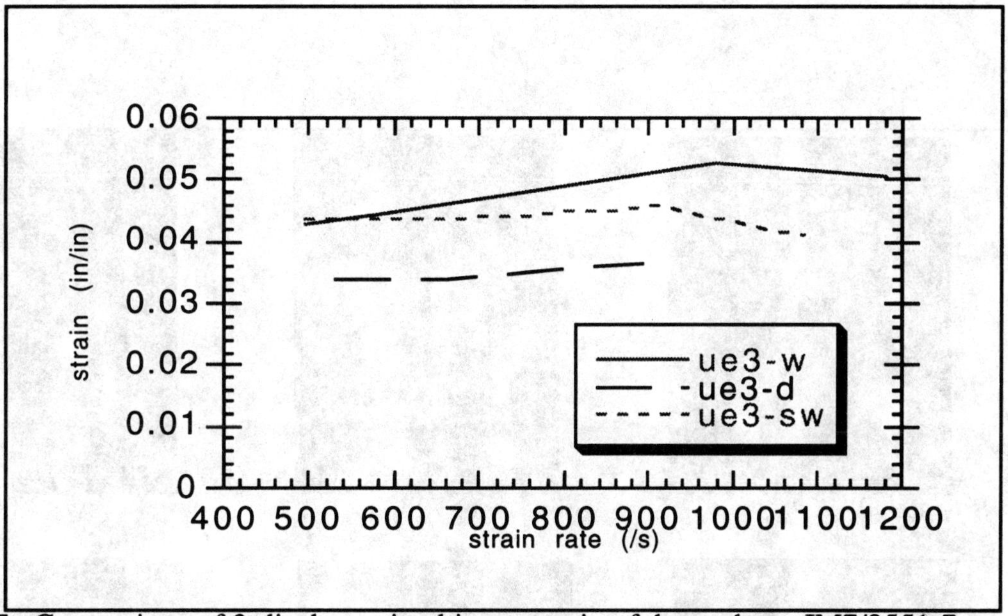

Figure 7. Comparison of 3-dir.dynamic ultimate strain of dry and wet IM7/8551-7 composite.

Figure 5-7 show the strain vs strain rate for dry (d), wet (w), and saturated wet (sw) samples. In the fiber and thickness direction, the wet samples which represent the specimens that were tested after 64 days of moisture absorption show the highest strain values, while in the transverse direction, the saturated wet samples which are at full saturation exhibit the highest strains. Figure 8 show the scanning electron microscope pictures of tested samples. The wet specimens show cleaner fiber surfaces but more broken pieces of the matrix than the dry specimens. This means that the water has weakened the strength of the interface region.

CONCLUSIONS

This study is unique because it is the first investigation of the effect of moisture on high strain rate properties of composites. This study includes 5 duplicates of dry and 3 duplicates of wet and saturated specimens, for a total 96 specimens. Moisture is usually taken to have a deteriorating effect on composite material properties. In high strain rate study (from 200 - 1200/s), this is not found to be the case. Contrarily, both the strength and strain of the IM7/8551-7 graphite/epoxy composite have increased due to the moisture intake. The stress in the fiber direction has increased by 0-25% and the transverse and thickness directions, by 0-20%, and the 1-direction strain has increased by 0-25% and the 2-direction strain has increased by 10-30% over strain rate range used. This can be attributed to the plasticization of the matrix, and leads to the conclusion that the matrix plays a major role in the high strain rate properties of the composite, because clearly if the matrix is not a major factor, the moisture tests should not have resulted in higher values than the dry ones.

Another observation is that even though the wet specimens show more clean fiber surfaces, which means the interface region has been weakened, the strength and strain of the wet specimen are higher than the dry counterparts for the same strain rate. Therefore, it can be concluded that, the fiber matrix interface is not a major factor in determining the high strain rate properties of the composite. Further moisture tests on other material systems are suggested to be performed to expand these conclusions to all types of composites.

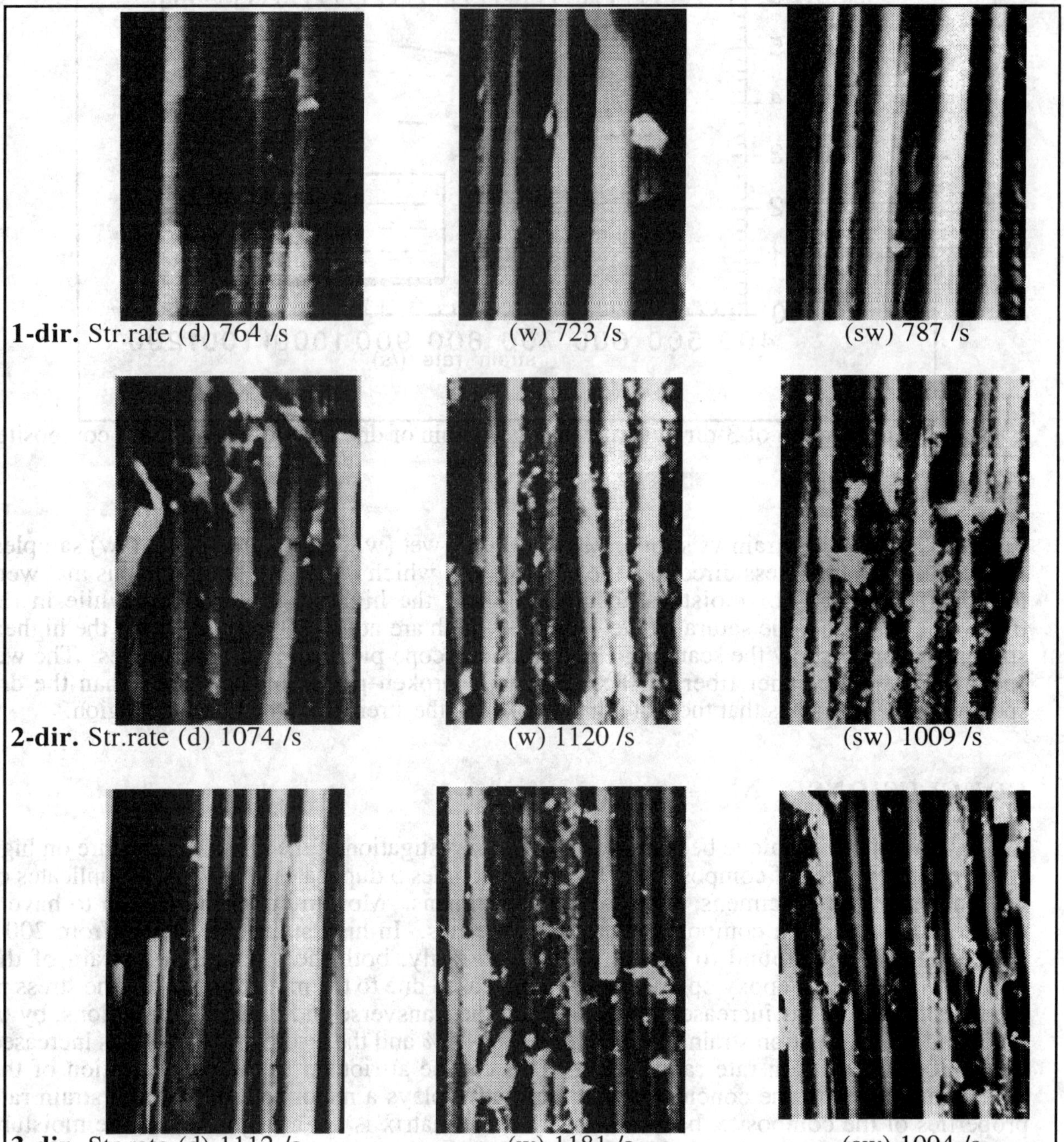

Figure 8. Scanning electron microscope pictures of dry, wet and saturated wet tested specimens in longitudinal, transverse and thickness directions, magnif.x640.

REFERENCES

Barker, A.J. and Balasundaram, V. 1987. "Compression Testing of Carbon Fiber-reinforced Plastics Exposed to Humid Environments," Composites. Vol. 18. No. 3. pp. 217-226.

Duffy, J., Campbell, J.D., and Hawley, R.H., "On the Use of Torsional Split Hopkinson Bar to Study Rate Effects up to 1500 s^{-1}." J. Appl. Mech. 38, 83-91.

Kolsky, H. 1949. "An investigation of the Mechanical Properties of Materials at Very High Rates of Loading." Proc. Phys. SOC. B. 62, 676-701.

Nicholas, T. 1971. "Tensile Testing of Materials at High Rates of Strain." Exp. Mech. 21, 177-185, 1980.

Powers, B.M. and J.R. Vinson. 1995. "High Strain Rate Effects on Materials in Sandwich Construction," Proceedings of the Third International Conference on Sandwich Constructions.

Powers, B.M., J.R. Vinson, I.W. Hall, and V. Nardone. 1995. "High Strain Rate Mechanical Properties to Silicon Carbide Reinforced 2080 Aluminum Metal Matrix Composites," Proceedings of the 10th International Conference on Composite Materials.

Powers, B.M., J.R. Vinson, and I.W. Hall. 1995. "High Strain Rate Mechanical Properties of IM7/8551-7 Graphite Epoxy Composites," Proceedings of the 10th Technical Conference of the American Society for Composites.

Powers, B.M., J.R. Vinson, M. Wardle, and B. Scott. 1995. " High Strain Rate Effects on Two AS4 Graphite Fiber Polymer Matrix Composites," ASME Aerospace Division Bound Volume AD-48, "High Strain Rate Effects on Polymer, Metal, and Ceramic Matrix Composites and Other Advanced Materials."

Powers, B.M., J.R. Vinson, I.W. Hall and R.F. Hubbard. 1995. "High Strain Rate Mechanical Properties of Cycom 5920/1583," Proceedings of the 36th AIAA/ASME/ASCE/AHS/ASC Structures, SDM Conference.

Powers, B.M., J.R. Vinson, M. Wardle, and B. Scott. 1996. "High Strain Rate Effects on Two Graphite Fiber K3B Polyimide Matrix Composites," Proceedings of the 37th AIAA/ASME/ASCE/AHS/ASC Structures, SDM Conference.

Shen, C.H. and Springer, G.S. 1981. "Effects of Moisture and Temperature on the Tensile Strength of Composite Materials," Environmental Effects on Composite Materials, Ed. G.S.Springer. Ann Arbor, Michigan. pp. 79-93.

Silverman, E.M., Wiacek, C.R. and Griese, R.A. 1992. "Characterization of IM7 Graphite/Thermoplastic Polyetheretherketone (PEEK) for Spacecraft Structural Applications," Composite Materials: Testing and Design (Tenth Volume), ASTM STP 1120, Glenn C. Grimes, Ed., American Society for Testing and Materials, Phladelphia, pp. 118-130.

Woldesenbet, E. and J.R. Vinson. 1997. "Effect of Specimen Geometry in High Strain Rate Testing of Graphite/Epoxy Composite," to be presented at the 38th AIAA/ASME/ASCE/AHS/ ASC Structures, SDM Conference.

Woldesenbet, E. and J.R. Vinson. 1997. "Effect of Fiber Orientation on High Strain Rate Properties of Graphite/Epoxy Composite," to be presented at the 12th Annual Technical Conference of the American Society for Composites (ASC), Dearborn, MI.

Zukas, J. 1982. (ed.), Impact Dynamics, John Wiley and Sons, New York, N.Y.

A Viscoplasticity Model for High Strain Rate Characterization of Polymeric Composites

S. V. THIRUPPUKUZHI* AND C. T. SUN**

ABSTRACT

The present study aims at characterizing the nonlinear rate dependent behavior of unidirectional S2Glass/8553-40 and woven 7781/F155 E glass fabric material systems in the strain rate regime of 0.0001/s to 1000/s. The objective is to develop a viscoplasticity constitutive model to predict the time dependent response of these material systems. Uniaxial tension tests under various constant total strain rates were conducted on a servohydraulic testing machine on off axis coupon specimens. Based on these tests, a three parameter viscoplasticity model capable of handling multiaxial loading conditions was developed for these composites. High strain rate experiments were conducted on a Split Hopkinson Pressure Bar. The proposed viscoplasticity model is shown to adequately model the material response of the composite laminates under various strain rate loadings.

INTRODUCTION

High strain rate characterization of composites has been a subject of great interest among researchers primarily because of the need to understand their behavior in loading situations where they are subjected to a spectrum of strain rates. With increasing use of composites in aerospace and marine applications, problems such as bird hits of composite aircraft wings, blast loading of a composite submarine hull and ballistic impact of composite armored vehicles are all situations which demand an understanding of the constitutive behavior of composites in order to predict their behavior. One of the most challenging issues is developing appropriate constitutive relations for the various strain rate regimes where there are different physically dominant mechanisms.. Various techniques have been employed over the years to generate different strain rates. Conventional servohydraulic testing machines have been employed to obtain strain rates in the order of 0-10/S, bar (Hopkinson) impact experiments have been used to generate

* Srikanth V. Thiruppukuzhi, Graduate student, School of Aeronautics and Astronautics, Purdue University, West Lafayette, IN 47907

** C.T. Sun, Professor, School of Aeronautics and Astronautics, Purdue University, West Lafayette, IN 47907

strain rates of 10-10000/S while plate impact experiments have been used to obtain strain rates above 10000/S.

The low strain rate response of fiber reinforced composites has been amply characterized by Sun and Chen (1989), Gates and Sun (1991), Yoon and Sun (1991), Chen and Sun (1993) and by many other researchers, but their behavior and characterization under dynamic loading is limited. Most work in dynamic characterization has been devoted to lateral impact testing with little effort on constitutive modeling. In this work a three parameter viscoplasticity model is proposed for a unidirectional and a woven composite material system based on inplane characterization. It is found to satisfactorily predict the material response of composite laminates, in the strain rate range from 0.0001/S to 1000/S.

EXPERIMENTAL PROCEDURES AND RESULTS

Uniaxial tension tests were conducted on a closed-loop servo-hydraulic MTS 20 kip machine on two different material systems, namely unidirectional S2Glass/8553 and woven Eglass/7781, in the strain rate regime of 0.0001/s to 1/s. Off axis coupon specimens were cut at angles of 0°, 15°, 30°, 45°, 60°, and 90°on a FLOW^{TM} waterjet cutting machine. A ply level characterization was used in characterizing the material behavior by testing laminates having the same lay up through the thickness $[\theta]_n$. Details of the off axis testing can be found in the paper by Thiruppukuzhi and Sun (1996). Based on the off axis testing, a one parameter plastic potential function proposed by Sun and Chen (1989), is found to adequately characterize the inplane behavior. This was derived on the assumption that there is no plastic deformation along the fiber direction, and is given by the following equation

$$f(\sigma_{ij}) = \frac{1}{2}\left(\sigma_{22}^2 + 2a_{66}\sigma_{12}^2\right) \tag{1}$$

From the offaxis testing, a unique relationship as below, can be established between the effective stress and the effective plastic strain, by choosing the parameter a_{66}.

$$\bar{\varepsilon}^p = A\left(\bar{\sigma}\right)^n \tag{2}$$

Details on the determination of a_{66} .can be found in the above references. Figure 1 shows a plot of the effective stress-effective plastic strain for the unidirectional material system. A similar approach is adopted in modeling the woven material system. It is found that no parameter is needed to characterize the inplane behavior of woven composites (Thiruppukuzhi, Sun, 1996). The plastic potential function for woven composites in plane stress is given by

$$f\left(\sigma_{ij}\right) = \sigma_{12}^2 \tag{3}$$

A dual power law is adopted in modeling the effective stress-effective plastic strain behavior, and when the effective plastic strain exceeds 6%, a second power law is used:

$$\bar{\varepsilon}^p = \begin{cases} A_1\left(\bar{\sigma}\right)^{n_1} & \bar{\varepsilon}^p \le 0.06 \\ A_2\left(\bar{\sigma}\right)^{n_2} & \bar{\varepsilon}^p > 0.06 \end{cases} \tag{4}$$

VISCOPLASTICITY MODELING OF POLYMER COMPOSITES

The question of rate influence is one of the central problems in the dynamic characterization of materials. It has been known for years that under dynamic loading, the behavior of materials is significantly different static. Numerous experimental papers have shown the effect of strain rate on the stress/strain behavior and the failure strength of different material systems. There is no longer a one-to-one correspondence between stress and strain as prescribed by a finite stress-strain relation. One of the challenging issues in developing constitutive equations for predicting rate dependent behavior of polymer composites is the fact that while constant total strain rate experiments are conducted, one has to deal with variable plastic strain rates.

The viscoplasticity model proposed here is the three parameter total form of the constitutive equation, and, as the name suggests, the constitutive equation is an absolute relationship among the three variables: effective dynamic stress, accumulated effective plastic strain, and effective plastic strain rate. The general functional form of the constitutive equation is

$$\dot{\bar{\varepsilon}}^p = B\left(\bar{\varepsilon}^p\right)^q \left(\bar{\sigma}\right)^p \qquad (5)$$

This is a viscoplasticity model without a yield surface. This implies that the material has a viscoplastic response corresponding to every state of loading and that there is no "quasistatic" response for the material. The functional form that is adopted for determining the viscoplasticity parameters as well as numerical implementation is

$$\bar{\varepsilon}^p = \chi\left(\dot{\bar{\varepsilon}}^p\right)^m \left(\bar{\sigma}\right)^n \qquad (6)$$

The parameters χ , m, and n are the three parameters determined from the experiments. Experiments show that master curves similar to figure 1 are obtained at different strain rates for the same value of a_{66} . This implies that the orthotropy parameter a_{66} in the plastic potential function can be assumed to be independent of strain rate. It is found that power laws similar to equation (2) can model the effective stress-effective plastic strain behavior at higher strain rates. The exponent n in the power law corresponding to effective stress $\bar{\sigma}$ is found to be a constant for all the strain rates, and hence the same value is retained in the viscoplasticity model. However, the amplitude A which is a constant for a given constant total strain rate, is a rate dependent coefficient, and is made a function of the effective plastic strain rate $\dot{\bar{\varepsilon}}^p$. Again a power law is chosen to get the rate dependent coefficient A as a function of the effective plastic strain rate along the entire loading path and is expressed as

$$A = \chi\left(\dot{\bar{\varepsilon}}^p\right)^m \qquad (7)$$

Theoretically, two constant effective plastic strain rate experiments are sufficient to determine the two viscoplasticity parameters, χ and m in this model. It is observed that in both unidirectional and woven polymer composites, uniaxial constant total strain rate experiments do not result in constant effective plastic strain rates. However, experimental data obtained from constant total strain rate experiments indicate that for both the material systems, the effective plastic strain rate reaches a steady state value and lies in a narrow bandwidth, after an initial rise time lasting 10-30 % of the loading duration.. Thus, the data corresponding to the initial rise time are truncated in determining the viscoplasticity parameters. As will be seen later, this does not influence the model predictions as the response of the material is dominated by the elastic behavior during the initial part of the loading. Figure 2 shows amplitude A as a function of the effective plastic strain rate for different off axis orientations on a log-log scale for the unidirectional S2glass composite material system. The viscoplasticity parameters χ and m are then determined from these plots as the intercept and the slope respectively. The values of these parameters for the two material systems are reported in table 1 at the end of the text. It is to be noted that the viscoplasticity parameter χ is the intercept corresponding to unit effective plastic strain rate. The important and powerful feature to be realized in the model discussed above, is the fact that the constitutive equations for the orthotropic material systems for all possible orientations and those under multiaxial stress state are captured in the form of a single scalar equation relating effective stress, accumulated effective plastic strain, and effective plastic strain rate.

LAMINATE MODELING AND PREDICTIONS

In problems of homogeneous deformation since the material response is the same at every point in the continuum, a single differential equation is sufficient to model the entire material response. However, in practice most of the structures are usually subjected to a nonhomogeneous state of stress and more complex loading situations. In particular, composite laminates can generate complex states of stress.

Incremental equations for orthotropic viscoplasticity are formulated, based on the above viscoplasticity model, for a plate element in plane stress. Finite element modeling is done using the finite element code ABAQUS. The incremental equations for viscoplasticity are integrated into ABAQUS through user material subroutine UMAT and other user interfaces. The rate form of the constitutive equations are integrated in time using explicit time integration to predict the time dependent material response. The laminate is discretized using a 4 noded, layered shell element with reduced integration S4R. Based on an additive decomposition of the total strain rate in terms of the elastic and viscoplastic components, the rate form of the constitutive equations are obtained as

$$\begin{Bmatrix} \dot{\varepsilon}_{11} \\ \dot{\varepsilon}_{22} \\ \dot{\gamma}_{12} \end{Bmatrix} = \begin{bmatrix} S_{11} & S_{12} & 0 \\ S_{12} & S_{22} + \dfrac{9a_{22}^2\sigma_{22}^2}{4H_p\bar{\sigma}^2} & \dfrac{9a_{22}a_{66}\sigma_{22}\sigma_{12}}{2H_p\bar{\sigma}^2} \\ 0 & \dfrac{9a_{22}a_{66}\sigma_{22}\sigma_{12}}{2H_p\bar{\sigma}^2} & S_{66} + \dfrac{9a_{66}^2\sigma_{12}^2}{H_p\bar{\sigma}^2} \end{bmatrix} \begin{Bmatrix} \dot{\sigma}_{11} \\ \dot{\sigma}_{22} \\ \dot{\sigma}_{12} \end{Bmatrix} \tag{8}$$

where H_P is a rate dependent plastic modulus and the parameter a_{22} assumes a value of 1.0 and 0.0, respectively, for the unidirectional and woven composite material systems.

Various symmetric laminates of both unidirectional and woven composite material systems were investigated in tension and under various strain rate loading conditions ranging from 0 - 1/S. Figures 3-6 show typical predictions for laminates with different stacking sequences. In general, the model predictions are found to be within 5%-10% of the experimental results.

HIGH STRAIN RATE EXPERIMENTS AND MODELING

Thick section composite laminates, approximate cubes of 9mm edge, were investigated under dynamic compressive loading on a Split Hopkinson Pressure Bar (SHPB). The details of the SHPB technique can be found in Nicholas (1982), Weeks (1995), and in a wealth of other references. Balanced symmetric laminates were investigated so as to generate homogeneous deformation within the laminate. Off axis unidirectional laminates cannot be investigated in a similar manner because the shear - extension coupling under dynamic loading produces a non-uniform deformation. This implies that tests such as those mentioned above cannot be used directly in the determination of viscoplasticity parameters at high strain rates. Tests are therefore conducted on balanced symmetric laminates to validate the viscoplasticity model for high strain rate loading. Figure 7 shows a simple schematic of the SHPB, and figure 8 shows typical signals recorded by the three strain gages on the pressure bars, obtained under dynamic loading. Figures 9 and 10 show typical strain rate history and the stress strain curves, respectively, obtained for the $[\pm 45]_{16S}$ unidirectional S2Glass laminate, under a stress pulse of 120 μS. It is to be noted that the curves corresponding to Test 2 were obtained by pulse shaping, (using softer material at the striker - incident bar interface) which gives stress pulses with larger rise times by eliminating high frequency components.

In order to predict the behavior of these laminates under high strain rate loading, complete three dimensional incremental equations for viscoplasticity were formulated and implemented through the user material subroutine UMAT in ABAQUS based on the 3-D plastic potential function for a transversely isotropic solid introduced by Chen and Sun (1993),

$$f(\sigma_{ij}) = \frac{1}{2}\left[(\sigma_{22} - \sigma_{33})^2 + 4\sigma_{23}^2 + 2a_{66}\left(\sigma_{13}^2 + \sigma_{12}^2\right)\right] \tag{9}$$

Formulation and modeling were verified for quasistatic compression of the thick section laminates in compression under low strain rate loading. In modeling the behavior of the thick section laminates under high strain rate loading, eight noded brick elements of ABAQUS, C3D8 with reduced integration were employed. Exploiting symmetry conditions, one eighth of the specimen was modeled using 2592 solid elements with 32 elements through the thickness, one element modeling each layer. The strain rate history recorded from the experiments were used in simulating the actual loading condition on the laminate. Figure 11 shows a typical prediction of the 3D finite element for the $[\pm 45]_{16S}$ laminate and its comparison to the experimental results. One aspect to be noted in the comparison is that the model predictions of the stress-strain curve are much smoother than those from the experiment. This is due to the fact that while the strain rate history itself is oscillatory, the displacement history applied on the finite element model is obtained by integrating the above and is much smoother.

CONCLUSIONS

The nonlinear rate dependent behavior of unidirectional S2Glass and the woven E-Glass fabric were investigated in the strain rate range from 0.0001/S to 1000/S. A three parameter viscoplasticity model was proposed based on inplane characterization. Incremental equations for viscoplasticity were formulated based on the one parameter plastic potential function, and this was used in predicting the behavior of laminates with a variety of stacking sequences and those under a range of strain rate loading conditions. The performance of the model is in general satisfactory in predicting the stress-strain behavior to within an accuracy of 5%-10%. The results for the unidirectional S2glass composite show moderate strain rate sensitivity. Additional tests are needed on different laminate stacking sequences to establish strain rate sensitivity for these material systems.

ACKNOWLEDGMENT

This work was supported by the Office of Naval Research through grant No. N00014-96-1-0822 to Purdue University. Dr. Y. D. S. Rajapakse was the technical monitor.

REFERENCES

Chen, J.L. and Sun, C.T.,1993 " A Plastic Potential Function Suitable for Anisotropic Fiber Composites," *Journal of Composite Materials*, Vol. 27, No.14, pp. 1379-1390.

Gates, T.S. and Sun, C.T. 1991. "Elastic/Viscoplastic Constitutive Model for Fiber Reinforced Thermoplastic Composites, " *AIAA Journal,* 29 :457-463.

Nicholas, T. 1982. "Material Behavior at High Strain Rates," Chapter 8, *Impact Dynamics*, Zukas, J.A., et al, John Wiley and Sons, New York, NY.

Sun, C.T. and Chen, J.L. 1989. "A Simple Flow Rule for Characterizing Nonlinear behavior of Fiber Composites," *Journal of Composite Materials,* 23 :1009-1020.

Thiruppukuzhi, S.V. and Sun, C.T.,1996 "High Strain Rate Characterization of Unidirectional Fiber and Woven Glass Composites," *American Society of Composites*, Atlanta, Georgia, pp.201-211.

Weeks, C.A.1995. " Nonlinear Rate Dependent Response of Thick-Section Composite Laminates," *Ph.D Dissertation*, Purdue University, West Lafayette, Indiana.

Yoon, K.J. and Sun, C.T. 1991. "Characterization of Elastic-Viscoplastic Properties of an AS4/PEEK Thermoplastic Composite, " *Journal of Composite Materials*, 25 :1277-1298.

Table 1 Material parameters for the unidirectional and the woven glass material systems for the total form of constitutive equation.

χ	1.045E-11(MPa)$^{-n}$
m	-0.14
n	4.0

χ_1	5.1E-10(MPa)$^{-n1}$
m_1	-0.1127
n_1	4.05
χ_2	2.192E-07(MPa)$^{-n2}$
m_2	-0.0766
n_2	2.75

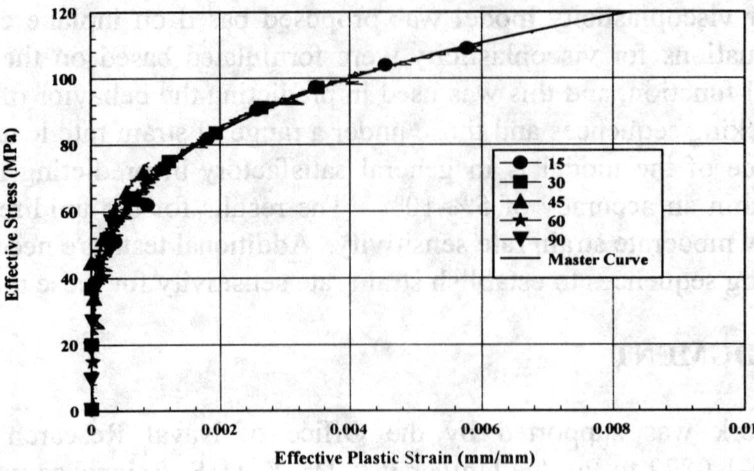

Figure 1 Effective stress-effective plastic strain curves for the unidirectional S2Glass at 0.0001/S.

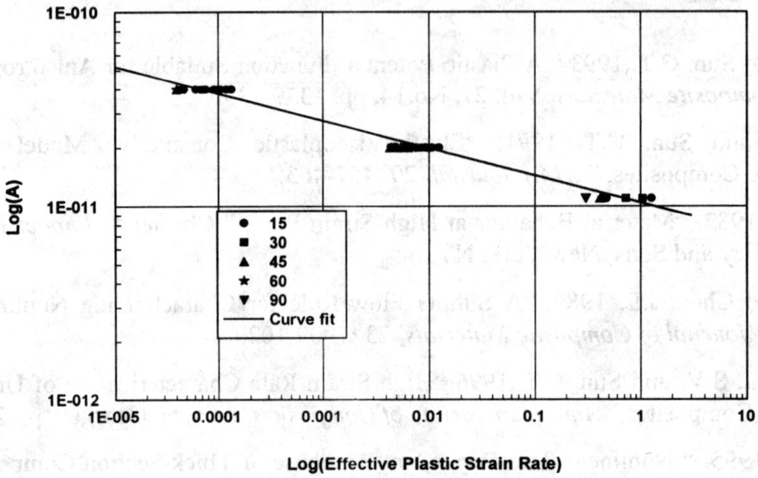

Figure 2 Material parameters of the total form of the constitutive equation for the unidirectional S2Glass material system.

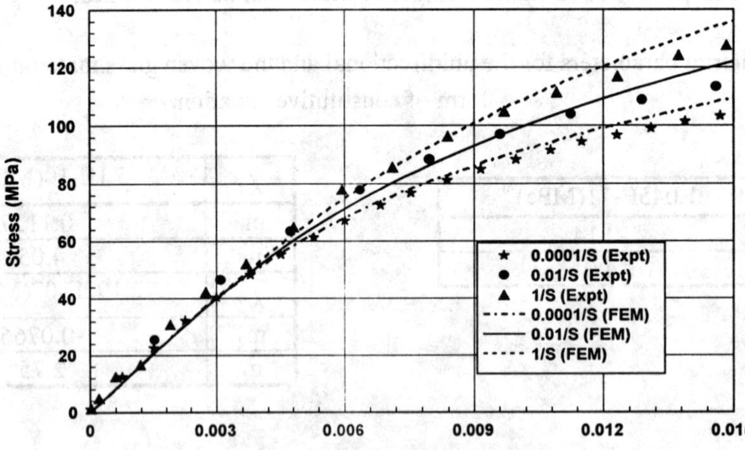

Figure 3 Finite Element predictions of stress-strain curves of the $[\pm 45]_{3S}$ unidirectional S2Glass laminate at various strain rates.

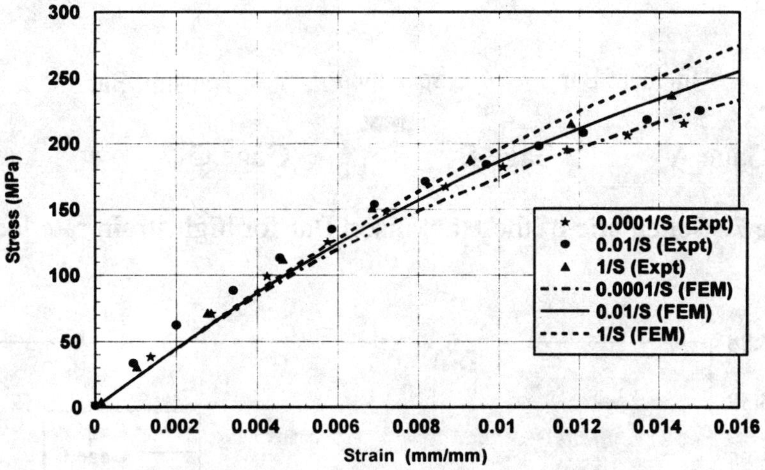

Figure 4 Finite lement predictions of stress-strain curves of the $[+15/-75]_{3S}$ unidirectional S2Glass laminate at various strain rates.

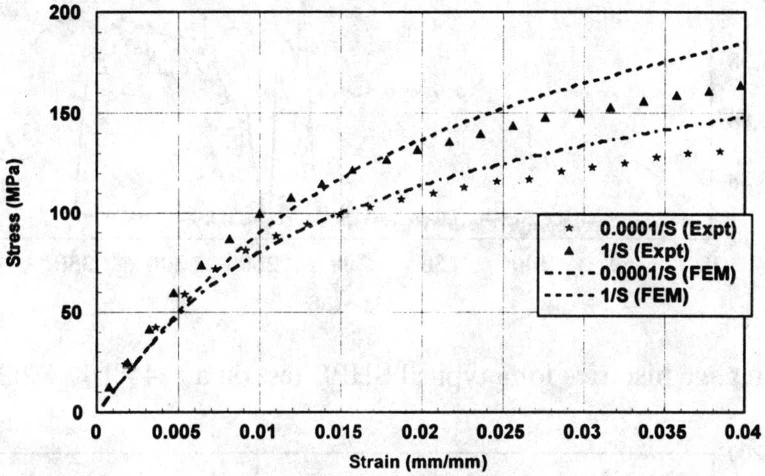

Figure 5 Finite element predictions of stress-strain curves of the $[\pm 45]_{3S}$ woven EGlass Laminate at various strain rates.

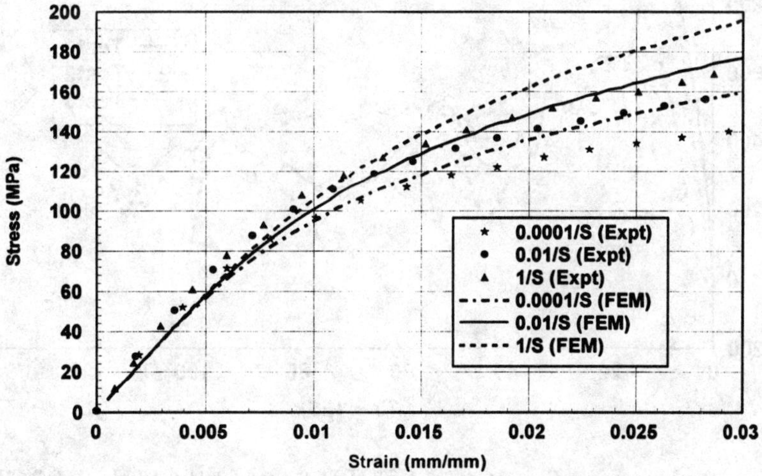

Figure 6 Finite lement predictions of stress-strain curves of the $[+30/-60]_{3S}$ woven EGlass laminate at various strain rates.

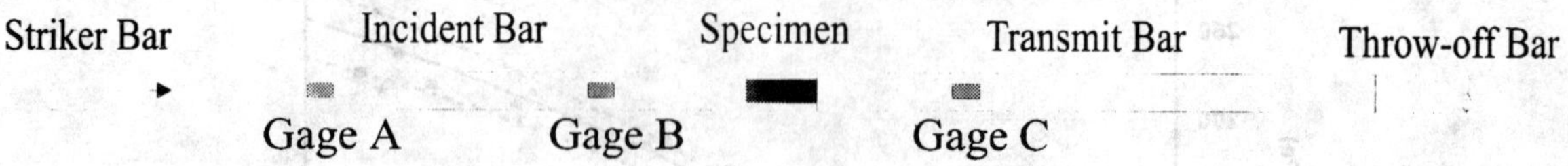

Figure 7 Schematic of the Hopkinson Bar for high strain rate loading.

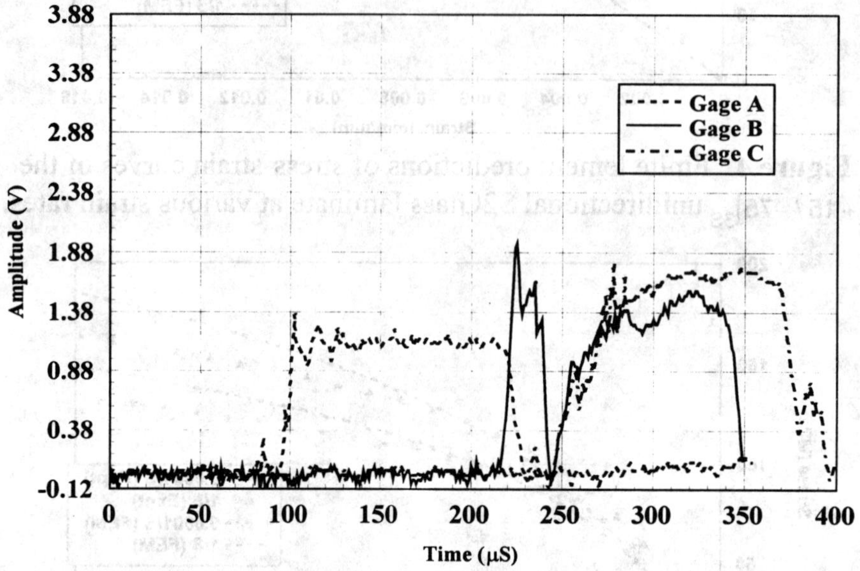

Figure 8 Strain gage histories for a typical SHPB test on a $[\pm 45°]_{16s}$ S2Glass laminate.

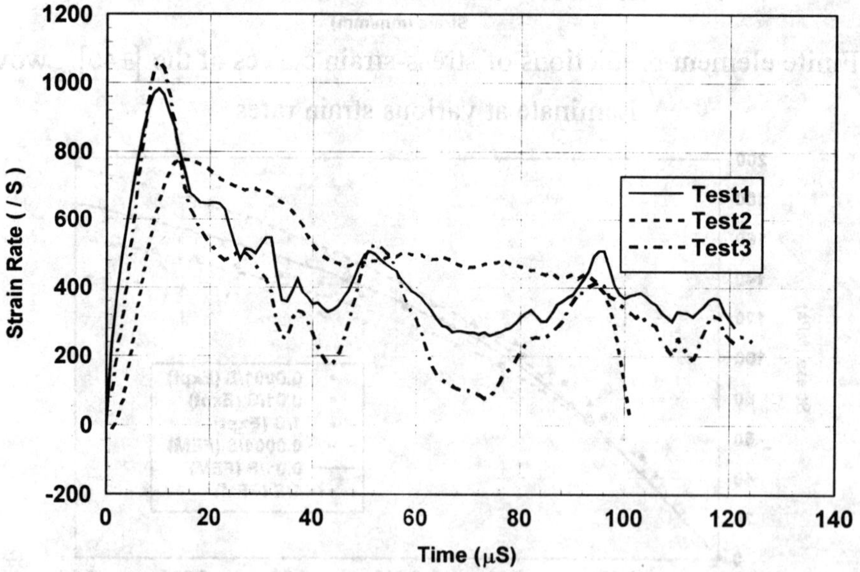

Figure 9 Strain rate histories for a typical SHPB test on a $[\pm 45°]_{16s}$ S2Glass laminate.

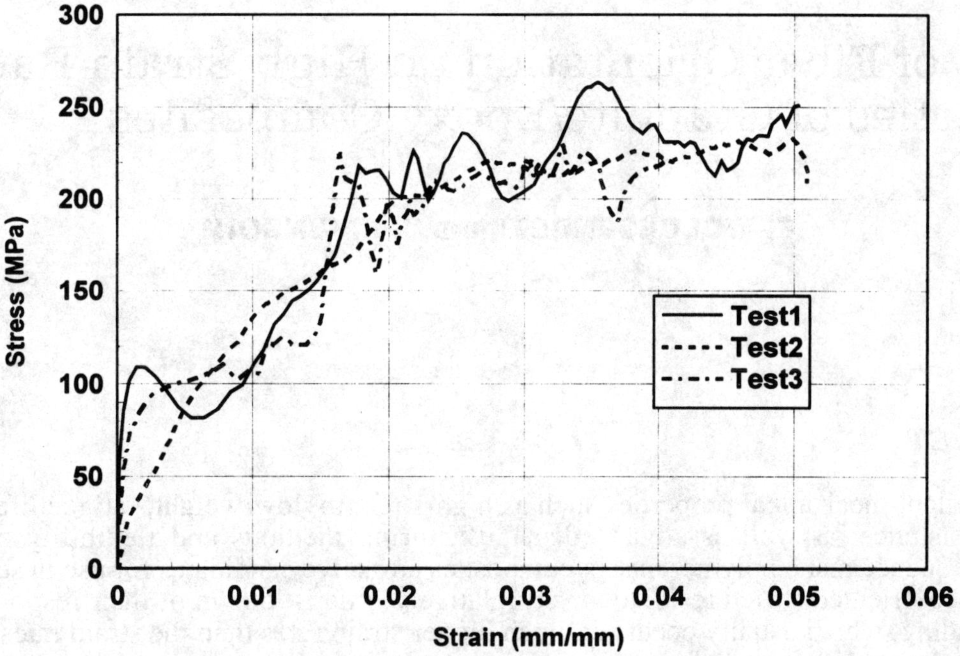

Figure 10 Stress-strain curves for a typical SHPB test on $[\pm 45°]_{16s}$ S2Glass laminate.

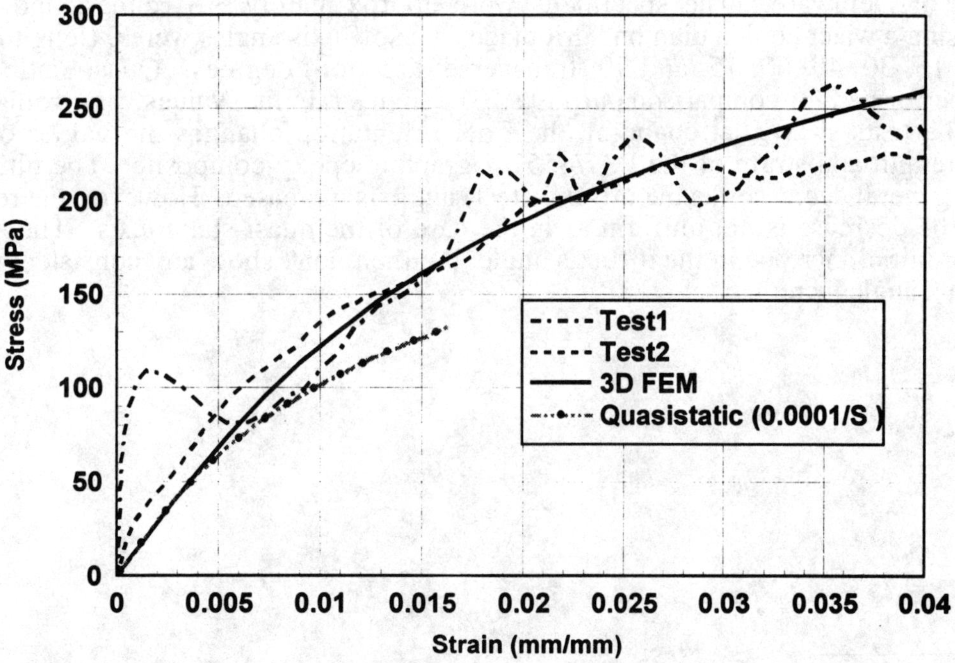

Figure 11 Finite element predictions of stress-strain curves of SHPB tests on the $[\pm 45]_{16S}$ unidirectional S2Glass laminate.

Effect of Fiber Orientation on High Strain Rate Properties of Graphite/Epoxy Composites

E. WOLDESENBET[1] AND J. R. VINSON[2]

ABSTRACT

Excellent mechanical properties such as high-strength, low-weight, fatigue life, and impact resistance, as well as advanced manufacturing methods and flexibility of the stacking sequence make polymer matrix composites attractive candidates for use in several performance-oriented structures. However, relatively little is known of their response to impact loading, which usually occurs at much higher strain rates than the strain rates used to measure the quasi-static mechanical properties of materials.

In this study, the split-Hopkinson bar is used to study the effect of fiber orientation on the compressive dynamic properties of a unidirectional IM7/8551-7 graphite/epoxy composite. Flat panels were fabricated at the Center of Composite Materials (CCM) of the University of Delaware. The specimens were approximately 3/8" cubes, and were prepared using a water cooled diamond grit blade. The off-axis angles were 0 (longitudinal direction), 15, 30, 45, 60, 75, and 90 (transverse direction) degrees. Quasi-static tests were also performed for comparison purposes. The high strain rate values vary from 250 - 1100/s. The tests show that changing the fiber orientation changes the values of the ultimate strength and strain of the IM7/8551-7 graphite/epoxy composite. The ultimate strength in general decreases as the off-axis (test) angle is increased. However, the results show that the decrease is not uniform as in the case of the quasi-static tests. The strain increase for initial increase in the off-axis angle, but then don't show any consistent trend for additional angle increments.

[1]Doctoral Candidate, Department of Mechanical Engineering, University of Delaware.

[2]H. Fletcher Brown Professor of Mechanical & Aerospace Engineering, University of Delaware.

INTRODUCTION

At present, a program is underway at the University of Delaware under the Office of Naval Research sponsorship (Dr. Y.D.S. Rajapakse) that involves three tasks :

(1) the high strain rate testing in compression and tension of various polymer matrix composite (and other) materials using the existing Split Hopkinson Pressure Bar facility, and correlation and statistical analysis of the dynamic material properties obtained;

(2) examination of the samples tested, using optical and scanning electron microscopy to characterize the deformation and fracture processes; and

(3) the evaluation of the suitability of current models, such as Johnson-Cook, Ferilli-Armstrong, Steinberg-Guinan-Lund, and Bodner-Partom, to describe the strength, deformation and failure of these materials at high strain rates and the modification of these models, or whenever necessary, the development of new models.

The present paper concentrates on Task 1 and 2.

The Split Hopkinson Pressure Bar facility at the University of Delaware is described in Figure 1 and the related Lagrangian diagram is shown in Figure 2. The Split Hopkinson Pressure Bar (SHPB) has been accepted as the test method for high strain rates since Kolsky first presented the test methodology for compression in 1949, (Kolsky, 1949). The method was developed based on wave propagation theory in elastic bars and the interaction between a stress wave and a short specimen. After Kolsky's introduction, the method was expanded to tensile testing (Nicholas, 1980) and shear testing (Duffy et al., 1971). Impact is initiated by releasing nitrogen gas from a pressurized chamber which causes the specimen to impacted between the two bars. The striker bar is displaced on Teflon rings through the barrel where it strikes the incident bar. The velocity of the striker bar V_{sb} before impact is measured using two infrared beams set four inches apart. At impact, the incident bar receives an elastic compressive stress wave with a wave velocity C_o and a wave shape $\varepsilon_I(t)$ and is displaced along its length axis. This bar is guided by Teflon bearings which are bolted to a steel I-beam and secured on a reinforced concrete base to minimize any vibration. When the wave reaches the bar 1 and specimen interface, a portion of the incident wave is reflected back into bar 1 as a tensile wave with a wave shape $\varepsilon_r(t)$.

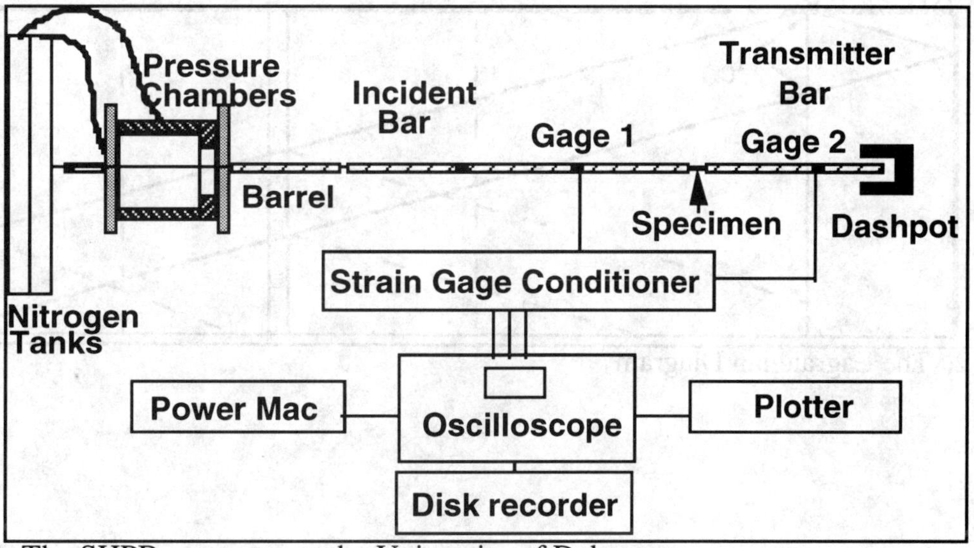

Figure 1. The SHPB apparatus at the University of Delaware

The remaining portion of the wave is transmitted into the specimen as a compressive wave. The wave transmitted into the specimen travels through the specimen and reaches the specimen and bar 2 interface, where a
portion of the wave is reflected back into the specimen. The rest is transmitted into bar 2 as a compressive wave with a wave shape $\varepsilon_t(t)$. This bar is also displaced along its length axis, and rests when the other end of it reaches a dash pot.

The initial stress wave in the specimen undergoes numerous internal reflections during the test, because the specimen length is short, and therefore, the wave reflections in the specimen are not shown in the Lagrangian diagram of the split-Hopkinson bar. The stress distribution is assumed to be uniform, i.e., the specimen is assumed to be in a uniform state of stress, due to the numerous reflections. In addition, the stress pulses or waves are assumed to undergo minimal dispersion, and the bars remain elastic and the ends of both the incident and transmitted bars in contact with the specimen remain flat. The Lagrangian diagram shows that there is a characteristic time window corresponding to the duration of the wave pulses. The wave pulse time window of the University of Delaware SHPB is approximately 290 microseconds. This means that the specimen must fail within 290 microseconds after the start of the test.

Strain gages are mounted on bars 1 and 2 equidistant from the specimen interfaces and are connected to an oscilloscope. Strain gages are mounted on bars 1 and 2 equidistant from the specimen interfaces and are connected to an oscilloscope. The first strain gage mounted on bar 1 also acts as a trigger for the oscilloscope. The oscilloscope records the strain gages output as voltage versus time graph. Using this data, stress vs strain curves can be generated for different strain rates. Zukas has shown that for many metals, the mechanical properties vary significantly with strain rate, (Zukas, 1982).

This new data will add to what is now available for other materials. At Delaware, material systems tested to date include glass/epoxy, glass/polyester, graphite/epoxy, carbon/metal matrix, and carbon/ceramic matrix, and the most recent papers regarding these materials are given in the Reference, (Powers and Vinson, 1995; Powers et al., 1995, Powers et al.,1996).

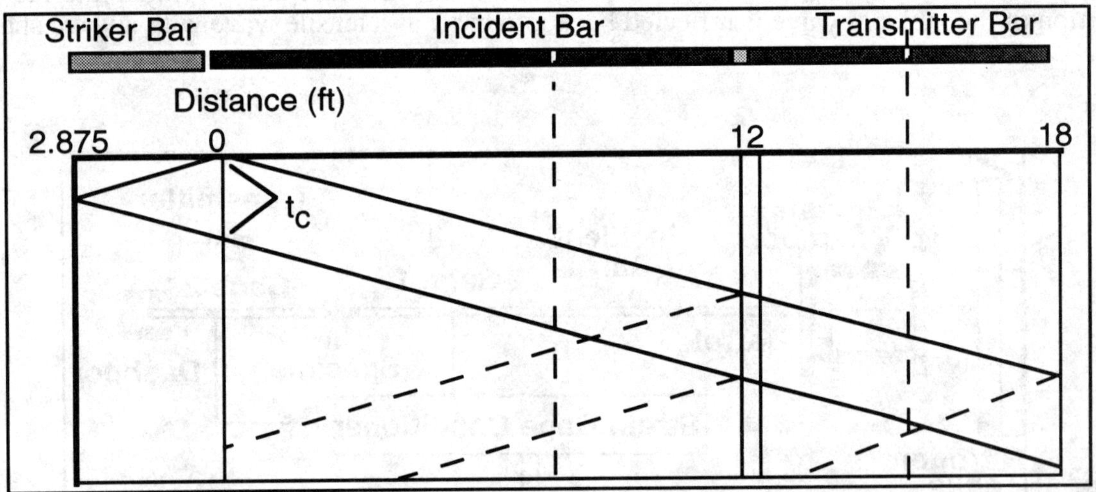

Figure 2. The Lagrangian Diagram

Some researchers, trying to compare the strain rate sensitivities of the fibers to the matrix, have performed tests at off-axis angles(Staab and Gilat, 1995). Using this kind of study to compare the strain rate sensitivities may lead to a wrong conclusion, because the fiber orientation can lead to a different mode of failure which doesn't account for the individual material properties of the fiber or the matrix. However, this study is very useful in understanding composite dynamic behavior loaded in different directions from the longitudinal or transverse directions.

SAMPLE PREPARATION

The IM7/8851-7 carbon/epoxy specimens were made at the University of Delaware from continuous fiber unidirectional pre-impregnated tape. The tape was 16 in. wide and was cut into 48 in. strips. A 75 ply unidirectional laminate was fabricated by hand lay-up to achieve the desired thickness after processing. The laminate was processed in a Thermal Equipment Corporation computer controlled autoclave. The finished laminate was then sectioned using a water-cooled masonry saw equipped with a diamond grit blade, which usually produces no delamination and rough surfaces. Square/rectangular samples cut out of the sectioned pieces were used to study the effect of varying the off-axis angle, zero angle is when the fibers are directed in the testing or loading direction. The specimens were approximately 3/8" cubes, and three sample per angle per strain rate were tested for a total of more than 105 specimens. Recently, a study was conducted using a graphite/epoxy composite that examined both various L/D ratios from 0.5 to 2.0 as well as a comparison between square/rectangular and cylindrical specimen geometry (Woldesenbet and Vinson, 1997). Within the range of variables examined, statistically, no significant differences could be found in the material properties and this provided the basis for making cubic samples with the specific dimension used in this study.

RESULTS & DISCUSSION

Figure 3 shows the compressive ultimate stress vs fiber orientation angle. The static values decrease uniformly as the off-axis angle increases. However, all the high strain rate values show a dip at 60 degrees. s0, s4, s6, and s8 represents strain rate values of quasi-static, 400, 600, and 800 (/s) respectively.

The fact that the stress values at 60 degrees are lower than the values at other angles might indicate that the 60 degrees plane could be the plane of maximum dynamic shear stress. For a quasi-static test, the shear stress has a maximum value at a plane of 45 degrees, according to Equation (1). The strain rate sensitivity of the composite also decreases as the off-axis angle increases. Some researchers have interpreted this and concluded that the matrix is less strain rate sensitive. However, the decrease in the strain rate sensitivity of the composite as the off-axis angle increases could be because of change of mode of failure. In a quasi-static study of the orientation dependence of ultimate stress, longitudinal or shear fractures can occur at different angles of orientation (Hull, 1981). In addition, fiber buckling and fiber kinking can be additional cause of failure. Therefore, this result should not be interpreted to define the strain rate sensitivity of the individual components of the composite.

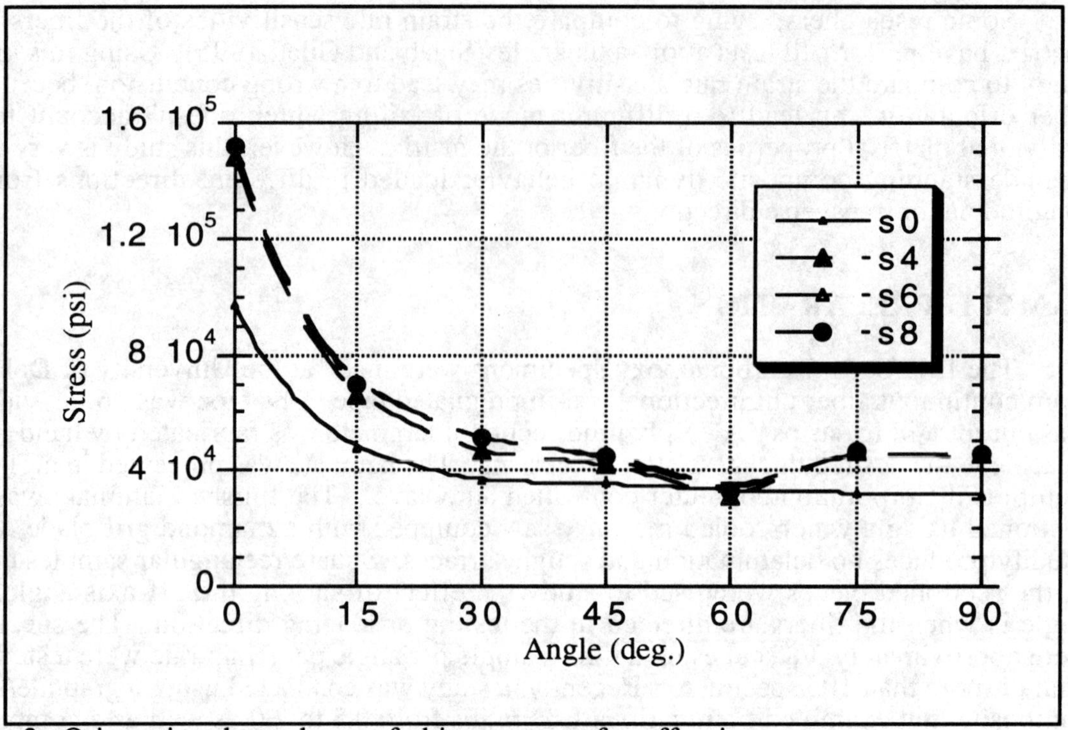

Figure 3. Orientation dependence of ultimate stress for off-axis tests.

In quasi-static studies, the in-plane shear stress can be determined by testing off-axis specimen at different angles, where there is relatively little interaction between the longitudinal stress and the shear stress (Hull, 1981). The in-plane shear stress can be calculated using

$$\tau = \sigma_\theta \sin\theta \cos\theta \qquad (1)$$

Takeda and Wan used this equation to obtain the shear stress in their impact study (Takeda and Wan, 1995). However, this equation does not take the strain rate sensitivity into account, and can lead to some error depending on the orientation dependence of the strain rate sensitivity of the composite. Therefore, Equation (1) has to account for strain rate sensitivity of the composite and should be modified as

$$\tau = \sigma_\theta \sin\theta \cos\theta * F(\dot{\varepsilon},\theta) \qquad (2)$$

where $F(\dot{\varepsilon},\theta)$ is a function that account for orientation dependence of strain rate sensitivity. Determination of $F(\dot{\varepsilon},\theta)$ is a subject of future study. However, the importance of this function is obvious by looking at Figure 3, and seeing, for example, the difference between the static and the high strain rate values at 0 degree is 50 ksi, while at 45 degrees it is only 8 ksi.

Figure 4 shows the compressive ultimate strain vs fiber orientation angle for different strain rates. The strain values seem to increase uniformly until 45 degrees. After 45 degrees, there is no uniform strain variation as the off-axis angle increases. However, looking at the strain values at 60 degrees closely, they are lower than the strain values at 45 and 75 degrees. This might indicate that the 60 degrees plane or orientation is a plane of maximum strain. Figures 5 - 9, show scanning electron microscope pictures of the fracture surfaces of specimens at different angles.

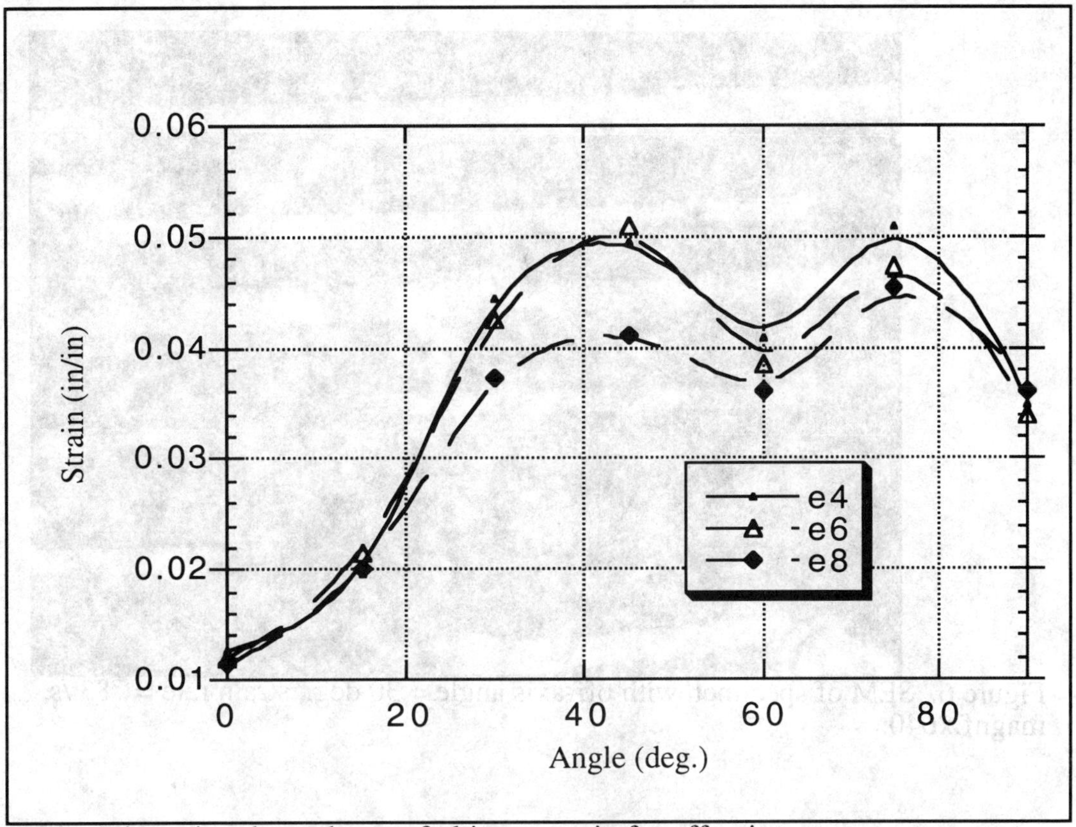

Figure 4. Orientation dependence of ultimate strain for off-axis tests.

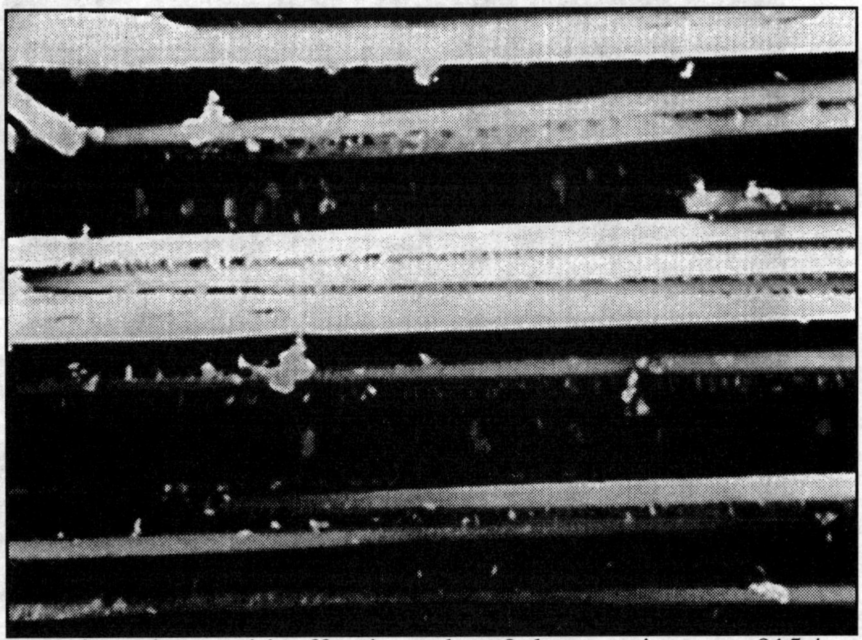
Figure 5. SEM of specimen with off-axis angle = 0 deg., strain rate = 815 /s, magnf.x640.

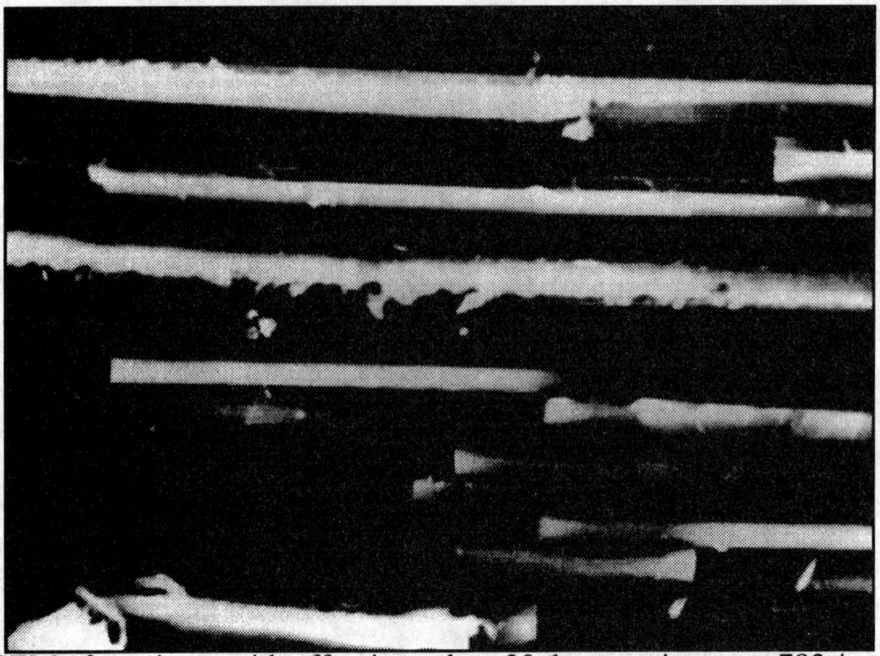
Figure 6. SEM of specimen with off-axis angle = 30 deg., strain rate = 783 /s, magnf.x640.

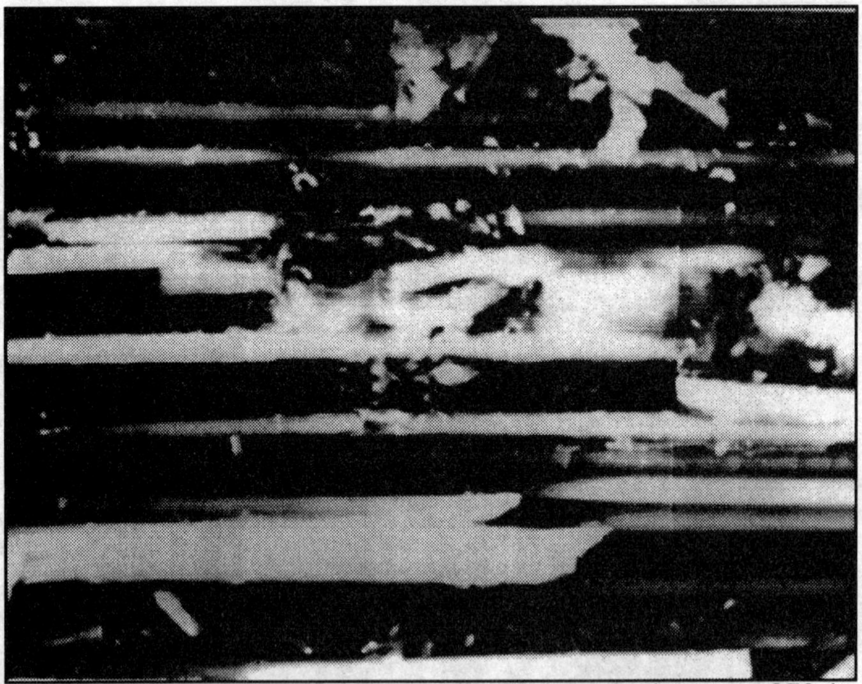
Figure 7. SEM of specimen with off-axis angle = 60 deg., strain rate = 879 /s, magnf.x640.

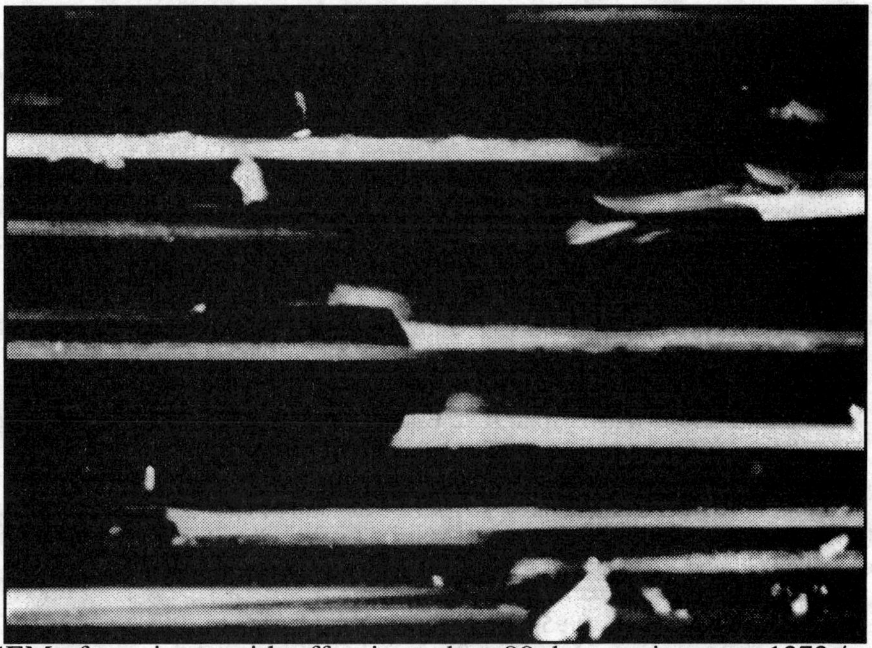

Figure 8. SEM of specimen with off-axis angle = 90 deg, strain rate = 1272 /s, magnf.x640.

CONCLUSIONS

Changing the fiber orientation changes the values of the ultimate strength and strain of the IM7/8551-7 graphite/epoxy composites. A total of 105 specimens, 3 duplicates for each test conditions, were used in this study. Clearly, the ultimate strength of the material becomes strain rate insensitive as the off-axis testing angle is increased. Particularly, at an angle of 60 degrees, the high strain rate strength values are lower or equal to the static strength value, and the graphs of the stress values show a dip at 60 degrees. The scanning electron microscope was used to investigate the fracture surfaces of the tested specimen. The result of the SEM did not provide a lot of information because there are no means of stopping the test right after reaching the ultimate stress. However, looking at the samples under the SEM and comparing the pictures at different off-axis angles, the fibers of the specimen tested at an angle of 60 and 90 degrees show cleaner surfaces than at other orientations, where there are flake-shape hackles.

The dynamic in-plane shear stress of a composite cannot be obtained from the static property equation. In obtaining the dynamic in-plane shear stress from off-axis tests, the static equation has to be modified to include the effect of strain rate and fiber orientation. The reason why there is an obvious reduction of the strength values at 60 degrees and a corresponding strain rate insensitivity is not yet clear. However, the function that would account for the effect of strain rate on the orientation dependence of the dynamic strength described in the results section can be the key to understanding and solve this problem.

REFERENCES

Duffy, J., Campbell, J.D., and Hawley, R.H., "On the Use of Torsional Split Hopkinson Bar to Study Rate Effects up to 1500 s^{-1}." J. Appl. Mech. 38, 83-91.

Hull, D. 1981. An Introduction to Composite Materials, Cambridge University Press, Cambridge.

Kolsky, H. 1949. "An investigation of the Mechanical Properties of Materials at Very High Rates of Loading." Proc. Phys. SOC. B. 62, 676-701.

Nicholas, T. 1971. "Tensile Testing of Materials at High Rates of Strain." Exp. Mech. 21, 177-185, 1980.

Powers, B.M. and J.R. Vinson. 1995. "High Strain Rate Effects on Materials in Sandwich Construction," Proceedings of the Third International Conference on Sandwich Constructions.

Powers, B.M., J.R. Vinson, I.W. Hall, and V. Nardone. 1995. "High Strain Rate Mechanical Properties to Silicon Carbide Reinforced 2080 Aluminum Metal Matrix Composites," Proceedings of the 10th International Conference on Composite Materials.

Powers, B.M., J.R. Vinson, and I.W. Hall. 1995. "High Strain Rate Mechanical Properties of IM7/8551-7 Graphite Epoxy Composites," Proceedings of the 10th Technical Conference of the American Society for Composites.

Powers, B.M., J.R. Vinson, M. Wardle, and B. Scott. 1995. " High Strain Rate Effects on Two AS4 Graphite Fiber Polymer Matrix Composites," ASME Aerospace Division Bound Volume AD-48, "High Strain Rate Effects on Polymer, Metal, and Ceramic Matrix Composites and Other Advanced Materials."

Powers, B.M., J.R. Vinson, I.W. Hall and R.F. Hubbard. 1995. "High Strain Rate Mechanical Properties of Cycom 5920/1583," Proceedings of the 36th AIAA/ASME/ASCE/AHS/ASC Structures, SDM Conference.

Powers, B.M., J.R. Vinson, M. Wardle, and B. Scott. 1996. "High Strain Rate Effects on Two Graphite Fiber K3B Polyimide Matrix Composites," Proceedings of the 37th AIAA/ASME/ASCE/AHS/ASC Structures, SDM Conference.

Staab, G.H. and Gilat, A. 1995. "High Strain Rate Response of Angle-Ply Glass/Epoxy Laminates,"J. of Comp. Matls., vol. 29, 1308-1320.

Woldesenbet, E. and J.R. Vinson. 1997. "Effect of Specimen Geometry in High Strain Rate Testing of Graphite/Epoxy Composite", to be presented at the 38th AIAA/ASME/ASCE/AHS/ ASC Structures, SDM Conference.

Zukas, J. 1982. (ed.), Impact Dynamics, John Wiley and Sons, New York, N.Y.

Analytical and Experimental Techniques for Improvement and Optimization of Damping in Advanced Composite Materials—A Review

I. C. FINEGAN AND R. F. GIBSON

ABSTRACT

Vibration damping is essential to the achievement of high performance goals in advanced engineering systems. Damping is a significant dynamic parameter for vibration and noise control, dynamic stability, fatigue endurance and impact resistance. Accurate information on damping of structural materials is essential in the design of structures and machines. This paper summarizes some recent analytical and experimental results regarding the improvement and optimization of damping in polymer composite materials. A summary of methods for controlling damping in composites at the macromechanical and micromechanical levels is presented in the first part of the review. The second part consists of a review of the experimental techniques for measuring damping in composite materials. The basic objective of the paper is to present the most recent results and solutions for damping improvement and optimization, and to offer recommendations for future research.

INTRODUCTION

Damping of vibrations has become an important requirement in the design of automotive and aerospace structures. Design considerations involve minimization of resonance amplitudes and extending the fatigue life of structures subjected to near resonant vibrations under suddenly applied forces. In particular, polymer composites have generated increased interest in the development of damped structural materials because of their low density and excellent stiffness and damping characteristics.

From previous research, it appears that an increase in damping will cause decreased stiffness and strength. At the same time, from previous literature, the improvement of damping can be achieved by active and passive controls. Active damping control requires sensors and actuators, a source of power, and a compensator which gives good performance in vibrating conditions. Passive damping control consists of the use of structural modifications, damping and isolation. Passive damping typically requires high loss viscoelastic or fluid materials and

---

Ioana C. Finegan, General Motors R&D Center, Physics and Physical Chemistry Department, Warren, MI 48090.
Ronald F. Gibson, Department of Mechanical Engineering, Advanced Composites Research Laboratory, Wayne State University, Detroit, MI 48202.

thermal control. Material damping can contribute to the passive control system by using the inherent capacity of the material to dissipate vibrational energy. Passive damping contributes more effectively to the improvement of the reliability of structures than active damping, because of reduced system complexity.

At the macromechanical level, research has emphasized the study of constituent layer properties and orientations, interlaminar effects, vibration coupling, surface attachments and damping treatments, cocured damping layers and hybridization of laminae. It has been shown that all of these parameters may have a significant influence on the attainment of improved damping characteristics.

At the micromechanical level, recent work has been oriented to the improvement of damping by optimizing the fiber orientation, fiber aspect ratio, fiber spacing, fiber/matrix interphase effects, fiber surface treatments, fiber coatings, fiber and matrix properties, and fiber hybridization. Some new results indicate that good damping characteristics can be achievable in a whisker or microfiber composite. Fiber hybridization has been shown to lead to stiff, light weight, highly damped composite structures.

Significant progress has been made also in the characterization of damping in composites and their constituents. Most of the experimental techniques for measuring damping are based on the use of vibration tests, particularly those involving flexural, extensional, or torsional vibrations.

IMPROVEMENT AND OPTIMIZATION OF DAMPING

The objective of designing composite structures with improved or optimized damping characteristics has led to the development of mechanical theories for the modeling of composite damping at the micromechanical and macromechanical levels. High damping of the composite materials combined with high stiffness and strength can be realized by control of the geometrical and mechanical properties of their constituents.

DAMPING AT THE MACROMECHANICAL LEVEL

At the macromechanical level, research has emphasized the study of constituent layer properties and orientations, interlaminar effects, vibration coupling, surface attachments and damping treatments, cocured damping layers and hybridization of laminae. It has been shown that all of these parameters may have a significant influence on the attainment of improved damping characteristics.

- **Effects of Interlaminar Stress and Lamina Orientation on Damping.**

Interlaminar stresses arise generally at the lamina interfaces, near free edges in angle-ply laminates. The existence of interlaminar stresses means that part of the total energy dissipation is due to interlaminar damping. The results presented in Figures 1 and 2 are for graphite/epoxy angle ply laminated composites under extensional vibrations and were predicted by using a three dimensional finite element/strain energy technique. Figure 1 [Hwang and Gibson, 1992] shows that there exists an optimal fiber orientation and an optimal laminate width-to-thickness ratio (w/t) for maximizing the contribution of interlaminar damping. It was also shown by Hwang and Gibson [1992], that a maximum contribution of interlaminar damping occurs at a fiber orientation of approximately 15^0 for graphite/epoxy laminates with $n = 1$, 3 and 7. The interlaminar damping in angle ply laminates is important only when the laminate is thick. Adams [1986] analyzed the influence of fiber orientation on the damping in the case of laminated beams and plates. Beams and plates are often made from a succession of laminae. If the elastic modulus and damping are known for unidirectional materials, it is possible to calculate the overall damping of the laminated beam or plate. Based on this approach, Adams developed a theoretical model to predict damping of beams and plates laminated from various types of fibers at various orientations.

- **Vibration Coupling Effects on Damping**

Vibration coupling in composites is influenced by parameters such as fiber orientation, stacking sequence and vibration modes of interest. Two of the major coupling modes are the coupling between bending and twisting in symmetric laminates, and the coupling between bending and extension in anti-symmetric laminates [Hwang and Gibson, 1992]. Coupling effects tend to increase the vibration damping. Such effects were predicted for graphite/epoxy by using a three dimensional strain energy/ finite element model. Hwang and Gibson, [1992] showed the contribution of damping due to bending-extensional coupling as a function of fiber angle. The maximum contribution of normal coupling-energy dissipation to damping was found at a fiber orientation between 30^0 and 45^0. It was concluded that the coupling between bending and extension has a significant effect on the total damping of anti-symmetric graphite/epoxy laminated composites.

- **Effects of Surface Attachments and Damping Treatments**

A passive method to increase the damping capacity of composite structures involves the use of surface damping treatments. This can be achieved by the application of damping tapes to the surfaces of the structures after fabrication. The damping tape is a viscoelastic material sandwiched between the base structure and a thin constraining layer. The damping will be improved due to the fact that the vibration energy will be dissipated by shearing motion of the viscoelastic layer as the base structure vibrates in flexure. The treatments can be applied as a single layer to one or both sides of the structures, providing in this way high damping capabilities. It was shown by Barrett [1991], that the application of constrained viscoelastic layer damping tapes is efficient in optimizing the damping, because most of the energy dissipation is due to shear deformations and almost none is due to dilatational deformations. Another approach to increasing the damping was to optimize the length of the constraining layers. Analytical and experimental results presented in Figure 3 [Mantena, et al. 1991] shows that an optimum tape-to-beam length ratio of 0.5 will lead to maximum damping. At the same time, the system damping will be higher for the case where the tape is fixed at the root than the case where the tape is free at the root, for all the tape-to-beam length ratios. Thus, significantly higher overall damping has been obtained by fixing the constrained layer tape at the clamped end. This means that the amount of damping tape, and, consequently, the amount of added weight, required to produce an improvement in damping can be minimized.

- **Cocured Damping Layers for Improving damping**

Recently it has been shown that the use of cocured damping layers is an effective way of obtaining high damping in composite structures. Although the passive damping methods presented previously increased the damping, these solutions generally added weight to the structures. The advantage of using composites cocured with viscoelastic materials is that no extra constraining layer is needed and the weight added to the system in this case is less than the weight added to the system by surface damping treatments with metal constraining layers. Rotz and Barrett, [1991], showed that damping can be designed into a structure right from the beginning rather than adding surface damping treatments later. Stiffness and damping materials were laid up in the composite structure to study the cocuring process, and Figure 4 [Rotz and Barrett, 1991] shows the proposed damped hat-stiffened structure. Note that the damping layers are placed at the laminate middle surfaces where the shear strains would be largest. Static and dynamic experimental tests have conducted on these panels for different prepreg tapes containing damping materials. Results show that the optimal design for the hat stiffened panels would be obtained by using ISD-112 polymeric damping material (manufacturer 3M Company) in the base panel (to maximize the shear damping) and AF-32 polymeric structural adhesive (manufacturer 3M Company) in the hat stiffener (to obtain good static properties while achieving desired level of damping).

- **Effects of Lamina Hybridization and Laminae Stacking Sequence on Laminate Damping**

Hybridization of laminates is a relatively new concept used in the design of high damping composite structures. The purpose of hybridization is to extend the concept of tailoring properties to suit the requirements. In general, it is difficult for a material to possess both high damping properties and high rigidity. It was shown by Adams and Zimmerman, [1986] that hybrid laminates consisting of polyethylene SPECTRA fibers and graphite fibers, have the qualities of stiff, highly damped, impact resistant and light weight composite materials. This hybridization was suggested by the fact that polyethylene fiber composites have better damping and impact resistance than graphite fiber composites, but low stiffness, while graphite composites have excellent stiffness, but low damping capacity and low impact resistance. It was proved from the experimental results that the position of the flexible layers is an important factor which influences the loss factor of the composite structure. A higher flexural damping was obtained by designing layups having more plies of polyethylene SPECTRA near the outer surfaces than layups having more graphite plies near the outer surfaces. The higher damping is the result of the product of the high loss factor (of SPECTRA plies) and the higher strain energy generated at the outer surfaces in flexural vibrations. It appears that further work is needed in order to determine the best locations for damping layers in laminates. Although the shear deformations tend to be the largest near the middle surface, the total strain energy tends to be largest near the outer surfaces.

DAMPING AT THE MICROMECHANICAL LEVEL

Micromechanical damping analysis deals with the prediction of composite damping properties from the corresponding damping properties and geometries of the constituent materials. The various geometric and material parameters used in estimating the damping of the composite systems are: fiber aspect ratio, fiber orientation, fiber interaction, fiber/matrix interphase effects, fiber surface treatments, fiber coatings, fiber and matrix properties, and fiber hybridization.

- **Effects of Fiber Aspect Ratio and Fiber Orientation**

Two primary sources contribute to the damping or energy dissipation in undamaged fiber reinforced polymer materials: (1) the viscoelastic behavior of the polymer matrix and interphase materials and (2) the friction at the interface caused by relative motion between fiber and matrix. The latter effect appears to be significant only in composites with poor interface bonds. Suarez, et al. [1986], and Gibson, et al. [1982], showed that the use of discontinuous fiber reinforcement with a stiffness mismatch between fiber and matrix will lead to the improvement of internal damping in fiber reinforced polymer composite materials by increasing the shear deformation near the fiber ends.

It was shown by Suarez, et al. [1986] that, for unidirectional short fiber composites, damping increases with decreasing fiber aspect ratio. Analytical and experimental results have been obtained to support these conclusions. Figure 5 illustrates the influence of fiber aspect ratio on the damping (loss factor) for unidirectional discontinuous fibers. Experimental research has not been done on damping for fiber aspect ratios (l/d) lower than about 100 due to the difficulties in fabricating aligned discontinuous fiber composites with such low l/d ratios. This is still a research area which could yield significant improvements in damping of composites.

- **Effects of Fiber Interaction on Damping**

The influence of fiber interaction on damping has been studied by analyzing the effect of the size of the fiber end gap on damping. The fiber end gap in this case is the distance between two adjacent fiber ends in a discontinuous aligned fiber composite. These studies have been presented by using experimental work and an analytical approach based on the strain energy finite element method [Hwang and Gibson, 1992]. Some of the resulting damping data for various fiber end gap sizes for a boron/epoxy composite are shown in Figure 6 [Hwang and Gibson, 1992]. The conclusion was that the composite damping generally increases with increasing fiber end gap size. The damping increases because of the stress transfer through a shear mechanism at the fiber matrix interface near the fiber ends.

- **Fiber/Matrix Interphase Effects**

In general, a complex state of stress exists in the fiber/matrix interphase region due to geometric and material discontinuities between the fiber and matrix materials. One study of the effects of such states of stress on composite damping was made by using a micromechanical strain energy/finite element analysis [Hwang and Gibson, 1992]. The micromechanical models used for these analyses were two - dimensional finite - element models with the interphase (not a separate fiber coating) as the third phase between the fiber and the matrix. Damping under longitudinal and transverse normal loading as well as in - plane and out - of - plane shear loading was studied. Two - dimensional finite element models for characterizing the interphase contribution on damping were reported also by Chaturvedi and Tzeng in 1991. A three dimensional model has been developed for laminate macromechanical analysis [Hwang and Gibson 1991]. Both micromechanical and macromechanical modeling of composite damping with the strain energy finite element method have been reviewed by Hwang and Gibson [1992]. Three dimensional finite element models have been used by Caruso and Chamis [1986], to study elastostatic micromechanical behavior of composites without interphase. They did not relate their work to the damping characteristics of the composite material, however. Friend, et al. [1992] used two models to analyze the dynamic properties of boron-aluminum, borsic-aluminum and silicon carbide-aluminum. The two models consisted of a two phase material having perfect bonding between fiber and matrix, and a three phase material where the effect of the interphase was considered. The dynamic shear properties, including shear storage modulus, shear loss factor, and shear loss tangent were analytically calculated by using the correspondence principle, and measured experimentally. One important conclusion was that the interphase contributed significantly to the damping. Dong and Gauvin [1993], showed that the performance of composite materials can be controlled by the properties of an interfacial region which can be a two-dimensional interface or a three-dimensional region (interphase).

- **Fiber Coating Effects**

Since shear deformations are so important in damping of viscoelastic polymers, it has been shown that the interphase damping of fiber reinforced polymer composites leads to significant improvement in damping because of the high shear strain energy stored in the interphase [Gibson, 1992]. The high shear strain in the region of the fiber/matrix interface is due to the mismatch between the fiber and matrix properties. Thus, improvement of damping can be achieved by putting a high damping material (fiber coating) in this region of high shear strain. To achieve a highly damped composite structure, both a high damping material and a high shear strain energy are needed - either one alone is not enough. Finegan and Gibson, [1997] proposed and presented analytical and experimental models to study the influence of fiber coating on the damping of the composite material. Details were presented by Finegan, [1997].

EXPERIMENTAL TECHNIQUES FOR MEASURING DAMPING IN COMPOSITE MATERIALS

Experimental methods to characterize damping in composite structures can be categorized as either free vibration of forced vibration methods. In the forced vibration technique, the excitation may be sinusoidal, random or impulsive, and the response may be analyzed in the time domain or the frequency domain. In the forced vibration techniques the frequency of oscillation of the exciter in the testing machine may or may not be equal to a resonant frequency of the specimen.

For the case of forcing frequency tuned to the resonant frequency of the specimen, the relationship between the input and the response takes on a special form, presented as the so-called resonant dwell method [Gibson and Plunkett, 1977; Gibson, Yau and Riegner, 1982]. Guild and Adams [1981] described an apparatus for the measurement of specific damping capacity in free-free flexure with excitation at the ends of the beam. Measurements where made for only the fundamental flexural mode of vibration, the stress level being varied by varying the energy input per cycle. Ni and Adams [1984] used a similar concept and developed a technique to measure the flexural Young's modulus and the specific damping capacity. Wren and Kinra [1988] presented an experimental procedure for the determination of structural damping in a material specimen at low frequencies and low strain amplitudes. Kinra and Wren [1992] described an experimental technique in which damping was defined as the phase angle by which the stress leads the strain. In their technique for measuring material damping a homogeneous strain field, namely, a uniform axial stress was used.

One of the most widely used techniques is the impulse-frequency response method, which is based on impulse excitation of the specimen and curve-fitting to the frequency response spectrum [Suarez, et al. 1984, and Suarez and Gibson, 1987]. An electromagnetic hammer with a force transducer attached to its tip provides the excitation for the beam. The response is sensed by a noncontacting eddy current proximity transducer, which provides input to the Fast Fourier transform (FFT) analyzer. The complex modulus (storage modulus and loss factor) are computed by a desktop computer interfaced with the FFT analyzer. Several variations on the technique have been used to test specimens in flexural, extensional and torsional modes of vibration. Recently Crane and Gillespie [1992] proposed a testing procedure which minimizes the effect of environmental sources of energy dissipation and provides a more accurate measure of the material loss factor. The procedure was also based on the impulse-technique. The authors conducted a study to determine explicitly the contribution of the aerodynamic damping and frictional losses at the clamped region of the beam. It was shown that aerodynamic damping could be minimized if the beam tip displacements were kept less than 30% of the specimen thickness. The frictional losses at the clamped region of the beam were minimized by using fiberglass fabric-reinforced Teflon sheet as clamping material. The effect of vibration amplitude on damping loss factor was also characterized in the study.

CONCLUSIONS AND DISCUSSION

A review of the most recent analytical and experimental results regarding improvement and optimization of the damping in polymer matrix composite materials with high performance fibers has been presented. Many advances regarding the improvement of damping have been reported and continue to appear in the literature. Much research has been reported in finite element modeling of damping , theoretical optimization studies of dynamic properties, the use of constrained layer damping in composite structures, and the use of hybrid materials. There are still areas which need attention. From the review of the literature, the following conclusions and recommendations on specific areas where future research should be directed can be drawn:

MACROMECHANICAL LEVEL

Cocuring of embedded layers of damping materials within the composite is a recent and effective method for fabricating highly damped composite structures, while minimizing the added weight. Analytical and experimental work has been done in modeling the system damping and resonance frequencies of composite plates with single and multiple viscoelastic layers. Combining the geometry of the composite with the cocuring of damping material is an interesting concept for obtaining high damping and stiffness in the structure. Thus far, such solutions as hat-stiffened panels with embedded viscoelastic layers have only been investigated experimentally; analytical investigations are also needed.

Since, in the literature it has already been proved that the optimization of damping is possible by using different laminate geometries, a new composite design could be a combination of hybrid geometry with hybrid material.

Hybridization of laminates is a relatively new concept used in the design of high damping composite structures. Experimental work has been done by combining polyethylene SPECTRA fiber reinforcement plies which have good damping capacities with graphite fiber reinforced plies which have high stiffness. Good damping and stiffness characteristics have been reported. Continuation of this research by experimenting with a composite hybrid obtained by lamina hybridization (macromechanical level) together with fiber hybridization (micromechanical level), and a classification of the compatible constituent materials that can lead to high damping and stiffness is also needed. There is still much potential to be explored in the area of hybrid composites.

MICROMECHANICAL LEVEL

It was shown that damping in discontinuous-aligned fiber composites increases at lower fiber aspect ratios. Experimental research has not been done on damping for fiber aspect ratios (l/d) lower than about 100 due to difficulties in fabricating aligned discontinuous fiber composites with such low l/d ratios. Since damping increases at lower fiber aspect ratios, it seems that near optimum damping could be obtained in a whisker or microfiber composite. It is known that whiskers are particles with a very low fiber aspect ratio. This is still an open research area which could yield significant improvements in damping of composites. Investigation of other micromechanical techniques for improving damping include studies of the effects of fiber orientation, fiber interaction, and fiber/matrix combinations.

Analytical/experimental efforts have been reported on the contribution of the fiber coating on the composite damping. It was shown that coating the fibers with a high damping polymer brings new solutions in the improvement of damping. This research can be continued and by studying different types of high damping polymers as fiber coatings and proposing the optimum polymer coating and optimum coating thickness. Also the manufacturing techniques for coating fibers with high damping polymers need to be investigated.

REFERENCES

Adams, D. F., and Zimmerman, R. S., 1986, "Static and Impact Performance of Polyethylene Fiber/Graphite Fiber Hybrid Composites", Proc. 31th National SAMPE Symposium, pp. 1456-1468.

Barrett, D. J., 1991, "Damped Composite Structures", Composite Structures, Vol. 18, pp. 283 -294.

Caruso, J. J., and Chamis, C. C., 1986, "Assessment of Simplified Composite Micromechanics Using Three-Dimensional Finite Element Analysis", J. of Composite Materials, Vol. 8, No. 3, pp. 77 - 83.

Crane, R. M., and Gillespie Jr. J. W., 1992, "A Robust Testing Method for Determination of the Damping Loss Factor of Composites", J. of Composites Technology & Research Vol. 14, No. 2, pp. 79-79.

Chaturvedi, S. K., and Tzeng, G. Y., 1991, "Microm4echanical Modeling of Material Damping in Discontinuous Fiber: Three-Phase Polymer Composites", Composites Engg., Vol. 1, pp. 49 - 60.

Dong, S., and Gauvin, R., 1993, "Application of Dynamic Mechanical Analysis for the Study of the Interfacial Region in Carbon Fiber/Epoxy Composite Materials", Polymer Composites, Vol. 14, No. 5, pp. 414 - 420.

Friend, R. D., Kennedy, J. M., and Edie, D. D., 1992, "Evaluating the Fiber-Matrix Interphase in Metal Matrix Composites Using Dynamic Mechanical Analysis", Proc. of the Damping of Multiphase Materials Symposium, ASM Materials Week, Chicago, IL, 2-5, Nov. 1992, pp. 123 - 135.

Finegan, I. C., and Gibson, R. F., 1997, "Analytical and Experimental Characterization of Damping and Stiffness in Polymer Composites Having Coated Fibers as reinforcement", to be presented at the International Mechanical Engineering Congress and Exposition, The 1997 Winter Annual Meeting of the ASME, November 16-2, Dallas, Texas.

Finegan, I. C., 1997, "*Analytical and Experimental Study of Improvement of Damping at the Micromechanical Level in Polymer Composite Materials by the Use of Special Fiber Coatings*", Wayne State University, Detroit, Michigan, Ph. D. Dissertation, 1997.

Gibson, R. F., 1992, "Damping Characteristics of Composite Materials and Structures", J. of Materials Engineering and Performance, Vol. 1, pp. 11 - 20.

Gibson, R. F., and Plunkett, R., 1977, "A Forced-Vibration Technique for Measurement of Material Damping", Experimental Mechanics, Vol. 11, No. 8, pp. 297 - 302.

Gibson, R. F., Rao, V. S., and Mantena, R., 1987, "Vibration Damping Characteristics of Highly Oriented Polyethylene Fiber Reinforced Epoxy Composites", Proc. 32nd International SAMPE Symposium, pp. 231- 245.

Gibson, R. F., Chaturvedi, S. K., and Sun, C. T., 1982, "Complex Moduli of Aligned Discontinuous Fiber Reinforced Polymer Composites", J. Materials Science, Vol. 17, pp. 3499 - 3509.

Gibson, R. F., Yau, A., and Riegner, 1982, "An Improved Forced Vibration Technique for Measurement of Material Damping", Experimental Techniques, Vol. 6, No. 2, pp. 10-14.

Guild, F. J. and Adams, R. D., 1981, "A New Technique for the Measurement of the Specific Damping Capacity of Beams in Flexure", J. Phys E: Sci Instr., Vol. 14, pp. 355 - 363.

Hwang, S. J. and Gibson, R. F., 1992, " The Use of Strain Energy - Based Finite Element Techniques in the Analysis of Various Aspects of Damping of Composite Materials and Structures", J. of Composite Materials, Vol. 26, No. 17, pp. 2585 -2605.

Hwang, S. J., and Gibson, R. F., 1991, "Effects of Three Dimensional States of Stress on Damping of Laminated Composites", Composites Scxience and Technology, Vol 41, pp., 379 - 393.

Kinra, V. K., and Wren, G. G., 1992, "Axial Damping in Metal-Matrix Composites. I. A New Technique for Measuring Phase Difference to 10^{-4} Radians", Experimental Mechanics, Vol. 32, No. 2, pp. 172 - 178.

Mantena, R. P., Gibson, R. F. and Hwang, S. J. 1991, "Optimal Constrained Viscoelastic Tape Lengths for Maximizing Damping in Laminated Composites", AIAA Journal, Vol. 29, No. 1, pp. 1678 - 1685.

Ni, R. G., and Adams, R. D., 1984, "Experimental Method for Obtaining the Dynamic Mechanical properties of laminae for predicting the Stiffness and Damping of Laminated Plates and Beams", Composites, Vol. 15, No. 3, pp. 193 - 199.

Olcott, D. D., Rotz, C. A., and Barrett, D. J., 1990, "Improved Damping in Composite Tubes through Stress Coupling and Co-cured Damping Layers", Proc. 23rd Int. SAMPE Tech. Conference, pp. 373-387.

Rotz, C. A., and Barrett, D. J., 1991, "Cocured Damping Layers in Composite Structures", Proc. of 23rd International SAMPE Technical Conference, Oct. 21-24, pp.352-363.

Suarez, S. A., Gibson, R. F., and Deobald, L. R., 1984, "Random and Impulse Techniques for Measurement of Damping in Composite Materials", Experimental Techniques, Vol. 8, No.10, 19-24.

Suarez, S. A., and Gibson, R. F., 1987, "Improved I
mpulse-Frequency Response Techniques for measurement of Dynamic Mechanical properties of Composite Materials", J. of Testing and Evaluation, Vol.15, No. 2, 114-121.

Suarez, S. A., Gibson, R. F., Sun, C. T. and Chaturvedi, S. K., 1986, " The Influence of Fiber Length and Fiber Orientation on Damping and Stiffness of Polymer Composite Materials", Experimental Mechanics, Vol. 26, No. 2, 175 - 184.

Wren, G. G., and Kinra, V. K., 1988, "An Experimental Technique for Determining a Measure of Structural Damping", J. of Testing and Evaluation, Vol. 16, No. 1, pp. 77-85.

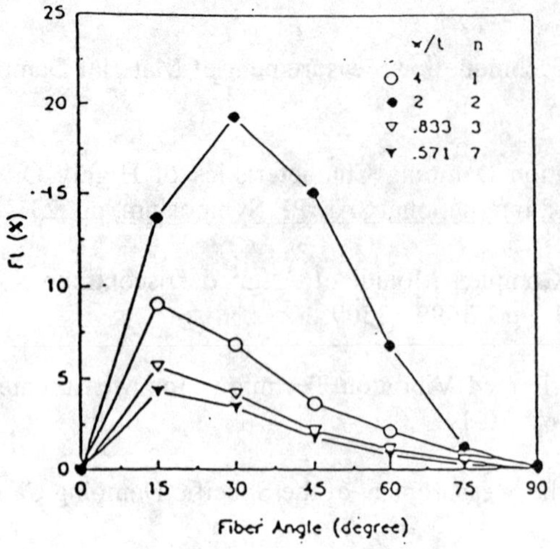

Figure 1 Contribution of interlaminar damping Ft as a function of fiber orientation for [n($\pm\theta$]s, graphite/epoxy laminates under uniaxial extension. [From Hwang and Gibson, 1992].

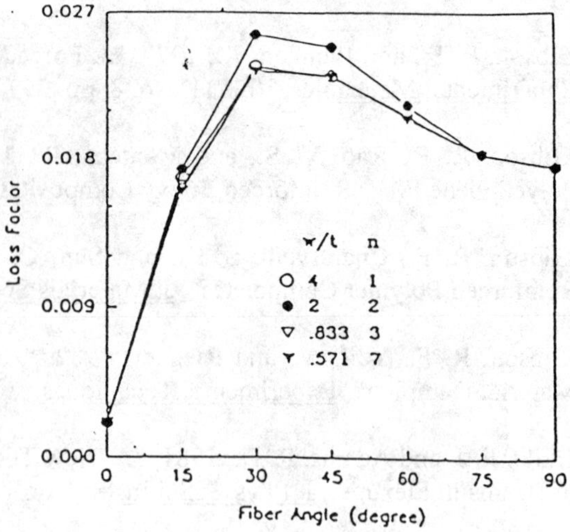

Figure 2 Variation of loss factor with fiber orientation for [n($\pm\theta$]s, graphite/epoxy laminates under uniaxial extension. [From Hwang and Gibson, 1992].

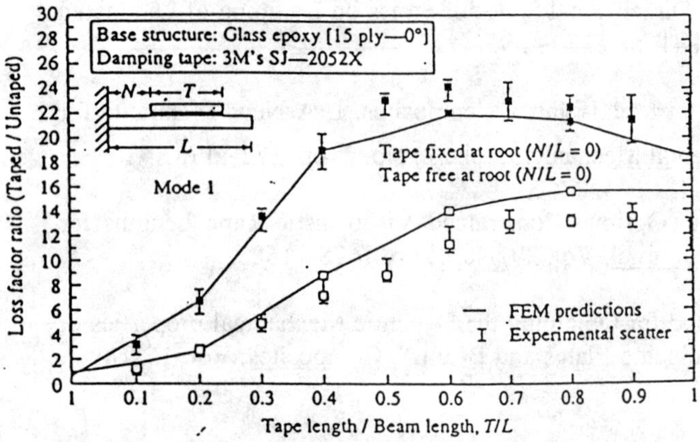

Figure 3 Measured and predicted damping for unidirectional glass/epoxy beam with constrained viscoelastic layer damping tapes of different lengths and tape end fixity conditions. [From Mantena et. al, 1991]

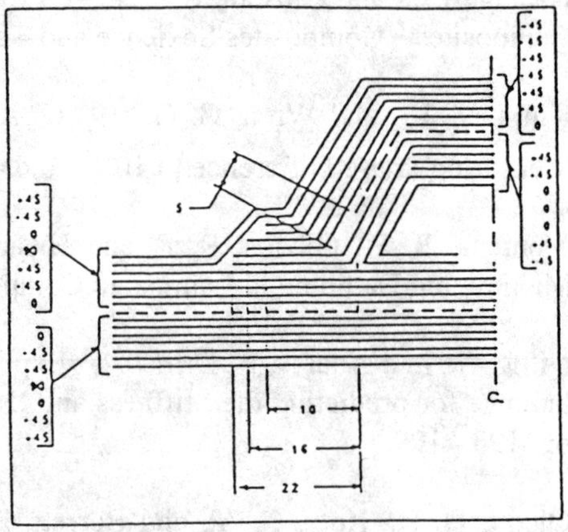

Figure 4 Layup diagram for the hat-stiffened panels. The dashed lines represent the damping layers. The linear dimensions are in mm. [From Rotz and Barrett, 1991]

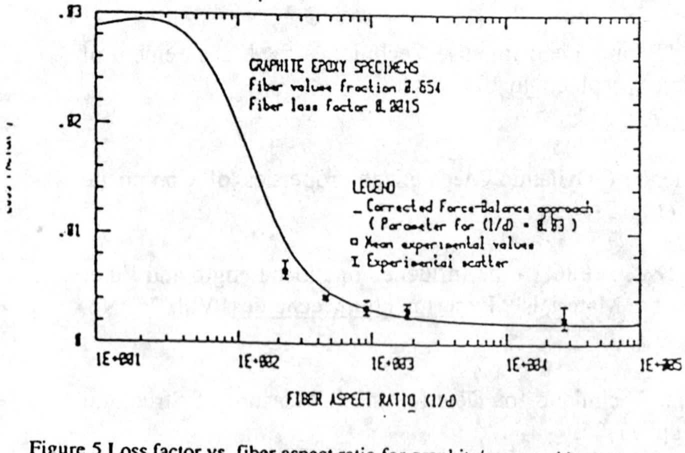

Figure 5 Loss factor vs. fiber aspect ratio for graphite/epoxy with curve fitting. [From Suarez, Gibson, Sun, and Chaturvedy, 1986]

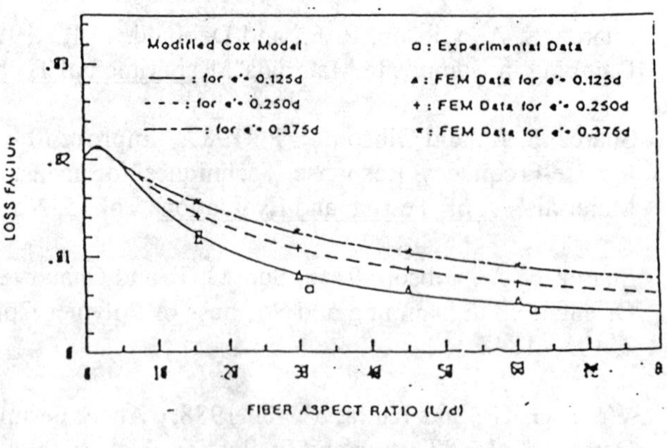

Figure 6 Variation of composite loss factor with fiber aspect ratio (l/D) for discontinuous fiber composites with different fiber end gap sizes (where e'=one half of fiber end size =e/2) [From Hwang and Gibson, 1992]

Micromechanical Analysis of Viscoelastic Damping in Woven Fabric-Reinforced Polymer Matrix Composites

H. GUAN AND R. F. GIBSON

ABSTRACT

This paper describes the development of a closed-form model and a finite element model for studying viscoelastic damping in woven fabric-reinforced polymer matrix composites. The closed-form model was obtained by first deriving an elastic solution based on the mechanics of materials theory, then applying the Elastic-Viscoelastic Correspondence Principle to the elastic solution. The damping loss factors were determined from the complex moduli. Three-dimensional finite element models with various mesh densities were constructed based on the actual geometry of a plain weave fabric-reinforced composite unit cell. Three different loading cases were designed to simulate extensional, shear and bending deformations of the unit cell. The strain energies stored in the fiber and matrix elements for each loading case are conveniently determined from the finite element analysis. A strain energy formulation based on earlier work by Ungar and Kerwin was used to calculate the damping loss factors for each loading case. Experiments were conducted using an impulse-frequency response technique to measure damping in beam samples made of plain weave E-glass fabric-reinforced vinyl ester resin matrix composite. The predictions from both analytical models show reasonably good agreement with experimental data.

INTRODUCTION

Damping in composites involves a variety of energy dissipation mechanisms that depend on vibration parameters such as frequency, amplitude, state of stress, and environmental conditions such as temperature and moisture. In fiber-reinforced polymers, the most important damping mechanisms include viscoelastic behavior of fiber and/or matrix materials, thermoelastic damping due to cyclic heat flow from regions of compressive stress to regions of tensile stress, coulomb friction due to slip in unbonded regions of the fiber/matrix interface, and energy dissipation occurring at sites of cracks and/or delaminations. Although viscoelastic damping is not the only damping mechanism

---

Huimin Guan, Ford Motor Company, PDC, MD299, P.O.Box 2053, Dearborn, MI 48121
Ronald F. Gibson, Mechanical Engineering, Wayne State University, Detroit, MI 48202

in fiber-reinforced polymer composites, it does appear to be the dominant mechanism in undamaged polymer composites vibrating at small amplitudes. Therefore, in the present study, only linear viscoelastic damping is considered. There are two basic approaches to the development of analytical models for predicting linear viscoelastic damping in composites. The first approach involves the use of the Elastic-Viscoelastic Correspondence Principle in combination with elastic solutions from mechanics of materials or elasticity theory. An analytical model developed by Gibson, Chaturvedi and Sun [1982] for predicting the damping of aligned discontinuous fiber composites, an analytical model developed by Sun, Wu and Gibson [1985] for predicting the damping of randomly oriented short-fiber composites, and an analytical model for predicting damping of laminated composites [Sun, Wu and Gibson, 1987] are examples of this approach. The second approach is based on the use of a strain energy formulation which relates the total damping in the structure to the damping of each element and the fraction of the total strain energy stored in that element. For example, Hwang and Gibson [1987] have applied the strain energy-based finite element approach for micromechanical analysis of damping in discontinuous fiber composites. This approach was also used to study the effects of the fiber/matrix interphase on composite damping properties [Gibson, Hwang and Kwak, 1991], and the effects of interlaminar stresses on damping of laminated composites [Hwang and Gibson, 1991]. Despite of the fact that analytical closed-form models as well as finite element models have been developed to study the various aspects of damping in composites at both micromechanical and macromechanical levels, the composite materials studied so far are limited to unidirectional or short fiber composites, and laminates based on these composites, and apparently no attempt has been made to explore the linear viscoelastic damping behavior of woven fabric-reinforced polymer matrix composites (WFC). In this study, two analytical models will be developed for studying damping in WFC. The first analytical model is a closed-form solution which involves the use of the Elastic-Viscoelastic Correspondence Principle in conjunction with the elastic solution from the mechanics of materials theory. The second model is a three-dimensional finite element implementation of a strain energy formulation. In order to validate the analytical models, experiments were conducted to determine the damping loss factors of E-glass/vinylester composite beam samples. The results obtained from experiments were then compared with the predictions from both analytical models. Finally, parametric studies were conducted to show the effects of fiber loss factor and fiber volume fraction on the damping properties of the composite.

MICROMECHANICAL CLOSED-FORM SOLUTION

Micromechanical damping analysis deals with the prediction of composite damping properties from the corresponding damping properties and geometries of the constituent materials. The derivation of the closed-form solution consists of two steps. The first step is to obtain the elastic closed-form solution based on the mechanics of materials theory. The second step is to convert the elastic solution into the corresponding viscoelastic solution by using the Elastic-Viscoelastic Correspondence Principle. The closed-form elastic solution is based on the "mosaic model" developed by Ishikawa and Chou [1983], where the basic assumption is that the classical lamination theory is valid for every infinitesimal piece of a WFC unit cell. Thus, the force-deformation relationships associated with a WFC unit cell can be written in the form:

$$\begin{Bmatrix} N_i \\ M_i \end{Bmatrix} = \begin{bmatrix} \overline{A}_{ij} & \overline{B}_{ij} \\ \overline{B}_{ij} & \overline{D}_{ij} \end{bmatrix} \begin{Bmatrix} \varepsilon_j^0 \\ \kappa_j \end{Bmatrix} \tag{1}$$

where N_i = in-plane forces $\quad M_i$ = moments

ε_j^0 = midplane strains $\quad \kappa_j$ = curvatures

$\overline{A}_{ij}$ = extensional stiffnesses of a WFC unit cell

$\overline{B}_{ij}$ = coupling stiffnesses of a WFC unit cell

$\overline{D}_{ij}$ = bending stiffnesses of a WFC unit cell

The upper and lower bounds of stiffnesses of a WFC unit cell were derived by Ishikawa and Chou [1983]. Based on the theorem of minimum potential energy and the theorem of least work, the upper and lower bounds of stiffnesses of the unit cell were determined from the stiffnesses of a cross-ply laminate.

The extensional, coupling and bending stiffnesses of a cross-ply laminate were given by:

$$\left(A_{ij}, B_{ij}, D_{ij}\right) = \sum_{k=1}^{2} \int_{h_{k-1}}^{h_k} \left(1, z, z^2\right) (\overline{Q}_{ij})_k \, dz \qquad \left(i, j = 1, 2, 6\right) \tag{2}$$

where $\overline{Q}_{ij}$ = stiffness constants of a unidirectional lamina

For a specially orthotropic lamina, as in the case here:

$$\overline{Q}_{ij} = \begin{bmatrix} E_{11}/D_\nu & \nu_{12}E_{22}/D_\nu & 0 \\ \nu_{21}E_{11}/D_\nu & E_{22}/D_\nu & 0 \\ 0 & 0 & G_{12} \end{bmatrix} \tag{3}$$

where

$$D_\nu = 1 - \nu_{12}\nu_{21} \qquad \nu_{21} = \frac{\nu_{12}E_{22}}{E_{11}}$$

Here, E_{11} and E_{22} are the longitudinal and transverse Young's moduli, respectively, G_{12} is the in-plane shear modulus, and ν_{12} denotes the major Poisson's ratio.

The prediction of the four independent moduli $E_{11}, E_{22}, G_{12}, \nu_{12}$, is based on the mechanics of materials models for an orthotropic continuous fiber-reinforced lamina. The matrix and the fibers are assumed to be linearly elastic and isotropic. For longitudinal modulus and Poisson's ratio, the "rules of mixtures" based on elementary mechanics of materials models are good enough for design use.

$$E_{11} = E_f v_f + E_m v_m \tag{4}$$

$$\nu_{12} = \nu_f v_f + \nu_m v_m \tag{5}$$

where E_f, E_m = longitudinal modulus of fiber and matrix, respectively.

v_f, v_m = fiber and matrix volume fraction, respectively.

ν_f, ν_m = Poisson's ratio of fiber and matrix, respectively.

However, improved mechanics of materials models are required for transverse modulus and shear modulus in order to obtain sufficient accuracy for design purposes. A set of equations for E_{22} and G_{12} has been derived by Chamis [1984]:

$$E_{22} = \frac{E_m}{1-\sqrt{v_f}\,(1-E_m/E_f)} \tag{6}$$

$$G_{12} = \frac{G_m}{1-\sqrt{v_f}\,(1-G_m/G_f)} \tag{7}$$

where G_f, G_m = shear modulus of fiber and matrix, respectively.

Based on Eqs. (2)-(7), the upper and lower bounds of stiffnesses of a WFC unit cell were obtained from the elastic properties and fiber volume fractions of the constituent materials, in this case, the fiber and matrix materials. Finally, the elastic solution was obtained by defining the stiffnesses of a WFC unit cell as the average of the upper and lower bound values:

$$\overline{A}_{ij} = \frac{1}{2}(\overline{A}_{ij}^{U} + \overline{A}_{ij}^{L}) \qquad \overline{B}_{ij} = \frac{1}{2}(\overline{B}_{ij}^{U} + \overline{B}_{ij}^{L}) \qquad \overline{D}_{ij} = \frac{1}{2}(\overline{D}_{ij}^{U} + \overline{D}_{ij}^{L}) \tag{8}$$

The viscoelastic solution was obtained by applying the Elastic-Viscoelastic Correspondence Principle to the elastic solution. For sinusoidal forces and deformations, Eqs. (1) can be rewritten as:

$$\begin{Bmatrix} \tilde{N}_i(t) \\ \tilde{M}_i(t) \end{Bmatrix} = \begin{bmatrix} \overline{A}_{ij}^*(\omega) & \overline{B}_{ij}^*(\omega) \\ \overline{B}_{ij}^*(\omega) & \overline{D}_{ij}^*(\omega) \end{bmatrix} \begin{Bmatrix} \tilde{\varepsilon}_j^0(t) \\ \tilde{\kappa}_j(t) \end{Bmatrix} \tag{9}$$

where t = time ; ω = frequency

$\overline{A}_{ij}^*$ = complex extensional stiffnesses of a WFC unit cell

$\overline{B}_{ij}^*$ = complex coupling stiffnesses of a WFC unit cell

$\overline{D}_{ij}^*$ = complex bending stiffnesses of a WFC unit cell

and the corresponding inverted force-deformation relationships are:

$$\begin{Bmatrix} \tilde{\varepsilon}_i^0(t) \\ \tilde{\kappa}_i(t) \end{Bmatrix} = \begin{bmatrix} \overline{a}_{ij}^*(\omega) & \overline{b}_{ij}^*(\omega) \\ \overline{b}_{ij}^*(\omega) & \overline{d}_{ij}^*(\omega) \end{bmatrix} \begin{Bmatrix} \tilde{N}_j(t) \\ \tilde{M}_j(t) \end{Bmatrix} \tag{10}$$

where $\overline{a}_{ij}^*(\omega)$ = complex extensional compliances of a WFC unit cell

$\overline{b}_{ij}^*(\omega)$ = complex coupling compliances of a WFC unit cell

$\overline{d}_{ij}^*(\omega)$ = complex bending compliances of a WFC unit cell

For plain weave fabric-reinforced composite, the effective complex extensional, shear and flexural moduli are defined as:

$$\overline{E}_e^*(\omega) = \frac{1}{h\overline{a}_{11}^*(\omega)} = \overline{E}_e^{'}(\omega) + i\overline{E}_e^{"}(\omega)$$
$$\overline{G}^*(\omega) = \frac{1}{h\overline{a}_{66}^*(\omega)} = \overline{G}^{'}(\omega) + i\overline{G}^{"}(\omega) \qquad (11)$$
$$\overline{E}_f^*(\omega) = \frac{12}{h^3\overline{d}_{11}^*(\omega)} = \overline{E}_f^{'}(\omega) + i\overline{E}_f^{"}(\omega)$$

where $\overline{E}_e^{'}(\omega)$; $\overline{E}_e^{"}(\omega)$ = extensional storage modulus and loss modulus, respectively.
$\overline{G}^{'}(\omega)$; $\overline{G}^{"}(\omega)$ = shear storage modulus and loss modulus, respectively.
$\overline{E}_f^{'}(\omega)$; $\overline{E}_f^{"}(\omega)$ = flexural storage modulus and loss modulus, respectively.
h = unit cell thickness

The viscoelastic damping loss factors of the composite are defined as:

$$\eta_e(\omega) = \frac{\overline{E}_e^{"}(\omega)}{\overline{E}_e^{'}(\omega)} \qquad \eta_s(\omega) = \frac{\overline{G}^{"}(\omega)}{\overline{G}^{'}(\omega)} \qquad \eta_f(\omega) = \frac{\overline{E}_f^{"}(\omega)}{\overline{E}_f^{'}(\omega)} \qquad (12)$$

where $\eta_e(\omega)$, $\eta_s(\omega)$, $\eta_f(\omega)$ = extensional, shear and flexural damping loss factor, respectively.

STRAIN ENERGY-BASED FINITE ELEMENT ANALYSIS

Three-dimensional finite element models with various mesh densities were created based on the actual geometry of a plain weave fabric-reinforced composite unit cell. A preprocessor PDGS/FAST, and a solver MSC/NASTRAN were used to construct the finite element models and carry out the calculations. A strain energy formulation was used in conjunction with finite element analyses to determine the viscoelastic damping loss factors of the composite.

Figure 1 shows a cross-sectional view of a plain weave fabric-reinforced composite unit cell, which consists of fill and warp yarns woven in a simple over-under-over pattern. In a balanced weave, as was modeled in this study, warp and fill yarns were assumed to be identical. The actual geometrical configuration and fiber volume fractions of the composite

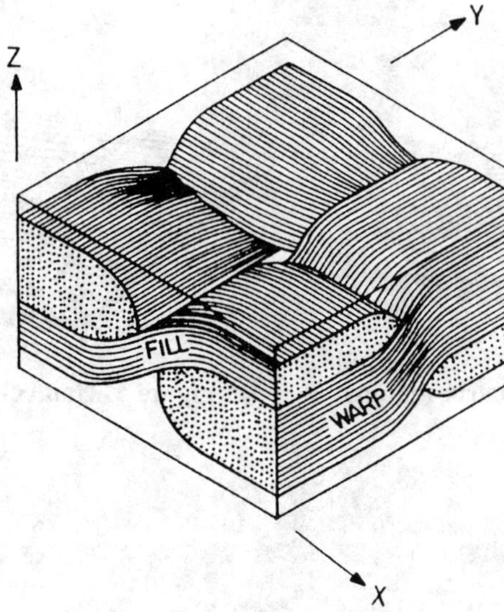

Figure 1 A cross-sectional view of a WFC unit cell (plain weave)

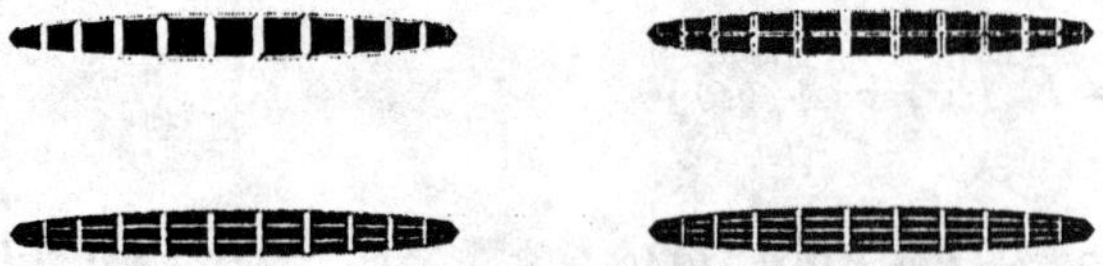

Figure 2 Different yarn cross section meshes

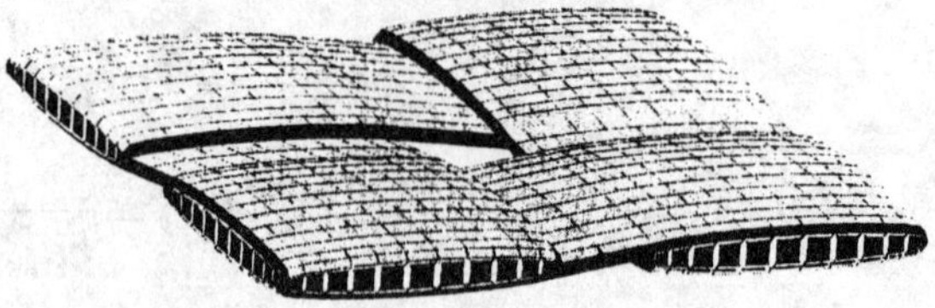

Figure 3 The modeling of the yarns

unit cell were determined from experiments [Guan, 1997]. The finite element modeling process started with the mesh generation of a typical yarn cross section. A yarn is a resin impregnated fiber bundle which contains thousands of filaments, but because of the limitation of finite element model size, it is impossible to model all the fiber filaments in a yarn. In order to conduct a convergence study, four yarn cross section meshes were generated with increasing mesh densities. As shown in Fig. 2, the first mesh has only one fiber element placed in the yarn thickness direction, the second mesh has two fiber elements placed in the yarn thickness direction, the third mesh has three fiber elements placed in the yarn thickness direction, and the last mesh has four fiber elements placed in the yarn thickness direction. In all four meshes, the fiber elements are surrounded by matrix elements, and no two fiber elements are directly connected to each other. The gaps between fiber elements are assumed to be approximately equal in order to represent the uniform distribution of the fiber filaments in a yarn. Once the yarn cross section mesh was generated, the whole yarn was modeled by dragging the cross section along the entire length of the yarn (Fig. 3). The drag path has to follow the surface of the intersecting yarns to make sure that all yarns match perfectly. Since the yarns are identical, once one yarn is modeled, others can be modeled by translating, rotating or reflecting the first yarn. The next step is to model the pure matrix regions in the unit cell. The pure matrix regions can be divided into eight pieces, four pieces on the top and four pieces at the bottom. Each piece covers a quarter of the top or bottom area, as shown in Fig. 4. Finally, the complete unit cell was assembled by putting yarns and pure matrix regions together, as shown in Fig. 5.

Figure 4 The modeling of the pure matrix regions

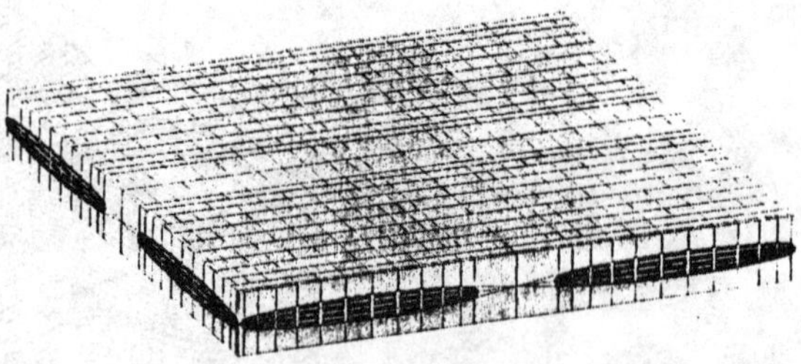

Figure 5 Unit cell #4, 23504 elements

The fiber and matrix materials used in the analyses are E-glass and vinyl ester resin, respectively. The elastic as well as damping properties of the fiber and matrix materials were obtained from experiments [Guan, 1997]. Three loading cases were designed to investigate the response of the unit cell under extensional, shear and flexural deformation. The strain energies stored in the unit cell model for each loading case were calculated by NASTRAN. The use of a strain energy relationship to calculate damping loss factors was first presented by Ungar and Kerwin [1962]. Ungar and Kerwin found that for an arbitrary system of linear viscoelastic elements, the system loss factor can be expressed as a summation of the products of the individual element loss factors and the fraction of the total strain energy stored in each element:

$$\eta = \frac{\sum_{i=1}^{N} \eta_i W_i}{\sum_{i=1}^{N} W_i} \qquad (13)$$

where η_i = loss factor for the ith element in system
W_i = strain energy stored in the ith element at maximum vibratory displacement
N = total number of elements in system

The Ungar-Kerwin equation is ideally suited for finite element implementation in the analysis of complex structures. In the finite element implementation the element index " i " in Eq. (13) refers to the element number, N refers to the total number of finite elements, and the strain energy terms, W_i, are determined from the finite element analysis for the loading case of interest.

Application of Eq. (13) to the unit cell with extensional loading yields:

$$\eta_e = \frac{\eta_e^m \sum_{i=1}^{K} W_i^m + \eta_e^f \sum_{i=1}^{L} W_i^f}{\sum_{i=1}^{N} W_i} \qquad (14)$$

where η_e = extensional loss factor of the composite
η_e^m, η_e^f = extensional loss factor of the matrix and fiber material, respectively.
W_i^m, W_i^f = strain energy stored in the ith matrix and fiber element, respectively.
W_i = strain energy stored in the ith element
K, L = total number of matrix and fiber elements, respectively.
N = total number of elements in the model = K+L

Similarly, for shear loading:

$$\eta_s = \frac{\eta_s^m \sum_{i=1}^{K} W_i^m + \eta_s^f \sum_{i=1}^{L} W_i^f}{\sum_{i=1}^{N} W_i} \qquad (15)$$

where η_s = shear loss factor of the composite

η_s^m, η_s^f = shear loss factor of the matrix and fiber material, respectively.

For flexural loading:

$$\eta_f = \frac{\eta_f^m \sum_{i=1}^{K} W_i^m + \eta_f^f \sum_{i=1}^{L} W_i^f}{\sum_{i=1}^{N} W_i} \quad (16)$$

where η_f = flexural loss factor of the composite

η_f^m = flexural loss factor of the matrix material = η_e^m

η_f^f = flexural loss factor of the fiber material = η_e^f

RESULTS AND DISCUSSION

An impulse-frequency response technique was used to conduct free-free beam extensional and cantilever beam flexural tests [Guan, 1997]. The composite beam samples were made of plain weave E-glass fabric preforms and vinyl ester resin. The damping properties of the composite predicted with both analytical models and measured in the experiments are presented in Figs. 6-7. The effects of finite element mesh density on the analysis results are demonstrated by comparing the predictions from four finite element models with different number of elements. As shown in Figs. 6 and 7, the predictions of extensional and flexural loss factors from the finite element models approach the corresponding experimental minimum as the number of elements in the model increases. It was also observed that the predicted extensional loss factor is closer to the experimental results than the predicted flexural loss factor. Overall, the comparison of the analytical and experimental results in Figs. 6-7 indicates that the predictions from both analytical models show reasonably good agreement with experimental data. The discrepancies between the predictions from the closed-form model and experimental data could be partially caused by the simplifying assumptions used in the derivation of the closed-form model. Due to the limitation of the finite element model size, the fiber filaments in the finite element model are much larger than those in the real composite. As a result, the fiber distribution in the model is much less uniform than that in the real composite, and this could be partially responsible for the discrepancies between the predictions from the finite element model and the experimental data.

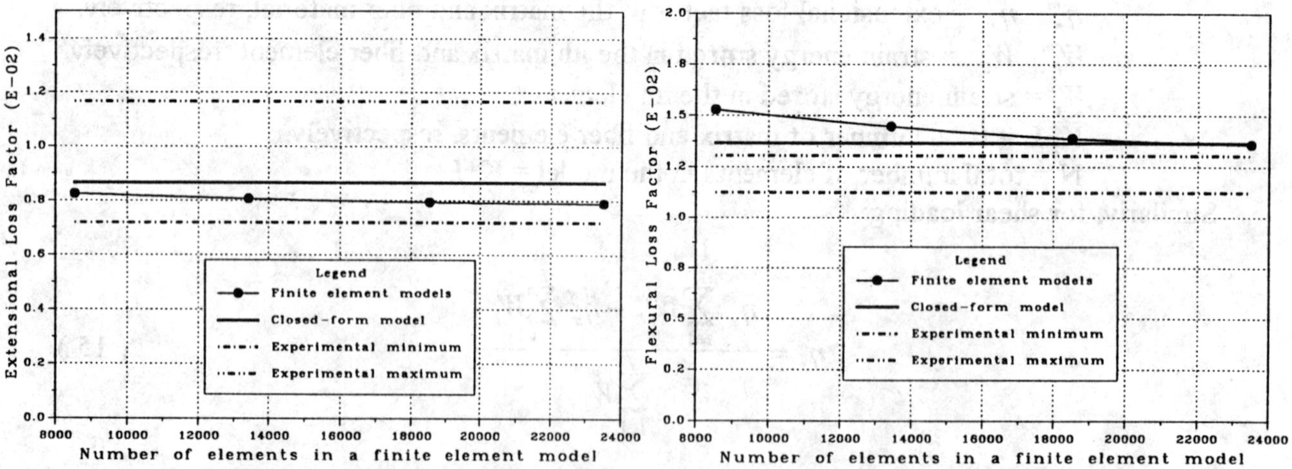

Figure 6-7 Analytical and experimental results for extensional and flexural loss factors

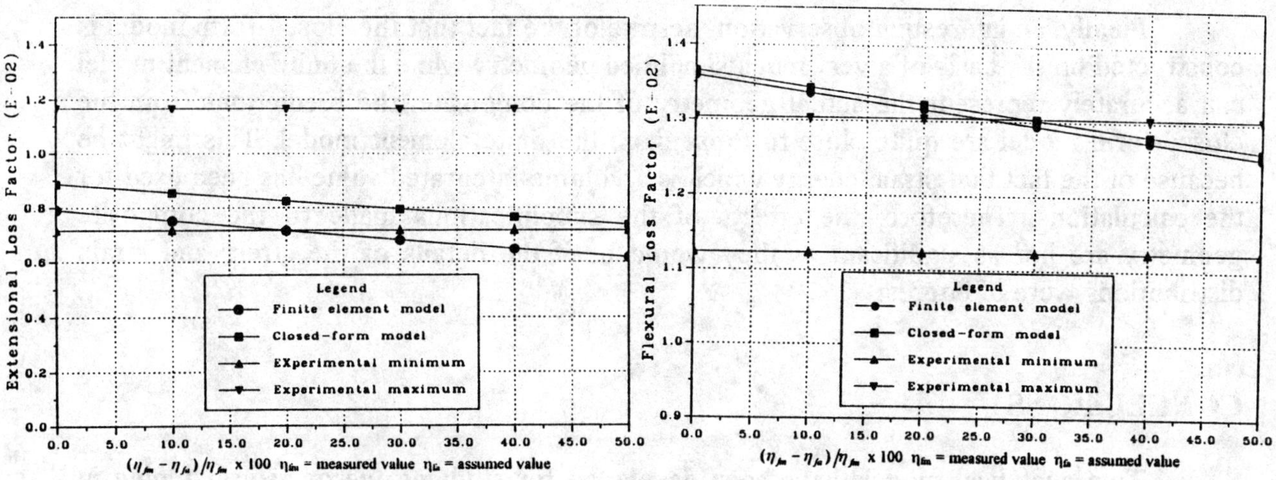

Figure 8-9 The effects of the fiber loss factor on the predictions of extensional and flexural loss factors.

Since both analytical models use the experimental data for the constituent materials as input, the discrepancies between the predictions and the experimental results could also be partially caused by the experimental errors in constituent material properties. For example, the measured fiber loss factor is probably higher than the actual fiber loss factor, and the measured local (yarn) fiber volume fraction is believed to be lower than its actual value. In order to estimate the effects of such experimental errors on the predictions from both analytical models, a sensitivity study of those parameters was conducted. The results were recorded in Figs. 8-11. In Figs. 8 and 9, the predicted composite extensional and flexural loss factors are described as the functions of the input fiber loss factor, which varies from the measured value down to 50% of that value. The results show that the predictions of the composite extensional and flexural loss factors decrease as the input fiber loss factor decreases. In Figs. 10 and 11, the predicted composite extensional and flexural loss factors are described as the functions of the local (yarn) fiber volume fraction which varies from the measured value up to 150% of that value. The results show that the predictions of the composite extensional and flexural loss factors decrease as the input local (yarn) fiber volume fraction increases. The parametric study indicates that the predictions of the composite loss factors from both analytical models can be improved by using a reduced fiber loss factor and/or a increased local fiber volume fraction as input.

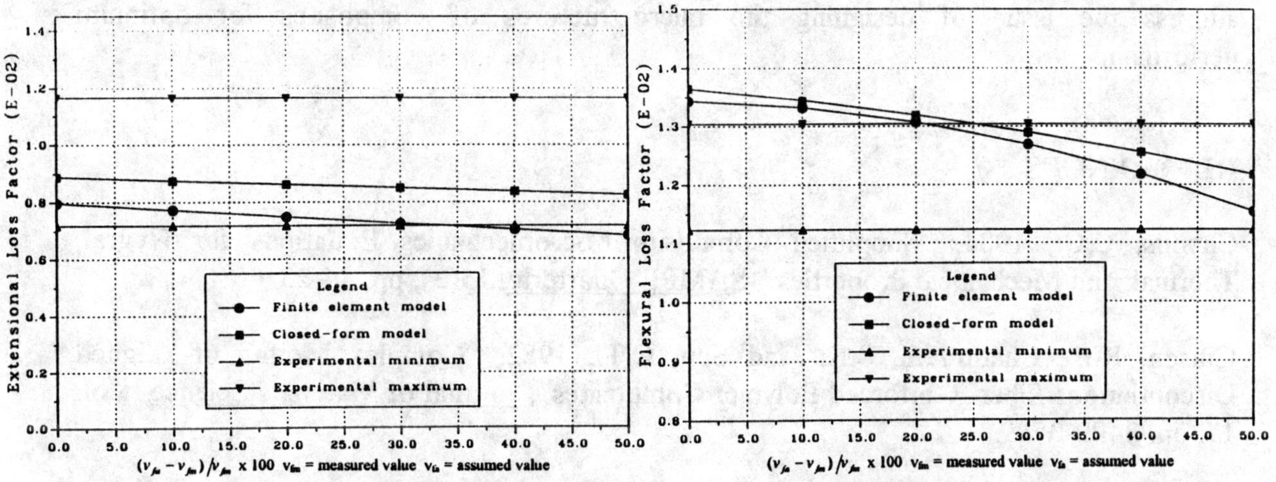

Figure 10-11 The effects of the local fiber volume fraction on the predictions of extensional and flexural loss factors.

Finally, an interesting observation, despite of the fact that the closed-form model is constructed on the basis of a very much simplified geometry while the finite element model can accurately represent the actual geometry of the composite, the predictions from the closed-form model are quite close to those from the finite element model. This might be because of the fact that strain energy which is a volume-integrated value has been used for the calculations. Therefore, the effects of the simplifications made to the composite geometry are not as significant as they would be if the details of the stress and strain distributions were of interest.

CONCLUSIONS

Two analytical models have been developed for studying viscoelastic damping in WFC. The predictions from both analytical models show reasonably good agreement with the experimental results. The closed-form model was constructed by the formulations derived from the mechanics of materials theory and the Elastic-Viscoelastic Correspondence Principle. The closed-form model can be a time and cost efficient tool for evaluating various design proposals and providing design recommendations at a early design stage. However, some accuracy has to be sacrificed due to the assumptions used to simplify the composite geometry. Three-dimensional finite element models with various mesh densities were constructed based on the actual geometry of a plain weave fabric-reinforced composite unit cell. Strain energies determined from finite element analyses were used in a strain energy formulation based on earlier work by Ungar and Kerwin [1962] to calculate the damping loss factors of the composite. The convergence study showed that the predictions from the finite element models become more accurate as the number of elements in the model increases. The finite element model may be best used in a late design stage for verifying and improving the original design. It was also observed in the parametric studies that the predictions of the composite loss factors from both analytical models can be improved by using a reduced fiber loss factor and/or a increased local fiber volume fraction as input. The methodology developed here can also be used to create analytical models for studying other types of structural composites. Because of the innumerable potential combinations of the geometry and material systems in the fabric reinforced composites, an analytical study seems to be the most practical approach to address the issue of designing the microstructures of composites for optimum performance.

REFERENCES

Chamis, C.C., 1984, "Simplified Composite Micromechanics Equations for Hygral, Thermal, and Mechanical Properties", SAMPE Quarterly, 15(3), pp. 14-23.

Gibson, R.F., Chaturvedi, S.K., and Sun, C.T., 1982, "Complex Moduli of Aligned Discontinuous Fiber-Reinforced Polymer Composites", Journal of Materials Science, Vol. 17, pp. 3499-3509.

Gibson, R.F., Hwang, S.J., and Kwak, H., 1991, "The effects of the fiber/matrix interphase on composite damping properties", Proceedings of the 36th International SAMPE Symposium, Vol. 36, pp. 592-606.

Guan, H., 1997, Micromechanical Analysis of Viscoelastic Damping in Woven Fabric-Reinforced Polymer Matrix Composites, Ph.D. dissertation, Wayne State University, Detroit, Michigan.

Hwang, S.J., and Gibson, R.F., 1987, "Micromechanical Modeling of Damping in Discontinuous Fiber Composites Using a Strain Energy/Finite Element Approach", Journal of Engineering Materials and Technology, Vol. 109, pp. 47-52.

Hwang, S.J., and Gibson, R.F., 1991, "The Effects of Three-Dimensional States of Stress on Damping of Laminated Composites", Composites Science and Technology, Vol. 41, pp. 379-393.

Ishikawa, T., and Chou, T.-W., 1983, "One-Dimensional Micromechanical Analysis of Woven Fabric Composites", AIAA Journal, Vol. 21, No. 12, pp. 1714-1721.

Sun, C. T., Wu, J. K., and Gibson, R.F., 1985, "Prediction of Material Damping in Randomly Oriented Short-Fiber Polymer Matrix Composites", Journal of Reinforced Plastics and Composites, Vol. 4, pp. 262-272.

Sun, C. T., Wu, J. K., and Gibson, R.F., 1987, "Prediction of Material Damping of Laminated Polymer Matrix Composites", Journal of Material Science, Vol. 22, pp. 1006-1012.

Ungar, E.E., and Kerwin, E.M., 1962, "Loss Factors of Viscoelastic Systems in Terms of Strain Energy", Journal of the Acoustical Society of America, Vol. 34(2), pp. 954-958.

Analytical Modeling of Damping at Micromechanical Level in Coated Fiber Polymer Composites

I. C. FINEGAN AND R. F. GIBSON

ABSTRACT

Theoretical models using theory of elasticity, finite element, and mechanics of materials formulations to characterize the damping loss factor and stiffness of coated fiber composites are described. Explicit micromechanics equations are presented in order to define the damping and stiffness of unidirectional coated fiber composites under longitudinal normal (LN), transverse normal (TN), transverse shear (TS) and longitudinal shear (LS) loading conditions. Since shear deformation is essential for viscoelastic damping in polymers, and large shear strains are generated near the fiber/matrix interface in composites, it is shown that the use of special viscoelastic polymer fiber coating is an effective way to improve damping in such composites. A strain energy approach was implemented in both theory of elasticity and finite element formulations in order to analyze the damping in the coated fiber composites. The models, based on a "representative volume element" or repeating element of a unidirectional continuous fiber composite, are used to study damping and stiffness under LN, TN, TS, and LS loading. In the theory of elasticity formulation, the stress and strain distributions inside the fiber, coating, and matrix for the four loading conditions are used to derive the strain energy in the component materials which, in turn, is used to derive the damping and stiffness of the coated fiber composite. The finite element formulations uses two and three finite element micromechanical models. Of particular interest are hybrid composites having a mixture of coated and uncoated fibers, since coating all the fibers would cause unacceptable reductions in other important properties like strength and stiffness. In order to analyze hybrid composites, two and three dimensional multi-cell finite element models have been developed. The mechanics of materials formulation is based on the equations developed by using the "subregion method" and the complex modulus notation used to define damping and stiffness of composites.

INTRODUCTION

Polymer composites have generated increased interest due to their low density and excellent stiffness and damping characteristics. The polymeric materials possess a large variety

---

Ioana C. Finegan, General Motors R&D Center, Physics and Physical Chemistry Department, Warren, MI 48090

Ronald F. Gibson, Department of mechanical Engineering, Advanced Composites Research Laboratory, Wayne State University, Detroit, MI 48202

of properties which are beneficial in many applications. Because of their dissipative properties, polymers are employed to damp the motions of mechanical and structural systems which are subjected to vibrations.

For analytical and computational convenience, these materials are usually idealized, with good accuracy to be viscoelastic. The viscoelastic damping of polymer composites arises from relaxation and recovery of the polymer network after it has been deformed. Strong correlations are noticed between vibration frequency effects and molecular motions within the material.

Shear deformation is an important factor in viscoelastic damping due to the fact that the relative motion of the polymer chains is driven by shear. Any design approach that increases the amount of shear deformation in the polymer matrix of composite material has the potential of improving the damping of the structure. Much of the strain energy of shearing is converted into heat and thereby provides a means for dissipating the energies of shock and vibration.

An important aspect of composite materials is the interface between the fiber and matrix. Recent research has been concerned with the complex state of stress which exists in the fiber/matrix interface region due to the geometric and material discontinuities between the fiber and the matrix materials. In some polymer composites such as graphite/epoxy, a third phase called the interphase is developed near the fiber/matrix interface during fabrication when some of the hardener molecules diffuse into the fiber sizing [Drzal, et al. 1983]. Analytical techniques for the investigation of the effects of this third phase on the damping and stiffness properties have been presented [Gibson, et al., 1991]. To simplify the analysis, the interphase is often assumed to have properties which are intermediate between those of the fiber and those of the matrix. Since shear deformations are so important in damping of viscoelastic polymers, it has been shown that the interphase damping of fiber reinforced polymer composites may lead to significant improvements in damping because of the high shear strain stored in the interphase region [Gibson et al., 1991]. The high shear strain in the region of the fiber/matrix interface is due to the mismatch between the fiber and matrix properties. Thus, a basic premise of this paper is to improve damping by putting a high damping material (fiber coating) in this region of high shear strain. To achieve a highly damped composite structure, both a high damping material and a high shear strain energy are needed-either one alone is not enough. To the best of the knowledge of the authors, no analytical work has been reported on the effect of viscoelastic fiber coatings on the damping of composites.

In this paper, a strain energy approach is implemented in finite element, theory of elasticity and mechanics of materials formulations to analyze the damping of the coated fiber composite.

STRAIN ENERGY METHOD APPLIED TO COMPOSITES AT MICROMECHANICAL LEVEL

The strain energy approach was implemented in both the finite element and theory of elasticity formulations to analyze the damping of the coated fiber composites. The strain energy method provides a convenient theoretical basis for the prediction of structural damping in composite structures. Micromechanical damping analysis is based on a "representative volume element" (RVE), or repeating element consisting of a single fiber embedded in a cylinder of matrix material. The strain energy approach was extended to analyze the damping of a three phase composite, the third phase being the interphase [Gibson, et. al, 1991, Chaturvedi and Tzeng, 1991]. Although the volume fraction of the interphase layer is generally low, the contribution to damping due to the interphase layer may be significant due to the high shear strain energy stored in this region. In this paper, the strain energy method was extended to analyze the damping of a three phase composite, the third phase being the fiber coating material. The fiber coating material has properties independent of those of the fiber and those of the matrix and plays an important role in energy dissipation of the coated fiber reinforced composite due to the high shear strain energy stored in the coating material.

$$\eta = \frac{\eta_f W_f + \eta_c W_c + \eta_m W_m}{W_f + W_c + W_m} \tag{1}$$

where: η_f = fiber loss factor; η_c= fiber coating loss factor; η_m = matrix loss factor W_f = strain energy in fiber at maximum vibratory displacement; W_c= strain energy in coating at maximum vibratory displacement; W_m = strain energy in matrix at maximum vibratory displacement.

The strain energy method is used to find both closed-form theoretical relationships and approximate numerical relationships between material properties and design parameters such as fiber orientation, coating thickness and coating volume fraction.

THEORY OF ELASTICITY FORMULATION

Analysis of the damping of the coated fiber composites requires knowledge of the strain energy distribution in the fiber, coating and matrix. The strain energy was expressed in terms of the stress and strain distribution inside the fiber, coating and matrix. The mathematical micromechanical models based on elasticity theory formulation generate theoretical stress and strain distributions inside the fiber, coating, and matrix for the four loading conditions: longitudinal normal (LN), transverse normal (TN), transverse shear (TS) and longitudinal shear (LS) loading.

The representative volume element (RVE) for the coated fiber composite is represented (Figure 1) by a set of four concentric cylinders representing the various phases (fiber, coating, matrix and equivalent composite). All the phases are considered homogeneous, elastic and isotropic, except the equivalent composite, which occupies the infinite region extending outside the matrix [Theocratis and Demakos, 1995]. The equivalent composite substitutes the surrounding bulk of the composite having properties matched to the average composite. All the concentric cylinders are considered to be geometrically similar, then the assemblage of cylinders are considered to have identical elastic properties as a single composite cylinder. This hypothesis gives closed form results for determining the longitudinal normal, and longitudinal shear properties. For the transverse normal and transverse shear properties, the effective properties to be determined are assigned to the infinite transversely isotropic material surrounding the concentric composite cylinders [Theocratis and Demakos, 1995 and Sullivan and Hashin, 1992]. This hypothesis gives closed form results for determining the longitudinal normal, and longitudinal shear properties. The concentric cylinder model is a so-called self-consistent model, since it is not dependent on any particular fiber packing array. The cylindrical coordinate system (r,θ,z) employed has the z as the axial coordinate and $r-\theta$ as the transverse plane. Three concentric cylinders represent the fiber, coating and matrix having radii r_1, r_2 and r_3 respectively. The three domains are denoted by 1, 2, 3 respectively. For the purpose of the strain energy analysis, the same hypotheses used in the finite element micromechanical models are used again.

In order to determine the stress distribution in the three constituent materials (fiber, coating and matrix), the generalized displacement field for each constituent is derived. The displacement field is determined for the proposed micromechanical model (Figure 1) from the equilibrium equations, constitutive relations and strain-displacement relations for each of the constituent materials, under the four loading conditions: longitudinal normal, transverse normal, transverse shear and longitudinal shear. The stress and strain distributions are then used to derive the strain energy distribution, the damping and the stiffness of the coated fiber composite for the loading cases mentioned above.

In order to determine the loss factor and storage modulus of the coated fiber composite under longitudinal normal, transverse normal, transverse shear, and longitudinal shear loading, an initial homogeneous surface displacement condition is applied in the longitudinal, transverse normal, transverse shear, and longitudinal shear directions respectively. The initial homogeneous surface displacement under the four loading conditions are:

Longitudinal normal loading (LN): A uniaxial strain, ε_1, along the axis of the fiber is applied to the micromechanical RVE in order to define the longitudinal loss factor and elastic modulus of the coated fiber composite. The uniaxial strain was applied by imposing:

$$\begin{aligned} u_r &= u(r) \\ u_\theta &= 0 \\ u_z &= \varepsilon_1 z \end{aligned} \tag{2}$$

where: ε_1 = constant and in the "z" direction according to Figure 1.

Transverse normal loading (TN): A radial stress of magnitude "σ_0" is applied at infinite radius to the equivalent composite (Figure 1) [Theocratis and Demakos, 1995]. The axial stresses remain constant in each phase (fiber, coating, matrix), so that the axial strains in the model remain equal to zero. Thus, the plane strain condition is assumed.

$$\begin{aligned} \sigma_r^{comp} &= \sigma_0 \\ \varepsilon_z &= 0 \end{aligned} \tag{3}$$

Transverse shear loading (TS): A displacement field associated with pure transverse strain is applied to the micromechanical RVE in order to define the transverse shear loss factor and storage modulus of the coated fiber composite. The applied transverse shear strains are [Figure 1]:

$$\begin{aligned} u_1 &= 0 \\ u_2 &= \gamma_{23}^0 x_3 \\ u_3 &= \gamma_{23}^0 x_2 \end{aligned} \tag{4}$$

where γ_{23}^0 = constant transverse shear applied to the composite assemblage.

Longitudinal shear loading (LS): In order to simulate the longitudinal shear loading case, the following displacement field is assumed at the boundary of the composite cylinder assemblage (Figure 1):

$$\begin{aligned} u_r^{(3)} &= 0 \\ u_\theta^{(3)} &= 0 \\ u_z^{(3)} &= \gamma_{12}^0 r_3 \cos\theta \end{aligned} \tag{5}$$

The above mentioned homogeneous displacements simulating the four loading conditions (LN, TN, TS, and LS) are replaced in the constitutive relations for transversely isotropic materials along principal material axes in polar cylindrical coordinates represented by [Sullivan and Hashin, 1990]. The set of equations for the constitutive relations, strain-displacement and stress equilibrium equations in polar coordinates are applicable for each of the constituents (fiber, coating, matrix) of the representative volume element presented in Figure 1. The generalized displacement for each constituent is derived by substituting the constitutive relations and the strain-displacement relations into the stress equilibrium equations and by solving the governing field equations for u_r, u_θ, u_z subject to the boundary conditions corresponding to four loading conditions: LN, TN, TS, LS. Details of this calculation are shown by Finegan, 1997.

FINITE ELEMENT FORMULATION

The micromechanical model used in the finite element analysis is based on a repeating single cell of matrix material containing a single coated fiber, subjected to four types of loading LN, TN, TS and LS. A coating layer of a high damping polymer is applied on the fibers. Because of the symmetry of the models, it is only necessary to consider one quarter domain for the finite element analysis (Figures 2 and 3).

Of particular interest are hybrid composites having a mixture of coated and uncoated fibers, since coating all of the fibers would cause unacceptable reductions in other important properties like strength and stiffness. The hybrid composites appear to be more practical than composites having all fibers coated. In order to simulate such composites, multi-cell models were required. Examples of two and three dimensional multi-cell finite element models and the corresponding mesh patterns are presented in Figures 4 and 5, respectively.

Details on the finite element formulation are presented by Finegan and Gibson, 1997 and Finegan, 1997.

MECHANICS OF MATERIALS FORMULATION

The set of linear elastic mechanics of materials equations for defining the storage modulus under four loading conditions (LN, TN, TS, and LS) was developed by Hopkins and Chamis, [1988] by using the so-called "subregion method". These equations are converted to the case of linear viscoelastic material under oscillatory loading by the use of the elastic-viscoelastic correspondence principle [Gibson, 1994]. Based on this principle, the elastic normal and shear moduli are replaced by he corresponding complex moduli as shown:

$$E_z^{*(comp)} = v_m E_z^{*(m)} + v_f \left\{ \left[1 - \left(\frac{D_f}{D_c} \right)^2 \right] E_z^{*(c)} + \left(\frac{D_f}{D_c} \right)^2 E_z^{*(f)} \right\} \tag{6}$$

where: $E_z^{*(comp)}, E_z^{*(f)}, E_z^{*(c)}, E_z^{*(m)}$ = complex longitudinal normal modulus for composite, fiber, coating, and matrix respectively defined as:

$$\begin{aligned} E_z^{*(comp)} &= E_z^{'(comp)} + iE_z^{''(comp)} \\ E_z^{*(f)} &= E_z^{'(f)} + iE_z^{''(f)} \\ E_z^{*(c)} &= E_z^{'(c)} + iE_z^{''(c)} \\ E_z^{*(m)} &= E_z^{'(m)} + iE_z^{''(m)} \end{aligned} \tag{7}$$

The loss factors under LN loading condition was determined by separating the real and imaginary parts in the equations defining the complex moduli.

$$\eta_z = \frac{E_z^{''(comp)}}{E_z^{'(comp)}} \tag{8}$$

where: η_z = loss factor of the composite under LN loading.

$E_z^{''(comp)}$ = normal and shear loss modulus of the composite under LN loading.

$E_z^{'(comp)}$ = normal and shear storage modulus of the composite under LN loading.

In a similar way the complex modulus for TN, TS, and LS loading conditions were determined by using Hopkins and Chamis, [1988] equations. The loss factor under TN, TS and LS loading conditions were defined by using equations similar to equation(6). Details on the calculations are presented by Finegan, [1997].

RESULTS AND DISCUSSIONS

Parametric studies were done with particular emphasis on the effects of coating thickness and coating volume fraction on the composite damping and stiffness. The materials used in the analysis are PVC coated copper wires embedded in an epoxy matrix (26AWGC and 28AWGC), hybrid composites having a mixture of coated and uncoated copper wires (26AWG in an epoxy matrix, and copper wires without coating (26AWGU) embedded in an epoxy matrix. The material properties used as input in the analytical models and experimental testing are listed by Finegan, [1997].

The following results are presented for the PVC coated copper wire composites. Figure 6 shows the variation of the composite longitudinal loss factor with coating volume fraction. Since the longitudinal properties of the composite are fiber dominated, an increase in the loss factor with the coating volume fraction can be observed. Figure 7 shows the variation of the composite longitudinal storage modulus with coating volume fraction. As expected, the trend of the storage modulus is opposite to the trend of the loss factor, i. e, as the coating volume fraction increases the composite storage modulus decreases, and the composite loss factor increases.

Similar analysis was done for PVC coated copper wires under transverse normal loading [Finegan and Gibson, 1997].

A comparison between the damping and stiffness of the composite having all fibers coated, hybrid composites and composites having all fibers uncoated is made in Figures 8 and 9.

Figure 9 shows the variation of the composite longitudinal damping loss factor with fiber volume fraction for the cases of a composite having all fibers coated, a hybrid composite and a composite having uncoated fibers as reinforcement. It can be seen that a hybrid composite having a combination of coated and uncoated fibers offers a good compromise regarding the improvement of damping and the corresponding reduction of stiffness. A similar analysis was done for composite transverse loss factor and stiffness, and presented by Finegan and Gibson, [1997].

It is shown in the above figures that reasonably good agreement is obtained for damping and stiffness results obtained by using experimental, finite element, elasticity, and mechanics of materials formulations. Details on the experimental characterization of coated fiber composites is presented by Finegan and Gibson, 1997. Also, the longitudinal damping loss factor of the PVC coated copper wire composite increases with increasing coating volume fraction. The longitudinal stiffness of the PVC coated copper wire composite decreases with increasing coating volume fraction.

The transverse damping loss factor of the PVC coated copper wire decreases as the coating volume fraction increases due to the fact that transverse damping is matrix dominated and (v_c + v_m) of the 26AWGC is less than (v_c + v_m) of the 28AWGC. The transverse stiffness of the PVC coated copper wire composite is increased by increasing the coating volume fraction, i.e it follows an opposite trend as compared to the transverse damping, as expected [Finegan, 1997].

The differences between the experimental results and the analytical results are believed to be partly due to the fact that the samples used in the experiments had many irregularities which were introduced during manufacturing. Nonuniformities in the fiber distribution and in the matrix volume fraction create differences between the geometry of the coated composites used in the analytical models and the geometry of the samples used in the experimental results.

CONCLUSIONS

From the analysis of the manufactured coated fiber composites, good agreement was obtained by using experimental, finite element, elasticity, and mechanics of materials formulations. The analysis was carried out for the longitudinal and transverse damping and stiffness.

An important conclusion is that the volume fractions of the coating and matrix both play a significant role in the improvement of damping of the overall composite. From Figure 8 it can be seen that by increasing the fiber volume fraction and decreasing the volume fractions of both the coating and matrix, the damping of the composite decreases accordingly. If the fiber volume fraction of the composite remains constant (Figure 6), and only the volume fraction of the coating increases, the damping of the composite also increases. Since the loss factor of the PVC coating is higher than the loss factors of the matrix and fiber respectively (η_c =0.220, η_m=0.0125, η_f = 0.0047) the damping of the composite is increasing. Also, we still get improved damping by using matrix material instead of coating since the damping loss factor of the matrix is higher than the damping loss factor of the fiber (η_m=0.0125 > η_f= 0.0047).

The finite element, elasticity, and mechanics of materials analytical micromechanical models have been proven to be reasonably accurate in predicting the dynamic properties of coated fiber composites. The analytical micromechanical models can be adjusted to study various commonly used fiber matrix combinations and coating materials in order to obtain optimized damping and stiffness properties for the composite structures.

REFERENCES

Chaturvedi, S. K., and Tzeng, G.Y., 1991, "Micromechanical Modeling of Material damping in Discontinuous Fiber Three - Phase Polymer Composites", Composites Engg., Vol. 1, pp. 49 - 60.

Drzal, L. T., Rich, M. J., Koenig, M. F., and Lloyd, P. F., 1983, "Adhesion of Graphite Fibers to Epoxy Matrices: II. The Effect of Fiber Finish", J. Adhesion, Vol. 16, pp. 133-152.

Finegan, I. C., and Gibson, R. F., 1997, "Analytical and Experimental Characterization of Damping at Micromechanical Level in Coated Fiber Polymer Composites", to be presented at the 12th Annual Technical Conference, American Society for Composites, October 6-8, 1997, Dearborn, Michigan.

Finegan, I. C., 1997, "*Analytical and Experimental Study of Improvement of Damping at the Micromechanical Level in Polymer Composite Materials by the Use of Special Fiber Coatings*", Wayne State University, Detroit, Michigan, Ph.D. Dissertation, 1997.

Gibson, R. F., Hwang, S. J., and Kwak, H., 1991, "Micromechanical Modeling of Damping in Composites Including Interphase Effects", Proc. of 36th International SAMPE Symposium, Vol. 1, pp. 592 - 606.

Hopkins, D. A., and Chamis, C. C., 1988, "A Unique Set of Micromechanics Equations fr High-Temperature Metal Matrix Composites", Testing Technology of Metal Matrix Composites, ASTM STP 964, American Society for Testing and Materials, pp. 159-176.

Sullivan, B. J., and Hashin, Z., 1990, "Determination of Mechanical Properties of Interfacial Region Between Fiber and Matrix in Organic matrix Composites", Controlled Interphases in Composite Materials, H. Ishida Ed., New-York, NY: Elsevier.

Theocratis, P. S. and Demakos, C. B., 1995, "Effect of Coatings With Smoothly Changing Transverse Rigidities on the Stiffness of Encapsulated Anisotropic Fiber Composites", Composite Science and Technology, Vol. 54, pp. 23 - 34.

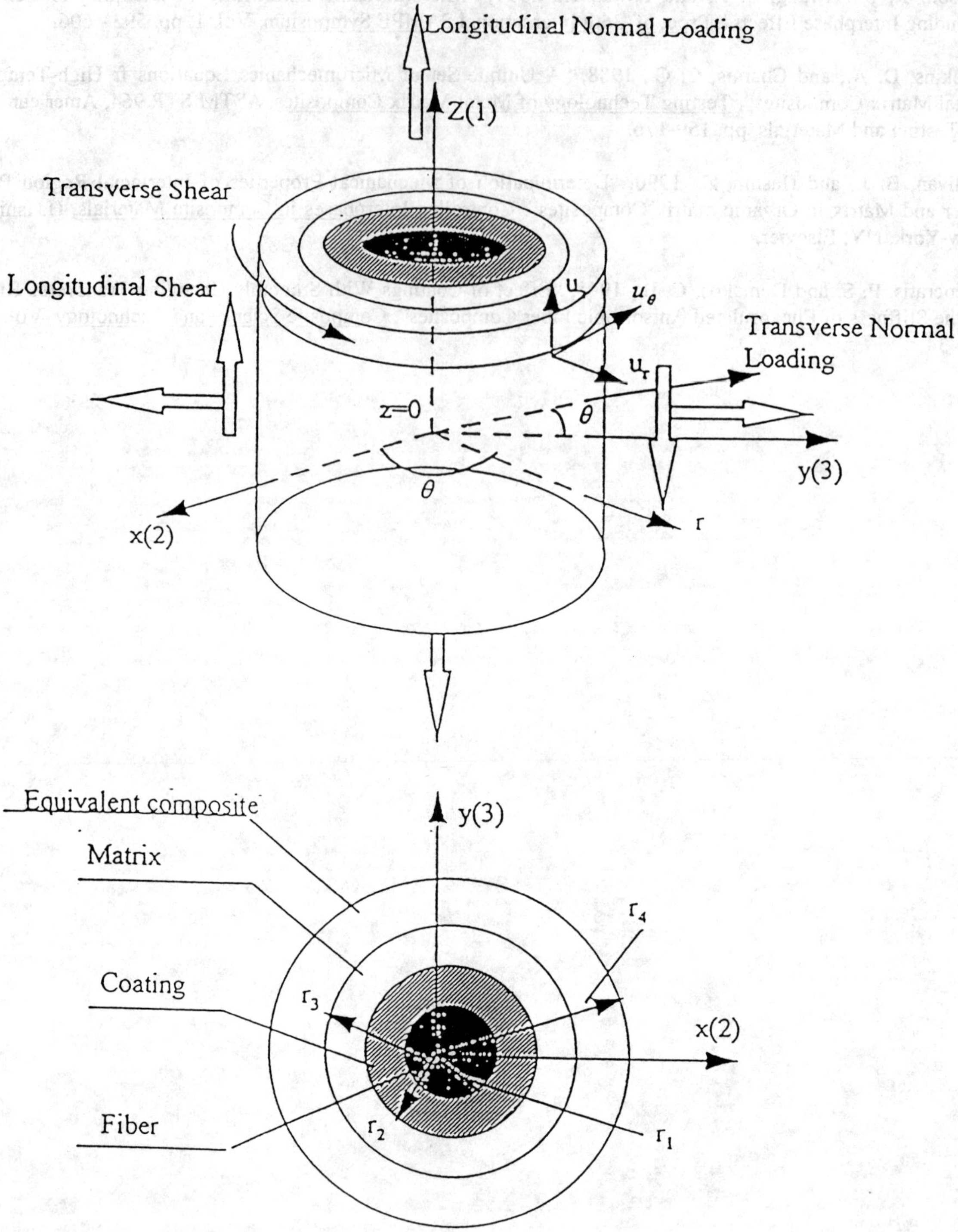

Figure 1 Representative Volume Element
Cylindrical Coordinate System
(Theory of Elasticity Model)

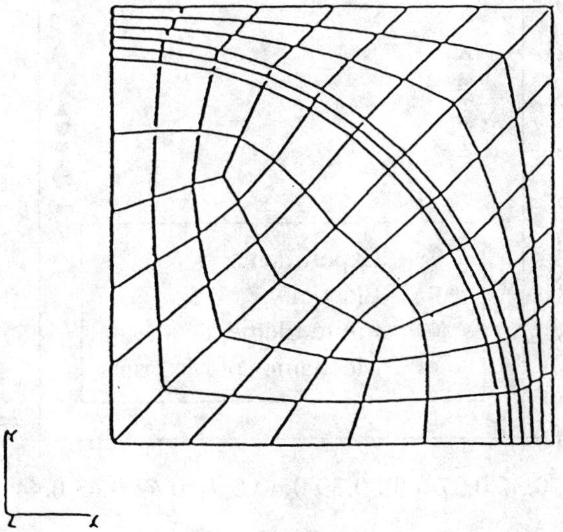

Figure 2 2D Unit-Cell Finite Element Model

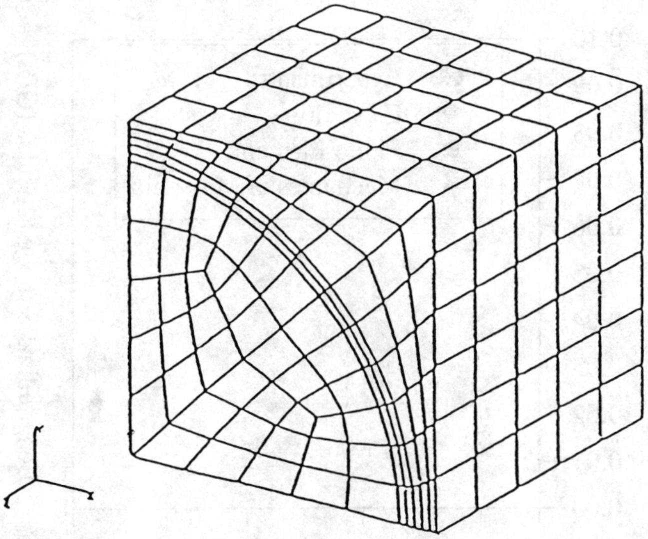

Figure 3 3D Unit-Cell Finite Element Model

Figure 4 2D Multi-Cell Finite Element Model

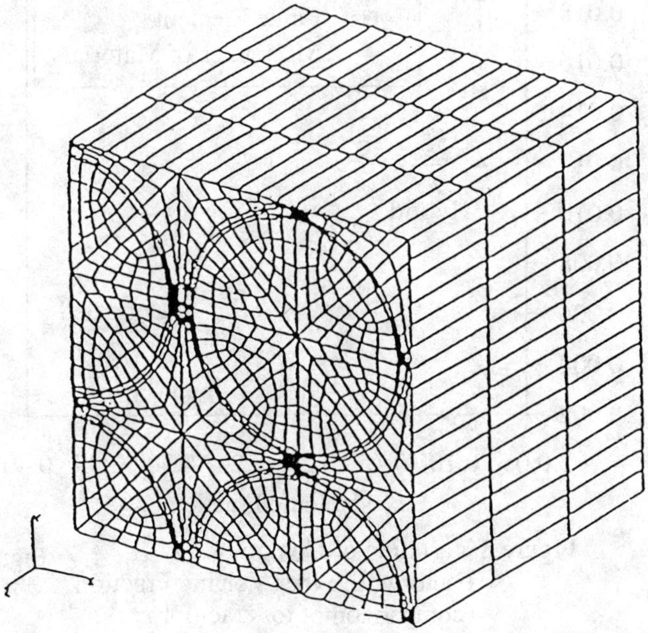

Figure 5 3D Multi-Cell Finite Element Model

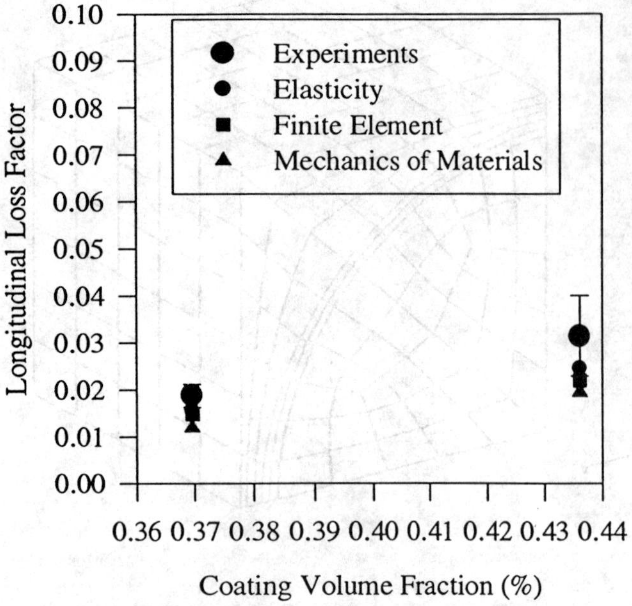

Figure 6 Variation of Composite Longitudinal Loss Factor with Coating Volume Fraction (composite reinforced with PVC coated copper wires)

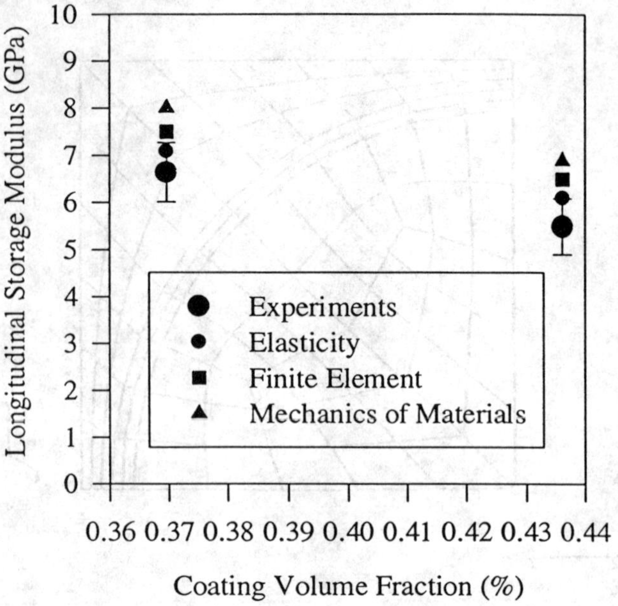

Figure 7 Variation of Composite Longitudinal Storage Modulus with Coating Volume Fraction (composite reinforced with PVC copper wires)

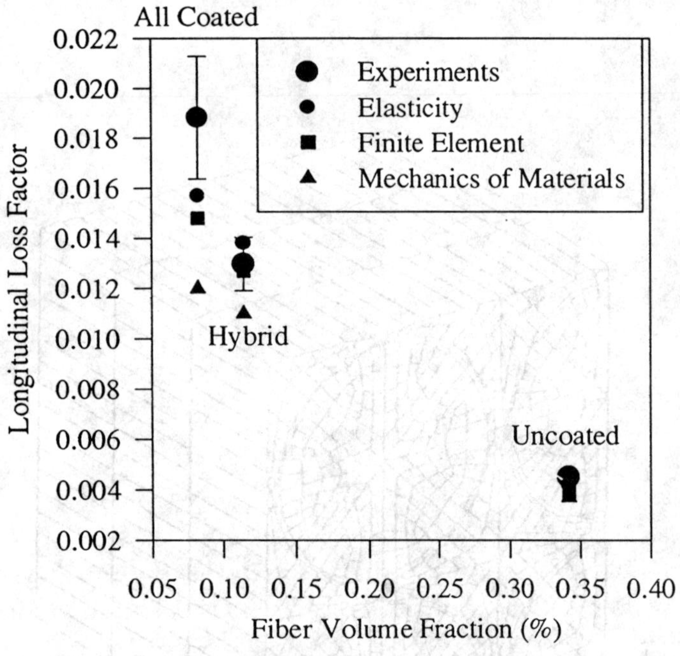

Figure 8 Variation of Composite Loss Factor with Fiber Volume Fraction (comparison of loss factor for composites having all fibers coated, hybrid and uncoated)

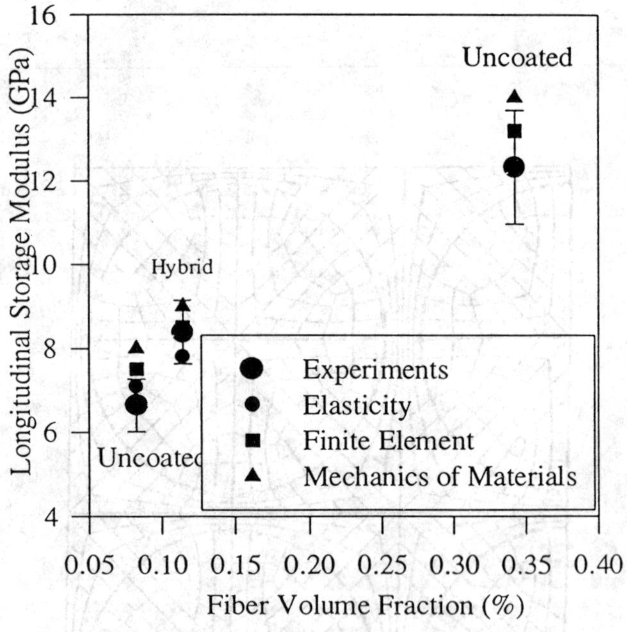

Figure 9 Variation of Composite Longitudinal Storage Modulus with Fiber Volume Fraction (comparison of storage modulus for composites having all fibers coated, hybrid and uncoated)

Use of Filament Wound Composites to Exploit the Lateral Confining Effect in Brittle Materials Subjected to Dynamic Loading

D. S. ADAMS* AND C. T. SUN†

ABSTRACT

Filament wound advanced composite materials are used to produce lateral confinement of cylindrical glass specimens subjected to axial quasi-static and dynamic high strain rate loading. A brief analysis is performed to identify the role of key parameters on the ultimate strength of brittle materials confined in this fashion. Experimental results are presented to demonstrate the potential performance and weight savings over traditional confinement techniques. Attention is given to the development of this technology as a means of abrading armor piercing projectiles for use in light armor applications.

INTRODUCTION

Brittle materials are frequently used in engineering applications where their hardness and high compressive strength can improve, and in some cases make possible, a structure's performance. One of the most dramatic examples of this is the use of ceramic materials in the fabrication of light weight armors to provide high ballistic resistance against long rod penetraor projectiles. Their relatively low density and high hardness make them a particularly attractive option for this application; however, ceramics are also quite brittle which limits their performance.

A popular technique to combat brittle failure modes is to provide a lateral confinement and an associated pre-stress pressure. This adds a hydrostatic component which shifts the deviatoric plane further from the origin producing higher strength in the brittle base material. A process using filament winding on a large scale is now being used on large structures, such as bridges, to reinforce concrete columns (Hoppel, Bogetti and Gillespie, 1995). This has been most popular in areas where earthquakes are common and

* Douglas S. Adams, Graduate Student, School of Aeronautics and Astronautics, Purdue University, West Lafayette, IN 47907

† C. T. Sun, Professor, School of Aeronautics and Astronautics, Purdue University, West Lafayette, IN 47907

a higher degree of damage tolerance is required. It has also been shown to be effective in bolstering aging concrete structures to prolong their service life. However, these studies have dealt exclusively with concrete whereas the remainder of this discussion will center on glass and ceramics but on a much smaller scale than is typical of civil structures.

Previous research reporting on the confining effect in homogeneous brittle materials has centered on the technique of shrink fitting a metal collar around the material (Chen and Ravichandran, 1996a, 1996b; Lankford, 1993, 1994; Shockey et al., 1990). This collar then produces a pre-loaded state of stress as well as additional confinement under axial loading by restricting the Poisson's expansion of the core material. This paper summarizes current research on a new technique that uses filament wound advanced composites to exploit the confining effect in brittle materials for improved strength and ballistic resistance. The central purpose of this study is to demonstrate the effectiveness and flexibility of the filament winding technique and to showcase its advantages over the traditional shrink fit approach.

Ultimately, the goal is to apply this technique to ceramic materials; however, the hardness and high strength of ceramics make them difficult to test. Since the interest is in demonstrating the filament wound confining effect, rather than the performance of the base material, the research discussed here focuses on the behavior of glass. This reduces the material cost, the demands on equipment, and allows visual observation of failure progression in the test specimens. There is no loss in generality in this as glass behaves in a brittle fashion and the results can be directly implied for the behavior of ceramics and other brittle materials.

SPECIMEN FABRICATION

The core material of the specimens used in this study was an 11 *mm* nominal diameter annealed soda-lime glass. The wrapped specimens were wound with a 1-K tow of Toray T300 pan based graphite fibers on an EnTec PW-10 filament winding machine. The matrix used was a two part room temperature curing (RTC) epoxy with a Shell EPON® 862 resin and EPI-CURE® 3234 curing agent. The recommended post cure cycle was not used in the composite processing. This was to avoid potential problems associated with thermal expansions in the heating cycle that arise due to differences in the coefficients of thermal expansion which may relax the room temperature pre-stress pressure.

All wound specimens were made with four hoop-wrapped plys which gave a nominally 0.3 *mm* thick wrap. The reported graphite properties include: a modulus of 230 *GPa*, an ultimate strength of 3530 *MPa*, and a nominal fiber diameter of 7 μm. The unwrapped and wrapped glass rods were then cut using a diamond saw into samples approximately 12 *mm* long. The ends of these cylinders were then polished to remove any major surface flaws and to bring the ends closer to parallel. The importance of the polishing step cannot be overemphasized as it can have an order of magnitude effect on the ultimate load. The shape and approximate specimen dimensions are shown in Figure 1.

PARAMETRIC ANALYSIS

Before proceeding further it is helpful to first examine the basic mechanics involved in the loading to understand the importance of some of the controlling variables. If we assume an idealized state of uniaxial stress, shown in Figure 2, then we can apply Hooke's law in cylindrical coordinates to draw some approximate results concerning the confinement effect. Briefly, we assume the wrap thickness is much smaller than the core diameter ($t_w << r_o$) and take the stress and strain in the wrap as

$$\sigma_\theta^w = +\frac{pr}{t} \tag{1}$$

$$\varepsilon_\theta^w = \frac{\sigma_\theta^w}{E_\theta^w} \tag{2}$$

where the transverse stiffness of the hoop wrap is assumed negligible and the hydrostatic compressive stress p is taken as positive. For the isotropic core we have

$$\varepsilon_\theta^c = \frac{1}{E_c}\left[\sigma_\theta^c - \nu_c\left(\sigma_r^c + \sigma_z^c\right)\right] \tag{3}$$

with appropriate elastic constants. Substituting the applied compressive stresses as shown in Figure 2

$$\begin{aligned} \sigma_z^c &= -\sigma_z^o \\ \sigma_\theta^c &= -p \\ \sigma_r^c &= -p \end{aligned} \tag{4}$$

and enforcing the compatibility condition at $r = r_o$ of

$$\varepsilon_\theta^w\Big|_{r_o} = \varepsilon_\theta^c\Big|_{r_o} \tag{5}$$

gives the relation between hydrostatic stress and applied stress of

$$\frac{pr_o}{t_w E_\theta^w} = \frac{1}{E_c}\left[-p(1-\nu_c) + \nu_c\sigma_z^o\right] \tag{6}$$

where σ_z^w is assumed negligible. This can then be rearranged to give the hydrostatic pressure as a function of the applied axial stress given in (7).

$$p = \frac{\nu_c}{1 - \nu_c + \frac{r_o}{t_w}\frac{E_c}{E_\theta^w}} \sigma_z^o \tag{7}$$

From this highly simplified result we see that higher t_w / r_o and E_θ^w / E_c ratios increase the expansion confining pressures for a given core Poisson's ratio. This follows intuition as we would expect stiffer and thicker wraps to produce a stronger lateral confinement. This result is plotted in Figure 3 in a broader manner for several Poisson ratios. As a more specific example, if we assume values of $E_\theta^w = 200$ *GPa*, $E_c = 69$ *GPa*, $\nu_c = 0.23$, $r_o = 5.5$ *mm*, and $t_w = 0.3$ *mm*, we find that $p = 0.032\sigma_z^o$ or 3.2% of the applied axial stress. Thus, while the behavior will enhance the material performance, we must rely on the fiber pretension for the majority of the initial confining stress. The current specimens have a small filament tension of ~10 *N* on a 1-K tow which, using Eqn. 1 produces a 14.2 *MPa* hydrostatic pressure with only a 0.3 *mm* thick wrap. Heavier tension and thicker wraps can be used to drive this much higher but with a minimal weight penalty as compared with conventional metal sleeves.

Finally, with the result in (7), and using the principal of linear superposition, we can then write an expression for τ_{max} as a function of the pre-stress pressure p_o and the pressure produced during loading p given in (8).

$$\tau_{max} = \frac{1}{2}\left(1 - \frac{p}{\sigma_z^o} - \frac{p_o}{\sigma_z^o}\right)\sigma_z^o \tag{8}$$

Thus the maximum shear stress behaves in a linear fashion with increasing axial stress. The pre-stress acts to reduce τ_{max} by a constant value and the confinement pressure acts to reduce the slope of the τ_{max} versus σ_z^o response.

EXPERIMENTAL PROCEDURES

Bare and wrapped specimens were tested both quasi-statically and dynamically to measure and observe the composite's confining effect. The quasi-static tests were done using a tool steel cylindrical indenter and base plate as shown in Figure 4. A hemispherical alignment fixture was included to produce minimally eccentric loading. The tests were performed inside a small plexiglass fixture with a steel base to contain the fragmenting glass; this was especially important as a properly prepared specimen explodes violently at its ultimate load. All quasi-static tests were performed at a strain rate of 0.0001 s^{-1} with the strain measurements estimated by subtracting the fixture compliance from the original data.

The high strain rate dynamic tests were done on a Split Hopkinson Pressure Bar (SHPB) which is functionally illustrated in Figure 5. The stress and strain rate in each specimen were calculated using the reduced expressions for homogeneous deformation

$$\dot{\varepsilon}(t) = \frac{-2c_o}{L}\varepsilon_r(t) \tag{9}$$

$$\sigma(t) = E\frac{A_o}{A}\varepsilon_t(t) \tag{10}$$

where E, A_o, and c_o are the modulus, area, and wave speed in the bars respectively, L and A are the specimen length and area, ε_r is the reflected pulse measured at Gage A, and ε_t is the transmitted pulse measured at Gage B. The sample strain is then found by numerically integrating the strain rate history (Follansbee, 1985).

Other authors have discussed difficulties associated with constructing accurate strain histories for brittle materials using this technique and recommend using strain gages mounted directly to the specimen (Nemat-Nasser, Isaacs and Starrett, 1991; Takeda and Wan, 1995). The rough surface of the graphite wrap complicates this measurement in practice; thus, all strain data reported herein is found using the described integration method. Protecting the SHPB from damage is another common issue. For this test series, the ends of the stainless steel incident and transmission bars were shielded from the test specimens by a pair of 3.66 *mm* thick hardened tool steel caps. These were made with the same diameter as the rods to provide identical impedance's to minimize their influence on the measurements.

EXPERIMENTAL RESULTS

Some typical quasi-static stress-strain plots for the bare and wrapped specimens are shown in Figure 6. The average σ_u values obtained from these tests were 890 *MPa* and 1170 *MPa* for the unwrapped and wrapped specimens respectively. It is likely that bending compliance in the upper test fixture produced uneven stresses in the specimens at loads above 70 *kN* and may have lead to premature failures. Thus, σ_u levels for the wrapped specimens may actually be well below the specimen capabilities. This compliance is evident in Figure 6 at specimen stress levels above 700 *Mpa*. Future tests are planned using a stiffer fixture and smaller samples to eliminate this feature of the data.

Several specimens were unloaded after reaching an applied load of 70 *kN*. The bare glass specimens cracked and showed heavy damage during the unloading process. However, the wrapped specimens had only minimal damage and retained essentially the same features as before the test. This improvement in wrapped specimens was both uniform and dramatic.

All dynamic testing was performed on a 12.7 *mm* diameter split Hopkinson pressure bar. A representative incident and reflected pulse from gages A and B are plotted in Figure 7 with compression taken as positive. Sample dynamic stress-strain and strain rate-time curves for both wrapped and unwrapped specimens are plotted in Figures 8 and 9 respectively. At lower stress levels the wrapped specimens showed little or no damage while the bare specimens fractured in all cases similar to the quasi-static testing results.

Unfortunately, the latest specimens were too large to reach adequate stress levels using the available bars. Larger diameter bars and smaller diameter specimens will be used in future tests to rectify this by providing much higher stress levels.

CONCLUSIONS

Lateral confinement using wound pre-stressed graphite composites was found to provide improved strength in brittle materials. Preliminary analysis indicates that the induced confinement stress state in the core material is a function of the wrap thickness, modulus, strength, and pre-tension. The high strength and modulus of graphite fibers allows improved performance over conventional shrink-fit specimens with lower added weight and higher potential confinement for a given thickness.

ACKNOWLEDGMENTS

This work was supported by the U.S. Army Research Office under grant number DAAH04-94-G-0181 to Purdue University. Additional support was provided by Torayca Corporation which supplied the graphite fibers and Shell Chemical Company which supplied the resin and curing agents.

REFERENCES

Chen, W., and G. Ravichandran. 1996a. "Static and Dynamic Compressive Behavior of Aluminum Nitride Under Moderate Confinement," *J. Am. Ceramic Soc.*, 79(3):579-584

Chen, W., and G. Ravichandran. 1996b. "An Experimental Technique for Imposing Dynamic Multiaxial-compression with Mechanical Confinement," *Experimental Mechanics*, 36(2):155-158

Follansbee, P. S. 1985. "The Hopkinson Bar," *Metals Handbook Ninth Edition*, 8:198-203

Hoppel, C. P. R., T. A. Bogetti, and J. W. Gillespie, Jr. 1995. "Design and Analysis of Composite Wraps for Concrete Columns," *Army Research Laboratory Report*, ARL-TR-794

Lankford, J. 1993. "Dynamic Compressive Failure of Brittle Materials Under Hydrostatic Confinement," *Experimental Techniques in the Dynamics of Deformable Solids*, ASME, 165:1-10

Lankford, J. 1994. "Utilization of the Split Hopkinson Pressure Bar Under Hydrostatic Confining Pressure to Characterize the Compressive Behavior of Ceramics and Ceramic Composites," *Mechanical Testing of Ceramics and Ceramic Composites*, ASME, 197:1-11

Nemat-Nasser, S., J. B. Isaacs, and J. E. Starrett. 1991. "Hopkinson Techniques for Dynamic Recovery Experiments," *Proc. R. Soc. Lond.*, A, 435:371-391

Shockey, D. A., A. H. Marchand, S. R. Skaggs, G. E. Cort, M. W. Burkett, and R. Parker. 1990. "Failure Phenomenology of Confined Ceramic Targets and Impacting Rods," *Int. J. Impact. Eng.*, 9:263-275

Takeda, N. and L. Wan. 1995. "Impact Compression Damage Evolution in Unidirectional Glass Fiber Reinforced Polymer Composites," *High Strain Rate Effects on Polymer, Metal and Ceramic Matrix Composites and other Advanced Materials*, ASME, 48:109-113

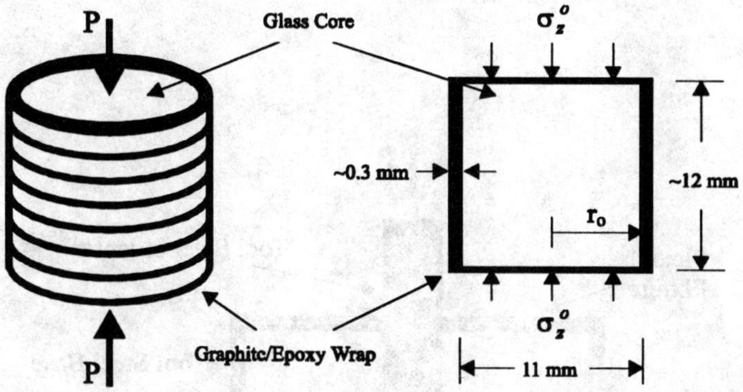

Figure 1: Wrapped specimen dimensions and applied stress.

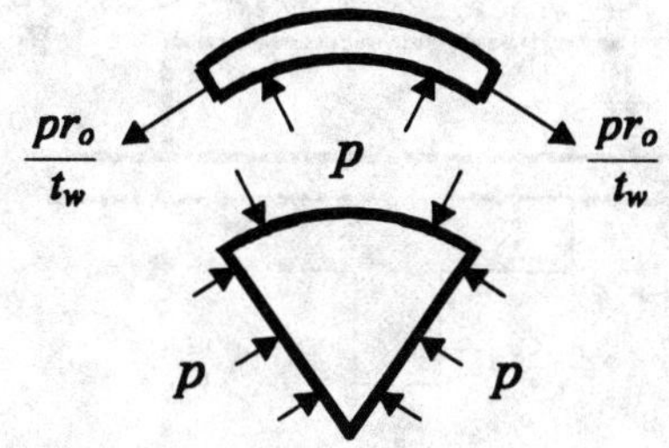

Figure 2: Free body diagram and internal stress state for wrapped specimens.

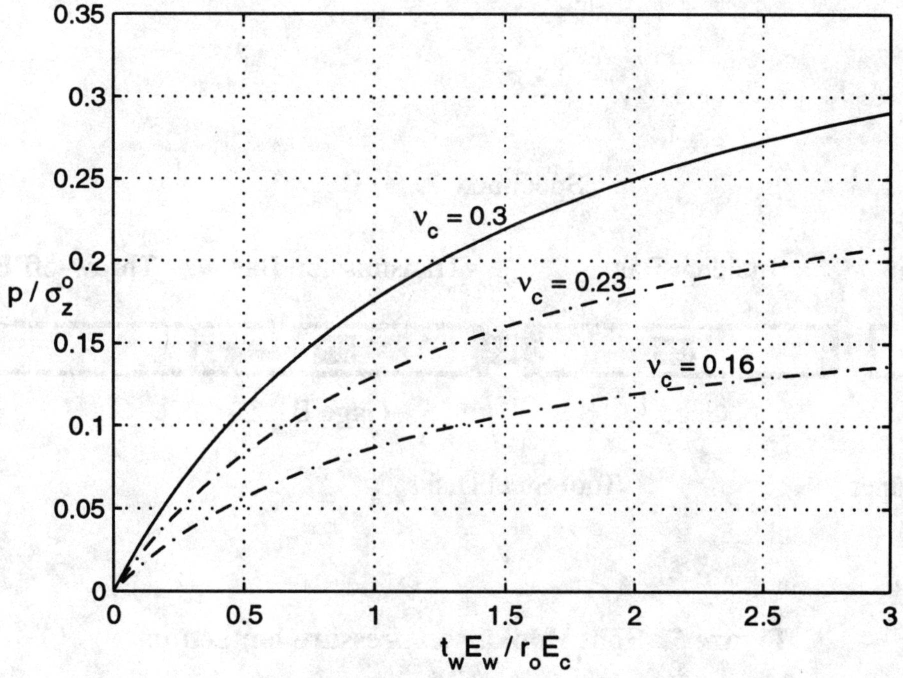

Figure 3: Normalized analytic confining pressure variable behavior.

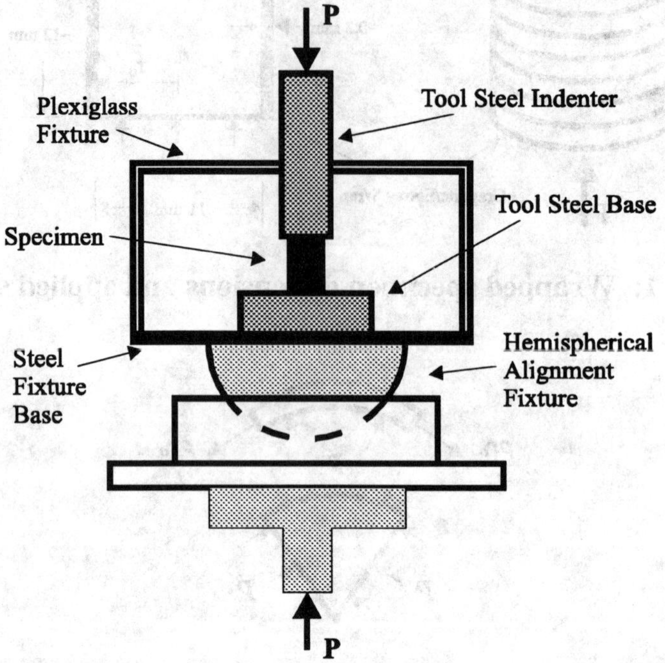

Figure 4: Quasi-static compression testing fixture.

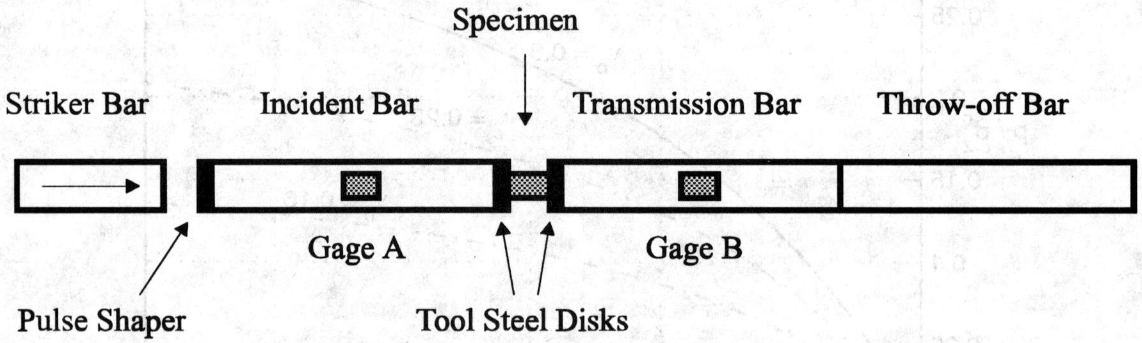

Figure 5: Split Hopkinson pressure bar setup.

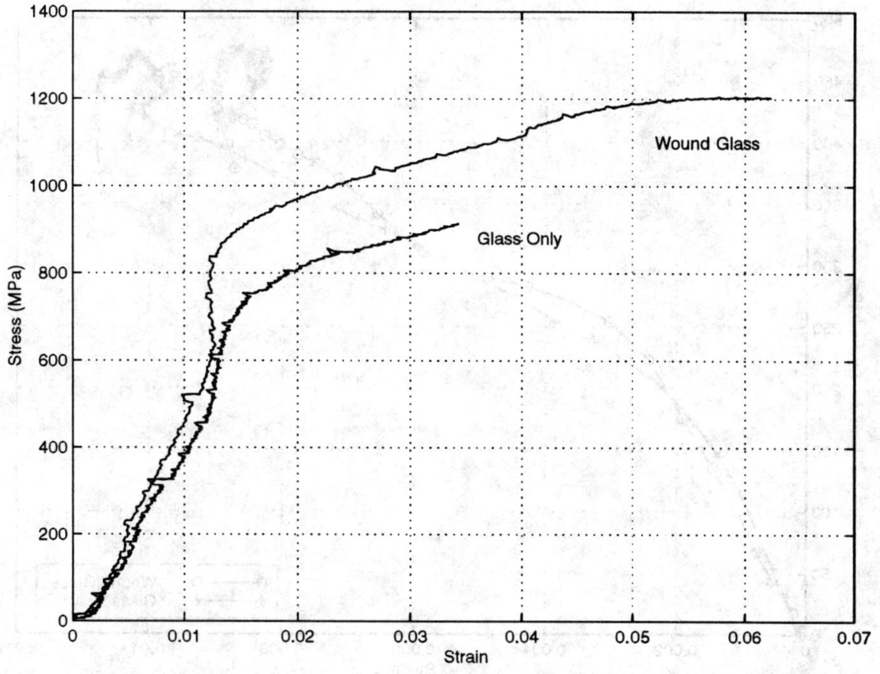

Figure 6: Typical quasi-static compression stress-strain results for wound and unmodified glass specimens.

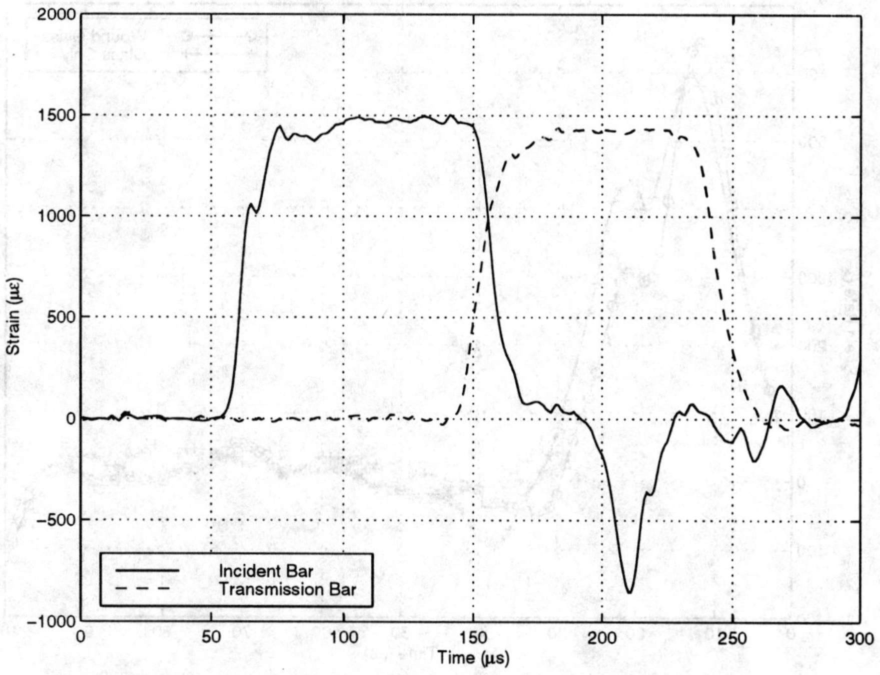

Figure 7: Typical SHPB incident and transmission bar strain gage measurements.

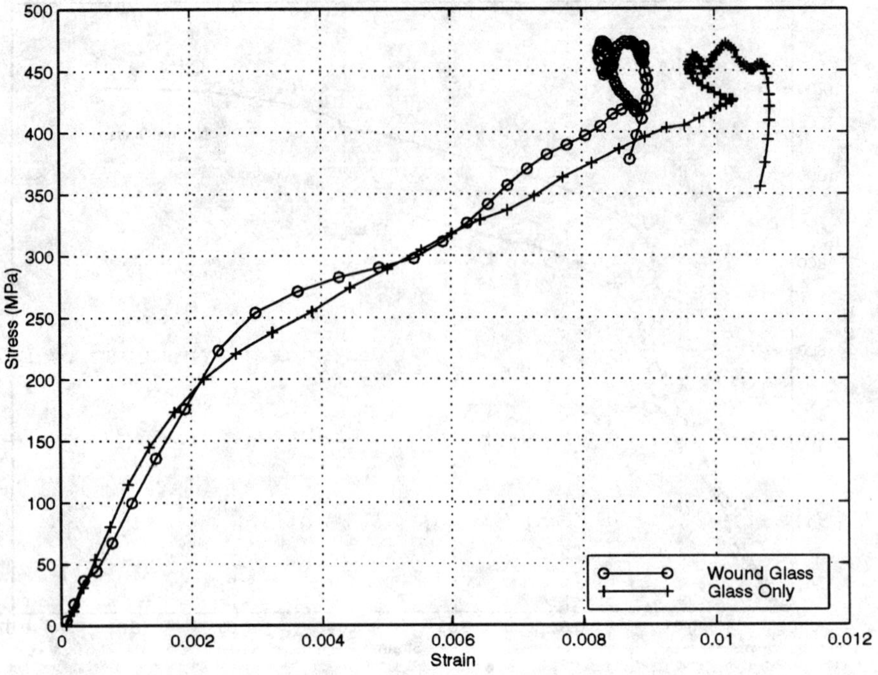

Figure 8: SHPB dynamic stress-strain results for wound and unmodified glass.

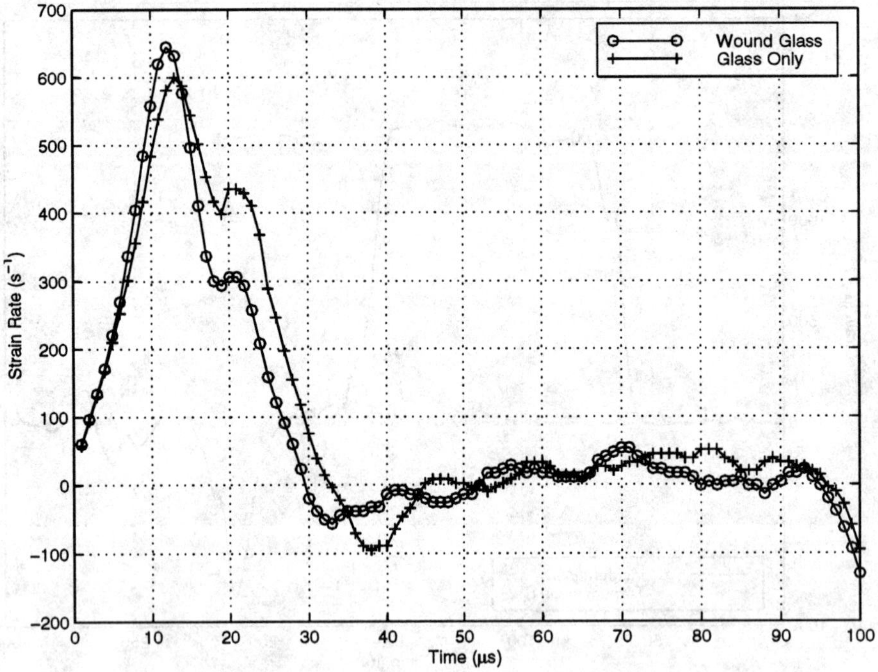

Figure 9: SHPB strain rate for wound and unmodified glass.

Wave Propagation in Composite Laminates Using a Stress Based Discrete Layer Plate Model

T. S. BUTALIA AND W. E. WOLFE

ABSTRACT

A stress based discrete layer plate model, that assumes a linear distribution of in-plane stresses through the lamina, is developed to predict transverse stresses in composite laminates. The model allows for time varying mass density of a layered monoclinic material and makes no assumptions on through the thickness distribution of the displacements. The governing equations are solved by using the finite element method for the spatial domain and Wilson's step forward integration scheme for the time domain. Stresses obtained using the proposed numerical model are verified by comparing the results with those available in literature for forced vibration response of isotropic and cross-ply laminated composite plates.

INTRODUCTION

Composite materials have been used in structures since the beginning of recorded civilization. Examples of mud and straw bricks used by early Incan and Egyptian societies to the concrete and plywood used by modern societies are well known. The advent of resin matrix composite materials has lead to the application of laminated composites in a wide variety of structures, particularly in the aerospace and automobile industries. This material is attractive to design engineers because of its high strength to weight ratio, superior corrosion resistance and the material properties that can be tailored through variations in fiber orientations, lamina thicknesses, and stacking sequences.

Under service conditions, structures made of composite materials are often subjected to dynamic loads. In particular, the performance of composite plates under low velocity impact (wheel loads, equipment drops, rock and debris impact) necessitates a detailed study. A large number of monographs and survey / review papers have been published on shear deformation theories and dynamics of composite plates (see Reddy, 1985,1990a,1990b; Bert, 1985a,1985b; Noor and Burton, 1989; Sandhu and Wolfe, 1991; Mallikarjuna and Kant, 1993; Reddy and Robbins, 1994; Noor et al., 1996). Dynamic plate models may be broadly classified as either displacement based or stress based models. They can be further be subdivided into smeared and discrete layer theories. The displacement based models are

Tarunjit S. Butalia, William E. Wolfe, The Ohio State University, 470 Hitchcock Hall, 2070 Neil Avenue, Columbus, Ohio 43210, USA

capable of predicting accurate global response (i.e., natural frequencies and deflected shape) but their ability to predict stresses, particularly interlaminar transverse stresses, has been poor. Schoeppner, 1991 presented a stress based discrete layer model for dynamic analysis of laminated plates. The model presented in this paper was developed by Butalia, 1996 and is an extension of Schoeppner's model to include the effect of time varying mass density so as to predict accurate interlaminar transverse stresses in composite laminates.

DEVELOPMENT OF NUMERICAL MODEL

The derivation of the model employs the standard index notation, i.e. Latin indices take on the range of values from 1 to 3, while the Greek indices take on values of 1 and 2.

EQUATIONS OF MOTION FOR A SINGLE LAMINA

Consider a rectangular lamina of thickness h bounded by $x_1=\pm a$, $x_2=\pm b$, and $x_3=\pm h/2$. The equilibrium equations can be written as

$$\sigma_{ij,j} + f_i - \dot{p}_i = 0 \qquad (1)$$

Separating the in-plane and out-of-plane equations gives

$$\sigma_{\alpha\beta,\beta} + \sigma_{\alpha 3,3} + f_\alpha - \dot{p}_\alpha = 0 \qquad (2)$$

$$\sigma_{\alpha 3,\alpha} + \sigma_{33,3} + f_3 - \dot{p}_3 = 0 \qquad (3)$$

where σ_{ij} is the symmetric Cauchy stress tensor, f_i is the body force per unit volume and $\dot{p}_i$ is the momentum density of the medium. The momentum density term can be expressed in terms of the medium mass density (ρ) and displacement (u_i) as

$$\dot{p}_i = (\rho\, \dot{u}_i)^{\cdot} = \dot{\rho}\ \dot{u}_i + \rho\, \ddot{u}_i \qquad (4)$$

The $\dot{\rho}$ term represents the rate of change of mass density with time. For small strains, this term can be expressed in terms of rate of change of volumetric strain as (Hiremath, 1987):

$$\dot{\rho} = -\rho\, \dot{\varepsilon}_{jj} = -\rho\, \dot{u}_{j,j} \qquad (5)$$

Substituting Equation (5) into (4) yields

$$\dot{p}_i = -\rho\ \dot{u}_{j,j}\ \dot{u}_i + \rho\, \ddot{u}_i = \rho\,(\ddot{u}_i - \dot{u}_{j,j}\ \dot{u}_i) \qquad (6)$$

The $\dot{u}_{j,j}\ \dot{u}_i$ term in Equation (6) is the non-linear contribution due to the compressibility of the material.

For a linear elastic monoclinic material (i.e. one plane of symmetry about the $x_3=0$ plane), the constitutive law may be written in an expanded form as

$$\sigma_{\alpha\beta} = E_{\alpha\beta\gamma\delta}\, \varepsilon_{\gamma\delta} + E_{\alpha\beta 33}\, \varepsilon_{33}\ ,\quad \sigma_{\alpha 3} = \sigma_{3\alpha} = 2\, E_{\alpha 3\beta 3}\, \varepsilon_{\beta 3}\ ,\quad \sigma_{33} = E_{33\gamma\delta}\, \varepsilon_{\gamma\delta} + E_{3333}\, \varepsilon_{33} \qquad (7)$$

or conversely as

$$\varepsilon_{\alpha\beta} = S_{\alpha\beta\gamma\delta}\, \sigma_{\gamma\delta} + S_{\alpha\beta 33}\, \sigma_{33}\ ,\quad \varepsilon_{\alpha 3} = \varepsilon_{3\alpha} = 2\, S_{\alpha 3\beta 3}\, \sigma_{\beta 3}\ ,\quad \varepsilon_{33} = S_{33\gamma\delta}\, \sigma_{\gamma\delta} + S_{3333}\, \sigma_{33} \qquad (8)$$

where E_{ijkl} and S_{ijkl} are the components of the rate independent isothermal elasticity and compliance tensors respectively.

For small deformations, the symmetric strain tensor may be written as

$$\varepsilon_{ij} = u_{(i,j)} = (u_{i,j} + u_{j,i})/2 \qquad (9)$$

Integrating kinematic relations (Equations (9)) along the lamina axes and substituting constitutive law (Equation (8)) yields the three displacement components as

$$u_1(x_1,x_2,x_3,t) = u_1(-a,x_2,x_3,t) + \int_{-a}^{x_1} [S_{11\alpha\beta}\,\sigma_{\alpha\beta} + S_{1133}\,\sigma_{33}]\,d\eta_1$$

$$u_2(x_1,x_2,x_3,t) = u_2(x_1,-b,x_3,t) + \int_{-b}^{x_2} [S_{22\alpha\beta}\,\sigma_{\alpha\beta} + S_{2233}\,\sigma_{33}]\,d\eta_2$$

$$u_3(x_1,x_2,x_3,t) = u_3(x_1,x_2,-\frac{h}{2},t) + \int_{-h/2}^{x_3} [S_{33\alpha\beta}\,\sigma_{\alpha\beta} + S_{3333}\,\sigma_{33}]\,d\eta_3 \qquad (10)$$

The three dimensional spatial problem is reduced to two dimensions by integrating over the thickness of the lamina. Substituting Equations (6) and (10) into Equations (2) and (3) and using Equation (8) gives

$$\sigma_{\alpha\beta,\beta} + \sigma_{\alpha 3,3} + f_\alpha - \rho\,[\ddot{u}_\alpha - \dot{u}_\alpha\, S_{kk\gamma\delta}\,\dot{\sigma}_{\gamma\delta} - \dot{u}_\alpha\, S_{kk33}\,\dot{\sigma}_{33}\,] = 0 \qquad (11)$$

$$\sigma_{\alpha 3,\alpha} + \sigma_{33,3} + f_3 - \rho\,[\ddot{u}_3 - \dot{u}_3\, S_{kk\gamma\delta}\,\dot{\sigma}_{\gamma\delta} - \dot{u}_3\, S_{kk33}\,\dot{\sigma}_{33}\,] = 0 \qquad (12)$$

Integrating Equations (11) and (12) and the first moment of Equation (11) over thickness of the lamina yields

$$N_{\alpha\beta,\beta} + (\sigma^+_{\alpha 3} - \sigma^-_{\alpha 3}) + F_\alpha - \int_{-h/2}^{h/2} \rho\,[\ddot{u}_\alpha - \dot{u}_\alpha\, S_{kk\gamma\delta}\,\dot{\sigma}_{\gamma\delta} - \dot{u}_\alpha\, S_{kk33}\,\dot{\sigma}_{33}\,]\,dx_3 = 0 \qquad (13)$$

$$V_{\alpha,\alpha} + (\sigma^+_{33} - \sigma^-_{33}) + F_3 - \int_{-h/2}^{h/2} \rho\,[\ddot{u}_3 - \dot{u}_3\, S_{kk\gamma\delta}\,\dot{\sigma}_{\gamma\delta} - \dot{u}_3\, S_{kk33}\,\dot{\sigma}_{33}\,]\,dx_3 = 0 \qquad (14)$$

$$M_{\alpha\beta,\beta} + \frac{h}{2}(\sigma^+_{\alpha 3} + \sigma^-_{\alpha 3}) - V_\alpha - \int_{-h/2}^{h/2} \rho\,[\ddot{u}_\alpha - \dot{u}_\alpha\, S_{kk\gamma\delta}\,\dot{\sigma}_{\gamma\delta} - \dot{u}_\alpha\, S_{kk33}\,\dot{\sigma}_{33}\,]\,x_3\,dx_3 = 0 \qquad (15)$$

The "+" and "-" superscripts denote the value of the variable at top ($x_3 = h/2$) and bottom ($x_3 = -h/2$) surface of the lamina. $N_{\alpha\beta}$ and $M_{\alpha\beta}$ denote the in-plane stress and moment resultants while V_α represent the transverse shear stress resultants. The body force per unit volume is assumed to be constant over the lamina thickness as

$$F_i \equiv \int_{-h/2}^{h/2} f_i \quad dx_3 = h\,f_i \qquad (16)$$

Assuming a control mass, the mass density at any instant is defined completely in terms of the initial mass density (ρ_o) and the volumetric strain (ε_{jj}) as

$$\rho = \rho_o / 1 + \varepsilon_{jj} \qquad (17)$$

For small strains ($\varepsilon_{ij} << 1$), the mass density can be approximated as

$$\rho = \rho_o\,(1 - \varepsilon_{jj}) \qquad (18)$$

Hence Equations (13) - (15) take the form of the lamina equations of motion as

$$N_{\alpha\beta,\beta} + (\sigma^+_{\alpha 3} - \sigma^-_{\alpha 3}) + F_\alpha + P_\alpha - \rho_o \int_{-h/2}^{h/2} \ddot{u}_\alpha\,dx_3 = 0 \qquad (19)$$

$$V_{\alpha,\alpha} + (\sigma^+_{33} - \sigma^-_{33}) + F_3 + P_3 - \rho_o \int_{-h/2}^{h/2} \ddot{u}_3\,dx_3 = 0 \qquad (20)$$

$$M_{\alpha\beta,\beta} + \frac{h}{2}(\sigma^{+}_{\alpha 3}+\sigma^{-}_{\alpha 3}) - V_\alpha + R_\alpha - \rho_o \int_{-h/2}^{h/2} \ddot{u}_\alpha\, x_3\, dx_3 = 0 \tag{21}$$

where

$$P_\alpha = \rho_o \int_{-h/2}^{h/2} \varepsilon_{jj}\ \ddot{u}_\alpha dx_3 + \rho_o \int_{-h/2}^{h/2} (1 - \varepsilon_{jj})\ \dot{u}_\alpha\ (S_{kk\gamma\delta}\ \dot{\sigma}_{\gamma\delta} + S_{kk33}\ \dot{\sigma}_{33})\ dx_3 \tag{22}$$

$$P_3 = \rho_o \int_{-h/2}^{h/2} \varepsilon_{jj}\ \ddot{u}_3 dx_3 + \rho_o \int_{-h/2}^{h/2} (1 - \varepsilon_{jj})\ \dot{u}_3\ (S_{kk\gamma\delta}\ \dot{\sigma}_{\gamma\delta} + S_{kk33}\ \dot{\sigma}_{33})\ dx_3 \tag{23}$$

$$R_\alpha = \rho_o \int_{-h/2}^{h/2} \varepsilon_{jj}\ \ddot{u}_\alpha\ x_3\ dx_3 + \rho_o \int_{-h/2}^{h/2} (1 - \varepsilon_{jj})\ \dot{u}_\alpha\ (S_{kk\gamma\delta}\ \dot{\sigma}_{\gamma\delta} + S_{kk33}\ \dot{\sigma}_{33})\ x_3\ dx_3 \tag{24}$$

P_i and R_α are the non-linear pseudo body force term contributions due to the compressibility of the monoclinic material.

Defining in-plane weighted displacements as

$$(\tilde{u}_i , \bar{u}_i) \equiv \int_{-h/2}^{h/2} (1, \frac{2x_3}{h})\ \frac{2}{h}\ u_i\, dx_3 \tag{25}$$

Equations (19) - (21) can then be written in the form

$$N_{\alpha\beta,\beta} + (\sigma^{+}_{\alpha 3} - \sigma^{-}_{\alpha 3}) + F_\alpha + P_\alpha - (\rho_o\, h/2)\ \ddot{\tilde{u}}_\alpha = 0 \tag{26}$$

$$V_{\alpha,\alpha} + (\sigma^{+}_{33} - \sigma^{-}_{33}) + F_3 + P_3 - (\rho_o\, h/2)\ \ddot{\tilde{u}}_3 = 0 \tag{27}$$

$$M_{\alpha\beta,\beta} + (h/2)\ (\sigma^{+}_{\alpha 3} + \sigma^{-}_{\alpha 3}) - V_\alpha + R_\alpha - (\rho_o\, h^2/4)\ \ddot{\bar{u}}_\alpha = 0 \tag{28}$$

Equations (26) - (28) represent the generalized equations of motion for a lamina. These equations account for the time varying density of the lamina using a pseudo body force approach. Up to this point, no displacement or stress assumptions have been made.

CONSISTENT STRESS FIELD FOR THE LAMINA

Reissner, 1947,1950 assumed a linear distribution of in-plane stresses across the thickness of a plate. In the current formulation, the in-plane stress distribution is assumed to be linear across thickness of each lamina as

$$\sigma_{\alpha\beta} = (N_{\alpha\beta}/h) + (12\, M_{\alpha\beta}/h^3)\ x_3 \tag{29}$$

Integrating Equation (29) through the lamina thickness and using Equations (6) and (18) yields

$$\sigma_{\alpha 3} = \sigma^{-}_{\alpha 3} - \int_{-h/2}^{x_3} \sigma_{\alpha\beta,\beta}\, d\eta_3 - \int_{-h/2}^{x_3} f_\alpha\, d\eta_3 + \rho_o \int_{-h/2}^{x_3} \ddot{u}_\alpha\, d\eta_3 - \rho_o \int_{-h/2}^{x_3} \varepsilon_{jj} \ddot{u}_\alpha\, d\eta_3 - \rho_o \int_{-h/2}^{x_3} (1-\varepsilon_{jj})\dot{\varepsilon}_{jj}\ \dot{u}_\alpha\, d\eta_3 \tag{30}$$

Substituting Equation (29) in (30), using Equations (19) and (21), and simplifying gives the transverse shear stress as

$$\sigma_{\alpha 3} = (\sigma^{+}_{\alpha 3} - \sigma^{-}_{\alpha 3})\frac{x_3}{h} + \frac{(\sigma^{+}_{\alpha 3} + \sigma^{-}_{\alpha 3})}{4}\ (\frac{12\, x_3^2}{h^2} - 1) + \frac{3}{2h}\ V_\alpha\, (1 - \frac{4\, x_3^2}{h^2}) - \frac{\rho_o}{2}\{\ \ddot{\tilde{u}}_\alpha\, (x_3 + \frac{h}{2})$$

$$+ \ddot{u}_\alpha\left(\frac{3\,x_3^2}{h}-\frac{3\,h}{4}\right)\} + \rho_o \int_{-h/2}^{x_3} \ddot{u}_\alpha\, d\eta_3 - \rho_o \int_{-h/2}^{x_3} \varepsilon_{jj}\, \ddot{u}_\alpha\, d\eta_3 - \rho_o \int_{-h/2}^{x_3} (1-\varepsilon_{jj})\, \dot{\varepsilon}_{jj}\, \dot{u}_\alpha\, d\eta_3$$
$$+\frac{1}{h}\int_{-h/2}^{x_3} P_\alpha\, d\eta_3 + \frac{12}{h^3}\int_{-h/2}^{x_3} R_\alpha\, \eta_3\, d\eta_3 \tag{31}$$

The transverse normal stress is evaluated by integrating Equation (3) over the thickness of the lamina and substituting Equations (6) and (18) in it as

$$\sigma_{33} = \sigma_{33}^- - \int_{-h/2}^{x_3} \sigma_{\alpha 3,\alpha}\, d\eta_3 - \int_{-h/2}^{x_3} f_3\, d\eta_3 + \rho_o \int_{-h/2}^{x_3} \ddot{u}_3\, d\eta_3 - \rho_o \int_{-h/2}^{x_3} \varepsilon_{jj}\, \ddot{u}_3\, d\eta_3 - \rho_o \int_{-h/2}^{x_3} (1-\varepsilon_{jj})\, \dot{\varepsilon}_{jj}\, \dot{u}_3\, d\eta_3 \tag{32}$$

The expanded form of Equations (31) and (32) were studied to look at the relative contributions of the terms. By selecting the lamina thickness, h, to be small, the contributions due to $O(h^1)$ and higher terms can be neglected. This gives rise to the transverse normal stress as

$$\sigma_{33} = \frac{(\sigma_{33}^+ + \sigma_{33}^-)}{4}\left(\frac{12\,x_3^2}{h^2}-1\right) + \frac{(\sigma_{33}^+ - \sigma_{33}^-)}{4}\left(\frac{40\,x_3^3}{h^3}-\frac{6\,x_3}{h}\right) + \frac{3\,N_{33}}{2\,h}\left(1-\frac{4\,x_3^2}{h^2}\right)$$
$$+\frac{15\,M_{33}}{h^2}\left(\frac{2\,x_3}{h}-\frac{8\,x_3^3}{h^3}\right) \tag{33}$$

and the transverse shear stress as

$$\sigma_{13} = (\sigma_{13}^+ - \sigma_{13}^-)\frac{x_3}{h} + \frac{(\sigma_{13}^+ + \sigma_{13}^-)}{4}\left(\frac{12\,x_3^2}{h^2}-1\right) + \frac{3}{2h}V_1\left(1-\frac{4\,x_3^2}{h^2}\right) + \frac{\rho_o}{h}\int_{-a}^{x_1} S_{1133}\{[\frac{h}{4}(\ddot{\sigma}_{33}^+ + \ddot{\sigma}_{33}^-)$$
$$-\frac{1}{2}\ddot{N}_{33}]\left[\frac{4\,x_3^3}{h^2}-x_3\right] + \left[\frac{h^2}{12}(\ddot{\sigma}_{33}^+ - \ddot{\sigma}_{33}^-) - \ddot{M}_{33}\right]\left[\frac{30\,x_3^4}{h^4}-\frac{9\,x_3^2}{h^2}+\frac{3}{8}\right]\}d\eta_1$$
$$-\frac{\rho_o}{h}\left(x_3+\frac{h}{2}\right)\int_{-h/2}^{h/2} \ddot{u}_1(-a,x_2,x_3,t)\, dx_3 - \frac{\rho_o}{h}\left(\frac{6\,x_3^2}{h^2}-\frac{3}{2}\right)\int_{-h/2}^{h/2} \ddot{u}_1(-a,x_2,x_3,t)\, x_3\, dx_3$$
$$+\rho_o \int_{-h/2}^{x_3} \ddot{u}_1(-a,x_2,\eta_3,t)\, d\eta_3 + \frac{1}{h}\int_{-h/2}^{x_3} P_1\, d\eta_3 + \frac{12}{h^3}\int_{-h/2}^{x_3} R_1\, \eta_3\, d\eta_3 - \rho_o \int_{-h/2}^{x_3} \varepsilon_{jj}\, \ddot{u}_1\, d\eta_3$$
$$-\rho_o \int_{-h/2}^{x_3} (1-\varepsilon_{jj})\, \dot{\varepsilon}_{jj}\, \dot{u}_1\, d\eta_3 \tag{34}$$

The expression for σ_{23} can be obtained by changing the subscript "1" to "2" and the limits of in-plane integration from "-a to x_1" to "-b to x_2".

GOVERNING EQUATIONS FOR A LAMINA

Consider an N layered laminate with k=1 and k=N representing the top and bottom laminae, respectively. For the k-th lamina, the top and bottom surface are represented by $x_3 = h_k/2$ and $-h_k/2$, respectively. Introducing the following generalized displacement definitions

$$(\tilde{g}, \bar{g}, \hat{g}, \bar{\bar{g}}, \hat{\hat{g}}) \equiv \int_{-h_k/2}^{h_k/2} g\left(1, \frac{2x_3}{h}, \frac{4x_3^2}{h^2}, \frac{8x_3^3}{h^3}, \frac{16x_3^4}{h^4}\right)\frac{2}{h}\, dx_3 \tag{35}$$

$$\bar{v}_\rho^{(k)} \equiv \frac{\tilde{u}_\rho^{(k)}}{2}, \quad \bar{\phi}_\rho^{(k)} \equiv \frac{3\,\tilde{u}_\rho^{(k)}}{h_k}, \quad \bar{v}_3^{(k)} \equiv \frac{3}{4}(\tilde{u}_3 - \hat{u}_3)^{(k)}, \quad \bar{\phi}_3^{(k)} \equiv (5\,\hat{u}_3 - \tilde{u}_3)^{(k)} \tag{36}$$

and eliminating the time derivatives using Gurtin's non-degenerate bilinear mapping, the self adjoint form of the field equations for the k-th lamina can be expressed in operator form as

$$[A]^{(k)}\{u\}^{(k)} + [B]^{(k)}\{\sigma\}^{-(k)} + [C]^{(k)}\{\sigma\}^{+(k)} + [D_u]^{(k)}\{F\}^{(k)} + \{H\}^{(k)} + \{E_u\}^{(k)} + \{Z_u\}^{(k)} = 0 \tag{37}$$

where $[A]^{(k)}$, $[B]^{(k)}$, $[C]^{(k)}$ and $[D_u]^{(k)}$ are the field operator matrices for the k-th layer (k = 1, 2, . . . N). $\{F\}^{(k)}$ is the vector of generalized body forces, $\{H\}^{(k)}$ is the vector of pseudo body force terms, $\{E_u\}^{(k)}$ is the vector of in-plane boundary terms and $\{Z_u\}^{(k)}$ is the vector of explicit initial conditions arising from Gurtin's convolution bilinear mapping. The vectors of (20N+3) field variables are defined as:

$$\{u\}^{(k)^T} = [\;\bar{v}_\gamma^{(k)}\;\; \bar{\phi}_\gamma^{(k)}\;\; \bar{v}_3^{(k)}\;\; \bar{\phi}_3^{(k)}\;\; \bar{u}_3^{(k)}\;\; N_{\alpha\beta}^{(k)}\;\; N_{33}^{(k)}\;\; M_{\alpha\beta}^{(k)}\;\; M_{33}^{(k)}\;\; V_\gamma^{(k)}\;] \tag{38}$$

$$\{\sigma\}^{\pm(k)^T} = [\;\sigma_{\gamma 3}^{\pm(k)}\;\; \sigma_{33}^{\pm(k)}\;] \tag{39}$$

The interface displacement and traction continuity equations for the k-th lamina can be written as

$$[\Lambda]^{(k)}\{\sigma\}^{-(k-1)} + [\bar{B}]^{(k)}\{u\}^{(k)} + [\Xi]^{(k)}\{\sigma\}^{-(k)} + [\bar{C}]^{(k+1)}\{u\}^{(k+1)} + [\bar{\Lambda}]^{(k+1)}\{\sigma\}^{-(k+1)}$$

$$+ [D_s]^{(k)}\{F\}^{(k)} + [\bar{D}_s]^{(k+1)}\{F\}^{(k+1)} + \{\bar{H}\}^{(k)} + \{\bar{\bar{H}}\}^{(k+1)} + \{E_s\}^{(k)} + \{Z_s\}^{(k)} = 0 \tag{40}$$

The detailed form of operators in Equations (37) and (40) are presented in Butalia, 1996.

The number of field variables for the governing function $\Omega(u,\sigma)$ of the self-adjoint problem (Equations (37)-(40)) was (20N+3). Specializations of the governing function that allowed the number of field variables to be reduced to 10N+3 were: (1) derivatives on generalized stresses were eliminated, (2) self-adjoint terms of interface continuity equations were eliminated, (3) constitutive law was exactly satisfied, (4) body forces were neglected (pseudo body forces were accounted for), (5) there were no existing jump discontinuities, and (6) the acceleration terms in constitutive law and interface continuity equations were neglected. The reduced field variables are:

$$\{u\}^{(k)^T} = [\;\bar{v}_\gamma^{(k)}\;\; \bar{\phi}_\gamma^{(k)}\;\; \bar{v}_3^{(k)}\;\; \bar{\phi}_3^{(k)}\;\; \bar{u}_3^{(k)}\;] \tag{41}$$

$$\{\sigma\}^{\pm(k)^T} = [\;\sigma_{\gamma 3}^{\pm(k)}\;\; \sigma_{33}^{\pm(k)}\;] \tag{42}$$

It was verified that vanishing of the Gateaux differential (Butalia, 1996) of the governing function $\Omega(u,\sigma)$ for all $\{u\}^{(k)}$ and $\{\sigma\}^{\pm(k)}$ leads to the field equations, boundary and initial conditions, and continuity as well as jump conditions. Spatial discretization of the governing equations was done using the Finite Element Method. Minimization of the governing function with respect to the field variables (refer Equations (41) and (42)) gives rise to

$$[M]\{\ddot{U}\} + [K]\{U\} = \{R\} - \{\bar{R}\} \tag{43}$$

where [M] and [K] are the global mass and stiffness matrix, respectively, and $\{U\}$ represents the nodal value field variable vector. $\{R\}$ and $\{\bar{R}\}$ denote the applied nodal load and pseudo body force vectors, respectively. The non-linearity in the problem was linearized at every time step. At each time step, the problem was first solved by neglecting the pseudo body force terms. Using the results obtained, vector $\{\bar{R}\}$ was calculated and the updated problem was solved again for $\{U\}$, $\{\dot{U}\}$ and $\{\ddot{U}\}$. $\{\bar{R}\}$ was recalculated and this iterative procedure was repeated until the required level of convergence was achieved for that time step.

The nine noded Heterosis element was used for the spatial discretization. Lagrangian interpolation was employed for $\bar{v}_\gamma^{(k)}$, $\bar{\phi}_\gamma^{(k)}$ and $\sigma_{\gamma 3}^{\pm(k)}$ field variables, while the Serendipity interpolation was used for the $\bar{v}_3^{(k)}$, $\bar{\phi}_3^{(k)}$, $\bar{u}_3^{(k)}$ and $\sigma_{33}^{\pm(k)}$ variables. Wilson's β–γ–θ step-forward time integration scheme was employed for the time domain solution.

NUMERICAL RESULTS

DYNAMIC RESPONSE OF AN ISOTROPIC PLATE

Consider a simply supported rectangular plate (a/b= $\sqrt{2}$, h/b=0.2) as shown in Figure 1. The top surface of the plate is subjected to a suddenly applied uniformly distributed load on a square (c/b=0.4) area at the center of the plate as :

$$q = q_o(x,y)\, H(t), \qquad \text{where } q_o(x,y) = \begin{cases} 1 & \text{for } |x| \le 0.2 \text{ and } |y| \le 0.2 \\ 0 & \text{for } |x| > 0.2 \text{ or } |y| > 0.2 \end{cases} \tag{44}$$

where H(t) is the Heaviside unit step function. Lee and Reismann, 1969 solved this problem within the framework of classical three dimensional elasticity to obtain a series solution. A finite element dynamic analysis was carried out using the following values: a = $\sqrt{2}$, b = 1, c = 0.4, h = 0.2, E = ρ = 1, ν = 0.3, β = 0.3025, γ = 0.6, θ = 1. The time step Δt was chosen to be 0.05 seconds which is one-fourth of the time it takes for the stress wave to travel from the top to the bottom of the plate. The spatial distribution of the suddenly applied loading was approximated by replacing it with an equivalent concentrated load suddenly applied at the center of the plate (i.e. F=0.16H(t) applied at x=0, y=0, z=h). The plate thickness was discretized into four layers of equal thickness. The generalized time (T) history for the generalized central displacement (W) is shown in Figure 2. The results for a 4x4 mesh with constant as well as variable mass density are compared with the data presented by Lee and Reismann, 1969. Inclusion of variable mass density in the model results in a larger central displacement and a small reduction in the time period of vibration as compared to the model in which the mass density is assumed constant over time. However, it needs to be observed that the approximation of an equivalent central concentrated load will give rise to a higher central displacement and lower time period of vibration in comparison to a uniformly distributed load applied on a square (c/b=0.4) area at the center of the plate. The time history for the transverse normal stress (σ_{33}) at x = 0, y = 0, z = h/4, obtained from the finite element model is shown in Figure 3. The results presented are for a 4x4 mesh with a four layer thickness discretization. Time histories for the constant mass density and variable mass density model are presented. Closer inspection of Figure 3 shows the superiority of the variable mass density model. For the variable mass density model, the compressive wave reaches z = h/4 at t=0.15 second, as would be expected. The compressive wave reaches the bottom free surface of the plate at t = 0.20 seconds and reflects back as a tensile wave and reaches the z = h/4 location at t=0.25 seconds. This behavior is modeled well by the variable mass density model, while the constant mass density model indicates a higher wave speed.

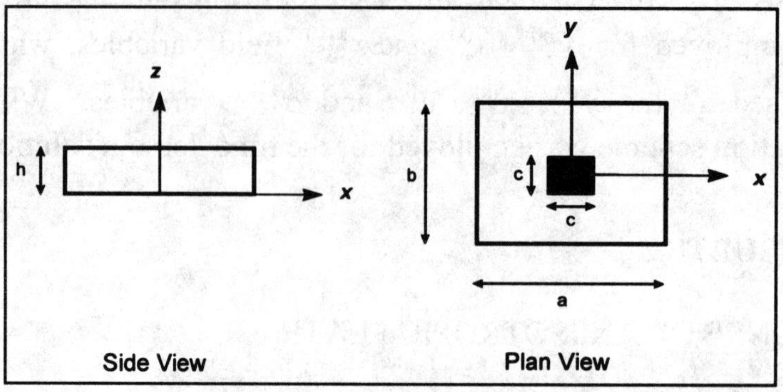

Figure 1. Plate Geometry

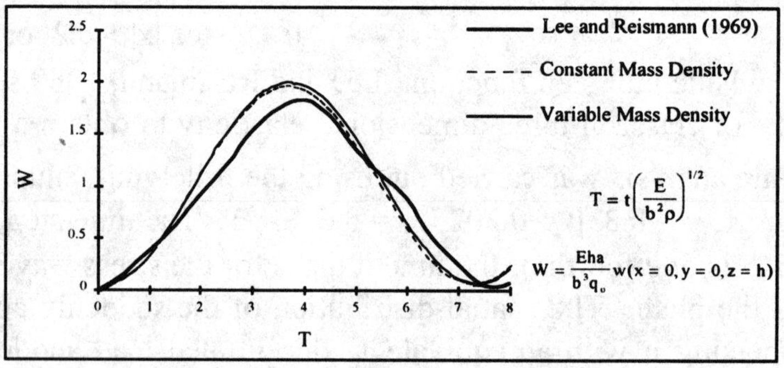

Figure 2. Central Displacement Time History for Isotropic Plate

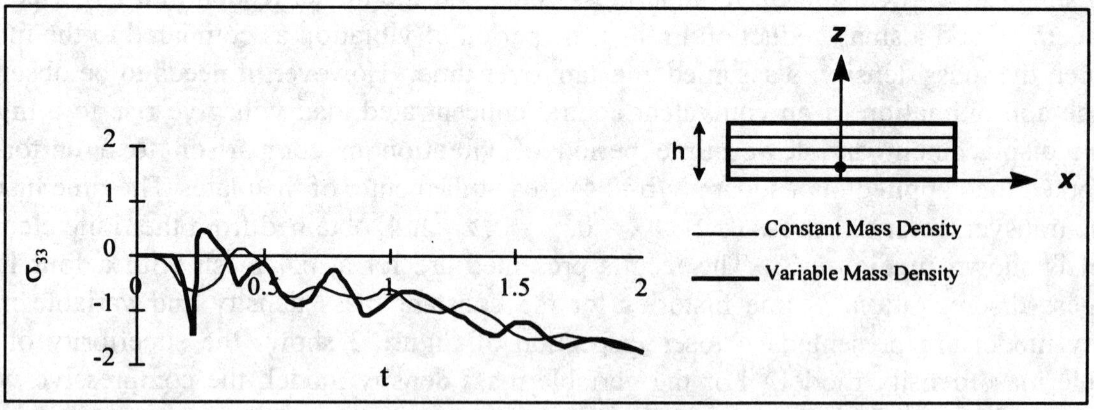

Figure 3. Transverse Normal Stress Time History for Isotropic Plate

DYNAMIC RESPONSE OF CROSS-PLY LAMINATED PLATE

Consider a clamped [0°/90°] cross-ply laminated square plate (a/b=1) as shown in Figure 1. The plate dimensions and material properties are: a=b= 7 in., h=0.1 in., c=1 in., E_1=19.52 x 10^6 psi, E_2= E_3=1.404 x 10^6 psi, G_{13}=G_{23}=G_{12}=0.65 x 10^6 psi, ν_{12}= 0.264, ν_{23}=ν_{13}=0.435, ρ=22.33 lb-sec^2/in.^{2}. The characteristic time (T), defined as the time taken for a longitudinal plane wave to propagate along the fiber direction from the center of the plate to the edge of the laminate, was calculated to be 3.74 x 10^{-3} seconds. A distributed loading was applied on a square (c/b=1/7) area at the center of the plate. The applied loading was a triangular pulse in time with a peak load of q_o at time τ = 0.5T. The loading may be represented as:

$$q(x,y,t) = \bar{q}(x,y)\ f(t) \qquad \text{where } \bar{q}(x,y) = \begin{cases} 1 & \text{for } |x| \le 0.5 \text{ and } |y| \le 0.5 \\ 0 & \text{for } |x| > 0.5 \text{ or } |y| > 0.5 \end{cases} \tag{45}$$

$$\text{where } f(t) = \begin{cases} q_o(2t/\tau) & \text{for } t \le \tau/2 \\ 2\,q_o\,(1-(t/\tau)) & \text{for } \tau/2 \le t \le \tau \\ 0 & \text{for } t \ge \tau \end{cases} \tag{46}$$

A 4x4 mesh discretization was employed for one quarter of the plate ($0 \le x \le 3.5$ and $0 \le y \le 3.5$). The 0° and 90° layers were subdivided into two sub-layers each with a total of four sub-layers through the laminate thickness. The time step was chosen to be .01T which is approximately one-third of the time its takes for the stress wave to traverse one sub-layer of thickness 0.025 inches. The parameters chosen for the time domain integration are β = 0.3025, γ = 0.6 and θ = 1. Results for the transverse normal stress time history at the 0°/90° interface at the center of the plate (x = 0, y = 0, z = 0.5h) are shown in Figure 4. The time histories for the stress based model are presented for the two cases of constant and variable mass density and are compared with the three dimensional finite element solution presented by Lee et al., 1984. Only the variable mass density model predicts the wave reaching the 0°/90° interface at approximately t/T = 0.05. All three cases predict the tensile wave reflected from the free surface of the plate (z = 0) reaching the 0°/90° interface at approximately t/T = 0.16. For the constant mass density model, the peak compressive and tensile stress values are 71 % and 46% respectively, of the values reported by Lee. However, the variable mass density model predicts peak compressive and tensile stress values which are 92% and 88% respectively, of the values reported by Lee. In general, the variable mass density stress based model predictions agree well with the three dimensional finite element results.

CONCLUSIONS

A stress based discrete layer model which accounts for variable mass density of composite laminates has been presented. The transverse stresses, which are known to be a major cause of delamination in composite laminates under low velocity impact, can now be accurately calculated.

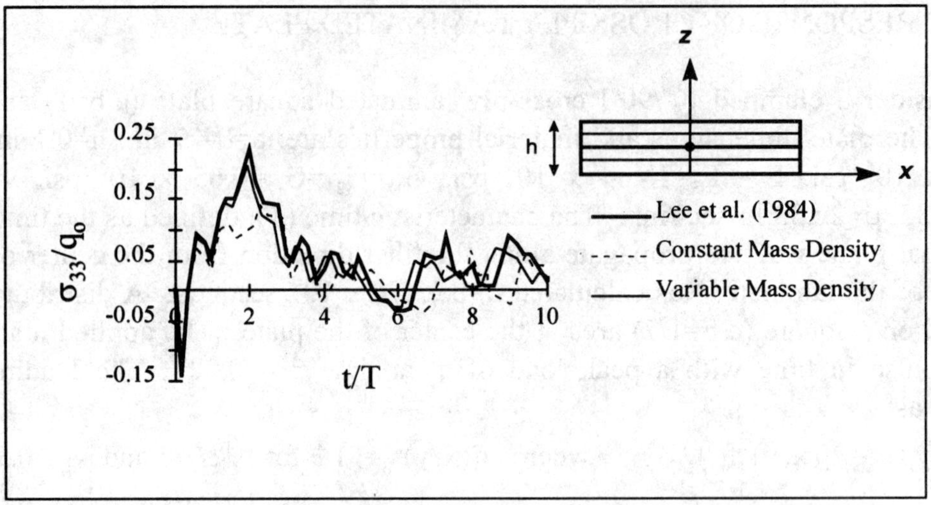

Figure 4. Transverse Normal Stress Time History for Cross-Ply Laminated Plate

REFERENCES

Bert, C.W. 1985a. "Dynamic Behavior of Composites: An Overview," *Proceedings of Society for Experimental Mechanics Conference*, 747-751.

Bert, C.W. 1985b. "Research on Dynamic Behavior of Composite and Sandwich Plates - IV," *Shock and Vibration Digest*, 17(11):3-15.

Butalia, T.S. 1996. "Dynamic Response of Advanced Composite Plates Subjected to Low Velocity Impact," Ph.D. Dissertation, The Ohio State University, Columbus, Ohio.

Hiremath, M.S. 1987. "A Dynamical Theory of Wave Propagation in Fluid - Saturated Solids," Ph.D. Dissertation, The Ohio State University, Columbus, Ohio.

Lee, Y.C. and H. Reismann. 1969. "Dynamics of Rectangular Plates," *International Journal of Engineering Science*, 7:93-113.

Lee, J.D., S. Du, and H. Liebowitz. 1984. "Three - Dimensional Finite Element and Dynamic Analysis of Composite Laminate Subjected to Impact," *Computers and Structures*, 19(5/6):807-813.

Mallikarjuna and T. Kant. 1993. "Critical Review and Some Results of Recently Developed Refined Theories of Fiber-Reinforced Laminated Composites and Sandwiches," *Composite Structures*, 23(4):293-312.

Noor, A.K. and W.S. Burton. 1989. "Assessment of Shear Deformation Theories for Multilayered Composite Plates," *Applied Mechanics Review*, 42(1):1-13.

Noor, A.K., W.S. Burton, and C.W. Bert. 1996. "Computational Models for Sandwich Panels and Shells," *Applied Mechanics Review*, 49(3):155-199.

Reddy, J.N. 1985. "A Review of the Literature on Finite Element Modeling of Laminated Composite Plates," *Shock and Vibration Digest*, 17(4):3-8.

Reddy, J.N. 1990a. "On Refined Theories of Composite Laminates," *Meccanica*, 25:230-238.

Reddy, J.N. 1990b. "A Review of Refined Theories of Laminated Composite Plates," *Shock and Vibration Digest*, 22(7):3-17.

Reddy, J.N. and D.H. Robbins. 1994. "Theories and Computational Models for Composite Laminates," *Applied Mechanics Review*, 47(6):147-169.

Reissner, E. 1947. "On Bending of Elastic Plates," *Quarterly of Applied Mathematics*, 5:55-68.

Reissner, E. 1950. "On a Variational Theorem in Elasticity," *Journal of Mathematical Physics*, 29:90-95.

Sandhu, R.S. and W.E. Wolfe, 1991. "Modeling Cumulative Damage Processes in Composite Laminates," Report # WL-TR-91-3024, Flight Dynamics Directorate, Wright Laboratory, WP Air Force Base, Ohio.

Schoeppner, G.A. 1991. "A Stress Based Theory Describing the Dynamic Behavior of Laminated Plates," Ph.D. Dissertation, The Ohio State University, Columbus, Ohio.

Free Vibration of Thin-Walled Closed-Section Composite Beams with Optimum and Near-Optimum Coupling

D. S. DANCILA, E. A. ARMANIOS AND W. K. LENTZ

ABSTRACT

An investigation of the free vibration characteristics for a set of extension-twist coupled thin-walled single-cell composite beams with optimal and near-optimal coupling is performed by using an analytical model, a finite element simulation and testing. It is found that, despite the relatively large beam slenderness ratio, the applicability of the analytical model is limited to the first bending modes, for which excellent agreement with the finite element and experimental results is obtained. For analytically predicted higher order modes increasingly significant in-plane warping is shown by the corresponding finite element results and an increasingly large variance is observed between the two set of results. The occurrence of shell modes, which cannot be captured by the analytical model, is also shown by the finite element results. Test data confirm finite element calculations as well as low modes predicted by the analytical method. The sensitivity of natural frequencies to lay-up perturbation angle is established over a frequency range of interest, providing insight into the modal characteristics' robustness for the global and the local optima, respectively.

INTRODUCTION

Composite materials play an increasingly important role in structural applications, in general, and in aeronautical applications, in particular. Their intrinsically anisotropic mechanical properties represent a challenge for analysis and design. When properly used, however, the anisotropic response of the material allows the introduction of structural coupling effects that result in improved performance. Among the numerous types of coupling possible, bending-twist coupling and extension-twist coupling are particularly relevant for lifting surface applications, where even relatively small changes in angle of attack could result in significant changes in airloads.

D. Stefan Dancila, Graduate Research Assistant, School of Aerospace Engineering, Georgia Institute of Technology, Atlanta, Georgia 30332-0150.
Erian A. Armanios, Professor, School of Aerospace Engineering, Georgia Institute of Technology, Atlanta, Georgia 30332-0150.
W. Karl Lentz, Research Engineer, Scaled Composites, Hangar 78 Airport, 1624 Flight Line, Mojave, California 93501-1663.

Weight penalties stemming from performance demands drive the design of aeronautical structures towards minimum-weight solutions, for which thin-walled constructions are suitable candidates. The use of composite materials in thin-walled structural applications leads to a more challenging analysis problem compared to the case where isotropic materials are used. The difficulties with properly accounting for deformation modes and coupling effects in such systems are well documented and a number of modeling approaches have been reported.

An asymptotic variational approach was applied by Badir (1992) to characterize the static response of closed-cell thin-walled composite beams. By the additional use of Hamilton's principle, Armanios and Badir (1995) investigated the dynamic response of thin-walled closed-cell composite beams. As an application for the general free vibration equations developed in that work, two lay-up configurations, namely the Circumferential Uniform Stiffness (CUS), producing extension-twist coupling, and the Circumferential Antisymmetric Stiffness (CAS), producing bending-twist coupling, were examined and model predictions were compared with published results. A further refinement of the solution methods for these two cases was presented by Dancila and Armanios (1995) and offered more insight into the solution process and the effects of coupling.

The problem of optimal extension-twist coupling in thin-walled closed-cell composite beams, with hygrothermal and frequency constraints, was investigated by Lentz, Armanios and Badir (1996) and Lentz (1997). A constrained optimization technique was used to determine CUS lay-ups that maximize extension-twist coupling while minimizing beam deformations due to hygrothermal effects and having the natural frequencies of the beam placed outside a prescribed frequency range. The results indicated the presence of both global and local optima satisfying the imposed constraints. The robustness of the respective optima was investigated by separately considering the effects of assumed distributions in material properties and the effects of lay-up angle perturbations around optimal values. An experimental verification was performed, by manufacturing and statically testing a set of CUS beams with lay-ups corresponding to the global optimum, a local optimum, and their respective perturbations.

TABLE I - BEAM GEOMETRIC CHARACTERISTICS AND MATERIAL PROPERTIES

Beam length	441.8 10^{-3}m
Exterior width	32.2 10^{-3} m
Exterior height	24.1 10^{-3}m
Ply thickness	0.668 10^{-3}m
Number of plies	8
E_{11}	135.6 GPa
E_{22}	9.96 GPa
G_{12}	4.2 GPa
ν_{12}	0.3
α_1	4.34 10^{-6} C^{-1}
α_2	37.0 10^{-6} C^{-1}
β_1	0
β_2	5.56 10^{-3} ε/% weight
ρ	1415 kg/m^3

TABLE II - LAYUP DESCRIPTION AND LABELING

Lay-up Description	Lay-up	Layup label
Global Optimum -3°	[$-81.33_3/25.13_5$]	GM
Global Optimum	[$-78.33_3/28.13_5$]	G
Global Optimum +3°	[$-75.33_3/31.13_5$]	GP
Local Optimum -3°	[34.84_8]	LM
Local Optimum	[37.84_8]	L
Local Optimum +3°	[40.84_8]	LP

In this work an investigation of the free vibration characteristics for a subset of these CUS beams is performed to determine the sensitivity of modal characteristics with respect to lay-up angle perturbations. This is accomplished through a comparison between analytical predictions, finite element calculations and test data.

BEAM SET CHARACTERISTICS

The six straight, rectangular cross section, thin-walled beams investigated in this work share identical geometric characteristics and material properties, shown in Table I, and they represent a subset of the CUS beam set manufactured and statically tested by Lentz (1997). The lay-ups of the six beams were identified in that work as the global optimum, a local optimum and their respective perturbations by ±3° for all plies, respectively. The corresponding stacking sequences are shown in Table II, together with a labeling scheme to be used in this work. The value of 3° for the perturbation angle is considered in the present work to be representative of the degree of variability associated with manufacturing tolerances.

TABLE III - ANALYTICAL PREDICTIONS FOR NATURAL FREQUENCIES FOR GM, G, AND GP CONFIGURATIONS

Mode	GM		G	GP	
	Hz.	% Diff.	Hz	Hz	% Diff.
VB1	124.4	7.60	115.6	108.4	-6.19
VB2	779.5	7.60	724.4	679.6	-6.19
VB3	2182.6	7.60	2028.5	1902.9	-6.19
HB1	156.5	7.60	145.5	136.5	-6.19
HB2	980.9	7.60	911.6	855.2	-6.19
HB3	2746.5	7.60	2552.6	2394.5	-6.19
TE1	998.5	-4.17	1041.9	1094.7	5.07
ET1	4107.5	4.10	3945.6	3764.5	-4.59

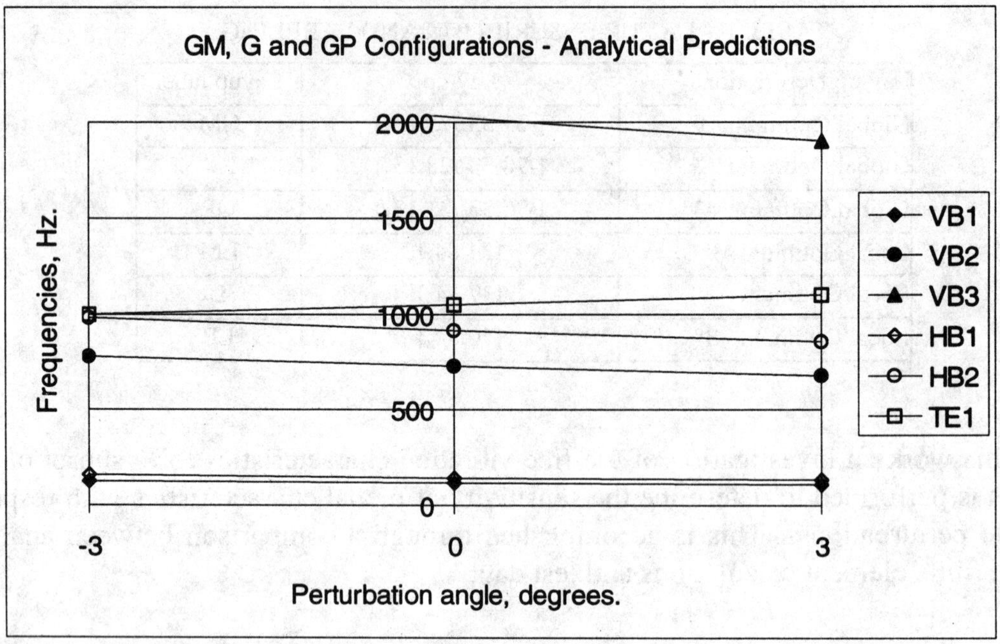

Figure 1. Analytical predictions for GM, G and GP configurations

A cantilevered support configuration is considered. Experimentally, this support condition is achieved by potting one end of each beam in an epoxy block and clamping the resulting base block to a stiff, massive support structure.

ANALYTICAL PREDICTIONS

The analytical model developed by Armanios and Badir (1995) and the solution procedure refined by Dancila and Armanios (1995) are applied for the prediction of dynamic characteristics. In the interest of brevity, neither the system of equations nor the solution

TABLE IV- ANALYTICAL PREDICTIONS OF NATURAL FREQUENCIES FOR LM, L AND LP CONFIGURATIONS

Mode	LM		L	LP	
	Hz.	% Diff.	Hz	Hz	% Diff.
VB1	103.5	4.87	98.7	94.8	-3.97
VB2	648.5	4.87	618.4	593.9	-3.96
VB3	1815.9	4.87	1731.5	1662.9	-3.96
HB1	129.4	4.87	123.4	118.5	-3.96
HB2	810.8	4.87	773.2	742.5	-3.96
HB3	2270.4	4.87	2164.8	2079.1	-3.96
TE1	1078.0	-2.51	1105.7	1131.3	2.31
ET1	3036.9	11.70	2718.8	2442.6	-10.16

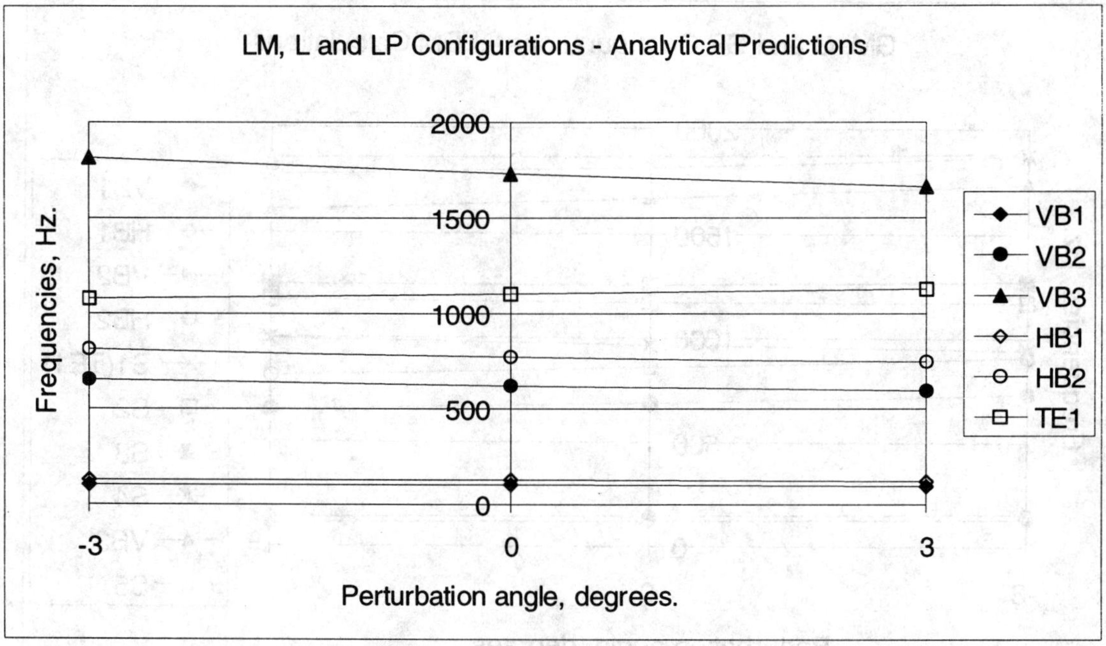

Figure 2. Analytical predictions for LM, L and LP configurations

procedure will be repeated here. The reader is directed to the respective references for full details. It is felt that two important aspects should be emphasized, however. Firstly, among the fundamental assumptions of the model are the one of in-plane nondeformability of the cross section and those of high beam slenderness and low wall-thickness-to-beam-length ratio. For the beams under consideration the values of these ratios are larger than 10 and 600, respectively. Secondly, the natural frequencies for the extension-twist coupled vibrations belong to two families, one for twist-dominated vibrations and the other one for extension-

TABLE V- FEM CALCULATION OF NATURAL FREQUENCIES FOR GM, G AND GP CONFIGURATIONS

Mode	GM		G	GP	
	Hz.	% Diff.	Hz	Hz	% Diff.
VB1	123.1	7.23	114.8	108	-5.92
HB1	155	7.34	144.4	135.7	-6.02
VB2	730.9	6.17	688.4	652.3	-5.24
HB2	908.5	6.00	857.1	813.3	-5.11
S1(TE1)	941.5	-2.75	968.1	994.7	2.75
S2	1165.6	-0.57	1172.3	1177.9	0.48
S3	1269.9	1.20	1254.8	1253.2	-0.13
S4	1869.5	3.19	1811.7	1783	-1.58
VB3	1733.7	3.93	1668.2	1607	-3.67
S5	1877.2	0.67	1864.7	1855.9	-0.47

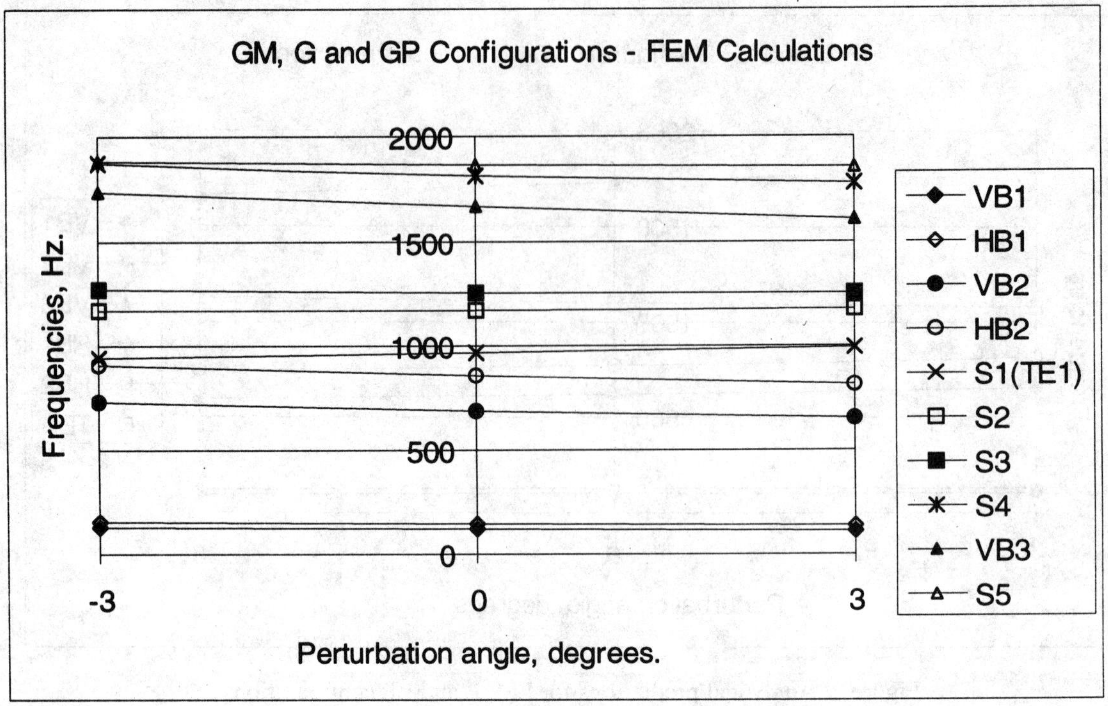

Figure 3. FEM calculations for GM, G and GP configurations

dominated vibrations. A criterion for discriminating between the two types of natural frequencies is provided in Dancila and Armanios (1995).

GLOBAL OPTIMUM AND GLOBAL NEAR-OPTIMUM CONFIGURATIONS

The analytical results for the natural frequencies of free vibration for beams GM, G and GP are shown in Table III and Fig.1. Included are the natural frequencies for the first

TABLE VI- FEM CALCULATION OF NATURAL FREQUENCIES FOR LM, L AND LP CONFIGURATIONS

Mode	LM		L	LP	
	Hz.	% Diff.	Hz	Hz	% Diff.
VB1	101.8	4.73	97.2	93.4	-3.91
HB1	128.2	4.82	122.3	117.4	-4.01
VB2	613.6	4.37	587.9	566.4	-3.66
HB2	769	4.40	736.6	709.4	-3.69
S1	891.3	-5.86	946.8	1004.4	6.08
S2(TE1)	1011.5	-4.65	1060.8	1114	5.02
S3	1172.1	-4.21	1223.6	1285.6	5.07
S4	1424.2	-0.52	1431.6	1456.6	1.75
VB3	1493.5	1.86	1466.2	1437.6	-1.95
HB3			1785.3	1773.8	-0.64

three bending modes in each plane, labeled HB*i* and VB*i*, respectively, and the first natural frequencies for torsionally-dominated coupled vibrations, labeled TE1, and for axially-dominated vibrations, labeled ET1. Also included in Table III are the sensitivities of each natural frequency to the perturbation in lay-up angle, expressed as a percentile difference with respect to the Global Optimum counterpart.

The sensitivity of all bending modes is 7.60% for a decrease in lay-up angle by 3° and -6.19% for an increase in lay-up angle by 3°. The extension-twist coupled modes show a reduced sensitivity of only 4-5%. While the bending natural frequencies are more sensitive to negative perturbation angles, the opposite is true for the extension-twist ones. With the exception of TE1, all natural frequencies decrease with respect to an increase in lay-up perturbation angle.

LOCAL OPTIMUM AND LOCAL NEAR-OPTIMUM CONFIGURATIONS

The analytical results for the natural frequencies of free vibration for beams LM, L and LP are shown in Table IV and Fig.2. As for the Global Optimum case, included are the natural frequencies for the first three bending modes in each plane and the first natural frequencies for torsionally-dominated and axially-dominated coupled vibrations. The same mode labeling convention is used, and the sensitivities of each natural frequency to the perturbation in lay-up angle are included.

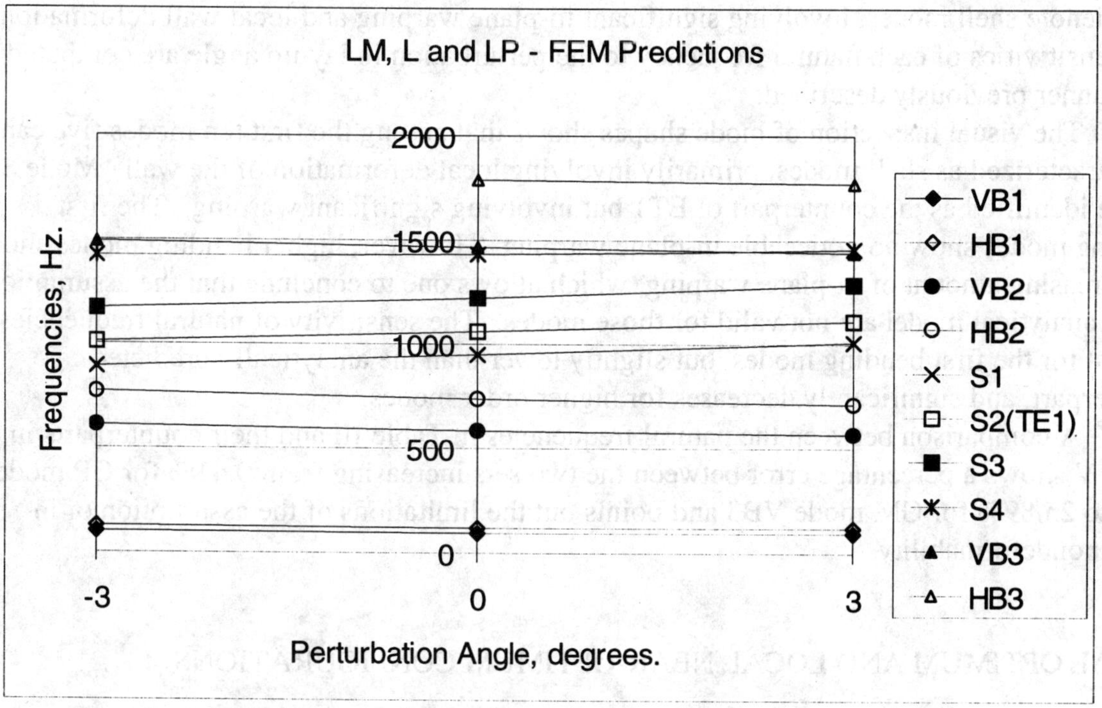

Figure 4. FEM calculations for LM, L and LP configurations

It can be seen that the bending modes appear in this case to be more robust with respect to lay-up angle perturbations, with sensitivity values representing only 64% of the Global Optimum counterparts. The same observation holds true for the ET1 mode, for which the values reduce to 60% and 46% for negative and positive perturbation angles, respectively. However, mode TE1 shows a significant increase in sensitivity, reaching 285% of its counterpart for negative perturbation angles and 221% for positive ones.

FINITE ELEMENT CALCULATIONS

A finite element analysis is performed to obtain the modal characteristics of the six beams by using the commercial finite element code ABAQUS. The finite element model contains 1260 quadrilateral reduced integration shell elements, designated as S4R, resulting in a total of 7668 variables and a wavefront size of only 124. By restraining all six degrees of freedom for the nodes located at one beam end, cantilevered boundary conditions are imposed. The ten lowest natural frequencies and mode shapes are sought. A postprocessor is used to display the computed mode shapes and visually identify the corresponding modes of deformation.

GLOBAL OPTIMUM AND GLOBAL NEAR-OPTIMUM CONFIGURATIONS

The finite element results for the natural frequencies for beams GM, G and GP are shown in Table V and Fig.3. The labeling convention previously used is extended by using S*i* to denote shell modes, involving significant in-plane warping and local wall deformations. The sensitivities of each natural frequency to the perturbation in lay-up angle are computed in the manner previously described.

The visual inspection of mode shapes shows that among the first ten modes five can be characterized as shell modes, primarily involving local deformation of the wall. Mode S1 can be identified as the counterpart of ET1 but involving significant warping. The first bending modes show no noticeable in-plane warping. However, higher bending modes show an increasing amount of in-plane warping, which allows one to conclude that the assumptions of the analytical model are not valid for those modes. The sensitivity of natural frequencies is highest for the first bending modes, but slightly lower than the analytically predicted counterpart, and significantly decreases for higher order modes.

A comparison between the natural frequencies in Table III and their counterparts in Table V shows a percentage error between the two sets increasing from 0.41% for GP mode VB1 to 25.89% for GM mode VB3 and points out the limitations of the assumption of in-plane nondeformability.

LOCAL OPTIMUM AND LOCAL NEAR-OPTIMUM CONFIGURATIONS

The finite element results for the natural frequencies of free vibration for beams LM, L and LP are shown in Table VI and Fig.4.

For this set one can identify four shell modes among the first ten calculated. The same observation regarding the presence of in-plane warping in higher order bending modes holds true in this case. For the first bending modes the sensitivity of the natural frequencies has values close to those predicted analytically for their counterparts and decreases for higher order bending modes. The sensitivity of the shell modes, for which there is no analytically computed counterpart, is in this case higher compared to the shell modes of the GM, G and GP lay-up.

A comparison between the natural frequencies in Table IV and their counterparts in Table VI shows a percentage error between the two sets increasing from 0.87% for L mode HB1 to 21.59% for LM mode VB3.

EXPERIMENTAL DETERMINATIONS

An experimental investigation is performed to identify the modal characteristics of the six beams by using a GenRad 2515 Test System and the PC-based STAR Modal Analysis software. Each of the six beams has one end potted in an epoxy block, thereby creating a cantilevered boundary condition at that end. The base block is in turn clamped to a massive, stiff support structure. A set of 20 nodal locations is selected on the surface of the mounted beam and a miniature piezoelectric accelerometer is placed at one selected location. An instrumented hammer, provided with a miniature piezoelectric force transducer, is used to apply impulse excitation at selected nodal locations. Both the force transducer and the accelerometer are connected to the GenRad system through signal amplifiers and the respective signals are acquired as time series and processed by the use of Fourier transforms to obtain Frequency Response Functions (FRF) for each pair of excitation/response locations. The FRFs are further processed to extract the modal characteristics of the structure by using the STAR software package. The processing is based upon a global curve fitting algorithm that identifies modal parameters by minimizing, in a least square sense, the error in fitting all provided FRFs.

TABLE VII- COMPARISON OF EXPERIMENTAL AND FEM NATURAL FREQUENCIES FOR CONFIGURATION L

Mode	FEM	Experimental	
	Hz	Hz	% Diff.
VB1	97.2	100.6	3.50
HB1	122.3	117.8	-3.68
VB2	587.9	584.8	-0.53
HB2	736.6	713.4	-3.15
S1	946.8	905.4	-4.37
S2(TE1)	1060.8	1040.0	-1.96
S3	1223.6	1160.0	-5.20
S4	1431.6	1300.0	-9.19
VB3	1466.2	1390.0	-5.20
HB3	1785.3	1530.0	-14.30

In order to obtain accurate results, one basic requirement of the testing method described is that the mass of the accelerometer be negligible with respect to the mass of the test structure. In the case of composite thin-walled beams, however, this requirement is difficult to meet even when using miniature accelerometers. The expected negative influence of the accelerometer mass was confirmed during preliminary testing. An appropriate choice of accelerometer location creates a tradeoff between the transducer signal-to-noise ratio and the lumped-mass effect on the structure.

The natural frequencies determined experimentally for beam L together with the percentage variance from their finite element counterparts are shown in Table VII. It can be observed that a good correlation is obtained for all modes except the last.

CONCLUSIONS

The present investigation has successfully determined a range of values and an upper bound for the sensitivity of natural frequencies to lay-up perturbation angles for the CUS beam set evaluated. While the L configuration is more robust than the G configuration for bending modes, the opposite is true for extension-twist coupled modes. Applicability limitations for the analytical model to composite beams with geometries similar to the one described can be determined. Careful choice of experimental modal analysis configuration has been shown to allow the accurate determination of natural frequencies for thin-walled beams.

ACKNOWLEDGMENTS

This work was supported by the National Rotorcraft Technology Center under Grant NCC2-945 and by the Georgia Space Grant Consortium. This support is gratefully acknowledged. The authors would also like to acknowledge the assistance of Mr. Cameron Coates with running finite element analyses and performing the experiments.

REFERENCES

Armanios, E.A., and Badir, A. M. 1995. "Free Vibration Analysis of Anisotropic Thin-Walled Closed-Section Beams," *AIAA Journal,* 33(10):1905-1910.

Badir, A. M. 1992. "Analysis of advanced thin-walled composite structures," Ph.D. Thesis, School of Aerospace Engineering, Georgia Institute of Technology, Atlanta, Georgia.

Dancila, D. S. and Armanios, E. A. 1995. "Influence of Coupling on the Free Vibration of Anisotropic Thin-Walled Closed-Section Beams" *Proceedings of the 36th AIAA/ASME/ASCE/AHS/ASC Structures, Structural Dynamics, and Materials Conference* and *AIAA/ASME Adaptive Structures Forum,* 4:2106-2115.

Lentz, W. K, Armanios, E. A. and Badir, A. M. 1996. "Constrained Optimization of Thin-Walled Composite Beams with Coupling," *Proceedings of the 37th AIAA/ASME/ASCE/AHS/ASC Structures, Structural Dynamics, and Materials Conference,* 4:2326-2334.

Lentz, W. K. 1997. " Optimum Coupling in Thin-Walled, Closed Section Composite Beams," Ph.D. Thesis, School of Aerospace Engineering, Georgia Institute of Technology, Atlanta, Georgia.

The Significance of Anisotropic Viscoelastic Poisson Ratio Time Dependencies

H. H. HILTON[1] AND S. YI[2]

ABSTRACT

Although viscoelastic moduli may be linear, Poisson ratios (PR) are always nonlinear functions of pairs of normal strains. This is equally true for physical PR defined in the real time space and for pseudo PR differently derived in the Fourier transform (FT) domain. It is shown analytically that only if anisotropic or isotropic viscoelastic moduli and relaxation or creep functions are characterized by identical time functions in all directions and stresses are constants or at most temporal and spatial separable functions, then corresponding PR must be time independent. Under all other conditions PR are proven to be time, stress and thermal expansion dependent through time integrals, although physical and pseudo PR are shown to be functionally unrelated. The consequences of PR nonlinearities are that their uniaxially determined values are not applicable to other uniaxial loadings with different time histories or to multiaxial loadings and thermal expansions, if the latter are present. Similarly, isotropic PR cannot generally be determined solely from viscoelastic Young's and shear moduli, even for linear materials. Consequently, viscoelastic material property characterization in terms of PR is not unique and viscoelastic responses are best described in terms of creep or relaxation functions. Anisotropic and isotropic viscoelastic PR time effects are investigated analytically and evaluated numerically.

INTRODUCTION

The phenomenon of deformations normal to the loading axis resulting from 1-D forces, known as the PR effect (Poisson, 1829), has received extensive attention in the theories of elasticity and plasticity. Historically, being deeply rooted in the usage of elastic and plastic metals, PR have enjoyed the advantage of relatively easy experimental determinations. Electric strain gauges can be used to great advantage on metals, because their resistance wires have the same elastic moduli as steel and aluminum. However, when applied to "softer" (less "rigid") materials, such as polymers, rubbery materials or composites, strain gauge moduli are five to six orders of magnitude higher than those of the tested materials. Indeed, here these gauges are a detriment to such experiments since they serve as reinforcements to the actual material. Under these conditions such

[1]Harry H. Hilton, University of Illinois, 104 S. Wright St., Urbana, IL 61801-2935 USA
[2]Sung Yi, Nanyang Technological University, Singapore 2263, Republic of Singapore

strain readings are the values specific to gauge wire displacements and do not properly reflect true actual strain magnitudes in the specimen.

By experimental necessity, multiaxial polymer material property determinations are conducted on thin strips or high aspect ratio thin plates under uniaxial loadings and multiaxial stress fields. While accurate measurements of strains and/or deformations in the directions of 1-D loads are generally achievable, because of the specimen's relatively large dimensions in the loaded direction, this facility is lost in the other orientations because of their relatively small deflections. These inherent problems pose limitations and make accurate experimental measurements of PR, moduli and of the associated multiaxial strains extremely difficult even under the simplest service conditions and nigh impossible during manufacture (cure) when material phase shifts and high temperature gradients are ever present.

These difficulties are further compounded in viscoelastic materials by the fact that all moduli and PR are time dependent material properties (Gurtin and Sternberg, 1963; Hilton, 1964a; Hilton and Dong, 1964b) and for anisotropic viscoelasticity they additionally take on distinct time functions in different directions. As a consequence of the operational restrictions associated with strain measurements along and particularly normal to the 1-D loading axis, many experimental researchers have assumed identical time functions for all anisotropic viscoelastic moduli, thus, as will be shown subsequently, reducing the attendant PR to constants. Such assumptions severely cloud the proper characterization of real viscoelastic materials and leave analysts in the dark regarding the true behavior of anisotropic moduli or compliances and relaxation and/or creep functions.

For a comprehensive and rigorous treatment of anisotropic elasticity, which forms the background for anisotropic viscoelasticity, the reader is referred to (Ting, 1996).

ANALYSIS

Consider a general linear anisotropic viscoelastic material in a Cartesian coordinate system $x = x_k$ with $k = 1, 2, 3$ where, under proper conditions, the Fourier transforms (FT) of the constitutive relations may be written in the form (Hilton and Dong, 1964a, 1964b)

$$\overline{\epsilon}_{kl}(x,\omega) = \overline{\overline{C}}_{klmn}(x,\omega)\,\overline{\overline{\sigma}}_{mn}(x,\omega) + \overline{\overline{q}}_{kl}(x,\omega)\,\overline{\overline{\mathcal{A}\mathcal{T}}}(x,\omega) \tag{2.1}$$

where $\overline{\overline{E}}_{kl\underline{mn}}$ and $\overline{\overline{C}}_{kl\underline{mn}}$ are respectively complex moduli and compliances with interdependences $\overline{\overline{E}}_{klmn} = 1/\overline{\overline{C}}_{klmn}$, q_{kl} are thermal expansion functions, $\mathcal{A}$ is the coefficient of thermal expansion, $\mathcal{T}$ the temperature w. r. t. a reference temperature T_o where the strains are zero and with the total temperature defined by $T = T_o + \mathcal{T}$. Repeated indices, unless underlined, indicate summations. The overscores indicate FT of time functions. (See (Yi *et al.* 1995, 1996, 1997) for a discussion on conditions when FT are applicable.)

For any general loading and/or thermal state, viscoelastic PR are definable in the real time domain in the same physical sense as the original elastic ones (Poisson, 1829), i. e. the ratio of normal strain pairs in mutually perpendicular directions, to whit

$$\nu_{kl}(x,t) \;=\; -\frac{\epsilon_{\underline{ll}}(x,t)}{\epsilon_{\underline{kk}}(x,t)} \qquad \text{for} \qquad k \neq l \tag{2.2}$$

with no summation over the underlined indices and with the understanding that the above physical definition of ν_{kl} mandates that

$$\overline{\overline{\nu}}_{kl}(x,\omega) \;\neq\; -\frac{\overline{\overline{\epsilon}}_{\underline{ll}}(x,\omega)}{\overline{\overline{\epsilon}}_{\underline{kk}}(x,\omega)} \qquad \text{and} \qquad \nu_{kl} \;\neq\; \nu_{lk} \qquad \text{for} \qquad k \neq l \tag{2.3}$$

where $\overline{\overline{\nu}}^{\mathbf{a}}_{kl} = -\overline{\overline{\epsilon}}_{\underline{ll}}/\overline{\overline{\epsilon}}_{\underline{kk}}$ is an alternate definition of a pseudo PR (Hilton and Yi, 1997). The elastic/viscoelastic analogy does not apply to PR including those for isotropic materials, even though it holds individually for strain, stress and displacement tensors. This is due to the fact that ν_{kl} in Eqs. (2.2) are related nonlinearly to pairs of normal strain components, including cases when materials obey linear constitutive relations.

Substituting general viscoelastic constitutive relations into (2.2), yields

$$\nu_{kl}(x,t) \;=\; -\frac{\int_0^t C_{\underline{ll}mn}(x,t,t')\,\sigma_{mn}(x,t')\,dt' \;+\; \int_0^t q_{\underline{ll}}(x,t,t')\,\mathcal{A}T(x,t')\,dt'}{\int_0^t C_{\underline{kk}op}(x,t,t')\,\sigma_{op}(x,t')\,dt' \;+\; \int_0^t q_{\underline{kk}}(x,t,t')\,\mathcal{A}T(x,t')\,dt'} \tag{2.4}$$

$$\text{for } k \neq l$$

Each viscoelastic PR, then, is not only dependent on the compliances (or moduli) but also on all nonvanishing stress tensor components (including shears) and temperature histories. A similar phenomenon has been noted by (Hilton and Piechocki, 1962) for viscoelastic shear centers. In the isotropic case, Eqs. (2.4) reduce to similar expressions with only two Cs and a single q function making ν_{kl} still both stress and thermal expansion time history dependent. PR for nonlinear viscoelastic materials remain prescribed by Eqs. (2.2).

Anisotropic and isotropic viscoelastic moduli are expressible in Prony series in the time domain, such that for linear materials

$$E_{klmn}(x,t) \;=\; \sum_{p=0}^{N^{klmn}} \widetilde{\phi}_{\underline{klmnp}}(x)\,\exp\left\{-\int_0^t \frac{dt'}{\tau_{\underline{klmnp}}[x,T(x,t'),M(x,t'),\alpha(x,t',T)]}\right\} \tag{2.5}$$

where $\phi_{klmn}(x,t)$ are anisotropic nonhomogeneous relaxation functions. The coefficients $\widetilde{\phi}_{klmnp}$, relaxation times $\tau_{klmnp}(x,t)$ and the number of series terms N^{klmn} are all distinct directional material property parameters. The $\widetilde{\phi}_{klmnp}$ are measures of moduli magnitudes (vertical shifts in Fig. 1), while the relaxation times determine the relative shifts on the horizontal time scale.

The simplest anisotropic model is one where amplitude and time shift multipliers are introduced. This characterization encompasses identical N, $\widetilde{\phi}_p$ and τ_p values in all directions, while being subjected to directionally sensitive distinct amplitudes and time dependences through orientationally influenced shift parameters a_{klmn}, and is represented by

$$E_{klmn}(x,t) = A_{\underline{klmn}} \sum_{p=0}^{N} \tilde{\phi}_p(x) \exp\left\{ - a_{\underline{klmn}} \int_0^t \frac{dt'}{\tau_p[x, T(x,t'), M(x,t'), \alpha(x,t',T)]} \right\} \tag{2.6}$$

Similarly, elastic PR can be derived from elastic constitutive relations equivalent to (2.1), except that the elastic compliances C^e_{klmn} are time independent.

DISCUSSION

The proper definition of viscoelastic PR is the clue to the discussion that follows. Aside from what has been called in this paper the physical (ν_{kl}, Eqs. (2.2)) and the pseudo ($\nu^{\mathbf{a}}_{kl}$) PR, there remains the issue of loading and thermal expansions time histories to be settled. Additionally, the question of whether either the proper or the pseudo definition is limited to uniaxial stresses at isothermal conditions needs to be explored. If the former is true, then as will be shown subsequently, strain data from multiaxial stress experiments cannot be used for uniaxial load PR determinations because the viscoelastic ν_{kl} and $\nu^{\mathbf{a}}_{kl}$ are load history dependent. Or conversely, uniaxially determined viscoelastic ν_{kl} and $\nu^{\mathbf{a}}_{kl}$ are not universal and exportable to other multiaxial load histories. Linear compliances, moduli and creep and relaxation functions are stress independent and, hence, load history insensitive, but PR by their very nature are nonlinear functions of pairs of perpendicular tensile or compressive strains. Even for uniaxial stress fields, additional complications arise in the experimental determination and definition of both isotropic and anisotropic ν_{kl} and $\nu^{\mathbf{a}}_{kl}$ if thermal expansions are present.

It can be seen from Eqs. (2.2) for the physical Poisson ratios ν_{kl} and for the pseudo $\nu^{\mathbf{a}}_{kl}$ that these two definitions are incompatible with each other. Consider the simplest possible set of conditions for an isothermal uniaxial time independent loading with compliances obeying identical time functions. Then from Eq. (2.4), one obtains

$$\nu_{kl}(x,t) = \nu^e_{kl}(x) = -\frac{C^o_{\underline{ll}11}(x)}{C^o_{\underline{kk}11}(x)} \tag{3.1}$$

and

$$\overline{\overline{\nu}}^{\mathbf{a}}_{kl}(x,\omega) = \nu^e_{kl}(x) \Longrightarrow \nu^{\mathbf{a}}_{kl}(x,t) = \nu^e_{kl}(x)\,\delta(t) \tag{3.2}$$

where $\delta(t)$ is the Dirac delta function.

Extensive tables of the dependence of the physical, pseudo and elastic PR on time and loadings are given by (Hilton and Yi, 1997). Several important observations on the true physical viscoelastic ν_{kl} are immediately evident from the preceding analysis, such as:

(A) Because of their nonlinear relations to strain tensors, neither physically defined isotropic nor anisotropic viscoelastic PR can be formulated in terms of the elastic/viscoelastic integral transform analogy, although moduli remain stress and displacement independent for linear materials.

(B) Both isotropic and anisotropic νs depend on material properties, loading (stress) and thermal expansion histories, but under some special circumstances these history functions factor out of the ν_{kl} expressions.
(C) By virtue of Eqs. (2.4), anisotropic νs can be time independent if and only if all moduli (compliances, relaxation functions) obey the same time functions in all directions, $\mathcal{T} = 0$ and the stress tensors are constants or are separable into $\sigma_{kl}(x,t) = g(t)\,\sigma^*_{kl}(x)$. This compliance class represents a highly restrictive set of conditions (Hilton, 1996).
(D) In multiaxial loading, isotropic νs can be time independent if and only if G and K have identical time functions, $\mathcal{T} = 0$, the temperature $T = T(t)$ only, and the stress tensors are separable *à la* (C) above. Separation of variables here can only be achieved if $\mathcal{T} = 0$ (Hilton and Russell, 1961).
(E) In uniaxial loading, isotropic νs are time independent if and only if $G/K =$ const. (same time functions) and $\mathcal{T} = 0$.

However, important observations and conclusions can be gleaned from analytic examinations of anisotropic and isotropic PR as shown in Figs. 2 - 4. More complicated relations then the isotropic ones for $\overline{\overline{\nu}}_{kl}$ in terms of the total set of $\overline{\overline{E}}_{klmn}$ can be derived for general anisotropic viscoelastic states as defined by Eqs. (2.1), which now additionally involve the shear moduli. In any case, viscoelastic PR depend on moduli function fractions as seen from Eqs. (2.4). If all moduli $E_{klmn}(x,t)$ and relaxation functions $\phi_{klmn}(x,t)$ have identical time functions, then all $\nu_{kl}(x,t)$ are time independent. This is due to the time function transforms in each complex moduli simply factoring and then dividing out of the expressions for $\overline{\overline{\nu}}$ in Eqs. (2.4). It has been noted by Hilton (1964a, 1996) that such constant viscoelastic PR values are physically true only for incompressible materials when $\nu(t) = .5$. Under all other circumstances constant ν values represent extremely restrictive conditions for real materials.

Under uniaxial loading ($\sigma_{11} \neq 0$), the FT of isotropic νs in terms of the two moduli is

$$\overline{\nu}(x,\omega) = \int_{-\infty}^{\infty} \underbrace{\left\{ \frac{\int_0^t [.5\,C_G(x,t,t') - C_K(x,t,t')]\;\sigma_{11}(x,t')\,dt'}{\int_0^t [C_G(x,t,t') + C_K(x,t,t')]\;\sigma_{11}(x,t')\,dt'} \right\}}_{=\,\nu(x,t)} \exp(-\imath\,\omega\,t)\;dt \quad (3.3)$$

where the shear and bulk compliances are $\overline{\overline{C}}_G = 1/\overline{\overline{G}}$ and $\overline{\overline{C}}_K = 1/\overline{\overline{K}}$ and $\imath = \sqrt{-1}$. Isotropic compliances can be expressed in Prony series similar to (2.5). Consider the simplest model given by Eqs. (2.6) which for $\sigma_{11} =$ const still yields a time dependent ν.

If $G(x,t)$ and $K(x,t)$ are expressed in terms of identical time functions, say $G(x,t) = G_o(x)f(t)$ or $G(x,t) = G_o f(x,t)$ and $K(x,t) = K_o(x)f(t)$ or $K(x,t) = K_o f(x,t)$, then after the inner integrations the inverses of the product functions $\overline{\overline{f}}\;\overline{\overline{\sigma}}_{11}$ factor, then $\nu(x,t) = \nu^e(x)$ where ν^e is the time independent elastic (instantaneous) PR. Similarly, the anisotropic νs of the previous section would reduce to their elastic counterparts, with $\nu_{kl}(x,t) = \nu^e_{kl}(x)$ for $k \neq l$, when the moduli have the same time functions in all directions, such that $E_{klmn}(x,t) = E^o_{klmn}(x)f(t)$ or $E_{klmn}(x,t) = E^o_{klmn}f(x,t)$.

Fig. 2 is a plot of isotropic C_G, C_K and ν vs. log time. All curves represent a real material and are modeled with $N = 29$. The C_G and C_K are normalized w. r. t. their own elastic instantaneous (maximum) values. Numerical values are $C^G_\infty/C^G_o = C^K_\infty/C^K_o = 1.5$, $C^K_o/C^G_o = .01$ and $a = .01$. It can be readily seen that ν is time independent only in those t regions where C^K_o/C^G_o or C^K_∞/C^G_∞ are constants. This condition indicates that both changes in shape and in volume must proceed either in an elastic manner or in a fully relaxed fashion for $\nu(t)$ to remain constant.

Fig. 3 and 4 respectively depict the effects of G_o/K_o and of a on PR. It is seen the larger the K_o compared to G_o the less time dependent ν becomes and approaches the incompressible value of 1/2. The influence of decreasing values of a is increase the relaxation times τ^K_p compared to τ^G_p and to widen the number of decades over which PR are time dependent.

The relative influence of Gs and Ks on isotropic PR of Eqs. (3.3) and in Figs. 3 and 4 are identical to that of $C_{\underline{ll}11}$ and $C_{\underline{kk}11}$ on anisotropic ν_{kl} of Eqs. (2.4) for $\mathcal{AT} = 0$ and uniaxial stress fields ($\sigma_{11} \neq 0$), with G and K defined in a manner similar to (2.5).

One must now ask what kind of real materials admit constant PR to characterize their viscoelastic behavior. One obvious answer is all incompressible isotropic materials do, where $\nu(t) = .5$. For isotropic materials,it is required that G and K obey the same time functions so that $G(x,t)/K(x,t) = G_o(x)/K_o(x)$ This relation, however, is not satisfied by real viscoelastic materials, except for the incompressible ones or where $\nu \simeq .4999^+$ and $K_o >> G_o$. For most materials K_o is only $> G_o$ but additionally real viscoelastic materials, such as high temperature metals, polymers, composites, etc., exhibit creep and relaxation rates for volume changes which are much slower than those contributing to shape changes, indicating $\tau^K_p > \tau^G_p$. Consequently, on a time plot $C_K(t)$ curves are always shifted in the direction of larger t when compared to $C_G(t)$. Conversely, since $\omega \sim 1/t$, $\overline{\overline{C}}_K(\omega)$ lags behind $\overline{\overline{C}}_G(\omega)$ on a frequency ω plot.

Similarly, in the anisotropic case Eqs. (2.4) again mandate that in order to achieve constant $\nu_{kl}(t)$ values all pertinent $E_{klmn}(t)$ must be in either their elastic or fully relaxed ranges as $t \to 0$ or $t \to \infty$.

Physically, constant viscoelastic ν_{kl} conditions over the entire time space can be approached if real materials such as, for instance, composites have fibers which either are elastic or have extremely slow creep rates with relaxation times orders of magnitude larger compared to the viscoelastic polymer matrix. The values of the E_{ijkl} in different orientations are then predominately dictated by fiber stiffnesses and directional arrangements, while the polymer matrix remains isotropically viscoelastic with identical constitutive relations in all directions.

The analysis indicates that isotropic viscoelastic materials require two moduli obeying different time functions, while anisotropic materials need as few as 9 and as many as 21 time dependent moduli for complete property specification. Additionally and independently, one thermal expansion function is called for in isotropic viscoelasticity and one to six are needed for anisotropic modeling.

Ultimately, it must be remembered that PR do not contribute anything additional to material characterization that cannot be achieved totally by moduli, compliances, or creep and relaxation functions. The utility of the PR because of their nonlinear behavior is, therefore, severely limited. Consequently, moduli or compliances, or creep or relaxation functions should be the vehicle of choice for describing viscoelastic responses.

Isotropic and anisotropic viscoelastic analyses in the present paper demonstrate the close relationship between viscoelastic moduli and PR, particularly in regard to their time and stress dependences. No experimental data for non-metallic materials seems to exist describing complete viscoelastic multiaxial experimental strain or displacement measurements necessary for proper PR determinations. The works of (Beckwith, 1980, 1984; Bogetti and Gillespie, 1991; Crossman *et al.* 1978; Flaggs and Crossman, 1981; Halpin and Pagano, 1968; Lin and Hwang, 1989; Tuttle and Brinson, 1985a, 1985b, 1986; and Yang *et al.*, 1983), among others, are typical examples of experimental investigations reporting on anisotropic moduli from which, because of inherent 2-D and 3-D experimental complexities it is difficult to deduce the total anisotropic modulus and PR picture. Additionally, an extensive collection of isotropic viscoelastic modulus data has been presented and analyzed by (Nashif *et al.*, 1985).

(Levitsky & Shaffer's, 1974) and (Shaffer and Levitsky's, 1974) interpretations of responses for elastic thermo-setting materials use PR which are not defined according to the present multiaxial analysis. Subsequently, (Levitsky and Shaffer, 1975) employed a single Maxwell viscoelastic model with constant bulk modulus and a PR defined by an essentially equivalent time dependent elastic ν, which is inapplicable for their 2-D time dependent stress states. Finally, (Kim and White, 1996, 1997) assumed a constant PR $\nu_o(t)$ during AS4/3501-6 composite cure and then calculated a relaxation modulus based on $E(t) = 2(1 + \nu_o)G(t)$, which essentially begs the question regarding appropriate viscoelastic values of ν and renders the calculation of $E(t)$ from experimentally determined $G(t)$ or vice-versa highly questionable for their multiaxial time dependent stress fields. It is unfortunate that these five publications chose an involvement with ill defined PR, because their analyses and interpretations of experimental data could be carried out in terms of moduli without ever appealing to any PR concepts. As a matter of fact, viscoelastic material characterization is best defined in terms of moduli or compliances, or creep or relaxation functions, which for linear materials do not bear the nonlinear and load-time history burden of PR. (Hwang, 1990) has reported viscoelastic experimental results for IM7/8551-7 graphite/epoxy composites indicating that in the fiber dominated direction the compliances C_{1111} and PR ν_{12} are time independent, while the compliances C_{2222} and C_{2323} exhibit distinct time dependent responses.

CONCLUSIONS

The isotropic and anisotropic viscoelastic analyses demonstrate the close relationship between viscoelastic moduli and PR, particularly in regard to the latter's time dependence and nonlinear functionality w. r. t. stress fields, thermal expansions and time histories. Uniaxially determined PR are not exportable to other uniaxial conditions with different loading time sequences nor are they applicable to multi-axial stress fields. PR add little to viscoelastic isotropic and anisotropic material characterization that is not already encompassed by moduli, compliances, and creep and relaxation functions. Consequently, any one of the latter should be the characterization model of choice, particularly because of the inherent nonlinear nature of PR. No experimental data for non-metallic materials seems to exist describing multiaxial experimental strain or displacement measurements necessary for proper PR determinations. It is, therefore, recommended that multiaxial experiments and data collection be undertaken on viscoelastic materials in order to correctly catalogue isotropic and anisotropic deterministic and stochastic material property parameters.

REFERENCES

Beckwith, S. W. 1980. "Creep evaluation of a glass/epoxy composite," *SAMPE Quarterly*, 11:8-15.

Beckwith, S. W. 1984. "Viscoelastic creep behavior of filament-wound case materials," *Journal of Spacecraft and Rockets*, 21:546-552.

Bogetti, T. A. and J. W. Gillespie, Jr. 1991. "Two-dimensional cure simulation of thick thermosetting composites," *Journal of Composite Materials*, 25:239-273.

Crossman, F. W., Mauri, R. E. and Warren, W. J. 1978. "Moisture altered viscoelastic response of graphite/epoxy composite," *Advanced Composite Materials: Environmental Effects, ASTM STP 658*, J. R. Vinson, ed., 205-220, ASTM, Philadelphia.

Flaggs, D. L. and F. W. Crossman. 1981. "Analysis of viscoelastic response of composite laminates during hygrothermal exposure," *Journal of Composite Materials*, 15:21-40.

Gurtin, M. E. and E. Sternberg. 1963. "A reciprocal theory of anisotropic viscoelastic solids," *Journal of the Society for Industrial Applied Mathematics*, 11:607-613.

Halpin, J. C. and N. J. Pagano. 1968. "Observations on linear anisotropic viscoelasticity," *Journal of Composite Materials*, 2:68-80.

Hilton, H. H. and H. G. Russell. 1961. "An extension of Alfrey's analogy to thermal stress problems in temperature-dependent linear viscoelastic media," *Journal of Mechanics and Physics of Solids*, 9:152-164.

Hilton, H. H. and J. J. Piechocki. 1962. "Shear center motion in beams with temperature-dependent linear elastic or viscoelastic properties," *Fourth U. S. National Congress of Applied Mechanics*, 2:1279-1289, ASME, New York.

Hilton, H. H. 1964a. "An introduction to viscoelastic analysis," *Engineering Design for Plastics*, E. Baer, ed., 199-276, Reinhold Publishing Corp., New York.

Hilton, H. H. and S. B. Dong. 1964b. "An analogy for anisotropic, nonhomogeneous, linear viscoelasticity including thermal stresses," *Developments in Mechanics* 58-73, Pergamon Press, New York.

Hilton, H. H. 1996. "On the inadmissibility of separation of variable solutions in linear anisotropic viscoelasticity," *Mechanics of Composite Materials and Structures*, 3:97-100.

Hilton, H. H. and S. Yi. 1997. "The significance of (an)isotropic viscoelastic Poisson ratio stress and time dependencies," *T. R. UILU ENG 97-0503*, University of Illinois at Urbana-Champaign.

Hwang, I. H. 1990. "Thermo-viscoelastic behavior of composite materials," *Ph. D. Thesis*, University of Washington, Seattle.

Kim, Y. K. and S. R. White. 1996. "Stress relaxation behavior of 3501-6 epoxy resin during cure," *Polymer Engineering and Science*, 36:2852-2862.

Kim, Y. K. and S. R. White. 1997. "Process-induced stress relaxation analysis of AS4/ 3501-6 laminate," *Journal of Reinforced Plastics and Composites*, 16:2-16.

Levitsky, M. and B. W. Shaffer. 1974. "Thermal stresses in chemically hardening elastic media with application to the molding process," *ASME Journal of Applied Mechanics* **41**:647-651.

Levitsky, M. and B. W. Shaffer. 1975. "Residual thermal stress in a solid sphere cast from a thermosetting material," *ASME Journal of Applied Mechanics*, 42:651-655.

Lin, K. Y. and I. H. Hwang. 1989. "Thermo-viscoelastic analysis of composite materials," *Journal of Composite Materials*, 23:554-569.

Nashif, A. D., D. I. G. Jones and J. P. Henderson. 1985. *Vibration Damping.* John Wiley & Sons, New York.

Poisson, S. D. 1829. "Mémoire sur l'équilibre et le mouvement des corps élastiques," *Mémoires de l'Académie Royal des Sciences de l'Institut de France*, 8:357-570, 623-627.

Shaffer, B. W. and M. Levitsky. 1974. "Thermoelastic constitutive equations for chemically hardening materials," *ASME Journal of Applied Mechanics*, 41:652-657.

Ting, T. C. T.. 1996. *Anisotropic Elasticity - Theory and Applications.* Oxford University Press, New York.

Tuttle, M. E. and H. F. Brinson. 1985a. "Accelerated viscoelastic characterization of T300/5208 graphite-epoxy laminates," *NASA Contractor Report 3871.*

Tuttle, M. E. and H. F. Brinson. 1985b. "Resistance fail strain gage technology as applied to composite materials," *NASA Contractor Report 3872.*

Tuttle, M. E. and H. F. Brinson. 1986. "Prediction of long term creep compliance of general composite laminates," *Experimental Mechanics*, 26:89-102.

Yang, P., L. A. Carlsson and S. S. Sternstein. 1983. "Dynamic-mechanical response of graphite/epoxy composite laminates and neat resin," *Polymer Composites*, 4:104-112.

Yi, S., H. H. Hilton and M. F. Ahmad. 1995. "Curing process induced viscoelastic residual stresses in polymer matrix laminated composites," *Proceedings ASME Symposium on Design and Manufacture of Composites*, MID-69-1:65-76.

Yi, S., H. H. Hilton, M. F. Ahmad and K. H. Pierson. 1996. 'Cure cycle simulations of composites with temperature and cure dependent anisotropic viscoelastic properties," *Proceedings 37th AIAA/ASME/ASCE/AHS/ASC Structures, Structural Dynamics and Materials Conference, AIAA Paper 96-1577.*

Yi, S. and H. H. Hilton. 1997. "Thermo-rheological behavior of polymer matrix composites during curing processes," *Proceedings 38th AIAA/ASME/ASCE/AHS/ASC Structures, Structural Dynamics and Materials Conference, AIAA Paper 97-1326.*

Zinoviev, P. A. and Y. N. Ermakov. 1994. *Energy Dissipation in Composite Materials*, Technomic Publishing Co., Lancaster PA.

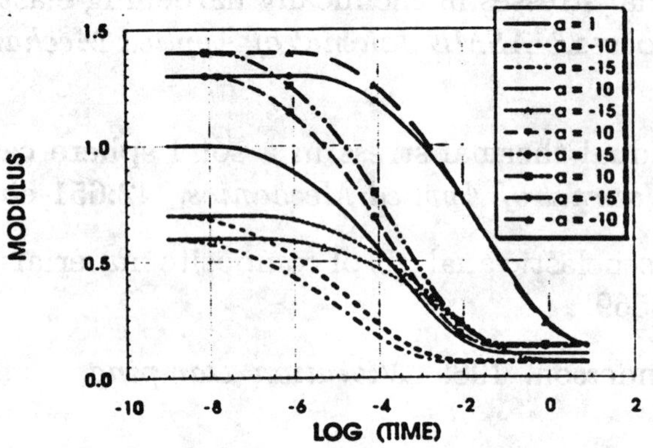

Fig. 1 EFFECTS OF E_0 AND TIME SHIFTS ON MODULI

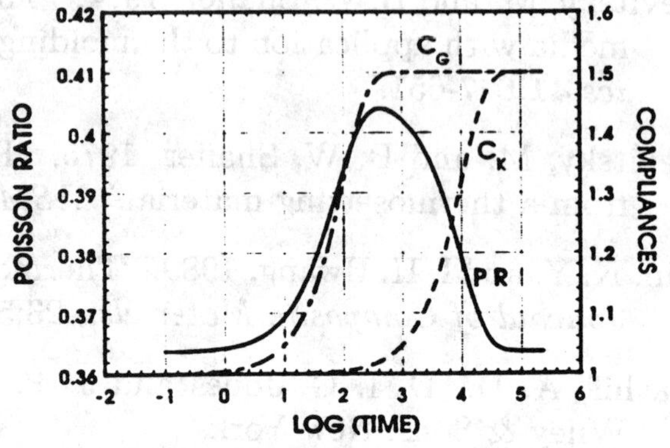

Fig. 2 POISSON RATIO & COMPLIANCES

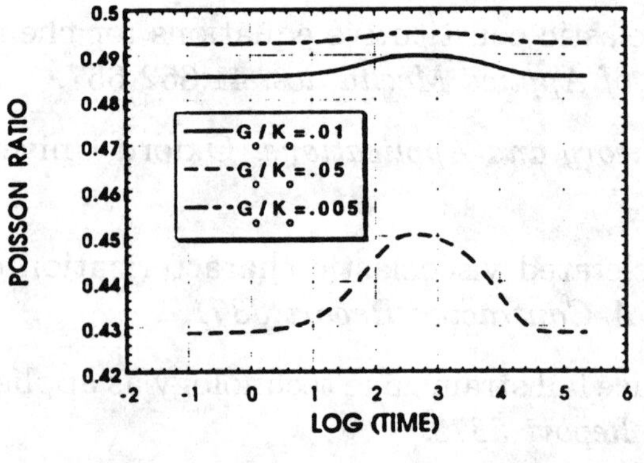

Fig. 3 G_o/K_o EFFECTS ON POISSON RATIO

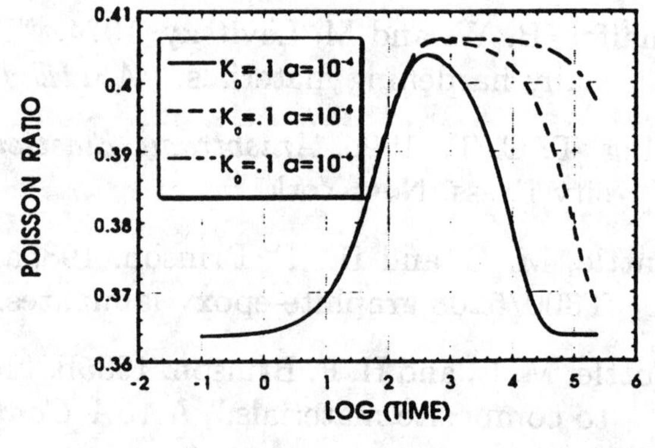

Fig. 4 EFFECT OF a ON POISSON RATIO

Viscoelastic Response of Composite Overwrapped Cylinders

J. T. TZENG

ABSTRACT

Creep and stress relaxation behavior of composite overwrapped cylinders are investigated based on anisotropic viscoelasticity. The analysis accounts for ply-by-ply variations of material properties, ply orientations, and temperature gradients through the thickness of cylinders. Currently, commercial finite element packages such as ABAQUS, ANSYS, and DYNA3D are not suitable for the viscoelastic analysis of composite cylinders because of the lack of anisotropic viscoelastic elements. However, fiber reinforced composite materials generally illustrate extreme anisotropy in viscoelastic behavior. The viscoelasticity exists mainly in matrix dominant properties such as transverse and shear while the fiber dominant properties behave more like elastic mediums. The study of thermal and mechanical stress relaxation in an overwrapped composite cylinder shows that the viscoelastic characteristics of composite cylinders are quite different from those of isotropic cylinders. Poisson effect alone from the creep behavior in the transverse properties of composite can result in a drastic change of stress and strain state of the cylinder.

INTRODUCTION

Composite materials are currently used for light-weight pressure vessels and high efficient rotors as an energy storage device. For both applications, pre-stresses are built in during fabrication of the cylinders through a "press-fit" procedure to enhance the mechanical performance. Composite overwrap pressure vessels are prestressed; therefore, the liner is in a state of compression (hoop) and the composite overwrap is in tension. For the other application, the rotors are subjected to a radial compression prior to operation. Accordingly, centrifugal force resulting from the rotation of rotors during operation generates tensile stresses in the radial and circumferential directions. Since the composite rotors are mainly reinforced in circumference (filament wound cylinders), the radial tensile is critical to the ultimate performance of the rotors. Accordingly, it is essential to design and build the rotors with radial pre-compression. However, the polymer matrix composites generally creep in long period of time, especially at elevated temperatures [1]. The associated stress relaxation in the composite will result in the loss of the pre-stresses and potentially lead to the failure of the rotors. The objective of this investigation is to develop an analytical method to study the viscoelastic behavior of the thick-walled composite cylinders. The analysis can be applied to the design of flywheel machinery and overwrapped composite pressure vessels.

Dr. Jerome T. Tzeng, U. S. Army Research Laboratory, Weapons and Materials Directorate, Attn:AMSRL-WM-PD, Aberdeen Proving Ground, Md 21005-5066.

To date, activities in the research of viscoelasticity have mainly concerned isotropic materials including the studies by Muki and Sternberg [2]; Schapery [3]; Williams [4]; and Christensen [5]. These basic theories of viscoelasticity were then extended to the area of heterogeneous and anisotropic materials for a variety of applications. Hashin [6] used the effective relaxation moduli and creep compliances to define the macroscopic viscoelastic behavior of linear viscoelastic heterogeneous media and its implementation in viscoelastic modeling. The general formulation of linear viscoelastic boundary value problems of composite materials, including the thermal viscoelastic problems for thermorheologically simple materials, and the applications of the correspondence principle were examined by Schapery [7]. Rogers and Lee [8] investigated the viscoelastic behavior of an isotropic cylinder.

VISCOELASTIC FORMULATION

In the following research, the linear quasi-static viscoelastic behavior of a thick laminated composite cylinder with an elevated temperature change is studied. The analysis accounts for ply-by-ply variation of properties, temperature changes, and fiber orientations. The thick cylinder is assumed to be in the absence of thermomechanical coupling and to be in the state of generalized plane strain such that all the stress and strain components are independent of the axial coordinate (Tzeng and Chien [9,10]). Moreover, due to the nature of axisymmetry, all the stress and strain components are also independent of the circumferential coordinate. The mechanical responses of this thick composite cylinder will, therefore, only have to satisfy the governing equation in the radial direction. Invoking the Boltzmann superposition integral for the complete spectrum of increments of anisotropic material constants with respect to time, the thermoviscoelastic constitutive relations of the anisotropic composite cylinder can be derived in integral forms. Since the thick composite cylinder is subjected to a constant elevated temperature change, and boundary conditions are all independent of time, formulations of the linear thermal viscoelastic problem can have forms identical to those of the corresponding linear thermoelastic problem by taking advantage of the elastic-viscoelastic correspondence principle. In other words, all of these integral constitutive equations reduce to the algebraic relations, which are very similar to those developed for thermoelastic media when they are Laplace transformed by means of the rule for convolution integrals. The thermoelastic analysis can thus be used to derive the transformed thermal viscoelastic solutions in the frequency domain.

The Boltzmann superposition integral of stress σ_{ij} $(i, j = 1, 2, 3)$ and strain ε_{ij} $(i, j = 1, 2, 3)$ relation for an isothermal viscoelastic problem with a constant temperature increase)T and the thermal expansion coefficient α_{kl} is

$$\sigma_{ij}(t) = \int_0^t C_{ij}^{kl}(T,t-\tau)\,\frac{\partial \varepsilon_{kl}(\tau)}{\partial \tau}\,d\tau - \beta_{ij}(T,t)\Delta T, \tag{1}$$

where $C_{ij}^{kl}(T,t)$ is the relaxation modulus dependent on temperature T and time t,

$$C_{ij}^{kl}(T(t),t) = C_{ij}^{kl}(T_0,\lambda(t))$$

$$\lambda(t) = \frac{t}{a_T\,(T(t))},$$

Here, T_0 is the base temperature, and a_T is the temperature shift factor. $\beta_{ij}(T,t)$ is given by $B_{ij}(T,t) = C_{ij}^{kl}(T,t)\bullet\alpha_{kl}$. It is often desirable to use the inverse form of the constitutive relation (1),

$$\varepsilon_{ij}(t) = \int_0^t A_{ij}^{kl}(T,t-\tau)\frac{\partial\sigma_{kl}(\tau)}{\partial\tau}d\tau + \psi_{ij}(T,t)\Delta T, \tag{2}$$

where $R_{ij}(T,t)$ is the tensor product of the creep compliance $A_{ij}^{kl}(T,t)$ and the thermal creep coefficient ϕ_{kl}. Since the elevated temperature change)*T* is constant above some reference value in time, the relaxation moduli and creep compliances are evaluated at that reference temperature regardless of whether or not the material is thermorheologically simple by employing the temperature shift-factor.

The Laplace transform of a function $f(t)$ is defined as

$$\bar{f} = \bar{f}(s) = \int_0^\infty e^{-st} f(t)dt, \tag{3}$$

where s is the Laplace transform variable. Applying (3) with the convolution rule to (1) and (2) reduces the integral constitutive equations to the following algebraic relations:

$$\bar{\sigma}_{ij} = \tilde{C}_{ij}^{kl}\tilde{\varepsilon}_{kl} - \tilde{\beta}_{ij}\frac{\Delta T}{s}, \tag{4}$$

and the inverse form

$$\tilde{\varepsilon}_{ij} = \tilde{A}_{ij}^{kl}\bar{\sigma}_{ij} + \tilde{\psi}_{ij}\frac{\Delta T}{s}, \tag{5}$$

respectively, where $\tilde{C}_{ij}^{kl} = s\bar{C}_{ij}^{kl}$, $\tilde{A}_{ij}^{kl} = s\bar{A}_{ij}^{kl}$, $\tilde{\alpha}_{ij} = s\bar{\alpha}_{ij}$, $\tilde{\phi}_{ij} = s\bar{\phi}_{ij}$, $\tilde{\beta}_{kl} = \tilde{C}_{ij}^{kl} \bullet \tilde{\alpha}_{ij}$, and $\tilde{\psi}_{kl} = \tilde{A}_{ij}^{kl} \bullet \tilde{\phi}_{ij}$. Furthermore, it can be shown that $\left[\tilde{A}_{ij}^{kl}\right] = \left[\tilde{C}_{ij}^{kl}\right]^{-1}$.

Consider a filament-wound axisymmetric thick composite cylinder consisting of N layers with the axial coordinate z, the radial coordinate r, and the circumferential coordinate 2, as shown in Figure 1(a). The composite cylinder has the inner radius a, the outer radius b, and the length L. Accordingly, there is a corresponding thermoelastic problem with the transformed displacement components $\bar{u}$, $\bar{v}$, and $\bar{w}$ in the axial direction, the circumferential direction, and the radial direction, respectively, in each layer. The axisymmetric character of the thick composite cylinder along with the assumption of the state of generalized plane strain leads to a simplified displacement field, which reflects the circumferential independence and only radial dependence of $\bar{w}$,

$$\bar{u}(r,\theta,z) = \bar{u}(r,z),\ \ \bar{v}(r,\theta,z) = \bar{v}(r,z),\ \ and\ \ \bar{w}(r,\theta,z) = \bar{w}(r). \tag{6}$$

Since each layer of the thick laminated cylinder is cylindrically monoclinic in reference to the global coordinates, there is no coupling between transverse shears and other deformations. Accordingly, the vanishing shear traction boundary conditions and interface continuity conditions generate zero out-of-plane shear tractions and shear strains for each layer. Moreover, owing to the absence of torsional deformation, the transformed displacement components $\bar{u}$ and $\bar{v}$ become

$$\bar{u} = \bar{\varepsilon}^{0} z \text{ and } \bar{v} = 0, \tag{7}$$

where the constant quantity $\bar{\varepsilon}^{0}$ has the physical interpretation of transformed axial strain of a layer. In fact, $\bar{\varepsilon}^{0}$, according to the present formulation, also represents the transformed axial strain of the entire composite cylinder. The calculation of $\bar{\varepsilon}^{0}$ requires the knowledge of end boundary conditions and will be given later. Likewise, solving for $\overline{w(r)}$ requires the information of transformed strain components, the constitutive equations, as well as the equilibrium equations.

The previously transformed displacement field gives the transformed strain components in cylindrical coordinates

$$\bar{\varepsilon}_{rr} = \frac{d\overline{w(r)}}{dr}, \quad \bar{\varepsilon}_{\theta\theta} = \frac{\overline{w(r)}}{r}, \quad \bar{\varepsilon}_{zz} = \frac{d\bar{u}}{dz} = \bar{\varepsilon}^{0}, \text{and} \tag{8}$$
$$\bar{\varepsilon}_{\theta r} = \bar{\varepsilon}_{zr} = \bar{\varepsilon}_{z} = 0.$$

The unabridged form of the constitutive equation (4) for each layer in cylindrical coordinates with the radial coordinate *r* normal to the plane of symmetry is expressed as

$$\begin{Bmatrix} \bar{\sigma}_{zz} \\ \bar{\sigma}_{\theta\theta} \\ \bar{\sigma}_{rr} \\ \bar{\sigma}_{\theta r} \\ \bar{\sigma}_{zr} \\ \bar{\sigma}_{z\theta} \end{Bmatrix} = \begin{bmatrix} \tilde{C}_{11} & \tilde{C}_{12} & \tilde{C}_{13} & 0 & 0 & \tilde{C}_{16} \\ \tilde{C}_{12} & \tilde{C}_{22} & \tilde{C}_{23} & 0 & 0 & \tilde{C}_{26} \\ \tilde{C}_{13} & \tilde{C}_{23} & \tilde{C}_{33} & 0 & 0 & \tilde{C}_{36} \\ 0 & 0 & 0 & \tilde{C}_{44} & \tilde{C}_{45} & 0 \\ 0 & 0 & 0 & \tilde{C}_{45} & \tilde{C}_{55} & 0 \\ \tilde{C}_{16} & \tilde{C}_{26} & \tilde{C}_{36} & 0 & 0 & \tilde{C}_{66} \end{bmatrix} \begin{Bmatrix} \bar{\varepsilon}_{zz} \\ \bar{\varepsilon}_{\theta\theta} \\ \bar{\varepsilon}_{rr} \\ \bar{\varepsilon}_{\theta r} \\ \bar{\varepsilon}_{zr} \\ \bar{\varepsilon}_{z\theta} \end{Bmatrix} - \frac{\Delta T}{s} \begin{Bmatrix} \tilde{B}_{zz} \\ \tilde{B}_{\theta\theta} \\ \tilde{B}_{rr} \\ 0 \\ 0 \\ \tilde{B}_{z\theta} \end{Bmatrix}. \tag{9}$$

Furthermore, from the previous discussions, it can be shown that two of the three equilibrium equations are satisfied automatically. The only nontrivial equilibrium equation is the one in the radial direction:

$$\frac{\partial \bar{\sigma}_{rr}}{\partial r} + \frac{\bar{\sigma}_{rr} - \bar{\sigma}_{\theta\theta}}{r} = 0. \tag{10}$$

Substituting (6), (7), and (8) into (9), the transformed stress components $\bar{\sigma}_{rr}$ and $\bar{\sigma}_{\theta\theta}$ are obtained in terms of the transformed radial displacement $\bar{w}$. Incorporating the resulting $\bar{\sigma}_{rr}$ and $\bar{\sigma}_{\theta\theta}$ functions with (10) gives a nonhomogeneous Euler differential equation of $\bar{w}$ for a layer

$$r^2 \frac{d^2\bar{w}}{dr^2} + r\frac{d\bar{w}}{dr} - \bar{\lambda}^2 \bar{w} = \frac{r}{\tilde{C}_{33}} \left[\frac{\Delta T}{s} (\tilde{B}_{rr} - \tilde{B}_{\theta\theta}) - (\tilde{C}_{13} - \tilde{C}_{12}) \bar{\varepsilon}^{o} \right], \tag{11}$$

where

$$\bar{\lambda}^2 = \frac{\tilde{C}_{22}}{\tilde{C}_{33}} \tag{12}$$

Solving (11) for $\bar{w}$ yields

$$\bar{w} = \bar{A}_1 r^{\lambda} + \bar{A}_2 r^{-\lambda} + \tilde{w}_p , \tag{13}$$

where

$$\tilde{w}_p = \tilde{f}_1 \bar{\varepsilon}^o r + \tilde{f}_3 r$$
$$\tilde{f}_1 = \frac{\tilde{C}_{12} - \tilde{C}_{13}}{\tilde{C}_{33} - \tilde{C}_{22}}$$
$$\tilde{f}_3 = \frac{\tilde{S}}{\tilde{C}_{33} - \tilde{C}_{22}} \tag{14}$$
$$\tilde{S} = \frac{\Delta T}{s}\left[\tilde{B}_{rr} - \tilde{B}_{\theta\theta}\right]$$

and $\bar{A}_1$ and $\bar{A}_2$ are coefficients to be determined from boundary and continuity conditions.

Finally, it is understood that the initial condition of the original thermoviscoelastic problem is displacement-free state of rest. The boundary conditions are of free tractions and, hence, of free transformed tractions on both inner and outer circular surfaces:

$$\bar{\sigma}_{rr} = \bar{\sigma}_{\theta r} = \bar{\sigma}_{zr} = 0 \quad \text{at } r = a, b . \tag{15}$$

On both end surfaces, stress resultants are zero:

$$\sum_{k=1}^{N} \int_{r_i}^{r_o} \bar{\sigma}_{zz} r dr = \bar{\sigma}_{zr} = \bar{\sigma}_{z\theta} = 0 \quad \text{at } z = 0, L , \tag{16}$$

where r_i and r_o are inner and outer radii, respectively, of the kth layer. The continuity conditions at each interface between two adjacent layers require continuous radial traction and continuous radial displacement at any instant as shown in Figure 1(b). Thus, when written in the transformed form, they become

$$\bar{\sigma}_{rr,o}^{(k)} - \bar{\sigma}_{rr,i}^{(k+1)} = 0 \tag{17}$$

and

$$\bar{w}_{,o}^{(k)} = \bar{w}_{,i}^{(k+1)} , \tag{18}$$

where $k = 1,\ldots,N - 1;$ and subscripts i and o denote inner and outer surfaces, respectively.

Accordingly, the formulation accounts for ply-by-ply variations of material properties and temperature change. The matrix form numerical solution procedure with parallel computing techniques resolved the complexity and time-consuming calculation procedures in Laplace transform of a multilayered composite cylinder [10].

RELAXATION OF THERMAL STRESSES

The time-dependent thermal viscoelastic behavior of a 100-layer AS-4/3502 graphite epoxy composite cylinder subjected to a temperature increase $\Delta T = 150°$ C is examined. The composite cylinder has an inner radius $a = 3.5$ in, an outer radius $b = 4.1$ in, and a thickness of each layer $h = 6.0 \times 10^{-3}$ in. Stacking sequence is given as $[0/30/60/90]_{25}$ from inside out with the 0° direction coinciding with the axis of the cylinder. The creep properties of AS-4/3502 graphite epoxy and AS-4/PEEK thermoplastic laminate with a fiber volume fraction of 0.67 were investigated at different temperatures by Kim and Hartness [11]. Figure 2 show the transverse and shear compliance of the composite at various temperatures. Increase of compliance with time due to creep behaviors of material was found at elevated temperatures. In this study, the least-squares curve fitting was used to express the creep compliances from the original AS-4/3502 data at $T = 150°$ C in a power law form

$$S_{22}(t) = \left[1.7051\,(t)^{0.1954} + 1\right] S^0_{22} \tag{19}$$

and

$$S_{66}(t) = \left[11.3076\,(t)^{0.2771} + 1\right] S^0_{66}\,, \tag{20}$$

where

$$S^0_{22} = 7.5328 \times 10^{-7}\ /\ \text{psi}$$

$$S^0_{66} = 1.3834 \times 10^{-6}\ /\ \text{psi}\ .$$

The compliance in the fiber direction, $S_{11} = 5.9 \times 10^{-8}$ / psi, and Poisson's ratios, $\nu_{12} = \nu_{13} = 0.3,\ \nu_{23} = 0.36$, are assumed to be time independent. Thermal expansion coefficients of the composites in three principal directions are $\alpha_{11} = -0.5 \times 10^{-6}$ /°C and $\alpha_{22} = \alpha_{33} = 40.0 \times 10^{-6}$ /°C, where the negative value indicates shrinkage with temperature increase.

Figures 3 and 4 show radial displacement and radial stress profiles across the thickness of the cylinder at three instants, t = 10^{-3} min, t = $10^{4.5}$ min, and t = 10^{12} min, respectively. The radial tractions and displacements satisfied the continuity conditions at the interfaces of layers at all instants. The "saw" shaped radial stress distribution is resulted from the discontinuity of material properties and various fiber orientations. The radial displacement, $w(t)$, reaches a steady state at $t = 10^{12}$ min due to a long-term creep behavior. The radial displacement of most layers reaches a constant value, except at the innermost and outermost portions of the cylinder. The free traction boundary at the surface of cylinders causes the gradients in the radial displacements. A similar phenomena is also observed in the radial stresses showing a constant level of stress distribution in the inner portion of cylinder away from boundary layers at $t = 10^{12}$ min. This long-term creep characteristic reflects the proposed power law form, (19) and (20), of the creep compliances. The hoop stress, $\sigma_{\theta\theta}(t)$ through the thickness of the cylinder are illustrated at three instants in Figures 5(A), 5(B), and 5(C), respectively. There exist two distinct values (discontinuity) of $\sigma_{\theta\theta}(t)$ across each interface of two adjacent layers due to the various fiber orientations through the thickness.

There is a trend showing the creep behavior of the hoop stress profiles from $t = 10^{-3}$ min to $t = 10^{12}$ min. The hoop stresses in 60° and 90° layers show a fairly steep gradient across the cylinder thickness initially. However, the gradient gradually disappears as time approaches to infinity.

RELAXATION OF MECHANICAL STRESSES

In the following study, a calculation was performed to simulate a composite cylinder pressed fit on a rigid steel mandrel as shown in a schamatic of Figure 6. The 0.6 inch thick composite cylinder has a layup construction of $[90/90/90/0]_{25}$, 75% in hoop and 25% in axial direction. The contact pressure is 1,000 psi at the interface of composite and mandrel, which resulting from the initial interference. The assembly is then heated and held at 93°C constantly for creep study. In modeling of the experiment, the inner radius of the composite overwrap was held constantly to allow creep and stress relaxation of the cylinder occurring at 93°C. The AS4/epoxy composite material properties were obtained from the reference report [11]. The compliance in the fiber direction is also assumed to be elastic as discussed in the previous section. Power equations were obtained by least square curve-fit technique for transverse and shear compliance with a unit of /psi as follows:

$$S_{22}(t) = 7.553 \times 10^{-7} \ (t)^{0.0175} \tag{21}$$

and

$$S_{66}(t) = 1.383 \times 10^{-6} \ (t)^{0.06} \tag{22}$$

Figure 7 and 8 illustrate the creep behavior of radial displacement and relaxation of radial stresses as a function of time, respectively. As shown in the plots, a significant loss of preload was in the composite cylinder constructed with this specific material and layup, subjected to an elevated temperature. Tests have been performed using similar basis graphite/epoxy composite materials, and about 20% of preload was lost in a few weeks at elevated temperatures. In general, the rate of loss of preload increases as the temperature increases. The stress relaxation in hoop-wound fiber layers results mainly from Poisson's effects of creep in the transverse direction since the properties is assumed to be elastic in the fiber direction. The stresses will not diminish completely due to elastic properties in the fiber direction. However, the reduction and redistribution of hoop stresses will cause a decrease of the radial compression in the cylinders in a long period. As the radial displacement approaches to a uniform state through the thickness, the radial stress will diminish completely depending upon the material properties and temperatures.

CONCLUSIONS

An analysis has been developed to study the thermal viscoelastic behavior of thick-walled laminated composite cylinders with ply-by-ply variation of anisotropic viscoelastic properties. The relaxation of stresses and creep behaviors are properly determined in a cylinder subjected to a uniform temperature change. The extreme anisotropy of viscoelastic behavior of composite results in very interesting while important long-term characteristics of cylinders, which cannot be modeled using currently available finite element codes. Creep and stress relaxation exists in the fiber direction even though the fiber dominant properties are elastic. This is mainly due to the contribution of the Poisson's effects of viscoelastic transverse and shear properties. Furthermore, the analysis is extended to study the viscoelastic behavior of composite overwrapped cylinders with built-in mechanical preloads. The relaxation of preload in the composite is critical to the service life cycle of applications such as pressure vessels and composite rotors which are designed and built with a preload to achieve desired mechanical performance.

REFERENCES

1. Ferry, J. D., Viscoelastic Properties of Polymer, John Wiley & Sons, Inc., New York, NY, 1961.
2. Muki, R., and E. Sternberg. "On Transient Thermal Stresses in Viscoelastic Materials With Temperature Dependent Properties." Journal of Applied Mechanics, pp. 193B207, 1961.
3. Schapery, R. A. "Application of Thermodynamics to Thermomechanical, Fracture, and Birefringent Phenomena in Viscoelastic Media." Journal of Applied Physics, vol. 35, no. 5, pp. 1451B1465, 1964.
4. Williams, M. L. "Structural Analysis of Viscoelastic Materials." AIAA Journal, vol. 2, no. 5, pp. 785B808, 1964.
5. Christensen, R. M. Theory of Viscoelasticity, An Introduction, 2nd edition, New York: Academic Press, Inc., 1982.
6. Hashin, Z. "Viscoelastic Behavior of Heterogeneous Media." Journal of Applied Mechanics, vol. 32, pp. 630B636, 1965.
7. Schapery, R. A. "Stress Analysis of Viscoelastic Composite Materials." Journal of Composite Materials, vol. 1, pp. 228B267, 1967.
8. Rogers, T. G., and E. H. Lee. "The Cylinder Problem in Viscoelastic Stress Analysis." Quarterly of Applied Mathematics, vol. 22, pp. 117B131, 1964.
9. Tzeng, J. T., and L. S. Chien. "A Thermal/Mechanical Model of Axially Loaded Thick-Walled Composite Cylinders." Journal of Composites Engineering, vol. 4, no. 2, pp. 219B232, 1994.
10. Chien, L. S., and J. T. Tzeng. "A Thermal Viscoelastic Analysis for Thick-walled Composite Cylinders." Journal of Composite Materials, vol. 29, no. 4, pp. 525-548, 1995.
11. Kim, R. Y., and J. T. Hartness. "Time-dependent Response of AS-4/PEEK Composite." Proceedings of the 19th International SAMPE Technical Conference, pp. 468-475, October 1987.

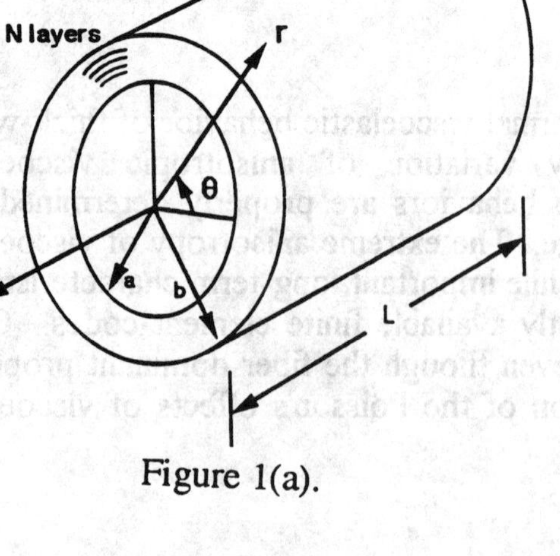

Figure 1(a).

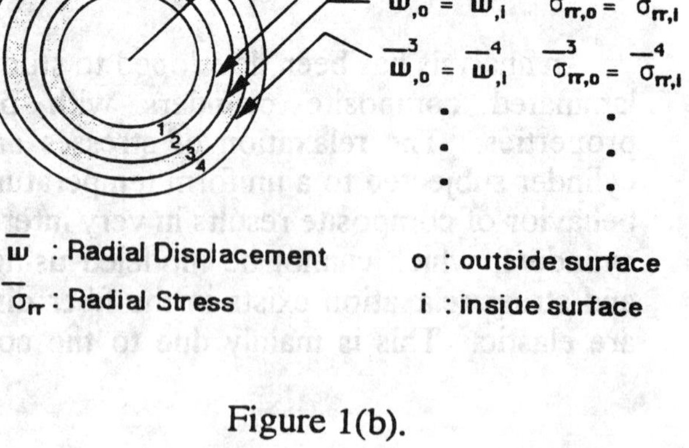

Figure 1(b).

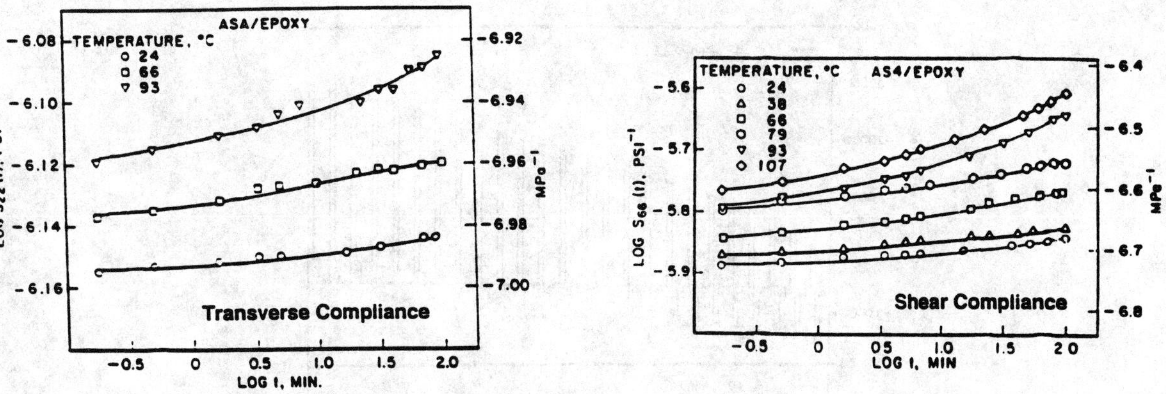

Figure 2. Time dependent transverse and shear compliances at various temperature (Kim and Hartness, 19 International SAMPE conference [11])

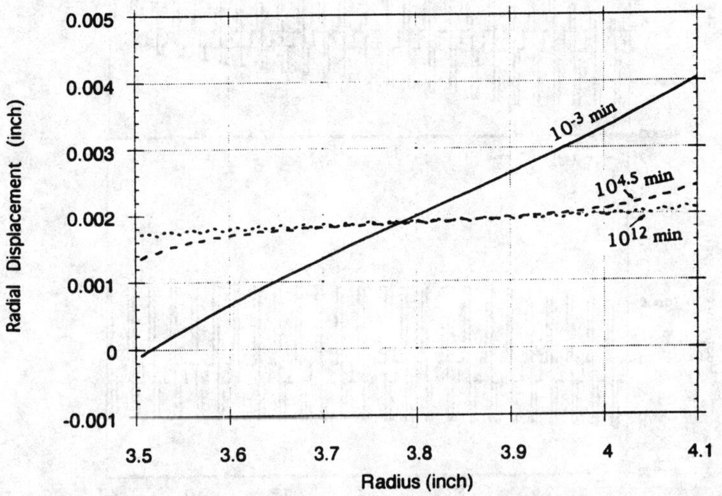

Figure 3. Radial displacement profiles across the thickness of the cylinder at three different instants: $t = 10^{-3}$ min, $t = 10^{4.5}$ min, and $t = 10^{12}$ min.

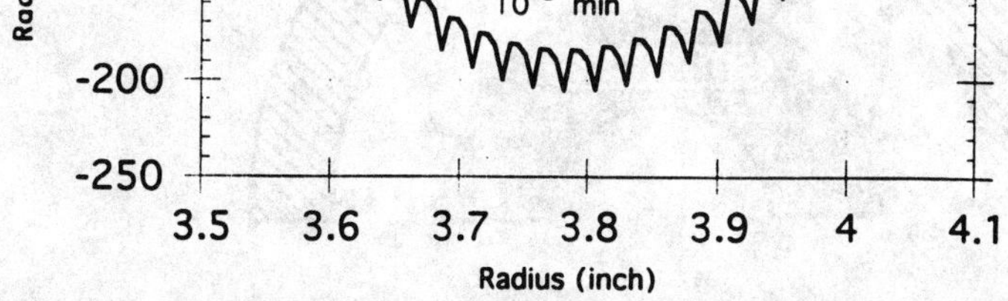

Figure 4. Radial stress profiles across the thickness of the cylinder at three different instants: $t + 10^{-3}$ min, $t = 10^{4.5}$ min, and $t = 10^{12}$ min.

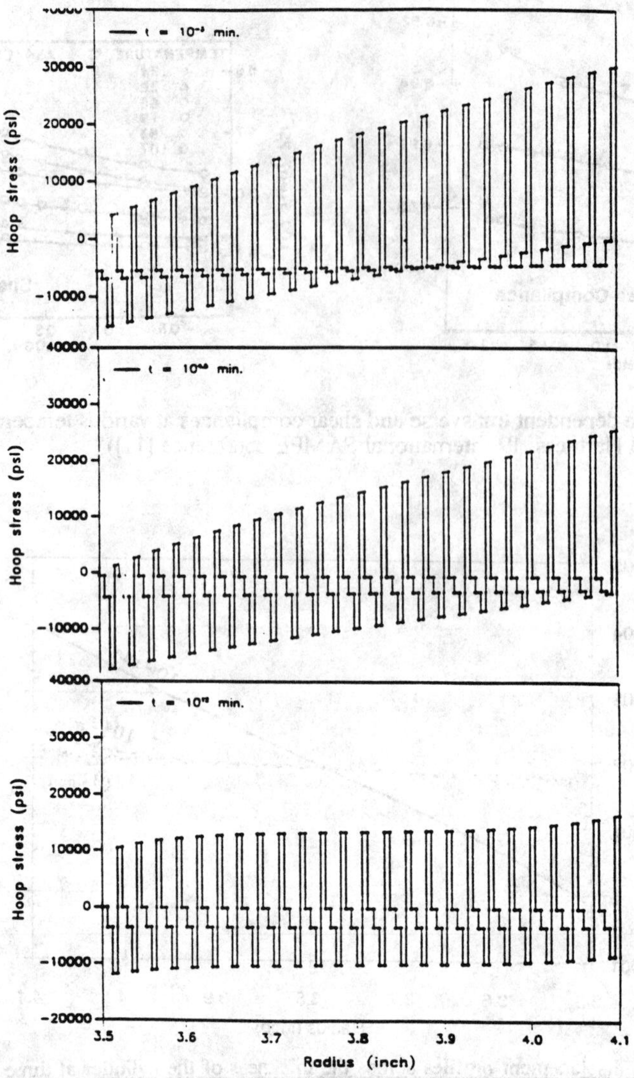

Figure 5. Hoop stress profiles across the thickness of the cylinder at three different instants: (A) $t = 10^{-3}$ min, (B) $t = 10^{4.5}$ min, and (C) $t = 10^{12}$ min.

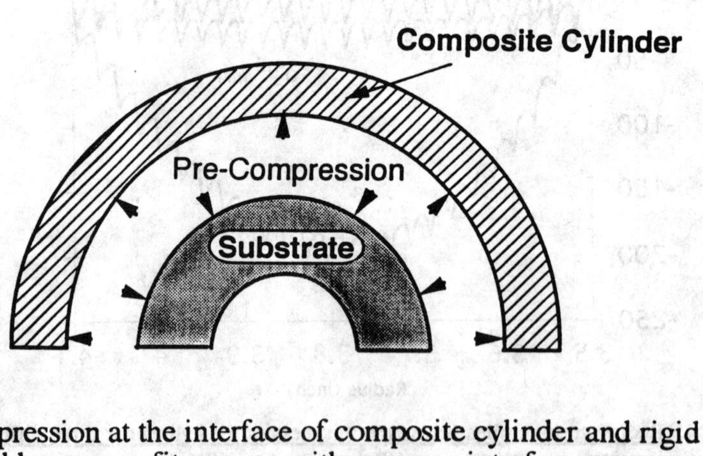

Figure 6. Pre-compression at the interface of composite cylinder and rigid mandrel can be achieved by a press-fit process with a proper interference

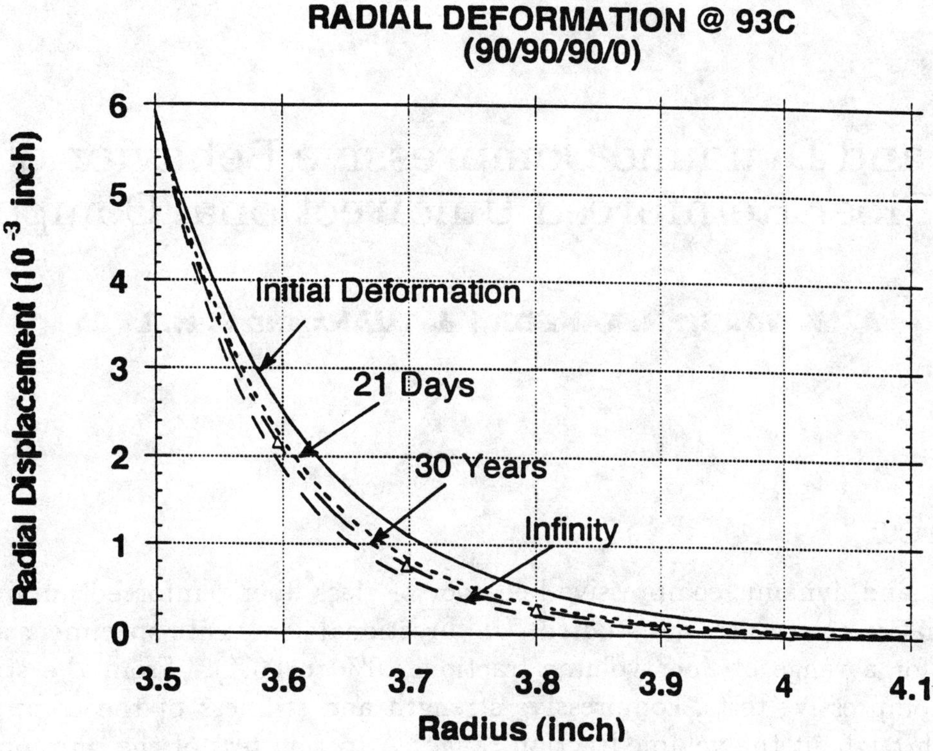

Figure 7. Creeping with time of the radial displacement through the thickness of a composite overwrapped cylinder

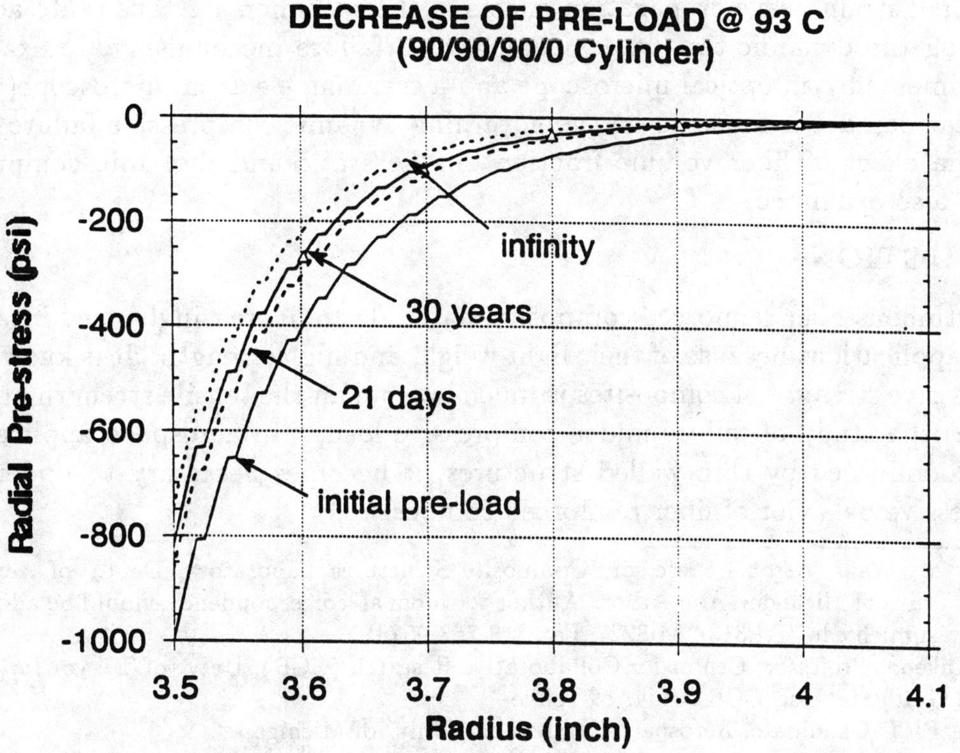

Figure 8. Relaxation with time of radial stress through the thickness of a composite overwrapped cylinder

Static and Dynamic Compressive Behavior of Glass Fiber Reinforced Unidirectional Composites

A. M. WAAS,* N. TAKEDA,† J. YUAN‡ AND S. H. LEE§

ABSTRACT

Static and dynamic compressive behavior of glass fiber reinforced unidirectional polymer matrix composites are studied. Unidirectional composite specimens are manufactured for a range of fiber volume fractions (0% to 60%). From the static and dynamic compressive tests, compressive strength and stiffness of the composite are obtained throughout the volume fraction range. A torsion test of the pure polymer is performed and in combination with the uniaxial compression test it is found that J2 incremental plasticity theory is a suitable constitutive description of the pure polymer material. Impact tests are performed via the SHPB (Split Hopkinson Pressure Bars) apparatus. In order to optimize the accuracy of dynamic behavior measured in the range of small strain, wave propagation effects in the specimen are taken into account in processing the dynamic test data. Compression failure mechanisms are examined (postexperiment) by an optical microscope and a scanning electron microscope(SEM). Matrix shear damage is found to be the governing dynamic compressive failure mechanism. The effect of fiber volume fraction on the static and dynamic compressive strength is also examined.

INTRODUCTION

Continuous fiber reinforced composite materials are increasingly used in various structural applications because of their light weight and high strength. It is known that the compressive strength of composites is much lower than the tensile strength and this necessitates the study of failure under compressive loads. In aerospace applications, which are dominated by thin walled structures, it becomes necessary to understand the compressive behavior of fiber reinforced composites.

---

*Anthony M. Waas, Assoc. Professor, Composite Structures Laboratory, Dept. of Aerospace Engineering, Univ. of Michigan, Ann Arbor. Author to whom all correspondence should be addressed. (e-mail : dcw@umich.edu, Tel:313-764-8227, Fax:313-763-0578)

†Nobuo Takeda, Professor, Center for Collaborative Research (CCR), Univ. of Tokyo, Japan

‡J. Yuan, Graduate Asst.,CCR, Univ. of Tokyo

§S.H. Lee, Ph.D Candidate, Aerospace Engineering, Univ. of Michigan

The main mechanisms of polymer matrix unidirectional composite failure under compression are elastic microbuckling, inelastic microbuckling, fiber crush and matrix failure[1]. Elastic microbuckling dominates low volume fraction situations while at higher volume fractions, the other mechanisms dictate failure. Tensile behavior is dominated by the much higher tensile strength of fibers, while geometrical instability of misaligned fibers, in a polymer matrix that has a progressively decreasing shear stiffness, is thought to be the main cause for the lowered compressive strength of the composite[2,3]. Much of this understanding of the compressive failure mechanism has been obtained on the basis of static experimental data. Relatively little work has been done to investigate the dynamic compressive behavior of composites[4,5].

In this study, glass fiber reinforced unidirectional polymer matrix composite specimens are subjected to static and dynamic compression loading. The specimens are of a circular cylinder shape with diameter of approximately 6.4 mm. Continuous E-glass fibers of 24.1 μm diameter (Certainteed R099-625 E-glass fiber) are contained in the correct amount of volume fraction in an epoxy vinyl ester resin (Dow Derakane 411-C50). The volume fractions of the specimens range from 10% to 60%. Specimens of the pure matrix material without fibers are also manufactured under the same process.

From the static and dynamic compression tests, compressive strength, stiffness and strains at failure are measured throughout the entire volume fraction range. For the pure matrix specimens, stress-strain data are obtained from static compressive tests and shear stress-shear strain data are obtained from a static torsion test. The unidirectional stress-strain data and the torsion test stress-strain data are examined in conjunction with J2 deformation theory and incremental theory of plasticity to establish an inelastic constitutive description for the pure matrix[6].

Dynamic compression tests are performed using the SHPB (Split Hopkinson Pressure Bars) test. Because the measurement of the dynamic behavior by the SHPB test is more accurate in the range of large strain than in the range of small strain[7], a new data processing method which takes into account wave propagation effects in the specimen is proposed in order to optimize the accuracy of the measured dynamic behavior. To check this procedure, the specimen strain is also measured directly from a strain gage mounted on the specimen surface. The failed specimens are examined using an optical microscope and a scanning electron microscope (SEM). The compression failure of this unidirectional composite is found to be dictated by matrix shear damage and fiber/matrix splitting.

EXPERIMENTS

Glass fiber reinforced vinyl ester unidirectional composite specimens are prepared with different fiber volume fractions. The fiber volume fractions range from 10% - 60%, in increments of 10%. The specimens are subjected to static and dynamic compression tests. Additionally, a torsion test of the pure matrix material is also performed.

SPECIMEN PREPARATION

The specimens consist of continuous E-glass fibers of 24.1 μm diameter (Certainteed R099-625 E-glass fiber) held in a vinyl ester resin (Dow Derakane 411-C50).

The Young's modulus of the E-glass fiber is approximately 72 GPa[8,9].

To make the specimens, an exact amount of the continuous E-glass fibers is put inside a glass tube with an internal diameter of 6.8 mm and 300 mm length. The vinyl ester resin, mixed with 1% weight of initiator (Benzoyl Peroxide), is injected into the glass tube under pressure (70 - 140 Pa) in an in-house fabricated set up. The resin is circulated in the glass tube so that all the fibers are adequately wetted. This results in a void free composite sample. After the tubes are filled, they are placed in a furnace and cured under 80°C for 2 hours. Subsequently, the specimens are cut by a diamond saw, which results in the cutting section to be flat and smooth.

STATIC COMPRESSION TESTS AND STATIC TORSION TESTS

Static compression tests are performed in a hydraulically actuated automated loading frame. The specimens of length 63.5 mm are fit into two separate holders which have a 25.4 mm length and a larger than 6.4 mm circular opening. The gage length of the specimen is 12.7 mm. When the specimens are compressed, the holders are restricted by side walls from lateral movements. The tests are performed under displacement control with a head speed of 0.0381 mm/sec. Loads are measured by a load cell at the top of the specimen and back to back strain gages are attached at the center of the specimen in the longitudinal direction to measure the compressive strain and any bending strain. A total of 24 specimens are tested (3 each of 0%, 10%, 20%, 30% volume fractions, 4 each of 40%, 50%, 60% volume fractions). From the test data, the compressive stress (σ) - strain (ϵ) curves are obtained.

A torsion test of the pure resin is also performed. This test is done to obtain the shear stiffness of the pure resin. A pure resin specimen is held by two grips at each end of the specimen. While one end is fixed, the other one is rotated by a small measurable angle. A strain gage is attached at the center of the specimen at 45° with respect to the axial direction of the specimen. The shear stress (τ) - shear strain (γ) curve of the pure resin is obtained from this test.

DYNAMIC COMPRESSION TESTS

A SHPB (Split Hopkinson Pressure Bars) system is used for the dynamic compression test. A schematic diagram of the SHPB is shown in Figure 1. The input and output bars are fabricated from hardened SUS-304 steel with 16 mm diameter and 1000 mm length. Specimens have 4.4 - 5.5 mm length. A total of 22 specimens are tested (2 each of 30%, 60% Vf, 3 each of 10%, 20%, 50% Vf, 4 each of 40% Vf, 5 of the pure resin). For the pure matrix, two different strain rates ($\bar{\dot{\epsilon}} = 670$ /sec, $\bar{\dot{\epsilon}} = 1510$ /sec) are used for the tests.

In the conventional data processing of the SHPB test, the incident and the transmitted waves, $\epsilon_I(t)$ and $\epsilon_T(t)$, are shifted by times of T_1 and T_2 (see Figure 2). After time shifting, the stress and strain of the specimen is calculated by the following formulae :

$$\sigma_S(t) \quad = \quad \frac{A_B}{A_S}\, E_B\, \epsilon_T(t)$$

$$\epsilon_S(t) = 2\,\frac{C_0}{l_s}\int_0^t [\epsilon_I(\tau) - \epsilon_T(\tau)]\,d\tau, \qquad (1)$$

where the subscript S and B represent the specimen and the bar, respectively. A, l and E represent the cross sectional area, the length and the Young's modulus, respectively. For impact tests of composites in the small strain range (less than 2%), the shift time of $T_1 + T_2$ has a great influence on the final results. The shift time is dependent on the nature of wave propagation in the bars and the specimen as well as the connecting condition between the specimen and the bars. To avoid any artificial effect, we propose the following supplemental method for determining the value of shift time.

For the SHPB test of a linear elastic specimen, the transmitted stress wave, $\sigma_T(t)$, can be determined by a simple one dimensional elastic wave propagation analysis as

$$\begin{aligned} \sigma_T(t) &= \left(\frac{1-F}{1+F}\right)^2 \sigma_T\left(t - \frac{2l_s}{c_s}\right) + \frac{4F}{(1+F)^2}\,\frac{A_B}{A_S}\,\sigma_I\left(t - \frac{l_s}{c_s}\right) \\ F &= \frac{A_B\,\rho_B\,c_b}{A_S\,\rho_S\,c_S} \end{aligned} \qquad (2)$$

In the actual SHPB test, the early stages of transmitted wave propagation is determined by the elastic behavior of the specimen, hence, equation (2) can be used to determine the shift time by the following operation (Figure 2). Presume a fictitious wave velocity for the specimen and then calculate the transmitted stress wave according to equation (2). Comparing the predicted transmitted wave with the measured one while using a suitable shift time of $T_1 + T_2$, a good fit can be obtained when the fictitious wave velocity of the specimen is quite near the actual one. The shift time needed to meet this condition is determined and used for subsequent calculation. After obtaining the shift time, the stress-strain curve is calculated using equation (1). A comparison of the strain-time history between that directly measured from the specimen and that predicted by equation (1) is found to be in very good agreement. Within the effective measuring range of the strain gage (about 1 % in the test), the above correlation lends confidence to the proposed shift time determination method. More details about this procedure are given in [4].

RESULTS

From the compression tests, $\sigma - \epsilon$ curves of the glass fiber reinforced vinyl ester resin composites for various fiber volume fractions are obtained. Important mechanical properties of the composites (compressive strength, longitudinal stiffness, failure strain) can be determined from the data. The $\tau - \gamma$ curve obtained from the torsion test provides information about the shear stiffness of the vinyl ester resin as a function of shear stress. These data are brought to conformity via the use of J2 plasticity theory[6].

COMPRESSION TEST RESULTS

Figure 3 shows typical $\sigma - \epsilon$ curves obtained from the static tests throughout the range of fiber volume fraction. From these curves, the compressive strength and the strains corresponding to this maximum stress as well as the axial stiffnesses are obtained. Compressive strength and corresponding strains are listed in Table 1. The strains for the pure resin specimens are not the strains at the maximum stresses, because the strain gages failed themselves before reaching the maximum stresses. Young's modulus of the pure resin material is measured as 3.69 GPa. Using this measurement the stiffness of the composite material in the axial direction E_{11} can be calculated via the rule of mixtures (E-glass fiber : E=72.4 Gpa). A comparison of the test results and the calculation are provided in Table 2.

Typical stress-strain curves obtained from the SHPB test are shown in Figure 4. From the $\sigma - \epsilon$ curves, dynamic compressive strength, strains and stiffnesses are obtained. Strain rates are also obtained from the $\epsilon - t$ data that is captured via an oscilloscope. Measured values are listed in Table 3 and Table 4. For the 60% Vf specimens, the impact forces are not large enough to cause compressive failure and this is to be noted.

In Figure 5, static and dynamic compressive strength as a function of fiber volume fraction is displayed. Both the static and dynamic compressive strength is seen to first increase with increasing fiber volume fraction after which a plateau region is experienced and further increase occurs beyond 40% Vf. The dynamic compressive strength follows a similar trend and is larger than the static case by a factor of 1.7. The static $\sigma - \epsilon$ curves (Figure 3) show that for all 6 fiber volume fractions the composite $\sigma - \epsilon$ curve is linear upto failure. On the other hand, in the dynamic compressive test, the $\sigma - \epsilon$ curves (Figure 4) show that the maximum stress points and fracture points are different. This shows that the mechanism of failure in the static case persists for a very short duration, too small to be captured by the data collection rate of the static tests. For low fiber volume fractions (10%, 20%, 30%), it can be seen that the dynamic $\sigma - \epsilon$ relation is non-linear before the maximum compressive stress is reached. Furthermore, the maximum compressive stresses are almost the same for this range of fiber volume fraction. For higher fiber volume fractions (40%, 50%, 60%), the dynamic $\sigma - \epsilon$ relation is linear up to the maximum stress points. Notice that there is a "jump" in the maximum stress value between the 30% and 40% fiber volume fraction specimens.

The static and dynamic strains corresponding to the maximum stress are tabulated in Tables 1 and 3 respectively. They are found to decrease as the fiber volume fraction increases. The dynamic strain to failure is larger than the static case by a factor of 1.7. For high volume fractions, both the static and dynamic cases tend to asymptote to a constant value, independent of fiber volume fraction.

In Table 2, the axial stiffnesses from the static and dynamic tests and the rule of mixtures calculation are shown. Static stiffnesses are in the range of 0.83 - 1.13 of the dynamic stiffness value. The rule of mixtures results are seen to be in very good agreement with the measured values.

The specimens are inspected under an optical microscope after the impact tests. From the viewing direction of the impact surface, the microdamage region looks black and is found to be concentrated in fiber-rich area and along the fiber direction. The microdamage region always has a distinctive border on one side and a vague border in the other side, moreover, part of the distinct border becomes a microcrack. Some specimens have little splitting along the fiber direction near the circular boundary of the specimen. SEM observation of the splitting region indicates that the matrix between the fibers have sustained inelastic shear failure. What is clear from the post experiment examination is that the mechanism of failure in the dynamic case includes inelastic matrix shear damage, from which we can conjecture splitting of the specimen.

TORSION TEST RESULTS

The $\sigma - \epsilon$ curve of the vinyl ester resin under compression and the $\tau - \gamma$ curve of the vinyl ester resin, are both found to be non-linear. The two curves are examined in conjunction with J2 plasticity theory. The $\sigma - \epsilon$ data is fitted with the Ramberg-Osgood relation,

$$\epsilon = \frac{\sigma}{E} + (\frac{\sigma}{B})^n.$$

For such a proposal, it is found that $E = 3.6GPa$, $B = 0.136GPa$, and $n = 8$.

From J2 deformation theory, the following $\tau - \gamma$ relation is obtained[6].

$$\gamma = \frac{\tau}{G} + (\frac{\tau}{A})^n$$

where $G = 1.3GPa$, $A = 0.073GPa$, and $n = 8$. It is found that the deformation theory underestimates the shear stiffness beyond $\gamma \simeq 0.0325$.

J2 incremental plasticity theory is checked with the use of ABAQUS, a non-linear finite element commercial code. A cylinder of 6.35 mm diameter and 40.64 mm length is modeled by 15-noded solid elements. The $\sigma - \epsilon$ data from the compression test is used as input for the plastic material property data card. While one end of the cylinder is fixed, the other end is rotated by a prescribed angle as is done in the test. The ABAQUS static analysis, which includes nonlinear geometry calculates the torsional moment at the fixed end. Two cases are considered in this analysis. One in which the cylinder is under pure torsion with no restraint on the axial deformation. The other restricts the cylinder from axial deformation. In the second case, the cylinder shows a slightly higher stiffness in the non-linear region due to the axial constraint that induces axial compression. Comparison of the various results shows that the J2 incremental theory provides a reasonable model for the vinyl ester resin.

CONCLUDING REMARKS

In this experimental study, static and dynamic compression tests are performed with glass fiber reinforced vinyl ester resin unidirectional composites to study the effect of fiber volume fraction and strain rate on compressive failure. A new data processing method for obtaining the $\sigma - \epsilon$ relation in SHPB test is proposed. This method is found to determine relatively accurate dynamic behavior in the range of small strain for the SHPB test.

Dynamic compressive strength and equivalent strain are about 1.7 times higher than the static compressive test over the entire range of fiber volume fractions examined.

Dynamic stiffnesses are similar to static stiffnesses and the measured stiffnesses are in the range of 0.84 - 1.03 of the calculated results based on the simple rule of mixtures.

It is found that failure strain approaches a constant value for higher volume fraction, 0.034 for dynamic compression and 0.016 for static compression. Matrix shear damage is found in the failed specimen upon observation under a SEM.

The compressive strength of the composite is clearly a function of fiber volume fraction and this dependence was found to persist in a similar way for both static and dynamic data, with the ratio of dynamic to static strength equal to 1.7. In the strain rate regime examined, the compressive $\sigma - \epsilon$ behavior of the pure resin shows no strain rate dependence and is adequately described by J2 incremental theory of plasticity.

References

[1] Jelf, P. M., and Fleck, N. A., *J. Comp. Mater.* Vol. 26, No. 18, pp. 2706-2726, 1992.

[2] Schultheisz, C. R., and Waas, A. M., "Compressive Failure of Composites, Part 1 : Testing and Micromechanical Theories" *Prog. Aerospace Sci.*, Vol. 32, pp 1-42, 1996.

[3] Waas, A. M., and Schultheisz, C. R.,, "Compressive Failure of Composites, Part 2 : Experimental Studies" *Prog. Aerospace Sci.*, Vol. 32, pp 43-78 1996.

[4] Yuan, J, Hiramatsu, M., Takeda, N. and Waas, A. M., "Influence of Fiber Volume Fraction on the Dynamic Compressive Behavior of Unidirectional Composite" National Japan Symposium on Strength of Aerospace Structures, July 1997, Osaka, Japan.

[5] Lankford, J., "Composites" Part A 28A, pp 215-222, 1997.

[6] Bisplinghoff, R. L., and Mar, J. W., "Statics of Deformable Solids" *Addison-Wesley*, pp. 206-227.

[7] Zhao, H., and Gray, G., *Int. J. Solids Structures*, Vol. 33, No. 23, pp 3363-3375, 1996.

[8] Jones, R. M., "Mechanics of Composite Materials" *Scripta book comp.*, p 3.

[9] Gibson, R. F., "Principles of Composite Material Mechanics" *McGraw-Hill*, p 8.

Table 1 : Maximum stress and strain results of static tests

Vf (%)	Stress[MPa]/Strain				
	No.1	No.2	No.3	No.4	Average
0.0	92.8/0.0465	82.5/0.0482	83.4/0.0518		86.2/0.0488
10.0	275.2/0.0231	339.8/0.0337	329.9/0.0333		315.0/0.03
20.0	366.5/0.0304	355.7/0.0207	464.8/0.0288		395.7/0.0266
30.0	499.0/0.0268	567.7/0.0251	486.0/0.0229		517.6/0.0250
40.0	374.0/0.0138	654.0/0.0220	451.9/0.0163	458.2/0.0178	484.5/0.0175
50.0	376.2/0.0117	627.2/0.0185	684.7/0.0184	516.6/0.0164	551.2/0.0162
60.0	658.8/0.0127	863.1/0.0217	545.3/0.0137	599.2/0.0149	666.6/0.0157

Table 2 : Stiffness results of static tests

Vf (%)	Axial stiffness E_{11} [MPa]					
	No.1	No.2	No.3	No.4	Average	Theory
0.0	4044.4	3654.2	3380.4		3693.0	
10.0	11912.2	11395.0	10705.6		11337.6	10562.8
20.0	13651.0	16973.8	16088.6		15571.1	17432.6
30.0	19293.6	23581.0	22668.5		21847.7	24302.4
40.0	28359.6	32510.1	33652.1	26817.0	30334.7	31172.2
50.0	32582.7	35517.8	32279.8	32158.6	33134.7	38042.0
60.0	46071.4	40120.3	52143.5	44373.2	45677.1	44911.8

Table 3 : Maximum stress and strain results of dynamic tests

Vf (%)	Stress[MPa]/Strain				
	No.1	No.2	No.3	No.4	Average
0.0	235/0.0720	233/0.0690			234/0.0700
0.0	000/0.0790	000/0.0820	000/0.0800		249/0.0800
10.0	549/0.0573	571/0.0612	529/0.0594		549/0.0593
20.0	634/0.0430	602/0.0422	604/0.0430		613/0.0427
30.0	628/0.0338	612/0.0359			620/0.0348
40.0	835/0.0325	837/0.0329	801/0.0350	839/0.0318	828/0.0330
50.0	946/0.0343	890/0.0318	894/0.0370		910/0.0344
60.0	893/0.0254	919/0.0242			906/0.0248

Vf	Stiffness[GPa]/Strain rate[/sec]				
(%)	No.1	No.2	No.3	No.4	Average
0.0	4.02/649	4.25/685			4.13/670
0.0	4.39/1420	4.08/1620	4.18/1490		4.22/1510
10.0	11.7/1079	10.5/1453	10.7/1399		10.9/1310
20.0	16.3/1055	16.2/1089	15.6/1265		16.0/1130
30.0	21.5/959	21.8/1054			21.7/1010
40.0	29.1/880	29.6/860	28.9/878	30.1/854	29.4/870
50.0	31.9/911	31.9/852	32.0/858		31.9/870
60.0	43.6/687	43.5/576			43.5/630

Table 4 : Stiffness and strain rate results of dynamic tests

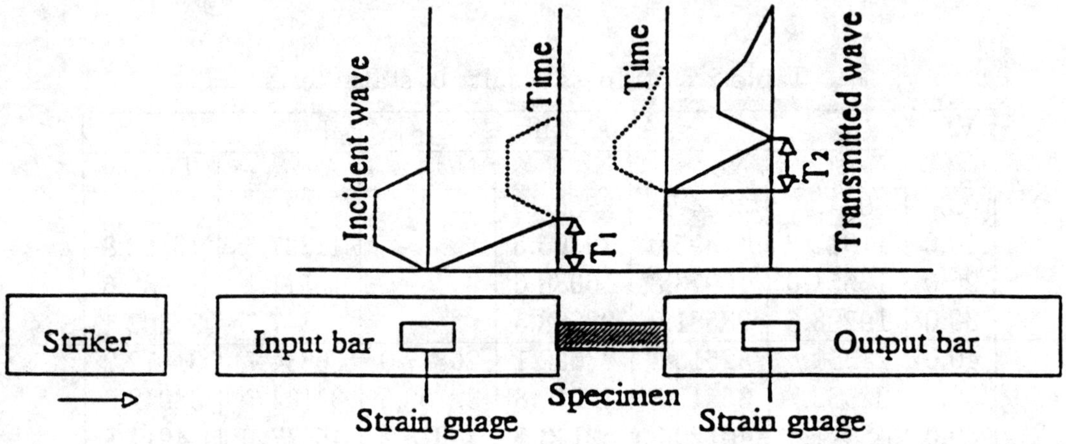

Figure 1: Schematic of SHPB test

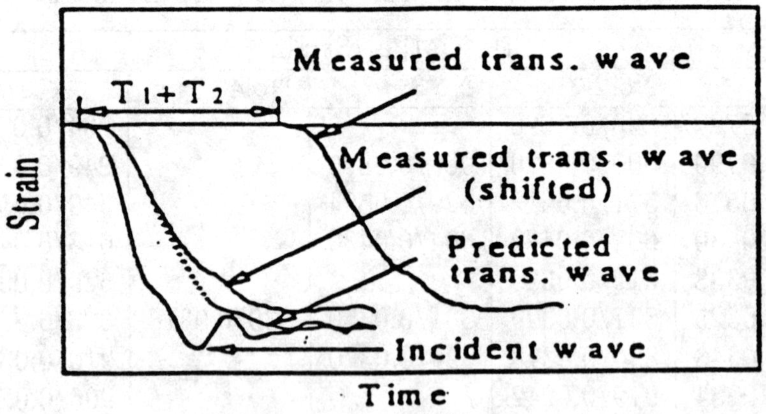

Figure 2: Method for determining shift time

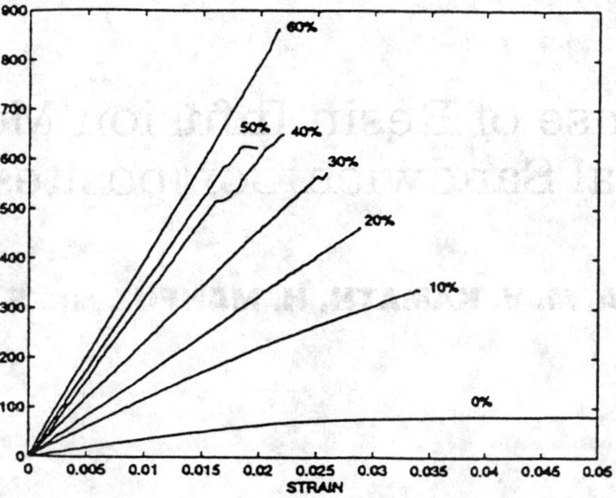

Figure 3: Typical $\sigma - \epsilon$ curves of static compression test

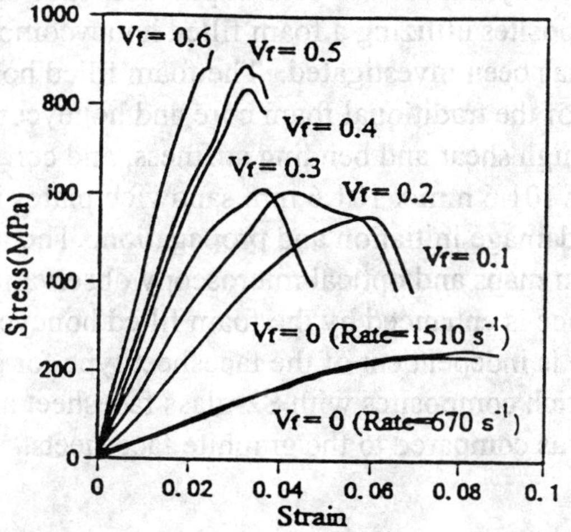

Figure 4: Typical $\sigma - \epsilon$ curves of dynamic compression tests

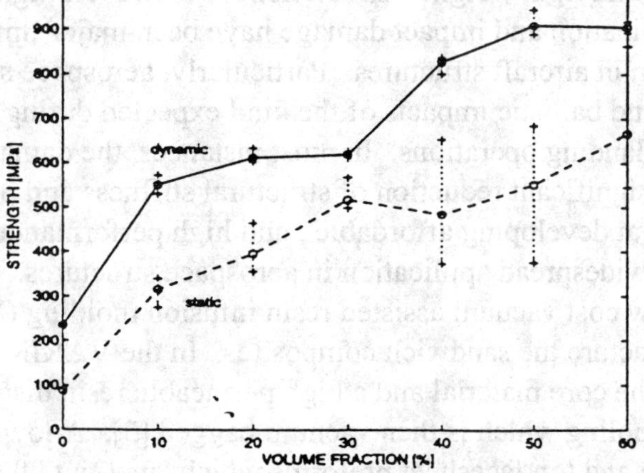

Figure 5: Compressive strength of static/dynamic tests

Impact Response of Resin Infusion Molded Unconventional Sandwich Composites

U. K. VAIDYA, M. V. KAMATH, H. MAHFUZ AND S. JEELANI

ABSTRACT

In this study the low-velocity impact and post-impact response of low-cost resin infusion molded sandwich composites utilizing a foam filled honeycomb core with graphite and S2-glass fabric facesheets has been investigated. The foam filled honeycomb core provides combined advantages of the traditional foam core and honeycomb sandwich composites in that it possesses high shear and bending stiffness, and cell wall stability. The low velocity impact response of 101.6 mm x 101.6 mm sandwich plates is studies at five energy levels representative of damage initiation and propagation. The low velocity damage is correlated to ultrasonic C-scan maps and optical microscopy observations. The results indicate that the damage tolerance is enhanced by the foam filled honeycomb core and that load required to initiate damage is independent of the facesheet type for any specific core/ facesheet thickness. The sandwich composites with S2-glass facesheet are found to be possess more damage tolerance as compared to the graphite facesheets.

INTRODUCTION

A number of advanced aircraft components utilize composite sandwich constructions due their design flexibility and light weight characteristics. However, high processing costs and susceptibility to delamination and impact damage have been major limiting factors in widespread implementation in aircraft structures. Particularly, aerospace structures are subjected to low velocity and ballistic impacts of the kind expected during aircraft maintenance, take-off and landing operations. In most instances, the damage is externally visible, however it causes significant reduction of structural stiffness and performance [1-2]. The current work is aimed at developing affordable , and high performance sandwich constructions that possess widespread application in aerospace structures. The cost issue is addressed by utilizing a low cost vacuum assisted resin infusion molding (VARIM) co-injection process to manufacture the sandwich composites. In the VARIM process, dry reinforcement along with the core material and a high permeable resin distributed layer is oriented in a single sided tooling which is then vacuum bagged [3]. A low viscosity resin is infused through the bottom and top facesheet preforms which "wet-out" the reinforcement and polymerizes. For sandwich composites, the process is particularly attractive as both the top and bottom facesheets can be co-infused simultaneously.

U.K.Vaidya, M.V.Kamath, H.Mahfuz and S.Jeelani, Center for Advanced Composites, Tuskegee University, Tuskegee, AL 36088

Sandwich composites with unique core constructions such as foam with honeycomb cells and open Z-fiber cores are of interest for providing open cell as well as enhanced damage tolerance benefits. There are few reported results available for these relatively new core types [4-6]. The work presented here is on sandwich composites that posses traditional honeycomb cell construction [7] with filled foam cells such as shown in Fig.1. The foam filled honeycomb core offers combined advantages of the foam and the honeycomb core because the cell walls provide the strengths and the foam provides a large surface area [8]. Honeycomb cell walls are further reinforced by the foam fill. The increased surface area allows stress forces to dissipate a over a greater area than offered by honeycomb alone. This core construction has the potential to absorb greater impacts by transmitting forces to adjoining cells. The effect is greater resistance to shear force perpendicular to the sandwich (breaking), increased moment (less bending) and better dampening of shock waves along the surface (less vibrations). When used in sandwich construction along with woven fabric reinforcement, they possess the capability to further enhance impact energy dissipation. Of particular importance is that the damage tolerance is designed into the sandwich composite by using a low-cost process.

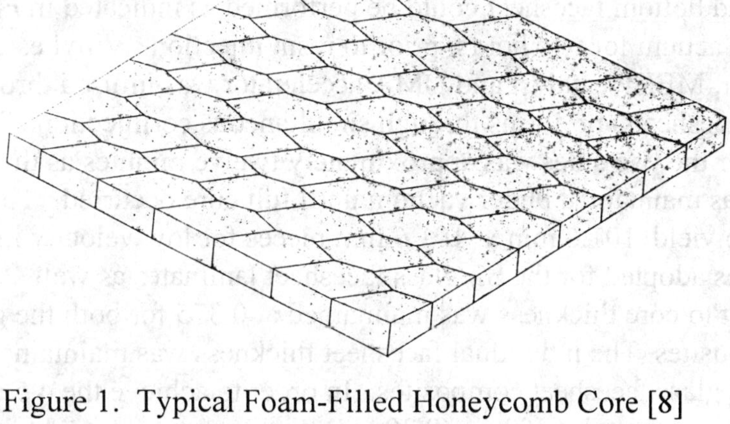

Figure 1. Typical Foam-Filled Honeycomb Core [8]

Two fabric facesheet types including woven plain weave graphite and woven 2x2 twill weave S2-glass and a foam filled honeycomb core have been considered. Low velocity impact studies have been conducted at five energy levels indicative of damage initiation and progression. The impact test results have been inspected in light of optical microscopy and ultrasonic nondestructive evaluation. Compression-After-Impact (CAI) performance has been studied to determine residual properties of the sandwich composites following low-velocity testing. Acoustic emission testing conducted during the CAI tests have also been presented.

FABRICATION OF SANDWICH COMPOSITES

Foam filled honeycomb core; Style 20, 2050R, 8 lbs/ft^3, 12 mm thick (Supplier : WESKOR) with phenolic reinforced Kraft paper honeycomb filled with polyurethane foam. Two layers of plain weave graphite fabric were bonded to the core using a microballoon mixed vinyl ester 350 resin system in a compression molding machine. After pre-compression, the panel was laid up in a single sided aluminum tooling between fourteen layers each (top and bottom) of plain weave graphite fabric preform, a layer of porous teflon and a highly permeable membrane and bagged in a vacuum bag film. The resin injection lines

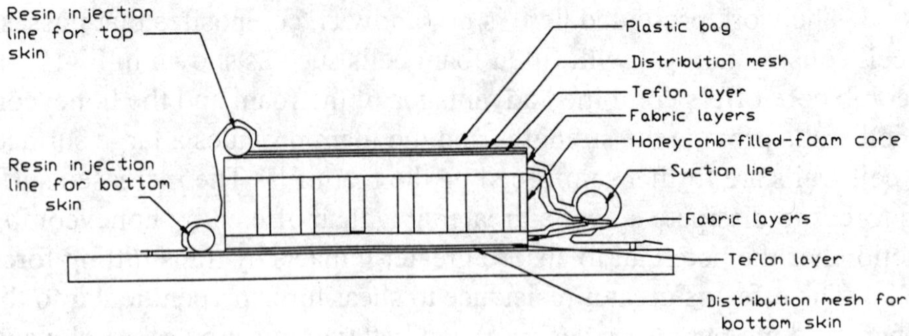

Figure 2. Schematic of the Resin Infusion Molding Process for Fabricating the Sandwich Composites

were inserted both at the top and bottom facesheet preforms so that simultaneous resin infusion of the top and bottom facesheet could be performed as indicated in Fig.2. The part was debulked under vacuum for two hours, prior to resin injection. Vinyl ester 350 resin with CoNap promoter, MEKP catalyst and DMA accelerator was infused through the top and bottom facesheets (facesheets) preforms through simultaneous co-injection. The in-plane and transverse fill time for the two sides was approximately twelve minutes as the resin wets the preform. The part was maintained under vacuum until full cure occurred. The resulting panel was trimmed to yield 101.6 mm x 101.6 mm pieces for low velocity impact testing. Similar procedure was adopted for the S2-glass facesheet laminates as well. The ratio between the facesheet to core thickness was maintained at 0.375 for both the graphite and glass sandwich composites. The individual facesheet thickness was maintained at 4.5 mm in both the graphite and glass facesheet composites. In order to achieve the 4.5 mm facesheet thickness, sixteen layers of graphite fabric (0.30 mm thick) were comparable to eight layers of S2-glass (0.58 mm thick) fabric. Hence, the graphite facesheets were constructed from sixteen plies, while the S2-glass facesheets were constructed from eight plies.

EXPERIMENTAL

A 8210 Dynatup low velocity impact tester equipped with a 1600 kgs load cell was used in conducting the impact tests. The 101.6 mm x 101.6 mm plate sample was placed in the fixture and clamped from all sides using pneumatic actuation. A few dummy tests were run to establish the energy levels at which the samples could be impacted. It was determined that the threshold initiation energy should exceed 8-10 J. The samples were then impacted at 11, 20, 28, 33 and 40 J energy levels. The force-time and load-energy curves were obtained using the software accompanying the Dynatup instrument.

Ultrasonic C-scan testing was performed in a pulse-echo immersion mode on a Testech system using a 2.25 MHz transducer. The top and bottom facesheets had to be individually scanned, as the ultrasonic attenuation was excessively high in the core. Pre-impact and post-impact C-scans were obtained for all the specimens. Microstructural studies were performed using a optical microscope.

RESULTS AND DISCUSSION

Low Velocity Impact Testing

Table 1 represents the nomenclature adopted in identifying the impact energy levels. Typical force-time curves and load-energy curves for the graphite facesheet composite samples at the indicated energy levels are shown in Figs. 3 and 4. For the 4.5 mm facesheet thickness graphite samples, the load at which the damage initiation occurs was found to be 5800 N. The damage initiation load is seen to be independent of the energy levels at which the samples were impacted. A load drop was obtained at 5800 N, which corresponds to the damage initiated in the top facesheet accompanied by local cell wall crushing and localized debonding of the cell/facesheet, beyond which energy continues to be absorbed by the specimen. Figures 5 and 6 represent two predominant damage mechanisms at the facesheet and the core. Figure 5 shows the fiber breakage, localized delamination at the location of the impact in the facesheet and localized wrinkling of a single core cell wall just below the impact point. At lower impact energy levels; A-C these phenomena are limited to the top facesheet only. Figure 6 represents that neighboring cells around the impact location undergo cell wall buckling, and core crushing. The foam cells appear to expand in dimension following the impact event and are seen as large black spots in the foam core in Fig.6. Additional damage (delamination and matrix cracking) is indicated by the further load drops in the force-time curves following the initiation of damage. The damage is concentrated at the core and occurs by cell wall buckling, interfacial debonding between the cell wall and the foam and through cell wall expansion due to the impact event. Herup [2] reported similar observations for conventional honeycomb composites. Although the overall nature of the damage remains similar, in this work, the damage initiation energy is much higher when the cells are reinforced with foam. As the energy increases from 11-40 J (impact velocity 1.3-2.4 m/s), the pulse width of the force-time curve increases from 5.8 msec to 9.5 msec. The time taken by the striker to rebound is longer as the impact energy increases. At energy levels of D and E, in addition to localized delamination in the top facesheet, and fiber breakage in the pockets of the weave, bottom facesheet delamination occurs. The top facesheet delamination is more extensive, and as can be seen in Fig. 7, multiples delamination span through the weave pockets of the facesheet. The bottom facesheet failure does not exhibit delaminations as in the top facesheet, but the failure is primarily by debonding of the core-facesheet, and cracking at the backside of the facesheet as indicated in Fig.8. The bottom facesheet failure is identified as a second load step in the force-time history at 2.5 msec, and is only characteristic for high energy levels of D and E. Samples A, B, and C did not exhibit bottom facesheet damage as evidenced through the force-time, load energy-history and corresponding C-scans. It appears that once the sandwich plate goes into bending under the impact load, the cells at the bottom facesheet debond at distributed locations, hence the damage zone is greater on the bottom facesheet, while it is more concentrated on the top facesheet.

Table 1: Impact Energy Levels Adopted in Testing the Sandwich Composites

Nomenclature	Impact Energy (J)
A	11
B	20
C	28
D	33
E	40

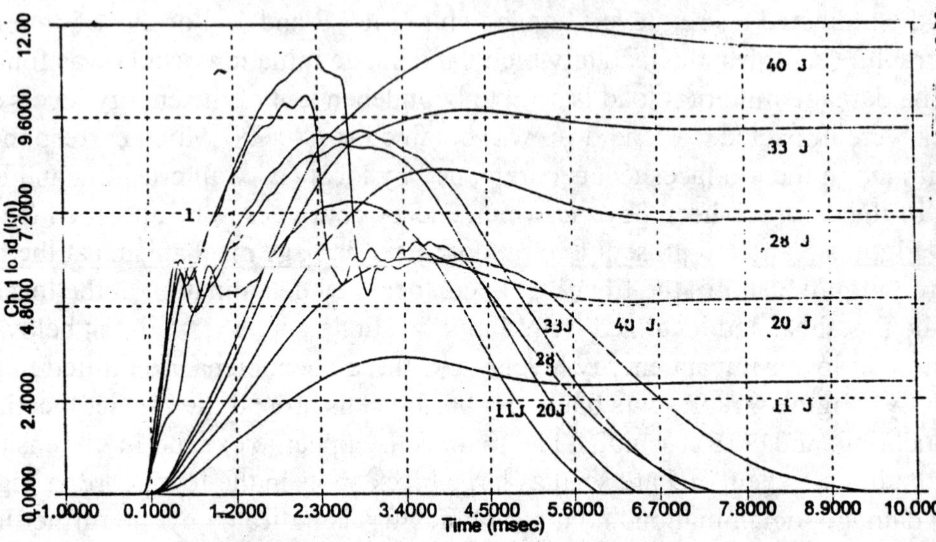

Figure 3. Typical force-time curves 16-ply woven graphite/vinyl ester foam filled honeycomb core sandwich composites

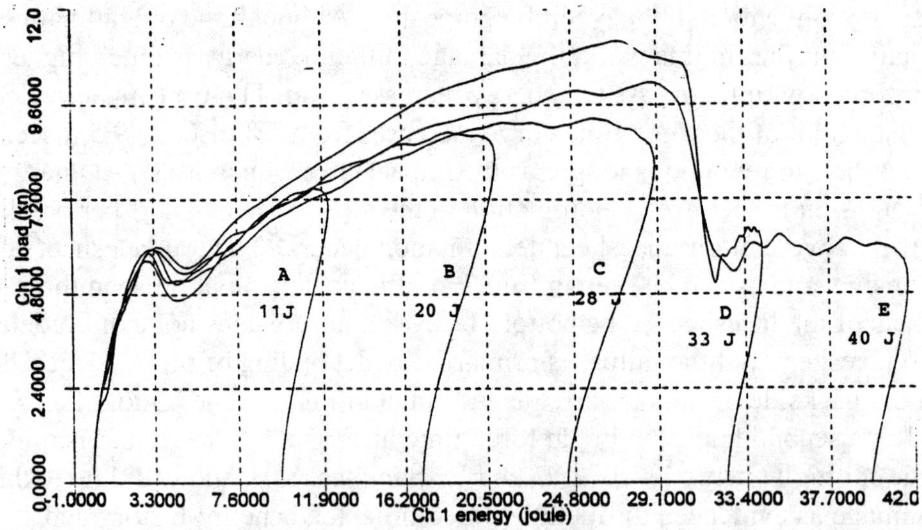

Figure 4. Typical load-energy curves 16-ply woven graphite/vinyl ester foam filled honeycomb core sandwich composites

Figure 5. Top Facesheet and Core Damage for Impact *Level A* (16-ply graphite /vinyl ester facesheets)

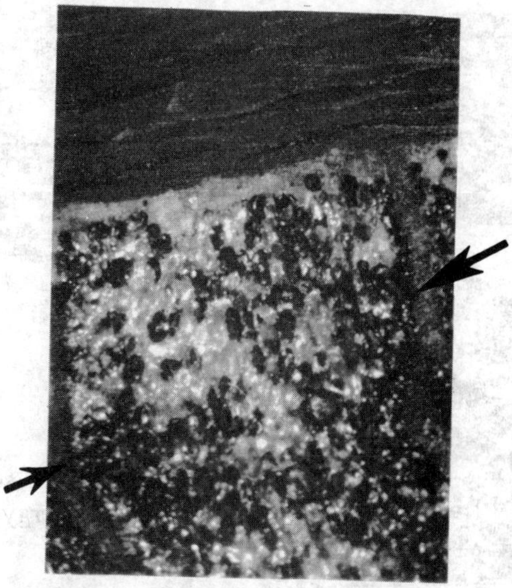

Figure 6. Core Damage in surrounding cells for Impact *Level A* (16-ply graphite/ vinyl ester facesheets)

Figure 7. Top Facesheet and Core Damage for Impact *Level D* (16-ply graphite/ vinyl ester facesheets)

Figure 8. Bottom facesheet failure for Impact *Level D* (16-ply graphite/vinyl ester facesheets)

For the pre-impact scans, the process was found to yield uniform and satisfactory quality of the laminate. Figure 9 represents the damage zones at 11, 20, 28, 33 and 40 J respectively for the C-scans taken on the top facesheet of the 16 layer graphite facesheet composites. The damage zone is observed to follow the energy input into the specimen, for higher energy values, the damage zone is greater. Typically, a four lobed elliptical damage zone develops away from the impact location. The damage is limited to 2-3 cell dimensions. Bottom facesheet scans were repeated at these energy levels as well. Only samples impacted at energy levels of D and E showed bottom facesheet damage, away from the impact zone.

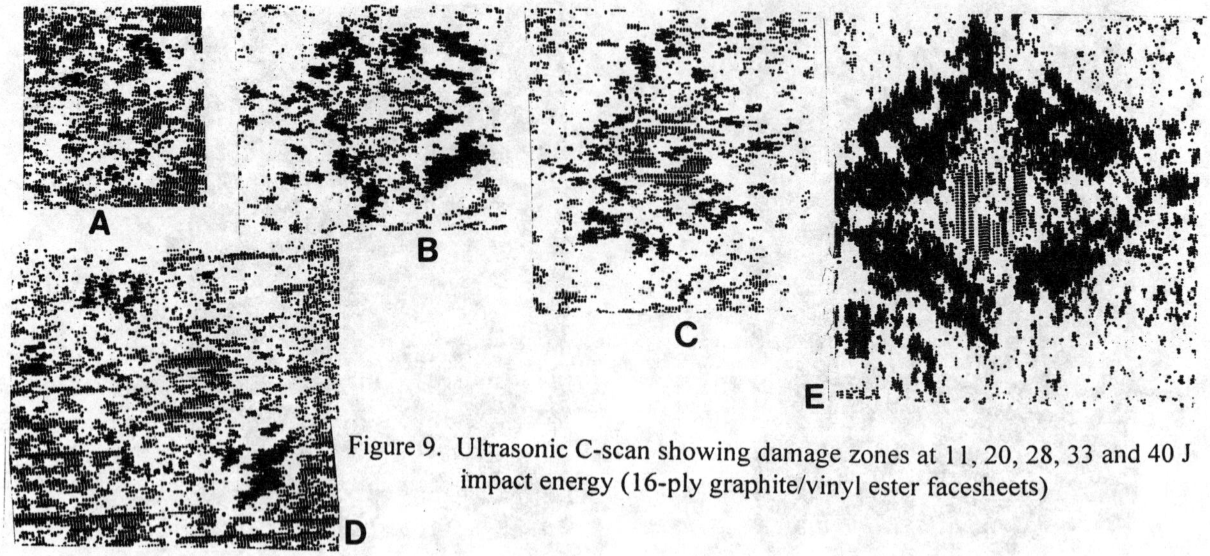

Figure 9. Ultrasonic C-scan showing damage zones at 11, 20, 28, 33 and 40 J impact energy (16-ply graphite/vinyl ester facesheets)

Figure 10 represents the force-time curves for the S2-glass facesheet composites impacted at the same energy levels. Interestingly the load at which damage initiation occurred in the S2-glass facesheet sandwich composites was also found to be 5800 N, indicating that the first damage is most likely the localized cell wall buckling/ crushing, accompanied by the localized interfacial debond of the core-facesheet wall for the specific ratio of facesheet/core thickness. Figures 11 a and b represent the micrographs for the D level tested S2-glass composite. These composites do not show extensive localized facesheet damage/delamination as observed in the graphite facesheets. The damage initiation load of 5800 N was independent of the energy levels and facesheet type at which the samples were impacted. As the energy increases from 11-40 J, the pulse width of the force-time curve increases from 6.0 msec to 8.5 msec. This indicates that the time taken by the striker to rebound is longer as the impact energy increases. The glass facesheet composites did not exhibit any bottom facesheet failure, as can be evidenced from both the load - energy and force- time curves.

Compression-After-Impact Studies

Following low velocity impact and nondestructive evaluation, the specimens were subjected to compression-after-impact testing in a SACMA recommended plate fixture [9]. The specimens were end loaded in a 110 kips MTS load frame. A 150 kHz acoustic emission (AE) sensor was fed to a AE LOCAN AT data acquisition system via a preamplifier [10]. Static compressive load was applied until failure at a crosshead rate of 1.27 mm/min. The failure loads based on impact levels for 16-ply graphite composites tested was B: 14,400 lbs; C: 13,920 lbs and E : 13, 800 lbs. For the S2-glass composites it was : B:14,300 lbs and C: 13,500 lbs. With the exception of two samples, which exhibited end crushing, the remaining samples failed in their gage section. Typically, the failure location coincided with the low velocity impact location as shown in Fig.12. The damage was found to be severe on the low velocity impact side facesheet, and little or no evidence of back facesheet failure due to CAI was evident. The damage state and failure loads were in close accordance with the impact energies that the specimens had suffered. Furthermore, the interface between the core and the

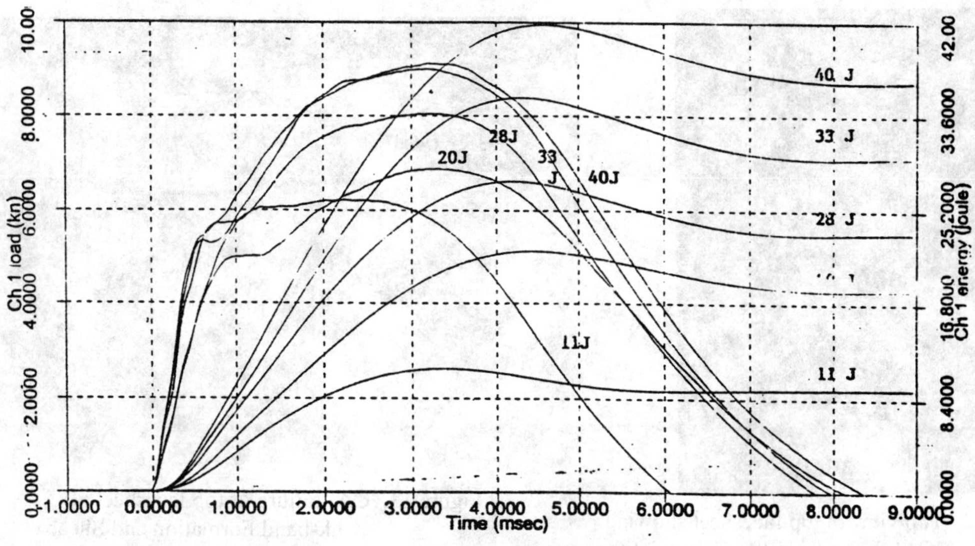

Figure 10. Typical load-energy curves 8-ply facesheet woven S2-glass/vinyl ester foam filled honeycomb core sandwich composites

Figures 11a and b. Top Facesheet and Core Damage for Impact *Level D* (8-ply S2-glass/vinyl ester facesheets)

facesheets remained intact, except for core shearing and core crushing in the vicinity of the damage location as shown in Fig.13 on the impact side facesheet. The impact side facesheet failed by kink-band formation and shear instability of the facesheet, showing a fork in the laminate at the impact location , as is also illustrated in Fig.13. The 8-ply S2-glass facesheet composite, showed higher degree of compliance, and tendency towards end crushing in comparison to the graphite facesheet composites.

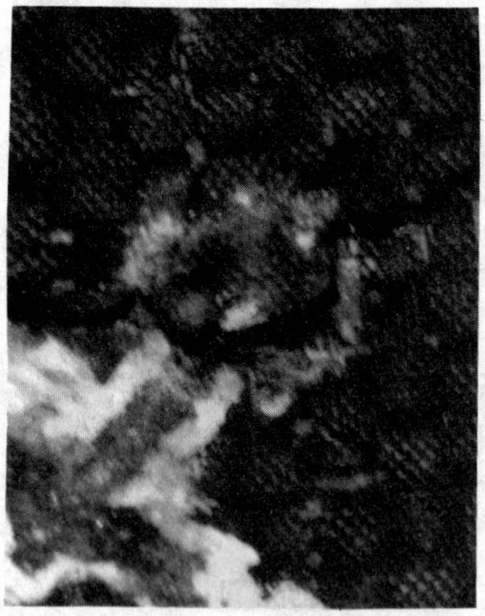

Figure 12. Top view of top facesheet showing CAI & failure in the vicinity of the impact location

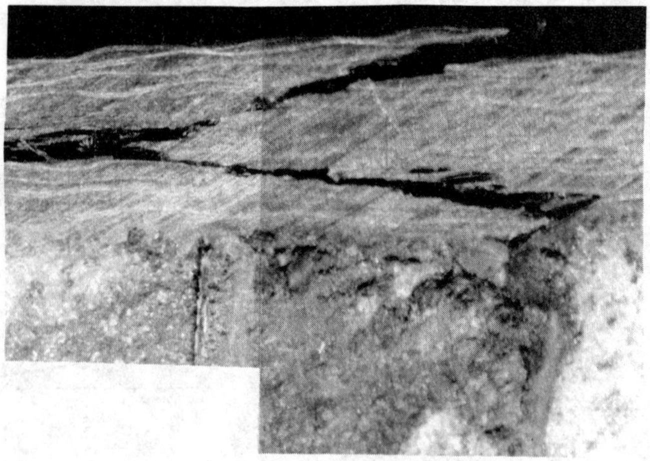

Figure 13. CAI failure of top facesheet : Kink-band Formation and Shear Instability of the facesheet

Acoustic Emission Testing

Figure 14 shows the parametric distribution of energy and amplitude of AE activity for the 16-ply facesheet graphite composites. The high energy events in the initial loading occur during the occurrence of full contact of the loading device to the specimen. With subsequent loading, high energy (>1500) and high amplitude (>80 dB) events occur coinciding with facesheet failure. It is also seen that a stream of low energy events occur throughout the loading period that are characteristic of slow matrix cracking, fiber kinking, kink-band formation and fiber filament level fracture. Beyond 42 sec of the time axis, three groups of events are seen to occur collectively, small energy (<600), low amplitude (40-60 dB) that are attributed to fast developing kink-bands and matrix cracks, medium energy (600-1200), medium amplitude (60-80 dB) that are attributed to frictional energy dissipation due to fiber kinking and matrix facesheet cracking , and high energy (>1500), high amplitude (80-100 dB) events that are due to fiber fracture and delamination around the kink-bands.

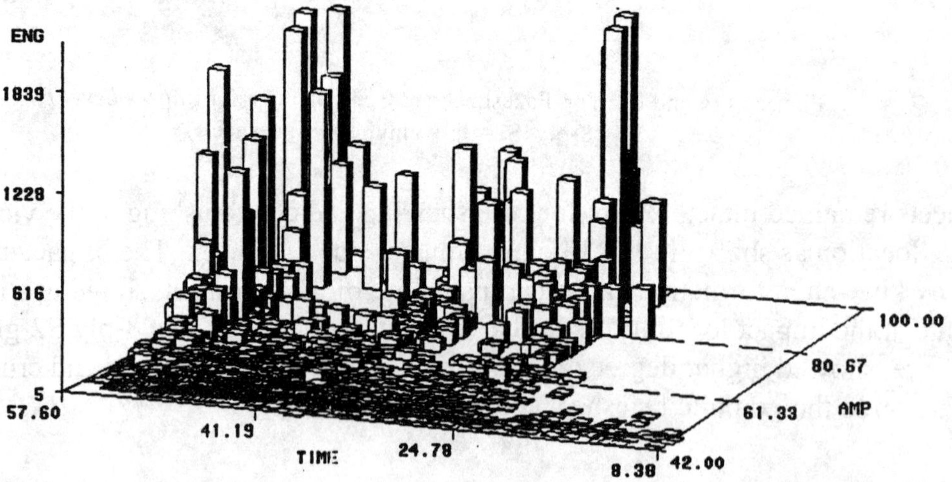

Figure 14. Parametric distribution of energy and amplitude of AE activity for the 16-ply facesheet graphite sandwich composite

CONCLUSIONS

A low-cost resin infusion molding process has been effectively adopted in developing foam filled honeycomb core sandwich composites. The damage initiation and containment mechanisms under low velocity impact for these composites are more attractive when compared to either foam or honeycomb core used alone. Within the energy levels tested, the damage in the core did not exceed past two and/or three cell dimensions. The damage initiation load is found to depend upon the ratio of facesheet/core thickness and occurs predictably by core related effects such as core crushing, cell wall buckling and localized debonding of the facesheet and the core. Localized top facesheet damage always accompanied the core related damage events. The S2-glass facesheet composites exhibit higher damage tolerance (minimal top facesheet delamination) as compared to the graphite facesheet sandwich composites. The force-time and energy-time histories could accurately predict distinct steps of top-facesheet as well as bottom facesheet failure.

REFERENCES

1. Abrate, S. 1991. "Impact on Laminated Composite Materials," *Applied Mechanics Review*, Vol.44, No.4, April, 155-190.
2. Herup,E. 1996. "Low Velocity Impact on Composite Sandwich Plates," Ph.D. Dissertation, Air Force Institute of Technology, AFIT/DS/ENY/96-11, Dayton OH, July.
3. *Reinforced Plastics*, Dec. 1994, Vol.38, No.12, Elsevier Publishers.
4. Wu, C.L. and Sun, C.T. 1996. "Low Velocity Impact Damage in Composite Sandwich Beams," *Composite Structures*, 24, 21-27.
5. Weeks, C.A. and Sun, C.T. 1994. "Multi-Core Composite Laminates," *Journal of Advanced Materials*, April, 28-37.
6. Wu, C.L., Week,C.A and Sun, C.T. 1995. " Improving Honeycomb-Core Sandwich Structures for Impact Resistance," *Journal of Advanced Materials*, 41-47.
7. Honeycomb Sandwich Constructions : *Hexcel Corporation Technical Literature*.
8. *WESKOR Foam Filled Honeycomb*, Westwind Composites Inc., Houston, TX, Company Literature.
9. Mallick, P.K. 1995. *Fiber Reinforced Composites*, Marcel Dekker Inc. Second Edition.
10. Vaidya, U.K. 1993. *Nondestructive Evaluation of Graphite Based Composites using Acoustic and Vibration Techniques*., Ph.D Dissertation, Auburn University.

ACKNOWLEDGMENTS

The support provided by Wright Laboratories, Flight Dynamics Directorate and Mr.Forrest Sandow, Technical Monitor under the Grant No. F22615-96-C-2200 is highly appreciated. The foam filled honeycomb core samples provided by Westwind Composites, Inc. is also appreciated.

Low-Velocity Impact of Multi-Reciprocal Braided E-Glass/Epoxy Composites

E. KWAN, H. WANG, X. R. XIAO, S. V. HOA AND H. HAMADA

ABSTRACT

A new braiding technique called the Multi-Reciprocal (MR) braiding method was used to fabricate laminated structural preforms. The MR braiding process provides a tortuous interface by reversal of braiding direction at a given distance called pitch. The composites made by this type of fabric is expected to exhibit a higher resistance to impact-induced delamination. In this work, the impact resistance of the laminated composites made of MR fabrics was studied by drop-weight impact test and compression-after-impact (CAI) test. The damage behaviour of MR composites with different pitch length was characterized and compared with that of the conventional braided laminate. The mechanism of enhanced resistance to impact-induced delamination was also discussed.

INTRODUCTION

Among the damage modes caused by impact, delamination is the most detrimental to the structural performance of composite laminates. The residual stiffness, strength and fatigue properties of impacted laminates have been found to decrease as the delamination size increases. Therefore, many studies have been conducted to improve the delamination resistance of composite materials in terms of reduced delamination for a given level of impact. The methods proposed so far include modifying the resin system for higher interlaminar toughness (Evans and Masters, 1987), through-the-thickness stitching (Mignery, Tan and Sun, 1985) interleaving (Sun and Rechak, 1988) and hybridization (Bhatia, 1982). In these cases, one wishes to provide a toughened interface or transverse reinforcement to the laminate in order to suppress or limit the delamination growth during impact.

Recent developments in textile technology have given rise to novel methods of fabricating structural preforms (dry fiber skeleton with no matrix material). The dry preforms can then be impregnated with a suitable resin, by resin transfer molding (RTM) for example, to produce three-dimensional structural parts. This offers additional approach to achieving enhanced damage tolerance. A technique developed by Hamada et al. (Hamada, et al, 1992; Nakai, Masui and Hamada, 1996) uses a modified braiding process, called the Multi-Reciprocal (MR) braiding method, for the fabrication of laminated structural preforms. In contrast to the conventional braiding method, which provides a straight and continuous interface between the layers of fabric,

Eddy Kwan, Bombardier Inc. Canadair Defense System Devision, Mirabel, Quebec, Canada
H. Wang, Institute de Materiaux Industriels, National Resarch Council, Montreal, Quebec, Canada
X.R. Xiao and S.V. Hoa, Concordia Center for Composites, Montreal, Quebec, Canada
H. Hamada, Kyoto Institute of Technology, Matsufasaki, Sakuo-ku, Kyoto, Japan

the MR braiding process provides a tortuous interface by reversal of braiding direction at each of a given distance (the pitch length as discussed later). The braided laminate as such and later impregnated with resin is expected to exhibit a higher resistance to impact-induced delamination, because the damage growth can be arrested by the reversed fiber tow at the pitch lines.

In this work, the impact resistance of laminated composites made of glass fiber MR fabrics and an epoxy resin was studied by drop-weight impact test and compression-after-impact (CAI) test. The damage behavior of MR composites with different pitch length was characterized and compared with that of the conventional braided laminate. The mechanism of enhanced resistance to impact-induced delamination was also discussed.

MANUFACTURE OF CONVENTIONAL AND MULTI-RECIPROCAL BRAIDED PREFORMS

Conventional braided preforms were fabricated using the Joint Braiding machine supplied by Muratec. The machine uses the traditional Maypole braiding technique. This process typically allows the axial yarns to retain a primarily straight orientation, while the braider yarns have a high degree of undulation as they travel over and under the axial yarns. This type of interlacing allows the preforms to be removed from the braiding mandrel, handled and manipulated before resin impregnation. No axial yarns were added. All conventional braided fabrics have a nominal thickness of 2.9 mm. This thickness was obtained by passing the preform through the braiding machine three times, braiding over the previously braided layers. So a total of three layers was braided. An aluminum cylindrical tube mandrel was used for this and the later discussed multi-reciprocal types specimens. The cylindrical mandrel has an outside diameter of 21.5 mm and a length of 25 mm. Fig. 1a shows the structure of a 2-D conventional braided fabric.

The MR braided preforms were manufactured using the "Multi-Reciprocal Braiding Process" [5-6]. This type of preforms can be fabricated using the same braiding machine as for the conventional braided fabrics, as illustrated in Fig. 1b. Initially, the first three layers are braided by normal reciprocal process over part A (The reversal point has been determined). Then at the third layer and at the end of part A, continue braiding to form the first layer of part B. This process is repeated until the desired length is obtained. The distance from each reversal point to the next is called the pitch. It is expected that with the reversed fiber tow at the pitch lines, the growth of delamination can be suppressed.

Two kinds of MR fabrics were fabricated using the constant pitch length of 25 mm and 12.5 mm respectively. After braiding, both conventional and the MR fabrics were cut longitudinally, removed from the mandrel, and border stitched to maintain handleability. The resulting flat pieces of fabric were later placed in molds and impregnated using resin transfer molding (RTM) process.

The fabrics used in the current work were fabricated from E-glass rovings, 1100 tex supplied by Nitto Boseki Co. of Japan. The glass fiber rovings were braided into dry preforms, which were subsequently impregnated with a Bisphenol A-Type liquid epoxy resin DER 324 provided by Dow Chemical with curing agent Ancamine 2167. The recommended curing agent to resin ratio was 28% by weight. Table 1 summarizes the physical characteristics of the braided composites.

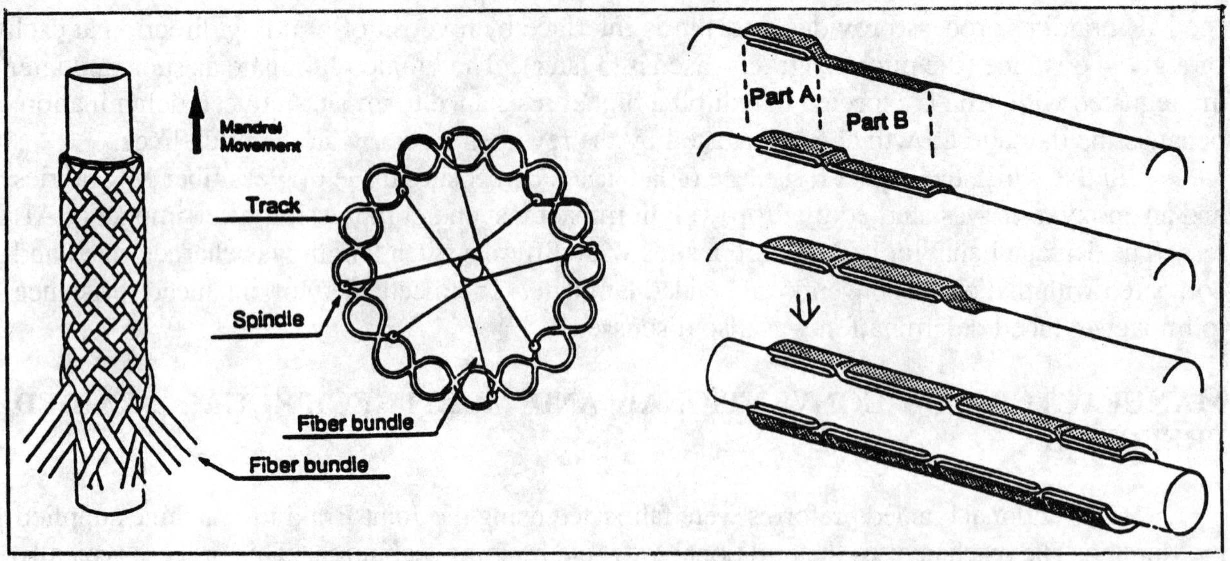

Figure.1a Conventional braiding process and structure Figure. 1b Multi-Reciprocal braiding process

IMPACT TEST AND DAMAGE ASSESSMENT

Impact tests were conducted using a Dynatup 8200 drop weight tester equipped with GRC 730-I instrumented impact test data system. It is capable to provide a digitized record of the impact event such as force history, impact velocity and impact energy. Each specimen was firmly clamped to a fixture which provides a circular test window of 38 mm in diameter. An impactor with the hemispherical nose of 12.7 mm in diameter and the total mass of 1.27 kg was used in this experiment. The impact energy was control by releasing the crosshead at a specific height above the center of the specimen determined from calibration prior to testing. During testing, since all impacts were sub-penetration type, the impactor would rebound. In order to avoid repeated strike, the impactor was captured during its rebound. Impact velocity was chosen between 1.89 m/s to 3.65 m/s, while the energy ranges from 2 J to 8 J.

After impact, pictures were taken for each specimen using high density fiber optic backlighting to record the internal damage. Surface damage was also recorded using the same method. Many studies have shown that, under low-velocity impact, the damage growth in laminated composites is very close to that induced by equivalent static indentation load. Therefore, some specimens were subjected to indentation test using the same fixture and an indentor of the same geometry as the impactor. The damage extension in such specimens was also documented with backlight assisted photography and, in selected cases, by cross-section examination.

COMPRESSION AFTER IMPACT (CAI) TEST

After impact testing, the specimens were compression tested using a MTS servohydraulic test machine to determine the residual compression strength. The specimen was set between the grips and clamped near both ends, leaving a test section of about 25.4 mm which contained the impact damage at the mid length. The axial load was transmitted through the ends of the

Table I. Physical Characteristics of Braided Composite Specimens

Type	Fiber Wt. (%)	Braiding Angle (degree)	Thickness (mm)	Width (mm)
CONV (conventional)	61	60-65	3.2	66
MR25 (pitch = 25 mm)	46.4	50-55	3.2	66
MR12 (pitch = 12.5 mm)	48.9	50-55	3.2	66

specimen and, up to compression failure at the damage site, no global buckling type of failure was observed.

RESULTS AND DISCUSSION

IMPACT RESPONSE

Fig. 2a shows the typical impact force history of a MR12 specimen impacted at 5 J. The transferred energy from the impactor to the specimen is also shown in Fig. 2b. For the subpenetration impact, the energy basically increases with the impact force during the loading stage while it decreases in the unloading stage, indicating that part of the elastic strain energy stored in the specimen was transferred back to the impactor during rebound. The total energy at the end of impact is that absorbed by the specimen due to damage creation and, for a minor amount, in terms of vibration and fixture deformation. Therefore the total energy, as summarized in Table 2, will be approximately used as a parameter for comparing the impact resistance of composites.

Table II Energy and peak force values recorded during impact test of the three types of specimen

Type	Impact Energy (J)	Peak Force F_p (kN)	Total Energy (J)	Max Delamination Area (mm^2)	Normalized Delamination Area
CONV	5.386	2.925	3.738	299.38	0.264
	8.298	3.912	6.017	598.07	0.5274
MR25	4.143	2.808	3.003	221.13	0.195
	8.299	4.344	5.686	624.38	0.5506
MR12	5.389	2.974	3.888	324.32	0.286
	8.202	4.123	5.745	405.61	0.3577

DAMAGE MORPHOLOGY

Each type of the specimens was impacted at one of the three energy levels: 2 J, 5 J and 8 J. At the lowest energy level, no visible damage was detected. At 5 J and 8 J energy level, a crush zone at the impact site and some backface matrix cracking were observed for all specimens. A whitened area was also visible as a result of light reflection at internal damage sites. The internal damage appears as a dark area on the photographs taken using strong backlighting. Fig.

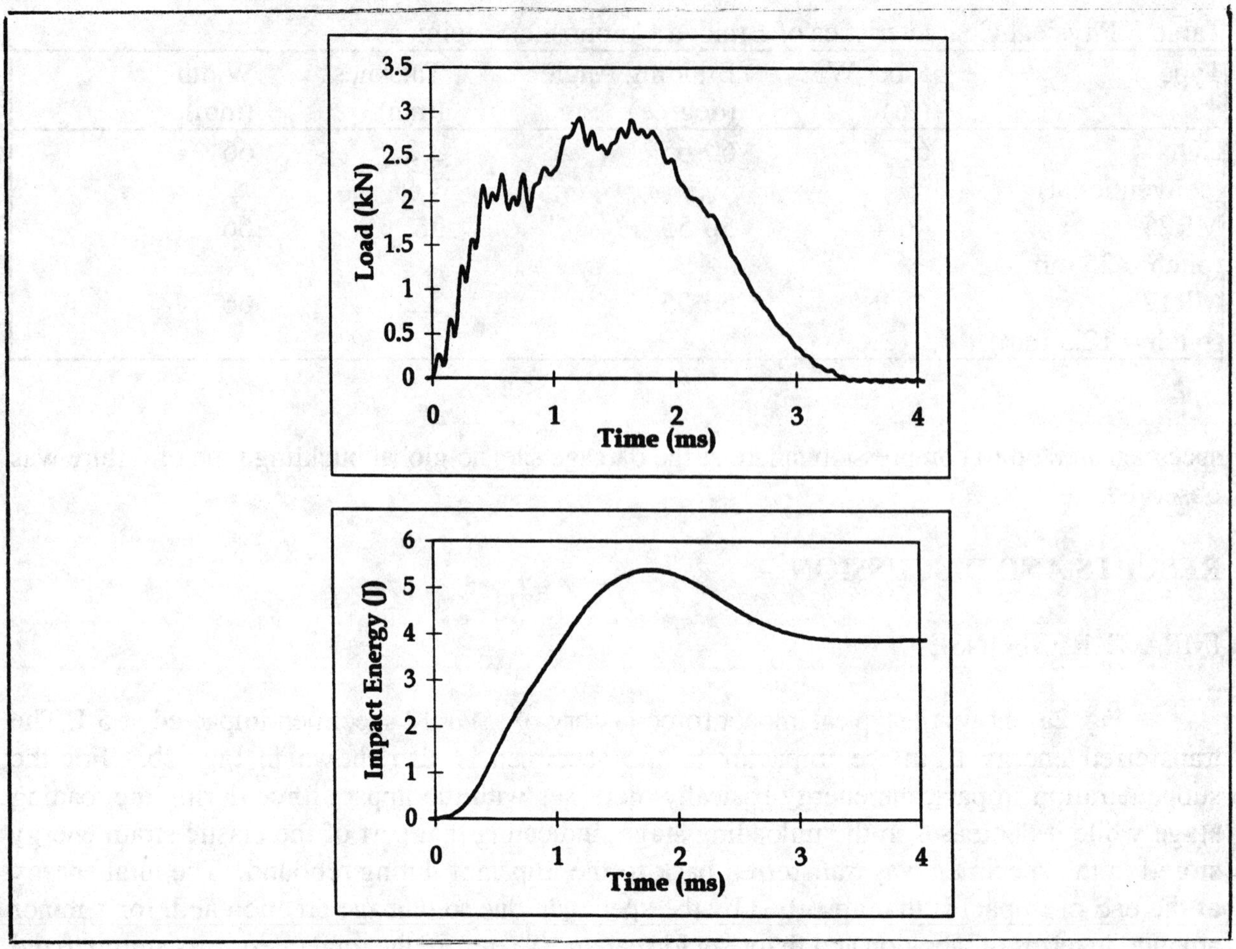

Figure.2 Typical force versus time curve and the corresponding energy versus time for MR12 specimen impacted at 5J

3 shows such photos for specimens impacted at different impact energy levels. It is interesting to observe the different shape and size of the whitened area depending on the types of braids. For the conventional type of specimens, the whitened area was almost a circular shape and its diameter increases with impact energy. The whitened area of the MR25 specimens was similar to that of the conventional specimens at about the same impact energy level. For the MR12 specimens, the whitened area was circular at about 5 J but became distorted and elongated parallel to the pitch line at about 8 J; in other words, the damage did not extend beyond the pitch line and was forced to grow sidewise. This demonstrates the effect of MR braided structure that the reversed fiber tow can restrict the damage within the pitch. Before delamination extends to the pitch line of the MR braid, however, the MR braids behave like the conventional braids, as in the cases of MR25 up to 8 J and MR12 at 5 J. The damage area for specimens subjected to static indentation tests was similar to the corresponding impact case, as shown in Fig.3d.

To determine the relation between the whitened area and the extend of internal damage, two specimens after identical static indentation loading were cross sectioned and inspected under an optical microscope. It was observed that the whitened area corresponds well to the extension of delaminations.

Fig.5 shows the mosaic pictures of these cross sections. The upper picture belongs to the specimens with the MR12 braid and the lower one the conventional braid. As seen, the braid/matrix interface delamination and fiber tow split occurred in both specimens, but the extension of the delamination was distinctly different. For the conventional braid, delaminations at the interface extended up to four braids from the indentation site. For the MR12 specimen, the entire length of interface delaminations is about four braids, which is equal to the pitch distance. As seen, in MR12 specimen, the interface delamination followed the reversed fiber tow and folded back at the pitch line. This crack reverse process requires much higher energy and hence the interface delamination became restricted at the vicinity of the pitch line. The same mechanism also restricts the fiber tow split to extend beyond the pitch line. Since the damage behavior under low velocity impact is very close to that under static loading, the whitened area on impact damaged specimens may also be assumed to be the delaminations. The delamination area was measured and normalized to the area exposed to impact, and is shown in Table 2. Fig. 4 compares the results of the different types of braided specimens. It shows that all the specimens behaved the same up to about 5 J. At 8 J, the increase of delamination area leveled off for the MR12 type but continued along the same line for the other types of braided specimens. This again demonstrates the damage restriction effect of the MR structure.

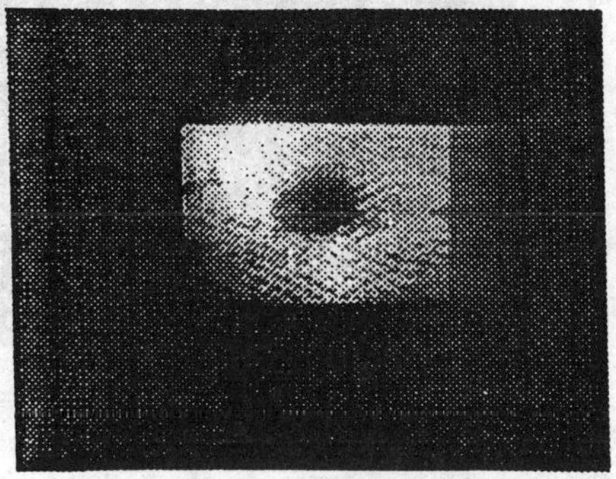

Impact Energy = 5 J

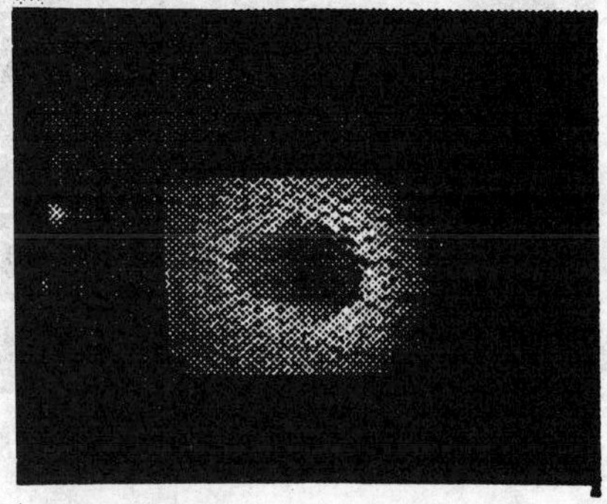

Impact Energy = 8 J

Figure. 3a Delamination shape of CONV specimens after impact

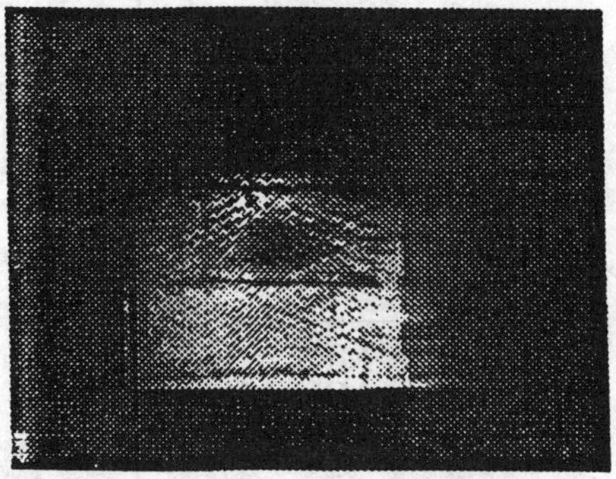

Impact Energy = 4 J

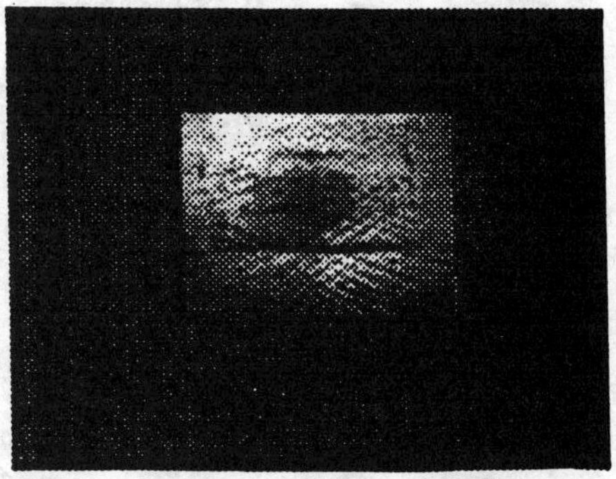

Impact Energy = 8 J

Figure. 3b Delamination shape of MR25 specimens after impact

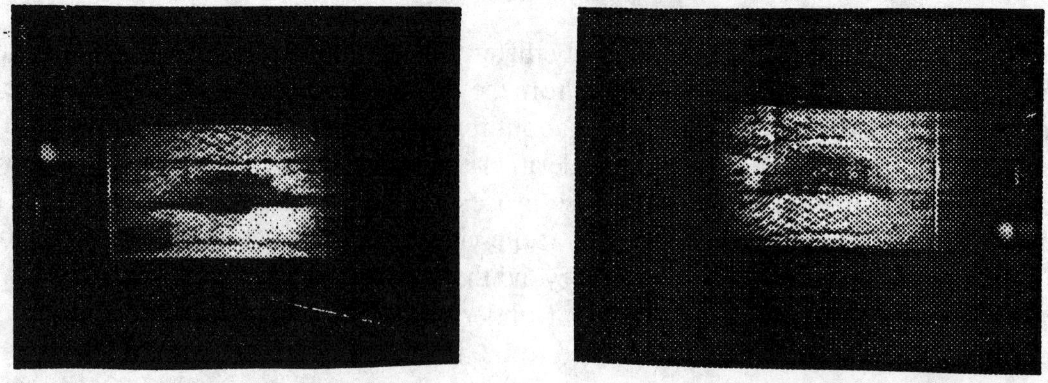

Impact Energy = 8J **Impact Energy = 5J**

Figure. 3c Delamination shape of MR12 specimens after impact

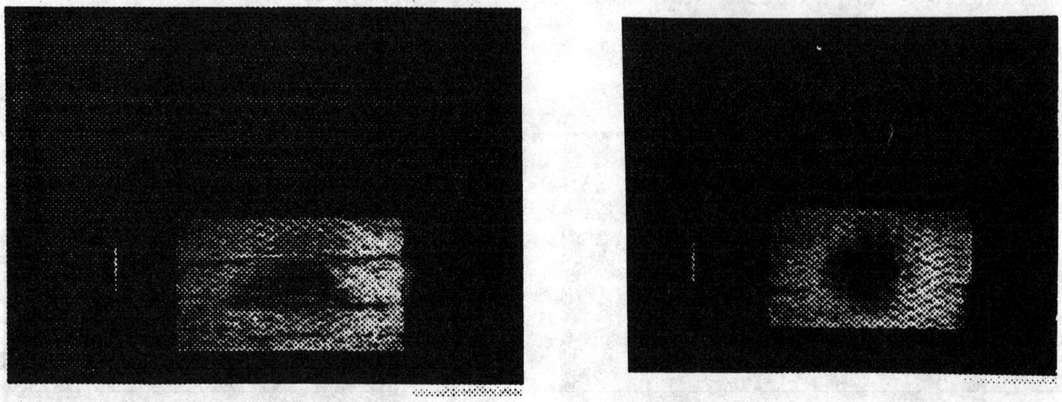

Figure. 3d Delamination shape of MR12 and CONV specimens after static indentation loading

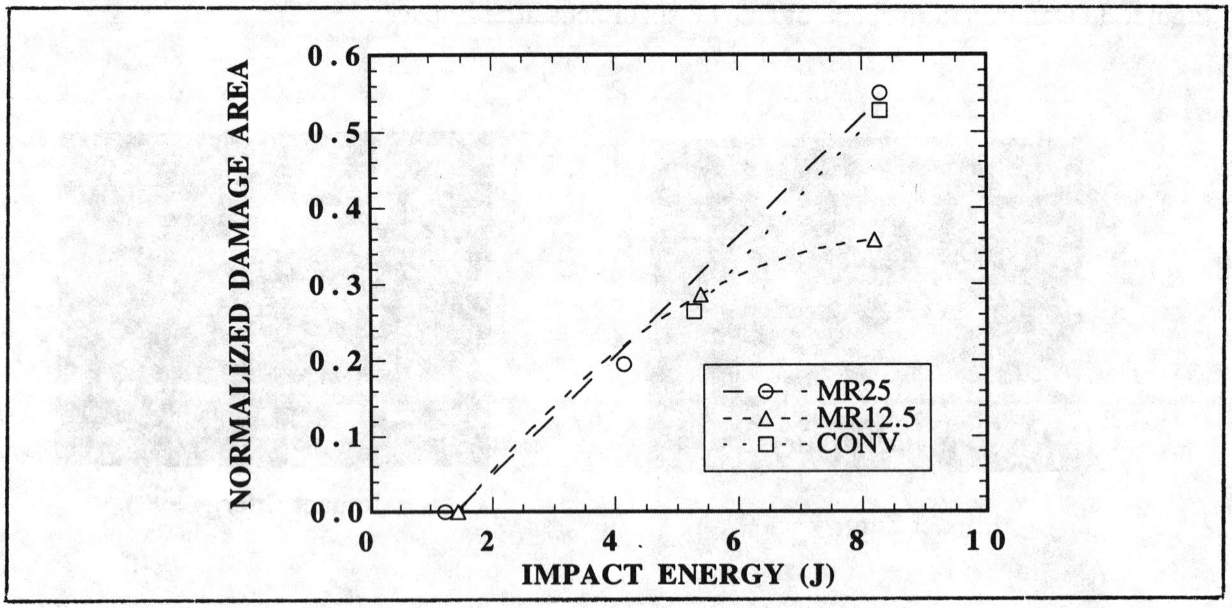

Figure.4 Comparison of damage area

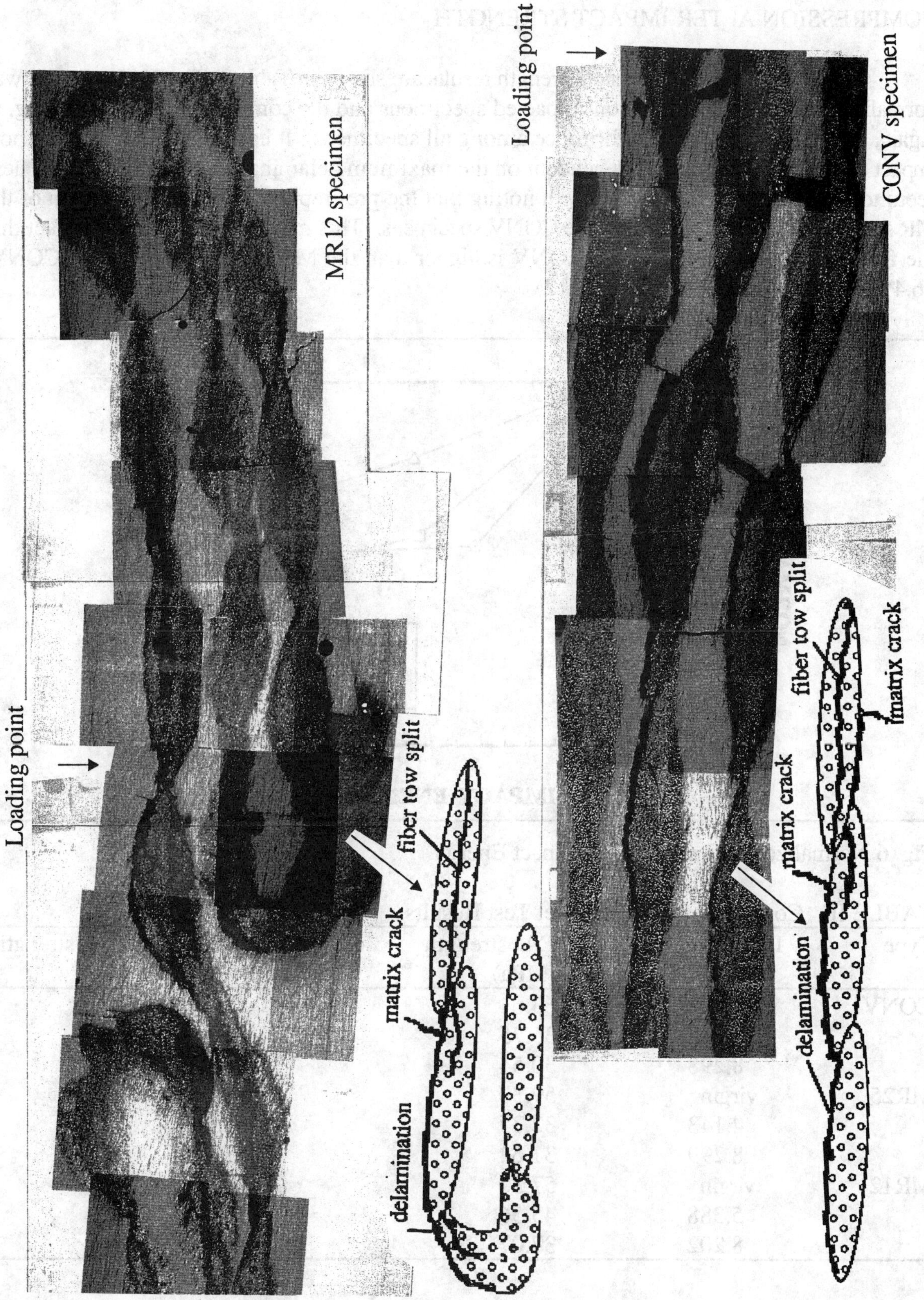

Figure. 5 Internal damage of MR12 and CONV type specimens after static indentation loading

COMPRESSION AFTER IMPACT STRENGTH

The post impact compression strength results are shown in Table 3. The CAI strength was normalized to that obtained on non-impacted specimens and the comparison is shown in Fig. 6. Again, MR12 shows the best performance among all specimens. It has been shown that the post impact compression strength is dependent on the maximum delamination area. The results here seem to support this finding. It is worth noting that the pre-impact compression strength of the MR specimens is lower than that of the CONV specimens. This may be attributed to the fact that the fiber weight percentage of the CONV is higher than the MR specimens (61% for CONV, 46.4% for MR25 and 48.9% for MR12).

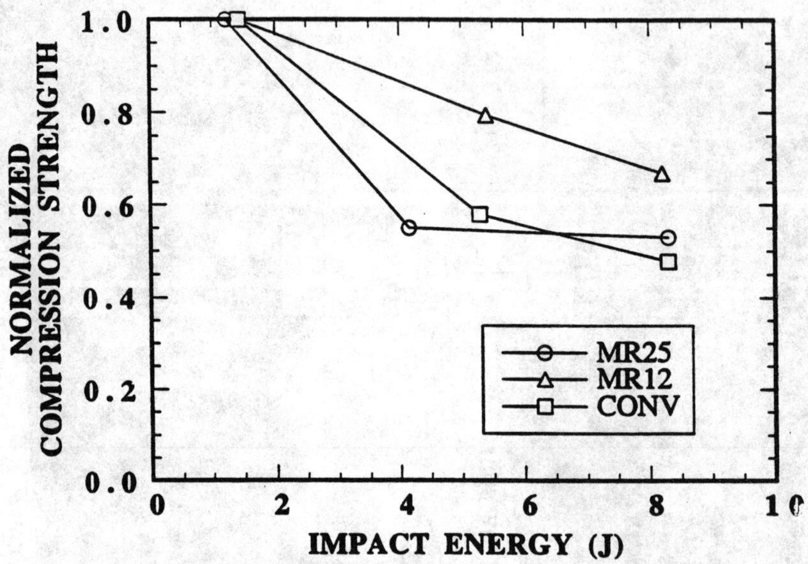

Fig. 6 Normalized CAI strength vs Impact Energy

TABLE III. Compression After Impact Test Results

Type	Impact Energy (J)	CAI strength (MPa)	Normalized CAI strength (%)
CONV	virgin	63.64	100
	5.268	36.75	57.7
	8.298	30.54	48.0
MR25	virgin	59.30	100
	4.143	32.75	55.6
	8.299	31.64	53.4
MR12	virgin	57.30	100
	5.388	45.56	79.6
	8.202	37.07	35.3

CONCLUSION

The current work has demonstrated that an improvement in the overall impact resistance can be achieved by using the multi-reciprocal braided composites. The results also suggest that the pitch (distance between the reversal points) influences the effect of the reinforcement of the MR fabrics. This is shown by comparing the damage size and normalized CAI strength between the MR25 and MR12 specimens. The cross-sectional inspection reveals that the increase in impact resistance of MR12 specimen can be attributed to the crack reverse mechanism at the fabric reversing points. This crack reverse process absorbs more energy and hence reduces the total damage area. Consequently, the MR12 specimen demonstrates the highest CAI strength. But there is a disadvantage in the MR type composites: their non-impact compression strength is lower than that of the conventional braided composites. This can be explained by the lower fiber volume fraction of the MR specimens since the MR braiding process requires the folding of the layers.

ACKNOWLEDGMENT

This work is supported by National Science and Engineering Research Council of Canada (NSERC) through an operating grant and Fonds pour la Formation de Chercheurs et L'Aide a la Recherche (FCAR) through a collaborative travel grant. We would also like to thank Mr. S.R. Jarugu and Mr. I.Al-Hmouz of Concordia Center for Composites for their assistance in composite fabrication and mechanical testing.

REFERENCES

R.E. Evans and J.E. Masters, "A New Generation of Epoxy Composites for Primary Structural Applications: Materials and Mechanics", Toughened Composites, ASTM STP 937, pp. 413-436. 1987.

L.A. Mignery, T.M. Tan and C.T. Sun. "The Use of Stitching to Suppress Delamination in Laminated Composites", in Delamination and Debonding of Materials, ASTM STP 876. W.S. Johnson, ed. pp. 371-385, 1985.

C.T. Sun, and S. Rechak, "Effect of Adhesive Layers on Impact Damage in composite Laminates", in Composite Materials: Testing and Design, (Eight Conference), ASTM STP 972, J.D. Whitcomb, ed. 1988.

N.M. Bhatia, "Strength and Fracture Characteristics of Graphite-Glass Intraply Hybrid Composites", Composite Materials: Testing and Design (Sixth Conference). ASTM STP 787, pp 183-199, 1982.

H. Hamada, A. Fujita, A. Nakai, A. Yokoyama and T. Uozumi, "New Fabrication System for Thick Composites - A Multi-Reciprocal Braiding System", 1992

A. Nakai, M. Masui and H. Hamada, "Fabrication and Mechanical Properties of Multi-Reciprocal Braided Composite Tube", Advanced Composites Letters, Vol.5, No.3 1996.

Impact Performance of Fiber Reinforced Composite Laminated Plates, and the Influences of Relative Fiber Content and Density

L. A. RUHALA AND R. S. ENGEL

ABSTRACT[1]

Fiber reinforced composite laminated plates (FRPs) have many potential applications in the area of automotive structural crashworthiness. Among these, a FRP may be used as a side door liner to absorb and dissipate energy during a side impact, thus allowing less door intrusion into the passenger compartment, potentially reducing occupant injury. As a first step in the design of such a plate, the authors have studied the effects of relative fiber content on the impact properties of laminated composite plates with a particular emphasis on the damage mode. That is, delamination, matrix cracking and fiber breakage. E-glass fiber/epoxy plates with fiber weight percents ranging from 50 to 80 percent were studied at impact energies, ranging from 5J to 40J.

INTRODUCTION

Impacts are generally categorized as low, high or hyper velocity, but there is not a clear transition between categories and authors disagree on their definition (Richardson and Wisheart, 1996). Low velocity impact events may be treated as quasi-static, since the contact time is long when compared with the period of the fundamental mode of vibration of the target composite laminate (Sjöblom, Hartness and Cordell, 1988; Shivakumar, Elber and Illg, 1985). The upper limit can vary from one to tens of m/s depending on the target stiffness, material properties and the impactor's mass and stiffness.

High-velocity impact response is dominated by stress wave propagation through the material, in which the structure does not have time to respond, leading to very localized damage. Boundary condition effects can be ignored because the impact event is over before the stress waves have reached the edge of the structure. In low velocity impact, the dynamic structural response of the target material is highly important since the contact duration is long

[1] Laura A. Ruhala, Ph. D. Candidate, The Pennsylvania State University, 227 Hammond Building, University Park, PA 16802
Renata S. Engel, Associate Professor, The Pennsylvania State University, 227 Hammond Building, University Park, PA 16802

enough for the entire structure to respond to the impact and hence, more energy is absorbed elastically.

An impact event may be classified as low velocity if it has a maximum velocity of 10 m/s, (Cantwell and Morton, 1991) or the event may even be classified as low velocity with maximum impact speeds of less than 100m/s (Abrate, 1991). The impacts conducted for this study had a maximum speed of 4 m/s, well within the definition of low-velocity.

Composite laminated E-glass fiber / epoxy plates were constructed using the hand lay-up method. By holding the amount of fiber constant and varying the amount of resin, the fiber weight percents of the plates ranged from 50 to 80 percent. The plates were then fixtured to minimize slippage and impacted in a drop-tower impact machine. The height at which the tup was released varied, allowing impact energies to range from 5J to 40J. The impact events were photographed with a high-speed video camera. A light-box was also used for analysis.

This paper will describe in further detail the fabrication and impacting of the plates. The types of plate damage will be shown, as well as the effect of plate penetration on the energy absorbed by the curves. The plates will be characterized by the amount of energy they absorb. Finally, the effect of relative fiber content on the impact properties will be discussed.

PROCEDURE

FABRICATE PLATES

The composite plates were fabricated using a hand lay-up method in a cross-ply configuration of $[0/90/0/\overline{90}]_s$. E-glass fibers (Anchor, Ancaref 5275) in mats bound together with a thin, randomly applied, polyester binder, were placed into a 30.5cm x 45.7cm mold, then saturated by four layers of epoxy. The epoxy consisted of Dow 383 resin and Huntsman Jeffamine T403 cure, mixed in a ratio of 45:100 parts cure to resin. The weight of fiber in each of the sheets remained constant, while the weight of resin was varied to change the relative fiber weight percent of the sheet (W_f). Nominal values of fiber volume fractions (V_f) were chosen as 40, 60, and 80, and the appropriate amount of resin was applied for each according to Eq. (1). The subscripts m and f designate the matrix and fiber, respectively, W is the weight, V is the volume, and ρ is the density.

$$W_m = \frac{V_m \rho_m W_f}{V_f \rho_f} \quad (1)$$

The mold was then heated (66°C) and compressed (138 kPa) in a programmable Tetrahedron smart press for 5 hours. As the Dow 383 / Jeffamine T403 epoxy cures, the viscosity initially falls from 1100 cp to 450 cp before increasing to 5500 cp when cured. That initial decrease in viscosity aggravated the problem of epoxy leakage from the mold during cure. Therefore, a "pre-cure" of 90 minutes at 38°C was performed on the epoxy. Also, the pressures during cure were varied in an attempt to control epoxy leakage.

Each composite sheet was cut with a standard bandsaw into 12.7cm x 12.7cm specimen plates. Two sheets were fabricated for each nominal V_f value, with each sheet yielding six plates. Therefore, each nominal V_f produced twelve plates. The sheets labeled "A" had the most resin, and therefore the lowest fiber weight percent. The sheets labeled "C" had the least amount of resin, and thus the highest fiber weight percent. The sheets labeled "B" had an intermediate amount of resin, and thus an intermediate fiber weight percent. The plates were identified in the following manner: "letter - sheet number - plate number", i.e. A-2-4 represented the fourth plate in the second sheet that has the lowest weight percent of fiber.

There were clear visible differences between the A, B, and C plates. The A plates appeared rigid while the C plates resembled stiffened fabric. The B plates fell between the A and C plates in their appearance.

The characteristics of the sheets varied significantly due to varying pressures during the cure cycle (in an attempt to control epoxy leakage). The sheets numbered "1" were what the authors considered of higher quality and uniformity, while the sheets numbered "2" were of lesser quality and uniformity. The differences between the sheets were most apparent in the B plates.

Following ASTM standards D 3171-76, Procedure A, D792-91 and D2734-94, acid digestion was performed on the plates to determine V_f and void content. Correction for glass fiber weight change by the digesting medium, 70% nitric acid, was not needed since weight loss was 0.4%, which the standard considers insignificant.

IMPACT TEST

"Slippage" is defined as a relative motion between the fixturing and the plates, which results in frictional energy absorption. Under impact loading, slippage seldom occurs in rigid plates. Thus, impact test standards such as ASTM D3029 and D5420 for flat, rigid plastic plates do not require a specific level of clamping pressure. However for fabric-like specimens, such as the C plates, slippage becomes more likely as the penetrator strikes and pulls the plate out from between the clamping fixtures due to the membrane tension. A specialized set of clamping fixtures developed by Walsh were used to provide the high clamping pressures necessary (Walsh, 1996). In his work, Walsh determined a critical clamping force; when this critical clamping force was met and exceeded, the impact energy was found to be independent of the clamping force, indicating that no slippage had occurred. The plates tested in this study were clamped into the specialized Walsh fixture, with clamping forces well above the critical level.

The plates were impacted using a Dynatup Model 8250 drop tower, complete with a Dynatup GRC 730-I instrumented impact test data system (see Figure 1). An instrumented 1.27cm diameter tup was used. The crosshead weight was set at 5.4kg. The heights from which the penetrator and crosshead were dropped varied from 9.4cm to 75.4cm, allowing the energy of the impact to vary from 5J to 40J, respectively. A minimum of two plates of the same type were tested for each energy level. The impact machine was also employed to remove the tup from the plates after impact, thereby avoiding additional damage to the plates from bending or twisting. A single-impact-rebound brake was added to the system to prevent multiple impacts.

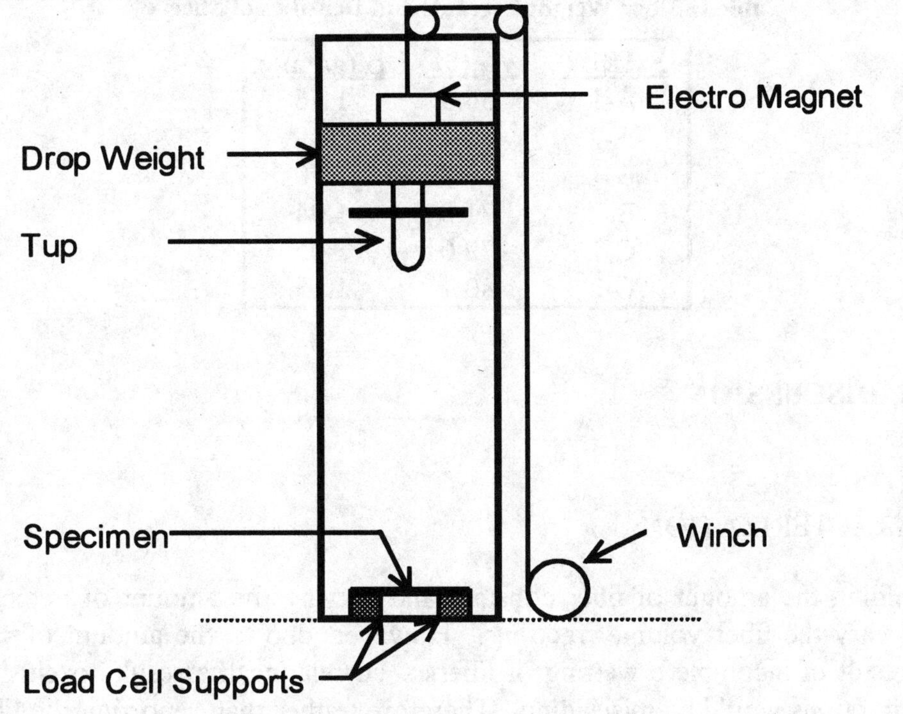

Figure 1: Schematic of Impact Test.

The Dynatup GRC 730-I instrumentation worked in conjunction with the drop tower to provide load, acceleration, velocity, and displacement data. Data counts stored on the transient recorder board at a frequency of 40 s^{-1} were converted to an array of load values and associated times. Using Newton's Second Law, acceleration data was obtained from the force data. The trapezoidal rule was then used to generate arrays of velocity and deflection values. Finally, the energy absorbed was calculated assuming conservation of total energy of the tup/specimen system throughout the test.

Whenever performing impact tests, the effects of inertial loads, harmonic oscillations, and stress waves must be addressed (Strait, 1989). The latter are associated primarily with ballistic tests. The maximum impact velocity encountered in this work was 4 m/s, many magnitudes lower than what would be considered ballistic. Thus, harmonic oscillations and stress wave effects were not a concern with these impacts. Inertial loads are a more common problem in impact tests, especially when testing heavy specimens of low strength material at high velocities. The inertial load is the load required to accelerate the plate to the velocity of the tup and may be large enough to be detected by the instrumented tup. A small inertial load may be mistakenly interpreted as incipient damage while a large inertial load may be interpreted as the peak load. No attempt was made to identify inertial effects.

A high-speed camera (Kodak-Eastman Spin Physics model SP2020, series II) recorded each of the impacts at 2000 frames/second. The video proved useful when studying the rebound (stored energy) results. A light-box was also used in the evaluation of the impacted plates, and some damage that did not show readily during normal viewing, was clear when surveyed on the light-box.

Table I: Fiber Weight Percents and Densities of Sheets

SHEET	W_f (%)	ρ (g/cm^3)
A-1	54.3	1.56
A-2	50.4	1.49
B-1	66.8	1.30
B-2	71.0	1.04
C-1	79.0	1.05
C-2	80.1	1.15

RESULTS & DISCUSSION

PLATE CHARACTERIZATION

By holding the amount of fiber constant and varying the amount of epoxy, an effort was made to vary the fiber volume fractions. However, due to the amount of voids in the plates – the result of incomplete wetting of fibers – correlating the results exclusively to the fiber volume fractions would be misleading. Therefore, rather than reporting the fiber volume fractions, the fiber weight percents and densities of the plates are being reported since they are independent of the voids, but still reflect the influence of the voids on plate characteristics (see Table I).

As Table I indicates, the fiber weight percents increase from the A to the B to the C plates. That is expected since the amount of resin used during fabrication decreased from A to B to C, so the relative amount of fiber weight increased.

Note, also, that the densities of the plates decrease from A to B to C. The density of the fiber is 2.14 g/cm^3 whereas the density of the resin is only 1.15 g/cm^3, so it is expected that the plate densities would increase as the fiber weight percents increase. The fact that the opposite occurs is indicative of the amount of voids in the plates. Recall that C plates have a stiffened fabric texture whereas the A plates are rigid composites. The density variation between plates of the same type was within 10% for A and C, but B-2 had significantly more voids than B-1, as demonstrated by the 20% reduction in density.

IMPACT EVENTS AND ASSOCIATED DAMAGE

It is necessary to understand the sequence of events occurring during an instrumented impact test, and how those sequences are reflected in the data. The sequence differs, depending on whether or not the plate is penetrated. Both situations are described below.

Plate Penetration

The tup is dropped from a specified height. When the tup contacts the plate, the plate first deforms elastically and the force data, as measured on the load cell, increases linearly with time. The load increases until permanent deformation occurs. This incipient damage is usually in the form of matrix cracking and delaminations, and the energy required for this damage represents a relatively small fraction of the total energy required for penetration. The load data reflects incipient damage with a slight drop in the load or a reduction in the slope of the force curve. At this point, the plate is still able to resist the forward movement of the tup and the force begins to rise again. When the force curve reaches a maximum, fiber fracture begins, most notably on the tension face of the plate, and unstable propagation of matrix cracks and delaminations occurs. The load then drops off rapidly, although some smaller peaks and plateaus are often observed as a result of the arrest of cracks and delaminations as the tup passes through the plate. That damage allows large amounts of energy to be absorbed during this time.

Plate Non-Penetration

This is identical to the above situation through incipient damage. However, when the plate is not penetrated, it maintains some stored energy: the plate transfers that energy back to the tup and causes it to start rebounding. This is indicated in the energy absorption curve by a reduction of the energy absorbed. On the energy absorbed vs. time graph, the stored energy of the part is indicated as the difference between maximum energy absorbed and final energy absorbed.

Types of Damage

The major types of damage that occurred include matrix cracking, delamination, denting, and fiber breakage. Photographs of plates A-1-3, B-1-3, and A-2-6 are examples of each type of damage and are shown in Figures (2-4). All of the plates showed a degree of delamination. Figure 3 shows a typical delamination pattern. Both Figures 2 and 3 are consistent with x-ray

Figure 2: Matrix cracking on plate A-1-3 (20J).

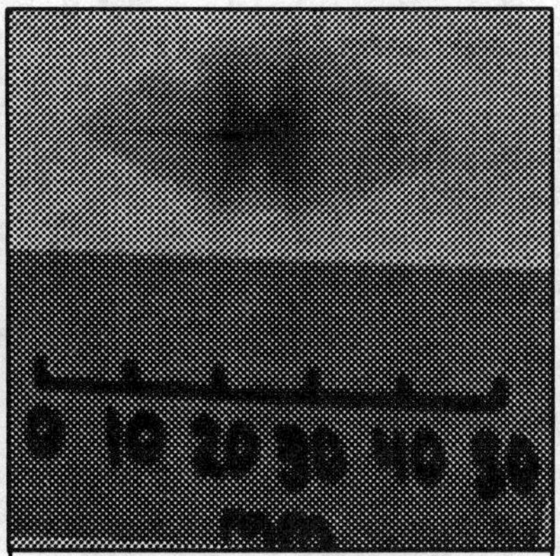

Figure 3: Delamination in B-1-3 (20J).

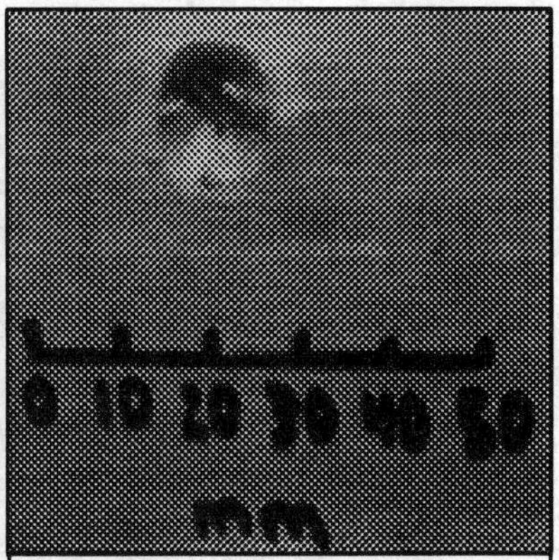

Figure 4: Denting & fiber breakage in A-2-6 (40J).

Table II: Damage in Representative Plates.

ENERGY	PLATE	VISIBLE DAMAGE
5J	C-2-4	dent
15J	A-1-5	matrix crack and very narrow dent
	A-2-3	light dent
	B-1-4	matrix crack & small dent
	B-2-3	dent
	C-1-6	big dent, top layers penetrated
20J	A-1-3	matrix crack
	A-1-4	matrix crack & dent
	B-1-1	dent
	B-1-3	dent
	C-1-5	nearly penetrated
	C-2-3	deep dent
40J	A-2-6	penetrated
	B-2-6	penetrated

and c-scan images which document composite laminate impact damage (Finn and Springer, 1991).

A total of 36 plates were impacted; Table II indicates the energy used for impact and describes the light-box viewed visible damage of representative plates. The remaining figures are the associated data for these representative plates.

THE EFFECTS OF RELATIVE FIBER CONTENT

Figures 5 and 6 show graphs of load versus time for representative A, B, and C plates impacted at 15J and 40J, respectively. In Figure 5, note that the duration of the 15J impact increases from approximately 9ms for the A and B plates to 11ms for the C plate. Also, the A, B, and C plates reach peak loads of 4400, 5000 and 2900 N, respectively. The peak width supports the observation that the damage was more localized in the rigid composites (A and

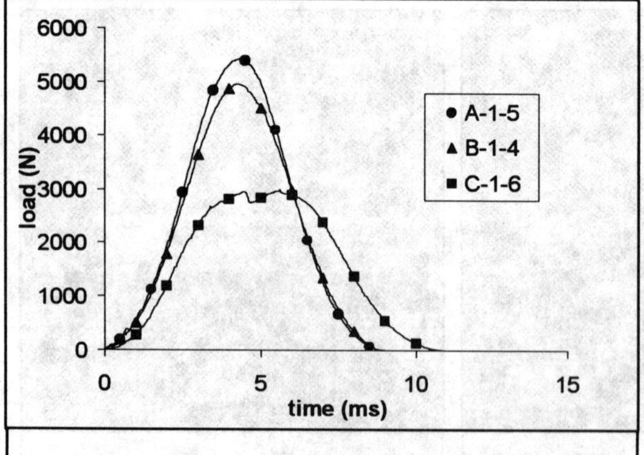

Figure 5: Load vs. time for representative plates at nominal impact energy of 15J.

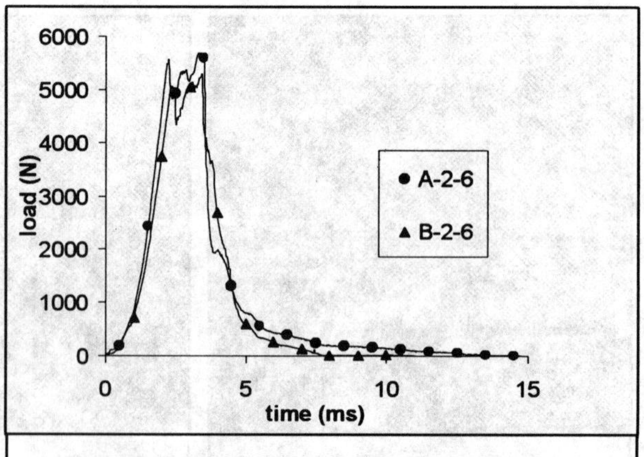

Figure 6: Load vs. time for representative plates at nominal impact energy of 40J.

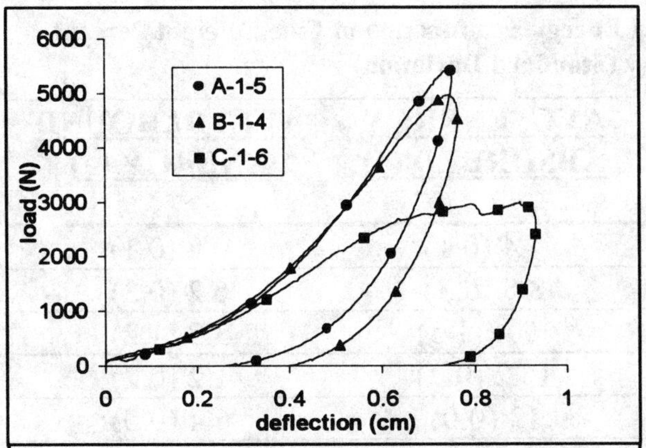

Figure 7: Load vs. deflection for representative plates at nominal impact energy of 15J.

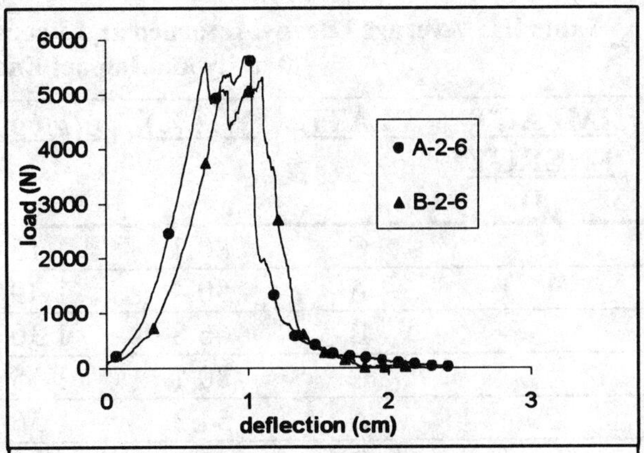

Figure 8: Load vs. deflection for representative plates at nominal impact energy of 40J.

B) than the stiffened fabric (C). Damage, in the form of fiber breakage was also apparent in C, and not A or B, type plates. The smoothness of the load-time data indicates minimal damage to plates A and B, while plate C has sustained damage. Figure 6, at 40J, indicates quite a bit of damage, and both of those plates were, in fact, penetrated.

When reviewing the load versus deflection graphs (Figures 7 and 8), the 15J plates indicate much less deflection than the 40J plates, both at peak load and when the load finally reaches zero, the deflection at which the tup is no longer in contact with the plate. The shape of the curve illustrates the rebound (Figure 7) and penetration (Figure 8) events of the impact. Also, the stiffened fabric plate deforms more and shows significantly less recovery than either A or B type plates.

The effects of the fiber weight percent are clearly indicated when viewing plots of energy absorbed (E_a) versus time (Figures 9 and 10). The energy data is smoother than the load data as a result of the integration performed when converting load to energy. In Figures 9 and 10, at impact energies of 15J and 20J, respectively, the A and B plates are not penetrated, while the C plates are penetrated. As described above, non-penetration is

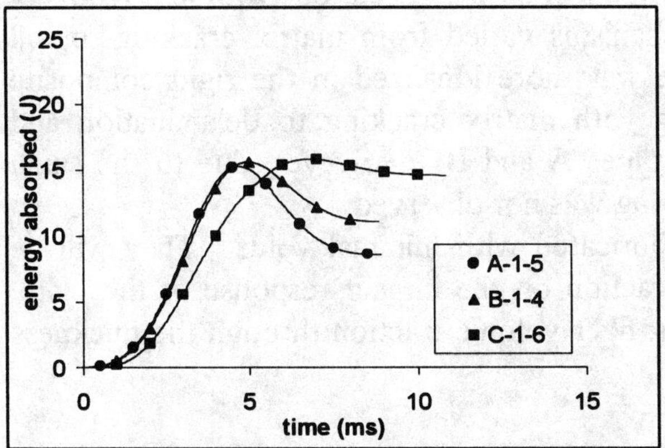

Figure 9: Energy vs. time for representative plates at nominal impact energy of 15J.

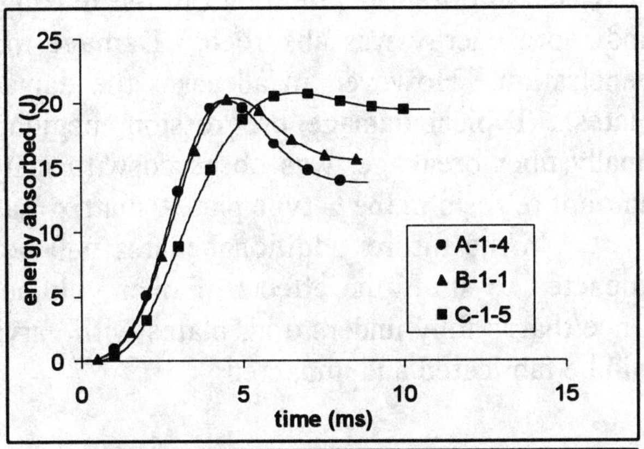

Figure 10: Energy vs. time for representative plates at nominal impact energy of 20J.

Table III: Average Energy Absorbed and Rebound Energy as a function of Fiber Weight Percent, Density and Impact Energy (Standard Deviation)

IMPACT ENERGY (J)	PLATE	W_f (%)	ρ(g/cm^3)	AVG. ENERGY ABSORBED (J)	AVG. REBOUND ENERGY (J)
5	C	80.1	1.15	4.4 (0.1)	0.8 (0.1)
15	A	50.4	1.49	8.8 (0.3)	6.2 (0.3)
	B	66.8	1.30	12.4 (1.5)	3.9 (1.2)
	C	80.1	1.15	15.2 (0.7)	1.2 (0.2)
20	A	54.3	1.56	13 (0.0)	6.4 (0.0)
	B	66.8	1.30	15.9 (1.1)	5.3 (0.3)
	C	79.0	1.05	19.6 (0.1)	1.0 (0.1)
40	A	50.4	1.49	30.5 (4.5)	0.0 (0.0)
	B	71.0	1.04	38 (3.0)	0.0 (0.0)

indicated by the decrease in absorbed energy after reaching a peak, whereas when the plate is penetrated, no energy is returned to the tup. In the A and B plates, not all of the impact energy is dissipated through permanent deformation; the plates act as springs and store the remaining energy which later causes the tup to rebound. However, all of the impact energy is absorbed as permanent deformation in the C plates. They are flexurally less stiff plates, and are penetrated by the tup. Table III indicates the average amount of energy absorbed and associated standard deviations, and the average rebound energy for each plate type. Variations between plate qualities "1" and "2" are not distinguished. Significant differences in energy absorbed are observed when relative fiber content is increased. The stiffened fabric absorbs more energy, but is more susceptible to penetration.

CONCLUSIONS

As the fiber weight percents increased, for the same energy level, the impact events increased in duration. In addition, the maximum deflection increased, but with less recovery and more energy was absorbed. Damage mechanisms varied from matrix cracking to full penetration. However, in all cases the damage was more localized in the rigid composite plates. Typical damage progression, beginning with matrix cracking to delamination and finally fiber breakage, was observed with plate types A and B. However, due to the small amount of resin in the C type plates, matrix cracking was not observed.

In the future, additional plates will be fabricated with minimal voids. They will be impacted to study the effects of fiber volume fraction on the impact response of the plate. Once that is fully understood, plates with varying fiber volume fraction through the thickness will be fabricated and impacted.

ACKNOWLEDGEMENTS

The authors wish to thank the following: the National Science Foundation Graduate Fellowship Program, Instron Dynatup in Santa Barbara, CA, Medinah Salaam for her assistance in obtaining density information, and Jim Sabo at Penn State's Applied Research Laboratory, for use of his impact facilities.

REFERENCES

Abrate, S. 1991. "Impact on laminated composite materials," *Appl. Mech. Review,* 44(4):155-190.

Cantwell, W. J. and Morton, J. 1991. "The impact resistance of composite materials – a review," *Composites,* 22(5):347-362.

Finn, S. R. and Springer, G. S. 1991. *Composite Plates Impact Damage: An Atlas.* Technomic Publishing Co., Lancaster, PA.

Richardson, M. O. W. and Wisheart, M. J. 1996. "Review of low-velocity impact properties of composite materials," *Composites Part A*, 27A:1123-1131.

Shivakumar, K. N., Elber, W. and Illg, W. 1985. "Predictions of low-velocity impact damage in thin circular laminates," *AIAA J.,* 23(3):442-449.

Sierakowski, R. L. and Chaturvedi, S. K. 1997. *Dynamic loading and characterization of fiber-reinforced composites*:30.

Sjöblom, P. O., Hartness, J. T. and Cordell, T. M. 1988. "On low-velosicy impact testing of composite matcrials," *J. of Composite Materials,* 22:30-52.

Strait, L. H. 1989. "The effects of seawater immersion on the impact resistance of polymer matrix composites," Master of Science thesis, The Pennsylvania State University, University Park, PA.

Walsh, T. F. 1996. "Penetration Failure Mechanisms of Woven Textile Composites – an Experimental and Numerical Investigation," Master of Science thesis, The Pennsylvania State University, University Park, PA.

Impact Resistance Modeling of Hybrid Laminated Composites[1]

A. ABATAN, H. HU AND D. OLOWOKERE

ABSTRACT This paper discusses the analytical and numerical results for the impact responses of two types of plates with equal aerial weights. These plates are the hybrid laminated composite plates and the homogeneous metallic plates. The loadings and the boundary conditions for the two types of plates are the same and an impact load was applied at the center of each plate, respectively. An analytical solution based on Fourier expansion was developed to get the impact responses of the plates. The finite element method (FEM) was used to perform the numerical analyses to verify analytical solutions. The response patterns from the results were studied comparatively and a preliminary conclusion made on impact resistance ratio between the hybrid laminates with proper interfacial bonding and that of equivalent homogeneous metallic plates. The effect of different shapes of pulse on the laminated composites was also examined. Three loading cases were considered, each being a function of time. The effect of different pulse shapes on the impact response was evaluated. Finally, a solution based on shear deformation theory was developed, and the comparison between classical plate theory results and shear deformation theory results indicates that for the problem examined classical plate theory predicts relatively accurate results.

INTRODUCTION

Laminated composite materials are used extensively in the various fields of structural engineering because of the light weight, high stiffness, high strength, and cost competitiveness due to recent advancement in material processing and manufacturing technology. The most common composite structural elements are formed into plates and shells. Therefore, understanding the dynamic response of composite laminates subjected to impact is important for structural design and assessment of damage resistance. Impact events occur in a wide variety of circumstances. Typical examples include a weight drop and a bullet impacting a target made of composite material. These impact events produce time-varying loads on a structure that can result in compromised properties as a result of damage. As a first step to

---

[1] Work supported by NASA Contracts NAS3-27785 and NCC3-552
Ayo Abatan and Hurang Hu, Engineering Department, Clark Atlanta University, Atlanta, GA 30314
David Olowokere, Civil & Environmental Engineering, University of Alabama at Birmingham, AL 35294

understanding the damage resistance issue in composite laminates, an accurate prediction of the transient response during an impact event is necessary.

It is desirable for simple analytic solutions to be available for impact response of composite laminates. This would allow the structural analyst to perform parametric studies and establish proper scaling laws. The analysis of dynamic loadings on laminated composite plates has undergone considerable development recently. Sun and Chattopadhyay (1975) and Dobyns (1981) used the plate equations to analyze a simply supported plate subjected to center impact. First order shear deformation plate theory was used by Christoforou and Swanson (1991), and Prasad et al (1993). Several investigators used variational methods of approximation. The Rayleigh-Ritz method was applied to determine the impact response of laminated plates by Cairns and Lagace (1989), and Qian and Swanson (1990).

In this study, the Navier type solution is obtained for a rectangular plate with four edges simply supported subjected to a center impact load. The impact responses of two types of plates with equal aerial weights were examined. The results were compared with the finite element analysis results. The closed-form solution is then applied to two other impact loading cases and the effect of different shapes of the pulse on deflections investigated. Finally, the solution is obtained using shear deformation theory, and the results are compared with classical plate theory results.

ANALYTICAL SOLUTION BASED ON CLASSICAL PLATE THEORY

THEORY

Assume an impact force P(x,y,t) is exerted over the top surface of a rectangular laminated plate with four edges simply supported, as shown in Fig. 1.

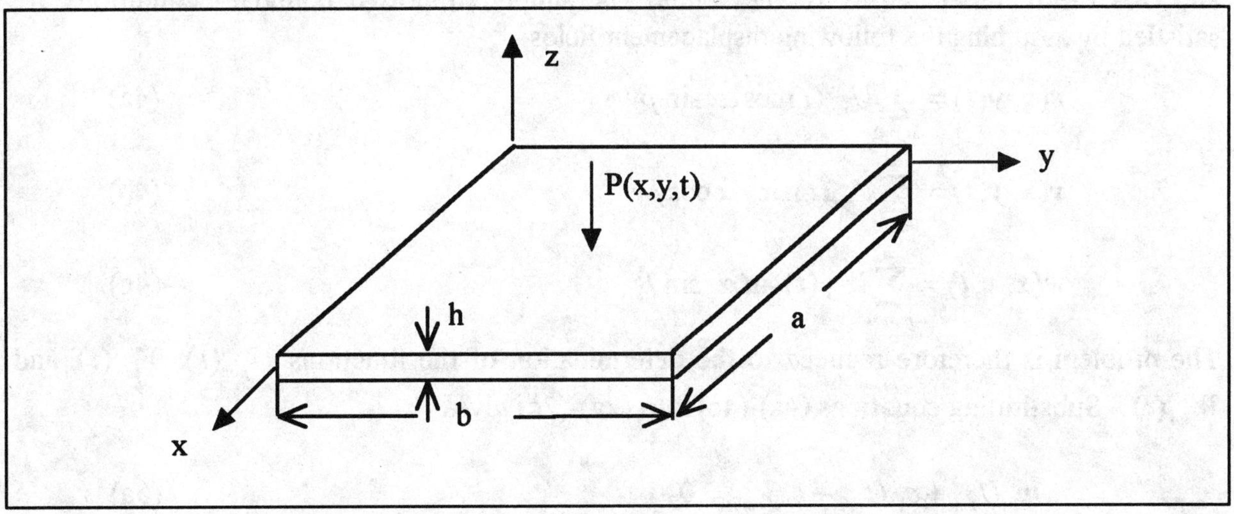

FIG. 1 LAMINATED PLATE UNDER AN IMPACT LOAD

The load may be expanded using double Fourier series as

$$P(x,y,t)=\sum_{m,n=1}^{\infty}P_{mn}(t)\sin\alpha x\sin\beta y \tag{1}$$

where $\alpha = m\pi/a$, $\beta = n\pi/b$, and $P_{mn}(t)$ are load coefficients which can be found from the load P(x,y,t). The boundary conditions may be expressed as $v = w = \frac{\partial w}{\partial y} = 0$ at x = 0 and a; $u = w = \frac{\partial w}{\partial x} = 0$ at y = 0 and b. Consider the laminates where $A_{16} = A_{26} = D_{16} = D_{26} = 0$, $B_{ij} = 0$, the governing equations in terms of the midplane displacements reduce to (Reddy and Miravete, 1995)

$$A_{11}\frac{\partial^2 u}{\partial x^2}+(A_{12}+A_{66})\frac{\partial^2 v}{\partial x\partial y}+A_{66}\frac{\partial^2 u}{\partial y^2}=I_0\frac{\partial^2 u}{\partial t^2} \tag{2a}$$

$$A_{66}\frac{\partial^2 v}{\partial x^2}+(A_{12}+A_{66})\frac{\partial^2 u}{\partial x\partial y}+A_{22}\frac{\partial^2 v}{\partial y^2}=I_0\frac{\partial^2 v}{\partial t^2} \tag{2b}$$

$$D_{11}\frac{\partial^4 w}{\partial x^4}+2(D_{12}+2D_{66})\frac{\partial^4 w}{\partial x^2\partial y^2}+D_{22}\frac{\partial^4 w}{\partial y^4}= \tag{2c}$$

$$P-I_0\frac{\partial^2 w}{\partial t^2}+I_2\frac{\partial^2}{\partial t^2}\left(\frac{\partial^2 w}{\partial x^2}+\frac{\partial^2 w}{\partial y^2}\right)$$

where I_0 and I_2 are the mass inertias defined as

$$(I_0, I_2)=\int_{-h/2}^{h/2}(1,z^2)\rho dz \tag{3}$$

and ρ is the material density of the solids. The simply supported boundary conditions are satisfied by assuming the following displacement fields:

$$u(x,y,t)=\sum_{m,n=1}^{\infty}U_{mn}(t)\cos\alpha x\sin\beta y \tag{4a}$$

$$v(x,y,t)=\sum_{m,n=1}^{\infty}V_{mn}(t)\sin\alpha x\cos\beta y \tag{4b}$$

$$w(x,y,t)=\sum_{m,n=1}^{\infty}W_{mn}(t)\sin\alpha x\sin\beta y \tag{4c}$$

The problem is therefore reduced to the determination of the functions $U_{mn}(t)$, $V_{mn}(t)$ and $W_{mn}(t)$. Substituting equations (4a)-(4c) into (2a)-(2c) gives:

$$m_{11}\ddot{U}_{mn}+c_{11}U_{mn}+c_{12}V_{mn}=0 \tag{5a}$$

$$m_{11}\ddot{V}_{mn}+c_{22}V_{mn}+c_{12}U_{mn}=0 \tag{5b}$$

$$m_{33}\ddot{W}_{mn}+c_{33}W_{mn}=P_{mn}(t) \tag{5c}$$

where $\ddot{U}$ denotes the double derivatives with respect to time t, and

$$c_{11} = A_{11}\alpha^2 + A_{66}\beta^2$$
$$c_{22} = A_{66}\alpha^2 + A_{22}\beta^2$$
$$c_{12} = (A_{12} + A_{66})\alpha\beta$$
$$c_{33} = D_{11}\alpha^4 + 2(D_{12} + 2D_{66})\alpha^2\beta^2 + D_{22}\beta^4$$
$$m_{11} = I_0, \qquad m_{33} = I_0 + I_2(\alpha^2 + \beta^2)$$

With initial displacements and initial velocities assumed to be zero, $U_{mn} = V_{mn} = 0$, and

$$W_{mn}(t) = \frac{1}{\Omega_{mn} m_{33}} \int_0^t P_{mn}(\tau) \sin \Omega_{mn}(t - \tau) d\tau \tag{6}$$

where Ω_{mn} is the vibration frequency of the plate given by

$$\Omega_{mn} = \left(c_{33} / m_{33}\right)^{1/2} \tag{7}$$

On substituting $W_{mn}(t)$ back in equation (4c), deflection w(x,y,t) can be determined. The strains and stresses in the laminate can be computed by using the strain-displacement relation and the constitutive relation, respectively.

NUMERICAL RESULTS

Consider an impact load of rectangular wave as shown in Fig. 2. From equation (6), the deflection is found to be

$$w(x, y, t) = \sum_{m,n=1,3,}^{\infty} \frac{4f}{ab\Omega_{mn}^2 m_{33}} (\cos \Omega_{mn} t - 1) \sin \alpha x \sin \beta y \tag{8}$$

where f is the maximum value of the load (ref. Fig. 2). It is noted that the maximum deflection will occur in the center of the plate, i.e. at x = a/2 and y = b/2.

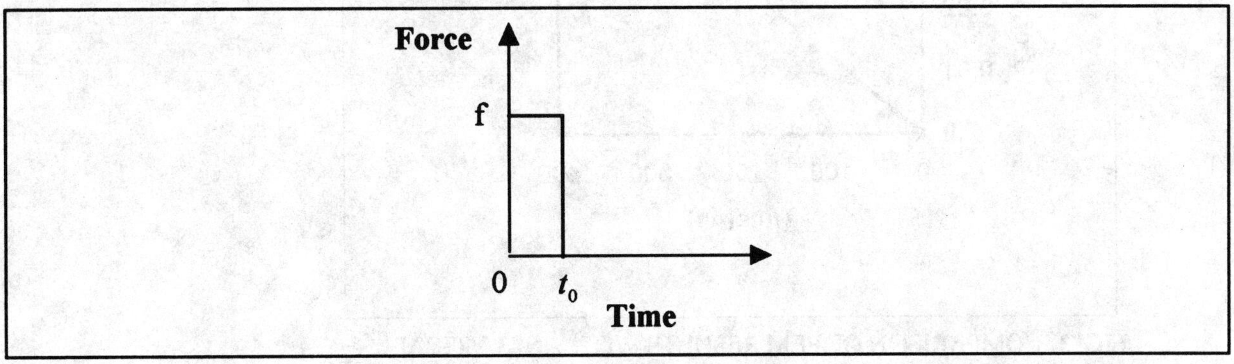

FIG. 2 AN IMPACT LOAD-TIME RELATION

Three plates with equal areal weight of 3 psf were analyzed. One is a homogeneous metallic plate with thickness 0.075″ (1.905mm), the second is a three layer laminate with stacking sequence [metal/polymer/metal], and the third is a 21 layer laminate with stacking sequence [metal/polymer/.../metal]. For all the plates, the dimensions in the x-y plane are 7″ × 7″ (0.1778m × 0.1778m). The material properties of the metal and polymer are listed in Table 1.

TABLE 1. MATERIAL PROPERTIES FOR METAL AND POLYMER

Property	E(msi)	ν	$\rho(lb \cdot \sec^2 / in^4)$
Metal	**28.0**	**0.3**	**7.254E-4**
Polymer	**1.0**	**0.2**	**9.067E-5**

In order to ensure the reliability of analytical results, we performed finite element analyses, and compared the finite element results with the analytical analysis results. The finite element code ANSYS was used to perform the numerical analyses using eight-noded layered shell element (ANSYS User's Manual, 1994). Due to the symmetry, only a quarter of the plate was discretized and analyzed. The finite element results for all three plates are very close to the analytical results. The comparison of the two results for the 21 layer laminated plate is shown in Fig. 3. Excellent agreement was observed. Therefore, the analytical results are reliable and accurate.

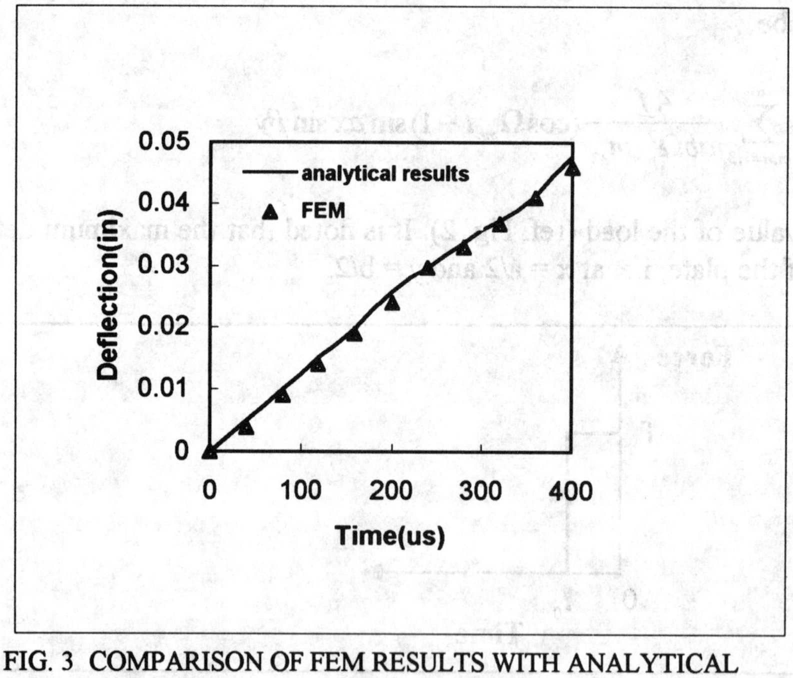

FIG. 3 COMPARISON OF FEM RESULTS WITH ANALYTICAL RESULTS FOR 21 LAYER LAMINATE

The deflections at plate center versus time for all three plates are shown in Fig. 4. The maximum deflections of the three plates are 0.082", 0.038", and 0.048", respectively. It is noted that the deflection of the metallic plate is much higher than that of the laminated plates, and the deflections of the 21 layer laminate is higher than that of the 3 layer laminate. It appears that the reason for this is because the 3 layer laminate has less metal (and therefore more polymer) than the 21 layer laminate. For plates with equal areal weight, considering the ratio of metal thickness over the total thickness as a variable, then the deflections and stresses will be proportional to that ratio. For homogeneous metallic plate which has the smallest volume, the deflections and stresses will be the largest; for homogeneous polymer plate which has the largest volume, the deflections and stresses will be the smallest. For all the laminates consist of the two materials with equal areal weight, the deflections and stresses will fall in between.

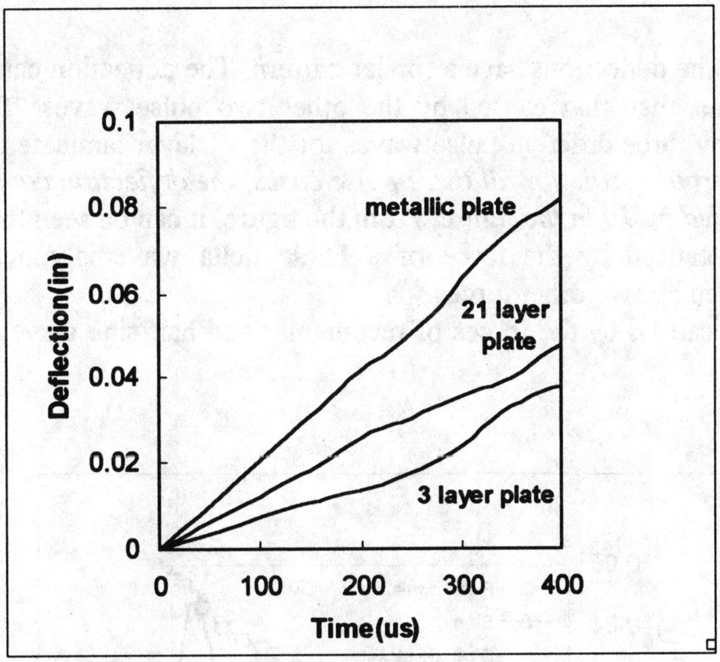

FIG. 4 DEFLECTIONS OF THREE PLATES WITH EQUAL AREAL WEIGHT

EFFECT OF DIFFERENT SHAPES OF THE PULSE ON THE DEFLECTIONS

The effect of the different shapes of the pulse with the same impact duration on the deflections of the plate was examined. The impact load is assumed to vary in time as a rectangular wave, a half sine wave, and a Dirac delta function, respectively. The rectangular wave is shown in Fig. 2. For a half sine wave, the impact force may be expressed as

$$P = f \sin \frac{t}{t_0} \pi \tag{9}$$

where t_0 is the duration of the impact. The deflections at x = a/2 and y = b/2 is found to be

$$w(t)=\sum_{m,n=1,3,}^{\infty}\frac{4f}{ab\Omega_{mn}^2 m_{33}}\frac{\Omega_{mn}}{(\pi/t_0-\Omega_{mn})(\pi/t_0+\Omega_{mn})}(\Omega_{mn}\sin\frac{t}{t_0}\pi-\frac{\pi}{t_0}\sin\Omega_{mn}t) \quad (10)$$

For a pulse of Dirac delta function

$$P=\delta(t-t_0/2) \quad (11)$$

The deflections at the center of the plate is found to be

$$w(t)=\begin{cases}0 & t<t_0/2\\ \sum_{m,n=1,3,}^{\infty}-\frac{4\Omega_{mn}}{ab\Omega_{mn}^2 m_{33}}\sin[\Omega_{mn}(t-t_0/2)] & t\geq t_0/2\end{cases} \quad (12)$$

For all three plates, the deflections have a similar pattern. The deflection caused by Dirac delta wave is much higher than that caused by the other two pulse waves. The comparison of deflections caused by three different pulse waves for the 21 layer laminate is shown in Fig. 5. *In order to have a proper scale for all three pulse cases, the deflection caused by Dirac delta wave has been divided by 10 in the figure*. From the figure, it can be seen that:

1. The deflection caused by the force of a Dirac delta wave is much higher that the deflections caused by two other forces.
2. The deflections caused by the forces of rectangular and half sine waves are close to each other.

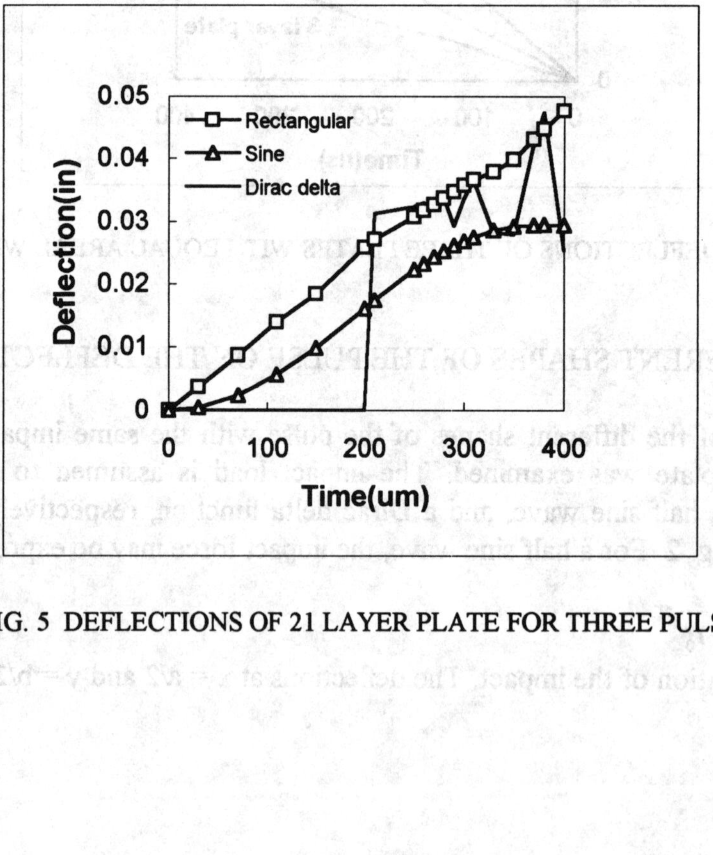

FIG. 5 DEFLECTIONS OF 21 LAYER PLATE FOR THREE PULSES

ANALYTICAL SOLUTION BASED ON SHEAR DEFORMATION THEORY

THEORY

The shear deformation is allowed by introducing planar rotations as independent variables. Let ψ_x and ψ_y be the shear rotations at the plate midplane in the x and y directions, then the governing equations of motion for the special orthotropic plates under consideration are (Dobyns, 1981):

$$D_{11}\frac{\partial^2\psi_x}{\partial x^2}+(D_{12}+D_{66})\frac{\partial^2\psi_y}{\partial x\partial y}+D_{66}\frac{\partial^2\psi_x}{\partial y^2}-kA_{55}(\psi_x+\frac{\partial w}{\partial x})=\frac{\rho h^3}{12}\frac{\partial^2\psi_x}{\partial t^2} \quad (13a)$$

$$D_{22}\frac{\partial^2\psi_y}{\partial y^2}+(D_{12}+D_{66})\frac{\partial^2\psi_x}{\partial x\partial y}+D_{66}\frac{\partial^2\psi_y}{\partial x^2}-kA_{44}(\psi_y+\frac{\partial w}{\partial y})=\frac{\rho h^3}{12}\frac{\partial^2\psi_y}{\partial t^2} \quad (13b)$$

$$kA_{55}(\frac{\partial\psi_x}{\partial x}+\frac{\partial^2 w}{\partial x^2})+kA_{44}(\frac{\partial\psi_y}{\partial y}+\frac{\partial^2 w}{\partial y^2})+P=\rho h\frac{\partial^2 w}{\partial t^2} \quad (13c)$$

where k is a shear correction factor which is taken to be $\pi^2/12$. The simply supported boundary conditions may be expressed as $w=\frac{\partial\psi_x}{\partial x}=0$ at $x=0$ and a; $w=\frac{\partial\psi_y}{\partial y}=0$ at $y=0$ and b. The solution is based on expansions of the loads, displacements and rotations in Fourier series which satisfy the specified boundary conditions. Assume the following functions:

$$\psi_x(x,y,t)=\sum_{m,n=1}^{\infty}A_{mn}(t)\cos\alpha x\sin\beta y \quad (14a)$$

$$\psi_y(x,y,t)=\sum_{m,n=1}^{\infty}B_{mn}(t)\sin\alpha x\cos\beta y \quad (14b)$$

$$w(x,y,t)=\sum_{m,n=1}^{\infty}W_{mn}(t)\sin\alpha x\sin\beta y \quad (14c)$$

Substituting equations (1) and (14a)-(14c) into (13a)-(13c), and neglecting in-plane and rotary inertia gives

$$d_{11}A_{mn}(t)+d_{12}B_{mn}(t)+d_{13}W_{mn}(t)=0 \quad (15a)$$
$$d_{12}A_{mn}(t)+d_{22}B_{mn}(t)+d_{23}W_{mn}(t)=0 \quad (15b)$$
$$d_{13}A_{mn}(t)+d_{23}B_{mn}(t)+d_{33}W_{mn}(t)=P_{mn}(t)-\rho h\ddot{W}_{mn}(t) \quad (15c)$$

where

$$d_{11}=D_{11}\alpha^2+D_{66}\beta^2+kA_{55}$$
$$d_{12}=(D_{12}+D_{66})\alpha\beta$$
$$d_{13}=kA_{55}\alpha$$
$$d_{22}=D_{66}\alpha^2+D_{22}\beta^2+kA_{44}$$

$$d_{23} = kA_{44}\beta$$
$$d_{33} = kA_{55}\alpha^2 + kA_{44}\beta^2$$

From (15a) and (15b) gives

$$A_{mn}(t) = \frac{d_{12}d_{23} - d_{13}d_{22}}{d_{11}d_{22} - d_{12}^2} W_{mn}(t) \tag{16a}$$

$$B_{mn}(t) = \frac{d_{12}d_{13} - d_{11}d_{23}}{d_{11}d_{22} - d_{12}^2} W_{mn}(t) \tag{16b}$$

Substituting (16a) and (16b) into (15c) gives the general solution of $W_{mn}(t)$ as

$$W_{mn} = \frac{1}{\omega_{mn}\rho h}\int_0^t P_{mn}(\tau)\sin\omega_{mn}(t-\tau)d\tau \tag{17}$$

where

$$\omega^2_{mn} = \frac{d_{11}d_{22}d_{33} + 2d_{12}d_{13}d_{23} - d_{33}d_{12}^2 - d_{22}d_{13}^2 - d_{11}d_{23}^2}{\rho h(d_{11}d_{22} - d_{12}^2)} \tag{18}$$

Comparing equation (17) with equation (6), it is clear that the deflections for the two theories have the same expression. Any one can be found from another by using the following correspondence:

Classical plate theory Ω_{mn} and m_{33} $\leftrightarrow$ shear deformation theory ω_{mn} and ρh

NUMERICAL RESULTS

Using shear deformation theory formulation, the deflections for all three plates under three different pulse waves were determined. For the 21 layer laminate, Fig. 6 shows the comparison of deflections at plate center between classical plate theory (CPT) and shear deformation theory (SDT) under three different pulses. *Note that in the figure the values of deflection caused by Dirac delta were divided by 10*. The comparison of maximum deflections between CPT and SDT is given in Table 2.

TABLE 2. COMPARISON OF MAXIMUM DEFLECTIONS BETWEEN CPT AND SDT

	Rectangular	Sine	Dirac delta
SDT (in)	**0.0482**	**0.0296**	**0.5035**
CPT (in)	**0.0477**	**0.0295**	**0.4736**

For pulses of rectangular wave and sine wave, the difference between SDT results and CPT results is about 1%. For Dirac delta function, the difference is about 6%. It appears that the Dirac delta function is not a good model for impact loading and for the problem discussed, classical plate theory predicts relatively accurate results.

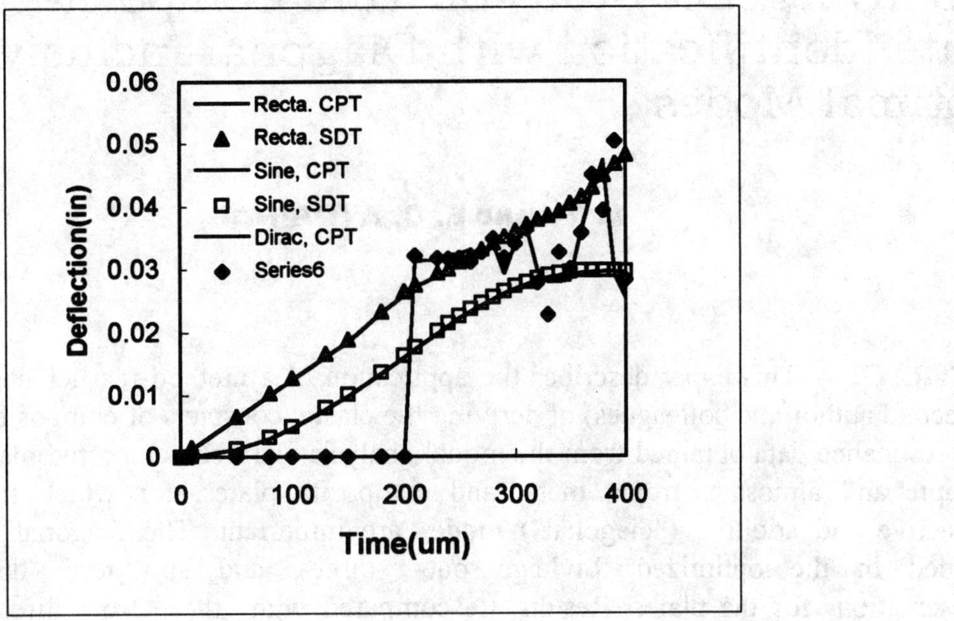

FIG. 6 COMPARISON OF DEFLECTIONS BETWEEN CPT AND SDT FOR THREE PULSES

CONCLUSION

An analytical solution was developed to investigate the impact response of hybrid laminates. The solution is simple and easy to apply for other loadings. Comparison with the finite element results shows that the analytical solution for the problem examined is accurate. With the same impact duration and maximum load, the effect of pulse shape on impact response is small. For the problems examined, classical plate theory predicts accurate results.

REFERENCES

ANSYS User's Manual, Swanson Analysis Systems, Inc., 1994.

Cairns, D.S. and Lagace, P.A. 1989. "Transient Response of Graphite/Epoxy and Kevlar/ Epoxy Laminates Subjected to Impact," AIAA J., **27**:1590-1596.

Christoforou, A.P. and Swanson, S.R.1991. "Analysis of Impact Response in Composite Plates," Int. J. Solids Structures, **27**:161-170.

Dobyns, A.L. 1981. "Analysis of Simply-Supported Orthotropic Plates Subjected to Static and Dynamic Loads," AIAA J., **19**: 642-650.

Qian, Y. and Swanson, S.R. 1990. "A Comparison of Solution Techniques for Impact Response of Composite Plates," Composite Struct., **14**:177-192.

Prasad, C.B., Ambur, D.R. and Starnes, J.H. 1993. "Response of Laminated Composite Plates to Low-Speed Impact by Airgun-Propelled and Dropped Weight Impactors," 34th AIAA Struct. Structural Dyn and Mat Conf, La Jolla CA, Part 2, pp. 887-900.

Reddy, J.N. and Miravete, A. 1995. Practical Analysis of Composite Laminates, CRC Press, New York.

Sun, C.T. and Chattopadhyay, S.1975. "Dynamic Response of Anisotropic Plates Under Initial Stress Due to Impact of Mass," J. Appl. Mech., **42**: 693-698.

A Convergence Study on Thin Composite Plate Identification with Diagonal-Inclusive Optimal Modes

L. YU AND E. O. AYORINDE

ABSTRACT This paper describes the application of a method (earlier developed by the second author and colleagues) of deriving the elastic constants of composite materials from resonance data obtained from the modal analysis of freely-supported plates, to thin isotropic and almost isotropic metal and composite plates, for which the coupled subtractive and additive ("diagonal") modes are important. The diagonal modes are included in the optimized Rayleigh one-, three-, and six- term displacement representations for the plates. Results are compared with those from direct Rayleigh expansions.

INTRODUCTION

Increasing industrial and military use of composite materials has made rapid, inexpensive determination of their elastic constants important. This impacts design, characterization and damage analysis of these materials.

Several numerical/experimental methods of obtaining the dynamic elastic constants have been proposed (as reviewed by Ayorinde and Gibson, 1993). A few of these are the approaches of Zelenev and Electrova (1973), DeWilde, et al. (1984), and Frederiksen (1997). Ayorinde and colleagues developed a method based on the use of a potential-energy-optimized modal representation (Kim and Dickinson, 1985) of the transverse displacement of a completely free plate in Rayleigh's method (Ayorinde, Gibson and Wen, 1993; Ayorinde and Gibson, 1993, 1995; Ayorinde 1995; Ayorinde and Yu, 1997). The last reference dwelt at some length on diagonal modes. These are composite modes made up of the sums and the differences of pure, indicially inverse modes such as (i,j) and (j,i), i.e. (i,j±j,i) modes. These modes occur in square samples when i and j are different and either both even or odd. This paper looks at the use of various versions of the optimized tranaverse vibration expression in the dynamic analysis. The sensitivity of the vibration modes to the elastic constants is also taken into account.

Lin Yu and Emmanuel O. Ayorinde, Advanced Composites Research Laboratory, Wayne State University, Detroit, MI 48202

In this study the diagonal modes are included in the analytical derivation of the frequency equation for the square, isotropic or almost isotropic plates by using the optimized three-mode Rayleigh formulation. The first logical step is the "forward" problem of estimating frequencies from the plate vibration model and known elastic constants because the accuracy of the solution to this problem provides an indication of the feasibility and relative ease of solution of the "inverse" material identification problem. The best mode set for each desired constant is automatically selected from the input experimental modal data, with the aid of sensitivity analysis.

ANALYSIS

The free transverse vibration of a symmetrically laminated rectangular plate at equilibrium in the xy-plane may be rendered as

$$\begin{aligned} &D_{11}\frac{\partial^4 w}{\partial x^4}+4D_{16}\frac{\partial^4 w}{\partial x^3\partial y}+2(D_{12}+2D_{66})\frac{\partial^4 w}{\partial x^2\partial y^2}\\ &+4D_{26}\frac{\partial^4 w}{\partial x\partial y^3}+D_{22}\frac{\partial^4 x}{\partial y^4}+\rho h\frac{\partial^2 w}{\partial t^2}=0 \end{aligned} \tag{1}$$

where the plate transverse deflection is w(x,y,t), x and y are the orthogonal plate coordinates, D_{ij} is the standard extension-bending stiffness of the classical lamination theory, ρ is the plate density, h is the plate thickness and t represents time. For specially orthotropic plates, $D_{16} = D_{26} = 0$, $D_{11} = D_x$, $D_{22} = D_y$, $D_{12} = \nu_{xy} D_y$ and $D_{66} = D_{xy}$, where $D_x = E_x h^3 / 12(1-\nu_{xy}\nu_{yx})$, $D_y = E_y D_x / E_x$, $D_{xy} = h^3 G_{xy} / 12$. The equilibrium equation for the isotropic plate is obtained from Eq.(1) simply by replacing ν_{xy} and ν_{yx} with ν, D_x and D_y with $D = Eh^3 / 12(1-\nu^2)$, and $2D_{xy} = 2(3-\nu)D$, where E is the Young's modulus and ν is the Poisson's ratio.

The deflection expression for transverse, six-mode representation displacement for a diagonally-symmetrical mode may be written as

$$\begin{aligned} W(x,y) = A\{&(\theta_i\phi_j - c\theta_i\phi_n - d\theta_m\phi_j - e\theta_i\phi_q - f\theta_m\phi_n - g\theta_p\phi_j)\\ \pm &(\theta_j\phi_i - c\theta_n\phi_i - d\theta_j\phi_m - e\theta_q\phi_i - f\theta_n\phi_m - g\theta_j\phi_p)\} \end{aligned} \tag{2}$$

where the mode shape of interest is the (i, j)th, A, c, d, e, f and g are constants, θ is a function of x and ϕ is a function of y. $\theta_i(x)$ and $\phi_j(y)$ are the appropriate i-th and j-th beam mode shapes, m=i+1, p=i+2, n=j+1, q=j+2 for dissimilar beam end and m=i+2, p=i+4, n=j+2, q=j+4 for the case where the conditions are the same at each end.

Within the representation of Equation (2), the maximum kinetic and potential energy expressions for square isotropic plates, respectively, are given by

$$T_{\max} = A^2 \rho h\omega^2 \left(1+c^2+d^2+e^2+f^2+g^2\right) X_i Y_j \tag{3}$$

$$\max = \frac{A^2 H\pi^4}{a^4}\Big\{C_{ij}+c^2C_{in}+d^2C_{mj}+e^2C_{iq}+f^2C_{mn}+g^2C_{pj}-2cE_{ij}-2dE_{ji}-2eE_{ij*}-2fQ_{ji}-2gE_{ji*}$$
$$+2cdF_{ij}+2ceE_{in}+2cfE_{ni}+2cgF_{i*j}+2deF_{ij*}+2dfE_{mi}+2dgE_{jm}+2efF_{in}+2egF_{i*j*}+2fgF_{mj}$$
$$\pm(\overline{C}_{ij}+c^2\overline{C}_{in}+d^2\overline{C}_{mj}+e^2\overline{C}_{iq}+f^2\overline{C}_{mn}+g^2\overline{C}_{pj}-2c\overline{E}_{ij}-2d\overline{E}_{ji}-2e\overline{E}_{ij*}-2f\overline{Q}_{ij}-2g\overline{E}_{ij*}$$
$$+2cd\overline{F}_{ij}+2ce\overline{E}_{in}+2cf\overline{E}_{ni}+2cg\overline{F}_{i*j}+2de\overline{F}_{ij*}$$
$$+2df\overline{E}_{mj}+2dg\overline{E}_{jm}+2ef\overline{F}_{in}+2eg\overline{F}_{i*j*}+2fg\overline{F}_{mj})\Big\}X_iY_j \tag{4}$$

where $X_i = \int_0^a \theta_i^2 dx,\ \ Y_j = \int_0^a \phi_j^2 dy$, and

$$C_{ij} = G_i^4 + G_j^4 + 2[\nu H_i H_j + (1-\nu)J_i J_j] \qquad \overline{C}_{ij}=\nu(K_{ij}^2+K_{ji}^2)+2(1-\nu)M_{ij}^2$$
$$E_{ij} = -\nu(K_j+L_j)H_i + 2(1-\nu)J_i M_j \qquad \overline{E}_{ij}=\nu(K_{ij}K_{in}+K_{ji}K_{ni})+2(1-\nu)M_{ij}M_{in}$$
$$E_{ji} = -\nu(K_i+L_i)H_j + 2(1-\nu)J_j M_i \qquad \overline{E}_{ji}=\nu(K_{ji}K_{mj}+K_{ij}K_{mj})+2(1-\nu)M_{ji}M_{mj}$$
$$F_{ij} = \nu(K_iK_j+L_iL_j)+2(1-\nu)M_iM_j \qquad \overline{F}_{ij}=\nu(K_{ij}K_{mn}+K_{ji}K_{nm})+2(1-\nu)M_{ij}M_{mn}$$
$$Q_{ij} = \nu(K_jL_i+K_iL_j)+2(1-\nu)M_iM_j \qquad \overline{Q}_{ij}=\nu(K_{in}K_{mj}+K_{ni}K_{jm})+2(1-\nu)M_{in}M_{jm} \tag{5}$$

The integrals G_i, H_i, J_i, K_i, L_i, M_i and K_{ij}, M_{ij} are based on normal-mode free-free beam characteristic functions, and were given in an earlier paper (Ayorinde and Gibson, 1993). In the above equations, m = i+2. The y-direction integrals, such as G_j, H_j, etc. are similarly obtained with appropriate indices. The constants with the overbar refer to the diagonal modes. The asterisked ones refer to a six-mode displacement expression.

$$K_{i*} = (a^2/\pi^2)\int_0^a \theta_i\theta_p''\,dx \Big/ \int_0^a \theta_i^2\,dx \tag{6}$$

where p = i + 1. Values of c, d, e, f, and g come from potential energy optimization with respect to these constants. This is equivalent to solving the matrix equation

$$\begin{bmatrix} C_{in}\pm\overline{C}_{in} & F_{ij}\pm\overline{F}_{ij} & E_{in}\pm\overline{E}_{in} & E_{ni}\pm\overline{E}_{ni} & F_{i*j}\pm\overline{F}_{i*j} \\ & C_{mj}\pm\overline{C}_{mj} & F_{ij*}\pm\overline{F}_{ij*} & E_{mj}\pm\overline{E}_{mj} & E_{jm}\pm\overline{E}_{jm} \\ & & C_{iq}\pm\overline{C}_{iq} & F_{in}\pm\overline{F}_{in} & F_{i*j*}\pm\overline{F}_{i*j*} \\ & Sym & & C_{mn}\pm\overline{C}_{mn} & F_{mj}\pm\overline{F}_{mj} \\ & & & & C_{pj}\pm\overline{C}_{pj} \end{bmatrix} \begin{Bmatrix} c\\ d\\ e\\ f\\ g \end{Bmatrix} = \begin{Bmatrix} E_{ij}\pm\overline{E}_{ij}\\ E_{ij}\pm\overline{E}_{ij}\\ E_{ij*}\pm\overline{E}_{ij*}\\ Q_{ij}\pm\overline{Q}_{ij}\\ E_{ji*}\pm\overline{E}_{ji*} \end{Bmatrix} \tag{7}$$

The resulting frequency equation is

$$\frac{\rho h a^4}{\pi^4} = \frac{H}{\omega^2(1+c^2+d^2)} * \Big[(C_{ij} \pm \bar{C}_{ij}) + c^2(C_{in} \pm \bar{C}_{in}) + d^2(C_{mj} \pm \bar{C}_{mj}) + e^2(C_{iq} \pm \bar{C}_{iq})$$
$$+ f^2(C_{mn} \pm \bar{C}_{mn}) + g^2(C_{pj} \pm \bar{C}_{pj}) - 2c(E_{ij} \pm \bar{E}_{ij}) - 2d(E_{ji} \pm \bar{E}_{ji}) - 2e(E_{ij*} \pm \bar{E}_{ij*}) \quad (8)$$
$$- 2f(Q_{ij} \pm \bar{Q}_{ij}) - 2g(E_{ji*} \pm \bar{E}_{ji*}) + 2cd(F_{ij} \pm \bar{F}_{ij})] + 2ce(E_{in} \pm \bar{E}_{in}) + 2cf(E_{ni} \pm \bar{E}_{ni})$$
$$+ 2cg(F_{i*j} \pm \bar{F}_{i*j}) + 2de(F_{ij*} \pm \bar{F}_{ij*}) + 2df(E_{mj} \pm \bar{E}_{mj}) + 2dg(E_{jm} \pm \bar{E}_{jm})$$
$$+ 2ef(F_{in} \pm \bar{F}_{in}) + 2eg(F_{i*j*} \pm \bar{F}_{i*j*}) + 2fg(F_{mj} \pm \bar{F}_{mj})\Big]$$

where $H = \nu_{xy} D_y + 2D_{xy} = D$ (9)

A residual found from the difference between experimental and theoretical frequency values (Ayorinde and Gibson, 1993) is then used as the objective function in the optimal search for the elastic constants.

SENSITIVITY ANALYSIS

The sensitivity of a modal frequency to an elastic constant, defined as

$$S_{\omega/E_x} = \frac{\partial\omega/\partial E_x}{\omega/E_x} \quad (10)$$

Sensitivity formulation for the diagonal modes may be obtained appropriate modification of those derived earlier for proper modes (Ayorinde, 1995). One then obtains,for example for E_x,

$$\partial \bar{E}_{ij}/\partial E_x = 2(2M_{ij}M_{in} - K_{ij}K_{in} - K_{ji}K_{ni})\partial(D_{xy}/H)/\partial E_x \quad (11)$$

METHOD

Sensitivity analysis is integrated into the inverse program for automatic selection of the best set of experimental modes to back out elastic constants based on the experimental modal data. Several solution approaches are possible. Here we use the lowest 10 frequencies as the input to the inverse program. Experimentally, these should be more accurately obtained than the higher modes.

We first solve the forward problem of using prescribed elastic constants to obtain plate natural frequencies, with 1-term, 3-term and 6-term displacement expressions including diagonal modes. Equation 8 is solved for frequencies using the corresponding modal indices. An aluminum plate measuring 254 mm × 254 mm × 3.16 mm and having a mass/unit volume density of 2770 kg/m^3, and an 1850 kg/m^3 SMC plate measured 306 mm × 305.6 mm × 2.682 mm., made of randomly oriented chopped glass fibers in a polyester matrix, are used. Experimental results (the plate of Deobald and Gibson, 1988), and the 36-term (6 × 6) and 64-term (8×8) Rayleigh-Ritz method are also included for comparison. The best four modes (having the highest four values of sensitivity) to each elastic constant are are then used to obtain the elastic constants.

RESULTS

Table 1 shows first 10 frequencies of the SMC plate with F-F-F-F boundary conditions obtained by using the static values of material properties. Table 2 shows elastic constants results both with and without the diagonal mode analysis.

Table 1 Natural frequencies of square SMC plate with completely-free boundary conditions

Model Indices	Natural Frequencies, Hz: Exptl.	1-mode with diag.	3-mode with diag.	6-mode with diag	36-term R.-R.[a]	64-term R.-R.[a]
2,2	54.4[-2.1] [b]	58.59[5.5]	55.92[0.7]	55.65[0.20]	55.56[0.04]	55.54
3,1-1,3	78.8[-2.75]	81.73[-0.86]	81.35[0.40]	81.10[0.09]	81.22[0.23]	81.03
1,3+3,1	99.2[0.90]	99.02[0.71]	98.53[0.21]	98.40[0.08]	98.47[0.15]	98.32
2,3	138.4[-3.42]	149.69[4.46]	145.15[1.29]	144.82[1.06]	143.50[0.14]	143.30
3,2	138.4[-3.42]	149.69[4.46]	145.15[1.29]	144.82[1.06]	143.50[0.14]	143.30
1,4	242.4[-2.76]	250.17[0.36]	249.98[0.28]	249.83[0.22]	249.64[0.15]	249.27
4,1	242.4[-2.76]	250.17[0.36]	249.98[0.28]	249.83[0.22]	249.64[0.15]	249.27
3,3	254.0[-3.02]	276.23[5.47]	264.19[0.87]	262.86[0.37]	263.34[0.55]	261.90
2,4-4,2	263.4[-7.81]	285.86[0.06]]	285.83[0.05]	285.75[0.02]	285.77[0.03]	285.70
4,2+2,4	306.6[-2.72]	325.23[3.26]	315.58[0.20]	314.77[-0.06]	315.37[0.13]	314.96

[a] From Deobald and Gibson (1988)
[b] [ ] contain % difference from 64-term Rayleigh-Ritz data

Table 2. Elastic constants of Aluminum from modal data of square plate

	Static	Inverse 3-term	Inverse 6-term	Modes used	Residual 3-term	Residual 6-term
Ex,Gpa	72.40	72.58	72.58	(3,1-1,3) (4,1) (4,2+2,4) (3,2) .	0.4447×10^{-2}	0.3089×10^{-2}
Ey,Gpa	72.40	70.38	71.03	(1,4) (1,3+3,1) (2,4-4,2) (2,3)	0.1848×10^{-3}	0.1815×10^{-3}
Gxy,Gpa	28.00	25.55	25.80	(2,2) (3,3) (2,3) (3,2)	0.2983×10^{-3}	0.7492×10^{-3}
ν_{xy}	0.33	0.32	0.33	(1,3+3,1) (4,2+2,4) (1,4) (4,1)	0.6478×10^{-3}	0.6137×10^{-3}

Table 3 Elastic constants of SMC from modal data of square plate

	Static	Inverse 3-term	6-term	Modes used	Residual 3-term	6-term
Ex,Gpa	16.20	15.90	15.90	(3,1-1,3) (4,1) (4,2+2,4) (3,2)	0.4577×10^{-2}	0.3730×10^{-2}
Ey,Gpa	16.20	13.80	13.94	(1,4) (1,3+3,1) (2,4-4,2) (2,3)	0.1510×10^{-3}	0.1001×10^{-3}
Gxy,Gpa	6.35	6.010	6.122	(2,2) (3,3) (2,3) (3,2)	0.1861×10^{-3}	0.6187×10^{-3}
v_{xy}	0.275	0.270	0.275	(1,3+3,1) (4,2+2,4) (1,4) (4,1)	0.5571×10^{-2}	0.5555×10^{-2}

DISCUSSION

From Tables 1, it is seen that the first 10 resonance frequencies of the SMC plate predicted by the optimized 1-mode, 3-mode and 6-mode Rayleigh method inclusive of diagonal modes compare well with those from the 36-term and 64-term Rayleigh-Ritz methods. 6-mode results with the diagonal modes deviate by only about 1% or less from the 64-term Rayleigh-Ritz results, while the 3-mode results deviate by about 1.3 % or less. Even for the 1-term Rayleigh results, the differences from the 64-term Rayleigh-Ritz results are generally below 6 percent. It seems that the optimal three-term approximation gives results comparable to the normal 36-term Rayleigh expansion, while the optimal six-term compares favorably with the 64-term Rayleigh expansion.

The inverse problem results are in Tables 2 and 3, for square aluminum plate and the SMC plate, respectively. The results are consistent with all the earlier results obtained by the previous works of Ayorinde and Gibson.

It may be seen from these tables that, the inverse results from the 6-term Rayleigh method are consistently better than those of the 3-term "inverse" result, but the improvement is not remarkable, and takes almost twice the CPU time. Therefore, the optimal 3-term Rayleigh displacement expression could satisfy the requirements of accurate and quick identification of elastic constants.

Table 1 shows that the experimental frequency data for SMC are lower than the theoretical values predicted from the static properties. This would lead to lower elastic constants than the static values as shown in Table 3. In addition, the relatively large difference between the predicted E_x and E_y of SMC plate may be caused by the slightly orthotropic behavior of the sample plate due to the peculiarities of mold flow during fabrication. In any case, only one Young's modulus value was given by the industrial supplier of the SMC plate, and the transverse value was taken to be the same on the basis of assumed isotropy.The Poisson's ratios for both of the plates are well predicted with lower residuals

CONCLUSIONS

Our results suggest that using a 3-mode vibration displacement function including the diagonal modes, for a completely-free square isotropic plate, as opposed to a 3-mode function without diagonal modes yields better results for the predicted elastic constants. Therefore in general, for enhanced accuracy, it is desirable to include the diagonal modes in the elastic identification of materials by methods involving explicit mode selections. Although there is accuracy improvement in the elastic identification exercise, going from use of the 1-mode to the 6- mode deflection representation, the improvement from the 3-mode to 6-mode is minimal and comes at a hefty price in computing resources

ACKNOWLEDGMENTS

The authors acknowledge the financial and other support of the Office of Naval Research and Program Director Dr. Yapa D.S. Rajapakse, and Wayne State University's Institute of Manufacturing Research and its director Prof. Robert Thomas for this research.

REFERENCE

Ayorinde, E. O. and Gibson, R. F. (1993). "Elastic constants of orthotropic composite materials using plate resonance frequencies, classical lamination theory and an optimized three-mode Rayleigh formulation," *Composites Engineering*, Vol.3, No. 5, pp.395-407

Ayorinde, E. O., Gibson, R. F. and Wen, Y. -F. (1993). "Elastic constants of isotropic composite materials from plate vibration test data," *Composite Materials: Testing and Design(Eleventh Volume)*, ASTM STP 1206, E. T. Camponeschi, Jr., Ed., American Society for Testing and Materials, Philadelphia, pp.150-161

Ayorinde, E. O. (1995). "On the sensitivity of derived elastic constants to the utilized modes in the vibration testing of composite plates," NCA-Vol. 20/AMD-Vol. 211, *Innovative Processing and Characterization of Composite Materials*, ASME.

Ayorinde, E. O. and Gibson, R. F. (1995). "Improved method for in-situ elastic constants of isotropic and orthotropic composite materials using plate modal data with trimodal and hexamodal Rayleigh formulations," *Journal of Vibrations and Acoustics (Transactions of the ASME)*, Vol. 117, April, pp.180-186

Ayorinde, E. O. and Yu, Lin (1997). "On the Use of Diagonal Modes in the Elastic Identification of Thin Plates", submitted to the *Journal of Vibartion and Acoustics, Trans ASME.*

DeWilde, W.P., Narmon, B., Sol, H. and Roovers, M. (1984). Determination of the material constants of an anisotropic lamina by free vibration analysis. *Proc. 2nd Modal Analysis Conf.*, Vol. I, pp. 44-49, Orlando, FL.

Frederiksen, P. S. (1997). "Numerical studies for the identification of orthotropic elastic constants of thick plates," *European Journal of Mechanics A - Solids*, Vol. 16, No. 1, pp. 117 - 140.

Kim, C. S. and Dickinson, S. M. (1985). "Improved approximate expressions for the natural frequencies of isotropic and orthotropic rectangular plates," *Journal of Sound and Vibration*, Vol. 103, No. 1, pp.142-149

Zelenev, Yu. V. and Electrova, L.M. (1973). Determination of the dynamic parameters of polymer plates. *Sov. Phys. Acoust.* 18(3), 339-341.

Static and Low-Velocity Impact Response Characteristics of Pultruded Hybrid Glass-Graphite/Epoxy Composite Beams

M. V. S. L. N. KOWSIKA AND P. R. MANTENA

ABSTRACT

The influence of hybridization on the crashworthiness and energy absorption characteristics of pultruded glass-graphite/epoxy composite beams was investigated. Low-velocity drop weight instrumented impact tests were conducted on these hybrid composites to determine the load-deformation behavior for evaluating the impact performance in terms of the ductility index, damage initiation, propagation and total absorbed energies. Three-point static flexural tests were also conducted to compare the static load-deformation characteristics with that of the dynamic low-velocity impact tests. The behavior under both static as well as dynamic loading conditions was simulated using finite element modeling procedures to identify the failure mechanisms for optimizing the performance of pultruded hybrid composites.

Experimental results show that the load and strain to failure of all-graphite/epoxy, all-glass/epoxy and other hybrid composites obtained from impact tests are significantly higher as compared to the static test data. The load bearing capability of composites after damage initiation (which is dictated by the ductility and failure index) has shown marked improvement for the graphite outside hybrids when compared with the all-graphite, all-glass and glass-outside hybrids. The high strain to failure glass fibers absorb considerably higher energy before ultimate failure compared to the brittle graphite fibers, as a result, the fiber content and geometric placement of each type of fiber significantly influenced the energy absorption characteristics of hybrids. Results indicate that the energy absorption behavior of pultruded hybrids predicted using finite element modeling is in close agreement with the behavior characterized from experimental impact tests. The effectiveness of hybridization to improve the impact performance of composites was demonstrated.

INTRODUCTION

Fiber-reinforced composite materials offer a number of distinct advantages in terms of specific strength and stiffness, corrosion resistance and fatigue performance and are currently used in a wide variety of low and high technology applications. Structural members are prone

Murthy V. S. L. N. Kowsika and P. Raju Mantena, Department of Mechanical Engineering, The University of Mississippi, University, MS 38677

to low-velocity impact loads and it is necessary to understand their response in order to optimize their impact performance while designing structural members for automotive, aerospace and other engineering applications. In these applications apart from analyzing the delamination fracture toughness, the energy absorption behavior before and after damage initiation as well as the total absorbed energy are considered during the material selection process.

Significant research has been conducted to tailor the energy absorption capability of composites by utilizing high strain to failure fibers such as Kevlar® and Spectra®, modifying the formulation of the matrix material (e.g., elastomers) and the interface adhesion properties between fiber and matrix (sizing or chemical treatment of the fiber surface). Innovative designs to manufacture 3-D composites by stitching, weaving and braiding are also considered to improve the impact performance of composites. Although significant improvements have been made, the benefits of hybridization, i.e., using two or more types of reinforcing fibers in a common matrix system, have not been widely investigated. An increase in the failure strain of carbon fiber reinforced composites was observed by hybridizing with glass fibers accompanied by a synergistic effect on the tensile, compressive and flexural properties (Kretsis, 1987). The effective utilization of low cost and high strain to failure E-glass fibers along with high cost and low strain to failure graphite fibers to produce hybrid composites having good impact performance accompanied by a reduction in the cost of manufacturing has been investigated by Mallick and Broutman (1977). Compared to composites reinforced with single type of fibers, a very limited amount of research has been conducted to characterize the impact performance of hybrid composites. In this study, an attempt has been made to understand and optimize the impact performance in terms of energy absorption characteristics of pultruded unidirectional glass-graphite/epoxy hybrid composite beams. Emphasis is placed on the improvements and changes that can be obtained by hybridization when compared to the mono-fiber type composites.

Low-velocity impact tests were conducted using an instrumented drop weight impact testing machine (DYNATUP®, Model 8250). The drop-weight testing procedure typically replicates low-velocity impact conditions such as a dropped tool during maintenance, and is widely used to characterize the impact performance. The load and energy-deflection traces obtained from each test were used to judge the impact performance of various pultruded hybrid composites. Three-point static flexural tests were also performed on pultruded hybrids to correlate the load-deformation characteristics with similar traces obtained from low-velocity impact tests. The influence of strain rate on the mechanical behavior of these pultruded hybrid composites was ascertained from the data obtained from static and low-velocity impact tests. The failure strain of the pultruded all-graphite/epoxy and all-glass/epoxy composite samples were utilized for modeling the energy absorption behavior using finite element analysis.

The experimental procedure adopted for performing low-velocity impact tests resulted in the load-displacement and energy absorption behavior associated with the global response of hybrid composites. Small undulations and changes in the slope of the load-displacement curve indicate progressive failure of the composite, however, it is difficult to attribute these changes to a specific failure mechanism. Thorough understanding of the failure mechanisms is necessary to comprehend the behavior which leads to optimization of the impact performance. Most of the analytical studies reported primarily focused either on the contact force of the impactor or on the impact response of the plates without taking impact damage into consideration. Since the failure of a composite subjected to impact loads follows a complicated pattern, comprehensive evaluation of the damage mechanisms using analytical techniques are not available. When detailed stress distributions in the contact zone are required, analysis is usually performed using two or three dimensional elasticity models. Finite element analysis is usually performed to analyze the stress distribution in these cases. A finite element model to compute the transient dynamic stress by incorporating impact-induced delamination criterion apart from the matrix dominated ply failure criterion to predict the initial impact damage and the extent of delamination due to point loading is reported elsewhere (Hyung and Chang, 1992). Apart from few other citations very little research has been conducted for predicting the post failure behavior due to progressive ply failure under low-velocity impact conditions.

In this study, an attempt was made to simulate the behavior of some pultruded hybrids under low-velocity dynamic loading conditions using 3-D dynamic stress analysis. Since the dynamic analysis procedures are computationally more expensive and are difficult for simulating the post failure characteristics after initial ply failure, the validity of using static analysis procedures to analyze the energy absorption behavior under low-velocity impact conditions was investigated. For this purpose, a quasi-two-dimensional static stress analysis has been performed and the results obtained at the point of initial ply failure were compared with similar data obtained from dynamic analysis. This preliminary investigation indicated that the overall energy absorption behavior, stress and strain distribution obtained from static and dynamic analysis are comparable (Kowsika, 1997). An attempt was also made to perform modal analysis using finite element procedures on these composite models to derive the fundamental frequency of vibration. It has been reported that if the excitation frequency (due to external impact loads) applied to the structure is less than one-third the fundamental frequency of vibration of the structure, then inertia effects can be neglected and the response can be treated as quasi-static (Cook, 1989). The results from modal analysis indicated that the time period of the fundamental mode of vibration is less than one-third the time of application of the applied impact load (Kowsika, 1997). The inferences from these investigations indicate that the low-velocity test conditions employed in this study can be treated as quasi-static and therefore, static analysis models can be used for predicting the energy absorption characteristics of the pultruded hybrid composites under consideration. Based on this preliminary study, static analysis was performed on all the pultruded hybrid composite models to simulate the energy absorption behavior.

EXPERIMENTAL PROCEDURE

Uni-directional sandwiched composite products having a rectangular cross-section (25.4 x 3.2 mm thick) were manufactured using a commercial size pultrusion machine (PULSTAR 804) with AS4-W 12K and/or PPG 2001 #112 E-glass fibers in a Shell EPON 862/W/527 epoxy resin matrix. The cross-section of these pultruded products showing the fiber content of each type of fiber and the lay-up sequence is listed in Table I. In order to perform experimental tests, samples of length around 55 mm having a thickness of 3.2 mm were cut from the long pultruded products. The width of the sample was trimmed to 20 mm to conform to the width of the falling tup insert. It should be noted that the samples were stored in a vacuum chamber maintained at 10^{-2} torr for at least two days to remove moisture and to provide similar environmental history prior to testing. At least three samples from each type of pultruded composite were considered for each testing method to account for experimental scatter. The experimental procedure utilized for testing the samples to evaluate the mechanical behavior under low-velocity impact and three-point static flexural tests is outlined in the following discussion.

LOW-VELOCITY IMPACT TESTS

The impact tests were performed using a standard Charpy impact test fixture (ASTM E-23) in conjunction with an instrumented drop-weight impact test machine (DYNATUP®, Model 8250). The instrumented impact testing system consists of a drop weight test machine coupled with signal conditioning and processing units and associated software (GRC's Model 930-I) for data acquisition and analysis. The raw data obtained in terms of load-time trace for each sample is further processed to obtain the load and energy-deflection traces for characterizing the impact performance. The Charpy impact test fixture (ASTM E-23) which is typically used for metallic materials was considered for testing the pultruded composites due to specimen size limitations imposed by other recommended fixtures for testing composites. It should be noted that no machining was performed to introduce a notch in the specimen as this study is intended to characterize the influence of hybridization which is associated with material property variation through the thickness. A similar testing technique has been utilized by other researchers to characterize the impact performance of composites (Kakarala and Roche, 1987). The span length was set at 41 mm and the composite specimens were tested under simply supported boundary conditions according to the procedure mentioned above. Preliminary screening tests were performed to ascertain the mass and velocity (drop height) which collectively influence the incident kinetic energy of impact. In this study, a falling mass of 5.94 kilograms was utilized to impact the composite specimens at a velocity of 2.7 m/s which provided an incident impact energy of 21.7 J, which was marginally higher than that required to break the samples (especially, the all-glass/epoxy sample which exhibited high energy absorption). It should be noted that the orientation of the uni-directional reinforcing fibers is perpendicular to the line of contact of the applied impact load (tup) and that the sample generally undergoes flexural deformation. A typical load, energy-time trace indicating the various parameters of interest is shown in Figure 1. The first point in the load-deflection curve where there is considerable deviation in the slope of the curve (shown in Figure 1) is termed as the incipient point. A new parameter has been specifically coined for this study which is termed as failure index (FI) and is equivalent to the ratio of the energy absorbed after incipient damage (E_T-E_I) to the incipient energy (E_I). The incipient, initiation, propagation and total absorbed energies as well as the ductility and failure indices are utilized to judge the impact performance of various materials.

FLEXURAL STATIC TESTS

Three-point static tests were performed under displacement control using an in-house 22 kN (5 kip) MTS servohydraulic test machine. Even though the testing facility is equipped with fixtures to perform three-point flexural testing according to the standard specifications (ASTM D790), the standard fixture (ASTM D2344) which is intended for short-beam shear tests was utilized for flexural tests to replicate the testing parameters used for impact testing. It is difficult to accommodate specimens of span length 41 mm (which was considered for impact tests) using the standard fixtures available for three-point flexural testing (ASTM D790). On the other hand, the fixture for short beam shear tests (ASTM D2344) can be adjusted to meet the desired span length conditions and static testing was performed using this modified configuration. Static testing was performed by maintaining the displacement rate at 0.085 mm/sec (0.2 in/min) and loading was continued even after the maximum load to observe the post failure load-deformation characteristics by recording the traces until the load drops to 99% of the peak load. A computer attached to the testing equipment was utilized to control the instrumentation as well as to acquire the data for analyzing the load-deformation characteristics. From the load-displacement traces obtained from static tests, the strain to failure, peak load as well as the parameters mentioned above for low-velocity impact tests (ductility index, initiation energy, etc.,) were derived.

MODELING

Finite element modeling was performed to simulate the energy absorption behavior of pultruded hybrid composites under low-velocity impact loading conditions using static stress analysis procedures. The commercial finite element code ABAQUS® has been utilized for this purpose. Each of the composite models was discretized by using eight-noded shell (plate) elements. The finite element mesh was constructed only at the bottom face of the model (plane containing the width and the length) and the rest of the model was constructed by laying (building) twelve hypothetical layers one on top of each other to model the material property variation through the thickness. The composite models were constrained at either ends with simply supported boundary conditions and subjected to incremental displacement at the mid-span. It should be noted that instead of using a line or a pointed load (displacement), a displacement pattern was imposed for the nodes at the mid-span which conformed to the contour of the tup insert utilized for experimental tests. This was necessary to simulate similar loading conditions that the sample experiences when it is statically loaded against the tup insert. The model composite was displaced gradually and for each increment in displacement the response in terms of stress and strain distribution and the overall absorbed strain energy was noted.

For each increment in prescribed displacement, the maximum strain experienced by each hypothetical layer was compared with the failure strain of either glass/epoxy or graphite/epoxy composites (depending on the type of hybrid), to detect individual ply (or plies) failure. If the maximum strain experienced by each hypothetical layers is below the allowable strain, the overall strain energy absorbed by the composite as well as the prescribed deflection at the mid-span were noted and the analysis is continued for the next increment in displacement. If the maximum longitudinal strain of any particular layer was close to the allowable strain then the corresponding elastic constants (E_{11}, E_{22}, E_{33}, G_{12}, etc.) were reduced to very low values to incorporate failure of that particular layer in the model and the analysis was repeated with these modified parameters to simulate the post-failure characteristics. The contribution of the degraded ply to the overall energy absorption is fairly insignificant and can be considered as equivalent to incorporating a crack used for modeling ply failure by other analysis techniques. A schematic of the procedure adopted for predicting the energy absorption behavior of pultruded hybrids by modeling progressive failure of individual plies is shown in Figure 2.

Figure 2 indicates two energy displacement curves viz., one for the intact laminate (pultruded composite) upto the point of initial ply 1 failure and the other curve with degraded elastic moduli for ply-1 with loading (displacement at the mid-span) being continued until the failure of ply-2. It is clear that the energy absorbed upto the failure of ply-1 is E_A and the additional energy required to cause failure of ply-2 is $E_B - E^*_A$. The total energy required to cause failure of plies 1 and 2 is $E_A+E_B - E^*_A$. This procedure is repeated until failure of all the hypothetical layers under consideration. The overall energy absorption behavior is derived after obtaining the information regarding the deflection and the cumulative energy absorbed during the course of the analysis. It should be noted that the failure strain of the individual hypothetical layers under tension and compression were considered to be equivalent. Since the pultruded composites under investigation are symmetric about the neutral axis, while modeling failure of a particular ply subjected to tension the corresponding ply (above the neutral axis) undergoing compression was also considered to have failed. By incorporating the above method the energy absorption behavior before and after damage initiation was effectively modeled for all the pultruded hybrids under consideration. The procedure mentioned above to predict the energy absorption behavior of pultruded hybrids using static stress analysis was employed for two different values of failure strain of all-glass/epoxy and all-graphite/epoxy obtained from static and low-velocity impact tests. It should be noted that even though static analysis is used to model the energy absorption behavior under low-velocity impact conditions, it is necessary to incorporate the failure strain obtained from the higher strain rate impact tests to take into account the strain rate sensitivity of individual fibers into consideration.

RESULTS AND DISCUSSION

Three-point flexural static and low-velocity impact tests were conducted on pultruded hybrids to evaluate the influence of hybridization on the mechanical properties, especially, the impact performance. Finite element modeling was also performed to simulate the energy absorption behavior as a result of progressive ply failure until complete fracture. The results obtained from experimental and finite element modeling are listed in Table I. Typical load-deflection traces obtained from static and low-velocity impact tests for all-graphite/epoxy and all-glass/epoxy pultruded composites is shown in Figure 3. It is clear from Figure 3 that the failure strain of glass/epoxy composite is considerably higher than graphite/epoxy composite. Also, for the glass/epoxy composite the strain to failure obtained from static tests is comparatively lower than the low-velocity impact tests where as for graphite/epoxy composite a marginal change is noticed indicating that glass fibers are strain rate sensitive. An increase in the failure strain enhances the extent of flexural deformation resulting in an increase in the energy absorption capability. Correspondingly, the amount of energy absorbed to induce incipient damage for glass/epoxy and glass-outside hybrids is much higher than graphite/epoxy and graphite-outside hybrids (Table I). This effect is more pronounced under low-velocity impact conditions as the glass fibers are strain rate sensitive. The total energy absorbed by these composites also indicate a similar trend (Table I) mentioned above and that it is dependent on the fiber content of glass fibers. It is interesting to note that the graphite/epoxy composite exhibited a ductile failure under static loads whereas a brittle failure (indicated by a sudden drop in the load after reaching maximum load) is noticed under low-velocity impact loading conditions. However, a reverse trend is observed for the glass/epoxy composite. For graphite-outside hybrids, after failure of the low-strain to failure graphite layers the core comprising of glass fibers continue to withstand the load absorbing considerable energy before ultimate failure. The load-deflection traces for these hybrids show a very distinct hump in the propagation phase (Kowsika, 1997) indicating that the failure index of these graphite-outside hybrids is higher than other all-glass/epoxy and other pultruded composites. The failure index of pultruded hybrids obtained from experimental and numerical modeling is shown in Figure 4. It is clear from Figure 4 that a higher failure index is exhibited by graphite-outside hybrids whereas reversing the lay-up sequence results in considerable reduction. The overall energy absorption behavior obtained from low-velocity impact tests and finite element modeling is shown in Figure 5. It is clear from Figure 5 that the energy absorption behavior predicted by static analysis using the failure strain of individual layers (fibers) obtained from static tests and low-velocity impact tests formed the lower and upper bound respectively to the actual experimental behavior deduced from low-velocity impact tests. From Figures 4 and 5 as well as Table I it is clear that the results obtained from finite element modeling are in close agreement with the experimental observations. The energy absorption behavior for all-graphite/epoxy composite and glass-outside hybrids indicates a flat response in the propagation phase. On the other hand, all-glass and graphite-outside hybrids continue to absorb energy in the propagation phase. It is clear that by substituting high strain to failure glass fibers near the neutral region of an all-graphite/epoxy composite the propagation energy as well as the ductility and failure indices can be improved significantly.

CONCLUSIONS

The influence of hybridization on the static and low-velocity impact performance of pultruded composite beams has been investigated. Results indicate that the glass fibers are strain rate sensitive which accordingly showed higher energy absorption capability. The lay-up sequence and the fiber content of each type of fiber significantly influenced the damage initiation, propagation energies and the failure index of pultruded composites. The results obtained from finite element modeling using static stress analysis procedures are in close agreement with the experimental observations. These results demonstrate that hybridization, in general, can be employed to tailor the energy absorption behavior of composites.

ACKNOWLEDGMENTS

This research was supported by the National Science Foundation/EPSCoR (Grant No. EPS-9452857), the State of Mississippi, the University of Mississippi, and the Shell Development Co.

REFERENCES

Hyung, Y. C. and Chang, F. K., 1992, "A Model for Predicting Damage in Graphite/Epoxy Laminated Composites Resulting from Low-Velocity Point Impact," *Journal of Composite Materials*, Vol. 26, No. 14, pp. 2134-2169.

Kakarala, S. Norm and Roche, L. James, 1987, "Experimental Comparison of Several Impact Test Methods," Instrumented Impact Testing of Plastics and Composite Materials, ASTM STP 936, S. L. Kessler, G. C. Adams, S. B. Driscoll, and Dr. R. Ireland, Eds., ASTM, Philadelphia, pp. 144-162.

Kowsika, V. S. L. N. Murthy, 1997, "Impact Performance Characteristics and Modeling Failure Mechanisms of Pultruded Glass-graphite/epoxy Hybrid Composite Beams," Ph.D. Dissertation, The University of Mississippi.

Kretsis, G., 1987, "A Review of the Tensile, Compressive, Flexural and Shear Properties of Hybrid Fiber-Reinforced Composites," *Composites*, Vol. 18, No. 1, pp. 14-23.

Mallick, C. K. and Broutman, L.J., 1977, "Static and Impact Properties of Laminated Hybrid Composites," *J. of Testing and Evaluation*, Vol. 5, No. 3, pp. 190-200.

TABLE I. STATIC AND LOW-VELOCITY IMPACT DATA OF PULTRUDED HYBRIDS.

Cross-section	Fiber content (%)		Incipient energy (J)				Total energy (J)				Failure index			
			Experimental		Modeling		Experimental		Modeling		Experimental		Modeling	
	Graphite	Glass	Static	Impact	Static	Impact	Static	Impact	Static	Impact	Static	Impact	Static	Impact
	60	0	2.5	3.1	3.5	5.2	7.0	6.8	6.2	7.7	1.79	1.18	0.76	0.47
	0	60	4.9	10.7	5.3	12.0	6.0	16.7	9.2	19.8	0.22	0.55	0.73	0.65
	20	40	4.8	10.2	6.2	14.8	6.7	12.1	8.8	15.0	0.40	0.20	0.41	0.01
	30	30	4.4	10.7	6.7	15.9	6.7	11.7	6.9	16.0	0.53	0.09	0.04	0.01
	40	20	3.4	2.7	4.8	5.2	6.2	8.7	7.6	10.5	0.81	2.16	0.58	1.00
	30	30	3.2	3.2	4.7	5.1	6.3	11.5	8.8	12.8	0.96	2.64	0.89	1.49

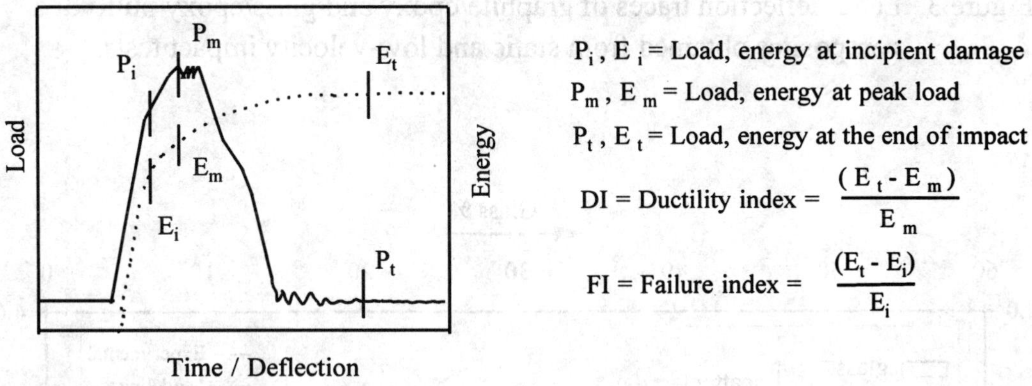

Figure 1. Schematic of the load, energy-time trace obtained from impact tests.

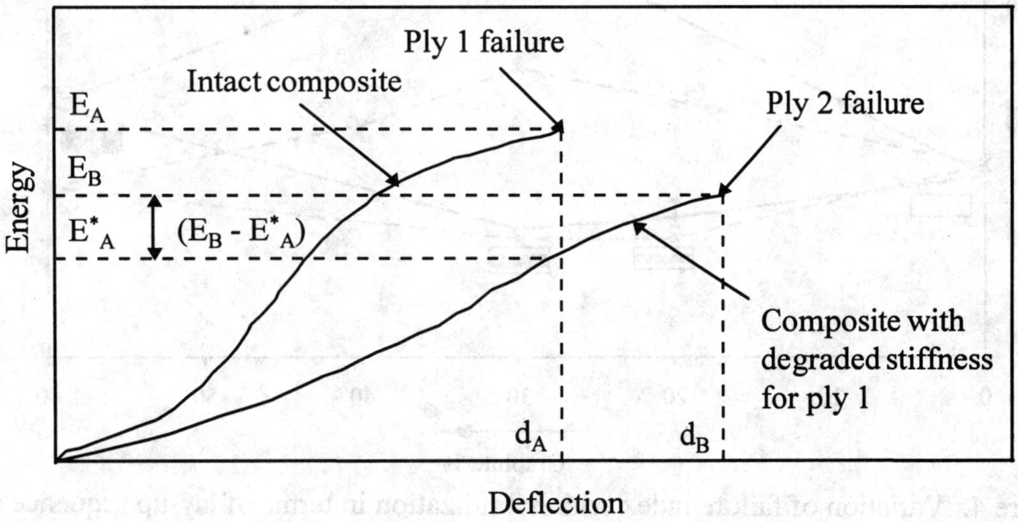

Figure 2. Schematic of the procedure used for predicting the energy absorption behavior.

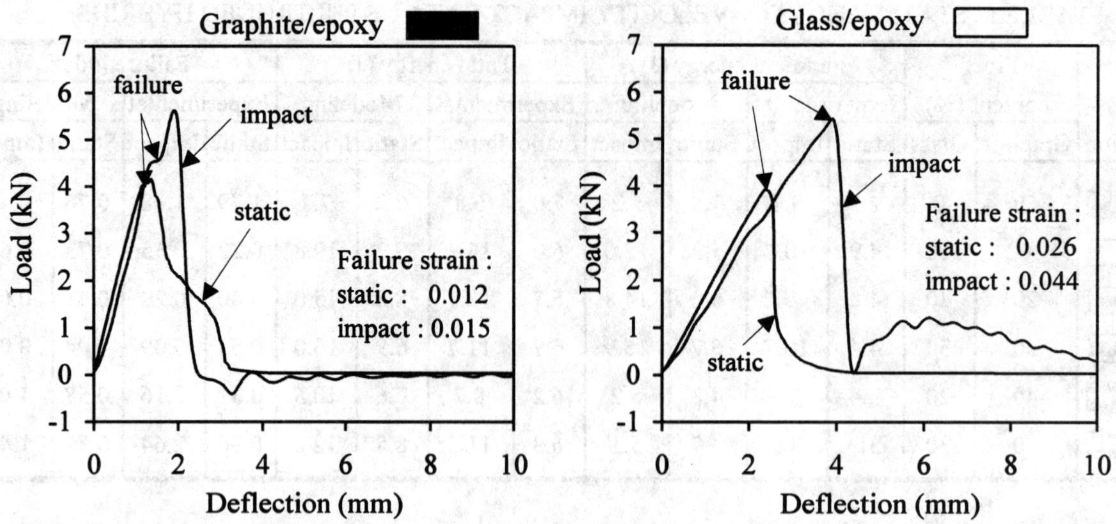

Figure 3. Load-deflection traces of graphite/epoxy and glass/epoxy pultruded composites obtained from static and low-velocity impact tests.

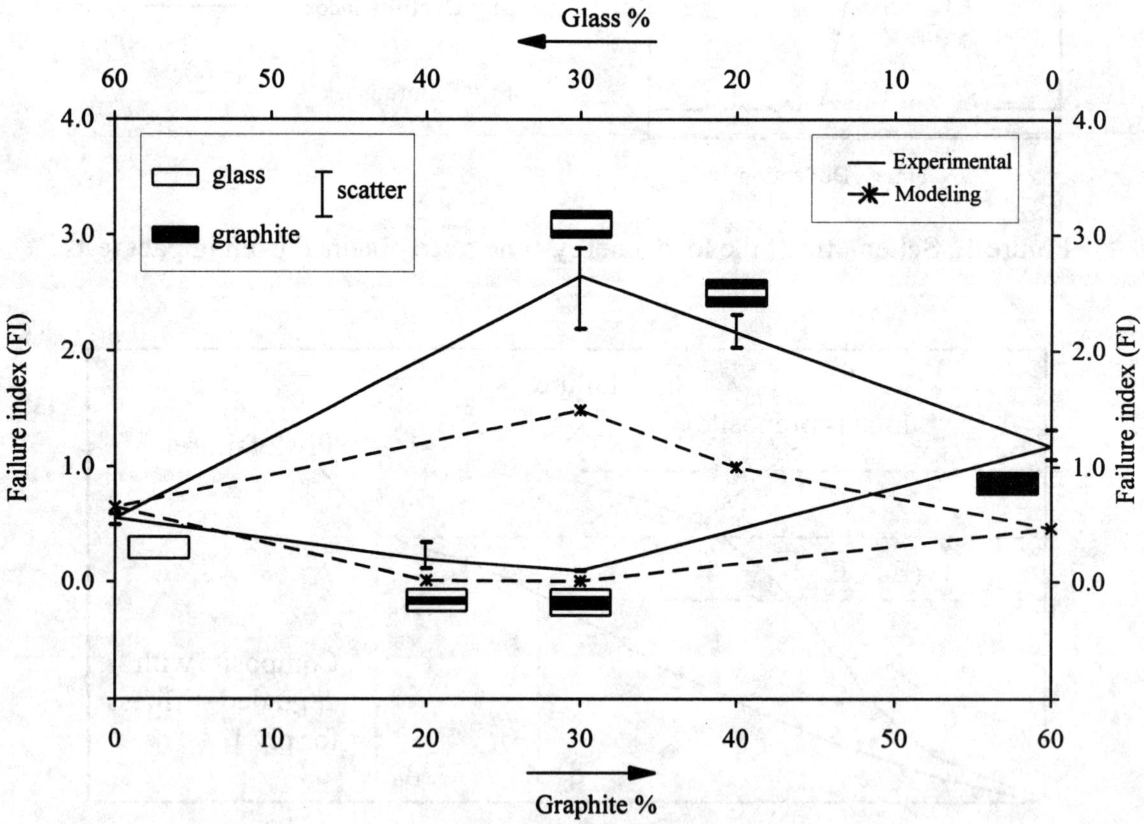

Figure 4. Variation of failure index with hybridization in terms of lay-up sequence and fiber content of each type of fiber obtained using impact tests and modeling.

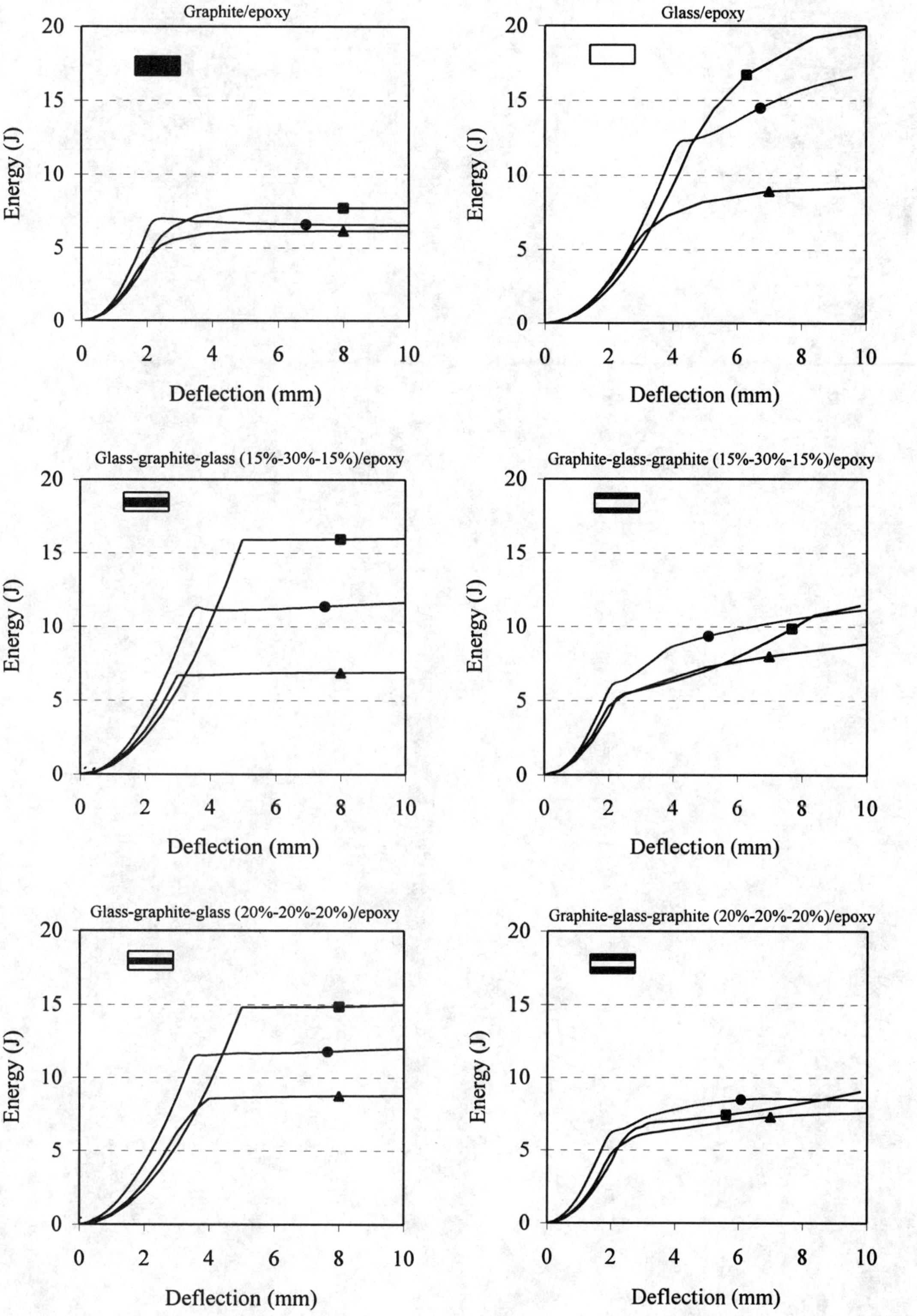

Figure 5. The energy absorption behavior of pultruded hybrid composites obtained from low-velocity impact tests (—●—) and finite element modeling based on failure strain derived from static (—▲—), impact (—■—) tests.

Education

Educating the Next Generation of Engineers in Composite Materials Structures

J. R. VINSON

ABSTRACT

Preparing the next generation of engineers to further increase the use of composites in every possible way is a challenge that can have significant effects on our civilization in the next century. Everywhere, composites are being used but the future opportunities are amazing. The key to success is through educating young engineers now.

INTRODUCTION

This paper represents the author's conclusions and opinions based upon teaching composite materials structures to students on average more then once per year since 1969. Not only has the author taught the course each year at the University of Delaware, but he has also given the course at Aberdeen Proving Ground, Maryland and in Buenos Aires, Argentina. In the process, two textbooks have been written and the later text is currently being revised for the second edition. The second text has also been translated into Japanese and Russian. More recently the course has been taught through video tape for distance learning.

Composite materials remains one of the most intriguing challenges and opportunities in modern day engineering science and technology. We still are scratching the surface on new materials, new manufacturing methods, new products and new uses for composites.

The education of engineers is essential, and fortunately it doesn't matter what the adjective is in front of the word engineer: mechanical, civil, manufacturing, materials, aerospace, marine or chemical - it just doesn't matter. Likewise, it doesn't matter whether the student is an undergraduate senior, graduate student, or an engineer in industry/government.

The course or courses needed to prepare the student simply builds upon the mathematics, mechanics and materials science at the undergraduate level in most engineering programs.

GOAL OF THE COURSE

The goal of the course in composite materials structures is to prepare the student to productively contribute to the design, analysis and development of products and structures totally or partially employing composite materials. Because the use of composite materials hopelessly intertwines materials-manufacturing-design-analysis, it is not worthwhile to separate out each of these categories. The course in composite structures should incorporate the elements given in the following sections to prepare the student to start a productive and rewarding career. The order is less important than the content.

H. Fletcher Brown Professor of Mechanical and Aerospace Engineering and the Center for Composite Materials, University of Delaware

COMPOSITE MATERIAL PROPERTIES

The student should be provided with the mechanical properties of numerous composite materials as well as conventional materials. These should include representative examples of graphite, Kevlar and glass reinforced polymeric matrix composites; unidirectional, cross-ply, angle ply and quasi-isotropic composites; and a few metal matrix and ceramic matrix materials. The student should have a homework problem involving the above materials as well as several aluminum and steel alloys by perhaps forming a graph of ultimate strength to density ratio as a function of modulus of elasticity to density ratio. By doing this early on, the student sees the fantastic potential of the use of composite materials compared to conventional materials, and can see the relative merits of various fibers, various stacking sequences, and various matrix materials.

Obtaining adequate and sufficient material property data is always a difficult time consuming process. The students should be urged to collect material property data on a prepared form in a notebook or stored in a computer which they will add to through their career. Because there is far more variability in composite material property data (compared to metallic alloy properties) it is important for the student to always record the complete source of the data, for referral when necessary.

MANUFACTURING METHODS

Manufacturing methods of producing composite structures, products and components are so numerous and change so rapidly, it is impossible to dwell on this topic too intensely. By default one can only briefly explain processes such as hand lay up, pultrusion, RTM, etc. so that the student knows the terminology. Of course, a second course dealing strictly with manufacturing processes is desirable but usually difficult for faculty with their university facilities to provide.

READING SOURCES

It is important to provide the student with a fairly complete list of textbooks on composite materials. Equally important is providing a complete list of journals dealing with composite materials. The latter is somewhat difficult because while it is obvious that the Journal of Composite Materials should be on the list, it is not so obvious that "Professional Boat Builder" should be, yet it is very instructive in the uses of composite materials.

It must be emphasized to the student that in this rapidly growing and changing area, the only way to attempt to keep current is to read or comb through numerous journals each month.

ANISTROPIC ELASTICITY

As a precursor to studying lamina and laminate theory and understanding the peculiar behavior of composite materials either micromechanically or macromechanically, basic anisotropic elasticity should be studied for the student to appreciate just what the basic concepts are in a material point, a control element, and the concepts associated with differing properties in different directions.

LAMINA AND LAMINATE THEORY

From anisotropic elasticity it is then important to introduce the generally agreed to terminology associated with material axes (1, 2 and 3) and structural axes (x, y and z) and the laminate terminology associated with the A, B, D stiffness matrix for composites. It is important for the student to fully understand bending-stretching coupling, stretching-shearing coupling, bending - shearing coupling, etc.

Even with composites associated with molding methods employing preforms, the standard methodology must be employed in order to build in (or avoid) various couplings to effect superior structures, products and components.

THERMAL CONSIDERATIONS

All manufacturing processes of composite materials involve thermal processes, it is even more important for engineers involved with composites to understand thermal effects than it is for engineers associated with conventional materials. To not be aware of the thermal effects can and has produced structures emanating from the autoclave to resemble pretzels, or for sandwich plates which have cores and faces of radically different coefficients of thermal expansion to develop face wrinkles during manufacturing which negated the usefulness of the structure and brought about the consternation and puzzlement of the manufacturer.

HYGROTHERMAL (MOISTURE) CONSIDERATIONS

For polymer matrix composites it is absolutely necessary for the student to understand the two primary effects of moisture-aggravated by temperature- hence the term hygrothermal. Not only does moisture cause expansion of the polymeric matrix analogous to thermal expansion, but it also lowers the glass transition temperature. Between the two the effects can be catastrophic. One example is a 500 fiberglas pound rotary fan designed to rotate at 1260 RPM which was designed to operate at ambient temperature and humidity because the manufacturer was unfamiliar with hygrothermal effects. However, with its operation at over 180°F and saturated moisture it failed during operation within a few months and the $25,000 fan closed down a 25 million dollar facility for months - all due to hygrothermal effects.

PLATE AND PANEL STRUCTURES

Unlike basic mechanics of materials courses it is more relevant and convenient to teach plates and panels prior to teaching beam-column-rod thcory. This is duc to thc fact that onc is employing two dimensional laminate theory with its A-B-D stiffness matrix as the integrated stress strain relations or the constitutive equations, which were developed earlier in the course.

It is important to teach plate theory, not merely the two-dimensional plane stress and plane strain mechanics. Because composite materials have such superior strength and stiffness, usually or quite often one uses very thin walled structures. When these are two dimensional and described best in a Cartesian coordinate system, they constitute a plate or panel; in a curvilinear coordinate system they comprise a shell; and of course in a one dimensional planar setting they constitute a beam.

Of course it makes sense to first deal with static loading problems. Therefore Navier and Levy solutions should be presented for classical plate theory. Then, transverse shear deformation (TSD) effects should be presented to show that with many composites TSD effects are important but do present complications in analyses and design. The student should also be made aware that in preliminary design in which there is a group effort, classical theory should be used to minimize turn around time. TSD effects resulting in a more complicated analysis can and should be used near to final design decisions to insure structural integrity of the structure or product.

One cogent example illustrates the necessity of the wise use of anisotropic plates. Consider a pultruded composite side rail channel of an extension ladder. All of the glass fibers are in the length direction. The flanges of the side rail are long, clamped on one edge, free on the outer edge and either clamped or simply supported at the ends. Through rather concentrated loads at the hard points between components of the extension sections, the load can be concentrated near or at the free edges. The result is substantial bending of the flange in the plane normal to the fibers, where both flexural stiffness and strength are matrix not fiber dominated. The result is unanticipated failure.

EIGENVALUE PROBLEMS - BUCKLING & NATURAL VIBRATION

Next the concept of the eigenvalue problems should be treated in the context of plates and panels.

Natural vibrations should be treated in some detail. The student must realize that every elastic body theoretically has an infinite number of natural frequencies (mathematically) but usually only the lowest few are of structural importance. These natural frequencies are the "signature" of the material-structural body, and when possible they must be avoided when the structure is dynamically loaded. Loads which coincide with natural frequencies can cause failure, large amplitude non-linear behavior and/or fatigue problems. Oscillating loads below the fundamental frequency can be treated quasi-statically, with no dynamic load factor. Students must realize that dynamic impact loads can easily cause a dynamic load factor of 2 in stresses and deformations, which can have catastrophic effects if unanticipated and not provided for in the design.

Ninety percent of the problems caused by dynamic loading that the writer has seen are easily diagnosed forensically by merely calculating the natural frequencies - for structures varying from composite satellite structures to composite extension ladders.

BUCKLING (ELASTIC STABILITY)

The other eigenvalue problem - that of buckling is a first cousin to the natural vibration problem, but only the lowest buckling load (lowest eigenvalue) is of practical importance, because the structure always changes shape significantly or fails almost without exception (some structures such as spars intentionally undergo elastic buckling at maximum loads - but this is only one example). Usually buckling must be avoided particularly with many composites whose stress-strain behavior is linear to failure and therefore unforgiving.

SANDWICH STRUCTURES

More and more sandwich construction is being used because a good sandwich construction can have a flexural stiffness hundreds of times greater than using the two faces together as a monolithic structure, and can have stresses dozens of times less than the outer fiber bending stresses if the faces were joined together with no core - and for very little more weight.

There will be insufficient time to dwell on sandwich structures too deeply in a one semester course but basic things such as means to calculate the A-B-D matrix for the sandwich, calculating maximum face stresses and core shear stresses are necessary.

Foam or solid core sandwich as well as honeycomb sandwich should be covered. Truss core and web core sandwich should be mentioned but not studied due to time constraints for the course.

Very importantly, the concept that with a monocoque construction discussed earlier only overall buckling can occur. However, in a sandwich, core shear instability and face wrinkling can and do occur, and for honeycomb core sandwich, monocell buckling or face dimpling can occur. Thus, design and analysis become more complicated - the price one pays for going to a better structure!

BEAMS - COLUMNS - RODS

After the preceding, this topic is simple; also the student is familiar with the topic from earlier courses. In general it is easy to show the student that fibers should always be in the length direction, and using the appropriate modulus - all of the isotropic solutions and equations previously studied can be used.

However, if coupling terms exist such as $( )_{16}$ and $( )_{26}$ terms in the constitutive relations, no matter how much the structure looks like a beam, it is not a beam - it is a "skinny" plate.

ENERGY METHODS

Although in elasticity theory one may employ Minimum Potential Energy (MPE), Minimum Complementary Energy or Reissner's Variational Theorem, only the former is practical. It is far easier to guess at trial functions for the displacements in a structure rather than guess at a trial stress functions.

Likewise it is important for the student to realize that solving differential equation sets with associated boundary conditions is digital - you either have a solution or you don't. However with MPE you can always get a solution - you cannot be stopped.

Furthermore with anisotropoic material properties, transverse shear deformation effects, thermal and hygrothermal effects, any structure incorporating these effects should be attacked by MPE, because trying to get an exact analytical solution is well nigh impossible.

For instance with thermal and hygrothermal effects, in many cases homogeneous boundary conditions become non-homogeneous, thus negating separation of variables - so at the minimum a lengthy transformation of dependent variables is necessary before one even starts to solve the problem.

COMPOSITE MATERIAL SHELLS

In a one semester course only cylindrical shells should be discussed in detail. Shell theory, in general, is sufficiently complex that it has to be approached and studied separately.

But, it is necessary that the student - through cylindrical shell study - learn that the shell behaves drastically different from beams and plates. The shell wants to act like a membrane with no bending effects everywhere except:

- near to a boundary
- near to a structural discontinuity
- near to a load discontinuity.

There, a bending boundary layer exists in which the bending stresses superimposed upon the membrane stresses reach values several times the easily calculated membrane stress value -with catastrophic results if they are unanticipated.

These concepts are necessary to infuse into the student, who likely will be designing, analyzing, or manufacturing thin walled curved composite structures or products. For instance, for many often treated loads, outside of the bending boundary layer, no interlaminar shear stresses will exist between plys and no shear stresses will occur in adhesive bonded joints. Not only is this important for assessing structural integrity, but also shows that detailed c-scan or x-ray examination of these regions is unnecessary.

STRENGTH AND FAILURE THEORIES

A brief introduction is all that is necessary here - simply for the student to be able to use the Tsai-Wu, Hoffmann and/or other criteria for sizing structures to avoid failure.

JOINING

Although composite structures, suitably designed, may contain only one tenth or one one hundredth the parts comprising a conventional structure - that is the good news - the bad news is that the only practical joining methods are adhesive bonding and mechanical fastening.

Mechanical fastening usually involves fiber cutting and stress concentrations, while adhesive bonding involves adhesives capable of carrying only 2000-3000 psi in shear, and nearly impossible to inspect after the joint is fabricated. The student must learn of the peculiar distribution of shear stresses and peel stresses in the bonded joint, and must learn that beyond a certain bond line length, the maximum stresses do not diminish. So neither method presents a happy picture and so must be discussed in sufficient detail to enable the student to approach the real world with confidence but concern.

FINAL REMARKS

The student must be encouraged to understand that upon completion of the course he/she knows more about composite structures than most other engineers - but we all need to continually learn more. Only by reading journal articles, attending conferences and building up a group of knowledgeable colleagues to periodically contact can any of us perform the job we would like to do.

The student must walk away from the course feeling confident, motivated and ready to take on the world.

I have intentionally not given any references because to do it justice and be fair, it could eclipse the length of the paper itself.

FOLLOW ON COURSES

If possible, the course discussed herein, perhaps given in the Fall of the academic year, should be followed by a design course and/or an experimental course in composite materials.

The design course that seems to work best is one in which the class is divided into teams which take on the design of a particular structure or product chosen either from a list provided by the professor or developed by the team themselves.

The following tips are given

- Each team selects its own leader
- The professor grades each team by awarding points.
- The elected leader grades each of his/her team members by distributing these points.
- At every class meeting several students are selected on the spot by the professor to give a progress report on his and his teams work - the persons chosen to speak are given no warning. (This provides much needed impromptu speech experience).

The experimental course should be one incorporating as many ASTM and other test procedures as possible to determine composite physical properties as well as fiber volume fraction, etc. Formal lab reports may or may not be required of all tests. This practical experience is most valuable.

An Overview of Composites Education at Georgia Tech

W. S. JOHNSON

ABSTRACT

This paper gives a brief overview of the composite materials and structures educational program at Georgia Tech. The Composites Education and Research Center is the umbrella organization on campus to consolidate and promote the composite materials and structures educational and research activities. This paper will review the current course offerings and the undergraduate and graduate composite certificate programs available.

INTRODUCTION

Composite materials offer exciting advantages over more traditional metallic materials. Their stiffness-to-weight and strength-to-weight ratios, when reinforced by modern stiff/high strength fibers, can be superior to any other known structural material. Modern advanced composites are a success story from the viewpoint of their many applications, ranging from golf club shafts and tennis rackets to advance aircraft and space vehicles. However the story could and should be better. More widely diverse applications of composite materials are hindered by the costs of production, the quality of manufacturing, questions of long term durability and the lack of ability to accurately predict life. The Composites Education and Research Center at Georgia Tech is poised to help resolve these issues by (1) Educating the scientist and engineers of the future in such a manner that they are equipped to tackle these problems and (2) Conducting the relevant research to solve many of the unanswered problems that are holding back even wider spread use of modern composite materials. We, at Georgia Tech, feel that the enhanced use of modern composites is a key ingredient to the national economic competitiveness of the United States. The Composites Education and Research Center (CERC) was formed in May, 1991, to promote interdisciplinary research and education in composite materials and structures. Prof. W. Steven Johnson currently serves as the Director of the Center. Although this center is not old compared to some others in the U.S., in a short time Georgia Tech has developed a very impressive, diverse, interdisciplinary group of faculty supported by excellent equipment and facilities. This has evolved to point were Georgia Tech is among the very best universities in

---

W. Steven Johnson, Professor, School of Materials Science and Engineering, Georgia Tech, Atlanta, GA 30332-0245

terms of composites education and research. The Center's faculty are drawn from six schools in the College of Engineering. Composites research is spread out across the campus and has been on going for a number of years. However, with the addition of several high profile faculty in the past few years, adding to the growth in professional recognition of some of the younger faculty, the reputation of Georgia Tech in the area of composite materials research has soared. Indeed, the number of involved faculty, added to the breadth and scope of the research (everything from new constituent development, fabrication, manufacture, testing, and modeling, to design), has placed Tech among the top tier in the nation in this important area of education and research.

The Schools of Aerospace, Civil and Environmental, Chemical, Materials Science and Engineering, Mechanical, and Textile and Fibers Engineering have active research programs in composites. The Composites Education and Research Center (CERC) is one of the multidisciplinary centers that operate through the College of Engineering at Georgia Tech. CERC's mission is to provide a forum for multidisciplinary, yet cohesive, education and research opportunities in composites for engineering undergraduate and graduate students. In all, forty faculty members from the participating departments are affiliated with CERC. Such large numbers of faculty with expertise in composites is unique among universities around the world and makes it possible for the Georgia Tech composites program to be very complete in all aspects of this critical technology.

EDUCATIONAL PROGRAMS

CERC is responsible for teaching the following four senior level undergraduate elective courses which are crosslisted in all six participating schools. Graduate students are allowed to count three 4000 level courses toward their degree requirements. Therefore there is normally a significant number of graduate students in these 4000 level composite courses.

4791 - Mechanical Behavior of Composites
4792 - Mechanics of Composites
4793 - Composite Materials and Processing
4794 - Laboratory in Composites Manufacturing and Testing

In addition to the above cross-listed courses, individual schools offer additional courses that deal at least in part with composites. Undergraduate students who take three of the above four courses plus two others from a separate list of courses are awarded an undergraduate *Multidisciplinary Certificate in Composites Engineering*. This certificate is available to students from all participating schools. Completion of the Undergraduate Certificate requirements implies that the student is familiar with the basic concepts of composites, the terminology, common constituents, and manufacturing procedures.

The following cross-listed graduate courses are also offered through CERC

7791 - Damage and Failure in Composites
7792 - Mechanics of Composites
7793 - Manufacturing of Composites

These courses form the core requirement for a Graduate Multidisciplinary Certificate Program in Composites Engineering. Upon the completion of these three core courses, plus six credit hours (2 courses) from a selected list, a student may obtain a *Graduate Certificate in Composites Engineering* under this program. The 4000 level courses serve as pre-requisites for the 7000 level courses. Thus, students completing the Graduate Certificate Program have considerable advanced knowledge in the field and are well prepared to tackle difficult research and applied problems dealing with composite materials and structures. Approximately a dozen certificates are awarded each year for the undergraduate and graduate levels combined.

DESCRIPTION OF COMPOSITE COURSES

The Composites Education and Research Center is responsible for overseeing the interdisciplinary composite courses. Below is a short description of each of the seven courses currently taught on the quarter system at Georgia Tech. The 4000 level courses are essentially senior level with background courses in basic mechanics. All of the composite courses described below are cross-listed between Aerospace Engineering, Chemical Engineering, Civil and Environmental Engineering, Materials Science and Engineering, Mechanical Engineering, and Textile and Fiber Engineering.

Mechanical Behavior of Composites (4791) - This course introduces the wide variety of fiber and matrix materials available by relating the material structure to mechanical properties. Next some simple micromechanical models for stiffness and strength of unidirectional continuous and discontinuous reinforced materials are covered. Typical stress-strain behavior as a function of constituent properties and lay-up are discussed and described with simple rule-of-mixtures equations. Tension, compression, fatigue, fracture and notched performance are discussed and standard testing techniques are reviewed. Adhesive bonding and mechanical fasteners are discussed for composite structural applications. Lastly, examples of current research problems and issues involving composite material applications are presented. This class is three hours of lecture per week for three credits.

Fundamentals of Fiber-Reinforced Composites (4792) - The course starts by defining fiber and matrix materials, advantages and uses of filamentary composites. Next a review of deformable body theory is given which includes stress, strain, equilibrium and compatibility. Constitutive laws and material property characterization methods (micro- and macro-mechanics) and lamiae and laminates (traditional lamination theory). Special laminate construction is covered with applications to beams, thin tubes, thin-walled box beams, wide beams and columns. This class is three hours of lecture per week for three credits.

Composite Materials and Processes (4793) - Basic principles of selecting component materials and manufacturing composites are presented. Most common manufacturing practices are covered. Polymer, metallic, and ceramic matrix systems are considered. This class is three hours of lecture per week for three credits.

Laboratory in Composite Manufacturing (4794) - Covers major manufacturing and testing techniques involved in composites. Analysis, planning, and design of practical parts is also covered. Students make parts by the following techniques: hand lay-up, filament winding, powder prepregging of thermoplastics, thermal forming, roll consolidation, injection molding and extrusion. The parts in many cases are inspected by NDE. This class involves two hours of lecture and three hours of lab each week for three credits.

Damage and Failure Analysis of Composites (7791) - Provides students with the knowledge of the recently developed methods for analysis of damage and failure of fiber reinforced composite material systems. This course covers mechanisms of toughening, fatigue damage, stiffness, strength and crack growth of short fiber reinforcement. For continuous fiber reinforcement, mechanisms of multiple cracking, delamination, fiber pullout and breakage, and mircobuckling are discussed. Various methods of analysis such as micromechanics, continuum damage models, statistical theories and phenomenological failure theories are covered. Attention is also given to woven and braided composites. Three hours of lecture for three graduate credits.

Mechanics of Composites (7792) - The purpose is to teach mechanics of composites to graduate students at a level required for doctorial research in composites. This course will build on the knowledge gained by the student in 4792. Topics to be covered include the following: basic theory of anisotropic elasticity, hygothermal constitutive equations, residual stresses, 3-D effects in laminate analysis, transverse shear deformation in thick composites, local effects due to free edges and ply terminations, fracture of composites, joints, hybrid and sandwich construction. Three hours of lecture for three graduate credits.

Manufacturing of Composites (7793) - This course presents major manufacturing techniques for metal, ceramic and polymer matrix composites. Modeling of processes is covered with emphasis on the fundamental mechanisms and effects. For ceramic matrix composites CVI, resin infiltration, direct metal oxidation, and reaction sintering processes are modeled. For metal matrix composites squeeze casting, foil-tape lay-up, hot pressing, whisker dispersion and powder metallurgy are covered. Composite and coating processing using CVD and CVI which includes an introduction to the equipment and the modeling process. Injection, transfer and compression molding processes for polymer matrix composites are also modeled. Three hours of lecture for three graduate credits.

A new interdisciplinary cross-listed course is currently being developed for the 4000 series entitled *Design of Composite Structures*. This new course will be structured to include a number of team projects and students will get experience in using composite computational codes as well as finite element based codes. Use of electronic media in the course development will take advantage of CAD/CAM visualization capabilities.

This past academic year 1996-97 the following number of students (*) completed the following courses: 4791 (34), 4792 (51), 4793 (63), 4794(9), 7791 (13), 7792 (12), and 7793 (13).

ADDITIONAL COMPOSITE COURSES

In addition to the multi-discipline cross listed courses described above, there exists several specialized courses that are very related to the composites field offered by the various schools in the College of Engineering at Georgia Tech. Some of these courses are as follows:

AE 4120 Introduction to Composite Aerospace Structures
CHE 6755 Polymer Structure and Mechanical Properties
CHE 6756 Mechanical Properties of Polymers
ME 7224 Micromechanics of Composite Materials
TEX 4600 High Performance Fibers
TEX 6100 Advanced Fiber Science
TEX 7220 Fiber Mechanics
TEX 7311 Polymer Degradation

GRADUATE SCHOLARSHIPS

CERC is in its fourth year of administering a 5-year $555K grant from NSF on Graduate Research Traineeship in Composites. Nine Ph.D. students are in this program. Two Ph.D.s graduated this past year. Four or five more are expected to graduate by September '97. CERC offers these fellowships to students with outstanding academic achievements.

In addition, Graduate Research Assistantships are also available through various government and industry funded research projects being directed by the CERC faculty, and through the participating schools. Opportunities are also available to conduct graduate research under the co-advisement of faculty members from different schools, making the program truly multidisciplinary.

COMPOSITES RESEARCH AT GEORGIA TECH - *A MULTIDISCIPLINARY APPROACH*

Resaerch, at both the graduate and undergraduate levels, is considered to be a very important part of the educational process. The inter-relationships between the various elements of research required for the reliable design of composites structural components is complex. It is apparent that a complete solution requires expertise from several disciplines including basic Materials Science, Manufacturing Technology, Testing and Evaluation, Mechanics and Design. The composites faculty organized under the Composites Education Research Center at Georgia Tech has extensive depth in all these areas. The research teams working in these areas often belong to different schools and units within Georgia Tech. Due to the diverse background of the participating faculty, the testing and analysis resources, and the interdisciplinary teaming available, we feel that Georgia Tech is capable of conducting composites research from the material synthesis, manufacture, characterization, and modeling, through product design and implementation. This scope of these capabilities makes Georgia Tech rather unique among academic institutions. This past year (1996/97) more than $5 million dollars was expended on sponsored composites research at Georgia Tech.

RESEARCH IN THE CLASSROOM

Essentially every professor that teaches one of our composites courses has active research programs. It is expected that he integrates the latest concerns and findings of the composites community into his lectures.

Dozens of undergraduates are employed by faculty to assist in composite research programs. In addition, many seniors choose composite materials and structures related projects for their senior design requirements.

FACILITIES FOR COMPOSITES RESEARCH AND EDUCATION

A short description of some of the facilities that are used for composites research and education are presented in this section.

High Performance Fiber Formation Laboratory:

Broad range of facilities for formation of carbon and ceramic fibers from polymer derived precursors exist in the School of Textile and Fiber Engineering. Equipment includes solution spinning,small scale melt spinning, solution mixers, low (up to 300° C), intermediate (up to 1200° C) and high (up to 2000° C) temperature tubular furnaces for continuous thermo-chemical conversion of pre-ceramic fibers to ceramic fibers under controlled time-temperature-stress-environment fields are available. Carbon, silicon carbide, boron nitride, and silica sol-gel fibers have been produced. Current research also includes developing processing routes for silicon carbide/nitride and mullite fibers.

Polymer Composites Processing Laboratory:

The Polymer Composites Processing Laboratory consists of two laboratories in the Manufacturing Research Center. The first is 5,000 sq. ft. of high bay space which contains industrial scale processing equipment. This includes three injection molding machines, two filament winding machines, single and twin screw extrusion systems, a powder coating line for continuous fiber tows, vacuum and standard hot presses, resin transfer molding and reaction injection molding systems, a pultruder and various ovens and sample preparation and testing equipment. The second area consists of 800 sq. ft. of space for wet chemistry and ultrasonic non-destructive testing. The laboratory is utilized by the Georgia Tech community for support of composites research and education.

Ceramic Composites Processing Laboratory:

The Ceramic Composites Processing Laboratory located in the Baker Building includes equipment for fabrication of ceramic composites and measurement of their physical properties. Fiber coating facilities include a chemical vapor deposition (CVD) system capable of continuous coating multifilament tows with a wide variety of metal and ceramic materials. The coater includes provisions for spreading and vibrating the tow to prevent bonding between filaments and a gas-tight fiber enclosure allowing operation at reduced pressures. Up to 500 meters of fiber can be coated in a continuous run. Composite fabrication equipment includes a filament winder and equipment for preform lay-up. For composite densification we have inert and reactive gas sintering furnaces and a modern, instrumented chemical vapor infiltration (CVI) furnace system. A UNIX-based DEC station 3100 workstation is used for process modeling and simulation. Extensive networking within and beyond the Georgia Tech campus provides access to additional mainframe and supercomputer facilities. State-of-the-art visual data analysis software allows effective display and presentation of multi-dimensional and multi-variate model results.

Mechanical Properties Research Laboratory:

The Mechanical Properties Research Laboratory (MPRL) located in the Bunger Henry Building and Manufacturing Research Center (MaRC) buildings is the nucleus of all experimental fracture, fatigue, creep and deformation behavior research at the Georgia Institute of Technology. It is equipped for a wide variety of investigations involving mechanical testing and structural characterization, and is the core facility for mechanical testing and quantitative evaluation of damage in composites. The laboratory currently has eleven servohydraulic and two screw driven test machines. Two of the servo-hydraulic machines are tension-torsion test machine. Seven dead-weight creep machines are also available for sustained load testing. Temperature capabilities range from room temperature to 1200° C. One servohydraulic machine is equipped with vacuum capability of 10^{-8} torr at temperatures up to 800° C. Several scanning and transmission electron microscopes are available at Georgia Tech. A facility for quantitative image analysis exists within Materials Science and Engineering, including both automatic and semi-automatic image analysis equipment. The Zeiss Videoplan System, coupled with a Zeiss ICM 405 inverted light microscope, handles all functions of quantitative microscopy on a semi-automated basis. There are also three Thermotron programmable aging ovens for temperatures up to 350°C. Each has 12 cubic feet of storage space.

Advanced Composite Structures Laboratory:

The facilities in this laboratory located in the School of Aerospace Engineering include an autoclave, a hot compression molding press, a filament winding machine for manufacturing of composites, two servohydraulic materials testing machines and a miniature tensile tester with acoustic emission, x-ray radiography, acousto-untrasonics, ultrasonic c-scan and microscope-videoscope systems.

SUMMARY

Georgia Tech has a very active education and research program in the area of composite materials and structures. There are 40 faculty involved in the Composites Education and Research Center. The Composites Education and Research Center is the umbrella organization on campus to consolidate and promote the composite materials and structures educational and research activities. Currently there are 7 multi-discipline cross-listed courses in composites that served 195 attendees this past year (1996/97). Another composite course that addresses structural design is being developed. Georgia Tech offers both an undergraduate and graduate certificate in Composites Engineering. Each year approximately a dozen certificates are awarded. Georgia Tech takes great pride in the true interdisciplinary nature of their composites education and research programs and feels as if it is among the best in the world.

Recent Changes in the Sophomore Engineering Curriculum at Texas A&M University

D. H. ALLEN, K. L. E. HELMS AND D. C. LAGOUDAS

ABSTRACT

Over the past decade Texas A&M University has implemented under National Science Foundation funding two new unified sophomore curricula in the Dwight Look College of Engineering. The first of these was funded for a period of five years, but continues today under the title ENGR 20X Program. The second was funded as the Foundation Coalition with six other universities. The authors have taught one of these courses (multiple times) in both programs, and this course deals with the mechanics and thermodynamics of deformable bodies. The essential change implemented in these two new sequences of courses (as opposed to more traditional courses) is that a more general framework based on continuum mechanics is first taught to all of the students, regardless of their engineering disciplines. Then this framework is utilized to solve a variety of problems within specialized disciplines such as strength of materials, heat transfer, electromagnetism, and fluid mechanics. In addition, the coalition program has adopted a teaming concept that more effectively utilizes computer hardware and software. These new educational programs are reviewed herein.

INTRODUCTION

In recent years there have been several efforts to overhaul the educational process at all levels of education. This has not gone without notice at the collegiate level in engineering colleges across the nation. Indeed, several new efforts have been made to revitalize the approach to engineering education in the United States. One such effort has been underway at Texas A&M University for the past ten years. The authors have been involved in this program for about seven years. During that time two new curricula have been introduced in the College of Engineering at Texas A&M University. These curricula are college wide, and are focused primarily on the first two years of undergraduate study. The intent of these new

---

David H. Allen*, Kayleen L.E. Helms, Dimitris C. Lagoudas, Center for Mechanics of Composites, Texas A&M University, College Station, TX 77843-3141

*Author to whom all correspondence should be directed.

programs has been to streamline the educational process through course synergism, more fundamental approaches to teaching, and adoption of advanced tools such as computer workstations, which encourage team activities. The second of these programs, called the Coalition Program, is in the process of developing a new sequence of core courses for the College of Engineering. This program, funded by the National Science Foundation, funds Texas A&M University and six other universities.

One of the objectives for the Foundation Coalition is to design an entire four year engineering curriculum. Starting in the Fall of 1994, each university within the foundation coalition began to design a sophomore year curriculum to fit their particular needs. The Texas A&M Sophomore Coalition Team was assembled during Fall 1994, and started meeting in late November. The team consisted of faculty from engineering, mathematics, physics, and English. In addition to the proposed instructors, there were other interested faculty contributing to the discussions. Almost two months were spent reviewing curricula and trying to establish the needs of the College of Engineering (COE), which includes nine departments and 12 programs. Finally, the committee decided to base the curriculum on a set of courses, previously as part of the ENGR 20X NSF funded program, which based the instruction on using conservation principles as the guiding philosophy. Once that decision was made, the following were the guidelines in trying to devise the content of the courses:

1. Provide a core program based on conservation principles that would be taken by all sophomore engineering students.
2. Integrate the courses.
3. Insure that the courses would be teachable.
4. Involve teaming and active learning within the program.
5. Employ technology in the courses.
6. Build the courses on the freshman coalition year.

With these concepts in mind, a team of faculty from engineering, mathematics, physics, and English created the following five three-hour courses, which were taught in the Fall of 1995 (Engr 1-3) and Spring of 1996 (Engr 4&5). The outline of these five courses is given below:

ENGR 211 - Conservation Principles and the Structure of Engineering

This first course deals with the conservation laws and their applications to macroscopic-sized systems. For example, the applications are to rigid body mechanics, flow of fluids through pipes, thermodynamics, and electrical circuits. The course presents the foundation and structure of the conservation principles for lumped or macroscopic systems, with applications in solid and fluid mechanics, electrical science, thermodynamics and heat transfer. The unified theme of conservation emphasizes its fundamental role in understanding and predicting the behavior of simple engineering systems. The course has built upon the skills developed by the students in teaming and collaborative learning, by assigning team projects early in the semester. A simple structural analysis software, Visual Analysis, has been introduced and used to carry out static structural analysis of trusses and frames. All notes developed for the course have been electronically submitted to the students via a local network, while a web page has been established and used as an electronic board to post the various activities of the class.

ENGR 212 - Conservation Principles and the Structure of Engineering-Thermo/Fluid Dynamics

The thermo/fluid dynamics course deals with the conservation laws and the unifying framework in mass, energy, enthalpy, and entropy. This course covers what a conservation law means. ENGR 212 creates an outline of which properties are conserved and which are not, how each is accounted for, and why these laws are so important. Thermodynamic and fluid property relationships are introduced for use with the conservation laws. These laws demonstrate how all of the thermodynamic properties of a system can be known if just two independent properties of that system are known. Applications of these laws are carried out for heat transfer, solid mechanics, and fluid mechanics applications. A simple equation solving program, EES (Engineering Equation Solver), allows students to carry out the theoretical thermodynamic analysis of such systems. This course compares different types of processes, such as steady state and non-steady state. ENGR 212 also investigates the ideal and non-ideal thermodynamic performance cycles of equipment such as turbines, boilers, pumps, and condensers.

ENGR 213 - Properties of Materials

The atomic structure of gases and liquids must be understood in order to predict the properties of materials. ENGR 213 presents a basis for understanding the behavior of matter through molecular and atomic theories. The purpose of this course is to introduce some basic principles, which are useful in modeling the structure of materials and describing /predicting their properties. This materials course investigates the dependence of macroscopic physical properties on chemical structure (bonding), crystal structure, and microstructure. The course starts from the basic particles, atoms and interatomic bonds, and moves up to the mechanical structure of matter such as the different type of crystal lattices. The ENGR 213 course introduces the role of conservation principles and the Second Law of Thermodynamics in determining the physical properties. These properties are then utilized in constitutive relationships or phase transformation equations. Mechanical properties and electrical conductivity are studied in relation to thermal properties. Structures are studied to predict such mechanical properties as elasticity/plasticity or ductility versus tensile strength. "Hands-on" teaming labs are incorporated into this course to re-enforce the material learned in class and to provide a physical example of actual materials testing.

ENGR 214 - Conservation Principles of Continuous Media

This course further develops the material presented in ENGR 211 and 212, by examining the application of conservation principles to continuous media, for which internal spatially variable changes are important in describing the system behavior. Thus, this course deals with areas such as the mechanics of solids and fluids and heat transfer. The students learn a unified approach to solving problems in the above areas, by selecting the appropriate conservation law(s), constitutive equations, geometric relations and boundary conditions. They realize that in order to evaluate the stress in a beam, the velocity in a simple flow, or the temperature in a heat conduction problem, they have to deal with vectors and concepts from calculus, linear algebra and differential equations. ENGR 214, therefore, is the course that brings together and integrates knowledge from undergraduate mathematics courses and ENGR 211 and ENGR 212 (for the basic conservation laws and material properties).

Emphasis is given to the formulation of problems, and explicit solutions are given for simple structural elements and simple fluid flows, as well as 1-D heat conduction problems. MAPLE is used to facilitate calculus and linear algebra, while Visual Analysis is used for structural analysis. FEMAP is introduced as a demonstration tool for the solution of more complicated 2-D and 3-D problems using finite element analysis. Team projects in heat transfer and structural analysis outside of class are planned, while team problems and quizzes are given during the recitation period.

The class notes are available through the local network and homework solutions are posted electronically. In the current web page, among other group activities and announcements, profiles of all students are available with their individual pictures. One key observation about the sophomore class is that students are not afraid to ask questions and to actively participate in the classroom, debating issues and answering questions. We believe that this is a result of their training during the freshman year and the first part of their sophomore year. A complete effectiveness assessment of the course is performed at the end of the sophomore year and comparisons are made against a similar group of students from the classical curriculum.

ENGR 215 - Principles and Applications of Electrical Engineering

The sophomore curriculum includes a separate electrical engineering course. The textbook is "Principles and Applications of Electrical Engineering" by Giorgio Rizzoni, which is intended for non-EE majors. This course replaces the more traditional required sophomore circuits course for EE majors and the traditional circuits/electronics/instrumentation service course required for most non-EE engineering degrees. Both of these goals must be achieved with only three credit hours! This constraint requires less emphasis on rigorous proofs and drill exercises. Slight adjustments in required junior level EE circuits and electronics courses can easily make up for any deficiencies in the depth of this introduction. We feel that these limitations will be compensated through integration with other courses, teaming experience, and extensive use of computer software. Students exploit computer and electronics experience that is not contained in the traditional sophomore circuits course. Further, this course will give the non-EE better preparation than the traditional service course.

The software specific to this course is "Electronics Workbench". The circuit simulation has all of the power of SPICE, the international standard, with the convenience of mouse operation. Students can very quickly test and adjust their intuitive understanding and analyses of a large variety of circuits (active-passive, linear-nonlinear, digital-analog, transient-steady state, etc.). In addition, the students will get hands-on hardware experience with demonstrations and team projects. One project will be the development of a small signal equivalent circuit model of a loudspeaker from electrical measurements on an actual loudspeaker.

EXAMPLES TAKEN FROM COURSE FOUR

The authors of the current paper have been primarily involved in the construction and teaching of the fourth course in the coalition sequence. The example shown in Table I demonstrates the unified approach to teaching where it can be seen that 1-D heat transfer and stress analysis of uniaxial bars are accomplished with identical mathematical models.

TABLE I

Heat Transfer

Equilibrium: $\frac{dQ}{dx} = -\rho\Phi \qquad Q = q_x A$

Constitution: $q_x = -k\gamma$

Kinematics: $\gamma = \frac{dT}{dx}$

Unknowns: Q, q, γ, T

Governing Differential Equation: $\frac{d}{dx}\left[kA \frac{dT}{dx} \right] + \rho\Phi = 0$

Where: **Q** is the heat transfer rate in units of [energy/ time]
q_x is the heat flux in units of [energy/ (time*area)]
γ is the temperature gradient in units of [degrees/ length]
T is the temperature in units of [degrees]
ρ is the density in units of [mass/ length^3]
ϕ is the heat source in units of [energy/ mass]
A is the cross sectional area in units of [length^2]
k is the thermal conduction coefficient in units of [energy/ (time*length*temperature)]

Uniaxial Bar

Equilibrium: $\frac{dP}{dx} = -p_x \qquad P = T_{xx} A$

Constitution: $T_{xx} = E\,\varepsilon_{xx}$

Kinematics: $\varepsilon_{xx} = \frac{du}{dx}$

Unknowns: $P, T_{xx}, \varepsilon_{xx}, u$

Governing Differential Equation: $\frac{d}{dx}\left[EA \frac{du}{dx} \right] + p_x = 0$

Where: **P** is the internal resultant in units of [force]
T_{xx} is the axial stress in units of [force/ area]
ε_{xx} is the axial strain in units of [length/ unit length]
u is the displacement in units of [length]
p_x is the external distributed load in units of [force/ length]
A is the cross sectional area in units of [length^2]
E is the Young's Modulus in units of [force/ area]

TABLE II

ENGR 489
Special Topics: Principles of Engineering IV
(Conservation Principles for Continuous Media)
Spring 1997

D. H. Allen

Nature, Scope, and Content of the Course: This course provides engineering students with a unified understanding of the handful of fundamental conservation laws and the second law of thermodynamics, which build a framework upon which our description of science and engineering is based, for *continuous media*. The first part of the course presents the fundamental concepts and illustrative examples of their uses. The latter part of the course provides a more complete examination of the manner in which these principles are applied to selected, traditional areas of engineering.

Required Course Text and Material: *Conservation Principles for Continuous Media*, C.J. Glover and H.L. Jones, 4th Edition, McGraw-Hill Custom Publishing Division, 1997. Available at local bookstores.

Course Prerequisites: ENGR 201 or equivalent; MATH 251, co-requisite: MATH 308; ENGR 202 or equivalent.

Grading Policy: Final course grades will be determined as follows:

Homework	25%
Recitation Quizzes (about 4)	25%
Midterm Exam	25%
Final Exam	25%
	100%

References: The references given in the text provide additional examples and perspectives.

Homework: Assignments generally will be made weekly and will be due one week later. Problem solutions themselves can be worked out with other students and this is frequently desirable. The write-up and explanation, however must be an individual effort on the part of each student. Certainly, in working together, discussion of concepts must occur (or should occur) but each person's write-up must be his/her own effort. This will help students prepare for the tests. All solutions should be clearly written, and should include descriptions or other narrative information, in the form of complete sentences.

Tests: There will be about four quizzes during the semester, given during the recitation period (see the accompanying syllabus and schedule). Additionally, there will be *comprehensive mid-term and final exams.* It is the very critical that this course is remembered from a comprehensive, broad perspective.

Instructor: David H. Allen. Professor, Department of Aerospace Engineering.
Office: 736A H.R. Bright Building Phone: (409) 845-1669 e-mail: dhallen@aero.tamu.edu

TABLE III

ENGR 489
Special Topics: Principles of Engineering IV
(Conservation Principles for Continuous Media)
Spring 1997

D. H. Allen

Part I- PERSPECTIVES

Chapter 1: Introduction
Chapter 2: Review: Macroscopic Analysis

Part II- THE FUNDAMENTAL LAWS

Chapter 3: The Conservation of Mass
Chapter 4: The Conservation of Charge
Chapter 5: The Conservation of Momentum

Rec Quiz I: Chs. 3-4 Wed., Jan. 29

Chapter 6: The conservation of Angular Momentum
Chapter 7: The conservation of Energy
Chapter 8: The Second Law of Thermodynamics

Rec Quiz II: Chs. 5-8 Wed., Feb. 19

Part III: APPLICATIONS

Chapter 11: Heat Transfer

Lecture Mid-Term: Chs. 1-8, 11 Wed., Mar. 19

Chapter 9: Stress, Strain, and Deformation in Solids

Rec Quiz III: Chs 9-10 Wed., Apr. 2

Chapter 10: Analysis of Linearly Elastic Solids

Rec Quiz IV: Ch 10 Wed., Apr 23

Chapter 12: Fluid Mechanics

FINAL EXAM: Comprehensive - Tuesday, May 6
10:30 am - 12:30 pm

Tables II and III show a typical course syllabus, and course outline for ENGR 214, respectively.

CONCLUSION

A sophomore core engineering program consisting of five 3-hour credit courses, for the COE at Texas A&M, has been designed. It is based on a previous NSF program that used conservation principles as the basis for its development. We have taken that basis and adjusted the curriculum to fit the reduced number of credits, and the increased use of technology, teaming, and collaborative learning as part of a long range plan to redesign the entire four-year curriculum.

ACKNOWLEDGEMENT

The support of the National Science Foundation is gratefully acknowledged.

REFERENCES

A National Action Agenda for Engineering Education, American Society for Engineering Education, 1987.

Smith, K. A., The Craft of Teaching Cooperative Learning: An Active Learning Strategy, 1989 Frontiers in Education Conference, ASEE, pp. 188-192, 1989.

Michaelsen, L. K. and Black, R. H., Building Learning Teams: The Key to Harnessing the Power of Small Groups in Higher Education, Growth Partners, Oklahoma City, OK, 1994.

Johnson, R. T., Johnson, D. W., and Smith, K. A., Cooperative Learning: An Active Learning Strategy for the College Classroom, Baylor Educator, pp. 11 - 16, Winter 1990.

Stice, J. E., A First Step Toward Improved Teaching, Engineering Education, vol. 66, no. 5, pp. 394 - 398, Feb., 1976.

Joints and Fasteners

Damage Initiation and Progression for Basic Laminates with a Hole: Part I—Unidirectional and Angle-Ply

J. R. SCHAFF[1] AND R. KIM[2]

ABSTRACT

The initiation and progression of damage for simple composite laminates with open holes under tension loading was recorded and observations were correlated with results from experimental measurements, from two-dimensional elasticity theory and from three-dimensional spline variational theory. During an incremental series of monotonic loadings, strain measurements were taken in close proximity to the hole boundary while radiographic images were captured following each loading increment. This information provided a direct comparison between the state of damage and the state of strain. Three laminate stacking sequences were considered: $[0_8]_T$, $[90_8]_T$, and $[\pm 45]_{2S}$. As expected, the initiation of damage and laminate fracture occurred almost simultaneously for the $[90_8]_T$ laminate. Transverse cracking occurred prior to specimen failure for the $[0_8]_T$ and $[\pm 45]_{2S}$ laminates and is evident in the measured strain behavior. Overall, theoretical strains obtained from spline variational theory agree well with experimental strain measurements at low loadings prior to damage initiation.

INTRODUCTION

The occurrence of holes in load bearing composite structures is pervasive in the aerospace industry as a result of the use of mechanical fasteners in assembly and the use of cutouts to accommodate wiring and hydraulic lines. The design of laminates with holes or with fasteners in composite structures is largely based on experimental test data (Shyprekevich, 1995). The most widely utilized analysis methods are those based on the two-dimensional elasticity solution provided by Lekhnitskii (1968) for an infinite orthotropic plate with an open hole. These methods, including the model of Garbo and Ogonowski (1981), are summarized by Synder *et al.* (1990). In an effort to model the fastener, Lekhnitskii assumes a cosine distribution of radial tractions and a contact zone size. Although the aforementioned analyses are useful design tools, a more rigorous stress analysis method and failure criterion is needed to predict the strength of laminates with holes and with fasteners.

Recently, Iarve (1996;1995) developed a three-dimensional stress analysis method for a composite laminate with an open hole or an elastic inclusion. This method is based on the spline variational theory and was verified through comparisons of in-plane and interlaminar stresses with results obtained from an asymptotic solution. Results presented herein were

1-United Technologies Research Center, M/S 129-73, East Hartford, CT 06108
2-University of Dayton Research Institute, 300 College Park Ave., Dayton, OH 45469

determined using Spline Variational Elastic Laminate Technology (SVELT), a dedicated computer package which implements the analysis described by Iarve (1996;1995).

A number of experimental studies have documented the progression of damage for composite laminates with open holes. Shalev and Reifsnider (1990) presented observations on the onset of delamination while transverse cracking was the focus of works by Joshi and Huang (1994) on quasi-isotropic laminates and Reddy et al. (1990) on cross-ply laminates. Lessard and Chang (1991) discuss the development of damage in laminates containing open holes for tension. In these works, observations were not made in a building block fashion for both unidirectional and basic laminates.

In this two-part work, quasi-static loading experiments were conducted on basic laminates to document the initiation and growth of damage and to record changes in the strain field. The objective of this work was the development of an experimental database from which three-dimensional stress analysis methods with discrete damage and associated failure criterion are later investigated and verified. In part I of this work, three IM7/5250-4 laminates, $[0_8]_T$, $[90_8]_T$, and $[\pm 45]_{2S}$ are incrementally loaded to failure. For each loading increment axial strain was recorded in close proximity to the hole and the specimen was x-rayed. Acoustic emission data was collected during loading and compared to strain measurements and radiographic images to identify significant damage events. Additional analysis was conducted by comparing experimental strain data to results obtained from elasticity theory and from spline variational theory. Part II of this work is in progress and considers more complex laminates.

EXPERIMENT

A series of tension tests were conducted to document the initiation and propagation of damage in basic composite laminates with a centrally located circular hole. Of particular interest is the response of IM7/5250-4. Testing was conducted in two phases: monotonic loading of unnotched tensile coupons to failure and incremental loading of open hole specimens to failure. The purpose of the first phase is to obtain mechanical properties including stiffness and strength in the transverse and longitudinal directions. The purpose of the second phase is to document the progression of damage and the corresponding strain response. Three laminate orientations were considered: $[0_8]_T$, $[90_8]_T$, and $[\pm 45]_{2S}$. Each laminate was cured according to the manufacturer's recommended cure cycle and was post cured at 227°C [440°F] for a duration of 5 hours and 40 minutes. The average fiber volume was reported as 62% based on procedures established in ASTM 3171-76, and the void content was determined to be less than 1%.

For the monotonic loading phase, the nominal tensile coupon dimensions were 2.54 cm wide, 30.48 cm long with a tab length of 2.54 cm for the $[90_8]_T$ and the $[\pm 45]_{2S}$ laminate. A variation of this specimen was used for the $[0_8]_T$ laminate where the coupon width was 1.27 cm. The average ply thickness was determined to be 0.134 mm [0.00527 in.]. A total of six specimens were tested for $[0_8]_T$ and $[90_8]_T$ laminates while three replicate tests were conducted for the $[\pm 45]_{2S}$ laminate. Measurements of axial and transverse strains were recorded for each specimen using strain gages during loading. Results of the monotonic tensile testing is summarized in Table I.

TABLE I-SUMMARY OF MECHANICAL PROPERTIES FOR BASIC LAMINATES

Laminate	Modulus, GPa	Strength, MPa	Strain to failure,%
$[0_8]_T$	151	2640	1.5
$[90_8]_T$	10.7	68.9	0.68
$[\pm 45]_{2S}$	21.4	208	>3.0

For the incremental loading phase, an open hole tension specimen is utilized with the following specimen dimensions: L=30.84 cm, W=7.62 cm, tab length =2.54 cm, and hole diameter, D=1.27 cm. A longer tab length, 3.81 cm, is selected for the $[0_8]_T$ laminate to avoid tab debonding. Prior to initial loading, no transverse cracks were observed upon radiographic inspection of each specimen. Each specimen was monotonically loaded to a value P1 and then unloaded. Axial strains were recorded during the test using up to 6 strain gages while acoustic emission measurements were monitored to identify potential damage events. Typically, two strain gages were employed in the far field region and four strain gages were arranged to record the strain intensity approaching the hole edge perpendicular to the loading direction. The load P1 was either a predetermined value based on assumed ply behavior, usually well below the anticipated transverse cracking stress, or a value corresponding to a significant acoustic event after which the specimen was unloaded. Following loading of the specimen to a value, P1, the specimen was x-rayed in the unloaded condition to document evidence of damage. This process was repeated for increasing loads P2, P3, etc. until failure.

A description of the drilling procedure is provided to emphasize the importance of minimizing drill induced damage. A carbide tipped steel drill bit with an operating speed of 800 rpm was selected for specimen fabrication procedures. This type of bit is specially designed for cutting holes for layered media such as circuit boards. Surface damage was minimized by using glass/epoxy laminates on the top and bottom surface of the specimen prior to drilling. This drilling process produced holes that were absent of transverse cracks for nine out of ten holes. It is important to note that drilling induced damage cannot be totally eliminated. Utilizing similar procedures, Schaff, et. al. (1996) showed that damage exists in the form of chipping and is visible in the surface ply around the hole periphery extending away from the edge a distance of one ply thickness. These regions were rarely more than a few fiber diameters in depth through-the-thickness of the ply and typically occurred where fibers approach the edge at a large angle.

RESULTS AND DISCUSSIONS

ANALYSIS

SVELT is based on the spline approximation of displacements *u*, *v*, and *w* in terms of unknown spline coefficients and sets of basic spline functions. Displacements, interlaminar tractions, and prescribed tractions are used to construct a system of equations by applying the minimum potential energy principle. After displacement boundary conditions are enforced, the remaining set of equations can then be solved for the unknown spline coefficients. Strains are determined from displacement relations and are continuous within a ply. The spline geometry utilized to obtain the present results is given by n′=3, k′=72, m′=22, m'_o=16, and q=1.4 according to the convention described by Schaff et al. (1996).

Two-dimensional results were obtained using the elasticity solution for an orthotropic plate with a circular hole (Lekhnitskii, 1968). All stress analysis results were obtained using the following properties: E_{11}=151 GPa, E_{22}=E_{33}=10.7 GPa, G_{12}=G_{23}=5.9 GPa, G_{13}=3.26 GPa, n_{12}=n_{23}=0.309, n_{13}=0.449, $\alpha_{11}=0.61\text{x}10^{-6}$/°C and $\alpha_{22}=25.2\text{x}10^{-6}$/°C.

$[90_8]_T$ LAMINATE

The $[90_8]_T$ laminate was subjected to two loading increments: 23.7 MPa [3.44 ksi] and 32.7 MPa [4.74 ksi]. The last loading corresponds to the failure load. It is apparent from acoustic emission sensors and strain measurements that initiation of a transverse crack and specimen failure occurred simultaneously. Inspection of the post-failed specimens indicates net tension failure perpendicular to the loading axis in the hole ligament. The location of the failure is described by transverse cracks at approximately 87° and 270° clockwise from the top of the specimen along the loading direction, hereafter referred to as the reference direction. Figure 1 shows the stress-strain curves obtained from four strain gages (e1, e2, e3, and e4) in close proximity to the hole and for the far field strain gage (e5). Gage e1 is inside the hole and e2, e3, and e4 are centered at distances 0.127cm [0.05 in.], 0.381cm [0.15 in.], and 0.889cm [0.35 in.] away from the hole edge perpendicular to the loading direction. Strains, e2 to e5 are linear up to failure whereas strain on the hole edge, e1, is initially linear and exhibits slightly nonlinear behavior at approximately 0.6% strain. This value of strain is close to the failure strain of the unnotched tensile coupons, 0.68 %. The strain recorded from gage e1 is 0.84% at failure which is 25% greater than the strain to failure reported in Table I.

In Figure 2, the axial strain from SVELT and from experiment (e2, e3, and e4) are plotted perpendicular to the direction of loading and along a radial path beginning at the hole edge and extending to 12.5T_{lam} where T_{lam} is 0.114cm [0.045 in]. Comparisons are made for applied stress values of 10.2 MPa [1.48 ksi], 15.3 MPa [2.22 ksi], 20.4 MPa [2.96 ksi], and 23.8 MPa [3.45 ksi]. In general, good agreement between SVELT and experiment is observed. Strain concentration factors at the hole edge can be determined by the strain ratio e1/e5 and compared to those determined by SVELT and Lehknitskii. The range of values of e1/e5 at low loadings is calculated as 2.54-2.65 which agrees well with those obtained from SVELT and elasticity, that is, 2.66 and 2.54 respectively.

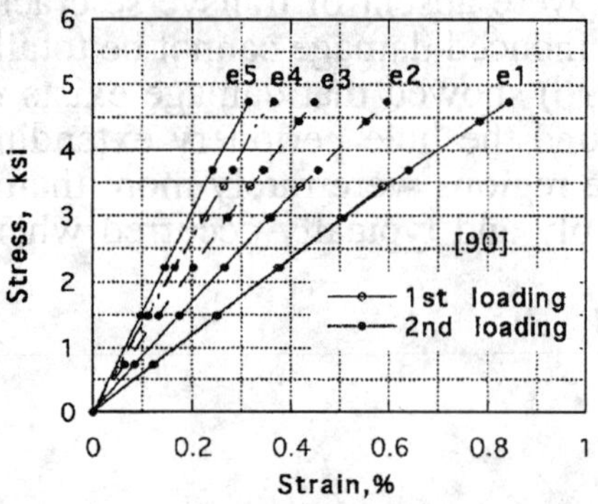

Figure 1. Stress-strain curves from gages e1-e5 for $[90_8]_T$ Laminate.

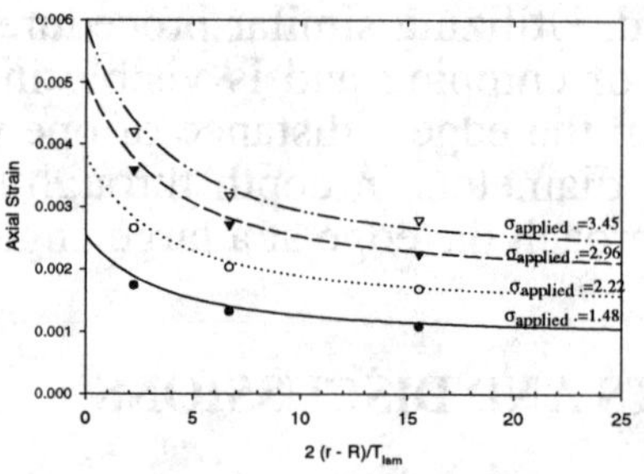

Figure 2. Comparison of strain measurements with axial strain from SVELT.

$[0_8]_T$ LAMINATE

The $[0_8]_T$ specimen was subjected to nine load steps where each step is defined by the following peak values of applied stress: 183.9 MPa, 229.8 MPa, 270.7 MPa, 332.0 MPa, 510.7 MPa, 766.1 MPa, 1277 MPa, 1787 MPa and 2119 MPa. Laminate failure occurred in net tension during the ninth load step at an applied stress of 2119 MPa. A description of the onset of damage and the subsequent growth of damage is discussed with the aid of radiographic images and stress strain curves for selected loadings. In Figures 3a through 3e, radiographic images following load steps 1, 2, 3, 4 and 8 are presented as a direct indication of the damage state. The $[0_8]_T$ laminate used the same arrangement of strain gages as the $[90_8]_T$ specimen except two strain measurements were recorded in the far field region. Inspection of Figures 3, show two far field strain gages at the bottom of the specimen where gage e5 is centered and gage e6 is closer to the edge. In Figures 3, a strain gage wiring post is positioned close the hole and 135° clockwise from the reference direction. The experimentally measured stress-strain results are shown for load steps 1, 2, 3, 4, 5, and 9 in Figures 4a through 4f, respectively. Results are provided for all gages in close proximity to the hole, e1 through e4, and gages in the far field region, e5 and e6. Gage e1 recorded for only the first two load steps due to inadequate strain gage capacity.

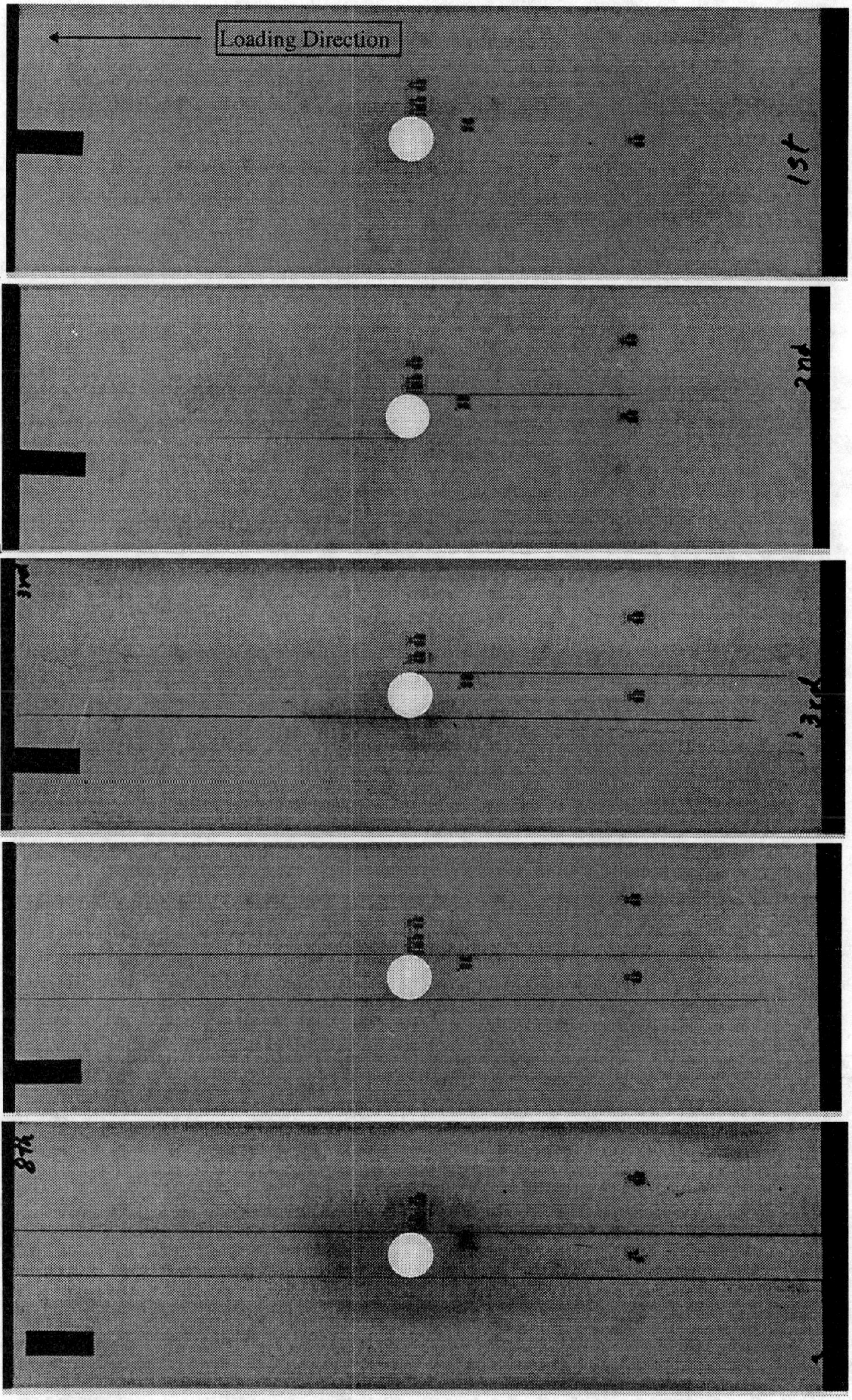

3a) 183.9 MPa

3b) 229.8 MPa

3c) 270.7 MPa

3d) 332.0 MPa

3e)1787MPa

Figures 3a-e. Radiographic images of $[0_8]_T$ specimen for load steps 1-4 and 8.

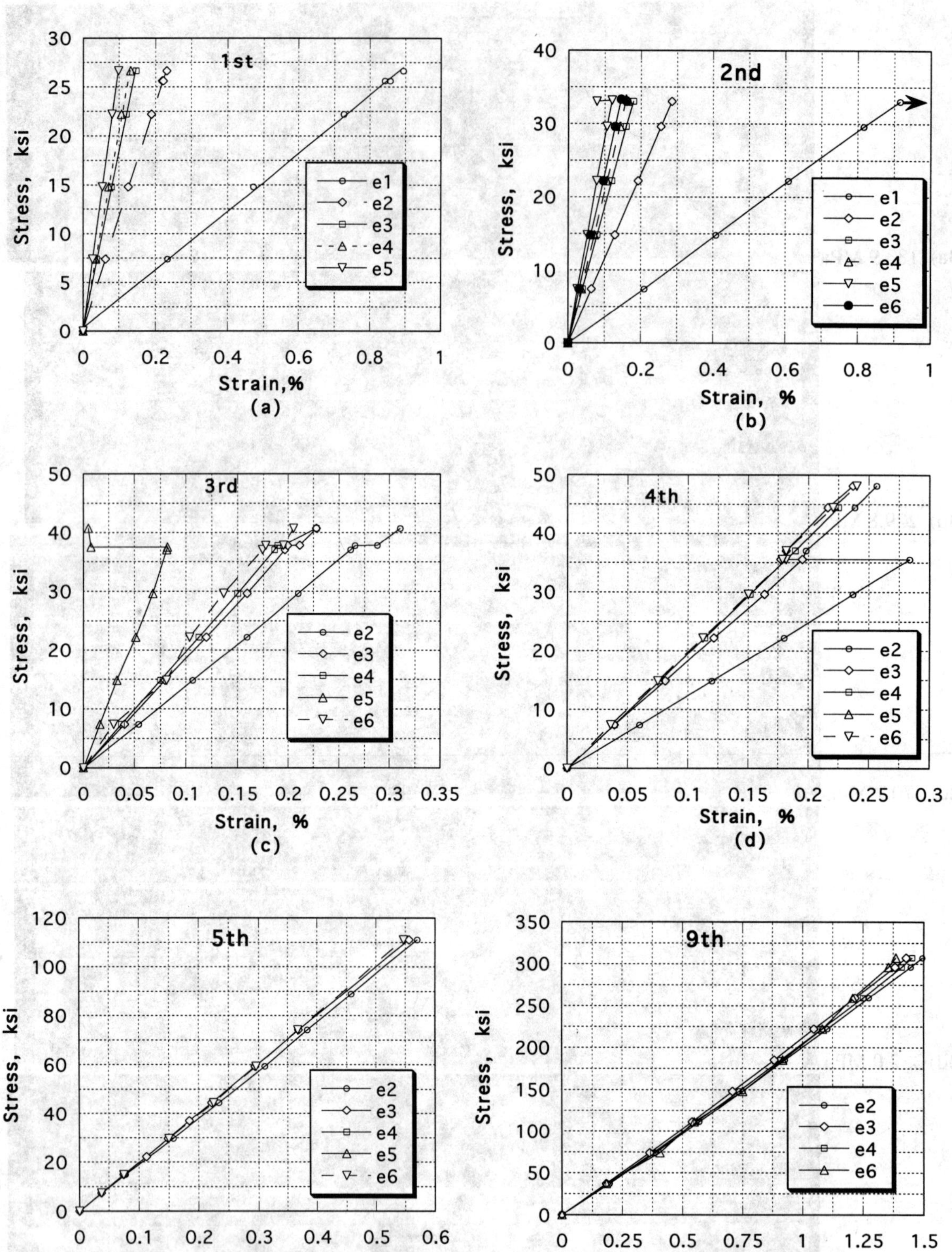

Figures 4. Stress strain curves for gages e1-e6 for $[0_8]_T$ laminate at load steps: a) 183.9 MPa, b) 229.8 MPa, c) 270.7 MPa, d) 332.0 MPa, e) 510.7 MPa, and f) 2119 MPa.

The first transverse crack emitted from the hole edge approximately 277° clockwise from the reference direction. Inspection of Figure 3a reveals that a transverse crack 6 mm in length appeared after the first loading step, 183.9 MPa. It is speculated that the crack initiated at a stress of 174 MPa based on the first acoustic emission peak. In general, the effect of the transverse crack was not evident in the measured strain data because the crack and the strain gages are located on opposite sides of the hole. However, the strain gage e1 showed a slight increase at 176 MPa which coincided with the first acoustic peak, 174MPa.

The second transverse crack occurred at the end of load step 2 and is shown in Fig. 3b. For this load step, the specimen was unloaded at the first signal of activity from the acoustic emission sensor. Strains e1 and e5 show erratic behavior in Fig. 4b at the end of the 2nd loading. Inspection of the x-ray radiograph for load step 3 exhibits the development of the third crack at 263° clockwise from the reference direction and extends within 1cm of the end tab. Figure 3c also shows the onset of this new crack and the growth of two previous transverse cracks to the tab region. During the third load step the strain e1 exceeded the gage capacity at an applied stress of 257.9 MPa. Almost simultaneously, a drop of strain e5 to near zero indicates stress relaxation in the center strip portion bounded by transverse cracks, that is, the load carried by the mid-strip was transferred to the edge ligaments. Furthermore, the load transfer is also detected by the increase of strains e2, e3, e4, and e6 as shown in Fig. 4c. The final transverse crack developed near strain gage e2 at approximately 83° clockwise from the reference direction during loading step 4 at a stress of 245.1 MPa.

Fig. 5 displays the experimental stress strain curves for gage e2 for load steps 1-9. Note that the e2 strain behavior is nearly identical after load step 4. The average moduli obtained from e2 for load steps 5-7 were determined to be within 1% of the experimental value from unnotched coupons. This implies that the strain concentration is diminished in close proximity to the hole boundary. The net stress based on ligament area at failure, 2543 MPa, agrees closely with the average static strength of the unnotched coupons, 2643 MPa.

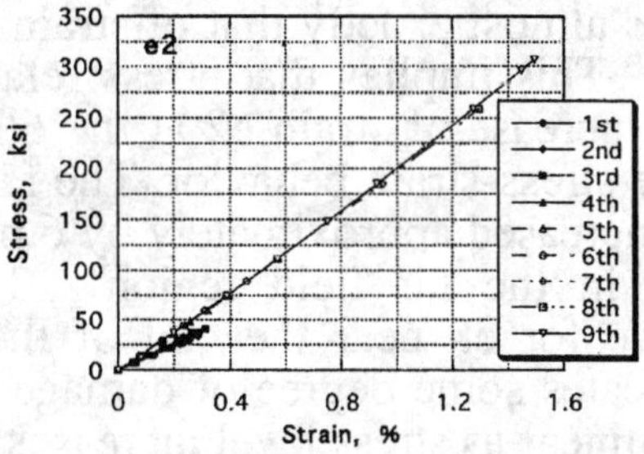

Figure 5. Experimental strain measurements from gage e2 for loadings 1- 9.

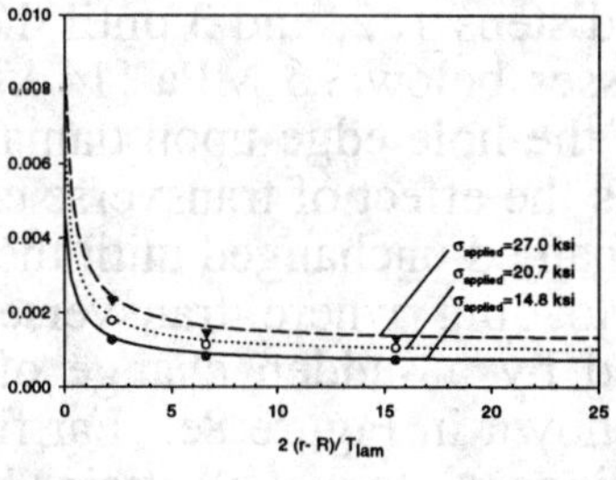

Figure 6. Comparison of experimental and theoretical strains for $[0_8]_T$ laminate.

In a similar fashion as the $[90_8]_T$ laminate, the experimental and theoretical strain results are compared in Figure 6. Results are shown for applied stresses of 102 MPa [14.8 ksi], 143 MPa [20.7 ksi], and 186 MPa [27 ksi]. As expected SVELT slightly overpredicts the experimental strains, however, results agree within 5% for all cases. Note that the $[0_8]_T$ laminate produced a higher stress concentration compared the $[90_8]_T$ laminate. Values of experimental strain concentration factors ranged from 5.92-9.2 at low loadings. Results are reported as 6.825 and 6.78 based on SVELT and Lekhnitskii's solution, respectively.

$[\pm 45]_{2S}$ LAMINATE

The $[\pm 45]_{2S}$ laminate was tested for a series of six incremental loadings corresponding to the following values of applied stress: 51.07 MPa, 76.61 MPa, 86.83 MPa, 114.9 MPa, 140.5 MPa and 158.3 MPa. On the sixth loading the laminate failed along the fiber direction similar to a typical unnotched $[\pm 45]_s$ tensile coupon. Figures 7a and 7b display radiographic images of the specimen following load steps 4 and 5, respectively, and Figures 8a through 8f show stress-strain curves for load steps 1 through 6. Note that the strain e1 is not shown after the third loading because of damage on the gage site.

The onset of cracking as determined from radiographic images was observed after the 4th loading. Unlike the previous laminates, no single initial crack was identified. Rather several transverse cracks in 45° and -45° plies were visible in Figure 7a which describes the initial damage state after the fourth loading. Transverse cracks appear in close proximity to the strain gages originating approximately 75° clockwise from the reference direction in the 45° ply and 105° in the -45° ply. The longest crack appears in the -45° ply and emanates a distance 7.2mm from the hole at 285° clockwise from the reference direction. The first significant acoustic emission event was recorded at 109.4 MPa and further peaks continued until specimen unloading at 114.9 MPa. The cracks grew in succeeding loads while multiple new cracks developed as shown in Figure 7b.

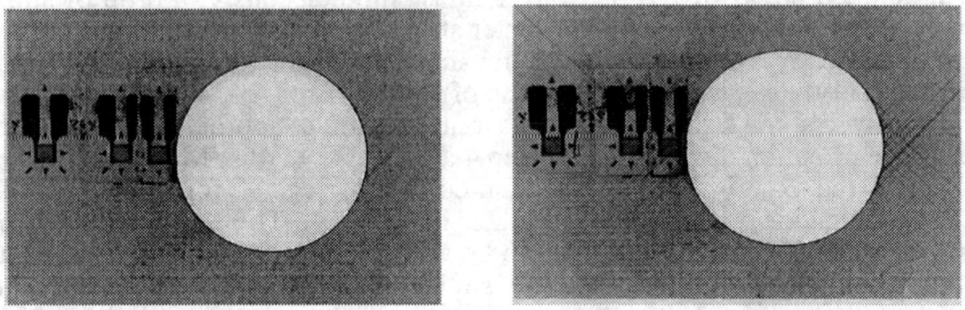

Figures 7. Radiographs of $[\pm 45]_{2S}$ laminate for stress of a) 114.9 MPa and b) 140.5 MPa.

Upon examination of strain measurements in Figures 8, a decrease in strain e1 is evident between load steps 1, 2, and 3 until the strain behavior is almost exactly that of strain e2 for applied stresses below 96 MPa [14 ksi] in load step 4. This implies that stress relaxation occurred on the hole edge upon damage initiation. An increase of strain e2 at the 6th load demonstrates the effect of transverse crack growth on the stress-strain behavior. The far field strain e5 remained unchanged until the 5th loading and increased approximately by 6 percent at the 6th loading where transverse cracks occurred in the far field region. This is demonstrated by a sudden change of stress-strain curve for e5 near the end of the fifth loading as shown in Figure 8e. Far field strain, e5 indicates some degree of damage at 5th and 6th loadings. Note that all strains become highly nonlinear as stress level increases.

The net stress at failure was 205.5 MPa and compared well with the unnotched tensile strength, 207.8 MPa. Good agreement between experimentally measured strains and SVELT is shown in Figure 9 for applied stresses of 27.4 MPa, 41.0 MPa, and 54.7MPa. The strain concentration factor was calculated in the initial linear region of stress strain curves both experimentally and theoretically. The experimentally determined average was 2.23 based on data from gage e1 and e5. Calculations from SVELT and Lekhnitskii resulting in values of 2.20 and 1.84, respectively.

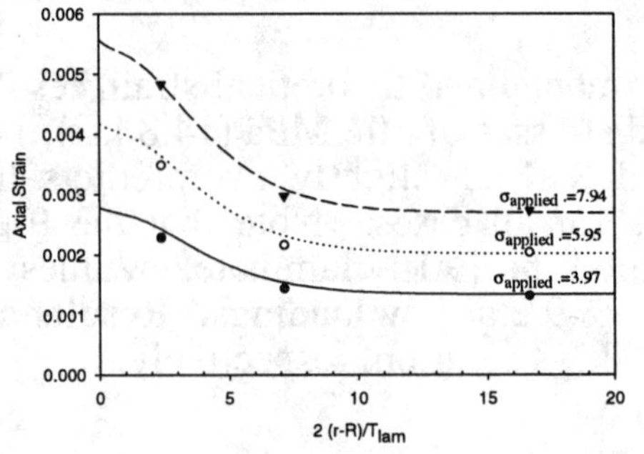

Figure 9. Experimental and theoretical strains for $[\pm 45]_{2S}$ laminate.

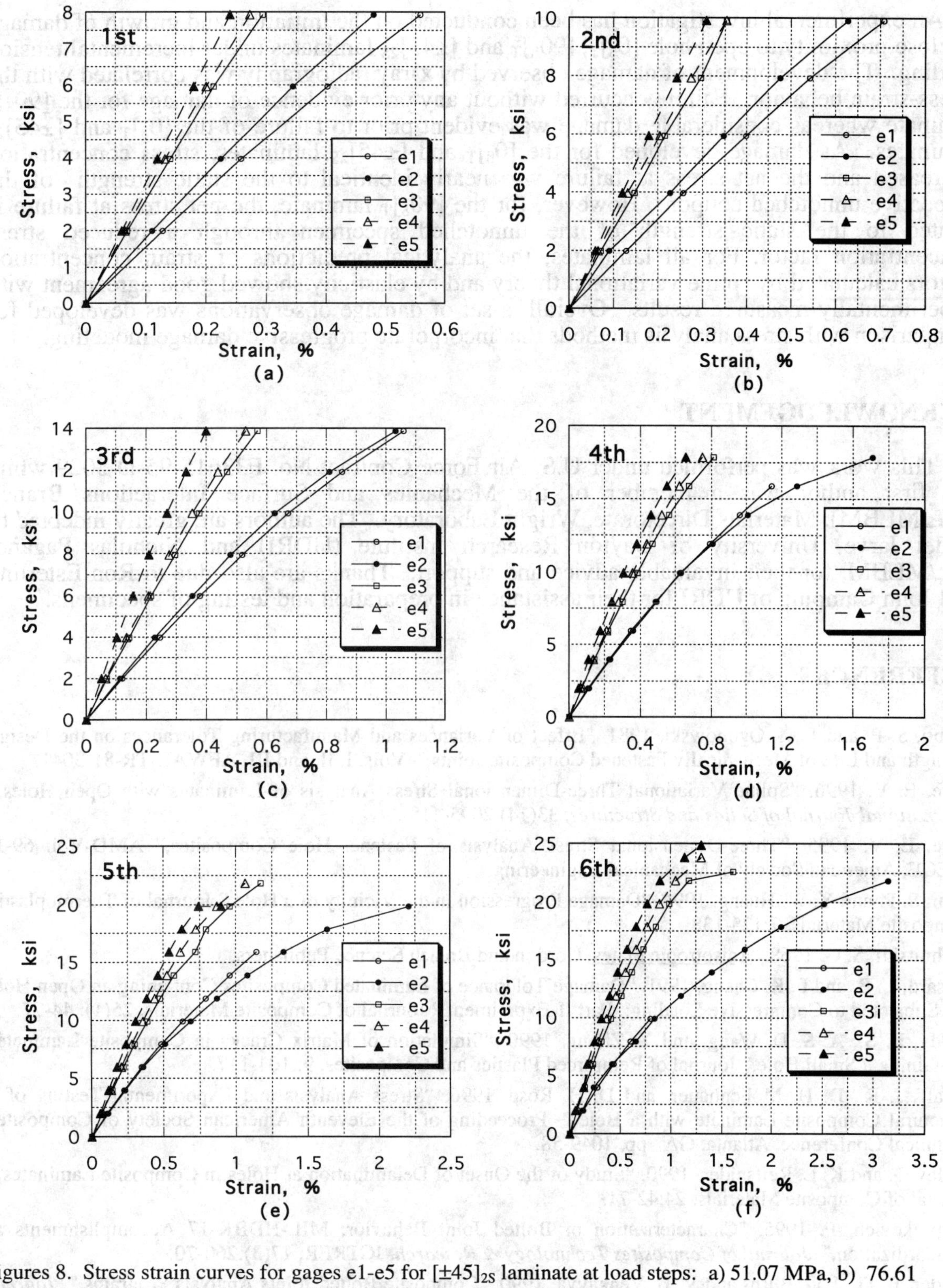

Figures 8. Stress strain curves for gages e1-e5 for $[\pm 45]_{2S}$ laminate at load steps: a) 51.07 MPa, b) 76.61 MPa, c) 86.83 MPa, d) 114.9 MPa, e) 140.5 MPa, and f) 158.3 MPa.

SUMMARY

An experimental investigation has been conducted on the initiation and growth of damage in close proximity to open hole $[0_8]_T$, $[90_8]_T$ and $[\pm 45]_{2S}$ laminates under incremental tension loading. The development of damage observed by x-ray radiography was correlated with the stress-strain behavior. Failure occurred without any prior evidence of damage for the $[90_8]_T$ laminate whereas considerable damage was evident prior to failure of the $[0_8]_T$ and $[\pm 45]_{2S}$ laminates. As damage developed for the $[0_8]_T$ and $[\pm 45]_{2S}$ laminates, stress concentration decreased and the net stress at failure was nearly identical to the static strength of the respective unnotched coupon. However, for the $[90_8]_T$ laminate, the net stress at failure is related to the static strength of the unnotched specimen through a reduced stress concentration factor. For all laminates, the analytical predictions of strain concentration factors calculated by spline variational theory and by elasticity showed good agreement with experimentally measured results. Overall, a set of damage observations was developed for comparison with stress analysis methods that incorporate progressive damage modeling.

ACKNOWLEDGEMENT

This work was performed under U.S. Air Force Contract No. F33615-95-D-5029 while the first author was a member of the Mechanics and Surface Interactions Branch (WL/MLBM), Materials Directorate, Wright Laboratory. The authors are greatly indebted to Endel Iarve, University of Dayton Research Institute (UDRI) and Nicholas Pagano, WL/MLBM, for their invaluable advice and support. Thanks are also due to Ron Esterline and John Camping of UDRI for their assistance in preparation and testing of specimens.

REFERENCES

Garbo, S. P. and J. M. Ogonowski. 1981. "Effect of Variances and Manufacturing Tolerances on the Design Strength and Life of Mechanically Fastened Composite Joints," Vols. I, II, and III, AFWAL-TR-81-3041.

Iarve, E. V. 1996. "Spline Variational Three-Dimensional Stress Analysis of Laminates with Open Holes," *International Journal of Solids and Structures*, 33(14):2095-118.

Iarve. E. V. 1995. "Three-Dimensional Stress Analysis of Fastener Hole Composites," AMD-Vol. 69-1, IMECE, American Society of Mechanical Engineering.

Joshi, S. P. and Y. W. Huang. 1994. "Damage Progression in the Vicinity of a Hole," Journal of Thermoplastic Composite Materials, 7:125-138.

Lekhnitskii, S. G. 1968. Anisotropic Plates, Gordon and Breach Science Publishers.

Lessard, L. B. and F. K. Chang. 1991. "Damage Tolerance of Laminated Composites Containing an Open Hole and Subjected to Compressive Loading: Part II-Experiment," Journal of Composite Materials, 25(1): 44-77.

Reddy, E. S., A. S. D. Wang, and Y. Zhong. 1990. "Simulation of Matrix Cracks in Composite Laminates Containing a Small Hole," Journal of Reinforced Plastics and Composites, 9: 104-117.

Schaff, J. R., D. H. Mollenhauer, and D. H. Rose. 1996. "Stress Analysis and Experimental Testing of a Structural Composite Laminate with a Hole," Proceeding of the Eleventh American Society of Composites Technical Conference, Atlanta, GA., pp. 1049-58.

Shalev D. and K. L. Reifsnider. 1990. "Study of the Onset of Delamination at Holes in Composite Laminates," Journal of Composite Materials, 24:42-71.

Shyprekevich, P. 1995. "Characterization of Bolted Joint Behavior: MIL-HDBK-17 Accomplishments at Standardization," *Journal of Composites Technology & Research*, JCTRER, 17(3):260-70.

Snyder, B. D., J. G. Burns and V. B. Venkayya. 1990. "Composite Bolted Joints Analysis Programs," *Journal of Composites Technology & Research,* JCTRER, 12(1):41-51.

Criteria for the Design of Marine Tee-Joints

J. BISH AND K. T. KEDWARD

ABSTRACT

The design of polymer matrix composite joints is strongly influenced by the presence of out-of-plane or through-the-thickness stresses. These stresses can be critical, since out-of-plane interlaminar tension and shear strengths, as well as moduli, are dominated by the strength and stiffness of the polymer matrix constituents as well as the fiber matrix interface. To aid in the design of these structures, this paper develops a methodology for predicting the structural adequacy of composites in the presence of through-the-thickness loadings. Specifically, this work details guidelines for the design of resin fillet tee-joints for use in marine applications.

A combination of finite element analyses and analytical approximations are used to create these practically oriented design guidelines. A parametric study, utilizing finite element models, was conducted examining the effects of changing various material and geometric variables on the stresses and failure modes of the tee-joint structure. In conjunction with this examination, the tee-joint structure was sectioned in order to facilitate the creation of analytical approximations to predict the failure stresses in the various sections, see Figure 1. The created sections approximate generic configurations seen in various structural design problems. These include a double lap joint, a curved beam, and a doubler termination or single lap joint.

Some important design guidelines have been successfully created, whereas other features are more challenging and represent important areas for future research. Changes in the tee-joint design configuration precipitate a wide variation of structural behavior covering a wide range of responses. Finite element methods are shown to be valuable in guiding the analytical effort by defining suitable boundary conditions at junctions of the subsections.

INTRODUCTION

Tee-joints are a form of connection where the two joined members are perpendicular to each other as illustrated in Figure 1. These joints occur in many areas of structural design, for instance in marine structures bulkhead to hull connections and at the intersection of interior and exterior bulkheads.

Jack Bish and Keith Kedward, UCSB Dept. of Mech. Engr., Santa Barbara, Ca. 93106

Previously reported research on this type of joint has mainly focused on finite element analysis of specific geometries with verification from experimental programs, an analytical approach appears to have been rarely attempted. Four reasons for this omission are given by Hawkins et al., 1993. First, a lack of continuity of the reinforcing fibers across the joint, due to the fact that typically the boundary angles can only be fabricated following the cure of the parent laminates. Second, the strength of the connections is dependent upon the relatively low tensile and shear strengths of the adhesive. Third, there are no continuous fibers extending through successive plies to arrest cracks or delaminations from propagating rapidly through the various members.

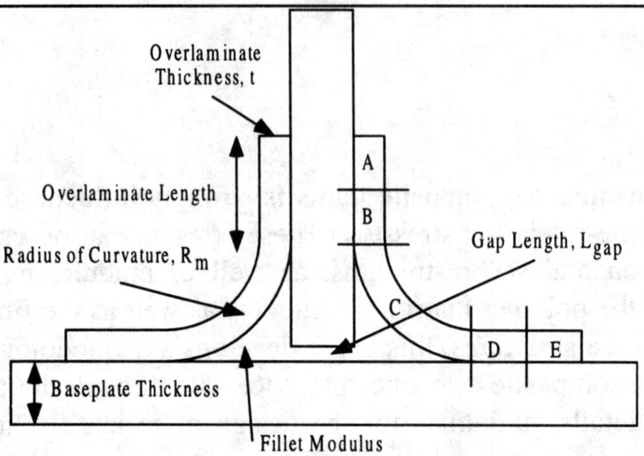

Figure 1: Tee-joint geometry with key design variables and an illustration of the various sections. The section titles are as follows: Section A double lap, Section B vertical doubler, Section C curved region, Section D interior horizontal overlaminate and Section E exterior horizontal overlaminate.

Fourth, the tendency of the boundary angles to peel and delaminate under load emphasizes the irregularities and imperfections of the laminate, compared with ductile characteristics of homogenous metallic materials. Although, these factors illustrate the difficulty in analytically modeling the entire joint as a complete structure, insight into the joint can be gained by examining the contributing pieces of the structure. Verification of these simplified expressions will be conducted through the use of finite element analysis.

The parametric, finite element study will examine the effects of several geometric and material variables on the tee-joint structure and the subsequent failure modes. Typically in this kind of structure with a complex geometry, failure occurs in the out-of-plane direction due to the relatively low strength of polymeric, composite systems in this material direction. These failure modes are seen in various areas such as curved sections, and the termination points of doubler regions to name a few seen in a tee-joint structure. The variables examined, in Bish, 1997, are gap length, resin modulus, R_m/t ratio, span length, baseplate thickness, overlaminate length and the inclusion of tapers. An example of the design guidelines examining the R_m/t ratio is included in this work.

The objective of this study is to produce a set of design guidelines which will give the designer insight into the failure modes of a tee-joint based upon several geometric and material properties. This study does not preclude the use of finite element studies, but will allow the designer to make intelligent design choices before utilizing computer modeling.

Thus, reducing the amount of design iterations and computer time necessary for the design of composite tee-joints.

LITERATURE REVIEW

Currently, tee-joints utilized in marine construction are designed by either extensive finite element analysis or broad guidelines that do not lend themselves to optimizing the joint or produce significant insight into the rationale for their various design rules. The included literature review examines previous work on tee-joints which attempt to aid in the production of more efficient tee-joints.

A series of papers examining the behavior of composite tee-joints has been published by R. A. Shenoi and others from the University of Southampton, Department of Ship Science. The first in the series, Shenoi and Violette, 1990, details an experimental test series conducted on fiberglass tee-joints. The tests series examines five different geometries under compressive loading applied at the top of the centerplate. The geometries include three different fillet radii, a foam pad and a triangular foam insert. The foam pad and triangular foam insert are placed at the base of the joint and replace the curved beam regions by straight sections. Unfortunately, the tests described utilize only one sample per configuration which calls into question the validity of the work. In fact in two of the tests, a lateral displacement was observed which caused the crosshead of the test machine to pivot resulting in an applied moment. But, the following conclusions are reached based upon the results of the conducted tests. First, the finite element results closely approximate the experimental results. Also, an analytical tool, based upon beam theory and implemented as a comprehensive group of computer algorithms, closely approximates the behavior of the joints. The described analytical tool is not given, but the comment is made that the model needs to be improved by reducing the number of approximations and increasing the simplicity so that the model can be used in practical engineering situations.

The next journal articles in this series, Shenoi and Hawkins, 1992 and Hawkins et al., 1993, detail an experimental investigation into tee-joints utilizing a baseline model designed according to current naval design compared to several specimen in which the geometry is altered to determine the affects on tee-joint strength and stiffness. The article begins with a discussion of current naval practices focusing on the lack of design requirements or modeling procedures. The paper continues with a discussion on the inability of traditional analytical methods to model this system. This inability is caused by several factors including the discontinuous nature of the lay-up within the boundary angles, the presence of imperfections caused during manufacture and the variations in material properties from one section to another. To alleviate the problems caused by the lack of rigorous design guidelines, the authors propose to carry out a set of experiments to identify the variables which most critically influence tee-joint design. These variables are similar to those shown in Figure 1. The major non-geometric parameter examined was the use of compliant matrix materials to help alleviate the peel stress magnitude at the resin overlaminate interface.

The test series presented in the paper, Shenoi and Hawkins, 1992, consists of seven joint configurations utilized to determine the effects of varying geometry and resins. The first two specimens are baseline tests constructed according to current naval standards. One of these samples used polyester resin while the other utilized a more compliant urethane-acrylate resin mix. The other five configurations were designed to test the effects of altering the fillet radius, the gap thickness, and tapering the edge of the web. Four samples of each configuration were tested with a load applied at 45° to the horizontal at the top of the centerplate. The following observations were made about the effects of altering the various parameters. Increasing the gap between the center and baseplates allowed greater deflections for a given load (100%) and a greater load potential (50%). The numbers in parentheses are the percentage change between tests specimens where the variable of interest was altered as given in the article. These numbers are only a relative change dependent upon the amount each variable was altered between the configurations. Tapering the edge of the center plate or web reduced the stiffness by a small amount (10%) while increasing the failure load (>50%). The addition of an overlaminate with no change in the fillet radius increased stiffness (50%) and failure load (100%). Increasing the radius of the fillet resulted in a higher failure load (150%), by decreasing the magnitude of the out-of-plane stress in the curved region, but did not change the stiffness of the specimen.

A finite element analysis examined the configurations used in the experimental phase of the study. In this study, the tee-joint was modeled using a coarse mesh with two elements across the resin fillet and the thickness of the boundary elements and one element in the center and baseplates. This mesh is coarse and omits important factors including the bond line between the overlaminates and the center and baseplates. The results of the numerical study were not successful in modeling the loads and deflection of the actual tee-joints. This fact was assumed to be caused by the inadequate and incomplete nature of the material data. Corrections to these values result in accurate finite element predictions as compared to the experimental evaluation.

Using the results of the finite element analysis in conjunction with the results of the experimental series, a few more configurations were examined. From this data, a study of efficiency was conducted. The desired efficiency was defined as the ability of a design configuration to withstand as large a load and as high a deflection with as low internal stresses as possible. The authors found that the most efficient tee-joint design is one utilizing a large radius flexible resin fillet with an overlaminate of minimal thickness, just sufficient to withstand the membrane tensile loads.

Further works in this topic by these authors include an examination of the fatigue characteristics of FRP tee-joints, Shenoi, Read and Hawkins, 1995, and a treatise on the aspects of out-of-plane joints in marine structures, Junhou and Shenoi, 1996 which includes work by Shenoi and others on tee-joints. The main importance of these previous studies to this work is the correlation between finite element analysis predictions and the experimentally determined failure modes.

MATERIAL PROPERTIES

The material systems examined in this program are typical of those currently considered for marine structures. Although the applicability of the model is to a large range of materials, this research will utilize various E-glass/vinylester laminates, most specifically in the form of a [0° / 90°] fiber orientation, and vinylester fillers. This choice of material differs from that adopted for investigation in most of the existing literature on the design of composite joints where the focus is typically on aerospace applications involving carbon fiber/epoxy laminates. This difference is due to the specific requirements associated with the design of marine structures. Unlike in aerospace engineering where the prime consideration is performance and the driving consideration is a high specific stiffness and/or strength, marine hull structures tend to be driven by durability and economic considerations.

In the tee-joint structure, the material properties depend upon the fiber orientations in the various sections. In the baseplate and horizontal overlaminates, when utilizing a [0° / 90°] E-glass laminate, the fibers are aligned along the length and width of the tee-joint. This results in the lower strength material properties being aligned perpendicular to the pull off load applied in the following finite element and simplified analytical models. Due to this weakness, the peel stress generated at the termination of the horizontal overlaminate becomes a concern to the designer. In the centerplate and vertical overlaminates, the fibers are aligned along the height and width of the tee-joint. This alignment results in an area of concern at the termination of the vertical overlaminate. In the curved region of the tee-joint structure, the fibers are aligned in the circumferential direction along the bend and through the width of the structure. This configuration results in an out-of-plane, radial stress generated in the curved region which is another potential failure zone. There are methods to increase the out-of-plane strengths of these composite systems, such as textile composites which include a three dimensional lay-up and/or stitches, but these are predominantly used in the aerospace industry and will not be addressed in this research.

The laminate properties utilized in this article have been documented in other work, Mallick, 1993, and others. The material properties are representative of an E-glass/resin system with a 60% fiber volume with a $[0° / 90°]_{12s}$ stacking sequence. The material properties for this laminate are given as: $E_1 = E_2 = 20.7$ GPa, $E_3 = 10.35$ GPa, $\nu_{12} = .2$, $\nu_{23} = \nu_{13} = .3$, $G_{12} = 5.52$ GPa and $G_{23} = G_{13} = 4.14$ GPa.

There are a wide range of resin materials which can be used in the construction of the fillet area of tee-joints. A few of these resins are epoxies, polyesters, vinylesters and urethane-acrylate. The material choice is based upon a variety of physical properties, such as environmental degradation, creep, fatigue life, and impact and shock loads. An extensive review of these factors is included in Obst and Gillespie, 1996. This review shows vinylesters to fare better under the various conditions seen by the in situ marine structures and is the material considered in this research. The properties of a typical vinylester are as follows: $E = 3.38$ GPa, $\sigma_{t,u} = 76 - 83$ MPa, and $\varepsilon_{ult} = 5\text{-}6\%$.

ANALYTICAL APPROXIMATIONS

The divided tee-joint structure has been examined to gain insight into the behavior of the complete tee-joint structure. This analysis has proven difficult due to the factors mentioned earlier and the changing boundary conditions present between the sections as the various design variables are altered. The only area that has been extensively examined is Section A, the double lap section. Approximations utilizing several mechanics based derivations have been created for the other sections, but these do not accurately predict the stresses in the tee-joint. The other approximations only give some insight into the trends in the tee-joint. These analytical approximations are examined in depth in Bish, 1997.

An analytical approximation is derived for Section A. One method of approximating the response of this type of joint under tensile loading conditions, as seen in this section of the

tee-joint, is by utilizing beam on elastic foundation analysis as detailed by Hetenyi, 1946 and others. By making various assumptions, a simplified expression for the maximum peel stress can be derived based upon geometric and material parameters. The derivation is as follows with the notation for the system shown in Figure 2.
This approximate method for modeling Section A of the divided tee-joint examines the shear deformation of the outer adherend. This need is dictated by the large anisotropy of many composite joints through the thickness where the properties are matrix dominated in comparison to the in-plane, fiber dominated properties.

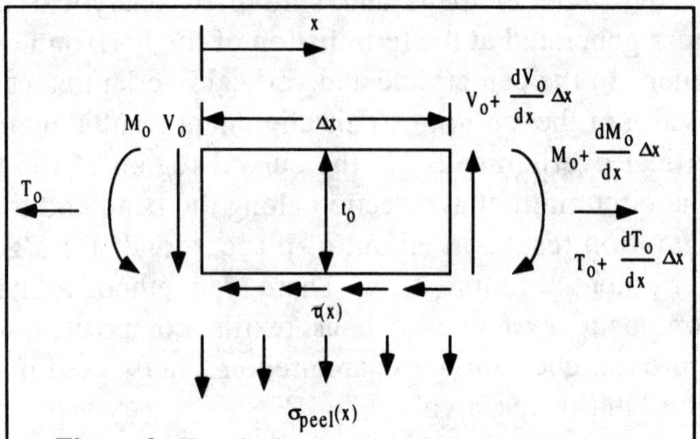

Figure 2: Free body diagram of a double lap joint.

An equation modeling the shear deformation of the adherend is derived in the following manner. The deflection, w, of the outer adherend which is undergoing elastic bending due to transverse shearing is given by

$$\frac{d^2w}{dx^2} = B\frac{dQ}{dx} - CM \tag{1}$$

where the constants are given by the following expressions

$$B = \frac{6}{5t_oG_{13}}, \text{ and } C = \frac{1}{D} = \frac{12(1-\nu_{12}\nu_{21})}{t_o^3E_1} \tag{2,3}$$

The peel stress is given by

$$\sigma = E_a\frac{w}{t_a} = Aw \tag{4}$$

for a thin adhesive layer. Combining eqns (1) and (4) results in,

$$\frac{d^4\sigma}{dx^4} = A\left[B\frac{d^3Q}{dx^3} - C\frac{d^2M}{dx^2}\right]. \tag{5}$$

The equilibrium equations for the outer adherend are

$$\frac{dQ}{dx} = \sigma, \text{ and } \frac{dM}{dx} = Q - \tau\frac{t_o}{2} \tag{6,7}$$

for vertical equilibrium and for moment equilibrium, respectively. The shear stress in the adhesive layer can be assumed to be perfectly plastic near the ends of the joint where the peel stress is highest. If this assumption is made, then the shear stress is a constant in this region and eqn. (7) reduces to

$$\frac{d^2M}{dx^2} = \frac{dQ}{dx} = \sigma. \tag{8}$$

Combing this equation, (8) with eqn. (5) yields the governing differential equation of the moment in the adherend,

$$\frac{d^6M}{dx^6}+E\frac{d^4M}{dx^4}+F\frac{d^2M}{dx^2}=0 \tag{9}$$

where for simplicity E = -AB and F = AC. The solution to eqn. (9) is found from the following characteristic equation,

$$\lambda^4+E\lambda^2+F=0. \tag{10}$$

For the case of adherends with high through-thickness anisotropy, for which $E^2 > 4F$, the roots of eqn. (10) are given by the following equations

$$\lambda_{1,2}=\pm\left[\frac{-E+\sqrt{E^2-4F}}{2}\right]^{\frac{1}{2}}=\pm a,\ \text{and}\ \lambda_{3,4}=\pm\left[\frac{-E-\sqrt{E^2-4F}}{2}\right]^{\frac{1}{2}}=\pm b \tag{11,12}$$

As in the previous cases, the portion of the joint under investigation is near the end of the joint where x = 0 and only the decaying part of the solution is retained.

$$M(x)=C_1e^{-ax}+C_2e^{-bx} \tag{13}$$

Utilizing the following boundary conditions zero moment and shear at x = 0, the solution has the following form

$$M(x)=\frac{\tau_p t_o}{2(a-b)}\left(e^{-ax}-e^{-bx}\right). \tag{14}$$

The peel stress can be found from this relation by utilizing eqn. (8),

$$\sigma_{peel}(x)=\frac{\tau_p t_o}{2(a-b)}\left(a^2e^{-ax}-b^2e^{-bx}\right). \tag{15}$$

The peak peel stress in the adhesive layer is at x = 0 and is given by

$$\sigma_{peel,\max}=\tau_p\frac{t_o}{2}(a+b). \tag{16}$$

This analytical approximation is compared to finite element models, see Figure 3, and has been shown to be resonable for thin adherends, corresponding to thin overlaminates in the tee-joint structure, and high loading levels where the ahdesive layer is undergoing plastic shear deformation.

DESIGN GUIDELINES

This section of the dissertation details a group of design guidelines based upon the analysis presented in the previous sections. Graphical representations of the affects of altering various parameters are created to allow the designer to determine the relative effect of changing the tee-joint geometry. The tee-joint modified in this section is the generic tee-joint shown in Figure 4 and is constrained by rollers located 12.7 mm from the termination of the horizontal overlaminates. In each of the following sections, only one variable is modified while the rest of the structure remains fixed.

The effects of changing each tee-joint parameter are examined by focusing on a group of variables which are critical to determining the failure mechanisms of the structure. A graphical representation of these variables is included in Figure 4 which corresponds to the notation included in the following list of variables. These variables are as follows:

i) vertical displacement - Point A on the following figure,

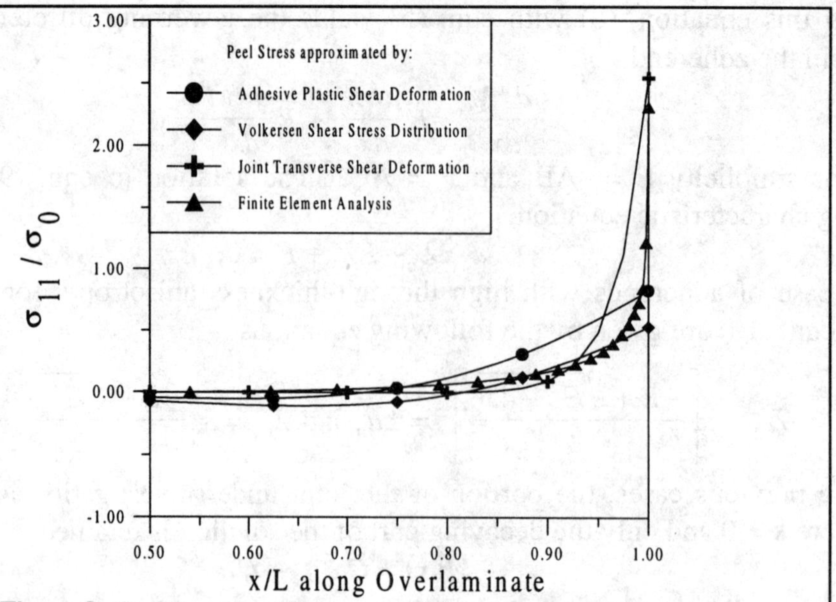

Figure 3: Peel stress profiles for a tee-joint constructed from a Carbon/epoxy laminate compared to the analytical expressions.

ii) the peel stress in the adhesive at the termination of the overlaminate - Point B,
iii) vertical stress in the baseplate directly under the centerplate - Point C,
iv) the peak radial stress in the curved region of the tee-joint - Point D,
v) the peak circumferential stress also in the curved region of the structure - Point E,
vi) the peel stress in the adhesive at the termination of the vertical overlaminate - Point F.

The vertical displacement is measured from a node on the line of symmetry of the tee-joint and 12.7 mm from the load application point, at point A. This distance is far enough from the load application point, so that there is no local displacement due to the load. The peel stress at the termination of the horizontal overlaminate is computed at the corner node, Point B, and is assumed to be in an element in the adhesive layer. The same node will have a different peel stress depending upon the material properties of the element chosen, in this case the node can be in either a baseplate (E-glass/epoxy) or an adhesive (vinylester) element. The vertical stress in the baseplate directly under the centerplate is computed for a node in a baseplate element at Point C. This stress is a compressive stress for most of the cases shown here and will be for joints with thin baseplates. As in the case of the peel in the horizontal overlaminate, the peel in the vertical overlaminate is measured at Point F and taken from the corner node in an element in the adhesive layer. All of the variables shown in Figure 4 have been examined for each of the tee-joint parameters, see Figure 1, and are included in Bish, 1997. One example is illustrated and discussed here.

The R_m/t ratio is the variable examined in this work. This is a measure of the mean radius of the curved region as a function of the thickness of the overlaminate. In this case, the thickness of the overlaminate has been altered which directly affects the stresses in the curved region and both the critical peel areas. Here, the peel stress between the horizontal overlaminate and the baseplate decreases as the thickness of the overlaminate decreases. The data point shown at R_m/t = 17.5 in Figure 5 is negative due to the interaction between the constraints and the peel stress. With a thin overlaminate, the peel stress decreases to such a low level that the local compression due to the rollers is large enough to dominate and a compressive stress is generated at an area which previously has been under a tensile stress. The other variables also change as the R_m/t ratio is altered, see Figure 5. An extensive discussion of the mechanisms for the changes are included in Bish, 1997.

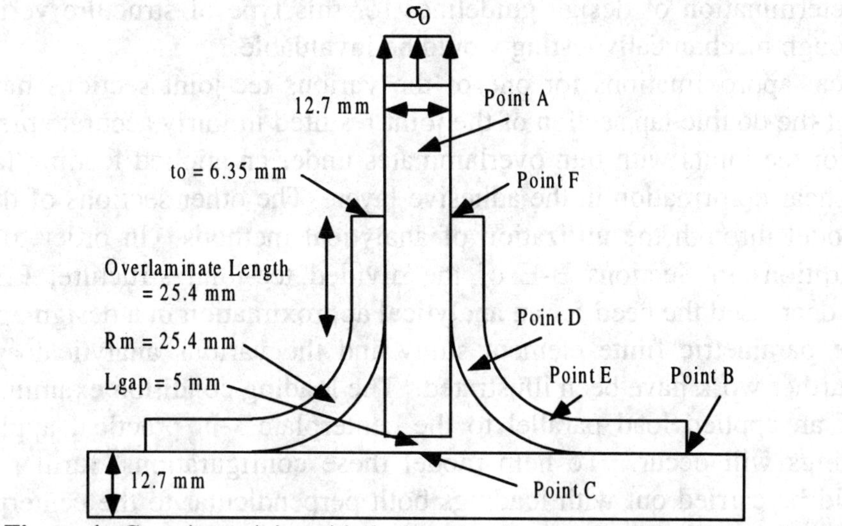

Figure 4: Generic tee-joint with the baseline dimensions utilized throughout section 5.0 and the location of the various variables utilized in this study.

CONCLUSIONS

In order to obtain a better understanding of the parameters affecting the design of resin fillet tee-joints for marine applications, finite element analyses and analytical approximations have been conducted.

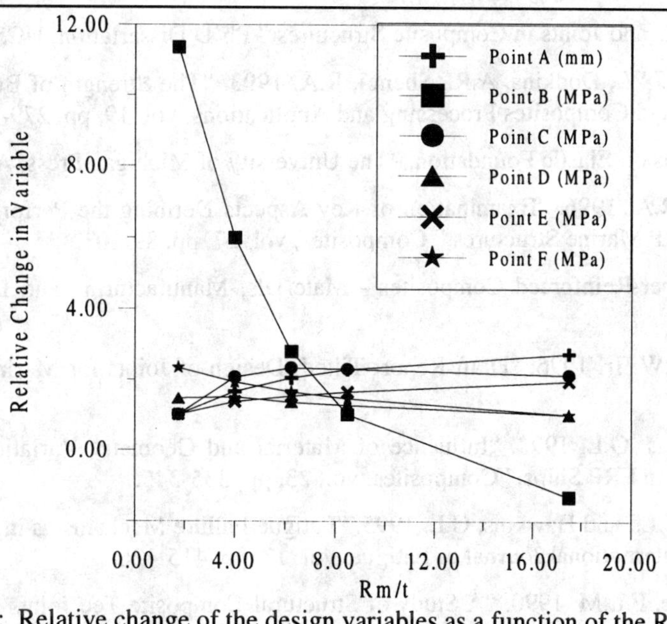

Figure 5: Relative change of the design variables as a function of the R_m/t ratio.

The finite element analyses were utilized as a parametric study to help determine a set of design guidelines for the construction of marine tee-joints. A set of material and geometric parameters were chosen and varied. An example of these design guidelines has been illustrated previously. It is recommended that an experimental test series verify the findings of this study. Although previous authors have concluded that FE analysis can be accurately used in the determination of design guidelines for this type of structure verification of the guidelines through mechanically testing would be invaluable.

An analytical approximations for one of the various tee-joint sections has been shown. The analysis of the double lap section of the joint resulted in fairly accurate predictions of the peel stresses for tee-joints with thin overlaminates under an applied loading large enough to cause plastic shear deformation in the adhesive layer. The other sections of the joint proved difficult to model through the utilization of analytical methods. In order to find pertinent boundary conditions in Sections B-E of the divided tee-joint structure, FE analysis was needed which decreased the need for an analytical approximation in a design application.

In both the parametric finite element study and the various analytical approximations, examples of further work have been illustrated. The loading condition examined in this study is the case of an applied load parallel to the centerplate. In practical applications, more complex loadings will occur. To help model these configurations, further finite element analyses should be carried out with loadings both perpendicular to the centerplate and at 45 degrees.

ACKNOWLEDGMENTS

Funding for this research was supplied by the Navy Surface Warfare Center through the University of Delaware, Center for Composite Materials.

REFERENCES

Bish, J. 1997. "Transitions and Joints in Composite Structures," Ph'D Dissertation, UCSB.

Hawkins, G.L., Holness, J.W., Dodkins, A.R., Shenoi, R.A. 1993. "The Strength of Bonded Tee-Joints in FRP Ships," Plastics, Rubber and Composites Processing and Applications, vol. 19, pp. 279-284.

Hetenyi, M. 1946. "Beams on Elastic Foundation," The University of Michigan Press, An Arbor.

Junhou, P. and Shenoi, R.A. 1996. "Examination of Key Aspects Defining the Performance Characteristics of Out-of-Plane Joints in FRP Marine Structures," Composites, vol. 27, pp. 89-103.

Mallick, P.K. 1993. "Fiber-Reinforced Composites - Materials, Manufacturing and Design," Marcel Dekker, Inc., New York.

Obst, A. and Gillespie, J.W. Jr. 1996. "Draft Report Tilted: Design of Joints for Marine Structures," Presented to NSWC.

Shenoi, R.A. and Hawkins, G.L. 1992. "Influence of Material and Geometry Variations on the Behaviour of Bonded Tee Connections in FRP Ships," Composites, vol. 23, pp. 335-345.

Shenoi, R.A., Read, P.J.C.L., and Hawkins, G.L. 1995. "Fatigue Failure Mechanisms in Fibre-Reinforced Plastic Laminated Tee Joints," International Journal of Fatigue, vol. 17, pp. 415-426.

Shenoi, R.A. and Violette, F.L.M. 1990. "A Study of Structural Composite Tee Joints in Small Boats," Journal of Composite Materials, vol. 24, pp. 644-666.

Fastener Design for Transversely Loaded Composite Plates

D. A. POLK, J. B. LIGON AND I. MISKIOGLU

ABSTRACT

The fastener design for the transfer of concentrated transverse (pull-out) loads to glass fiber reinforced thermoset polymers was investigated. The elastic material properties, void content, and glass content of the composite were determined and a finite element model was used to analyze and compare the performance of the various washer designs for reducing the stress and strain levels near the edge of the washer. Experimental studies were conducted to verify the finite element model.

INTRODUCTION

Glass fiber reinforced composite panels have a wide range of current and potential applications in industry. An understanding of how composite panels perform under various loading and fastening conditions is essential to proper design of the attachment that will transfer the concentrated load to the composite body.

Many investigations have been done on the ability of composite panel-fastener interfaces to withstand bearing (in-plane) loads where failure can occur by bearing failure, shearout failure, cleavage-tension failure, and tension failure (Hart-Smith, 1980). Cloud et al., (1987, 1990) studied similar effects with both a pin and a bolt-washer joint. They studied the effects of initially low and high axial bolt preloads with various washer designs on the failure and strain levels of the composite under in-plane loads. The results were that higher preloads were the most desirable for preventing failure due to in-plane loads. In the area of transverse loading of composite panels (Fig. 1), Hughes and Sikarskie (1993) employed classical mechanics and experimental methods, and Dykstra et al., (1996) used finite element and experimental methods to predict the behavior of virgin as well as repaired composite panels under transverse (pull-out) loading.

In the above mentioned transverse loading studies, the majority of the panels tested failed or started to fail in the region near the edge of the washer. Hence, in the present study, it

D. M. Polk, General Motors Corporation Engineering Development Center, Flint, MI 48557
J.B. Ligon, ME-EM Department, Michigan Technological University, Houghton, MI, 49931
I. Miskioglu, ME-EM Department, Michigan Technological University, Houghton, MI, 49931

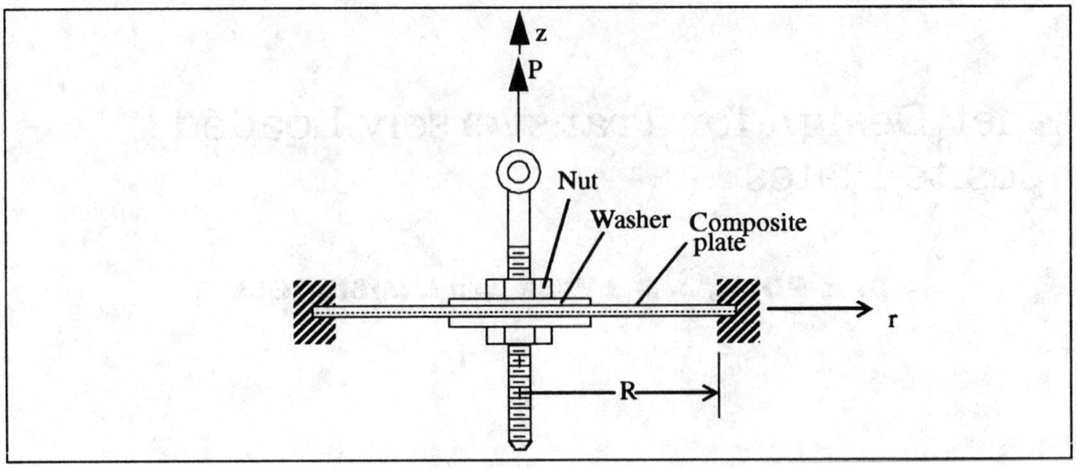

Figure 1. Circular plate subjected to transverse load P.

was proposed to optimize the design of the washer used in the fastening system to better utilize the strength of the composite panel. The effect that the design of a washer has on failure can be directly applied to fastener design and finite element studies for transverse loads or can serve as a good starting point for analyzing plates under combined loads (such as moments, in-plane loads, and torque). The design variables evaluated in order to reduce the stress concentration at the fastener were shoulder radius, diameter, and thickness (in order to change the washer flexibility) of the washer. Two specific types of washers were studied: a thick steel washer with a shoulder radius and a thin flexible washer made from spring steel. These two types of washers will hereafter be referred to as rigid washers and flexible washers.

Glass fibre reinforced plates that were used throughout this study were provided by the Michigan Materials and Processing Institute (MMPI). The material is a polyurethane SRIM reinforced with Owens Corning M8610 continuous strand fiberglass mat (E-glass, 2oz/ft^2).

In the numerical part of the study a nonlinear finite element analysis (FEA) of the transverse loading problem was conducted to analyze and compare the performance of rigid and flexible washers. The model accounted for both material property nonlinearities and geometric nonlinearities inherent to the problem. Also, the interface between the circular panel and the washer was modeled to study the slipping taking place between the washer and the composite plate.

On the experimental side of this study, circular plates were transversely loaded to failure. Applied load and the resulting center deflection of the plate along with the in-plane strains (measured by strain gages) at specified points of the plate were measured in a typical test. All test results were compared with the finite element analysis results

MATERIAL CHARACTERIZATION

Simple tension tests in accordance with ASTM Standard 3039 were performed to determine the elastic modulus and failure stress of the composite material. Specimens for the tension tests were prepared and conditioned according to ASTM Standard D 5229, procedure C.

Also percent composition of the constituents was determined in accordance with ASTM Standard D 2584 and the specific gravity of the composite was determined by ASTM D 792-91. A summary of the properties determined is presented in Table 1. A statistical analysis of all accumulated data showed that no distinction could be made between the moduli measured from samples taken along different orientations of a plate.

TABLE 1. MATERIAL PROPERTY DATA

Modulus of Elasticity (Msi)	1.30
Failure Stress (ksi)	23.3
Specific Weight (lb/ft^3)	95.8
Weight Percent Glass	37.8
Volume Percent Void	3.59

FINITE ELEMENT ANALYSIS

For the finite element analysis in this study, ABAQUS (1992) software was used. The IDEAS(1991) software was used for the pre and post processing Since the plate geometry, boundary conditions and loading were all axisymmetric, the FEA model was simplified using these axisymmetric properties. Thin shell elements were used since the panel was essentially a thin plate. Figure 2 shows the actual dimensions of the test plate which were also used in the finite element model. To model the plate, linear axisymmetric shell elements were used with a higher concentration of elements in the vicinity of the washer area. The flexible washer was likewise generated with axisymmetric shell elements using the same concentration of elements as the plate model. The rigid washer was modeled geometrically without the need for nodes and elements.

The axisymmetric model is defined by radial(r) and out of plane(z) dimensions of the plate as shown in Fig.1. There are three degrees of freedom for this model: the radial direction, the z direction, and the rotation in the r-z plane (ϕ).The outer edge of the plate was clamped, so the r, z, and ϕ directions were all initially restricted from movement. Since the loading of the plate was strictly transverse (in the z direction), the center bolted joint region was constrained from rotation about the z axis. Hence the 'base' boundary conditions applied were completely clamped at the 8-inch diameter outside perimeter, and restrained in all directions except in the direction of loading at the center bolt and washer. The loading of the model was achieved by defining a displacement associated with the center portion of the washer.

The interface between the flexible washer and the plate is represented by "slide line elements" (ABAQUS, 1992). These elements allow sliding movement between the plate and the washer interface, so that as the plate deforms, the washer may slide relative to the plate surface. A friction coefficient of 0.17 was determined experimentally and used for the interface between the washer and the composite plate. A more detailed discussion of the modeling of both flexible and rigid washers can be found in the MS Thesis of Polk (1995).

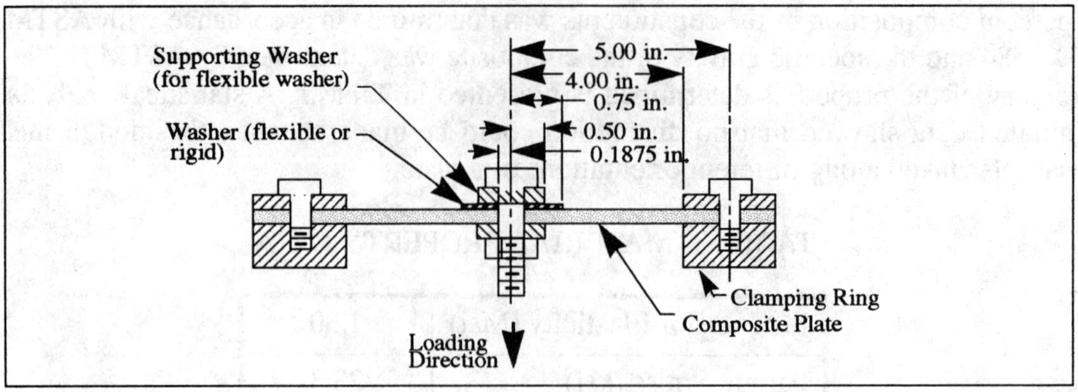

Figure 2. The dimensions of the test plate.

Experimental results were utilized to refine the 'base' boundary conditions. The comparisons between the experimental and FEA results using the 'base' and 'refined' boundary conditions are shown on Fig. 3. The 'refined' boundary conditions were used in the subsequent numerical experiments performed to determine the best type of the washer design.

The comparison of the radial strains for the rigid and flexible washers are given in figures 4, 5 and 6. Figures 4 and 5 compare washers of three and four different thicknesses at two different deflections. The 0.114 inch thick washer represents a rigid washer with a "stock manufactured shoulder radius. The 0.032 inch thick flexible washer performed best for minimizing the maximum strain observed for a center plate deflection of 0.065 inches. For plates with the larger deflection of 0.40 inches, the 0.042 inch thick washer produced the minimum strain.

Figure 6 compares rigid washers with varying shoulder radii. Although a rigid washer with a 3 inch shoulder radius reduced the strain experienced by the plate the same amount as the 0.032 inch thick and 0.042 inch thick flexible washers, the flexible washer design is more

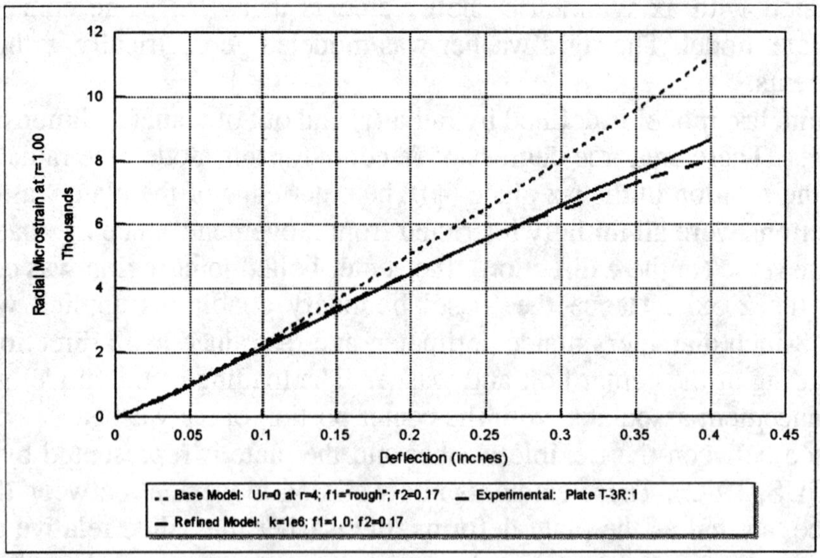

Figure 3. Refined FEA model compared o base model and experimental; Washer thickness of 0.032 in and a plate thickness of 0.1105 in.

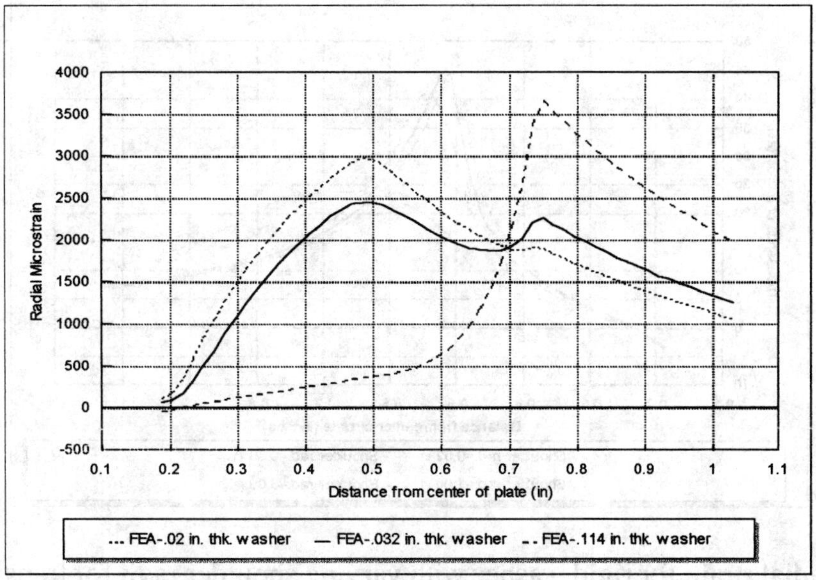

Figure 4. Radial strain for transverse plate deflection of 0.065 inches.

advantageous since it offers reduced weight and ease of manufacturability. A flexible washer is also less likely to be assembled incorrectly as opposed to a shaped rigid washer. Therefore the experimental work focussed on the flexible washer concept.

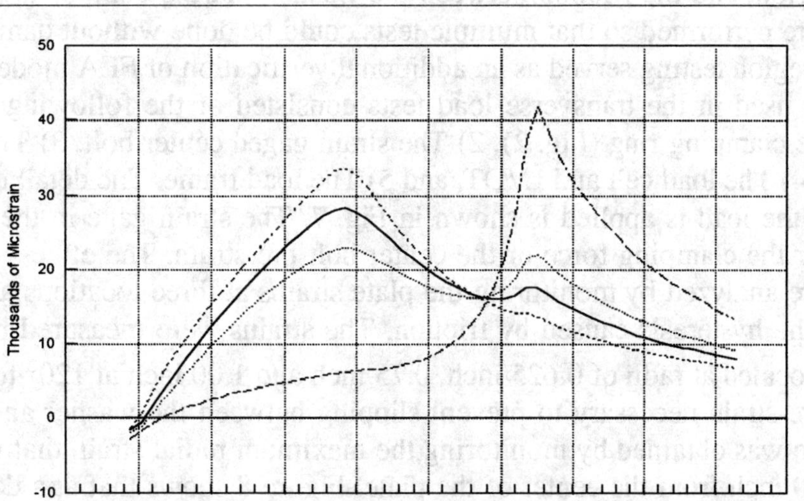

Figure 5. Radial strain for transverse plate deflection of 0.4 inches.

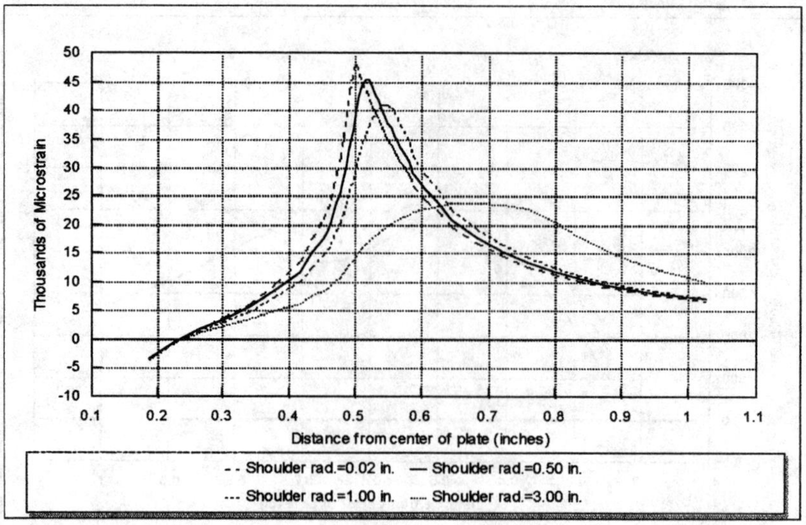

Figure 6. Radial strain for rigid washer with varying shoulder radii for transverse plate deflection of 0.4 inches.

EXPERIMENTAL TRANSVERSE LOADING OF CIRCULAR PLATES

The transverse loading tests were performed primarily to verify the finite element model. The tests focussed on three areas of interest: The linear stress-strain region, the nonlinear stress-strain region, and the friction between the washer and the composite plate. The linear region tests were performed so that multiple tests could be done without damage to the plate. The nonlinear region testing served as an additional verification of FEA model.

The fixture used in the transverse load tests consisted of the following assemblies/sub-systems: 1) The clamping ring (Fig. 2), 2) The strain gaged center bolt, 3) The hydraulic ram and controller, 4) The load cell and LVDT, and 5) The load frame. The detail of the center bolt through which the load is applied is shown in Fig. 7. The strain gage in the center bolt was used to monitor the clamping force or the center bolt pre-strain. The effects of the pre-strain on the bolt were analyzed by monitoring the plate strains at three locations and by observing the change in the hysteresis caused by friction. The strains were measured by three element strain rosettes located at radii of 0.625 inch, 0.75 inch and 1.00 inch at 120° to each other.

The bolt pre-strain necessary to prevent slipping between the washer and the plate for a given deflection was obtained by monitoring the maximum radial strain that was reached at a distance of 1.00 inch from the center of the plate. Figure 8 shows that at a deflection of 0.04 inches, the radial strain reached a maximum value at a bolt pre-strain of 1100 $\mu\varepsilon$. This would indicate that no slippage occurred beyond this pre-strain since the strain should reach a maximum when the plate is not allowed to slip and relax part of the in-plane deformation.

Radial and tangential strains calculated from the strain gage rosette data obtained at the three locations mentioned above and their comparisons with the FEA results for 0.065 inch center deflection are shown in Figures 9 and 10. These results indicate that the finite element

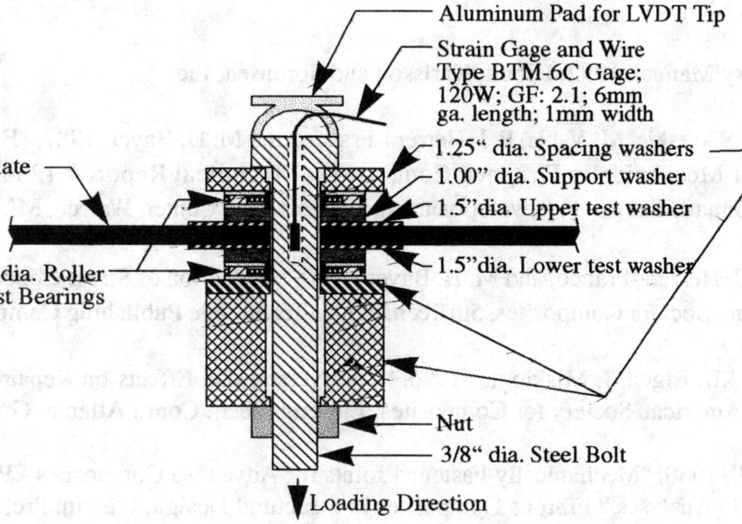

Figure 7. Detail of the joint assembly.

model predicted which washer parameters performed the best to reduce the strains in the plane of the plate near the washer edge. Good correlation were found at radii from the plate center of 0.75 inch and 1.0 inch. Results for large deflections are shown in Fig. 11.

CONCLUSIONS

An axially symmetric finite element model was used for the fastener design in a transversely loaded glass fiber reinforced thermoset polymer composite plate. The model accounted for the material and geometric nonlinearities and slipping between the washer and the composite plate. Washer parameters predicted by this model were found to be in good agreement with the experimental results. It was found that both rigid washers with specific shoulder radii as well as flexible steel washers can be optimized to reduce the stresses in the composite plate near the edge of the washer. However, a flexible washer is the recommend choice since it is not only lighter in weight but easier to manufacture and assemble correctly.

ACKNOWLEDGMENTS

Authors would like to sincerely acknowledge the collaboration and efforts of M. Vable and D.L. Sikarskie, and the financial and technical support of Chrysler Corp., General Motors Corp. and the State of Michigan through the Michigan Materials and Processing Institute during the course of this study.

REFERENCES

ABAQUS Theory Manual, 1992, Hibbit, Karlsson and Sorensen, Inc.

Cloud, G. L., D. Sikarskie, M. Vable, P. J. Herrera-Franco, and M. H. Bayer, 1987, "Experimental and Theoretical Investigation of Mechanically Fastened Composites," Technical Report # 12844 of the U. S. Army Tank-Automotive Command Research, Development, and Engineering Center, Warren, MI.

Cloud, G. L., P. J. Herrera-Franco, and M. H. Bayer, 1990, "Reduction of Stress Concentrations at Bolted Joints in FGRP," Proc. Am. Soc. for Composites, 5th Tech. Conf., Technomic Publishing Company, Lancaster, PA.

Dykstra, D. R., J.B. Ligon, I. Miskioglu, 1996, "Transverse Load Effects on Repaired Composite Panels", Proceedings of the American Society for Composites, Eleventh Tech. Conf., Atlanta, GA: 1059-1068.

Hart-Smith, L. J., 1980, "Mechanically Fastened Joints for Advanced Composites - Phenomenological Considerations and Simple Analyses," Fibrous Composites in Structural Design, Plenum Press, New York: 543-573.

Hughes C. F. and D. L Sikarskie, 1993, "Concentrated Load Transfer on Composite Plates," Proc. Am. Soc. for Composites, 8th Tech. Conf., Cleveland, OH: 943-952.

IDEAS Users Guide, 1991, Structural Dynamics Research Corporation.

Polk, D. M., 1995, "Fastener design for Transversely Loaded Composite Plates," Master of Science Thesis, Michigan Technological University.

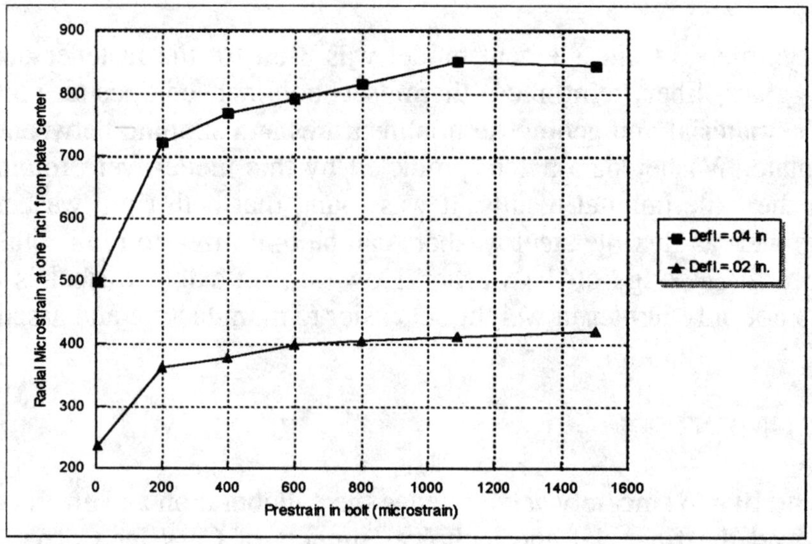

Figure 8. Effect of pre-strain in the center bolt on radial strain in the plate.

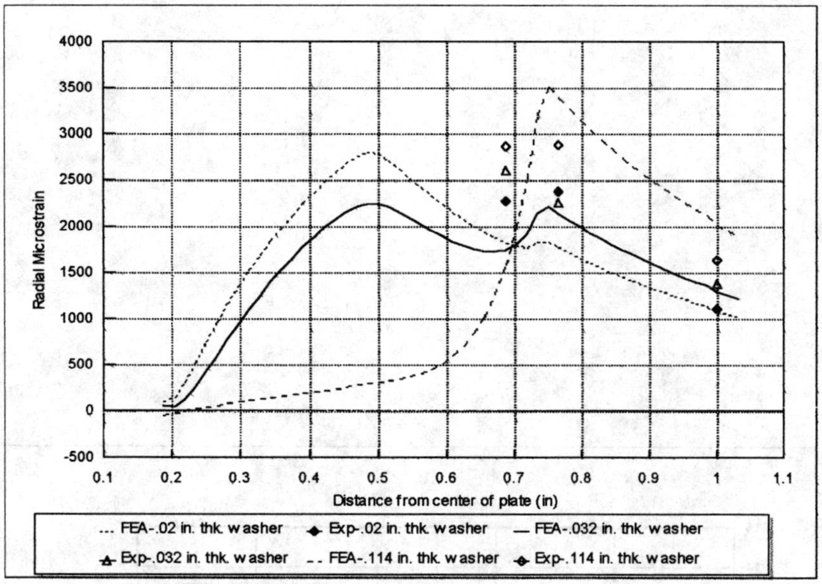

Figure 9. Radial strains at the three rosette locations for 0.065 inch center deflection.

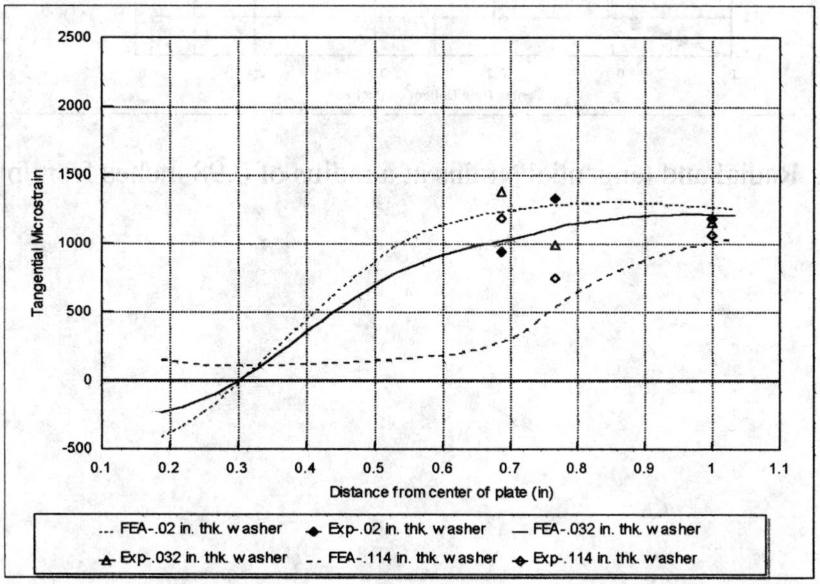

Figure 10. Tangential strains at the three rosette locations for 0.065 inch center deflection.

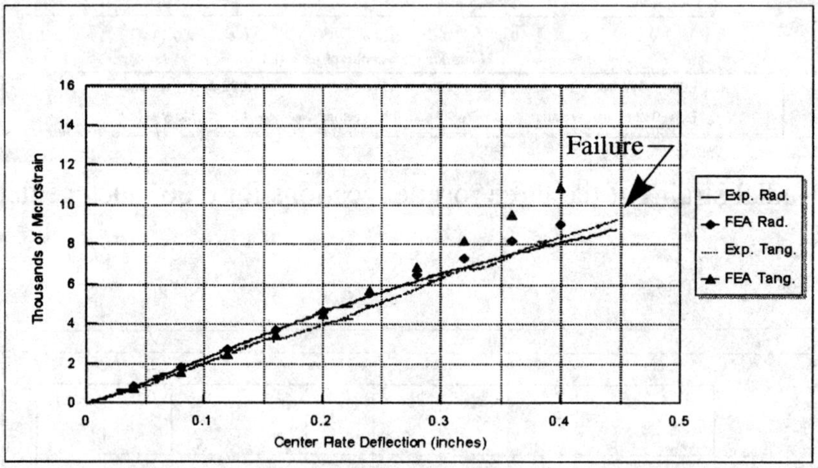

Figure 11. Radial and tangential strains at a radius of 0.98 inches from plate center.

Stresses around a Bi-Axially Loaded Fastener Hole in a Laminate of Finite Geometry

Y. XIONG AND C. POON

ABSTRACT

The stress analysis of a bi-axially loaded fastener hole in a laminate of finite width has been conducted using a complex variational approach developed at the Institute for Aerospace Research. The formulation of the complex variational approach developed is described in this paper. In this approach, an infinitely stiff fastener is assumed to be in contact with one half of the hole edge. The fastener load is in equilibrium with the applied loads along the exterior edges and a cosine distribution over the contact region is assumed. The stresses around a single fastener hole under bi-axial loading conditions were determined. Various anisotropic laminates were analyzed. The effects of the bi-axial loading ratios on the joint performance were examined. Results from the analytical approach are presented and discussed in this paper.

INTRODUCTION

Mechanically fastened composite joints are important components in aircraft structures for assembly of parts as well as for bolted patch repair of damaged composite components. The design of mechanically fastened joints represents one of the most challenging tasks for composite structures. To investigate the failure characteristics of the joints, stresses around the fastener hole need to be determined. Extensive work has been devoted to mechanically fastened composite joints but the majority of the work dealt with uni-directional loading conditions, as cited by Oplinger, 1996. In most of the actual cases, such as in battle damage repair of composite panels using metallic patches with mechanical fasteners, the loading conditions around each fastener hole will not be uni-directional. The stress analysis of a bi-axially loaded joint is more difficult due to its complexity than that of a uni-directionally loaded joint.

The closed-form analytical approaches following Lekhnitskii, 1968 (such as the work by de Jong, 1982; and Zhang and Ueng, 1984) have been used to deal with a single fastener joint under uni-axial loading conditions. Superposition procedures have to be employed if bi-axial loads are considered. In addition, in these approaches, an infinite plate assumption is adopted. That is, finite width correction is required when the diameter of the fastener hole is not small compared with the plate size (Garbo and Ogonowski, 1981). The finite element methods can be used for any joint configurations under general loading conditions, (see, e.g., Waszczak and Cruse, 1971; Crews et al, 1981; and Chang et al, 1984), but the numerical procedures are not

Structures, Materials and Propulsion Laboratory, Institute for Aerospace Research, National Research Council Canada, Ottawa, CANADA K1A 0R6

efficient for parametric studies and design optimization.

Research has been conducted at the Institute for Aerospace Research of the National Research Council Canada (IAR/NRCC) to develop efficient approaches for stress analysis and failure prediction of single- and multi-fastener joints, Xiong and Poon, 1994. A complex variational approach is developed in this paper for calculating the stresses around a single fastener joint of finite geometry under bi-axial loads. The approach is based on a variational formulation involving complex variables. The stresses and displacements in the joined members of the joint are represented using two complex stress potentials. The boundary effects are taken into account by the variational statement with the stress potentials being used as the trial functions. As a result, the approach can be used for a single joint with finite geometry without the requirement for a finite width correction. This paper presents some results of stress analyses of single fastener joints under bi-axial loads using the approach developed.

STATEMENT OF PROBLEM

The problem under study is a rectangular laminated plate of length L and width W with a centrally located hole with a radius of a, as shown in Figure 1. This is a two-dimensional problem of an anisotropic plate under bi-axial loads, which are generally distributed along its four external edges. These given loads are shown in barred symbols in Figure 1. The central hole is subjected to a bearing load by the fastener, which is in equilibrium with the external loads. In addition, the following assumptions are made in the analysis:

(i) The pin is infinitely rigid and frictionless;

(ii) The pin load is in a cosine distribution over one half of the hole edge; and

(iii) The center of all the distributed normal loads is located at the mid-span of the respective edges.

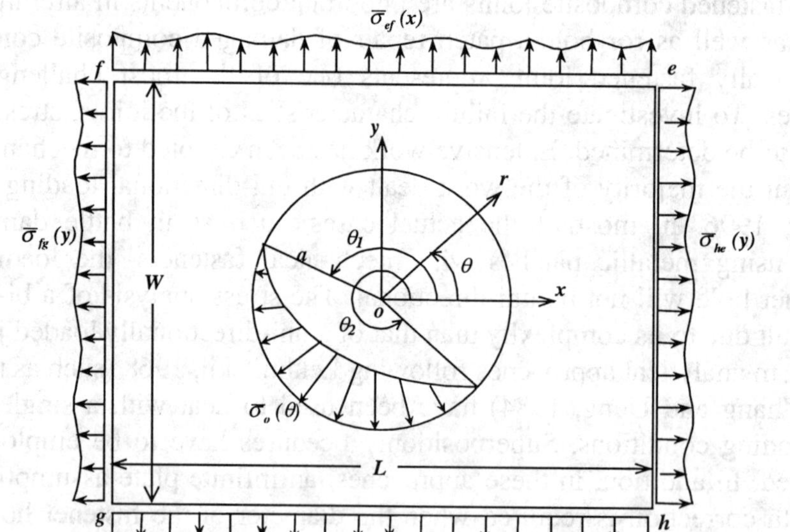

FIGURE 1. GEOMETRY CONFIGURATION OF A SINGLE JOINT

A global co-ordinate system, *x-y*, is employed which is located at the centre of the plate. The corresponding stress and displacement components in the plate are denoted as σ_x, σ_y, τ_{xy}, u, and v, respectively. For the purpose of convenience, a polar system, $r-\theta$, is used on the periphery of the hole with the *r*-axis being the outer normal direction and $\theta = 0$ being the *x*-direction. Using the coordinate transformations, the stress and displacement components in the polar system can be derived, which are denoted as σ_r, σ_θ, $\tau_{r\theta}$, u_r, and u_θ, respectively.

The problem is to find solutions for the stress and displacement components, which satisfy the equilibrium and compatibility equations in the plate. In addition, the solutions must satisfy the boundary conditions along all external and internal edges. The boundary conditions for the exterior edges can be expressed as:

$$\begin{aligned} &\sigma_y(x,\frac{W}{2}) = \bar{\sigma}_{ef}(x), \qquad &&\sigma_x(-\frac{L}{2},y) = \bar{\sigma}_{fg}(y), \\ &\sigma_y(x,-\frac{W}{2}) = \bar{\sigma}_{gh}(x), \qquad &&\sigma_x(\frac{L}{2},y) = \bar{\sigma}_{he}(y) \end{aligned} \tag{1}$$

The fastener load is applied against the hole causing one half of the hole edge to be in compression. Thus the boundary conditions along the hole edge can be written as:

$$\begin{aligned} &\sigma_r(a,\theta) = \tau_{r\theta}(a,\theta) = 0, \qquad && \theta_2 - 2\pi \le \theta \le \theta_1 \\ &\begin{cases} \sigma_r(a,\theta) = \bar{\sigma}_o(\theta), \\ \tau_{r\theta}(a,\theta) = 0, \end{cases} && \theta_1 \le \theta \le \theta_2 \end{aligned} \tag{2}$$

where the two angles θ_1 and θ_2 define the contact region between the fastener and the hole edge as $\theta_1 - \theta_2$ which is assumed to be a half circle in the present work. The compression by the fastener over the contact region is assumed to be in a cosine distribution as:

$$\bar{\sigma}_o(\theta) = -\frac{2P_f}{\pi a}\cos\left(\frac{\theta-\theta_1}{\theta_2-\theta_1}\pi + \frac{\pi}{2}\right), \qquad \theta_1 \le \theta \le \theta_2 \tag{3}$$

where P_f is the fastener load. Based on the equilibrium consideration with the external applied loads, the fastener load and the contact region can be determined for various situations, as:

$$\begin{aligned} &P_f = \int_{-L/2}^{L/2} [\bar{\sigma}_{ef}(x) - \bar{\sigma}_{gh}(x)]dx, \qquad && \text{if} \quad \int_{-L/2}^{L/2} [\bar{\sigma}_{ef}(x) - \bar{\sigma}_{gh}(x)]dx > 0 \\ &\theta_1 = \pi,\ \theta_2 = 2\pi, && \qquad\ \ \int_{-W/2}^{W/2} [\bar{\sigma}_{he}(y) - \bar{\sigma}_{fg}(y)]dy = 0 \end{aligned} \tag{4}$$

$$\begin{aligned} &P_f = \int_{-L/2}^{L/2} [\bar{\sigma}_{gh}(x) - \bar{\sigma}_{ef}(x)]dx, \qquad && \text{if} \quad \int_{-L/2}^{L/2} [\bar{\sigma}_{ef}(x) - \bar{\sigma}_{gh}(x)]dx < 0 \\ &\theta_1 = 0,\ \theta_2 = \pi, && \qquad\ \ \int_{-W/2}^{W/2} [\bar{\sigma}_{he}(y) - \bar{\sigma}_{fg}(y)]dy = 0 \end{aligned} \tag{5}$$

$$P_f = \int_{-W/2}^{W/2} [\bar{\sigma}_{he}(y) - \bar{\sigma}_{fg}(y)]dy, \quad \theta_1 = \pi/2, \ \theta_2 = 3\pi/2, \quad \text{if} \quad \int_{-L/2}^{L/2} [\bar{\sigma}_{ef}(x) - \bar{\sigma}_{gh}(x)]dx = 0, \ \int_{-W/2}^{W/2} [\bar{\sigma}_{he}(y) - \bar{\sigma}_{fg}(y)]dy > 0 \tag{6}$$

$$P_f = \int_{-W/2}^{W/2} [\bar{\sigma}_{fg}(y) - \bar{\sigma}_{he}(y)]dy, \quad \theta_1 = -\pi/2, \ \theta_2 = \pi/2, \quad \text{if} \quad \int_{-L/2}^{L/2} [\bar{\sigma}_{ef}(x) - \bar{\sigma}_{gh}(x)]dx = 0, \ \int_{-W/2}^{W/2} [\bar{\sigma}_{he}(y) - \bar{\sigma}_{fg}(y)]dy < 0 \tag{7}$$

$$P_f = \sqrt{\{\int_{-L/2}^{L/2} [\bar{\sigma}_{ef}(x) - \bar{\sigma}_{gh}(x)]dx\}^2 + \{\int_{-W/2}^{W/2} [\bar{\sigma}_{he}(y) - \bar{\sigma}_{fg}(y)]dy\}^2},$$
$$\theta_1 = \pi/2 + \arctan\{\frac{\int_{-L/2}^{L/2} [\bar{\sigma}_{ef}(x) - \bar{\sigma}_{gh}(x)]dx}{\int_{-W/2}^{W/2} [\bar{\sigma}_{he}(y) - \bar{\sigma}_{fg}(y)]dy}\}, \quad \theta_2 = \theta_1 + \pi, \tag{8}$$
$$\text{if} \quad \int_{-L/2}^{L/2} [\bar{\sigma}_{ef}(x) - \bar{\sigma}_{gh}(x)]dx \neq 0, \ \int_{-W/2}^{W/2} [\bar{\sigma}_{he}(y) - \bar{\sigma}_{fg}(y)]dy > 0$$

$$P_f = \sqrt{\{\int_{-L/2}^{L/2} [\bar{\sigma}_{ef}(x) - \bar{\sigma}_{gh}(x)]dx\}^2 + \{\int_{-W/2}^{W/2} [\bar{\sigma}_{he}(y) - \bar{\sigma}_{fg}(y)]dy\}^2},$$
$$\theta_1 = -\pi/2 + \arctan\{\frac{\int_{-L/2}^{L/2} [\bar{\sigma}_{ef}(x) - \bar{\sigma}_{gh}(x)]dx}{\int_{-W/2}^{W/2} [\bar{\sigma}_{he}(y) - \bar{\sigma}_{fg}(y)]dy}\}, \quad \theta_2 = \theta_1 + \pi, \tag{9}$$
$$\text{if} \quad \int_{-L/2}^{L/2} [\bar{\sigma}_{ef}(x) - \bar{\sigma}_{gh}(x)]dx \neq 0, \ \int_{-W/2}^{W/2} [\bar{\sigma}_{he}(y) - \bar{\sigma}_{fg}(y)]dy < 0$$

COMPLEX VARIATIONAL APPROACH

In the theory of the two dimensional anisotropic elasticity, two complex stress potentials, φ_1 and φ_2, are employed to derive the stress and displacement components. In doing so, the equilibrium and compatibility equations are automatically satisfied in the domain of the plate. The stress components take the following form:

$$\sigma_x = 2\,\mathrm{Re}[\mu_1^2 \frac{d\varphi_1(z_1)}{dz_1} + \mu_2^2 \frac{d\varphi_2(z_2)}{dz_2}]$$
$$\sigma_2 = 2\,\mathrm{Re}[\frac{d\varphi_1(z_1)}{dz_1} + \frac{d\varphi_2(z_2)}{dz_2}] \tag{10}$$
$$\tau_{xy} = -2\,\mathrm{Re}[\mu_1 \frac{d\varphi_1(z_1)}{dz_1} + \mu_2 \frac{d\varphi_2(z_2)}{dz_2}]$$

and the corresponding displacement components are:

$$u = 2\,\mathrm{Re}[p_1\,\varphi_1(z_1) + p_2\,\varphi_2(z_2)], \qquad v = 2\,\mathrm{Re}[q_1\,\varphi_1(z_1) + q_2\,\varphi_2(z_2)] \tag{11}$$

where $z_k = x + \mu_k\, y$ $(k = 1, 2)$ are the coordinate variables, μ_k are the two distinct complex roots with positive imaginary part of the characteristic equation of the laminate, and p_k and q_k are two pairs of complex constants.

Since the stress and displacement components expressed in terms of the two stress potentials as in Eqs. (10) and (11) satisfy the equilibrium and compatibility equations in the domain of the plate, the remaining task is to find appropriate stress potentials so that the respective boundary conditions can be satisfied. To this end, a variational formulation is established below with the energy functional written as:

$$
\begin{aligned}
\Pi = \frac{1}{2}\int_{\Omega}\Big\{ & A_{11}\Big(\frac{\partial u}{\partial x}\Big)^2 + 2A_{12}\frac{\partial u}{\partial x}\frac{\partial v}{\partial y} + A_{22}\Big(\frac{\partial v}{\partial y}\Big)^2 \\
& + 2\Big(A_{16}\frac{\partial u}{\partial x} + A_{26}\frac{\partial v}{\partial y}\Big)\Big(\frac{\partial u}{\partial y} + \frac{\partial y}{\partial x}\Big) + A_{66}\Big(\frac{\partial u}{\partial y} + \frac{\partial v}{\partial x}\Big)^2\Big\}\, dxdy \\
& - \int_{x_e}^{x_f} \overline{\sigma}_{ef}(x)\, v\Big(x, \frac{W}{2}\Big)\, dx - \int_{y_f}^{y_g} \overline{\sigma}_{fg}(y)\, u\Big(-\frac{L}{2}, y\Big)\, dy \\
& + \int_{x_g}^{x_h} \overline{\sigma}_{gh}(x)\, v\Big(x, -\frac{W}{2}\Big)\, dx + \int_{y_h}^{y_e} \overline{\sigma}_{he}(y)\, u\Big(\frac{L}{2}, y\Big)\, dy \\
& + \int_{\theta_1}^{\theta_2} \overline{\sigma}_o(\theta)\, u_r(a,\theta)\, a\, d\theta
\end{aligned}
\tag{12}
$$

where Ω denotes the domain of the plate. It can be shown, after cumbersome mathematical manipulations, that the vanishing condition of the energy functional in Eq. (12) is equivalent to the boundary conditions. That is, the minimisation of the energy functional results in the correct solution of the problem under study. This condition can be written in a variational formulation, which involves boundary integrations only, as:

$$
\begin{aligned}
\delta\Pi = & \int_{x_e}^{x_f} \Big[\sigma_y\Big(x, \frac{W}{2}\Big) - \overline{\sigma}_{ef}\Big]\, \delta v\, dx + \int_{y_f}^{y_g} \Big[\sigma_x\Big(-\frac{L}{2}, y\Big) - \overline{\sigma}_{fg}\Big]\, \delta u\, dy \\
& - \int_{x_g}^{x_h} \Big[\sigma_y\Big(x, -\frac{W}{2}\Big) - \overline{\sigma}_{gh}\Big]\, \delta v\, dx - \int_{y_h}^{y_e} \Big[\sigma_x\Big(\frac{L}{2}, y\Big) - \overline{\sigma}_{he}\Big]\, \delta u\, dy \\
& + \int_{\theta_2 - 2\pi}^{\theta_1} \{\sigma_r(a,\theta)\, \delta u_r + \tau_{r\theta}(a,\theta)\, \delta u_\theta\}\, a\, d\theta \\
& + \int_{\theta_1}^{\theta_2} \{[\sigma_r(a,\theta) + \overline{\sigma}_o(\theta)]\, \delta u_r + \tau_{r\theta}(a,\theta)\, \delta u_\theta\}\, a\, d\theta \\
= & \; 0
\end{aligned}
\tag{13}
$$

Now the problem is to select the appropriate stress potentials, φ_1 and φ_2, as the trial functions for the variational formulation derived. The selection of trial functions is an important step in the variational methods for accurate numerical calculations. For the convergence and efficiency of calculations in the complex theory of two-dimensional elasticity, the well-known Laurent series is often used for the stress potentials. In this work, functions in the form of a truncated Laurent series with conformal mapping are employed for the stress potentials, φ_1 and φ_2, as:

$$\varphi_1 = C_{10} \ln \zeta_1 + \sum_{n=1}^{N} (C_{1n} \zeta_1^n + D_{1n} \zeta_1^{-n})$$
$$\varphi_2 = C_{20} \ln \zeta_2 + \sum_{n=1}^{N} (C_{2n} \zeta_2^n + D_{2n} \zeta_2^{-n}) \quad \textbf{(14)}$$

where the undetermined constants C_{1n}, D_{1n}, C_{2n}, and D_{2n} are, in general, complex and the mapping functions are:

$$\zeta_k(x, y) = \frac{z_k \pm \sqrt{z_k^2 - a^2(1+\mu_k^2)}}{a(1 - i\mu_k)}, \qquad k = 1, 2 \quad \textbf{(15)}$$

where i is the imaginary unit and the sign of the square root terms is chosen such that the hole of radius a is mapped to a unit circle. Noting that the stress potentials in Eq. (14) involve logarithm terms, the following conditions to ensure single-valued displacement components are imposed, as in Xiong and Poon, 1994:

$$\mathrm{Im}(p_1 C_{10} + p_2 C_{20}) = 0, \qquad \mathrm{Im}(q_1 C_{10} + q_2 C_{20}) = 0 \quad \textbf{(16)}$$

The complex constants in the trial functions are to be determined by the variational formulation derived and the number of constants is $2N$+1. Once these constants are computed, the stress and displacement components can be calculated using Eqs. (10) and (11). The determination of the stress distributions along the fastener hole edge as well as in its surrounding area can then be used for the prediction of joint failure.

RESULTS AND DISCUSSION

To verify the analytical approach developed, several cases involving a laminate with a single fastener hole have been analysed. The problem is a square AS4/3501-6 laminate with a centrally located hole. The plate edge length is $L = W = 152.4\,\text{mm}$. The mechanical properties of each ply are: $E_{11} = 140\,\text{GPa}$, $E_{22} = 8.2\,\text{GPa}$, $G_{12} = 6.2\,\text{GPa}$, and $\nu_{12} = 0.3$. The layer thickness is 0.127 *mm*. For the sake of simplicity, uniform normal loads are applied along the exterior edges.

In the first case, an analysis was conducted to examine the convergence characteristics of the approach by changing the number of terms in the stress potentials. The lay-up of the laminate considered is quasi-isotropic $[45/0/\text{-}45/90]_{5S}$. The diameter of the hole is $d = 50.8\,\text{mm}$ and the four edge loads are $\bar{\sigma}_{ef} = 2\,\text{Pa}$, $\bar{\sigma}_{fg} = 1\,\text{Pa}$, $\bar{\sigma}_{gh} = 1\,\text{Pa}$, and $\bar{\sigma}_{he} = 2\,\text{Pa}$. The peak stress components along the hole edge calculated versus the number of terms in each of the stress potentials are presented in Table 1. It is shown that the calculations are stable and converged when the total number of terms in each of the stress potentials is 11 or more. Therefore, in the following examples, the number of terms in each of the stress potentials was chosen by setting $N = 6$. It is also shown in Table 1 that the fastener load is relatively insensitive to the number of terms in the stress potentials as can be seen from the relative error, ε, between the calculated fastener load and the exact one which is in equilibrium with the external edge loads.

TABLE 1. STRESS COMPONENTS AROUND HOLE EDGE (PA)

N	2	3	4	5	6
$(\sigma_\theta)_{max}$	7.78	7.83	7.82	7.96	7.97
$(\sigma_r)_{max}$	-5.03	-5.45	-5.46	-5.40	-5.38
$(\tau_{r\theta})_{max}$	-0.73	-0.21	-0.18	0.10	0.10
P_f (N)	0.217	0.217	0.217	0.217	0.216
ε	0.68%	0.67%	0.67%	0.65%	0.06%

In the second case, the hoop and bearing stress distributions along the fastener hole edge in two laminates with the same geometry and loading conditions but different lay-ups, are shown in Figure 2. The first lay-up is a quasi-isotropic as $[0/90/45/-45]_{6s}$, and the other is an anisotropic as $[45/45/-45/-45/0/0/45/-45/90/90/45/-45/90/90/45/-45/0/0/45/-45/90/0/0/0]_s$ which represents a typical lay-up of a wing skin panel in an F/A-18 aircraft. It is seen that the laminate lay-up has effects on both the stress distributions and the locations of stress concentrations in joints, even for the same applied and fastener loads.

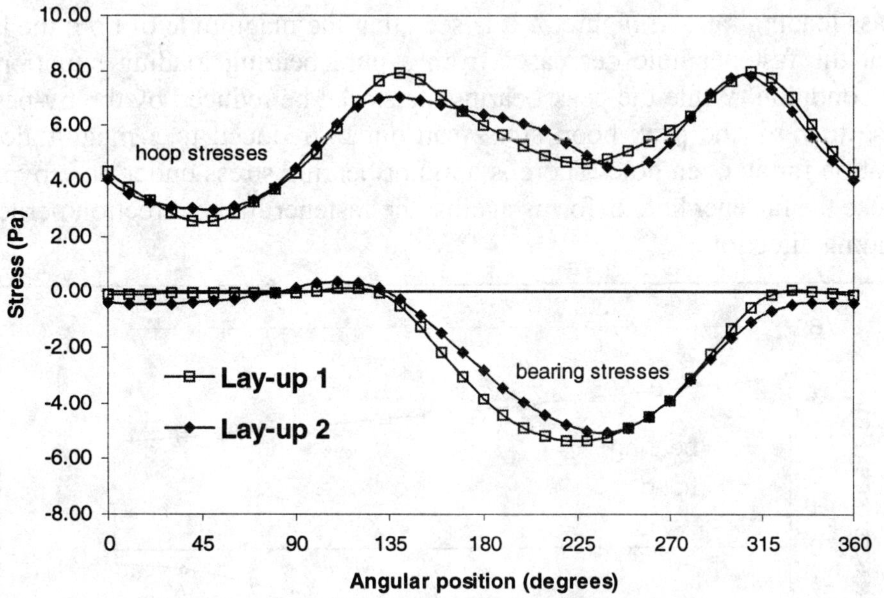

FIGURE 2. HOOP AND BEARING STRESSES IN TWO JOINTS

In the third case, the effect of the bi-axial loading ratio on the stress concentrations is examined. In this case, only the two edges, *e-f* and *h-e*, are loaded and the applied loads vary such that they result in a fastener load with a fixed magnitude but different directions. The geometry of the joint is the same as in the above cases and the lay-up is [45/45/-45/-45/0/0/45/-

45/90/90/45/-45/90/90/45/-45/0/0/45/-45/90/0/0/0]$_S$. The two peak stresses calculated are presented versus the bi-axial loading ratio in Table 2. It is seen that while the peak hoop stress is not very sensitive to the bi-axial loading ratio, significant change of the peak bearing stress is observed when the bi-axial loading ratio increases. The anisotropic lay-up causes the unsymmetric changes of the peak stresses with the bi-axial loading ratio.

TABLE 2. STRESS COMPONENTS AROUND HOLE EDGE (PA) UNDER VARIOUS BI-AXIAL LOADING RATIOS

$\overline{\sigma}_{he} / \overline{\sigma}_{ef}$	0	0.27	0.58	1.0	1.73	3.73	∞
α *	180°	195°	210°	225°	240°	255°	270°
$(\sigma_\theta)_{max}$	5.33	4.87	4.10	4.07	4.69	5.30	5.55
$(\sigma_r)_{max}$	-4.15	-4.10	-3.98	-3.79	-3.53	-3.23	-2.98

* α is the direction angle of the fastener load with respect to *x*-direction

In the last case, the effect of the bypass loading ratio on the stress concentrations is examined. The geometry and lay-up of the joint are the same as in the third case discussed above. The applied loads in this case are $\overline{\sigma}_{ef} = 1\,\text{Pa}$, $\overline{\sigma}_{fg} = \overline{\sigma}_{he} = 0.5\,\text{Pa}$, and $0 \le \overline{\sigma}_{gh} \le \overline{\sigma}_{ef}$. The bypass loading ratio is defined as $\overline{\sigma}_{gh} / \overline{\sigma}_{ef}$. The two peak stresses calculated are presented versus the bypass loading ratio in Figure 3. It is seen that the magnitude of both the bearing and hoop stresses at the fastener hole decreases from a pure bearing loading situation to a pure bypass loading condition. While the peak bearing stress can be reduced by the by-pass load to a magnitude close to zero, the peak hoop stress can not be reduced to a magnitude below the stress concentration for an open hole. There is a minor bearing stress under pure bypass loading condition because the fastener hole deforms against the fastener in the direction perpendicular to the primary loading direction.

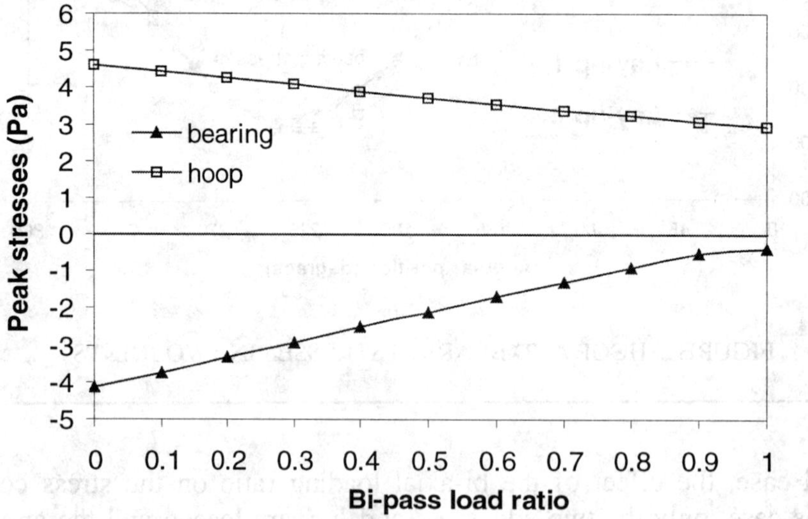

FIGURE 3. EFFECTS OF BYPASS LOADING RATIO ON PEAK STRESSES

CONCLUSIONS

An analytical method based on a complex variational formulation has been developed for the stress analysis of finite composite plates containing a single fastener hole. The variational formulation derived can deal with joints of finite geometry under bi-axial and bypass loading conditions without the requirement for finite width correction factors. The overall stress distributions in the joint as well as the fastener load can be determined. Solution procedures for the stress and displacement components have been implemented into a computer code written in FORTRAN, which can be executed in a PC environment. As follow-on work, the computer program, along with a bi-axial test program, will be used to develop appropriate failure criteria for a fastener loaded hole with bi-axial and bypass stresses in its surrounding area.

Results of the example cases discussed show the effectiveness of the complex variational approach developed. The following conclusions can be drawn:

(1) The accuracy of the calculated stress concentrations relies on the number of terms in the stress potentials but the fastener load is relatively insensitive to this number. This feature will be useful in analysing multi-fastener joints where the determination of fastener loads is the major concern;

(2) The laminate lay-up affects both the stress distribution and the location of stress concentrations along the fastener hole edge for the same applied loads;

(3) The bi-axial loading ratio has a more significant effect on the bearing stress than on the hoop stress; and

(4) The bypass load reduces the magnitude of both the bearing and hoop stresses.

ACKNOWLEDGEMENT

This work has been carried out under IAR Program 3G3, Aerospace Structures, Structural Dynamics and Acoustics, Project/Subproject SAA-00 Computational Structures Technology and Project/Subproject JGM-02 Mechanical Joining of Composite Structures.

REFERENCES

Chang, Fu-Kuo, Scott, R. A., and Springer, G. S. 1984. "Strength of Bolted Joints in Laminated Composites," *AFWAL-TR-84-4029.*

Crews, Jr., J. H., Hong, C. S., and Raju, I. S. 1981. "Stress-Concentration Factors for Finite Orthotropic Laminates with a Pin-Loaded Hole," *NASA Technical Paper 1862.*

De Jong, Th. 1977. "Stresses Around Pin-Loaded Holes in Elastically Orthotropic or Isotropic Plates," *Journal of Composite Materials*, **11**: 313-331.

Garbo, S. P. and Ogonowski, J. M. 1981. "Effect of Variances and Manufacturing Tolerances on the Design Strength and Life of Mechanically Fastened Composite Joints, I. Methodology Development and Data Evaluation," *AFWAL-TR-81-3041.*

Lekhnitskii,, S. G. 1968. "Anisotropic Plates," Gordon and Breach Science Publishers, New York.

Oplinger, D. W. 1996. "Bolted Joints in Composite Structures – an Overview," *AGARD Conference Proceedings 590*: Bolted/Bonded Joints in Polymeric Composites, NATO, France.

Waszczak, J. P. and Cruse, T. A. 1971. "Failure Mode and Strength Predictions of Anisotropic Bolt Bearing Specimens," *Journal of Composite Materials*, **5:** 421-425.

Xiong, Y. and Poon, C. 1994. "A Design Model for Composite Joints with Multiple Fasteners," *IAR-AN-80, NRC No. 32165*, National Research Council Canada, Ottawa.

Zhang, K. D. and Ueng, C. E. S. 1985. "Stresses Around a Pin-Loaded Hole in Orthotropic Plates with Arbitrary Loading Direction," *Composite Structures*, **3:** 119-143.

Experimental Study of Ceramic Matrix Composites at High Temperatures

M. XIE AND N. D. FREY

ABSTRACT

Test system design and general techniques for high temperature testing of ceramic matrix composite materials are discussed in the present study. Detailed design of furnace, strain extensometer and grip are presented, as well as a creep rupture test system and a thermal mechanical fatigue test system. Typical experimental data obtained are also presented for static, fatigue and creep tests of ceramic matrix composite materials at high temperatures.

INTRODUCTION

Advanced ceramic matrix composite (CMC) materials offer great potential in many engineering applications. Recently, they have been used to manufacture high temperature structural components. For example, acoustic treatment of jet engine exhaust nozzle to reduce noise can be achieved by attaching a layer of acoustic liners on the nozzle side walls. These acoustic liners are multi-rib stiffened open boxes made of ceramic matrix composite materials (Xie, et al., 1996).

However, as a relatively new high temperature structural material, the stress-strain response, and damage and failure mechanism of ceramic matrix composite materials under complex loading conditions are still not well understood. The difficulties of high temperatures testing of ceramic composites arise from the requirement of special fixtures. For example, the CMC acoustic liners mentioned above have an operational temperature range of about 1000°F to 1900°F. To test CMC materials at this temperature range, heating and gripping of the specimens and strain measurements are much more difficult than ambient temperature testing. Ceramic matrix composites also tend to require more delicate testing considerations than polymer matrix composites and metal matrix composites because of the brittle nature of the materials.

In the present study, test system design and general techniques for high temperature testing of ceramic matrix composite materials will be discussed, as well as the design of some experimental fixtures, such as furnace and extensometers. Typical experimental data obtained by using these test fixtures will then be presented for static, fatigue and creep tests of ceramic matrix composite materials at high temperatures.

Ming Xie and Norman D. Frey, AdTech Systems Research, Inc., 1342 N. Fairfield Road, Beavercreek, OH 45432

DESIGN OF HIGH TEMPERATURE TEST SYSTEMS

The most important parts of a high temperature mechanical test system include heating source to elevate and maintain the test temperature, strain measurement device to measure the deformation in the test specimen, and gripping device to hold the test specimen and apply the mechanical loads in high temperature environments.

A furnace is usually used to generate and maintain high temperature environment. There are different types of furnace for high temperature mechanical testing. Except some integrated units (testing machine and furnace are integrated together) and controlled atmosphere unit, most of them are conventional split furnaces which can be separated or opened to allow loading and unloading the specimen (Buddery, 1988). Many important factors are involved in designing a furnace, for example, the choice of heater materials and insulation materials, and temperature control. Many conductive materials have been used as heater materials including silicon carbide, carbon and platinum alloys. Temperature control should be adaptive and the thermocouple positions inside the furnace are also very important to the accurate temperature control. There is no standard furnace design for high temperature CMC testing at this time.

The design of a small furnace, with dimensions of about 4x4x4 inches, is shown in Figure 1. The furnace shell is made of stainless steel. The insulation material inside the split furnace has a thickness of about 0.5 inches, leaving an interior volume of about 3x3x3 inches. This size of the furnace can accommodate a specimen of about 6 inches or longer, which is commonly used in mechanical testing. The central split design allows easy specimen loading and unloading. Four heating elements are used in the furnace, which are made of silicone carbide and with a nominal resistance of 60 ohms at ambient temperature. These heating elements have a capability of raising their own temperature to 1800°F in about 15 seconds and have a maximum temperature of about 3000°F. Of course, the heating rate for test specimen depends on specimen material. Usually it takes about 20 minutes to reach the test temperature and another 20 minutes to reach thermal equilibrium. The furnace has been used for high temperature testing up to about 2200°F. This kind of silicone carbide resistance heater has been used for many high temperature ceramic composite tests, including continuous creep rupture test at 1880°F for more than 1200 hours (50 days) with no heater failure.

The furnace is controlled by a programmable temperature controller, which has a microprocessor that improves its reliability and accuracy. A thermocouple inserted into the furnace provides the input to the controller, and the controller adjusts the voltage output to the heating elements accordingly to control the temperature in the furnace. The thermocouple should be close to the center of the furnace to maintain an accurate temperature control. Temperature overshoot during heat-up usually cannot be avoided, as well as temperature fluctuation during testing. In the present study, a 1% overshoot is often observed unless the set point on the controller be manually adjusted during heat-up. For this reason, a fuzzy logic temperature controller has also been used. Employing fuzzy logic, this controller virtually eliminates the temperature overshoot and effectively suppresses the temperature fluctuation due to external disturbances.

This furnace design also proves to be able to generate a uniform temperature distribution inside the furnace. A ceramic composite specimen was used to calibrate the furnace and seven (7) thermocouples were placed on different locations inside the furnace on the specimen, some of them being embedded inside the specimen. The temperature

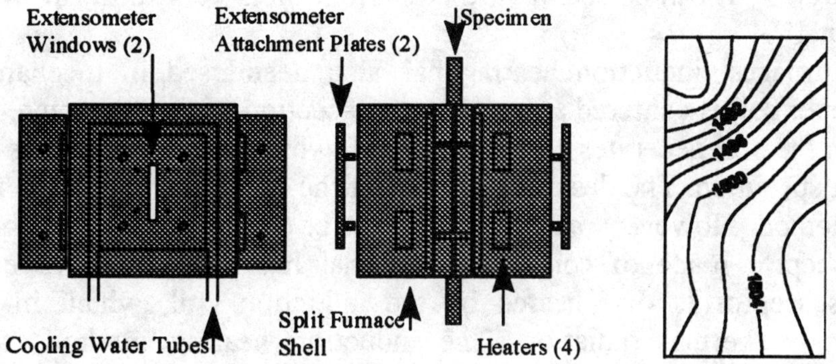

Figure 1. Furnace Design and Interior Temperature Distribution

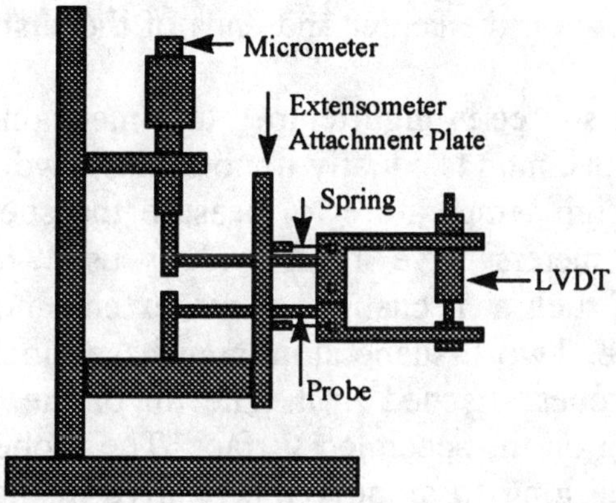

Figure 2. Extensometer Design and a Calibration Device

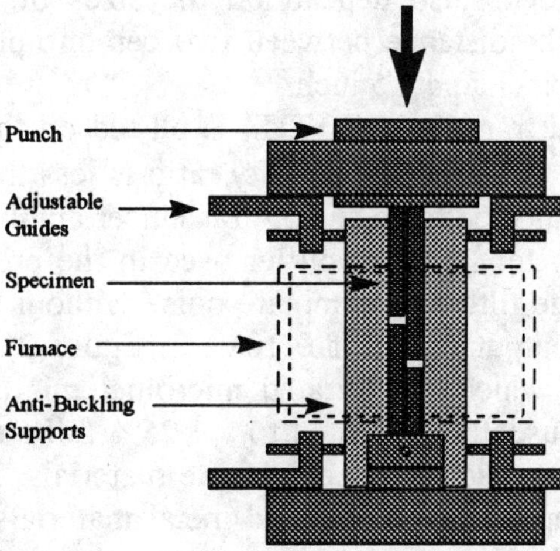

Figure 3. A Double Notch Shear Test Fixture

distribution obtained from the thermocouples is also shown in Figure 1 for a nominal testing temperature of 1500°F. It can be seen that the distribution is very uniform, with a maximum deviation of about 1%.

Besides furnaces, induction heating has also been used in mechanical testing. A water-cooled copper coil is centered around the gage-section of the specimen. The alternating electric current in the coil generates a magnetic field, which in turn generates electric current in the conducting specimen. The design flexibility of the induction coil is the main advantage of this heating method. However, many ceramic matrix composites are not good conductive materials. A susceptor made of conductive material has to be used to enclose the test specimen. The susceptor is first heated by the induction coil, which in turn heats the specimen inside by thermal radiation. The induction heating method is usually more expensive than a furnace, mostly because of the cost of the induction power supply. The accuracy of temperature control by induction heating is not better than furnace heating. For example, a study was performed (Liu, 1988) to investigate the temperature profile inside a pair of SiC clamshell susceptors, which are about the same size as the above mentioned furnace. Temperature measurements were made at 800C, 1000C and 1200C, and the temperature differences between the center and ends of the susceptor ranged from about 1% to 3%.

Besides the heating source in high temperature mechanical testing, the measurement of deformation in the test specimen is equally important. However, conventional strain gages can not be used at such high temperature to measure the specimen deformation. Instead, mechanical strain extensometers have been widely used because they can be made economically. A design of such a mechanical strain extensometer is shown in Figure 2, as well as a calibration device. Two L-shape aluminum base blocks are hinged together, each block having a ceramic probes attached to it. The tip of the ceramic probe can be either conical or chisel, depending on the specimen surface. The probes should be short and the L-shape base block should be long to amplify the relative displacement between the tips of probes. However, in order not to apply too much bending to the specimen, the base blocks also have to be light enough. Besides these general guidelines, the actual length of the probes and the actual size of the base blocks also depend on the sizes of the furnace and the specimen. Two gage lengths, i.e., the distance between two ceramic probes, have been for this extensometer design, namely, 1 inch and 0.5 inch.

A linear variable differential transducer (LVDT) is placed on the base block, which has a displacement range of ±0.005 inch and its linearity ratio is less than 0.25% of the full range output. The LVDT is then connected to an analog transducer amplifier which interfaces with a computer data acquisition system. The amplifier used in the present study is a self-contained unit which has multi-stage filters to eliminate noise without sacrificing response. This amplifier has a DC voltage output range of 0-10V. In figure 2, the extensometer is loaded on a calibration device, on which a calibrated micrometer is used to calibrate the extensometer. The extensometer is usually calibrated to a 1.25% full strain range for a 10V signal level, which is suitable for ceramic matrix composite materials. This design has also been utilized to measure other composite materials and metal materials using LVDTs with greater displacement ranges. After the extensometer is loaded on the specimen, it is held on a brass attachment plate by four springs, while the plate being attached to the furnace shell as shown in Figure 1. The plate also serves as a heat shield to assure that the LVDT operates at an acceptable temperature. As shown in Figure 1, a brass tube is also wrapped around the edges of the plate with cooling water continuously running inside.

This extensometer design can be used for static and creep tests without any limitation. However, for fatigue test, it can only be used under a certain frequency. This frequency limitation is due, in part, to the length and low spring-applied contact force of the probes. The present design has been used for up to 0.5 Hz fatigue test with good performance. To measure strain at a higher frequency, the length of the probes should be as short as possible to improve the frequency response. A dummy extensometer can be placed on the other side of the specimen to balance the bending moment and this arrangement allows higher contact force be applied to the specimen to prevent the possible probe slippage. The extensometer can also be placed on the edge of the specimen, instead of on the face, to take advantage of the increased stiffness on that direction.

The grip is also an integrated part of the high temperature test system. In high temperature tests, including static, fatigue and creep tests, the common practice is that only the central portion of the coupon specimen is placed inside the furnace, because the commonly used hydraulic grips that hold the specimen are not able to endure the high temperature. However, many experimental studies reveal that, for some ceramic matrix composites, even coupon specimen with a central hole or dogbone specimen failed near the edge of the furnace instead of the center of the specimen. This is also observed in the present study. The reason could be due to the thermal gradient at the edge of the furnace which caused material degradation at a lower temperature. If the whole specimen and the grips can be put inside the furnace, the specimen will have uniform temperature gradient and thus the simulated high temperature environmental effect will be more realistic.

This so-called hot grip technology has been used to test non-coupon types of ceramic material specimen, for example, by Carroll, et al (1989) and Ohji and Yamauchi (1992). The grips were made of SiC Material and the specimen had to be machined to special requirement by the grips. However, for ceramic matrix composite materials, the machining is very difficult, especially for textile composites. A set of hot grips was designed in the present study and was made of Inconel 718. Conventional saw teeth wedges were enclosed in a pair of bolted matching sleeve plates. Pin loaded extension bars were used to connect the hot grips to hydraulic grips outside the furnace. This hot grip design did not succeed, the specimen often slipping out of the grips during loading because of the deformation of the sleeve plates. More grip design will be discussed later for a creep rupture test system.

Another concern for compression or reversed cyclic tests for CMC materials is the buckling of the specimen. An anti-buckling fixture should be used in these kinds of tests and the fixture should be able to function properly in high temperature environment. A fixture design for high temperature double notch shear test is shown in Figure 3. A double notch shear CMC specimen is sandwiched between two anti-buckling support plates made of silicon nitride. All the remaining parts are made of Inconel 718 except the punch which is also made of silicon nitride. Two pairs of adjustable guides, located at the top and the bottom of the fixture, respectively, are used to clamp and center the sandwiched specimen. Another pair of adjustable guides is placed at the bottom of the fixture to center the specimen in the width direction. This double notch shear fixture has been used successfully in many high temperature experimental studies of bonded interface between ceramic matrix composites. The furnace used for this fixture has to be smaller than the one mentioned before, because of the size of the double notch shear specimen which is usually 3.125 inch long. Only two heating elements, instead of four, are needed for this smaller furnace.

All the above mentioned high temperature test fixtures have been used extensively in the laboratory. In fact, they are integrated into various high temperature test systems in the present study, for example, a creep rupture test system and a thermal mechanical fatigue test system.

Long-term creep testing requires that the test system is reliable and accurate. All aspects of the test system must be chosen carefully to obtain valid results. The design of a direct loading creep frame is shown in Figure 4. The direct loading mechanism, or the so-called dead-weight loading, is better than lever arm loading, since the applied load can be accurately controlled. In a lever arm frame, after the load is applied and the specimen sets-in the grips, the lever arm position may change which will in turn change the applied load. This actually happened several times during the present study, in which two commercial lever arm frames and one direct loading frame were used. It is also easier in a direct loading frame to achieve better alignment of the specimen and grips, which reduces the bending moment on the specimen. As shown in Figure 4, the specimen and grips are sandwiched between a swivel joint and a flexible chain, and this arrangement allows the specimen to be self aligned. If a dummy extensometer were placed on the other side of the specimen to balance the spring load, the bending effect on the specimen would be further minimized.

Commercially available wedge coupling grips were modified in the present study to integrate into the creep frame. In the original grip design, specimen alignment within the coupling wedges could not be maintained. When the collar was tightened on the U-shaped sleeve, the specimen and wedges would shift within the sleeve. Moreover, the collars had to be moved along the sleeves with shock loading and the impact could cause damage to the brittle ceramic matrix composites. The grips did not provide adequate gripping during the trial tests, the CMC specimens often slipping out of the grips. In the present study, an alignment fixture was designed and made. A threaded hole was drilled at the end of the coupling sleeve to allow applying pressure on the wedges by a screw. A procedure for assembling the specimen and grips is established in which the collar is tightened on the sleeve by shock loading first. The specimen and wedges are then assembled into the sleeve in the alignment fixture, with a layer of sand paper between the CMC specimen and the saw teeth wedges. A screw is then inserted into the sleeves to push the wedge into the tightened end and the screw remains inside the sleeve during the test. A water-cooled brass coil is wrapped around the grips to reduce the thermal expansion and therefore minimize the chance of specimen slipping. The present grip design performs very well during long term creep testing without noticeable slippage.

Another innovative design in the present study is a thermal mechanical fatigue test system. Spectrum thermal cyclic loading and spectrum mechanical cyclic loading can be applied at the same time in this test system. The system setup and a typical loading spectrum are shown in Figures 5 and 6, respectively. The system is integrated into a commercial test frame with a controller. The spectrum mechanical cyclic loading is relatively easy to implement through the test frame controller and an automatic data acquisition system. The difficulty is with the temperature spectrum which requires the furnace, as well as the specimen in the furnace, to be very responsive to the input temperature spectrum. The ceramic matrix composite materials usually have relatively low heat conductivity constants and therefore the CMC specimens may not follow the temperature change fast enough as required by the spectrum input.

As shown in Figure 5, a pair of air driven actuators is attached to the furnace shell. When the solid state relay, controlled by the computer, outputs a voltage to the 4-way air control solenoid, compressed air is allowed to pass the solenoid and flow to the actuators. The split design furnace shell is then opened or closed by the air driven actuators, depending

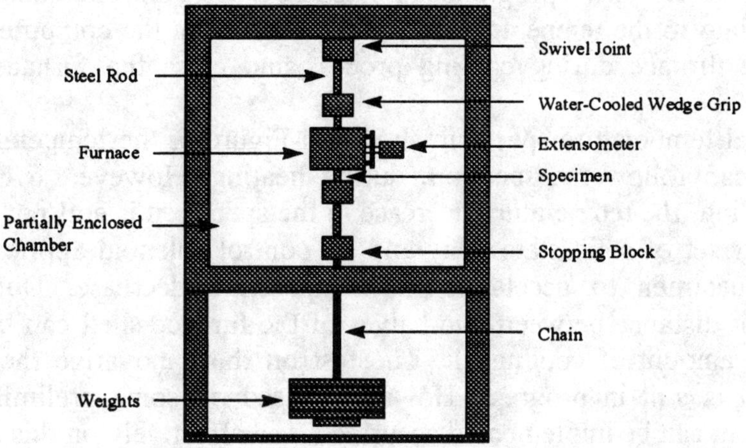

Figure 4. A Direct Loading Creep Test System

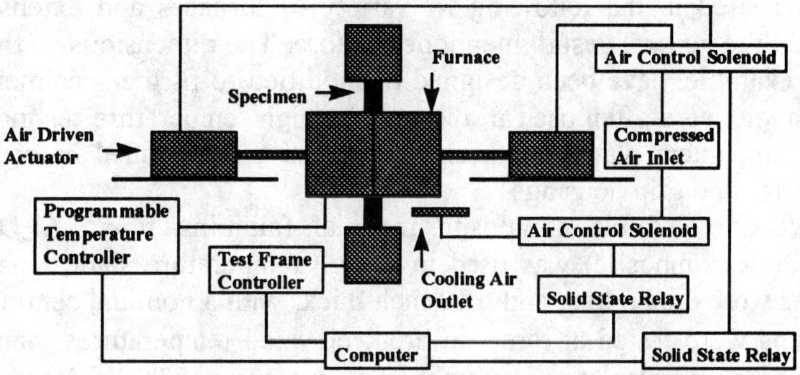

Figure 5. A Thermal Mechanical Fatigue Test System

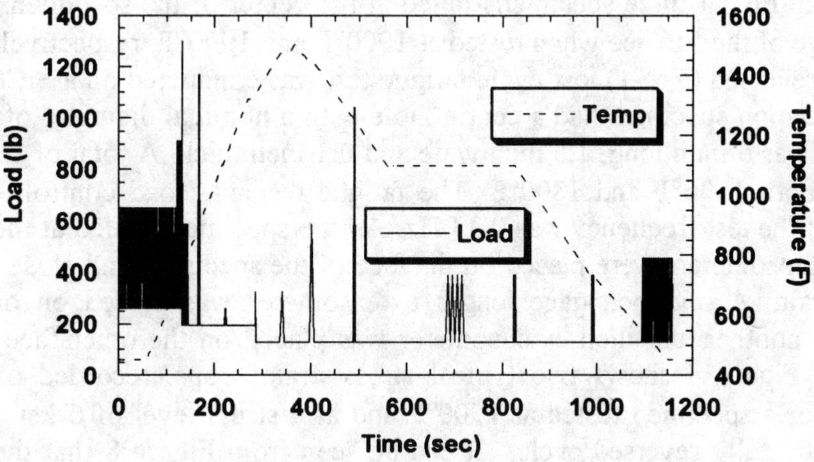

Figure 6. A Typical Temperature and Mechanical Loading Spectrum

on which way the air flows. A programmable temperature controller controls the furnace temperature according to the temperature spectrum input from the computer. The air driven actuators open the furnace during cooling process and close the furnace during heating process.

For a typical temperature spectrum shown in Figure 6, the temperature in the CMC specimen usually can follow the spectrum during heating. However, even the furnace is opened during cooling, the temperature decrease in the specimen is still not fast enough. For this reason, another set of solid state relay and air control solenoid applies compressed air directly on the specimen to accelerate the temperature decrease. During the system calibration, the split distance between two halves of the furnace shell can be determined, as well as the output amount of cooling air. The test on this innovative thermal mechanical fatigue test system is still in progress. However, based on some preliminary results, the temperature spectrum can be implemented accurately and effectively on this test system.

TESTING OF CERAMIC COMPOSITES AT HIGH TEMPERATURES

Some typical experimental results for high temperature testing of ceramic matrix composites are presented in the following. A variety of furnaces and extensometers have been used, based on the general design mentioned before. The dimensions of the furnace and extensometer, for example, have been designed or modified to fit the specimen dimensions. Cooling water systems were often used in the present high temperature testing, not only on the extensometer attachment plate and mechanical grips as mentioned before, but also on hydraulic wedge grips and grip housings.

An Oxide/Oxide ceramic composite material (alumina fiber (Al_2O_3) reinforced aluminosilicate matrix composite) was used in a high temperature static tensile test. The dogbone specimens were 6 inch long and 0.12 inch thick, with a nominal central width of 0.4 inch. The specimens were tested at three different elevated temperatures, namely, 1000°F, 1800°F and 1900°F. It was also tested at ambient temperature (72°F) for comparison. Two specimens were tested at each temperature, except that three were tested at ambient temperature. The average failure strength and ultimate strain, with little scatter, are shown in Figure 7. The strain was measured by a one-inch gage length extensometer. When tested at ambient temperature, all three specimens failed at the center of the specimen. However, two failed at the edge of the furnace when tested at 1000°F and 1900°F, respectively.

A fully reversed (R= -1) low cycle fatigue test was conducted on a SiC/SiC composite material. The coupon specimen had a central hole with a nominal diameter of 0.75 inch. The specimen itself was 6 inch long, 1.5 inch wide and 0.1 inch thick. A total of seven specimens were tested at both 1500°F and 1800°F. The fatigue test was load controlled with various stress levels and the test frequency was 0.33 Hz. All the specimens failed at the holes. During the test, two extensometers were placed on the face of the specimen and close to the edges to monitor the strain. A one-inch gage length extensometer was placed on one face of the specimen while another half-inch extensometer was placed on the other face to balance the bending effect. Figure 8 shows two typical stress-strain loops, recorded on the one-inch extensometer, for a specimen tested at 1500°F and at a stress level of 6 ksi. This specimen failed after 18406 fully reversed cycles. It can be seen from Figure 8 that the extensometer followed the specimen deformation quite well under fully reversed loading at this frequency.

The direct loading creep frame designed in the present study was proved to be very reliable. A typical creep curve, generated on this direct loading frame, is shown in Figure 9.

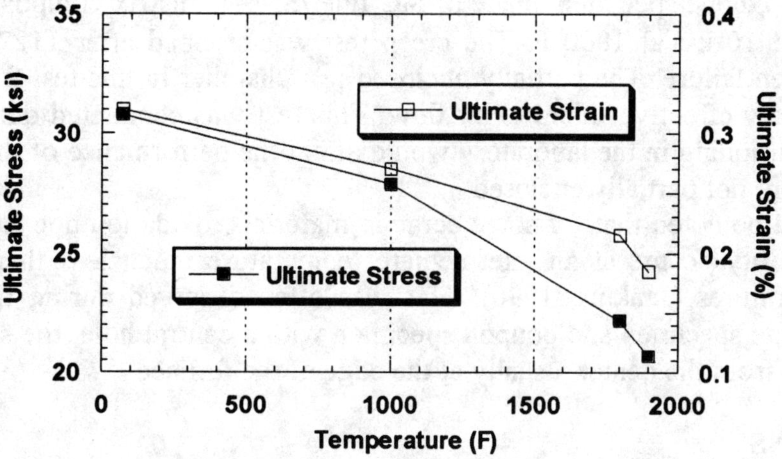

Figure 7. Failure Strength and Ultimate Strain of an Oxide/Oxide CMC Material

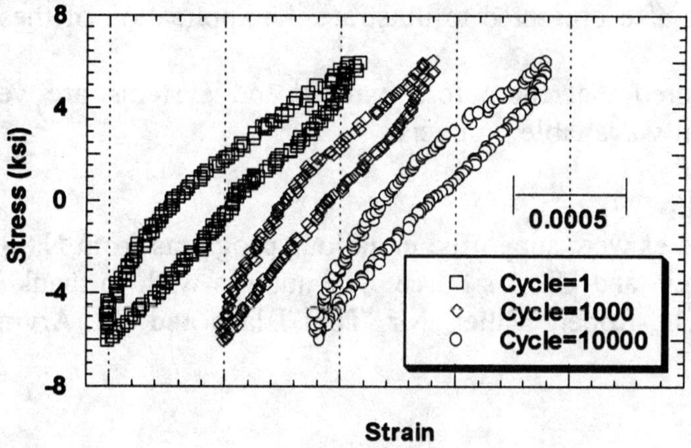

Figure 8. Stress-Strain Curves of a SiC/SiC CMC Material at 1500°F

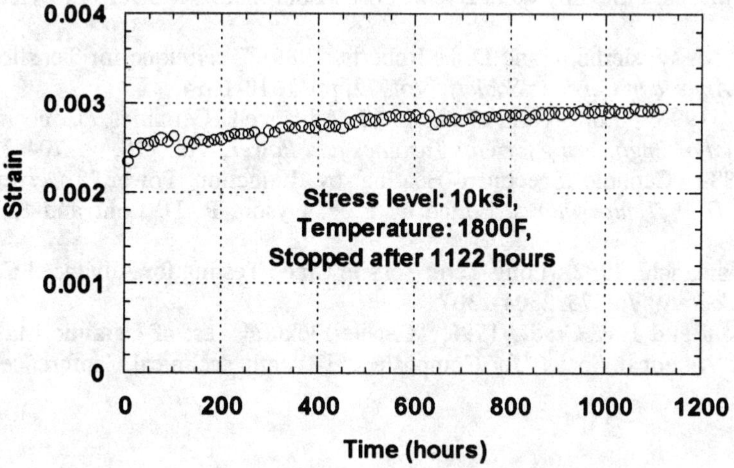

Figure 9. Creep Curve for a CMC Material at 1800°F

A 6 inch long dogbone specimen, made of SiC fiber/Si-N-C matrix composite, was loaded at a stress level of 10 ksi at 1800°F. The creep test was stopped after 1122 hours (47 days) without specimen failure. The partially enclosed test chamber in this test frame, as shown in Figure 4, is a very effective shield to air flow. This test was conducted during summer time and the air conditioning in the laboratory would affect the performance of the extensometer if the chamber were not partially enclosed.

It should be noted that for some ceramic materials, oxidation due to low-temperature chemical instability occurs at an intermediate temperature much less than the upper limit elevated temperatures (Jenkins, 1996). This was often observed during the present study. Even for dogbone specimen and coupon specimen with a central hole, the specimen failed at a location away from the center, usually at the edge of the furnace.

CONCLUSIONS

Various aspects of high temperature test system are discussed in the present study. High temperature furnace, extensometer and grip designs are presented in details. They are integrated into two high temperature test systems, namely, a creep rupture test system and a thermal mechanical fatigue test system. Some typical experimental results for ceramic matrix composite materials are also presented to illustrate the application of these test fixtures and system.

It should be noted that these test fixtures and systems are very cost effective, compared to commercially available products.

ACKNOWLEDGEMENTS

These experimental studies were supported in various programs from NASA Lewis Research Centers, Pratt & Whitney, and US Air Force. The authors wish to thank Dr. Joseph Grady, Mr. Michael Verrilli, Dr. Robert Miller, Mr. Tom Blase and Dr. Arvind Nagar for their support.

REFERENCES

Buddery, J. H., 1988, "Furnace Design and Temperature Control," *Mechanical Testing of Engineering Ceramics at High Temperatures*, Edited by B. F. Dyson, R. D. Lohr and R. Morrell, Elsevier Applied Science, London.

Carroll, D. F., S. M. Wiederhorn, and D. E. Roberts, 1989, "Technique for Tensile Creep Testing of Ceramics," *Journal of the American Ceramic Society*, Vol. 72, pp. 1610-1614.

Jenkins, M. G., 1996, "Long-Term Testing of Advanced Ceramics: Concerns, Insights, and Recommendations," *Journal of Engineering for Gas Turbines and Power*, Vol. 118, pp. 704-710.

Liu, K. C., 1988, "Ceramic Specimen Heating by Induction Power," *Mechanical Testing of Engineering Ceramics at High Temperatures*, Edited by B. F. Dyson, R. D. Lohr and R. Morrell, Elsevier Applied Science, London.

Ohji, T. and Y. Yamauchi, 1992, "Long-Term Tensile Creep Testing for Advanced Ceramics," *Journal of the American Ceramic Society*, Vol. 75, 2304-2307.

Xie, M., S. R. Soni and J. E. Grady, 1996, "Tensile-Flexural Test of Ceramic Matrix Composite T-Joints," Proceedings of the American Society for Composites, Eleventh Technical Conference, Atlanta, GA, pp. 398-406.

On the Elastic Behavior of a Cross-Ply Composite Pin-Joint with Clearance Fits

F. L. DI SCALEA, F. CAPPELLO AND G. L. CLOUD

ABSTRACT

The effect of pin/hole clearance fits on the elastic membranal behavior of a *cross-ply* fiber glass-reinforced epoxy laminate plate loaded in tension through a pin is examined by a finite element nonlinear contact analysis. No *a priori* assumption is made on the distribution of the contact pressure on the surface of the hole. Friction effects are not included. The numerical results correlate well with Hertzian theory of contact. It is demonstrated that clearance increases the peak compressive stress in the plate up to a factor of 3000% with respect to the perfect-fit case. Clearance also shifts the location of the peak tensile stress towards the pin/hole contact arc, and decreases its magnitude by 4% with respect to the perfect-fit case. The degree of clearance does not significantly influence the stress distribution in the plate outside the contact arc.

INTRODUCTION

Among the different techniques for joining structural members, mechanical fastening through a pin is a common choice because of low cost, simplicity, and facilitation of disassembly for repair. In order to predict failure, it is necessary to determine the stress distribution in the vicinity of the boundary of the hole. For anisotropic materials, this distribution is strongly dependent on several parameters such as material properties, laminate lay-up, ply orientation, and fit, whether clearance or interference, between the pin and the hole.

Pin/hole clearance is usually present in at least one member of a pin-joint in order to permit articulation of the joint. Using the Finite Element Method (FEM) with incremental loading and iterative procedures to account for the nonlinearity in the stress/load relation, earlier researchers (Rowlands et al., 1982) found that clearance significantly increases the maximum radial contact stress in a spruce joint.

Francesco Lanza di Scalea and Gary L. Cloud, Materials Science and Mechanics Department, Michigan State University, A304 Engineering Building, East Lansing, Michigan 48824-1226
Francesco Cappello, Dipartimento di Meccanica ed Aeronautica, Universita' degli Studi di Palermo, Viale delle Scienze, 90128 Palermo, Italy

Other studies (Pradhan and Ray, 1984) examined the effect of clearance in isotropic materials and several fiber-reinforced plastic composites using FEM with an initial distribution of radial stress imposed on the edge of the hole as a boundary condition for the analysis. It was concluded that, for small holes or large width plates, the maximum circumferential stress decreases when the pin/hole contact angle decreases for highly anisotropic plates. For plates of lower degree of anisotropy, the same stress increases when the contact angle decreases. However, for large holes or small width plates, the maximum circumferential stress increases when the contact angle decreases, regardless of the degree of anisotropy of the plate. In another study (Eriksson, 1986), FEM was used with an incremental procedure to study the effect of clearance in a highly anisotropic T300/914C graphite/epoxy laminate. The conclusion was that, for different degrees of clearance, the peak radial stress on the edge of the hole is about the same in magnitude, but different in location. However, at a small distance from the edge of the hole, both magnitude and location of this peak stress are influenced by the degree of clearance. The complex variable approach with an iterative method to determine the arc of contact between the pin and the hole was used to study $[0/\pm45/90]_s$ and $[0_2/\pm45]_s$ laminates (Hyer, Klang and Cooper, 1987). It was demonstrated that the assumption of an initial cosinusoidal distribution of radial stress in the contact region of the hole leads to erroneous results. It was also concluded that clearance increases the bearing and circumferential stress concentration factors and changes the location of these stresses with respect to the perfect-fit case, leading to a smaller load capacity of the plate.

The results obtained in previous work are limited in that there has not been a comprehensive study of laminates in practical use. Specifically, it appears that the effects of high degrees of clearance on the elastic behavior of *cross-ply* laminates have never been investigated. Additionally, earlier results do not always coincide and have rarely been validated by a second method of investigation. The present study is considered a further contribution to the literature because it provides a detailed understanding of the effect of pin/hole clearance in a *cross-ply* fiber glass-reinforced epoxy laminate loaded in tension through the pin. The investigation is performed by using a FEM frictionless analysis with an incremental-iterative procedure to account for the nonlinear contact problem, without any *a priori* assumption on the radial stress distribution on the surface of the hole. Predictions of Hertz' theory of contact are presented for validation of the numerical results.

CASES STUDIED

The object of the study was the determination of the effects of pin/hole clearance fits on the stress distribution in a $[0/90/0/90/0/90/\overline{0}]_s$ fiber glass-reinforced epoxy (R1500/1581) *cross-ply* laminate loaded in tension through the pin. The 0° plies were aligned with the direction of application of the load. Figure 1 shows the geometry, loading condition, boundary conditions, nomenclature, and definition of the problem. The principal dimensions of the plate are presented in Table I.

The membranal behavior of the laminate in the (x,y) plane was described by the following effective mechanical properties, which were determined experimentally in a previous work (Herrera Franco, 1985):
axial Young's modulus: E_y= 21980 N/mm^2;
transverse Young's modulus: E_x=21251 N/mm^2;

shear modulus: G_{xy}=2620 N/mm^2;
Poisson's ratio: ν_{xy}=0.11.

The pin was a steel one, having E=208000 N/mm^2 and ν=0.28. A tensile load P=120N, giving a far-field tensile stress σ_y=0.66 N/mm^2 in the plate, was considered throughout the work. The study encompassed two degrees of pin/hole clearance (diametral clearance λ=0.37 and λ=0.055) and a perfect-fit case (λ=0).

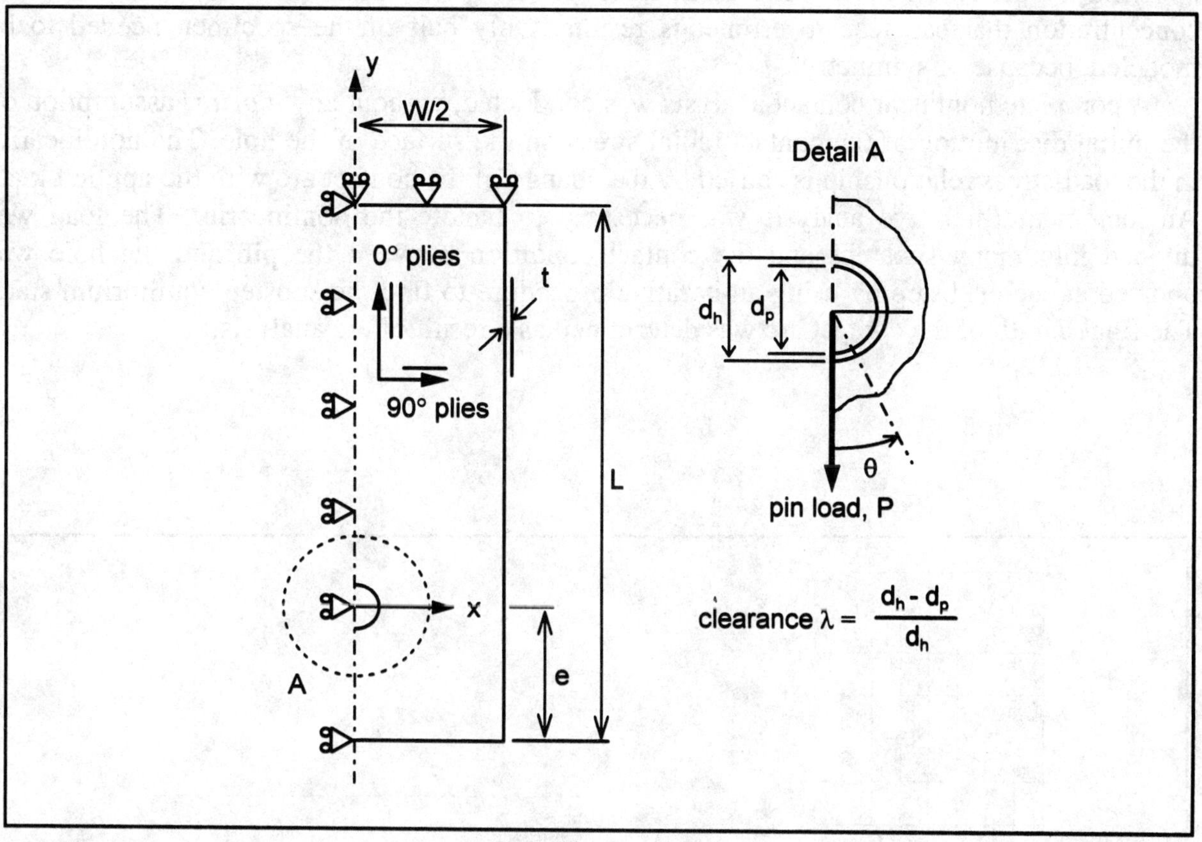

Figure 1. Geometry, loading condition and notation for the $[0/90/0/90/0/90/\bar{0}]_s$ FGRP pin-loaded composite plate.

TABLE I- PRINCIPAL DIMENSIONS OF THE PLATE

length, L	203.2 mm (8.0 inch)
thickness, t	3.556 mm (0.14 inch)
hole diameter, d_h	6.35 mm (0.25 inch)
edge dimension, e/d_h	4.0
width, w/d_h	8.0

ANALYSIS

FEM was used to investigate the three cases mentioned above. Models of the rectangular pin-loaded plate were generated and analyzed using ANSYS ver.5.3. Figure 2 shows the mesh employed for the plate and a detail of the mesh in the pin/hole contact area. The plate was loaded along its axis of symmetry (see Figure 1), by applying a tensile load distribution on an inner area of the pin covering about 80% of the whole pin cross-section. This solution was chosen in alternative to earlier authors' choices (e.g. Grant, Smart and Stanley, 1994) of applying the load only to the central node of the pin, which produces a local stress concentration that can lead to erroneous results. Only half of the specimen needed to be modeled, because of symmetry.

A complete nonlinear contact analysis was conducted, without any *a priori* assumption on the initial distribution of the contact radial stress on the surface of the hole. The nonlinearity in the load-stress relationship is caused by the change of the contact arc with the applied load. An incremental-iterative analysis was necessary to handle the nonlinearity. The load was divided into many substeps and the contact condition between the pin and the hole was updated at each substep by using an iterative procedure to find the substep equilibrium state. The final length of the contact arc was determined as a result of the analysis.

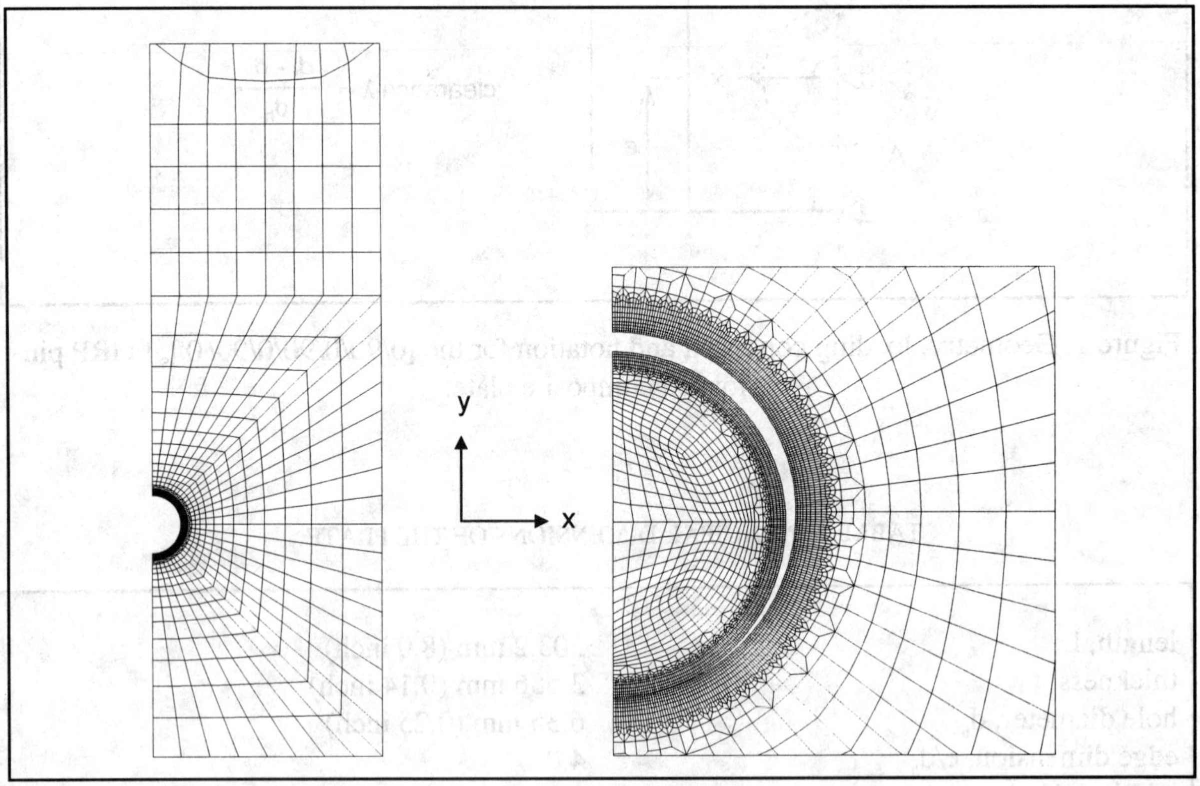

Figure 2. Mesh employed for the plate and a detail of mesh in the pin/hole contact area.

The mesh consisted of four-noded plane-stress finite elements for the plate and the pin, with 2-D point-to-surface contact elements for the pin/hole interface. The choice of a plane-stress analysis derived from the consideration that the thickness of the plate was small compared to its in-plane dimensions (see Table I). By using the contact finite elements, the contact condition could be simulated even when the exact location of the contact arc was not known *a priori*.

In order to check the adequacy of the mesh refinement for the high spatial gradients arising near the hole, the continuity of the resulting strain and stress distributions at the interface of the finite elements was examined, and found to be satisfactory. Also, the equilibrium condition at the pin-hole interface was checked by observing equal radial stress in the pin and the plate at the hole boundary. Friction was not included in this study. The material was considered to behave elastically.

RESULTS

Distributions of the in-plane stress components are obtained along the edge of the hole and along the bearing direction below the hole. The stress values are normalized by the average bearing stress $\sigma_b = P/d_h t$, where P, d_h and t are shown in Figure 1.

Figures 3, 4 and 5 show the FEM results for the distribution of the cartesian components of stress σ_x, σ_y and τ_{xy} along the edge of the hole, for the cases of clearance λ=0.37, clearance λ=0.055, and perfect-fit (λ=0) respectively.

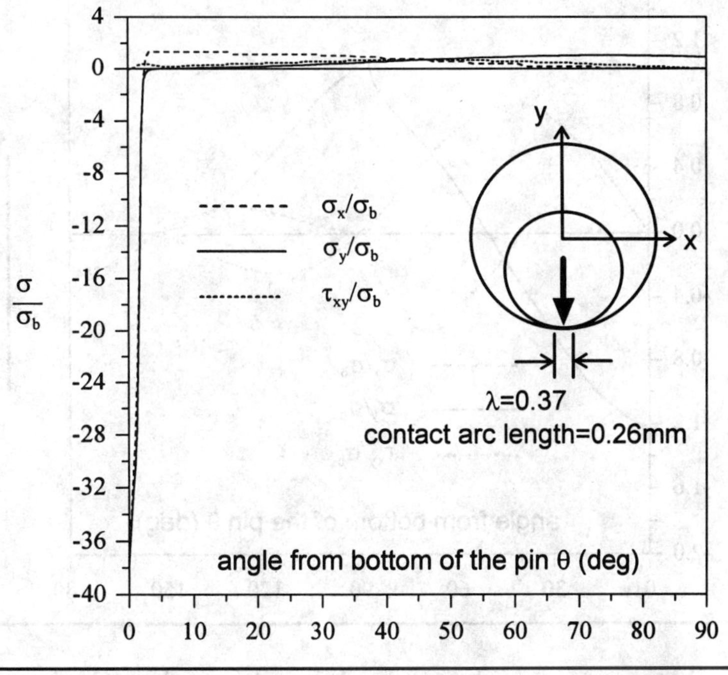

Figure 3. Normalized stress σ_x, σ_y and τ_{xy} around the edge of the hole for clearance λ=0.37.

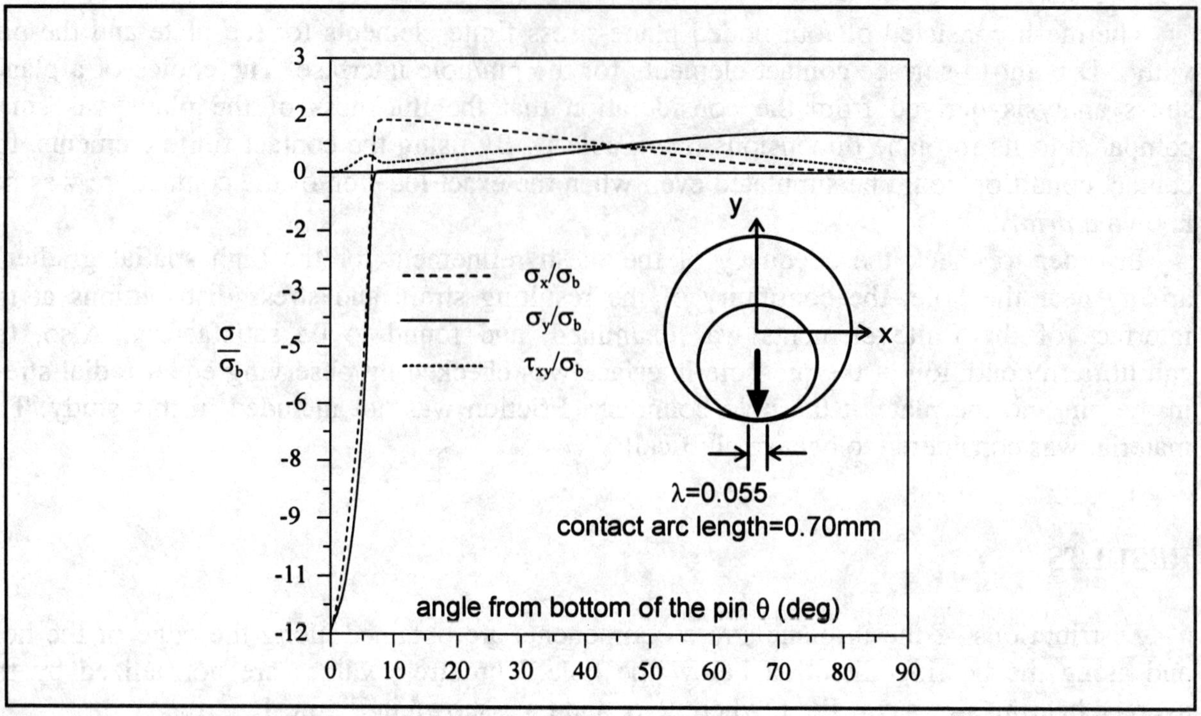

Figure 4. Normalized stress σ_x, σ_y and τ_{xy} around the edge of the hole for clearance λ=0.055.

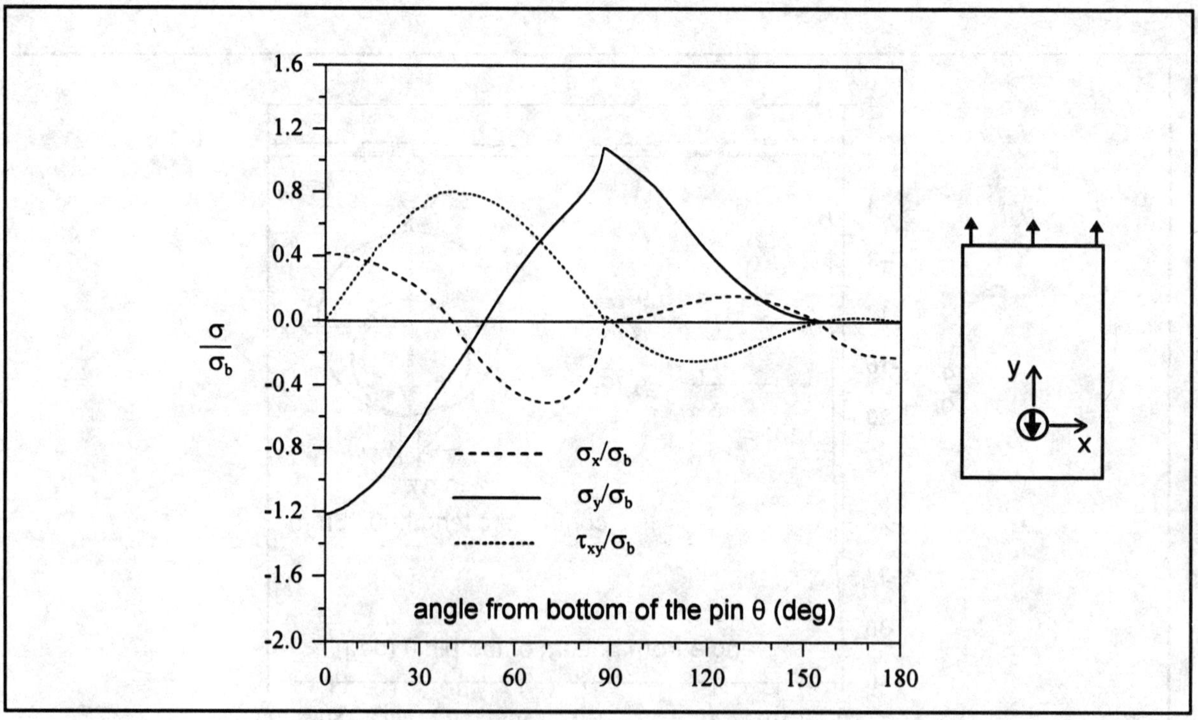

Figure 5. Normalized stress σ_x, σ_y and τ_{xy} around the edge of the hole for perfect-fit (λ=0).

Due to the small contact arc, only $0° \leq \theta \leq 90°$ is shown in the plots of Figures 3 and 4, the stress values in the rest of the edge of the hole being negligible. For the perfect-fit case, results are presented for $0° \leq \theta \leq 180°$, because of the symmetry of the plate. The plots in Figure 5 are in general agreement with those obtained by earlier authors (Pradhan and Ray, 1984; De Jong, 1977). It is shown in Figures 3 and 4 that highly concentrated compressive stress arises in the pin-hole contact arc. Table II indicates the magnitude and angular position of the peak tensile and compressive stress σ_{ymax} and σ_{ymin} respectively, along the edge of the hole for the three cases examined. The increase in peak compressive stress σ_{ymin} with respect to the perfect-fit case is shown to be about 3000% for $\lambda=0.37$ and 900% for $\lambda=0.055$. Also, both magnitude and position of the peak tensile stress σ_{ymax} seem to be substantially independent of the level of clearance. For very small contact arcs (high degrees of clearance), this observation is consistent with the St. Venant principle, which states that different statically equivalent force distributions produce the same effects far from the area of application of the force. It is also observed that clearance shifts the position of σ_{ymax} towards the contact area and decreases its magnitude by only about 4%, with respect to the perfect-fit case.

TABLE II- EFFECT OF CLEARANCE: MAGNITUDE AND POSITION OF NORMALIZED PEAK TENSILE AND COMPRESSIVE STRESS σ_y ALONG THE EDGE OF THE HOLE

	maximum tensile stress σ_{ymax}/σ_b		maximum compressive stress σ_{ymin}/σ_b	
	magnitude	position θ	magnitude	position θ
clearance $\lambda=0.37$	1.05	74°	-38.5	0°
clearance $\lambda=0.055$	1.05	73.8°	-11.85	0°
perfect-fit ($\lambda=0$)	1.09	87.9°	-1.22	0°

Validation of the numerical results was provided by Hertz' theory of contact (Love, 1920), which yields a closed-form solution for isotropic materials, obtained with the assumptions that the strains are within the elastic limit and no frictional forces are present. The application of Hertz's theory to the pin-loaded composite plate here examined seemed meaningful because of the very low degree of anisotropy of the laminate (axial to transverse elastic stiffness ratio $E_y/E_x \sim 1$), and the high degrees of clearance (small contact arcs) examined. This comparison also provided a method to test the refinement of the finite element mesh in the contact area. In calculating Hertz' solution, the average of the laminate's

E_x and E_y was considered as Young's modulus of the plate. The length of the contact arc for λ=0.37 is equal to 0.26mm from the FEM analysis (see Figure 3) and 0.25mm according to Hertz. For λ=0.055, it is 0.70mm from FEM (see Figure 4) and 0.68mm according to Hertz. The maximum pressures calculated numerically along the edge of the hole in the bearing direction are basically coincident with the results of Hertzian theory.

Figure 6 shows the distribution of stress σ_x and σ_y along the bearing direction (direction x=0) for the two levels of clearance. In the figure, good agreement is shown with the results from Hertzian theory, represented with continuous lines. Some discrepancy between numerical and theoretical results can be attributed to the anisotropy of the FEM model and to the restrictive approximation of "semi-infinite" bodies in the theory, which does not account for the actual dimensions of the component. The plots in Figure 6 also show that the degree of clearance does not have any significant effect on the stress distribution at a distance from the edge of the hole of about 0.2 times the radius r_h, the plots relative to λ=0.37 and λ=0.055 being practically coincident thereafter. This behavior is consistent with the St. Venant principle.

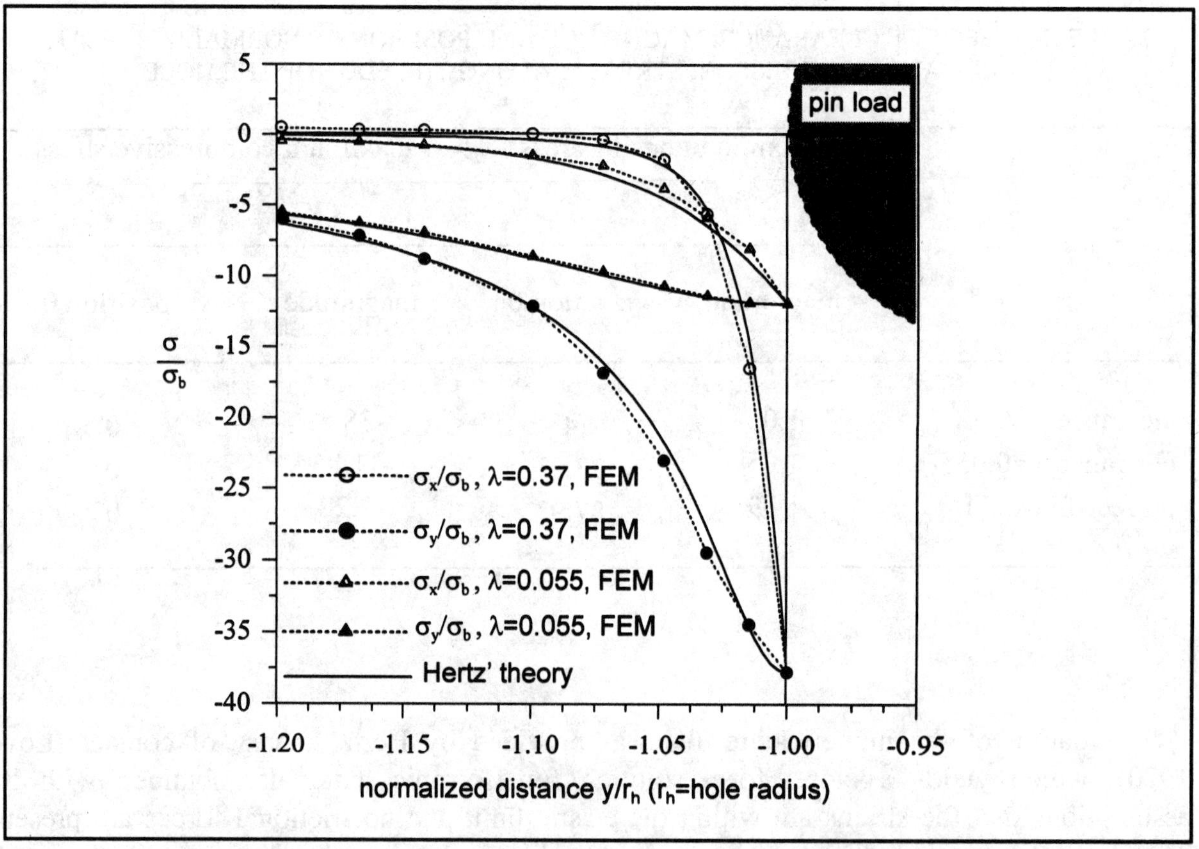

Figure 6. Comparison between FEM and Hertzian theory: normalized stress σ_x and σ_y close to the hole along the bearing direction x=0, for clearance λ=0.37 and λ=0.055. Plotted symbols are nodal solutions from FEM.

DISCUSSION

It has been showed that the magnitude and position of the peak tensile stress σ_{ymax} around the edge of the hole are not significantly influenced by the level of clearance (see Table II). Owing to St. Venant principle, it appears that this conclusion is generally true for high ratios of degree of clearance/applied load, such as the ones examined in this work, in which cases the pin-hole contact arc is a very small portion of the hole. Due to the high compressive stress concentration arising in the bearing direction of the plate (Figures 3 and 4, Table II), clearance fits should be avoided.

As said earlier, the stress-load relation in pin-loaded holes is generally nonlinear because of the change in contact arc with the applied load. Yet, for isotropic pin-loaded plates of different dimensions and pin fits, an earlier work (Grant, Smart and Stanley, 1994) concludes that in the case of clearance, the only significant nonlinearities occur over the contact arc. No evidence exists which would lead to different conclusions for anisotropic materials. Thus, although the present study involves only one level of load, some of the results can be extended to different load levels.

The stress distributions presented are relative only to the *cross-ply* fiber glass-reinforced epoxy laminate examined in the present study; the mechanical behavior of other laminates can be very different. Although failure criteria and their applications have not been the object of this study, the stress distributions presented can be useful for predicting failure modes of the joint.

CONCLUSIONS

1. FEM nonlinear contact analysis is a powerful technique to simulate the elastic behavior of anisotropic pin-loaded plates, for which theoretical closed-form solutions are either nonexistent or very complex.
2. Hertz' contact theory provides a good estimate of stresses where clearance fits are involved.
3. Clearance fits in pin-loaded FGRP *cross-ply* laminates should be avoided, as they cause large compressive stresses in the bearing area below the hole.
4. For high ratios degree of clearance/applied load, the stress distribution outside the pin/hole contact arc is not significantly influenced by the degree of clearance used.

ACKNOWLEDGEMENTS

Portions of this research were supported by the Research Excellence Fund of Michigan and the U.S. Army Tank Automotive Command, Warren, Michigan.

REFERENCES

De Jong, T. 1977. "Stresses around pin-loaded holes in elastically orthotropic or isotropic plates," *J. of Composite Materials*, 11: 313-331.

Eriksson, L. I. 1986. "Contact stresses in bolted joints of composite laminates," *Composite Structures*, 6(1-3): 57-75.

Grant, R. J., J. Smart, and P. Stanley. 1994. "A parametric study of the elastic stress distribution in pin-loaded lugs," *J. of Strain Analysis*, 29(4): 299-307.

Herrera Franco, P. J. 1985. "A Study of Mechanically Fastened Composite Using High Sensitivity Interferometric Moiré Technique," Ph.D. thesis, Michigan State University, East Lansing, Michigan.

Hyer, M. W., E. C. Klang, and D. E. Cooper. 1987. "The effects of pin elasticity, clearance and friction on the stresses in a pin-loaded orthotropic plate," *J. of Composite Materials*, 21(3): 190-206.

Love, A. E. H. 1920. "Theory of elasticity," (3rd edition), Cambridge University Press.

Pradhan, B. and K. Ray. 1984. "Stresses around partial contact pin-loaded holes in FRP composite plates," *J. of Reinforced Plastics and Composites*, 3(1): 69-84.

Rowlands, R. E., M. U. Rahman, T. L. Wilkinson, and Y. I. Chiang. 1982. "Single- and multiple-bolted joints in orthotropic materials," *Composites*, 13(3): 273-279.

Evaluation of Electron-Beam Cured Resins for Repair of Composite Structures

K. R. ULECK* AND A. J. VIZZINI†

ABSTRACT

An experimental study was conducted to compare electron beam curing with conventional thermoset curing for composite repair applications. To compare the strength of repairs made with electron beam cured resins to conventional thermally cured epoxy resins, two versions of double strap joints using composite adherends were run. The first version employed a pre-cured patch or doubler, and the second version employed a cocured patch. Results show that the electron beam cured resins compare favorably to conventional systems when used for constructing cocured patches, but do not work well for adhesive bonding of precured patches.

COMPOSITE REPAIR

Methods for repair of damaged composite structures must be developed and refined as advanced composite materials are used more frequently on aircraft. The current lack of mature composite repair technology is often seen as a critical disadvantage of composite structures because of the associated costs in down time, materials, and labor. To decrease these costs and to make composite materials more attractive, simple repair procedures must be developed that minimize the facilities, skill level and time required to make an acceptable repair [1].

Adhesive repairs tend to be lighter than mechanically fastened patches. Indeed this benefit is important in balanced structures such as rotor blades. A drawback of traditional mechanical joining techniques used for repairing metal structures, such as bolting and riveting, is that they induce stress concentrations which can lead to crack development and premature failure when used on composites. Adhesive bonding provides smooth load transfer minimizing stress concentrations [2]. Additionally, bonded joints can be made to have a smooth, sealed surface, thereby minimizing aerodynamic disturbances and prohibit water ingress.

Major obstacles to adhesive repairs are material handling and storage requirements, surface preparations, adhesive preparation and application. Traditional adhesive systems have limited shelf life, and many require cold storage. This not only necessitates freezer space, but presents logistical problems for transporting these frozen materials [2, 3]. Another disadvantage of current resin systems is that improper proportioning or mixing of two-part systems can occur.

* Kevin R. Uleck, Undergraduate Research Assistant, Dept. of Aerospace Eng., University of Maryland, College Pk, MD 20742.

† Anthony J. Vizzini, Associate Professor, Dept. of Aerospace Eng., University of Maryland, College Pk, MD 20742 member ASC.

Radiation Curing

Radiation curing is the use of electron beam (EB) or ultraviolet (UV) radiation as an energy source to induce the rapid conversion of reactive liquids to solids [4]. Ultraviolet radiation is primarily used for curing inks, paints, and thin coatings, and is recently seeing use in filament winding as a staging method. Electron beam radiation can penetrate much deeper than UV, and therefore can be used for laminating and adhesive bonding.

In electron-beam curing, electrons produced from a hot filament and cathode are accelerated and concentrated with high voltage coils, then directed into the uncured resin causing direct initiation of the cross-linking reaction [5]. This can be compared to conventional thermally activated resin systems, where the catalyst becomes active when heated thus initiating polymerization.

Advantages Electron beam curing has numerous advantages over conventional methods that make it attractive for repair applications such as ambient temperature processing, controllable cure, single component materials, and extensive cross linking [4].

Ambient Temperature Processing: Electron beam curing involves little or no part heat up. This provides several advantages. First, it reduces the complexity of the repair because no external heating required. Second, no thermally induced stresses are produced because there is no appreciable part heat up [3, 6]. Third, no special high temperature tooling is required, making repair procedures simple and quick.

Controlled but Rapid Cure: Electron beam curing resins do not cure except in the presence of electron radiation, therefore they have an indefinite shelf life and can be stored at room temperature. They also have an indefinite pot life which allows the technician to take time and care to properly construct a repair, and then cure it in seconds upon irradiation [6].

Single Component Materials: Electron beam curing resins are supplied as a single, ready-to-use component. This eliminates error due to improper measuring and mixing of two-part systems and simplifies repair procedures.

Extensive Cross Linking: Electron beam curing resins exhibit a high degree of cross linking. This results in high thermal stability, good chemical resistance, and low moisture absorption [7].

Approach

The first step in a repair procedure is to assess whether the repair will be permanent for the life of the aircraft, or temporary to enable the aircraft to complete its mission or to fly to another facility. Next, the requirements of the repair are determined. For aircraft, these generally fall into three categories: aerodynamic restoration, environmental protection, or load bearing. Once these are determined, a repair is designed, often with the help of computer codes, to recover the performance of the composite and prevent further damage accumulation.

Figure 1 shows techniques for a patch type repair of a typical sandwich structure. The patch can be constructed of either pre-cured material that is bonded onto the structure (a,b) or plies laid up and cocured onto the structure (c,d).

EXPERIMENTAL TEST PROGRAM

Guidelines for repair designs and procedures using current materials and technology have been established by previous researchers. The main factors affecting repair performance are:

1. Substrate construction (reinforcement, matrix, core material, layup)

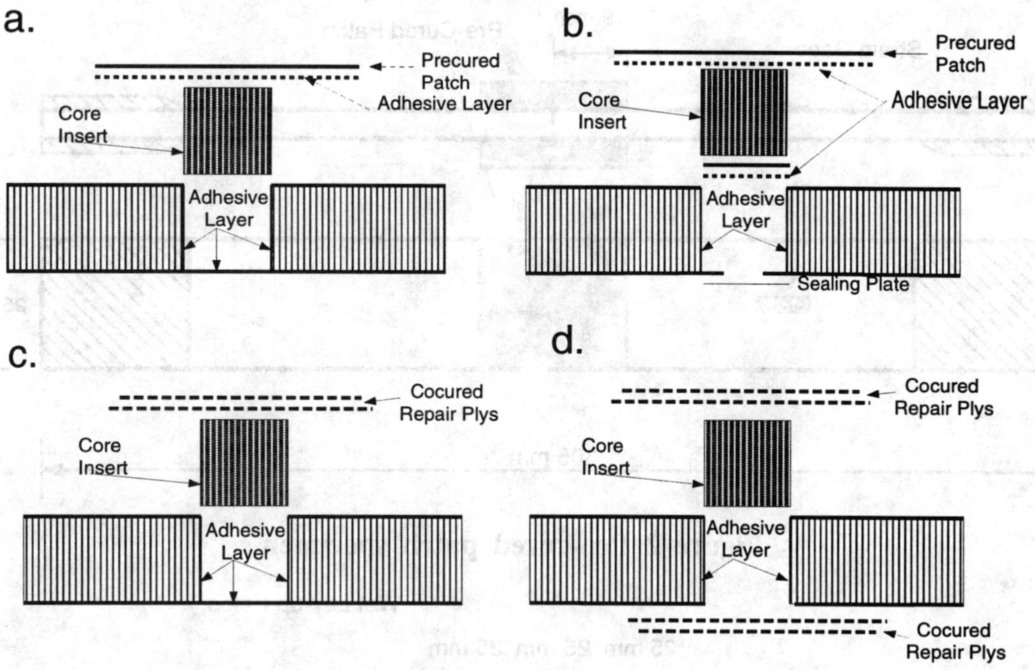

Figure 1 Typical sandwich structure repairs [8]

2. Substrate surface finish and geometry
3. Repair design (patch type, geometry)
4. Repair materials (resin, adhesive, reinforcement, layup)
5. Repair procedure (conditions, cure cycle, surface preparation)
6. Service environment

It is the aim of this research to compare repairs made with electron-beam cured resins to repairs made with conventional resin systems to determine the utility of electron-beam curing technology for composite structure repair. To make an accurate assessment and comparison of overall performance, a simple test that minimizes process and design variables was desired.

Existing repair procedures were surveyed to determine common structural features and critical parameters of bonded repairs [9]. The lap joint is a fundamental structural component of repairs, present in various configurations. Additionally, two different repair methods were identified; those constructed using pre-cured patch material (Figure 1a, b) and those where the patch is cocured onto the structure (Figure 1c, d).

The test method selected for this study, ASTM D-3528B, uses double lap joints which eliminate bending and minimizes the peel stress. It was modified for use with composites and to more accurately represent a repair scenario. Two specimen types were tested to represent the two primary repair methods. The first used pre-cured patches bonded to the adherends (Figure 2) to test electron-beam curable resins when used as an adhesive. The second used patches constructed through wet layup and then cocured onto the adherends (Figure 3) to test electron-beam curable resins when used for laminating. Note that these test specimens are designed so that failures occur at the joint, not in the adherends.

Material Selection

The following adherend material and resin systems were selected based on a survey of past studies, industry use, and availability.

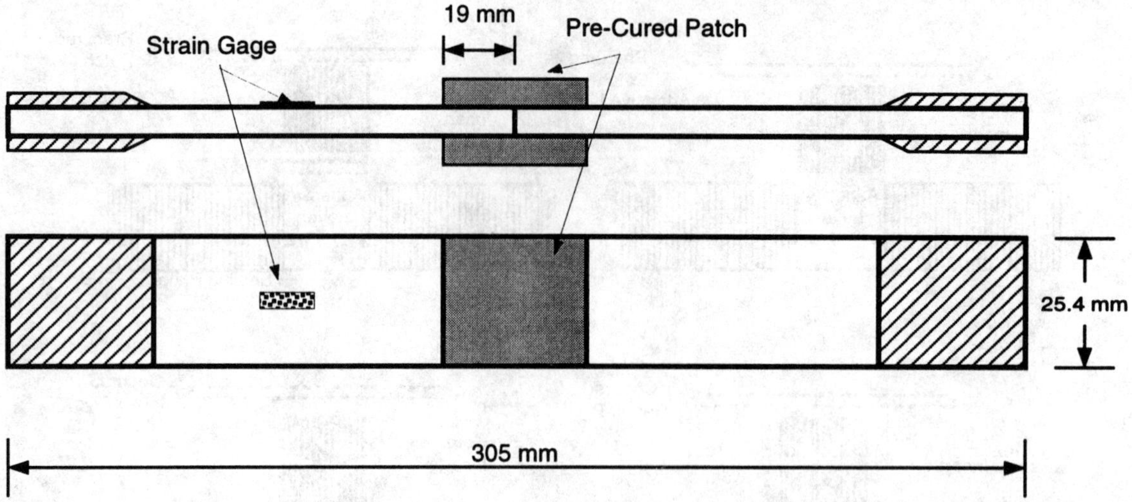

Figure 2 Pre-cured patch specimen

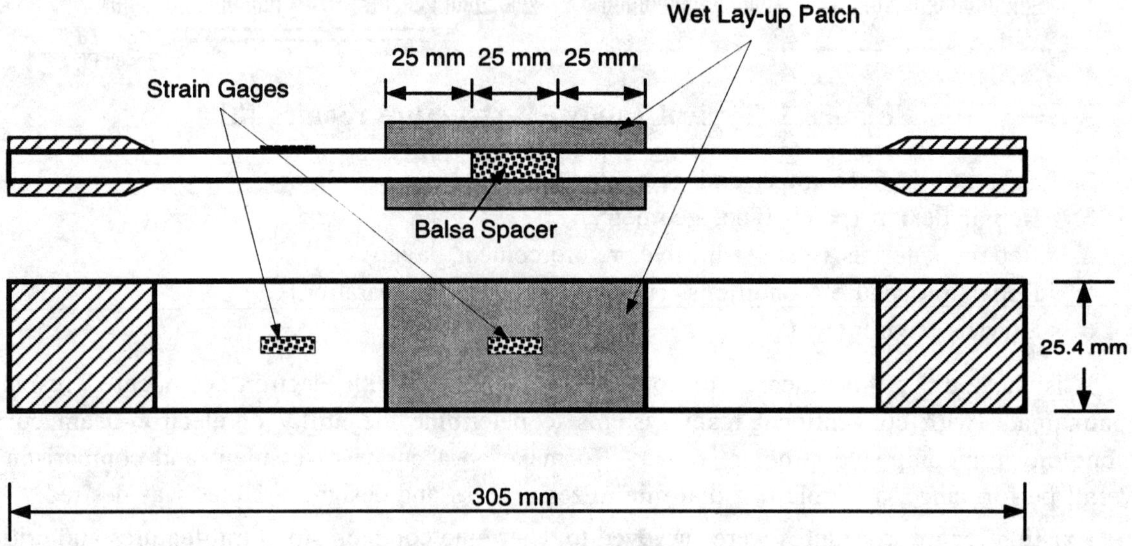

Figure 3 Cocured patch specimen

Adherends: *S2/E773FR* $[0°]_8$: This typical glass/epoxy composite is found in a wide variety of applications and was selected as the primary adherend material for the first phase of testing. The [0°] layup was chosen to ensure that failures take place in the patch/adherend bond, and the thickness was chosen to be typical of thin face sheet stiffness.

S2/PPS $[0°]_8$: PPS (Polyphenylene Sulfide) thermoplastic is traditionally difficult to bond to due to its highly crystalline structure and few sites for hydrogen bonding. Again, layup was chosen to ensure that failures take place in the patch/adherend bond.

AS4/3501–6 $[0°]_{12}$: One series of tests was conducted to test determine the effect of adherend stiffness on bond strength for joints using pre-cured patches. The thickness was chosen to be similar to the glass/epoxy specimens.

Adhesives (Pre-cured patch specimens): *American Cyanamid FM-123:* Common in manufacturing for secondary bonding, this adhesive requires an autoclave cure cycle. Specimens using pre-cured patches made with this were tested to represent an upper-limit ideal bond.

Dexter Hysol EA 9309: A two part epoxy paste adhesive widely used for secondary bonding and repair. Cure is seven days at room temperature.

DAMALIC Formulation #1001: A specially formulated electron-beam curable acrylate-based resin made by DAMALIC corporation for this study.

Laminating Resins (Cocured patch specimens) *Dexter Hysol EA 9396*: This special two part epoxy cures in seven days at room temperature. It was chosen because it was used in a recent study by Sikorsky Aircraft Company where it was cured at an elevated temperature [8].

Dexter Hysol EA 9390: This two-part epoxy cures in 220 minutes at 93 °C. It was also used in the Sikorsky study.

DAMALIC Formulation #1002, 1003: Specially formulated electron-beam curable vinyl ester resins made by DAMALIC corporation for this study.

Processing and Handling Comparison

Radiation curable resins offer many advantages over these conventional systems making them attractive for repair applications. Several key processing advantages are highlighted below in Table 1.

Table 1 Comparison of Adhesives and Laminating Resins

	Cure Time	Cure Temp	Components	Shelf Life
Adhesives				
DAMILIC #1001	Minutes	Ambient	One	Infinite
Dexter-Hysol EA 9309	7 Days	Ambient	Two	1 Year
American Cyanamid FM-123	2 Hours	121 °C[a]	One	1 Year[b]
Laminating Resins				
DAMILIC #1002, #1003	Minutes	Ambient	One	Infinite
Dexter-Hysol EA 9396	7 days	Ambient	Two	1 Year
Dexter-Hysol EA 9390	4 hours	93 °C	Two	1 Year

[a] High pressure required

[b] Refrigeration required

Specimen Manufacture and Testing

Test panels were formed by joining two pre-tabbed adherend panels with the appropriate lap joint, then vacuum bagging and curing. All specimens were cured under vacuum only, with the exception of specimens with FM-123 film adhesive which were autoclave cured under total pressure of 241 kPa. After the lap joints were cured, five individual specimens were machined on a mill equipped with diamond grit tooling and water irrigation. Each set of five specimens constituted a series as described in Table 2. Completed specimen dimensions are shown in Figures 2 and 3. All specimens were instrumented with one foil resistance strain gage to measure adherend elongation. The cocured patch specimens also received an additional gage in the unsupported patch section, as shown in Figure 3, to measure patch elongation.

Table 2 Test Variation Matrix

	Glass/Epoxy	Glass/PPS	Graphite/Epoxy
Precured Patch Specimens			
Dexter Hysol EA 9309	Test Series E	Test Series H	——
American Cyanimid FM-123	Test Series D	——	——
DAMALIC #1001	Test Series F, G	Test Series I	Test Series J
Cocured Patch Specimens			
Dexter Hysol EA 9390	Test Series C	Test Series L	——
Dexter Hysol EA 9396	Test Series A	——	——
DAMALIC #1002 and #1003	Test Series B	Test Series K	——

All electron-beam cures took place at Irradiation Industries Incorporated in Gaithersburg, Maryland. The vacuum bagged test panels were placed on a conveyer system which passed them under the scanned electron beam. Electron energy was 2.8 MeV, and dose rates varied from 40 kGy to 950 kGy per pass based on accelerator settings and integrated continuous dosimetry. The total dose the specimens received was typically 240 kGy.

All testing was performed at ambient conditions on an uniaxial hydraulic test frame with computer data acquisition sampling at a rate of 0.2 s. Specimens were loaded to failure under a constant stroke rate of 0.0423 mm/s giving an equivalent strain rate of 167 μstrain/min for a 254 mm test section.

RESULTS

The parameter used to compare joint performance is *joint efficiency*. This is the percentage of original laminate strength recovered by the joint, and is defined as the strength of the joint divided by the original strength of the adherend alone.

Adhesive Tests (Precured Patch)

In general, precured patch joints using EB cured resin (series F, I) were approximately half as strong as the joints using EA 9309 (series E, H) as shown in Figure 4. In all cases, failure occurred at the patch/adherend bond, however, the electron-beam cured joints exhibited a brittle failure different from the failure of the EA 9309 joints.

During testing of the radiation cured specimens, cracking was heard, thought to indicate brittle failure of the bond. This is visible in Figure 5 as the "steps" in the load-stroke plot. Two factors contribute to this failure mechanism: the brittleness of the electron-beam cured resin, and the thinness of the bond line. The bond lines of the electron-beam cured joints and thickness of the EA 9309 joints were 0.05 mm and 0.13 mm, respectively. This reduced thickness of the bond line, increases the shear stress, thus decreasing the strength of the electron-beam cured joint.

To verify the brittle failure hypothesis, a series of electron-beam adhesive tests was conducted using graphite/epoxy adherends and patches (series H). The graphite/epoxy specimens (series H) failed at higher joint stress than the glass/epoxy specimens (series F, G), 220 MPa and 196 MPa, respectively. This confirms the trend that for extensible materials, the brittleness of the electron-beam cured adhesive used in this study reduces the joint strength.

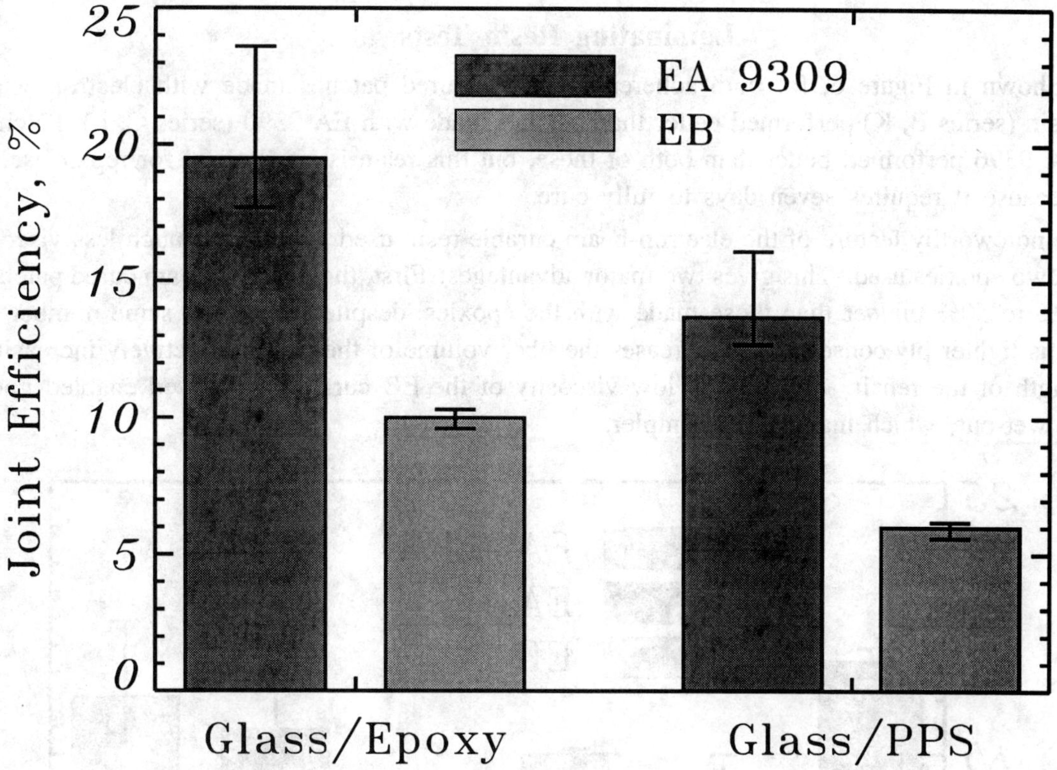

Figure 4 Comparison of pre-cured patch characteristics

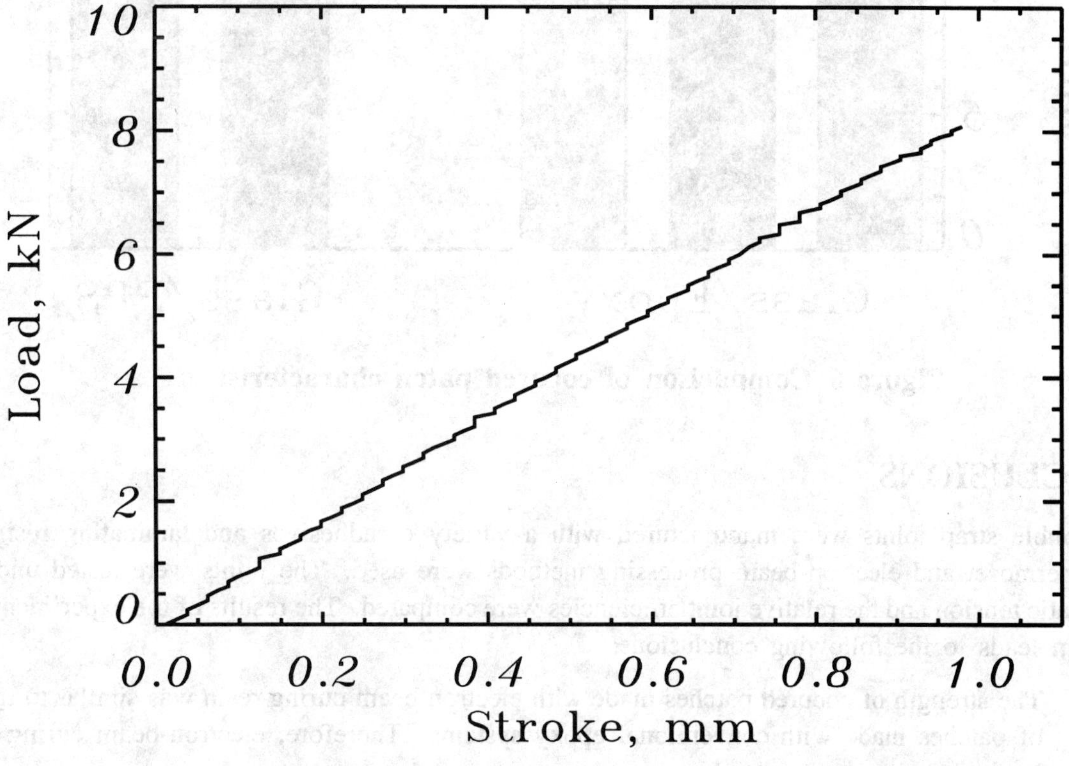

Figure 5 Typical load-stroke plot of glass/epoxy precured patch specimen

Laminating Resin Tests

As shown in Figure 6, for both adherend types, cocured patches made with electron-beam cured resin (series B, K) performed better than patches made with EA 9390 (series C, L). Patches using EA 9396 performed better than both of these, but this resin is impractical for repair use as tested because it requires seven days to fully cure.

One noteworthy feature of the electron-beam curable resin used is that it is much less viscous than the two epoxies used. This gives two major advantages: First, the electron-beam cured patches were 30% to 50% thinner than those made with the epoxies, despite having the same number of plies. This tighter ply consolidation increases the fiber volume of the patch, effectively increasing the strength of the repair. Second, the low viscosity of the EB curable resin used enabled good material wet-out, which made layup simpler.

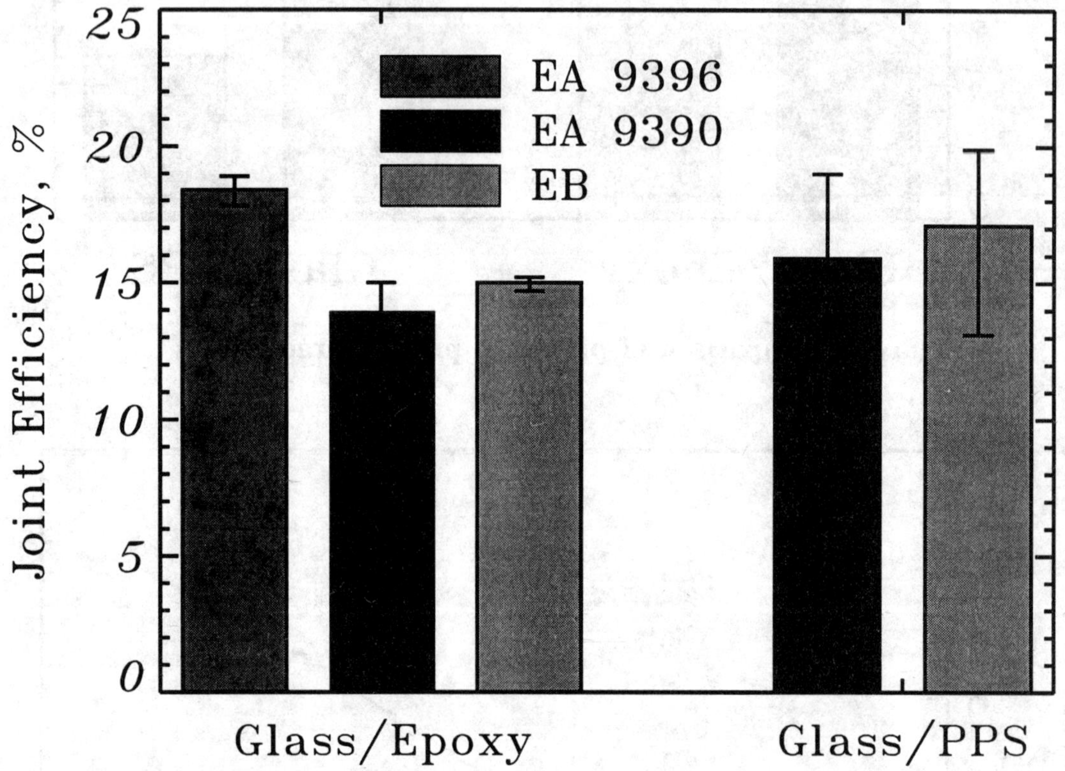

Figure 6 Comparison of cocured patch characteristics

CONCLUSIONS

Double strap joints were manufactured with a variety of adhesives and laminating resins. Both thermoset and electron-beam processing methods were used. The joints were tested under quasistatic tension and the relative joint efficiencies were compared. The results of the experimental program leads to the following conclusions:

1. The strength of cocured patches made with electron beam curing resin was similar to that of patches made with conventional epoxy systems. Therefore, electron-beam curing of laminated repairs is feasible.
2. However, the electron beam curing resin tested did not compare favorably to conventional epoxy systems for adhesive bonding of pre-cured patches. Therefore, further resin

development is necessary before electron-beam curing can be considered for adhesively bonded repairs.

ACKNOWLEDGMENTS

The authors would like to thank Dr. Walter Chappas and Dr. Barbara Devney of the DAMALIC Corporation and Mike Lilly of Irradiation Industries Incorporated.

References

1. J. D. Labor, "Composite Repair Concepts For Depot Level Use" *Proceedings of the 26th International SAMPE Symposium*, 1981, pp. 118–128
2. R. Stone, "Field-Level Repair Materials and Processes" *Proceedings of the 28th National SAMPE Symposium*, April 1983, pp. 87–99
3. J. Scott, C. Mashiba, "Improved Resins for Wet Layup Repair of Advanced Composite Structure" *Proceedings of the 16th National SAMPE Technical Conference*, October 1984, pp. 110–116
4. A. Pincus, "Radiation Curing Adhesives Give Cure and Performance Benefits" *Adhesives Age*, May 15, 1995, pp. 16–17.
5. T. Triplett, "Rad-Cure Market Heats Up" *Industrial Paint and Powder*, June 1995, pp. 16–18.
6. J. Lacey and A. Keough, "Radiation Curing: A Discussion of Advantages, Features and Applications" Association for Finishing Processes of SME, Dearborn MI, July 1981
7. C. J. Janke, R. Norris, K. Yarborough, "Critical Parameters For Electron Beam Curing of Cationic Eopxies and Property Comparison of Electron Beam Cured Cationic Eopxies Versus Thermal Cured Resins and Composites" *Proceedings of the 42nd International SAMPE Symposium*, Anaheim, CA, May 1997, pp. 477-486.
8. A. Dobyns, E. Rosenzweig, C. Francis, "Development of Repair Methodology for the MH-53E Composite Sponson" *Proceedings of the 11th ASC Technical Conference*, Atlanta, GA, October 1996, pp. 111-120.
9. J. D. Labor, "Repair of Composites" FAA Handbook on Manufacture and Repair of Fiber-Reinforced Composites, July 1987, DOT/FAA/CT-87/9

Static and Fatigue Properties of Adhesive Joints in SRIM-SRIM Composites

S. K. MAZUMDAR AND P. K. MALLICK

ABSTRACT

This paper presents static and fatigue properties of adhesively bonded joints in a random continuous strand mat reinforced structural reaction injection molded (SRIM) composite. Both static and fatigue tests were conducted using single lap joints between SRIM substrates. The effects of joint design parameters, such as lap length, adhesive thickness, surface preparation and recessing on the static strength and fatigue life were investigated. The effect of water absorption on the static and fatigue properties is also investigated.

INTRODUCTION

Polymer matrix composites with randomly oriented fibers are used in many structural applications outside the aerospace industry. Examples of such composites are the sheet molding compound composite (SMC) and the structural reaction injection molded composite (SRIM). With increasing use of these materials in automotive and durable goods industries, it is becoming important to develop design data, including the joint design data, for these materials. Many experimental and analytical studies on the static behavior and fatigue behavior of adhesive joints in laminated composites are available in the literature. Similar studies on random fiber SMC and SRIM composites are rare.

Dillard and Spinu (1994) studied the durability of aluminum-SMC joints that were exposed to an environmental cycle or to fixed conditions of temperature and humidity, with and without load. In their study, the metal surface treatment proved to be significantly important for the durability of the SMC / urethane adhesive/ aluminum specimens. Krause, Holubka and Chun (1986) examined the effects of moisture, temperature, loading environment, and exposure time on the fatigue resistance of adhesive joints in random fiber SMC adherends. Urethane as well as epoxy/polyamide adhesives were used. Exposure to water was found to cause significant reductions in the fatigue resistance of these joints.

Holubka, Chun, Krause and Shyu (1986) investigated the effects of various corrosive

S.K. Mazumdar and P.K. Mallick, Department of Mechanical Engineering, University of Michigan-Dearborn, 4901 Evergreen Road, Dearborn, MI 48128

environments on the mechanism of failure in adhesive bonds between steel and SMC adherends. Three to four orders of magnitude reduction in fatigue life of adhesive bonds were observed for bonds exposed to environment prior to fatigue testing.

Recently, Mazumdar and Mallick (1997) conducted experimental investigation on the performance of SMC-SMC adhesive bonded joints. They determined the effects of various joint design parameters, such as lap length, bond thickness and surface preparation, on the static and fatigue behavior of adhesively bonded single lap joints using SMC as the substrate material. The fatigue strength was approximately 50% of the static strength. Effect of water exposure on the static and fatigue performance of SMC-SMC adhesive bonded joints were determined and had negligible effect on the joint performance.

The objective of this study is to examine the effects of joint parameters, such as lap length, adhesive thickness and surface characteristics, on the static and fatigue performance of adhesive bonded single lap joints in SRIM-SRIM composites. The effect of recessing on the static strength of SRIM- SRIM joints is considered. The effect of moisture absorption on the static and fatigue behavior of SRIM-SRIM joints is also examined.

EXPERIMENTS

CHARACTERIZATION OF SRIM PLATES

SRIM plates, 635 mm x 635 mm in size, were supplied by the Automotive Composites Consortium (ACC). The SRIM material was identified as DOW 364 U750 CSM, which contains randomly oriented continuous E-glass strand mat in a polyisocyanurate resin. The nominal fiber content was 40 percent by weight. The plate thickness was between 3.12 and 3.35 mm. Each plate had one direction marked with an arrow indicating direction of the glass roll. Specimens for adhesively bonded joints were cut perpendicular to the arrow direction.

Tensile properties of SRIM plates were determined along the arrow direction as well as transverse to the arrow direction. The average tensile strength in the arrow direction is 172 ± 4.8 MPa and the average tensile modulus is 8.06 ± 0.12 GPa. Tensile strength and modulus transverse to the arrow direction are 213 ± 7 MPa and 9.32 ± 0.17 GPa, respectively.

A major factor in developing good adhesive joint strength is the surface characteristics of the adherends, which includes both surface roughness and surface chemistry. The as-molded SRIM surface was very rough. The microscopic examination of the SRIM surface showed a large number of surface pores, deep voids and exposed fibers.

SPECIMEN PREPARATION

Single lap joints with various adhesive thickness and lap lengths were prepared using a reformulated epoxy adhesive, also supplied by ACC. A few specimens were prepared with recessed adhesive layer (Fig. 1). The adhesive was applied in the lap area and was cured in an air circulating oven at 121°C for 45 minutes following the procedure recommended by the adhesive manufacturer. Lap lengths of 12.7, 25.4 and 38.1 mm were used and adhesive thickness of either 0.762, 1.27 or 2.54 mm was obtained. The lap width was 25.4 mm. The joining surfaces were wiped with a clean dry cloth and acetone cleaned before applying the adhesive. Care was taken to wipe off excess adhesive from the sides of the joint.

Before applying the adhesive, SRIM coupons were dried to remove the absorbed moisture. Presence of moisture in the plate can cause voids at the adhesive-substrate interfaces when the adhesive is cured at 120°C. For drying, SRIM coupons were placed at 80 °C inside the oven and subsequently the oven temperature was increased by 10°C every 15 minutes until 110°C. The SRIM coupons were kept for half an hour at 110°C and then slowly cooled to the room temperature. Lap joints were prepared immediately following the drying operation. In addition to the dried specimens, several tests were conducted with the as-received material (i.e., without drying).

STATIC AND FATIGUE TESTS

Both static and fatigue tests were conducted at room temperature in a computer controlled MTS servohydraulic testing machine. Specimens were placed in the grips of the testing machine with the outer 25.4 mm of each end clamped between the jaws. A special fixture for gripping the specimen was designed to make sure that the loading direction coincides with the center line of the adhesive.

Except for a few cases, all static tests were conducted at a cross head speed of 1.27 mm per min. The peak load, representing the failure of the joint, was recorded for each case and the lap shear value was calculated by dividing the peak load with the lap area. The lap area is the product of lap width and lap length. Tension-tension cyclic fatigue tests were conducted in a load-controlled mode at 10 Hz. The load ratio (P_{min}/P_{max}) was +0.1.

RESULTS

STATIC TEST RESULTS

Single lap joint specimens were prepared according to the ASTM standard D 1002 -72 with a 12.7 mm lap length and 0.762 mm adhesive thickness. A typical load-displacement curve obtained in a single lap shear test is shown in Fig. 2. The load-displacement curve is slightly non-linear and shows small drops in load before reaching the maximum value at which the joint failed. It is possible that these drops occur due to initiation of internal damages. The average failure load for the four specimens tested at a cross head speed of 1.27 mm/min. was 3.4 ± 0.67 kN and the average lap shear value was 10.55 ±2.07 MPa. The failure mode in all four specimens was found to be 100% adhesive failure as shown in Fig. 3. Adhesive failure is defined as the failure at the interface of the adhesive and the substrate.

The data in Table 1 also show that both static failure load and lap shear value are higher for the 2.54 mm adhesive thickness than for the 0.762 mm adhesive thickness. The failure load increases with increasing lap length; however, the lap shear value, obtained by dividing the peak load with the lap area, decreases with increasing lap length.

To examine the effect of cross head speed, single lap joints were tested at cross head speeds of 1.27 mm/min. and 12.7 mm/min. The joint strength is 24 percent higher at the higher cross head speed. Similar effects were observed with SMC-SMC joints (Mazumdar and Mallick, 1997).

TABLE I - STATIC TEST RESULTS OF SRIM-SRIM SINGLE LAP JOINTS

Condition	Lap Length (mm)	Adhesive Thickness (mm)	Cross head Speed (mm/min.)	Failure Load (kN)	Lap Shear Value (MPa)
Dried	12.7	0.762	1.27	3.4±0.67	10.55±2.07
	12.7	2.54	1.27	3.87±0.08	12.0±0.25
	38.1	0.762	1.27	4.8±0.39	4.96±0.4
	63.5	0.762	1.27	10.29±1.12	6.38±0.69
	12.7	0.762	12.7	4.24±0.21	13.13±0.66
As-received	12.7	0.762	1.27	2.91±0.55	9.03±1.72
	12.7	1.27	1.27	3.74±0.4	11.61±1.23

A few adhesive joints were prepared with the as-received SRIM substrates. The average failure load of the jointed specimens with the as-received SRIM substrates was 2.91 ± 0.55 kN, whereas with the dried SRIM substrates, the average failure load was 3.4 ± 0.67 kN. The lower joint strength with the as-received substrates was due to the presence of a greater number of pores in the adhesive, which results from moisture desorption from the as-received SRIM surface during the adhesive curing stage.

The above test results were obtained with an adhesive which was of very thick consistency and did not flow much. The adhesive consistency was much thinner in another batch of adhesive, which was supplied after the first batch was exhausted. A few static test specimens were prepared with 12.7 mm lap length and 0.762 mm adhesive thickness using this second batch of adhesive. The adhesive was cured in the manner described earlier. The average static failure load for these specimens was 5.12 ± 0.36 kN, which is almost 50 percent higher than that obtained with the first batch of adhesive. It is possible that the second batch of adhesive penetrates more into the pores, voids and roughness on the SRIM surface due to its lower viscosity and creates a better mechanical bond.

RECESSED BONDING

The static test data discussed so far included a continuous layer of adhesive. To examine the effect of recessing in SRIM-SRIM joints, specimens with both continuous bonding and recessed bonding were prepared with a total (gross) lap length of 63.5 mm, as shown in Fig. 1. The recessed joints had either a single recess of 38.1 mm at the center, or two recesses, each 12.7 mm long, between the adhesive layers. The adhesive thickness was 0.762 mm in all cases. The average failure load for the continuous adhesive layer with a 63.5 mm lap length was 10.29 ± 1.12 kN. With a single recess, the average failure load was 6.2 ± 1.21 kN, and with two recesses, the average failure load was 12.85 ± 0.46 kN. The lap shear values calculated on the basis of the net bonded area are 6.38 ± 0.69 MPa for the continuous bonding, 9.61 ± 1.88 MPa for the single recessed bonding and 8.57 ± 0.47 MPa for the double recessed bonding. Thus, even with lesser amount of adhesive used, higher lap shear values are obtained with the recessed bonding compared to the continuous bonding. Similar results were also observed with the SMC-SMC joints (Mazumdar and Mallick, 1997).

FATIGUE TEST RESULTS

Specimens for fatigue tests were prepared with dried as well as as-received SRIM coupons. With dried SRIM coupons, three different groups of fatigue specimens were prepared. They were (i) 12.7 mm lap length and 0.762 mm adhesive thickness, (ii) 12.7 mm lap length and 2.54 mm adhesive thickness, and (iii) 38.1 mm lap length and 0.762 mm adhesive thickness. Fatigue specimens with as-received SRIM substrates were prepared with 12.7 mm lap length and 0.762 mm adhesive thickness. The fatigue test results are discussed below.

(a) 12.7 mm lap length and 0.762 mm adhesive thickness

The test results for this group of specimens are shown in Fig. 4. The maximum fatigue load applied was varied between 45.9 % and 65 % of the average peak load in static tests. At the maximum fatigue load equal to 45.9 %, failure did not occur for 4 x 10^6 cycles and tests were discontinued after 4 x 10^6 cycles. The maximum fatigue load at 4 x 10^6 cycles is estimated as 1.56 kN (45.9 % of the static failure load). Failure in all these samples was 100% adhesive.

(b) 12.7 mm lap length and 2.54 mm adhesive thickness

For this group of specimens with 2.54 mm adhesive thickness, the static failure load was higher than that of the previous group with 0.762 mm adhesive thickness. The fatigue test results, shown in Fig. 4, indicates that the fatigue S-N curve is significantly lower for the 2.54 mm adhesive thickness than for the 0.762 mm adhesive thickness. The maximum fatigue load at 4 x 10^6 cycles is estimated as 1.25 kN (32.2 % of the static failure load), which is almost a 20 % decrease in the fatigue load as compared to 0.762 mm adhesive thickness. Out of the five specimens tested in this group, two did not fail for 5 x 10^6 cycles. In the three specimens that failed , two had nearly 100% adhesive failure accompanied by light fiber tear, and one had 100% deep fiber tear (Fig. 3).

(c) 38.1 mm lap length and 0.762 mm adhesive thickness

Fatigue test results for this group of specimens are also plotted in Fig. 4. The fatigue S-N curve is slightly higher for the 38.1 mm lap length compared to that for the 12.7 mm lap length. The maximum fatigue load at 4 x 10^6 cycles is estimated as 2.41 kN (50.1 % of static failure load), which is higher than that for the 12.7 mm lap length.

A few fatigue specimens with 12.7 mm lap length and 0.762 mm adhesive thickness were prepared with as-received SRIM coupons. The average static failure load for this group of specimens was 2.91 ± 0.55 kN, which is 14.4 % less than the average static failure load measured with the dried specimens. Fatigue test results for the dried and as-received specimens are compared in Fig. 4. There is no appreciable difference in the fatigue performance of these two types of specimens. The maximum fatigue load for this group of specimens at 4 x 10^6 cycles is estimated as 1.44 kN , which is 49.6 % of the static failure load. The failure mode in all as-received (undried) specimens was of 100% adhesive failure type.

The above tests were performed with SRIM surfaces cleaned with a cloth and then wiped with acetone before applying the adhesive. In another set of experiments, the joining surface areas were first lightly sanded with a 100 grit aluminum oxide paper before adhesive bonding. Single lap joints with 12.7 mm lap length and 0.762 mm adhesive thickness were then prepared using the second batch of adhesive. The average static failure load for the sanded specimens was 5.17 $\pm$ 0.35 kN, which is the same as that obtained for the unsanded specimens. Failure mode for the sanded specimens were adhesive failure combined with fiber tear. Results on fatigue tests of sanded specimens are shown in Fig. 5 . The maximum fatigue load at 4 x 10^6 cycles is around 30 % of the static failure load. Failure mode in sanded specimens was dominated by fiber tear. For unsanded specimens, adhesive failure was predominant.

EFFECT OF WATER ABSORPTION

In order to determine the effect of water absorption on the joint strength, single lap joints with 12.7 mm lap length and 0.762 mm adhesive thickness were prepared using the second batch of adhesive. Specimens were kept immersed in tap water at room temperature for 214 hr. (8.9 days). During this time, specimens were periodically taken out of the water, wiped and then weighed in a balance with an accuracy of 0.1 mg. To calculate the amount of water absorbed by the adhesive, unbonded SRIM coupons were also kept in the water and weighed in the similar way. After 214 hours of water exposure, specimens were taken out from the water, wiped, weighed and coated with a moisture resistant silicone coating. Both static and fatigue tests were performed with these specimens.

Water absorption (in weight percent) is shown as a function of the square root of time (in hours) in Fig. 6. The amount of water absorption did not reach the equilibrium level in 214 hours. In 214 hours of water immersion, the SRIM-SRIM specimens absorbed 0.59 $\pm$ 0.017% of water. However, when the water absorption in the adherends (which was determined from the unbonded specimens) is taken into account, it is observed that the water absorption in the adhesive is negligible. It is concluded that the water absorption in the bonded specimens is mainly due to the water absorption in the adherends.

Static test results for single lap joint specimens after water immersion show a small drop in the failure load. The average static failure load after 214 hours of water exposure was 4.89 $\pm$ 0.43 kN compared to 5.12 $\pm$ 0.36 kN for the unexposed specimens prepared with the same batch of adhesive. The failure mode for both unexposed and exposed specimens was close to 100% adhesive failure with light fiber tear. The effect of water exposure on the fatigue behavior of SRIM - SRIM single lap joints is shown in Fig. 7. The maximum fatigue load at 4 x 10^6 cycles is approximately 35 % of the static failure load, whereas for the unexposed specimens, it is approximately 45.9%. The failure mode in the water exposed joints was a combination of adhesive failure and fiber tear.

CONCLUSIONS

Following are the principal conclusions drawn from the static and fatigue tests performed on SRIM-SRIM adhesive bonded joints:

1. Increasing the adhesive thickness increases the static failure load as well as the lap shear value of single lap joints. Increasing the lap length increases the static failure load; however, the lap shear value decreases.
2. The static failure is 24.4% higher at a cross head speed of 12.7 mm/min. than at 1.27 mm/min.

3. The lap shear values calculated on the basis of the net bonded area is higher for the recessed bonding with gaps than for the continuous bonding.
4. Lower viscosity adhesive gave 50% higher static failure load than higher viscous adhesive, possibly due to its ability to fill the pores and voids present on the SRIM surface.
5. The fatigue load at 4 x 10^6 cycles is estimated as 45.9 % of the peak static load for dried SRIM specimens with 12.7 mm lap length and 0.762 mm adhesive thickness. The fatigue S-N curve is slightly higher for the greater lap length, but significantly lower for the greater adhesive thickness.
6. Drying the SRIM substrates prior to adhesive bonding improves the load carrying capacity of the single lap SRIM-SRIM joints. Drying also improves the fatigue performance.
7. Sanding the SRIM surface prior to bonding gave almost the same static failure load as that without sanding ; however, sanding produced a lower maximum fatigue load, which is only 30% of the static failure load.
8. Water exposure seems to lower both static and fatigue performance of SRIM-SRIM joints.

ACKNOWLEDGMENTS

Authors are thankful to the Michigan Materials and Processing Institute and the Automotive Composites Consortium for their financial support of this project. Authors are also thankful to Dr. Jessica Schroeder of General Motors Research Center for valuable discussions and encouragement during this project.

REFERENCES

Dillard, J.G. and I.M Spinu. 1994. "Durability in Automotive Adhesive Bonding," 39th Int. SAMPE Symposium, p. 3143.

Holubka, J.W., W. Chun., A.R. Krause and J. Shyu. 1986. "Effects of Corrosive Environments on the Locus and Mechanism of Failure of Adhesive Joints," ACS Symposium Series 322, p. 194, Polymeric Materials for Corrosion Control, R.A. Dickie and F.L. Floyd, Eds., American Chemical Society, Washington, DC.

Krause, A.R., J.W. Holubka and W. Chun. 1986. "The Effect of Adhesive Chemistry on the Fatigue Resistance of Adhesive Bonds in Service Environment," Advanced Composites: The Latest Developments, ASM/ESD 2nd Conference on Advanced Composites.

Mazumdar, S.K. and P.K Mallick. 1997. "Static and Fatigue Behavior of Adhesive Joints in SMC-SMC Composites", submitted to Polymer Composites.

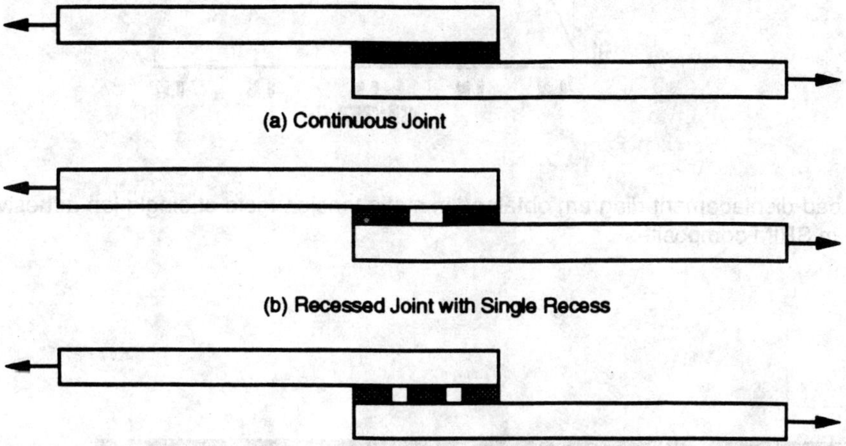

(a) Continuous Joint

(b) Recessed Joint with Single Recess

(c) Recessed Joint with Double Recesses

Figure 1: Continuous and recessed single lap joints between SRIM substrates

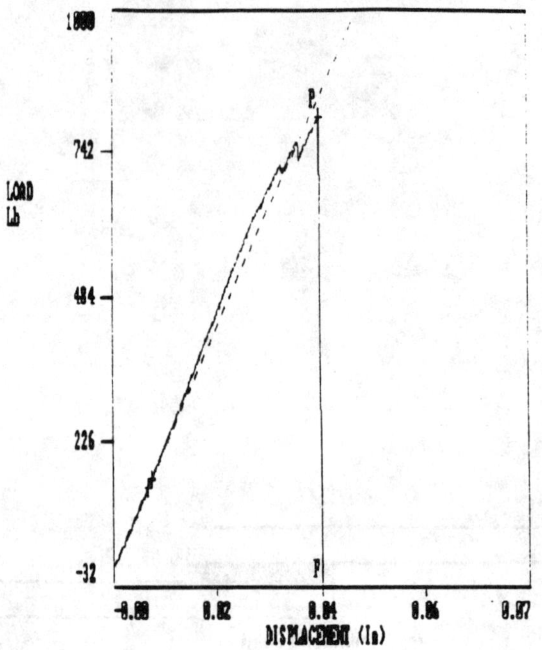

Fig. 2: Load-displacement diagram obtained in static tension tests of single lap adhesive joints in SRIM composites

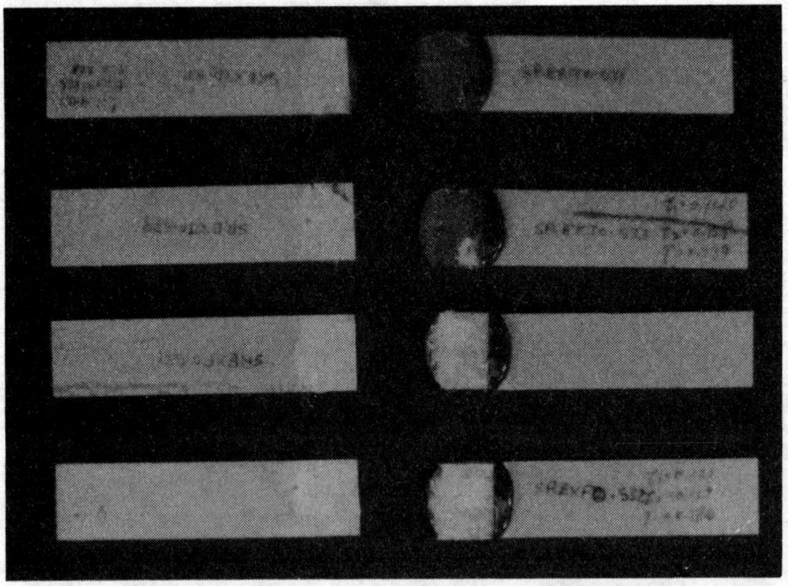

Fig. 3: Failure modes observed in static and fatigue tests of single lap adhesive joints in SRIM composites (Top: adhesive failure, Bottom: fiber tear)

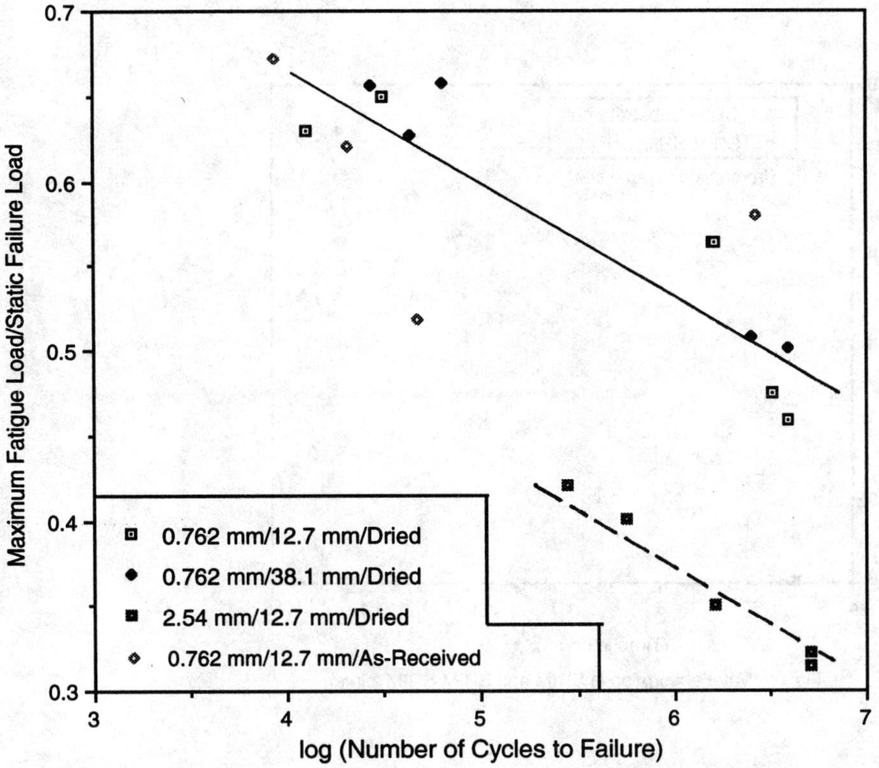

Fig. 4: Tension-tension fatigue data for SRIM-SRIM single lap joints (R = 0.1, Frequency = 10 Hz., RT)
(The first number in the nomenclature is the adhesive thickness and the second number is the lap length)

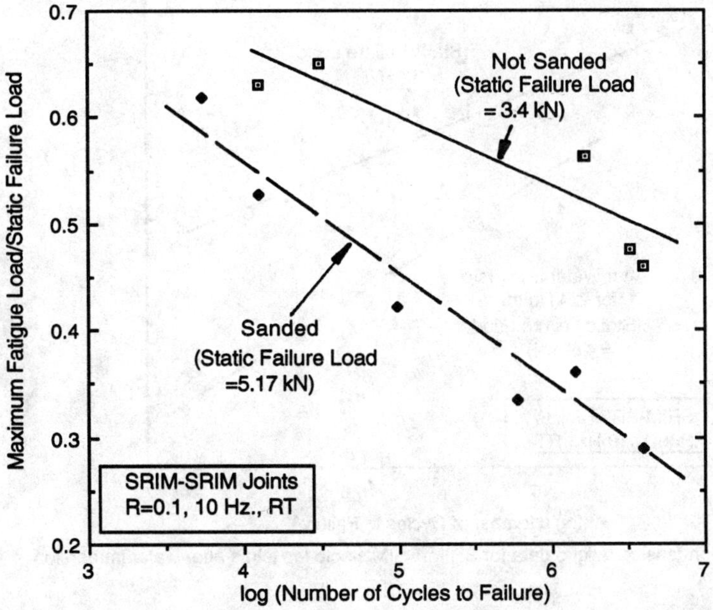

Fig. 5: Comparison of tension-tension fatigue data for sanded and unsanded SRIM-SRIM single lap joints

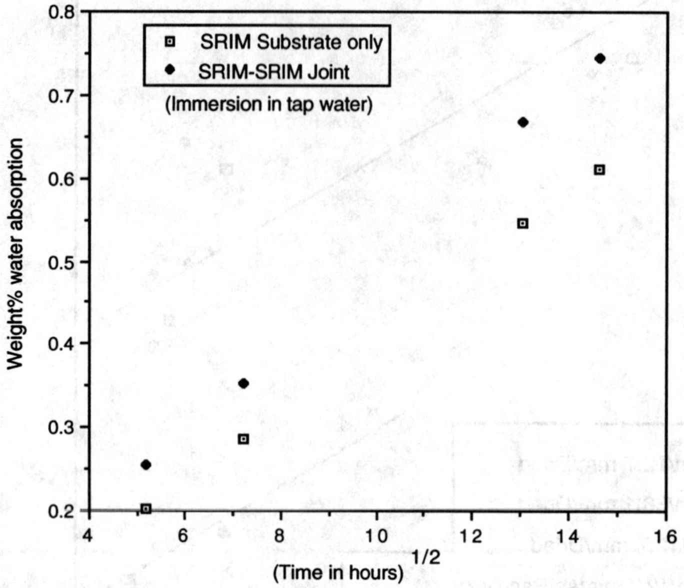

Fig. 6: Water absorption in SRIM and SRIM-SRIM joints

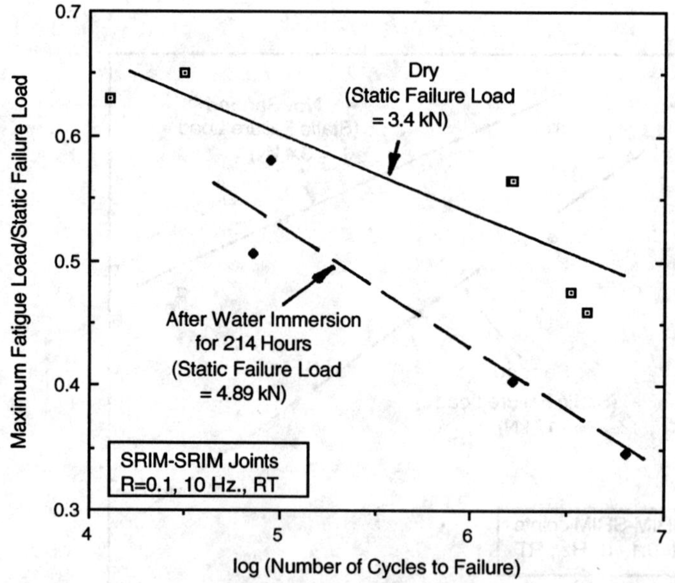

Fig. 7: Tension-tension fatigue data for SRIM-SRIM single lap joints after water immersion

On the Effect of Interference Fits in Composite Pin-Joints

F. L. DI SCALEA, F. CAPPELLO AND G. L. CLOUD

ABSTRACT

The Finite Element Method is used to determine the effect of pin/hole interference-fits on the elastic membranal behavior of an orthotropic plate loaded in tension through a pin. A complete contact analysis is performed to handle the nonlinear contact problem. The study encompasses two levels of interference-fits, two degrees of anisotropy of the plate, and two load levels. Friction effects are not included. Experimental results from Digital Speckle Pattern Interferometry are also presented. It is shown that the stress changes in the plate vary linearly with load when no pin/plate separation occurs. It is also shown that interference-fits can have beneficial effects on the static and fatigue behavior of the plate.

INTRODUCTION

Pin/hole interference-fits are generally used in pin-loaded joints to lower the stress concentration factors and to reduce the magnitude of the local oscillatory stress component in the case of oscillatory loads, thereby increasing the fatigue life of the joint. Although this effect has been studied extensively for isotropic materials, the literature concerning interference-fit pin-loaded holes for composite materials is very limited, and this is the main reason why interference-fits have not been widely used when anisotropy is involved.

Sendeckyj and Richardson (1974) observed beneficial effects of interference-fit pins on the fatigue life of graphite/epoxy laminates. Ramamurthy (1989) determined the behavior of an interference-fit pin in an orthotropic plate subjected to both pull and push-type pin loads, using the *inverse technique* in a FEM analysis. He found nonlinearities in the load transfer mechanism of the joint, caused by the change of the pin/hole contact arc with the applied load. Using Digital Speckle Pattern Interferometry (Jones and Wykes, 1983), Cloud and Lanza di Scalea (1997) observed a reduction in change of stress due to the applied load when interference-fit pins are used in a cross-ply glass/epoxy laminate. Their results suggest that interference-fits increase the fatigue life of the joint.

Francesco Lanza di Scalea and Gary L. Cloud, Materials Science and Mechanics Department, Michigan State University, A304 Engineering Building, East Lansing, Michigan 48824-1226
Francesco Cappello, Dipartimento di Meccanica ed Aeronautica, Universita' degli Studi di Palermo, Viale delle Scienze, 90128 Palermo, Italy

In order to gain a better knowledge of the effect of interference in pin-loaded composites, more study and experimental data are needed. Moreover, the stress distribution near the hole is strongly influenced by several parameters such as material properties, laminate lay-up, and ply orientation; therefore a dedicated study has to be performed for each single laminate. The present study is considered a further contribution to the literature because it provides a detailed understanding of the effect of pin/hole interference on the stress distribution in orthotropic pin-loaded plates. Two levels of interference, two degrees of anisotropy, and two load levels are examined by using a FEM frictionless analysis with an incremental-iterative procedure to account for the nonlinear contact problem. No *a priori* assumption on the radial stress distribution on the surface of the hole is made. Results from Digital Speckle Pattern Interferometry are also presented for validation of the numerical predictions.

CASES STUDIED

Figure 1 shows the geometry, loading condition, boundary conditions, nomenclature, and definition of the problem. The principal dimensions of the plate are presented in Table I. Table II shows the effective mechanical properties for the two degrees of anisotropy examined. The plate was considered orthotropic. The steel pin had Young's modulus $E=208000$ N/mm^2 and Poisson's ratio $\nu=0.28$. The material was considered to behave elastically throughout the study.

The study encompassed two levels of load, and two levels of pin/hole interference-fit, in addition to the pin/hole perfect-fit case, namely:

1. pin load P=60N (far-field $\sigma_y=0.33$N/mm^2) and P=120N (far-field $\sigma_y=0.66$N/mm^2);
2. diametral interference I=0.0002 and I=0.0004; where $I=(d_p-d_h)/d_h$, being d_p and d_h the diameters of pin and hole respectively (see Figure 1);
3. perfect-fit (I=0).

TABLE I- PRINCIPAL DIMENSIONS OF THE PLATE

length, L	203.2 mm (8.0 inch)
thickness, t	3.556 mm (0.14 inch)
hole diameter, d_h	6.35 mm (0.25 inch)
edge dimension, e/d_h	4.0
width, w/d_h	8.0

TABLE II- EFFECTIVE MECHANICAL PROPERTIES OF THE ORTHOTROPIC PLATES STUDIED

properties	material A	material B
E_x [N/mm^2]	21250	24200
E_y [N/mm^2]	21980	102000
G_{xy} [N/mm^2]	9720	111000
ν_{xy}	0.11	0.3

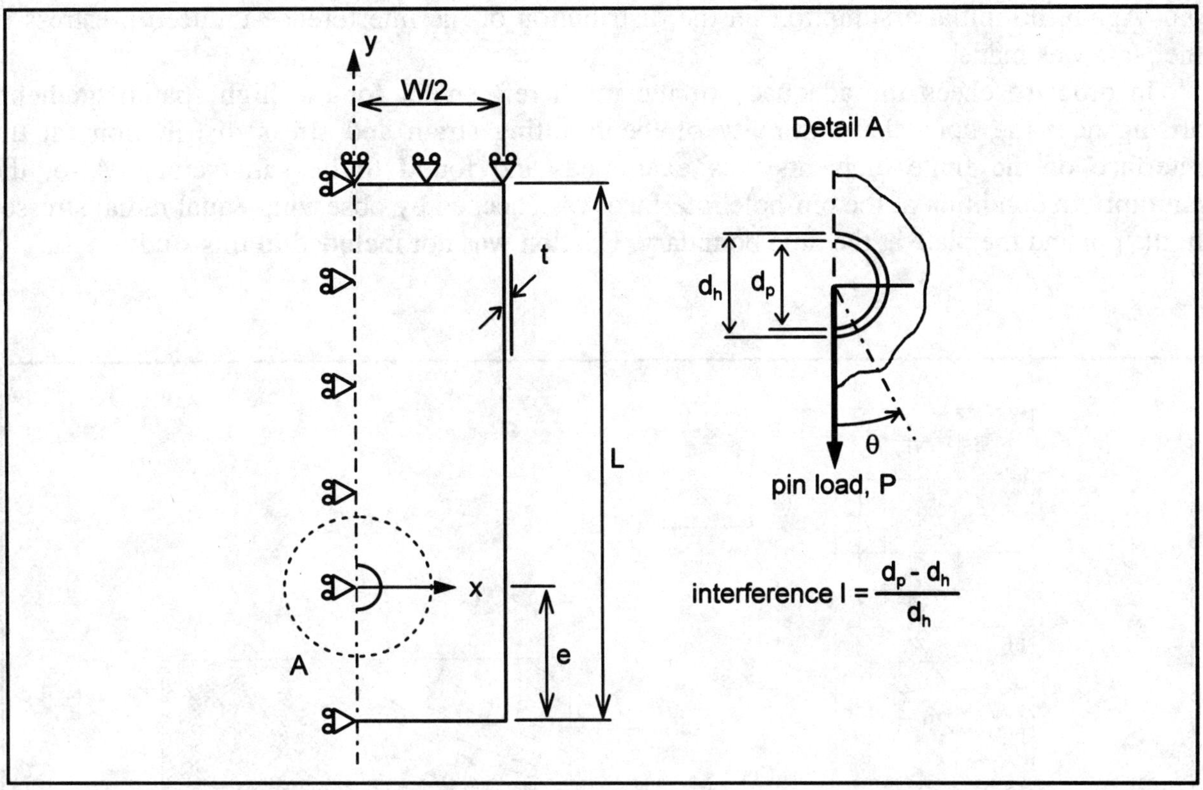

Figure 1. Geometry, loading condition, nomenclature, and definition of the problem.

NUMERICAL ANALYSIS

FEM was used as main tool for the investigation. Models of the rectangular pin-loaded plate were generated and analyzed using ANSYS ver.5.3. Figure 2 shows the mesh employed for the plate and a detail of the mesh in the pin/hole contact area. The plate was loaded along its axis of symmetry (see Figure 1), by applying a tensile load distribution on an inner area of the pin covering about 80% of the whole pin cross-section. This solution was chosen in alternative to earlier authors' choices (e.g. Grant, Smart and Stanley, 1994) of applying the load only to the central node of the pin, which produces a local stress concentration that can lead to erroneous results. Only half of the specimen needed to be modeled, because of symmetry.

A complete nonlinear contact analysis was conducted, without any *a priori* assumption on the initial distribution of the contact radial stresses on the surface of the hole. The nonlinearity in the load-stress relationship is caused by the change of the contact arc with the applied load. An incremental-iterative analysis was necessary to handle the nonlinearity. The load was divided into many substeps and the contact condition between the pin and the hole was updated at each substep by using an iterative procedure to find the substep equilibrium state. The final length of the contact arc was determined as a result of the analysis.

The mesh consisted of four-noded plane-stress finite elements for the plate and the pin, with 2-D point-to-surface contact elements for the pin/hole interface. By using the contact finite elements, we could simulate the pin/hole interference by simply modeling an oversized pin. Again, no initial assumption on the distribution of the interference-induced prestress in the plate was made.

In order to check the adequacy of the mesh refinement for the high spatial gradients arising near the hole, the continuity of the resulting strain and stress distributions at the interface of the finite elements was examined, and found to be satisfactory. Also, the equilibrium condition at the pin-hole interface was checked by observing equal radial stresses in the pin and the plate at the hole boundary. Friction was not included in this study.

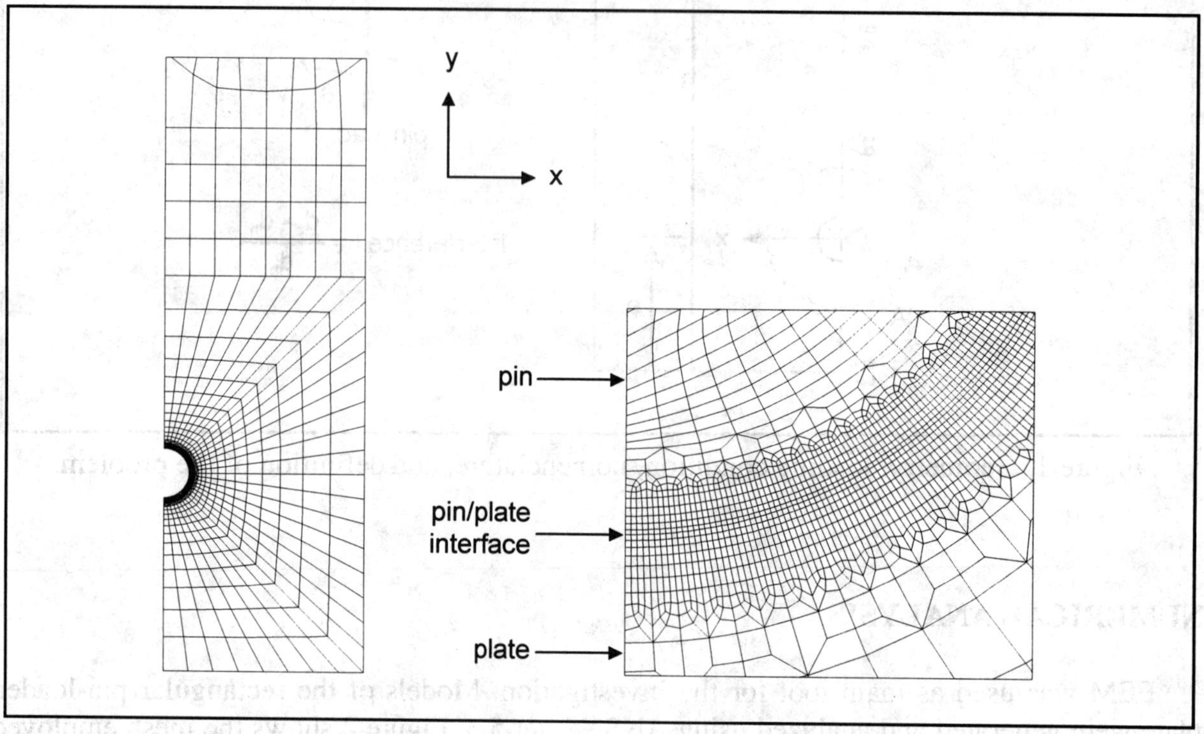

Figure 2. Mesh employed for the plate and a detail of mesh in the pin/hole contact area.

EXPERIMENTS

A *cross-ply* fiber glass-reinforced (R1500/1581) pin-loaded plate was tested using dual-beam illumination Digital Speckle Pattern Interferometry (DSPI) (Jones and Wykes, 1983; Cloud, 1995) with a phase-stepping technique (Creath 1985). DSPI is an accurate, high resolution and high sensitivity technique which allows noncontact whole-field deformation measurements at an object's surface.

The plate tested had the same mechanical properties as material A of the FEM analysis (see Table II). The case of pin/hole perfect-fit and the two cases of interference-fit mentioned above were investigated, for comparison with the numerical results. Geometry and loading condition of the specimen were also the same as the FEM analysis. The in-plane displacement components d_x and d_y at the surface of the specimen were measured by DSPI.

The displacement plots were fitted with least-square polynomials, which were eventually differentiated along the x and y directions in order to calculate the in-plane strain components $\varepsilon_x=\partial d_x/\partial x$, $\varepsilon_y=\partial d_y/\partial y$ and $\gamma_{xy}=\partial d_x/\partial y+\partial d_y/\partial x$. The in-plane stress components σ_x, σ_y and τ_{xy} were determined from the well-known stress-strain relations for an orthotropic laminate (Lekhnitskii, Tsai and Cheron, 1968), the laminate's mechanical properties being known. The steel pins were considered infinitely rigid.

For details on the technique and the experimental procedure, the reader is referred to two previous works (Cloud and Lanza di Scalea, 1996 and 1997). A difference with these previous works is that in the present study the steel pins were lubricated with colloidal graphite grease, for comparison of the results with the frictionless numerical analysis.

RESULTS AND DISCUSSION

Distributions of the in-plane stress components are obtained along critical locations of the plate, namely the edge of the hole, the net-section, and the bearing direction. Along the edge of the hole, the discussion is limited to $0° \leq \theta \leq 180°$, because of symmetry. Since in the experimental tests the interference-fit specimen had the pin already installed in its initial "undeformed" configuration, the experimental results presented for the interference-fit cases are representative of the *change* in stress $\Delta\sigma$ caused by the applied load.

EFFECT OF LOAD

The FEM result for the polar components of stress σ_r and σ_θ along the edge of the hole for the two levels of interference-fit are presented in Figures 3 and 4 for the two degrees of anisotropy examined. Results encompass two load levels, in addition to the zero load case. For the zero load case (interference-induced prestress), the stress distribution is almost constant around the hole for the small anisotropy case (material A), whereas it is highly variable with the angular position for the high anisotropy case (material B). An effect of the application of the load to the interference-prestressed plate is an increase of σ_r in the lower quarter of the hole, and a decrease of it in the upper quarter. The stiffer material B exhibits higher stresses than material A everywhere in the hole. A $\sigma_r=0$ indicates pin/plate separation occurring at the highest load for material A at small interference. This separation causes change in contact arc, and thus nonlinearity in the load transfer mechanism. Everywhere else around the hole, the changes of stresses vary linearly with load. The position of both radial and circumferential stresses is substantially independent of the load level. Similar observations were made for isotropic materials (Grant, Smart and Stanley, 1994). Thus, only the results at maximum load are presented when discussing the effect of interference-fits.

It is expected that this linear behavior generally occurs for high ratios interference/load, which do not exhibit pin/plate separation. Additionally, nonlinearity can be caused by friction effects (not considered in the present analysis), because of change in slip/adherence regions with load.

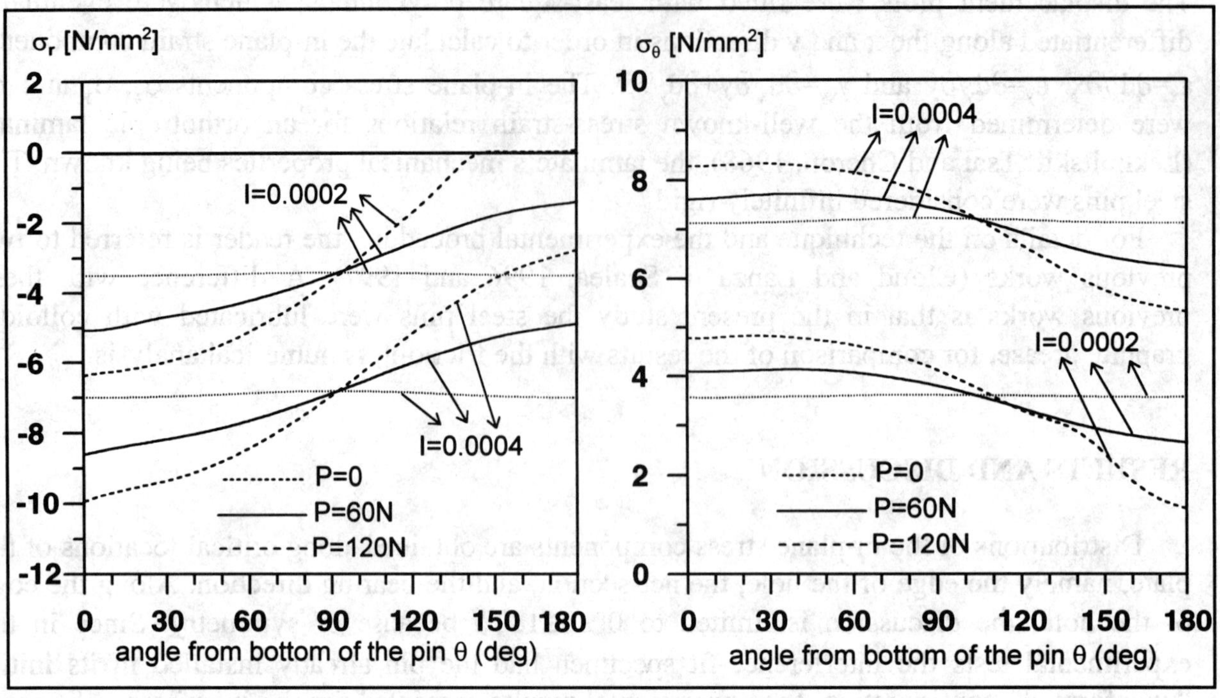

Figure 3. Material A: stresses σ_r and σ_θ around the hole for different levels of interference and load

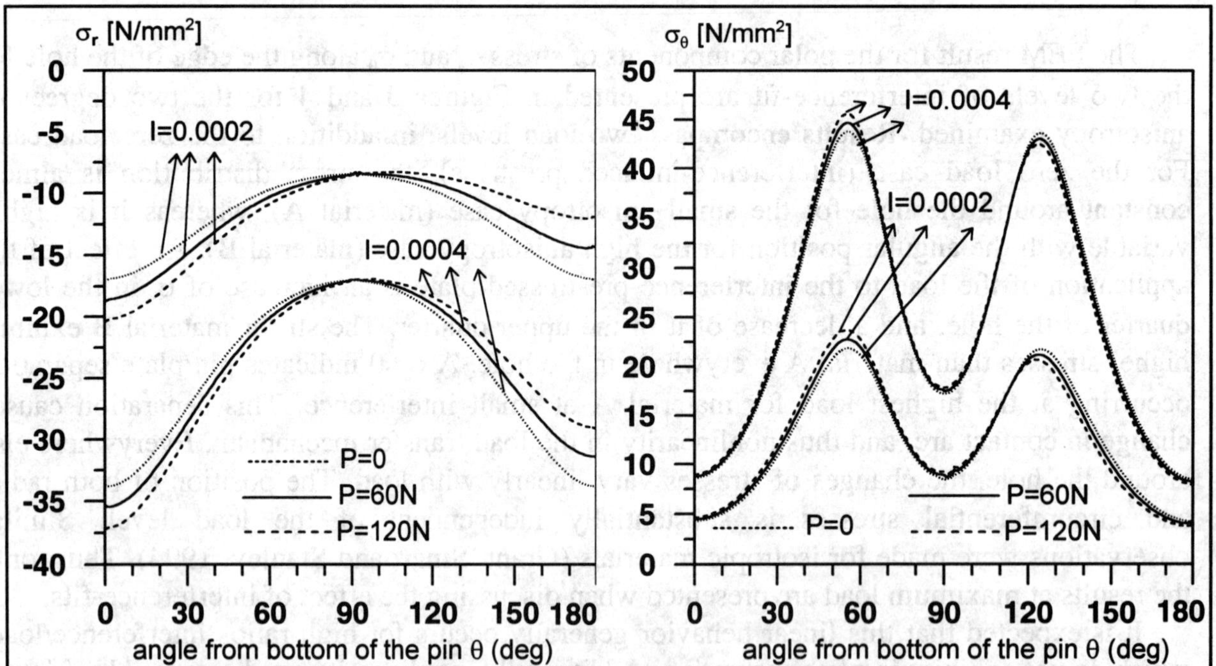

Figure 4. Material B: stresses σ_r and σ_θ around the hole for different levels of interference and load.

EFFECT OF INTERFERENCE-FITS

Figure 5 shows the FEM results for the circumferential stress σ_θ along the edge of the hole for the different degrees of interference-fit and anisotropy. The stresses are normalized by the average bearing stress $\sigma_b = P/d_h t$ (P, d_h and t as in Figure 1). The low interference in material A exhibits a peak circumferential stress about 20% smaller than the perfect-fit case. This decrease in peak stress is not observed in the stiffer material (material B), where the interference stresses are always higher than the perfect-fit ones.

It is interesting to observe the *change* $\Delta\sigma$ in stress values, relative to the unloaded condition, produced by the applied load in the interference-fit cases. In fact, $\Delta\sigma$ can be looked at as the alternating component of the stress for a case of a pulsating load. By subtracting the plots relative to the final (loaded) and initial (unloaded) conditions for the interference-fit cases in Figure 5, it is easy to see that $\Delta\sigma_\theta$ is smaller than the corresponding σ_θ for the perfect-fit case along most of the edge of the hole, for both materials A and B.

Table III shows the magnitude of peak circumferential stress $\sigma_{\theta max}$ and stress *change* $\Delta\sigma_{\theta max}$ relative to unloaded pin, for the cases studied. It is observed that $\sigma_{\theta max}$ is strongly dependent on material properties and degree of anisotropy. Alternatively, $\Delta\sigma_{\theta max}$ is only slightly influenced by material properties and interference level.

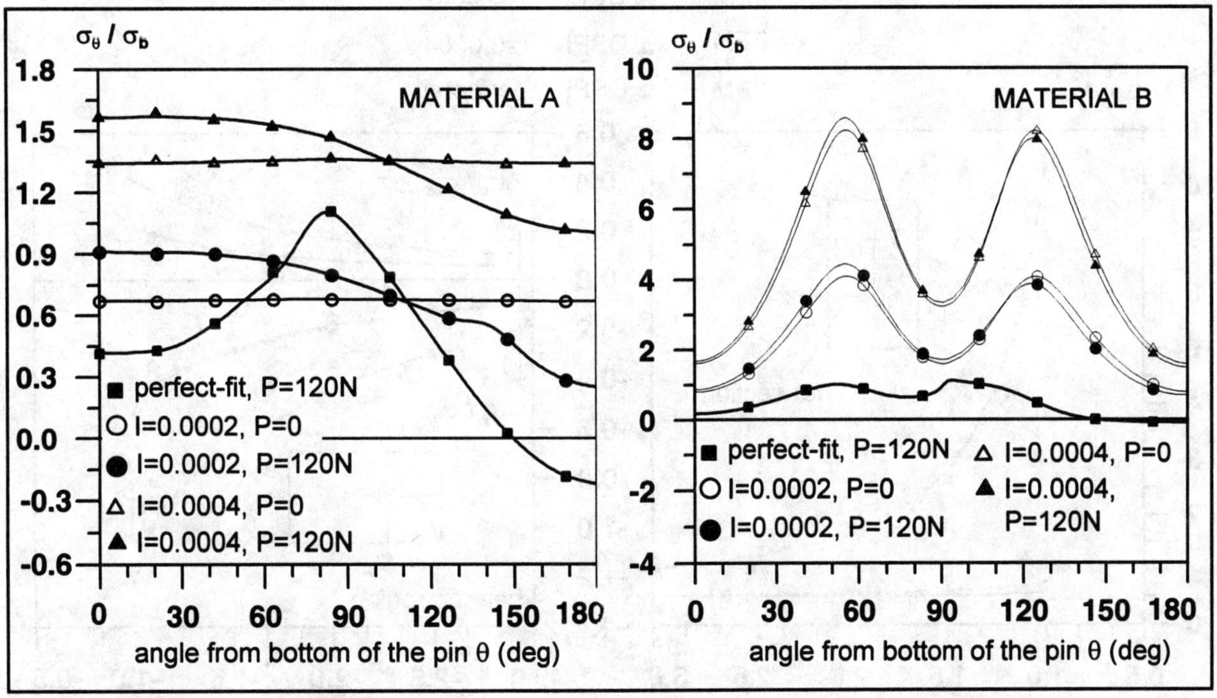

Figure 5. Materials A and B: normalized stress σ_θ around the hole for the perfect-fit and the two interference-fit cases (symbols used for line identification only).

TABLE III- EFFECT OF INTERFERENCE: MAGNITUDE OF NORMALIZED PEAK CIRCUMFERENTIAL STRESSES $\sigma_{\theta max}$ AND STRESS CHANGES $\Delta\sigma_{\theta max}$ AROUND THE HOLE

	$\sigma_{\theta max}/\sigma_b$		$\Delta\sigma_{\theta max}/\sigma_b$	
	material A	material B	material A	material B
perfect-fit (I=0)	1.2	1.1	1.2	1.1
interference I=0.0002	0.9	4.4	0.4	0.4
interference I=0.0004	1.6	8.5	0.3	0.4

In Figure 6 numerical as well as DSPI experimental results are presented, relative to the distribution of stress change $\Delta\sigma$ along two critical directions of the material A plate, namely the net-section (y=0) and the bearing direction (x=0). The perfect-fit case and the two interference cases are considered. The reference stress in the first graph of Figure 6 is the far-field tensile stress σ_N=P/wt (P, w and t as in Figure 1). The general trend of the plots of Figure 6 shows good agreement between FEM and DSPI. The mismatch is probably due to possible presence of residual stresses and/or mechanical damage induced in the specimen by the drilling of the hole. A substantial decrease in $\Delta\sigma$ is observed in both the numerical and the experimental plots when an interference-fit is used.

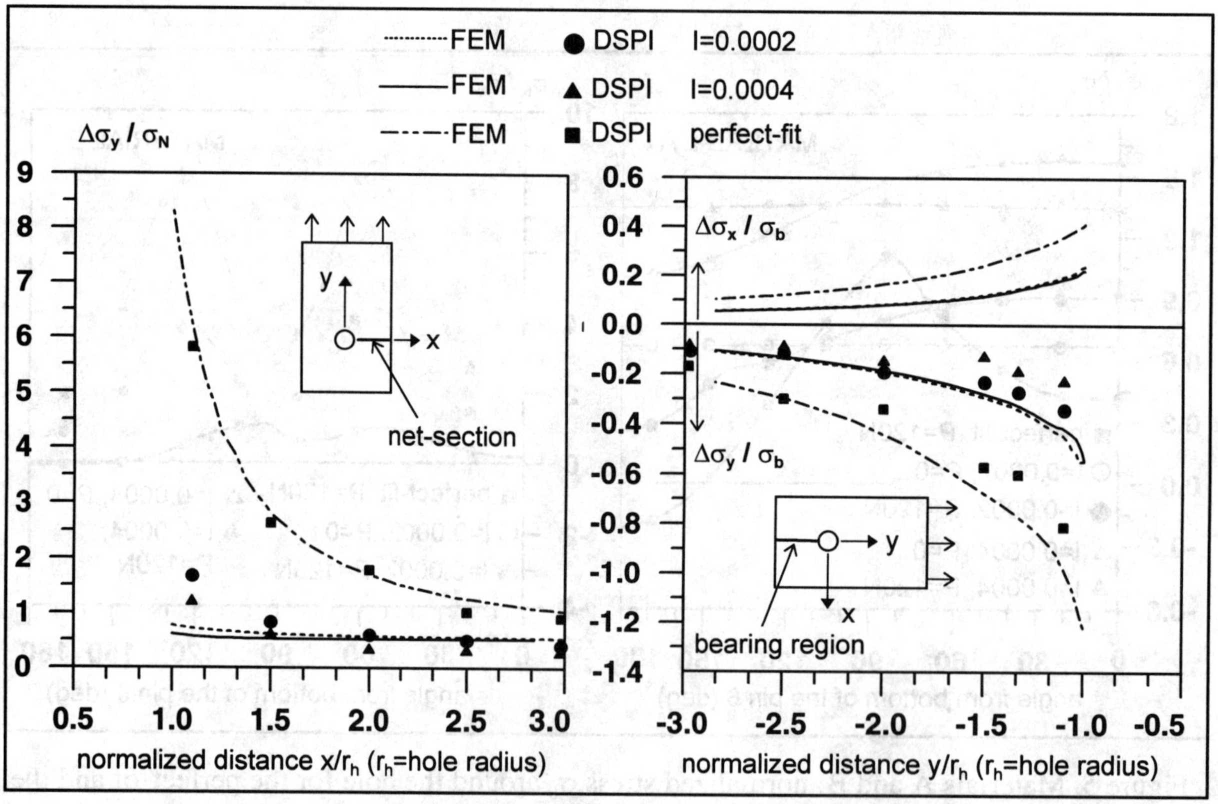

Figure 6. Material A: normalized stress changes $\Delta\sigma$ along net-section and bearing direction of the plate (results from FEM and DSPI - DSPI results for $\Delta\sigma_x$ not available).

CONCLUSIONS

- FEM nonlinear contact analysis is a well suited technique for the study of interference-fits in orthotropic pin-loaded plates, for which theoretical solutions are very difficult to obtain.
- Digital Speckle Pattern Interferometry provides a valid comparison with the numerical analysis.
- For high ratios interference-fit/load (when no pin/plate separation occurs), the stress changes in the plate vary linearly with load.
- For a static load, pin/hole interference-fits have a beneficial effect on plates with relatively low stiffness. For these materials, it exists an optimum level of interference which produces maximum decrease in peak circumferential stress in the plate.
- For a pulsating load, the results suggest that interference-fits can increase the fatigue life of the joint, in that they lower the alternating component of the stress at critical points in the plate.

ACKNOWLEDGEMENTS

Portions of this research were supported by the Research Excellence Fund of Michigan and the U. S. Army Tank Automotive Command, Warren, Michigan.

REFERENCES

Cloud, G. L. 1995. "Optical methods of engineering analysis," Cambridge University Press.

Cloud, G. L. and F. Lanza di Scalea. 1996. "Determination of stress around a single pin-loaded hole in fiber-reinforced plastic using phase-stepped digital speckle pattern interferometry," Proceedings of the XXV AIAS National Conference and International Conference on Material Engineering, Gallipoli, Italy: 319-326.

Cloud, G. L. and F. Lanza di Scalea. 1997. "Stress near interference-fit composites fasteners by speckle interferometry," Abstract Proceedings of the SEM Spring Conference on Experimental Mechanics, Bellevue, Washington: 150-151.

Creath, K. 1985. "Phase-shifting speckle interferometry," *Applied Optics*, 24(18): 3053-3058.

Grant, R. J., J. Smart, and P. Stanley. 1994. "A parametric study of the elastic stress distribution in pin-loaded lugs," *J. of Strain Analysis*, 29(4): 299-307.

Jones, R. and C. Wykes. 1983. "Holographic and Speckle Interferometry," 1st edition, Cambridge University Press.

Lekhnitskii, S. G., S. W. Tsai, and T. Cheron. 1968. "Anisotropic Plates," Gordon and Breach Science Publishers.

Ramamurthy, T. S. 1989. "Recent studies on the behavior of interference fit pins in composite plates," *Composite Structures*, 13(2): 81-99.

Sendeckyj, G. P. and M. D. Richardson. 1974. "Fatigue behavior of a graphite-epoxy laminate loaded through an interference-fit pin," Proceedings of 2nd Air Force Conference on Fibrous Composites in Flight Vehicle Design, Tech. Rep. AFFDL-TR-74-103, Air Force Flight Dynamics Lab, WPAFB, Ohio.

A Simplified Method for the Geometrically Nonlinear Analysis of the Single Lap Joint

F. E. PENADO

ABSTRACT

A simplified method for the geometrically nonlinear analysis of the single lap joint is presented. The method consists of finding first the force/moment boundary conditions at the joint overlap ends. This is achieved by solving the differential equations that relate the resulting moment to the applied load and the unknown transverse deflection. The solution for the deflection is eventually reduced to the solution of a system of six linear equations in six unknown coefficients. Once the force/moment boundary conditions for a given load are known, a linear finite element analysis is used to find the joint stresses using only the overlap region of the joint. Since in many practical joints adhesive fillets exist at the ends of the overlap, the existence of these fillets is taken into account in the formulation. Also, to allow for various support conditions of the grips of the testing machine, solutions are presented for both simply supported and clamped-clamped end supports. Additional boundary conditions on the linear finite element model using multipoint constraints are also given that allow consideration of only one half of the overlap length in the analysis. This is possible due to the inherent conditions of antisymmetry that exist between the top and bottom of the joint overlap. The accuracy of the simplified method was verified by comparison of results against a full size geometrically nonlinear finite element model, and it was found that the results were very close, within 3%. Overall, the method results in considerable savings in both solution time and modeling effort, while maintaining accuracy.

INTRODUCTION

Because of its simplicity and ease of manufacture, the single lap joint is widely used in the characterization of adhesively bonded joints with composite adherends. In order to determine joint strength parameters at the adhesive/adherend interface, some form of analysis of the joint must be performed. This is particularly true if a fracture mechanics approach such as the generalized intensity factor method (Gradin, 1982; Groth, 1988), which requires a detailed analysis for strength parameter determination, is to be used subsequently to design a

F. Ernesto Penado, Associate Professor, Mechanical Engineering Department, Northern Arizona University, P. O. Box 15600, Flagstaff, AZ 86011-5600

composite structure. Because of the complexities in the analysis, a finite element model is normally employed to determine the necessary strength parameters from the measured load at failure of the joint. A difficulty that is encountered when performing the analysis is due to the geometric nonlinearities that arise due to joint rotation and transverse deflection, which lead to cumbersome and time-consuming finite element models. The effect of these nonlinearities can be significant, and it is shown that, for typical failure loads, the error resulting from using a linear model instead of a nonlinear model can be near 50%.

METHOD OF ANALYSIS

Consider the single lap joint shown in Figure 1. Due to the eccentricity in the load path, a moment is created in the overlap region which is dependent on both the magnitude of the applied load per unit length, P, and the joint rotation. Since the amount of rotation also depends on the magnitude of P, the relationship between the load and rotation is nonlinear. Furthermore, the relationship between P and the resulting stresses is also nonlinear.

Now consider the static equilibrium of the single lap joint shown in Figures 2 and 3. In Figure 2 the joint has been divided into three regions with origins denoted by x_1, x_2, and x_3. The moment M_0 depends on the support conditions of the grips of the testing machine and will be found later. From overall equilibrium, the unknown end shear V_0 can be expressed in terms of the end moment M_0 as follows

$$\sum M_A = 0 = -2M_0 + 2V_0(\ell + c) - P(t + t_c) \;\Rightarrow\; V_0 = P\alpha + \frac{M_0}{(\ell + c)} \tag{1}$$

$$\text{where } \alpha = \frac{1}{2}\frac{(t + t_c)}{(\ell + c)}$$

Also, from equilibrium of the center portion of the joint in Figure 3a, Q and M are related by the expression

$$Q = \frac{1}{2(c + e)}\left[(t + t_c)P - 2M\right] \tag{2}$$

Now, for each region, let M_1, M_2 and M_3 represent the moments per unit length, and w_1, w_2 and w_3 be the deflections (positive downward). Then, after using (1) we get

$$M_1 = P\left[\left(\alpha + \frac{M_0}{P(\ell + c)}\right)x_1 - w_1\right] - M_0 \qquad 0 \le x_1 \le (\ell - e) \tag{3}$$

$$M_2 = P\left[\left(\alpha + \frac{M_0}{P(\ell + c)}\right)(\ell - e + x_2) - w_2 - \bar{y}\right] - M_0 \qquad 0 \le x_2 \le e \tag{4}$$

$$M_3 = P\left[\left(\alpha + \frac{M_0}{P(\ell + c)}\right)(\ell + x_3) - w_3 - \frac{1}{2}(t + t_c)\right] - M_0 \qquad 0 \le x_3 \le c \tag{5}$$

where $\bar{y}$ is the distance from the centroid of the cross-section of the adherend to the centroid of the cross-section of the adherend-fillet combination (see Figure 2b). $\bar{y}$ can be found by using the equivalent width technique (Young, 1989, p. 117). The result is

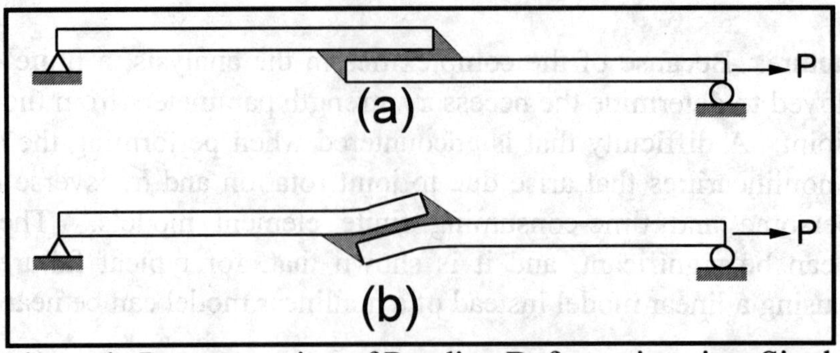

Figure 1. Schematic Representation of Bending Deformations in a Single Lap Joint. (a) Undeformed Joint. (b) Deformed Joint.

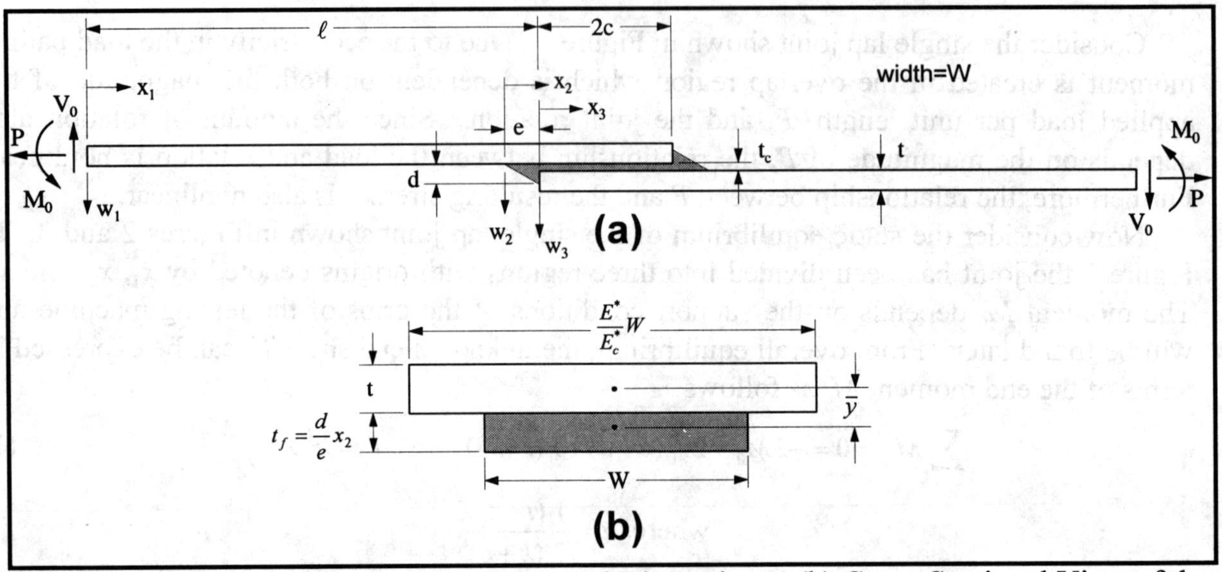

Figure 2. Single Lap Joint. (a) Geometry and Dimensions. (b) Cross-Sectional View of the Joint in the Fillet Region (region between x_2 and x_3).

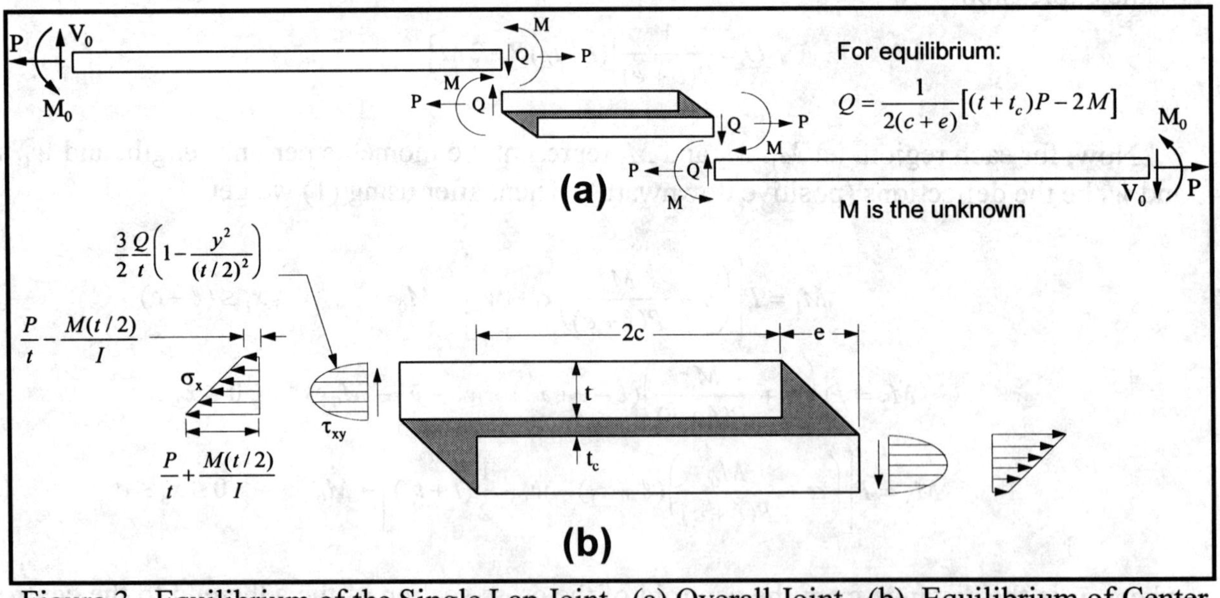

Figure 3. Equilibrium of the Single Lap Joint. (a) Overall Joint. (b) Equilibrium of Center Portion with Edge Loads Substituted by Statically Equivalent Edge Stresses.

$$\bar{y} = \frac{\left(\frac{d}{e}\right)\left(\frac{t}{2}\right)x_2 + \frac{1}{2}\left(\frac{d}{e}\right)^2 x_2^2}{\left(\frac{E^*}{E_c^*}\right)t + \left(\frac{d}{e}\right)x_2} \qquad (6)$$

where, assuming that both the adherends and adhesive behave as wide beams,

$E^* = \frac{12D_{11}}{t^3}$ =effective flexural modulus of adherend laminate, which must be symmetric (Swanson, 1997, p. 195)

$E_c^* = \frac{E_c}{1-\nu_c^2}$

D_{11}=first entry of the laminate bending stiffness matrix (Gibson, 1994, p. 206)

E_c, ν_c=Young's modulus and Poisson's ratio of adhesive

It can be seen in (3)-(5) that the moment-load relationship is nonlinear, since the moments M_1, M_2 and M_3 depend not only on the magnitude of the applied load P, but also on the deflections w_1, w_2 and w_3, which also depend on the magnitude of the load. Additional relations between the moment and deflection can be can be obtained by writing the moment-curvature relations for each region:

$$(EI)_1 \frac{d^2 w_1}{dx_1^2} = -M_1\,; \qquad 0 \le x_1 \le (\ell - e) \qquad (7)$$

$$(EI)_2 \frac{d^2 w_2}{dx_2^2} = -M_2\,; \qquad 0 \le x_2 \le e \qquad (8)$$

$$(EI)_3 \frac{d^2 w_3}{dx_3^2} = -M_3\,; \qquad 0 \le x_2 \le c \qquad (9)$$

where the moments M_1, M_2 and M_3 are given by equations (3)-(5), and

$$(EI)_1 = \frac{E^* W t^3}{12} \qquad (10)$$

$$(EI)_3 = \frac{E^* W}{12}\left[(2t + t_c)^3 - (1 - E_c^* / E^*) t_c^3\right] \qquad (11)$$

and $(EI)_2$ was obtained as the equivalent EI for a composite strip (Young, 1989, p. 118) with $t_f = (d/e)x_2$ (see Figure 2b) as follows:

$$(EI)_2 = \frac{Wt^2 E^*\left(\frac{d}{e}\right)}{12\left(\frac{d}{et}x_2 + \frac{E^*}{E_c^*}\right)}\left[4x_2 + 6\left(\frac{d}{et}\right)x_2^2 + 4\left(\frac{d}{et}\right)^2 x_2^3 + \frac{E_c^*}{E^*}\left(\frac{d}{et}\right)^3 x_2^4 + \frac{E^*}{E_c^*}\left(\frac{et}{d}\right)\right] \qquad (12)$$

The solutions to equations (7) and (9) are of the form (Goland and Reissner, 1944)

$$w_1 = A_1 \cosh(u_1 x_1) + B_1 \sinh(u_1 x_1) + \alpha\, x_1 + M_0\left[\frac{x_1 - (\ell + c)}{P(\ell + c)}\right]; \qquad 0 \le x_1 \le (\ell - e) \qquad (13)$$

$$w_3 = A_3 \cosh(u_3 x_3) + B_3 \sinh(u_3 x_3) + \alpha\,(\ell + x_3) - \frac{1}{2}(t + t_c) + M_0\left[\frac{x_3 - c}{P(\ell + c)}\right]; \qquad 0 \le x_3 \le c \quad (14)$$

where $u_i = \sqrt{\dfrac{P}{(EI)_i}}; \qquad i = 1,3$

The solution to equation (8) is not as simple due to the variable coefficients. In view of the power form of $(EI)_2$, it seems reasonable to seek the solution to (8) in a convergent power series of the form

$$w_2 = \sum_{n=0}^{\infty} a_n x_2^n + M_0\left[\frac{x_2 - (c + e)}{P(\ell + c)}\right] \qquad (15)$$

where the coefficients a_n are to be determined from the governing differential equation (8) and the boundary conditions.

Substituting (15) into (8) and equating like powers of x_2 yields, after some algebra, the following recurrent relations for the a_n's:

for $n \le 4$

$$-k_9 a_0 + 2k_6 a_2 = -K_4 \qquad (16)$$

$$-k_8 a_0 - k_9 a_1 + 2k_5 a_2 + 6k_6 a_3 = -K_3 \qquad (17)$$

$$-k_7 a_0 - k_8 a_1 + (2k_4 - k_9)a_2 + 6k_5 a_3 + 12k_6 a_4 = -K_2 \qquad (18)$$

$$-k_7 a_1 + (2k_3 - k_8)a_2 + (6k_4 - k_9)a_3 + 12k_5 a_4 + 20k_6 a_5 = -K_1 \qquad (19)$$

$$(2k_2 - k_7)a_2 + (6k_3 - k_8)a_3 + (12k_4 - k_9)a_4 + 20k_5 a_5 + 30k_6 a_6 = 0 \qquad (20)$$

and for $n \ge 5$

$$\begin{aligned} &k_1(n-3)(n-4)a_{n-3} + [k_2(n-2)(n-3) - k_7]a_{n-2} + [k_3(n-1)(n-2) - k_8]a_{n-1} \\ &\quad + [k_4 n(n-1) - k_9]a_n + k_5(n+1)n a_{n+1} + k_6(n+2)(n+1)a_{n+2} = 0 \end{aligned} \qquad (21)$$

where

$$k_1 = \left(\frac{d}{e}\right)c_4; \quad k_2 = \left(\frac{E^*}{E_c^*}\right)tc_4 + \left(\frac{d}{e}\right)c_3; \quad k_3 = \left(\frac{E^*}{E_c^*}\right)tc_3 + \left(\frac{d}{e}\right)c_2; \quad k_4 = \left(\frac{E^*}{E_c^*}\right)tc_2 + \left(\frac{d}{e}\right)c_1; \quad k_5 = \left(\frac{E^*}{E_c^*}\right)tc_1 + \left(\frac{d}{e}\right)c_0; \quad k_6 = \left(\frac{E^*}{E_c^*}\right)tc_0; \quad k_7 = 12\left(\frac{P}{E^*}\right)\left(\frac{d}{e}\right)^2\frac{1}{t}$$

$$k_8 = 24\left(\frac{P}{E^*}\right)\left(\frac{d}{e}\right)\left(\frac{E^*}{E_c^*}\right); \quad k_9 = 12\left(\frac{P}{E^*}\right)\left(\frac{E^*}{E_c^*}\right)^2 t; \quad k_{10} = -6\left(\frac{P}{E^*}\right)\left(\frac{d}{e}\right)^3\frac{1}{t}; \quad k_{11} = -6\left(\frac{P}{E^*}\right)\left(\frac{d}{e}\right)^2\left(1 + \frac{E^*}{E_c^*}\right); \quad k_{12} = -6\left(\frac{P}{E^*}\right)\left(\frac{d}{e}\right)\left(\frac{E^*}{E_c^*}\right)t$$

$$c_0 = Wt^3\left(\frac{E^*}{E_c^*}\right); \quad c_1 = 4\beta; \quad c_2 = 6\left(\frac{\beta}{t}\right)\left(\frac{d}{e}\right); \quad c_3 = 4\left(\frac{\beta}{t^2}\right)\left(\frac{d}{e}\right)^2; \quad c_4 = W\left(\frac{d}{e}\right)^4\left(\frac{E_c^*}{E^*}\right)\frac{1}{t}; \quad \beta = Wt^2\left(\frac{d}{e}\right)$$

$$K_1 = \alpha k_7 + k_{10}; \quad K_2 = \alpha(\ell - e)k_7 + \alpha k_8 + k_{11}; \quad K_3 = \alpha(\ell - e)k_8 + \alpha k_9 + k_{12}; \quad K_4 = \alpha(\ell - e)k_9; \quad \alpha = \frac{1}{2}\frac{(t + t_c)}{(\ell + c)}$$

It can be seen in equations (16)-(21) that there are two more unknowns than equations and the system is overdetermined. Therefore, the solution to the nonhomogeneous system is of the form (e.g., O'Neil, 1987, p. 493)

$$\begin{Bmatrix} a_0 \\ a_1 \\ a_2 \\ \vdots \\ a_N \end{Bmatrix} = \begin{Bmatrix} a_0^{(p)} \\ a_1^{(p)} \\ a_2^{(p)} \\ \vdots \\ a_N^{(p)} \end{Bmatrix} + A_2 \begin{Bmatrix} a_0^{(1)} \\ a_1^{(1)} \\ a_2^{(1)} \\ \vdots \\ a_N^{(1)} \end{Bmatrix} + B_2 \begin{Bmatrix} a_0^{(2)} \\ a_1^{(2)} \\ a_2^{(2)} \\ \vdots \\ a_N^{(2)} \end{Bmatrix} \tag{22}$$

where the $a_i^{(p)}$, i=0,...N correspond to the particular solution of the system, $a_i^{(1)}$ and $a_i^{(2)}$ are homogeneous solutions, A_2 and B_2 are arbitrary constants, and N is the number of terms in the truncated series. The $a_i^{(p)}$, $a_i^{(1)}$ and $a_i^{(2)}$ coefficients are found by using row reduction. Substituting (22) into (15) yields

$$w_2 = \sum_{n=0}^{N} a_n^{(p)} x_2^n + A_2 \sum_{n=0}^{N} a_n^{(1)} x_2^n + B_2 \sum_{n=0}^{N} a_n^{(2)} x_2^n + M_0 \left[\frac{x_2 - (c+e)}{P(\ell + c)} \right] \tag{23}$$

where the convergent series has been truncated to N terms. It was found that N=10 produced adequate convergence (5 or 6 significant figures). A_2 and B_2, along with A_1, B_1, A_3 and B_3 in equations (13) and (14) can be found from the boundary conditions, which are discussed next.

BOUNDARY CONDITIONS

The unknown constants A_i, B_i (i=1,2,3) in equations (13), (14) and (23) must be found by considering the boundary conditions. Two cases will be considered, depending on the support conditions of the grips of the testing machine: simply supported and clamped-clamped end supports (Figure 4).

SIMPLY SUPPORTED ENDS

This is true in situations where the grips of the testing machine allow rotation of the specimen ends (Figure 4a). The boundary conditions to be satisfied are:

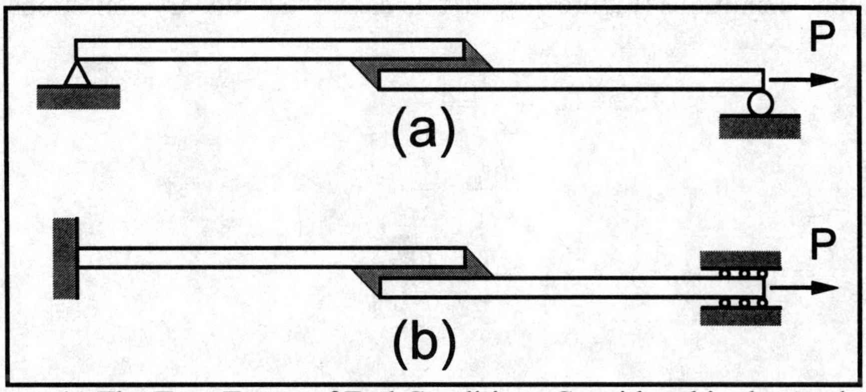

Figure 4. The Two Types of End Conditions Considered in the Analysis. (a) Simply Supported. (b) Clamped-Clamped.

$$@x_1 = 0: \qquad M_0 = 0,\ w_1 = 0,\ \frac{d^2 w_1}{dx_1^2} = 0 \ \textit{(automatically satisfied)} \tag{24}$$

$$@x_1 = (\ell - e),\ x_2 = 0: \qquad w_1 = w_2,\ \frac{dw_1}{dx_1} = \frac{dw_2}{dx_2} \tag{25}$$

$$@x_2 = e,\ x_3 = 0: \qquad w_2 = w_3,\ \frac{dw_2}{dx_2} = \frac{dw_3}{dx_3} \tag{26}$$

$$@x_3 = c: \qquad w_3 = 0 \tag{27}$$

Substituting the expressions for the displacements, equations (13), (14) and (23) into the above boundary conditions, the following system of equations in the unknown coefficients is obtained:

$$A_1 = 0 \tag{28}$$

$$\begin{bmatrix} \sinh[u_1(\ell-e)] & -a_0^{(1)} & -a_0^{(2)} & 0 & 0 \\ u_1\cosh[u_1(\ell-e)] & -a_1^{(1)} & -a_1^{(2)} & 0 & 0 \\ 0 & \sum_{n=0}^{N} a_n^{(1)} e^n & \sum_{n=0}^{N} a_n^{(2)} e^n & -1 & 0 \\ 0 & \sum_{n=1}^{N} n a_n^{(1)} e^{n-1} & \sum_{n=1}^{N} n a_n^{(2)} e^{n-1} & 0 & -u_3 \\ 0 & 0 & 0 & \cosh(u_3 c) & \sinh(u_3 c) \end{bmatrix} \begin{Bmatrix} B_1 \\ A_2 \\ B_2 \\ A_3 \\ B_3 \end{Bmatrix} = \begin{Bmatrix} a_0^{(p)} - \alpha(\ell - e) \\ a_1^{(p)} - \alpha \\ -\sum_{n=0}^{N} a_n^{(p)} e^n - \alpha c \\ -\sum_{n=1}^{N} n a_n^{(p)} e^{n-1} + \alpha \\ 0 \end{Bmatrix} \tag{29}$$

where e is the distance from the x_2 to the x_3 axes (see Figure 2), and not the natural log base.

It was verified analytically that in the limit when d=0 (no fillet), the series for w_2, equation (23), becomes

$$w_2 = a_0^* \cosh(u_1 x_2) + a_1^{**} \sinh(u_1 x_2) + \alpha(\ell - e + x_2) + M_0 \left[\frac{x_2 - (c+e)}{P(\ell + c)} \right] \tag{30}$$

where a_0^* and a_1^{**} are arbitrary constants. It can be seen that equation (30) is of the same form as equation (13), as would be expected if there were no fillet.

CLAMPED-CLAMPED ENDS

When the gripping machine is rigid, the supports can be better approximated by assuming clamped-clamped conditions (Figure 4b). In this case, the boundary conditions to be satisfied are:

$$@x_1 = 0: \qquad w_1 = 0,\ \frac{dw_1}{dx_1} = 0 \tag{31}$$

$$@x_1 = (\ell - e),\ x_2 = 0: \qquad w_1 = w_2,\ \frac{dw_1}{dx_1} = \frac{dw_2}{dx_2} \tag{32}$$

$$@x_2 = e,\ x_3 = 0: \qquad w_2 = w_3,\ \frac{dw_2}{dx_2} = \frac{dw_3}{dx_3} \tag{33}$$

$$@x_3 = c: \qquad w_3 = 0 \tag{34}$$

Substituting the expressions for displacements, equations (13), (14) and (23) into the above boundary conditions, we obtain the following system of equations:

$$M_0 = A_1 P \tag{35}$$

$$\begin{bmatrix} \dfrac{1}{\ell+c} & u_1 & 0 & 0 & 0 & 0 \\ \cosh[u_1(\ell-e)] & \sinh[u_1(\ell-e)] & -a_0^{(1)} & -a_0^{(2)} & 0 & 0 \\ u_1\sinh[u_1(\ell-e)] & u_1\cosh[u_1(\ell-e)] & -a_1^{(1)} & -a_1^{(2)} & 0 & 0 \\ 0 & 0 & \sum\limits_{n=0}^{N} a_n^{(1)} e^n & \sum\limits_{n=0}^{N} a_n^{(2)} e^n & -1 & 0 \\ 0 & 0 & \sum\limits_{n=1}^{N} n a_n^{(1)} e^{n-1} & \sum\limits_{n=1}^{N} n a_n^{(2)} e^{n-1} & 0 & -u_3 \\ 0 & 0 & 0 & 0 & \cosh(u_3 c) & \sinh(u_3 c) \end{bmatrix} \begin{Bmatrix} A_1 \\ B_1 \\ A_2 \\ B_2 \\ A_3 \\ B_3 \end{Bmatrix} = \begin{Bmatrix} -\alpha \\ a_0^{(p)} - \alpha(\ell-e) \\ a_1^{(p)} - \alpha \\ -\sum\limits_{n=0}^{N} a_n^{(p)} e^n - \alpha c \\ -\sum\limits_{n=1}^{N} n a_n^{(p)} e^{n-1} + \alpha \\ 0 \end{Bmatrix} \tag{36}$$

VERIFICATION OF THE MODEL

The simplified model was verified by comparing against a full size geometrically nonlinear finite element model. The mesh used for the full model is shown in Figure 5a. Simply supported end conditions were assumed for simplicity, since it was found that simply supported or clamped-clamped ends produced almost identical results.

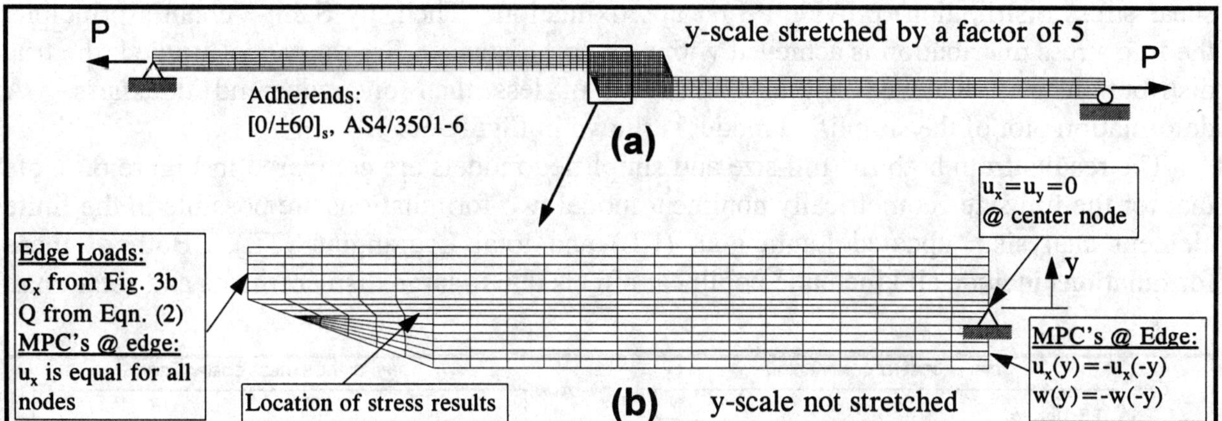

Figure 5. Full Size and Simplified FE Models Used in the Analysis. (a) Full Size Geometrically Nonlinear Mesh. (b) Simplified Linear Model of the Joint Region Showing Loads and Boundary Conditions Used. (c) Deformed Shape of the Simplified Model.

The adherends consisted of a $[0/\pm60]_s$ laminate with plies made of AS4/3501-6 carbon/epoxy. The adhesive was epoxy. The material properties used in the model were:

Adherends: E_1=20.0 msi, E_2=1.30 msi, ν_{12}=0.30, G_{12}=1.0 msi, ply thickness=0.0052 in.

Adhesive: E=0.50 msi, ν=0.35

The finite element analysis was conducted using COSMOS/M, and the model consisted of 960 8-noded plane strain elements and 3,087 nodes. For the adherends, six rows of elements were used through the thickness—each row corresponding to each of the plies and having the corresponding orthotropic material properties. The adhesive was assumed isotropic. The applied load, P, was increased from 0 to 6,000 lb using automatic stepping (which selects the optimum step increment) and resulted in a total of 18 incremental steps. The following dimensions were used for the joint (see Figure 2):

ℓ=3.5 in., e=0.0824 in., d=0.0412 in., $2c$=0.5 in., t=0.0312 in., t_c=0.010 in.

The simplified model consisted of half of the center region of the full model, had a total of 300 elements and 959 nodes, and is shown in Figure 5b (see also Figure 3b). Only half of the center region is needed in the simplified model because of the special kind of antisymmetry that the joint possesses. This antisymmetry can be invoked by using the multiple point constraints (MPC's) shown in Figure 5b. In addition to displacement boundary conditions, the simplified model also had the edge loads shown in Figure 3a. The introduction of the edge loads into the FE model can be facilitated by the use of statically equivalent edge stresses, as shown in Figure 3b. It should be pointed out that with laminated composites the interply stresses are not necessarily continuous; however, a linear distribution for the normal stresses and a parabolic distribution for the shear stresses (or even constant shear stress distribution) provide a first approximation. Then, by Saint-Venant's principle, the true stress distribution is achieved within a short distance. For the model studied, the true distribution was achieved within a distance of less than one adherend thickness. A deformation plot of the simplified model is shown in Figure 5c.

The results from both the full size and simplified models are compared in Figure 6. Note that for the full size geometrically nonlinear model two formulations are possible in the finite element analysis: updated Lagrangian (UL) and total Lagrangian (TL). Both of these formulations include all kinematic nonlinear effects due to large displacements and rotations.

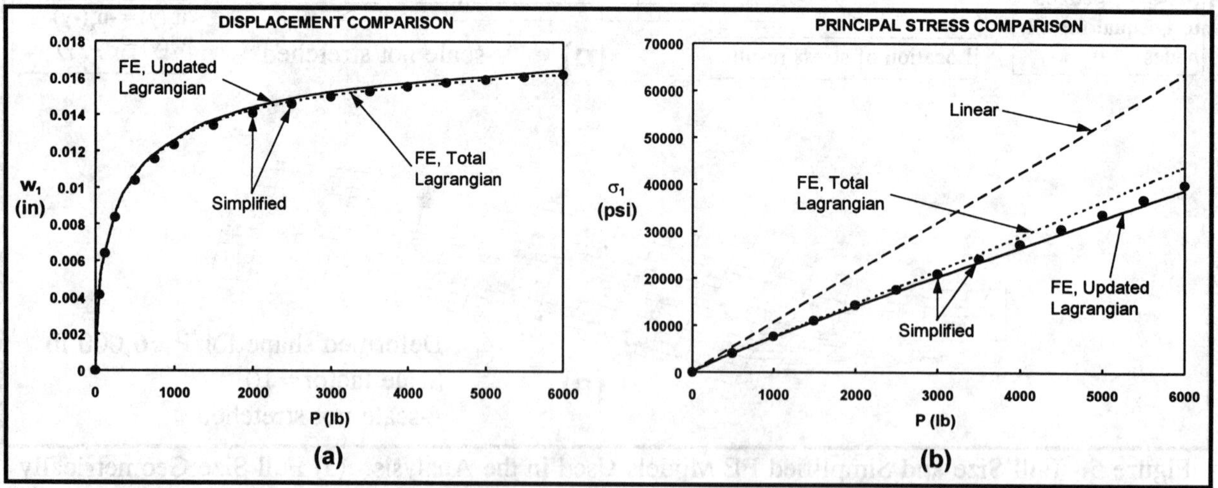

Figure 6. Comparison of Displacement and Principal Stress Results. (a) Displacement w_1 at $x_1=(\ell-e)$ vs. Load. (b) Maximum Principal Stress in the Adhesive vs. Load.

The difference lies in the fact that in the TL formulation all static and kinematic variables are referred to the initial configuration, while in the UL formulation the same variables are referred to the last calculated configuration. Both configurations were used in the full model.

The deflection w_1 at a location $x_1=(\ell-e)$ is compared in Figure 6a. This deflection is important in the calculation of the edge moment from Equation (3). As can be seen in Figure 6a, the load-deformation relationship is highly nonlinear. Also, there is very close agreement between the simplified and full size models—the simplified model is within 2% of the updated Lagrangian formulation.

The maximum element principal stress in the adhesive[†] (see Figure 5b for location) is compared in Figure 6b. The linear solution for the full size model is also shown for comparison. It can be seen that there is close agreement between the simplified and full size geometrically nonlinear models—the simplified model is within 3% of the updated Lagrangian formulation. In contrast, the assumption of linearity would lead to an answer that is 64% higher at P=6,000 lb, and about 49% higher at P=1,500 lb (a typical failure load).

CONCLUSIONS

A simplified model for the geometrically nonlinear analysis of the single lap joint has been presented. The model can be expected to produce accurate results while using linear analysis of a small portion of the overall finite element mesh. This can be advantageous when a fine mesh of the overlap region is needed (e.g., in fracture mechanics analysis of failure) since the size of the FE model becomes relatively small, high element aspect ratios in the adherends are eliminated, and the complexities of a nonlinear analysis are avoided.

REFERENCES

Gibson, R. F. 1994. *Principles of Composite Material Mechanics*, McGraw-Hill, New York.

Goland, M. and Reissner, E. 1944. "The Stresses in Cemented Joints," *Journal of Applied Mechanics*, 11:A17-A27.

Gradin, P. A. 1982. "A Fracture Criterion for Edge-Bonded Bimaterial Bodies," *Journal of Composite Materials*, 16:448-456.

Groth, H. L. 1988. "Stress Singularities and Fracture at Interface Corners in Bonded Joints, *International Journal of Adhesion and Adhesives*, 8(2):107-113.

O'Neil, P. V. 1987. *Advanced Engineering Mathematics*, Second Edition, Wadsworth Publishing Company, Belmont, CA.

Swanson, S. R. 1997. *Introduction to Design and Analysis with Advanced Composite Materials*, Prentice Hall, New Jersey.

Young, W. C. 1989. *Roark's Formulas for Stress and Strain*, Sixth Edition, McGraw-Hill, New York.

[†] A stress singularity exists at the adhesive-adherend interface corner, and theoretically the maximum principal stress is infinite. However, due to the finite element discretization, the singularity disappears but the maximum stress depends on the mesh size used.

Analysis of a Composite Notched-Lap Joint Using Higher Order Plate Theory

J. M. WHITNEY

ABSTRACT

A stress analysis of a notched-lap joint is developed in conjunction with a higher order plate theory in which transverse shear and normal strain are included. In order to physically capture the proper free-edge effects, each component of the joint is considered as a separate plate with appropriate interface continuity conditions enforced. A sufficient number of boundary conditions are available with this approach to allow free-edge conditions to be satisfied. Effect of the adhesive layer is ascertained by comparing stress distributions for the case where the adhesive layer is included in the model versus the case where the adhesive layer is neglected. Intensity of the transverse normal stress is significantly increased by including the adhesive layer in the model.

INTRODUCTION

Because of the weight introduced by the extensive use of mechanical fasteners, there is an increasing interest in the use of adhesive bonding for joining structural parts. In the area of composite materials technology this involves joining composites to other composites and/or composites to metals. In addition to the conventional structural need to join parts, bonding technology has become a key player in the area of composite material repair.

An important issue in adhesive bonding technology is determining the shear strength of a bonded joint. Various configurations for determining shear strength of an adhesive have evolved over the years. Among the various methods developed, those using a tensile type specimen for introducing shear load have become very popular. Of course the most famous of these is the single-lap joint. Because of the bending introduced into the single-lap specimen, many researchers have favored other geometries such as a notched joint, ASTM D3165. This specimen consists of two composites bonded together with notches cut in the top and bottom as shown in Fig.1. Unlike a single lap configuration, this geometry allows for an external axial load to be applied along the centerline of the specimen.

From a stress analysis standpoint the free-edge zone near the boundaries of a notched-lap joint, where peak stresses occur, causes considerable difficulty. In particular, many approaches (including finite elements) are unable to satisfy the proper free-edge boundary conditions. Finite element models which approach the correct boundary conditions require more complex, higher order elements.

In the present paper a stress analysis of a notched joint is developed in conjunction with a higher order plate theory in which transverse shear and normal strain are included. In order to physically capture the proper free-edge effects, each component of the joint is considered

---

James M. Whitney is a Professor in the Department of Civil & Environmental Engineering and Engineering Mechanics, University of Dayton, 300 College Park, Dayton, OH 45469-0243, USA

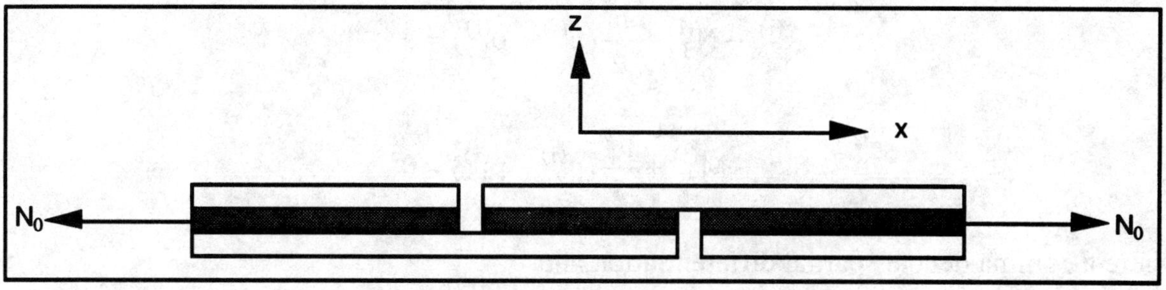

Figure 1. Notched joint geometry.

as a separate plate with appropriate interface continuity conditions enforced. A sufficient number of boundary conditions are available with this approach to allow free-edge conditions to be satisfied.

ANALYSIS

The present stress analysis is confined to the center region (gage section) of the notched joint configuration shown in Fig. 1. By taking advantage of anti-symmetry conditions, it is only necessary to model the upper portion of the center region. The composite and adhesive are modeled as two separate plates under cylindrical bending joined together by appropriate traction and displacement continuity conditions. The global laminated plate model developed by Pagano and Soni, 1983, are utilized in the plate analysis. This model is based on the following displacement field:

$$u^{(i)} = u_0^{(i)}(x) + z_i \psi_x^{(i)}(x), \; v_i = 0 \qquad (i = 1, 2)$$
$$w^{(i)} = w_0^{(i)}(x) + z_i \psi_z^{(i)}(x) + \frac{z_i^2}{2} \phi^{(i)} \tag{1}$$

where u and w denote displacements in the x and z directions, respectively. The superscripts 1 and 2 denote composite and adhesive, respectively. A cursory examination of eqs. (1) reveal that they contain both transverse shear and normal strain components. Constitutive equations in terms of force and moment resultants along with equilibrium equations and proper boundary conditions are developed from the theory of minimum potential energy.

For the case under consideration, the following five equilibrium equations are applicable to both the composite and adhesive:

$$N_{1,x}^{(i)} + (q_2^{(i)} - q_1^{(i)}) = 0$$
$$M_{1,x}^{(i)} - Q_5^{(i)} + \frac{h_i}{2}(q_2^{(i)} + q_1^{(i)}) = 0 \tag{2}$$
$$Q_{5,x}^{(i)} + (p_2^{(i)} - p_1^{(i)}) = 0$$

$$R^{(i)}_{5,x} - N^{(i)}_3 + \frac{h_i}{2}(p^{(i)}_2 + p^{(i)}_1) = 0$$

$$S^{(i)}_{5,x} - M^{(i)}_3 + \frac{h_i^2}{8}(p^{(i)}_2 - p^{(i)}_1) = 0 \tag{2}$$

where a comma denotes partial differentiation and

$$(N^{(i)}_k, M^{(i)}_k) = \int_{-h/2}^{h/2} \sigma_k(1, z_i)dz_i \quad (k = 1,3)$$

$$(Q^{(i)}_5, R^{(i)}_5, S^{(i)}_5) = \int_{-h/2}^{h/2} \tau^{(i)}_{xz}(1, z_i, \frac{z_i^2}{2})dz_i \tag{3}$$

In addition, the interlaminar shear and normal stresses at the top of the ith plate are denoted by $q^{(i)}_2$ and $p^{(i)}_2$, respectively, while the corresponding stress at the bottom are denoted by $q^{(i)}_1$ and $p^{(i)}_1$.

Constitutive relations are of the form

$$N^{(i)}_k = A^{(i)}_{k1}u^{(i)}_{0,x} + A^{(i)}_{k3}\psi^{(i)}_z$$

$$M^{(i)}_k = D^{(i)}_{k1}\psi^{(i)}_{x,x} + D^{(i)}_{k3}\phi^{(i)} \quad (k = 1, 3)$$

$$Q^{(i)}_5 = A^{(i)}_{55}(\psi^{(i)}_x + w^{(i)}_{0,x}) + \frac{D^{(i)}_{55}}{2}\phi^{(i)}_{,x} \tag{4}$$

$$R^{(i)}_5 = D^{(i)}_{55}\psi^{(i)}_{z,x}$$

$$S^{(i)}_5 = \frac{D^{(i)}_{55}}{2}(\psi^{(i)}_x + w^{(i)}_{0,x}) + \frac{H^{(i)}_{55}}{4}\phi^{(i)}_{,x}$$

where

$$(A^{(i)}_{kj}, D^{(i)}_{kj}) = \int_{-h/2}^{h/2} C^{(i)}_{kj}(1, z_i^2)dz \quad (k, j = 1,3)$$

$$H^{(i)}_{55} = \int_{-h/2}^{h/2} C^{(i)}_{55}z_i^4 dz$$

and $C^{(i)}_{kj}$ are elastic stiffnesses of the composite and adhesive. Substituting eqs. (4) into eqs (2), we obtain 5 linear differential equations with constant coefficients in terms of the displacements $u^{(i)}_0, w^{(i)}_0$, $\psi^{(i)}_x$, $\psi^{(i)}_z$, and $\phi^{(i)}$, for the composite and adhesive, respectively. These 10 equations can be further reduced by considering continuity and conditions.

The solution involves superposition of two problems in order to preserve anti-symmetric requirements at the joint centerline. Nomenclature for the first problem is shown in Fig. 2.

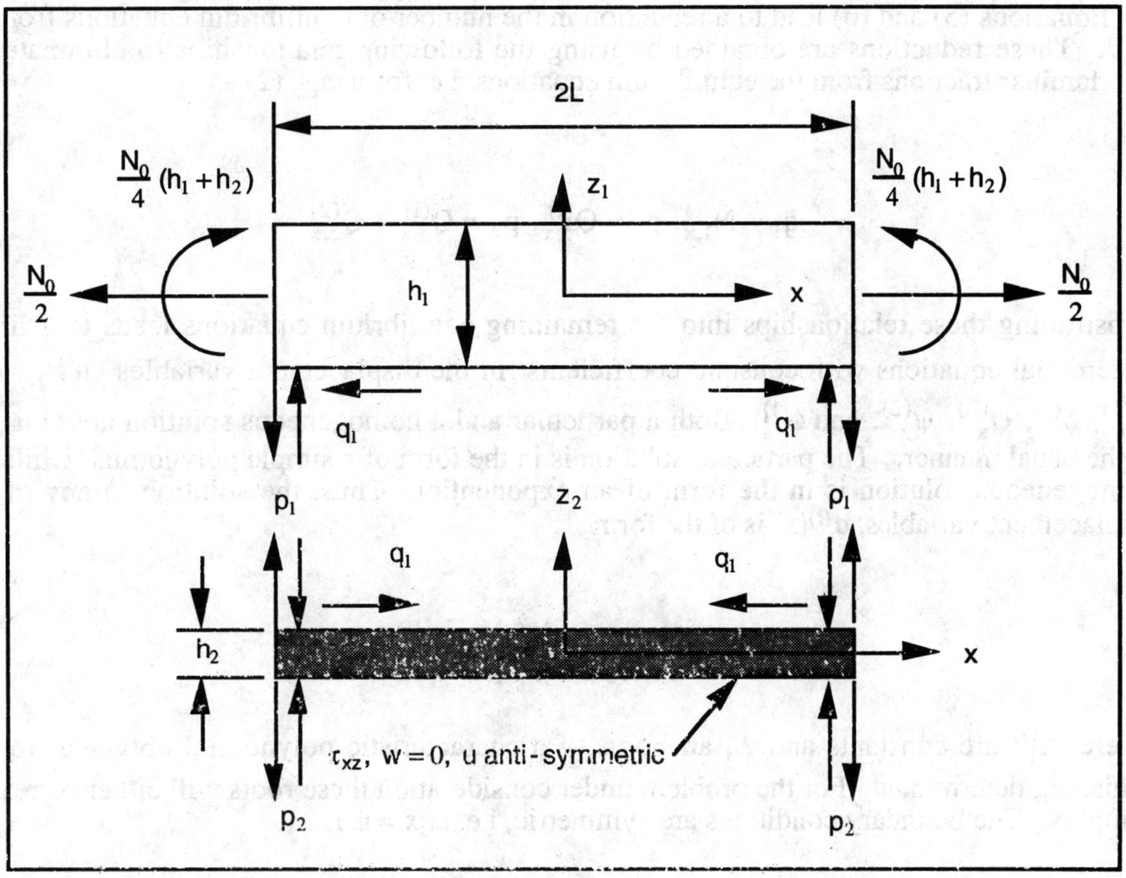

Figure 2. Nomenclature for problem 1.

Both the transverse deflection and interlaminar shear stress vanish along the centerline of the adhesive. In addition the displacement u must be anti-symmetric with respect to x at the center of the adhesive. The shear stress at the interface between the composite and adhesive is denoted by q_1, while the interlaminar normal stresses at the composite/adhesive interface and the center of the adhesive are denoted by p_1 and p_2, respectively. Vanishing of w at the center of the adhesive leads to the relationship

$$w_0^{(2)} = \frac{h_2}{2}\psi_z^{(2)} - \frac{h_2^2}{8}\phi^{(2)} \tag{5}$$

Continuity of u and w at the composite/adhesive interface lead to the following relationships

$$u_0^{(2)} = u_0^{(1)} - \frac{h_1}{2}\psi_x^{(1)} - \frac{h_2}{2}\psi_x^{(2)}$$

$$w_0^{(1)} = \frac{h_1}{2}\psi_z^{(1)} + h_2\psi_z^{(2)} - \frac{h_2^2}{8}\phi^{(2)} \tag{6}$$

Equations (5) and (6) lead to a reduction in the number of equilibrium equations from 10 to 7. These reductions are obtained by using the following relationships to eliminate the interlaminar tractions from the equilibrium equations, i.e. from eqs. (2)

$$q_1 = N^{(1)}_{1,x},\ p_1 = Q^{(1)}_{5,x},\ p_2 = Q^{(1)}_{5,x} + Q^{(2)}_{5,x} \tag{7}$$

Substituting these relationships into the remaining equilibrium equations leads to 7 linear differential equations with constant coefficients in the displacement variables $u^{(1)}_0$, $\psi^{(1)}_x$, $\psi^{(1)}_z$, $\phi^{(1)}$, $\psi^{(2)}_x$, $\psi^{(2)}_z$, and $\phi^{(1)}$. Both a particular and a homogeneous solution are obtained in the usual manner. The particular solution is in the form of a simple polynomial, while the homogeneous solution is in the form of an exponential. Thus, the solution to any of the displacement variables, $d^{(i)}(x)$ is of the form

$$d^{(i)}(x) = \sum_j A^{(i)}_j e^{\lambda_j x} + d^{(i)}_p(x) \tag{8}$$

where $A^{(i)}_j$ are constants and λ_j are roots of a characteristic polynomial obtained from a vanishing determinant. For the problem under consideration these roots will either be real or complex. The boundary conditions are symmetric, i.e. at $x = \pm L$

$$N^{(1)}_1 = \frac{N_0}{2},\ M^{(1)}_1 = -\frac{N_0}{4}(h_1 + h_2)$$

$$Q^{(1)}_5,\ N^{(2)}_1,\ M^{(2)}_1,\ Q^{(2)}_5,\ q_1 = 0 \tag{9}$$

Nomenclature for the second problem is shown in Fig. 3. In this case both the interlaminar normal stress and the displacement u vanish at the centerline of the adhesive, while the displacement w must be anti-symmetric with respect to x. Vanishing of u at the center of the adhesive leads to the following relationship

$$u^{(2)}_0 = \frac{h_2}{2}\psi^{(2)}_x \tag{10}$$

The following displacement relations are obtained by considering continuity at the composite/adhesive interface

$$u^{(1)}_0 = \frac{h_1}{2}\psi^{(1)}_x + h_2\psi^{(2)}_x \tag{11}$$

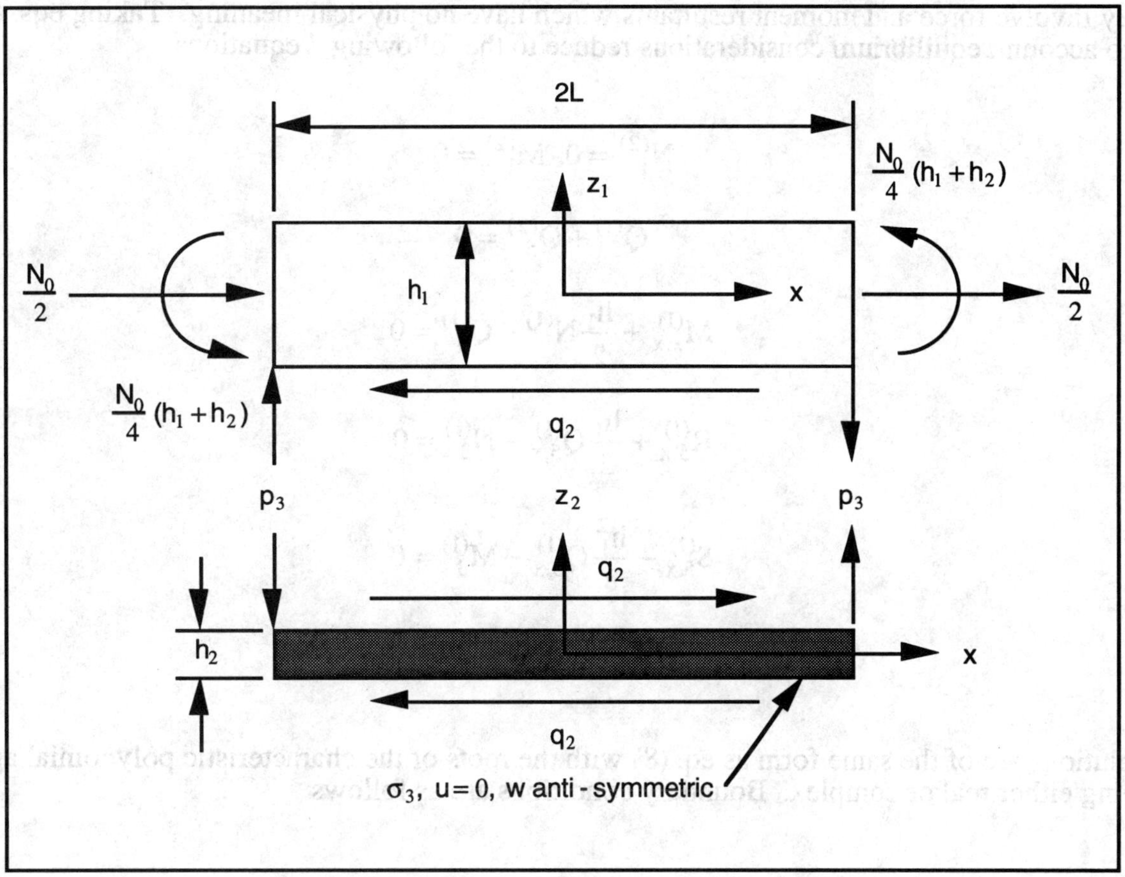

Figure 3. Nomenclature for problem 2

$$w_0^{(2)} = w_0^{(1)} - \frac{h_1}{2}\psi_z^{(1)} + \frac{h_1^2}{8}\phi^{(1)} - \frac{h_2}{2}\psi_z^{(2)} - \frac{h_2^2}{8}\phi^{(2)} \tag{11}$$

From equilibrium consideration, eqs. (2), the composite/adhesive interface tractions are as follows

$$q_2 = N_{1,x}^{(1)},\ p_3 = Q_{5,x}^{(1)} \tag{12}$$

Substituting these relationships into the remaining equilibrium equations, taking into account eqs. (10) and (11), leads to 8 linear differential equations with constant coefficients in the 7 variables $w_0^{(1)}$, $\psi_x^{(1)}$, $\psi_z^{(1)}$, $\phi^{(1)}$, $\psi_x^{(2)}$, $\psi_z^{(2)}$, $\phi^{(2)}$. Thus, we have one too many equations for the number of variables. This difficulty is due to the assumption that the shear traction at the top and bottom of the adhesive layer are equal, which leads to the vanishing of $N_1^{(2)}$. In order to alleviate this problem, we assume that $\sigma_1^{(2)} = 0$. This implies that $M_1^{(2)} = 0$. The last two equilibrium relations in eqs. (2) are dropped with respect to the adhesive layer, as they involve force and moment resultants which have no physical meaning. Taking eqs. (12) into account, equilibrium considerations reduce to the following 7 equations

$$N_1^{(2)} = 0, \; M_1^{(2)} = 0$$

$$Q_5^{(1)} + Q_5^{(2)} = 0$$

$$M_{x,x}^{(1)} + \frac{h_1}{2} N_1^{(1)} - Q_5^{(1)} = 0$$

$$R_{5,x}^{(1)} + \frac{h_1}{2} Q_{5,x}^{(1)} - N_3^{(1)} = 0 \tag{13}$$

$$S_{5,x}^{(1)} - \frac{h_1^2}{8} Q_{5,x}^{(1)} - M_3^{(1)} = 0$$

$$Q_5^{(1)} + h_2 N_{1,x}^{(1)} = 0$$

Solutions are of the same form as eq. (8) with the roots of the characteristic polynomial again being either real or complex. Boundary conditions are as follows

$$\underline{\text{at } x = L}: \quad N_1^{(1)} = \frac{N_0}{2}, \; M_1^{(1)} = -\frac{N_0}{4}(h_1 + h_2), \; q_3 = 0$$

$$\underline{\text{at } x = -L}: \quad N_1^{(1)} = -\frac{N_0}{2}, \; M_1^{(1)} = \frac{N_0}{4}(h_1 + h_2), \; q_3 = 0 \tag{14}$$

Note that equilibrium automatically assures that $Q_5^{(1)}$, $Q_5^{(2)}$, $N_1^{(2)}$, $M_1^{(2)}$ vanish at $x = \pm L$.

If the adhesive layer is neglected we obtain the free-body diagrams illustrated in Figs. 4 and 5 for problems 1 and 2, respectively. The solution procedure is the same as for the previous case with some simplification due to the presence of the composite layer only. Details will not be presented here. A similar analysis was performed by Whitney, 1989.

NUMERICAL RESULTS

Consider a notched joint with the composite composed of 0° unidirectional material with the following ply properties

$$\frac{E_L}{E_T} = 14, \; \frac{E_3}{E_T} = 1, \; \frac{G_{LT}}{E_T} = \frac{G_{L3}}{E_T} = 0.533, \; \frac{G_{T3}}{E_T} = 0.385, \; \nu_{LT} = \nu_{L3} = 0.3, \; \nu_{T3} = 0.5 \tag{15}$$

where E_L, E_T, and E_Z are the elastic moduli parallel to the fibers, transverse to the fibers,

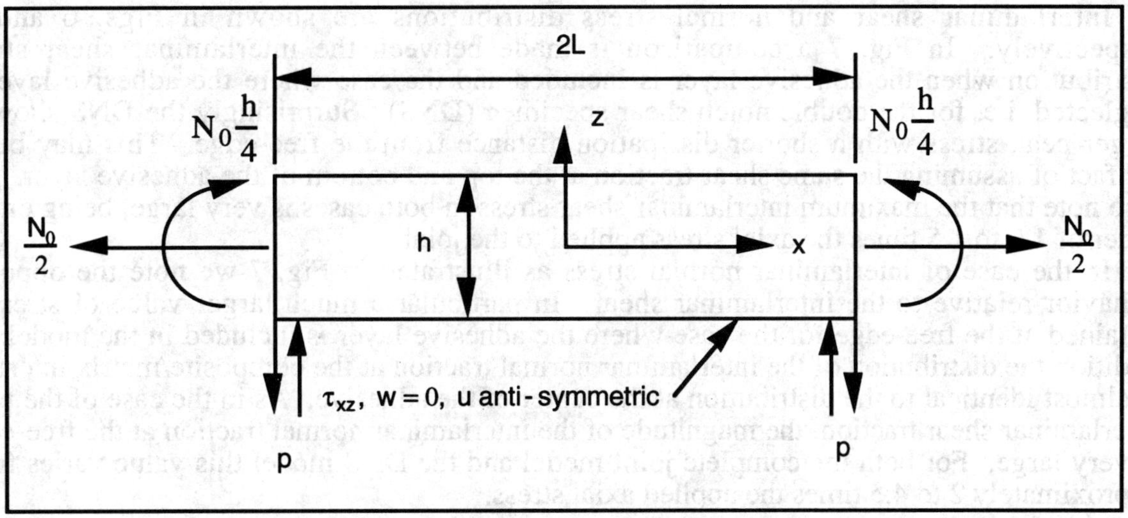

Figure 4. Nomenclature problem 1 - adhesive layer neglected.

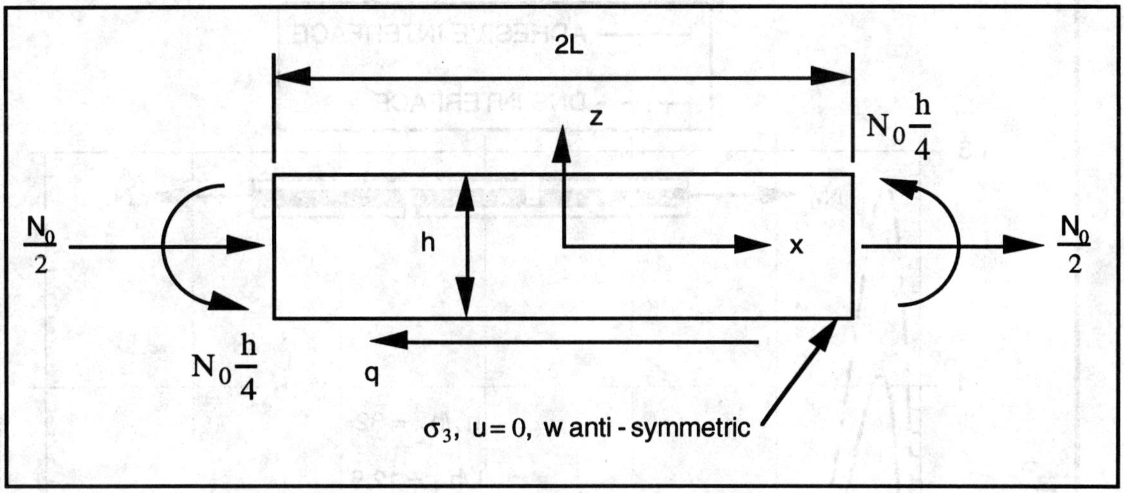

Figure 5. Nomenclature problem 2 - adhesive layer neglected.

and through-the-thickness, respectively, while G_{LT}, G_{L3}, and G_{T3}, denote shear moduli relative to the L-T plane L-3 plane, and T-3 plane, respectively. The Poisson's ratio ν_{LT} is determined from a uniaxial tensile test parallel to the fibers while measuring contraction transverse to the fibers. Analogous definitions are attached to ν_{L3} and ν_{T3}. These properties are typical of current state-of-the-art graphite/epoxy unidirectional composites. Denoting properties of the isotropic adhesive layer with a subscript A, we assume

$$\frac{E_A}{E_T} = 0.333, \; \nu_A = 0.35 \tag{16}$$

Interlaminar shear and normal stress distributions are shown in Figs. 6 and 7, respectively. In Fig. 7 a comparison is made between the interlaminar shear stress distribution when the adhesive layer is included and the case where the adhesive layer is neglected, i.e. for the double notch shear specimen (DNS). Surprisingly the DNS shows a larger peak stress with a shorter dissipation distance from the free-edge. This may be an artifact of assuming the same shear traction at the top and bottom of the adhesive layer. We also note that the maximum interlaminar shear stress in both cases is very large, being on the order of 1.0 to 1.5 times the axial stress applied to the joint.

In the case of interlaminar normal stress as illustrated in Fig. 7 we note the opposite behavior relative to the interlaminar shear. In particular a much larger value of stress is obtained at the free-edge for the case where the adhesive layer is included in the model. In addition the distribution of the interlaminar normal traction at the composite/matrix interface is almost identical to the distribution at the center of the adhesive. As in the case of the peak interlaminar shear traction, the magnitude of the interlaminar normal traction at the free-edge is very large. For both the complete joint model and the DNS model this value varies from approximately 2 to 4.5 times the applied axial stress.

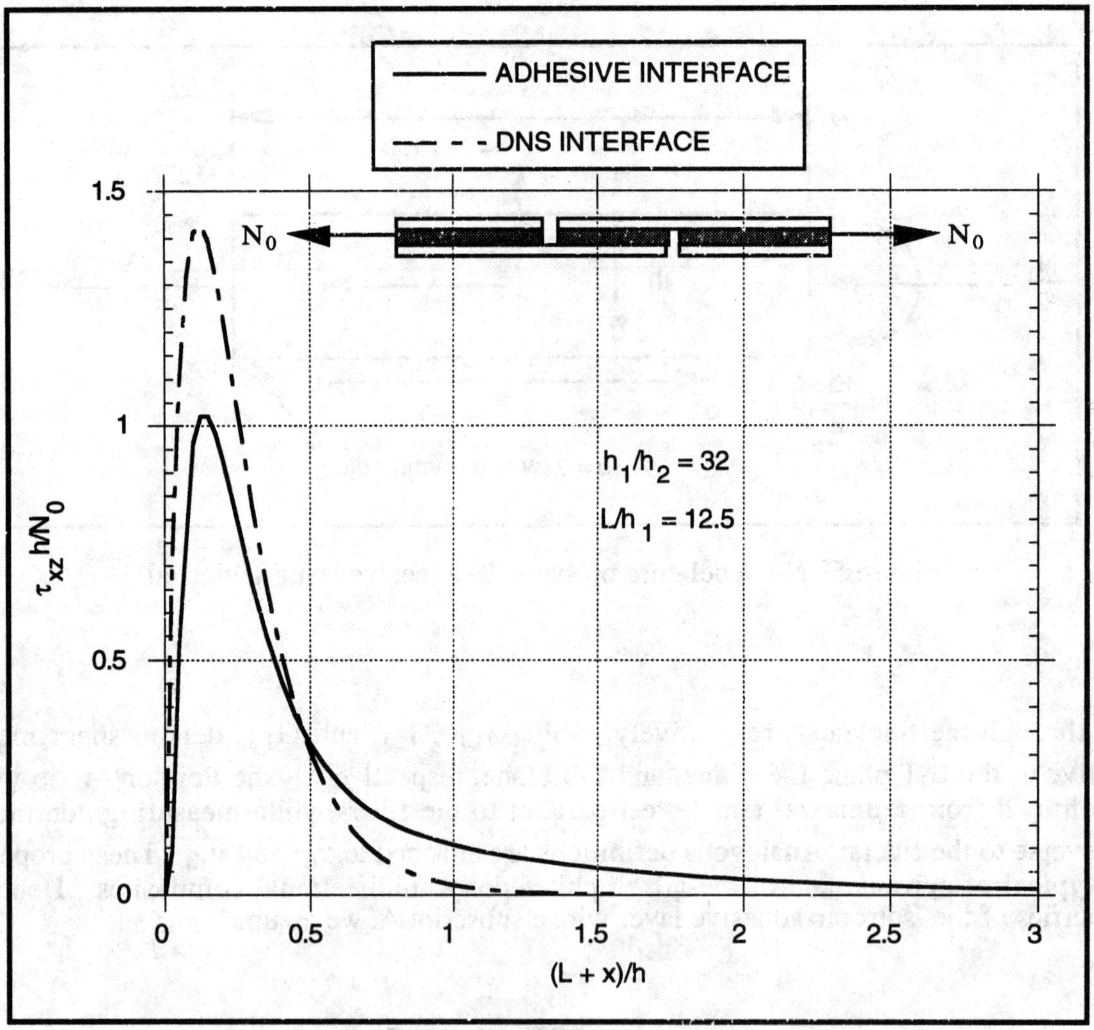

Figure 6. Interlaminar shear stress distribution.

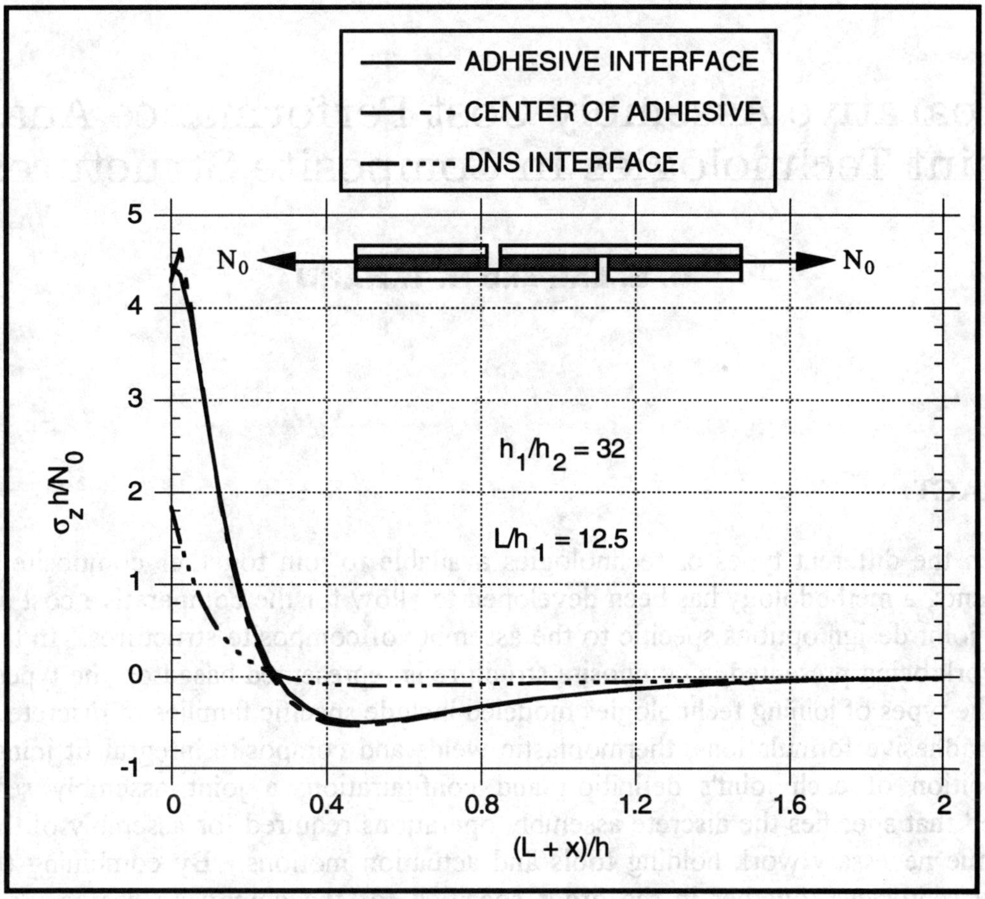

Figure 7. Interlaminar normal stress distribution.

ACKNOWLEDGMENT

This work was sponsored by the Materials Directorate of the Air Force Wright Laboratory under a contract AF33615-95-C-5029 with the University of Dayton Research Institute.

REFERENCES

Pagano, N. J. and S. R. Soni, S. R., 1983, "Global-Local Laminate Variational Model," *International Journal of Solids and Structures*, Vol. 19, No. 3, pp. 207-228.

Whitney, J. M., 1989, "Stress Analysis of the Double Notch Shear Specimen," *Proceedings of the American Society for Composites, Fourth Technical Conference*, Technomic Publishing Co., Lancaster, PA, pp. 325-333.

Comparative Assembly Cost Performance Analysis of Joint Technologies in Composite Structures

D. E. LEE AND H. T. HAHN

ABSTRACT

Given the different types of technologies available to join together composite structural components, a methodology has been developed to allow for the comparative cost analysis of multiple joint design options specific to the assembly of composite structures. In the context of the work being presented, a composite structure is represented based on the types of joints used. The types of joining technologies modeled include specific families of discrete fasteners, bonding adhesive formulations, thermoplastic welds and composite integral fit joints. Using the definition of each joint's definition and configuration, a joint assembly sequence is generated that specifies the discrete assembly operations required for assembly of the joint as well as the necessary work holding tools and actuation motions. By combining these joint assembly sequences together in the order specified for the composite structure's assembly plan, the overall assembly operation sequence is then evaluated based on available simulation models of the discrete operations needed and the tooling and actuation used. This approach to analyzing the assembly of a composite structure was selected because of the need to capture the time and cost of operations required to simply prepare a specific joint in the structure for assembly and because of the differences in assembly sequences and tooling requirements between different joint technologies such as discrete fasteners or integral fit joints.

1. INTRODUCTION

In seeking to compare different technologies to used to join composite components together, differing factors drive the cost of utilizing one joint over another. Focusing specifically on the time and cost of assembling different joining technologies, a methodology is being developed to estimate the cost of assembling individual composite joints as well as providing a mechanism by which to compare one technology over another. While the primary goal of this research has been to provide an effective comparison between different types of joints in composite structures, the results are intended to be broadly applicable to the many technologies used in joining mechanically assembled products.

David E. Lee and H. Thomas Hahn, Integrated Manufacturing Engineering Program for Advanced Transportation Systems, UCLA School of Engineering and Applied Science, Box 951597, Los Angeles, CA 90095-1597. *david@ime.ucla.edu* & *hahn@seas.ucla.edu; http://ime.ucla.edu/~david*

1.1 JOINING TECHNOLOGIES IN COMPOSITES

Two principal classes of joining technologies have predominated in composite structures: a) adhesive bonding and b) discrete mechanical fasteners (Matthews, 1987). In addition to these technologies, two other joining methods have been investigated for applications in composite materials: c) thermoplastic welding and d) integral fit joints (Lee and Hahn, 1997b) (Figure 1).

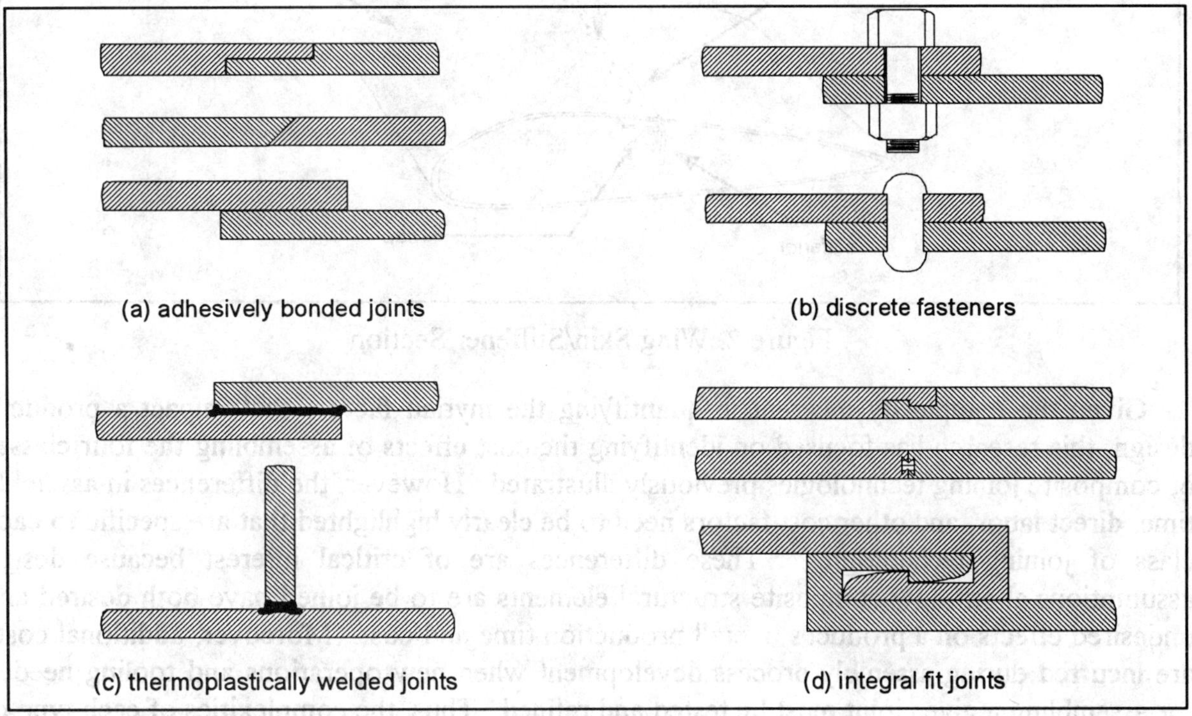

Figure 1. Composite Structural Joints

With each of these joining technologies, different performance factors specific to a product design affect the use and application of each joint in its structure. These include design specific metrics such as the types and magnitudes of forces transmitted through the joint, joint failure characteristics and the effects of thermally-induced joint expansion/ contraction behaviors. Beyond these, performance factors related to other stages in a product's life cycle (Section 1.2) are of interest.

For instance, in the case of a aircraft wing's skin/stiffener joint (Figure 2), concerns related to the assembly and maintenance of the composite structure must be addressed during product design. In the assembly of the skin/stiffener joint, access to the underside of the cap may be required depending on the type of fasteners used. Additionally, given a long extent of the wing section, tooling may be necessary to support the skins, stiffener and cap both for spatially positioning each of these structural elements as well as needing to withstand the applied assembly forces. In the case of maintaining the wing, regular access to service wiring or fuel lines may preclude the use of one type of fastener over another. Moreover, the acceptable repair operations that will be necessary to the outer skin sections may dictate one joint design over another.

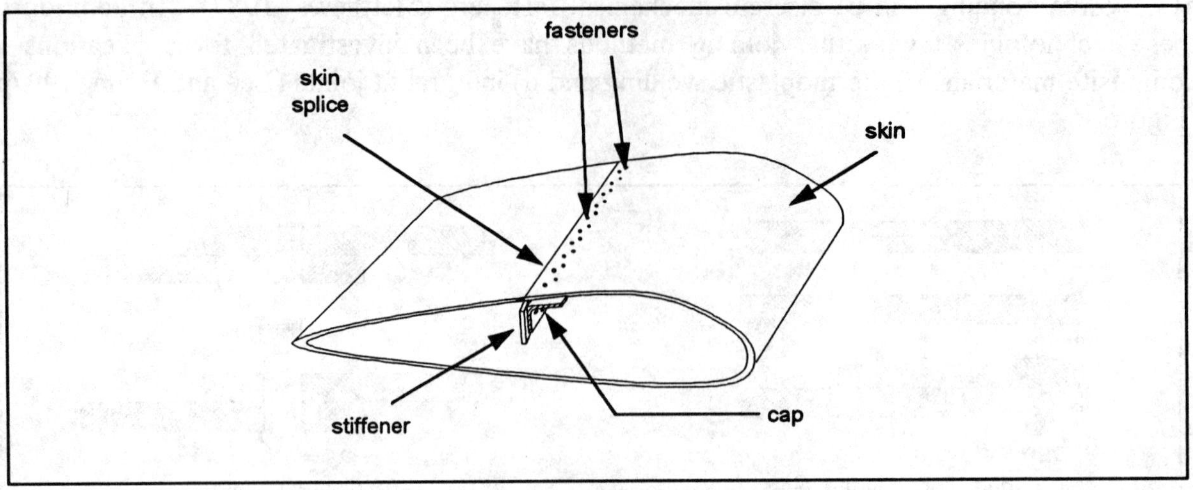

Figure 2. Wing Skin/Stiffener Section

Given the complexity inherent in quantifying the myriad factors that impact a product's design, this research has focused on identifying the cost effects of assembling the four classes of composite joining technologies previously illustrated. However, the differences in assembly time, direct labor, and other cost factors need to be clearly highlighted that are specific to each class of joining technologies. These differences are of critical interest because design assumptions about how composite structural elements are to be joined have both desired and undesired effects on a product's overall production time and cost. Moreover, additional costs are incurred during assembly process development when new operations and tooling needed for assembling a given joint must be tested and refined. Thus, the complexities of each type of joint provide a joint-centered view of both product and process development.

1.2 PRODUCT AND PROCESS DEVELOPMENT

Central to this joint-centered view of assembly cost analysis is the assumption that both a product and its manufacturing processes undergo specific developmental cycles which are directly coupled by the joining technologies that are used (Figure 2). That individual products are subject to a development process that involves specific stages of evolution - product design data, manufacturing information, testing evaluation, systems use and maintenance as well as final disposal - has been published earlier (National Research Council, 1991; Lee and Melkanoff, 1993). What has not been focused on are the development refinement cycles required to implement and validate the manufacturing processes necessary to fabricate a product and the processes needed to assemble each joints (Lee and Hahn, 1996a).

1.3 PRODUCT AND PROCESS ASSEMBLY COST MODELING

Each cost element that comprises the total cost of a product (Figure 4) can be treated as a linear sum of the costs that are either dependent or independent on the volume manufactured.

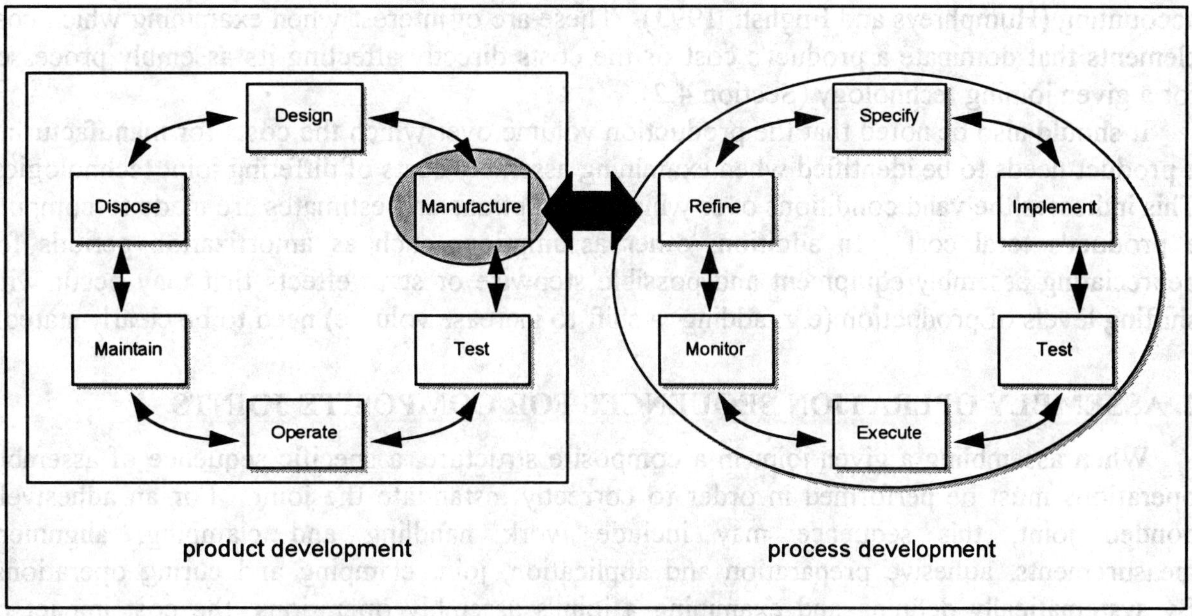

Figure 3. Interaction Between Product and Process Development Cycles

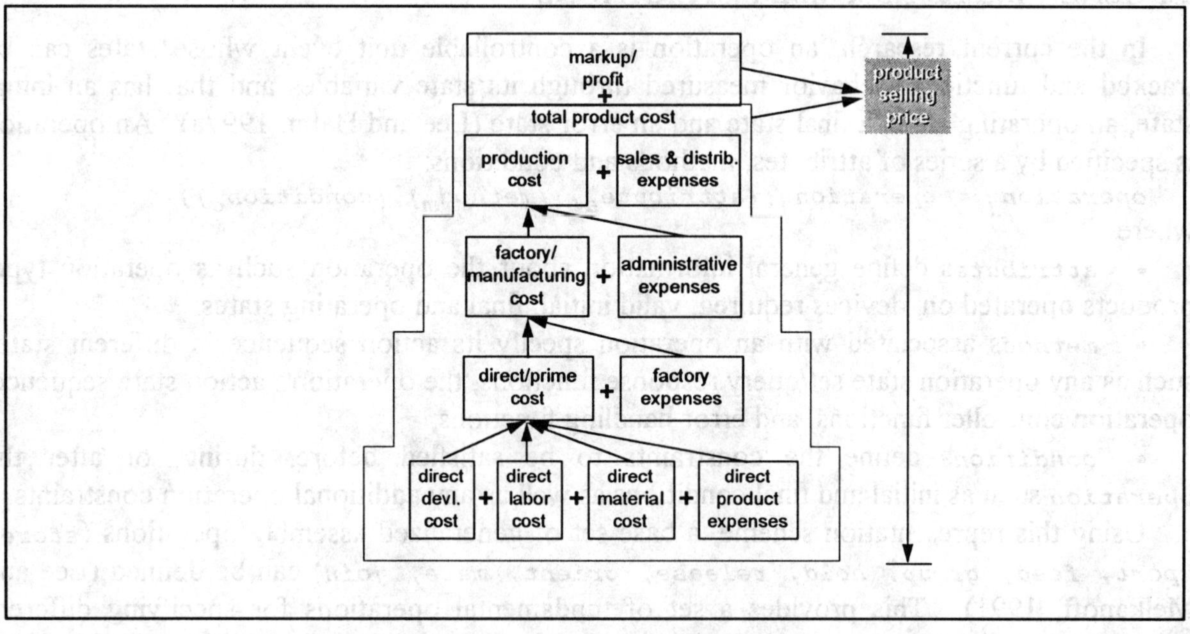

Figure 4. Product Cost Elements

This indicates that there are a set of costs that are invariant over the number of products manufactured (i.e. the fixed costs per product such as development costs, overhead, etc.) as well as those costs directly affected by the production volume (i.e. variable costs such as labor and material) (Creese, Adithan and Pabla, 1992). Further, a cost can be classified as being direct (attributable specific to a product) or indirect (attributed to many products). Each cost element can be additionally accounted for and classified into specific categories such as labor cost, material cost, development cost, sales expenses, etc. in an approach known as cost accounting (Humphreys and English, 1993). These are of interest when examining which cost elements that dominate a product's cost or the costs directly affecting its assembly processes for a given joining technology (Section 4.2).

It should also be noted that the production volume over which the costs for manufacturing a product needs to be identified when examining assembly costs of differing joint technologies. This indicates the valid conditions over which assumptions and estimates are made to compute a product's total cost. In addition, other assumptions such as amortization periods for depreciating assembly equipment and possible stepwise or state effects that may occur with shifting levels of production (e.g. adding a shift to increase volume) need to be clearly stated.

2. ASSEMBLY OPERATION SEQUENCES FOR COMPOSITE JOINTS

When assembling a given joint in a composite structure, a specific sequence of assembly operations must be performed in order to correctly instantiate the joint. For an adhesively bonded joint, this sequence may include work handling and clamping, alignment measurements, adhesive preparation and application, joint clamping and curing operations. By systematically defining and examining a joint's assembly operations, the cost impact of each discrete step of the assembly sequence can thus be identified and measured.

2.1 GENERALIZED ASSEMBLY OPERATIONS

In the current research, an operation is a controllable unit event whose states can be tracked and functional behavior measured through its state variables and that has an initial state, an operating state, a final state and an error state (Lee and Hahn, 1997a). An operation is specified by a series of attributes, methods and conditions:

$$operation_j = operation_j(\{attribute_a\},\{method_m\},\{condition_c\})$$

where

- *attributes* define general information about the operation such as operation type, products operated on, devices required, valid initial, final and operating states;
- *methods* associated with an operation specify its action sequence in different states such as any operation state set/query response functions, the operation's action state sequence, operation controller functions, and error handling functions;
- *conditions* define the constraints to be satisfied before, during, or after the *operation* such as initial and final conditions as well as any additional operation constraints.

Using this representation scheme, a base set of generalized assembly operations (*store, xport, feed, group, hold, release, orient, mate, join*) can be defined (Lee and Melkanoff, 1991). This provides a set of fundamental operations for specifying different assembly operation sequences required to instantiate a given joint. In addition, this identifies an object-oriented framework for adding definitions of new generalized operations to handle new types of components as well as a scheme in which non-assembly operations such as measuring or marking can be incorporated into an assembly sequence.

2.2 COMPOSITE ASSEMBLY OPERATION SEQUENCES

Using the identified generalized assembly operations, an assembly operation sequence can be defined as those operations, when constrained to follow in a required sequence, necessary to complete a single joint. Moreover, the assembly of a common family of joining technologies can be represented as a sequence of assembly operations that is parametrized by the common traits. Such traits might include discrete hole sizing values for fasteners and assembly force values for integral fit joints. This would allow for common assembly operation sequences to be tailored for either a specific geometric setup or generic for each type of joint.

For the wing skin/stiffener example, assembly operation sequences can be evaluated for four different joint designs (Figure 5) based on the joining technologies previously identified. In the case of the thermoplastically welded joint, the assembly operation sequence can be broken down into a series of distinct material handling, surface preparation, alignment measurement, welding, cooling and testing operations (Figure 6). For each of the other joint designs, similar specialized assembly operation sequences can also be generated.

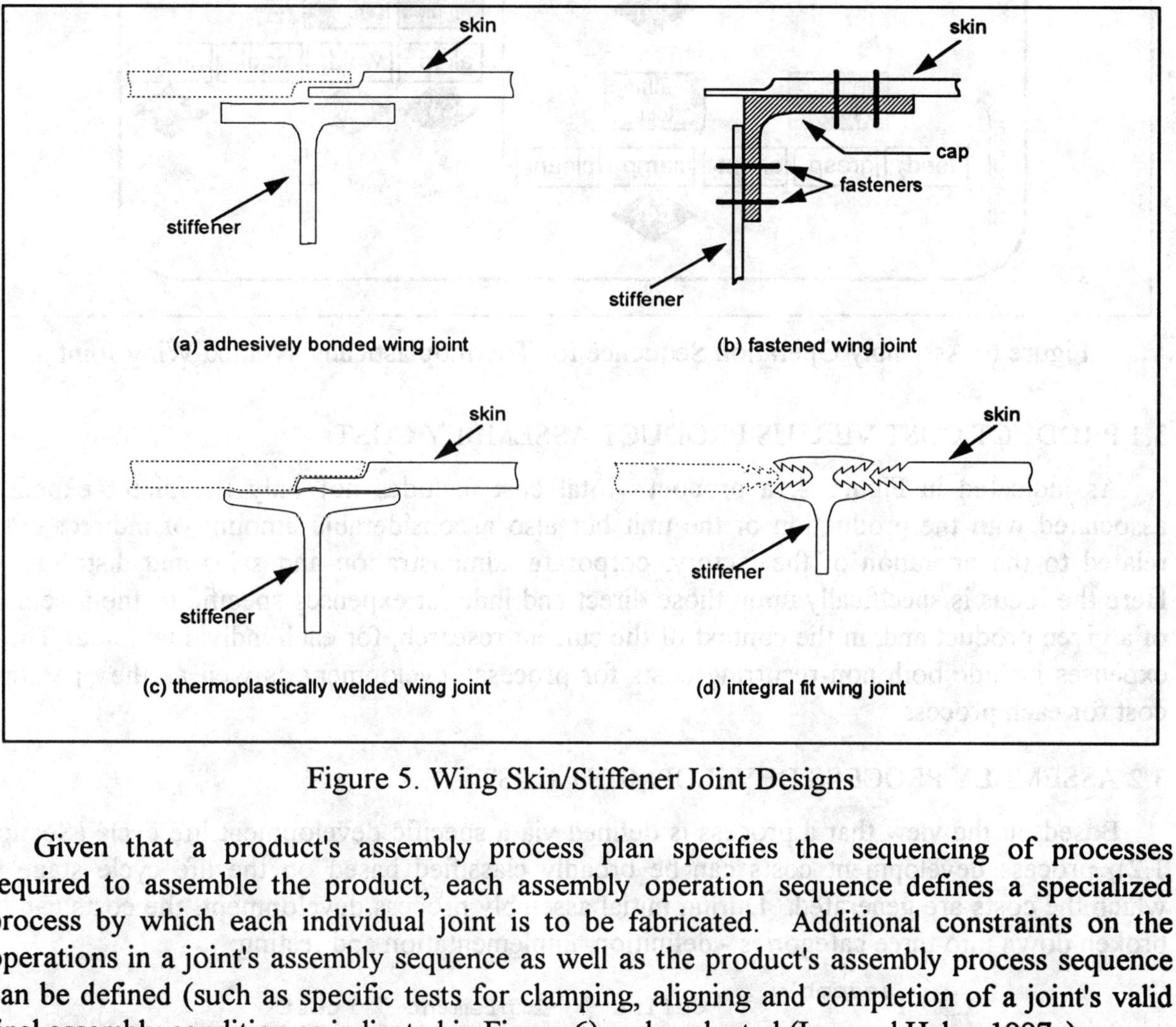

Figure 5. Wing Skin/Stiffener Joint Designs

Given that a product's assembly process plan specifies the sequencing of processes required to assemble the product, each assembly operation sequence defines a specialized process by which each individual joint is to be fabricated. Additional constraints on the operations in a joint's assembly sequence as well as the product's assembly process sequence can be defined (such as specific tests for clamping, aligning and completion of a joint's valid final assembly condition as indicated in Figure 6) and evaluated (Lee and Hahn, 1997a).

3. A COST MODEL FOR THE ASSEMBLY OF COMPOSITE STRUCTURES

In order to identify the primary cost factors in assembling specific types of joints in composite structures, a formalized cost model is needed that allows for assessing both the cost of direct assembly expenses as well as development costs for integrating together sequences of assembly operations.

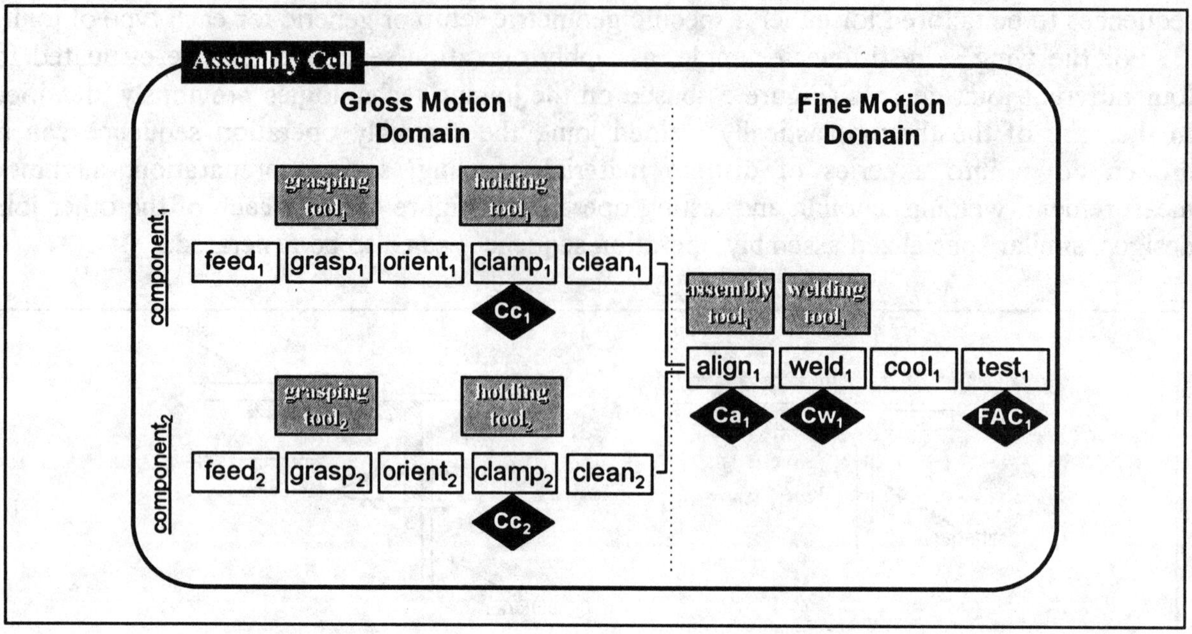

Figure 6. Assembly Operation Sequence for Thermoplastically Welded Wing Joint

3.1 PRODUCT COST VERSUS PRODUCT ASSEMBLY COST

As indicated in Figure 4, a product's total cost includes not only the direct expenses associated with the production of the unit but also a considerable amount of indirect costs related to the operation of the factory, corporate administration and sales and distribution. Here the focus is specifically upon those direct and indirect expenses specific to the assembly of a given product and, in the context of the current research, for each individual joint. These expenses include both non-recurring costs for process development as well as the operating cost for each process.

3.2 ASSEMBLY PROCESS DEVELOPMENT COST

Based on the view that a process is defined via a specific development life cycle (Section 1.2), process development costs can be broadly classified based on the life cycle stage in which the costs are generated. During initial assembly process development, the costs can be broken down into three categories - definition, implementation and testing:

$$c_i^{assembly} = c_{define} + c_{implement} + c_{test} \quad (1)$$

In the case where an assembly process relies upon both predefined generic operations as well as requires the development of other operations specific to the tooling and motions required for the process, the development cost can be redefined based on the development costs of each of the individual operations and the cost of integrating an operation sequence together:

$$c_i^{assembly} = r_1(\Sigma c_{op-exist}) + r_2(\Sigma c_{op-new}) + r_3(\Sigma c_{integrate}) \quad (2)$$

Here, additional weighting factors (r_1, r_2, r_3) are included that assign risk toward the definition, implementation and testing of the operation sequence. Note that while an existing operation's development cost may be much less than a new operation, it is not zero because of the need to tailor and test the existing operation in the sequence context.

3.3 ASSEMBLY PROCESS OPERATING COST

When performing the necessary assembly operation sequence that defines a given assembly process and that will create a specific joint, the assembly process operating cost can be treated to be an element of a product's direct cost - a product's assembly cost. In Figure 4, a product's direct cost is the sum of direct expenses and costs requited to fabricate the product:

$$c_d^{product} = c_{develop} + c_{labor} + c_{material} + c_{expenses} \tag{3}$$

However, some costs are amortized over the entire production volume (e.g. the product's development costs) while aggregating processing costs into a general expenses category. From a cost analysis perspective, this makes it difficult to analyze the cost drivers affecting the assembly of a given joint. Thus, a product's assembly cost can be defined as the cumulative sum of the operating costs of all required assembly processes. For a given assembly process, its operating cost is given by:

$$c_o^{assembly} = c_{develop} + c_{startup} + c_{execute} + c_{refine} + c_{decomm} \tag{4}$$

In this case, the amortized process development cost is identical to $c_i^{assembly}$ and includes a per process execution costs of process startup, process refinement or improvement, and eventual costs of decommissioning the process as well as the direct execution cost of running the assembly operation sequence.

3.4 PREVIOUS WORK

In the area of general assessment of a product's assembly costs, the work known as Design for Assembly (DFA) as defined by Boothroyd, Dewhurst and Knight (1994) has focused on estimating theoretical minimum handling and assembly times for a given product assembly configuration and then ascribing costs per handling and assembly (insertion) operation. Identifying the costs of developing an assembly process or addressing the operations specific to the assembly of composite structures are not addressed.

In the area of cost engineering, methods that focus on cost accounting and estimation primarily for construction has been published (Humphreys and English, 1993). Specific to cost engineering in manufacturing, Creese, Adithan and Pabla (1992) focuses on addressing cost analysis issues in metal manufacturing but provides cost models for metal welding processes. Sims (1995) provides a time- and facilities-based standard costing system that moves away from indirect cost distribution approaches that simply assign overhead costs using production volume and toward a model that directly assigns these costs based specifically on the facilities used in the manufacture of a product. Lotter (1989) provides an operating cost model for different types of assembly workcells but focuses on workcell infrastructure costs such as power consumed, space costs, etc.

4. COMPARATIVE ASSEMBLY COST PERFORMANCE ANALYSIS

Because different types of joining technologies are used in composite structures, providing a methodology by which the costs of assembling specific joint design options is necessary to evaluate the effectiveness of one design over another.

4.1 BREAK-EVEN COSTS

Assuming that a joint's assembly costs can be divided into a set of fixed costs (costs that are invariant with the volume produced) and those that vary as a linear function of produced volume, a joint's break-even cost can be identified by:

$$c_{total} = c_{fixed} + (N * c_{variable}) \tag{5}$$

In this context, c_{total} identifies the total cost for producing N joints, c_{fixed} and $c_{variable}$ are the fixed costs and variable costs and N is the number of joints to be produced. From this, a comparative break-even analysis can be generated for two alternative joint designs by matching total costs ($c_{total}{}^{a} = c_{total}{}^{b}$) and then for a given common production volume N, the break-even point is given by:

$$c_{break\text{-}even}(N) = [c_{fixed}{}^{b} - c_{fixed}{}^{a}] \;/\; [c_{variable}{}^{a} * c_{variable}{}^{b}] \tag{6}$$

Here, after the break-even point, the joint with the lower cost for a smaller volume costs more to produce than the other with higher initial cost.

4.2 CRITICAL OPERATION AND CRITICAL JOINT COSTING

In addition to identifying the break-even costs for different joints and comparing joint costs, critical operations in an assembly operation sequence can be identified based on different factors. These include overall cost, tooling cost and operation execution time among others. Moreover, because some joints must be established prior to others when assembling a composite structure, the cost of using one joining technology over another in a critical joint can be evaluated to determine if there are assembly operation sequencing constraints that can be performed prior to the assembly of a critical joint. This allows for assessing implicit operation parallelism such as simultaneously mounting multiple components (Figure 6).

4.3 COMPARATIVE JOINT ASSEMBLY COSTING

Given the basis of using generalized assembly operations as a methodology for representing the steps necessary for assembling a given joint in a composite structure, the costs for each operation as well as the tooling required to assemble the joint can be explicitly called out as part of the overall cost evaluation of the joint. In performing comparative joint assembly costing, similar generalized operations can be identified across different types of joining technologies. Moreover, the costs inherent to a certain type of technology can be isolated and understood. For the different joint designs proposed for the wing's skin/stiffener joint, the requisite assembly operations and tools demonstrate the similarities and differences in specific cost factors when assembling each of the joints (Table 1).

Table 1. Comparative Assembly Costing for Skin/Stiffener Joints

joint type	*component count*	*fixturing/ workholding*	*assembly operations*	*assembly time*	*assembly cost*
adhesive bonding	2 primary parts	1 assembly jig, 2 work locating templates, 1 adhesive application template, 6 clamps	mounting of parts, align parts, apply adhesive, mate parts, cure adhesive, detach	mount: 10 min, align: 15 min, adhesive: 5 min, mate: 10 min, cure: 120 min, detach: 10 min (2 foot section)	assembly: $210 support ops: $10 parts: $1500 Total: $1720
fasteners	3 primary parts, 24 fasteners/foot	1 assembly jig, 2 hole drilling templates, 2 work locating templates	mounting of primary parts, align parts, drill holes, insert/cinch fasteners, detach	mount: 30 min, align: 20 min, drill: 90 min, fasten: 60 min, detach: 10 min (2 foot section)	assembly: $200 support ops: $150 parts: $2500 Total: $2850
plastic welding	2 primary parts	1 assembly jig, 2 work locating tools, 1 welding unit, 6 clamps	mounting of parts, align joints, thermoplastically weld, cool, detach	mount: 10 min, align: 15 min, weld: 40 min, cool: 20 min, detach: 10 min (2 foot section)	assembly: $300 support ops: $75 parts: $2700 Total: $3075
integral fit	2 primary parts	1 assembly jig, 2 clamps, 1 work support tool	mounting of primary parts, align parts, join, detach	mount: 10 min, align: 10 min, join: 10 min, detach: 10 min (2 foot section)	assembly: $50 support ops: $0 parts: $2000 Total: $2050

REFERENCES

Boothroyd, G., Dewhurst, P., and Knight, W., 1994. Product Design for Manufacture and Assembly, Marcel Dekker, New York, NY.

Creese, R. C., Adithan, M., and Pabla, B. S., 1992. Estimating and Costing for the Metal Manufacturing Industries, Marcel Dekker, New York, NY.

Humphreys, K. K., and English, L. M., 1993. Project and Cost Engineers' Handbook, Marcel Dekker, New York, NY.

Lee, D. E., and Hahn, H. T., 1996a. "A Coordinated Product and Process Development Environment for Design for Assembly", *Proceedings, 1996 ASME Design for Manufacturing Conference (CD-ROM)*, (96-DETC/DFM-1300), Irvine, CA, ASME Press, New York, NY.

Lee, D. E., and Hahn, H. T., 1996b. "Assembly Modeling and Analysis of Integral Fit Joints for Composite Transportation Structures", *Proceedings, 1996 ASME Flexible Assembly Conference (CD-ROM)*, (96-DETC/FAS-1362), Irvine, CA, ASME Press, New York, NY.

Lee, D. E., and Hahn, H. T., 1997a. "Generic Modular Operations for Virtual Manufacturing Process Engineering", *Proceedings, 1997 ASME Computers in Engineering Conference (CD-ROM)*, (DETC97/CIE-4309), Sacramento, CA, ASME Press, New York, NY.

Lee, D. E., and Hahn, H. T., 1997b. "Integral Fit Joint Technologies for Composites", to be published in *Journal of Thermoplastic Composite Materials*.

Lee, D. E. and Melkanoff, M. A., 1993. "Issues in Product Life Cycle Engineering Analysis", *Advances in Design Automation - 1993: Vol. 1*, ASME Design Automation Conference, Albuquerque, NM, pp. 75-86.

Lotter, B., 1989. Manufacturing Assembly Handbook, Butterworths, London, UK.

Matthews, F. L., ed. 1987. Joining Fibre-Reinforced Plastics, Elsevier Applied Science, London, UK.

National Research Council, 1991. Improving Engineering Design: Designing for Competitive Advantage, the Committee on Engineering Design Theory and Methodology and the Manufacturing Studies Board, National Academy Press, Washington, D.C.

Sims, E. R., Jr., 1995. Precision Manufacturing Costing, Marcel Dekker, New York, NY.

Scaling Effects of Composite Thickness on Mechanical Fastening

D. LIU, B. RAJU, J. YOU AND G. CLOUD

ABSTRACT

Mechanical joints are suitable for joining composite laminates, especially thick-section composites. In most studies on mechanical joints, the interactions between pin diameter (D) and composite dimensions such as width (W) and distance from hole center to composite end (e), i.e. W/D and e/D ratios, were the primary concerns. However, as composite thickness increases, the contact interaction between the pin and composite becomes more complicated. This study investigated the mechanical joint efficiency of a composite material made of woven glass fabric and phenolic matrix. Sixteen joint configurations based on four composite thicknesses and four pin diameters were examined. Both experimental method and finite element analysis were performed in this study. Results showed that thick composites with small pins and thin composites with large pins had lower efficiencies of joint stiffness and joint strength than those having similar dimensions between pin diameter and composite thickness.

INTRODUCTION

There are obvious advantages in utilizing very thick composites for the bodies or hulls of combat vehicles. The joining of these components to substructure or other components must be, to a large degree, by mechanical fastening in order to gain high load-carrying capability and ease of structural disassembly. However, the testing of thick composites can become very costly as the thickness rapidly increases. Accordingly, it will prove to be very valuable, if the conclusions obtained from the analysis of relatively thin composites can be extended to predict the behaviors of very thick composites, i.e. if the thickness scaling can be established. This study is to determine the scaling effect of thickness on mechanical fastening of thick composites.

D. Liu, J. You, and G. Cloud, Dept. of Materials Science and Mechanics, Michigan State University, East Lansing, MI 48824
B. Raju, U.S. Army - TARDEC, Warren, MI 48397

Improvements in our understanding of the mechanics of composite materials, and advances in the technologies of making and using them, are greatly extending the range of potential applications of these materials. Until recently, applications of composite materials have tended to be limited to thin shell structures such as aircraft fuselages or pressure vessels. There is increasing interest in utilizing composites for thick section structures, including submarine hulls and combat vehicle bodies. The extension of these applications from thin shells to thick sections is, however, not trivial. In fact, it requires careful modifications in almost every aspect of composite technology, e.g. manufacturing, testing, analysis, and design [1]. Taking composite joining as an example, the design of a practical adhesive joint may turn out to be impractical if the composite thickness becomes too large. If mechanical fastening is to be used, the change from quasi-two-dimensional thin shells to three-dimensional thick sections can greatly alter the contact mechanisms between the fastener and the composite; subsequently the stress distributions in the composite will be different from those expected on the basis of conventional design procedures, experiments, or analyses. As a consequence, the failure mechanism may change, and unexpected failures or expensive over-design might result. That is to say, unwise extension of thin-wall design procedures to thick sections can cause the design process to lose its major significant assets, those being predictability and control.

The design of fastening systems for composites has tended to be done on a case-by-case basis, with great reliance on testing. A reason is that the experiential base, derived from mechanics analysis, controlled experiments, and field experience, is not yet large enough, especially for new applications of thick composites. This project is one component of an effort to advance our knowledge of mechanical fastening of composites so that joint design is more effective by virtue of being more scientific and less phenomenological.

Study of mechanical fastenings for composites with various thicknesses would be very tedious if no "unified rule" is included or developed. Alternatively, the study may be very efficient if some fundamental knowledge for thin composite mechanical fastening can be "transformed into" that for thick composite mechanical fastening. In other words, it will prove to be very useful if the fundamentals of a scaling effect of thickness can be extracted from the existing information for thin composite fastening and some extra, but limited, studies on thick composite fastening. Although there have been already some studies on the scaling effect of composite materials, they are more focused on the scaling effect of in-plane dimensions rather than on the effect of thickness change.

Scaling effect has long been considered as a very important study for engineering designs [2]. If scaling effect is fully understood, many experimental and computational investigations may be saved. However, the conventional studies on scaling effect have been limited to two-dimensional structures. As more thick-section composites are used in engineering applications, thickness effect can no long be neglected. This study presents a unique investigation on thickness scaling which is not available in the literature.

Thick-section composites are suitable for combat vehicles due to their high weight saving and impact resistance. This study investigates the joining mechanism for thick-section composites and is critically important to composite combat vehicle designs.

INVESTIGATIONS

1. EXPERIMENTAL INVESTIGATION

1.1 Composite Material

In studying the scaling effect, double-lap and single-pin joints of various hole sizes and thicknesses were investigated. Both experimental and computational investigations were performed. The composite material of interest was made of phenolic matrix and S-2 glass fabrics of plane-weave, i.e. (0/90). A 45° difference of fiber orientation was imposed in laying up the fabrics alternatively, resulting in composite laminates of quasi-isotropic stacking sequence, i.e. $[(\underline{0/90})/(\underline{45/-45})]_{ns}$ where n is layer number and s stands for symmetry with respect to midplane.

1.2 Thickness Effect

The contact mechanism between the steel pin and the composite material was of primary concern in studying the thickness effect. Four composite thicknesses, 8-ply, 24-ply, 40-ply, and 80-ply with the ply thickness of 0.01", were investigated. Given a constant hole size, the load-bearing capacity of composite laminates should be linearly proportional to the thickness provided the contact mechanisms for the composite laminates with various thicknesses are identical. However, if the contact mechanisms are different, the stress distributions around the hole edges would not be the same.

1.3 Hole Size Effect

Another study on contact mechanism was to investigate pins with various rigidities, e.g. various pin sizes. Although the pin size effect for thin composites has been very well established, it was primarily a study on in-plane (two-dimensional) effect instead of thickness (three-dimensional) effect. Since the pin size affects the contact mechanism which subsequently affects the stress distribution around the hole edge, pins (and also holes) with diameters of 1/8", 1/4", 1/2", and 1" were investigated.

1.4 Hole Drilling

In order to reduce the possible damage around the hole edge during machining, carbide drills were used. For small holes such as 1/8" and 1/4", drilling was done in one process. However, for large holes like 1/2" and 1", a smaller hole was first created before using the carbide drills for final sizing.

Figure 1 shows the schematic diagrams of a double-lap, single-pin joint from front view and side view. Although there are four thicknesses and four hole sizes, the width and the distance between the hole center and the specimen edge are kept constant, i.e. 4" and 3", respectively.

In experimental investigations, assemblies of composite laminates with various thicknesses and pin sizes were loaded up to failure in order to evaluate the stiffness and strength of the mechanical fasteners. This investigation played an important role in mechanical fastening designs as it addressed the important issue regarding the interaction between the pin size and the composite thickness.

2. COMPUTER SIMULATION

2.1 Finite Element Mesh

In computational study, a commercial finite element program, namely ABAQUS, was utilized in parameter analysis. Due to the symmetry of joint geometry, loading condition, and boundary condition, only a quarter part of each three-dimensional structure was modeled. In addition, a contact interface was established in between the pin and the composite laminate. Consequently, as many as 16,000 finite elements were used in computer simulations.

2.2 Contact Stresses

Although experimental technique gave both stiffness and strength of the mechanical fasteners, it was not able to offer the contact stresses between the pin and the composite. However, the finite element method was able to present all stresses required for joining analysis.

RESULTS

1. EXPERIMENTAL RESULTS

1.1 Strengths

Sixteen cases were investigated both experimentally and computationally. Table 1 gives the experimental results of the strengths of the sixteen mechanical fasteners. It can be concluded that thick plates with small holes and thin plates with large holes seem to give lower strengths.

1.2 Damage Mode

Since the composite material under investigation was woven-fabric laminate, it was difficult to drill a hole through the laminate thickness without causing any delamination. Before being tested, each specimen was examined by X-ray radiography. After testing, bearing damage with local delamination was also found as the only damage mode.

2. COMPUTATIONAL RESULTS

In order to understand the interaction between the pin size and the composite thickness, stress distributions on the contact surfaces were examined. Based on the contact stresses obtained from computer simulation and the bearing strength of composite laminate, i.e. 22.21 ksi, the analytical strengths for individual cases were calculated. The results are also shown in Table 1. Figure 2 shows the comparsion between the experimental and computational results. Apparently, the trend of the analytical results seem to match with those from experiments except in the case of 80-ply laminate with 0.125" hole.

CONCLUSIONS

The study of scaling effect for composite thickness provided some basic understanding of the use of mechanical fasteners in joining thick composites. Based on experimental and computational studies, it is found that the joining strengths for all sixteen cases are not proportional to the bearing area. In other words, both the joint stiffnesses per unit bearing area and the joint strengths per unit bearing area are different for different cases. Theses results indicate that there exist different contact mechanisms for the cases investigated. That is to say that thickness scaling effects do exist in mechanical joints.

REFERENCES

1. Mechanics of Thick Composites, ed. Y. D. S. Rajapakse, AMD-Vol. 162, The American Society of Mechanical Engineers, 1993.

2. Workshop on Scaling Effects in Composite Materials and Structures, NASA Langley Research Center, Hampton, VA, November 15-16, 1993, Report No. 94-6, Institute for Mechanics and Materials, University of California, San Diego, CA.

Table 1: Joint strengths from both experiments and computations.

unit: lbs

	8-ply	24-ply	40-ply	80-ply
D/8	214 (178)*	541 (602)	800 (641)	643 (1169)
D/4	346 (359)	1499 (1078)	2135 (1778)	2383 (1948)
D/2	423 (707)	2396 (2249)	4288 (3871)	7808 (7928)
D	762 (1393)	4313 (4498)	7063 (7750)	12416 (13328)

* computational results in parentheses

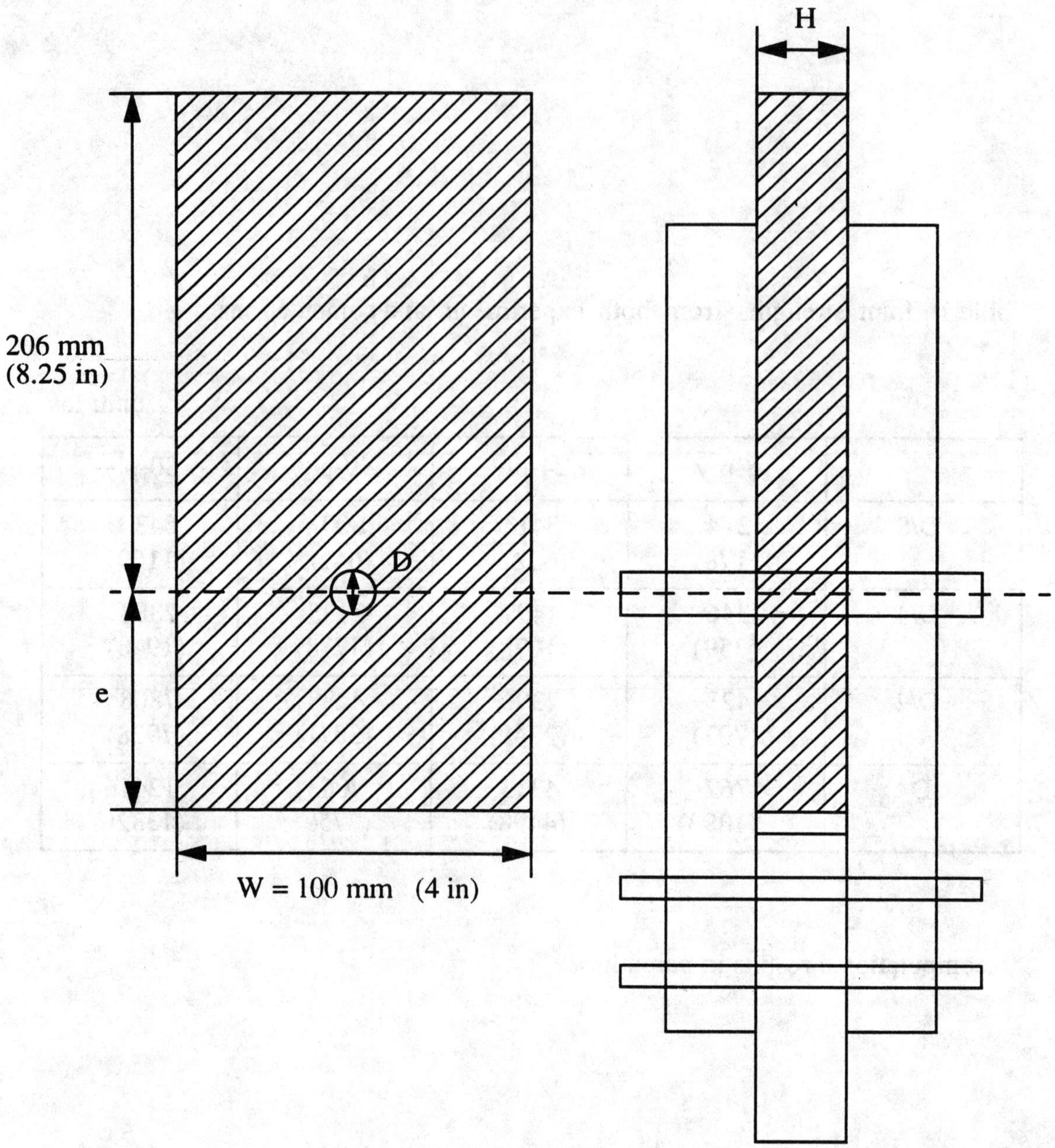

Fig. 1 - Schematic diagram for double-lap, single-pin joint.

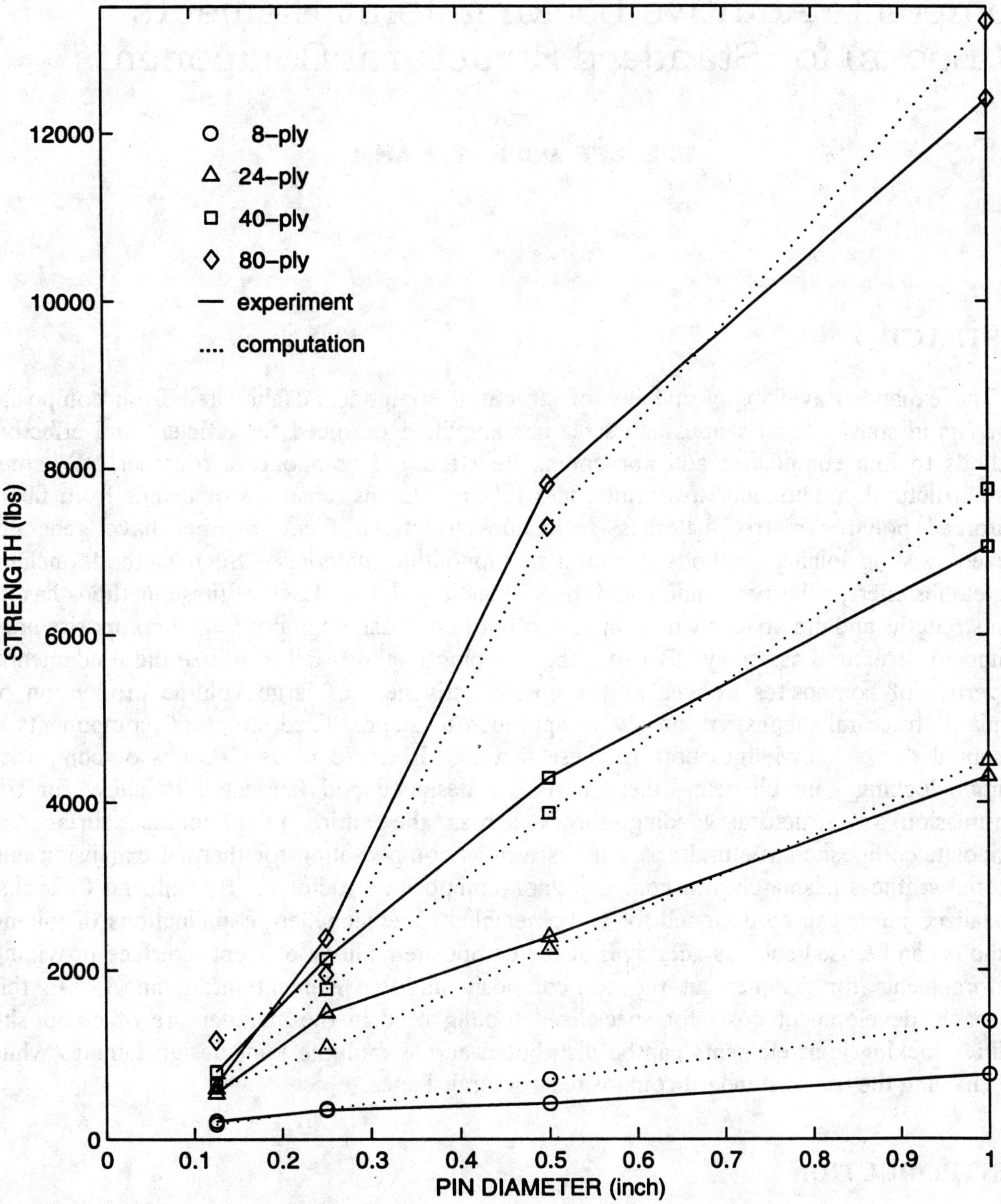

Fig. 2 – Comparison of joint strengths.

Composite Additive Locking Joint Elements (C-Locks) for Standard Structural Components

D. E. LEE AND H. T. HAHN

ABSTRACT

The expanded availability and use of structural components fabricated from composite materials in standardized shapes and sizes has amplified the need for efficient and effective methods to join composites and non-composite structural components together. Whether these structural components are formed into tubes, columns, channels or beams from fiber-reinforced polymer matrix materials, techniques to attach them together have generally utilized existing joining methods common to monolithic materials. Such methods include discrete fasteners, adhesive bonding and thermoplastic welding. Each of these methods has its own strengths and drawbacks when applied to both composite/composite and composite/non-composite structural assembly. Because these joining methods fail to utilize the fundamental properties of composites as well as the cost-effectiveness of large volume production of standard structural shapes, an innovative approach to standardized structural components is presented that acknowledges both of these factors. Referred to as C-Locks or composite additive locking joint elements, these parts are designed and fabricated to allow for the transmission of structural loading forces across the entire joint contact surface in composite/composite structural assembly as well as compensating for thermal expansion and material stiffness mismatches in composite/non-composite structures. By utilizing C-Locks, specialized joints can be designed for and assembled together where combinations of joining methods can be used such as adhesives at the composite/additive element interface providing reinforcements for fasteners at the non-composite/additive element interconnect. In this approach, development costs for specialized tooling used in the manufacture of composite additive locking joint elements can be distributed across multiple joint design families while still enabling the use of standard composite structural shapes.

1. INTRODUCTION

Given the ability to design and manufacture structural components with specific directional material properties that satisfy multiple loading criteria using composite materials, the use of standardized structural components fabricated from fiber reinforced composites

David E. Lee and H. Thomas Hahn, Integrated Manufacturing Engineering Program for Advanced Transportation Systems, UCLA School of Engineering and Applied Science, Box 951597, Los Angeles, CA 90095-1597. *david@ime.ucla.edu* & *hahn@seas.ucla.edu; http://ime.ucla.edu/~david*

might be viewed as a less than optimal application of composites. However, a large number of applications exist which do not require uniquely designed fiber arrangement sequences or complex component geometries in most structural members and instead focus on utilizing the strength/weight performance and the durability properties of composites. In addition, design standards are available focusing on the use of standardized structural components and that allow an engineer to apply straightforward analytical tools to evaluate the loading performance of structures built from these standard shapes.

1.1 STANDARD STRUCTURAL COMPONENTS

When referring to standard structural components this research has focused primarily on structural elements that have a constant two-dimensional cross section along a principal axis. These elements may be solid material across the cross section (Figure 1a), utilize thin walls to form a closed section (Figure 1b) or an open section (Figure 1c). Each element has material properties that can vary both in cross section and along its principal axis but maintains constant geometric properties along both. In addition, standard components used principally in reinforced concrete structures are also considered (Figure 1d) (MacGregor, 1994).

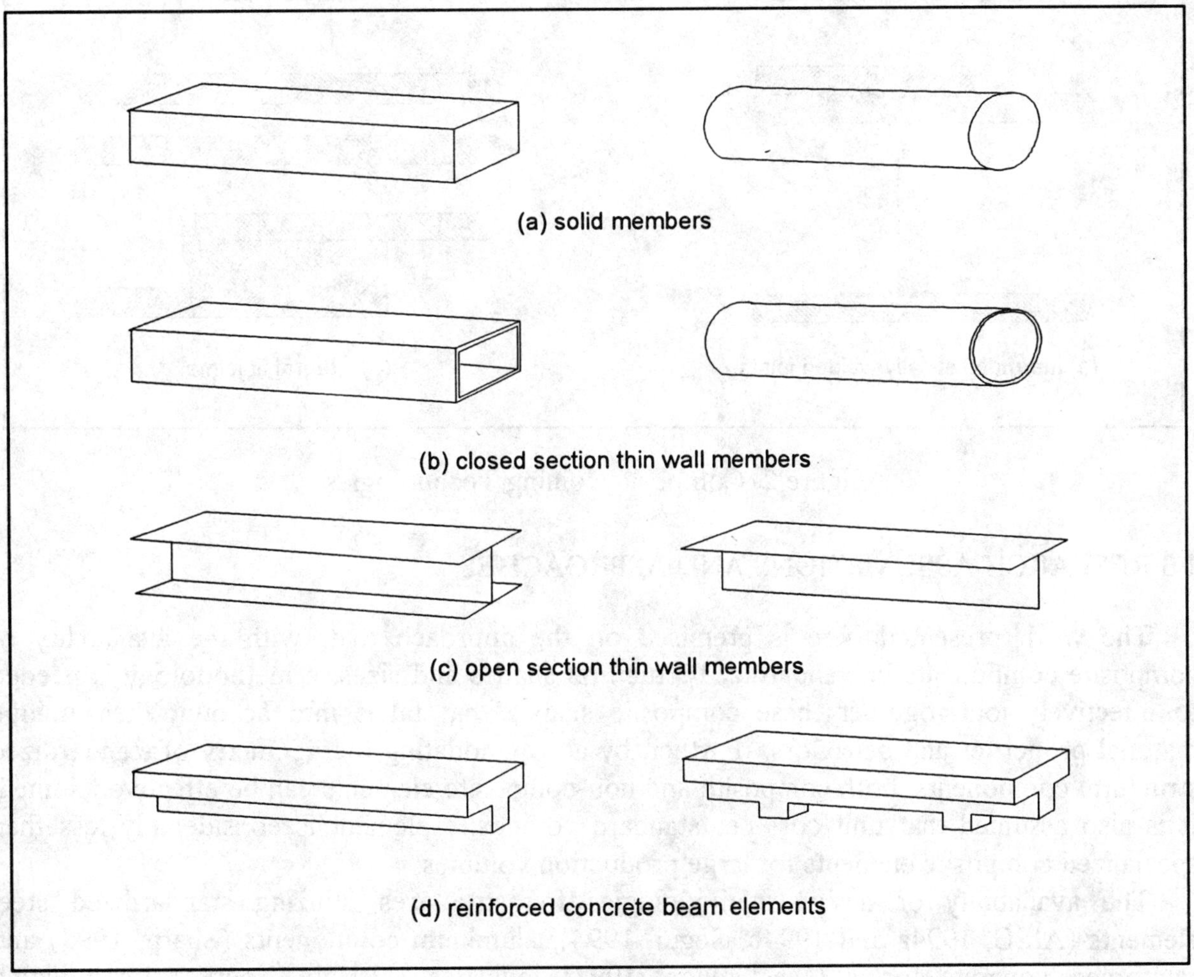

Figure 1. Standard Structural Components

1.2 JOINING TECHNOLOGIES FOR COMPOSITES

The primary types of joining technologies used in composite structures have been adhesive bonding (Figure 2a) and discrete mechanical fasteners (Figure 2b) (Matthews, 1987, Messler, 1993). In addition, two other methods have been investigated for use in joining composite components: thermoplastic welding (Figure 2c) (Grimm, 1990, Howie, Gillespie and Smiley, 1993) and integral fit joints (Figure 2d) (Lee and Hahn, 1997).

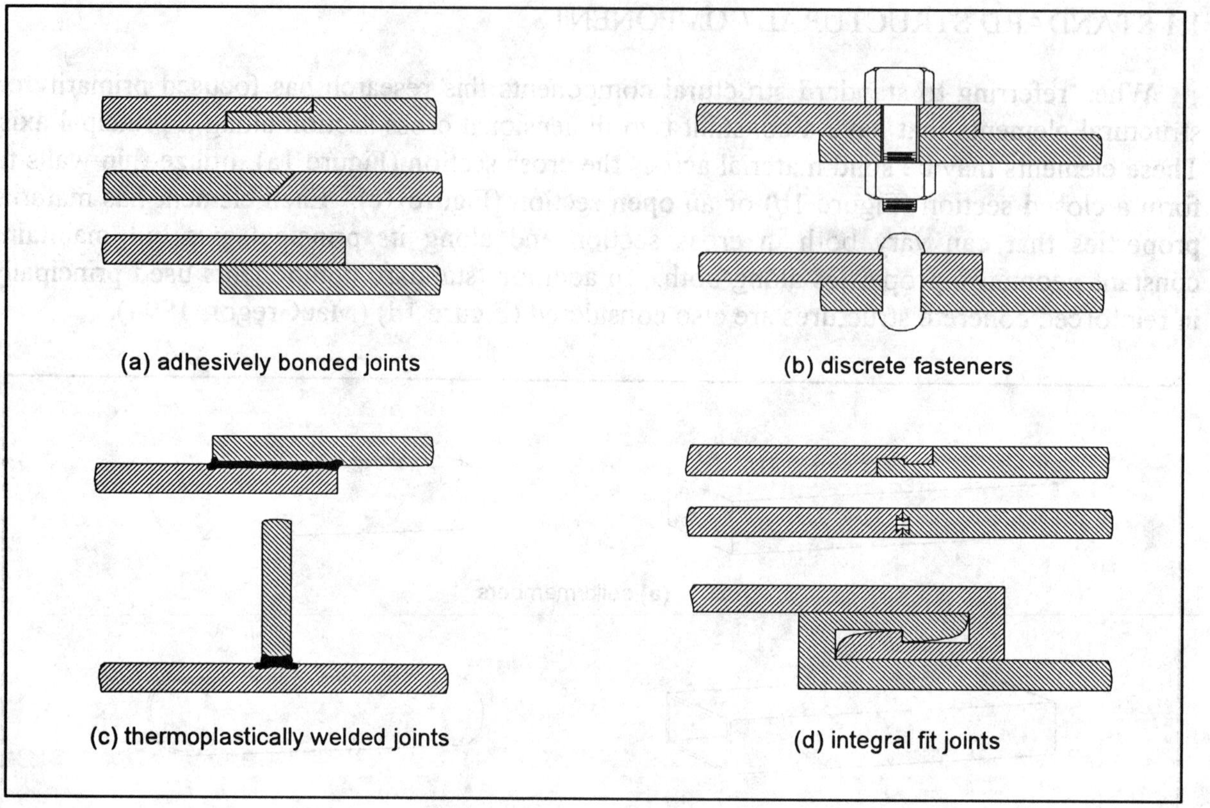

Figure 2. Composite Joining Technologies

1.3 RESEARCH ASSUMPTIONS AND APPROACHES

The work presented here is premised on the approach that, with the availability of composite components in standardized structural shapes and sizes, a methodology is needed to effectively join together these composite shapes that takes into account their unique material properties and behaviors. Further, by accommodating the geometry of standardized structural components, both composite and non-composite elements can be effectively joined. It is also assumed that unit cost per standard composite element is considerably less than specialized composite elements for large production volumes.

The availability of design methodologies for structures utilizing standardized steel elements (AISC, 1994a and 1994b, Segui, 1994), aluminum components (Sharp, 1993) and reinforced concrete forms (MacGregor, 1997) indicates a sizable market for utilizing standardized shapes in composites and methods for joining them together if specific material properties required by these design methods can be readily provided for composite components. This would provide structure designers with the data necessary to analyze the performance and cost-effectiveness of composite structural elements as compared to currently available materials as well as provide opportunities for using these composite elements in applications suited to the load bearing capabilities of the components.

Because of the focus of this research is specifically on joining standardized composite shapes, it differs somewhat in its approach as compared to our previous work that has addressed joining technologies for composite components in advanced transportation systems (Lee and Hahn, 1996). The joint elements discussed in this paper are primarily intended for structural elements used in civil infrastructure applications such as buildings, bridges and heating/ventilation systems. However, it is our intent to use the results presented here for these applications in structural frameworks for transportation systems.

2. JOINING COMPOSITE STRUCTURAL ELEMENTS

As noted previously, composite structural elements provide an opportunity to expand the use of composite materials in applications that have been previously been dominated by monolithic materials or reinforced concrete. However, to use these elements, new methods for joining together standardized composite shapes are needed that differ from existing joint technologies that take into account the poor bearing stress performance of fiber reinforced composites and the need to transfer concentrated forces and moments across the joint interface without overly disrupting the fiber architecture of a composite structural element.

2.1 EXISTING APPROACHES TO JOINING STRUCTURAL ELEMENTS

In steel frame structures, the principal joining methods have been to use discrete fasteners (such as structural bolts) and welds to join one structural element to another via a connector element such as shear plates, corner angle plates, and flange stiffeners (AISC, 1994b). For non-structural joints, aluminum structures have been bonded using adhesives but the principal types of structural joints in aluminum have also been discrete fasteners and welds (Sharp, 1993). In the case of reinforced concrete structures, structural elements may be cast in place so that steel reinforcements are tied together prior to emplacing the concrete or formed and cured off-site and then placed in position with a curing layer of concrete in between elements (MacGregor, 1997).

For interfacing composite structural elements or composite/non-composite elements together, these identified joining technologies may not all be viable. In the case of using discrete fasteners, localized disruptions of the fiber architecture are incurred that lower the moment response of the composite members. For thermoplastically formed composite structural elements, welding shapes together may be quite appropriate but the joint may not have the load bearing capacity of a similar steel weld nor are the fiber architectures of the two elements linked to transfer the applied load. Using adhesives to join composites is a tested process but incurs joint loading modes such as peel stresses. For reinforced concrete, methods of tying steel rebar with random mat fibers may be challenging at least and the alkali content of the concrete may have a negative effect on the fibers used in a composite element.

2.2 COMPOSITE INTERLOCKING JOINT ARCHITECTURES

Rather than using existing approaches to joining composite structural elements, two other methods can be examined: interlocking joint architectures and additive locking joint elements (C-Locks). In the case where the composite structural elements can be designed and manufactured to fit a specific application or where the elements are designed to be interconnected, interlocking joint architectures for composites can be used. When loading on members is primarily axial, a system of composite struts that fit end-to-end might be appropriate (Figure 3a). If loads are applied transversely to the members, an arrangement similar to the interlocking beam and column elements in Figure 3b can be applied. Each of these architectures expands the load bearing area of the joint by the design of the common joint interfacing as well as avoiding the use of fasteners and/or adhesives in maintaining the joint connection. These joint architectures are examples of composite integral fit (CIF) joints (Lee and Hahn, 1997).

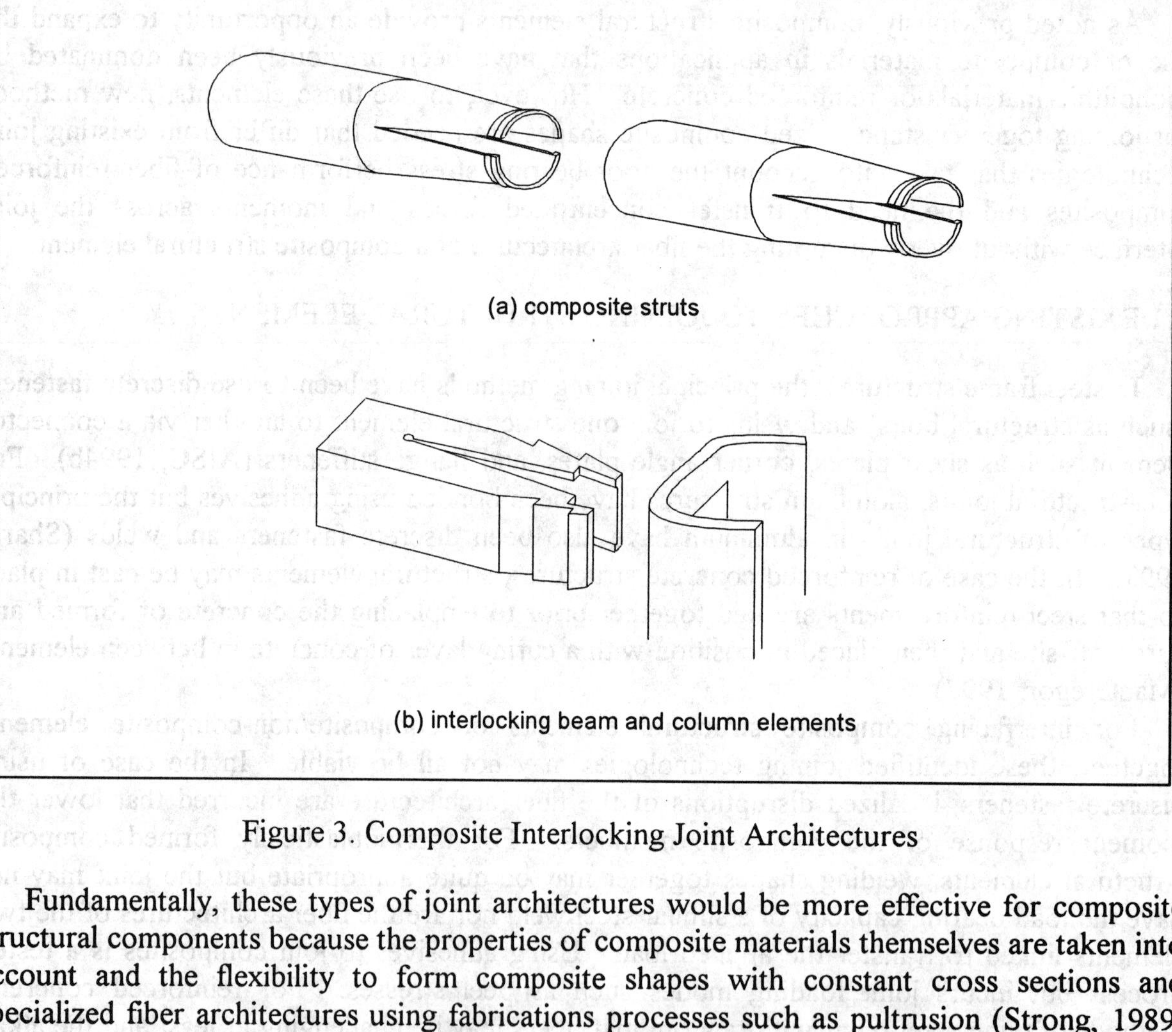

(a) composite struts

(b) interlocking beam and column elements

Figure 3. Composite Interlocking Joint Architectures

Fundamentally, these types of joint architectures would be more effective for composite structural components because the properties of composite materials themselves are taken into account and the flexibility to form composite shapes with constant cross sections and specialized fiber architectures using fabrications processes such as pultrusion (Strong, 1989, Larock, Hahn, and Evans, 1989). However the development cost of taking an interlocking joint architecture to full scale production and the lack of familiarity and experience with these joining approaches indicates the need to enhance the use of standardized composite shapes.

3. COMPOSITE ADDITIVE LOCKING JOINT ELEMENTS (C-LOCKS)

Rather than attempting to bring an interlocking joint system to market and having to compete with the experience of designers in applying standard structural shapes, an innovative approach referred to as composite additive locking joint elements or C-Locks is introduced. These joint elements serve as the joint interface between pairs of composite structural members or pairs of composite/non-composite members and seek to reduce or eliminate the need to use fasteners or adhesives in the joint region.

3.1 CLASSIFYING C-LOCKS

C-Locks can be classified based on multiple criteria. The criteria selected here include: a) the types of loads principally handled in the joint region; b) the shape types of elements being joined; c) the profile of the C-Lock interface; d) the types of materials being joined; e) the material composition of the C-Lock itself. These criteria are primarily identified based on joining the standard components illustrated in Figure 1 - constant cross-section axial members and are demonstrated in the following sections.

In the case of the principal load types handled, these include:

- shear
- tension
- compression
- torsion
- multi-axial

With respect to element shape types joined, the pairings are:

- same pairs (i.e. elements have the same profile and geometric dimensions)
- similar pairs (i.e. elements have the same profile but different dimensions)
- dissimilar pairs (i.e. elements have different profiles)

For C-Lock interface profiles, these are:

- intraprofile (i.e. fits within the cross section profile)
- profile (i.e. matches the cross section profile)
- extraprofile (i.e. extends around the cross section profile)

Looking at the element material types joined, these include:

- composite/composite (which can be specialized further (e.g. thermoplastic/ thermoset, etc.)
- composite/non-composite
- non-composite/non-composite

Finally, C-Locks can be classified based on the type or types of composite materials used in the additive element.

3.2 C-LOCKS FOR CLOSED SECTION THIN WALL MEMBERS

For closed section thin wall members such as rectangular cross section composite tubes, C-Locks that allow the joining of beam elements (Figure 4a), beam and column elements (Figure 4b) and column and footing elements (Figure 4c) can be identified. For both the beam/beam and beam/column joints, the C-Locks are intraprofile elements - they fit within the profile of the tube cross section and serve to handle shear loads that occur between the structural elements. Additional features illustrated include tapered entry interfaces, raised locking lips for maintaining C-Lock/structural element contact and extended contact surfaces for bearing loads and/or bonding. The column/footing C-Lock is an extraprofile compressive fitting that resists torsional loading and provides distributed contact load surfaces.

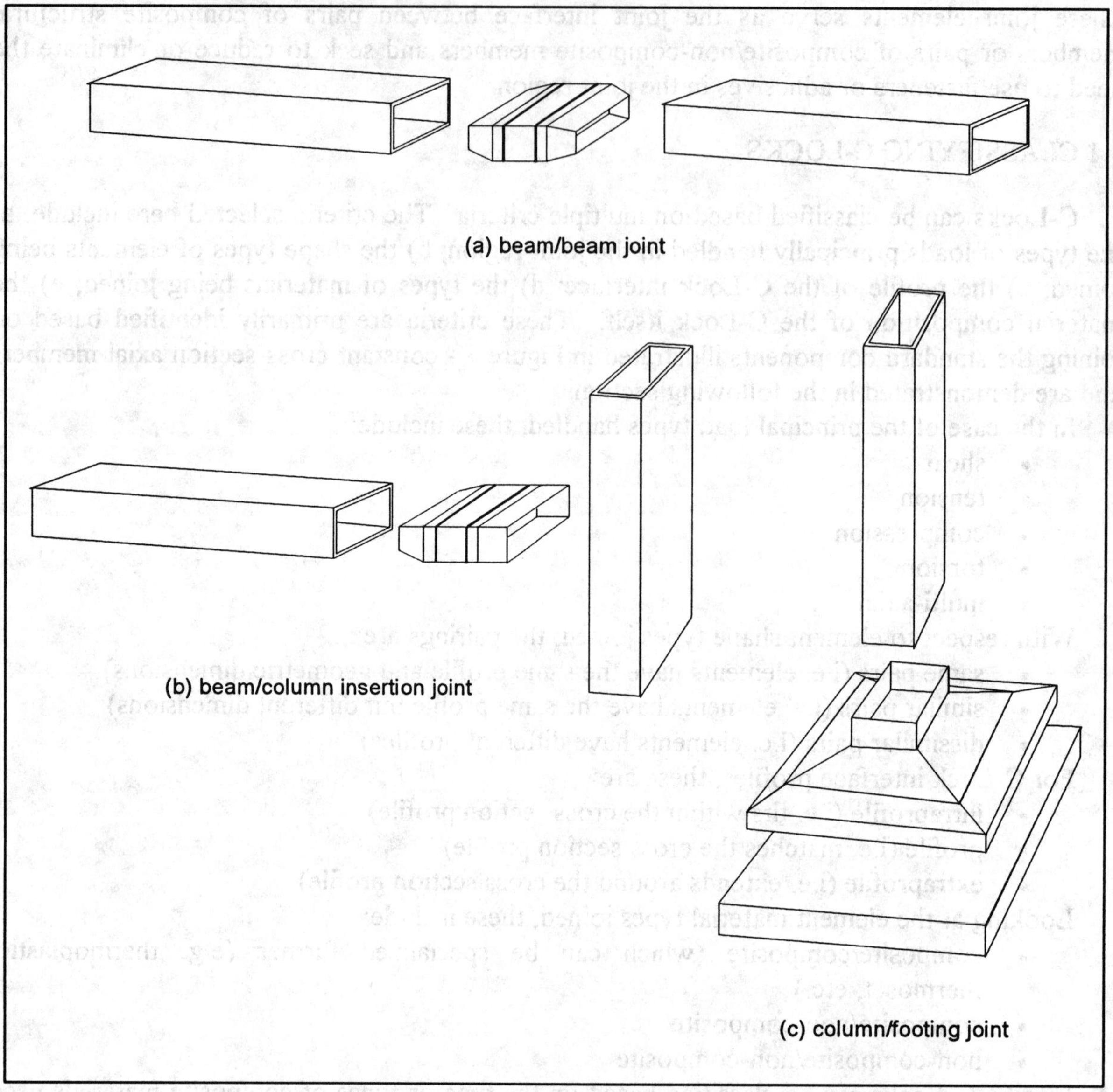

Figure 4. Rectangular Section Tube Intraprofile C-Locks

3.3 C-LOCKS FOR OPEN SECTION THIN WALL AND SOLID MEMBERS

For open section thin wall members and solid members, C-Locks that are extraprofile elements can be used. In the case of common structural element such as an I-beam, a C-Lock can transmit flexural loads between the flange elements in shear and resist torsional loads as in Figure 5a. For the beam/column extraprofile joint in Figure 5b, the C-Lock accepts the shear load imposed by the beam and transmits it to the column elements that are tied together with the same C-Lock. Internal to each of these C-Locks are tapers on the all sides to aid in joint assembly as well as maintain joint contact. In the case of the column/footing joint, the C-Lock is thickened at the flange/lock contact surfaces.

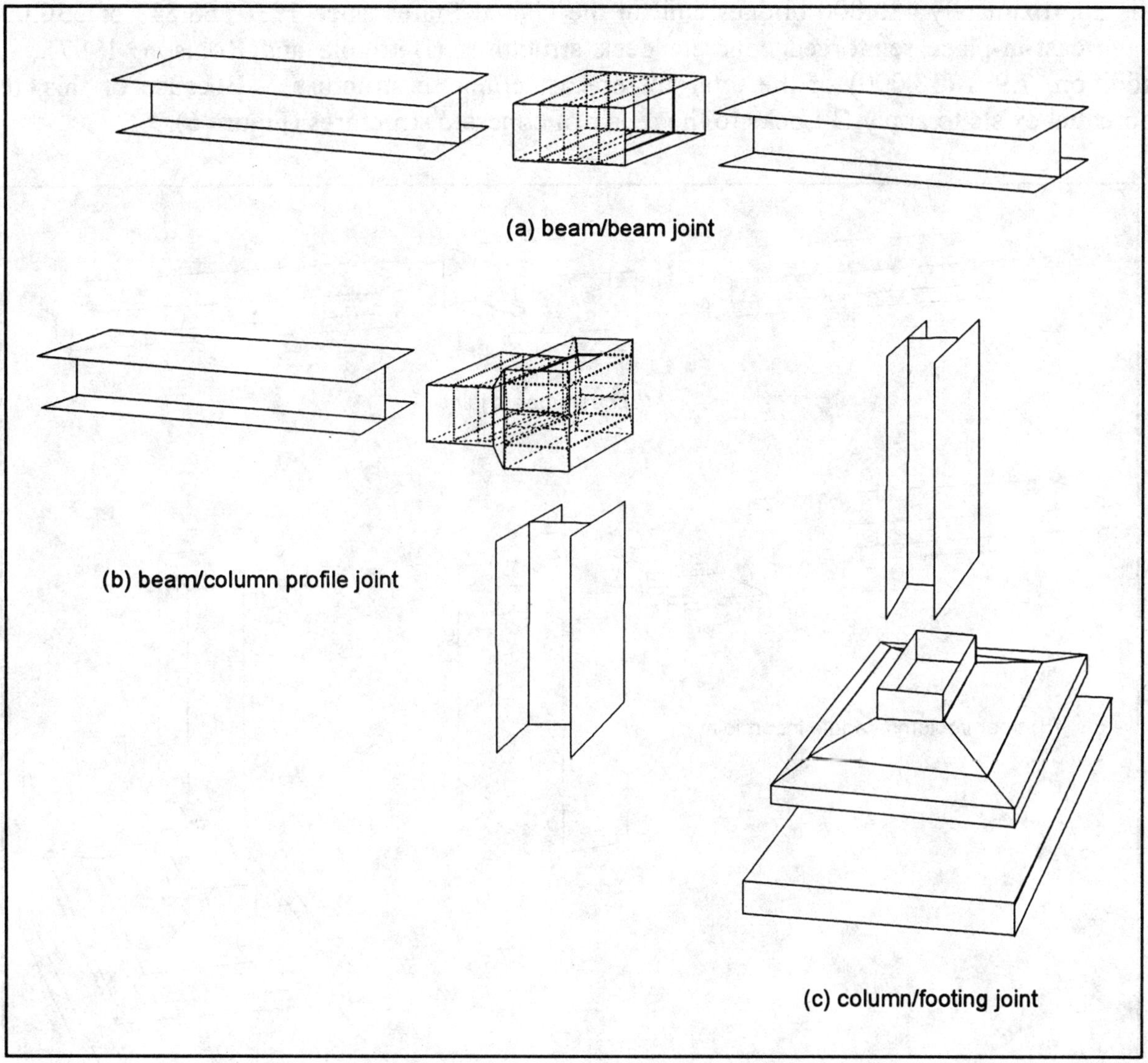

Figure 5. I-Beam Extraprofile C-Locks

4. APPLICATIONS OF C-LOCKS

Because this research has primarily focused on standard structural shapes of constant cross section as primarily used in civil infrastructure designs, there are multiple applications involved in infrastructure repair where C-Locks can be utilized and evaluated. Also, the issues surrounding joining of dissimilar materials and overall life cycle cost of using these elements are considered.

4.1 USING C-LOCKS IN INFRASTRUCTURE REPAIR APPLICATIONS

Because of the ubiquitous application of reinforced concrete in infrastructure projects, the use of C-Locks to repair and rehabilitate these structures can and should be considered. Of the approximately 480,000 bridges built in the United States since 1930, 68.8% or 330,000 have cast-in-place reinforced concrete deck structures (Bettigole and Robison, 1997). In addition, 7.9% (38,000) of the total are precast concrete structures. Because of this, the potential exists to apply C-Locks to the repair of concrete structures (Figure 6).

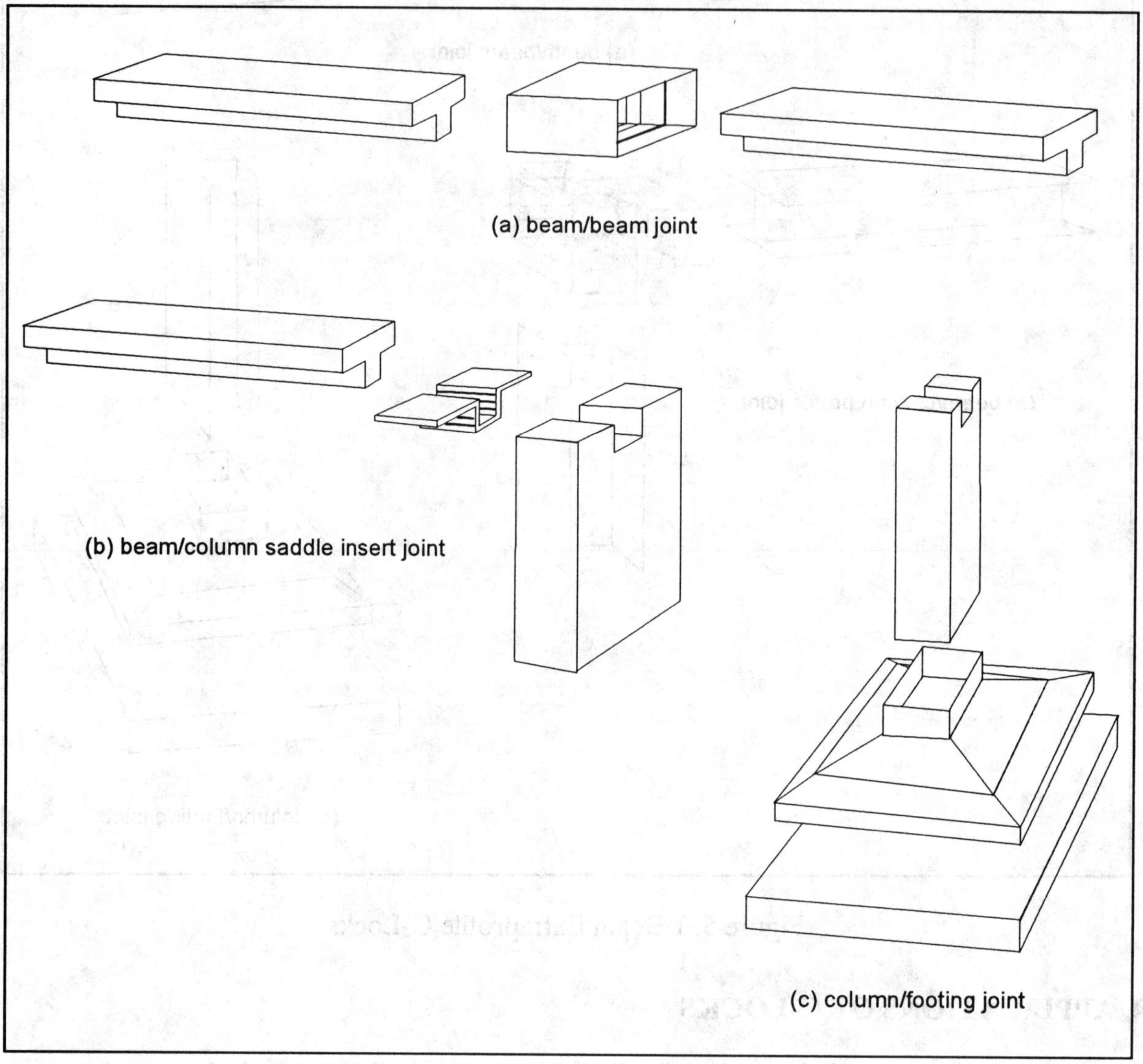

Figure 6. Applying C-Lock Elements to Reinforced Concrete Structural Components

In addition, given the availability of standard structural shapes for construction applications such as I-beams, angles, and tubes (AISC, 1994a), C-Locks might be used in new designs and could be applied in the repair of seismically ruptured welds in over 150 steel framed buildings generated as a result of the Northridge 1994 earthquake (Willman, 1996).

4.2 JOINING DISSIMILAR MATERIALS USING C-LOCKS

As indicated in the classification of C-Locks, the material types of the standard shapes to be joined by a C-Lock can and should be identified. Given the differences in coefficients of thermal expansion (CTE), C-Locks can be designed to accommodate differences in CTE between dissimilar materials. Additionally, the fabrication of a C-Lock can incorporate wear or impact surfaces such as Teflon™ or Kevlar™ panels for specialized applications.

4.3 LIFE CYCLE COSTING OF C-LOCKS

Because the structures in which C-Locks may be applied have operational lives from 30 to 100 years, the issue of the cost of using C-Locks in a structure's overall life-cycle cost must be considered (Lee and Melkanoff, 1993, Bettigole and Robison, 1997). This may involve accelerated testing procedures to assess multi-axial loading of a joint (and subject to different environmental conditions for exposed structures such as bridges). Issues of servicing and maintaining a C-Lock in the field as well as repair and replacement are of concern for structures that will exist for many years to come.

ACKNOWLEDGMENTS

This research has been supported by the Integrated Manufacturing Engineering (IME) Program for Advanced Transportation Systems at UCLA under sponsorship by the U.S. Department of Transportation and the Industrial Affiliates of the IME Program.

REFERENCES

AISC, 1994a. Manual of Steel Construction: Load and Resistance Factor Design (LRFD), v. 1: Structural Members, Specifications, & Codes, 2nd Ed., American Institute of Steel Construction, Chicago, IL.

AISC, 1994b. Manual of Steel Construction: Load and Resistance Factor Design (LRFD), v. 2: Connections, 2nd Ed., American Institute of Steel Construction, Chicago, IL.

Bettigole, N., and Robison, R., 1997. Bridge Decks, ASCE Press, New York, NY.

Howie, I., Gillespie, J. W., Jr., and Smiley, A. J., 1993. "Resistance Welding of Graphite-Polyarylsulfone/ Polysulfone Dual-Polymer Composites", *Journal of Thermoplastic Composite Materials*, v. 6, pp. 205-225.

Larock, J. A., Hahn, H. T., and Evans, D. J., 1989. "Pultrusion Processes for Thermoplastic Composites", *Journal of Thermoplastic Composite Materials*, v. 2, pp. 216-229.

Lee, D. E., and Hahn, H. T., 1996. "Assembly Modeling and Analysis of Integral Fit Joints for Composite Transportation Structures", *Proceedings, 1996 ASME Flexible Assembly Conference (CD-ROM)*, (96-DETC/FAS-1362), Irvine, CA, ASME Press, New York, NY.

Lee, D. E., and Hahn, H. T., 1997. "Integral Fit Joint Technologies for Composites", to be published in *Journal of Thermoplastic Composite Materials.*

Lee, D. E. and Melkanoff, M. A., 1993. "Issues in Product Life Cycle Engineering Analysis", *Advances in Design Automation - 1993: Vol. 1*, ASME Design Automation Conference, Albuquerque, NM, pp. 75-86.

MacGregor, J., G., 1997. Reinforced Concrete: Mechanics and Design, Prentice Hall, Upper Saddle River, NJ.

Matthews, F. L., ed. 1987. Joining Fibre-Reinforced Plastics, Elsevier Applied Science, London, UK.

Messler, R. W., Jr., 1993. Joining of Advanced Materials, Butterworths-Heinemann, Boston, MA.

Sharp, M. L., 1993. Behavior and Design of Aluminum Structures, McGraw-Hill , New York, NY.

Strong, A. B., 1989, Fundamentals of Composites Manufacturing, SME Press, Dearborn, MI.

Willman, D., 1996. "Weld Metal Tests Stir Steel Building Concerns", *Los Angeles Times*, Monday, December 30, 1996, pp. A1, A8-A9.

Manufacturing

Rapid Process Development for Resin Transfer Molding

M. J. TARI AND H. T. HAHN

ABSTRACT

An expensive and time consuming aspect product development manufacturing the prototype of the design concept. The goal of current research is to devise a methodology for the rapid process development of resin transfer molding (RTM) in order to reduce product development cycle time and lower the cost of developing composite parts. Towards this goal an initial methodology has been developed and implemented in the manufacturing of a composite I-beam. Since tooling costs comprise a substantial portion of the cost associated with RTM, the methodology employs rapid prototyping (RP) techniques to fabricate the molds. Furthermore, the methodology also utilizes simulation tools to determine both processing conditions and optimum gate locations. Finally, the total hours to manufacture the first acceptable I-beam were quantified for comparison with future studies.

INTRODUCTION

Many studies have been undertaken to investigate the RTM process. Some (Gebart, 1992; Parnas and Salem, 1993; Wang) have focused on predicting and measuring the permeability of the fiber mats associated with RTM, while others (Hayward and Harris, 1990; Lundstrom and Gebart, 1993) have quantified the effects of RTM process variables on the quality of the final parts. Still more studies (Young, 1994; Lin and Hahn, 1996) have concentrated on the modeling and simulation of resin flow during RTM mold filling as well as the local variations in preform permeability due to mat deformation and edge effects (Fong and Lee, 1994; Hammami et al., 1995).

Although the information from the above mentioned work has been useful, there is still a desire to reduce the cost and time associated with developing an RTM process for a new product. The goal of current research is to devise a methodology for the rapid process development of resin transfer molding (RTM) in order to reduce product development cycle time and lower the cost of developing composite parts. Towards this goal, an initial

---

Michael J. Tari and H. Thomas Hahn, Mechanical and Aerospace Engineering Department, School of Engineering and Applied Science, University of California Los Angeles, Los Angeles, CA 90095-1597

methodology has been developed and implemented in the manufacturing of a composite I-beam. Since tooling costs comprise a substantial portion of the cost associated with RTM a rapid prototyping (RP) technique known as laminated object manufacturing (LOM) was employed to fabricate the I-beam mold. LOM builds parts by sequentially bonding together layers of paper coated with thermoplastic. Furthermore, the simulation program RTMSIM, developed at UCLA by Lin and Hahn (1996), was used to determine optimum gate locations. Finally, the total hours to manufacture the first acceptable I-beam were quantified for comparison with future studies.

THE METHODOLOGY

As can be seen in Figure 1 the entire procedure starts with the product specifications. These include elementary detail such as the geometry and surface finish of the product as well as profound as its load bearing and fire safety requirements. Once these specifications are established, the designer can determine appropriate fiber/resin combinations along with the fiber volume fraction necessary to achieve the desired performance. It should also be noted that it has been assumed that material properties such as fiber mat permeability and resin viscosity, heat of reaction and cure temperature are available. If they are not, experiments should be performed to determine them. The initial mold design can be made at this point since the geometry of the product is known. However optimum gate location should be determined using simulation which takes place later in the process development cycle. Also, if a high permeability layer (HPL) is going to be incorporated into the lay-up, the initial mold design should account for its thickness.

The properties of the chosen fiber mat will dictate both the driving pressure necessary to pump resin through the mold, as well as the clamping pressure required to create the appropriate fiber volume fraction. The driving pressure will also be affected if an HPL is incorporated into the lay-up. The properties of the resin will also affect the process parameters through the mold temperature necessary to bring about a suitable resin viscosity as well as complete curing.

Once the process variables, resin properties and initial mold design have been determined they must be checked to insure they are compatible with the RP device (in this case the LOM machine) being used to fabricate the mold. For example a mold produced by the LOM machine cannot withstand high forces which act parallel to the individual sheets of paper which make it up. To correct such a problem one can either change the design of the mold or lower the driving pressure. However, reducing the driving pressure may entail going back and choosing a different fiber mat, so that the fiber volume fraction can also be reduced. This new choice of fiber mat may, in turn, dictate some change in the type of resin used. Another situation may be that the resin's exothermic reaction produces to a temperature that is high enough to weaken the bonding between the layers of paper in an LOM mold. Again, going back and choosing a new resin may entail choosing a new fiber mat as well as new process parameters. Finally, if the mold is too large to be fabricated in one piece, a new arrangement, where pieces are fastened together to achieve desired size, may have to be designed.

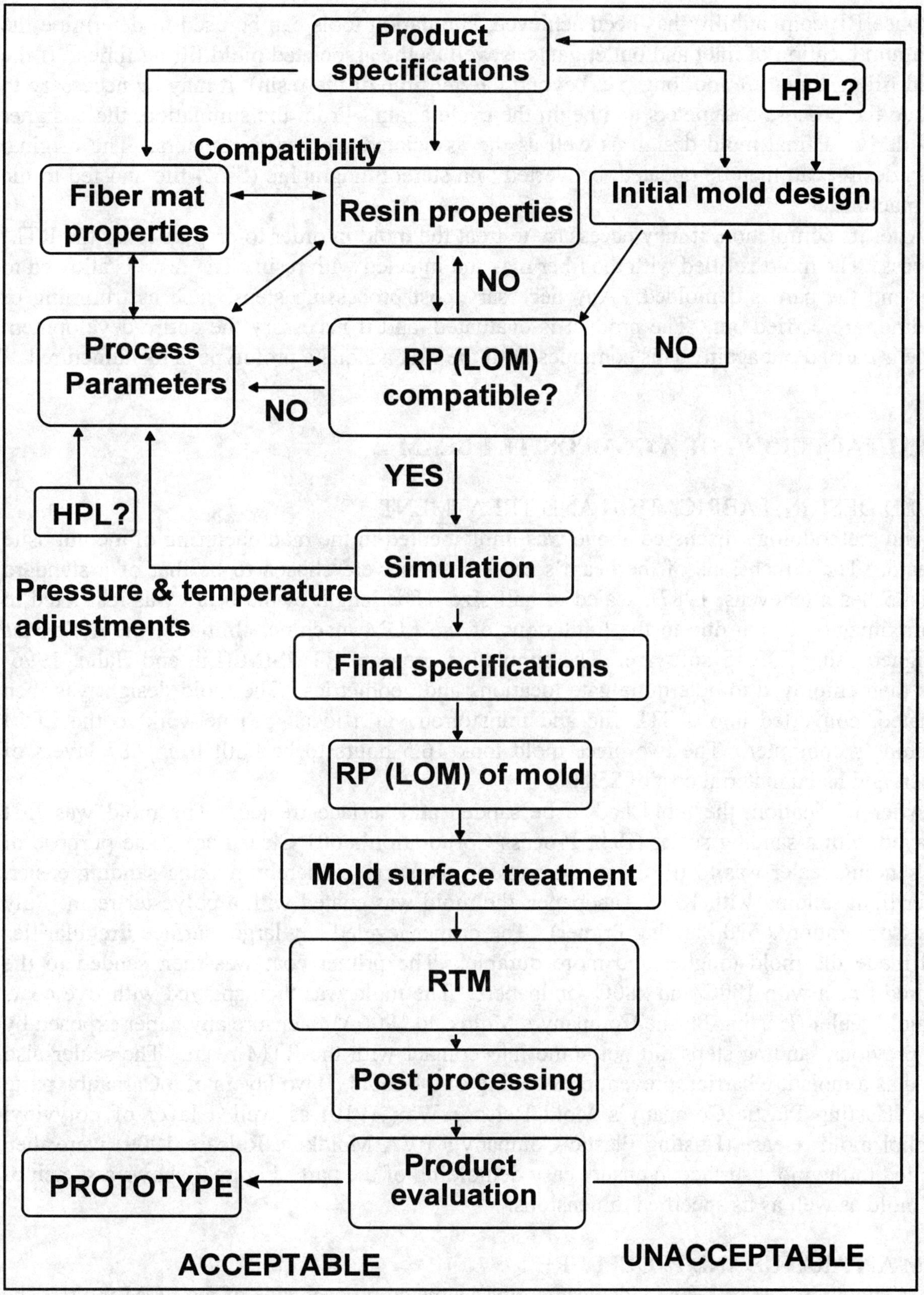

Figure 1. Flowchart of Methodology

Once RP compatibility has been achieved, simulation tools can be used to determine the optimum location of inlet and outlet gates as well as the associated mold filling times. If the mold filling times are too long (i.e. beyond the gel time of the resin), it may be necessary to update the process parameters and begin the cycle again. From the simulation, the designer will derive a final mold design as well as the associated process parameters. The original mold design can then be updated, converted to a stereolithographic (STL) file and fed to the RP machine.

After its completion, it may necessary to treat the mold in order to prepare it for the RTM process. The mold is filled with the fiber mat and injected with resin. The resin is allowed to cure and the part is demolded. Any necessary post processing steps, such as trimming or sanding, are carried out. The product is evaluated, and if necessary the entire development cycle is carried out again. This continues until the first adequate prototype is manufactured.

MANUFACTURING OF A COMPOSITE I-BEAM

MOLD DESIGN, FABRICATION AND TREATMENT

The methodology discussed above was implemented in the manufacturing of a composite I-beam. The dimensions of the beam's cross section were chosen to be that of a standard W6x25 beam (Stevens, 1987), scaled to half size. The length of the beam was restricted to approximately 25 cm due to the limitations of the LOM machine. Initially, the mold was designed using IDEAS software. The simulation program RTMSIM (Lin and Hahn, 1996) was then employed to determine gate locations and geometries. The mold design was then updated, converted into a STL file and transferred via a local area network to the LOM machine's computer. The two-piece mold took 16.5 hours to be built from 687 layers of paper, and had a material cost of $58.00.

After fabrication, the mold had to be sanded and surface treated. The mold was first sprayed with a sanding sealer (Behr Process Corporation's 601 Clear Lac). The purpose of the sanding sealer was to fill the micro-gaps on the surface, thereby making sanding easier. After light sanding with 180C sandpaper, the mold was coated with a polyester resin (Poly Lux Corporation's 670 Sanding Primer). The primer leveled any larger surface irregularities and made the mold tougher and more durable. The primer coat was then sanded to the desired finish with 180C and 280C sandpapers. The mold was then sprayed with five coats of mold sealer (Hasting Plastic Company's Moldkote 1906A) to ensure any paper exposed by the previous sanding steps did not come into contact with the RTM resin. The sealer also acted as a moisture barrier, preventing swelling of the mold. Two layers of a Carnauba paste wax (Hasting Plastic Company's Mold Release Wax 1910) as well a layer of polyvinyl alcohol mold release (Hasting Plastic Company's PVA Moldkote Release 1909) were then applied to the mold surface to ensure easy demolding of the part. Figure 2 shows a sketch of the mold as well as its specified dimensions.

RTM APPARATUS AND PROCEDURE

As shown in Figure 3, one video camera was situated on each side of the LOM mold. The mold was fitted with a clear polycarbonate top and bottom to facilitate recording of the flow front during mold filling. Polycarbonate was chosen since it could withstand the exotherm

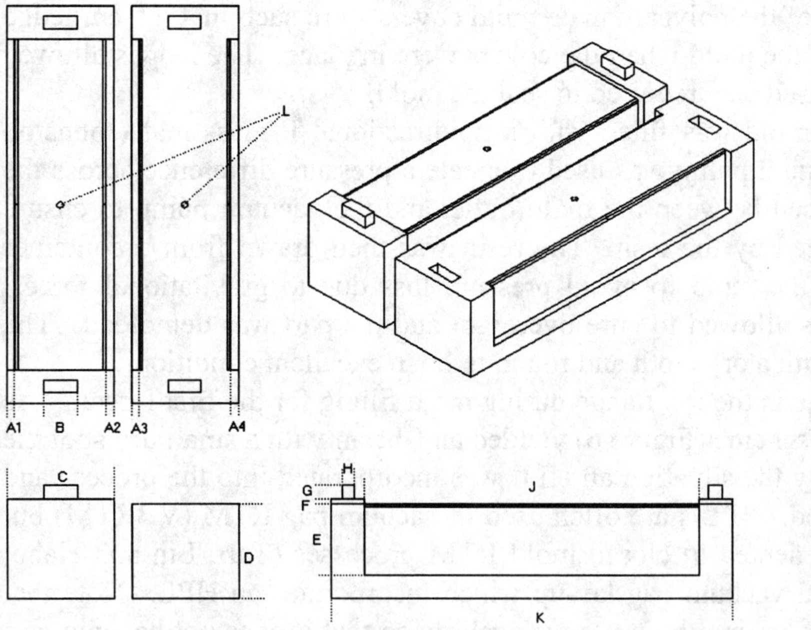

Dimension	Specified (mm)
A1	5.791
A2	5.791
A3	5.791
A4	5.791
B	24.928
C	80.899
D	72.136
E	54.864
F	4.064
G	10.160
H	9.400
J	254.000
K	304.800
L (diameter)	5.000

Figure 2. Mold Dimensions

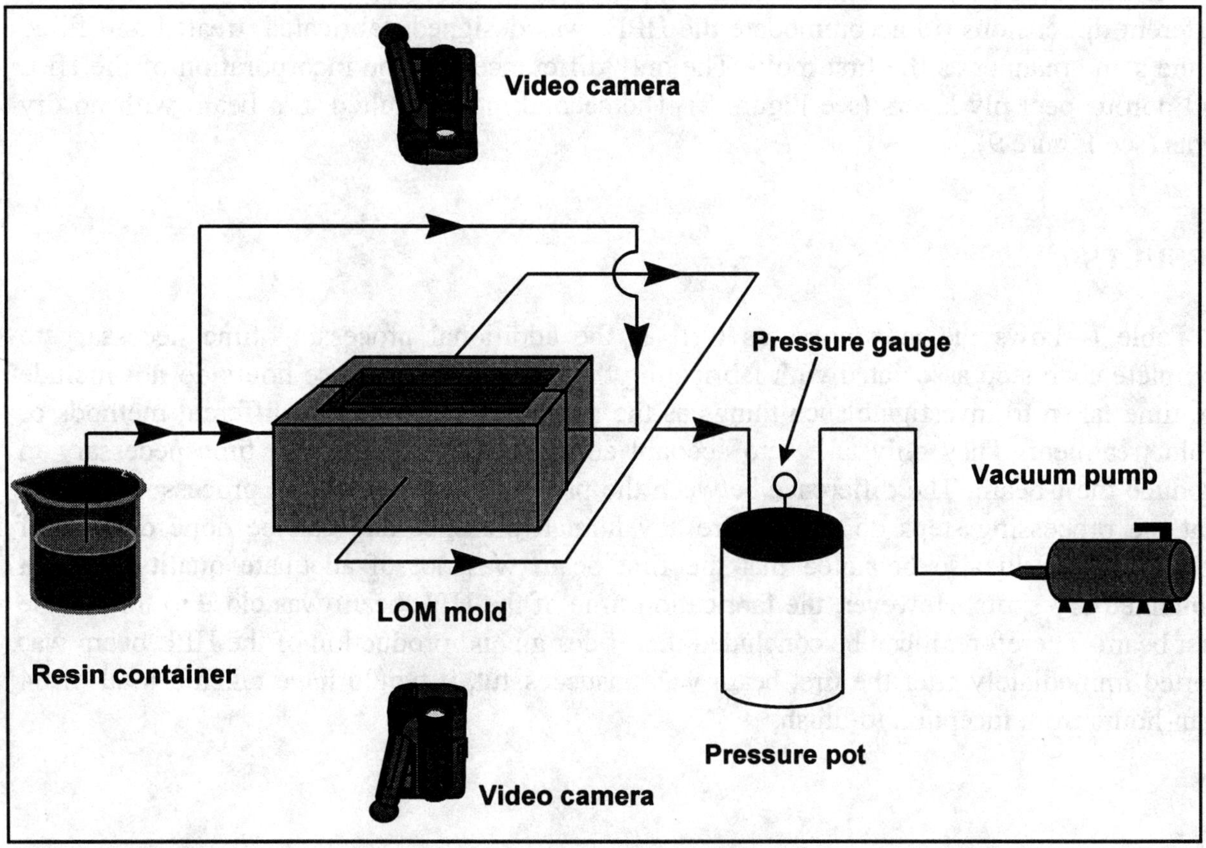

Figure 3. RTM Apparatus

(~100 C) in the vinylester resin (Dow Chemical's Derakane 450-311 PA) without losing its rigidity. The length and width of the polycarbonate mold covers were such that a 1 cm ledge existed around the periphery of the mold when the covers were in place. The ledges allowed for easy placement of the oil-based clay required to seal the mold.

As shown in Figure 4, the mold was filled with a unidirectional E-glass mat (Johnston Composite's ET 0800). A vacuum pump was used to create a pressure difference across the mold. A pressure pot was placed between the mold outlet and the vacuum pump to ensure the pump would not be damaged by the resin. The resin was then drawn from a container located at the same height as the mold to avoid pressure loss due to gravitational forces. After mold filling the resin was allowed to cure overnight and the part was demolded. The LOM mold was then cleaned with a dry cloth and found to be in excellent condition.

Figure 5 show the flow fronts in the top flange during mold filling for the first I-beam. As can be seen the initial injection scheme (Figure 6) yielded an I-beam with a small dry spot the center of the flange. To remedy the situation an HPL was incorporated into the process and an end to end injection was used. HPL's are often used in vacuum bag RTM (VBRTM) but can also be successfully implemented in closed mold RTM processes (Tari, Lin and Hahn, 1997). Figure 7 shows a typical vacuum bag lay-up which incorporates an HPL. Note that the HPL is separated from the fiber mat by a porous peel ply so that it does not become part of the final product. HPL's work by creating a transverse flow in the fiber mat thus facilitating mold filling. Tari, Lin and Hahn (1997) showed that incorporating an HPL into closed mold RTM can reduce mold filling time by up to 90%. A second mold with slightly different dimensions (to accommodate the HPL) was designed, fabricated, treated and filled in the same manner as the first mold. The only difference was the incorporation of the HPL and porous peel ply layers (see Figure 8). The second mold resulted in a beam with no dry spots (see Figure 9).

RESULTS

Table I shows the man hours as well as the additional processing time necessary to complete each step associated with fabricating the first I-beam. These hours do not include the time taken to investigate such things as the material properties and different methods of mold treatment. They only take into account actual work done on, and time necessary to produce the I-beam. The difference between the preparation time and the processing time is that the processing steps do not require any human presence and can be done during off hours. It should also be noted that the first beam was not of adequate quality since it contained dry spots. However, the fabrication time of the HPL beam was close to that of the first beam. Therefore, it can be concluded that if design and production of the HPL beam was started immediately after the first beam was unsuccessful, it would have taken a total of 64 man-hours from inception to finish.

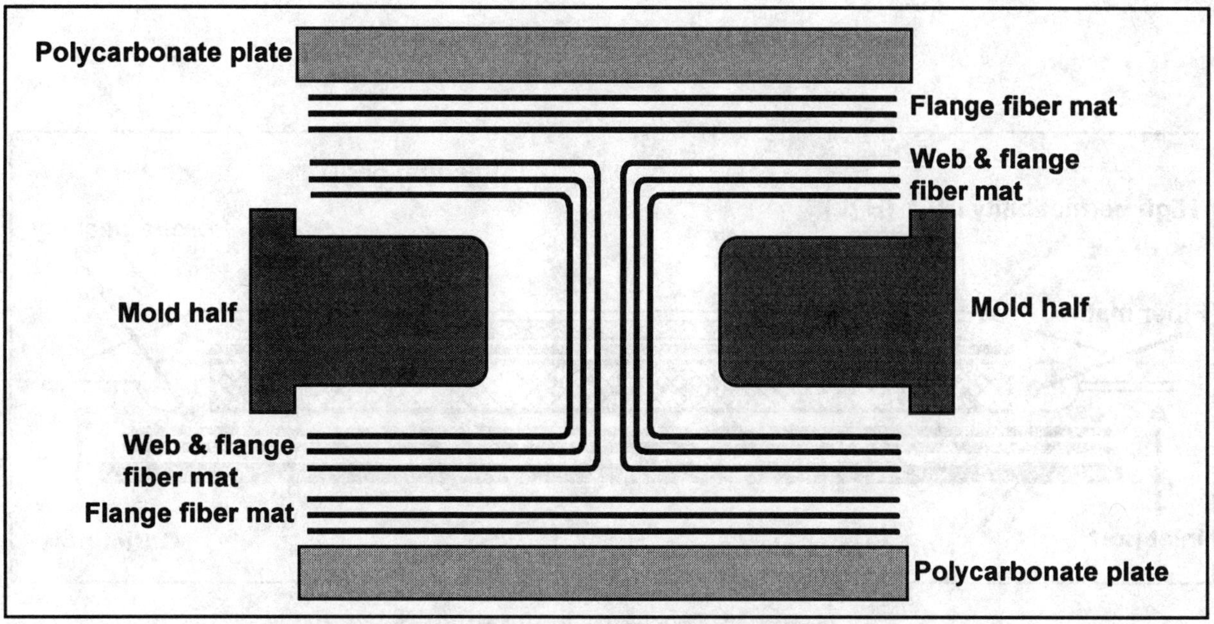

Figure 4. Cross Section of the LOM Mold Filled with Fiber Mat (Exploded View)

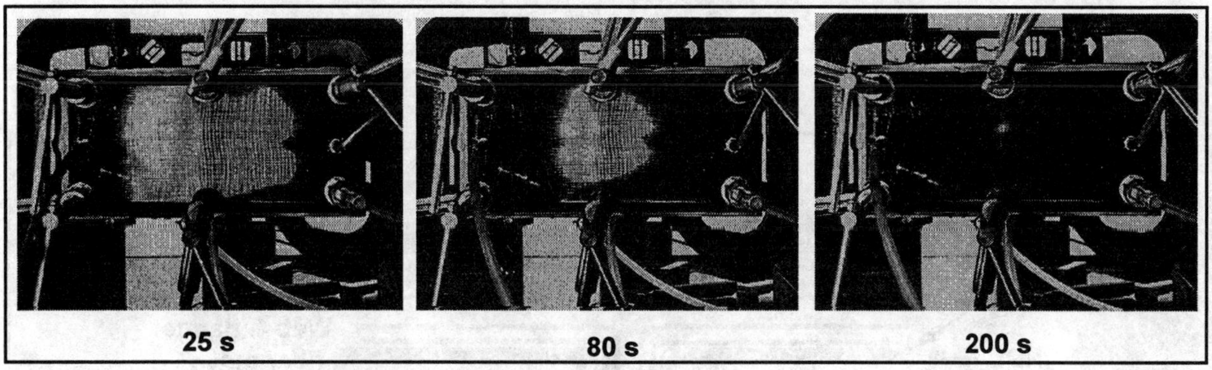

Figure 5. Mold Filling for the First I-beam

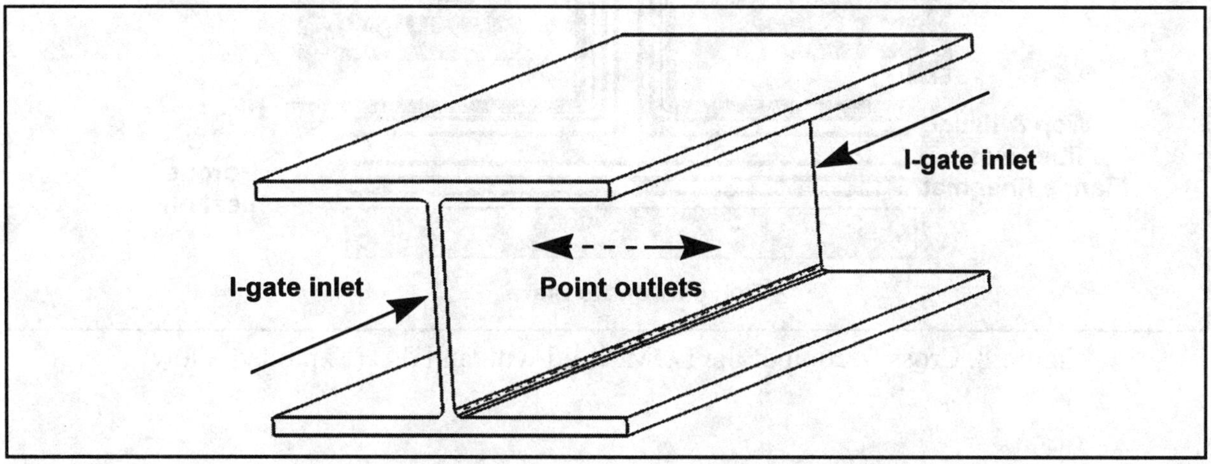

Figure 6. Injection Scheme for the First I-beam

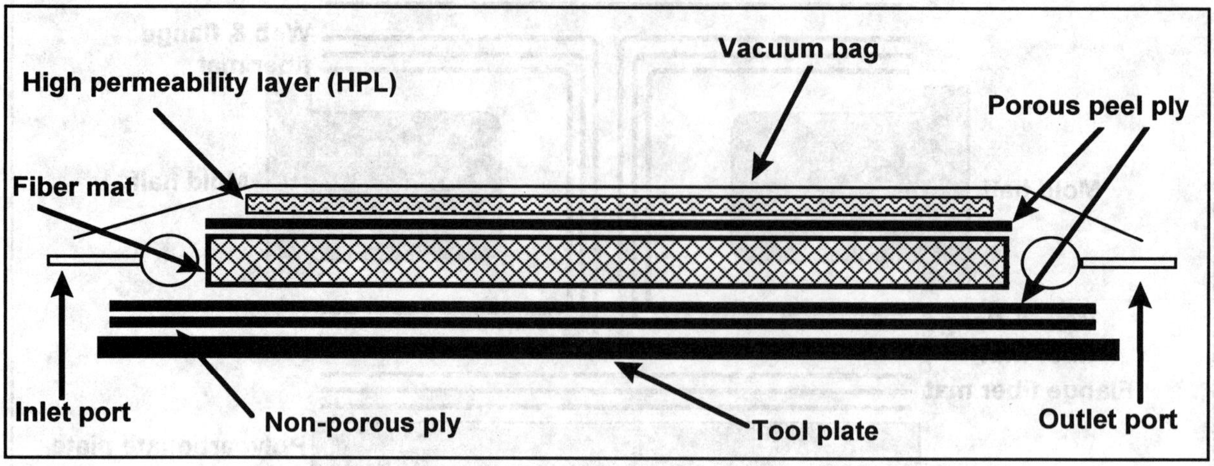

Figure 7. Vacuum Bag RTM Lay-up

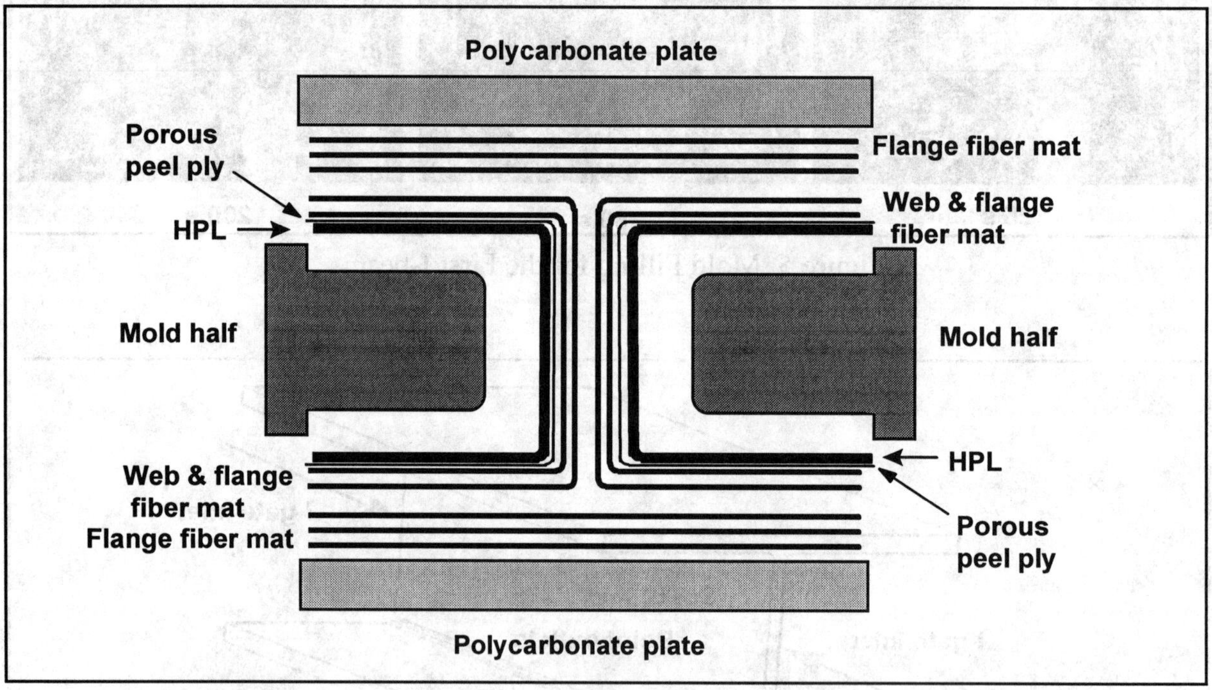

Figure 8. Cross Section of the LOM Mold with an HPL (Exploded View)

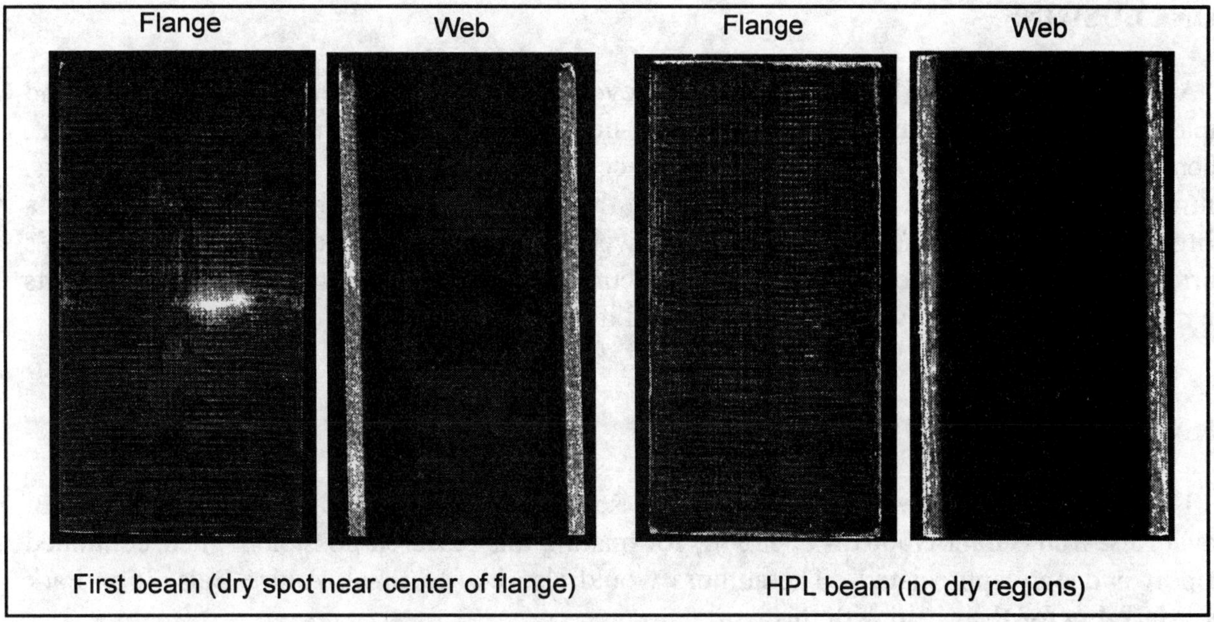

Figure 9. Completed Beams

TABLE I-TIME ASSOCIATED WITH FABRICATING THE FIRST I-BEAM

Methodology step	Preparation time (man-hours)	Processing time (hours)
Mold design	4	
Simulation (meshing and mold filling)	12	4
LOM processing	3	16.5
Mold surface treatment	4	8
RTM: fiber preparation, mold filling and curing	8	8
Demolding	1	
Total	32	36.5

CONCLUSIONS

A methodology for the rapid process development of RTM has been devised and implemented in the fabrication of a composite I-beam. The current work has also shown that, when properly treated, an LOM mold can be successfully incorporated into the RTM process. Future studies will involve the application of the methodology to the manufacturing of a prototype snowboard. The total man hours as well as the cost associated with the prototype's fabrication will be recorded and compared to current industry numbers. Using these results the methodology will be evaluated and modified if necessary.

AKNOWLEDGEMNET

The authors would like to thank Yapa D. S. Rajapakse, Scientific Officer, and the Office of Naval Research (Grant N00014-92-J1846) for making this research possible. Their continued support is deeply appreciated. The authors would also like to thank André Bals, Joon Park and Mark Lin for their help with this work.

REFERENCES

Fong, L. and Lee, L.J., "Preforming Analysis of Thermoformable Fiber Mats-Preforming Effects on Mold Filling," *Journal of Reinforced Plastics and Composites*, Vol. 13, July 1994, pp. 637-663

Gebart, B.R. "Permeability of Unidirectional Reinforcements for RTM," *Journal of Composite Materials*, Vol. 26, No. 8, 1992, pp. 1100-1133.

Hammami, A., Gauvin, R., Trochu, F., Touret, O., and Ferland P., "Analysis of the Edge Effect on Flow Patterns in Liquid Composite Molding," *Proceedings of The Tenth International Conference on Composite Materials*, Whistler, Canada, August, 1995 Vol. 3, pp. 277-284

Hayward, J.S., and Harris, B., "Effect of Process Variables on the Quality of RTM Moldings," *SAMPE Journal*, Vol. 26, No. 3, May/June, 1990 pp. 39-46.

Lin M.Y., and Hahn, H.T., "A Robust and Efficient Approach for RTM Simulation," *1996 ASME International Mechanical Engineering Congress and Exhibition*, November, 1996, MD-Vol. 74, pp.75-76.

Lundstrom, T.S. and Gebart, B.R., "Influence from Different Process Parameters on Void Formation in Resin Transfer Molding," Submitted to *Polymer Composites* 1993

Parnas, R.S., Salem, A.J., " A Comparison of the Unidirectional and Radial In-Plane Flows of Fluids Through Woven Composite Reinforcements," *Polymer Composites*, October 1993, Vol. 14, No. 5 pp. 383-394.

Stevens, K.S., *Statics and Strength of Materials*, Second Edition, Prentice-Hall Inc., 1987

Tari, M.J., Lin, M.Y., Hahn, H.T., "Resin Transfer Molding with High Permeability Layers," *Proceedings of the SME Composites '97 Manufacturing and Tooling Conference and Tabletop Exhibits*, January, 1997, pp. 177-187.

Young, W.B., "Three Dimensional Non-isothermal Mold Filling Simulations in Resin Transfer Molding," *Polymer Composites*, Vol. 15, No. 2, 1994, pp. 118-127

Stochastic Analysis of the Resin Transfer Molding Process

S. PADMANABHAN AND R. PITCHUMANI

ABSTRACT

Resin Transfer Molding (RTM) offers the potential to manufacture composite materials of complex geometries cost-effectively, and at high throughputs. However, strong variabilities in the associated process variables such as the preform permeability hamper effective control of the process and pose a stiff challenge to the commercial manufacture of quality composites by this process. Although a number of process simulation models have already been developed for RTM, a systematic analysis of the process parameter uncertainties has been the subject of little attention, and forms the focus of this study.

This paper presents a methodology for the systematic analysis of process parameter variabilities towards robust quality manufacturing of composites using liquid molding techniques. A stochastic model is developed to study the uncertainties in the resin transfer molding process. The uncertainties in the process inputs are quantified by assigning probability distribution functions to them, and the stochastic model is used to obtain the corresponding process output uncertainties. Parametric studies are carried out to examine the effects of the nature and magnitude of the input uncertainties on the output uncertainties.

INTRODUCTION

Resin Transfer Molding (RTM) is one of the liquid composite molding techniques that is increasingly being used in the commercial manufacture of composites. In the basic process a catalyzed resin is pumped into a closed mold cavity which has a pre-placed fiber stack called the *preform*. After the preform is saturated with the resin, the mold is subjected to high temperatures to initiate complex cross-linking chemical reactions in the resin. These reactions, called *cure*, lead to an increase in the viscosity of the resin, and eventually to the formation of the structurally hard composite product. The major physical phenomena involved in the process are the fluid mechanics of the resin permeating the preform, and the heat transfer and kinetics associated with the curing of the resin. Air entrapment, mobility, and dissolution control the formation and growth of voids and the void fraction of the composite. These phenomena are in turn influenced

Srinivasan Padmanabhan and Ranga Pitchumani, Composites Processing Laboratory, Department of Mechanical Engineering, University of Connecticut, U-139, Storrs, CT 06269-3139

by the rheological characteristics and the kinetic properties of the resin, and the microstructure of the preform.

A thorough understanding of these process phenomena is essential for designing molds, locating and sizing injection ports and vents, choosing injection pressures for complete permeation, and determining mold fill times. Many experimental investigations of the resin flow have been reported in literature (Bruschke and Advani, 1989; Chen, Macosko, and Davis, 1995a), and flow models have been developed to simulate the flow of the resin through the preform (Cai, 1992; Coulter and Guceri, 1989; Li and Gauvin, 1991). Associated with the permeation of the preform by the resin is the entrapment of air pockets which leads to the formation of voids in the composite product. Several models have been developed to study void formation based on the concept of two levels of flow: flow in the gaps between/around the fiber tows, and flow into the fiber tows (Kardos, Dave, and Dudukovic, 1988; Parnas and Phelan, 1991; Chan and Morgan, 1992; Lundstrom, 1997).

The flow dynamics of the resin and the formation of voids in the preform depend on such process variables as the viscosity, the kinetic parameters, and the density of the resin, and the permeability of the preform. These variables are highly sensitive to the operating conditions and exhibit considerable uncertainties during a process run, and from one processing run to another. For example, the disorder in the preform microstructure leads to impreciseness in the quantification of the permeability, which in turn leads to unpredictable mold filling and void formation. Such parameter uncertainties cause unpredictable variations in the quality of the products, and thus hinder reliable process design and control, which are critical to the commercial manufacture of composites. All of the aforementioned models are *deterministic* in the way they treat the process variables, and as a result, are of little practical importance in the face of process uncertainties.

A systematic *stochastic analysis* of the effects of such parameter variabilities on the process performance forms the focus of this paper. Stochastic analysis refers to the study of the effects of the input uncertainties on the variabilities in the output parameters, and has previously been used in hydrology and petrology (Campbell and Rao, 1987), and in safety assessment of technological systems (Apostolakis, 1990; Diwekar and Rubin, 1991). In the present context, stochastic process simulation can be used to study the variabilities in the processing times and the product quality caused by the interaction of the input uncertainties. The stochastic model, in principle, requires performing multiple deterministic simulations using combinations of input parameters chosen by means of an appropriate sampling of their respective distributions. The deterministic model thus forms the foundation on which the stochastic analysis is built.

In this study, a deterministic process model is developed to simulate the two-dimensional flow of the resin and the void formation. Uncertainties in the permeabilities, the resin viscosity and the inlet pressure are quantified by assigning appropriate probability distributions to them, and samples are chosen from these distributions using the Latin Hypercube Sampling (LHS) procedure (Iman and Shortencarier, 1984). The outputs of the stochastic model are probability distributions of the mold-filling time and the composite void content. The effects of the mean and variance of the input parameters on the mean and variance of the output parameters are determined, and the most important parameters with respect to controlling the quality of the product are highlighted.

DETERMINISTIC MODEL

Figure 1(a) shows a schematic of the resin transfer molding process as considered in this study. As illustrated in the Figure, the resin enters a rectangular mold through the inlet slot, and

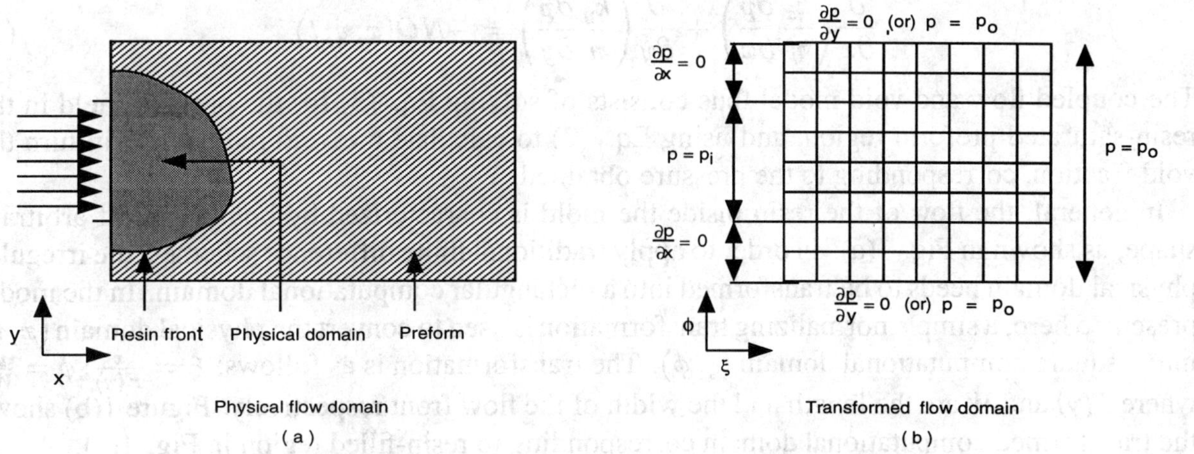

Figure 1: (a) The physical domain, and (b) the transformed domain

advances through the preform at a rate that is governed by the injection pressure, the permeability of the preform, and the viscosity of the resin. The two-dimensional flow of the resin through the porous preform can be described using Darcy's law, which for a Cartesian *x-y* coordinate system, is written as follows:

$$u = \frac{k_x}{\eta}\left(\frac{\partial p}{\partial x}\right); v = \frac{k_y}{\eta}\left(\frac{\partial p}{\partial y}\right) \tag{1}$$

where k_x and k_y are the principal permeabilities of the porous medium, η is the viscosity of the resin, p is the pressure inside the flow domain, and u and v are the velocities in the x and y directions, respectively. The coordinate system (x, y) is so chosen that they coincide with the principal directions of the porous medium. The pressure field in the resin-filled region of the preform is obtained using the velocity components u and v, given by Eq. (1), in the continuity equation for two-dimensional incompressible flow, modified to include a mass sink term that arises from the void model described below.

The void model used is similar to that developed by Parnas and Phelan (1991). In their model, the fiber bundles are treated as flow sinks which absorb the resin with a strength that is proportional to the radial pressure gradient within the fiber bundles. The mass sink term in the continuity equation is obtained as the product of the sink strength, $Q(x, y, t)$, and the sink density, N. The sink strength is given by $Q = 2\pi r v_r$, where r is the instantaneous radius of the voids inside the fiber bundles, and v_r is the velocity in the radial direction. The velocity v_r is given by a modified form of the Darcy's law written in the radial coordinates:

$$v_r = -\frac{k_b}{\eta}\left(\frac{p_o \frac{r^2}{R^2} - p}{r \ln \frac{r}{R}}\right) \tag{2}$$

where k_b is the fiber bundle permeability, R is the fiber bundle radius, and p_o is the initial pressure in the unsaturated fiber bundles, and is usually equal to the outlet pressure. The governing equation for the process is obtained by substituting the expressions for the velocity components and the sink strength in the continuity equation:

$$\frac{\partial}{\partial x}\left(\frac{k_x}{\eta}\frac{\partial p}{\partial x}\right)+\frac{\partial}{\partial y}\left(\frac{k_y}{\eta}\frac{\partial p}{\partial y}\right)=-NQ(x,y,t) \tag{3}$$

The coupled flow and void model thus consists of solving Eq. (3) for the pressure field in the resin-saturated preform region, and using Eq. (2) to calculate the void radius, and in turn the void fraction, corresponding to the pressure obtained.

In general, the flow of the resin inside the mold is characterized by boundaries of arbitrary shape, as shown in Fig. 1(a). In order to apply traditional finite difference schemes, the irregular physical domain needs to be transformed into a rectangular computational domain. In the model presented here, a simple normalizing transformation is used to convert the physical domain (x, y) into a square computational domain (ξ, ϕ). The transformation is as follows: $\xi = \frac{x}{F(y)}$; $\phi = \frac{y}{W}$, where *F(y)* and *W* are the length and the width of the flow front respectively. Figure 1(b) shows the transformed computational domain corresponding to resin-filled region in Fig. 1(a).

The boundary conditions are presented in Fig. 1(b), and are stated as follows: At the inlet, the pressure is constant, and is equal to the injection pressure i.e., $p = p_i$. At the resin front, the pressure is equal to the outlet pressure, i.e., $p = p_o$, while along the impermeable mold walls, there is no resin flow normal to the mold walls, i.e., $\frac{\partial p}{\partial n} = 0$, where *n* is the normal vector.

In order to solve for the pressure field, the governing equation, Eq. (3), is transformed to the computational coordinates. A control volume based finite difference method (Patankar, 1980) is used to discretize Eq. (3), and the set of linear equations so obtained is solved using the numerical method of Successive Over Relaxation (SOR). Darcy's law is used to calculate the resin front velocities from the computed pressure distribution, and the front is advanced to its next location inside the mold. The simulation is stopped when the resin fills the mold completely. The corresponding time constitutes the mold fill time, and the void fraction at that instant is considered to be the final void fraction in the composite.

STOCHASTIC MODEL

Stochastic modeling of a physical process refers to the simulation of associated phenomena for special, but randomly selected stochastic instances. A deterministic process model often forms the foundation on which the stochastic framework is built. The uncertain input parameters of a stochastic model are assumed to be random variables that are described by probability distribution functions, instead of the point values that would be assigned to them in conventional, deterministic models. An appropriate sampling technique is used to choose parameter values from these distribution functions. Corresponding to these input distribution functions, the stochastic model generates distribution functions for the output parameters. In the present context, the anisotropic preform permeabilities (k_x, k_y), the resin viscosity (η), and the injection pressure (p_i) form the uncertain input variables, and the mold fill time (τ_{fill}) and the void fraction (v_f) are the process outputs.

The input variables are characterized by the distribution functions listed in Table I. Uncertainty analyses conducted in the fields of hydrology and petrology have indicated that the hydraulic conductivity of soil follows a lognormal distribution (Campbell and Rao, 1987; Jury, Russo, and Sposito, 1987). Owing to the similarity in the flow physics in the RTM process to that in these fields, a lognormal distribution is assumed for the preform permeabilities in this study. The principal permeabilities (k_x and k_y) are assumed to be spatially variable throughout the mold. The longitudinal permeability, k_x, is assumed to vary about a fixed mean, while the transverse

TABLE I—PROBABILITY DISTRIBUTIONS

Parameter	Distribution
Permeability, k_x	Lognormal
Permeability, k_y	Lognormal
Viscosity, η	Lognormal
Inlet Pressure, p_i	Normal

TABLE II—DETERMINISTIC MODEL VALIDATION

p_i (psi)	k_x ($\times 10^{-10} m^2$)	τ_{model} (sec)	$\tau_{Cai,1992}$ (sec)
12	1.196	1354.9	1375.7
16	3.16	384.6	390.5
20	4.75	204.7	207.8
25	4.73	163.7	167.1

permeability, k_y, is assumed to vary about a location-dependent mean that is higher near the edges of the mold, and decreases rapidly away from the edges, towards the centerline. Such a trend for k_y is corroborated by the experimentally observed phenomenon of preferential flow along the mold walls. The viscosity (η) of the resin is also assumed to follow a lognormal distribution. This is justified by the fact that at low viscosities and high normalized variances, the distribution would be constrained by a lower bound of zero for the viscosity, and hence the distribution develops a "skew" that is characteristic of lognormal distributions. The injection pressure (p_i) is assumed to follow a normal distribution, as the normalized variances on the pressure are typically smaller than those on the resin viscosity.

The uncertain input variables are sampled using the Latin Hypercube Sampling (LHS) technique (McKay, Conover, and Beckman, 1979). LHS is a stratified sampling technique, and works by dividing the range of each input variable into *m* equal probability intervals, where *m* is the number of samples desired. One value is selected for the input parameters from each interval on their respective distributions, and the values are combined in a random manner to form *m* vectors of the input parameters. The stochastic model performs the deterministic simulation for each of the *m* input parameter combinations, and generates *m* output vectors. The *m* output vectors are then used to generate the mean and variance of the output distribution functions.

A numerical code developed by Iman and Shortencarier (1984) and implementing the Latin Hypercube Sampling (LHS) procedure was used to sample from the input distributions. The optimum number of samples for the stochastic analysis was chosen by doing a "convergence test" that consisted of studying the output mean and variance as a function of the number of samples, and choosing the smallest number of samples that gave a fairly good convergence of the mean and the variance. An optimum number of samples of 250 was chosen on the basis of less than 1% change in the mean and about 3% change in the variance.

RESULTS AND DISCUSSION

Since the deterministic model is the basis for the stochastic analysis, the validation of the deterministic model is discussed. The validation is presented with respect to the analytical model developed by Cai (1992)for the mold fill time in the case of one-dimensional flow of the resin. For the purposes of the validation, the preform permeability was considered to be isotropic. Table II shows the values of the fill time for different injection pressures and fiber bed permeabilities. It is evident from this comparison that the deterministic model is accurate enough to base the stochastic analysis upon.

Two sets of studies were performed to stochastically analyze the flow of the resin through a rectangular mold with a line inlet, for a range of process parameter values. The studies include: *(a)* fixing the mean, μ, of each input parameter and changing the variance, σ^2, and *(b)* fixing

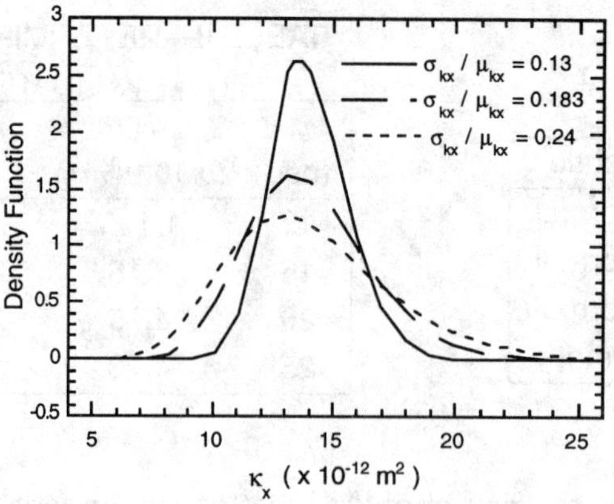

Figure 2: Input distribution for fiber bed permeability

the variance, σ^2, of each input parameter and changing the mean, μ. In each case, the effects of these variations on the fill time (τ) and the void fraction (v_f) were examined. For the fixed mean runs, three different variances were investigated for each input parameter, while for the fixed variance runs, three different mean values were considered for each input variable. Further, while studying the effect of the uncertainties in one parameter, the uncertainties in the other parameters were maintained constant, i.e., their probability distributions are fixed. A typical set of input distribution functions used for the fixed mean runs is shown in Fig. 2, which illustrates three different variances for a mean fiber bed permeability of $1.4 \times 10^{-11}\ m^2$.

As explained previously, the stochastic model generates the distribution functions for the fill time and the void fraction. The distributions of both these parameters were generally observed to follow a lognormal trend, as shown in Fig. 3. In this Figure, the values of the fill time are plotted against their probability density functions so that the total area under the curve is one. The data points denote the stochastic simulation results, while the solid line denotes the lognormal fit made through the data. The high degree of correlation between the data and the fit is noteworthy. Similar results are obtained for the void fraction also. The lognormal fit is characterized by its mean, μ, and the standard deviation, σ, which are the parameters in terms of which the outputs are analyzed. Further, a normalized standard deviation, σ/μ, is used in the presentation of the results.

The first set of studies investigate the effects of changing the input parameter variances, keeping their mean values fixed. Figures 4–6 show the results obtained from these studies. In all these Figures, the normalized standard deviation of the input parameters is plotted on the x-axis, and the normalized standard deviations of the fill time and the void fraction are plotted on the y-axis.

The effects of the uncertainties in the injection pressure are illustrated in Fig. 4. As seen from this Figure, the variabilities in the injection pressure have a greater effect on the void fraction than on the fill time. For the same change in the pressure, the normalized standard deviation of the void fraction shows a steeper slope than that of the fill time. Furthermore, the slope of the line for the fill time is less than unity, thereby indicating that the standard deviation of the fill time increases at a much slower rate than the standard deviation of the pressure itself. The magnitude of the variabilities in the void fraction are seen to be equal to the uncertainties in the pressure.

Figure 5 shows the results of changing the variances of the preform permeability. It is clear from this Figure that as the variance of the permeability increases, the variance of the fill time

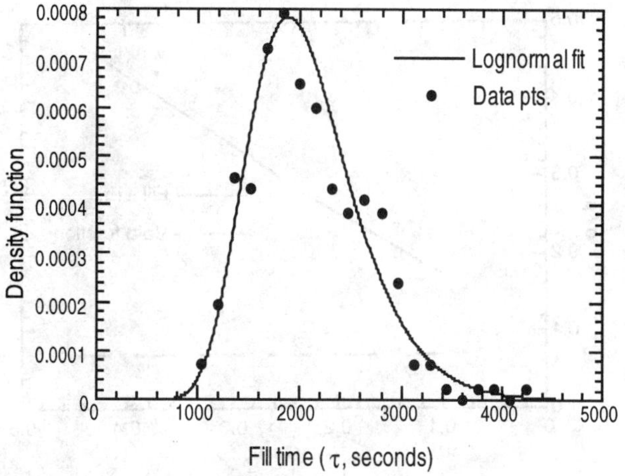

Figure 3: Lognormal fits for the output parameters

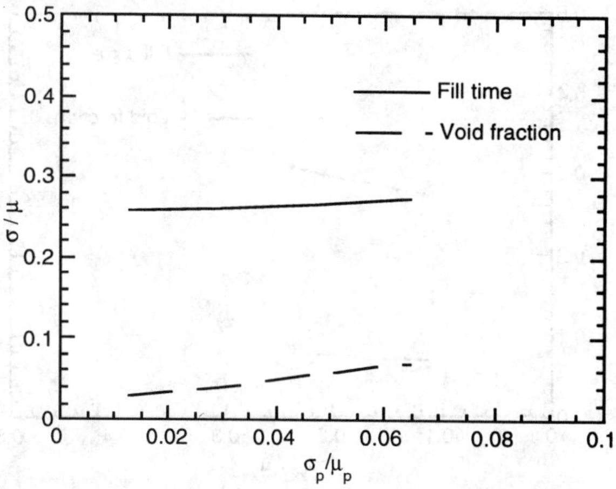

Figure 4: Fixed mean runs for pressure, μ_{p_i}=9.3 *bar*

also increases, but at a slower rate. However, the effect of the preform variability on the variance of the void fraction is much smaller. This may be attributed to the fact that the fiber bed permeability only marginally affects the penetration of the resin into the fiber bundles, and hence the void fraction, by way of changing the pressure distribution inside the resin-filled region.

The effects of the variabilities in the viscosity on the output uncertainties are shown in Fig. 6, which indicates that of all the uncertain input parameters considered in this study, viscosity variations have the greatest effect on the fill time. This plot also shows that the standard deviation of the fill time increases non-linearly with the variance of the viscosity. This suggests that controlling the viscosity variabilities through careful metering and mixing of the resin and initiators as well as the tight control of the injection temperature might be vital for robust processing. It is further seen in Fig. 6 that viscosity variations have little effect on the uncertainties in the void fraction. This is perhaps the result of the simplifying assumptions made in the void model, that the penetration of the resin into the fiber bundles is driven only by the pressure differential across the cross section of the bundles. A more comprehensive void model might show a greater dependence of the void fraction on the viscosity.

The second set of studies focus on the effects of varying the mean values of the input parameters, while maintaining their variances constant. Three different mean values were chosen for the preform permeability, the injection pressure and the viscosity. This set of studies, in combination with the one described earlier, gives comprehensive information on the effects of changing both the mean and the variance of the inputs.

Figures 7(a) and 7(b) present the effects of the variabilities in the permeability on the distributions of the fill time and the void fraction, respectively. These Figures show the variations in the mean value, the standard deviation, and the normalized standard deviation of the fill time and the void fraction. As expected, the mean fill time decreases with increasing mean permeability. Associated with this decrease in the mean fill time is a decrease in the uncertainty on the fill time. The standard deviation decreases faster than the mean, thereby causing the normalized standard deviation to decrease. As with the case of the fixed mean runs investigated in the first set of studies, the void fraction is relatively unaffected by the uncertainties in the preform permeability.

The effects of increasing the mean values while maintaining a fixed variance on the injection pressure are depicted in Figures 8(a) and 8(b). These Figures show that the uncertainty in the

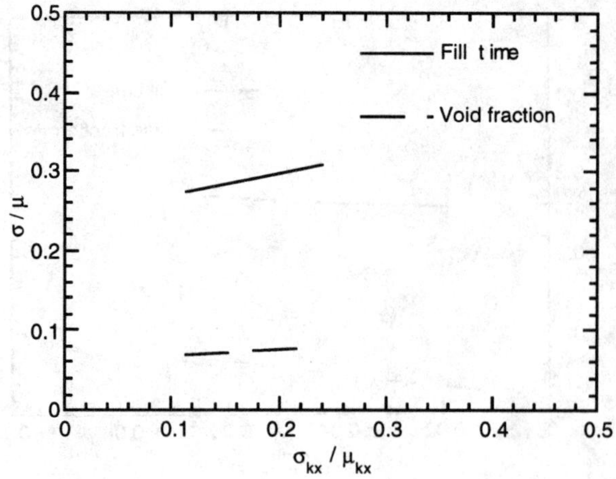

Figure 5: Fixed mean runs for permeability, μ=1.4 × 10^{-11} m^2

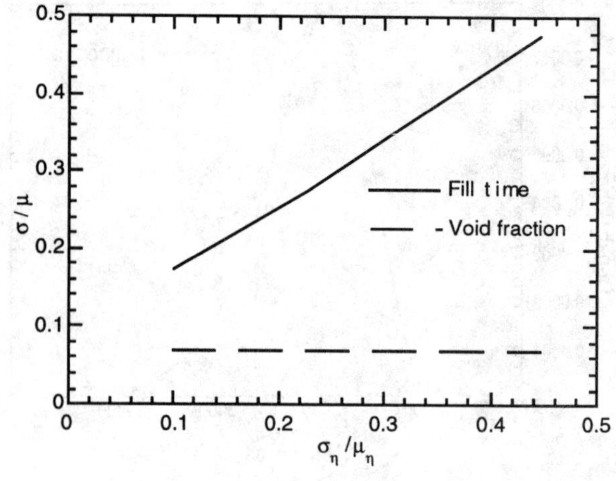

Figure 6: Fixed mean runs for viscosity, μ=2.05 *Pa.s*

fill time, as characterized by its standard deviation, decreases, as the mean injection pressure increases. Further, the decrease in the standard deviation is seen to be larger than that in the mean. As a result, the normalized standard deviation decreases. The effect of the mean injection pressure on the void fraction is similar. As the mean pressure increases, the mean void radius decreases, and the standard deviation decreases at a faster rate, whereby, the normalized standard deviation also decreases.

Figures 9(a) and 9(b) illustrate the effects of varying the mean viscosity, for a constant variance. Increasing the mean viscosity decreases the velocity of the resin flow, and causes the mean fill time to increase. As seen from the Figures, the standard deviation on the fill time also increases with the mean, although the magnitude of the change is smaller. The values of the normalized standard deviation drop as a result, implying that the severity of the effect of varying the input mean viscosities on the output variabilities decreases with increasing viscosities.

It can be inferred from these results that the uncertainties in the viscosity of the resin have the greatest effect on the fill time. Hence, controlling the viscosity of the resin by suitably controlling the mixing dynamics and the temperatures would lead to better control on the processing times. This result is of considerable significance, since minimizing variabilities in the process time decreases production costs, leading to more efficient manufacturing procedures. The studies conducted further demonstrated that the uncertainties in the injection pressure influence the variabilities on the void fraction of the composite the most. The simplicity of the void model might be the cause for the relative insensitivity of the void fraction to changes in the preform permeability and the viscosity of the resin.

CONCLUSIONS

A stochastic analysis approach to characterize and study the variabilities in the RTM process is presented. Though a number of models have been developed to describe the process, a systematic analysis of the variabilities in the process has not been performed before. The stochastic analysis approach has the potential to be a valuable tool in the quality control of commercial composites, and could lead to better and more reliable designs, and a tighter control on the processing times. This approach provides a powerful tool for minimizing the variabilities in the process, particularly in processes with a number of non-linear effects. In conjunction with

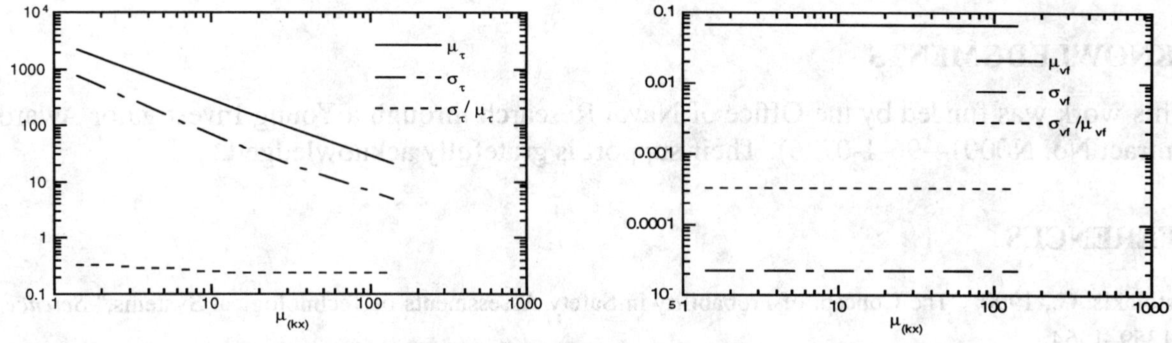

(a) (b)

Figure 7: Effect of varying the mean permeabilities, for a fixed variance, on (a) the fill time, (b) the void fraction

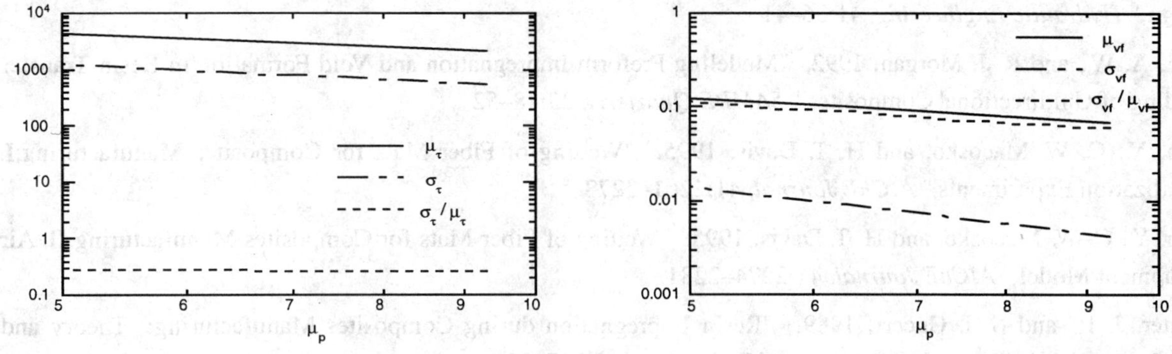

(a) (b)

Figure 8: Effect of varying the mean pressures, for a fixed variance, on (a) the fill time, (b) the void fraction

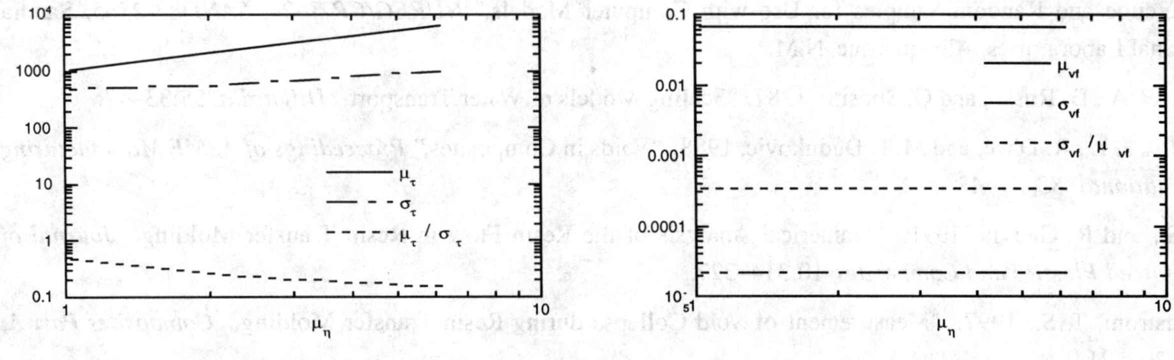

(a) (b)

Figure 9: Effect of varying the mean viscosities for a fixed variance on (a) the fill time, (b) the void fraction

an optimizer, stochastic analysis can be used to identify process parameter spaces that lead to minimum output variabilities.

ACKNOWLEDGMENTS

This work was funded by the Office of Naval Research through a Young Investigator Award (Contract No. N00014-96-1-0726). Their support is gratefully acknowledged.

REFERENCES

Apostolakis, G., 1990. "The Concept of Probability in Safety Assessments of Technological Systems," *Science,* 250:1359–1364.

Bruschke, M. V., and S. G. Advani, 1989. "A Numerical Simulation of Resin Transfer Mold Filling Process," ANTEC '89, 1769–1773.

Cai, Z., 1992. "Simplified Mold Filling Simulation in Resin Transfer Molding," *J. Composite Materials,* 26:52–75.

Campbell, J. E., and B. S. R. Rao, 1987. "An Uncertainty Analysis Methodology Applied to Sheetpile Cofferdam Design," *Hydraulic Engineering,* 41:36–41.

Chan, A. W., and R. J. Morgan, 1992. "Modeling Preform Impregnation and Void Formation in Resin Transfer Molding of Unidirectional Composites," *SAMPE Quarterly,* 23:48–52.

Chen, Y., C. W. Macosko, and H. T. Davis, 1995. "Wetting of Fiber Mats for Composites Manufacturing:I. Visualization Experiments," *AIChE Journal,* 41:2261–2273.

Chen, Y., C. W. Macosko, and H. T. Davis, 1995. "Wetting of Fiber Mats for Composites Manufacturing:II. Air Entrapment Model," *AIChE Journal,* 41:2274–2281.

Coulter, J. P., and S. I. Guceri, 1989. "Resin Impregnation during Composites Manufacturing: Theory and Experimentation," *Composites Science and Technology,* 35:317–330.

Diwekar, U. M., and E. S. Rubin, 1991. "Stochastic Modeling of Chemical Processes," *Computers and Chemical Engineering,* 15(2):105–114.

Iman, R. J., and M. J. Shortencarier, 1984. "A FORTRAN77 Program and User's Guide for Generation of Latin Hypercube and Random Samples for Use with Computer Models," *NUREG/CR-3624, SAND83-2365,* Sandia National Laboratories, Albuquerque, N.M.

Jury, W. A., D. Russo, and G. Sposito, 1987. "Scaling Models of Water Transport," *Hilgardia,* 55:33–47.

Kardos, J. L., R. Dave, and M. P. Dudukovic, 1988. "Voids in Composites," *Proceedings of ASME Manufacturing International '88,* 41–48.

Li, S., and R. Gauvin, 1991. "Numerical Analysis of the Resin Flow in Resin Transfer Molding," *Journal of Reinforced Plastics and Composites,* 10:314–327.

Lundstrom, T. S., 1997. "Measurement of Void Collapse during Resin Transfer Molding," *Composites Part A,* 28A:201–214.

McKay, M. D., W. J. Conover, R. J. Beckman, 1979. "A Comparison of Three Methods for Selecting Values of Input Variables in the Analysis of Output from a Computer Code," *Technometrics,* 21:239–245.

Parnas, R. S., and F. R. Phelan, 1991. "The Effects of Heterogeneities in Resin Transfer Molding Preforms on Mold Filling," *SAMPE Quarterly,* 22:53–60.

Patankar, S. V., 1980. "Numerical Heat Transfer and Fluid Flow," McGraw-Hill, New York.

Lumped Mass Lagrangean Method for Heat and Species Transport in Resin Transfer Molding

M. Y. LIN AND H. T. HAHN

ABSTRACT

A first order, lumped mass Lagrangean method is introduced as an unconditionally stable scheme that can be easily implemented for solving convection dominant conservation laws. The key to this algorithm is the consistent definition of the nodal Courant number and the associated nodal flux. This upwind biased scheme does not produce spurious oscillations in regions of sharp gradient. The consistent Courant number can be obtained explicitly without the costly Fourier analysis. The design of this algorithm guarantees unconditional stability and optimal time stepping. When tested over a 2-D non-uniform mesh, the proposed method has been found to be in excellent agreement with semi-analytical solutions. Simulation result of mold filling with power law fluids shows significant difference from an earlier result published by other researchers.

INTRODUCTION

Considerable amount of work on resin transfer molding (RTM) simulation has already been accumulated in the literature. In general, isothermal flow, which can be modeled by Darcy's Law, has been extensively exploited with various degrees of success. However, methods for modeling heat transfer and cure kinetics remain in doubt. According to Tucker (Tucker, 1996), careful attention should be given to the problem of distorted, inaccurate solutions which are obtained in spite of the limitations of the numerical schemes used. Because "the existence of a robust computer code with color graphics does not guarantee accurate results," Tucker called for rigorous validation of models by analytical solutions. Tucker's argument is supported in part by the fact that some researchers have reported difficulties in controlling truncation errors (Young, 1994).

BACKGROUND

Solutions of convection-dominated problems can be unstable when a conventional numerical scheme such as the Galerkin's method is applied. The following is a brief review of the three most successful finite element methods (Donea and Quartapelle, 1992). The first

Mark Lin and H.Thomas Hahn, Mechanical and Aerospace Engineering Department, School of Engineering and Applied Science, University of California Los Angeles, CA 90095-1597

family of approaches is the characteristic or semi-Lagrangean method. The advantage of this method is that along the characteristic lines, the governing partial differential equations can be transformed into ordinary differential equations which are considerably easier to solve. Avariation of this method has been found to be accurate and efficient enough to be used in meteorological forecast where a large amount of data is processed at a time. This method is unconditionally stable. Nonetheless, for RTM applications the characteristic lines will be undefined across the flow front, where a different treatment may be required.

The other two families of numerical schemes mentioned came about to address a serious shortcoming of the classical Galerkin method. Basically, the simple Euler explicit time stepping scheme negates the high spatial accuracy achieved by the discretization process. The Taylor-Galerkin approaches use higher order terms in the Taylor series expansion of the time derivative to obtain third or even fourth order temporal accuracy. These approaches are only conditionally stable, but are very accurate. The Least Square or Petrov-Galerkin schemes achieve unconditional stability by applying the variational principle to both the temporal and the spatial discretizations to obtain a compatible level of accuracy. These schemes perform poorly when Courant numbers exceed unity. Recall that for one dimensional problems the CFL condition requires that the Courant number be less or equal to one. Let C be the wave speed, than the Courant number is:

$$\left|\frac{C\Delta t}{\Delta x}\right| \leq 1 \tag{1}$$

What also makes the Courant number important is the fact that for most algorithms, the solution is most accurate when the Courant number is unity. Therefore, the general goal is to maximize the Courant number without exceeding unity. However, despite these important properties, the Courant number is very difficult to evaluate explicitly in multi-dimension problems with a non-uniform mesh. Often some approximation based on velocity and element size is used. Sometimes, empirical formulas have been tried (Lin, Lee and Liou, 1991). These schemes do not offer the optimal time step and do not offer complete safeguard against violation of the CFL condition.

In the field of finite difference methods, the essentially non-oscillating (ENO) finite difference scheme was developed to capture the motion of steep gradients in the flow without incurring the smoothing effect or spurious oscillations (Harten, Engquist, Osher and Chakravarthy, 1987). Despite its effectiveness in many applications, how ENO can be implemented in a finite element method is not immediately apparent and remains a ground for future research. Nevertheless there are two important concepts in the ENOS schemes that set the foundation for the algorithm to be presented below. First, high order polynomial interpolation should not be used across a region of steep gradients. This is the main source of spurious oscillations in the results. Second, convection terms can be solved separately from diffusion terms. In this paper, convection terms will be solved by Lagrangean method while diffusion terms will be solved by the well tested Crank-Nicholson method.

GOVERNING EQUATIONS FOR PMR-15 RTM SYSTEMS

This non-Newtonian fluid problem was first modeled by Chan and Hwang using the least square method(Chan and Hwang, 1993). Crosslinking and conversion are assumed to take place mostly after the mold is filled. The variables of interest are the location of the flow front, the pore pressure field, and the fiber and resin temperatures. The mass conservation of a liquid resin flowing through a fiber preform can be modeled with Darcy's Law:

$$\frac{\partial s}{\partial t} + \nabla \cdot \mathbf{q} = \frac{\partial s}{\partial t} + \nabla \cdot \phi\, \mathbf{v} = \frac{\partial s}{\partial t} - \nabla \cdot \left(\frac{1}{\mu} \mathbf{K} \cdot \nabla p \right) = 0 \tag{2}$$

where $\mathbf{q}$ is defined as the volumetric flow rate per unit area. ϕ is the porosity of the fiber preform and $\mathbf{v}$ is the average resin velocity. $\mathbf{K}$ is the permeability tensor and μ the resin viscosity (Bear, 1988). The saturation factor s is analogous to the fill factor of the finite element control volume method(Young,1994). The power law model for the resin viscosity is given by an empirical relationship:

$$\mu = \mu_\infty\, exp\left(\frac{\Delta E}{RT}\right)\left(\frac{|\mathbf{v}|}{H}\right)^{N-1} \tag{3}$$

The energy balance equation for the fiber preform is:

$$\frac{\partial T_f}{\partial t} - \kappa_f \nabla^2 T_f = \frac{ha(T - T_f)}{(1-\phi)\rho_f C p_f} \tag{4}$$

where ρ is the resin density, C_p is the resin thermal capacitance, and κ is the scaler thermal conductivity constant. The heat transfer coefficient between the fiber and the resin is give as a linear function of the resin velocity: $h = 100|\mathbf{v}|$ W/m²·K. Other parameters are listed in Table I. The energy balance equation for the PMR-15 resin is:

$$\frac{\partial T}{\partial t} + \mathbf{v} \cdot \nabla T - \kappa\, \nabla^2 T = \frac{ha(T_f - T)}{\phi \rho C p} + \frac{h_{mold}\, a_{mold}\,(T_{mold} - T)}{\phi \rho C p} \tag{5}$$

where the heat transfer coefficient between the mold and the resin is a square root function of the resin velocity: $h_{mold} = 1700\sqrt{|\mathbf{v}|}$ W/m²·K. The injection pressure is 500KPa and the injection temperature is 310K. The initial temperature of the fiber and the mold is 380K. It is assumed that no cross linking takes place before the mold is filled and that imidization occurs at temperature higher than 380K.

VALIDATION PROBLEM

First, the solution algorithm is presented and validated with a degree of cure balance equation. The diffusion effect and the viscosity change are assumed to be insignificant, so that a semi-analytical solution in 2-D can be obtained. Conservation form of Darcy's law(Eq.2) still applies. The isothermal degree of cure balance law is the 2-D extension to a 1-D benchmark problem given by Tucker (Tucker,1996).

$$\phi \frac{\partial \alpha}{\partial t} + \mathbf{q} \cdot \nabla \alpha = \phi (0.001 + 5\alpha)(1-\alpha)^2 \qquad (6)$$

where α is given as the *intrinsic phase average* degree of cure. It can be shown that for axi-symmetric isothermal flow a semi-analytical solution of Eq.6 exists.

SOLUTION PROCEDURE

In this section, an 'upwinding' scheme is developed by modifying the Galerkin method. Even though only Eq.6 is used, the same procedure can be applied to heat transfer equations as well. Because of the novelty of the approach, an outline of the derivation process is necessary. First the classical Galerkin method is formulated. For one-dimensional problems with a uniform mesh the Galerkin method becomes identical to the central difference CFD method. Therefore, the central difference CFD method is a special case of the Galerkin method. The approach is to modify the Galerkin method to be a general case of the first order upwind scheme CFD method, which is known to be stable provided the CFL condition is satisfied. The modification follows the observation made by comparing the forms of the Galerkin method, the central difference CFD method and the first order upwind CFD method. The consistent Courant number for this method will be constructed. Because the first order upwind scheme CFD method can be shown to be a lumped mass Lagrangean method, the resulting FEM algorithm is called the Lumped Mass Lagrangean FEM method.

Let α be approximated by the nodal values $\bar{\alpha}_i$ and the interpolation functions H_i over a finite element domain Ω so that

$$\alpha = \sum_i H_i \bar{\alpha}_i \qquad (7)$$

Galerkin's method requires that

$$\int_\Omega \left[\phi \frac{\partial \alpha}{\partial t} + \mathbf{q} \cdot \nabla \alpha - \phi (0.001 + 5\alpha)(1-\alpha)^2 \right] H_i \, d\Omega = 0 \qquad (8)$$

Applying the Euler explicit time stepping method to (8) leads to the following matrix equation:

$$\mathbf{F}\left(\bar{\alpha}^{t+\Delta t} - \bar{\alpha}^t\right) + \Delta t \cdot \mathbf{Q}\bar{\alpha}^t = \mathbf{R} = \int_0^{\Delta t} \int_\Omega f(\bar{\alpha}) \mathbf{H} \, d\Omega \, d\tau \qquad (9)$$

where $\bar{\alpha}$ is the vector of nodal values and the time step indices are in superscript. $f(\bar{\alpha})$ is the right hand side of Eq.(6). $\mathbf{H}$ is the vector of shape functions. $\mathbf{Q}$ is the convection matrix with components written as

$$\mathbf{Q} = \int_\Omega \mathbf{H} \cdot \mathbf{q} \cdot \nabla \mathbf{H}^T \, d\Omega \qquad (10)$$

Because of the following conforming shape function property,

$$\sum_j H_j = 1 \tag{11}$$

F is the 'lumped' mass matrix with the following diagonal elements

$$F_i = \int_\Omega \phi H_i d\Omega = \sum_{j=1}^{n} \int_\Omega \phi H_i H_j d\Omega \tag{12}$$

Equation (12) implies that the diagonal terms F_i can be considered to be a *nodal volume* defined by the shape function H_i. Therefore, Eq.(9) can be rearranged in the form of a conservation law with respect to the nodal volumes. For example, for nodal volume i,

$$F_i(\overline{\alpha}_i^{t+\Delta t} - \overline{\alpha}_i^{t}) + \sum_j \Delta t \cdot Q_{ji}\overline{\alpha}_j^{t} + \sum_k \Delta t \cdot Q_{ik}\overline{\alpha}_k^{t} + \Delta t \cdot Q_{ii}\overline{\alpha}_i^{t} = R_i \tag{13}$$

where Q_{ji} is always negative and represent the coupling effect between an upstream node j and node i. Q_{ik} is always positive and represent the coupling effect between node i and an downstream node k. Q_{ii} is the diagonal coefficient and carries no physical interpretation of interest.

Consider a one-dimensional problem in x, with <u>scalar</u> mass flow rate per unit area q, and uniformly spaced mesh points, Q_{ii} becomes zero in Eq.(13). Q_{ji} and Q_{ik} becomes $q/2$ and -$q/2$ respectively. In this case, Eq.(13) is identical to the central difference finite difference scheme commonly found in the literature and given by Tucker (Tucker, 1996):

$$\frac{\overline{\alpha}_i^{t+\Delta t} - \overline{\alpha}_i^{t}}{\Delta t} + \frac{q}{2} \cdot \frac{\overline{\alpha}_j^{t} - \overline{\alpha}_k^{t}}{\Delta x} = 0 \tag{14}$$

where q is positive when the flow is from node j to node i and to node k. In fact, for two-dimensional problems over a uniform mesh it can be shown that both methods are closely related. As discussed previously, both the central difference CFD method and the Galerkin method is unsuitable for this problem. Again, the objective is to find the generalized FEM equivalent to the stable first order upwinding finite difference scheme discussed by Tucker:

$$\frac{\overline{\alpha}_i^{t+\Delta t} - \overline{\alpha}_i^{t}}{\Delta t} + q \cdot \frac{\overline{\alpha}_j^{t} - \overline{\alpha}_i^{t}}{\Delta x} = 0 \tag{15}$$

where the solution is based on values solely at the upstream nodes. It can be observed that this equation is a Lagrangean equation based on the nodal value in the center of a cell spanned by Δx. When the Courant number $q\Delta t/\Delta x$ is equal to one,

$$\overline{\alpha}_i^{t+\Delta t} = \overline{\alpha}_j^{t} \tag{16}$$

In other words, the degree of cure at node i has been entirely replaced by the degree of cure at node j.

Even though Eq.(15) is a simple and popular CFD method for stabilizing convection dominant problems, it is one of the few CFD schemes without a FEM equivalent or counterpart. The derivation of such a scheme is not straight forward, and would be presented in a backward manner, with the expected form of the discretized matrix equation presented first. In the discretized domain defined with respect to the nodal volume the following form of conservation law is expected:

$$F_i(\bar{\alpha}_i^{t+\Delta t} - \bar{\alpha}_i^t) + \sum_j \Delta t \cdot \bar{q}_{ji}\bar{\alpha}_j^t + \sum_k \Delta t \cdot \bar{q}_{ik}\bar{\alpha}_i^t = R_i \qquad (17)$$

where $\bar{q}_{ji}$ is the *nodal volume flux* from node *j* upstream to node *i* and is always negative. $\bar{q}_{ik}$ is the nodal volume flux from node *i* downwind to node *k* and is always positive. Unlike Eq.(13) the lower case *q* is used because these are no longer simply matrix coefficients for coupling effects, but values with physical interpretations as mass flux in the discretized domain.

For the remainder of this discussion, the subscript *j* will be reserved for upstream nodes and the subscript *k* will be reserved for downstream nodes. Subscript *m* will imply every node in the discretized volume. To satisfy mass conservation

$$\sum_j \left|\bar{q}_{ji}\right| = \sum_k \left|\bar{q}_{ik}\right| \qquad (18)$$

$$\sum_j \bar{q}_{ji} + \sum_k \bar{q}_{ik} = 0 \qquad (19)$$

except for unsaturated nodes. $\bar{q}_{ji}$ and $\bar{q}_{ik}$ can not be determined explicitly. Furthermore, because the central difference scheme in multi-dimensions requires the diagonal terms as shown in Eq.(13) there is no simple relationship between the Q_{im} coefficients and the $\bar{q}_{ji}$ coefficients as in the one-dimensional case. However, by comparing the convection coefficients of Eq.(13) and Eq.(14), it can be presumed that the sum of the magnitudes of Q_{im} is also equal to the total mass flow rate. This is partly supported by the fact that

$$\sum_m Q_{im} = 0 \qquad (20)$$

as in the case of Eq.(14). This leads to the following assumption about the net nodal volume flux

$$\sum_m \left|Q_{im}\right| = \sum_j \left|\bar{q}_{ji}\right| = \sum_k \left|\bar{q}_{ik}\right| = \sum_m \left|\int_\Omega H_i \underline{\mathbf{q} \cdot \nabla H_m} \, d\Omega\right| \qquad (21)$$

This assumption is complemented by the fact that the last term of Eq.(21) can be interpreted as the integral of convection effects(terms that are underlined) within the nodal volume H_i which was defined earlier. The validation of this assumption lies of course mainly on the numerical results to be presented. Because the solution of Eq.(17) requires only the knowledge of the sum of $\bar{q}_{ik}$ and not the individual components, all that remains to do is the determination of $\bar{q}_{ji}$. Assuming that the magnitude of the coupling effect by an upstream node is proportional to the total magnitude of the coupling effect from all the upstream nodes, the convection coefficients should obey the following ratio

$$\frac{\bar{q}_{ji}}{\sum\left|\bar{q}_{ji}\right|}=\frac{Q_{ji}}{\sum\left|Q_{ji}\right|}=\frac{\int_{\Omega} H_i(q\cdot\nabla)H_j d\Omega}{\sum_j\left|\int_{\Omega} H_i(q\cdot\nabla)H_j d\Omega\right|} \quad (22)$$

The solution of Eq.(17) is now straightforward. The proposed algorithm has the CFL property (Anderson, 1988) that exact answers are obtained whenever the Courant number is equal to 1. For multi-dimension problems the equivalent Courant number, ν, for node *i* is

$$\nu_i=\frac{\Delta t\cdot\sum_j\left|\bar{q}_{ji}\right|}{F_i}=\frac{\Delta t\cdot\sum_k\left|\bar{q}_{ik}\right|}{F_i} \quad (23)$$

Because the Courant number can be calculated explicitly for each node, the maximum Δt that does not violate the CFL condition can be determined. For the formulation to be consistent with the explicit integration method used to solve Darcy's flow, the coefficients with unsaturated nodes are excluded in the evaluation of Eq.(22).

RESULTS OF THE VALIDATION PROBLEM

Eq.6 is now solved with the proposed algorithm in a radial flow problem. Explicit time integration is used to solve Darcy's law and to update the flow front represented by the partially saturated region. The geometry is defined by a finite element mesh with 176 bilinear quadrilateral elements composing of 207 nodes. The inlet radius is 2 cm while the mold radius is 6 cm. Thc injcction pressure is 100MPa and k/μ is 1.0e-8 m2/Pa-s. The porosity is 0.5. The final distribution of α as a function of *r* for this problem is plotted in Figure 1 with the semi-analytical solution. The agreement between the semi-analytical solution and the simulation result is very good.

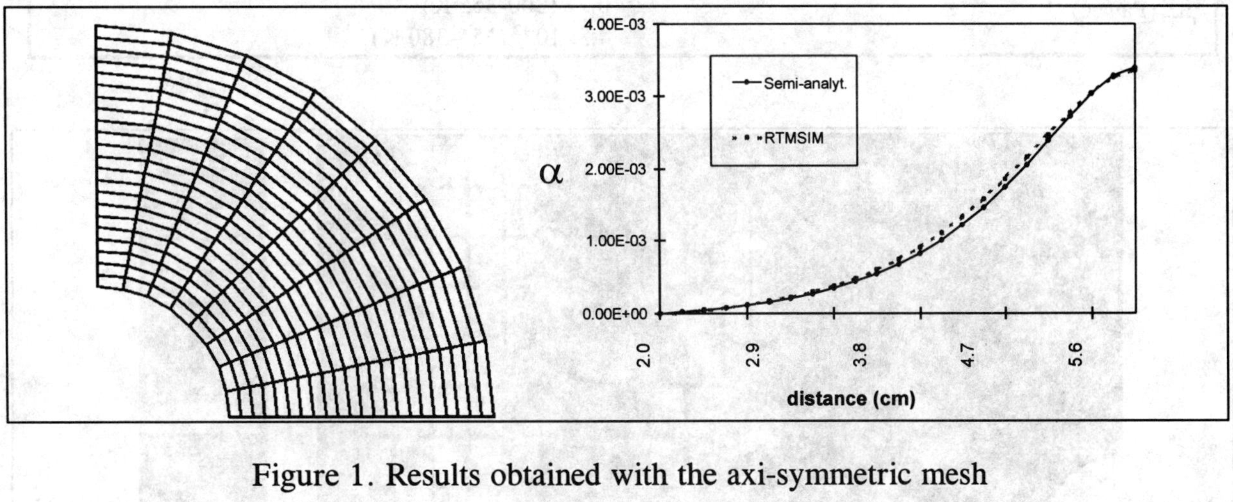

Figure 1. Results obtained with the axi-symmetric mesh

RESULTS OF THE PMR-15 RTM SIMULATION

An iteration procedure is used to solve the coupled equations. Convergence is set to be less or equal to 0.0001 of the residual of the velocity norms. Solution of Eq.2 has been checked with 2-D analytical solutions for power law fluids. The flow front progression is illustrated in Figure 2. The resin pore pressure and temperature at 3 different times are graphed in Figure 3 and Figure 4 respectively. The mold filling time obtained is 2628 sec, which is more than five times longer than the filling time given by the results obtained by Chan and Hwang using the least squares method. The present result is supported by the validation of convective equation solution described in the earlier part of the paper. If the present result is correct, than it is doubtful that cure kinetics can be neglected as originally assumed. Therefore, this exercise underlines the fact that much work remains in this particular subject area.

Table I Parameters of the PMR-15 RTM system

mold dimension	0.1x0.05x0.003m
injection port radius	0.005m
a_{mold}	2/0.003
fiber density ρ_f	1760 kg/m^3
fiber heat capacity C_{p_f}	1000 J/kg·K
κ_f	0.8 W/m·K
a	2×10^4m^{-1}
porosity ϕ	0.7
resin density ρ	940 kg/m^3
resin heat capacity C_p	800 J/ kg·K
power law constant N	0.12
κ	0.2 W/m·K
ΔE (kJ/mol)	14.7 (300-355 K) 87.4 (355-380 K)
μ_∞ (Pa·sec)	0.22 (300-355 K) 4.3×10^{-12} (355-380 K)

Figure 2 Flow front progression of PMR-15 resin

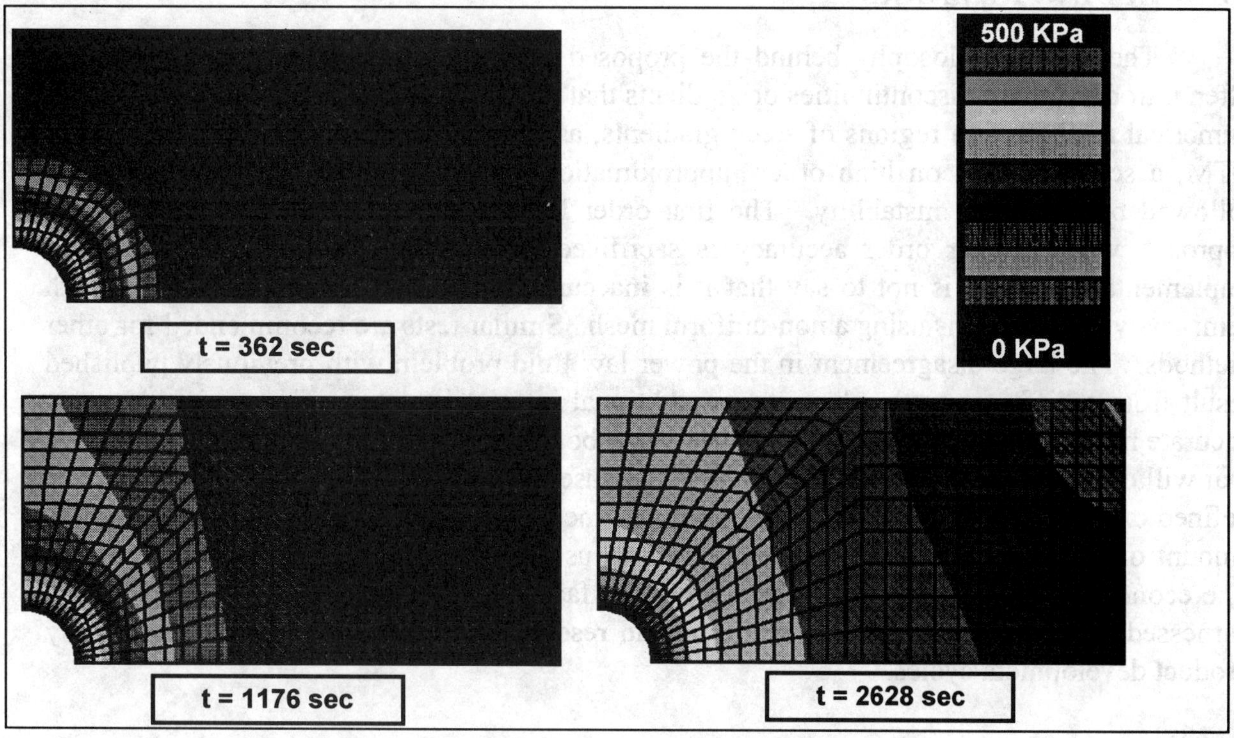

Figure 3 Pore pressure contour of PMR-15 resin

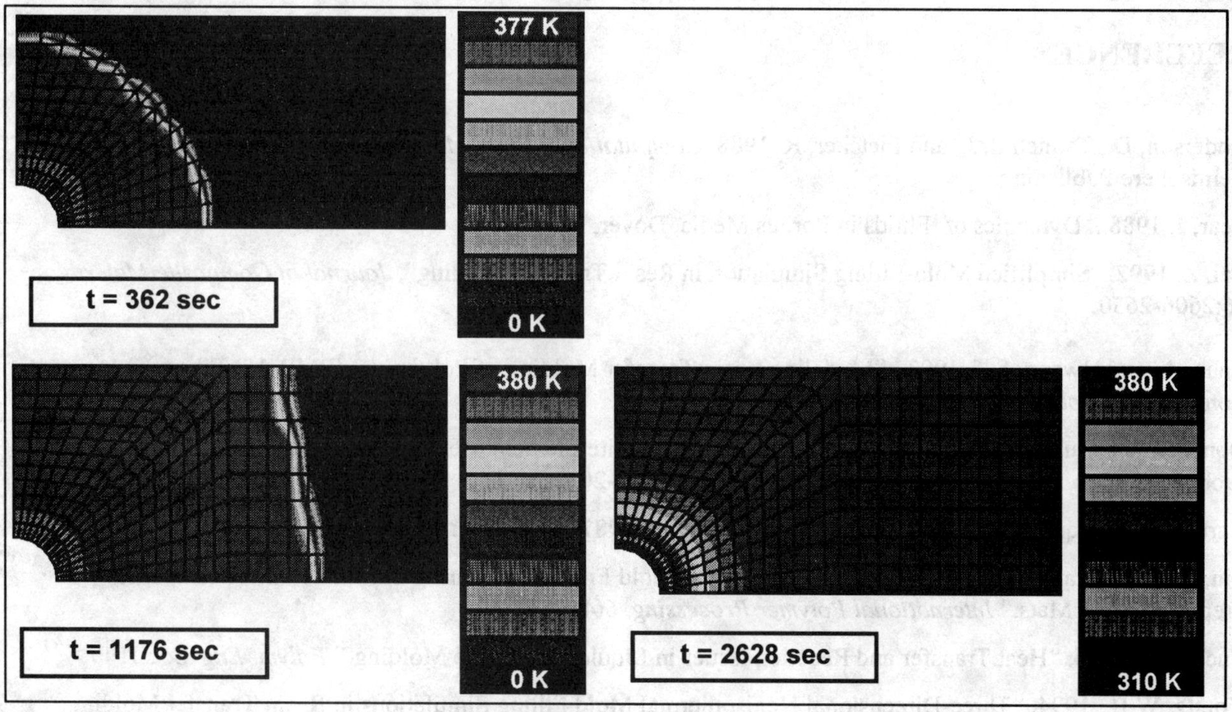

Figure 4 Temperature field of PMR-15 resin

CONCLUDING REMARKS

The design philosophy behind the proposed algorithm is that real world problems often introduce sharp discontinuities or gradients that are difficult to model with conventional numerical methods. In regions of steep gradients, as often encountered at the flow front in RTM, a scheme based on high order approximation usually leads to spurious oscillation followed by numerical instability. The first order Lagrangean approach is a 'brute force' approach where higher order accuracy is sacrificed for stability, robustness and ease of implementation. This is not to say that it is inaccurate because it has been validated with semi-analytical solutions using a non-uniform mesh. Similar tests are recommended for other methods. The large disagreement in the power law fluid problem with previously published result from a different approach warrants additional interest in determining which the most accurate method. Again, there are no conceivable boundary conditions or material properties that will destabilize the proposed algorithm. Because the consistent Courant numbers can be defined explicitly in this approach, the optimum time step is always used. Consequently, the amount of knowledge and experience required to use this algorithm effectively is minimal. The economy of this approach can be seen in the fact that more computer resources can be harnessed for less price every day while human resources remain to be expensive in many product development cycles.

ACKNOWLEDGEMENTS

This work is supported by the Office of Naval Research under ONR Grant N00014-92-J1846. Sincere appreciation is extended to Yapa D.D. Rajapakse, Scientific Officer

REFERENCES

Anderson, D., Tannehill, J., and Pletcher, R. 1988. *Computational Fluid Mechanics and Heat Transfer*, Hemisphere Publishing.

Bear, J. 1988... Dynamics of Fluids in Porous Media, Dover, New York.

Cai, Z. 1992. "Simplified Mold-Filling Simulation in Resin Transfer Molding", *Journal of Composite Materials* 26:2606-2630.

Chan, A. and Hwang, S.T. 1993., "Modeling Resin Transfer Molding of Polyimide (PMR-15)/Fiber Composites", *Polymer Composites*, 14(6):524-528.

Donea, J. and Quartapelle. L.1992. "An Introduction to Finite Element Methods for Transient Advection Problems," *Comput. Meth. in App. Mech. and Eng.* 95. 169-203

Harten, A., Engquist, S., Osher, S., and Chakravarthy, S. 1987.. *J. Comput. Phys.* 71, 231.

Lin, R., Lee, J., and Liou, M. 1991. "Non-isothermal Mold Filling and Curing Simulation in Thin Cavities with Preplaced Fiber Mats," *International Polymer Processing,* 6(4):356-369

Tucker, C. 1996. "Heat Transfer and Reaction Issues in Liquid Composite Molding," *Polym. Eng. Sci.* 1. 17.

Young, W.B.. 1994. "Three-Dimensional Nonisothermal Mold Filling Simulations in Resin Transfer Molding," *Polymer Composites* 15(2):118-127

Assessment of Thermal Spray SRIM Composites

O. O. OCHOA, J. H. BARRON AND B. W. HILL

ABSTRACT

Material and mechanical properties of Structural Reaction Injection Molded (SRIM) composites manufactured with the Thermal Spray (TS) preforming process are investigated analytically and experimentally. Flexure, compressive, and tensile coupon tests are conducted to obtain respective moduli and strength. In addition to classical flat coupons, hat beams are tested to assess structural response. The elastic constants obtained from the coupon tests were used in the finite element models of the hat beam flexure tests which provided good agreement with the experimental load-deflection response.

INTRODUCTION

The development of inexpensive manufacturing options for composite components with primary load carrying ability is one of the most important requirements for engaging high volume industrial applications. Injection molding of fiber reinforced preforms offers promising savings where the flexibility of mixing multiple continuous and discontinuous fibers in the preforms offers a wide range of loading options. The present program motivation is focused on establishing the impact of maintaining multiple reinforcement architecture on load capability when manufactured by SRIM.

The Thermal Spray preforming process developed by Dow Chemical is used to place the E-glass fiber reinforcement in the preform construction (Barron and Entringer, 1994). This approach requires a glass roving chopper and a thermal spray gun. Glass roving is cut to specific length in the chopper while the thermal spray gun is used to spray a thermoplastic binder. Chopped glass fibers and binder are sprayed concurrently onto a perforated metal screen which serves as the template for the preform shape. The binder then cools and solidifies, allowing the preform to retain the shape of the template. Vacuum is applied to the back of the metal screen to help direct the binder and glass toward the template, and help compact and cool the preform as the binder solidifies. Then the preform is removed from the

Ozden O. Ochoa, Center for Mechanics of Composites, College Station, TX 77843
James H. Barron, The Dow Chemical Company, 2301 N. Brazosport Blvd., Freeport TX 77541
Bryan W. Hill, The Dow Chemical Company, 2301 N. Brazosport Blvd., Freeport, TX 77541

screen and placed in a mold for the injection molding procedure (Howell, Camargo, and Bityk, 1989).

Two distinct reinforcement architectures are utilized in the preform. The first group has only Thermal Spray (TS) reinforcement (discontinuous, random E-glass fibers) with glass content maintained at 30, 40, 50, or 60 weight percent. The second group is a three-layer sandwich configuration where the core is TS and the outer layers (skins) are unidirectional (UNI) roving fabric. Total weight percent glass content for the TS/UNI architectures is maintained at 40 percent. A schematic of the TS/UNI sandwich panel is presented in Figure 1. Flat panels and hat-shaped beams with the aforementioned reinforcements are manufactured with the SRIM procedure. Flat preforms are easier to make in comparison to the hat beams. The placement of the fibers and the square cross section of the beams with sharp corners interferes with the flow of the resin (Liou et. al., 1989).

TESTING

Material properties are obtained by testing coupons cut from the flat panels. Tensile, flexure, and compression coupon tests are performed according to ASTM composite laminate test standards.

COUPON TESTS

Tensile testing is performed according to ASTM standard D-3039 on 48 coupons of 1" x 10.5" in size. Crosshead displacement is set at 0.10 in/min and an extensometer is used to measure strain within the 2 inch gage region. The loading is tracked with an 10,000-lb cell on an Instron Model 1127. The coupons are obtained from nine TS and six TS/UNI architecture panels.

ASTM Standard D-790, Method I, Procedure A is used for flexure testing 1" x 3" rectangular coupons with 0.125" nominal thickness. This is a three-point testing procedure with the lower supports 2" apart on the coupon. Crosshead displacement is set at 0.05 in/min. Similar to the tensile tests, nine TS and six TS/UNI panels are used for the coupons.

An Iitri test fixture is used with ASTM standard D-3410 for compression tests. The coupons dimensions are 0.5" x 5.5" x 0.125". Crosshead displacement is maintained at 0.05 in/min. Strain is measured with an 0.125" square strain gage.

HAT BEAM FLEX TESTS

Hat beams are studied with the four point flexure testing to characterize their load-deflection response. These beams are fabricated with the same TS and TS/UNI architectures used in the coupon tests, and are tested in "open" and "restrained" configurations. UNI reinforcement is aligned axially along the length of the beam for all TS/UNI architectures. The fixture presented in Figure 2 is designed to accommodate the geometry of the beams. It provides a twenty one inch span to support the beams, and the upper loading bars are separated seven inches. Loading bars of 1" diameter are used to contact the beams. The

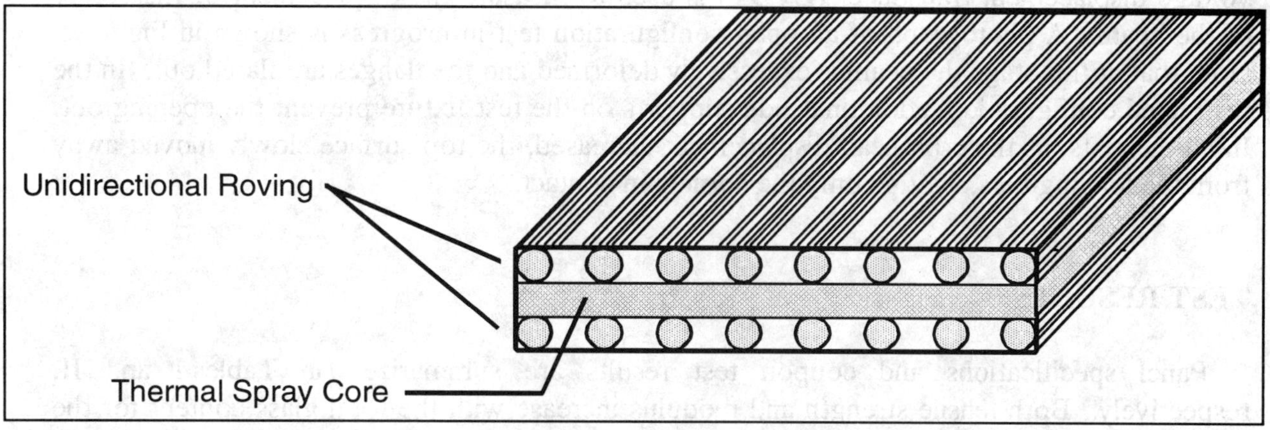

Figure 1 - Sandwich Structure of TS/UNI Architecture

Figure 2 - Hat Beam Experiment Test Fixture: Open Configuration

downward displacement rate of the upper loading bars is 0.25 in/min during testing. A linear voltage displacement transducer (LVDT) is used to measure the displacement of the center of the beam. A photograph of an open configuration test in progress is shown in Figure 2. Note that at this stage the beam is excessively deformed and the flanges are flared out. In the restrained configuration, additional side supports on the test fixture prevent the opening out. In all the tests, as the crosshead displacement increased, the top surface slowly moved away from the loading bars, leaving only the corners in contact.

TEST RESULTS

Panel specifications and coupon test results are summarized in Tables I and II, respectively. Both tensile strength and modulus increase with the total glass content for the TS reinforcement. Tensile strength with thirty percent glass reinforcement is about 17 ksi, whereas for higher reinforcement content it approaches 30 ksi. Similarly, tensile modulus values range from 1.0 Msi to 1.7 Msi as a function of increasing glass content.

For the TS/UNI composites, similar observations hold as a function of increasing UNI roving. The longitudinal tensile strength in the sandwich construction reached to 45.2 ksi (average) and modulus increased to 2.2 Msi for architectures near 20% UNI glass content. On the other hand, as UNI glass is added there is less fiber oriented in the transverse direction, so a reduction in the transverse properties occur. For this same reinforcement, transverse tensile strength is about 12 ksi and the transverse modulus is about 0.8 Msi.

The flex test results are summarized in Tables III and IV. Within the TS grouping, the increased glass content governs the performance improvement as before. Once UNI rovings are added as skin, the response is dramatically improved since the highest bending strain is experienced by the outer roving layers. The transverse properties of the TS/UNI continue to decrease in comparison to the TS configuration. It is also noted that the coefficient of variation (CV) of the measured properties is lower for TS/UNI groups reinforcement (3.9 %). The TS panels with 40% glass content had flexure strengths of approximately 36 ksi and a CV of 9.0%.

Compression strength for the TS architectures varies from 19-26 ksi as the glass content is increased and the modulus range is 1.1-2.0 Msi respectively. The longitudinal compression strength of the TS/UNI coupons approaches 41 ksi while remaining at 16 ksi in the transverse direction. Modulus in the longitudinal direction is as high as 2.4 Msi while the transverse modulus is no lower than 1.1 Msi. In comparison, the average TS strength at 40% weight glass content is 21.7 ksi and average TS modulus is 1.4 Msi.

Tests results show that the hat beam performance is dependent upon the total TS glass content of the reinforcement. Load versus mid-beam deflection curves of 36% TS glass content test beams are presented in Figure 3. As mentioned earlier, loading capacity of beams in the restrained tests is higher than beams tested in the open configuration. Maximum load is achieved near mid center displacement of 0.25" (0.20”-open) for the restrained tests. Testing of the TS/UNI composites reveal that the amount of UNI roving in the reinforcement has little effect on beam performance; results for beams with 20% UNI glass are displayed in Figure 4. All load-deflection curves are essentially collinear and reach highest loading when the mid-beam displacement reaches 0.3”. This indifference to UNI content implies that the UNI rovings are not effectively oriented to carry significant loading.

TABLE I - TS COMPOSITE TENSILE TEST RESULTS

Plaque Number	Total Wt. %	UNI Wt. %	Thickness (inch)	Coupons Tested	Strength, ksi ksi	Strength, ksi CV, %	Modulus, Msi Msi	Modulus, Msi CV, %
142003	30.8	0.0	0.1245	3	17.0	5.0	1.02	9.5
142002	31.3	0.0	0.1299	6	17.8	7.7	0.97	13.7
143003	39.4	0.0	0.1274	6	20.7	12.4	1.34	12.4
143001	40.9	0.0	0.1243	6	20.6	8.9	1.20	9.1
143002	41.4	0.0	0.1335	9	19.5	6.2	1.40	10.6
144002	45.7	0.0	0.1441	6	23.1	9.2	1.46	7.2
144004	48.3	0.0	0.1267	6	28.1	5.4	1.46	8.5
145003	52.7	0.0	0.1320	3	29.7	1.9	1.74	6.4
145002	53.1	0.0	0.1285	6	29.1	8.0	1.68	15.5

TABLE II - TS/UNI COMPOSITE TENSILE TEST RESULTS

			LONGITUDINAL					
Plaque Number	Total Wt. %	UNI Wt. %	Thickness (inch)	Coupons Tested	Strength, ksi ksi	Strength, ksi CV, %	Modulus, Msi Msi	Modulus, Msi CV, %
143003	39.4	0.0	0.1274	6	19.5	6.2	1.40	10.6
143001	40.9	0.0	0.1243	6	20.7	12.4	1.34	12.4
143002	41.4	0.0	0.1335	9	20.6	8.9	1.20	9.1
151002	44.6	6.7	0.1248	3	28.6	4.2	1.46	5.2
151001	46.6	7.0	0.1237	3	28.3	0.6	1.45	1.6
152002	41.8	14.6	0.1273	3	35.7	0.8	1.82	5.8
152001	42.9	15.0	0.1228	3	42.4	2.7	1.85	5.1
153002	42.6	20.9	0.1260	3	47.4	3.7	2.17	3.7
153001	43.1	21.1	0.1283	3	42.9	3.2	2.15	2.5
			TRANSVERSE					
Plaque Number	Total Wt. %	UNI Wt. %	Thickness (inch)	Coupons Tested	Strength, ksi Avg.	Strength, ksi CV, %	Modulus, Msi Avg.	Modulus, Msi CV, %
143003	39.4	0.0	0.1274	6	19.5	6.2	1.40	10.6
143001	40.9	0.0	0.1243	6	20.7	12.4	1.34	12.4
143002	41.4	0.0	0.1335	9	20.6	8.9	1.20	9.1
151002	44.6	6.7	0.1293	3	18.9	5.8	1.14	2.7
151001	46.6	7.0	0.1280	3	20.3	4.5	1.50	13.4
152002	41.8	14.6	0.1310	3	15.8	11.6	1.17	6.2
152001	42.9	15.0	0.1263	3	15.6	6.4	1.21	4.9
153002	42.6	20.9	0.1248	3	12.1	5.4	1.02	5.4
153001	43.1	21.1	0.1323	3	11.7	8.7	0.82	3.3

TABLE III - TS COMPOSITE FLEXURE TEST RESULTS

Plaque Number	Total WT %	UNI Wt. %	Thickness (inch)	Coupons Tested	Strength, ksi ksi	Strength, ksi CV, %	Modulus, Msi Msi	Modulus, Msi CV, %
142003	30.8	0.0	0.1286	14	28.3	6.7	1.02	7.3
142002	31.3	0.0	0.1278	6	30.4	6.2	0.98	5.7
143003	39.4	0.0	0.1258	24	37.9	8.5	1.38	8.0
143001	40.9	0.0	0.1232	22	36.1	9.2	1.39	11.9
143002	41.4	0.0	0.1332	8	34.8	9.2	1.18	6.6
144002	45.7	0.0	0.1478	23	39.3	9.7	1.39	10.1
144004	48.3	0.0	0.1279	24	39.8	5.7	1.60	7.0
145003	52.7	0.0	0.1383	27	42.3	7.7	1.63	9.6
145002	53.1	0.0	0.1291	23	43.8	8.2	1.80	7.8

TABLE IV - TS/UNI COMPOSITE FLEXURE TEST RESULTS

			LONGITUDINAL					
Plaque Number	Total Wt. %	UNI Wt. %	Thickness (inch)	Coupons Tested	Strength, ksi ksi	Strength, ksi CV, %	Modulus, Msi Msi	Modulus, Msi CV, %
143003	39.4	0.0	0.1258	24	37.9	8.5	1.38	8.0
143001	40.9	0.0	0.1232	22	36.1	9.2	1.39	11.9
143002	41.4	0.0	0.1332	8	34.8	9.2	1.18	6.6
151001	43.6	6.5	0.1268	3	44.8	7.0	1.77	4.4
151002	44.6	6.7	0.1288	3	47.6	6.2	1.84	3.7
152002	41.8	14.6	0.1305	3	60.2	3.2	2.63	3.3
152001	42.9	15.0	0.1257	3	58.6	2.2	2.49	2.6
153002	42.6	20.9	0.1245	3	73.5	2.1	3.28	0.1
153001	43.1	21.1	0.1318	3	67.9	2.4	3.11	1.6
			TRANSVERSE					
Plaque Number	Total Wt. %	UNI Wt. %	Thickness (inch)	Coupons Tested	Strength, ksi Avg.	Strength, ksi CV, %	Modulus, Msi Avg.	Modulus, Msi CV, %
143003	39.4	0.0	0.1258	24	37.9	8.5	1.38	8.0
143001	40.9	0.0	0.1232	22	36.1	9.2	1.39	11.9
143002	41.4	0.0	0.1332	8	34.8	9.2	1.18	6.6
151001	43.6	6.5	0.1288	3	30.3	3.2	0.98	1.4
151002	44.6	6.7	0.1315	3	31.6	6.1	1.03	6.0
152002	41.8	14.6	0.1309	3	24.4	14.6	0.91	10.2
152001	42.9	15.0	0.1264	3	23.7	6.5	0.92	5.6
153002	42.6	20.9	0.1275	3	17.8	3.7	0.81	1.5
153001	43.1	21.1	0.1303	3	17.4	10.1	0.79	8.6

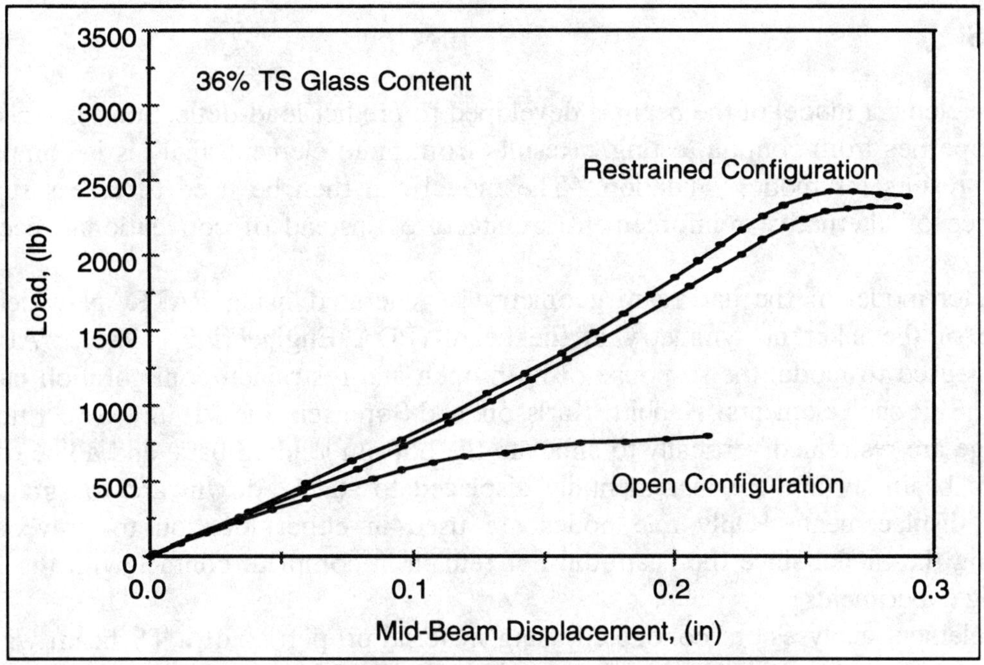

Figure 3 - TS Hat Beam Load v. Mid-Beam Deflection
Open and Restrained Configuration Tests

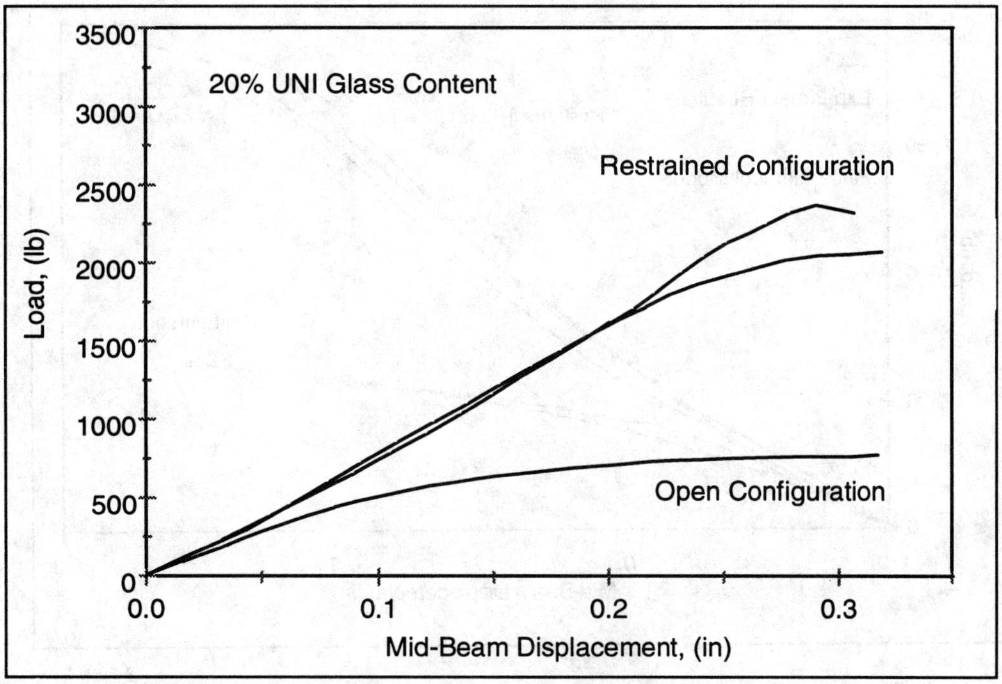

Figure 4 - TS/UNI Hat Beam Load v. Mid-Beam Deflection
Open and Restrained Configuration Tests

MODELS

A finite element model of the beam is developed to predict load-deflection response based on the properties from coupon testing. Results from finite element analysis is compared to beam test results for model validation. The model can then be used to assess structural performance of alternative reinforcement architectures instead of conventional mechanical testing.

A quarter-model of the hat beam geometry is generated using PATRAN which takes advantage of the inherent symmetry in the beam (PDA Engineering, 1988). ABAQUS software is used to model the response of both open and restrained configuration hat beam tests using S8R shell elements (Habbitt, Karlsson, and Sorensen, Inc., 1992). Select nodes on beam flange are restrained vertically to simulate the bottom loading bars; and a line of nodes on the top beam surface are incrementally displaced to -0.25" downward to simulate the crosshead displacement. Only five nodes are used in either location to prevent over-constraining the model since the beam did not remain in complete contact with the loading bars during experiments.

Finite element analyses are conducted using material properties of a TS beam with 28% total glass content, and for a TS/UNI beam with 20% UNI glass and 20% TS glass Beams with TS reinforcement use a single layer shell element with 0.125" thickness and isotropic properties. Beams with TS/UNI reinforcement are treated as three layer laminates. The TS core is considered to be an isotropic layer with 0.075" thickness, while each UNI layer is considered to be orthotropic with 0.025" thickness.

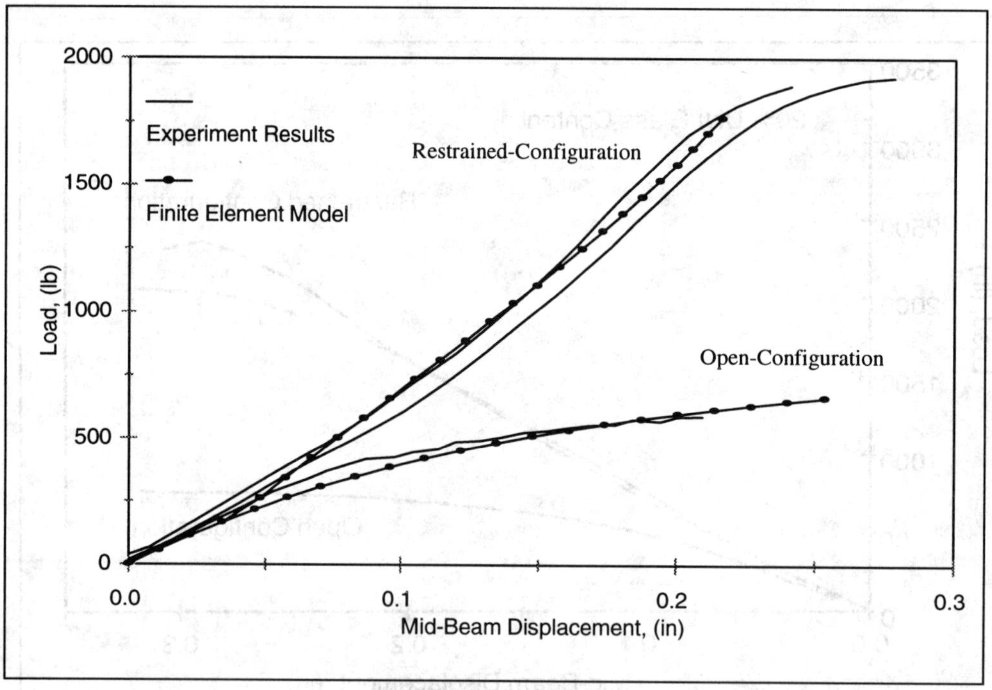

Figure 5 - Comparison of TS Beam FEA Results to Test Results
Load V. Mid-Beam Displacement

Load-deflection results of the open and restrained configuration are plotted in Figure 5 and are compared with TS beam experiments. As illustrated, the open-configuration FEA predicts the initial loading of the experiments well, but is about 50 pounds lower from 0.05" to 0.09". The restrained-form model initially under-predicts the loading of the two experiment beams, but follows the two curves better at higher loading. This initial error is likely due to a small gap which was present between the flanges of the beams and the side supports during actual testing.

CONCLUSION

ASTM tensile, flexure, and compressive tests are conducted to establish a database for thermal spray panels with multiple reinforcement options. Hat beams were fabricated using the same panel architectures and tested in four point bending fixture to determine the load-deflection response. Properties obtained from coupon testing were used in the finite element model of the hat beam to predict and compare its structural response in the experiments.

The corresponding TS strengths and moduli increased with increasing glass content. For TS/UNI architectures, increases in the longitudinal strengths and moduli and the decrease in the transverse values are attributed to the addition of UNI rovings. Similarly, TS hat beam flex tests in the open or restrained configuration carried higher loads for higher glass content. However, hat beams with the TS/UNI reinforcement are indifferent to the changes in the UNI content as observed from both open and restrained configuration tests. Cross-comparison of TS and TS/UNI beam results suggest that a beam with 40% TS glass content has a higher load capacity than a TS/UNI beam at 40% total glass content. Finite element analysis of the beams reasonably predicts the TS beam experiments and the TS/UNI beam experiments for the open configuration test.

REFERENCES

Barron, J. H., and Entringer, J., "Thermal Spray Preforms: An Economical Route to SRIM Composites," *SAE International Congress and Exposition,* Society of Automotive Engineers, Detroit, 1994.

Howell, T. B., Camargo, R. E., and Bityk, D. A., "Improving the Processing Characteristics of Structural RIM Systems," *Proceedings of the Fifth Annual ASM/ESD Advanced Composites Conference,* ASM International, Metals Park, 1989.

Liou, et. al, "Mold Fillinand Analysis of Structural Reaction Injection Moulding (SRIM) and Resin Transfer Molding (RTM)," *Proceedings of the Fifth Annual ASM/ESD Advanced Composites Conference,* ASM International, Metals Park, 1989.

PATRAN Plus, PDA Engineering, Costa Mesa, CA, 1988.

ABAQUS, Version 4.9, Habbitt, Karlsson, and Sorensen, Inc., Providence, RI, 1992.

Permeability Measurement of Advanced Fiber Architecture Preforms

A. C. LOOS, J. C. FINGERSON, K. W. FURROW AND T. W. KNOTT

ABSTRACT

A test fixture was designed, fabricated and used to measure in-plane permeabilities of fiber preforms by both the advancing front and steady-state techniques. Advancing front and steady-state permeabilities were determined for an E-glass fabric and compared. Permeability values obtained by the two techniques were the same. The steady-state permeabilities of a multiaxial warp knit fabric were measured at fiber volume fractions ranging from 55 to 65 percent. These preforms were highly compacted due to through-the-thickness stitching. Off-axis measurements were taken to determine the principal permeability axes and principal permeabilities. The principal axes for permeability were found to be rotated 13° from the stitching direction and varied slightly with fiber volume fraction.

1. INTRODUCTION

Resin transfer molding (RTM) has become a popular cost-effective processing technique for the manufacture of primary composite structural components for the aerospace and automotive industries. The RTM process lends itself the use of near net shape textile preforms manufactured through a variety of automated textile processes such a knitting and braiding. Often, these advanced fiber architecture preforms have through-the-thickness stitching for improved damage tolerance and delamination resistance. The challenge presently facing RTM is to refine the process to ensure complete infiltration and cure of a geometrically complex shape preform with the high fiber volume fraction needed for structural applications. An accurate measurement of the permeability of fabric prefoms used in the RTM resin infiltration process is critical for the successful modeling of the process. Small changes in permeability can result in very different infiltration behavior and times. In

---

A. C. Loos, Professor, Virginia Tech, Department of Engineering Science and Mechanics, Blacksburg, VA 24061-0219

John C. Fingerson, former graduate student, currently Process Engineer, Hewlett Packard, Corvallis, OR 97333

Keith W. Furrow, Research Associate, Department of Engineering Science and Mechanics, Blacksburg, VA 24061-0219.

Tamara W. Knott, Research Associate, Department of Engineering Science and Mechanics, Blacksburg, VA 24061-0219.

this investigation, test methods that can be used to measure the compaction behavior and permeabilities of high fiber volume, advanced fiber architecture textile preforms were developed. These preforms are often highly compacted due to through-the-thickness stitching. A test fixture has been designed, fabricated and used to measure in-plane permeabilities by both the advancing front and steady-state techniques. The fixture was used to measure the steady-state permeabilities of multiaxial warp knit fabric at fiber volume fractions ranging from 55 to 65 percent Off axis permeability measurements of warp knit fabric preforms were taken to determine principal axes and principal permeabilities as a function of fiber volume fraction.

2. PERMEABILITY

Darcy's law (Darcy, 1856) empirically describes flow through porous media and is written as

$$q_i = -\frac{1}{\eta} S_{ij} P_{,j} \qquad i,j \rightarrow 1,2,3 \tag{1}$$

The superficial or filter velocity (q_i) is proportional to the pressure gradient ($P_{,j}$) across the porous media. The proportionality constants are the fluid viscosity (η) and the experimentally determined permeability (S_{ij}). Darcy's law is valid when the flow process is quasi-static, the material system behaves as a continuum, the fluid is Newtonian and the flow is in the laminar regime.

For in-plane flow through a thin preform the pressure gradient in the *z*-direction is zero and Darcy's law simplifies to

$$\begin{bmatrix} q_x \\ q_y \end{bmatrix} = -\frac{1}{\eta} \begin{bmatrix} S_{xx} & S_{xy} \\ S_{xy} & S_{yy} \end{bmatrix} \begin{bmatrix} \partial P/\partial x \\ \partial P/\partial y \end{bmatrix} \tag{2}$$

The permeability (S_{ij}) is a second order tensor and transforms according to:

$$S'_{ij} = a_{ik}\, a_{jl} S_{kl} \qquad i,j,k,l \rightarrow 1,2 \tag{3}$$

where S'_{ij} is the transformed permeability, S_{ij} is the original permeability and a_{ik} , a_{jl} are the direction cosines. The principal permeability as a function of fiber volume fraction can be found by solving the following system of equations:

$$\left(S_{ij} - S\delta_{ij}\right) n_j = 0 \qquad i,j \rightarrow 1,2 \tag{4}$$

$$n_i n_i = 1 \tag{5}$$

where S is the principal permeability and n_i is the principal direction vector. The angle of rotation to the principal permeability (γ) can be computed from the eigenvector as follows:

$$\gamma = \tan^{-1}(n_2/n_1). \tag{6}$$

Permeability is the resistance to flow through porous media. Factors affecting preform permeability include through-the-thickness stitching, interply interfaces, fiber orientation, local changes in fiber volume fraction and wetting characteristics. The complex fiber architectures of textile preforms make analytical permeability predictions difficult. As a result the permeability is often measured directly. Two common methods for measuring the permeability of fabrics and preforms are the steady-state and the advancing front methods.

The advancing front method measures the permeability of a dry preform to an infiltrating fluid. This method accounts for the effects of surface tension between the fluid and fiber. The dry preform is placed in the test fixture and compressed to the desired fiber volume fraction. Then a fluid of known viscosity is injected at constant flow rate into the perform. The position of the flow front, with respect to an initial flow front location, and the pressure at the initial flow front location are measured as a function of time. By plotting pressure as a function of flow front position, the permeability can be determined from the slope of the data and the one-dimensional form of Darcy's law:

$$q = -\frac{S}{\eta}\frac{dP}{dx} \tag{7}$$

Full advancing front permeability characterization requires a new preform at each volume fraction

The steady-state method measures the permeability of a saturated preform under constant flow rate conditions. The preform is compressed to the desired fiber volume fraction and saturated with a fluid of known viscosity. As fluid flows through the preform, the pressure drop across the preform is recorded. The permeability of the preform is then determined from the one-dimension form of Darcy's law (7). Full steady-state characterization requires only one preform for the entire range fiber volume fractions.

3. PERMEABILITY MEASUREMENT SYSTEM

The permeability measurement system was composed of a permeability fixture, a constant flow rate pump and a data acquisition system. Detailed drawings of the permeability fixture shown in figure 1 can be found in Fingerson, Loos and Dexter, 1995. The fixture was designed utilizing a pre-manufactured die set. An LVDT mounted to the die set measured mold height to determine fiber volume fraction. The mold cavity is 17.78 cm long by 15.32 cm wide. The preform measures 15.2 cm long by 15.32 cm wide. The extra cavity length forms an inlet manifold allowing even inlet pressure across the face of the preform. A unique O-ring seat design eliminated the leakage path located between a standard O-ring and the preform.

Figure 1. In-Plane Permeability Fixture

Five pressure transducers were mounted in the fixture-one at the inlet, three located in line beneath the preform and one at the outlet (figure 2). The transducers have small 3.6 mm diameter tips for minimal disruption of the flow pattern. The transducers beneath the preform detect flow front location for advancing front measurements while the inlet transducer is used for steady-state permeability measurements (Fingerson, Loos and Dexter, 1995).

A real time data acquisition and data reduction system was comprised of a DAQ board, signal conditioning and multiplexing hardware and LabVIEW data acquisition soft ware. The constant flow rate pump was controlled through a serial interface using LabVIEW. The inputs were inlet pressure, three intermediate pressures, mold displacement and load. These inputs along with a known flow rate and a known viscosity allowed direct calculation of permeability, fiber volume fraction and compaction pressure.

4. MATERIALS

Two types of reinforced textile preforms were characterized during the course of this investigation. The first was 162 E-glass, a plain weave, bi-axial fiberglass fabric with an areal weight per ply of 392 g/m^2. This material was used in the comparison of the steady state and advancing front techniques. The second material was a multiaxial warp knit AS4 carbon fiber preform (Dexter, Palmer and Hasko, 1993; Hinrichs et al., 1994; Furrow, 1996) composed of a 9-ply subgroup with a $[\pm 45, 0_2, 90]_s$ layup which was knitted together with

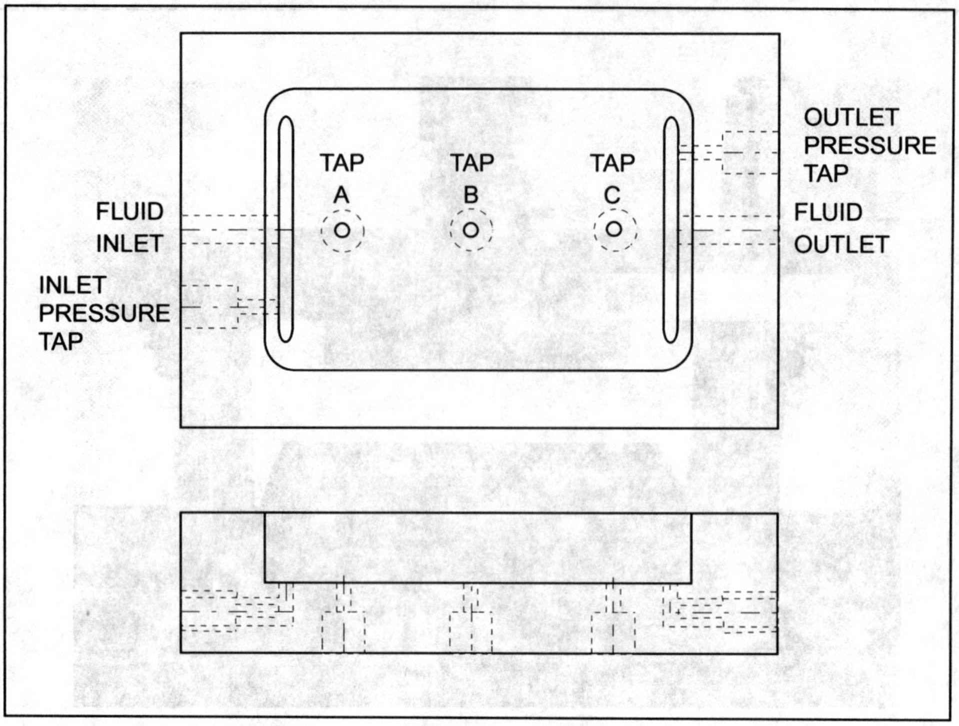

Figure 2. In-Plane Fixture Mold Cavity

polyester thread. The subgroups were stitched through-the-thickness along the 0° direction. The stitch spacing and pitch were 0.476 cm and 0.317 cm, respectively. The areal weight of the 9-ply subgroup is 1425 g/m^2.

5. RESULTS

5.1 COMPARISON OF STEADY STATE AND ADVANCING FRONT

Pressure gradient versus flow rate curves generated with the fixture described in this study are linear and the intercepts pass through the origin as evident in figure 3. This verifies that the one-dimensional form of Darcy's law (7) can be used to determine permeability. Under steady-state conditions the permeability at each fiber volume fraction of interest is determined from the slope of a linear least squares fit to the pressure versus flow rate data as

$$S = \frac{slope\ \eta}{A}, \tag{8}$$

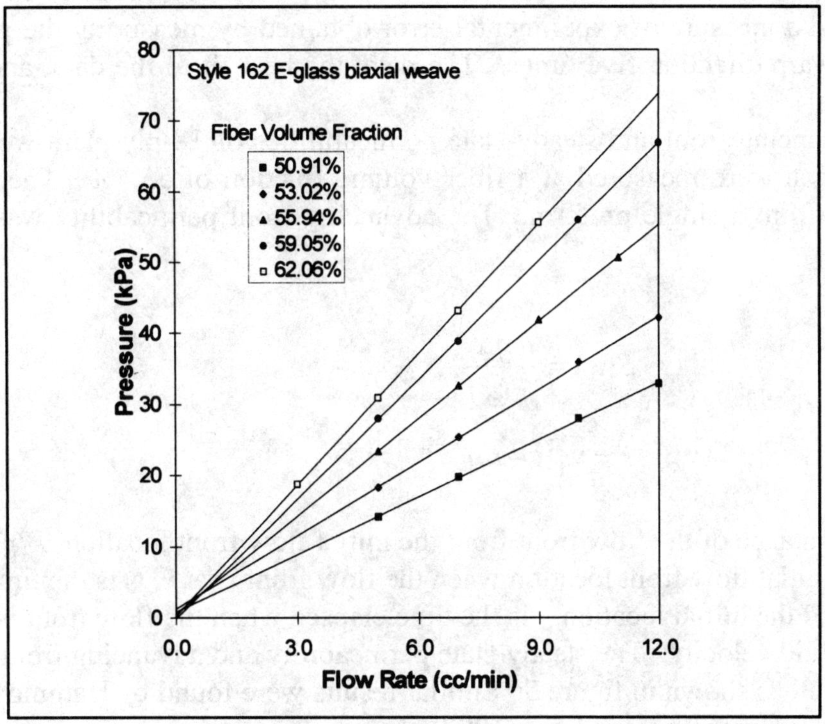

Figure 3. Pressure versus Flow Rate for E-glass 162.

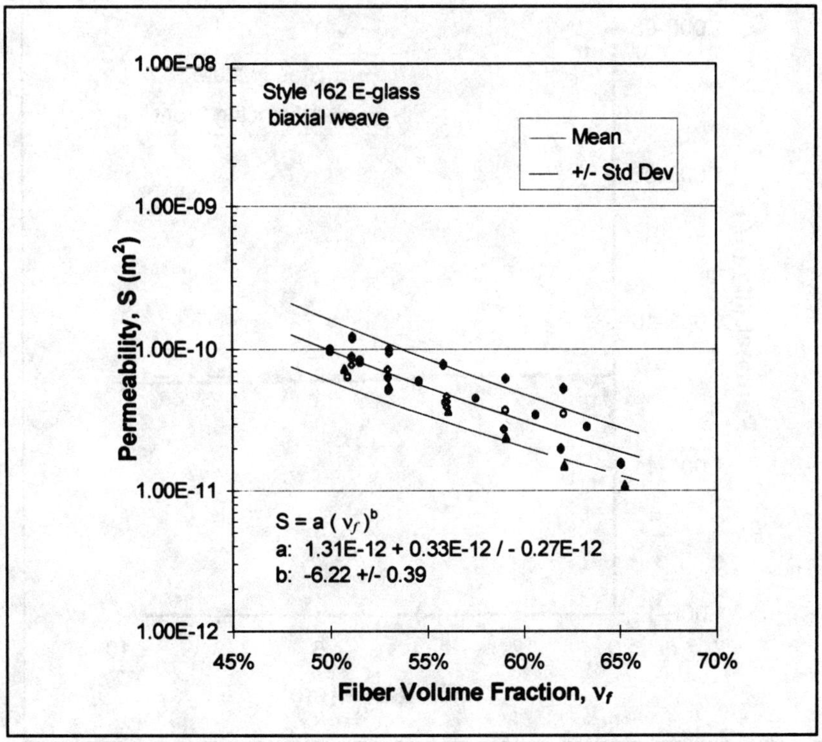

Figure 4. Repeatability of E-glass Warp Direction Permeability.

where A is the cross sectional area normal to the flow. The repeatability of measurements was verified and a measure of experimental error obtained by measuring the permeability of E-glass in the warp direction five times. The data, the best fit to the data, and the error are shown in Figure 4

The advancing front and steady-state permeabilities of 18-ply plain weave E-glass in the warp direction were measured at a fiber volume fraction of 55.4%. The permeabilities were measured from a single preform. The advancing front permeability was calculated as follows:

$$S(x) = \frac{q \eta x}{P(x)} \tag{9}$$

$$x = v_x (t - t_o) \tag{10}$$

where x is the distance of the flow front from the initial flow front location, $P(x)$ is the pressure at the initial flow front location when the flow front is at x, t_0 is the time when the flow front passed the initial location, t is the time elapsed when the flow front is at x, and v_x is the interstitial fluid velocity. The steady-state permeability and advancing front permeability data compare well as shown in figure 5. Similar results were found by Hammond and Loos (Hammond, 1993; Hammond and Loos, 1997).

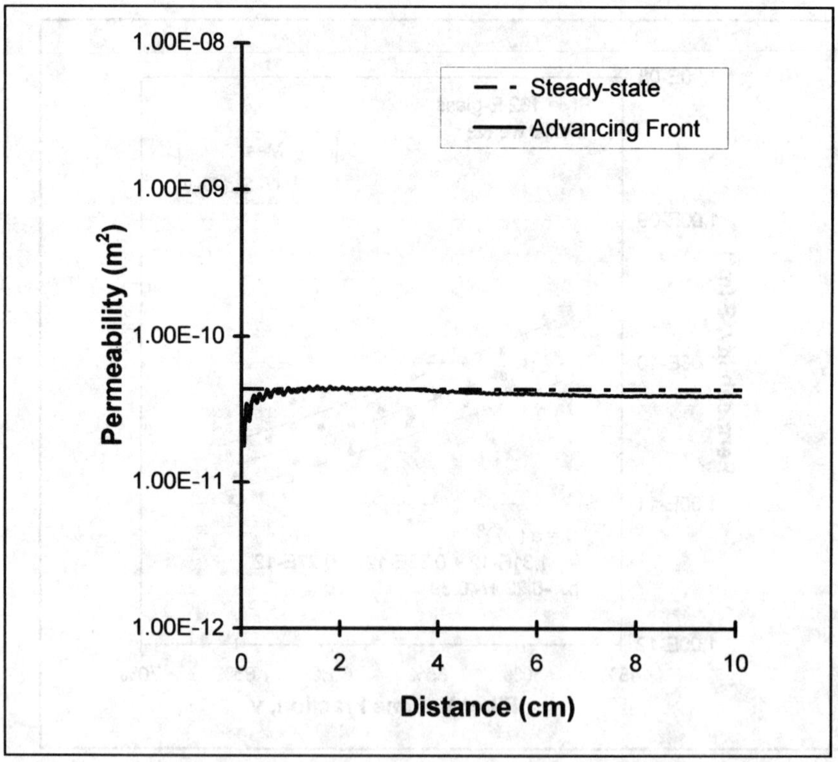

Figure 5. Advancing Front and Steady-State Permeability of 18-ply E-glass in the Warp Direction at 55.4% Fiber Volume Fraction.

5.2 OFF AXIS

In this study, the permeability was measured parallel to the stitching direction (S_{xx}), perpendicular to the stitching direction(S_{yy}) and at 45° degrees to the stitching direction. (see figure 6) When the axes are rotated 45°, the transformation equation for S'_{xx} simplifies to

$$S'_{xx} = S_{xy} + \frac{1}{2}(S_{xx} + S_{yy}) \tag{11}$$

and the permeability tensor component S_{xy} can be computed. Using off-axis permeability measurements (see figure 6) and S_{xy} of the warp knit preform, the principal axes were calculated from equation 6 to be 13° from the stitching direction (figure 7) and varied slightly with fiber volume fraction. The principal permeability values are plotted in figure 8 as a function of fiber volume fraction. They are found to be nearly equal to the permeability values measured parallel and perpendicular to the stitching. Rotation of the principal axes from the stitching direction could be due to experimental error since the principal permeabilities lie within one standard deviation of the power law regression model fit for the permeabilities calculated parallel and perpendicular to the stitching. This method of computing the principal permeabilities and principal directions has the advantage of allowing the direction to be measured as a function of fiber volume fraction, and all results are obtained from three tests in a single fixture.

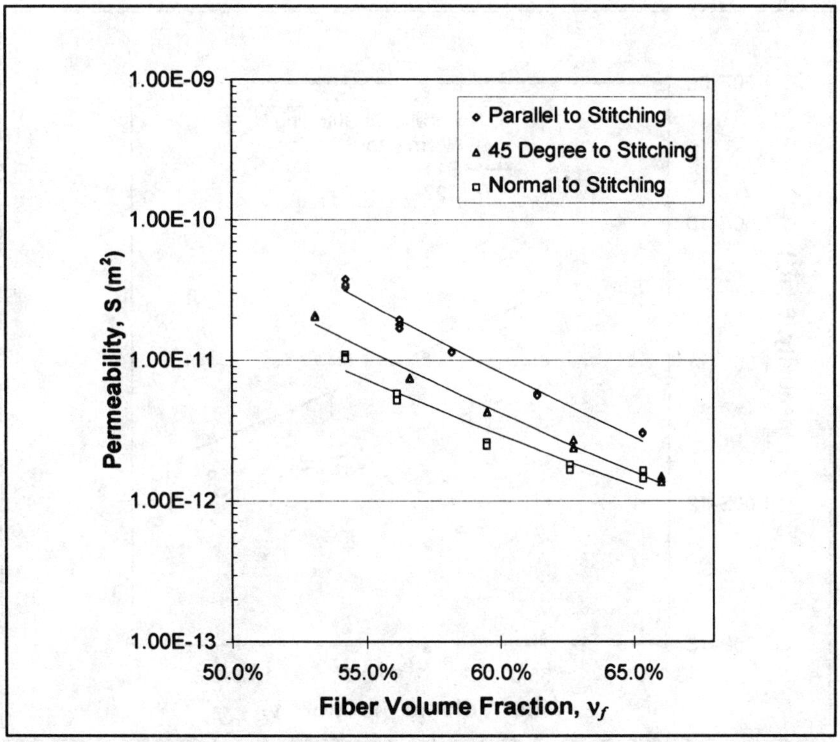

Figure 6. Permeability data and power-law regression fits for 81 ply multiaxial warp knit preform.

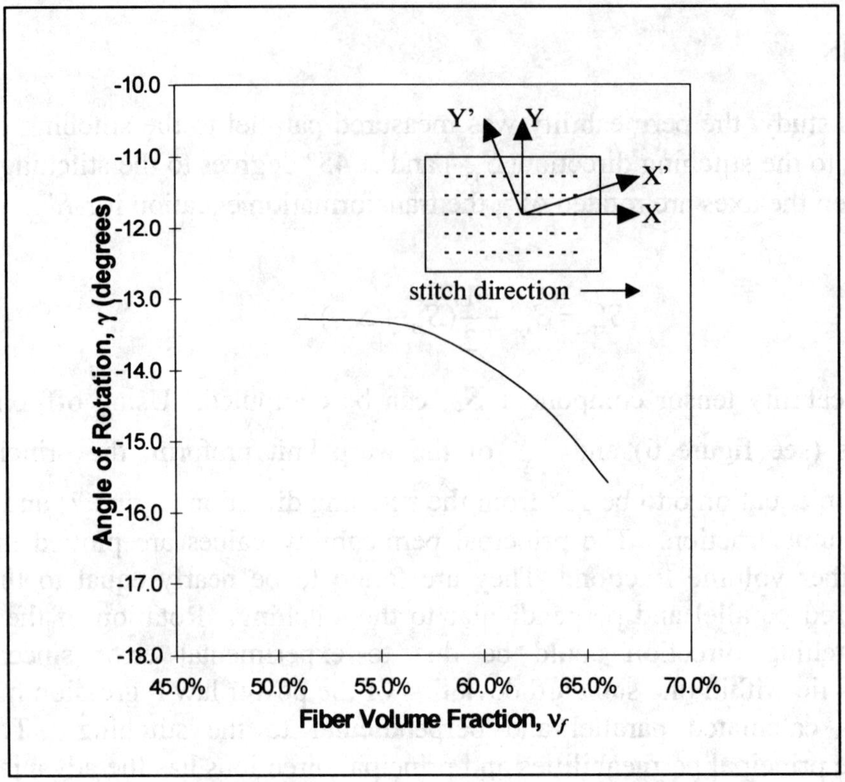

Figure 7. Angle of rotation from original coordinate axes aligned with preform stitching to principal permeability axes.

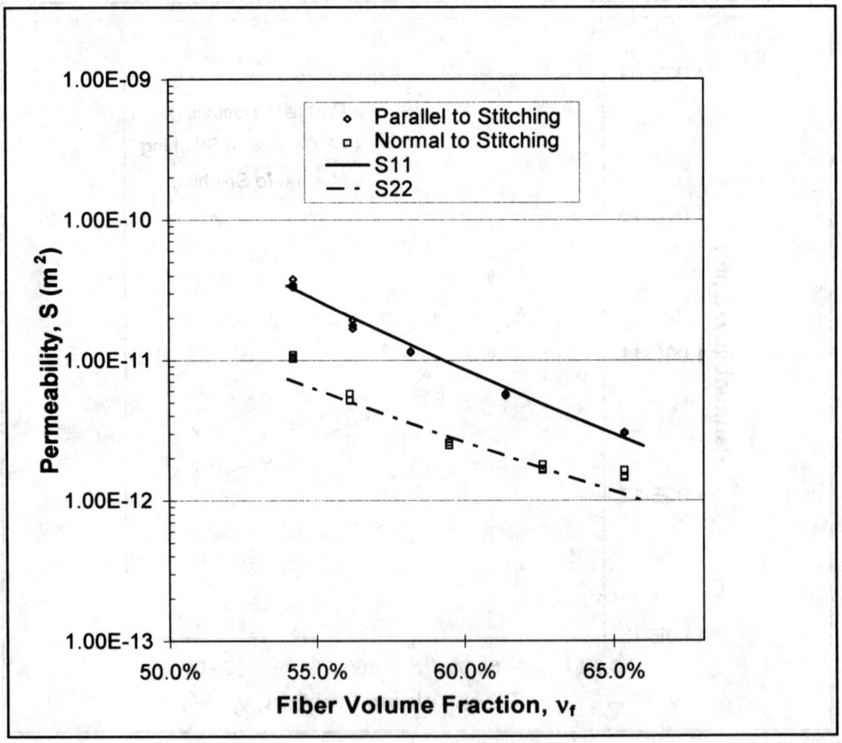

Figure 8: Principal permeabilities and permeability data measured parallel and normal to the stitching direction

6. CONCLUDING REMARKS

A system capable of real-time, in-plane advancing front permeability as well as steady-state permeability measurements of high fiber volume fraction preforms has been developed. The advancing front permeability has been shown to closely match the steady-state permeability for Style 162 E-glass at a fiber volume fraction of 55 percent. Finally, a method has been developed to measure the principal directions and permeabilities as a function of fiber volume fraction of a preform. Results obtained from an 81-ply multiaxial warp knit preform showed the principal axes are nearly aligned with the directions parallel and normal to the stitching.

7. ACKNOWLEDGMENTS

This research was supported by the NASA-Virginia Tech Composites Program, Grant No. NAG1-343, and by NASA Grant No. NAG1-1881; Mr. H. Benson Dexter, grant monitor. This work was also supported by McDonnell Douglas Corporation, McDonnell Douglas Aerospace, Advanced Systems & Technology; Mr. Kim Rohwer, technical monitor.

8. REFERENCES

Darcy, H. 1856. Les Fontaines Publiques de la Villa de Dijon. Dalmount.

Dexter, H. Benson; Raymond J. Palmer, Gregory H. Hasko. 1993. "Mechanical Properties and Damage Tolerance of Multiaxial Warp Knit Structural Elements," Fourth NASA/DOD Advanced Composites Technology Conference, NASA CP-3229, Salt Lake City, UT.

Fingerson, John C., Alfred C. Loos, and H. Benson Dexter. 1995. " Verification of a Three-Dimensional Resin Transfer Molding Process Simulation Model," Virginia Tech Center for Composite Materials and Structures: CCMS-95-10, Blacksburg, VA.

Furrow, Keith, W. 1996. "Material Property Evaluation of Braided and Braided/Woven Wing Skin Blade Stiffeners," NASA Contractor Report 198303, Hampton, VA

Hammond, V. H. 1993. "Verification of a Two-Dimensional Infiltration Model for the Resin Transfer Molding Process." Master's Thesis, Virginia Polytechnic Institute and State University, Blacksburg, VA.

Hammond, V. H., A. C. Loos. 1997. "The Effects of Fluid Type and Viscosity on the Steady-State and Advancing Front Permeability Behavior of Textile Preforms," Journal of Reinforced Plastics and Composites, Vol. 16, no. 1/1997, pp. 50- 72.

Hinrichs, Steve; Raymond J. Palmer, Aaron Ghumman; Jerry Deaton, Keith W. Furrow, Larry C. Dickinson. 1994. "Mechanical Property Evaluation of Stitched/RFI Composites." Fifth NASA/DOD Advanced Composites Technology Conference, NASA CP-3294, Seattle, WA.

Wrinkle-Free Preforms for Fiber Composite Structures

R. E. ROBERTSON, T.-J. CHU, M. PARK AND R. J. GERARD

ABSTRACT

This report is concerned with the mapping of a flat fabric, woven or engineered, with reinforcing fibers in two directions, into pyramids and pyramidal depressions, regular or truncated, single pyramids or pairs. The mapping is not difficult, but the results are not intuitive. The collected results reported here have not been reported elsewhere, yet they have application to the general preforming problem. The sharp pyramidal edges result in regions of distinct deformation in the fabric, which allows a critical comparison with experiment.

INTRODUCTION

The handling of the fibers has always been the most costly and difficult part of manufacturing fiber composite structures. One of the attractive approaches to handling the fibers is to assemble them into a mat or fabric and then to form the mat or fabric into the shape of the structure to create a preform. This is especially attractive for structures consisting of large contoured surfaces without openings or with only small openings. The use of fabrics has the advantage over mats that a high density of fibers can be obtained in the completed composite structure. A disadvantage of using fabrics is that they are not very deformable. In general, they can be neither lengthened nor shortened along the fiber axes of the fabric without either cutting or folding them or allowing wrinkling to occur. Cutting generally adds costs and entails other issues, and folding and wrinkling almost always should be avoided. Wrinkling lowers the strength and stiffness of the fabric across the wrinkle, it can severely lower resin permeability during liquid molding, and it looks bad. This means that, except for some local fiber sliding, the fabric can only be deformed in shear, by stretching along the bias or diagonal of the fabric. But even this deformation is limited.

As seen in Figure 1, the area of the square or rectangular repeating unit decreases as the fabric is stretched along the bias. For a plain-weave fabric the unit area must remain at least as large as the cross sectional area of the fibers passing through it. For satin-weave and engineered fabrics, the limiting unit area can be smaller. But in any case, the area must remain finite, and even when the area is allowed to collapse to zero, the resulting deformation is equivalent to an elongation along the bias of 41%. (If the sides of the square repeating unit are assumed to be equal to one, the diagonal is of length $\sqrt{2}$ before deformation and to 2 after the deformation.)

Richard E. Robertson, Tun-Jen Chu, Min Park, and Robert J. Gerard,
Macromolecular Science & Engineering Center and Department of Materials Science & Engineering, The University of Michigan, Ann Arbor, MI 48109-2136

In the following, we summarize our work in shaping a fabric over a single regular or truncated pyramid and over a pair of regular or truncated pyramids (Robertson et at., 1997). The fabric and the crossings of the two sets of fibers are like a fishing net, with knots at the crossings and the cords between the knots pulled taut. This analogy was previously used to study the shaping of a two-dimensional fabric into a hemisphere (Robertson et al., 1981) and into a cone with a rounded top (Robertson, Hsiue and Yeh, 1984).

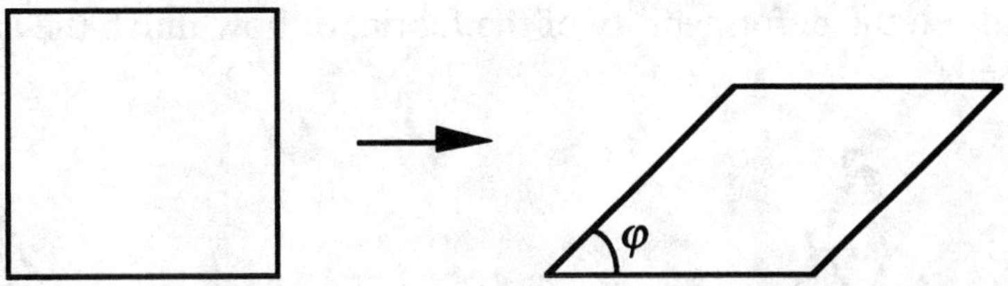

Figure 1. Deformation of the repeating unit of a plain weave fabric when stretched along the bias.

In this work, we are largely concerned with deformation maps; i.e., where should the deformation be located in the shaped fabric, and what magnitude and orientation should the deformation have. These deformation maps show what is possible, but not how it can be attained.

COVERING OF A REGULAR FOUR-SIDED (SQUARE-BASE) PYRAMID

FIBERS PARALLEL WITH BASE DIAGONALS

The plan-view of a regular four-sided pyramid and one of the faces are shown in Figure 2. Let us assume that a plain weave fabric is oriented so that before shaping over the pyramid the fibers are parallel to the base diagonals. On shaping, then, fibers run along the edges of the pyramid faces. The fibers at the two edges of each face would have been perpendicular to each other in the flat fabric. Though the fitting of the fabric to the pyramid surface turns out to be relatively simple, it is not intuitive that it can be done without wrinkling. The solution is quickly obtained with the fishing net analogy.

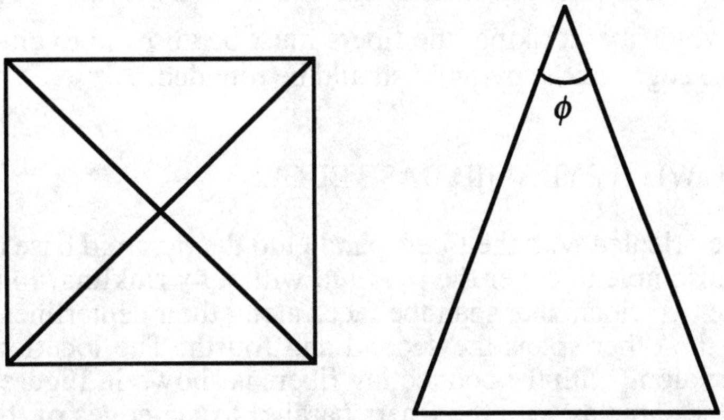

Figure 2. Plan-view of a regular four-sided pyramid and one of its faces.

Using the fishing net analogy, a row of knots with the knots evenly spaced are placed along each of the face edges. With one of the knots placed at the pinnacle of the pyramid, the fishing net is fitted to the pyramid by pulling it down over each of the faces. The locations of knots covering the upper portion of a typical pyramidal face are shown in Figure 3(a). This pattern is repeated over the entire face with $\varphi = \phi$, where ϕ is the included angle of the face at the top. The pattern found is unique; i.e., there is no other pattern that covers each face of the pyramid with the fibers along the face edges without wrinkling. The theoretical limit to the angle φ is zero, at which point the diamond is collapsed and the pyramid has degenerated into a straight line segment. Whether a given fabric can be wrapped over a pyramid with a finite angle ϕ, however, depends on the deformability of the fabric, or how much the area of the repeating unit can be decreased.

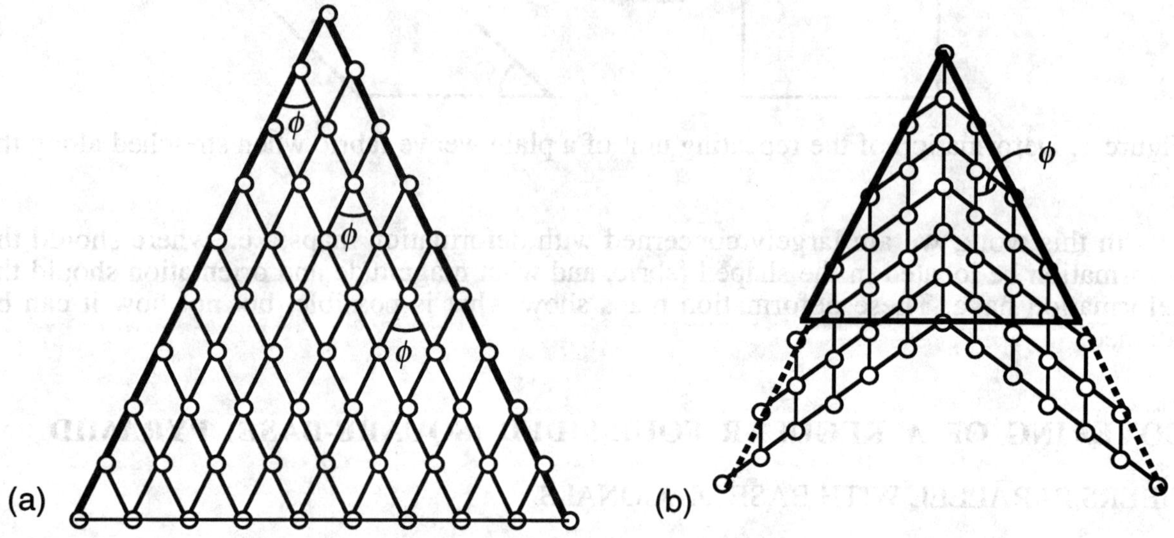

Figure 3. The covering of a pyramidal face by a fabric oriented with the principal fibers (a) along the face edges and (b) along the face bisector.

Though the fibers on the faces of the pyramid are all straight, they undergo a discontinuous change in direction at the edges. There are both the nearly 90° change in direction made by the cross fibers when they change faces and the $(180° - 2\phi)$ change in direction due to the fitting to each face. The two result in a change in direction given by $(180° - 2\phi')$, where $\sin \phi' = (\sin\phi)/\sqrt{2}$. To avoid fiber kinking, the fibers must be allowed to change direction over a small distance and the edges of the pyramid should be rounded.

FIBERS PARALLEL WITH PYRAMID BASE EDGES

A piece of fabric oriented with the fibers parallel to the pyramid base edges before shaping over the pyramid is also able to cover the pyramid without wrinkling. In this case, the knot at the apex connects the two fibers that span the faces along their centerlines. One fiber spans the first and third face, the other spans the second and fourth. The locations of the knots on a single pyramidal face along with the connecting fibers is shown in Figure 3(b). The pattern of knots is essentially the same as with the fibers parallel to the edges of the face, consisting of diamonds with the smaller included angle equal to ϕ. The difference is that the fibers across the face change direction abruptly, by the angle $(180° - 2\phi)$, at the center of the face, instead of at the edges. When projected into the plane of the base, all fibers are parallel with one or the other edge of the base.

COVERING OF A TRUNCATED FOUR-SIDED (SQUARE-BASE) PYRAMID

The plan-view of a regular truncated (flat) four-sided pyramid and one of its side faces are shown in Figure 4.

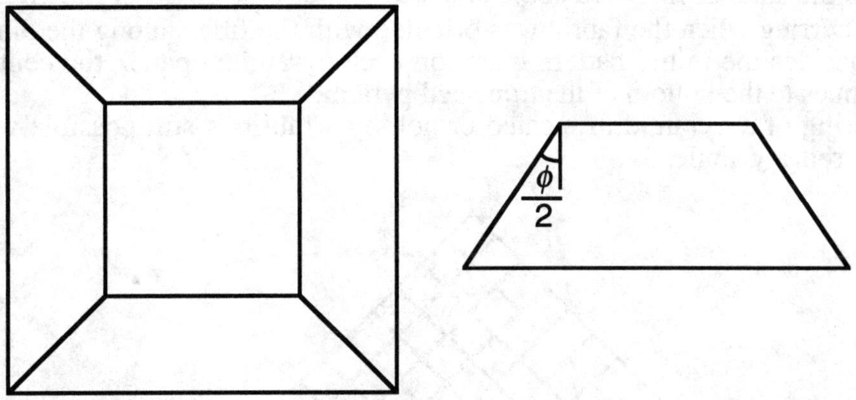

Figure 4. Plan-view of a regular truncated (flat-top) four-sided pyramid and one of its side faces.

FIBERS ALONG THE TOP DIAGONALS AND SIDE-PANEL EDGES

An example of the covering of a truncated pyramid by a fabric oriented with fibers along the diagonals of the flat top and the edges of the side panels is shown in Figure 5.

Three fabric deformations result, which are highlighted by the use of shadings. The unshaded region is the square repeating unit of the original fabric. The darker shaded region is the pattern that results when the pyramid is not truncated. The lightly shaded region along the edges is intermediate between these two. As with the previous symmetrical coverings, this also is unique. There is no other way to arrange the fibers (and knots) and have the pattern be symmetrical and have the same pattern on each face.

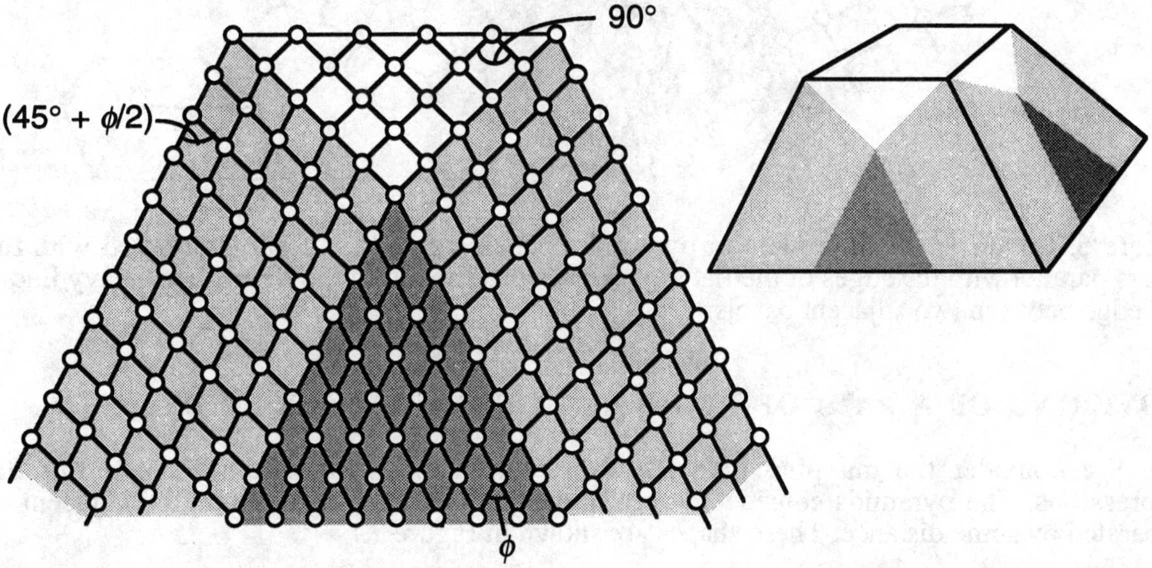

Figure 5. Side panel of a tall truncated four-sided pyramid after wrapping with a fishing net.

FIBERS PARALLEL WITH EDGES OF FLAT TOP AND BASE

Consider now the orientation of the fabric with the fibers parallel with the edges of the flat top and base. In the corner region of Figure 6, the repeating unit of the fabric is deformed into diamonds with the smaller included angle equal to ϕ, the same angle as shown in Figure 3(a). Unlike that occurring when the fabric was oriented with the fibers along the diagonal, where two patterns besides the initial pattern occur on each pyramidal panel, the central pattern in Figure 6 continues to the bottom of the truncated pyramid.

The covering of a pyramid, truncated or not, by a fabric is still possible with the fabric oriented at an arbitrary angle.

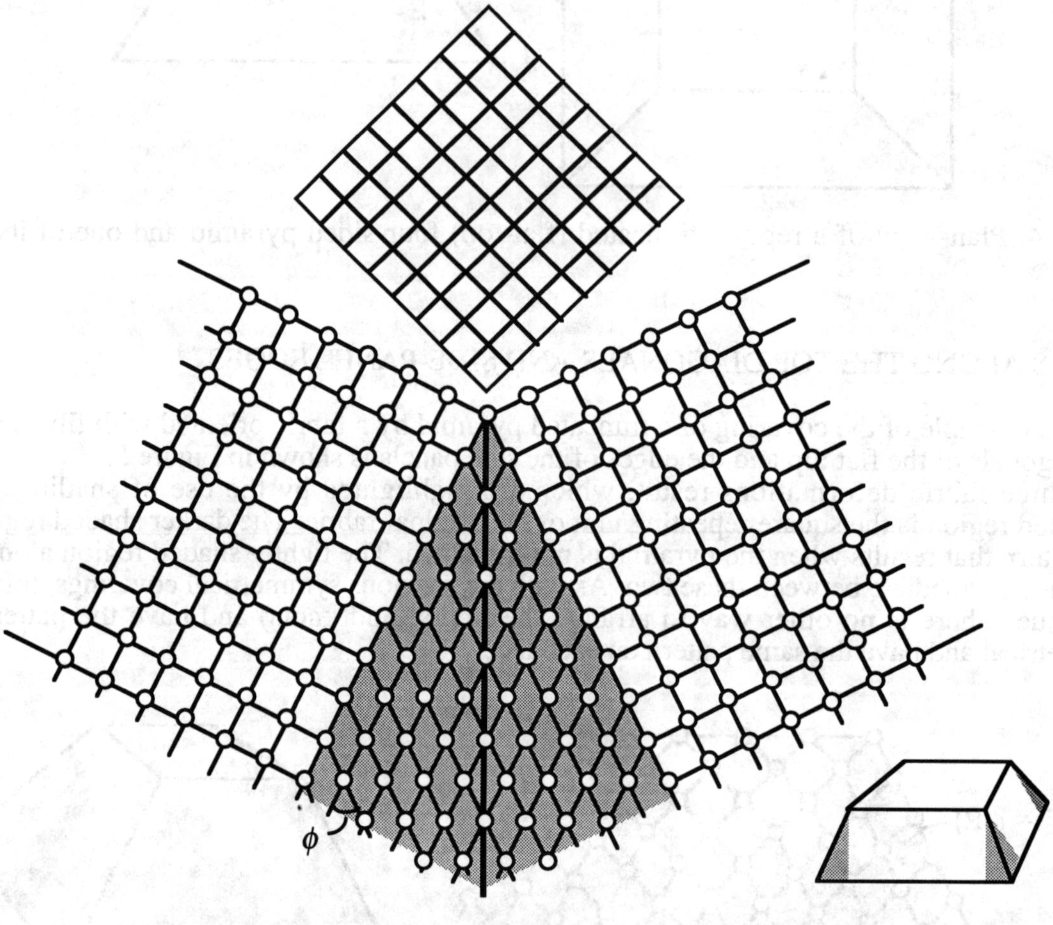

Figure 6. Covering of the corners of pyramidal panels when the fabric is oriented with the fibers parallel with the edges of the flat top, as shown at the top of the figure. The heavy line is the edge between two adjacent panels.

COVERING OF A PAIR OF REGULAR FOUR-SIDED PYRAMIDS

We consider the mapping of a flat fabric onto a pair of pyramids or pyramidal depressions. The pyramids considered are regular four-sided, and they may be adjacent or separated by some distance. These shapes are shown in Figure 7.

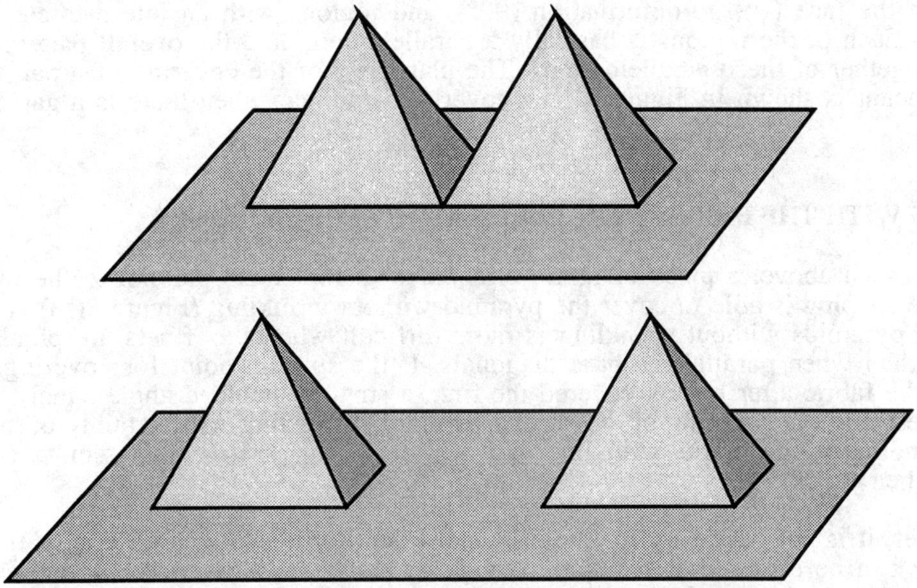

Figure 7. Possible arrangements of a pair of regular, four-sided pyramids.

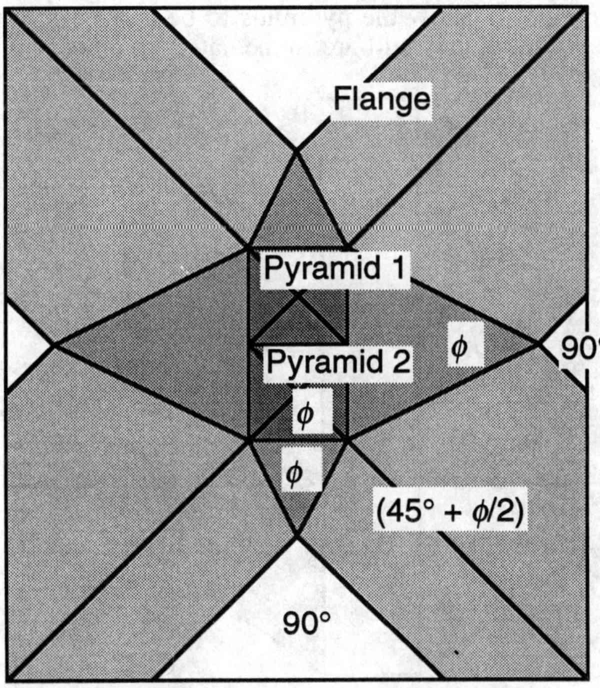

Figure 8. Deformation zones in a fabric covering a pair of adjacent four-sided-pyramids and a flat surrounding flange.

COVERING WITH THE FIBERS PARALLEL WITH THE FACE EDGES

A plain weave fabric oriented with the fibers parallel to the pyramidal face edges (pyramid diagonals) was previously shown to be able to cover a regular four-sided pyramid without wrinkling. Although the covering of these regular pyramids themselves is straight forward, with the deformation being the same everywhere, this is not so for the attached flange. Rather, the flange consists of regions of three different degrees of fabric deformation: the deformation occurring in the face (ϕ), no deformation (90°), and regions with an intermediate deformation (45° +ϕ/2)). Each of the regions is basically a parallelogram, and the overall pattern arises from the fitting together of these parallelograms. The plan view of the covering of a pair of pyramids that are adjacent is shown in Figure 8. The covering is similar when there is a gap between the pyramids.

COVERING WITH THE FIBERS PARALLEL WITH THE BASE EDGES

As was seen above, a piece of fabric oriented with the fibers parallel to the pyramid base edges before shaping is able to cover the pyramid without wrinkling (Figure 3(b)). The covering of a pair of pyramids without wrinkling is more difficult when the fibers are parallel with the base edges than when parallel the base diagonals. If the starting point for covering the second pyramid is the fabric after having covered the first, a smaller included angle equal to $\varphi = (3\phi - 180°)$ will result in the covering of the second pyramid. Wrinkling will certainly occur when $\varphi < 0$. But depending on the fabric, wrinkling may still occur for positive angles of φ. (One notices that $\varphi > 0$ when $\phi > 60°$.)

However, it is not necessary for the principal fibers to be centered on one of the pyramids. Rather, they can be centered at the symmetry point half-way between the pyramids. The only thing that happens is that the deformation in the fabric on each of the faces will be different.

The plan view of the full covering of both pyramids and surrounding flange is shown in Figure 9. The fabric on the pyramidal faces has been darkened slightly over that in the flange for the same degree of deformation, to allow the pyramids to be more easily seen, though since there are several patterns on the pyramid, they still are immediately visible.

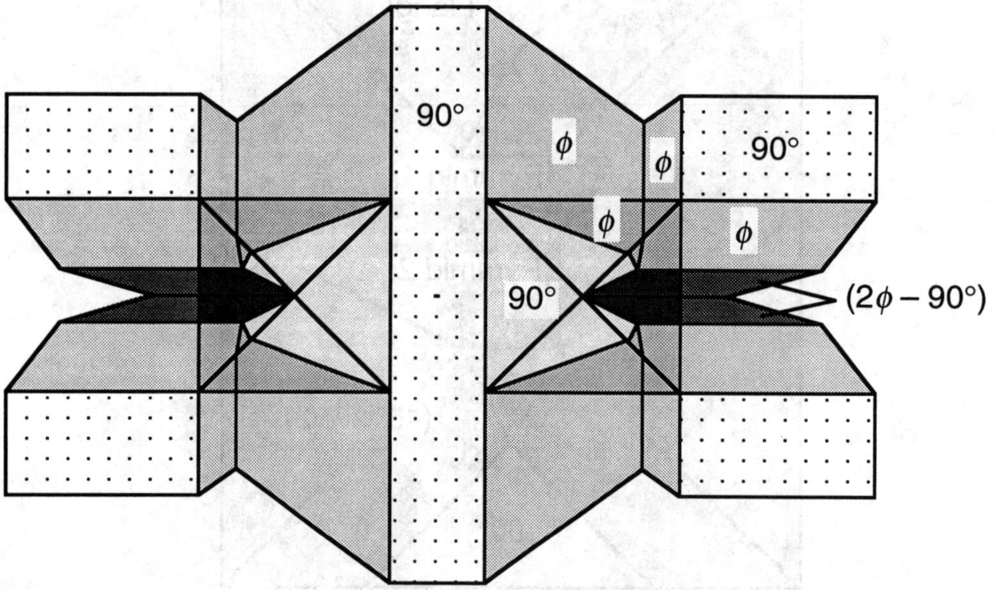

Figure 9. Plan view of fabric covering a pair of pyramids with the fabric centered at the symmetry point between the pyramids.

COVERING OF A PAIR OF TRUNCATED FOUR-SIDED PYRAMIDS

Finally, we consider the mapping of a flat fabric onto a pair of truncated pyramids. The pyramids may be close to one another or not. These shapes are shown in Figure 10.

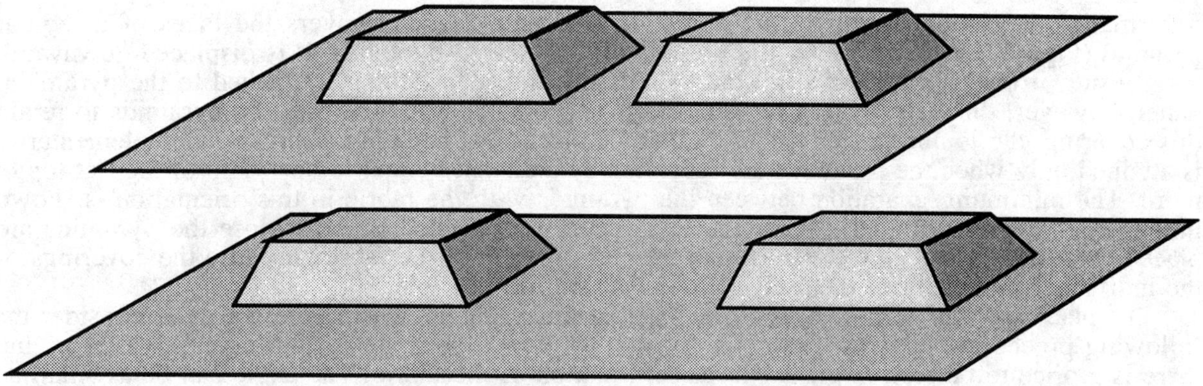

Figure 10. Possible shapes and arrangements of a pair of pyramids.

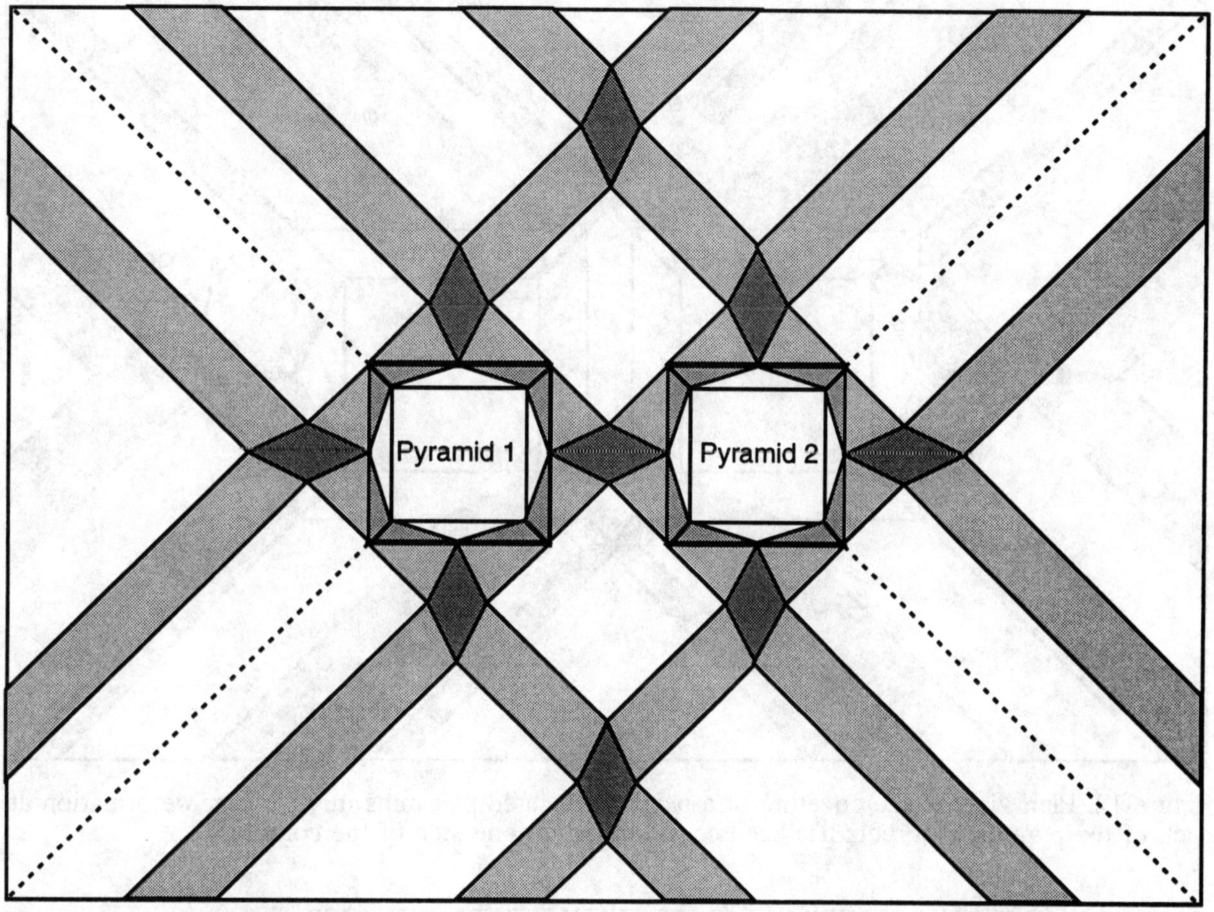

Figure 11. Deformation zones in the flat flange surrounding a pair of flat-topped four-sided pyramids placed as close together as possible without wrinkling when covered with a ±45° fabric in a most uniform deformation mode.

COVERING WITH THE FIBERS PARALLEL TO THE TOP DIAGONALS

A fabric oriented with the fibers parallel with the diagonals of the flat top of a truncated pyramid was previously shown to be able to cover the pyramid without wrinkling (Figure 5). The pyramid has a triangular (actually the upper half of a diamond) zone where the fabric deformation is characterized by the angle ϕ. This zone is what covers the faces of a regular pyramid (Figure 3(a)), and while it is present with truncated pyramids, it is displaced downward. As a result, it may not appear at all on a short pyramid. If a flange is attached to the pyramidal panel, however, this deformation region always appears. To allow both of the pyramids to retain this covering, the joining edges where the two connect must be complementary. Complementarity is attained only when the pyramids are separated by a distance equal to the width of the flat top or more. The minimum separation between the pyramids with the fabric in this orientation is shown in the plan view in Figure 11. Although this covering is limited by how close the pyramids are placed together, it does give the most uniform deformation of the fabric, because the coverings of the individual pyramids are themselves the most uniform possible.

To place the pair of pyramids closer together than that shown in Figure 11, we consider the following procedure, for which there is no limit on how close together the pyramids can be, but there is a price to be paid in greater local deformation of the fabric. The angle can be as small as $(2\phi - 90°)$. The procedure involves moving the deformation in the two pyramidal panels that face each other around to the side panels. An example is shown in Figure 12.

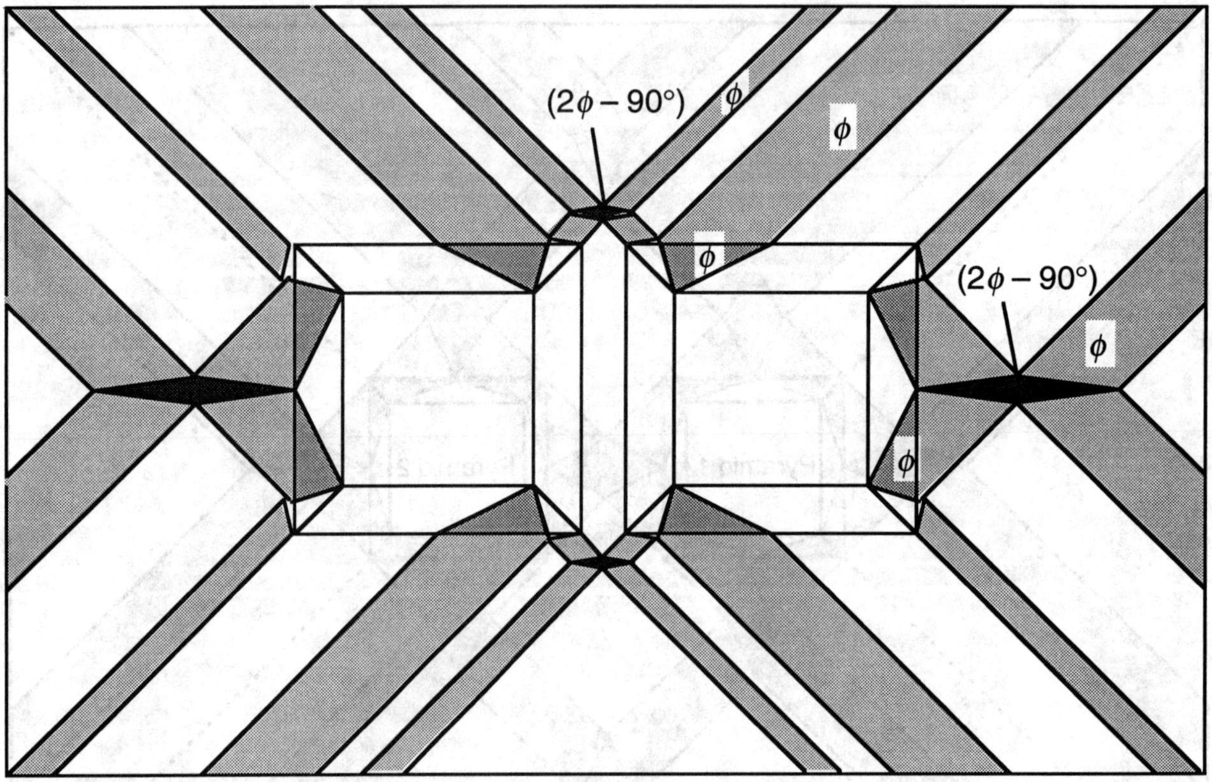

Figure 12. Plan view of the covering of a pair of truncated pyramids in which the deformation at each of the pyramidal corners has been concentrated on one side of the corner.

COVERING WITH THE FIBERS PARALLEL WITH THE BASE EDGES

Consider now the covering that gives the most uniform distribution of deformation (and the least maximum deformation) with the fibers parallel with the base edges of the fabric. The two pyramids are given the same covering.

The plan view of the covering of a pair of truncated pyramids with the fabric centered at the symmetry point between the pyramids and oriented with the fibers parallel with base edges is shown in Figure 13. Before covering, the fabric was rectangular. Although this covering has the most uniform distribution of deformation (and the least maximum deformation), the maximum deformation is still seen to be $(2\phi - 90°)$.

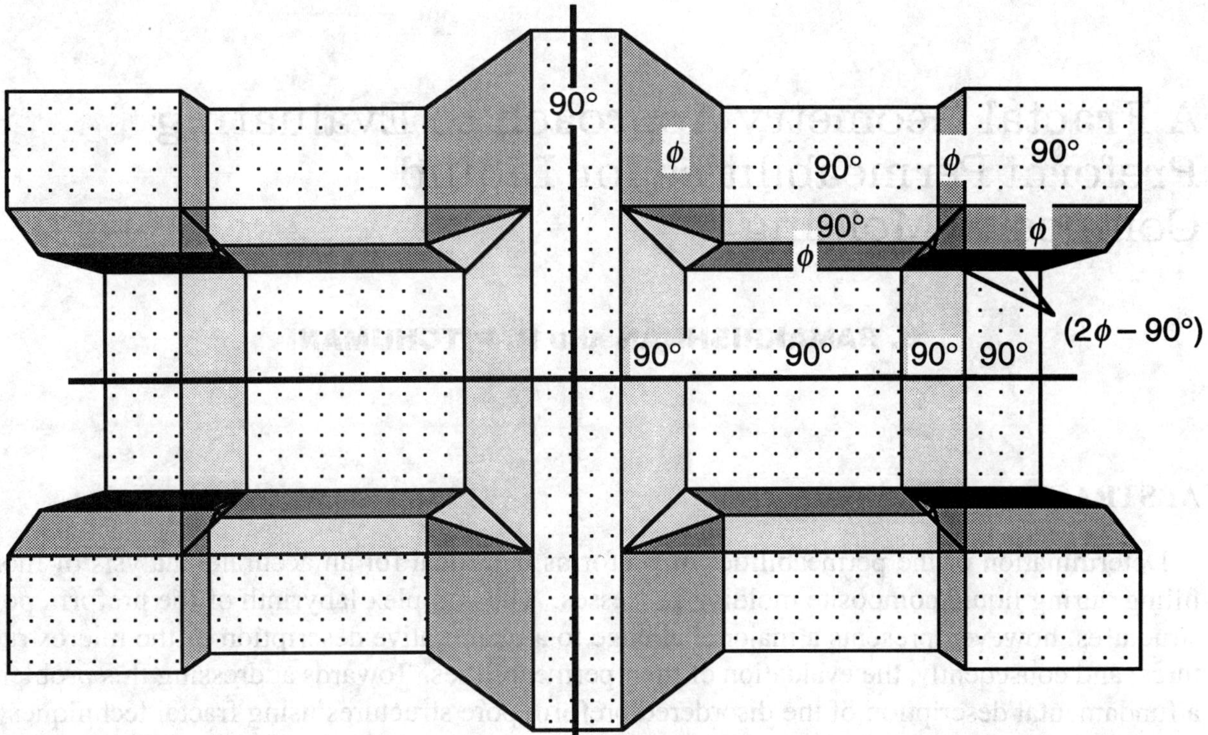

Figure 13. Plan view of fabric covering a pair of truncated pyramids with fabric centered at the symmetry point between the pyramids and oriented with fibers parallel with base edges.

Conclusions

The above calculations show that a single pyramid, regular or truncated, and a pair of pyramids can in principle be covered with a two-dimensional fabric without wrinkling. The technique used was the computation of the location of knots of a fishing net analogy to the fabric, with taut strings of fixed length between the knots.The patterns of knots were unique; no other patterns symmetrically arrayed on each face would have covered the pyramids examined without wrinkling. Of the two fabric orientations considered for the pairs of pyramids, that having the fibers parallel with the face edges (base diagonals) is more likely to be accomplished without wrinkling than that having the fibers parallel with the base edges. The minimum deformed angle arising with the fibers parallel with the face edges is ϕ, whereas, the minimum deformed angle with the fibers parallel with the base edges is $(2\phi - 90°)$, and $\phi \geq (2\phi - 90°)$ since $\phi \leq 90°$.

REFERENCES

Robertson, R.E., E.S. Hsiue, E.N. Sickafus and G.S.Y. Yeh (1981), *Polym. Compos.*, **2**, 126.

Robertson, R.E., E.S. Hsiue and G.S.Y. Yeh (1984), *Polym. Compos.*, **5**, 191.

Robertson, R.E., T.-J. Chu, M. Park and R. J. Gerard (1997), to be published.

A Fractal Geometry Approach to Evaluating Preform Permeabilities for Liquid Composite Molding

B. RAMAKRISHNAN AND R. PITCHUMANI

ABSTRACT

Determination of the permeabilities of preforms is critical for an accurate analysis of mold filling during liquid composite molding processes. The complex labyrinth of the preform pore structures, however, presents a major challenge to a quantitative description of the microstructures, and conseqently, the evaluation of their permeabilities. Towards addressing this problem, a fundamental description of the disordered preform pore structures using fractal techniques is presented. A fractal permeation model is developed which relates the preform permeabilities to the actual microstructures in terms of two fractal dimensions—one relating the size of the capillary flow pathways to their population, and the other describing the tortuosity of the capillary pathways. The analytical model predictions are validated by experimentally-determined permeabilities for a wide range of volume fractions and injection pressures. The model development, experimental studies and the model validation are presented and discussed.

INTRODUCTION

Liquid composite molding, including techniques such as resin transfer molding and its variants, offers potential for affordable fabrication of high-performance products of complex shapes and sizes. The process begins with the stacking of fibrous reinforcement units to obtain a porous net-shaped structure termed a *preform*. The preform is placed into a closed mold cavity, and subsequently infiltrated by injecting a reactive polymeric resin into the mold. The resin-saturated preform is then exposed to high temperatures which initiate and sustain a cross-linking polymerization reaction (referred to as the *cure*) of the resin and transform the fiber-resin mixture in the mold cavity to a structurally hard composite product.

An important—and challenging—problem in the analysis and practical design of liquid molding process is the description of the flow of the resin in the porous preform. A thorough understanding of the flow behavior is essential for properly designing molds, locating and sizing injection ports and vents, designing injection pressures for complete permeation, and determining mold fill times. Towards this end, several theoretical and experimental investigations have

---

Bhaskar Ramakrishnan and Ranga Pitchumani, Composites Processing Laboratory, Department of Mechanical Engineering, University of Connecticut, U-139, Storrs, CT 06269-3139

been reported in the literature. A partial list of some recent studies are as follows: Bruschke and Advani (1989); Dave (1990); Kim *et al.* (1991); Karbhari, *et al.* (1993); Lin, *et al.* (1993); and Ranganathan, *et al.* (1994). A common thread weaving all the theoretical approaches is the use of Darcy's law (Happel and Brenner, 1986) to relate the fluid flow to the pressure gradient, using the fluid viscosity and the permeability of the porous medium.

A key parameter in the Darcian flow description is the preform permeability, for which closed-form expressions have been derived as a function of the fiber diameter, fiber arrangement, and fiber volume fraction. The theoretical derivations follow one of two approaches: One approach uses a capillary model with or without a correction factor to account for the effect of capillary tortuosity on the pressure gradient (Carman, 1979; Advani, *et al.*, 1994), while another approach analyzes flow past a bank of cylinders to arrive at the effective permeabilities (Sangani and Acrovis, 1982; and Bruschke and Advani, 1993).

The theoretical approaches, however, model the preform microstructure using idealized arrangements of fibrils and fiber bundles, which are far removed from the actual microstructures, which exhibit considerable disorder in the distribution of the fibrils and fiber bundle. Furthermore, the preform consists of capillary pathways of cross sectional sizes spanning several length scales ranging from nominal inter-fibril spacings within a fiber bundle (typically on the order of a micron) to representative inter-bundle distances (on the order of a fraction of a millimeter) at the upper end. As a result, the use of a single diameter capillary in the capillary-based permeability models is inappropriate. A fundamental description of the disordered preform structures towards the evaluation of the flow permeabilities is therefore of critical importance and is the focus of this paper.

In this paper, the preform pore microstructures are characterized using two fractal dimensions (Mandelbrot, 1983; Feder, 1988), one relating the size of the capillary pathways for resin flow to their population in a preform cross section, and the other describing the tortuosity (the extent of convolutedness) of the capillary pathways. The fractal description of the pore structures is used to construct a flow model which yields an analytical expression for the permeability in terms of the fiber volume fraction in the preform and the two fractal dimensions. A methodology for the evaluation of the fractal dimensions of actual microstructures is presented, along with experimental validation of the fractal permeation model for the case of knit fiber preform architectures, over a wide range of processing conditions. A noteworthy distinction of the present work is that while fractal dimensions have been used in literature for characterizing material microstructures as an objective by itself (for example, Kindo, *et al.*, 1989), this paper presents a practical use of these geometric descriptors in the evaluation of a physical property.

FRACTAL DESCRIPTION OF PREFORM MICROSTRUCTURES

The disordered nature of preform microstructures, coupled with the presence of multiple length scales for resin flow, suggests the existence of fractal structure (Mandelbrot, 1983) formed by the pores in the preform. Fractal structures are characterized by scale-dependent measures and are described by fractal dimensions, which is the exponent of a power-law scale relationship between the measure and the length scale used for the measurement. Fractal structures of engineering relevance often exhibit the scaling relationship in a statistical sense (Voss, 1985)and it is this concept which is used in the description of the preform microstructures.

Consider a representative cubic preform volume of size L_o, as shown in Fig. 1. The preform can be envisioned to be made up of capillaries of various sizes ranging from nominal inter-fibril spacings at the lower end to typical inter-bundle separations at the upper end. Let the size of a

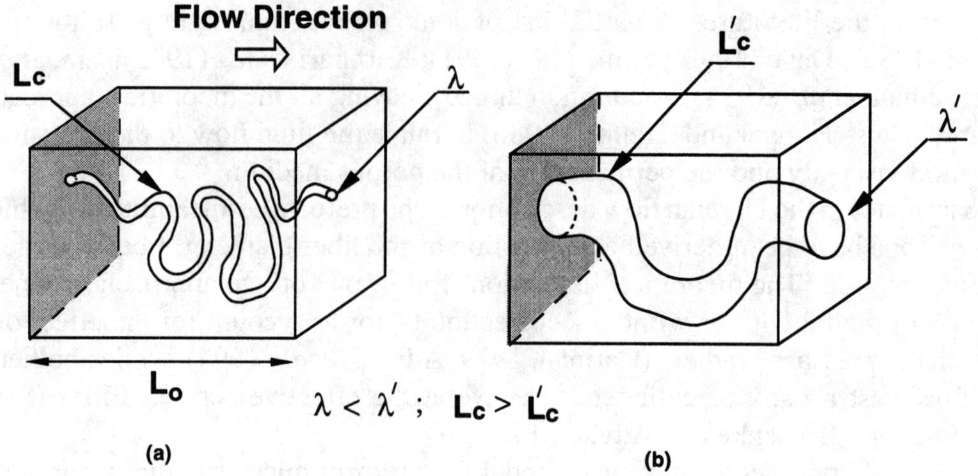

Figure 1: Schematic of a representative preform volume composed of tortuous λ-sized capillaries. Sketches *(a)* and *(b)* illustrate the decrease in tortuosity with increasing capillary size.

generic capillary be λ, and its tortuous length within the preform volume be $L_c(\lambda)$, as illustrated schematically in Fig. 1(a). Owing to the tortuous nature of the capillary, its end-to-end length, $L_c(\lambda)$, is greater than the side of the cubic volume, L_o. Furthermore, the extent to which a capillary is convoluted within the preform volume, i.e., its tortuosity, depends on the capillary size, λ; the smaller the capillary size, greater is its tortuous length, and vice-versa, as illustrated schematically in Figs. 1(a) and 1(b). Thus, the capillary exhibits a fractal tortuosity with respect to the capillary size, λ, and its length (expressed in a dimensionless form by scaling with respect to the cube dimension) is given by

$$\frac{L_c(\lambda)}{L_o} = \left(\frac{L_o}{\lambda}\right)^{d_T - 1} . \tag{1}$$

In Eq. (1), d_T is the fractal dimension of the capillary path, and lies in the range $1 < d_T < 2$. Higher values of d_T within this range correspond to a highly tortuous capillary, while $d_T = 1$ denotes a straight capillary path.

Equation (1) describes the convolutedness of the capillary pathways and is one of the relationships characterizing the fractal pore structure. It is also essential to determine the number of capillary pathways corresponding to every pore size, λ. This is obtained by considering a cross sectional plane of the preform volume normal to the overall flow direction, as shown by the shaded planes in Figs. 1(a) and 1(b). The capillary pathways intersecting the plane range in cross sectional sizes from a minimum value of λ_{min} to a maximum value of λ_{max}. The population of the intersecting pores in the cross section exhibits the general trend that $N(L \geq \lambda)$, which denotes the total number of pores of size exceeding or equal to a value λ, increases as the pore size, λ, decreases. This scaling relationship of the cumulative pore population in a cross section may be expressed mathematically as

$$N(L \geq \lambda) = \left(\frac{\lambda_{max}}{\lambda}\right)^{d_N} \tag{2}$$

where d_N is the fractal dimension of the intersecting pore cross sections with a plane normal to the flow direction. Since d_N defines a fractal surface in a two-dimensional plane, its value lies in the range $1 < d_N < 2$.

The number of pores whose size falls within the infinitesimal range λ to $\lambda + d\lambda$ is obtained by differentiating the cumulative pore size distribution, Eq. (2), with respect to λ. The resulting expression may be written as

$$dN = -d_N \frac{d\left(\frac{\lambda}{\lambda_{max}}\right)}{\left(\frac{\lambda}{\lambda_{max}}\right)^{d_N+1}}. \tag{3}$$

Note that the negative sign in the above equation reflects a decrease in the pore population with increasing pore size. The number-size relationship given by Eq. (3), together with the tortuosity equation, Eq. (1), forms the basis for the development of the permeability model discussed below.

FRACTAL PERMEATION MODEL

Permeation of resin through the preform during the process principally depends upon the preform permeability, the driving pressure differential for the flow, and the fluid viscosity. The preform permeability, in turn, is a function of the preform geometry, namely the preform architecture, which is quantified in terms of two fractal dimensions d_T and d_N, and the fiber volume fraction, v_f. In this section, we seek to develop an analytical model relating the permeability to the three geometric preform parameters. The overall approach to the model development consists of *(a)* determining the total volumetric flow rate, Q, through the representative preform volume, and *(b)* using Darcy's law, to determine the effective permeability, κ.

For the purpose of the model development, the overall resin flow is considered to be in the direction as shown in Fig. 1(a). The corresponding permeability would be a component of the permeability tensor along the direction of flow. The other principal permeabilities may be evaluated following a similar procedure as that presented here. Since the flow of resin through the preform results from flow through several tortuous capillary pathways of various sizes, as mentioned before, the overall volumetric flow rate, Q, through the preform is a cumulative of the flows through all the individual capillaries. The flow rate through a single λ-sized capillary, $q(\lambda)$, is given by the Hagen-Poiseulle equation (Denn, 1980),

$$q(\lambda) = g_q \frac{\Delta p}{L_c(\lambda)} \frac{\lambda^4}{\mu}, \tag{4}$$

where Δp is the pressure differential applied to drive the flow, μ is the viscosity of the resin permeating through the preform, and g_q is a geometric factor based on the shape of the capillary cross section (for a circular cross section, $g_q = \pi/128$, with λ as the cross sectional diameter).

The net flowrate through all the capillaries whose cross sectional size lie in the interval λ to λ to $d\lambda$ is given by the product of the number of such capillaries, $dN(\lambda)$ [Eq. (3)], and the individual capillary flowrate, $q(\lambda)$ [Eq. (4)]. Further, the total volumetric flowrate, Q, through the representative preform volume is obtained by integrating the flowrate contribution by the pores in every $d\lambda$ interval, over the entire range of prevalent pore sizes, λ_{min} to λ_{max}, in the preform. The preform permeability, κ, is determined from the flow rate, Q, and using Darcy's law as

$$\kappa = \frac{\mu L_o Q}{\Delta p A_o} = \frac{\mu}{L_o \Delta p} \int_{\lambda_{min}}^{\lambda_{max}} q(\lambda) dN(\lambda) = g_q \lambda_{max}^2 \beta^{1+d_T} \left[1 + d_N \left(1 - \frac{d_N\left(1 - \alpha^{3+d_T-d_N}\right)}{3 + d_T - d_N}\right)\right] \tag{5}$$

In the above equation, A_o $(= L_o^2)$ is the cross sectional area of the representative preform volume normal to the overall flow direction. Note that the permeability is a function of only the geometrical parameters, namely α, β, and λ_{max}, and the two fractal dimensions, d_T and d_N. The determination of α, β, and λ_{max} are discussed in the remainder of this section, while the methodology for evaluating the fractal dimensions forms the subject of the next section.

The dimensionless parameters α and β are determined from consideration of the overall pore volume fraction in the preform, which may be evaluated as the cumulative area fraction of the pores in a preform cross section. Following the steps similar to those used to arrive at the total flow rate, Q, above, the pore volume fraction, v_p, which equals $1 - v_f$, where v_f is the fiber volume fraction in the preform may be evaluated as follows

$$v_p = 1 - v_f = \frac{1}{L_o^2}\int_{\lambda_{min}}^{\lambda_{max}} g_v \lambda^2 dN(\lambda) = g_v \beta^2 \left[1 + d_N \left(\frac{1-\alpha^{2-d_N}}{2-d_N}\right)\right]. \tag{6}$$

As α approaches zero, the situation wherein the size of the smallest pores is negligible or the maximum pore is very large, the pore volume fraction in the preform, v_p, approaches the maximum achievable value of g_v. Using this condition in Eq. (6) yields the following expression for β in terms of the fractal dimension d_N as $\beta = \sqrt{\left(\frac{2-d_N}{2}\right)}$. Equation (6) can be rewritten to obtain α as a function of the fiber volume fraction and the fractal dimension, d_N, by substituting the expression for β.

The maximum pore size, λ_{max}, appearing in the permeability expression [Eq. (5)], corresponds to the pore space formed between fiber bundles in the preform. For the purpose of estimating λ_{max}, the maximum pore space may be considered to result from one of two ordered bundle arrangements in the preform, as explained below. When the number of fiber mat layers stacked inside the mold cavity is the minimum required to completely fill the mold cavity thickness, the preform cross section is likely to exhibit an in-line arrangement of the bundles within and across fiber layers [Fig. 2(a)]. On the other hand, for higher fiber volume fractions, which require a dense stacking of the fiber mat layers within the mold cavity, a staggered arrangement of the bundles with respect to bundles in adjacent fiber layers results [Fig. 2(b)]. The maximum pore area, $A_{p,max}$, in each of these cases is the difference between the (rectangular or triangular) unit cell area, and the total bundle area enclosed within the unit cells, as identified by the lightly-shaded regions in Figs. 2(a) and 2(b). The maximum pore size, λ_{max}, is then given by $\lambda_{max} = \sqrt{\frac{A_{p,max}}{g_v}}$.

Introducing the expressions for α, β, and λ_{max} in Eq. (5), the permeability emerges as a function of the fiber volume fraction, v_f, the architectural parameters of the fiber mats, and the fractal dimensions, d_T and d_N.

EVALUATION OF d_T and d_N

TORTUOSITY DIMENSION, d_T

It may be recalled that the tortuosity dimension, d_T, represents the extent of convolutedness of the capillary pathways for resin flow through the preform. Here, we propose a simple technique based on analysis of two-dimensional microstructural images of the preform cross section through the thickness and parallel to the flow direction. The proposed method for the measurement of d_T is based on the correlation between the tortuosity and the contiguity of the

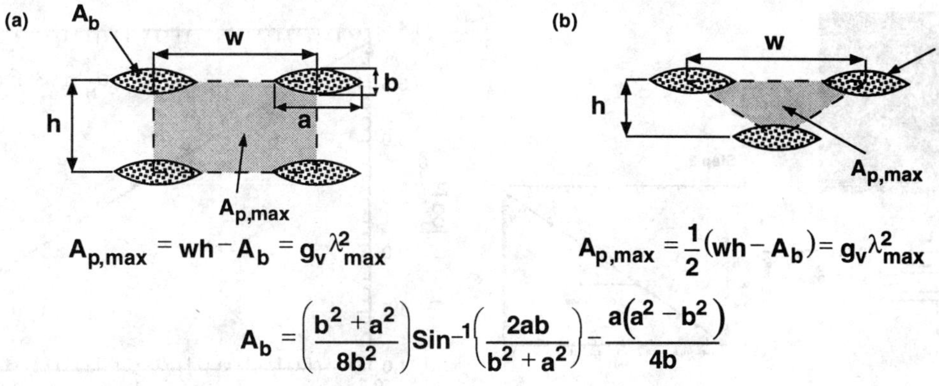

Figure 2: Illustration of the *(a)* Rectangular, and *(b)* Staggered arrangement of the fiber bundles considered in the estimation of the maximum pore size, λ_{max}.

porous regions as seen in the two-dimensional cross section. It may be realized that if the pore contiguity is high, the flow paths are more or less straight, which implies a smaller tortuosity for the capillary pathways, i.e., $d_T \approx 1$. Conversely, a reduced contiguity of the pores corresponds to a greater physical barrier presented by the fibers to the flow, which results in the flow taking on a more tortuous path, and a higher value of d_T. It follows, therefore, that the determination of d_T reduces to the problem of measuring the "contiguity" of the pores in the preform microstructure.

Contiguity of the fibers in the cross section of a continuous fiber-reinforced composite was studied by Pitchumani and Yao (1991), and Pitchumani (1994) in the context of evaluating the transverse thermal conductivities of the composite. They proposed a contiguity fractal dimension, d_C, whose values lie in the range 0 (least contiguous) to 1 (highly contiguous). This suggests that the tortuosity dimension, d_T, may be obtained from the measured values of the pore contiguity dimension, d_C, using the simple correlation: $d_T = 2 - d_C$.

The steps involved in the measurement of the pore contiguity dimension, and in turn the tortuosity dimension, are summarized below. A pictorial illustration of the steps on an actual preform cross section is presented in Fig. 3. *Step (1)*: The preform cross section is discretized into a number of square grid cells, whose size serves as the resolution for the measurements. *Step (2)*: The entire cross section is systematically scanned by placing a measuring line segment of length L on every grid cell, with the line oriented along the principal flow direction. Each time a reference point on the line (for example, its left end) falls on a porous region, the total length of the intercepts between the line and the pores is recorded. An arithmetic average of all the recorded total intercepted lengths is the mean intercepted length, L_p. In Fig. 3, the measuring line is shown as a narrow rectangle for the purpose of clarity. *Step (3)*: Step (2) is repeated for several values of the measuring length L in the range of 2–64 pixels, and a log-log plot of L_p against L is constructed from the measured values. The slope of the best-fit straight line through the points is the contiguity dimension, d_C, from which the tortuosity dimension, d_T, is obtained as $2 - d_C$. Note that the data points in Fig. 3 closely follow a straight line, which confirms the fractal nature of the microstructure.

AREA DIMENSION, d_N

The area dimension may be obtained using the box-counting method (Mandelbrot, 1983; Feder, 1988). In this method, the pore areas in the cross section under analysis are completely

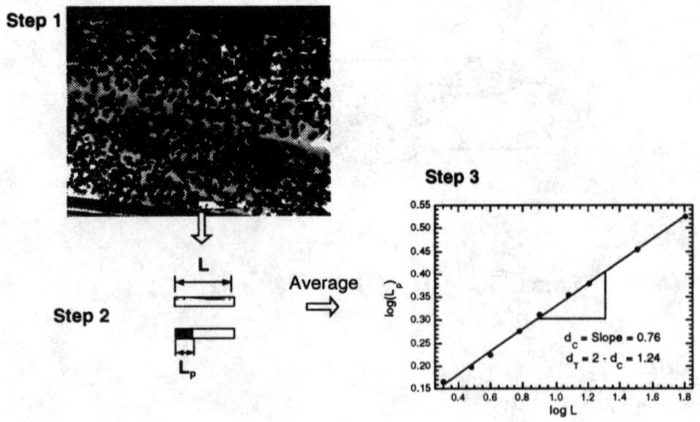

Figure 3: Schematic illustration of the procedure for evaluation of the tortuosity dimension, d_T.

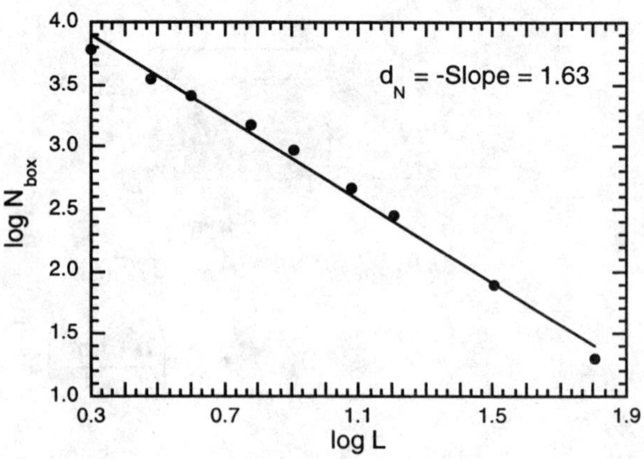

Figure 4: Area fractal dimension of the microstructure in Fig. 3, evaluated using the box counting method.

covered using boxes of size L, and the number of such L-sized boxes required to cover the pore areas is recorded as $N(L)$. This step is repeated for various box sizes in the range $2 \leq L \leq 16$, and a log-log plot of $N(L)$ versus L is constructed as shown in Fig. 4. The magnitude of the slope of a linear fit through the data gives the area dimension, d_N.

EXPERIMENTAL STUDIES

Experimental studies were carried out with the objectives of *(a)* investigating the effects of the process and material parameters on the permeability of preforms for an actual resin and catalyst system, and *(b)* validating the theoretical permeability model. The experimental validation studies were conducted using a lab-scale resin transfer molding (RTM) setup, shown schematically in Fig. 5. The main components include a pressure tank that houses the catalyzed resin, and a mold in which the preform is placed. The mold consists of a sandwich construction of a bottom aluminum base plate, a middle Plexiglas spacer, and a top Plexiglas plate. The aluminum base plate comprises of a resin inlet port which discharges into a concave runner along the width of the mold, as illustrated in Fig. 5. The runner ensures an uniform, essentially one-dimensional permeation of the preform along the mold length. As shown in the Fig. 5, the aluminum mold additionally incorporates a cavity that houses a strip heater, for curing the resin. The cavity is closed by a copper plate that houses the thermocouples and ensures uniform surface heating of the composite. The middle spacer, placed between the aluminum base plate and the Plexiglas top plate, determines the size of the mold cavity which was kept constant in this study at a value of 6 in. $\times$ 3 in. $\times$ 0.125 in. The overall sandwich setup was secured by six hexagonal bolts.

A knitted glass fabric, A1110, supplied by the Collins Craft Company was used in the studies. The fiber mats have an isotropic architecture in the plane of the fabric. The permeation studies were conducted using an epoxy resin, Epon-815, catalyzed with Epicure 3274, both supplied by the Shell Chemical Company. A noteworthy distinction of the present study is the use of an actual resin for the permeability measurements, whereas the studies in the literature use simulated fluids such as corn syrup and motor oil, among others. The resin selection for the study was based on its low viscosity which permitted investigating a wide range of fiber volume fractions using relatively low injection pressures (12 psi for a fiber volume fraction of 38% to 75 psi for the 58% fiber volume fraction). The flow rates for each combination of the volume

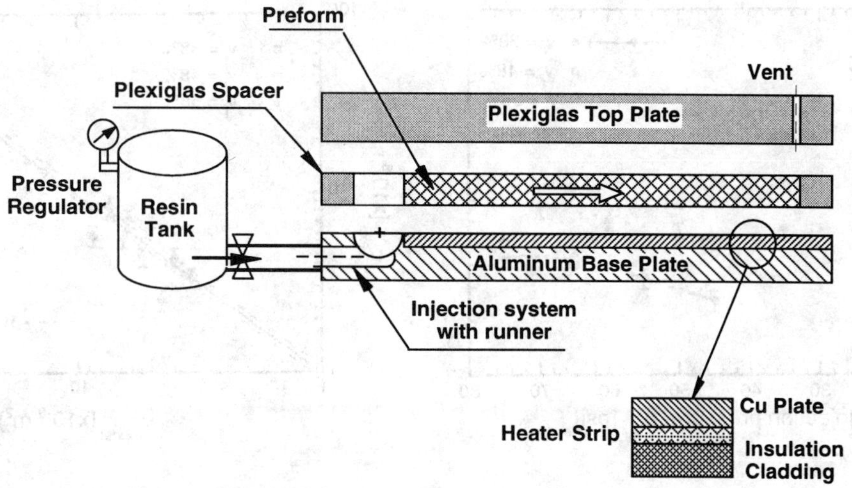

Figure 5: Schematic of the experimental setup used for the permeation studies.

fraction and injection pressures were measured by timing the flow of the resin as it exited through the two vents into a collection beaker. The preform permeabilities were calculated using the measured flow rates in Darcy's law.

Apart from providing the permeability data representative of those encountered in realistic processing environments, the use of a catalyzed resin in the experiments also provided for preserving the relative fiber arrangement in the composite for image analysis. A consequence of the isotropic fabric architecture used in the study is that the preform cross sections are identical along the planes normal to, and parallel to the principle flow direction. This simplifies the image analysis procedure, wherein the same cross section may be used for obtaining both the fractal dimensions, d_T and d_N. Specimens were cut along the width of the rectangular plate samples fabricated, which provided for analysis of the microstructures on planes normal to the flow direction. The polished specimens were analyzed under a magnification of $10\times$, by considering four adjacent fields of view on the microstructure. The microstructures were captured into a 8100/100AV Power PC Macintosh computer using a Scion image grabber through a Hitachi CCD camera mounted on a Nikon Optiphot microscope. The image analysis software, NIH Image, was used to evaluate the fractal dimensions of the captured digitized images using the procedure described previously.

RESULTS AND DISCUSSION

Figure 6 shows the variation of the experimentally-determined permeabilities and the model-predicted permeabilities, plotted along the ordinate, with the injection pressure plotted along the abscissa, for varying fiber volume fractions. The open symbols in the plot represent the model-predicted values which are seen to be in close agreement with the experimentally-measured values of permeabilities, denoted by solid symbols. The lines through the solid symbols represent smooth fits of the data and serve to elucidate the general trend in the permeability variation. The error bands on the measured permeabilities varies from 14 percent for a 0.38 fiber volume fraction case, to 12 percent for a 0.58 fiber volume fraction case. The increase in the fiber volume fraction causes a decrease in the permeability of the preform, as seen in Fig. 6, owing to the decrease in the available flow volume for the resin. Moreover, the capillary tortuosity increases with increasing fiber volume fraction (and therefore the fluid flow path is more convoluted),

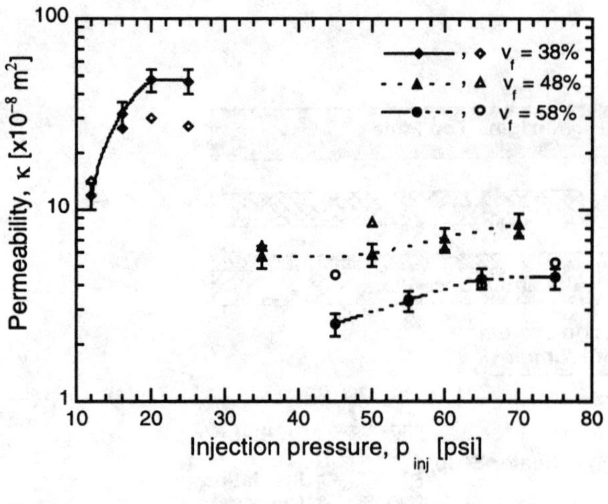

Figure 6: Effects of the fiber volume fraction and the injection pressure on the permeablities.

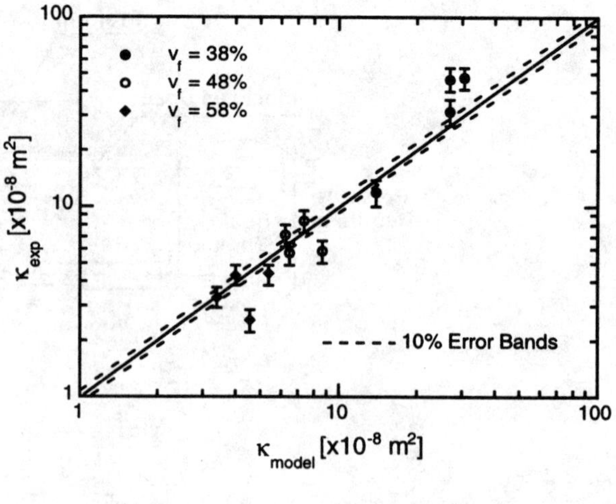

Figure 7: Comparison between the predicted and the measured permeability values.

which further contributes to the decrease in the preform permeability. It is also observed from Fig. 6 that the permeability increases with the injection pressure. This might be attributed to the dilation of the pores in the preform, due to increased fluid pressure, which leads to a reduced flow resistance and therefore, an increase in the permeability.

A quantitative comparison between the experimentally-determined permeabilities and the model-predicted values is presented in Fig. 7. The plot encompasses all the cases studied, and covers a wide range of injection pressures and fiber volume fractions. The solid line diagonal to the plot frame represents the case of perfect agreement between the experimental data and the model predictions, while the dashed lines indicate the 10 percent error bands. As seen in Fig. 7, most of the values lie within a 10 percent of the experimental data. The excellent agreement is especially remarkable in light of the fact that unlike other models reported in the literature (for example, Carman, 1979) the fractal model does not involve any empirical constants, and, for the first time, accounts for the pressure-dependence on the permeability.

The foregoing results suggest that the fractal dimensions provide an effective way of describing the disordered preform microstructures, for the permeability evaluation. The work represents a valuable contribution of fundamental and practical significance. Studies on various preform architectures, and determination of the anisotropic permeabilities using the fractal technique presented in this paper, are presented elsewhere.

CONCLUSIONS

A novel approach using the tool of fractal dimensions was presented for characterization of the disordered pore structures of preforms used in liquid composite molding processes. The fractal pore description was used to obtain an analytical expression for the permeabilities of the preform. Experimental validation of the fractal model was conducted over a wide range of injection pressures and fiber volume fractions. Excellent agreement was demonstrated between the predicted and the measured permeability values. The theoretical model was also shown to capture the pressure-dependence of the permeability accurately.

ACKNOWLEDGMENTS

The authors gratefully acknowledge the financial support for the project provided by National Science Foundation through Grant No. DMI-9522801. The authors also wish to thank the Collins Craft Company and the Shell Chemical Company for their gracious donation of the materials used in the study.

REFERENCES

Advani, S. G., M. V. Bruschke, and R. S. Parnas. 1994. "Resin Transfer Molding Transfer Phenomena in Polymeric Composites," in *Flow and Rheology in Polymer Composites Manufacturing*, S. G. Advani, ed., Elsevier, 10:465–515.

Bruschke, M. V., and S. G. Advani. 1989. "A Numerical Simulation of Resin Transfer Mold Filling Process," *ANTEC '89*, 1769–1773.

Carman, P. C. 1979. *Flow of Gases through Porous Media,* Academic Press, New York.

Darcy, H. 1856. *Les fontaines publiques de la ville de Dijon,* Dalmont, Paris.

Dave, R. 1990. "A Unified Approach to Modeling Resin Flow During Composite Processing," *J. Composite Materials,* 24:22–41.

Denn, M. M. 1980. *Process Fluid Mechanics,* Prentice-Hall, New Jersey.

Feder, J. 1988. *Fractals,* Plenum, New York.

Happel, J., and H. Brenner. 1986, *Low Reynolds Number Hydrodynamics, Mechanics of Fluids and Transport Process,* Martinus Nejhoff, Boston.

Karbhari, V. M., D. A. Steenkamer, and G. R. Palmese. 1993. "Effect of Preform Architecture on Flow and Processing in Resin Transfer Molding," SAE Paper 930166, Presented at the SAE Int. Congress and Expo.

Kim, Y. R., S. P. McCarthy, J. P. Fanucci, S. C. Nolet, and C. Koppernaes. 1991. "Resin Flow Through Fiber Reinforcements During Composites Processing," SAMPE *Quarterly,* April 1991, 16–22.

Kindo, K., et al. 1989. "High Magnetic Field Effect in Martensite Transformation," *Physica B,* 155:207–210.

Lin, R. J., L. J. Lee, and M. J. Liou. 1993,"Mold Filling and Curing Analysis in Liquid Composite Molding," *Polymer Composites*, 14(1):71–81.

Mandelbrot, B. B. 1983. *The Fractal Geometry of Nature,* W. H. Freeman, New York.

Pitchumani, R., and S. Yao. 1991. "Correlation of Thermal Conductivities of Unidirectional Fibrous Composites Using Local Fractal Techniques," *Trans. ASME, Journal of Heat Transfer,* 113(4):788–796.

Pitchumani, R. 1994. "A Fractal Geometry Approach to Modeling Disordered Composite Microstructures and Their Properties," in *Proceedings of the 9th American Society for Composites Conference,* Newark, DE, 807–817.

Ranganathan, S., S. G. Advani, G. M. Wise, F. R. Phelan, and R. S. Parnas. 1994. "A Numerical and Experimental Study of the Permeability of Fiber Preforms," in *Proceedings, 10th ASM/ESD Advanced Composites Conf.*, 309–319.

Sangani, A. S., and A. Acrivos. 1982. "Slow Fast Periodic Arrays of Cylinders with Application to Heat Transfer," *International Journal of Multiphase Flow,* 8(3):193–201.

Voss, R. F. 1985. "Random Fractals: Characterization and Measurement," in *Scaling Phenomena in Disordered Systems*, R. Pynn and Skjeltorp, eds., NATO ASI Series B, Plenum, New York, 133:1–11.

Simulation and Sensor System for an Injected Pultrusion Process

S. KOMMU, M. AL-HAYAN, J. L. KARDOS AND B. KHOMAMI

ABSTRACT

Injected Pultrusion is an attractive process for high volume, high performance, and low cost manufacturing of continuous fiber reinforced composites. The process involves injection of a polymeric resin (usually thermosetting) into a die through which fibers are pulled at a constant velocity. In this study we have developed a set of governing equations for conservation of mass, momentum, chemical species and energy for the injected pultrusion process using a local volume averaging approach. The utility of this simulation model in the design and optimization of the IP process has been demonstrated through a number of example simulations. It has been shown that with the aid of the simulation model one can effectively design the die geometry (i.e., optimum location of injection port, optimum shape of the injection section) as well as optimize the processing conditions (i.e., injection pressure or flow rate, fiber pull speed, wall temperature). Specifically, it has been shown that tapering the die in the injection section aids in quick attainmenzt of a flat flow front leading to uniform wetting of fibers and minimize formation of voids. However, it is also shown that a 3-dimensional solution of the pressure evolution equation is required for taper angles greater than 10 degrees. A dielectric-based sensing system is described that can be used to verify the model and potentially to optimize and control the process on-line.

INTRODUCTION

Fiber-reinforced polymer matrix composites are manufactured by a variety of processing methods including Resin Transfer Molding (RTM), Pultrusion, Autoclaving, Compression Molding and Filament Winding. Injected Pultrusion (IP) is currently being developed for the manufacture of high performance composites having constant cross-sectional area normal to one principal symmetry direction. It is a continuous composite manufacturing process which evolved as a combination of Resin Transfer Molding (RTM) and Pultrusion processes. To our knowledge, the IP process was originally developed by Hexcel (Blosser, Florio and Donti, 1994), and more recently by Owens Corning (Anonymous, 1996).

S. Kommu, M. Al-Hayan, J.L. Kardos and B. Khomami, Department of Chemical Engineering, and the Materials Research Laboratory, Washington University, St. Louis, MO 63130-4899

In IP, the concept of liquid injection molding has been incorporated into the conventional pultrusion process. The process (See Figure 1) involves injection of a polymeric resin into a heated die through which fibers are pulled at a desired velocity. The resin cures and solidifies around the fibers and the solid composite is drawn out of the die (Blosser, Florio and Donti, 1994). Carbon and glass fibers or preforms are commonly used as reinforcement in this process, especially for their applications in automotive, aerospace, sporting goods and civil structures industries. The main advantage of this process over other composite manufacturing processes is it's high volume and high quality production of composites having constant cross-sectional area at low cost .

For efficient fabrication of better quality composite parts, process optimization along with proper design are essential. However, it is prohibitively expensive to conduct experiments required for determining the optimum parameters for die design and for optimizing the IP process parameters. This motivated us to construct a reliable computer model which can simulate and optimize the IP process.

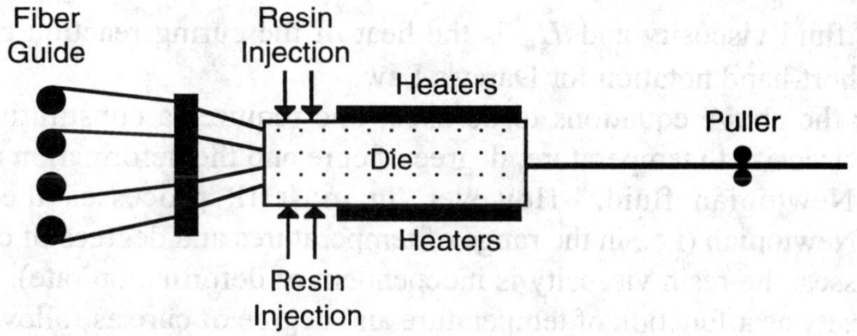

FIGURE 1: A Schematic Representation of the Injection Pultrusion Process

PROBLEM FORMULATION

GOVERNING EQUATIONS

The first and most important step for computer simulation of IP is development of a set of balance equations for transport of mass, momentum and energy in the moving porous fiber medium. It is extremely difficult to model macroscopic transport of these quantities based on microscopic transport models that account for variation of velocity and temperature as well as other quantities of interest past individual solid objects. Hence, volume averaging plays a central role in our development of field equations for various quantities of interest.

(Khomami, 1997) has recently derived the basic balance equations for transport of mass, momentum and energy for the IP process. These equations are,

Mass Conservation : $$\nabla.\langle\vec{\upsilon}_r\rangle=0 \qquad (1)$$

Cured Species Mass Conservation: $$\varepsilon_r \frac{\partial \langle \alpha \rangle^r}{\partial t} + \langle \vec{\upsilon}_r \rangle . \vec{\nabla} \langle \alpha \rangle^r = \varepsilon_r f(\langle \alpha \rangle^r, \langle T_r \rangle^r) \quad (2)$$

Conservation of Momentum: $$\langle \vec{\upsilon}_r \rangle = -\frac{1}{\mu} \underline{\underline{S}} \cdot \vec{\nabla} \langle P_r \rangle^r + \varepsilon_r \langle \vec{\upsilon}_f \rangle^f \quad (3)$$

Conservation of Energy:

$$\left\{ \varepsilon_r (\rho\, c_p)_r + \varepsilon_f (\rho\, c_p)_f \right\} \frac{\partial \langle T \rangle}{\partial t} + \left\{ (\rho\, c_p)_r \langle \vec{\upsilon}_r \rangle + (\rho\, c_p)_f \langle \vec{\upsilon}_f \rangle \right\} \cdot \vec{\nabla} \langle T \rangle = \\ \vec{\nabla} \cdot \left\{ (\varepsilon_r k_r + \varepsilon_f k_f) \cdot \vec{\nabla} \langle T \rangle \right\} + \varepsilon_r \rho_r H_R f\left(\langle \alpha \rangle^r, \langle T_r \rangle^r \right) + \mu \langle \vec{\upsilon}_r \rangle \cdot \underline{\underline{S}}^{-1} \cdot \langle \vec{\upsilon}_r \rangle \quad (4)$$

In the above equations, $\vec{\upsilon}$, P, α, ρ, c_p, k and T are the velocity, pressure, degree of cure, density, heat capacity, thermal conductivity and temperature, respectively. $f\left(\langle \alpha \rangle^r, \langle T_r \rangle^r\right)$ represents the rate of production of the cured species per unit volume, $\underline{\underline{S}}$ is the permeability tensor, μ is the fluid viscosity and H_R is the heat of the curing reaction per unit volume. Equation 3 is a short-hand notation for Darcy's Law.

In order for the above equations to be used, one requires a constitutive equation that relates the resin viscosity to temperature, degree of cure and the deformation rate tensor if the resin is a non-Newtonian fluid. However, in most IP processes the resins can be characterized as Newtonian (i.e., in the range of temperatures and degrees of cure encountered in most IP processes, the resin viscosity is independent of deformation rate). Hence, one can express the viscosity as a function of temperature and degree of cure as follows,

$$\langle \mu \rangle^r = f\left(\langle \alpha \rangle^r, \langle T_r \rangle^r \right) \quad (5)$$

DISCRETIZATION TECHNIQUE

A solution to the above set of non-linear partial differential equations in arbitrary domains cannot be obtained analytically, hence a numerical approach to the problem is required. A number of different numerical techniques such as finite differences with boundary fitted coordinates (Coulter and Guceri, 1988), finite elements with a moving grid (Dupret and Vandenschuven, 1988) or finite element control volume techniques can be used (Lin, Lee and Lion, 1991). Among these techniques the finite element control volume method (FE/CV) offers the most flexibility in terms of dealing with complex geometries and variable physical properties, as well as being more cost efficient than traditional finite element techniques with moving grids (Lin, Lee and Lion, 1991). Hence, in this study we have used this technique to discretize our governing equations. The implementation of this technique is described in detail elsewhere (Kommu, Khomami and Kardos, 1997).

Figure 2 is a schematic diagram of the IP process geometry. The x-y plane is the plane of the die cavity and z is the thickness direction of the die. The characteristic length, width and thickness of the die can be represented by L, W and H. In most IP processes cavities are thin, i.e., H<<L. Hence, based on a scaling argument, one can show that the pressure is mainly a function of x and y and the continuity and momentum balance equations can be solved in 2-D. An order-of-magnitude analysis of equation (4) shows that for typical resins used in the IP process, the heat transfer due to convection is dominant in the x-y plane compared to conduction, but for thin cavities heat transfer due to conduction is dominant along the z direction. Thus, the energy balance equation must be solved in 3-dimensions. Moreover, since the degree of cure is a sensitive function of the temperature in the resin phase, the chemical species balance equation (i.e., equation 2) must also be solved in 3-dimensions. The details of the numerical solution techniques are described elsewhere (Kommu, Khomami and Kardos, 1997).

An iterative scheme is employed to solve the above coupled system of non-linear equations. At time zero the initial temperature and degree of cure distributions are known. Using Equation (5) the viscosity of the resin can then be calculated. Next, the pressure distribution in the cavity is evaluated. In turn, the velocities are computed using equation 3. Utilizing Equation 5, equations (2) and (4) are solved to yield the temperature which in turn affects the viscosity of the resin in the flow domain. With the new viscosity information, the pressure solution is obtained and the iterations continue at the same time level until a converged viscosity is achieved. Upon convergence, the fill factors are updated using the calculated flow rates (Eqn. 3). The complete procedure is repeated for subsequent time steps until the die is filled and steady state is achieved all along the die.

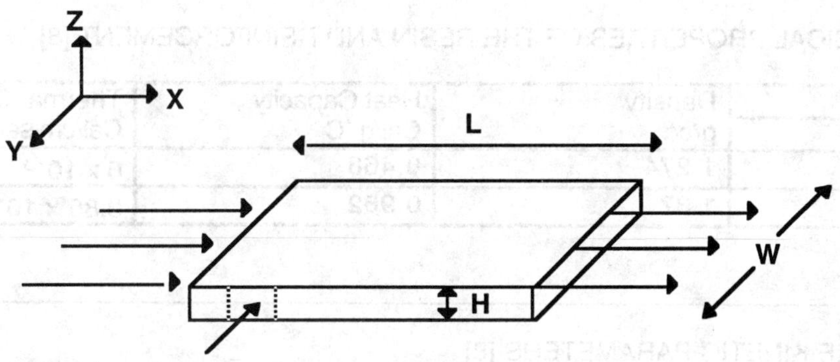

FIGURE 1: A Schematic Representation of a Typical Injection Pultrusion Die

SIMULATION OF THE IP PROCESS

TYPICAL 3-DIMENSIONAL NON-ISOTHERMAL SIMULATIONS

A 3-dimensional non-isothermal simulation of a typical IP process has been performed. The density, specific heat and thermal conductivity of the resin and reinforcement used in the simulations are given in Table 1. The cure kinetics for a typical high performance epoxy resin are as follows (Lee, Loos and Springer, 1982):

$$f\left(\langle\alpha\rangle^r, \langle T\rangle\right) = \left(K_1 + K_2 \langle\alpha\rangle^r\right)\left(1 - \langle\alpha\rangle^r\right)\left(B - \langle\alpha\rangle^r\right) \qquad \langle\alpha\rangle \leq 0.3 \qquad (11)$$

$$f\left(\langle\alpha\rangle^r, \langle T\rangle\right) = K_3\left(1 - \langle\alpha\rangle^r\right) \qquad \langle\alpha\rangle > 0.3 \qquad (12)$$

Where $K_i = A_i \exp\left(-\Delta E_i / R\langle T\rangle\right), i = 1,2,3$ and B is a constant that is determined through a fitting procedure. Table 2 summarizes the values used in the simulation. The viscosity of the resin is given by (Lee, Loos and Springer, 1982),

$$\langle\mu\rangle^r = \mu_\infty \exp\left(U/R\langle T\rangle + W\langle\alpha\rangle^r\right) \qquad (13)$$

where $\mu_\infty = 7.9 \times 10^{-14}\ Pa \cdot s, U = 9 \times 10^4\ J/mol, R = 8.314\ J/mol^\circ K$ and $W = 14.1$. The heat of reaction H_R for this resin is 473.6 J/g (Lee, Loos and Springer, 1982).

The permeability of the continuous carbon fibers as a function of porosity has been obtained from (Skartsis, Khomami and Kardos, 1992),

$$K_{yy} = 6.4 \times 10^{-11} \times 10^{(4.75\,\varepsilon_r - 4.275)} \qquad K_{xx} = \frac{\varepsilon_r^2}{\left(1 - \varepsilon_r^2\right)} \times 4 \times 10^{-12} \qquad (14, 15)$$

In all the simulations, the porosity has been assumed to be 0.5.

TABLE I: PHYSICAL PROPERTIES OF THE RESIN AND REINFORCEMENT [8]

Properties / Materials	Density g/cc	Heat Capacity Cal/g °C	Thermal Conductivity Cal/cm sec °C
Resin	1.274	0.468	6×10^{-3}
Fiber	1.37	0.962	3.85×10^{-4}

TABLE II: CURE KINETIC PARAMETERS [8]

Parameters	Value
A_1	2.101×10^9 min^{-1}
A_2	-2.014×10^9 min^{-1}
A_3	1.960×10^5 min^{-1}
ΔE_1	$8.07 \times 10^4\ J/mol$
ΔE_2	$7.78 \times 10^4\ J/mol$
ΔE_3	$5.66 \times 10^4\ J/mol$
B	0.47

The finite element mesh used in the computations is shown in the top of Figure 3. In the bottom of Figure 3 a typical flow front progression for the IP process is shown for isothermal filling with a low fiber pull speed. Although the fibers are fully wetted prior to the exit and a flat flow front is obtained quickly providing uniform curing of the resin, significant leakage exists from the back of the die as indicated by the large spacing between flow fronts near the back. To avoid this undesirable leakage, the fiber pull speed was significantly increased. The result of this simulation is shown on top of Figure 4. As seen in this figure, the selected pull speed is high enough to prevent any back flow, however, it has resulted in incomplete wetting of the fibers. This is highly undesirable since it will result in a part with a void in its center. Using the simulation code, one can obtain the optimum pull speed for a given set of operating parameters. At the optimum speed the fibers are fully wetted, a flat flow front is obtained in the curing region, and there is no leakage from the back (very closely spaced flow fronts in the rear of the die are indicative of no back flow). The flow front progression for this optimum condition is shown in the bottom of Figure 4. Sample temperature profiles and degrees of cure in the curing region for this optimum condition are shown in Figure 5. As depicted in the top of the figure, the wall temperature in the curing section was set at 160°C and a parabolic temperature profile is observed in the thickness direction (z-direction) at the midpoint of the x-y plane, indicating diffusion dominated heat transfer. Along the flow direction (the x-direction) the temperature at the midpoint of the z-y plane, is raised from 80°C, which is the temperature of the entering resin, to 160°C at the exit. Clearly along the flow direction heat transfer is convection-dominated. The bottom of Figure 5 shows that the cure profiles are very similar to the temperature profiles. At the inlet to the curing section the degree of cure is zero, while at the exit the desired value of 50 percent cure is achieved. In this particular simulation the goal was to obtain a void-free part with a 50 percent degree of cure at the exit. Specifically, by varying the injection pressure, fiber pull speed, temperature of the wall and length of the curing section, we were able to obtain the desired part quality. It should be noted that the simulation code can be used to address other important issues such as determining the optimum taper angle which is used in many IP processes in order to aid in rapid attainment of a flat flow front. In what follows, we briefly address this issue.

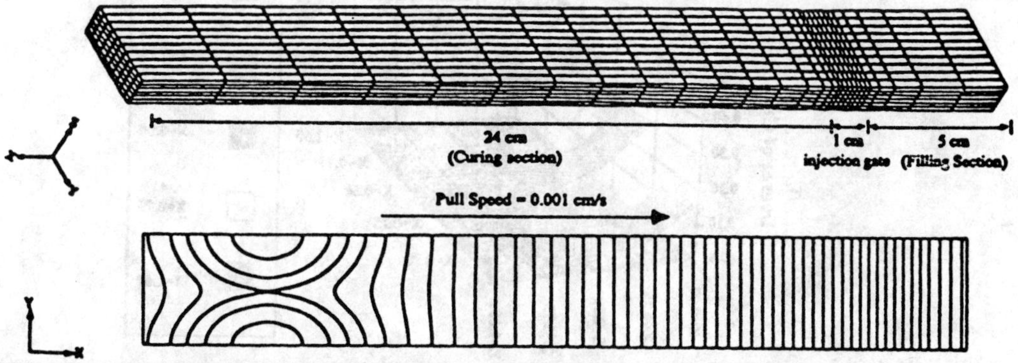

FIGURE 3: Sample isothermal simulation of the IP process
Top: Finite Element Mesh. Total length 30cm $(0 \le x \le 30)$, total height 1cm $(0 \le z \le 1)$, total width 3cm $(0 \le y \le 3)$. Fluid is injected from both sides (through the xz plane) through a 1cm x 1cm square. Bottom: Flow front progression, $K_{xx} = 2K_{yy}$; $K_{xy} = 0$

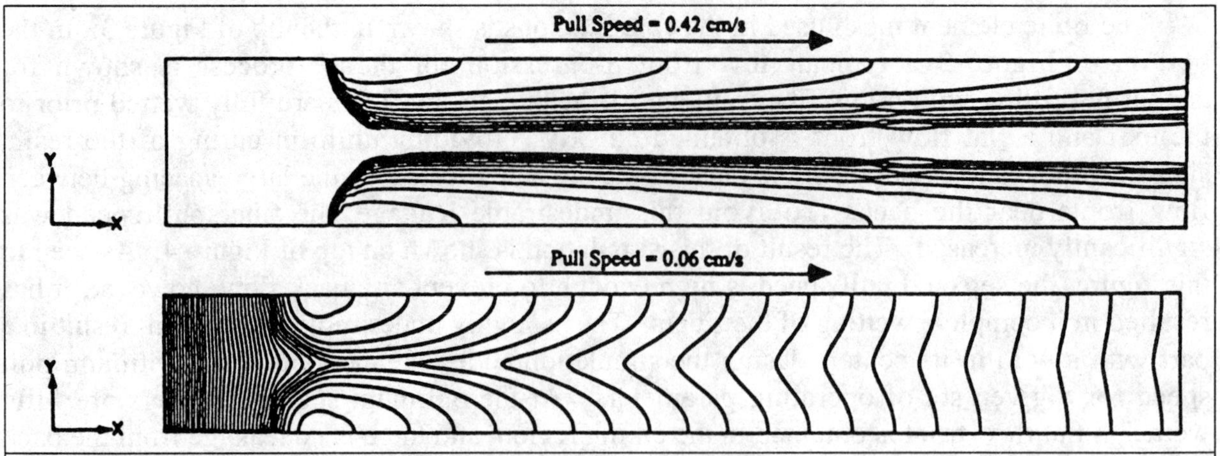

FIGURE 4: Sample Isothermal simulation of the IP process.
Top: Flow front progression, $K_{xx} = 2K_{yy}; K_{xy} = 0$; Bottom: Flow front progression, $K_{xx} = 2K_{yy}; K_{xy} = 0$

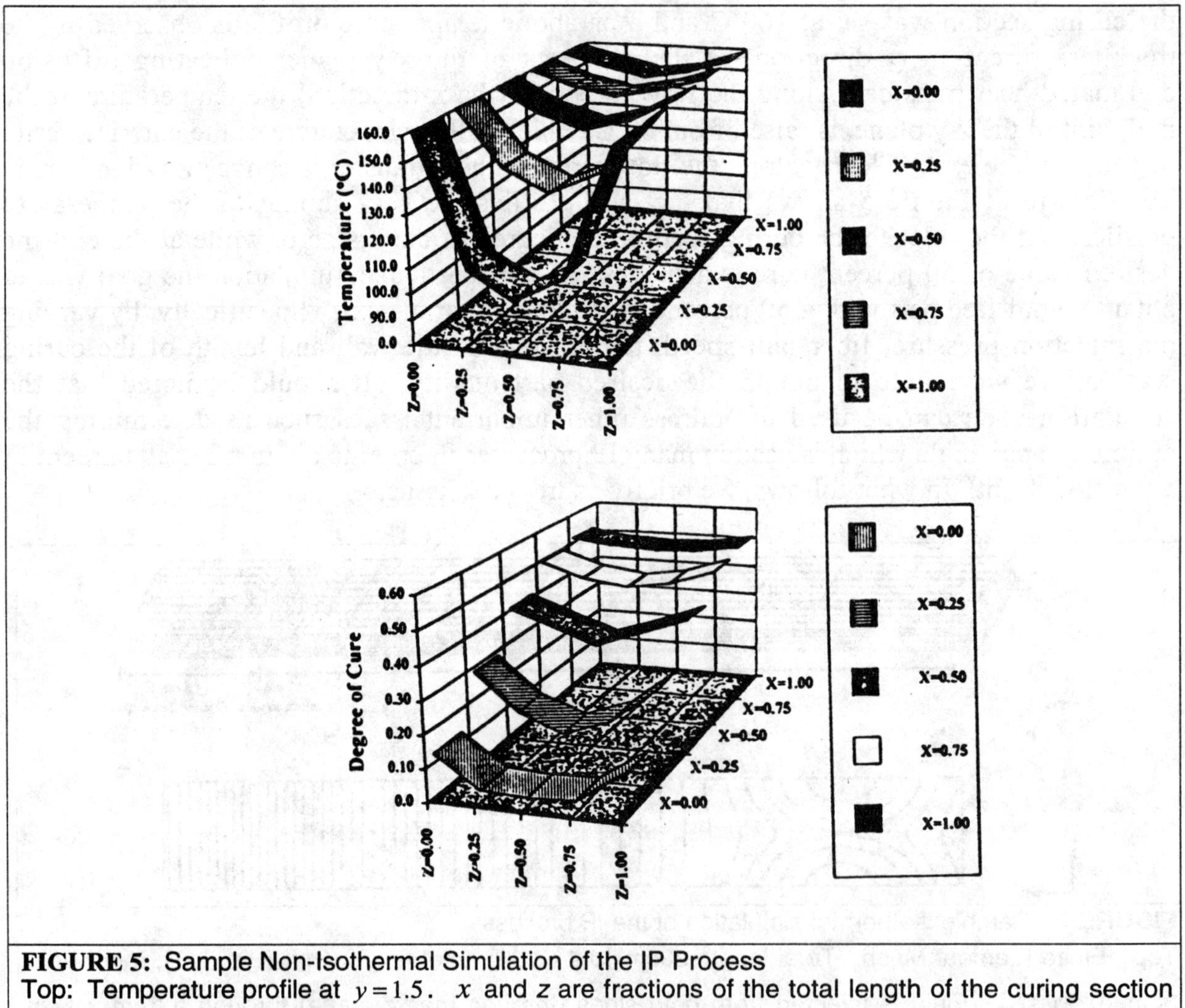

FIGURE 5: Sample Non-Isothermal Simulation of the IP Process
Top: Temperature profile at $y = 1.5$. x and z are fractions of the total length of the curing section and the channel height respectively (i.e., $x = 0$ denotes the entrance to the curing section). Bottom: Degree of cure profile at $y = 1.5$

EFFECT OF TAPER ANGLE

As mentioned earlier, based on an ordering argument it can be shown that for thin cavities, the pressure gradient in the thickness direction is negligible compared to the pressure gradient in the plane of the cavity. However, when one considers tapering of the test section in the flow direction of the injection section of the die, the magnitude of the pressure gradient in the z-direction becomes more significant as the taper angle is increased. The magnitude of this gradient can be approximated as follows,

$$\frac{\partial \langle P_r \rangle^r / \partial z}{\partial \langle P_r \rangle^r / \partial y} \simeq \frac{\Delta P_z W}{\Delta P_y H} = \frac{H}{W} \frac{S_{yy}}{S_{zz}} \tag{16}$$

where S_{yy} and S_{zz} are the components of the permeability tensor in the y and z directions, respectively. Therefore,

$$\frac{\partial \langle P_r \rangle^r}{\partial z} = \frac{H}{W} \frac{S_{yy}}{S_{zz}} \frac{\partial \langle P_r \rangle^r}{\partial y} \tag{17}$$

Using this equation, the importance of the pressure gradient in the z-direction can be estimated. To investigate the effect of taper angle on the pressure profile in the filling section, we have performed a number of simulations utilizing the 2-dimensional pressure evolution equations as well as a three-dimensional pressure field. Equation (17 was used to approximate the pressure gradient along the z-direction. The results of these simulations show that the pressure distribution in the injection section is accurately predicted by the 2-dimensional simulations for taper angles below 10 degrees. However, for higher taper angles, a full three dimensional simulation of the pressure field is required. Additionally, as expected, increasing the taper angle will result in more rapid attainment of a flat flow front. In turn, this leads to minimization of void formation, which is detrimental to the total quality of the part.

PRELIMINARY ON-LINE SENSING WITH THE FDEMS SYSTEM

Two preliminary on-line runs were performed with the Hexcel CRTM™ IP process using an in-situ Frequency-Dependent Electro-Magnetic Sensing (FDEMS) system (Hewlett-Packard 4192 A LF impedance analyzer, 5-1.3x10^7 hz) on two different-sized C-section parts. The vinyl ester resin system was used for both parts and the DekDyne Dapton EFEMS 750 sensors were fed into the die between plies at desired locations. This type of sensor provides complex permittivity information relatable to the resin cure state as a function of time and position along the die. The sensor is inert and viable at temperatures up to 400°C and pressures up to 1000 psi.

The C-section part is shown in cross section in Figure 6. The sensor was placed in the middle of the web as shown. The normalized ionic conductivity is shown in Figure 7 as a function of time and position along the die. The initial dip in the log (ε"ω) values indicates consolidation of the glass plies. At the die length fraction of 0.15, the log (ε"ω) curve rapidly increases along the die and the viscosity decreases. At about halfway down the die, the reaction begins (at about 85°C). The minimum in viscosity (maximum in log (ε"ω)) occurs at 85% of the die length, followed by the main exothermic reaction taking off. Gelation is attained at about 95% of the die length.

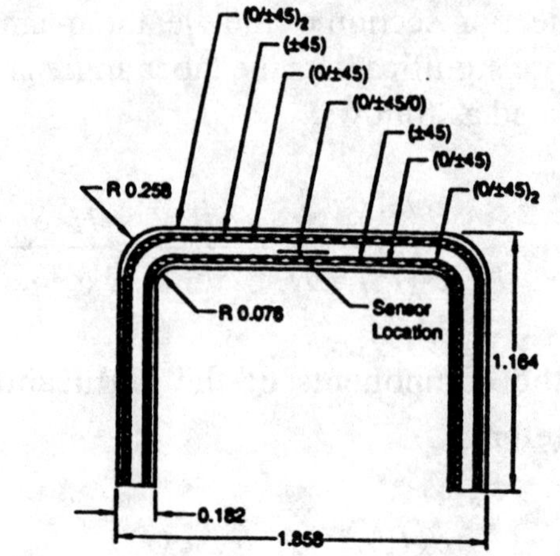

FIGURE 6: Cross Section of Test Beam

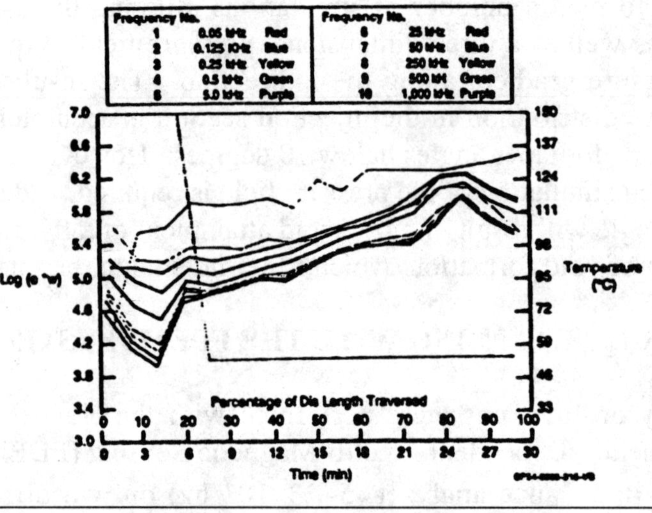

FIGURE 7: Ionic Conductivity as a Function of Time/Position

It is hard to determine the point of minimum viscosity for this particular resin, as was shown in the off-line FDEMS experiments and the rheological data. However, the least we can say is that it is in the range of 50-100°C. At about 60-65% down the die, the main exothermic cure reaction takes place, as indicated by the sharp drop in (ε"ω) values by more than two orders of magnitude. However, due to the high temperature caused by the exothermic reaction, some of the uncured resin becomes less viscous causing a subsequent rise in ($\varepsilon''\omega$) values. As the resin starts to cool, and more specifically at a resin temperature of 125°C (which is coincidentally the glass transition temperature), the values of ($\varepsilon''\omega$) drop sharply again (as the reaction overwhelms the temperature effect) to the final level and remain constant there, indicating that the final cure state is reached.

ACKNOWLEDGMENTS

This work was supported in part by the Advanced Research Project Agency under ARPA Order No. 9284, Affordable Polymeric Composite Materials Synthesis and Processing Program. We are indebted to Dr. John Florio at Hexcel Corp. for his help on the sensing measurements.

REFERENCES

Anonymous, Composites Technology, Sept./ Oct. 1996, p. 7.

R. Blosser, J. Florio Jr. and P. Donti, "Continuous Resin Transfer Molding of High Quality, Low Cost, Constant Cross Section, Composite Structural Elements", Proceedings of the 39th International SAMPE Symposium, Pages 1961-1972, April 11-14, Anaheim, CA (1994).

J.P. Coutler and S.I. Guceri, "Resin Impregnation During Manufacturing of Composite Materials Subjected to Prescribed Injection Rate", J. Reinf. Plast. Compos., **7**, 200 (1988).

F. Dupret and L. Vandenschuren, "Calculations of the Temperature Field in Injection Molding", AIChE J., **34**, 1959 (1988).

B. Khomami, "A Unified Approach to Modeling Transport of Heat, Mass and Momentum in Processing of Polymer Matrix Composite Materials", in Processing of Continuous Fiber Reinforced Composites; Editors, R.S. Dave and A.C. Loos, Hanser Publishers, Munich (to appear, 1997).

S. Kommu, B. Khomami and J.L. Kardos, Modeling of Injected Pultrusion Processes: A Numerical Approach", Polymer Composites, in press (1997).

W.J. Lee, A.C. Loos and G.S. Springer, "Heat of Reaction, Degree of Cure, and Viscosity of Hercules 3501-6", *J. of Comp. Matl.*, **16**, 510 (1982).

R. Lin, L.J. Lee and M.J. Lion, "Non-Isothermal Mold Filling and Curing Simulations in Thin Cavities with Pre-Placed Fiber Melts", Int. Polym. Process, **6**, 356 (1991).

L. Skartsis, B. Khomami and J.L. Kardos, "Resin Flow Through Fiber Beds During Composite Manufacturing Processes. Part II: Numerical and Experimental Studies of Newtonian Flow Through Ideal and Actual Fiber Beds", Poly. Engrg. Sci., **32**, 231 (1992).

Intelligent Process Control for the Microwave Curing of Composites

R. HAWKINS, M. C. HAWLEY, T. LENZ, J. K. McDOWELL, Y. QIU AND J. STICKLEN

ABSTRACT

In current research within the Composite Materials and Structures Center, Michigan State University, one thrust is aimed at developing microwave curing to achieve higher throughput, lower product cost and higher energy efficiency, relative to autoclave curing. Microwave heating may be characterized as single-mode or multi-mode. Single-mode heating sets up a single standing wave in the cavity so that the energy is highly focused and heating is efficient, however the heating pattern is generally uneven (Asmussen, 1987). In multi-mode heating, the microwave energy is shared by multiple modes which have different heating patterns, and therefore heating is not efficient and although more uniform than single mode heating, not acceptably uniform. To achieve uniform and efficient heating using single-mode microwave technology therefore requires switching between different heating modes in "real-time." Power application and mode selection in real-time curing must be carefully monitored and adjusted to minimize thermal gradients. The Intelligent Process Control Architecture (IPCA) was designed to specifically address these problems. IPCA is a control architecture for dynamic processes based on the Generic Task approach to knowledge-based systems, with extensions that provide a complete architecture supporting control system synthesis and real-time execution. IPCA interacted with an external controller which provides sensor monitoring and low-level mode tuning. The IPCA system was able to achieve thermal gradients lower than 10°C using a mechanically tuned cavity. The current thrusts of our research are on propelling the intelligent microwave curing technology further towards commercial viability by upgrading the system from a mechanically tuned method to a frequency tuned method, expanding the material domain coverage, and substantially extending the intelligent planning component by using learning methods to investigate curing without invasive sensing.

R. Hawkins and J. Sticklen, Department of Computer Science, Michigan State University, East Lansing, MI, 48824-1226

M.C. Hawley, T. Lenz, J.K. McDowell, and Y. Qiu, Dept. of Chemical Engineering, Michigan State University, East Lansing, MI, 48824-1226

INTRODUCTION

To date, the bulk of thermoset composite materials have been autoclave cured. Some alternatives to curing in an autoclave include microwave curing, E-Beam curing, and photo initiated curing. In the Microwave Research Group at Michigan State University, microwave curing technology is being studied with major goals of the research being: achieving higher throughput, lower final product cost, and higher energy efficiency, all relative to autoclave curing. Microwave heating can be either single-mode or multi-mode in nature. Home microwave ovens are an example of a multimode microwave chamber; such devices produce "cold spots" in the items being heated even when using turntables to rotate the items. This unevenness in multimode heating is due to the interaction between the multiple heating modes which are set up in the oven, which disperses the EM radiation throughout the oven. The modes produce heating patterns which are largely unfocused, with both hot and cold spots being produced. Single-mode heating, however, sets up a single standing wave in the cavity so that the energy is highly focused and heating is efficient. However the heating pattern is generally more uneven across the item than in the multimode setup. The single-mode method offers an attractive alternative to conventional heating for curing composite materials efficiently in a short time, because of the potentially higher heating rate. However, large thermal gradients across the structure are a normal by-product of single mode curing. High thermal gradients negatively impact final material properties, and are thus undesirable. To minimize thermal gradients, different modes can be selected throughout the cure cycle so that heating is uniform across the structure. Using this technique, microwave energy can be applied when and where it is needed for optimal heating and the control of material properties, provided the proper mode is applied at appropriate times during cure. Mode selection can be achieved either through mechanically altering the dimensions of the microwave cavity, thus altering the standing wave mode, or through manipulation of the frequency of the microwave power source.

Heating with single-mode microwave technology therefore requires switching between different heating modes in real-time. Power application and mode selection in real-time curing must be carefully monitored and adjusted to minimize thermal gradients. The Intelligent Process Control Architecture (IPCA) was developed in our laboratory in previous work of Decker (Decker, 1995) and Adegbite (Adegbite, 1995), and was designed to specifically address this problem. IPCA is a control architecture for dynamic processes based on the Generic Task(GT) approach to knowledge-based systems, with extensions that provide a complete architecture supporting control system synthesis and real-time execution. IPCA interacted with an external controller which provided sensor monitoring and low-level mode tuning. The IPCA system was able to achieve thermal gradients lower than 10°C using a mechanically tuned cavity.

Having shown broad proof of concept in the first generation system (IPCA), our current emphasis is on the transition from a mechanical method to a frequency based method of tuning modes in the microwave cavity (Wei, 1996). Frequency-based tuning provides access to more modes, almost instantaneous mode tuning, and enhanced reliability. The first and second items (more modes and very rapid tuning) heighten the importance of reliable computer control. In addition to extending the IPCA architecture to variable frequency

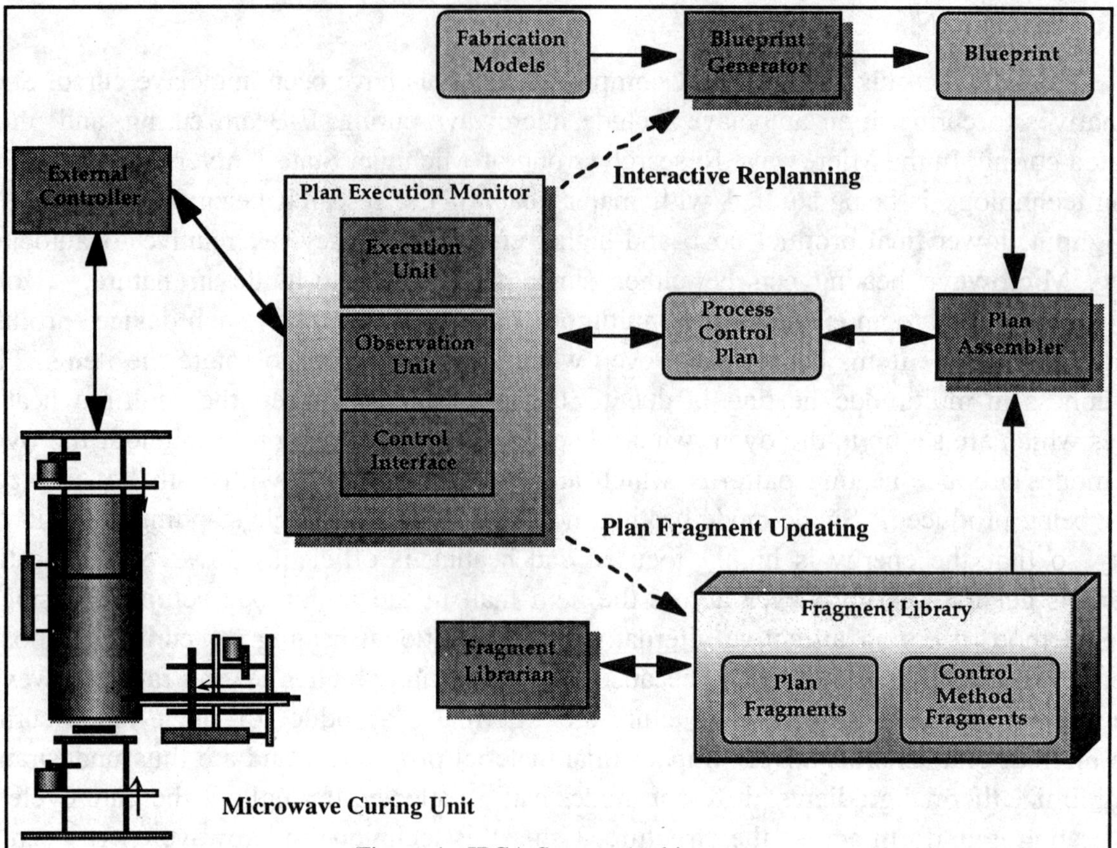

Figure 1: IPCA System Architecture

tuning, we also will extend to cover wider ranges of material samples, and to less reliance on invasive sensing.

IPCA: THE FIRST GENERATION

The IPCA control architecture is rooted in the Generic Task (GT) approach to knowledge-based systems, which emphasizes similarity in reasoning, knowledge acquisition and knowledge organization among certain types of problem solving tasks. IPCA expanded on the GT base by providing a complete architecture supporting knowledge-based control system synthesis and real-time execution. IPCA incorporated three primary components: a Blueprint Generator, a Plan Assembler, and a real time reactive Plan Execution Monitor (Figure 1).

The Blueprint Generator produces a specification for a complex process control plan which is designed for a specific execution environment. The Blueprint Generator's problem solving mechanism is based on a generic task known as *Routine Design*, which emphasizes the commonality in design problems which do not require "innovation". The Plan Assembler implements each blueprint by combining pre-compiled plans and control method fragments and assembling them to form an executable process control plan. The Plan Execution Monitor incrementally determines the execution of appropriate fragments from the process control plan based on sensory input (temperatures, reflected power, mode setting) gathered from real-time process monitoring, reactively selecting an appropriate path through the process

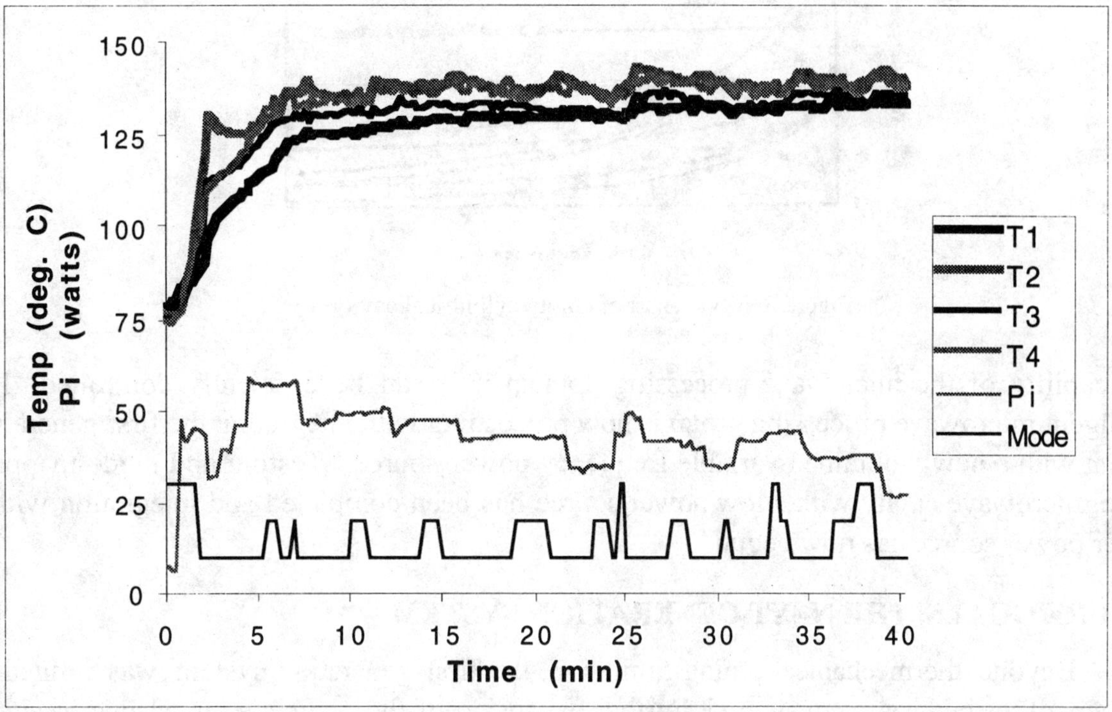

Figure 2: Thermal Profile

control plan structure. The Plan Execution Monitor selects the appropriateness of a code fragment using a generic task construct: the *Sponsor/Selector* decision mechanism.

Experiments were performed on a carbon-fiber reinforced DGEBPA prepreg. Thermal gradients across the part were reduced to below 10 degrees C. (Figure 2) This represented a 66% reduction over previous automated processing methods. (Fellows, 1991) These results demonstrated that IPCA effectively controlled the microwave curing apparatus, resulting in high-quality, high-speed composites processing for the relatively simple planar material samples used to date.

CURRENT STATUS: FREQUENCY BASED TUNING

Theoretically, there are two factors that determine the resonant modes in a microwave cavity: the microwave frequency, and the dimensions of the cavity. Thus two methods can be used to select a mode in a cylindrical cavity, mechanically modifying the length of the microwave cavity, or electronically altering the frequency of the microwave power source (Qiu, 1996). As seen in figure 3, the number of resonant modes available when tuning by varying frequency is greater than those available when changing cavity length at a fixed frequency and cavity diameter. Because of the nearly instantaneous mode switching that frequency based tuning will provide, the process of power application and mode selection in real-time curing must be even more carefully monitored and adjusted to further minimize thermal gradients(Qiu, 1996). Frequency-based tuning provides access to more modes, almost instantaneous mode tuning, and enhanced reliability, and will clearly enhance the commercial

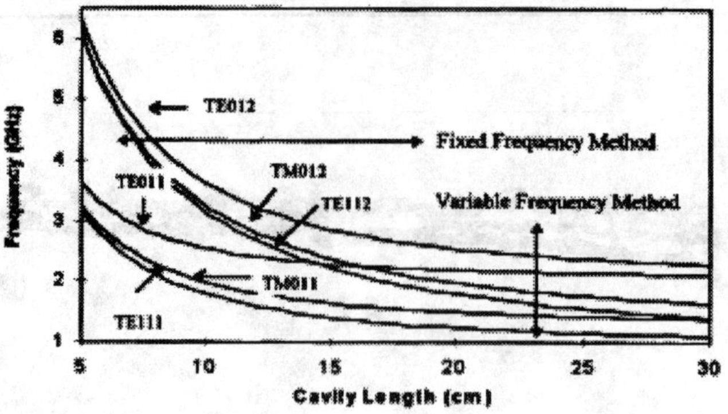

Figure 3: Mode chart of empty cylindrical cavity

applicability of the microwave processing domain if it can be efficiently controlled. The intelligent microwave processing group is now proceeding with integrating the first generation system with a newly obtained variable frequency power source. Testing and mode-mapping of the microwave cavity with a low power source has been completed and integration with a higher power source has now begun.

FUTURE GOALS: THE NEXT GENERATION SYSTEM

Beyond the mechanical tuning limitation, the first generation system was limited in numbers of testbed cases considered. Further expansion of the planner's knowledge base and the library of plan and control fragments will be incorporated to cover a larger array of composite material systems. More advanced control and planning fragments are also being investigated to try to deal with more complex fiber and part geometries, which tend to cause perturbations in the microwave EM field, making heating patterns far more unpredictable. In the first generation testbed of IPCA, 1", 2" and 3" square 12 ply unidirectional samples were used for proof of concept. For the next generation, the array of part sizes and shapes will be expanded, and non-unidirectional fiber orientations will be investigated.

One of the main thrust of the new research will be to experimentally characterize processing parameter effects. The first generation of IPCA monitored processing interactively through the use of invasive thermocouples to sense part temperature. Ideally, for commercialization, it would be desirable to develop a profile for a particular "case" (part geometry, material system, part size, etc.) through preliminary runs with sensing, then be able to use case data with confidence to produce parts without sensing. Experiments will be run both with sensors and then without followed by comparisons of part cure quality to determine the viability of this option and to determine what parameters (part placement, geometry, material composition, etc.) affect processing, and to what degree.

Also, we are investigating interactive replanning between the execution monitor and the plan generator to enable the system to replan in response to radical changes in operating conditions. Finally, we are investigating the addition of learning mechanisms to automatically update the planning knowledge-base through experience of past curing solutions will further enhance the system. Curing profiles and mode maps for particular "cases" will be modified automatically from experience. In addition, mode maps for new cases will be automatically generated with these. (Dashed lines in Figure 1) .

ACKNOWLEDGMENTS

We would like to acknowledge the contributions of Dave Decker and Bill Punch of the Intelligent Systems Laboratory, and Val Adegbite of the Composite Materials and Structures Center at Michigan State University. In addition, we gratefully acknowledge the continuing financial support of the National Science Foundation Center for Low-Cost, High-Speed Polymer Composites Processing at Michigan State University and its member companies and the generous equipment donations of Apple Computer.

REFERENCES

Adegbite,V.A., Automation and Control of the Microwave Processing of Composite Materials, Ph.D. Dissertation, Dept. of Chem. Eng., Michigan State University, 1995.

Asmussen, J., H.H. Lin, et al. "Single Mode or Controlled Multimode Microwave Cavity Applicators for Precision Material Processing." In Review of Scientific Instruments, pgs. 1477-1486, 1987

Decker,D.B., IPCA: An Intelligent Control Architecture Based on the Generic Task Approach to Knowledge Based Systems, Ph.D. Dissertation, Dept. of Comp. Sci., Michigan State University, 1995

Fellows, L.A. Microwave Processing of Polyimide Composites, M.S. Thesis, Dept. of Chem. Eng., Michigan State University, East Lansing, 1991.

Qiu, Y., Adegbite, V. and Hawley, M., "Intelligent Microwave Processing Using Variable Frequency Method: Preliminary Results", In Advanced Composites Conference and Exposition, 155-166. Dearborn, Michigan, 1995.

Wei, J.B., Shidaker, T. and Hawley, M. "Recent Progress in Microwave Processing of Polymers and Composites" In Trends in Polymer Science, Jan. 1996, V. 4, No. 1, pgs. 18-24

In-situ Monitoring of Graphite/Epoxy Cure Using Optical Fiber and Ultrasonic Sensors

J.-Y. CHEN,[a] S. V. HOA,[a] C.-K. JEN[b] AND H. WANG[b]

ABSTRACT The cure process of graphite/epoxy composites has been monitored by fiber-optic and ultrasonic sensors simultaneously for the first time. The results have shown that: both the fiber-optic and ultrasonic sensors can sense the gelation but only ultrasonic monitoring can sense the end-of-cure. This suggests that ultrasonic monitoring may be an effective way to in-situ assess the state-of-cure. In addition, through the ultrasonic monitoring we have also found that the "degree of cure" measured by DSC technique does not accurately reflect the mechanical property development during cure. Significant changes in the modulus still exist when the "degree of cure" indicated by DSC is almost fully developed.

INTRODUCTION

Advanced composites can be made of continuous, fiber-reinforced, thermosetting resin matrix prepreg materials. They are fabricated by laminating multiple plies into the desired shape and then cured in an autoclave with simultaneous application of proper heat and pressure. The quality of the final part is directly affected by the cure time, temperature, and pressure. Thus, the knowledge of cure process is very important in order to obtain fully cured and high quality composites.[1] In-situ sensors capable of monitoring the change of physical and chemical properties during cure process are therefore desirable. Several technologies have been investigated to monitor the cure of epoxy and among the most promising ones are thermal DSC, fiber-optic, and ultrasonic techniques.

Traditionally, the "degree of cure" (DOC) of composites, measured by using thermal DSC technique, is used to assess the state-of-cure. [2] Several observations on the cure kinetics of AS4/3501-6 prepregs, have been made. Figure 1 shows the DOC as a function of cure time with different cure temperatures. As cure temperature increased, the DOC increased and the time for completion of the cure reaction decreased. At the cure temperature of 176°C and 186°C, cure reaction was almost fully completed after 60 minutes, while at 146°C cure, the DOC only reached around 0.7 after 120 minutes. The DOCs measured at 156°C and 166°C fall between these extremes. However, even though the chemical reaction was completed, the mechanical properties of composites seemingly were not well developed. The critical objective of the process engineer is to know when the material has been "fully processed", thus, the "DOC" may not be a unique criteria to evaluate the end-of-cure. [3]

---

[a] J.-Y. Chen, S.V. Hoa, Dept. of Mech.Eng., Concordia Univ., Montreal, Quebec, Canada H3G 1M8
[b] C.-K. Jen and H. Wang, IMI, National Research Council, Boucheville, Quebec, Canada J4B 6Y4

Recently, by using fiber-optic strain gauges, strain measurements [4,5] may provide us with an opportunity for in-situ cure monitoring studies because during cure process, significant residual stresses in composites are introduced. [6-9] Usually, the primary causes of residual stress are thermal stresses due to the different coefficients of thermal expansion of the constituents, and curing stresses which result from chemical shrinkage of polymer matrix material during processing. Measurements of strains and temperature, combined with some computational model, can provide us valuable information on the state of cure. Because fiber-optic sensors can be incorporated into the component during the manufacturing process, for example in prepregs, they are suitable for in-situ cure monitoring in composites. Furthermore, the dimensions and orientations of the optical fiber sensors within prepregs can be selected to have a minimal impact on the mechanical properties of the host composite.

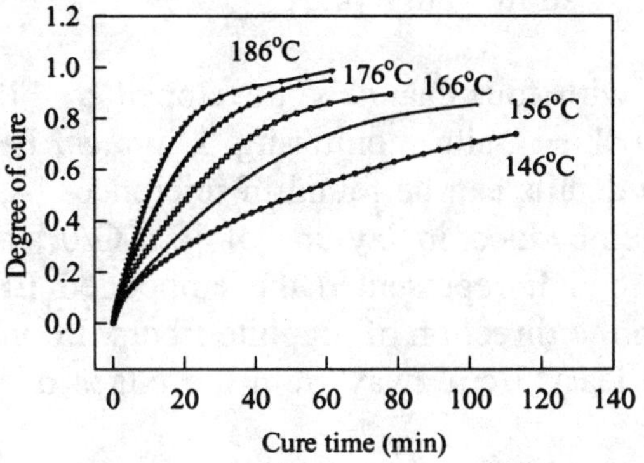

Figure 1 The degree of cure *vs.* cure time for AS4/3501-6 prepregs with different cure temperatures

The use of ultrasonics in cure monitoring was reported early in 1952 by Sofer and Hauser. [10] Using a pulse transmission method with longitudinal waves of about 2 MHz, they found that the longitudinal wave velocity and attenuation vary significantly with the cure time of the resin. The velocity increases monotonically with time whereas the attenuation peaks are located near the gelation point. Since then, ultrasonic techniques for cure monitor have been reported in numerous literatures [11-17] because it can provide a direct and nondestructive measure of the elastic properties during cure process. There are a number of limitations though, specifically, commonly used piezoelectric ultrasonic transducers (UTs) can only be operated continuously at an upper temperature of about 60°C. This limitation presents difficulties when applying ultrasonic techniques in actual processing environment which has a temperature of 176°C. However, the above limitations can be overcome by using a buffer rod together with a cooling system.

In this paper, the Extrinsic Fabry-Perot Interferometric (EFPI) fiber-optic and ultrasonic sensors are investigated as potential candidates to monitor cure processes of composite materials in an autoclave. Ultrasonic sensors are used to measure the change of viscoelastic properties, while the EFPI sensors are used to detect the evolution of local strain caused by temperature, chemical shrinkage, or process induced residual stresses.

EXPERIMENTAL SETUP

Our in-situ cure monitoring system consists of fiber-optic and ultrasonic sensors, and associated electronics. It is incorporated into an autoclave as shown in Figure 2. Two pieces of aluminum (Al) plates with the thickness of around 1 cm were used as the mold and the materials were Hercules AS4/3501-6 prepregs.

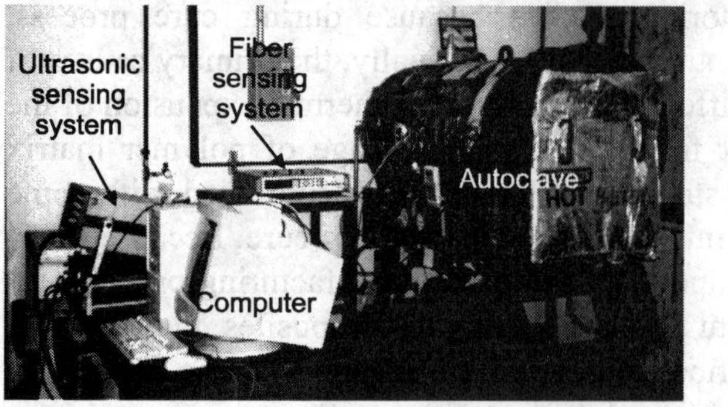

Figure 2 Actual in-situ monitor system

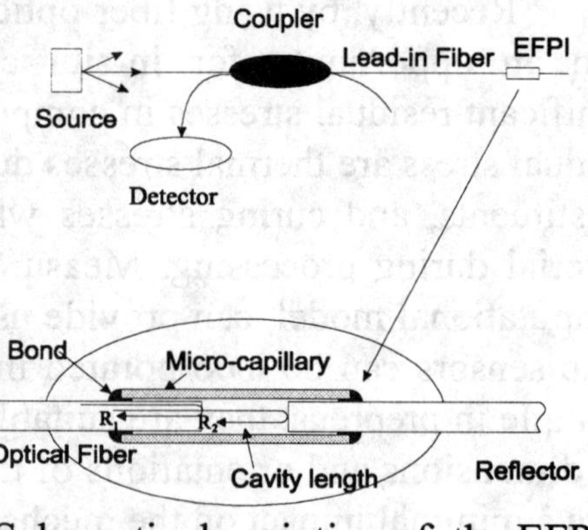

Figure 3 Schematic description of the EFPI strain gauge [4,5]

FIBER-OPTIC SENSOR

A WFS-100 EFPI fiber-optic sensor system with four channels, developed by FISO Technologies, Inc. in Quebec, Canada, was used for this in-situ monitoring. A typical EFPI sensing mechanism is illustrated in Figure 3 and the details can be found in references [4,5]. In our experiments, the EFPI strain gauges were embedded in lay-ups of $[0_8/SG/0_8]$ and $[90_3/SG1/90_3/0_2/SG2/0_2/90_6]$ composite panels, where SG*i* represented the embedded EFPI strain gauges. All EFPI strain gauges were aligned in the direction of graphite fibers. In order to avoid any edge effect, the EFPIs were placed at least 5 cm away from the edges of the panel.

ULTRASONIC SENSOR

The ultrasonic sensing system was composed of high performance 5 MHz broadband longitudinal piezoelectric UTs, couplants, clad buffer rods [18] and air cooling systems which surrounded the UTs, as shown in Figure 4. This system can be operated in either pulse/echo or/and through-transmission mode. The clad buffer rods consisted of aluminum core and a thermal sprayed aluminum cladding. The length of each buffer rod was 17 cm and the width was less than 1.5 cm. These buffer rods were screwed into the mold, with the end face flush with the mold inner surface as indicated in Figure 5. The principle of the ultrasonic guidance in clad buffer rods is like the clad optical fibers in which the energy is guided in the core, and thus the threading in the cladding region does not disturb the guided ultrasonic energy inside the core. The variation of the amplitude of the reflected ultrasonic signals at the rod/composite interface and those of the amplitude and time delay of signals traversing back and forth through the composite thickness enable us to monitor the cure. The 100 psi compressive air cooling allows the cooling of the UTs from 190°C down to 60°C which is the maximum allowable temperature for the UTs used. The 17 cm clad buffer rod length is chosen so that the temperature gradient is not large and the thermal characteristics of the threaded buffer rod end are the same as those of the mold. Ultrasonic waveforms were recorded with a sampling rate of 100 MHz in every 5 or 10 minutes period during the entire cure process using a computerized system consisting of an ultrasonic MP270 pulser, MR106 receiver, LeCroy 9400 dual 125 MHz digital oscilloscope and a PC Pentium®. The whole data acquisition and signal processing was handled by computer programs written in LabVIEW ® software.

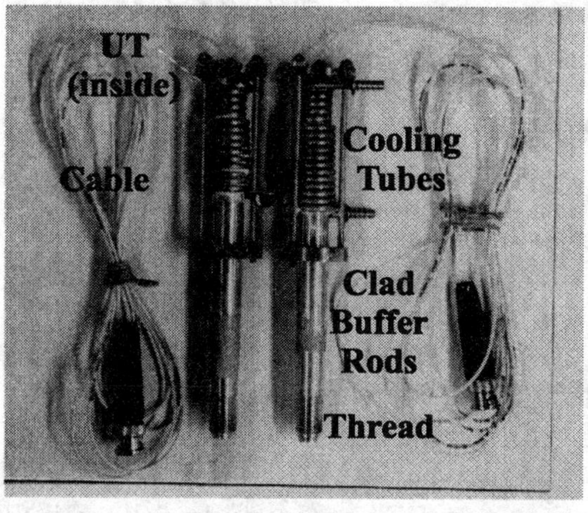

Figure 4 Clad Al buffer rods

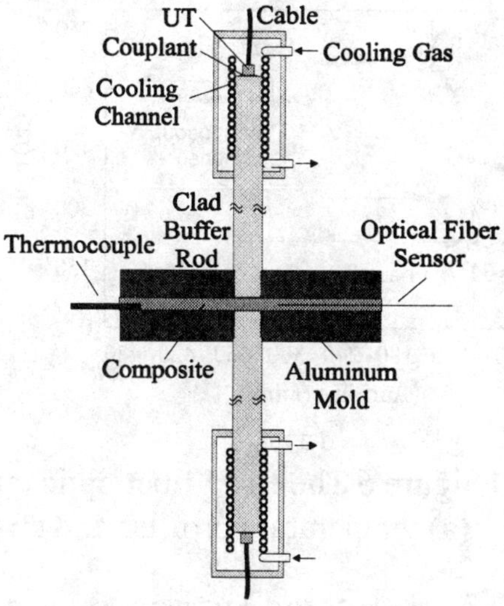

Figure 5 Illustration of cure monitoring using optic fiber and ultrasonic sensors simultaneously

PANEL PROCESSING

The panels of carbon fiber reinforced epoxy were laid up manually from the same lot of AS4/3501-6 prepregs. We used the commercially recommended cure cycle indicated as follows: the prepreg was debulked under vacuum while the temperature was raised to 116°C; after one hour at 116°C (phase I or 1st dwelling) the temperature was raised to the cure temperature of 176°C, and held there for 2 hrs (phase II or 2nd dwelling); the autoclave gas pressure was kept constant at 85 psi during the entire cure processing to secure prepregs compaction and good thermal conduction; the cured panel was then slowly cooled to room temperature. This cure cycle is indicated in Figure 6. The temperature of the panel was monitored by a K-type thermocouple embedded into the panel and close to the location of embedded fiber-optic or/and ultrasonic sensors.

FIBER-OPTIC IN-SITU CURE MONITORING

Figure 6(a) presents a typical result of the strain in-situ sensing for composites with the lay-up sequence of $[0_8/SG/0_8]$ cured at 176°C for 2 hrs. The corresponding plot of strain vs. panel temperature is shown in Figure 6(b). This result shows that during the ramp-up from room temperature to just below 176°C, the strain increased at a linear rate of 4.55 με/°C which was slightly higher than the thermal expansion coefficient of 4.40 με/°C of the bare EFPI fiber-optic strain gauge. It indicates a negligible load transfer between the embedded fiber-optic strain gauge and the laminate at the early stage of processing. This is to be further proved by the following EFPI fiber-optic in-situ monitoring on cross-ply panels. After that, the strains were built up, presumably due to the chemical shrinkage caused by crosslinking of the polymer matrix and thermal expansion. These strains appeared to build up slowly as the composite was cured. During cool-down, initially, thermal contraction rapidly increased at a rate of 11.28 με/°C. After a while, the strain bounced back a little bit (D_1), then thermally induced strain slowly changed at a rate of -0.09 με/°C which was close to the reported longitudinal thermal expansion coefficient of the unidirectional laminate. [19]

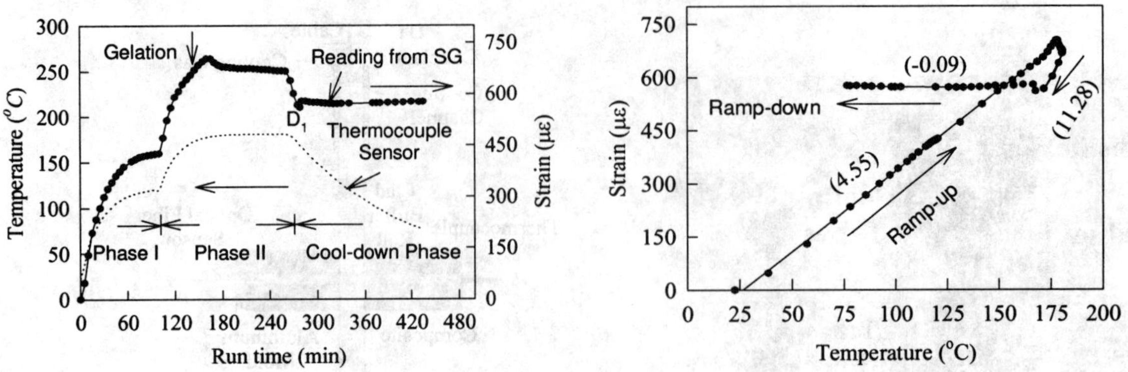

(a) (b)

Figure 6 The EFPI fiber optic cure monitoring for $[0_{16}]$ cured at 176°C
(a) temperature profiles and strain readings (b) strain vs. temperature

In the case of $[90_3/SG1/90_3/0_2/SG2/0_2/90_6]$ panel, two SGs successfully followed the cure cycle. The experimental results indicate four regions of interest which can be identified in Figure 7. Initially, from room temperature to around 170°C, both SGs have almost the same rate of thermal expansion regardless of their direction of alignment in the laminate. This proves again that during the initial curing stage, the load transfer between the fiber-optic gauges and the laminate was negligible. Then, above 170°C due to the appearance of gelation of prepregs, the bonding between the gauges and the laminate was strong enough to effectively transfer load, the readings from two SGs started to be separated and followed the local strain changes where they were embedded. Over the entire cure reaction, two SGs showed the strains induced from both chemical shrinkage and thermal expansion of the panel. After cure phase II, the panel began cool-down. Initially, again, the thermal contraction showed a rapid accumulation at a rate of 3.23 and 22.93 με/°C, respectively. Until around 150°C, small dips (D_1s) in the strain curve appeared as shown in Figure 7, and then the variation of thermal strain with cooling temperature kept the rate of 0.26 and 3.87 με/°C, respectively. These thermal expansion coefficients reasonably agree with the calculated values of 0.72 and 4.74 με/°C, respectively, using classical lamination theory (CLT).

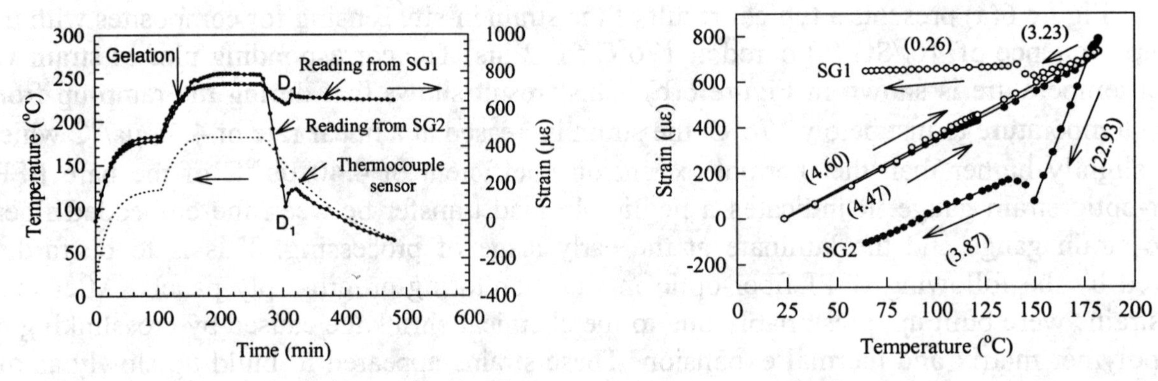

(a) (b)

Figure 7 The EFPI fiber optic cure monitoring for $[90_3/SG1/90_3/0_2/SG2/0_2/90_6]$ laminate cured at 176°C (a) temperature profiles and strain readings (b) strain vs. temperature

The above observations indicate that the EFPI can sense the gelation, but cannot determine the end-of-cure. Furthermore, the appearance of a dip during the cool-down phase of cure cycle may very well be related to the detachment of the laminate from Al mold plates near the sensor location. This means that prior to detaching, the panel followed the change of thermal shrinkage of the Al mold plate, whose thermal expansion coefficient was much higher than that of composite panel. After detaching, the panel followed its own thermal shrinkage. Such conclusion will be further verified by ultrasonic in-situ monitoring.

ULTRASONIC IN-SITU CURE MONITORING

When using ultrasonics for cure monitoring two parameters are typically employed to infer the state-of-cure of a material: the ultrasonic (1) velocity or time delay and (2) attenuation. The physical material properties which are useful for determining the state-of-cure, such as moduli and viscosity can be subsequently deduced from this information by using, e.g., a viscoelastic analytical model. By monitoring the evolution of the ultrasonic time delay and attenuation in a composite laminate as a function of cure time, we can understand the cure process qualitatively and quantitatively.

PULSE/ECHO MODE

In the pulse/echo mode, the ultrasonic wave traveling through the material is reflected from the back side of the laminate to the originating surface where it is detected by the same UT that generates the pulses.

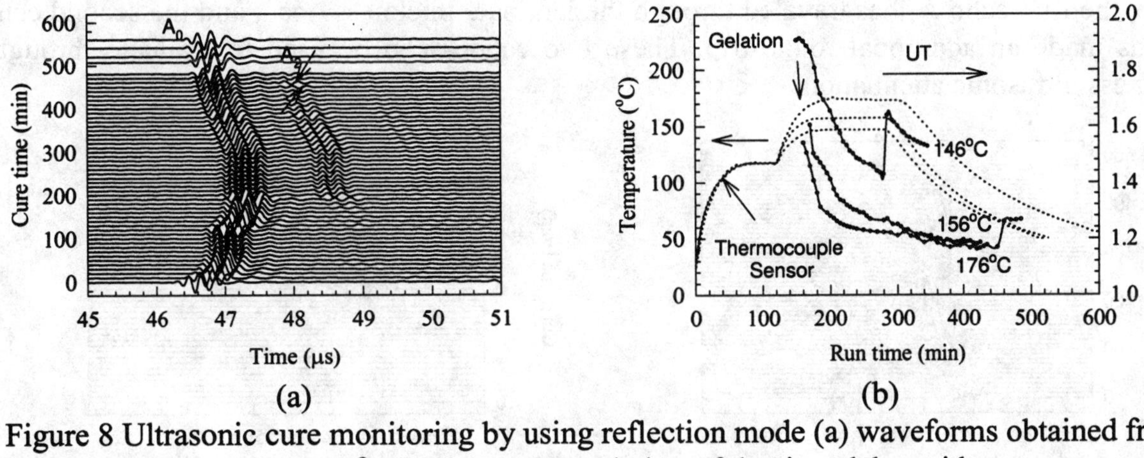

Figure 8 Ultrasonic cure monitoring by using reflection mode (a) waveforms obtained from the $[0_{16}]$ panel cured at 176°C for 2 hrs. (b) variation of the time delay with cure temperature

The results for the $[0_{16}]$ panels obtained during the cure monitoring, as given in Figure 8, show that during the initial stage the ultrasonic waves reflected from the back surface of the laminate were too weak to be recorded using the current setup. After reaching the gelation condition, ultrasonic signals were recorded. Although ultrasonic dispersion induced by absorption and scattering losses caused the signals to have a slight distortion, the time delay between the first echo A_0 reflected from clad Al buffer rod/top surface of laminate interface and the second echo A_2 reflected from the back surface of the laminate can be determined using a cross correlation algorithm. The sharp decay in the time delay, in the temperature range above phase I, could be attributed to the gelation of prepregs. When the progress of chemical crosslinking reaction initiated, a continuous decrease in the time delay was observed. The effects of the state of cure on the time delay can be, therefore, described as a continuous decrease of the time delay as the material changes from a liquid, through a gel, to a solid due to increasing stiffness of composite materials. After cure phase II, the panel started to cool down, and the time delay decreased. After a certain time, a sudden rise in the time delay was found. This discontinuity results from the increase of thickness caused by the sudden detachment of the laminate from the Al mold plate at the sensing region. This interpretation is supported by the observation in Figure 8(a) of a sudden phase change of 180° and an increase in amplitude of the echo reflected from the back surface of the laminate. This increase is confirmed by the change of the interface condition from the laminate/Al mold plate to laminate/air.

Also from Figure 8(b), increasing cure temperature from 146°C to 176°C shifts the curve of the time delay vs. run time to a well distinct lower value, which indicates that ultrasonic in-situ cure monitoring can monitor the change induced by different curing temperatures.

TRANSMISSION/REFLECTION MODE

In the transmission mode, the ultrasonic wave passes through the laminate and is detected by another UT located at the opposite side of the mold. Meanwhile, two UTs can still conduct reflection mode measurements individually. For these measurements of the $[0_{16}]$ panel, the cure time for temperature 176°C was arbitrarily extended from 2 to 3 hrs., in order to observe the end-of-cure. Figure 9 shows the obtained waveforms. In the transmission mode, the first echo A_1 has traveled through the laminate thickness once, and the second echo A_3 has made an additional round-trip. These two echoes can be used to evaluate through-thickness ultrasonic attenuation.

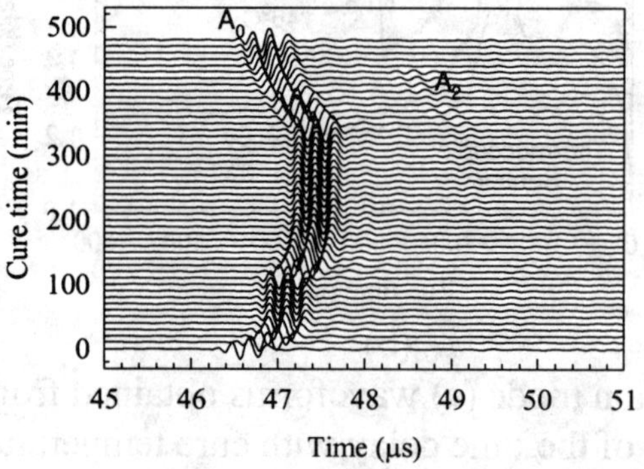

(a)

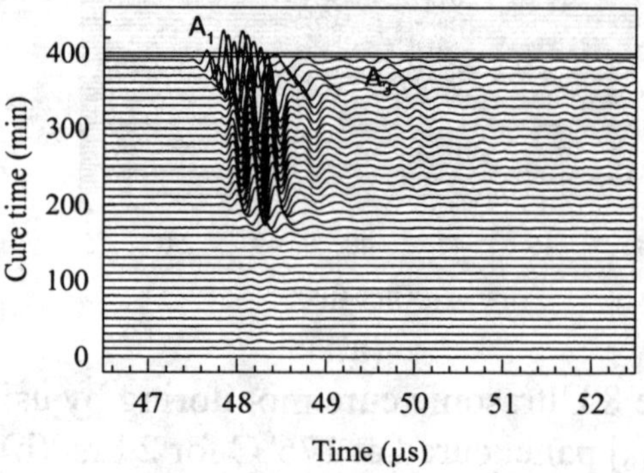

(b)

Figure 9 Ultrasonic waveforms obtained from (a) reflection and (b) transmission mode

The variations of the time delay vs. run time from both top UT and bottom UT exhibit the same trend, as shown in Figure 10. After curing for 2 hrs, the time delay reaches a plateau, which suggests that the end-of-cure has been reached and the extra hour of curing does not make more contribution to the improvement of the stiffness of composites. Here, again, as the laminate cooled down, the time delay decreased. After a certain time, because of detaching of laminate from the top Al mold plate at the sensing region, the signal from the top

UT completely disappeared, while the time delay obtained from the bottom UT showed a discontinuity. For the transmission mode, the shape of the time delay variation is the same as those obtained from reflection mode, as indicated in Figure 10.

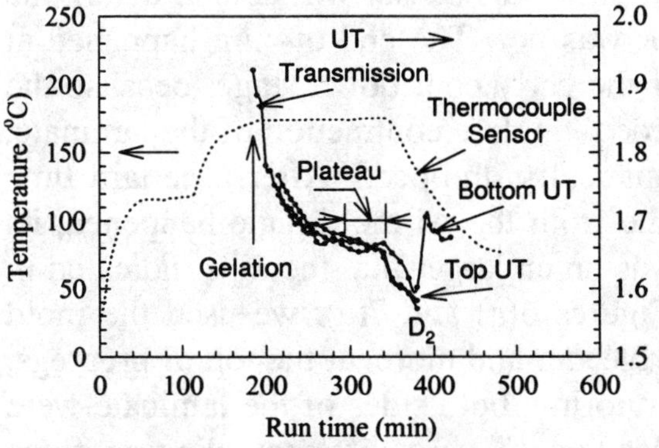

Figure 10 Time delay vs. run time obtained from reflection/transmission mode

Figure 11 presents the variation of ultrasonic attenuation of the prepregs vs. run time during the cure, where ultrasonic attenuation α is defined as $(10/h)\log(A_3/A_1)$. A_1 and A_3 are the amplitude of these pulses indicated in Figure 9(b), and h is the thickness of the laminate. The variation in Figure 11 represents the basic features of viscoelastic behavior of epoxy resin as it cures [20]. First, the ultrasonic attenuation drops dramatically just after gelation, indicating the rapid rise of viscosity due to accelerated crosslinking at the early stage of cure reaction. Then, the attenuation gradually decreases because, as crosslinking proceeds, the material behaves more as an elastic medium, with reduced absorption of ultrasound energy. The plateau is reached when the constant viscoelastic properties of the composite have been developed at the end of cure. Finally, the attenuation further decreases as the material is cooled down. It has been observed by Dynamic Mechanical Analyzer (DMA) measurement that the thermoelastic damping of the cured composite decreases with increasing temperature.[3]

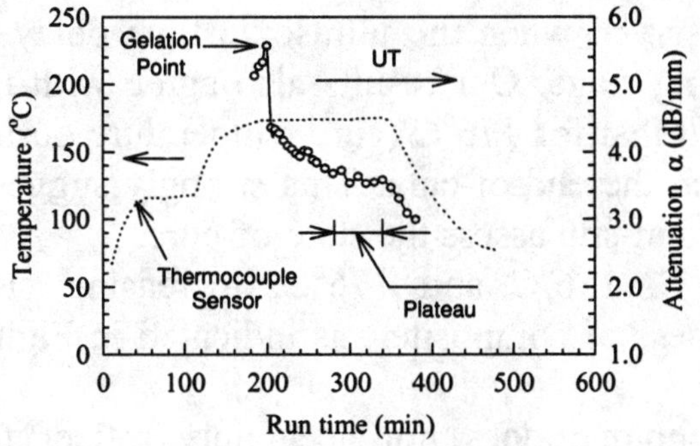

Figure 11 Ultrasonic attenuation vs. run time

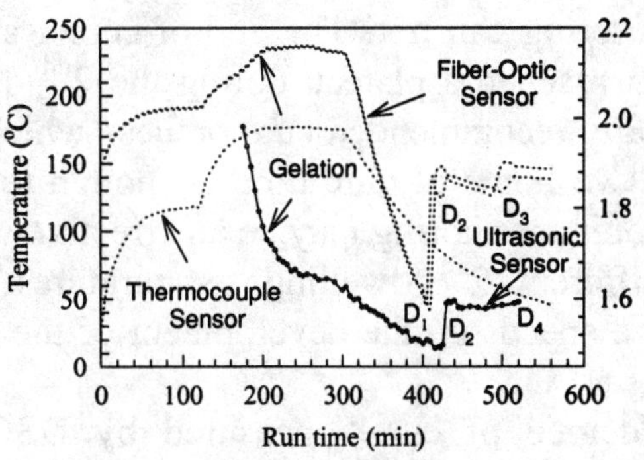

Figure 12 Results of simultaneous fiber-optic and ultrasonic cure monitoring

SIMULTANEOUS FIBER-OPTIC AND ULTRASONIC MEASUREMENT

In order to compare the fiber-optic and ultrasonic monitoring capabilities, we have carried out experiments on panel with lay-ups of $[90_6/0_2/SG1, SG2/0_2/90_6]$ in which both of them were operated simultaneously. The locations of the sensing region were different but close to each other. The 2nd dwelling temperature and cure time were 176°C and 2hrs respectively. One result is given in Figure 12. It can be seen that both sensors can indicate the appearance of gelation of the composite. Furthermore, ultrasonic sensor was able to determine the end-of-cure, although the EFPI fiber-optic sensor was not. The end-of-cure happened at the ultrasonic time delay reaching a plateau. During the early cool down stage, because the laminate was attached to the Al mold plate, this impeded the free contraction of the laminate, the thermal strain detected by the EFPI sensor dramatically dropped. After a certain time (D_1), the sudden detaching of one side of the laminate from the Al mold plate happened, its thermal strain sharply bounced back. This dip (D_1) was much larger that the dips indicated in Figures 6(a) and 7(a). It is due to the fact that in Figures 6(a) and 7(a), we used the mold release agent at bottom and accessories like peel ply, bleeder and more, at the top of prepregs, while for simultaneous fiber-optic and ultrasonic monitoring, both sides of the laminate were directly in contact with the two Al mold plates, leading to very different clinging force between the laminate and the mold. Meanwhile, ultrasonic time delay appeared a discontinuity (D_2). Once more, when the detaching of another side of the laminate happened, ultrasonic sensor totally lost its signals (D_4), while the readings from the EFPI fiber-optic sensor experienced another small bouncing (D_3).

CONCLUSION

The use of the EFPI fiber-optic and ultrasonic sensors to monitor the process state of composites was investigated and compared to the DSC monitor. The results have shown that:
(1) Both the EFPI fiber-optic and ultrasonic sensors can sense the gelation of composite, and the detachment of composites from Al mold plate. In particular, the sudden decrease in the ultrasonic attenuation curve gives us a clear indication of the gelation. Also the timing information of the detachment may provide us the information concerning the development of residual stress during the cool-down phase.
(2) In our example, ultrasonic monitoring can sense the end-of-cure, while the EFPI fiber-optic monitoring can not. The end-of-cure was observed when the ultrasonic time delay or attenuation reached a plateau during the 2nd dwelling stage. Our results also agree with the commercially recommended cure process which said that for 176°C (cure temperature during phase II), two hours of cure time is enough to reach the end-of-cure. This strongly suggests that ultrasonic monitoring may be an effective way to in-situ assess the state-of-cure.
(3) For different 2nd dwelling temperatures 146°C, 156°C and 176°C, increasing cure temperature speeds up the development of the stiffness of composites, as indicated in Figure 8(b).
(4) The "degree of cure" measured by DSC technique does not accurately reflect the mechanical property development during cure. Significant changes in the modulus still exist when the "degree of cure" indicated by DSC is almost fully developed.

ACKNOWLEDGMENT

The authors wish to thank FISO Technologies Inc., Quebec, for loaning the WFS-100 Fiber-optic sensors, H.Hebert of IMI, NRC for his cooling system design and S.-S.L.Wen of McGill University, Montreal, for writing LabVIEW® software. The financial support from a strategy and an operating grant of Natural Sciences and Engineering Research Council of Canada is also acknowledged.

REFERENCES

(1) G.C.Sih, "*Towards decision making in knowledge-based systems for composite cure*", in "Advanced technology for design and fabrication of composite materials and structure", G.C.Sih, A.Carpenteri, and C.Surace (eds.), Kluwer Academic Publishers, The Netherlands, 1994:1-14

(2) W.I.Lee, A.C.Loos, and G.S.Springer, "*Heat of reaction, degree of cure, and viscosity of Hercules 3501-6 resin*", J.Composite Materials, Vol.16, 1982:510-520

(3) J.-Y.Chen, *Ph.D dissertation*, Concordia University, 1997

(4) C.M.Lawrence, D.V.Nelson, J.R.Spingarn, and T.E.Bennett, "*Measurement of process-induced strains in composite materials using embedded fiber optic sensors*", SPIE, Vol.2718, 1996:60-68

(5) C. Belleville and G. Duplain, "White-light interferometric multimode fiber-optic strain sensor", Optics Letters, Vol.18(1), 1993:78-80

(6) P.A. Crosby, G.R.Powell, G.F.Fernando, C.M.France, R.C. Spooncer, and D.N. Waters, "*In situ cure monitoring of epoxy resins using optical fiber sensors*", Smart Mater. Struct., Vol.5, 1996: 415-428

(7) J.Dorighi, S.Krishnaswamy, and J.Achenbach, "*A fiber optic ultrasonic system to monitor the cure of epoxy*", Res. Nondestr. Eval., Vol.9, 1997:13-24

(8) S.S.J.Roberts and R.Davidson, "*Cure and fabrication monitoring of composite materials with fiber-optic sensors*", Composites Sci. & Tech., Vol.49, 1993:265-276

(9) R.G.May, J.M. Sanderson, and R.O. Claus, "*Combined fiber optic strain sensor and composite cure monitor for smart structure applications*", SPIE, Vol.2191, 1994:46-57

(10) G.A. Sofer and E.A. Hauser, "*A new tool for determining the stage of polymerisation of thermosetting polymer*", J.Polymer Sci., Vol.8(6), 1952:611-620

(11) T.M.Whitney and R.E.Green, Jr., "*Cure monitoring of carbon epoxy composites: an application of resonant ultrasound spectroscopy*", Ultrasonics, Vol.34, 1996:347-353

(12) W.P.Winfree, and F.R.Parker, "*Measurement of the degree of cure in epoxies with ultrasonic velocity*", Review of progress in QNDE, Vol.5B, 1986:1055-1061

(13) F.R.Parker and W.P.Winfree, "*Acoustic characterization of composite cure*", Review of Progress in QNDE, Vol.5B, 1986:1063-1067

(14) S.R.White and P.T.Mather, "*Ultrasonic and thermal cure monitor of an epoxy resin*", 36th International SAMPE Symposium, April, 1991:15-18

(15) T.Saliba, S.Saliba, J.Lanzafame, and L.Gudeman, "In-situ cure monitoring of advanced composite materials", 37th International SAMPE Symposium, March, 1992:9-12

(16) P.J.Biermann, J.H.Cranmer, C.A.Lebowitz and L.M.Brown, "*End-of-cure sensing using ultrasonics for autoclave fabrication of composites*", SPIE, Vol.2948, Scottsdale, Arizona, 1996: 72-83

(17) S.Chiou, P.Kukuchek, D.Echternach, G.Carman, and L.Lai, "*Processing monitoring of carbon/phenolic composites using smart sensors*", ibid., 1994:338-348

(18) C.-K. Jen and J.-G. Legoux, "*High performance clad metallic buffer rods*", Proc. IEEE Ultrasonics Symp., 96CH35993, Nov., 1996:771-776

(19) S.W.Tsai and H.T. Hahn, "Introduction to composite materials", Technimic, 1980

(20) R.G.May and R.O.Claus, "In-situ fiber-optic sensor for composite cure monitoring through characterization of resin viscoelasticity, " SPIE, Vol.2948, 1996:24-34

Thermal Control System for Thick Composite Laminates Based on Forecasting

D. D. SHIN, L. LAI AND H. T. HAHN

ABSTRACT

An expert-system-based thermal control system for thick composite laminates was developed by incorporating forecasting ability to control the curing process. This ability enables the system to predict future cure conditions at designated forecast time so that the system can respond in real-time to avoid any undesirable conditions that may arise in the future. A 1-dimensional heat transfer model along with kinetic model for Hercules 3501-6 epoxy resin was utilized to determine the temperature profile within the laminate. By incorporating a forecasting ability, the system was able to derive efficient cure cycles without sacrificing the quality of the final part. The system was also able to derive cure cycles for laminate thickness of 3 cm. The quality was ensured by maintaining uniform temperature distribution and minimizing exotherms. As a result, the degree of cure was also uniform through the thickness of the laminate.

INTRODUCTION

An intelligent composite processing using an expert-system-based (ESB) cure control system has been a viable replacement to the manufacturer's recommended cure cycle. The major draw back with the manufacturer's cure cycle is the lack of specific guidelines to cure composite structures with varying geometry, and it does not guarantee quality final parts. An ESB system may devise a different cure path depending on the material property or curing conditions. The advantage of the ESB cure control system is its potential to produce quality composite structures by custom controlling the cure cycle depending on the geometry avoiding any undesirable conditions that may appear during cure.

The first of the ESB cure control systems was known as Quality Process Automation (QPA) which used qualitative knowledge based rules to control the curing process (LeClair and Abrams, 1988; Manzini and Roehl, 1990). The generic knowledge based rules were written to accommodate different resins systems such as 1034C and Hercules 3501-6. The basic control strategy of this system was to compact and cure the laminate as rapidly as possible using the data collected from thermocouples and dielectric sensor. Their system was

Daniel Shin and H. Thomas Hahn, Mechanical and Aerospace Engineering Department, School of Engineering and Applied Science, University of California Los Angeles, CA 90095-1597
Larry Lai, Hughes Space and Communications, P.O. Box 92919, Los Angeles, CA, 90009

able to cure 256 ply laminate using 1034C based resin while handling only 56 ply thick laminate for 3501-6 resin. Although improvements in cure time were observed, the system was only capable of making decisions based on the immediate and past time frames.

Others such as Johnson and Roberts have relied heavily on dielectric sensors to control the cure process (Johnson and Roberts, 1989). Their control strategy was based on an event based method which adapted cure cycles to account for rheological variations observed by the system rather than forcing part rheology to strictly conform to a predetermined profile. Their system separated the cure process into three separate time frames: minimum viscosity, gelation, and end-of cure. The time events were recognized by determining the first and the second derivative of the ionic conductivity measured using dielectric sensors. The maximum temperature rate was applied throughout the cure cycle to reduce the cure time. This system was ideal for curing moderately thin polymide based laminates.

Sensor based cure control systems did not always provide adequate control for the cure process. To accommodate this deficiency, others have tried to incorporate mathematical models to accurately evaluate the cure process. Springer and Ciriscioli were among the first to develop an off-line system to evaluate its cure control performance which was known as SIMULATE (Ciriscioli, Springer, and Lee, 1991). The core of this system was based on rule based expert system which maintained the following: reasonably uniform temperature distribution, minimal exotherm, fully compacted laminate, minimal void content, complete and uniform cure reached in shortest time, and reduced residual stress. In particular, the exotherm was minimized by decreasing the autoclave temperature if the acceleration of the mid-temperature became greater than zero. The online system known as SECURE used sensors to collect data on pressure, temperature, and ionic conductivity of the resin. The system was able to minimize exotherms for laminates fabricated up to 1000 plies for Fiberite T300/976.

Recent development in knowledge based cure control system was made by Pillai et. al. (Pillai et al., 1994). Their system derived an optimal off-line cure cycle using an algorithm called Local Criterion Optimization. The derived cure cycle was then applied on-line by process trend analysis where the cure cycle was treated as a collection of desired trends rather than time-stamped data. The system was tested with 2.54 cm thick CYCOM-4102 glass polyester prepreg. The control strategy was to force the derived off-line cure cycle on-line. Although their on-line system was able to follow the derived cure cycles, discrepancies existed between the temperature profiles obtained from the simulation and the experiment.

Significant strides have been made to develop better intelligent cure control systems, yet no system has the ability to forecast cure conditions on-line. It is clear that using future cure conditions to control the curing process would be advantageous since it may prevent any undesired cure conditions before they occur. The objective of this research is to develop a cure strategy for an expert-system-based thermal control system for thick composites laminates. It is impractical to use optimization methods on-line to minimize cure time due to prohibitively long computational time. However, forecasting approach improves cure control by providing predictions which can be used in combination with sensor data. The quality is ensured by maintaining uniform temperature distribution and minimizing exotherms.

THERMAL CONTROL SYSTEM

The implementation of the entire control operation is based on EXSYS Professional©, an expert system development package. The expert system serves as a tool for the closed loop control process. It also collects necessary data which it passes to the decision maker. The decision maker is a central brain of the expert system which consists of an expert system shell and knowledge based rules. The expert system follows appropriate rules and resolves any conflicting decisions for proper control action. It consists of three main functional groups: present state simulation model, future state simulation model, and an expert system consisting of knowledge based rules and a command language that controls the entire process.

Simulation

The state of the internal laminate is determined by simulating existing thermal and cure kinetic models. The 1-dimensional thermal behavior of the composite laminate has been modeled using conductive heat transfer through the thickness with internal heat generation

$$\frac{\partial T}{\partial t} = \left(\frac{k}{\rho C_p}\right)\frac{\partial^2 T}{\partial x^2} + \left(\frac{H_R \rho_m (1 - V_f)}{\rho C_p}\right)\frac{d\alpha}{dt} \tag{1}$$

where the parameters k, C_p, ρ_m, and ρ are thermal conductivity, heat capacity, resin density, and composite density, respectively (Loos and Springer, 1983). The simulation is based on hot press curing where excess resin from the laminate is squeezed out from the side. Since bleeders are not applied on top of the laminate and the bagging materials are relatively thin, it is assumed that no heat is lost between the platen and the laminate surface. The consolidation is assumed to be uniform throughout the cure cycle, and a constant fiber volume fraction is used. The density of the fiber is determined by the relation

$$\rho = v_f \rho_f + (1 - v_f)\, \rho_m \tag{2}$$

where ρ is the density of the composite, ρ_f is the density of the fiber, ρ_m is the density of the resin, and v_f is the fiber volume fraction. The thermal conductivity of the composite is determined by

$$k = \frac{k_f k_m}{k_m v_f + k_m (1 - v_f)} \tag{3}$$

where k_m and k_f are resin and fiber conductivity, respectively. The specific heat of the composite is

$$C_p = \frac{C_f \rho_f v_f + C_m \rho_m (1 - v_f)}{\rho} \tag{4}$$

where C_f and C_m are specific heat of the fiber and resin, respectively.

The degree of cure and viscosity are based on Hercules 3501-6 epoxy resin. The kinetic equations for this resin system is based on the work by Lee, et. al. (Lee, Loos, and Springer, 1982).

$$K_1 = A_1\, exp\left(\frac{-E_1}{RT}\right)$$
$$K_2 = A_2\, exp\left(\frac{-E_2}{RT}\right)$$
$$K_3 = A_3\, exp\left(\frac{-E_3}{RT}\right)$$

$$\frac{d\alpha}{dt} = (K_1 + K_2\alpha)(1-\alpha)(B-\alpha) \qquad \alpha \le 0.3$$
$$\frac{d\alpha}{dt} = K_3(1-\alpha) \qquad \alpha > 0.3 \qquad (5)$$

where α is degree of cure. The variable K_x is obtained by an Arrhenius type equation derived from the relationship between gel time and temperature. Table 1 shows kinetic parameters used for the above equations.

Forecasting Strategy

The forecasting capability foresees the future cure condition at a specified forecast time so that necessary measures can be implemented by the system to avoid unwanted conditions before they occur. Forecasting is performed at each time interval using a simulation model. The forecasting simulation model is an exact replica of the present state simulation model with an exception of applying different boundary conditions. The boundary condition at the forecast time is determined by projecting the temperature ramp rate prescribed by the knowledge base to that time, Figure 1. The initial thermal and degree of cure distribution necessary to solve the forecast model are the last values calculated by the present state simulation. Thus the forecast model picks up where the present state simulation left off.

A restriction is imposed on the boundary condition so that the forecast surface temperature never exceeds the prescribed maximum dwell temperature. If in any case the surface temperature reaches the dwell temperature before the forecast time is reached, then it remains constant at the dwell temperature. Often, the middle temperature exceeds the dwell temperature while the forecasting is in progress. The forecasting simulation is programmed so that it terminates when the maximum over shoot is reached even when the prescribed forecast time is not. The maximum over shoot is the peak temperature reached at the middle of the laminate before it starts to decrease, Figure 1. A similar restriction is imposed for temperature overshoot occurring under the isothermal dwell temperature. If the forecast middle temperature exceeds the surface temperature by more than 11°C then the forecasting simulation terminates. This is important since the expert system does not consider the temperature history but considers only the final temperatures provided by the forecast simulation model.

Knowledge Base Expert System

Episodes

To minimize conflicts among rules, the EXSYS shell allows the operator to define command language to control the execution of knowledge based rules. The command language allows the operator to combine knowledge based rules under various episodes. This system categorizes the entire knowledge based rules into two episodes. All the variables used in the episodes and rules are defined in Appendix 1. The knowledge base is defined such that the variables starting with prefix Temp- or Alpha- corresponds to present time state. The variables starting with prefix T- or A- corresponds to future time state. The first episode is defined as

Episode #1
IF [Alphadelta] < 0.02 & [Tmid] < [Isotemp] Then Rules 1-2

If the conditions of the first episode are met, then rules 1 and 2 are applied. The primary concern of this episode is to reach high temperature quickly yet trying to maintain reasonably uniform temperature throughout the initial stage of the cure cycle. The concern for temperature overshoot is ignored since it is unlikely to occur when the change in degree of cure is less than 2%. The second episode is defined as

Episode #2
IF [Alphadelta] ≥ 0.02 Then Rules 3-6

In this episode, the temperature is assumed to have reached reasonably high enough to cause possible temperature over shoots. Thus the primary concern for this episode is to eliminate any potential exotherms by applying rules 3 to 6.

Rules

The knowledge based rules provide important curing actions. The rules were derived so that they would be as general as possible. The majority of the rules are based on temperature since the main objective is to minimize potential exotherms and to maintain uniform temperature distribution so that uniform degree of cure can be achieved. Furthermore, the rules do not deal with variables regarding pressure and viscosity. Since the material is an epoxy based resin, it does not produce any byproducts during cure nor any evidence exists that proves applying higher pressure at low viscosity produces higher consolidated laminates. The variables that must be assigned prior to the start of the simulation are thickness of the laminate, temperature ramp rate, maximum isothermal dwell temperature, and forecasting time. Once these variables are defined, the system derives custom cure cycles pertaining to given constraints. However, certain aspect of the manufacturer's recommended cure cycle is maintained such as 177°C the maximum isothermal dwell temperature.

The purpose of Rule 1 is to reach high temperature as quickly as possible.

RULE: 1
IF: [Alphasurf] < 0.02 & [Tempmid] < [Isotemp]
THEN: [Delta] is given the value [Ramp]

Rule 2 pertains to the middle temperature lag with respect to the surface. Although the system tries to maintain uniform temperature by decreasing the temperature ramp rate, no attempts are made to strictly enforce the uniform temperature at the first episode.

RULE: 2
IF: [Tempsurf]-[Tempmid] > 11

THEN: [Delta] is given the value [Ramp]/2

Rule 3 deals with middle temperature lag and overshoot. If overshoot is detected then heat is not applied.

RULE: 3
IF: [Tempsurf] - [Tempmid] < 11 & [Tempsurf] > [Tempmid]
THEN: [Delta] is given the value [Ramp]
ELSE: [Delta] is given the value 0

Rule 4 prevents to minimize potential temperature overshoots by observing the forecast middle temperature.

RULE: 4
IF: [Tmid] - [Tsurf] < 11 & [Tempmid] - [Tempsurf] < 11 & [Tmid] < [Isotemp]
THEN: [DELTA] is given the value [Ramp]

Rule 5 compares the rate of the forecast middle temperature to the prescribed maximum temperature ramp rate.

RULE: 5
IF: ([Tmid] - [Tempmid])/[ET] > [Ramp]
THEN: [Delta] is given the value 0

Rule 6 prevents any temperature from exceeding the prescribed dwell temperature. Allowing the surface temperature to exceed the dwell temperature would result in possible exotherms.

RULE: 6
IF: [Tmid] > [Isotemp] OR [Tempsurf] ≥ [Isotemp]
THEN: [Delta] is given the value 0

This end of cure cycle is reached when the degree of cure throughout the laminate becomes 99% or higher. Usually full degree of cure is reached at the middle of the laminate first because of the excess heat generation in the middle.

RESULTS AND DISCUSSION

The cure simulations were performed using manufacturer's cure cycles for a 3 cm thick laminate, Figure 2. This cure cycle was used as a base-line to compare cure cycles derived using the thermal control system.

The conditions used to derive cure cycles for the thermal control system were 4.44°C/min at 10 minute forecast time and 2.22°C/min at 5 minute forecast time, Figures 3 and 4. The system virtually eliminated exotherms while maintaining temperature difference below the allowable limit for both conditions. Furthermore, the derived cure cycle showed a drastic reduction in cure time. A total of 77 minutes were reduced from the manufacturer's cure cycle while maintaining uniform temperature gradient and degree of cure, Figure 4. Although faster than 4.44°C/min could had been employed by the simulation, this value was chosen due to the limitations of the hot press. The figures also indicate that longer forecast time was needed when faster temperature ramp was prescribed.

CONCLUSION

The cure cycles derived from the thermal control system were clearly an improvement over the manufacturer's suggested cure cycle. Any exotherms observed from the manufacturer's cure cycle were virtually eliminated using the system, and shorter cure time was observed for 3 cm thick laminates. The final degree of cure throughout the laminate was also uniform and reached 99%. The forecasting capability of the system was proven to be very effective.

ACKNOWLEDGMENT

The present paper is based on work supported by the Office of Naval Research under Grant N00014-92-J-1846. Sincere appreciation is extended to Yapa D. S. Rajapakse, Scientific Officer. Appreciation is also extended to Evan Kempner for assisting with the simulation code.

REFERENCES

Ciriscioli, P.R., Springer, G.S., and Lee, W.I., "An Expert System for Autocalve Curing of Composites", *Journal of Composite Materials*, Vol. 25, December 1991, pp. 1542-1587.

Hojjati, M. and Hoa, S.V., "Curing Simulation of Thick Thermosetting Composites", *Composites Manufacturing*, vol. 5, No. 3, 1994, pp. 159-169.

Johnson, S.A. and Roberts, N.K., "Production Implementaion of Fully Automated, Closed Loop Cure Control for Advanced Composite Structures", *34th International SAMPE Symposium*, May 8-11, 1989, pp. 373-384.

LeClair, S.R., and Abrams, F.L., "Qualitative Process Automation", *Proceedings of the 27th Conference on Decision and Control*, December, 1988, pp. 558-562.

Lee, W.I., Loos, A.C., and Springer, G.S., "Heat of Reaction, Degree of Cure, and Viscosity of Hercules 3501-6 Resin", *Journal of Composite Materials*, vol. 12, November 1982, pp. 510-520.

Loos, A.C. and Springer, G.S., "Curing of Epoxy Matrix Composites", *Journal of Composite Materials*, vol. 17, no. 2, 1983, pp.135-169.

Manzini, R.A. and Roehl, E.A., "Flexible Control of an Organic Matrix Composite Cure Process Using Object-Oriented Control Concept", *American Control Conference Proceeding,* 1990, pp. 1980-1985.

Pillai, et.al., "Implementation of Model-Based Optimal Temperature Profiles for Autoclave Curing of Composites Using a Knowledge-Based System", *Industrial Engineering and Chemistry Research*, vol.33, 1994, pp. 2243-2452.

Saliba, T.E., Quinter, S.R., and Abrams, F.L., "Expert Model Process Control of Composite Materials in a Press", *36th International SAMPE Symposium*, April 15-18, 1992, pp.1487-1497.

Table 1a. Kinetic parameters for epoxy resin [Lee, Loos, and Springer, 1982]

Heat of reaction (J/g) H_R	473.6
B	0.47
Pre-exponential factor (min^{-1}) A_1	$2.101x10^9$
Pre-exponential factor (min^{-1}) A_2	$-2.014x10^9$
Pre-exponential factor (min^{-1}) A_3	$1.960x10^5$
Activation energy (J/mol) E_1	80700
Activation energy (J/mol) E_2	77800
Activation energy (J/mol) E_3	56600

Table 1b. Prepreg properties for Hercules AS4/3501-6 [Hojjati and Hoa, 1994]

Resin density (g/cm^3) ρ_m	1.26
Fiber density (g/cm^3) ρ_f	1.79
Specific heat of resin ($J/cm^{3o}K$) C_m	1.5876
Specific heat of fiber ($J/cm^{3o}K$) C_f	1.27448
Thermal conductivity of resin (W/cm^oK) k_m	0.00167
Thermal conductivity of fiber (W/cm^oK) k_f	0.26

Appendix 1

Variable Definitions of Knowledge Based Rules:	
[Tempsurf]	Present state surface temperature
[Tempmid]	Present state middle temperature
[Tsurf]	Forecast surface temperature
[Tmid]	Forecast middle temperature
[Alphasurf]	Simulated degree of cure at surface
[Alphamid]	Simulated degree of cure at middle of laminate
[Delta]	Temperature ramp rate into the simulation

Input Parameters:	
[Thick]	Thickness of the laminate
[Ramp]	Temperature ramp rate
[ET]	Forecast time interval
[Isotemp]	Isothermal cure temperature

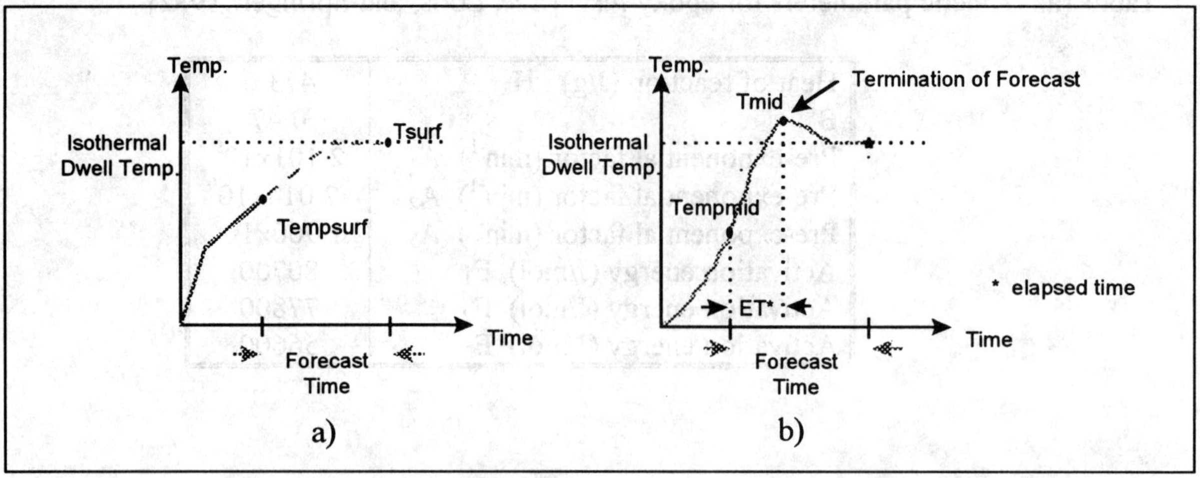

Figure 1. a) Once forecast surface temperature reaches isothermaldwell temperature, it remains constant until the end of forecast time. b) Forecasting simulation terminates upon maximum temperature observed at the middle of the laminate regardless of complete forecast time reached.

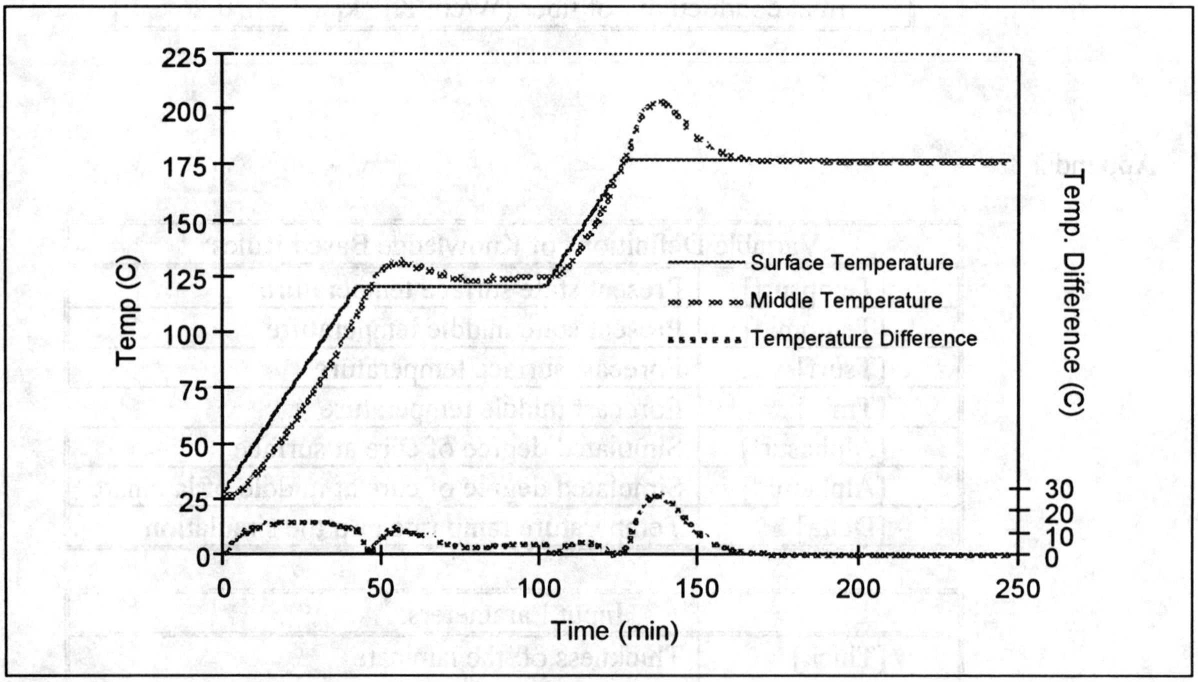

Figure 2. Manufacturer's cure cycle for 3 cm thick laminate. Cure time was 247 minutes and maximum temperature reached upto 204 °C.

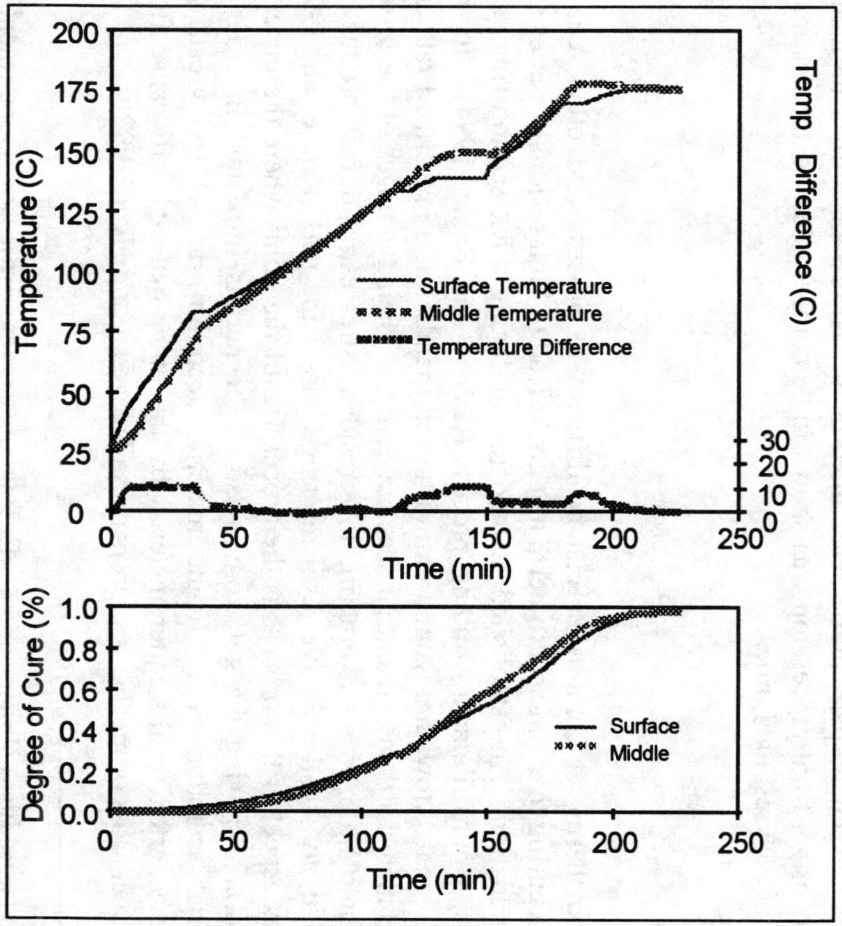

Figure 3. Cure cycle derived for 3 cm thick laminate using thermal control system. Cure time was 226 minutes and maximum temperature reached was 179 °C. 5 minute forecast time and 2.22 °C/min ramp rate are applied. Uniform degree of cure reached.

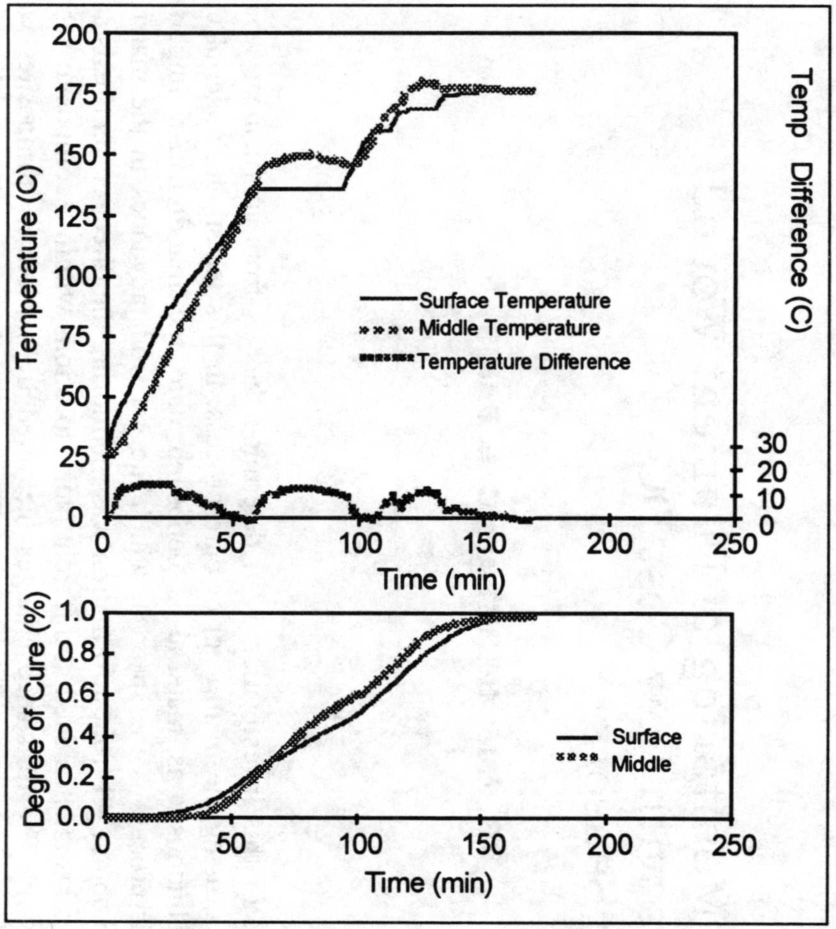

Figure 4. Cure cycle derived for 3 cm thick laminate using thermal control system. Cure time was 170 minutes and maximum temperature reached was 180 °C. 10 minute forecast time and 4.44 °C/min ramp rate are applied. Uniform degree of cure reached.

In-situ Gel Monitoring of Filament Wound Composites with Low Cost DC Resistance Sensors

G. P. McKNIGHT AND C. E. BAKIS

ABSTRACT

The purpose of this investigation is to determine the gel front in filament wound composites produced using the *in-situ* gel or cure process. In this technique, material cure state increases continuously as material is deposited onto the mandrel. An important characteristic of this process is the speed at which the gel front advances in the material. The method used to obtain gel front information was monitoring the direct current resistance of several low cost sensors embedded at different radial locations within a composite during the winding process. The resistance changes measured by sensors in composites were compared with similar measurements in neat resin. The results of this investigation show that the gel front in a filament wound part can be determined using DC resistance measurements. Furthermore, the rate of gel front procession was shown to be comparable to estimates made from measurements of neat resin gel times.

INTRODUCTION

Thick section thermoset composites can be difficult to process and often require lengthy and complex curing to avoid an exotherm that could lead to high residual stresses and possible matrix degradation. An established method for producing thick section rings and cylinders is the *in-situ* cure filament winding process (Gabrys and Bakis, 1994). In this technique resin-impregnated tows are wound onto a heated mandrel at a prescribed rate and tension. The wound material then cures at an accelerated rate. The advantages of the *in-situ* cure technique include little or no exotherm, low residual stresses, and fast production resulting from a relatively small effective resin reacting zone in the part (Kim et al., 1995, 1996). An important aspect of this process is the procession of the point where the material has gelled (gel front). Knowing the gel front speed in the part is important in order to optimize the material accretion rate. If the material accretion rate is too great, the compressive stresses exerted on the inner material by the outer material will cause fiber waviness and subsequent loss of mechanical properties (Telegadas and Hyer, 1992).

G. P. McKnight, undergraduate student and C. E. Bakis, associate professor, Department of Engineering Science and Mechanics, 227 Hammond Bldg., The Pennsylvania State University, University Park, PA 16802

Additionally, an excessive material accretion rate could lead to an exotherm within the part and loss of mechanical properties due to matrix degradation and high residual stresses. Economy dictates that parts be made as quickly as possible and thus the accretion rate should be as fast a possible while still producing high quality parts. Previous estimates of the gel front have been made assuming the resin on the mandrel will behave as neat resin would at the same temperature (Gabrys and Bakis, 1994).

The objective of this study is to determine the rate of gelation in a filament wound part using DC resistance measurements. Previous work has shown that DC resistance measurements provide similar information on gelation as the relatively costly dielectric techniques (Lee and Rice, 1996). The technique employed here involves placing a voltage across a pair of electrodes embedded within a filament wound part and monitoring the resistance between the electrodes. As the cure state in the resin progresses, sensor resistance increases in concert with the bulk resistivity of the resin. A specific increase in resistance can be related to gelation of the resin and thus the time to gelation for specific radial position can be determined. Placing sensors at different radial positions enables the monitoring of gel point at different positions within the part and thus create a profile of gel point versus time in the part.

EXPERIMENTAL METHODS

The procedures given below are separated into two distinct areas: using DC resistance measurements to determine gelation of neat epoxy resin and using DC resistance sensors as *in-situ* gel monitoring sensors for the fiber reinforced epoxy filament winding process.

USING DC RESISTANCE MEASUREMENTS TO DETERMINE GELATION

Sensor Design

In order to correlate DC resistance with gelation, resistance and gel point of neat resin at different isothermal temperatures were recorded by a series of experiments. Resistance sensors, designed for use in both neat resin as well as in a filament wound part, were fabricated by hand and consisted of two 5-mm-square electrodes made from 50-μm-thick copper foil. Polyimide-coated copper lead wires were soldered to the back of the plates for electrical signal transmission. The electrodes were then separated by wrapping the each one with a 0.254-mm-thick glass fiber mat (0.0035 g/cm^2 areal density). The fabric served both to insulate the electrodes and provide a porous, rigid, separation layer between the electrodes that would allow resin infiltration yet maintain a constant distance between the plates. The total sensor thickness including the glass overwrap was 2.54 mm. The sensor design is shown in Figure 1.

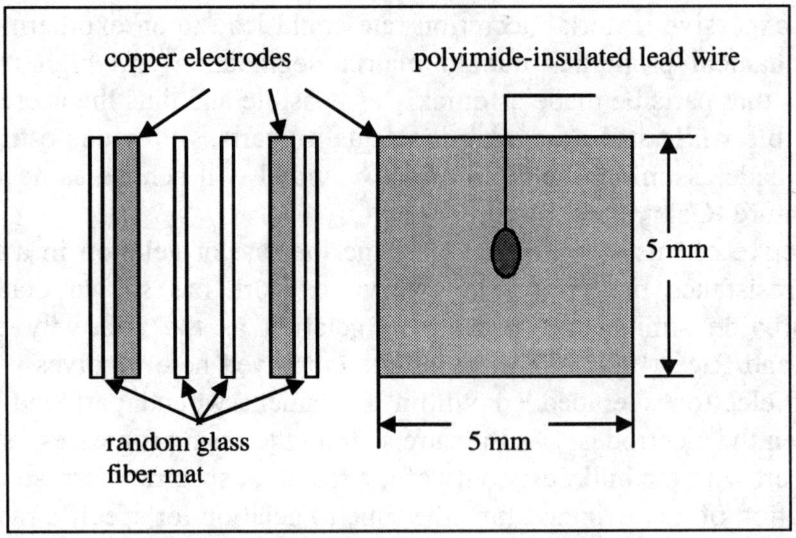

Figure 1. DC resistance sensor design.

Data Acquisition

Initial experiments determined the requirements of the data acquisition system designed for this study. The epoxy resin used in the experiments had a very high resistivity (sensor resistance of 1 MΩ and higher) and temperature dependency. In addition it was found to be necessary to apply a voltage to the electrodes for only short periods of time to avoid an upward drift of resistance with time. Thus the system had to be capable of switching a voltage on and off, recording the temperature along with DC resistance at every sensor position, and filtering out unwanted noise in the signal encountered due to the high resistance measured.

The data acquisition system used a microcomputer with a National Instruments AT-MIO 16-E data acquisition (DAQ) card and a program written in LabView™. Voltage across a dummy resistor in a potentiometer circuit was recorded and used to calculate the resistance of the sensor. The value of the dummy resistor used in these experiments was 1 MΩ and was selected for the highest sensitivity possible over the entire range of cure states measured. The temperature at each sensor was recorded by the DAQ card through the use of an Analog Devices AD594 thermocouple cold junction compensator and amplification integrated circuit. The system required precise timing due to the transient nature of the measured resistance. In addition it was necessary to sample and average the resistance and temperature data to increase the signal to noise ratio (S/N). The entire system used shielded cables tied to their own electrical ground to further improve the S/N. The timing details included a 100-ms delay between voltage turn-on and a 30-sample scan of the voltage across the dummy resistor at 150 Hz. The voltage was then turned off and the thermocouple voltage was scanned 30 times at 1000 Hz. The temperature was calculated and then there was a one second delay before the next sensor was scanned.

Characterization of Epoxy with Isothermal Neat Resin Experiments

Experiments to determine the relationship between resin resistivity and gelation were performed using an aluminum mold shown in Figure 2. The 2.54-cm-thick upper plate contained six 2.25-cm-diameter through-holes that were taped shut to hold 2 ml of resin each. Each of the six resin wells were covered using an aluminum foil coated, wooden lid. The upper plate was then clamped to the lower 0.65-cm-thick plate and both were heated by a 500-W ceramic heater bolted to the lower plate. Three of the resin wells contained resistance sensors and thermocouples. The remaining three wells were used to determine the gel point of the resin. The resin was considered gelled at the time at which a wooden probe (4.76 mm X 0.80 mm X 11.4 cm) could be inserted into the resin and pulled out with no resin adhering to the probe. The resistance values recorded by the three sensors placed in the wells were averaged at the time when the other three wells had reached gelation. These three wells reached gelation within a few minutes of each other, and an average of these times was used for an overall gel time. The change in resistance thus measured was considered valid for the particular temperature used in the experiment. The temperature control of the system was accurate to ±2°C.

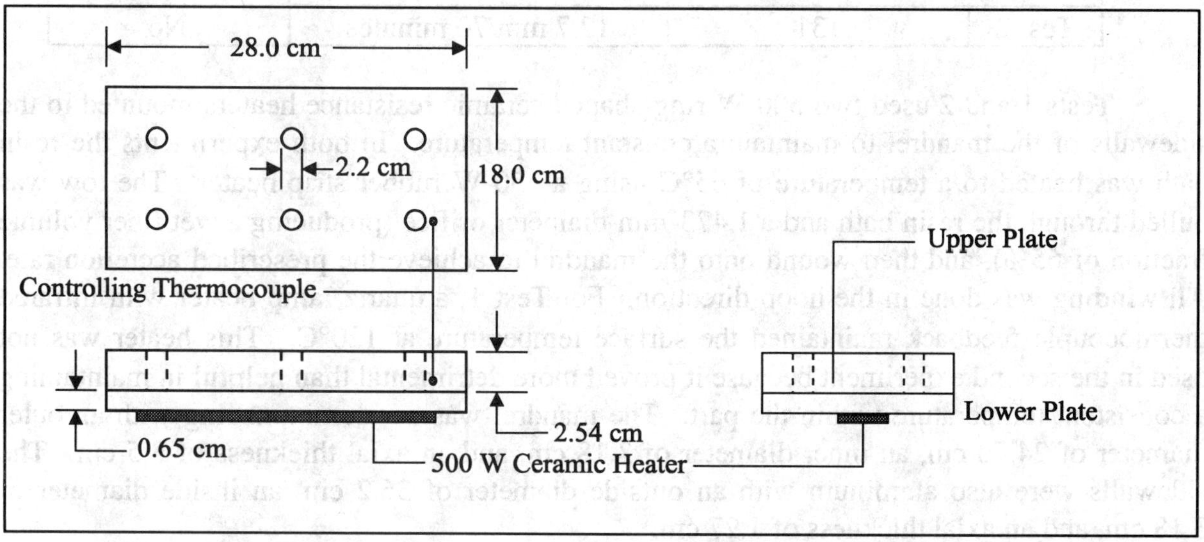

Figure 2. Schematic illustration of heated mold used for neat resin characterization

To make comparisons between experiments with different temperatures and sensor geometries, two scaling operations were performed. The first was to normalize the resistance values recorded just after mixing to unity and reference all subsequent values to this beginning resistance, which accounted for slight variations in sensor geometry. The second was a temperature compensation operation done to compare experiments at different temperatures. The relationship between temperature and resistance for freshly mixed resin was determined at approximately zero degree of cure, $\alpha = 0$, by quickly ramping down the temperature of the resin batch while recording the resistance. This was found to be linear within the region of interest for this resin system. This information was then used to reference all values to a common starting temperature, 120°C, for the resin system used in this study. Thus, all resistance values were referenced to a common condition of $\alpha = 0$ and T = 120°C.

FILAMENT WINDING

The filament winding done in these experiments very closely follows the procedures given by Gabrys and Bakis (1994). The filament winder used for these experiments was a McClean Anderson 3-axis winder. The materials used were Owens Corning 1998 tex type 30 E-glass filament tows and Shell Epon 9404 epoxy resin and 9470 hardening agent mixed in a 100:34 ratio by weight. In addition 0.3% by volume BYK 501 air release agent was added to decrease foaming in the resin bath. Two separate experiments at different operating temperatures and material accretion rates were performed, henceforward referred to as Test 1 and Test 2. The details of these experiments are shown in Table I.

Table I. Details of filament winding experiments.

Test	Mandrel Temp. (°C)	Material Accretion Rate	Surface Heater
Test 1	120	12.7 mm/60 minutes	Yes
Test 2	131	12.7 mm/70 minutes	No

Tests 1 and 2 used two 500-W ring-shaped ceramic resistance heaters mounted to the sidewalls of the mandrel to maintain a constant temperature. In both experiments the resin bath was heated to a temperature of 65°C using a 150-W rubber strip heater. The tow was pulled through the resin bath and a 1.473-mm-diameter orifice (producing a wet fiber volume fraction of 65%), and then wound onto the mandrel to achieve the prescribed accretion rate. All winding was done in the hoop direction. For Test 1, a quartz lamp heater with infrared thermocouple feedback maintained the surface temperature at 120°C. This heater was not used in the second experiment because it proved more detrimental than helpful in maintaining a consistent temperature within the part. The mandrel was an aluminum ring with an outer diameter of 24.76 cm, an inner diameter of 3.18 cm, and an axial thickness of 3.5 cm. The sidewalls were also aluminum with an outside diameter of 36.2 cm, an inside diameter of 3.18 cm, and an axial thickness of 1.27 cm.

SENSOR INSERTION

Six resistance sensors and thermocouples were inserted into the part starting at the mandrel surface and then every 6.4 mm thereafter. The sensors were inserted at evenly spaced angular positions to minimize disturbances to the thermal kinetics of the system. In addition, 36-gage type-J thermocouple wire was used to provide as little disturbance to the system as possible. The resistance sensors and thermocouples were placed into the part by stopping the mandrel and placing the sensor under the filament tow and then continuing winding. The pressure of the tow was sufficient to keep the sensor and thermocouple firmly in place. The lead wires for the sensor and thermocouple were then inserted through a hole in the side wall and connected to a sixteen-conductor IER VL16R slip ring which was, in turn, connected to the power and data acquisition electronics. The overall equipment setup is shown in Figure 3. The sensors required 2-3 minutes to insert into the part and connect to the slip ring terminals. In all, sensors were placed at radial positions of approximately 2.5, 6, 12, 19, 25, and 32 mm as measured from the mandrel surface. After the sensor located at 32 mm was inserted, another 7 mm of material was added and then the entire part was allowed to cure while rotating until the outer surface appeared gelled.

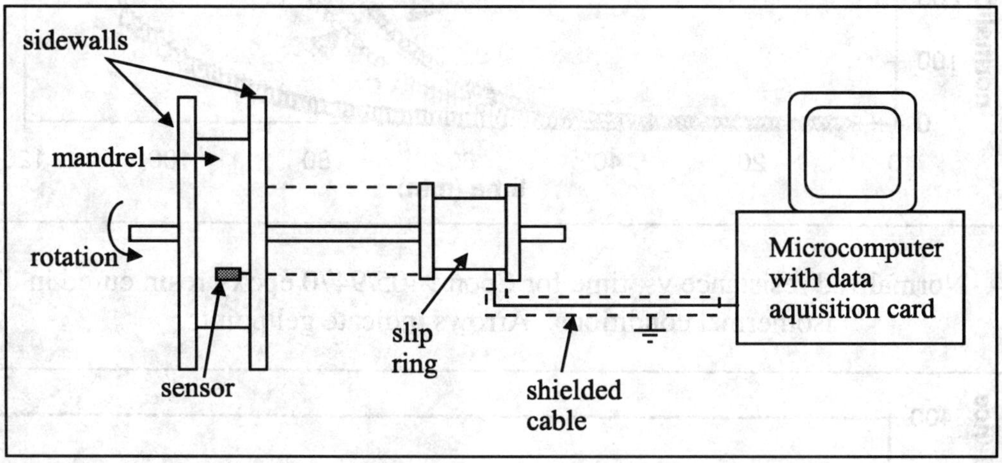

Figure 3. Schematic illustration of overall filament winding and electronics setup.

RESULTS AND DISCUSSION

USING DC RESISTANCE MEASUREMENTS TO DETERMINE GELATION

The resin characterization tests proved that DC resistance measurements could be used as a measurement of gelation for the Epon 9405/9470 epoxy system studied. Experiments revealed that there was a consistent increase in resistance of over two orders of magnitude accompanying gelation. The actual sensor DC resistance measured varied widely with temperature but in general started near 1 MΩ for freshly mixed resin and rose to over 500 MΩ at cure states past gelation. Three characterization experiments were run in isothermal conditions at 112, 124, and 130°C. The results of these tests were normalized and temperature compensated as described previously to reference all trials to a common condition of $\alpha = 0$, and T = 120°C. The average of the three resistance sensors placed in the aluminum mold at each of the three different temperatures is shown in Figure 4. The change in normalized resistance measured at gelation for each of these trials is presented in Figure 5.

As can be noted from Figure 4, the maximum sensitivity of the technique is at points near gelation, where the sensor resistance climbs exponentially. This indicates that DC resistance measurements may be best used as a gel point monitor. Other parts of the cure cycle do exhibit a change in resistance but not nearly to the extent observed near gelation.

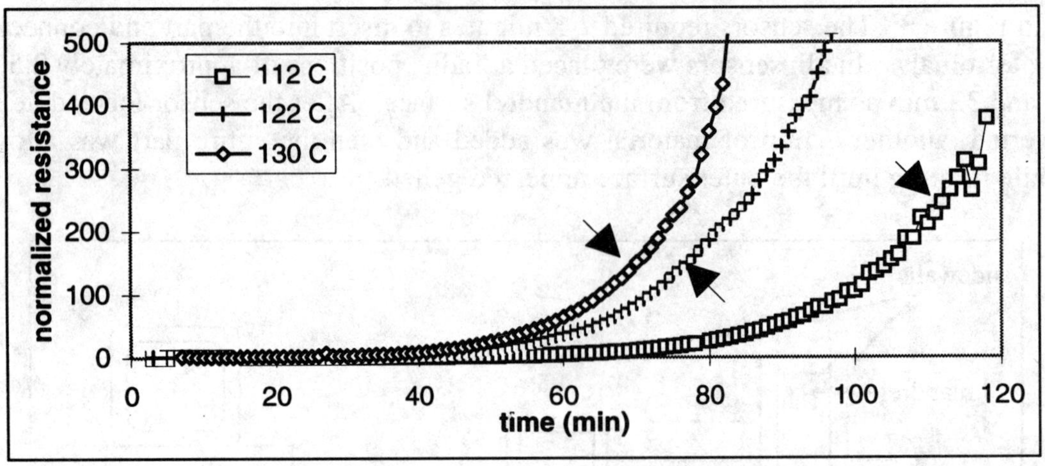

Figure 4. Normalized resistance vs. time for Epon 9405/9470 epoxy resin cured in different isothermal conditions. Arrows indicate gel points.

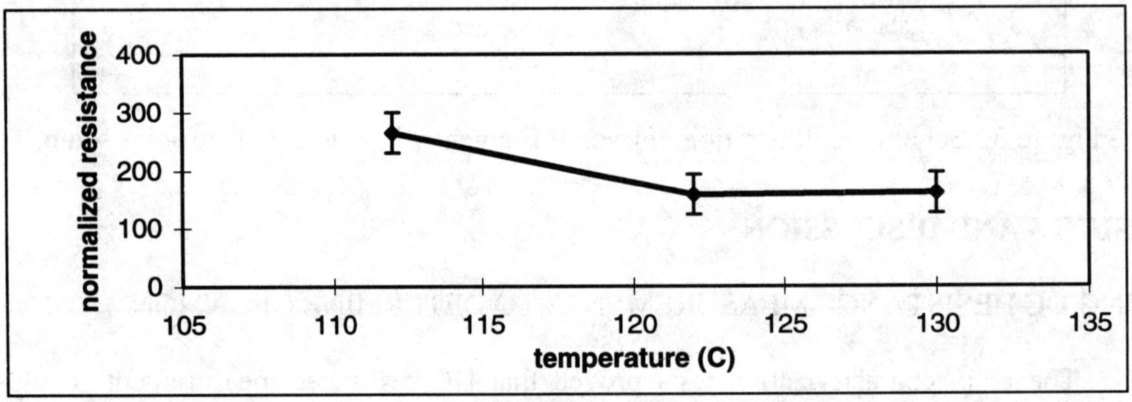

Figure 5. Change in normalized resistance accompanying gelation at different temperatures. Error bars reflect difficulty in assigning exact gelation times as well as variation within individual trials.

IN-SITU GEL MONITORING IN FILAMENT WINDING

Two separate experiments were conducted to determine the gel front characteristics within a filament wound part using the *in-situ* cure method. The results of these experiments indicate that DC resistance measurements can be used as an *in-situ* gel monitor for the filament winding process. The resistance recorded at each of the sensors in these tests were normalized and temperature compensated at a time just after insertion as was done for the neat resin testing. The normalized, temperature compensated resistance values for embedded sensors in each of the experiments is graphed against time in Figures 6 and 7.

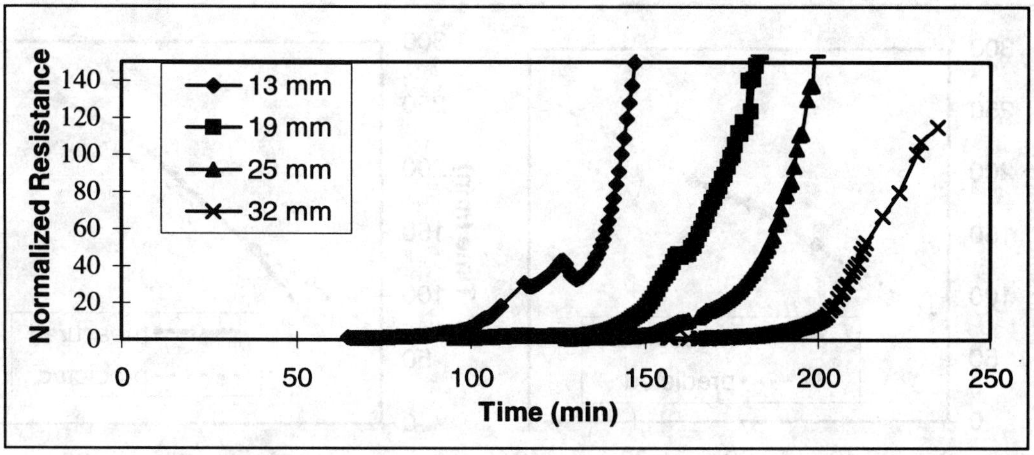

Figure 6. Normalized, temperature-compensated sensor resistance vs. time for sensors at different radial positions measured from the mandrel surface in Test 1.

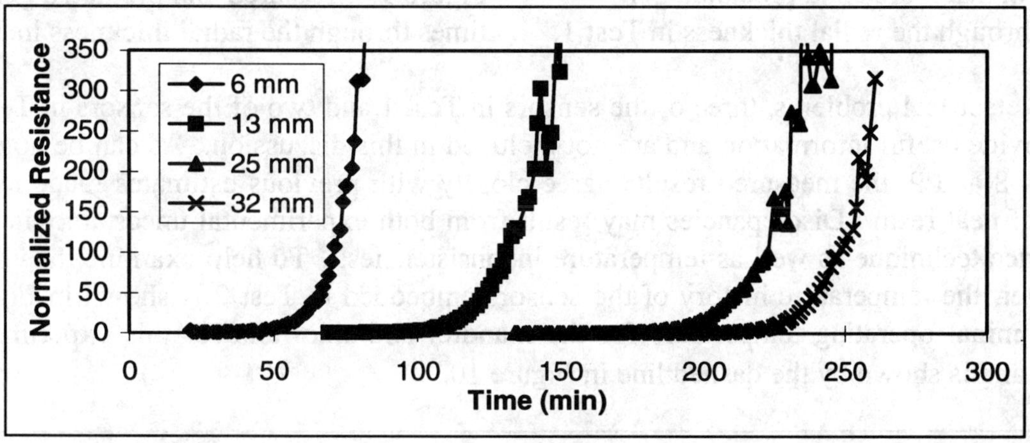

Figure 7. Normalized, temperature-compensated sensor resistance vs. time for sensors at different radial positions as measured from the mandrel surface in Test 2.

The resistance data presented above can be combined with the neat resin characterization (Figure 5) and the recorded temperature history of the sensors to produce the time at gelation for the radial position of each of the sensors. This information can then be compared to previous estimates made using neat resin gel testing to compare estimated gel front speed with the measured speed. The estimated gel time can be determined by Equation (1).

$$t_{gel,x} = \frac{x}{R} + t_{gel,neat} \tag{1}$$

Where $t_{gel,x}$ is the time (min.) to gelation for a point at x mm radial position as measured from the mandrel surface, x is the radial position (mm) in question, R is the radial accretion rate in mm/min , and $t_{gel,\ neat}$ is the neat resin time (min.) to gelation at the nominal temperature. The results of these calculations are presented in Figures 8 and 9 for Tests 1 and 2, respectively.

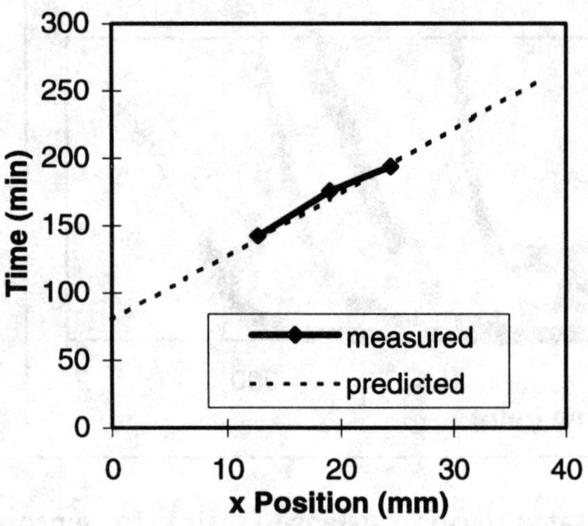

Figure 8. Measured and predicted gel times through the radial thickness in Test 1.

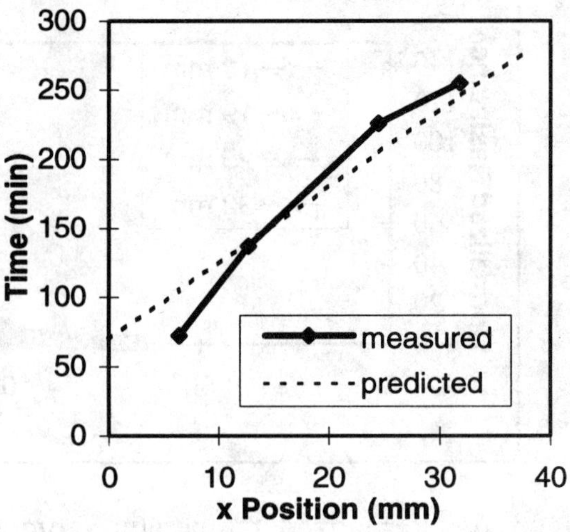

Figure 9. Measured and predicted gel times through the radial thickness in Test 2.

Due to electrical problems, three of the sensors in Test 1 and two of the sensors in Test 2 did not provide useful information and are not included in this discussion. As can be noted from Figures 8 and 9, the measured results agree closely with previous estimates made using gel times of neat resin. Discrepancies may result from both experimental uncertainty in the DC resistance technique as well as temperature inconsistencies. To help examine the effects of the latter, the temperature history of the sensors embedded in Test 2 is shown in Figure 10. The nominal operating temperature for the mandrel and sidewalls for this experiment was 131°C and is shown by the dashed line in Figure 10.

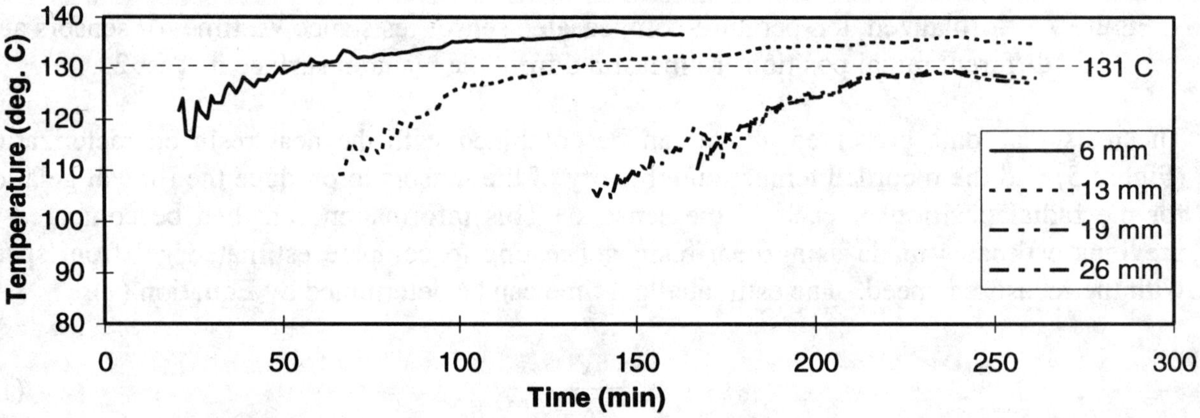

Figure 10. Temperature history of resistance sensors for Test 2.

The sensor located at 6 mm was above the nominal temperature for an extended period of time before gelation and should thus gel at a faster rate than if it was at the nominal temperature. This is reflected in Figure 9 where the measured gel time is less than the predicted value. Similarly, the sensors located at 19 and 26 mm are below the nominal temperature for the entire test and thus would be expected to gel at a slower rate than predicted. Again, this is reflected in Figure 9 where the measured gel times are greater than their predicted values.

From these results, it is concluded that the most important factor in maintaining a constant rate of gel front procession in the *in-situ* filament winding process is maintaining a constant temperature within the part. This can be challenging, depending on the heating arrangement, part geometry, and the amount of heat evolved in the reaction. In addition, the evolving thermal properties of the material system, including fiber and matrix, are important.

CONCLUSIONS

This study has confirmed that low cost DC resistance measurements can be used to measure gelation in a neat epoxy resin system. It was necessary to normalize the resistance measurements to compensate for the effects of temperature and geometry of the sensor. As used here, the technique is most sensitive at cure states at or near gelation. Most importantly the normalized, temperature-compensated DC resistance technique was found to be effective in measuring the procession of the resin gel front in an *in-situ* cured filament wound composite.

The results of this study confirm that radial gel rate in the *in-situ* cure filament winding process is dependent on the neat resin gel time at the operating temperature. Hence, the results support previous techniques used to obtain estimates of the gel rate performed by Gabrys and Bakis (1994).

ACKNOWLEDGMENTS

The authors thank Ryan Emerson of the Pennsylvania State University Composites Manufacturing Technology Center, C.W. Lee of the University of Dayton Research Institute, and Bruce Fink of the University of Delaware Center for Composite Materials for their input.

REFERENCES

Gabrys, C.W. and C.E. Bakis. 1994. "Fabrication of Thick Filament Wound Carbon epoxy Rings Using *in-situ* Curing: Manufacturing and Quality," Proc. American Society for Composites 9th Technical Conference, Sept. 20-22, 1090-1097.

Kim, C., H. Teng, C. L. Tucker, III, and S. R. White. 1995. "The Continuos Curing Process Thermoset Polymer Composites. Part 1: Modeling and Demonstration," *J. Composite Materials*, 29:1222-1253.

Kim, C., and S. R. White. 1996. "The Continuous Curing Process for Thermoset Polymer Composites, Part 2: Experimental Results for a Graphite/Epoxy Laminate," *J. Composite Materials*, 30:627-647.

Lee, C. W., and B. Rice. 1996. "Direct Current Based Resin State Sensors," 41st International SAMPE Symposium, March 24-28, 1511-1517.

Telegadas, H. K., and M. W. Hyer. 1992. "The Influence of Layer Waviness on the Stress State of Hydrostatically Loaded Cylinders: Failure Predictions," *J. Reinforced Plastics and Composites*, 11:127-145.

A Bending Model for Prepreg Material

H. LI, T. G. GUTOWSKI AND S.-B. SHIM

ABSTRACT

Forming of advanced composites has shown strong potential as a low cost substitute for hand lay-up. However, existing forming processes are limited by a variety of failure modes, such as laminate wrinkling, fiber misalignment, and thickness variation. The difficulty of finding solutions to these problems is related to the incomplete knowledge of fundamental deformation mechanisms, especially compressive stress build-up in composites during forming. In this paper, based on experimental observations on a three point bending test of an aligned continuous fiber prepreg material Toray T800H/3900-2, an innovative conceptual model is introduced. This model gives reasonable solutions to some length scaling problems and may lead to a broader understanding on the forming mechanisms.

INTRODUCTION

Most forming models for advanced composites treat the prepreg material as a highly viscous and homogeneous fluid with reinforcement in the fiber direction. Kinematic constraints are imposed on the deformation of such a material by the assumptions of incompressibility and fiber inextensibility (Rogers and O'Neill, 1991; Martin et al., 1995). Although these models are successful to illuminate some of the many important aspects of the material behavior, the corresponding analyses of forming often turn out to deviate from the experimental results significantly. Even for the simple bending of a laminated prepreg beam, the models predict pure viscous shear deformation, which features a singular distribution of normal stress on the cross section and very high stress concentration in the surface layers. The resulting computations for constant deformation rate can neither describe the temporal development of transverse load, nor provide the correct scaling laws for forming load versus the beam length and thickness.

Some other models emphasize the material inhomogeneity and incorporate layer slip mechanisms (Tam and Gutowski, 1989). When a prepreg or a laminate is assumed to consist of elastic and viscous layers, the viscoelastic transient behavior of the material during bending is well depicted. However, an equilibrium state dominated by viscous shear mechanism is induced by an implicit assumption that the elastic layers can bear very high axial normal stress by tensile or compressive elastic deformation. This leads to a singular

---

H. Li, T.G. Gutowski, S.-B. Shim, Laboratory for Manufacturing and Productivity,
Massachusetts Institute of Technology, Room 35-005, 77 Massachusetts Avenue, Cambridge, MA 02139

normal stress distribution on the cross section (Figure 1) and scaling laws similar to that in the reinforced fluid model.

The discrepancies between the analytical solutions of these models and experimental results are nontrivial. The most important reason might be that these models have not successfully incorporated the significant difference between the material responses to tensile and compressive loads along the fiber direction.

The purpose of the work reported in this paper is to develop a new model which includes the above consideration. Comparison is made between the proposed model and the existing ones with experimental data for three point bending of beams with different length, thickness, and under different deformation rate and temperature.

THREE POINT BENDING OF SINGLE LAYER PREPREG BEAMS

In the existing models, inextensibility of the fibers and incompressibility of the material are commonly used as two fundamental assumptions to provide kinematic constraints to deformation. These, in combination with the availability of "ideal" shear mechanisms, lead to a shear beam in the three point bending problem.

However, one important feature of the material is missing: Since a fiber is long and thin, under forming conditions it has very limited resistance to axial compressive stress before transverse deformation occurs. This transverse deformation can be minor waviness, fiber spreading and bunching-up, which are admissible in a certain range. It can also be large scale in-plane fiber buckling and misalignment, excessive thickness variation and laminate wrinkling, which are the failure modes that we are concerned for.

There is evidence (Figure 2) that during bend forming of a cross-plied laminate, very little, if any, longitudinal through-the-thickness shear is induced within each ply and the plane cross section hypothesis approximately holds. This means that macroscopically the bending of a single layer prepreg may be achieved by normal strain in the fiber direction instead of shear. Perhaps a more important implication of this observation is that within a multi-ply laminate, where ideal inter-ply shear is often large and impossible to achieve for double-curvature parts (Li et al., 1995), compressive stress may build up in a similar way.

The three point bending test of aligned prepreg beams was used to study the constitutive behavior of the material. The prepreg material tested was Toray T800H/3900-2. Referring to Figure 1, the distance between the two supports was 2 to 4 inches and the punch was centered between the supports. The samples were made of a single layer prepreg (thickness=0.009 inch). The width was 2 inches. The length was 12.5% longer than the distance between the supports, and was along the fiber direction. The test temperature was from 72 to 145 °F. While the experimental results at all these temperatures have shown applicability to the theory proposed in this paper, due to the difficulty of collecting data at higher temperatures, the discussion below is based on the more reliable data at room temperature (72 °F). Figure 3 shows typical test results for different displacement rates.

The relationship between the drape force F and the displacement rate at the center of the beam $\dot{\Delta}$ had been investigated according to a shear beam model and is described as follows.

Suppose that the beam length is $2a$, along which is the fiber direction. In addition, define the sample width as constant b, the thickness as constant h, and the longitudinal through-the-thickness shear as γ. Assume a power-law stress-strain relationship for the shear:

$$\tau = m\dot{\gamma}^n \tag{1}$$

where m and n are non-Newtonian viscosity coefficient and exponent respectively.

From equilibrium and (1) it can be shown that (Neoh, 1992):

$$F_v \approx 2bhm\dot{\gamma}^n \tag{2}$$

where F_v is the crosshead load when the viscous shear is fully developed, which is indicated by a leveling off of the load-deflection curve. The longitudinal through-the-thickness shear rate along the beam can be expressed by:

$$\dot{\gamma} \approx \frac{\dot{\Delta}}{a} \tag{3}$$

where it is assumed that the deflection Δ is small relative to the length $2a$. In our experiments, $0 \le \frac{\Delta}{2a} \le 0.15$.

If one changes the deflection rate $\dot{\Delta}$ and the beam length $2a$ while keeping all other experimental parameters constant, it should be expected that F_v has the following relationship with $\dot{\Delta}$ and a:

$$F_v \propto \left(\frac{\dot{\Delta}}{a}\right)^n \tag{4}$$

However, test results show that the scaling law (4) cannot be justified (Figure 4).

Now, assume that the plane cross section hypothesis approximately holds for the prepreg beam under bending. It should be noticed that the normal stress on the compressive side of the cross section may be very complicated in nature. However, for simplicity the prepreg can be modeled as a viscoelastic material which has the following unique properties:

When an element of the material is under tensile stress in the fiber direction, the elastic component (fiber) dominates because of its extremely high tensile modulus; when the element is under compressive stress in the fiber direction, the response is viscoelastic initially and then primarily viscous after yielding because the elastic component (fiber) has a very low resistance to the compressive load due to slenderness of the fibers. This is shown by the fact that fiber waviness can be observed on the compressive side of the prepreg under bending (Figure 5).

According to the above assumptions, for steady state analysis the compressive response is viscous. This is expressed by:

$$\sigma_v = m\dot{\varepsilon}^n = m\left[\frac{\partial}{\partial t}\left(y\frac{\partial^2 v}{\partial x^2}\right)\right]^n \tag{5}$$

where y is the distance to neutral axis for any point under compressive normal stress on the cross section and v is the deflection of the beam. The normal stress distribution is as shown in Figure 6. The neutral axis is far from the geometrical center of the cross section area because of the significant difference between the magnitude of the tensile and compressive stresses. Hence the moment equilibrium about the neutral axis gives:

$$\int \sigma_v \, y dA \approx \frac{F}{2}x \qquad \text{for } 0 \le x \le a \tag{6}$$

where A is the cross section area. Please notice that the contribution of the tensile stress to the left-hand side of (6) is omitted because it is much less than the contribution of the compressive stress.

Substitute (5) into (6):

$$\int_0^h m\left[\frac{\partial}{\partial t}\left(y\frac{\partial^2 v}{\partial x^2}\right)\right]^n by\,dy \approx \frac{F}{2}x \tag{7}$$

Since $F = F(t)$, the curvature of the beam can be expressed as:

$$\frac{\partial^2 v}{\partial x^2} \approx f(t)x^{\frac{1}{n}} \tag{8}$$

where $f(t)$ is a function of time only.

The boundary conditions are:

$$\begin{cases} v(0) = 0 \\ v(a) = -\Delta \\ \dfrac{\partial v}{\partial x}(a) = 0 \end{cases} \tag{9}$$

From (8) and (9) the deflection of the beam can be solved as:

$$v = \Delta\left(\frac{n}{n+1}\frac{x^{\frac{1}{n}+2}}{a^{\frac{1}{n}+2}} - \frac{2n+1}{n+1}\frac{x}{a}\right) \quad \text{for } 0 \le x \le a \tag{10}$$

The measured beam shape can be fit into a curve expressed by the above equation with n=0.24 (Figure 7).

By substituting v into (7), it can be deduced that while keeping all experimental parameters other than $\dot{\Delta}$ and a constant, the following scaling law should hold:

$$F_v \propto \frac{1}{a}\left(\frac{\dot{\Delta}}{a^2}\right)^n \tag{11}$$

The experimental data for Toray T800H/3900-2 material fit this relationship very well with n=0.243 (Figure 8 compared with Figure 4).

THREE POINT BENDING OF MULTI-PLY PREPREG BEAMS

Now that the conceptual model introduced above has given a reasonable length scaling law to the three point bending of beams made of a single layer of prepreg, further interests may include the length scale of multi-ply beams with wider range of length and the effects of inter-ply stresses.

The three point bending tests were carried out for beams made of up to 8 layers of the Toray T800H/3900-2 prepreg, all with fiber orientation along the beam length. The beam length ranges from 2 to 18 inches. Let the number of layers be N so that the beam thickness is

$$H = Nh \tag{12}$$

Suppose that the plane cross section hypothesis still holds for the laminate, which means that the through-the-thickness shear is zero. From equations (7) and (10) it can be shown that

$$F_v \propto H^{2+n} \tag{13}$$

while keeping all experimental parameters except H constant. Thus

$$F_v \propto N^{2+n} \qquad (14)$$

since h is also a constant for one kind of prepreg.

The experimental results shown in Figure 9 apparently disagree with the thickness scaling law (14). However, it should not be surprising. Figure 2 has clearly shown that in a formed 90° bend it is relatively easier for the inter-ply shear to occur even when there is no evidence that within individual plies through-the-thickness shear is induced. It is logical to believe that inter-ply shear would help alleviate compressive stress build-up within the laminate.

Although the distribution of normal stress on a cross section is not clear for the multi-ply beam, it is interesting to notice that for the same thickness, the length scale in equation (11) still holds (Figure 10) for a much wider range of length, while the shear beam models cannot provide the right length scale at all. This suggests that the beam cannot be in a pure shear status. In other words, normal strain and stress are built up through the thickness, at least within individual plies.

By now, it should be noticed that when a multi-ply laminate with different fiber orientations in different layers is formed into complex shape with double curvature, the inter-ply shear required may have a very complicated distribution throughout the laminate and can be larger in orders of magnitude compared with the inter-ply shear under single curvature bending (Li et al., 1995). In such cases, compressive stress may be induced due to the lack of inter-ply shear. To study stress build-up in forming with such degree of complexity, the three point bending test is not enough . Nevertheless the model developed in this work has provided an innovative view of fiber deformation in forming and a start point for further analysis.

CONCLUSIONS

A conceptual model has been presented for aligned continuous fiber composites forming. The consequence of this model on the three point bending test is a non-singular axial stress distribution on the cross section of the beam at dynamic steady state. The stress distribution is asymmetric about the neutral axis and a corresponding power law relationship gives solution to some length scaling problems which is impossible to get by the earlier models. The nature of the compressive response is still not clear and the degree of complexity is expected to increase when dealing with multi-ply laminate deformation. The most significant meaning of the model is to reveal that while composites incompressibility and fiber inextensibility supply kinematic constraints for ideal fiber mapping, the difference between the tensile and compressive resistance in fiber direction of prepreg materials may play an important role in actual fiber deformation.

REFERENCES

Li, H., Dillon, G., Chey, S. and Gutowski, T. 1995. "A Kinematic Approach for the Diaphragm Forming of Advanced Composites", *Proceedings of the 40th International SAMPE Symposium*, 1595-1602.

Martin, T.A., Bhattacharyya, D. and Collins, I.F. 1995. "Bending of Fibre Reinforced Thermoplastic Sheets", *Composites Manufacturing*, 6: 177-187.

Neoh, E.T. 1992. "Drape Properties of Thermosetting Prepregs", M.S. Thesis, Department of Mechanical Engineering, M.I.T.

Rogers, T.G. and O'Neill, J.M. 1991. "Theoretical Analysis of Forming Flows of Fibre-Reinforced Composites", *Composites Manufacturing*, 2: 153-160.

Tam, A.S. and Gutowski, T.G. 1989. "Ply-Slip During the Forming of Thermoplastic Composite Parts", *Journal of Composite Materials*, 23: 587-605.

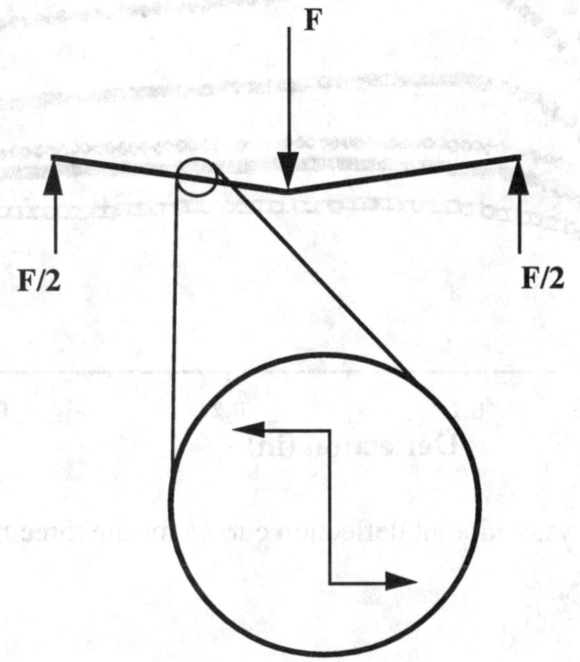

Figure 1. Singular distribution of normal stress on the cross section of a shear beam.

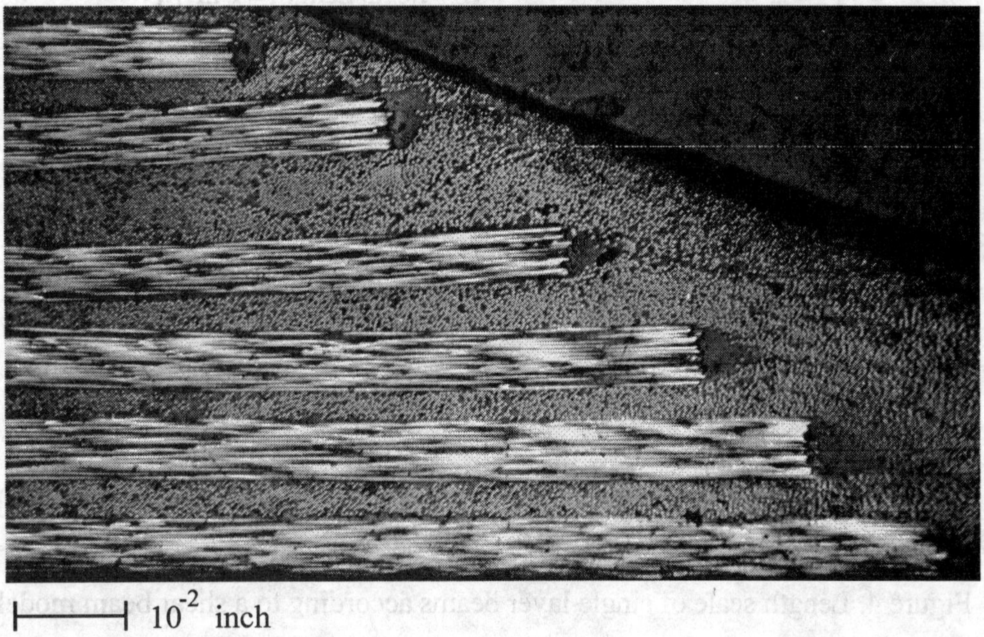

Figure 2. Photomicrograph of the end section of a formed APC-2 90° bend. (Courtesy of Dr. C. M. ÓBrádaigh, University College, Galway, Ireland.)

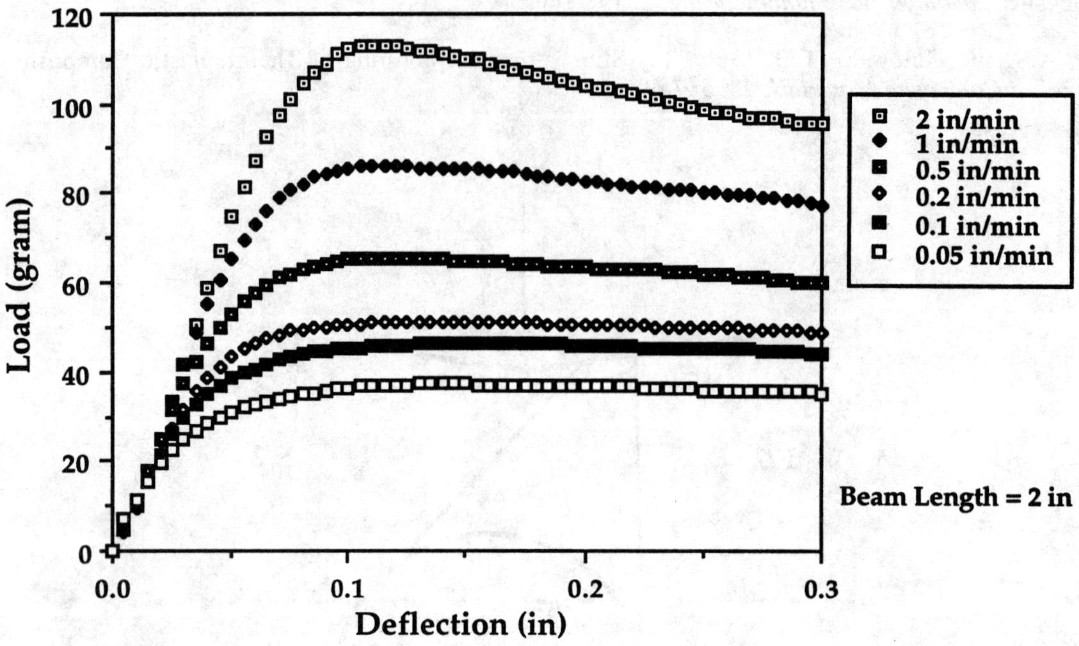

Figure 3. Typical load vs. midpoint deflection curves for the three point bending of prepreg beams.

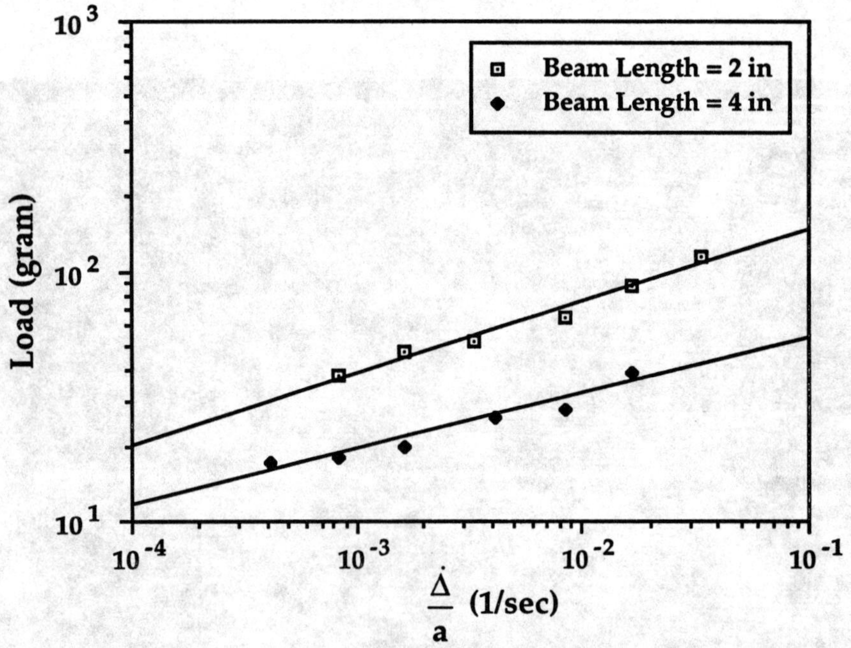

Figure 4. Length scale of single layer beams according to a shear beam model.

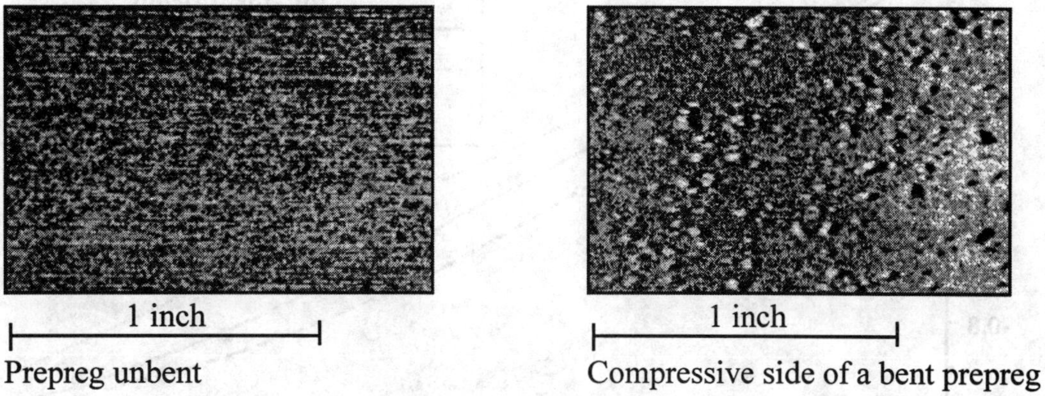

Figure 5. Fiber waviness on the compressive side of a prepreg under bending.

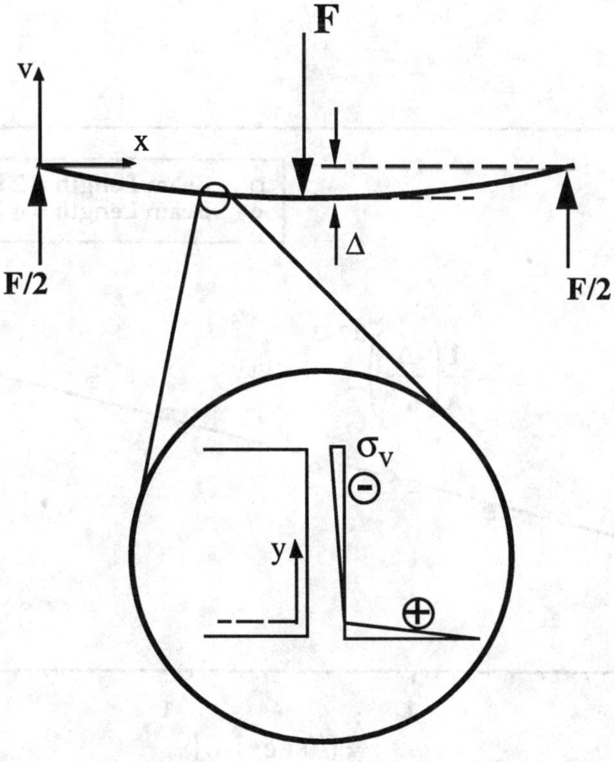

Figure 6. Normal stress distribution at steady state according to the proposed model.

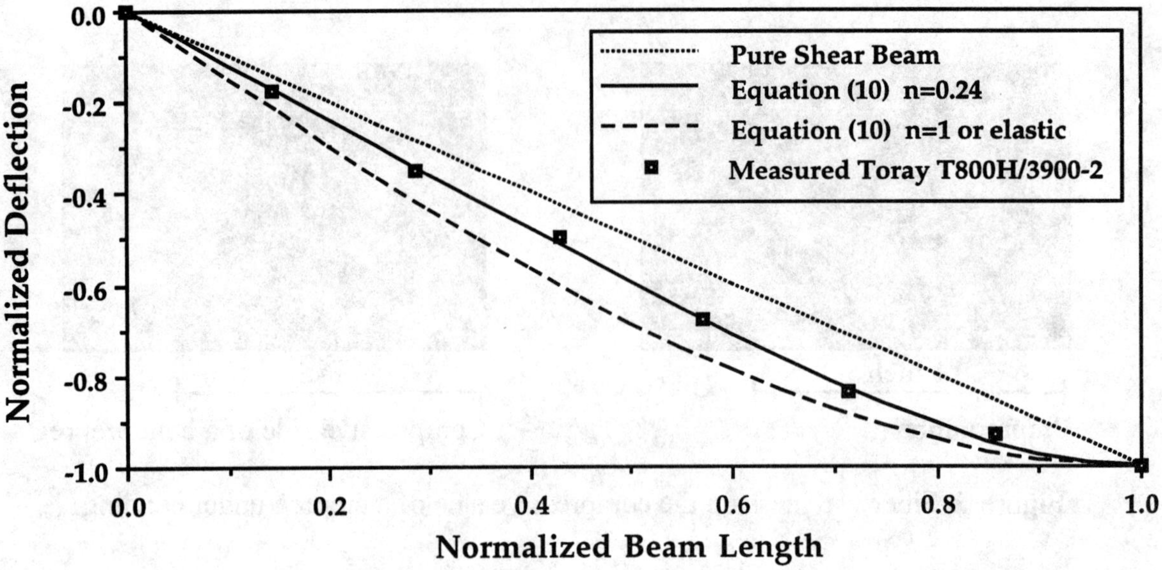

Figure 7. Deflection curves of the beam predicted by different models vs. the measurement.

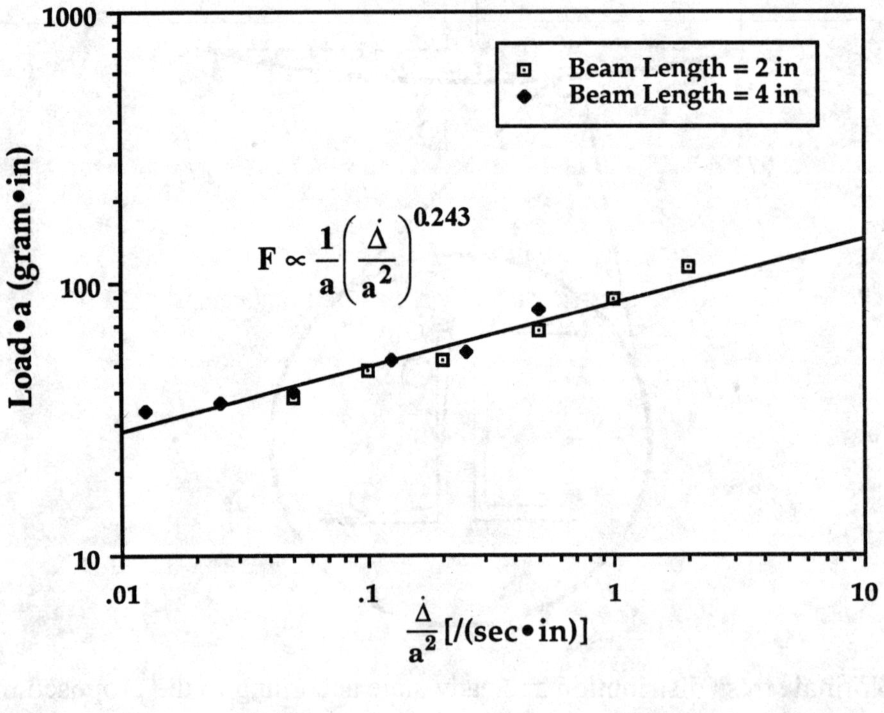

Figure 8. Length scale of single layer beams according to the proposed model.

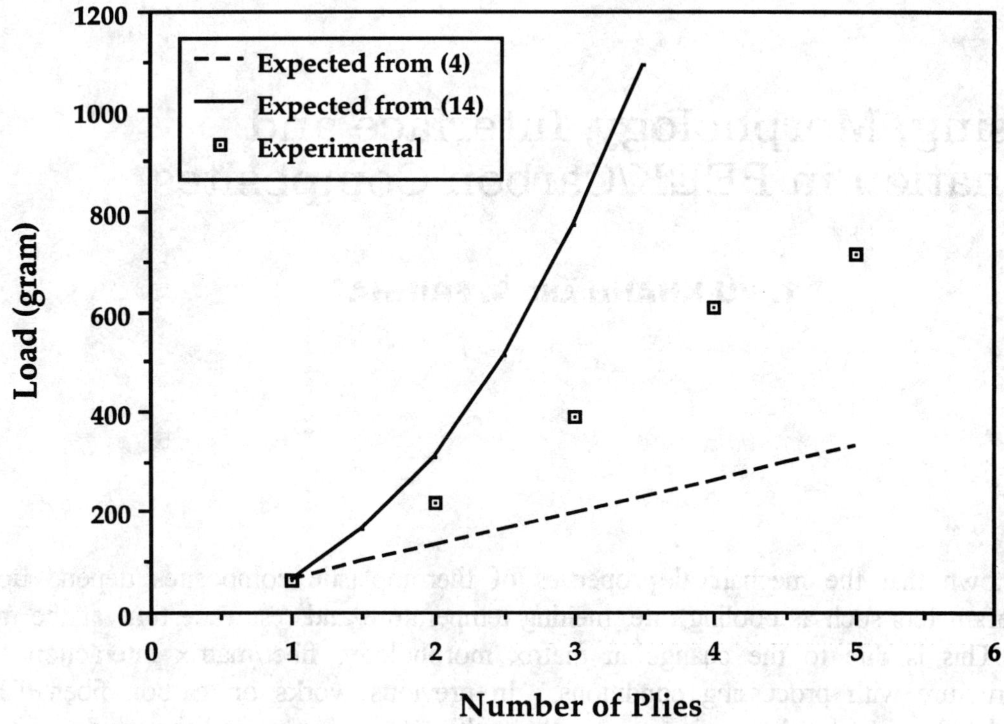

Figure 9. Thickness scale of multi-layer beams (Length = 2 inches).

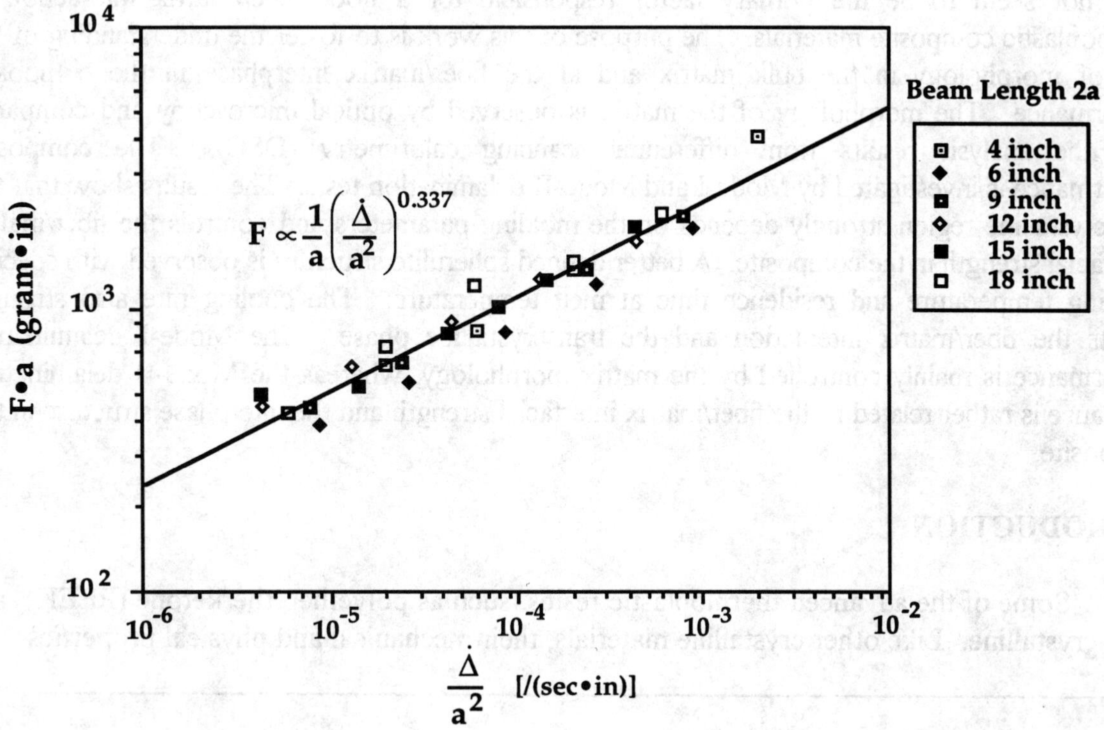

Figure 10. Length scale of multi-layer beams (4 plies) according to the proposed model.

Processing, Morphology, Interface and Delamination in PEEK/Carbon Composites

T. VU-KHANH AND S. FRIKHA

ABSTRACT

It is well known that the mechanical properties of thermoplastic composites depend upon processing parameters such as cooling rate, melting temperature and residence time at the melt temperature. This is due to the change in matrix morphology, fiber/matrix interaction and interphase structure with processing conditions. In previous works on carbon fiber/PEEK composite, it has been found that an increase in the melting temperature and the residence time improves the short-beam-shear strength of the composite, and increasing the cooling rate results in a decrease in this strength. Crystallinity at the interface appeared to be efficient only if there is a strong physico-chemical interaction between the matrix and the fiber. Moreover, transcrystallinity does not seem to be the primary factor responsible for a good fiber/matrix interaction in thermoplastic composite materials. The purpose of this work is to foster the understanding of the role of morphology in the bulk matrix and at the fiber/matrix interphase in the composite performance. The morphology of the matrix is observed by optical microscopy and compared with the analysis results from differential scanning calorimetry (DSC). The composite performance is investigated by Mode-I and Mode-II delamination tests. The results show that the transcrystalline region strongly depends on the molding parameters and controls the fiber/matrix interfacial strength in the composite. A better defined spherulite structure is observed with specific molding temperature and residence time at melt temperature. The cooling rate also strongly affects the fiber/matrix interaction and the transcrystalline phase. The Mode-I delamination performance is mainly controlled by the matrix morphology, whereas the Mode-II delamination resistance is rather related to the fiber/matrix interfacial strength and the interphase structure in the composite.

INTRODUCTION

Some of the advanced thermoplastic resins, such as polyetheretherketone (PEEK), are semi-crystalline. Like other crystalline materials, their mechanical and physical properties

---

- Toan Vu-Khanh, Université de Sherbrooke, Faculté des sciences appliquées / Genie mécanique, 2500 boul. de l'Université, Sherbrooke, Québec, Canada, J1K 2R1 (E-mail: toan.vu-khanh@sca.usherb.ca)
- Slim Frikha, Former Ph.D. student at Ecole Polytechnique, C.P. 6079, Montréal, Québec, Canada H3C 3A7

are related to the crystalline structure of the polymer. Consequently, some processing conditions may result in crystalline morphology that is detrimental to the mechanical performance of the material.

The effect of processing conditions on the crystalline morphology [1-6] and mechanical properties [7-10] of PEEK resin and PEEK matrix composites has received much attention. It has been recognised that the molding conditions play a key role in the mechanical performance of PEEK/carbon composites. In previous work [9-11], the effects of such parameters as molding temperature, cooling rate, and residence time on the microstructure and performance of these materials have been investigated. Molding temperature has been found to strongly influence the mechanical properties of the composite. It has also been shown [5, 9] that the temperature corresponding to the onset of crystallization of the PEEK matrix in APC-2 composite varies with the molding temperature. A plot of the onset temperature of crystallization as a function of molding temperature in the range of 365°C to 460°C displays three regions: a steadily decreasing region, a plateau region and another decreasing region. Correspondingly, the matrix-dominated mechanical property, as measured by the tensile strength of $[+/-45°]_s$ samples, increases with the molding temperature, attains a plateau region, and then decreases. The plateau region probably corresponds to the melting temperature range in which all the crystalline entities are eliminated. At lower temperatures, these entities remain in the melt and play a nucleating role [12]. At higher molding temperatures, matrix degradation occurs, resulting in a decrease in the crystallization temperature, the heat of crystallization and the overall rate of crystallization. Previous works [9-11] have also suggested that in the APC-2 prepreg, good fiber/matrix bonding is always present, but in the commingled fabric, this interaction must be created during the molding process. Fiber/matrix adhesion in the commingled system depends on molding temperature and residence time at melt temperature, as well as cooling rate during the consolidation process. An optimum combination of molding temperature and residence time must be respected to insure good interfacial performance in the composite. A transition between cohesive and adhesive fracture has been observed when the cooling rate increases from 30°C/min to 71°C/min for the commingled system. While fiber/matrix interaction is always present in the composite made from the prepreg form, the interfacial strength in this system has also been found to be significantly affected by the cooling rate. There is a critical cooling rate above which, no fiber/matrix interaction is created.

This paper presents the effects of molding parameters (temperature, residence time at melt temperature, and cooling rate) on Mode-I and Mode-II delaminations. The role of transcrystallinity in the complex mechanism of fiber/matrix interaction is studied.

EXPERIMENTAL

The commingled systems consist of hybrid yarns of intimate blends of continuous thermoplastic PEEK filaments (15OG) with AS-4 graphite filaments. The non-woven unidirectional fabric NCS-1057, made from sized AS-4 fibers commingled with PEEK (15OG) filaments, and the non-woven square symmetric bi-directional (0/90) fabric NCS-1025, consisting of unsized AS-4 fibers commingled with PEEK (15OG) filaments, were obtained from BASF Structural Materials, Inc. The APC-2 prepreg tapes consisting of "VICTREX" PEEK reinforced with unidirectional AS-4 graphite fibers were obtained from ICI.

To properly compare the performance of the NCS-1025 fabric with that of the APC-2 composite, square-symmetric cross-ply laminates containing the same volume fraction of 0° and 90°, were compression molded. Square plaques 150 mm x 150 mm x 5 mm thick were compression molded in a matched mold and pressed in a programmable Wabash press permitting reproducible molding cycles. The plaques were molded between 365°C and 455°C; the temperature of the melt was measured with a thermocouple inserted in the laminates. Heating was done without pressure and when the molding temperature was reached, high pressure was applied. Various cooling rates were obtained by transferring the hot mold to a second press set at different temperatures or by modifying the cooling system of the first press. The cooling rate reported was calculated from the time required to cool the melted sample from 400°C to 100°C, assuming that the PEEK crystallization under 100°C is negligible [1].

The Mode-I delamination fracture toughness was measured by using double cantilever beam (DCB) specimens as described in [13]. The mode-II delamination tests were performed using a precracked cantilever beam as described in [14]. The precrack was introduced by placing a strip of polyimide film 50 μm thick in the midplane of the laminates. The tests were carried out on an Instron machine, Model 1025, at a cross-head speed of 10 mm/min. After testing, the crack length was measured on the fractured specimens with a Vernier. The G_I and G_{II} values were calculated from the relations based on beam theory [13, 14].

RESULTS AND DISCUSSIONS

MOLDING TEMPERATURE

Figure 1 shows the Mode-I critical strain energy release rate, G_{Ic}, at crack initiation for the two composites NCS-1025 and APC-2 at various molding temperatures. The fracture energy at initiation continuously increases with molding temperature up to about 420°C and then decreases. The effect is similar to that on the tensile strength of $[+/-45°]_s$ samples and could be explained, in the same manner, by the unfavorable effect of the residual crystalline entities on the morphology of the PEEK matrix [9]. The DSC analysis of the evolution of the temperature at onset of crystallization of the PEEK matrix, T_o, in the composite as a function of melting temperature, T_m also revealed three distinct regions [5, 9]. The first region, which corresponds to the decrease of T_o with T_m at low melting temperature, was associated with the decrease of the number of remaining crystalline entities in the melt when T_m increases [12]. The second region, a plateau region, corresponds to the melting condition in which all of the pre-existent crystalline entities are destroyed. At high melting temperature, there is a third region where T_o decreases as T_m increases, due to the degradation of the PEEK matrix. It should be noted that while T_o strongly depends on T_m in the low temperature range, the heat of crystallization does not seem to be influenced by T_o. A decrease in the heat of crystallization is observed only in the region of matrix degradation. This result indicates that the melting temperature affects the number and the size of the crystalline entities but does not alter the level of crystallinity of the composite. The optimum Mode-I delamination resistance would correspond to a melting condition in which all the crystalline germs are destroyed. At higher temperature, the matrix degradation can explain the reduction in the Mode-I delamination resistance. The result suggests therefore that the Mode-I delamination is strongly influenced by the property of the matrix. It is also interesting to note that a higher initiation energy corresponds to a more stable crack propagation. Table 1 gives the number of observed unstable crack growths amongst the tested samples at different temperatures.

Figure 2 shows the Mode-II delamination energy at initiation, G_{IIc}, of the two materials as a function of temperature. In this case, fracture is always unstable. Up to about 420°C, the G_{IIc} value of the NCS-1025 composite increases with temperature before but the APC-2 composite exhibits a constant G_{IIc} that corresponds to the maximum G_{IIc} value observed in the NCS-1025 composite. Above about 420°C, the G_{IIc} value in both composites decreases with temperature. The result confirms the effect observed on the short beam shear strength measured on these composites [9]. Under shear stress, fracture would occur at fiber matrix interface and would be controlled by the fiber-matrix adhesion. In the APC-2 prepreg system, the fiber-matrix interaction exists before molding but in the commingled system such as NCS-1025, this interaction is created during molding, resulting in a significant effect of molding temperature on the G_{IIc} value.

RESIDENCE TIME AT MELT TEMPERATURE

In order to investigate the effects of residence time, in the molding cycle, the plaques were heated up to 400°C at 20°C/min. At this temperature they were maintained for different intervals and then cooled down at a cooling rate of 20°C/min. Figures 3 and 4 show respectively the Mode-I and Mode-II delamination initiation energies of the two composites as a function of residence time at 400°C. It can be seen that in both cases the delamination resistance is affected by the residence time at the melt temperature of 400°C. The delamination energies at initiation in both Mode-I and Mode-II increase as the residence time at 400°C is prolonged, attains a maximum and then decreases. The maximum in the G_{Ic} value is observed between 5 and 20 min and, within the experimental error, seems to occur at a shorter residence time for the NCS-1025 composite. Examination of the fracture behavior reveals that as the residence time increases, the Mode-I crack propagation becomes more and more unstable as shown in Table 2. The G_{IIc} value is also found to be slightly higher for the APC-2 system but its maximum corresponds to about the same residence time and fracture is constantly unstable in both systems.

In order to verify the effect of the residence time at melt temperature on the PEEK matrix morphology, the PEEK filaments were blended with some carbon fibers, heated between two microscope glass lamella and maintained at 400°C for different residence times and cooled down at the same cooling rate as that of the molding cycle (i.e. 20°C/min). The matrix morphology was then observed by transmission microscopy under polarized light. Figure 5 gives a qualitative view of the morphology of the PEEK matrix after three different residence times at 400°C. It can be seen that after a residence time of 20 min, the fiber-matrix interphase region shows a large transcrystalline structure whereas after a residence time of 1 min or 80 min, the transcrystalline morphology is almost inexistent or not easily observed. A more careful measurement of the width of transcrystalline region is shown in Figure 6. Even though the result shows a very large scatter, it can be seen that the transcrystalline region grows with increasing residence time up to about 20 min and then decreases. In previous work [10, 11], it has been found that the fiber-matrix interaction is a result of several complex processes that are time and temperature dependent. The interphase crystallinity is essential to insure good fiber-matrix adhesion but it is not the sole parameter responsible for the creation of fiber-matrix interaction. The observed change in the transcrystalline at the interphase in Figure 5 and 6 does not contradict the above hypothesis but suggests that the transcrystalline structure is only formed under specific physico-chemical state of interaction between the fibers and the matrix. It has been also found that the crystallinity of the PEEK matrix drops from about 35% to 30% when the residence time at 400°C is above 10 min This drop does not prevent the development of a large transcrystalline region observed for the residence time of 20 min. This favorable state for transcrystallinity formation also seems to be the optimum condition in terms of interfacial adhesion in this composite.

It is worth noting that the crystalline morphology of the matrix significantly changes with the residence time as shown in Figure 5. The spherulites seem to be larger and better defined for the residence time corresponding to a larger transcrystalline region. This change in the matrix morphology is probably responsible for the variation in Mode-I delamination initiation energy as shown in Figure 3. The results also suggest that the PEEK matrix is quite unstable in the region of its molding temperatures. Its morphology always evolves with the residence time at the melt temperature and care must be taken to insure a proper molding cycle, leading to a better performance. As discussed above, the drop in crystallinity above 10 min at 400°C suggests a possible degradation of the polymer. However, this does not prevent the formation of well-defined transcrystalline and spherulitic structures.

COOLING RATES

Another major processing parameter that can affect the composite performance is the cooling rate since it is well known that this parameter strongly affects crystallinity. To analyse the effect of cooling rate on the composite morphology, similar samples for transmission microscopy observation as described above were heated up to 400°C. After a residence time of 5 min at this temperature, the samples were cooled down at different rates, ranging from 1°C/min to 80°C/min. Figure 7 shows such effect of cooling rate on the morphology of the PEEK matrix. As it can be expected, increasing the cooling rate results in a reduction in the formation of crystalline structure. The sizes of the spherulites in the bulk matrix and of the transcrystalline region around the fibers strongly diminish from 1°C/min to 80°C/min. Figure 8 shows that, as the cooling rate increases, the width of the transcrystalline region decreases from about 40 μm to about 5 μm. The most severe drop occurs between 1°C/min and 20°C/min with a reduction in the transcrystalline region from 40 μm to lest than 10 μm.

Figures 9 and 10 show the effects of cooling rate on the Mode-I and Mode-II delamination initiation energy of the two systems. The Mode-I delamination energies at initiation, G_{Ic}, in both composites slightly increase from 1°C/min to 20°C/min and the remains relatively constant as the cooling rate augments. Since the toughness of semi-crystalline thermoplastics tends to increase when the crystallinity diminishes, the reduction of crystallinity could be responsible for the variation of G_{Ic} value with cooling rate in both materials (Figure 9). The result also confirms the dominant role of the matrix in Mode-I delamination resistance. In terms of Mode-II delamination resistance, the initiation energy, G_{IIc}, in both materials continuously decreases with increasing cooling rate. The result suggests that the shear mode of delamination is rather controlled by the fiber-matrix interfacial strength. As demonstrated above, a larger transcrystalline region usually results in a better interfacial strength. Consequently, the reduction of this region with the cooling rate is probably responsible for decrease in the G_{IIc} value.

CONCLUSION

It has been shown that the Mode-I and Mode-II delamination resistance in both the APC-2 and the NCS-1025 PEEK/carbon fiber composites are strongly affected by processing conditions. The Mode-I delamination energy is controlled by the matrix property whereas the Mode-II delamination rather depends on the fiber-matrix adhesion. There is an optimum region of molding temperature that results in a better Mode-I delamination energy. Below this zone the crystalline germs in the PEEK matrix are not completely destroyed, leading to a lower toughness of the matrix. Above this zone, the matrix degradation results in a reduction of the Mode-I fracture energy. A variation in the residence time results in a significant change in the matrix morphology but does not affect the Mode-I delamination energy. There exists an optimum residence time that leads to a better Mode-II delamination resistance. The higher Mode-II fracture energy is attributed to the presence of the transcrystalline region around the fibers. The transcrystalline structure is created at a specific physico-chemical state of the fiber-matrix interaction. The cooling rate must be also slow enough to allow the formation of the transcrystalline region around the fibers to insure a better Mode-II delamination resistance. However, the higher crystallinity at low cooling rate also results in a slight decrease in the Mode-I delamination energy.

REFERENCES

1. Blundell, D. 1. and B. N. Orborn. 1983. *Polymer*, 24:953-958.
2. Cebe, P. 1989. *SPE ANTEC*, pp. 1413-1415.
3. Osttherg, G. M. and J. C. Seferis. 1987. *J. Applied Polymer Science*, 33:29-39.
4. Bassett, D. C., R. H. Olley and I. A. M. Al Raheil, 1988. *Polymer*, 29:1745-1754.
5. Denault J., Vu-Khanh T., 1992. Polym. Compos., 13: 361-371.
6. Olley, R. H., D. C. Bassett and D. 1. Blundell, 1986. *Polymer*, 27:344-348.
7. Talbott, M. F. and G. S. Springer. 1987. *J. Compos. Mater.*, 21:1056-1081.
8. Lustiger, A., F. S. Uralil and G. M. Newaz. 1991. *Polym. Compos*, 11(1):65-75.
9. Vu-Khanh, T. and J..Denault. 1991. *J. Thermoplastic Compos. Mater.*, 4:363-376.
10. Vu-Khanh, T. and J..Denault. 1993. *J. Thermoplastic Compos. Mater.*, 12:916-931.
11. Denault J., Vu-Khanh T., "Fiber/Matrix Interaction in Carbon/PEEK Composites", Journal Thermoplastic Compos. Mater., Vol. 6, pp. 190-204, 1993.
12. Caramaro L., B. Chabert, J. Chauchard, T. Vu-Khanh. 1991. *Polym. Eng. Sci.*, 31:1279-1285.
13. Xiao X.R., T. Vu-Khanh. 1993. *J. Thermoplastic Compos. Mater.*, 6:323-341,
14. Vu-Khanh T., 1987. *Polym. Compos.*, 8:331-341.

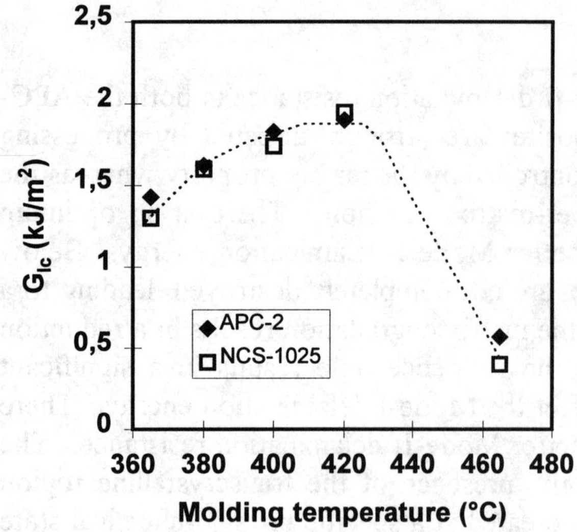

Figure 1: Effect of molding temperature on Mode-I delamination initiation energy

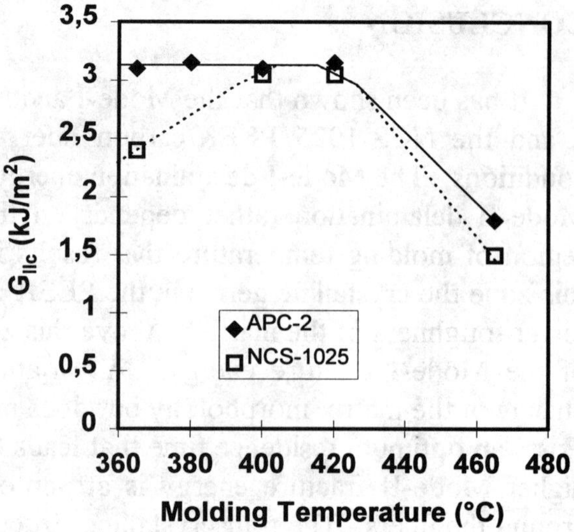

Figure 2: Effect of molding temperature on Mode-II delamination initiation energy

TABLE I: EFFECT OF MOLDING TEMPERATURE ON UNSTABLE FRACTURE

Molding Temperature (°C)	Number of Unstable Fracture/ Number of Tested Samples
365	4/4
380	4/4
400	2/4
420	2/4
455	4/4

TABLE II: EFFECT OF RESIDENCE TIME ON UNSTABLE FRACTURE

Residence Time at 400°C (min)	Number of Unstable Fracture/ Number of Tested Samples
1	1/4
5	2/4
20	4/4
60	4/4

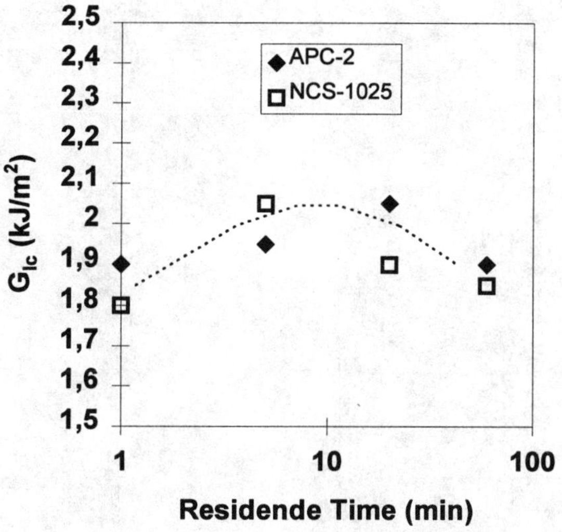

Figure 3: Effect of residence time on Mode-I delamination initiation energy

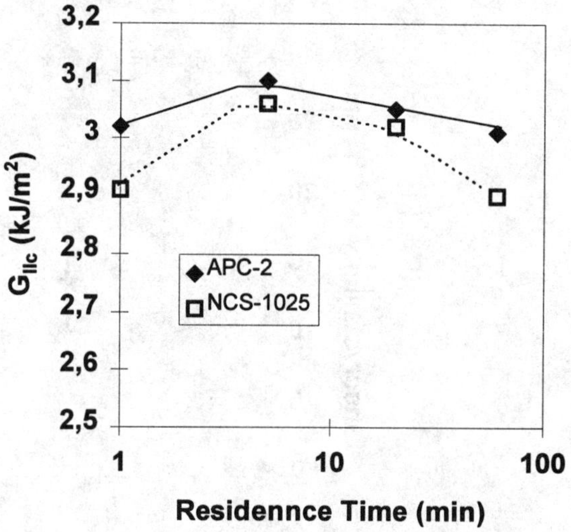

Figure 4: Effect of residence time on Mode-II delamination initiation energy

(a)

(b)

(c)

Figure 5: Morphology of the PEEK matrix after a residence time at 400°C for:
(a) 1 min
(b) 20 min
(c) 80 min

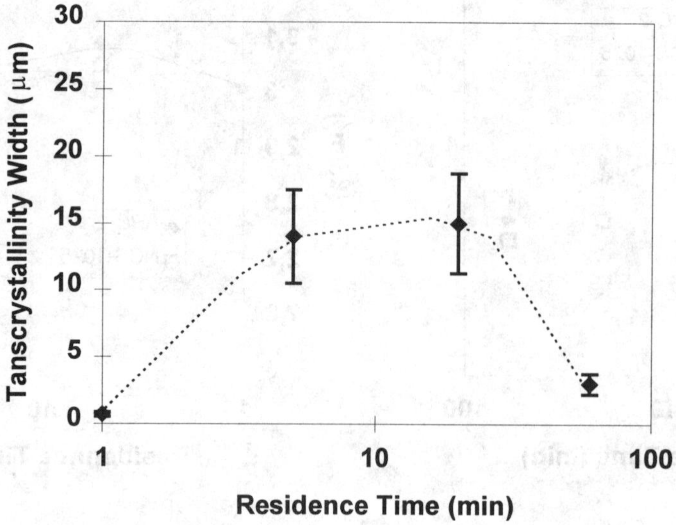

Figure 6: Variation of the width of transcrystalline region with residence time at 400°C

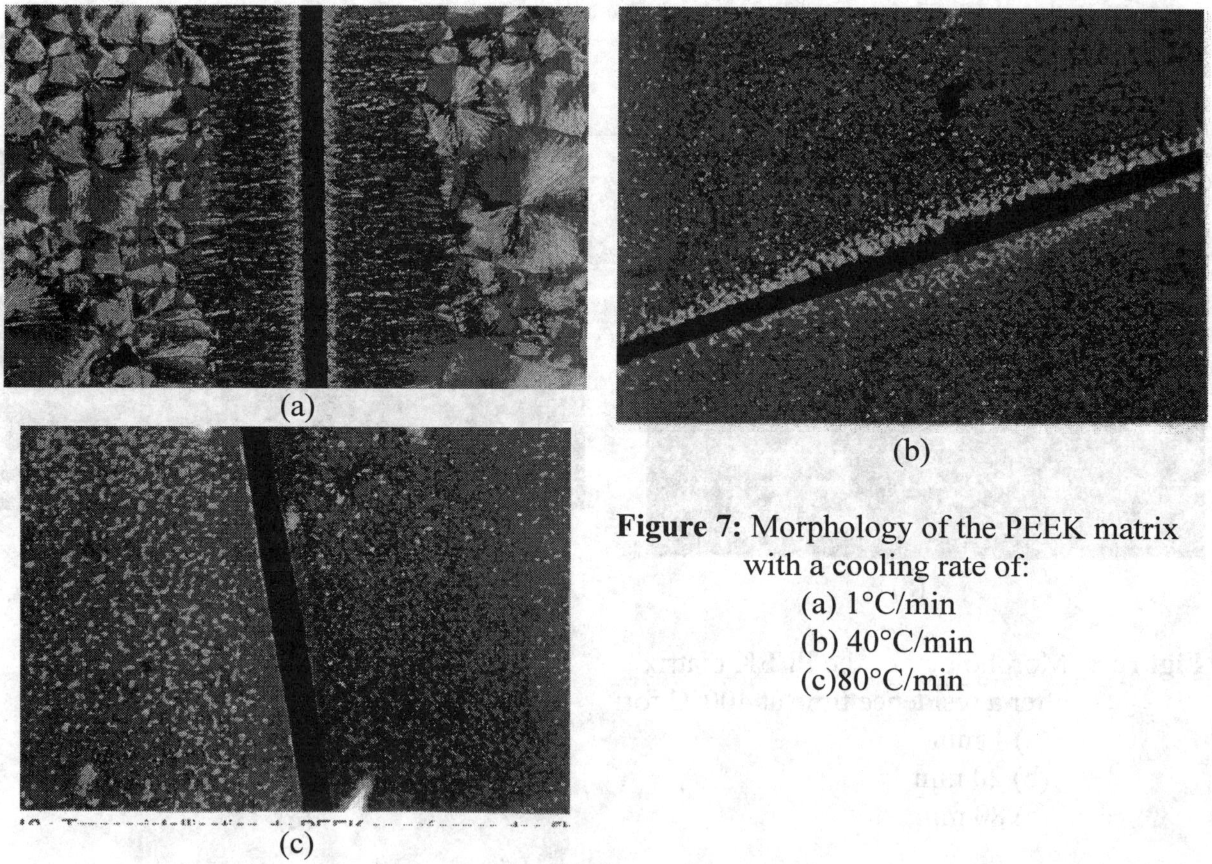

Figure 7: Morphology of the PEEK matrix with a cooling rate of:
(a) 1°C/min
(b) 40°C/min
(c)80°C/min

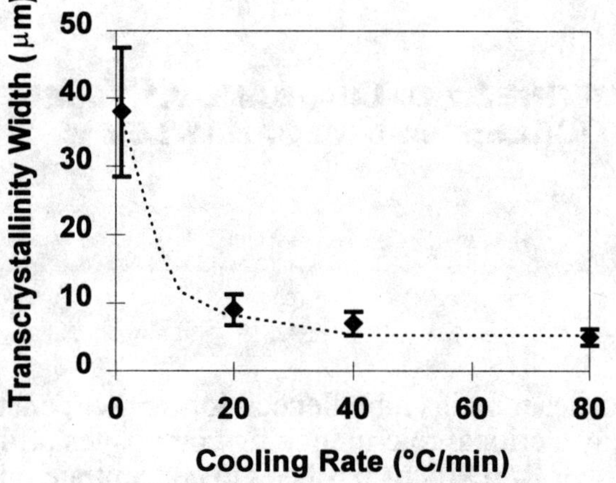

Figure 8: Variation of transcrystallinity width with cooling rate

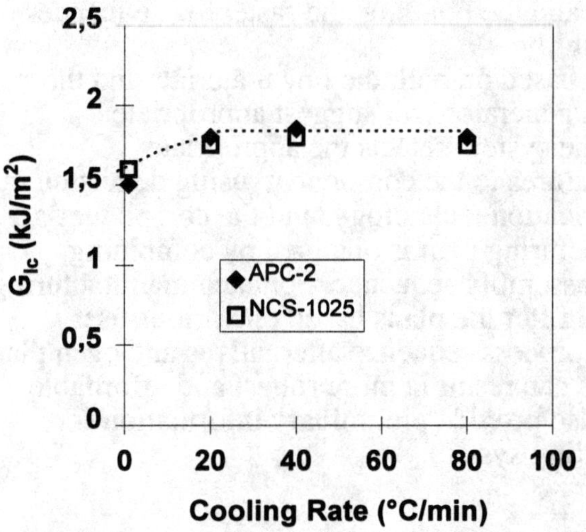

Figure 9: Effect of cooling rate on Mode-I delamination initiation energy

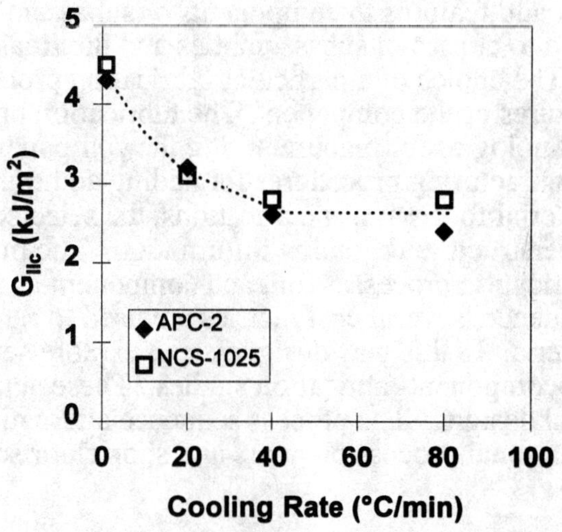

Figure 10: Effect of cooling rate on Mode-II delamination initiation energy

Design of Conceptual Manufacturing Sequences for Composite Assemblies

J. K. McDOWELL, I. MARTINEZ,* O. LUKIBANOV,* T. LENZ, A. KAMEL,* J. STICKLEN* AND M. C. HAWLEY

ABSTRACT

The manufacturing of composites remains a bottleneck for market penetration. Although process technology advances play a critical role in making composites affordable, changes in design practice are also critical. Our R&D focus has been to concentrate on conceptual design support. The basic theme is determining a preliminary manufacturing sequence for a product as it is conceptualized without waiting for the final prints.

Starting with what we refer to as a conceptual composite assembly, the goal is generating a manufacturing process sequence to produce the composite product. The conceptual composite assembly product design contains information about: configuration of the assembly and its components; internal assembly/component and external assembly/superstructure connections; raw materials; functional satisfaction of the design requirements; and partial geometries (dimensions and features). The manufacturing process sequence includes: the composite fabrication processes used to produce the components; the machining and post-process steps that add features to components or subassemblies; and the bonding and fastening sequences used to construct subassemblies and the final assembly.

The choice of a particular fabrication process is based on both the raw materials and the features of the component. The fabrication process generator will suggest appropriate technologies for manufacturing the component. The system selects the appropriate manufacturing procedures for adding additional features to the component using details on raw material forms and specifications, the selected fabrication technology, and the component's superstructure & joining information. The manufacturing plan is obtained by combining fabrication processes for each component and the assembly sequence. For each manufacturing sequence, several cost metrics are used to rank and filter the plans based on various user criteria. In this way designers can explore several process sequence alternatives and even plan subcomponent fabrication studies. These activities can result in more robust and affordable final designs. The process sequence alternatives also provide preliminary information for traditional process planners and shop floor scheduling systems.

INTRODUCTION

Composite materials hold considerable promise as a material of the future. Unfortunately, their successful use in aerospace applications has not directly transferred to the durable goods markets, most significant of which is the automotive industry. Much of this can be attributed to the lack of low-cost, high speed processes comparable to metals manufacturing. We believe

Department of Chemical Engineering, Department of Computer Science*, all at Michigan State University, E. Lansing, MI 48824 there is also substantial ground to make up in the design practices that utilize composites. Composites offer considerable design flexibility as well as demand an appreciation of their complexity. This is a double edged sword that we believe can be successfully navigated using advanced computing technology, most notably intelligent decision support software.

The manufacturing of composites remains a bottleneck for market penetration. Though process technology advances play a critical role in making composites affordable, changes in design practice are also critical. The traditional "build to print" approach, where an artifact is first "designed" and then passed to a subcontractor for manufacturing, contributes much to the cost of the final composite product. And though themes like Concurrent Engineering (CE) and Design for Manufacturing (DFM) have made impacts, there still remains much to be done in this arena.

THE DESIGN ACTIVITY IN COMPOSITES

It is critical to remember that the promise of polymer matrix composites as an advanced material comes at the price of significantly more design complexity. This increase in design complexity has two serious consequences. The first is that in an effort to control the complexity, only simple geometries are used for composites (e.g. skins and stiffened panels). The second is the ingenuous reproduction of complex metal geometries in composites. This direct replacement of materials fails to consider the anisotropic nature of composites. The first underutilizes composites and the other, in an attempt to broaden their use, uses them incorrectly.

To effectively use composites, the designer must consider the interacting dimensions of materials, part/assembly and the manufacturing processes. This puts an increased burden on the designer and requires iterative considerations of these issues as the design evolves and becomes more and more concrete. This is very different from the linear approach to the design activity often employed for metal products. Many in that design community are now embracing the themes of concurrent engineering and integrated product/process development. These efforts are showing impressive gains through shorter design times, improved product quality, and better management of development costs.

INTELLIGENT DECISION SUPPORT

A range of opportunities to aide the design process using advanced computing techniques exits. The structural analysis community continues to expand their capability to model composites with ever increasing fidelity. Computing techniques based on artificial intelligence and knowledge-based systems also have considerable in the design activity surrounding composites. Knowledge-based systems techniques enable the effective capture and reuse of design expertise and heuristics. These systems can also suggest design alternatives in materials, structural concepts, or processing technologies. Knowledge-based systems support the designer by taking on more routine, mundane tasks and allowing the designer to be innovative and creative.

Our decision support vision does not see a software panacea for the composites design problem. Rather, we envision a design activity where reasoning engines and computing engines work in concert with the designer. This facilitates an exploration of a wide design space and a subsequent concentration on an affordable and effective combined materials, structures and process solution. A number of traditional computing tools and intelligent decision support systems assist the designer in exploring the design space as well as evaluating alternatives.

PREVIOUS WORK IN INTELLIGENT DECISION SUPPORT FOR COMPOSITES

Intelligent decision support system prototypes already exist for process design and material system design. These systems complement the current work described in this paper in manufacturing sequence design. COMADE (COmposite MAterial DEsigner) [Lenz, et al, 1994], is a third generation prototype for polymer composite material systems design. Considering the performance requirements and the environmental conditions for composite part/assembly, COMADE generates multiple designs for the matrix material, any necessary chemical agents, the fiber type and the fiber length. COMADE can generate over 1000 different material system designs. COFATE (COmposite FAbrication Technology Evaluator) [Moy, et al, 1994], is a second generation prototype for polymer composite fabrication technology selection. COFATE evaluates 16 different fabrication technologies based on qualitative economics, abstractions concerning the part/assembly geometry and other processing issues.

These intelligent decision support systems provide considerable assistance in the composites design process. COMADE provides choices for composite material systems and COFATE suggests how to manufacture the composite assembly or portions of the assembly. This assistance was demonstrated in MADEFAST, a collaboration design experiment, where COMADE and CTechSel [McDowell, et al, 1996] (the first generation process selector) were used in the design of a polymer composite seeker casing [Lenz, et al, 1994]]. MADEFAST did not involve any intelligent decision support for assembly design, either at the conceptual or detailed phases, and once the process technology was selected, design of the process parameters was straightforward. An overview of the evolution of COMADE and MADEFAST can be found in a recent publication [Lenz, et al, 1996].

CURRENT FOCUS IN INTELLIGENT DECISION SUPPORT FOR COMPOSITES DESIGN - THE NEXT LEVEL

The next level of activity is characterized by providing decision support to the general problem of converting metal assemblies into composites. Here, issues of geometry and the behavior and function of the structure seem to dominate. There is also a need for rough costing information early in the design process. A number of traditional computational tools are available for analyzing composite structures based on the finite element method, the boundary element method, or finite difference method. Tools to support the early stages of design where the concept for the composite assembly is being formed are, however, nearly nonexistent. Our R&D activity to address the development of intelligent decision support tools for this problem is embodied in two related efforts. The first effort takes a metal assembly as a starting point and produces a preliminary design result called a conceptual composite assembly. The second effort takes that conceptual composite assembly design and produces concepts for the manufacturing plans to produce that composite product. Each problem solving activity is expected to produce multiple results and the generation of evaluation metrics is critical for the designer to interpret the results.

The use of metals follows riles of practice consistent with homogeneous materials and typical metals processing techniques. The geometric features of a metal part are heavily influenced by these factors. With such homogeneous materials, it is the geometry that then determines the mechanical properties of the structure. Specific geometric features are used to provide strength, stiffness and weight reduction. As composites are anisotropic, they are quite different from homogeneous metals. Composites have internal details that contribute to directional mechanical properties. Therefore, when redesigning a metal assembly to composites, a designer should not fixate on gross metal features.

The redesign of a metal part using polymer composites can have the following possible goals: 1) further reduce the weight of the part, and 2) reduce the cost of fabrication with lower tooling cost molding and part count reduction. Critical issues in the redesign of metals to polymer composites include:

- The redesigned artifact must reproduce the equivalent functionality.
- The redesign artifact must fit into the surrounding superstructure. It is NOT necessary to reproduce all the original geometry.
- There is a focus on pre-geometric and partial geometric features with the goal of leveraging the use of polymer composites.

It should be noted that a polymer composite design that attempts to reproduce all the geometric features of the metal component could lead to a project failure, a part that could not satisfy all the functionality, or a part that could not be affordably fabricated.

TASK STRUCTURE FOR CONCEPTUAL DESIGN - OVERVIEW

The process for conceptual design begins with the existing metal assembly, its geometry, materials, and knowledge about how the components and features of the metal assembly provide the functionality to achieve various goals. A reverse engineering process abstracts the knowledge about the metal assembly into two related knowledge models. The first knowledge model represents the functional aspects of the metal assembly. The second knowledge model represents the configuration of the metal assembly and shows the relationships among the components and features in the metal assembly. The configuration knowledge model also shows how the assembly connects to the larger system that surrounds it. The composite assembly must reproduce the functionality of the existing metal assembly and it must fit properly in the existing superstructure. The next step replaces portions of the metal assembly with components made of composites. The goals of this process are to reduce the cost of manufacturing the assembly and/or reduce the assembly's weight. In the context of the load paths in the assembly and the number of fasteners used, the assembly is retrofitted with composite and, where possible, segments are consolidated to eliminate fasteners. The conceptual composite assembly then goes through a process of criticism and modification. This stage eliminates incompatible configurations, insures that loadings are qualitatively satisfied and identifies areas which contribute to downstream costs. The final problem solving step recovers important and relevant geometry information from the original metal assembly. This task structure and related problem solving is described in [McDowell, et al, 1995] and [McDowell, et al, 1997]. The final design result is then a conceptual composite assembly that includes configuration and functional knowledge about the assembly as well as partial geometric information. The conceptual composite assembly forms a natural basis for detailed product design but also provides a starting point to designing the manufacturing process.

CoMAP: COMPOSITE MANUFACTURING ASSEMBLY PLANNER

THE PROBLEM SOLVING ACTIVITY

The focus of this work is to concentrate on conceptual design for manufacturing. The basic theme is to begin to determine the manufacturing sequence for a product as it is being conceptualized. Starting with a conceptual composite assembly (described in the previous section), the goal is to generate the manufacturing process sequence that would produce the composite assembly artifact. The conceptual composite assembly product design contains: configuration information about the assembly and its components, how they are connected to each other and to the surrounding superstructure; functional information about how the assembly and its components satisfy the design requirements; and partial geometric information on dimensions and features of the assembly and its components. The manufacturing process sequence includes: the composite fabrication processes used to produce the components; the machining and post-process steps that add features to components or subassemblies; and the bonding and fastening sequences used to construct subassemblies and the final assembly.

Figure 1 summarizes these notions and in a graphical form presents the activities that must take place in the problem solving architecture called CoMAP, which stands for Composite Manufacturing Assembly Planner. Using CoMAP, designers can explore several process sequence alternatives and even plan subcomponent fabrication studies resulting in a more robust and affordable final designs. These process sequence alternatives also provide preliminary information for traditional process planners and shop floor scheduling systems.

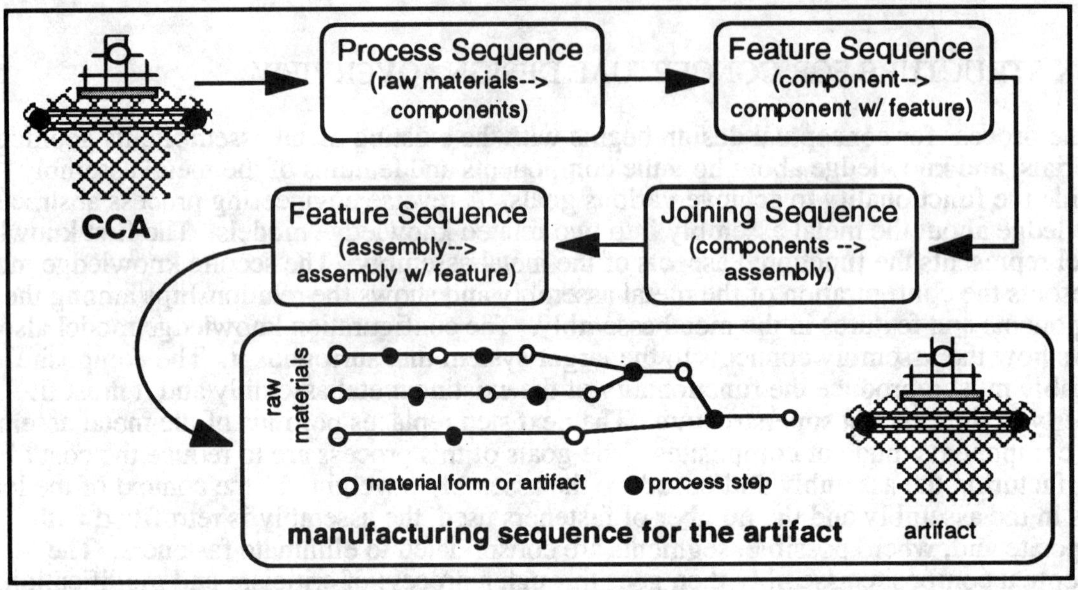

Figure 1. Expected stages for problem solving activity.

A TESTBED - THE COMPOSITE BONDED WINGBOX

A testbed example was designed that would represent the output of the metal-to-composites redesign system and yet exercise all aspects of the manufacturing sequence system. A composite bonded wing box (CBWB) patterned after Air Force's bonded wing structure is being used as a testbed. An inventory of the composite components is provided in Table 1.

Table 1. Composite Bonded Wing Box (CBWB) Components.

Component Name	Number Needed	Component Name	Number Needed
stringer	10	rear spar stiffener	2
upper skin	1	upper rib chord	3
lower skin	1	lower rib chord	3
front spar web	1	rib web	2
front spar stiffener	2	rib web w/access	1
rear spar web	1	access door	1

There are enough individual composite components in the CBWB to explore a wide variety of assembly sequences as well as component generation and feature generation.

CoMAP's PROBLEM SOLVING SEQUENCE

Table 2 outlines the problem solving sequence used by CoMAP to arrive at a set of conceptual manufacturing sequences for the given conceptual composite assembly.

Table 2. Problem Solving Sequence for CoMAP.

- Design Model Analysis and Manufacturing Model Generation.
- Fabrication Technology Selection.
- Assembly Sequence Generation.
- Manufacturing Sequence Generation.
- Cost Metrics Determination.

The operations described in Table 2 do not necessarily proceed in sequence. At many points in the problem solving, the user may change aspects of the design and repeat several of the operations.

The first stage of problem solving in CoMAP involves changing the configuration model of the conceptual composite assembly from a design centered model to a manufacturing centered model. This transformation involves converting design features to manufacturing features and may involve the conversion of features to components or components to features. The result of this analysis is the *assembly hierarchy*, which represents the organized constraints that apply to assembling the components in the composite assembly. The assembly hierarchy is used to generate the assembly sequences.

The configuration model contains knowledge about how the components in the composite assembly are connected to each other as well as how they connect to the overall superstructure. The configuration model also identifies the nature of the connection or joining in terms of the type of join. Components can be mechanically fastened, bonded, fastened & bonded, co-cured or just in contact with each other. This knowledge from the configuration model is important in the assembly operations that follow the production of the individual components in the assembly.

The next stage in CoMAP problem solving is fabrication technology selection. The production of each component is determined by an intelligent decision support system pattered after the COFATE system described in [Moy, et al, 1994]. The inputs for process selection will come from the composite assembly configuration model as well directly from the designer. The designer will specify a number of inputs related to the production scenario he would like to explore. Like COFATE, the process selector is likely to suggest more than one feasible processing technology for each component considered. Table 3 presents some possibilities for the components in the CBWB assembly.

The list in Table 3 is only one of many possible combinations of processing technologies to produce the components of the CBWB assembly. For example, the upper and lower skins could be produced using a fiber placement machine rather than by hand layup. Alternative process technologies are also possible for the other components in Table 3.

Table 3. Component Fabrication Possibilities.

Component Name	Process Technology
stringer	pultrusion
upper skin	prepreg hand layup w/autoclave cure
lower skin	prepreg hand layup w/autoclave cure
front spar web	prepreg hand layup w/autoclave cure
front spar stiffener	pultrusion
rear spar web	prepreg hand layup w/autoclave cure
rear spar stiffener	pultrusion
upper rib chord	RTM
lower rib chord	RTM
rib web	compression molding
rib web w/access	compression molding
access door	compression molding

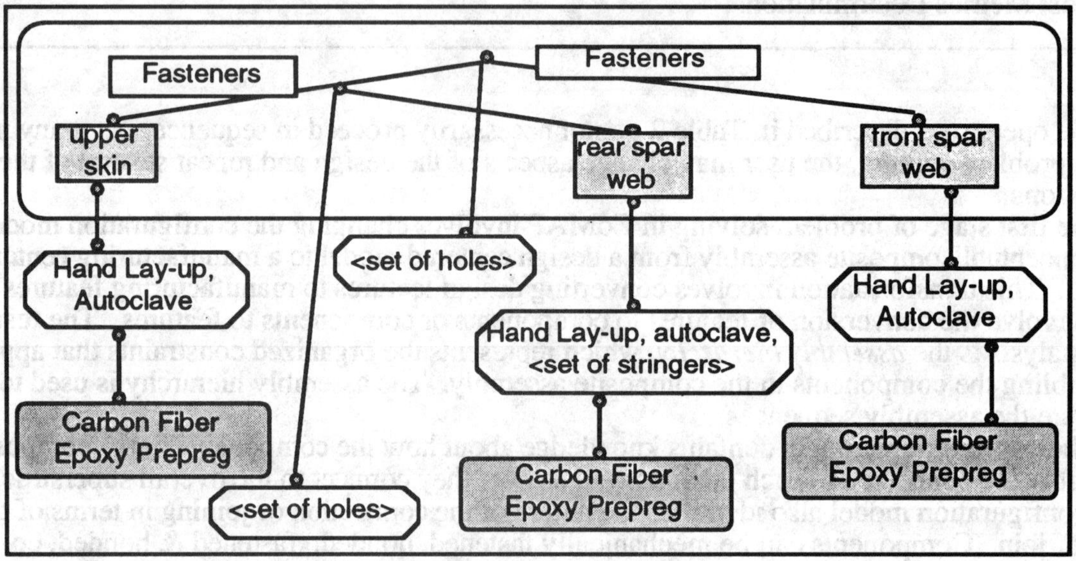

Figure 2. Process sequence for assemblies and components.

Figure 2 shows the conceptual process sequence for a portion of the CBWB. Table 3 identifies these components (upper skin, rear spar web and the front spar web) as being produced using hand layup with autoclave curing. Identifying the process determines the starting materiel form (a prepreg), the component fabrication process steps and the necessary tooling and process equipment. This information is retrieved from a process sequence library and instantiated for the components.

At least one conceptual process sequence is defined for each component in the composites assembly. In most cases there will be multiple conceptual process sequences for a component given that more than one candidate process technology could be recommended and variations within a single process technology are possible.

COST METRICS FOR THE MANUFACTURING SEQUENCE

The manufacturing sequence then includes the process sequences for all the components in the assembly, the assembly sequence for combining components into subassemblies and subassemblies into the complete assembly and any post-processing operations to add features to components, subassemblies or the final assembly. As stated before, it is very likely that the designer will be presented with several design alternatives to manufacturing this composite assembly. Table 4 presents some of the preliminary cost metrics that CoMAP can use to rank order these alternatives and provide some tradeoffs to the designer at this early stage.

Table 4. Possible Cost Metrics.

Number and Types of Process Steps
Number and Types of Process Equipment
Number of Primary Tools
Number of Secondary Tools and Fixtures

In addition to the items in Table 4, CoMAP is also expected to provide criticism of the various manufacturing sequences based on knowledge about processing composites in an actual manufacturing environment. These would be more pragmatic issues, related to machining composites, bonding and fastening composites that are not touched on in the generation of the conceptual manufacturing sequences.

SUMMARY/CONCLUSIONS

This paper presents a unique approach to providing concepts for manufacturing sequences for a composite assembly, early in the design phase. The system CoMAP is described in terms of the major task elements involved as well as the inspirations for many of the features of this problem solving activity. These manufacturing sequences are determined from a conceptual composite assembly design which represents the functional and configurational aspects of the assembly along with partial geometric information. These manufacturing sequence concepts can then be reviewed under a number of cost metrics. They can also be used to guide shape detailed design by suggesting useful subcomponent fabrication studies and preliminary design of the manufacturing process.

ACKNOWLEDGMENTS

This research is supported by the Defense Advanced Research Projects Agency's (DARPA) Rapid Design Exploration and Optimization (RaDEO) Program under the direction of Mr. K. Lyons and by the Research Excellence Fund of the State of Michigan. The principal investigators are also supported by the National Science Foundation's Rapid Prototyping Initiative and by the S/I/U CRC on Low-Cost High-Speed Polymer Composites Processing at Michigan State University. The Intelligent Systems Laboratory is also supported by Apple Computer through its generous equipment donations. The authors would also like to thank John Scanlon, Gary Wigell and Gary Franck Jr. of Auto-Air Composites for their valuable insights in composites design and manufacturing.

All correspondence concerning this paper should be addressed to Dr. James K. McDowell, Michigan State University, 2100 Engineering Bldg., East Lansing, MI 48824. phone (517) 353-5010, FAX (517) 432-1634, email mcdowelj@cps.msu.edu

REFERENCES

Lenz, T., McDowell, J.K., Sticklen, J., and Hawley, M.C., "An Integrated Approach to the Design and Fabrication of a Composite Seeker Casing," MSU Processor, NSF S/I/U Center for Low Cost, High Speed Polymer Composites Processing, Michigan State University, Vol. 1, Issue 5, 1994.

Lenz, T., McDowell, J.K., Moy, B., Sticklen, J., and Hawley, M.C., "Intelligent Decision Support for Polymer Composite Material Design in an Integrated Design Environment," Proceedings of the American Society of Composites 9th Technical Conference, Newark, Delaware, September 1994, pp 685-691.

Lenz, T., J. K. McDowell, A. Kamel, J. Sticklen, & M. C. Hawley. The Evolution of a Decision Support Architecture for Polymer Composites Design. IEEE Expert, 11(5), 1996, 77-83.

McDowell, J.K, T.J. Lenz, M.C. Hawley, J. Sticklen, J.F. Scanlon and G.A. Wigell, "Virtual Prototyping for Composites," Proceedings of the American Society of Composites 10th Technical Conference, 1995, pp. 12-21.

McDowell, J.K., Kamel, A., Sticklen, J., and Hawley, M.C., "Integrating Material/Part/Process Design for Polymer Composites. A Knowledge-Based Problem Solving Approach," Journal of Thermoplastic Composite Materials , Vol. 9, July 1996, pp 218-238.

McDowell, J.K., J. Sticklen, Y. Lin, K. Zhou and M. C. Hawley with J. F. Scanlon and G. A. Wigell, "Conceptual Design of Composite Assemblies in the Redesign of Metal Assemblies." appears in the Proceedings of ACCE'97, Advanced Composites Conference & Exposition, held in Detroit, MI, April 1997.

Moy, B., McDowell, J.K., Lenz, T., Sticklen, J., and Hawley, M.C., "Expansion of an Intelligent Decision Support System for Process Selection and Process Design in Polymer Composites," Proceedings of the 26th International SAMPE Technical Conference, Atlanta, Georgia, October 1994, pp 162-171.

Microwave Processing of Thick-Section Polymer Composites

E. T. THOSTENSON AND T.-W. CHOU

ABSTRACT

In the manufacture of thick-section thermosetting composite laminates, the low thermal conductivity of most polymers and the exothermic nature of the curing process combine to create complex temperature distributions within the laminate. These thermal gradients result in non-uniform curing and high residual stresses. Microwaves deliver energy to materials through the interaction of electromagnetic fields at a molecular level. The result is heat generation throughout the volume of the material. Volumetric heating can reduce thermal gradients, decrease processing times, and save energy due to the more efficient transfer of energy. Thus, microwave processing is a desirable technique for manufacturing thick-section composites.

Microwave heating has been examined numerically and experimentally as an alternative to conventional thermal processing techniques. A numerical simulation has been developed to predict the one-dimensional transient temperature profile of the composite laminate during both microwave and conventional heating. Numerical and experimental results are presented for a glass/epoxy laminate with a thickness of 25 mm (1 inch). Results of the numerical simulation and experimental work show that it is possible to cure thick composites uniformly and eliminate temperature excursions due to chemical reactions during the cure cycle. In addition, through continuous wave feedback control of the microwave power it is possible to monitor the cure behavior of the composite laminate.

INTRODUCTION

In the manufacturing of composite materials, the processing cycle must be designed to optimize the composite properties. The flow and cure behavior of the resin, thermal properties of the constituent materials, thermal history of the manufactured composite, and the composite microstructure all contribute to the final quality of the composite material. In many composite materials applications, thick-section composites are required to meet the requirements of the imposed loading conditions. For thick laminates, the processing difficulties are often magnified. The proper manufacturing of these laminates requires careful design of the cure cycle to maintain a uniform temperature profile to avoid material degradation and thermal stresses.

Erik Thostenson and Tsu-Wei Chou, Center for Composite Materials and Department of Mechanical Engineering, University of Delaware, Newark, DE 19716

In traditional autoclave processing, materials must be heated at a slow rate to allow for the conduction of heat through the thickness of the material. As a result, the processing times for thick laminates are often excessive due to the low thermal conductivity of the polymer matrix. In addition, the exothermic chemical reaction of thermosetting resins generates heat that is slow to dissipate through conduction. The consequence is a complex temperature distribution within the laminate that can give rise to non-uniform cure, thermal degradation, and process induced stresses.

Unlike traditional autoclave processing, which relies on heat transfer through convection and conduction, microwave processing supplies energy directly to the material through interaction with the microwave field. This results in heating throughout the material thickness. Thus, in microwave processing, thermal gradients and processing times can be reduced through the volumetric deposition of energy. In addition, because heating is instantaneous when the microwaves are turned on, the internal temperature excursion that often occurs due to the exothermic chemical reaction can be minimized by turning off the microwave field and allowing heat to dissipate to the unheated cavity.

Energy transfer at the molecular level can also have some additional benefits. Many researchers report non-thermal effects that are not simply a result of the more uniform and rapid heating that is possible with microwaves. These non-thermal effects have been broadly termed "microwave effects." Some of the "microwave effects" that have been reported by researchers includes accelerated reaction rates and lower activation energies of thermosetting resins (Marand, Baker, Graybeal, 1992). Thus, it may be possible to reduce the processing time through more uniform, volumetric heating and through accelerating chemical reactions with microwaves.

Although there are clearly advantages to using microwaves for processing of materials, the apparent simplicity of microwave heating can be misleading. Microwave processing is a dynamic process where the ability of the material to absorb microwave energy, the state of cure, and the temperature are interrelated. It is important to accurately model the microwave curing process in order to optimize the manufacturing of thick-section composites. Therefore, it is important to understand the fundamentals of the microwave/materials interactions and electromagnetic fields in addition to the other relevant process variables. A brief introduction to microwave/materials interactions and microwave processing equipment is followed by the experimental and numerical work on microwave processing of thick-section composites.

MICROWAVE/MATERIALS INTERACTIONS

Energy transfer occurs through the interaction of microwaves, the portion of the electromagnetic spectrum with wavelengths between one millimeter and one meter, with electric dipoles within the material. The dielectric properties determine the ultimate effect of the electromagnetic field, and for heat to be generated within a material, the microwaves must be able to penetrate the material and transmit energy. The dielectric constant (ε') and the dielectric loss factor (ε'') describe the dielectric response. They are commonly expressed in terms of the complex dielectric constant (ε^*)

$$\varepsilon^* = \varepsilon' - i\varepsilon'' \qquad (1)$$

or the loss tangent:

$$\tan\delta = \frac{\varepsilon''}{\varepsilon'} \qquad (2)$$

Dipole polarization is considered to be the dominant polarization mechanism for energy transfer at microwave frequencies (Mijovic and Wijaya, 1990). Local charge moves in response to an applied electric field, and motion of bound charge results in polarization. The motions of the dipoles are restricted, and the time lag between the applied field and the motion of the dipoles is due to dissipation of energy as heat. This phenomenon is known as dielectric relaxation. Dielectric relaxation is analogous to relaxation in mechanical systems and electrical circuits.

For thin materials, where the electric field can be considered constant across the thickness, the following approximate equation for power, P, per unit volume can be used:

$$P = 2\pi f \varepsilon'' E^2 \tag{3}$$

where f is the microwave frequency and E is the electric field strength.

For thick materials, the electric field decreases as energy is absorbed through the depth. Penetration depth is defined as the distance from the sample surface where the electric field is $(\exp)^{-1}$ of the value of the surface. Beyond the penetration depth, heat generation due to microwaves is minimal. By assuming that the dielectric constant of free space is ε° and c is the speed of light, the penetration depth, d, is given by the following equation (Mijovic and Wijaya, 1990):

$$d = \frac{c\varepsilon^{o}}{2\pi f \varepsilon''} \tag{4}$$

Equations (3) and (4) give insight as to which materials are suitable for microwave processing. Materials with high loss factors have penetration depths that approach zero and are considered reflectors. Materials that have low loss factors have penetration depths which approach infinity, and they absorb very little power.

Depending on the chemical structure, thermosetting resins couple with microwaves more readily when they are in their liquid state. As crosslinking occurs, the dielectric loss decreases because the dipoles become more restricted (Martinelli, Rolla and Tombari, 1985).

MICROWAVE APPLICATORS

The microwave applicator, or cavity, transmits the electromagnetic energy to the material being heated. The geometry of the applicator determines the distribution of the electromagnetic field. Because non-uniform electromagnetic fields result in non-uniform heating, the design of microwave applicators is important for microwave processing. Single mode and multi-mode resonant cavities are the most common applicators used for materials processing due to their high field strengths.

The single mode cavity is designed specifically to support one resonant mode. In general, these types of cavities have an area within the cavity that has a "hot spot" where the electromagnetic field strength is high. Because these cavities can be characterized using analytical techniques, the location and area of high field strength is known. From this knowledge, the object to be heated can be placed in the cavity to obtain optimum coupling with the microwaves. Because these cavities are designed to support one resonant mode, the size of the cavity is on the order of one wavelength (12.2 cm at the common industrial frequency of 2.45 GHz). Single mode cavities are usually designed for a specific application. They have been used for dielectric properties measurement (Guillon, 1990), continuous pultrusion (Ghaffariyan, and Metheven, 1991), and thermoplastic tape placement (Lind et al., 1990).

Because of their non-uniform fields and small size, single mode applicators are unable to heat large or irregularly shaped objects.

Multi-mode cavities are large enough to support several different resonant modes. The presence of multiple modes results in a more uniform distribution of electromagnetic fields within the applicator, but the uniformity of electromagnetic fields is still a concern. Often there are multiple hot spots within multi-mode cavities. The presence of the different hot spots can result in non-uniform heating of large objects. The uniformity of the electromagnetic fields can be enhanced by increasing the size of the applicator because the number of resonance modes increases with the size of the cavity. To further enhance the uniformity of the microwave field, mode stirrers are used. Mode stirrers rotate within the microwave cavity and constantly reflect and redistribute the electromagnetic waves. In addition, the uniformity can be enhanced by adding multiple microwave inputs within a cavity. Different inputs are capable of exciting different resonant modes. The uniformity of the microwave fields can be further enhanced by processing at higher frequencies or by varying the frequency of the microwave output. Both methods increase the number of possible resonant modes within the applicator, but operating at higher frequencies reduces the depth of penetration because, in Equation (4), the penetration depth is inversely proportional to the microwave frequency.

EXPERIMENTAL

To investigate the processing of thick-section composites using microwave heating, 25 mm (1 inch) thick laminates were processed in a large multi-mode microwave furnace. The microwave system, located at the University of Delaware Center for Composite Materials, is shown in Figure 1 (Model 101

Figure 1: 6 kW, 2.45 GHz, multi-mode microwave processing system.

Figure 2: Experimental set-up for microwave processing of polymer composites through vacuum assisted resin transfer molding or hand lay-up/vacuum bag molding.

Microwave Furnace, Microwave Materials Technologies, Inc., Knoxville, TN). The applicator is a cylindrical multi-mode cavity with an internal volume of 500 liters. The large cavity is equipped with a mode stirrer and multiple microwave inputs to enhance the uniformity of the microwave field. The system is run by a computer-based control and data acquisition system with a PID controller, and the microwave source generates up to 6 kW of microwave power at a frequency of 2.45 GHz. The output power of the microwave source can be varied continuously from 0 to 6 kW.

The polymer composite laminates were manufactured using a Vacuum Assisted Resin Transfer Molding (VARTM) technique. The experimental set-up is shown in Figure 2. The resin used for the experimental investigation of microwave curing was a bisphenol-F/epichlorohydrin epoxy resin (Shell Epon 862) mixed 100/26.4 parts by weight with an aromatic diamine curing agent (Shell Epi-Cure W). The fiber preform consisted of 42 layers of plain weave E-glass fabric. The preform was placed in the microwave, hermetically sealed in a vacuum bag, and drawn under a vacuum to compact the layers together. The materials used for the tooling and sealing of the vacuum bag were chosen so that the microwave energy absorbed by the process materials was minimal. Once a high vacuum was achieved, the resin was drawn into the preform under vacuum. To achieve infiltration of the preform, the resin was heated to 50°C to reduce the viscosity. Once the preform was fully infused, the door to the microwave was closed and the temperature cycle was started. The vacuum pressure was applied throughout the cure cycle.

The temperature cycle used for the microwave curing process was designed to mimic a typical autoclave ramp and soak cycle. The heating rate was set to be 5°C/min during heat up and cool down and the soak temperature was set to be 165°C. Feedback control was accomplished through embedding a shielded thermocouple within the composite laminate. The temperature was maintained at the soak temperature for 150 minutes. Although it is possible to uniformly and to rapidly heat materials using microwaves, the heating rate was selected for comparison with autoclave curing. The heating rate is limited only by the available microwave power, but non-uniformity in the microwave fields is more significant at higher powers. Therefore, it is likely that heating rates below 10°C/min will be required so that areas of the laminate within regions of higher field strength do not become overheated.

NUMERICAL SIMULATION

A numerical model of the microwave curing process was developed to facilitate a parametric investigation of the process. The simulation numerically solves the one-dimensional transient heat conduction equation:

$$\rho C_p \frac{\partial T}{\partial t} = k_{33} \frac{\partial^2 T}{\partial z^2} + \dot{Q}'''_{gen} + \dot{Q}'''_{mw} \tag{5}$$

where ρ, C_p and k_{33} are the density, specific heat, and thermal conductivity in the z-direction, respectively. The term $\dot{Q}'''_{gen}$ is the volumetric heat generation due to the exothermic chemical reaction of the resin, and the term $\dot{Q}'''_{mw}$ is internal heat generation due to the absorbed microwaves. Equation (5) was solved numerically by using an implicit Crank-Nicolson finite difference solution method (Poulikakos, 1994). To simulate the processing conditions within the autoclave, the $\dot{Q}'''_{mw}$ term is dropped and the boundary conditions are modified.

CURE KINETICS

The first internal heat generation term in Equation (5) is the instantaneous heat generation per volume of material due to the exothermic chemical reaction of the resin. This quantity is dependent on the cure rate of the resin. The cure kinetics can be examined experimentally using a Differential Scanning Calorimeter (DSC). The degree of cure is defined using DSC by assuming that the exothermic heat is liberated only by the crosslinking reaction. The internal heat generated as a function of the degree of cure, α, and the total heat of reaction, H_r, is:

$$\dot{Q}'''_{gen} = \rho H_r \frac{d\alpha}{dt} \tag{6}$$

The cure kinetics of bisphenol-A epoxy resin cured with a trimellitic anhydride has been investigated by Woo and Seferis (1990), and the expression for cure rate was determined to be:

$$\frac{d\alpha}{dt} = A e^{\left(-E_a/RT\right)} (1-\alpha)^n \tag{7}$$

where A is the pre-exponential coefficient, E_a is the activation energy, R is the universal gas constant, and n is the order of the reaction. The reaction was determined to follow second order cure kinetics, and the values for the activation energy and pre-exponential coefficient were determined experimentally (Woo and Seferis, 1990).

MICROWAVE HEATING

The interaction of microwaves with anisotropic composite materials has been investigated by Lee and Springer (1984a,b). They have developed a model that determines the transmitted and reflected energy by a single anisotropic ply. The difference between the transmitted and reflected energy is the energy generated by microwaves within that ply. As the microwave penetrates a laminated composite, the energy absorbed and reflected in each ply can be calculated based on the energy transmitted from the previous ply in the laminate. The Lee and Springer model (1984a) has been adapted for this work, and the incident microwave is assumed to be isotropic.

BOUNDARY CONDITIONS

For the simulation of autoclave curing, the top and bottom boundary conditions were assumed to be at the temperature setpoint of the autoclave. In microwave curing, the cavity is heated only by heat lost through convection from the surface of the material being heated. The microwave used in the experimental work is water cooled and it is assumed that the conditions within the microwave are isothermal at 25°C. The surface of the microwave processed laminate is subjected to free convection. The value of $(h/k)_{eff}$ is assumed to be 0.1 cm^{-1} for free convection within the microwave cavity.

RESULTS AND DISCUSSION

The results of the numerical simulation for the center temperature of a 25 mm (1 inch) thick laminate are presented in Figure 3. For the autoclave cure cycle, there is a significant lag between the center temperature of the laminate and the autoclave temperature during the temperature ramp. After the soak temperature is reached, the chemical reaction continues to generate heat, and the result is a significant temperature excursion from the desired temperature. The peak internal temperature of the autoclave processed laminate was 240°C. In the microwave processed laminates, however, the temperature overshoot was virtually eliminated. There was also no lag in temperature during the initial temperature ramp for the microwave processed laminates.

Figure 4 presents the results of the composite manufactured within the microwave. There was no lag during the temperature ramp, and the exothermic overshoot was completely eliminated. The results of the numerical simulation compare well with the experimental results. Although different resin systems were used for the experimental and numerical work, the ability to eliminate the temperature excursions were illustrated in both.

The microwave power is controlled continuously throughout the process through feedback control from temperature measurement. Most microwave systems use an on/off control for programmed power and temperature. For example, if the microwave is set to output 50% power, the microwave operates at full power for 50% of the time. With continuous control of the power output, some information is given

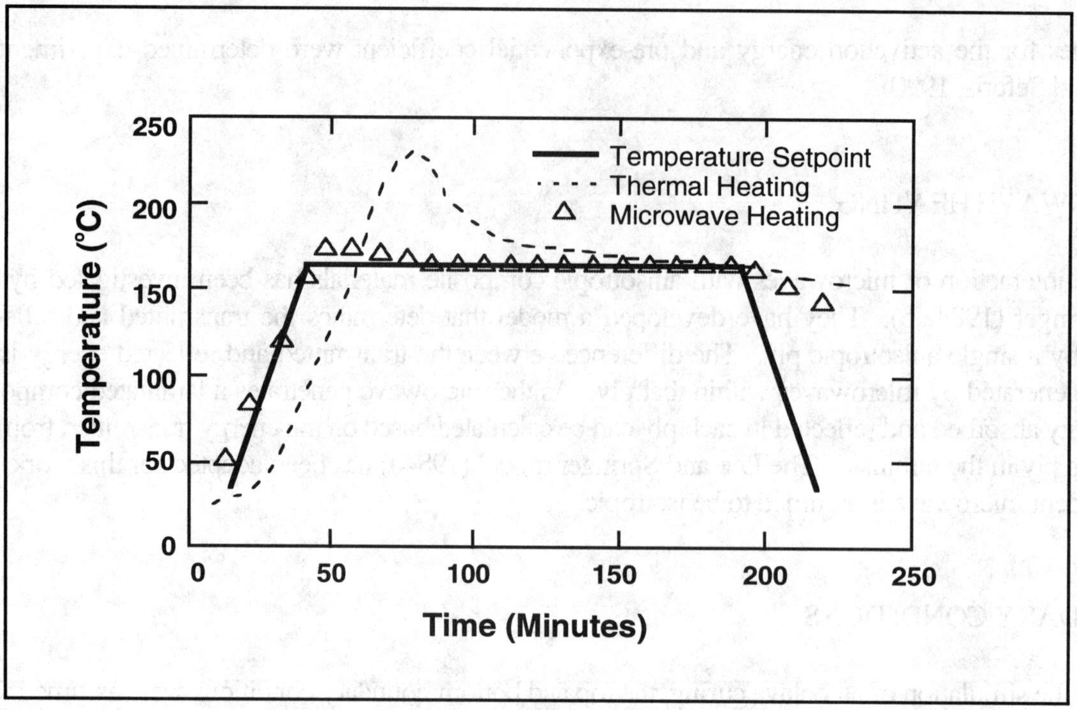

Figure 3: Results of the numerical simulation for the center temperature history of a 25 mm (1 inch) thick laminate for microwave and thermal curing.

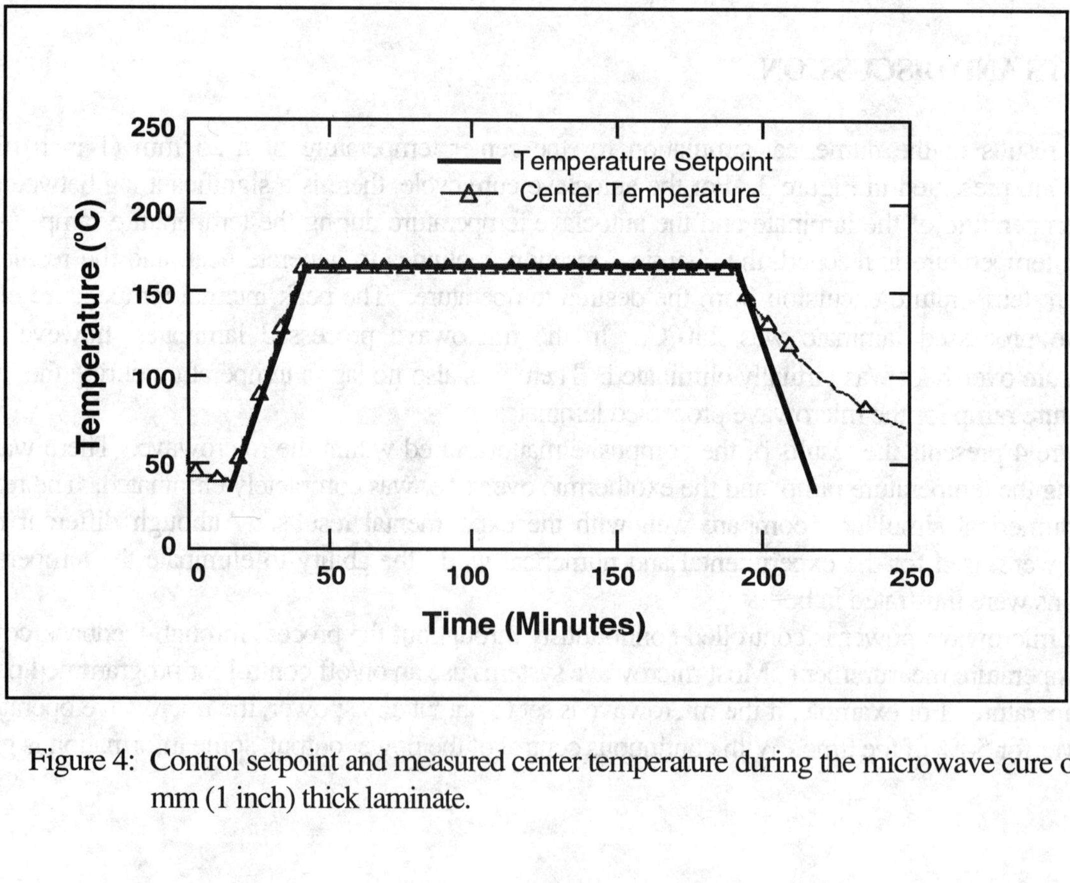

Figure 4: Control setpoint and measured center temperature during the microwave cure of a 25 mm (1 inch) thick laminate.

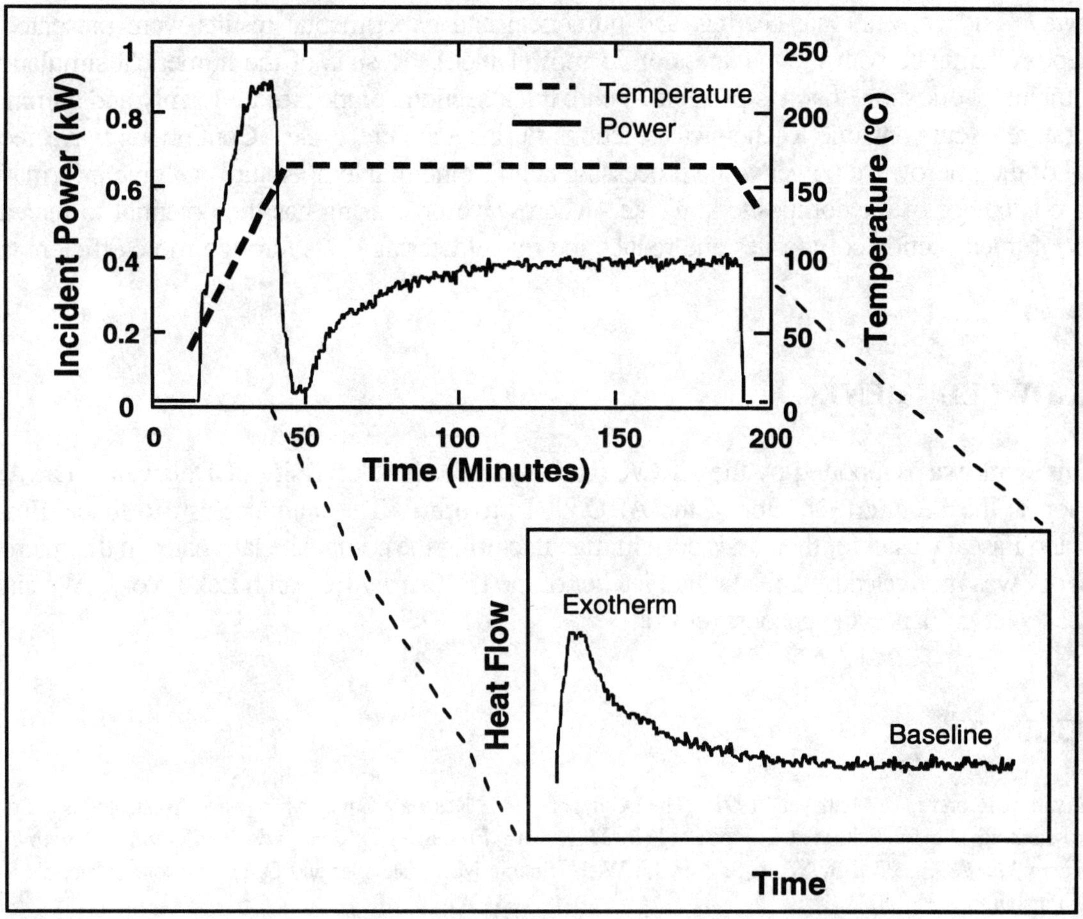

Figure 5: Microwave power applied by the feedback controller throughout the cure cycle

on the progression of the chemical reaction within the laminate. In Figure 5, the incident power is plotted throughout the cure cycle. During the temperature ramp, the power increases rapidly at the beginning of the heating cycle, but as the temperature approaches the soak temperature, the power begins to level off. This indicates that the reaction is beginning to generate additional heat. Once the soak temperature is reached, the microwave power decreases significantly. During the isothermal portion of the cure cycle, the microwave is adding only enough power to balance heat lost due to convection and conduction. If the isothermal portion of the power curve is inverted (Figure 5) it resembles an isothermal DSC thermogram. There is an initial peak where the crosslinking reaction is generating heat. As the chemical reaction progresses, the heat generated begins to decline, and the heat flow approaches the baseline when the chemical reaction is complete. This information may be used to determine when the laminate has been fully cured.

CONCLUSIONS

Microwave processing has been examined numerically and experimentally as an alternative to conventional autoclave processing. A finite difference numerical simulation has been developed to predict the one-dimensional transient temperature profile of the composite laminate during both microwave and conventional heating, and numerical and experimental results were presented for a glass/epoxy laminate with the thickness of 25 mm (1 inch). Results of the numerical simulation and experimental work show that it is possible to cure thick-section composites uniformly and eliminate the temperature excursions due to chemical reactions during the cure cycle. Continuous wave feedback control of the microwave power while processing at an isothermal temperature can give information on the cure behavior of the composite laminate. Microwave processing has the potential to increase the quality of thick-section composites and reduce the manufacturing costs through more efficient transfer of energy.

ACKNOWLEDGMENTS

This work was supported by the ARO/URI program at the University of Delaware. Dr. Andrew Crowson is the Program Director of the ARO/URI program. The authors wish to thank Emanuele Gillio and Pascal Casari for their assistance in manufacturing the composite laminates in the microwave. This work was motivated by Dr. Martha Fletcher of the U.S. Army Research Laboratory. We sincerely appreciate her guidance and encouragement.

REFERENCES

Ghaffariyan, S. R. and J. M. Metheven. 1991. "The Design of a TM_{10} Resonant Cavity Microwave Applicator as a Preheating and Crosslinking Die for Pultruded Composites," in *Microwave Processing of Materials II, Materials Research Society Proceedings Vol. 189*. Edited by W. B. Snyder, Jr., W. H. Sutton, M. F. Iskander and D. L. Johnson. Materials Research Society, Pittsburgh. 135-140.

Guillon, P. 1990 "Microwave Techniques for Measuring Microwave Dielectric Properties," in *Microwave Processing of Materials IV, Materials Research Society Proceedings Vol. 347*. Edited by M. F. Iskander, R. J. Lauf and W. H. Sutton. Materials Research Society, Pittsburgh. 203-208.

Lind, A. C., L. N. Medgyesi-Mitschang, J. E. Kurz, H. F. McKinney and F. C. Wear. 1990. "Microwave Heating for Manufacturing Carbon-Fiber Thermoplastics," in *Microwave Processing of Materials II, Materials Research Society Proceedings Vol. 189*. Edited by W. B. Snyder, Jr., W. H. Sutton, M. F. Iskander and D. L. Johnson. Materials Research Society, Pittsburgh. 461-467.

Marand, E., H. R., Baker, J. D. Graybeal. 1992. "Comparison of the Reaction Mechanisms of Epoxy Resins Undergoing Thermal and Microwave Cure from In-Situ Measurements of Microwave Dielectric Properties and Infrared Spectroscopy," *Macromolecules*, 25:2243-2252.

Martinelli, M., P. A. Rolla and E. Tombari. 1985. "A Method for Dynamic Dielectric Measurements at Microwave Frequencies: Applications to Polymerization Process Studies," *IEEE Transactions on Instrumentation and Measurement*, 34(3):417-421.

Mijovic, J. and J. Wijaya. 1990. "Review of Cure of Polymers and Composites by Microwave Energy," *Polymer Composites*, 11(3):184-190.

Poulikakos, D. 1994. *Conduction Heat Transfer*, Prentice-Hall, Englewood Cliffs, NJ.

Woo, E. M. and J. C. Seferis. 1990. "Cure Kinetics of Epoxy/Anhydride Thermosetting Matrix Systems," *Journal of Applied Polymer Science*, 40:1237-1256.

Microwave and Conventional Joining of Ceramic Composites Using Spin-On Materials

K. N. SEIBER, K.-Y. LEE AND E. D. CASE

ABSTRACT[1]

Polycrystalline alumina, glass ceramic composites reinforced by mica platelets (MaCor™), and alumina/zirconia composites were joined using a commercial spin-on material which is also used as a passivation layer in semiconductor processing. The film thickness was controlled by varying spin speed, spin time, and temperature for heat treatment. Curing a coated specimen at 200°C produced a film thickness ranging from about 0.2 micron to about 0.6 micron. Joining was performed in both a conventional electrical resistance furnace and a single-mode 2.45 GHz cylindrical microwave cavity. Effects of film thickness, joining temperature, and heating time on the joining of the materials will be discussed.

INTRODUCTION

Ceramic components with complex geometric shapes are difficult to process. Intricately shaped components are of interest in high temperature heat exchanger and engine applications. Complex structures, however, can be fashioned by joining ceramic components of simpler shape. Joining simpler ceramic subcomponents assists quality control as well, by allowing flawed subcomponents to be rejected prior to joining.

Using conventional (radiant) heating, ceramic components have been brazed via oxide or oxinitride glasses, slurries of metal particles, and metal layers (Sandage et al., 1996). Oxide glasses have been shown to join alumina (Moorhead, 1987; Zdaniewski, Shah, and Kirchner, 1987) and oxinitride glasses have been used to join silicon nitride (MeCartney, Sinclair, and Loehman, 1985). Diffusion bonding has also been utilized to join ceramic components under high applied pressures and temperatures (Santella, 1992).

Ceramic-ceramic joining using the microwave system typically involves externally applied pressures. An applied load of 3 MPa has been utilized to join alumina (in air) and silicon nitride (in nitrogen gas) (Silberglitt et al., 1991), while loads ranging from 0.6 MPa to 9 MPa have been used to join alumina, mullite, and silicon nitride. (Fukushima,Yamanaka, and Matsui, 1990). The objectives of this study were to 1) join the subcomponents with the use of a spin-on film and 2) to use ambient or low externally applied pressures in the joining process. A low external

---

Kiersten N. Seiber, Ki-Yong Lee and Eldon D. Case, Materials Science and Mechanics Department, Michigan State University, East Lansing, MI 48824

pressure, provided by dead weight loading, avoided the use of loading systems (such as those using a piston and load cell) which can be awkward and interfere with the microwave fields. In this study, a total of 15 pairs of MaCor™ specimens, five pairs of alumina specimens and 1 pair of alumina/zirconia composite specimens were joined.

EXPERIMENTAL PROCEDURE

MaCor™, polycrystalline alumina specimens, and alumina/zirconia particulate composite specimens were joined in this study. MaCor™ is a machineable glass ceramic reinforced by mica platelets (Corning code 9658). The MaCor™ specimens were cut from as received billets of the material into 1 cm x 1 cm components using a low speed diamond saw.

To fabricate the alumina specimens, about 2 grams of alumina powder was hard-die pressed at about 32 GPa from Sumitomo AKP-30 powders, resulting in a disc-shaped powder compact of about 2.2 cm diameter and 2 mm thick. The specimens were sintered at 1550°C for 20 minutes using a 2.45 GHz single-mode microwave cavity (Lee et al., 1996; Lee, Case, and Reinhard, 1997; Lee, Case, and Asmussen, 1997). The average grain size of the sintered alumina specimens was about 8.6 µm after multiplying the average intercept size by a stereographic factor of 1.5 (Fullman, 1953). The density determined by Archimedes technique was about 3.806 g/cm^3, which is about 95% of theoretical density of alumina (National Bureau of Standards, 1959).

For alumina/zirconia particulate composite specimens, AKP30 alumina powder was mixed with 15wt% of zirconia powder (Fisher Scientific Co.). The mixture was ball-milled with alumina grinding media in a plastic container for 48 hours.

Approximately 2 grams of the Al_2O_3/15wt% ZrO_2 powder mixture (pressed using conditions identical to those used for the alumina compacts) was sintered at 1550°C for 20 minutes in the microwave cavity.

Using an automatic polisher (LECO Corporation, St. Joseph, MI), alumina and alumina/zirconia particulate specimens were polished using a series of grits; 25, 17, 10, 6 and 1 µm diamond paste. MaCor™ specimens were polished using 17, 10, and 1 µm diamond paste. After polishing, selected specimens of MaCor™, alumina, and alumina/zirconia were coated with silica film by placing ten drops of the silica film (Silicafilm, Emulsitone Co., Whippany, NJ) onto the specimens and spinning the specimens on a substrate spinner. The specimens were spun for 20 minutes at spinning rates ranging from 500 rpm to 2000 rpm. Curing the film in a conventional oven for 20 minutes at 200°C yielded a high purity silica coating.

Notches were made in five MaCor™ specimens and four alumina specimens with a stationary sonic mill (Sonic-Mill, Rio Grande Jewelers Supply Inc., Albuquerque, NM). A razor blade tool was made by silver soldering a commercial razor blade to a standard flat sonic-mill screw. The sonic mill vibrated the razor blade tool in a slurry of boron carbide particles, which in turn vibrated the boron carbide particles contacting the razor blade tool. The specimen was cut (notched) by the vibration of the boron carbide media itself.

A portion of the specimen containing the notch was then sectioned and reserved for scanning electron microscope (SEM) analysis, in order to examine the notch prior to joining (Figure 1). The remaining section was used in the joining experiment.

Microwave heating was done in a single-mode, 2.45 GHz microwave cavity using the TM_{111} mode (Lee et al., 1996; Lee, Case, and Reinhard, 1997; Lee, Case, and Asmussen, 1997).

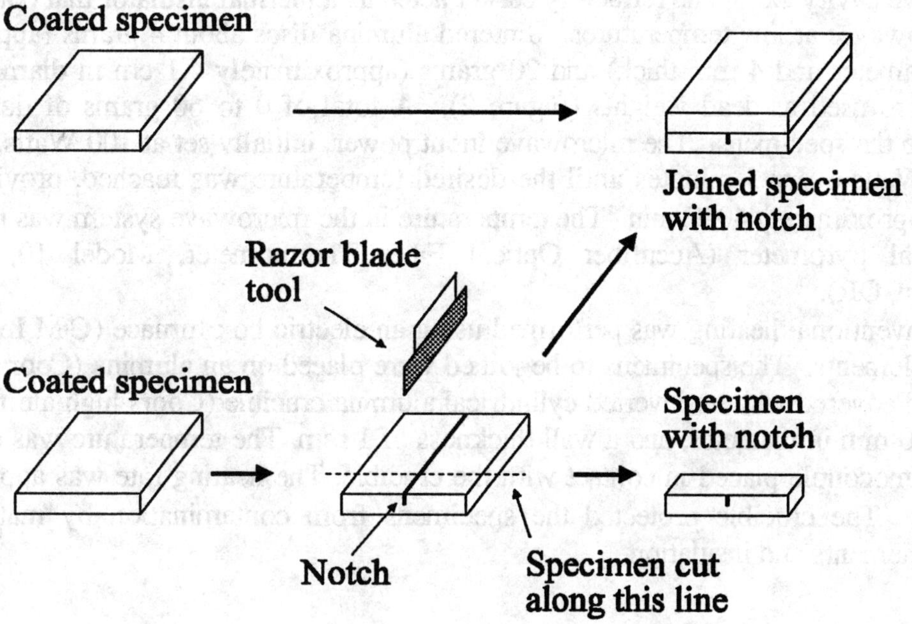

Figure 1. Schematic showing notched specimen preparation.

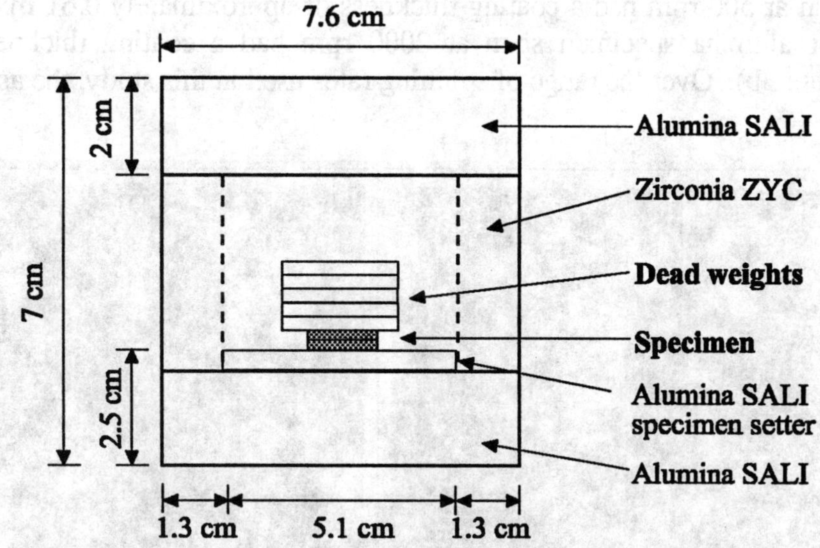

Figure 2. Schematic for refractory casket used for microwave heating, showing dead weights placed on the specimen for joining.

The specimens were positioned in a refractory casket (Figure 2) that was centered along the microwave cavity axis. The refractory casket acted as a thermal insulator that coupled well with the microwaves at low temperatures. Sintered alumina discs about 4 grams (approximately 1.8 cm in diameter and 4 mm thick) and 20 grams (approximately 4.1 cm in diameter and 4 mm thick) were used as dead weights (Figure 2). A total of 0 to 60 grams of dead weight was applied to the specimens. The microwave input power, initially set at 100 Watts, was increased by 100 Watts every 3 minutes until the desired temperature was reached, providing a heating rate of approximately 40°C/min. The temperature in the microwave system was measured using an optical pyrometer (Accufiber Optical Fiber Thermometer, Model 10, Luxtron Co., Beaverton, OR).

Conventional heating was performed using an electric box furnace (C·M Inc.) with $MoSi_2$ heating elements. The specimens to be joined were placed on an alumina (Coors high alumina) setter and covered with an inverted cylindrical alumina crucible (Coors high alumina) 60 mm in height, 10 mm in diameter, and a wall thickness of 1 mm. The temperature was sensed by a K-type thermocouple placed in contact with the crucible. The heating rate was approximately 30-40°C/min. The crucible protected the specimens from contamination by material from the furnace elements and insulation.

RESULTS AND DISCUSSION

Four MaCor™ specimens and four alumina specimens were coated using spinning rates between 500 rpm and 2000 rpm. The specimens were then cured at 200°C for 20 minutes. The silica film thickness was a function of the spinning rate, such that the coating thickness decreased with increasing spinning rate (Figures 3 and 4, Table I). For example, an alumina specimen spun at 500 rpm had a coating thickness of approximately 0.61 microns (Figure 3a), while another alumina specimen spun at 2000 rpm had a coating thickness of about 0.20 microns (Figure 3b). Over the range of spinning rates used in this study, the as-cured silica film

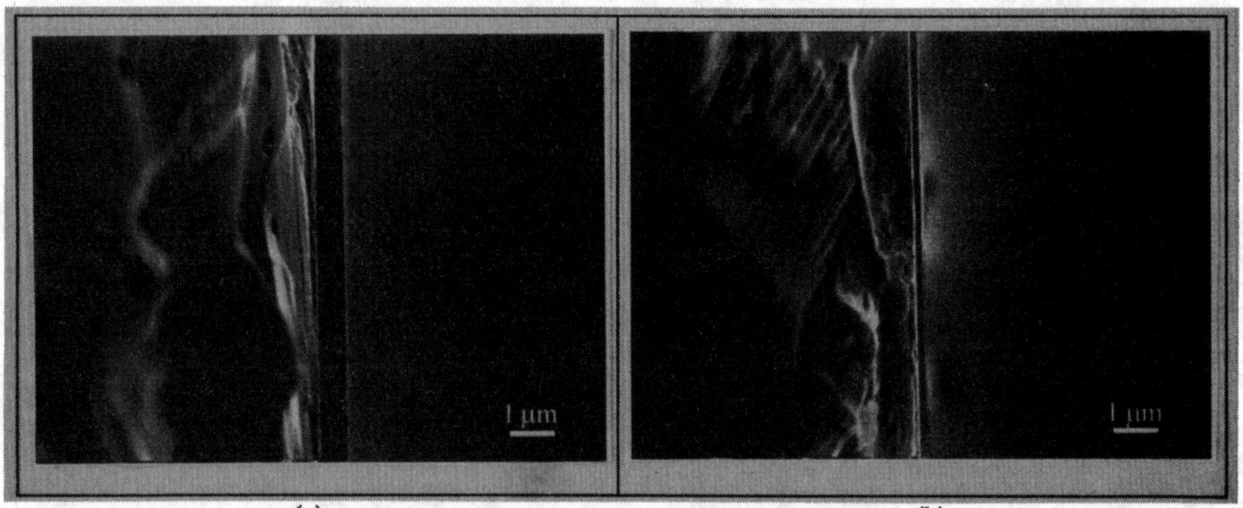

(a) (b)

Figure 3. SEM images of silica film spun on alumina specimens at (a) 500 rpm and (b) 2000 rpm.

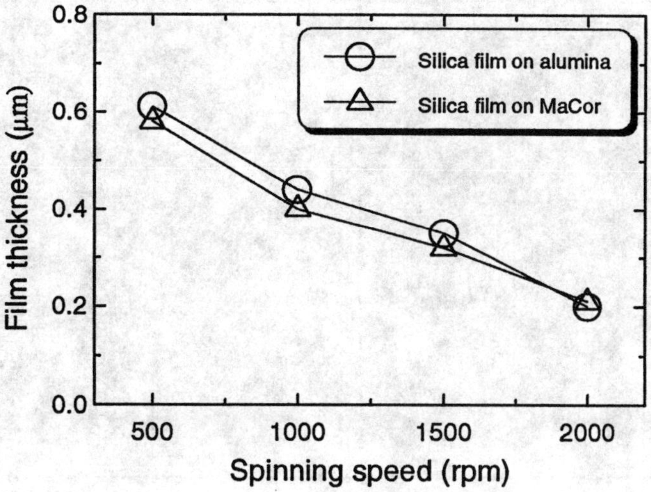

Figure 4. Silica film thickness as a function of spinning speed.

TABLE I. THICKNESS OF THE AS-CURED SILICA FILM AS A FUNCTION OF SPINNING SPEED.

Spinning speed	500 rpm	1000 rpm	1500 rpm	2000 rpm
Silica film on alumina (µm)	0.61	0.44	0.35	0.20
Silica film on MaCor™ (µm)	0.58	0.40	0.32	0.21

thicknesses for the MaCor™ and alumina specimens agreed to within ± 9% (TABLE I).

Ten pairs of MaCor™ specimens were joined in the microwave cavity using 20 gram dead weight loading at temperatures of 1050°C and 1075°C with an approximate joining time of 20 minutes. Conventionally, five pairs of MaCor™ specimens were joined at 1050°C with no externally applied load and with 20 gram dead weight loading, using an approximate joining time of 20 minutes. Initial results indicated that heating times shorter than 20 minutes may not be sufficient to join MaCor™ specimens. For example, one pair of MaCor™ specimens was tested with a joining temperature of 1050°C, a joining time of 10 minutes, and a 20 gram dead weight loading. The 10 minute heating time failed to join the MaCor™ specimens.

The width and depth of notches in the MaCor™ specimens changed by less than 5.7 percent during joining (Figure 5, Table II), while for the alumina specimens the notch width and depth changed by less than 4.1 percent (Figure 6, Table II). Note that the dimensional stability observed for the notch geometries holds for a range of notch sizes (Table II).

For five alumina and for five MaCor™ specimens, the fraction of the interface that included pores was evaluated by SEM observations, where F_P is defined as

$$F_P = \frac{\text{total length of porous regions along specimen interface}}{\text{total length of specimen interface}} \quad (1)$$

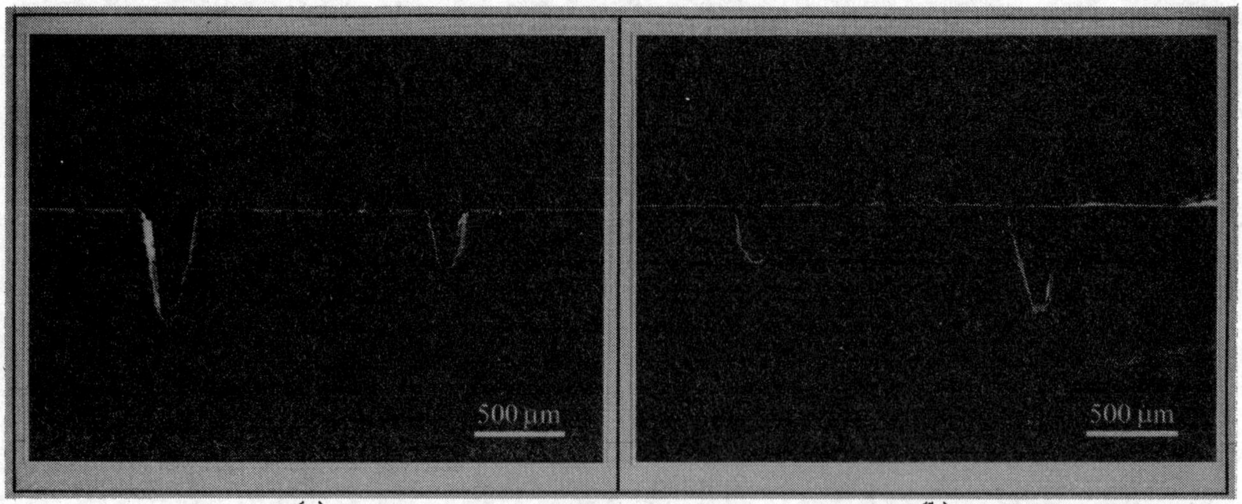

(a) (b)

Figure 5. SEM images of notch made in MaCor™ specimen (a) before and (b) after joining.

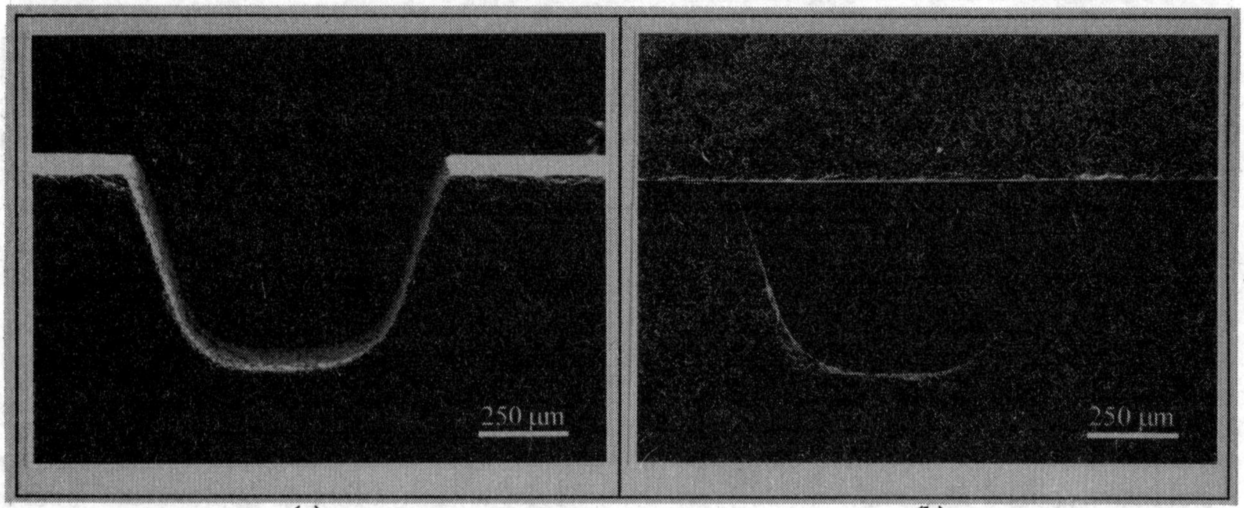

(a) (b)

Figure 6. SEM images of notch made in alumina specimen (a) before and (b) after joining.

TABLE II. DIMENSIONS OF NOTCHES BEFORE JOINING AND AFTER JOINING.

Material	MaCor™				Alumina	
Dimension	Depth (µm)	Width (µm)	Depth (µm)	Width (µm)	Depth (µm)	Width (µm)
Before joining	619	321	331	228	531	907
After joining	585	312	314	215	528	870
% difference	-5.5	-2.8	-5.1	-5.7	-0.6	-4.1

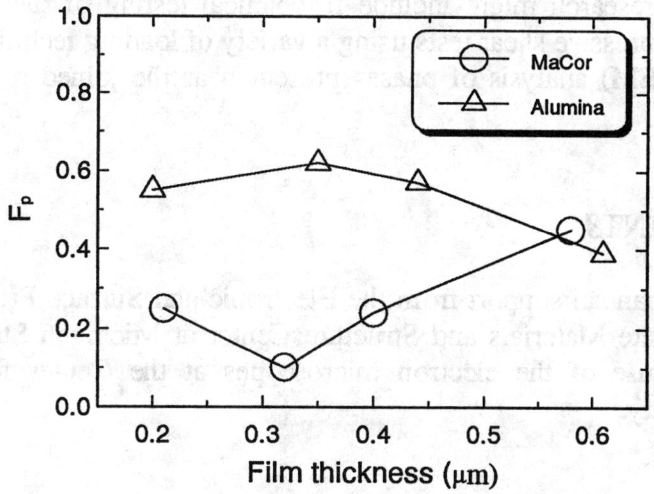

Figure 7. Total fraction of pores along the joint interface in joined MaCor™ and alumina specimens as a function of film thickness.

The fraction of F_P is a function of the silica film thickness (Figure 7) for both the MaCor™ and alumina specimens. However, the functional dependence of F_P is quite different for alumina and MaCor™ specimens (Figure 7). One pair of microwave heated, polished, and coated MaCor™ specimens was compared to a conventionally heated, polished and coated MaCor™ specimen. Both the microwave and the conventionally heated specimens were processed at a maximum temperature of 1050°C for 20 minutes, with a 20 gram dead-weight load. The coatings for both specimens were spun at 2000 rpm and cured for 20 minutes at 200°C. The F_P values were 0.14 and 0.25 for the conventional and microwave heated specimens, respectively.

SUMMARY AND CONCLUSIONS

A total of five pairs of alumina specimens, fifteen pairs of MaCor™ specimens and one pair alumina/zirconia particulate composite specimens were successfully joined with low externally applied pressure. The joining temperature for alumina was 1575°C with a joining time of 20 minutes in the microwave system. The MaCor™ specimens were joined using both conventional and microwave heating techniques for approximately 20 minutes at 1050°C and 1075°C. The quality of the joins was assessed using F_P (equation 1). Figure 7 shows the functional dependence of F_P on the film thickness.

Notches in MaCor™ specimens changed dimension by less than 5.7% in width and depth during joining (Figure 5, Table II). Alumina specimens changed by less than 4.1% in width and depth (Figure 6, Table II). Notch shape retention indicates that ceramic components with intricate features, such as small holes and channels, could maintain their complex geometry during joining. Also for the MaCor™ specimens, the constancy of notch geometry indicates a lack of wide-scale viscous flow.

Additional work needs to be done to compare the quality of the joins as a function of the join temperature, time, coating thickness, and heating mode (microwave versus conventional heating). Also, future research might include mechanical testing of the joined region using techniques, such as compressive shear tests using a variety of loading techniques. Transmission electron microscopy (TEM) analysis of phases present near the joined region should also be investigated.

ACKNOWLEDGEMENTS

We acknowledge the financial support from the Electronic and Surface Properties of Materials Center and the Composite Materials and Structure Center at Michigan State University. We also acknowledge the use of the electron microscopes at the Center for Electron Optics, Michigan State University.

REFERENCES

Fukushima, H., T. Yamanaka, and M. Matsui. 1990. "Microwave Heating of Ceramics and Its Application to Joining," *J. Mater. Res.*, 5 (2): 397-405, 84.

Fullman, R. L. 1953. "Measurements of particle sizes in opaque bodies," *Trans. Met. Soc.* AIME 197(3): 447-452.

Lee, K. Y., E. D. Case, J. Asmussen, Jr., and M. Siegel. 1996. "Binder Burnout in a Controlled Single-Mode Microwave Cavity," *Scripta Materialia*, 35(1):107-111.

Lee, K. Y., E. D. Case, and D. Reinhard. 1997. "Microwave Joining and Repair of Ceramics and Ceramic Composites," *Ceramic Eng. and Sci. Proc.* vol. 18.

Lee, K. Y., E. D. Case, and J. Asmussen, Jr. 1997. "Microwave Binder Burn-out for Batch Processing of Al_2O_3, Al_2O_3/SiC Platelet, and Al_2O_3/ZrO_2 Particle Powder Compacts," *Cer. Trans.*, 80:539-546.

Mecartney, M. L., R. Sinclair, and R. E. Loehman. 1985. "Silicon Nitride Joining," *J. Amer. Ceram. Soc.*, 68:472-488.

Moorhead, A. J. 1987. "Direct Brazing of Alumina Ceramics," *Adv. Ceram. Mater.* 2:159-166.

National Bureau of Standards (U.S.). 1959. *Circ. 539*, 9:3.

Sandage, K. H., H. J. Schmutzler, R. Wheeler, and H. L. Fraser.1996. "Mullite Joining by Oxidation of Malleable, Akaline-Earth-Metal-Bearing Bonding Agents," *J. Am. Ceram. Soc.*, 79(7): 1839-50.

Santella, M. 1992. "A Review of Techniques for Joining Advanced Ceramics," *Am. Ceram. Soc. Bull.*, 71:947-54.

Silberglitt, R., D. Palaith, W. M. Black, H. S. Sa'adaldin, J. D. Katz and R. D. Blake. 1991. "Investigation of interlayer Materials for the Microwave Joining," *Ceram. Trans.*, 21:473-480

Zdaniewski, W. A., P. M. Shah, and H. P. Kirchner, 1987. "Crystallization Toughening of Ceramic Adhesives for Joining Alumina," *Adv. Ceram. Mater.*, 2:204-208.

Discontinuously Reinforced Al-SiC and Al-AlN Composites Formed by a Low Cost Chemically Activated Wetting Technique

M. V. KEVORKIJAN

ABSTRACT

This paper deals with an attempt to rationalize the interfacial behavior between a particulate ceramic reinforcement and molten metal at different interfacial chemical compositions. The ultimate objective is, however, to develop a predictive model which could be used to model the interfacial behaviour of different MMCs mainly from a knowledge of a few basic thermodynamic parameters, such as ΔG of formation of the compound. Using the model developed, several particularly exothermic chemical reactions localized near the ceramic/liquid metal interface were theoretically and experimentally explored as promoters for wetting and bonding in Al-SiC and Al-AlN DRMMCs.

INTRODUCTION

To date, it is impossible to predict the effectiveness and the kinetics of the wetting of ceramic species introduced into molten metal using a set of fundamental thermodynamic and kinetic parameters for the observed interfacial chemical reaction. For example, there is no theoretical explanation of which (and even if) energy changes that occur during interfacial chemical reaction (exothermic or endothermic) are (more) effective in activation of wetting. Even in very recent reviews of this subject (Li,1994) it is possible to find only a basic thermodynamic scenario that the negative change of Gibbs free energy combined with an existing gradient of a chemical potential at the interface is the only "driving force" for chemically activated wetting. Referring to this it is important to note that there is still no derivation of some fundamental correlation, based on a clear process mechanism, between the energy change introduced by an interfacial chemical reaction and the change in solid/liquid interfacial energy, γ_{sl}

Kevorkijan Varuzan, Alpha and Omega in Composites, Lackova 139, 2341 Limbus, Slovenia, EU

The equilibrium wetting state between a liquid, solid, and gaseous environment is most often described by equation for the mechanical equilibrium of surface tension

$$\gamma_{lv}\cos\theta = \gamma_{sv} - \gamma_{sl} \tag{1}$$

The γ_{lv}, γ_{sv} and γ_{sl} are the surface energies of the three different interfaces; namely the liquid/solid, the solid/vapor, and the liquid/vapor interfaces. θ is the contact or wetting angle. According to Eq. 1, the system minimizes its free energy by maximizing the low-surface-energy interfacial area at the expense of high-energy interfacial areas, consistent with the constraints imposed by the gross geometry of the system. If $\gamma_{sv}>\gamma_{sl}$, then θ will be less than 90°, and the liquid can reduce the energy of the system by spreading, thereby creating more solid/liquid and solid/vapor interfaces at the expense of the solid/vapor interface. The liquid is then said to wet the solid. If $\gamma_{sv}<\gamma_{sl}$, then the system can reduce its free energy by reducing its solid/liquid and liquid/vapor interfacial area and increasing its solid/vapor interface area. The liquid is then said to be in a nonwetting condition.

It is unfortunate that specific values of γ_{sv} and γ_{sl} cannot presently be measured, or at least measured easily and reliably, but liquid surface energies γ_{lv} and contact angles are the only experimentally observable variables of the wetting phenomenon. The other important quantity which usually describes the wetting properties of solid ceramic materials by liquid metals is the work of adhesion W; this is the work per unit area which should be provided to the system in order to separate reversibly a solid/liquid interface to create a solid/vapor interface and a liquid/vapor interface:

$$W = \gamma_{sv} + \gamma_{lv} - \gamma_{sl} \tag{2}$$

Taking into account Eq. (1) , the work of adhesion W can then be written as:

$$W = \gamma_{lv}(1 + \cos \Theta) \tag{3}$$

According to Eq. (3), it is deduced that the wettability is determined by the adhesion energy W between the solid and the liquid contacting phases and by the cohesion energy $W_{coh} = 2\,\gamma_{lv}$ of the liquid phase. However, wettability can also be measured by the work necessary to immerse a solid in a liquid (Delannay, Froyen, Deruyttere, 1987) This method is more appropriate to measurements made on particulate materials. Note that the work of immersion is the work done on the system in exchanging a solid/vapor interface for a solid/liquid interface:

$$W_i = \gamma_{sl} - \gamma_{sv} \tag{4}$$

The "driving force for wetting" , D_f, may thus be defined as (Delannay, Froyen, Deruyttere, 1987):

$$D_f = \gamma_{sv} - \gamma_{sl} = -W_i \quad (5)$$

When $D_f > \gamma_{lv}$, $\theta = 0$ and the liquid spreads spontaneously on the solid surface. Several attempts have been also directed to correlate chemically activated wetting with the substantial lowering of the interfacial energy caused by the interfacial chemical reaction (Naidich, 1981; Laurent, 1987; Aksay, Hoge and Pask, 1974). This theoretical approach, which is still not generally accepted, is based on the assumption that only the interfacial energy,γ_{sl}, could be influenced by the chemical reaction localized at the interfacial region. The maximal lowering of γ_{sl} due to the free energy released by the interfacial chemical reaction is said to be given by the following differential equation (Aksay, Hoge, Pask, 1974):

$$\gamma_{sl} = \gamma_{sl}^{\circ} + d(\Delta G_R)/(d\Omega_{sl} \cdot dt) \text{ (with } \Delta G_R < 0 \text{)} \quad (6)$$

where γ_{sl} is the dynamic interfacial energy at time t and γ_{sl}° is the initial interfacial energy at the time t = 0 . The above equation takes into account the contribution to wetting of the free energy ΔG_R of the reaction per unit area of interface ($d\Omega_{sl}$) and per unit of time (dt) , and as a result, the importance of the wetting kinetics comes into consideration. As proposed by (Aksay, Hoge, Pask, 1974), in many cases, the formation of new compounds at the metal/ceramic interface takes place, and the contribution of the change $\Delta\gamma_{sl}$ in the interfacial energy resulting from the replacement of the initial solid/liquid interface with at least one new interface after reaction has to be added to Eq. (6):

$$\gamma_{sl} = \gamma_{sl}^{\circ} + d(\Delta G_R)/(d\Omega_{sl} \cdot dt) - \Delta\gamma_{sl} \quad (7)$$

Let us recall that γ_{sl} in Eq. (7) represents the nonequilibrium interfacial energy of the new ceramic phase formed *in situ* at the interface by interfacial chemical reaction. In this way, the formation of a new surface, $\Delta\Omega$, in liquid metal, which is the main prerequisite for the successful introduction of ceramic particles into a melt, is thermodynamically possible only if $\Delta G < 0$. Moreover, the energy of the interface created in liquid metal by introduction of ceramic particles into a melt has to be equal to the energy created in the system by the interfacial chemical reaction

$$\Delta G = \Delta\Omega \cdot \gamma_{sl} \quad (8)$$

In this way, the number of ceramic dispersoids which form a stable dispersion in liquid metal is represented by the simple relation:

$$\Delta G = m \cdot S \cdot \gamma_{sl} \quad (9)$$

where m is the mass of ceramic particles successfully introduced into the melt and S is the specific surface area of ceramic powder.

The mass of ceramic particles which forms a stable dispersion in liquid metal may then be expressed by the following relation:

$$m = \Delta H_r / S\gamma_{S\text{-}l} \qquad (10)$$

This means that: (i) an exothermic interfacial chemical reaction; and (ii) the decreasing of the interfacial energy caused by the formation of a new ceramic phase at the interface quantitatively determines the number of ceramic particles which will be stable dispersed in the liquid metal. Generally, in this thermodynamic scenario, the energy released by the decrease in solid/liquid interfacial energy caused by an (exothermic) interfacial reaction will be consumed for the formation of the new interface in the system. Under these conditions the dispersion of ceramic particles in liquid metal should be stable, because there is no total energy change in the system. The main assumption (and a possible source of criticism) in the model presented is that a thermodynamically favorable chemical reaction ($\Delta G << 0$) located at the interface should *necessarily* result in a reduction of interfacial energy.

Nevertheless, the main purpose of this simple thermodynamic model is to be used as a means of correlating the critical concentration of ceramic particles into a liquid metal with the available data on the ceramic/molten metal interface.

The introduction of ceramic particles into a melt

The introduction of particulate into the melt can be expressed in terms of thermodynamics. For a spherical particle, (Neuman, Van Oss and Szekely, 1973) calculated the work of immersion for each step and found that the total work of immersion per unit area is then equivalent to the difference between the initial unwetted state, γ_{sv}, and the final immersed state, γ_{sl}:

$$W_{tot} = \gamma_{sl} - \gamma_{sv} = -\gamma_{lv}\cos\theta = W_i \qquad (11)$$

The conditions for spontaneous wetting can be expressed in terms of the surface energies, i.e., if the solid-vapor surface energy γ_{sv}, exceeds the solid liquid surface energy γ_{sl}, the work of immersion becomes negative. In this way, the lowering of the solid-liquid interface energy γ_{sl} or/and increasing the solid-vapor interface energy γ_{sv}, caused by a change of the surface composition (one can name this procedure " γ - engineering "), seems the most reasonable approach to immersion of particles into a melt.Assuming that the suspension of immersed particles will be thermodynamically stable when the energy change caused by formation of a new ceramic/liquid metal interface is equal or even less than the energy released by an interfacial chemical reaction (note that under these conditions the total energy change in the system is zero or, in the ideal case, negative), one can derive the criterion for successful and thermodynamically stable immersion of ceramic particles into a melt:

$$\gamma_{sl} - \gamma_{sv} = - \gamma_{lv}\cos\theta \le k \cdot \Delta G_r / S \tag{12}$$

where k is the dimensionless system dependent constant, S is the surface area of the ceramic reinforcement, and ΔG_r is the free energy change for the interfacial chemical reaction. Note that the term $\gamma_{lv}\cos\theta$ has to be determined at equilibrium conditions, i.e. when the interfacial reaction proceeded completely. The empirical constant k appears to relate to the free energy yield in the system during the interfacial reaction. Taking into account that the surface area of the immersed ceramic reinforcement can be expressed as $S = S^\circ \cdot m$ where S° is the specific surface are of the ceramic phase and m is the mass of ceramic reinforcement which is successfully introduced into a melt, the expression (12) may be written as:

$$\gamma_{sl} - \gamma_{sv} = - \gamma_{lv}\cos\theta \le k \cdot \Delta G_r / S = k \cdot \Delta G_r / S^\circ \cdot m \tag{13}$$

Finally, the portion of ceramic reinforcement which may be introduced into a melt according to reactive wetting is defined by the following formula:

$$m = k \cdot \Delta G_r / S^\circ \gamma_{lv}\cos\theta \tag{14}$$

SURFACE MODIFICATION STRATEGIES

The improvement of wetting by the reaction of liquid silicon with carbon

Recently, (Kevorkijan, 1995; Kevorkijan and Sustarsic, 1996; Kevorkijan and Sustarsic, 1997; Kevorkijan, 1996) have reported an improvement in wetting of carbon coated SiC particles by an Al-40 wt. % Si alloy in a technical vacuum or even in a protective atmosphere. In the same time, an other research group has presented results on the spontaneous infiltration of carbon containing SiC preform by Al-40 wt% Si alloy in an open air atmosphere, without the needing for external pressure or control of the atmosphere (Xi and Yang, 1996). In both cases, the "driving force" for wetting was the interfacial reaction between Si and C. However, the applicability of this technique is limited on DRMMCs with a silicon-rich aluminum matrix and a ceramic reinforcement which is chemically stable in contact with carbon.

The improvement of wetting by the reaction of TiO_2 - Al - C

The metallothermic-reduction reaction, which has been used in the preparation of various metals and alloys for several decades, is generally an exothermic process which is self-sustained by the heat liberated. Recently, the synthesis of TiC-Al_2O_3

composites by self-propagating high-temperature synthesis has drawn much attention as an alternative to conventional processes because of its simplicity and energy efficiency (Choi and Rhee, 1995). The reaction can be described as follows:

$$3TiO_2 + 4Al + 3C \rightarrow 3TiC + 2Al_2O_3 \qquad (15)$$

Note that $\Delta H^{\circ}_{f,298.15K} = -1076.6$ kJ/mol and $\Delta G^{\circ}_{f,298.15K} = -1038.9$ kJ/mol.

EXPERIMENTAL PROCEDURE

Two types of β-SiC powders produced by low temperature carbothermic reduction (AP-1 consisting of fine particles with a diameter range between 2-12 μm, and AP-2 consisting of coarser particles with about 20μm mean diameter), commercially available M-Grade ART SiC whiskers and laboratory scale prepared SiC platelets (size range: 30-50 μm, thickness: 3-5 μm, aspect ratio 8-10, purity: max.1000 ppm of metallic impurities, particulate content 5-10 wt. %, oxygen 1.1 wt. % and free carbon 0.53 wt. %) were used in this study. Commercially available A-500 grade ART AlN powder with two selected fractions of particles A-500/1 and A-500/2 with average particle size near 10 and 20 μm, respectively, were also used as reinforcement. Note that the selected AlN fractions were comparable in size to the AP-1 and AP-2 SiC powders. Standard 356-T6 alloy with a nominal composition of 7 wt. % of Si and 0.3 wt. % of Mg was used in all experiments. SiC reinforcements used in this study were surface treated with a carbon layer by the following procedures: Phenolic resin and SiC reinforcements were first wet mixed in acetone. Then the acetone was evaporated with constant stirring in order to maintain the homogeneous distribution of species. The dry mixture was then pyrolyzed in flowing argon for 4h at 450°C to permit the formation of an amorphous carbon layer on the surface of the SiC reinforcements; Some of the SiC reinforcements were coated with a carbon layer doped with TiO_2 as follows: Phenolic resin, SiC reinforcements and TiO_2 powder (99.9% pure, rutile, average particle size 1 μm) were wet-milled in ethanol for 24 h using alumina milling media. Using the same procedure as just described, SiC reinforcement were covered with an amorphous carbon layer doped with TiO_2. AlN powders were also covered with a carbon layer doped with TiO_2 by following the above procedure. Note that in a separate set of experiments, the wetting tendency of the as- received SiC and AlN reinforcements were also evaluated. Both treated and untreated SiC reinforcements (except SiC reinforcement doped with TiO_2 + C) were first dispersed into the Si-Al melt at 1150°C in vacuo. The composition of the melt was usually 60 wt. % Si - 40 wt. % Al. The SiC reinforcement was introduced into the melt through a ceramic tube using a flow of carrier gas (argon with some additives). When a sufficient amount of SiC reinforcement had been incorporated into the Si - Al melt, an Al alloy (Al-4wt%Mg) and pure Al were added at a controlled rate and under vigorous mixing conditions in vacuo, or under a protective atmosphere, in order to achieve the 356-T6 alloy composition. SiC and AlN particles doped with TiO_2 + C were dispersed directly into the Al alloy melt in an argon atmosphere. The apparatus consisted of a graphite crucible surrounded by an induction heating coil. The mixing assembly consisted of a DC variable speed motor, the spindle and the dispersion impeller having blades angled from about 15° to about 45° from a line perpendicular to the shaft. The crucible was provided with a protective cover and an inert gas or vacuum protection chamber.

RESULTS AND DISCUSSION

Table I summarizes the upper experimentally monitored concentrations of SiC and AlN reinforcement which result in their stable dispersion in the melt.
In this way, the efficiency of the direct immersion of SiC and AlN reinforcements into an Al-Si melt was quantitatively compared with the potential of the new processing route.
The preliminary results in Table I show , as expected, that only a very limited level of uncoated fine SiC particulate could be successfully introduced into the melt. Rejection occurs spontaneously at a particulate concentration between 2-3 wt. %. More promising results were obtained with fine SiC particles previously oxidised in air (at 1200°C for 1 h). In that case, 6-9 wt. % of fine SiC particles could be successfully incorporated into the matrix prior to their spontaneous rejection. However, note that oxidation of fine SiC particles at 1200°C for 1 h changes their chemical composition dramatically. Thus, the wetting of oxidised fine SiC particles in an Al melt was rather governed by the wetting tendency of the equivalent amorphous SiO_2 particulate. Using the exothermic reaction between carbon, previously deposited on the surface of SiC particles, and silicon in the melt, a high portion of ceramic filler could be sucessfully immersed, Table I. For example, it was found that only 10-15 wt. % uncovered fine SiC particles could be successfully incorporated into a Si melt without significant rejection. In contrast, under the same experimental conditions more than 30 wt. % fine SiC particles covered by a carbon layer were successfully introduced into a Si melt, in accordance with the exothermic reaction between the carbon layer and silicon in the melt.

In the second processing route used in this work, approximately 25 vol. % of fine SiC particles or SiC reinforcement with high aspect ratio were successfully introduced **directly** into the melt by the exothermic self-sustaining chemical reaction between the melt and a mixture of TiO_2 and carbon deposited on the surface of the SiC particles. Moreover, it was found that covering other ceramic dispersoids (e.g. AlN) with a TiO_2/C layer also resulted in an enhancement of the wetting of this species with an Al alloy melt.

This clearly confirms that rapid, exothermic interfacial (i.e. exergonic) chemical reaction can drive the immersion of ceramic particles into a melt. However, it is important to note that the exothermic nature of an interfacial chemical reaction (or, more generally speaking, the negative change of its free energy) is only one aspect in determining of their effectiveness in wetting improvement. The other parameters, especially the kinetic aspect of the selected interfacial reaction, should be also considered.

TABLE I- THE CONCENTRATION OF CERAMIC REINFORCEMENT WHICH LEADS TO THE SPONTANEOUS REJECTION OF CERAMIC DISPERSOIDS FROM THE MELT

SiC reinforcement

Introduction of SiC phase directly to an Al-Si melt				Immersion of SiC phase to a Si-Al melt			
Type of reinforcement				Type of reinforcement			
AP-1	AP-2	whiskers	platelets	AP-1	AP-2	whiskers	platelets
Coating SiC with a carbon layer							
/	/	/	/	40-50* 10-20**	50-60 10-25	50-60 10-25	50 10-20
*concentration in Si-Al melt, **concentration in final eutectic composition							
Coating SiC with a carbon layer doped with TiO_2							
18	25	23	23	/	/	/	/
As-received SiC phase							
1-2	2-3	1	1	20-30 5-8	25-35 5-7	25-35 5-6	20-30 3-5

AlN reinforcement

Type of reinforcement		Type of reinforcement	
AlN A-500/1	AlN A-500/2	AlN A-500/1	AlN A-500/2
Coating AlNwith a carbon layer doped with TiO_2			
20	26	/	/
As received AlN phase			
3-4	3-5	/	/

To confirm this, one can compare the upper level of successfully immersed particles achieved by the Si + C and $3TiO_2 + 3C + 4Al$ interfacial reactions (see Table I), with their free energy change and reaction enthalpy. An inspection of the data presented in Table 1 shows that, in practice, a one order of magnitude higher reaction enthalpy and free energy change for the aluminothermic reaction ($3TiO_2+3C+4Al$) does not result in a proportionally higher level of successfully immersed ceramic particles into the melt.

It seems that, as well as kinetics, other parameters such as the efficiency of the system in using the free energy released by the interfacial reaction for the formation of a new interface, are also operative.

CONCLUSION

An attempt was made to rationalize the interfacial behavior of particulate ceramic reinforcement with a molten metal at different interfacial chemical compositions. The main objective was to develop a predictive model which could be used to model the interfacial behavior of different DRMMCs, mostly from a knowledge of the reaction enthalpy and the free energy change of a selected interfacial reaction.

Using this simple model, several exothermic chemical reactions which are localized near the ceramic/liquid metal interface were thermodynamically explored as a "driving force" for reactive wetting in Al-SiC and AlN-Al DRMMCs.

Two processing routes for introducing larger amounts of fine SiC particles into an Al alloy melt without rejection of the reinforcement, and without unwanted chemical reactions between the matrix and the reinforcement were described. Both processes are based on an exothermic interfacial reaction, i.e. on the chemically activated wetting of ceramic dispersoids in a melt.

The collected experimental data clearly confirm that rapid, exothermic interfacial (i.e. exergonic) chemical reaction can drive the immersion of ceramic particles into a melt. However, it is important to note that the exothermic nature of the interfacial chemical reaction (or, more generally speaking, the negative change of its free energy) is only one aspect in determining their effectiveness in wetting improvement. The other parameters, especially the kinetic aspect of the selected interfacial reaction, should also be considered.

To confirm this, the upper level of immersed particles successfully achieved by Si + C and $3TiO_2 + 3C + 4Al$ interfacial reactions (see Table I) was correlated by the free energy change and reaction enthalpy of the selected interfacial reactions. The analysis performed showed that, in practice, a one order of magnitude higher reaction enthalpy and free energy change for the aluminothermic reaction ($3TiO_2$+3C+4Al) does not result in a proportionally higher level of stable immersed ceramic particles into the melt.

This indicates that, besides the kinetics, other parameters such as the efficiency of the system in using the free energy released by the interfacial reaction for the formation of a new interface are also operative.

In this context, further research on this topic by the scientific community should be encouraged because of the potential technological importance of the findings.

REFERENCES

Li, J. G. 1994. " Wetting of Ceramic materials by Liquid Silicon, Aluminum and Metallic Melts Containing Titanium and Other Reactive Elements: A Review," *Ceram. Int.,* 20: 391-412

Delannay, F., Froyen, L. and Deruyttere, A. 1987. " The Wetting of Solids by Molten Metals and its Relation to the Preparation of Metal- Matrix Composites," *J. Mater. Sci.* 22:1-16.

Naidich, J. V. 1981. " The Wettability of Solids by Liquid Metals," *Progress in Surface and Membrane Science*, 14: 353-483.

Laurent, V.1988. " Mouillabilite et reactivite dans les systemes composites metal-ceramique: Etude du couple Al/SiC," Ph. D Thesis, INP, Grenoble, France.

Aksay, I. A, Hoge, C. E. and Pask. 1974.: " Wetting under Chemical Equilibrium and Nonequilibrium Conditions ," *J. Phys. Chem.,* 78(12):1178-83.

Neumann, W. A., van Oss C. J. and Szekely, J. 1973: " Thermodynamic of Particle Engulfment," *Kolloid-Z. u. Z. Polymere*, 251: 415-423.

Mortensen, A. 1991. " Solidification Processing of Reinforced Metals ," Metal Matrix Composites-Processing, Microstructure and Properties, 12. Riso International Symposium on Materials Science, Roskilde, Denmark:101-122.

Kevorkijan, M. V. and Sustarsic, B. 1995 " The introduction of fine SiC particles into a molten Al alloy matrix: application to MMCs preparation via a foundry route," Proceedings of the 10th. Annual Meeting of ACS:569-581.

Kevorkijan, V.1996. "Stabilization of fine SiC particles dispersed in a molten Al alloy: application to composite casting," *Ceram. Trans.,* 79:183.

Kevorkijan, M. V. and Sustarsic, B. 1996. "The New Production Technology for Discontinuously Reinforced Al-SiC Composites," *Key Engineering Materials* 127:471-478.

Kevorkijan, M. V. 1996. "The New Production Technology for Discontinuously Reinforced Al-SiC Composites: Future Prospects for Commercial Development," Procccedings of the 11th. Annual Meeting of ACS:717-728.

Xi, X. M. and Yang, X. F.1996: " Spontaneous Infiltration of Aluminum-Silicon Alloy into Silicon Carbide Preforms in Air," *J. Am. Ceram. Soc.,* 79(1): 102 - 108.

Choi, Y. and Rhee, S. W.1995. "Reaction of TiO_2-Al-C in the Combustion Synthesis of TiC-Al_2O_3 Composite," *J. Am. Ceram. Soc.,* 78(4):986-992.

Some General Considerations about the "Ideal" Ceramic Reinforcement for Castable Particulate Reinforced Aluminium Composites for Structural Applications

M. V. KEVORKIJAN

ABSTRACT

An overview of the current understanding of mechanical properties of discontinuously reinforced metal matrix composites (DR MMCs) is presented and the "ideal" ceramic reinforcement for light metal composites based on Al alloys is identified.Three main problems which must be solved to facilitate the expanding application of DR MMCs with light alloys matrix metals are: (i) superior cost performances - in Al based DR MMCs the quality/cost ratio should be above one, (ii) Active implementation of environmental concerns- DR MMCs should be easy recyclable, and (iii) the improved ductility - the ductility of the composite should follow the ductility of unreinforced matrix. It is almost postulated that market competitive DR MMCs should be defect free castable and re-castable. It seams that only in this way the most inexpensive processes for production of final components could be developed. Referring to this, the cost effective preparation of molten metal dispersion with at least 20 vol% of fine (less than 10 μm) ceramic particles produced *in situ* or directly introduced into a melt becomes the key technological request.

In this article, several new preparation routes for DR MMCs based on chemically improved wetting behavior of ceramic dispersoids in molten aluminum alloys and chemically activated stabilization of dispersoids in melt have been discussed. The intrinsic economy of these methods and some new market opportunities for DR MMCs have been also presented. Finally, a further selling of the concept of MMCs based on the improvement of the price/quality ratio and the important automotive market pull for the improved properties of DR MMCs were introduced.

INTRODUCTION

In recent years, a considerable development has taken place in the synthesis of discontinuously reinforced cast composite materials (DR MMCs) with tailored properties because of potential improvements in mechanical properties such as modulus, 0.2% proof stress, hardness and tensile strength. Such improvements are, however, usually accompanied by reduction in both tensile ductility and fracture toughness.

Kevorkijan Varuzan, Alpha and Omega in Composites, Lackova 139, 2341 Limbus, Slovenia, EU

Moreover, other technical problems arise when one intends to produce a ***castable*** DR MMCs for structural application: the matrix-reinforcement reactivity, the increased viscosity of a melt, the machinability of the cast ingot to the final shape at a reasonable price, problems of successful and cost effective recycling, etc.

In order to illustrate the complexity of an intrinsic economy of DR MMCs, cost factors for an extruded wrought automotive component are listed in Table I.

TABLE I - COST FACTORS FOR AN EXTRUDED WROUGHT AUTOMOTIVE COMPONENT

Finished component cost	**100%**
Extrusion billet cost	**18%**
Material	15%
Sawing	3%
Extrusion cost	**30%**
Extrusion speed	7%
Die cost	10%
Die wear	8%
Recoveries	5%
Cutting	**8%**
Machining	**15%**
Bending and forming	**5%**
Welding	**4%**
Surface finishing - anodising or painting	**1%**
Transportation	**9%**
Recycling	**10%**

Source: Alpha and Omega in Composites, 1996

The key solution for improving the quality and the economy of cast DR MMCs simultaneously is the creating of new casting techniques to reduce casting defects and to disperse fine ceramic particles in a melt, or the development of new ceramic reinforcements with superior characteristics such as: (i) spontaneous wetting and (ii) chemical stability in the melt

Let now try to answer the main question: Which is or which will be the "perfect" ceramic reinforcement for cast DR MMCs? The following criteria should be considered: (i) the application of the composite, (ii) the matrix alloy composition, (iii) the cost of reinforcement, and (iv) the ratio between the extra cost and the achieved improvements in commercial properties of composite. Commercial properties of composite are a very complex term consisting of mechanical properties, corrosion resistance, tribological behavior, formability of final parts, and the ability to recycle the scrap and/or the final components when their useful life is over.

APPLICATION OF THE COMPOSITE

For a structural application, the modulus, strength, and density of the composite become important. Usually, a high modulus and low density are required. Particle shape may also play an important role, since angular particles can act as local stress raisers reducing ductility. In order to improve the basic mechanical properties such as modulus of elasticity, strength and ductility, it seams the most important to find a way of introducing large amounts (20-30 vol. %) of fine ceramic particles (less than 10 μ m) directly into an Al alloy melt without rejection of the reinforcement and without excessive unwanted chemical reactions. However, as reported, finer particles increase the viscosity of the melt, making processing difficult (Lloyd, 1994). Based on these investigations, it was postulated that a large amount of fine particulate reinforcement is generally difficult to incorporate using classical foundry processes. Several new chemically activated foundry methods have been suggested in order to overcome the major problem: the extreme difficulty of simultaneously obtaining sufficient wetting of the reinforcement by the molten matrix metal without incurring excessive or disastrous reactivity between the components, (Lloyd, 1994). However, all of these methods are based on improved wetting between ceramic dispersoids and melt caused by interface chemical reaction. Thus, the challenge is to control the interfacial reaction to the correct level to obtain MMCs with high(er) level of fine ceramic particles in matrix and to prevent, in the same time, a larger degradation of, at least, key properties of the composite material. However, this is no an easy task. The main problem involves the natural dichotomy between the wettability of the reinforcement and the excessive reactivity with it. On this way, the choice of ceramic reinforcement for DR MMCs involves compromises between cost, wettability, excessive reactivity, and effects of interfacial reactions on the properties and heat treatment responses of the matrix itself.

To date, there is no cost effective foundry method able to introduce 20-30 vol % of fine (less than 10 μm ceramic particles) into Al alloy melt. Some very recent foundry improvements have been suggested in Al-SiC system based on simultaneous changing of the composition of the Al alloy during the introduction of SiC particles incorporated in an alloy consisting of 60% Si and 35% Al and 5% Mg to the final matrix composition with 7-12% Si, 1-3% Mg and 85-92 Al (Kevorkijan, 1996a) In spite of some reported improvements, this technique seems no flexible enough for other important systems. However, further understanding of the phenomenon of the wetting and the stabilisation of ceramic dispersoids in Al alloy melt (Kevorkijan, 1997) combined with tremendous improvements in fluidity of the dispersion, which may be caused by the introduction of selected alloying additives into a melt, could result in requested cost effective and castable DR MMCs.

If the composite to be used in thermal management applications, the coefficient of thermal expansion and thermal conductivity are dominant. The coefficient of thermal expansion is generally important because it influences the composite strength. The ability to tailor properties, such as thermal conductivity and coefficient of thermal expansion, for example, provides the basis for applications demanding dimensional tolerance and stability. The thermal conductivity is reduced by the presence of ceramic elements. In the same way, the thermal expansion coefficients of aluminum alloys can be partly adjusted by addition of ceramic elements. However, the thermal properties such as thermal conductivity and coefficient of thermal expansion are not so easily predictable. In these cases, the rule of mixture predictions can usually provide an initial " ball-park " estimate..

The first applications for which DR MMCs have been investigated are related to wear resistance. As reported earlier, in the case of DR MMCS, the mechanical properties are not significantly altered (as in continuous-fibres composites), but the tribological properties (wear, friction, galling) show marked improvements. It was found that soft, solid lubricant particles, such as graphite and mica, and probably fly ash, improve the sizure resistance of aluminium alloys, while hard particles such as SiC, Al_2O_3, WC, TiC, ZrO_2, SiO_2, B_4C, TiB_2 greatly improve the resistance to abrasion of aluminium alloys (Rohatgi, Asthana, Das, 1986).

It is generally recognised that larger particles are more effective against abrasive wear than smaller ones (Rohatgi, Asthana, Das, 1986).

According to (Rohatgi, Asthana, Das, 1986), the ideal solution would involve the analysis of stress distribution in a tribocomponent followed by selective reinforcement of the regions of maximum stress concentration. This reinforcement would be accomplished through the addition of high-strength and high-modulus whiskers or short fibres with a sufficiently high aspect ratio. Finally, dispersed graphite or other solid lubricant particles would be added to regions subjected to sliding contact.

In many applications the corrosion behavior of DR MMCs is important. On the other hand, there is very little published work on the corrosion behavior of DR MMCs. In spite of the lack of published data, all performed studies have stated that DR MMCs suffer more severe attack than the respective non-reinforced metals. Several reasons have been given for the degradation including galvanic corrosion between reinforcement and matrix, galvanic corrosion between intermetallic compounds and matrix, preferential attack because of carbon diffusion from the reinforcement into the matrix, and an increase in the number of anodic sites on the interfaces . In general, the corrosion resistance of the metallic phase in DR MMCs is due to a passive film created on a less noble metal. Hence, the corrosion resistance of DR MMCs can be influenced by the ceramic reinforcement in several ways. The presence of a particles in the surface introduce a break in the passive film, and the break could be an initiation point for corrosion attack. As mentioned, an electrical conducting particle can act as a cathode, creating local galvanic micro elements, which increase the metal solution. Since many ceramic particles, such as Al_2O_3, are insulators, they would not be expected to affect corrosion behavior directly. SiC, however, is a conductor and changes in corrosion behavior need to be assessed. Silicon carbide particles, if not covered by an insulating surface layer, operate as local cathodes. The locally produced hydroxyl ions give rise to an increase in the pH-value of the environment near the particles. This change in pH is a destabilizing factor for the passive film on the aluminum. However, even in the case of insulating reinforcing particles, the composite microstructure is different of that of the matrix as a result of interface reactions and other processing, and this can modify the corrosion response.

THE MATRIX ALLOY COMPOSITION

The matrix alloy should be chosen only after giving a careful consideration to its chemical compatibility with the reinforcement, to its ability to wet the reinforcement, to its own characteristically properties (the most critical is ductility), and the processing behavior (the key requirement is sufficient fluidity).

In past, commercial DR MMCs producers have decided to use commercial wrought or casting alloy compositions as matrix material. To day, the matrix is selected following the above mentioned criteria.

On this way, it is necessary to answer on the following fundamental question: Is it a better approach to select the ceramic reinforcement and after that to look for a compatible matrix, or to select the matrix and than to define the adequate ceramic reinforcement? It seems that both approaches are reasonable. To date, the final selection of one of the variables - the matrix or the reinforcement, introduces very serious limitations in the selection of the second. Moreover, note that by selecting the matrix and the reinforcement, the technological ability for the production of a requested composite are also defined, almost in all.

COST OF REINFORCEMENT

It is well known that the main reason for using particles as the reinforcement in MMCs is to reduce the cost of the composite. Referring to the cost, one can recognize three categories of ceramic reinforcement:

Fillers which are more expensive as the matrix. To date, in this class are all commercially available synthetic ceramic powders with well defined morphology and chemical composition. The two reinforcements receiving the most attention are SiC and Al_2O_3. However, very recently, AlN has also introduced as cost effective performance material for a broad array of filled light metal. Other to costly reinforcements like zircon, alumina, silica, titanium carbide, silicon nitride, boron nitride and other, are still used in lab scale.

Fillers which cost is comparable with the matrix cost. The representatives of this class are coarser powders usually prepared by milling and some purification of natural products like graphite (carbon), shell char, clay, mica, talc etc.

Fillers which are less expensive as the matrix. Some waste by-products like fly ash , slag, sludge waste from high speed steels etc. Note that the introduction of these fillers results in composites which are cheaper as an unreinforced matrix.

Of course, the above listed status is subject to rapid changes. Large efforts are invested and also planning to be applied in further improvement of economy of ceramic powders preparation routes. Nevertheless, it is unrealistic to expect that in next one or two decades the synthetic ceramic fillers will become cheaper as the light metal matrix. Hence, the further selling of the concept of MMCs will be strictly on the value added basis i.e. on the improvement of the price/ quality ratio. However, there must be a market pull for the improved properties before anyone will agree to pay for them. If the final users dwell on the price first, there will be little incentive for developing innovative solutions.

PRICE FOR THE INCREASING PERFORMANCES

The whole DR MMCs business philosophy is build up on the value added basis. In other words, the introduction of carefully selected ceramic reinforcement into a light alloy matrix should result in a profitable material. The possible target prices for the increasing performances of DR MMCs concerning different market segments are indicated in Table II. Note that automotive - the largest potential market for DR MMCs is traditionally interested in very low value of improved performance. The problem is that none of the available DR MMC's production technologies can meet these requirements. The present competition of various DR MMCs production methods is demonstrated in Table III.

TABLE II - VALUE OF PERFORMANCE VERSUS COST FOR DIFFERENT MMCs:

Market segment	Value of Improved Performance ($/kg)	Potential Market Volume (tones/year)	Reinforcement Quality Level
Space	2500-12000	2000	A
Aircraft	500-1800	20000	B
Military vehicles	15-200	350000	C,D,E
Industrial	10-120	400000	C,D,E
Leisure	10-15	20000	E,F
Automotive	10-15	15000000	E,F

A-Continuous fiber; B-Whiskers/PM; C-Particulate/PM; D-Staple/Pressure Cast; E-Whiskers/Squeeze Casting; F-Particulate/Ingot
Source: Advanced Refractory Technologies

As evident, all potential production methods are still operated in no added value region. However, it is also important to note that, according to the very low production cost, foundry methods have the largest potential shear of added value. The problem is how to meet the market specifications at the same target price or, on the other words, how to improve the quality of cast DR MMCs inside the same cost segment?

TABLE III- CURRENT QUALITY AND COST LEVEL FOR DR MMCs PRODUCED BY DIFFERENT LIQUID METAL ROUTES AND BY POWDER METALLURGY

	Foundry Methods	Squeeze Casting	in situ	Infiltration	PM
Target Quality (For Structural Application in Automotive Industry)	100	100	100	100	100
Target Price (For Structural Application in Automotive Industry)	3-4	3-4	4-6	6-8	20-40
Current Quality Level	60-80	70-80	40-90	80-90	>100
Current Price Level ($/kg) (For non-shaped ingot)	6	5-6	10	10-15	>30
Current Cost of Production ($/kg) (For non-shaped ingot)	4-5	3.5-5	5-7	10-15	>20

Source: Alpha and Omega in Composites, 1996

Unfortunately, there is no single, well defined type of market specification. Market requirements are sprayed up on better structural and thermal performances, cost effective machining, ability to recycle or reclaim the individual components of the composite, etc. The situation is somewhat complicated by the fact that many of these requirements are opposite and exclude each other. It is evident that the introduction of discontinuous reinforcement into a light metal matrix can improve the mechanical properties of matrix very modestly in comparison with continuous reinforcement. However, very often in some composite applications, the mechanical properties are not critical. Addition of SiC particles to aluminum alloy can reduce the coefficient of thermal expansion of the alloy, still maintaining the thermal conductivity equivalent of the matrix alloy. As recognized many times in practice, low coefficient of thermal expansion and high thermal conductivity is an attractive combination for applications requiring dimensional stability.

In many cases, the wear resistance is the most critical composite specification. Because of their excellent wear resistance, DR MMCs are candidates for bearings, bushings, cylinder liners, and break rotors. Frequently, high wear resistance is required only at the surface of the component, and the ability to prepare a composite with some gradient of hardness and wear resistance could be the most competitive solution for a particular final user.

To date, it is well established that the cast DR MMCs will be the most competitive on the automotive market. The real cast systems are, unfortunately, faraway from these expectations. Very recent investigation of the influence of silicon carbide particle size, aspect ratio and volume fraction on room temperature tensile properties of DR MMC showed that, according to the strengthening mechanism, higher volume fraction and smaller particle size are beneficial (Kevorkijan, Sustarsic 1996). On the other hand, because silicon carbide particles were introduced and dispersed into molten aluminium alloy by a mechanical stirring method, adding particles resulted in casting defects which deteriorate the properties of the composite. In this way, the damage accumulation processes caused by cast and extrusion defects, limit further improvement of the mechanical properties of cast MMCs based on a high reinforcement aspect ratio. Because of that, the use of costly reinforcement like SiC platelets or whiskers in cast MMCs seems unreasonable. However, the same study has shown that the introduction of fine (less than 10 μm) rounded-off SiC particles combined with significant reduction in cast and extrusion defects could result in much better mechanical properties of Al-SiC composites. It was also recommended that the key factor which needs to be controlled is the fluidity of the melt and not the shape and type of reinforcement. Of course, the immersion of fine ceramic particles directly into a melt by mechanical stirring is not the only possible way for production of castable DR MMCs. Moreover, it seems that the potential of direct mechanical or even chemically activated immersion of ceramic particles in molten metal has already been explored. Most of these procedures are operated by ceramic particles in the 10-20 μm size range. On the contrary, it was shown (R. M. Aikin, 1990) that finer particles can be successfully produced *in situ* into a melt and this might be a new effective direction for the preparation of cast DR MMCs with superior mechanical properties.

SOME GENERAL COMMENTS

Following different criteria and accumulated data one can conclude that DR MMCs will become an engineering material of general use. At the end of 1970s, the automobile and transport industries all around the world had a very strong interest in and high hopes for DR MMCs, due to their superior specific strength and rigidity over conventional aluminum alloys. After two decades of development, the first DR MMCs application, an engine connecting-rod is ready to market. Also, major applications in the automobile industry, such as brake rotors and drive shafts, are now in the late stages of development, and other engineering applications are already well advanced. DR MMCs technology is now recognized in the world-wide industry and research groups in this field are also able to introduce remarkable applications of these materials to the market.

Three main problems must be solved in order to facilitate the expanding application of DR MMCs with light alloys matrix metals. These are:

Superior cost performances - in Al based DR MMCs the quality/cost ratio should be more than one (the values between 1.3 - 1.5 are strongly requested). That means DR MMCs should be a class of high profitable materials with, at least, 30 -50 % of added value.

Active implementation of environmental concerns - DR MMCs should be the easy recyclable or, at least, reclaimable materials.

The improved ductility - the ductility of the composite should be near to the ductility of the unreinforced matrix.In this moment, the solving of these problems is impossible. However, many researchers and managers firmly believe that solutions will be found if the same amount of energy is devoted to solving these problems as that which was expended during the last two decades of DR MMCs research and development.

It is also necessary to point out that the long history of steel could learn us how difficult is to optimise the microstructure of advanced materials especially if their consist from several phases. According to the some estimations, the final optimization of the DR MMCs will take next 20-30 years.

To date, it's to early to predict what the " perfect " solution will be. It seems that DR MMCs should be defect free castable and re-castable - on this way the most inexpensive processes for production of final components could be established. Referring to this, the cost effective preparation of dispersion with at least 20 vol% of fine (less than 10 μm) ceramic particles produced *in situ* or introduced directly into a melt becomes as, probably, the key technological request. Three main questions - big 3Q's - are:

How to improve the wetting behavior between ceramic particles and molten alloy in order to make their immersion much more easier and cost effective?

How to stabilize fine ceramic dispersoids into dispersion and to prevent their agglomeration, settling and rejection after a successful immersion?

How to improve the fluidity of dispersion of small ceramic particles? As reported, a sufficient fluidity is required to enable ingot casting directly from the composite processing reactor and, also, the casting fluidity after remelting of composite billets, is necessary for the secondary shape casting of composite parts.

The "Big 3 Q's " are fundamental questions and the answers will be found through the opening of the new frontiers in the knowledge of wetting, surface engineering and the interfacial chemistry. It is demonstrated very recently (Kevorkijan, 1997) that the wetting of ceramic particles by the molten metal could be improved by introducing very exothermic interface reactions which provide sufficient energy for the wetting activation. The same author suggested that, in principle, the stabilization of ceramic particles previously dispersed in molten metal could be promoted by some endothermic interfacial reactions which binds a part of intrinsic energy of ceramic disperoids. The author demonstrated some progress made on this way in the Al-SiC system (Kevorkijan, 1996b) However, the effectiveness of the offered solution is still very close to the existing commercial foundry procedures which operate with particles in size 10 to 20 μm. The improvements in fluidity could be, probably, achieved using new alloying elements or additives or, even, increasing the melt temperature (the rising of working temperature could also enhance the wetting tendency between particles and melt, but, unfortunately, in the same time, also the reactivity between particles and molten alloy). However, it seams that the type and the shape (aspect ratio) of discontinuous reinforcement don't play the dominant role in further improvements of mechanical properties of DR MMCs but mostly the size and the volume fraction of successfully incorporated particles into a melt.

CONCLUSION

The further commercialization of DR MMCs will be strictly on the value added basis. To be successful on market, DR MMCs should become a class of high profitable materials. With only modest improvements in most mechanical properties and even by a worse ductility as in unreinforced matrix, the producers of DR MMCs now offer to final users preferably superior wear resistance and some flexibility in tailoring of thermal properties of DR MMCs and still not profitable concept for their future structural applications.In order to achieve this end, castable, defect free DR MMCs with at least 20 vol% of fine (less than 10 μm) ceramic particles in matrix should be prepared in cost effective way.In the present paper, some general recommendations concerning this aim have been suggested. It was estimated that the type and the shape (aspect ratio) of ceramic reinforcement will not play the dominant role in further improvement of mechanical properties of cast DR MMCs but their size, homogeneous distribution and volume fraction.

On the contrary, the immersion of fine particles into a melt or their *in situ* formation as well as their stability into dispersion will be mostly governed by the surface chemistry of ceramic dispersoids and interfacial chemical reactions between particles and melt.However, prior to take all advantages of discontinuous reinforcement of light metals on high profitable way, further, very fundamental knowledge, concerning the wetting of ceramic particles by the molten metal and the fluidity of dispersions, should be developed.

REFERENCES

Lloyd, J. D.D. 1994."Particle Reinforced Aluminum and Magnesium Matrix Composites," *Intern. Materials Reviews*, 39: 1-23.

Kevorkijan, M. V. 1996a. " The New Production Technology for Discontinuously Reinforced Al-SiC Composites - Future Prospects for Commercial Development," *Proc. ACS.*,11: 717-728.

Kevorkijan, M. V. 1997. "The Improvement of Wetting in AlN_p-Al_l System Achieved by an Exothermic Interface Reaction," *The Abstract Book of 99. AcerS Annual Meeting, paper C-21-97*: 308

Rohatgi, P. K., Asthana, R. and Das, S. 1986. " Solidification, Structures, and Properties of Cast Metal-Ceramic Particle Composites," *Intern. Metals Reviews*, 31(3): 115 - 139.

Kevorkijan, M. V and Sustarsic, B.. 1997 " A New Production Technology for Discontinuously Reinforced Al-SiC Composites," *Key Engineering Materials*, 127:471-478.

Aikin, R. M. 1990. "The Mechanism of Dispersion Strengthening and Fracture in Al-based XD™ Alloys," *Contract Report 4276, NASA*.

Kevorkijan, M. V. 1996b "Stabilization of Fine SiC Particles Dispersed in a Molten Al Alloy: Application to Composite Casting," *Ceram. Trans.* 79:183-190.

Development of Non-Oxide Ceramic Matrix Composites Based on Reactive Infiltration and *In situ* Growth of AlN Bonding Phase

M. V. KEVORKIJAN

ABSTRACT

Novel ceramic composites (CMCs) were prepared by reactive infiltration of porous (50-70% T.D.) ceramic preforms and *in situ* growth of an AlN bonding phase (10-35 vol%) from molten Al alloys. The continuity of the forming phases results in a very favourable balance of both physical and mechanical properties. Moreover, because the prepared composites have a continuous Al phase (5-15 vol. %), they exhibit some traditional macroscopic plasticity, despite being approximately 70 vol. % ceramic. Preforms with optimal porosity and permeability were prepared by conventional ceramic processes using commercially available AlN, SiC and Si_3N_4 powders. Different non-oxide species were selected in order to demonstrate the flexibility of the preparation route. Wetting-assisted pressureless infiltration of Al into the preforms was accomplished in nominally pure N_2 or under an argon atmosphere at temperatures between 850°C and 1200°C. The AlN bonding phase was prepared by displacement reactions between Al and selected commercially available nitrides (Si_3N_4, Mg_3N_2,, Zn_3N_2, 325 Mesh powders) previously deposited on the surface of the porous ceramic body. Composite growth rates between 0.4 - 0.6 mm/min. were observed and correlated with the interfacial reaction enthalpy change. Initial characterization of selected mechanical properties (hardness, Young's modulus, compressive and flexural strength, toughness) of the composites with the same ratio of forming phases was also performed.

INTRODUCTION AND THEORETICAL BACKGROUND

The need for high strength, lightweight, and high stiffness materials has, in recent years, attracted much interest in the development of manufacturing processes for metal matrix composites (MMCs). The most important limitation encountered in fabricating these materials is that molten aluminium alloys normally do not wet the ceramic reniforcement and it is difficult to infiltrate ceramic preforms with molten aluminum alloys.

Kevorkijan Varuzan, Alpha and Omega in Composites, Lackova 139, 2341 Limbus, Slovenia, EU

In many instances the properties of a reinforced metal have been shown to provide a performance advantage over a monolithic metal, but the high cost of producing the composite has prohibited widespread commercial use. One way to reduce the cost is to obtain near-net-shape composites through rapid pressureless infiltration. Recently, it has been recognized (Kevorkijan, 1996) that pressureless infiltration can be achieved by utilizing certain exothermic chemical reactions that enhance wetting between the ceramic fillers and the molten metal alloy:

$$\mathbf{Si_3N_4 + 4Al \rightarrow 4AlN + 3Si \qquad \Delta H^{\circ} = -248.1\ kJ/mol}$$

$$\mathbf{Zn_3N_2 + 2Al \rightarrow 2AlN + 3Zn \qquad \Delta H^{\circ} = -457.1\ kJ/mol}$$

$$\mathbf{Mg_3N_4 + 4Al \rightarrow 4AlN + 3Mg \qquad \Delta H^{\circ} = -511.2\ kJ/mol}$$

These displacement reactions are the basis for the present work in which porous ceramic preforms with Si_3N_4, Mg_3N_4 and Zn_3N_2 layer-coated surfaces were spontaneously infiltrated with molten Al alloy. It must be noted that a chemically assisted wetting of a porous ceramic network by molten alloy, which is the key factor in rapid spontaneous infiltration, is based on the exothermic nature of these interfacial reactions. This contribution is effective for wetting only if the reactions are really localized near the solid-liquid interface and if they proceed quickly.

Assuming that a rapid exothermic interfacial reaction can lead to pressureless infiltration of molten alloy into a preform, it was theoretically predicted that the infiltration length, L can be expressed as (Kevorkijan, 1996):

$$\mathbf{L = (1/\varepsilon)\, k^{3/4}\, \rho\, [\, S_s(1-\varepsilon)/\mu\,]^{1/2}\, \Delta H^{1/2}\, t^{1/2}} \qquad \mathbf{(1)}$$

where ε is the porosity of the ceramic preform, k is the intrinsic permeability, S_s is the surface area per unit mass of porous preform, ρ is the theoretical density of the ceramic phase, μ is the viscosity of the molten alloy, ΔH° is the standard reaction enthalpy for above listed interfacial reactions and t is the time of infiltration.

The aim of this work was threefold: (i) to demonstrate that exothermic interfacial reaction is the key factor in spontaneous infiltration of a ceramic preform by Al alloy; (ii) to confirm the validity of Eq. 1 by using interfacial chemical reactions with different standard reaction enthalpies; and (iii) to present the processing details and results of initial property measurements of CMCs prepared by this route.

EXPERIMENTAL PROCEDURE

The AlN, SiC or Si_3N_4 preforms were typically fabricated using traditional ceramic green forming techniques. Pressing, tape casting, injection molding, casting, and green machining were all successfully used to make preforms. The preforms were impregnated by a selected nitride layer using a slurry of nitride particles in acetone.

Infiltration was performed by directly dipping an impregnated preform into the Al alloy bath in a nominally pure nitrogen atmosphere. The process was accomplished at temperatures of 850 - 950°C. The impregnated preform was dipped and held beneath the surface of the molten alloy using quartz rods. The alloy bath temperature was then increased to 1200°C quickly (in less than 2 min). After about 10-20 s, the preform sank to the bottom of the alloy bath. After 100 min of infiltration at 1200°C, the alloy bath was cooled to 900°C before the sample was removed from the alloy.

Characterization of materials

The length of infiltration was determined by using optical microscopy, SEM and EDS. Vickers hardness was measured on all samples with a 10-kg load, following the procedure outlined in ASTM Standard E 92. Measurements of elastic properties by an ultrasonic pulse echo technique in accordance with ASTM Standard D 2845, compressive strength per ASTM Standard E 9, and flexural strength in four-point bending following MIL-STD-1942A were made on selected materials.

Results - the cause of spontaneous infiltration

Spontaneous infiltration is probably caused by the negative capillary pressure generated at the infiltration front as a result of the enhanced wetting between the molten alloy and the solid preform. This enhanced wetting can be attributed to the exothermic interfacial reaction between the Al alloy and the surface dopants (Si_3N_4, Mg_3N_4 or Zn_3N_2) in the preform, i.e. by local increase in the temperature. It is well known that the wetting angle between a molten Al alloy and various non-oxide ceramics preforms decreases substantially as the temperature increases beyond approximately 1000°C in a very low oxygen pressure environment.

The requirements for spontaneous infiltration, which include the use of a magnesium containing alloy and a nitrogenous atmosphere, have been studied by many researchers (e. g. see in Aghajanian et al., 1991) It was demonstrated that no infiltration occurred without both magnesium in the alloy and a nitrogen atmosphere. In a more recent paper (Schiroky et al., 1997) the same authors offered the explanation that Mg reacts with the nitrogen process atmosphere to form a coating of magnesium nitride (Mg_3N_2) on the ceramic reinforcement. When molten aluminum comes into contact with Mg_3N_4, this compound is reduced to form AlN, with Mg dissolving into and re-evaporating from the alloy. The authors concluded that the displacement reaction between Mg_3N_4 and Al ***induces*** wetting for spontaneous infiltration.

In contrast, the present investigation has shown that spontaneous infiltration can performed without addition of magnesium (i.e. by using Zn_3N_2 or Si_3N_4 as a surface dopant). This indicates that the local increase of temperature (the temperature gradient) caused by an exothermic interfacial reaction is the ***real driving force*** for spontaneous infiltration. The result was confirmed experimentally using different interfacial reactions to cause spontaneous infiltration. The accumulated data are listed in Table I. A higher infiltration rate was achieved by applying an interfacial reaction with a higher standard reaction enthalpy. In this way, keeping other experimental parameters and the processing time constant, a more exothermic interfacial reaction resulted in a greater depth of infiltration as evident in Table I. Moreover, it was found that some comparable exothermic interfacial reactions which are operative without magnesium (e.g. the reaction between Zn_3N_2 and Al alloy) have a similar beneficial effect on the infiltration kinetics as Mg_3N_4. This result was recently additionally confirmed using some extremely exothermic interfacial reactions (Kevorkijan, 1997):

$$\mathbf{3TiO_2 + 4Al + 3C \rightarrow 3TiC + 2\,Al_2O_3 \qquad \Delta H^\circ = -\,1038.9\ kJ/mol\,.}$$

The same interfacial reactions are also used in our laboratory to prepare cast MMCs based on different non-oxide ceramic particles, previously covered by a nitride or TiO_2+C layer.

TABLE I - EFFECT OF THE STANDARD REACTION ENTHALPY OF SELECTED INTERFACIAL REACTIONS ON THE INFILTRATION LENGTH OF POROUS CERAMIC PREFORMS (EXPERIMENTAL CONDITIONS: INFILTRATION TIME: 100 MIN, TEMPERATURE: 1200°C , ATMOSPHERE: NITROGEN)

Preform	Interfacial reaction	ΔH (kJ/mol)	Infiltration length (mm)
AlN	$Si_3N_4 + 4Al \rightarrow 4AlN + 3Si$	- 248.1	40.4
AlN	$Zn_3N_2 + 2Al \rightarrow 2AlN + 3Zn$	- 457.1	55.1
AlN	$Mg_3N_4 + 4Al \rightarrow 4AlN + 3Mg$	- 511.2	58.2
AlN	$3TiO_2 + 4Al + 3C \rightarrow 3TiC + Al_2O_3$*	- 1038.9	82.1
SiC	$Si_3N_4 + 4Al \rightarrow 4AlN + 3Si$	- 248.1	40.8
SiC	$Zn_3N_2 + 2Al \rightarrow 2AlN + 3Zn$	- 457.1	55.7
SiC	$Mg_3N_4 + 4Al \rightarrow 4AlN + 3Mg$	- 511.2	58.6
SiC	$3TiO_2 + 4Al + 3C \rightarrow 3TiC + Al_2O_3$	- 1038.9	83.0
Si_3N_4	$Zn_3N_2 + 2Al \rightarrow 2AlN + 3Zn$	- 457.1	54.9
Si_3N_4	$Mg_3N_4 + 4Al \rightarrow 4AlN + 3Mg$	- 511.2	58.4
Si_3N_4	$3TiO_2 + 4Al + 3C \rightarrow 3TiC + Al_2O_3$	- 1038.9	83.0

***The reaction was performed at 1200°C in an argon atmosphere**

CONCLUSION

CMCs made up of primarily of AlN were produced using a novel infiltration and reaction approach.

The present investigation showed that the spontaneous and rapid infiltration of different non-oxide porous ceramic preforms can be performed without addition of magnesium if an exothermic interfacial reaction improves the wettability, giving rise to a modification of metal-ceramic interfacial chemistry. Based on this, one can conclude that the local increase of temperature caused by an exothermic interfacial reaction is the real driving force for spontaneous infiltration.

The same approach can be successfully used to produce castable discontinuously reinforced MMCs by direct immersion of ceramic particles into an Al alloy melt.

Initial characterization results demonstrate that composites produced in this manner are light, hard, stiff, and strong.

REFERENCES

Kevorkijan, M. V.. 1996.: "Discontinuous Al-SiC Composites Formed by a Low Cost Chemically Activated Infiltration Technique," *Metals, Alloys, Technologies,* 30(6):565-572.

Aghajanian, M. K., Rocazella, M. A., Burke, J. T. and Keck, S. D. 1991. " The Fabrication of Metal Matrix Composites by a Pressureless Infiltration Technique ," *J. Mater. Sci.*, 26:447-454.

Schiroky, G. H., Miller, D. V., Aghajanian, M. K. and Fareed, A. S.1997 " Fabrication of CMCs and MMCs using Novel Processes," *Key Engineering Materials* , 127:141-152.

Kevorkijan, M. V. 1997."The Improvement of Wetting in AlN_p-Al_l System Achieved by an Exothermic Interface Reaction," *The Abstract Book of the 99. Annual Meeting of ACerS, abstract C-021-97*:308.

Experimental Characterization of Orientation Formation in a Mold Cavity with Sudden Contraction

K. A. OLIVERO AND M. C. ALTAN

ABSTRACT

Orientation formation of short fibers suspended in a flow through a planar mold cavity is experimentally analyzed. The suspension of corn syrup and nylon fibers is injected into the mold at constant flow rate through a narrow, planar inlet gate. The mold cavity contains a sudden expansion following the inlet gate, and a three to one sudden contraction downstream. Orientation data is obtained using computerized image analysis on photographs of thirteen zones of interest near the sudden contraction. The photographs are taken through the transparent top mold wall, thus collecting average planar orientation data through the mold thickness. Orientation distribution histograms are generated from the individual fiber orientations to graphically depict orientation microstructure. These histograms reveal bimodal fiber orientations in nearly all zones of interest. Experiments performed at different volume flow rates and viscosities also confirmed the bimodal orientation distribution observed through the top mold wall. Orientation results are observed to be more random compared to those predicted using Jeffery's theory (He et al., 1997). Fiber size or wall effects may contribute to the formation of more random orientation field.

INTRODUCTION

Injection molding is among the most commonly used methods to manufacture short-fiber-reinforced composite materials. During injection molding processes, the flow field in the mold cavity causes the fibers to translate and rotate resulting in a complex orientation field in the molded part. The mechanical properties of the molded part are highly dependent on orientation field, thus models to characterize this flow induced orientation field are needed. Accurate experimental characterization of orientation field is also necessary to assess the predictive capability of these models.

Several experimental studies on the orientation formation of short fibers in a viscous flow have been conducted. Initial experimental studies observe the motion of individual fibers traveling through an experimental flow configuration. For example, Murty and Modlen (1977) and Lee and George (1978) investigated the flow kinematics, rotation of fibers, and formation

---

Kristian A. Olivero and M. Cengiz Altan, School of Aerospace and Mechanical Engineering, University of Oklahoma, Norman, OK, 73019.

of stagnant eddies in convergent channels. Vincent and Agassant (1985) studied fiber rotation in a diverging channel flow. Experimental observation of individual fibers is useful for comparison with theory based on the dynamics of a single fiber, but may not be adequate to construct a detailed description of orientation field formed by a large number of fibers.

Bay and Tucker (1992a) performed an experimental study of orientation field in center-gated disk and film-gated strip mold cavities. Their study involved the injection molding of nylon 6/6 reinforced with 43 wt% glass fibers. Sections were cut from the molded parts and polished. Cross sections of fibers on the polished faces were digitized and analyzed to obtain orientation data for each fiber. From this data, the second-order moment of the orientation probability density function (i.e., second-order orientation tensor), S_{ij}, was calculated as described by Bay and Tucker (1992b) at each location. The second-order orientation tensor is commonly used to represent both experimental and theoretical microstructure, and is shown to completely characterize orientation field in dilute, non-Brownian suspensions of spheroidal particles (Rao, 1996). Although the process used to obtain the data provided by Bay and Tucker (1992a) is rather involved, the results presented are comprehensive and useful for comparison with theoretical models.

In addition to experimental studies, numerical prediction of orientation field in injection molded parts is studied by a number of researchers. Theoretical studies to predict orientation field are mostly based on the rotation of a hydrodynamically isolated ellipsoidal particle immersed in a viscous Newtonian flow (Jeffery, 1922). Initially, Altan et al. (1990) presented numerical prediction of fiber orientation in Hele-Shaw flows, thus simulating orientation formation in injection molding. Some of the other studies predicting orientation structure in injection molding are published by Matsuoka et al. (1990), Bay and Tucker (1992c), De Frahan et al. (1992), Freidrichs et al. (1993), Chung and Kwon (1995), Altan and Rao (1995) and Chung and Kwon (1996).

In this study, orientation formation in a steady Newtonian flow through a planar mold cavity containing a three to one sudden contraction is experimentally studied. Computerized image analysis is performed on photographs of thirteen zones of interest near the sudden contraction. Orientation results for nine zones are presented in the form of orientation distribution histograms. Histograms for different filling conditions are also presented to depict the effect of Reynolds number on the orientation field.

EXPERIMENTAL SETUP AND DATA ANALYSIS

Filling experiments are performed using the experimental setup shown in Fig. 1. The suspension is injected into a planar mold cavity using a Masterflex I/P 73 peristaltic pump. The microprocessor controlled pump can be set to provide constant flow rates ranging from 0.25 and 8.11 L/min for water. Experiments are performed at pump setting of 0.25, 0.50, and 1.00 L/min for water, which yield 10-50% lower flow rates due to the higher viscosity of the corn syrup suspension used in this study. The flow oscillations caused by the peristaltic pump are minimized by an in-line pulse dampener. Tygon tubing having an inside diameter of 9.5 mm (3/8 inch) is used throughout the setup. The suspension is stored in a feed reservoir and pumped through the experimental mold cavity. After flowing through the cavity, the suspension is returned to the reservoir to facilitate continuous operation at steady conditions.

For each experiment, the corn syrup is diluted with water to a selected viscosity ranging from 300-2000 cp. Viscosity measurements are made with a Brookfield DVII+ viscometer. A specific gravity cup with 0.5% accuracy is used to obtain the density of the corn syrup and fibers. The density of the undiluted corn syrup and fibers are found to be 1400 kg/m^3 and 770 kg/m^3, respectively. Short fibers are prepared by hand cutting longer nylon fibers with 0.17 mm diameter to an average length of 1.55 mm. This leads to an average aspect ratio of 9.14. Suspensions are prepared having less than 0.2% fibers by weight (0.36% by volume). The fibers are slowly mixed with the corn syrup solution to ensure uniform dispersity.

Three different Reynolds numbers for the experimental conditions can be defined. At the fiber length scale, two microscale Reynolds numbers are defined in terms of the fiber diameter, D, and average fiber length, L_{ave}

$$Re_D = \frac{\rho V_{max} D}{\mu} \quad \text{and} \quad Re_L = \frac{\rho V_{max} L_{ave}}{\mu} \tag{1}$$

where V_{max} is the maximum midplane velocity at a particular zone of interest The minimum and maximum microscale Reynolds numbers for the experimental conditions are Re_{Lmax}=0.004 and Re_{Dmin}=0.0002 downstream of the sudden contraction. These values are much less than unity, justifying comparisons with the numerical orientation predictions based on Jeffery's theory. The third macroscale Reynolds number is defined in terms of the flow rate Q and half gap width h.

$$Re = \frac{\rho Q}{\mu h} \tag{2}$$

The experimental conditions lead to a macroscopic Reynolds number in the range of 0.12 to 0.28. These values are less than unity, thus indicating the presence of Hele-Shaw flow.

The planar mold cavity is formed by placing an aluminum spacer plate between two sheets of plexiglass. The spacer plate, and thus the cavity, has a thickness of 6.35 mm (¼ inch). The suspension is injected into the cavity through a 25.4 mm (1 inch) wide inlet gate, specially designed to achieve a planar velocity profile. After the inlet gate, the suspension undergoes a sudden expansion to 228.6 mm (9 inch) wide mold. At 279.4 mm (11 inch) downstream, a three to one sudden contraction down to 76.2 mm (3 inch) takes place. The suspension exits the mold through a gate identical to the inlet gate. The fibers are buoyant and rise to the top wall in a horizontally positioned mold. Thus, the entire mold is aligned vertically to minimize gravity effects on orientation field. The mold geometry, dimensions, and the locations of the thirteen zones of interest are shown in Fig. 2.

Data is collected after the mold is completely filled and steady conditions are established. Photographs of fibers in thirteen 25.4 mm (1 inch) square zones of interest near the sudden contraction are taken through transparent mold walls. The photos are digitized and computerized image analysis is performed on each picture to obtain the planar orientation of each fiber. Figure 3 is an example of a digitized photograph of fibers in zone 12. In Hele-Shaw flows, fibers have predominantly inplane orientations, thus the planar orientation data may be adequate to describe the orientation structure. The photographs contain fibers throughout the mold thickness, leading to orientation results averaged through the mold cavity. Several photographs of each location are superimposed to reduce statistical variations by increasing the total number of fibers in each zone. Typically three photographs of each zone are used yielding a total of 200-300 fibers.

The second-order orientation tensor, S_{ij}, is currently used most often to characterize orientation field. In this study, S_{ij} is found to be inadequate to represent the experimental data. Reasons for this include averaging microstructure through the mold thickness and bimodal orientation distributions. Rather than utilizing a higher order orientation tensor, fiber orientation data is presented in the form of radial orientation distribution histograms. The circumfrencial axis represents orientation angles (0° indicating the direction of the symmetry axis of the mold) and the radial axis represents the number of fibers in each orientation interval. The orientation angles are divided into 30° intervals. Smaller intervals give jagged curves and larger intervals do not give enough resolution to examine orientation details. The two ends of each fiber are indistinguishable, thus, the histograms have angular symmetry with a period of 180°.

RESULTS

Figures 4-12 are radial orientation distribution histograms for nine zones of interest near the sudden contraction (results for zones 4, 5, 9, and 10 are omitted due to mold symmetry). The location of each zone is shown in Fig. 2. Zones 1-10 are upstream of the sudden contraction and zones 11-13 are downstream of the contraction. Results in Figs. 4-12 are obtained using a suspending fluid with viscosity μ=1200 cp injected at a flow rate of $Q=7.5x10^{-6}$ m^3/s. In each zone, orientation results are compared to orientation predictions given by He et al. (1997). He et al. numerically solved the Hele-Shaw flow for the same mold geometry. Using this flow the orientation field was calculated based on Jeffery's theory and the average orientation structure through the thickness was given at thirteen zones of interest. It is also illustrative to see the direction perpendicular to the local flow on orientation histograms. Thus, the preferred fiber orientation based on Jeffery's theory and the direction perpendicular to the flow are superimposed on each orientation histogram.

Orientation data for zone 1 is shown in Fig. 4. A bimodal distribution is clearly observed with peaks near the predicted preferred orientation and in the direction perpendicular to the local flow. Figure 5 indicates higher degree of fiber alignment in zone 2. Bimodal alignment peaks are more pronounced compared to zone 1. The first orientation peak is near the calculated preferred orientation. The second peak is around perpendicular to the flow direction. Orientation data from zone 3 is shown in Fig. 6. This zone is on the symmetry axis, and fibers are observed to align predominantly perpendicular to the flow direction. Orientation results for zone 6, shown in Fig. 7, depict fiber alignment primarily in the predicted preferred orientation. The orientation distribution, however, has a smaller second alignment peak near perpendicular to the flow direction. Zone 7 is closer to the symmetry axis than zone 6. Figure 8 shows more distinct bimodal distribution in this zone compared to zone 6. Alignment peaks are observed in the predicted preferred orientation and near perpendicular to the flow. Figure 9 shows the orientation histogram for zone 8, which is on the symmetry axis just prior to the sudden contraction. Although Jeffery's theory predicts alignment in the flow direction in zone 8, fibers are seen to align predominantly perpendicular to the local flow. Figure 10 shows orientation distribution downstream of the sudden contraction in zone 11. Fibers are mostly aligned perpendicular to the flow direction, but this alignment is not as much as that of zone 8. A small alignment peak in the flow direction (i.e., theoretical preferred orientation) is apparent. The orientation histogram for zone 12 is shown in Fig. 11. The histogram depicts a somewhat random orientation distribution with a slight alignment peak along the predicted orientation (i.e., in the flow direction). Fibers are more aligned in the flow direction than those in zone 11, which is consistent with the theory. Zone 13, shown in Fig. 12, is the final zone before the exit gate. As predicted by the theory, the fibers are primarily aligned in the flow direction; however a smaller second orientation peak exists perpendicular to the local flow direction.

The experimental data for zones 1-3 and zones 6-8 indicate increasing degree of alignment in the direction predicted by Jeffery's theory with increasing distance from the symmetry axis. In zones 3 and 8 (on the mold's symmetry axis) the fibers are aligned predominantly perpendicular to the flow, with low degree of alignment in the predicted preferred orientation direction. In zones 2 and 7 (slightly away from the mold's symmetry axis) fibers are mostly aligned perpendicular to the flow, with distinct alignment peaks in the predicted orientation directions. In zones 1 and 6 the degree of alignment in the predicted direction is even stronger.

The theory predicts increasing fiber alignment in the flow direction as the fibers move through the sudden contraction along the symmetry axis. Zones 3 and 8 show fiber alignment predominantly perpendicular to the flow direction. In zone 11, however, an alignment peak in the flow direction becomes evident. As fibers move through zones 12 and 13, the alignment in the flow direction increases as predicted by the Jeffery's theory.

Figures 8, 13, and 14 show orientation distributions in zone 7 for different Reynolds numbers. Data from each experiment show similar bimodal peaks. The first peak is observed along the predicted preferred orientation, whereas the second peak is perpendicular to the local flow.

CONCLUSIONS

Orientation formation in short-fiber suspensions injected into a planar mold cavity is experimentally studied. Planar fiber orientation field is characterized at nine zones of interest near the sudden contraction. Bimodal orientation distributions are observed in almost all zones, even under different flow conditions. Fibers tend to align both in the direction predicted by Jeffery's theory and in the direction perpendicular to the local flow. Large fiber size or wall effects may cause the observed alignment perpendicular to the local flow.

REFERENCES

Altan, M. C., Subbiah, S., Güçeri, S. I., and Pipes, R. B., 1990, "Numerical Prediction of Three-Dimensional Fiber Orientation in Hele-Shaw Flows," Polymer Eng. Sci., Vol. 30(14), pp. 848-859.

Altan, M. C., and Rao, B. N., 1995, "Closed-Form Solution for the Orientation Field in a Center-Gated Disk," J. Rheology, Vol. 39(3), pp. 581-599.

Bay, R. S., and Tucker, C. L., 1992a, "Fiber Orientation in Simple Injection Moldings. Part II: Experimental Results," *Polymer Composites*, Vol. 13, pp. 332-341.

Bay, R. S., and Tucker, C. L., 1992b, "Stereological Measurement and Error Estimates for Three-Dimensional Fiber Orientation," *Polymer Engr. Sci.*, Vol. 32(4), pp. 240-253.

Bay, R. S., and Tucker, C. L., 1992c, "Fiber Orientation in Simple Injection Moldings. Part I: Theory and Numerical Methods," *Polymer Composites*, Vol. 13, pp. 317-331.

Chungh, S. T., and Kwon, T. H., 1995, "Numerical Simulation of Fiber Orientation in Injection Molding of Short-Fiber-Reinforced Thermoplastics," *Polymer Engr. Sci.*, Vol. 35(7), pp. 604-618.

Chungh, S. T., and Kwon, T. H., 1996, "Coupled Analysis of Injection Molding Filling and Fiber Orientation, Including In-Plane Velocity Gradient Effect," *Polymer Composites*, Vol. 17(6), pp. 859-872.

De Frahan, H. H., Verleye, V., Dupret, F., and Crochet, M. J., 1992, "Numerical Prediction of Fiber Orientation in Injection Molding," *Polymer Eng. Sci.*, Vol. 32(4), pp. 254-266.

Friedrichs, B., Güçeri, S. I., Subbiah, S., and Altan, M. C., 1993, "A Numerical Approach to Short Fiber Reinforced Reaction Injection Molding (RRIM)," *J. Mat. Processing & Manuf. Sci.*, Vol. 1, pp. 331-349.

He, J., Olivero, K. A., and Altan, M. C., "Orientation Formation in Planar Mold Filling: Theory and Numerical Predictions," to be published, Proceedings of the International Mechanical Engineering Congress and Exposition, ASME, Dallas, TX, 1997.

Jeffery, G. B., 1922, "The Motion of Ellipsoidal Particles Immersed in a Viscous Fluid," *Proc. Royal Soc.*, Vol. A102, pp. 161-179.

Lee, W., and George, H. H., 1978, "Flow Visualization of Fiber Suspensions," *Polymer Eng. Sci.*, Vol. 18(2), pp. 146-156.

Murty, K. N., and Modlen, G. F., 1977, "Experimental Characterization of the Alignment of Short Fibers During Flow," *Polymer Engr. Sci.*, Vol. 17(12), pp. 848-853.

Rao, B. N., 1996, "Rheology and Flow of Fiber Suspensions in Composite Materials Manufacturing," Ph. D. Thesis, University of Oklahoma, Norman, OK.

Matsuoka, T., Takabatake, J., Inoue, Y., and Takahashi, H., 1990, "Prediction of Fiber Orientation in Injection Molded Parts of Short-Fiber-Reinforced Thermoplastics," *Polymer Engr. Sci.*, Vol. 30(16), pp. 957-966.

Vincent, M., and Agassant, J. F., 1985, "Experimental and Theoretical Study of Short Fiber Orientation in Diverging Flows," *Rheol. Acta*, Vol. 24, pp. 603-610.

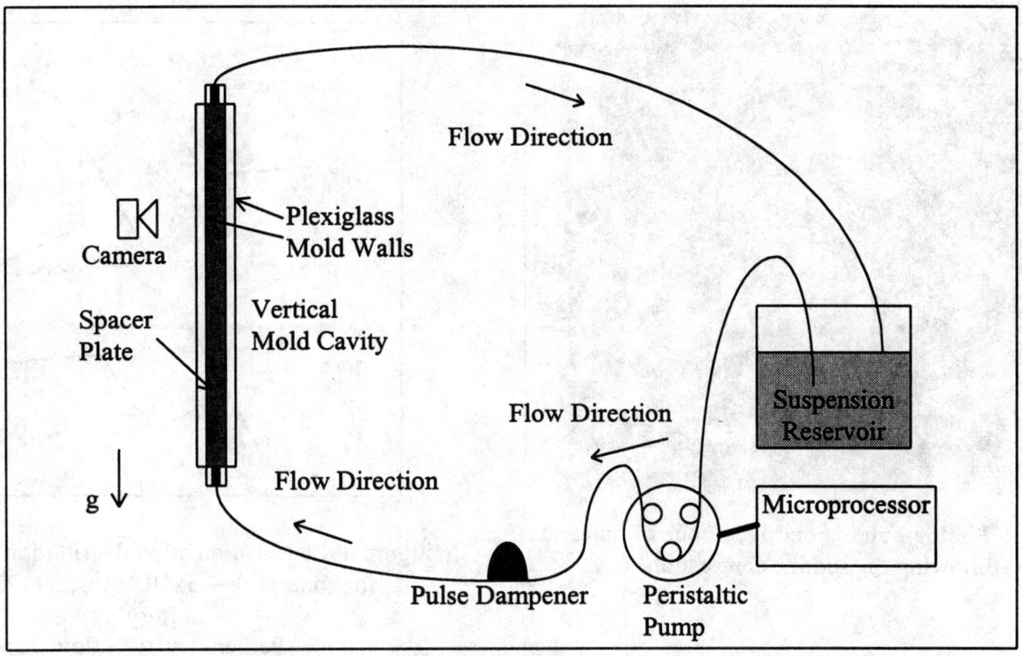

Figure 1: Experimental molding setup used to collect fiber orientation data.

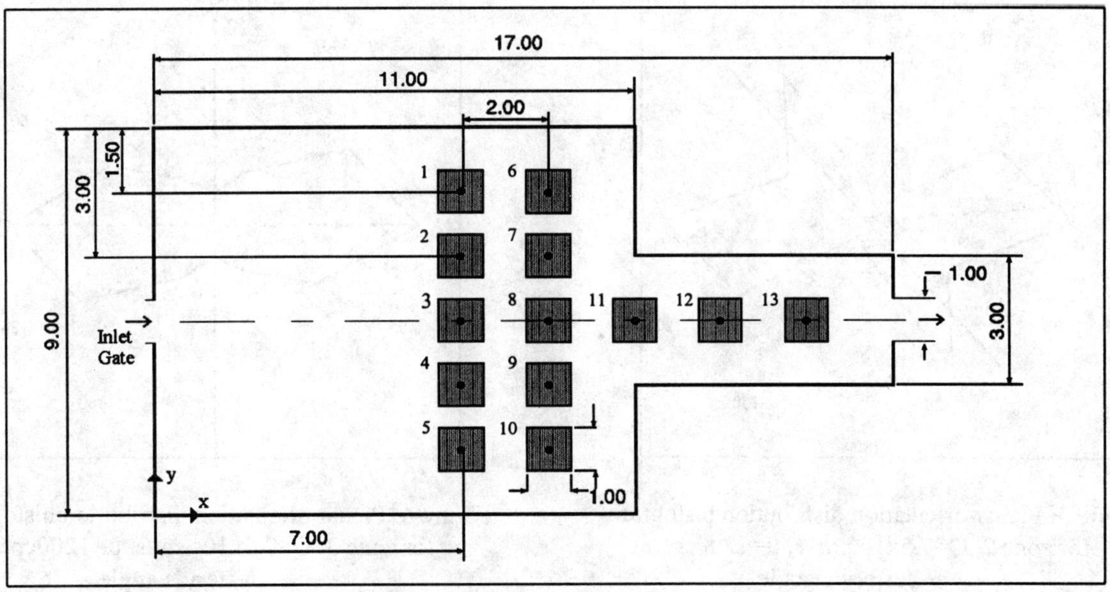

Figure 2: Mold geometry, dimensions (in inches), and locations of interest.

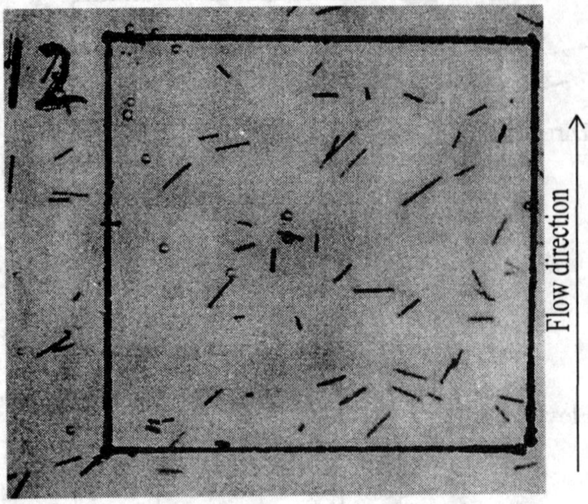

Figure 3: Photograph of zone 12, a zone of interest following the sudden contraction.

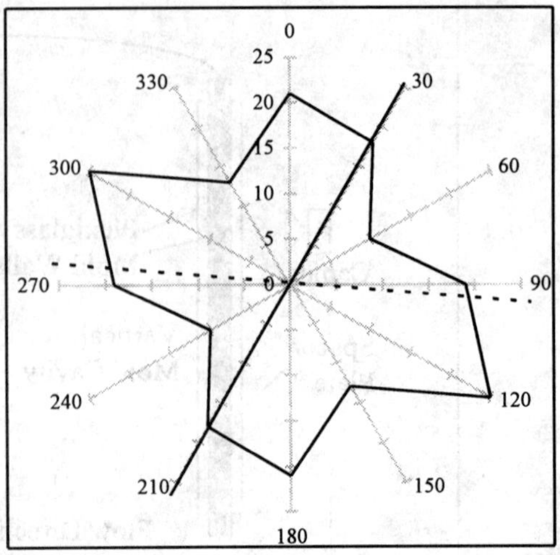

Figure 4: Planar orientation distribution histogram for zone 1, Q=7.5x10^{-6} m^3/s, μ=1200cp.
——— Jeffery's angle
- - - - Perpendicular to flow direction

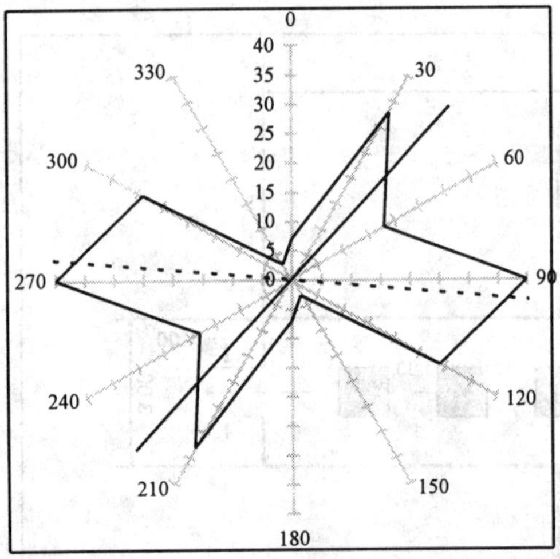

Figure 5: Planar orientation distribution histogram for zone 2, Q=7.5x10^{-6} m^3/s, μ=1200cp.
——— Jeffery's angle
- - - - Perpendicular to flow direction

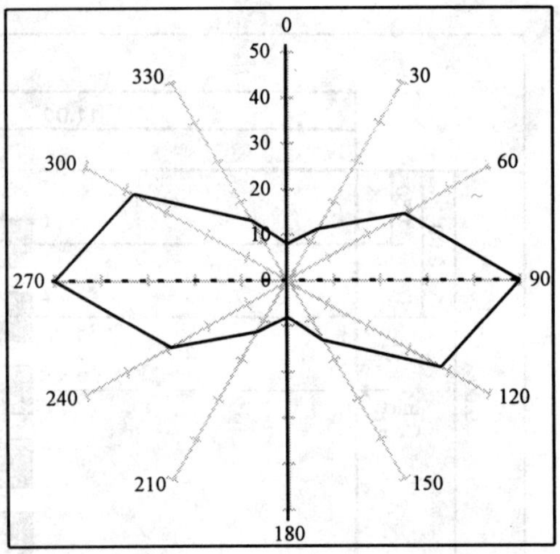

Figure 6: Planar orientation distribution histogram for zone 3, Q=7.5x10^{-6} m^3/s, μ=1200cp.
——— Jeffery's angle
- - - - Perpendicular to flow direction

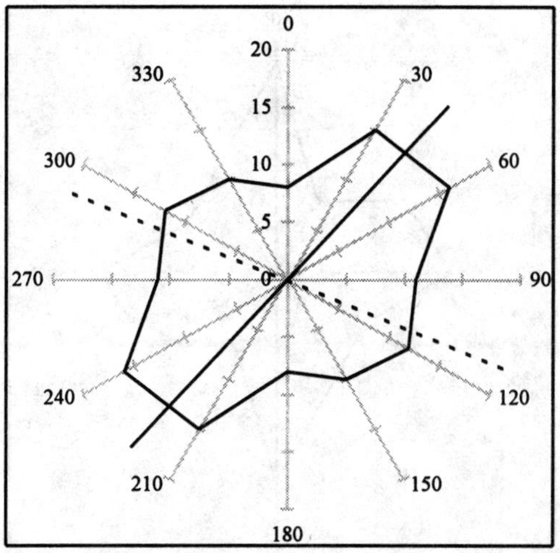

Figure 7: Planar orientation distribution histogram for zone 6, Q=7.5x10^{-6} m^3/s, μ=1200cp.
—— Jeffery's angle
- - - - Perpendicular to flow direction

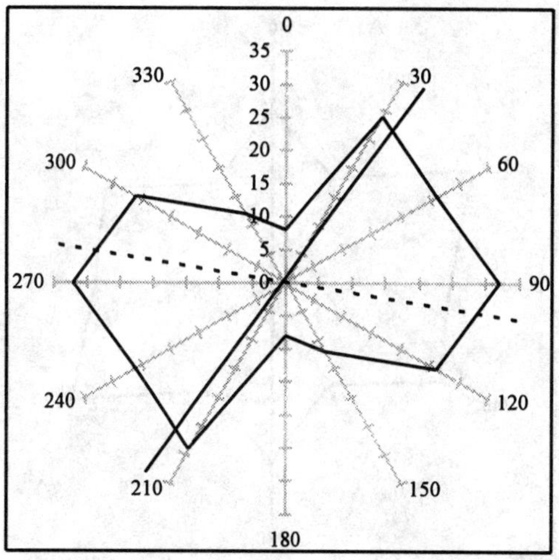

Figure 8: Planar orientation distribution histogram for zone 7, Q=7.5x10^{-6} m^3/s, μ=1200cp.
—— Jeffery's angle
- - - - Perpendicular to flow direction

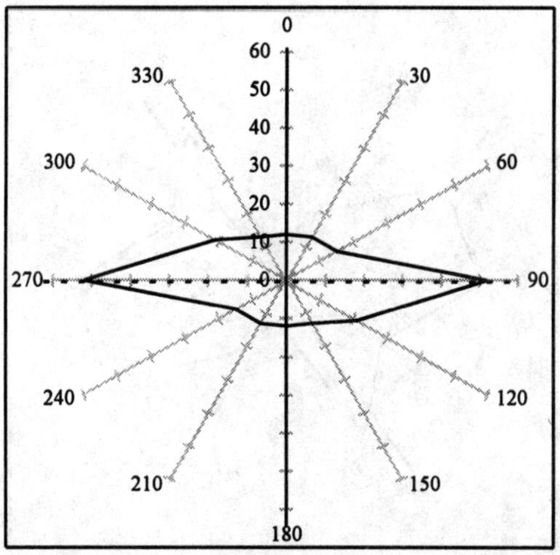

Figure 9: Planar orientation distribution histogram for zone 8, Q=7.5x10^{-6} m^3/s, μ=1200cp.
—— Jeffery's angle
- - - - Perpendicular to flow direction

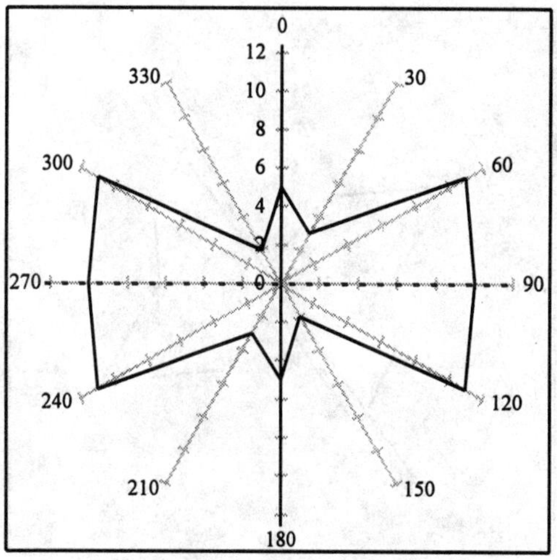

Figure 10: Planar orientation distribution histogram for zone 11, Q=7.5x10^{-6} m^3/s, μ=1200cp.
—— Jeffery's angle
- - - - Perpendicular to flow direction

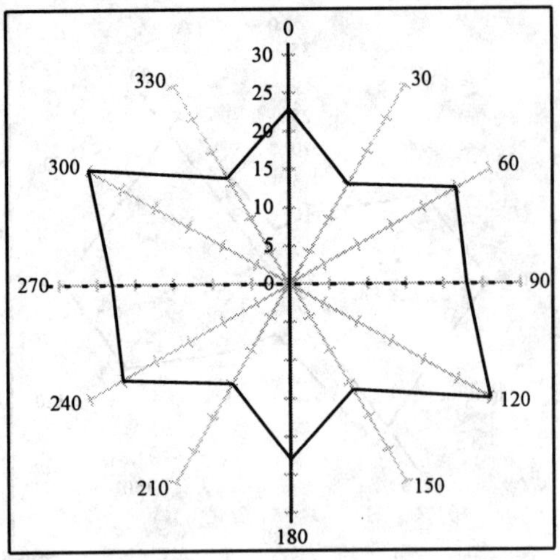

Figure 11: Planar orientation distribution histogram for zone 12, Q=7.5x10^{-6} m^3/s, μ=1200cp.
—— Jeffery's angle
- - - - Perpendicular to flow direction

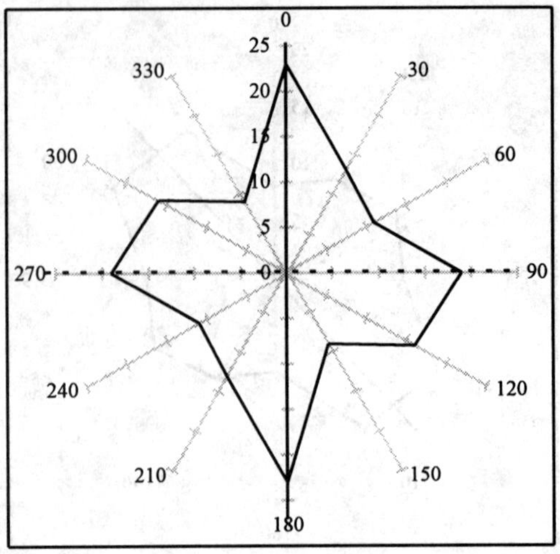

Figure 12: Planar orientation distribution histogram for zone 13, Q=7.5x10^{-6} m^3/s, μ=1200cp.
—— Jeffery's angle
- - - - Perpendicular to flow direction

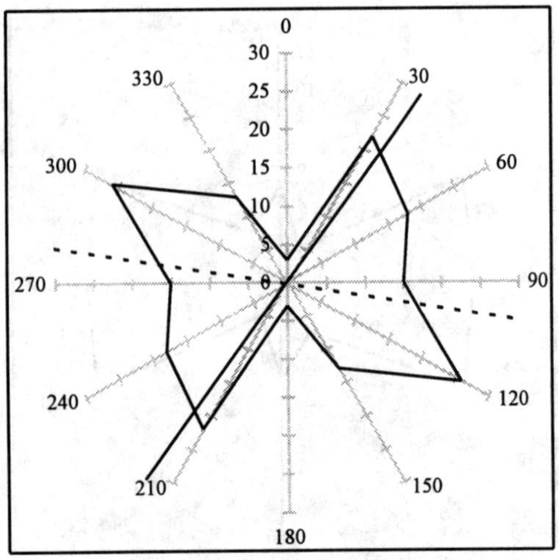

Figure 13: Planar orientation distribution histogram for zone 7, Q=4.2x10^{-6} m^3/s, μ=2000cp.
—— Jeffery's angle
- - - - Perpendicular to flow direction

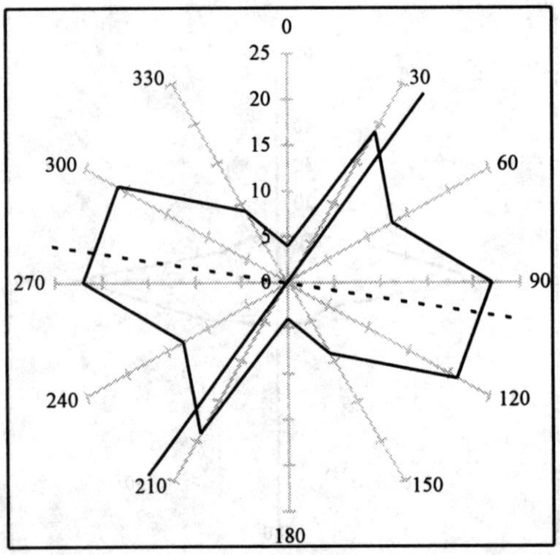

Figure 14: Planar orientation distribution histogram for zone 7, Q=3.3x10^{-6} m^3/s, μ=2000cp. —— Jeffery's angle
- - - - Perpendicular to flow direction

Recycling Process for Automotive Plastics and Composites*

R. E. ALLRED, T. J. DOAK AND A. B. COONS

ABSTRACT

The use of engineering plastics and composites in automotive applications is increasing to achieve weight reductions and the accompanying performance increases. Those materials are difficult to recycle using conventional technologies because they are contaminated with metals and paints or are cross-linked polymer structures. A novel tertiary recycling process is being developed as an economical means for recycling scrap automotive plastics and composites. Early development work on this process for recycling scrap aerospace composites and electronics shows that it can convert a variety of polymers and composites into low molecular weight hydrocarbons at temperatures near 300°C. The hydrocarbons produced can then be reused as chemicals, fuels, or monomers. Metal, glass, fillers, and fibers are separated from the hydrocarbons during the process and can be reclaimed for reuse. Initial feasibility studies have shown that the proposed catalytic conversion process is effective for recycling plastics and composites used in automotive parts. This process does not require presorting of plastic wastes and allows fibers and fillers to be reclaimed for reuse. Conversion products from recycling sheet molding compounds (SMC), auto shredder residue (ASR), and scrap mixtures of automotive plastic body panels show that commercial recycling plants based on the catalytic conversion process should be an economical alternative to landfilling.

INTRODUCTION

Goals for the Next Generation Vehicles (NGV) program include increasing fuel economy, reducing emissions, and increasing recyclability while maintaining the comfort, capacity, safety, and price of current vehicles [1].

Ronald E. Allred, Thomas J. Doak, and Andrew B. Coons, Adherent Technologies, Inc., 11208 Cochiti SE, Albuquerque, NM 87123, (505) 298-8662.

*This work supported by the National Science Foundation under Grant DMI-9660673.

Achievement of these goals requires that the weight of current vehicles must drop substantially. Weight reduction will be most easily achieved through increased use of engineering plastics and polymer matrix composites [2-5]. Those materials, however, are the most difficult to recycle by conventional means, and they represent the majority of automotive materials that are currently not recycled [1,6,7].

The difficulty in recycling automotive plastics and composites is due to both technical and economic limitations with current recycling approaches. Automakers are meeting the recycling challenge by designing for recyclability and fabricating multiple parts from single, recyclable plastics [6]. That approach provides short- and medium-term gains in recycling content, but may be too limiting for the NGV because of the large weight reductions required. To achieve the NGV goals, designers will need the full complement of plastic and composite materials properties available, which requires the development of alternative recycling technologies for those materials. Thermoset plastics and composites in particular are difficult to recycle using conventional technology [6].

The majority of work on automotive composites recycling that has appeared in the literature is concerned with grinding, chipping, or flaking the composite into suitable size to be used as filler in new molded composite parts [8-11]. The largest firm using those processes, Phoenix Fiberglas, closed its doors in 1996. Their primary feedstock was automotive sheet molding compound (SMC). Their departure leaves a large void to be filled for recycling of automotive composites, while the use of SMC composites is predicted to rise by 43% in 1997 [12].

Tertiary recycling, conversion of plastics into low molecular weight hydrocarbons for reuse as chemicals or fuels, represents a potentially economically viable means for recycling waste automotive plastics and composites [13-18]. Adherent Technologies, Inc., has been developing a low-temperature, catalytic conversion tertiary recycling process for organic materials (such as electronics assemblies and cross-linked polymers and composites) that do not lend themselves to recycling with conventional technologies. Initial results show that all types of plastics and composites, thermosets as well as thermoplastics, can be converted in high yields to valuable hydrocarbon products with this novel catalytic conversion process [19-23]. This process also performs equally well with complex mixtures of multiple plastics, metals, and inorganics such as found in aircraft structures and electronic assemblies [24,25]. Conversion times are rapid, and the process is closed and, thus, nonpolluting. Fillers, fibers, and impurities are separated from the hydrocarbons for reuse or disposal. The resultant hydrocarbon products represent potential chemical feedstocks or fuel sources.

The feasibility studies conducted by Adherent Technologies show that a significant opportunity exists to recycle scrap automotive plastics and composites using a related version of this catalytic conversion technology. Successful implementation of this technology will allow the extensive use of engineering plastics and polymer composites in NGVs while simultaneously meeting recycling goals and conserving national resources. This paper summarizes our initial feasibility studies on recycling scrap automotive plastics and composites. Results show that SMC, thermoplastic composites, mixtures of plastic automotive body panels, and auto shredder residue (ASR) are all readily recycled using the catalytic conversion process. Fibers and fillers are reclaimed in a form for reuse while the polymers are converted into low molecular weight mixtures for use as chemical feedstocks or fuels.

EXPERIMENTAL PROCEDURES

Materials

Scrap automotive SMC, reinforced polypropylene (RPP), and mixed body panel parts were obtained from local repair shops and size reduced at Saturn-Granutech (Grand Prairie, TX). Additional SMC and ASR materials were supplied by Philip Environmental (Detroit, MI) from their automotive recycling subsidiaries.

Recycling Procedures

The recycling trials were conducted in our custom-made bench-scale reactor. A schematic of the reactor used is given in Figure 1. The reactor is a continuous-feed screw kiln type with a 3-cm-diameter auger drive for feedstock material transport. The reactor is heated by an electric tube furnace.

Conversion times in the system shown in Figure 1 were nominally 30 minutes at temperatures between 200 and 450°C. As the polymeric materials depolymerize, the hydrocarbon conversion products are removed as a gas and condensed in a series of condensers. Uncondensible gases travel through the vacuum pump and are incinerated. If useable as fuel, the uncondensible gases would be used to heat a gas-fired reactor in a commercial system.

Hydrocarbon Characterization

Hydrocarbon conversion products were analyzed using gas chromatography-mass spectroscopy (GC-MS) techniques in a Shimadzu QP-5000 instrument with a Restek RTX-502.2 column optimized for volatile separation. The system carrier gas was high-purity helium at a flow rate of 1.0 sccm. Sample fractions were diluted to 1-2% in methylene chloride or methanol before introduction of 1.0 μl into the GC. The GC is programmed for a ramp rate of 5°C/min. from 40 to 250°C where the temperature is held 20 min. to allow high boiling point compounds to elute. The MS is programmed to scan from a mass/charge ratio (M/Z) of 15 to 450 at a rate of 1,000 scans/sec. Peak spectra are matched against the Wiley Library that contains 139,859 spectra of known compounds. Peak retention times are also compared to pure chemical standards as a reference.

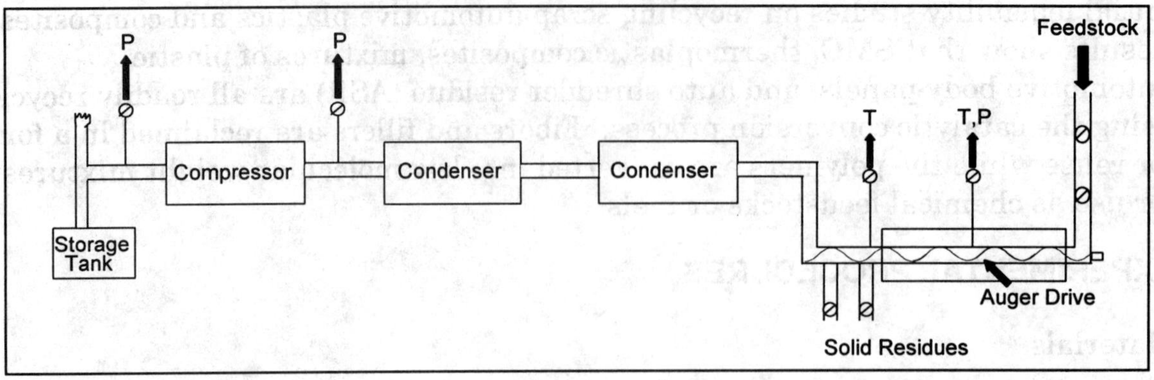

Figure 1. Schematic of bench-scale tertiary recycling system

Fiber Characterization

Samples of the reclaimed carbon fibers were examined using an ISI Model SIIIA scanning electron microscope (SEM). The fibers were placed on aluminum specimen holders using adhesive carbon tape and gold coated for observation. Observations were made regarding the presence or absence of residual resin material and the physical condition of the fiber surfaces.

Single fiber tensile tests were performed on recycled fibers recovered from SMC and RPP composites. Virgin glass fibers were tested as controls. Twenty-five of each of the control fibers and recycled fibers were tested in accordance with ASTM D3379-75 using an Instron Model 1101 mechanical testing machine with a crosshead speed of 0.5 cm/min.

RESULTS AND DISCUSSION

Process Development

Eighteen processing trials were conducted on the scrap automotive materials. To determine the basic parameters for catalytic conversion, chamber temperatures ranged from 200°C to 450°C with a relatively constant vacuum of between 18 and 21 inches Hg. Each sample weighed 100 g and had a chamber residence time of 30 minutes.

The catalytic conversion results showed that the recycling process under investigation can be implemented over a wide range of process conditions with conversion kinetics mainly affected by temperature. A process temperature of 300°C was found to be near optimum for all the automotive scrap feedstocks. Solid products from recycling at 300°C appeared free from all organic materials. Conversion product yields for the various scrap automotive materials are given in Table I. Liquid yields are understated in Table I due to the small sample sizes used. The losses are due to coating of the reactor piping and condensing system walls. In a commercial system operating at steady state, the liquid/gas ratio would increase. The solid residue amounts given in Table I appear to be in the correct range for these types of feedstocks.

TABLE I. PRODUCT YIELDS FOR CATALYTIC CONVERSION OF AUTOMOTIVE SCRAP MATERIAL

Scrap Material	Solid Residue (%)	Liquid Product (%)	Gaseous Product (%)
SMC	82	2	16
ASR	39	35	26
RPP	37	18	45
Mixed Auto Body Panels	31	46	23

Characterization of Reclaimed Glass Fibers

Photographs of a reinforced polypropylene bumper and the reclaimed glass fibers after tertiary recycling are shown in Figure 2 and 3, respectively.

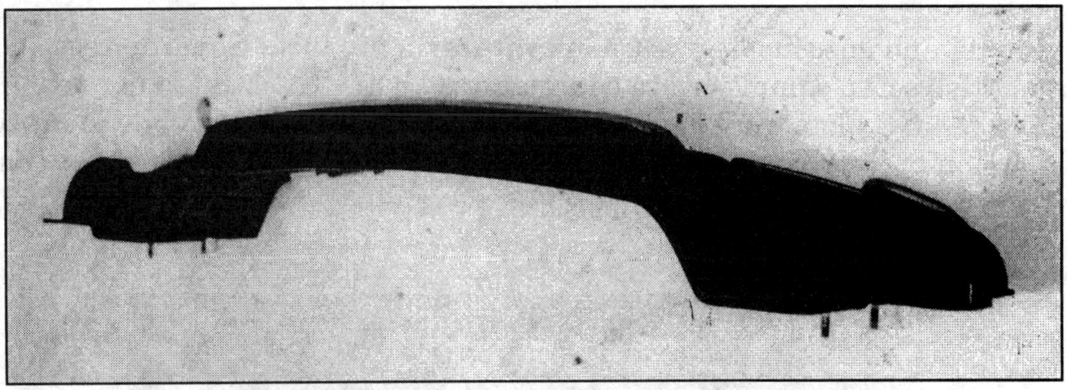

Figure 2. Reinforced polypropylene bumper

Figure 3. Reclaimed glass fibers after tertiary recycling

Glass fibers reclaimed from SMC appeared visually similar to those shown in Figure 3 except that they were mixed with filler particles. The glass fibers reclaimed from SMC and RPP using catalytic conversion were characterized for fiber surface texture and tensile strength. An SEM photomicrograph of the glass fibers reclaimed from SMC is shown in Figure 4.

As seen in Figure 4, the surface texture of the reclaimed glass fibers is smooth with no evidence of pitting or other surface attack. There is also little residual resin evident remaining on the glass fiber surface.

Single fiber tensile tests revealed that the reclaimed glass fibers had lost a significant amount of strength (≈50%) compared to literature values. This is in contrast to what has previously been observed for reclaimed carbon fibers from carbon/epoxy aircraft components [22,24]. The sensitivity of glass fiber surfaces is well known and is the reason that coupling agents and sizings are applied immediately after spinning glass fibers during their manufacture. The reclaimed glass fibers will probably find maximum value as milled fibers for use in injection molding compounds.

Characterization of Hydrocarbon Products

Liquid and gaseous hydrocarbon conversion products from the recycled automotive scrap samples were characterized using GC-MS techniques. Peaks seen in the GC plots were identified from their mass spectra. The identified conversion product compounds from SMC with concentrations greater than 2% are given in Table II.

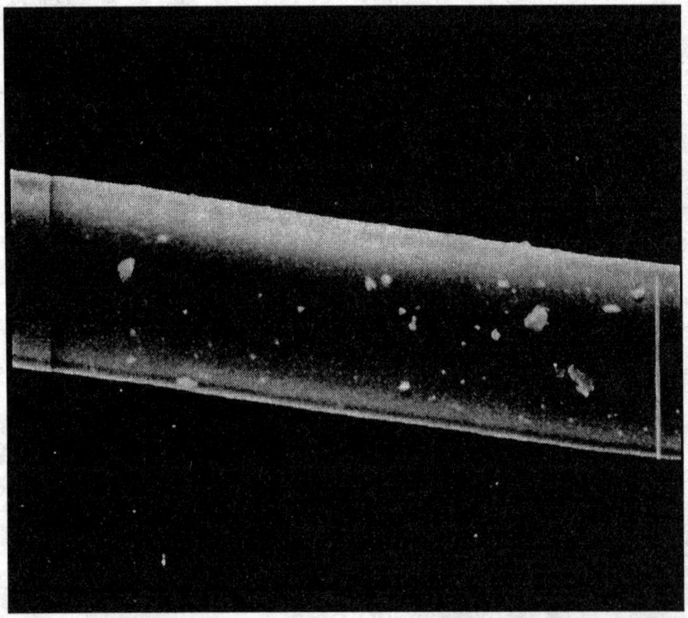

Figure 4. Appearance of glass fiber reclaimed from SMC (2500X)

The results given in Table II show that styrene constitutes the majority of the liquid conversion product from tertiary recycling of SMC. Separation of the styrene by distillation for use in new SMC formulations should prove economically viable and add value to the catalytic conversion recycling process. The balance of the liquid hydrocarbon conversion products can be separated if economically feasible or used as fuels.

The gaseous compounds from catalytic conversion of SMC (Table III) show that acetone is the major component. An optimized condensing system would reclaim the acetone and acetaldehyde as liquids. Otherwise, the gaseous conversion products should provide a good fuel for heating the recycling reactor.

The major liquid products from catalytic conversion of the mixed body panels were styrene, α-methyl styrene, and ethyl benzene, which would be expected from ABS. Various nitriles and aliphatic hydrocarbons were also seen. The RPP

TABLE II. HYDROCARBON CONVERSION PRODUCTS FROM SMC COMPOSITES

Compound	Retention Time (minutes)	Concentration (%)
Allyl Alcohol	4.18	2.1
Ethylene Glycol	8.79	15.0
Styrene	15.54	57.8
α-Methyl Styrene	18.66	6.0
Benzoic Acid	25.18	2.6
Other Hydrocarbons	various	16.5
	Total	**100.00**

TABLE III. GASEOUS CONVERSION PRODUCTS FROM SMC COMPOSITES

Compound	Retention Time (minutes)	Concentration (%)
1-Propene	1.91	6.6
Acetaldehyde	2.55	6.3
Acetone	4.32	81.6
Other Hydrocarbons	various	5.5
	Total	**100.00**

bumper material yielded a wide variety of branched aliphatic hydrocarbons that should find value as a high octane fuel. Cracking of the polypropylene also yields a high volume of gaseous conversion products (Table I), primarily methane, propene, and pentane. The high volume of gas is not unexpected from the instability of the polypropylene chain once a free radical is formed.

Liquid hydrocarbon conversion products from ASR with concentrations greater than 2% are given in Table IV.

Because of the wide variety of plastics in ASR, a large number of conversion products is expected. Results given in Table IV show the expected large number of compounds; however, the majority are common industrial chemicals that have value if separated or can be used as high BTU fuels.

Gaseous conversion products from ASR consisted primarily of propene (44%), acetaldehyde (17%), and acetone (36%). Again, an optimized condensing system would reclaim the acetaldehyde and acetone as liquids for reuse, leaving the propene and remaining hydrocarbons for use as fuel to heat the reactor.

The solid residue from catalytic conversion of ASR also contains numerous valuable components. Ferrous metals accounted for 70% of the solid residue, while copper, tin, and lead composed the majority of the remaining material along with smaller amounts of char, glass, and antimony. Separation of those metals for reuse with conventional extractive metallurgy techniques should also prove economical and provide further improvements in the overall economics of the catalytic conversion recycling process.

TABLE IV. LIQUID HYDROCARBON CONVERSION PRODUCTS FROM ASR

Compound	Retention Time (minutes)	Concentration (%)
Acetic Acid	4.93	3.5
Acetol	7.93	2.7
Ethylene Glycol	8.68	7.3
1-(1-Methylethoxy)-2-propanone	9.28	2.3
Phenol	18.70	14.1
Hexanoic Acid	20.75	18.3
Benzoic Acid	24.89	11.8
Tridecene	26.80	4.1
$C_{12}H_{24}$	26.96	12.6
Tetradecene	29.68	3.5
Phthalic Anhydride	31.59	7.2
Other Hydrocarbons	various	12.6
	Total	**100.00**

SUMMARY AND CONCLUSIONS

Initial studies of tertiary recycling of scrap automotive materials using a catalytic conversion process show that the process is effective for all types of polymer-based automotive scrap including SMC, ASR, RPP, and mixtures of body panels. Conversion products include valuable hydrocarbon chemicals and fuels, fibers and fillers, and metals. Once the polymer fraction is removed by the catalytic conversion process, separation of the remaining solid residue into its individual components for reuse is straightforward using technologies developed by the mining industry. By obtaining value from each component in the scrap, the economics of recycling automotive materials is maximized. Implementation of this process will save valuable petroleum and strategic metal resources that are currently being landfilled.

REFERENCES

1. D. L. Illman, "Automakers Move Toward New Generation of 'Greener' Vehicles," *Chemical & Engineering News*, August 1, 1994, pp. 8-16.
2. M. C. Gabriele and R. Monks, "New Automotive Plastics Materials and Technology," *Plastics Technology*, Apr 1993, pp. 67-80.
3. "SMC Use is Surging in Automotive Sector," *Modern Plastics*, Feb 1994, p. 15.
4. I. E. Poston, "Polymer Composites for Automotive Applications," *Proc. ANTEC 96*, Soc. Plastics Engr., May 1996.
5. J. A. Grande, "Plastics help define GM's electric car performance," *Modern Plastics*, Feb 1997, pp. 62-65.
6. H. Hock and M. A. Allen Jr., "Preliminary Study of the Recovery and Recycling of Automotive Plastics," in *Automobile Life Cycle Tools and Recycling Technologies*, SAE Special Publications, 1993, pp. 59-71.
7. P. Mapleston, "Auto Sector's Recycling Goals Keep Plastics on Hot Seat," *Modern Plastics*, May 1995, pp. 48-58.
8. A. Weaver, "Recycling in Action," *Reinforced Plastics*, Feb 1992, pp. 32-33.
9. "Recycled SMC Now in Appearance Parts," *Plastics Technology*, Oct 1993, p. 94.
10. J. Makower, "Recycled Parts Make a Heap of Difference to New Cars," United Press Syndicate, March 28, 1994.
11. T. Inoh, T. Yokoi, K. Sekiyama, N. Kawamura, and Y. Mishima, "SMC Recycling Technology," *J. Thermoplastic Composite Materials, 7*, 1, 1994, pp. 42-55.
12. "SMC Use Grows in Autos and Trucks," Modern Plastics, Apr 1997, p. 13.
13. R. D. Leaversuch, "Chemical Recycling Brings Real Versatility to Solid-waste Management," *Modern Plastics*, July 1991, pp. 40-43.
14. P. Mapleston, "Chemical Recycling May Be an Option to Meet Mandated Reclaim Levels," *Modern Plastics*, Nov 1993, pp. 58-61.
15. K. Fouhy, I. Kim, S. Moore, and E. Culp, "Plastics Recycling's Diminishing Returns," *Chemical Engineering*, Dec 1993, pp. 30-33.

16. R. Leaversuch, "Industry Backs Pyrolysis as a Recycling Option," *Modern Plastics*, Jan 1994, p. 93.
17. G. A. Mackey, "Thermolysis, Chemolysis, and Gasification as Advanced Recycling Technology for Waste Plastic," 208th American Chemical Society Meeting, Washington, DC, Aug 1994.
18. M. Day, Z. Shen, and J. D. Cooney, "Pyrolysis: An Acceptable Solution to the Auto Shredder Residue Problem?," Proc. Auto Recycle '94, Dearborn, MI, November 8-9, 1994.
19. R. E. Allred and R. M. Salas, "Chemical Recycling of Scrap Composites," Proc. AIA, SACMA, and NASA Joint Conference on Environmental, Safety, and Health Considerations--Composite Materials in the Aerospace Industry, Phoenix, AZ, October 20-21, 1994.
20. R.E. Allred, R. M. Salas, and B.W. Gordon, "Tertiary Recycling Process for Plastics, Composites, and Electronic Materials," *Proc. 40th Intl. SAMPE Symp. and Exhib.*, Anaheim, CA, May 8-11, 1995.
21. R. E. Allred, "Recycling Process for Scrap Composites and Prepregs," *Proc. 41st Intl. SAMPE Symp. and Exhib.*, Anaheim, CA, March 24-28, 1996, pp. 21-31.
22. R. E. Allred, A. B. Coons, and R. Simonson, "Properties of Carbon Fibers Reclaimed from Composite Manufacturing Scrap by Tertiary Recycling," *Proc. 28th Intl. SAMPE Tech. Conf.*, Seattle, WA, November 4-7, 1996, pp. 139-150.
23. R. E. Allred, T. J. Doak, L. D. Busselle, B. W. Gordon, L. A. Harrah, and A. E. Hoyt, "Catalytic Conversion Process for Recycling Navy Shipboard Plastic Wastes," *Polymeric Materials Science and Engineering, 76*, American Chemical Society, Washington, DC, pp. 578-579.
24. R. E. Allred, T. J. Doak, A. B. Coons, G. C. Newmeister, and R. C. Cochran, "Tertiary Recycling of Cured Composite Aircraft Parts," *Proc. Composites '97 Manufacturing and Tooling Conf.,* Society of Manufacturing Engineers, Dearborn, MI, 1997, pp. EM97-110 to EM97-110-17.
25. R. E. Allred and L. D. Busselle, "Economical Recycling Process for Mixtures of Electronic Scrap," *Proc. 1997 IEEE Intl. Symp. on Electronics and the Environment*, IEEE, Piscataway, NJ, 1997, pp. 115-120.

Manufacturing and Property Analysis of a Novel Class of 3-D Woven Composites

S. P. YUSHANOV, A. E. BOGDANOVICH AND M. H. MOHAMED

ABSTRACT

Structural and elastic properties of a novel class of 3-D woven orthogonally reinforced composites are studied. Comparative analysis of various theoretical approaches for elastic properties prediction of spatially reinforced composites is carried out. Theoretical predictions are compared with available experimental data. The effect of local yarn waviness on the elastic response of 3-D woven composites is studied using developed novel probabilistic theory.

INTRODUCTION

Mohamed and Zhang (1992) developed a new process for forming variable cross-sectional shaped three-dimensional fabrics. This method of weaving 3-D net shapes utilizes different weft yarn insertion from at least one side of the warp layers for selectively inserting weft yarns into different portions of the fabric cross-sectional profile defined by the warp yarn layers. If inserted from both sides of the warp yarn layers, the weft yarns may be inserted simultaneously or alternately from each side of the warp yarn layers. The vertical yarns, which are part of the warp, are then inserted into fabric by reciprocating a plurality of harnesses which separate the vertical yarn systems as required by the shape of the three-dimensional fabric being formed. The structure could be either 3-D orthogonal, angle interlock, warp interlock, or a combination of all three in the same piece. A detailed characterization of the structural parameters of 3-D orthogonal woven preforms is provided using approach developed by Gu and Mohamed (1993).

Various approaches have been developed for the elastic property predictions of textile composites. All of them are purely deterministic. The state-of-the-art and bibliography on elastic response of textile composites can be found in Bogdanovich and Pastore (1996). In the present study, the Modified Matrix Method (MMM) originated by Abolinsh (1966) is further generalized for accounting anisotropic properties of the reinforcement elements. Mathematical algorithm includes several successive steps corresponding to each of the spatial reinforcement directions. In each step, the composite is considered as continuous macro-homogeneous media in the sense of unidirectional macro-mechanics. As a result, the close-form relations are

---

S. P. Yushanov and A. E. Bogdanovich, AdTech Systems Research, Inc., Beavercreek, OH 45432-2968.
M. H. Mohamed, College of Textiles, North Carolina State University, Raleigh, NC 27695-8301.

obtained for all components of the compliance tensor for 3-D composite. Theoretical predictions obtained with this Generalized Modified Matrix Method (GMMM) are compared with the other existing theoretical approaches and available experimental data for the XYZ composite system. This comparison shows better predictive capabilities of the GMMM compared to the other analyzed theories.

A novel probabilistic theory of elastic response of composite materials with local waviness of a spatial reinforcement is outlined. The theory is based on the local spatial averaging of the reinforcement paths and allows one to obtain full set of the mean values of elastic characteristics. The effect of local yarn waviness on the elastic characteristics of 3-D woven composites is studied.

3-D WOVEN MATERIAL ARCHITECTURE

To describe the macroscopic properties of woven composite, the unit cell of the fabric is introduced, as shown in Figure 1. The repeat of this unit cell will generate the whole woven structure. To calculate fiber volume fraction of woven preform, it is necessary to take into account that the number of the weft layers is one more than that of warp layers: if there are n layers of warps, then the number of weft layers will be $n+1$. Therefore, the thickness , h, of the fabric is equal to n thicknesses of the unit cell plus thickness of the weft yarn:

$$h = nZ_z + F_z \tag{1}$$

Since this volume contains n layers of warp yarns, $n+1$ layers of weft yarns, and Z yarns having volume hS_Z, fiber volume fraction of the fabric, V_f, is calculated as

$$V_f = \frac{nS_W}{hF_y} + \frac{(n+1)S_F}{hW_x} + \frac{S_Z}{W_x F_y} \tag{2}$$

In the right-hand side of equation (2), the first, second, and third terms represent fiber volume fractions of fabric along *x*, *y*, and *z* directions, respectively.

The following parameters of a fabric are usually specified by direct measurements: n - number of warp layers; S_W, S_F, and S_Z - cross-sectional areas of warp, weft and Z yarns; W_y and $\rho_W = 1/F_y$ - warp width and reed density; $\rho_F = 1/W_x$ - pick density. These parameters

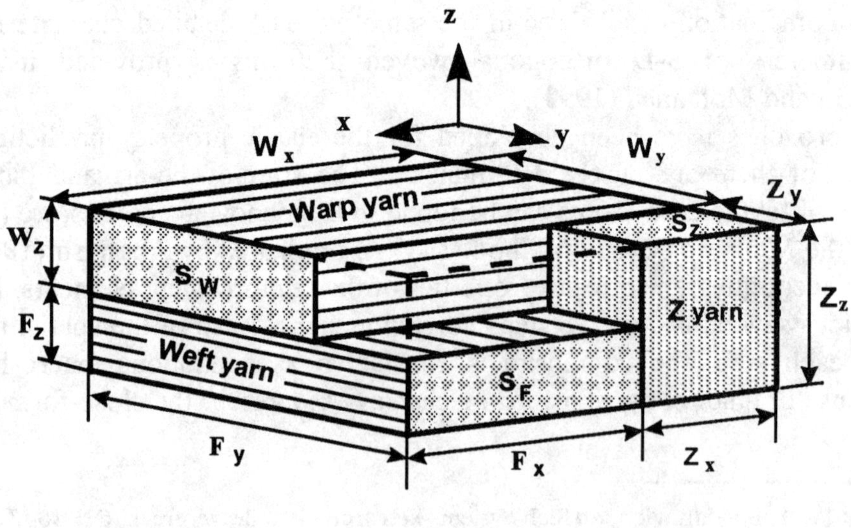

Figure 1. Unit cell of an orthogonal woven fabric.

supplemented by some obvious relations following from the unit cell geometry define uniquely parameters F_y and W_x entering into (2). In such a way, if parameters n, S_W, S_F, S_Z, W_y, ρ_W and ρ_F are prescribed, then micro-structure of the fabric is completely defined and, therefore, fiber volume fraction in each reinforcement direction can be calculated.

ELASTIC PROPERTY PREDICTION

The GMMM is adopted to analyze the elastic properties of orthogonally reinforced composites. Clear physical assumptions, simplicity of realization, and good predicting capability are the main advantages of this approach. Originally, the MMM was proposed to predict the elastic constants of the material reinforced in two orthogonal directions (Abolinsh 1966). Later on, Tarnopolskii *et al.* (1973) used this approach to analyze materials reinforced with a system of three straight, mutually orthogonal fibers. In their study, fibers were assumed isotropic. Assessment of the effect of fiber anisotropy has become of considerable importance due to the widespread use of carbon and graphite fibers which are highly anisotropic. The effect of the fiber anisotropy on the elastic response of the uniaxially reinforced composites was studied by Dean and Turner (1973), Kriz and Stinhcomb (1979), Kawabata (1990). However, there is a lack of study related to the effect of the fiber anisotropy on the elastic response of 3-D woven composites. In the present study, the MMM is generalized to account for fiber anisotropy when predicting elastic properties of 3-D reinforcement composites.

Elastic characteristics of 3-D orthogonally reinforced composites are evaluated in several successive steps corresponding to each of the reinforcement directions. In each step, the composite is considered as continuous macro-homogeneous media in the sense of unidirectional macro-mechanics. First step represents generalization of the micromechanics developed by Abolinsh (19965). Fibers along i direction are embedded in an isotropic matrix. In the second step, fibers along j direction are embedded in transversely isotropic homogeneous matrix with the plane of isotropy j-k. Finally, in the third step, fibers along k direction are embedded into homogeneous orthotropic matrix which has two mutually orthogonal planes of isotropy i-k and j-k. Performing derivations according these successive steps, one obtains the recursive equations for the evaluation of the elements of the composite compliance matrix of 3-D reinforced composite as following

$$S_{kk} = \frac{S_{kk}^{(j,i)}}{1+\left(E_L^{f(k)} S_{kk}^{(j,i)} - 1\right)V_f^{(k)}};\; S_{jk} = \frac{\left(1-V^{(k)}\right)S_{jk}^{(j,i)} - V_f^{(k)} \nu_{LT}^{f(k)} S_{kk}^{(j,i)}}{1+\left(E_L^{f(k)} S_{kk}^{(j,i)} - 1\right)V_f^{(k)}} \tag{3}$$

The compliances of the 2-D reinforced composite entering in equations (3) are evaluated as

$$S_{kk}^{(j,i)} = \frac{\left[b^{(i)}V_f^{(j)} + E_L^{f(j)} S_{jj}^{(i)}\left(1-V_f^{(j)}\right)\right]\left[1+\left(E_L^{f(j)} S_{jj}^{(i)} - 1\right)V_f^{(j)}\right] - \left(S_{jk}^{(i)} E_L^{f(j)} + \nu_{LT}^{f(j)}\right)^2 \left(1-V_f^{(j)}\right)V_f^{(j)}}{\left[1+\left(E_L^{f(j)} S_{jj}^{(i)} - 1\right)V_f^{(j)}\right]E_L^{f(j)}}$$

$$S_{ji}^{(j,i)} = \frac{\left(1-V_f^{(j)}\right)S_{jk}^{(i)} - V_f^{(j)} \nu_{LT}^{f(j)} S_{kk}^{(i)}}{1+\left(E_L^{f(j)} S_{jj}^{(i)} - 1\right)V_f^{(j)}} \tag{4}$$

Equations (4) contain compliances of unidirectional fiber reinforced composite:

$$S_{jj}^{(i)} = \frac{\left[b^{(i)}V_f^{(i)} + n^{(i)}\left(1-V_f^{(i)}\right)\right]\left[1+\left(n^{(i)}-1\right)V_f^{(i)}\right] - \left(n^{(i)}\nu^m - \nu_{LT}^{f(i)}\right)^2\left(1-V_f^{(i)}\right)V_f^{(i)}}{\left[1+\left(n^{(i)}-1\right)V_f^{(i)}\right]E_L^{f(i)}} \tag{5}$$

$$S_{jk}^{(i)} = -\frac{\left[\nu_{TT}^{f(i)}b^{(i)}V_f^{(i)} + n^{(i)}\nu^m\left(1-V_f^{(i)}\right)\right]\left[1+\left(n^{(i)}-1\right)V_f^{(i)}\right] + \left(n^{(i)}\nu^m - \nu_{LT}^{f(i)}\right)^2\left(1-V_f^{(i)}\right)V_f^{(i)}}{\left[1+\left(n^{(i)}-1\right)V_f^{(i)}\right]E_L^{f(i)}}$$

where $n^{(i)} = E_L^{f(i)}/E^m$ and $b^{(i)} = E_L^{f(i)}/E_T^{f(i)}$. The shear moduli are calculated as

$$G_{kj} = G_{kj}^{(j,i)}\frac{G_L^{f(k)}\left(1+V_f^{(k)}\right)+G_{kj}^{(j,i)}\left(1-V_f^{(k)}\right)}{G_L^{f(k)}\left(1-V_f^{(k)}\right)+G_{kj}^{(j,i)}\left(1+V_f^{(k)}\right)}, \quad G_{kj}^{(j,i)} = G_L^{(i)}\frac{G_L^{f(j)}\left(1+V_f^{(j)}\right)+G_L^{(i)}\left(1-V_f^{(j)}\right)}{G_L^{f(j)}\left(1-V_f^{(j)}\right)+G_L^{(i)}\left(1+V_f^{(j)}\right)} \tag{6}$$

$$G_L^{(i)} = G^m\frac{G_L^{f(i)}\left(1+V_f^{(i)}\right)+G^m\left(1-V_f^{(i)}\right)}{G_L^{f(i)}\left(1-V_f^{(i)}\right)+G^m\left(1+V_f^{(i)}\right)} \tag{7}$$

In equations (3)-(7), $i,\ j,\ k \in \{1,\ 2,\ 3\}$, and $i \neq j \neq k \neq i$. Further, E^m, G^m and ν^m are elastic modulus, shear modulus and Poisson's ratio of matrix material. Notations $E_L^{f(i)}$ $E_T^{f(i)}$, $G_L^{f(i)}$, and $\nu_{LT}^{f(i)}$, $\nu_{TT}^{f(i)}$ denote elastic modulus, shear modulus, and Poisson's ratios of fibers in the i-direction. Subscripts "L" and "T" denote the longitudinal and transverse directions related to the fiber, respectively. Finally, $V_f^{(i)}$ is fiber volume fraction in the i th - direction.

In the paper by Tarnopolskii *et al.* (1973), transverse shear modulus of unidirectional fiber reinforced composite is used instead of the longitudinal shear modulus. We suggest that using longitudinal shear modulus given by equation (7) is more consistent with the adopted methodology. Also, a mistake has been found in the formula for the compliance component $S_{kk}^{(j,i)}$ in the paper by Tarnopolskii *et al.* (1973): for the case of isotropic fiber, the term $\left(S_{jk}^{(i)}E_f^{(i)} + \nu_f^{(j)}\right)$ entering in the expression for $S_{kk}^{(j,i)}$ should read $\left(S_{jk}^{(i)}E_f^{(i)} + \nu_f^{(j)}\right)^2$.

To validate the developed GMMM analysis, a comparison of the experimental data by Zhigun *et al.* (1973) for 3D orthogonal Glass Reinforced Plastic (GRP) with various theoretical predictions has been carried out. Fiber volume fractions of GRP are: $V_f^{(1)}$=23.48%, $V_f^{(2)}$=32.39%, $V_f^{(3)}$=3.10%, and the total fiber volume fraction $\sum_i V_f^{(i)}$ =59.0%. Elastic characteristics of the constituents are: $E_f = 73.1\ GPa$, $E_m = 3.3\ GPa$, $\nu_f = 0.25$, $\nu_m = 0.35$. The results of the comparison are presented in Table 1. Theoretical predictions based on the other theoretical models, namely, results obtained by Lagzdin' *et al.* (1989) using Stiffness Averaging Method (SAM) and Compliance Averaging Method (CAM), results obtained by using MMM in the form of equations (1)-(7) by Tarnopolskii *et al.* (1973), and results obtained using Pagano's approach (1988) are also presented. The results presented in Table I show that compliance and stiffness averaging methods provide lower and upper bounds for the elastic moduli. Pagano's model gives best prediction for E_3 but too high estimation for E_1 and E_2. The present method gives the best predictions for E_1 and E_2 but somewhat lower value for E_3. Although the present analysis, as well as stiffness and compliance averaging methods give good predictions of the shear moduli, the results from present analysis appear closer to the experimental data. Pagano's model gives too high values of the shear moduli. All

TABLE I - ELASTIC PROPERTIES OF THE XYZ GRAPHITE COMPOSITE

Property	Experimental		Theoretical				
	Range	Average	GMMM	MMM	CAM	SAM	Pagano's approach
E_1, GPa	21.0-24.0	22.0	21.7	21.3	15.2	24.5	25.32
E_2, GPa	28.0-30.0	29.0	27.3	26.7	18.2	29.7	30.76
E_3, GPa	13.5-15.5	14.0	10.4	8.66	11.0	13.5	14.75
ν_{12}	0.128-0.136	0.13	0.122	0.114	0.13	0.12	-
ν_{21}	0.165-0.174	0.17	0.15	0.139	-	-	0.13
ν_{31}	0.130-0.150	0.14	0.178	0.141	0.29	0.20	0.29
ν_{32}	0.122-0.140	0.128	0.134	0.107	0.23	0.16	0.30
G_{12}, GPa	3.4-4.3	3.86	3.572	3.49	4.1	4.1	4.15
G_{13}, GPa	3.3-4.1	3.60	3.570	2.85	3.80	3.90	4.33
G_{23}, GPa	3.1-4.0	3.50	3.577	3.02	3.90	3.90	4.43

TABLE II - THE SPECIFICATION OF THE FABRIC MATERIAL

Description	Characteristic, Unit	Value
Number of warp layers	n	5
Warp yarn 12K	S_W, mm^2	0.449
Warp width	W_y, mm	1.397
Weft yarn 2 × 6K	S_F, mm^2	0.449
Z yarn 3K	S_Z, mm^2	0.11225
Reed density	ρ_W, $dents/cm$	5.51
Pick density	ρ_F, $picks/cm$	5.51

theories predict quite well Poisson's ratios in the plane normal to the thickness direction, but only present method provides results which are in a good agreement with the experimental data for the other two Poisson's ratios, ν_{13} and ν_{23}. The CAM and SAM approaches as well as Pagano's model significantly overestimate (by factor of 2) these elastic characteristics.

COMPARISON OF THE THEORETICAL PREDICTIONS AND EXPERIMENTAL DATA FOR 3-D WOVEN COMPOSITES

The characteristics of 3-D woven preform (Celion® G30-500 carbon fibers) are given in Table II. Those are adopted from Gu and Mohamed (1993). Using these data, fiber volume fractions for each reinforcement direction are calculated by equation (2): $V_f^{(x)} = 0.223$,

$V_f^{(y)} = 0.273$, $V_f^{(z)} = 0.034$ for the number of warp layers $n = 5$. Elastic characteristics of the fibers are (Fiberite®, 1995.): $E_f = 220\ GPa$, $\nu_f = 0.3$. Assumed packing density of fiber bundles =0.65. Elastic properties of the matrix are: $E_m = 3.45\ GPa$, $\nu_m = 0.35$. Experimental specimens have been manufactured and tested at NCSU. Measured tensile elastic moduli in the warp and weft directions along with the theoretical predictions based on GMMM are presented in Table III. Reasonable agreement is achieved between the measured and predicted moduli. For comparison, numerical predictions based on the finite element analysis (FEA) due to Gu and Mohamed (1993) are also presented in Table III. It is seen that closed-form solution (3) - (8) fits experimental data result than the finite element analysis.

TABLE III. ELASTIC CONSTANTS OF 3-D WOVEN COMPOSITE

Property	Experiment	Theory (GMMM)	FEA
E_{11}, GPa	55	54.1	53.1
E_{22}, GPa	72	64.4	62.5

STOCHASTIC THEORY AND EFFECT OF YARN LOCAL WAVINESS

Yarn local waviness is almost inevitable in the manufacturing processes of textile composites, see Brandt *et al.*, 1996 for example. It has been known for a long time that even small reinforcement waviness unidirectional composites may result in a significant change of elastic characteristics (Bolotin, 1966). To account for the yarn waviness, a novel probabilistic theory of the elastic response of composites with local reinforcement waviness has been developed. The theory considers composite materials with continuous-fiber reinforcements, where the spatial reinforcement path is treated as a random vector function characterized in 3-D space by its mean value and covariance matrix. By introducing local stochastic basis at any point along the reinforcement path and further applying the series expansion technique, the mean value and covariance of elastic characteristics of the composite material are expressed in terms of the mean value and covariance of the reinforcement path. The theory assumes that the deflections of the reinforcement path from the "perfect" one are rather small compared to the characteristic scale of its mean value. The full description of the developed theory and its various applications will be published elsewhere. Here, we recapitulate one particular application of the theory.

When applied to the unidirectional composite with reinforcement along the axis *1* and random deflection of the yarn in the planes *1-2* and *1-3*, the theory gives the following final result for the mean stiffnesses:

$$\left\langle C_{ij}^{(1)} \right\rangle \approx C_{ij}^{0} + \sigma_{\theta_{12}^{(1)}}^{2} C_{ij}' + \sigma_{\theta_{13}^{(1)}}^{2} C_{ij}'' \tag{8}$$

where $\sigma_{\theta_{ij}^{(k)}}$ are the standard deviations of the deflection angle in the plane *i-j* for the case of reinforcement in the *k*-direction. The terms of order $O(\sigma_\theta^4)$ and higher are neglected in equation (8). The non-zero coefficients C_{ij}' and C_{ij}'' are derived as

$$\begin{aligned} &C'_{11}=C''_{11}=2\left(-C^0_{11}+C^0_{12}+2C^0_{66}\right); && C'_{22}=C''_{33}=2\left(-C^0_{22}+C^0_{12}+2C^0_{66}\right)\\ &C'_{44}=C''_{44}=-\left(C^0_{22}-C^0_{23}-2S^0_{66}\right)/2; && C'_{55}=C''_{66}=\left(C^0_{22}-C^0_{23}-2C^0_{66}\right)/2\\ &C'_{66}=C''_{55}=C^0_{11}+C^0_{22}-2C^0_{12}-4C^0_{66}; && C'_{12}=C''_{13}=C^0_{11}+C^0_{22}-2C^0_{12}-4C^0_{66}\\ &C'_{13}=C''_{12}=C^0_{23}-C^0_{12}; && C'_{23}=C''_{23}=C^0_{12}-C^0_{23} \end{aligned} \tag{9}$$

where S^0_{ij} are stiffnesses of a unidirectional composite with straight yarns which are calculated by conventional relations. Application of stiffness averaging procedure for 3-D woven orthogonally reinforced composite with account for yarn waviness results in the following expression for total mean stiffness

$$\langle C_{ij}\rangle=\sum_{k=1}^{3}\mu^{(k)}\langle C^{(k)}_{ij}\rangle;\quad \mu^{(k)}=V^{(k)}_f\Big/\sum_{i=1}^{3}V^{(i)}_f \tag{10}$$

where $V^{(k)}_f$ is the volume fraction of the reinforcement yarns in the k-direction; $\langle C^{(k)}_{ij}\rangle$ is the mean stiffness matrix of unidirectional composite reinforced in the k-direction, which is evaluated with account for the local yarn waviness. For $k=1$, the mean stiffnesses are evaluated according to relations (8) and (9). The values of $\langle C^{(2)}_{ij}\rangle$ are obtained from (8), (9) by the cyclic permutation of the indices: 2 for 1, 3 for 2, 1 for 3, 5 for 4, 6 for 5, and 4 for 6. Finally, stiffnesses $\langle C^{(3)}_{ij}\rangle$ are obtained from (8), (9) by the following cyclic permutation of the indices: 3 for 1, 1 for 2, 2 for 3, 6 for 4, 4 for 5, and 5 for 6.

Simulations were carried out to study the effect of yarn waviness on the elastic response of 3-D woven composite. Since experimental data concerning quantification of yarn waviness are not available, only parametric study has been performed. Illustrative example correspond to 3-D woven composite with number of weft layers $n=10$. Fiber volume fractions for each reinforcement direction calculated by equation (2) are: $V^{(1)}_f=0.250$, $V^{(2)}_f=0.256$, $V^{(3)}_f=0.0317$. Figure 2 illustrates the case where only the warp direction yarns have waviness; the weft and Z-direction yarn are straight. In this case the non-zero standard deviations of the deflection angle are: $\sigma_{\theta^{(1)}_{12}}=\sigma_{\theta^{(1)}_{13}}=\sigma_\theta$. The results show considerable drop (about 40% at $\sigma_\theta=20^\circ$) of E_{11} and less pronounced drop of E_{22} and E_{33}. When considering shear moduli, Figure 2 (b), G_{12} increases 2.2 times at $\sigma_\theta=20^\circ$; the values of G_{13} are slightly growing with increasing waviness, and G_{23} practically does not change. As seen in Figure 2 (c), Poisson's ratio ν_{12} increases by factor of 1.8 at $\sigma_\theta=20^\circ$; the values of ν_{13} are slightly higher with the waviness, while drop of the values of ν_{23} is about 30%.

Since the reinforcement fractions of the warp and weft yarns are practically the same, the effect of the weft yarn waviness on the elastic characteristics of 3-D woven composite is similar to those shown in Figure 2, only the indices 1 and 2 interchange.

Figure 3 shows the effect of the Z-yarn waviness on the elastic characteristics of 3-D woven composite. The non-zero standard deviations of the deflection angle are: $\sigma_{\theta^{(3)}_{31}}=\sigma_{\theta^{(3)}_{32}}=\sigma_\theta$. It is seen that the mean values of the elastic characteristics related to the Z-direction change significantly with the increase of the standard deviation of the deflection angle of the Z-yarn. Namely, mean elastic modulus $<E_{33}>$ decreases, while mean shear moduli, $\langle G_{12}\rangle$, $\langle G_{23}\rangle$ and mean Poisson's ratios, $\langle \nu_{13}\rangle$, $\langle \nu_{23}\rangle$ increase. Since volume fraction of the Z-yarns is low, the elastic characteristics related to the *1*- and *2*- directions change less significantly.

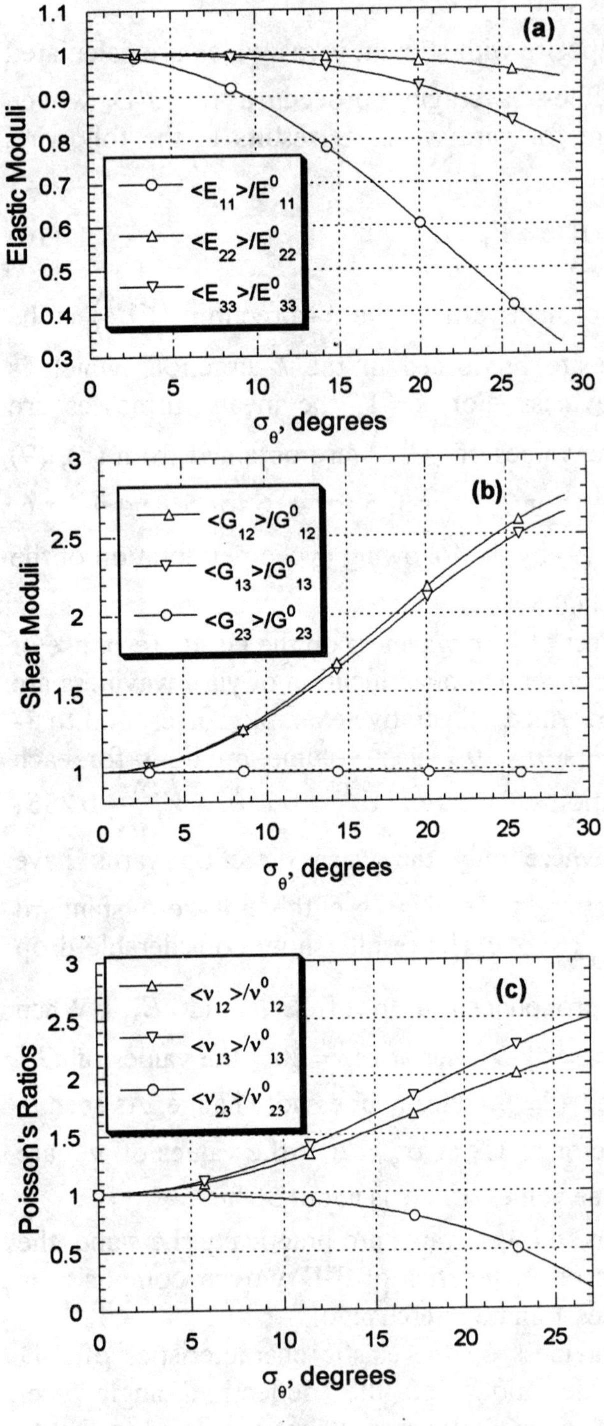

Figure 2 - Dependencies of the elastic constants of 3-D woven composite on the standard deviations of the deflection angles of the warp yarn.

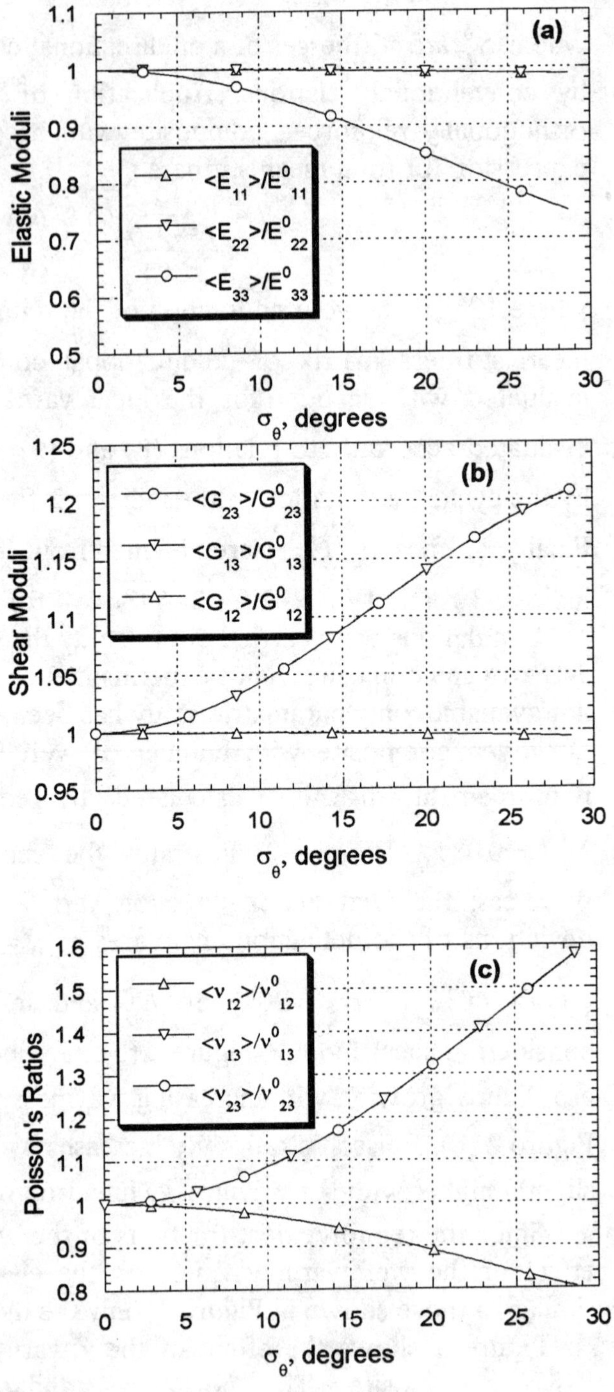

Figure 3 - Dependencies of the elastic constants of 3-D woven composite on the standard deviations of the deflection angles of the Z-yarn.

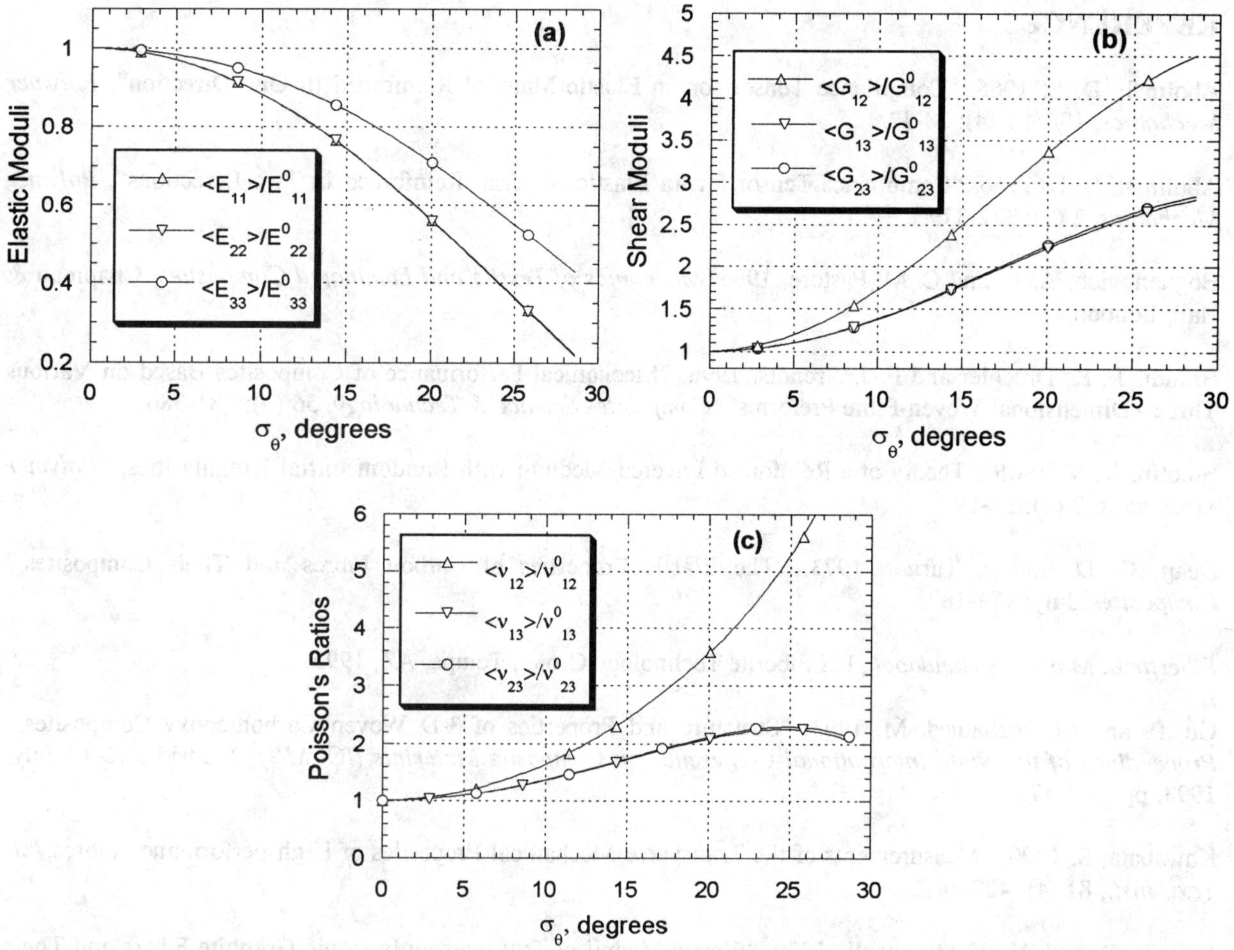

Figure 4 - Dependencies of the elastic constants of 3-D woven composite on the standard deviations, σ_θ , of the deflection angles of the warp, weft and Z-yarn.

Figure 4 illustrates the case when waviness of all three sets of yarns is taken into account: $\sigma_{\theta^{(1)}_{12}} = \sigma_{\theta^{(1)}_{13}} = \sigma_{\theta^{(2)}_{21}} = \sigma_{\theta^{(2)}_{23}} = \sigma_{\theta^{(3)}_{31}} = \sigma_{\theta^{(3)}_{32}} = \sigma_\theta$ and the rest of standard deviations are zero. A kind of combined effect is observed. Most distinctive feature, if comparing Figure 4 with Figures 2 and 3, is that all elastic moduli decrease while all shear moduli and Poisson's ratios increase. Naturally, changes in the elastic characteristics are more significant when accounting for the waviness of yarns in all three direction.

CONCLUSION

Structural and elastic characteristics of a novel class of 3-D woven composites were studied. A good agreement between theoretical and experimental predictions was achieved. The effect of local yarn waviness on the full set of elastic characteristics of 3-D woven composite was analyzed using a novel developed stochastic theory.

ACKNOWLEDGMENT - This work was financially supported by Office of Naval Research through Grant No. N00014-95-1-0642, Dr. Yapa D. S. Rajapakse is the grant monitor.

REFERENCES

Abolinsh, D. S. 1965. "Compliance Tensor for an Elastic Material Reinforced in One Direction", *Polymer Mechanics*, 1965, 1 (4): 28-32.

Abolinsh, D. S. 1966. "Compliance Tensor for an Elastic Material Reinforced in Two Directions", *Polymer Mechanics*, 2 (3): 372-379.

Bogdanovich, A. E. and C. M. Pastore. 1996. *Mechanics of Textile and Laminated Composites*, Chapman & Hall, London.

Brandt, J., K. Drechler and F.-J. Arendts. 1996. "Mechanical Performance of Composites Based on Various Three - Dimensional Woven-Fibre Preforms". *Composites Science & Technology*, 56 (3): 381-386.

Bolotin, V. V. 1966. "Theory of a Reinforced Layered Medium with Random Initial Irregularities," *Polymer Mechanics*, 2 (1): 11-19.

Dean, G. D. and P. Turner. 1973. "The Elastic Properties of Carbon Fibres and Their Composites," *Composites*, July: 174-180.

Fiberite®, Materials Handbook, ICI Fiberite Technology Group, Tempe, AZ, 1995.

Gu, P. and M. Mohamed, M. 1993. "Structure and Properties of 3-D Woven Carbon/epoxy Composites," *Proceedings of the Ninth International Conference on Composite Materials (ICCM/9)*, Madrid , 12-16 July, 1993, pp. 568-575.

Kawabata, S. 1990. "Measurements of the Transverse Mechanical Properties of High-performance Fibres," *J. Text. Inst.*, 81 (4): 432-447.

Kriz, R. D. and W. W. Stinhcomb. 1979. "Elastic Moduli of Transversely Isotropic Graphite Fibers and Their Composites," *Experimental Mechanics*, February: 41-49.

Lagzdin', A. Z., V. P. Tamuzh, G. A. Teters, and A. F. Kregers. 1989. *Orientation Averaging Method in Mechanics* of *Materials* (in Russian), Riga, Zinatne, 1989.

Mohamed, M. H. and Z.-H. Zhang. 1992. "Method of Forming Variable Cross-Sectional Shaped Three-Dimensional Fabrics," *U.S. Patent No. 5085252.*

Pagano, N. 1988. "Theory of an Anisotropic Elastic Continuum," *Journal of Composite Materials*, 22 (1): 1-15.

Tarnopol'skii, Y. M., V. A. Polyakov, and I. G. Zhigun. 1973. "Composite Materials Reinforced With a System of Three Straight, Mutually Orthogonal Fibers, I: Calculation of Elastic Characteristics," *Polymer Mechanics* No. 5, 853-850.

Zhigun, I. G., I. G, Dushin, V. A. Polyakov, and V. A. Yakushin. 1973. "Composite materials reinforced with a system of three straight, mutually orthogonal fibers, 2: Experimental study," *Polymer Mechanics*, 6 (December): 895-900.

Manufacturing and Testing of Thermoplastic Composite High Load Hinge

M. HOU,[1] L. YE[2] AND Y.-W. MAI[2]

ABSTRACT

A composite hinge has been designed to replace an existing metallic hinge on an aircraft aileron. The new design aims to reduce the manufacturing cost while maintaining the operating loads and conditions. Thermoplastic composite hinges were produced from solvent impregnated fiber fabric reinforced polyetherimide prepreg. To get fully consolidated thermoplastic composite parts with high quality, optimum forming conditions such as forming temperature, holding time and forming pressure were determined by an impregnation model. The final void content in the thermoplastic composite hinge was less than 0.2%. A special test rig was designed to test the hinges under the worst loading condition expected during operations. Composite hinges with different fiber orientation and flange/web thickness ratio were tested up to failure. All hinges behaved linearly up to the onset of the first crack growth in the tensile section. After this event, non-linear behavior was observed. The main failure mode observed in all hinges appeared to be delamination in the tensile section. The load at which first delamination started was found to be in the range of 15-17 kN, depending on the fiber orientation and flange/web thickness ratio in the hinges, and catastrophic failure was not observed up to the specified maximum load of 25 kN. The ultimate load expected during the operation of the metallic hinge was met by the thermoplastic hinges.

INTRODUCTION

Continuous fiber reinforced thermoplastic composites have recently been introduced as structural materials for high performance aerospace and industrial applications. High strain to failure, increased fracture toughness, better impact tolerance, short processing cycle time, infinite shelf life of prepreg, recyclability and repairability are some of the reasons cited for their growing popularity (Offringa, 1996; Harper, 1992). The most important advantage of thermoplastics may lie in their potential for rapid, low-cost, mass production of reinforced composites. To assess the potential for the use of advanced thermoplastic composites in

[1] Cooperative Research Centre for Advanced Composite Structures Ltd. 361 Milperra Road, Bankstown, NSW 2200, AUSTRALIA
[2] Centre for Advanced Materials Technology, Department of Mechanical and Mechatronic Engineering, The University of Sydney, NSW 2006, AUSTRALIA

spacecraft, the Cooperative Research Center for Advanced Composite Structures (CRC-ACS), Australia, conducted a project on manufacturing technology of thermoplastic composites. This paper outlines the current status of manufacturing and testing of thermoplastic composite high load hinges for a civil aircraft.

The metallic hinge of an aircraft aileron was used as the baseline design for a composite equivalent part. The principal objective was to reduce the manufacturing cost of the hinge while maintain the operational loads and conditions. The overall geometry of the composite hinge was dictated by the aileron's front spar and also the position of the fork fitting which connects the hinge to the wing. To aid in defining the optimised geometrise, the actual metallic hinge was analysed with finite element analysis so as to acquire an understanding on load paths within the hinge (Rispler et al., 1997). The load cases applied to the aileron in the analysis were obtained from the aircraft manufacturer and were combined to give worst loading conditions on the aileron, as shown in Figure 1a. The values of these critical loadings includes a safety factor of 1.5. After comparing a number of aspects in terms of manufacturing cost it was decided to further develop the single web hinge design (Figure 1b), which can reduce the number of parts in the hinge assembly. This leads to reduced manufacturing cost and weight.

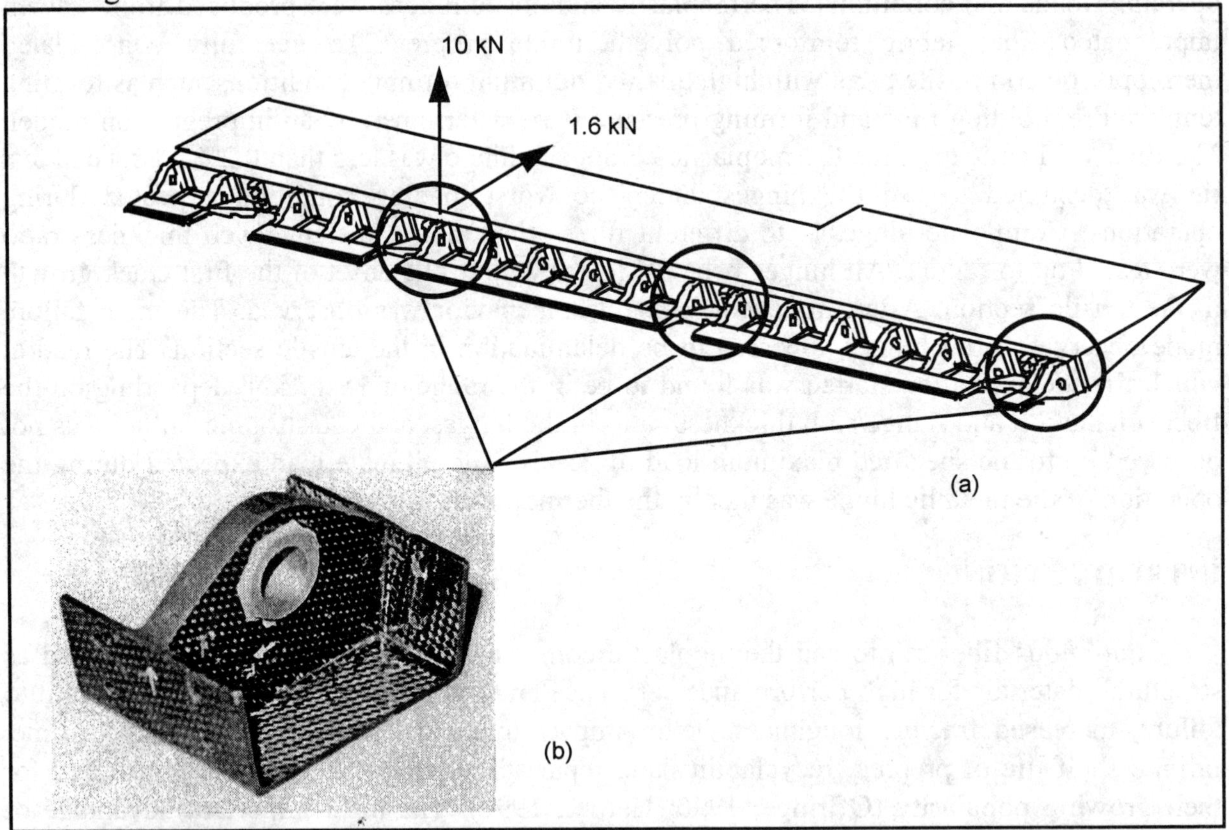

FIGURE 1: LOCATION AND LOADING CONDITIONS OF HINGE

MANUFACTURING PROCEDURE OF COMPOSITE HINGE

Currently, there are several methods being used in the manufacturing of thermoplastic composites (Hou et al., 1995). Each method has some specific limitations on the part shape that can be produced. Based on the geometry of the hinge and the availability of manufacturing facilities, compression molding process was selected as the first step to manufacture the high load hinge. Because of the complex geometry of the hinge, the co-consolidation technique was applied to manufacture the hinge, in which several components with simple geometry were first compression molded individually, and then these components were co-consolidated together. The manufacturing process is schematically shown in Figure 2. In this approach, the composite preform (prepreg) was preheated first in a heating oven, when the desired temperature was reached, the material was transferred into a hot forming mold and a compression pressure was applied to form and consolidate the composite material. Then the composite part was cooled down under pressure. It should be noted that to match the consolidation requirement, ie. to provide a sufficient consolidation pressure on the web of the hinge in the co-consolidation step, a special mold system was designed.

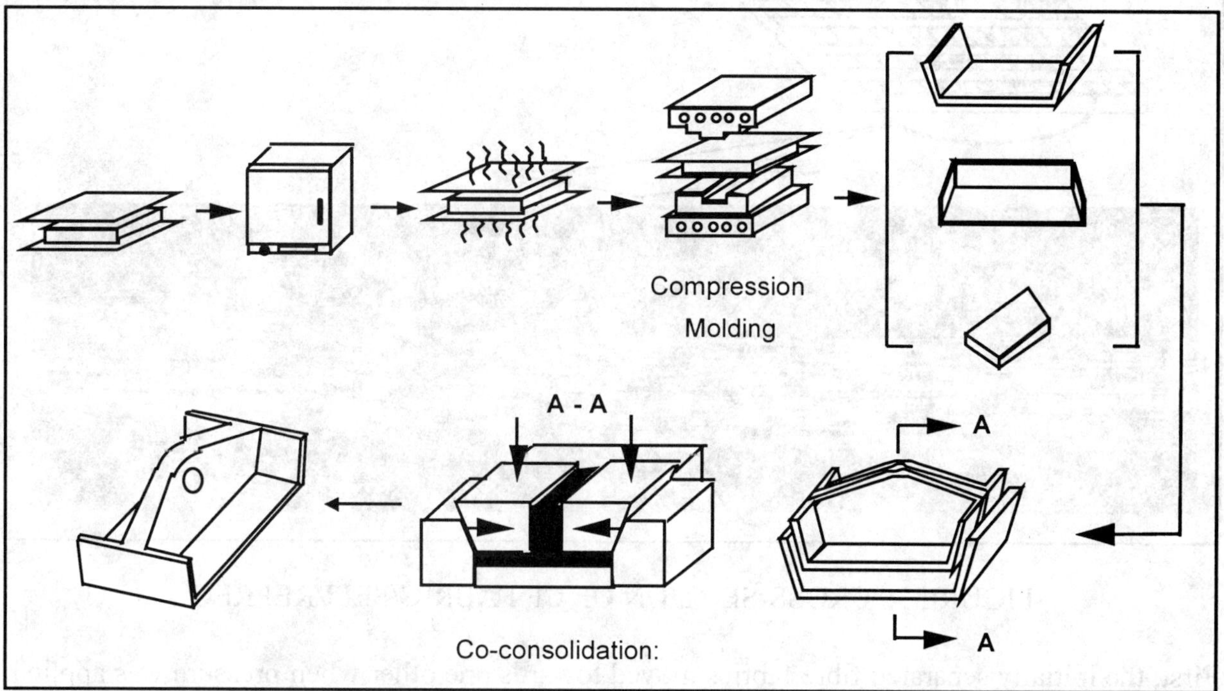

FIGURE 2: MANUFACTURING PROCEDURE OF COMPOSITE HINGE

The material used in this study was the solvent impregnated carbon fibre fabric reinforced polyetherimide (Ultem 1000) prepreg supplied by Ten Cate Advanced Composites, the Netherlands. The carbon fabric has a 5H satin woven construction. The prepreg has a resin content of 44.1 wt %. PEI is an amorphous polymer and has a glass transition temperature of about 217°C (Fines, 1988). During the solvent impregnation process some air was entrapped in the PEI polymer, which formed many tiny air bubbles on the surface of the prepreg. The cross-section micrograph of the CF/PEI prepreg shows that most of the PEI matrix (about 60%) remains outside the warp/weft fibre bundles (Figure 3).

To get the optimum processing conditions, e.g. forming temperature, holding time and forming pressure, the impregnation/consolidation mechanism was first studied using small scale compression molding experiments, in which the consolidation of CF fabric/PEI prepreg was performed using a small steel mold with a rectangular cavity (100 mm x 80 mm) and a laboratory hot press. Various processing conditions, e.g. two temperature levels of 280°C and 310°C, four holding times of 5, 10, 20 and 30 minutes, and three compression pressures of 0.5, 2 and 4 MPa, were applied in the consolidation of the laminates. The finished composite panels were cross-sectioned perpendicular to one fiber direction, and then polished. The impregnation status was examined by reflection light microscopy. Micrographs in Figure 4 illustrate some typical cross-sections of composite laminates at different stages of consolidation.

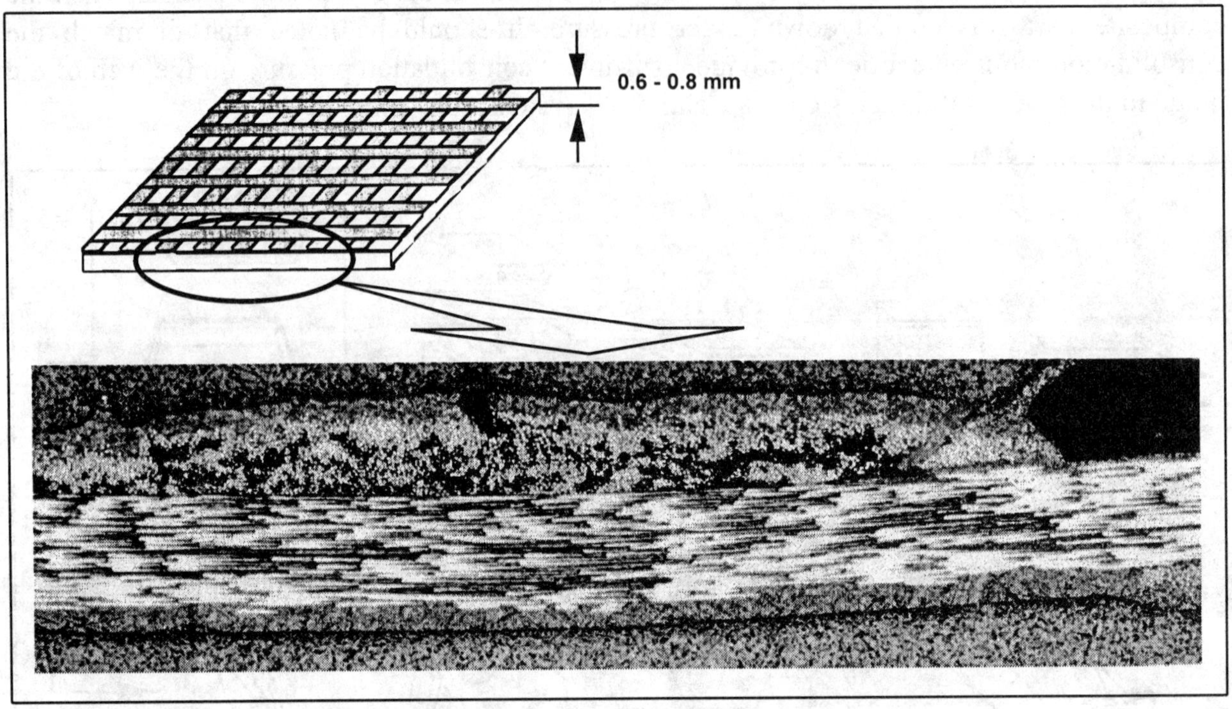

FIGURE 3: CROSS-SECTION OF CF FABRIC/PEI PREPREG

First, the initially separated fibre fabrics moved towards one other when pressure was applied. When the temperature was increased above T_g, the matrix of the fabric melted first. Then intimate contact occurred and the flow of the matrix removed the entrapped air at the prepreg surface and allowed local redistribution of the matrix and formation of a resin rich interlayer which permitted bonding of different layers, as shown in Figure 4a. Coalescence between the melting matrix layers then occurred and the space between adjacent fabrics was filled by the matrix. Meanwhile, significant compaction of fibre tows (warp/weft) took place, and fibres moved towards each other under the applied pressure; the cross-section of PEI impregnated fibre bundles was further compacted. By the end of this period, the boundary of each of the fibre tows could be clearly identified (Figure 4b). In the second step, with elapse of holding time, the applied pressure drove the matrix layer from both sides of the prepreg further into the warp and weft fibre bundles thus pushing the air between them away and reducing the voids (Figure 4c). This process continued until the weft/warp fibre bundle was completely impregnated and full consolidation was achieved. In this final condition, the distance between fibres within each bundle became a minimum and the fibre distribution in bundles was uniform (Figure 4d). These results show that the inward impregnation of molten matrix from polymer rich layers to the fibre bundles dominates the consolidation procedure of the CF/PEI composite. As an approximate approach, it is assumed that the impregnation of each individual fibre bundle is dominated by the flow of the molten polymer matrix layer belonging to this bundle. Based on this main mechanism and several basic assumptions, a qualitative model has been developed to describe the consolidation quality of this material (Hou et al., 1997). According to this model and a predetermined critical void content (X_{VC}), for instance, X_{VC}=0.5% for the consolidation quality, the relationship between the processing temperature, applied pressure and holding time can be illustrated as a processing window (Figure 5) for the compression molding of CF fabric/PEI composite. In Figure 5, two processing temperatures (280°C and 310°C) were selected, and the shadowed area represents the processing window. The possible combinations of appropriate processing variables to be read from this diagram will lead to a laminate in which the void content is less than the specified critical value (X_{VC} =0.5%).

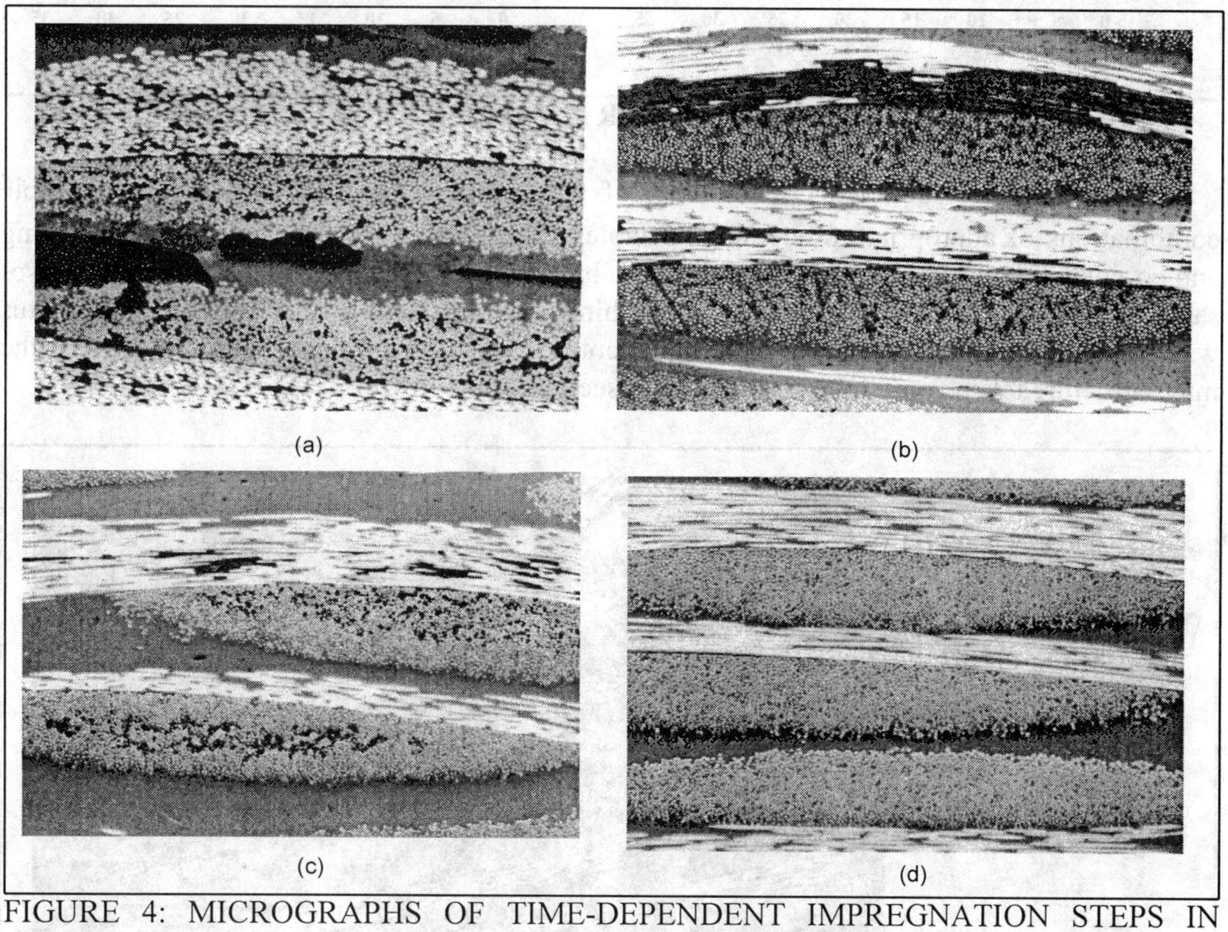

FIGURE 4: MICROGRAPHS OF TIME-DEPENDENT IMPREGNATION STEPS IN CF/PEI LAMINATES. (PROCESSING TIME WAS INCREASED STEPWISE FROM A TO D)

Based on the established optimum processing window, several prototypes of high load hinges with different fiber orientation and flange/web thickness ratio were produced from the

CF/PEI composite prepreg. Composite hinges were trimmed to the required dimensions and holes for the fastening bolts were drilled (Figure 1b). The main bush hole (load pick-up point) was drilled with a hole saw. The bush was then bonded with BR95 paste adhesive and left to cure at room temperature for over 24 hours.

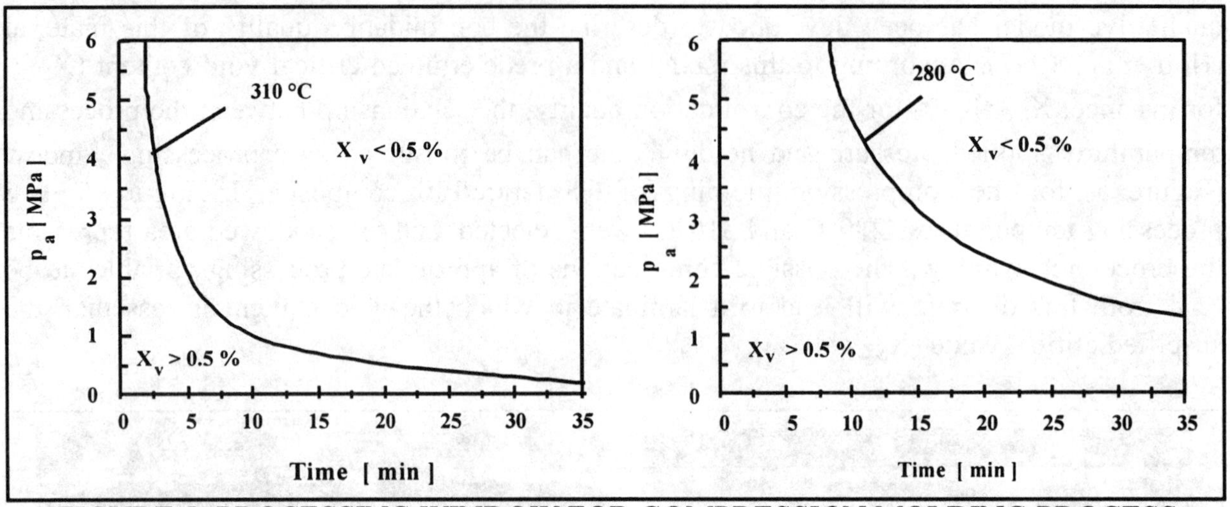

FIGURE 5: PROCESSING WINDOW FOR COMPRESSION MOLDING PROCESS

According to a specified critical level of void content (X_{VC} =0.2%) several possible combinations of appropriate processing variables were chosen for the compression molding and co-consolidation procedures to produce the high load hinges. The microstructure study of samples taken from different sections of the hinges reveals that a fully impregnation status was achieved in these composite components. As an example, Figure 6 shows the microstructure taken from the web/flange intersection of the hinge.

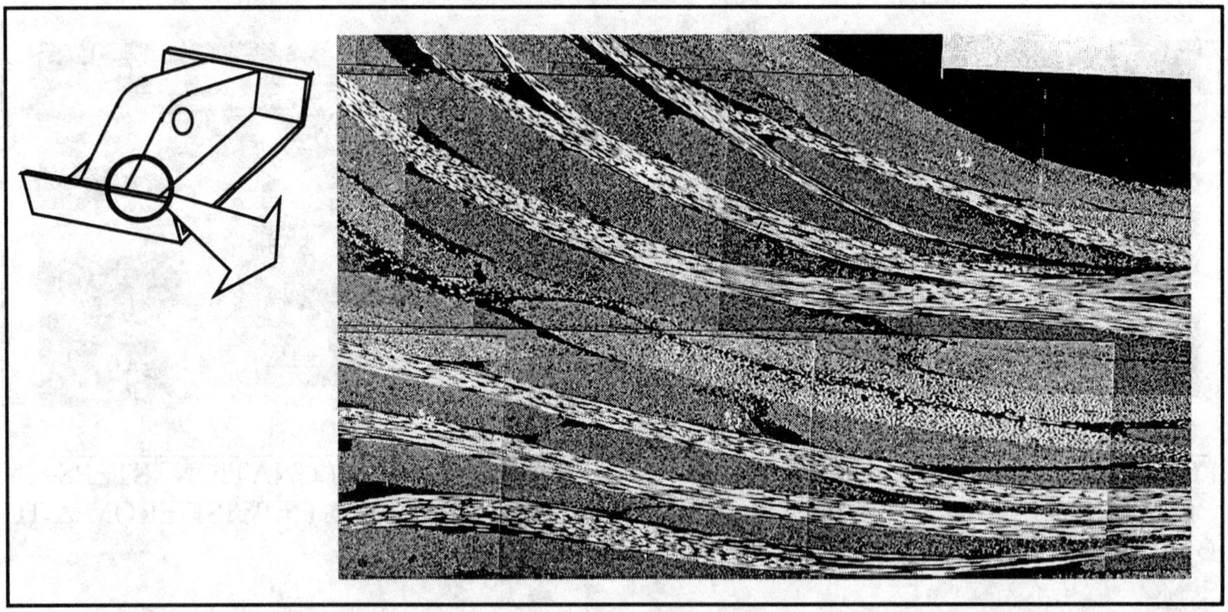

FIGURE 6: MICROSTRUCTURE OF WEB/FLANGE INTERSECTION OF HINGE

TESTING OF COMPOSITE HINGE

A special purpose test rig was designed to test the hinge specimens under the worst loading case expected during operating conditions. The tests were carried out on an INSTRON 1195 machine under displacement control at a cross-head rate of 0.2 mm/min. The load-displacement curve was obtained from a computer-based data acquisition system and the test was filmed with a SVHS video camera so as to be able to determine crack initiation/propagation. The test set-up consisted of a hollow steel beam with two machined steel plates matching the profile of the front spar. These parts were bolted together to a larger steel plate fitted to the INSTRON testing machine. A schematic of the test set-up and an actual photograph of the loaded hinge specimen are shown in Figure 7.

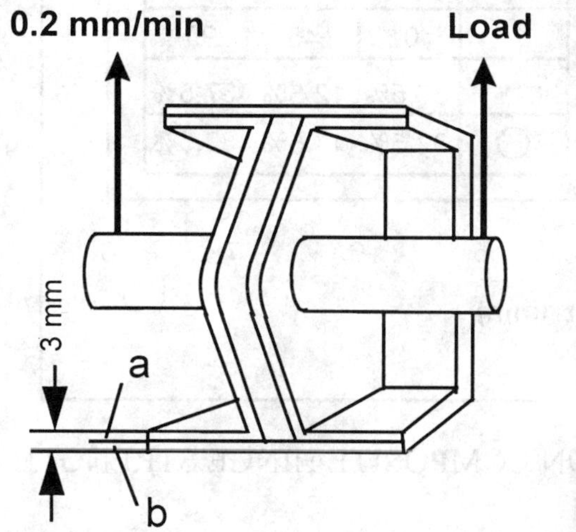

FIGURE 7: TEST SET-UP

Figure 8 shows that test results of two composite hinges with the same flange/web thickness ratio (a/b=5/5) but different fiber orientation. Both hinges behaved linearly up to the onset of the first crack growth in the bottom section the hinge being tested. After this event, non-linear behaviour was observed in both specimens. No catastrophic failure was observed up to the specified maximum test load of 25 kN. After the load was removed, only small delaminations could be seen in the hinges. Although the ultimate load expected during the operation of the hinge was met by both hinges, the hinge with more fiber oriented at in the ± 45 direction (17.5%) started delamination at a higher load of 15.5 kN. The result indicated that the fiber orientation has no or negligible effect on the globe stiffness but can influence the delamination load of the hinges. The main failure mode observed in both hinges appeared to be delamination in the joint section (as shown in Fig. 9), while no failure indication was observed in the compression side. The primary cause for delamination is due to excessive stress in the through-thickness direction in a composite laminate. These large stresses in the tensile section of the hinge are associated with the out-of-plane deformation of the web/flange intersection, tending to pull away from the center line of the web.

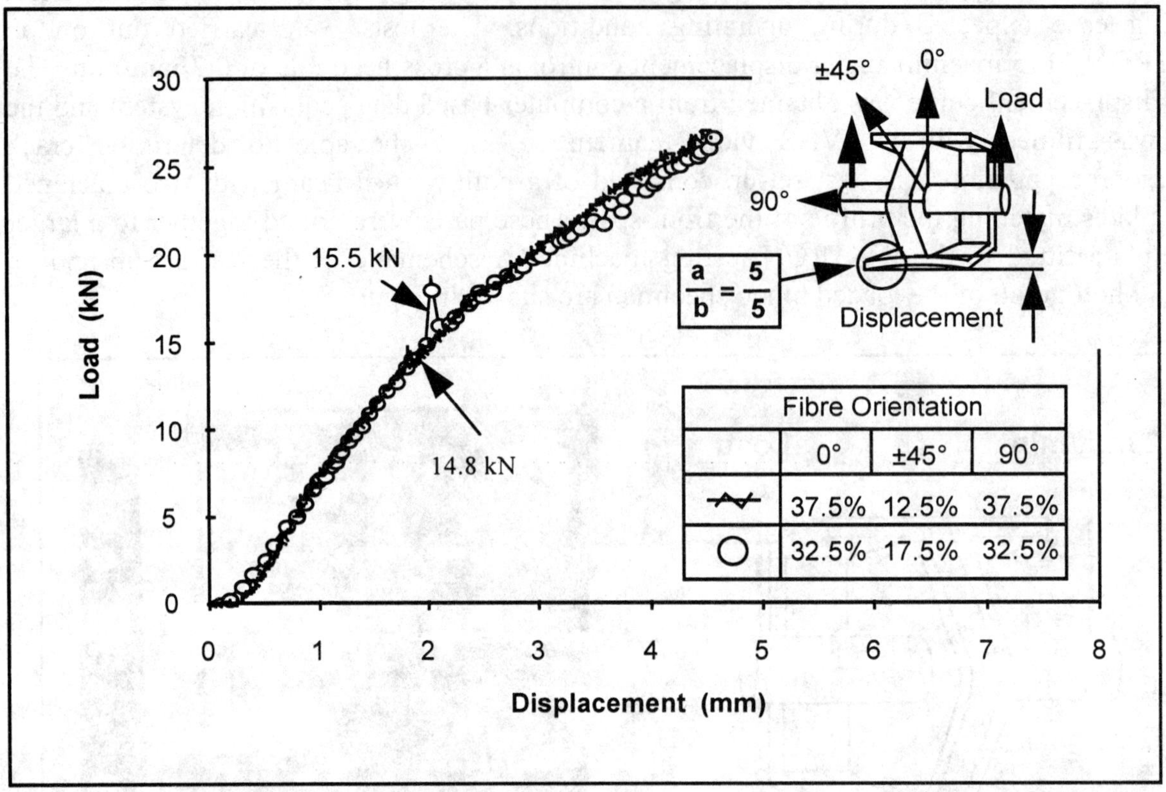

FIGURE 8: EFFECT OF FIBER ORIENTATION ON COMPOSITE HINGE STRENGTH

The effect of flange/web thickness ratio on the hinge strength was studied using composite hinges with the same fiber orientation. The flange is made up of 10 layers of CF fabric/PEI prepreg and has a total thickness of 3 mm. To obtain continuous fibre alignment in the outer side of the hinge flange as well as in the web/flange intersection, these 10 layers of prepreg were divided into two thickness groups, namely thickness **a** and thickness **b** (as indicated in Fig. 7). Group **a** corresponds to the build up around the web/flange intersection while group **b** forms the outer side of the flange. Two kinds of thermoplastic hinges with the same fibre orientation but with two different flange/web thickness ratios of **a/b** = 7/3 and **a/b** = 5/5, respectively, were manufactured and tested. The results are shown in Fig. 10. As expected, the hinge with **a/b** = 7/3 delaminated at a higher load level (17.8 kN) than that of **a/b** = 5/5 (15.5 kN). The reason is believed to be due to the fact that with more continuous fibre oriented in the load direction, the secondary tensile stresses within the matrix can be effectively reduced in the web/flange intersection.

CONCLUSIONS

A CF/PEI composite hinge has been designed to replace an existing metallic hinge on an aircraft aileron. The manufacturing cost can significantly be reduced by adopting the single

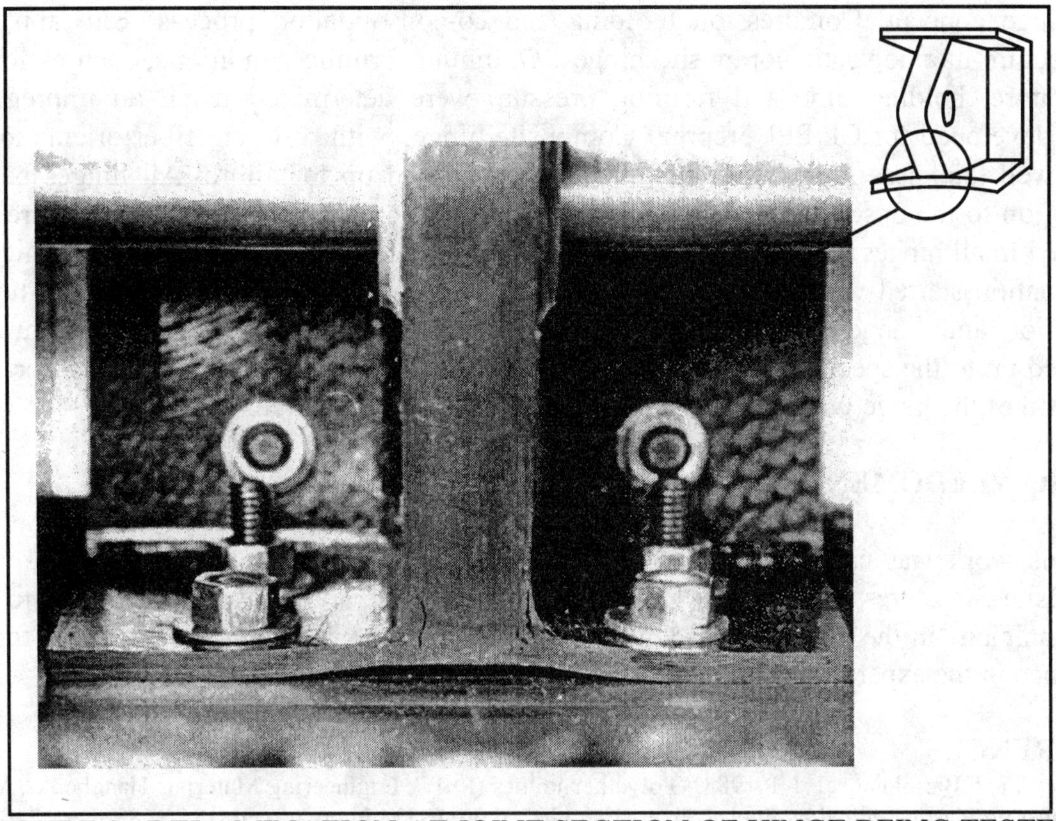

FIGURE 9: DELAMINATION AT JOINT SECTION OF HINGE BEING TESTED

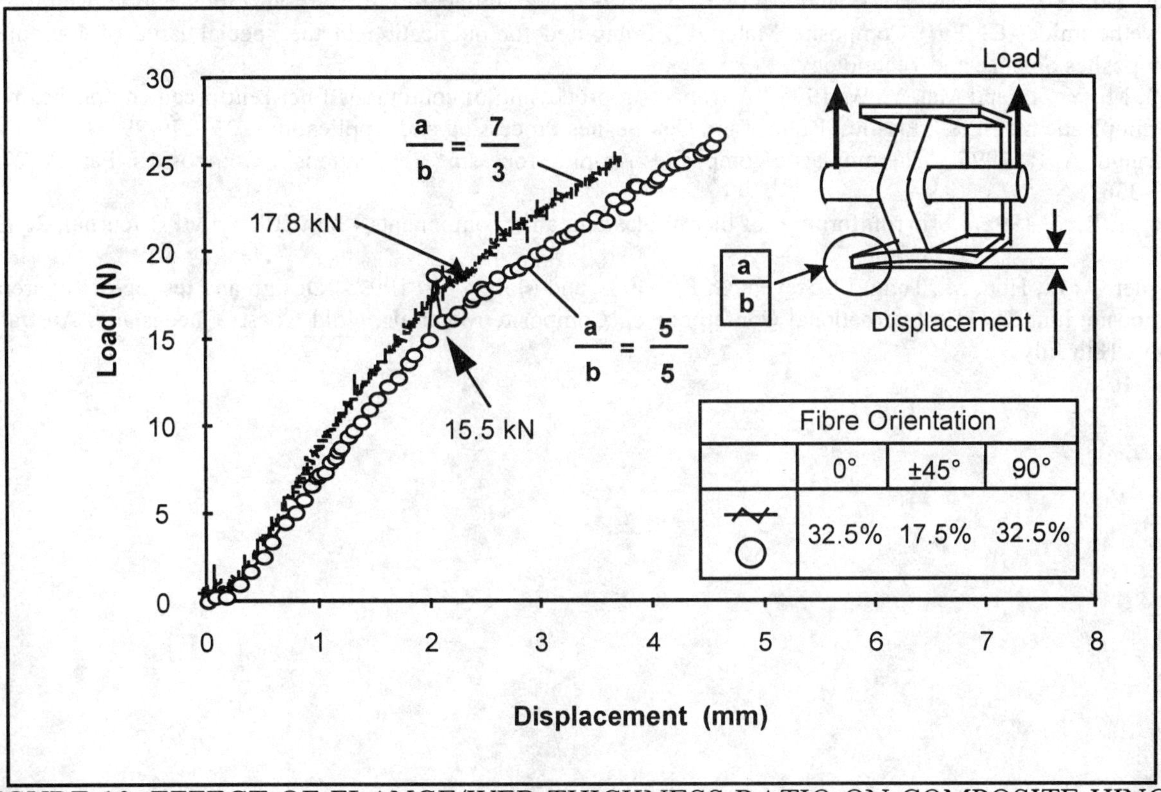

FIGURE 10: EFFECT OF FLANGE/WEB THICKNESS RATIO ON COMPOSITE HINGE STRENGTH

web hinge concept. Compression molding and co-consolidation process were applied to manufacture thermoplastic composite hinges. Optimum forming conditions, such as forming temperature, holding time and forming pressure were determined using an impregnation model developed for CF/PEI prepreg. Composite hinges with different fiber orientation and flange/web thickness ratio were manufactured and tested up to failure. All hinges behaved linearly up to the onset of the first crack growth in the joint section. The main failure mode observed in all hinges appeared to be delamination in the joint section. The load at which first delamination started was found to be in the range of 14.8-17.8 kN, depending on the fiber orientation and flange/web thickness ratio in the hinges and no catastrophic failure was observed up to the specified maximum load of 25 kN. The ultimate load expected during the operation of the hinge was met by the thermoplastic composite hinges.

ACKNOWLEDGMENTS

This work was carried out as part of the Cooperative Research Centre for Advanced Composite Structures (CRC-ACS) Research Programs. The authors would also like to thank the technicians in the Materials Laboratory of CAMT at the University of Sydney, for their assistance in the experiments.

REFERENCES

Fines, R. E. and Bartolomucci, J. P.1988. "Polyether-imides (PEI)", Engineering Materials Handbooks, Volume 2, Engineering Plastics, edited by Epel, J. N. et al. Published by ASTM International, 156-158

Hou, M., Ye, L., Lee, H. J. and Mai, Y. W. 1997. "Manufacture of Carbon Fibre Fabric Reinforced Polyetherimide (CF/PEI) Composite Material", Submitted for publication in the special issue of Journal of Composites Science and Technology

Hou, M., Ye, L. and Mai, Y. W. 1995 "Advances in processing of continuous fiber reinforced composites with thermoplastic matrices", Plastics, Rubber and Composites Processing and Applications, 23: 279-293

Offringa, A. R. 1996. "Thermoplastic composites - rapid processing applications", Composites: Part A, 27A: 329-336

Harper, R. C. 1992. "Thermoforming of thermoplastic matrix components - Part II", SAMPE Journal, 28 (3): 9-17

Rispler, A. R., Hou, M., Tong, L., Steven, G. P., Ye, L. and Mai, Y. W. 1997. "Design and testing of an aircraft composite hinge", 11th International Conference on Composite Materials, Gold Coast, Queensland, Australia, 14th - 18th July

Demonstration of a Dynamic Polymer Composite Connector

F. STOLL, R. F. GIBSON, W. ZHAO, S. YANG, N. HUSAIN AND D. BETTINGER

ABSTRACT

The paper introduces the dynamic polymer composite (DPC) connector concept for creating structural joints of tubular composite members. DPC connectors are fiber-reinforced tubular elements featuring pretensioned circumferential fibers and a thermoplastic matrix. Application of heat softens the thermoplastic, allowing the circumferential fibers to clamp the connector onto the tubes to be joined. A laboratory demonstration of the DPC connector concept is described. General design relationships are presented to relate the basic performance measures to material properties and prestress levels. The need to maximize certain properties of the prestressing fiber and the thermoplastic matrix is discussed.

INTRODUCTION

Fiber-reinforced composite materials have been shown to offer large improvements in structural performance over metals, particularly in weight-critical applications. However, the high cost of traditional high-performance composites has prevented them from being adopted in all but a small fraction of potential applications. These high costs arise not only from the cost of raw materials and starting material forms, but also because of the difficulties of joining composite structural components. The development of low-cost joining techniques will lower the cost of composites technology, thereby helping make the widespread use of composites a reality.

This paper introduces the dynamic polymer composite (DPC) connector concept for creating structural composite joints. The patented DPC technology (Bettinger, 1996) uses the time-dependent response of viscoelastic polymers to produce a controlled dimensional change in a composite material. DPC connectors have three basic constituents: 1) a thermoplastic matrix, possibly including fillers or short-fiber reinforcement; 2) longitudinally oriented fiber reinforcement to supply bending strength and stiffness to the joint; and 3) circumferentially

---

F. Stoll, Univ. of Dayton Research Inst., 300 College Park Ave., Dayton, OH 45469-0110
R.F. Gibson, W. Zhao, S. Yang and N. Husain, Mechanical Engineering. Dept., Wayne State Univ., 5050 Anthony Wayne Dr., Detroit, MI 48202
D. Bettinger, The Technology Partnership, 8030 Coventry, Grosse Ile., MI 48138-1119

oriented fibers which are prestressed in tension. These constituents are illustrated in Figure 1. Because of the tensile prestress in the circumferential fibers, the thermoplastic matrix is prestressed in circumferential compression. Upon installation, heat is applied to warm and soften the thermoplastic matrix, causing the material to relax, thus allowing the prestressed fibers to contract and clamp the connector onto the tubes to be joined. The connector accomodates diameter variations and surface irregularities, and conforms to surface features to create mechanical locking. Adhesive is used to improve load transfer between the tubular members and the connector or to seal connections for pipe applications.

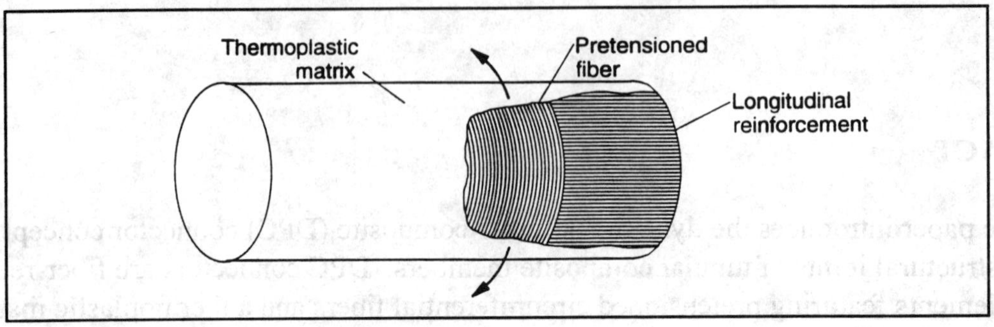

Figure 1. Constituents of a dynamic polymer composite (DPC) connector.

A Phase I Small Business Innovative Research (SBIR) program was recently performed (Bettinger and Gallagher, 1997) by The Technology Partnership (TTP) with assistance from the University of Dayton Research Institute (UDRI) and the Advanced Composites Research Laboratory (ACRL) of Wayne State University. As a part of the program, proof-of-concept DPC connectors were designed and fabricated and were used to join sections of fiberglass pipe. Other program activities included the identification of candidate materials, the development of a one-dimensional axisymmetric analysis code, and the analysis of the connector installation process with a commercial finite element code. Work has recently begun on a Phase II SBIR program to mature DPC connector technology and lay the groundwork for technology commercialization. Included in this paper are a discussion of the Phase I SBIR connector demonstration, a presentation of key design relationships to relate the basic performance measures to material properties and prestress levels, and an outline of plans for the Phase II SBIR program.

CONNECTOR DEMONSTRATION PROGRAM

The goal of the Phase I demonstration was to show that the connector could contract and securely clamp tubular members to create a joint, though not necessarily with optimal materials or fabrication techniques. Structural performance of the connector after installation was not a priority.

DESIGN

A commercially manufactured fiberglass pipe was selected for joining. The pipe featured a smooth, controlled-dimension inner diameter (ID), but a variable outer surface quality with a ±12.5% tolerance on the wall thickness. This resulted in a wide range for the outer diameter (OD) which, ideally, the connector design would accommodate.

While a DPC connector will generally feature fibers imbedded in a thermoplastic matrix, the challenges of producing such a design at a specified diameter while also obtaining the required prestress state were prohibitive for the Phase I SBIR program. For the demonstration, both the longitudinal reinforcing fiber and the circumferential prestressing fibers were applied to the outside of a plain thermoplastic tube. The fibers were left bare for the demonstration because of concerns that a stiff binder/matrix material would inhibit the contraction during connector installation.

A one-dimensional axisymmetric analysis method was used to establish the cross-sectional design. The analysis method, written in FORTRAN and implemented on a personal computer, uses a numerical procedure to compute the elasticity solution for a multi-layer tubular cross section. Some details of the procedure have been reported (Stoll, 1996). The analysis simulates the sequential application of multiple fiber layers, computes the contraction potential and the maximum available clamping pressure, and computes the stress state of the final joint. Thermal expansion was not included in the analysis. Prestress levels were selected such that creep/relaxation effects would be small over a several day period preceding installation. The analysis of the final design indicated that the full OD range of the fiberglass pipe could be accommodated by the design of the demonstration connector.

FABRICATION AND INSTALLATION

The demonstration connector was fabricated at UDRI. Commercial "off-the-shelf" thermoplastic tubing was selected as the starting material for the connector core. The required size for the core was obtained by starting with an oversized tube, machining a longitudinal slice from the pipe, and clamping and bonding the cut surfaces using a solvent welding compound. The longitudinal reinforcing fabric was mounted using only a non-structural flexible adhesive, and, therefore, its only true function was to distribute the load between the circumferential fiber and the thermoplastic tube. Filament winding of an untwisted yarn was used to create the prestress. There was difficulty in achieving and maintaining the target level of prestress during winding. The yarn experienced breakage at irregular but fairly frequent intervals when the tension was set near the desired value. It is suspected that the yarn was suffering some damage as it was pulled off of the spool due to the large pay-out tension in comparison with the low yarn tension in the spool. This caused friction as the tensioned yarn was stretched across the yarn on the spool. In future efforts special consideration must be given to the pay-out tensioning system to circumvent this problem.

Two lengths of fiberglass pipe were prepared for joining, and the ends of the two pieces were centered inside the connector. The initial fit was distinctly loose, and the pipes could be visibly rocked inside the connector. Thermocouples were mounted to the inside wall of the fiberglass pipe, the connector/pipe interface, and the outside wall of the connector for temperature monitoring and control. An electrical-resistance heating blanket was wrapped around the connector. The outside of the connector was heated to a target temperature which was then maintained until the temperature at the connector inner wall reached a temperature which assured that the thermoplastic was sufficiently softened to allow contraction. The connected joint was inspected following cooling. The pipe sections were securely clamped by the connector, with no sign of motion or looseness detectable during hand-loading of the joint. The fiber wraps were slightly loose following cooling, a fact attributed to a difference in the thermal expansion coefficients for the circumferential fiber and the thermoplastic. A photograph of the final pipe connection is presented in Figure 2.

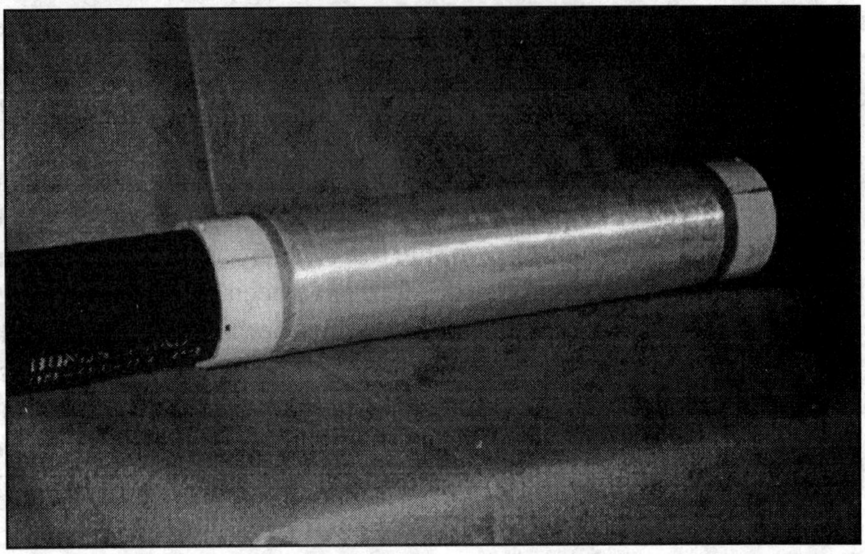

Figure 2. Demonstration joint connecting two sections of nominal four-inch fiberglass pipe.

FINITE ELEMENT ANALYSIS

Finite element (FE) simulations of the connector installation were performed by ACRL using both two-dimensional axisymmetric models and three-dimensional models. The three-dimensional model, which used three symmetry planes, is depicted in Figure 3. Solid elements were used for the pipe and thermoplastic, and shell elements were used for the thin fiber layer. Thermal expansion effects were included. Special contact elements were used to simulate the contact condition as the connector contracts to clamp the fiberglass pipe. No viscoelastic material model was used, but the heat-softening of the thermoplastic was modeled by using a temperature-dependent modulus. The temperature-dependent modulus of thermoplastic was determined experimentally at ACRL using an impulse-frequency-response vibration test method on cantilever-beam thermoplastic specimens enclosed in a temperature-controlled oven.

FE results agreed qualitatively with all observations of the demonstrator. Measurements of material response during the demonstration were inadequate for allowing quantitative comparisons. The ability of the connector to clamp the fiberglass pipe for the demonstrator geometry was demonstrated analytically. Residual stresses were computed to be well below material strength limits and were uniform along the connector length except near the ends. The analytical results provide confidence in the FE simulations for future work, where the flexibility of FE analysis will be exploited to study not only processing phenomena, but also local stress concentrations and the response to mechanical loading. A viscoelastic material model will be incorporated to account for the influence of inelastic material behavior on connector performance.

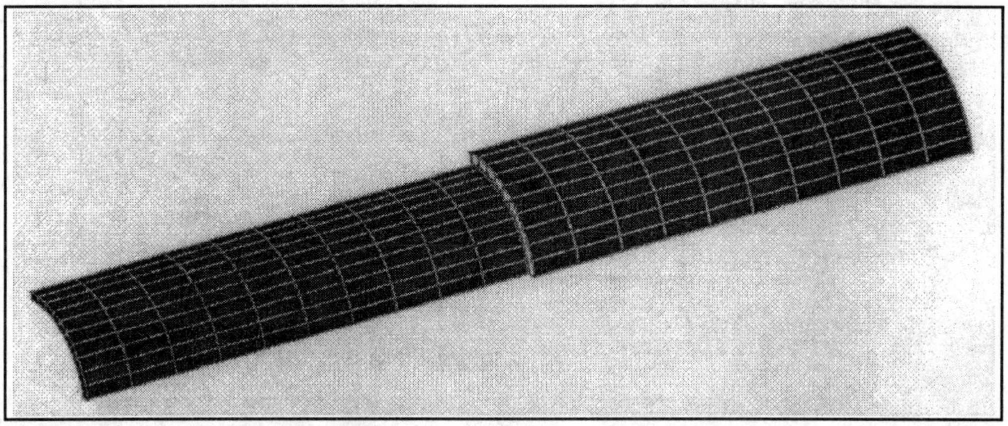

Figure 3. Finite element model of connector and pipe with three symmetry planes.

GENERAL DESIGN RELATIONSHIPS

REQUIRED CONTRACTION

Potential DPC connector applications vary widely with respect to the dimensional precision and alignment of the tubular components to be joined. The elastic strain limits of the prestressing fibers will impose limits on the range of geometry to which a DPC connector can conform. Two sources of data are examined here to characterize the diameter contraction required of the connectors. The maximum-tolerance clearance for a Class 1 - Loose Fit given by the Limits of Cylindrical Fits (American Standard ASA B4a, 1925) is plotted versus nominal diameter in Figure 4. This clearance can be interpreted as a conservative measure of the clearance required for fit-up. The second curve in Figure 4 indicates the variability in a commercially available fiberglass pipe. The pipe ID is closely controlled, but the tolerance on the wall thickness is ±12.5%. It can be seen that the percent variation increases with decreasing diameter. The percent diameter contraction available from a DPC connector approximately equals the prestrain in the circumferential fibers. For the smaller diameter range, the >5% diameter variation pushes the limits of elastic strain available with current high-performance fibers. However, for diameters greater than about 2 inches, the diameter variability can be accommodated rather easily. For tubular members with more highly controlled outer surface dimensions, the contraction requirements on the connector are greatly reduced.

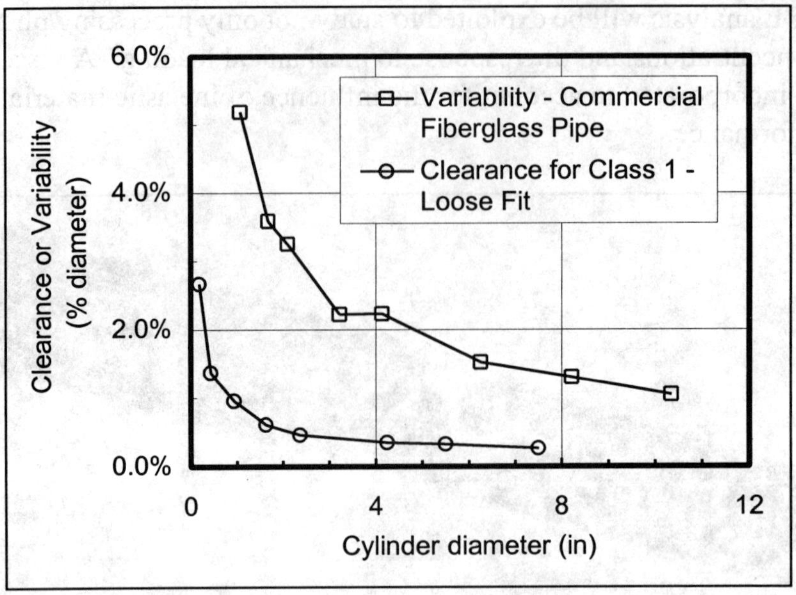

Figure 4. Example curves for cylindrical-fit clearance and diameter variability versus diameter.

THIN-WALLED CONNECTOR ANALYSIS RESULTS

A simplified thin-walled analysis is used here to illustrate several design relationships. Assume that a tube of circumferentially oriented fibers has an initial diameter of D_f when unstressed, and a thermoplastic tube has an initial diameter of D_p when unstressed, where $D_f < D_p$. The two tubes are forced together to a new composite diameter, D_c, where $D_f < D_c < D_p$. The fibers are now in tension and the thermoplastic tube is now in circumferential compression. Assuming linearly elastic material behavior, the circumferential strains in the fibers and thermoplastic, ε_f and ε_p, respectively, are given by

$$\begin{aligned} \varepsilon_f &= (D_c - D_f)/D_f > 0 \quad \text{Tension} \\ \varepsilon_p &= (D_c - D_p)/D_p < 0 \quad \text{Compression} \end{aligned} \tag{1}$$

Consider the contraction potential of the connector, ΔD_{max}, defined as the maximum connector diameter reduction which can be obtained through the contraction process. This corresponds to case where the thermoplastic has lost all stiffness and the circumferential fibers have contracted to their initial state:

$$\frac{\Delta D_{max}}{D_c} = \frac{D_c - D_f}{D_c} = \varepsilon_f \frac{D_c}{D_f} \approx \varepsilon_f \tag{2}$$

where the last approximation is made because the ratio D_c/D_f will be close to unity for high-performance structural fibers. This equation indicates that the relative contraction potential of the connector is directly related to the elastic strain in the prestressed fiber. For the case where high contraction potential is required, a high-elastic-strain fiber must be used and must be prestressed to the highest level practical.

Let t_f be the equivalent solid thickness of the fiber layer and t_p be the thickness of the thermoplastic tube. Equilibrium of internal forces in the connector requires that

$$\frac{t_f}{t_p} = \frac{-\sigma_p}{\sigma_f} \tag{3}$$

For a maximum-contraction connector with a thin wall, materials will be stressed near their limits with respect to strength and allowable creep. The ratio $-\sigma_p/\sigma_f$ and, therefore, the ratio t_f/t_p could be on the order of 1/100 for a structural fiber used with an unreinforced thermoplastic.

The theoretical maximum clamping pressure, P_{max}, is computed by assuming that there is no change in connector diameter, but the thermoplastic relaxes and the fiber force is balanced by the interface pressure acting on the inside wall of the connector. The fiber stress is related to the circumferential stress in the thermoplastic before connector installation, σ_p, through equation (3), providing:

$$P_{max} = \frac{2\sigma_f t_f}{D_c} = \frac{-2\sigma_p t_p}{D_c} \tag{4}$$

It is shown that σ_p limits both the amount of circumferential fiber (equation (3)) and the maximum clamping pressure (equation (4)). This gives a motivation for identifying and/or developing thermoplastic materials (with embedded longitudinal fibers) which maximize the allowable level of prestress.

CONTRACTION AMPLIFICATION DUE TO THICK-WALL EFFECT

If the prestressed circumferential fibers are concentrated at the outer diameter of the connector, as occurs with filament winding, there is an enhancement to the contraction of a DPC connector over that predicted by equation (2). Assume that for a thick-walled connector the cross-sectional area of the thermoplastic tube remains constant before, during, and after installation, implying that there is no flow of the thermoplastic along the length of the connector. This causes the inside diameter (ID) to contract at a greater rate than the outside

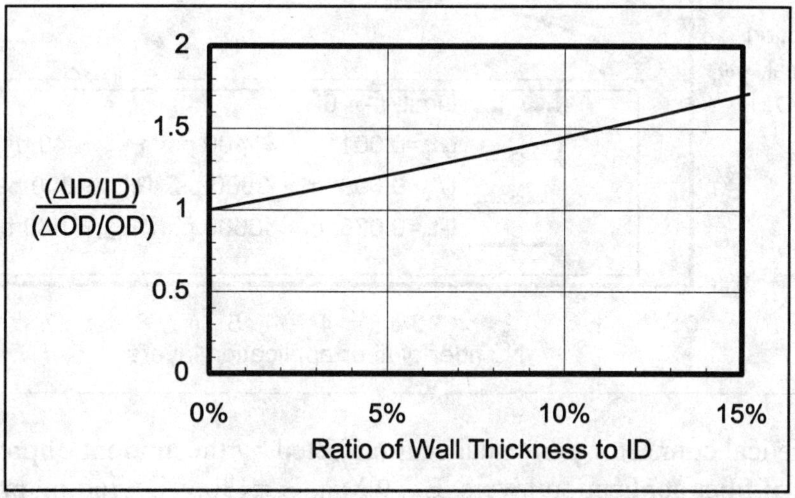

Figure 5. Amplification of the relative contraction at the inside connector wall compared to the outside wall for the case of circumferential fibers concentrated at the outer diameter.

diameter (OD), where the fibers are, thereby amplifying the relative contraction available from the fiber prestrain. The ratio of ΔID/ID to ΔOD/OD for small diameter changes is plotted as a function of the wall-thickness-to-ID ratio in Figure 5. Considering the strain limits of conventional structural fibers, this thick-wall effect may be important to obtain the contraction required for some applications.

EFFECTS OF CIRCUMFERENTIAL FILAMENT WINDING ON PERFORMANCE

Filament winding is a straightforward method to produce the circumferential stresses in DPC connectors. The process gradually deforms the thermoplastic tube as winding progresses, and the question arises as to what will be the stress state and performance potential of the connector. A study was performed to investigate the sensitivity of the contraction to the number of fiber application layers and to the prestress level in the thermoplastic material. The one-dimensional analysis method was used. Simulations were performed for a starting tube with a wall-thickness/ID ratio of 6% and prestressing fiber applied under tension at 3.3% strain. The thermoplastic was modeled with a modulus, E_p, of 430 ksi and a Poisson's ratio of 0.30, while the fiber was modeled with a modulus, E_f, of 9 Msi. Computations were performed for three different values of equivalent total fiber thickness, and for four different values of the number of winding passes used to apply the fiber. Contraction potential, maximum clamping pressure (P_{max}), and circumferential stress in the thermoplastic were computed and are summarized in Figure 6. Also included is the theoretical performance limit obtained using a vanishingly small amount of prestressing fiber.

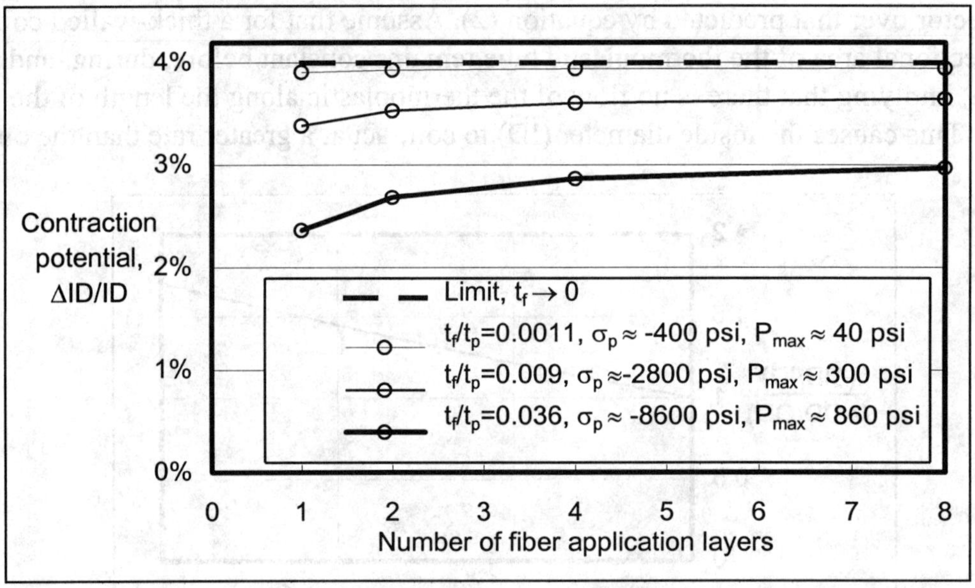

Figure 6. Theoretical connector performance as affected by the amount of prestressing fiber and the number of fiber application layers, E_f= 9 Msi, E_p=430 ksi, winding prestrain=3.3%, wall-thickness/ID = 6% .

Due to the amplification effect described earlier, the 4.0% theoretical limit of contraction is greater than the fiber prestrain value of 3.3%. There is a penalty to the contraction performance as the total amount of fiber is increased. This is because the tube OD contracts progressively during fiber application, unloading the prestressing fibers already in place. P_{max} increases with an increase in the amount of fiber, but this also increases the σ_p, which must be kept low enough to avoid excessive creep before installation of the connector. There is also some increase in the contraction potential with the number of winding passes used to apply the total amount of fiber. This effect increases with an increase in the amount of fiber.

PHASE II ACTIVITIES

A two-year Phase II SBIR program to mature DPC connector technology was initiated in April 1997. Some anticipated activities of the program are noted here:

- Identify optimal material systems for several specific DPC connector applications.
- Characterize fibers and thermoplastics with respect to creep/relaxation, accounting for the specific stress/temperature loadings in the DPC connector application.
- Develop a computer-based design method based on the cross-sectional analysis method.
- Perform structural tests of DPC connector joints.
- Perform finite element analysis of structurally loaded DPC connector joints; verify strength prediction methodology.
- Develop fabrication techniques for the economical manufacturing of DPC connectors.

CONCLUDING REMARKS

The dynamic polymer composite (DPC) connector concept uses circumferentially prestressed fibers in conjunction with a reinforced thermoplastic matrix to produce a low-profile cylindrical unit which contracts with the application of heat to structurally join cylindrical composite pipes and tubes. The DPC connector concept was successfully demonstrated in a Phase I SBIR program. The ability to analytically simulate the mechanisms of DPC connectors was demonstrated using both two- and three-dimensional finite element analyses and a one-dimensional cross section analysis. Fundamental design relationships were presented to relate the basic connector performance measures to material properties. Common structural fibers will provide adequate contraction for many larger-diameter applications. Fibers exhibiting greater elastic strain may be required for some small-diameter applications. The allowable compressive stress in the thermoplastic matrix was shown to limit both the amount of prestressing fiber which can be used, and the maximum clamping pressure which can be generated.

ACKNOWLEDGMENTS

This work was supported under a USAF Phase I SBIR program, Contract Number F33615-96-C-9622, awarded to The Technology Partnership. The USAF technical monitor was Mr. Vincent Johnson, Manufacturing Technology Directorate, Wright Laboratory.

Additional funding for Wayne State University was provided by the Michigan Materials and Processing Institute.

REFERENCES

Bettinger D.S., 1996. "Dynamic Polymer Composites," U.S. Patent No. 5,552,197, issued Sept. 3, 1996.

Bettinger D.S., and Gallaghar, J., 1997. "DPC Connectors for Aforadable Composite Structures," AF96-177 SBIR Phase I Final Report, Contract No. F33615-96-C-5622.

Stoll, F., 1996. "Analysis of Dynamic Polymer Composite Connectors," Rept. No. UDR-TR-96-75, The University of Dayton Research Institute, Dayton, OH.

Fabrication, Analysis, and Experimentation of a Practically Constructed Laminated Composite I-Beam under Pure Bending

J. A. DRUMMOND AND W. S. CHAN

ABSTRACT

A practically constructed I-beam with AS4/3501 graphite/epoxy is tested and analyzed. The I-beams having a layup sequence of $[45/\text{-}45/0_3/\text{-}45/45]_T$ with various flange widths varying from 3/4 to 2-1/2 inches and the web height at 1 inch are fabricated. The beams are tested under 4-point bending producing results for the pre-buckled stiffness, buckling moment, and ultimate moment.

Results indicate that wider flange beams exhibit buckling at lower moments than narrower flange beams. The flange width also affects the buckling mode while wider flanged beams shows larger edge displacements and lower wave values than narrower flanged beams. The ultimate moment capability appears to be only slightly increased by increased stiffness and typically shows that increasing the flange width does not contribute significantly the ultimate strength of the beam. For wider flanged beams delamination occurs at the flange and then propagates to the web causing final failure. However, narrower flanged beams fails due to fiber/matrix crushing at the flange /web intersection.

INTRODUCTION

Future developments in fiber reinforced composite structural components must be optimally designed and cost effectively produced with clearly defined performance characteristics. At that point, structures can be designed by selection and use of components similar to the methodology used with isotropic counterparts. In this paper practical design and manufacturability aspects are applied to a classic structural shape, the I-beam.

Barbero, et al.(1993) discussed a closed form solution of the stiffness and the shear correction factor for thin walled structures having a continues cross section, open or closed,

J. A. Drummond, former graduate student, currently working at Lockheed Martin Vought Systems, Dallas, Texas

Wen S. Chan, Professor, Department of Mechanical and Aerospace Engineering, University of Texas at Arlington, P.O. Box 19023, Arlington, TX 76019

made from flat sub-sections. No experimental work and discussion of buckling effects were included. Morton and Webber, (1994) including buckling and a shear correction factor for the flange, developed an iterative procedure for designing a composite I-beam. However, no failure at the web/flange in their beams were discussed. An I-beam under compression was analytically and experimentally investigated by Parnas, et al (1995) for its non-linear behavior. No flange/web buckling and out-of plane stiffness of I-beam were discussed.

ANALYSIS AND PRACTICAL DESIGN

BENDING STIFFNESS PRE-BUCKLED BEAM

The equivalent stiffness matrices of an I-beam given by Tsai (1987) are;

$$[\bar{A}]^* = [\bar{A}]^*_{web} + 4[\bar{A}]^*_{spandrel} + \sum_{i=1} b_i[A]_i \quad (1)$$

$$[\bar{B}]^* = \sum_{i=1} b_i([B]_i + z_i[A]_i) \quad (2)$$

$$[\bar{D}]^* = [\bar{D}]^*_{web} + 4[\bar{D}]^*_{spandrel} + \sum_{i=1} b_i([D]_i + 2z_i[B] + (z_i)^2[A]) \quad (3)$$

where b is given by figure 1, and [A], [B], and [D] refer to the axial, coupling, and bending stiffness of the flanges, respectively.

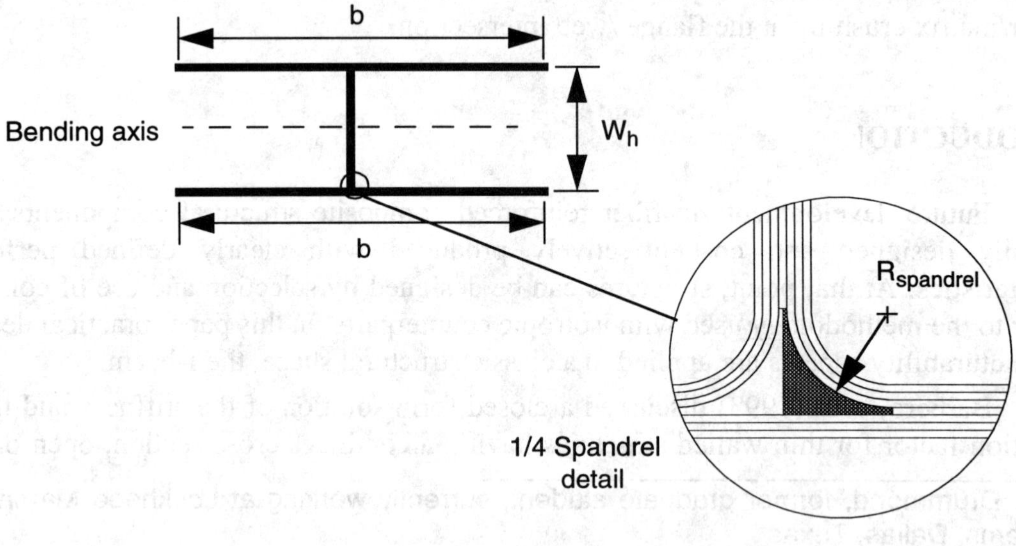

Figure 1. Simplified cross section of I-beam showing exploded view of spandrel detail

The non-conventional components defined in above three equations are defined below as;

$$[\bar{A}]^*_{web} = \begin{bmatrix} \bar{A}^*_{11_{web}} & 0 & 0 \\ 0 & 0 & 0 \\ 0 & 0 & 0 \end{bmatrix} \quad \text{and} \quad [\bar{A}]^*_{spandrel} = \begin{bmatrix} \bar{A}^*_{11_{spandrel}} & 0 & 0 \\ 0 & 0 & 0 \\ 0 & 0 & 0 \end{bmatrix} \tag{4}$$

$$\bar{A}^*_{11_{spandrel}} = \left(1 - \frac{\pi}{4}\right) r^2_{spandrel} (\bar{Q}_{11}) \tag{5}$$

$$\bar{A}^*_{11_{web}} = W_h \sum_{n=1}^{plies} (\bar{Q}_{11})_i t_{ply,\,i} \tag{6}$$

$$[\bar{D}]^*_{web} = \begin{bmatrix} \bar{D}^*_{11_{web}} & 0 & 0 \\ 0 & 0 & 0 \\ 0 & 0 & 0 \end{bmatrix} \quad \text{and} \quad [\bar{D}]^*_{spandrel} = \begin{bmatrix} \bar{D}^*_{11_{spandrel}} & 0 & 0 \\ 0 & 0 & 0 \\ 0 & 0 & 0 \end{bmatrix} \tag{7}$$

$$\bar{D}^*_{11_{web}} = \frac{1}{12} W h^3 \sum_{i=1}^{plies} (\bar{Q}_{11})_i t_{plyi} \tag{8}$$

$$\bar{D}^*_{11_{spandrel}} = z^2 \bar{A}^*_{11_{spandrel}} + Q_{11}\left(1 - \frac{5\pi}{16}\right) r^4_{spandrel} \tag{9}$$

For a symmetric I-beam, the bending moment resultant, [M] can be written as

$$\overline{M_x}^* = \overline{D_{11}}^* \kappa_x + \overline{D_{12}}^* \kappa_y + \overline{D_{16}}^* \kappa_{xy} \tag{10}$$

Conventionally, the equivalent bending stiffness along the longitudinal direction can be represented in familiar notation as

$$EI_{equiv} = \overline{D_{11}}^* \tag{11}$$

Equation 11 has been given by (1993) as the equivalent bending stiffness for a laminated I-beam cross section. This equation states that determination of the bending stiffness, about the axis shown in figure 1, is based entirely upon the stiffness in the longitudinal direction and the geometric properties of the cross section. This expression, equation 11, totally disregards the effect of the induced curvatures, κ_y and κ_{xy} due to any anti-symmetric, unbalanced, or Poisson's ratio effects of the structure.

To account for these effects Chan and Chou, (1995) suggested using the reciprocal of the entire flexibility of the I-beam as shown below:

$$EI_{equiv} = \frac{1}{\bar{d}^*_{11}} \tag{12}$$

FINITE ELEMENT MODELING

A three-dimensional ANSYS model was developed to evaluate the bending stiffness of I-beam and compared to that of the analytical results. The element SOLID73 in ANSYS was used to model the half of the beam because of its symmetry across the length. Symmetric constraint was enforced at one end and the axial tension and compression were applied to the lower and upper flanges, respectively at the other end of the beam to simulate the bending moment.

PRACTICAL LAYUP CONFIGURATION

Practical construction of the composite I-beam dictates the usage of negative mandrels to form the interior 'c' shape of the I-beam. A quick sketch of this junction will show that the wrapping of the mandrels will form a void at the web-flange intersection. To fill the spandrel created 0° plys are selected as a filler. The desired layup sequence of both flange and web is $[45,-45,0,0,0,-45,45]_T$. To accomplish continuity from the web to the flange sections, the outer three layers on each side of the web are extended to form the first three layers of the flange. The remaining flange sections can be applied to the outer surfaces of the flange to complete the flange layup. This will also assure fiber continuity between flange half sections. The additional 0° ply in the web lacks continuity with the flange section but since the ply is 0° the argument for continuity is pointless.

With the beam having web-flange continuity, the next concern is the layup sequence for the flanges and web. The symmetry of the web is sought in the design which will eliminate the configuration shown in figure 2a. The remaining option is shown in figure 2b. As can be seen by following the plies that turn from the web, two of the flanges will be symmetric and two of the flanges will be anti-symmetric. It is not possible to make all of the flanges and web asymmetric and balanced at the same beam.

TEST SPECIMEN FABRICATION AND PREPARATION

An experimental program was conducted to investigate the integrity of the predicted pre-buckled bending stiffness, nonlinear performance, and failure character of the laminated I-beam. Seven beams were fabricated from AS4/3501 carbon fiber/epoxy prepreg having a layup sequence of $[45,-45,0,0,0,-45,45]_T$ with various flange widths varying from 3/4 to 2-1/2 inches and 1 inch web height. To construct a co-cured beam, several sub-components, as shown in figure , are needed. To fill the spandrel, the cross sectional area created by the mandrel radii, as shown in figure 1, was calculated. Since the thickness of the pre-preg was known, the width of 0° pre-preg to be used in the fill could be calculated.

Before testing crush blocks are mounted in-between flanges at the ends of the beam and load pads are mounted at the outside flange areas for both ends and sides of all beams. The end tab arrangement creates essentially two planes, parallel to the cross section, to which moment will be applied when the beams are placed in four point bending. The remaining test section and bending arm are 6 inches and 3 inches, respectively, for all beams tested. The beams were tested in an MTS servohydraulic testing machine. Loading is applied in load control and applied to the specimen with a four point bend fixture having the lower (and wider) two-point set fixed and the upper (and narrower) two-point set connected by a pinned rocker allowing rotational freedom of the upper test fixture along the direction parallel to the pins.

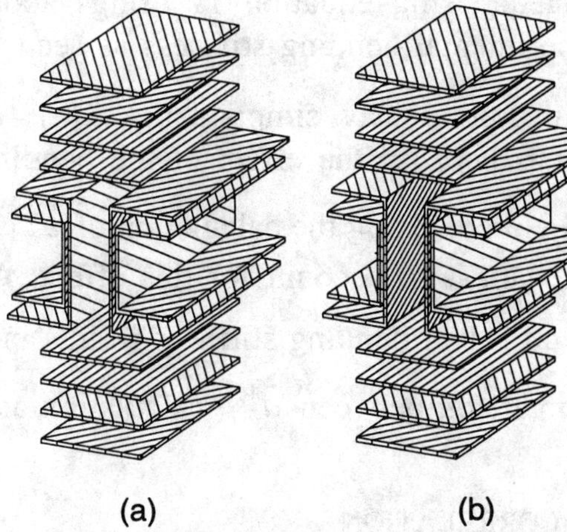

Figure 2. View of an I-beam showing (a) flange symmetric and web anti-symmetric, and (b) opposing flange symmetric anti-symmetric and web symmetric

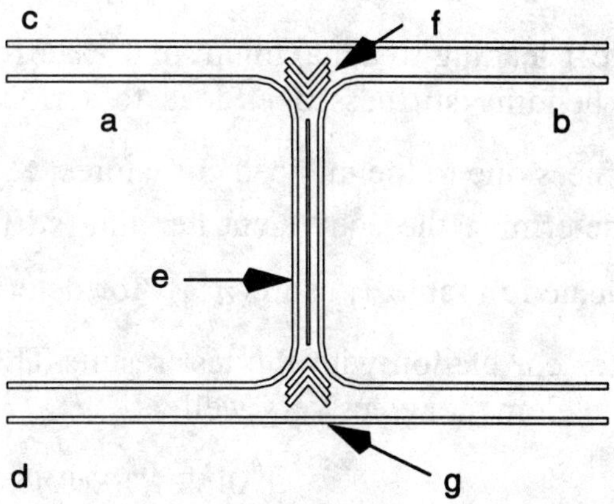

Figure 3. seven subcomponents needed for practical construction of laminated I-beam.

For data acquisition, 4 sets of strips strain gages were mounted across the width of the flange in four locations. Three of the locations are on the top side (compressive loading) at 1/4, 1/2, and 3/4 length of the beam. The remaining strip set was located at the middle-bottom side (tensile loading) of the beam.

RESULTS

LINEAR PERFORMANCE

The experimental bending stiffness for the beam is determined from equation 13. The results of these calculations appear in table 1, along with the results obtained from analytical solution.

$$EI = y\frac{M}{\varepsilon} \tag{13}$$

From the experimental data, a linear interpolation is made for the moment versus strain data from the strain gages located nearest to the web on the top and bottom flange at middle of the beam. These two points closely represent the deformation in the longitudinal direction of the beam that occurs to the entire cross section neglecting any effects due to flange deflection or end effects. The linear interpolation is made by analyzing consecutive sets of data points until the highest correlation coefficient can be found using a given set of points, which allows for elimination of points in the initial loading region and points in the nonlinear region of the behavior.

The experimental to calculated bending stiffness comparison using equation 12 typically have 4% or less error for every beam tested. Using equation 11 brings about relatively larger error between the calculated and experimental bending stiffness as seen in table 1 leading to a maximum of 15% error. As explained previously, simply using $\overline{D}^*_{11}$ for the bending stiffness does lead to serious error. This is because the effect of the bending stiffness due to the induced curvatures, κ_x and κ_{xy} is ignored. Using the calculation of $\overline{d}^*_{11}$, to determine the equivalent bending stiffness of the beam is to account for this effect. As indicated in table 1, using $\overline{d}^*_{11}$ for determining the equivalent bending stiffness results in a better correlation with the test results The results comparison between $\overline{d}^*_{11}$ calculation and finite element agrees very well.

TABLE 1: BENDING STIFFNESS CORRELATION

Flange Width (in.)	Analytical			EI_{exp} (kip-in^2)	% Error Compared with Experiment		
	$\overline{D}^*_{11}$ (kip-in^2)	$1/\overline{d}^*_{11}$ (kip-in^2)	FEM, EI (kip-in^2)		$\overline{D}^*_{11}$	$1/\overline{d}^*_{11}$	FEM
2.460	697	610	608	587	+15.7	+3.7	+3.6
2.195	611	537	555	530	+13.3	+1.3	+4.7
1.936	568	499	503	480	+15.5	+3.9	+4.8
1.471	462	409	411	424	+8.2	-3.7	-3.1
1.257	397	353	368	364	+8.2	-3.2	+1.1
1.141	393	350	345	360	+8.2	-3.0	-4.2
0.774	288	260	273	256	+11.0	+1.5	+6.6

NONLINEAR INITIATION MOMENT

The nonlinear initiation moment is determined based upon on visual inspection of the moment-strain results to locate the bifurcation point and on statistically search for the point at which the linear response is lost.

Figure 4 represents a typical load strain curve for all the beams tested. The bifurcation points seen in this figure occur due to nonlinear loading effects in the beam. Dissection of the bifurcation point allows for clear determination of the global transition from linear to nonlinear behavior but lacks sensitivity to subtle changes in the behavior occurring in the beam.

In order to quantify less obvious changes of behavior in the beam a linear interpolation was made for a consecutive series which considered the highest correlation coefficient in search of the experimental stiffness of the beam. The highest point of loading seen in the series is taken as the point of initiation of nonlinear behavior. This criterion allows for detection subtle higher order effects that are occurring in the beam such as limited web buckling. The result indicates that as the flange width increases the moment applied to create the bifurcation point decreases.The results obtained by the statistical analysis give similar results in some cases but tended to diverge in other cases as also seen in figure 5. The statistical analysis survey shows that there is a maximum point of linear behavior which then decreases significantly with an increase or decrease in flange width.

The various dimensions and material properties will determine the amount and location of the buckling phenomena. With a significantly large flange its easy to imagine that buckling would occur in the flange segments first. However, having a significantly smaller flange would provide the converse in which the buckling would initiate in the compressive half or the web.

For all the beams tested, it is observed that slight buckling effects occurred in the web but that the flange buckling predominated the behavior of the beam. As loading increases, the web sections is thought to deflect slightly until the flange constrains the motion. At wider flange widths, seen in figure 5, the flange buckling occurs before the web begins to buckling. When the flanges buckle, their contribution to stiffness is reduced which then results in a change in the centroid of the beam as detected by the statistical analysis. For example, the widest flange beam (2.460"), the web section probably still had not undergone buckling before the bifurcation point but immediately after the bifurcation point the statistical analysis showed a change in the centroid.

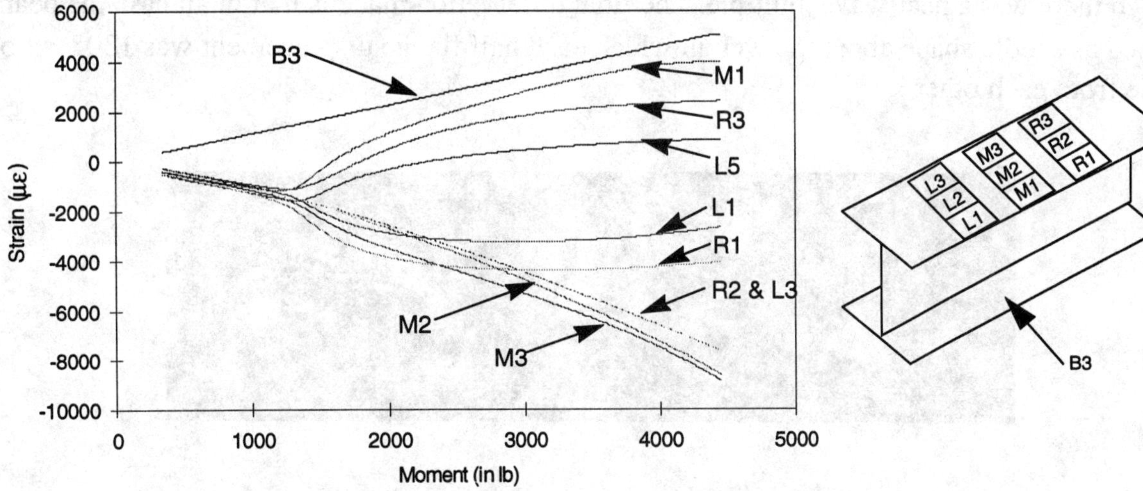

Figure 4. Strain moment results for beam #3, flange width 1.936 inches

From figure 5, at flange widths of 1.257" and below their is a clear trend between flange width and buckling initiation moment taken from the statical analysis. This trend is provided that the flange segments had not buckled thus disguising the actual web buckling. Although the effects of the web buckling are small, they are still present and may contribute to the overall performance of the structure. In some cases the point at which the bifurcation point occurred was not the same as that seen in the statistical analysis. Conception of the bifurcation point as the initiation point of nonlinear effects could then lead to error in the linear analysis.

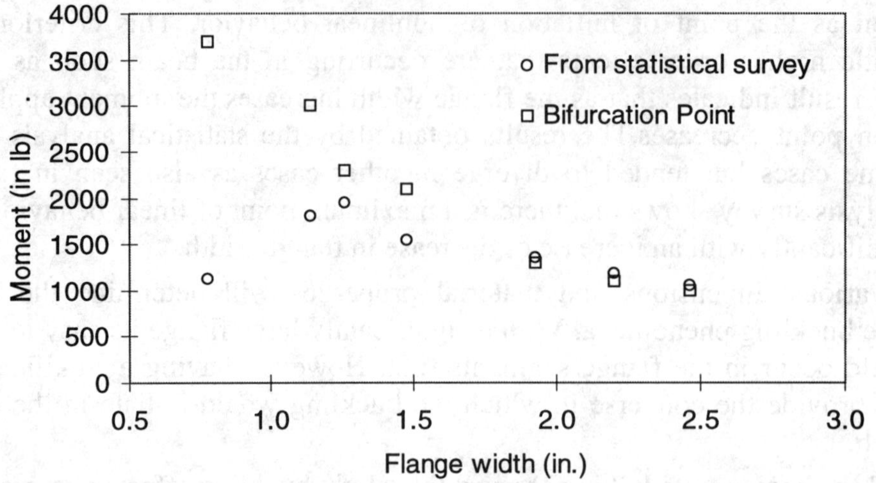

Figure 5. Buckling initiation moment as a function of flange width.

BUCKLED SHAPE

Figure 6 shows a representative case of the periodic displacements. Three observations were made for the beams tested in this study. The first observation is that as the flange width decreases the buckling mode increases. The second observation is that in some cases the mode appeared to be even multiples of half waves while in other cases it appeared as though there was a near wave multiple. The final observation made is that in all case the beam formed as saddle shape about the web in which each half flange displacement was 180° out of phase from each other.

Figure 6. Beam #3 under pure bending, flange width 1.936 inches.

ULTIMATE MOMENT

Table 2 shows the failure moment of each beam tested. The experimental results of the bending stiffness is also included. Results show that the ultimate load carrying capability of the beam may not be directly associated with the stiffness but more dependent upon the buckling mode. Overall, all the beams tested failed at significantly lower loads than predicted by linear behavior predictions.

TABLE 2: ULTIMATE MOMENT RESULTS FOR BEAMS TESTED

Beam	Flange Width (in)	EI_{equiv} (klb in^2)	Failure Moment (in lb)
1	2.460	587	4775
2	2.195	530	4935
3	1.936	480	4445
4	1.471	424	3566
5	1.257	364	4412
6	1.141	360	4082
7	0.774	256	4098

FAILURE CHARACTERIZATION

The observation of the failure reveals that for the widest flange (width is greater than 2 inches), the beams experience less shear web damage and more severe delamination at the flange. However, for the shorter flanges, at widths less than 1.4 inches, extensive shear web damage was observed.

Although beams 5, 6, and 7 showed nonlinear behavior in the strain moment results the failure is thought to originate from material failure alone. Buckling was observed in testing but most likely that buckling slightly changed the stiffness. Previously failure initiation was believed to come from delamination/separation of flange components for the

wide flange beams. For the short flange beams the failure initiation is seen as material failure seen as crushing in the compressive flange.

Figure 7. Specimen after failure, beam #3 flange width 1.936 inches.

CONCLUSIONS

A laminated I-beam was fabricated, analyzed, and test. The beams were tested in 4-point bending to evaluate the overall performance. The following are concluded in this work:

1) The preferred configuration, web symmetric, dictates the occurrence of two opposing symmetric flanges and two opposing antisymmetric flanges.
2) The equivalent bending stiffness evaluated by finite element method and the proposed analytical method based upon the reciprocal of the entire flexibility gives a less than 4% difference compared to the experimental results.
3) The bifurcation point was seen to be a function of the flange width. With a decrease in flange width the bifurcation moment increases as well as the buckling mode present in the flange is increased.
4) The global effects of the antisymmetric effects somewhat unclear. The usage of all symmetric flanges was not possible so no comparison could be made. The local effects appear to be minimum showing that stress distributions do not vary significantly from symmetric to anti symmetric flanges. Delamination initiation failure appeared to occur in both symmetric and antisymmetric flanges.
5) Wider flange beams appear to fail under the initiation of delamination due to excessive buckling displacements. The narrower flange beams appear to fail entirely from material failure associated with crushing in the compressive flange.
6) Buckling contribution essentially dictates failure for wider flange beams, resulting in beam failure at loads lower than predicted by linear analysis.

ACKNOWLEDGMENT

The authors would like to thank Mr. C.Y. Lin, a graduate student for performing the finite element analysis.

REFERENCES

Barbero, E. J., Lopex-Anido, R., and Davalos, J. F.,"On the Mechanics of Thin-Walled Laminated Composite Beams", *J. Composite Materials*, 27 (1993) 806-829.

Chan, W. S. and Chou, C. J., "Effects of delamination and ply fiber waviness on effective axial and bending stiffness in composite laminates", *Composite Structures*, 30 (1995) 299-306.

Morton, S. K., and Webber, J. P. H., "Optimal Design of a composite I-beam", *Composite Structures*, 28 (1994) 149-168.

Parnas, L., Armanios, E. A., Sriram, P., and Rehfield, L., "Postbuckling and Crippling of I-Section Composite Stiffener", *J. Aerospace Engineering*, 8 (1995) 32-42.

Tsai, S., W., Composite Design, 3rd Ed., *Think Composites*, Dayton OH (1987).

Development of Homogeneous and Composite Low Frequency Acoustic Absorbers

J. LEE AND E. O. AYORINDE

ABSTRACT Good sound absorption is presently achievable in the high frequency region, but absorption at low frequencies has presented problems. This has resulted in the use of alternative methods such as active control, noise cancellation, etc. in this area. This paper relates a development exercise for passive, low frequency acoustic absorbers made from homogeneous and composite materials. Effects of material type, absorber architecture, void geometry and other parameters are investigated. Experimental results are obtained by impedance tube measurements.

INTRODUCTION

Unwanted noise is a problem in several situations, including vehicle and aircraft compartments, workplace, domestic and sports spaces and vicinities of highways and airports. Apart from this, noise must be controlled in many machines and equipment. Research into active noise control, absorbent solids, and cavity acoustics (e.g. Schilling(1996), Grinfeld and Norris(1996), Craggs and Ayorinde(1990)) and use of local materials(e.g. Sangolola and Ayorinde (1988)) is still active. Besides re-designing the structure, which is often not practicable because it is primarily intended for another specific (load-bearing, decor, mechanical motion, etc.) purpose, noise is generally controlled in two ways - passive (judicious design using appropriate noise-absorbing materials), and active (use of noise monitoring, control and feedback circuits for regulation). Active noise control is much more expensive and complicated than passive, but, beside re-design, it is the only choice presently available for the low-frequency range (the more "bass" noises, <= 500 Hz). Low-frequency noise is unavoidable. Noise control is very important in various sectors of the economy. For example, the noise aspect of vehicle performance has in fact become a strong selling point. The development of viable low-frequency passive absorbers therefore has very significant economic appeal.

Jaesuk Lee and Emmanuel O. Ayorinde, Advanced Composites Research Laboratory, Wayne State University, Detroit, MI 48202

The two main technical approaches to noise control currently are the active (through the so-called "noise cancellation{ method whereby the signal-profile opposite of monitored noise is generated and fed back to the noise to cancel it) and the passive (in which proper sizes and configurations of materials possessing suitable acoustic

absorption properties are deployed to adequately absorb the noise). There is a progressive increase in the use of composite materials in various industries. They, and other materials are used in various primary and covering or cladding applications, in which acoustic properties may become important. Although specific acoustic absorption property values of claddings designed for acoustic treatment of conventional metallic materials are generally available from manufacturers, factors that govern acoustic absorption in composites have not been adequately studied. Some scattered experimental findings, mainly in the course of other work, have shown that for composites, these properties are strongly influenced (Koval,1980) by stacking order and other geometrical or material properties. This work is aimed at investigating the acoustic absorption properties of composite materials at low frequencies, with a view to generating designs for low-frequency-efficient acoustic absorbers.

ANALYSIS

Modeling of the acoustic transmission process in materials has been a continuing exercise. A few examples are the works of Rettinger (1936), Zwikker and Kosten (1949), Maa (1975) and Beyene and Burdiso (1996). Absorbent materials may be flexible or non-flexible, porous or impervious. Each is generally modeled differently, and refinements to existing models continue. This paper focuses on solid porous absorbent materials. The absorption coefficient for a solid porous material may be written as (Maa, 1975).

$$\alpha = \frac{4r}{(1+r)^2 + (2\pi gy - ctg 2\pi y)^2},$$

where

$$y = fD / c,$$

$$g = mc / D,$$

$$r = \frac{32\mu}{pc}\frac{t}{d^2}\left[\sqrt{1+\frac{x^2}{32}}+\frac{\sqrt{2}}{8}x\frac{d}{t}\right],$$

$$m = \frac{t}{pc}\left[1+\frac{1}{\sqrt{3^2+\frac{x^2}{2}}}+0.85\frac{d}{t}\right],$$

$$x = \sqrt{\frac{\varpi}{\mu}}\frac{d}{2}. \tag{1}$$

where μ is the kinematic viscosity of air.

ϖ is the circular frequency and f is frequency in Hertz.
C is the velocity of sound in air
D is thickness of air gap behind the panel and t is thickness of material
p is hole area per unit area of panel and d is the diameter of hole.

EXPERIMENTATION

The impedance tube (Bruel & Kjaer) was used for the experiments. Figures 1 shows a schematic setup for measuring several acoustic indices. The complex reflection coefficient, R, is calculated from the overall frequency response function, H_1, the frequency response function associated with the incident component, H_i, and the frequency response function associated with the reflected component, H_r, as

$$R = \left(\frac{H_1 - H_i}{H_r - H_1} \right) e^{j2k(1+s)} \tag{2}$$

where k = wave number, l is the distance between the first microphone location and the front of the sample, and s is the spacing between the microphones. The normalized impedance ratio and the absorption coefficient can then respectively be obtained as

$$\frac{z}{\rho c} = \frac{1 + R}{1 - R} \tag{3}$$

$$\alpha = 1 - |R|^2 \tag{4}$$

A large number of samples of dedicated acoustic absorbent composite materials, and other samples for fabricating composite panels, but not presently being used for acoustics, have been collected from different manufacturers in industry. These were tested for acoustic absorption, specific acoustic impedance and acoustic reflection. Figure 2 shows a typical result for one of these materials. Examination of the absorption coefficient plot shows that the materials tend to be effective (high values of absorption) at some high frequencies, but are acoustically weak at low frequencies. Microstructures (e.g. Figure 3) of the materials were studied to reveal what contribution to their acoustics characteristics is from their geometry and physical micro-construction. The logic of such a study is that, apart from the quality of sound conduction along the fibers themselves, the goodness of an acoustic absorbent ultimately depends on how many "helmholtz resonators" it effectively forms, and how efficient these are. The "helmholtz resonator" is equivalent to a narrow-mouthed blind alley where sound gets trapped and is internally reflected and absorbed very many times until only very little escapes back. An acoustical absorbent of this type may be conceived as an n-tuple arrangement where each absorbent layer has tiny holes of diameter "d" modeled at an average spacing "b" resulting in a hole-area-per-unit-panel-area "p", and with an air-gap of thickness "D". The parameters can be varied. Thus a good analytical model can be built for practically any configuration. In addition to analytical modeling, actual samples can be tested. The procedure included (i) acoustic testing of samples, (ii) micrographic analysis of the samples, (iii) development of computer program for configurational- vs.- acoustic analyses of samples, and (iv) correlation of experimental and analytical results.

RESULTS

Figures 2 and 3 show the absorption curve and microstructure respectively for a TTI 19.5 oz/sq. yd. MC8-4625B 100% aramid material. Figure 4 shows the experimental and simulation results for a basalt material, indicating that it is possible to obtain reasonable analytical models for acoustic models. Figures 5 to 8 show parametric study results for some of the important variables, for a three-layer construction, keeping overall construction thickness constant, Figure 5 shows that the thinner perforated material absorbs better at low frequencies. Figure 6 shows that hole diameter is critical, and must be optimized for the particular construction. Below and above an optimum hole size range, the absorption deteriorates sharply. Figure 7 shows that generally, higher hole perforation density leads to higher acoustic absorption, in the range investigated. Figure 8 shows that the wider the air gap the better, up to some optimum region after which performance deteriorates.

CONCLUSION

A model for acoustic absorption has been practically verified. Some of the factors that govern acoustic absorption at low frequencies have been investigated. Further investigation of the controlling parameters is being done.

References:

Beyene, S. and Burdiso, R.A., "A new hybrid passive/active noise absorption system", *Journal of the Acoustical Society of America,* Vol. 101, No. 3, March 1997, pp.1512-1515.

Craggs, A. & Ayorinde, E.O., "Isochronous oscillators: Their importance in low frequency sound transmission in passenger vehicles", *Journal of Sound and Vibration*, Vol 140 No. 2, 1990, pp. 360-363.

Grinfeld, M.A., and Norris, A.N., "Acoustoelasticity theory and applicatios for fluid-saturated porous-media", *Journal of the Acoustical Society of America*, Vol. 100 No. 3, 1996, pp. 1368-1374.

Koval, R.L., "Sound transmission into a laminated composite cylindrical shell", *Journal of Sound and Vibration*, (1980) Vol. 71, No. 4, pp. 523-530.

Maa Dah-you, "Theory and design of microperforated panelsound absorbing constructions", *Scientia Sinica* XVIII, 1975, pp 55-71.

Rettinger, M,.The theory of sound absorption of porous materials, flexible and nonflexible, *Journal of Acoustical Society of America* Volume 8, pp 53-59, 1936.

Sangolola, B.A. & Ayorinde, E.O., "Sound absorption properties of local wood panelling products", *Nigerian Journal for Technological Development*, Vol. 1 No. 2, 1988, pp. 157-162.

Schilling, R.J., et al., "Active noise control using a dual feedforward approach", *International Journal of Systems Science*, Vol. 27 No. 10, Oct 1996, pp. 905-915.

Zwikker, C. and Kosten, C.W., *Sound Absorbing Material*, (Elsevier, Amsterdam, 1949) pp. 5-18.

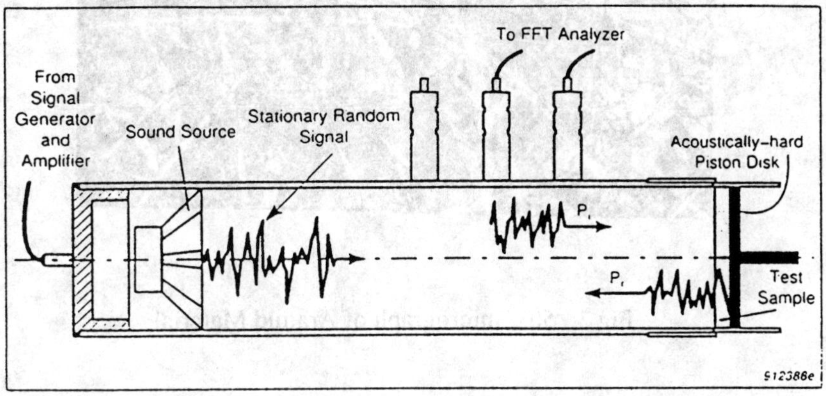

Fig. 1 Cut-away diagram of the impedance measurement tube, showing the incident and reflected components of the stationary-random signal

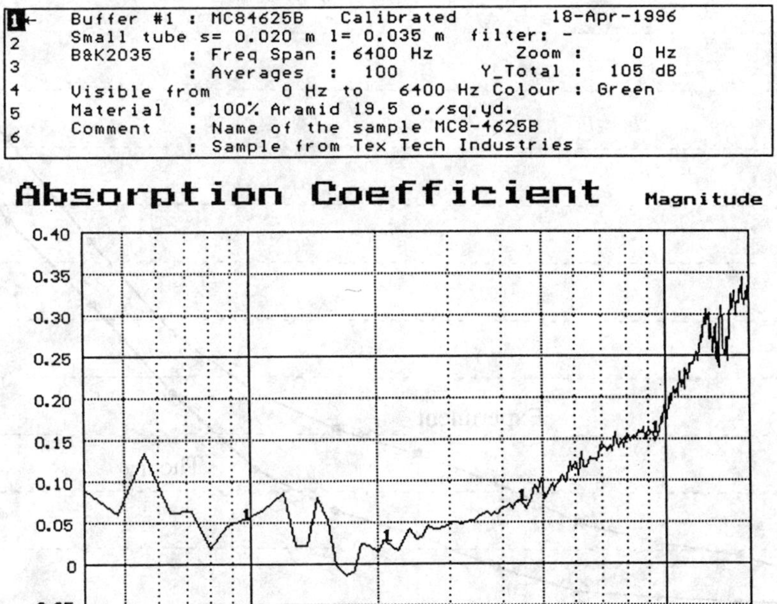

Fig.2. Absorption coefficient of Aramid Material

Fig.3. 50x micrograph of Aramid Material

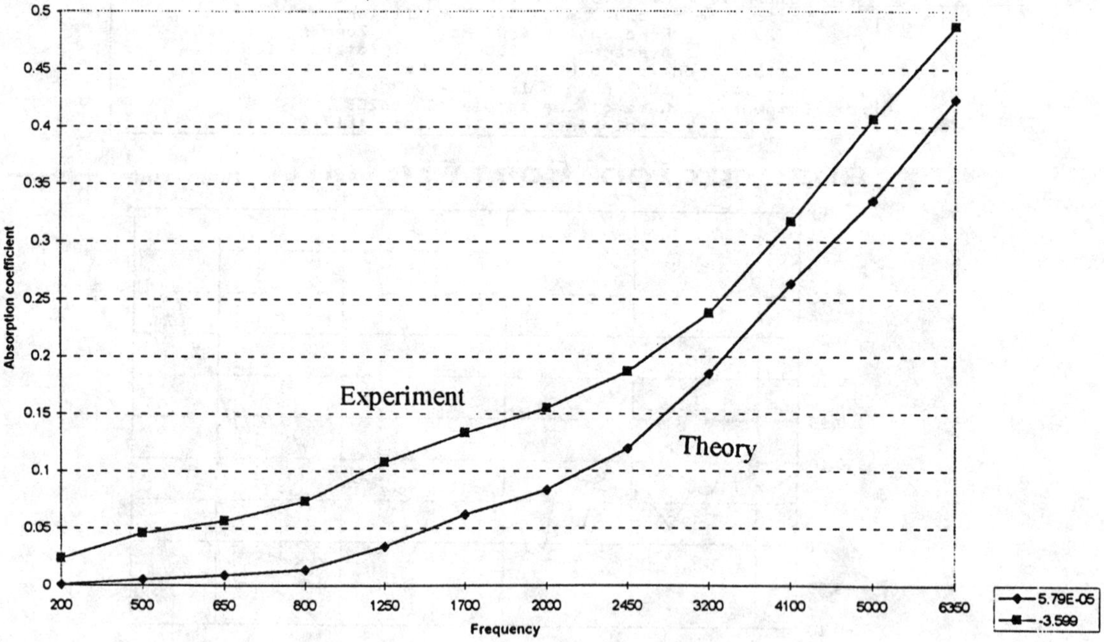

Fig.4. Comparison of experiment and theory (Basalt 1000gm/mxm, 1/4 inch)

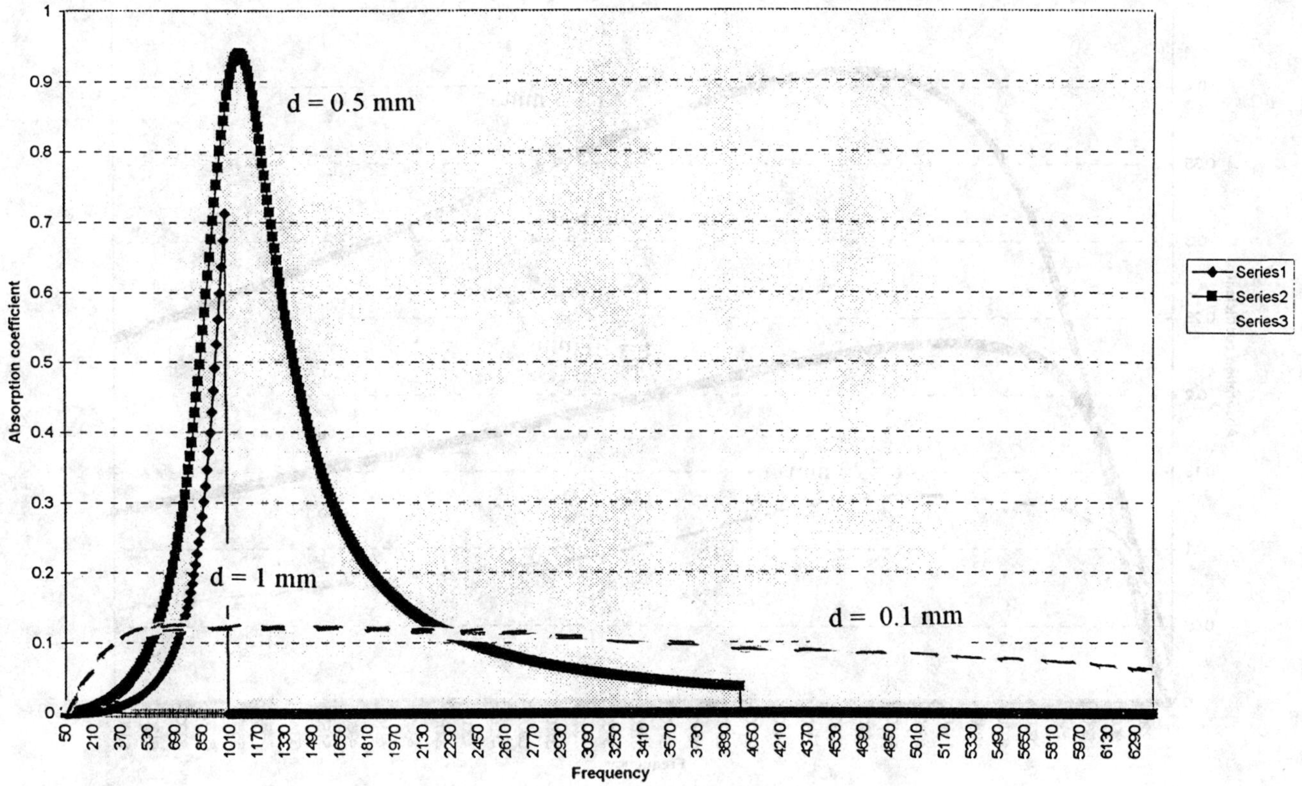

Fig.5. Parameter study for hole diameter

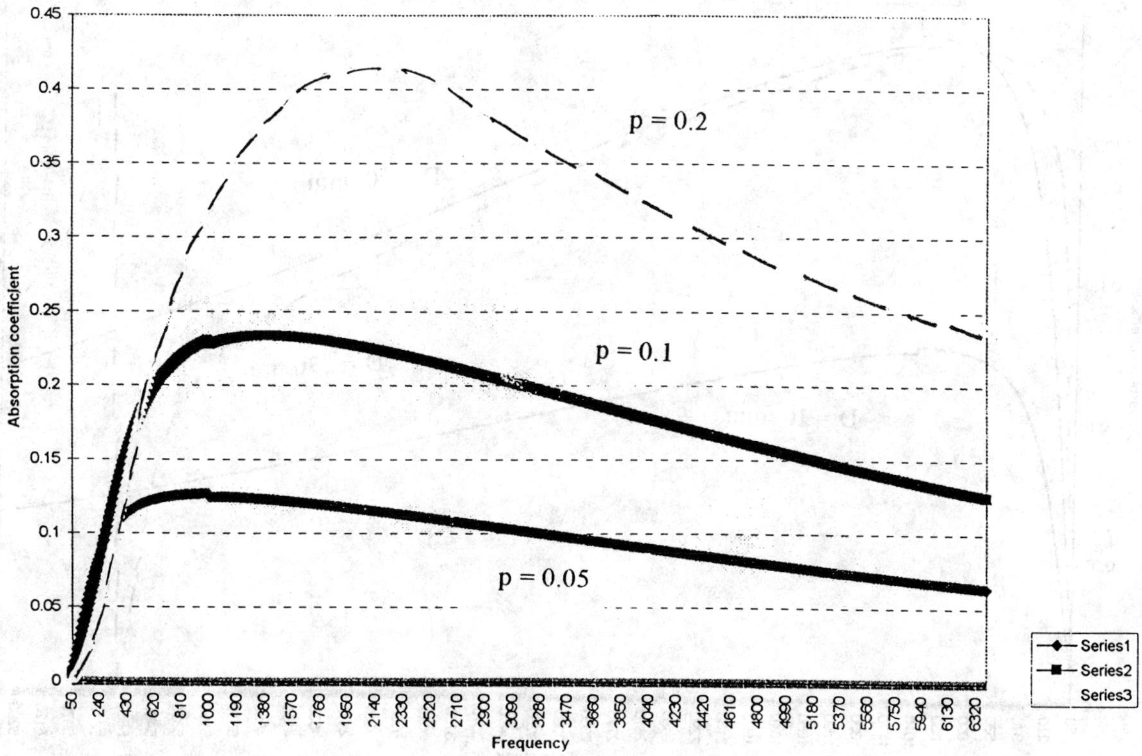

Fig.6. Parameter study for perforation ratio

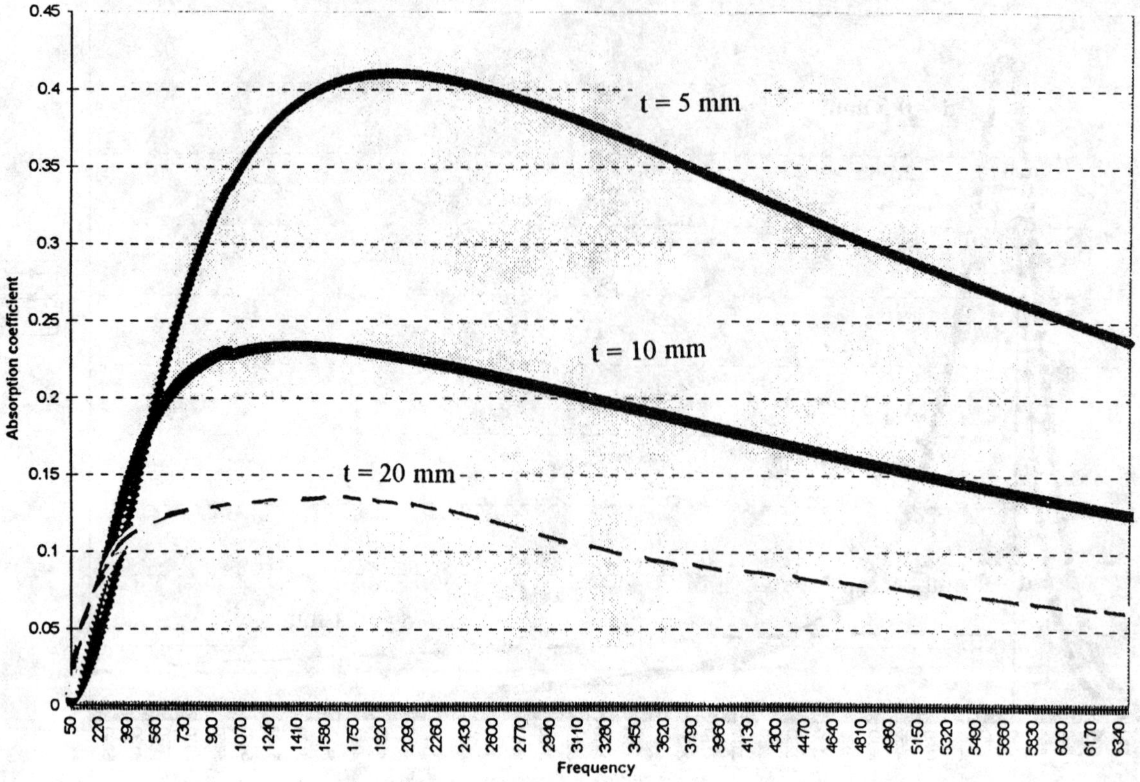

Fig.7. Parameter study for thickness

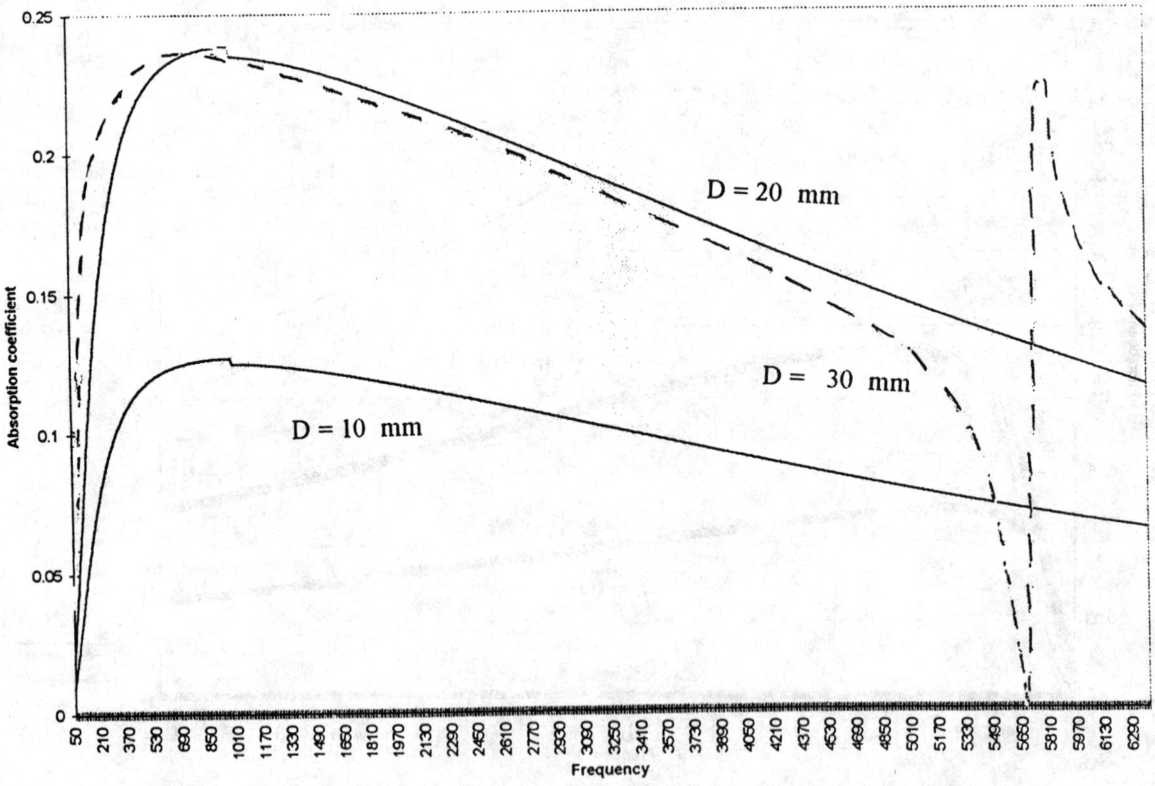

Fig.8. Parameter study for Air gap(D)

Application of Coherency Point in a Hybrid Model of Phase Transformation

O. J. ILEGBUSI AND M. D. MAT

ABSTRACT

This paper reports on the progress made in the development of a hybrid model of the mushy region in phase change problems. The mushy region is modeled as a non-Newtonian semi-solid medium below the coherency point and a porous medium thereafter. The validity of the model is tested on the solidification of a binary alloy in a 2-D rectangular cavity. The results are comparable with those of the conventional porous medium model.

INTRODUCTION

Phase transformation is inherent in many materials processing operations including casting, welding, and crystal growth. The process may occur at a constant temperature as in pure materials or over a temperature range (mushy zone) in alloy systems.

Extensive research (Mehrabian et al, 1970; Rappaz, 1989) has demonstrated the crucial role of fluid flow in the mushy and liquid regions on the final product quality. The major inhomogeneity in a casting system is the macrosegregation which occurs during the solidification as a result of differences in the component solubilities in the solid and liquid phases. The solute enrichment or depletion of the liquid occurs in the two-phase, mushy-region while solutally and thermally induced buoyancy forces can drive the liquid through and from the mushy region. The alloy component may thus be redistributed into segregates or regions that are rich or lean in a particular constituent.

Due to the difficulty of obtaining exact solutions to phase change problems, attention has been focused on the application of numerical techniques. Such techniques can be categorized into multiple region and continuum formulations (Brent et al., 1988). The multiple-region formulation involves separation of the domain into discrete regions and solution of transport equations for each region. Two adjoining regions are coupled through appropriate boundary conditions (Sparrow et al., 1977; Szekely and Jassal, 1978; Kroeger and Ostrach, 1989). A major shortcoming of this approach is the need to track interfaces that are generally unknown functions of space and time. Such problem is exercabated by the highly irregular interfaces in alloy solidification.

Olusegun J. Ilegbusi, Associate Professor
Mahmut D. Mat, Graduate student
Department of Mechanical, Industrial and Manufacturing Engineering
Northeastern University, Boston, MA 02115

the highly irregular interfaces in alloy solidification.

The alternative continuum or single domain approach has received considerable attention in recent years (Bennon and Incorpera, 1987; Voller and Prakash, 1987; Brent et al., 1988). It eliminates the need for separate treatment of each phase or region by establishing conservation equations that are valid in all regions.

While continuum models have been successfully used to simplify complex equations and represent the solidification process, none has adequately addressed the critical issue of fluidity of the mushy region. Some models consider a linearly increasing liquid viscosity with solid fraction (Gratling, 1980). A major drawback of such models is that the viscosity becomes infinitely large when the solid fraction approaches unity and a fix is often required for numerical stability. The most common approach is to consider the mushy region as a porous medium that obeys the Darcy law (Bennon and Incorpera, 1987; Voller and Prakash, 1987). The velocity arising from the momentum equations thus approaches zero when solidification is completed. This model assumes stationary solid phase and is therefore inadequate to represent free-floating equiaxed grain structure, rheocasting, compocasting and DC casting systems.

In a recent paper, Ilegbusi and Mat (1997) presented an alternative model that addresses the limitations of the porous-medium methodology. The proposed approach is a hybrid model which assumes that the mushy region behaves as a non-Newtonian semi-solid slurry at low solid fractions $f_s < f_{tr}$ and a porous medium at higher fractions $f_s > f_{tr}$. The transition solid fraction f_{tr} was chosen as the equivalent packing density of solid spheres based on Frankel and Acrivos (1967). The objective of the present paper is to further examine the choice of f_{tr}. Specifically, f_{tr} is here chosen as the solid fraction (f_{coh}) at the coherency point. The coherency point represents the temperature at which the dendrites first form an interconnected network, and is in general dependent on the alloy composition and process conditions. The non-Newtonian sub-model assumes a power-law constitutive behavior for the semi-solid slurry. The standard Darcy-law formulation is applied beyond f_{coh}. The predictions are compared with the conventional porous medium model and the effect of f_{coh} on the solidification process is investigated in the range $0.2 \le f_{coh} \le 0.625$.

FORMULATION

The solidification of a generic binary alloy is considered in a two-dimensional cavity as shown in Fig. 1. Initially, the cavity is filled with super heated alloy at temperature T_i. Solidification is initiated by maintaining the left wall at $T_{cold} < T_e$ (eutectic temperature) and the right wall at $T_{hot} > T_e$ while the top and bottom walls are insulated. The thermosphysical properties of the binary alloy are given in Table 1. By defining the following scaling relations,

$$\tau = \frac{t}{H^2/\alpha} \quad \bar{x} = \frac{x}{H} \quad \bar{y} = \frac{y}{H} \quad \bar{\rho} = \frac{\rho}{\rho_{ref}}$$

$$\bar{u} = \frac{u}{\alpha/H} \quad \bar{v} = \frac{v}{\alpha/H} \quad \bar{p} = \frac{p}{\alpha^2\rho_{ref}/H^2} \quad \theta = \frac{T - T_{ref}}{T_{hot} - T_{cold}}$$

where ρ_{ref} is the alloy density at a reference temperature T_{ref}, and α is the thermal diffusivity, the non-dimensionalized continuum conservation equations in the system can be

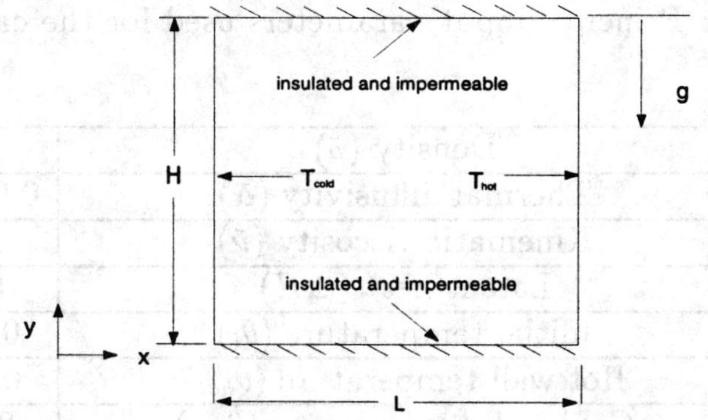

Figure 1: Schematic sketch of the test problem

expressed as:

Continuity

$$\frac{\partial \bar{u}}{\partial \bar{x}} + \frac{\partial \bar{v}}{\partial \bar{y}} = 0 \tag{1}$$

x-momentum

$$\frac{\partial}{\partial \tau}(\bar{\rho}\bar{u}) + \frac{\partial}{\partial \bar{x}}(\bar{\rho}\bar{u}^2) + \frac{\partial}{\partial \bar{y}}(\bar{\rho}\bar{u}\bar{v}) = Pr\Big[\frac{\partial}{\partial \bar{x}}(\bar{\mu}\frac{\partial \bar{u}}{\partial \bar{x}}) + \frac{\partial}{\partial \bar{y}}(\bar{\mu}\frac{\partial \bar{u}}{\partial \bar{y}})\Big] - \frac{\partial \bar{p}}{\partial \bar{x}} + S_x \tag{2}$$

y-momentum

$$\frac{\partial}{\partial \tau}(\bar{\rho}\bar{v}) + \frac{\partial}{\partial \bar{x}}(\bar{\rho}\bar{u}\bar{v}) + \frac{\partial}{\partial \bar{y}}(\bar{\rho}\bar{v}^2) = Pr\Big[\frac{\partial}{\partial \bar{x}}(\bar{\mu}\frac{\partial \bar{v}}{\partial \bar{x}}) + \frac{\partial}{\partial \bar{y}}(\bar{\mu}\frac{\partial \bar{v}}{\partial \bar{y}})\Big] - \frac{\partial \bar{p}}{\partial \bar{y}} + S_y + S_B \tag{3}$$

Energy

$$\frac{\partial}{\partial \tau}(\bar{\rho}\bar{h}) + \frac{\partial}{\partial \bar{x}}(\bar{\rho}\bar{u}\bar{h}) + \frac{\partial}{\partial \bar{y}}(\bar{\rho}\bar{v}\bar{h}) = \frac{\partial}{\partial \bar{x}}(\bar{\alpha}\frac{\partial \bar{h}}{\partial \bar{x}}) + \frac{\partial}{\partial \bar{y}}(\bar{\alpha}\frac{\partial \bar{h}}{\partial \bar{y}}) + S_h \tag{4}$$

where S_x, S_y are the Darcy source terms, and S_B and S_h are the buoyancy and enthalpy source terms respectively. Assuming Boussinesq fluid,

$$S_B = Pr Ra \bar{\rho}\bar{h} \tag{5}$$

where Ra $(=\rho\beta\Delta h H^3/\nu\alpha c_P)$ is the Rayleigh number. S_h represents the liberation of latent heat and can be expressed as (Voller and Prakash, 1987):

$$S_h = \frac{\partial \bar{\rho}}{\partial \tau} + \frac{\partial}{\partial \bar{x}}(\bar{\rho}\bar{u})\bar{\Delta H} + \frac{\partial}{\partial \bar{y}}(\bar{\rho}\bar{v})\bar{\Delta H} \tag{6}$$

where $\bar{\Delta H}(= \Delta H/(h_{hot} - h_{cold})$ is the dimensionless latent heat.

The solid fraction f_s is calculated from the linear relation:

$$f_s = \begin{cases} 0 & \theta \geq \theta_L \\ \frac{\theta_L - \theta}{\theta_L - \theta_E} & \theta_L > \theta \geq \theta_E \\ 1 & \theta < \theta_E \end{cases} \tag{7}$$

Table 1: Principal input parameters used for the calculation

Density ($\bar{\rho}$)	1
Thermal diffusivity ($\bar{\alpha}$)	0.001
Kinematic viscosity ($\bar{\nu}$)	1
Latent heat (ΔH)	5
Initial temperature (θ_i)	0.5
Hot wall temperature (θ_{hot})	0.5
Cold wall temperature (θ_{hot})	-0.5
Liquidus temperature (θ_l)	0.1
Eutectic temperature (θ_e)	0.5
Rayleigh number ($Ra = g\beta\Delta h H^3/\nu\alpha c_p$)	10^4

where θ_L and θ_E are the dimensionless liquidus and eutectic temperatures respectively.

HYBRID MODEL

Following Ilegbusi and Mat (1997), the mushy region is assumed to behave like a semi-solid slurry at low solid fractions ($f_s < f_{tr}$) and a porous medium at high solid fractions ($f_s \geq f_{tr}$). The transition solid fraction f_{tr} is here assumed to be the fraction solid at the dendritic coherency point. The need for this "two-layer" or hybrid approach is strongly supported by recent experimental results on equiaxed solidification (Arnberg et al., 1993). It was found that there exists a critical solid fraction f_{coh} at which dendrites start to impinge and form a network of solid skeleton. This fraction, corresponding to the dendritic coherency point, also represents a dramatic transition in the rheological behavior of the alloy.

One of the major issues in the hybrid model is the determination of the coherency point. Studies on coherency solid fraction are limited and there are discrepancies in the available data. It generally falls in the range $0.15 < f_{coh} < 0.65$ (Arnberg et al., 1993). Such scatter has been attributed to the poor resolution in the experimental devices, the various interpretations of coherency point, and the strong influence of composition and process conditions (Arnberg et al., 1993).

In this paper the effect of coherency point on the flow and thermal field is investigated by parametrically varying f_{coh} in the range $0.2 \leq f_{coh} \leq 0.625$. The upper limit is deliberately chosen here to accord with previous studies on the viscosity of particulate suspensions (Frankel and Acrivos, 1967). Such studies indicate that beyond f_s=0.625 which corresponds to the mean volume fraction of solid in a randomly-packed bed of uniform spheres, the viscosity of suspensions increase exponentially.

Model for $f_s < f_{coh}$

Below the transition solid fraction, the viscosity of the semi-solid mushy region is assumed to obey a power law relation (Joly, 1974):

$$\mu = m|\frac{1}{2}(\Delta : \Delta)^{1/2}|^{n-1} \tag{8}$$

where Δ is the rate of deformation tensor, m and n are empirical coefficients defined as:

$$m = exp(9.783f_s + 1.435) \tag{9}$$

$$n = \begin{cases} 0.1055 + 0.41f_s & f_s < 0.30 \\ -0.308 + 1.78f_s & f_s \geq 0.30 \end{cases} \tag{10}$$

Although these relations were originally developed for Sn-Pb alloys (Joly, 1974), their validity for other alloy systems has been demonstrated in Ilegbusi (1996).

Model for $f_s \geq f_{coh}$

Above the solid fraction at the coherency point, the dendritic network is assumed to behave like a porous medium. The momentum source terms S_x and S_y can thus be expressed using the Darcy law as (Brent et al., 1988):

$$S_x/\bar{u} = S_y/\bar{v} = \gamma H^2/\rho_{ref} \tag{11}$$

where γ is the permeability of the mushy region which can be expressed using the Carman-Kozeny relation as:

$$\gamma = K_o \frac{f_s^2}{(1-f_s)^3} \tag{12}$$

where the empirical constant K_o depends on the microstructure and the solid fraction.

NUMERICAL METHOD

The governing equations are solved numerically with a fully implicit finite-domain scheme embodied in the PHOENICS code (Rosten and Spalding, 1986). The source terms are implemented using the technique suggested by Bennon and Incorpera (1987). The viscosity of the mushy region is first calculated from a Newtonian fluid assumption to obtain a deformation tensor and later updated with the formulation in Eq. (8). A 40x40 fixed grid structure with τ = 0.0025 time increment is employed in all computations. The results are found to be grid independent in both spatial and temporal coordinates through systematic grid refinement. Each calculation requires 250 iterations per time step for convergence and takes about 5 minutes of CPU time on a Sun SPARC-4 workstation.

RESULTS

Figure 2 shows the evolution of the velocity field, streamline pattern and thermal field when f_{coh}=0.625. The mushy region is bounded by the solidus and liquidus isotherms. The isotherms are plotted with ten equal increments between θ_{hot} and θ_{cold}. The lateral temperature gradient induces a counter-clockwise buoyancy flow in the liquid phase and a much weaker flow is induced in the mushy zone. The velocity field in turn affects the

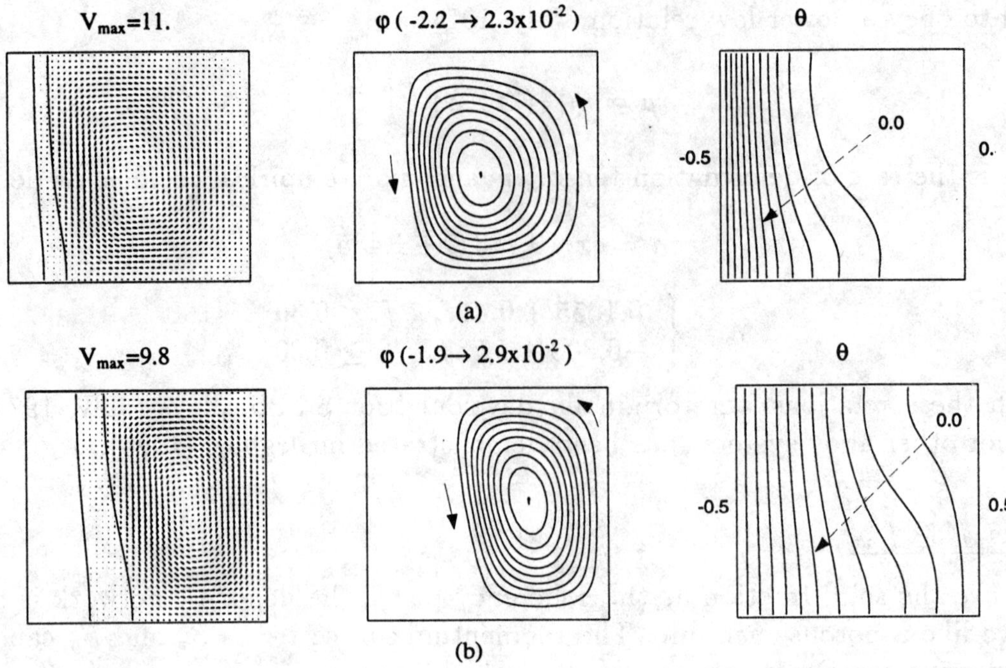

Figure 2: Flow field and temperature profiles for f_{coh}=0.625 at (a) τ=0.5, (b) τ=2.0

thermal field, resulting in non-uniform distribution of the isotherms in the liquid. A comparison of Figs. 2a and 2b shows that the solidification front moves into the flow field and the mushy region widens with time. The width of the mushy region is larger at the bottom part of the cavity because of the cold fluid being directed rightwards while the return flow from the right wall delays solidification at the upper region. As expected, pure diffusion occurs in the solid, producing large uniform temperature gradient, while thermal convection in the liquid phase enhances mixing with smaller thermal stratification.

Figure 3 shows the calculated viscosity profile in the mushy region. The viscosity reaches a maximum value in the vicinity of the transition solid fraction f_s=0.625. The maximum viscosity contours appear as "islands" in the mushy zone. Since the fluidity scales inversely with the viscosity, such result indicates that the hybrid model indeed has the potential to represent inter-dendritic flow.

Figures 4 and 5 compare the predicted velocity and temperature profiles respectively with the result based on the conventional porosity model of Voller and Prakash (1987) at three horizontal lines between the cold and hot surfaces. The agreement between the two model is generally satisfactory. There is however some discrepancy in the magnitudes of the flow velocity, especially at $\bar{y}$=0.25 (near the bottom surface) where the mushy zone is widest. This result confirms the need for accurate representation of the flow as it affects the mushy size and the solidification pattern.

The effect of the solid fraction at the coherency point (f_{coh}) on the solidification process is presented in Figs. 6-8. Three values of f_{coh} are considered: 0.2, 0.4 and 0.625. Figure 6 shows the vertical velocity at $\bar{y}$=0.25 (near the bottom wall) and τ=0.5. The magnitude of the velocity in the mushy zone increases with the value of f_{coh}. This result is consistent with the inherent presumption of the hybrid model that the semi-solid behavior

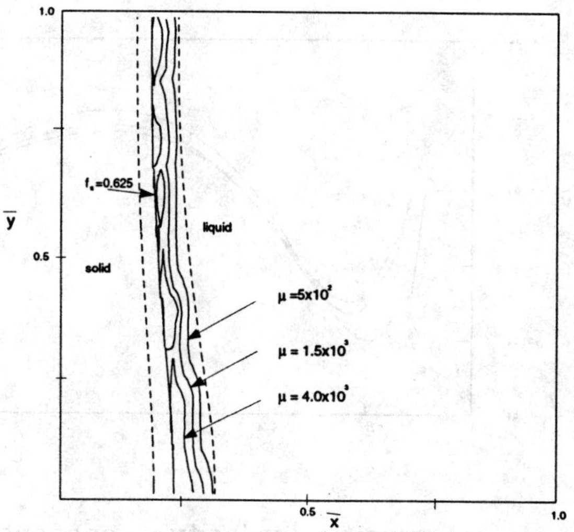

Figure 3: Viscosity distribution in the mushy region

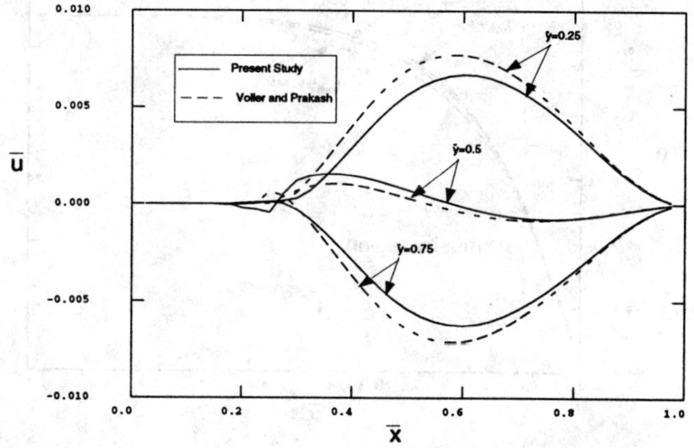

Figure 4: Comparison of hybrid model with that of porous medium model

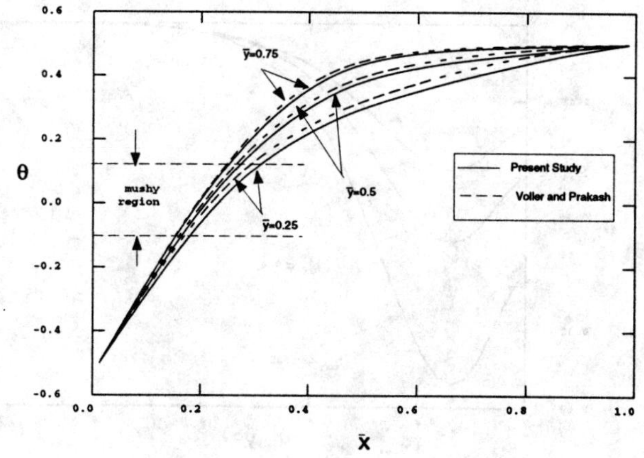

Figure 5: Comparison of hybrid model with that of porous medium model

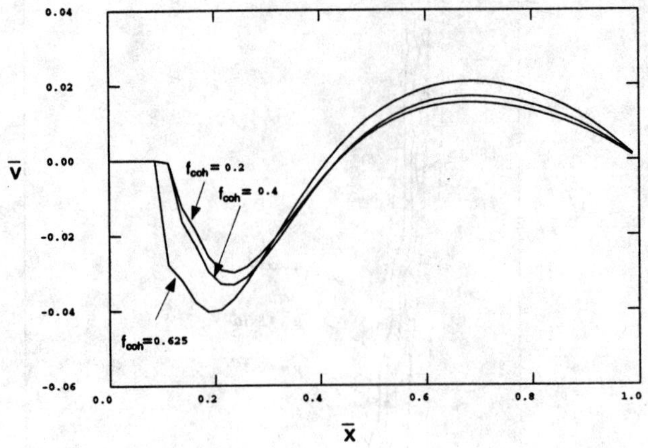

Figure 6: Effect of f_{coh} on vertical velocity component at $\bar{y}$=0.25 (τ=0.5)

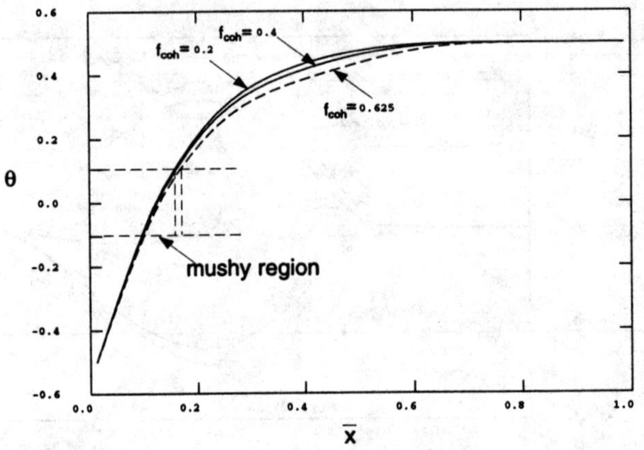

Figure 7: Effect of f_{coh} on temperature profile at $\bar{y}$=0.25 (τ=0.5)

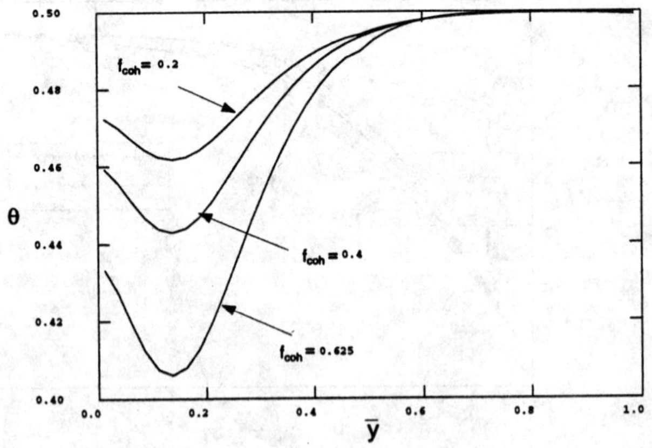

Figure 8: Effect of f_{coh} on temperature profile at $\bar{x}$=0.5 (τ=0.5)

occurs over a wider range of mushy region as f_{coh} increases. This trend is clearly evident on the corresponding temperature profiles presented in Fig. 7.

The vertical temperature profiles at τ=0.5 and $\bar{x}$=0.5 (mid width) are presented in Fig. 8. It is seen that the choice of f_{coh} significantly affects the temperature, especially in the bottom part of the cavity where the mushy zone is widest. This trend is consistent with the enhanced flow in that region as f_{coh} increases.

CONCLUSION

A hybrid mathematical model has been used to represent the fluidity of the mushy region in a generic phase-change problem. The model assumes the mushy region behaves like a semi-solid slurry below the coherency point and as a porous medium thereafter.

The predictions for a binary alloy in a 2-D cavity compare favorably with the results of the conventional porous-medium model. The predicted maximum viscosity isopleths and correspondingly, minimum fluidity in the mushy zone, appear as "islands". This result suggests that the hybrid model indeed has the potential to represent inter-dendritic flow. The limited results also demonstrate the suitability of the hybrid model to represent a range of microstructural morphologies including dendritic equiaxed free-floating grains.

The choice of the transition (coherency) solid fraction f_{coh} has profound effect on the details of the velocity and thermal field and size of the mushy zone. The flow velocity, thermal gradient and width of the mushy zone increase with the coherency solid fraction.

Despite its promise, the model needs to be further developed on several fronts. First, additional experimental data is required on the coherency solid fraction for a range of alloys. It is also important to re-examine carefully the choice of f_{coh} as the transition solid fraction. Perhaps the networked dentrites at $f_s = f_{coh}$ are still too weak to form a rigid porous medium and a certain function of f_{coh} is more appropriate. This paper has provided a first attempt at addressing this issue through a parametric study of the effect of f_{coh} on the solidification phenomena. The appropriate rheological behavior of the semi-solid slurry for $f_s < f_{coh}$ needs to be refined beyond the formulation that was originally established for a specific alloy. It is also possible that, like many semi-solid suspensions, the mushy zone exhibits thixotropy to which the present formulation is not valid. This issues will be addressed in subsequent publications.

Acknowledgment

The financial support of the NSF is gratefully acknowledged.

REFERENCES

Arnberg, L., Chai, G., and Backerud, L., 1993. "Determination of dendritic coherency in solidifying melts by rheological measurements," Mater. Sci. Engn., A173:101-103.

Bennon, W.D., and Incorpera, F.P. 1987. "Numerical analysis of binary solid-liquid phase change using a continuum model," Numer. Heat Transfer 13:277-296.

Brent, A.D., Voller, V.R., and Reid, K.J. 1988."The enthalpy porosity technique for modelling convection-diffusion phase change:application to the melting of a pure metal," Numer. Heat Transfer 13:297-318.

Frankel, N.A, and Acrivos, A., 1967. "On the viscosity of a concentrated suspension of solid spheres," Chem. Engn. Sci., 22:847-853.

Gartling, D.K., 1980. "Finite element analysis of convective heat transfer problems with change of phase," In Numerical Methods in Thermal Problems (Eds. Lewis, R.W., and Morgan, K.) Pineridge Press, Swansea, 4:595-609.

Ilegbusi, O.J., 1996. "Application of a time-dependent constitutive model to rheocast systems," J. Mater. Engn. Performance, 5:11-123.

Ilegbusi, O.J., and Mat, M.D., 1997. "A hybrid model of the mushy region in phase-change problems," J. Materials Processing & Manufacturing Science (Accepted for publication).

Joly, P.A., 1974. Ph.D. thesis, Department of Materials Science and Engineering, M.I.T.

Kroeger, P.G., and Ostrach, S. 1974. "The solution of a two dimensional freezing problem including convection effects in the liquid region," Int. J. Heat Mass Transfer, 17:1191-1207.

Mehrabian, R., Keane, M., Flemings, M.C. 1970. "Interdendritic fluid flow and macrosegregation; influence of gravity," Met. Trans. B1:1209-1220.

Rappaz, M., 1989. "Modeling of macrostructure formation in solidification processes," Int.Mater. Rev., 34:93-123.

Rosten, H., and Spalding, D.B., 1986. *PHOENICS Beginner's Guide and User's Manual*, CHAM Limited (UK) Technical Report, TR/100.

Sparrow, E.M., Patankar, S.V., and Ramadhyani S. 1977. "Analysis of melting in the presence of natural convection in the melt region," J. Heat Transfer 99:520-526.

Szekely, J., and Jassal, A.S. 1978. "An experimental and analytical study of the solidification of a binary dendritic system," Met. Trans. 9B:389-398.

Voller, V.R., and Prakash, C., 1987. "A fixed grid numerical modeling methodology for convection-diffusion mushy region phase-change problems," Int. J. Heat Mass Trans., 30:1709-1719.

Development of a Constitutive Model for Semi-Solid Metal Alloy Slurries at High Solid Fractions

O. J. ILEGBUSI AND M. D. MAT

ABSTRACT

A mathematical model is proposed to represent the thixotropic behavior of semi-solid slurries at high solid fractions ($f_s > 0.6$) as occurs in dispersion-strengthened metal-matrix composite processing. The model assumes the slurry behaves as a viscoplastic porous medium saturated with liquid phase and hardened or softened, depending on the shear rate and solid fraction. The model is validated by comparison of its predictions with a recent set of experimental data on Sn-Pb alloys.

INTRODUCTION

Processing of materials in the semi-solid state has become a potentially attractive means of producing near-net parts from segregation-prone alloys and for controlling the microstructure. It involves partially solidifying a melt or partially melting a solid to produce solid particles in a liquid matrix. The solid fraction may vary over a wide range depending on the application and the microstructure can be dendritic as in conventional solidification or globular as a consequence of vigorous stirring during the solidification. The stirred slush becomes thixotropic and its viscosity decreases with the stirring speed and time.

Although semi-solid processing has existed as a net shape forming technology for many years, the flow behavior of semi-solid metal alloys is still poorly characterized at high solid fractions. This is the subject of the present investigation.

The constitutive behavior of semi-solid alloys is very complex, as it depends on the solid fraction, the morphology of the solid phase and thermochemical history. As a result, most previous works on the constitutive behavior of semi-solid slurries have been limited to low solid fractions. In that range, the solid particles are not fully connected and the slurry rheology is history-dependent. The flow resistance however increases dramatically as the solid fraction reaches a critical value representing the development of an interconnected network, and is highly dependent on the particle morphology and the degree of agglomeration between the particles.

Olusegun J. Ilegbusi, Associate Professor
Mahmut D. Mat, Graduate student
Department of Mechanical, Industrial and Manufacturing Engineering
Northeastern University, Boston, MA 02115

The constitutive behavior of semi-solid slurries has been widely studied. Brown (1989) introduced an internal variable or structural variable to represent the degree of agglomeration among the solid particles, fluid phase viscosity, particle size morphology and distribution of particle sizes. The extensive set of experimental data of Joly and Mehrabian (1976) with rounded particles showed that the flow resistance after some transience would continue to decrease with increase in shear stress due to the breaking of the particle agglomerates. Kattamis et al. (1991) used a Coutte type viscometer to demonstrate that increasing the cooling rate results in smaller particle size and higher apparent viscosity for a given solid fraction.

The kinetics of agglomeration and disagglomeration of the primary phase particles in semi-solid Al-6%Si alloy reinforced with SiC particles was studied by Mada and Ajerch (1990). An analytical model was developed based on the analogy of dissociation and restructuring with chemical reaction.

Several models have been proposed to represent slurry viscosity at high solid fractions (Nguyen et al., 1992; Gunesekera, 1992; Wang and Wang, 1992; Hernandez et al., 1992; Martin et al., 1995). Most of these models consider the slurry as a viscoplastic porous medium saturated with a liquid phase and impose a power-law dependence of the strain rate on shear stress. Thus a constant slurry viscosity is predicted after the flow field becomes fully developed. However, such trend contradicts recent experimental evidence (Martin et al., 1995) that the slurry exhibits thixotropic (history-dependent) behavior even at high solid fractions. Therefore, the objective of the present study is to develop a constitutive model which accounts for this thixotropy at high solid fractions. The model is based on the internal-variable concept of Brown (1989), with modification for the high solid loading. It is assumed that thixotropy occurs due to the structural change under shear.

The paper is divided into four sections of which this introduction is the first. Next, the mathematical formulation of the model is presented, followed by a brief description of the numerical solution procedure. The results are then presented before the concluding remarks.

FORMULATION

Consider two concentric cylindrical vessels simulating a Couette rheometer, Fig.1. The slurry consisting of Sn-15%wtPb alloy is contained in the annulus while the outer cylinder is rotated around the stationary inner cylinder to provide a shear force. This system mirrors experimental set up of Martin et al. (1995) to afford direct comparison of the numerical results with the experimental data. The principal input parameters used in the computation are presented in Table 1.

This problem may be stated mathematically in terms of the axial, radial and azimuthal components of the equation of motion, together with that for mass conservation. Assuming cylindrical symmetry, the governing equations can be expressed as:

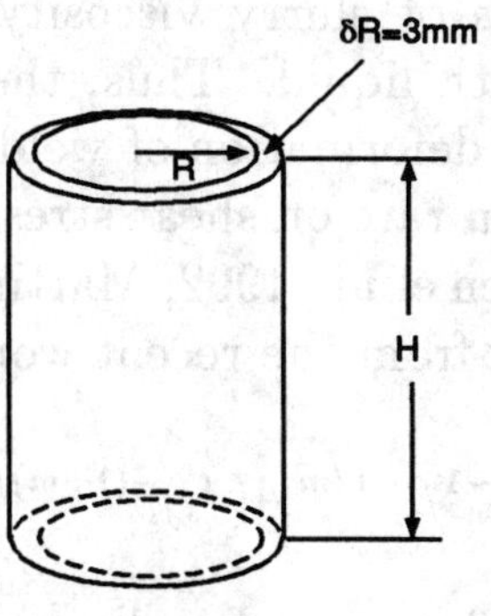

Figure 1: Schematic sketch of the problem considered

Table 1: Principal input data used in the computations

Height of cylinder (H)	0.06 m
Outer radius of cylinder (R_{out})	0.023 m
Inner radius of cylinder (R)	0.02 m
Density of slurry (ρ)	8107 kg/m^3
Ref. viscosity of slurry (μ)	0.04 Pa.s

Mass Conservation

$$\frac{\partial w}{\partial t} + \frac{1}{r}\frac{\partial rw}{\partial r} = 0. \tag{1}$$

Axial Momentum

$$\frac{\partial}{\partial t}(\rho w) + \nabla.(\rho \mathbf{u} w) = -\frac{\partial p}{\partial z} - (\nabla.\tau)_z \tag{2}$$

Radial Momentum

$$\frac{\partial}{\partial t}(\rho v) + \nabla.(\rho \mathbf{u} v) = -\frac{\partial p}{\partial r} - (\nabla.\tau)_r \tag{3}$$

Azimuthal Momentum

$$\frac{\partial}{\partial t}(\rho u) + \nabla.(\rho \mathbf{u} u) = -(\nabla.\tau)_\theta \tag{4}$$

CONSTITUTIVE RELATIONSHIP

The parameter τ in Eqs. 2-4 is the shear stress tensor through which the constitutive relationship is represented. In general, the stress components can be expressed in terms of the slurry apparent viscosity μ, and the shear rate in tensorial notation as (Ilegbusi, 1996):

$$\tau_{ij} = -\mu(\frac{\partial U_i}{\partial x_j} + \frac{\partial U_j}{\partial x_i}) \tag{5}$$

where subscripts i and j refer to the coordinate directions and U and x represent the generalized velocity and spatial coordinates respectively.

Most of the existing models of slurry viscosity at high solid fractions assume a viscoplastic medium saturated with liquid. Thus, the flow resistance is principally due to the thermally activated plastic deformation of welds between the solid particles and a power-law dependence of the strain rate on shear stress is usually assumed (Hernandez et al., 1992; Gunesekara, 1992; Nguyen et al., 1992; Martin et al., 1995). A useful relationship can be derived for slurry viscosity from the recent work of Martin et al. (1995) thus:

$$\mu = 2\gamma^{(1-m)/m}\alpha^{-1}\beta^{-1/m}A(f_s)^{-(1+m)/2m}exp(\frac{-Q}{RT})^{-1/m} \tag{6}$$

where γ is the shear rate, α and β are empirical constants that depend on the material, Q is the activation energy, R is the universal gas constant, T is the temperature, m is the power law exponent with a value of 5.7 established by Nguyen et al. (1992) for Sn-15wt%Pb alloys, and $A(f_s)$ is a function of the solid fraction and microstructure. Martin et al. (1995) proposed the following functional relationship for globular microstructure:

$$A(f_s) = \frac{3.0}{f_s^{15}} \tag{7}$$

While Eq. (6) may provide a basis for representing slurry behavior, it has a major drawback that the shear stress calculated with it is invariant with the shear rate at constant temperature and solid fraction. Such trend contradicts the recent experimental result of Martin et al. (1995) in which the slurry exhibits thixotropic behavior even at high solid fractions. To adequately represent such behavior, it is assumed here that the microstructure continues to evolve even at high solid loading and that thixotropy results from this structural evolution. In analogy to the work of Brown (1989) at low solid fraction, we propose adding a thixotropic term to Equation (6) of the form:

$$\mu_{th} = S(\gamma, f_s)\gamma^{(m-1)/m} \tag{8}$$

such that the effective viscosity μ_{eff} at high solid loading is:

$$\mu_{eff} = \mu + \mu_{th} \tag{9}$$

In Equation (8), $S(\gamma, t)$ is an internal variable that represents the structural evolution. Evidence for the form of $S(\gamma, t)$ may be provided by the limited data of Martin et al. (1995) at high solid fraction in which there occurs an initial exponential shear stress-shear strain dependence and a subsequent decay of the shear stress with strain after a threshold strain. The data also suggests that this critical strain may be dependent on the shear rate. While additional data will be needed to confirm these conclusions, we propose the following functional relationship for S:

$$S(\gamma, t) = H(\gamma)exp(-ct) \tag{10}$$

where $H(\gamma)$ is a shear-rate dependent function and c is a parameter used to express the agglomeration and disagglomeration of particles depending on the shear rate relative to the critical value γ_{crit}. This parameter may be expressed as:

$$c = d(\gamma_{crit} - \gamma) \tag{11}$$

Table 2: The values of the model constants used in the computations

Constant	α (MPa^{-1})	β	d (s)	e (s^{-1})	m	Q (kJ/kg)	R (kJ/kg.K)
Values	0.03	$4.5x10^{12}$	0.1	140.0	5.7	257	8.13

where d is an empirical coefficient to be determined by systematic comparison of predictions with the experimental data (described in the following section).

Equations (10) and (11) in effect imply that if $\gamma < \gamma_{crit}$, the slurry continues to agglomerate while disagglomeration occurs thereafter. The disagglomeration of the slurry after a critical strain as observed by Martin et al. (1995) can be accommodated by making the coefficient d strain dependent. It can also be deduced from the data of Martin et al. (1995) that γ_{crit} is a power-law function of the solid fraction. This functional relationship is here represented as:

$$\gamma_{crit} = e f_s^m \tag{12}$$

where m is the power-law exponent and e is an empirical constant, both to be determined by systematic comparison of prediction with the experimental data. The values of the model constants based on the limited experimental data are given in Table 2.

BOUNDARY CONDITIONS

A zero slip condition is imposed on the velocity components on all the solid surfaces. The fluxes of all variables are neglected at the free surface.

SOLUTION PROCEDURE

The governing equations are solved with a modified version of the PHOENICS computational package (Rosten and Spalding, 1986) for which a new routine was added to represent the non-Newtonian slurry behavior. The velocities are first calculated from a Newtonian fluid formulation to obtain a preliminary deformation tensor. Next, the viscosity is calculated using the proposed formulation in Eq. 9. The transport equations are then solved with the new viscosity at the same time step until the solution converges. The numerical procedure uses an implicit finite-domain formulation of the equations and the SIMPLER algorithm (Patankar and Spalding, 1972). The results are verified for numerical accuracy through systematic refinement of the temporal and spatial grids until the results become grid independent. Typically, 15x20x20 grids in the r, θ and z directions and 0.01 time step are required for grid independence.

RESULTS

Figure 2a shows the velocity profile at mid height for a shear rate γ=0.1 s^{-1} after a shear strain of 5. The corresponding radial variation of velocity at $z = H/2$ is presented in Fig. 2b. The distribution is essentially laminar as evidenced by the approximately

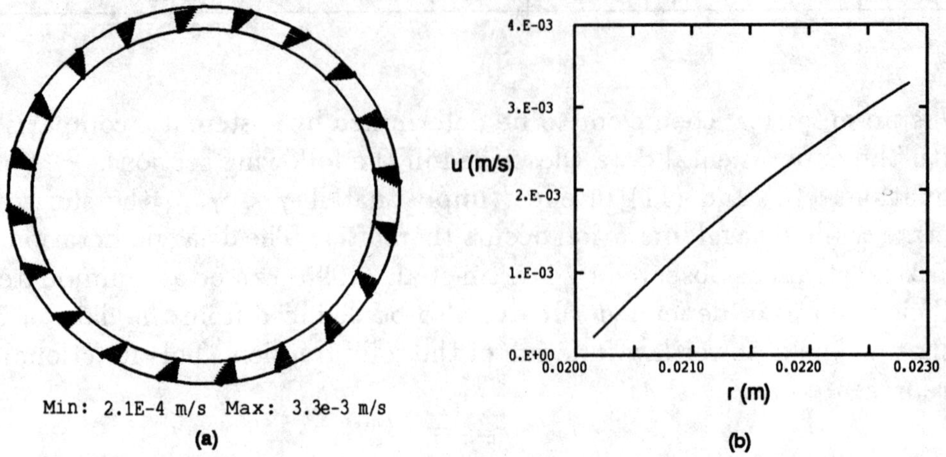

Figure 2: Velocity profile (a) and radial variation of the velocity (b) at z=H/2 for γ=0.1s^{-1}

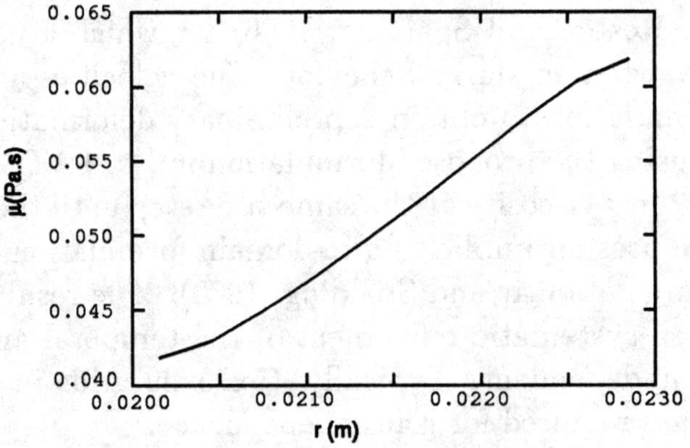

Figure 3: Radial variation of viscosity at z=H/2 for $\gamma = 0.1s^{-1}$ and f_s=0.63.

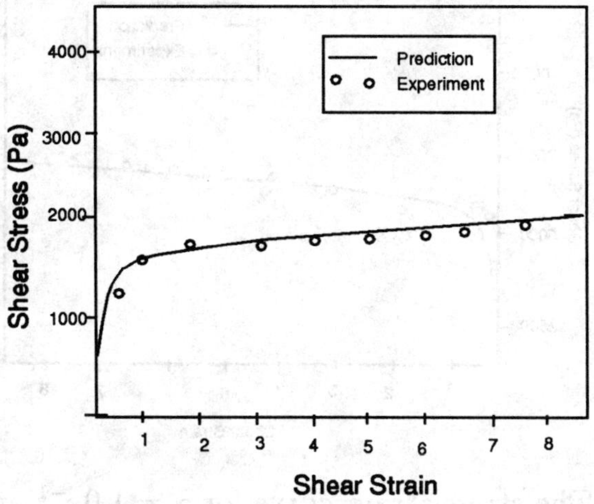

Figure 4: The stress-strain curve for $\gamma = 0.1s^{-1}$ and f_s=0.63.

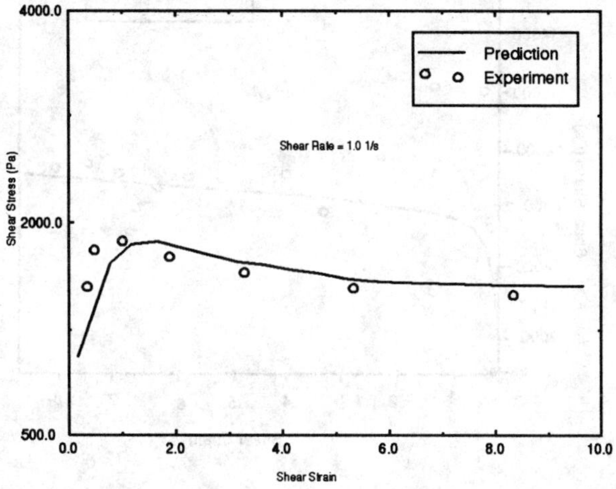

Figure 5: The stress-strain curve for $\gamma = 1.0s^{-1}$ and f_s=0.63.

linear variation between the cylinders. This result may be associated with the assumed no-slip conditions at the walls. While this assumption may not be valid at high solid loading due to slippage, the present result may be considered adequate since slippage can be minimized in a real situation by putting grooves on the rheometer walls.

Figure 3 shows the variation of the viscosity of the slurry along the radial direction at $z = H/2$. The non-Newtonian behavior is quite evident as the viscosity decreases towards the stationary inner cylinder (r=0.02 m) where the shear stress is highest.

Figure 4 compares the predicted shear stress-strain relation with the experimental data of Martin et al. (1995) for a Sn-15%Pb alloy with $\gamma = 0.1\ s^{-1}$, and f_s=0.63. The agreement between the predictions and measurement is quite good over the entire range. The shear stress increases with the shear strain. The variation is initially rapid until a strain of about 1.0 and stress of 1800 Pa, and subsequently approaches a steady state. This result provides a strong support for the exponential term in the proposed model Eq. (10).

Figure 5 shows the shear stress response of the slurry at a higher shear rate ($\gamma =$

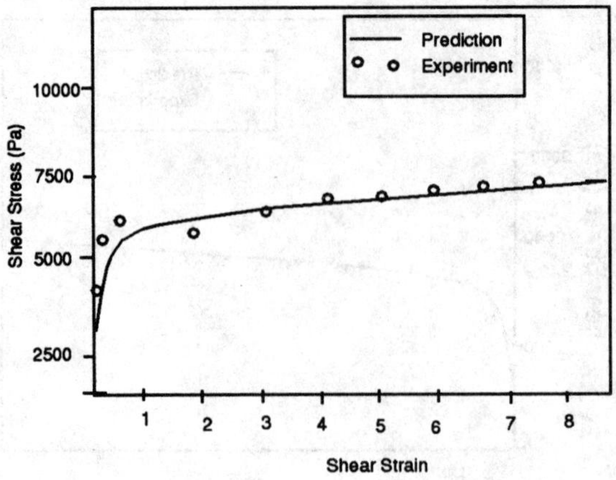

Figure 6: The stress-strain curve for $\gamma = 1.0s^{-1}$ and f_s=0.7.

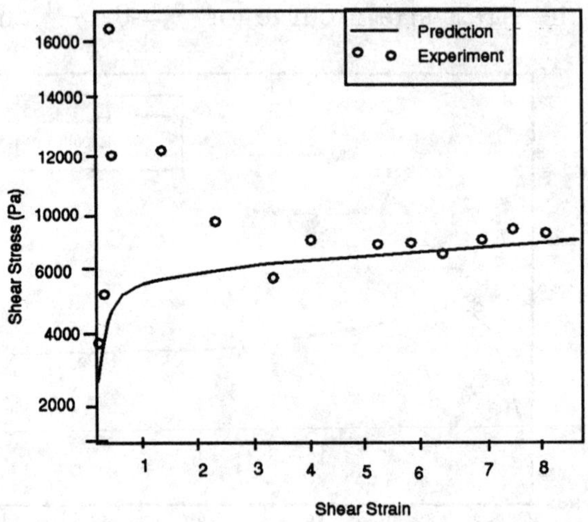

Figure 7: The stress-strain curve for $\gamma = 10.0s^{-1}$ and f_s=0.7.

$1.0s^{-1}$). Again, the experimental data of Martin et al. (1995) is adequately represented by the proposed model. The shear stress initially increases until a critical shear strain is reached, after which it decreases. This trend may be attributed to the breakdown of the solid structure or rupture of liquid welds between solid particles at the higher shear rate.

Figures 6 and 7 show the shear stress-strain relation at a higher solid fraction, $f_s = 0.7$. Figure 6 compares the prediction and the experimental data for $\gamma = 1\ s^{-1}$ and Fig. 7 compares the results at $\gamma = 10.0\ s^{-1}$. The shear stress in both cases continuously increases with the shear strain. Since the shear stress does not decay at any stage of the strain, the results suggest that for this high solid fraction the critical shear stress may be higher than that observed in Figs. 4 and 5 for a lower solid fraction f_s=0.63. The large discrepancy between the predicted shear stress and the experimental data at low strains in Fig. 7 may be due to other physical mechanisms in play at that range such as a positive shear sensitivity of the materials at high solid fraction (McLelland et al., 1992).

The calculations were carried beyond the experimental information to predict the slurry behavior at yet higher solid fraction and shear rates. Typical results are presented in Fig. 8 for $\gamma = 20.0s^{-1}$ and $\gamma = 50s^{-1}$ for f_s=0.8. Although such high solid loading

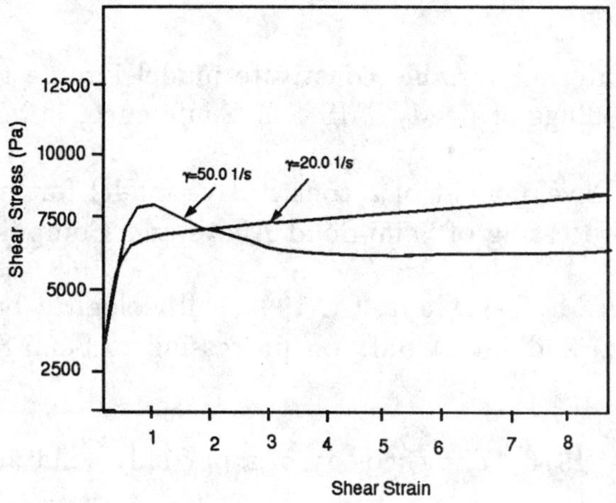

Figure 8: The stress-strain curve for f_s=0.8 and γ=20.0 s^{-1}, γ=50 s^{-1}.

may be difficult to achieve in practice, the results are quite instructive. There is indeed a breakdown of the structure at high shear rates even for very high solid loading, as evidenced by the subsequent decrease in the shear stress after a critical maximum value.

CONCLUSION

A mathematical model is presented to represent the rheological behavior of semi-solid slurries at high solid fraction. The model is based on the internal-variable concept in which thixotropy is associated with the structural change of the slurries under shear. The main proposal is that the slurry behavior at high solid fraction exhibits a shear decay with strain after a critical shear stress that depends on the solid fraction. This decay may be attributed to the deformation of liquid welds under shear.

The model predictions agree quite well with the available experimental data for Sn-15%wtPb alloy at solid fractions of 0.63 and 0.7. The model is also applied beyond the available experimental data for a system at higher solid fraction of 0.8 and higher shear rate of 20 s^{-1} and 50 s^{-1}. The shear stress-shear strain curve exhibits a continuously exponential behavior at low shear rates and low-to-moderate solid fractions. The stress decays after a critical shear stress for higher solid fractions. Additional experimental data is needed to further develop and validate the model. Such data will provide evidence for relating the model constants to the spatial and temporal distribution of the microstructure. Furthermore, the model should be applied to practical systems such as mold filling of semi solid slurries involving a wide range of solid fractions.

Acknowledgment

The financial support of the NSF is gratefully acknowledged.

REFERENCES

Brown, S.B., 1989. "An internal variable constitute model for the thixotropic behavior of metal semi-solid slurries," Proceedings of ASM/TMS Fall Conference, Indianapolis, Indiana.

Gunasekera, J.S., 1992. "Development of a constitutive model for mushy (semi-solid) materials," Proc. 2nd Int. Conf. on processing of Semi-Solid Alloys and Composites, p.211-222, Boston, MA.

Hernandez, P.V., Chaze, A.M., Levaillant, C., 1992. "Rheological behavior of Zn-Sn alloy in the solidification range," Proc. 2nd Int. Conf. on processing of Semi-Solid Alloys and Composites, p.316-327, Boston, MA.

Joly, P.A., and Mehrabian, 1976 "The rheology of a partially solid alloy", J. Mater. Sci. 11:1393-1418.

Kattamis, T.Z., and Piccone, T.J., 1991. "Rheology of semi-solid Al-4,5% Cu -1,5%Mg alloy," Mater. Sci. &Eng, A131:265-272.

Mada, M., and Ajerch, F., 1990. "Thixotropic effects in semi-solid Al-6%Si alloy reinforced with SiC particles," Metal &Ceramic Matrix Composites: Processing, Modeling & Mechanical behavior, p.337-350; The Minerals, Metals and Materials Society.

Martin, C.L., Brown, S.B., Favier, D., and Suery, M., 1995. " Shear deformation of high solid fraction (> 0.60) semi-solid Sn-Pb under various structures", Mater. Sci.& Eng., A202:11-122.

McLelland, A.R.A., Henderson, N.G., Atkinson, H.V., and Sellars, C.M. 1992. "Semi-solid Processing and microstructure in MMCs based on hypereutectic Al/Si alloys,"Proc. 2nd Int. Conf. on processing of Semi-Solid Alloys and Composites, p.290-295, Boston, MA.

Nguyen, T.G., Suery, M., and Favier, D. 1992. "Mechanical behavior of semi-solid alloys under drained compression with lateral pressure," Proc. 2nd Int. Conf. on processing of Semi-Solid Alloys and Composites, p.296-304, Boston, MA.

Patankar, S.V. and Spalding, D.B. 1972. "A calculation procedure for heat, mass and momemtum transfer in three-dimensional parabolic flows," Int. J. Heat Mass trans. 15:1787-1806.

Rosten, H., and Spalding, D.B., 1986. *PHOENICS Beginner's Guide and User's Manual*, CHAM Limited (UK) Technical Report, TR/100.

Wang, S.P., and Wang, K.K., 1992. "Die casting of semi-solid metals,"Proc. 2nd Int. Conf. on processing of Semi-Solid Alloys and Composites, p.336-345, Boston, MA.

Effects of Forming-Induced Evolution of Microstructure in Plain Weave Composites

T. M. McBRIDE, J. CHEN AND L. LAM

ABSTRACT

When forced to conform to multiply curved surfaces, woven fabrics may undergo large shear or trellis deformation. As a result, the fibrous microstructure will evolve considerably. In formed fabric reinforced composites then, microstructural features will vary slowly in a macroscale, in addition to the rapid microscale variations present in planar fabric reinforced composite geometries. When considering plain weave composites, a unit cell may be identified as being bounded by four adjacent, interlaced yarn segments. The configuration of the unit cell, which may involve dimensions such as yarn width, yarn spacing or the fiber volume fraction within a yarn, will vary slowly from cell to cell within a ply of a formed woven fabric composite. In this work, we investigate the relationship between the level of trellis deformation in plain weave fabric and the corresponding configuration of its microstructure. Even for simple part geometries, the unit cell fiber volume fraction can vary throughout a ply by a factor of two for certain fabrics. The fiber volume fraction within yarns may also vary by more than 15% as a result of forming induced evolution of microstructure. At a given level of trellis deformation, expressions are presented which describe the current microstructural geometry as a function of parameters which may be easily measured from the undeformed fabric. This information, together with existing kinematic models which yield yarn paths on the tool surface, may be used to map the distribution of microstructure throughout a multiply curved component. In doing so, the unit cell microstructure over which we average to obtain effective properties is defined and the resulting slow variation of these averaged properties (composite stiffness or preform permeability) in the macroscopic scale of a fabric composite ply with multiple curvature may be investigated. Further, in sheet forming of dry fabric, the continuously evolving microstructure generates strong deformation dependence of the average fabric response.

T.M. McBride, Texas Instruments Incorporated, Attleboro, MA

Julie Chen, University of Massachusetts-Lowell, Dept. of Mechanical Engineering, Lowell, MA

Lisa Lam, Parametric Technology Corporation, Waltham, MA

INTRODUCTION

Exact analysis of physical processes in heterogeneous domains is rarely achievable. This hindrance is largely due to rapid variations of the constants in the governing equations. If however, we limit our concern to events occuring over length scales much larger than the microstructure, approximations may be sought which serve to describe the average, overall or macroscopic behavior. That is, instead of solving the original equations with rapidly oscillating coefficients, we apply homogenized equations whose coefficients (the effective coefficients) do not oscillate rapidly (see e.g. Bakhvalov and Panasenko, 1989; Sanchez-Palencia and Zaoui, 1987). These averaged equations follow from the idealization that the microstructure is periodic and are defined over a unit cell with periodic boundary conditions.

This idea is central to many existing approaches to mechanical analysis of fibrous composites. In predicting effective mechanical constants of textile composites, additional simplifications are typically made in the solution to the unit cell problem. Often, this involves the assumption that the state of stress or strain is uniform throughout the cell. The predicted average properties under such assumptions have been shown to agree favorably with more detailed finite element simulation and with experimental data (see e.g. Naik, Ifju and Masters, 1994; Marrey and Sankar, 1995; Cox and Flanagan, 1997). These ideas are also used in analysis of mold filling in resin transfer molding operations where application of Darcy's Law for flow through porous media results from homogenization of the Stokes equation (Sanchez-Palencia and Zaoui, 1987). In this case, the averaged coefficients are the effective permeabilities (see e.g. Phelan, 1991; Gong, 1992; Calado and Advani, 1996).

In these various averaging procedures, the effective properties depend on the arrangement of microstructure, which for textile materials can be prohibitively complex. Most of the results found in the literature, for effective thermo-mechanical constants or effective permeabilities, have been limited to textile structures in the planar, undeformed state. One of the commonly cited advantages of using textile reinforcements is related to their drapeability, enabling parts with multiply curved geometries to be formed in resin transfer molding operations. When formed into three-dimensional geometries however, the basic textile structure will vary slowly in the macroscopic scale. This effect on average properties has not received considerable attention in the past.

MICROSTRUCTURE EVOLUTION

Kinematic models, based on the pin-joint or trellis assumptions long used in the apparel industry (Weissenberg, Chadwick and Shorter, 1949), have been applied to describe the fitting of fabric to general surfaces. Applications to composite forming include simple fitting to hemispherical surfaces (Robertson et al, 1981) and CAD based simulation of fitting to more general surfaces (CDT, 1997). The fitting or mapping models are based on the assumption that the material behaves as a pin-jointed net consisting of two families of inextensible cords. Deformation occurs via trellising, rotation of the initially orthogonal cords about pinned joints. Figure 1 shows typical results for a quadrant of a hemispherical mold. The angle θ is referred to as the yarn angle. It measures the acute

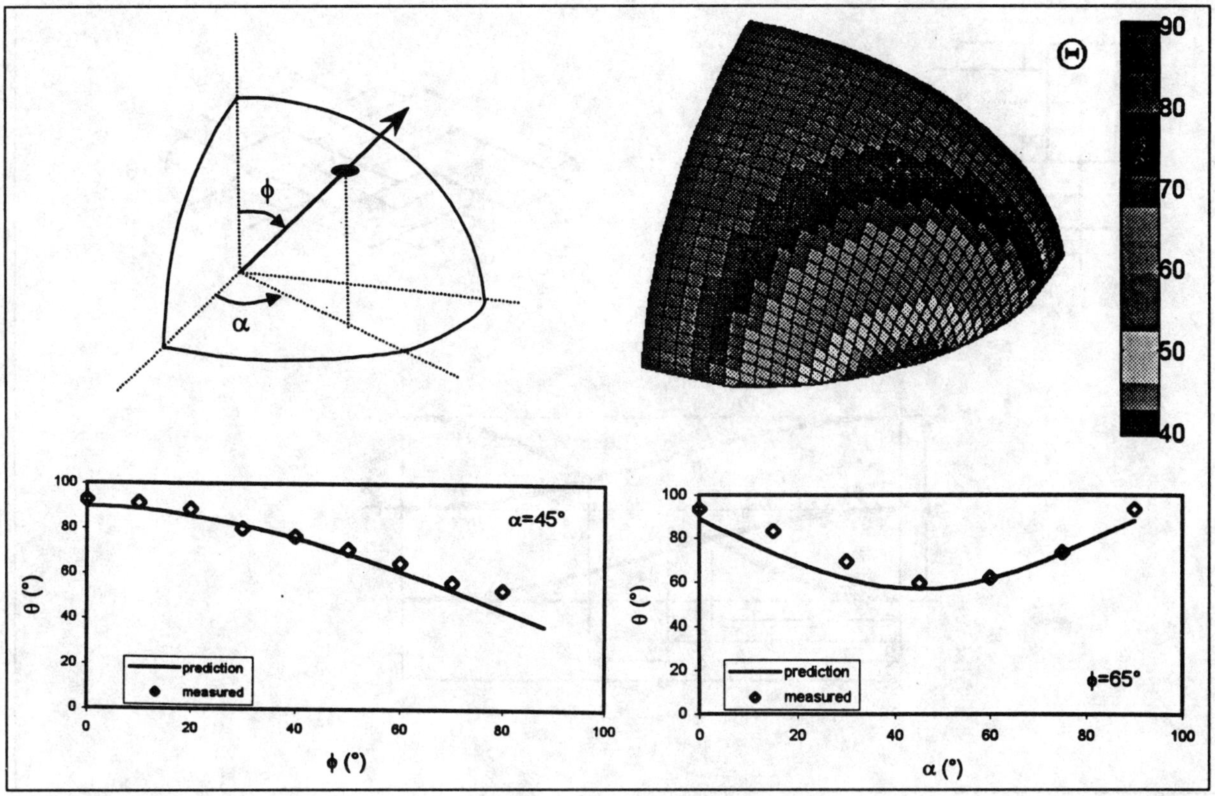

Figure 1: Predicted Yarn Paths and Yarn Angles for Hemisphere

angle between warp and weft yarns and varies widely for even this simple mold geometry. While such models do accurately predict yarn paths for woven materials, they omit description of the three-dimensional nature of the fabric microstructure.

It is proposed that the configuration of the plain weave unit cell may be described by a set of four cosinusoids (see equations 1-4 and Figure 2). These expressions depend on the fabric thickness $2h$, the yarn width tw and the yarn spacing s. Here, we investigate the dependence of each of these dimensions on the yarn angle θ. In doing so, the continuous evolution of microstructure during fabric trellising may be defined. Note that the yarn width tw is related to the dimension w by $tw(\theta) = w(\theta)\sin\theta$.

$$y_1(x,\theta) = \frac{h(\theta)}{2}\left[\cos\frac{\pi x}{s(\theta)} + 1\right] \qquad (0 < x < s(\theta)) \tag{1}$$

$$y_2(x,\theta) = \frac{h(\theta)}{2}\left[\cos\frac{\pi x}{s(\theta)} - 1\right] \qquad (0 < x < s(\theta)) \tag{2}$$

$$y_3(x,\theta) = -h(\theta)\cos\left[\frac{\pi x}{\psi(\theta)}\right] \qquad (0 < x < \frac{w(\theta)}{2}) \tag{3}$$

$$y_4(x,\theta) = -h(\theta)\cos\left[\frac{\pi(x-(s(\theta)-\psi(\theta)))}{\psi(\theta)}\right] \qquad (s(\theta) - \frac{w(\theta)}{2} < x < s(\theta)) \tag{4}$$

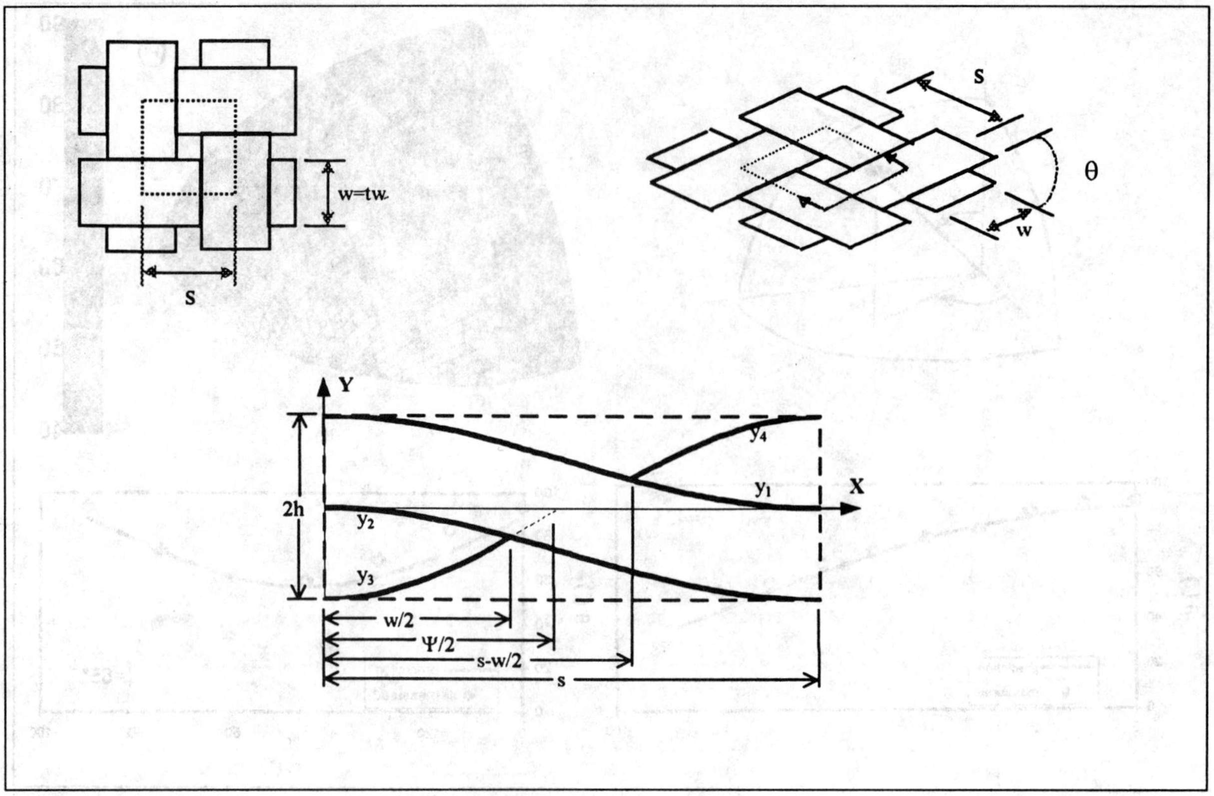

Figure 2: Plain Weave Unit Cell Geometry

ψ is defined as the half period of y_3 so that equating $y_3(x = w(\theta)/2)$ with $y_2(x = w(\theta)/2)$ yields:

$$\psi(\theta) = \frac{\pi w(\theta)}{2 \arccos \left[\sin^2 \left(\frac{\pi w(\theta)}{4s(\theta)} \right) \right]} \tag{5}$$

A set of balanced plain weave fabrics were extended in the bias direction from their initial configuration in which warp and weft yarns are orthogonal, and the dimensional changes of the unit cell microstructure were observed (McBride and Chen, 1997). Consistent with the pin-joint assumptions, the yarn spacing was found to be approximately independent of the yarn angle ($s(\theta) = s_o$). The subscript o refers to the initial fabric configuration for which $\theta = \pi/2$. Further, the fabric thickness did not change as the fabric trellised ($h(\theta) = h_o$). Recalling that we are focusing on dry fabric, this latter result should not be applied to coated, or preimpregnated fabrics, where the incompressibility of the matrix material causes thickness changes with trellising. The yarn width, however, was found to decrease substantially with trellis deformation. The empirical expression in equation 6 was developed to accurately describe the dependence of yarn width on the yarn angle as shown in Figure 3. The term $(s_o - w_o)/\psi_o$ in the exponent may be considered as a measure of fabric tightness and is nonnegative. For full cover fabrics, with no gaps between parallel yarns, this term is zero. For looser fabrics, the value of

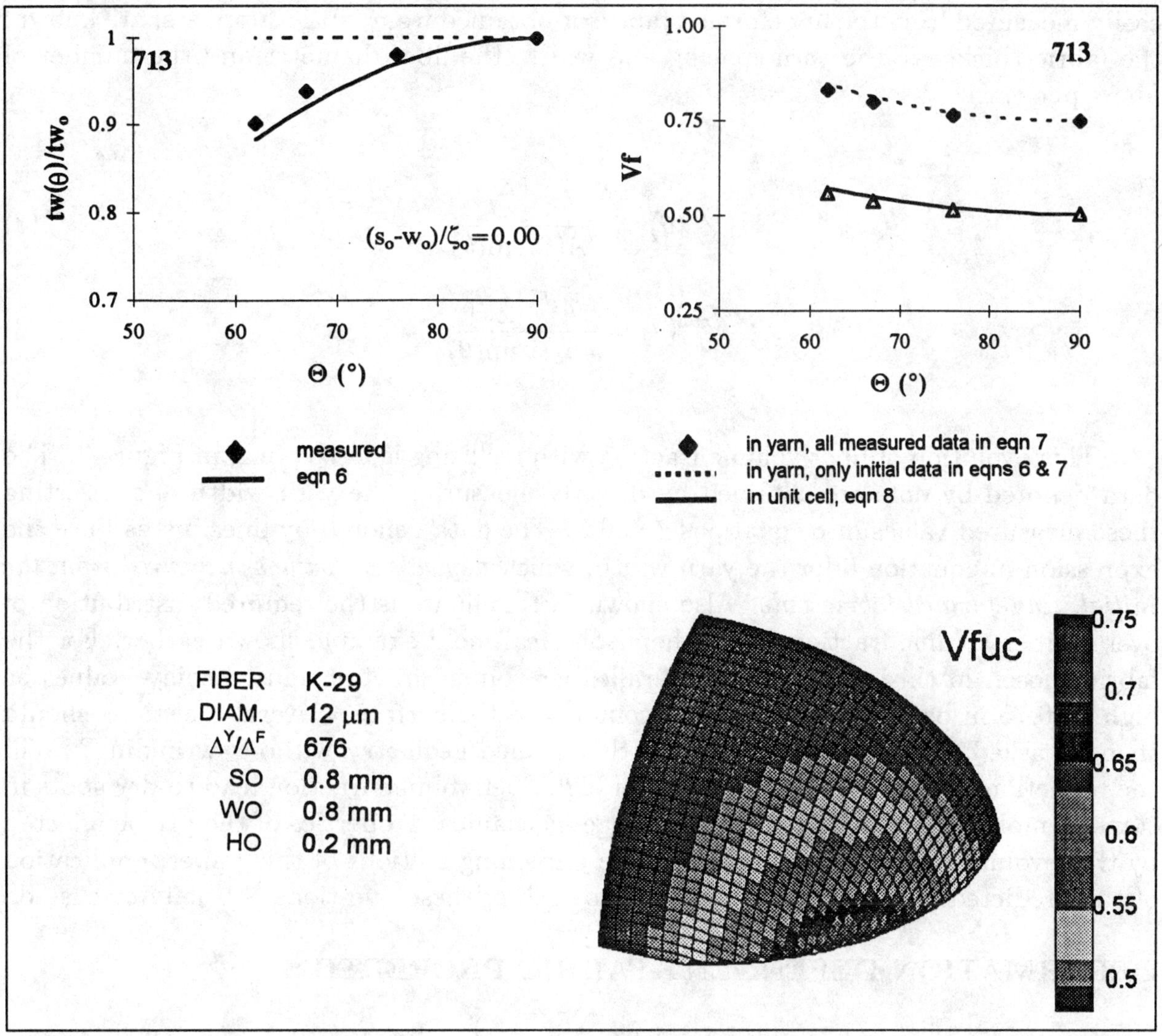

Figure 3: Evolution of Microstructure–Yarn Width and Fiber Volume Fractions

this term will be larger, but remains much less than one for fabrics of interest in textile composites, where high fiber volume fractions are desired. This expression has the advantage of being defined in terms of initial fabric parameters only.

$$tw(\theta) = tw_o(\sin(\theta))^{1 - \frac{s_o - w_o}{\psi_o}} \qquad (6)$$

We define $2A(\theta)$ as the area exposed by cutting a warp yarn parallel to a weft yarn or vice versa (i.e. $A(\theta) = \int_0^{w(\theta)/2}[y_3(x) - y_2(x)]dx$). If the fiber diameter (d), and the fiber (Δ^f) and yarn (Δ^y) denier (or alternatively, the number of fibers per yarn) are known, the fiber volume fractions in the yarns and the overall fiber volume fraction of the unit cell may then be defined as in equations 7 and 8. To summarize then, the evolution of microstructure may be described provided that the following parameters, which may be easily measured from the undeformed fabric or obtained from fabric suppliers, are known: the fabric thickness, the yarn spacing and width, the fiber diameter and the number of fibers per yarn.

$$V_f^y(\theta) = \frac{\pi d^2 \left[\frac{\Delta^y}{\Delta^f}\right]}{8A(\theta)\sin(\theta)} \tag{7}$$

$$V_f^{uc}(\theta) = \frac{\pi d^2 \left[\frac{\Delta^y}{\Delta^f}\right]}{4h_o s_o \sin(\theta)} \tag{8}$$

This evolution of fiber volume fraction with trellising is also shown in Figure 3. The data denoted by dots are obtained by directly measuring the yarn width and inserting these measured values into equations 7 and 8. The data denoted by lines instead use the expression in equation 6 for the yarn width, which *depends on values measured from the initial, undeformed fabric only.* Also shown in this figure is the required distribution of overall fiber volume fraction for the hemispherical mold example shown earlier. For the fabric chosen in this example, the overall fiber volume fraction must achieve values as high as 0.75 in order for the fabric to conform to the surface. Several questions should then be asked: Can this fabric achieve this formed geometry without wrinkling? Will the reduction of permeability in regions of high fiber volume fraction lead to dry spots in transfer molded parts? How will the average mechanical properties of the ply be affected by this evolution of microstructure? In the remaining sections of this paper, application of the predicted microstructure evolution to each of these questions is briefly discussed.

DEFORMATION DEPENDENT FABRIC PROPERTIES

In the transverse direction, aligned fiber yarn bundles have been modeled as nonlinear elastic springs with finite compressibility. Cai and Gutowski give functional expressions for average transversely isotropic yarn compliances in terms of the current fiber volume fraction within the yarn (Cai and Gutowski, 1992). While their work was motivated by the deformation behavior of lubricated fiber bundles, we have extended it in describing the behavior of dry aligned fiber yarns. Inelastic effects due to fiber-fiber friction and fiber sliding are not characterized and have been neglected. From the results described in the above section, the current fiber volume fraction in the yarns may be related to the level of trellis deformation in the fabric, as measured by the yarn angle θ. At a given level of fabric deformation then, the distribution of material properties within the fabric is known. One may then average over a unit cell of this microstructure to determine the current effective fabric properties.

To do this, we seek the incremental constitutive relation $< \Delta\sigma > = C^{\star}_{eff} < \Delta\epsilon >$ describing the average nonlinear response of the plain weave fabric. $C^{\star}_{eff}$ is the current effective constitutive tensor of the fabric. At any increment of fabric deformation, the current yarn stiffness and the current geometry of the yarns within the unit cell are updated. $C^{\star}_{eff}$ is then obtained by orientation averaging (see e.g. Naik, Ifju and Masters,

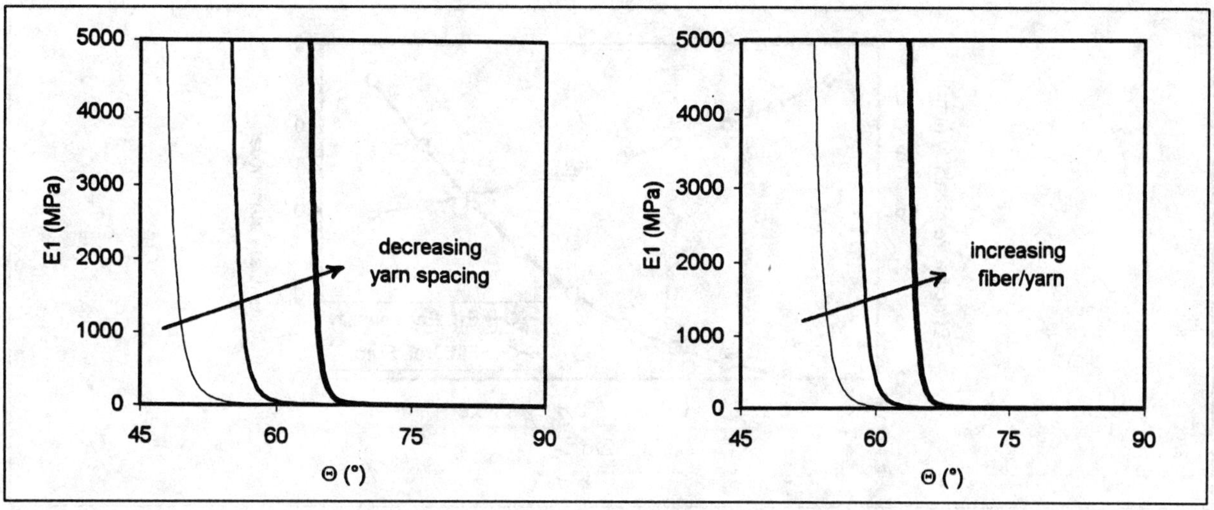

Figure 4: Rapid Stiffening in Trellis Deformation–Effect of Initial Fabric Architecture

1994; Cox and Flanagan, 1997) over this current unit cell configuration.

This approach has been used to investigate limits to achievable trellis deformation in plain weave fabrics. Analagous to forming limits for sheet metal, fabrics display limiting forming behavior through wrinkling. This model provides deformation dependent average fabric properties which may be used to model large deformation fabric forming operations. Because the model incorporates features of the fabric architecture, it may be used as a design tool to anticipate defects such as wrinkling and therefore to guide fabric selection. Typical results for the bias modulus (in direction bisecting the yarn angle) as a function of trellis deformation are shown in Figure 4. As examples, parametric variation of the yarn spacing and number of fibers per yarn is seen to affect the location of rapidly stiffening behavior.

SLOW VARIATIONS IN EFFECTIVE PREFORM PERMEABILITY

In regions of the preform for which the local fiber volume fractions have evolved to high levels as a result of forming induced deformations, decreased permeabilities may result in dry spots. Design tools which are capable of anticipating such defects can serve to greatly reduce development costs (see e.g. Baichen, Bickerton and Advani, 1996). The permeability depends strongly on the local arrangement of fiber, so that in complex geometry preforms, the effect of forming induced microstructure evolution should be significant. To illustrate this point, consider transverse flow through a square array of solid, aligned cylinders (Phelan, 1991). Figure 5 shows the relationship between fiber spacing, fiber volume fraction and effective transverse permeability. The capability of this work in describing the distribution of fiber volume fraction throughout multiply curved preforms can be used to characterize the correponding slow variation in effective permeability.

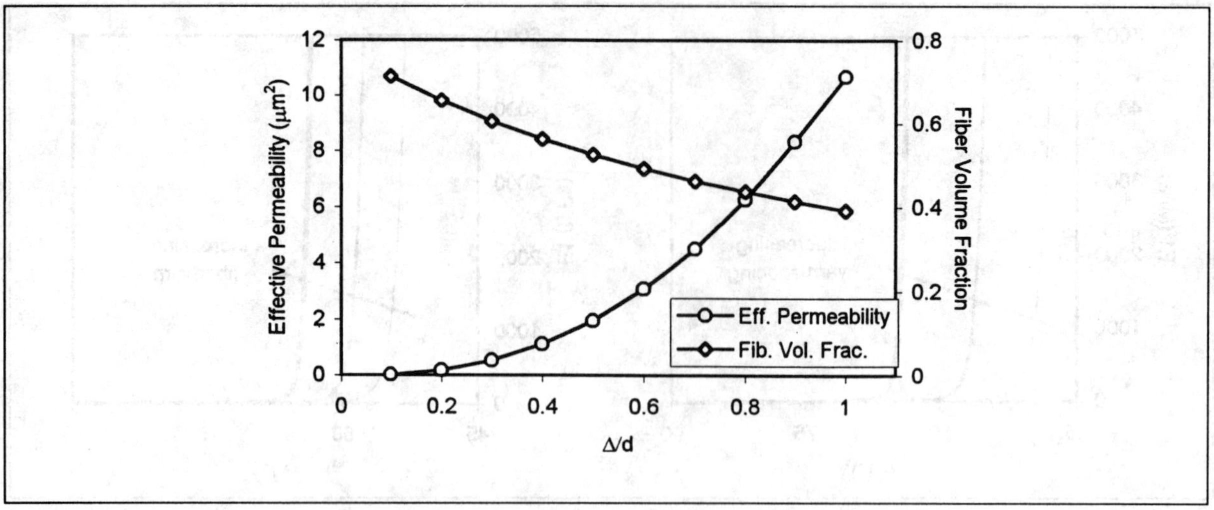

Figure 5: Slow Variation in Average Permeability

SLOW VARIATIONS IN EFFECTIVE COMPOSITE PROPERTIES

The effect of forming induced microstructure evolution should also be considered when determining effective thermo-elastic properties of fabric composite lamina. Consider a plain weave composite unit cell oriented such that the 1 and 2 axes lie in the plane of the fabric and the 1 axis bisects the acute angle between yarns. The orientation averaging procedure referred to earlier is used to predict the effective ply properties (see e.g. Naik, Ifju and Masters, 1994; Cox and Flanagan, 1997). The unit cell consists of impregnated yarn regions (transversely isotropic with axial direction tangent to yarn centerline) and neat resin regions (isotropic). The fiber and resin material properties shown in Table 1 were used. Also shown are the expressions used for the transversely isotropic properties of the impregnated yarn/resin regions. In Figure 6, two cases are given for the dependence of average fabric properties on the yarn angle θ. Each point corresponds to a particular unit cell configuration as measured by the yarn angle. The point at $\theta = \pi/2$ corresponds to a fabric composite in which the yarns are in their orthogonal, undeformed state. Points for which $\theta < \pi/2$ represent composites in which the fabric has been trellised prior to the resin impregnation. In one case (filled symbols), changes in yarn cross section are neglected. In the other case (open symbols), the evolution of yarn cross section as described earlier is included. It is seen that for smaller yarn angles, the effect of local changes in yarn cross section (and therefore in yarn fiber volume fraction) can be significant. In the case for which changes in yarn cross section are included, the yarn fiber volume fraction increased from 0.64 at $\theta = 90^o$ to 0.9 at $\theta = 40^o$. In both cases, the overall fiber volume fraction increases from 0.48 at $\theta = 90^o$ to 0.75 at $\theta = 40^o$. This latter effect explains the increase in $E2$ for smaller yarn angles.

TABLE 1

Resin Properties:	Fiber Properties:
$E_r = 3.45$ GPa $G_r = 1.28$ GPa	$E_{lf} = 191.9$ GPa $E_{tf} = 18.7$ GPa $G_{lf} = 11.3$ GPa $G_{tf} = 7.4$ GPa $\nu_{ltf} = 0.19$

Trans. Iso. Impregnated Yarn Properties	
$E_l = E_{lf}V_f^y + E_r(1 - V_f^y)$	$E_t = \frac{E_r}{1-\sqrt{V_f^y}(1-\frac{E_r}{E_{tf}})}$
$G_l = \frac{G_r}{1-\sqrt{V_f^y}(1-\frac{G_r}{G_{lf}})}$	$G_t = \frac{G_r}{1-\sqrt{V_f^y}(1-\frac{G_r}{G_{tf}})}$
$\nu_{lt} = \nu_{ltf}V_f^y + \nu_r(1 - V_f^y)$	$\nu_{tt} = \frac{E_t}{2G_t} - 1$

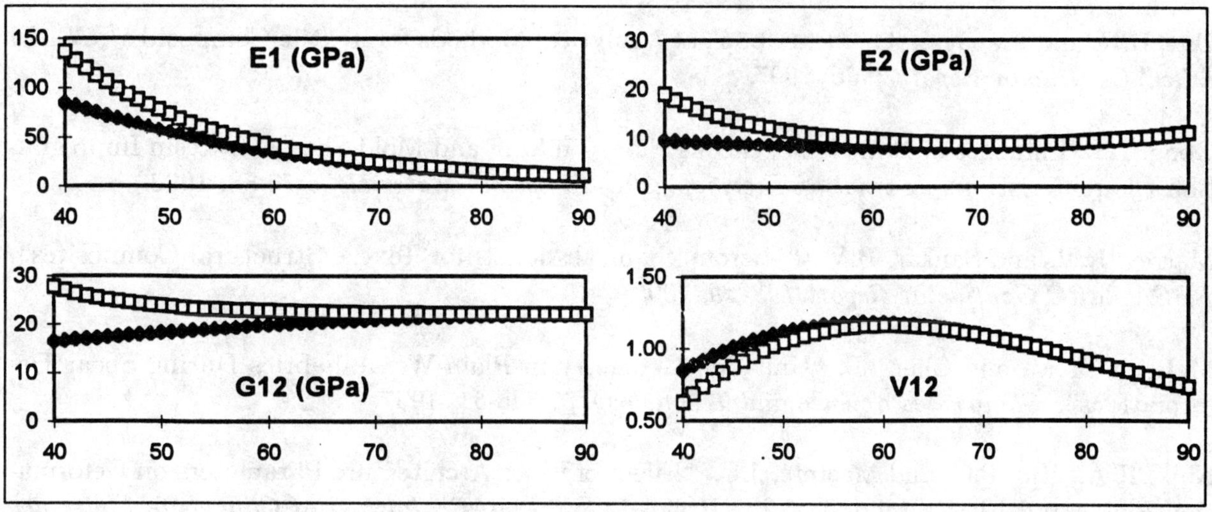

Figure 6: Slow Variation in Average Fabric Composite Ply Properties

SUMMARY

In forming fabric composites with multiple curvature, the evolution of fabric microstructure is often significant and may be predicted. Its effect is evident in the strong deformation dependence of the average fabric response to trellising. The predicted evolution of microstructure may also be used in defining local problems for determination of effective preform permeabilities or effective mechanical properties of formed fabric composite plies. While this investigation is limited to plain weave fabrics, it illustrates an additional complexity in analysis of textile composite forming processes which may be significant.

REFERENCES

Baichen, L., Bickerton, S. and Advani, S.G., "Modeling and Simulation of Resin Transfer Molding (RTM)-Gate Control, Venting and Dry Spot Prediction", *Composites Part A: Applied Science and Manufacturing*, 27A(2):135-41, 1996.

Bakhvalov, N. and Panasenko, G., "Homogenisation: Averaging Processes in Periodic Media", Kluwer, 1989.

Cai, Z. and Gutowski, T.G., "The 3-D Deformation Behavior of a Lubricated Fiber Bundle", *Journal of Composite Materials*, 26(8):1207-37, 1992.

Calado, V.M.A. and Advani, S.G., "Effective Average Permeability of Multi-Layer Preforms in Resin Transfer Molding", *Composites Science and Technology*, 56:519-31, 1996.

Composite Design Technologies (CDT), *FiberSim, version 2.0, release date January 22, 1997*, Waltham, MA.

Cox, B.N. and Flanagan, G., "Handbook of Analytical Methods for Textile Composites", *NASA LaRC Contractor Report 4750*, 1997.

Gong, H., "Influence of Non-Homogeneous Fiber Preform and Mold Shape on Resin Impregnation Characteristics", *Journal of Thermoplastic Composite Materials*, 5:76-88, 1992.

Marrey, R.V. and Sankar, B.V., "Micromechanical Models for Textile Structural Composites", *NASA LaRC Contractor Report 198229*, 1995.

McBride, T.M. and Chen, J., "Unit-Cell Geometry in Plain-Weave Fabrics During Shear Deformations", *Composites Science and Technology*, 57:345-51, 1997.

Naik, R.A., Ifju, P.G. and Masters, J.E., "Effect of Fiber Architecture Parameters on Deformation Fields and Elastic Moduli of 2-D Braided Composites", *Journal of Composite Materials*, 28(7):656-81, 1994.

Phelan, F.R., "Modeling of Microscale Flow in Fibrous Porous Media", *Advanced Composite Materials: New Developments and Applications Conference Proceedings*, 175-85, 1991.

Robertson, R.E., Hsiue, E.S., Sickafus, E.N. and Yeh, G.S.Y., "Fiber Rearrangements During the Molding of Continuous Fiber Composites: I. Flat Cloth to a Hemisphere", *Polymer Composites*, 2(3):126-31, 1981.

Sanchez-Palencia, E. and Zaoui, A., editors, "Lecture Notes in Physics No. 272, Homogenization Techniques for Composite Media", Springer-Verlag, 1987.

Weissenberg, K., Chadwick, G.E. and Shorter, S.A., "A Trellis Model for the Application and Study of Simple Pulls in Textile Materials", *Journal of the Textile Institute*, 40(2):108-60, 1949.

Nondestructive Evaluation

High Temperature Moiré Interferometry and Its Applications to Composite Laminates

X. HE, D. ZOU AND S. LIU

ABSTRACT

Laminated composites have been widely used in both military and commercial applications. For instance, Boeing 777 has used toughened composites for its primary structures. Recently, high density and low cost electronics required the introduction of organic laminates (printed wiring boards) which consist of fiber reinforced layers, conductor layers, power planes, and thermal conducting vias, forming some kind of hybrid composites.

One critical problem for all these complex composite systems is due to the significant thermal mismatch between various materials, which produces significant thermal residual stresses within layers and along interfaces. There is a great need for measuring the thermal stress induced during the fabrication and in operation.

However, there exists difficulties in measuring the stresses involved during all these processes. The first problem is that high temperature is needed due to the enhancing curing temperature of some resins. Secondly, the thin layers exist in these special laminates and deformation is so small that the deformation cannot be resolved. Thirdly, the real time capability is needed to resolve the refined local/micro deformation field, while being able to monitoring the global/macro deformation field. The latter capability is useful for many applications, such as crack tip deformation field, the interface and interphase fields, and advanced smart materials.

In this study, several advances in high temperature moiré interferometry and a multi-function macro-micro moiré interferometer (4M) are reported. High temperature grating replication technique was developed up to 200°C, the curved surface deformed grating replication technique was realized, high temperature resistant gratings which can work at the temperature of 1000°C was developed, which can be useful for high temperature composites. Nano-scale deformation resolution is reported
to be able to resolve nano-scale deformation field by a PZT driven phase shifting system.

The above techniques were successfully applied in the study of several structures. For instance, the technique has been used in the study of the interlaminar thermal shear strains of C/BMI composite laminate with complex-ply. The experimental results showed that unlike the case of the cross-ply laminate, the interlaminar shear strains of C/BMI complex-ply laminate are not only at the corner of the free boundary region but also at the middle of the boundary region, and the largest interlaminar shear strains are in the +45/-45 interface.

The standard moiré interferometry method was applied in the thermo-mechanical analysis of microelectronic packages, with both high temperature grating replication and real time measurement

---

Xiaoyuan He, Daqing Zou, and Sheng Liu, Dept.of Mechanical Engineering, Wayne State University, Detroit, MI 48202

techniques. The grating replication technique of moiré interferometry at the temperature up to 200 °C was presented for the coefficient of thermal expansion (CTE) measurement of metal ball grid array (BGA) package. Considering the complicated and large thermal deformation of electronic packages, gratings of 600 l/mm was used instead of traditional gratings of 1200 l/mm in the thermal displacement measurement of the flip-chip (BGA) package cross-section. Thermal deformation of the packaging systems was measured and the effects of the encapsulation, soldering, and geometry on the deformation were investigated. The real time measurement of moiré interferometry was also applied with the aid of a vacuum thermal chamber and a computer imaging analysis system. Thermal deformation of electronic package as a function of temperature was analyzed. CTE of a molding compound, a highly filled system, was measured from room temperature to 260 °C.

Finally, 4M has been used for deformation analysis of a corner and interface of a plastic IC packaging. In addition, a new generation flip-chip bonded onto the hybrid laminate (PWB) was analyzed. Global deformation (warpage) of the device and local detailed deformation around the tiny solder balls, woven layers, and interfaces are reported.

Thermal-Wave Imaging for NDE of Composites

X. HAN, L. D. FAVRO AND R. L. THOMAS

ABSTRACT

We describe the application of infrared (IR) thermal wave imaging for the nondestructive evaluation (NDE) of composites. The technique uses pulsed surface heating (with flashlamps) and fast, synchronous IR imaging of the surface temperature to form images of subsurface defects in composite materials. We present experimental examples of the detection of typical defects.

1. INTRODUCTION

Pulse-echo thermal wave imaging is accomplished using high-power flash lamps as a pulsed heat source, an infrared (IR) video camera as a detector, and image processing hardware and software to display the results. A schematic diagram is shown in Fig. 1. When the energy from the flashlamps is absorbed by the sample, a thermal wave pulse is launched from its surface, and propagates into the material as a time-dependent temperature variation (the thermal wave). If the thermal wave encounters a subsurface defect, it gets reflected and propagates back to the surface, thereby changing the rate of cooling of the surface. The largest change in the cooling rate occurs with a time delay determined by the thermal wave transit time from the surface to the defect and back. This return of the thermal wave signal is monitored by means of the IR video camera, which records the time-dependent surface temperature distribution. The images from the camera are processed in the computer which controls the system. In this paper we present examples of the application of thermal wave imaging of defects in composites.

2. THERMAL WAVE IMAGES OF COMPOSITES

In Figs. 2a - 2d, we show an example of thermal wave imaging of impact damage on a multi-ply graphite-fiber-reinforced epoxy laminate. The impact point was somewhat to the left of the center of the image. The four thermal wave images are at successively later times

---

Xiaoyan Han, L.D. Favro, and R.L. Thomas
Department of Physics and Institute for Manufacturing Research
Wayne State University, Detroit, MI 48202, USA

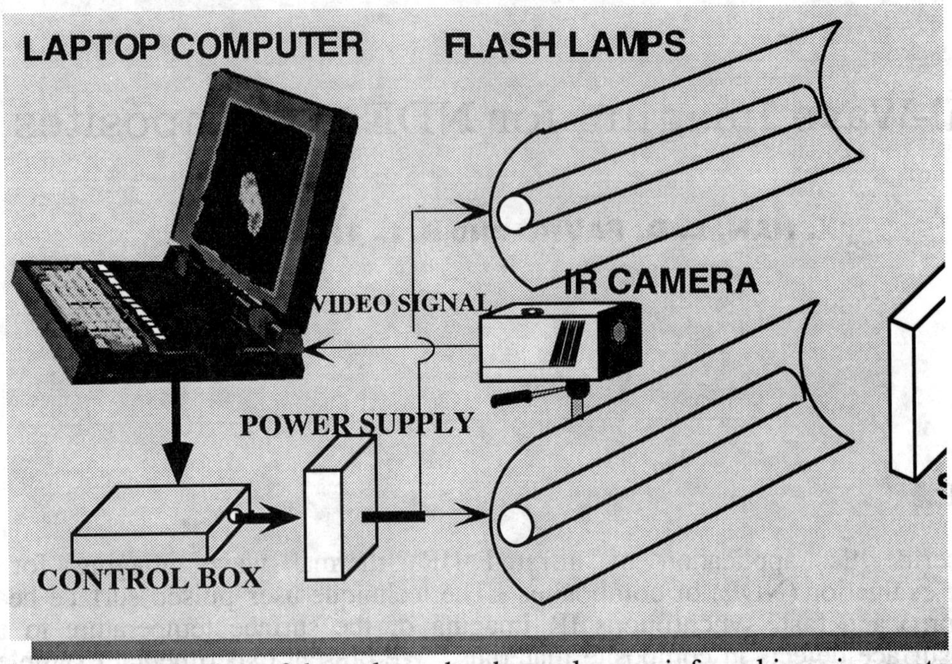

Fig. 1. Schematic diagram of the pulse-echo thermal wave infrared imaging system.

after the flash. First, in the image in Fig. 2a (shortest thermal wave propagation time), among the prominent features are the graphite fiber tows in the top plies. It can be seen that the tows are nonuniformly distributed, and are oriented at 45° angles with respect to one another. In addition, a bright feature is seen, oriented along one of the tow directions, radiating vertically down towards the lower left of the image from the impact point. This feature is a delamination plies which are close to the top surface of the sample. Also, there is an indication in Fig. 1a of a second delamination, one ply deeper, which is rotated 45° with respect to the first. In the next image, Fig. 2b, a third and a fourth delamination are visible, symmetrically oriented about the impact point, corresponding to delaminations between successively deeper pairs of plies, and rotated sequentially by 45° . In Fig. 2c, the pattern has blurred, and in the latest image of this set, Fig. 2d, a still deeper delamination appears in the image, extending vertically upward to the upper left corner. This delamination is between plies which are close to the opposite face of the laminated plate. These images were all obtained from a single flash, and were selected from a larger set of thirteen images. They were chosen to illustrate the ability of thermal wave imaging to determine not only the presence of delaminations, but also to give depth information. As can be seen from Figs. 2a-2d, the images become more blurred as the thermal waves propagate more deeply into the material. This is the result of lateral heat diffusion, and is a well-known phenomenon in thermal wave imaging.

Another example of thermal wave imaging of defects in composites is presented in Fig. 3. In this example, the material is a boron-fiber-reinforced multi-ply laminate, bonded to an aluminum aircraft skin. In this sequence of thermal wave images, taken at progressively later gate times (in frames: 1f = 16.7 msec), several intentionally placed subsurface delaminations are evident. For example, there is a near subsurface feature, located in the left center of the images, which peaks in contrast at approximately at gate 5 (115 frames). Deeper features in the bottom and the center of the images peak at much later times. Once again, this illustrates the ability of thermal wave imaging to discriminate among defects at different depths.

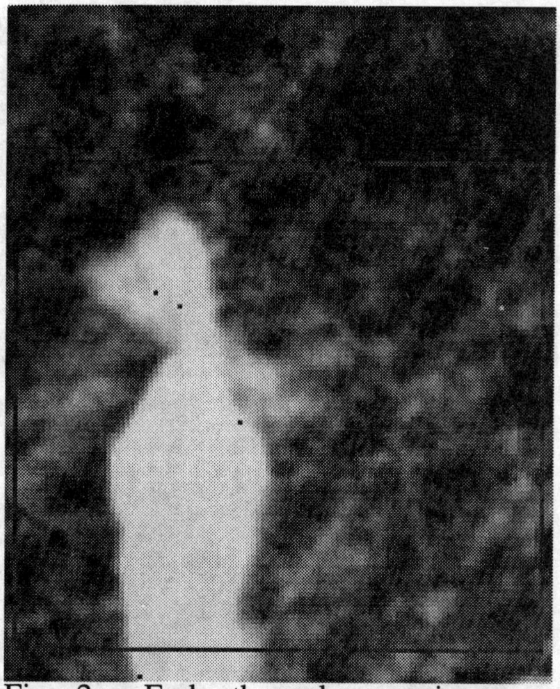

Fig. 2a Early thermal wave image, showing the graphite fiber tows, and two subsurface delaminations, one of which extends vertically downward to the edge of the specimen

Fig. 2b Later thermal wave image, showing the defects from Fig. 2a, together with two additional delaminations radiating from the impact point at 45° intervals in the form of a "flower", from pressively deeper defects.

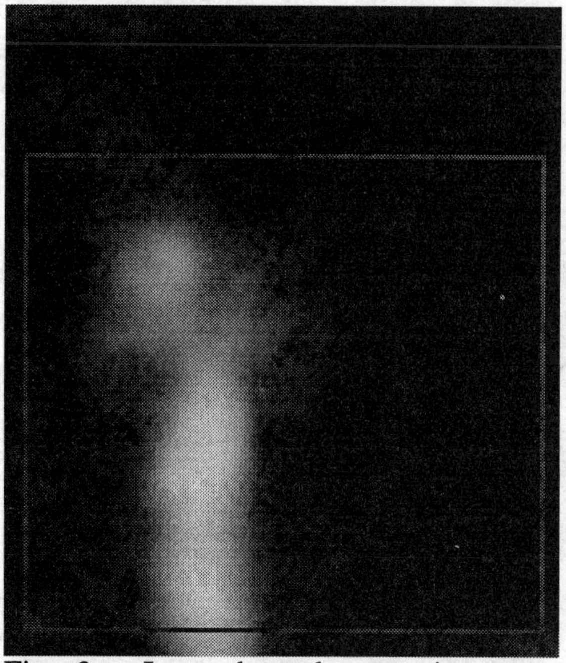

Fig. 2c Later thermal wave image, showing the defects from Figs. 2a and 2b, with blurring from lateral heat diffusion.

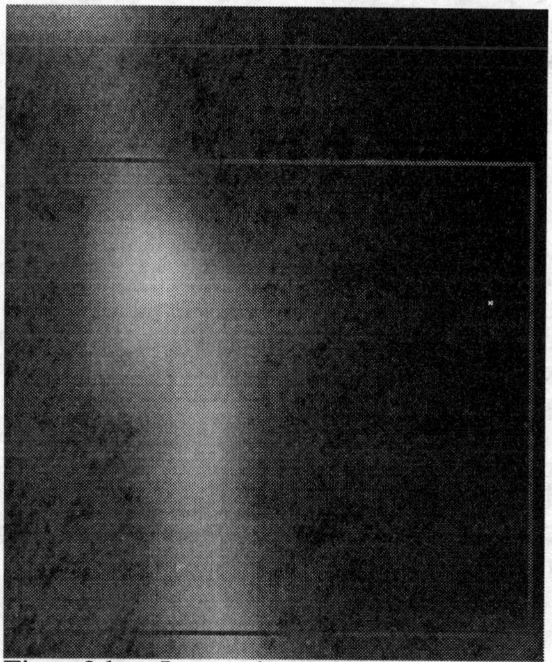

Fig. 2d Later thermal wave image, showing the defects from Figs. 2a, 2b and 2c, plus a delamination which runs to the upper left edge of the sample, located close to its opposite face.

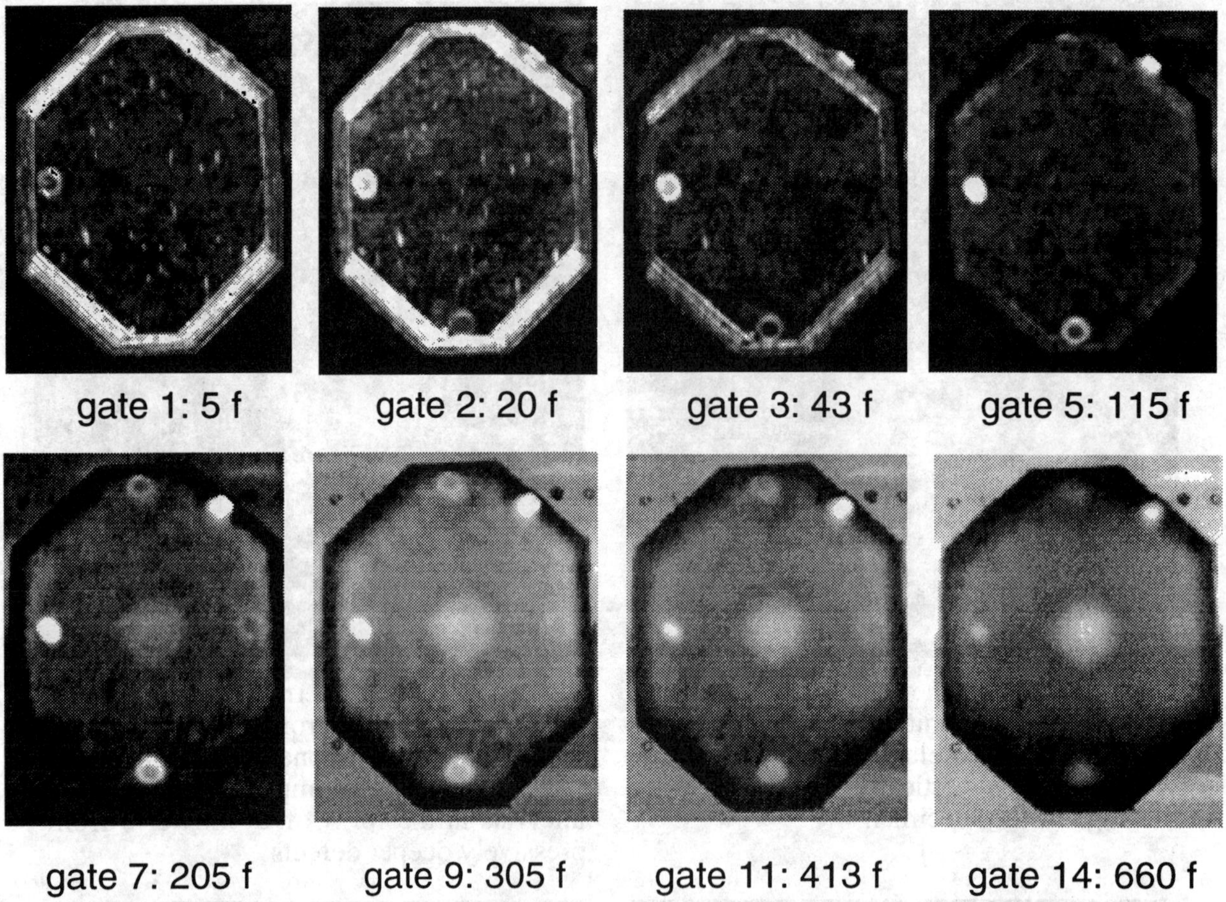

Fig. 3 Sequence of thermal wave images at progressively later gate times (in frames: 1f = 16.7 msec). The near subsurface feature in the left center of the image peaks approximately at gate 5, whereas the deeper features in the bottom and the center of the images peak at much later times.

Figure 4 is a thermal wave image of a composite test sample, with a graphite-fiber-reinforced epoxy skin over a honeycomb core. The five more or less circular shaped bright spots in the upper left of the image are the result of intentional implants to simulate skin to core disbonds. The remaining irregular bright areas are the result of unintentional defects. The honeycomb core structure is clearly imaged through the composite skin. Another important type of subsurface defect in composite materials used on aircraft is that of fluid intrusion. For example, if water manages to penetrate the composite, the weight of the structure can increase significantly, and also, the inevitable freezing temperatures at high altitude can lead to damage. This fluid intrusion is readily seen by thermal wave imaging, as shown in Fig. 5. The dark areas in this figure are areas in which the honeycomb core contains water.

3. CONCLUSIONS

We have demonstrated that thermal wave imaging is an effective nondestructive inspection technique for the detection of subsurface defects in a variety of composite structures. The method is contactless, rapid, and has the capability of discriminating among defects located at different depths beneath the surface of the structure.

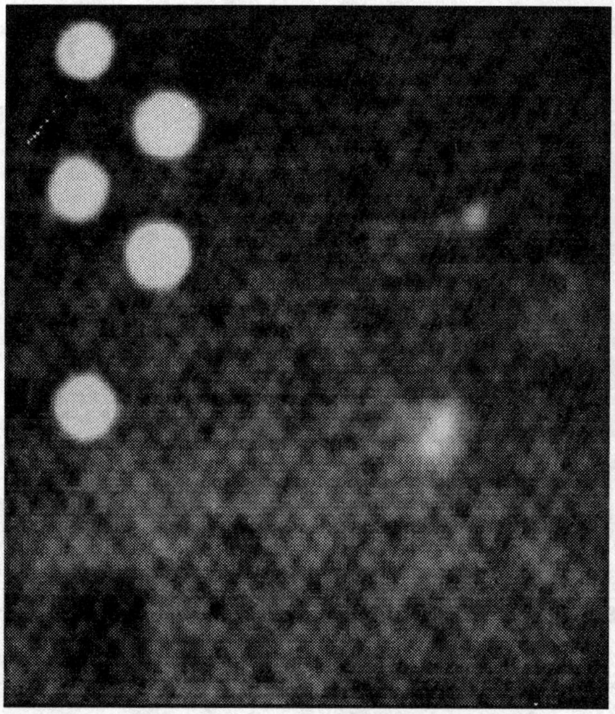

Fig. 4 Thermal wave image of a honeycomb core under a composite skin, showing various defects.

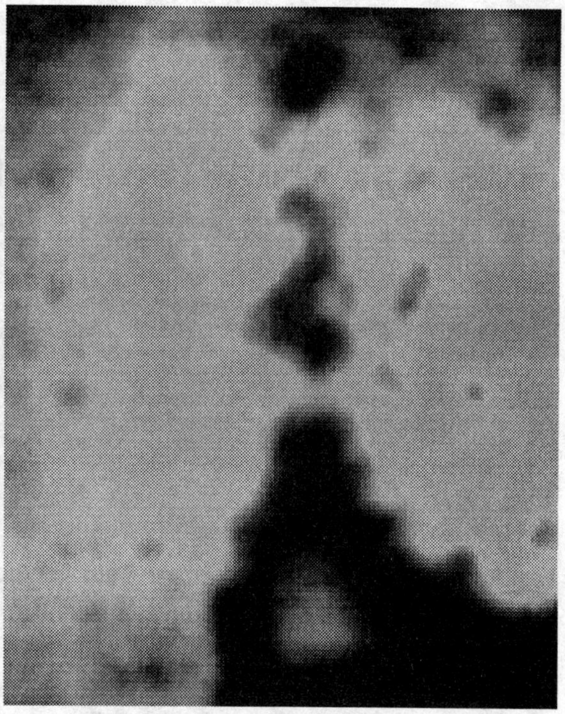

Fig. 5 Thermal wave image of water intrusion (darker regions) in a composite aircraft structure.

4. ACKNOWLEDGMENTS

This work is sponsored by the FAA-Center for Aviation Systems Reliability, operated by Iowa State University and supported by the Federal Aviation Administration Technical Center in Atlantic City, New Jersey, under grant number 94-G-011, by AFOSR under Grant No. F49620-93-1-0428, and by the Institute for Manufacturing Research.

Advanced Ultrasonic Testing of Joints in Multi-Layered Composite Materials with Irregular Structures

E. G. NESVIJSKI

ABSTRACT

The last several years have been critical for composite materials as along with their advantagesand growing field of applications, there are a lot of practical evidence of their "illness", failure, and construction damage. That is why future application of these materials is closely connected with optimization of their structures and creation of special means of monitoring of their behavior for forecast of safety and durability. One of the new and promising methods of nondestructive testing is ultrasonic harmonic waves method. This method proved to be very sensitive to structural changes in materials with regular structures. Harmonic waves method uses wave phase characteristics which can be measured with high accurate. Possibility of application of this method for composite materials with irregular structures as well as joints of materials with distinctive properties is the aim of this work as it gives opportunity to characterize a wider class of composites such as ceramics, concrete, and metal-ceramic-plastic composites. Stochastic approaches to the analysis of distribution of mechanical properties are used. Results of mathematical simulation of ultrasonic testing of joints in multi layered composites are demonstrated. Analysis of sensitivity of the method is made and some practical applications as well as schematic design of ultrasonic instrumentation are presented in this work.

INTRODUCTION

The most advanced fields of engineering and technology where composites are playing today a decisive role are shipbuilding, aviation and space. The main types of composites which are used there are laminates, matrix composites, and "sandwich" composites. The last type are presented by synthetic resin reinforced by fibres. As rule such structures may be considered as regular structures. During their service time significant irregularities are brought about due to

---

Edouard G. Nesvijski, Federal University of Santa Maria, Campus Universitario, Santa Maria, CEP 97119, RS, Brazil

variety of impacts, such as multi directional mechanical loads, temperature affects, osmos as well as due to violations of technological manufacturing process. All these results in the structural irregularities of composite material. That is why regular monitoring of their state is an important problem of prediction and prevention of failure. In a number of case it is necessary to have a reliable instrumentation for evaluation of emergency situation and recommendation for maintenance and repairing. At present NDT methods are providing new possibilities to solve these problems. A new advanced ultrasonic method for testing of composites is presented in this work.

HARMONIC WAVES METHOD

Usage of harmonic ultrasonic waves with fixed frequency is a basis of this method. The main informative parameter here is the phase characteristics of the waves. Thus, harmonic wave may be described as the following expression for amplitude $\mathbf{U(t) = A\cos(2\pi ft + \varphi)}$, where **A** is amplitude, **f** - frequency, φ - phases of the wave (Firestone, 1945). It is proposed to use this method for monitoring of multi layered composite materials attracting physical process of harmonic wave difraction on the layers with different acoustic properties (Krautkramer and Krautkramer). Integration of phase shift $\Delta\varphi$ may be observed during propagation of the wave through multi layered structure. Special models may be taken for description of the physical nature of this method (Baranov, 1990).

MODEL FOR MULTI LAYERED COMPOSITE WITH REGULAR STRUCTURE

We may represent multi layered composite as a two-component medium with senquentionally interlacing layers. Propagation of acoustic waves in such a medium can be considered on the base of the following scheme (Nesvijski, 1982)

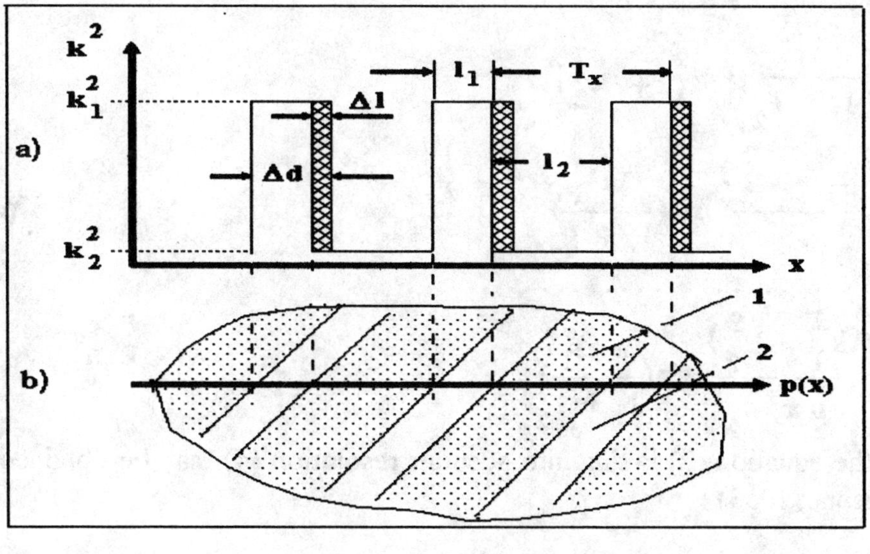

Figure 1.
a - analytical model for calculation of the acoustic field in the two-component medium; b - physical model for multi layered composite in two-component structure: 1 - matrix, 2 - filler.

There are repetition of two components with period **T** on coordinate **x.** Their elastic properties are represented by square wave number $\mathbf{k_{1,2}}$, where $\mathbf{i}$ = 1,2. The discrete structure of the medium creates difractional-interferential acoustic field due to reflection and difraction of harmonic waves on the borders between layers (Brekhovskih, 1989). Mathematical model of waves propagation in such a medium may be described by wave equation with periodical sequence for the wave number $\mathbf{k_{1,2}}$

$$\frac{\partial^2 p}{\partial x^2} + k_i^2 (T_x) p = 0 \qquad (1)$$

where **p** - pressure in the harmonic wave; $\mathbf{T_x}$ - period of layers interlace. Equation (1) may be transformed into Hill equation (Bondarenko, 1936) with periodical coefficient $\theta(\mathbf{T_x})$

$$\frac{\partial^2 p}{\partial x^2} + \theta(T_x) p = 0 \qquad (2)$$

Resolute $\theta(\mathbf{T_x})$ into Fourier series for the case presented in the Figure 1

$$\theta_n(T_x) = \{\frac{l_1}{l_1 + l_2}\}[k_1^2 + k_2^2] + \Sigma \frac{2(k_1^2 - k_2^2)}{n\pi} \sin(\frac{n\pi l_1}{l_1 + l_2}) \cos(\frac{2xn\pi}{l_1 + l_2}) \qquad (3)$$

where $\theta(\mathbf{T_x})$ - amplitude of the series members

$$\left.\begin{aligned} \theta_0(T_x) &= \{\frac{l_1}{l_1 + l_2}\}[k_1^2 + k_2^2] \\ \theta_1(T_x) &= \frac{2(k_1^2 - k_2^2)}{\pi} \sin(\frac{\pi l_1}{l_1 + l_2}) \\ &\cdots\cdots\cdots\cdots \\ \theta_n(T_x) &= \frac{2(k_1^2 - k_2^2)}{n\pi} \sin(\frac{n\pi l_1}{l_1 + l_2}) \end{aligned}\right\} \qquad (4)$$

Solution of the equation (2) taking into account resolution (4) may be obtained using Flocke method (Sansone, 1953) :

$$p = \exp(\mu x) \sum b_m \exp(2imx) \quad (5)$$

in the form of infinite series with uncertain coefficients bm. It is possible to use Witteker method (Witteker, 1953) for substitution of (5) into (2) and equation of coefficients of members with exponent for resolution **exp(2iμx)** into series (3) in accordance with the **θ(T_x)** problem condition the following equation system may be received

$$(\mu + 2ni)\, b_m + \sum \theta_n(T_x)\, b_{(m-n)} = 0 \quad (6)$$

The solution of the system (5) as it is known can be transformed into the equation where the roots satisfy the original equation

$$\Delta(0) \sin\left(\frac{\pi \theta_n}{2}\right) = \sin\left(\frac{\pi \mu i}{2}\right) \quad (7)$$

where **Δ(0)** - determinant of the system (5) with μ = 0

$$\Delta(0) = \begin{vmatrix} \cdots & 1 & -\dfrac{\theta_1}{4^2 - \theta_0} & -\dfrac{\theta_2}{4^2 - \theta_0} & -\dfrac{\theta_3}{4^2 - \theta_0} & \cdots \\ \cdots & -\dfrac{\theta_1}{2^2 - \theta_0} & 1 & -\dfrac{\theta_2}{2^2 - \theta_0} & -\dfrac{\theta_3}{2^2 - \theta_0} & \cdots \\ \cdots & -\dfrac{\theta_2}{\theta_0} & -\dfrac{\theta_1}{\theta_0} & 1 & -\dfrac{\theta_1}{\theta_0} & \cdots \end{vmatrix} \quad (8)$$

It is possible to find the approximate value of **Δ(0)** with a limited number of members, for example - three,

$$\Delta(0) = 1 + \frac{\theta_1^2}{\theta_0(4-\theta_0)^2} + \frac{\theta_1^2\theta_2}{\theta_0(4-\theta_0)^2} + \frac{\theta_1\theta_0}{\theta_0(4-\theta_0)^2} + \frac{\theta_1^2}{\theta_0(4-\theta_0)^2} \quad (9)$$

Value of the characteristic parameter m can be calculated from (7) taking into account (4) and (9). **μx** is in accordance with the number **m** of the layer. Thus, phase characteristic φ can be described by the following formula

$$\varphi = \frac{2}{\pi} \arcsin \{[\Delta (0)^{0.5} \sin (\frac{\pi \theta_0^{0.5}}{2})]\} \quad (10)$$

It may be inferred from (10) that the phase characteristic of the harmonic waves in the multi layered medium is directly connected with the structural parameters of the medium: elastic constants, periodical sequence of layers and their quantity (Nesvijski and Nogin, 1981).

MODEL FOR MULTI LAYERED COMPOSITE WITH IRREGULAR STRUCTURE

For the mediums which can be represented by the models for multi layered composites with irregular structure where irregularity is due to the random thickness of layers or random character of elasto-plastic properties of the layers materials, the following differential equation can be used for description of the harmonic waves propagation :

$$\frac{\partial^2 p}{\partial x^2} + k_i^2 n^2(z)(T_x) p = - \frac{\delta(z - z_0)\delta(x)}{2x\pi} \quad (11)$$

where **p** - pressure of the wave and a function of the integral acoustic field created by dispersion on irregularities, **$\delta(z-z_0)$, $\delta(x)$** - Dirac function describing difraction effects on irregularities:

$$p(x,z) = \Sigma g_n(x,z) \quad (12)$$

As a result we get the model of acoustic field as follows

$$p(x,z) = \sum_{m=0}^{\infty} \frac{ik}{4\pi} \int_0^{\infty} I_0(k,x)\,[(R_1R_2)^m\, U(k,z)\, D(k,z) + \cdots$$
$$\cdots + (R_1R_2)^{m-n}\, U(k,z)\, D(k,z)\,]\, dx\, dz \quad (13)$$

where

$$R_1 = -\frac{U(k,0)}{D(k,0)} \quad ; \quad R_2 = -\frac{\frac{\partial}{\partial x}[D(k,-h)]}{\frac{\partial}{\partial z}[U(k,-h)]} \quad (14)$$

and **U(k, z)** and **D(k,z)** are particular solutions of (11), where members $\mathbf{(R_1\ R_2)^m}$,**U(z)** and $\mathbf{D(z_0)}$ describe the wave reflected **m**-times on the border between layers with thickness **h.** Solution of (14) regarding phase characteristics of the wave is build as numerical solution of the reverse problem. Algorithm of the solution is a combination of analytical transformations and computer simulation (for example, Monte Carlo method) of random components. Stochastic character of the solution is determined by the chosen probability distributions for layers thickness and physical properties of the composite components (Nesvijski, 1983,1984).

COMPUTER SIMULATION RESULTS

Special algorithms were worked out for calculation of acoustic fields in multi layered composite materials with regular and irregular structures. For calculations according to the above shown formulas computer program was created in MATCAD 6.0 PLUS. There are some examples of graphs received with the help of this program below. Dependency of the harmonic wave phase φ on the variation of the layer thickness $\mathbf{\Delta l_2}$ of the one of the components of the two-component medium is shown in the Figure 2.

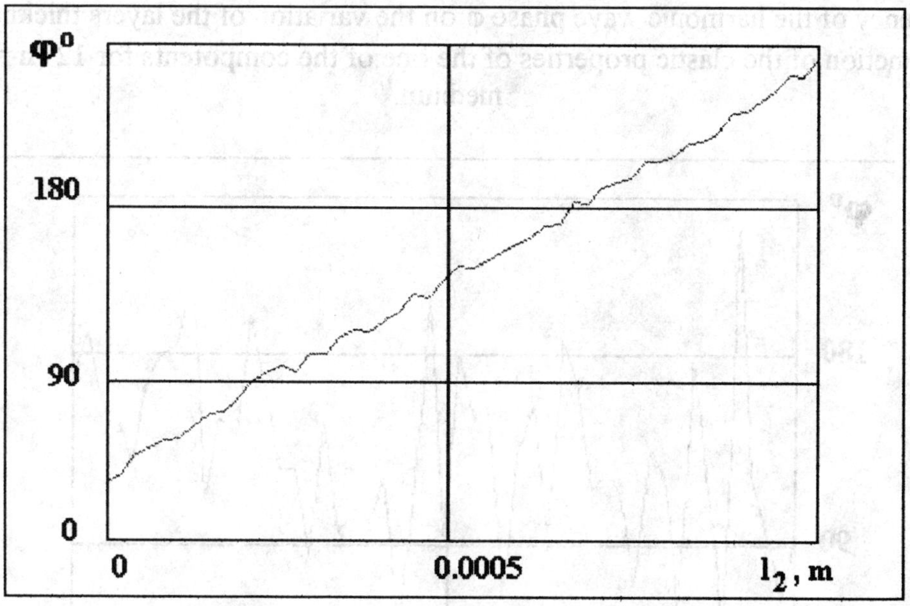

Figure 2.
Dependency of the harmonic wave phase φ on the variation of the layers thickness $\mathbf{l_2}$ and random function of the layers thickness $\mathbf{l_1}$ for five (**n**=5) layered medium.

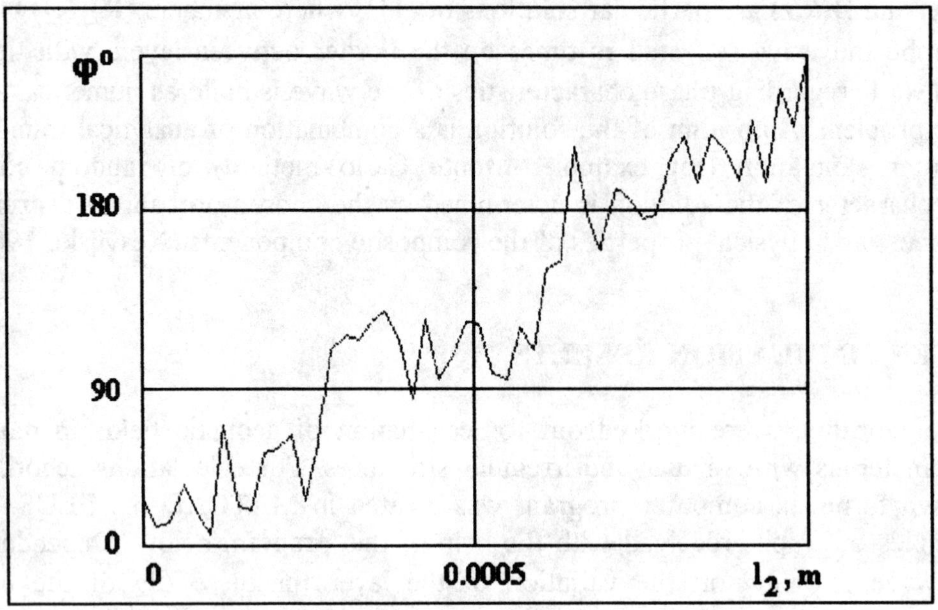

Figure 3.
Dependency of the harmonic wave phase φ on the variation of the layers thickness l_2 and random function of the elastic properties of the one of the compotents for 12 (n=12) layered medium.

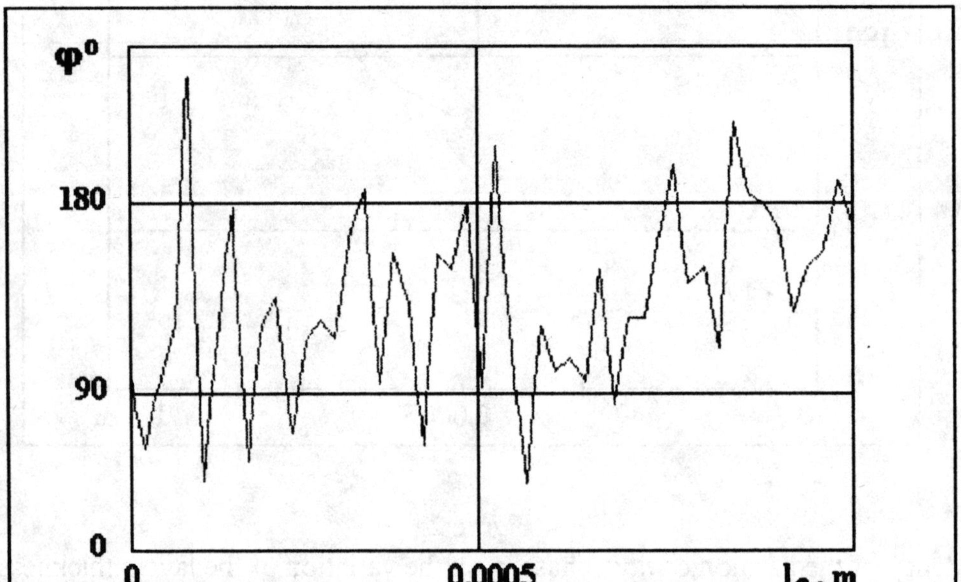

Figure 4.
Dependency of the harmonic wave phase φ on the variation of the layers thickness l_2 and random function of the layers thickness l_1 and of the elastic properties of the one of the compotents for 16 (n=16) layered medium.

CONCLUSIONS

The advanced ultrasonic testing method for multi layered composite materials presented in this work is based on the high sensitivity of the harmonic wave phase characteristic to the variation of the structure and properties of the components of multi layered materials. Regular as well as irregular structures of composite materials which are often found in practical application in multi layered composites are considered in this work. The build mathematical models for anaylysis of waves propagation in such mediums give a reliable picture of practical application of this method. It seems to be possible to use this method for long time monitoring of the composite structures in service or for their testing during inspections. This method may be also developed as phase spectral method for ultrasonic phase frequency scanning of the testing materials and structures. It may be based on modern measuring instruments using computerization and neural network.

REFERENCES

Baranov, V.M. 1990. "Acoustics Measurements in Nuclear Power," Energoatomizdat, Moscow, Russia.

Brekhovskikh, L. M. and Godin, O.A. 1989. "Acoustics of Layered Media", Nauka, Moscow, Russia.

Bondarenko, G.V. 1936. "Hill Equation and Its Application in the Area of Mechanical Vibrations", Academy of Sciences of the USSR, Kharkov.

Firestone, F.A. 1945. "Supersonic Reflectoscope an Instrument for Inspecting the Interior of Solid Parts by Sound Waves," *J. Acoust. Soc. Am.* 17 (1945) 287- 289.

Krautkramer, J. and Krautkramer, H. 1986. "Werkstoffprufung mit Ultraschall", Springer-Verlag , Berlin (West) Heidelberg.

Nesvijski, E.G. and Noguin, S.I. 1980. "Application of Monochromatic Oscillation for Testing of the Cracking Process in Reinforced Concrete Structures," Quality Control of Reinforced Concrete Structures, Stroyizdat, Moscow, Russia.

Nesvijski, E.G. and Noguin, S.I. 1981. "Digital Ultrasonic Phase Device for Testing of Anchors' Length of Pre-Stressed Constructions," Report at the All Union Conference on Experimental Researches of Engineering Structures, Tallinn, Estonia.

Nesvjski, E.G. 1982. "Usage of the Phase Method for Appraisal of Development of Micro Cracking Defects," *Defektoscopy*, 6, Academy of Sciences of the USSR, Moscow, Russia.

Nesvijski, E.G. 1984. "On Sensitivity of the Harmonic Waves Method to the Process of Cracking of Materials," *Defektoscopy*, 1, Academy of Sciences of the USSR, Moscow, Russia.

Sansone, Jh. 1953. "Simple Differential Equations," Inostrannaya Literatura, Moscow, Russia.

Wittecker, R.G. and Watson, Jh. 1953. "Course of Modern Analysis", V.2, Inostrannaya Literatura, Moscow, Russia.

Nondestructive Detection of Defects and Environmentally Induced Degradation in Adhesively Jointed Composite Structures

S. YANG, L. GU AND R. F. GIBSON

ABSTRACT

Continued efforts on the application of modal damping and frequency measurements for nondestructive and rapid screening of adhesively bonded joints in composite structures are summarized. Particular emphasis here is on the use of nondestructive modal damping measurements to determine the effects of moisture absorption on the mechanical behavior of adhesively bonded composite sandwich beam specimens. Besides the moisture absorption investigation, the same nondestructive dynamic testing technique was demonstrated by the investigation of the structural integrity of real automotive parts. Sections of floor panels of a composite pick up box with and without defects were tested, and the defective sample was found to have increased damping and reduced frequencies relative to the control sample.

INTRODUCTION

In recent years, adhesive bonding has been increasingly used in joining of composite material components. Compared with traditional fasteners that usually result in reduced structural integrity due to stress concentrations around holes, adhesively bonding has the potential advantage of maintaining the integrity of structures or components. Other advantages of adhesive bonding over traditional fastening include high strength-to-weight ratio, better design flexibility, smooth appearance, uniform load distribution, and ease of processing. With the increasing use of adhesive bonding, considerable efforts have been put into developing new methods and techniques for the characterization of the mechanical behavior of adhesively bonded joints. Nondestructive testing and evaluation by various means, such as ultrasonic and thermographic techniques, has been extensively explored.

In the past two years or so, the authors have been conducting research on the development of a new, fast, economical, and simple evaluation method for checking the structural integrity of adhesively bonded joints in composites by the use of nondestructive

---

Shuo Yang, Lan Gu, and Ronald F. Gibson, Advanced Composites Research Laboratory, Mechanical Engineering Department, Wayne State University, Detroit, MI 48202

modal damping measurements. The method is based on the fact that some modal vibration parameters, e.g., modal damping, are very sensitive to microstructural changes, damage and degradation. Impulsive excitation of the structure and subsequent analysis of the frequency response curve are used. Details can be found in several previous published articles (Yang et al., 1996; 1997), where the application of modal damping factors and frequencies for screening out specimens with void and unprepared adherend surface defects were described.

In this paper, the recent application of modal damping and frequency measurements in the detection of environmentally induced degradation in adhesively bonded joints in composite structures is reported. The feasibility of this method was demonstrated by inducing moisture absorption effects in a simulated laboratory environment. As proof of the usefulness of the newly developed method for testing not only small laboratory made specimens but full scale components, tests of two sections of floor panels of a composite pick up box are also described here.

MOISTURE ABSORPTION TESTS

It is a well known fact that moisture absorption degrades the mechanical properties of polymers by plasticization, which causes a reduction in the glass transition temperature and softening of the polymer (Gibson, 1994). Therefore, the objective of this research was to demonstrate the use of nondestructive modal vibration measurements to characterize such degradation in adhesively bonded polymer matrix composites.

DESCRIPTION OF SPECIMENS

The sandwich beam samples used here are described in Table 1 and Figure 1. These samples were made from Vetrotex CertainTeed U750 continuous E-glass fiber strand mat reinforced Dow MM 364 polyisocyanurate SRIM composite panel bonded with Ashland Pliogrip polyurethane adhesive. Samples 1 and 2 were made with completely prepared (i.e., sanded and cleaned) adherend surfaces, Samples 3 and 4 were made with 27% unprepared adherend surfaces, while Samples 5 and 6 were made with 40% unprepared adherend surfaces. The samples with partially unprepared adherend surfaces were prepared in such a way that a portion of the adherend surfaces was kept unsanded, while the remaining portion was sanded carefully by using 100 grit sand paper.

EXPERIMENTAL PROCEDURE

In order to obtain data over the full range of moisture contents from fully dry to fully saturated, the samples were first oven dried until no further moisture loss was observed. Samples were weighted on a microbalance at various times during the oven drying cycle. Figure 2 shows moisture loss curves during the drying process. After being totally dried, the samples were immersed in a distilled water bath at room temperature to allow moisture absorption until the samples were fully saturated with moisture. During this soaking process, the samples were removed from the water bath at given time intervals, weighed on a microbalance, and the corresponding damping loss factors and vibration frequencies were nondestructively measured and recorded.

Damping loss factors and vibration frequencies were measured by the combination of the half power band width method and the impulse-frequency response techniques. The impulse-frequency response method was used to excite the vibration of the specimen, and the half power band width method was applied to measure the damping loss factor from the first mode of the resonant peak in the frequency response spectrum. Figure 3 shows the experimental set up for measuring modal damping loss factors and frequencies from free-free beam flexural vibration tests. The details on the experimental technique, instrumentation, and data acquisition were described in several previous publications (Yang et. al., 1996; 1997).

RESULTS AND DISCUSSION

The moisture diffusion process to achieve the fully saturated condition in the specimens was very slow. After immersion in water for about 2,643 hours, the samples appeared to have reached full saturation. Table 2 shows the modal parameter variation at various times during the moisture saturation process for Samples 1 and 2, Table 3 shows the corresponding results for Samples 3 and 4, and Table 4 shows the results for Samples 5 and 6. Figure 4 displays the percent weight gain due to moisture absorption, Figure 5 shows the first mode damping loss factor, Figure 6 shows the first mode frequency, and Figure 7 gives the percentage gain of the first mode damping, all versus soaking time for the above six samples. The concluding point is that both vibration frequency and damping loss factor can be used to identify the global integrity and quality of adhesively bonded composites under various degrees of moisture-induced degradation. As shown in the figures and tables, Samples 1 and 2 with completely prepared adherend surface usually have the highest vibration frequencies and lowest damping loss factors, while Samples 5 and 6 with the largest unprepared surfaces generally have the lowest vibration frequencies and highest damping loss factors. It is also interesting to note that, although damping is more sensitive to moisture content for all samples, the frequency seems to be more sensitive to the amount of unprepared adherend surface

FLOOR PANEL TESTS

DESCRIPTION OF SAMPLES

The possibility of full scale component testing by modal parameter measurements was investigated in preliminary tests. Two corrugated composite floor panel samples were provided by Ford Motor Company. Those samples were made as part of the Automotive Composites Consortium focal project on the composite pick up box. These samples are much larger and have much more complicated geometry than the beam samples that have been used in the previous tests. Figure 8 shows the geometry and dimensions of the samples. The samples are made from three layers: a corrugated SRIM composite top panel, an epoxy adhesive layer, and a flat SRIM composite bottom layer. The composite used in both panels is Vetrotex CertainTeed U-750 continuous strand mat reinforced Dow MM 364 polyisocyanurate resin manufactured by the SRIM process. The adhesive is B. F. Goodrich 654 ETG epoxy resin. One of the samples, which is free of defects, is called the "control sample", while the other, which has various defects in the adhesive layer, is called the "test sample". The defects in the adhesive layer of the "test sample" include a Teflon inclusion, a steel sheet inclusion, and a gap in the adhesive.

EXPERIMENTAL PROCEDURE

As with the free-free beam flexural vibration test described above, the free-free plate flexural vibration test was conducted to nondestructively measure the modal damping loss factors and frequencies of these samples in their first three flexural vibration modes. The vibrations were impulsively excited at both the top panel and the bottom panel. Figure 9 shows the schematic set up for such a plate vibration test.

RESULTS AND DISCUSSION

The experimental results are tabulated in Table 5 for the "control sample" and in Table 6 for the "test sample. The corresponding bar charts are shown in Figure 10 for the "control sample" and Figure 11 for the "test sample", respectively. By comparing the results, it can be seen that the "control sample" and the "test sample" show distinct differences in their modal damping and frequencies. The differences in their damping loss factors, however, are much greater than the corresponding differences in their frequencies. Compared with the damping loss factors of the "control sample", the damping loss factor of the "test sample" increased about 40.48%, 22.54%, and 36.17%, respectively; while the corresponding frequencies decreased about 4.29%, 7.05%, and 21.91%, respectively; for modes 1, 2, and 3. From these comparisons a conclusion consistent with the previous conclusion regarding the beam samples can be drawn: nondestructively measured modal damping and frequencies can be used to screen the integrity of adhesively bonded composite structures with large dimensions and complicated geometries.

SUMMARIES

Nondestructive modal damping and vibration frequency measurements were used to identify moisture induced degradation of adhesively bonded composite laboratory specimens and full scale automotive composite parts with various kinds of manufacturing induced defects. Preliminary experimental results show that (1) this approach can be effectively used to monitor both the manufacturing defects and degradation induced by environmental factors, and (2) damping loss factors seem to be more sensitive to structural integrity than vibration frequencies in certain circumstances.

ACKNOWLEDGMENT

The authors gratefully acknowledge the financial support from the Michigan Materials and Processing Institute, and Ford Motor Comapny, and the technical advice and support of Tom Brimhall of Ford Motor Company, the industry project monitor. We also thank Wolfgang Oplesch of Ashland Chemical for providing the adhesives and Jessica Schroeder of General Motors for providing the SRIM materials for the beam samples.

REFERENCES

Gibson, R. F., 1994, Principles of Composite Material Mechanics, Chapter 5, McGraw Hill, New York.

Yang, S., Gibson, R. F., Gu, L., and Chen, W.-H., 1996, "Modal Parameter Analysis of Degraded Adhesively Bonded Composite Beams by Finite Element Numerical Simulation," Proceedings of the 14th International Modal Analysis Conference, Dearborn, MI, pp. 1342-1349.

Yang, S,. Gu, L., and Gibson, R. F., 1997, "Nondestructive Detection of Weak Joints in Adhesively Bonded Composite Structures," Proceedings of the 12th Annual Advanced Composites Conference & Exposition, ESD-The Engineering Society, Detroit, MI, pp. 349-366.

TABLE 1. DESCRIPTION OF SAMPLE ADHEREND SURFACE CONDITION

Sample No.	Adherend Surface Condition
1-2	Complete Prepared Surface
3-4	27% Unprepared Surface
5-6	40% Unprepared Surface

TABLE 2. MODAL PARAMETER VARIATION AT VARIOUS TIME FOR SAMPLES 1 AND 2

Step	Change in Frequency (%)		Change in Damping (%)		Change in Weight (%)	Soaking Time (Hrs)
	Mode 1	Mode 2	Mode 1	Mode 2		
2	-0.73	-0.51	5.03	5.30	0.29	87
4	-1.96	-2.84	20.04	12.25	0.57	193
7	-6.42	-7.62	32.70	22.76	0.84	458
10	-8.08	-10.17	65.10	42.15	1.12	892
14	-0.10	-0.05	63.64	49.62	1.42	1677
17	-0.51	-0.39	73.42	53.53	1.58	2499
18	-0.51	-0.42	73.42	54.11	1.59	2643

TABLE 3. MODAL PARAMETER VARIATION AT VARIOUS TIME FOR SAMPLES 3 AND 4

Step	Change in Frequency (%)		Change in Damping (%)		Change in Weight (%)	Soaking Time (Hrs)
	Mode 1	Mode 2	Mode 1	Mode 2		
2	-3.78	-3.09	2.89	2.89	0.32	87
7	-8.16	-7.89	25.45	14.59	0.72	458
10	-10.47	-10.20	57.98	30.37	0.98	892
14	-0.93	-0.43	55.80	35.78	1.30	1677
17	-1.28	-0.53	62.28	47.16	1.36	2499
18	-0.86	-0.56	63.07	39.57	1.36	2643

TABLE 4. MODAL PARAMETER VARIATION AT VARIOUS TIME FOR SAMPLES 5 AND 6

Step	Change in Frequency (%)		Change in Damping (%)		Change in Weight (%)	Soaking Time (Hrs)
	Mode 1	Mode 2	Mode 1	Mode 2		
2	4.54	-2.86	2.66	2.79	0.35	87
7	-6.55	-7.31	16.55	20.52	0.77	458
10	-2.89	-10.91	52.42	32.79	1.06	892
14	0.98	0.14	43.83	47.49	1.38	1677
17	-0.07	-0.30	55.76	46.67	1.46	2499
18	-0.16	-0.30	55.98	46.79	1.47	2643

TABLE 5. EXPERIMENTAL RESULTS FOR THE COMPOSITE PICK UP BOX "CONTROL SAMPLE"

Mode	No. of Tests	Frequency (Hz)		Damping Loss Factor			
		Std. Dev	Mean	Max (10^{-3})	Min (10^{-3})	Std. Dev (10^{-3})	Mean (10^{-3})
1	8	0.06	133.24	13.15	10.73	0.12	11.98
2	8	0.11	312.36	11.53	11.33	0.01	11.40
3	8	3.24	691.82	18.90	18.90	0.17	20.93

TABLE 6. EXPERIMENTAL RESULTS FOR THE COMPOSITE PICK UP BOX "TEST SAMPLE"

Mode	No. of Tests	Frequency (Hz)		Damping Loss Factor			
		Std. Dev	Mean	Max (10^{-3})	Min (10^{-3})	Std. Dev (10^{-3})	Mean (10^{-3})
1	8	0.42	127.64	23.29	12.43	0.34	16.83
2	8	0.21	290.33	14.77	13.24	0.06	13.97
3	8	1.51	540.34	30.88	25.73	0.19	28.50

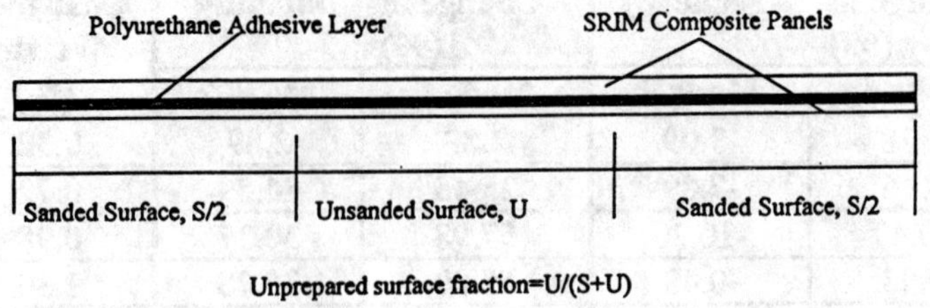

Figure 1. Composite sandwich beam with prepared and unprepared adherend surface

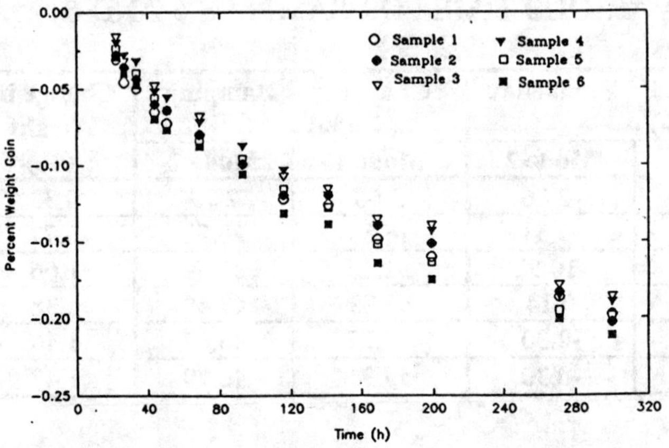

Figure 2. Weight loss during drying in an oven at 65°C

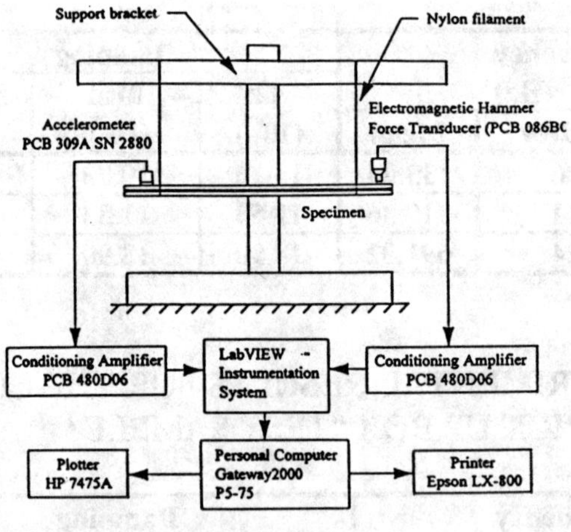

Figure 3 Experimental set up of free-free beam flexural vibration test.

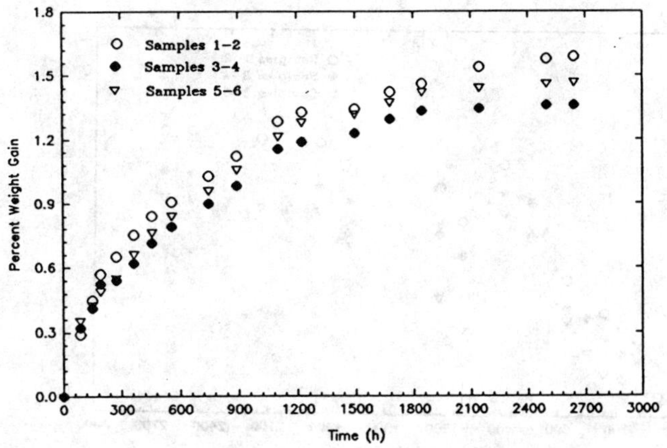

Figure 4. Percent weight gain versus soaking time: all samples in distilled water at 20°C

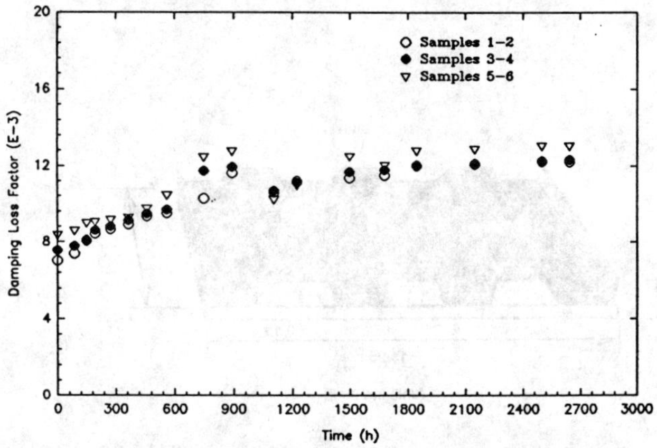

Figure 5. First mode damping loss factor versus soaking time: all samples in distilled water at 20°C

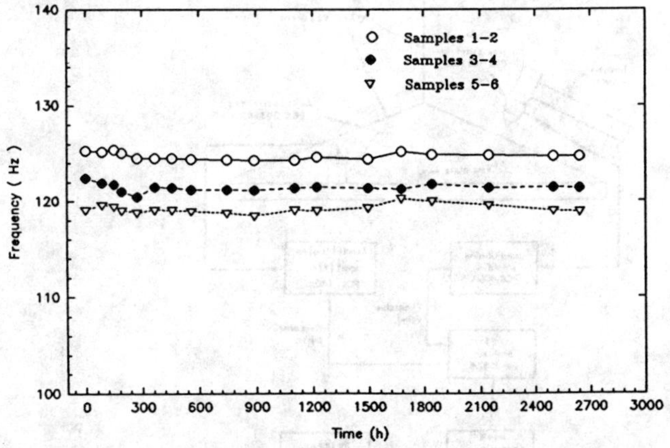

Figure 6. First mode vibration frequency versus soaking time: all samples in distilled water at 20°C

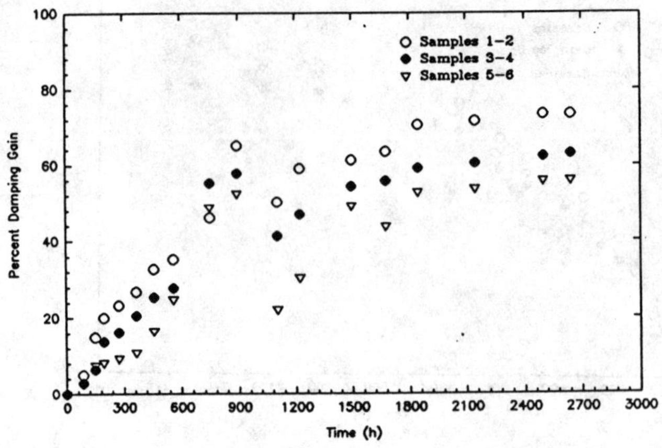

Figure 7. First mode damping gain versus soaking time: all samples in distilled water at 20°C

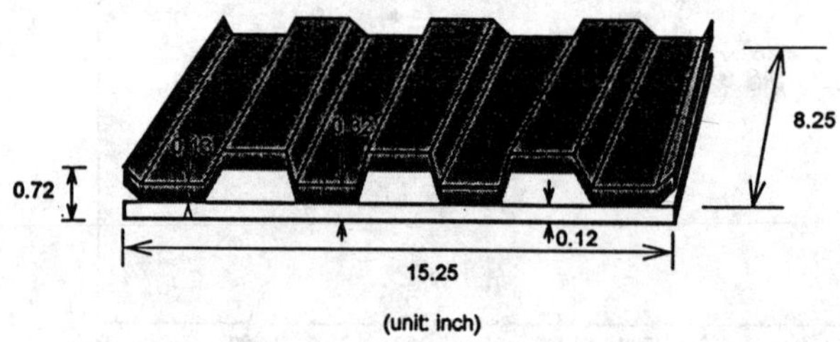

Figure 8. Schematic of floor panel section of composite pick up box

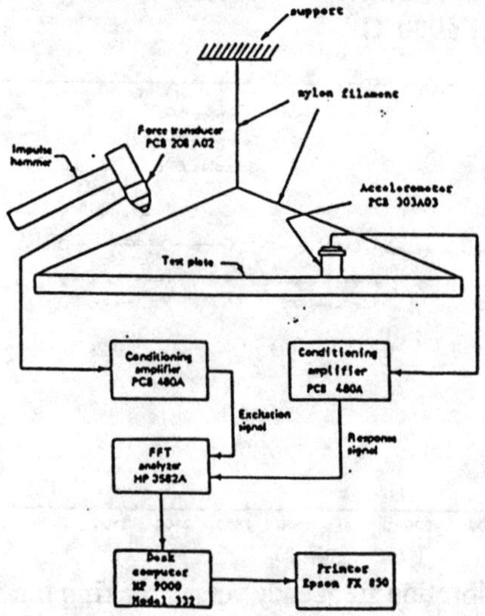

Figure 9. Experimental set up of free-free plate flexural vibration test

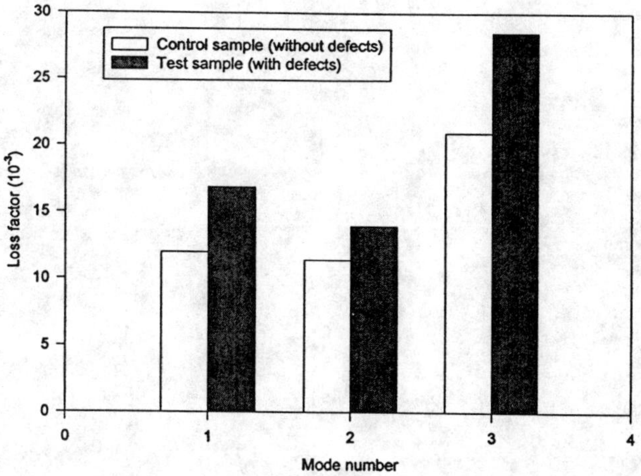

Figure 10. Damping loss factors versus mode numbers for "control sample" and "test sample"

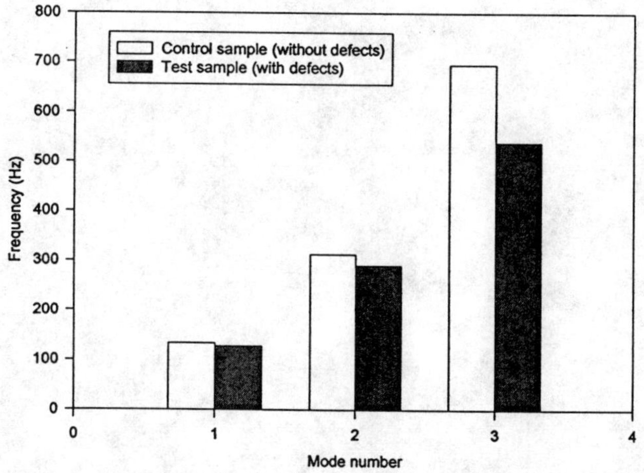

Figure 11. Vibration frequencies versus mode numbers for "control sample" and "test sample"

Thermal and Mechanical Characterization

Measured and Predicted Changes in Laminate CTE Due to Microcracking

R. Y. KIM, A. S. CRASTO AND G. A. SCHOEPPNER

ABSTRACT

Strain gages were employed to measure changes in the coefficient of thermal expansion (CTE) of a graphite/epoxy (AS4/3501-6) cross-ply laminate with physical damage and the results compared with analytical predictions. Refinements of this standard measurement technique coupled with the use of ultra-low-expansion titanium silicate as a reference material allowed the accurate determination of very low CTEs. Longitudinal and transverse CTEs of a unidirectional composite were first measured over the temperature range of -150°F to 250°F. Specimens of a $[0_2/90_2]_S$ laminate were then stressed in uniaxial tension with incrementally increasing loads to produce microcracking within the 90° plies. The laminate CTE was measured in the same location after each incremental loading, and the corresponding crack density (in the vicinity of the strain gage) was determined by microscopic examination of a polished specimen edge. CTE variation with crack density was predicted using an axisymmetric formulation of a large-radius, hollow, layered-cylinder model which is equivalent to the flat laminate formulation. The experimental results compared fairly well with the analytical predictions.

INTRODUCTION

Composites are steadily replacing metals in weight-critical spacecraft structures primarily because of their high specific strengths and moduli. Many space structures require not only high stiffness but precision alignment and dimensional stability as well, which makes the low CTE of carbon fiber composites attractive. Composite laminates can be tailored to obtain a near-zero CTE which is essential to maintaining the dimensional stability of composite structures in a space thermal environment. However, laminate CTE can change in service due to physical damage or material property changes associated with exposure to temperature fluctuations, outgassing, atomic oxygen, radiation, etc. Therefore, optimum selection and application of composites in space structures requires an understanding of how the service environment influences dimensional stability. For this study we have selected to investigate the influence of a common form of physical damage, viz. microcracking in the off-axis plies, on changes in CTE (and hence dimensional stability) of composite laminates. A number of studies have been reported in this area [1-3], but the influence of microcracking on changes in laminate CTE is still not fully understood.

Ran Y. Kim, Allan S. Crasto, Senior Research Engineers, University of Dayton Research Institute, 300 College Park, Dayton, OH 45469-0168; Gregory A. Schoeppner, Materials Engineer, WL/MLBM, Wright-Patterson AFB, OH 45433-7750.

The main objective of this study, therefore, is to investigate the influence of transverse ply cracks on the variation in in-plane CTE of a cross-ply laminate. A conventional aerospace graphite/epoxy, AS4/3501-6, was selected over candidate composite materials for space structures since it exhibits characteristics ideal for a generic study of this nature: it is well-characterized, exhibits consistent properties with little variation and is not cost-prohibitive. Transverse ply cracks were introduced in tensile coupons of cross-ply laminates by axial loading in increments. After each incremental loading the specimen's CTE was measured with strain gages and correlated with the measured crack density. These experimental correlations were compared with those from an analytical model.

MEASUREMENT OF COMPOSITE CTE

A number of techniques have been employed to measure composite CTE including dilatometry, thermomechanical analysis, laser interferometry [4] and strain gages [5]. Each technique has its advantages and disadvantages. A strain gage has the disadvantage of covering a relatively small area of the composite; however, thermal strains from a number of gages can be monitored simultaneously in a single run to provide an average CTE of the composite part. This technique is inexpensive and readily accessible in most test laboratories. In addition it is particularly conducive to *in situ* measurements on selected sites of the composite, e.g. for correlation with localized environmental or mechanical damage. However, appropriate care must be taken to ensure accuracy by eliminating artifacts associated with the strain measurement.

In this study composite CTEs were measured using strain gages in conjunction with a computer-controlled temperature chamber and data acquisition system. Ultra-low-expansion titanium silicate (with a maximum thermal strain of 10 ppm over the temperature range of interest) was used as the reference material. WK-series 350Ω strain gages from Micro-Measurements, Inc., with a gage length of 0.25 in., were employed. Heat generated in the strain gage from the excitation current, which gives rise to an apparent strain, was minimized by reducing the input voltage to the strain gage circuit to one volt. Two strain gages, one on the reference material and one on the test specimen, were taken from the same manufacturing lot to assure a close match of their apparent strain outputs. The strain gages were mounted with a high-temperature adhesive, cured as recommended by the manufacturer, and connected as adjacent arms of the bridge circuit.

The measurement system employed can accommodate eight channels; one was used for temperature, one for time, and the remaining six for collecting thermal strains. A software program controlled the chamber temperature, allowing prescribed ramps to a series of temperatures, holds for a fixed period of time to allow the specimen temperature to equilibrate, and continuous acquisition of temperature and strain data. Data were collected during both heating and cooling cycles. A specimen was cycled between -150°F and 250°F at a rate of 5°F/min and held for 15 minutes at each increment of 40°F of temperature. Thermal strains recorded at the end of each holding period were utilized for determination of CTE. The resolution of strain and temperature measurement with this experimental setup is better than 1×10^{-6} and 1°F, respectively. The correction for transverse sensitivity of the strain gage was made from the information supplied by the manufacturer. Since the longitudinal thermal strain is two orders of magnitude smaller than the transverse thermal strain of a unidirectional composite, the transverse sensitivity correction is significant. Furthermore, the transverse sensitivity of the WK series gages employed is approximately 4%, which is an order of magnitude greater than other types of gages.

The accuracy and reliability of this measurement technique was verified with two materials of known CTE, aluminum (Al-2024-T4) and titanium (Ti-6Al-4V) alloys. The average measured CTEs for these materials are 12.73×10^{-6}/°F and 4.88×10^{-6}/°F for the aluminum and titanium alloys, respectively, which compare very well with the corresponding published handbook values of 12.9×10^{-6}/°C and 4.9×10^{-6}/°F. The excellent reproducibility of the measurement technique was demonstrated by a variation of less than one percent in the CTE between consecutive measurements for four strain gages. Although strain gages on the test specimen and reference specimen may be from the same manufacturing lot, this does not guarantee that their electrical properties will be identical. To examine this effect four strain gages were mounted on the reference material, titanium silicate. Two gages were used as reference gages while the other two were used as test gages, with one reference gage and one test gage forming a half-bridge. Assuming the titanium silicate to have homogeneous thermal expansion, any output from this arrangement is considered apparent strain due to differences in electrical properties between the reference gage and test gage. Figure 1 shows the result of measurements with this setup, with various combinations of the four strain gages as test and reference gages. The measured thermal strain over the test temperature range varies from near-zero to $50x10^{-6}$ which corresponds to a maximum apparent CTE of $0.125x10^{-6}$/°F.

Lamina longitudinal and transverse CTEs are fundamental thermal properties of a composite material and are used to calculate the thermal properties of various laminate orientations. To measure these CTEs a rectangular coupon, 4"x1", was sectioned from an $[0]_{8T}$ composite panel, and one longitudinal and one transverse strain gage were mounted on each face. Thermal strains from all four gages were measured simultaneously in the temperature range of -150°F to 250°F. Because of the considerable scatter of the very low longitudinal CTE, a total of three specimens were tested in this manner for this property. Figures 2 and 3 show some of the typical strain data recorded, with the longitudinal thermal strain corrected for the transverse sensitivity of the strain gages. The CTE was obtained from a linear fit to the data over the entire test temperature range. The longitudinal CTE (of six specimens) ranged from 0.15 to 0.31 $x10^{-6}$/°F, with an average value of $0.22x10^{-6}$/°F, while the transverse CTE ranged from 13.99 to $14.71x10^{-6}$/°F, with an average value of 14.37 $x10^{-6}$/°F. The longitudinal CTE data have a wider scatter than the transverse data. The source of this scatter could be apparent strains from a mismatch in electrical properties of the two strain gages, and material variations within the laminate which arise from local variations in ply thickness, fiber alignment and fiber volume.

The CTEs of a $[0_2/90_2]_S$ laminate were determined from a 6"x12" panel. Three locations on the panel were selected for CTE measurements, and back-to-back axial and transverse gages were mounted at each location. Figure 4 shows typical axial and transverse thermal strains as functions of temperature. Excellent reproducibility with this measurement technique is demonstrated from a comparison of data from two consecutive runs. The CTEs were determined from a linear fit to the data in Figure 4 over the test temperature range. The axial CTE (six measurements at three back-to-back locations) ranges from 1.39 to 1.59 $x10^{-6}$/°F, with an average value of 1.47 $x10^{-6}$/°F, while the transverse CTE ranges from 1.44 to 1.73 $x10^{-6}$/°F, with an average value of $1.59x10^{-6}$/°F. The corresponding coefficients of variation are 5.6% and 6.8%, respectively. The CTE predicted by the model used in this study is $1.464x10^{-6}$/°F, calculated from the average values of the longitudinal and transverse CTEs of the $[0]_{8T}$ composite. This is in excellent agreement with the value of $1.456x10^{-6}$/°F obtained from laminated plate theory.

INFLUENCE OF MICROCRACKING ON LAMINATE CTE

ANALYTICAL BACKGROUND

The analytical model used in this study, developed by Schoeppner and Pagano [6], employs the variational theorem by Reissner [7] to predict the stress fields and energy release rates for cracked flat laminates containing orthotropic layers. The model can be used to examine the initiation, propagation and interaction of damage for thermoelastic problems in flat laminates under generalized plane strain conditions. A two-dimensional (2-D) representative volume element for uniform transverse crack spacing in a graphite-epoxy $[0_2/90_2]_S$ laminate was used to model the change in the axial coefficient of thermal expansion (CTE) α_x as a function of crack density. Figure 5 illustrates the analytical model for the $[0_2/90_2]_S$ laminate. For the unit or periodic cell of length L, it is assumed that transverse cracks extend the full depth of the 90° layers in the cross-ply laminates and that the cracks are normal to the layer interface. It is assumed that lamina interfaces are perfectly bonded and that the traction free cracks are equally spaced. Furthermore, free edge effects are neglected in this 2-D model. The displacements U, V, and W correspond to the Cartesian x, y, and z coordinates, respectively.

The calculation of α_x requires the solution to two boundary value problems. The first boundary value problem is used to determine the effective axial complaince S_{xx} and has the following boundary conditions: $U(0,z)=\tau_{xz}(0,z)=0$ for the 0° plies; $\sigma_x(0,z)=\tau_{xz}(0,z)=0$ for the 90° plies; $U(L,z)=1$; $\tau_{xz}(L,z)=0$; and $\sigma_z(x,\pm T/2)=\tau_{xz}(x,\pm T/2)=0$. These boundary conditions result in $\overline{\sigma}_y=\overline{\sigma}_z=0$. The average axial stress $\overline{\sigma}_x$ and average axial strain $\overline{\varepsilon}_x$ are then related by $\overline{\varepsilon}_x=S_{xx}\overline{\sigma}_x$, where

$$\overline{\sigma}_x=\frac{1}{LT}\int_{-T/2}^{T/2}\int_0^L \sigma_x(x,z)dxdz \qquad \overline{\varepsilon}_x=\frac{1}{LT}\int_{-T/2}^{T/2} U(L,z)dz$$

The second boundary value problem required to calculate α_y has the following set of boundary conditions: $U(0,z)=\tau_{xz}(0,z)=0$ for the 0° plies; $\sigma_y(0,z)=\tau_{xz}(0,z)=0$ for the 90° plies; $U(L,z)=\tau_{xz}(L,z)=0$; $\sigma_z(x,\pm T/2)=\tau_{xz}(x,\pm T/2)=0$; with a unit temperature change. These boundary conditions also result in $\overline{\sigma}_y=\overline{\sigma}_z=0$. Then we get

$$\overline{\varepsilon}_x=0=S_{xx}\overline{\sigma}_x+\alpha_x\Delta T$$

where ΔT is the unit temperature change. Knowing S_{xx}, the effective axial CTE α_x can be determined. The undamaged effective CTE was determined in a similar manner by analyzing a unit cell without a transverse crack. For calculation of the effective compliance and effective axial CTE, a 12 layer model was used.

EXPERIMENTAL RESULTS

Cross-ply $[0_2/90_2]_S$ laminates were used to investigate the change in laminate CTE due to microcracking brought about by mechanical loading. Specimens 6" long and 1" wide were prepared from a $[0_2/90_2]_S$ panel such that the fibers in the surface layers were parallel to the specimen length. Specimen edges were ground and then polished to enhance their microscopic examination. An axial and a transverse strain gage (each 0.25 inches long) were mounted on the specimen and the initial CTE in the axial direction measured prior to loading, as described earlier. Specimens were then loaded to a level at which transverse cracks initiated under the strain gage, as determined by monitoring acoustic emission during loading. The polished edges of the specimen were examined under a microscope to determine the nature of the transverse cracking and determine the crack density below the strain gage. As defined by the analytical model, only transverse cracks which extended over the entire specimen width as well as the thickness of the 90-degree plies were counted. After the crack density was determined, the specimen was subjected to another increment of loading to obtain new cracks. Figure 6 shows a plot of thermal strain versus temperature for a representative specimen, for a range of crack densities. The applied stress, crack density and corresponding CTE are also shown in the figure. This incremental loading was repeated a number of times until the crack density reached saturation, i.e. when additional load produced no further increase in crack density. A total of four specimens were tested in this manner.

COMPARISON BETWEEN PREDICTION AND EXPERIMENT

The variation in CTE as a function of crack density for a $[0_2/90_2]_S$ laminate is plotted in Figure 7. The lines in the plot are theoretical predictions obtained from unidirectional composite properties considering the scatter in the CTEs α_x and α_y; the solid line represents the average value, and the two dashed lines represent the upper and lower bounds calculated from the largest and smallest values of the longitudinal CTE determined from $[0]_{8T}$ composites, as detailed earlier. The symbols in the plot are the experimental data from the four test specimens. The experimental CTE data tend to deviate from the analytical bounds in the region where the crack density reaches a limiting value, as seen in Figure 7. As the crack density approaches the limiting value, a number of partial cracks appear near the major transverse crack, as seen in Figure 8. These partial cracks were not included in the analysis. The variations in modulus and Poisson's ratio with crack density for a typical specimen are shown in Figure 9. A sharp drop in the values of both these properties at a crack density of 40 in^{-1} demonstrates the influence of the partial cracks on the observed correlations. In addition further opening of the existing cracks may also contribute to the observed deviation between experiment and prediction. It should be noted that the large variation of Poisson's ratio with increasing crack density has not been considered in the analysis.

SUMMARY AND CONCLUSIONS

Strain gages provide an inexpensive yet accurate means of measuring very low CTEs (and CTE changes) of composite laminates provided care is taken to minimize extraneous sources of error. This technique was employed to determine lamina and laminate CTEs of a graphite/epoxy composite system, AS4/3501-6. Lamina thermal strains of this composite material varied linearly with temperature over the range studied. A significant scatter in the longitudinal CTE appears to be due to material variability and apparent strains from the measurement system. Transverse cracking in a $[0_2/90_2]_S$ laminate, induced by mechanical loading, is reflected in a change in laminate CTE. The analytical model employed to predict this variation in CTE due to transverse cracking demonstrates great potential. The use of strain gages provides the flexibility to correlate localized damage with changes in composite CTE. This work is being extended to investigate variations in laminate CTE due to thermally-induced microcracking.

ACKNOWLEDGMENTS

The assistance of John Camping and Ron Esterline of the University of Dayton Research Institute is greatly appreciated, for automation of the CTE measurement technique and for conducting the CTE measurements, respectively. This work was performed under U.S. Air Force Contract No. F33615-95-D-5029.

REFERENCES

1. C. H. Park and H. L. McManus, "Thermally Induced Damage in Composite Space Structures: Predictive Methodology and Experimental Corelation," Proc. of American Society of Composites, 9, (1994), pp. 161-168.

2. H. L. McManus, D. E. Bowles, and S. S. Tompkins, "Prediction of Thermal Cycling Induced Matrix Cracking," J. Reinforced Plast. Compos., 15, (1996), pp. 124-139.

3. R. Y. Kim and A. S. Crasto, "Dimensional Stability of Composites in Space: CTE Variations and Their Prediction," Proc. Int. Conf. Compos. Mater. ICCM-10, Vol. IV, (1995), p. 513.

4. S. S. Tompkins, D.E. Bowles and W. R. Kennedy, "A Laser Interferometric Dilatometer for Thermal Expansion Measurements of Composites," Experimental Mechanics, 26, (1986), pp. 1-6.

5. Micro-Measurements Group, "Measurement of Thermal Expansion Coefficient Using Strain Gages," Tech. Note No. TN-513, Micro-Measurements Group, Raleigh, NC (1986).

6. G. A. Schoeppner. and N. J. Pagano, "Stress Fields and Energy Release Rates in Cross-Ply Laminates," International Journal of Solids and Structures (*acceptedMar. 1997*).

7. E. Reissner, "On a Variational Theorem in Elasticity," Journal of Mathematics and Physics, 29, (1950), pp. 90-95.

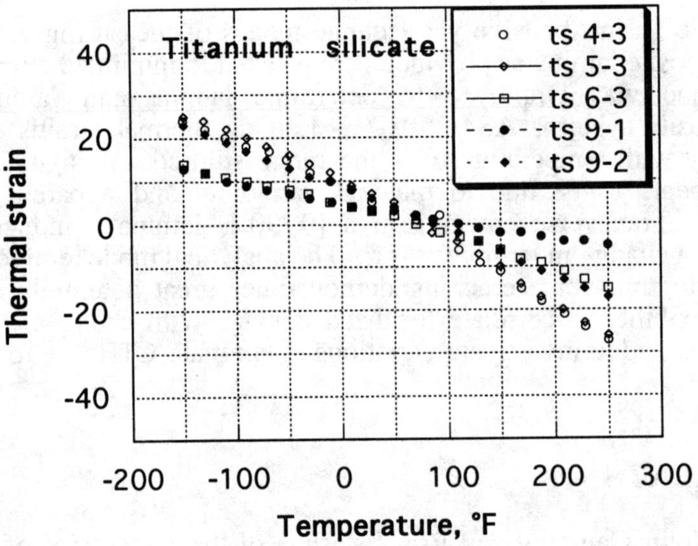

Figure 1. Apparent thermal strain as a function of temperature.

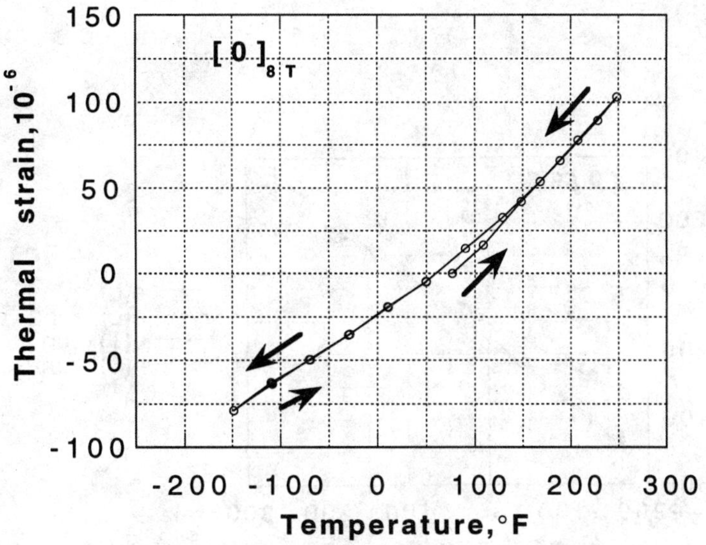

Figure 2. Thermal strain of a $[0]_{8T}$ through heating and cooling cycles.

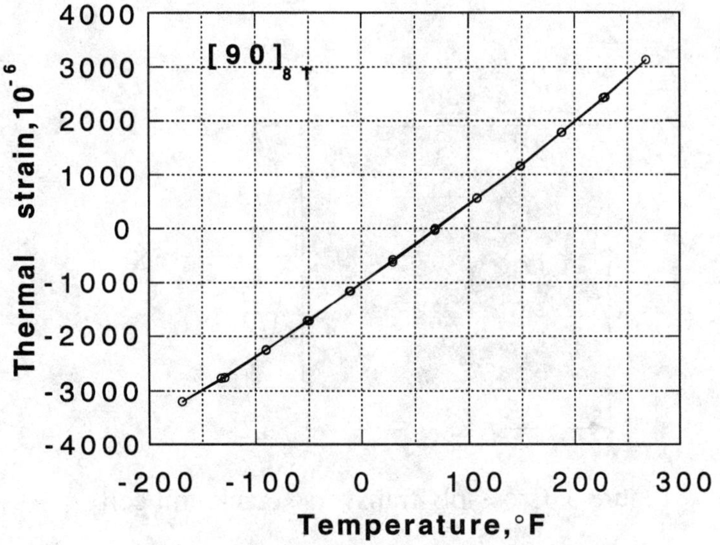

Figure 3. Thermal strain of a $[90]_{8T}$ laminate through heating and cooling cycles.

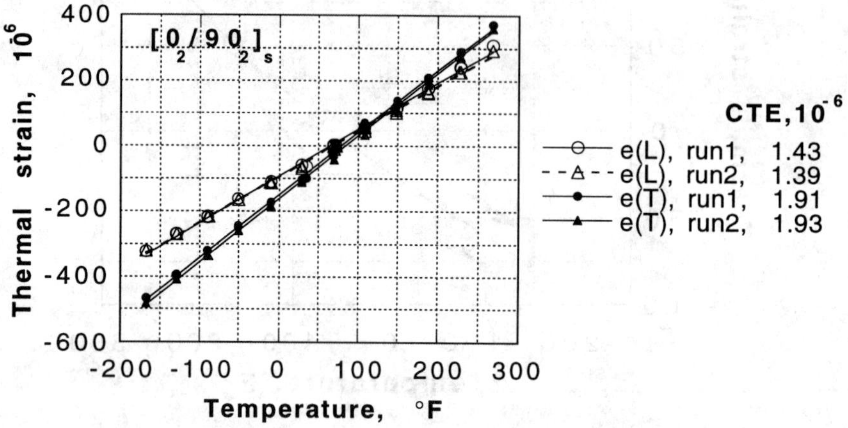

Figure 4. Variation in CTE of a $[0_2/90_2]_S$ laminate in two normal directions. (L: parallel to fiber in the surface ply; T: perpendicular to fiber in the surface ply)

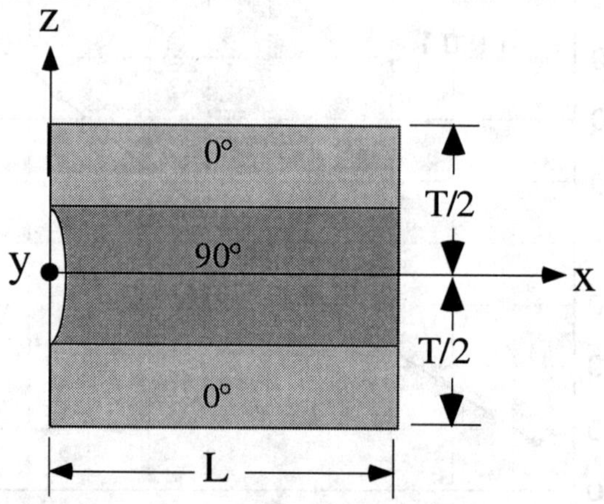

Figure 5. Cross-ply transverse crack unit cell

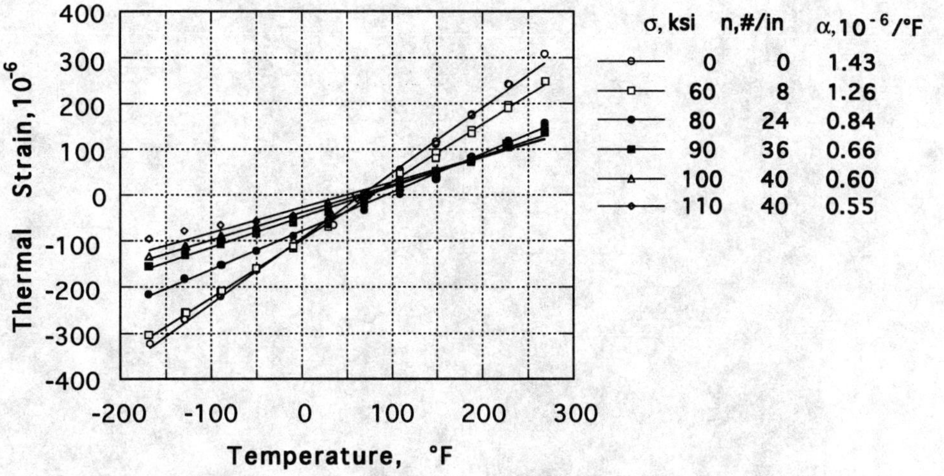

Figure 6. Thermal strain as a function of temperature and crack density for a $[0_2/90_2]_s$ laminate.

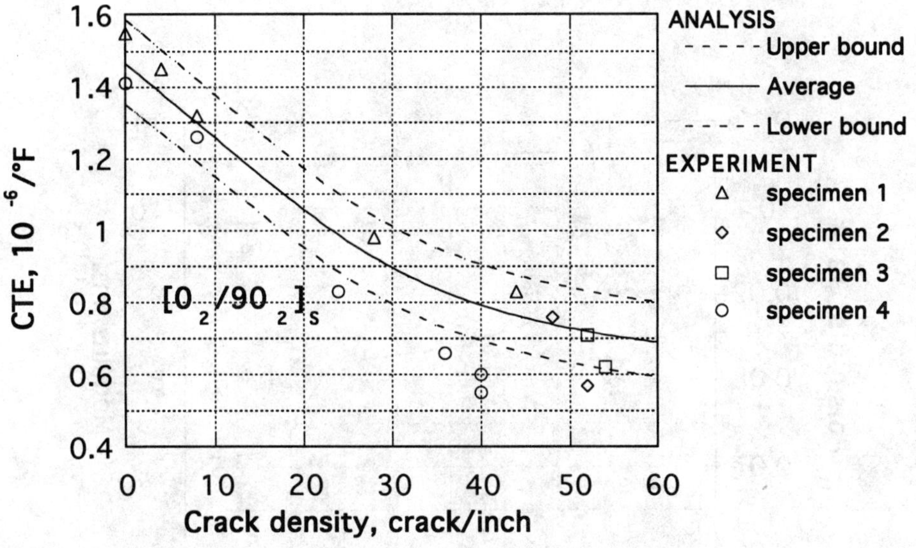

Figure 7. Comparison of experimental and predicted CTEs.

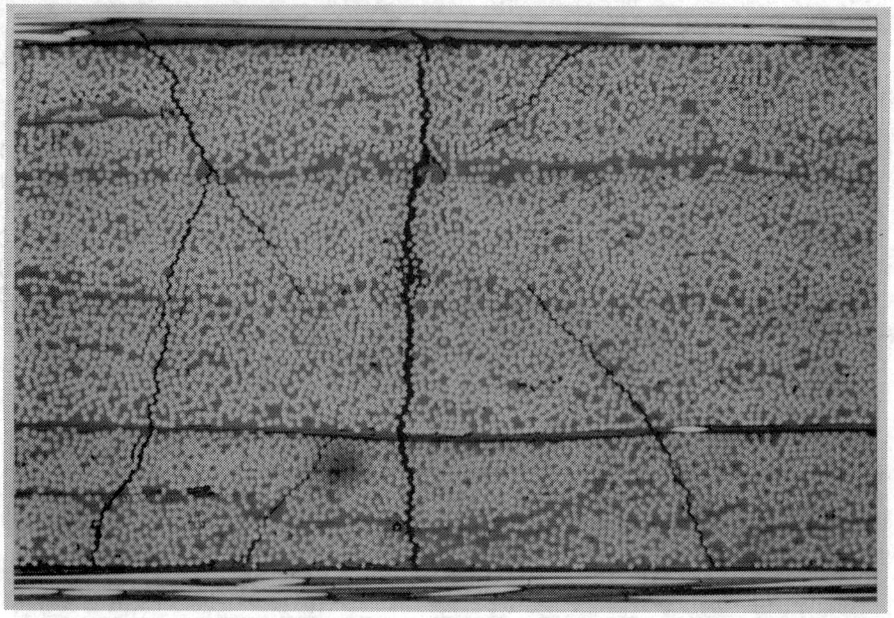

Figure 8. Micrograph showing transverse cracks in a $[0_2/90_2]_S$ laminate with secondary partial cracks at/near the state of crack saturation.

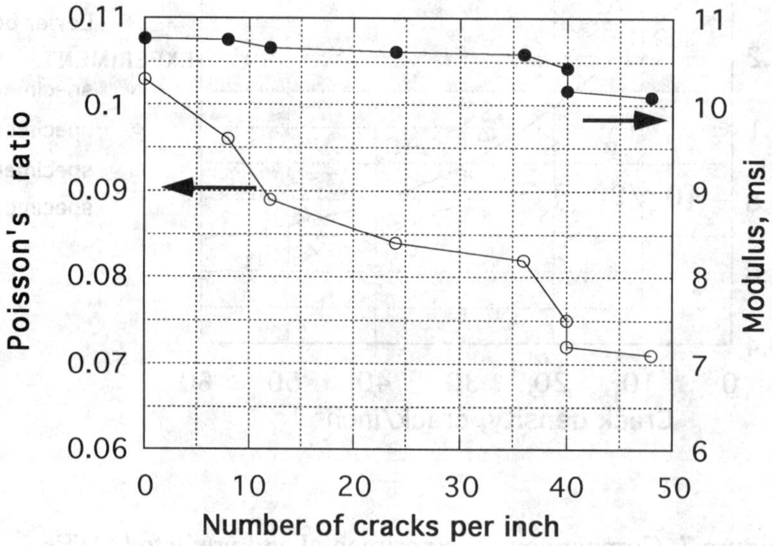

Figure 9. Variation of $[0_2/90_2]_S$ laminate elastic properties with crack density.

Hygrothermal Durability and Thermal Aging Behavior Prediction of High Temperature Polymer Matrix Composites and Their Resins

E. E. SHIN, R. J. MORGAN, J. ZHOU, J. LINCOLN, R. JUREK AND D. B. CURLISS

ABSTRACT

Prediction of the cure induced T_g increases and associated matrix and composite mechanical property deterioration of BMI systems in real service environment is attempted by network structure interrelations with mechanical and thermal properties as a function of composition (initial monomer ratio) and time-temperature cure cycles. Tensile and flexural properties of four BMI compositions at six different cure cycles (or degree of cure) have been determined at three temperatures, 23°C, 177°C, and 250°C and correlated to T_g and density of the systems.

Systematic studies on hygrothermal durability of bismaleimide (BMI) and various polyimide (PI) - carbon fiber composites and neat resins were conducted. The combined effects of moisture and thermal exposures such as hygrothermal spikes up to 250°C and hygrothermal aging under various time-temperature-moisture conditions including an accelerated aging at saturated steam environment; 160°C and 110 psi, on microscopic damage, polymer matrix physical structural state and residual properties of those composites and neat resins are presented. The onset of blistering in moist K3B PI- C fiber composites occurred at 229 °C during the thermal spiking. It is evident that the hygrothermal performance stability is one of the prime guidelines in future aerospace applications, especially for PI composites. The physical and chemical structural state of PI matrices such as K3B and AFR700B as a function of hygrothermal exposure are discussed in terms of hydrolytic chemical degradation, moisture vapor-induced physical damage and molecularly locked-in water. Those structural states are characterized by systematic weight monitoring, ^2H NMR, and various thermal, mechanical property measurements using D_2O water environment.

EXPERIMENTAL

MATERIALS

The BMI resins utilized was the BMPM/DABPA (BMI 5292, Ciba-Geigy) system. The PI resins studied were Avimid K3B (DuPont) formed from an aromatic diethyl ester diacid and an aromatic diamine in N-methyl -2-pyrrolidone (NMP) and AFR700B supplied from Wright Laboratory, AFML. The BMI neat resin sheets of about 2 mm thick were cast in clamped two tempered glass plates. PI sheets were compression-molded with fully imidized powder; 10 mm

E.Eugene Shin, Roger J. Morgan, Jiming Zhou, Jason Lincoln, Robert Jurek, Advanced Materials Engineering Experiment Station (AMEES), Michigan State University, 2203 Eastman Avenue, Midland, MI 48640
David B. Curliss, Wright Laboratory, Materials Directorate, WL/MLBC, Wright-Patterson AFB, OH 45433-7750

thick K3B plaque consolidated at 350°C in a deep metal mold and about 2 mm thick AFR700B plates cured at 382°C. Composites were fabricated with IM-7 carbon fibers. BMI 5292-IM7 composites were cured at 177°C for 1 hr, then post-cured for 2 hrs at 200°C and for up to 6 hrs at 250°C. K3B-IM7 composites were consolidated at 360°C using autoclave vacuum bagging process at Boeing Commercial Airplane Group, Seattle.

METHODS

T_g of BMI resins as a function of composition and cure time-temperature cycle were determined by a modulated DSC (TA Instrument) under N_2. In addition, DMA or TMA was also used to measure T_g of composite specimens and some resins. Density was measured in METTLER electronic balances with a resolution of 0.0001 g using the Archimedean Principle. TGA with N_2 was used to monitor weight loss of BMI resin during cure cycles. Tensile and flexure tests were performed in Instron and MTS test frame, respectively, using machine cross-head speed following the standard test conditions.

RESULTS AND DISCUSSION

BMI NETWORK STRUCTURE INTERRELATIONS WITH MECHANICAL, THERMAL AND PHYSICAL PROPERTIES

Based on our previous durability studies (Morgan, 96; Shin, 95b;96), the most likely potential, critical aging mechanisms involve cure advancement of the system with associated T_g increases. The BMI cure reactions are incomplete after standard composite fabrication and post-cure procedures and further cure can occur in service environment exposure conditions resulting in T_g increases and associated composite mechanical property deteriorations and enhanced microcrack development. It was reported (Loechett, 92; Morgan, 97) that real time T_g increased for commercial BMI-carbon fiber composites (BASF 5260) of up to 70°C after isothermal exposure in the 150-200°C for up to 8,000 hrs. Others (Li, 90; Pederson, 95; Burchan, 95) also reported considerable mechanical property deterioration of similar BMI composites after isothermal aging as a function of time, temperature and ply depth. Thus, it is important to study the relationship between the BMI molecular network structure and composite processing, service environment exposure and mechanical property relations.

Recently, we have initiated a systematic study of the effects of composition (four systems) and cure conditions (six different cure cycles) on the mechanical, thermal and physical properties of BMI resins (Morgan, 96a). The overall properties of cured BMI neat resins are summarized in Table I in terms of composition, cure conditions, and test temperatures. In case of mechanical properties, only flexural properties are reported since tensile properties showed similar behavior.

Figure 1 shows T_g, density, and their changes as a function of cure conditions. As expected, T_g of all systems increased significantly with higher cure temperatures and/or longer cure time which had been suggested in our previous studies of cure reaction mechanisms and kinetics (Morgan, 96;97). It should be noted that after extended cure, considerable further T_g increases up to about 45°C from standard cured conditions were observed in all compositions. However, at the standard cure temperature regions, the most T_g increase is observed in system 4 (DABPA rich) which is supportive of our and others (Mijovic, 96) claim on the ether cross-linking reaction via hydroxyl dehydration as a main cause of further increase in T_g of the BMI system. This is also supported by the % wt loss data even though the change is rather small. Interestingly enough, density of the BMI system decreases with increasing degree of cure or higher cross-linking density. This phenomenon has been confirmed experimentally using a Positron Annihilation technique (PALS) counting total size of micro or nano pores formed in cross-linked network system, such as epoxies (Yee, 97), i.e., increasing total amount of pores with higher cross-linking density. Note also that the density value directly relates with BMI composition. This density change along with weight loss data suggests that there may not be large volumatric shrinkage occurred with further cross-linking development.

Overall mechanical property deterioration with more extended cure conditions in terms of modulus, strain at failure and ultimate strength is shown in Figures 2 thru 4. It is certain that the property degeneration with cure advancement is closely manifested by density change even though the data were somewhat scattered.

Combining these results with the ongoing fracture surface analysis and the BMI cure reaction characteristics and kinetics information obtained from the other parallel task in our AFOSR program (Morgan and Shin et al, 96; 97; Choi,96), we will develop a BMI network structure model and the relationship between the network structure, resin mechanical response and T_g which will be utilized to develop long-term predictions of BMI thermal and mechanical properties.

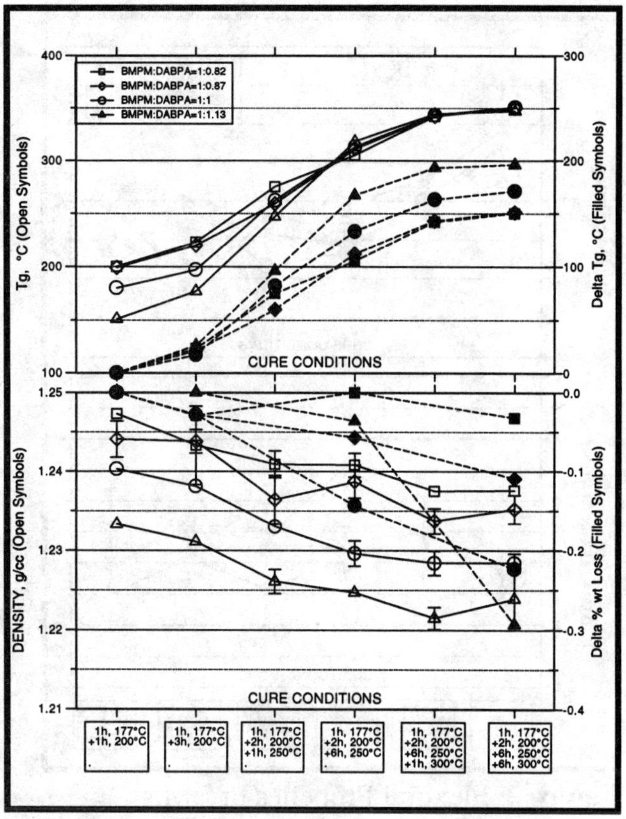

Figure 1 T_g and density Changes of BMIs

TABLE I - OVERALL MECHANICAL, THERMAL AND PHYSICAL PROPERTIES

CURE & POSTCURE CONDITIONS		Flexural Properties	BMI NEAT RESIN COMPOSITIONS BY BMPM:DABPA RATIO (molar)															
			SYSTEM 4 (1:1.13)				SYSTEM 1 (1:1)				SYSTEM 2 (1:0.87)				SYSTEM 3 (1:0.82)			
			Test Temp., °C			T_g, °C/ Density g/cc	Test Temp., °C			T_g, °C/ Density g/cc	Test Temp., °C			T_g, °C/ Density g/cc	Test Temp., °C			T_g, °C/ Density g/cc
			25	177	250		25	177	250		25	177	250		25	177	250	
LOW TEMP. CURE	1h @ 177°C + 1h @ 200°C	E (Mpa)	4454	-	-	151/ 1.2333	4368	-	-	180/ 1.2400	4369	-	-	199/ 1.2441	4122	-	-	200/ 1.2473
		σ_S (Mpa)	131	-	-		150	-	-		124	-	-		149	-	-	
		ε_f (%)	3.14	-	-		3.80	-	-		2.70	-	-		3.77	-	-	
	1h @ 177°C + 3h @ 200°C	E (Mpa)	4410	-	-	177/ 1.2311	4009	-	-	197/ 1.2381	3925	-	-	220/ 1.2438	3927	-	-	223/ 1.2433
		σ_S (Mpa)	178	-	-		159	-	-		173	-	-		168	-	-	
		ε_f (%)	4.50	-	-		4.34	-	-		4.97	-	-		4.63	-	-	
STANDARD TEMP. CURE	1h @ 180°C + 2h @ 200°C + 1h @ 250°C	E (Mpa)	3967	-	-	247/ 1.2261	3665	2527	537	262/ 1.2330	3732	2580	459	259/ 1.2364	3702	-	-	275/ 1.2409
		σ_S (Mpa)	162	-	-		143	101	n/a		169	110	n/a		199	-	-	
		ε_f (%)	4.49	-	-		4.09	7.64	n/a		4.65	8.93	n/a		7.10	-	-	
	1h @ 180°C + 2h @ 200°C + 6h @ 250°C	E (Mpa)	3886	2234	1603	318/ 1.2247	3573	2432	1965	313/ 1.2296	3779	2431	1894	311/ 1.2387	3794	2531	1781	305/ 1.2408
		σ_S (Mpa)	152	94	59		113	100	75		164	114	67		157	120	63	
		ε_f (%)	4.55	5.51	5.38		3.15	4.91	5.55		5.10	7.41	7.38		4.77	6.64	8.59	
HIGH TEMP. CURE	1h @ 180°C + 2h @ 200°C + 6h @ 250°C + 1h @ 300°C	E (Mpa)	3443	2468	1759	344/ 1.2215	3965	2415	1628	343/ 1.2284	3511	2810	1962	342/ 1.2337	3553	2728	1986	342/ 1.2375
		σ_S (Mpa)	137	110	79		138	108	80		156	131	88		150	123	90	
		ε_f (%)	4.37	5.99	7.71		4.13	6.39	7.80		5.20	7.23	8.52		4.99	6.78	8.47	
	1h @ 180°C + 2h @ 200°C + 6h @ 250°C + 6h @ 300°C	E (Mpa)	3988	2072	1602	347/ 1.2239	3846	2416	1723	351/ 1.2284	3960	2344	1760	350/ 1.2352	3741	2334	1860	350/ 1.2376
		σ_S (Mpa)	137	78	74		107	106	79		133	108	87		132	106	92	
		ε_f (%)	4.02	4.47	7.14		2.98	5.95	6.63		3.87	6.00	6.65		4.27	5.76	7.65	

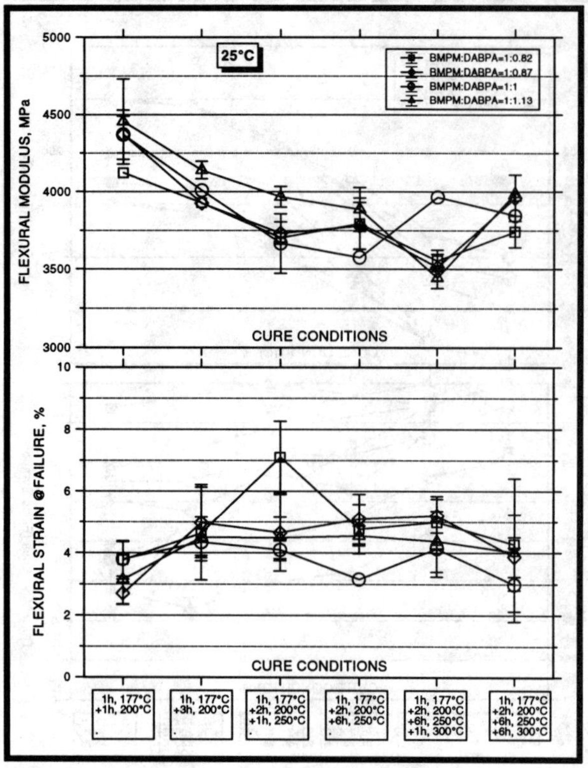

Figure 2 Flexural Propeties of BMIs

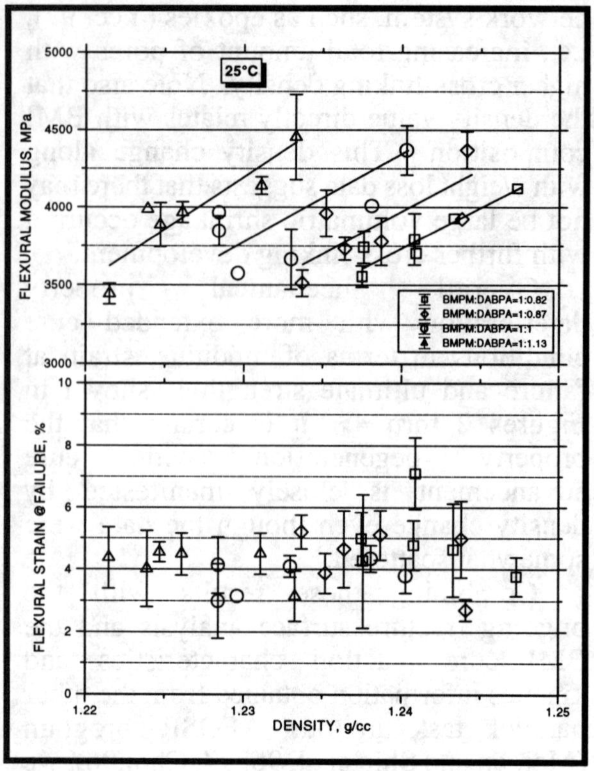

Figure 3 Density Dependency of BMIs.

HYGROTHERMAL DURABILITY

Recently, hygrothermal performance of high temperature polymer matrix composites has been great concerns in aerospace applications since hygrothermal exposure of BMI- and PI-carbon fiber composites has resulted in reported matrix blistering and composite delamination and permanent "dry" T_g decreases up to 80°F to 150°F after hygrothermal spiking or aging as a results of possible matrix physical and chemical modifications (Cornelia, 95; 96; Curliss, 96; Hoogstenden, 95; Morgan, 96; Price, 95; Rice, 96; Shin, 93; 95b; 96; Thorp, 96). Curliss (96a) reports significant modification of the NMR spectra of hygrothermal exposed PIs that are subsequently dried. These disturbing reported permanent T_g decreases have been attributed to possible imide ring hydrolysis at high temperatures with reversion to a polyamic acid linkage, which in itself is susceptible to chain scission via hydrolysis (Morgan, 96b). However, the higher T_g decreases reported would require 30-40% formation of polyamic acid linkages. It was also reported (Cornelia, 95;96) a broad range of Tg knockdown of

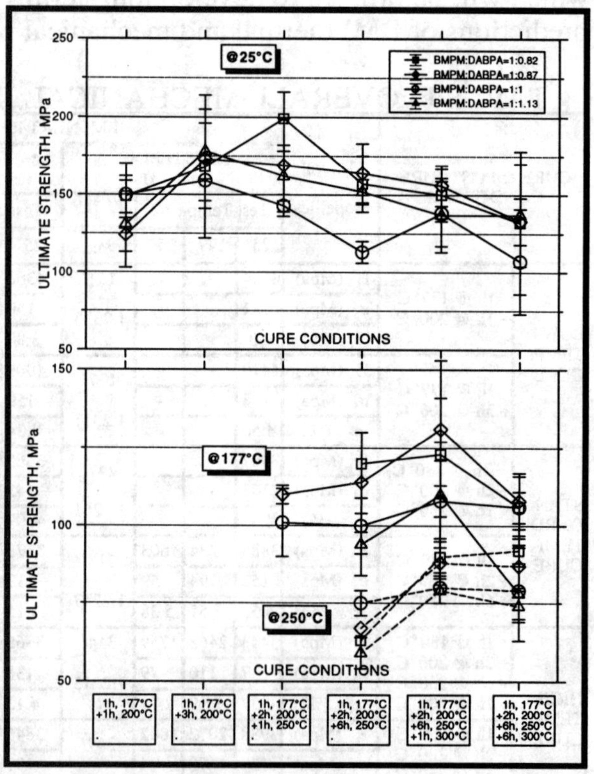

Figure 4 Flexural Strength of BMI Systems at Various Temperatures.

various PI resin systems as a function of moisture content, i.e., "dry T_g -wet T_g " up to about 300 to 350°F at about 2 wt% moisture content. This suggests that an alternative mechanism of the reported "dry" T_g decreases is a result of moisture molecular "lock-in." At higher temperatures new matrix H-bonding sites become available to moisture as the molecular structure of the matrix expands with temperature. Upon relatively rapid cool-down water molecules can become locked into molecular cages and can only diffuse out upon subsequent exposure to high temperatures. Such water molecules can not be released at low temperatures but upon release at higher temperatures act as plasticizers and lower the measured matrix T_g. It is crucial to identify most possible mechanisms of the T_g decreases and their contributions.

Lately, we have been conducting two hygrothermal durability programs, so-called, hygrothermal spiking and accelerated hygrothermal aging in pressure-bomb. The experimental details of those programs and some preliminary results were discussed in our previous paper (Shin, 96; Morgan, 96a). In this paper we report the latest analyses of the test results.

Effects of Hygrothermal Spiking

Both dry and wet samples were thermal spiked from 30°C to 250°C up to 12 cycles. The most dramatic result observed was severe blistering of moisture saturated K3B-carbon fiber at 229°C at the first spike. Similar type of blistering was also observed in K3B neat resin after several more thermal spikes. However, all other specimens including dry K3B and all BMI resins and composites, wet or dry, showed no damage or structural morphological changes. When the trapped moisture vapor pressure exceeds the local polymer matrix yield stress during dynamic hygrothermal exposure matrix cavitation will occur. Such matrix cavitation and associated delamination depends on a series of variables such as (i) previous humidity-time-temperature exposure and associated moisture concentration profiles; (ii) component thickness; and (iii) rapid heat-time service environment exposures, such as thermal spikes. However, BMI-carbon fiber composites exhibit no evidence of blistering for a similar thermal spike. A thermoset is more resistant to cavitation and associated macro blistering than a thermoplastic because thermoset cavitation requires rupture of covalent crosslinked molecular segments.

Moisture weight gain behavior was monitored during the entire thermal spiking exposure, Figure 5. The exposure cycle consisted of three steps, i.e., (1) initial saturation at 30°C, 90%RH, (2) absorption and desorption during thermal spiking, and (3) desorption during dry conditioning at 80°C in vacuum. Both [0/90] and [0] BMI composite laminates were used to identify the effects of microcracks on moisture absorption behavior. The cross-ply laminate always

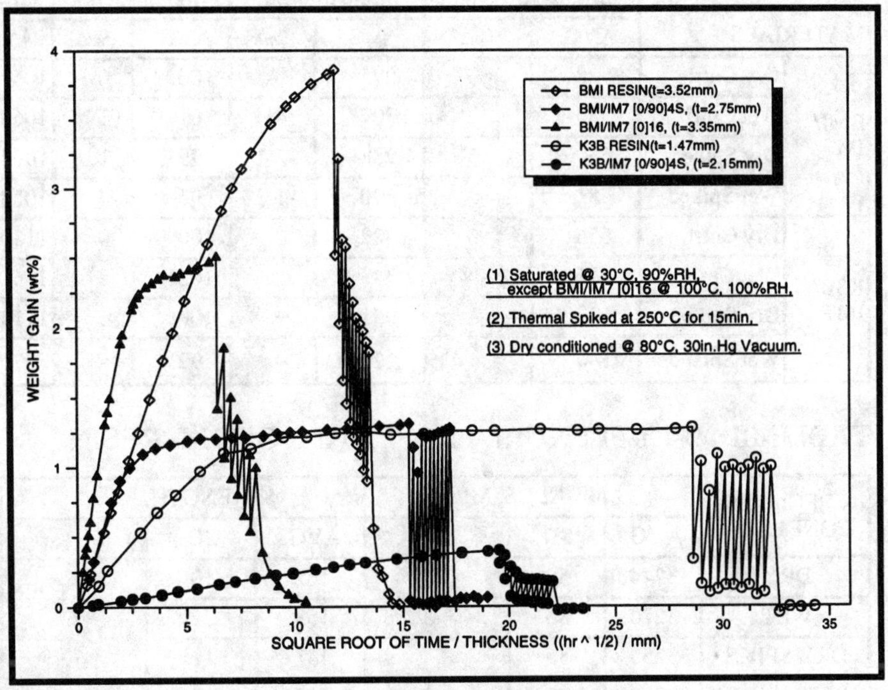

Figure 5 Overall Moisture Absorption-Desorption Behavior.

contained cure-induced microcracks while the unidirectional laminate was completely microcrack-free. At the initial wet conditioning the BMI composite and K3B resin reached the equilibrium moisture saturation in a relatively short exposure time per thickness while K3B composite and BMI resin were still absorbing moisture even after the extended exposure time. During thermal spiking, BMI resin, BMI unidirectional composite, and K3B composite behaved similar, i.e., (i) the moisture content decreased gradually at the early spikes until leveling off at about 50w% loss and (ii) the amount of water in and out during thermal spiking was about 30 - 40% of maximum moisture content. On the other hand, both cross-ply BMI composites and K3B resin always maintained a high moisture content and almost 80 to 90w% gain and loss during spiking which implies high diffusion rate of the systems. In the cross-ply BMI composites the microcracks are the main cause of the high diffusivity while the intergranular boundary effect is suspected for K3B resin since the resin plaque was fabricated by compression molding of a fully imidized K3B powder. It was also observed that the diffusivity of the K3B composites was almost doubled after the spiking exposures because of the hygrothermal-induced damage; micro-cavitation, blistering and delamination. Interestingly enough, there are always some residual moisture left in all BMI/IM7 composite regardless of drying time while BMI resin, K3B resin and composites came back to the original dry weight. This is most likely due to that the wet T_g of the BMI composites is higher than the spiking temperature so that some lock-in moisture can not diffuse out at lower drying temperature.

Tables II and III summarize the residual mechanical property changes before and after the thermal spike exposure. All mechanical test specimens including wet-control samples for the residual property measurement were re-dried after thermal spikes prior to mechanical testing. Most property degradation was observed in wet spiked K3B/IM7 composites obviously due to micro-cavitation and blistering. There was no considerable property change observed from other conditions except some improvements in strength and toughness were seen from BMI/IM7 composites after wet exposure and thermal spikes probably due to plasticization effect by locked

TABLE II. RESIDUAL MECHANICAL PROPERTIES OF COMPOSITES

PROPERTIES		OHC				4-PT FLEXURE					
		Stiffness, MPa/mm		Strength, MPa		Modulus, GPa		Strength, MPa		Strain @ failure, %	
MATERIALS		AVG.	%Δ	AVG.	%Δ	AVG.	%Δ	AVG.	%Δ	AVG.	%Δ
BMI/ IM7	Dry Cntrl	528	0	295	0	100	0	1054	0	1.06	0
	Wet Cntrl	530	0.4	307	4	101	1	1069	1.4	1.09	2.8
	Dry Spiked	560	6	288	-2.4	99	-1	1064	1	1.31	24
	Wet Spiked	481	-9	290	-1.7	96	-4	1058	0.4	1.32	25
K3B/ IM7	Dry Cntrl	656	0	322	0	100	0	1130	0	1.21	0
	Wet Cntrl	611	-7	343	7	102	2	1161	5.3	1.28	6
	Dry Spiked	630	-4	351	9	100	0	1113	1	1.19	-2
	Wet Spiked	4340	-34	229	-29	92	-8	571	-48	0.71	-41

TABLE III. 3-PT FLEXURE PROPERTIES OF BMI RESIN

SPECIMEN CONDITION	MODULUS, MPa			STRENGTH, MPa			STRAIN AT FAILURE, %		
	AVG.	SD	%Δ	AVG.	SD	%Δ	AVG.	SD	%Δ
DRY	3244	59	0	162	10	0	5.19	0.46	0
WET	3270	86	0.8	153	22	-6	4.87	0.95	-6
DRY SPIKE	3255	35	0.3	141	11	-13	4.41	0.45	-15
WET SPIKE	3312	23	2.1	125	14	-23	3.75	0.42	-28

in moisture during thermal spiking. However, it is of interest to note that the residual flexure properties of BMI neat resin after thermal spikes were considerably lowered, i.e., BMI neat resin was already damaged by moisture exposure alone or dry thermal spike alone, and further damaged by combination of moisture and thermal spikes. This suggests possible changes of hygrothermal degradation mechanisms between neat resin and composites, for example, the role of fiber-matrix interface or interphase under hygrothermal exposures.

Dry T_g changes of both neat resins and composites were determined after various exposures. BMI systems show most changes, i.e., dry T_g drops up to 5°C after wet exposure probably due to residual moisture, but increases about 10°C after spikes mostly due to cure advancement. However, it should be noted that no T_g changes in K3B systems was observed from wet and spike exposures even though density of K3B composite dropped more than 40% due to hygrothermal induced micro-cavitation and blistering. This suggests that the hygrothermal spiking does not cause any chemical de-polymerization in K3B systems.

Effects of Hygrothermal Aging

An accelerated hygrothermal aging experiment followed by a systematic drying program as a function of time and temperature has been developed to identify the mechanisms of the permanent "Dry" T_g drop in the hygrothermal exposed PI as discussed earlier. Pressurized steam in a sealed stainless steel pressure bomb was employed to accelerate aging and to obtain uniform moisture distribution thru the thickness of test specimens in a reasonably short exposure time. However, all test specimens were kept in Teflon container in pressure bomb to avoid possible galvanic attack. DMA (51 mm×9.8 mm×1.5 mm) and density specimen (10 mm×10 mm×10 mm) were prepared from K3B resin plaque. The steam pressure was about 110 psi at 160°C. The samples were aged for 40 days, then weight loss, T_g and density were monitored as a function of drying temperature and time.

Figure 6 illustrates overall weight changes through the entire aging process. The weight change data plotted in this figure was normalized by the final weight after the additional drying at 210°C for 10 days in vacuum oven. It should be noted that when dried at lower temperatures, such as 80°C and 130°C, there were still some residual moisture remained in the system regardless of drying time. From this data, it can be concluded that the water molecules locked into molecular network structure at high temperatures can only diffuse out upon subsequent exposure to the same high temperatures or higher.

During drying density of the specimens was monitored as a function of time and temperature, Figure 7. Interestingly enough, the data indicated three physical transitions; (i) initial decrease of density at all

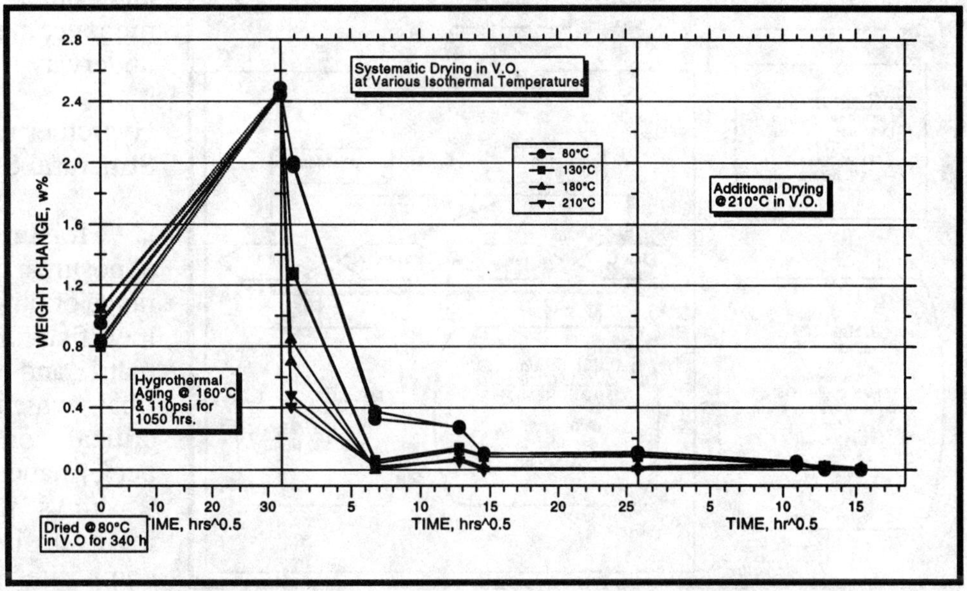

Figure 6 Moisture Absorption-Desorption Data for K3B Resin.

temperatures which is mainly contributed by the loss of moisture, (ii) sharp drop of density at the very early drying at 210°C due to hygrothermal induced micro-cavitation which was discussed in thermal spiking program (iii) density re-increase after longer drying time caused by densification thru physical annealing process, especially at higher temperatures.

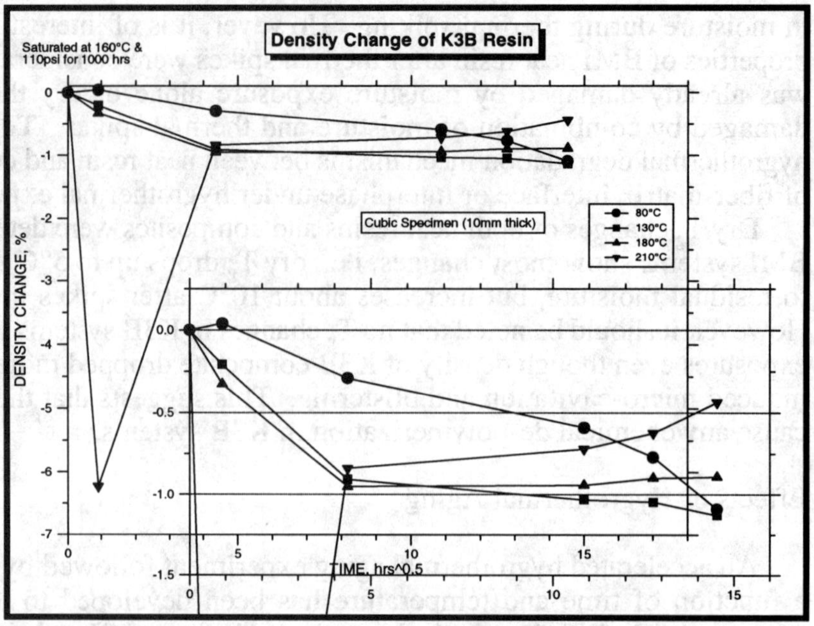

Figure 7 Density Changes of K3B Resin

Another systematic studies of "pressure-bomb" hygrothermal aging experiment is being conducted jointly with Wright Patterson Air Force Materials Laboratory in order to investigate high temperature, moisture-induced degradation mechanisms of PI matrices including both K3B and AFR700B. The overall program plan is summarized in Figure 8. D_2O water was used instead of H_2O for ^2H NMR studies to distinguish between bonded and non-bonded hydration (hydrolysis and exchange). We will report controlled isothermal temperature-time weight loss and other physical property measurements presently underway.

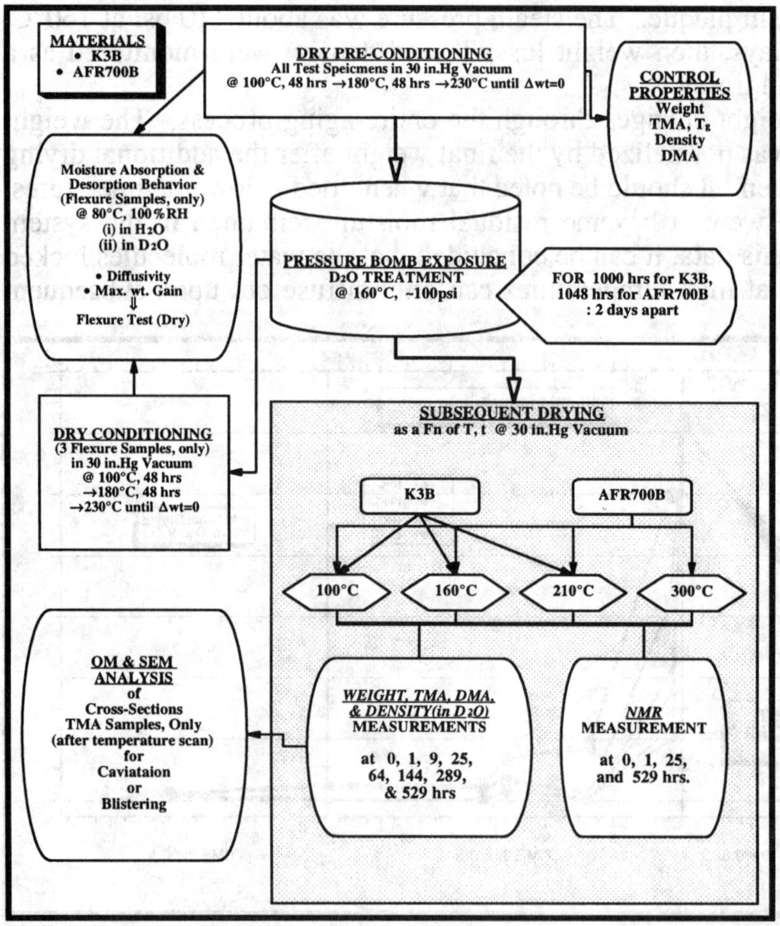

Figure 8 Hygrothermal Aging Program Plan

Hygrothermal Induced Physical Structural State

Moisture-temperature-time exposures can lead through a range of physical structural state modifications of the polymer matrix and its interface with the fiber which will modify the critical composite mechanical performance threshold. Figure 9 shows future requirements involved in high temperature composite applications. Efforts have been made to develop the physical state phase diagram as a function of various hygrothermal durability variables and materials variables. Figure 10 is a typical 2-D moisture concentration-temperature plots showing various hygrothermal induced polymer physical structural phase, such as homogeneous elastic swelling and inhomogeneous elastic cavitation, both of which are reversible upon removal of moisture, moisture molecular "lock-in" as discussed earlier, plastic swelling or blistering, and densification etc. Various modified phase diagram will be developed based on our hygrothermal durability program results.

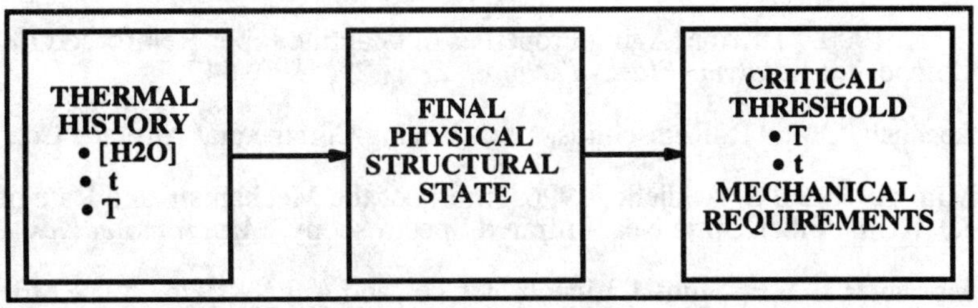

Figure 9 High Temperature Composite Future Requirements

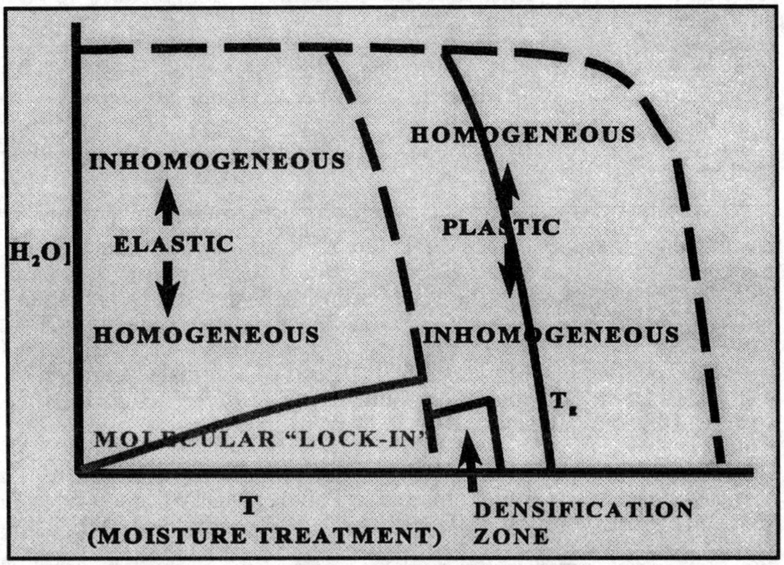

Figure 10 Hygrothermal Induced Physical Structural State Phase Diagram

ACKNOWLEDGMENT

The authors thank Air Force Office of Scientific Research (AFOSR), sponsored by Dr. Charles Lee, and State of Michigan Research Excellence Funds for financial support and encouragement of this work.

REFERENCES

Burchan, L.J., R.F. Eduljee and J.W. Gillespie, 1995, "Investigation of the Microcracking Behavior of Bismaleimide Composites During Thermal Aging," *Polymer Composites,* 16, 507.

Choi, J.K., A. Lee, E. Shin, and R. Morgan, 1996, "Cure Kinetics and Network Formation of Bismaleimide (BMI) Resins," *Proceedings of 28th Central Region, ACS Meeting,* June 1996, Dayton, Ohio, p III-A1.

Cornelia D., 1995, "BMI Mechanical Properties and Performance," *Proc. Of High Temple Workshop XV,* Santa Fe, New Mexico, January 16-19, Paper C-1.

Cornelia, D., 1996, "Long Term Hygrothermal Durability of Composites," *Proc. Of High Temple Workshop XVI,* Orange Beach, Alabama, January 29-February 1, Paper W-1.

Curliss, D.B., 1996a, "Solid-State NMR Investigation of Processing and Environmental Durability of High-Temperature Polyimides," *Proc. Of 28th Central Region ACS Meeting,* Dayton OH, Paper III-C-1~13.

Curliss, D. B., 1996b, "Hygrothermal Studies on Fluorinated Polyimides-Model Compound Studies of Cure and Degradation", *Proceedings of the 28th International SAMPE Technical Conference*, November 4-7, Seattle, WA. Vol. 28, pp

Hoogstenden, B and R. Kuhbander, 1995, "Effect of Thermal Cycling Upon AFR 700B Composites," *Proc. Of High Temple Workshop XV,* Santa Fe, New Mexico, January 16-19, Paper U-1.

Li, L., 1990, "Thermal Aging Properties of Graphite Fiber Reinforced PEEK and Graphite Fiber Reinforced BMI Composites," *Polym.-Plast. Technol. Eng.*, 29(5&6), 549.

Loechett, C. and R. Rothschilds, 1992, Boeing Commercial Airplane Co., Seattle, WA, (Personal Communication)

Mijovic, J. and S. Andjelic, 1996, "Study of the Mechanism and Rate of Bismaleimide Cure by Remote in-Situe Real Time Fiber Optic Near-Infrared Spectroscopy," *Macromolecules,* 29, 239.

Morgan R.J., E.E. Shin, J. Lincoln, J. Choi, and A. Lee, 1996a, "The Structure-Property Relations of Bismaleimide Composite Matrices and Durability of High Temperature Polymer Matrix Composites," *Proceedings of the 28th International SAMPE Technical Conference*, November 4-7, Seattle, WA. Vol. 28, pp 213-223.

Morgan, R.J., 1996b, "Thermal Characterization of Composites," in *Thermal Analysis of Polymers,* 2nd Edition, Ed. E.Turi, Academic Press, Chap. 9.

Morgan R.J., E.E.Shin, B.A.Rozenberg and, A. Jurek, 1997, "Characterization of the Cure Reactions of Bismaleimide Composite Matrices", *Polymer*, Vol. 38, No. 3, p639.

Pederson, C.L., J.W. Gillespie, R.L. McCullough, R.J. Rothschilds and S.L. Stanek, 1995, *Polymer Composites,* 16, 154.

Price, W.A., B.P.Rice, A.S.Crasto, K.A.Thorp, K.M.Johnson, and D. Cornelia, 1995, "Hygrothermal Aging of Imide Composites," *Proc. Of High Temple Workshop XV,* Santa Fe, New Mexico, January 16-19, Paper S-1.

Rice, B.P., 1996, "Recent Studies Concerning Hydrolytic Degradation of High Temperature Polyimides," *Proc. Of High Temple Workshop XVI,* Orange Beach, Alabama, January 29-February 1, Paper U-1.

Shin, E.E. and Roger J. Morgan, 1993, "Effects of Thermal Cycling on Mechanical Properties of High Temperature Bismaleimide(BMI)-carbon Fiber Composites", *Conference Proceedings of ANTEC '93 on Plastic Engineering*, Society of Plastics Engineers, Vol II, May, p1357.

Shin E.E., R.Jurek, L.T.Drzal, R.J.Morgan, J.K.Choi, and A.Lee, 1995a, "Durability and Critical Fundamental Aging Mechanisms of High Temperature Polymer Matrix Carbon Fiber Composites: PART I," *Proceedings of the ASME Materials Division,* MD-Vol. 69-1, 1995 IMECE, ASME P183-189.

Shin E.E., C.Dunn, E.Fouch, R.J.Morgan, M.Wilenski, and L.T.Drzal, 1995b, "Durability and Critical Fundamental Aging Mechanisms of High Temperature Polymer Matrix Carbon Fiber Composites: PART II," *Proceedings of the ASME Materials Division,* MD-Vol. 69-1, 1995 IMECE, ASME P191-200.

Shin E.E., R.J. Morgan, M. Wilenski, J. Zhou, J. Lincoln, and L.T. Drzal, 1996, "The Effects of Processing Conditions and Service Environmental Exposures on High Temperature Polymer Matrix Carbon Fiber Composites," *Proceedings of the 28th International SAMPE Technical Conference*, November 4-7, Seattle, WA, Vol.28, pp 225-235.

Thorp, K.E. and A.S.Crasto, 1996, "Hygrothermal Cycling of AFR700B," *Proc. Of High Temple Workshop XVI,* Orange Beach, Alabama, January 29-February 1, Paper V-1.

Yee, A. and C.Soles, 1997, (Personal Communication). U. of Mich., Ann Arbor

Modeling the High-Temperature Degradation of Graphite/Epoxy

L. K. CREWS AND H. L. MCMANUS

ABSTRACT

In order to predict the response of graphite/epoxy composites exposed to fires or other severe high-temperature environments, modeling of the matrix material degradation and prediction of degraded composite properties must be addressed. Previous work is reviewed. An Arrhenius reaction model is proposed to model matrix degradation, and a semi-empirical micromechanics-based model is used for composite properties. Thermogravimetric analysis (TGA) tests are used to determine the reaction constants. The Arrhenius reaction model correlates well with data and existing temperature-dependent models. Tensile strength of axial, transverse, and $[\pm 45]_s$ graphite/epoxy laminates exposed to various temperatures is tested. Testing is carried out both at-temperature and after cooling to room temperature. Results are presented along with a summary of prior work. A preliminary composite property model that includes both temperature and permanent degradation effects is presented. Results are reasonable, although more data is needed.

INTRODUCTION

Advanced composite materials are vulnerable to fires and other severe high temperature environments. While graphite fibers are relatively unaffected by heat, polymer matrices lose properties at temperatures only slightly above their cure temperature. In order to be able to predict the behavior of composites exposed to high temperature environments, one must consider all aspects of their response, including the thermal response, degradation of the matrix, the thermomechanical response of the composite, damage, and failure. The flowchart in Figure 1 illustrates the approach. Each box represents a model which uses the output of the other models and information about the environment and applied loads as input. These models are often implemented as computer codes or subroutines. For example, a thermal response code predicts the temperature distribution within the structure. A chemistry code then determines the degree to which the matrix has degraded based on the temperatures calculated by the thermal response code. The degradation state and the temperature distribution, along with mechanical loads, are used to predict the thermomechanical response. The codes continue to build on each other in this fashion. The flowchart contains feedback loops that represent the fact that mechanical damage may change the interaction of the structure with the environment.

This paper addresses the issue of how best to model the thermal degradation of epoxy and its effects on the mechanical properties of graphite/epoxy composites, the areas highlighted in Figure 1. Here, degradation is defined as a permanent change in the material due to chemical

Lauren K. Crews, Massachusetts Institute of Technology, Room 41-219, Cambridge, MA 02139
Hugh L. McManus, Room 33-311, Cambridge, MA 02139

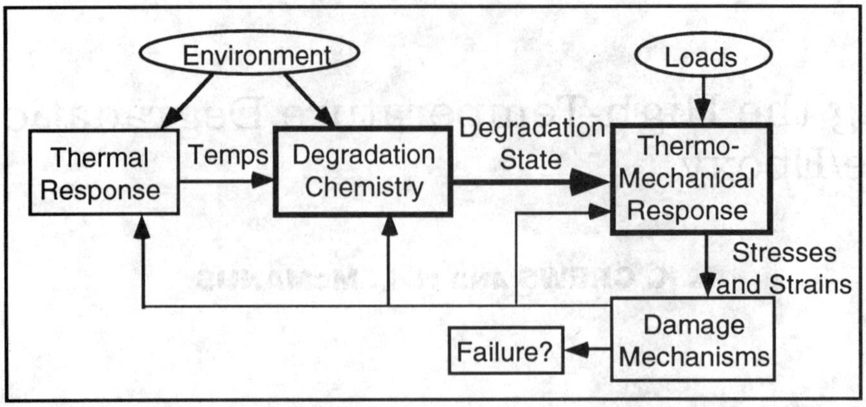

Figure 1. Comprehensive modeling flowchart

reactions. Degradation chemistry models are reviewed, and a model developed and quantified using TGA tests. The model predicts matrix degradation at a given location in terms of a parameter α, called degradation state, which represents the relative mass loss out of an infinitesimal volume centered at the given location. This mass-loss parameter is selected because it is easily measured in uniformly heated specimens; it is *assumed* that it is a useful metric of chemical degradation. The first step in determining the thermomechanical response is to determine the material properties as functions of temperature and α. Here, limited tensile strength data is collected. Additional data is also assembled from the literature. Strength is correlated with the temperature and degradation state of the specimens, and with models available in the literature.

BACKGROUND

Previous investigators have developed models of the thermal response, reaction chemistry, and thermomechanical response of composites, including the accumulation of damage and failure, with varying levels of complexity and accuracy. Pering, Farrell, and Springer (1980), Griffis, Masumura, and Chang (1981), Griffis et al. (1986), Henderson and Wiecek (1986), and Milke and Vizzini (1991) each developed increasingly complex models of the thermal behavior of composites. In general, the thermal response of fire-exposed composites is well understood, although it is complicated by the anisotropy and inhomogeneity of the material and the lack of agreement as to the value of material properties such as transverse thermal conductivity.

Two models have been used to track the reaction chemistry of the degrading epoxy. Griffis, Masumura, and Chang (1981), Chen, Sun, and Chang (1985), Griffis et al. (1986), and Milke and Vizzini (1991) all accounted for the degradation of epoxy by using temperature-dependent material property data originally collected by Menousik and Monin (1979) and later modified slightly by Griffis, Masumura, and Chang (1981). This data shows sharp transitions in the specific heat, density, and thermal conductivity at temperatures corresponding with the onset and completion of the degradative reactions. Figure 2 shows the specific heat data used in these studies. The spike in the data, between 343°C and 510°C, represents the additional heat absorbed by the epoxy as it degrades. The spike corresponds to a linear change in the density of the epoxy from 1.60 g/cm^3 to 1.15 g/cm^3 (Griffis, Masumura, and Chang, 1981).

Pering, Farrell, and Springer (1980) used an Arrhenius-type reaction model of the matrix degradation. The model used mass loss as an indicator of degradation state, and the rate of degradation was dependent on both temperature and time of exposure. Reaction constants were fit to mass loss data measured from quasi-isotropic graphite/epoxy panels exposed to gas flame burners.

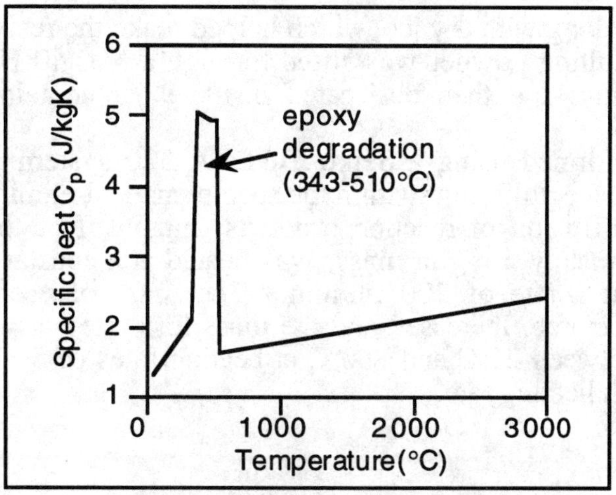

Figure 2. Temperature-dependent specific heat data
(from Griffis, Masumura, and Chang, 1981)

In modeling the thermomechanical response, investigators have primarily used temperature-dependent mechanical material properties in conjunction with models of structural response. Chen, Sun, and Chang (1985) modeled stiffness, Poisson's ratio, and coefficients of thermal expansion using the rule of mixtures in conjunction with a glass transition temperature of 200°C for epoxy. Strengths were modeled by assuming that the strength of the epoxy decreases linearly to zero between 200°C and 260°C, bringing transverse and shear strengths to zero and diminishing the axial tensile strength. Ha and Springer (1987) measured the properties of neat resin at various temperatures, fit an exponential expression to describe property loss as a function of temperature, and then used rule of mixtures models and Tsai-Hahn models to determine composite properties at various temperatures based on temperature-dependent epoxy properties and constant fiber properties. McManus and Chamis (1996) developed a model using both temperature and degradation state to determine changes in matrix strength and stiffness properties. These properties were in turn used to find composite properties using micromechanics. For matrix-dominated properties such as transverse strength, the model predicts properties with the form

$$\frac{P}{P_o} = \sqrt{\frac{T_g - T}{T_g - T_o}}\left(1 + s\alpha^n\right) \tag{1}$$

where P is the property of interest, T_g is the glass transition temperature, T is the temperature at which the property is being calculated, α is the mass loss-based degradation state, s and n are fit parameters, and the subscript o refers to reference values.

EXPERIMENTAL WORK

TGA TESTS

Thermogravimetric analysis (TGA) tests were performed on powdered neat 3501-6 epoxy resin. The resin was cured into plaques approximately 2 mm thick by placing pellets of frozen, uncured 3501-6 epoxy into a pan and heating in a 180°C oven in air for eight hours at ambient pressure. Heat first caused a decrease in viscosity, allowing the resin to flow. The resin then gelled and cured. The thermal treatment of the epoxy plaques was equivalent to a standard cure and postcure of the graphite/epoxy laminates studied. The resin plaques were ground in a standard coffee grinder, along with dry ice, which helped make the resin more brittle and hence easier to grind. The resulting powder was sifted through a No. 40 U.S.A. Standard Testing Sieve (425 micron grating) and then desiccated inside a jar containing anhydrous calcium sulfate.

TGA tests were performed using a Seiko TG/DTA 320 system. The use of powdered specimens ensured thermal equilibrium within the specimens and eliminated the possibility that transport effects (e.g. diffusion of reaction products) might affect the results. Each small sample, initially approximately 5 mg in mass, was heated at a constant rate inside a chamber flushed with nitrogen at a rate of 200 ml/min. The mass of each sample was recorded continuously as the temperature increased and the mass loss rate was calculated continuously. Tests were performed between 25°C and 800°C at heating rates of 5, 10, and 15°C/min. One sample was tested at each heating rate.

STRENGTH TESTS

At-temperature and residual (i.e. post cooldown) tensile strength tests were performed on AS4/3501-6 graphite/epoxy coupons with lay-ups of $[0]_4$, $[90]_{12}$, and $[\pm 45]_s$. The coupons were cut from panels vacuum-bag cured in an autoclave for 2 hours at 180°C and 85 psi, and air postcured at the same temperature for eight hours. An Instron 8500 mechanical testing machine was used to apply the loads, and load, stroke, and temperature data was collected at 2 Hz. Coupons were 5 cm wide and 71 cm long and had 5 cm by 7.5 cm glass loading tabs bonded to each end. The unusual length was necessary because a 46 cm-tall ATS series 2961 split cylinder furnace surrounded the coupons, which had to be gripped above and below the furnace, outside of the heated zone. The specimens, initially held only in the top grip of the testing machine, were free to expand as they heated. The heating rate was approximately 15°C/min. When the test temperature was reached, the specimens were soaked at the test temperature for ten minutes. For at-temperature strength tests, the bottom loading tab of the specimen was gripped after the soak period, and the specimen was loaded in tension to failure. Loading was performed under stroke control at a rate of 2 mm/min. For residual strength tests, the furnace was turned off after the soak period. The specimen cooled to room temperature at approximately 20°C/min. The bottom loading tab was gripped and the specimen was loaded to failure under stroke control at a rate of 2 mm/min.

RESULTS

TGA TESTS

Reaction rate results as a function of temperature for different heating rates are shown in Figure 3. The rate of change in α was calculated by dividing the raw mass loss rate data by the total mass loss of each specimen:

$$\alpha = \frac{\Delta m}{\Delta m_{max}} = \frac{m_i - m}{m_i - m_f} \tag{2}$$

$$\frac{d\alpha}{dt} = \frac{dm/dt}{m_i - m_f} \tag{3}$$

where α is the degradation state metric, ranging from 0 to 1; m refers to mass; and subscripts i and f indicate initial and final quantities, respectively. The mass of unreactive char is m_f.

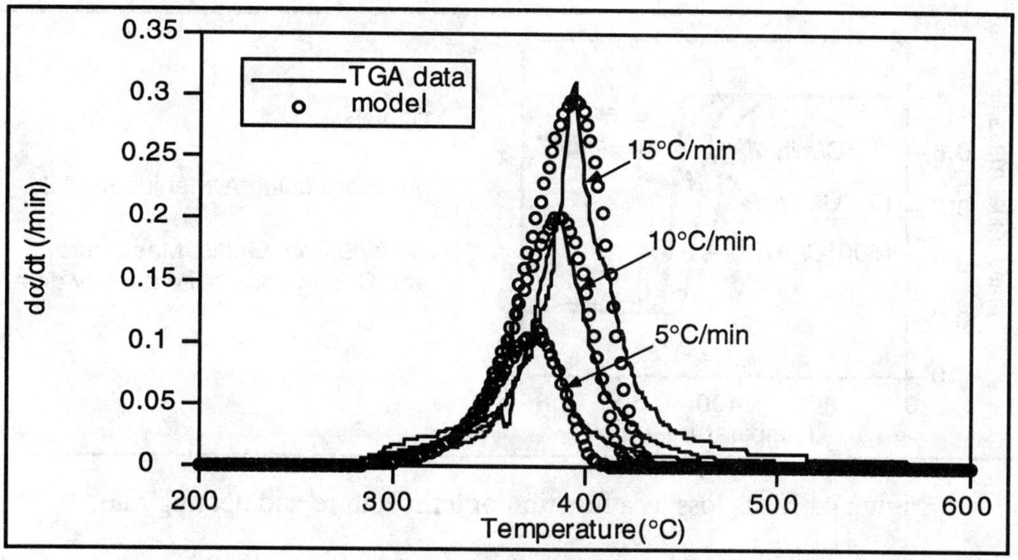

Figure 3. Reaction rate data and model

The shift in the peaks of the reaction rate curves in Figure 3 down and towards lower temperatures as the heating rate decreases is characteristic of a reaction which can be modeled using an Arrhenius equation of the form

$$\frac{d\alpha}{dt} = k(1-\alpha)^n \exp\left(\frac{-E}{RT}\right) \tag{4}$$

where k is the rate constant, n is the reaction order, E is the activation energy of the reaction, R is the universal gas constant (8.314 J/molK), and T is absolute temperature. Eq. (4) was fit to the three data curves in Figure 3. Kissinger's method (Cunningham, 1997) was used to evaluate the reaction constants based on the location of the reaction rate peaks. The absolute temperature T corresponding with the reaction rate peak was determined for each TGA test, and $\ln(Q/T^2)$ was plotted against (1/T), where Q refers to heating rate in (°C/min). The slope and intercept of the plot were used to determine activation energy and rate constant respectively. The calculated constants were activation energy E = 185 kJ/mol, and rate constant k = 2.34×10^{14}/min, with an assumed reaction order n = 1. The results of the model are superimposed on the TGA reaction rate data in Figure 3. The fit is not perfect, but the model captures the reaction rate peak magnitudes and the peak shifts.

Figure 4 offers a direct comparison of the mass loss data used by many previous investigators (assumed to be rate independent) with the mass loss results generated using the Arrhenius reaction constants determined above. The mass loss data was derived from the composite density data used by Griffis, Masumura, and Chang (1981) assuming a matrix volume fraction of 35% and no mass loss from fibers. The heating rate employed in the experiments from which the density data was obtained was unspecified, however the heat source was a laser, hence one may assume a high heating rate. The rate-based curves were generated, using the Arrhenius constants determined above, by integrating Eq. (6):

$$\alpha = \int \frac{\partial \alpha}{\partial t} dt = \int Q \frac{\partial \alpha}{\partial t} dT \tag{5}$$

$$\frac{m}{m_i} = 1 - \left(1 - \frac{m_f}{m_i}\right) \int_{T_i}^{T_f} Qk(1-\alpha)^n \exp\left(\frac{-E}{RT}\right) dT \tag{6}$$

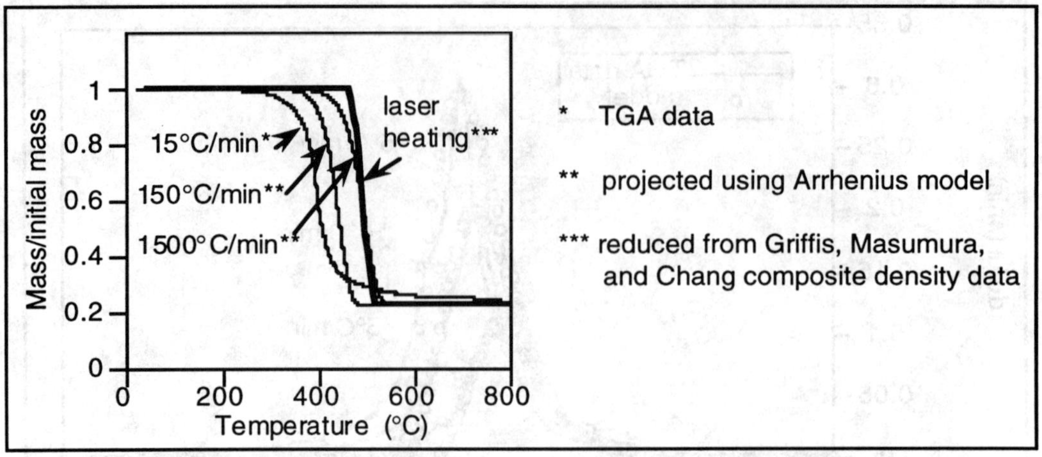

Figure 4. Mass loss as a function of temperature and heating rate

where Q represents heating rate (in °C/min). The average residual mass percentage (i.e. the char yield), m_f/m_i, was measured in the TGA experiments as 23%.

STRENGTH TESTS

The at-temperature and residual strengths measured in the current experiments are plotted, along with results from previous investigators, in Figures 5, 6, and 7. Previous data is taken from Greszczuk (1984) (as reported in Griffis et al. (1986)), Weems and Rini (1985), Hinger (1986), Ha and Springer (1987), and Ottaviano and Yeh (1996). The two symbols used to display the current data (by Crews) represent the two different test conditions: white triangles indicate at-temperature strengths, while black triangles indicate residual strengths. All other authors reported at-temperature strength except Ottaviano and Yeh, who reported residual strength.

Although all available data are shown here as functions of temperature, these plots are somewhat misleading. First, much of the data shown was collected at temperature, so both permanent effects of degradation and reversible effects of temperature are present. Also, material, specimen preparation, thermal environment (e.g. heating rate and soak time), and test technique all varied between investigations. Hence, maximum temperature reached is not the sole determinant of material degradation and material properties.

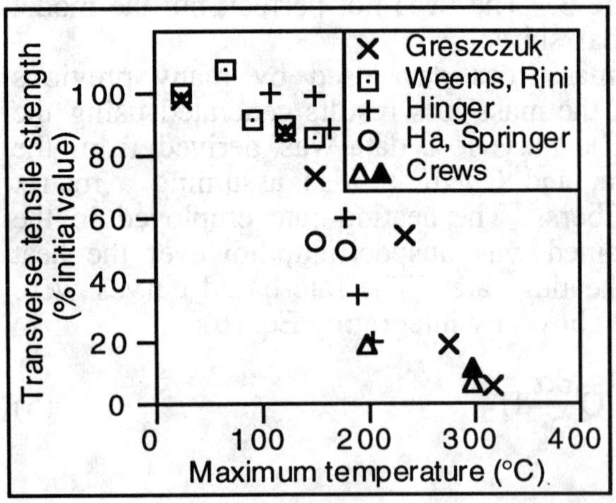

Figure 5. Transverse strength as a function of temperature

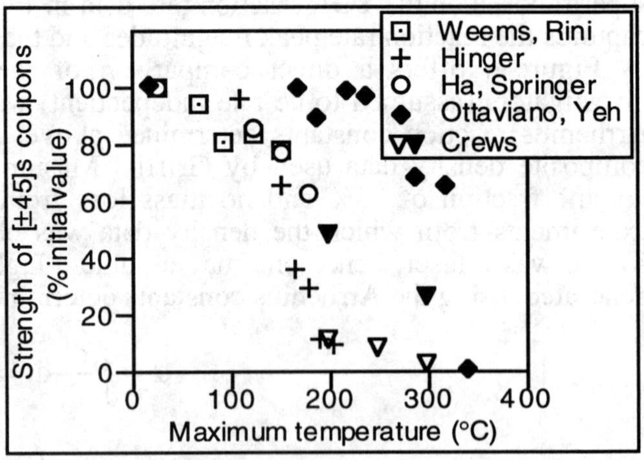

Figure 6. $[\pm 45]_s$ tensile strength as a function of temperature

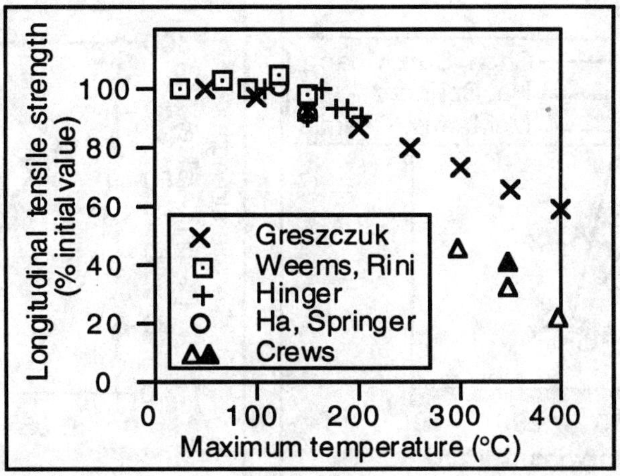

Figure 7. Longitudinal strength as a function of temperature

In an effort to determine if degradation state might be a useful metric for evaluating permanent strength loss, the strengths measured in the current investigation, along with those of Ottaviano and Yeh (1996), were plotted as a function of degradation state, α. Degradation state was calculated by using measured time vs. temperature curves for each specimen and integrating Eq. (4). The values of the final degradation states spanned several orders of magnitude, from 3.3×10^{-8} to 1. The results are plotted in Figures 8 and 9. The notation is similar to that used previously; white symbols denote at-temperature strengths, while darkened symbols denote residual strengths.

CORRELATION WITH MODELS

Available material property models were compared with available data. Figure 10 compares the transverse tensile strength data from Figure 5 with models from Chen, Sun, and Chang (1985), Ha and Springer (1987), and McManus and Chamis (1996). Values for the Chen, Sun, and Chang model were taken directly from their paper. The Ha and Springer model for transverse tensile strength employed a Tsai-Hahn micromechanics expression with two fitting parameters, whose values were taken from their paper. The McManus and Chamis curve in Figure 10 was generated using Eq. (1) assuming T_g=240°C and α=0.

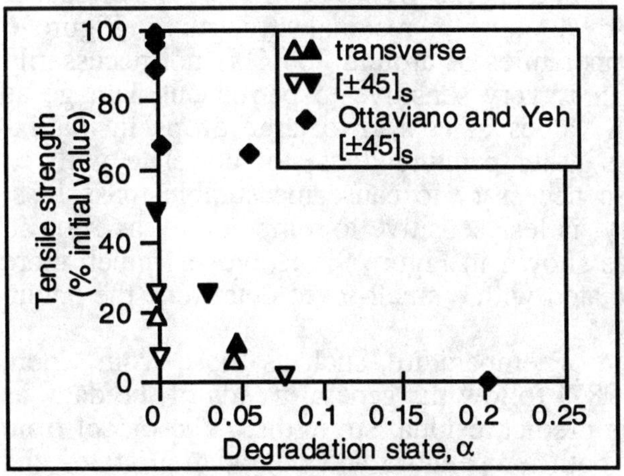

Figure 8. Matrix-dominated strength as a function of degradation state

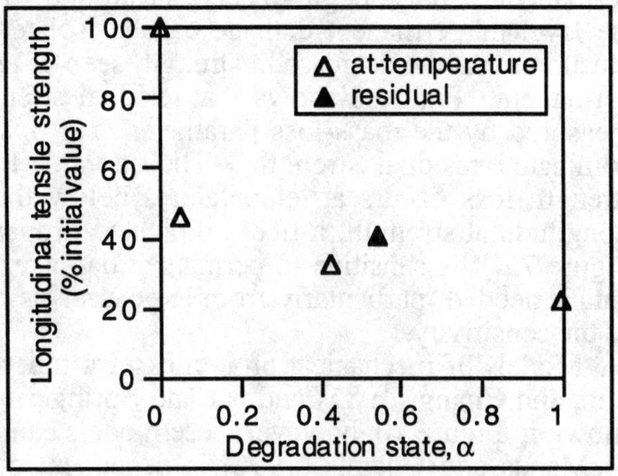

Figure 9. Fiber-dominated strength as a function of degradation state

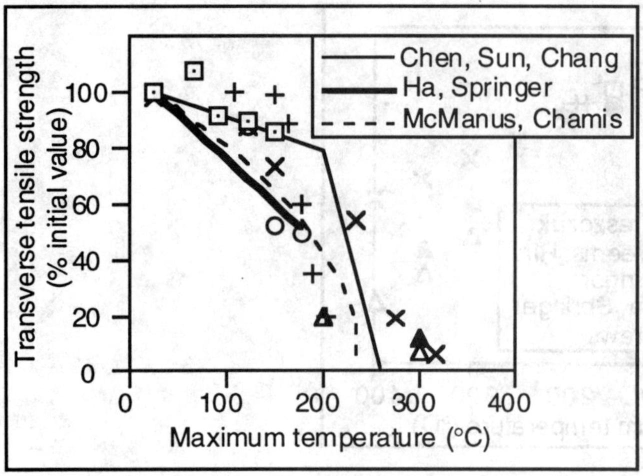

Figure 10. Comparison of transverse tensile data with available temperature-based models

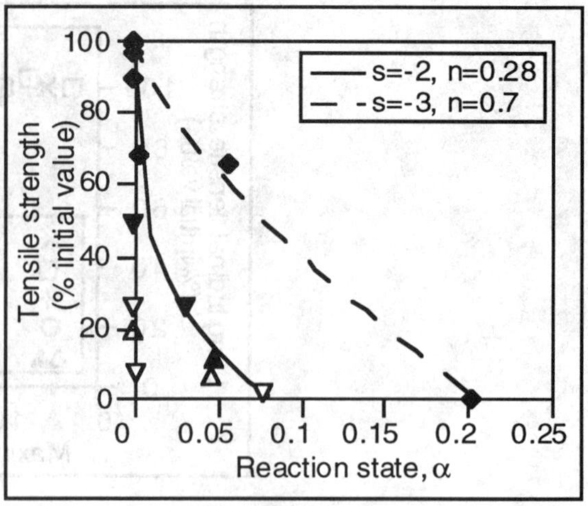

Figure 11. Comparison of matrix-dominated strength data and degradation state-based model

Figure 11 compares the tensile strength data plotted in Figure 8 with the McManus and Chamis model. Eq. (1) was used to generate the curves for the model. Temperature was taken to be room temperature, and various values were used for the fitting parameters s and n.

DISCUSSION

The correlation between the Arrhenius reaction-dependent model and observed TGA results, as shown in Figure 3, indicates that the permanent degradation of epoxy as measured by mass loss can be reasonably modeled by a single Arrhenius equation. This is useful because in addition to predicting the temperature dependence of the degradation, the model can capture the effects of varying heating rates and soak times.

Mechanical material properties depend on both ambient temperature and the level of permanent damage. Figures 5 and 6 show that matrix-dominated properties such as transverse strength and the strength of $[\pm45]_s$ laminates (related to shear strength) are very sensitive to temperature. The at-temperature strengths, dependent on temperature and the level of permanent damage, begin to decrease around 150°C and approach zero at 300°C. The effects of the level of permanent damage can be isolated by looking at residual strengths. Figure 6 illustrates that the decrease in strength seen at temperatures of around 200°C is not necessarily permanent. Figure 8 shows that residual strength is very sensitive to permanent damage as measured by the mass-loss parameter, α. Small values of α lead to large drops in matrix-dominated residual strength. The strength loss at-temperature does not correlate with α; strength loss occurs at temperatures below those necessary to cause measurable mass loss. Longitudinal strength, a fiber-dominated property, is less sensitive to temperature, as seen in Figure 7. It is sensitive to permanent damage, as shown in Figure 9. However, much more data is needed, particularly from exposures associated with a small α, to determine the nature of the sensitivity.

Models of mechanical properties as a function of temperature, such as those from Chen, Sun, and Chang (1985) and Ha and Springer (1987) follow the general trends of the data, as shown in Figure 10, however these models cannot predict residual strengths. The model from McManus and Chamis (1996) is dependent on both temperature and degradation state, and hence may be useful for predicting both at-temperature and residual mechanical properties. Figure 11 shows two possible curve fits to the matrix-dominated tensile strength data using the

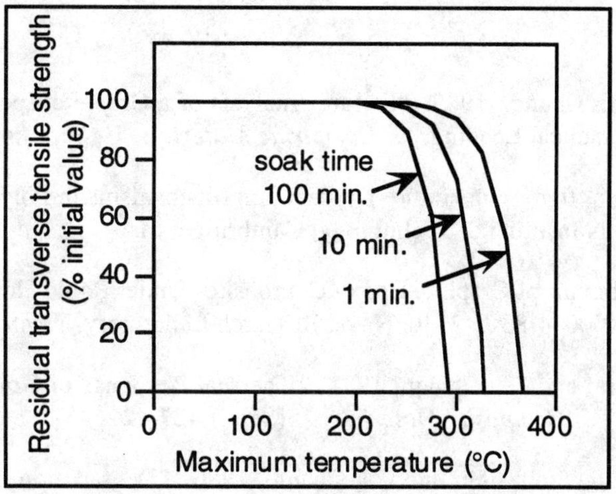

Figure 12. Predicted residual transverse tensile strength

McManus and Chamis model. More data is necessary to determine if this model is truly appropriate, and if so, to accurately determine the fit constants.

An accurate material property model based on temperature and degradation state will allow one to predict the at-temperature and residual material properties of composites exposed to a wide variety of high temperature environments. The results of the best-effort model generated here are shown in Figure 12, which was generated by using the Arrhenius reaction model developed above to calculate the degradation states of specimens exposed to different temperatures for given soak times, and using the McManus and Chamis (1996) model with s= -2, n=0.28 to determine residual strengths. This is a very preliminary example, but it represents a goal of this modeling effort. It is interesting to note that despite the relative crudeness of the fits used to generate the models, the predictions are reasonable, and consistent with existing data (e.g. the residual strength data in Figures 5 and 6).

CONCLUSIONS

- Reaction rate data collected from TGA tests has enabled the development of a simple thermal degradation model which predicts degradation (as indicated by mass loss) as a function of thermal history.
- Mechanical properties are dependent on both temperature and the level of permanent degradation. Most published data was taken at the exposure temperature, and thus includes both temperature and degradation effects. More residual strength data is needed to isolate the dependence of material properties on degradation state.
- A proper model of mechanical properties must include the effects of both temperature and degradation state. The McManus and Chamis (1996) model has this form. It gives surprisingly reasonable predictions using parameters fit to existing data. However, more data is necessary to determine if the model is appropriate and, if so, to accurately determine the values of the model parameters.
- A verified material property model will allow the user to determine residual and at-temperature mechanical properties for specified thermal histories. This is a key step in developing the ability to predict the effects of high temperature exposure on composites.

ACKNOWLEDGMENTS

This work was sponsored by the National Science Foundation under NSF Grant 9257612-CMS, and made use of MRSEC Shared Facilities supported by NSF award DMR-9400334.

REFERENCES

Chen, J.K., C.T. Sun, and C.I. Chang. 1985. "Failure Analysis of a Graphite/Epoxy Laminate Subjected to Combined Thermal and Mechanical Loading," *J. Composite Materials*, 19(Sept.):408-23.

Cunningham, Ronan. 1997. "High Temperature Degradation Mechanisms in Polymer Matrix Composites," Master's thesis, Massachusetts Institute of Technology, Cambridge, MA.

Greszczuk, L.B. 1984. "Behavior of Graphite Epoxy Composites Under Rapid Heating, Rapid Loading," Final Report, NRL Contract No. N00014-83-C-2330, Naval Research Laboratory, Washington, DC.

Griffis, C.A., R.A. Masumura, and C.I. Chang. 1981. "Thermal Response of Graphite Epoxy Composite Subjected to Rapid Heating," *J. Composite Materials*, 15(Sept.):427-42.

Griffis, C.A., J.A. Nemes, F.R. Stonesifer, and C.I. Chang. 1986. "Degradation in Strength of Laminated Composites Subjected to Intense Heating and Mechanical Loading," *J. Composite Materials*, 20(May):216-35.

Ha, S.K. and G.S. Springer. 1987. "Mechanical Properties of Graphite Epoxy Composites at Elevated Temperatures," presented at the 6th International Conference on Composite Materials and 2nd European Conference on Composite Materials, London, England.

Henderson, J.B. and T.E. Wiecek. 1987. "A Mathematical Model to Predict the Thermal Response of Decomposing, Expanding Polymer Composites," *J. Composite Materials*, 21(April):373-93.

Hinger, J. 1985. "Temperature Effects on Graphite/Epoxy," Master's thesis, Massachusetts Institute of Technology, Cambridge, MA.

Menousik, J.F. and D.L. Monin, D.L. 1979. "Laser Thermal Modeling of Graphite Epoxy," Naval Weapons Center Technical Memorandum 3834.

McManus, H.L. and C.C. Chamis. 1996. "Stress and Damage in Polymer Matrix Composites Due to Material Degradation at High Temperatures," NASA Technical Memorandum 4682.

McManus, H.L., and G.S. Springer. 1992a. "High Temperature Thermomechanical Behavior of Carbon-Phenolic and Carbon-Carbon Composites, I. Analysis," *J. Composite Materials*, 26(2): 206-29.

McManus, H.L., and G.S. Springer. 1992b. "High Temperature Thermomechanical Behavior of Carbon-Phenolic and Carbon-Carbon Composites, II. Results," *J. Composite Materials*, 26(2):230-55.

Milke, J.A. and A.J. Vizzini. 1991. "Thermal Response of Fire-Exposed Composites," *J. Composites Technology and Research*, 13(3):145-51.

Ottaviano, M.J., and J. Yeh. 1996. "Fire Damage to Composite Structures," Final Report for 16.622 Project, Department of Aeronautics and Astronautics, Massachusetts Institute of Technology, Cambridge, MA.

Pering, G.A., P.V. Farrell, and G.S. Springer. 1980 "Degradation of Tensile and Shear Properties of Composites Exposed to Fire or High Temperature," *J. Composite Materials*, 14(Jan.):54-68.

Weems, D.B., and L.S. Rini. 1984. "Tensile Behavior of Graphite/Epoxy at Elevated Temperatures," TELAC Report 84-11, Department of Aeronautics and Astronautics, Massachusetts Institute of Technology, Cambridge, MA.

Moisture Diffusion in a Graphite/ Epoxy Composite

T. A. BOGETTI,[1] C. P. R. HOPPEL,[1] J. F. NEWILL,[1] J. ELWOOD[2] AND J. W. GILLESPIE, JR.[2]

ABSTRACT

The objective of this work was to characterize the moisture diffusion rates of a graphite epoxy material fabricated from two different quality grades of pre-preg. The material investigated was a T650/1914 graphite epoxy system. One pre-preg for this material was considered to be of a high quality grade (HQ), typically used in high performance applications, while the second pre-preg was representative of a lower quality pre-preg (LQ), which would more likely be used in high volume production applications. The material system fabricated from the HQ pre-preg exhibited significantly lower diffusion rates than that from the LQ pre-preg ($3.74\text{x}10^{-6}$ in.^{2}/sec. versus $6.05\text{x}10^{-6}$ in.^{2}/sec., respectively). The differences in observed diffusion rates between the two were attributed to the much lower void volume content of the material made from the HQ pre-preg than that of the LQ pre-preg (0.65% versus 1.86%, respectively). Due to the unique configuration of the moisture specimens used in this study, the diffusion rates measured were characteristic of in-plane laminate direction diffusion, as opposed to the typically reported through-the-thickness values.

INTRODUCTION

Moisture absorption is a significant design consideration for polymer matrix composites. In composites, water can cause weight gain, hygrothermal expansion, and degradation of material properties [Pipes et al., 1976; VanLandingham et al., 1995; Hoppel and Pangborn, 1994]. In general, polymers are hygroscopic in nature and therefore will absorb moisture over time. The moisture diffused into the polymer matrix disrupts the local structure and causes the polymer to expand. Low levels of water in the polymer can cause a reduction in the glass transition temperature (T_g) and the elastic modulus of the matrix [McCrum et al., 1988]. Higher levels of moisture can cause cracks and delaminations to

[1] T. A. Bogetti, C.P.R. Hoppel, J.F. Newill, Army Research Laboratory, WMRD, APG, MD 21005
[2] J. Elwood, J. W. Gillespie, University of Delaware, CCM, Newark, DE 19716

form in the composite. Moisture-induced expansion can also cause significant problems where close structural tolerances are required.

Moisture diffusion into a composite structure is generally described by Fick's second law (Equation 1) [Pipes et al., 1976].

$$D\frac{\partial^2 M}{\partial z^2} - \frac{\partial M}{\partial t} = 0 \qquad (1)$$

In Equation 1, D is a diffusion constant, M is the moisture content, t is time, and z is the depth of moisture penetration. The diffusion constant, D, is an Arrhenius function of the temperature, with diffusion increasing with increasing temperature [Springer, 1988]. Equation 1 is generally accepted as a reasonable approximation for diffusion of moisture in graphite epoxy exposed to humid air [Springer, 1988].

While the intrinsic moisture diffusion characteristics of a composite will depend on the particular system, other factors can have a profound effect on moisture diffusion as well. Significant deviations from Fick's law can occur at high temperatures, when materials are immersed in liquids, or when diffusion occurs through cracks or voids in the material [Springer, 1988; Cai and Weitsman, 1994]. Immersion of the composite in liquids reduces the activation energy of diffusion, causing moisture to be absorbed at a higher rate than predicted by Equation 1 [Woo and Piggott, 1987]. Cracks and voids facilitate moisture diffusion by increasing exposed surface area and acting as easy paths for moisture transport.

Eventually a maximum moisture content for a given surrounding humidity level is reached. The relation between the maximum moisture constant and the external humidity level is given by Equation 2.

$$M_m = a(RH)^b \qquad (2)$$

where, M_m is the maximum moisture content, a and b are empirically determined material constants, and RH is the surrounding relative humidity level [Tsai, 1988].

The objective of this work was to characterize the moisture diffusion rates of a graphite epoxy material which was fabricated from two different quality grades of pre-preg. The material investigated was a T650/1914 graphite epoxy material system. Qualitatively, the first pre-preg was considered to be of a high quality grade (HQ), typically used in high performance applications, while the second pre-preg was representative of a lower quality material (LQ), which would more likely be used in high volume production applications. The HQ pre-preg has a smoother or flatter surface than the LQ pre-preg. These differences stem from quality control standard and the rates at which fiber impregnation was done during the pre-pregging process.

This study was motivated by interest in the moisture diffusion behavior of a graphite epoxy material system that is used in an Army structural application. To gain the most realistic moisture diffusion rate data as possible, the specimens used in the characterization were machined directly from as-manufactured components of the structure.

Coupon sized specimens were machined from the structures and standard moisture diffusion rate characterization techniques were applied to identify differences in the moisture absorption behavior of the structure as a function of the pre-preg quality used. Significant differences in moisture absorption behavior were observed which led to further physical inspection of the material microstructure. Photomicrographs identified differences in material void content, which ultimately helped to explain the two distinct moisture diffusion rates.

It is interesting to note that due to the unique configuration of the moisture specimens used in this study, the diffusion rates measured were due to in-plane laminate direction diffusion as opposed to the typically reported through-the-thickness values. The in-plane direction is the x-y plane on Figure 1. Details of the experimental procedure and findings are summarized below.

EXPERIMENTAL PROCEDURE

SPECIMEN CONFIGURATION

Specimens were machined directly from as-manufactured composite structures. Machining was done using a water cooled circular diamond saw. For data reduction purposes, it is sufficient to note that specimen dimensions were nominally 0.9 cm wide by 12.2 cm long by 0.254 cm thick. The specimens were machined such that the significant moisture diffusion occurred through the in-plane laminate direction as shown in Figure 1.

A total of twelve specimens were machined for this study, six from each of the prepreg types (i.e., HQ and LQ). One of the HQ specimens, however, fractured during machining and was not included in the study.

Three thickness measurements across the length of each specimen were taken with a pair of digital calipers. The average value was used for data reduction purposes. No significant thickness variation over the length of any of the specimens was observed. All specimen weights were taken with a Mettler PM400 scale and recorded prior to the dry-out procedure.

DRY-OUT

Specimens were dried in a Fisher Scientific Isotemp Vacuum Oven, Model 282A for 14 days at 60°C (140°F) under a vacuum of 0.1 inch of mercury. Weight loss during dry-out was monitored daily for the first week then every other day for the second week. Specimens were allowed to cool down to room temperature before measurements were taken to minimize any thermal effects. Weight loss monitoring indicated that full dry-out was achieved at approximately 7 days.

MOISTURE CONDITIONING

After the dry-out period, specimens were environmentally conditioned in a Fisher Scientific 10" Desiccator, Model 683. A relative humidity level of 89% was maintained in the desiccator with a solution of potassium nitrate and water, as described in ASTM E104-85 [1991]. Aluminum stands were used to hold the specimens upright in order to maximize the surface area exposure to the environment. The desiccator was placed in a stainless steel container which was maintained at a constant 40°C / 104°F. The temperature of the container was held constant with a Micro DIN microprocessor-based temperature controller and an Omegalux Silicone Rubber Laminated Heating Blanket, No. 10960.

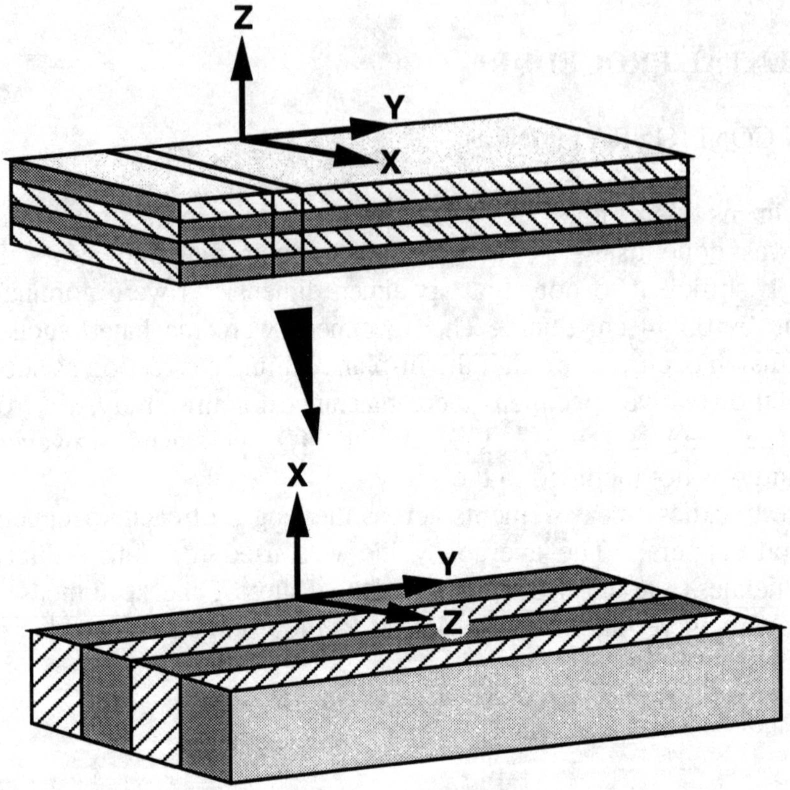

Figure 1. Moisture test specimen configuration. (Note that the test coupon is cut so that diffusion occurs primarily in the in-plane (x,y) direction.)

All the specimen weights were monitored and recorded during the moisture conditioning period. Measurements were taken every day for the first week and every other day thereafter. All specimens were found to become fully saturated within 50 days of the start of the conditioning.

VOID VOLUME ANALYSIS

Void volume analysis on the material was also performed. Two representative specimens from each prepreg type were used. Specimens were first dried, sectioned appropriately, potted and then polished. Photomicrographs were taken over a large sampling area of each of the polished specimen surfaces. A computer assisted digital enhancing technique was employed in the photomicroscopy. Void volume fractions were then obtained from each of the samples.

RESULTS AND DISCUSSION

The test specimen were conditioned for a 50 day period, a point at which all samples appeared to have become fully saturated. Moisture-induced weight gain (recorded in terms of weight percent) is plotted as a function of time for all specimens in Figure 2. From each of these curves, a diffusion rate constant was calculated according to Equation 3, as specified by ASTM Standard D 5229/D5229M-92 [1992].

$$D = \pi\left(\frac{h}{4M_m}\right)^2\left(\frac{M_2 - M_1}{\sqrt{t_2} - \sqrt{t_1}}\right)^2 \tag{3}$$

where h is the specimen thickness, M_m is the moisture saturation level and the quantity in the second set of parenthesis is the slope of the initial linear region on a plot of moisture weight gain versus the square root of time. This initial linear region is normally taken to be 60% of the time required for saturation. Diffusion rate coefficients for each of the samples represented in Figure 2 are plotted in Figure 3. The average diffusion rate coefficients, moisture saturation levels, and their respective standard deviations are given in Table 1 for the two materials.

A void volume analysis was conducted to help explain the differences in diffusion rate coefficient and moisture saturation level between the two material systems. The average void volume fractions and standard deviations for the material systems are listed in Table 1.

Table 1. Diffusion Rate Constants and Moisture Saturation Levels

Material System	Diffusion Rate Coefficient (in^2/sec.)	Standard Deviation	Moisture Saturation Level	Standard Deviation	Void Volume Fraction	Standard Deviation
LQ	6.05×10^{-6}	0.31×10^{-6}	0.80%	0.07	1.86%	0.34
HQ	3.74×10^{-6}	0.34×10^{-6}	0.66%	0.11	0.65%	0.13

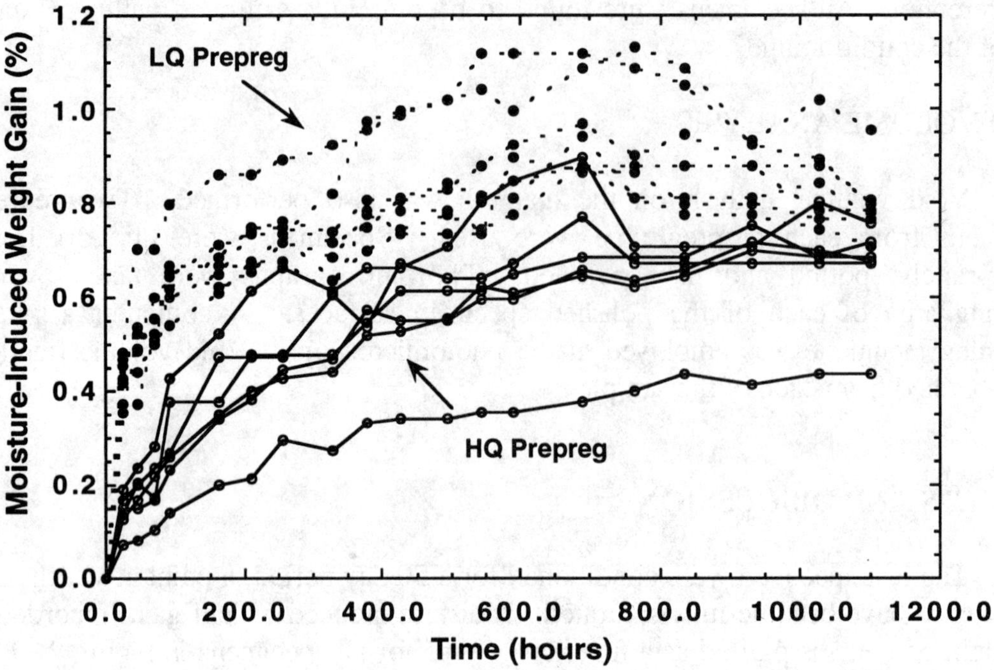

Figure 2. Moisture weight gain versus time.

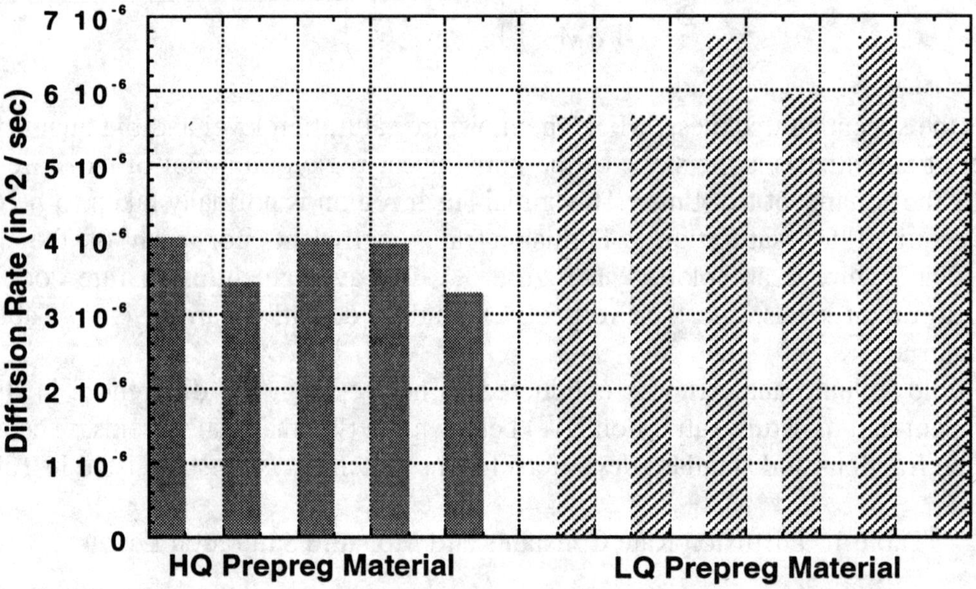

Figure 3. Diffusion rates for the two material systems.

In these two material systems, the only known difference was in the flatness of the initial pre-preg. Both materials contained the same resin system, sizing and fibers. The processing cycle for the material was not changed. However, the flatter (HQ) material consolidated better during the mold cycle giving much lower void volume fraction material. This difference in void volume fraction appears to be the mechanism for the lower diffusion rate in the HQ material system. The voids may facilitate the moisture diffusion by increasing the amount of exposed surfaces within the material. The higher moisture saturation levels for the LQ material may also indicate that moisture can accumulate at the voids in the material.

CONCLUSIONS

The objective of this work was to characterize the moisture diffusion rates of a graphite epoxy material fabricated from two different quality grades of pre-preg. The material system fabricated from the flatter HQ pre-preg exhibited significantly lower diffusion rates than that of the LQ pre-preg (3.74×10^{-6} in.^{2}/sec. versus 6.05×10^{-6} in.^{2}/sec., respectively). Saturation levels were also found to be lower in the HQ material. The differences in measured diffusion rates and saturation levels between the two was explained by closer examination of the material microstructure wherein a void volume content analysis was conducted. The much lower void volume content of the HQ material helped to explain the observed differences in moisture response between the two materials.

Based on the results of this limited study, a more thorough investigation is being initiated. In this future study, processing conditions and pre-preg quality will be varied to create a wider range of material void volume fractions, thus producing a better determination of the relationship between void volume fraction and diffusion rate. The materials will also be tested at a wider range of temperatures and humidity levels.

REFERENCES

ASTM D 5229/D 5229M - 92, "Standard Test Method for Moisture Absorption Properties and Equilibrium Conditioning of Polymer Matrix Composite Materials," *American Society for Testing and Materials, Volume D*, Philadelphia, 1992.

ASTM E 104-85 (Reapproved 1991), "Standard Practice for Maintaining Constant Relative Humidity by Means of Aqueous Solutions," *American Society for Testing and Materials, Volume E*, Philadelphia, 1991.

Cai, L. W. and Y. Weitsman, "Non-Fickian Moisture Diffusion in Polymeric Composites," *Journal of Composite Materials*, V. 28, n.2, pp. 130-154, 1994.

Hoppel, C. P. R., and R. N. Pangborn, "The Evaluation of Fatigue Damage in Short Fiber-Reinforced Styrene-Maleic Anhydride," *Cyclic Deformation, Fracture, and Nondestructive Evaluation of Advanced Materials: Second Volume, ASTM STP 1184*, M.R. Mitchell and O. Buck, Eds., American Society for Testing and Materials, Philadelphia, pp. 278-300, 1994.

McCrum, N. G., C. P. Buckley, and C. B. Bucknall, *Principles of Polymer Engineering*, Oxford University Press, New York, 1988.

Pipes, R. B., J. R. Vinson, and T. W. Chou, "On the Hygrothermal Response of Laminated Composite Systems", *Journal of Composite Materials*, V. 10, pp. 129-148, April 1976.

Springer, G. S., "Environmental Effects,", published in *Composites Design, 4th Edition*, S.W. Tsai, ed., Think Composites: Dayton, Ohio, 1988.

Tsai, S. W., *Composites Design, 4th Edition*, Think Composites: Dayton, Ohio, 1988.

VanLandingham, M. R., R. F. Eduljee, and J. W. Gillespie, Jr., "Hygrothermal Effects on the Material Properties and Behavior of Thermoplastic Polyimide Composites, *Proceedings of the American Society for Composites Tenth Technical Conference,* Technomic, pp. 525-536, 1995.

Woo, M. and M. R. Piggott, "Water Absorption of Resins and Composites: I. Epoxy Homopolymers and Copolymers," *Journal of Composites Technology & Research*, V. 9, No. 3, pp. 101-107, Fall 1987.

Measurement of Residual Stress in Laminated Composites

P. G. IFJU, B. C. KILDAY, X. NIU AND S.-C. LIU

ABSTRACT

A non-destructive method to measure the residual stresses in laminated composites was developed. The method incorporates the use of moiré interferometry (a laser based optical method) and replication of diffraction gratings on the surface of laminates in the autoclave during cure. This allows the strains due to resin shrinkage and thermal expansion to be referenced back to the stress free state. For thermoset matrix composites, the stress free state exists just prior to matrix solidification. To determine the residual stress in multidirectional laminates, the strain difference from cure temperature to room temperature was first measured. Then, by doing the same for a unidirectional panel, that was cured simultaneously with the multidirectional laminate, the free thermal expansion of the unidirectional lamina could be defined for the multidirectional case. Using laminate theory, the residual stresses in each ply of the multidirectional laminate were calculated accordingly. The method was evaluated by measuring the residual stresses in balanced symmetric cross-ply, unbalanced symmetric cross-ply and symmetric angle-ply AS4/3501-6 graphite/epoxy laminates. To determine the repeatability of the method, five separate autoclave runs were performed, each with the three multidirectional panels, and the unidirectional panel.

INTRODUCTION

With the ever increasing use of fiber reinforced composites, comes a need to fully characterize their mechanical properties. Residual stresses due to elevated temperature can degrade the load carrying capacity of composites by consuming a significant portion of the ultimate strength [1]. Residual stresses are present on two scales of the composite; the fiber scale, where resin shrinkage and the mismatch of thermal expansion of the resin and matrix leads to stresses on the micromechanics level, and the ply scale, where adjacent plies mutually constrain each other. This work is concerned with residual stresses on the ply scale of the composite.

Peter G. Ifju, Brian C. Kilday, Xiaokai Niu, Shao-Chun Liu
Aerospace Engineering, Mechanics and Engineering Science Department
University of Florida, Gainesville, Florida 32611-6250

There has been a great deal of research conducted to develop methods to predict, characterize and measure residual stresses in laminated composites. The experimental methods used to determine residual stress can be divided into two categories, destructive methods which rely on cutting, drilling or sectioning specimens to elicit and measure stress relaxation, and non-destructive methods, which reveal residual stresses by measuring temperature dependent geometry or incorporation of probes in the composite.

Destructive methods, as the name implies, gather material information on the composite by way of physically damaging the specimen. An example of such a method is the hole drilling method [2-4]. This method has been used extensively for measuring the residual stresses in isotropic materials such as metals. By introducing a hole (free edge) into a residually stressed body, the residual stresses relax around the vicinity of the hole. The strains due to relaxation are measured via a strain gage or full-field method [5]. An elasticity solution has been developed in order to relate the strain release to the far field residual stress. The hole drilling method has also been used with orthotropic materials [6-8]. It has yet to be used for laminated composites however, since the strain is only relaxed over a length scale equivalent to a few ply thicknesses, rendering strain gages useless due to spatial resolution constraints.

Cutting or sectioning methods, much like the hole drilling method, create a free edge in the composite to elicit a stress relaxation. Sectioning, in combination with strain gages or moiré interferometry has been used by numerous investigators [9-13]. Much like the hole drilling method, sectioning is destructive and further testing on the specimen is not possible.

Another destructive method called the first ply failure method has been used to determine residual stresses in laminated composites [1,14,15]. By the assumption that moisture absorption is analogous to thermal expansion, the use of hygroscopic expansion has been used to nullify cure induced residual stresses. Testing to failure of the composites can then lead to knowledge of the preexisting residual stresses.

For obvious reasons, non-destructive methods are generally more favorable over destructive methods. Analysis to determine the temperature dependent warpage of nonsymmetric laminates has been developed [16-22]. Additionally, predictive methods, verified by experiment, have been used to determine the after cured shape of composite structures. Although some of these tests are specific to nonsymmetric laminates, they have been shown to be useful.

Embedded sensors such as strain gages and fiber optics [23,24] have been incorporated into laminates during the curing process. These methods require sophisticated temperature compensation and have proved to be difficult to perform. Additionally, questions have risen as to whether the sensors actually measure the residual stress or rather the sensors effect on the residual stress of the laminate.

The X-Ray diffraction method [25-27] has been used to measure stresses in metal parts. By incorporating metal particles into the matrix of laminated composites, the atomic spacing of the metal can be used to measure the residual stress. As with other imbedded sensors, the metal particles influence the stress state as well as alter the natural thermal expansion characteristics.

Currently there is a need for the development of a method to measure residual stresses in multidirectional laminates. Ideally, a suitable method would be non-destructive, applicable to any symmetric ply lay-up, able to measure the influence of matrix shrinkage and temperature effects separately, and for long term testing and combined loading. This paper outlines a novel method that incorporates these virtues.

EXPERIMENTAL METHODOLOGY

The experimental method of moiré interferometry [28] was used to measure the strains due to residual stresses. The methodology relies on determining the strain on the composites at room temperature as they relate to the stress free temperature. The stress free state is that at which the matrix is a liquid and cannot transmit stress to the fibers of the composite. The stress free temperature is the temperature at which the matrix transforms from a liquid to the gel state and solidifies. A method of replicating a diffraction grating on the face of the composite at the stress free temperature was developed, as illustrated in Fig. 1. This replication process relies on the same mechanism of matrix solidification as that which leads to the development of stresses in the composite. This forms the basis of the key assumption in the present experimental methodology. At the time and temperature at which the grating is attached to the panel, the matrix solidifies and the composite makes the transition from the stress free state to that which stresses can exist. Initially, matrix shrinkage causes stress and later, after the panel is brought to room temperature, thermal expansion mismatch in ply orientation contributes.

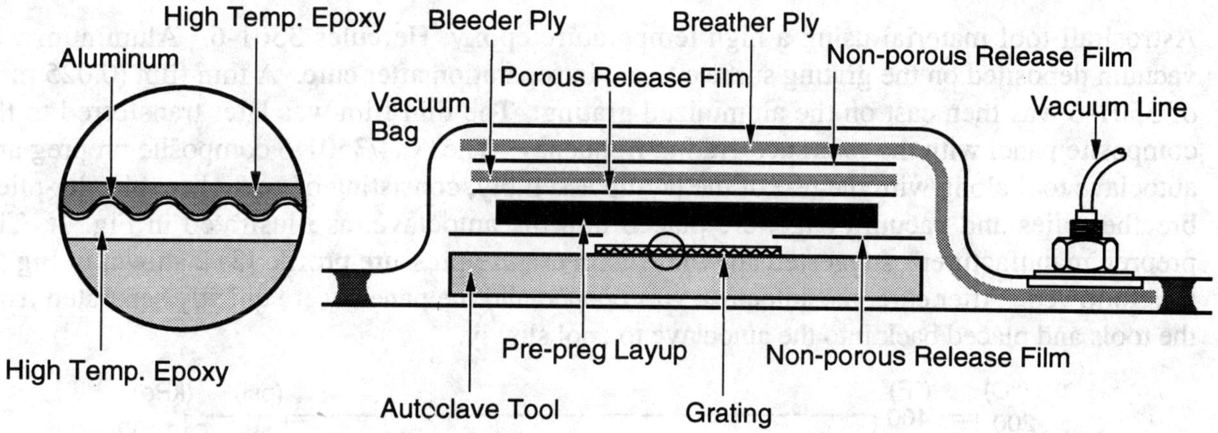

Figure 1. Method of attaching the moiré grating to the composite in the autoclave.

Moiré interferometry, schematically illustrated in Fig 2, was used for this technique because of some of its beneficial attributes. The method is characterized by high displacement and strain sensitivity, high spatial resolution, and high signal to noise ratio. It is a non contacting, full-field method that can measure both the normal and shear strains. One trait that was taken advantage of in this work is that the optical system, known as the interferometer, can be tuned to a specific grating frequency known as a reference. The reference grating can be removed from the interferometer and replaced with a deformed grating such as a specimen grating, and the absolute strain difference can be determined [29,30]. In this work, the reference grating was the grating that was used to emboss the specimen grating in the autoclave and the specimen grating was the embossed grating on the composite panel. The reference grating was backed by a ultra low expansion, high modulus material called Astrositall. Over the temperature and pressure range of the autoclave the reference had minimal dimensional change. The reference expansion, although small, as is discussed in the next section, does not contribute to the residual stress calculations.

The following procedure was developed to replicate diffraction gratings on the composite panels during autoclave cure. A reference grating was first replicated on the

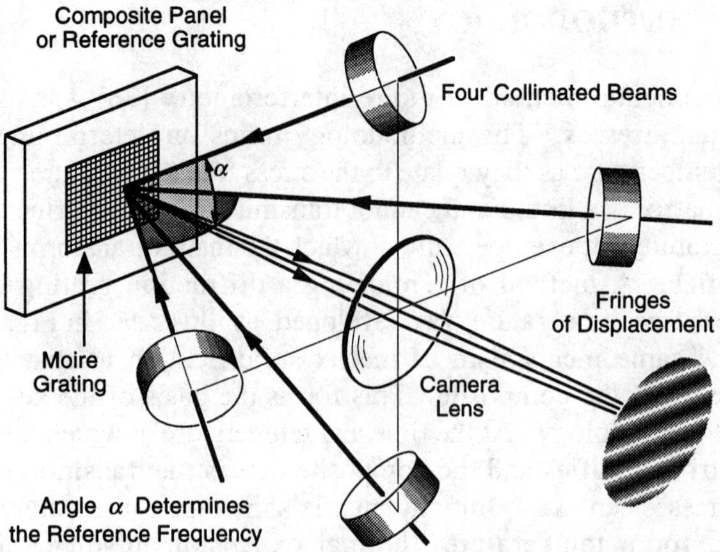

Figure 2. Schematic of moiré interferometry.

Astrositall tool material using a high temperature epoxy, Hercules 3501-6. Aluminum was vacuum deposited on the grating surface to aid in separation after cure. A thin film (0.025 mm) of 3501-6 was then cast on the aluminized grating. The thin film was later transferred to the composite panel with the reference grating frequency. The AS4/3501-6 composite prepreg and autoclave tool along with the rest of the lay-up assembly, consisting of peel plies, bleeder plies, breather plies and vacuum bag were placed into the autoclave, as illustrated in Fig. 1. The prepreg manufacturer's suggested autoclave temperature/pressure profile [31], shown in Fig. 3, was followed. After cure, the autoclave was opened and the panels were quickly separated from the tools and placed back into the autoclave to cool slowly.

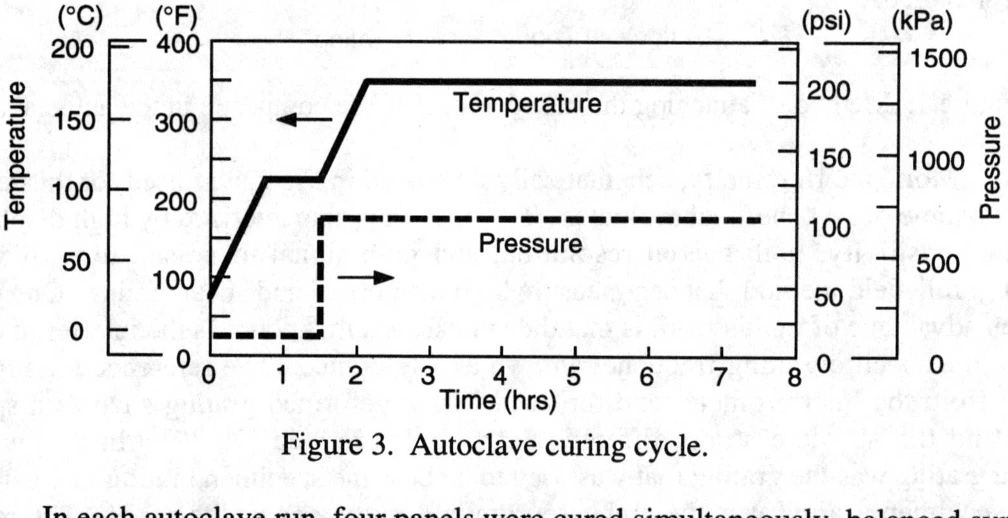

Figure 3. Autoclave curing cycle.

In each autoclave run, four panels were cured simultaneously, balanced symmetric cross-ply $[0_2/90_2/0_2/90_2]_S$, unbalanced symmetric cross-ply $[0_3/90/0_3/90]_S$, symmetric angle-ply $[0_2/45_2/0_2/45_2]_S$ and a unidirectional $[0_{16}]_T$. The three multidirectional lay-ups were selected in order to evaluate the method on panels with markedly different behavior. All four panels were placed in the same vacuum bag. The unidirectional panel served to provide free thermal expansion information for lamina properties. A total of five separate autoclave runs were performed in order to determine repeatability.

Once the panels cooled to room temperature and were removed from the autoclave the moiré interferometer was tuned to the frequency of the diffraction grating on the tool for each of the panels. The tools were then substituted with the specimens and the moiré fringe patterns were recorded. Fringe gradients ($\partial u/\partial x, \partial v/\partial y, \partial u/\partial y+\partial v/\partial x$) were measured to determine the average normal and shear strains over a square area of 12.7 x 12.7 mm in the center of the panels. Examples of the moiré interferometry fringe patterns are presented in Figs. 4-7.

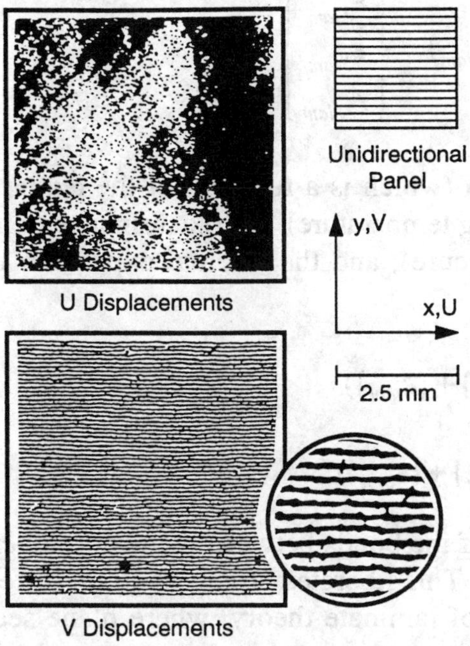

Figure 4 Moiré interferometry fringe patterns for the unidirectional panel.

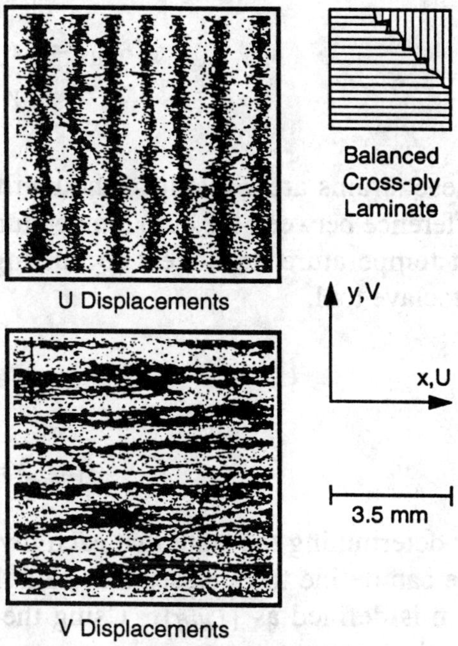

Figure 5 Moiré interferometry fringe patterns for the balanced cross-ply laminate.

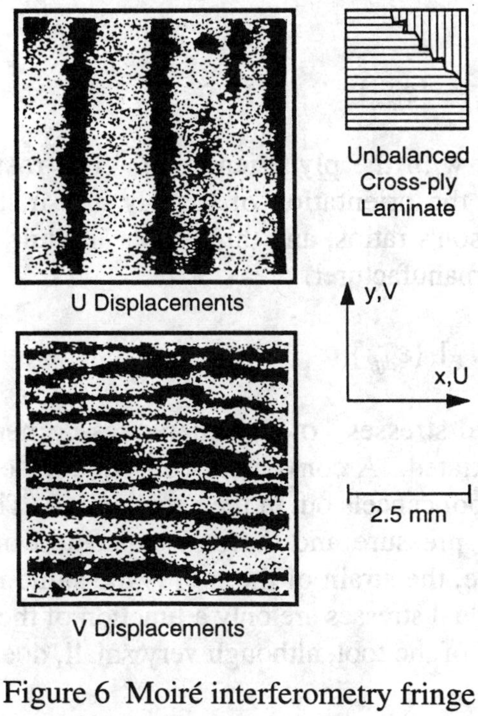

Figure 6 Moiré interferometry fringe patterns for the unbalanced cross-ply laminate.

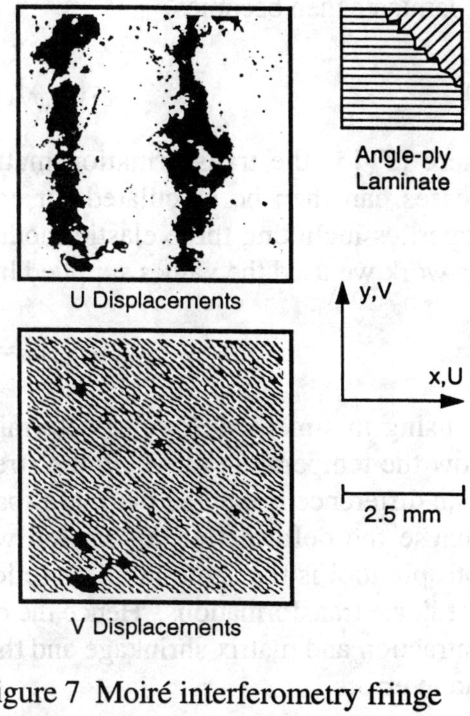

Figure 7 Moiré interferometry fringe patterns for the angle-ply laminate.

ANALYSIS

In the experimental methodology, moiré interferometry was used to measure the normal and shear strains on the faces of unidirectional $\{\varepsilon_{uni}\}$ and multidirectional $\{\varepsilon_{lam}\}$ laminates.

$$\{\varepsilon_{uni}\} = \begin{Bmatrix} \varepsilon_{1_{uni}} \\ \varepsilon_{2_{uni}} \\ \gamma_{12_{uni}} \end{Bmatrix} \qquad \{\varepsilon_{lam}\} = \begin{Bmatrix} \varepsilon_{x_{lam}} \\ \varepsilon_{y_{lam}} \\ \gamma_{xy_{lam}} \end{Bmatrix}$$

These strains are a result of the thermal contraction (which is a function of the temperature difference between the cure temperature and operating temperature), matrix shrinkage (which is not temperature dependent, occurring only during cure), and the thermal expansion of the autoclave tool.

$$\{\varepsilon_{uni}\} = (\{\varepsilon_{uni}^{T}\} + \{\varepsilon_{uni}^{S}\} + \{\varepsilon_{tool}\})$$

$$\{\varepsilon_{lam}\} = (\{\varepsilon_{lam}^{T}\} + \{\varepsilon_{lam}^{S}\} + \{\varepsilon_{tool}\})$$

By determining the state that each ply would reach if the bonds between them were liberated, one can define the stresses that held them together. This liberated strain, or strain difference term is defined as $\{\varepsilon_{diff}\}$. Using the assumptions of laminate theory, where plane sections remain plane, for symmetric lay-ups, the strain on the surface matches that at any point in the interior, as long as the location is a few ply thicknesses away from the free edge. In this study the location of measurement was in the center of the panel. The strain difference in each ply of the laminate then becomes

$$\{\varepsilon_{diff}\}_k = [T_k]^{-1} \cdot \{\varepsilon_{uni}\} - \{\varepsilon_{lam}\}$$

where $[T_k]$ is the transformation matrix associated with the ply orientation. The residual stresses can then be calculated for each ply using the orientation of the ply and material properties including three elastic moduli, three Poisson's ratios, and three shear moduli. For this work we used the values supplied by the prepreg manufacturer.

$$\{\sigma_{res}\}_k = [T_k]^{-1} \cdot [Q] \cdot [T_k] \cdot \{\varepsilon_{diff}\}_k$$

By using this methodology to determine the residual stresses $\{\sigma_{res}\}_k$, there was no need to know the temperature at which the stresses were initiated. A convenient consequence of the strain difference is that the thermal expansion of the tool cancels out in the calculations. This is because the deformation associated with autoclave pressure and thermal expansion of the isotropic tool is the same in all directions. Therefore, the strain of the tool is the same in all coordinate transformations. Hence the calculated residual stresses are only a function of thermal contraction and matrix shrinkage and the deformation of the tool, although very small, does not lead to errors.

RESULTS AND DISCUSSION

Figure 4-7 shows typical moiré interferometry fringe patterns. Normal and shear strains from the five autoclave runs are presented in Table I along with the coefficient of variation for the four types of panels. The repeatability was typically around 10% except for those with strain measurements less than 350 $\mu\varepsilon$. The repeatability for these cases was not as faithful. However, since the average measured values were low, the error in stress calculations was minimal. It can be seen that the largest strain was found in the unidirectional composite in the direction transverse to the fibers. The smallest strain was found in the unidirectional panel parallel to the fiber direction. This was expected. Note that there was no measurable shear strain in any of the laminates except for the angle-ply.

Table I Residual strains for the five autoclave runs.

Laminate	Strain	Run 1	Run 2	Run 3	Run 4	Run 5	Ave.	COV
Unidirectional $[0_{16}]_T$	ε_x	17.77	28.16	16.40	22.51	8.24	18.62	39.8
	ε_y	4950	4998	5206	5287	5092	5106	2.75
Balanced Cross-ply $[0_2/90_2/0_2/90_2]_S$	ε_x	333.6	362.3	399.2	415.9	365.1	375.2	8.67
	ε_y	343.1	304.8	373.2	416.4	462.5	380.0	16.22
Unbalanced Cross-ply $[0_3/90/0_3/90]_S$	ε_x	173.6	191.4	207.8	229.5	203.1	201.1	10.27
	ε_y	706.8	775.1	874.9	867.4	878.4	820.5	9.34
Angle-ply $[0_2/45_2/0_2/45_2]_S$	ε_x	153.1	151.7	194.9	194.9	230.6	185.0	17.9
	ε_y	2946	3314	3535	3582	3091	3293	8.38
	γ_{xy}	-2636	-3357	-3221	-3124	-3003	-3068	8.94

values in $\mu\varepsilon$

From the average strain information and the laminate theory analysis, the stresses were calculated for each ply as shown in Table II. The material properties for the lamina were taken from the literature and were not measured for this study.

For the symmetric balanced cross-ply, it was expected that the normal stress along the fibers of the 0° plies would be equal and opposite to the normal stress transverse to the fibers in the 90° plies. This must be the case if equilibrium is to be satisfied. From these results, the values were calculated to be -38.03 MPa and 43.24 MPa. The trend is evident, although equilibrium is not strictly satisfied. Likewise, for the symmetric, unbalanced cross-ply, there are 3 times as many 0° plies as 90° plies, therefore the normal stress parallel to the fibers in the 0° plies should be the opposite sign and one third of the stress perpendicular to the fiber in the 90° plies. The stresses are -14.22 MPa and 43.64 MPa respectively. Along the same line, the shear stresses in the angle-ply laminate in the 0° and 45° plies should be equal and opposite to satisfy equilibrium. These are 21.17 MPa and -19.01 MPa respectively.

Table II Residual Stresses in the three multidirectional laminates

Balanced Cross-ply

$[0_2/90_2/0_2/90_2]_S$

	σ_x	σ_y	τ_{xy}
0° plies	-38,03	43.22	0
90° plies	43.24	-38.71	0
0° plies	-38,03	43.22	0
90° plies	43.24	-38.71	0

MPa

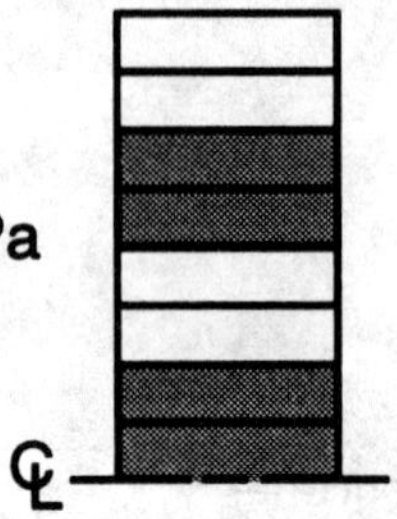

Unbalanced Cross-ply

$[0_3/90/0_3/90]_S$

	σ_x	σ_y	τ_{xy}
0° plies	-14.22	39.58	0
90° plies	43.64	-101.6	0
0° plies	-14.22	39.58	0
90° plies	43.64	-101.6	0

MPa

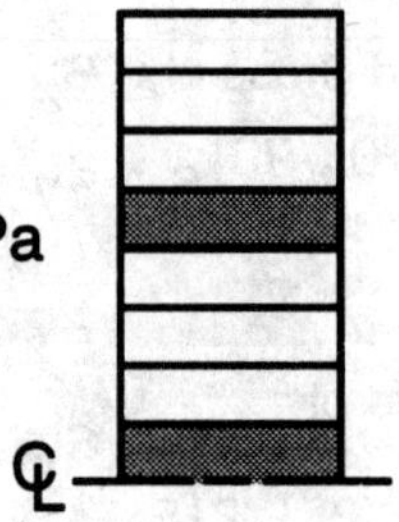

Angle-ply

$[0_2/45_2/0_2/45_2]_S$

	σ_x	σ_y	τ_{xy}
0° plies	-18.28	16.51	21.17
45° plies	19.11	-23.84	-19.01
0° plies	-18.28	16.51	21.17
45° plies	19.11	-23.84	-19.01

MPa

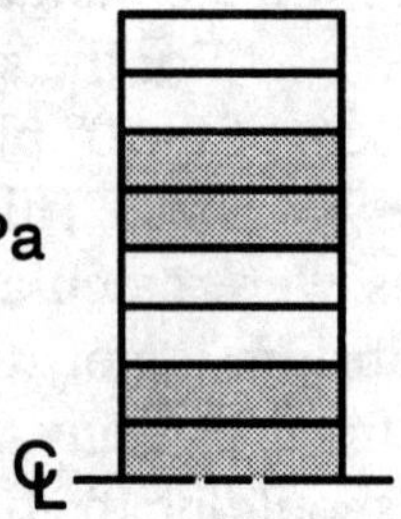

These results are encouraging and the small inconsistencies can be attributed to a number of factors. The material properties that we used in the calculations were from the literature, and it would be more appropriate to measure them for our specific laminates. Additionally, it is believed that the linear elastic assumption of laminate theory leads to some error, and that it would be more appropriate to use viscoelastic material properties.

The results for these tests are encouraging and it is believed that the method could be used to determine the residual stresses in any thin multidirectional laminate with any fiber/thermoset matrix combination. Extension to thick laminates, where pronounced through-the-thickness residual stress variations are present, has yet to be attempted. Additionally, the use of this method for thermoplastic matrix composites is foreseen.

FUTURE WORK

The next step in our investigation is to develop the capability to determine the individual contributions of matrix shrinkage and thermal expansion mismatch to the residual stress. This will be accomplished by heating the panels to the cure temperature in a special interferometer/thermal chamber. Preliminary results show that at this temperature all of the residual strains do not relax out, therefore, the remaining strains are due to resin shrinkage.

By establishing this capability to determine the strain with reference to the stress free state, we have essentially obtained the capability to determine the "absolute stresses" in the composite. This can be used with combined thermal, mechanical, and machining stresses to determine the absolute stresses with knowledge of the residual stress.

CONCLUSIONS

A non-destructive method to measure the residual stress in composites has been developed. The method was used to determine the residual stresses in three different multidirectional laminate lay-ups of AS4/3501-6 graphite /epoxy composite. It was found that the repeatability of the method was reasonable.

ACKNOWLEDGMENTS

The authors would like to acknowledge the National Science Foundation for sponsorship of this research under the Career program CMS 9502678.

REFERENCES

[1] Pagano, N. J., Hahn, H. T., "Evaluation of Composite Curing Stresses," *Composite Materials: Testing and Design (Fourth Conference),* ASTM STP 617, pp. 317-329 (1977).

[2] "Determining Residual Stress by the Hole-Drilling Strain Gage Method" ASTM Standard E8376-81, Philadelphia, (1981).

[3] "Measurement of Residual Stresses by the Hole-Drilling Strain-Gage Method," *Tech Note TN-503-3*, Measurements Group, pp. 1-15 (1988).

[4] Rendler, J. J., and Vigness, I "Hole-Drilling Strain Gage Method of Measuring Residual Stresses," *Proc. SESA*, Vol. XXIII, No.2, pp. 577-586, (1966).

[5] Nicoletto, G., "Moiré Interferometry Determination of Residual Stresses in the Presence of Gradients," *Experimental Mechanics,* Vol. 31, No. 3, pp. 252-256 (1991).

[6] Bert, C. W., Thompson, G. L., “A Method for Measuring for Measuring Planar Residual Stresses in Rectangularly Orthotropic Materials,” *J. Composite Materials,* Vol. 2, No. 2, pp. 244-253 (1968).
[7] Lake, B. R., Appl, F. J., Bert. C. W., “An Investigation of the Hole-drilling Technique for Measuring Planar Residual Stress in Rectangularly Orthotropic Materials,” *Experimental Mechanics,* Vol. 10, June, pp. 233-239 (1970).
[8] Schajer, G. S., Yang, L., “Residual-stress Measurement in Orthotropic Materials Using the Hole-drilling Method,” *Experimental Mechanics,* Vol. 34, No. 4, pp. 324-333 (1994).
[9] Lee, J., Czarnek, R., “Measuring Residual Strains in Composites Laminates Using Moiré Interferometry,” *Proceedings of the 1991 SEM Conference on Experimental Mechanics,* Milwaukee, WI, pp. 405-415, (1991).
[10] Lee, J., Czarnek, R., Guo, “Interferometric Study of Residual Strains in Thick Composites,” *Proceedings of the 1989 SEM Conference on Experimental Mechanics, Cambridge, MA,* pp. 356-364, (1989).
[11] Gascoigne, H. E.,“Residual Surface Stresses in Laminated Cross-ply Fiber-epoxy Composite Materials.,” *Experimental Mechanics,* Vol. 34, No.1, pp. 27-36 (1994).
[12] Joh, D., Byun, K. Y., Ha, J., “Thermal Residual Stresses in Thick Graphite/Epoxy Composite Laminates-Uniaxial Approach,” *Experimental Mechanics,* Vol. 33, No.1, pp. 70-76 (1993).
[13] Sunderland, P., Yu, W. J., Månson, J. E., "A Technique for the Measurement of Process-Induced Internal Stresses in Polymers and Polymer Composites,” *Proceedings of ICCM-10,* Vol. 3, pp. 125-132, (1995).
[14] Hahn, H. T., “Residual Stresses in Polymer Matrix Composite Laminates,” *J. Composite Materials,* Vol. 10, , pp. 265-277 (1976).
[15] Kim, R. Y., Hahn, H. T., “Effect of Curing Stresses on the First Ply-failure in Composite Laminates,” *J. Composite Materials,* Vol. 13, January, pp. 2-16 (1979).
[16] Jain, L. K., Mai, Y. W., “On Residual Stress Induced Distortions during Fabrication of Composite Shells,” *Proceedings of the American Society for Composites,* Tenth Technical Conference, pp. 261-270 (1995).
[17] Jeronimidis, G., Parkyn, A. T., "Residual Stresses in Carbon Fiber-Thermoplastic Matrix Laminates,” *J. Composites Materials,* Vol. 22, May, pp. 401-415 (1988).
[18] Månson, J. E., Seferis, “Process Simulated Laminate (PSL)” A Methodology to Internal Stress Characterization in Advanced Composite Materials,” *J. Composite Materials,* Vol. 26, No. 3, pp.405-431 (1992).
[19] Hyer, M. W., “Calculation of Room-Temperature Shapes of Unsymmetric Laminates,” *J. Composite materials,* Vol. 15, July, pp. 296-310 (1981).
[20] Kim, K. S., Hahn, H. T., "Residual Stress Development during Processing of Graphite/Epoxy Composites,” *Composites Science and Technology,* Vol. 36, pp. 121-132 (1989).
[21] Kim, K. S., Hahn, H. T., Croman, R. B., "The Effect of Cooling Rate on Residual Stresses in a Thermoplastic Composite,” *J. Composites Technology & Research,* Vol. 11, No. 2, pp. 47-52 (1989).
[22] Fakuda, H., Takahashi, K., Toda, S., “Thermal Deformation of Anti-Symmetric Laminates At Cure,” *Proceedings of ICCM-10,* Vol 3, pp. 141-148, (1995).
[23] Unger, W. J., Hansen, J. S., “The effect of Thermal Processing on Residual Strain Development in Unidirectional Graphite Fibre Reinforced PEEK,” *J. Composite Materials,* Vol. 27, No. 1, pp. 59-81 (1993).
[24] Carman, G. P., Sendechyj, G. P., “Review of the Mechanics of Embedded Optical Sensors,” *J. Composites Technology & Research,* JCTRER, Vol. 17, No. 3, pp.183-193, (1995).
[25] Smith, W. F., *Foundations of Materials Science and Engineering, Second Edition,* McGraw Hill, Inc, New York, pp.102-113, (1993).
[26] Predecki, P., Barrett, C. S., “Stress Measurement in Graphite/Epoxy Composites by X-Ray Difftraction from Fillers,” *J. Composite Materials,* Vol. 13, pp. 61-71 (1979).
[27] Fenn, R. H., Jones, A. M., Wells, G. M., “X-Ray Diffraction Investigation of Triaxial Residual Stresses in Composite Materials,” *J. Composite Materials,* Vol. 27, No. 14, pp. 1338-1351 (1993).
[28] Post, D., Han, B. and Ifju, P. G., "High Sensitivity Moire: Experimental Analysis for Mechanics and Materials," Springer Verlag, New York, NY, (1994).
[29] Post, D. and Wood, J. D., "Determination of Thermal Residual Strains by Moire Interferometry," *Experimental Mechanics*, Vol. 29, No. 3, pp. 318-322 (1989).
[30] Post, D., Wood, J. D. and Han, B., "Thermal Stresses in a Bimaterial Joint: An Experimental Analysis," *ASME J. Applied Mechanics,* Vol. 61, No. 1, pp 192-198 (1994).
[31] Boll, D. J., “Curing of 3501-6 Composite Systems,” *Hercules,* February, (1992).

Occupation of Interstitial Positions and Diffusion of Carbon at the Interface in Cu-C Composites

S. DORFMAN AND D. FUKS

ABSTRACT

Model of the interstitial solid solution was used to choose whether octahedral or tetrahedral interstitial positions are occupied by the fibre atoms dissolved in the matrix. On the example of Cu-C composite material we show the way to calculate the occupancy of the interstitial positions and its temperature dependence. The results obtained on the basis of non-empirical calculations predict the preferable occupation of octahedral positions till T~1200K. This confirms the structure of the interstitial solid solution. In the framework of this model we calculate the diffusion barrier for carbon diffusion in copper.

INTRODUCTION

The problem of stability of copper-carbon solid solutions is significant for the develpement of modern metal matrix composites with the aim of adding improved mechanical properties to the excellent electrical and thermal conductivity properties of copper (see, e.g. [1, 2]). Of particular interest is the possibility of developing improved high temperature strength. Addition of particulate graphite to copper matrices permits considerable improvements in tribological properties. Thus Cu-carbon composites see regular application in sliding electrical contacts applications.

Carbon is insoluble in copper up to very high temperatures, its solubility does not exceed 0.02 at%. Thus a problem of wettability of carbon fibers by copper arises. This wettability is extraordinary small [3] and does not allow to fabricate the composite material. So the interface bonding in copper-carbon composites is extremely weak [4]. Formation of very dilute Cu-C interstitial solid solution at the interface may be diffusion-controlled. To predict the diffusion behaviour of carbon atoms we have to study the structure and interatomic interactions in such alloys.

Simon Dorfman, Department of Physics, Technion, 32000 Haifa, Israel

David Fuks, Materials Engineering Department, Ben-Gurion University of the Negev, P.O.Box 653, 84105 Beer Sheva, Israel.

In order to write the expression for the energy of the interstitial solution one may use the pertubation series on pseudopotentials in reciprocal space (PSP RS - method). It is evident that this approximation will only roughly describe the basic particularities of the electronic spectrum of the solutions. This is especially true for the solutions, where transition metals are chosen as alloying elements. But, as thermodynamic values are always obtained as a result of averaging over the spectrum, they are less sensitive to its particularities than, for example, optical characteristics[5]. Making use of PSP RS method we carried out calculations of potetials of Cu-Cu, Cu-C, and C-C interactions in Cu-C alloy. The use of these potentials allow us to study the occupation of tetrahedral and octahedral interstitial positions in fcc Cu by C atoms. Exploring this result and the same interactive potentials we calculate the height of the diffusion barrier for the carbon atom which pass between the nearest interstitial positions and predict the diffusivity of carbon in copper. The distinguishing feature of our study is accounting for the nonuniform static displacements which occur in interstitial solid solutions. In contradiction to the traditional calculations of the diffusion barier in the rigid lattice we show an influence of the displacements on the value of the barrier.

Displacements depend not only on relative "geometric" sizes of the atoms of the impurity and those of the host matrix, but also on the electron charge redistribution in the vicinity of the impurity. Thus, both a geometrical factor and a charge transfer have to be accounted for in the framework of the same formalism.

In our study we show how nonuniform static displacements, which are local effect near the interstial-type impurity atom, influence on the jump of this atom across the saddle point and change the shape of lnD vs 1/T dependence.

OCCUPATION OF INTERSTITIALS POSITIONS

Let us assume that interstitial atoms C occupy both octahedral and tetrahedral positions in the lattice. The number of octahedral positions is M_o and the number of the tetrahedral positions is M_t. It means that the total number of intestitial positions is $M = M_o + M_t$. The n atoms C are placed in the interstial positions in the following manner: n_o atoms in octahedral positions and n_t atoms ($n_t = n - n_o$) in tetrahedral positions. The part of the total energy of the crystal depending on the the number of interstitial atoms may be written as

$$E = n_o u_o + n_t u_t, \tag{1}$$

in assumption that C atoms do not interact with each other. Here u_o and u_t are the energies of of C atoms in the octahedral and tetrahedral positions correspondly. The number of different permutations of C atoms on the interstial positions is

$$L = \frac{M_o!}{n_o!(M_o - n_o)!} \cdot \frac{M_t!}{n_t!(M_t - n_t)!}, \tag{2}$$

while the entropy of the system, S, is

$$S = klnL, \tag{3}$$

where k is the Boltzmann constant. Substituting $n_t = n - n_o$ and making use of the Stirling formula we obtain the free energy $F = E - TS$ in the form

$$F = n_o u_o + (n - n_o)u_t - kT\{ln M_o! - n_o(ln n_o - 1) - (M_o - n_o)[ln(M_o - n_o) - 1] + ln M_t! - (n - n_o)[ln(n - n_o) - 1] - (M_t - n + n_o)[ln(M_t - n + n_o) - 1]\}. \quad (4)$$

Applying the equilibrium conditions $\frac{\partial F}{\partial n_o} = 0$ the following equation for the equilibrium numbers n_o and n_t may be obtained

$$\frac{(M_t - n_t)(n - n_t)}{(M_t - n + n_o)n_o} = \frac{(M_o - n_o)n_t}{(M_t - n_t)n_o} = e^{\frac{u_o - u_t}{kT}}. \quad (5)$$

It is easy to see that when the temperature increases the system of C-atoms seeks to the uniform distribution of C-atoms on on the interstitial positions. In this case (at $T \to \infty$) we obtain

$$\frac{n_t}{n_o} = \frac{M_t}{M_o}. \quad (6)$$

To clarify the temperature dependence of n_o and n_t Eq.(5) may be solved giving

$$n_o = \frac{M_o + n - M_t\mu - n\mu \pm \sqrt{(M_o + n + M_t\mu - n\mu)^2 - 4M_o n(1 - \mu)}}{2(1 - \mu)}. \quad (7)$$

Here we use the equality $n_t = n - n_o$ and notified $\mu = exp[(u_o - u_t)/kT]$. Using the definition of partial concentrations of atoms in octahedral and tetrahedral interstial positions $c_o = n_o/n$ and $c_t = n_t/n$ the following result may be obtained

$$c_o = \frac{\alpha + \beta \pm \sqrt{(\alpha + \beta)^2 - 4\beta}}{2\beta}, \qquad c_t = 1 - c_o. \quad (8)$$

Here $\alpha = 1 + \frac{M_t}{M_o}\mu$; $\beta = \gamma(1 - \mu)$; and $\gamma = n/M_o$. For the relatively small n this result has to transform to the Boltzmann distribution. This condition leads to the consideration that only the minus sign before the square root in (7) has to be left and in this case from (5) it follows

$$c_o = \frac{1}{1 + \frac{M_t}{M_o}\mu}; \quad c_t = \frac{1}{1 + (\frac{M_t}{M_o}\mu)^{-1}}. \quad (9)$$

When the value γ is small the value of β is also small and the square root may be expanded in Taylor's series. Then from (7) in the limiting case $\gamma \to 0$ we get $c_o = 1/\alpha$ which coinsides with (8).

In the case of the fcc lattice with N lattice sites $M_o = N$ and $M_t = 2N$. Thus the final expressions for the concentrations c_o and c_t have the forms

$$c_o = \frac{1}{1 + 2e^{\frac{u_o - u_t}{kT}}}; \qquad c_t = \frac{1}{1 + \frac{1}{2}e^{\frac{u_t - u_o}{kT}}}. \quad (10)$$

NONUNIFORM DISPLACEMENTS IN INTERSTITIAL SOLID SOLUTIONS

For the interstitial dilute copper-carbon solid solution in the copper-rich region we calculate quasielastic Kanzaki forces and nonuniform static displacements of copper atoms. These displacements are considered for the first and the second coordination shells of the intersite carbon atom. In our study we use the microscopic phenomenological theory of the interstitial solid solutions [6]. It was shown that Fourier transforms $\vec{\epsilon}(\vec{k})$ of nonuniform static displacements of the host atoms are

$$\epsilon_i(\vec{k}) = G^{ij}(\vec{k}) \sum_{p=1}^{\nu} \bar{F}_p^j(\vec{k}) c_p(\vec{k}), \tag{11}$$

where $G^{ij}(\vec{k})$ is defined by

$$G^{ij}(\vec{k}) A^{jl}(\vec{k}) = \delta_{il}.$$

Here $A^{jl}(\vec{k})$ is the dynamic matrix in the reciprocal space. This matrix for fcc structure is presented in Table 1. The displacements in the real space are obtained with the Fourier transformation of $\epsilon_i(\vec{k})$.

The matrix of the force constants on the first coordination shell has the following symmetry (see for example Ref.[7])

$$\left\{ \begin{array}{ccc} \beta & \gamma & 0 \\ \gamma & \beta & 0 \\ 0 & 0 & \alpha \end{array} \right\}. \tag{12}$$

The force constants α, β, and γ may be calculated from the experimetally measured elastic constants C_{11}, C_{12}, and C_{44}, according to the following relations [7]

$$\begin{aligned} \alpha &= 1/2(C_{11}/2 - C_{44}); \\ \beta &= -1/4C_{11}; \\ \gamma &= -1/4(C_{12}/2 - C_{44}). \end{aligned} \tag{13}$$

All these equations (13) are written in the lattice parameter units.

Performing the Fourier transformation (4) for $A^{ij}(\vec{R})$ we obtain the dynamic matrix in the reciprocal space. Matrix elements $A^{ij}(\vec{k})$ are presented in Table 1. The Fourier transform of the vector of Kanzaki forces is expressed in terms of perfect lattice elastic constants for a pure matrix. When the vector of Kanzaki forces occupied only octahedral positions it has the form [6]

$$\vec{F}(\vec{k}) = -i\frac{a^2}{2}(C_{11} + 2C_{12}) \cdot \hat{u} \cdot \left(sin\frac{k_x a}{2}, sin\frac{k_y a}{2}, sin\frac{k_z a}{2} \right), \tag{14}$$

where $\hat{u} = a^{-1}(da/dc)$. a is the lattice parameter in an alloy. c is the atomic fraction of the interstitial atoms in the alloy.

Table 1: Components of the dynamic matrix $A_{ij}(\vec{k})$.

A_{xx}	$4\cdot(2\cdot\beta+\alpha)-4\cdot\beta\cdot cos\left(k_x\cdot\frac{a}{2}\right)\cdot\left[cos\left(k_y\cdot\frac{a}{2}\right)+cos\left(k_z\cdot\frac{a}{2}\right)\right]-$ $-4\cdot\alpha\cdot cos\left(k_y\cdot\frac{a}{2}\right)\cdot cos\left(k_z\cdot\frac{a}{2}\right)$
A_{yy}	$4\cdot(2\cdot\beta+\alpha)-4\cdot\beta\cdot cos\left(k_y\cdot\frac{a}{2}\right)\cdot\left[cos\left(k_x\cdot\frac{a}{2}\right)+cos\left(k_z\cdot\frac{a}{2}\right)\right]-$ $-4\cdot\alpha\cdot cos\left(k_x\cdot\frac{a}{2}\right)\cdot cos\left(k_z\cdot\frac{a}{2}\right)$
A_{zz}	$4\cdot(2\cdot\beta+\alpha)-4\cdot\beta\cdot cos\left(k_z\cdot\frac{a}{2}\right)\cdot\left[cos\left(k_x\cdot\frac{a}{2}\right)+cos\left(k_y\cdot\frac{a}{2}\right)\right]-$ $-4\cdot\alpha\cdot cos\left(k_x\cdot\frac{a}{2}\right)\cdot cos\left(k_y\cdot\frac{a}{2}\right)$
A_{xy}	$2\cdot\gamma\cdot sin\left(k_x\cdot\frac{a}{2}\right)\cdot sin\left(k_y\cdot\frac{a}{2}\right)$
A_{xz}	$2\cdot\gamma\cdot sin\left(k_x\cdot\frac{a}{2}\right)\cdot sin\left(k_z\cdot\frac{a}{2}\right)$
A_{yz}	$2\cdot\gamma\cdot sin\left(k_y\cdot\frac{a}{2}\right)\cdot sin\left(k_z\cdot\frac{a}{2}\right)$

DIFFUSION OF INTERSTIAL ATOMS

Let us assume that we have a disordered interstitial alloy A-C and atoms C occupy the interstitial positions. In the case of fcc lattice that we are going to study there are one octahedral and two tetrahedral interstitial sites per atom. We shall consider only octahedral positions for interstitial atoms. This consideration corresponds to the results of the calculations for occupation probabilities for Cu-C solid solution, which will be discussed in the next section. These octahedral positions are also forming the fcc lattice and in the case of absolutely disordered solid solution they are energetically equivalent. All interstitial atoms are situated in the same mean-field lattice potential. This potential is formed by A-type atoms distribute on the fcc lattice sites and on C-type atomson the interstitial sites.

Taking into account only first and second nearest neighbours we obtain the energies of C atom in the octahedral position $\bar{u}_i$ and in the saddle point $\bar{u}_s$

$$\bar{u}_i = -\{6\cdot v_{ac} + 8\cdot v'''_{ac}\},$$
$$\bar{u}_s = -\{2\cdot v'_{ac} + 4\cdot v''_{ac}\}. \tag{15}$$

Where $v_{ac} = V_{ac}(R_1)$, $v'_{ac} = V_{ac}(R'_1)$, $v''_{ac} = V_{ac}(R'_2)$, and $v'''_{ac} = V_{ac}(R_2)$ are the values of the interatomic interaction energies determined from the values of interatomic interaction potentials V_{ac} between atoms of types A and C. $R_1 = a/2$ and $R_2 = a\sqrt{3}/2$ are the radii of the first and the second coordination shells for the C atom in the interstitial position and $R'_1 = a\sqrt{2}/4$ and $R'_2 = a\sqrt{6}/4$ are the same for the atom C in the saddle point. Substitution of $R_1+\epsilon_1$ in R_1 and similar substitutions in R_2, R'_1, and R'_2 allow to account nonuniform displacements in the interstitial solid solution.

The height of potential barrier is

$$\Delta U = |\bar{u}_i - \bar{u}_s| = = |6v_{ac} + 8v'''_{ac} - 2v'_{ac} - 4v''_{ac}|. \tag{16}$$

The diffusion coefficient of C atoms that are situated in the octahedral interstices of fcc lattice may be obtained by substituting this expression into Arrhenius-type formulae

$$D \sim exp(-\frac{\Delta U}{kT}). \tag{17}$$

Now we have to turn to the study of the changes of pre-exponential factors caused by alloying. The Gibbs free energy of migration includes entropy term and the pressure-dependent term. The last can be neglected because we are studying the diffusion process at the athmospheric pressure that in our units is approximately zero. The entropy of migration in a binary alloy A-C may be calculated, according to [9], with our values for the migration energy ΔU.

$$S_{mig} = \beta \frac{\Delta U}{T_m}, \tag{18}$$

where $\beta \simeq 0.35$ [9] and T_m is the melting point. Following relation may be written now

$$D \simeq a^2 exp\left[\frac{(1-\beta\tau)}{kT}\Delta U\right], \tag{19}$$

here $\tau = T/T_m$.

RESULTS ADND DISCUSSION

Our mean-field calculations in the framework of local density approximation with the semiempirical pseudopotentials [5] show the completely different character of C-C interaction in the matrix of diamond and in the Cu matrix (see Fig.1). In the Cu matrix the C-C pair potential has a very strong repulsive tendency and in the diamond lattice the behaviour of the pair potential is traditional. The obtained results have proved the statement: *the interactions between the same atoms in the different environment are completely different.*

The interatomic potential presented in Fig. 1 was used to calculate the carbon atom energies in different interstitial positions u_o and u_t in Cu matrix. In these calculations we take into account the interaction in the first and the second coordination shell. The energies u_o and u_t were substituted in Eq. (10) and the temperature dependence of the relative concentrations c_o and c_t were obtained. They are plotted in Fig. 2.

Our results show clearly that the probability fo find carbon atoms in octahedral positions is much higher then in tetrahedral ones. This statement will be true up to a temperature $T \sim 1000K$. Thus we are convinced to use the model of the interstitial diffusion of carbon atoms in Cu-C solid solution with the carbon atoms randomly situated in octahedral positions. This model is true at leat till $T \sim 1000K$. Proving the carbon occupation of the octahedral interstitial positions in copper it would be worth to calculate nonuniform static displacements of copper atoms in the vicinity of impurity according to the scheme given in the Section 3.

Using the elastic constants from [8] we obtained the matrix of the force constants.

$$\left\{\begin{array}{ccc} 3.0 & 3.52 & 0 \\ 3.52 & 3.0 & 0 \\ 0 & 0 & -0.215 \end{array}\right\} \tag{20}$$

Table 2: Phonon frequencies of copper in Brillouin zone points. $\vec{k}$-points are in $2\pi/a$.

$\vec{k}$-points	frequencies, $10^{12}\ s^{-1}$	
	theory	experiment (taken from Ref.[12])
(001)	7.672	7.40
	5.525	4.88
$\left(\frac{1}{2},\frac{1}{2},\frac{1}{2}\right)$	6.880	7.56
	4.526	3.55
$\left(\frac{1}{2},\frac{1}{2},0\right)$	6.195	6.51
	5.425	5.33
	4.526	3.40

The averaged value of the atomic volume may be obtained from the Zen's law. In our calculations we assumed $c = 0.02\ at\%$ and obtained $\hat{u} = 0.271$. The atomic volume of copper and carbons were taken from [10], where these volumes were extrapolated from the experimental data to the zero temperature. The elements of the dynamic matrix and the Fourier transform of the vector of Kanzaki forces were calculated in the special points of the Brillouin zone [11]. The quality of the dynamic matrix calculations may be checked after carrying out the diagonalization of this matrix and comparison of the obtained phonon frequencies in some Brillouin zone points with the experimental data (see Table 2). The average error in phonon frequency calculations performed in the suggested procedure is about 13.5%. This result encouraged us to use the calculated values of the matrix elements A_{ij} and the Fourier transform of the vector of Kanzaki forces in the calculations of the nonuniform displacements of copper atoms in dilute Cu-C solid solutions.

The interstitial carbon atom was situated in the point (0,0,0), while copper atoms were located in points of types $0.5{\cdot}a(1,0,0)$ and $0.5 \cdot a(1,1,1)$. This points correspond to the first and the second nearest neighbors (NN) atoms respectively. The results of our calculations carried out in accordance with Eq. (11) show that copper atoms of the first coordination shell move to the carbon atom and the second NN copper atoms move away from the carbon. We discuss here the motion relatively to the positions in the uniformly distorted lattice. The values of the displacements on the first and the second coordination shells are $\epsilon_1 = 0.98\ a.u.$ and $\epsilon_2 = 0.27\ a.u.$ For ϵ_1' and ϵ_2' we obtained much smaller values (0.03 and 0.01 a.u., correspondingly). Using the expressions for R_i and R_i' and potentials given in Fig. 1 we calculated Δu values in the rigid lattice and in the lattice with displacements. Δu for the rigid lattice is 0.99 eV and for the displaced lattice is 1.09 eV. The ratio of diffusion coefficients in the displaced and rigid lattice at the temperatures about 800K is

$$\zeta = \frac{D_{displ}}{D_{rigid}} = exp\left[\frac{1-\beta_T}{kT}(\Delta u_{rigid} - \Delta u_{displ})\right]$$

At the temperature 800 K $\zeta = 3.17$. Thus displacements lead to the drastic decreasing of diffusivity in Cu-C interstitial solid solution.

Acknowledgements

We highly appreciate the support of this research by the Grant # 94-44/2 from the United States-Israel Binational Science Foundation (BSF), Jerusalem, Israel. S.D. acknowledges the Israel Ministry of Absorption for the additional support.

References

[1] X.Kuang, G.Caronuto, Z.Zhou, and L.Nicolais, Science and Engineering of Composite Materials, **50**, 9 (1996).

[2] S. Dorfman and D. Fuks, Composites **27**A, 697 (1996).

[3] D.A. Mortimer and M.J.Nicholas, Journal of Mater. Sci. **5**, 149 (1970).

[4] S.J. Sun and M.D. Zhang, Journal of Mater. Sci. **26**, 5762 (1991).

[5] R.J.Elliott, J.A.Krumhansl, and P.L.Leath, Rev. of Modern Phys. **46**, 465 (1978).

[6] A.G.Khachaturyan, *Theory of Structural Transformations in Solids*, (John Wiley, New York, 1983).

[7] G. Leibfried and W. Ludvig, Solid State Physics **12**, 41 (1961).

[8] Ch. Kittel, *Introduction to Solid State Physics*, (Wiley,New York, 1976).

[9] P. G. Shewmon, *Diffusion in Solids*, (McGraw-Hill, New York, 1962).

[10] G.B.Bachelet, D.R.Hamman, and M.Schlüter, Phys. Rev. B**26**, 4199 (1982).

[11] D.J.Chadi and M.L.Cohen, Phys. Rev. B**8**, 5747 (1973).

[12] N.Singh, N.S.Banger, and S.P.Singh, Phys. Rev. B**38**, 7415 (1988).

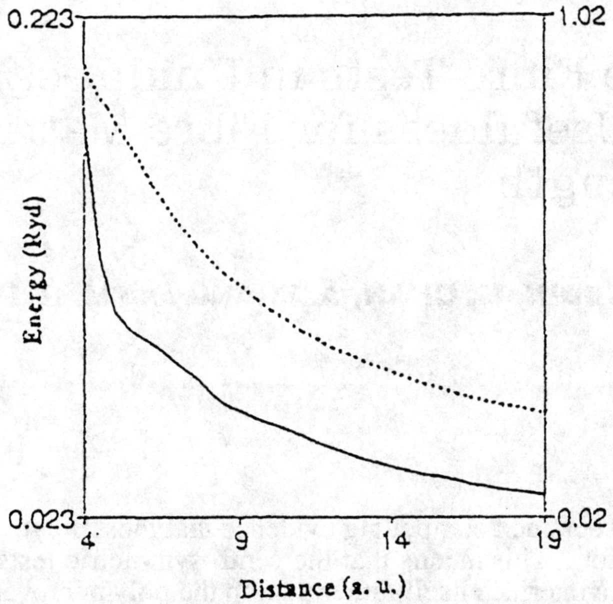

Figure 1: The effective pair potentials of Cu-C, the wide line, and of C-C, the dotted line, in Cu-based metal matrix composite. The left Y-axis is energies for Cu-C potential and the right one is energies for C-C potential. Energies are in Ry.. Distances (X-axis) are in atomic units.

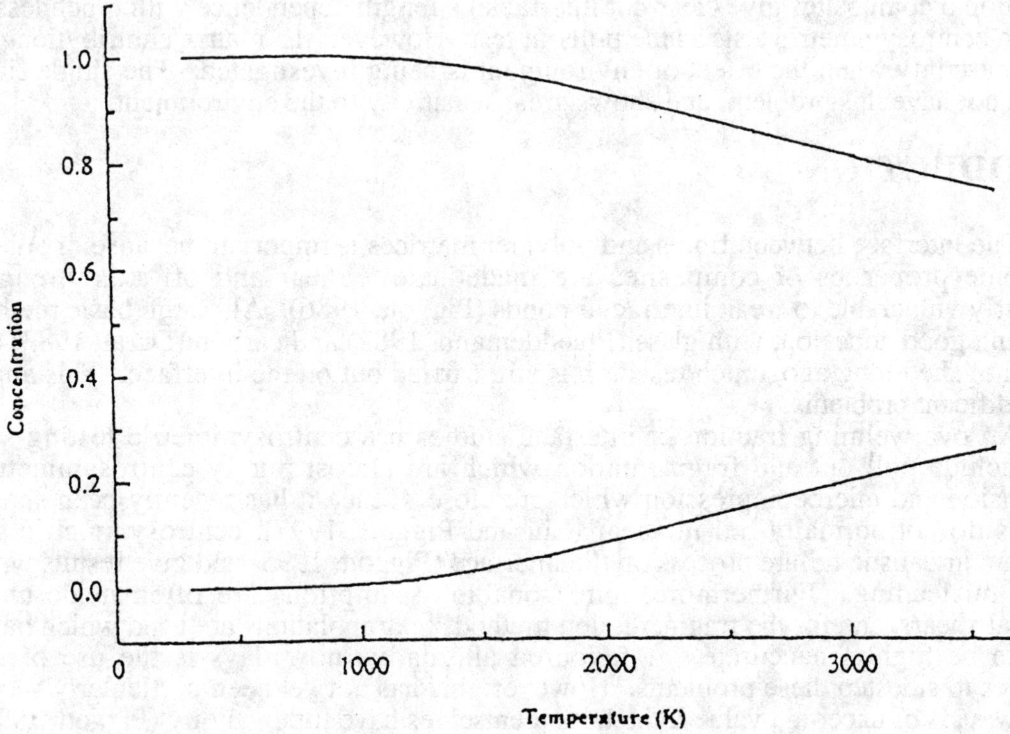

Figure 2: Temperature dependences of partial concentrations of carbon atoms occupying octahedral (upper curve) and tetrahedral (lower curve) interstitial positions in copper

Normal to the Fibre Tests in Unidirectional Composites: Usefulness for Fibre Matrix Interface Strength

M. A. ALIMUDDIN, M. CHAN, A. ZHANG AND M. R. PIGGOTT

ABSTRACT

We have recently obtained compelling evidence that most polymers do not fail in shear. Instead, they fail in tension. This means that the centrosymmetric tests for interface strength impel the failure along the interface itself rather than in the polymer, even when the interface is extremely strong. The actual mechanism of failure is tensile and is unduly influenced by surface roughness, friction, and cure shrinkage pressure. High apparent interface strengths can be obtained even when adhesion is very poor, for example when the fibres are coated with release agent. These problems aply to all the tests currently lused; i.e. pull out microtension, fibre fragmentation and microcompression. Results from tests on the transverse strength of unidirectional composites give clear-cut interfacial strength dependence with much less scatter than such centrosymmetric tests as the pull out test. However, the matrix contribution presents some uncertainty when the effect of environment is being investigated. The single fibre peel test does not have this problem, and shows great sensitivity to the environment.

INTRODUCTION

The interface between fibres and polymer matrices is important because, if adhesion is poor, some properties of composites are inadequate. Shear and off-axis strengths are particularly vulnerable to weak interfacial bonds (Piggott, 1986). Although basic methods for developing good adhesion, with glass (Plueddemann, 1982) and carbon (Drzal, 1983) at least, were established long ago, much research is still carried out on the interface. This appears to have significant problems.

An overwelming fraction of interface studies use centrosymmetric testing systems. These include pull out and fragmentation which are almost purely centrosymmetric, and microtension and microcompression which are close. Since it has recently been shown that polymers do not normally fail in shear (Liu and Piggott, 1997), centrosymmetric systems impose an unrealistic failure process on the interface (Piggott, 1995) and give results which can be very misleading. Furthermore, unreasonable assumptions are often made about the interfacial shears, and for the fragmentation methods, extrapolations are used which have been shown to be highly inaccurate. A favoured alternative nowadays is the use of fracture mechanics to sidestep these problems. However, this has not yet been particularly successful, and anyway, is of uncertain value if the tests themselves have little validity (Piggott, 1997).

This paper describes the results of some attempts to get away from the centrosymmetric tests.

M.R. Piggott, Department of Chemical Engineering and Applied Chemistry, University of Toronto, Toronto, Ontario M5S 3E5, Canada

TRANSVERSE TESTS

Since the transverse strength is one of the most significantly affected properties, this has been examined to see what it can reveal about the interface strength.

When a unidirectional composite is broken in such a way that the fracture surface is parallel to the fibres, we can use an approximate expression for the transverse strength, σ_{2u}, i.e.

$$\sigma_{2u} = A_i \sigma_{iu} + B A_m \sigma_{mu} \tag{1}$$

Here A_i is the projected area fraction of interface normal to the stress, σ_{2u}, and A_m is that for the matrix ($A_i + A_m = 1$). B, a constant which takes into account the stress concentrating effect of the fibres, is expected to be ≤ 1.0. σ_{mu} is the matrix ultimate tensile strength, and σ_{iu} that of the interface.

TABLE 1 - RESULTS FROM INTERFACE TESTS AND OFF-AXIS TESTS FOR CARBON FIBRE COMPOSITES

Curing Agent Agent	σ_{mu}[1] (MPa)	σ_{2u}[2] 9MPa)	V_m[3]	A_m[4]	τ_{iu}[5] (MPa)
AS1-Epoxy with TETA[6]	61 ± 5	30 ± 7	0.60	0.55	52 ± 8
AS1-Epoxy with Anchor 1171[7]	49 ± 3	23 ± 3	0.60	0.32	71 ± 9
AS4 -PEEK	103[8]	91± 3	0.39	> 0.99	85 ± 29

Notes:

1. Polymer matrix tensile strength
2. Strength normal to the fibres
3. Volume fraction of polymer
4. Area fraction of polymer in fracture surface produced by off axis tensile test
5. Debonding force/embedded area of fibre in pull out test
6. EPON 815 cured at 120° for 4hr; 12 phr of curing agent used.
7. EPON 815 cured at 50°C for 2h then 150°C for 2 h with 5 phr curing agent
8. Data from Chang and Lees (1988)

Data on σ_{mu}, σ_{2u} and A_m is given in Table 1 for three different systems. For the uncoated AS1 carbon fibres in epoxy, A_m is a little less than V_m with the TETA curing agent, and little more than half V_m with Anchor 1171. Both polymers appear to be adhering only moderately well, since σ_{2u} is only about 80% of $V_m \sigma_{mu}$. With AS4 carbon-PEEK (APC 2), on the other hand, $\sigma_{2u} \cong 0.9\ \sigma_{mu}$. Furthermore, Am > 0.99, i.e. almost no bare fibre can be seen, in marked contrast with the epoxy cases: see fig. 1. Evidently in this case $B \cong 0.9$.

Fig. 1. Fracture surfaces of well adhering carbon-1171 epoxy (at left) and carbon-PEEK (at right). Fibres are 8 μm diameter.

Fig. 2. Fracture surfaces of poor fibres are 8 μm diameter adhering carbon-epoxy (at left); (fibres coated with release agent) and carbon PEEK after 300 h in 90°C water (at right). Fibres are 8 μm diameter.

When the fibres were coated with release agents, σ_{2u} fell drastically, to less than 0.25 $V_m\sigma_{mu}$ (table 2). The amount of bare fibre increased very greatly, and misaligned fibres could be plucked out, leaving channels and tunnels; see fig. 2. When the APC 2 was immersed in water at 90°C for 300h, there was no significant loss of adhesion, table 2. Again, almost no bare fibre regions were visible on the fracture surface; see fig. 2.

TABLE 2- RESULTS FROM INTERFACE TESTS AND OFF-AXIS TESTS WHEN ADHESION HAS BEEN IMPAIRED

Curing Agent Agent	σ_{mu}[1] (MPa)	σ_{2u}[2] 9MPa)	V_m[3]	A_m[4]	τ_{iu}[5] (MPa)
AS1-Epoxy with TETA[6]	61 ± 5	6.3 ± 0.7	0.59	0.32	39 ± 6
AS1-Epoxy with Anchor 1171[7]	49 ± 3	7 ± 1	0.60	0.21	44 ± 8
AS4 -PEEK	-	85 ± 6	0.39	> 0.99	69 ± 24

Notes:

1. Polymer matrix tensile strength.
2. Strength normal to the fibres
3. Volume fraction of polymer
4. Area fraction of polymer in fracture surface produced by off axis tensile test
5. Debonding force/embedded area of fibre in pull out test
6. EPON 815 cured at 120° for 4hr; 12 phr of curing agent used. Release agent coating was undiluted Frekote 44-NC.
7. EPON 815 cured at 50°C for 2h then 150°C for 2 h with 5 phr curing agent; release agent was Dow Corning D.C. 20, from a 10% solution in tolune 10 parts and alcohol 80 parts.

These results should be contrasted with the pull out shear strengths, τ_{iu}, given in Tables 1 and 2. With the epoxies, these could be greater than σ_{mu}, and did not reflect the drastic loss of adhesion shown by the transverse tests. Even τ_{iu} for the release agent coated AS1 fibre in the Anchor 1115 cured epoxy was not significantly different from the matrix strength. It should be emphasised that the τ_{iu} values were simply estimated from the force balance, i.e. the debonding force F_A, equal to the product of the mean shear stress and the total immersed interfacial area, neglecting the fibre embedded end, i.e.

$$F_A = \pi d L \tau_{iu} \tag{2}$$

where d is the fibre diameter and L the embedded length.

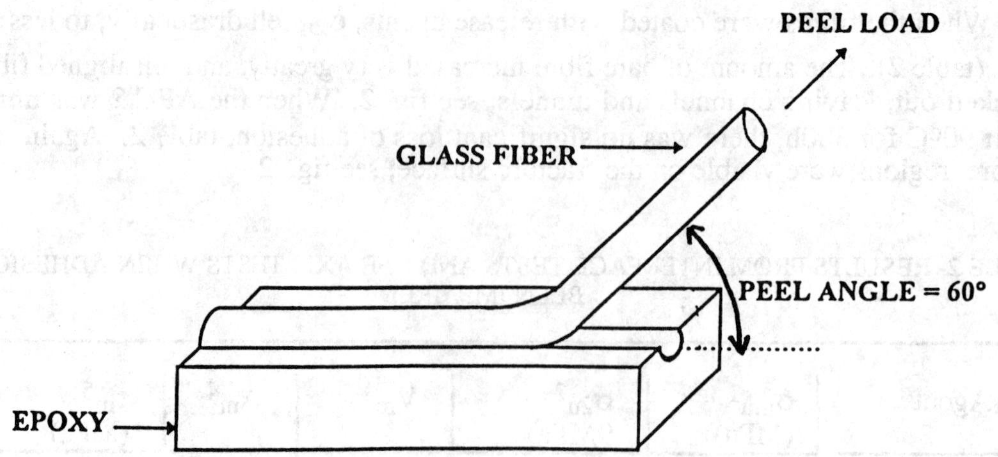

Fig. 3. Single fibre 60º peel test.

The situation was quite different with PEEK. This is a ductile thermoplastic. Clearly $\sigma_{iu} \geq \sigma_{mu}$, since $\sigma_{2u} = \sigma_{mu}$, and no significant amount of fibre could be seen in the fracture surface (fig. 1). The pull out values are very variable (the coefficient of variation (cv) = 35%) compared with σ_{2u} (cv = 6%) but are close enough to σ_{mu} to appear to be in possible agreement. However, with such large scatter, they can hardly be regarded as useful. σ_{2u} shows that there is probably no adhesion loss after water immersion. This is confirmed by the fractographic evidence (fig. 2). Again, this seems more informative than the pull out result where the large scatter of results makes any conclusion uncertain. (The τ_{iu} value given was for 100h immersion, for 1000h the result was 56 ± 12 MPa. No result was available for 300h.)

FIBRE PEEL TEST

Fibres were embedded in epoxy in such a way that they were "floating" parallel to the polymer surface, with only about one half of their diameter submerged. This was done by allowing the resin to gel slightly, then placing the fibre on the surface. Then the sample was put in the oven at 60ºC for 2h, during which the fibre sank the required amount and the resin hardened. The resin used was epoxy EPON 815 with 12% TETA as curing agent. It was held in a polyethylene capsule. The samples were checked in a microscope, and only those which had more than 80% of their diameter showing, as seen from above, were used for testing.

With equipment presently available, it was not possible to measure the force required for the perpendicular peel, since this was too small. Instead, with glass, it was carried out at 60ºC peel angle; see fig. 3. (With carbon, the force was too small at any useful angle.)

Glass adhesion to epoxy was very susceptible to the environment. At room temperature, only four hours were required for the adhesion to disappear completely in the presence of water. Other environments acted more slowly, but still only a few hours were required for very large losses of adhesion to occur: see fig. 4.

DISCUSSION

The transverse tests show clearly the importance of fibre-matrix adhesion. These direct composite tests appear to be much more useful guides to reductions in adhesion than pull out tests. Furthermore, some of the actual mean pull out shear stresses appear to be unrealistic, although based on simple force balances.

The apparent agreement between transverse strength and interface strength described by Madhukar and Drzal (1991a, 1991b) may possibly be fortuitous. Fibre fragmentation was used for the interface evaluation and τ_{iu} was calculated from

$$\tau_{iu} = \sigma_{fu} d/2L_c \tag{3}$$

with σ_{fu}, the fibre strength, obtained by extrapolation. The extrapolation error can be as much as 50% (Dai and Piggott, 1993). Furthermore, the shear stress at the interface is not uniform over the fibre critical length L_c, as assumed when equation 3 is used. There is now a very large body of laser RAMAN data which clearly demonstrates that the shear stress has large peaks at the ends of approximately constant shear stress region. Theory indicates that these peaks can be three to four times the value of the mean stress (Piggott, 1991). Finally, when the transverse strength was measured in flexure, instead of tension, the agreement was not very good.

The preliminary results of the peel test are very promising. The test reveals environmental effects with glass very quickly indeed, compared with pull out tests at least (Chua et al, 1992). This test could be regarded as an interfacial tensile test. Ideally, 90° peel should be used rather than 60°, so that the failure is entirely opening mode. We are at the moment exploring this possibility with more sensitive equipment.

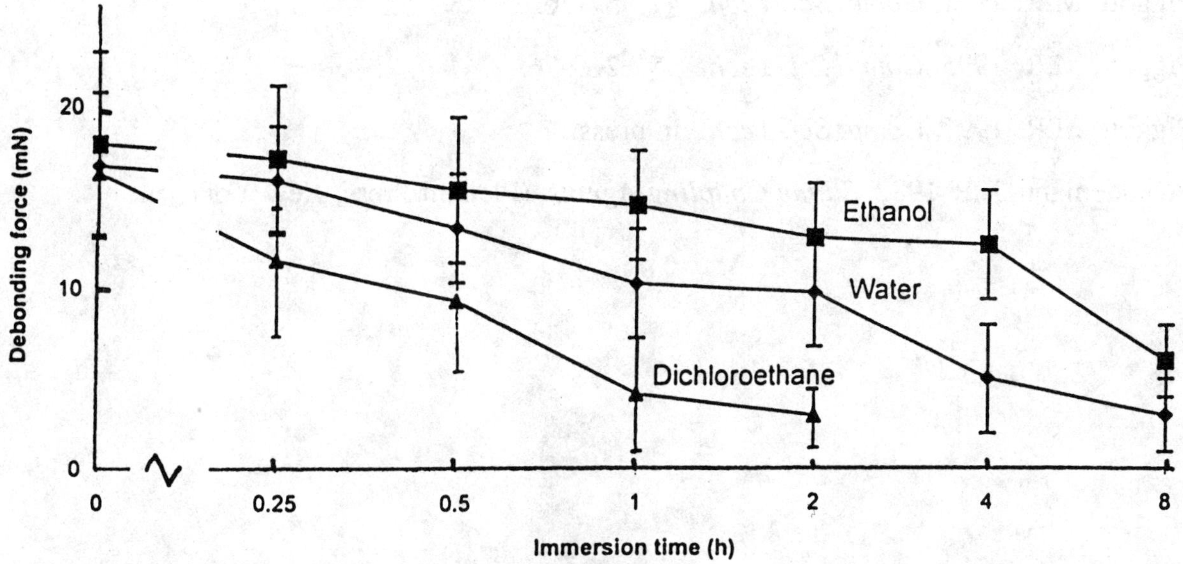

Fig. 4. Results for single glass fibres peeled from epoxy, with various liquids on polymer surface.

CONCLUSIONS

Transverse tests appear to be much more useful guides to interface degradation than pull out tests, while any agreement between fibre fragmentation and transverse strength should probably be regarded as purely fortuitous. The most sensitive transverse test appears to be the single fibre peel test. This responds very quickly to environments such as water, in the case of glass at least. It is recommended that transverse testing be examined further and that centrosymmetric tests be phased out.

ACKNOWLEDGEMENTS

The authors are grateful to United States Air Force Office of Scientific Research for financial support which made this work possible.

REFERENCES

Chang, I.Y. and Lees, J.K. 1988. *J. Thermoplastic Comp. Mater.*, 1, 272-96.

Chua, P.S., S.R. Dai and M.R. Piggott. 1992. *J. Mater Sci.* 27, 911-24.

Dai, S.R. and M.R. Piggott. 1993. *Comp. Sci. Tech.,* 40, 81-7.

Drzal, L.T. 1983. *SAMPE J.,* 19, (5), 7 -13.

Liu, K. and M.R. Piggott. 1997. *Polymer Eng. Sci..,* in press; two papers.

Madhukar, M.S. and L.T. Drzal. 1991a. *J. Comp. Mater.* 25, 932-51.

Madhukar, M.S. and L.T. Drzal 1991b. *J. Comp. Mater.* 25, 958-97.

Piggott, M.R. 1982. *Polymer Composites,* 3, 179-88.

Piggott, M.R. 1991. *Comp. Sci. Tech.* 42, 57-76.

Piggott, M.R. 1995. *Comp. Sci. Tech.* 55, 269-76.

Piggott, M.R. 1997. *Comp. Sci. Tech.* in press.

Plueddemann, E.P. 1982. *Silane Coupling Agents*, (Plenum Press, New York,), 1.

Mechanics of Stochastic Fibrous Networks

A. M. SASTRY, X. CHENG AND C. W. WANG

ABSTRACT

Point-bonded fibrous networks are used in a variety of engineering applications, including reinforcement for polymeric composites, paper, and electrochemical substrates, among many others. Understanding of the nature of the deformation of these irregular microstructures is key to developing useful manufacturing guidelines for their cost-effective production. The current approach allows such analysis, through use of a stochastic approach in constructing the network. Resulting network properties are compared with those in more regular arrays. Ultimately, determination of local deformations in these kinds of networks will be used to predict local deformation due to flow front progression in fibrous preforms for production of polymeric composite materials.

INTRODUCTION

Various approaches have been used in the modeling of the deformation processes and effective stiffnesses of nonwovens and bonded nonwoven fibrous materials. The majority of classic approaches comprise conservation of strain energy analyses (e.g. Hearle and Newton, 1968; Hearle, 1980). These analyses revolve around calculation of the strain energy associated with affine deformations of unit cells of known orientation relative to the direction of applied load. The assumption of affine deformation, or deformation in which the strains at all scales of interest are geometrically similar, neglects effects such as microbuckling of fibers and random fiber slack. Other approaches, (e.g. Backer and Petterson, 1960) employ averagings of orientation of fibers to derive elastic moduli of bonded networks. While affine unit cell approaches and their extensions can provide satisfactory prediction of elastic moduli of short fiber nonwovens, they amount to highly averaged, continuum predictions.

A.M. Sastry, X. Cheng, and C.W. Wang
Department of Mechanical Engineering and Applied Mechanics, 2350 Hayward Street, The University of Michigan, Ann Arbor, Michigan 48109-2125

Homogenization approaches (e.g. Gudes and Kikuchi, 1991) have also been developed to address the problem of optimization of structures using smoothed continuum constitutive rules calculated from known properties of distinct phases.

Current approaches in analysis of fibrous media can be thus categorized roughly into three types of models: 1) continuum approaches, relying upon macroscopic experiments to provide properties of interest, 2) micromechanical approaches using regular unit cells and constituent properties to model macroscopic properties (including both strength-of-materials approaches and elasticity approaches) and 3) homogenization approaches, where microstructural features are assessed and smoothed to produce local continuum properties.

Full optimization of these materials requires that models incorporate the specific microstructures obtainable, along with incorporation of their specified or anticipated statistical distributions in the component. Directed fiber approaches allow production of fibrous preforms, for example (Fig. 1), with continuously varying fiber orientation, to within $\pm 5^{o}$ precision (e.g. Gerard and Jander, 1993).

Cost control in production of composites reinforced with these materials requires that analysis be detailed enough to 1) determine the effect of continuous changes in microstructure on these properties (e.g. the change in permeability with a 10% change in local orientation of fibers), and 2) determination of sensitivity to these properties to production specifications (e.g. the determination of angle sensitivity and fiber laydown, and thus production precision).

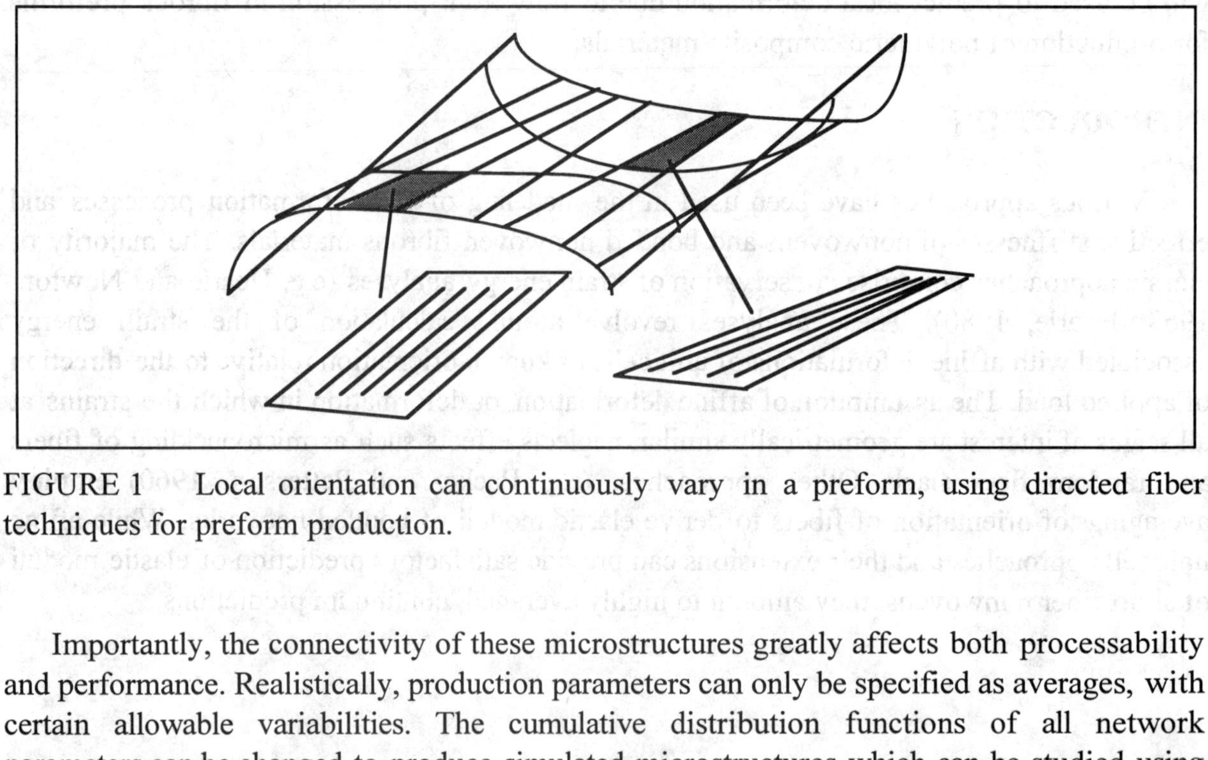

FIGURE 1 Local orientation can continuously vary in a preform, using directed fiber techniques for preform production.

Importantly, the connectivity of these microstructures greatly affects both processability and performance. Realistically, production parameters can only be specified as averages, with certain allowable variabilities. The cumulative distribution functions of all network parameters can be changed to produce simulated microstructures which can be studied using the current approach. The current approach employed stochastic microstructures for study of the material properties.

METHODOLOGY

Stochastic fiber networks were generated to understand and predict local deformations in point-bonded, fused fiber networks, addressing fundamental questions about the drapability and processability of these structures. The main approach involved four steps:

1) Networks were initially constructed by selecting distribution functions for six main variables: fiber volume fraction within a unit cell, fiber orientation distribution, fiber length distribution, fiber centerpoint (x- and y-) location, and fiber aspect ratio (L/d). Data were generated randomly according to these parameters, and fibers were "placed" in the unit cell, as in Fig. 2a. The intersection points were calculated, and thereafter assumed to be rigid bonds.

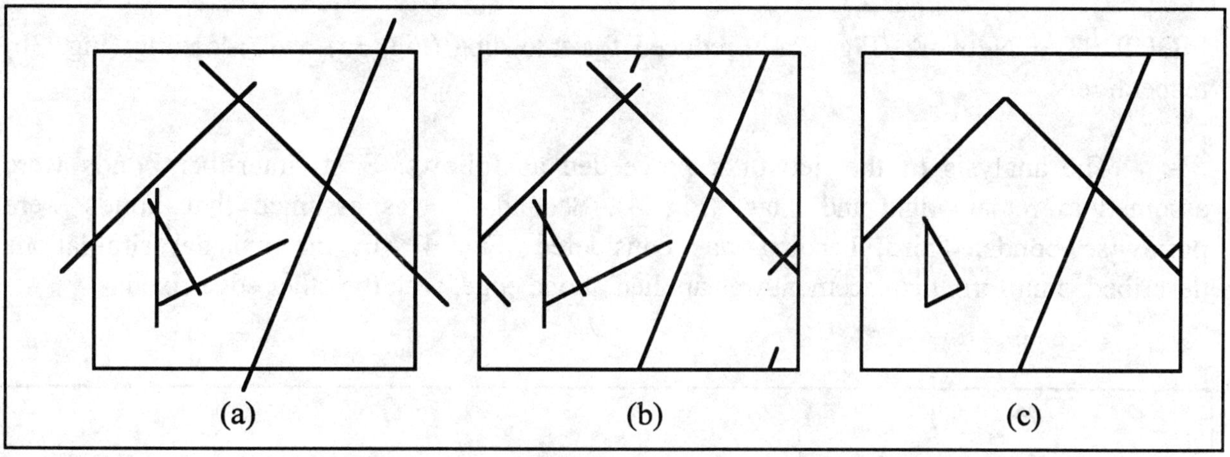

FIGURE 2 Network construction technique; square cells of unit area.

2) Periodic boundary conditions were imposed (Fig. 2b). In this way, the initial volume fraction specified at the outset was imposed exactly for the unit cell, as ends were "wrapped" back into the cell. Also, periodic boundary conditions allowed analysis without spurious edge effects.

3) Non-load-bearing segments were removed. This included ends of overlapping fibers which did not intersect one of the load boundaries, and structures within the cell which did not span the boundaries (Fig. 2c), such as the triangle shown. Finally, a load direction was chosen (structures studied here were placed in uniaxial tension), and the networks were reduced accordingly, as in Figs. 3a and 3b, for x- and y-direction loading, respectively.

Using this methodology, several types of networks were constructed. By the technique described, the controllable parameters in the simulated networks were identical to those adjustable in a manufacturing process. The great majority of the generated structures were highly indeterminate (mechanisms, if bending were not considered to prevent unrestricted rotations about joints).

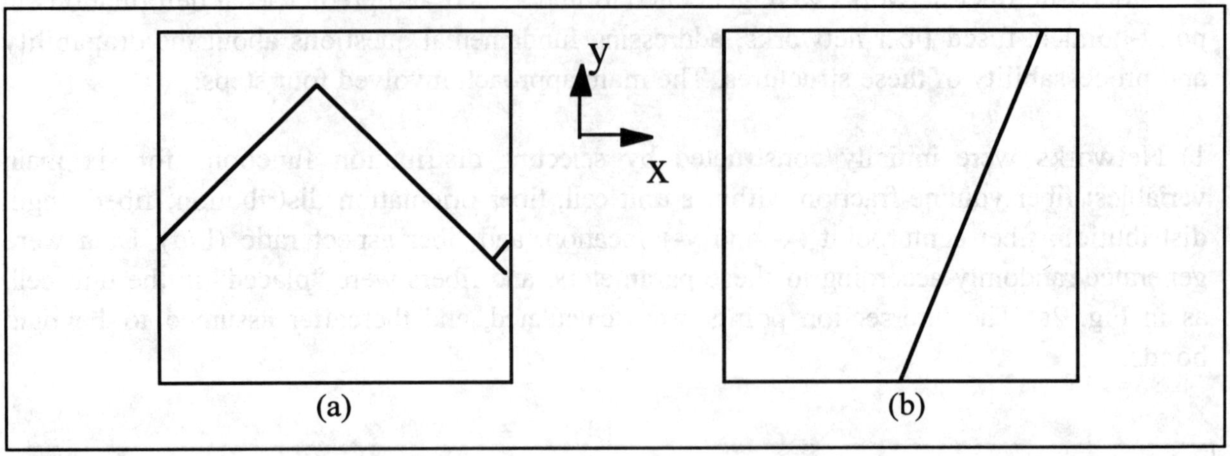

FIGURE 3 Network (Figs. 2a-c) reduced for x-loading (Fig. 3a) and y-loading (Fig. 3b) respectively.

The analysis of the networks proceeded as follows. First, interfiber bonds were assumed to remain rigid and intact (Fig. 4). Second, it was assumed that bonds were pointwise bonds. Third, bending was considered (Fig. 4). In the uniaxial simulations described, a uniform displacement was applied at one edge, with the other edge fixed.

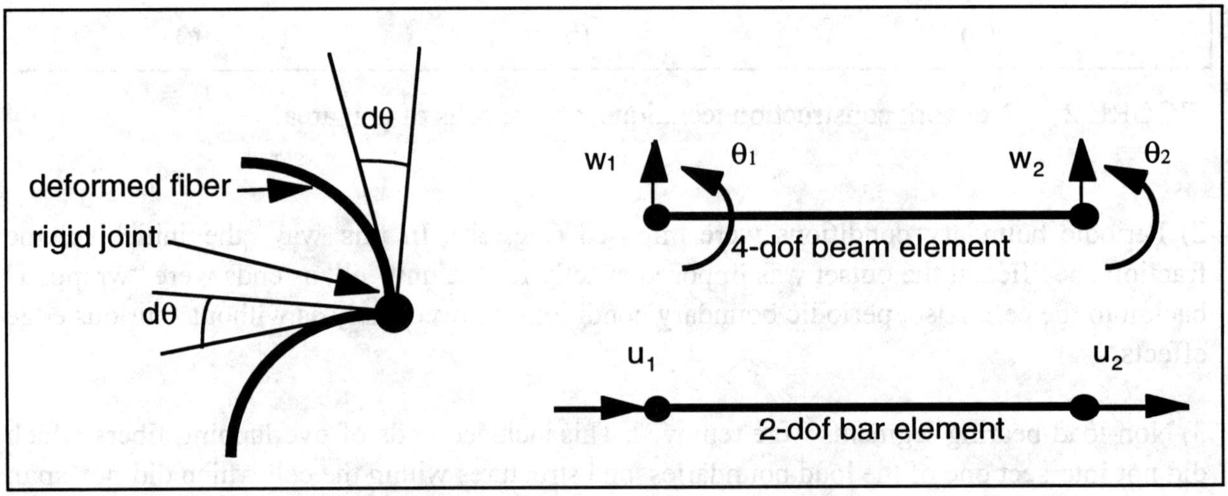

FIGURE 4 Network loading scheme. Point bonds are assumed rigid; segment geometry can be altered through moment of inertia and aspect ratio.

The mechanical analysis proceeds from a calculation of potential energy in the random structure (e.g. Cook *et. al*, 1989). The general expression for potential energy in a linear elastic body is

$$\Pi_p = \int_V (1/2\{\varepsilon\}^T[E]\{\varepsilon\} - \{\varepsilon\}^T[E]\{\varepsilon_o\} + \{\varepsilon\}^T\{\sigma_o\})dV - \int_V \{u\}^T\{F\}dV - \int_S \{u\}^T\{\Phi\}dS - \{D\}^T\{P\} \quad (1)$$

where terms are defined

Πp...potential energy
{u}...displacement field
{ε}...strain field
[E]...stiffness matrix
{F}...body forces
{Φ}...surface tractions
{D}...nodal d.o.f. of the structure
{P}...externally applied loads to d.o.f.
S,V...surface area, volume of structure
$\{\varepsilon_o\}$, $\{\sigma_o\}$...initial values of {ε}, {σ}

For a system of discrete elements (in this case, the segments between nodes), (1) becomes

$$\Pi_p = 1/2 \sum_{n=1}^{numel} \{d\}_n^T[k]_n\{d\}_n - \sum_{n=1}^{numel} \{d\}_n^T\{r_e\}_n - \{D\}^T\{P\} \quad (2)$$

where

{P}...externally applied loads at nodes
[k]...element stiffness
[d]...transformation matrix (each "element" is oriented at a different angle)

Making Πp stationary with respect to small changes in displacement, we obtain

$$[K]\{D\} = \{R\} \quad (3)$$

where

[K]...element stiffness
{D}...nodal displacements; here, u, v, θ
{R}...externally applied loads at nodes, here, at boundary nodes

A schematic of the deformation of fibers about a rigid bond is shown at left in Fig. 4; the elements shown at right (uniform, and lying along the x-axis) were superposed to obtain the stiffness matrix:

$$[K'] = \int_0^L [B]^T EI[B]dx = \frac{AE}{L}\begin{bmatrix} 1 & 0 & 0 & -1 & 0 & 0 \\ 0 & 0 & 0 & 0 & 0 & 0 \\ 0 & 0 & 0 & 0 & 0 & 0 \\ -1 & 0 & 0 & 1 & 0 & 0 \\ 0 & 0 & 0 & 0 & 0 & 0 \\ 0 & 0 & 0 & 0 & 0 & 0 \end{bmatrix} + \frac{EI}{L^3}\begin{bmatrix} 0 & 0 & 0 & 0 & 0 & 0 \\ 0 & 12 & 6L & 0 & -12 & 6L \\ 0 & 6L & 4L^2 & 0 & -6L & 2L^2 \\ 0 & 0 & 0 & 0 & 0 & 0 \\ 0 & -12 & -6L & 0 & 12 & -6L \\ 0 & 6L & 2L^2 & 0 & -6L & 4L^2 \end{bmatrix} \quad (4)$$

where (T) is the transformation matrix. Edge displacements (representing uniaxial tension for a displacement-controlled experiment) were applied to the boundary nodes the cell. A unit dummy load technique was used to solve for the applied displacement boundary conditions.

Several cases were investigated, to demonstrate the high variance in the behavior of these structures, owing to their stochastic construction.

RESULTS

Table 1 gives a summary of the structures simulated. Five cases were studied, to assess both the effect of bending on stochastic networks, as controlled by fiber aspect ratio (where higher aspect ratio fibers produce networks whose deformation is more bending-dominated), and also to assess the effect of fiber length-to-unit cell ratio on the networks (it was not the aim to determine a "percolation point," for short fiber materials, or a point at which the volume fraction was sufficiently high to produce a load-bearing network). Five networks were generated and reduced for each case. In case C, no load-bearing structures were generated for any of the five data (because of the high aspect ratio, the volume fractions studied dictated only between one and three fibers in each case), so a plot is not included.

TABLE I—SIMULATION PARAMETERS

Case	Fiber Length	Fiber Aspect Ratio	Orientation Distribution
A	1	0.01	uniform, 0–90 degrees
B	1	0.02	uniform, 0–90 degrees
C	1	0.1	uniform, 0–90 degrees
D	1.5	0.01	uniform, 0–90 degrees
E	1.5	0.02	uniform, 0–90 degrees
F	1.5	0.1	uniform, 0–90 degrees

Several plots of normalized effective network modulus versus initial (Fig. 2b) volume fraction are shown in Figs. 5 and 6. A significant portion of the stochastic networks generated produced no load-bearing structures, resulting in zero normalized moduli.

DISCUSSION/FUTURE WORK

The nonaffine deformations within the structures produced moduli outside of the bounds predictable by strength-of-materials approximations, or by use of models for analysis of regular microstructures. The computed moduli in all cases, for example, were substantially less than that predicted via simple rule-of-mixtures, due to the significant role of bending in the deformation. Little difference was found in the structural properties of networks generated with staple lengths of 1–1.5.

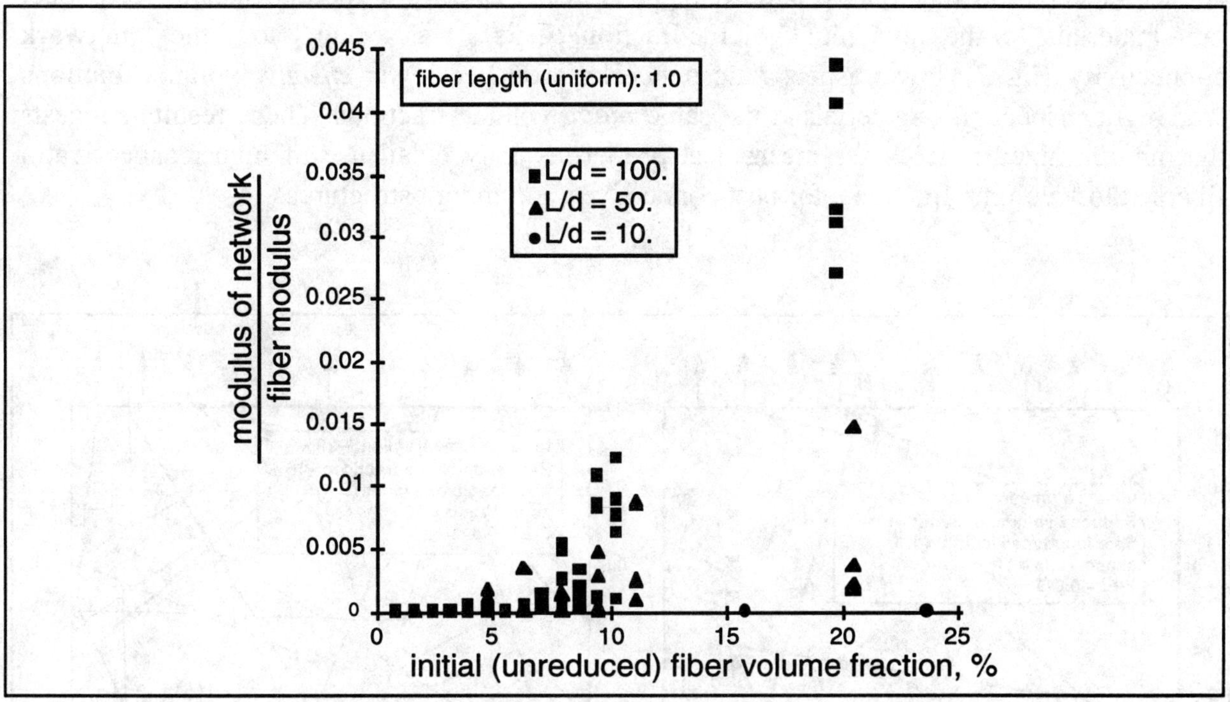

FIGURE 5 Normalized network moduli, for fiber length = cell length.

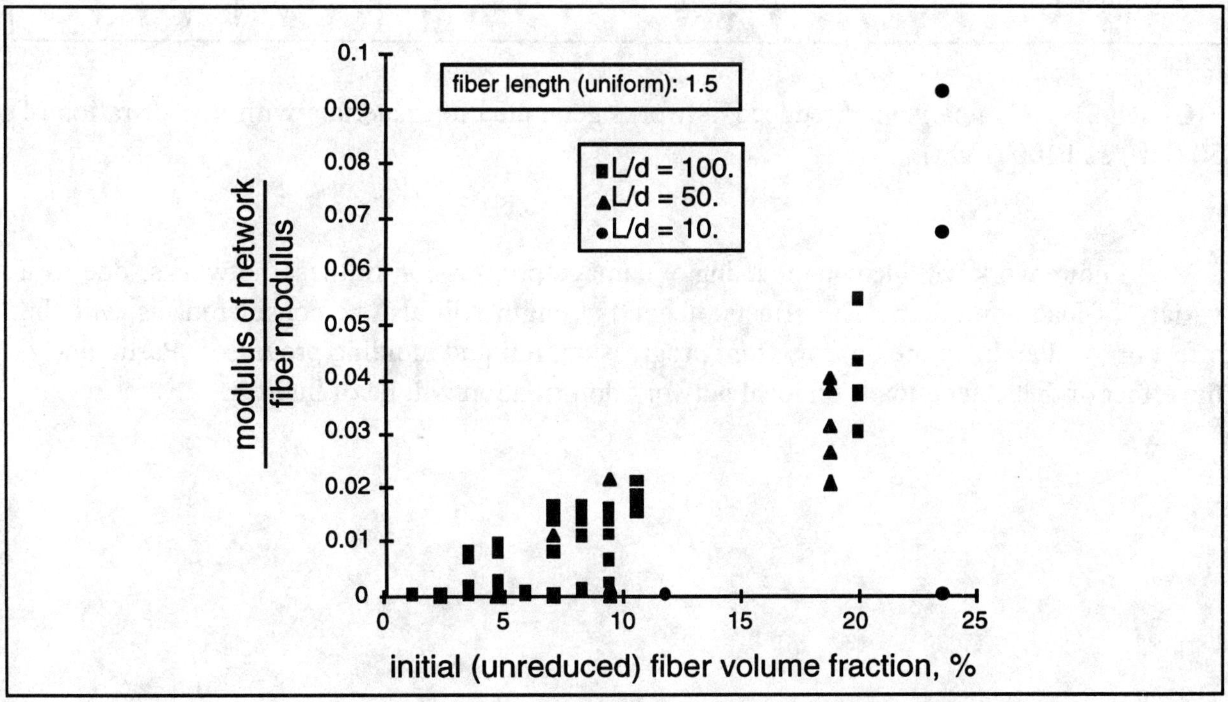

FIGURE 6 Normalized network moduli, for fiber length = 1.5 cell length.

Higher aspect ratio materials produced higher effective network moduli than lower aspect ratio materials, for the same initial volume fraction. This was due to the network connectivity (Fig. 7): lower aspect ratio networks produced lower *effective* volume fractions than higher aspect ratio materials at the same *initial* volume fractions. These results suggests that an optimized process for mechanical properties may entail use of higher aspect ratio fibers at low volume fractions, for best connectivity in random structures.

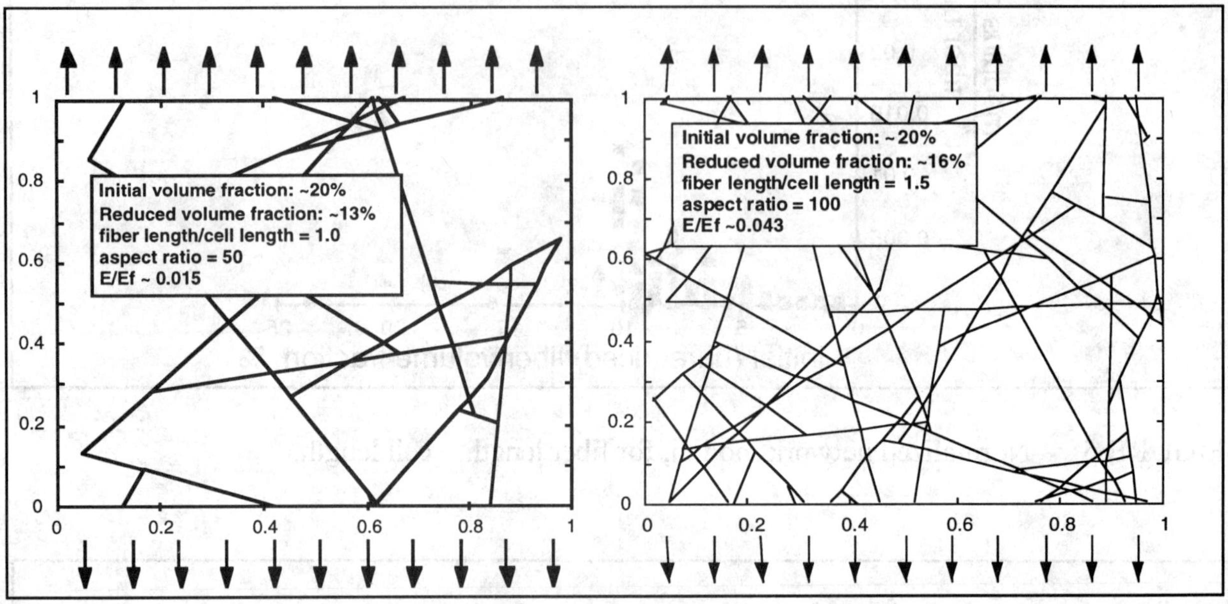

FIGURE 7 Comparison of reduced networks generated for materials with aspect ratios of 50 (left) and 100 (right).

Future work will include modeling of damage progression in these networks, due to a variety of load conditions. The effects of bond strength will also be considered, as will the effect of local architecture on flow front progression in liquid molding processes. Particularly, the effect of distributed loads on local network deformation will be of interest.

REFERENCES

Backer, S., and D. R. Petterson. 1960. "Some Principles of Nonwoven Fabrics," *Textile Research Journal* **30**:704-711.

Cook, R. D., D. S. Malkus, M. E. Plesha. 1989. "Concepts and Applications of Finite Element Analysis," John Wiley & Sons, New York.

Gerard, J., and M. Jander. February, 1993.*48th Annual Conference, The Composites Institute*, The SPI.

Gudes, J. M. and N. Kikuchi. 1990. "Preprocessing and Post Processing for Material Based on the Homogenization Method with Adaptive Finite Element Methods," *Computer Methods in Applied Mechanics and Engineering* **83**:143.

Hearle, J. W. S. 1980. "The mechanics of Dense Fibre Assemblies," in *The Mechanics of Flexible Fibre Assemblies*, 51-86, J.W.S. Hearle, J.J. Thwaites and J. Amirbayat, eds., Sijthoff and Noordhoff, New York.

Hearle, J. W. S., and A. Newton. 1968. "Part XV: The Application of the Fiber Network Theory, " *Textile Research Journal*, **1**:343-351.

ACKNOWLEDGMENTS

This work was supported in part by the Lawrence Berkeley Laboratories, through the Exploratory Research Program of the U.S. Department of Energy, Dr. Albert Landgrebe and Dr. Kim Kinoshita, program managers, and is gratefully acknowledged. Insightful technical discussions with Professor James R. Barber at the University of Michigan are also gratefully acknowledged.

Influence of Edge Effects in Estimating Interfacial Normal Strength in Model Unidirectional Composites

G. P. TANDON,[1] R. Y. KIM,[2] S. G. WARRIER[3] AND B. S. MAJUMDAR[3]

ABSTRACT

The present study is aimed at designing and evaluating some simple test methods for estimating the normal strength of the fiber-matrix interface. In order to remove the influence of the free surfaces from tests on traditional straight-sided specimens, two alternate geometries, namely, embedded fiber and cruciform design, are considered. These model specimens are incrementally loaded in tension to failure and the fractured halves examined to identify the location and extent of failure. Analytical modeling of the specimens is also done using 3-D finite element analysis and is coupled with local asymptotic analysis for calculating the stress concentration at the fiber matrix interface and at fiber corners, respectively. Results demonstrate the importance of stress field singularities in controlling damage initiation in straight-sided samples and the success of cross-shaped design in forcing debond initiation in the interior of the sample (region free of stress singularities).

INTRODUCTION

Failure modes of fiber reinforced composites is governed in part by the transfer of stress between fiber and the matrix. This transfer occurs across the interface between the components, and the properties of this interface, therefore, will effect the properties of the composite. Unlike the interfacial shear strength, the measurement of the interfacial normal strength (Tandon and Kim, 1997) has, to our knowledge, received limited attention. Therefore the present study is aimed at designing and evaluating some simple test methods for estimating the normal strength of the fiber-matrix interface.

Interfacial normal strength is commonly determined by testing straight-sided single fiber composite specimens (Fig. 1) and interpreting debond initiation from the beginning of non-linearity or first occurrence of acoustic emission on the stress strain curve. In these straight-sided specimens, the free surface is a favored site for interface debonding because stresses are singular at that location. In order to remove the influence of the free surfaces from the test, two alternate geometries can be considered. The first design involves covering the ends of the fiber by embedding the fiber (Hu et al., 1995; Tandon et al, 1996)) completely inside the matrix (Fig. 2), whereas, the second design is of a composite sample in the shape of a cruciform or cross (Gundel, et al., 1995) with an extremely large width in the gage section (Fig. 3). With the cross-shaped geometry, the central portion of the fiber/matrix interface is highly stressed under tensile loading, thereby forcing debond initiation and propagation in the center, away from the free edges.

Materials Directorate, USAF Wright Laboratory, Wright-Patterson AFB, OH 45433
[1] AdTech Systems Research Inc. [2] Univ. of Dayton Research Inst. [3] Universal Energy Systems Inc.

In this study, model single fiber composite specimens were made in the three geometrical configurations, namely, i) straight sided specimen with fiber ends exposed, ii) straight-sided specimens with fiber ends covered and iii) cross-shaped design. These model specimens are incrementally loaded in tension to failure and the fractured halves examined to identify the location of failure initiation and the extent of failure while strain and acoustic emission activity is continuously monitored. The three specimen designs have also been analyzed using 3-D finite element methods to estimate the interfacial stresses in terms of constituent material properties. Finally, some failure criterias (Whitney and Nuismer, 1974; Folias, 1991) have been employed in conjunction with the experimental observations to estimate the normal bond strength of the interface. Results indicate good correlation between experiments and the analytical predictions for test configurations considered in this study and demonstrate the importance of stress field singularities in controlling damage initiation in straight-sided samples. It is further shown that the cross-shaped design is successful in forcing debond initiation in the interior of the sample (region free of stress singularities) and therefore the results produced by testing of such specimens may be capable of providing the true transverse response of the fiber-matrix interface.

SPECIMEN DESIGN , FABRICATION AND TEST METHODS

Single fiber composite tests are increasingly recognized as a most useful tool to study the mechanical/chemical interactions at the fiber-matrix interface region. Single fiber specimen are easier to make compared to unidirectional composites avoiding the difficulty of fiber alignment in unidirectional composites. Further, single fiber composites help to focus on localized failure, such as interface debonding, avoiding the distraction of randomness in failure and the neighboring effect of other fibers in unidirectional composites. In this work, model single fiber composite specimens were made in the three geometrical configurations, namely, i) straight sided specimen with fiber ends exposed, ii) straight-sided specimens with fiber ends covered and iii) cross-shaped design.

The composite specimens were cast in rectangular and cruciform shaped silicone rubber molds for straight-sided and cross-shaped samples, respectively. The specimen cross-sectional dimensions are shown in Figures 1-3. The samples were 0.198" (or 5.03 mm) thick for all three geometries considered. The composite reinforcement consisted of aluminum rods, 0.092" (or 2.34 mm) in diameter, while the matrix was an epoxy resin (Epon 828 from Shell Chemical Co.) cured with a polyetheramine (Jeffamine D-230 from Texaco, In.) for 3 days at ambient temperature. The rod surfaces were merely cleaned with acetone, which enabled the epoxy to strongly bond with the reinforcement. Rods of predetermined length were approximately positioned in the mold and epoxy resin cast around them. Curing at ambient temperature eliminated thermal residual stresses which are otherwise induced from the mismatch in coefficients of thermal expansion of fiber and matrix.

The free edges of the straight-sided specimen with exposed fiber ends were progressively ground and then polished using successively smaller diameter alumina polishing powder (final size was 0.3 micrometer) in order to enhance the microscopic image for crack detection. The free edges of the cruciform specimen and straight-sided samples with embedded fiber ends were also polished after curing. Strain gauges (gauge length 3.2 mm) were then mounted on the flat surface in the center and along the edges of the specimen perpendicular to the fiber to monitor strain during loading. An acoustic emission transducer was also attached to the specimen to monitor failure events. For straight-sided samples with fiber ends exposed, the specimen was loaded to either a prescribed level slightly higher than the expected interface debonding level or up to the first occurrence of acoustic emission on the stress strain curve. This load was held during application of a fluorescent penetrant to the polished surface, and then unloaded for microscopic examination to capture all failure events which occurred. Straight sided samples with fiber ends embedded and cruciform shaped specimen were also loaded in tension, under displacement control, to failure and the fractured halves examined to identify the locus of failure.

EXPERIMENTAL RESULTS

In Table I, we have listed the applied stress level at debond initiation obtained either from first occurrence of acoustic emission or beginning of non-linearity in the stress-strain curve for the three specimen designs. The values listed in this table are the average of six specimens for each configuration considered.

TABLE I: Applied Stress At Debond Initiation For Model Composites

Model Single Fiber Specimen	Average Applied Stress at Debond Initiation	Standard Deviation (%)
Straight-sided with fiber ends exposed	1207 psi (8.3 MPa)	14.1
Straight-sided with fiber end embedded	1787 psi (12.3 MPa)	4.6
Cruciform (Cross-shaped)	2832 psi (19.5 MPa)	6.7

Figures 4a and 4b are the photo micrographs showing the front and back views of the straight-sided exposed fiber specimen at the first occurrence of acoustic emission on the stress-strain curve, respectively. The portion of the interface which was observed to be debonded under the microscope is indicated by arrows. Notice the symmetric nature of the debonded interface, with respect to the horizontal loading axis, observed both in front (Fig 4a) and back (Fig 4b) of the specimen. Subsequent loading resulted in the debond crack branching into the matrix leading to final failure as shown in the front and back of the specimen in Figs 4c and 4d, respectively. Figure 5 is the photo micrograph of the failure surfaces of straight sided specimen with fiber ends covered. It is seen that failure initiates at the fiber corner and propagates over some distance along the fiber length before branching into the matrix. However, for cruciform (or cross-shaped) specimen, it is observed that failure initiates in the center (region free of stress singularities) away from the fiber ends, as seen in the photo micrograph in Figure 6. The debond propagates for some distance along the length of the reinforcement before branching into matrix.

NUMERICAL ANALYSIS

In this work, model single fiber composite specimens have been analyzed using 3-D finite element analysis employing the ANSYS code. The model was simplified by using symmetry planes so that only an eighth of the total specimen was required to be modeled. These planes of symmetry were constrained by symmetry boundary conditions, whereas the outer surfaces were free to move in all directions. The aluminum reinforcement and the epoxy matrix were treated as three dimensional eight node brick elements with the following elastic properties (handbook values).

Material	E(GPa)	ν
Aluminum	72.4	0.32
Epoxy	2.9	0.35

Figure 7 shows the radial stress at the fiber-matrix interface in the loading direction as a function of distance along the length of the reinforcement for the three specimen geometries considered. Analysis reveals that the radial stress at the interface is the dominant stress component and is maximum in the loading direction. The local stresses are represented by the stress concentration factor (SCF), defined as the ratio of the local stress to the applied stress. The distance is normalized with respect to the radius of the aluminum rod. The SCF value is 1.2 in the center of the cruciform specimen and remains reasonably constant (within

10 percent) over two-thirds of the loading region. The SCF gradually decreases at locations further away from the center of specimen approaching zero value before increasing marginally near the free surface. For straight-sided samples, the SCF at the center of the specimen is approximately 1.3 for both embedded and exposed fiber ends. However, as the free surface is approached for exposed rod or the edge of the embedded rod is approached, a sharp increase in the SCF is observed, although the stresses in the latter are significantly lower than the former.

The results in the previous sections for the specimens tested indicated that only in the case of the cruciform specimen did interface failure occur far from the ends (where a factor of 1.2 is more appropriate due to partial loading of the wings), and that cracking initiated at the fiber corners in both the straight sided and embedded fiber specimens. The likely explanation for this behavior is the high radial stresses that occur at the fiber corners for both the exposed-fiber and embedded-fiber specimens, as seen in Figure 7. The lower far-field applied stress at debonding for these specimens, compared with the cruciform geometry, are consistent with this local elevation of radial stress at the interface. Additional insight regarding the stresses at the fiber corners is provided in the next section.

Figure 8 shows the transverse strain at the free surface of the cruciform specimen as a function of the distance along the length of the aluminum rod at locations which represent the positions of the strain gages in the experiments. The strain values have been normalized with respect to the applied stress level and measured matrix modulus. As seen in Figure 8, the experimentally measured values of strain were within 10 percent of the calculated value at several locations along the length of the aluminum rod. Excellent match between strain calculations and measurements has also been obtained previously for metal matrix composite cruciform shaped specimens (Warrier et al., 1996).

SINGULAR STRESS FIELDS AT INTERFACE CORNERS

Bogy's (1971) analysis for edge-bonded elastic wedges provides an elegant methodology for analyzing the stress concentration at the fiber corners. Using Airy's stress function approach, Bogy (1971) showed that the stresses at the corner are actually singular, and can be described in the form $\sigma_{ij} \sim r^{-\lambda}$. Here λ is the order of the singularity and is a function of the geometry and the elastic constants of the constituent materials, and r is the radial distance of a material point from the interface corner. Using Mellin's transform, and following Bogy's formulation, one obtains an 8x8 determinant, D, for both the exposed fiber and embedded fiber problem. The dominant singularity, λ, is equal to -s-2, where s is the eigenvalue in the domain $-2<\mathrm{Re}(s)<-1$ for D=0.

Table II: Order of Stress Singularity for Straight-sided specimens

Model Single Fiber Specimen	λ (asymptotic analysis)	λ (FEM analysis)
Straight-sided with fiber ends exposed	0.275	0.2748
Straight-sided with fiber end embedded	0.210	0.2024

Utilizing the elastic constants for epoxy matrix and aluminum fiber, the order of the singularity obtained from the asymptotic approach is listed in Table II. Notice that the order of the singularity is indeed smaller for the embedded fiber, indicating that it is beneficial to embed the fiber in order to reduce the tendency towards interface failure from the fiber ends. The strength of the singularity from finite element analysis is next obtained by plotting the radial stress distribution at the fiber-matrix interface in the neighborhood of the singularity from the numerical solution on a double logarithmic plot. The slope and the intercept from that plot averaged over a distance $.007\ R < r < 0.1\ R$, where R is the fiber radius, provide the values of λ and k (see Eq 1). Although there are small differences between the two approaches in the order of the singularity, the agreement may be considered to be good as seen in Table II.

CHARACTERISTIC FAILURE DISTANCES FROM THE STRAIGHT SIDED SPECIMENS WITH EXPOSED AND EMBEDDED FIBERS

The singular stress profiles for the straight sided exposed-fiber and embedded fiber specimens indicate that there are two ways any useful data can be obtained from such specimens. As an extreme case, the singular stress profile can be used to imply that a crack already exists at the fiber ends at the interface, so that the maximum load in the tensile test simply reflects the attainment of criticality based on the fracture toughness of the interface. The problem with this approach is that it requires some method of determining the initial crack length. The alternate approach is to assume that the failure of the interface is governed by a probability distribution, whereby failure occurs when area integrated probability of failure reaches a critical value. Therefore, even though stress can become infinite over an infinitesimal distance, the probability of failure will remain finite, being determined by the integrated probability of failure. In this study, the focus here is to determine a characteristic distance for the exposed and embedded fiber specimens so that the apparent strength of the interface from those specimens is equal to that from the cruciform specimen.

To do this, we express the interface stress close to the fiber ends in the form:

$$\sigma_{rr}/\sigma_a = k\,(r/R)^{-\lambda} \tag{1}$$

where we now identify r with the length of the fiber (z) starting from the fiber end and k and λ are the quantities determined from finite element analysis. Integrating, the average stress concentration factor, SCF_{avg}, over a normalized distance z_{avg}/R is then given by

$$SCF_{avg} = \frac{\sigma_{rr}^{avg}}{\sigma_a} = \frac{k}{(1-\lambda)\left(\frac{z_{avg}}{R}\right)^{\lambda}} \tag{2}$$

Note that the fitted parameters, λ and k, in Eqns (1) and (2) vary with the averaging distance over which they are evaluated, as seen in Table III. Moreover, the values of λ reported in Table III, differ somewhat from the values obtained from asymptotic analysis (Table I). The use of the fitted parameters which vary with averaging distance is justified in lieu of Eq (2) where stress averaging is done over the same distance over which these parameters have been evaluated. It is further observed that small variations in the quantities λ and k obtained from FEM solution, have negligible influence on the stress averaging given by Eq 2.

The tensile strength of the interface using the cruciform specimen results can be estimated by multiplying the externally applied stress with the stress concentration factor at the interface and is calculated as 3398 psi (=2832 x 1.2) [or 23.4 MPa]. This value is significantly higher than the value of 1750 psi (= 1207 x 1.45) [or 12.1 MPa] reported earlier (Tandon and Kim, 1997) which was based on 2-D concentric cylinder model (Pagano and Tandon, 1988) neglecting the end effects (Kroupa and Ashbaugh, 1995; Folias, 1989) and size (specifically thinness) of the samples. As seen in Fig 7, the radial stress at the interface settles down to approximately 1.3 times the far field stress (σ_a) at a sufficient distance from the fiber end for straight sided specimens, as shown from 3-D finite element analysis. This value (1.3) is less than the value (1.45) determined using 2-D concentric cylinder analysis and can be ascribed to stress relaxation due to proximity of the specimen faces. As shown in Table III, the stress concentration factor calculated at the exposed fiber end (for the three averaging distances considered) for straight-sided specimens are higher than the value of

1.45 used for earlier predictions, which resulted in lower estimates of the bond strength of the interface. On the other hand, we have demonstrated that the cross-shaped design is successful in forcing debond initiation in the interior of the sample (region free of stress singularities) and therefore the results produced by testing of such specimens may be capable of providing the true transverse response of the fiber-matrix interface. Comparison of the interface strength value (3398 psi or 23.4 MPa) with the predicted strengths in Table III suggests that $0.5 < z_{avg}/R < 1$ is a reasonably good distance over which stresses should be averaged to obtain an average strength of the interface that is representative of the material for straight-sided specimen with fiber ends embedded. However, for straight-sided specimen with fiber ends exposed, the predicted strength values are lower when averaged over the same distance. One of the possible causes for lower strength values is machining induced damage during specimen preparation prior to testing. For exposed fiber specimen, the radial stress at the interface peaks at the free edge whereas, for embedded specimen, it is at the fiber corner in the interior of the sample. These are precisely the locations where failure was found to initiate during testing of straight-sided specimen. Thus, surface damage induced during sample preparation will facilitate failure for exposed fiber specimen and will have negligible influence on specimen with fiber ends embedded.

Table III: Predicted bond strength from straight-sided specimen

Distance	Straight-sided with fiber ends exposed				Straight-sided with fiber ends embedded			
$(\frac{z_{avg}}{R})$	λ	k	$\frac{\sigma_{rr}^{avg}}{\sigma_a}$	Bond Strength*	λ	k	$\frac{\sigma_{rr}^{avg}}{\sigma_a}$	Bond Strength+
0.5	0.2415	1.310	2.04	2462 Psi (16.98 MPa)	0.2136	1.350	1.99	3556 Psi (24.52 MPa)
1	0.2188	1.386	1.77	2136 Psi (14.73 MPa)	0.1987	1.402	1.75	3127 Psi (21.56 MPa)
2	0.2074	1.414	1.55	1871 psi (12.9 MPa)	0.1843	1.391	1.50	2681 Psi (18.48 MPa)

* Based on an average applied stress at debonding of 1207 psi
\+ Based on an average applied stress at debonding of 1787 psi

CONCLUDING REMARKS

Careful design and testing of model composites appears to be useful in estimating the normal strength of the fiber-matrix interface in unidirectional fiber composites. Two simple specimen geometries, namely, embedded fiber and cruciform design, have been considered to remove the influence of the free surfaces from tests on traditional straight-sided specimens. Results indicate good correlation between experiments and the analytical predictions and demonstrate the importance of stress field singularities in controlling damage initiation in straight-sided samples. It is further shown that only the cross-shaped design is successful in forcing debond initiation in the interior of the sample (region free of stress singularities) and therefore the results produced by testing of such specimens may be capable of providing the true transverse response of the fiber-matrix interface.

ACKNOWLEDGMENTS

The authors would like to thank Bill Ragland and Ron Trejo of the University of Dayton Research Institute for specimen preparation and conducting the necessary mechanical testing, respectively. This research was performed under U. S. Air Force Contract No. F33615-95-C-5041.

REFERENCES

Bogy, D.B. 1971. "Two edge bonded elastic wedges of different materials and wedge angles under surface tractions", *J. Appl. Mech.*, pp 377-386

Folias, E. S. 1989. " On the Stress Singularities at the intersection of a Cylindrical Inclusion with the Free Surface of a Plate ", *Int J Fracture*, 39: 25-34

Folias, E. S. 1991. " On the Prediction of Failure at a Fiber/Matrix Interface in a Composite Subjected to a Transverse Tensile Load ", *Journal of Composite Materials.*, 25: 869-886

Gundel, D. B., B. S. Majumdar and D. B. Miracle. 1995. " Evaluation of the Transverse Response of Fiber-Reinforced Composites Using a Cross-Shaped Sample Geometry, " *Scripta Metallurgica et Materialia,* 33: 2057-2065

Kroupa, J. L., and N. E. Ashbaugh. 1995. " Stress-Free Edge Effects on the Transverse Response of a Unidirectional Metal Matrix Composite, " *Composites Engineering*, 5: 569-582

Pagano, N. J., and G. P. Tandon. 1988. " 'Elastic Response of Multidirectional Coated Fiber Composites, " *Composites Science and Technology*, 31: 273-293

Shoufeng, H., P. Karpur,, T. E. Matikas, L. Shaw, and N. J. Pagano. 1996. " Free Edge Effect on Residual Stresses and Debond of a Composite Fiber/Matrix Interface ", to appear in *Mechanics of Composite Materials and Structures*

Tandon, G. P., R. Y. Kim and R. E. Dutton. 1996. " Micromechanical Edge Effects in Glass Matrix Composites ", presented at ASME IMECE, Atlanta, GA, Nov 17 - 22, 1996

Tandon, G. P. and R. Y. Kim. 1997. " Test Methods for Estimation of Interfacial Normal Strength in Unidirectional Fiber Reinforced Composites, " to appear in *Journal of Reinforced Plastics and Composites*

Warrier, S. G., D.B. Gundel, B.S. Majumdar and D.B. Miracle. 1996. " Interface Effects on the Micromechanical Response of a Transversely Loaded Single Fiber SCS-6/Ti-6Al-4V Composite, " *Metall. Mater. Trans.*, vol. 27A, pp. 2035-2043.

Whitney, J. M. and R. J. Nuismer. 1974. " Stress Fracture Criteria for Laminated Composites Containing Stress Concentrations, " *Journal of Composite Materials.*, 8: 253-265.

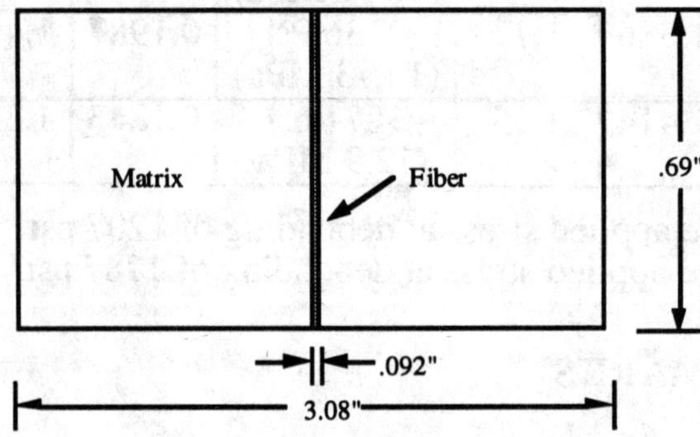

Fig 1: Straight-sided specimen with fiber ends exposed

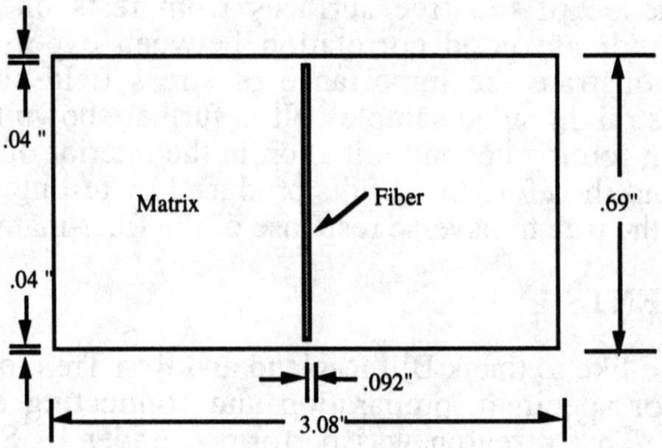

Fig 2: Straight-sided specimen with fiber ends embedded

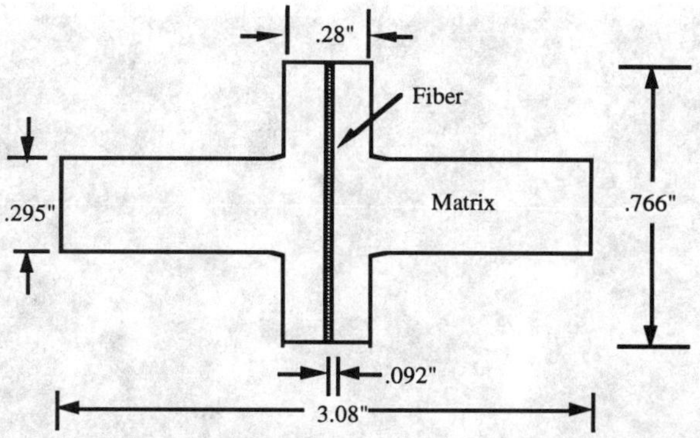

Fig 3: Cruciform (or cross-shaped) specimen

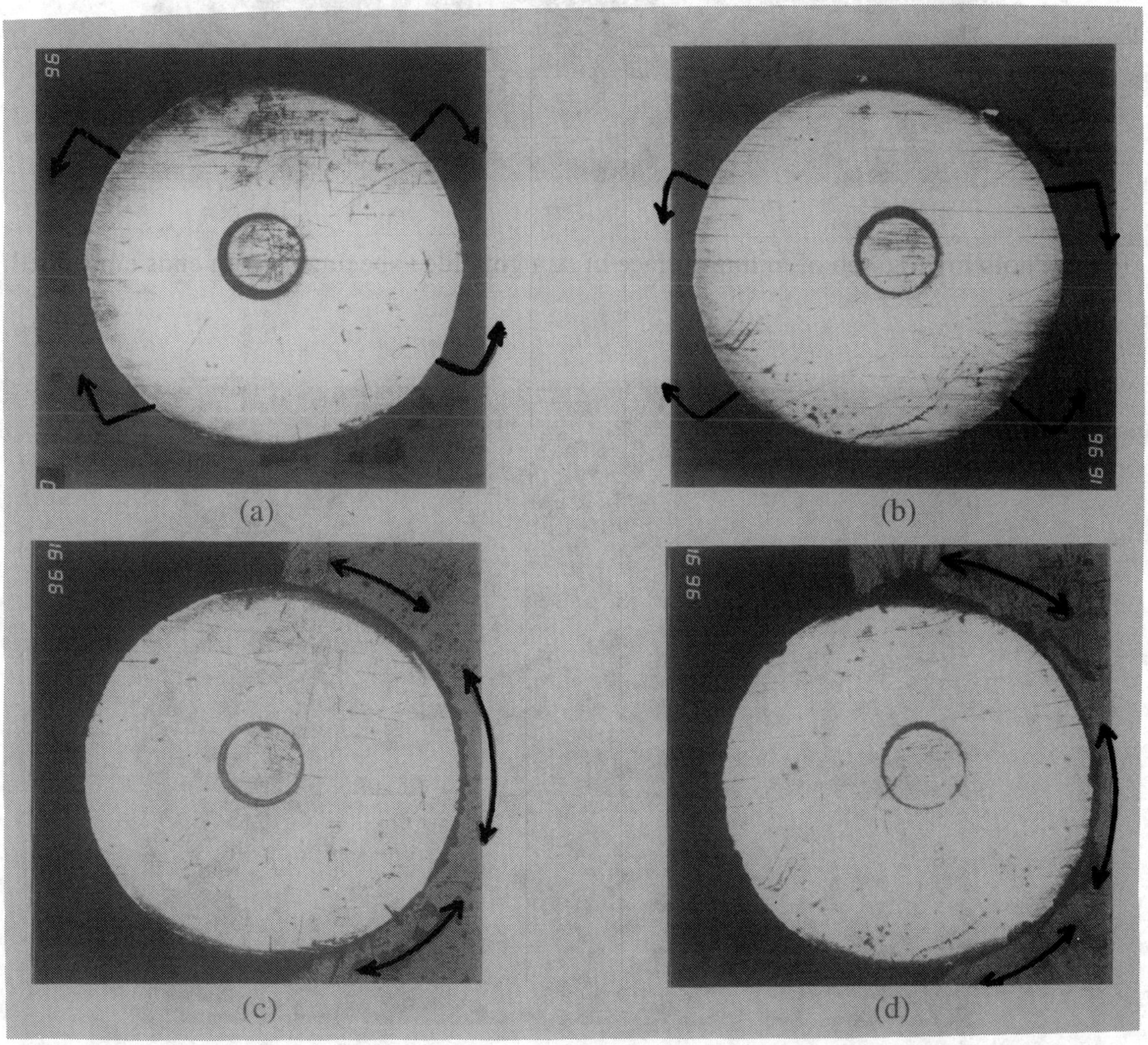

Fig 4: Photo micrographs of aluminum rod specimen showing front and back views at the first occurrence of acoustic emission on the stress-strain curve (4a and 4b) and at failure (4c and 4d), respectively.

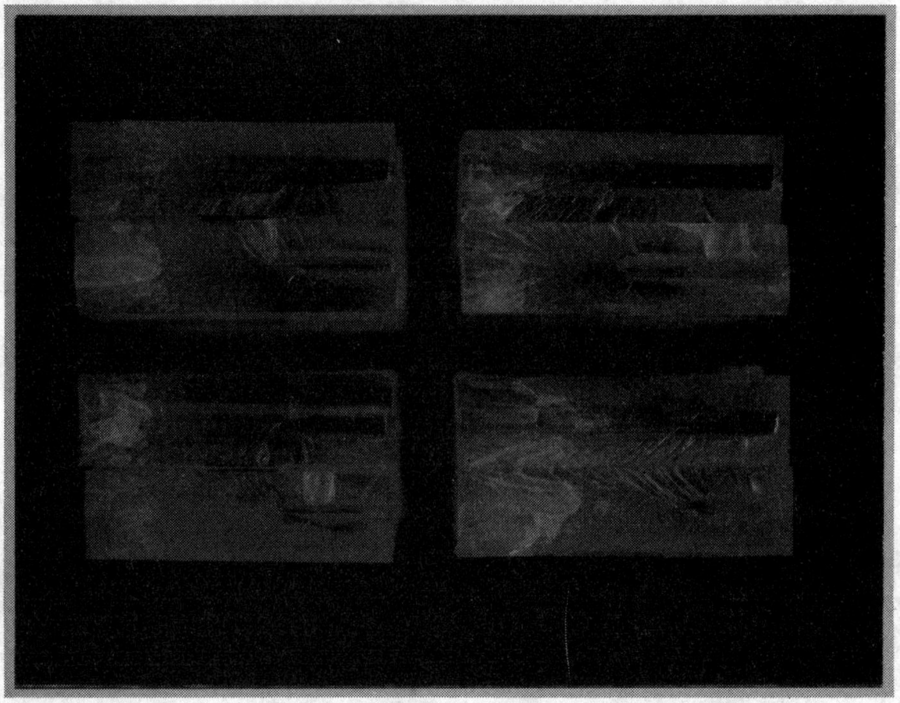

Fig 5: Photo micrograph of failure surface of straight-sided specimens with ends embedded

Fig 6: Photo micrograph of failure surface of cruciform (or cross-shaped) specimens

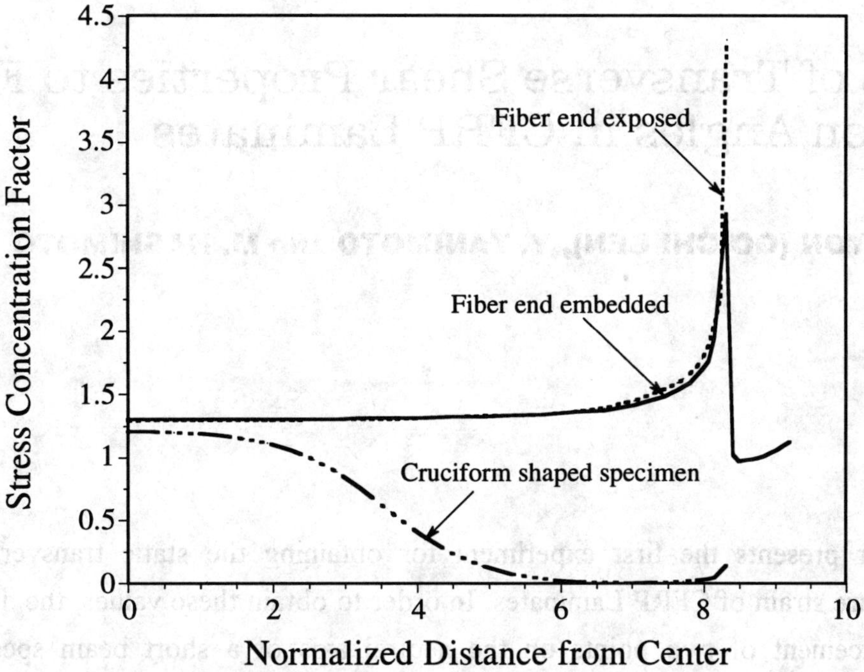

Fig 7: Variation of radial stress with normalized distance from center of the specimen along the length of the aluminum rod

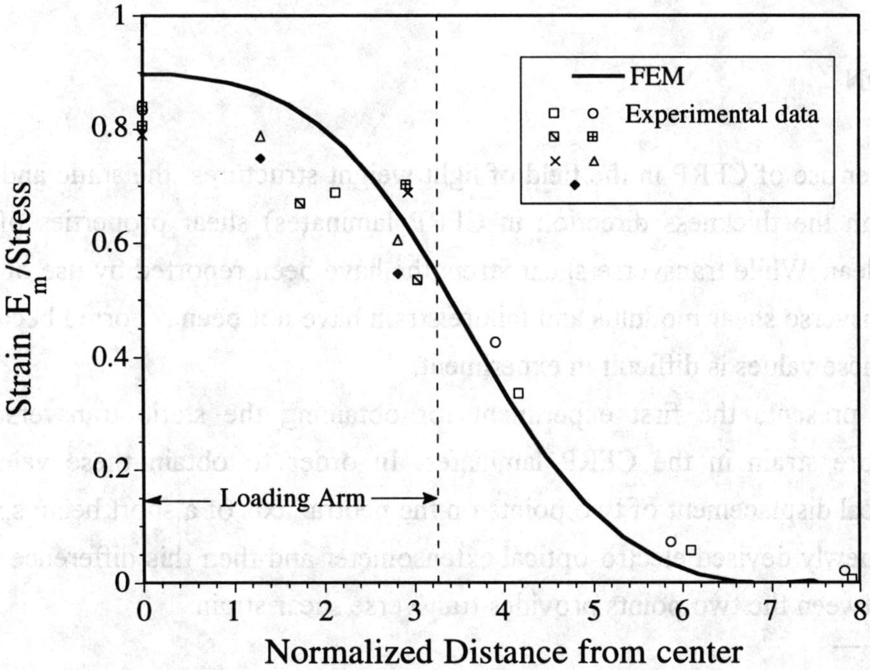

Fig 8: Variation of transverse strain with normalized distance from center of the specimen along the length of the aluminum rod

Relations of Transverse Shear Properties to Fiber Orientation Angles in CFRP Laminates

O-IL BYON (GOICHI BEN), Y. TANIMOTO AND M. HASHIMOTO

ABSTRACT

This paper presents the first experiment for obtaining the static transverse shear modulus and failure strain of CFRP Laminates. In order to obtain these values, the difference in vertical displacement of two points on the neutral axis of a short beam specimen is measured by a newly devised electro-optical extensometer. The transverse isotropy is first confirmed by our proposed method. Furthermore, the experimental results of transverse shear modulus and failure strain are compared with ones based on classical lamination theory and a three dimensional FEM code, respectively.

INTRODUCTION

For the wider use of CFRP in the field of light weight structures, the static and impact transverse (through the thickness direction in CFRP laminates) shear properties of CFRP should be made clear. While transverse shear strengths have been reported by use of a short beam test, the transverse shear modulus and failure strain have not been reported because the measurement of these values is difficult in experiment.

This paper presents the first experiment for obtaining the static transverse shear modulus and failure strain in the CFRP laminates. In order to obtain these values, the difference in vertical displacement of two points on the neutral axis of a short beam specimen is measured by a newly devised electro-optical extensometer and then this difference divided by the distance between the two points provides transverse shear strain.

O-Il Byon(Goichi Ben) Department of Mechanical Engineering, College of Industrial Technology, Nihon University, 1-2-1, Izumicho, Narashino ,Chiba 275 Japan.

Yasuhiro Tanimoto and Miho Hashimoto, Graduate School, Nihon University

First, the usefulness of our proposed method is confirmed by epoxy resign test. The variations of transverse shear modulus and failure strain are measured when the fiber orientation angle of unidirectional laminates is changed from 0° to 90° and the first experimental confirmation of the transverse isotropy is provided by our method. After the experimental modulus is compared with the analytical one based on classical lamination theory, it is shown that the experimental value agrees with the analytical one. Furthermore, the stress distribution at failure is discussed by using a three dimensional FEM and it is compared with the result obtained from the detection of flaws by the ultrasonic technique.

EXPERIMENT

Unidirectional CFRP laminated plates each having length of 200mm, width of 100mm, and thickness of 3mm are made by a hot press instrument and they have seven kinds of fiber orientation angles, $0°_{24}$, $15°_{24}$, $30°_{24}$, $45°_{24}$, $60°_{24}$, $75°_{24}$, and $90°_{24}$,respectively. Each angle of laminated plate is cut into five specimens each having length of 21mm and width of 10mm as shown in Fig.1. The distance of the supports is 15mm which is 5 times the thickness (JIS, 1991).

A conventional electro-optical extensometer is used to measure the difference in vertical displacement of two points on the vertical axis. In this paper, a new extensometer is devised that can measure the difference in vertical displacement of two points on the horizontal axis. In order to measure vertical displacements on the neutral axis, the specimen is

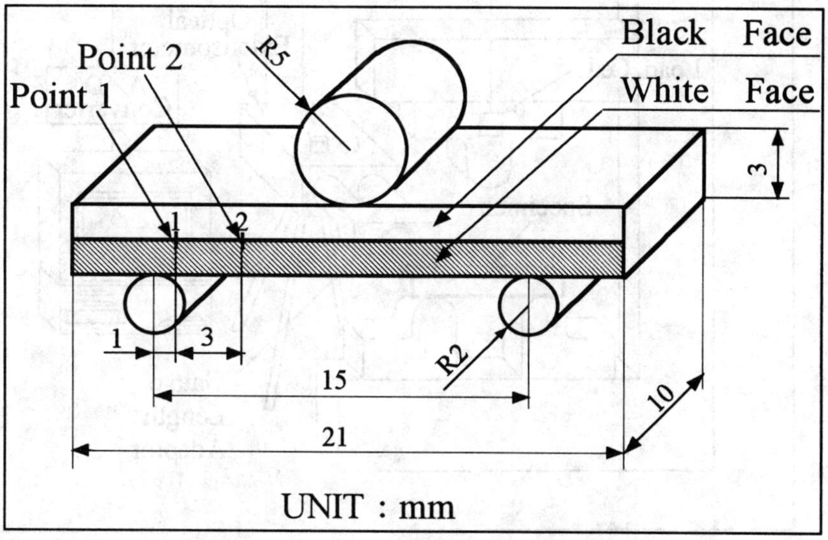

Figure 1. Test Specimen

accurately divided into halves through the thickness direction and one of them is painted white with correction fluid. After the effects of the position of gauge points and the gauge length on measuring the transverse strain are examined, the gauge length is determined to be 3mm and gauge point 1 is 1mm from the left support and gauge point 2 is 4mm from the left support(Byon and Tanimoto 1996). In other words, each vertical displacement of the points crossing the gauge line and the black/white boundary line is measured.

Fig.2 shows the experimental apparatuses for the static loading. The gauge length adapter is attached to the front of two lenses of the electro-optical extensometer and this extensometer faces the direction of beam axis. The static load is measured by a load cell and the average transverse shear stress τ_{ts} and the transverse shear strength F_{ts} are obtained by the following equations.

$$\tau_{ts} = \frac{P}{2bh}, \quad F_{ts} = \frac{3P_{\max}}{4bh} \tag{1}$$

in which, P is the load within the elastic limit, P_{max} is the maximum load at failure, b and h are the width and the thickness of specimen. Since the vertical displacements of w_1 and w_2 measured by the electro-optical extensometer include the flexural displacements of w_{b1} and w_{b2}, the transverse strain γ_{ts} is calculated by the following equations.

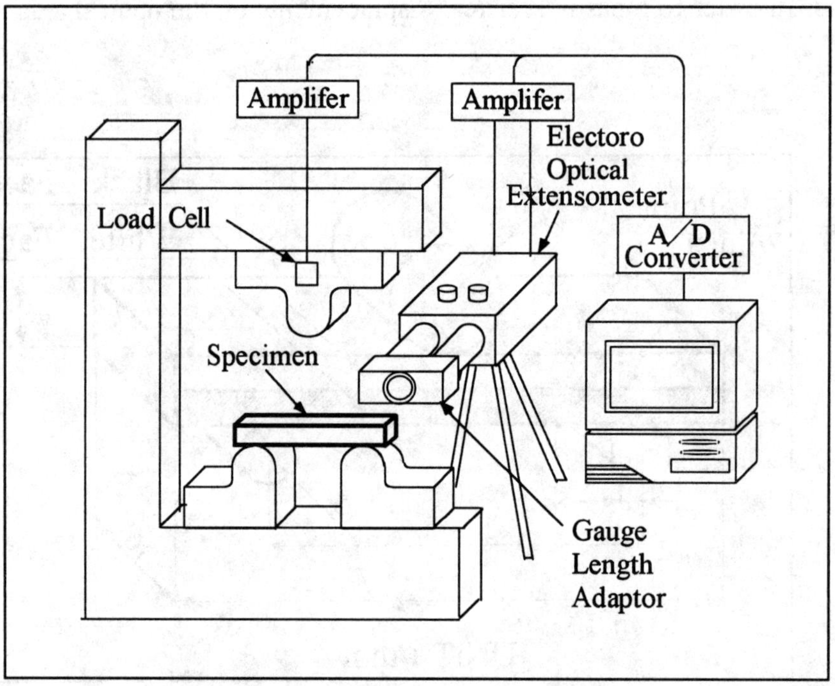

Figure 2. Apparatus for Test

Table 1. Variation of Flexural Modulus to Fiber Orientation Angle in Unidirectional CFRP Laminates

Fiber Orientation Angle θ (deg)	Flexural Modulus E(GPa)
$[0°]_{24}$	118.1
$[15°]_{24}$	84.9
$[30°]_{24}$	34.1
$[45°]_{24}$	18.5
$[60°]_{24}$	12.1
$[75°]_{24}$	11.0
$[90°]_{24}$	10.2
Epoxy Resin	3.8

$$\gamma_{ts} = \frac{(w_2 - w_{b2}) - (w_1 - w_{b1})}{x_2 - x_1}, w_{b1} = \frac{Px_1}{12\overline{\overline{EI}}}\left(\frac{3}{4}L^2 - x_1^{\ 2}\right), w_{b2} = \frac{Px_2}{12\overline{\overline{EI}}}\left(\frac{3}{4}L^2 - x_2^{\ 2}\right) \quad (2)$$

where, L is the distance (15mm) between the two supports, x_1 and x_2 are the distances(=1mm and 4mm) from the left support and $\overline{\overline{EI}}$ is the flexural rigidity of the specimen. The strain gauge is pasted to the center of back face in the specimen and the flexural modulus of specimen E is also measured together with the vertical displacements. Table 1 shows the experimental relation of fiber orientation angle to the flexural modulus. After amplifying the experimental data, converting from analogue data to digital and inputting it into a personal computer, the relation of the average transverse shear stress and shear strain and the transverse shear strength are calculated. Then the transverse shear modulus is obtained by the initial slope of τ_{ts}-γ_{ts} figure.

ANALYTICAL EQUATIONS

Let axes 1 and 2 be the directions of the fibers and normal of unidirectional laminates and x, y be the axial and the width directions of the beam specimen. When the x axis is rotated clockwise from the 1 axis by θ, the transverse shear modulus is given as follows(Vinson and Sierakowski, 1986).

$$G_{ts} = \frac{G_{23}G_{31}}{G_{31}\sin^2\theta + G_{23}\cos^2\theta} \quad (3)$$

where, G_{31} and G_{23} are the transverse shear modulus of 0° and 90° specimens. In the case of the symmetric laminates, the transverse shear modulus is obtained by the following equations

$$G_{ts} = \frac{A_{44} \cdot A_{55} - A_{45}{}^{2}}{A_{44}h}$$

$$A_{ij} = \frac{5}{4}\sum_{k=1}^{N}\left(\overline{Q_{ij}}\right)_{k}\left[h_k - h_{k-1} - \frac{4}{3}\left(h_k{}^{3} - h_{k-1}{}^{3}\right)\frac{1}{h^2}\right] \quad (i,j=4,5) \tag{4}$$

where, N is the number of lamination and the transformed reduced stiffness $\overline{Q_{ij}}$ is also defined as follows

$$\begin{aligned} \overline{Q_{44}} &= G_{23}\cos^2\theta + G_{31}\sin^2\theta \\ \overline{Q_{45}} &= \left(G_{31} - G_{23}\right)\cos\theta\sin\theta \\ \overline{Q_{55}} &= G_{23}\sin^2\theta + G_{31}\cos^2\theta \end{aligned} \tag{5}$$

THREE DIMENSIONAL FINITE ELEMENT ANALYSIS

In order to examine the stress distribution and the effect of cross elasticity on the deformation of the off-axis specimens, PAFEC, a three dimensional finite element code, is used. The specimen is divided into 22, 10 and 12 elements in the axial, width and thickness directions, respectively. The total number of elements is 2640. The static load is assumed to be distributed over a region of 1.0mm around the load point. The displacement of the load

Table 2. Mechanical Properties of Unidirectional CFRP Laminates

Modulus and Poisson' ratio	Unit	Value
E_x	GPa	118.1
E_y	GPa	10.2
E_z	GPa	10.2
G_{xy}	GPa	4.61
G_{zx}	GPa	8.23
G_{yz}	GPa	3.50
ν_{xy}	—	0.30
ν_{yx}	—	0.026
ν_{yz}	—	0.37
ν_{zy}	—	0.37
ν_{xz}	—	0.30
ν_{zx}	—	0.026

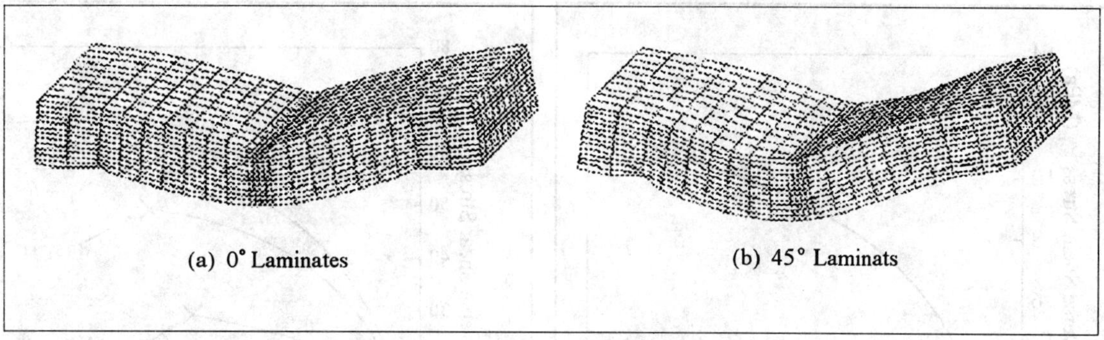

Figure 3 Deformation of CFRP Laminates obtained by 3D FEM

point is prevented in the axial and width directions and the displacement in all three dimensions at the supports is also prevented.

The mechanical properties of the CFRP used in the FEM analysis are listed in Table 2. The values of Gzx and Gyz are explained in the next section. The other values are provided by the fiber maker.

Figs 3 (a) and (b) show the deformation for the 0° and 45° specimens, respectively. In the case of the 0° specimen, the shear deformation mainly appears in the central plane through the thickness direction, but the 45° specimen shows torsional deformation with shear and flexural deformations owing to the effect of cross elasticity. In order to remove this effect from the experimental results, both fiber orientation angles of $+\theta$ and $-\theta$ are tested for the 15°, 30°, 45°, 60° and 75° specimens and the average value of the positive and negative cases is used for obtaining the transverse shear strain and modulus.

RESULTS AND DISCUSSIONS

TRANSVERSE SHEAR STRIN AND STRESS

In order to show the validity of our method, the specimens made of epoxy resin used as the matrix of the CFRP are tested by the same method described above. Fig.4 shows the relation of transverse shear stress to transverse shear strain in epoxy, the maximum transverse shear strain of 2.34% and the maximum transverse shear strength of 10.7MPa. The transverse shear modulus obtained from the initial slope in this figure is 1.45GPa and this value agrees well with the value(=1.38Gpa) calculated from Young's modulus (=3.8GPa) and Poisson's ratio(=0.38).

Fig.5 shows the relation of the transverse shear stress to shear strain from the beginning

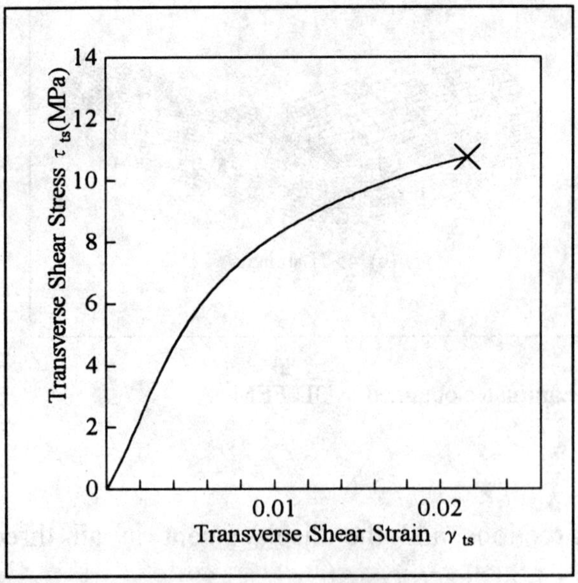

Figure 4. Relation of Transverse Shear Stress to Shear Strain in Epoxy Resin

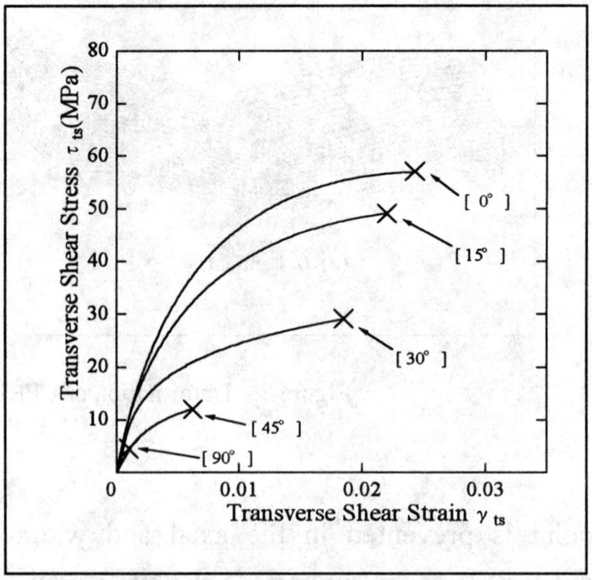

Figure 5. Relations of Transverse Shear Stress to Shear Strain in Unidirectional CFRP Laminates

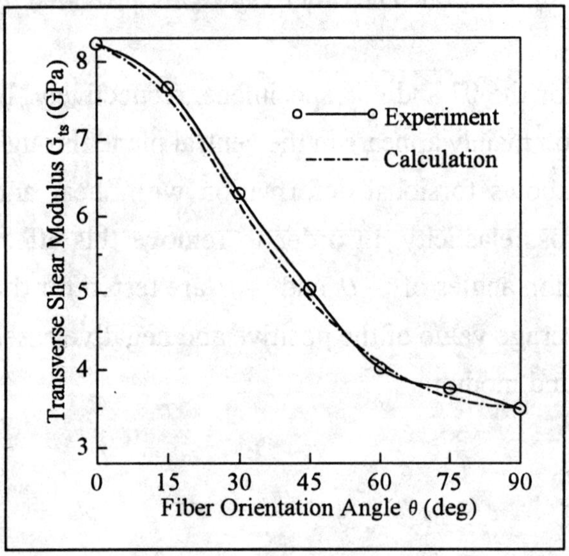

Figure 6. Variation of Transverse Shear Modulus to Fiber Orientation Angle

Table 3. Conparision of Transverse Shear Modulus in Quasi-isotropic Laminates

	Transverse Shear Modulus
Present Method	4.75(GPa)
Analytical value	4.70(GPa)

to failure for the CFRP specimens. All cases are nonlinear and even the curve of the 0° specimen is also remarkably nonlinear. The initial slope of the curve decreases as the fiber orientation angle increases.

TRANSVERSE SHEAR MODULUS

Fig.6 shows the change of the transverse shear modulus G_{ts} to the fiber orientation θ.

The modulus decreases with increasing θ and it agrees well with the one calculated by equation (3). Consequently, the transverse shear modulus of the unidirectional laminate having any fiber orientation angle can be calculated by using the values of the 0° and 90° laminates.

Table 3 shows that the experimental transverse shear modulus of the quasi-isotropic laminates agrees well with the value calculated by equation (4) and that the transverse shear modulus of the symmetric laminates having any stacking sequence can be also obtained by using classical lamination theory.

TRANSVERSELY ISOTROPY

Since the assumption of transverse isotropy through the thickness direction in the unidirectional laminates is often used, this assumption is confirmed by the use of experimental result of the 90° specimen. The experimental transverse shear modulus of 3.50GPa in the 90° specimen is nearly the same as the value(=3.72GPa) calculated from the equation, $G_{yz} = E_Z / 2(1+\upsilon_{yz})$, using the values of $E_Z = 10.2\,\text{GPa}, \upsilon_{yz} = 0.37$ given in Table 2. This is the first confirmation of the validity of transverse isotropy assumption.

TRANSVERSE SHEAR FAILURE STRAIN AND STRENGTH

The relations of the transverse shear failure strain and strength to the fiber orientation angles are shown in Fig. 7. Both curves have maximums at θ=0° and decrease nonlinearly

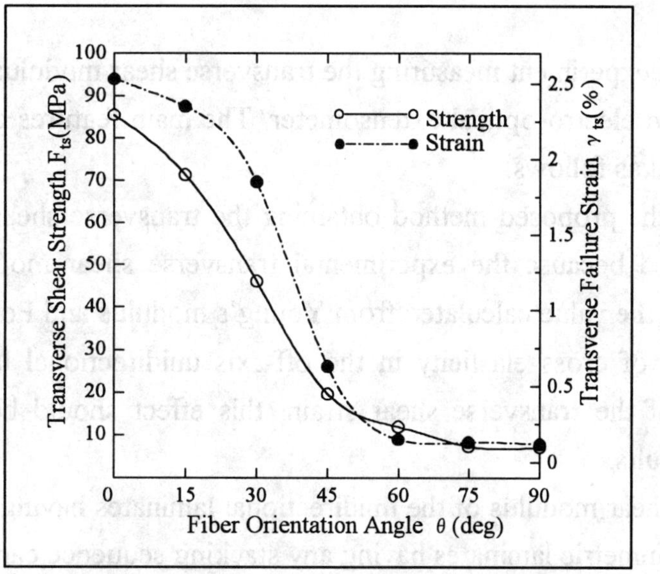

Figure 7. Variations of Transverse Shear Strength and Failure Strain to Fiber Orientation Angle

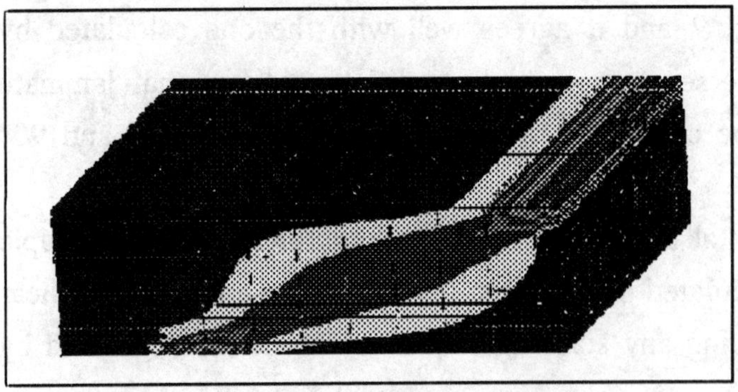

Figure 8. Transverse Shear Stress Distribution in 0° Laminates obtained by 3D FEM

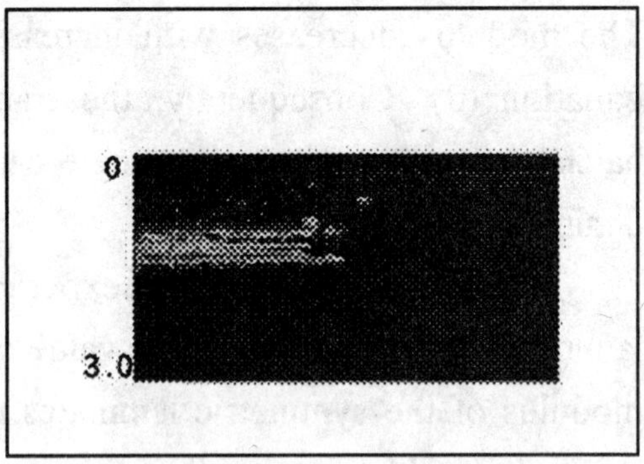

Figure 9. Damage Distribution in 0° Laminates obtained by Ultrasonic Inspection

with increasing θ. Both have an inflection point near 30° and the failure mode is expected to change from the transverse shear failure to the flexural failure at this angle. The failure modes of the θ=0° and 15° specimens are confirmed to be transverse shear by the ultrasonic detection of flaws. In the cases of the over 45° specimens, the failure mode is the flexural one and the 30° specimen shows a combination of these two modes.

Fig.8 shows the distribution of the transverse shear stress τ_{ts} in the 0° specimen obtained by the three dimensional FEM code. Although the stress concentration appears near the support and the load point, this distribution is similar to the one obtained by the ultrasonic detection of flaws as shown in Fig. 9.

CONCLUSIONS

This is the first experiment measuring the transverse shear modulus and failure strain by using a newly devised electro-optical extensometer. The main features obtained through this study are summarized as follows:

(1) The validity of the proposed method obtaining the transverse shear modulus and strain can be confirmed because the experimental transverse shear modulus of epoxy resin agrees well with the value calculated from Young's modulus and Poisson's ratio.

(2) Since the effect of cross elasticity in the off-axis unidirectional laminates influences measurements of the transverse shear strain, this effect should be removed from the experimental results.

(3) The transverse shear modulus of the unidirectional laminates having any fiber orientation angle and the symmetric laminates having any stacking sequence can be calculated by the values of the 0° and 90° laminates and classical lamination theory.

(4) This is the first confirmation of the validity of the transverse isotropy assumption.

(5) The unidirectional laminates having the fiber orientation angle of over 30° are subjected to shear, flexure and torsion and these specimens are not broken by pure transverse shear. Therefore, the fiber orientation angle of the unidirectional laminate is restricted to less than 30° when the short beam method is used in order to obtain the transverse shear strength and failure strain.

(6) Using this method, the impact transverse shear modulus and failure strain are being now examined and they will be reported in the near future.

This paper was summarized during the period when the primary author was a visiting professor at the Center for Aerospace Structure, University of Colorado. The author wishes to thanks members of Nihon University and Prof. K.C.Park, University of Colorado, who made this sabbatical possible.

REFERENCES

JIS(Japanese Industrial Standard),1991,"Testiny method for apparent interlamina shear strength of carbon fiber reinforced plastics by three point loading method" (JIS K 7078).

O-Il Byon(Goichi Ben) and Yasuhiro Tanimoto,1996, "Transvers Shear Properties of CFRP Laminates under Static and Impact Loading(in Japanese)",Proceeding of the 74th JASME Fall Annual Meeting, Vol. 2,p123-124.

J.R.Vinson and R.L.Sierakowski,1986, "The Behavior of Structures Composed of Composite Materials", Martinus Nijhoff Pub.

Author Index